French Dictionary of Telecommunications
Dictionnaire anglais des télécommunications

Routledge French Dictionary of Telecommunications
Dictionnaire anglais des télécommunications

French–English/English–French
Français–Anglais/Anglais–Français

London and New York

First published 1997
by Routledge
11 New Fetter Lane, London EC4P 4EE

Simultaneously published in the USA and Canada by Routledge
29 West 35th Street, New York, NY 10001

Typeset in Monotype Times, Helvetica Neue and Bauer Bodoni by Routledge
Printed in England by TJ International Ltd, Cornwall

Printed on acid-free paper

British Library Cataloguing-in-Publication Data
A catalogue record for this book is available from the British Library

Library of Congress Cataloguing-in-Publication Data
Applied for

ISBN: 0-415-13348-3

French Dictionary of Telecommunications
Dictionnaire anglais des télécommunications

Project Manager/Gestion de projet
Neil Clements

Managing Editor/Coordination éditoriale
Sinda López

Editorial/Rédaction
Justine Bird Lisa Carden Cécile Guinard Janice McNeillie Robert Timms

Programme Manager/Direction de collection
Elizabeth White

Marketing
Vanessa Markey Rachel Miller

Systems/Informatique
Omar Raman Simon Thompson

Production
Michelle Draycott Nigel Marsh Joanne Tinson

Terminologist/Terminologue
Stewart Wittering

Specialist Consultants/Spécialistes consultés
Ismaïla Biaye
Golden Gate University

Josette Mell
France Télécom

Réjane G. Amery Marie Désy-Field Catherine Maryan-Green

Janet Ormrod
École Nationale Supérieure des Télécommunications de Bretagne

Chris Sharp

Roland Ginguay

Lexicographers & Proofreaders/Lexicographie & Correction des épreuves
Nicola Cooke Anne Convery Alison Crann Hazel Curties Stephen Curtis Gemma Marren
Jeremy Munday Mary Rigby Maggie Seaton Robert Timms

Keyboarders/Clavistes
Emmanuelle Bels Beatriz Fernández Ilona Lehmann Rosa Gálvez López David Morgan
Roger Pena Muiño Nathalie Pernstich Fabienne Rangeard Debbie Thomas

Acknowledgements

We are also particularly indebted to the following organizations for their assistance during the compilation of this dictionary:

Remerciements

Nous tenons aussi particulièrement à exprimer nos remerciements aux organismes qui nous ont apporté leur concours lors de la rédaction de ce dictionnaire:

Belgacom Ecole Nationale Supérieure des Télécommunications Edusoft Ericsson Language Services
Eutelsat Evercom France Telecom Intelmatique Groupement Français de l'Industrie de l'Information
Nokia France SA Nortel Matra Cellulaire Novamédia Réseaux & Télécoms Softbank
Secrétariat à la Recherche et à l'Espace TRT Lucent Technologies

Contents/Table de matières

Foreword/Préface

Telecommunications is one of the key industries propelling us forward to new levels of global business and multi-regional integration. It covers a growing range of subjects and encompasses many new technologies. Because of its relevance to so many scientific and commercial fields, those serving the translation needs of a large number of industries will increasingly have recourse to a bilingual work comprising a wide range of telecommunications terms.

The international telecommunications industry is regulated by various entities including the Federal Communications Commission (FCC) in the US, the Canadian Radio-Television and Telecommunications Commission (CRTC), the Office of Telecommunications (OFTEL) in the UK and the European Commission (EC) throughout Europe. In France, Switzerland and Belgium, regulatory authorities include the *Ministre délégué à la poste aux télécommunications et à l'espace*, the Telecommunication Regulatory Service, the *Office fédéral de la communication (OFCOM)*, the Belgian Institute for Post & Telephony (BIPT) and the *Cabinet du ministre de l'économie et des télécommunications* respectively. Standards are determined by a number of bodies including the International Standards Organisation (ISO), the International Telecommunications Union (ITU), the International Electrotechnical Commission (IEC), the American National Standards Institute (ANSI), the Canadian General Standards Board (CGSB) and the Canadian Standards Association (CSA). Amongst others, the British Standards Institution (BSI) deals with standards in the UK, the French Standards Association (AFNOR) in France, the Belgian Standardisation Institute (BSI) in Belgium, the *Schweizerischen Normen-Vereinigung (SNV)* in Switzerland and the European Telecommunications Standards Institute (ETSI) throughout Europe. Notwithstanding the distinctions between European and US telecommunications and television standards, differences

Les télécommunications font partie de ces industries clés qui nous projettent vers les nouveaux horizons du commerce mondial et de l'intégration multirégionale. Ce secteur, qui englobe un nombre croissant de sujets et comprend une variété de technologies nouvelles, touche particulièrement aux très nombreux domaines scientifiques et commerciaux. Ainsi, cet ouvrage bilingue dans lequel figure une grande partie de la terminologie des télécommunications, s'avère un outil indispensable pour tous ceux qui servent les besoins en traduction d'une multitude d'industries.

Parmi les organismes réglementant l'industrie internationale des télécommunications figurent la Commission fédérale des communications (FCC) aux États-Unis, le Conseil de la radiodiffusion et des télécommunications canadiennes (CRTC) au Canada, l'*Office of Telecommunications (OFTEL)* au Royaume-Uni, et la Commission Européenne (CE) en Europe. En France, en Suisse et en Belgique, les institutions chargées de cette réglementation sont respectivement le Ministre délégué à la poste, aux télécommunications et à l'espace, le Service de la Réglementation des Communications (SRT), l'Office fédéral de la télécommunication (OFCOM), l'Institut belge des services postaux et des télécommunications (IBPT) et le Cabinet du Ministre de l'économie et des télécommunications. Les normes sont définies par un certain nombre d'organismes tels que l'Organisation internationale de normalisation (ISO), l'Union internationale des télécommunications (UIT), la Commission électrotechnique internationale (CEI), l'Institut américain de normalisation (ANSI), l'Office des normes générales du Canada (ONGC) ainsi que l'Association canadienne de normalisation (CSA). On peut citer également l'Institut britannique de normalisation (BSI) au Royaume-Uni, l'Association française de normalisation (AFNOR) en France, l'Institut belge de normalisation (IBN) en Belgique, la *Schweizerischen Normen-Vereinigung (SNV)* en Suisse et l'Institut européen des normes de télécommunications (ETSI) en Europe.

in regional terminology in this area are relatively limited in English or French, as the field is still dynamic and technical frames of reference tend, for example, to be the same in France as Canada and in the UK as the US.

International business depends increasingly on the speed and reliability of communication, which nowadays means reliance on modern telecommunications tools including the Internet, electronic commerce, teleconferencing and telecommuting facilities. Financial institutions, such as brokerage houses, are repositioning themselves in the marketplace and emphasizing the product differentiation made possible by their technological astuteness and their leadership role in telecommunications. The universal tool of the Internet has dramatically increased our ability to exchange information and to process business transactions efficiently. Countries are investing in their telecommunications infrastructure, and the ingenious use of such technology as Electronic Data Interchange enables the clearing of cargo ships through a busy port in a matter of hours instead of days. It is possible to earn a university degree entirely via interactive video and audio conferencing in the comfort of your own office or home, or equally to telecommute to work. We are limited only by our imagination as to how we can use this technology to enhance the quality of our life. Meanwhile, we can best benefit from it by mastering its language and adapting to how it works.

Due to the fast pace of technological change in the field of telecommunications, there has been an exponential growth in technical jargon over the past ten years. New words are added as a result of the merging of communications technology and computing in entertainment, education and business. Unusually among specialized technical fields, communications technology is so ubiquitous that business people are frequently confronted with choices and even purchase decisions that require some basic familiarity with the new language of cyberspace. Increasing numbers of people will soon need to be conversant with search engines, web browsers and home pages to locate and download publications, including those no longer available in print. A computer and an Internet account may be used to access information, or neighbourhood libraries will provide access. The challenge may be in mastering the new vocabulary and adapting to addresses with no street names but unique combinations of characters

En dépit des normes distinctes établies en Europe et aux États-Unis en matière de télécommunications et de télévision, les différences terminologiques entre les régions sont assez limitées en anglais ou en français, étant donné qu'il s'agit d'un secteur en plein dynamisme dont les systèmes techniques de référence sont généralement les mêmes en France qu'au Canada, et les mêmes au Royaume-Uni qu'aux États-Unis.

Le commerce international est de plus en plus tributaire de la rapidité et de la fiabilité de la communication, ce qui signifie qu'il nous faut compter sur les moyens modernes de télécommunications dont nous disposons à l'heure actuelle, comme par exemple, Internet, le commerce électronique, la téléconférence et le télétravail. Quant aux institutions financières, comme les maisons de courtage, elles sont en train de revoir leur position sur le marché et, s'aidant de leur sagacité en matière de technologie et de leur rôle de leader dans la sphère des télécommunications, mettent l'accent sur la différenciation de produits. Grâce à l'Internet, cet outil universel, nous avons considérablement amélioré notre système d'échange d'informations et notre capacité à traiter de manière efficace des opérations commerciales. Alors que les pays cherchent à mettre en valeur l'infrastructure de leurs télécommunications, l'utilisation ingénieuse d'une technologie telle que l'échange de données informatisées permet d'autoriser l'entrée ou la sortie des cargos dans un port encombré en l'espace de quelques heures, alors qu'auparavant, cette procédure nécessitait plusieurs jours. Il est tout à fait possible d'obtenir un diplôme universitaire par la vidéo interactive et l'audioconférence sans quitter son bureau ou son lieu de résidence, ou bien encore de travailler chez soi (télétravail à domicile). Nous n'avons que notre imagination pour apprendre à utiliser cette technologie qui nous permettra d'améliorer notre qualité de vie. Pour l'heure, nous pouvons en tirer les meilleurs avantages en maîtrisant sa langue et en nous adaptant à son fonctionnement.

Dans le secteur des télécommunications, le jargon technique a connu une croissance exponentielle ces dix dernières années en raison de la rapidité des bouleversements technologiques. La technologie des communications et l'informatique, que ce soit dans les domaines des loisirs, de la formation ou des affaires ont fusionné, ce qui a permis d'introduire de nouveaux termes. A la différence des autres disciplines techniques spécialisées, les télécommunications sont omniprésentes si bien que les acteurs du monde des affaires se trouvent

lacking geographic or cultural symbolism instead.

There is ample evidence that telecommunications plays a central role in the growth and expansion of international business. Many more firms and countries are able to participate in today's global marketplace thanks to the accessibility of communications technology. With this ease of access comes an increased reliance on language, context and accuracy, whether it is for electronic commerce, international financial transactions or technical, professional and business documentation. Routledge's valuable programme of practical specialized bilingual dictionaries is filling a critical need in creating reference publications for translators, researchers and professionals throughout the French- and English-speaking world.

As much as telecommunications is crucial to the development of modern business, it remains a medium for information exchange, and it is what flows through it that has value. What bilingual dictionaries provide is essential to clear and accurate communications: they play a key role in the provision of reliable reference information for the accurate and professional translation of technical vocabulary.

The problem for professional translators, scholars, writers, engineers and business people involved in telecommunications and other areas of information technology is the lack of a reliable professional reference source able to address effectively in both English and French the scope of this ever-expanding field. This dictionary seeks to provide you with the tools essential to bridge the gap between technical jargon and user-accessible international business language. It contains approximately 30,000 entries in each language, covering not only terms used in Telephony & Telegraphy, Television & Video, Radio-based Technologies and Signalling, but also developments in related and fast-moving fields including Computing, Electronics and Testing Procedures. By making available an up-to-date tool for accurate, intelligent and useful information exchange to and from two major languages at the forefront of Telecommunications terminology, it fulfils a valuable and timely role in bilingual specialist publishing in this field.

The effective use of such a carefully developed bilingual dictionary will be of great value in facilitating and enhancing communication and business transactions between anglophone and francophone cultures. It represents an important element for information exchange on the global

fréquemment face à des choix et, parfois même, à des décisions professionnelles nécessitant une connaissance de base de la nouvelle langue du cyberspace. Il deviendra bientôt indispensable à la plupart d'entre nous de connaître les moteurs de recherche, les explorateurs Web et les pages d'accueil afin de rechercher et de télécharger des publications, y compris celles dont la version imprimée n'est plus disponible. Pour avoir accès à l'information, différents moyens sont à notre disposition: un ordinateur et un compte sur le réseau Internet, ou encore les bibliothèques de quartier. S'il y a un défi à relever, c'est peut-être celui de maîtriser le nouveau vocabulaire et de l'adapter à des adresses, constituées non pas de noms de rue, mais de combinaisons uniques de caractères dénuées de tout symbolisme géographique ou culturel.

Il n'est plus à démontrer que les télécommunications jouent un rôle essentiel dans le développement du commerce international. Et c'est parce que la technologie des communications est devenue accessible qu'un nombre croissant d'entreprises et de pays peuvent jouer un rôle sur le marché mondial. A cette facilité d'accès s'ajoute la capacité, toujours plus grande, de pouvoir se référer à une base linguistique et contextuelle précise, qu'il s'agisse de commerce électronique, d'opérations financières internationales, ou encore de documentation technique, professionnelle ou commerciale. L'unique programme lancé par Routledge prévoyant la publication de dictionnaires bilingues spécialisés répond à un besoin critique de créer des ouvrages de référence pouvant s'adresser à l'ensemble des traducteurs, des chercheurs et des professionnels, de langues française et anglaise.

Si les télécommunications sont essentielles au développement du monde moderne des affaires, elles restent surtout un moyen destiné à l'échange d'informations dont la valeur dépend du contenu mis en circulation. Les dictionnaires bilingues sont essentiels à la clareté et la précision de la communication et, en fournissant une information de référence de qualité, ils jouent un rôle majeur dans la traduction professionnelle du vocabulaire technique.

Les traducteurs professionnels, les enseignants, les rédacteurs, les ingénieurs et les hommes et femmes d'affaires travaillant dans les télécommunications et les autres domaines de l'informatique ne disposent pas toujours d'une source de référence sérieuse pouvant offrir un large éventail de termes anglais et français relatifs à cette technologie qui évolue sans cesse. Ce dictionnaire vise à vous fournir les outils essentiels permettant de rapprocher le

information superhighway, and without it you may still be searching for the on-ramp.

Prof. Ismaila Biaye, Director, Technology Management Programs
Golden Gate University, San Francisco

jargon technique et la langue internationale des affaires. On y recense environ 30 000 entrées dans chaque langue s'appliquant non seulement à des domaines tels que la téléphonie & la télégraphie, la télévision & la vidéo, les technologies radio et la structure & traitement des signaux, mais qui tiennent également compte de l'évolution de disciplines en constant développement comme l'informatique, l'électronique et les tests. Cet outil moderne auquel chacun peut avoir recours, permet d'échanger des informations de manière précise, intelligente et utile, entre deux langues majeures, le français et l'anglais (dans les deux directions), qui sont également à l'avant-garde de la terminologie des télécommunications. Ainsi, tous les ingrédients sont réunis pour offrir à cet ouvrage une place de choix dans le monde de l'édition des dictionnaires spécialisés bilingues.

Ce dictionnaire bilingue, élaboré avec le plus grand soin, se révèlera à l'usage un atout précieux en facilitant et en améliorant la communication et les opérations commerciales entre les cultures anglophone et francophone. Sans cet outil indispensable à l'échange de données sur les autoroutes mondiales de l'information, vous continuerez peut-être à chercher la voie d'accès.

Les éditeurs

Introduction/Introduction

Following on from the programme of general technical and business dictionaries launched in October 1994, the *French Dictionary of Telecommunications* is the latest in a series of specialist dictionaries to be published by Routledge. Given the growing importance of telecommunications in everyday life and the fact that English and French as a language combination has become increasingly important over recent years, this dictionary will be a valuable resource for specialist translators, terminologists and experts working in all fields of telecommunications. Approximately 35,000 entries in each language cover not only terminology used in traditional disciplines, such as Telephony & Telegraphy and Radio-based Technologies, but also terms drawn from developing areas such as Electronics and Computing.

It would not have been possible to compile this dictionary within a realistic timescale, and to the standard achieved, without the use of a custom-designed database. The database used is a relational one: term records for each language are held in separate files, with further files consisting only of link records. Links between terms within a single-language file represent cross-references, themselves of a wide variety of types: spelling variants, geographical variants and abbreviations.

The content of the database for this dictionary was compiled and vetted by a team of English and French terminologists with current practical experience of a narrowly-defined subject area within telecommunications in order to ensure its currency, the accuracy of translations and the adequacy of coverage. Finally, each language file was reviewed by editors to ensure coverage of North American terms and spelling variants, which are clearly labelled and distinguished.

The compilation and editing of the database of terms was, however, only the first stage in the making of the dictionary. Within the database, the distinction between source and target languages is not meaningful, but for this printed dictionary it was necessary to format the data to produce

Le *Dictionnaire anglais des télécommunications* est le dernier d'une série de dictionnaires spécialisés bilingues que Routledge publie dans le cadre d'un programme lancé en octobre 1994, et dont les premières publications sont les dictionnaires de technique générale et les dictionnaires des affaires, du commerce et de la finance. Non seulement les télécommunications jouent un rôle croissant dans notre vie quotidienne, mais la combinaison de langues anglais et français, de plus en plus utilisée aujourd'hui, fait de ce dictionnaire un outil précieux pour les traducteurs techniques, terminologues et spécialistes travaillant dans le domaine des télécommunications. Les 35 000 entrées dans chaque langue englobe une terminologie relative à des disciplines telles que la téléphonie & télégraphie et les technologies radio, mais également à des secteurs qui évoluent sans cesse comme l'électronique et l'informatique.

La sortie de ce dictionnaire dans les délais et conformément aux critères prévus imposait la compilation d'une base de données spécifique qui est de type relationnel: dans chaque langue, les termes sont stockés dans des fichiers séparés, tandis que d'autres fichiers ne se composent que de liens. Les liens entre les différent termes d'un même fichier de langue représentent des renvois, eux-mêmes de différents types: variantes orthographiques, variantes géographiques et abréviations.

La base de données a été compilée et approuvée par une équipe de terminologues anglais et français, ayant une connaissance pratique approfondie des domaines s'appliquant aux télécommunications afin de garantir l'actualité et la précision des explications, ainsi que l'étendue des sujets traités. Enfin, chaque fichier de langue a été contrôlé par des lexicographes des pays concernés, chargés de veiller au respect des variantes géographiques et orthographiques. Dans chaque langue, les formes et l'orthographe (anglaises, américaines ou canadiennes) sont clairement identifiées.

La création et la vérification de la base de données terminologique n'ont cependant représenté que la

separate French–English and English–French sections. The data was processed by a further software module to produce two alphabetic sequences, of French headwords with English translations and vice-versa, each displaying the nesting of compounds, ordering of translations, style for cross-references of different types, and other features according to the algorithm.

At this stage the formatted text was edited by a team of experienced English and French lexicographers whose task it was to eliminate duplication or inconsistency; edit the contextual information; and to remove terms that were too general for inclusion in a specialist dictionary.

The editorial team

première étape de la constitution du dictionnaire. La distinction entre les langues cible et source n'étant pas significative au niveau de la base de données, nous avons dû, pour la version imprimée, mettre les données en forme pour obtenir des parties français-anglais et anglais-français séparées. Les données ont alors été traitées par un logiciel permettant de produire deux séquences alphabétiques, des mots-clés français avec leurs traductions anglaises et vice-versa avec, dans chaque cas, indication de l'imbrication des composants, de l'ordre des traductions, du style des renvois de différents types, et d'autres fonctions, selon l'algorithme.

A ce stade, le texte mis en forme a été vérifié par une équipe de lexicographes français et anglais expérimentés, qui avaient pour mission d'éliminer les doublons et les incohérences, de vérifier les informations et les explications contextuelles, et de supprimer les termes trop généraux pour être inclus dans un dictionnaire technique.

Les éditeurs

Features of the dictionary/ Caractéristiques du dictionnaire

The following text extracts illustrate the principal features of the dictionary. For a more detailed explanation of each of these features and a full guide to using the dictionary, see pages xix–xxi.

Les éclatés ci-contre mentionnent les principales caractéristiques de ce dictionnaire. Pour une explication détaillée de chacune de ces caractéristiques et un guide complet de l'utilisation du dictionnaire, reportez-vous aux pages xxiii–xxv.

A colon introduces a compound where the headword has no technical value

Subject area labels given in alphabetical order show appropriate translation

Genders are indicated at French nouns

Compound terms are nested alphabetically at the first element

Cross-references from abbreviations to their full form are shown for both French and English

British English and American English spelling variants are given in full and labelled accordingly

Superscript numbers denote homographs

PABX *abrév* (*autocommutateur privé*) APP SERV, COMMUT, CONTRÔLE, TÉL PABX (*private automatic branch exchange*); **~ à intégration de services** *m* APP SERV service integration PABX; **~ sans fil** *m* PÉRIPH cordless PABX

package *m* DONNÉES package

packetage: **~ standard** *m* DONNÉES standard packages

PAD *abrév* DONNÉES, TÉL (*assemblage-désassemblage de paquets*) PAD (*packet assembly-disassembly facility*), (*assembleur-désassembleur de paquets*) PAD (*packet assembler-disassembler*)

pagaille *f* ÉLECTRON chaos

page *f* DONNÉES, INFORMAT page, PÉRIPH type page; **~ d'accueil** DONNÉES opening screen, welcome page, INFORMAT front page, home page, welcome page; **~ à l'écran** DONNÉES screen page; **~-écran** TV screen page; **~ suivante** TÉL form feed (*FF*); **~ télétex** TÉL teletex page; **~ de titre** INFORMAT title page, cover page; **~ Web** DONNÉES, INFORMAT Web page

pages: **~ jaunes** *f pl* TÉL yellow pages

pagination *f* GÉNÉRAL, INFORMAT *traitement de texte* page numbering, pagination, paging

paginer *vt* GÉNÉRAL, INFORMAT number pages, page, paginate

paiement: **~ électronique** *m* APP SERV electronic payment; **~ à l'émission** *m* TV pay-per-view (*PPV*); **~ par émission** *m* TV pay-per-view (*PPV*)

pair[1] *adj* GÉNÉRAL, INFORMAT even

pair[2] *m* INFORMAT peer

pairage *m* TV pairing, twinning

paire *f* GÉNÉRAL pair; **~ de bornes** ALIM ÉLEC terminal pair; **~ en câble** TRANSMIS cable pair; **~ de service** TÉL administrative pair; **~ en simplex** TÉL simplex pair; **~ symétrique non-blindée** ÉLECTRON, TRANSMIS unshielded twisted pair (*UTP*); **~ torsadée** ÉLECTRON, TRANSMIS twisted pair; **~ torsadée blindée** ÉLECTRON, TRANSMIS shielded twisted pair (*STP*); **~ torsadée non-blindée** ÉLECTRON, TRANSMIS unshielded twisted pair (*UTP*)

PAL *abrév* (*logique à matrice programmable*) INFORMAT, TV PAL (*programmable array logic*)

palette *f* DONNÉES palette; **~ en bois** COMPOSANTS wooden paddle; **~ de couleurs** INFORMAT color scheme (*AmE*), colour scheme (*BrE*); **~ graphique** TV graphics package; **~ de jeu** INFORMAT paddle; **~ d'outils** DONNÉES, INFORMAT tool palette, toolbox

palier *m* DONNÉES, GÉNÉRAL level, RADIO plateau; **~ de comptage** TÉL metering rate

panel *m* GÉNÉRAL representative sample

paneuropéen *adj* GÉNÉRAL pan-European

panier: **~ de terminaison de ligne** *m* TRANSMIS line-terminating magazine

panne[1]: **en ~** *adj* TÉL, TESTS faulty; **~ facultative** *adj* ÉLECTRON fault-tolerant

panne[2] *f* GÉNÉRAL failure, hitch, breakdown, fault; **~ affectant les fonctions** TESTS function-disturbing fault; **~ d'alimentation** ALIM ÉLEC mains failure, power failure, power outage; **~ avec dégradation des fonctions** TESTS function-degrading fault; **~ complète** TÉL, TESTS function-preventing fault; **~ corrigeable** TESTS *sûreté de fonctionnement* correctable fault; **~ corrigée** TESTS *sûreté de fonctionnement* corrected fault

Deux-points indique un mot dépourvu de valeur technique et introduit un mot composé

Domaines en ordre alphabétique indiquent la traduction appropriée à chaque domaine

Genre de chaque nom indiqué pour le nom français

Termes composés imbriqués en ordre alphabétique sous le premier élément

Abréviation du terme donnée à l'entrée pour la forme intégrale

Informations complètes sur les variantes américaines et anglaises suivies de codes géographiques

Les numéros en exposant indiquent les homographes

biconical: ~ **aerial** *n* RADIO antenne biconique *f*; ~ **antenna** *n* RADIO antenne biconique *f*

bid *n* GEN tentative de prise *f*, RADIO offre *f*, soumission *f*, tentative *f*; ~ **for a line** TELEP demande de ligne *f*

Les numéros en exposant indiquent les homographes — *Superscript numbers denote homographs*

bidirectional[1] *adj* TRANS, TV bidirectionnel

bidirectional[2]: ~ **diode thyristor** *n* ELECT, PARTS thyristor diode bidirectionnel *m*, thyristor monté en diode bidirectionnel *m*; ~ **network** *n* TV réseau bidirectionnel *m*; ~ **transistor** *n* ELECT, PARTS transistor bidirectionnel *m*; ~ **triode thyristor** *n* ELECT, PARTS thyristor monté en triode bidirectionnel *m*, thyristor triode bidirectionnel *m*

bidirectionally: ~-**predictive coded picture** *n* (*b-picture*) COMP, TV image à codage prédictif bidirectionnel *f* (*image-b*)

Deux-points indique un mot dépourvu de valeur technique et introduit un mot composé — *A colon introduces a compound where the headword has no technical value*

bids: ~ **per circuit per hour** *phr* TELEP tentatives de prise par circuit et par heure

bifilar: ~ **winding** *n* POWER enroulement bifilaire *m*

bilateral: ~ **closed user group** *n* (*BCUG*) TELEP groupe fermé d'usagers bilatéral *m*; ~ **closed user group call indicator** *n* TELEP indicateur d'appel de groupe fermé d'usagers bilatéral *m*; ~ **closed user group with outgoing access** *n* (*BCUGOA*) TELEP groupe fermé d'usagers bilatéral avec accès sortant *m*; ~ **control** *n* TELEP, TRANS commande bilatérale *f*

Genre de chaque nom indiqué pour la traduction française — *Genders are indicated at French translations*

bi-level: ~ **image** *n* TELEP image à deux niveaux *f*

bilinear *adj* PER APP *scanner* bilinéaire

billed: ~ **services** *n pl* TELEP services facturés *m pl*

Domaines en ordre alphabétique indiquent la traduction appropriée à chaque domaine — *Subject area labels given in alphabetical order show appropriate translation*

billing *n* APPL, DATA, RADIO facturation *f*, TELEP facturation *f*, taxation *f*; ~ **address** TELEP destination des factures téléphoniques *f*; ~ **center** *AmE*, ~ **centre** *BrE* APPL centre de facturation *m*; ~ **information** TELEP renseignements tarifaires *m pl*

bimetal *n* PARTS bimétal *m*; ~ **spring** CONTR *regulating device* ressort bimétallique *m*; ~ **springset** CONTR groupe de ressorts bimétalliques *m*

Termes composés imbriqués en ordre alphabétique sous le premier élément — *Compound terms are nested alphabetically at the first element*

binary: ~ **coding** *n* SIG codage binaire *m*; ~ **counter** *n* ELECT *logic elements* compteur binaire *m*; ~ **data** *n pl* COMP, DATA, TRANS données binaires *f pl*; ~ **digit** *n* COMP, DATA, TRANS bit *m*, chiffre binaire *m*; ~ **error rate** *n* (*BER*) COMP, DATA, TRANS taux d'erreurs binaires *m* (*TEB*); ~ **figure** *n* COMP, TRANS chiffre binaire *m*; ~ **file** *n* COMP, DATA fichier binaire *m*; ~ **file transfer** *n* (*BFT*) COMP, DATA transfert de fichiers binaires *m* (*BFT*); ~ **image** *n* DATA image binaire *f*; ~ **modulation** *n* DATA modulation binaire *f*; ~ **multiplier** *n* DATA multiplicateur binaire *m*; ~ **numeral** *n* COMP, DATA nombre binaire *m*; ~ **phase-shift keying** *n* (*BPSK*) RADIO, SIG modulation par déplacement de phase bivalente *f* (*MDPB*); ~ **rate** *n* COMP, DATA, TELEP, TRANS débit binaire *m*; ~ **search** *n* DATA recherche binaire *f*; ~ **search procedure** *n* DATA procédure de recherche binaire *f*, procédure de recherche dichotomique *f*; ~ **sequence** *n* DATA séquence binaire *f*; ~ **serial signaling rate** *AmE*, ~ **serial signalling rate** *BrE* *n* SIG débit binaire série *m*; ~ **signal** *n* SIG signal binaire *m*; ~ **synchronous communication** *n* (*BSC*) DATA transmission binaire synchrone *f*; ~ **system** *n* COMP système binaire *m*; ~ **tariff system** *n* TELEP système de tarification binaire *m*

Informations complètes sur les variantes américaines et anglaises suivies de codes géographiques — *British English and American English spelling variants are given in full and labelled accordingly*

bind *n* DATA rattachement *m*

binomial: ~ **distribution** *n* GEN *statistics* distribution binomiale *f*

Abréviation du terme donnée à l'entrée pour la forme intégrale — *Cross-references from abbreviations to their full form are shown for both the French and English translation*

BIOS *abbr* (*basic input/output system*) DATA BIOS (*système d'exploitation des entrées/sorties*)

BIP *abbr* (*bit interleaved parity*) DATA parité d'entrelacement des bits *f*

Using the dictionary

Range of coverage

This is a specialist French–English/English–French dictionary that covers the whole range of telecommunications and the scientific knowledge that underlies it. It contains a broad base of terminology drawn from traditional areas of technology such as Parts & Materials, Radio-based Technologies and Power Systems, but also includes the vocabulary of newly prominent subject areas such as Computing, Testing Procedures and Terminal & Peripheral Apparatus.

Selection of terms

We have aimed to include the essential vocabulary of each subject area, and the material has been checked by leading subject experts to ensure that both the English and the French terms are accurate and current, that the translations are valid equivalents, and that there are no gaps in coverage.

Placement of terms

All terms are ordered alphabetically. This is also the policy for hyphenated compounds.

Stoplists

All terms are entered at their first element except where that element is an article, preposition, conjunction, pronoun or other delexicalized word. In such cases, the term is entered at its next valid element.

Open compounds consisting of more than one element are listed at their first element. In these nested listings, the simple form is replaced by a swung-dash (~). For example:

orbital: ~ **decay** *n* RADIO dégénérescence d'une orbite *f*; ~ **elements** *n pl* RADIO éléments d'une orbite *m pl*; ~ **guidance** *n* RADIO guidage orbital *m*; ~ **inclination control** *n* RADIO correction d'inclinaison de l'orbite *f*

When this first element is itself a headword with a technical sense, compounds follow the simple form in alphabetical order. For example:

collecteur *m* ALIM ELEC, ELECTRON collector, commutator; ~ **annulaire de mise à la terre** *m* ALIM ELEC earth ring collector (*BrE*), ground ring collector (*AmE*); ~ **commun** *m* ELECTRON common collector

If the first element is not itself translated, a colon precedes the compounds. For example:

bec: ~ **à déclic** *m* ELECTRON trigger nozzle

Ordering of terms

All terms, including hyphenated compounds, are nested according to part of speech. For example:

magneto[1]: ~**-optic** *adj* (*MO*) COMP, ELECT magnéto-optique (*MO*)
magneto[2]: ~**-optic effect** *n* ELECT effet magnéto-optique *m*; ~**-optical device** *n* DATA appareil magnéto-optique *m*

Within nests, articles, pronouns and prepositions are ignored in determining the sequence of nested open compounds. For example:

indépendance: ~ **du code utilisé** *f* TRANSMIS code transparency; ~ **des données** *f* ORDINAT data independence; ~ **de la séquence des bits** *f* ORDINAT bit sequence independence

Abbreviations and acronyms written in upper case appear after entry terms of the same form written in lower case. For example:

ns *abbr* (*nanosecond*) TEST ns (*nanoseconde*)
NS *abbr* (*network service*) DATA, TELEP service de réseau *m*

Terms containing figures and symbols are alphabetized according to the usual expansion when written out in full. For example:

USB *abbr* (*upper sideband*) RADIO, TRANS BL sup. (*bande latérale supérieure*)
USB1 *abbr* (*unscrambled binary ones*) DATA uns binaires non brouillés *m pl*
use *n* GEN utilisation *m*, *working* fonctionnenment *m*

Homographs

Every term is accompanied by a label indicating its part of speech. For a complete list of these labels and their expansions, please see pages xxvii.

When terms beginning with the same element fall into two or more part-of-speech categories, the different categories are distinguished by a raised number immediately following the headword. The sequence is adjective, adverb, noun, verb, followed by less frequent parts of speech. For example:

fail[1]: **~-safe** *adj* TEST à sûreté intégrée, de sécurité intrinsèque
fail[2]: **~-safe** *n* TELEP protection contre les dérangements *f*; **~-safe concept** *n* TEST concept de sécurité intrinsèque *m*; **~ verdict** *n* DATA échec verdict *m*
fail[3] *vi* COMP, GEN, TELEP, TEST, TRANS échouer

Ordering of translations

Every term is accompanied by one or more labels indicating the technological area in which it is used. For a complete list of these labels and their expansions, please see pages xxvii–xxviii.

Where the same term is used in more than one technological area, multiple labels are given as appropriate. These labels appear in alphabetical order. Where a term has the same translation in more than one technological area, this translation is given after the sequence of labels. For example:

volume *n* COMP, ELECT, POWER, RADIO volume *m*

Where a term has different translations according to the technological area in which it is used, the appropriate translation is given after each label or set of labels. For example:

discriminateur *m* COMMUT, DONNÉES discriminator, ÉLECTRON discriminator, restrictor, TÉL call-barring equipment

Supplementary information

The gender is given for every French noun term in the French–English section as well as for every French translation that is a noun in the English–French section. In the case of compound terms this is the gender of the noun head. For example:

analyse *f* PÉRIPH scanning; **~ de l'acheminement** TÉL route analysis; **~ d'arbre de défaillance** TESTS fault-tree analysis; **~ d'arbre des pannes** TESTS fault-tree analysis

In many cases additional data is given about a term in order to show how it is used. Such contextual information can be:

(a) the typical subject or object of a verb, for example:

interrompre *vt* ALIM ÉLEC *courant* cut off, break

(b) typical nouns used with an adjective, for example:

lisible *adj* GÉNÉRAL, INFORMAT *fichier, données* readable, legible

(c) words indicating the reference of a noun, for example:

intersection *f* ÉLECTRON *élements logiques* conjunction

(d) a narrower subject area than indicated by the label, for example:

epoxide *n* PARTS *plastics* epoxyde *m*

When various different translations apply in the same subject area, contextual information is also used to show which translation is appropriate in different circumstances. For example:

fault *n* POWER, TELEP, TRANS dérangement *m*, *in machine* anomalie *f*, défaut *m*, *mistake* faute *f*, *breakdown* panne *f*

Cross-references

All terms not universally applicable are accompanied by a geographical label. In the French–English section of the dictionary coverage includes Canadian terms. In the English–French section, British, American and Canadian terms are covered, and these are differentiated by regional labels. In the case of lexical variants, full information including translations and cross-references to other regional forms is given at each entry. For example:

railroad: **~ radio system** *n AmE* (*cf railway radio system BrE*) RADIO système de radio communicationspour les chemins de fer *m*, système radio ferroviaire *m*
railway: **~ radio system** *n BrE* (*cf railroad radio system AmE*) RADIO système de radiocommunications pour les chemins de fer *m*, système radio ferroviaire *m*

Geographical variants, both spelling and lexical, are given in full when they are translations. For example:

bouche: **~ d'incendie** *f* COMPOSANTS fire hydrant (*BrE*), fireplug (*AmE*)

Both abbreviations and their full forms are entered in the main body of the dictionary in alphabetical sequence. Full information – including translations and cross-references to the full form or abbreviation – is given.

Where an abbreviation in the source language translates to an abbreviation in the target language, the full form in each case is given after the abbreviation in brackets and italics. For example:

MPIL *abbr* (*maximum permissible interference level*) DATA NMAB (*niveau maximal admissible de brouillage*)

If a source language abbreviation does not translate to a target language abbreviation, the translation of the abbreviation is in roman.

For example:

MAN *abbr* (*metropolitan area network*) TRANS réseau de zone urbaine *m*

Comment utiliser ce dictionnaire

Domaines traités

Ce dictionnaire est un ouvrage spécialisé français–anglais/anglais–français qui couvre l'ensemble des télécommunications et des connaissances scientifiques sur lesquelles elles reposent. Il comporte une importante base terminologique extraite des disciplines traditionnelles, comme les composants & matériaux, les technologies radio et les alimentations électriques, ainsi que des nouvelles technologies comme l'informatique, les procédures d'essai et les équipements terminaux & périphériques.

Sélection des termes

Les termes qu'il contient se rapportent au vocabulaire de base de chaque discipline, et ont été minutieusement vérifiés par des spécialistes de haut niveau, chargés de veiller à l'exactitude et à l'actualité des termes anglais et français, à la parfaite correspondance des termes, et à la cohérence de l'ensemble.

Classement des termes

Tous les termes sont classés par ordre alphabétique. Les termes composés apparaissent sous leur premier élément, quelle que soit la structure sémantique de l'ensemble.

Termes ignorés

Tous les termes apparaissent sous leur premier élément, sauf dans le cas où cet élément est un article, une préposition, une conjonction, un pronom ou tout autre terme depourvu de valeur lexicale. Lorsque de tels cas se présentent, le terme apparaît sous l'élément significatif le plus proche.

Les locutions ou les syntagmes figurent sous le premier élément. Dans les listes imbriquées, la forme simple est remplacée par un 'tilde' (~). Exemple:

orbital: ~ **decay** *n* RADIO dégénérescence d'une orbite *f*; ~ **elements** *n pl* RADIO éléments d'une orbite *m pl*; ~ **guidance** *n* RADIO guidage orbital *m*; ~ **inclination control** *n* RADIO correction d'inclinaison de l'orbite *f*

Lorsque ce premier élément est un mot-vedette ayant un sens technique, les mots composés suivent la forme simple et sont classés par ordre alphabétique. Exemple:

collecteur *m* ALIM ELEC, ELECTRON collector, commutator; ~ **annulaire de mise à la terre** *m* ALIM ELEC earth ring collector (*BrE*), ground ring collector (*AmE*); ~ **commun** *m* ELECTRON common collector

Si le premier élément n'est pas traduit, ses composants sont précédés de deux-points. Exemple:

bec: ~ **à déclic** *m* ELECTRON trigger nozzle

Ordre des termes

Les locutions et termes comportant des traits-d'union sont classés selon leur catégorie grammaticale. Exemple:

magneto^1: **~-optic** *adj* (*MO*) COMP, ELECT magnéto-optique (*MO*)
magneto^2: **~-optic effect** *n* ELECT effet magnéto-optique *m*; **~-optical device** *n* DATA appareil magnéto-optique *m*

Les articles, prépositions et pronoms ne sont pas pris en compte lors du tri. Exemple:

indépendance: ~ **du code utilisé** *f* TRANSMIS code transparency; ~ **des données** *f* ORDINAT data independence; ~ **de la séquence des bits** *f* ORDINAT bit sequence independence

Les abréviations et acronymes en majuscules figurent après les termes de même forme en minuscules. Exemple:

ns *abbr* (*nanosecond*) TEST ns (*nanoseconde*)
NS *abbr* (*network service*) DATA, TELEP service de réseau *m*

Les termes contenant des chiffres ou des symboles sont classés comme s'ils étaient écrits en lettres. Exemple:

USB *abbr* (*upper sideband*) RADIO, TRANS BL sup. (*bande latérale supérieure*)
USB1 *abbr* (*unscrambled binary ones*) DATA uns binaires non brouillés *m pl*
use *n* GEN utilisation *m*, *working* fonctionnement *m*

Homographes

Chaque terme est accompagné d'une abréviation indiquant sa fonction grammaticale. La liste complète de ces abréviations figure page xxvii.

Un terme ayant plusieurs fonctions grammaticales figure sous une entrée séparée pour chaque fonction. Ces entrées se distinguent par un chiffre en exposant qui suit immédiatement le mot-vedette. La séquence est la suivante: adjectif, adverbe, nom, verbe, suivie des fonctions moins fréquentes. Exemple:

fail^1: **~-safe** *adj* TEST à sûreté intégrée, de sécurité intrinsèque
fail^2: **~-safe** *n* TELEP protection contre les dérangements *f*; **~-safe concept** *n* TEST concept de sécurité intrinsèque *m*; **~ verdict** *n* DATA échec verdict *m*
fail^3 *vi* COMP, GEN, TELEP, TEST, TRANS échouer

Ordre des traductions

Chaque terme est accompagné d'une ou plusieurs abréviations indiquant le domaine auquel il se rapporte. La liste complète de ces abréviations figure pages xxvii–xxviii.

Lorsqu'un terme est utilisé dans plusieurs disciplines, il est suivi des abréviations appropriées, en ordre alphabétique. Lorsqu'un terme est traduit de la même manière dans différentes disciplines, la traduction figure après la liste des abréviations. Exemple:

volume *n* COMP, ELECT, POWER, RADIO volume *m*

Lorsqu'un terme peut se traduire de différentes façons, selon les disciplines, sa traduction figure après chaque abréviation ou série d'abréviations. Exemple:

discriminateur *m* COMMUT, DONNÉES discriminator, ÉLECTRON discriminator, restrictor, TÉL call-barring equipment

Informations complémentaires

Le genre de chaque nom est indiqué dans la partie français–anglais, ainsi que pour chaque traduction française s'il s'agit d'un nom dans la partie anglais–français. Pour les expressions ou termes composés, le genre est celui du terme dans son ensemble (c'est-à-dire de son mot-vedette). Exemple:

analyse *f* PÉRIPH scanning; **~ de l'acheminement** TÉL route analysis; **~ d'arbre de défaillance** TESTS fault-tree analysis; **~ d'arbre des pannes** TESTS fault-tree analysis

Dans de nombreux cas, des informations complémentaires expliquent l'utilisation du terme concerné. Exemple:

(a) le sujet ou l'objet typique d'un verbe:

interrompre *vt* ALIM ÉLEC *courant* cut off, break

(b) les noms typiques utilisés avec un adjectif:

lisible *adj* GÉNÉRAL, INFORMAT *fichier, données* readable, legible

(c) les mots indiquant la référence d'un nom:

intersection *f* ÉLECTRON *élements logiques* conjunction

(d) un domaine plus limité que celui indiqué par l'abréviation:

epoxide *n* PARTS *plastics* epoxyde *m*

Lorsqu'un terme peut donner lieu à différentes traductions, des informations contextuelles indiquent également la traduction appropriée. Exemple:

fault *n* COMP, ELECT défaut *m*, POWER, TELEP, TRANS dérangement *m*, *in machine* anomalie *f*, défaut *m*, *mistake* faute *f*, *breakdown* panne *f*

Renvois

Tous les termes non universels sont accompagnés d'une mention géographique. Dans la partie français–anglais, les variantes linguistiques (entre le français de France et du Canada) sont toujours indiquées. Dans la partie anglais–français, les variantes linguistiques (entre l'anglais britannique, l'anglais américain et l'anglais canadien) sont également accompagnées de la mention géographique qui convient.

En cas de variantes lexicales, les informations les concernant et notamment les traductions et renvois à d'autres formes régionales sont fournies systématiquement à chaque entrée. Exemple:

railroad: **~ radio system** *n AmE* (*cf railway radio system BrE*) RADIO système de radiocommunications pour les chemins de fer *m*, système radio ferroviaire *m*
railway: **~ radio system** *n BrE* (*cf railroad radio system AmE*) RADIO système de radiocommunications pour les chemins de fer *m*, système radio ferroviaire *m*

Les variantes géographiques, tant orthographiques que lexicales, sont indiquées intégralement lorsqu'il s'agit de traductions. Exemple:

bouche: ~ **d'incendie** *f* COMPOSANTS fire hydrant (*BrE*), fireplug (*AmE*)

Les abréviations et leurs formes intégrales figurent dans le corps principal du dictionnaire, classées par ordre alphabétique. Les informations complètes – et notamment leurs traductions et les renvois à leur abréviation ou à leur forme intégrale, le cas échéant – sont indiquées.

Lorsque qu'une abréviation dans la langue source est traduite par une abréviation dans la langue cible, la forme développée est indiquée dans chaque cas entre parenthèses en italique après l'abréviation. Exemple:

MPIL *abbr* (*maximum permissible interference level*) DATA NMAB (*niveau maximal admissible de brouillage*)

Si une abréviation dans la langue source n'est pas traduite par une abréviation dans la langue cible, la traduction de l'abréviation est en caractères romains. Exemple:

MAN *abbr* (*metropolitan area network*) TRANS réseau de zone urbaine *m*

Abbreviations used in this dictionary/Abréviations utilisées dans ce dictionnaire

Parts of speech/Catégories grammaticales

abbr	abbreviation	abréviation
abrév	abréviation	abbreviation
adj	adjective	adjectif
adv	adverb	adverbe
f	feminine	féminin
f pl	feminine plural	féminin pluriel
loc	locution	phrase
m	masculine	masculin
m pl	masculine plural	masculin pluriel
n	noun	nom
n pl	noun plural	nom pluriel
phr	phrase	locution
pref	prefix	préfixe
prep	preposition	préposition
vi	intransitive verb	verbe intransitif
v réfl	verbe réfléchi	reflexive verb
vt	transitive verb	verbe transitif

Geographic codes/Codes géographiques

AmE	American English	anglais américain
BrE	British English	anglais britannique
Can	Canadian English/French	anglais/français canadien

Subject area labels/Domaines

APPL	Applications & Services	Applications & services
CLIM	Climate Control	Climatisation
COMP	Computing	Informatique
CONTR	Control Technology	Contrôle & automatisation
DATA	Data	Structure & traitement des données
ELECT	Electronics	Electronique
GEN	General Telecommunications	Général
PARTS	Parts and Materials	Composants & matériaux
PER APP	Terminal & Peripheral Apparatus	Équipements terminaux & périphériques
POWER	Power Systems	Alimentations électriques
RADIO	Radio-based Technologies	Technologies radio

SIG	Signalling	Structure & traitement des signaux
SWG	Switching	Commutation
TELEP	Telephony & Telegraphy	Téléphonie & télégraphie
TV	Television & Video	Télévision & vidéo
TEST	Testing Procedures	Procédures d'essai
TRANS	Transmission	Transmissions

ALIM ÉLEC	Alimentations électriques	Power Systems
APP SERV	Applications & services	Applications & Services
CLIMAT	Climatisation	Climate Control
COMMUT	Commutation	Switching
COMPOSANTS	Composants & matériaux	Parts & Materials
CONTRÔLE	Contrôle & automatisation	Control Technology
DONNÉES	Structure & traitement des données	Data
ÉLECTRON	Électronique	Electronics
GÉNÉRAL	Télécommunications générales	General Telecommunications
INFORMAT	Informatique	Computing
PÉRIPH	Équipements terminaux & périphériques	Terminal & Peripheral Apparatus
RADIO	Technologies radio	Radio-based Technologies
SIGNAL	Structure & traitement des signaux	Signalling
TÉL	Téléphonie & télégraphie	Telephony & Telegraphy
TESTS	Procédures d'essai	Testing Procedures
TRANSMIS	Transmissions	Transmission
TV	Télévision & vidéo	Television & Video

Registered trademarks

Note that neither the presence nor the absence of the symbol ® should be regarded as affecting the legal status of any trademark.

Marques déposées

Veuillez noter que l'absence ou la présence de la mention ® est sans effet sur le statut légal d'une marque déposée, quelle qu'elle soit.

FRANÇAIS–ANGLAIS
FRENCH–ENGLISH

A

AA *abrév* (*assistance annuaire*) APP SERV, TÉL DQ (*directory enquiries BrE*), directory information (*AmE*), directory assistance

AAMT *abrév* (*appareil automatique de mesure de transmission*) COMMUT ATME (*automatic transmission-measuring equipment*)

AARE *abrév* (*unité de données du protocole d'application réponse*) DONNÉES, TÉL AARE (*A-associate response*)

AARQ *abrév* (*unité de données du protocole d'association demande*) DONNÉES, TÉL AARQ (*A-associate request*)

abaissement: **~ de fréquence** *m* TRANSMIS frequency down-conversion; **~ de la tension** *m* ALIM ÉLEC step-down transformation

abaisser *vt* GÉNÉRAL lower, INFORMAT *performances* downgrade; **~ par transformateur** ALIM ÉLEC *tension* step down

abaisseur: **~ de fréquence à double changement de fréquence** *m* TRANSMIS double-conversion converter

abandon *m* INFORMAT abort

aberration: **~ optique** *f* TRANSMIS optical aberration; **~ sphérique** *f* ÉLECTRON, TRANSMIS spherical aberration

abonné *m* DONNÉES subscriber, user, GÉNÉRAL user, INFORMAT, RADIO, TÉL, TV subscriber, user; **~ à bord** RADIO *service maritime mobile* on-board subscriber; **~ d'un commutateur privé** TÉL PBX subscriber; **~ itinérant** RADIO roamer, roaming subscriber, TÉL roamer; **~ prioritaire** TÉL priority subscriber; **~ professionnel** RADIO business subscriber; **~ au téléphone** TÉL telephone subscriber

abonnement *m* GÉNÉRAL subscription, PÉRIPH line rental, RADIO, TÉL, TV subscription; **~ annuel** APP SERV, GÉNÉRAL annual subscription; **~ mensuel** APP SERV, GÉNÉRAL, INFORMAT monthly subscription; **~ multizone** RADIO multizone subscription; **~ principal** TÉL basic subscription; **~ téléphonique** TÉL telephone subscription

abonner: **s'~** *v réfl* GÉNÉRAL subscribe

aboutement *m* TESTS patch-out

abouter *vt* GÉNÉRAL butt-joint

aboutir *vi* TÉL, TRANSMIS terminate; **~ à un poste** TÉL, TRANSMIS *pour un appel* terminate to a station

aboutissement: **~ de l'appel** *m* TÉL completion of call

abrasif[1] *adj* GÉNÉRAL abrasive

abrasif[2] *m* GÉNÉRAL abrasive

ABRT *abrév* (*unité de données du protocole d'application A-ABANDON*) DONNÉES ABRT (*A-ABORT application protocol data unit*)

absence: **~ d'alarme** *f* TESTS no-alarm; **~ d'échec** *f* TESTS no-failure

absorption *f* GÉNÉRAL absorption; **~ acoustique** TÉL acoustic absorption; **~ d'énergie** ÉLECTRON energy absorption; **~ hyperfréquence** RADIO microwave absorption; **~ résonnante** ÉLECTRON resonant absorption; **~ d'un signal** SIGNAL consumption of a signal

accélérateur *m* ÉLECTRON accelerator, INFORMAT booster; **~ promoteur** COMPOSANTS *matières plastiques* accelerator, promotor

accélération: **~ radiale** *f* INFORMAT radial acceleration

acceptation *f* INFORMAT acceptance; **~ d'adresses étendues** DONNÉES long-address acceptance; **~ d'appel** TÉL call acceptance, call accepted; **~ automatique d'appels** TÉL automatic call acceptance; **~ de débordement** COMMUT, TÉL overflow accept (*OA*); **~ de la sélection rapide** TÉL fast-select acceptance; **~ de la taxation à l'arrivée** APP SERV, TÉL collect-call acceptance (*AmE*), collect-calling acceptance (*AmE*), reverse-charge acceptance (*BrE*), reverse-charging acceptance (*BrE*); **~ de la taxe à l'arrivée** APP SERV, TÉL collect-call acceptance (*AmE*), collect-calling acceptance (*AmE*), reverse-charge acceptance (*BrE*), reverse-charging acceptance (*BrE*); **~ de transaction pour remise** DONNÉES input transaction accepted for delivery (*ITD*)

accepté *adj* GÉNÉRAL approved

accepteur *m* DONNÉES, ÉLECTRON *semiconducteurs* acceptor; **~ de l'association** DONNÉES association responder

accès[1] *m* GÉNÉRAL access, RADIO port; **~ d'administration** DONNÉES administration port; **~ aléatoire** INFORMAT random access; **~ aléatoire à la mémoire** INFORMAT memory random access; **~ d'arrivée** DONNÉES inlet access; **~ de base** DONNÉES, TÉL *ISDN*, TRANSMIS basic access; **~ de base RNIS** TÉL basic rate interface; **~ à des bases de données distantes** DONNÉES remote database access (*RDA*); **~ bidirectionnel à l'opérateur** TÉL operator office (*OO*); **~ clandestin** INFORMAT piggyback entry; **~ code multiple** DONNÉES multiple-code access; **~ conditionnel** TV conditional access; **~ au débit primaire** TÉL primary rate access (*PRA*); **~ du débit primaire du RNIS** TÉL ISDN primary rate access; **~ à la demande** DONNÉES access on demand, semipermanent access; **~ de départ** COMMUT outlet access; **~ de dépôt indirect** DONNÉES indirect submission port; **~ direct** DONNÉES, INFORMAT, PÉRIPH direct access; **~ direct à l'arrivée** DONNÉES, TÉL direct dialing-in (*AmE*), direct dialling-in (*BrE*); **~ direct à l'interurbain par numérotation** TÉL trunk dialing (*AmE*), trunk dialling (*BrE*); **~ direct à la mémoire** INFORMAT, PÉRIPH direct memory access (*DMA*); **~ direct à un poste** TÉL direct inward dialing (*AmE*) (*DID*), direct inward dialling (*BrE*) (*DID*); **~ direct au réseau** TÉL direct outward dialing (*AmE*) (*DOD*), direct outward dialling (*BrE*) (*DOD*); **~ à distance** DONNÉES, TÉL remote access; **~ du document distant** DONNÉES, TÉL remote document access; **~ entrant** DONNÉES incoming access (*IA*); **~ d'entrée** PÉRIPH line-in port; **~-fiable-imposé-mts** DONNÉES mts-forced-reliable-access; **~-fiable-ms** DONNÉES ms-reliable-access; **~-fiable-mts** DONNÉES mts-reliable-access; **~ aux fichiers** DONNÉES, INFORMAT file access; **~ homologué** INFORMAT registered access; **~ instantané** INFORMAT, TRANSMIS instantaneous access; **~ intégré** TÉL integrated access; **~ interdit** TÉL access barred, TRANSMIS access denied; **~ limité** GÉNÉRAL limited access; **~ en mode ascendant** DONNÉES Dial-up Internet Protocol, Dial-up IP; **~ multiple** GÉNÉRAL multiple access; **~ multiple aléatoire** INFORMAT random multiple access; **~ multiple avec affectation à la demande** SIGNAL demand-assignment multiple access (*DAMA*); **~ multiple avec écoute de la porteuse et détection de collisions** (*CSMA/CD*) RADIO, SIGNAL, TÉL, TRANSMIS carrier-sense multiple access with collision detection (*CSMA/CD*); **~ multiple par détection de la porteuse** (*AMDP*) RADIO, SIGNAL, TÉL, TRANSMIS carrier-sense multiple access (*CSMA*); **~ multiple par différence de code** (*AMDC*) RADIO, SIGNAL, TÉL, TRANSMIS code-division multiple access (*CDMA*); **~ multiple par impulsions d'adresse** (*AIMA*) RADIO pulse-address multiple access (*PAMA*); **~ multiple par répartition de fréquence** (*AMRF*) RADIO, SIGNAL, TÉL, TRANSMIS frequency-division multiple access (*FDMA*); **~ multiple par répartition en fréquences orthogonales** (*AMRFO*) RADIO, SIGNAL, TÉL, TRANSMIS orthogonal frequency-division multiple access (*ODFMA*);

~ **multiple à répartition dans le temps** (*AMRT*) RADIO, SIGNAL, TÉL, TRANSMIS time-division multiple access (*TDMA*); ~ **numérique intégré** (*IDA*) DONNÉES integrated digital access (*IDA*); ~ **aux octets de surdébit** DONNÉES overhead access (*OHA*); ~ **par carte intelligente** APP SERV smartcard access; ~ **par carte à mémoire** APP SERV smartcard access; ~ **par carte à microcircuit** APP SERV smartcard access; ~ **par carte à microprocesseur** APP SERV smartcard access; ~ **par mot-clé** INFORMAT keyword access; ~ **primaire** DONNÉES primary rate interface, TÉL primary access, primary rate interface; ~ **primaire RNIS** DONNÉES, TÉL primary rate interface ISDN; ~ **privé par réseau commuté** COMMUT private dial-up port; ~ **rapide** INFORMAT fast access; ~ **de recherche** DONNÉES retrieval port; ~ **au réseau** DONNÉES, RADIO, TRANSMIS network access; ~ **au réseau AUTOVON** TÉL AUTOVON trunk access; ~ **restreint** TÉL restricted access; ~ **au RNIS** TÉL ISDN access; ~ **sécurisé** DONNÉES secure access; ~ **semipermanent** DONNÉES semipermanent access; ~ **séquentiel** DONNÉES sequential access, INFORMAT sequential access, serial access; ~ **aux services supplémentaires** TÉL access to supplementary services; ~ **sortant** DONNÉES outgoing access (*OA*); ~ **spécialisé** TRANSMIS dedicated port; ~ **spécialisé pour les liaisons de données** DONNÉES datalink specialized access; ~ **au surdébit** DONNÉES overhead access (*OHA*); ~ **au tarif de base** TÉL basic rate access (*BRA*); ~ **de voie de service téléphonique** TÉL, TRANSMIS voice-order wire port

accès[2]: **à ~ multiple** *loc* RADIO multiple-access

accessibilité *f* GÉNÉRAL accessibility, TÉL availability

accord *m* INFORMAT *musique* chord; ~ **d'antenne** RADIO antenna tuning; ~ **de basse automatique** INFORMAT auto bass chord; ~ **cadre** GÉNÉRAL memorandum of understanding (*MOU*), outline agreement, framework agreement; ~ **de la CAF** RADIO AFC hold-in; ~ **d'échange** GÉNÉRAL communications agreement; ~ **électronique** ÉLECTRON electronic tuning; ~ **de fréquence** RADIO frequency tuning; ~ **optique** TRANSMIS optical tuning; ~ **par réluctance** RADIO permeability tuning, reluctance tuning; ~ **par varactor** ÉLECTRON varactor tuning; ~ **par variation de perméabilité** RADIO permeability tuning; ~ **en phase** ÉLECTRON phase tuning; ~ **silencieux** RADIO squelch

accorder *vt* GÉNÉRAL tune, RADIO *autorisation* grant, TÉL *autorisation* hook up, TV *autorisation* grant

accords: ~ **de prix** *m pl* APP SERV price-fixing

accouplement: ~ **d'arbres** *m* COMPOSANTS shaft coupling; ~ **à baïonnette** *m* ÉLECTRON bayonet coupling

accréditation *m* TESTS *qualité* accreditation

accrochage *m* COMPOSANTS singing, ÉLECTRON capture, GÉNÉRAL *acoustique* feedback instability, RADIO squealing; ~ **automatique** RADIO automatic acquisition; ~ **de cible** RADIO target acquisition

accroissement: ~ **du trafic** *m* TÉL traffic growth

accumulateur *m* ALIM ÉLEC accumulator, battery, storage battery, ÉLECTRON battery; ~ **négatif** ALIM ÉLEC negative battery; ~ **au plomb** ALIM ÉLEC lead accumulator, lead storage battery; ~ **positif** ALIM ÉLEC positive battery; ~ **rechargeable** ALIM ÉLEC rechargeable battery

accumulation: ~ **des porteurs** *f* COMPOSANTS, ÉLECTRON hole storage; ~ **des porteurs de charge** *f* COMPOSANTS, ÉLECTRON carrier storage; ~ **des trous** *f* COMPOSANTS, ÉLECTRON hole storage

accusé: ~ **de réception** *m* (*AR*) DONNÉES, SIGNAL, TÉL acknowledgement (*ACK*); ~ **de réception d'appel** *m* TÉL call acknowledgement; ~ **de réception électronique** *m* RADIO handshake, handshaking; ~ **de réception de message d'entrée** *m* DONNÉES input message acknowledgement (*IMA*); ~ **de réception négatif** *m* (*NACK*) RADIO negative acknowledgement (*NACK*); ~ **de réception non-numéroté** *m* TÉL unnumbered acknowledgement (*UA*); ~ **de réception technique** *m* DONNÉES technical acknowledgement

accuser: ~**-réception de** *vt* APP SERV, DONNÉES, TÉL *signal* acknowledge

achat: ~ **à domicile** *m* GÉNÉRAL, TÉL, TV armchair shopping

acheminement *m* SIGNAL, TÉL, TRANSMIS routing; ~ **adaptatif** SIGNAL, TÉL, TRANSMIS adaptive path-routing, adaptive routing; ~ **automatique sur voie secondaire** SIGNAL, TÉL, TRANSMIS automatic alternative routing (*AAR*); ~ **détourné** SIGNAL, TÉL, TRANSMIS alternative routing, second-attempt routing; ~ **détourné automatique** SIGNAL, TÉL, TRANSMIS automatic alternative routing (*AAR*); ~ **direct** SIGNAL, TÉL, TRANSMIS direct routing; ~ **hiérarchique** SIGNAL, TÉL, TRANSMIS hierarchical routing; ~ **des messages de signalisation** SIGNAL, TÉL, TRANSMIS signaling-message routing (*AmE*), signalling-message routing (*BrE*); ~ **normal** SIGNAL, TÉL, TRANSMIS normal routing; ~ **normal de signalisation** SIGNAL, TÉL, TRANSMIS normal routing of signaling (*AmE*), normal routing of signalling (*BrE*); ~ **par voie détournée** SIGNAL, TÉL, TRANSMIS second-attempt routing; ~ **réseau** DONNÉES, SIGNAL, TÉL, TRANSMIS network routing; ~ **retour** COMMUT, SIGNAL, TÉL, TRANSMIS reverse routing; ~ **section par section** SIGNAL, TÉL, TESTS, TRANSMIS link-by-link traffic routing; ~ **sémaphore** SIGNAL, TÉL, TRANSMIS signaling routing (*AmE*), signalling routing (*BrE*); ~ **de la signalisation** SIGNAL, TÉL, TRANSMIS signaling routing (*AmE*), signalling routing (*BrE*); ~ **de signalisation de secours** SIGNAL, TÉL, TRANSMIS alternative routing of signaling (*AmE*), alternative routing of signalling (*BrE*); ~ **du trafic** SIGNAL, TÉL, TESTS, TRANSMIS traffic routing; ~ **en transit** SIGNAL, TÉL, TRANSMIS transit routing

acheminer *vt* ÉLECTRON dispatch, GÉNÉRAL, TÉL, TRANSMIS route

ACID *abrév* (*atomicité, cohérence, isolation et durabilité*) DONNÉES ACID (*atomicity, consistency, isolation and durability*)

ACN *abrév* (*Association canadienne de normalisation*) GÉNÉRAL CSA (*Canadian Standards Association*)

acoustique[1] *adj* GÉNÉRAL acoustic

acoustique[2] *f* GÉNÉRAL acoustics; ~ **de salle** TÉL room acoustics

acousto-optique *adj* ÉLECTRON acousto-optical (*AO*)

acquéreur *m* TÉL subscriber equipment keeper

acquisition *f* GÉNÉRAL acquisition, capture; ~ **en boucle ouverte** ALIM ÉLEC, CONTRÔLE open-loop acquisition; ~ **de connaissance** APP SERV, INFORMAT knowledge acquisition; ~ **et réception en régime permanent** DONNÉES acquisition and steady-state reception (*ASSR*); ~ **et synchronisation de trame** *f pl* TRANSMIS frame acquisition and synchronization; ~ **en mode de recherche** DONNÉES search mode acquisition (*SMA*); ~ **d'un repère terrestre** RADIO earth acquisition; ~ **vidéo** INFORMAT video acquisition

ACT *abrév* (*affaiblissement de conversion transversale*) DONNÉES, TÉL TCL (*transverse conversion loss*)

actif *adj* GÉNÉRAL active, INFORMAT hot, TÉL asserted

action *f* GÉNÉRAL action; ~ **automatique** DONNÉES auto-action; ~ **automatique générale** DONNÉES general auto-action; ~ **en deux étapes** CONTRÔLE two-step action; ~ **d'entité fonctionnelle** DONNÉES functional entity action (*FEA*); ~ **par échantillonnage et maintien** CONTRÔLE sample and hold action; ~ **par étapes** CONTRÔLE step action; ~ **par modulation** CONTRÔLE modulating action; ~ **par tout ou peu** CONTRÔLE high-low action; ~ **en plusieurs étapes** CONTRÔLE multistep action; ~ **progressive** CONTRÔLE *commande automatique* continuous action; ~ **proportionnelle** CONTRÔLE proportional action; ~ **réciproque** GÉNÉRAL interplay; ~ **tunnel** COMPOSANTS tunnel action

actionnement *m* GÉNÉRAL activation, INFORMAT actuation; **~ du crochet commutateur** COMMUT, TÉL hook flash
activateur *m* RADIO activating unit
activation *f* GÉNÉRAL activation, initialization, INFORMAT, RADIO, TESTS activation
activer *vt* GÉNÉRAL activate, initialize, trigger, INFORMAT, RADIO, TESTS, activate; **~ par une impulsion** ÉLECTRON strobe
activité: **~ décentralisée** *f* COMMUT, SIGNAL decentralized operation; **~ optique** *f* TRANSMIS optical activity
actualités *f pl* DONNÉES news
adaptateur *m* (*HDR*) ÉLECTRON header (*HDR*), GÉNÉRAL adaptor; **~ d'anneau** DONNÉES ring adaptor; **~ coupleur** COMPOSANTS gang-mountable adaptor; **~ EGA** (*EGA*) INFORMAT enhanced graphics adaptor (*EGA*); **~ d'essai** TESTS test adaptor; **~ d'impédance** ÉLECTRON balun adaptor; **~ intégré** INFORMAT integrated adaptor; **~ d'interface** INFORMAT, TÉL interface adaptor; **~ de jonction** INFORMAT, TÉL interface adaptor; **~ de ligne** DONNÉES, INFORMAT data adaptor unit; **~ de mesure** (*HDR*) ÉLECTRON header (*HDR*); **~ monocartouche** INFORMAT single-cartridge adaptor; **~ de périphérique** INFORMAT device adaptor; **~ RNIS** TÉL ISDN adaptor; **~ secteur** GÉNÉRAL AC power adaptor; **~ de support** ÉLECTRON, TÉL medium adaptor (*MA*); **~ de terminal** DONNÉES, PÉRIPH, TÉL terminal adaptor (*TA*); **~ de terminal pour le RNIS-LB** TÉL B-ISDN terminal adaptor (*TA-LB*); **~ de test** TESTS test adaptor; **~ à traversée de cloison** ÉLECTRON bulkhead feed-through adaptor
adaptation *f* ÉLECTRON, GÉNÉRAL matching; **~ de conduit d'ordre inférieur** TÉL, TRANSMIS lower-order path adaptation (*LPA*); **~ de conduit d'ordre supérieur** TÉL, TRANSMIS higher-order path adaptation (*HPA*); **~ défectueuse** GÉNÉRAL, TRANSMIS mismatch; **~ d'impédance** RADIO impedance matching; **~ de jonction** GÉNÉRAL adaptation unit; **~ de section** DONNÉES section adaptation (*SA*)
addition: **~ d'harmoniques** *f* TRANSMIS harmonic addition
additionneur *m* DONNÉES, ÉLECTRON adder; **~ analogique** ÉLECTRON analog adder, summer; **~ complet** ÉLECTRON full adder; **~ parallèle** ÉLECTRON parallel adder; **~ en série** ÉLECTRON serial adder
adhérence *f* ÉLECTRON *soudage, collage* adhesion,
adhérer *vi* ÉLECTRON *soudage, collage* adhere
adhésif: **~ anaérobique** *m* ÉLECTRON anaerobic adhesive
administrateur: **~ de réseau** *m* INFORMAT network administrator
administration *f* GÉNÉRAL administration; **~ de destination** GÉNÉRAL country of destination, TÉL administration of destination; **~ d'origine** GÉNÉRAL country of origin, TÉL administration of origin; **~ de réseau** DONNÉES, GÉNÉRAL, TRANSMIS network management; **~ des télécommunications** TÉL telecommunications administration; **~ terminale** TÉL terminal administration; **~ de transit** TÉL, TRANSMIS transit administration
admittance *f* ALIM ÉLEC admittance; **~ en circuit ouvert** ALIM ÉLEC open-circuit admittance; **~ en court-circuit** ALIM ÉLEC short-circuit admittance; **~ effective d'entrée** ALIM ÉLEC, effective input admittance; **~ effective de sortie** ALIM ÉLEC effective output admittance; **~ électrique** ÉLECTRON electrical admittance; **~ d'électrode** ÉLECTRON electrode admittance; **~ d'entrée** ALIM ÉLEC, input admittance; **~ de sortie** ALIM ÉLEC, output admittance
adr: **~ tél.** *abrév* (*adresse télégraphique*) TÉL cable address, telegraphic address
adressage *m* DONNÉES, INFORMAT addressing; **~ calculé** DONNÉES, INFORMAT hash coding, hash function, hashing; **~ différé** INFORMAT deferred addressing; **~ direct** INFORMAT direct addressing; **~ de l'enregistreur** COMMUT register addressing; **~ immédiat** INFORMAT immediate addressing; **~ implicite** INFORMAT implied addressing; **~ indirect** INFORMAT indirect addressing; **~ de position active** (*PAP*) TÉL active-position addressing; **~ à progression automatique** INFORMAT one-ahead addressing; **~ symbolique** INFORMAT symbolic addressing; **~ virtuel** INFORMAT virtual addressing
adresse *f* DONNÉES directory entry, GÉNÉRAL address, TÉL *address* call destination; **~ abrégée** TÉL shortened address; **~ absolue** INFORMAT absolute address; **~ d'acheminement** COMMUT, DONNÉES routing address; **~ d'acheminement retour** COMMUT, DONNÉES reverse routing address; **~ d'appel** RADIO call address; **~ de base** INFORMAT base address; **~ de bloc logique** INFORMAT logical block address (*LBA*); **~ CD-ROM** INFORMAT CD-ROM address; **~ du correspondant** DONNÉES party address; **~ du demandeur** TÉL calling-party address; **~ de destination** TÉL destination address; **~ différentielle** INFORMAT self-relative address; **~ électronique** DONNÉES e-mail address, electronic address; **~ E-mail** DONNÉES e-mail address, electronic address; **~ d'équipement terminal de traitement de données** DONNÉES, PÉRIPH, TÉL data terminal equipment address; **~ figurant dans l'annuaire** DONNÉES directory entry; **~ générique** DONNÉES generic address; **~ indexée** INFORMAT indexed address; **~ indirecte** INFORMAT indirect address; **~ d'installation** TÉL installation address; **~ d'instruction** INFORMAT instruction address; **~ de la ligne appelante** TÉL calling-party address; **~ N** DONNÉES N address; **~ N d'appelant** DONNÉES calling-N address; **~ N d'appelé** DONNÉES called-N address; **~ N en réponse** DONNÉES responding-N address; **~ de NSAP** DONNÉES N-service access-point address; **~ physique** (*AP*) TÉL physical address; **~ du point d'accès au service N** DONNÉES N-service access-point address; **~ de point d'accès au service de réseau** DONNÉES network service access-point address; **~ de point d'accès du service de transport** TÉL, TRANSMIS transport-service access-point address; **~ du point d'accès à des services N** DONNÉES N-service access-point address; **~ du point de rattachement du sous-réseau** DONNÉES subnetwork point of attachment address; **~ de présentation** DONNÉES presentation address; **~ de réacheminement** TÉL redirection address; **~ réelle** INFORMAT effective address, real address; **~ relative** INFORMAT relative address; **~ de réseau** DONNÉES network address; **~ au sol** GÉNÉRAL ground address; **~ de sous-réseau** DONNÉES subnetwork address; **~ symbolique** INFORMAT symbolic address; **~ système** (*AS*) TÉL system address (*SA*); **~ télégraphique** (*adr tél.*) TÉL cable address, telegraphic address; **~ translatable** INFORMAT, TÉL relocatable address; **~ d'usager** TÉL user address; **~ virtuelle** INFORMAT virtual address
adresses: **~ multiples** *f pl* TÉL multiaddress calling
adsorption *f* GÉNÉRAL *physique* adsorption
AECHO *abrév* (*affaiblissement d'écho*) TÉL echo loss, pulse echo attenuation, pulse echo return loss
AEN *abrév* (*affaiblissement équivalent pour la netteté*) TÉL articulation reference equivalent loss, equivalent articulation loss
aérographe *m* GÉNÉRAL airbrush
aérosol *m* RADIO aerosol
AFB *abrév* (*amplificateur à faible bruit*) ÉLECTRON, RADIO, TRANSMIS LNA (*low-noise amplifier*)
affaiblir[1] *vt* ALIM ÉLEC, INFORMAT, RADIO, TRANSMIS attenuate
affaiblir[2]: **s'~** *v réfl* ALIM ÉLEC die away, die down
affaiblissement *m* ALIM ÉLEC attenuation, GÉNÉRAL loss, INFORMAT attenuation, decay, RADIO attenuation, loss, radio fade, TRANSMIS attenuation, loss; **~ adaptable** TÉL adaptive attenuation; **~ d'adaptation** TÉL, TRANSMIS return loss; **~ d'adaptation optique** TÉL, TRANSMIS optical return loss (*ORL*); **~ atmosphérique par temps clair** RADIO clear weather attenuation; **~ en bande passante** TRANSMIS pass band attenuation; **~ de blocage** TÉL suppression loss; **~ en**

boucle ouverte ALIM ÉLEC, CONTRÔLE open-loop loss (*OLL*); ~ **composite** TÉL, TRANSMIS composite loss; ~ **de conversion transversale** (*ACT*) DONNÉES, TÉL transverse conversion loss (*TCL*); ~ **copolaire** RADIO co-polar attenuation; ~ **de courbure** TRANSMIS bending loss; ~ **en demi-boucle** TÉL semiloop loss; ~ **de diffusion** RADIO scattering loss; ~ **de distorsion harmonique d'ordre n** TÉL nth order harmonic distortion; ~ **dû à l'absorption par ciel clair** RADIO clear-sky absorption loss; ~ **dû à l'effet d'écran du bâtiment** RADIO building penetration loss; ~ **dû au pointage d'antenne** RADIO *satellite maritime* antenna-pointing loss; ~ **d'écho** (*AECHO*) TÉL echo loss, pulse echo attenuation, pulse echo return loss; ~ **de l'écho à la réception** TÉL listener echo loss, receive echo loss; ~ **d'emplacement** RADIO site attenuation; ~ **entre antennes isotropes** RADIO basic transmission loss; ~ **entre bornes d'antennes** RADIO system loss; ~ **d'équilibrage pour l'écho** TÉL echo balance return loss; ~ **équivalent pour la netteté** (*AEN*) TÉL articulation reference equivalent loss, equivalent articulation loss; ~ **d'espace libre** RADIO, TRANSMIS *d'une liaison radioélectrique* free-space basic transmission loss, free-space transmission loss; ~ **en espace libre** RADIO, TRANSMIS free-space attenuation; ~ **géométrique** RADIO spreading loss; ~ **global** RADIO, TV *d'une liaison radioélectrique* total loss; ~ **d'intermodulation** TRANSMIS intermodulation attenuation; ~ **intrinsèque de raccordement** TRANSMIS intrinsic junction loss; ~ **linéique** RADIO, TRANSMIS attenuation constant; ~ **minimal admissible** TRANSMIS minimum net loss; ~ **modal différentiel** TRANSMIS differential modal attenuation, differential mode attenuation; ~ **en mode différentiel** TRANSMIS differential mode attenuation; ~ **optique** DONNÉES, TRANSMIS optical loss; ~ **par rapport à l'espace libre** RADIO, TRANSMIS *d'une liaison radioélectrique* loss relative to free space; ~ **par traitement non linéaire** (*ATNL*) TÉL nonlinear processing loss; ~ **de la porteuse** TRANSMIS carrier rejection; ~ **de propagation** RADIO, TRANSMIS *d'une liaison radioélectrique* basic transmission loss; ~ **propre** TRANSMIS inherent loss; ~ **d'un récepteur** RADIO receiver rejection; ~ **à la réception** TÉL receive loss; ~ **de référence** TRANSMIS basic loss; ~ **de réflexion** TÉL, TRANSMIS return loss; ~ **de réflexion monofréquence** RADIO, SIGNAL, TÉL single-frequency return loss; ~ **de réflexion transversale** TÉL transverse return loss (*TRL*); ~ **du signal** SIGNAL signal attenuation; ~ **du signal copolaire** RADIO co-polar attenuation; ~ **des sons** TÉL sound attenuation; ~ **de stabilité** TÉL stability loss; ~ **supplémentaire** RADIO overfading; ~ **sur la fréquence conjuguée** RADIO image rejection ratio; ~ **sur fréquence intermédiaire** RADIO intermediate frequency rejection ratio; ~ **sur la fréquence parasite** RADIO spurious response rejection ratio; ~ **sur les irrégularités** TÉL regularity loss; ~ **sur un trajet** RADIO, TÉL path attenuation; ~ **du système** GÉNÉRAL system loss; ~ **du trajet a-t-b** RADIO path a-t-b, TÉL path a-t-b, transmission loss of a-t-b, TRANSMIS transmission loss of a-t-b; ~ **transductique** TÉL transducer loss; ~ **de transmission** TÉL, TRANSMIS transmission loss; ~ **de transmission directe** TÉL, TRANSMIS through-transmission loss; ~ **de transmission normalisé** (*ATN*) TRANSMIS normalized transmission loss (*NTL*); ~ **de transmission de référence minimal admissible** (*ATRMA*) TRANSMIS minimum permissible basic transmission loss (*MPBTL*); ~ **de transmission pour un trajet radioélectrique** TRANSMIS ray-path transmission loss

affaiblisseur *m* ALIM ÉLEC, ÉLECTRON, INFORMAT, RADIO, TRANSMIS attenuator; ~ **à absorption** TRANSMIS absorptive attenuator; ~ **étalonné** TRANSMIS calibrated attenuator; ~ **fixe** ALIM ÉLEC pad, TRANSMIS fixed attenuator, pad; ~ **en guide d'ondes** RADIO, TRANSMIS waveguide attenuator; ~ **à guillotine** RADIO guillotine attenuator; ~ **hyperfréquence** RADIO microwave attenuator; ~ **à lame** RADIO flap attenuator; ~ **de niveau** INFORMAT level attenuator; ~ **de niveau de sortie** INFORMAT output-level attenuator; ~ **non réciproque** RADIO isolator; ~ **optique** TRANSMIS optical attenuator; ~ **de perte d'insertion** TRANSMIS insertion loss pad; ~ **à piston** RADIO piston attenuator; ~ **réactif** RADIO reactive attenuator; ~ **résistif** ÉLECTRON resistive attenuator; ~ **rotatif** RADIO rotary attenuator

affaissement *m* GÉNÉRAL collapse

affectation *f* GÉNÉRAL, INFORMAT allocation, assignment; ~ **aléatoire** TÉL random allotment; ~ **binaire adaptative** INFORMAT adaptive bit allocation; ~ **de la commande de souffle** INFORMAT breath-controller assign; ~ **à la demande** RADIO demand assignment; ~ **dynamique des ressources** INFORMAT dynamic resource allocation; ~ **d'espace** GÉNÉRAL space allocation; ~ **de mémoire** INFORMAT storage allocation; ~ **de la molette de modulation** INFORMAT modulation-wheel assign; ~ **de paramètre** INFORMAT parameter assign; ~ **de la pédale de soutenu** INFORMAT sustain-pedal assign; ~ **de la pression après enfoncement** INFORMAT after-touch assign; ~ **de sortie** INFORMAT output assign; ~ **de touches** INFORMAT key assign; ~ **de valeur** INFORMAT value assignment

affichage *m* DONNÉES, ÉLECTRON display, GÉNÉRAL read-out, INFORMAT display, read-out, visual read-out, PÉRIPH, RADIO display; ~ **alphagéométrique** TÉL alphageometric display; ~ **alphanumérique** DONNÉES, ÉLECTRON alphanumeric display; ~ **analogique** DONNÉES, ÉLECTRON analog display; ~ **de la cause du renvoi automatique** TÉL call forward reason display; ~ **du coût de la communication** APP SERV charge advice; ~ **à cristaux liquides** DONNÉES, ÉLECTRON, liquid crystal display (*LCD*); ~ **à diodes électroluminescentes** DONNÉES, ÉLECTRON LED display, INFORMAT LED read-out; ~ **de l'état de la station de référence** RADIO reference-status display; ~ **en inverse** INFORMAT reverse display; ~ **lumineux** INFORMAT lighted display, RADIO illuminated display; ~ **matriciel** ÉLECTRON matrix display; ~ **à la mise sous tension** ALIM ÉLEC, PÉRIPH power-on display; ~ **numérique** DONNÉES, ÉLECTRON digital display, digital read-out, numerical display, RADIO numerical display, TÉL digit display; ~ **du numéro appelé** TÉL called-number display; ~ **du numéro composé** TÉL dialed-number display (*AmE*), dialled-number display (*BrE*); ~ **du numéro demandeur** TÉL call display, caller display; ~ **à sept segments** CONTRÔLE seven-segment display

afficher *vt* GÉNÉRAL display, INFORMAT read out, show

afficheur: ~ **alphanumérique** *m* TV alphanumeric display; ~ **de caractères** *m* ÉLECTRON character display

affinage *m* GÉNÉRAL finishing off, refinement

affiner *vt* GÉNÉRAL finish off, refine

affirmation: ~ **de valeur d'attribut** *f* DONNÉES attribute-value assertion (*AVA*)

affluent *m* TRANSMIS tributary (*TR*)

AFNOR *abrév* (*Association française de normalisation*) GÉNÉRAL AFNOR (*French National Standards Association*)

agence: ~ **commerciale** *f* RADIO local sales office

Agence: ~ **spatiale européenne** *f* RADIO European Space Agency (*ESA*)

agent *m* GÉNÉRAL agent; ~ **de durcissement** COMPOSANTS hardening agent, setting agent; ~ **d'imprégnation** GÉNÉRAL impregnating agent; ~ **de système d'annuaire** (*DSA*) DONNÉES directory system agent (*DSA*); ~ **télématique** DONNÉES, TÉL telematic agent (*TLMA*); ~ **de traitement d'appel** DONNÉES call control agent (*CCA*); ~ **de transfert de messages** DONNÉES, TÉL message transfer agent (*MTA*); ~ **usager d'annuaire** APP SERV, DONNÉES, TÉL, directory user agent; ~ **utilisateur** (*UA*) APP SERV, DONNÉES, TÉL user agent (*UA*); ~ **utilisateur d'annuaire** APP SERV, DONNÉES, TÉL directory user agent; ~ **utilisateur EDI** APP SERV, DONNÉES, TÉL EDI user agent (*EDI-UA*); ~ **utilisateur de système de messagerie de personne à personne** APP SERV, DONNÉES,

RADIO, TÉL interpersonal messaging system user agent (*IPMS UA*); **~ utilisateur du système de messagerie vocale** APP SERV, DONNÉES, TÉL voice messaging system user agent (*VMGS-UA*)

agile: **~ en fréquence** *adj* RADIO dynamic

AGP *abrév* (*alarme de groupe primaire*) TRANSMIS CGA (*carrier group alarm*)

agrément *m* DONNÉES agreement, TESTS type approval

agrémenter *vt* DONNÉES *web page* embellish

aide: **~ à la programmation graphique** *f* INFORMAT graphic programming aid

aigrette *f* ALIM ÉLEC brush discharge

aiguillage *m* TRANSMIS feeding cable

aiguille *f* INFORMAT needle

aiguiller *vt* TÉL direct

aiguilleur *m* TV switcher

aigus *m pl* TRANSMIS treble

AIMA *abrév* (*accès multiple par impulsions d'adresse*) RADIO PAMA (*pulse-address multiple access*)

aimant *m* ÉLECTRON magnet; **~ de focalisation** ÉLECTRON focusing magnet; **~ permanent** COMPOSANTS, ÉLECTRON permanent magnet; **~ de réenclenchement** COMMUT resetting magnet; **~ de relais** COMPOSANTS, ÉLECTRON relay magnet; **~ de remise au repos** COMMUT restoring magnet

aimantation *f* ÉLECTRON magnetization; **~ résiduelle** ÉLECTRON residual magnetism

AIR *abrév* (*Association internationale de retransmission*) RADIO IAB (*International Association of Broadcasting*)

aire: **~ de cohérence** *f* TRANSMIS coherence area, coherent area; **~ de couverture** *f* RADIO coverage area; **~ de propagation** *f* ÉLECTRON, GÉNÉRAL *physique* area of propagation, propagation area; **~ de recouvrement** *f* RADIO overlap area

AISM *abrév* (*Association internationale de signalisation maritime*) RADIO IALA (*International Association of Lighthouse Authorities*)

ajustage *m* ALIM ÉLEC *des composants*, ÉLECTRON trimming, TÉL *métrologie* adjustment; **~ d'appoint** ALIM ÉLEC, ÉLECTRON trimming; **~ au laser** ÉLECTRON laser trimming; **~ par étincelage** ÉLECTRON high-voltage pulse trimming

alarme *f* DONNÉES, TÉL, TESTS alarm; **~ anti-effraction** TESTS *sécurité* burglar alarm; **~ de blocage** TÉL blocking alarm; **~ de coupure de courant** ALIM ÉLEC power-off alarm; **~ cyclique** TESTS cyclic alarm; **~ de dérangement** TÉL failure alarm; **~ de distribution d'énergie** ALIM ÉLEC power distribution alarm; **~ externe** RADIO external alarm; **~ de fusible** TESTS fuse alarm; **~ de fusible d'alvéole** ALIM ÉLEC, COMPOSANTS, TÉL shelf fuse alarm; **~ de fusible de travée** ALIM ÉLEC, COMPOSANTS suite fuse alarm; **~ de groupe primaire** (*AGP*) TRANSMIS carrier group alarm (*CGA*); **~ indépendante** PÉRIPH independent alarm; **~ à indicateur visuel** TESTS visual alarm; **~ indicatrice d'unité défectueuse** TÉL unit-indicating alarm; **~ majeure** TÉL urgent alarm, TESTS major alarm; **~ de manque de tension** ALIM ÉLEC undervoltage alarm; **~ mineure** TESTS minor alarm; **~ neutralisée** TRANSMIS alarm inhibited; **~ de panne de courant** ALIM ÉLEC mains-failure alarm; **~ de perte du mot unique** DONNÉES unique-word loss alarm; **~ de service** TÉL service alarm; **~ sonore** TESTS audible alarm; **~ de surtension** ALIM ÉLEC overvoltage alarm; **~ de TEB élevé** DONNÉES high-BER alarm; **~ urgente** TÉL urgent alarm; **~ vers l'arrière** TESTS backward alarm; **~ visuelle** TESTS visual alarm

aléatoire *adj* GÉNÉRAL random

alerte *f* RADIO, TESTS *service maritime mobile* alerting; **~ de détresse** RADIO distress alerting; **~ navire-navire** RADIO ship-to-ship alerting; **~ navire-terre** RADIO ship-to-shore alerting; **~ passive** RADIO passive alerting; **~ terre-navire** RADIO shore-to-ship alerting

alerter *vt* TÉL alarm, alert

alésage *m* COMPOSANTS boring, *fabrication* broaching, reaming

aléser *vt* COMPOSANTS bore, *fabrication* broach, ream

ALGOL *m* (*langage algorithmique*) INFORMAT ALGOL (*algorithmic language*)

algorithme *m* INFORMAT algorithm; **~ à arbre recouvrant** ÉLECTRON spanning-tree algorithm (*STA*); **~ à deux porteurs** INFORMAT, RADIO, TRANSMIS dual-carrier algorithm; **~ MPEG** DONNÉES motion picture experts group algorithm, MPEG algorithm; **~ de Rivest, Shamir et Adleman** TÉL Rivest, Shamir, Adleman algorithm (*RSA*); **~ à un seul porteur** INFORMAT, RADIO, TRANSMIS single-carrier algorithm

algorithmique *adj* INFORMAT algorithmic

alias *adv* GÉNÉRAL alias

aligné[1] *adj* INFORMAT, RADIO, SIGNAL, TÉL, TRANSMIS *fibres optiques* aligned

aligné[2]: **~ sur les octets** *loc* INFORMAT byte-aligned

alignement *m* INFORMAT, RADIO, SIGNAL, TÉL, TRANSMIS *fibres optiques* alignment; **~ de caractères** TÉL character alignment; **~ de descente à amplitude de référence nulle** RADIO null-reference glide path; **~ initial** TÉL initial alignment; **~ initial des trames sémaphores** SIGNAL initial signal unit alignment, initial signing unit alignment; **~ d'octets** TÉL octet alignment; **~ des trames sémaphores** SIGNAL signal unit alignment

aligner *vt* INFORMAT, RADIO, SIGNAL, TÉL, TRANSMIS *fibres optiques* align

aligneur *m* TÉL aligner

alimentation *f* ALIM ÉLEC feed, feeding, power, power supply, power supply unit, power unit, COMPOSANTS feeding, ÉLECTRON feed, power supply, INFORMAT feed, feeding; **~ d'antenne** RADIO aerial feed, antenna feed; **~ axiale** RADIO vertex feed; **~ de bâti** ALIM ÉLEC rack power; **~ CC** ALIM ÉLEC DC power; **~ en col de cygne** RADIO gooseneck feed, swan feed; **~ en courant** ALIM ÉLEC current feed; **~ en courant individuel** ALIM ÉLEC, TÉL split feed, individual current feed; **~ Cutler** RADIO Cutler feed; **~ directe** RADIO direct feed; **~ électrique** ALIM ÉLEC power supply; **~ électrique de secours** ALIM ÉLEC emergency power supply; **~ en énergie** ALIM ÉLEC current feed; **~ en énergie électrique** ALIM ÉLEC power feed; **~ feuille à feuille** PÉRIPH single-page feeding, single-sheet feeding; **~ frontale** RADIO front feed; **~ hors service** ALIM ÉLEC power-off condition; **~ informatisée** ALIM ÉLEC computerized power supply; **~ ininterrompue** ALIM ÉLEC no-break power; **~ non-interruptible** (*UPS*) ALIM ÉLEC uninterruptible power supply (*UPS*); **~ du papier** PÉRIPH paper feed; **~ par batterie** ALIM ÉLEC battery power supply; **~ par fentes** RADIO Cutler feed; **~ par piles** ALIM ÉLEC battery power supply; **~ périscopique** RADIO beam waveguide; **~ des porteuses** TRANSMIS carrier supply; **~ sans coupure** (*UPS*) ALIM ÉLEC uninterrupted power supply (*UPS*); **~ de secours** ALIM ÉLEC power supply unit, reserve power, standby power, standby power supply, TÉL backup power supply; **~ secteur** ALIM ÉLEC mains connection, INFORMAT AC power

alimenté *adj* ALIM ÉLEC, powered; **~ par batterie** ALIM ÉLEC battery-powered; **~ par énergie solaire** ALIM ÉLEC solar-powered; **~ par piles** ALIM ÉLECbattery-powered

alimenter *vt* ALIM ÉLEC power, feed, INFORMAT, RADIO, TV feed; **~ en retour** INFORMAT feed back

allée *f* COMMUT, COMPOSANTS, TÉL aisle; **~ latérale** TÉL side aisle; **~ principale** COMMUT main aisle

allocation: **~ dynamique de fréquences** *f* GÉNÉRAL dynamic allocation of frequency, RADIO dynamic channel allocation; **~ de fréquence** *f* RADIO frequency allocation, frequency assignation; **~ d'itinéraire** *f* COMMUT path allocation; **~ de spectre** *f* RADIO spectrum allocation

allumage *m* ÉLECTRON firing, firing time, ignition

allumer: **s'~** *v réfl* INFORMAT light up

alphabet *m* DONNÉES alphabet; **~ international de référence** (*IRA*) DONNÉES, TÉL international reference alphabet (*IRA*); **~ international n° 5** DONNÉES, TÉL international alphabet no. 5 (*IA5*); **~ télégraphique** DONNÉES, TÉL telegraphic alphabet; **~ télégraphique international n° 2** DONNÉES, TÉL international telegraphic alphabet no. 2 (*ITA2*)
alphagéométrique *adj* TV alphageometric
alphamosaïque *adj* TV alphamosaic
alphaphotographique *adj* TV alphaphotographic
alt. *abrév* (*alternative*) DONNÉES *Internet* alt. (*alternative*)
alternat[1] *adj* TRANSMIS half-duplex, simplex
alternat[2] *m* TÉL alternate transmission
alternat[3]: **à l'~** *loc* RADIO with alternate handling
alternateur *m* ALIM ÉLEC AC generator, alternating current generator; **~ à champ tournant** ALIM ÉLEC, RADIO, TÉL rotating field alternator
alternative *f* (*alt.*) DONNÉES *Internet* alternative (*alt.*)
alterné *adj* RADIO alternated
altimètre *m* TÉL, TESTS *indicateurs* altimeter; **~ radioélectrique** RADIO, TÉL radio altimeter
altitude: **~ de l'apogée** *f* RADIO altitude of the apogee
alvéole *m* ÉLECTRON duct, tube, INFORMAT pit, RADIO cell, TRANSMIS duct, tube
AMDC *abrév* (*accès multiple par différence de code*) RADIO, SIGNAL, TÉL, TRANSMIS CDMA (*code-division multiple access*)
AMDP *abrév* (*accès multiple par détection de la porteuse*) RADIO, SIGNAL, TÉL, TRANSMIS CSMA (*carrier-sense multiple access*)
âme *f* GÉNÉRAL, TRANSMIS cable core; **~ en forme** COMPOSANTS shaped conductor; **~ rétreinte** TRANSMIS compacted conductor; **~ segmentée** ÉLECTRON milliken conductor
améliorer *vt* GÉNÉRAL enhance, upgrade
American: **~ Standard Code for Information Interchange** *m* (*ASCII*) DONNÉES, INFORMAT American Standard Code for Information Interchange (*ASCII*)
aminoplaste *m* COMPOSANTS *matières plastiques* aminoplastic
amorçage *m* ALIM ÉLEC striking, *d'un parafoudre* spark-over, COMPOSANTS singing, ÉLECTRON ignition, striking, primer ignition, INFORMAT boot-up, bootstrap, computer bootstrapping, TÉL howling
amorce *f* ALIM ÉLEC stub cable, DONNÉES leader, INFORMAT bootstrap, TRANSMIS pigtail; **~ de bande** DONNÉES leader; **~ de bande magnétique** DONNÉES leader, magnetic-tape leader; **~ de rupture** TESTS indication of fracture
amorcer *vt* INFORMAT boot, boot up, bootstrap
amortir *vt* GÉNÉRAL amortize, TRANSMIS damp
amortissement *m* GÉNÉRAL amortization, INFORMAT decay, TRANSMIS damping; **~ critique** TÉL *acoustique* critical damping
amortisseur: **~ de barre de sélection** *m* COMMUT selecting bar damper; **~ de bruit** *m* TÉL, TRANSMIS silencer
ampérage *m* ALIM ÉLEC amperage
ampère *m* ALIM ÉLEC ampere; **~-tour** ALIM ÉLEC ampere turn
ampèremètre *m* TESTS *dispositif de mesure* ammeter; **~ thermique** TESTS hot-wire ammeter
ampliop *m* (*amplificateur opérationnel*) ÉLECTRON, TRANSMIS op amp (*operational amplifier*)
amplificateur: **~ asymétrique** *m* ALIM ÉLEC single-ended amplifier; **~ audio** *m* PÉRIPH, TÉL voice amplifier; **~ bas niveau** *m* ÉLECTRON, SIGNAL small-signal amplifier; **~ basse fréquence** *m* PÉRIPH, TÉL voice amplifier; **~ de cadre** *m* TESTS loop amplifier; **~ à champs croisés** *m* TRANSMIS crossed-field amplifier; **~ de clavier** *m* INFORMAT keyboard amplifier; **~ commandé par tension** *m* ALIM ÉLEC voltage-controlled amplifier; **~ à contre-réaction** *m* ÉLECTRON negative-feedback amplifier; **~ à correction aval** *m* ÉLECTRON feedforward amplifier; **~ à déclenchement périodique** *m* ÉLECTRON gated amplifier; **~ à découpage** *m* ÉLECTRON chopper amplifier; **~ déphaseur** *m* ÉLECTRON paraphrase amplifier; **~ différentiel** *m* TRANSMIS differential amplifier; **~ dynamique numérique** *m* INFORMAT digital dynamic amplifier; **~ d'entrée** *m* ÉLECTRON input amplifier; **~ à faible bruit** *m* (*AFB*) ÉLECTRON, RADIO, TRANSMIS low-noise amplifier (*LNA*); **~ faible puissance** *m* ÉLECTRON low-frequency amplifier; **~ FI** *m* RADIO, TRANSMIS IF amplifier; **~ à fibre dopée erbium** *m* TRANSMIS erbium-doped fiber amplifier (*AmE*), erbium-doped fibre amplifier (*BrE*); **~ de filtre actif** *m* ÉLECTRON filter amplifier; **~ final** *m* ÉLECTRON final amplifier; **~ à fréquence intermédiaire** *m* RADIO, TRANSMIS intermediate frequency amplifier; **~ de fréquences radioélectriques** *m* TÉL RF amplifier; **~ à gain élevé** *m* ALIM ÉLEC, DONNÉES, INFORMAT, TRANSMIS power amplifier (*PA*); **~ à grille à la masse** *m* ÉLECTRON earthed-grid amplifier (*BrE*), grounded-grid amplifier (*AmE*); **~ à haute fréquence** *m* TÉL RF amplifier; **~ à haute puissance** *m* ALIM ÉLEC, ÉLECTRON high-power amplifier (*HPA*); **~ hyperfréquence** *m* RADIO microwave amplifier; **~ inverseur** *m* ALIM ÉLEC inverting amplifier; **~ à large bande** *m* ÉLECTRON wideband amplifier, TRANSMIS broadband amplifier; **~ à laser semiconducteur** *m* ÉLECTRON semiconductor laser diode amplifier (*SLDA*); **~ de ligne** *m* TV line amplifier; **~ linéaire** *m* TRANSMIS linear amplifier; **~ magnétique** *m* ALIM ÉLEC transductor amplifier; **~ de mesure** *m* ÉLECTRON measuring amplifier; **~ non symétrique** *m* ALIM ÉLEC single-ended amplifier; **~ numérique variable** *m* INFORMAT variable digital amplifier; **~ à ondes progressives** *m* ÉLECTRON, TRANSMIS traveling-wave amplifier (*AmE*) (*TWA*), travelling-wave amplifier (*BrE*) (*TWA*); **~ opérationnel** *m* (*ampliop*) ÉLECTRON, TRANSMIS operational amplifier (*op amp*); **~ optique** *m* DONNÉES, TRANSMIS optical amplifier; **~ paramétrique** *m* TRANSMIS parametric amplifier (*paramp*); **~ de petits signaux** *m* ÉLECTRON, SIGNAL small-signal amplifier; **~ psophométrique** *m* TÉL noise-measuring amplifier; **~ de puissance** *m* ALIM ÉLEC, DONNÉES, INFORMAT, TRANSMIS power amplifier (*PA*); **~ de puissance linéaire** *m* TRANSMIS linear power amplifier; **~ de puissance transistorisé** *m* ALIM ÉLEC, COMPOSANTS solid-state power amplifier (*SSPA*); **~ push-pull** *m* ÉLECTRON push-pull amplifier; **~ reflex** *m* ÉLECTRON reflex amplifier; **~ régulateur** *m* ÉLECTRON regulating amplifier; **~ de relais** *m* ALIM ÉLEC relay driver; **~ RF** *m* TÉL RF amplifier; **~ à seconde FI** *m* ALIM ÉLEC second IF amplifier; **~ à semiconducteurs** *m* ÉLECTRON semiconductor amplifier; **~ séparateur** *m* ÉLECTRON buffer amplifier; **~ sommateur-pondérateur** *m* ALIM ÉLEC summing-scaling amplifier; **~ de sommation et pondération** *m* ALIM ÉLEC summing-scaling amplifier; **~ télégraphique** *m* TÉL telegraphic magnifier; **~ de tension** *m* ALIM ÉLEC voltage amplifier; **~ à transistors** *m* ÉLECTRON solid-state amplifier; **~ à tube à ondes progressives** *m* ÉLECTRON, TRANSMIS traveling-wave tube amplifier (*AmE*) (*TWTA*), travelling-wave tube amplifier (*BrE*) (*TWTA*); **~ à variation temporelle** *m* INFORMAT time-variant amplifier
amplification *f* ALIM ÉLEC, ÉLECTRON, INFORMAT, RADIO, TRANSMIS amplification; **~ de courant** ALIM ÉLEC current amplification; **~ FI** RADIO, TRANSMIS IF amplification; **~ à fréquence intermédiaire** RADIO, TRANSMIS intermediate frequency amplification; **~ non linéaire** TRANSMIS nonlinear amplification; **~ d'onde** ÉLECTRON wave amplification; **~ optique en ligne** TRANSMIS on-line optical amplification; **~ de puissance** ALIM ÉLEC, DONNÉES, INFORMAT, TRANSMIS power amplification
amplifier *vt* ALIM ÉLEC, ÉLECTRON, INFORMAT, RADIO, TRANSMIS amplify
amplitude *f* GÉNÉRAL amplitude, INFORMAT depth; **~ de base** RADIO basic amplitude; **~ de crête** TÉL *d'écho élémentaire* peak amplitude; **~ crête à crête** GÉNÉRAL, RADIO peak-to-peak amplitude; **~ de décision** TÉL decision value; **~ de**

déplacement ÉLECTRON displacement amplitude; **~ d'une impulsion** TÉL pulse magnitude, TRANSMIS pulse amplitude; **~ relative** TÉL *d'écho élémentaire* relative amplitude; **~ du signal** SIGNAL signal amplitude; **~ du spectre** RADIO spectrum amplitude; **~ virtuelle de décision** TÉL virtual decision value

AMRF *abrév* (*accès multiple par répartition de fréquence*) RADIO, SIGNAL, TÉL, TRANSMIS FDMA (*frequency-division multiple access*)

AMRFO *abrév* (*accès multiple par répartition en fréquences orthogonales*) RADIO, SIGNAL, TÉL, TRANSMIS ODFMA (*orthogonal frequency-division multiple access*)

AMRT *abrév* (*accès multiple à répartition dans le temps*) RADIO, SIGNAL, TÉL, TRANSMIS TDMA (*time-division multiple access*); **~ à faible trafic** *f* RADIO thin-route TDMA

A-N *abrév* (*analogique-numérique*) DONNÉES, ÉLECTRON, INFORMAT, SIGNAL, TÉL A-D (*analog-to-digital, analog-digital, analog/digital*)

analogique *adj* GÉNÉRAL analog

analogique-numérique *adj* (*A-N*) DONNÉES, ÉLECTRON, INFORMAT, SIGNAL, TÉL analog-to-digital (*A-D*), analog-digital (*A-D*), analog/digital (*A-D*)

analyse *f* PÉRIPH scanning; **~ de l'acheminement** TÉL route analysis; **~ d'arbre de défaillance** TESTS fault-tree analysis; **~ d'arbre des pannes** TESTS fault-tree analysis; **~ de bruit** TÉL noise monitoring; **~ du champ lointain** RADIO, TRANSMIS far-field analysis; **~ du champ proche** RADIO, TRANSMIS near-field analysis; **~ du cheminement critique** GÉNÉRAL critical-path analysis; **~ des contraintes** TESTS stress analysis; **~ dans le domaine fréquentiel** ALIM ÉLEC, ÉLECTRON, RADIO frequency-domain analysis; **~ de la dimension temporelle** DONNÉES time-domain analysis; **~ documentaire** DONNÉES abstracting; **~ des données multidimensionnelles** DONNÉES multidimensional data analysis; **~ des erreurs** APP SERV error analysis; **~ de faisceau** TÉL route analysis; **~ du flux de données** DONNÉES data-flow analysis; **~ fonctionnelle** INFORMAT, TESTS systems analysis; **~ harmonique** ÉLECTRON harmonic analysis; **~ de l'image** TV image analysis; **~ du langage naturel** DONNÉES natural-language analysis; **~ de longueur de numéro** TÉL analysis for number length; **~ de maintenance** TESTS maintenance analysis; **~ des modes de panne** TESTS fault-mode analysis (*FMA*); **~ nodale** TRANSMIS nodal analysis; **~ des pannes** TESTS fault analysis; **~ par boucles** ÉLECTRON mesh analysis; **~ par nud** TRANSMIS nodal analysis; **~ de Pareto** TESTS Pareto analysis; **~ de la parole** TÉL speech analysis; **~ de protocole** DONNÉES protocol analysis; **~ de risque** TESTS hazard analysis; **~ du risque** TESTS risk analysis; **~ de signaux** SIGNAL signal analysis; **~ spectrale** TESTS spectral analysis; **~ statistique** ÉLECTRON advanced statistical analysis; **~ syntaxique** TÉL syntactic analysis; **~ des tarifs** APP SERV tariff analysis; **~ de taxation** APP SERV charging analysis; **~ taxation** APP SERV call-charging analysis; **~ de trafic** DONNÉES, TÉL, TESTS traffic analysis

analyser *vt* INFORMAT scan

analyseur *m* COMMUT scanner unit, ÉLECTRON analyser (*BrE*), analyzer (*AmE*), INFORMAT scanner, SIGNAL analyser (*BrE*), analyzer (*AmE*); **~ de catégorie** TÉL category analyser (*BrE*), category analyzer (*AmE*); **~ de débordement** TÉL analysis meter; **~ de distorsion de signaux** TESTS, signal distortion analyser (*BrE*) (*SDA*), signal distortion analyzer (*AmE*) (*SDA*); **~ de faisceaux hertziens** TESTS microwave link analyser (*BrE*), microwave link analyzer (*AmE*); **~ d'image** TV image analyser (*BrE*), image analyzer (*AmE*); **~ de liaison en mode paquet** DONNÉES burst-mode link analyser (*BrE*) (*BMLA*), burst-mode link analyzer (*AmE*) (*BMLA*); **~ de liaisons hyperfréquences** TESTS microwave link analyser (*BrE*), microwave link analyzer (*AmE*); **~ de lignes** TESTS line analyser (*BrE*), line analyzer (*AmE*); **~ logique** DONNÉES logic analyser (*BrE*), logic analyzer (*AmE*); **~ d'ondes** TESTS wave analyser (*BrE*), wave analyzer (*AmE*); **~ de réseau** INFORMAT network computer; **~ de réseaux** TESTS network analyser (*BrE*), network analyzer (*AmE*); **~ de signal** INFORMAT, SIGNAL signal analyser (*BrE*), signal analyzer (*AmE*); **~ de spectre** TESTS spectrum analyser (*BrE*), spectrum analyzer (*AmE*); **~ syntactique** TÉL syntactic analyser (*BrE*), syntactic analyzer (*AmE*); **~ de taxes téléphoniques** TÉL telephone charge analyser (*BrE*), telephone charge analyzer (*AmE*); **~ de trafic** DONNÉES, TÉL, TESTS traffic analyser (*BrE*), traffic analyzer (*AmE*); **~ de trafic téléphonique** TESTS telephone traffic analyser (*BrE*), telephone traffic analyzer (*AmE*)

analyste: **~ en informatique** *m* GÉNÉRAL EDP analyst

anamorphose *f* GÉNÉRAL anamorphosis

ancien: **~ indicatif** *m* TÉL *dans le cas d'un changement d'indicatif* old code

ange *m* RADIO radar angel

angle: **~ d'admission** *m* TRANSMIS acceptance angle; **~ d'arrivée** *m* TRANSMIS angle of arrival; **~ d'azimut** *m* RADIO *radar* angle of azimuth, azimuth; **~ de Brewster** *m* TRANSMIS Brewster's angle; **~ de champ** *m* RADIO field of vision (*FOV*); **~ critique** *m* TRANSMIS critical angle; **~ de départ** *m* TRANSMIS angle of departure; **~ de dépression** *m* RADIO angle of depression; **~ d'élévation** *m* RADIO, TRANSMIS *radar* angle of elevation; **~ d'empiétement** *m* ÉLECTRON overlap angle; **~ d'évitement** *m* RADIO avoidance angle; **~ exocentrique** *m* RADIO exocentric angle; **~ géocentrique** *m* RADIO geocentric angle; **~ de gisement** *m* RADIO *radar* angle of azimuth, azimuth; **~ de Hall** *m* ÉLECTRON Hall angle; **~ d'incidence** *m* TRANSMIS angle of incidence; **~ d'inclinaison** *m* RADIO angle of tilt; **~ de lancement** *m* TRANSMIS launch angle; **~ d'ouverture** *m* TRANSMIS *fibres optiques* acceptance angle; **~ par rapport à l'axe** *m* RADIO angle from the axis; **~ de phase** *m* ALIM ÉLEC, INFORMAT phase angle; **~ de pointage** *m* RADIO pointing angle; **~ de rayonnement** *m* ÉLECTRON radiation angle, TRANSMIS radiation angle, output angle; **~ de rotation** *m* TRANSMIS *fibres optiques* angle of rotation; **~ de site** *m* RADIO angle of elevation, elevation angle, TRANSMIS angle of elevation, angle of site; **~ au sommet de l'alvéole** *m* RADIO vertex angle; **~ de sortie** *m* ÉLECTRON, TRANSMIS output angle, radiation angle; **~ de strabisme** *m* RADIO squint, squint angle; **~ topocentrique** *m* RADIO topocentric angle; **~ de transit** *m* ÉLECTRON transit angle; **~ de visée** *m* RADIO angle of sight

animation: **~ comportementale** *f* INFORMAT behavioral animation (*AmE*), behavioural animation (*BrE*); **~ faciale** *f* INFORMAT facial animation; **~ par images-clés** *f* INFORMAT keyframe animation, keyframing; **~ sonore** *f* INFORMAT sound animation

anisochrone *adj* TRANSMIS anisochronous

anisotrope *adj* TRANSMIS anisotropic

anneau[1] *m* DONNÉES, ÉLECTRON ring; **~ d'ancrage** TRANSMIS *pose de câbles* anchoring bracket; **~ de fixation** COMPOSANTS fixing ring; **~ de garde** TÉL guard ring; **~ guide-fils** COMPOSANTS jumper-wire guide, jumper-wire eyelet; **~ hybride** RADIO hybrid ring, hybrid circulator, rat race; **~ à jeton** DONNÉES, INFORMAT, TRANSMIS token ring; **~ passe-jarretières** COMMUT jumper ring

anneau[2]: **à ~ double** *loc* ÉLECTRON dual-ring

annexe *adj* INFORMAT ancillary

annonce *f* PÉRIPH, TÉL outgoing message (*OGM*); **~ de durée réglable** PÉRIPH *répondeur* variable-length OGM, variable-length outgoing message; **~ enregistrée** TÉL recorded announcement (*RAN*), recorded message; **~ enregistrée générale** TÉL general recorded announcement; **~ enregistrée générale négative** TÉL general negative-recorded announcement; **~ enregistrée générale positive**

TÉL general positive-recorded announcement; **~ enregistrée spécifique** TÉL specific recorded announcement; **~ enregistrée spécifique négative sans information** TÉL specific negative-recorded announcement without information; **~ enregistrée spécifique positive avec information supplémentaire** TÉL specific positive-recorded announcement with supplementary information; **~ enregistrée spécifique positive sans information** TÉL specific positive-recorded announcement without information; **~ numérique** TÉL digital announcement; **~ parlée** *Can* TÉL recorded announcement (*RAN*)

annotateur *m* TÉL recording operator

annotation *f* INFORMAT, TÉL annotation; **~ vocale** INFORMAT, TÉL voice annotation

annuaire *m* DONNÉES, ÉLECTRON directory, TÉL telephone book, telephone directory; **~ électronique** DONNÉES, TÉL electronic directory; **~ interne** DONNÉES office directory; **~ papier** TÉL paper directory; **~ téléphonique** TÉL telephone directory, telephone book

annulateur: **~ de paquets** *m* DONNÉES burst negator

annulation *f* GÉNÉRAL deletion; **~ d'écho** ÉLECTRON, RADIO, TRANSMIS echo cancelation (*AmE*), echo canceling (*AmE*), echo cancellation (*BrE*), echo cancelling (*BrE*), echo suppression; **~ d'enregistrement** INFORMAT record cancel; **~ d'une localisation** RADIO location cancelation (*AmE*), location cancellation (*BrE*); **~ sélective** RADIO selective cancelation (*AmE*), selective cancellation (*BrE*)

annuler *vt* GÉNÉRAL delete, INFORMAT undo; **~ une inscription** INFORMAT unsubscribe

annuleur: **~ d'écho** *m* ÉLECTRON, RADIO, TRANSMIS echo canceler (*AmE*), echo canceller (*BrE*); **~ de perturbation** *m* RADIO interference canceler (*AmE*), interference canceller (*BrE*); **~ de transpolarisation** *m* RADIO cross-polarization canceler (*AmE*), cross-polarization canceller (*BrE*)

anode *f* ÉLECTRON anode, plate

anomalie *f* GÉNÉRAL fault; **~ de fonctionnement** TESTS service irregularity, service incident

ANSI *abrév* (*Institut américain de normalisation*) GÉNÉRAL ANSI (*American National Standards Institute*)

ant. *abrév* (*antenne*) INFORMAT AE (*aerial*), RADIO AE (*aerial*), ANT (*antenna*), radio mast

ante: **~-serveur** *m* DONNÉES, INFORMAT back-end server

antémémoire *f* DONNÉES, INFORMAT cache memory, cache storage; **~ de disque** DONNÉES, INFORMAT disk caching

antenne *f* (*ant.*) INFORMAT aerial (*AE*), RADIO aerial (*AE*), antenna (*ANT*), radio mast; **~ acoustique** RADIO acoustic antenna; **~ adaptative** RADIO adaptive antenna; **~ à alimentation déphasée** RADIO antenna fed out of phase; **~ à alimentation focale** RADIO focal-fed antenna; **~ en anneau** RADIO ring antenna; **~ apériodique** RADIO aperiodic aerial, aperiodic antenna; **~ en arête de poisson** RADIO fishbone antenna; **~ autoportante** RADIO self-supporting antenna; **~ autoporteuse** RADIO self-supporting antenna; **~ d'azimut** RADIO azimuth antenna, azimuth aerial; **~ à balayage** RADIO scanning aerial, scanning antenna, sweep aerial, sweep antenna; **~ à balayage électronique** RADIO phased-array antenna; **~ biconique** RADIO biconical antenna; **~ cadre** RADIO loop aerial, loop antenna; **~ champ E** RADIO E-field antenna; **~ cierge** RADIO polyrod, dielectric rod radiator; **~ cigare** RADIO cigar aerial, cigar antenna; **~ classique** TV television aerial; **~ coaxiale** RADIO coaxial aerial, coaxial antenna; **~ co-implantée** RADIO co-located antenna; **~ collective** RADIO, TV collective aerial, common aerial, community aerial, community antenna; **~ communautaire** RADIO, TV community aerial, community antenna; **~ à compensation électronique** RADIO electronically-despun antenna; **~ à contre-rotation électronique** RADIO electronically-despun antenna; **~ à cornet** RADIO horn antenna; **~ en cornet** RADIO horn antenna; **~-cornet** RADIO flare antenna; **~ à correction par balayage** RADIO scanning-corrected aerial, scanning-corrected antenna; **~ coudée** RADIO bend aerial; **~ cylindre parabolique** RADIO parabolic cylinder aerial, parabolic cylinder antenna; **~ en D** RADIO cheese antenna, hogtrough antenna; **~ dièdre** RADIO corner aerial, corner antenna, TESTS dihedral antenna; **~ diélectrique** RADIO dielectric aerial, dielectric radiator, dielectric antenna; **~ dipôle** RADIO dipole aerial, dipole antenna; **~ directionnelle** RADIO beam antenna; **~ directive** RADIO directional aerial, directional antenna, beam aerial, beam antenna, unidirectional aerial, unidirectional antenna; **~ à disque** RADIO disc antenna (*BrE*), disk antenna (*AmE*); **~ à double faisceau** DONNÉES dual-beam antenna; **~ à double polarisation** RADIO dual-polarized antenna; **~ à double réflecteur** RADIO double-reflector aerial, double-reflector antenna; **~ de doublet** RADIO dipole aerial, dipole antenna; **~ émettrice** RADIO transmit aerial, transmit antenna; **~ d'émission** RADIO transmit aerial, transmit antenna; **~ d'émission à faisceau étroit** RADIO narrow-beam transmit aerial, narrow-beam transmit antenna; **~ d'émission du faisceau européen** RADIO eurobeam transmit aerial, eurobeam transmit antenna; **~ d'émission du satellite** RADIO satellite transmit aerial, satellite transmit antenna; **~ encastrée** RADIO suppressed aerial, suppressed antenna, flush antenna; **~ escamotable** RADIO retractable aerial, retractable antenna; **~ en espace libre** RADIO free-space antenna; **~ extérieure** RADIO outdoor aerial, outdoor antenna; **~ à faisceau en cosécante carrée** RADIO co-secant-squared antenna; **~ à faisceau étalé** RADIO fan-beam aerial, fan-beam antenna; **~ à faisceau fixe** RADIO fan-beam aerial, fan-beam antenna; **~ à faisceau formé** RADIO shaped-beam aerial, shaped-beam antenna; **~ à faisceau modelé** RADIO shaped-beam aerial, shaped-beam antenna; **~ à faisceau orientable** RADIO steerable-beam aerial, steerable-beam antenna; **~ à faisceau-crayon** RADIO pencil-beam aerial, pencil-beam antenna; **~ à ferrites** RADIO ferrite antenna; **~ à fente** RADIO slot aerial, slot antenna; **~ à fentes croisées** RADIO crossed-slot aerial, crossed-slot antenna; **~ fictive** RADIO artificial aerial, artificial antenna, dummy aerial, dummy antenna; **~ à fil** RADIO wire aerial, wire antenna; **~ filaire** RADIO wire aerial, wire antenna; **~ à fils inclinés** RADIO V-shaped aerial, V-shaped antenna; **~ fouet** RADIO whip aerial, whip antenna; **~ fouet quart d'onde** RADIO quarter-wave whip aerial, quarter-wave whip antenna; **~ fouet en tourniquet** RADIO turnstile whip; **~ à gain élevé** RADIO high-gain aerial, high-gain antenna; **~ à grand gain** RADIO high-gain aerial, high-gain antenna; **~ Gregory excentrée** RADIO offset Gregory antenna; **~ à guide à fentes** RADIO waveguide aerial, waveguide antenna; **~ à guide d'ondes fendu** RADIO slotted-waveguide aerial, slotted-waveguide antenna; **~ en hélice** RADIO helical aerial, helical antenna, helix aerial, helix antenna; **~ hélicoïdale** RADIO corkscrew aerial, corkscrew antenna, helical aerial, helical antenna; **~ hyperfréquence** RADIO microwave aerial, microwave antenna; **~ incorporée** RADIO integral antenna; **~ individuelle** TV individual aerial, individual antenna; **~ isotrope** RADIO isotropic aerial, isotropic antenna, isotropic radiator; **~ à lentille** RADIO lens aerial, lens antenna; **~ à lobes multiples** RADIO multilobe aerial, multilobe antenna; **~ log-périodique** RADIO log-periodic aerial, log-periodic antenna; **~ en losange** RADIO rhombic antenna; **~ monopôle** RADIO monopole aerial, monopole antenna; **~ multibande** RADIO multifrequency aerial, multifrequency antenna; **~ multifaisceau** RADIO multispot aerial, multispot antenna, multiple-beam aerial, multiple-beam antenna; **~ multifréquence** RADIO multifrequency aerial, multifrequency antenna; **~ en nappe** RADIO flat-top antenna; **~ à nappe horizontale** RADIO flat-top antenna; **~ normale économique** RADIO economic standard antenna; **~ normale minimale** RADIO minimum-standard aerial, minimum-standard antenna; **~ à noyau magnétique** RADIO ferrite antenna; **~ omnidirectionnelle** RADIO omnidirectional aerial, omnidirectional antenna; **~ à ondes progressives** RADIO, TRANSMIS traveling-wave aerial (*AmE*), traveling-wave antenna (*AmE*), travelling-wave aerial (*BrE*), travelling-wave antenna (*BrE*); **~ à optique Cassegrain** RADIO Cassegrain antenna; **~ orientable** RADIO rotatable aerial, rotatable antenna, steerable aerial, steerable antenna; **~ orientable en tous sens** RADIO full-steering capability aerial, full-steering capability antenna; **~ à ouverture** RADIO aperture aerial, aperture antenna; **~ à ouverture alimentée de guide d'ondes** RADIO waveguide-fed aperture aerial, waveguide-fed aperture antenna; **~ à ouverture rayonnante** RADIO aperture aerial, aperture antenna; **~ panneau** RADIO bow-tie aerial, bow-tie antenna; **~ parabolique** RADIO dish antenna, parabolic antenna, dish, parabolic aerial; **~ parabolique avec alimentation décalée** RADIO offset parabolic aerial, offset parabolic antenna; **~ parabolique à faisceau formé** RADIO shaped-beam offset parabolic aerial, shaped-beam offset parabolic antenna; **~ à périodicité logarithmique** RADIO log-periodic aerial, log-periodic antenna; **~ périscope** RADIO periscope aerial, periscope antenna; **~ à plan de sol** RADIO ground plane antenna; **~ plane** RADIO flat antenna; **~ plaquée** RADIO suppressed aerial, suppressed antenna; **~ à polarisation horizontale** RADIO horizontal antenna; **~ de poursuite** RADIO tracker, tracking aerial, tracking antenna; **~ quasi-équidirective** RADIO quasi-omnidirectional antenna; **~ quasi-omnidirective** RADIO quasi-omnidirectional antenna; **~ radar** RADIO radar aerial, radar antenna; **~ radioastronomique** RADIO radio astronomical antenna; **~ rayonnante** RADIO slot antenna, slot aerial; **~ à rayonnement longitudinal** RADIO endfire array aerial, endfire array antenna; **~ à rayonnement transversal** RADIO broadside array aerial, broadside array antenna; **~ à rayonnement vers l'arrière** RADIO backfire aerial, backfire antenna; **~ de réception** RADIO receive aerial, rceeive antenna, receiving aerial, receiving antenna; **~ de réception par satellite** RADIO satellite receiver aerial, satellite receiver antenna; **~ de réception de télévision seulement** TV television receive-only aerial, television receive-only antenna, TVRO aerial, TVRO antenna; **~ réceptrice** RADIO receive aerial, receive antenna; **~ rectiligne** RADIO rectilinear aerial, rectilinear antenna; **~ redresseuse** RADIO rectenna, rectifying aerial, rectifying antenna; **~ à réflecteur** RADIO reflecting aerial, reflecting antenna, reflector aerial, reflector antenna; **~ à réflecteur cylindrique** RADIO cylindrical reflector aerial, cylindrical reflector antenna; **~ à réflecteur et source frontale** RADIO front-fed reflector aerial, front-fed reflector antenna; **~ à réflecteur grégorien** RADIO Gregorian reflector aerial, Gregorian reflector antenna; **~ à réflecteur à grille** RADIO gridded reflector aerial, gridded reflector antenna; **~ à réflecteur parabolique** RADIO dish aerial, dish antenna, parabolic reflector aerial, parabolic reflector antenna; **~ à réflecteurs multiples** RADIO multiple-reflector aerial, multiple-reflector antenna; **~ repliée** RADIO bow-tie aerial, bow-tie antenna; **~ réseau** RADIO array aerial, array antenna; **~ réseau à commande de phase** RADIO phased-array antenna; **~ réseau plan** RADIO planar-array antenna; **~ réseau planaire** RADIO planar-array antenna; **~ réseau rectiligne** RADIO linear-array antenna; **~ rhombique** RADIO rhombic antenna; **~ rideau** RADIO aerial curtain, antenna curtain; **~ satellite** RADIO, TV satellite aerial, satellite antenna; **~ sphérique** RADIO spherical aerial, spherical antenna; **~ spirale** RADIO spiral antenna; **~ à strabisme** RADIO squint aerial, squint antenna; **~ de télécommande** RADIO command antenna; **~ de télévision** TV television aerial, television antenna; **~-tige** RADIO rod antenna; **~ de toit** TV rooftop aerial, rooftop antenna; **~ toroïdale** RADIO toroidal aerial, toroidal antenna; **~ de transmission** RADIO transmitting aerial, transmitting antenna; **~ en trèfle** RADIO cloverleaf antenna, conformal antenna, conformal aerial; **~ unipolaire** RADIO unipole aerial, unipole antenna; **~ en V incliné** RADIO sloping V aerial, sloping V antenna; **~ Yagi** RADIO Yagi aerial, Yagi antenna

antennes: **~ de balayage dos-à-dos** *f pl* RADIO back-to-back scanners; **~ paraboliques dos-à-dos** *f pl* RADIO back-to-back parabolic aerials, back-to-back parabolic antennas

anti-ballottant *adj* RADIO anti-sloshing

antibrouillage *m* ÉLECTRON interference suppression, RADIO interference suppression, anti-jamming (*AJ*), TESTS disturbance suppression

anti-chocs: **~ en mousse** *m* ÉLECTRON shock-absorbing foam

antidéflagrant *adj* TRANSMIS explosion-proof

anti-effraction *adj* TESTS *sécurité* burglar-proof

antifading *m* ÉLECTRON automatic volume control (*AVC*)

antilocal *adj* TÉL anti-sidetone

Antiope® *m Fra* APP SERV Ceefax® (*BrE*)

antiparasitage *m* ÉLECTRON, RADIO interference suppression, TESTS disturbance suppression

antireflet *m* DONNÉES anti-reflection coating

antiréflexion *adj* (*AR*) GÉNÉRAL anti-reflection (*AR*)

antirepliement *adj* INFORMAT, TRANSMIS anti-aliasing

antisismique *adj* TESTS earthquake-resistant

antistatique *adj* ÉLECTRON anti-static

AP *abrév* (*adresse physique*) TÉL physical address

APDU *abrév* (*unité de données de protocole d'application*) DONNÉES APDU (*application protocol data unit*)

aperçu *m* INFORMAT preview

apoapside *m* RADIO apoapsis

apoastre *m* RADIO apoapsis

apogée *m* RADIO apogee

appareil *m* INFORMAT apparatus, device, system; **~ d'affichage** DONNÉES display device; **~ à aimant mobile** ALIM ÉLEC moving-magnet instrument; **~ d'alarme anti-agression** TESTS *sécurité* assault alarm device; **~ analogique** GÉNÉRAL analog device; **~ à annonce enregistrée** PÉRIPH, TÉL recorded announcement machine; **~ arythmique** TÉL start-stop apparatus; **~ automatique de mesure de transmission** (*AAMT*) COMMUT automatic transmission-measuring equipment (*ATME*); **~ de balayage** PÉRIPH, RADIO scanner; **~ à cadre mobile** ALIM ÉLEC moving-coil instrument; **~ à cadre mobile et à aimant permanent** ALIM ÉLEC moving-coil permanent magnet instrument; **~ de commande de tempo** INFORMAT tempo controller; **~ à commutation vocale** COMMUT voice switching equipment; **~ de contrôle** TESTS monitor; **~ de contrôle de bruit** TESTS noise monitor; **~ d'éclairage** ALIM ÉLEC lighting fixture, light fitting, lighting fitting; **~ émetteur** PÉRIPH, TRANSMIS transmitting equipment; **~ enregistreur** TESTS recording instrument; **~ d'entrée** DONNÉES input device; **~ d'essai** PÉRIPH, TESTS tester, testing device, testing equipment; **~ externe** INFORMAT external device; **~ à fer mobile** ALIM ÉLEC moving-iron instrument; **~ indicateur** TESTS indicating-measuring instrument; **~ magnéto-optique** DONNÉES magneto-optical device; **~ de mesure** TESTS measuring instrument; **~ de mesure analogique** TÉL *métrologie*, TESTS analog measuring instrument; **~ de mesure enregistreur** TESTS recording instrument; **~ de mesure indicateur** TESTS indicating-measuring instrument; **~ de mesure intégrateur** TESTS integrating-measuring instrument; **~ de mesure de puissance optique** TRANSMIS optical power meter; **~ de mesure témoin** TESTS figure of merit meter; **~ de prise de courant** ALIM ÉLEC current collector; **~ programmable** ÉLECTRON programmable device; **~ récepteur** INFORMAT, TÉL receiving device; **~ de repérage de microformes** DONNÉES microform-retrieval terminal; **~ de réserve** GÉNÉRAL stand-by set; **~ de saisie**

INFORMAT grabber; ~ **de signalisation sonore** INFORMAT beeper; ~ **de suivi de batterie** ALIM ÉLEC battery monitor; ~ **de surveillance** TESTS monitor; ~ **de surveillance de bruit** TESTS noise monitor; ~ **de surveillance du taux d'erreur** TÉL error rate monitor; ~ **de surveillance de transmission en mode continu** TESTS continuous-mode transmission monitor; ~ **de télécommunications pour les sourds** (*ATS*) TÉL telecommunications device for the deaf (*TDD*); ~ **de téléphone** PÉRIPH, TÉL telephone attachment; ~ **téléphonique** PÉRIPH, TÉL telephone set, telephone instrument; ~ **téléphonique à clavier** PÉRIPH, TÉL keypad telephone, keyset telephone, push-button set, push-button telephone; ~ **téléphonique à mémoire** PÉRIPH, TÉL telephone set with memory; ~ **téléphonique mobile** PÉRIPH, TÉL table instrument, table set; ~ **terminal sur réseau** TRANSMIS network terminating unit (*NTU*)

appareillage *m* INFORMAT equipment; ~ **embarqué** RADIO on-board apparatus

appariage: ~ **des bandes de fréquence** *m* RADIO frequency band pairing

apparition: ~ **d'un dérangement** *f* TÉL failure occurrence; ~ **progressive** *f* INFORMAT fade-in

appel[1] *m* DONNÉES *datacom* calling, TÉL *sonnerie* ringing, *communication* call, call-in software; ~ **aide-mémoire** APP SERV reminder call; ~ **annoté** TÉL record call; ~ **d'arrivée** DONNÉES, RADIO incoming call (*IC*), TÉL in call, incoming call (*IC*); ~ **d'arrivée DAA** TÉL ACD in call; ~ **assisté par opérateur** TÉL operator-assisted call (*OAC*); ~ **en attente** TÉL waiting call, call waiting (*CW*); ~ **automatique** DONNÉES automatic call, automatic calling, TÉL automatic call, automatic calling, auto-dialing (*AmE*), auto-dialling (*BrE*), direct-dial call; ~ **avec assistance** TÉL operator-assisted call; ~ **avec préavis** TÉL person-to-person call, personal call; ~ **avec tarification au demandé** TÉL freephone call (*BrE*), toll-free call (*AmE*); ~ **ayant abouti** TÉL successful call; ~ **branché par l'opérateur** TÉL operator-connected call; ~ **cadencé** TÉL interrupted ringing; ~ **connecté par l'opérateur** TÉL operator-connected call; ~ **de courant** ALIM ÉLEC current rush, current surge; ~ **direct** DONNÉES, TÉL direct call; ~ **direct de l'extérieur** TÉL direct inward dialing (*AmE*) (*DID*), direct inward dialling (*BrE*) (*DID*); ~ **écoulé** TÉL trunk-routed call; ~ **efficace** TÉL completed call; ~ **de l'entité d'application** DONNÉES, RADIO, TÉL application-entity invocation (*AEI*); ~ **entrant** DONNÉES, RADIO, TÉL incoming call (*IC*); ~ **entre standardistes** TÉL attendant-to-attendant calling; ~ **d'essai abonné à abonné** TÉL end-to-end test call; ~ **d'essai du type abonné à abonné** TESTS test call of the subscriber-to-subscriber type; ~ **à frais virés** *Can* TÉL collect call (*AmE*), reverse-charge call (*BrE*), transfer-charge call; ~ **général** TÉL general call; ~ **global** RADIO global call; ~ **gratuit** TÉL freephone call (*BrE*), toll-free call (*AmE*); ~ **de groupe** TÉL PBX call, group number call; ~ **importun** TÉL abusive call, malicious call, nuisance call; ~ **inefficace** TÉL ineffective call, unsuccessful call; ~ **en instance** TÉL call waiting (*CW*); ~ **intérieur** TÉL inland call; ~ **interurbain** TÉL toll call (*AmE*), trunk call (*BrE*); ~ **local** TÉL city call, local call; ~ **longue distance** TÉL long-distance call, toll call (*AmE*), trunk call (*BrE*); ~ **malveillant** TÉL abusive call, malicious call; ~ **manuel** TÉL manual ringing, manual calling; ~ **non accepté** TÉL call not accepted; ~ **numéro vert** *Fra* TÉL freephone call (*BrE*), toll-free call (*AmE*); ~ **d'offres** GÉNÉRAL request for proposal (*RFP*); ~ **par sonnerie codée** TÉL coded ringing; ~ **par tonalité** RADIO, TÉL tone call; ~ **en PCV** *Fra* TÉL collect call (*AmE*), reverse-charge call (*BrE*); ~ **perdu** COMMUT, TÉL, TRANSMIS lost call; ~ **prioritaire** TÉL priority call; ~ **à prix partagé** TÉL local charge-rate call; ~ **de procédure** RADIO procedure call; ~ **en provenance de l'extérieur du réseau** TÉL off-net to on-net call; ~ **en provenance du réseau privé** TÉL on-net to off-net call; ~ **réacheminé** TÉL redirected call; ~ **réservé** APP SERV, TÉL booked call; ~ **sans numérotation** DONNÉES hot line, TÉL nondialed connection (*AmE*), nondialled connection (*BrE*); ~ **sans réponse** TÉL unanswered call; ~ **sélectif** DONNÉES, INFORMAT polling, RADIO polling, selective calling, TÉL selective ringing; ~ **sélectif numérique** RADIO digital selective call (*DSC*); ~ **semi-sélectif** TÉL coded ringing; ~ **en série** TÉL serial call; ~ **de service** TÉL service call; ~ **sortant** DONNÉES, RADIO, TÉL outgoing call; ~ **suivant** TÉL follow-on call; ~ **sur carte de crédit** TÉL credit card call; ~ **sur ligne occupée** TÉL busy occurrence; ~ **téléphonique** TÉL phone call, telephone call; ~ **de TPSU** DONNÉES TPSU invocation, transaction processing service user invocation (*TPSUI*); ~ **urgent** TÉL urgent call

appel[2]: ~ **par appel** *loc* RADIO on a per-call basis

appelant *m* RADIO originator, TÉL calling party, caller

appelé *m* TÉL called party

appeler[1] *vt* GÉNÉRAL call, RADIO initiate, *personne* page, TÉL *au téléphone* call, make a call to, phone, telephone, ring (*BrE*), *au télé-appel* page, initiate, invoke; ~ **à l'interphone** TÉL call over the intercom; ~ **par numéro vert** TÉL call on a freephone number (*BrE*), call toll-free (*AmE*); ~ **qn en PCV** TÉL call sb collect (*AmE*), make a reverse-charge call to sb (*BrE*)

appeler[2] *vi* GÉNÉRAL call, TÉL make a call, originate a call, phone, telephone, ring (*BrE*); ~ **en PCV** TÉL make a collect call (*AmE*), call collect (*AmE*), make a reverse-charge call (*BrE*), reverse the charges (*BrE*)

appellation: ~ **AE** *f* DONNÉES application-entity title; ~ **AP** *f* DONNÉES application process title; ~ **d'entité d'application** *f* DONNÉES application-entity title; ~ **de processus d'application** *f* DONNÉES application process title

appels[1]: ~ **destinés à des groupes de navires** *m pl* TÉL group calls; ~ **répétés** *m pl* TÉL repeat calls

appels[2]: ~ **en attente** *loc* PÉRIPH calls waiting; ~ **en instance** *loc* PÉRIPH calls waiting; ~ **interdits** *loc* TÉL calls barred; ~ **par seconde** *loc* GÉNÉRAL calls per second (*CS*)

application *f* GÉNÉRAL application; ~ **active** INFORMAT active application; ~ **de communication** (*CA*) DONNÉES communication application (*CA*); ~ **distribuée** INFORMAT distributed application; ~ **hypermédia** INFORMAT hypermedia application; ~ **hypertexte** INFORMAT hypertext application; ~ **d'impression à durée critique** APP SERV time-critical printing application; ~ **informatique** APP SERV, DONNÉES, INFORMAT software application; ~ **locale** (*LA*) TÉL local application (*LA*)

applications: ~ **de câblage vertical en immeuble** *f pl* TRANSMIS intrabuilding riser applications

appliquer[1] *vt* ALIM ÉLEC *tension*, INFORMAT apply; ~ **une impulsion d'horloge à** ÉLECTRON clock; ~ **une polarité négative à** ALIM ÉLEC connect negative; ~ **un traitement antistatique à** INFORMAT destaticize

appliquer[2] *vi* TÉL map

appréciation *f* GÉNÉRAL valuation; ~ **de la netteté** RADIO articulation score

apprentissage: ~ **analogique** *m* INFORMAT analogical learning

apprêt *m* GÉNÉRAL primer

apprêter *vt* ÉLECTRON, GÉNÉRAL prime

approximation: ~ **asymptomatique** *f* GÉNÉRAL asymptomatic approximation; ~ **linéaire** *f* ÉLECTRON linear approximation; ~ **de Tchebychev** *f* TRANSMIS Chebyshev approximation

apside *f* RADIO apoapsis

APT *abrév* (*Télécommunauté Asie-Pacifique*) RADIO APT (*Asia-Pacific Telecommunity*)

aptitude: ~ **du processus** *f* TESTS process capability; ~ **au service** *f* APP SERV, TÉL serviceability; ~ **au stockage** *f* TESTS storability; ~ **au traitement** *f* TESTS process capability; ~ **au transport** *f* TESTS transportability

AR *abrév* DONNÉES (*accusé de réception*) ACK (*acknowledgement*), GÉNÉRAL (*antiréflexion*) AR (*antireflection*), SIGNAL, TÉL (*accusé de réception*) ACK (*acknowledgement*)
arbitre: **~ de bus** *m* INFORMAT bus arbitrator
arborescence *f* DONNÉES, ÉLECTRON, INFORMAT, TRANSMIS tree and branch structure, tree structure; **~ de relais** COMPOSANTS relay tree
arbre *m* COMPOSANTS *génie mécanique* axle, shaft, DONNÉES, ÉLECTRON, INFORMAT tree; **~ d'appel** DONNÉES calling tree; **~ balancé** DONNÉES B-tree; **~ comportemental** DONNÉES behavior tree (*AmE*), behaviour tree (*BrE*); **~ de décision** DONNÉES decision tree; **~ de défaillance** TESTS fault tree; **~ de dénomination** DONNÉES naming tree; **~ d'informations d'annuaire** (*DIT*) DONNÉES, TÉL directory information tree (*DIT*); **~ local** DONNÉES local tree; **~ de nom hiérarchique d'enregistrement** DONNÉES registration-hierarchical name tree; **~ des pannes** TESTS fault tree; **~ racine** DONNÉES root tree
arc *m* ALIM ÉLEC arc; **~ électrique** ALIM ÉLEC electric arc; **~ de service** RADIO service arc; **~ de visibilité** ÉLECTRON visible arc
arcanson *m* ÉLECTRON colophony
Archie *m* DONNÉES *Internet* Archie
architecture *f* INFORMAT architecture; **~ arborescente** DONNÉES, INFORMAT tree and branch architecture; **~ en arbre** DONNÉES, INFORMAT tree and branch architecture; **~ à cœur réparti spécialisé** COMMUT function division system architecture; **~ en couches** INFORMAT layered architecture; **~ DCA** (*DCA*) DONNÉES document content architecture (*DCA*); **~ distribuée** COMMUT distributed architecture; **~ EISA** (*EISA*) INFORMAT extended industry standard architecture (*EISA*); **~ en étoile** TRANSMIS, TV star architecture; **~ de la gestion de réseaux** TRANSMIS network management architecture; **~ de mémoire** INFORMAT memory architecture; **~ parallèle** TRANSMIS parallel architecture; **~ de processeur** DONNÉES processor architecture; **~ de protocole télématique** (*TPA*) TÉL telematic protocol architecture (*TPA*); **~ de réseau ouverte** DONNÉES open network architecture; **~ de réseau de transport d'Ericsson** TRANSMIS Ericsson transport network architecture (*ETNA*); **~ de systèmes distribués** INFORMAT distributed-system architecture (*DSA*); **~ de systèmes ouverts** TÉL, TRANSMIS open systems architecture (*OSA*); **~ systolique** APP SERV, TRANSMIS systolic architecture
archivage *m* DONNÉES, INFORMAT, TV archiving; **~ des messages** DONNÉES message storing
archival *m* DONNÉES archivage
archive *f* DONNÉES, TÉL storage file, TV archive material
archiver *vt* DONNÉES, INFORMAT archive, TÉL put on file
archives *f pl* INFORMAT, TV archives
ardoise: **~ électronique** *f* INFORMAT notepad computer
argenter *vt* COMPOSANTS silver-plate; **~ par électrolyse** COMPOSANTS silver-electroplate
argenture *f* COMPOSANTS silver-plating; **~ électrolytique** COMPOSANTS silver-electroplating
argument *m* DONNÉES argument; **~ de paramètre** TÉL parameter argument; **~ de paramètre composé** TÉL compound parameter argument; **~ de paramètre simple** TÉL simple parameter argument; **~ du périgée** RADIO argument of perigee
armature *f* ALIM ÉLEC *de condensateur* plate, INFORMAT key signature; **~ d'un condensateur** ALIM ÉLEC capacitor plate
armer *vt* RADIO set, TRANSMIS *câble* armor (*AmE*), armour (*BrE*)
armoire *f* COMMUT cabinet, INFORMAT rack, TRANSMIS cabinet; **~ d'alimentation** ALIM ÉLEC power rack; **~ à batteries** ALIM ÉLEC battery rack; **~ de câblage** ÉLECTRON wiring closet; **~ à câbles** TRANSMIS *pose de câbles* cable cabinet; **~ de commande** TRANSMIS cabinet; **~ de contrôle** CONTRÔLE control cabinet; **~ de conversion** ALIM ÉLEC converter cabinet; **~ de convertisseur** ALIM ÉLEC converter rack; **~ de dérivation** ALIM ÉLEC cable distribution cabinet, distribution cabinet; **~ de distribution** ALIM ÉLEC distribution rack; **~ de distribution par câble** ALIM ÉLEC cable distribution cabinet; **~ murale** TÉL wall cabinet; **~ quartier** TV roadside cabinet, streetside cabinet; **~ de répartiteur** TÉL cross-connect cabinet; **~ satellite** RADIO satellite closet; **~ à sept logements** COMPOSANTS, ÉLECTRON seven-slot rack mounting assembly; **~ de sous-répartiteur** TRANSMIS distribution cabinet; **~ de sous-répartition** TRANSMIS subdistribution cupboard, TV roadside cabinet, streetside cabinet; **~ standard** ÉLECTRON component housing; **~ de télécommunications** ÉLECTRON telecommunications closet (*TC*)
armoires: **~ batterie** *f pl* TRANSMIS battery cabinets; **~ de sous-répartition** *f pl* TRANSMIS roadside furniture, streetside furniture
armure *f* COMPOSANTS armor (*AmE*), armour (*BrE*), armoring (*AmE*), armouring (*BrE*); **~ en acier** COMPOSANTS steel armor (*AmE*), steel armour (*BrE*); **~ de clé** INFORMAT key signature; **~ externe** TRANSMIS cable armor (*AmE*), cable armour (*BrE*); **~ en feuillard métallique** COMPOSANTS steel tape armor (*AmE*), steel tape armour (*BrE*)
arpège *m* INFORMAT arpeggio
arpégiateur *m* INFORMAT arpeggiator
arrangement: **~ simplifié par rapport au modèle de Clos** *m* TÉL, TRANSMIS sub-Clos arrangement
arrangeur *m* INFORMAT arranger; **~ automatique** INFORMAT automatic arranger, auto-arranger
arrêt *m* ALIM ÉLEC shutdown, GÉNÉRAL cutoff, INFORMAT standstill, TÉL *du courant d'appel* trip, tripping; **~ automatique** INFORMAT automatic cutoff, auto-stop, automatic stop; **~ automatique de sonnerie** TÉL automatic ring trip, automatic ring tripping; **~ bouclé** INFORMAT loop stop; **~ de chronologie** ÉLECTRON hold; **~ de courant d'appel** TÉL ring trip, ring tripping; **~ dispositif** TÉL device stop; **~ immédiat de l'appel** TÉL ring trip, ring tripping; **~ non programmé** TESTS unscheduled stop; **~ programmé** TESTS scheduled stop; **~ de sonnerie** TÉL ring trip, ring tripping; **~ sur image** TV freeze-frame, still frame, stop frame; **~ de surtension** ÉLECTRON surge arrester; **~ d'urgence** GÉNÉRAL emergency tripping device
arrêter *vt* GÉNÉRAL cut off, switch off, turn off, TÉL trip; **~ le courant d'appel** TÉL trip the ringing
arriéré *m* INFORMAT backlog
arrière: **d'~-plan** *adj* GÉNÉRAL, INFORMAT background
arséniure: **~ de gallium** *m* COMPOSANTS gallium arsenide (*GaAs*)
artère *f* TÉL traffic route; **~ principale** TÉL backbone route; **~ de transmission** TRANSMIS transmission highway
articulation *f* INFORMAT joint
arythmique *adj* GÉNÉRAL start-stop
AS *abrév* (*adresse système*) TÉL SA (*system address*)
ascenseur *m* INFORMAT *Macintosh* scroll box
ASCII *abrév* (*American Standard Code for Information Interchange*) DONNÉES, INFORMAT, TÉL ASCII (*American Standard Code for Information Interchange*)
ASE *abrév* (*élément de service d'application*) DONNÉES, TÉL ASE (*application service element*); **~ d'utilisateur** *m* DONNÉES, TÉL user-ASE
ASIC *abrév* (*circuit intégré spécifique*) COMPOSANTS, ÉLECTRON ASIC (*application-specific integrated circuit*)
ASN *abrév* (*Association suisse de normalisation*) GÉNÉRAL Swiss Standards Association
ASP *abrév* (*primitive de service abstraite*) DONNÉES ASP (*abstract service primitive*)
aspiration: **~ à vide** *f* ÉLECTRON vacuum suction
ASP(N) *abrév* (*primitive de service abstraite (N)*) DONNÉES ASP(N) (*abstract (N) service primitive*)

assemblage *m* GÉNÉRAL assembling, assembly; ~ **de câble** COMPOSANTS, GÉNÉRAL, TRANSMIS cable assembly; ~ **de câble optique** TRANSMIS optical cable assembly; ~**-désassemblage de paquets** (*PAD*) DONNÉES, TÉL packet assembly-disassembly facility (*PAD*); ~**-désassemblage de paquets pour la télécopie** DONNÉES, TÉL facsimile packet assembly-disassembly; ~ **séquentiel** INFORMAT sequential assembling

assembleur *m* INFORMAT assembler, assembly program, TRANSMIS jointer; ~ **de câbles** TRANSMIS *outil* cable jointer; ~ **croisé** INFORMAT cross assembler; ~**-désassembleur de paquets** (*PAD*) DONNÉES, TÉL packet assembler-disassembler (*PAD*)

assertion: ~ **de valeur d'attribut** *f* DONNÉES attribute-value assertion (*AVA*)

asservissement *m* SIGNAL compelled signaling (*AmE*), compelled signalling (*BrE*); ~ **maître-esclave** CONTRÔLE master-slave control; ~ **numérique** ÉLECTRON digital feedback

assignation *f* GÉNÉRAL assignment; ~ **de fréquence** RADIO frequency assignment

assistance: ~ **annuaire** *f* (*AA*) APP SERV, TÉL directory assistance, directory enquiries (*BrE*) (*DQ*), directory information (*AmE*); ~ **du standardiste** *f* TÉL attendant recall

assistant: ~ **en micro-informatique** *m* GÉNÉRAL PC support engineer; ~ **numérique personnel** *m* INFORMAT personal digital assistant

assisté: ~ **par ordinateur** *loc* INFORMAT computer-aided, computer-assisted

association *f* DONNÉES association, INFORMAT relationship; ~ **abstraite** DONNÉES abstract association; ~ **d'application** DONNÉES application association; ~ **N** DONNÉES N-association

Association: ~ **canadienne de normalisation** *f* (*ACN*) GÉNÉRAL Canadian Standards Association (*CSA*); ~ **européenne de constructeurs d'ordinateurs** *f* (*ECMA*) INFORMAT European Computer Manufacturers Association (*ECMA*); ~ **française de normalisation** *f* (*AFNOR*) GÉNÉRAL French National Standards Association (*AFNOR*); ~ **internationale de normalisation des cartes à mémoire pour PC** *f* (*PCMCIA*) INFORMAT Personal Computer Memory Card International Association (*PCMCIA*); ~ **internationale de retransmission** *f* (*AIR*) RADIO International Association of Broadcasting (*IAB*); ~ **internationale de signalisation maritime** *f* (*AISM*) RADIO International Association of Lighthouse Authorities (*IALA*); ~ **suisse pour certificats d'assurance de la qualité** *f* (*SQS*) GÉNÉRAL Swiss Association for Quality Assurance Certificates (*SQS*); ~ **suisse de normalisation** *f* (*ASN*) GÉNÉRAL Swiss Standards Association

associé: **non** ~ *m* TÉL nonassociated mode

assombrir *vt* TV darken

assourdir *vt* INFORMAT mute

assurance: ~ **qualité** *f* APP SERV, TESTS quality assurance

astatique *adj* GÉNÉRAL *physique* astatic

asymétrie *f* ALIM ÉLEC, INFORMAT unbalance

asymétrique *adj* ALIM ÉLEC unbalanced, GÉNÉRAL asymmetrical, INFORMAT unbalanced

asynchrone *adj* DONNÉES asynchronous, ÉLECTRON unclocked, INFORMAT, TRANSMIS asynchronous

atelier: ~ **d'énergie** *m* ALIM ÉLEC power room; ~ **de réparation** *m* TÉL repair shop

Atelier: ~ **européen pour les systèmes ouverts** *m* (*EWOS*) GÉNÉRAL European Workshop for Open Systems (*EWOS*)

atmosphère *f* GÉNÉRAL atmosphere; ~ **radioélectrique normale** RADIO standard radio atmosphere; ~ **de référence pour la réfraction** RADIO reference atmosphere for refraction

ATN *abrév* (*affaiblissement de transmission normalisé*) TRANSMIS NTL (*normalized transmission loss*)

ATNL *abrév* (*affaiblissement par traitement non linéaire*) TÉL nonlinear processing loss

atomicité *f* DONNÉES atomicity; ~, **cohérence, isolation et durabilité** (*ACID*) DONNÉES atomicity, consistency, isolation and durability (*ACID*)

atonal *adj* INFORMAT atonal

ATRMA *abrév* (*affaiblissement de transmission de référence minimal admissible*) TRANSMIS MPBTL (*minimum permissible basic transmission loss*)

ATS *abrév* (*appareil de télécommunications pour les sourds*) TÉL TDD (*telecommunications device for the deaf*)

attache *f* COMPOSANTS clip; ~ **autoblocante** COMPOSANTS tie wrap, TY wrap; ~ **de conducteur** TRANSMIS cable lug; ~**-fil** COMPOSANTS wire fastener; ~ **de profilé** COMPOSANTS bar holder; ~**-ressort** COMPOSANTS spring holder

attacher *vt* ÉLECTRON clip

attaches *f pl* COMPOSANTS fasteners; ~ **de câble** COMPOSANTS cable tie

attaque *f* GÉNÉRAL, INFORMAT attack; ~ **chimique** GÉNÉRAL etching

attaquer *vt* GÉNÉRAL, INFORMAT attack; ~ **chimiquement** GÉNÉRAL etch

attaqueur: ~ **de relais** *m* ALIM ÉLEC relay driver

atteinte: ~ **de but** *f* RADIO hit

attendez *loc* TÉL stand by

attente[1] *f* INFORMAT, RADIO stand-by; ~ **après numérotation** RADIO, TÉL post-dialing delay (*AmE*), post-dialling delay (*BrE*); ~ **connexion** TÉL connect hold; ~ **d'enregistrement** INFORMAT record stand-by; ~ **de message** TÉL await message; ~ **de message de service** TÉL wait-for-overhead message; ~ **musicale** TÉL courtesy music, music-on-hold system

attente[2]: **d'**~ *loc* ALIM ÉLEC stand-by; **en** ~ *loc* GÉNÉRAL standby

atténuateur *m* ALIM ÉLEC attenuator, pad, INFORMAT attenuator, fader, RADIO, TRANSMIS attenuator, pad; ~ **à absorption** TRANSMIS absorptive attenuator; ~ **étalonné** TRANSMIS calibrated attenuator; ~ **fixe** ALIM ÉLEC pad, TRANSMIS fixed attenuator, pad; ~ **en guide d'ondes** RADIO, TRANSMIS waveguide attenuator; ~ **à guillotine** RADIO guillotine attenuator; ~ **hyperfréquence** RADIO microwave attenuator; ~ **à lame** RADIO flap attenuator, vane attenuator; ~ **de niveau** INFORMAT level attenuator; ~ **de niveau de sortie** INFORMAT output-level attenuator; ~ **optique** TRANSMIS optical attenuator; ~ **de perte d'insertion** TRANSMIS insertion loss pad; ~ **à piston** RADIO piston attenuator; ~ **réactif** RADIO reactive attenuator; ~ **résistif** ÉLECTRON resistive attenuator; ~ **rotatif** RADIO rotary attenuator; ~ **unidirectionnel** RADIO isolator

atténuation *f* ALIM ÉLEC attenuation, INFORMAT, RADIO attenuation, loss, TRANSMIS attenuation, damping; ~ **d'affaiblissement d'équilibrage** *m* TRANSMIS balance-return loss attenuation; ~ **d'équilibrage** TRANSMIS balance-return attenuation; ~ **à fréquence intermédiaire** RADIO IF rejection, intermediate-frequency rejection; ~ **géométrique** RADIO spreading loss; ~ **en mode différentiel** TRANSMIS differential-mode attenuation; ~ **de niveau de sortie** INFORMAT output-level attenuation; ~ **numérique** TRANSMIS digital loss; ~ **de propagation** RADIO, TÉL path attenuation; ~ **de tension** ALIM ÉLEC voltage attenuation; ~ **vers l'arrière** TRANSMIS backward attenuation

atténuer *vt* ALIM ÉLEC, INFORMAT, RADIO, TRANSMIS attenuate

atterrissage: ~ **de câble** *m* TRANSMIS cable landing

attribuer *vt* ALIM ÉLEC apply, ÉLECTRON allot, INFORMAT allocate, apply, assign, RADIO allocate, assign, allot, TÉL allot, *classes de service* assign, TRANSMIS allocate

attribut *m* DONNÉES, GÉNÉRAL attribute; ~ **d'accès** DONNÉES access attribute; ~ **défini de domaine** TÉL domain-defined attribute, defined domain attribute (*DDA*); **~ pour un ensemble de valeurs** DONNÉES set-valued attribute; **~ de fichier** DONNÉES, INFORMAT file attribute; **~ général** DONNÉES general attribute; **~ à plusieurs valeurs** DONNÉES multivalued attribute; **~ à valeur unique** DONNÉES single-valued attribute; **~ vidéo** TV video attribute

attribution *f* RADIO allocation, TÉL *classes de service* assignment; **~ aléatoire** TÉL random allotment; **~ du bruit** TRANSMIS noise allocation; **~ de capacité** RADIO allotment of capacity; **~ dynamique de mémoire** INFORMAT dynamic storage allocation; **~ dynamique des ressources** INFORMAT dynamic resource allocation, dynamic allocation; **~ d'espace** GÉNÉRAL space allocation; **~ de fréquence** RADIO frequency allocation, frequency allotment, frequency assignment; **~ de mémoire** INFORMAT storage allocation; **~ de l'orbite** RADIO assignment of orbit; **~ de répéteurs** DONNÉES, ÉLECTRON, RADIO transponder allotment; **~ de valeur** INFORMAT value assignment

attributs: **~ de fichier BFT** *m pl* TÉL BFT file attributes; **~ du nœud** *m pl* INFORMAT node attributes; **~ du secteur** *m pl* INFORMAT sector attributes

audibilité *f* GÉNÉRAL clarity, *acoustique* audibility, INFORMAT, RADIO audibility

audible *adj* GÉNÉRAL, INFORMAT, RADIO audible

audio[1] *adj* INFORMAT audio; **~-numérique** INFORMAT audio-digital

audio[2]: **~-messaging interchange specifications** *m* TÉL audio-messaging interchange specifications; **~-numérique multicanaux** *m* (*MADI*) INFORMAT multichannel audio-digital interface (*MADI*); **~-vidéographie** *f* APP SERV audio-videotex; **~-vidéotexte** *m* APP SERV audio-videotex

audioconférence *f* RADIO, TÉL audio conference

audiofréquence *f* RADIO, TÉL audio frequency (*AF*)

Audio-Interchange: **~ File Format** *m* DONNÉES, INFORMAT Audio-Interchange File Format

audiovisuel[1] *adj* TV audiovisual

audiovisuel[2]: **~ interactif** *m* TV interactive audiovisual

audit *m* GÉNÉRAL audit; **~ de qualité** APP SERV, TESTS quality audit; **~ de sécurité** DONNÉES security audit

augmentation *f* INFORMAT enhancement; **~ de capacité** GÉNÉRAL capacity increase

augmenter *vt* INFORMAT enhance; **~ par transformateur** ALIM ÉLEC *tension* step up

authentification *f* DONNÉES authentication, SIGNAL authentication, authentication procedure; **~ de l'entité homologue** DONNÉES peer-entity authentication, peer-identity authentication; **~ ferme** DONNÉES strong authentication; **~ de l'origine de données** DONNÉES data-origin authentication; **~ simple** DONNÉES simple authentication

auto-amorçage *m* ÉLECTRON self-boot

autocommutateur *m* COMMUT switcher, DONNÉES central switching office, TÉL automatic exchange, automatic telephone exchange, switcher; **~ électronique à commande centralisée** COMMUT centralized control system; **~ à intégration voix-données** COMMUT integrated voice-data switch; **~ multiservice** COMMUT multiservice automatic exchange; **~ numérique** COMMUT digital switch; **~ numérique à intégration de services** COMMUT integrated digital services exchange; **~ privé** (*PABX*) APP SERV, COMMUT, CONTRÔLE, TÉL private automatic branch exchange (*PABX*), private automatic exchange (*PAX*); **~ privé à programme enregistré** APP SERV, COMMUT, CONTRÔLE, TÉL stored-program control (*AmE*) (*SPC*), stored-programme control (*BrE*) (*SPC*); **~ privé réparti** APP SERV, COMMUT, CONTRÔLE, TÉL distributed PBX; **~ privé pour le RNIS-LB** APP SERV, COMMUT, CONTRÔLE, TÉL broadband ISDN PBX, TÉL B-ISDN PBX; **~ privé sans fil** APP SERV, COMMUT, CONTRÔLE, TÉL wireless PBX; **~ privé voix-données** APP SERV, COMMUT, CONTRÔLE, TÉL integrated voice-data PABX; **~ à quatre fils** COMMUT four-wire switch; **~ radiotéléphonique** RADIO radio exchange; **~ de rattachement** COMMUT, TÉL local switch; **~ RNIS** COMMUT ISDN switch; **~ temporel** COMMUT time switch; **~ visiophonique** COMMUT videophone switching system

autocommutateurs *m pl* COMMUT electronic automatic exchanges

autoconfiguration *f* ÉLECTRON auto-installation

autocorrection *f* DONNÉES, INFORMAT, SIGNAL, TÉL auto-correction

autodécharge *f* ALIM ÉLEC self-discharge

auto-échauffement *m* ALIM ÉLEC self-heating

auto-enregistrement *m* DONNÉES self-registration

auto-identification *f* RADIO self-identification

auto-inductance *f* ALIM ÉLEC self-inductance

automate: **~ à états finis** *m* DONNÉES finite automation

automatique[1] *adj* GÉNÉRAL, INFORMAT, RADIO, TÉL automatic

automatique[2]: **~ international** *m* TÉL international direct dialing (*AmE*) (*IDD*), international direct dialling (*BrE*) (*IDD*), international direct distance dialing (*AmE*) (*IDDD*), international direct distance dialling (*BrE*) (*IDDD*), international subscriber dialing (*AmE*), international subscriber dialling (*BrE*)

automatisation *f* CONTRÔLE automatization, GÉNÉRAL automation; **~ de conception électronique** COMPOSANTS electronic design automation (*EDA*)

automatiser *vt* CONTRÔLE automatize, GÉNÉRAL automate

automode *m* DONNÉES auto-mode

automodulation *f* TRANSMIS auto-modulation; **~ de phase** TRANSMIS self-phase modulation (*SPM*)

autonome *adj* DONNÉES off-line, stand-alone, ÉLECTRON self-contained, INFORMAT autonomous, off-line, self-contained, stand-alone, TÉL autonomous, off-line

auto-optimisation *f* ÉLECTRON self-optimization

autorégénération *f* ALIM ÉLEC self-healing

autoriser *vt* GÉNÉRAL license

autorisation *f* COMPOSANTS, CONTRÔLE, DONNÉES clearance, GÉNÉRAL authorization, INFORMAT clearance, *action de donner un permis* licensing, TÉL clearance; **~ d'accès par opérateur** TÉL operator authorization; **~ d'exploitation** TV operating authorization

autorité *f* GÉNÉRAL authority; **~ d'adressage de réseau** DONNÉES network-addressing authority; **~ de certification** TESTS certification authority; **~ chargée de la comptabilité** TÉL accounting authority; **~ de dénomination** DONNÉES naming authority

autoroute: **~ de l'information** *f* TRANSMIS information highway

auto-shunt *m* COMPOSANTS, ÉLECTRON thinnet tap

autosynchronisation *f* TRANSMIS auto-synchronization

autotarification *f* APP SERV self-metering

autotransformateur *m* ALIM ÉLEC autotransformer, single-winding transformer

auxiliaire *adj* GÉNÉRAL backup, ancillary

aval: **en ~ de** *loc* GÉNÉRAL downstream

avalanche *f* ÉLECTRON *semiconducteurs* avalanche

avance: **~ pas à pas** *f* GÉNÉRAL stepping; **~ de phase du signal transmis** *f* TRANSMIS transmitted phase advance; **~ rapide** *f* TÉL fast-forward playback

avancement: **~ d'un interligne** *m* DONNÉES line-feed character; **~ du papier** *m* PÉRIPH sheet feed

avant: **~-première** *f* TV movie preview

avertir *vt* RADIO alert

avertissement *m* TÉL alert; **~ automatique** DONNÉES auto-alert; **~ radio** RADIO radio warning; **~ radioélectrique** RADIO radio warning

avertisseur *m* RADIO detector; **~ lumineux** CONTRÔLE warning lamp, warning light, TESTS flashing light; **~ radio** RADIO bleeper; **~ sonore** RADIO audible alarm

avis *m* GÉNÉRAL notification, notice; **~ de chaîne de service** DONNÉES service string advice; **~ de modification** GÉNÉRAL change notice (*CN*); ~ **de non-réception** DONNÉES, TÉL nonreceipt notification (*NRN*), nonrecipient notification (*NRN*); ~ **de non-remise** DONNÉES, TÉL nondelivery notification (*NDN*); ~ **de réception** DONNÉES, TÉL receipt notification (*RN*); ~ **de service** DONNÉES service notification (*SN*), TÉL service advice, service notification (*SN*); ~ **de service en retour** DONNÉES, TÉL return-service advice; ~ **de service à la suite** DONNÉES, TÉL follow-on service advice; ~ **de service taxé** DONNÉES, TÉL paid-service advice; ~ **de taxation** APP SERV advice of charge, charge advice, charging information

axe *m* COMPOSANTS *génie mécanique* axle, shaft, spindle, ÉLECTRON route, INFORMAT axis; ~ **de chrominance** TV chrominance axis; ~ **de faisceau** RADIO *antennes* beam axis; ~ **du faisceau** RADIO beam center (*AmE*), beam centre (*BrE*); ~ **de faisceau électromagnétique** RADIO beam axis; ~ **de la fibre** TRANSMIS fiber axis (*AmE*), fibre axis (*BrE*), optic axis, optical axis; ~ **des gains** ÉLECTRON gain scale; ~ **de luminance** INFORMAT luminance axis; ~ **optique** TRANSMIS optic axis, optical axis; ~ **de référence d'espacement** INFORMAT character-spacing reference line; ~ **de roulis** RADIO roll axis; ~ **de symétrie** RADIO center line (*AmE*), centre line (*BrE*); ~ **de temps** TESTS, TRANSMIS time axis; ~ **de visée** RADIO boresight, TRANSMIS line of sight

axial *adj* GÉNÉRAL axial

axiome *m* DONNÉES axiom

azimut *m* RADIO azimuth; ~ **du grand arc** RADIO long-path bearing; ~ **inverse** RADIO reciprocal bearing

B

babillard *m Can* DONNÉES, INFORMAT, TÉL notice board; **~ électronique** *Can* DONNÉES, INFORMAT, TÉL bulletin board (*BB*), bulletin board system (*BBS*), notice board

bac *m* TÉL subrack; **~ à compartiments** ALIM ÉLEC *batteries* block container, multiple container; **~ d'éléments** ALIM ÉLEC *batteries* battery cell container, cell container; **~ de mélange** ÉLECTRON mixing tray; **~ à papier** PÉRIPH paper tray

bague *f* COMPOSANTS, TRANSMIS ferrule; **~ collectrice** ALIM ÉLEC slip ring; **~ d'espacement** COMPOSANTS spacing ring; **~ filetée** COMPOSANTS, ÉLECTRON threaded bushing; **~ de fixation** COMPOSANTS fixing ring; **~ de retard** COMPOSANTS *relais* delay slug, slug

baguette *f* ÉLECTRON rod, TRANSMIS preformed fiber (*AmE*), preformed fibre (*BrE*)

baie *f* TÉL bay, rack

baïonnette: **~ à** *adj* ALIM ÉLEC, COMPOSANTS, ÉLECTRON bayonet

baisse: **~ de tension** *f* ALIM ÉLEC brownout

bakélite *f* COMPOSANTS *matières plastiques* bakelite

BAL *abrév* (*boîte aux lettres*) DONNÉES mailbox

baladeuse *f* GÉNÉRAL inspection lamp

balai *m* COMMUT wiper

balance: **~ droite-gauche** *f* INFORMAT left-right balance

balayage *m* ÉLECTRON, INFORMAT, PÉRIPH, RADIO scanning, scan; **~ aléatoire** TV stroke writing; **~ avec effet de loupe** RADIO expanded sweep; **~ avec entrelacement** TV interlaced scanning; **~ en basse fréquence** INFORMAT low-frequency sweep; **~ cavalier** INFORMAT directed scan, directed-beam scan, vector scanning; **~ conique** RADIO conical scan; **~ conique avec nutation** RADIO conical scanning with nutation; **~ découpé** TV dot interlace scanning, dot interlace sweep; **~ dilaté** RADIO expanded sweep; **~ élargi** RADIO expanded sweep; **~ entrelacé** TV interlaced scanning; **~ de fréquence** RADIO, TRANSMIS frequency scan, frequency sweep; **~ horizontal** INFORMAT line scan, TV horizontal sweep; **~ intermédiaire** RADIO interscan; **~ de ligne** TV horizontal sweep, line sweep; **~ ligne par ligne** INFORMAT progressive scanning, sequential scanning; **~ par électrons lents** ÉLECTRON low-velocity scanning; **~ par électrons rapides** ÉLECTRON high-velocity scanning; **~ par lignes** INFORMAT line scan; **~ par lignes contiguës** INFORMAT progressive scanning; **~ du rayon laser** ÉLECTRON laser beam scanning; **~ sans entrelacement** INFORMAT progressive scanning, sequential scanning; **~ sectoriel** RADIO sector scanning; **~ séquentiel** TÉL sequential scanning; **~ de trames** INFORMAT field scan, vertical scan, TV field sweep; **~ vectoriel** INFORMAT vector scanning

balayer *vt* ÉLECTRON sweep, scan, INFORMAT scan, PÉRIPH, RADIO scan, sweep

balayeur: **~ de fréquence** *m* RADIO, TRANSMIS frequency scanner

balisage: **~ ésotérique** *m* DONNÉES esoteric marking; **~ manuel** *m* DONNÉES manual marking

balise *f* RADIO beacon; **~ embarquée** RADIO on-board beacon; **~ passive** RADIO radar reflector buoy; **~ prédéfinie** DONNÉES predefined marker; **~ radar** (*racon*) RADIO radar beacon (*racon*); **~ de radiodétection** (*racon*) RADIO radar beacon (*racon*)

ballast *m* ALIM ÉLEC ballast; **~ capacitif à crête d'amorçage** ALIM ÉLEC lead-peak ballast

ballon: **~ de transmission de données** *m* DONNÉES, TRANSMIS data transmission balloon

banc: **~ de broches décalé** *m* TÉL slipped bank; **~ de contacts** *m* ALIM ÉLEC contact bank; **~ de contrôle** *m* TESTS check-out equipment; **~ d'essai** *m* ALIM ÉLEC test bed, test bench, TESTS test bed, test bench, test stand, proving stand; **~ d'essai de ligne** *m* TESTS line test set; **~ de filtrage** *m* INFORMAT filterbank; **~ de filtres** *m* INFORMAT filterbank; **~ de filtres d'analyse** *m* INFORMAT analysis filterbank; **~ optique** *m* TRANSMIS optical bench; **~ de test** *m* ALIM ÉLEC, TESTS test bed, test bench; **~ de test et de mesure** *m* TESTS test and measurement set

bancs: **~ de broches alignés** *m pl* TÉL straight bank

bande[1] *f* APP SERV tape, GÉNÉRAL ribbon, INFORMAT band, tape; **~ atténuée** TRANSMIS stop band, suppressed frequency band; **~ audio** INFORMAT audio tape; **~ audionumérique** (*DAT*) INFORMAT, TV digital audio tape (*DAT*); **~ audionumérique à tête rotative** (*R-DAT*) DONNÉES, INFORMAT rotary-head digital audio tape (*R-DAT*); **~ de base** RADIO, TRANSMIS, TV baseband; **~ basse** TRANSMIS low-pass band; **~ C** RADIO C-band; **~ CB** RADIO citizens' band (*CB*); **~ chauffante** COMPOSANTS, ÉLECTRON heat strip; **~ de conduction** ÉLECTRON conduction band; **~ de connexion** ALIM ÉLEC, COMPOSANTS strap; **~ de contact** RADIO finger strip, finger stock; **~ de court-circuit** ALIM ÉLEC, COMPOSANTS jumper; **~ critique** TÉL critical band; **~ effective globale de bruit** TRANSMIS effective overall noise band; **~ d'énergie** COMPOSANTS energy band; **~ à entraînement décalé vers l'avant** INFORMAT advance-feed tape; **~ étiquetée de désignation transparente** COMPOSANTS, ÉLECTRON transparent designation labeling strip (*AmE*), transparent designation labelling strip (*BrE*); **~ étroite** TRANSMIS narrow band; **~ d'extension** RADIO expansion band; **~ de fréquence assignée** RADIO assigned-frequency band; **~ de fréquences** RADIO band, frequency band; **~ des fréquences** RADIO frequency range; **~ de fréquences vidéo** TRANSMIS, TV videofrequency band; **~ des fréquences vocales** TRANSMIS speech band, voice-frequency band; **~ de garde** RADIO guard band; **~ d'incertitude des fréquences** TÉL frequency-uncertainty band; **~ interdite** COMPOSANTS energy gap; **~ intermédiaire satellite** RADIO intermediate satellite band; **~ large** TRANSMIS wideband; **~ latérale** (*BL*) RADIO, TÉL, TRANSMIS sideband (*SB*); **~ latérale avec porteuse** RADIO, TÉL, TRANSMIS sideband with carrier; **~ latérale indépendante** (*BLI*) RADIO independent sideband (*ISB*); **~ latérale inférieure** TRANSMIS lower sideband (*LSB*); **~ latérale résiduelle** (*BLR*) RADIO, TRANSMIS, TV residual sideband (*RSB*); **~ latérale restante** (*BLR*) RADIO, TRANSMIS, TV vestigial sideband (*VSB*); **~ latérale supérieure** (*BL sup.*) RADIO, TRANSMIS upper sideband (*USB*); **~ latérale unique** (*BLU*) RADIO, TRANSMIS single sideband (*SSB*); **~ magnétique** DONNÉES, INFORMAT magnetic tape; **~ magnétique sonore** DONNÉES, INFORMAT audio tape; **~ mère** TV master tape; **~ occupée** RADIO occupied band; **~ passante** DONNÉES, ÉLECTRON bandwidth, INFORMAT, RADIO, TÉL, TRANSMIS pass band, TV bandwidth; **~ passante à la demande** INFORMAT bandwidth on demand; **~ passante limitée** DONNÉES limited bandwidth; **~ perforée à alignement frontal** INFORMAT advance-feed tape; **~ pilote** DONNÉES format tape, PÉRIPH carriage-control tape; ~ **de radiodiffusion** RADIO broadcasting band; ~ **relais** ÉLECTRON tag strip; ~ **de satellite** RADIO satellite band; ~ **de sauvegarde** INFORMAT backing tape, backup tape, TESTS backup tape; ~ **de sécurité** APP SERV clear area, clear band; ~ **son** INFORMAT audio tape, soundtrack, TV soundtrack; ~ **sonore** INFORMAT audio tape; ~ **de synchronisation** RADIO lock-in range; ~ **de valence** COMPOSANTS valence band; ~ **vidéo** TV video tape, videotape; ~ **vidéofréquences** TRANSMIS, TV videofrequency band; ~ **vocale** DONNÉES voice band

bande[2]: **dans la** ~ *loc* SIGNAL, TRANSMIS in-band; **en ~ de base** *loc* RADIO, TRANSMIS, TV baseband; **à ~ latérale unique** *loc* (*BLU*) RADIO, TRANSMIS single-sideband (*SSB*)

bandeaux *m pl* INFORMAT bands

banque *f* DONNÉES, INFORMAT, TV bank; ~ **de données** DONNÉES data bank; ~ **d'images** TV image bank

barème *m* APP SERV scale of charges

Bark *m* INFORMAT Bark

barre: ~ **d'alimentation** *f* ALIM ÉLEC, INFORMAT, TÉL busbar; ~ **d'alimentation principale** *f* ALIM ÉLEC main bus; ~ **bus à courant continu** *f* ALIM ÉLEC DC busbar; ~ **bus en cuivre** *f* ALIM ÉLEC copper busbar; ~ **de défilement** *f* INFORMAT scroll bar; ~ **de distribution d'énergie** *f* ALIM ÉLEC, INFORMAT, TÉL busbar; ~ **d'énergie** *f* ALIM ÉLEC power busbar; ~ **d'espacement** *f* INFORMAT, PÉRIPH space bar; ~ **d'essai** *f* COMMUT, TESTS test bar; ~ **d'essai d'alimentation** *f* ALIM ÉLEC power-test bus; ~ **guide-fil** *f* COMPOSANTS wire guide beam; ~ **d'impression** *f* TÉL print bar, type bar; ~ **d'interconnexion d'éléments** *f* ALIM ÉLEC cell interconnection bar; ~ **de levage de verticale** *f* COMMUT lifting bar; ~ **de masse** *f* ALIM ÉLEC, ÉLECTRON earth bar (*BrE*), ground bar (*AmE*); ~ **de menu** *f* INFORMAT menu bar; ~ **de métallisation** *f* ÉLECTRON plating bar; ~ **de mise à la terre** *f* ALIM ÉLEC, ÉLECTRON earth bar (*BrE*), ground bar (*AmE*); ~ **neutre** *f* ALIM ÉLEC neutral bar; ~ **oblique** *f* INFORMAT oblique stroke, slant, slash; ~ **oblique inverse** *f* INFORMAT inverse solidus, back slash; ~ **omnibus** *f* ALIM ÉLEC, INFORMAT, TÉL busbar; ~ **omnibus d'alimentation** *f* ALIM ÉLEC busbar, power busbar, INFORMAT, TÉL busbar; ~ **omnibus de distribution** *f* ALIM ÉLEC busbar, distribution busbar, INFORMAT, TÉL busbar; ~ **omnibus principale** *f* ALIM ÉLEC main busbar; ~ **positive** *f* ALIM ÉLEC positive bar; ~ **de régulation de puissance** *f* ALIM ÉLEC power control bus; ~ **de sélection** *f* COMMUT, COMPOSANTS selecting bar; ~ **de sélection d'aiguillage de niveau** *f* COMMUT level-selecting horizontal bar; ~ **de sélection de niveau de déclenchement** *f* COMMUT level-selecting bar; ~ **de sélection numérique** *f* COMMUT numerical horizontal, numerical selecting bar; ~ **de soutien** *f* COMPOSANTS stay bar; ~ **de support** *f* COMPOSANTS supporting bar; ~ **de titre** *f* INFORMAT title bar

barreau: ~ **d'échelle** *m* TRANSMIS *pose de câbles* cable-run rung

barres: ~ **de court-circuit** *f pl* ÉLECTRON shorting bars

barrette *f* ALIM ÉLEC, COMPOSANTS jumper, strap, TESTS *outils* barrette file; ~ **à bornes** ÉLECTRON connection strip, tag strip; ~ **de connexion** ÉLECTRON connection strip; ~ **à contacts lyre** ÉLECTRON fork contact unit; ~ **à douilles** ALIM ÉLEC sleeve contact unit

barrière: ~ **de potentiel** *f* COMPOSANTS potential barrier; ~ **de vapeur** *f* CLIMAT vapor barrier (*AmE*), vapour barrier (*BrE*)

bartering *m* TV bartering

bas *m* INFORMAT bottom, low; ~ **débit** GÉNÉRAL low speed; ~ **médium** INFORMAT low medium

bascule *f* ÉLECTRON flip-flop (*FF*); ~ **adressable** ÉLECTRON *éléments logiques* addressable latch; ~ **bistable** ÉLECTRON flip-flop (*FF*); ~ **bistable à une entrée** ÉLECTRON single-control bistable trigger circuit; ~ **de changement de programme** INFORMAT program switch flip-flop; ~ **déclenchée** COMPOSANTS, ÉLECTRON triggered flip-flop; ~ **déclenchée par front d'impulsion** ÉLECTRON edge-triggered bistable; ~ **de dérivation** COMMUT branch flip-flop; ~ **à entrée de forçage** COMPOSANTS, ÉLECTRON steered flip-flop; ~ **esclave** COMPOSANTS slave flip-flop; ~ **instable** ÉLECTRON multivibrator (*MV*); ~ **maître-esclave** ÉLECTRON master-slave flip-flop; ~ **monostable** COMPOSANTS, ÉLECTRON single-shot flip-flop; ~ **de Schmitt** ÉLECTRON Schmitt limiter, Schmitt trigger; ~ **de temporisation** COMPOSANTS, TÉL time supervision flip-flop

basculement *m* ALIM ÉLEC toggling, COMMUT side shifting, toggling, ÉLECTRON toggling, GÉNÉRAL changeover, INFORMAT toggling, TÉL cut-over, toggling

basculer *vt* ALIM ÉLEC, COMMUT, ÉLECTRON toggle, INFORMAT flip, toggle, TÉL toggle

base[1]: **de** ~ *adj* RADIO baseline

base[2] *f* ÉLECTRON *semiconducteurs* base; ~ **améliorée** RADIO augmented baseline; ~ **commune** ÉLECTRON common base; ~ **de connaissances** APP SERV, INFORMAT knowledge base; ~ **de données** DONNÉES, INFORMAT database; ~ **de données et d'analyse** *m* TÉL subscriber and analysis database (*ML TR*); ~ **de données de gestion** DONNÉES, INFORMAT management information base (*MIB*); ~ **de données graphiques** DONNÉES, INFORMAT graphics database; ~ **de données hors ligne** DONNÉES, INFORMAT off-line database; ~ **de données en ligne** DONNÉES, INFORMAT on-line database; ~ **de données multimédia** DONNÉES, INFORMAT multimedia database; ~ **de données numériques** DONNÉES, INFORMAT numerical database; ~ **de données off-line** DONNÉES, INFORMAT off-line database; ~ **de données relationnelles** (*BDR*) DONNÉES, INFORMAT relational database (*RDB*); ~ **de données sémantiques** DONNÉES, INFORMAT semantic database; ~ **de données temporelles** DONNÉES, INFORMAT temporal database; ~ **de données des terrains et superficies** DONNÉES, INFORMAT terrain and surface database; ~ **de données des terrains et surfaces** DONNÉES, INFORMAT terrain and surface database; ~ **de faits** INFORMAT facts base; ~ **d'impulsion** TÉL pulse base; ~ **d'information** DONNÉES information base; ~ **d'informations de gestion de sécurité** DONNÉES security management information base (*SMIB*); ~ **de rattachement** RADIO home base; ~ **temporelle** ÉLECTRON, INFORMAT time base; ~ **de temps** ÉLECTRON, INFORMAT time base; ~ **de temps déclenchée** ÉLECTRON triggered time-base; ~ **de temps pour les éléments de signal** SIGNAL signal-element timing; ~ **de textes** DONNÉES, INFORMAT corpus; ~ **de transistor** COMPOSANTS, ÉLECTRON transistor base

basse: ~ **fréquence** *f* (*BF*) COMMUT, RADIO low frequency (*LF*); ~ **résistance nuque-terre** *f* TESTS low B to earth, low ring to ground; ~ **résistance pointe-nuque** *f* TESTS low A to B, low tip to ring; ~ **résistance pointe-nuque-terre** *f* TESTS low A to B to earth; ~ **résistance pointe-terre** *f* TESTS low A to earth, low tip to ground; ~ **tension** *f* ALIM ÉLEC low voltage (*LV*); ~ **vitesse** *f* GÉNÉRAL low speed

basses *f pl* GÉNÉRAL bass; ~ **fréquences** GÉNÉRAL bass

bassin: ~ **d'audience** *m* TV franchise area

bâti *m* ALIM ÉLEC *pour équipements électroniques*, TÉL rack; ~ **d'antenne** RADIO antenna structure, antenna supporting structure; ~ **pour appareils divers** COMPOSANTS, TÉL miscellaneous apparatus rack (*MAR*); ~ **autonome** TÉL stand-alone rack; ~ **de base** COMPOSANTS basic rack; ~ **pour batterie** ALIM ÉLEC battery rack; ~ **câblé** COMPOSANTS, TÉL wired rack; ~ **de commutateurs** COMMUT, TÉL selector rack, switch rack; ~ **de compteurs** TÉL meter rack; ~ **de contrôle** TESTS test rack; ~ **de convertisseur** ALIM ÉLEC converter rack; ~ **de distribution** ALIM ÉLEC distribution rack; ~ **à double face** TÉL double-sided rack; ~ **d'énergie** ALIM ÉLEC power rack; ~ **d'équipement d'alarme** TÉL alarm equipment rack; ~ **d'essai** TESTS test rack; ~ **étroit** TÉL slim rack; ~ **d'extension** COMPOSANTS continuation rack, TÉL extension rack; ~ **à une face** TÉL single-sided rack; ~ **de**

matériel de base COMPOSANTS basic equipment rack; **~ non câblé** COMPOSANTS, TÉL unwired rack; **~ de sélecteurs** COMMUT, TÉL selector rack, switch rack

bâtonnet *m* COMPOSANTS, INFORMAT stroke

batterie *f* ALIM ÉLEC, ÉLECTRON battery, INFORMAT cluster; **~ d'accumulateurs** ALIM ÉLEC accumulator, accumulator battery; **~ anodique** ALIM ÉLEC anode battery, plate battery; **~ auxiliaire** ALIM ÉLEC booster battery; **~ de central** TÉL exchange battery; **~ centrale** TÉL central battery, common battery; **~ chargée** ALIM ÉLEC fully-charged battery; **~ de conversation** RADIO, TÉL talk battery; **~ à électrolyte gélifié** ALIM ÉLEC gel battery; **~ électronique** INFORMAT drum machine; **~ flottante** ALIM ÉLEC floating battery; **~ locale** TÉL local battery (*LB*); **~ longue durée** ALIM ÉLEC long-life battery; **~ négative** ALIM ÉLEC negative battery; **~ d'ordinateurs** INFORMAT computer bank; **~ ouverte** ALIM ÉLEC open battery; **~ à pile sèche** ALIM ÉLEC dry-cell battery; **~ au plomb** ALIM ÉLEC lead accumulator, lead storage battery; **~ positive** ALIM ÉLEC positive battery; **~ rechargeable** ALIM ÉLEC rechargeable battery; **~ de redresseurs à semiconducteurs** COMPOSANTS semiconductor rectifier stack; **~ scellée** ALIM ÉLEC sealed battery; **~ de secours** PÉRIPH backup battery, battery backup; **~ solaire** ALIM ÉLEC solar battery; **~ tampon** ALIM ÉLEC buffer battery; **~ terminale** ALIM ÉLEC end cell

battre: **~ la mesure** *vt* INFORMAT beat time

baud *m* INFORMAT baud

bavardage *m* TÉL chat

bavure *f* COMPOSANTS blanking burr, burr

BB: **~ Edit** *m* DONNÉES BB Edit

BBS *abrév* (*système de babillard électronique*) DONNÉES, INFORMAT, TÉL BBS (*bulletin board system*)

Bcc *abrév* (*blind carbon copy*) DONNÉES *courrier électronique* Bcc (*blind carbon copy*)

BCS *abrév* (*système de câblage BULL*) DONNÉES, INFORMAT BCS (*BULL cabling system*)

BDR *abrév* (*base de données relationnelles*) DONNÉES, INFORMAT RDB (*relational database*)

bec: **~ à déclic** *m* ÉLECTRON trigger nozzle

Bel *m* RADIO *acoustique* Bel

BER *abrév* (*règles de codage de base*) DONNÉES, INFORMAT BER (*basic encoding rules*)

BF *abrév* (*basse fréquence*) COMMUT, DONNÉES, INFORMAT, RADIO, TÉL LF (*low frequency*)

BFO *abrév* (*oscillateur de battement de fréquence*) ÉLECTRON, TRANSMIS BFO (*beat-frequency oscillator*)

BFR *abrév* (*borne fixed radio*) RADIO fixed radio beacon

BFT *abrév* (*transfert de fichiers binaires*) TÉL BFT (*binary file transfer*)

biais *m* INFORMAT, TESTS *qualité* bias

bibliothèque *f* DONNÉES, INFORMAT library; **~ de base** INFORMAT basic software package; **~ des comportements par défaut** DONNÉES default library; **~ de données** DONNÉES library of data; **~ d'images** INFORMAT clip-art library; **~ d'installation** GÉNÉRAL installation library; **~ de liens dynamiques** (*DLL*) TÉL dynamic-link library (*DLL*); **~ de modules de test** TESTS test-step library; **~ de programmes** INFORMAT program library; **~ de signaux** COMMUT signal library

bidirectionnel *adj* GÉNÉRAL two-way, both-way, INFORMAT bidirectional, duplex, RADIO, TRANSMIS, TV bidirectional, two-way; **~ simultané** RADIO, TÉL duplex

bien: **~ reçu** *loc* DONNÉES successfully-received, ACK, Roger

bifibus *m* TV two-wire bus

bifilaire *adj* TV two-wire

bifurquée *f* TÉL forked working

bilan: **~ de liaison** *m* ALIM ÉLEC, RADIO, TRANSMIS link budget; **~ de puissance** *m* ALIM ÉLEC, RADIO, TRANSMIS power budget; **~ de puissance d'une liaison** *m* ALIM ÉLEC, RADIO, TRANSMIS link power budget; **~ thermique** *m* ÉLECTRON heat budget

bilanmètre *m* TESTS net radiometer

bilatéral: **~ simultané** *adj* RADIO, TÉL duplex, TRANSMIS full-duplex

bilatérale *f* TÉL two-way simplex, two-way simplex connection

bilinéaire *adj* DONNÉES bilinear

bimétal *m* COMPOSANTS *métallurgie* bimetal

binaires: **~ brouillés** *m pl* DONNÉES scrambled binary ones (*SB1*)

BIOS *abrév* (*système d'exploitation des entrées/sorties*) TESTS BIOS (*basic input/output system*)

bip *m* PÉRIPH tone, RADIO *appareil* beep, bleep, bleeper tone pager, *son* beep, bleep, TÉL pip; **~ sonore** INFORMAT beep sound, PÉRIPH tone; **~ vocal** RADIO tone-plus-voice paging

bipper *m* RADIO alert-only pager, beeper, bleep, bleeper

biréfringence *f* DONNÉES, TRANSMIS *fibres optiques* birefringence

bistabilité: **~ optique** *f* TRANSMIS optical bistability

bistable[1] *adj* ÉLECTRON *éléments logiques* bistable

bistable[2]: **~ maître-esclave** *m* ÉLECTRON master-slave flip-flop

bistandard *adj* GÉNÉRAL dual-standard

bit[1] *m* DONNÉES, INFORMAT, TRANSMIS binary digit, bit; **~ auxiliaire** DONNÉES housekeeping bit; **~ de bourrage** DONNÉES dummy bit; **~ de commande/réponse** TÉL command/response bit (*C/R*); **~ de continuation** TÉL M-bit; **~ de contrôle** DONNÉES check bit, RADIO parity bit; **~ de données appliqué** DONNÉES, TÉL applied data bit; **~ de données appliqué en brouillage** DONNÉES, TÉL applied data-bit scrambling process; **~ de drapeau** DONNÉES flag bit; **~ drapeau** DONNÉES tag bit; **~ erroné** INFORMAT erroneous bit; **~ factice** DONNÉES dummy bit; **~ de fin** TÉL final bit; **~ d'identification de la priorité** DONNÉES priority-indication bit; **~ indicateur vers l'arrière** INFORMAT, TÉL backward indicator bit; **~ indicateur vers l'avant** INFORMAT, TÉL forward indicator bit; **~ d'information** DONNÉES, TÉL information bit; **~ d'interrogation** DONNÉES poll bit; **~ d'invitation** TÉL poll bit; **~ d'invitation à émettre/fin** TÉL P/F; **~ M** TÉL M-bit; **~ de parité** DONNÉES parity bit; **~ de plus faible poids** DONNÉES least significant bit (*LSB*); **~ de plus fort poids** DONNÉES most significant bit (*MSB*); **~ pseudo-aléatoire** DONNÉES pseudorandom bit; **~ de service** DONNÉES housekeeping bit, INFORMAT service bit, TÉL overhead bit; **~ de servitude** DONNÉES overhead bit; **~ de signalisation** SIGNAL, TRANSMIS signaling bit (*AmE*), signalling bit (*BrE*); **~ de signe** INFORMAT sign bit; **~ significatif** DONNÉES significant bit; **~ de stop** DONNÉES stop bit; **~ suivant transmis** TRANSMIS next transmitted bit; **~ supplémentaire** DONNÉES overhead bit; **~ de synchronisation** TÉL synchronization bit; **~ de traçage** COMMUT tracer bit; **~ utile** DONNÉES information bit; **~ de verrouillage de trame** TÉL framing bit

bit[2]: **~ à bit** *loc* DONNÉES bit-by-bit

bitimbral *adj* INFORMAT bi-timbral

Bitnet *m* DONNÉES Bitnet

bits[1]: **~ différents** *m pl* DONNÉES bit disagreements; **~ transmis précédemment** *m* TRANSMIS earlier-transmitted bits; **~ transmis précédemment en brouillage** *m* TÉL earlier-transmitted bits scrambling process

bits[2]: **~ par couleur** *loc* DONNÉES bits per color (*AmE*), bits per colour (*BrE*); **~ par pouce** *loc* (*bpp*) DONNÉES bits per inch (*bpi*); **~ par seconde** *loc* (*b/s*) DONNÉES bits per second (*bps*)

bivocal *m* TÉL speech plus duplex, speech-plus-duplex equipment

BL *abrév* INFORMAT (*bus local*) local bus, RADIO, TÉL, TRANSMIS (*bande latérale*) SB (*sideband*); **~ sup.** (*bande latérale supérieure*) RADIO, TRANSMIS USB (*upper sideband*)

blanc *m* ÉLECTRON light, INFORMAT white, RADIO look-through, TÉL white; **~ de l'image** INFORMAT picture white; **~ des lettres automatique** TÉL unshift-on-space; **~ nominal** TÉL nominal white

BLI *abrév* (*bande latérale indépendante*) RADIO ISB (*independent sideband*)

blind: **~ carbon copy** *m* (*Bcc*) DONNÉES *courrier électronique* blind carbon copy (*Bcc*)

blindage *m* ALIM ÉLEC shield, COMPOSANTS shielding, TRANSMIS cable armor (*AmE*), cable armour (*BrE*); **~ antimagnétique** DONNÉES anti-magnetic shield; **~ Beldfoil en aluminium-polyester** ÉLECTRON Beldfoil aluminium-polyester shield (*BrE*), Beldfoil aluminum-polyester shield (*AmE*); **~ à tresse en cuivre étamé** COMPOSANTS, ÉLECTRON tin-plated copper braid shield

blindé *adj* GÉNÉRAL armored (*AmE*), armoured (*BrE*)

bloc *m* GÉNÉRAL block; **~ accus** ÉLECTRON accumulator module; **~ d'alimentation** ALIM ÉLEC power pack, power supply, power supply unit; **~ alvéolé** COMPOSANTS, TRANSMIS multiple-duct block; **~ de bobinage** COMPOSANTS, ÉLECTRON yoke assembly; **~ de canalisations monotubulaire** COMPOSANTS, TRANSMIS single-duct block; **~ de canalisations multitubulaire** COMPOSANTS, TRANSMIS multiple-duct block; **~ capteur solaire** RADIO sun sensor assembly (*SSA*); **~ de circuits téléphoniques** COMMUT, ÉLECTRON, TÉL, TRANSMIS trunk circuit block; **~ codé** DONNÉES code block; **~ de commande de transfert** COMMUT transfer-control block; **~ de conduite unitaire** COMPOSANTS single-duct block; **~ de descripteur de commande** DONNÉES command descriptor block (*CDB*); **~ de données** DONNÉES data block; **~ de données conservé** DONNÉES *procédure de traduction du CTP* retained-data block; **~ d'enfichage par impact** ÉLECTRON punch-down block; **~ erroné** DONNÉES erroneous block (*EB*); **~ erroné consécutif** DONNÉES consecutive error block (*CEB*); **~ d'essai** TESTS test block; **~ fonctionnel** COMMUT function block, TÉL *en LDS* functional block; **~ de joncteurs** COMMUT junctor block; **~ de liaison montante** RADIO uplink block; **~ de mémoire** DONNÉES, INFORMAT memory bank; **~ de mémoires de données** DONNÉES, INFORMAT data store block; **~ de mémoires de programmes** DONNÉES, INFORMAT program store block; **~ de messages de CRC** TÉL CRC message block (*CMB*); **~ mot** DONNÉES word block; **~-notes électronique** INFORMAT notepad computer; **~ numérique** DONNÉES, TÉL digital block; **~ de paramètres spécifiques** TÉL parameter block; **~ primaire** TÉL primary block; **~ de raccordement** DONNÉES, ÉLECTRON tag block; **~-source** ÉLECTRON source block; **~ de télécommande** CONTRÔLE command unit, remote-control unit (*RCU*); **~ de test** TESTS test block; **~ de test de route** TÉL route test block; **~ de texte** INFORMAT text block; **~ de touches** INFORMAT keypad; **~ de transfert** COMMUT transfer block

blocage *m* COMMUT, GÉNÉRAL blocking, RADIO barring, TÉL barring, blocking; **~ automatique** RADIO muting, squelch, TÉL muting; **~ d'un canal sémaphore** SIGNAL signaling-link blocking (*AmE*), signalling-link blocking (*BrE*); **~ externe** SIGNAL external blocking; **~ de faisceau** TÉL route blocking; **~ interne** COMMUT internal blocking; **~ d'une liaison de signalisation** SIGNAL signaling-link blocking (*AmE*), signalling-link blocking (*BrE*); **~ de ligne à distance** CONTRÔLE remote blocking; **~ partiel de l'accès à l'interurbain** COMMUT conditionally toll-denied (*CTD*); **~ de retour** TÉL return block; **~ de la zone de Fresnel** TRANSMIS Fresnel zone blockage

blocs: **~ éventuellement supprimables** *m pl* DONNÉES optionally-droppable blocks; **~ non supprimables** *m pl* DONNÉES nondroppable blocks

bloqué *adj* COMPOSANTS tightened, ÉLECTRON cut-off, inhibited, RADIO barred, TÉL barred, blocked, busied out

bloquer *vt* COMPOSANTS tighten, ÉLECTRON cut off, inhibit, RADIO bar, TÉL bar, block

BLR *abrév* RADIO, TRANSMIS, TV (*bande latérale résiduelle*) RSB (*residual sideband*), (*bande latérale restante*) VSB (*vestigial sidebande*)

BLU *abrév* RADIO, TRANSMIS (*bande latérale unique*) SSB (*single sideband*)

bobinage *m* ALIM ÉLEC, COMPOSANTS winding

bobine *f* ALIM ÉLEC coil, COMPOSANTS spool, DONNÉES reel; **~ d'absorption** ALIM ÉLEC drainage coil; **~ d'arrêt** ALIM ÉLEC choke; **~ de charge** TRANSMIS loading coil; **~ de charge d'antenne** RADIO antenna-loading coil; **~ de choc** ALIM ÉLEC choke, choking coil; **~ de compensation** ALIM ÉLEC peaking coil; **~ de concentration** ÉLECTRON focusing coil; **~ de crête** ALIM ÉLEC peaking coil; **~ de déviation** COMPOSANTS, ÉLECTRON deflection yoke, yoke coil; **~ d'électro de barre de sélection** COMMUT selecting-bar coil; **~ d'électro-aimant** ÉLECTRON iron-core coil, magnet coil; **~ exploratrice** RADIO search coil; **~ de guidage** TÉL guide spool; **~ d'inductance** ALIM ÉLEC inductance coil, inductor; **~ d'induction** ÉLECTRON induction coil; **~ isolante** ALIM ÉLEC bobbin; **~ de maintien** COMPOSANTS holding coil; **~ à noyau de fer** ÉLECTRON iron-core coil; **~ oscillatrice** RADIO oscillator; **~ de papier** PÉRIPH reel of paper; **~ Pupin** TRANSMIS loading coil; **~ de relais** ALIM ÉLEC relay coil; **~ de relèvement** ALIM ÉLEC peaking coil; **~ thermique** ALIM ÉLEC heat coil

bobiner *vt* ALIM ÉLEC, COMPOSANTS wind

bogue *f* DONNÉES bug

boîte: **~ de câble à une seule paire** *f* COMPOSANTS single-pair cable box; **~ de charge** *f* TRANSMIS loading-coil pot; **~ de bobine de charge** *f* TRANSMIS loading-coil case; **~ de commande** *f* CONTRÔLE control box; **~ de commutation** *f* COMMUT switch unit; **~ de dérivation** *f* ALIM ÉLEC, TRANSMIS distribution box; **~ de dialogue** *f* INFORMAT dialog box; **~ de distribution** *f* ALIM ÉLEC, TRANSMIS distribution terminal, distribution box, terminal box; **~ de distribution et de traitement des signaux faibles** *f* GÉNÉRAL low-level signaling device (*AmE*), low-level signalling device (*BrE*); **~ de division** *f* ALIM ÉLEC, ÉLECTRON, GÉNÉRAL, TRANSMIS cable box, junction box; **~ à écho** *f* RADIO echo box; **~ d'essai** *f* TESTS test box; **~ d'extrémité** *f* COMPOSANTS termination joint box; **~ d'interconnexion optique** *f* ÉLECTRON lightguide interconnection unit (*LIU*); **~ de jonction** *f* ALIM ÉLEC, ÉLECTRON, GÉNÉRAL, TRANSMIS cable box, junction box; **~ de jonction de télécommande** *f* CONTRÔLE command interface unit; **~ à lettres** *f* APP SERV, INFORMAT text mailbox; **~ aux lettres** *f* (*BAL*) DONNÉES mailbox; **~ aux lettres électronique** *f* APP SERV, INFORMAT electronic mailbox; **~ à lettres vocale** *f* APP SERV, TÉL voice mailbox; **~ à plaques** *f* PÉRIPH plate block; **~ de raccordement** *f* ALIM ÉLEC, ÉLECTRON, GÉNÉRAL, cable box, junction box, TÉL block terminal, connecting box, connection box, terminal box, TRANSMIS junction box, cable box; **~ de résistance à décades** *f* ALIM ÉLEC decade resistance box; **~ à rythmes** *f* INFORMAT drum machine, rhythm programmer; **~ de test** *f* TESTS *pour cartes imprimées* test box

boîtes: **~ et coffrets de raccordement** *f pl* COMMUT junction boxes and cabinet

boîtier *m* COMPOSANTS case, cover, housing, DONNÉES box, ÉLECTRON *semiconducteurs* can, case, breakout box, GÉNÉRAL case, PÉRIPH remote control; **~ d'abonné** TV subscriber box; **~ de dérivation** COMPOSANTS, ÉLECTRON branching block; **~ à deux rangées de broches** (*boîtier DIP*) DONNÉES dual-in-line package (*DIP*); **~ DIP** (*boîtier à deux rangées de broches*) DONNÉES DIP (*dual-in-line package*); **~ de distribution** COMPOSANTS, ÉLECTRON distribution block; **~ d'écouteur** PÉRIPH *postes téléphoniques*, TÉL receiver case; **~ d'enrobage** GÉNÉRAL potting boot; **~ d'épissures** ÉLECTRON splice closure; **~ externe** DONNÉES add-on unit; ~ **partageur de ressources** DONNÉES, INFORMAT shared-resource device; ~ **plat** ALIM ÉLEC flat-pack; ~ **protecteur** COMPOSANTS protective cover; ~ **de protection** COMPOSANTS protective case; ~ **de raccordement** TV connection box; ~ **à une rangée de connexions** (*boîtier SIP*) DONNÉES single in-line package (*SIP*); ~ **de récepteur** PÉRIPH *postes téléphoniques*, TÉL receiver case; ~ **SIP** (*boîtier à une rangée de connexions*) DONNÉES SIP (*single in-line package*); ~ **SOIC** ÉLECTRON small-outline integrated circuit (*SOIC*); ~ **de test** TESTS test box; ~ **des têtes optiques** TRANSMIS head housing

bond *m* RADIO hop

booléen *adj* DONNÉES, INFORMAT Boolean

booter *vt* INFORMAT boot, boot up, bootstrap

bord *m* ÉLECTRON edge; ~ **de carte** ÉLECTRON card edge; ~ **recourbé** INFORMAT edge curl; ~ **de segment** INFORMAT stroke edge

bords: ~ **recourbés** *m pl* INFORMAT lipping

bordure: ~ **du filtre** *f* ÉLECTRON filter flank

borne *f* ALIM ÉLEC, COMPOSANTS terminal, GÉNÉRAL beacon, PÉRIPH terminal; ~ **d'anode** ÉLECTRON *semiconducteurs* anode terminal; ~ **d'appel d'urgence** TÉL emergency call box; ~ **d'appels simultanés** TÉL simultaneous-call terminal; ~ **de base** ÉLECTRON *semiconducteurs* base terminal; ~ **de cathode** ÉLECTRON cathode terminal; ~ **de connexion** INFORMAT connecting terminal; ~ **pour connexion enroulée** COMPOSANTS wire-wrap terminal, wrap post; ~ **d'élément** ALIM ÉLEC terminal clip; ~ **de l'émetteur** COMPOSANTS emitter terminal; ~ **d'entrée de la ligne** PÉRIPH line-in port; ~ **de fixation** ÉLECTRON clamp terminal; ~ **fixe radio** (*BFR*) RADIO fixed radio beacon; ~ **de gâchette** ÉLECTRON gate terminal; ~ **de Hall** ÉLECTRON Hall terminal; ~ **interactive** INFORMAT touch-screen kiosk, TV interactive terminal; ~ **de mise à la terre** ALIM ÉLEC earth terminal (*BrE*), ground terminal (*AmE*); ~ **multimédia** TV multimedia kiosk; ~ **négative** ALIM ÉLEC battery negative, negative terminal; ~ **neutre** ALIM ÉLEC neutral terminal; ~ **positive** ALIM ÉLEC battery positive, positive terminal; ~ **à poussoir** INFORMAT push-button terminal; ~ **de raccordement** ALIM ÉLEC terminal, ÉLECTRON connector terminal, INFORMAT connection terminal; ~ **de sortie** ALIM ÉLEC output terminal, INFORMAT output jack; ~ **supérieure** TRANSMIS upper bound; ~ **de tension de batterie** ALIM ÉLEC battery voltage terminal; ~ **terminale** TÉL circuit terminal; ~ **à vis** ALIM ÉLEC screw terminal

bornes: ~ **d'appels** *f pl* TRANSMIS call-binding posts; ~ **d'appels d'urgence** *f pl* PÉRIPH emergency call stations

bosse *f* ALIM ÉLEC *circuits imprimés* bump

Bottin® *m* TÉL telephone book, telephone directory

bouchon: ~ **d'insertion de câble double** *m* ÉLECTRON dual-purpose wire insertion cap

bouclage *m* ÉLECTRON loopback, INFORMAT looping, word wrapping, wraparound, TÉL, TRANSMIS looping; ~ **d'anneau** ÉLECTRON ring wrap

boucle[1] *f* ALIM ÉLEC loop circuit, loop, INFORMAT loop, TRANSMIS loop, loopback; ~ **d'abonné** RADIO local loop; ~ **d'accès rapide** DONNÉES rapid-access loop, revolver loop, revolver; ~ **anticontraintes mécanique** ÉLECTRON stress relief loop; ~ **d'asservissement** CONTRÔLE *commande automatique* control loop; ~ **audio** TÉL audio loop; ~ **de communication multi-usage** TÉL multiuse communication; ~ **pour contrôle de continuité** TÉL check loop; ~ **à contrôle d'itération** ÉLECTRON kick-the-dog loop; ~ **de Costas** TÉL Costas loop; ~ **de courant** ALIM ÉLEC current loop; ~ **fermée** ÉLECTRON closed loop; ~ **fermée de réaction** ÉLECTRON feedback closed loop; ~ **d'induction magnétique** ÉLECTRON magnetic-induction current loop; ~ **de mesure** TESTS measuring loop, test loop; ~ **minigauge** TÉL minigage loop (*AmE*), minigauge loop (*BrE*); ~ **numérique** TRANSMIS digital loop; ~ **optique** TRANSMIS fiber-optic loop (*AmE*), fibre-optic loop (*BrE*); ~ **ouverte** ALIM ÉLEC, INFORMAT, TRANSMIS open loop; ~ **à phase asservie** ÉLECTRON phase-locked loop (*PLL*); ~ **principale de gestion d'événements** INFORMAT main event loop; ~ **de protection contre les erreurs** TÉL error-control loop; ~ **de la qualité** TESTS quality loop, quality spiral; ~ **de réaction** ÉLECTRON, INFORMAT feedback loop; ~ **de régulation** CONTRÔLE *commande automatique* control loop; ~ **à retard de phase** ÉLECTRON delay-lock loop; ~ **de retour de l'information** RADIO feedback loop; ~ **de rétroaction** ÉLECTRON, INFORMAT feedback loop; ~ **sans fin** INFORMAT endless loop; ~ **simple** INFORMAT basic loop; ~ **unigauge** TÉL unigage loop (*AmE*), unigauge loop (*BrE*); ~ **à verrouillage de phase** ÉLECTRON phase-locked loop (*PLL*); ~ **à verrouillage de phase optique** ÉLECTRON optical phase-locked loop (*OPLL*); ~ **verrouillée** ÉLECTRON locked loop; ~ **vidéo** TV video loop

boucle[2]: **en** ~ **fermée** *loc* ÉLECTRON closed-loop

bouclé *adj* ÉLECTRON *circuit* looped

boucler *vt* ÉLECTRON loop, TRANSMIS terminate, loop; ~ **un circuit** ÉLECTRON connect in a loop, loop a circuit

boule: ~ **de commande** *f* INFORMAT rolling ball, trackball; ~ **roulante** *f* INFORMAT trackball

bourdonnement *m* GÉNÉRAL hum

bourrage *m* COMMUT traffic padding, DONNÉES traffic padding, stuffing, GÉNÉRAL filler, INFORMAT *cartes, imprimantes* jam

bourre: ~ **de coton** *f* COMPOSANTS *matériaux* cotton waste

bout: **de** ~ **en bout** *loc* GÉNÉRAL end-to-end; **au** ~ **du fil** *loc* TÉL on the line

bouton *m* CONTRÔLE knob, GÉNÉRAL button, INFORMAT box; ~ **d'appels en série** TÉL serial-call key; ~ **d'arrêt** INFORMAT stop button; ~ **à bascule** ALIM ÉLEC rocker button; ~ **de commande** CONTRÔLE control knob, INFORMAT activate button; ~ **de conversation** RADIO press-to-talk button, push-to-talk button; ~ **du cylindre** TÉL platen knob; ~ **d'éjection** DONNÉES, INFORMAT eject button; ~ **d'expansion** INFORMAT expansion button; ~ **lumineux** CONTRÔLE luminous button; ~ **de mise à la terre** COMMUT earth button (*BrE*), ground button (*AmE*), earthing button (*BrE*), grounding button (*AmE*); ~ **de navigation** DONNÉES, INFORMAT navigation button; ~ **de note** INFORMAT annotation button; ~**-poussoir** CONTRÔLE plunger key, pushbutton; ~ **de réglage** CONTRÔLE control knob; ~ **de réglage fin** ÉLECTRON fine switch; ~ **de réglage grossier** ÉLECTRON coarse switch; ~ **de réglage de volume** CONTRÔLE volume control; ~ **de remise en position initiale** CONTRÔLE reset button, restoring button; ~ **de remise à zéro** CONTRÔLE reset button, restoring button; ~ **de restauration** CONTRÔLE reset button, restoring button

bpp *abrév* (*bits par pouce*) DONNÉES bpi (*bits per inch*)

bracelet: ~ **antistatique** *m* ÉLECTRON anti-static wristband

branché[1] *adj* ÉLECTRON, INFORMAT connected

branché[2]: ~ **sur le secteur** *loc* ALIM ÉLEC mains-connected

branche *f* ALIM ÉLEC, DONNÉES branch; ~ **de câble** TRANSMIS cable branch; ~ **commune** ALIM ÉLEC mutual branch, TRANSMIS common branch; ~ **d'une transaction** DONNÉES transaction branch

branchement *m* DONNÉES, ÉLECTRON, GÉNÉRAL, INFORMAT connection, TRANSMIS branch joint, branch-off, distribution point (*DP*); ~ **d'abonné** RADIO loop; ~ **clandestin** TÉL tapping, telephone tapping, wiretapping; ~ **de liaison** INFORMAT interface connection; ~ **optique** TRANSMIS optical branching, optical tapping

brancher *vt* ALIM ÉLEC, COMMUT switch on, ÉLECTRON plug in, switch on, INFORMAT hook up, switch on, PÉRIPH, RADIO switch on; ~ **en parallèle** ALIM ÉLEC connect in parallel, parallel up, ÉLECTRON connect in parallel; ~ **en série** ALIM ÉLEC connect in series

bras: ~ **articulé** *m* ÉLECTRON swing-arm feature; ~ **porte-balais** *m* COMMUT wiper arm; ~ **porte-contacts** *m* COMMUT

wiper arm; ~ **de réactance** *m* RADIO nondissipative stub, stub
brasabilité *f* COMPOSANTS solderability
brasable *adj* COMPOSANTS solderable
brasage *m* ÉLECTRON soldering; ~ **en phase vapeur** ÉLECTRON condensation soldering, vapor-phase soldering (*AmE*), vapour-phase soldering (*BrE*); ~ **tendre** ÉLECTRON soldering; ~ **tendre en grande série** ÉLECTRON mass soldering; ~ **tendre à la machine** ÉLECTRON automatic soldering, machine soldering; ~ **tendre par refusion** ÉLECTRON reflow soldering; ~ **tendre simultané** ÉLECTRON mass soldering; ~ **tendre au trempé** ÉLECTRON dip soldering
braser *vt* ÉLECTRON solder
brassage *m* ÉLECTRON cross-connect, patching; ~ **de modes** TRANSMIS mode mixing; ~ **modulaire** ÉLECTRON modular patching
brasseur: ~ **ATM** *m* COMMUT ATM switch; ~ **de modes** *m* TRANSMIS mode mixer, mode scrambler; ~ **de réseau à haut débit ATM** *m* COMMUT high bit-rate ATM network mixer
brasure *f* COMPOSANTS soldered joint
brevet: ~ **en cours** *m* ÉLECTRON patent pending
bribe *f* DONNÉES chip; ~ **de parole** RADIO talk spurt, TÉL talk spurt, voice spurt
bride: ~ **de connexion** *f* ALIM ÉLEC, COMPOSANTS jumper, strap; ~ **de dissipation de chaleur** *f* COMPOSANTS cooling flange; ~ **de fixation** *f* COMPOSANTS fixing bracket, mounting bracket; ~ **lisse** *f* RADIO plain coupling flange, plain connector; ~ **de masse** *f* ALIM ÉLEC earth clamp (*BrE*), ground clamp (*AmE*); ~ **de retenue** *f* COMPOSANTS strain-relief mounting bracket, ÉLECTRON sheath retention clamp
brillance *f* INFORMAT brightening
brin *m* COMPOSANTS, ÉLECTRON strand
Broadcast: ~ **Vidéo U-Matic** *m* (*BVU*) TV Broadcast Video U-Matic (*BVU*)
brochage *m* ÉLECTRON pin-out
broche[1] *f* COMPOSANTS pin, tag, ÉLECTRON pin, socket, GÉNÉRAL drift, driftpin; ~ **d'assemblage** GÉNÉRAL drift, driftpin; ~ **de contact** COMPOSANTS, ÉLECTRON contact pin, plug pin; ~ **enrouleuse** COMPOSANTS wrapping pin; ~ **de fiche mâle** COMPOSANTS, ÉLECTRON contact pin, plug pin; ~ **de raccordement** ÉLECTRON connecting pin, TÉL block terminal; ~ **à souder** COMPOSANTS soldering lug, soldering tag
broche[2]: ~ **à broche** *loc* COMPOSANTS, ÉLECTRON pin-to-pin
brouillage *m* DONNÉES jamming, scrambling, ÉLECTRON jamming, GÉNÉRAL *interférence* disturbance, INFORMAT jamming, scrambling, RADIO scrambling, jamming, breakthrough, radio-frequency interference (*RFI*), interference, TÉL, TV scrambling; ~ **accepté** RADIO accepted interference; ~ **admissible** RADIO, TRANSMIS permissible interference; ~ **atmosphérique** TRANSMIS precipitation static, static; ~ **du canal adjacent** RADIO, TRANSMIS adjacent-channel interference (*ACI*); ~ **dans le même canal** RADIO, TRANSMIS co-channel interference; ~ **dans la même voie** RADIO co-channel interference; ~ **de diversion** RADIO confusion jamming, deception jamming; ~ **dû à une bande latérale** RADIO, TÉL, TRANSMIS sideband interference; ~ **dû aux éclipses du soleil** RADIO eclipse interference; ~ **dû aux harmoniques** RADIO harmonic interference; ~ **dû à la présence du soleil** RADIO solar interference; ~ **électromagnétique** ÉLECTRON electromagnetic interference (*EMI*), TESTS electromagnetic disturbance, electromagnetic interference (*EMI*); ~ **global** RADIO aggregate interference; ~ **intentionnel** ÉLECTRON jamming, RADIO radio-frequency interference (*RFI*), jamming; ~ **intersymbole** INFORMAT, TÉL, TRANSMIS intersymbol interference (*ISI*); ~ **monofréquence** RADIO, SIGNAL, TÉL single-frequency interference; ~ **en palissade** ÉLECTRON railings; ~ **par balayage** ÉLECTRON, RADIO sweep jamming; ~ **par feuilles réfléchissantes** RADIO chaff jamming; ~ **par leurres** RADIO chaff jamming, window jamming; ~ **par paillettes** RADIO window jamming; ~ **préjudiciable** RADIO harmful interference; ~ **quasi-impulsif** RADIO quasi-impulsive interference; ~ **des radiocommunications** RADIO radio jamming; ~ **sélectif** ÉLECTRON, RADIO selective jamming, spot jamming; ~ **sur la voie commune** RADIO, TRANSMIS co-channel interference; ~ **tolérable** RADIO acceptable interference; ~ **transmis par rayonnement** ÉLECTRON radiated interference
brouillard: ~ **protection** *m* ÉLECTRON jabber protection
brouillé *adj* DONNÉES scrambled, jammed, ÉLECTRON jammed, INFORMAT scrambled, jammed, *estompé* blurred, RADIO, TÉL jammed, scrambled, TV scrambled
brouiller *vt* DONNÉES scramble, jam, ÉLECTRON jam, GÉNÉRAL *interférer* disturb, INFORMAT scramble, jam, RADIO, TÉL jam, scramble, TV scramble
brouilleur *m* DONNÉES scrambler, ÉLECTRON jammer, GÉNÉRAL encrypting equipment, INFORMAT, RADIO, TÉL jammer, scrambler, TV scrambler; ~ **intentionnel** ÉLECTRON jammer, RADIO jammer, jamming transmitter; ~ **de modes** TRANSMIS mode mixer, mode scrambler; ~ **numérique linéaire de la voix** TÉL linear-digital voice scrambler; ~ **de parole** RADIO, TÉL speech scrambler; ~**-répéteur** ÉLECTRON repeater-jammer
broutage: ~ **longitudinal** *m* TÉL, TRANSMIS longitudinal judder; ~ **transversal** *m* TÉL, TRANSMIS transverse judder
B-routeur *m* DONNÉES B-router
brucelles *f pl* COMPOSANTS wire tweezers, GÉNÉRAL picking pliers
bruit *m* GÉNÉRAL, RADIO, TÉL noise; ~ **acoustique** DONNÉES, TÉL acoustic noise; ~ **additif** TÉL additive noise; ~ **d'agitation thermique** COMPOSANTS, ÉLECTRON thermal agitation noise; ~ **aléatoire** ÉLECTRON, RADIO, TÉL random noise; ~ **aléatoire uniforme** ÉLECTRON, RADIO, TÉL uniform spectrum random noise; ~ **d'ambiance** INFORMAT, TÉL ambient noise; ~ **ambiant** INFORMAT, TÉL ambient noise; ~ **d'amorçage** ÉLECTRON primer noise; ~ **artificiel** TESTS artificial noise; ~ **atmosphérique** RADIO atmospheric noise; ~ **d'avalanche** ÉLECTRON *semiconducteurs* avalanche noise; ~ **à bande étroite** TRANSMIS narrow-band noise; ~ **blanc** RADIO, TÉL white noise; ~ **de brouillage** RADIO interference noise; ~ **coloré** RADIO colored noise (*AmE*), coloured noise (*BrE*); ~ **dynamique** TÉL dynamic noise; ~ **en 1/f** ÉLECTRON noise 1/f; ~ **erratique** ÉLECTRON, RADIO, TÉL random noise; ~ **erratique triangulaire** ÉLECTRON, RADIO triangular random noise; ~ **de fond** ÉLECTRON ground noise, GÉNÉRAL background noise, hum, INFORMAT grass, background noise, basic noise, noodle, PÉRIPH background noise, background mush, RADIO grass, TÉL background noise, basic noise; ~ **galactique** RADIO galactic noise; ~ **gaussien** TÉL, TRANSMIS Gaussian noise; ~ **granulaire** TÉL granular noise; ~ **de granulation** TRANSMIS speckle noise; ~ **de grenaille** ÉLECTRON, TÉL shot noise; ~ **impulsif** TÉL impulse noise, TRANSMIS burst noise; ~ **industriel** ÉLECTRON, RADIO man-made noise; ~ **d'intensité** ÉLECTRON relative-intensity noise (*RIN*); ~ **d'intermodulation** TRANSMIS intermodulation noise; ~ **interne** TRANSMIS intrinsic noise; ~ **ionique** ÉLECTRON ion noise; ~ **à large bande** DONNÉES wideband noise, INFORMAT broadband noise; ~ **de ligne** INFORMAT line noise, TÉL circuit noise; ~ **microphonique** ÉLECTRON microphone noise, microphonic noise, microphonism; ~ **modal** TRANSMIS modal noise, speckle noise; ~ **de modulation** TRANSMIS modulation noise; ~ **à modulation de fréquence** TRANSMIS frequency-modulation noise; ~ **naturel** ÉLECTRON natural noise; ~ **non gaussien** TÉL, TRANSMIS non-Gaussian noise; ~ **numérisé** DONNÉES digital noise; ~ **optique** DONNÉES optical noise; ~ **parasite** INFORMAT chatter, TÉL background noise, extraneous noise; ~ **de partition des modes** TRANSMIS mode-partition noise; ~ **photonique** ÉLECTRON, TRANSMIS photon noise; ~ **des photons**

ÉLECTRON, TRANSMIS photon noise; ~ **propre** INFORMAT, TÉL basic noise; ~ **pseudo-aléatoire** DONNÉES, TRANSMIS pseudorandom noise; ~ **de quantification** ÉLECTRON, TRANSMIS quantizing noise, quantization noise; ~ **quantique** ÉLECTRON, TRANSMIS quantum noise; ~ **radioélectrique** RADIO radio noise; ~ **de répartition** ÉLECTRON partition noise; ~ **résiduel** INFORMAT residual noise; ~ **rose** INFORMAT pink noise; ~ **de salle** TÉL room noise; ~ **de scintillation** TV flicker noise; ~ **de sélecteurs** COMMUT selector noise; ~ **de surface** TÉL scratching noise; ~ **de tacheture** TRANSMIS modal noise, speckle noise; ~ **de télégraphe** TÉL telegraphic noise; ~ **thermique** COMPOSANTS, ÉLECTRON thermal noise, thermal agitation noise; ~ **de traînage** RADIO trailing noise; ~ **transmis par rayonnement** ÉLECTRON radiated noise; ~ **de transmodulation** RADIO cross-modulation noise; ~ **de tube** ÉLECTRON tube noise; ~ **d'une voie au repos** TRANSMIS idle-channel noise

bruitage *m* INFORMAT sound effect

brûlure: ~ **d'écran** *f* ÉLECTRON screen burn; ~ **de mire** *f* ÉLECTRON raster burn

b/s *abrév* (*bits par seconde*) DONNÉES bps (*bits per second*)

BS *abrév* (*station de base*) RADIO, TÉL BS (*base station*)

BSI *abrév* (*Institut britannique de normalisation*) GÉNÉRAL BSI (*British Standards Institute*)

BT *abrév* (*classe de transfert de masse de document*) TÉL BT (*document bulk transfer class*)

BTM *abrév* (*classe de transfert de masse et de manipulation de document*) TÉL BTM (*document bulk transfer and manipulation class*)

bulles: ~ **d'aide** *f pl* INFORMAT balloon help

bulletin *m* TÉL bulletin; ~ **d'information** INFORMAT newsletter

bureau[1] *m* DONNÉES desktop, INFORMAT desk, office, desktop; ~ **mobile** RADIO mobile office; ~ **de poste** TÉL post office (*PO*); ~ **public de télécopie** TÉL public facsimile bureau; ~ **télégraphique** TÉL telegraphic office (*TO*); ~ **du trafic** RADIO traffic bureau; ~ **virtuel** GÉNÉRAL virtual office

bureau[2]: **de** ~ *loc* INFORMAT desktop

bureautique[1] *adj* DONNÉES office-level

bureautique[2] *f* DONNÉES, GÉNÉRAL office automation

bureaux: ~ **connectés** *m pl* TÉL linked offices

bus *m* ALIM ÉLEC, COMMUT, INFORMAT, TÉL bus; ~ **d'alimentation** ALIM ÉLEC, INFORMAT, TÉL bus; ~ **de courant continu** ALIM ÉLEC direct current bus (*DC bus*); ~ **d'entrée/sortie** COMMUT input/output bus; ~ **d'interface universel** DONNÉES general-purpose interface bus (*GPIB*); ~ **local** (*BL*) INFORMAT local bus; ~ **de mémoire de programme** COMMUT, INFORMAT program store bus; ~ **optique** TRANSMIS optical-data bus; ~ **passif** ALIM ÉLEC passive bus; ~ **de station multiprocesseur** TÉL multiprocessor station bus; ~ **de test et d'exploitation** TÉL, TESTS test and operation bus; ~ **de transfert** COMMUT transfer bus

buse *f* CLIMAT spout, nozzle, COMPOSANTS spout, RADIO nozzle; ~ **de pulvérisation** CLIMAT, COMPOSANTS spray nozzle

butée: ~ **de cadran** *f* PÉRIPH *postes téléphoniques*, TÉL finger stop; ~ **inférieure** *f* COMPOSANTS bottom stud

butineur *m* DONNÉES browser

BVU *abrév* (*Broadcast Vidéo U-Matic*) TV BVU (*Broadcast Video U-Matic*)

bypass *m* INFORMAT bypass

byte *m* DONNÉES, INFORMAT octet

C

CA *abrév* ALIM ÉLEC (*courant alternatif*) AC (*alternating current*), DONNÉES (*communication acceptée*) CA (*call accepted*), INFORMAT, SIGNAL (*communication accepté*) CA (*call accepted*), TÉL (*application de communication*) CA (*communication application*)

cabine *f*: **~ à cartes** *f* TÉL card-operated payphone, cardphone; **~ à pièces** *f* TÉL coin-operated payphone; **~ publique** *f* TÉL call box, public call box; **~ téléphonique** *f* TÉL call box, phone box, telephone booth, telephone box, telephone kiosk; **~ téléphonique publique** *f* TÉL call box, public call box

cabinet: **~ d'expertise** *m* GÉNÉRAL consultancy

câblage *m* COMPOSANTS, ÉLECTRON, TV, TRANSMIS cabling, stranding, wiring; **~ aérien** ALIM ÉLEC, TRANSMIS overhead cabling; **~ d'alimentation** COMPOSANTS harness; **~ de bâti** TÉL rack cabling; **~ broche à broche** COMPOSANTS, ÉLECTRON pin-to-pin wiring; **~ en couches concentriques** TRANSMIS concentric stranding; **~ en cuivre** TRANSMIS copper cabling; **~ direct** COMPOSANTS direct wiring; **~ effiloché** ÉLECTRON frayed wiring; **~ électrique** ÉLECTRON wiring harness; **~ externe** ÉLECTRON external cabling; **~ en fil nu** TÉL bare wiring; **~ fixe** TÉL permanent wiring; **~ interne** ÉLECTRON internal wiring; **~ libre** TRANSMIS loose cable structure; **~ des locaux** TÉL premises wiring; **~ à paires torsadées** COMPOSANTS, ÉLECTRON, TRANSMIS twisted-pair cabling; **~ par paires** GÉNÉRAL paired cable; **~ point à point** COMPOSANTS point-to-point wiring; **~ standardisé** COMPOSANTS, RADIO, TÉL uniform wiring; **~ transversal** TRANSMIS cross-stranding

câble[1] *m* GÉNÉRAL cable, cord, wire, TV cable; **~ aérien** ALIM ÉLEC, TRANSMIS aerial cable, overhead cable; **~ d'alimentation** ALIM ÉLEC, DONNÉES, ÉLECTRON, INFORMAT power cable, RADIO, TRANSMIS feeder; **~ amorce** ALIM ÉLEC stub cable, ÉLECTRON pigtail; **~ amorce d'épissage** COMPOSANTS splicing pigtail; **~ amorce orientable** ÉLECTRON swivel stub; **~ anti-feu** ÉLECTRON plenum-rated cable; **~ armé** ÉLECTRON, INFORMAT, TRANSMIS armored cable (*AmE*), armoured cable (*BrE*); **~ autoporté** TRANSMIS self-supporting cable, self-carrying cable; **~ autoporteur** TRANSMIS self-carrying cable, self-supporting cable; **~ de batterie** ALIM ÉLEC battery cable; **~ blindé** ALIM ÉLEC shielded cable shielded cable, screened cable, ÉLECTRON, INFORMAT armored cable (*AmE*), armoured cable (*BrE*), TÉL loaded cable, screened cable, TRANSMIS armored cable (*AmE*), armoured cable (*BrE*), loaded cable; **~ en canalisation** GÉNÉRAL duct cable; **~ à ceinture** TRANSMIS belted cable; **~ à champ radial** GÉNÉRAL radial-field cable; **~ à charge continue** TRANSMIS krarup cable; **~ chauffant** ALIM ÉLEC heating cable; **~ chauffant pour tuyauterie** TRANSMIS heat trace cable; **~ classique à fibres libres** TRANSMIS loose-tube cable; **~ coaxial** GÉNÉRAL pipe cable, TRANSMIS coax, coaxial cable; **~ coaxial fin** TRANSMIS thin coaxial cable; **~ coaxial gros** TRANSMIS thick coaxial cable; **~ à cur rainuré** TRANSMIS slotted-core cable; **~ de commande** CONTRÔLE, TRANSMIS control cable; **~ à conducteurs métalliques** TRANSMIS metallic conductor cable; **~ de contrôle** TRANSMIS control cable; **~ de cuivre** TRANSMIS copper cable; **~ cylindrique rainuré** TRANSMIS slotted-core cable; **~ de dérivation** ÉLECTRON break-out cable; **~ à deux conducteurs** ÉLECTRON, TRANSMIS two-conductor cable; **~ directement enterré** ÉLECTRON direct-buried cable; **~ de distribution** ALIM ÉLEC distribution cable, ÉLECTRON fan-out cable; **~ de distribution secondaire** TRANSMIS local distribution cable; **~ domestique** ALIM ÉLEC house-wiring cable; **~ double** TRANSMIS twin AX; **~ à écran** ALIM ÉLEC screened cable, shielded cable, TÉL screened cable; **~ enfichable** ÉLECTRON plug-in cable; **~ enterré** ALIM ÉLEC underground cable; **~ équipé** TRANSMIS cable assembly; **~ étanche** ALIM ÉLEC weather-sealed cable; **~ Ethernet noir** ÉLECTRON thin Ethernet; **~ externe** ÉLECTRON external cable; **~ à fibres enrobées** TRANSMIS tight-jacketed cable; **~ à fibres optiques** TRANSMIS optical cable, optical fiber cable (*AmE*), optical fibre cable (*BrE*); **~ à gainage serré** TRANSMIS tight-jacketed cable; **~ de garde** ALIM ÉLEC earth wire (*BrE*), ground wire (*AmE*); **~ haute fréquence** RADIO radio-frequency cable; **~ horizontal de suspension** TRANSMIS suspension strand; **~ intempérisé** COMPOSANTS weather-sealed cable; **~ inter-sites** TRANSMIS intersite cable; **~ interurbain** TRANSMIS trunk cable; **~ à isolant plastique** TRANSMIS plastic-insulated cable; **~ à isolant synthétique** TRANSMIS plastic-insulated cable; **~ isolé** ALIM ÉLEC, ÉLECTRON insulated cable; **~ de jonction** ÉLECTRON crossover patch cable; **~ krarupisé** TRANSMIS krarup cable; **~ de liaison** TRANSMIS connecting cable; **~ à masse** TRANSMIS paper-insulated mass-impregnated cable; **~ à matière non-migrante** TRANSMIS paper-insulated non-draining cable; **~ métallique** TRANSMIS metal conductor cable; **~ de mise à la terre** ALIM ÉLEC earth cable (*BrE*), ground cable (*AmE*); **~ monoconducteur** ALIM ÉLEC single-core cable; **~ monofibre** ALIM ÉLEC single-fiber cable (*AmE*), single-fibre cable (*BrE*); **~ multiconducteur** TRANSMIS multiconductor cable; **~ multifibre** TRANSMIS multifiber cable (*AmE*), multifibre cable (*BrE*); **~ multipaire** TRANSMIS, TV multipair cable; **~ multiple** COMMUT multiple cable; **~ multiple vertical** COMMUT vertical multiple cable; **~ à neutre périphérique** TRANSMIS concentric neutral cable; **~ non chargé** TRANSMIS unloaded cable; **~ non pupinisé** TÉL nonloaded cable, TRANSMIS unloaded cable; **~ omnibus** ALIM ÉLEC bus cable; **~ optique** TRANSMIS optical cable, optical fiber cable (*AmE*), optical fibre cable (*BrE*); **~ optique assemblé** TRANSMIS optical cable assembly; **~ optique à cœur rainuré** TRANSMIS slotted-core optical fiber cable (*AmE*), slotted-core optical fibre cable (*BrE*); **~ à paires** ÉLECTRON twin conductor cable, GÉNÉRAL paired cable; **~ à paires câblées en étoile** TRANSMIS quad-pair cable; **~ à paires symétriques** ÉLECTRON twin cable, TRANSMIS symmetrical-pair cable, twin cable; **~ à paires torsadées** ÉLECTRON, TRANSMIS twisted-pair cable; **~ à perte de couplage par gradation** TRANSMIS grading-coupling loss cable; **~ plat** ÉLECTRON flat cable, TRANSMIS ribbon cable; **~ plenum** ÉLECTRON plenum-rated cable; **~ porteur** ALIM ÉLEC suspension strand, TRANSMIS messenger cable, messenger wire, bearer cable; **~ à porteur central** TRANSMIS central suspension cable; **~ pressurisé** ALIM ÉLEC, TRANSMIS pressurized cable; **~ principal** ÉLECTRON trunk cable; **~ prolongateur** ALIM ÉLEC extension cable; **~ de protection** TRANSMIS protection cable; **~ pupinisé** TÉL, TRANSMIS loaded cable; **~ à quarte à paires combinables** TRANSMIS multiple twin cable; **~ à quartes en étoile** TRANSMIS quad cable; **~ à quatre fils** ÉLECTRON four-wire cable; **~ de raccordement** TRANSMIS connecting cable, connection cable, jumper cable, splicer cable; **~ rainuré** COMPOSANTS, TRANSMIS grooved cable; **~ rempli** TRANSMIS filled cable; **~ rempli de gelée imperméable**

TRANSMIS jelly-filled cable; ~ **de réseau** TRANSMIS network cable; ~ **ruban** TRANSMIS ribbon cable; ~ **de signalisation** SIGNAL signaling cable (*AmE*), signalling cable (*BrE*); ~ **soufflé** TRANSMIS blown fiber (*AmE*), blown fibre (*BrE*); ~ **souple** ALIM ÉLEC flexible cable; ~ **sous gaine non-métallique** TRANSMIS nonmetallic sheathed cable (*NMSC*); ~ **sous plomb** ALIM ÉLEC lead-sheathed cable; ~ **sous pression** ALIM ÉLEC, TRANSMIS pressurized cable; ~ **sous pression gazeuse** TRANSMIS internal gas pressure cable; ~ **sous-marin** TRANSMIS submarine cable; ~ **souterrain** TRANSMIS underground cable; ~ **à structure à fibres libres** TRANSMIS loose-structure cable; ~ **à structure à fibres lisses** TRANSMIS loose-structure cable; ~ **à structure ruban** TRANSMIS ribbon cable; ~ **supraconducteur** TRANSMIS superconductor cable; ~ **de télécommunications** TRANSMIS telecommunications cable; ~ **terrestre** TRANSMIS land cable; ~ **tréplomb** TRANSMIS separately sheathed cable; ~ **triax** TRANSMIS triax, triax cable, triaxial cable; ~ **triaxial** TRANSMIS triax, triax cable, triaxial cable; ~ **à tubes assemblés** TRANSMIS loose-tube cable; ~ **unifilaire** ÉLECTRON single-conductor cable

câble[2]: **~-diffusion** *f* TV *cable television* cable TV

câble[3]: **à ~ connecté** *loc* ÉLECTRON rack-mounted

câblé: ~ **en étoile** *adj* ALIM ÉLEC star-wired

câbler *vt* TV wire

câblerie *f* GÉNÉRAL, TRANSMIS cable factory

câbles: ~ **pré-équipés** *m pl* TRANSMIS preassembled cables; ~ **réparation** *m pl* TRANSMIS repairing cables

câblo[1]: **~-diffusion** *f* TV cable television, cable TV

câblo[2]: **~-distributeur** *m* TV cable operator; **~-opérateur** *m* TV cable operator

câblodistribution *f* TV cabled distribution

cabochon *m* ALIM ÉLEC lamp lens

cadence *f* GÉNÉRAL pace, INFORMAT pace, rate, swing, RADIO cadence; ~ **de commutation** COMMUT, TESTS switching rate; ~ **de création** ÉLECTRON generation rate; ~ **des évanouissements** RADIO fading rate; ~ **d'impulsions** APP SERV impulse rate

cadencer *vt* DONNÉES run, set

cadenceur *m* ÉLECTRON clock oscillator

cadrage: ~ **précis** *m* DONNÉES precision framing

cadran *m* INFORMAT, PÉRIPH, TÉL dial; ~ **alpha** INFORMAT alpha dial; ~ **d'appel** PÉRIPH *postes téléphoniques* dial; ~ **à clavier** APP SERV, PÉRIPH keypad, keyset; ~ **de commande** INFORMAT alpha dial; ~ **de syntonisation** ÉLECTRON tuning position meter

cadre *m* APP SERV, COMPOSANTS frame, framework, TV frame; ~ **d'étiquette** GÉNÉRAL label frame, designation frame; ~ **multiplage** COMMUT multiple frame; ~ **de page** DONNÉES, INFORMAT page frame; ~ **pour peigne** TRANSMIS formboard base; ~ **de sélection** INFORMAT selection box; ~ **de travail** INFORMAT framework

cadrer *vt* INFORMAT justify; ~ **à droite** DONNÉES, INFORMAT right-justify; ~ **à gauche** DONNÉES, INFORMAT left-justify

CAF *abrév* (*commande automatique de fréquence, contrôle automatique de fréquence*), ÉLECTRON, TÉL, TRANSMIS AFC (*automatic frequency control*)

CAG *abrév* (*commande automatique de gain, contrôle automatique de gain*), ÉLECTRON, TÉL, TRANSMIS AGC (*automatic gain control*); ~ **avec précompensation** *m* TRANSMIS feedforward AGC; ~ **à contre-réaction** *m* TRANSMIS feedback AGC

cahier: ~ **des charges** *m* GÉNÉRAL schedule

CAI *abrév* (*interface radio commune*) RADIO CAI (*common air interface*)

CAIG *abrév* (*commande automatique instantanée de gain, contrôle automatique instantanée de gain*) ÉLECTRON, TÉL, TRANSMIS IAGC (*instantaneous automatic gain control*)

caisse: ~ **de groupement** *f* ALIM ÉLEC *batteries* battery tray; ~ **intermédiaire** *f* TÉL coin buffer

caisson *m* ALIM ÉLEC *batteries* battery tray

calcul: ~ **d'adresse** *m* INFORMAT address generation; ~ **d'échelle linéaire** *m* DONNÉES linear-scaling calculation

calculateur: ~ **pilote** *m* INFORMAT controlling computer

calé *adj* TV in-register

cale: ~ **de guide d'ondes** *f* COMPOSANTS, RADIO, TRANSMIS waveguide shim; ~ **mince de raccordement** *f* COMPOSANTS, RADIO, TRANSMIS waveguide shim

calfeutrage: ~ **antibrouillage RF** *m* RADIO RFI caulking

calibrage *m* TESTS gaging (*AmE*), gauging (*BrE*)

calibre *m* COMPOSANTS, GÉNÉRAL jig, TESTS gage (*AmE*), gauge (*BrE*); ~ **de filetage** TESTS thread gage (*AmE*), thread gauge (*BrE*); **~à mâchoires** TESTS snap gage (*AmE*), snap gauge (*BrE*)

CAM *abrév* (*carte à mémoire*) APP SERV smartcard, swipe card, DONNÉES memory card, ÉLECTRON chip card, INFORMAT smartcard, TÉL swipe card

cambreur: ~ **de lames de contact** *m* COMPOSANTS, TESTS spring bender; ~ **de ressort** *m* COMPOSANTS, TESTS spring adjuster, spring bender; ~ **de ressort porte-contact** *m* TESTS contact spring adjuster, contact spring bender

came: ~ **d'impulsions** *f* PÉRIPH *postes téléphoniques* pulse cam disc (*BrE*), pulse cam disk (*AmE*), pulse disc (*BrE*), pulse disk (*AmE*)

camembert *m* INFORMAT pie chart

caméra *f* TV camera; ~ **vidéo** TV video camera

caméscope *m* TV camcorder

camouflage: ~ **contre la radiodétection** *m* RADIO radar camouflage; ~ **radar** *m* RADIO radar camouflage

CAMR *abrév* (*Conférence administrative mondiale des radiocommunications*) GÉNÉRAL World Administrative Radiocommunications Conference

CAN *abrév* (*convertisseur A-N, convertisseur analogique-numérique*) DONNÉES, ÉLECTRON, INFORMAT, SIGNAL, TÉL ADC (*A-D converter, analog-to-digital converter*)

canal *m* DONNÉES, INFORMAT, RADIO, TRANSMIS, TV channel; ~ **d'accès** TÉL access channel; ~ **adjacent** RADIO, TÉL adjacent channel; ~ **amorphe** TRANSMIS amorphous channel; ~ **d'appel** RADIO *mobicom* calling channel; ~ **audio** TV audio channel; ~ **avec évanouissement Rice** ÉLECTRON Rician-fading channel; ~ **B** DONNÉES, TRANSMIS *information* B-channel; ~ **de base** INFORMAT basic channel; ~ **bifréquence** RADIO dual-frequency channel; ~ **à code d'accès** TV limited-access channel, restricted channel; ~ **de commande** CONTRÔLE control channel (*CC*); ~ **de commande intégré** DONNÉES embedded control channel (*ECC*); ~ **de communication** DONNÉES, INFORMAT pipe; ~ **de communication de données** DONNÉES, TRANSMIS data-communication channel (*DCC*); ~ **contrapolaire** TRANSMIS cross-polar channel; ~ **de contrôle** CONTRÔLE, TRANSMIS control channel (*CC*); ~ **de contrôle commun** CONTRÔLE, TRANSMIS common control channel; ~ **D** DONNÉES, TÉL D-channel; ~ **à 2,048 Mbit/s tramé** DONNÉES framed 2.048 Mbit/s channel; ~ **discret** TRANSMIS discrete channel; ~ **disponible** RADIO idle channel; ~ **de distribution** DONNÉES distribution channel; ~ **d'émission** TRANSMIS transmit channel; ~ **d'enregistrement** TÉL recording channel; ~ **d'entrée** ÉLECTRON, INFORMAT input channel; ~ **d'entrée/sortie** INFORMAT input/output channel; ~ **E/S** INFORMAT I/O channel; ~ **d'exploitation intégré** DONNÉES embedded operations channel (*EOC*); ~ **de fréquence** RADIO frequency channel; ~ **de Gauss** DONNÉES Gaussian channel; ~ **de gestion et de retard** TRANSMIS control-and-delay channel; ~ **de groupement de signalisation** SIGNAL signaling grouping channel (*AmE*) (*SGC*), signalling grouping channel (*BrE*) (*SGC*); ~ **image** TV picture channel; ~ **interurbain** TRANSMIS trunk channel; ~ **interurbain intermédiaire** TRANSMIS intermediate trunk channel; ~ **à large bande** TRANSMIS wideband channel; ~ **linéaire** TRANSMIS linear channel; ~ **local** TV local channel; ~ **de maintenance** TRANSMIS maintenance channel; ~ **de maintenance des exploitants de réseau** TRANSMIS network-operators' maintenance channel (*NOMC*); ~ **MCS** TÉL MCS channel; ~ **de migration simple** ÉLECTRON straightforward migration path; ~ **mosaïque** TV preview channel; ~ **MPEG** TV MPEG channel, motion picture experts group channel; ~ **à multiplexage dans le temps** COMMUT, TÉL time-division highway; ~ **à multiplexage dans le temps en commutation** COMMUT, TÉL time-division highway in switching; ~ **multiplexeur** TRANSMIS multiplexer channel; ~ **à ondes multiples** RADIO multiray channel; ~ **optionnel** TV premium channel; ~ **premier-adjacent** RADIO first-adjacent channel; ~ **privé MCS** TÉL MCS private channel; ~ **radio** RADIO radio channel; ~ **de radiodiffusion** RADIO broadcast channel; ~ **radioélectrique** RADIO radio-frequency channel, RF channel; ~ **radioélectronique** RADIO channel, radio-frequency channel, RF channel; ~ **de Rayleigh** DONNÉES, TÉL, TRANSMIS Rayleigh channel; ~ **de réception** INFORMAT receive channel, receiving channel, reception channel, TRANSMIS receive channel; ~ **de recherche de personnes** TRANSMIS paging channel; ~ **RF** RADIO channel, RF channel; ~ **satellite de TV** TV satellite television channel; ~ **sémaphore** SIGNAL, TÉL signaling channel (*AmE*), signalling channel (*BrE*), signaling link (*AmE*), signalling link (*BrE*); ~ **sémaphore actif** SIGNAL active signaling link (*AmE*), active signalling link (*BrE*); ~ **sémaphore à l'état actif** SIGNAL active signaling link (*AmE*), active signalling link (*BrE*); ~ **sémaphore inactif** SIGNAL inactive signaling link (*AmE*), inactive signalling link (*BrE*); ~ **sémaphore normal** SIGNAL regular signaling link (*AmE*), regular signalling link (*BrE*); ~ **sémaphore de secours** SIGNAL reserve signaling link (*AmE*), reserve signalling link (*BrE*); ~ **son** TV sound channel; ~ **de sortie** INFORMAT output channel; ~ **spécialisé pour la signalisation** SIGNAL dedicated signaling channel (*AmE*), dedicated signalling channel (*BrE*); ~ **de surcharge** TRANSMIS overload channel; ~ **thématique** TV theme channel; ~ **de trafic** DONNÉES, RADIO traffic channel; ~ **de traitement transactionnel** DONNÉES transaction processing channel; ~ **de transmission** RADIO, TRANSMIS transmitting channel, transmit channel, transmission channel, TV transmission channel; ~ **de transmission de données** DONNÉES, TRANSMIS data-communication channel (*DCC*); ~ **virtuel** SIGNAL, TRANSMIS virtual channel (*VC*); ~ **virtuel de signalisation** SIGNAL, TRANSMIS signaling virtual channel (*AmE*) (*SVC*), signalling virtual channel (*BrE*) (*SVC*); ~ **vocal libre** RADIO idle voice channel

canalisation *f* COMPOSANTS conduit, ÉLECTRON, TRANSMIS conduit, duct

caniveau *m* ÉLECTRON troughing; ~ **de câble** TRANSMIS cable duct

canon: ~ **à électrons** *m* ÉLECTRON electron gun; ~ **immergé** *m* ÉLECTRON immersed gun; ~ **d'injection de magnétron** *m* ÉLECTRON magnetron injection gun; ~ **de lecture** *m* ÉLECTRON reading gun; ~ **de Pierce** *m* ÉLECTRON Pierce gun

canules: ~ **d'application** *f pl* ÉLECTRON dispensing tips

CAO *abrév* (*conception assistée par ordinateur*) COMPOSANTS EDA (*electronic design automation*), INFORMAT CAD (*computer-aided design*)

capacité *f* COMMUT size, ÉLECTRON capacitance, capacitor, GÉNÉRAL capacity, INFORMAT ability, SIGNAL size; ~ **d'absorption d'humidité** GÉNÉRAL moisture-absorbing capacity; ~ **d'acquisition** PÉRIPH scanning rate, scanning capability; ~ **attribuée d'office** TRANSMIS automatically allocated capacity; ~ **du boîtier** ÉLECTRON case capacitance; ~ **de canal** TRANSMIS channel capacity; ~ **du canal** RADIO channel loading; ~ **des canaux** TRANSMIS channel size; ~ **de charge** ÉLECTRON load capacitance, load capacity; ~ **de communication** GÉNÉRAL communication capability; ~ **de correction des paquets d'erreurs** DONNÉES error burst-correcting capability; ~ **différentielle** ÉLECTRON small-signal capacitance; ~ **de diffusion** COMPOSANTS diffusion capacitance; ~ **d'écoulement du trafic** DONNÉES traffic-handling capacity, TÉL traffic performance, trafficability performance, traffic-handling capacity; ~ **effective** TÉL effective capacity; ~ **effective d'entrée** ALIM ÉLEC effective input capacitance; ~ **effective de sortie** ALIM ÉLEC effective output capacitance; ~ **électrique** ÉLECTRON electrical capacitance; ~ **d'électrode** ÉLECTRON electrode capacitance; ~ **entre électrodes** ÉLECTRON interelectrode capacitance; ~ **d'entrée** ÉLECTRON input capacitance; ~ **fictive** TRANSMIS dummy capacitance; ~ **de fonctionnement en période d'éclipse** RADIO eclipse capability; ~ **interélectrode** ÉLECTRON interelectrode capacitance; ~ **de jonction** ÉLECTRON junction capacitance; ~ **de jonction de l'émetteur** COMPOSANTS emitter depletion-layer capacitance; ~ **de la mémoire interne** INFORMAT internal-memory capacity; ~ **mémoire maximale** PÉRIPH maximum memory capacity; ~ **de mise hors tension** ALIM ÉLEC switch-off capability; ~ **parasite** ALIM ÉLEC stray capacitance; ~ **de servitude de trame** DONNÉES frame overhead; ~ **de sortie** ÉLECTRON output capacitance; ~ **de stockage** DONNÉES storage capacity; ~ **du support** TRANSMIS bearer capacity; ~ **support** TÉL, TRANSMIS bearer unit, bearer capability (*BC*); ~ **terminale** RADIO *d'antenne* capacity top; ~ **thermique** COMPOSANTS, ÉLECTRON thermal capacitance; ~ **de trafic** TÉL, TRANSMIS traffic capacity; ~ **de transmission** DONNÉES transmission capability; ~ **utile** RADIO throughput, thruput (*AmE*)

capacités: ~ **de fonctionnement locales** *f* (*CLF*) COMMUT local functional capabilities (*LFC*); ~ **multimédia** *f pl* INFORMAT multimedia capabilities

capacitif *adj* ÉLECTRON capacitive

capillaire: ~ **en verre** *m* ÉLECTRON glass capillary

capot *m* COMPOSANTS case, cover, hood, lid; ~ **d'antenne** RADIO radar dome, radome, shroud; ~ **antipoussière** COMPOSANTS dust cover; ~ **d'extrémité de rangée** TÉL row-end cover; ~ **d'insonorisation** COMPOSANTS, TÉL sound-attenuating cover; ~ **protecteur** COMPOSANTS protective case, protective cover

capsule: ~ **d'écouteur** *f* PÉRIPH *télex*, SIGNAL, TÉL receiver inset; ~ **émettrice** *f* PÉRIPH *postes téléphoniques* transmitter inset; ~ **de microphone** *f* PÉRIPH *postes téléphoniques*, SIGNAL, TÉL transmitter cap; ~ **réceptrice** *f* PÉRIPH *télex*, SIGNAL, TÉL receiver inset

capter *vt* RADIO pick up

capteur *m* ÉLECTRON, GÉNÉRAL, RADIO sensor, TESTS probe; ~ **CCD** (*capteur de dispositif à couplage de charge*) PÉRIPH CCD detector (*charge-coupled device detector*); ~ **de chaleur** TV heat detector; ~ **de champ lumineux** TRANSMIS *fibres optiques* bright-field sensor; ~ **de dispositif à couplage de charge** (*capteur CCD*) PÉRIPH charge-coupled device detector (*CCD detector*); ~ **d'effraction** TV heat detector; ~ **à fibres optiques** TRANSMIS optical fiber sensor (*AmE*), optical fibre sensor (*BrE*); ~ **d'images** PÉRIPH, TV image sensor; ~ **infrarouge** TRANSMIS infrared detector; ~ **d'intensité** TRANSMIS intensity sensor; ~ **de lumière** TV light detector; ~ **matriciel** PÉRIPH matrix detector; ~ **de phase** TRANSMIS phase sensor; ~ **solaire de première approximation** RADIO coarse sun sensor; ~ **de température** CLIMAT temperature sensor, CONTRÔLE temperature sensor, thermometer probe

capture *f* GÉNÉRAL capture; ~ **vidéo** INFORMAT video capture, TV frame grabbing

caractère *m* APP SERV, DONNÉES, INFORMAT character, TÉL code character, character, TESTS characteristic; ~ **accusé de réception négatif** TÉL negative acknowledge character; ~ **d'annulation de bloc** INFORMAT block-cancel character; ~ **binaire de signe** INFORMAT sign-binary character; ~ **de bourrage** DONNÉES stuffing character; ~ **de changement**

de ligne DONNÉES line-feed character; **~ de code normal** DONNÉES shift-in character; **~ de code spécial** DONNÉES shift-out character; **~ de commande** DONNÉES, INFORMAT, TÉL control character; **~ de commande d'appel** TÉL call control character; **~ de commande d'impression** DONNÉES print-control character; **~ de contrôle d'erreurs** RADIO error-check character; **~ de correction** TÉL correction character; **~ de début de texte** (*STX*) DONNÉES, INFORMAT start of text (*STX*), start-of-heading character (*STX*); **~ d'échappement** DONNÉES escape character; **~ d'effacement** INFORMAT deletion character; **~ espace** DONNÉES space character (*SP*); **~ d'espacement** DONNÉES space character (*SP*); **~ d'espacement arrière** INFORMAT backspace character (*BS*); **~ d'exécution** TÉL execution character; **~ final** DONNÉES final character; **~ de générique** INFORMAT joker; **~ générique** INFORMAT wildcard character; **~ graphique** DONNÉES, TÉL graphic character, graphical character; **~ inactif** TÉL inactive character; **~ de mise en page** DONNÉES, TÉL format effector (*FE*); **~ nul** (*NUL*) DONNÉES null character (*NUL*); **~ de présentation** DONNÉES, INFORMAT format effector (*FE*), layout character, TÉL format effector (*FE*); **~ de rejet** DONNÉES reject character; **~ de remplacement** (*SUB*) DONNÉES, TÉL substitute character (*SUB*); **~ de remplissage** DONNÉES fill character, filler, filler character, gap character; **~ de répétition** TÉL continuation character; **~ de retour arrière** INFORMAT backspace character (*BS*); **~ de retour à la ligne** DONNÉES new-line character (*NL*), character return (*CR*); **~ de séparation d'informations** DONNÉES information-separator character; **~ de signe** INFORMAT sign character; **~ spécial** DONNÉES special character; **~ de substitution** (*SUB*) DONNÉES, TÉL substitute character (*SUB*); **~ suite** TÉL continuation character; **~ de synchronisation** (*SYN*) DONNÉES synchronization character (*SYN*), synchronous idle character (*SYN*); **~ de tabulation horizontale** DONNÉES, RADIO horizontal tabulation character (*HT*); **~ de tabulation verticale** INFORMAT vertical tabulation character (*VT*); **~ télégraphique** TÉL code character

caractères[1]: **~ redéfinissables dynamiquement** *m pl* INFORMAT dynamically-redefinable characters (*DRCS*)

caractères[2]: **~ par pouce** *loc* (*cpp*) INFORMAT characters per inch (*cpi*)

caractéristique: **~ de communication virtuelle** *f* DONNÉES, TÉL virtual-call characteristic; **~ de conduction directe** *f* ÉLECTRON forward characteristic; **~ de destruction** *f* TÉL destruction characteristic of protector; **~ directe** *f* ÉLECTRON forward characteristic; **~ dynamique** *f* INFORMAT dynamic characteristic; **~ d'émission** *f* RADIO, TRANSMIS transmit characteristic; **~ fonctionnelle** *f* GÉNÉRAL functional characteristic, functional feature; **~ de fonctionnement** *f* TÉL operating characteristic, working characteristic, TESTS performance characteristic; **~ fréquentielle** *f* RADIO frequency characteristic; **~ optique** *f* TRANSMIS optical characteristic; **~ phase-amplitude** *f* TRANSMIS phase-amplitude characteristic; **~ de réception** *f* RADIO receive characteristic; **~ de réponse à un échelon** *f* ÉLECTRON, SIGNAL, TRANSMIS square-wave response characteristic; **~ du réseau** *f* TÉL network parameter; **~ de sensibilité énergétique monochromatique relative** *f* ÉLECTRON relative spectral sensitivity characteristic; **~ de sensibilité spectrale relative** *f* ÉLECTRON relative spectral sensitivity characteristic; **~ du signal** *f* SIGNAL signal characteristic; **~ spectrale** *f* ÉLECTRON spectral characteristic; **~ spectrale de rendement quantique** *f* ÉLECTRON quantum-efficiency spectral characteristic; **~ statique** *f* ÉLECTRON static characteristic; **~ temporelle** *f* ÉLECTRON, INFORMAT, TESTS, TRANSMIS time characteristic; **~ tension-courant d'anode-cathode** *f* ÉLECTRON *semiconducteurs* anode-cathode characteristics; **~ thermique** *f* COMPOSANTS, ÉLECTRON thermal characteristic; **~ de transfert** *f* ÉLECTRON transfer characteristic; **~ de transfert de signal lumineux** *f* ÉLECTRON light-signal transfer characteristic; **~ de transmission** *f* RADIO, TRANSMIS transmission characteristic, transmit characteristic

caractéristiques: **~ de bobinage** *f pl* ALIM ÉLEC winding data; **~ en charge** *f pl* ALIM ÉLEC load function; **~ dans le domaine fréquentiel** *f pl* RADIO frequency-domain characteristics; **~ dans le domaine temporel** *f pl* TESTS time domain characteristics; **~ électriques** *f pl* TRANSMIS electrical data; **~ d'enroulement** *f pl* ALIM ÉLEC winding data; **~ fréquencielles** *f pl* RADIO frequency-domain characteristics; **~ générales** *f pl* INFORMAT *en LDS* general parameters; **~ des lobes latéraux** *f pl* RADIO side-lobe performance; **~ de propagation** *f pl* RADIO propagation properties, TÉL propagation performance; **~ du signal** *f pl* SIGNAL character of signal; **~ de sortie** *f pl* ÉLECTRON output characteristics

carcasse: **~ de bobine** *f* ALIM ÉLEC coil former

carcinotron *m* ÉLECTRON carcinotron

carte *f* DONNÉES, ÉLECTRON, INFORMAT card; **~ accélératrice** DONNÉES accelerator board; **~ d'affectation** DONNÉES assignment map; **~ d'alarme** APP SERV alarm card; **~ audio** INFORMAT audio card, sound board; **~ auxiliaire** ALIM ÉLEC *circuits imprimés* auxiliary board; **~ CGA** INFORMAT CGA card; **~ de circuit imprimé** INFORMAT circuit board; **~ à circuits** ÉLECTRON circuit board; **~ à circuits imprimés** ÉLECTRON integrated circuit panel assembly (*ICPA*), printed circuit board (*PCB*); **~ à cur métallique** ÉLECTRON metal-core board; **~ de communication** DONNÉES communication card; **~ de compression audio** DONNÉES audio compression board; **~ de compression de données** DONNÉES data-compression board; **~ de compression d'images** DONNÉES, TV image compression board; **~ de connexion** ALIM ÉLEC paddle board; **~ de connexion de câble** TRANSMIS *mise à l'essai* cable connection board (*CCB*); **~ de contrôle** CONTRÔLE control card; **~ de données** DONNÉES, INFORMAT data card; **~ enfichable** ÉLECTRON plug-in board, plug-in chip card; **~ en-tête** INFORMAT header card; **~ équipée** ÉLECTRON printed board assembly; **~ équivalente** ÉLECTRON replacement board; **~ d'essai** ÉLECTRON, TESTS test board; **~ de l'étendue** TESTS range chart; **~ Ethernet** DONNÉES ethernet card; **~ d'extension** DONNÉES add-on board, ÉLECTRON extension board, INFORMAT add-in board, expansion card, upgrade board; **~ fille** ÉLECTRON, INFORMAT daughterboard; **~ flexo-rigide** ÉLECTRON rigid flex card; **~ graphique couleur** (*CGA*) INFORMAT colour graphics adaptor (*BrE*) (*CGA*); **~ imprimée** ÉLECTRON printed board; **~ imprimée double face** ÉLECTRON double-sided printed board; **~ imprimée équipée** ÉLECTRON printed board assembly; **~ imprimée souple** ÉLECTRON flexible printed board; **~ intelligente** APP SERV, INFORMAT smartcard; **~ d'interface de ligne** COMMUT line interface board (*LIB*); **~ d'interface S** TÉL S-universal interface card; **~ de ligne analogique** COMMUT analog line board (*ALB*); **~ de ligne intérieure** TÉL line circuit pack; **~ de ligne numérique** COMMUT digital line board (*DLB*); **~ magnétique** DONNÉES magnetic card; **~ à mémoire** (*CAM*) APP SERV smartcard, swipe card, DONNÉES memory card, ÉLECTRON chip card, INFORMAT smartcard, TÉL swipe card; **~ mémoire** DONNÉES memory board; **~ mémoire de programme** RADIO code memory board; **~ mère** INFORMAT motherboard; **~ à microcircuit** APP SERV smartcard, ÉLECTRON chip card, INFORMAT smartcard; **~ à microprocesseur** APP SERV smartcard, ÉLECTRON chip card, INFORMAT smartcard; **~ modem** DONNÉES modem card; **~ nue** ÉLECTRON printed board; **~ pastel** DONNÉES phonecard, TÉL charge card, phonecard; **~ PC de communication** INFORMAT communication control card; **~ périphérique** INFORMAT peripheral card; **~ à piste magnétique** ÉLECTRON magnetic-strip card; **~ de production** ÉLECTRON product board; **~ à puce** ÉLEC-

TRON chip card; ~ **de remplacement** ÉLECTRON replacement board; ~ **à somme cumulée** TESTS *qualité* cumulative sum chart; ~ **sur bus local** INFORMAT local bus card; ~ **de test** *f* ÉLECTRON, TESTS test board; ~ **de traitement** DONNÉES, INFORMAT processing card; ~ **de traitement de la parole** APP SERV speech-processing card; ~ **vidéo** INFORMAT video digitizer, video card, video board, TV frame grabber

cartouche *f* GÉNÉRAL cartridge; ~ **amovible** GÉNÉRAL removable cartridge; ~ **de bande magnétique** DONNÉES magnetic-tape cartridge; ~ **d'exécution** INFORMAT performance cartridge; ~ **de programmes** INFORMAT program cartridge; ~ **quart de pouce** (*QIC*) TV quarter-inch cartridge (*QIC*); ~ **RAM** DONNÉES RAM cartridge

CAS *abrév* (*fonction commutateur d'accès aux services*) COMMUT, TÉL service access switch function

cascader *vi* ÉLECTRON cascade

case: ~ **de commande** *f* INFORMAT button; ~ **d'option** *f* INFORMAT check box, radio button; ~ **par défaut** *f* INFORMAT default button; ~ **zoom** *f* INFORMAT zoom box

casque *m* TÉL headphones, headset; ~ **à deux écouteurs indépendants** TÉL split headphones; ~ **d'écoute** TÉL headphones, receiver headset; ~ **téléphonique** TÉL headset

cassette *f* INFORMAT, TV cassette; ~ **compacte numérique** (*cassette DCC*) INFORMAT, TV digital compact cassette (*DCC*); ~ **DCC** (*cassette compacte numérique*) INFORMAT, TV digital compact cassette (*DCC*); ~ **d'épanouissement** COMPOSANTS splice tray; ~ **numérique** INFORMAT digital cassette; ~ **vidéo** TV videocassette

cassure *f* TRANSMIS break

catalogue *m* DONNÉES, INFORMAT, TÉL catalog (*AmE*), catalogue (*BrE*); ~ **des défaillances** TÉL failure catalog (*AmE*), failure catalogue (*BrE*); ~ **des erreurs** TÉL error catalog (*AmE*), error catalogue (*BrE*); ~ **en ligne** DONNÉES on-line catalog (*AmE*), on-line catalogue (*BrE*); ~ **des pannes** TÉL error catalog (*AmE*), error catalogue (*BrE*), failure catalog (*AmE*), failure catalogue (*BrE*); ~ **photo CD** INFORMAT photo CD catalog (*AmE*), photo CD catalogue (*BrE*)

cataloguer *vt* DONNÉES, INFORMAT, TÉL catalog (*AmE*), catalogue (*BrE*)

cathode *f* COMPOSANTS, ÉLECTRON cathode; ~ **chaude** ÉLECTRON hot cathode; ~ **à chauffage direct** COMPOSANTS, ÉLECTRON directly-heated cathode; ~ **à chauffage indirect** COMPOSANTS, ÉLECTRON indirectly-heated cathode; ~ **à couche d'oxydes** COMPOSANTS, ÉLECTRON oxide-coated cathode; ~ **liquide** COMPOSANTS, ÉLECTRON pool cathode; ~ **en tungstène thorié** COMPOSANTS, ÉLECTRON thoriated-tungsten cathode

CAV *abrév* (*commande automatique de volume*) ÉLECTRON, TÉL, TRANSMIS AVC (*automatic volume control*)

cavalier *m* ALIM ÉLEC strap, U-link, COMPOSANTS strap, INFORMAT jumper

cavité *f* ÉLECTRON, TRANSMIS cavity; ~ **optique** TRANSMIS optical cavity; ~ **optique résonnante** ÉLECTRON, TRANSMIS cavity resonator; ~ **résonnante** ÉLECTRON, TRANSMIS resonant cavity, cavity resonator

CB *abrév* (*bande publique*) RADIO CB (*citizens' band*)

CC *abrév* ALIM ÉLEC (*courant continu*) DC (*direct current*), COMMUT, TÉL (*central de commutation, centre de commutation*) switching center (*AmE*), switching centre (*BrE*), switching exchange, exchange

CCAL *abrév* (*coupleur central d'alarmes*) CONTRÔLE central alarm coupler

CCE *abrév* (*code correcteur d'erreurs*) DONNÉES, INFORMAT, SIGNAL, TÉL ECC (*error-correcting code*)

CCIR *abrév* (*Comité consultatif international des radiocommunications*) RADIO IRCC (*International Radio Consultative Committee*)

CCITT *abrév* (*Comité consultatif international télégraphique et téléphonique*) TÉL ITTCC (*International Telegraph and Telephone Consultative Committee*)

CCL *abrév* COMMUT (*contrôle de charge de ligne*) LLC (*line load control*), GÉNÉRAL (*centre de construction de lignes*) external plant center (*AmE*), external plant centre (*BrE*), TÉL (*contrôle de charge de ligne*) LLC (*line load control*)

CCM: ~ **d'entrées** *m* (*centre de commutation d'entrées mobile*) COMMUT gateway MSC (*gateway mobile switching center AmE, gateway mobile switching centre BrE*); ~ **nominal** *m* (*centre de commutation nominal mobile*) RADIO anchor MSC (*anchor mobile switching center AmE, anchor mobile switching centre BrE*); ~ **de rattachement** *m* (*centre de commutation de rattachement mobile*) RADIO anchor MSC (*anchor mobile switching center AmE, anchor mobile switching centre BrE*)

CCN *abrév* (*centre de coordination national*) TESTS NCC (*national control center AmE, national control centre BrE*)

CCS *abrév* (*copie conforme sélective*) TÉL BCC (*blind courtesy copy*)

CCTM *abrév* (*centre de commutation de téléphonie mobile*) RADIO MTSO (*mobile telephone switching office*)

C^3 *abrév* (*C-cube*) INFORMAT C^3 (*C-cube*)

C-cube *m* (*C^3*) INFORMAT C-cube (*C^3*)

CCV *abrév* (*commutateur à circuits virtuels*) COMMUT VCS (*virtual-circuit switch*)

CD *abrév* ALIM ÉLEC (*centre de distribution*) distribution center (*AmE*), distribution centre (*BrE*), INFORMAT (*disque compact*) CD (*compact disc*), TV (*centre de distribution*) distribution center (*AmE*), distribution centre (*BrE*); ~ **multicouche** *m* DONNÉES multilayer CD

CD-Audio *m* INFORMAT CD-Audio

CDE *abrév* (*code détecteur d'erreurs, code détection d'erreur*) DONNÉES, INFORMAT, SIGNAL, TÉL EDC (*error-detecting code, error-detection code*)

CD-I *abrév* (*disque compact interactif*) INFORMAT CD-I (*compact disc interactive*)

CD-R *abrév* (*disque compacte à lecture seule*) INFORMAT CD-R (*compact disc-read only*)

CD-ROM *abrév* (*disque optique numérique*) INFORMAT CD-ROM (*compact disc read-only memory*)

CD-RTOS *abrév* INFORMAT CD-RTOS (*CD-ROM real-time operating system*)

CDV *abrév* (*disque compact vidéo*) TV video CD

CD-WO *abrév* (*disque compacte à écriture seule*) INFORMAT CD-WO (*compact disc-write once*)

CE *abrév* COMMUT (*centre d'enregistrement*) recording center (*AmE*), recording centre (*BrE*), TÉL (*centre d'enregistrement*) recording center (*AmE*), recording centre (*BrE*)

CED *abrév* (*centre d'enregistrement à distance*) RADIO remote operating center (*AmE*), remote operating centre (*BrE*)

CEI *abrév* (*Commission électrotechnique internationale*) GÉNÉRAL IEC (*International Electrotechnical Commission*)

ceinture: ~ **de câble** *f* ALIM ÉLEC, COMPOSANTS tie wrap, TY wrap

cellulaire *adj* ÉLECTRON, RADIO, TÉL, TV cellular

cellule *f* DONNÉES, ÉLECTRON, INFORMAT, RADIO cell; ~ **ATM** DONNÉES, INFORMAT ATM cell; ~ **de base** RADIO base unit (*BU*); ~ **cible** RADIO target cell; ~ **électrolytique** ÉLECTRON electrolytic cell; ~ **de filtrage** ÉLECTRON filter section; ~ **galvanique** ALIM ÉLEC galvanic cell; ~ **invalidée** INFORMAT invalid cell; ~ **de maintenance** DONNÉES maintenance entity; ~ **de mémoire** DONNÉES storage cell, storage element; ~ **OAM de la couche physique** RADIO physical-layer operations and maintenance cell (*PL-OAM cell*); ~ **parapluie** RADIO umbrella cell; ~ **photoconductrice** ÉLECTRON photoconductive cell; ~ **photoélectrique** ÉLECTRON photoelectric cell; ~ **photosensible** DONNÉES, ÉLECTRON photosensitive cell; ~ **photovoltaïque** ÉLECTRON photovoltaic cell; ~ **de pluie** RADIO rain cell; ~ **sectorielle**

RADIO sector cell; ~ **standard** ÉLECTRON standard cell; ~ **valide** INFORMAT valid cell

CEN *abrév* (*Comité européen de normalisation*) GÉNÉRAL CEN (*European Committee for Standardization*)

CENELEC *abrév* (*Comité européen de normalisation électrotechnique*) GÉNÉRAL CENELEC (*European Committee for Electrotechnical Standardization*)

central: ~ **d'abonné** *m* COMMUT, TÉL private branch exchange (*PBX*); ~ **d'arrivée** *m* COMMUT, TÉL terminating exchange; ~ **automatique** *m* TÉL automatic exchange, automatic telephone exchange, unattended exchange, unmanned exchange; ~ **automatique local** *m* COMMUT community dial office (*AmE*) (*CDO*), unit automatic exchange (*UAX*); ~ **automatique privé** *m* APP SERV private automatic exchange (*PAX*); ~ **automatique rural** *m* COMMUT rural automatic exchange; ~ **autonome** *m* COMMUT stand-alone exchange; ~ **auxiliaire** *m* COMMUT dependent exchange, satellite exchange; ~ **auxiliaire discriminateur** *m* COMMUT discriminating satellite exchange; ~ **à cadran** *m* COMMUT rotary exchange; ~ **de commutation** *m* (*CC*) COMMUT, TÉL switching center (*AmE*), switching centre (*BrE*), switching exchange, exchange; ~ **de départ** *m* COMMUT originating exchange; ~ **de destination** *m* TÉL called exchange, destination exchange, destination office; ~ **discriminateur** *m* COMMUT discriminating satellite exchange; ~ **électronique** *m* COMMUT electronic exchange; ~ **électronique à relais à lames souples** *m* COMMUT reed-relay electronic exchange; ~ **interurbain** *m* COMMUT, TÉL, TRANSMIS trunk exchange; ~ **local** *m* COMMUT local exchange (*LE*), TÉL local central office, local exchange (*LE*); ~ **manuel privé** *m* COMMUT, TÉL private manual exchange; ~ **d'origine** *m* COMMUT home exchange; ~ **pas à pas** *m* TÉL step-by-step office; ~ **principal** *m* RADIO host exchange, host office; ~ **privé** *m* COMMUT private branch exchange (*PBX*), TÉL private branch exchange (*PBX*), private exchange; ~ **public** *m* TÉL public exchange; ~ **régional** *m* TÉL regional center (*AmE*) (*RC*), regional centre (*BrE*) (*RC*); ~ **rural** *m* TÉL rural exchange; ~ **sans personnel** *m* TÉL unattended exchange, unmanned exchange; ~ **satellite** *m* COMMUT satellite exchange, TÉL dependent exchange, subordinate exchange; ~ **serveur** *m* COMMUT serving exchange; ~ **de télécommunications** *m* TÉL telecommunications exchange; ~ **téléphonique** *m* COMMUT, TÉL central office, switching center (*AmE*), switching centre (*BrE*), switching exchange, telephone exchange; ~ **téléphonique automatique** *m* COMMUT, TÉL automatic telephone exchange; ~ **télésurveillé** *m* TÉL remote-controlled exchange; ~ **télex** *m* COMMUT, TÉL telex exchange, telex switching center (*AmE*), telex switching centre (*BrE*); ~ **temporel** *m* COMMUT time-division exchange; ~ **terminal** *m* TÉL, TRANSMIS terminal exchange; ~ **de transit** *m* COMMUT, DONNÉES, TÉL transit exchange; ~ **de transit urbain** *m* TÉL tandem exchange; ~ **tributaire** *m* COMMUT dependent exchange; ~ **type** *m* TÉL type exchange; ~ **urbain** *m* COMMUT, TÉL local exchange (*LE*)

centrale *f* GÉNÉRAL center (*AmE*), centre (*BrE*); ~ **électrique** ALIM ÉLEC power plant, electric power station; ~ **d'orientation** RADIO attitude control unit

centralographe *m* TÉL centralograph

centre: ~ **d'accès** *m* COMMUT gateway exchange; ~ **d'alarme** *m* TÉL alarm center (*AmE*), alarm centre (*BrE*); ~ **d'analyse du réseau** *m* TÉL network analysis point; ~ **d'assistance opérateur** *m* TÉL operator service center (*AmE*), operator service centre (*BrE*); ~ **automatique de commutation de messages** *m* TÉL automatic message switching center (*AmE*), automatic message switching centre (*BrE*); ~ **de calcul** *m* INFORMAT computing center (*AmE*), computing centre (*BrE*); ~ **du cœur** *m* TRANSMIS core center (*AmE*), core centre (*BrE*); ~ **de collecte de données** *m* DONNÉES data-collection center (*AmE*) (*DCC*), data-collection centre (*BrE*) (*DCC*); ~ **de commande de rétablissement du service** *m* TÉL restoration control point (*RCP*); ~ **de communication de données** *m* COMMUT, DONNÉES, ÉLECTRON data-switching exchange (*DSE*); ~ **de communication de groupement** *m* COMMUT group switching center (*AmE*) (*GSC*), group switching centre (*BrE*) (*GSC*); ~ **de commutation** *m* (*CC*) COMMUT, TÉL switching center (*AmE*), switching centre (*BrE*), switching exchange, exchange; ~ **de commutation à bord du satellite** *m* COMMUT, RADIO satellite switching center (*AmE*) (*SSC*), satellite switching centre (*BrE*) (*SSC*); ~ **de commutation de circuits** *m* COMMUT circuit-switching center (*AmE*), circuit-switching centre (*BrE*); ~ **de commutation de circuits virtuels** *m* COMMUT, TRANSMIS virtual-circuit switching node; ~ **de commutation de données** *m* COMMUT, DONNÉES, ÉLECTRON data-switching exchange (*DSE*); ~ **de commutation de groupement** *m* COMMUT group switching center (*AmE*) (*GSC*), group switching centre (*BrE*) (*GSC*); ~ **de commutation interurbaine** *m* COMMUT trunk switching center (*AmE*), trunk switching centre (*BrE*); ~ **de commutation local** *m* COMMUT, TÉL local central office; ~ **de commutation maritime** *m* COMMUT, TÉL maritime switching center (*AmE*) (*MSC*), maritime switching centre (*BrE*) (*MSC*); ~ **de commutation de messages** *m* COMMUT message-switching center (*AmE*), message-switching centre (*BrE*); ~ **de commutation mobile** *m* COMMUT, TÉL mobile switching center (*AmE*) (*MSC*), mobile switching centre (*BrE*) (*MSC*); ~ **de commutation numérique** *m* COMMUT digital switching center (*AmE*), digital switching centre (*BrE*); ~ **de commutation numérique du réseau principal** *m* COMMUT digital main network switching center (*AmE*), digital main network switching centre (*BrE*); ~ **de commutation de paquets** *m* COMMUT, DONNÉES, TÉL, TRANSMIS packet-switching exchange (*PSE*); ~ **de commutation du service maritime par satellite** *m* COMMUT, RADIO maritime satellite switching center (*AmE*) (*MSSC*), maritime satellite switching centre (*BrE*) (*MSSC*); ~ **de commutation de services mobiles** *m* COMMUT, TÉL mobile services switching center (*AmE*) (*MSC*), mobile services switching centre (*BrE*) (*MSC*); ~ **de commutation de téléphonie mobile** *m* (*CCTM*) RADIO mobile telecommunication switching office (*MTSO*); ~ **de commutation de transit** *m* COMMUT, DONNÉES, TÉL transit switching center (*AmE*), transit switching centre (*BrE*); ~ **de construction de lignes** *m* (*CCL*) GÉNÉRAL external plant center (*AmE*), external plant centre (*BrE*); ~ **continental** *m* COMMUT continental exchange; ~ **de contrôle des satellites** *m* RADIO satellite control center (*AmE*) (*SCC*), satellite control centre (*BrE*) (*SCC*); ~ **de coordination national** *m* (*CCN*) TESTS national control center (*AmE*) (*NCC*), national control centre (*BrE*) (*NCC*); ~ **directeur** *m* COMMUT controlling exchange; ~ **de direction des opérations** *m* TESTS operations control center (*AmE*) (*OCC*), operations control centre (*BrE*) (*OCC*); ~ **de distribution** *m* (*CD*) ALIM ÉLEC, TV distribution center (*AmE*), distribution centre (*BrE*); ~ **d'enregistrement** *m* (*CE*) COMMUT, TÉL recording center (*AmE*), recording centre (*BrE*); ~ **d'enregistrement à distance** *m* (*CED*) RADIO remote operating center (*AmE*), remote operating centre (*BrE*); ~ **d'enregistrement de la taxation** *m* TÉL charging-recording center (*AmE*), charging-recording centre (*BrE*); ~ **d'essai** *m* TESTS test center (*AmE*), test centre (*BrE*); ~ **d'étalonnage** *m* TESTS *métrologie* calibration center (*AmE*), calibration centre (*BrE*); ~ **d'exploitation** *m* GÉNÉRAL, TÉL, TV operating center (*AmE*), operating centre (*BrE*), operational center (*AmE*), operational centre (*BrE*), operations center (*AmE*), operations centre (*BrE*); ~ **d'exploitation et de maintenance** *m* TÉL, TESTS operation and maintenance center (*AmE*), operation and maintenance centre (*BrE*); ~ **d'exploitation et de maintenance de**

réseau *m* TESTS network operation and maintenance center (*AmE*), network operation and maintenance centre (*BrE*); **~ d'extrémité** *m* TÉL, TRANSMIS terminal exchange; **~ de facturation** *m* APP SERV billing center (*AmE*), billing centre (*BrE*); **~ de gaine** *m* TRANSMIS *fibres optiques* cladding center (*AmE*), cladding centre (*BrE*); **~ géographiquement dispersé** *m* COMMUT geographically-distributed exchange; **~ de gestion des dérangements** *m* TÉL fault center (*AmE*), fault centre (*BrE*); **~ de gestion, d'exploitation et de maintenance** *m* TÉL, TESTS operation, administration and maintenance center (*AmE*) (*OAMC*), operation, administration and maintenance centre (*BrE*) (*OAMC*); **~ de gestion de réseau** *m* TESTS network control center (*AmE*) (*NCC*), network control centre (*BrE*) (*NCC*), network management center (*AmE*) (*NMC*), network management centre (*BrE*) (*NMC*); **~ de gestion du réseau** *m* TÉL network management point; **~ d'impureté** *m* ÉLECTRON impurity center (*AmE*), impurity centre (*BrE*); **~ d'interconnexion mural** *m* ÉLECTRON wall-mountable interconnect center (*AmE*) (*WIC*), wall-mountable interconnect centre (*BrE*) (*WIC*); **~ intercontinental de transit** *m* TÉL, TV intercontinental transit exchange; **~ international** *m* COMMUT international exchange; **~ international automatique** *m* COMMUT, TÉL international switching center (*AmE*) (*ISC*), international switching centre (*BrE*) (*ISC*); **~ international de commutation** *m* COMMUT, TÉL international switching center (*AmE*) (*ISC*), international switching centre (*BrE*) (*ISC*); **~ international de commutation de données** *m* COMMUT, DONNÉES international data switching exchange (*IDSE*); **~ international de maintenance de commutation** *m* (*CIMC*) TESTS international switching maintenance center (*AmE*) (*ISMC*), international switching maintenance centre (*BrE*) (*ISMC*); **~ international de maintenance de transmission** *m* (*CIMT*) TESTS international transmission-maintenance center (*AmE*) (*ITMC*), international transmission-maintenance centre (*BrE*) (*ITMC*); **~ international de transit principal** *m* COMMUT main international switching center (*AmE*), main international switching centre (*BrE*), main international-trunk switching center (*AmE*), main international-trunk switching centre (*BrE*); **~ interurbain** *m* COMMUT, TÉL, TRANSMIS trunk exchange; **~ de maintenance des accès clients** *m* TESTS customer-access maintenance center (*AmE*) (*CAMC*), customer-access maintenance centre (*BrE*) (*CAMC*); **~ maritime** *m* TÉL maritime center (*AmE*), maritime centre (*BrE*); **~ maritime station côtière** *m* TÉL maritime center shore station (*AmE*), maritime centre shore station (*BrE*); **~ maritime station terrienne côtière** *m* TÉL maritime center shore station (*AmE*), maritime centre shore station (*BrE*); **~ mixte urbain et de transit** *m* COMMUT combined local and transit exchange; **~ mobile de transmissions** *m* TRANSMIS message center vehicle (*AmE*) (*MCV*), message centre vehicle (*BrE*) (*MCV*); **~ national** *m* COMMUT national exchange; **~ national des applications super-informatiques** *m* (*NCSA*) DONNÉES National Center for Supercomputing Applications (*AmE*) (*NCSA*); **~ nodal** *m* TÉL tandem exchange; **~ nodal automatique** *m* TÉL tandem exchange; **~ d'opérateurs** *m* TÉL operator center (*AmE*), operator centre (*BrE*); **~ d'opérations** *m* RADIO, TESTS operations center (*AmE*), operations centre (*BrE*); **~ d'opérations des engins spatiaux de télécommunications** *m* RADIO communications spacecraft operation center (*AmE*) (*COMSOC*) communications spacecraft operation centre (*BrE*), (*COMSOC*); **~ pour les essais** *m* TESTS test center (*AmE*), test centre (*BrE*); **~ pour les essais de la signalisation de ligne** *m* COMMUT testing point, SIGNAL, TESTS testing point, testing point line signaling (*AmE*), testing point line signalling (*BrE*); **~ pour les essais de transmission** *m* COMMUT, SIGNAL testing point, TESTS testing point, testing point transmission; **~ primaire** *m* TÉL primary center (*AmE*), primary centre (*BrE*); **~ radiophonique international** *m* (*CRI*) RADIO international sound-program center (*AmE*) (*ISPC*), international sound-programme centre (*BrE*) (*ISPC*); **~ radiophonique national** *m* (*CRN*) RADIO national sound-program center (*AmE*), national sound-programme centre (*BrE*); **~ de rattachement** *m* COMMUT serving exchange, TV connection center (*AmE*), connection centre (*BrE*); **~ de recherche documentaire** *m* (*IRC*) DONNÉES information retrieval center (*AmE*) (*IRC*), information retrieval centre (*BrE*) (*IRC*); **~ de recherche sur les communications** *m* GÉNÉRAL communications research center (*AmE*) (*CRC*), communications research centre (*BrE*) (*CRC*); **~ régional** *m* TÉL regional center (*AmE*) (*RC*), regional centre (*BrE*) (*RC*); **~ de reportage** *m* TV outside location; **~ RNIS** *m* COMMUT ISDN exchange; **~ satellite** *m* COMMUT remote switching unit (*RSU*); **~ satellite numérique** *m* TÉL subscriber digital-access unit; **~ du service d'abonnés** *m* TÉL subscriber's service center (*AmE*), subscriber's service centre (*BrE*); **~ de service vidéotex** *m* APP SERV, TÉL, TV videotex service center (*AmE*), videotex service centre (*BrE*); **~ de la surface de référence** *m* TÉL, TRANSMIS reference-surface center (*AmE*), reference-surface centre (*BrE*); **~ tandem** *m* TÉL tandem exchange; **~ de taxation** *m* APP SERV charging center (*AmE*), charging centre (*BrE*); **~ télécommandé** *m* COMMUT remotely-controlled exchange; **~ de télécommunications** *m* TÉL telecommunications exchange; **~ télévisuel international** *m* TV international television center (*AmE*) (*ITC*), international television centre (*BrE*) (*ITC*), international television program center (*AmE*) (*ITPC*), international television programme centre (*BrE*) (*ITPC*); **~ télévisuel national** *m* TV national television center (*AmE*) (*NTC*), national television centre (*BrE*) (*NTC*); **~ terminal** *m* TÉL, TRANSMIS terminal exchange; **~ terminal international** *m* TÉL terminal international center (*AmE*), terminal international centre (*BrE*), terminal international exchange; **~ terminal national** *m* TÉL terminal national center (*AmE*), terminal national centre (*BrE*); **~ pour les tests de la transmission** *m* TESTS testing point transmission; **~ de traitement de l'information** *m* (*CTI*) DONNÉES data-processing center (*AmE*), data-processing centre (*BrE*); **~ de transit** *m* COMMUT trunk transit exchange, subzone center (*AmE*), subzone centre (*BrE*), transit exchange, DONNÉES, TÉL transit exchange; **~ de transit international** *m* TRANSMIS gateway; **~ de transit principal** *m* COMMUT main trunk switching center (*AmE*), main trunk switching centre (*BrE*); **~ de transit privé** *m* COMMUT hub; **~ de transmissions** *m* TRANSMIS communications center (*AmE*), communications centre (*BrE*); **~ utilisant un système de signalisation sur voie** *m* COMMUT common channel exchange

centres: **~ de maintenance de la transmission pour la ligne** *m pl* TÉL transmission maintenance points for international line

centreur: **~ d'impédance** *m* RADIO compensating network

CEPT *abrév* (*Conférence européenne des administrations des postes et des télécommunications*) GÉNÉRAL ECPT (*European Conference of Postal and Telecommunications Administrations*)

céramique[1] *adj* COMPOSANTS ceramic

céramique[2] *f* COMPOSANTS ceramics

cerclage *m* TV leafleting

cercle: **~ fermé** *m* ALIM ÉLEC feeding loop; **~ primitif de référence** *m* GÉNÉRAL pitch circle; **~ de qualité** *m* TESTS quality circle

certificat *m* GÉNÉRAL certificate; **~ d'assurance qualité** TESTS quality certificate; **~ d'assurance qualité SQS** TESTS SQS quality certificate; **~ de conformité** TESTS certificate of compliance, certificate of conformance, certificate of conformity; ~ **d'étalonnage** TESTS *métrologie* calibration certificate; ~ **d'homologation** GÉNÉRAL certificate of approval; ~ **SQS** TESTS SQS quality certificate

certification *f* GÉNÉRAL certification

césure *f* DONNÉES hyphenation

CFAO *abrév* (*conception et fabrication assistées par ordinateur*) INFORMAT CAD/CAM (*computer-aided design and computer-aided manufacturing*)

CFNR *abrév* (*conduit fictif numérique de référence*) RADIO, TÉL, TRANSMIS HRDP (*hypothetical reference digital path*)

CFR *abrév* (*circuit fictif de référence*) RADIO, TÉL, TRANSMIS HRC (*hypothetical reference circuit*)

CG *abrév* (*convertisseur de gestion*) APP SERV charge-metering converter

CGA *abrév* (*carte graphique couleur*) INFORMAT CGA (*color graphics adaptor AmE, colour graphics adaptor BrE*)

chaînage: ~ **avant** *m* DONNÉES forward chaining; ~ **dynamique** *m* DONNÉES dynamic chaining; ~ **statique** *m* DONNÉES static chaining

chaîne *f* COMMUT chain, DONNÉES string, TV channel; ~ **d'appel** TÉL call string; ~ **de caractères** DONNÉES, INFORMAT character string; ~ **centrale de connexion** COMMUT, TÉL switching complex; ~ **de connexion de données** DONNÉES data connection; ~ **décodée** INFORMAT decoded stream; ~ **élémentaire** INFORMAT elementary stream; ~ **généraliste** TV general entertainment channel; ~ **image** TV picture channel; ~ **d'information** TV information channel; **internationale** TÉL international chain; ~ **en option** TV premium channel; ~ **d'origine** INFORMAT source stream; ~ **de pilotage** RADIO attitude control system; ~ **de qualité** APP SERV quality chain; ~ **de relais** COMPOSANTS relay chain; ~ **secondaire** DONNÉES substring; ~ **séquentielle** DONNÉES sequencing system; ~ **de télévision** TV television channel; ~ **de télévision à péage** TV pay television network, pay channel; ~ **de texte** TÉL text string; ~ **de transmission** TÉL, TRANSMIS transmission link; ~ **unitaire** DONNÉES unit string; ~ **vide** DONNÉES null string

chaînon: ~ **de voie de communication** *m* RADIO, TÉL, TRANSMIS transmission line, transmission link; ~ **de voie de transmission** *m* RADIO, TÉL, TRANSMIS transmission line, transmission link

chaleur: ~ **spécifique** *f* GÉNÉRAL specific heat

chambre: ~ **anéchoïque** *f* TESTS *acoustique* anechoic chamber, anechoic room, echoless chamber; ~ **d'épissage** *f* ÉLECTRON splice chamber; ~ **à grenailles** *f* COMPOSANTS *microphones* carbon-granule chamber; ~ **sourde** *f* TESTS *acoustique* anechoic chamber, anechoic room, dead room, echoless chamber, free-field room; ~ **souterraine** *f* TRANSMIS underground chamber, cable chamber, cable cellar, cable vault

champ *m* GÉNÉRAL field; ~ **en l'absence d'absorption** RADIO unabsorbed field strength; ~ **acoustique** TÉL acoustic field; ~ **d'adresse** DONNÉES, TÉL address field (*AF*); ~ **aléatoire** RADIO random field; ~ **de bruit** TRANSMIS noise field; ~ **de classe de surcharge** TÉL overload class field; ~ **coercitif** ÉLECTRON coercive field; ~ **de commande UIH** TÉL UIH control field; ~ **croisé** TÉL crossed field; ~ **de données du paquet** DONNÉES burst data section; ~ **de données d'usager** DONNÉES user data field; ~ **électrique** ÉLECTRON electric field; ~ **électromagnétique** ÉLECTRON electromagnetic field; ~ **évanescent** ÉLECTRON, TRANSMIS evanescent field; ~ **exploré** TÉL scanning field; ~ **de fuite** ÉLECTRON leakage field; ~ **d'identificateur de protocole** DONNÉES protocol identifier field; ~ **d'indicateur de longueur** DONNÉES length-indicator field; ~ **d'induction** RADIO induction field; ~ **d'informations télécopie** DONNÉES, INFORMAT facsimile information field (*FIF*); ~ **lointain** RADIO, TRANSMIS far field; ~ **lu** TÉL scanning field; ~ **magnétique** ÉLECTRON magnetic field; ~ **magnétisant** ÉLECTRON magnetizing field; ~ **minimal utilisable** RADIO minimum usable power flux-density, minimum usable field strength; ~ **de mode** TRANSMIS mode field; ~ **observé** RADIO field of vision (*FOV*); ~ **proche** RADIO, TRANSMIS near field; ~ **de recherche** DONNÉES search field; ~ **de répartition** ALIM ÉLEC connection field; ~ **réservé** ÉLECTRON reserved field (*RES*); ~ **réverbéré** TÉL reverberated field; ~ **des services inter-réseaux** TÉL network utility field; ~ **sombre** TRANSMIS dark field; ~ **statique** ÉLECTRON static field; ~ **de tolérance de gaine** TRANSMIS cladding tolerance field; ~ **tournant** ALIM ÉLEC, RADIO, TÉL rotating field; ~ **utilisable** RADIO usable field strength; ~ **utilisable de référence** RADIO reference-usable field strength, reference-usable power flux density; ~ **de visée** INFORMAT aiming circle, aiming field, aiming symbol; ~ **de visibilité** RADIO field of vision (*FOV*); ~ **de vision** RADIO field of vision (*FOV*)

champmètre *m* TESTS field intensity meter, field strength meter

changement *m* INFORMAT alteration, change; ~ **de commande** CONTRÔLE control change; ~ **de disque** INFORMAT disk change; ~ **de fréquence** RADIO frequency changing, frequency conversion, TRANSMIS, TV frequency conversion; ~ **d'horaire** PÉRIPH clock change; ~ **de ligne** COMMUT, DONNÉES, INFORMAT, RADIO, TÉL line feed (*LF*); ~ **de mode** PÉRIPH, TRANSMIS mode jumping

changer *vt* INFORMAT alter; ~ **de mode** PÉRIPH, TRANSMIS switch mode

changeur: ~ **abaisseur de fréquence** *m* RADIO down-converter; ~ **élévateur de fréquence** *m* DONNÉES, RADIO up-converter; ~ **de fréquence** *m* GÉNÉRAL frequency converter, frequency changer

chanter: ~ **en playback** *vt* TV mime

chantier: ~ **d'installation** *m* GÉNÉRAL installation site

chaque: ~ **mot deux fois** *loc* RADIO words twice

charge *f* ALIM ÉLEC charge, power, COMPOSANTS filler, ÉLECTRON charge, power, GÉNÉRAL load; ~ **d'accumulateur** ALIM ÉLEC battery charging; ~ **adaptée** RADIO matched load; ~ **additionnelle** ALIM ÉLEC additional load; ~ **automatique de batterie** ALIM ÉLEC automatic battery charging; ~ **de la base** ÉLECTRON *semi-conducteurs* base charge; ~ **de batterie** ALIM ÉLEC battery charging; ~ **de biberonnage** ALIM ÉLEC *batteries* boost charge, quick charge, rapid charge; ~ **du canal** RADIO channel loading; ~ **des canaux** TRANSMIS channel loading; ~ **capacitive** ÉLECTRON capacitive load; ~ **capacitive terminale** RADIO top capacitive loading; ~ **de compensation** ALIM ÉLEC trickle charge; ~ **continue** GÉNÉRAL *câbles* continuous loading, TRANSMIS krarup loading; ~ **diélectrique** ÉLECTRON dielectric charge; ~ **à l'eau** RADIO water load; ~ **d'égalisation** ALIM ÉLEC equalizing charge; ~ **électrique** ÉLECTRON electrical charge; ~ **d'électron** ÉLECTRON electron charge; ~ **d'entretien** ALIM ÉLEC floating charge, trickle charge; ~ **d'espace** ÉLECTRON space charge; ~ **d'essai** TESTS test load; ~ **fictive** RADIO dummy load, artificial load; ~ **inductive** ÉLECTRON inductive load; ~ **d'intensité de trafic** COMMUT, TÉL traffic intensity load; ~ **lente** ALIM ÉLEC trickle charge; ~ **d'une ligne** RADIO loading; ~ **liquide** RADIO water load; ~ **de maintien** ALIM ÉLEC trickle charge; ~ **maximale** ALIM ÉLEC full load; ~ **minimale** ALIM ÉLEC, TÉL base load; ~ **nominale** ÉLECTRON nominal load; ~ **optimale** RADIO optimum load; ~ **par cartouche** INFORMAT cartridge loading; ~ **rapide** ALIM ÉLEC *batteries* boost charge, quick charge, rapid charge; ~ **réactive** ÉLECTRON reactive load; ~ **résistive** ALIM ÉLEC, ÉLECTRON resistive load; ~ **de rupture** COMPOSANTS, TESTS breaking strength, ultimate strength; ~ **de service** ALIM ÉLEC normal charging; ~ **au sol** GÉNÉRAL floor load; ~ **à la sortie** ALIM ÉLEC output load; ~ **de sortie** ALIM ÉLEC output load; ~ **de trafic** COMMUT, TÉL traffic load, traffic loading; ~ **utile** ÉLECTRON, GÉNÉRAL payload

chargé[1] *adj* INFORMAT loaded

chargé[2]: **~ directement en parallèle** *loc* ÉLECTRON jammed; **~ par cartouche** *loc* INFORMAT cartridge-loaded

chargement: **~ automatique** *m* INFORMAT autoload; **~ de cartouche** *m* INFORMAT loading cartridge; **~ frontal** *m* GÉNÉRAL front loading; **~ rapide** *m* INFORMAT quick load; **~ de système** *m* INFORMAT boot-up; **~ unitaire** *m* INFORMAT loading single

charger *vt* ALIM ÉLEC, APP SERV charge, INFORMAT boot, boot up, bootstrap, load; **~ directement en parallèle** DONNÉES, ÉLECTRON, INFORMAT jam; **~ en formation** ALIM ÉLEC form charge

chargeur *m* ALIM ÉLEC battery charger, charger, charging set; **~ de batterie** ALIM ÉLEC battery charger, charging set; **~ de compensation** ALIM ÉLEC trickle charger; **~ de documents** DONNÉES document loader; **~ d'entretien** ALIM ÉLEC trickle charger; **~ lent** ALIM ÉLEC trickle charger; **~ de maintien** ALIM ÉLEC trickle charger; **~ de modulateur** ÉLECTRON modulator charger; **~ monobatterie** ALIM ÉLEC single-compartment charger; **~ monocellulaire** ALIM ÉLEC single-compartment charger

chariot: **~ à barres** *m* TÉL type-bar carriage; **~ à jeu de barres** *m* TÉL type-bar carriage; **~ porte-batteries** *m* ALIM ÉLEC battery cart, battery trolley

charrue: **~ enfouisseuse** *f* GÉNÉRAL *génie industriel extérieur* TRANSMIS cable layer, cable plough (*BrE*), cable plow (*AmE*); **~ rigoleuse** *f* GÉNÉRAL *génie industriel extérieur*, TRANSMIS cable layer, cable plough (*BrE*), cable plow (*AmE*); **~ sous-marine** *f* GÉNÉRAL, TRANSMIS submarine cable layer, submarine cable plough (*BrE*), submarine cable plow (*AmE*)

châssis *m* COMPOSANTS frame, framework, mounting frame, ÉLECTRON chassis, TÉL chassis, frame; **~ d'alvéole** TÉL shelf chassis; **~ de cartes** COMPOSANTS *equipment practice* board frame; **~ de montage** COMPOSANTS mounting frame; **~ de relais** TÉL relay frame

chef: **~ de chantier d'installation** *m* GÉNÉRAL installation site manager; **~ des essais** *m* TESTS testing manager; **~ de montage** *m* COMPOSANTS construction manager; **~ de poste** *m* TÉL chief operator, chief supervisor; **~ de surveillance** *m* TÉL chief operator, chief supervisor

chemin: **~ d'accès** *m* INFORMAT access path, path name; **~ de câble** *m* TRANSMIS *pose de câbles* cable riser, cable runway, cable run; **~ de câble d'alimentation** *m* ALIM ÉLEC, DONNÉES, ÉLECTRON, INFORMAT, TÉL power cable chute; **~ de câble de bus** *m* ALIM ÉLEC bus cable chute; **~ de câble principal** *m* COMMUT main chute; **~ de câble de signalisation** *m* TÉL signal cable chute; **~ de câbles** *m* TRANSMIS cable trough, *pose de câbles* chute; **~ de cartes** *m* INFORMAT card path; **~ critique** *m* GÉNÉRAL critical path; **~ de dérivation** *m* DONNÉES derivation path; **~ de navigation** *m* DONNÉES, INFORMAT navigation path; **~ de navigation hypermédia** *m* INFORMAT hypermedia navigation path; **~ optimal** *m* TRANSMIS optimal path; **~ optique** *m* TRANSMIS optical path; **~ parcouru** *m* INFORMAT *Internet* chosen path

cheminée: **~ d'ascension** *f* TRANSMIS *pose de câbles* cable runway

chercheur: **~ d'appel** *m* COMMUT, DONNÉES, INFORMAT, RADIO, TÉL line finder (*LF*), TÉL call finder, line finder (*LF*); **~ d'émetteur de code** *m* TÉL code-sender finder, sender finder; **~ d'enregistrement** *m* COMMUT register finder; **~ de ligne** *m* COMMUT, DONNÉES, INFORMAT, RADIO, TÉL line finder (*LF*), line feed (*LF*), call finder; **~ de ligne appelante** *m* COMMUT, DONNÉES, INFORMAT, RADIO, TÉL line finder (*LF*); **~ d'opérateur** *m* TÉL access matrix; **~ radiogoniométrique** *m* RADIO radiogoniometer

chevauchement *m* GÉNÉRAL overlapping, overlap

chevaucher: **se ~** *v réfl* INFORMAT, RADIO, TÉL overlap

chez: **~ l'émetteur** *loc* TÉL at the near end

chiffon: **~ non pelucheux** *m* ÉLECTRON crocus cloth, lint-free cloth

chiffraison *f* TESTS scale numbering

chiffre *m* DONNÉES, INFORMAT digit, RADIO cipher, TÉL, TRANSMIS digit; **~ binaire** INFORMAT, TRANSMIS binary digit, binary figure; **~ de bourrage** DONNÉES stuffing digit; **~ des centaines** INFORMAT hundreds digit; **~ des dizaines** INFORMAT tens digit; **~ facultatif spécifique au réseau** TÉL optional network-specific digit (*ONSD*); **~ hexadécimal** INFORMAT hexadecimal digit; **~ d'identification de nationalité** TÉL national identification digit (*NID*); **~ inférieur** DONNÉES, INFORMAT subscript; **~ de langue** TÉL language digit; **~ de signe** INFORMAT sign digit; **~ supérieur** DONNÉES, INFORMAT superscript; **~ supplémentaire** TÉL extra digit; **~ x** TÉL code x

chiffré *adj* DONNÉES, INFORMAT, TRANSMIS, TV encoded

chiffrement *m* GÉNÉRAL encryption, *brouillage* scrambling, encipherment, encoding; **~ de bout en bout** DONNÉES end-to-end encipherment; **~ à clé publique** RADIO public-key encryption; **~ de liaison** GÉNÉRAL link encryption; **~ de liaison par liaison** DONNÉES link-by-link encipherment; **~ section par section** DONNÉES link-by-link encipherment

chiffrer *vt* GÉNÉRAL encode, encipher, encrypt, *brouiller* scramble

chiffres: **~ d'acheminement** *m pl* COMMUT routing digits; **~ appelés** *m pl* TÉL digits dialed (*AmE*), digits dialled (*BrE*); **~ d'identification de voie d'acheminement** *m pl* TÉL route information digits; **~ de taxation** *m pl* TÉL fee digits

chloroprène *m* COMPOSANTS chloroprene

chlorure: **~ de polyvinyle** *m* (*PVC*) ÉLECTRON polyvinyl chloride (*PVC*)

choc ALIM ÉLEC, ÉLECTRON shock; **~ acoustique** *m* ÉLECTRON acoustic shock; **~ électrique** ALIM ÉLEC electric shock, surge, ÉLECTRON surge

chromaclé *f* INFORMAT chromakey

chromaticité *f* TRANSMIS, TV chromaticity

chromatique *adj* TRANSMIS, TV chromatic

chrominance *f* GÉNÉRAL, TV chrominance

chronocode *m* DONNÉES, INFORMAT time code

chronogramme *m* GÉNÉRAL timing diagram, timing chart

chronométrer *vt* DONNÉES, INFORMAT, RADIO, TÉL, TESTS time

chronomicromètre *m* TESTS time interval meter (*TIM*)

chute *f* GÉNÉRAL drop; **~ du palier** TÉL negative tilt, pulse droop; **~ de température** CLIMAT temperature drop; **~ de tension** ALIM ÉLEC voltage drop; **~ de tension directe** ÉLECTRON forward voltage drop; **~ de tension du tube** ÉLECTRON tube voltage drop

CI *abrév* COMPOSANTS, ÉLECTRON (*circuit imprimé*) PC (*printed circuit*), printed wiring, (*circuit integré*) IC (*integrated circuit*), chip, RADIO (*conversation impossible*) CI (*conversation impossible*); **~ à transistors** *m* COMPOSANTS, ÉLECTRON transistor array IC

cible *f* GÉNÉRAL target; **~ de mémoire** ÉLECTRON storage target; **~ passive** RADIO skin target; **~ répondeuse** RADIO *radar* active target

CIG *abrév* (*identification de l'abonné appelant*) DONNÉES, TÉL CIG (*calling-subscriber identification*)

CIG-DTC *abrév* (*identification de l'abonné demandeur-DTC*) DONNÉES, TÉL CIG-DTC (*calling-subscriber identification-DTC*)

CIL *abrév* (*ligne d'identification de l'appel*) TÉL CIL (*call-identification line*)

CIMC *abrév* (*centre international de maintenance de commutation*) TESTS ISMC (*international switching maintenance center AmE, international switching maintenance centre BrE*)

CIMT *abrév* (*centre international de maintenance de transmission*) TESTS ITMC (*international transmission-maintenance center AmE, international transmission-maintenance centre BrE*)

cintrage *m* ALIM ÉLEC *circuits imprimés* bow, COMPOSANTS bend

cintrer *vt* ALIM ÉLEC *circuits imprimés* bow, COMPOSANTS bend

CIRC *abrév* (*code Reed-Solomon à entrelacement croisé*) INFORMAT CIRC (*cross-interleaved Reed-Solomon code*)

circonscription *f* ÉLECTRON region, TÉL local exchange area; **~ administrative** TÉL administration district; **~ tarifaire** APP SERV charging area; **~ de taxe** TÉL tariff area

circuit[1] *m* GÉNÉRAL circuit; **~ d'abonné** TÉL subscriber circuit; **~ d'accès** TÉL access circuit; **~ d'accord** ÉLECTRON tuner circuit, tuning circuit **~ accordé** ALIM ÉLEC tuned circuit; **~ d'affaiblissement d'adaptation** TRANSMIS best possible circuit; **~ d'alarme** TÉL alarm trunk, alarm circuit, TESTS alarm circuit; **~ d'alarme de dérangement** TÉL failure alarm circuit, fault alarm circuit; **~ d'alignement** ÉLECTRON clamp, clamping circuit; **~ d'alimentation** ÉLECTRON feed system; **~ amplificateur** ALIM ÉLEC amplifier circuit; **~ amplifié** TRANSMIS amplified circuit; **~ analogique** INFORMAT, TRANSMIS analog circuit; **~ d'annonces parlées** TÉL announcement trunk; **~ antilocal** TÉL anti-sidetone circuit; **~ antirésonnant** ALIM ÉLEC, ÉLECTRON parallel resonant circuit; **~ approprié** TÉL earth-return phantom circuit (*BrE*), ground-return phantom circuit (*AmE*); **~ approprié de fantôme** TÉL earth-return double phantom circuit (*BrE*), ground-return double phantom circuit (*AmE*); **~ d'arrivée** TÉL, TRANSMIS incoming trunk circuit (*ITC*); **~ asynchrone** DONNÉES asynchronous circuit; **~ d'attaque** ALIM ÉLEC driving circuit, ÉLECTRON driver, driving stage; **~ d'attaque de bus** ÉLECTRON *éléments logiques* bus driver; **~ automatique international** TÉL international automatic circuit; **~ auxiliaire** ALIM ÉLEC auxiliary circuit; **~ avec affectation à la demande** ÉLECTRON demand-assignment circuit; **~ avec tolérance de transmission de données** DONNÉES permissive circuit; **~ basse fréquence** RADIO *mobicom* audio-frequency circuit; **~ bidirectionnel** ÉLECTRON two-way circuit; **~ bifilaire** ÉLECTRON, TÉL, TRANSMIS two-wire circuit; **~ bipolaire** ÉLECTRON *éléments logiques* bipolar circuit; **~ bouchon** ÉLECTRON tank circuit, RADIO wave trap; **~ bouclé** ALIM ÉLEC ring main; **~ en boucle fermée** ÉLECTRON closed-loop circuit; **~ à cellules prédéfinies** ÉLECTRON standard cell circuit; **~ de charge** ÉLECTRON load circuit; **~ de codage** SIGNAL coding circuit; **~ collecteur de données discrètes** DONNÉES, INFORMAT discrete-data collector; **~ combinant** TÉL side circuit; **~ combinatoire** ÉLECTRON combinational circuit, combinatorial circuit; **~ combiné** TÉL, TRANSMIS phantom circuit; **~ de commande** CONTRÔLE control circuit, ÉLECTRON driving stage, *éléments logiques* bus driver, driver, TESTS override circuit; **~ d'un commutateur privé raccordé à un circuit de transmission** TÉL PBX trunk circuit; **~ de commutation d'antennes** RADIO antenna-switching circuit; **~ commuté** ÉLECTRON, TRANSMIS switched circuit; **~ de compensation** ÉLECTRON compensation circuit, RADIO compensating network; **~ de compensation de charge** RADIO absorber circuit, absorption circuit; **~ à composants discrets** ÉLECTRON discrete circuit; **~ de comptage** ÉLECTRON counting circuit; **~ de concentration de trafic** TÉL release trunk; **~ de conférence** TÉL conference circuit; **~ de connexion** COMMUT junctor, trunk, COMPOSANTS, DONNÉES junctor, TÉL junctor, trunk, TRANSMIS junctor; **~ de connexion intracentral** COMMUT intraoffice junctor circuit; **~ de connexion local** COMMUT own-exchange supervisory circuit; **~ continental** TÉL continental circuit; **~ de contournement** INFORMAT bypass; **~ de conversation** TÉL, TRANSMIS *postes téléphoniques* speech circuit; **~ de correction d'impulsions** TRANSMIS pulse-correction circuit; **~ à couplage de charge** DONNÉES charge-coupled device circuit; **~ à courants porteurs** TRANSMIS carrier circuit; **~ de décision** TÉL decision circuit; **~ de déclenchement** ÉLECTRON trigger circuit; **~ à déclenchement par boucle** TÉL loop-start trunk; **~ de démarrage** ALIM ÉLEC start-up circuit; **~ de départ** COMMUT, TÉL outgoing trunk; **~ déphaseur** ALIM ÉLEC phase-shifting circuit; **~ de dérivation** ALIM ÉLEC, ÉLECTRON bridge circuit; **~ détecteur avertisseur** TÉL alerting detection device; **~ de différenciation** ALIM ÉLEC differentiating circuit; **~ différentiel** TÉL telephone hybrid; **~ à diodes** ÉLECTRON diode circuit, diode tree; **~ de données** DONNÉES, TRANSMIS data circuit; **~ d'égalisation** TRANSMIS equalizer circuit; **~ égaliseur** TRANSMIS equalizer circuit; **~ électrique** ALIM ÉLEC, ÉLECTRON electrical circuit; **~ électronique** ÉLECTRON electronic circuit; **~ à émetteur commun** ÉLECTRON earthed emitter circuit (*BrE*), grounded emitter circuit (*AmE*); **~ d'enregistrement et de raccordement** TÉL recording-completing trunk; **~ entrant** COMMUT, ÉLECTRON, TÉL incoming circuit, incoming trunk, TRANSMIS incoming trunk circuit (*ITC*); **~ entre centres locaux** TÉL interexchange trunk (*BrE*), interoffice trunk (*AmE*); **~ entre centres de transit** TÉL intertandem trunk; **~ entre groupes marqueurs** TÉL intermarker group trunk (*IM*); **~ entre installations de raccordement** (*IFL*) TRANSMIS interfacility link (*IFL*); **~ d'entrée** ÉLECTRON input circuit; **~ d'entretien** ALIM ÉLEC, ÉLECTRON keep-alive circuit; **~ d'équivalence** ÉLECTRON, INFORMAT IF-AND-ONLY-IF element; **~ équivalent de Thévenin** ALIM ÉLEC Thévenin equivalent circuit; **~ d'essai** TÉL, TESTS test circuit; **~ d'essai de transmission** TÉL, TESTS, TRANSMIS transmission test trunk (*TTT*); **~ ET** ÉLECTRON AND circuit; **~ d'excitation** ALIM ÉLEC energizing circuit; **~ d'exclusion** ÉLECTRON NOT-IF-THEN element; **~ fantôme** TÉL, TRANSMIS phantom circuit; **~ fictif de référence** (*CFR*) RADIO, TÉL, TRANSMIS hypothetical reference circuit (*HRC*); **~ fictif de référence pour système de terre** RADIO, TÉL, TRANSMIS terrestrial hypothetical reference circuit; **~ fictif de référence pour la téléphonie** TÉL hypothetical reference circuit for telephony; **~ fictif de référence pour transmissions radiophoniques** RADIO, TRANSMIS hypothetical reference circuit for sound program transmissions (*AmE*), hypothetical reference circuit for sound programme transmissions (*BrE*); **~ de garde** TÉL stopper circuit, guard circuit; **~ générateur de parité** COMMUT parity-generating circuit; **~ à grande impédance** ÉLECTRON high-impedance circuit; **~ de grande longueur** ÉLECTRON, TÉL long circuit; **~ à haut débit** TRANSMIS megastream circuit; **~ d'horloge** GÉNÉRAL timing circuit; **~ hybride** ÉLECTRON hybrid circuit; **~ identificateur** ÉLECTRON identifier circuit, recognition circuit; **~ d'identification** ÉLECTRON identifier circuit, recognition circuit; **~ d'identité** ÉLECTRON, TÉL identity element, identity gate; **~ à impédance optimale** TRANSMIS best possible circuit; **~ imprimé** (*CI*) COMPOSANTS, ÉLECTRON printed circuit (*PC*), printed wiring; **~ imprimé multicouche** ÉLECTRON multilayer printed circuit; **~ INCLUSIF OU** INFORMAT, ÉLECTRON INCLUSIVE-OR element, INCLUSIVE-OR circuit; **~ d'inclusion** ÉLECTRON IF-THEN element; **~ inductif** ÉLECTRON inductive circuit; **~ d'intégration** ÉLECTRON integrating circuit; **~ intégré** (*CI*) COMPOSANTS, ÉLECTRON integrated circuit (*IC*), chip; **~ intégré à connexions courtes** ÉLECTRON small-outline integrated circuit (*SOIC*); **~ intégré à couches épaisses** ÉLECTRON thick-film integrated circuit, thick-layer integrated circuit; **~ intégré à effet de champ** ÉLECTRON field-effect integrated circuit; **~ intégré à effet Hall** ÉLECTRON Hall integrated circuit; **~ intégré à grande échelle** ÉLECTRON, TÉL, TRANSMIS large-scale integration circuit; **~ intégré à grande vitesse** ÉLECTRON high-speed integrated circuit; **~ intégré hybride** ÉLECTRON hybrid integrated circuit; **~ intégré linéaire** ÉLECTRON linear integrated circuit; **~ intégré logique** PÉRIPH integrated office system; **~ intégré pour**

micro-ondes ÉLECTRON microwave integrated circuit (*MIC*); **~ intégré monolithique** COMPOSANTS, ÉLECTRON monolithic integrated circuit (*MIC*); **~ intégré monolithique hyperfréquence** ÉLECTRON monolithic microwave integrated circuit (*MMIC*); **~ intégré numérique** ÉLECTRON digital integrated circuit (*digital IC*); **~ intégré optique** ÉLECTRON integrated optical circuit (*IOC*); **~ intégré spécifique** (*ASIC*) ÉLECTRON application-specific integrated circuit (*ASIC*); **~ intégré à très grande échelle** ÉLECTRON, TÉL, TRANSMIS very large-scale integration circuit; **~ intégré à très grande vitesse** (*CITGV*) ÉLECTRON very high-speed integrated circuit (*VHSIC*); **~ intercentral** TÉL interchange trunk, interoffice trunk; **~ d'interception** TÉL intercept trunk; **~ d'intercommunication entre positions** TÉL interposition trunk; **~ d'interconnexion** *f* TÉL interconnection circuit; **~ intercontinental** TÉL intercontinental circuit; **~ intercontinental de transit** TÉL intercontinental transit circuit; **~ d'interface** ÉLECTRON interface circuit; **~ d'interface de ligne** COMMUT line interface circuit; **~ d'interface de ligne d'abonné** COMMUT, TÉL subscriber line interface circuit (*SLIC*); **~ d'interface de transmission** COMMUT, TÉL line interface circuit (*LIC*); **~ international** TÉL international circuit; **~ international loué** TÉL international leased circuit; **~ interne** ÉLECTRON internal circuitry; **~ interposition** TÉL interposition trunk; **~ interurbain** COMMUT trunk circuit, TÉL intertoll trunk, trunk circuit, trunk telephone circuit; **~ interurbain primaire** TÉL primary interlock trunk; **~ intracentral** TÉL interchange trunk, intraoffice trunk; **~ inverse** ÉLECTRON reverse-current circuit; **~ inverseur** ÉLECTRON NOT circuit, inverter circuit; **~ de jonction** COMMUT, ÉLECTRON, TÉL junction circuit, tie trunk, trunk circuit, trunk, TRANSMIS interchange circuit; **~ à large bande** TRANSMIS wideband circuit; **~ libre** TRANSMIS idle circuit; **~ de ligne** COMMUT, TÉL line interface circuit (*LIC*), line circuit; **~ de ligne d'abonné** COMMUT, TÉL subscriber line circuit (*SLC*); **~ linéaire** ÉLECTRON, TRANSMIS linear circuit; **~ logique** DONNÉES, ÉLECTRON logic circuit; **~ logique majoritaire** DONNÉES, ÉLECTRON majority gate; **~ logique programmable** DONNÉES, ÉLECTRON programmable logic circuit; **~ logique programmable effaçable électriquement** DONNÉES, ÉLECTRON erasable programmable logic device (*EPLD*); **~ loué** TÉL, TRANSMIS leased circuit; **~ loué à temps partiel** TÉL, TRANSMIS part-time leased circuit; **~ LSI** ÉLECTRON, TÉL, TRANSMIS LSI circuit; **~ magnétique** ÉLECTRON magnetic circuit; **~ de maintien** ÉLECTRON holding circuit; **~ majoritaire** ÉLECTRON majority element, majority gate; **~ maritime par satellite** RADIO maritime satellite circuit; **~ de masse** ÉLECTRON earth loop (*BrE*), ground loop (*AmE*); **~ à matrice de portes** ÉLECTRON gate array circuit; **~ matriciel** ÉLECTRON matrix circuit; **~ mélangeur** ÉLECTRON combinatorial circuit; **~ de mémoire** ÉLECTRON memory circuit; **~ de messages enregistrés** TÉL recorded announcement trunk; **~ de mesure** TESTS measuring circuit; **~ métallique** TÉL metallic circuit; **~ de mise en forme d'impulsions** ÉLECTRON pulse-shaping circuit; **~ mixte** TRANSMIS both-way circuit; **~ monolithique** ÉLECTRON monolithic circuit; **~ à montage en surface** COMPOSANTS, ÉLECTRON surface-mounted device (*SMD*); **~ moyen débit** APP SERV kilostream circuit; **~ MSI** TÉL MSI circuit; **~ multi-étage** ÉLECTRON multistage circuit; **~ multipoint simple** TÉL simple multipoint circuit; **~ national** TÉL national circuit; **~ NI** ÉLECTRON NOR circuit, NOR gate; **~ NON** ÉLECTRON NOT circuit, inverter circuit, NOT element, NOT gate; **~ NON-ET** ÉLECTRON NAND circuit; **~ non inductif** ALIM ÉLEC noninductive circuit; **~ nonlinéaire** TRANSMIS nonlinear circuit; **~ NON-OU** ÉLECTRON NOR circuit, NOR gate; **~ non réciproque** ÉLECTRON nonreciprocal circuit; **~ non utilisé** TRANSMIS idle circuit; **~ numérique** TRANSMIS digital path, digital circuit; **~ optique intégré** ÉLECTRON integrated optical circuit (*IOC*); **~ d'optique intégré** ÉLECTRON, TRANSMIS integrated optical circuit; **~ optoélectronique intégré** ÉLECTRON, TRANSMIS integrated optoelectronic circuit; **~ oscillant** ALIM ÉLEC oscillatory circuit, ÉLECTRON oscillatory circuit, tank circuit; **~ OU** ÉLECTRON OR circuit, OR gate; **~ ouvert** ALIM ÉLEC open circuit; **~ en parallèle** ÉLECTRON parallel circuit; **~ parallèle** ÉLECTRON parallel circuit; **~ de parole** TÉL, TRANSMIS speech circuit; **~ passif** ÉLECTRON passive circuit; **~ permanent** TÉL permanent circuit; **~ en pont** ALIM ÉLEC, ÉLECTRON bridge circuit; **~ de poursuite** ÉLECTRON tracker; **~ à précharge** ÉLECTRON ratioless circuit; **~ principal** TRANSMIS backbone circuit; **~ de prise** TÉL seizure circuit; **~ de protection** TESTS protection circuit; **~ purement résistif** ALIM ÉLEC purely-resistive circuit; **~ à quatre fils** TÉL four-wire circuit; **~ de radiocommunication** RADIO radio circuit; **~ radiophonique** TV sound-program circuit (*AmE*), sound-programme circuit (*BrE*); **~ radiophonique international** TV international sound-program circuit (*AmE*), international sound-programme circuit (*BrE*); **~ radiophonique international multiple** TV international multiple-destination sound-program circuit (*AmE*), international multiple-destination sound-programme circuit (*BrE*); **~ à rapport de résistances** ÉLECTRON ratio circuit; **~ RCD** ÉLECTRON RCD circuit; **~ réciproque** ÉLECTRON reciprocal circuit; **~ de reconnaissance** ÉLECTRON identifier circuit, recognition circuit; **~ de récupération de rythme** ÉLECTRON, TRANSMIS timing-recovery circuit; **~ de réfrigération** CLIMAT refrigeration circuit; **~ de refroidissement** GÉNÉRAL cooling system; **~ résistif** ÉLECTRON resistive circuit; **~ résonnant** ALIM ÉLEC, ÉLECTRON resonant circuit; **~ résonnant parallèle** ALIM ÉLEC, ÉLECTRON parallel resonant circuit; **~ résonnant série** ALIM ÉLEC, ÉLECTRON series resonant circuit; **~ de rétablissement du rythme** ÉLECTRON clock-recovery circuit; **~ à retard** ÉLECTRON delay element, delay circuit; **~ à retardement** ÉLECTRON time-delay circuit; **~ à retour de courant** ÉLECTRON reverse-current circuit; **~ à retour par la terre** TRANSMIS earth-return circuit (*BrE*), ground-return circuit (*AmE*); **~ sans diviseur** ÉLECTRON ratioless circuit; **~ sans perte** ÉLECTRON zero loss circuit; **~ de secours** TÉL reserve circuit, TESTS fall-back circuit; **~ semi-duplex** TÉL half-duplex circuit; **~ séquentiel** DONNÉES, ÉLECTRON sequential circuit; **~ série** ÉLECTRON series circuit; **~ de service** TÉL speaker circuit, TRANSMIS engineering service circuit (*ESC*); **~ de service direct** TÉL direct service circuit; **~ de service multiterminal** TÉL multiterminal service circuit; **~ de service omnibus** TÉL omnibus service circuit; **~ de signalisation** SIGNAL, TÉL, TRANSMIS signal circuit, signal path; **~ simplex** ÉLECTRON simplex circuit; **~ sortant** COMMUT, ÉLECTRON, TÉL outgoing circuit, outgoing trunk; **~ de sortie** ÉLECTRON output circuit; **~ souple** ALIM ÉLEC flexstrip; **~ spécialisé** TÉL leased circuit; **~ stabilisateur de tension** ÉLECTRON voltage-stabilizing circuit; **~ stéréo** INFORMAT stereophonic circuit; **~ superfantôme** TRANSMIS double-phantom circuit, superphantom circuit; **~ superposé** TÉL superposed circuit; **~ synchrone** TRANSMIS synchronous circuit; **~ tampon** DONNÉES, INFORMAT buffer circuit; **~ de télécommunication** RADIO, TÉL telecommunication circuit; **~ de télégraphie harmonique** TÉL, TRANSMIS voice-frequency telegraph circuit; **~ télégraphique** TÉL, TRANSMIS telegraphic circuit; **~ télégraphique fantôme** TÉL, TRANSMIS phantom telegraph circuit; **~ télégraphique superfantôme** TÉL, TRANSMIS double-phantom balanced telegraph circuit, superphantom balanced telegraph circuit; **~ de téléphonie-données à l'alternat** TÉL, TRANSMIS alternate voice-data circuit (*AVD*); **~ téléphonique** TÉL, TRANSMIS telephone circuit, telephone connection; **~ téléphonique**

international TÉL, TRANSMIS international telephone circuit; **~ téléphonique interurbain** TÉL, TRANSMIS trunk telephone circuit; **~ télévisuel** TV television circuit; **~ télévisuel international** TV international television circuit; **~ télévisuel international multiple** TV international multiple-destination television circuit; **~ de temporisation** ÉLECTRON timer, timing circuit; **~ temporisé** ÉLECTRON time-delay circuit; **~ terminal** ALIM ÉLEC, TÉL terminal circuit; **~ terminal de central** COMMUT exchange terminal circuit (*ETC*); **~ terrestre spécialisé** TÉL, TRANSMIS dedicated terrestrial circuit; **~ terrestre du système maritime** TÉL, TRANSMIS maritime terrestrial circuit; **~ de test** TESTS test circuit; **~ de traduction** TÉL translation circuit; **~ de trafic** TÉL traffic circuit; **~ de transmission** TRANSMIS trunk circuit; **~ de transmission de données** TRANSMIS data circuit; **~ à très haut niveau d'intégration** ÉLECTRON, TÉL, TRANSMIS very large-scale integration circuit; **~ à très haute intégration** ÉLECTRON very high-speed integrated circuit (*VHSIC*); **~ triphasé** ALIM ÉLEC three-phase circuit; **~ type quatre fils** TÉL four-wire-type circuit; **~ de type téléphonique** RADIO, TÉL telephone-type circuit, telephone-type connection; **~ ULSI** ÉLECTRON ULSI circuit, ultralarge scale integration circuit; **~ unifilaire** COMPOSANTS, ÉLECTRON single-wire circuit; **~ de vérification de signal** SIGNAL, TESTS signal test circuit; **~ à verrouillage** ÉLECTRON latch circuit; **~ de verrouillage** ÉLECTRON clamping circuit; **~ virtuel** DONNÉES, TRANSMIS virtual circuit (*VC*); **~ virtuel commuté** (*CVC*) COMMUT, ÉLECTRON switched virtual circuit (*SVC*), SIGNAL signaling virtual channel (*AmE*) (*SVC*), signalling virtual channel (*BrE*) (*SVC*), switched virtual circuit (*SVC*), TRANSMIS switched virtual circuit (*SVC*); **~ virtuel permanent** COMMUT, ÉLECTRON, SIGNAL, TRANSMIS permanent virtual circuit (*PVC*); **~ virtuel sur canal D** TÉL D-channel virtual circuit; **~ VLSI** ÉLECTRON, TÉL, TRANSMIS VLSI circuit

circuit[2]: **en ~ fermé** *loc* ALIM ÉLEC closed-circuit, TÉL loaded, TV closed-circuit; **à ~ ouvert** *loc* ALIM ÉLEC unloaded; **~ interurbain libre** *loc* TÉL trunk free; **~ pris** *loc* TÉL trunk seized

circuits *m pl* ÉLECTRON circuitry; **~ de câblage des locaux** TÉL premises wiring; **~ à commande unique** ÉLECTRON ganged circuits; **~ électroniques** ÉLECTRON electronic circuitry; **~ informatiques** INFORMAT computer circuitry; **~ numériques** INFORMAT digital circuitry; **~ tous occupés** TÉL all-paths busy (*APB*), all-trunks busy (*ATB*); **~ prédiffusés** ÉLECTRON gate array

circulariser *vt* RADIO *orbite*, TRANSMIS circularize

circulateur *m* RADIO, TRANSMIS circulator; **~ hyperfréquence** RADIO, TRANSMIS microwave circulator; **~ optique** TRANSMIS optical circulator

circuler *vi* ALIM ÉLEC, CONTRÔLE flow

CIRD *abrév* (*code d'identification de réseau de données*) DONNÉES, TÉL DNIC (*data-network identification code*)

CIRM *abrév* (*Comité international radio-maritime*) RADIO CIRM (*International Marine Radio Association*)

CIRT *abrév* (*code d'identification de réseau télex*) DONNÉES, TÉL telex-network identification code

CISE *abrév* DONNÉES CISE (*complex instruction set equipment*)

CITGV *abrév* (*circuit intégré à très grande vitesse*) ÉLECTRON VHSIC (*very high-speed integrated circuit*)

clair[1]: **en ~** *adv* TÉL uncoded, TV in clear

clair[2]: **en ~** *loc* TÉL uncoded, TV in clear

clapet: **~ à bille** *m* COMPOSANTS *tubes électroniques* ball valve; **~ coupe-feu** *m* COMPOSANTS fire valve

claquage *m* ALIM ÉLEC, COMPOSANTS, ÉLECTRON *de fusible* blowing, *de semiconducteur* electric breakdown; **~ du diélectrique** ALIM ÉLEC dielectric breakdown, COMPOSANTS electric breakdown, ÉLECTRON dielectric breakdown; **~ disruptif** ALIM ÉLEC disruptive breakdown; **~ par avalanche** ÉLECTRON *de semiconducteur* avalanche breakdown; **~ par emballement thermique** COMPOSANTS, ÉLECTRON thermal breakdown

claquement: **~ de manipulation** *m* RADIO, TÉL key click

claquer *vi* ALIM ÉLEC, COMPOSANTS, ÉLECTRON *fusible* blow, *semiconducteur* break down

classe *f* GÉNÉRAL class, TESTS grade; **~ d'associations** DONNÉES association class; **~ de changement de tarif** COMMUT tariff-switching class; **~ de commande** TÉL command class; **~ de débit** DONNÉES throughput class, thruput class (*AmE*); **~ d'émission** RADIO class of emission; **~ erreur de protocole** DONNÉES protocol error class; **~ d'exploitation asymétrique en mode réponse normal** DONNÉES unbalanced operation normal response mode class (*UNC*); **~ fonctionnelle** (*FC*) TÉL functional class (*FC*); **~ fonctionnelle A** (*FCA*) TÉL functional class A (*FCA*); **~ fonctionnelle B** (*FCB*) TÉL functional class B (*FCB*); **~ de HDLC** TÉL class of HDLC; **~ d'objet** DONNÉES object class; **~ d'objet géré** DONNÉES managed-object class (*MOC*); **~ d'opérations** DONNÉES operation class; **~ de protocole** DONNÉES protocol class; **~ ressource indisponible** DONNÉES resource-unavailable class; **~ service-ou-option indisponible** DONNÉES service-or-option class not-available; **~ service-ou-option non mis en œuvre** DONNÉES service-or-option class not-implemented; **~ situation normale** DONNÉES normal-situation class; **~ de tarif** TÉL tariff class; **~ de transfert de masse de document** (*BT*) TÉL document bulk transfer class (*BT*); **~ de transfert de masse et de manipulation de document** (*BTM*) TÉL document bulk transfer and manipulation class (*BTM*)

classement *m* TÉL grading; **~ des défaillances** TESTS classification of failures

classer *vt* TÉL grade

classification *f* GÉNÉRAL classification; **~ prioritaire** TÉL priority classification

classifier *vt* GÉNÉRAL classify

clavier *m* GÉNÉRAL keyboard, PÉRIPH keyboard, push-button dial, TÉL keysender; **~ à action directe** PÉRIPH sawtooth keyboard; **~ alphanumérique** INFORMAT alphanumeric keyboard; **~ dynamique** INFORMAT dynamic keyboard; **~ d'enregistrement** INFORMAT recording keyboard; **~ externe** INFORMAT external keyboard; **~ des fonctions** INFORMAT function keyboard; **~ inférieur** INFORMAT lower keyboard; **~ lumineux** RADIO illuminated keyboard; **~ nu** INFORMAT keyboard matrix, key matrix; **~ numérique** APP SERV, INFORMAT, PÉRIPH, TÉL keypad, keyset, numeric keypad, numeric keyset; **~ de numérotation** PÉRIPH, TÉL keypad, keyset; **~ perforateur** GÉNÉRAL keyboard perforator; **~ piézosensible** INFORMAT pressure-sensitive keyboard; **~ pondéré** INFORMAT weighted keyboard; **~ programmable** INFORMAT programmable keyboard; **~ à réponse sensitive** INFORMAT velocity-sensitive keyboard; **~ sans boîtier** INFORMAT keyboard matrix, key matrix; **~ supérieur** INFORMAT upper keyboard; **~ téléphonique** APP SERV keypad, keyset, PÉRIPH keypad, telephone keypad, *postes téléphoniques* keyset; **~ à verrouillage** INFORMAT, PÉRIPH shift-lock keyboard

clé *f* APP SERV, DONNÉES, GÉNÉRAL, INFORMAT key; **~ absolue** INFORMAT actual key; **~ d'autorisation** TÉL authorization key; **~ de déchiffrement** RADIO code key; **~ électronique** INFORMAT dongle; **~ de groupe** ÉLECTRON group key; **~ d'habitation** TÉL authorization key; **~ d'interrogation** INFORMAT polling key; **~ polygonale** COMPOSANTS, TESTS box spanner (*BrE*), box wrench (*AmE*); **~ publique** RADIO public key; **~ de recherche** INFORMAT search key

clés: **~ de protection** *f pl* ÉLECTRON offset key protection

CLF *abrév* (*capacités de fonctionnement locales*) COMMUT LFC (*local functional capabilities*)

clic *m* ÉLECTRON transient, INFORMAT click

cliché *m* PÉRIPH printing block; **~ de panne** TÉL fault pattern, fault syndrome; ~ **de production** ÉLECTRON production master; ~ **de production à images multiples** TV multiple-image production master; ~ **de production original** ÉLECTRON original production master

clicher *vt* INFORMAT dump

client *m* GÉNÉRAL customer, client; ~ **pilote** TESTS test customer

client/serveur *m* INFORMAT client/server

clignotement *m* INFORMAT, TÉL, TESTS blinking, flashing, winking

clignoter *vi* INFORMAT, TÉL, TESTS *indicateur* blink, flash, wink

clip *m* COMPOSANTS, ÉLECTRON clip; ~ **étamé à montage rapide** COMPOSANTS solder-plated quick clip; ~ **de verrouillage** ÉLECTRON locking clip

cliquer *vt* INFORMAT click; ~ **sur** INFORMAT click on

cliquet *m* COMMUT pawl

cloison *f* RADIO septum

clone *m* DONNÉES, INFORMAT clone

cloner *vt* DONNÉES, INFORMAT clone

clore *vt* INFORMAT log-out

CLUT *abrév* (*table de couleurs*) INFORMAT CLUT (*color look-up table AmE, colour look-up table BrE*)

CMF *abrév* (*compression-extension avec modulation de fréquence*) TRANSMIS CFM (*companding and frequency modulation*)

CMS *abrév* (*composant monté en surface*) COMPOSANTS, ÉLECTRON SMC (*surface-mounted component*)

C-n *abrév* (*conteneur-n*) APP SERV, DONNÉES, TÉL C-n (*container-n*)

CNA *abrév* (*convertisseur numérique-analogique, convertisseur N-A*) DONNÉES, ÉLECTRON, INFORMAT, SIGNAL, TÉL DAC (*digital-to-analog converter, digital-analog converter, D-A converter*)

CNC *abrév* (*concentration numérique des conversations*) TÉL, TRANSMIS DSI (*digital speech interpolation*)

CNE *abrév* (*concentrateur numérique éloigné*) TÉL remote digital concentrator

CNP *abrév* TÉL, TRANSMIS (*concentration numérique de la parole*) digital speech compression

CNS *abrév* (*concentrateur numérique satellite*) RADIO digital satellite concentrator

coax *m* COMPOSANTS, TRANSMIS coax, coaxial cable; ~ **fin** COMPOSANTS, TRANSMIS thin coax; ~ **gros** COMPOSANTS, TRANSMIS thick coax

coaxial[1] *adj* ÉLECTRON, RADIO, TRANSMIS, TV coaxial

coaxial[2]: ~ **carré** *m* TRANSMIS squareax

cocanal *m* RADIO co-channel; ~ **orthogonal** RADIO orthogonal co-channel

coche *f* ÉLECTRON blank

codage[1] *m* DONNÉES, INFORMAT, RADIO, SIGNAL, TÉL, TRANSMIS, TV coding, encoding; ~ **adaptif** SIGNAL adaptive coding; ~ **arithmétique** DONNÉES, INFORMAT arithmetic coding; ~ **auto-orthogonal par convolution** DONNÉES self-orthogonal convolutional coding; ~ **avec code bipolaire** TRANSMIS conversion to bipolar code; ~ **binaire** SIGNAL binary coding; ~ **en bipolaire** TRANSMIS conversion to bipolar code; ~ **bit à bit** DONNÉES bit-by-bit encoding; ~ **de canal entrelacé** TRANSMIS interleaved channel coding; ~ **chaîné** DONNÉES concatenated coding (*CC*); ~ **convolutif** INFORMAT convolutional coding; ~ **correcteur d'erreurs** DONNÉES error correction coding; ~ **correcteur d'erreurs sans circuit de retour** INFORMAT, TÉL forward error correction coding; ~ **à débit réduit** DONNÉES low-rate encoding (*LRE*); ~ **de détection d'erreurs** DONNÉES error-detection coding; ~ **direct** DONNÉES absolute encoding; ~ **pour la durée du traitement** INFORMAT run-length encoding (*RLE*); ~ **d'échantillonnage multiple** DONNÉES multiple subsampling encoding (*MUSE*); ~ **entropique** SIGNAL entropy coding; ~ **à faible débit** DONNÉES low-rate encoding (*LRE*); ~ **de fréquence** RADIO frequency encoding; ~ **Golay** DONNÉES Golay coding; ~ **d'image** DONNÉES image coding, INFORMAT, TÉL picture coding; ~ **d'impulsions** DONNÉES, TRANSMIS pulse coding; ~ **intertrame** DONNÉES interframe coding; ~ **intra-image** TV intraframe coding; ~ **intratrame** DONNÉES in-frame coding; ~ **à longueur variable** DONNÉES, INFORMAT, TÉL variable-length coding (*VLC*); ~ **non intratrame** INFORMAT non-intra coding; ~ **non uniforme** TÉL, TRANSMIS nonuniform encoding; ~ **numérique** DONNÉES digital coding, digital encoding; ~ **par blocs** TRANSMIS block coding; ~ **par découpage** TRANSMIS gap coding; ~ **par modulation de largeur** RADIO width coding; ~ **par transformation acoustique adaptive** INFORMAT adaptive transform acoustic coding (*ATRAC*); ~ **par transformée** DONNÉES, ÉLECTRON, INFORMAT, SIGNAL, TÉL transform coding; ~ **par transformée en cosinus discrète** DONNÉES, ÉLECTRON, INFORMAT, SIGNAL, TÉL discrete cosine transform coding; ~ **par transition** DONNÉES differential encoding; ~ **de la parole** INFORMAT, RADIO, TÉL, TRANSMIS speech coding, speech encoding; ~ **photographique** INFORMAT, TÉL photographic coding; ~ **prédicitif linéaire** DONNÉES lower-order path connection (*LPC*); ~ **prédictif** DONNÉES predictive coding; ~ **prédictif linéaire** DONNÉES linear predicting coding (*LPC*); ~ **progressif** INFORMAT, TÉL progressive coding; ~ **séquentiel** INFORMAT, TÉL sequential coding; ~ **séquentiel compatible avec le codage progressif** INFORMAT, TÉL progressive-compatible sequential coding; ~ **séquentiel à progression unique** INFORMAT, TÉL single-progression sequential coding; ~ **de signal composite** TRANSMIS composite signal coding; ~ **en sous-bandes** SIGNAL sub-band coding; ~ **à tonalités séquentielles** DONNÉES sequential tone coding; ~ **en treillis** DONNÉES trellis coding; ~ **uniforme** TÉL, TRANSMIS uniform encoding; ~ **de la voix** INFORMAT, RADIO, TÉL, TRANSMIS speech coding, vocoding

codage[2]: **à ~ convolutionnel différentiel** *loc* DONNÉES convolutionally and differentially encoded (*CDE*)

code[1]: **en ~** *adj* DONNÉES shift-in; **de ~ couleur** *adj* ÉLECTRON color-coded (*AmE*), colour-coded (*BrE*)

code[2] *m* COMMUT, DONNÉES, INFORMAT, SIGNAL, TÉL, TRANSMIS code; ~ **abrégé** TÉL short code; ~ **d'accès** DONNÉES, TV access code; ~ **d'accès à l'interurbain automatique** TÉL direct distance dialing access code (*AmE*), subscriber trunk dialing access code (*AmE*); ~ **alphabétique** INFORMAT alphabetic code; ~ **AMI** TRANSMIS AMI code, alternate mark inversion signal; ~ **d'arrêt** SIGNAL stop code; ~ **d'autorisation** TÉL authorization code; ~ **à barres** APP SERV, DONNÉES bar-code; ~ **baudot** INFORMAT baudot code; ~ **binaire haute densité** DONNÉES, ÉLECTRON, INFORMAT high-density binary code; ~ **binaire MIC** DONNÉES, ÉLECTRON, TÉL PCM binary code; ~ **binaire pur** DONNÉES, ÉLECTRON straight binary code; ~ **binaire réfléchi** DONNÉES, ÉLECTRON reflected binary code; ~ **binaire symétrique** DONNÉES, ÉLECTRON, TÉL, TRANSMIS symmetrical binary code; ~ **biphase** APP SERV *ISDN* biphase code; ~ **bipolaire** SIGNAL bipolar code; ~ **bipolaire avec substitution de trois zéros** DONNÉES, SIGNAL, TÉL bipolar code with three-zero substitution (*B3ZS*); ~ **bipolaire à haute densité d'ordre 2** (*HDB2*) DONNÉES, SIGNAL, TÉL high-density bipolar 2 code (*HDB2*); ~ **bipolaire à haute densité d'ordre 3** (*HDB3*) DONNÉES, SIGNAL, TÉL high-density bipolar 3 code (*HDB3*); ~ **de blocs cycliques** DONNÉES cyclic-block code; ~ **de blocs linéaires** SIGNAL linear-block code; ~ **de bruit pseudo-aléatoire** DONNÉES, TRANSMIS pseudorandom-noise code; ~ **de câble** TÉL cable code; ~ **de câble bivalent** TÉL, TRANSMIS double-current code, two-condition cable code; ~ **de câble trivalent** TÉL, TRANSMIS three-condition cable code; ~ **de canal sémaphore** SIGNAL signaling-link code (*AmE*), signalling-link code (*BrE*); ~ **de caractère** INFORMAT character code; ~ **à cinq**

moments TÉL five-unit code; ~ **CMI** TRANSMIS code-mark inversion (*CMI*); ~ **de commande** DONNÉES, TÉL command code; ~ **complémentaire** TRANSMIS complementary code; ~ **complet** INFORMAT block code; ~ **de compte** TÉL account code; ~ **confidentiel** GÉNÉRAL personal identity number (*PIN*), RADIO secret code; ~ **considéré** DONNÉES given code; ~ **continental** TÉL *telex* continent code; ~ **de contrôle d'erreurs** DONNÉES error protection code; ~ **convolutif** INFORMAT convolutive code; ~ **de convolution** INFORMAT convolution code, convolutional code; ~ **de convolution discontinu** INFORMAT punctured convolutional code; ~ **correcteur d'erreurs** (*CCE*) DONNÉES, INFORMAT, SIGNAL, TÉL error-correcting code (*ECC*, *error correction code*); ~ **couleur analogique** TÉL analog color code (*AmE*), analog colour code (*BrE*); ~ **couleurs** GÉNÉRAL color code (*AmE*), colour code (*BrE*); ~ **de créneau temporel** SIGNAL time slot code; ~ **cyclique** DONNÉES cyclic code; ~ **decimal codé binaire étendu** (*EBCDIC*) DONNÉES, TRANSMIS extended binary coded decimal interchange code (*EBCDIC*); ~ **détection d'erreur** (*CDE*) DONNÉES, INFORMAT, SIGNAL, TÉL error-detecting code (*EDC*), error detection code (*EDC*); ~ **deux sur cinq** DONNÉES two-out-of-five code; ~ **de diagnostic** TÉL diagnostic code; ~ **à disparité compensée** TÉL paired-disparity code; ~ **DNTX** SIGNAL DNTX code, do-not-transmit code; ~ **de données** DONNÉES data code, TÉL code; ~ **d'échappement** DONNÉES, TÉL escape code (*EC*); ~ **d'épellation** RADIO phonetic alphabet; ~ **d'erreur** TÉL error code, fault code; ~ **d'état et de commande** CONTRÔLE, DONNÉES status and control code; ~ **à faible herméticité** RADIO low-grade code; ~ **de fonction utilisé pour le lancement d'un appel** TÉL originating feature code; ~ **Golay de rendement 1/2** DONNÉES rate 1/2 Golay code; ~ **Gray** DONNÉES Gray code; ~ **d'habilitation** TÉL authorization code; ~ **de Hamming** DONNÉES Hamming code; ~ **Hollerith** INFORMAT Hollerith code; ~ **horaire** RADIO time code; ~ **d'identification** TÉL identification code; ~ **d'identification de l'autorité chargée de la comptabilité** TÉL accounting authority identification code (*AAIC*); ~ **d'identification du destinataire** DONNÉES, TÉL recipient identification code; ~ **d'identification de l'entreprise** PÉRIPH company code, company identity code; ~ **d'identification de réseau de données** (*CIRD*) DONNÉES, TÉL data-network identification code (*DNIC*); ~ **d'identification de réseau privé de données** (*PNIC*) DONNÉES, TÉL private data-network identification code (*PNIC*); ~ **d'identification de réseau télex** (*CIRT*) DONNÉES, TÉL telex-network identification code; ~ **d'identification de support** DONNÉES, TÉL bearer identification code; ~ **interprétatif** INFORMAT interpretive code; ~ **d'intervalle de temps** SIGNAL time slot code; ~ **à inversion de marque alternée** SIGNAL, TÉL, TRANSMIS alternate-mark inversion code; ~ **de liaison de signalisation** SIGNAL signaling-link code (*AmE*), signalling-link code (*BrE*); ~ **en ligne** DONNÉES, TÉL line code; ~ **linéaire** DONNÉES linear code; ~ **à longueur variable** DONNÉES, INFORMAT, TÉL variable-length code; ~ **machine** INFORMAT machine code; ~ **à moments** DONNÉES, TÉL equal-length code; ~ **Morse** RADIO, TÉL Morse code; ~ **Morse international** RADIO, TÉL international Morse code (*IMC*); ~ **à n éléments** TÉL n-unit code; ~ **à n éléments unitaires** TÉL n-unit code; ~ **à n moments** TÉL n-unit code, n-level code; ~ **natif** DONNÉES proprietary code; ~ **national de destination** DONNÉES, TÉL national destination code (*NDC*); ~ **ne pas émettre** SIGNAL DNTX code, do-not-transmit code; ~ **non binaire** DONNÉES nonbinary code; ~ **non-retour à zéro** (*code NRZ*) DONNÉES, TRANSMIS nonreturn to zero code (*NRZ code*); ~ **NRZ** (*code non-retour à zéro*) DONNÉES, TRANSMIS NRZ code (*nonreturn to zero code*); ~ **numérique** DONNÉES numerical code; ~ **opération** INFORMAT operation code; ~ **d'organisation** DONNÉES, INFORMAT organization code (*OC*); ~ **de paquet de référence principal** DONNÉES PRB code, primary reference burst code; ~ **de paquet de référence secondaire** DONNÉES secondary reference burst code; ~ **de point** DONNÉES, TÉL point code; ~ **du point de destination** DONNÉES, TÉL destination point code; ~ **du point d'origine** DONNÉES, TÉL originating point code (*OPC*); ~ **d'un point sémaphore** SIGNAL signaling-point code (*AmE*), signalling-point code (*BrE*); ~ **d'un point de signalisation** SIGNAL signaling-point code (*AmE*), signalling-point code (*BrE*); ~ **postal** INFORMAT post code (*BrE*), zip code (*AmE*); ~ **de priorité de traitement** DONNÉES, INFORMAT processing priority code; ~ **de priorité de transmission** DONNÉES, TRANSMIS transmission priority code; ~ **de protection contre les erreurs** DONNÉES error-protecting code, error protection code; ~ **4B3T** TRANSMIS code 4B3T; ~ **quinaire** DONNÉES two-out-of-five code; ~ **de redondance cyclique** (*CRC*) DONNÉES, TÉL cyclic redundancy code (*CRC*); ~ **redondant** SIGNAL redundant code; ~ **de Reed-Solomon** (*code RS*) INFORMAT Reed-Solomon code, RS code; ~ **Reed-Solomon à entrelacement croisé** (*CIRC*) INFORMAT cross-interleaved Reed-Solomon code (*CIRC*); ~ **de réponse** RADIO reply code, TÉL answer code; ~ **de réponse partielle** DONNÉES partial response code; ~ **de retour sur canal sémaphore normal** SIGNAL changeback code; ~ **RS** INFORMAT Reed-Solomon code, RS code; ~ **SDTNX** (*code sélectif ne pas émettre*) SIGNAL SDTNX code (*selective do-not-transmit code*); ~ **sélectif ne pas émettre** (*code SDTNX*) SIGNAL selective do-not-transmit code (*SDTNX code*); ~ **de sélection** TÉL selection code; ~ **de service** COMMUT, TÉL service code; ~ **de signal** SIGNAL signal code; ~ **de signalisation** SIGNAL signal code; ~ **de signalisation de commande** SIGNAL control signaling code (*AmE*), control signalling code (*BrE*); ~ **de signalisation de ligne** DONNÉES, SIGNAL line signaling code (*AmE*) (*LSC*), line signalling code (*BrE*) (*LSC*); ~ **de signaux vocaux** TRANSMIS speech code; ~ **de station de référence hors fonctionnement** RADIO inoperative reference station code; ~ **de synchronisation** DONNÉES, INFORMAT sync code, synchronization code; ~ **de synchronisation de trame** DONNÉES frame code; ~ **télégraphique** TÉL, TRANSMIS code, telegraphic code; ~ **télégraphique bivalent** TÉL, TRANSMIS two-condition telegraph code; ~ **télégraphique trivalent** TÉL, TRANSMIS three-condition telegraph code; ~ **télex de destination** DONNÉES, TÉL telex destination code (*TDC*); ~ **temporel** DONNÉES, INFORMAT time code; ~ **temporel d'intervalle vertical** DONNÉES, INFORMAT vertical interval time code (*VITC*); ~ **temporel longitudinal** (*CTL*) INFORMAT longitudinal time code (*LTC*); ~ **temporel SMPTE** INFORMAT, TV SMPTE time code; ~ **de tranche horaire** SIGNAL time-slot code; ~ **de transmission** TRANSMIS line code; ~ **de verrouillage** DONNÉES, TÉL interlock code (*IC*); ~ **de verrouillage de multitrame** TRANSMIS multiframe alignment code

codé[1] *adj* DONNÉES, INFORMAT, RADIO, SIGNAL, TÉL, TRANSMIS, TV coded, encoded

codé[2]: ~ **Manchester** *loc* INFORMAT Manchester-coded

codec *abrév* (*codeur/décodeur*) INFORMAT, TRANSMIS codec (*coder/decoder*); ~ **de loi mu** *m* TRANSMIS mu law codec; ~ **de sous-bande** *m* INFORMAT sub-band codec; ~ **vidéo** *m* INFORMAT video codec

coder *vt* DONNÉES, INFORMAT, RADIO, SIGNAL, TÉL, TRANSMIS, TV code, encode

codet *m* INFORMAT code element

codeur *m* DONNÉES, INFORMAT, RADIO, SIGNAL, TÉL, TRANSMIS coder, encoder; ~ **d'angle optique** ÉLECTRON optical shaft encoder; ~ **angulaire** DONNÉES angle encoder; ~ **de canal** RADIO channel coder; ~ **CELP** SIGNAL CELP coder; ~ **convolutionnel** INFORMAT convolutional encoder; **~-décodeur numérique** INFORMAT, TRANSMIS digital encoder-decoder; **~ à fréquences vocales** INFORMAT, RADIO, TÉL voice-operated coder; **~ à fréquences vocales activé par la hauteur du son** RADIO pitch-excited vocoder; **~ à fréquences vocales à canaux** RADIO channel vocoder; **~ à fréquences vocales à formant** RADIO formant vocoder; **~ à fréquences vocales LPC** RADIO linear predicting coding vocoder; **~ d'interface** TRANSMIS interface coder; **~ MICDA** TÉL ADPCM encoder; **~ multifréquence** TRANSMIS multiple-frequency encoder; **~ de parole** INFORMAT, RADIO, TÉL speech encoder; **~ prédictif linéaire** TÉL linear predictive vocoder

codeur/décodeur *m* (*codec*) INFORMAT, TRANSMIS coder/decoder (*codec*)

coefficient *m* GÉNÉRAL coefficient; **~ d'affaiblissement** RADIO, TRANSMIS attenuation coefficient; **~ d'affaiblissement d'image** TRANSMIS image attenuation coefficient; **~ d'amplification** ALIM ÉLEC mu factor, ÉLECTRON m-factor; **~ d'atténuation** RADIO, TRANSMIS attenuation coefficient; **~ de diffusion** RADIO scattering coefficient; **~ de dilatation linéaire** COMPOSANTS, ÉLECTRON coefficient of linear expansion, linear expansion coefficient; **~ de dilatation thermique** COMPOSANTS, ÉLECTRON thermal expansion coefficient; **~ de directivité d'antenne** RADIO antenna directivity factor; **~ de dispersion** TRANSMIS dispersion coefficient; **~ de dispersion de matériaux** TRANSMIS material dispersion coefficient; **~ de fiabilité** GÉNÉRAL confidence coefficient; **~ de Hall** ÉLECTRON Hall coefficient; **~ de multiplication d'erreurs** TÉL error multiplication factor; **~ d'occupation des sols** RADIO ground clutter coefficient; **~ de pratique expérimentale** TÉL practice factor, TESTS crew factor; **~ de qualité** RADIO figure of merit; **~ de réflexion** RADIO, TRANSMIS reflection coefficient; **~ de réflexion diffuse** RADIO, TRANSMIS diffuse reflection coefficient; **~ de réflexion spéculaire** RADIO, TRANSMIS specular reflection coefficient; **~ de remplissage** DONNÉES filling coefficient; **~ de température** GÉNÉRAL temperature coefficient; **~ de temps** TESTS time coefficient; **~ du temps de montée** GÉNÉRAL, INFORMAT, TÉL rise-time coefficient; **~ de transmission** TRANSMIS transmission coefficient; **~ d'utilisation** ALIM ÉLEC, RADIO duty cycle

coercivité *f* ÉLECTRON coercivity

cœur *m* INFORMAT heart, TRANSMIS core; **~ de chaîne** DONNÉES core of exchange, RADIO base unit (*BU*)

coffret *m* COMPOSANTS case, housing, cabinet, DONNÉES cabinet, box, INFORMAT box, cabinet, RADIO cabinet; **~ d'abonné** TV subscriber unit, set-top unit; **~ de batterie** ALIM ÉLEC battery case; **~ à cartouches** INFORMAT cartridge box; **~ de commande** CONTRÔLE control cabinet; **~ à disquettes** INFORMAT disk box; **~ de distribution** TRANSMIS distribution cabinet; **~ d'épissure locale** ÉLECTRON premises splice organizer (*PSO*); **~ d'épissure mural** ÉLECTRON wall-mountable splice center (*AmE*) (*WSC*), wall-mountable splice centre (*BrE*) (*WSC*); **~ d'essai** TESTS test case; **~ à fusibles** ALIM ÉLEC fuse box; **~ de protection** ÉLECTRON component housing; **~ de sous-répartition** TRANSMIS subdistribution cabinet

cofréquence *f* RADIO co-channel

cogniticien *m* INFORMAT knowledge engineer

cognitif *adj* GÉNÉRAL cognitive

cognitique *f* INFORMAT knowledge engineering

cohérence *f* DONNÉES, GÉNÉRAL, TRANSMIS coherence, consistency; **~ de fréquence** RADIO coherence of frequency; **~ d'une modulation** TÉL modulation coherence; **~ d'onde** ÉLECTRON wave coherence; **~ optique** TRANSMIS optical coherence; **~ partielle** TRANSMIS partial coherence; **~ de phase** RADIO, TRANSMIS phase coherence; **~ d'un sématème** TÉL modulation coherence; **~ spatiale** TRANSMIS spatial coherence; **~ temporelle** TRANSMIS temporal coherence

coiffe: **~ à détrompeur amovible** *f* ÉLECTRON removable keyed shroud; **~ insonorisante** *f* GÉNÉRAL noise source enclosure

coïncidence *f* TÉL coincidence, TV registration

coïndice *m* RADIO refractivity

col *m* ÉLECTRON neck

collage *m* ALIM ÉLEC bonding; **~ d'armature** ÉLECTRON *relais* armature stick

collecte: **~ de données** *f* DONNÉES data collection; **~ et visualisation d'alarmes** *f pl* TÉL alarm collection and display

collecteur *m* ALIM ÉLEC, ÉLECTRON collector, commutator; **~ annulaire de mise à la terre** ALIM ÉLEC earth ring collector (*BrE*), ground ring collector (*AmE*); **~ commun** ÉLECTRON common collector; **~ de données** DONNÉES data sink; **~ de messages** TRANSMIS message sink; **~ de terre** ALIM ÉLEC earth collector (*BrE*), ground collector (*AmE*), earth collector bar (*BrE*), ground collector bar (*AmE*); **~ de transistor** ÉLECTRON transistor collector

collectique *f* DONNÉES, INFORMAT groupware

collectivité: **~ locale** *f* TV local authority

coller *vt* INFORMAT paste

collerette *f* RADIO skirt

collier: **~ de câble** *m* TRANSMIS *pose de câbles* cable clamp; **~ de déviation** *m* COMPOSANTS, ÉLECTRON yoke assembly; **~ de terre** *m* ALIM ÉLEC earth clamp (*BrE*), ground clamp (*AmE*)

colliers: **~ d'acheminement** *m pl* ÉLECTRON routing wings

collimateur *m* TRANSMIS collimator

collimation *f* TRANSMIS collimation

collision *f* ÉLECTRON collision, RADIO clashing; **~ d'appel à l'interface ETTD/ETCD** TÉL call collision at the DTE/DCE interface; **~ d'appels** TÉL call collision; **~ d'électrons** ÉLECTRON electron collision; **~ de libération** TÉL clearing collision

colloque: **~ de reconnaissance** *m* DONNÉES, INFORMAT handshake, handshaking

côlon *m* INFORMAT colon

colonne: **~ de broches** *f* ÉLECTRON pin column; **~ montante** *f* ÉLECTRON riser shaft, riser, TRANSMIS *pose de câbles* cable riser; **~ positive** *f* ÉLECTRON positive glow

colophane *f* ÉLECTRON colophony

coloration *f* INFORMAT coloration

colorimétrie *f* GÉNÉRAL colorimetry

colorisation *f* TV colorization

ColorSence® *m* INFORMAT *Kodak* ColorSence®

COMA *abrév* (*ML COM de branche A*) TÉL COMA (*ML COM of A branch*)

COMB *abrév* (*ML COM de branche B*) TÉL COMB (*ML COM of B branch*)

combinaison *f* DONNÉES, INFORMAT, TÉL combination; **~ binaire** TÉL bit combination; **~ de circuits** TÉL, TRANSMIS phantoming; **~ d'éléments binaires** INFORMAT bit combination; **~ d'exécution** INFORMAT performance combination; **~ spéciale de caractères** DONNÉES special character combination

combinateur: **~ optique** *m* TRANSMIS optical combiner

combinatoire *adj* ÉLECTRON combinational (*AmE*), combinatorial (*BrE*)

combiné *m* PÉRIPH, TÉL handset; **~ d'essai** TÉL, TESTS test handset; **~ de radio mobile** RADIO cellular radio handset; **~ téléphonique** PÉRIPH, TÉL handset, telephone, telephone set; **~ téléphonique à écoute amplifiée** PÉRIPH, TÉL amplified handset; **~ téléphonique à mémoire** PÉRIPH, TÉL telephone set with memory; **~ de test** TÉL, TESTS test handset

combineur *m* ÉLECTRON combiner, combiner unit

comité *m* GÉNÉRAL committee; **~ d'application des recommandations techniques** GÉNÉRAL technical recommendations applications committee (*TRAC*); ~ **consultatif** GÉNÉRAL advisory committee

Comité: ~ **consultatif international des radiocommunications** *m* (*CCIR*) RADIO International Radio Consultative Committee (*IRCC*); ~ **consultatif international télégraphique et téléphonique** *m* (*CCITT*) TÉL International Telegraph and Telephone Consultative Committee (*ITTCC*); ~ **européen de normalisation** *m* (*CEN*) GÉNÉRAL European Committee for Standardization (*CEN*); ~ **européen de normalisation électrotechnique** *m* (*CENELEC*) GÉNÉRAL European Committee for Electrotechnical Standardization (*CENELEC*); ~ **international radio-maritime** *m* *CIRM* RADIO International Marine Radio Association (*CIRM*); ~ **international spécial des perturbations radioélectriques** *m* (*CISPR*) RADIO International Special Committee on Radio Interference (*CISPR*)

commande[1] *f* TÉL command, control; ~ **d'accès** INFORMAT access instruction; ~ **de l'accès conditionnel** RADIO conditional access control; ~ **d'accès au support** CONTRÔLE medium access control (*MAC*); ~ **alphanumérique** INFORMAT character instruction; ~ **asservissement** CONTRÔLE closed-loop control; ~ **d'attitude** RADIO attitude control; ~ **automatique** CONTRÔLE automatic control; ~ **automatique de fréquence** (*CAF*) ÉLECTRON, TÉL, TRANSMIS automatic frequency control (*AFC*); ~ **automatique de gain** (*CAG*) ÉLECTRON, TÉL, TRANSMIS automatic gain control (*AGC*); ~ **automatique instantanée de gain** (*CAIG*) ÉLECTRON, TÉL, TRANSMIS instantaneous automatic gain control (*IAGC*); ~ **automatique de puissance** ALIM ÉLEC automatic power control (*APC*); ~ **automatique de volume** (*CAV*) ÉLECTRON, TÉL, TRANSMIS automatic volume control (*AVC*); ~ **bilatérale** TÉL, TRANSMIS bilateral control; ~ **de blocage automatique** RADIO squelch control; ~ **de boucle** TV loopback command; ~ **en boucle ouverte** ALIM ÉLEC, CONTRÔLE open-loop control; ~ **de brillance** TV brilliance control; ~ **en cascade** INFORMAT cascade control; ~ **de clavier** INFORMAT keyboard controller; ~ **de contraste** TV contrast control; ~ **cyclique du gain** CONTRÔLE sensitivity time control (*STC*); ~ **de début de document** DONNÉES command document start; ~ **directe** COMMUT, INFORMAT direct control; ~ **à distance** CONTRÔLE, INFORMAT, PÉRIPH, RADIO, TÉL remote control; ~ **d'édition** INFORMAT format control; ~ **d'émission numérique** (*DTC*) DONNÉES digital-transmit command (*DTC*); ~ **et surveillance du système** TESTS control and system monitoring; ~ **d'expression** INFORMAT expression control; ~ **de flux** DONNÉES, TÉL flow control; ~ **de fonction non normalisée** DONNÉES, TÉL nonstandard facility command (*NSC*); ~ **de fonction non normalisée CIG-DTC** (*NSC-CIG-DTC*) DONNÉES, TÉL nonstandard facility command CIG-DTC (*NSC-CIG-DTC*); ~ **de fondu sonore** INFORMAT fader; ~ **de fréquence** RADIO frequency control; ~ **de gain** ÉLECTRON gain control; ~ **d'impression** INFORMAT print command; ~ **d'intensité** TV intensity control; ~ **d'intensité sonore** CONTRÔLE, INFORMAT volume control; ~ **invalide** INFORMAT invalid command; ~ **de liaison de données** DONNÉES data-link control (*DLC*); ~ **de liaison de données à haut niveau** (*HLDLC*) DONNÉES, TÉL high-level data link control (*HLDLC*); ~ **de liaison logique** CONTRÔLE logical link control (*LLC*); ~ **linéaire** CONTRÔLE linear control, INFORMAT linear control, linear slider, slider, slider control; ~ **locale** CONTRÔLE, TRANSMIS local control, single-ended control; ~ **locale et distante** CONTRÔLE, TRANSMIS double-ended control; ~ **de luminosité** INFORMAT, TV brilliance control; ~ **manuelle** CONTRÔLE manual control; ~ **manuelle d'émission** (*MOX*) DONNÉES manually-operated transmission (*MOX*); ~ **monocartouche** INFORMAT single-stroke control key; ~ **multipoint de demande de jeton** DONNÉES multipoint command token claim (*MCT*); ~ **multipoint d'imposition de visualisation** DONNÉES multipoint command visualization forcing (*MCV*); ~ **de niveau** INFORMAT level control; ~ **de niveau automatique** INFORMAT automatic level control; ~ **du niveau de bruit** TÉL noise level control; ~ **non valide** INFORMAT invalid command; ~ **numérique** DONNÉES, INFORMAT digital control, network connection (*NC*), numerical control (*NC*); ~ **opérateur** DONNÉES operator command; ~ **optimale** CONTRÔLE optimal control; ~ **d'orientation** CONTRÔLE orientation control, RADIO attitude control; ~ **d'orientation de faisceau** RADIO beam steering; ~ **de panoramique** INFORMAT pan control; ~ **par microprocesseurs** COMMUT microprocessor control; ~ **par modem** DONNÉES modem control; ~ **par programme enregistré** COMMUT, CONTRÔLE, TÉL stored-program control (*AmE*) (*SPC*), stored-programme control (*BrE*) (*SPC*); ~ **parlée** CONTRÔLE, PÉRIPH, TÉL speech control, voice control; ~ **par la voix** CONTRÔLE, PÉRIPH, TÉL speech control, voice activation, voice control; ~ **de phase** ÉLECTRON phase control; ~ **de polarisation** INFORMAT bias control; ~ **programmable** COMMUT, CONTRÔLE, TÉL programmable control stored-program control (*AmE*) (*SPC*), stored-programme control (*BrE*) (*SPC*); ~ **programmée** CONTRÔLE programmed control, TÉL programmed command; ~ **de puissance d'émetteur** RADIO transmitter-power control; ~ **relative** INFORMAT relative order; ~ **de serveur** DONNÉES server command; ~ **du sol** RADIO ground control; ~ **de souffle** INFORMAT breath control, breath controller; ~ **de stabilité verticale** TV vertical hold; ~ **de synchronisation** CONTRÔLE, INFORMAT timing control; ~ **de synchronisation verticale** TV vertical hold; ~ **du système de télécommunications** GÉNÉRAL communications system control (*CSC*); ~ **de tempo** INFORMAT tempo control; ~ **du temps de conduction** ÉLECTRON conduction-time control; ~ **en temps réel** INFORMAT real-time control; ~ **de tonalité** CONTRÔLE, INFORMAT tone control; ~ **de transfert** COMMUT transfer control; ~ **de transit numérique** SIGNAL digital transit command; ~ **de trémolo** INFORMAT tremolo control; ~ **unilatérale** TÉL, TRANSMIS unilateral control; ~ **vidéo de demande de gel de l'image** TV video command freeze-picture request (*VCF*); ~ **vidéo de demande de rafraîchissement rapide** TV video command fast-update request (*VCU*); ~ **vocale** CONTRÔLE, PÉRIPH, TÉL voice control; ~ **de voie** TRANSMIS channel control; ~ **de volume** CONTRÔLE, INFORMAT volume control; ~ **de volume général** PÉRIPH master volume control

commande[2]: **à ~ locale** *loc* CONTRÔLE, TÉL locally-controlled; **à ~ manuelle** *loc* CONTRÔLE manually-controlled; **à ~ par marqueur** *loc* COMMUT, CONTRÔLE marker-controlled; **à ~ vocale** *loc* CONTRÔLE, PÉRIPH, TÉL speech-controlled, voice-activated, voice-controlled, voice-operated

commandé: ~ **à distance** *loc* CONTRÔLE remote-controlled; ~ **localement** *loc* TÉL locally-controlled; ~ **à la main** *loc* CONTRÔLE manually-controlled; ~ **par l'abonné** *loc* TÉL subscriber-controlled; ~ **par événement** *loc* INFORMAT event-driven; ~ **par fréquence vocale** *loc* CONTRÔLE, PÉRIPH, TÉL speech-controlled, voice-activated, voice-controlled, voice-operated; ~ **par ordinateur** *loc* INFORMAT computer-controlled; ~ **par tension** *loc* ALIM ÉLEC, CONTRÔLE voltage-controlled; ~ **par touche** *loc* INFORMAT key-controlled; ~ **par la voix** *loc* CONTRÔLE, PÉRIPH, TÉL speech-controlled, voice-activated, voice-controlled, voice-operated; ~ **en tension** *loc* ALIM ÉLEC, CONTRÔLE voltage-controlled

commander *vt* APP SERV *communication* book, order, GÉNÉRAL control, TÉL *communication* book, order; ~ **à distance** CONTRÔLE, INFORMAT, PÉRIPH, RADIO, TÉL telecommand

commencement: ~ **de session** *m* DONNÉES, INFORMAT log-in

comment: ~ **me recevez-vous?** *loc* RADIO how do you read me?

commerce: ~ **électronique** *m* ÉLECTRON electronic commerce

commission: ~ **X12 de l'Institut américain de normalisation** *f* GÉNÉRAL American National Standards Institute committee 12 (*ANSI X12*)

Commission: ~ **électrotechnique internationale** *f* (*CEI*) GÉNÉRAL International Electrotechnical Commission (*IEC*); ~ **fédérale des communications** *f* (*FCC*) DONNÉES Federal Communication Commission (*FCC*)

communauté: ~ **de conception** *f* TÉL commonality; ~ **de répéteurs** *f* RADIO community of transponders

Communauté: ~ **régionale des communications** *f* GÉNÉRAL Regional Commonwealth in the Field of Communications (*RCC*)

communicant *adj* DONNÉES, GÉNÉRAL networked

communicateur *m* TÉL communicator

communication[1] *f* DONNÉES, GÉNÉRAL communication, RADIO communication, connection, TÉL call, connection; **en ~ avec** *loc* TÉL connected to; ~ **en attente** TÉL call waiting (*CW*); ~ **avec avis d'appel** (*communication AVP*) TÉL messenger call; ~ **avec préavis** (*communication PAV*) TÉL person-to-person call, personal call; ~ **AVP** (*communication avec avis d'appel*) TÉL messenger call; ~ **bidirectionnelle** RADIO, TÉL, TV two-way communication, both-way communication; ~ **bifurquée** TÉL forked working; ~ **bilatérale** TÉL two-way simplex, two-way simplex connection; ~ **continentale** TÉL continental connection; ~ **en cours** TÉL in-call, call in progress; ~ **en cours d'établissement** TÉL connection in progress, call in progress; ~ **défectueuse** TÉL, TESTS faulty call, faulty connection; ~ **de départ** DONNÉES, TÉL outgoing call; ~ **de données** DONNÉES, TRANSMIS data communication, data transmission; ~ **de données bas débit** DONNÉES, RADIO low bit data communication; ~ **échelonnée** TÉL echelon telegraphy; ~ **fictive de référence** TÉL hypothetical reference connection, hypothetical reference circuit (*hrc*); ~ **intercontinentale** TÉL intercontinental connection; ~ **internationale** TÉL international connection; ~ **interne** TÉL inside call; ~ **interrompue** TÉL split connection; ~ **interurbaine** TÉL toll call; ~ **libérée** DONNÉES call released; ~ **locale** TÉL local call; ~ **longue distance** TÉL long-distance call; ~ **monomédia** TV monomedia communication; ~ **non taxée à la durée** TÉL untimed call; ~ **non verbale** DONNÉES nonverbal communication; ~ **par impulsions météoriques** RADIO meteor-burst communication; ~ **par salves météoriques** RADIO meteor-burst communication; ~ **PAV** (*communication avec préavis*) TÉL person-to-person call, personal call; ~ **payable à l'arrivée** TÉL collect call (*AmE*), reverse-charge call (*BrE*), transfer-charge call; ~ **en PCV** TÉL collect call (*AmE*), reverse-charge call (*BrE*), transfer-charge call; ~ **personnelle** TÉL person-to-person call, personal call; ~ **point à multipoint** RADIO, TRANSMIS point-to-multipoint communication; ~ **point à point** RADIO, TRANSMIS point-to-point communication; ~ **point à zone** RADIO, TRANSMIS point-to-area communication; ~ **poste à poste** TÉL end-to-end communication; ~ **radioélectrique** RADIO radiocommunication; ~ **radiophonique internationale** TV international sound-program connection (*AmE*), international sound-programme connection (*BrE*); ~ **radiophonique internationale multiple** TV international multiple-destination sound-program (*AmE*), international multiple-destination sound-programme (*BrE*); ~ **semi-duplex** TÉL half duplex connection; ~ **en série** TÉL serial call; ~ **simplex** TÉL simplex circuit; ~ **sous-marine** RADIO, TÉL underwater communication; ~ **taxée à la durée** TÉL timed call; ~ **téléphonique** TÉL telephone call, telephone conversation, speech connection, phone call; ~ **téléphonique internationale** TÉL international telephone call, international telephone connection, international telephone conversation, international phone call; ~ **télétex** TÉL teletex call; ~ **télévisuelle internationale** TV international television connection; ~ **télévisuelle internationale multiple** TV international multiple-destination television connection; ~ **télex d'État** TÉL government telex call; ~ **télex privée ordinaire** TÉL ordinary private telex call; ~ **télex de service** PÉRIPH service telex call; ~ **à trois** PÉRIPH, RADIO, TÉL three-way call; ~ **unidirectionnelle** COMMUT, RADIO, TÉL, TV one-way communication, unidirectional communication; ~ **unilatérale** COMMUT, RADIO, TÉL, TV one-way communication, unidirectional connection; ~ **urbaine** TÉL local call; ~ **virtuelle** TÉL virtual call (*VC*), virtual communication; ~ **virtuelle circuit virtuel commuté** COMMUT virtual call switched virtual circuit

communication[2]: ~ **acceptée** *loc* (*CA*) DONNÉES call accepted (*CA*); ~ **établie** *loc* DONNÉES call connected (*CC*)

communications: ~ **avec les mobiles** *f pl* RADIO mobile communications; ~ **hyperfréquence** *f pl* RADIO microwave communications; ~ **micro-ondes** *f pl* RADIO microwave communications; ~ **sur place** *f pl* RADIO *service maritime mobile* on-scene communications (*OSC*)

communiqué *m* TÉL bulletin, communiqué

commutateur *m* ALIM ÉLEC, COMMUT, ÉLECTRON switch, switching unit, changeover switch, selector switch, RADIO switch driver, TÉL selector; ~ **d'abonnés** COMMUT, TÉL local exchange (*LE*), private branch exchange (*PBX*), switchboard, telephone switchboard; ~ **d'accès** COMMUT access selector, access switch; ~ **à accès flexible** COMMUT flexible access switch (*FAS*); ~ **à anneau** RADIO ring switch; ~ **d'arrivée** COMMUT, TÉL terminating exchange; ~ **automatique** COMMUT, TÉL auto-manual switchboard, automatic switchboard, automatic switching equipment; ~ **automatique rural** COMMUT, TÉL rural automatic exchange; ~ **à autonomie d'acheminement** (*GSC*) COMMUT group switching center (*AmE*) (*GSC*), group switching centre (*BrE*) (*GSC*); ~ **auxiliaire** COMMUT auxiliary switch, auxiliary switching unit; ~ **à bande étroite** COMMUT narrow-band switch; ~ **de bits** COMMUT bit switch; ~ **caractère par caractère** COMMUT character coupling switch; ~ **de caractères** COMMUT character switch; ~ **central manuel d'abonnés** COMMUT, TÉL private manual branch exchange (*PMBX*); ~ **de circuits** COMMUT circuit-switching unit, circuit switch, circuit-switched exchange; ~ **à circuits rapides** COMMUT fast circuit switch; ~ **à circuits virtuels** (*CCV*) COMMUT virtual-circuit switch (*VCS*); ~ **coaxial** TRANSMIS coaxial switched jumper; ~ **commandé par porteur** RADIO carrier-operated switch (*COS*); ~ **à commande vocale** COMMUT, COMPOSANTS voice-operated relay, voice-operated switch; ~ **à coulisse** COMPOSANTS slide switch; ~ **à couteau** ALIM ÉLEC knife switch; ~ **crossbar** COMMUT crossbar switch; ~ **déjouable** ALIM ÉLEC, COMMUT trickable switch; ~ **à déplacement de fibre** COMMUT moving fiber switch (*AmE*), moving fibre switch (*BrE*); ~ **de destination** TÉL called exchange, destination exchange, destination office; ~ **à deux fils** COMMUT two-wire switch, two-wire switching system, two-wire system; ~ **à dicorde** COMMUT double-cord switchboard; ~ **de données** COMMUT, DONNÉES, ÉLECTRON data switch; ~ **électronique** COMMUT electronic exchange, electronic switch, DONNÉES electronic switch, INFORMAT electronic exchange, electronic switch; ~ **électronique de messages** COMMUT electronic message switch; ~ **électro-optique** COMMUT electro-optic switch; ~ **d'éléments** ALIM ÉLEC cell switch; ~ **d'enveloppes** COMMUT envelope switch; ~ **en étoile** COMMUT star switch; ~ **de fonction** GÉNÉRAL function switch; ~ **en fréquence** COMMUT frequency-division switching system (*FDS system*); ~ **à glissière** COMPOSANTS slide switch; ~ **de guide d'ondes** COMMUT waveguide switch; ~ **holographique** COMMUT holographic exchange; ~ **hybride** COMMUT hybrid switch; ~ **à inertie** ALIM ÉLEC impact switch; ~ **interurbain** COMMUT toll switch; ~ **à lame** ALIM ÉLEC knife switch; ~ **à large bande** COMMUT broadband switch; ~ **de lignes** COMMUT line switch; ~ **local** COMMUT, TÉL local exchange (*LE*), local switch; ~ **en longueur d'onde** COMMUT wavelength division switch; ~ **manuel** COMMUT, TÉL telephone switchboard; ~ **manuel à clés** COMMUT, TÉL cordless switchboard; ~ **manuel à monocordes** COMMUT, TÉL single-cord switchboard; ~ **manuel privé** COMMUT, TÉL private manual branch exchange (*PMBX*); ~ **manuel sans cordons** COMMUT, TÉL cordless switchboard; ~ **manuel télégraphique** COMMUT, TÉL telegraphic switchboard; ~ **matriciel** COMMUT matrix switch; ~ **de messages** COMMUT message switch; ~ **de mobiles** RADIO cellular mobile exchange; ~ **multiplex** COMMUT multiple switchboard; ~ **numérique** COMMUT digital exchange; ~ **numérique intégré** COMMUT integrated digital exchange; ~ **n° sept** COMMUT no. seven exchange; ~ **d'octets** COMMUT byte switch; ~ **ll/12 GHz** RADIO e/t switch; ~ **optique** COMMUT, TRANSMIS optical switch, optical exchange; ~ **optique à coupleur directif** COMMUT, TRANSMIS directional coupler optical switch; ~ **optique intégré** COMMUT, TRANSMIS integrated optical switch; ~ **optomécanique** COMMUT mechanical optical switch; ~ **de paquets** (*CP*) COMMUT, DONNÉES, TÉL, TRANSMIS packet switch (*PS*); ~ **de paquets rapide** COMMUT, DONNÉES, TRANSMIS, TÉL fast packet switch (*FPS*); ~ **de paquets voix-données** COMMUT, DONNÉES, TÉL, TRANSMIS voice-data packet switch; ~ **pas à pas** COMMUT step-by-step selector; ~ **de passage à zéro** COMMUT zero crossing switch; ~ **de point de transfert des signaux** COMMUT, SIGNAL, TÉL, TRANSMIS signal transfer point switch; ~ **principal** COMMUT, TÉL main exchange; ~ **privé** COMMUT, TÉL private branch exchange (*PBX*); ~ **privé haut débit** COMMUT, TÉL high bit-rate private switch; ~ **privé nodal** COMMUT, TÉL, TRANSMIS tandem PBX (*tandem private branch exchange*); ~ **privé de terminaison** COMMUT, CONTRÔLE stored-program control exchange (*AmE*), stored-programme control exchange (*BrE*), TÉL terminating PBX (*terminating private branch exchange*); ~ **à programme enregistré** COMMUT, CONTRÔLE stored-program control exchange (*BrE*), stored-programme control exchange (*BrE*); ~ **public** COMMUT, TÉL public exchange, central exchange switch (*BrE*), central office switch (*AmE*), public telephone exchange (*BrE*); ~ **à quatre fils** COMMUT four-wire switch, four-wire switching system; ~ **de rattachement** COMMUT home switching center (*AmE*), home switching centre (*BrE*), host exchange, main exchange (*BrE*), serving exchange; ~ **pour renvoi sur le service de nuit** COMMUT night-service key; ~ **de réseau privé** COMMUT, TÉL private network exchange (*PNX*); ~ **rotatif** GÉNÉRAL, INFORMAT rotary switch; ~ **rural** COMMUT rural switch, TÉL rural exchange; ~ **S** (*commutateur spatial*) COMMUT, TÉL S switch (*space switch*); ~ **sans cordons** COMMUT, TÉL cordless switchboard; ~ **des sauts** COMMUT, RADIO hopping switch; ~ **secondaire** INFORMAT auxiliary switch; ~ **séquentiel** ALIM ÉLEC, COMMUT sequence switch; ~ **de soutenu** INFORMAT sustain switch; ~ **spatial** (*commutateur S*) COMMUT, TÉL space switch (*S switch*); ~ **spatial multiplex** COMMUT, TÉL multiplex space switch; ~ **statique** ALIM ÉLEC, COMMUT static switch; ~ **sur satellite** COMMUT, RADIO satellite switch; ~ **T** (*commutateur temporel*) COMMUT, DONNÉES TSI (*time-slot interchanger*), time switch; ~ **téléphonique** COMMUT, TÉL central office, exchange, switching center (*AmE*), switching centre (*BrE*), switching exchange, telephone exchange; ~ **téléphonique privé** COMMUT, TÉL private branch exchange (*PBX*); ~ **télex** COMMUT, TÉL telex exchange, telex switching center (*AmE*), telex switching centre (*BrE*); ~ **temporel** (*commutateur T*) COMMUT, DONNÉES time-slot interchanger (*TSI*), time switch; ~ **tout à relais** COMMUT, TÉL relay system, relay-type switch; ~ **de transit** COMMUT, TÉL tandem exchange, transit exchange, transit switching center (*AmE*), transit switching centre (*BrE*); ~ **de transit international** COMMUT, TÉL international gateway exchange, international transit exchange (*INTTR*); ~ **de transit urbain** COMMUT, TÉL junction tandem exchange, tandem exchange; ~ **universel** COMMUT universal switch; ~ **urbain** COMMUT central exchange, metropolitan switch, TÉL central switching unit; ~ **vidéo** COMMUT video switch

commutation *f* GÉNÉRAL switching; ~ **des appels en cours** COMMUT call-in-progress switching; ~ **automatique** TESTS automatic changeover, TV *pour circuits* automatic switching; ~ **automatique sur liaison de réserve** COMMUT automatic protection switching (*APS*); ~ **satellite** COMMUT, RADIO satellite switching; ~ **de circuits** COMMUT circuit switched (*CS*), TÉL circuit switching (*CS*); ~ **à commande vocale** COMMUT, CONTRÔLE voice-controlled operation, voice-operated switching, TÉL voice-controlled operation; ~ **à deux fils** COMMUT two-wire switching; ~ **de données** COMMUT, DONNÉES, ÉLECTRON data switching; ~ **écoute** RADIO speech switchover; ~ **électronique** COMMUT, INFORMAT electronic switching; ~ **de liaison de données** DONNÉES, ÉLECTRON data-link switching (*DLS*); ~ **de longueur d'onde** COMMUT wavelength switching; ~ **de messages** COMMUT message switching; ~ **normal-secours** TESTS redundancy switching; ~ **numérique** COMMUT digital switching; ~ **optique** COMMUT, TRANSMIS optical switching; ~ **de paquets** (*CP*) COMMUT, DONNÉES, TÉL, TRANSMIS packet switching (*PS*); ~ **de paquets rapide** COMMUT, DONNÉES, TÉL, TRANSMIS fast packet switching; ~ **par cellules** RADIO cell relay; ~ **par paquets** DONNÉES, TÉL packet-mode operation; ~ **par répartition dans le temps** COMMUT, TRANSMIS time-division switching; ~ **par répartition en fréquence** COMMUT, TRANSMIS frequency-division switching; ~ **par trames** TRANSMIS frame relay; ~ **parole** RADIO speech switchover; ~ **de point d'essai et signalisation inter-registre** COMMUT, TESTS testing point switching and interregister signaling (*AmE*), testing point switching and interregister signalling (*BrE*); ~ **de point de test et signalisation inter-registre** COMMUT, TESTS testing point switching and interregister signaling (*AmE*), testing point switching and interregister signalling (*BrE*); ~ **de protection automatique** COMMUT automatic protection switching (*APS*); ~ **à quatre fils** COMMUT four-wire switching; ~ **redondante** TESTS redundancy switching; ~ **séquentielle** COMMUT sequential switching; ~ **spatiale** COMMUT, TÉL space-division switching; ~ **spatiale multiplexée** COMMUT, TÉL spacial multiplex switching; ~ **sur les équipements de secours** TESTS redundancy switching; ~ **sur liaison de réserve** COMMUT, TÉL protection switching; ~ **sur liaison de secours** COMMUT, TÉL protection switching; ~ **de taxe** COMMUT tariff switching; ~ **temporelle** COMMUT, DONNÉES, TÉL, TRANSMIS time switching, time-division switching; ~ **temporelle asynchrone** INFORMAT cell switching; ~ **temporelle multiplexée** COMMUT time multiplex switching; ~ **de transit** COMMUT trunk switching; ~ **de voies** COMMUT channel switching; ~ **des voies entre cellules** RADIO intercell switching

compacter *vt* DONNÉES compress, condense

compagnon: ~ **silencieux** *m* INFORMAT lurker

comparaison: ~ **directe** *f* TESTS direct comparison; ~ **logique** *f* INFORMAT logical comparison; ~ **temporelle** *f* RADIO, TRANSMIS time comparison; ~ **de temps** *f* RADIO, TRANSMIS time comparison

comparateur *m* CONTRÔLE comparing element, ÉLECTRON comparator, comparison circuit; ~ **de courant cryogénique** ÉLECTRON cryogenic current comparator; ~ **de signaux** RADIO signal comparator

compartiment: ~ **pour carte imprimée** *m* COMPOSANTS board compartment

compatibilité: ~ **ascendante** *f* INFORMAT upward compatibility; ~ **de couche inférieure** *f* COMMUT, DONNÉES, TÉL low-layer compatibility (*LLC*); ~ **de couche supérieure** *f* COMMUT, DONNÉES, TÉL high-layer compatibility (*HLC*); ~ **descendante** *f* INFORMAT downward compatibility

compatible: ~ **avec les appareils auditifs** *loc* APP SERV, PÉRIPH hearing-aid compatible; ~ **ETD** *loc* TRANSMIS data-permissible, data-permissive; ~ **RNIS** *loc* RADIO ISDN compatible

compensateur: ~ **d'amplitude** *m* INFORMAT amplitude equalizer; ~ **de distorsion de phase** *m* ALIM ÉLEC, GÉNÉRAL phase equalizer; ~ **d'écho** *m* ÉLECTRON, TRANSMIS echo canceler (*AmE*), echo canceller (*BrE*); ~ **de phase** *m* ALIM ÉLEC, ÉLECTRON phase equalizer; ~ **de retard** *m* INFORMAT delay equalizer; ~ **synchrone** *m* COMPOSANTS, ÉLECTRON synchro capacitor, synchronous capacitor, synchronous condenser

compensation *f* GÉNÉRAL, TRANSMIS balancing; ~ **de dérive** TÉL drift compensation; ~ **de la distorsion caractéristique** DONNÉES, TÉL characteristic distortion compensation; ~ **d'écho** ÉLECTRON, TRANSMIS echo cancelation (*AmE*), echo cancellation (*BrE*); ~ **de luminance** TV luminance compensation; ~ **de mouvement** TV motion compensation; ~ **par dopage** COMPOSANTS doping compensation; ~ **de phase** ALIM ÉLEC, ÉLECTRON phase compensation, phase equalization

compensé *adj* GÉNÉRAL balanced

compilateur *m* INFORMAT, TÉL compiler

compiler *vt* INFORMAT, TÉL compile

complément: ~ **d'étude** *m* DONNÉES further study (*FS*); ~ **de ligne numérique** *m* TRANSMIS digital pad; ~ **du service** *m* APP SERV complementary service

complémenter *vt* ÉLECTRON, INFORMAT negate

compléments: ~ **de ligne R ou T plage du terminal** *m pl* TÉL R or T pads

complexe: ~ **d'antenne** *m* RADIO antenna farm

comportement *m* DONNÉES behavior (*AmE*), behaviour (*BrE*); ~ **de l'abonné** TÉL subscriber's dialing habits (*AmE*), subscriber's dialling habits (*BrE*); ~ **par défaut** DONNÉES default behavior (*AmE*), default behaviour (*BrE*); ~ **progressif** INFORMAT, TÉL progressive behavior (*AmE*), progressive behaviour (*BrE*); ~ **séquentiel** INFORMAT, TÉL sequence behavior (*AmE*), sequence behaviour (*BrE*)

composant *m* COMPOSANTS, ÉLECTRON component; ~ **actif** ÉLECTRON active component; ~ **d'alliage** COMPOSANTS *métallurgie* alloying constituent; ~ **électronique** COMPOSANTS, ÉLECTRON, INFORMAT electronic component; ~ **étalon** COMPOSANTS, ÉLECTRON standard component; ~ **fondamental** TESTS fundamental component; ~ **harmonique** TRANSMIS harmonic component; ~ **de haute qualité** COMPOSANTS high-grade component; ~ **imprimé** ÉLECTRON printed component; ~ **de luminance** INFORMAT luminance component; ~ **monté en surface** (*CMS*) COMPOSANTS, ÉLECTRON surface-mounted component (*SMC*); ~ **normalisé** COMPOSANTS, ÉLECTRON standard component; ~ **d'onde de surface** ÉLECTRON surface-wave component; ~ **d'ondulation** ALIM ÉLEC ripple component; ~ **ondulatoire** ALIM ÉLEC ripple component; ~ **optoélectronique** COMPOSANTS optoelectronic component; ~ **passif** COMPOSANTS, ÉLECTRON passive component; ~ **porteur** ÉLECTRON carrier component; ~ **en quadrature** RADIO quadrature component; ~ **de régulation** CONTRÔLE *commande automatique* control component; ~ **à semiconducteur** COMPOSANTS semiconductor device; ~ **de service** INFORMAT service component; ~ **de signal** SIGNAL signal component; ~ **soudé** ÉLECTRON flying lead, flying-lead component; ~ **spectral** COMPOSANTS, TRANSMIS spectral component; ~ **unique autonome** COMPOSANTS, ÉLECTRON single device housing; ~ **Z** RADIO Z component

composante *f* ÉLECTRON component; ~ **de chrominance** TV chrominance component; ~ **conditionnelle** COMPOSANTS, ÉLECTRON conditional component; ~ **continue** TÉL DC component; ~ **continue non significative** ÉLECTRON non-useful d.c. component; ~ **de couleur** TÉL color component (*AmE*), colour component (*BrE*); ~ **discrète** COMPOSANTS discrete component

composants[1]: **à** ~ **liés** *adj* ALIM ÉLEC, COMPOSANTS strap-down

composants[2]: ~ **de diaphonie inintelligible** *m pl* TÉL unintelligible crosstalk components; ~ **de diaphonie intelligible** *m pl* TÉL intelligible crosstalk components; ~ **extra-bandes neutres** *m pl* TÉL harmless out-of-band components; ~ **extra-bandes nuisibles** *m pl* TÉL harmful out-of-band components

composer *vt* TÉL dial

composeur *m* TÉL dialer (*AmE*), dialler (*BrE*); ~ **de rythmes** INFORMAT rhythm programmer; ~ **téléphonique** TÉL automatic dialer (*AmE*), automatic dialler (*BrE*); ~ **vocal** TÉL voice dialer (*AmE*), voice dialler (*BrE*)

compréhensif *adj* DONNÉES all-in-one

compresser *vt* DONNÉES compress, INFORMAT compact, TÉL, TRANSMIS compress

compresseur *m* DONNÉES compressor, INFORMAT compactor, TÉL, TRANSMIS compressor; **~-extenseur** DONNÉES, SIGNAL compander, compressor-expander

compression[1] *f* DONNÉES compressing, INFORMAT compaction, TÉL, TRANSMIS compressing, compression; ~ **adaptative** INFORMAT adaptive compression; ~ **audio** INFORMAT audio compression; ~ **des débits** DONNÉES bit-rate compression; ~ **des données** DONNÉES data compression; ~ **dynamique** INFORMAT dynamic compression; **~-extension** DONNÉES, INFORMAT, TÉL, TRANSMIS companding; **~-extension avec modulation de fréquence** (*CMF*) TRANSMIS companding and frequency modulation (*CFM*); **~-extension par blocs** INFORMAT block companding; **~-extension syllabique** INFORMAT, TÉL, TRANSMIS syllabic companding; ~ **de fichier** DONNÉES, INFORMAT file compression; ~ **d'image** DONNÉES, TV image compression; ~ **d'impulsions** ÉLECTRON pulse compression; ~ **inter-image** TV interframe compression; ~ **intra-image** TV intraframe compression; ~ **de la largeur de bande** DONNÉES bandwidth compression; ~ **des luminances** TÉL compression of the luminance range; ~ **numérique** TV digital compression; ~ **de signal** SIGNAL signal compression; ~ **sonore** TÉL speech compression; ~ **spatiale** INFORMAT spatial compression; ~ **de télécopie** TÉL facsimile compression; ~ **temporelle** DONNÉES, INFORMAT temporal compression, time compression; ~ **de temps** DONNÉES, INFORMAT time compression; ~ **vidéo** DONNÉES, TV video compression

compression[2]: **avec** ~ *loc* DONNÉES, TÉL, TRANSMIS compressed

comprimer *vt* DONNÉES compress, condense, INFORMAT squeeze, zip, TÉL, TRANSMIS compress; **~-étendre** DONNÉES, INFORMAT, TÉL, TRANSMIS compand

comptage *m* APP SERV charging, TÉL metering; ~ **aléatoire** TÉL random metering; ~ **d'appels** TÉL call metering; ~ **de conversations** TÉL call metering; ~ **au hasard** TÉL random metering; ~ **multiple** TÉL multimetering, repetitive metering; ~ **multiple aléatoire** TÉL random repetitive metering; ~ **simple** TÉL single-fee metering, unit-charge metering

compte: ~ **Internet** *m* INFORMAT Internet account; ~ **de justification de pointeur** *m* DONNÉES pointer justification count (*PJC*); ~ **maritime** *m* TÉL maritime account; ~ **de l'opérateur** *m* ÉLECTRON operator's cashflow; ~ **de taxe** *m* TÉL charge meter account; **~-temps** *m* TÉL running time meter

compteur *m* GÉNÉRAL, TÉL counter, meter; ~ **d'abonné** TÉL subscriber's meter, subscriber's call meter; ~ **d'appels** TÉL call register (*CR*), call counter, call meter; ~ **d'appels non**

aboutis TÉL overflow meter; ~ **asynchrone** TESTS ripple-through counter, asynchronous counter; ~ **binaire** ÉLECTRON *éléments logiques* binary counter; ~ **de blocages** TÉL blocking counter; ~ **des blocs bien reçus** TÉL block-acknowledged counter; ~ **de conversations** TÉL call counter, call meter, call register (*CR*); ~ **de débordements** TÉL overflow meter; ~ **à décades** ÉLECTRON decade counter; ~ **de défaillances** TÉL error counter, failure counter; ~ **des dizaines** ÉLECTRON tens counter; ~ **d'encombrements de faisceau** COMMUT, TÉL congested-route counter, route congestion counter; ~ **de fréquence** TESTS frequency counter, frequency meter; ~ **horaire** APP SERV, TESTS hour meter, time meter; ~ **horaire électronique** TESTS electronic time meter; ~ **imprimant télétaxe** TÉL telephone charge printer; ~ **d'impulsions** ÉLECTRON, SIGNAL pulse counter, TÉL pulse meter; ~ **à induction** ALIM ÉLEC induction meter; ~ **d'intervalles de temps** ALIM ÉLEC interval timer; ~ **de Johnson** ÉLECTRON Johnson counter; ~ **de messages reçus** PÉRIPH, TÉL message received counter; ~ **modulo-n** ÉLECTRON modulo-n counter; ~ **de niveau des prises** TÉL traffic-level counter; ~ **personnel** PÉRIPH, TÉL subscriber's private meter; ~ **de perturbations** TÉL disturbance counter; ~ **de prises** TÉL seizure counter; ~ **à rebours** PÉRIPH countdown timer; ~ **de règlement des comptes** TÉL accounting counter; ~ **de renvois** COMMUT redirecting counter; ~ **à source liquide** RADIO liquid-sample counter; ~ **de statistiques** TESTS statistics counter; ~ **de taxes** APP SERV call charge counter, DONNÉES message register, TÉL tariff pulse meter; ~ **de taxes d'abonné** PÉRIPH, TÉL subscriber's meter; ~ **télétaxe** TÉL subscriber's check meter; ~ **de temps de fonctionnement** TÉL running-time meter; ~ **de trafic** ÉLECTRON, TÉL, TESTS traffic meter; ~ **des trames I** (*IC*) TÉL I-frame counter (*IC*); ~ **des unités** ÉLECTRON units counter

concaténation *f* DONNÉES concatenation, chaining

concentrateur *m* DONNÉES hub, ÉLECTRON, TRANSMIS concentrator (*CON*), hub; ~ **d'abonnés** TÉL, TRANSMIS line concentrator; ~ **d'abonnés local** TÉL, TRANSMIS collocated concentrator; ~ **arborescent** ÉLECTRON tree concentrator; ~ **de câblage** ÉLECTRON wiring concentrator; ~ **de câblage en boucle** ÉLECTRON loop wiring concentrator (*LWC*); ~ **de central** TÉL exchange concentrator; ~ **de central distant** TÉL remote-exchange concentrator; ~ **de central local** TÉL co-located exchange concentrator; ~ **distant** TRANSMIS remote concentrator; ~ **de données** DONNÉES data concentrator; ~ **hôte** INFORMAT host concentrator; ~ **de lignes** TÉL, TRANSMIS line concentrator; ~ **de lignes d'abonnés** TÉL, TRANSMIS local line concentrator; ~ **numérique** TÉL digital concentrator; ~ **numérique éloigné** (*CNE*) TÉL remote digital concentrator; ~ **numérique local** TÉL local digital concentrator; ~ **numérique satellite** (*CNS*) RADIO digital satellite concentrator; ~ **à paires torsadées** COMPOSANTS, ÉLECTRON twisted-pair hub; ~ **en radiocommunication numérique** RADIO digital radio concentrator; ~ **satellite électronique** RADIO electronic satellite concentrator; ~ **satellite numérique** RADIO remote concentration unit (*RCU*); ~ **de voies** TRANSMIS channel concentrator; ~ **de voies asynchrones** TRANSMIS access concentrator

concentration *f* ÉLECTRON concentration, focusing, GÉNÉRAL concentration; ~ **des conversations avec affectation temporelle** TRANSMIS time assignment speech concentration; ~ **dans un étage de commutation** COMMUT concentration in a switching stage; ~ **d'erreurs** DONNÉES error density; ~ **ionique** ÉLECTRON ion concentration; ~ **magnétique** ÉLECTRON magnetic focusing; ~ **numérique des conversations** (*CNC*) TÉL, TRANSMIS digital speech interpolation (*DSI*); ~ **numérique de la parole** (*CNP*) TÉL, TRANSMIS digital speech compression; ~ **de vapeur** CLIMAT vapor concentration (*AmE*), vapour concentration (*BrE*)

concept *m* TESTS concept; ~ **d'aptitude** TESTS *sécurité de fonctionnement* ability concept; ~ **de capacité** TESTS *sécurité de fonctionnement* performance concept; ~ **des grandes artères** RADIO major-path concept; ~ **de sécurité intrinsèque** TESTS fail-safe concept

concepteur *m* GÉNÉRAL designer

conception: ~ **assistée par ordinateur** *f* (*CAO*), COMPOSANTS electronic design automation (*EDA*), INFORMAT computer-aided design (*CAD*); ~ **de base** *f* RADIO baseline design; ~ **des circuits** *f* GÉNÉRAL circuit design; ~ **et fabrication assistées par ordinateur** *f* (*CFAO*) INFORMAT computer-aided design and computer-aided manufacturing (*CAD/CAM*); ~ **logique** *f* DONNÉES logic design; ~ **d'unités** *f* COMMUT, PÉRIPH unit design

concordance *f* DONNÉES matching, ÉLECTRON registration

condensateur *m* ALIM ÉLEC, COMPOSANTS, ÉLECTRON capacitor, condenser; ~ **d'accélération** ALIM ÉLEC, COMPOSANTS speed-up capacitor; ~ **ajustable** COMPOSANTS, ÉLECTRON trimmer capacitor; ~ **céramique** COMPOSANTS, ÉLECTRON ceramic capacitor; ~ **à couche métallique** COMPOSANTS, ÉLECTRON metal-film capacitor; ~ **à couche métallisée** COMPOSANTS, ÉLECTRON metalized-film capacitor (*AmE*), metallized-film capacitor (*BrE*); ~ **à couche mince** COMPOSANTS, ÉLECTRON thin-film capacitor, thin-layer capacitor; ~ **à couche d'oxyde** COMPOSANTS, ÉLECTRON oxide capacitor; ~ **de couplage** ALIM ÉLEC, ÉLECTRON coupling capacitor; ~ **de découplage** ALIM ÉLEC, ÉLECTRON decoupling capacitor; ~ **à deux cages** ALIM ÉLEC, COMPOSANTS two-gang capacitor, two-gang condenser; ~ **électrochimique** COMPOSANTS electrochemical capacitor; ~ **à électrolyte solide** COMPOSANTS solid electrolyte capacitor; ~ **électrolytique** COMPOSANTS electrolytic capacitor; ~ **étalon** ALIM ÉLEC standard capacitor; ~ **étouffeur d'étincelles** COMPOSANTS spark-quenching capacitor; ~ **d'extinction des étincelles** COMPOSANTS spark-quenching capacitor; ~ **à film métallique** COMPOSANTS, ÉLECTRON metal-film capacitor; ~ **à film métallisé** COMPOSANTS, ÉLECTRON metalized-film capacitor (*AmE*), metallized-film capacitor (*BrE*); ~ **à film plastique** ALIM ÉLEC, COMPOSANTS plastic-film capacitor, plastic-foil capacitor; ~ **de filtrage** ALIM ÉLEC, COMPOSANTS filter capacitor, smoothing capacitor; ~ **à huile** COMPOSANTS, ÉLECTRON oil-filled capacitor; ~ **à jonction** ÉLECTRON junction capacitor; ~ **de lissage** ALIM ÉLEC, COMPOSANTS smoothing capacitor; ~ **au mica** COMPOSANTS, ÉLECTRON mica capacitor; ~ **au mica argenté** COMPOSANTS, ÉLECTRON silver mica capacitor; ~ **au mylar** ÉLECTRON mylar capacitor; ~ **de neutrodynage** ÉLECTRON neutralizing capacitor; ~ **NPO** ALIM ÉLEC NPO-type capacitor; ~ **au papier** ALIM ÉLEC, COMPOSANTS paper capacitor; ~ **au papier métallisé** ALIM ÉLEC, COMPOSANTS metalized-paper capacitor (*AmE*) (*MP capacitor*), metallized-paper capacitor (*BrE*) (*MP capacitor*); ~ **en parallèle** ALIM ÉLEC, COMPOSANTS shunt capacitor; ~ **pare-étincelles** COMPOSANTS spark-quenching capacitor; ~ **au tantale** COMPOSANTS, ÉLECTRON tantalum capacitor; ~ **de traversée** ALIM ÉLEC feedthrough capacitor; ~ **à trois cages** ALIM ÉLEC three-gang condenser; ~ **tubulaire** ALIM ÉLEC, COMPOSANTS tubular capacitor; ~ **variable** COMPOSANTS, ÉLECTRON variable capacitor; ~ **variable à air** COMPOSANTS, ÉLECTRON variable air capacitor

condition *f* GÉNÉRAL condition; ~ **A** TÉL condition; ~ **d'acceptation** TESTS *qualité* acceptance condition; ~ **acceptation d'appel** DONNÉES, TÉL call-accepted condition; ~ **de conformité dynamique** DONNÉES dynamic conformance requirement; ~ **de conformité statique** DONNÉES static conformance requirement; ~ **de dérangement** TÉL fault condition; ~ **initiale** ÉLECTRON initial condition; ~ **de refus** TESTS rejective condition; ~ **de régime permanent** ÉLECTRON, TRANSMIS steady-state condition; ~ **de rejet** TESTS rejective condition; ~ **de repos**

intercalaire entre les caractères TÉL intercharacter rest condition; ~ **Z** TÉL condition Z

conditions: ~ **ambiantes** *f pl* CLIMAT environmental requirements; ~ **d'exploitation** *f pl* TÉL operating conditions, operational conditions; ~ **de fonctionnement** *f pl* ALIM ÉLEC service conditions, ÉLECTRON operating conditions, hot conditions, operational conditions, TÉL operating conditions, operational conditions; ~ **d'injections d'équilibre** *f pl* ÉLECTRON steady-state launching conditions; ~ **de propagation en espace libre** *f pl* RADIO free-space conditions; ~ **de réception** *f pl* INFORMAT receive conditions, reception conditions; ~ **de référence** *f pl* TESTS reference conditions; ~ **de signalisation** *f pl* SIGNAL signaling conditions (*AmE*), signalling conditions (*BrE*); ~ **de transmission** *f pl* INFORMAT, TRANSMIS transmission conditions; ~ **de veille** *f pl* GÉNÉRAL stand-by conditions

conductance *f* ALIM ÉLEC, ÉLECTRON conductance; ~ **électrique** ÉLECTRON electrical conductance; ~ **d'électrode** ÉLECTRON electrode conductance

conducteur[1] *adj* ALIM ÉLEC, ÉLECTRON conducting, conductive

conducteur[2] *m* GÉNÉRAL, TÉL conductor; ~ **affleurant** ÉLECTRON flush conductor; ~ **annulaire commun de mise à la terre** ALIM ÉLEC collective earth system (*BrE*), collective ground system (*AmE*); ~ **central** TÉL inner conductor, TRANSMIS inner conductor, central conductor; ~ **coaxial** TRANSMIS concentric conductor; ~ **de courant de retour** ALIM ÉLEC current return conductor; ~ **creux** TRANSMIS hollow conductor; ~ **de cuivre** ÉLECTRON copper conductor; ~ **encastré** ÉLECTRON flush conductor; ~ **étamé** COMPOSANTS, ÉLECTRON tin-plated conductor; ~ **extérieur** TÉL, TRANSMIS outer conductor; ~ **imprimé** ÉLECTRON printed conductor; ~ **intérieur** TÉL, TRANSMIS inner conductor; ~ **isolé** COMPOSANTS insulated conductor, TRANSMIS *câbles* core; ~ **de masse** ALIM ÉLEC earth conductor (*BrE*), ground conductor (*AmE*), earthing conductor (*BrE*), grounding conductor (*AmE*); ~ **massif** COMPOSANTS solid conductor; ~ **métallisé** ÉLECTRON metal-clad conductor; ~ **monobrin en cuivre** COMPOSANTS solid copper-weld conductor; ~ **multifilaire** COMPOSANTS, ÉLECTRON multistrand wire, multiwire conductor; ~ **neutre** ALIM ÉLEC, GÉNÉRAL neutral conductor, neutral wire; ~ **de nuque** COMPOSANTS ring, TÉL B-leg; ~ **de phase** ALIM ÉLEC phase conductor; ~ **de pointe** TÉL A-leg; ~ **profilé** COMPOSANTS shape conductor; ~ **de retour** ALIM ÉLEC return conductor; ~ **à revêtement métallique** ÉLECTRON metal-clad conductor; ~ **sectoral** COMPOSANTS sector-shaped conductor; ~ **souple** ALIM ÉLEC flexible conductor; ~ **de terre** ALIM ÉLEC earth conductor (*BrE*), earthing conductor (*BrE*), ground conductor (*AmE*), grounding conductor (*AmE*); ~ **torsadé** COMPOSANTS, ÉLECTRON stranded conductor; ~ **unique** COMPOSANTS, ÉLECTRON single conductor, single wire; ~ **visuel électronique** INFORMAT electronic storyboard

conducteur[3]: ~ **sous tension** *loc* ALIM ÉLEC live wire

conductibilité *f* COMPOSANTS, ÉLECTRON conductivity: ~ **thermique** *f* COMPOSANTS, ÉLECTRON thermal conductivity

conduction *f* GÉNÉRAL conduction; ~ **extrinsèque** COMPOSANTS, ÉLECTRON extrinsic conduction; ~ **intrinsèque** COMPOSANTS, ÉLECTRON intrinsic conduction; ~ **par électrons** COMPOSANTS, ÉLECTRON electron conduction; ~ **type N** COMPOSANTS, ÉLECTRON N-type conduction

conduit *m* COMPOSANTS, ÉLECTRON, TRANSMIS conduit, duct, pipe; ~ **élevé** RADIO elevated duct; ~ **fictif numérique de référence** (*CFNR*) RADIO, TÉL, TRANSMIS hypothetical reference digital path (*HRDP*); ~ **hertzien numérique** RADIO, TÉL digital radio path; ~ **interne** ÉLECTRON inner duct; ~ **de ligne numérique** TÉL, TRANSMIS digital line link, digital line path; ~ **monotubulaire** COMPOSANTS, TRANSMIS single-duct block; ~ **multitubulaire** COMPOSANTS, TRANSMIS multiple-duct block; ~ **numérique** RADIO, TÉL, TRANSMIS digital path; ~ **au sol** RADIO surface duct; ~ **de surface** RADIO ground-based duct; ~ **troposphérique** RADIO tropospheric radioduct; ~ **virtuel** DONNÉES, TRANSMIS virtual path (*VP*); ~ **virtuel ATM** APP SERV ATM virtual path

conférence *f* RADIO, TÉL conference; ~ **établie par un poste** RADIO station conference; ~ **établie par standardiste** TÉL attendant-controlled conference; ~ **rendez-vous** APP SERV meet-me conference; ~ **téléphonique** APP SERV, TÉL phone conference, conference call, telephone conference; ~ **à trois** APP SERV, TÉL three-party conference; ~ **à trois en chaîne** TÉL three-way call chaining

Conférence: ~ **administrative mondiale des radiocommunications** *f* (*CAMR*) GÉNÉRAL World Administrative Radiocommunications Conference; ~ **européenne des administrations des postes et des télécommunications** *f* (*CEPT*) GÉNÉRAL European Conference of Postal and Telecommunications Administrations (*ECPT*); ~ **internationale des grands réseaux électriques à haute tension** *f* (*CIGRE*) RADIO International Conference on Large High Voltage Electric Systems (*CIGRE*)

configuration *f* GÉNÉRAL configuration; ~ **en anneau** TRANSMIS ring configuration; ~ **en boucle** TRANSMIS ring configuration; ~ **à écrans multiples** TV multiscreen configuration; ~ **d'électrode** ÉLECTRON electrode configuration; ~ **d'erreurs** DONNÉES error pattern; ~ **et gestion des données** *f pl* DONNÉES configuration and data management; ~ **en étoile** TRANSMIS star configuration; ~ **de fond** ÉLECTRON backplane; ~ **de l'impression** INFORMAT print settings; ~ **en matrice** ÉLECTRON matrix configuration; ~ **mémoire** INFORMAT memory configuration; ~ **multimédia** INFORMAT multimedia configuration; ~ **par bus** INFORMAT bus configuration; ~ **des perforations** COMPOSANTS, ÉLECTRON hole pattern; ~ **de relais pontés** TÉL bridged-relay configuration; ~ **du réseau** TÉL network configuration; ~ **du système** INFORMAT system configuration; ~ **totale** COMMUT total configuration; ~ **de type conférence** TV conferencing configuration

configurer *vt* INFORMAT configure

confirmation *f* (*conf*) GÉNÉRAL confirmation (*conf*); ~ **de connexion** SIGNAL connect confirm (*CC*); ~ **correspondante** DONNÉES matching order; ~ **de libération** TÉL clear confirmation, clearing confirmation; ~ **de message** DONNÉES message confirmation (*MCF*); ~ **primitive** (*conf*) DONNÉES, TÉL, confirmation (*conf*), primitive confirm (*conf*); ~ **de réception** DONNÉES, TÉL confirmation of receipt (*COR*); ~ **de remise** TÉL delivery confirmation; ~ **de remise de datagramme** DONNÉES, SIGNAL datagram delivery data confirmation

conflit *m* GÉNÉRAL contention; ~ **d'accès** GÉNÉRAL contention

conformateur *m* GÉNÉRAL shaper; ~ **d'impulsions** TÉL pulse shaper

conformité: ~ **induite** *f* DONNÉES dependent conformance

confort: ~ **téléphonique** *m* RADIO, TÉL phoning comfort

congestion: ~ **temporelle** *f* RADIO, TÉL time congestion

conique *adj* COMPOSANTS tapered, RADIO conical

conjoncteur *m* ALIM ÉLEC circuit closer, COMMUT, COMPOSANTS, PÉRIPH, TÉL telephone jack

conjonction *f* ÉLECTRON *éléments logiques* conjunction

connectable: ~ **par cavalier** *adj* ALIM ÉLEC strappable

Connect-Additional *m* TÉL Connect-Additional

connecté *adj* DONNÉES, INFORMAT, TÉL connected, on-line; ~ **à distance** TÉL remotely-connected; ~ **en permanence** GÉNÉRAL permanently connected; ~ **en série** GÉNÉRAL series-connected

connecter[1] *vt* ALIM ÉLEC switch, switch on, connect, wire; ~ **en boucle** ÉLECTRON connect in a loop; ~ **en dérivation** ALIM ÉLEC *électricité* bypass; ~ **directement** COMMUT put through, through-connect, TÉL put through; ~ **en guirlande** ÉLECTRON, INFORMAT daisy-chain; ~ **les jarretières** ÉLECTRON jumper up; ~ **en parallèle** ALIM ÉLEC, ÉLECTRON connect in parallel; ~ **en série** ÉLECTRON connect in series

connecter[2]: **se** ~ *v réfl* DONNÉES get connected; **se** ~ **à** *v réfl* RADIO log onto, log on, sign onto, sign on

connecteur *m* GÉNÉRAL connector, INFORMAT slot; ~ **blindé** ÉLECTRON shielded connector; ~ **de bord** ÉLECTRON edge connector; ~ **de bout de carte** ÉLECTRON edge connector; ~ **bout-à-bout** COMPOSANTS splice connector; ~ **de données universel** ÉLECTRON universal data connector (*UDC*); ~ **à double usage** ÉLECTRON dual-purpose connector; ~ **encartable** ÉLECTRON edge connector; ~ **enfichable** COMPOSANTS, ÉLECTRON plug-in connector; ~ **d'entrée** GÉNÉRAL in-connector; ~ **européen** COMPOSANTS European-type connector; ~ **d'extension** INFORMAT expansion slot; ~ **à faible perte** TRANSMIS low-loss connector; ~ **femelle** ÉLECTRON, INFORMAT, TESTS female plug; ~ **à fibres à faible perte** TRANSMIS low-loss fiber connector (*AmE*), low-loss fibre connector (*BrE*); ~ **à fibres optiques** TRANSMIS optical fiber connector (*AmE*), optical fibre connector (*BrE*); ~ **fixe à contacts lyre** COMPOSANTS fixed fork connector; ~ **fixe multibroche** COMPOSANTS fixed blade connector, fixed pin connector; ~ **hybride** TRANSMIS hybrid connector; ~ **interconnectable** TRANSMIS intermateable connector; ~ **interne** ÉLECTRON internal connecter; ~ **libre** COMPOSANTS, ÉLECTRON free connector; ~ **libre à contacts lyre** COMPOSANTS, ÉLECTRON free fork connector; ~ **libre à douilles** COMPOSANTS, ÉLECTRON free sleeve connector; ~ **libre multibroche** COMPOSANTS, ÉLECTRON free blade connector, free pin connector; ~ **mâle** ÉLECTRON, INFORMAT, TESTS male plug; ~ **multibroche** COMPOSANTS, ÉLECTRON multipoint connector; ~ **multicontact** COMPOSANTS, ÉLECTRON multipin connector; ~ **multifibre** TRANSMIS multifiber connector (*AmE*), multifibre connector (*BrE*); ~ **plat** GÉNÉRAL edge connector; ~ **série** INFORMAT serial port; ~ **de sortie** GÉNÉRAL out connector; ~ **spécial** COMPOSANTS special connector; ~ **de station en boucle** ÉLECTRON loop station connector (*LSC*); ~ **symétrique à lignes coaxiales** TRANSMIS coaxial push-pull connector; ~ **à vis** COMPOSANTS socket connector block

connecteurs: ~ **associés** *m pl* ÉLECTRON associated connectors; ~ **biconiques** *m pl* ÉLECTRON biconic connectors; ~ **homologues** *m pl* GÉNÉRAL mating connectors; ~ **interchangeables** *m pl* TRANSMIS interchangeable connectors; ~ **multimode** *m pl* TRANSMIS multiple-mode connectors; ~ **rainurés** *m pl* ÉLECTRON array connectors

connectez: ~ **le poste de données sur la ligne** *loc* DONNÉES connect-data set to line

Connect-Initial *m* TÉL Connect-Initial

connectique *f* INFORMAT, TRANSMIS interfacing

connectivité *f* INFORMAT connectivity

Connect-Response *m* TÉL Connect-Response

Connect-Result *m* TÉL Connect-Result

connexion *f* ALIM ÉLEC, COMPOSANTS wiring, DONNÉES, ÉLECTRON connection, INFORMAT connection, hook-up, TÉL connection, TRANSMIS *câbles* branch joint, connection; ~ **ATM** INFORMAT ATM connection; ~ **automatique en séquence** TÉL automatic sequential connection; ~ **en boucle** TÉL trombone connection, trombone loop connection; ~ **brasée** COMPOSANTS soldered connection; ~ **de canal virtuel** DONNÉES, TRANSMIS virtual-channel connection (*VCC*); ~ **de conduit d'ordre inférieur** TÉL, TRANSMIS lower-order path connection (*LPC*); ~ **de conduit d'ordre supérieur** TÉL, TRANSMIS higher-order path connection (*HPC*); ~ **de conduit virtuel** DONNÉES, TRANSMIS virtual-path connection (*VPC*); ~ **à la demande** DONNÉES dial-up connection; ~ **en dial-up** DONNÉES dial-up connection; ~ **dirigée** TÉL directed connection; ~ **à distance** ÉLECTRON remote log-in; ~ **à encliquetage** COMPOSANTS snap connection; ~ **enroulée** COMPOSANTS wire-wrap connection, GÉNÉRAL wrapped connection; ~ **entre couches** ÉLECTRON interlayer connection; ~ **étoilée** TRANSMIS star connection; ~ **à fil nu** TÉL bare-wire strap; ~ **inductive** ALIM ÉLEC impedance bond; ~ **informatique** DONNÉES data-processing connection; ~ **Internet PPP** DONNÉES PPP Internet connection; ~ **à intervalles de temps multiples** TÉL multislot connection; ~ **des jarretières** ÉLECTRON jumpering; ~ **lâche** ALIM ÉLEC loose connection; ~ **MCS** TÉL MCS connection; ~ **N** TÉL N-connection; ~ **numérique** TRANSMIS digital connection; ~ **par câble** TRANSMIS cable connection; ~ **par commutation de circuits** COMMUT circuit-switched connection; ~ **par enroulement** COMPOSANTS wire-wrapping; ~ **par fiche et jack** ALIM ÉLEC plug-and-jack connection; ~ **pas à pas** ALIM ÉLEC step-by-step connection; ~ **permanente** DONNÉES permanent connection; ~ **physique** TRANSMIS physical connection (*PhC*); ~ **point à multipoint** APP SERV, RADIO, TRANSMIS point-to-multipoint connection; ~ **de pointeur** COMMUT pointer connection; ~ **pour liaison de données** DONNÉES data-link connection (*DLC*); ~ **de présentation** DONNÉES presentation connection; ~ **progressive** ALIM ÉLEC step-by-step connection; ~ **radiophonique et télévisuelle** TV sound and television program connection (*AmE*), sound and television programme connection (*BrE*); ~ **rainurée** ÉLECTRON array connection; ~ **de réseau** DONNÉES, INFORMAT, TÉL network connection (*NC*); ~ **du réseau de terre radial** ALIM ÉLEC earth radial termination (*BrE*), ground radial termination (*AmE*); ~ **sauvage** INFORMAT wild connection; ~ **semi permanente** TÉL semipermanent connection; ~ **de session** DONNÉES session connection; ~ **soudée** COMPOSANTS soldered connection; ~ **superficielle** ÉLECTRON surface connection; ~ **TC additionnelle** TÉL additional TC; ~ **TC initiale** TÉL initial TC; ~ **de terre** ALIM ÉLEC earth connection (*BrE*), ground connection (*AmE*); ~ **de trajet de communication** TÉL connection communication path; ~ **de trajet virtuel** DONNÉES, TRANSMIS virtual-path connection (*VPC*); ~ **de transport** TÉL transport connection (*TC*); ~ **transversale** ALIM ÉLEC, ÉLECTRON clinched-wire through connection, through connection; ~ **transversale à fil** ALIM ÉLEC, ÉLECTRON wire through connection; ~ **à travers un central** COMMUT connection through an exchange; ~ **via modem** DONNÉES, ÉLECTRON dial-up; ~ **virtuelle** TÉL virtual connection; ~ **à vis** ALIM ÉLEC screwed connection; ~ **de voie virtuelle** DONNÉES, TRANSMIS virtual-channel connection (*VCC*), virtual-path connection (*VPC*); ~ **volante** ALIM ÉLEC, COMPOSANTS jumper, strap

connexité *f* INFORMAT, TRANSMIS, TV connectivity; ~ **numérique** TRANSMIS digital connectivity; ~ **numérique de bout en bout** ÉLECTRON end-to-end digital connection

Conseil: ~ **de la radiodiffusion et des télécommunications canadiennes** *m* (*CRTC*) RADIO, TÉL, TV Canadian Radio, Television and Telecommunications Commission (*CRTC*)

consignation: ~ **de reprise** *f* DONNÉES recovery log

console *f* INFORMAT console; ~ **d'affichage** INFORMAT display console; ~ **d'exploitation** TÉL operator console; ~ **de mélange** INFORMAT mixing console, mixing desk; ~ **de mixage** INFORMAT mixing console, mixing desk; ~ **à paiement** PÉRIPH pay console; ~ **de système** TÉL system console; ~ **de table** RADIO desk cassette, table cassette; ~ **télématique** DONNÉES communication console; ~ **de télétraitement** DONNÉES remote console; ~ **de visualisation** TV display console, display unit

consommation *f* GÉNÉRAL consumption; ~ **compteur** TÉL meter reading; ~ **de courant** ALIM ÉLEC current consumption, current drain; ~ **électrique** ALIM ÉLEC, ÉLECTRON, GÉNÉRAL, INFORMAT power consumption; ~ **d'énergie** ALIM ÉLEC energy consumption, power consumption, ÉLECTRON power consumption, GÉNÉRAL energy consumption, power consumption, INFORMAT power consumption;

~ **secteur** ALIM ÉLEC mains outlet, mains voltage outlet; ~ **téléphonique** TÉL telephone consumption

constante: ~ **d'affaiblissement** *f* RADIO, TRANSMIS attenuation constant; ~ **d'affaiblissement d'image** *f* TRANSMIS image attenuation constant; ~ **diélectrique** *f* ALIM ÉLEC dielectric constant; ~ **de diffusion** *f* COMPOSANTS diffusion constant; ~ **de dissociation électrolytique** *f* GÉNÉRAL electrolytic dissociation constant; ~ **d'erreur** *f* CONTRÔLE error constant; ~ **figurative** *f* INFORMAT figurative constant; ~ **d'un instrument de mesure** *f* TESTS instrument constant; ~ **de phase** *f* ÉLECTRON, GÉNÉRAL, RADIO, TRANSMIS phase coefficient, phase constant, phase-change coefficient; ~ **de propagation** *f* ÉLECTRON, RADIO, TRANSMIS propagation coefficient, propagation constant; ~ **temporelle** *f* ÉLECTRON, INFORMAT, TESTS, TRANSMIS time constant

constructeur *m* GÉNÉRAL manufacturer

construction: ~ **ATTACH** *f* DONNÉES ATTACH construct

consultable *adj* DONNÉES *Internet* accessible

consultation *f* INFORMAT browsing, RADIO interrogation; ~ **séparée des correspondants** TÉL call splitting

consulter *vt* RADIO interrogate

contact *m* ALIM ÉLEC, COMPOSANTS, ÉLECTRON contact; ~ **à accompagnement** ALIM ÉLEC, COMPOSANTS trailing contact; ~ **à action rapide** COMPOSANTS snap-action contact; ~ **auto-interrupteur** COMPOSANTS self-breaking contact; ~ **autonettoyant** COMPOSANTS self-cleaning contact; ~ **auxiliaire** ALIM ÉLEC auxiliary contact; ~ **à chevauchement** ÉLECTRON make-before-break contact; ~ **à double fermeture** COMPOSANTS, ÉLECTRON double-make contact; ~ **à double rupture** COMPOSANTS, ÉLECTRON double-break contact; ~ **à douille** COMPOSANTS sleeve contact; ~ **élastique** COMPOSANTS spring contact; ~ **d'extrémité de carte** DONNÉES finger, ÉLECTRON edge-board contact; ~ **à faible inertie** COMPOSANTS snap-action contact; ~ **de fin de papier** PÉRIPH paper-out switch; ~ **glissant** COMPOSANTS sliding contact, wiping contact, INFORMAT slide contact; ~ **imprimé** ÉLECTRON printed contact; ~ **imprimé d'extrémité** ÉLECTRON printed edge-connector; ~ **imprimé d'extrémité de carte** ÉLECTRON printed edge-board contact; ~ **d'interruption de tonalité** TÉL tone code sender; ~ **lâche** ALIM ÉLEC loose contact; ~ **à lames souples** COMPOSANTS reed contact; ~ **lyre** COMPOSANTS, ÉLECTRON fork contact; ~ **de maintien** COMPOSANTS holding contact; ~ **mâle** COMPOSANTS, ÉLECTRON pin contact, male contact; ~ **au mercure** COMPOSANTS, ÉLECTRON mercury contact; ~ **de numérotation** TÉL pulsing contact; ~ **ohmique** COMPOSANTS ohmic contact; ~ **de passage** COMPOSANTS passing contact; ~ **ponctuel** COMPOSANTS point contact; ~ **à pression** COMMUT, COMPOSANTS pressure contact; ~ **à pression directe** COMMUT, COMPOSANTS, TÉL butt contact; ~ **de relais** COMPOSANTS relay contact; ~ **sabre** COMPOSANTS, ÉLECTRON blade contact; ~ **scellé** COMPOSANTS reed; ~ **à semiconducteur** COMPOSANTS semiconductor contact; ~ **simple** COMPOSANTS single contact; ~ **à la terre** ALIM ÉLEC accidental earthing (*BrE*), accidental grounding (*AmE*)

contacteur *m* COMMUT, ÉLECTRON contact unit, contactor

contacts: ~ **doublés** *m pl* COMPOSANTS twin contacts; ~ **jumelés** *m pl* COMPOSANTS twin contacts

contenance: ~ **d'un câble** *f* TRANSMIS cable capacity

conteneur *m* APP SERV, DONNÉES, GÉNÉRAL, TÉL container (*CT*); ~-**n** (*C-n*) APP SERV, DONNÉES, TÉL container-n (*C-n*); ~-**n virtuel** APP SERV, DONNÉES, TÉL virtual container-n (*VC-n*); ~ **virtuel** (*CTV*) APP SERV, DONNÉES, TÉL virtual container (*VC*)

contexte *m* ÉLECTRON background; ~ **par défaut** DONNÉES default context; ~ **de présentation** DONNÉES presentation context; ~ **de taxation** TÉL call charge context

continu *adj* RADIO seamless

contour *m* GÉNÉRAL, INFORMAT *d'image* outline; ~ **de caractère** APP SERV character outline; ~ **de classe A** RADIO A-grade contour; ~ **de classe B** RADIO B-grade contour; ~ **de protection** TV protected contour

contournement *m* ALIM ÉLEC flashover, bypass, GÉNÉRAL bypass, bypassing

contourner *vt* ALIM ÉLEC flash over, *électricité* bypass, GÉNÉRAL bypass

contracter: **se** ~ *v réfl* COMPOSANTS shrink

contraction *f* COMPOSANTS reduction of area, shrinkage, shrinking

contrainte *f* COMPOSANTS stress, DONNÉES constraint; ~ **d'environnement** CLIMAT environmental constraints; ~ **de fonctionnement** TESTS functional stress; ~ **de licence** DONNÉES licence restriction (*BrE*), license restriction (*AmE*); ~ **modifiée** DONNÉES modified constraint

contraste *m* GÉNÉRAL contrast; ~ **accentué** TV accentuated contrast; ~ **de l'indice de réfraction** TRANSMIS refractive-index contrast; ~ **de luminance** TV luminance contrast; ~ **radiographique** RADIO radiographic contrast

contre: ~-**mesure électronique** *f* RADIO electronic countermeasure; ~-**pression** *f* GÉNÉRAL *physique* back pressure; ~-**réaction** *f* ALIM ÉLEC, CONTRÔLE negative feedback, DONNÉES, ÉLECTRON, RADIO feedback, negative feedback; ~-**vérification** *f* INFORMAT cross-check, TÉL, TESTS retest

contrefaire *vt* INFORMAT, TV pirate

contrefait *adj* INFORMAT, TV pirated

contribution: **à** ~ **de bruit nul** *loc* TRANSMIS noise-free

contributions: ~ **par satellite** *f pl* RADIO, TV satellite news gathering (*SNG*)

contrôlabilité *f* CONTRÔLE controllability

contrôle *m* GÉNÉRAL control, monitoring, inspection; ~ **d'accès** DONNÉES, TESTS *sécurité* access control; ~ **d'accès à l'anneau** ÉLECTRON ring-access control; ~ **d'accès au support** CONTRÔLE medium access control (*MAC*); ~ **d'affaiblissement** TRANSMIS attenuation control; ~ **automatique d'affaiblissement** TRANSMIS automatic attenuation regulation; ~ **automatique de l'altitude par radar de bord** RADIO radar altitude control mode; ~ **automatique de fréquence** (*CAF*) ÉLECTRON, TÉL, TRANSMIS automatic frequency control (*AFC*); ~ **automatique de gain** (*CAG*) ÉLECTRON, TÉL, TRANSMIS automatic gain control (*AGC*); ~ **automatique instantanée de gain** (*CAIG*) ÉLECTRON, TÉL, TRANSMIS instantaneous automatic gain control (*IAGC*); ~ **de bloc** INFORMAT block check; ~ **de bruit** TÉL noise monitoring; ~ **du câblage** COMPOSANTS wiring check; ~ **au casque** TÉL headphones monitoring; ~ **de charge de ligne** (*CCL*) COMMUT, TÉL line load control (*LLC*); ~ **de la circulation aérienne** RADIO air traffic control; ~ **de continuité à travers un commutateur** COMMUT cross-office check; ~ **en cours de fabrication** TESTS in-process inspection; ~ **de dépassement de capacité** INFORMAT overflow check; ~ **dynamique de la charge** COMMUT dynamic load control; ~ **d'enregistrement** RADIO playback, replay, reproduction; ~ **d'entrée** TESTS incoming inspection, receiving inspection; ~ **d'erreur sur en-tête** DONNÉES header error control (*HEC*); ~ **d'état** CONTRÔLE status monitoring, RADIO status check, TESTS status monitoring; ~ **d'exactitude** INFORMAT accuracy control; ~ **final** TESTS final inspection; ~ **de flux** CONTRÔLE, DONNÉES, TÉL flow control; ~ **de flux d'émission** TÉL, TRANSMIS transmit flow control; ~ **de flux générique** CONTRÔLE, TRANSMIS *à l'interface usager-réseau* generic flow control (*GFC*); ~ **de flux de trafic** DONNÉES, TÉL traffic flow control; ~ **de flux de trafic sémaphore** DONNÉES, SIGNAL signaling-traffic flow control (*AmE*), signalling-traffic flow control (*BrE*); ~ **de flux de transmission** TÉL, TRANSMIS transmit flow control; ~ **de fréquence** RADIO frequency control; ~ **des fréquences du clavier** TESTS dial test; ~ **d'imparité** DONNÉES, TÉL odd parity check; ~ **des impulsions du cadran** TESTS dial test; ~ **incorporé** TESTS built-in check; ~ **d'interruption** TÉL interruption control; ~ **de liaison synchrone de données** (*SDLC*) DONNÉES synchronous data link control (*SDLC*); ~ **de livraison** TESTS delivery inspection; ~ **local** TÉL local record; ~ **longitudinal par redondance** DONNÉES longitudinal redundancy check; ~ **de la mémoire** DONNÉES storage control; ~ **de niveau** RADIO level monitoring; ~ **normal** TESTS normal inspection; ~ **par attributs** TESTS inspection by attributes; ~ **par balance carrée** INFORMAT cross check; ~ **par caractère** TÉL character check; ~ **par cliquet** ÉLECTRON ratchet control; ~ **par échantillonnage** TESTS sampling inspection; ~ **par mesures** TESTS inspection by variables; ~ **par redondance** DONNÉES redundancy check; ~ **par relecture** INFORMAT read-back check; ~ **par retour de l'information** TÉL message feedback; ~ **par sondage** TESTS spot check; ~ **par tonalité pilote** TRANSMIS pilot-tone monitoring; ~ **par totalisation** INFORMAT summation check; ~ **de parité** DONNÉES, TÉL even parity check, parity check; ~ **des périphériques** DONNÉES, ÉLECTRON, INFORMAT device control; ~ **de pertinence** DONNÉES relevance feedback; ~ **à plusieurs variables** TESTS multivariable control; ~ **en première présentation** TESTS original inspection; ~ **de présence** INFORMAT null-data check; ~ **préventif** TESTS preventive inspection; ~ **de processus** CONTRÔLE process control, TESTS process control, process inspection; ~ **de programmation** INFORMAT desk checking; ~ **de qualité** APP SERV, TESTS quality audit, quality control; ~ **de réception** TESTS *qualité* acceptance inspection; ~ **de redondance cyclique** (*CRC*) DONNÉES, TÉL cyclic redundancy check (*CRC*); ~ **réduit** TESTS reduced inspection; ~ **renforcé** CONTRÔLE, TESTS tightened inspection; ~ **réparti** DONNÉES distributed control; ~ **de restriction d'accès au réseau** CONTRÔLE, TÉL trunk discrimination control; ~ **de routage** DONNÉES routing control; ~ **séquentiel** RADIO sequential monitoring; ~ **de surcharge** TÉL overload control; ~ **de système** INFORMAT system control; ~ **de télévision** TESTS, TV television monitoring; ~ **transversal par redondance** DONNÉES transverse redundancy check; ~ **de validité** CONTRÔLE, DONNÉES validity check; ~ **volant** TESTS patrol inspection

contrôler *vt* CONTRÔLE, TÉL monitor, TESTS inspect

contrôleur *m* GÉNÉRAL controller; ~ **de communications** DONNÉES data communications controller; ~ **d'écran** DONNÉES display board; ~ **d'entrée/sortie** INFORMAT input/output controller; ~ **d'E/S** INFORMAT I/O controller; ~ **de fréquence** RADIO frequency monitor; ~ **d'imprimante** DONNÉES, INFORMAT printer controller; ~ **de lecteur de disque** INFORMAT disk drive controller; ~ **à logique programmable** DONNÉES programmable logic controller (*PLC*); ~ **de niveau** TRANSMIS level monitor; ~ **de nœud terminal** TRANSMIS terminal node controller (*TNC*); ~ **de périphériques** DONNÉES, ÉLECTRON, INFORMAT device controller, peripheral controller; ~ **de polarisation** ÉLECTRON polarization controller (*PC*); ~ **secteur** ALIM ÉLEC, CONTRÔLE mains monitor, voltage monitor; ~ **de station de base** RADIO base station controller; ~ **de synthétiseur** INFORMAT synthesizer driver; ~ **de tension** ALIM ÉLEC, CONTRÔLE voltage monitor; ~ **de trafic** DONNÉES, TÉL traffic controller, traffic superintendent; ~ **universel** TESTS multimeter

convection *f* CLIMAT convection; ~ **libre** CLIMAT free convection; ~ **naturelle** CLIMAT free convection

convention: ~ **d'établissement** *f* TV network establishment agreement; ~ **d'exploitation** *f* TV network operating agreement

convergence *f* DONNÉES, TRANSMIS convergence; ~ **de transmission** TRANSMIS transmission convergence

conversation[1] *f* RADIO talk, TÉL call, talk; ~ **en clair** RADIO clear speech; ~ **en différé** DONNÉES indirect conversation; ~ **en direct** DONNÉES live conversation; ~ **mains libres** PÉRIPH, TÉL hands-free conversation; ~ **payable par un tiers** TÉL third-party billing call; ~ **personnelle automatique** TÉL automatic personal call; ~ **téléphonique** TÉL call, telephone call, telephone conversation; ~ **à trois** PÉRIPH, RADIO, TÉL three-way call, three-way calling, three-way conversation

conversation[2]: ~ **en cours** *loc* TÉL call in progress; ~ **impossible** *loc* (*CI*) RADIO conversation impossible (*CI*)

conversationnel *adj* INFORMAT interactive

converser *vi* RADIO, TÉL talk

conversion *f* DONNÉES conversion; ~ **de code** ÉLECTRON, TÉL, TRANSMIS code conversion; ~ **de données** DONNÉES data conversion; ~ **de fichier** DONNÉES, INFORMAT file conversion; ~ **de fréquence** RADIO, TRANSMIS, TV frequency conversion; ~ **hypermédia** INFORMAT hypermedia conversion; ~ **hypertexte** INFORMAT hypertext conversion; ~ **d'image** DONNÉES, TV image conversion; ~ **numérique-analogique** DONNÉES, INFORMAT, SIGNAL, TÉL digital-to-analog conversion; ~ **série-parallèle** DONNÉES serial-parallel conversion; ~ **de signaux** SIGNAL, TÉL signal conversion; ~ **de support** DONNÉES media conversion

convertir *vt* DONNÉES, INFORMAT convert; ~ **en bipolaire** TRANSMIS convert to bipolar code; ~ **en images point à point** INFORMAT rasterize; ~ **du vectoriel en bitmap** INFORMAT rasterize

convertisseur *m* ÉLECTRON, GÉNÉRAL, INFORMAT converter; ~ **d'alimentation** ALIM ÉLEC, DONNÉES, INFORMAT power converter; ~ **A-N** (*CAN, convertisseur analogique-numérique*) DONNÉES, ÉLECTRON, INFORMAT, SIGNAL, TÉL A-D converter (*ADC*); ~ **analogique** DONNÉES, ÉLECTRON, INFORMAT, SIGNAL, TÉL analog converter; ~ **analogique-numérique** (*CAN, convertisseur A-N*) DONNÉES, ÉLECTRON, INFORMAT, SIGNAL, TÉL analog-to-digital converter (*ADC*); ~ **de balayage** RADIO scan converter; ~ **de CC à CC** ALIM ÉLEC, DONNÉES DC-DC converter; ~ **de code** ÉLECTRON, INFORMAT, TÉL, TRANSMIS code converter; ~ **continu-continu** ALIM ÉLEC, DONNÉES DC-DC converter; ~ **de courant** ALIM ÉLEC, DONNÉES, INFORMAT power converter; ~ **courant continu** ALIM ÉLEC direct current converter; ~ **de données** DONNÉES, INFORMAT data converter; ~ **élévateur de fréquence** DONNÉES, RADIO up-converter; ~ **d'énergie** ALIM ÉLEC energy converter; ~ **de fréquence** ALIM ÉLEC, ÉLECTRON, TRANSMIS, TV frequency converter; ~ **de gestion** (*CG*) APP SERV charge-metering converter; ~ **hauteur-MIDI** INFORMAT pitch-to-MIDI converter; ~ **hauteur-tension** INFORMAT pitch-to-voltage converter; ~ **d'interface** DONNÉES interface converter; ~ **d'interface de signalisation** SIGNAL signal interface converter (*SIC*); ~ **N-A** (*CNA, convertisseur numérique-analogique*) DONNÉES, ÉLECTRON, INFORMAT, SIGNAL, TÉL D-A converter (*DAC*), digital-analog converter (*DAC*); ~ **numérique** DONNÉES, INFORMAT, SIGNAL, TÉL digital converter; ~ **numérique-analogique** (*CNA, convertisseur N-A*) DONNÉES, ÉLECTRON, INFORMAT, SIGNAL, TÉL digital-to-analog converter (*DAC*), digital-analog converter (*DAC*), D-A converter (*DAC*); ~ **parallèle-série** ÉLECTRON, TÉL parallel-to-serial converter, serializer; ~ **en pont** ALIM ÉLEC bridge converter; ~ **de protocole** DONNÉES protocol converter; ~ **série-parallèle** ÉLECTRON serial-to-parallel converter, TÉL serial-to-parallel converter; ~ **de signaux** SIGNAL, TÉL signal converter; ~ **survolteur** ALIM ÉLEC *électricité* booster converter; ~ **du système de télévision** TV television system converter; ~ **télégraphique** TÉL telegraphic converter; ~ **de tension** ALIM ÉLEC voltage converter

convivial *adj* GÉNÉRAL user-friendly

convivialité *f* GÉNÉRAL user-friendliness

coopérant *adj* DONNÉES, GÉNÉRAL networked

coordinateur: ~ **d'engagement** *m* DONNÉES commitment coordinator

copie *f* DONNÉES copy; ~ **carbone** (*Cc*) DONNÉES carbon copy

(*Cc*); ~ **aveugle** DONNÉES *Internet* blind copy; ~ **conforme sélective** (*CCS*) TÉL blind courtesy copy (*BCC*); ~ **d'écran** DONNÉES screen copy; ~ **locale** TRANSMIS local copy; ~ **de sécurité** DONNÉES, GÉNÉRAL, INFORMAT backup, backup copy
copier *vt* INFORMAT copy, pick up; ~**-coller** INFORMAT cut and paste
copolaire *adj* RADIO co-polarized, co-polar
coprocesseur *m* INFORMAT coprocessor
coproduction *f* TV coproduction
coque: ~ **arrière du réflecteur** *f* RADIO back dish
corbeille *f* DONNÉES basket
cordon *m* GÉNÉRAL, INFORMAT cord; ~ **d'alimentation** ALIM ÉLEC line cord, power cable, power cord, DONNÉES power cable, ÉLECTRON line cord, power cable, power cord, INFORMAT AC mains cord, line cord, power cable, power cord; ~ **d'appareil téléphonique** COMPOSANTS, TÉL telephone instrument cord; ~ **blindé** COMPOSANTS shielded cord; ~ **de brassage** ÉLECTRON patch cord; ~ **du combiné** PÉRIPH, TÉL handset cord; ~ **connecteur** TRANSMIS cord set; ~ **flexible** PÉRIPH, TÉL instrument cord; ~ **de montage** ÉLECTRON mounting cord; ~ **optique** TRANSMIS optical cord; ~ **prolongateur** ALIM ÉLEC extension cord; ~ **de raccordement** TRANSMIS patch cord; ~ **de raccordement coaxial** ÉLECTRON coarse jumper; ~ **secteur** INFORMAT line connector cord
corésident *m* DONNÉES, INFORMAT, TÉL terminate and stay resident (*TSR*)
cornet *m* RADIO horn, flare antenna, cornucopia; ~ **d'alimentation** ÉLECTRON feed horn; ~ **d'alimentation à surface cannelée** RADIO corrugated horn feed; ~ **à correction de phase** RADIO, TÉL phase-corrected horn; ~ **parabolique** RADIO hoghorn; ~ **rayonnant** RADIO radiating horn
cornières: ~ **de montage en retrait** *f pl* ÉLECTRON recessed access brackets
corps *m* COMPOSANTS, ÉLECTRON hood, GÉNÉRAL body, INFORMAT *typographie* font size; ~ **d'essai** DONNÉES, TESTS test body; ~ **fini** SIGNAL finite field; ~ **isolant** ALIM ÉLEC insulant; ~ **non conducteur** ÉLECTRON nonconductor; ~ **de palier** COMPOSANTS *génie mécanique* bearing housing, bearing shell; ~ **principal** RADIO primary body; ~ **de test** DONNÉES, TESTS test body
correcteur *m* ALIM ÉLEC equalizer (*EQL*), CONTRÔLE compensating element, equalizer (*EQL*), ÉLECTRON, TÉL, TRANSMIS equalizer (*EQL*); ~ **de forme d'onde** ÉLECTRON waveform corrector; ~ **de phase** ALIM ÉLEC, GÉNÉRAL phase equalizer
correction *f* GÉNÉRAL correction; ~ **active** TESTS active control; ~ **automatique** DONNÉES, INFORMAT, SIGNAL, TÉL auto-correct, automatic correction; ~ **automatique d'erreurs** DONNÉES, INFORMAT, SIGNAL, TÉL automatic error correction; ~ **de contour** INFORMAT contour correction; ~ **directe d'erreurs** DONNÉES, INFORMAT, SIGNAL, TÉL *satellite maritime* forward error correction; ~ **de distorsion** DONNÉES error correction; ~ **d'erreurs** DONNÉES, INFORMAT, RADIO, SIGNAL, TÉL error correction; ~ **d'erreurs sans voie de retour** DONNÉES, INFORMAT, RADIO, SIGNAL, TÉL forward error correction codex (*FECC*); ~ **immédiate** DONNÉES in-line recovery; ~ **d'impulsions** SIGNAL, TRANSMIS pulse correction; ~ **d'inclinaison de l'orbite** RADIO orbital inclination control; ~ **d'ombrage** TV shading correction; ~ **de synchronisme** TÉL synchronous correction; **mode avec ~ d'erreur** *m* TÉL error-correction mode, ECM
corrélateur *m* ÉLECTRON, GÉNÉRAL correlator
corrélation *f* GÉNÉRAL correlation; ~ **de formes** APP SERV, INFORMAT pattern matching; ~ **spatio-temporelle** COMMUT, DONNÉES space-time correlation; ~ **temporelle** COMMUT, DONNÉES time correlation
correspondance: ~ **télégraphique de service** *f* TÉL telegraphic service correspondence
correspondant *m* TÉL called party; ~ **extérieur** TÉL outside party
cosécante *f* RADIO cosecant
cosinus: ~ **de Lambert** *m* TRANSMIS cosine emission
COSPAS-SARSAT *abrév* (*Système international de satellites pour les recherches et le sauvetage*) GÉNÉRAL COSPAS-SARSAT (*International Satellite System for Search and Rescue*)
cosse *f* ÉLECTRON terminal connector, tag; ~ **de câble** TRANSMIS cable lug; ~ **à souder** COMPOSANTS soldering lug, soldering tag
cotation: ~ **du fournisseur** *f* TESTS supplier rating, vendor rating
cote: ~ **effective** *f* TESTS *qualité* actual dimension, observed dimension
couche *f* ÉLECTRON, GÉNÉRAL, INFORMAT layer; ~ **d'accumulation** ÉLECTRON *semiconducteurs* accumulation layer; ~ **d'adaptation ATM** DONNÉES, GÉNÉRAL, TÉL, TRANSMIS ATM adaptation layer (*AAL*); ~ **antireflets** TRANSMIS anti-reflection coating; ~ **d'appauvrissement** COMPOSANTS depletion layer; ~ **application** APP SERV *interconnexion de systèmes ouverts*, DONNÉES, TRANSMIS application layer; ~ **de barrage** ÉLECTRON *semiconducteurs* barrier layer; ~ **barrière** ÉLECTRON barrier layer; ~ **de bobinage** ALIM ÉLEC winding layer; ~ **conductrice** ÉLECTRON conducting layer, conductive layer; ~ **E sporadique** RADIO sporadic E layer; ~ **électronique** GÉNÉRAL electron shell; ~ **d'éléments de câblage** GÉNÉRAL layer of cabling elements; ~ **d'enroulement** ALIM ÉLEC winding layer; ~ **épitaxiale** COMPOSANTS epitaxial layer; ~ **externe** ÉLECTRON external layer; ~ **F** INFORMAT, RADIO F-layer; ~ **F1** INFORMAT, RADIO F1-layer; ~ **F2** INFORMAT, RADIO F2-layer; ~ **de faisceaux** GÉNÉRAL layer of units; ~ **de guidage** RADIO ducting layer; ~ **d'interconnexion de systèmes ouverts** INFORMAT open systems interconnection layer; ~ **interne** ÉLECTRON internal layer; ~ **intrinsèque** ÉLECTRON intrinsic layer; ~ **d'inversion** TRANSMIS inversion layer; ~ **ionosphérique** RADIO ionospheric layer; ~ **ISO** DONNÉES, INFORMAT OSI layer, OSI model layer; ~ **isolante** ALIM ÉLEC insulating layer; ~ **liaison** DONNÉES, TRANSMIS link layer; ~ **liaison de données** DONNÉES, TRANSMIS data-link layer; ~ **mince métallique** ÉLECTRON metallic thin film; ~ **N** DONNÉES, ÉLECTRON N-layer; ~ **paquet** DONNÉES, TÉL packet layer; ~ **physique** DONNÉES, TRANSMIS physical layer; ~ **de plastique protectrice** INFORMAT protective plastic coating; ~ **présentation** DONNÉES, TRANSMIS presentation layer; ~ **réseau** DONNÉES, TÉL, TRANSMIS network layer (*NL*); ~ **de réseau commuté** COMMUT, TRANSMIS, TV switched network layer; ~ **réseau d'interconnexion de systèmes ouverts** DONNÉES open systems interconnection network layer; ~ **réseau ISO** DONNÉES OSI network layer; ~ **réseau de transmission** TRANSMIS transmission layer; ~ **de services** TRANSMIS service layer; ~ **session** DONNÉES session layer, session level, TRANSMIS session level, session layer; ~ **de structure** COMPOSANTS, ÉLECTRON sheathing layer; ~ **superficielle** COMPOSANTS surface layer; ~ **transport** DONNÉES, TRANSMIS transport layer; ~ **transport d'interconnexion de systèmes ouverts** DONNÉES open systems interconnection transport layer; ~ **transport ISO** DONNÉES OSI transport layer
couches: ~ **croisées** *f pl* ÉLECTRON crossplay sheath
coude *m* ÉLECTRON knee point, RADIO bend; ~ **H** RADIO H-corner
coulage *m* COMPOSANTS plastic molding (*AmE*), plastic moulding (*BrE*)
coulée: ~ **de résine** *f* ÉLECTRON resin smear
couler: ~ **sous pression** *vt* COMPOSANTS die-cast

couleur *f* GÉNÉRAL color (*AmE*), colour (*BrE*); **~ achromatique** INFORMAT, TV achromatic color (*AmE*), achromatic colour (*BrE*); **~ additive** INFORMAT additive color (*AmE*), additive colour (*BrE*); **~ chromatique** TV chromatic color (*AmE*), chromatic colour (*BrE*); **~ primaire additive** INFORMAT additive primary color (*AmE*), additive primary colour (*BrE*); **~ primaire soustractive** INFORMAT subtractive primary color (*AmE*), subtractive primary colour (*BrE*); **~ sans tonalité chromatique** INFORMAT, TV achromatic color (*AmE*), achromatic colour (*BrE*)

coulisse *f* COMPOSANTS slide

couloir *m* COMMUT, COMPOSANTS aisle, RADIO corridor, lane, TÉL aisle; **~ intertravée** TÉL row aisle

coup: **~ de but** *m* RADIO hit; **~ de fil** *m* TÉL call, phone call, telephone call; **~ de foudre indirect** *m* ALIM ÉLEC indirect stroke; **~ parasite** *m* RADIO spurious count

coupé *adj* ALIM ÉLEC, RADIO cut-off

coupe[1]: **~-bande** *m* TRANSMIS band rejection filter, bandstop filter; **~-circuit** *m* ALIM ÉLEC fuse unit; **~-circuit à répétition** *m* ALIM ÉLEC repeater fuse; **~-gaine transversal** *m* TRANSMIS cable trimmer

coupe[2] *f* GÉNÉRAL cutting; **~ transversale** GÉNÉRAL *dessin technique* cross section

couper[1] *vt* ALIM ÉLEC *courant* trim, cut off, *électricité* open, break, ÉLECTRON trim, INFORMAT trim, cut, RADIO cut off, TÉL interrupt

couper[2]: **se ~** *v réfl* ALIM ÉLEC break

couplage *m* ÉLECTRON coupling; **~ acoustique** DONNÉES, TÉL acoustic coupling; **~ acoustique interactif** TÉL acoustic feedback; **~ d'antenne variable** RADIO variable antenna coupling; **~ capacitif** ÉLECTRON capacitive coupling; **~ entre étages** ÉLECTRON interstage coupling; **~ inductif** ALIM ÉLEC impedance bond; **~ de liaison** RADIO link coupling; **~ magnétique** ÉLECTRON magnet coupling; **~ de modes** TRANSMIS mode coupling; **~ mutuel** ÉLECTRON mutual coupling; **~ d'ondes** ÉLECTRON wave coupling; **~ par résistance** ÉLECTRON resistance coupling; **~ par résistance-capacité** (*couplage RC*) ÉLECTRON resistance-capacitance coupling (*RC coupling*); **~ parasite** ÉLECTRON parasitic coupling; **~ RC** (*couplage par résistance-capacité*) ÉLECTRON RC coupling (*resistance-capacitance coupling*)

couplé *adj* ÉLECTRON ganged

coupler *vt* INFORMAT couple, PÉRIPH intercouple

coupleur *m* ÉLECTRON coupler, flex-ferrule; **~ acoustique** DONNÉES, TÉL acoustic coupler; **~ alternatif** ALIM ÉLEC alternating current coupler; **~ central d'alarmes** (*CCAL*) CONTRÔLE central alarm coupler; **~ clos** TÉL closed coupler; **~ continu** ALIM ÉLEC direct current coupler (*DC coupler*); **~ directif** COMMUT, DONNÉES, TRANSMIS directional coupler; **~ directionnel** COMMUT, DONNÉES, TRANSMIS directional coupler; **~ en étoile** TRANSMIS star coupler; **~ femelle-femelle** ÉLECTRON, INFORMAT female-female adapter; **~ fileté** COMPOSANTS, ÉLECTRON threaded coupler; **~ de guides d'ondes** RADIO, TRANSMIS waveguide coupler; **~ Mach-Zehnder** TÉL Mach-Zehnder interferometer; **~ monomode** TRANSMIS single-mode coupler; **~ MRL** TRANSMIS WDM coupler; **~ multiplex principal** TÉL main multiplex coupler; **~ multiplex secondaire** TÉL secondary multiplex coupler; **~ optique** DONNÉES, ÉLECTRON, TRANSMIS optical coupler, optical fiber coupler (*AmE*), optical fibre coupler (*BrE*), optocoupler; **~ par répartition en longueur d'onde** TÉL, TRANSMIS wavelength division multiplex coupler; **~ satellite d'alarmes** TÉL alarm satellite coupler; **~ à sélection de mode** TRANSMIS mode-selective coupler; **~ de signalisation multiprotocole** SIGNAL multiprotocol signaling coupler (*AmE*), multiprotocol signalling coupler (*BrE*); **~ en T** COMPOSANTS, ÉLECTRON T-branching device, T-coupler; **~ téléphone-radio** RADIO, TÉL phone patch; **~ de traitement à signal vocal** TÉL voice-signal processing coupler; **~ en Y** ÉLECTRON, TRANSMIS Y-branching device, Y-coupler

coupure[1] *f* ALIM ÉLEC shutdown, interruption, GÉNÉRAL break, cut, cutoff, TESTS outage, TRANSMIS breakdown; **~ automatique d'alimentation** ALIM ÉLEC, INFORMAT auto power off; **~ automatique du laser** TRANSMIS automatic laser shutdown (*ALS*); **~ automatique de la radio** PÉRIPH, RADIO automatic radio mute; **~ de courant** ALIM ÉLEC power cut, mains failure; **~ de filtre** INFORMAT filter cutoff; **~ de microphone** PÉRIPH microphone cutoff; **~ de page-écran** INFORMAT clip; **~ secteur** ALIM ÉLEC power cut, mains failure

coupure[2]: **sans ~** *loc* TRANSMIS break-free

courant[1] *m* ALIM ÉLEC, ÉLECTRON, GÉNÉRAL current; **~ d'accrochage** ALIM ÉLEC latching current; **~ actif** ALIM ÉLEC operating current; **~ d'alimentation** ALIM ÉLEC feeding current, supply current; **~ d'alimentation du microphone** ALIM ÉLEC, TÉL microphone current feed; **~ d'alimentation du microphone à charbon** ALIM ÉLEC, TÉL carbon microphone current feed; **~ alternatif** (*CA*) ALIM ÉLEC alternating current (*AC*); **~ alternatif de décharge** ÉLECTRON alternating discharge current; **~ alternatif monophasé** ALIM ÉLEC single-phase alternating current; **~ anodique** ÉLECTRON anode current, plate current; **~ d'antenne** RADIO aerial current, antenna current; **~ d'appel** ALIM ÉLEC operate current, TÉL ringing current; **~ d'arc** ÉLECTRON arc current; **~ assigné** ALIM ÉLEC rated current; **~ d'attaque** ALIM ÉLEC driving current; **~ d'autodécharge** ALIM ÉLEC self-discharge current; **~ de base** ÉLECTRON *semiconducteurs* base current; **~ de bruit** TÉL noise current; **~ cathodique minimal** ÉLECTRON minimum cathode current, cathode-covering current; **~ de charge** ALIM ÉLEC load current, charging current ÉLECTRON charging current; **~ de charge flottante** ALIM ÉLEC float charging current; **~ de chauffage** ÉLECTRON filament current, heater current; **~ de chauffage au démarrage** ÉLECTRON filament starting current, filament surge current, heater starting current, heater surge current; **~ de choc de décharge** ÉLECTRON impulse discharge current; **~ en circuit fermé** ALIM ÉLEC closed-circuit current; **~ de collecteur** ÉLECTRON collector current; **~ de commande** ALIM ÉLEC driving current; **~ continu** (*CC*) ALIM ÉLEC direct current (*DC*); **~ continu permanent à l'état passant** ÉLECTRON *thyristors* continuous direct on-state current; **~ continu pulsé** ALIM ÉLEC pulsating DC current; **~ coupé limité** ALIM ÉLEC limiting current; **~ de coupure** ÉLECTRON cutoff current; **~ de coupure du collecteur** ÉLECTRON collector cutoff current; **~ de court-circuit** ALIM ÉLEC short-circuit current; **~ de court-circuit symétrique** ALIM ÉLEC symmetrical short-circuit current; **~ de décalage** ÉLECTRON offset current; **~ de déclin** ÉLECTRON transient decay current; **~ de défaut** ÉLECTRON fault current; **~ de démarrage du filament** ÉLECTRON filament starting current, heater starting current; **~ en dents de scie** ÉLECTRON sawtooth current; **~ de désamorçage par la gâchette** ÉLECTRON gate turn-off current; **~ direct** ALIM ÉLEC, ÉLECTRON forward current; **~ direct moyen** ALIM ÉLEC, ÉLECTRON mean forward current; **~ direct de pointe** ALIM ÉLEC, ÉLECTRON surge forward current; **~ direct de pointe récurrent** ALIM ÉLEC, ÉLECTRON repetitive peak forward current; **~ direct de surcharge prévisible** ALIM ÉLEC, ÉLECTRON overload forward current; **~ d'effluve** ÉLECTRON glow current of protector; **~ électrique** ALIM ÉLEC electric current; **~ électronique** ALIM ÉLEC, ÉLECTRON electron current, electron flow; **~ d'émetteur** ÉLECTRON emitter current; **~ d'émission électronique secondaire** ÉLECTRON secondary electron emission current; **~ d'entrée** ÉLECTRON input current; **~ à l'état bloqué** ÉLECTRON off-state current; **~ à l'état passant** ÉLECTRON on-state current; **~ faible** ALIM ÉLEC, ÉLECTRON, TRANSMIS light current; **~ de faible puissance** ALIM ÉLEC, ÉLECTRON, TRANSMIS light current; ~ **fantôme** ÉLECTRON sneak current; ~ **de filament** ÉLECTRON filament current, heater current; ~ **de filament au démarrage** ÉLECTRON filament starting current, filament surge current, heater starting current, heater surge current; ~ **de fonctionnement** ALIM ÉLEC operating current; ~ **fort** ALIM ÉLEC heavy current, power current; ~ **de Foucault** ALIM ÉLEC eddy current; ~ **fourni** ALIM ÉLEC current load; ~ **de fuite** ÉLECTRON fault current, leakage current; ~ **de fuite d'une grille** ÉLECTRON gate cutoff current; ~ **de fuite à la terre** ALIM ÉLEC earth leakage current (*BrE*), ground leakage current (*AmE*); ~ **de gâchette de déblocage** ÉLECTRON gate turn-on current; ~ **de gaine** TESTS sheath current; ~ **de gaz** ÉLECTRON gas current; ~ **de grille** ÉLECTRON grid current; ~ **inactif maximal** ÉLECTRON maximum non-operate current; ~ **induit** ALIM ÉLEC induced current; ~ **inverse** ÉLECTRON reverse current; ~ **inverse de la base** ÉLECTRON *semiconducteurs* base reverse current; ~ **inverse de blocage** ALIM ÉLEC, ÉLECTRON reverse blocking current; ~ **inverse d'électrode** ÉLECTRON inverse electrode current, reverse electrode current; ~ **inverse résistif** ÉLECTRON resistive reverse current; ~ **ionique** ÉLECTRON ionic current; ~ **d'isolement entre filament et cathode** ÉLECTRON heater-cathode insulation current; ~ **de libération** ALIM ÉLEC release current; ~ **de ligne** INFORMAT line current; ~ **magnétisant** ÉLECTRON magnetizing current; ~ **de magnétisation** ÉLECTRON magnetizing current; ~ **de maintien** ÉLECTRON hold current, holding current; ~ **maximal de non-attraction** ÉLECTRON maximum non-operate current; ~ **minimal d'attraction** ÉLECTRON minimum operate current; ~ **minimal de commande** ÉLECTRON minimum operate current; ~ **minimal de fonctionnement** ÉLECTRON hold current, holding current; ~ **de mouillage** ALIM ÉLEC, ÉLECTRON wetting current; ~ **nominal** ALIM ÉLEC rated current; ~ **nominal alternatif de décharge nominal** ALIM ÉLEC nominal alternating discharge current; ~ **nominal de choc de décharge** ALIM ÉLEC nominal impulse discharge current; ~ **de non-amorçage par la gâchette** ÉLECTRON gate non-trigger current; ~ **d'obscurité** ÉLECTRON electrode dark current, dark current, TRANSMIS dark current; ~ **par charge d'espace** ÉLECTRON space-charge limited current; ~ **photoélectrique** ALIM ÉLEC, ÉLECTRON, TRANSMIS light current, photocurrent, photoelectric current; ~ **de pic** ALIM ÉLEC peak point current; ~ **plaque** ÉLECTRON anode current, plate current; ~ **de pointe** ALIM ÉLEC surge current; ~ **de pointe du filament** ÉLECTRON filament surge current, heater surge current; ~ **de pointe récurrent à passant** ÉLECTRON repetitive peak onstage current; ~ **de polarisation** ALIM ÉLEC polarization current, ÉLECTRON bias current; ~ **porteur** TRANSMIS carrier; ~ **principal** ALIM ÉLEC main current; ~ **réactif** ALIM ÉLEC reactive current; ~ **de recouvrement inverse** ÉLECTRON reverse recovery current; ~ **réfléchi** TRANSMIS reflected current, return current; ~ **de repos** ÉLECTRON standing current; ~ **résiduel** ÉLECTRON cutoff current; ~ **résiduel de l'émetteur** ÉLECTRON emitter cutoff current; ~ **de retournement** ALIM ÉLEC breakover current; ~ **de saturation anodique** ALIM ÉLEC, ÉLECTRON voltage saturation current; ~ **de saturation par irradiation** ALIM ÉLEC, ÉLECTRON irradiation-saturation current; ~ **secteur** INFORMAT line current; ~ **de seuil** ALIM ÉLEC, ÉLECTRON, TRANSMIS threshold current; ~ **de signal** ÉLECTRON signal current; ~ **simple** ALIM ÉLEC, TRANSMIS single current; ~ **de sonde** ÉLECTRON probe current; ~ **de sonnerie** TÉL ring current; ~ **de sortie** ÉLECTRON output current; ~ **de substrat** ÉLECTRON substrate current; ~ **superficiel** ÉLECTRON surface current; ~ **de surcharge** ÉLECTRON inrush current; ~ **de surtension transitoire** ALIM ÉLEC surge current; ~ **total de réflecteur** ÉLECTRON total reflector current; ~ **de traînage** ÉLECTRON transient decay current; ~ **de transfert** ÉLECTRON transfer current; ~ **de transfert de l'électrode d'amorçage** ÉLECTRON starter transfer current; ~ **de travail** ALIM ÉLEC operating current; ~ **vagabond** ALIM ÉLEC stray current; ~ **de validation** ALIM ÉLEC enable current; ~ **à vide** ALIM ÉLEC open-circuit current

courant[2]: **en ~ formé** *loc* ALIM ÉLEC, TV closed-circuit

courbe *f* INFORMAT curve, TRANSMIS curve, *câbles* bend; ~ **d'amplitude** INFORMAT amplitude curve; ~ **d'attaque initiale** INFORMAT initial attack slope; ~ **de Bézier** INFORMAT Bézier curve; ~ **de câble** TRANSMIS cable bend; ~ **d'écho** TÉL echo curve; ~ **d'effet de masque** TÉL masking curve; ~ **d'efficacité** GÉNÉRAL operating characteristic curve (*OC curve*); ~ **d'enveloppe** INFORMAT envelope curve; ~ **exponentielle** INFORMAT exponential curve; ~ **exponentielle ascendante** INFORMAT rising exponential curve; ~ **exponentielle descendante** INFORMAT falling exponential curve; ~ **de fonctionnement** ÉLECTRON operating curve, operating line; ~ **de fréquence** TESTS frequency curve; ~ **de gamma** PÉRIPH gamma curve; ~ **gaussienne** GÉNÉRAL Gaussian curve; ~ **d'hystérésis** ÉLECTRON hysteresis curve; ~ **isosonique** TÉL loudness contour; ~ **linéaire** INFORMAT linear curve; ~ **linéaire ascendante** INFORMAT rising linear curve; ~ **linéaire descendante** INFORMAT falling linear curve; ~ **de M** RADIO M-curve; ~ **de phase** ÉLECTRON phase plot; ~ **de répartition de la résistance** ÉLECTRON taper curve; ~ **de réponse en fréquence** ÉLECTRON, RADIO frequency response curve; ~ **de tension d'amorçage au choc en fonction du temps** TÉL *de parafoudre* impulse sparkover voltage-time curve; ~ **de tension en fonction du courant de décharge** TÉL voltage discharge-current curve

courbes: ~ **de Bode** *f pl* ALIM ÉLEC, CONTRÔLE *commande automatique* Bode diagram

courbure *f* ALIM ÉLEC *circuits imprimés* bow, INFORMAT bend, RADIO *de la terre* curvature, TRANSMIS *câbles* bend; ~ **de câble** TRANSMIS cable bend

couronne: ~ **numérotée** *f* PÉRIPH *postes téléphoniques*, TÉL digit plate

courrier: ~ **électronique** *m* APP SERV, DONNÉES, INFORMAT electronic mail (*e-mail*); ~ **vidéo** *m* INFORMAT video mail

courroie: ~ **de commande** *f* COMPOSANTS driving belt; ~ **d'entraînement** *f* COMPOSANTS driving belt

court[1]: **~-circuit** *m* ALIM ÉLEC, ÉLECTRON, INFORMAT short, short circuit (*S/C*); **~-circuit franc** *m* ALIM ÉLEC hard shorting; **~-circuit partiel** *m* ALIM ÉLEC soft shorting

court[2]: **~-circuiter** *vt* ALIM ÉLEC, ÉLECTRON, INFORMAT short, short-circuit

courte: **à ~ distance** *loc* TRANSMIS short-haul

courtier *m* INFORMAT broker

co-usager *m* TRANSMIS joint user

coût *m* GÉNÉRAL cost; ~ **des défaillances externes** TESTS external failure cost; ~ **d'évaluation** TESTS *qualité* appraisal cost; ~ **de fabrication** GÉNÉRAL production cost; ~ **d'obtention de la qualité** TESTS producer's quality cost; ~ **de possession** GÉNÉRAL *qualité* cost of ownership; ~ **de prévention** TESTS prevention cost; ~ **de production** GÉNÉRAL cost of production, production cost; ~ **qualité** TESTS quality cost; ~ **qualité supporté par l'utilisateur** TESTS user's quality cost; ~ **système** GÉNÉRAL overheads; ~ **total de la communication** APP SERV, INFORMAT, RADIO, TÉL total charge

couteau: ~ **à dénuder** *m* COMPOSANTS stripper, stripping knife

couvercle *m* COMPOSANTS cover, lid; ~ **coulissant** COMPOSANTS sliding lid; ~ **en deux parties** COMPOSANTS, TÉL two-piece cover; ~ **à glissière** COMPOSANTS sliding lid; ~ **d'insonorisation** COMPOSANTS, TÉL sound-attenuating cover; ~ **monobloc** TÉL one-piece cover; ~ **à vis** COMPOSANTS screw cover, screw top

couverture *f* RADIO coverage; ~ **de chemin** ÉLECTRON raceway cover; **~ globale** RADIO global coverage; **~ hémisphérique** RADIO hemispherical coverage; **~ à l'intérieur des bâtiments** RADIO in-building coverage; **~ par faisceaux ponctuels** RADIO spot-beam coverage; **~ radioélectrique** RADIO radio coverage; **~ totale** RADIO blanket coverage; **~ vidéo en direct** TV live video coverage

CP *abrév* COMMUT, DONNÉES, TÉL, TRANSMIS (*commutateur de paquets*) PS (*packet switch*), (*commutation de paquets*) PS (*packet switching*)

cpp *abrév* (*caractères par pouce*) INFORMAT cpi (*characters per inch*)

crachement *m* TÉL crackle, crackling noise, line scratches

craquelure *f* TESTS crack

cratère *m* COMPOSANTS pit

crayon: **~ électronique** *m* DONNÉES electronic pen; **~ lecteur** *m* INFORMAT *de codes-barres* wand reader; **~ optique** *m* APP SERV, INFORMAT light pen, optical pencil

CRC *abrév* DONNÉES, TÉL (*code de redondance cyclique*) CRC (*cyclic redundancy code*), (*contrôle de redondance cyclique*) CRC (*cyclic redundancy check*)

création: **~ de l'arborescence** *f* INFORMAT, DONNÉES tree expansion

crédit: **~-bail** *m* TV leasing

créer *vt* INFORMAT devise, TV create; **~ une image miroir de** INFORMAT mirror

crénage *m* DONNÉES interletter spacing

créneau *m* RADIO, TV slot; **~ horaire** RADIO, TV time slot, timetable slot; **~ orbital** RADIO orbital slot; **~ temporel** COMMUT, RADIO, SIGNAL, TRANSMIS time slot; **~ temporel élémentaire justifiable** TÉL, TRANSMIS justifiable-digit time slot; **~ temporel de verrouillage de trame** TÉL, TRANSMIS frame-alignment time slot; **~ temporel de voie** TRANSMIS channel time slot

crénelage *m* INFORMAT aliasing

crépitement *m* TÉL crackling noise

crépiter *vi* TÉL crackle

crête[1] *f* ALIM ÉLEC, ÉLECTRON peak, GÉNÉRAL, INFORMAT apex, peak; **~ de longueur d'onde** GÉNÉRAL wavelength peak; **~ de tension** ALIM ÉLEC voltage peak

crête[2]: **~ à crête** *loc* ALIM ÉLEC peak-to-peak

creux *m* COMPOSANTS pit, ÉLECTRON trough, INFORMAT pit; **~ central d'indice** TRANSMIS index dip; **~ d'indice** TRANSMIS index dip

CRI *abrév* (*centre radiophonique international*) RADIO ISPC (*international sound-program center AmE, international sound-programme centre BrE*)

criquage *m* TESTS cracking

crique: **~ due à la tension** *f* COMPOSANTS stress crack

cristal: **~ germe** *m* ÉLECTRON seed crystal; **~ piézoélectrique** *m* ÉLECTRON piezoelectric crystal

cristaux: **~ liquides** *m pl* COMPOSANTS, ÉLECTRON liquid crystals; **~ liquides nématiques** *m pl* COMPOSANTS, ÉLECTRON nematic liquid crystals; **~ liquides en phase smectique** *m pl* COMPOSANTS, ÉLECTRON smectic liquid crystals

critère *m* TESTS criterion; **~ d'acceptation** TESTS *qualité* acceptance number; **~ de défaillance** TESTS failure criterion; **~ de qualité** TESTS quality requirement; **~ de rejet** TESTS nonacceptance number, rejection number

CRN *abrév* (*centre radiophonique national*) RADIO national sound-program center (*AmE*), national sound-programme centre (*BrE*)

crochet *m* ÉLECTRON dental pick; **~ commutateur** RADIO cradle, cradle switch, switch hook, TÉL cradle switch, switch hook

croisement: **~ au point nul** *m* ALIM ÉLEC, COMMUT zero crossing

croquis *m* GÉNÉRAL outline; **~ coté** ÉLECTRON outline drawing; **~ d'implantation** ÉLECTRON layout sketch

CRTC *abrév* (*Conseil de la radiodiffusion et des télécommunications canadiennes*) RADIO, TÉL, TV CRTC (*Canadian Radio, Television and Telecommunications Commission*)

cryogénie *f* GÉNÉRAL cryogenics

cryotron *m* ÉLECTRON cryotron

cryptage *m* DONNÉES, INFORMAT, RADIO, SIGNAL, TÉL, TV encryption, scrambling

crypter *vt* DONNÉES, INFORMAT, RADIO, SIGNAL, TÉL, TV encrypt, scramble

crypteur *m* DONNÉES encryptor, encrypting equipment, scrambler, INFORMAT scrambler, RADIO, SIGNAL, TÉL, TV encryptor, encrypting equipment, scrambler

cryptogramme *m* DONNÉES, RADIO, SIGNAL, TÉL, TV ciphertext, cryptogram

cryptographe *m* DONNÉES encrypting equipment, scrambler, INFORMAT scrambler, RADIO, SIGNAL, TÉL, TV encrypting equipment, scrambler

cryptographie *f* DONNÉES cryptanalysis, cryptography, scrambling, INFORMAT scrambling, RADIO, SIGNAL, TÉL, TV cryptanalysis, cryptography, scrambling

cryptophone *m* DONNÉES, INFORMAT, RADIO, SIGNAL, TÉL, TV scrambler

cryptophonie *f* INFORMAT, TÉL speech ciphering

CSE: **~ distant** *m* TÉL remote CSE

CSI-DIS *abrév* (*identification de l'abonné demandé-DIS*) DONNÉES, TÉL CSI-DIS (*called-subscriber identification-DIS*); **~ de fonction nonnormalisée** *f* (*NSF-CSI-DIS*) DONNÉES, TÉL nonstandard facility CSI-DIS (*NSF-CSI-DIS*)

CSMA/CD *abrév* (*accès multiple avec écoute de la porteuse et détection de collisions*) DONNÉES CSMA/CD (*carrier-sense multiple access with collision detection*)

CSN: **~ distant** *m* TÉL remote CSN; **~ local** *m* TÉL local CSN

CT2 *abrév* (*téléphone sans fil de deuxième génération*) PÉRIPH, TÉL CT2 (*cordless telephone 2*)

CTI *abrév* (*centre de traitement de l'information*) DONNÉES data-processing center (*AmE*), data-processing centre (*BrE*)

CTL *abrév* (*code temporel longitudinal*) INFORMAT LTC (*longitudinal time code*)

CTR *abrév* (*réponse pour continuer à corriger*) DONNÉES CTR (*response to continue to correct*)

CT1 *abrév* (*téléphone sans fil de première génération*) PÉRIPH, TÉL CT1 (*cordless telephone 1*)

CTV *abrév* (*conteneur virtuel*) APP SERV, DONNÉES, TÉL VC (*virtual container*)

cuirassé *adj* COMPOSANTS sheathed

cuivre *m* GÉNÉRAL copper; **~ au béryllium** COMPOSANTS *métallurgie* beryllium copper; **~ électrolytique** GÉNÉRAL electrolytic copper; **~ étamé** COMPOSANTS, ÉLECTRON tin-plated copper

culasse *f* COMPOSANTS, ÉLECTRON, TÉL relay yoke, yoke

culot *m* ALIM ÉLEC lamp cap; **~ à baïonnette** ALIM ÉLEC *de lampe* bayonet cap; **~ de lampe** ALIM ÉLEC lamp base; **~ de tube** COMPOSANTS, ÉLECTRON tube base, valve base; **~ à vis** COMPOSANTS screw base, screw cap

curseur[1] *m* INFORMAT cursor, slider, slider control; **~ clignotant** INFORMAT flashing cursor; **~ de défilement** INFORMAT scroll box; **~ de volume général** INFORMAT master volume slider; **~ de volume de rythme** INFORMAT rhythm volume slider

curseur[2]: **~ arrêté** *loc* INFORMAT COF (*cursor off*); **~ en marche** *loc* INFORMAT CON (*cursor on*)

cuve: **~ de stockage de câble** *f* TRANSMIS cable storage hold, cable tank

CVC *abrév* COMMUT, ÉLECTRON (*circuit virtuel commuté*) SVC (*switched virtual circuit*), SIGNAL SVC (*switched virtual circuit*), (*circuit virtuel commuté*) SVC (*signaling virtual channel AmE, signalling virtual channel BrE*), TRANSMIS (*circuit virtuel commuté*) SVC (*switched virtual circuit*)

cyber *adj* INFORMAT cyber

cybercafé *m* INFORMAT cybercafé
cyberculture *f* INFORMAT cyberculture
cybernaute *m* INFORMAT cybernaut
cybernétique *f* INFORMAT cybernetics
cyberspace *f* INFORMAT cyberspace
cyclage *m* TESTS cycling; ~ **thermique** CLIMAT, TESTS temperature cycling
cycle *m* GÉNÉRAL, INFORMAT cycle; ~ **asservi** SIGNAL compelled cycle; ~ **de chargement** ALIM ÉLEC load cycle; ~ **d'essai** TESTS testing cycle; ~ **d'horloge** INFORMAT clock cycle; ~ **d'hystérésis** ÉLECTRON hysteresis cycle, hysteresis loop
cycler *vt* TESTS cycle
cycles: ~ **par seconde** *loc* INFORMAT cycles per second
cyclique *adj* ÉLECTRON cyclical

D

DAA *abrév* TÉL (*distributeur d'appels automatique*) ACD (*automatic call distributor*), (*distribution automatique d'appels*) ACD (*automatic call distribution*)
DALP *abrév* (*dépôt axial latéral au plasma*) ÉLECTRON ALPD (*axial lateral plasma deposition*)
DAT *abrév* (*bande audionumérique*) INFORMAT, TV DAT (*digital audio tape*)
Data: **~ Discman®** *m* DONNÉES, INFORMAT Data Discman®
datagramme *m* DONNÉES datagram
date *f* GÉNÉRAL date; **~ de création** INFORMAT creation date; **~ de l'échange** DONNÉES interchange date; **~ et heure de dépôt** DONNÉES submission time; **~ et heure de transmission** DONNÉES date and time of transmission; **~-heure de création** DONNÉES creation time; **~ d'installation** GÉNÉRAL installation date; **~ julienne** RADIO Julian date; **~ julienne modifiée** (*DJM*) RADIO modified Julian date (*MJD*); **~ limite** INFORMAT deadline; **~ ordinale** GÉNÉRAL ordinal date; **~ ou heure de préparation** DONNÉES date or time of preparation
DAV *abrév* (*dépôt axial en phase vapeur*) ÉLECTRON VAD (*vapor axial deposition AmE, vapour axial deposition BrE*)
dB *abrév* (*décibel*) RADIO *acoustique* dB (*decibel*)
DBL *abrév* (*double bande latérale*) RADIO, TRANSMIS DSB (*double sideband*)
dBm *abrév* TÉL dBm (*decibels above one milliwatt*)
DBV *abrév* (*débit binaire variable*) DONNÉES, INFORMAT, TÉL, TRANSMIS VBR (*variable bit rate*)
DCA *abrév* (*architecture DCA*) DONNÉES DCA (*document content architecture*)
DCC *abrév* (*dispositif à couplage de charge*) ÉLECTRON CCD (*charge-coupled device*)
DCN *abrév* (*déconnexion*) DONNÉES DCN (*disconnect*)
DCX *abrév* (*fichier PCX multipage*) DONNÉES DCX (*multipage PCX file*)
DD *abrév* (*double densité*) INFORMAT DD (*double density*)
débit *m* DONNÉES throughput, thruput (*AmE*), GÉNÉRAL ratio, INFORMAT throughput, thruput (*AmE*); **~ binaire** DONNÉES, INFORMAT, TÉL, TRANSMIS binary rate, bit rate, data signaling rate (*AmE*), data signalling rate (*BrE*); **~ binaire de base** DONNÉES, INFORMAT, TÉL, TRANSMIS basic bit rate; **~ binaire composite** DONNÉES, INFORMAT, TÉL, TRANSMIS gross bit rate; **~ binaire constant** DONNÉES, INFORMAT, TÉL, TRANSMIS constant bit rate (*CBR*); **~ binaire cumulé d'un multiplex** TÉL multiplex aggregate bit rate; **~ binaire équivalent** TÉL equivalent bit rate; **~ binaire intermédiaire** DONNÉES intermediate data rate (*IDR*); **~ binaire série** SIGNAL binary serial signaling rate (*AmE*), binary serial signalling rate (*BrE*); **~ binaire de transmission** RADIO, TRANSMIS transmission bit rate; **~ binaire variable** (*DBV*) DONNÉES, INFORMAT, TÉL, TRANSMIS variable bit rate (*VBR*); **~ de bribes** DONNÉES chip rate; **~ des éléments** DONNÉES chip rate; **~ d'information garanti** DONNÉES committed information rate (*CIR*); **~ maximal de justification** TÉL maximum justification rate; **~ en mots** DONNÉES word rate; **~ nominal de justification** TÉL nominal justification rate; **~ numérique** TRANSMIS digit rate; **~ numérique de ligne** TRANSMIS line digit rate; **~ de la parole** TÉL speaking rate; **~ de service** INFORMAT service rate; **~ de symboles** DONNÉES, TÉL symbol rate; **~ de symboles nominal** DONNÉES, TÉL nominal symbol rate; **~ de transfert** DONNÉES, INFORMAT transfer rate; **~ de transfert d'information** DONNÉES, INFORMAT information transfer rate; **~ de transfert de l'information d'usager** DONNÉES, INFORMAT user information transfer rate
débitmètre *m* CONTRÔLE flowmeter
déblocage *m* RADIO, TÉL deblocking, unblocking; **~ d'un canal sémaphore** SIGNAL signaling-link unblocking (*AmE*), signalling-link unblocking (*BrE*); **~ du faisceau** ÉLECTRON unblanking; **~ d'une liaison de signalisation** SIGNAL signaling-link unblocking (*AmE*), signalling-link unblocking (*BrE*)
débloquer[1] *vt* RADIO, TÉL unblock
débloquer[2]: **se ~** *v réfl* ÉLECTRON turn on
débordement *m* GÉNÉRAL alternate, alternative routing, overflow, TÉL interflow, overflow; **~ automatique des appels en file d'attente** TÉL automatic call queue overflow; **~ de données de connexion** DONNÉES connect data overflow (*CDO*)
débranché *adj* ÉLECTRON, INFORMAT, PÉRIPH disconnected, off-hook, unplugged, TESTS disconnected, unplugged
débrancher *vt* ÉLECTRON disconnect, unplug, INFORMAT disconnect, unplug, trap, PÉRIPH, TESTS disconnect, unplug
début: **~ de la demande d'appel** *m* DONNÉES beginning of call demand; **~ d'impulsion** *m* INFORMAT beginning of pulse; **~ de message** *m* DONNÉES, PÉRIPH beginning of message (*BOM*)
décalage *m* INFORMAT lag, offset, shift, PÉRIPH shift, RADIO offset, TRANSMIS out-of-sync error, out-of-synchronization error; **~ angulaire** TÉL angular displacement; **~ de fréquence** RADIO, TRANSMIS frequency staggering, frequency offset; **~ de fréquence de faible précision** RADIO, TRANSMIS nonprecision frequency offset; **~ de fréquence normée** RADIO, TRANSMIS normalized frequency offset; **~ de fréquence porteuse** RADIO, TRANSMIS carrier-frequency offset; **~ de fréquence de précision** RADIO, TRANSMIS precision frequency offset; **~ de hauteur** INFORMAT key shift; **~ d'horloge** INFORMAT shift clock, TRANSMIS clock phasing; **~ image-son** INFORMAT picture-sound desynchronization, picture-sound lag; **~ logique** INFORMAT logic shift; **~ normé** RADIO, TRANSMIS normalized offset; **~ d'octave** INFORMAT octave shift; **~ par rapport à la fréquence centrale** RADIO, TRANSMIS displacement from center frequency (*AmE*), displacement from centre frequency (*BrE*); **~ parallèle** RADIO, TRANSMIS parallel displacement; **~ de phase** ALIM ÉLEC phase deviation, phase shift, ÉLECTRON phase shift, GÉNÉRAL phase deviation, RADIO, TRANSMIS phase deviation, phase shift; **~ de ton** INFORMAT key shift
décalé *adj* INFORMAT, RADIO off-time, offset
décaler *vt* INFORMAT, RADIO offset, rub out, lag
décentralisation *f* GÉNÉRAL distribution
décentraliser *vt* GÉNÉRAL, INFORMAT decentralize, distribute
décharge *f* ALIM ÉLEC discharge, flat; **~ auto-entretenue** ALIM ÉLEC, ÉLECTRON self-maintained discharge; **~ autonome** ALIM ÉLEC, ÉLECTRON self-maintained discharge; **~ en circuit ouvert** ALIM ÉLEC local action; **~ disruptive** ALIM ÉLEC sparking; **~ électrostatique** ÉLECTRON, TESTS electrostatic discharge (*ESD*); **~ du gaz** ÉLECTRON gas discharge; **~ luminescente** ÉLECTRON glow discharge; **~ non-autonome** ÉLECTRON non-self-maintained discharge; **~ spontanée** ALIM ÉLEC self-discharge
décharger *vt* ALIM ÉLEC discharge, TÉL decrease load, relieve; **~ un satellite** RADIO offload a satellite

déchiffrage *m* DONNÉES, GÉNÉRAL, INFORMAT decryption
déchiffrement *m* GÉNÉRAL deciphering, decipherment, INFORMAT, RADIO, SIGNAL, TV decoding
déchiffrer *vt* GÉNÉRAL decipher, INFORMAT, RADIO, SIGNAL, TV decode
déchirer *vt* TRANSMIS break
décibel *m* (*dB*) RADIO *acoustique* decibel (*dB*)
décideur *m* TV decision-maker
déciseconde: **~ sans erreur** *m* TÉL, error-free decisecond (*EFdS*)
décision: **~ douce** *f* DONNÉES, ÉLECTRON, SIGNAL soft decision; **~ ferme** *f* DONNÉES, ÉLECTRON, SIGNAL hard decision; **~ heuristique** *f* DONNÉES heuristic decision; **~ pondérée** *f* DONNÉES soft decision; **~ programmable** *f* DONNÉES soft decision; **~ souple** *f* DONNÉES soft decision
déclaration *f* DONNÉES, INFORMAT declaration, directive; **~ de conformité d'instance** GÉNÉRAL implementation conformance statement (*ICS*); **~ de conformité d'une instance de protocole** (*PICS*) DONNÉES protocol implementation conformance statement (*PICS*); **~ de conformité de mise en œuvre du protocole** (*PICS*) DONNÉES protocol implementation conformance statement (*PICS*); **~ de conformité d'objet géré** DONNÉES managed-object conformance statement (*MOCS*); **~ de conformité du système** DONNÉES system conformance statement (*SCS*); **~ en notation combinée arborescente et tabulaire** DONNÉES TTCN statement
déclasser *vt* GÉNÉRAL downgrade
déclenchement *m* ALIM ÉLEC tripping, ÉLECTRON triggering, tripping, INFORMAT activation, triggering, RADIO, TESTS triggering; **~ d'alarme** TRANSMIS alarm on; **~ d'un appel** TÉL call initiation; **~ par boucle** TÉL loop start; **~ du signal d'appel** TÉL ring start
déclencher *vt* ALIM ÉLEC *disjoncteurs* trip, ÉLECTRON trigger, trip, INFORMAT activate, trigger, RADIO, TESTS *alarme* trigger, set off
déclin *m* INFORMAT decay
décodage *m* INFORMAT, RADIO, SIGNAL, TV decoding; **~ algébrique** SIGNAL algebraic decoding; **~ à décision douce** DONNÉES, ÉLECTRON, SIGNAL soft-decision decoding; **~ à décision formelle** DONNÉES, ÉLECTRON, SIGNAL hard-decision decoding; **~ à décision programmable** DONNÉES *satellite maritime* soft-decision decoding; **~ à décision souple** DONNÉES soft-decision decoding; **~ local** ÉLECTRON feedback decoding; **~ séquentiel** TÉL sequential decoding; **~ à seuil** DONNÉES threshold decoding; **~ un sur cinq** ÉLECTRON one out of five decode; **~ de Viterbi** TÉL Viterbi decoding
décoder *vt* INFORMAT, RADIO, SIGNAL, TV decode
décodeur *m* INFORMAT decoder, RADIO decoder, descrambler, SIGNAL, TV decoder; **~ FEC de rapport 1/2** SIGNAL half-rate FEC decoder; **~ d'interface** TRANSMIS interface decoder; **~ en ligne** TRANSMIS line decoder; **~ MICDA** TÉL ADPCM decoder; **~ à vraisemblance maximale** RADIO maximum likelihood decoder
décollement: **~ interlaminaire** *m* ÉLECTRON delamination; **~ de pastille** *m* ÉLECTRON lifted land
décompacté *adj* DONNÉES, INFORMAT decompressed
décomposition *f* COMPOSANTS decomposition, GÉNÉRAL *d'une commande* breakdown
décompression *f* DONNÉES, INFORMAT decompression, unzipping, RADIO rarefaction; **~ de fichier** DONNÉES, INFORMAT file decompression
décomprimer *vt* DONNÉES, INFORMAT decompress, unzip
décompte *m* TÉL revenue-sharing arrangement
décompteur *m* PÉRIPH countdown timer
décondenser *vt* DONNÉES, INFORMAT unpack
déconnecté *adj* ALIM ÉLEC, COMMUT, ÉLECTRON switched off, INFORMAT off-hook, PÉRIPH, RADIO switched off, TÉL disconnected, off-hook
déconnecter *vt* GÉNÉRAL disconnect, switch off
déconnexion *f* (*DCN*) DONNÉES disconnect (*DCN*), INFORMAT disconnection; **~ de priorité** TÉL priority disconnection
découpage *m* INFORMAT clipping; **~ de la bande selon un code de brassage** DONNÉES rolling-code band splitting; **~ réception-émission** RADIO receiver-transmitter isolation; **~ du temps** INFORMAT time slicing; **~ en voies** TRANSMIS channeling (*AmE*), channelling (*BrE*)
découpe: **~ de montage** *f* ÉLECTRON mounting cut-out
découplage *m* ALIM ÉLEC, ÉLECTRON decoupling; **~ de polarisations croisées** TRANSMIS cross-polar discrimination
décourager *vt* INFORMAT deter
décrochage *m* APP SERV lifting, DONNÉES off-hooking, PÉRIPH, TÉL lifting, unhooking
décroché *adj* TÉL off-hook, off the hook
décrocher *vt* APP SERV lift, DONNÉES off-hook, PÉRIPH, TÉL lift, unhook
décroissance *f* GÉNÉRAL decrease, INFORMAT decay, RADIO roll-off
décrypter *vt* INFORMAT unscramble
DECT *abrév* (*service numérique européen de téléphonie sans fil*) ALIM ÉLEC DECT (*digital European cordless telephone service*)
défaillance *f* GÉNÉRAL breakdown, failure, TÉL, TESTS failure, outage; **~ aléatoire** TESTS random failure; **~ d'un canal sémaphore** SIGNAL signaling-link failure (*AmE*), signalling-link failure (*BrE*); **~ catastrophique** TESTS sudden total failure, catastrophic failure; **~ complète** TESTS complete failure; **~ critique** TESTS *sûreté de fonctionnement* critical failure; **~ cumulée** TESTS *sécurité de fonctionnement* accumulative failure; **~ d'un élément vital** TESTS single-point failure (*SPF*); **~ interne** TESTS internal damage, internal failure; **~ d'une liaison de signalisation** SIGNAL signaling-link failure (*AmE*), signalling-link failure (*BrE*); **~ majeure** TESTS major failure; **~ au niveau élémentaire** TESTS single-point failure (*SPF*); **~ non significative** TESTS nonrelevant failure; **~ par dérive** TESTS drift failure, gradual failure; **~ par usure** TESTS *sécurité de fonctionnement* ageing failure, wear-out failure; **~ par vieillissement** TESTS *sécurité de fonctionnement* ageing failure, wear-out failure; **~ partielle** TESTS partial failure; **~ pertinente** TESTS relevant failure; **~ précoce** TESTS early failure; **~ primaire** TESTS primary failure; **~ progressive** TESTS drift failure, gradual failure; **~ reproductible** TESTS reproducible failure; **~ du réseau** TÉL network failure; **~ de satellite** RADIO, TV satellite outage; **~ secondaire** TESTS secondary failure; **~ de signal** SIGNAL signal failure (*SF*); **~ significative** TESTS relevant failure; **~ soudaine** TESTS sudden failure; **~ soudaine et complète** TESTS sudden total failure; **~ sporadique** TESTS sporadic failure; **~ systématique** TESTS systematic failure; **~ technique** TESTS technical breakdown
défaut *m* GÉNÉRAL defect, fault, imperfection; **~ d'adaptation** GÉNÉRAL, TRANSMIS mismatch; **~ d'alignement** TRANSMIS alignment fault; **~ d'asservissement** INFORMAT mistracking; **~ de continuité** TESTS continuity fault; **~ d'enroulement** COMPOSANTS wrapping fault; **~ intermittent** TESTS intermittent fault; **~ d'isolement** ALIM ÉLEC insulation fault; **~ à la masse** ALIM ÉLEC earth fault (*BrE*), ground fault (*AmE*); **~ mineur** TESTS minor defect; **~ par le réseau** ÉLECTRON, TÉL network default (*ND*); **~ en réception à l'extrémité distante** SIGNAL far-end receive failure (*FERF*); **~ de rémanence** TV image retention fault; **~ à la terre** ALIM ÉLEC earth fault (*BrE*), ground fault (*AmE*); **~ de verrouillage trame** (*DVT*) DONNÉES out-of-frame
défectueux *adj* INFORMAT, TÉL, TESTS defective, faulty; **~ critique** TESTS *qualité* critical defective; **~ majeur** TESTS major defective; **~ mineur** TESTS minor defective

défectuosité: ~ **d'image** *f* TV picture blemish
défense: ~ **centrale** *f* TÉL central defence
défiché *adj* ÉLECTRON unplugged
défilement *m* INFORMAT scrolling
défini: ~ **par l'utilisateur** *adj* INFORMAT user-defined
définition *f* DONNÉES, ÉLECTRON, INFORMAT, TÉL, TRANSMIS, TV definition, resolution; ~ **à l'émission** TÉL, TRANSMIS transmission definition; ~ **de fichier** DONNÉES, INFORMAT file layout; ~ **générique** DONNÉES generic definition; ~ **d'image** TV picture definition; ~ **à la réception** TÉL reception definition
déflecteur *m* RADIO deflector
déformation *f* GÉNÉRAL distortion, garbling, INFORMAT warping, SIGNAL, TÉL pulse droop; ~ **faciale** INFORMAT facial warping; ~ **de forme libre** INFORMAT freeform deformation (*FFD*); ~ **plastique** COMPOSANTS plastic deformation
dégagement *m* ALIM ÉLEC clearance hole
dégager *vt* GÉNÉRAL disengage, TÉL clear
dégainer *vt* GÉNÉRAL remove the sheath from
dégauchir *vt* COMPOSANTS straighten
dégénérescence: ~ **d'une orbite** *f* RADIO orbital decay
dégiguer *vt* TRANSMIS dejitter
dégradation *f* RADIO, SIGNAL, TRANSMIS degradation, impairment; ~ **catastrophique** TÉL catastrophic degradation; ~ **de gain** RADIO gain degradation; ~ **progressive** RADIO, TRANSMIS graceful degradation, gradual degradation; ~ **du signal** SIGNAL signal degradation (*SD*), signal weakening
degré: ~ **de cohérence** *m* TRANSMIS coherence degree, degree of coherence; ~ **conventionnel de distorsion** *m* TÉL conventional degree of distortion; ~ **de distorsion arythmique** *m* TÉL degree of start-stop distortion; ~ **de distorsion arythmique globale** *m* TÉL degree of gross start-stop distortion; ~ **de distorsion arythmique au synchronisme** *m* TÉL degree of synchronous start-stop distortion; ~ **de distorsion d'essai normalisé** *m* TÉL degree of standardized test distortion; ~ **de distorsion isochrone** *m* TÉL degree of isochronous distortion; ~ **de distorsion parallèle anisochrone en avance** *m* TÉL degree of early anisochronous parallel distortion; ~ **de distorsion parallèle anisochrone en retard** *m* TÉL degree of late anisochronous parallel distortion; ~ **de distorsion propre** *m* TÉL degree of inherent distortion; ~ **de distorsion en service** *m* TÉL degree of distortion in service; ~ **de liberté** *m* GÉNÉRAL degree of freedom; ~ **de modulation** *m* TRANSMIS degree of modulation; ~ **d'occupation des voies** *m* RADIO channel occupancy rate
dégroupement: ~ **de la charge d'espace** *m* ÉLECTRON space-charge debunching; ~ **par répulsion mutuelle** *m* ÉLECTRON space-charge debunching
dégrouper *vt* INFORMAT ungroup
dehors: **en** ~ **de l'axe du faisceau principal** *adj* RADIO off-axis
DEL *abrév* (*diode électroluminescente, diode émettrice de lumière*) COMPOSANTS, ÉLECTRON, TÉL, TRANSMIS LED (*light-emitting diode*); ~ **à émission frontale** *f* COMPOSANTS, TRANSMIS Burrus diode; ~ **à émission longitudinale** *f* TRANSMIS edge-emitting LED
délai: ~ **d'acheminement** *m* DONNÉES routing delay, transmission delay; ~ **administratif** *m* TESTS *sécurité de fonctionnement* administrative delay; ~ **d'attente après numérotation** *m* RADIO, TÉL post-dialing delay (*AmE*), post-dialling delay (*BrE*); ~ **de conservation** *m* RADIO retention period; ~ **d'établissement de la communication** *m* TÉL call setup delay (*CSUD*); ~ **d'exécution** *m* INFORMAT turnaround time; ~ **haute tension** *m* RADIO high-voltage delay time, HV delay time; ~ **d'intervention** *m* GÉNÉRAL response time; ~ **d'inversion** *m* DONNÉES modem turnaround; ~ **logistique** *m* TÉL, TESTS logistic delay; ~ **moyen d'attente** *m* INFORMAT mean waiting time; ~ **moyen de rétablissement du service** *m* TESTS mean time to service restoration (*MTSR*); ~ **de raccordement** *m* TV connection delay; ~ **de recette technique** *m* GÉNÉRAL commissioning time; ~ **de réponse** *m* TÉL answering time, answering delay; ~ **de restitution** *m* TÉL restitution delay; ~ **rotationnel** *m* INFORMAT rotational delay; ~ **technique** *m* TESTS technical delay; ~ **de transmission de la demande** *m* TRANSMIS request transmission time; ~ **de transmission de paquets** *m* DONNÉES packet delay
délester *vt* TÉL decrease load, relieve
délimiteur *m* GÉNÉRAL, TESTS delimiter; ~ **arithmétique** INFORMAT arithmetic delimiter
délissage *m* RADIO desmoothing
demande[1]: **à la** ~ *adj* TV on-demand; ~**-réponse** *adj* RADIO request-response
demande[2] *f* DONNÉES, INFORMAT, RADIO, TÉL, TV request (*req*); ~ **d'acceptation de changement** DONNÉES ready-to-change request; ~ **d'accusé de réception** DONNÉES acknowledgement request; ~ **d'appel** TÉL call request; ~ **de communication** TÉL call request; ~ **de connexion** DONNÉES charge request, connection request; ~ **de connexion de réseau** DONNÉES N-connect request; ~ **de gel de l'image** TV freeze-picture request; ~ **d'identification** DONNÉES interrogation; ~ **latente** RADIO suppressed demand; ~ **de libération** TÉL clear request condition, clearing request; ~ **de libération émise par l'ETTD** TÉL DTE clear request; ~ **de libération forcée** SIGNAL compulsory clearing request; ~ **de ligne** TÉL bid for a line; ~ **de modification** DONNÉES connection request, GÉNÉRAL change request; ~ **de modification d'appel** (*MAD*) TÉL call-modification request (*CMR*); ~ **de page partielle** DONNÉES, TÉL partial-page request (*PPR*); ~ **partielle de valeurs d'attribut** DONNÉES partial attribute request; ~ **pour émettre** DONNÉES request to send (*RTS*); ~ **préalable** TÉL priority request; ~ **de prestation** GÉNÉRAL request for service; ~ **primitive** DONNÉES primitive request; ~ **de rafraîchissement rapide** INFORMAT fast update request (*FUR*); ~ **de renvoi** TÉL forwarding request; ~ **de service complémentaire** DONNÉES, TÉL facility request; ~ **de service supplémentaire** *m* DONNÉES, TÉL facility request; ~ **de trafic** RADIO traffic demand; ~ **de transfert de données** DONNÉES, TÉL request data transfer; ~ **de vidage** INFORMAT dump request
demandé *m* TÉL called party, called subscriber, callee
demander: ~ **à une unité de prendre le relais** *loc* INFORMAT shoulder-tap
demandeur *m* DONNÉES invoker, requestor, TÉL caller, calling party, calling subscriber, originator; ~ **d'association** DONNÉES association initiator
démarrage: ~ **à distance** *m* CONTRÔLE remote start; ~ **étoile-triangle** *m* ALIM ÉLEC star-delta starting; ~ **intempestif** *m* GÉNÉRAL false start; ~ **de la taxation** *m* APP SERV initial metering
déménagement *m* ALIM ÉLEC removal
déménager *vt* ALIM ÉLEC remove
demi[1]: ~**-onde** *adj* ALIM ÉLEC, ÉLECTRON, RADIO, TRANSMIS half-wave
demi[2]: ~**-additionneur** *m* ÉLECTRON half-adder; ~**-alternance** *f* ALIM ÉLEC half-cycle; ~**-grand axe** *m* RADIO semimajor axis; ~**-intensité** *f* TV half-intensity; ~**-période** *f* ALIM ÉLEC half-cycle; ~**-soustracteur** *m* ÉLECTRON half-subtracter; ~**-suppresseur d'écho** *m* TÉL half-echo suppressor; ~**-ton** *m* INFORMAT semitone, TÉL half-toning
démodulateur *m* ÉLECTRON, INFORMAT, RADIO, TRANSMIS demodulator; ~ **à abaissement de seuil** ÉLECTRON, RADIO threshold extension demodulator; ~ **de fréquence** TRANSMIS frequency demodulator; ~ **à seuil amélioré** ÉLECTRON, RADIO threshold extension demodulator; ~ **télégraphique** TÉL telegraphic demodulator

démodulation *f* ÉLECTRON, INFORMAT, RADIO, TRANSMIS demodulation; **~ cohérente** TRANSMIS coherent demodulation; **~ de fréquence** TRANSMIS frequency demodulation; **~ de phase** ÉLECTRON phase demodulation
démoduler *vt* ÉLECTRON, INFORMAT, RADIO, TRANSMIS demodulate
démontage *m* ALIM ÉLEC removal, GÉNÉRAL disassembly, dismantling, TÉL *d'un volume* unloading
démouillage *m* ÉLECTRON dewetting
démultiplexage *m* INFORMAT, RADIO, TÉL, TRANSMIS, TV demultiplexing
démultiplexeur *m* INFORMAT, RADIO, TÉL, TRANSMIS, TV demultiplexer; **~ numérique** TÉL digital demultiplexer; **~ d'onde guidée** TRANSMIS guided-wave demultiplexer (*GWD*); **~ optique** TV optical demultiplexer
déni *m* APP SERV *de service* denial
dénier *vt* APP SERV deny
densité: **~ acoustique** *f* TÉL sound density; **~ d'assemblage** *f* DONNÉES packing fraction; **~ de courant** *f* ALIM ÉLEC current density; **~ du courant de seuil** *f* ALIM ÉLEC, ÉLECTRON, TRANSMIS threshold current density; **~ d'énergie** *f* GÉNÉRAL energy density; **~ d'enregistrement** *f* DONNÉES packing density, recording density, ÉLECTRON packing density; **~ d'enregistrement binaire** *f* INFORMAT linear bit density; **~ flux** *f* ÉLECTRON flux density; **~ de flux magnétique** *f* ÉLECTRON magnetic-flux density; **~ d'intégration** *f* DONNÉES, ÉLECTRON packing density; **~ ionique** *f* ÉLECTRON ion concentration; **~ de piste** *f* INFORMAT track density; **~ de puissance** *f* ALIM ÉLEC power density; **~ spectrale** *f* DONNÉES, TRANSMIS spectral density; **~ spectrale d'éclairement** *f* GÉNÉRAL spectral irradiance; **~ spectrale d'éclairement énergétique** *f* GÉNÉRAL spectral irradiance; **~ spectrale de luminance** *f* TRANSMIS spectral radiance; **~ spectrale de luminance énergétique** *f* TRANSMIS spectral radiance; **~ surfacique** *f* ÉLECTRON flux density; **~ téléphonique** *f* TÉL telephone density; **~ de trafic** *f* RADIO, TÉL, TRANSMIS traffic density; **~ volumique de charge** *f* ALIM ÉLEC volume density of charge
dentures: **~ de masse** *f pl* ÉLECTRON earthing indents (*BrE*), grounding indents (*AmE*)
dénudage *m* COMPOSANTS stripping
dépanner *vt* DONNÉES debug, GÉNÉRAL, TESTS repair
dépaqueteur *m* DONNÉES, TÉL depacketizer
départ[1]: **de ~** *adj* GÉNÉRAL outgoing (*o/g*), TÉL originating, TRANSMIS outgoing (*o/g*)
départ[2]: **~ synchro** *m* (*départ synchronisé*) INFORMAT sync start, synch start
dépassement *m* TÉL overshoot; **~ de capacité** INFORMAT overflow; **~ de temps imparti** DONNÉES, ÉLECTRON, RADIO, TÉL time-out
dépasser: **~ sur le plan des performances** *loc* INFORMAT outperform; **~ le temps imparti** *loc* DONNÉES, ÉLECTRON, RADIO, TÉL time-out
dépendance: **~ de validation** *f* ÉLECTRON enable dependency
dépendant: **~ de la fonction** *adj* GÉNÉRAL function-dependent; **~ de la tension** *adj* ALIM ÉLEC, CONTRÔLE voltage-dependent
déperditeur: **~ de chaleur** *m* ALIM ÉLEC, ÉLECTRON heat sink
déphasage *m* ALIM ÉLEC, ÉLECTRON, RADIO, TRANSMIS phase shift; **~ angulaire** TÉL angular displacement; **~ arrière** ALIM ÉLEC lagging phase shift; **~ avant** ALIM ÉLEC leading phase shift; **~ linéique** ÉLECTRON, GÉNÉRAL *acoustique*, RADIO, TRANSMIS phase constant, phase coefficient, phase change coefficient; **~ linéique de propagation acoustique** GÉNÉRAL *acoustique* acoustic phase coefficient; **~ sur images** TRANSMIS image phase-change coefficient; **~ vers l'avant** TRANSMIS phase advance; **~ vers l'avant du signal transmis** TRANSMIS transmitted phase advance
déphaseur *m* ÉLECTRON, RADIO, SIGNAL phase changer, phase shifter; **~ de guide d'ondes** RADIO waveguide phase changer, waveguide phase shifter; **~ hyperfréquence** RADIO microwave phase changer; **~ rotatif** RADIO rotary phase changer, rotary phase shifter
dépister *vt* COMMUT trace
déplacement *m* ÉLECTRON shift, GÉNÉRAL displacement, INFORMAT, RADIO, TÉL, TRANSMIS shift; **~ en cours de communication** TÉL portability; **~ de fréquence** RADIO, TRANSMIS frequency displacement, frequency shift; **~ longitudinal** GÉNÉRAL longitudinal displacement; **~ mécanique du spot** ÉLECTRON mechanical spot displacement; **~ parallèle** TRANSMIS parallel displacement; **~ vertical** GÉNÉRAL vertical displacement
déplacer *vt* INFORMAT move, RADIO relocate; **~ vers le bas** INFORMAT move down; **~ vers le haut** INFORMAT move up
déploiement *m* GÉNÉRAL deployment, RADIO deployment, roll-out; **~ anticipé** GÉNÉRAL early deployment
déployer *vt* RADIO *antenne* roll out, unfurl
dépointage *m* RADIO pitch bias, depointing; **~ d'antenne** RADIO antenna swing off
dépolarisation *f* RADIO depolarization
dépose *f* ÉLECTRON removal
déposer *vt* ÉLECTRON remove, GÉNÉRAL demount
dépôt *m* COMPOSANTS, ÉLECTRON deposition, INFORMAT, TV deposit; **~ axial latéral au plasma** (*DALP*) ÉLECTRON axial lateral plasma deposition (*ALPD*); **~ axial en phase vapeur** (*DAV*) ÉLECTRON vapor axial deposition (*AmE*) (*VAD*), vapour axial deposition (*BrE*) (*VAD*); **~ chimique en phase vapeur** (*DCV*) ÉLECTRON vapor chemical deposition (*AmE*), vapour chemical deposition (*BrE*); **~ de garantie** TV deposit; **~ latéral au plasma** ÉLECTRON lateral plasma deposition; **~ par autocatalyse** ÉLECTRON autocatalytic plating; **~ par vaporisation** COMPOSANTS outer vapor deposition (*AmE*) (*OVD*), outer vapour deposition (*BrE*) (*OVD*); **~ en phase vapeur interne** ÉLECTRON inside vapor phase oxidation (*AmE*), inside vapour phase oxidation (*BrE*) (*IVPO*); **~ en phase vapeur de matériaux métal-organiques** ÉLECTRON, TRANSMIS metal-organic chemical vapor deposition (*AmE*), metal-organic chemical vapour deposition (*BrE*); **~ sans courant** ÉLECTRON electroless plating
dépupiniser *vt* GÉNÉRAL de-load
déquantification *f* INFORMAT dequantization
dérangement[1]: **en ~** *adj* TÉL out-of-action, TESTS out-of-order
dérangement[2] *m* ALIM ÉLEC fault, GÉNÉRAL disturbance, failure, TÉL, TESTS failure, fault; **~ dégradant les fonctions** TÉL function-degrading failure; **~ empêchant l'accomplissement des fonctions** TÉL, TESTS function-preventing failure; **~ intermittent** TÉL intermittent failure; **~ d'une ligne** TÉL line fault; **~ permanent** TÉL permanent failure; **~ permettant l'accomplissement des fonctions** TÉL, TESTS function-permitting failure; **~ de réception à l'extrémité** SIGNAL far-end receive failure (*FERF*); **~ du réseau** DONNÉES network failure; **~ simple** TESTS individual fault; **~ temporaire** TÉL, TESTS temporary fault, temporary malfunction
déranger *vt* GÉNÉRAL disturb
déréglage *m* INFORMAT detuning
déréglementation *f* GÉNÉRAL deregulation, TV liberalization
dérégulation *f* GÉNÉRAL, TV deregulation
dérivateur *m* RADIO hybrid ring
dérivation[1]: **en ~** *adj* TÉL bridged
dérivation[2] *f* ALIM ÉLEC bypassing, ÉLECTRON bridge, INFORMAT bypass, TRANSMIS branch-off; **~ automatique** ÉLECTRON automatic bypass; **~ et insertion** TRANSMIS drop and insert; **~ optique** TRANSMIS optical branching, optical tapping
dérivé *m* GÉNÉRAL derivative
dérive *f* GÉNÉRAL deviation, RADIO, TRANSMIS drift; **~ de**

courant ALIM ÉLEC current drift; ~ **de fréquence** ÉLECTRON, RADIO, TRANSMIS frequency drift; ~ **de fréquence normée** RADIO normalized frequency drift; ~ **des gyroscopes** RADIO gyroscopic drift; ~ **en longitude de l'orbite** RADIO orbit drift
dérivée *f* GÉNÉRAL derivative
dériver[1] *vt* GÉNÉRAL derive
dériver[2] *vi* GÉNÉRAL drift
dernier[1]: ~ **commutateur n° sept** *m* COMMUT no. seven exchange - last
dernier[2]: ~ **entré, premier sorti** *loc* (*LIFO*) INFORMAT last in, first out (*LIFO*)
dernière: ~ **localisation connue** *f* RADIO last known location; ~ **tentative** *f* TÉL last try
dérogation: ~ **après production** *f* TESTS concession; ~ **avant production** *f* TESTS deviation permit, production permit
déroulement *m* GÉNÉRAL, TESTS unwrapping
dérouler *vt* GÉNÉRAL pay out, unwind, unwrap, TESTS unwrap
dérouleur *m* INFORMAT streamer; ~ **de bande** APP SERV, INFORMAT, TV tape transport unit (*TTU*); ~ **de bande magnétique** TV magnetic-tape drive
dérouter *vt* RADIO redirect; ~ **sur** INFORMAT divert to
dérouteuse *f* DONNÉES routing device
désabonner: **se** ~ *v réfl* INFORMAT unsubscribe
désaccentuation *f* ÉLECTRON de-emphasis
désaccord *m* INFORMAT detune
désaccordage *m* INFORMAT detune
désaccorder *vt* INFORMAT detune
désactivation *f* DONNÉES, ÉLECTRON, INFORMAT, TÉL deactivation
désactiver *vt* DONNÉES, ÉLECTRON deactivate, INFORMAT deactivate, disable, TÉL deactivate
désadaptation *f* GÉNÉRAL, TRANSMIS mismatch
désaimantation *f* ÉLECTRON demagnetization, demagnetizing
désaimanter *vt* ÉLECTRON demagnetize
désalignement: ~ **angulaire des axes de fibres** *m* TRANSMIS *fibres optiques* angular displacement
désassemblage: ~ **de paquets** *m* DONNÉES, TÉL packet disassembly
désassembleur: ~ **de paquets** *m* DONNÉES, TÉL depacketizer
descente: ~ **d'antenne** *f* RADIO, TÉL antenna lead-in, down lead, lead-in
descripteur *m* INFORMAT descriptor
description: ~ **de bloc fonctionnel** *f* TÉL functional-block description; ~ **de cellule de maintenance** *f* DONNÉES maintenance cell description (*MCD*); ~ **du dispositif** *f* TÉL device description; ~ **d'exploitation** *f* TÉL operational description; ~ **fonctionnelle** *f* GÉNÉRAL *en LDS* functional description (*FD*); ~ **d'interfonctionnement** *f* GÉNÉRAL interwork description; ~ **de processus** *f* TÉL process description; ~ **de la signalisation** *f* SIGNAL signaling description (*AmE*), signalling description (*BrE*)
désembrouiller *vt* DONNÉES, RADIO, SIGNAL, TÉL, TV descramble
désensibilisation *f* ÉLECTRON desensitization
déséquilibré *adj* ALIM ÉLEC, COMMUT, ÉLECTRON, INFORMAT, TÉL unbalanced
déséquilibre *m* ALIM ÉLEC, COMMUT, ÉLECTRON, INFORMAT, TÉL unbalance; ~ **d'impédance** ÉLECTRON, TÉL impedance unbalance; ~ **de la répartition du trafic** COMMUT, TÉL traffic distribution imbalance, traffic imbalance, traffic load imbalance
désétamage *m* COMPOSANTS tin stripping
désignateur: ~ **de code international** *m* DONNÉES international code designator (*ICD*)
désignation *f* GÉNÉRAL, TÉL, TRANSMIS designation; ~ **de câble** TRANSMIS cable designation; ~ **de fonction** GÉNÉRAL function designation; ~ **de gabarit pour peigne** TRANSMIS formboard designation
désigner *vt* GÉNÉRAL, TÉL, TRANSMIS designate
désionisation *f* ÉLECTRON de-ionization
desserte *f* APP SERV serving, GÉNÉRAL service, servicing; ~ **aux abonnés** DONNÉES, INFORMAT, RADIO, TÉL subscriber service
desservir *vt* GÉNÉRAL serve, service
dessin: ~ **de câblage** *m* GÉNÉRAL *documentation* cabling plan; ~ **d'état de surface** *m* COMPOSANTS surface finish drawing; ~ **d'étiquetage** *m* GÉNÉRAL designation drawing; ~ **modèle** *m* ALIM ÉLEC *circuits imprimés* artwork master; ~ **de numérotage des broches** *m* GÉNÉRAL numbering drawing; ~ **de parcours de câble** *m* GÉNÉRAL *documentation* cable-way drawing; ~ **de pose** *m* GÉNÉRAL *documentation* cable drawing; ~ **de raccordement** *m* GÉNÉRAL terminal drawing; ~ **de trajet de câble** *m* GÉNÉRAL *documentation* cable-way drawing
dessinateur: ~ **projeteur** *m* GÉNÉRAL engineering draftsman
dessiner *vt* GÉNÉRAL, INFORMAT draw
dessouder *vt* COMPOSANTS, ÉLECTRON unsolder
destinataire *m* DONNÉES receiver, recipient, transmission recipient, GÉNÉRAL recipient, PÉRIPH, RADIO, TÉL receiver, TRANSMIS transmission recipient; ~ **de l'échange** DONNÉES, TÉL interchange recipient ~ **N** DONNÉES N-recipient
destination *f* DONNÉES, INFORMAT, TÉL destination; ~ **des factures téléphoniques** TÉL billing address
destiner *vt* RADIO, TÉL, TV target
désynchronisation: ~ **image-son** *f* INFORMAT picture-sound desynchronization, picture-sound lag
désynchronisé *adj* INFORMAT offbeat
détachement *m* GÉNÉRAL detachment
détacher *vt* DONNÉES unbind
détecter *vt* GÉNÉRAL detect, sense
détecteur *m* GÉNÉRAL detector, sensor; ~ **actif** RADIO active sensor; ~ **d'autocorrélation** RADIO auto-correlation detector; ~ **bifréquence** TESTS two-tone detector; ~ **de bruit** TÉL noise detector; ~ **de câble** TRANSMIS *génie industriel extérieur* cable locater; ~ **de champ lumineux** TRANSMIS *fibres optiques* bright-field detector; ~ **de champ sombre** TRANSMIS *fibres optiques* dark-field detector; ~ **de crête** GÉNÉRAL peak detector; ~ **de données dans la bande vocale** DONNÉES voice-band data detector; ~ **d'erreurs** DONNÉES error detector; ~ **à filtre et échantillonnage** ÉLECTRON filter-and-sample detector; ~ **de fissures** TESTS crack detector; ~ **de fuite à la terre** ALIM ÉLEC earth leakage guard (*BrE*), ground leakage guard (*AmE*); ~ **d'horizon terrestre** RADIO earth sensor; ~ **d'hyperfréquences** RADIO microwave detector; ~ **d'image** PÉRIPH, TV image sensor; ~ **d'interruption de la porteuse de données** DONNÉES, TRANSMIS data carrier failure detector; ~ **d'interruption de la voie de données** DONNÉES, TÉL data channel failure detector; ~ **d'intrusion** APP SERV intruder presence detector; ~ **linéaire** ÉLECTRON linear detector; ~ **de lumière** DONNÉES photodetector, TRANSMIS optical detector, photodetector; ~ **de mesure** TESTS measuring detector; ~ **de microcourbures** TRANSMIS microbend sensor; ~ **optique** TRANSMIS fiber-optic sensor (*AmE*), fibre-optic sensor (*BrE*), photodetector, TRANSMIS optical detector, photodetector; ~ **d'orientation** RADIO attitude sensor; ~ **de parole** RADIO, TÉL speech detector; ~ **passif** RADIO passive sensor; ~ **de perte de verrouillage de trame** TÉL loss of frame alignment detector; ~ **de phase** ÉLECTRON phase detector; ~ **photosensible** DONNÉES photodetector, TRANSMIS optical detector, photodetector; ~ **quadratique** ÉLECTRON square-law detector; ~ **de radar** RADIO radar warning receiver; ~ **de signal** SIGNAL signal tracer, signal detector; ~ **de signal de ligne reçu sur la voie de données** DONNÉES, TRANSMIS data channel received line signal detector; ~ **de signalisation** SIGNAL signaling detector (*AmE*), signalling detector (*BrE*); ~ **de température résistif** CLIMAT resistive temperature detector (*RTD*); ~ **de zéro** RADIO null detector

détection *f* CONTRÔLE, DONNÉES, ÉLECTRON, GÉNÉRAL, INFORMAT, RADIO, TÉL, TRANSMIS detection; **~ d'activité** INFORMAT active sensing; **~ automatique** RADIO *radar* automatic detection; **~ automatique d'erreurs** RADIO automatic error detection; **~ des câbles** TRANSMIS cable localization; **~ cohérente** TRANSMIS coherent detection; **~ de collision** ÉLECTRON collision detection (*CD*); **~ des contours** DONNÉES edge detection; **~ par crayon optique** INFORMAT light-pen detection; **~ de défaut** GÉNÉRAL fault detection; **~ des dérangements** GÉNÉRAL fault detection; **~ d'erreur** DONNÉES error detection, RADIO error check; **~ hétérodyne** ÉLECTRON heterodyne detection; **~ homodyne** TRANSMIS homodyne detection; **~ d'impulsions** SIGNAL pulse detection; **~ linéaire** ÉLECTRON linear detection; **~ de modification** CONTRÔLE, DONNÉES manipulation detection; **~ du mot unique** DONNÉES unique word detection; **~ de panne** TÉL fault tracing, TESTS error recognition; **~ par photostyle** INFORMAT light-pen detection; **~ de porteuse** TRANSMIS carrier detect (*CD*); **~ de porteuse de données** DONNÉES data carrier detection (*DCD*); **~ quadratique** ÉLECTRON square-law detection; **~ de la qualité du signal de données** DONNÉES, SIGNAL data signal quality detection; **~ du repère** DONNÉES detection of marker; **~ séquentielle** TESTS sequential sensing; **~ de signal** SIGNAL signal detection

détectivité *f* ÉLECTRON, TRANSMIS detectivity; **~ normalisée** ÉLECTRON, TESTS, TRANSMIS normalized detectivity, specific detectivity, D-star; **~ normée** ÉLECTRON, TESTS, TRANSMIS normalized detectivity, specific detectivity, D-star; **~ spécifique** ÉLECTRON, TESTS, TRANSMIS normalized detectivity, specific detectivity, D-star

détendeur *m* COMPOSANTS pressure-reducing valve

détente *f* ÉLECTRON, GÉNÉRAL expansion

détenteur: **~ d'une carte** *m* TÉL card holder

détérioration: **~ de la qualité** *f* TESTS deterioration of quality

détermination: **~ automatique** *f* RADIO *radar* automatic measurement; **~ de faisceau** *f* TÉL route determination; **~ d'itinéraire** *f* TÉL route determination; **~ de l'orbite** *f* RADIO orbit determination; **~ de statut** *f* INFORMAT set status; **~ de taxation** *f* TÉL tariff determination

déterminer: **~ l'origine d'un appel** *vt* TÉL trace a call; **~ l'origine d'une communication** *vt* TÉL trace a call

déterrement *m* APP SERV excavation damage, excavator damage

détourage *m* INFORMAT scissoring

détournement *m* TÉL detour

détourneur: **~ d'appels** *m* TÉL automatic call transfer unit, call diverter

détramage *m* DONNÉES deframing, descreening

détrompage *m* TÉL keying

détrompeur *m* ALIM ÉLEC key, ÉLECTRON key, polarizing slot

Deutsche Industrie-Normen *m* (*DIN*) GÉNÉRAL Deutsche Industrie-Normen (*DIN*)

deux: **à ~ paires de bornes** *adj* ÉLECTRON, TRANSMIS two-terminal-pair

deuxième: **~ amplificateur FI** *m* ALIM ÉLEC second IF amplifier; **~ canal adjacent** *m* RADIO second adjacent channel; **~ choix** *m* TÉL second choice; **~ écho** *m* RADIO second time around echo

devancer *vt* ÉLECTRON lead

développement *m* GÉNÉRAL development; **~ de l'arborescence** DONNÉES, INFORMAT tree expansion

développer *vt* GÉNÉRAL develop

devenir: **~ conducteur** *vi* ÉLECTRON turn on; **~ passant** *vi* ÉLECTRON turn on

déverminage *m* INFORMAT burn-in

déviateur: **~ d'appel** *m* TÉL call forwarder, call diverter

déviation *f* ALIM ÉLEC, ÉLECTRON, GÉNÉRAL, RADIO, TÉL, TESTS, TRANSMIS deflection; **~ de l'aiguille** TÉL needle deflection, pointer deflection; **~ de fréquence** RADIO frequency departure; **~ latérale** RADIO lateral deviation; **~ magnétique** ÉLECTRON magnetic deflection; **~ de phase** ALIM ÉLEC, GÉNÉRAL, RADIO, TRANSMIS phase deviation; **~ symétrique** ÉLECTRON symmetrical deflection

devise: **~ personnelle** *f* DONNÉES personal icon

dévolter *vt* ALIM ÉLEC step down

dézingage *m* COMPOSANTS zinc-stripping

DGAD *abrév* (*domaine de gestion d'administration*) DONNÉES, TÉL ADMD (*administration management domain*)

diac *m* ÉLECTRON diac

diagnostic *m* GÉNÉRAL, INFORMAT, TESTS diagnosis, diagnostics; **~ détaillé** GÉNÉRAL detailed diagnosis, fine diagnosis; **~ d'erreurs** DONNÉES error diagnosis; **~ de panne** TESTS fault diagnosis

diagramme *m* GÉNÉRAL diagram; **~ d'arbre de blocs** DONNÉES block tree diagram; **~ de charge** ALIM ÉLEC load function; **~ de corrélation** INFORMAT scatter diagram; **~ de courbe de charge** ALIM ÉLEC load curve diagram; **~ de couverture** RADIO *radar* coverage diagram; **~ de diffraction** RADIO, TRANSMIS diffraction pattern; **~ de diffraction en champ lointain** RADIO, TRANSMIS far-field diffraction pattern; **~ de diffraction en champ proche** RADIO, TRANSMIS near-field diffraction pattern; **~ de diffraction de Fraunhofer** TRANSMIS Fraunhofer diffraction pattern; **~ de diffraction de Fresnel** TRANSMIS Fresnel diffraction pattern; **~ de directivité d'antenne** RADIO antenna directivity diagram; **~ de directivité horizontal** RADIO horizontal directivity pattern; **~ de directivité vertical** RADIO vertical directivity pattern; **~ de fiabilité** TESTS reliability block diagram; **~ de fluence de signal** SIGNAL signal flow diagram; **~ de flux de contrôle** DONNÉES control flow diagram; **~ de fonctionnement** INFORMAT running diagram, working diagram, TÉL sequence chart; **~ de gain d'antenne** RADIO antenna gain pattern; **~ de gain de référence** RADIO antenna gain pattern, reference antenna pattern; **~ en huit** RADIO figure-of-eight diagram; **~ des lignes de force** ÉLECTRON flux pattern; **~ de Nyquist** CONTRÔLE Nyquist diagram; **~ en œil** ÉLECTRON, TRANSMIS eye diagram, eye pattern, eyeshape pattern; **~ de programmation** INFORMAT programming flow chart; **~ de rayonnement** RADIO, TRANSMIS radiation pattern; **~ de rayonnement d'antenne** RADIO antenna radiation pattern; **~ de rayonnement de champ lointain** RADIO, TRANSMIS far-field radiation pattern; **~ de rayonnement de champ proche** RADIO, TRANSMIS near-field radiation pattern; **~ de rayonnement à l'équilibre** TRANSMIS equilibrium radiation pattern; **~ séquentiel** TÉL sequence chart; **~ de structure** INFORMAT structure flow chart; **~ syntaxique** TÉL syntax diagram; **~ de système** INFORMAT system flow chart; **~ de transition d'état** TESTS state transition diagram

dialogue *m* DONNÉES dialog, talk, GÉNÉRAL interwork, interworking, INFORMAT, TV dialog; **~ homme-machine** INFORMAT man-machine dialog; **~ d'introduction** ÉLECTRON log-in dialog

dialoguer *vi* DONNÉES talk, GÉNÉRAL interwork, TV communicate

diamètre: **~ de l'âme** *m* TÉL, TRANSMIS *câble* core diameter; **~ du champ de mode** *m* TRANSMIS mode field diameter; **~ du cœur** *m* TÉL, TRANSMIS *fibres optiques* core diameter; **~ d'une couche** *m* GÉNÉRAL layer diameter; **~ extérieur** *m* GÉNÉRAL outer diameter; **~ d'un faisceau** *m* TRANSMIS *fibres optiques* beam diameter, beam width; **~ du faisceau guidé** *m* TRANSMIS mode field diameter; **~ de gaine** *m* TRANSMIS *fibres optiques* cladding diameter; **~ moyen du cœur** *m* TRANSMIS average core diameter; **~ moyen de la gaine** *m* TRANSMIS average cladding diameter; **~ moyen de la surface de référence** *m* TÉL average reference-surface diameter (*DRav*); **~ du noyau** *m* TÉL, TRANSMIS core diameter; **~ primitif** *m* GÉNÉRAL pitch diameter; **~ du spot** *m* TV spot size; **~ du stockage** *m* TRANSMIS stock diameter;

~ **sur flanc** *m* COMPOSANTS theoretical pitch diameter; ~ **de la surface de référence** *m* TÉL, TRANSMIS reference-surface diameter
diapason *m* GÉNÉRAL, INFORMAT tuning fork
diaphonie *f* TÉL, TRANSMIS crosstalk; ~ **entre les deux sens de transmission** TRANSMIS go to return crosstalk; ~ **inintelligible** TÉL unintelligible crosstalk
diaphragme *m* COMPOSANTS diaphragm, membrane
diaporama *m* INFORMAT slide show
diapositive *f* COMPOSANTS, INFORMAT slide; ~ **compilée** INFORMAT built slide; ~ **couleur** GÉNÉRAL color slide (*AmE*), colour slide (*BrE*)
dicorde *m* TÉL, TESTS cord circuit; ~ **avec point** TÉL bridge control cord circuit
didacticiel *m* DONNÉES, INFORMAT courseware, teachware, tutorial
diélectrique[1] *adj* COMPOSANTS, ÉLECTRON dielectric
diélectrique[2] *m* COMPOSANTS, ÉLECTRON dielectric; ~ **en rexolène cellulaire** ÉLECTRON cellular rexolene conductor
différé *adj* GÉNÉRAL, INFORMAT deferred; **en** ~ RADIO recorded, TÉL off-line, TV recorded
différence: ~ **entre échelles de temps** *f* INFORMAT, RADIO, TRANSMIS timescale difference; ~ **entre temps d'horloge** *f* GÉNÉRAL clock time difference; ~ **de fréquence** *f* RADIO frequency difference; ~ **de fréquence normée** *f* RADIO normalized frequency difference; ~ **de gain** *f* ÉLECTRON gain inequality; ~ **d'indice de réfraction** *f* TÉL ESI refractive index difference; ~ **de l'indice de réfraction** *f* TRANSMIS refractive-index contrast; ~ **de phase** *f* ALIM ÉLEC, ÉLECTRON phase difference; ~ **de potentiel** *f* ALIM ÉLEC potential difference, voltage (*V*); ~ **de potentiel entre deux terres** *f* ALIM ÉLEC earth potential difference (*BrE*), ground potential difference (*AmE*); ~ **de potentiel magnétique** *f* ÉLECTRON magnetic-potential difference
différentiation *f* TÉL differentiation
différentiel *adj* GÉNÉRAL differential
différés *m pl* ÉLECTRON deferrals
difficile: ~ **à exécuter** *adj* INFORMAT user-unfriendly; ~ **à utiliser** *adj* INFORMAT user-unfriendly
diffraction *f* TRANSMIS diffraction; ~ **multiple** TRANSMIS multiple diffraction; ~ **d'onde** ÉLECTRON wave diffraction
diffusé *adj* GÉNÉRAL broadcast, TÉL distributed, TV broadcast; ~ **par satellite** TV broadcast by satellite
diffuser *vt* GÉNÉRAL broadcast, TÉL distribute, TV broadcast; ~ **depuis un satellite** RADIO beam down
diffuseur *m* CLIMAT diffuser, TV broadcaster; ~ **d'appels** PÉRIPH message broadcaster; ~ **de messages** PÉRIPH message broadcaster
diffusiomètre *m* TESTS scatterometer; ~ **vecteur vagues** RADIO wavemeter
diffusion[1]: **en** ~ *adj* RADIO in broadcast mode
diffusion[2] *f* CLIMAT diffusion, GÉNÉRAL, TÉL, TRANSMIS broadcasting, *télex* broadcast (*BC*); ~ **différée** APP SERV, TRANSMIS store-and-forward transmission; ~ **de données** DONNÉES, TÉL data distribution; ~ **d'émissions de télévision par satellite** TV satellite television broadcasting; ~ **de fibrage** TRANSMIS fiber scattering (*AmE*), fibre scattering (*BrE*); ~ **ionosphérique** RADIO ionospheric scatter propagation, ionospheric scatter; ~ **de matériaux** RADIO material scattering; ~ **nonlinéaire** RADIO, TRANSMIS nonlinear scattering; ~ **de Raman** RADIO, TRANSMIS Raman scattering; ~ **Raman stimulée** RADIO, TRANSMIS stimulated Raman scattering (*SRS*); ~ **de Rayleigh** DONNÉES, TÉL, TRANSMIS Rayleigh scattering; ~ **restreinte** RADIO narrowcasting; ~ **stéréo** TV stereophonic broadcast; ~ **sur les précipitations** RADIO precipitation scatter; ~ **troposphérique** RADIO tropospheric scatter
digitalisation *f* GÉNÉRAL digitalization, digitization
digitalisé *adj* GÉNÉRAL digitalized, digitized
digitaliser *vt* GÉNÉRAL digitalize, digitize
dilatation *f* COMPOSANTS, ÉLECTRON, GÉNÉRAL expansion; ~ **linéaire** GÉNÉRAL linear expansion; ~ **thermique** COMPOSANTS, ÉLECTRON thermal expansion
dimension *f* DONNÉES, GÉNÉRAL, INFORMAT dimension, size; ~ **fractale** INFORMAT fractal dimension; ~ **temporelle** GÉNÉRAL time domain
dimensionnement *m* GÉNÉRAL, INFORMAT dimensioning, sizing
dimensionner *vt* GÉNÉRAL, INFORMAT size
diminuer[1] *vt* GÉNÉRAL decrease, INFORMAT decay
diminuer[2] *vi* GÉNÉRAL decrease, INFORMAT decay
diminution *f* GÉNÉRAL decrease, INFORMAT decay; ~ **de capacité** GÉNÉRAL capacity decrease (*CD*); ~ **de la qualité** TESTS quality degradation
DIN *abrév* (*Deutsche Industrie-Normen*) GÉNÉRAL DIN (*Deutsche Industrie-Normen*)
diode *f* COMPOSANTS, ÉLECTRON, TÉL, TRANSMIS diode; ~ **à avalanche** ÉLECTRON *semiconducteurs* avalanche diode; ~ **de Burrus** COMPOSANTS, TÉL, TRANSMIS Burrus diode; ~ **à capacité variable** COMPOSANTS, ÉLECTRON *semiconducteurs* capacitance diode, varactor; ~ **de commutation** COMMUT, COMPOSANTS, ÉLECTRON switching diode, switchover diode; ~ **à cristal** ALIM ÉLEC crystal diode; ~ **à double base** COMPOSANTS double-base diode; ~ **électroluminescente** (*DEL*) COMPOSANTS, ÉLECTRON, TÉL, TRANSMIS light-emitting diode (*LED*); ~ **électroluminescente à émission frontale** COMPOSANTS, TÉL, TRANSMIS surface-emitting light-emitting diode, *fibres optiques* Burrus diode; ~ **émettrice de lumière** (*DEL*) COMPOSANTS, ÉLECTRON, TÉL, TRANSMIS light-emitting diode (*LED*); ~ **à émission par la tranche** ÉLECTRON, TRANSMIS edge-emitting diode; ~ **de fixation de niveau** ÉLECTRON clamp diode, clamping diode; ~ **génératrice de bruit** ÉLECTRON noise-generator diode; ~ **idéale de bruit** ÉLECTRON ideal noise diode; ~ **à jonction** COMPOSANTS junction diode; ~ **laser** ÉLECTRON, TRANSMIS laser diode (*LD*); ~ **laser de Fabry-Pérot** TRANSMIS Fabry-Pérot laser diode (*FP laser diode*); ~ **laser à faible puissance** ÉLECTRON low-power laser diode; ~ **laser à injection** TRANSMIS injection laser diode (*ILD*); ~ **laser à rétroaction répartie** TRANSMIS distributed feedback laser diode (*DFB laser diode*); ~ **de niveau** ÉLECTRON clamp diode, clamping diode, RADIO DC clamp diode, DC restorer diode; ~ **planaire** COMPOSANTS planar diode; ~ **au plasma** ÉLECTRON plasma diode; ~ **à pointe** COMPOSANTS point-contact diode; ~ **à pointe d'or** COMPOSANTS gold-bonded diode; ~ **à porteurs chauds** ÉLECTRON hot carrier diode; ~ **de redressement à avalanche contrôlée** ÉLECTRON *semiconducteurs* controlled-avalanche rectifier diode; ~ **de redressement à semiconducteur** COMPOSANTS, ÉLECTRON semiconductor rectifier diode; ~ **de référence de tension** COMPOSANTS, ÉLECTRON voltage reference diode; ~ **régulatrice de tension** COMPOSANTS, ÉLECTRON voltage regulator diode; ~ **Schottky** COMPOSANTS Schottky diode; ~ **selon la deuxième approximation** COMPOSANTS, ÉLECTRON second approximation diode; ~ **à semiconducteur** COMPOSANTS, ÉLECTRON semiconductor diode; ~ **de signal** ÉLECTRON signal diode; ~ **au silicium** COMPOSANTS silicon diode; ~ **superluminescente** (*DSL*) COMPOSANTS, ÉLECTRON, TRANSMIS superluminescent LED (*SLD*), superluminescent light-emitting diode (*SLD*), superradiant diode (*SRD*); ~ **tunnel** COMPOSANTS, ÉLECTRON tunnel diode; ~ **unijonction** COMPOSANTS, ÉLECTRON unijunction diode; ~ **unitunnel** COMPOSANTS, ÉLECTRON unitunnel diode; ~ **varicap** COMPOSANTS, ÉLECTRON varactor; ~ **Zener** ÉLECTRON Zener diode
diplex *m* COMPOSANTS, RADIO, TÉL, TRANSMIS diplex
diplexeur *m* COMPOSANTS, RADIO, TÉL, TRANSMIS diplexer
direct *adj* GÉNÉRAL direct; **en** ~ DONNÉES live, TÉL on-line, INFORMAT, TV live
direction *f* RADIO, TÉL direction; ~ **de conversation** TÉL

direction of speech; **~ d'objectif** RADIO target direction; **~ de l'ombre** INFORMAT shadow offset; **~ des télécommunications** (*DT*) TÉL telecommunications board

directivité *f* RADIO directivity

discordance *f* ÉLECTRON mismatch

discret *adj* GÉNÉRAL discrete

discriminateur *m* COMMUT, DONNÉES discriminator, ÉLECTRON discriminator, restrictor, TÉL call-barring equipment; **~ de durée d'impulsion** RADIO pulse length discriminator; **~ de protocole** (*DP*) DONNÉES protocol discriminator (*PD*); **~ de transmission d'événement** DONNÉES event-forwarding discriminator

discrimination: **~ d'antenne** *f* RADIO antenna discrimination; **~ des appels** *f* TÉL call barring, call discrimination; **~ des messages** *f* INFORMAT message discrimination; **~ des messages de signalisation** *f* SIGNAL signaling-message discrimination (*AmE*), signalling-message discrimination (*BrE*); **~ par polarisation croisée** *f* RADIO, TÉL cross-polarization discrimination, TRANSMIS cross-polar isolation, cross-polarization discrimination; **~ de polarisation** *f* RADIO, TÉL, TRANSMIS cross-polarization discrimination, polarizing isolation

disharmonie *f* ÉLECTRON mismatch

disjoncteur *m* ALIM ÉLEC, ÉLECTRON circuit breaker; **~ à fusible** ALIM ÉLEC fuse switch; **~ du moteur** ALIM ÉLEC motor protector switch

disjonction *f* ÉLECTRON EXCLUSIVE-OR operation, non-equivalence, nonequivalence operation

disparité *f* GÉNÉRAL, TÉL disparity

disparition: **~ d'alarme** *f* TRANSMIS alarm off; **~ progressive** *f* INFORMAT, RADIO fade, fade-out, fading; **~ des signaux** *f* RADIO blackout

dispersion *f* GÉNÉRAL dispersion; **~ acoustique** TÉL acoustical dispersion; **~ anomale** TRANSMIS anomalous dispersion; **~ chromatique** TRANSMIS chromatic dispersion; **~ d'énergie** RADIO energy dispersal; **~ de guidage** RADIO, TRANSMIS waveguide dispersion; **~ de guide d'ondes** RADIO, TRANSMIS waveguide dispersion; **~ d'impulsion** ÉLECTRON, TÉL, TRANSMIS pulse broadening, pulse dispersion, pulse spreading; **~ de matériaux** TRANSMIS material dispersion; **~ d'ondes** ÉLECTRON wave dispersion; **~ de profil** TRANSMIS profile dispersion; **~ temporelle** ÉLECTRON, INFORMAT, TESTS, TRANSMIS time dispersion; **~ des temps** ÉLECTRON, INFORMAT, TESTS, TRANSMIS time dispersion; **~ des temps par trajets multiples** RADIO multipath time dispersion; **~ de vitesse de groupe** TRANSMIS group velocity dispersion (*GVD*)

Display: **~ PostScript** *m* DONNÉES Display PostScript

disponibilité *f* DONNÉES availability, ÉLECTRON uptime, GÉNÉRAL, TÉL, TESTS availability; **~ asymptotique** TÉL, TESTS *sécurité de fonctionnement* asymptotic availability; **~ d'un circuit** RADIO, TÉL circuit availability; **~ effective** TESTS operational availability factor; **~ indisponibilité** TÉL availability unavailability; **~ indisponibilité asymptotique** TÉL asymptotic availability unavailability; **~ instantanée** TÉL instantaneous availability, TESTS *sécurité de fonctionnement* availability function; **~ limitée** COMMUT limited availability; **~ moyenne** TÉL mean availability; **~ des programmes** TV program availability (*AmE*), programme availability (*BrE*); **~ du service** TÉL service availability performance; **~ totale** TÉL full availability

dispositif[1] *m* ÉLECTRON apparatus, device, GÉNÉRAL appliance, device, unit, INFORMAT device, TÉL device, feature; **~ à accès direct** INFORMAT random access device; **~ d'accord d'antenne** RADIO antenna tuning unit (*ATU*); **~ d'adaptation** ÉLECTRON matching device; **~ d'affichage** INFORMAT display device; **~ d'alimentation** INFORMAT, PÉRIPH feeder, feeding device; **~ analogique** INFORMAT analog device; **~ anti-chocs** TÉL anti-shock device; **~ antilocal** TÉL anti-sidetone device; **~ d'antiparasitage** TESTS noise suppression component, noise suppressor device; **~ avertisseur** TÉL alerting device; **~ de blocage d'entraînement** PÉRIPH transport locking device; **~ de bourrage** DONNÉES stuffing device; **~ de chiffrement** GÉNÉRAL encrypting equipment, scrambler; **~ de commande** CONTRÔLE control device, controlling device, operating device, GÉNÉRAL driver; **~ de commutation** ALIM ÉLEC, COMMUT, COMPOSANTS, TÉL switchgear, switching device; **~ de comptage** ÉLECTRON counting device; **~ de correction de synchronisation** INFORMAT timing correction device; **~ à couplage de charge** (*DCC*) ÉLECTRON charge-coupled device (*CCD*); **~ à courants porteurs** TRANSMIS carrier-current device; **~ en cours d'essai** TESTS device under test; **~ de cryptage** GÉNÉRAL encrypting equipment; **~ de décodage à décision formelle** DONNÉES, ÉLECTRON, SIGNAL hard-decision decoder; **~ de désignation** INFORMAT pointing device; **~ à effet Hall** ÉLECTRON Hall effect device; **~ émetteur** TÉL originating device; **~ d'enregistrement autonome** TÉL autonomous registration device; **~ d'entraînement** GÉNÉRAL driver; **~ d'entrée** INFORMAT input unit; **~ d'entrée vocale** INFORMAT voice input device; **~ d'entrée/sortie** INFORMAT input/output device (*I/O device*); **~ d'entrelacement** RADIO, TRANSMIS interleaver; **~ d'épissage** INFORMAT splicing facility; **~ d'essai** PÉRIPH, TESTS testing device; **~ d'essai automatique** TESTS routiner; **~ d'essai de liaison** COMMUT, TESTS link test equipment; **~ d'essai en mode paquets** DONNÉES, TESTS burst-mode tester; **~ d'essai de taxation** APP SERV, TESTS charging tester; **~ d'exclusion** TÉL queueing device; **~ d'extrémité de fibre optique** TRANSMIS fiber-optic terminal device (*AmE*), fibre-optic terminal device (*BrE*); **~ d'inhibition de temporisation** ÉLECTRON time-out disable facility; **~ à injection de charge** ALIM ÉLEC charge injection device (*CID*); **~ d'intégration électrique** ÉLECTRON electrical integration unit; **~ intermédiaire** INFORMAT intermediate equipment; **~ logique** DONNÉES, ÉLECTRON logic device, logic element; **~ de médiation** INFORMAT mediation device (*MD*); **~ de mise en attente** TÉL queueing device; **~ de mise à la terre** ALIM ÉLEC earthing arrangement (*BrE*), grounding arrangement (*AmE*); **~ non-conforme** TESTS nonconforming unit; **~ à ondes acoustiques de surface** COMPOSANTS, ÉLECTRON, TRANSMIS surface acoustic wave device (*SAW device*); **~ périphérique** DONNÉES, INFORMAT, PÉRIPH peripheral device; **~ de pointage** INFORMAT pointing device; **~ de préhension détecteur de contact optimal** CONTRÔLE rigidize-sensing grapple fixture; **~ de priorité** DONNÉES, TÉL override feature; **~ de protection anti-écho** RADIO echo-control device; **~ de protection anti-erreur** DONNÉES error-control device (*ECD*); **~ de protection du réseau** (*DPR*) TÉL network protection device (*NPD*); **~ de raccordement homologué** TÉL approved connecting agent; **~ récepteur** INFORMAT, TÉL receiving device; **~ de réduction d'écho** TÉL echo-control device; **~ de réglage silencieux** RADIO muting device; **~ de renvoi d'appel** TÉL call diverter; **~ de réserve** GÉNÉRAL standby set; **~ de saisie vocale** INFORMAT speech capture device; **~ de sécurité à clé** TESTS keyswitch; **~ à semi-conducteurs** COMPOSANTS, ÉLECTRON solid-state device; **~ de sortance en pyramide** ÉLECTRON fan-out device; **~ de sortie** INFORMAT output device, output unit; **~ de suralimentation** TÉL loop extender; **~ de surveillance de phase** ALIM ÉLEC phase monitor; **~ de télémesure** TESTS telemetering device; **~ de téléréunion** APP SERV meet-me bridge; **~ de temporisation** COMPOSANTS time-out device; **~ de traitement de paquets** DONNÉES, TÉL packet handler (*PH*); **~ de traitement de signaux** SIGNAL signal processor (*SP*); **~ de transfert à la chaîne** COMMUT bucket brigade device (*BBD*); **~ à transfert de charge** (*DTC*) ÉLECTRON charge transfer device (*CTD*); **~ transistorisé** COMPOSANTS, ÉLECTRON solid-state device;

~ **de verrouillage** ÉLECTRON latch lock, PÉRIPH *postes téléphoniques* locking device; ~ **de verrouillage de défilement** PÉRIPH transport locking device; ~ **de visée** RADIO boresight
dispositif[2]: ~ **en attente** *loc* TÉL device waiting (*DW*)
disque: ~ **d'azimut** *m* RADIO cursor; ~ **du cadran d'appel** *m* PÉRIPH, TÉL dial disk (*AmE*), finger wheel (*BrE*); ~ **à changement de phase** *m* DONNÉES phase-change disk; ~ **CI** *m* (*disque à circuits intégrés*) ÉLECTRON solid-state disc (*BrE*), solid-state disk (*AmE*); ~ **à circuits intégrés** *m* (*disque CI*) ÉLECTRON solid-state disc (*BrE*), solid-state disk (*AmE*); ~ **compact** *m* (*CD*) INFORMAT compact disc (*CD*); ~ **compact interactif** *m* (*DCI*) INFORMAT compact disc interactive (*CD-I*); ~ **compact multicouche** *m* DONNÉES multilayer compact disc, multilayer optical compact disc; ~ **compact réinscriptible** *m* DONNÉES, INFORMAT rewritable compact disc; ~ **compact vidéo** *m* TV video CD; ~ **compacte à écriture seule** *m* (*CD-WO*) INFORMAT compact disc-write once (*CD-WO*); ~ **compacte à lecture seule** *m* (*CD-R*) INFORMAT compact disc-read only (*CD-R*); ~ **de démarrage** *m* DONNÉES, INFORMAT start-up disk; ~ **dur** *m* INFORMAT hard disk; ~ **dur sur carte d'extension** *m* INFORMAT hard-disk card; ~ **magnétique** *m* DONNÉES, ÉLECTRON, INFORMAT magnetic disk; ~ **magnétique amovible** *m* DONNÉES, ÉLECTRON, INFORMAT removable magnetic disk; ~ **magnéto-optique** *m* DONNÉES, ÉLECTRON, INFORMAT magneto-optical disk, MO disk; ~ **mobile** *m* PÉRIPH *postes téléphoniques*, TÉL dial disk (*AmE*), finger wheel (*BrE*); ~ **numéroté** *m* PÉRIPH *postes téléphoniques* digit plate; ~ **optique** *m* DONNÉES, INFORMAT optical disc; ~ **optique numérique** *m* (*DON*) DONNÉES, INFORMAT compact disc read-only memory (*CD-ROM*), optical disc; ~ **optique réinscriptible** *m* DONNÉES, INFORMAT rewritable optical disk; ~ **partagé** *m* INFORMAT shared disk; ~ **à polir** *m* ÉLECTRON polishing jig; ~ **RAM** *m* DONNÉES, INFORMAT RAM disk; ~ **souple** *m* DONNÉES, INFORMAT diskette, flexible disk, floppy disk; ~ **virtuel en mémoire vive** *m* DONNÉES RAM disk
disquette *f* DONNÉES, INFORMAT diskette, flexible disk, floppy disk; ~ **de connexion** DONNÉES, INFORMAT connection disk; ~ **double face** INFORMAT reversible flexible disk
dissimulation *f* RADIO concealment
dissipation: ~ **d'électrode** *f* ÉLECTRON electrode dissipation; ~ **d'énergie** *f* ALIM ÉLEC, ÉLECTRON heat dissipation
dissociation: ~ **électrochimique** *f* GÉNÉRAL electrochemical dissociation; ~ **électrolytique** *f* GÉNÉRAL electrolytic dissociation
dissymétrie *f* GÉNÉRAL *de distribution* skewness
distance: ~ **de bord** *f* ÉLECTRON edge distance; ~ **de la cible** *f* RADIO *radar* slant range; ~ **de contournement** *f* ALIM ÉLEC flashover clearance; ~ **entre conducteurs** *f* ÉLECTRON conductor spacing; ~ **entre les stations de base** *f* RADIO *mobile* base station separation; ~ **de fuite** *f* ALIM ÉLEC creeping distance; ~ **de Hamming** *f* INFORMAT Hamming distance, SIGNAL signal distance; ~ **de l'horizon radioélectrique** *f* RADIO radio horizon distance; ~ **d'isolement** *f* ALIM ÉLEC flashover clearance; ~ **moyenne de réutilisation** *f* RADIO mean re-use distance; ~ **de propagation** *f* ÉLECTRON drive distance; ~ **radiale** *f* RADIO *radar* slant range; ~ **de réutilisation des fréquences dans la même voie** *f* RADIO co-channel reuse distance; ~ **du satellite** *f* RADIO satellite range; ~ **de saut** *f* RADIO, TÉL skip distance
distorsion *f* GÉNÉRAL distortion, TÉL clipping; ~ **d'affaiblissement** TÉL attenuation distortion, TRANSMIS attenuation distortion, nonuniform frequency response; ~ **d'amplitude** TRANSMIS amplitude distortion; ~ **arythmique** TÉL start-stop distortion; ~ **de balayage horizontal** TRANSMIS line non-linearity; ~ **biaise** ÉLECTRON, TÉL asymmetrical distortion, bias distortion; ~ **caractéristique** DONNÉES, TÉL characteristic distortion; ~ **chromatique** TV chromatic distortion; ~ **en coussin** TV pincushion distortion; ~ **cyclique** TÉL cyclic distortion; ~ **de demi-teinte** TV half-tone distortion; ~ **de déphasage** TRANSMIS phase-shift distortion; ~ **de la deuxième harmonique** ÉLECTRON second harmonic distortion (*2HD*); ~ **dissymétrique** TÉL asymmetrical distortion; ~ **à l'émission** TÉL, TRANSMIS transmitter distortion; ~ **fortuite** TÉL fortuitous distortion; ~ **géométrique** TV geometric distortion; ~ **harmonique** TRANSMIS harmonic distortion; ~ **d'impulsion** TÉL pulse distortion; ~ **individuelle** TÉL individual distortion; ~ **d'intermodulation** TRANSMIS intermodulation distortion; ~ **intramodale** TRANSMIS intramodal distortion; ~ **linéaire** TRANSMIS linear distortion; ~ **linéaire d'amplitude** TRANSMIS linear amplitude distortion; ~ **de luminance horizontale** INFORMAT line-time luminance distortion; ~ **de luminance verticale** INFORMAT field-time luminance distortion; ~ **modale** TRANSMIS intermodal distortion, modal distortion, mode distortion, multimode distortion; ~ **non-linéaire** ÉLECTRON, SIGNAL, TRANSMIS nonlinear distortion; ~ **de l'ordre d'une trame** INFORMAT field-time waveform distortion; ~ **de la parole** TÉL clipping, speech distortion; ~ **de phase** ÉLECTRON, TRANSMIS phase distortion; ~ **phase-amplitude** ÉLECTRON, TRANSMIS phase-amplitude distortion; ~ **phase-fréquence** ÉLECTRON, TRANSMIS phase-frequency distortion; ~ **de quantification** TÉL, TRANSMIS quantizing distortion; ~ **de repliement** INFORMAT, TRANSMIS *PCM* aliasing; ~ **en S** ÉLECTRON S-distortion; ~ **du signal** GÉNÉRAL, SIGNAL signal distortion, waveform distortion; ~ **de spot** ÉLECTRON spot distortion; ~ **télégraphique** TÉL telegraphic distortion; ~ **du temps de propagation de groupe** TRANSMIS group-delay distortion; ~ **trapézoïdale** ÉLECTRON trapezoidal distortion; ~ **de la troisième harmonique** ÉLECTRON third harmonic distortion (*3HD*); ~ **de vitesse de balayage de ligne** TRANSMIS line non-linearity
distortiomètre: ~ **à cadran** *m* TESTS signal distortion analyser (*BrE*) (*SDA*), signal distortion analyzer (*AmE*) (*SDA*)
distribué *adj* COMMUT, GÉNÉRAL, INFORMAT distributed; ~ **par satellite** TV broadcast by satellite
distribuer *vt* GÉNÉRAL dispatch, distribute
distributeur *m* GÉNÉRAL distributor; ~ **agréé** GÉNÉRAL authorized dealer; ~ **analyseur** ÉLECTRON scanner distributor; ~ **d'appels** TÉL call allotter, call distributor, call sequencer; ~ **d'appels automatique** (*DAA*) TÉL automatic call distributor (*ACD*); ~ **de chercheurs** TÉL selection allotter, selection allotter relay, selection allotter relay chain; ~ **d'émetteur multiple** TÉL multiple-transmitter distributor; ~ **de priorité aléatoire** TÉL random allotter, jumping allotter; ~ **rapide** COMMUT central pulse distributor; ~ **réception** RADIO receiver distributor; ~ **séquentiel d'appels** TÉL call routine sequencer, call sequencer; ~ **des trames** TRANSMIS frame distributor
distribution *f* GÉNÉRAL distribution; ~ **d'amplitude de bruit** TRANSMIS noise amplitude distribution (*NAD*); ~ **arborescente** TRANSMIS tree distribution; ~ **en arbre** TRANSMIS tree distribution; ~ **automatique d'appels** (*DAA*) TÉL automatic call distribution (*ACD*); ~ **à basse impédance** ALIM ÉLEC low-resistance distribution; ~ **binomiale** GÉNÉRAL *statistique* binomial distribution; ~ **de clés de chiffrement** RADIO encryption key distribution; ~ **de courant** ALIM ÉLEC current distribution; ~ **directe** RADIO direct distribution; ~ **électrique** ALIM ÉLEC power distribution; ~ **d'énergie** ALIM ÉLEC power distribution; ~ **en étoile** TRANSMIS star distribution; ~ **exponentielle** GÉNÉRAL exponential distribution; ~ **des fichiers reçus** (*DRF*) DONNÉES dispatch-received file (*DRF*); ~ **de fréquence du rythme** TRANSMIS clock-rate distribution; ~ **des fréquences** GÉNÉRAL frequency distri-

bution; ~ **gaussienne** GÉNÉRAL Gaussian distribution; ~ **de l'indice de réfraction** TRANSMIS refractive-index distribution; ~ **indirecte** RADIO indirect distribution; ~ **des messages** TÉL message distribution; ~ **des messages de signalisation** SIGNAL signaling-message distribution (*AmE*) signalling-message distribution (*BrE*); ~ **optique** TRANSMIS optical distribution; ~ **du paquet de référence en mode global** DONNÉES global reference burst distribution; ~ **du paquet de référence en mode séquentiel** DONNÉES sequential reference burst distribution; ~ **par câble** TRANSMIS cable distribution; ~ **de probabilité des amplitudes** (*DPA*) SIGNAL amplitude probability distribution (*APD*); ~ **des signaux** SIGNAL signal distribution; ~ **de signaux horloge** TRANSMIS clock distribution; ~ **de signaux de rythme** TRANSMIS clock-signal distribution

DIT *abrév* (*arbre d'informations d'annuaire*) DONNÉES, TÉL DIT (*directory information tree*)

divergence: ~ **de faisceau** *f* RADIO, TRANSMIS beam divergence; ~ **de sortie** *f* ÉLECTRON output divergence

diversification *f* ÉLECTRON chaos

diversion: ~ **par altération** *f* TÉL manipulative deception

diversité *f* GÉNÉRAL, RADIO diversity; ~ **d'acheminement** TRANSMIS routing diversity; ~ **à deux ondes** RADIO two-ray diversity; ~ **de fréquence** RADIO, TRANSMIS frequency diversity; ~ **à ondes multiples** RADIO multiray diversity; ~ **en réception** RADIO receiver diversity; ~ **temporelle** DONNÉES, RADIO time diversity

divertisseur: ~ **de surtension** *m* COMPOSANTS surge diverter

diviseur *m* ÉLECTRON, INFORMAT divider, splitter; ~ **audiofréquence** TRANSMIS audio splitter; ~ **de fréquence** RADIO, TRANSMIS frequency divider; ~ **de fréquence de trame** TV file divider; ~ **de puissance** ALIM ÉLEC, RADIO power divider; ~ **de tension** ALIM ÉLEC voltage divider

division *f* GÉNÉRAL split, splitting; ~ **de câble** TRANSMIS cable branching; ~ **de cellules** RADIO cell splitting; ~ **du clavier** INFORMAT key split; ~ **de l'échelle** TESTS scale division; ~ **syllabique** APP SERV syllabization

DJM *abrév* (*date julienne modifiée*) RADIO MJD (*modified Julian date*)

DLL *abrév* (*bibliothèque de liens dynamiques*) TÉL DLL (*dynamic-link library*)

DM *abrév* TÉL (*manipulation de documents*) DM (*document manipulation*), (*mode déconnecté*) DM (*disconnected mode*)

document *m* GÉNÉRAL document; ~ **de base** DONNÉES, INFORMAT source document; ~ **de câblage** COMPOSANTS wiring document; ~ **hypermédia** INFORMAT hypermedia document; ~ **hypertexte** INFORMAT hypertext document; ~ **interchange format** DONNÉES document interchange format; ~ **maître** INFORMAT master document; ~ **multimédia** DONNÉES multimedia document; ~ **SGML** INFORMAT SGML document; ~ **sonore** DONNÉES audio document; ~ **source** DONNÉES, INFORMAT document source, source document; ~ **source d'origine** DONNÉES, INFORMAT original document; ~ **de spécification des tests de conformité** DONNÉES conformance testing specification document; ~ **de structure** INFORMAT structure document; ~ **télétex** TÉL teletex document; ~ **vectoriel** DONNÉES vector document; ~ **vidéotex** INFORMAT videotex document; ~ **Web** DONNÉES, INFORMAT Web document

documentation *f* GÉNÉRAL documentation; ~ **du central** GÉNÉRAL exchange library; ~ **d'installation** GÉNÉRAL installation documentation, installation library

doigt: ~ **de guidage** *m* TÉL guide finger; ~ **d'indication de barre de sélection** *m* COMMUT, COMPOSANTS selecting finger

doigté *m* INFORMAT fingering

domaine *m* GÉNÉRAL domain, field; ~ **d'adressage du réseau** DONNÉES network addressing domain; ~ **de dénomination** DONNÉES naming domain; ~ **d'état** TÉL status field; ~ **fréquentiel** ALIM ÉLEC, ÉLECTRON, RADIO, TESTS, TRANSMIS frequency domain; ~ **de gestion** DONNÉES, TÉL management domain (*MD*); ~ **de gestion d'administration** (*DGAD*) DONNÉES, TÉL administration management domain (*ADMD*); ~ **de gestion privé** DONNÉES, TÉL private management domain (*PRMD*); ~ **globale d'adressage du réseau** DONNÉES global network addressing domain; ~ **d'informations de signalisation** TÉL signaling information field (*AmE*), signalling information field (*BrE*); ~ **MCS** TÉL MCS domain; ~ **de référence** TESTS reference range; ~ **de sélection du canal sémaphore** SIGNAL signaling-link selection field (*AmE*), signalling-link selection field (*BrE*); ~ **de sélection de liaisons de signalisation** SIGNAL signaling-link selection field (*AmE*), signalling-link selection field (*BrE*); ~ **de tolérance du cœur** TRANSMIS core tolerance field; ~ **de tolérance de gaine** TRANSMIS cladding tolerance field

domotique *f* APP SERV domotics, home automation

DON *abrév* (*disque optique numérique*) INFORMAT CD-ROM (*compact disc read-only memory*)

données *f pl* GÉNÉRAL data; ~ **d'acheminement** TÉL route-choice data, routing data; ~ **d'action atomique** DONNÉES atomic action data; ~ **alphabétiques** INFORMAT alphabetic data; ~ **analogiques** INFORMAT analog data; ~ **d'appel** TÉL calling-party data; ~ **ASCII** INFORMAT ASCII data; ~ **binaires** DONNÉES, INFORMAT binary data, TÉL BINARY-DATA-Message, TRANSMIS binary data; ~ **brutes** DONNÉES raw data; ~ **CD-I** DONNÉES, INFORMAT CD-I data; ~ **de commande** INFORMAT control data; ~ **compactées** DONNÉES compressed data; ~ **comprimées** DONNÉES compressed data; ~ **condensées de sous-paquet** DONNÉES condensed sub-burst data; ~ **confidentielles** DONNÉES, INFORMAT sensitive data; ~ **dans la bande** DONNÉES in-band data; ~ **décomprimées** DONNÉES, INFORMAT decompressed data; ~ **de description de tâches** TÉL task description data (*TDD*); ~ **émises** DONNÉES transmitted data; ~ **d'émission** DONNÉES, TRANSMIS transmit data; ~ **d'entrée** DONNÉES input data; ~ **d'état** DONNÉES, TESTS status data; ~ **d'exécution** DONNÉES, INFORMAT performance data; ~ **d'exploitation** DONNÉES, TÉL operational data; ~ **de fonction** DONNÉES function data; ~ **de gestion** DONNÉES management data; ~ **de gestion de réseau** INFORMAT network management data; ~ **à grande vitesse** DONNÉES high-speed data; ~ **d'image** DONNÉES, TV image data; ~ **infravocales** DONNÉES, RADIO data under voice (*DUV*); ~ **d'installation** GÉNÉRAL installation data; ~ **intrabandes** DONNÉES in-band data; ~ **liées** DONNÉES bound data; ~ **Manchester** INFORMAT Manchester data; ~ **numériques** DONNÉES, INFORMAT digital data, numeric data, numerical data; ~ **numérisées** DONNÉES digitized data; ~ **permanentes** DONNÉES master data; ~ **relatives au terrain** INFORMAT terrain data; ~ **de sauvegarde** INFORMAT backup information; ~ **semi-permanentes** TÉL semipermanent data; ~ **de séquence** DONNÉES, INFORMAT sequence data; ~ **de sortie** DONNÉES output data; ~ **statistiques** ÉLECTRON statistical frame traffic information; ~ **supravocales** DONNÉES, RADIO data above voice (*DOV*); ~ **de synchronisation** INFORMAT synchronization data; ~ **système** INFORMAT system data; ~ **techniques** GÉNÉRAL engineering data; ~ **temporaires** DONNÉES working data; ~ **de terrain** RADIO ground data; ~ **de transmission** DONNÉES, TRANSMIS transmit data; ~ **d'usager** DONNÉES, INFORMAT user data; ~ **utilisateur** DONNÉES, INFORMAT user data; ~ **utilisateur N** DONNÉES, TÉL N-user data; ~ **vidéo** DONNÉES, INFORMAT, TV video data

donneur *m* COMPOSANTS donor; ~ **d'ordre** GÉNÉRAL customer

doper *vt* COMPOSANTS, INFORMAT dope

dorsal *m* DONNÉES back-end processor

DOS: ~ **Microsoft®** *m* (*MS-DOS®*) INFORMAT Microsoft disk operating system® (*MS-DOS*)

doublage *m* GÉNÉRAL dubbing

double[1] *adj* GÉNÉRAL dual; **à ~ courant** TÉL double current; **en ~ précision** INFORMAT double-precision; **à ~ sens** RADIO, TÉL, TRANSMIS both-way, two-way

double[2]: **~ appel** *m* TÉL simultaneous call; **~ clic** *m* DONNÉES, INFORMAT double click; **~ courant** *m* TÉL double current; **~ dibit non-embrouillé** *m* DONNÉES unscrambled double dibit (*SI*); **~ image** *m* TV ghost; **~ sens** *m* TÉL two-way

double[3]: **~ bande latérale** *f* (*DBL*) RADIO, TRANSMIS double sideband (*DSB*); **~ connexion** *f* ÉLECTRON dual attachment; **~ densité** *f* (*DD*) INFORMAT double density (*DD*); **~ écriture** *f* INFORMAT write-through; **~ vérification** *f* INFORMAT double check

doublet *m* DONNÉES, INFORMAT two-bit byte, RADIO dipole, dipole aerial, dipole antenna; **~ demi-onde** RADIO half-wave dipole; **~ en onde entière** RADIO full-wave dipole; **~ replié** RADIO folded-dipole antenna; **~ symétrisé par coaxial fendu** RADIO slot-fed dipole

doubleur *m* ÉLECTRON doubler; **~ de fréquence** TRANSMIS frequency doubler; **~ de tension** ALIM ÉLEC voltage doubler

doublure *f* COMPOSANTS lining

DP *abrév* (*discriminateur de protocole*) TÉL PD (*protocol discriminator*)

DPA *abrév* (*distribution de probabilité des amplitudes*) SIGNAL APD (*amplitude probability distribution*)

DPR *abrév* (*dispositif de protection du réseau*) TÉL NPD (*network protection device*)

DQDB *abrév* INFORMAT DQDB (*distributed-queue dual bus*)

DQDBS *abrév* DONNÉES DQDBS (*distributed-queue double system*)

drag: **~ and drop** *vt* DONNÉES drag and drop

dragonne *f* ÉLECTRON belt loop

drapeau *m* DONNÉES flag, GÉNÉRAL switch indicator

drastique *adj* ÉLECTRON dramatic

drawmap *m* INFORMAT drawmap

DRF *abrév* (*distribution des fichiers reçus*) DONNÉES DRF (*dispatch-received file*)

driver *m* ÉLECTRON driver

droit: **~ d'accès** *m* TV connection charge; **~ de raccordement** *m* TV connection charge

DS *abrév* DONNÉES (*monocouche répartie*) DS (*distributed single-layer*), (*service d'annuaire*) DS (*directory services*)

DSA *abrév* (*agent de système d'annuaire*) DONNÉES DSA (*directory system agent*); **~ autonome** *m* DONNÉES stand-alone DSA; **~ centralisé** *m* DONNÉES centralized DSA; **~ coopérant** *m* DONNÉES cooperating DSA

DSE *abrév* (*méthode de test monocouche distribuée encastrée*) DONNÉES DSE (*distributed single-layer embedded test method*)

DSL *abrév* (*diode superluminescente*) COMPOSANTS, ÉLECTRON, TRANSMIS SRD (*superradiant diode*), SLD (*superluminescent LED, superluminescent light-emitting diode*)

DSP *abrév* (*processeur numérique de signaux*) DONNÉES DSP (*digital signal processor*)

DT *abrév* (*direction des télécommunications*) TÉL telecommunications board

DTC *abrév* DONNÉES (*commande d'émission numérique*) DTC (*digital-transmit command*), ÉLECTRON (*dispositif à transfert de charge*) CTD (*charge transfer device*)

duplex[1] *adj* GÉNÉRAL duplex

duplexage: **~ par division de fréquence** *m* TRANSMIS frequency-division duplexing

duplexeur *m* COMMUT, RADIO, TRANSMIS TR switch, transmit-receive switch, duplexer; **~ de radar** RADIO radar duplexer

durabilité *f* GÉNÉRAL durability

durée *f* GÉNÉRAL duration, time; **~ administrative pour la maintenance corrective** TÉL administrative time for corrective maintenance; **~ d'affectation des canaux** TRANSMIS channel allocation time; **~ d'analyse** RADIO scanning interval; **~ de l'appel** TÉL call duration; **~ d'attente de tonalité** TÉL dial tone delay; **~ de balayage** ÉLECTRON sweep time; **~ de bon fonctionnement** TESTS uptime; **~ de la charge** PÉRIPH charging period; **~ de la charge complète** PÉRIPH complete charging period; **~ de chauffage du filament** ÉLECTRON heater warm-up time; **~ de cohérence** TRANSMIS coherence time; **~ de la communication** TÉL call duration; **~ de commutation** COMMUT, TÉL, TESTS switching time; **~ de connexion** INFORMAT connect time; **~ de conservation** TESTS shelf life; **~ de conversation** RADIO, TÉL conversation time; **~ de la conversation** TÉL call duration; **~ cumulée** TESTS *sécurité de fonctionnement* accumulated time; **~ cumulée d'indisponibilité** TESTS *sécurité de fonctionnement* accumulated downtime; **~ de délai technique** TESTS technical delay time; **~ de démarrage d'un tube** ÉLECTRON tube starting time; **~ de désionisation** ÉLECTRON recovery period; **~ du diagnostic d'un dérangement** TÉL failure diagnosis time; **~ de disponibilité** TÉL uptime; **~ de disponibilité-indisponibilité** TÉL up-down time; **~ d'émission** RADIO airtime; **~ d'émissions sans effet** RADIO ineffective airtime; **~ d'essai** TESTS test duration; **~ d'établissement** TÉL setting-up time; **~ d'établissement d'une communication** TÉL call setup delay; **~ d'établissement d'une connexion transversale** TÉL through-connection delay; **~ d'exécution** INFORMAT run duration, run time; **~ de fonctionnement prévue** RADIO scheduled operating time; **~ de glissade** INFORMAT portamento time; **~ d'impulsion** DONNÉES, ÉLECTRON, RADIO, TÉL, TRANSMIS pulse duration, pulse length, pulse width (*PW*); **~ d'indisponibilité** TÉL downtime; **~ d'indisponibilité interne** TÉL internal downtime; **~ d'indisponibilité interne-externe** TÉL internal-external downtime; **~ d'interruption** TESTS downtime; **~ d'ionisation** ÉLECTRON ionization time; **~ jusqu'à la mi-valeur** TESTS time to half value; **~ de maintenance** TÉL, TESTS maintenance time; **~ de maintenance active** TÉL, TESTS active maintenance time; **~ de maintenance corrective** TÉL, TESTS corrective maintenance time; **~ de maintenance corrective active** TÉL, TESTS active corrective maintenance time; **~ de maintenance préventive** TÉL, TESTS preventive maintenance time; **~ de maintenance préventive active** TÉL, TESTS active preventive maintenance time; **~ de maintien de synchronisme de modem** DONNÉES modem holdover period; **~ maximum de l'annonce** TÉL maximum announcement duration; **~ maximum du message** TÉL maximum message duration; **~ à mi-crête** (*FDHM*) DONNÉES full-duration half-maximum (*FDHM*); **~ de mise en œuvre** CONTRÔLE setting-up time; **~ moyenne de conversation** TÉL average call duration; **~ moyenne de disponibilité** TESTS mean uptime; **~ moyenne d'indisponibilité** TESTS mean downtime; **~ moyenne de maintenance en homme-heure** TESTS mean maintenance man-hour; **~ moyenne des manœuvres** TÉL average operating time; **~ moyenne de panne** TESTS mean time to recovery (*MTTR*); **~ moyenne de réparation** TESTS mean repair time; **~ moyenne de vie** TESTS mean life; **~ moyenne de vie utile** TESTS mean life; **~ de notes** INFORMAT note length; **~ de numérotation** TÉL dialing period (*AmE*), dialling period (*BrE*), dialing time (*AmE*), dialling time (*BrE*); **~ d'occupation** TÉL holding time, occupation time, seizure time; **~ d'occupation moyenne par prise** TÉL mean holding time per seizure; **~ de pas** INFORMAT step length; **~ de perte du verrouillage de trame** TÉL out-of-frame alignment time; **~ de présélection** TÉL incoming response delay; **~ de relève d'un dérangement** TÉL failure correction time; **~ de réparation active** TÉL active repair time; **~ requise** TÉL, TESTS required time; **~ requise non-requise** TÉL required unrequired time; **~ de salve** DONNÉES burst duration; **~ sans panne** TESTS time to failure; **~ de sélection d'un commutateur** COMMUT switching delay, exchange call set-up delay; **~ de service** DONNÉES housekeeping lifetime; **~ du signal** SIGNAL signal duration; **~ de**

sonnerie TÉL ringing duration, TV ringing time; **~ statistique de préamorçage** ÉLECTRON statistical delay of ignition; **~ taxable** APP SERV, TÉL chargeable duration; **~ taxée** APP SERV, TÉL charged duration; **~ de test** TESTS test duration; **~ théorique** TÉL theoretical duration; **~ totale de démarrage** ÉLECTRON total starting time; **~ totale d'occupation** TÉL total holding time; **~ de trame** TRANSMIS frame length; **~ de vérification** TÉL check-out time; **~ de vie** RADIO, TESTS lifespan, lifetime; **~ de vie jusqu'à l'usure** TESTS wear-out life; **~ de vie mécanique** TESTS mechanical life; **~ de vie à poste** RADIO station-keeping lifetime; **~ de vie en stockage** TESTS shelf life; **~ de vie utile** TÉL, TESTS operational lifetime, useful life

durée²: **~ d'occupation moyenne par prise** *loc* TÉL mean holding time per seizure

dureté: **~ par rayage** *f* COMPOSANTS scratch hardness; **~ aux rayonnements** *f* COMPOSANTS radiation hardness; **~ superficielle** *f* COMPOSANTS surface hardness

DVT *abrév* (*défaut de verrouillage trame*) DONNÉES out-of-frame

dynamique¹ *adj* GÉNÉRAL dynamic

dynamique² *f* GÉNÉRAL dynamics

dynamo: **~ génératrice de courant continu** *f* ALIM ÉLEC DC generator

dysfonctionnement *m* TESTS faulty working

E

EAO *abrév* (*enseignement assisté par ordinateur*) INFORMAT CAL (*computer-assisted learning*)
EB *abrév* (*élément binaire*) DONNÉES, TRANSMIS bit
EBCDIC *abrév* (*code decimal codé binaire étendu*) DONNÉES, TRANSMIS EBCDIC (*extended binary coded decimal interchange code*)
éblouissement *m* TV glare
ébonite *f* COMPOSANTS ebonite
écart *m* COMPOSANTS, CONTRÔLE manipulated variable, GÉNÉRAL deviation, RADIO separation, TÉL, TESTS, TRANSMIS deviation, TV spacing; **~ d'affaiblissement** TRANSMIS attenuation deviation; **~ angulaire en orbite** RADIO orbit gap; **~ de capacité** TÉL difference of capacity; **~ diaphonique** TRANSMIS signal-to-crosstalk ratio (*SCR*); **~ entre porteuses** TRANSMIS carrier separation; **~ équivalent** TÉL equivalent resistance error; **~ équivalent corrigé** TÉL corrected equivalent resistance error; **~ de fréquence normée** RADIO normalized frequency departure; **~ moyen** TESTS *qualité* mean deviation; **~ observé** TESTS *qualité* observed deviation; **~ réel** TESTS *qualité* actual deviation; **~ de régulation** CONTRÔLE system deviation; **~ sur le diamètre de l'âme** TÉL, TRANSMIS core diameter deviation; **~ sur le diamètre du cœur** TÉL, TRANSMIS core diameter deviation; **~ sur le diamètre de la gaine** TÉL cladding surface diameter deviation; **~ sur le diamètre du noyau** TÉL, TRANSMIS core diameter deviation; **~ sur la surface de référence** TÉL, TRANSMIS reference-surface diameter deviation; **~ systématique** TESTS systematic variation; **~ de temps** ÉLECTRON, INFORMAT, TESTS, TRANSMIS time displacement; **~ type** GÉNÉRAL standard deviation, root mean square (*rms*)
écartométrie *f* GÉNÉRAL deviation measurement
ECD *abrév* (*échec du conditionnement*) DONNÉES, TÉL FTT (*failure to train*)
échange *m* DONNÉES, GÉNÉRAL exchange, interchange, INFORMAT exchange, TV barter; **~ d'authentification** DONNÉES, SIGNAL authentication exchange; **~ de documents informatisés** (*EDI*) DONNÉES, INFORMAT, TÉL electronic data interchange (*EDI*); **~ de données** DONNÉES, INFORMAT data interchange; **~ de données commerciales** DONNÉES trade data interchange (*TDI*); **~ de données dynamiques** DONNÉES dynamic data exchange (*DDE*); **~ de données informatisées** (*EDI*) DONNÉES, INFORMAT, TÉL electronic data interchange (*EDI*); **~ de données informatisées pour l'administration, le commerce et le transport** (*EDIFACT*) DONNÉES electronic data interchange for administration, commerce and transport; **~ dynamique de données** DONNÉES dynamic data exchange (*DDE*); **~ entre intervalles de temps** COMMUT time slot interchange; **~ d'états du terminal de référence** RADIO reference terminal status interchange (*RTSI*); **~ d'identifications** (*XID*) TÉL exchange of identification (*XID*); **~ d'information** DONNÉES, INFORMAT information exchange; **~ ionique** ÉLECTRON ion exchange
échanger *vt* DONNÉES, GÉNÉRAL interchange, TÉL hold, TV barter
échanges: **~ entre matériels** *m pl* GÉNÉRAL interwork, interworking
échangeur *m* TÉL interchange unit program; **~ d'ions** ÉLECTRON ion exchanger
échantillon *m* GÉNÉRAL sample; **~ biaisé** TESTS *qualité* biased sample; **~ pour contrôle de livraison** TESTS shipping sample; **~ et mémorisation** CONTRÔLE sample and hold; **~ localisé** TESTS spot sample; **~ d'orientation spécifiée** TESTS directional sample, geometric sample; **~ de parole** TRANSMIS speech sample; **~ partiel** TESTS part sample; **~ reconstitué** TÉL reconstructed sample; **~ représentatif** TESTS representative sample; **~ simple au hasard** TESTS simple random sample; **~ stratifié** TESTS stratified sample
échantillonnage *m* GÉNÉRAL sampling; **~ aléatoire** ÉLECTRON random sampling; **~ en chaîne** TESTS chain sampling; **~ continu multiniveau** TESTS multilevel continuous sampling; **~ continu uniniveau** TESTS single-level continuous sampling; **~ pour contrôle de réception** TESTS *qualité* acceptance sampling; **~ et maintien** CONTRÔLE sample and hold; **~ en grappes** GÉNÉRAL cluster sampling; **~ imbriqué** TESTS nested sampling; **~ intégral** TRANSMIS integral sampling; **~ de matériaux en vrac** TESTS *qualité* bulk sampling; **~ multi-étage** TESTS multistage sampling; **~ multiple** TRANSMIS multiple sampling, multisampling; **~ de Nyquist** INFORMAT Nyquist sampling; **~ optimal** TRANSMIS optimal sampling; **~ progressif** TESTS sequential sampling; **~ simple** TESTS single sampling; **~ stratifié** TESTS stratified sampling; **~ successif partiel** TESTS skip lot sampling; **~ systématique** TESTS systematic sampling; **~ systématique périodique** TESTS periodic systematic sampling
échantillonner *vt* GÉNÉRAL sample
échantillonneur *m* GÉNÉRAL sampler; **~-bloqueur** CONTRÔLE sample and hold
échantillons: **~ par seconde** *loc* TRANSMIS samples per second
échappement *m* (*ESC*) DONNÉES, INFORMAT, TÉL escape (*ESC*); **~ dans les données** (*EDD*) DONNÉES escape in data (*EID*)
échauffement *m* CLIMAT, GÉNÉRAL temperature increase, temperature rise, warming; **~ propre** ALIM ÉLEC self-heating
échéancier *m* GÉNÉRAL *de remboursements* schedule
échec *m* DONNÉES, GÉNÉRAL, TÉL, TESTS, TRANSMIS failure; **~ d'appel** TÉL call failure; **~ du conditionnement** (*ECD*) DONNÉES, TÉL failure to train (*FTT*); **~ verdict** DONNÉES fail verdict
échelle: **~ d'affichage** *f* INFORMAT view scale; **~ de câbles** *f* TRANSMIS *pose de câbles* cable rack; **~ de dégradation à cinq notes** *f* TESTS five-point impairment scale; **~ fixe** *f* COMPOSANTS, TÉL supporting comb; **~ de force sonore** *f* TÉL loudness scale; **~ des gains** *f* ÉLECTRON gain scale; **~ de gris** *f* DONNÉES, TÉL gray scale (*AmE*), grey scale (*BrE*); **~ de hauteur** *f* TÉL pitch scale; **~ linéaire** *f* TESTS linear scale; **~ microtonale** *f* INFORMAT microtonal scale; **~ mobile** *f* INFORMAT sliding scale; **~ nonlinéaire** *f* TESTS nonlinear scale; **~ de nuances** *f* TÉL tone wedge; **~ numérique** *f* TESTS digital scale; **~ Pantone** *f* ÉLECTRON Pantone scale; **~ roulante** *f* COMPOSANTS ladder on rollers, rolling-track ladder; **~ à roulettes** *f* COMPOSANTS ladder on rollers, rolling-track ladder; **~ temporelle** *f* INFORMAT, RADIO, TRANSMIS timescale; **~ de temps** *f* INFORMAT, RADIO, TRANSMIS timescale; **~ de temps atomique** *f* RADIO atomic time scale; **~ de temps coordonné** *f* GÉNÉRAL coordinated time scale; **~ de temps en synchronisme** *f* RADIO timescale in synchronisation
échelon *m* TESTS scale division, TRANSMIS step; **~ de quantification** TRANSMIS quantization step; **~ unitaire de trame** TRANSMIS frame unit step; **~ unité** TÉL unit step

échelonné *m* TÉL echelon telegraphy

écho *m* GÉNÉRAL echo, RADIO echo, blip, *radar* angel; **~ arrière** RADIO *radar* back echo; **~ avant tour de terre** RADIO forward round-the-world echo; **~ de cible** RADIO target echo; **~ corrigé en amplitude** TÉL amplitude-corrected echo; **~ corrigé en phase** RADIO, TÉL phase-corrected echo; **~ distant** DONNÉES remotely-generated display; **~ élémentaire** TÉL elementary echo; **~ fantôme** RADIO *radar* angel; **~ fixe** RADIO permanent echo; **~ hors cycle** RADIO second time around echo; **~ latéral** RADIO side echo; **~ de leurres** RADIO chaff echo; **~ local** DONNÉES locally generated echo, INFORMAT local echo; **~ de mer** RADIO sea return, sea clutter; **~ de nuage** RADIO cloud return; **~ de paillettes** RADIO chaff echo; **~ parasite** RADIO unwanted echo, clutter; **~ permanent** RADIO permanent echo; **~ de précipitations atmosphériques** RADIO precipitation return; **~ radar** RADIO radar echo; **~ à la réception** TÉL receive end echo, listener echo; **~ de sol** RADIO ground return; **~ de terre** RADIO land return; **~ tour de terre** RADIO round-the-world echo; **~ utile** RADIO target echo; **~ de vagues** RADIO sea clutter, sea return; **~ vers l'amont** TÉL backward echo; **~ vers l'aval** TÉL forward echo

échomètre: **~ à impulsions** *m* TÉL pulse echo meter

échouer *vi* GÉNÉRAL, INFORMAT, TÉL, TESTS, TRANSMIS fail

ECL *abrév* (*logique à couplage par les émetteurs*) DONNÉES, ÉLECTRON ECL (*emitter-coupled logic*)

éclaboussure: **~ de soudure** *f* COMPOSANTS solder spatter, solder splash

éclairage *m* ALIM ÉLEC, INFORMAT, TV lighting; **~ ambiant** ALIM ÉLEC, INFORMAT ambient lighting; **~ par l'arrière** ALIM ÉLEC, INFORMAT backlighting; **~ de secours** ALIM ÉLEC emergency lighting

éclairé *adj* ALIM ÉLEC, INFORMAT illuminated; **~ par l'arrière** ALIM ÉLEC, INFORMAT backlit

éclairement: **~ énergétique** *m* TRANSMIS irradiance; **~ énergétique spectral** *m* GÉNÉRAL, TRANSMIS spectral irradiance; **~ spectral** *m* TRANSMIS spectral irradiance

éclairer *vt* ALIM ÉLEC, INFORMAT illuminate

éclatement *m* TÉL splitting

éclateur *m* ALIM ÉLEC, ÉLECTRON spark gap; **~ déclenché** ALIM ÉLEC, ÉLECTRON triggered spark gap

ECMA *abrév* (*Association européenne de constructeurs d'ordinateurs*) INFORMAT ECMA (*European Computer Manufacturers Association*)

économie: **~ d'échelle** *f* GÉNÉRAL economy of scale; **~ d'énergie** *f* GÉNÉRAL energy saving; **~ en mémoire** *f* DONNÉES bit efficiency; **~ du risque** *f* TESTS risk economy

économiseur: **~ d'écran** *m* INFORMAT screensaver

écoulement *m* GÉNÉRAL flow; **~ de données** DONNÉES traffic flow; **~ du trafic** DONNÉES, TÉL traffic flow, traffic handling

écouler[1] *vt* TÉL carry

écouler[2]: **s'~** *v réfl* ALIM ÉLEC, GÉNÉRAL flow

écoute *f* RADIO, TÉL, TRANSMIS listening, listening-in; **~ amplifiée** PÉRIPH amplified reception; **~ automatique des messages** INFORMAT auto-play; **~ clandestine** INFORMAT eavesdropping, TÉL tapping, telephone tapping, wiretapping; **~ collective** TÉL group listening; **~ collective sur postes distincts** TÉL group monitoring; **~ discrète** RADIO, TÉL observe; **~ indiscrète** RADIO, TÉL eavesdropping; **~ téléphonique filaire** TÉL tapping, telephone tapping, wiretapping

écouter *vt* PÉRIPH play back; **~ à** TÉL listen; **~ clandestinement** TÉL tap, wiretap

écouteur *m* INFORMAT, PÉRIPH, TÉL earphone, earpiece; **~ téléphonique** PÉRIPH, SIGNAL receiver inset, TÉL receiver inset, earphone; **~ témoin** PÉRIPH watch receiver, watch-keeping receiver

écran *m* DONNÉES, ÉLECTRON, GÉNÉRAL, INFORMAT, RADIO, TÉL, TV display, screen; **~ d'accueil** DONNÉES, INFORMAT welcome page; **~ acoustique** GÉNÉRAL baffle; **~ cathodique** COMPOSANTS, ÉLECTRON cathode-ray screen; **~ de contrôle des signaux** CONTRÔLE waveform monitor; **~ à cristaux liquides** ÉLECTRON, INFORMAT liquid crystal display (*LCD*); **~ de démarrage** INFORMAT start-up screen; **~ électrique** ALIM ÉLEC electric screen; **~ électroluminescent à couche mince** COMPOSANTS, ÉLECTRON thin-film electroluminescent display (*TFEL*); **~ fluorescent** ÉLECTRON, INFORMAT, TV luminescent screen; **~ graphique** INFORMAT graphics screen; **~ magnétique** ÉLECTRON magnetic shield; **~ métallisé** ÉLECTRON metalized screen (*AmE*), metallized screen (*BrE*); **~ multifréquence** TV multifrequency screen; **~ à plasma** ÉLECTRON, INFORMAT plasma panel, gas panel, plasma display, gas discharge display; **~ plat** PÉRIPH flat panel display, TV flat screen; **~ plat à matrice active** TV active-matrix flat screen; **~ plat de projection à cristaux liquides** DONNÉES LCD projection panel; **~ de protection** ALIM ÉLEC shield, COMPOSANTS shielding; **~ solaire** RADIO sunshield; **~ solaire sélectif en fréquence** RADIO frequency-selective sunshield; **~ sur âme** TRANSMIS conductor screen; **~ sur enveloppe isolante** ALIM ÉLEC core screen, *câbles* insulation screen; **~ surdimensionné** DONNÉES large screen display; **~ tactile** INFORMAT tactile screen, touch screen; **~ thermique** ÉLECTRON heat shield; **~ VGA** INFORMAT video graphics array (*VGA*); **~ de visualisation** INFORMAT, PÉRIPH, TV display screen, video display terminal (*AmE*) (*VDT*), video display unit (*BrE*) (*VDU*)

écrasement *m* ÉLECTRON, INFORMAT, TÉL, TRANSMIS crushing; **~ des blancs** INFORMAT, TÉL, TRANSMIS white crushing

écraser *vt* DONNÉES overwrite, ÉLECTRON, INFORMAT, TÉL, TRANSMIS crush

écrêtage *m* COMMUT, TÉL clipping, TRANSMIS peak limiting in quantizing, peak limiting

écrêter *vt* COMMUT clip

écriture: **~ différée** *f* INFORMAT write-back; **~ miroir** *f* DONNÉES mirror writing; **~ par étapes** *f* INFORMAT step write; **~ pas à pas** *f* INFORMAT step write; **~ en temps réel** *f* INFORMAT real-time writing

EDA *abrév* (*enregistrement des données d'appel*) APP SERV CDR (*call data recording*)

EDC *abrév* (*enregistrement des données de communications*) APP SERV CDR (*call data recording*)

EDD *abrév* (*échappement dans les données*) DONNÉES EID (*escape in data*)

EDI *abrév* DONNÉES, INFORMAT, TÉL (*échange de données informatisées*) EDI (*electronic data interchange*)

EDIFACT *abrév* (*échange de données informatisées pour l'administration, le commerce et le transport*) DONNÉES, TÉL EDIFACT (*electronic data interchange for administration, commerce and transport*)

édifice: **~ intelligent** *m* APP SERV intelligent building (*IB*)

éditer *vt* GÉNÉRAL, INFORMAT edit

éditeur *m* GÉNÉRAL, INFORMAT editor; **~ de liens** INFORMAT linkage editor, linker; **~ de texte** INFORMAT text editor

édition *f* GÉNÉRAL edition, issue, INFORMAT editing; **~ des alarmes** TESTS alarm printout; **~ de caractères** INFORMAT character generation; **~ de chaîne** INFORMAT chain edit; **~ de données** DONNÉES, INFORMAT data editing; **~ électronique** APP SERV, DONNÉES, INFORMAT desktop publishing (*DTP*), digital publishing, electronic publishing; **~ numérique** INFORMAT digital editing; **~ de schéma** GÉNÉRAL diagram issue; **~ en temps réel** INFORMAT real-time editing; **~ de texte** APP SERV, INFORMAT text editing

éditique *f* APP SERV, DONNÉES, INFORMAT desktop publishing

éducation: **~ assistée par ordinateur** *f* INFORMAT computer-based learning; **~ informatisée** *f* INFORMAT computer-based learning

effacement *m* GÉNÉRAL deletion, INFORMAT cleaning, erasing, PÉRIPH, RADIO erasure, TÉL *vidéo* blanking; **~ de canal** INFORMAT delete channel; **~ global** INFORMAT bulk erasing, master clear; ~ **mémoire** INFORMAT memory cleaning, memory erasure; ~ **sélectif** DONNÉES selective erasing

effacer *vt* GÉNÉRAL delete, INFORMAT wipe out, erase, rub, zap, *écran* clear, clean, PÉRIPH, RADIO erase

effectif: ~ **de l'échantillon** *m* TESTS sample size; ~ **moyen contrôlé** *m* TESTS *qualité* average sample number (*ASN*); ~ **moyen de l'échantillon** *m* TESTS *qualité* average sample number (*ASN*)

effectuer *vt* ÉLECTRON *pontage* connect, INFORMAT perform, RADIO, TÉL initiate; ~ **un appel** TÉL make a call; ~ **un appel de départ** TÉL originate a call; ~ **une inversion dans** ÉLECTRON, INFORMAT negate; ~ **une rotation** INFORMAT rotate

effet *m* DONNÉES, INFORMAT effect; ~ **acousto-optique** TRANSMIS *fibres optiques* acousto-optic effect; ~ **d'allongement** TV lingering effect; ~ **d'antenne** RADIO aerial effect, antenna effect; ~ **d'avalanche** ÉLECTRON *semiconducteurs* avalanche effect; ~ **de blocage** GÉNÉRAL blockage effect; ~ **de bords** ALIM ÉLEC fringe effect, fringing; ~ **de brouillard** INFORMAT fog effect; ~ **de capture** ÉLECTRON capture effect; ~ **de champ** DONNÉES, ÉLECTRON field effect; ~ **de charge de surface** ÉLECTRON surface-charge effect, S-effect; ~ **couronne** ALIM ÉLEC corona effect; ~ **de cumul** TRANSMIS *fibres optiques* accumulation effect; ~ **de décalage de phase** INFORMAT phase-shifting effect; ~ **de déphasage** INFORMAT phasing-type effect; ~ **Doppler** GÉNÉRAL Doppler effect; ~ **d'écho** INFORMAT, TÉL echo effect, TRANSMIS howling, TV multiple image, ghost; ~ **d'écran** GÉNÉRAL, RADIO, TV screen effect, screening effect, shielding effect; ~ **d'écran des ondes** RADIO wave shadowing effect; ~ **électroacoustique** ÉLECTRON acousto-electric effect; ~ **électro-optique** TRANSMIS electro-optic effect; ~ **d'entraînement de fréquence** ÉLECTRON frequency pulling effect; ~ **d'exécution** INFORMAT performance effect; ~ **Faraday** TRANSMIS Faraday effect; ~ **de glissement** INFORMAT glide effect, slide effect; ~ **Hall** ÉLECTRON Hall effect; ~ **Hall quantique** ÉLECTRON quantum Hall effect; ~ **Joule** GÉNÉRAL Joule effect; ~ **Kelvin** ALIM ÉLEC Kelvin effect, skin effect; ~ **Kerr** TRANSMIS Kerr effect; ~ **Larsen** GÉNÉRAL, TÉL acoustic feedback; ~ **local** TÉL sidetone; ~ **magnéto-optique** ÉLECTRON magneto-optic effect; ~ **de masque** TÉL masking, obstacle effect; ~ **de mèche** COMPOSANTS, ÉLECTRON wicking; ~ **microphonique** ÉLECTRON microphonic effect, microphonism, microphony; ~ **Miller** ÉLECTRON Miller effect; ~ **mosaïque** INFORMAT mosaic effect, pixellization (*BrE*), pixelization (*AmE*), pixelation; ~ **nocturne** RADIO night effect; ~ **non-linéaire** ÉLECTRON, TRANSMIS nonlinear effect; ~ **de nuit** *f* RADIO night effect, TV shading; ~ **d'ombrage** TV shading; ~ **Pockels** TRANSMIS Pockels effect; ~ **de peau** ALIM ÉLEC Kelvin effect, skin effect; ~ **pelliculaire** ALIM ÉLEC skin effect, Kelvin effect; ~ **perturbateur** GÉNÉRAL perturbing effect; ~ **photoconductif** ÉLECTRON photoconductive effect; ~ **photoélectrique** ÉLECTRON, TRANSMIS photoelectric effect; ~ **photoélectrique externe** ÉLECTRON, TRANSMIS external photoelectric effect, photoelectric emission; ~ **photoélectrique interne** ÉLECTRON, TRANSMIS internal photoelectric effect, photoconductivity; ~ **photoémissif** ÉLECTRON, TRANSMIS photoemissive effect; ~ **de photoémission** ÉLECTRON, TRANSMIS photoemissive effect; ~ **photovoltaïque** ÉLECTRON photovoltaic effect; ~ **de pincement** COMPOSANTS pinch effect; ~ **plastique** INFORMAT plastic effect; ~ **produit par les ondes pilotes** TRANSMIS pilot interference; ~ **propagé par les conducteurs** ÉLECTRON conducted effect; ~ **de queue de comète** INFORMAT comet tail effect; ~ **des radiations transitoires sur les circuits électroniques** ÉLECTRON transient radiation effect on electronics (*TREE*); ~ **des résidus de porteurs** TRANSMIS carrier-link interference; ~ **de retard** INFORMAT delay effect; ~ **rocky-point** ÉLECTRON rocky-point effect; ~ **S** ÉLECTRON S-effect; ~ **Schottky** ÉLECTRON Schottky effect; ~ **Seebeck** ÉLECTRON Seebeck effect, thermoelectric effect; ~ **sonore** INFORMAT, TÉL sound effect; ~ **de sous-quantification** INFORMAT posteriation, countouring; ~ **de surbrillance** INFORMAT spotlight effect; ~ **de surface** ÉLECTRON surface-charge effect; ~ **systématique** TESTS systematic effect; ~ **en temps réel** INFORMAT real-time effect; ~ **thermoélectrique** ÉLECTRON thermoelectric effect, Seebeck effect; ~ **Thomson** GÉNÉRAL Thomson effect; ~ **de trace** INFORMAT teletrack; ~ **de traînage dynamique** INFORMAT comet tail effect; ~ **de traînée** INFORMAT teletrack; ~ **tunnel** COMPOSANTS tunnel effect, ÉLECTRON tunneling (*AmE*), tunnelling (*BrE*); ~ **de variation de hauteur** INFORMAT pitchbend effect

efficacité *f* COMPOSANTS, ÉLECTRON, GÉNÉRAL effectiveness, efficiency, TÉL *de l'oreille artificielle* sensitivity; ~ **des canaux** TRANSMIS channel efficiency; ~ **du codage** SIGNAL code efficiency; ~ **du couplage** ÉLECTRON coupling efficiency; ~ **de détection en puissance** COMPOSANTS detector power efficiency; ~ **de détection en tension** COMPOSANTS detector voltage efficiency; ~ **de l'écran** ÉLECTRON screen efficiency; ~ **à l'émission** TÉL *d'un système téléphonique local* sending sensitivity; ~ **d'émission thermionique** ÉLECTRON thermionic emission efficiency; ~ **de l'emploi du spectre** RADIO spectrum efficiency; ~ **lumineuse d'une source de lumière** DONNÉES luminous efficacy of a source of light; ~ **à la réception** TÉL *d'un système téléphonique local* receiving sensitivity

effluve *f* ALIM ÉLEC corona

effort: ~ **de tension** *m* COMPOSANTS tensile stress; ~ **de traction** *m* COMPOSANTS tensile force; ~ **unitaire** *m* COMPOSANTS stress

EGA *abrév* (*adaptateur EGA*) INFORMAT EGA (*enhanced graphics adaptor*)

égal: **d'~ à égal** *loc* TRANSMIS peer-to-peer

égalisateur *m* ALIM ÉLEC, CONTRÔLE, ÉLECTRON, TÉL, TRANSMIS equalizer (*EQL*)

égalisation *f* ALIM ÉLEC, CONTRÔLE, ÉLECTRON, TÉL, TRANSMIS equalization; ~ **d'amplitude** TRANSMIS amplitude equalization; ~ **fixe des modems** TRANSMIS compromise equalization; ~ **de phase** ALIM ÉLEC, ÉLECTRON phase equalization; ~ **du temps de propagation** TRANSMIS propagation time equalization; ~ **du temps de propagation de groupe** TRANSMIS group-delay equalization; ~ **des trajets électriques** TRANSMIS electrical path equalization

égaliser *vt* ALIM ÉLEC, CONTRÔLE, ÉLECTRON, TÉL, TRANSMIS equalize

égaliseur *m* ALIM ÉLEC, CONTRÔLE, ÉLECTRON, TÉL, TRANSMIS equalizer (*EQL*); ~ **d'amplitude** TRANSMIS amplitude equalizer; ~ **avec rétroaction de décision** ÉLECTRON decision feedback equalizer; ~ **de phase** ALIM ÉLEC, GÉNÉRAL phase equalizer; ~ **de tension d'élément** ALIM ÉLEC cell voltage equalizer

EIAO *abrév* (*enseignement intelligemment assisté par ordinateur*) INFORMAT ITS (*intelligent tutoring systems*)

EISA *abrév* (*architecture EISA*) INFORMAT EISA (*extended industry standard architecture*)

éjecter *vt* DONNÉES *disque*, INFORMAT eject

élaguer *vt* GÉNÉRAL prune, crop

élagueur *m* RADIO defruiter

élargissement: ~ **d'impulsion** *m* ÉLECTRON, TÉL, TRANSMIS pulse broadening, pulse dispersion, pulse spreading, pulse widening; ~ **spectral** *m* RADIO spectrum expansion

élargisseur: ~ **d'impulsions** *m* ÉLECTRON, TÉL, TRANSMIS pulse stretcher

élasticité *f* GÉNÉRAL elasticity; ~ **aux erreurs** INFORMAT error resilience

élastomètre *m* TESTS elastometer

électret *m* ALIM ÉLEC, COMPOSANTS electret

électrisation *f* TESTS electrification

électroacoustique[1] *adj* TÉL electroacoustic
électroacoustique[2] *f* TÉL electroacoustics
électroaimant *m* ALIM ÉLEC, COMPOSANTS electromagnet; **~ d'ascension** TÉL stepping magnet; **~ de barre de sélection** COMMUT selecting bar damper; **~ de relais** COMPOSANTS, ÉLECTRON relay magnet; **~ de rotation** TÉL stepping magnet
électrode *f* ALIM ÉLEC, ÉLECTRON electrode; **~ d'amorçage** ÉLECTRON primer electrode, starting electrode; **~ de concentration** ÉLECTRON focusing electrode; **~ de déclenchement** COMPOSANTS, ÉLECTRON trigger electrode; **~ d'entretien** ALIM ÉLEC, ÉLECTRON keep-alive electrode; **~ de focalisation** ÉLECTRON focusing electrode; **~ de signal** ÉLECTRON signal electrode; **~ de terre** ALIM ÉLEC earth electrode (*BrE*), ground electrode (*AmE*)
électrodynamique[1] *adj* ÉLECTRON, TÉL electrodynamic
électrodynamique[2]: **~ quantique** *f* ÉLECTRON quantum electrodynamics (*QED*)
électroformage *m* ÉLECTRON electroforming
électroluminescence *f* ÉLECTRON, TRANSMIS electroluminescence
électrolyte *m* ÉLECTRON electrolyte
électromagnétique *adj* GÉNÉRAL electromagnetic
électromécanique[1] *adj* ALIM ÉLEC, COMPOSANTS, GÉNÉRAL electromechanical
électromécanique[2] *f* ALIM ÉLEC, COMPOSANTS, GÉNÉRAL electromechanics
électron *m* GÉNÉRAL electron; **~ de conduction** ÉLECTRON conduction electron; **~ interne** ÉLECTRON inner shell electron; **~ libre** ÉLECTRON conduction electron; **~ périphérique** ÉLECTRON outer electron; **~ positif** ÉLECTRON, GÉNÉRAL positive electron, positron; **~ de valence** ÉLECTRON conduction electron
électronicien *m* ÉLECTRON, GÉNÉRAL electronics technician
électronique[1] *adj* ÉLECTRON, GÉNÉRAL electronic
électronique[2] *f* ÉLECTRON, GÉNÉRAL electronics; **~ intégrée** ÉLECTRON integrated electronics; **~ moléculaire** ÉLECTRON molecular electronics; **~ de puissance** ALIM ÉLEC power electronics; **~ quantique** ÉLECTRON quantum electronics; **~ spatiale** ÉLECTRON astrionics
électron: **~-volt** *m* ÉLECTRON, GÉNÉRAL electron-volt
électro: **~-optique** *adj* COMMUT, ÉLECTRON, TRANSMIS electro-optic, electro-optical
électrostatique[1] *adj* ALIM ÉLEC, COMPOSANTS, ÉLECTRON, PÉRIPH, TÉL, TESTS electrostatic
électrostatique[2] *f* ALIM ÉLEC, COMPOSANTS, ÉLECTRON, PÉRIPH, TÉL, TESTS electrostatics
électrotechnique *f* ALIM ÉLEC, GÉNÉRAL electrical engineering
électrozinguer *vt* COMPOSANTS zinc-electroplate
élément *m* ALIM ÉLEC cell, DONNÉES chip, component, element, object, ÉLECTRON, GÉNÉRAL component, element, INFORMAT object, TÉL component, element; **~ d'accès au réseau** DONNÉES, TÉL network access component; **~ d'accès au réseau service** DONNÉES, TÉL network service access component; **~ additionnel** INFORMAT add-on unit; **~ d'alliage** COMPOSANTS *métallurgie* alloying constituent; **~ de batterie** ALIM ÉLEC battery cell; **~ binaire** (*EB*) DONNÉES, TRANSMIS bit; **~ de câblage** TRANSMIS cabling element; **~ de commutation** ÉLECTRON switching element; **~ conditionnel** DONNÉES conditional component, C component, ÉLECTRON, TÉL conditional component; **~ de connexion** DONNÉES connection element (*CE*); **~ de connexion d'accès** DONNÉES, TÉL access connection element (*ACE*); **~ de connexion de transit** TRANSMIS transit connection element (*TCE*); **~ constitutif** GÉNÉRAL component, component part; **~ à contact mâle** COMPOSANTS, ÉLECTRON blade contact unit, pin contact unit; **~ corps** INFORMAT body object; **~ de demande du service** TÉL service invocation component; **~ détectable** INFORMAT detectable element; **~ de diagnostic** TÉL diagnostic element (*DIAG*); **~ de données CC** (*élément de données en code caractère*) DONNÉES, TÉL coded character data element; **~ de données en code caractère** (*élément de données CC*) DONNÉES, TÉL coded character data element; **~ enfichable** COMPOSANTS, ÉLECTRON, INFORMAT plug-in module, plug-in unit (*PIU*), TÉL plug-in unit (*PIU*); **~ excité** RADIO driven element; **~ facultatif** DONNÉES optional component, O component; **~ graphique** COMMUT, INFORMAT, PÉRIPH, SIGNAL, TÉL display element, pictorial element (*PE*); **~ graphique de catégorie d'abonné** TÉL subscriber category PE; **~ graphique de catégorie d'équipement** TÉL terminal category PE; **~ graphique de circuit interurbain** TÉL trunk PE; **~ graphique d'élément de commande** TÉL control element PE; **~ graphique d'émetteur sémaphore** SIGNAL signaling sender PE (*AmE*), signalling sender PE (*BrE*); **~ graphique d'émetteur de signalisation** SIGNAL signaling sender PE (*AmE*), signalling sender PE (*BrE*); **~ graphique d'équipement terminal** PÉRIPH terminal equipment PE; **~ graphique de ligne d'abonné** TÉL subscriber line PE; **~ graphique de limite de bloc fonctionnel** TÉL functional-block boundary PE; **~ graphique de module de commutation** COMMUT switching module PE; **~ graphique de récepteur sémaphore** SIGNAL signaling receiver PE (*AmE*), signalling receiver PE (*BrE*); **~ graphique de récepteur de signalisation** SIGNAL signaling receiver PE (*AmE*), signalling receiver PE (*BrE*); **~ graphique de tableau commutateur** COMMUT switchboard PE; **~ graphique de taxation en cours** TÉL charging in progress PE; **~ graphique de temporisateur de contrôle d'un** TÉL time supervision of a process PE; **~ graphique de trajet de commutation** COMMUT switching path PE; **~ graphique de trajet de commutation établi** COMMUT connected switching path PE; **~ graphique de trajet de commutation réservé** TÉL reserved switching path PE; **~ identité** TÉL identity element (*ID*); **~ d'image** ÉLECTRON, INFORMAT, TV picture element (*pel, pixel*); **~ d'image variable** INFORMAT display foreground; **~ d'indication** TÉL indicating element; **~ d'information** TÉL information element (*IE*); **~ d'information de compatibilité de couche inférieure** (*LLC-IE*) COMMUT, TÉL low-layer compatibility information element (*LLC-IE*); **~ à lames souples** COMPOSANTS reed element; **~ logique** DONNÉES, ÉLECTRON logic element; **~ logique séquentiel** ÉLECTRON sequential logic element; **~ de mémoire** DONNÉES storage element; **~ de mémoire de données** COMMUT, DONNÉES data store element; **~ de modulation** TÉL, TRANSMIS modulation element; **~ de modulation restitution signal** SIGNAL modulation restitution signal element; **~ monostable** ÉLECTRON monostable element; **~ moteur** TESTS meter movement; **~ mural** ÉLECTRON wall mounting; **~ numérique** DONNÉES, TÉL, TRANSMIS digit; **~ numérique de bourrage** DONNÉES stuffing digit; **~ numérique justifiable** TÉL, TRANSMIS justifiable digit; **~ numérique de justification** TÉL justifying digit; **~ numérique de poids faible** DONNÉES least significant digit (*LSD*); **~ numérique de service** TÉL, TRANSMIS service digit; **~ numérique de service de justification** TÉL, TRANSMIS justification service digit, justification service unit; **~ de numéro aléatoire** TÉL random number element (*RAND*); **~ obligatoire** DONNÉES mandatory component, M component; **~ d'opposition** ALIM ÉLEC counter cell; **~ optique holographique** DONNÉES holographic optical element; **~ OU exclusif** ÉLECTRON EXCLUSIVE-OR element; **~ parasite** RADIO parasitic element; **~ de parité** INFORMAT parity digit; **~ passerelle de réseau** TRANSMIS gateway network element; **~ de pile alcaline** ALIM ÉLEC alkaline cell; **~ de réglage** CONTRÔLE *commande automatique* controlling element; **~ de réglage final** CONTRÔLE *commande automatique* final controlling element; **~ de remplacement à fusion** ALIM ÉLEC enclosed fuse-link;

~ de remplissage ÉLECTRON filling compound; **~ de réponse signée** TÉL signed response element; **~ de réseau** TRANSMIS network element (*NE*); **~ de réseau non-SDH** TRANSMIS non-SDH network element (*NNE*); **~ à retard** ÉLECTRON delay element; **~ sensible** TESTS sensing element; **~ de service** DONNÉES, TÉL, TRANSMIS service element; **~ de service d'application** (*ASE*) DONNÉES, TÉL application service element (*ASE*); **~ de service d'application de la gestion de réseau** DONNÉES network management application service element (*NM-ASE*); **~ de service d'application pour la gestion des systèmes** DONNÉES, INFORMAT systems management application service element; **~ de service d'application pour le traitement des transactions** DONNÉES transaction processing application service element (*TPASE*); **~ de service de commande d'association** DONNÉES, TÉL association control service element (*ACSE*); **~ de service commun d'informations de gestion** DONNÉES common management-information service element (*CMISE*); **~ de service de contrôle d'association** DONNÉES, TÉL association control service element (*ACSE*); **~ de service de dépôt de messages** DONNÉES message submission service element (*MSSE*); **~ de service de gestion de messages** DONNÉES message administration service element (*MASE*); **~ de service de message de personne à personne** APP SERV, DONNÉES, TÉL interpersonal message element of service (*IPMEOS*); **~ de service de recherche de messages** DONNÉES message retrieval service element (*MRSE*); **~ de service de remise de messages** DONNÉES message delivery service element (*MDSE*); **~ de service de système de traitement de messages** (*MHS-SE*) APP SERV, DONNÉES, RADIO, TÉL message handling system service element (*MHS-SE*); **~ de service du transfert, accès et gestion de fichiers** DONNÉES, INFORMAT file transfer access and management service element (*FTAMSE*); **~ de service de transfert de messages** DONNÉES, TÉL message transfer service element (*MTSE*); **~ à seuil** ÉLECTRON threshold element; **~ de signal** SIGNAL, TÉL signal element; **~ de signal télégraphique** SIGNAL telegraphic signal element; **~ signature** SIGNAL, TÉL signature element (*SIG*); **~ supplémentaire** INFORMAT add-on unit; **~ universel** INFORMAT universal element; **~ d'utilisateur** DONNÉES, TÉL user element (*UE*); **~ d'utilisation du réseau** DONNÉES, TÉL network utilization component; **~ d'utilisation du réseau service** DONNÉES, TÉL network service utilization component

éléments: **~ fonctionnels du sous-réseau** *m pl* DONNÉES subnetwork functionality; **~ non conformes** *m pl* TESTS nonconforming items; **~ d'une orbite** *m pl* RADIO orbital element; **~ de service d'application communs** *m pl* DONNÉES common ASEs; **~ de taxation** *m pl* DONNÉES charging data

éliminateur: **~ de brouillage** *m* RADIO wave trap; **~ de modem** *m* DONNÉES null modem; **~ de réponses parasites** *m* RADIO defruiter

élimination *f* ÉLECTRON, RADIO, TÉL suppression; **~ de bruit** RADIO noise suppression; **~ des échos parasites** RADIO clutter suppression; **~ des lignes cachées** INFORMAT hidden-line removal; **~ progressive** RADIO phasing-out; **~ de silence** TESTS, TRANSMIS silence elimination

éliminer *vt* ÉLECTRON suppress, GÉNÉRAL phase out, INFORMAT kill, RADIO delete, suppress, TÉL suppress

emballement: **~ thermique** *m* ÉLECTRON thermal runaway

embarqué *adj* INFORMAT on-board, RADIO vehicle-mounted

embase *f* COMPOSANTS mounting base, base

emboîter *vt* INFORMAT, TESTS nest

embosser *vt* GÉNÉRAL emboss

embouchure *f* INFORMAT aperture, PÉRIPH mouthpiece

embout *m* TÉL end part; **~ de câble** TRANSMIS cable cap

embouteiller *vt* DONNÉES, INFORMAT jam

embrayage: **~ électrique** *m* COMPOSANTS magnetic clutch

EMCN *abrév* (*équipement de multiplication de circuits numériques*) TÉL, TRANSMIS DCME (*digital circuit multiplication equipment*)

émetteur *m* GÉNÉRAL emitter, transmitter (*XMTR*); **~ d'alarme** TÉL alarm transmitter; **~ d'annonces** TÉL message announcement system; **~ asservi** RADIO slave transmitter; **~ d'auto-excitation** RADIO self-excited transmitter; **~ automatique d'appels** TÉL automatic call transmitter; **~ automatique d'indicatif** DONNÉES, TÉL answer-back unit; **~-balise** RADIO beacon transmitter; **~ basse puissance** RADIO low-power transmitter; **~ de brouillage** ÉLECTRON jammer, RADIO jammer, jamming transmitter; **~ brouilleur** RADIO interfering transmitter; **~ à clavier** APP SERV, PÉRIPH keyboard sender; **~ de code** TÉL code sender; **~ commun** ÉLECTRON common emitter; **~ à courte portée** RADIO short-range transmitter; **~ à cylindre** TÉL drum transmitter; **~ de détresse à faible puissance** RADIO low-power distress transmitter (*LPDT*); **~ esclave** RADIO slave transmitter; **~ à étincelles** RADIO spark transmitter; **~ faible puissance** RADIO low-power transmitter; **~ d'images de télévision** TV television picture transmitter; **~ d'impulsions de compteur** TÉL meter pulse sender; **~ d'informations** DONNÉES coded-data transmitter; **~ laser** TRANSMIS laser transmitter; **~ de localisation d'urgence** RADIO emergency locater transmitter (*ELT*); **~ lumière** TV light emitter; **~ de messages de tarification** TÉL tariff message sender; **~ de modem** DONNÉES modem transmitter; **~ multiple** RADIO multiple transmitter; **~ multiplex à répartition dans le temps** INFORMAT, TÉL, TRANSMIS time-division multiplex transmitter; **~-multiplexeur** TRANSMIS transmitter-multiplier; **~ à numérotation automatique** TÉL automatic numbering equipment; **~ optique** TRANSMIS optical transmitter, transmit fiber optic terminal device (*AmE*), transmit fibre optic terminal device (*BrE*); **~ à plat** PÉRIPH, TRANSMIS flat-bed transmitter; **~ radio** RADIO radio transmitter; **~ de radiodiffusion** PÉRIPH broadcast transmitter; **~ de radiodiffusion sonore** RADIO sound broadcast transmitter; **~-récepteur** (*TR*) DONNÉES transceiver, ÉLECTRON talk set, transmitter-receiver (*TR*), RADIO, TRANSMIS radio unit, transceiver, transmitter-receiver (*TR*); **~-récepteur asynchrone universel** (*UART*) RADIO universal asynchronous receiver-transmitter (*UART*); **~-récepteur combiné** DONNÉES, RADIO transceiver; **~-récepteur pour essais de continuité** TÉL continuity check transceiver; **~-récepteur multifréquence** SIGNAL, TÉL MF sender-receiver, multi-frequency sender-receiver; **~-récepteur portable à dos d'homme** RADIO man-pack transceiver; **~ secondaire** RADIO secondary transmitter; **~ de secours** RADIO emergency transmitter; **~ de signaux** SIGNAL, TRANSMIS signal sender; **~ de son de télévision** TV television sound transmitter; **~ télégraphique** TÉL telegraphic transmitter; **~ de télévision** TV television transmitter; **~ à transistor** ÉLECTRON transistor emitter; **~ de voie** TRANSMIS channel transmitter

émetteurs: **~ numériques** *m pl* DONNÉES digital transmitters

émettodyne *m* COMPOSANTS emitter follower

émettre *vt* DONNÉES, INFORMAT transmit, RADIO emit, transmit, *un appel* initiate, TÉL transmit, initiate, TRANSMIS send, transmit; **~ des impulsions** GÉNÉRAL, SIGNAL outpulse, pulse out; **~ un signal** DONNÉES, INFORMAT, RADIO, SIGNAL, TÉL signal

éminceur *m* COMPOSANTS, ÉLECTRON slicer

émission[1]: **à ~ par la tranche** *adj* TRANSMIS edge-emitting

émission[2] *f* DONNÉES sending, ÉLECTRON emission, GÉNÉRAL broadcast (*BC*), INFORMAT sending, RADIO, TRANSMIS emission, sending, transmission; **~ d'appel** TÉL call origination; **~ à bande latérale atténuée** RADIO, TRANSMIS vestigial-sideband transmission; **~ à bande latérale résiduelle**

RADIO, TRANSMIS vestigial-sideband emission; **~ à bande latérale unique** (*émission BLU*) RADIO, TRANSMIS single sideband emission (*SSB emission*); **~ BLU** (*émission à bande latérale unique*) RADIO, TRANSMIS SSB emission (*single sideband emission*); **~ à commande manuelle** (*MOX*) DONNÉES manually-operated transmission (*MOX*); **~ continue** RADIO continuous wave (*CW*); **~ d'électrons primaires** ÉLECTRON primary-electron emission; **~ d'électrons secondaires** ÉLECTRON secondary electron emission; **~ de fréquences étalon** RADIO standard-frequency emission; **~ hors bande** RADIO out-of-band emission (*OBE*); **~ d'impulsions** TÉL pulsing, impulsing; **~ d'impulsions par ouverture de boucle** TÉL loop pulsing, loop dialing (*AmE*), loop dialling (*BrE*); **~ indésirable** RADIO unwanted emission; **~ ionique secondaire** ÉLECTRON secondary ionic emission; **~ lambertienne** TRANSMIS Lambertian emission; **~ d'ondes pilotes de référence** RADIO, TRANSMIS reference-pilot generation; **~ par effet de champ** ÉLECTRON field-effect emission; **~ par satellite** RADIO, TV satellite broadcast; **~ parasite** ÉLECTRON stray emission; **~ photoélectrique** ÉLECTRON, TRANSMIS photoelectric emission; **~ à porteuse complète** RADIO full-carrier emission; **~ à porteuse réduite** RADIO reduced-carrier emission; **~ à porteuse supprimée** RADIO suppressed-carrier emission; **~ de signal** SIGNAL, TRANSMIS signal; **~ des signaux horaires** RADIO standard time-signal emission; **~ spontanée** ÉLECTRON, TRANSMIS spontaneous emission; **~ stimulée** ÉLECTRON, TRANSMIS stimulated emission; **~ sur chaîne câblée** TÉL cable program (*AmE*), cable programme (*BrE*); **~ en synchronisme** RADIO, TV simulcast; **~ par télécopie** TÉL facsimile transmission; **~ de télévision** TV television program (*AmE*), television programme (*BrE*); **~ de télévision commerciale par satellite** TV commercial satellite telecast; **~ de télévision par satellite** TV satellite telecast; **~ thermionique** ÉLECTRON thermionic emission; **~ utile** ÉLECTRON, RADIO, SIGNAL wanted emission

émission[3]: **~ en cours** *loc* RADIO transmit on

émissivité *f* ÉLECTRON, RADIO, TRANSMIS emissivity

émittance *f* ÉLECTRON, RADIO, TRANSMIS emittance; **~ énergétique spectrale** ÉLECTRON, TRANSMIS spectral radiant emittance

empaqueter *vt* DONNÉES bundle

empilable *adj* COMPOSANTS, GÉNÉRAL stackable

empilage *m* COMMUT spring set, COMPOSANTS stacking; **~ automoteur** COMMUT, COMPOSANTS self-operating spring set; **~ à deux temps** COMMUT, COMPOSANTS two-step contact spring set; **~ échelonné** COMMUT, COMPOSANTS sequence spring set

emplacement[1]: **à ~ interne** *adj* COMPOSANTS, ÉLECTRON slot-based

emplacement[2] *m* COMPOSANTS, ÉLECTRON, INFORMAT slot, RADIO location, site; **~ de carte** COMPOSANTS board position; **~ du hub** ÉLECTRON slot in the hub; **~ libre** TÉL empty space, vacant position; **~ libre pour extension** INFORMAT expansion slot; **~ orbital** RADIO orbital location; **~ du satellite** RADIO satellite location

emploi: **~ du temps** *m* GÉNÉRAL, RADIO schedule

empreinte *f* RADIO footprint; **~ d'un faisceau** RADIO beam area; **~ d'un faisceau de liaison de connexion** RADIO, TRANSMIS feeder-link beam area; **~ de faisceau du satellite** RADIO footprint; **~ de lobe d'antenne du satellite** RADIO footprint; **~ Torx** GÉNÉRAL Torx system; **~ vocale** INFORMAT, TÉL voice print

émulateur *m* DONNÉES, ÉLECTRON, INFORMAT emulator; **~ de protocole** DONNÉES protocol emulator (*PE*); **~ de terminal** DONNÉES terminal emulator

émulation *f* DONNÉES, ÉLECTRON, INFORMAT emulation; **~ de terminal** DONNÉES terminal emulation

émuler *vt* DONNÉES, ÉLECTRON, INFORMAT emulate

encapsulation *f* COMPOSANTS, ÉLECTRON, GÉNÉRAL, INFORMAT encapsulating, encapsulation; **~ plastique** ÉLECTRON plastic encapsulation

enceinte: **~ acoustique** *f* GÉNÉRAL *acoustique* baffle; **~ d'insonorisation** *f* TÉL, TRANSMIS silencer

enchaînement: **~ de carte** *m* PÉRIPH *publiphone* card follow-on; **~ de communication** *m* PÉRIPH *publiphone* follow-on call; **~ par volet** *m* TV wipe

enchaîner *vt* INFORMAT catenate, concatenate, chain

enchevêtrement *m* ÉLECTRON tangling

enclenchement *m* ALIM ÉLEC pull-in

enclencher *vt* COMPOSANTS snap in; **~ à zéro** GÉNÉRAL reset

encliqueter *vi* COMPOSANTS, ÉLECTRON snap together

encoche *f* GÉNÉRAL nick, RADIO notch; **~ de positionnement** ÉLECTRON locating notch

encodage *m* DONNÉES, INFORMAT, SIGNAL, TÉL, TRANSMIS coding, encoding

encoder *vt* DONNÉES, INFORMAT, TÉL, TRANSMIS encode

encodeur *m* DONNÉES, INFORMAT, TÉL, TRANSMIS encoder

encombrant *adj* COMPOSANTS bulky, cumbersome

encombré *adj* COMMUT, TÉL, TRANSMIS congested

encombrement *m* COMMUT congestion, DONNÉES blocking, TÉL, TRANSMIS congestion; **~ d'appel** TÉL call congestion; **~ d'appel perte** TÉL call congestion loss; **~ d'appel probabilité de perte** TÉL call congestion probability of loss; **~ de faisceau** TÉL route congestion; **~ d'itinéraire** TÉL route congestion; **~ radio** RADIO radio congestion; **~ de réception** TÉL reception congestion; **~ du réseau** DONNÉES network locking, TÉL, TRANSMIS network congestion; **~ du spectre** RADIO spectrum congestion

encombrer *vt* COMMUT congest, INFORMAT *disque dur* clutter, TÉL, TRANSMIS congest

endommagé *adj* TRANSMIS damaged

endonnager *vt* TRANSMIS damage

endommagement: **~ de facette** *m* TRANSMIS facet damage

enduit: **~ de facette** *m* TRANSMIS facet coating

énergie[1]: **à ~ solaire** *adj* ALIM ÉLEC solar-powered

énergie[2] *f* ALIM ÉLEC, COMPOSANTS, ÉLECTRON, GÉNÉRAL, RADIO, TRANSMIS energy, power; **~ d'activation** ÉLECTRON *semiconducteurs* activation energy; **~ électrique** ALIM ÉLEC electric power; **~ de fuite de pointe** ÉLECTRON spike leakage energy; **~ d'ionisation** ÉLECTRON ionization energy; **~ rayonnante** ÉLECTRON, TRANSMIS radiant energy; **~ solaire** ALIM ÉLEC solar power; **~ à zéro absolu** GÉNÉRAL zero point energy

enfichable *adj* ALIM ÉLEC plug-in, ÉLECTRON pluggable, INFORMAT plug-in

enficher *vt* ALIM ÉLEC, GÉNÉRAL plug in, add in

enfoncement *m* ÉLECTRON indentation; **~ de touche** INFORMAT key-on

engagement *m* DONNÉES commitment; **~, concurrence et rétablissement** DONNÉES commitment, concurrency and recovery (*CCR*); **~ de transaction** DONNÉES transaction commitment

engendrer *vt* GÉNÉRAL generate

énoncé *m* INFORMAT banner, utterance

énoncer *vt* INFORMAT utter

ENR *abrév* (*enregistrement de fréquence*) RADIO REC (*frequency record*)

enregistrable *adj* INFORMAT recordable, TÉL recordable, recording

enregistré *adj* PÉRIPH stored, RADIO registered, TÉL recorded

enregistrement *m* DONNÉES registration, recording, record, GÉNÉRAL recording, INFORMAT logging, PÉRIPH recording, RADIO registration; **~ analogique** GÉNÉRAL analog recording; **~ des appels** TÉL call logging; **~ automatique de diagrammes de rayonnement** RADIO automatic antenna-pattern recording; **~ automatique de la position des navires** APP SERV automatic location registration of ships; **~ avec retour à un niveau de référence** INFORMAT return-to-reference recording; **~ avec retour à zéro** INFORMAT return-to-zero recording; ~ **de chiffres** TÉL storing of digits, digit storing; ~ **de consignation** DONNÉES log record; ~ **de consignation de dommage** DONNÉES log-damage record; ~ **de consignation d'engagement** DONNÉES log-commit record; ~ **de consignation heuristique** DONNÉES log-heuristic record; ~ **de consignation prêt** DONNÉES log-ready record; ~ **digital** INFORMAT, PÉRIPH, TESTS digital recording; ~ **de données** DONNÉES data recording; ~ **des données d'appel** (*EDA*) APP SERV, DONNÉES call data recording (*CDR*); ~ **des données de communications** (*EDC*) APP SERV, DONNÉES call data recording (*CDR*); ~ **et retransmission** APP SERV, DONNÉES store and forward; ~ **de fréquence** (*ENR*) RADIO frequency record (*REC*); ~ **incrémentiel** TV incremental recording; ~ **initial** INFORMAT home record; ~ **magnétique** ÉLECTRON magnetic recording; ~ **en modulation de phase** DONNÉES phase modulation recording, phase encoding; ~ **multicanal** TV multichannel recording; ~ **non-retour à un niveau de référence** DONNÉES nonreturn to reference recording; ~ **non-retour à zéro** DONNÉES nonreturn to zero recording; ~ **numérique** INFORMAT, PÉRIPH, TESTS digital recording; ~ **optique** TRANSMIS optical recording; ~ **par paquets fixes** INFORMAT fixed packet recording; ~ **par paquets variables** INFORMAT variable-packet recording; ~ **par surimpression** INFORMAT overdub recording; ~ **par zone** INFORMAT zone bit recording (*ZBR*); ~ **de la parole** TÉL speech recording; ~ **pas à pas** INFORMAT step recording; ~ **des perturbations** TÉL disturbance recording; ~ **piste par piste** INFORMAT track-at-once recording; ~ **de la position** RADIO location registration; ~ **de la position de la station mobile** RADIO mobile location registration; ~ **potentiel de consignation** DONNÉES potential log record; ~ **sans retour à un niveau de référence** DONNÉES nonreturn to reference recording; ~ **sans retour à zéro** DONNÉES nonreturn to zero recording; ~ **du son** INFORMAT, RADIO, TV sound recording; ~ **sonore** INFORMAT audio recording; ~ **en temps réel** INFORMAT real-time recording; ~ **de trafic** TÉL traffic record, traffic recording; ~ **de transaction** DONNÉES transaction logging; ~ **unique** INFORMAT at-once recording; ~ **vidéo** TV video recording

enregistrer: ~ **sur bande** *vt* APP SERV, INFORMAT tape

enregistreur *m* APP SERV recorder, COMMUT register, PÉRIPH recorder, TÉL register, recorder, TESTS, TV recorder; ~ **d'arrivée** TÉL incoming register; ~ **automatique** INFORMAT logger; ~ **de bande vidéo** TV videotape recorder (*VTR*); ~ **de cassettes vidéo** TV video recorder, videocassette recorder (*VCR*); ~ **de conversation** TÉL call recorder; ~ **de départ** COMMUT originating register; ~ **de données** DONNÉES data logger; ~ **de données à cassette** DONNÉES, PÉRIPH data cassette recorder (*DCR*); ~ **d'impulsions** TÉL, TRANSMIS pulse recorder; ~ **international R2 d'arrivée** TÉL *système de signalisation* incoming international R2 register; ~ **international R2 de départ** TÉL *système de signalisation* outgoing international R2 register; ~ **local** COMMUT local register; ~ **de localisations** RADIO location register; ~ **de localisation visitée** RADIO visited location register; ~ **monovoie** TESTS single-channel recorder; ~ **multipiste** DONNÉES multitrack recorder; ~ **multiple** PÉRIPH multiple recorder; ~ **de niveau** TESTS, TRANSMIS level recorder; ~ **numérique** INFORMAT, PÉRIPH digital recorder; ~ **photo CD** INFORMAT PCD writer, photo CD writer; ~ **de position de rattachement** (*HLR*) RADIO home location register (*HLR*); ~ **de positions** RADIO location register; ~ **R2 d'arrivée** TÉL *système de signalisation* incoming R2 register; ~ **R2 de départ** TÉL *système de signalisation* outgoing R2 register; ~ **de séquences** INFORMAT sequence recorder; ~ **de signal et d'état** COMMUT signal and state recorder; ~**-traducteur** COMMUT register translator; ~ **de trafic** TÉL traffic recorder

enrichir *vt* COMPOSANTS, GÉNÉRAL, INFORMAT, RADIO enhance

enrichissement *m* COMPOSANTS, GÉNÉRAL, INFORMAT, RADIO enhancement; ~ **d'image** TV image enhancement

enroulé: ~ **en spirale** *adj* COMPOSANTS, ÉLECTRON spiral-wound

enroulement *m* ALIM ÉLEC winding, wrapping, COMPOSANTS winding; ~ **antagoniste** ALIM ÉLEC counteracting winding; ~ **auxiliaire** ALIM ÉLEC auxiliary winding; ~ **bifilaire** ALIM ÉLEC balanced opposing winding, bifilar winding; ~ **d'équilibrage** ALIM ÉLEC balance winding; ~ **d'excitation** COMPOSANTS operating winding; ~ **d'induit** ALIM ÉLEC armature winding; ~ **de maintien** COMPOSANTS holding winding; ~ **ondulé** INFORMAT cinching; ~ **opposé** ALIM ÉLEC counteracting winding; ~ **de phase** COMPOSANTS phase winding; ~ **primaire** COMPOSANTS primary winding; ~ **de relais** COMPOSANTS relay winding; ~ **secondaire** COMPOSANTS secondary winding; ~ **en spirale** ÉLECTRON helix spiral wrap; ~ **de travail** ALIM ÉLEC active winding

enrouler *vt* ALIM ÉLEC, COMPOSANTS wind, wrap

enrouleur: ~ **de papier** *m* TÉL paper winder; ~ **de prolongateur téléphonique** *m* COMPOSANTS, PÉRIPH telephone extension reel

enseignement: ~ **assisté par ordinateur** *m* (*EAO*) APP SERV computer-based training, INFORMAT computer-assisted learning (*CAL*), computer-assisted instruction, computer-based training (*CBT*); ~ **informatisé** *m* APP SERV, INFORMAT computer-based training; ~ **intelligemment assisté par ordinateur** *m* (*EIAO*) INFORMAT intelligent tutoring systems (*ITS*)

ensemble *m* TÉL cluster; ~ **des abonnés** RADIO, TÉL subscriber base; ~ **antenne-récepteur** RADIO antenna-receiver unit; ~ **des circuits** ÉLECTRON circuitry; ~ **complet de signaux d'entrée valides** DONNÉES complete valid input-signal set; ~ **de contextes définis** DONNÉES defined-context set (*DCS*); ~ **coupe-circuit de batterie** ALIM ÉLEC, COMPOSANTS battery fuse unit; ~ **d'équipements terminaux** TÉL terminal equipment package; ~ **de lancement** RADIO launch complex; ~ **des propositions conditionnelles multiples** DONNÉES set of alternatives; ~ **des propositions SI** DONNÉES set of alternatives; ~ **terminal** TÉL local end

entité: ~ **d'application** *f* TÉL application entity (*AE*)

entraînement: ~ **de bande magnétique** *m* INFORMAT tape drive; ~ **de fréquence** *m* ÉLECTRON pulling figure; ~ **du papier** *m* PÉRIPH paper feed; ~ **par picots** *m* PÉRIPH pin feed

entraîneur *m* COMPOSANTS dog, driver

entrance *f* ÉLECTRON fan-in

entrant *adj* GÉNÉRAL incoming (*I/C*)

entraxe *adj* GÉNÉRAL between centers (*AmE*), between centres (*BrE*)

entrée *f* DONNÉES entry, inlet access, input, ÉLECTRON input, GÉNÉRAL item, SIGNAL, TÉL, TRANSMIS entry, inlet, input; ~ **d'acceptation** TÉL acceptance input; ~ **analogique** INFORMAT analog input; ~ **asymétrique** INFORMAT unbalanced input; ~ **audio** ÉLECTRON audio input, INFORMAT audio in, audio input; ~ **de câble** TRANSMIS cable entry, cable inlet; ~ **contenu renvoyé** DONNÉES returned-content entry; ~ **de déclenchement** ÉLECTRON trigger input; ~ **de données** DONNÉES, INFORMAT data entry; ~ **de forçage** ÉLECTRON steered input, steering input; ~ **de gamme** DONNÉES entry level; ~ **de ligne** INFORMAT line input; ~ **en ligne** TÉL interruption; ~ **maîtresse** DONNÉES parent entry; ~ **message remis** DONNÉES delivered-message entry; ~ **du microphone** ÉLECTRON microphone input; ~ **mono** INFORMAT, PÉRIPH mono input; ~ **muette** DONNÉES blank entry; ~ **numérique** INFORMAT digital input; ~ **numérique directe** INFORMAT direct digital input; ~ **de paramètres avec format** INFORMAT format parameter input; ~ **à plusieurs bits** DONNÉES multibit input; ~ **principale** DONNÉES main entry; ~ **rapport remis** DONNÉES delivered-report entry; ~ **de réception double** RADIO dual front end; ~ **secondaire** DONNÉES, INFORMAT child-entry; ~ **de**

sélection TÉL selection input; **~ de la source** ÉLECTRON feed input; **~ stéréo** INFORMAT, PÉRIPH stereo input; **~ symétrique** INFORMAT balanced input; **~ de terre de signalisation** SIGNAL, TRANSMIS signal earth input (*BrE*), signal ground input (*AmE*); **~ en tiers** TÉL override, busy override, executive override; **~ en tiers du standardiste** TÉL attendant busy override, attendant override; **~ des unités** ÉLECTRON units input; **~ de validation** ÉLECTRON enable input; **~ vidéo** DONNÉES, TV video input

entrées: **~ de chargement parallèle direct** *f pl* ÉLECTRON jam inputs

entrée/sortie[1] *adj* (*E/S*) ÉLECTRON, INFORMAT, TÉL input/output (*I/O*)

entrée/sortie[2] *f* ÉLECTRON, INFORMAT, TÉL input/output (*I/O*); **~ virtuelle** INFORMAT virtual input/output (*VIO*)

entrefer *m* ALIM ÉLEC air gap, INFORMAT head gap

entrelacé *adj* RADIO, TÉL, TRANSMIS interlaced, interleaved, TV interlaced

entrelacement *m* RADIO, TÉL, TRANSMIS interlacing, interleave, interleaving, TV interlacing; **~ des blocs** RADIO block interleaving; **~ de câblage** TRANSMIS cable lacing; **~ de fréquences porteuses** TRANSMIS interleaved carriers; **~ d'impulsions** TÉL, TRANSMIS pulse interlacing; **~ des impulsions** TÉL, TRANSMIS pulse interleaving

entrelacer *vt* RADIO, TÉL, TRANSMIS interlace, interleave, TV interlace

entreprise: **~ de télécommunications** *f* GÉNÉRAL common carrier, TÉL network carrier, TRANSMIS common carrier, telecommunications common carrier; **~ de télécommunications de type 1** *f* TRANSMIS T1 carrier (*T1C*)

entrer[1] *vt* DONNÉES input, INFORMAT feed, input

entrer[2]: **~ en ligne** *vi* TÉL go into circuit

entretien *m* ÉLECTRON holding; **~ de l'image** TV image refreshing; **~ systématique** TESTS scheduled maintenance

entretoise *f* COMPOSANTS spacing block, spacer; **~ tubulaire** COMPOSANTS distance tube

entropie *f* GÉNÉRAL entropy; **~ moyenne par caractère** DONNÉES information rate per character, mean entropy per character; **~ à zéro absolu** GÉNÉRAL zero point entropy

enveloppe *f* COMMUT envelope, COMPOSANTS case, cover, housing; **~ céramique** ÉLECTRON ceramic encapsulation, ceramic encasement; **~ de refroidissement** COMPOSANTS cooling jacket; **~ de signal** SIGNAL signal envelope; **~ de timbre** INFORMAT timbre envelope; **~ verrouillée en permanence** TÉL permanently-locked envelope; **~ de volume** INFORMAT volume envelope

environnement *m* GÉNÉRAL environment

envoi *m* GÉNÉRAL transmission, sending, TÉL, TRANSMIS transmission; **~ en différé** DONNÉES deleted-time sending; **~ d'essai** DONNÉES, TESTS *Internet* test message; **~ par télécopie** TÉL fax transaction

épandeur: **~ de matériel d'étanchéité** *m* COMPOSANTS sealant applicator

épanouissement: **~ polaire** *m* ALIM ÉLEC pole shoe

EPD *abrév* (*équipement périphérique distant*) INFORMAT RPE (*remote peripheral equipment*)

épilogue *m* DONNÉES, TESTS postamble; **~ d'essai** TESTS test postamble; **~ de procédure** TÉL procedure epilogue; **~ de test** TESTS test postamble

épine: **~ dorsale en fibres optiques** *f* ÉLECTRON fiber-optic backbone (*AmE*), fibre-optic backbone (*BrE*)

épisseur *m* COMPOSANTS, GÉNÉRAL, TRANSMIS jointer, splicer; **~ de câbles** TRANSMIS cable jointer

épissure *f* COMPOSANTS, GÉNÉRAL, TRANSMIS joint, splice; **~ de câble** TRANSMIS cable joining, cable joint, cable splice; **~ en dérivation** TRANSMIS bridge splice; **~ droite** COMPOSANTS, ÉLECTRON straight joint, straight splice; **~ de fibres optiques** TRANSMIS optical fiber splice (*AmE*), optical fibre splice (*BrE*); **~ longitudinale** GÉNÉRAL linear joint; **~ mécanique** TRANSMIS mechanical splice; **~ optique** TRANSMIS optical fiber splice (*AmE*), optical fibre splice (*BrE*), optical splice; **~ par fusion** TRANSMIS fused-fiber splice (*AmE*), fused-fibre splice (*BrE*), fusion splice; **~ par soudage** TRANSMIS fused-fiber splice (*AmE*), fused-fibre splice (*BrE*), fusion splice; **~ de terminaison** COMPOSANTS termination joint

épissurer *vt* GÉNÉRAL splice

épissures: **~ rabattues** *f pl* ÉLECTRON configurations foldback

épitaxie *f* COMPOSANTS, ÉLECTRON epitaxy; **~ en phase vapeur de matériaux métal-organiques** ÉLECTRON metal-organic vapor phase epitaxy (*AmE*) (*MOVPE*), metal-organic vapour phase epitaxy (*BrE*) (*MOVPE*)

époxy *adj* COMPOSANTS, ÉLECTRON, TRANSMIS epoxy

époxyde *m* COMPOSANTS epoxide

EPR *abrév* (*exploitation privée reconnue*) APP SERV, DONNÉES recognized private operation, RPOA (*recognized private operating agency*)

épreuve: **à l'~ de l'effraction** *adj* TESTS *sécurité* burglar-proof; **à l'~ du feu** *adj* COMPOSANTS fireproof

EPROM *abrév* (*mémoire morte programmable électroniquement*) ÉLECTRON EPROM (*electrically programmable read-only memory*)

éprouvette *f* TESTS test coupon, test piece

équation: **~ exponentielle** *f* GÉNÉRAL exponential equation; **~ du radar** *f* RADIO radar equation; **~ de régression** *f* GÉNÉRAL, INFORMAT regression equation; **~ de Sellmeir** *f* TRANSMIS Sellmeir's equation

équilibrage *m* GÉNÉRAL balancing

équilibre: **~ chromatique** *m* TV chromatic balance; **~ des circuits réels** *m* ALIM ÉLEC longitudinal balance; **~ d'impédance** *m* ÉLECTRON, TÉL impedance balance; **~ de ligne** *m* TRANSMIS line balance; **~ des modes** *m* TRANSMIS steady-state condition, equilibrium mode distribution

équipement *m* GÉNÉRAL equipment; **~ d'abonné** COMPOSANTS, TÉL subscriber equipment; **~ d'accès externe** TÉL external access equipment (*EA*); **~ acoustique** INFORMAT audio equipment; **~ d'alimentation** ALIM ÉLEC power supply equipment; **~ annexe** INFORMAT accessory equipment, accessory unit; **~ d'antiparasitage** PÉRIPH interference suppression equipment; **~ en attente** TÉL standby equipment; **~ auxiliaire** TÉL accessory equipment; **~ bivocal** TÉL speech-plus-duplex equipment; **~ à bord** RADIO on-board equipment; **~ de brassage intermédiaire** SIGNAL intermediate cross-connect (*IC*); **~ de canaux** TRANSMIS channel equipment (*CE*); **~ de chiffrement** DONNÉES, SIGNAL encryption unit; **~ de circuit** TÉL circuit equipment, TRANSMIS trunk equipment; **~ de collecte de données** DONNÉES data-collection equipment; **~ de commande** CONTRÔLE control equipment, controlling equipment; **~ commun des terminaux de trafic** TÉL common traffic terminal equipment (*CTTE*); **~ de commutateur** COMMUT, TÉL switchboard equipment; **~ de commutation** COMMUT, TÉL server; **~ de comptage d'appels** APP SERV call-metering equipment; **~ de contrôle** TESTS check-out equipment; **~ de conversion de signaux** SIGNAL, TÉL signal conversion equipment; **~ destinaire** INFORMAT destination equipment; **~ en double** TÉL duplicate equipment, standby equipment; **~ duplex à pont** TÉL bridge duplex system; **~ dupliqué** TÉL duplicate equipment, standby equipment; **~ électronique** ÉLECTRON electronic equipment; **~ embarqué** RADIO on-board equipment; **~ d'entrée** ÉLECTRON entrance facility (*EF*); **~ d'essai** PÉRIPH, TESTS testing equipment, test equipment; **~ d'essai de câblage** TESTS wiring test set, wiring tester; **~ d'essai de faisceau** TÉL, TESTS traffic route tester; **~ d'essai d'itinéraire** TÉL, TESTS traffic route tester; **~ d'extension de canal** TRANSMIS channel extension equipment; **~ de facturation** APP SERV charging equipment; **~ de facturation**

des appels APP SERV call-charging equipment; **~ aux fréquences de la bande de base** RADIO baseband equipment; **~ à fréquences vocales** TRANSMIS audio equipment; **~ de gestion des systèmes** INFORMAT systems management equipment (*SME*); **~ d'interception** TÉL interception equipment; **~ d'interconnexion** COMMUT digital cross-connect equipment (*DXC*); **~ d'interface téléphonique** TÉL telephone interface equipment; **~ intermédiaire** INFORMAT, TÉL intermediate equipment; **~ lecteur optique de marques** DONNÉES mark scanning equipment; **~ de ligne d'abonné** DONNÉES, INFORMAT, RADIO, TÉL, TV subscriber's line equipment; **~ de ligne interautomatique** TRANSMIS tie-circuit interface; **~ local d'abonnés** PÉRIPH customer premises equipment (*CPE*); **~ de matériel de groupe primaire** TÉL, TRANSMIS group translating equipment (*GTE*); **~ de matériel de groupe secondaire** TÉL supergroup translating equipment; **~ de médiation** INFORMAT mediation device (*MD*); **~ de mesure** TESTS measuring equipment; **~ de mise en file d'attente** TÉL queueing equipment; **~ de modulation de groupe primaire** TÉL, TRANSMIS group translating equipment (*GTE*); **~ de modulation de groupe secondaire** TÉL supergroup translating equipment; **~ multiplex** TÉL, TRANSMIS multiplex equipment; **~ de multiplexage MIC** TÉL, TRANSMIS PCM multiplex equipment; **~ de multiplexage numérique** TÉL, TRANSMIS digital multiplex equipment; **~ de multiplication de circuits numériques** (*EMCN*) TÉL, TRANSMIS digital circuit multiplication equipment (*DCME*); **~ non dupliqué** TÉL nonduplicated equipment; **~ optoélectronique** ÉLECTRON optoelectronic equipment; **~ optoélectronique de réception** TRANSMIS fiber-optic receiver (*AmE*), fibre-optic receiver (*BrE*); **~ périphérique** DONNÉES peripheral equipment, GÉNÉRAL external equipment, peripheral equipment, INFORMAT peripheral equipment; **~ périphérique distant** (*EPD*) INFORMAT remote peripheral equipment (*RPE*); **~ de position** TÉL position set; **~ de pressurisation de câbles** GÉNÉRAL cable pressurization equipment; **~ de puissance** ALIM ÉLEC power equipment; **~ de raccordement téléphonique** TÉL telephone interface equipment; **~ de redémarrage automatique** CONTRÔLE automatic restart unit; **~ régulateur d'ambiance** CLIMAT climate control equipment; **~ de réponse** TÉL answering equipment; **~ de reproduction** RADIO maximum usable sensitivity; **~ pour réseaux de transmission** TRANSMIS transmission equipment; **~ de réserve** TÉL standby equipment; **~ RF associé à l'antenne** RADIO front-end equipment; **~ sans possibilité d'adressage au réseau** TÉL network nonaddressing equipment; **~ de secours** TÉL standby equipment; **~ de service des abonnés absents** APP SERV, TÉL absent-subscriber service equipment; **~ de signalisation de ligne** TÉL line signaling equipment (*AmE*), line signalling equipment (*BrE*); **~ de signalisation routière** SIGNAL traffic signaling equipment (*AmE*), traffic signaling installation (*AmE*), traffic signalling equipment (*BrE*), traffic signalling installation (*BrE*); **~ de soutien au sol** RADIO ground support equipment (*GSE*); **~ de standard** COMMUT, TÉL switchboard equipment; **~ de supervision** DONNÉES monitoring equipment; **~ de surveillance** DONNÉES housekeeping unit; **~ de taxation** APP SERV charging equipment; **~ de taxation des appels** APP SERV call-charging equipment; **~ de taxation automatique** APP SERV automatic charging equipment; **~ de taxation centralisée** APP SERV central-charging equipment; **~ de télécommunications** GÉNÉRAL telecommunications equipment; **~ de télétraitement de l'information** DONNÉES remote data unit (*RDU*); **~ de terminaison de circuit de données** (*ETCD*) DONNÉES, INFORMAT, PÉRIPH, TRANSMIS data circuit terminal equipment (*DCE*), data circuit terminating equipment (*DCTE*); **~ de terminaison de ligne** PÉRIPH, TÉL line-terminating equipment; **~ terminal** DONNÉES, INFORMAT, PÉRIPH, TÉL terminal equipment (*TE*), terminating equipment; **~ terminal demandé** TÉL called terminal; **~ terminal demandeur** TÉL calling terminal; **~ terminal de données** DONNÉES, PÉRIPH, TÉL data terminal equipment (*DTE*); **~ terminal à raccordement acoustique** TÉL acoustically-connected terminal equipment; **~ terminal de réseau** TÉL, TRANSMIS network terminal equipment (*NTE*); **~ terminal pour le RNIS-LB** TÉL B-ISDN terminal equipment; **~ terminal sémaphore** TÉL signaling terminal equipment (*AmE*) (*STE*), signalling terminal equipment (*BrE*) (*STE*); **~ terminal de signalisation** TÉL signaling terminal equipment (*AmE*) (*STE*), signalling terminal equipment (*BrE*) (*STE*); **~ terminal de traitement de données** (*ETTD*) DONNÉES, PÉRIPH data-processing terminal equipment; **~ terminal de traitement de données mobile** (*ETTD-MT*) RADIO mobile data-processing terminal equipment; **~ terminal de travée** TÉL suite supervision equipment, suite terminal equipment; **~ terminal de voie** TRANSMIS channel terminal; **~ de test** PÉRIPH, TESTS testing equipment; **~ de test incorporé** TÉL, TESTS built-in test equipment; **~ de tonalités et auxiliaires** TÉL auxiliary management software machine; **~ de traitement de l'image** DONNÉES, TV image-processing equipment; **~ de transmission plésiochrone** TRANSMIS plesiochronous transmission equipment; **~ de transposition de groupe secondaire** TÉL supergroup translating equipment; **~ univocal** TÉL speech plus simplex, speech-plus-simplex equipment; **~ d'usager** PÉRIPH customer equipment (*CEQ*); **~ de vérification de l'acheminement du trafic** TÉL, TESTS traffic route tester; **~ de vidéo** TV video equipment; **~ de vidéocommunication** TV videocommunication equipment; **~ de voie** TRANSMIS channel unit

équiper *vt* GÉNÉRAL, INFORMAT, TRANSMIS equip

équivalence: **~ logique** *f* ÉLECTRON, INFORMAT IF-AND-ONLY-IF operation

équivalent *m* ALIM ÉLEC, COMPOSANTS, ÉLECTRON, RADIO, TÉL equivalent; **~ d'un circuit téléphonique** TÉL telephone circuit loss; **~ de couplage du terminal** TÉL terminal coupling loss (*TCL*); **~ d'une liaison téléphonique** TÉL telephone circuit loss; **~ minimal net** TRANSMIS minimum net loss; **~ net** TRANSMIS net loss; **~ pondéré de couplage du terminal** TÉL weighted terminal coupling loss (*TCLw*); **~ de référence** TÉL reference equivalent; **~ de référence corrigé** (*ERC*) TÉL corrected reference equivalent (*CRE*)

ER *abrév* (*exploitation reconnue*) APP SERV, DONNÉES ROA (*recognized operating agency*)

ERC *abrév* (*équivalent de référence corrigé*) TÉL CRE (*corrected reference equivalent*)

erlang *m* ÉLECTRON, TÉL Erlang

erlangmètre *m* ÉLECTRON, TÉL, TESTS traffic meter

ERR *abrév* (*réponse pour fin de transmission*) DONNÉES, TÉL ERR (*response for end of transmission*)

erreur *f* GÉNÉRAL error; **~ d'adaptation** GÉNÉRAL, TRANSMIS mismatch; **~ d'ajustage** TÉL *métrologie* adjustment error; **~ aléatoire** DONNÉES, TÉL, TESTS random error; **~ d'alignement** RADIO, TÉL alignment error; **~ d'amplitude** INFORMAT amplitude error; **~ d'approximation** GÉNÉRAL approximation error; **~ d'assortiment** GÉNÉRAL, TRANSMIS mismatch; **~ binaire** TRANSMIS bit error; **~ de bloc à l'extrémité distante** DONNÉES far-end block error (*FEBE*); **~ de câblage** COMPOSANTS wiring error; **~ de calage** TESTS loop alignment error; **~ centrée** GÉNÉRAL *statistique*, INFORMAT balanced error; **~ de chargement** INFORMAT loading error; **~ de communication** GÉNÉRAL communication error; **~ de concentricité cœur-gaine** TÉL, TRANSMIS core-cladding concentricity error; **~ de concentricité cœur-surface de référence** TÉL, TRANSMIS core-reference surface concentricity error; **~ de directivité** RADIO squint angle, squint; **~ distante** DONNÉES remote error; **~ double** TÉL double error; **~ due à l'inertie** TESTS inertia error; **~ due à la température** CLIMAT, TESTS temperature error; **~ d'échantillonnage** TRANSMIS sampling error; **~ d'émetteur** RADIO, TRANSMIS transmitter error; **~ d'entrée** DONNÉES, INFORMAT input error; **~ d'espacement d'antenne** RADIO antenna spacing error, aerial-spacing error; **~ d'étalonnage** TRANSMIS calibration error; **~ d'étalonnage de gain** ÉLECTRON gain calibration error; **~ d'excentrement** RADIO pointer centering error (*AmE*), pointer centring error (*BrE*); **~ de facturation** APP SERV charging error; **~ de fonctionnement** INFORMAT operation error; **~ de fréquence** TÉL frequency uncertainty; **~ de guidage** INFORMAT mistracking; **~ humaine** TESTS human error; **~ instrumentale** TESTS instrumental error; **~ d'interpolation** TESTS interpolation error; **~ intrinsèque** TESTS intrinsic error; **~ irréparable** INFORMAT irrecoverable error; **~ de lecture** INFORMAT misread, reading error; **~ de linéarité** COMPOSANTS linearity error; **~ locale** RADIO site error; **~ maximum d'intervalle de temps** TESTS maximum time interval error (*MTIE*); **~ de mesure** TESTS measurement error; **~ de méthode** TESTS error of method; **~ moyenne** TESTS mean error; **~ nocturne** RADIO night error; **~ de nuit** RADIO night error; **~ numérique** TÉL digital error, TRANSMIS bit error; **~ de numérotation** APP SERV, PÉRIPH keying error, TÉL dialing error (*AmE*), dialling error (*BrE*); **~ d'observation** TESTS observation error; **~ octantale** RADIO octantal error; **~ de parallaxe** TESTS parallax error; **~ parasite** TESTS parasitic error; **~ de plage** RADIO swing error; **~ de pointage** RADIO mispointing; **~ de polarisation** RADIO polarization error; **~ de poursuite** RADIO tracking error; **~ de prédiction** INFORMAT prediction error; **~ de procédure** DONNÉES procedural error; **~ de programmation** DONNÉES, INFORMAT programming error; **~ de propagation** RADIO propagation error; **~ de propagation ionosphérique** RADIO ionospheric path error; **~ de protocole** TÉL protocol error; **~ quadrantale** RADIO quadrantal error; **~ de quantification** TRANSMIS quantization error; **~ réduite** TESTS reduced error; **~ relative** TESTS relative error; **~ relative maximum d'intervalle de temps** DONNÉES maximum relative time interval error (*MRTIE*); **~ résiduelle** DONNÉES, INFORMAT, TÉL residual error; **~ semi-circulaire** RADIO semicircular error; **~ simple** TESTS single error; **~ sur la durée** TÉL time interval error; **~ de synchronisation** TRANSMIS clock error; **~ systématique** TESTS systematic error; **~ de taxation** APP SERV charging error; **~ de test élémentaire** TESTS test case error; **~ de test élémentaire abstraite** DONNÉES, TESTS abstract test case error; **~ de test élémentaire exécutable** DONNÉES, TESTS executable test case error; **~ de transmission** TESTS, TRANSMIS transmission error; **~ de transposition** DONNÉES translation error; **~ type** GÉNÉRAL standard error; **~ type de polarisation** RADIO, TRANSMIS standard wave error; **~ de zéro** TESTS zero error

éruption: **~ due au contact avec les blocs de raccordement** *f* ÉLECTRON tag rash

E/S *abrév* (*entrée/sortie*) ÉLECTRON, INFORMAT, TÉL I/O (*input/output*)

ES *abrév* (*étalement du spectre*) RADIO SS (*spread spectrum*)

ESC *abrév* (*échappement*) DONNÉES, INFORMAT, TÉL ESC (*escape*)

ésotérique *adj* DONNÉES esoteric

espace: **~ arrière** *m* INFORMAT backspace (*BS*); **~ de bobinage** *m* ALIM ÉLEC winding space; **~ chromatique** *m* TV chromatic space, color space (*AmE*), colour space (*BrE*); **~ des couleurs** *m* TV chromatic space, color space (*AmE*), colour space (*BrE*); **~ de déclenchement** *m* ÉLECTRON starter gap, trigger gap; **~ de disque** *m* INFORMAT disk space; **~ d'échange** *m* DONNÉES open forum; **~ d'enroulement** *m* ALIM ÉLEC winding space; **~ entre blocs** *m* DONNÉES interblock gap, block gap; **~ entre enregistrements** *m* DONNÉES interrecord gap; **~ image** *m* INFORMAT image space; **~ d'interaction** *m* ALIM ÉLEC interaction gap; **~ interbloc** *m* DONNÉES interblock gap, block gap; **~ intertravée** *m* TÉL row spacing; **~ lointain** *m* RADIO deep space; **~ mémoire** *m* INFORMAT memory space; **~ Morse** *m* TÉL Morse space; **~ objet** *m* INFORMAT object space; **~ panoramique** *m* DONNÉES panoramic space; **~ principal** *m* ÉLECTRON main gap; **~ proche de la Terre** *m* RADIO near-Earth space; **~ réservé au corps** *m* INFORMAT body placeholder; **~ des résultats possibles** *m* TESTS sample space

espacement *m* GÉNÉRAL, INFORMAT, RADIO spacing; **~ d'air** ALIM ÉLEC insulation gap; **~ angulaire** RADIO angular separation; **~ d'antenne** RADIO aerial spacing, antenna spacing; **~ des bandes de radio fréquences** RADIO radio channel spacing; **~ des broches** ÉLECTRON pin spacing; **~ des canaux** RADIO channel separation, channel spacing; **~ des canaux radio** RADIO radio channel spacing; **~ de cartes** COMPOSANTS board spacing; **~ entre les antennes** RADIO aerial spacing; **~ entre canaux** RADIO channel separation, channel spacing; **~ entre impulsions** TÉL, TRANSMIS pulse spacing, pulse interval; **~ des fréquences** RADIO frequency spacing, frequency separation; **~ de garde** RADIO guard space; **~ des impulsions** TÉL, TRANSMIS pulse spacing; **~ modulaire** TÉL modular spacing; **~ des relais** TÉL relay spacing; **~ des répéteurs** TRANSMIS repeater spacing, repeater span; **~ de section** TÉL section spacing; **~ des trous** COMPOSANTS, ÉLECTRON hole spacing

esquisse *f* GÉNÉRAL outline

essai *m* TESTS test; **~ accéléré** TESTS *sécurité de fonctionnement* accelerated test; **~ d'alignement** TRANSMIS alignment check; **~ d'ambiance** CLIMAT, TESTS environmental test; **~ d'arrachement** COMPOSANTS, TESTS pulling test, stripping test; **~ automatique** TESTS self-test; **~ au banc** TESTS bench testing; **~ en boucle** TRANSMIS back-to-back test; **~ Brinell** COMPOSANTS *résistance* Brinell test; **~ de câblage** COMPOSANTS wiring check, TESTS wiring check, wiring test; **~ censuré** TESTS censored test; **~ de chaleur humide** TESTS damp heat testing; **~ au choc** TESTS shock test; **~ de chute** TESTS drop test; **~ en circuit ouvert** ÉLECTRON open-circuit test (*OC test*); **~ de la commutation** COMMUT, TESTS switching test; **~ de conformité** TESTS compliance test; **~ de continuité de bout en bout** TÉL end-to-end continuity test; **~ corona** TESTS corona arc testing; **~ en cours de fabrication** TESTS in-process testing; **~ en court-circuit** ÉLECTRON short-circuit test; **~ de criquage** TESTS cracking test; **~ de déroulement** TESTS unwrap test, unwrapping test; **~ destructif** DONNÉES, TESTS destructive testing; **~ de détermination** TESTS determination test; **~ de durée de vie** TESTS life testing; **~ d'écoute** RADIO listening test; **~ à l'égratignure** COMPOSANTS, TESTS scratch hardness test; **~ d'endurance** TESTS endurance test, fatigue test; **~ d'environnement** GÉNÉRAL environmental test; **~ d'étanchéité** ALIM ÉLEC insulation test, leakage test; **~ final** TESTS final test; **~ de fissuration** TESTS cracking test; **~ de fonctionnement** TESTS performance test, functional test; **~ en grandeur réelle** TESTS field test, field trial, full-scale test; **~ d'isolement** ALIM ÉLEC insulation test, leakage test; **~ en laboratoire** TESTS laboratory test, alpha trial; **~ de mailles** TESTS link test; **~ de mise en service** TESTS commissioning test; **~ de netteté des syllabes** TÉL, TESTS syllable articulation test; **~ de niveau de bruit** TÉL, TESTS noise test; **~ au niveau carte** TESTS *épreuves* board-level testing; **~ au niveau composants** TESTS component-level testing; **~ non destructif** DONNÉES, TESTS nondestructive testing; **~ d'ordre des phases** ALIM ÉLEC phase-sequence test; **~ par échantillonnage** TESTS sampling test; **~ par lots** COMPOSANTS batch testing, DONNÉES group testing; **~ par "tout ou rien"** TÉL, TESTS yes or no test; **~ partiel** TESTS partial test, subtest; **~ pilote** TESTS beta-trial, pilot trial, TRANSMIS pilot trial; **~ pratique** TESTS field test, field trial; ~ **préalable** TESTS pretest; ~ **préliminaire** TESTS preliminary test; ~ **préventif** TESTS routine test; ~ **de qualification** GÉNÉRAL, TESTS acceptance test; ~ **en radiofréquence** RADIO, TESTS RF test; ~ **de réception** GÉNÉRAL, TESTS acceptance test; ~ **de recette** GÉNÉRAL, TESTS acceptance test; ~ **de résistance** TESTS strength test, impact test; ~ **de résistance aux rayures** COMPOSANTS, TESTS scratch hardness test; ~ **de rigidité diélectrique** ALIM ÉLEC dielectric strength test; ~ **de rupture par traction** TESTS, TRANSMIS proof test; ~ **de satellite intégré** TESTS integrated satellite test (*IST*); ~ **aux secousses** COMPOSANTS, TESTS *résistance* bump test; ~ **de sélection** TESTS screening test; ~ **sémaphore** TESTS signaling test (*AmE*), signalling test (*BrE*); ~ **séquentiel** TESTS sequential test; ~ **de série** TESTS routine test; ~ **de la signalisation** TESTS signaling test (*AmE*), signalling test (*BrE*); ~ **de signalisation de ligne** TESTS line signaling test (*AmE*), line signalling test (*BrE*); ~ **au sol** TRANSMIS ground test; ~ **sous conditions ambiantes** CLIMAT environmental test; ~ **sous contrainte échelonnée** TESTS step-stress test; ~ **subjectif** TESTS subjective test; ~ **sur prélèvement** TESTS sampling test; ~ **systématique** TESTS routine test; ~ **de système** INFORMAT, TESTS system testing; ~ **de système intégré** TESTS integrated system test (*IST*); ~ **de taxation** APP SERV charging test; ~ **technique** TESTS trial; ~ **de téléphonie en bande de base** TESTS telephony baseband test; ~ **de tension** ALIM ÉLEC, TESTS voltage test; ~ **de tenue à la foudre** ALIM ÉLEC lightning impulse test; ~ **de traction** COMPOSANTS, TESTS tensile test; ~ **de trafic opérationnel** TESTS live traffic test; ~ **de trafic réel** TESTS live traffic test; ~ **de type** TESTS type test; ~ **de vibration** TESTS vibration test; ~ **aux vibrations en régime sinusoïdal** TESTS sine vibration test; ~ **à vide** ÉLECTRON open-circuit test (*OC test*); ~ **en vraie grandeur** TESTS field trial, field test, full-scale test

essais: ~ **d'ambiance** *m pl* CLIMAT, TESTS environmental testing; ~ **conjoints** *m pl* TESTS joint testing, joint tests; ~ **à la demande** *m pl* TESTS on-demand testing; ~ **d'endurance** *m pl* TESTS life testing; ~ **d'environnement** *m pl* CLIMAT, TESTS environmental testing; ~ **de fonctionnement** *m pl* TESTS functional testing

estampille *f* COMMUT, INFORMAT, TESTS time stamp; ~ **de décodage** INFORMAT decoding time stamp (*DTS*); ~ **de présentation** INFORMAT presentation time stamp (*PTS*)

estampiller *vt* COMPOSANTS stamp

estimation: ~ **de mouvement** *f* TV motion estimation; ~ **du risque** *f* TESTS risk assessment

estimé *adj* TESTS estimated

estimer *vt* TESTS estimate

estompé *adj* INFORMAT dimmed

établir *vt* DONNÉES, GÉNÉRAL, RADIO *communication*, TÉL *connexion* establish, set up; ~ **une communication** TÉL put through a call; ~ **une liaison** DONNÉES dial up

établissement *m* GÉNÉRAL establishment, setting-up, setup, RADIO, TÉL setting-up; ~ **d'appel** TÉL call establishment, call setup; ~ **de la CAF** RADIO AFC pull-in; ~ **de communication** RADIO setup, TÉL call connection, call setup, setup; ~ **d'une connexion sans émission** RADIO off-air call setup; ~ **de liaison** DONNÉES, INFORMAT handshake, handshaking; ~ **nonnormalisé** (*NSS*) DONNÉES, TÉL nonstandard setup (*NSS*); ~ **d'un organigramme** GÉNÉRAL flowcharting

étage *m* GÉNÉRAL stage; ~ **d'abonnés** COMMUT subscriber switch; ~ **d'amplificateur** ALIM ÉLEC, ÉLECTRON amplifier stage; ~ **d'amplification** ÉLECTRON amplification stage; ~ **d'attaque** ÉLECTRON driving stage; ~ **de brassage** TRANSMIS distribution stage; ~ **de chercheurs de ligne** TRANSMIS line-finder stage; ~ **de communication d'abonnés** COMMUT subscriber stage; ~ **de commutation** COMMUT, TÉL switching stage, switching point; ~ **de commutation distante** COMMUT, TÉL remote switching point, remote switching stage; ~ **de concentration** TRANSMIS concentration stage; ~ **à émetteur commun** ÉLECTRON earthed emitter stage (*BrE*), grounded emitter stage (*AmE*); ~ **d'entrée** ÉLECTRON input stage; ~ **d'expansion** TRANSMIS expansion stage; ~ **FI** ÉLECTRON, RADIO IF stage; ~ **final** RADIO, TÉL final stage; ~ **à fréquence intermédiaire** ÉLECTRON, RADIO intermediate frequency stage; ~ **de génération de fréquence** TRANSMIS frequency generating stage; ~ **mélangeur** RADIO mixer stage; ~ **pilote** ÉLECTRON driver stage; ~ **de puissance** ALIM ÉLEC, RADIO power stage; ~ **de sélecteur final** TÉL final selector stage; ~ **de sélecteurs** TÉL rank of selectors; ~ **de sélection** COMMUT selection stage, TÉL selector stage; ~ **de sélection final** COMMUT final selection stage; ~ **de sortie** ÉLECTRON, RADIO output stage; ~ **T** (*étage temporel*) COMMUT, TÉL T stage (*time stage*); ~ **temporel** (*étage T*) COMMUT, TÉL time stage (*T stage*); ~ **terminal** COMMUT terminating stage; ~ **de transit** TÉL transit stage

étagère *f* TÉL shelf; ~ **de câble** TRANSMIS cable shelf

étai *m* COMPOSANTS stay, guy, GÉNÉRAL brace

étalement *m* ÉLECTRON, RADIO spread; ~ **d'impulsion** ÉLECTRON, TÉL, TRANSMIS pulse broadening, pulse dispersion, pulse spreading; ~ **de la largeur de bande** RADIO, TRANSMIS bandwidth expansion; ~ **du spectre** (*ES*) RADIO spectrum spreading, spread spectrum (*SS*); ~ **du spectre hybride** RADIO hybrid spread spectrum; ~ **du spectre à sauts de fréquence** RADIO frequency-hopping spread spectrum; ~ **du spectre à séquence directe** RADIO direct-sequence spread spectrum; ~ **du temps de transit** ÉLECTRON transit time spread

étalon *m* GÉNÉRAL standard; ~ **de capacité** ÉLECTRON capacitance standard; ~ **de fréquence** RADIO frequency standard; ~ **international** TESTS international standard (*IS*); ~ **primaire de fréquence** RADIO primary frequency standard; ~ **secondaire de fréquence** RADIO secondary frequency standard; ~ **de temps** RADIO time standard; ~ **de travail** TESTS working standard

étalonnage *m* TESTS *métrologie* calibration

étalonner *vt* TESTS *métrologie* calibrate

étalons: ~ **atomiques de fréquence** *m pl* ÉLECTRON atomic frequency standards

étamage *m* COMPOSANTS tin-plating, tinning; ~ **électrolytique** COMPOSANTS tin-electroplating

étamé *adj* COMPOSANTS solder-plated

étamer *vt* COMPOSANTS tin-plate; ~ **à chaud** COMPOSANTS hot-dip tin; ~ **par électrolyse** COMPOSANTS tin-electroplate

étanche *adj* COMPOSANTS waterproof, watertight; ~ **à l'humidité** GÉNÉRAL moisture-proof; ~ **à la poussière** GÉNÉRAL dustproof

étanchéisation *f* COMPOSANTS waterproofing

étancher *vt* COMPOSANTS seal

étape: ~ **de communication interurbaine** *f* COMMUT trunk switching stage; ~ **de commutation interurbaine** *f* TÉL trunk switching stage; ~ **d'extension** *f* TÉL extension step

état[1]: **à** ~ **solide** *adj* COMPOSANTS solid-state

état[2] *m* GÉNÉRAL state, condition, INFORMAT report; ~ **absence de fréquence** ÉLECTRON, INFORMAT, RADIO, TÉL tone-off condition; ~ **activé** GÉNÉRAL activated condition; ~ **activité** COMPOSANTS, ÉLECTRON on-state; ~ **actuel de la technique** GÉNÉRAL current state of the art; ~ **d'alarme** TESTS alarm condition; ~ **d'alerte** TÉL state of alert; ~ **d'attente** INFORMAT waiting state, TESTS standby state; ~ **B** ÉLECTRON low, low level, L; ~ **bas** ÉLECTRON low, low level, L; ~ **des batteries** ALIM ÉLEC battery condition; ~ **bloqué** COMPOSANTS, ÉLECTRON off-state; ~ **bloqué dans le sens inverse** COMPOSANTS, ÉLECTRON reverse blocking state; ~ **de canal** TRANSMIS channel state; ~ **conducteur** ÉLECTRON conducting state, on-state; ~ **croisé** TRANSMIS *fibres optiques* cross state; ~ **dangereux** TESTS hazardous state; ~ **défavorable** RADIO *brouillage* adverse state; ~ **de dérangement** TÉL failure state, fault state; ~ **de disponibilité** COMMUT idle state, TÉL idle state, up-state, TESTS up-state; ~ **de l'entrée** DONNÉES entry status; ~ **d'équilibre** GÉNÉRAL steady state; ~ **final** DONNÉES final state; ~ **fini** DONNÉES, SIGNAL finite state; ~ **flottant** ÉLECTRON high-impedance state; ~ **de fonctionnement** TESTS operating state; ~ **H** ÉLECTRON high, H, high-level state; ~ **haut** ÉLECTRON high, high-level state; ~ **haute impédance** ÉLECTRON high-impedance state; ~ **hors circuit** TESTS disabled state; ~ **hors fonction** TESTS disabled state; ~ **hygrométrique** GÉNÉRAL moisture content; ~ **inactif** COMMUT, TÉL idle state; ~ **inactivité** ÉLECTRON off-state; ~ **d'indisponibilité** TÉL down-state; ~ **initial** DONNÉES, ÉLECTRON initial state; ~ **instable** GÉNÉRAL unstable state; ~ **libre** COMMUT idle state, DONNÉES free, TÉL idle state; ~ **limite** TESTS limiting state; ~ **logique** ÉLECTRON logic state; ~ **de marche** GÉNÉRAL operate condition, operated position; ~ **mésomorphe** ÉLECTRON mesomorph state; ~ **nématique** ÉLECTRON nematic phase; ~ **non dangereux** TESTS nonhazardous state; ~ **de non-fonctionnement** TÉL, TESTS nonoperating state; ~ **d'occupation** TÉL busy status; ~ **de panne** TÉL fault condition, fault state; ~ **passant** COMPOSANTS on-state, ÉLECTRON conducting state, on-state; ~ **de porte** ÉLECTRON gate condition; ~ **présence de fréquence** ÉLECTRON, INFORMAT, RADIO, TÉL tone-on condition; ~ **prêt à l'engagement** DONNÉES ready-to-commit state; ~ **de prise** TÉL seizure condition; ~ **de raccrochage** TÉL on-hook condition; ~ **de recherche** RADIO search condition; ~ **de repos** ALIM ÉLEC idle condition, rest state, COMPOSANTS off position, ÉLECTRON, TÉL, TRANSMIS rest state; ~ **de révision** GÉNÉRAL revision state; ~ **saturé** ÉLECTRON saturated state; ~ **du signal** GÉNÉRAL signal state; ~ **de signalisation** GÉNÉRAL signaling state (*AmE*), signalling state (*BrE*); ~ **significatif** SIGNAL, TÉL, TESTS significant condition; ~ **smectique** ÉLECTRON smectic state; ~ **stable** ÉLECTRON stable state; ~ **de surface brut de presse** ÉLECTRON plate finish; ~ **tampon** ÉLECTRON buffer state; ~ **de test** TESTS testing state; ~ **de test initial** DONNÉES initial testing state; ~ **de test au repos** TESTS idle testing state, TESTS idle state test; ~ **de test stable** TESTS stable testing state; ~ **de test transitoire** DONNÉES, TESTS transient testing state; ~ **de transition** TÉL state of transition, transitional state; ~ **transitoire** TÉL transitional state; ~ **vacant** TESTS free state, idle state; ~ **de veille** RADIO search condition; ~ **de verrouillage de trame** TRANSMIS frame-alignment state

états: ~ **auxiliaires** *m pl* TÉL auxiliary conditions

ETAV *abrév* (*événement téléphonique d'interfonctionnement vers l'avant*) TÉL FITE (*forward interworking telephony event*)

ETCD *abrév* (*équipement de terminaison de circuit de données*) DONNÉES, INFORMAT, PÉRIPH, TRANSMIS DCE (*data circuit terminal equipment*), DCTE (*data circuit terminating equipment*); ~ **en attente** *m* TÉL DCE waiting

éteindre[1] *vt* ALIM ÉLEC, COMMUT, ÉLECTRON, INFORMAT, PÉRIPH, RADIO switch off, turn off

éteindre[2]: **s'~** *v réfl* ALIM ÉLEC die away, die down, go out

étendre *vt* DONNÉES, INFORMAT, RADIO, TRANSMIS, TV extend

étendue *f* DONNÉES range, ÉLECTRON spread, GÉNÉRAL range, RADIO spread, range, TÉL range, TESTS range, scope, TRANSMIS range; ~ **d'audibilité** TÉL audibility range, range of audibility; ~ **de mesure** TESTS measuring range; ~ **à mi-crête** (*FWHM*) DONNÉES, ÉLECTRON, TRANSMIS full-width half-maximum (*FWHM*); ~ **optimale de l'échelle** TESTS optimum scale range; ~ **des relèvements** RADIO spread of bearings; ~ **de sensibilité auditive** TÉL range of audibility

ETF *abrév* (*modulation de huit à quatorze*) INFORMAT ETF (*eight to fourteen modulation*)

éthernet *adj* DONNÉES ethernet

Ethernet *m* DONNÉES Ethernet; ~ **noir** ÉLECTRON thin Ethernet

étincelle *f* ALIM ÉLEC spark

étiquetage *m* ÉLECTRON labeling (*AmE*), labelling (*BrE*), GÉNÉRAL designation, labeling (*AmE*), labelling (*BrE*), INFORMAT tagging

étiqueter *vt* ÉLECTRON label, GÉNÉRAL designate, label, INFORMAT tag

étiqueteur *m* INFORMAT tagger; ~ **grammatical** INFORMAT grammatical tagger

étiquette *f* ÉLECTRON label, GÉNÉRAL designation, label, INFORMAT label, tag, TRANSMIS tag; ~ **d'acheminement** TÉL routing label; ~ **de câble** TRANSMIS cable tag; ~ **de graduation** INFORMAT tick-mark label; ~ **de sécurité** DONNÉES security label; ~ **verte** APP SERV application appliance label; ~ **vierge** ÉLECTRON blank label

étirement: ~ **des lignes** *m* INFORMAT rubber banding

ETM *abrév* (*modulation de huit à dix*) INFORMAT ETM (*eight to ten modulation*)

étouffer *vt* TRANSMIS damp

étouffeur: ~ **d'étincelles** *m* ALIM ÉLEC spark quench, spark quencher, spark suppressor

étrier: ~ **de fixation** *m* COMPOSANTS fixing bracket; ~ **de montage** *m* COMPOSANTS mounting bracket, fixing bracket

ETTD *abrév* (*équipement terminal de traitement de données*) PÉRIPH data-processing terminal equipment; ~ **en attente** *m* TÉL DTE waiting; ~ **non-prêt automatique** *m* TÉL DTE-uncontrolled not-ready; ~ **non-prêt commandé** *m* TÉL DTE-controlled not-ready; ~ **occupé** *m* TÉL DTE busy

ETTD-MT *abrév* (*équipement terminal de traitement de données mobile*) RADIO mobile data-processing terminal equipment

étude: ~ **de cas** *f* GÉNÉRAL case study; ~ **de faisabilité** *f* GÉNÉRAL feasibility study; ~ **d'intensité de trafic** *f* TÉL busy study, traffic study; ~ **logique** *f* GÉNÉRAL logic design

étui *m* COMPOSANTS, INFORMAT, TÉL case, casing; ~ **souple** INFORMAT soft case

EU *abrév* (*Euronorm*) GÉNÉRAL EN (*Euronorm*)

EURESCOM *abrév* (*Union européenne de recherches et d'études stratégiques en télécommunications*) GÉNÉRAL EURESCOM (*European Institute of Research and Strategic Studies in Telecommunications*)

Euronorm *m* (*EU*) GÉNÉRAL Euronorm (*EN*)

EUTELSAT *abrév* (*Organisation internationale de télécommunications par satellites*) RADIO EUTELSAT

évaluation: ~ **du fournisseur** *f* TESTS supplier evaluation, supplier valuation, vendor appraisal, vendor assessment; ~ **des performances** *f* INFORMAT benchmark testing; ~ **de la qualité** *f* APP SERV, TESTS quality assessment; ~ **du risque** *f* TESTS risk evaluation; ~ **du vendeur** *f* TESTS supplier evaluation, supplier valuation, vendor appraisal, vendor assessment

évanouir: **s'~** *v réfl* INFORMAT, RADIO fade

évanouissement *m* INFORMAT fade, fade-out, fading, RADIO fade-out, fading, null, radio fade; ~ **brusque** RADIO radio fade-out; ~ **dû à la pluie** RADIO rain fading; ~ **dû à la propagation par trajets multiples** RADIO multipath fading; ~ **lent** RADIO log-normal shadowing; ~ **log-normal** RADIO log-normal fading; ~ **par interférence** RADIO flat fading; ~ **par trajets multiples** RADIO multipath fading; ~ **de Rayleigh** DONNÉES, TÉL, TRANSMIS Rayleigh fading; ~ **sélectif** RADIO selective fading; ~ **sur le trajet montant** RADIO uplink fade; ~ **uniforme** RADIO flat fading

événement *m* DONNÉES, ÉLECTRON, INFORMAT, RADIO, TÉL, TESTS event; ~ **d'envoi** DONNÉES send event; ~ **d'envoi implicite** DONNÉES implicit send event; ~ **de fin de temporisation** DONNÉES time-out event; ~ **imprévu** TESTS unforeseen event; ~ **invalide sur le plan de la sémantique** TESTS semantically invalid event; ~ **ionisant** ÉLECTRON, RADIO ionizing event; ~ **ionisant initial** RADIO initial ionizing event; ~ **de justification de pointeur** DONNÉES pointer justification event (*PJE*); ~ **non identifié** TESTS unidentified event; ~ **non qualifié** DONNÉES unqualified event; ~ **parasite sporadique** TESTS single event upset (*SEU*); ~ **qualifié** DONNÉES qualified event; ~ **de réception** DONNÉES receive event; ~ **de référence de couche paquets** DONNÉES, TÉL packet layer reference event (*PE*); ~ **relatif à la sécurité** TESTS security-related event; ~ **sinon** DONNÉES otherwise event; ~ **téléphonique d'interfonctionnement vers l'arrière** TÉL *maritime mobile* backward-interworking telephony event (*BITE*); ~ **téléphonique d'interfonctionnement vers l'avant** (*ETAV*) TÉL forward interworking telephony event (*FITE*); ~ **téléphonique de traitement des opérations de commutation à l'interface** TÉL *maritime mobile* switching process interworking telephony event (*SPITE*); ~ **de test** DONNÉES, TESTS test event; ~ **de test inopportun** DONNÉES, TESTS inopportune test event; ~ **de test invalide** DONNÉES, TESTS invalid test event; ~ **de test invalide sur le plan de la syntaxe** DONNÉES, TESTS syntactically-invalid test event; ~ **valide** DONNÉES, TESTS valid event

évolutivité *f* GÉNÉRAL scalability

EWOS *abrév* (*Atelier européen pour les systèmes ouverts*) GÉNÉRAL EWOS (*European Workshop for Open Systems*)

exact *adj* GÉNÉRAL accurate

exactitude *f* GÉNÉRAL accuracy

examen *m* GÉNÉRAL examination, TESTS investigation; ~ **de conformité statique** DONNÉES static conformance review

examiner *vt* GÉNÉRAL examine, TESTS investigate

excentrage: ~ **mécanique du spot** *m* ÉLECTRON mechanical spot misalignment

excès: ~ **de soudure** *m* ÉLECTRON excess solder

excitateur: ~ **de relais** *m* ALIM ÉLEC relay driver

excitation *f* ÉLECTRON *d'un aimant* excitation, operation; ~ **aléatoire** ÉLECTRON random excitation; ~ **ligne par ligne** ÉLECTRON line-sequential drive; ~ **par choc** ÉLECTRON impulse excitation, shock excitation, RADIO shock excitation; ~ **séquentielle** ÉLECTRON line-sequential drive

exciter *vt* ÉLECTRON excite

exclusion *f* ÉLECTRON exclusion; ~ **des postes** DONNÉES, PÉRIPH, TÉL privacy

excroissance *f* ÉLECTRON outgrowth

excursiomètre *m* TESTS frequency deviation meter

excursion: ~ **efficace multivoie** *f* TRANSMIS multichannel rms deviation, multichannel root mean square deviation; ~ **de fréquence** *f* GÉNÉRAL frequency swing, RADIO frequency deviation; ~ **de la fréquence de crête** *f* RADIO, TRANSMIS peak frequency deviation; ~ **de fréquence maximale** *f* RADIO, TRANSMIS peak frequency deviation

exécuter *vt* GÉNÉRAL, INFORMAT carry out, perform

exécution *f* INFORMAT play, running time, playback; ~ **automatique** INFORMAT automatic playback; ~ **pas à pas** INFORMAT step-by-step operation; ~ **polyphonique** INFORMAT polyphonic play; ~ **synchrone** INFORMAT synchronous execution; ~ **synchronisée** INFORMAT synchronized playback; ~ **de travail** INFORMAT job execution

exempt: ~ **de bruit de fond** *adj* INFORMAT noiseless; ~ **de droits** *adj* GÉNÉRAL free of charge, INFORMAT toll-free; ~ **d'erreur** *adj* INFORMAT bug-free

exigence *f* INFORMAT, TESTS requirement; ~ **de qualité** TESTS quality requirement

exigences: ~ **de fonctionnement échelonné** *f pl* COMMUT sequence requirements

expandeur *m* GÉNÉRAL, INFORMAT expander

expansion *f* GÉNÉRAL, INFORMAT expansion

expéditeur *m* DONNÉES sender, transmission sender, GÉNÉRAL dispatcher, INFORMAT sender, TÉL originator, sender, transmission sender, TRANSMIS transmission sender; ~ **de l'échange** DONNÉES, TÉL interchange sender

expéditeur/destinataire *m* DONNÉES, TÉL originator/recipient (*O/R*)

expérience: ~ **acquise en exploitation** *f* TÉL operational experience; ~ **d'exploitation** *f* TÉL operational experience; ~ **factorielle** *f* TESTS factorial experiment

expérimental *adj* GÉNÉRAL pilot

expérimentation *f* TESTS field test, field trial, trial

expérimenté: ~ **en grandeur nature piloté** *adj* GÉNÉRAL field-tested; ~ **sur le terrain** *adj* GÉNÉRAL field-tested

expérimenter: ~ **sur le terrain** *vt* GÉNÉRAL field-test

expert: ~ **du domaine** *m* INFORMAT domain expert

expiration: ~ **de la temporisation** *f* RADIO expiration of timer (*AmE*), expiry of timer (*BrE*)

exploitant *m* DONNÉES, GÉNÉRAL, RADIO, TÉL, TV operator; ~ **public** DONNÉES, TÉL public operator; ~ **de réseau** TÉL network operating company, network operator; ~ **de réseaux câblés de télévision** TV CATV operator; ~ **de réseaux interurbains** GÉNÉRAL carrier, long-distance carrier; ~ **de réseaux de télécommunications** GÉNÉRAL, TRANSMIS common carrier; ~ **de systèmes câbles** TV cable television operator; ~ **de télécommunications** GÉNÉRAL telecommunications operator

exploitation *f* CONTRÔLE, GÉNÉRAL operation, running, PÉRIPH working, RADIO, SIGNAL operation, TÉL operating, operation, running, system, TESTS, TRANSMIS operation; ~, **administration, maintenance et mise en service** TÉL, TESTS operation, administration, maintenance and provisioning (*OAMP*); ~ **en autonome** TÉL off-line working; ~ **avec mise en file d'attente des appels** TÉL call-queueing system; ~ **avec préparation** TÉL preparation operating; ~ **avec station de base commune** RADIO common base station working; ~ **à bandes croisées** TRANSMIS cross-band operation; ~ **à distance** DONNÉES, TESTS remote operation (*RO*); ~ **duplex** TRANSMIS full-duplex; ~ **et maintenance** DONNÉES operations and maintenance (*OM*), TÉL, TESTS operations and maintenance (*OM*), organization and maintenance; ~ **interurbaine automatique** PÉRIPH, TÉL automatic trunk working; ~ **limitée par diffraction** TRANSMIS diffraction-limited operation; ~ **locale** CONTRÔLE local operation; ~ **multirépéteur** RADIO multiple-transponder operation; ~ **par voies conjuguées** TÉL two-way simplex system; ~ **pilote** TÉL pilot operation; ~ **à porteuses distinctes** RADIO spaced carrier operation; ~ **privée reconnue** (*EPR*) APP SERV, DONNÉES recognized private operating agency (*RPOA*), recognized private operation; ~ **reconnue** (*ER*) APP SERV, DONNÉES recognized operating agency (*ROA*); ~ **sans attente** TÉL demand working; ~ **en service rapide** TÉL demand operating; ~ **simplex** TÉL simplex operation, simplex working; ~ **simultanée** TÉL simultaneous operation; ~ **sur deux fréquences** RADIO, SIGNAL two-frequency operation; ~ **sur une fréquence** RADIO, SIGNAL, TÉL single-frequency operation; ~ **sur une même voie** RADIO co-channel operation

exploiter *vt* CONTRÔLE, GÉNÉRAL, TÉL run, operate

explorateur *m* INFORMAT, PÉRIPH, RADIO scanner; ~ **Web** INFORMAT web browser

exploration *f* GÉNÉRAL, INFORMAT scanning, RADIO beam scanning, scanning, TÉL, TRANSMIS scanning; ~ **du champ proche** RADIO, TRANSMIS near-field scanning; ~ **conique** RADIO conical scanning with rotation; ~ **réduite** TÉL limited scanning; ~ **sectorielle** RADIO sector scanning; ~ **séquentielle** TÉL sequential scanning; ~ **de table** GÉNÉRAL table scanning

explorer *vt* GÉNÉRAL, INFORMAT, RADIO, TÉL, TRANSMIS scan

exponentiel *adj* ÉLECTRON, GÉNÉRAL, INFORMAT exponential

export *m* DONNÉES export

exportation *f* ÉLECTRON export

exposant: ~ **linéique de propagation** *m* ÉLECTRON, RADIO, TRANSMIS propagation coefficient, propagation constant; ~ **linéique de propagation longitudinale** *m* ÉLECTRON,

RADIO, TRANSMIS axial propagation coefficient; **~ de transfert sur images** *m* DONNÉES, TRANSMIS image transfer coefficient, image transfer constant

expression *f* DONNÉES, INFORMAT expression, phrase, phrasing, TÉL expression; **~ active** DONNÉES active expression; **~ arithmétique** INFORMAT *en LHM* arithmetic expression; **~ conditionnelle** DONNÉES conditional expression; **~ à valeur aléatoire** DONNÉES anyvalue expression

extenseur: **~ de ligne à phase variable** *m* RADIO line lengthener, line stretcher

extensibilité *f* INFORMAT, TÉL expandability, extensibility

extension *f* DONNÉES, GÉNÉRAL extension, SIGNAL expansion, extension, TÉL extension; **~ d'adresse de réseau** DONNÉES network address extension (*NAE*); **~ de code** TÉL code extension; **~ de code graphique** TÉL graphic code extension; **~ de signal** SIGNAL signal expansion, signal extension

extensomètre *m* COMPOSANTS strain gage (*AmE*), strain gauge (*BrE*)

extinction: **~ automatique** *f* ÉLECTRON automatic shut-off

extracteur: **~ de carte** *m* TESTS *outils* board extraction tool, board extractor; **~ de matrice** *m* ÉLECTRON matrix remover; **~ de modes de gaine** *m* TRANSMIS cladding mode stripper, mode stripper

extraction *f* INFORMAT retrieval, extraction, SIGNAL, TRANSMIS extraction; **~ de contour** INFORMAT contour extraction; **~ de signaux** SIGNAL, TRANSMIS signal extraction

extraire *vt* INFORMAT extract, RADIO withdraw

extrémité[1]: **à ~ en plastique** *adj* ÉLECTRON plastic-tipped

extrémité[2] *f* DONNÉES endpoint, ÉLECTRON edge, termination, TRANSMIS endpoint; **~ de connexion de canal virtuel** DONNÉES, TRANSMIS virtual-channel connection endpoint (*VCCE*); **~ de connexion de circuit virtuel** DONNÉES, TRANSMIS virtual-path connection endpoint (*VPCE*); **~ de connexion de trajet virtuel** DONNÉES, TRANSMIS virtual-path connection endpoint (*VPCE*); **~ de connexion de voie virtuelle** DONNÉES, TRANSMIS virtual-path connection endpoint (*VPCE*); **~ de fil** COMPOSANTS wire end; **~ de gaine** TRANSMIS cable butt; **~ réceptrice** TRANSMIS receiving terminal; **~ de section** TÉL section termination; **~ virtuelle** COMMUT virtual switching point; **~ virtuelle analogique** COMMUT virtual analog switching point; **~ de voie de transmission** DONNÉES, TRANSMIS transmission path endpoint (*TPE*)

EyePhone *m* INFORMAT EyePhone

F

F *abrév* (*farad*) ALIM ÉLEC F (*farad*)
FA *abrév* (*faux appel*) TÉL handset off hook
fabricabilité *f* DONNÉES workability
fabrication *f* COMPOSANTS manufacturing, processing; **~ assistée par ordinateur** (*FAO*) INFORMAT computer-aided manufacturing (*CAM*); **~ intégrée par ordinateur** (*FIO*) INFORMAT computer-integrated manufacturing (*CIM*)
face: **~ avant en retrait** *f* COMPOSANTS, ÉLECTRON tilted front; **~ clivée** *f* TRANSMIS cut face; **~ de la fibre** *f* TRANSMIS fiber end face (*AmE*), fibre end face (*BrE*); **~ des fibres à raccorder** *f* TRANSMIS fiber end face (*AmE*), fibre end face (*BrE*); **~ polaire** *f* TÉL pole face
facette *f* TRANSMIS facet
facile[1]: **~ d'emploi** *adj* INFORMAT easy-to-use
facile[2]: **~ à exécuter** *loc* INFORMAT, GÉNÉRAL user-friendly; **~ à utiliser** *loc* INFORMAT, GÉNÉRAL user-friendly
facilité: **~ d'entretien** *f* APP SERV, TÉL serviceability; **~ N** *f* TÉL N-facility
façonnage: **~ de câble** *m* TRANSMIS cable forming; **~ au tour** *m* COMPOSANTS turning
façonnier *m* INFORMAT service bureau
facsimile *m* GÉNÉRAL facsimile, fax
facteur *m* GÉNÉRAL factor; **~ d'absorption** RADIO absorption factor; **~ d'activité des signaux vocaux** TÉL speech activity factor; **~ d'affaiblissement** ALIM ÉLEC loss factor, TRANSMIS attenuation factor; **~ aléatoire** INFORMAT random factor; **~ ambiant** CLIMAT environmental factor; **~ d'amplification en tension** ALIM ÉLEC voltage amplification factor; **~ d'antenne** RADIO antenna factor; **~ d'aspect** RADIO, TÉL aspect ratio; **~ d'aspect de pixel** INFORMAT, RADIO, TÉL pel aspect ratio, pixel aspect ratio; **~ d'atténuation** TRANSMIS attenuation factor; **~ de bruit** RADIO noise factor, noise figure, spot noise factor, spot noise figure, TESTS, TRANSMIS noise factor, noise figure; **~ de bruit du récepteur** RADIO receiver-noise figure; **~ de câblage** TÉL take-up factor; **~ de charge** ALIM ÉLEC load factor; **~ clé** TV key factor; **~ de coopération** TÉL factor of cooperation; **~ de correction** TESTS *métrologie* correction factor; **~ de couplage** ÉLECTRON coupling factor; **~ de couverture** RADIO coverage factor; **~ de cylindre** TÉL drum factor; **~ de dispersion de profil** TRANSMIS profile-dispersion parameter; **~ de dissipation** ALIM ÉLEC dissipation factor; **~ de durée d'impulsions** TÉL pulse duty factor; **~ d'efficacité dans le temps** TÉL efficiency factor in time; **~ d'émission électronique secondaire** ÉLECTRON secondary electron emission factor; **~ d'émission secondaire d'électrons** ÉLECTRON secondary electron emission factor; **~ d'environnement** CLIMAT environmental factor; **~ de forme** TESTS form factor; **~ de gravure** ÉLECTRON etch factor; **~ G/T** GÉNÉRAL G/T; **~ d'irrégularité** TÉL peakedness factor; **~ de luminance** TV luminance factor; **~ de mérite** GÉNÉRAL G/T, RADIO figure of merit; **~ multiplicatif du rayon terrestre** RADIO effective Earth-radius factor; **~ de multiplication du gaz** ÉLECTRON gas multiplication factor; **~ d'ombre** RADIO shadow factor; **~ d'opérateur** RADIO operator factor; **~ de perte** ALIM ÉLEC loss factor; **~ de post-accélération** ÉLECTRON post-deflection acceleration factor; **~ de puissance** ALIM ÉLEC power factor; **~ pyramidal** ÉLECTRON fan-out; **~ Q** (*facteur de qualité*) ALIM ÉLEC Q factor (*quality factor*); **~ de qualité** (*facteur Q*) ALIM ÉLEC quality factor (*Q factor*); **~ de réflexion** CLIMAT, ÉLECTRON, RADIO, TÉL, TRANSMIS reflection factor, reflectance; **~ de réflexion complexe** TRANSMIS *fibres optiques* amplitude reflection factor; **~ de réflexion sur les irrégularités** TÉL irregularity reflection coefficient; **~ de réflexion transversale** TÉL transverse reflection factor; **~ de remplissage** DONNÉES packing factor; **~ de réponse** ÉLECTRON, TRANSMIS responsivity; **~ de sensibilité** RADIO sensitivity factor; **~ TIF** TÉL telephone interference factor; **~ de transmission** ÉLECTRON, TRANSMIS transmittance; **~ d'uniformité d'éclairement** ALIM ÉLEC illuminance uniformity ratio; **~ d'utilisation** ALIM ÉLEC load factor, RADIO duty cycle; **~ de vide** ÉLECTRON vacuum factor, gas-content factor
factice *adj* GÉNÉRAL dummy
factorielle *f* GÉNÉRAL factorial
factorisation *f* GÉNÉRAL factoring
facturation *f* GÉNÉRAL billing; **~ à l'arrivée** APP SERV, TÉL reverse charging; **~ à chaud** RADIO hot billing; **~ détaillée** TÉL detailed billing, itemized billing (*BrE*); **~ d'un tiers** TÉL third-party billing
facultatif *adj* GÉNÉRAL optional
faculté: **~ d'intervention** *f* TÉL forward transfer facility
faible[1] *adj* INFORMAT low
faible[2]: **~ déplacement** *m* RADIO narrow shift
faible[3]: **~ atténuation** *f* TÉL low loss; **~ dissipation** *f* ÉLECTRON low dissipation; **~ perte** *f* TÉL low loss; **~ perte d'insertion** *f* TRANSMIS low insertion loss; **~ perte d'isolement** *f* TÉL low insulation loss
faible[4]: **à ~ débit** *loc* TV low-bit-rate; **à ~ encombrement** *loc* ÉLECTRON quiet-front; **à ~ portée** *loc* TRANSMIS short-haul; **à ~ résistance** *loc* ALIM ÉLEC low-resistance
faisable *adj* GÉNÉRAL feasible
faisceau *m* COMPOSANTS conductor, harness, GÉNÉRAL beam, bundle, circuit group, fan-out, group, TRANSMIS *téléphonique* cabling unit, bundle, route, trunk group, group, *fibres optiques* beam; **~ abonné-téléphoniste** TÉL recording-completing trunk group; **~ d'acheminement unique** TÉL only-route circuit group; **~ d'alarme** TÉL alarm trunk group (*AL*); **~ d'âme** TRANSMIS conductor unit; **~ d'antenne du satellite** RADIO satellite spot beam; **~ bidirectionnel automatique** TÉL *téléphoniste* operator office trunk group; **~ de canaux sémaphores** SIGNAL signaling-link set (*AmE*), signalling-link set (*BrE*); **~ de circuits** TRANSMIS circuit route; **~ de circuits abonné-téléphoniste** TÉL recording-completing trunk group; **~ de circuits d'alarme** TÉL AL, alarm trunk group; **~ de circuits à débordement** TÉL primary high-usage trunk group; **~ de circuits entre centres locaux** TÉL interexchange trunk group (*BrE*), interoffice trunk group (*AmE*); **~ de circuits entre centres de transit** TÉL intertandem trunk group; **~ de circuits entre groupes marqueurs** TÉL intermarker trunk group (*IM*); **~ de circuits de faux appels** TRANSMIS permanent-signal trunk group circuit; **~ de circuits d'intercommunication entre positions** TÉL interposition trunk group; **~ de circuits intermédiaire à débordement** TÉL intermediate high-usage trunk (*IH*); **~ de circuits interurbain** TÉL intertoll trunk group (*IT*), trunk group; **~ de circuits interurbain primaire** TÉL primary intertoll trunk group; **~ de circuits intracentral** TÉL interchange trunk group (*BrE*), intraoffice trunk group (*AmE*); **~ de circuits local** TÉL interlocal trunk group; **~ de circuits mixte d'opératrice** TÉL operator office trunk group; **~ de circuits**

de transit local d'arrivée TRANSMIS tandem completing trunk; ~ **de couverture globale** RADIO global beam; ~ **débordant** TÉL high-usage circuit group; ~ **de dernier choix** COMMUT, ÉLECTRON, TÉL alternate final trunk group, last-choice circuit group; ~ **direct** TÉL transverse route; ~ **direct à débordement** TÉL primary high-usage trunk group; ~ **directionnel** RADIO directional beam; ~ **double** COMPOSANTS, ÉLECTRON double conductor, twin conductor, GÉNÉRAL twin conductor; ~ **à double sens** TRANSMIS both-way group; ~ **échelonné** TÉL echelon telegraphy; ~ **électromagnétique** RADIO beam, *antennes* electromagnetic beam; ~ **électronique** ÉLECTRON electron beam (*EB*); ~ **d'émission** RADIO down-link beam; ~ **d'énième choix** TÉL nth choice group; ~ **entrant** TRANSMIS incoming group; ~ **entre centres locaux** TÉL interexchange trunk group (*BrE*), interoffice trunk group (*AmE*); ~ **entre centres de transit** TÉL intertandem trunk group; ~ **équivalent** TÉL equivalent random circuit group; ~ **étroit** RADIO, TV spot beam; ~ **en éventail** ALIM ÉLEC fan-shaped beam; ~ **de faux appels** TRANSMIS permanent-signal trunk group circuit; ~ **de fibres** TRANSMIS fiber bundle (*AmE*), fibre bundle (*BrE*); ~ **filiforme** RADIO pencil beam, spot beam; ~ **final** TÉL final circuit group; ~ **formé** RADIO shaped beam; ~ **à fort trafic** TÉL high-usage circuit group; ~ **gaussien** TRANSMIS Gaussian beam; ~ **hertzien** RADIO microwave beam, microwave radio, radio relay system, TRANSMIS microwave relay links, radio relay system; ~ **hertzien à large bande** RADIO wideband radio relay system; ~ **hertzien numérique pour hiérarchie synchrone** (*HNS-FHN*) RADIO digital radio relay for synchronous hierarchy (*SDH-DRSS*); ~ **hertzien transhorizon** RADIO transhorizon radio relay system; ~ **d'intercommunication entre positions** TÉL interposition trunk group; ~ **intermédiaire à débordement** TÉL intermediate high-usage trunk (*IH*); ~ **interurbain** TÉL intertoll trunk group (*IT*), trunk group; ~ **interurbain primaire** TÉL primary intertoll trunk group; ~ **intracentral** TÉL interchange trunk group (*BrE*), intraoffice trunk group (*AmE*); ~ **laser** ÉLECTRON laser beam; ~ **latéral** RADIO direct beam; ~ **de liaisons de signalisation** SIGNAL signaling-link set (*AmE*), signalling-link set (*BrE*); ~ **de lignes** COMMUT line group; ~ **local** TÉL interlocal trunk group; ~ **lumineux** ÉLECTRON light beam; ~ **de messages enregistrés** TÉL announcement trunk group (*AN*); ~ **mixte** TRANSMIS both-way group; ~ **moléculaire** TV molecular beam; ~ **multiple** RADIO multiple beam; ~ **d'opératrice** TÉL operator junctor trunk group; ~ **optique** ÉLECTRON lightwave stream, optical beam; ~ **plat** RADIO paddle-shaped beam; ~ **ponctuel** RADIO pencil beam, spot beam; ~ **de premier choix** COMMUT, ÉLECTRON, TÉL first-choice circuit group, first-choice group; ~ **de remplacement** TÉL substitute route; ~ **de réseaux** TÉL network cluster; ~ **de routes sémaphores** SIGNAL signaling-route set (*AmE*), signalling-route set (*BrE*); ~ **de routes de signalisation** SIGNAL signaling-route set (*AmE*), signalling-route set (*BrE*); ~ **de secours** GÉNÉRAL emergency trunk group; ~ **sémaphore** SIGNAL signaling-link set (*AmE*), signalling-link set (*BrE*); ~ **de services auxiliaires** TÉL auxiliary services trunk group; ~ **sortant** TÉL outgoing group; ~ **de système de télégraphie harmonique** TÉL voice-frequency telegraph; ~ **totalement fourni** TÉL fully-provided circuit group; ~ **de transit local d'arrivée** TRANSMIS tandem completing trunk; ~ **de voies** TRANSMIS channel pool

fan: ~ **d'ordinateur** *m* INFORMAT computer buff

fanion: ~ **de nouvelles données** *m* DONNÉES new data flag (*NDF*)

fantôme *m* TV ghost

fantômiciel *m* INFORMAT vaporware (*AmE*), vapourware (*BrE*)

FAO *abrév* (*fabrication assistée par ordinateur*) INFORMAT CAM (*computer-aided manufacturing*)

farad *m* (*F*) ALIM ÉLEC farad (*F*)

fatigue *f* COMPOSANTS fatigue

fausse: ~ **alarme** *f* GÉNÉRAL false alarm; ~ **plaque** *f* COMPOSANTS dummy strip

faute *f* ALIM ÉLEC, TÉL, TESTS fault; ~ **de frappe** INFORMAT typing mistake, typo; ~ **de signalisation** TÉL signaling fault (*AmE*), signalling fault (*BrE*)

faux[1]: **en ~ appel** *adj* COMMUT, TÉL parked

faux[2]: ~ **appel** *m* (*FA*) GÉNÉRAL , TÉL, TRANSMIS wrong connection; ~ **démarrage** *m* GÉNÉRAL false start; ~ **écho** *m* RADIO false echo; ~ **numéro** *m* TÉL wrong number, wrong connection; ~ **panneau** *m* TÉL dummy front, dummy panel; ~ **plancher** *m* GÉNÉRAL, TRANSMIS *mise à l'essai* cable floor

fax *m* GÉNÉRAL fax

FC *abrév* (*classe fonctionnelle*) TÉL FC (*functional class*)

FCA *abrév* (*classe fonctionnelle A*) TÉL FCA (*functional class A*)

FCB *abrév* (*classe fonctionnelle B*) TÉL FCB (*functional class B*)

FCC *abrév* (*Commission fédérale des communications*) DONNÉES FCC (*AmE*) (*Federal Communication Commission*)

fcém *abrév* (*force contre-électromotrice*) ALIM ÉLEC cemf (*counter-electromotive force*)

fcm *abrév* (*force cyclomotrice*) RADIO cmf (*cyclomotive force*)

FDHM *abrév* (*durée à mi-crête*) DONNÉES FDHM (*full-duration half-maximum*)

fédérer *vt* ÉLECTRON combine

feedback *m* GÉNÉRAL feedback

fém *abrév* (*force électromotrice*) ALIM ÉLEC, ÉLECTRON emf (*electromotive force*); ~ **psophométrique** *f* (*force électromotrice psophométrique*) TÉL psophometric emf (*psophometric electromotive force*)

fendage: ~ **de la fibre** *m* TRANSMIS cleaving

fendoir *m* COMPOSANTS stroke-cleaving tool

fendon *m* TRANSMIS cleaver

fenêtrage *m* INFORMAT clipping, windowing

fenêtre *f* DONNÉES, INFORMAT, RADIO, TÉL, TRANSMIS, TV window; ~ **d'accusé de réception** TÉL acknowledgement window; ~ **d'affichage** INFORMAT display window; ~ **de Hann** TV Hann window; ~ **du mot unique** DONNÉES unique word window (*UW window*); ~ **spectrale** TRANSMIS spectral window; ~ **temporaire** INFORMAT temporary window; ~ **de transmission** TÉL, TRANSMIS transmission window

fenêtres: ~ **en cascade** *f pl* INFORMAT cascading windows

fente *f* COMPOSANTS, ÉLECTRON, INFORMAT, PÉRIPH, RADIO, TRANSMIS slot; ~ **d'insertion de disque** INFORMAT disk slot; ~ **d'introduction de pièces** PÉRIPH coin slot; ~ **longitudinale** TRANSMIS longitudinal slot; ~ **rayonnante** RADIO *antenne* slot radiator; ~ **transversale** TRANSMIS transverse slot

fermer *vt* ALIM ÉLEC, COMMUT, ÉLECTRON, INFORMAT, PÉRIPH close, switch off, RADIO close, switch off, withdraw

fermeture *f* RADIO *de service* withdrawal; ~ **à ressort** COMPOSANTS snap locking; ~ **de session** INFORMAT log-off

ferrite *f* ALIM ÉLEC, INFORMAT ferrite

ferrule *f* COMPOSANTS ferrule, ÉLECTRON tip

ferrure: ~ **d'assemblage** *f* COMPOSANTS fishplate

ferrures *f pl* COMPOSANTS fittings

FET *abrév* (*transistor à effet de champ*) COMPOSANTS, ÉLECTRON FET (*field effect transistor*)

feu: ~ **clignotant** *m* TV flashing light; ~ **fixe** *m* TV fixed light; ~ **scintillant** *m* TV flickering light

feuille *f* ALIM ÉLEC sheet, COMPOSANTS foil, DONNÉES leaf, ÉLECTRON, GÉNÉRAL sheet, INFORMAT sheet, childless node, tip node; ~ **d'arbre** DONNÉES, INFORMAT tree leaf; ~ **de calcul** INFORMAT spreadsheet; ~ **de collage** ALIM ÉLEC, ÉLECTRON bonding sheet; ~ **isolante** ALIM ÉLEC insulating sheet; ~ **de norme** GÉNÉRAL standard sheet; ~ **de spécification** DONNÉES data sheet; ~ **de travail** INFORMAT work sheet

feuilles: ~ **d'étiquettes adhésives** *f pl* ÉLECTRON fan-fold sheets
feuillet: ~ **magnétique** *m* ÉLECTRON, TÉL magnetic strip
feuilleteur *m* DONNÉES browser
FI *abrév* ALIM ÉLEC (*identificateur de format*) FI (*format identifier*), RADIO (*fréquence intermédiaire*) IF (*intermediate frequency*), TÉL (*identificateur de format*) FI (*format identifier*), TRANSMIS (*fréquence intermédiaire*) IF (*intermediate frequency*)
fiabilité *f* ÉLECTRON, GÉNÉRAL reliability, dependability, INFORMAT, TÉL, TESTS reliability; ~ **d'un équipement** ÉLECTRON, GÉNÉRAL, INFORMAT, TÉL, TESTS reliability performance
fiable *adj* GÉNÉRAL dependable, reliable
fibrage *m* TRANSMIS drawing process
fibre *f* GÉNÉRAL fiber (*AmE*), fibre (*BrE*); ~ **amincie par fusion** TRANSMIS fuse-tapered fiber (*AmE*), fuse-tapered fibre (*BrE*); ~ **amorce** TRANSMIS pigtail, launching fiber (*AmE*), launching fibre (*BrE*), dummy fiber (*AmE*), dummy fibre (*BrE*); ~ **à dispersion aplatie** TRANSMIS dispersion-flattened fiber (*AmE*), dispersion-flattened fibre (*BrE*); ~ **dopée lanthanide** TRANSMIS lanthanide doped fiber (*AmE*), lanthanide doped fibre (*BrE*); ~ **à gradient exponentiel** ALIM ÉLEC, ÉLECTRON, TRANSMIS power law index fiber (*AmE*), power law index fibre (*BrE*); ~ **à gradient d'indice** COMPOSANTS, ÉLECTRON, TRANSMIS graded-index fiber (*AmE*), graded-index fibre (*BrE*), grading-index fiber (*AmE*), grading-index fibre (*BrE*); ~ **à guidage faible** TRANSMIS weakly-guiding fiber (*AmE*), weakly-guiding fibre (*BrE*); ~ **d'injection** TRANSMIS dummy fiber (*AmE*), dummy fibre (*BrE*), launching fiber (*AmE*), launching fibre (*BrE*), pigtail; ~ **monomodale** COMPOSANTS, DONNÉES, TRANSMIS single-mode fiber (*AmE*), single-mode fibre (*BrE*); ~ **monomode** COMPOSANTS, DONNÉES, TRANSMIS single-mode fiber (*AmE*), single-mode fibre (*BrE*); ~ **multimodale** COMPOSANTS, DONNÉES, TRANSMIS multimode fiber (*AmE*), multimode fibre (*BrE*); ~ **multimode** COMPOSANTS, DONNÉES, TRANSMIS multimode fiber (*AmE*), multimode fibre (*BrE*); ~ **optique** COMPOSANTS, DONNÉES, INFORMAT, TRANSMIS optical fiber (*AmE*), optical fibre (*BrE*); ~ **optique monomode** COMPOSANTS, DONNÉES, TRANSMIS single-mode fiber (*AmE*), single-mode fibre (*BrE*); ~ **optique multimode à large cœur** *m* TRANSMIS large-core multiple-mode optic fiber (*AmE*), large-core multiple-mode optic fibre (*BrE*); ~ **optique à saut d'indice** COMPOSANTS, DONNÉES, TRANSMIS step index optical fiber (*AmE*), step index optical fibre (*BrE*); ~ **plastique** COMPOSANTS, DONNÉES, TRANSMIS all-plastic fiber (*AmE*), all-plastic fibre (*BrE*), plastic fiber (*AmE*), plastic fibre (*BrE*); ~ **à saut d'indice** COMPOSANTS, DONNÉES, TRANSMIS step index fiber (*AmE*), step index fibre (*BrE*); ~ **de silice** COMPOSANTS, DONNÉES, TRANSMIS silica fiber (*AmE*), silica fibre (*BrE*); ~ **silice-plastique** COMPOSANTS, DONNÉES, TRANSMIS PCS-fiber (*AmE*), PCS-fibre (*BrE*), plastic-clad silica fiber (*AmE*), plastic-clad silica fibre (*BrE*); ~ **silice-silicone** COMPOSANTS, DONNÉES, TRANSMIS PCS-fiber (*AmE*), PCS-fibre (*BrE*), plastic-clad silica fiber (*AmE*), plastic-clad silica fibre (*BrE*); ~ **tout silice** COMPOSANTS, DONNÉES, TRANSMIS all-silica fiber (*AmE*), all-silica fibre (*BrE*); ~ **unimodale** COMPOSANTS, DONNÉES, TRANSMIS single-mode fiber (*AmE*), single-mode fibre (*BrE*); ~ **de verre** COMPOSANTS glass fiber (*AmE*), glass fibre (*BrE*), ÉLECTRON fiberglass (*AmE*), fibreglass (*BrE*), TRANSMIS all-glass fiber (*AmE*), all-glass fibre (*BrE*), glass fiber (*AmE*), glass fibre (*BrE*)
fibrer *vi* TRANSMIS form the fiber (*AmE*), form the fibre (*BrE*)
fibres: **à** ~ **optiques** *adj* TRANSMIS fiber-optic (*AmE*) (*F/O*), fibre-optic (*BrE*) (*F/O*)
FIC *abrév* (*format intermédiaire commun*) INFORMAT CIF (*common intermediate format*)
ficeleuse *f* TRANSMIS cable lasher
fiche *f* ALIM ÉLEC, COMPOSANTS, ÉLECTRON, INFORMAT, RADIO, TESTS plug; ~ **banane** ÉLECTRON, TESTS banana plug; ~ **coaxiale** ÉLECTRON coaxial plug; ~ **de code** RADIO code plug; ~ **de connexion** TRANSMIS patch cord; ~ **à contacts lyre** COMPOSANTS, ÉLECTRON free fork connector; ~ **demi-grandeur** COMPOSANTS half-size plug; ~ **de dérangement** TÉL error report, trouble report, fault report; ~ **DIN** INFORMAT DIN plug; ~ **de distribution** ALIM ÉLEC distribution plug; ~ **à douilles** COMPOSANTS, ÉLECTRON free sleeve connector; ~ **femelle** GÉNÉRAL socket; ~ **mâle** GÉNÉRAL male plug; ~ **de mesure** TESTS measuring plug; ~ **multibroche** ÉLECTRON free blade connector, free pin connector; ~ **de pontage** ALIM ÉLEC, COMPOSANTS strapping plug; ~ **de strapage** ALIM ÉLEC, COMPOSANTS strapping plug; ~ **technique** DONNÉES data sheet; ~ **de travail** INFORMAT work sheet; ~ **tripolaire** COMPOSANTS three-conductor plug
fichier *m* DONNÉES, INFORMAT, TÉL, TESTS file; ~ **d'abonnés** COMMUT subscriber's store, TÉL directory store; ~ **d'aide** INFORMAT help file; ~ **d'archives** DONNÉES, INFORMAT archive file; ~ **binaire** DONNÉES, INFORMAT binary file; ~ **document** DONNÉES document file; ~ **de gestion** INFORMAT job-recovery control file, job-recovery file; ~ **historique des alarmes** TÉL alarm history file; ~ **image** DONNÉES image file; ~ **de liens** DONNÉES web file, INFORMAT link file, link map, web file; ~ **logique** DONNÉES logical file; ~ **PCX multipage** (*DCX*) DONNÉES multipage PCX file (*DCX*); ~ **de référence** DONNÉES master register; ~ **réparti** DONNÉES distributed file; ~ **de session** INFORMAT kill file; ~ **son** DONNÉES sound file; ~ **source** DONNÉES, INFORMAT file source; ~ **texte** DONNÉES text file; ~ **de travail** INFORMAT work file; ~ **verrouillé** INFORMAT locked file
fidélité *f* RADIO fidelity, resettability, TESTS precision
Fidonet *m* INFORMAT Fidonet
FIF *abrév* (*fonction d'interfonctionnement*) TÉL IWF (*interworking function*)
fil *m* ALIM ÉLEC, COMMUT lead, wire, COMPOSANTS conductor, lead, wire, ÉLECTRON lead, solder, wire, GÉNÉRAL lead, wire, *outils* edge, TÉL wire, leg, lead, TRANSMIS cable; ~ **d'acier** COMPOSANTS, ÉLECTRON, GÉNÉRAL steel wire; ~ **aérien** GÉNÉRAL open wire; ~ **à âme décapante** ÉLECTRON resin-core solder; ~ **d'appel** TÉL call wire; ~ **blindé** COMPOSANTS shielded wire; ~ **de bobinage** ALIM ÉLEC winding wire, ÉLECTRON magnet wire; ~ **à bobiner** ÉLECTRON magnet wire; ~ **à braser** COMPOSANTS soldering wire; ~ **de compteur** TÉL meter wire; ~ **de connexion** ÉLECTRON connecting wire, hook-up wire; ~ **de continuité d'écran** GÉNÉRAL drain wire; ~ **de conversation** COMPOSANTS speech wire; ~ **de cuivre** COMPOSANTS, ÉLECTRON, GÉNÉRAL copper wire; ~ **de déchirement** ÉLECTRON ripcord; ~ **dénudé** COMPOSANTS stripped wire; ~ **détecteur** COMPOSANTS sensor wire; ~ **électrique** GÉNÉRAL wire; ~ **émaillé** COMPOSANTS, ÉLECTRON, GÉNÉRAL enameled wire (*AmE*), enamelled wire (*BrE*); ~ **d'enroulement** ALIM ÉLEC winding wire; ~ **d'essai** COMMUT, TÉL test wire; ~ **étamé** COMPOSANTS, ÉLECTRON, GÉNÉRAL tin-plated wire; ~ **à flux incorporé** ÉLECTRON resin-core solder; ~ **guipé** COMPOSANTS, ÉLECTRON, TRANSMIS braided cable; ~ **isolé** ÉLECTRON insulated conductor; ~ **jarretière** TÉL jumper wire; ~ **de lecture** TÉL reading wire; ~ **de libération** COMMUT release wire; ~ **de masse** ALIM ÉLEC earth lead (*BrE*), earth wire (*BrE*), ground lead (*AmE*), ground wire (*AmE*); ~ **multibrin** COMPOSANTS, ÉLECTRON stranded conductor, stranded wire; ~ **négatif** TÉL negative wire; ~ **non isolé** ALIM ÉLEC *conducteurs électriques*, ÉLECTRON bare wire, uninsulated conductor; ~ **nu** ALIM ÉLEC *conducteurs électriques*, ÉLECTRON bare wire, uninsulated conductor, GÉNÉRAL open wire; ~ **nu**

aérien GÉNÉRAL open-wire line; **~ de nuque** (*fil R*) COMPOSANTS ring, TÉL ring wire (*R-wire*), B-leg; **~ de pointe** COMPOSANTS tip, TÉL A-leg; **~ positif** TÉL positive wire; **~ de prise** COMPOSANTS, TÉL seizure wire; **~ R** (*fil de nuque*) TÉL R-wire (*ring wire*); **~ recuit** COMPOSANTS annealed wire; **~ de retour** ALIM ÉLEC return wire; **~ rosette** COMPOSANTS, ÉLECTRON tinsel conductor; **~ à souder** COMPOSANTS soldering wire; **~ T** TÉL T wire; **~ de terre** ALIM ÉLEC earth wire (*BrE*), ground wire (*AmE*); **~ de test** COMMUT, TÉL test wire; **~ torsadé** COMPOSANTS, ÉLECTRON stranded conductor, twisted wire; **~ tressé** COMPOSANTS, ÉLECTRON, TRANSMIS braided cable; **~ volant** ALIM ÉLEC, COMPOSANTS jumper, strap

filament *m* COMPOSANTS, ÉLECTRON filament

file: **~ d'attente** *f* GÉNÉRAL queue; **~ d'attente d'appels** *f* TÉL call queue; **~ d'attente d'appels au décrochage** *f* TÉL off-hook queue (*OHQ*); **~ d'attente avec rappel automatique** *f* TÉL callback queue (*CBQ*); **~ d'attente limitée** *f* TÉL limited-waiting queue; **~ d'attente à serveurs multiples** *f* DONNÉES, INFORMAT multiple-server queue; **~ d'attente à un seul serveur** *f* DONNÉES, INFORMAT single-server queue; **~ d'attente temporelle** *f* DONNÉES, ÉLECTRON, RADIO, TÉL time queue; **~ variable** *f* TÉL variable queue

filin: **~ en nylon** *m* ÉLECTRON nylon binder

film *m* ALIM ÉLEC, COMPOSANTS film, foil, ÉLECTRON film, TV film, movie; **~ isolant** ALIM ÉLEC insulating foil; **~ plastique** COMPOSANTS plastic film, plastic foil; **~ de polissage** ÉLECTRON lapping film

fils[1] *m* DONNÉES child, INFORMAT son, child

fils[2] *m pl* COMPOSANTS wiring; **~ d'aramide** ÉLECTRON aramid yarn; **~ effilochés** ÉLECTRON frayed wiring; **~ de liaison à œillet** ÉLECTRON eyelet bond strap

filtrage *m* APP SERV pattern matching, DONNÉES filtering, ÉLECTRON filtering, trapping, INFORMAT filtering, pattern matching, TÉL *d'appels*, TRANSMIS, TV filtering; **~ adaptatif** TRANSMIS adaptive filtering; **~ des appels** TÉL call filtering, call screening; **~ des canaux** TV channel filtering; **~ des communications** TÉL call filtering, call screening; **~ hors voie** TRANSMIS out-of-channel filtering; **~ linéaire** TRANSMIS linear filtering; **~ logique** INFORMAT logic filtering; **~ multicadence** DONNÉES multirate filtering; **~ multidimensionnel** ÉLECTRON multidimensional filtering; **~ non linéaire** TRANSMIS nonlinear filtering; **~ numérique** DONNÉES, TRANSMIS digital filtering

filtré *adj* ALIM ÉLEC smooth

filtre *m* ALIM ÉLEC filter, APP SERV pattern, COMPOSANTS, DONNÉES filter, ÉLECTRON filter, trap, INFORMAT filter, pattern, RADIO, TRANSMIS filter; **~ d'aiguillage** ÉLECTRON branching bypass; **~ d'aiguillage directionnel** ÉLECTRON directional filter; **~ analogique** INFORMAT, TRANSMIS analog filter; **~ d'antenne** ÉLECTRON interference filter, RADIO wave trap, TRANSMIS interference filter; **~ antibrouillage** ÉLECTRON interference filter, RADIO wave trap, TRANSMIS interference filter; **~ antiparasite** ÉLECTRON interference filter, GÉNÉRAL parasitic stopper, INFORMAT line filter, TRANSMIS interference filter; **~ antirayonnement** DONNÉES anti-radiation filter; **~ antireflet** DONNÉES anti-glare filter; **~ antirepliement** TRANSMIS anti-aliasing filter; **~ de bandes** ÉLECTRON image parameter filter; **~ blanchissant** ÉLECTRON whitening filter; **~ à capacités commutées** COMPOSANTS, ÉLECTRON switched-capacitor filter; **~ commandé par tension** ALIM ÉLEC, INFORMAT voltage-controlled filter; **~ pour contrastes** TV *photographie* contrast filter; **~ coupe-bande** ÉLECTRON notch filter, TRANSMIS band elimination filter, band rejection filter, bandstop filter, notch filter; **~ coupe-bande à bande étroite** ÉLECTRON, TRANSMIS notch filter; **~ coupe-fréquence** ÉLECTRON, TRANSMIS notch filter; **~ coupleur** ÉLECTRON filter coupler; **~ à coupure brusque** ÉLECTRON, TRANSMIS notch filter; **~ décolorant** ÉLECTRON whitening filter; **~ en demi-treillis** ÉLECTRON, TRANSMIS half-lattice filter; **~ à déphasage minimal** TRANSMIS minimum-phase filter; **~ dérivé en m** RADIO m-derived filter; **~ dichroïque** ÉLECTRON, TRANSMIS dichroic filter; **~ dynamique numérique** INFORMAT digital dynamic filter; **~ en échelle** RADIO ladder type filter; **~ d'entrée** DONNÉES input filter; **~ FI** RADIO, TRANSMIS IF filter; **~ de fréquence** ÉLECTRON, RADIO frequency filter; **~ à fréquence intermédiaire** RADIO, TRANSMIS intermediate frequency filter; **~ d'impulsions** TÉL pulse corrector; **~ incorporé** ÉLECTRON built-in filter; **~ interdigital** ALIM ÉLEC interdigital filter; **~ interférentiel** ÉLECTRON, TRANSMIS interference filter; **~ en L** ÉLECTRON L-section filter; **~ de ligne** ALIM ÉLEC line filter; **~ linéaire** TRANSMIS linear filter; **~ de lissage** ALIM ÉLEC, COMPOSANTS smoothing filter; **~ de manipulation** RADIO keying filter; **~ miroir en quadrature** TRANSMIS quadrature mirror filter; **~ de mise en forme** DONNÉES shaping filter; **~ de mise en forme du signal** ÉLECTRON, SIGNAL signal-shaping filter; **~ de mode** TRANSMIS mode filter; **~ de mode à anneaux** RADIO ring-mode filter; **~ de mode à réflexion** RADIO reflection-mode filter; **~ de mode résonnant** RADIO resonant-mode filter; **~ monolithique** ÉLECTRON monolithic filter; **~ multibande** TRANSMIS multiband filter; **~ multivoie** TRANSMIS multichannel filter; **~ non linéaire** DONNÉES, TRANSMIS nonlinear filter; **~ non récursif** DONNÉES, TRANSMIS nonrecursive filter; **~ numérique** DONNÉES, INFORMAT, TRANSMIS digital filter; **~ numérique à mémoire infinie** DONNÉES recursive digital filter; **~ numérique multidimensionnel** DONNÉES multidimensional digital filter; **~ numérique variable** INFORMAT variable digital filter; **~ optique** TRANSMIS optical filter; **~ orangé** DONNÉES orange filter; **~ passe-bande** (*FPB*) ALIM ÉLEC, TRANSMIS bandpass filter (*BPF*); **~ passe-bas** ALIM ÉLEC, TRANSMIS low-pass filter; **~ passe-haut** ALIM ÉLEC, TRANSMIS high-pass filter; **~ passif** ÉLECTRON passive filter; **~ à phase asservie** ÉLECTRON phase-locked filter; **~ pondérateur** ÉLECTRON weighting filter; **~ à pondération** ÉLECTRON weighting filter; **~ psophométrique** ÉLECTRON weighting filter; **~ quadripolaire** ÉLECTRON four-pole filter; **~ à quartz** ÉLECTRON quartz filter, RADIO crystal filter; **~ récursif** DONNÉES, TRANSMIS recursive filter; **~ récursif non linéaire** DONNÉES, TRANSMIS nonlinear recursive filter; **~ réjecteur** ÉLECTRON, TRANSMIS notch filter; **~ de réjection** ÉLECTRON, TRANSMIS notch filter; **~ à réponse impulsionnelle finie** ÉLECTRON, TRANSMIS finite impulse response filter; **~ à réponse impulsionnelle finie ondulation** ÉLECTRON equiripple finite-impulse response filter (*EFIR*); **~ à réponse impulsionnelle infinie** ÉLECTRON, TRANSMIS infinite impulse response filter; **~ à réponse temporelle** ÉLECTRON time domain filter; **~ RIF** ÉLECTRON, TRANSMIS FIR filter; **~ RII** ÉLECTRON, TRANSMIS IIR filter; **~ rouge** ÉLECTRON red pass filter; **~ secteur** INFORMAT line filter; **~ sécuritaire** DONNÉES, INFORMAT firewall; **~ spectral** COMPOSANTS, TRANSMIS spectral filter; **~ temporel** ÉLECTRON time domain filter; **~ de touche** INFORMAT key filter; **~ transversal** PÉRIPH, TRANSMIS transversal filter; **~ à variation temporelle** INFORMAT time-variant filter; **~ de visualisation** INFORMAT viewing filter

filtrer *vt* ALIM ÉLEC smooth, ÉLECTRON trap

filtres: **~ en chaîne** *m pl* TRANSMIS multiple-filter chain

fin *f* RADIO termination, TÉL end; **~ de bande magnétique** DONNÉES magnetic-tape trailer; **~ de communication** DONNÉES, RADIO termination, TÉL end of communication (*EOC*), termination, TRANSMIS termination; **~ de course** ALIM ÉLEC limit switch; **~ de dialogue** TÉL end of dialog (*AmE*), end of dialogue (*BrE*); **~ d'impulsion** INFORMAT end of pulse; **~ de ligne** PÉRIPH end of line; **~ de message** DONNÉES, TÉL end of message (*EOM*); **~ de procédure** DONNÉES, TÉL end of procedure (*EOP*); **~ de réémission EOM** TÉL end of retransmission EOM; **~ de réémission**

EOP DONNÉES, TÉL end of retransmission EOP; **~ de réémission MPS** DONNÉES, TÉL end of retransmission MPS; **~ de réémission NULL** DONNÉES, TÉL end of retransmission NULL; **~ de réémission PRI-EOM** DONNÉES, TÉL end of retransmission PRI-EOM; **~ de réémission PRI-EOP** DONNÉES, TÉL end of retransmission PRI-EOP; **~ de réémission PRI-MPS** DONNÉES, TÉL end of retransmission PRI-MPS; **~ de retransmission EOM** DONNÉES, TÉL end of retransmission EOM; **~ de retransmission EOP** DONNÉES, TÉL end of retransmission EOP; **~ de retransmission MPS** DONNÉES, TÉL end of retransmission MPS; **~ de retransmission NULL** DONNÉES, TÉL end of retransmission NULL; **~ de retransmission PRI-EOM** DONNÉES, TÉL end of retransmission PRI-EOM; **~ de retransmission PRI-EOP** DONNÉES, TÉL end of retransmission PRI-EOP; **~ de retransmission PRI-MPS** DONNÉES, TÉL end of retransmission PRI-MPS; **~ de sélection** DONNÉES, TÉL end of selection (*EOS*); **~ de session** INFORMAT log-off; **~ de sortie** DONNÉES, TÉL end of output; **~ de transaction** DONNÉES, TÉL end of transaction (*EOT*); **~ de transmission** DONNÉES, TÉL end of transmission (*EOT*)

finesse: **~ d'exploration** *f* TÉL scanning density

fini: **~ de surface** *m* COMPOSANTS surface finish

finition *f* INFORMAT finish, smoothing

FIO *abrév* (*fabrication intégrée par ordinateur*) INFORMAT CIM (*computer-integrated manufacturing*)

fixation *f* COMPOSANTS holder, support, INFORMAT socket, TRANSMIS bracket; **~ de câble** TRANSMIS cable bracket; **~ des tarifs** APP SERV, TÉL tariff setting

fixer *vt* ÉLECTRON anchor

flash *m* GÉNÉRAL flash

FLCD *abrév* (*écran à cristaux liquides ferro-électriques*) INFORMAT FLCD (*ferro-electric liquid crystal display*)

flèche *f* GÉNÉRAL arrow, *physique* sag

flexibilité *f* GÉNÉRAL flexibility; **~ du réseau** ÉLECTRON backbone flexibility; **~ du segment fédérateur** ÉLECTRON backbone flexibility; **~ de la structure** ÉLECTRON backbone flexibility

flexible[1] *adj* GÉNÉRAL flexible

flexible[2] *m* COMPOSANTS flexible hose, flex

flexion *f* ÉLECTRON flex, INFORMAT bend

flot: **~ de données** *m* DONNÉES data stream; **~ de travaux** *m* INFORMAT job stream, run stream

flotte *f* RADIO mobile radio fleet, vehicle fleet

flou[1] *adj* INFORMAT blurred

flou[2] *m* RADIO flatness of bearing, TV fuzziness; **~ de mouvement** TV motion blur

fluctuation *f* ALIM ÉLEC, TESTS fluctuation; **~ de fréquence d'impulsion** RADIO chirping; **~ de longueur d'onde** TRANSMIS chirping; **~ de phase** TV jitter; **~ du temps de transit** ÉLECTRON transit time jitter; **~ de tension** ALIM ÉLEC, TESTS voltage fluctuation, voltage variation

fluorescence *f* RADIO, TRANSMIS fluorescence

flux *m* GÉNÉRAL flow, flux, power, current, stream; **~ de données** DONNÉES, INFORMAT data-flow; **~ électronique** ALIM ÉLEC electron current, electron flow; **~ d'électrons** ALIM ÉLEC electron current, electron flow; **~ énergétique** ÉLECTRON, TRANSMIS radiant flux; **~ lumineux** TRANSMIS irradiance; **~ lumineux au coude** ÉLECTRON knee sensitivity, knee-luminous flux; **~ de puissance de brouillage** TRANSMIS interfering power flux; **~ résineux** ÉLECTRON rosin flux; **~ de trafic** TÉL traffic relation; **~ du trafic** DONNÉES traffic flow, TÉL call flow, traffic flow

fluxmètre *m* TESTS fluxmeter

FM *abrév* GÉNÉRAL (*fréquence multiple*) MF (*multiple frequency*), RADIO (*modulation de fréquence*) FM (*frequency modulation*)

foncteur *m* INFORMAT functor

fonction[1]: **en ~ NON-ET** *adj* ÉLECTRON NANDed; **en ~ OU** *adj* ÉLECTRON ORed with

fonction[2] *f* GÉNÉRAL function, feature; **~ d'accord** INFORMAT tuning function, tune function; **~ d'alarme** TESTS alarm function; **~ d'application** DONNÉES application function; **~ d'application de gestion** DONNÉES management application function (*MAF*); **~ d'authentification** DONNÉES authentication function; **~ auxiliaire** DONNÉES housekeeping function; **~ de canal sémaphore** RADIO, SIGNAL signaling-link function (*AmE*), signalling-link function (*BrE*); **~ colorimétrique** INFORMAT colorimetric function; **~ de commande** CONTRÔLE control function (*CF*), DONNÉES, INFORMAT control function (*CF*), control operation, TÉL control function (*CF*); **~ de commande pour la présentation** TÉL presentation control function; **~ de communication de messages** DONNÉES message communication function (*MCF*); **~ commutateur d'accès aux services** (*CAS*) COMMUT, TÉL service access switch function; **~ de commutation** COMMUT, TÉL switching function; **~ comptage** ÉLECTRON counting function; **~ de contrôle** CONTRÔLE, DONNÉES, INFORMAT, TÉL control function (*CF*); **~ de contrôle d'association multiple** DONNÉES multiple association control function (*MACF*); **~ de contrôle d'association unique** DONNÉES single-association control function; **~ de convergence dépendant du sous-réseau** DONNÉES subnetwork dependent convergence function (*SNDCF*); **~ de correction** TESTS *métrologie* correction function; **~ de couche supérieure** TRANSMIS higher-level function (*HLF*); **~ courante** INFORMAT routine task; **~ de coût** CONTRÔLE performance index; **~ décalage** INFORMAT shift function; **~ de décision** TESTS decision function; **~ de densité de probabilité** DONNÉES probability density function (*PDF*); **~ de dépistage** COMMUT tracing function; **~ de Dirac** TÉL Dirac function; **~ directe** INFORMAT direct function; **~ de distribution cumulative** DONNÉES cumulative distribution function (*cdf*); **~ échelon** TÉL step function; **~ d'élément de réseau** TRANSMIS network element function (*NERF*); **~ EMCN** TÉL DCME function; **~ d'enregistrement automatique** RADIO roaming capability; **~ d'enregistreur** TÉL, TRANSMIS register function; **~ exponentielle** GÉNÉRAL exponential function; **~ de gestion d'équipement synchrone** APP SERV synchronous equipment management function (*SEMF*); **~ de la gestion de réseau** TRANSMIS network management function; **~ de gestion des systèmes** DONNÉES, INFORMAT systems management function; **~ hypermédia** INFORMAT hypermedia function; **~ d'interfonctionnement** (*FIF*) TÉL interworking function (*IWF*); **~ d'interfonctionnement télex-mode paquet** DONNÉES telex packet interworking function (*TPIWF*); **~ inverse** GÉNÉRAL inverse function; **~ de liaison de signalisation** RADIO, SIGNAL signaling-link function (*AmE*), signalling-link function (*BrE*); **~ liée à la connexion** DONNÉES, TRANSMIS connection-related function (*CRF*); **~ liée à la connexion sur les canaux virtuels** DONNÉES, TRANSMIS virtual-channel connection-related function (*VCCRF*); **~ liée à la connexion de conduit virtuel** DONNÉES, TRANSMIS virtual-path connection-related function (*VPCRF*); **~ liée à la connexion sur les voies virtuelles** DONNÉES, TRANSMIS virtual-channel connection-related function (*VCCRF*); **~ liée à une connexion de transit** DONNÉES, TRANSMIS transit connection-related function (*TCRF*); **~ de localisation** RADIO tracing function; **~ de maintenabilité** TESTS maintainability function; **~ de maintenance** GÉNÉRAL maintenance function; **~ marche-arrêt** PÉRIPH on-off function; **~ de médiation** INFORMAT mediation function (*MF*); **~ N** DONNÉES N function; **~ nonnormalisée** DONNÉES, TÉL nonstandard facility (*NSF*); **~ NON-OU** ÉLECTRON NOR operation; **~ de parité** TÉL parity function; **~ de Pierce** ÉLECTRON nondisjunction, NOR operation; **~ potentielle** ALIM ÉLEC potential function; ~ **primaire** DONNÉES primary function; ~ **rampe** TÉL ramp function; ~ **de rappel périodique** DONNÉES, TÉL phone-home feature; ~ **de recherche** INFORMAT locate function; ~ **de récurrence** DONNÉES, INFORMAT recursion, recursive function; ~ **récursive** DONNÉES, INFORMAT recursive function; ~ **de relais et de routage** COMMUT relaying and routing function; ~ **de renouvellement** TESTS renewal function; ~ **de répartition de la durée de vie** TESTS life-length distribution function; ~ **de répertoire N** DONNÉES N-directory-function; ~ **répondeur** PÉRIPH, TÉL answer machine function; ~ **de réponse en bande de base** APP SERV, CONTRÔLE, DONNÉES, TÉL, TRANSMIS baseband response function; ~ **de sauvegarde** GÉNÉRAL backup function; ~ **secondaire** DONNÉES secondary function; ~ **de secours** GÉNÉRAL backup function; ~ **de service** DONNÉES housekeeping function; ~ **de service sans connexion** APP SERV connectionless service function (*CLSF*); ~ **de seuil** ÉLECTRON threshold function; ~ **de suivi** RADIO tracing function; ~ **supplémentaire** INFORMAT additional function; ~ **de système d'exploitation** DONNÉES, INFORMAT, TÉL, TESTS operating system function (*OSF*); ~ **de traçage** COMMUT, TESTS tracing function; ~ **de transfert** APP SERV, CONTRÔLE, DONNÉES, TRANSMIS transfer function; ~ **de transfert en bande de base** APP SERV, CONTRÔLE, DONNÉES, TRANSMIS baseband transfer function; ~ **de transfert de modulation** ÉLECTRON modulation transfer function

fonctionnalité *f* DONNÉES functionality, INFORMAT feature, TESTS functionality; ~ **de confiance** DONNÉES, TESTS trusted functionality

fonctionnement *m* GÉNÉRAL function, operation, use; ~ **asynchrone** TÉL asynchronous operation; ~ **avec chevauchement** SIGNAL, TÉL overlap operation; ~ **de barre de sélection** COMMUT operation of horizontal, operation of selecting; ~ **à deux fréquences avec relais automatique** TÉL two-frequency operation automatic repeater mode; ~ **à deux fréquences sans relais** TÉL two-frequency operation non-repeater mode; ~ **en duplex** RADIO, TÉL duplex working, TRANSMIS full-duplex operation; ~ **en dupliqué** TÉL duplicate operation; ~ **échelonné** COMMUT sequence function; ~ **entièrement automatique** TÉL fully-automatic operation; ~ **d'horizontale** COMMUT operation of horizontal, operation of selecting; ~ **limité par l'affaiblissement** TRANSMIS attenuation-limited operation; ~ **limité par le bruit quantique** ÉLECTRON quantum-limited operation, quantum-noise-limited operation; ~ **limité par la distorsion** ÉLECTRON distortion-limited operation; ~ **limité par la largeur de bande** TRANSMIS bandwidth-limited operation; ~ **mains libres** RADIO hands-free operation; ~ **en mode asynchrone symétrique** TÉL set asynchronous balanced mode (*SABM*); ~ **en mode asynchrone symétrique étendu** TÉL set asynchronous balanced mode extended (*SABME*); ~ **en mode dégradé** RADIO, TRANSMIS graceful degradation; ~ **en mode de déplétion** COMPOSANTS depletion mode operation; ~ **en mode d'enrichissement** COMPOSANTS enhancement mode operation; ~ **en parallèle** COMMUT parallel operation; ~ **pas à pas** CONTRÔLE step operation; ~ **en push-pull** ÉLECTRON push-pull operation; ~ **redondant** TESTS standby working; ~ **séquentiel** INFORMAT sequential operation; ~ **à un seul porteur** INFORMAT, RADIO, TRANSMIS single-carrier operation; ~ **à une seule fréquence** RADIO, SIGNAL single-frequency operation; ~ **en simplex** TÉL simplex working; ~ **de transmission** TÉL, TESTS, TRANSMIS transmission performance

fonctions: ~ **complémentaires** *f pl* INFORMAT support activities; ~ **de gestion des canaux sémaphores** *f pl* SIGNAL signaling-link management functions (*AmE*), signalling-link management functions (*BrE*); ~ **de gestion des liaisons de signalisation** *f pl* SIGNAL signaling-link management functions (*AmE*), signalling-link management functions (*BrE*); ~ **de gestion du réseau sémaphore** *f pl* SIGNAL signaling-network management functions (*AmE*), signalling-network management functions (*BrE*); ~ **de gestion du réseau de signalisation** *f pl* SIGNAL signaling-network management functions (*AmE*), signalling-network management functions (*BrE*); ~ **de gestion des routes sémaphores** *f pl* SIGNAL signaling-route management functions (*AmE*), signalling-route management functions (*BrE*); ~ **de gestion des routes de signalisation** *f pl* SIGNAL signaling-route management functions (*AmE*), signalling-route management functions (*BrE*); ~ **de gestion de trafic sémaphore** *f pl* SIGNAL signaling-traffic management functions (*AmE*), signalling-traffic management functions (*BrE*); ~ **d'une IUT** *f pl* DONNÉES capabilities of an IUT; ~ **d'orientation des messages de signalisation** *f pl* SIGNAL signaling-message handling functions (*AmE*), signalling-message handling functions (*BrE*); ~ **du réseau sémaphore** *f pl* SIGNAL signaling-network functions (*AmE*), signalling-network functions (*BrE*); ~ **de réseau de signalisation** *f pl* SIGNAL signaling-network functions (*AmE*), signalling-network functions (*BrE*)

fond *m* COMPOSANTS unit, DONNÉES background, ÉLECTRON unit, INFORMAT background, bottom; ~ **d'écran** DONNÉES screen background; ~ **d'image fixe** INFORMAT display background; ~ **inversé** TV inverted background; ~ **de panier** COMPOSANTS back panel

fondamentale *f* ÉLECTRON *électronique* basic frequency

fondre[1] *vi* ALIM ÉLEC *fusibles* blow, melt, INFORMAT, RADIO fade

fondre[2]: **se** ~ *v réfl* INFORMAT coalesce

fondu *m* INFORMAT, RADIO, TV fading; ~ **enchaîné** INFORMAT fade, TV crossfade; ~ **à la fermeture** TV fade-out; ~ **à l'ouverture** TV fade-in; ~ **sonore** INFORMAT sound fade-out

force: ~ **d'adhérence** *f* ÉLECTRON peel strength; ~ **d'arrachement** *f* COMPOSANTS stripping force, ÉLECTRON pull-off strength, TESTS pulling force; ~ **contre-électromotrice** *f* (*fcém*) ALIM ÉLEC counter-electromotive force; ~ **cyclomotrice** *f* (*fcm*) RADIO cyclomotive force (*cmf*); ~ **électromagnétique** *f* ALIM ÉLEC, ÉLECTRON electromagnetic force; ~ **électromotrice** *f* (*fém*) ALIM ÉLEC, ÉLECTRON electromotive force (*emf*); ~ **électromotrice psophométrique** *f* (*fém psophométrique*) TÉL psophometric electromotive force (*psophometric emf*); ~ **d'empilage de lames de contact** *f* COMPOSANTS spring-set load; ~ **expansive** *f* GÉNÉRAL expansion force; ~ **de frappe** *f* INFORMAT key touch force; ~ **du signal** *f* SIGNAL signal strength

forfait: ~ **mensuel** *m* APP SERV, INFORMAT monthly subscription

formage: ~ **de faisceau** *m* ÉLECTRON, RADIO beam forming

formant *m* TÉL formant

format *m* DONNÉES, INFORMAT, PÉRIPH, TÉL, TV format; ~ **arythmique** DONNÉES start-stop format; ~ **asynchrone** DONNÉES asynchronous format; ~ **de document** DONNÉES document format; ~ **d'échange de documents numériques** DONNÉES digital document interchange format (*DDIF*); ~ **d'échange de fichiers JPEG** (*JFIF*) INFORMAT JPEG file interchange format (*JFIF*); ~ **d'échange de graphiques** (*GIF*) INFORMAT graphics interchange format (*GIF*); ~ **d'image** TV frame format; ~ **intermédiaire commun** (*FIC*) INFORMAT common intermediate format (*CIF*); ~ **international X.121** DONNÉES international X.121 format; ~ **libre** DONNÉES free format; ~ **non numéroté** TÉL unnumbered format; ~ **du numéro international pour la transmission de données** DONNÉES international data number format; ~ **ouvert d'échange de documents** DONNÉES open document interchange format (*ODIF*); ~ **du papier** PÉRIPH paper size, paper format; ~ **de paquet** TÉL packet format; ~ **du pixel** INFORMAT, RADIO, TÉL pel aspect ratio, pixel aspect ratio; **~ de référence des publications** DONNÉES reference publication format (*RPF*); **~ vertical** DONNÉES, TRANSMIS vertical format

formatage *m* DONNÉES, INFORMAT formatting

format: **~ A4** *m* GÉNÉRAL *documentation* A4 size

formation: **~ des blocs** *f* DONNÉES blocking; **~ de files d'attente** *f* TÉL queueing; **~ des signaux par modulation de fréquence** *f* RADIO frequency change signaling (*AmE*), frequency change signalling (*BrE*)

forme *f* APP SERV pattern, DONNÉES form, INFORMAT form, pattern, shape, RADIO form, shape; **~ de Backus-Naur** (*forme BNF*) DONNÉES Backus-Naur form (*BNF*); **~ BNF** (*forme normale de Backus*) DONNÉES BNF (*Backus Normal Form, Backus-Naur form*); **~ BNF étendue** DONNÉES *utilisée en notation TTCN* extended BNF; **~ libre** INFORMAT free form; **~ normale de Backus** (*forme BNF*) DONNÉES Backus Normal Form (*BNF*); **~ normalisée** INFORMAT normalized form; **~ d'onde** GÉNÉRAL waveform, wave shape; **~ d'onde de déclenchement** RADIO triggering waveform; **~ d'onde MIC** INFORMAT PCM waveform; **~ d'onde non sinusoïdale** INFORMAT non-sine-wave waveform

formulaire: **~ MOCS** *m* DONNÉES MOCS proforma; **~ PCIS** *m* DONNÉES PCIS proforma; **~ PICS** *m* DONNÉES PICS proforma; **~ PIXIT** *m* DONNÉES PIXIT proforma; **~ de réglage** *m* TRANSMIS line-up record

forum *m* DONNÉES forum

FOT *abrév* RADIO (*fréquence optimale de travail*) OWF (*optimum working frequency*), (*fréquence optimale de trafic*) OTF (*optimum traffic frequency*)

fouillis: **~ d'écho** *m* RADIO clutter

fournisseur *m* APP SERV, DONNÉES, INFORMAT, TÉL provider; **~ d'accès Internet** INFORMAT Internet Service Provider; **~ de l'ACSE** DONNÉES, TÉL ACSE provider; **~ connexion** DONNÉES connection provider; **~ d'information** DONNÉES information provider; **~ de RTSE** DONNÉES, TÉL RTSE provider; **~ du service abstrait mémoire de messages** DONNÉES, TÉL MS abstract service provider; **~ du service ACSE** DONNÉES, TÉL ACSE service provider; **~ du service CMISE** DONNÉES, TÉL CMISE service provider; **~ du service de contrôle d'association** DONNÉES, TÉL ACSE service provider; **~ de service MCS** DONNÉES, TÉL MCS provider; **~ du service de réseau** DONNÉES network service provider; **~ de service de traitement transactionnel** DONNÉES transaction processing service provider (*TPSP*); **~ de services** GÉNÉRAL service provider; **~ supérieur du service MCS** DONNÉES, TÉL top MCS provider; **~ du système** APP SERV system provider

fourniture *f* APP SERV, DONNÉES, INFORMAT, TÉL, TRANSMIS provision; **~ de réseaux ouverts** TRANSMIS open network provision (*ONP*)

foyer: **~ hi-tech** *m* TV hi-tech home

foyers: **~ cerclés** *m pl* TV homes passed

FPB *abrév* (*filtre passe-bande*) ALIM ÉLEC, TRANSMIS BPF (*bandpass filter*)

fraction: **~ de gel** *f* TRANSMIS freeze-out fraction; **~ de niveau de brouillage** *f* RADIO interference level contribution

fragilisation: **~ par l'hydrogène** *f* TESTS hydrogen embrittlement

fragment *m* INFORMAT chunk; **~ de savoir** INFORMAT knowledge chunk

frais: **~ d'abonnement** *m pl* DONNÉES, INFORMAT, RADIO, TÉL, TV subscription fee; **~ d'accès au réseau** *m* GÉNÉRAL access charge; **~ de communication** *m pl* APP SERV call charge; **~ généraux** *m pl* GÉNÉRAL overheads; **~ de location** *m* GÉNÉRAL rental charge; **~ de raccordement au réseau** *m* TÉL network connection fee; **~ de service composés** *m pl* APP SERV multielement service charge

frappe *f* INFORMAT key depression, stroke

fraude: **~ informatique** *f* INFORMAT computer fraud

frein *m* COMPOSANTS *génie mécanique*, GÉNÉRAL brake; **~ d'inertie** ALIM ÉLEC inertia damper

fréquence[1]: **à ~ vocale** *adj* PÉRIPH, TÉL tone-pulse

fréquence[2] *f* GÉNÉRAL frequency; **~ d'accrochage** ÉLECTRON capture frequency; **~ asservie** RADIO frequency lock; **~ assignée** RADIO assigned frequency; **~ audio** INFORMAT, TÉL audio frequency (*AF*); **~ d'autoneutrodynage** ÉLECTRON self-neutralization frequency; **~ de balayage** ÉLECTRON sweep frequency, RADIO swept frequency, sweep frequency; **~ de balayage horizontal** TV horizontal sweep frequency, line frequency, scanning line frequency; **~ de balayage vertical** TV field frequency; **~ de base** ÉLECTRON basic frequency, INFORMAT base frequency; **~ de battement** ÉLECTRON, TRANSMIS beat frequency; **~ caractéristique** RADIO characteristic frequency; **~ centrale** TRANSMIS center frequency (*AmE*), centre frequency (*BrE*); **~ de commande** CONTRÔLE control frequency; **~ de coupure** ÉLECTRON cutoff frequency; **~ critique** ÉLECTRON threshold frequency, TRANSMIS critical frequency, threshold frequency; **~ de découpage** RADIO quench frequency; **~ désignée** TRANSMIS designated frequency; **~ d'échantillonnage** TRANSMIS sampling frequency, sampling rate; **~ des éclairs** ALIM ÉLEC flashing rate; **~ étalon** RADIO standard frequency, frequency standard; **~ d'exploitation** RADIO operating frequency; **~ d'exploration horizontale** INFORMAT line frequency, TV horizontal line frequency; **~ fixe** RADIO fixed frequency; **~ de fonctionnement** RADIO operating frequency; **~ fondamentale** ÉLECTRON basic frequency; **~ d'hétérodyne** RADIO heterodyne frequency; **~ d'horloge** TRANSMIS clock rate; **~ idler** RADIO idler frequency; **~ d'image** INFORMAT picture frequency, TV picture frequency, frame frequency; **~ image** RADIO image frequency; **~ d'images-clés** INFORMAT keyframe rate; **~ inférieure** RADIO lower frequency; **~ infra-acoustique** GÉNÉRAL subaudio frequency; **~ infra-téléphonique** TÉL subtelephone frequency; **~ instantanée** TRANSMIS instantaneous frequency; **~ intercalée** RADIO interspersed frequency; **~ intermédiaire** (*FI*) RADIO, TRANSMIS intermediate frequency (*IF*); **~ d'interruption** TÉL rate of interruption; **~ latérale** RADIO side frequency; **~ de ligne** TV horizontal sweep frequency, line frequency; **~ de ligne de balayage** TV scanning line frequency; **~ maximale utilisable** (*MUF*) RADIO maximum usable frequency (*MUF*); **~ minimale utilisable** (*LUF*) RADIO lowest usable frequency (*LUF*); **~ de modulation** TÉL, TRANSMIS modulating frequency, modulation frequency; **~ moyenne de passage à zéro** GÉNÉRAL average crossing rate; **~ multiple** (*FM*) GÉNÉRAL, SIGNAL, TÉL multiple frequency (*MF*); **~ nominale** ALIM ÉLEC rated frequency; **~ normalisée** RADIO, TRANSMIS normalized frequency; **~ normée** TRANSMIS normalized frequency; **~ optimale de trafic** (*FOT*) RADIO optimum traffic frequency (*OTF*); **~ optimale de travail** (*FOT*) RADIO optimum working frequency (*OWF*); **~ d'oscillateur** ÉLECTRON oscillator frequency; **~ d'oscillateur libre** ÉLECTRON free-running oscillator frequency; **~ d'oscillateur pilote** ÉLECTRON master oscillator frequency; **~ d'oscillation** ÉLECTRON oscillation frequency; **~ des paquets de signaux** DONNÉES burst rate; **~ par largeur de classe** TESTS frequency density; **~ partagée** RADIO shared frequency; **~ pilote** TRANSMIS pilot frequency; **~ de points** TV dot frequency; **~ porteuse** TRANSMIS carrier frequency; **~ porteuse d'image** TV picture tone; **~ porteuse son** TV sound carrier frequency; **~ propre** ALIM ÉLEC natural frequency, ÉLECTRON free-running frequency, RADIO self-resonant frequency; **~ de rafraîchissement** DONNÉES, INFORMAT refresh rate; **~ de récurrence des impulsions** (*FRI*) ÉLECTRON, SIGNAL, TÉL, TRANSMIS pulse recurrence frequency (*PRF*); **~ de référence** RADIO, TRANSMIS reference frequency; **~ de répétition** TÉL repetition frequency, repetition rate; **~ de répétition des impulsions** (*FRI*)

ÉLECTRON, SIGNAL, TÉL, TRANSMIS pulse repetition frequency (*PRF*); ~ **du réseau** ALIM ÉLEC mains frequency; ~ **réservée** TRANSMIS dedicated frequency; ~ **résistive de coupure** ÉLECTRON resistive cut-off frequency; ~ **de résonance** ÉLECTRON, RADIO, TÉL resonance frequency, resonant frequency; ~ **du rythme** *m* TRANSMIS clock-rate distribution; ~ **du secteur** ALIM ÉLEC mains frequency, TV line frequency; ~ **de seuil** ÉLECTRON threshold frequency, TRANSMIS critical frequency, threshold frequency; ~ **de la sous-porteuse** RADIO, TÉL, TRANSMIS subcarrier frequency; ~ **sous-porteuse** RADIO, TÉL, TRANSMIS subcarrier frequency; ~ **supérieure** RADIO higher frequency; ~ **supra-acoustique** GÉNÉRAL super-audio frequency; ~ **supratéléphonique** TÉL, TRANSMIS super-telephone frequency; ~ **téléphonique** TÉL, TRANSMIS telephone frequency; ~ **de trame** DONNÉES frame rate, TV frame frequency; ~ **de travail** RADIO working frequency; ~ **ultrasonore** RADIO ultrasonic frequency; ~ **unique** GÉNÉRAL single tone; ~ **verticale** DONNÉES, INFORMAT refresh rate; ~ **vocale** (*FV*) GÉNÉRAL voice frequency (*VF*)

fréquencemètre *m* TESTS frequency counter, frequency meter; ~ **hétérodyne** TESTS heterodyne frequency meter; ~ **à lames vibrantes** TESTS reed frequency meter, vibrating-reed frequency meter

FRI *abrév* ÉLECTRON, SIGNAL, TÉL, TRANSMIS (*fréquence de récurrence des impulsions*) PRF (*pulse recurrence frequency*), (*fréquence de répétition des impulsions*) PRF (*pulse repetition frequency*)

frigoporteur *m* CLIMAT secondary refrigerant

frigorigène *m* CLIMAT refrigerant

frittage *m* COMPOSANTS sintering

friture *f* TÉL frying, crackle, crackling noise, line scratches

front: ~ **arrière** *m* COMPOSANTS, ÉLECTRON, GÉNÉRAL, TÉL trailing edge; ~ **arrière d'impulsion** *m* ÉLECTRON pulse trailing edge; ~ **avant** *m* COMPOSANTS, ÉLECTRON, GÉNÉRAL, TÉL leading edge; ~ **avant d'impulsion** *m* ÉLECTRON pulse leading edge; ~ **d'onde** *m* GÉNÉRAL wavefront

frontal[1]: **en** ~ *adj* RADIO add-on

frontal[2] *m* COMMUT front-end computer

frontière *f* GÉNÉRAL boundary, border; ~ **d'une cellule** RADIO cell border

FTAM *abrév* (*transfert, accès et gestion de fichiers*) DONNÉES, INFORMAT FTAM (*file transfer access and management*)

fuire: ~ **en surface** *vi* ALIM ÉLEC creep

fuite *f* ALIM ÉLEC, ÉLECTRON leakage, TÉL leak; ~ **en surface** ALIM ÉLEC creepage, surface leakage; ~ **à la terre** ALIM ÉLEC earth leakage (*BrE*), ground leakage (*AmE*)

full[1]: ~ **duplex** *m* RADIO duplex

full[2]: ~ **view** *f* INFORMAT full view

fuseau: ~ **z** *m* GÉNÉRAL z time

fusée: ~ **de lancement** *m* RADIO launch rocket

fusible *m* ALIM ÉLEC, COMPOSANTS, ÉLECTRON fuse wire, fuse; ~ **à action instantanée** ALIM ÉLEC, COMPOSANTS quick-acting fuse; ~ **d'alimentation** ALIM ÉLEC, COMPOSANTS feeding fuse; ~ **d'alvéole** ALIM ÉLEC, COMPOSANTS, TÉL shelf fuse; ~ **d'appareil** ALIM ÉLEC, COMPOSANTS equipment fuse; ~ **de batterie** ALIM ÉLEC, COMPOSANTS battery fuse; ~ **à cartouche** ALIM ÉLEC, COMPOSANTS grip-type fuse; ~ **cartouche** ALIM ÉLEC, COMPOSANTS cartridge fuse; ~ **de courant de fuite enfichable** ÉLECTRON push-on sneak current fuse; ~ **pour courants forts** ALIM ÉLEC, COMPOSANTS power fuse; ~ **de distribution** ALIM ÉLEC, COMPOSANTS distribution fuse; ~ **pour faible intensité** ALIM ÉLEC, COMPOSANTS fine-wire fuse; ~ **à fil** ALIM ÉLEC, COMPOSANTS wire fuse; ~ **de filtre** ALIM ÉLEC, COMPOSANTS filter fuse; ~ **fondu** ALIM ÉLEC, COMPOSANTS blown fuse; ~ **interrupteur** ALIM ÉLEC fuse switch; ~ **miniature** ALIM ÉLEC, COMPOSANTS miniature fuse; ~ **principal** ALIM ÉLEC, COMPOSANTS main fuse; ~ **sauté** ALIM ÉLEC, COMPOSANTS blown fuse; ~ **de secteur** ALIM ÉLEC, COMPOSANTS mains fuse; ~ **de travée** ALIM ÉLEC, COMPOSANTS suite fuse; ~ **à tube** ALIM ÉLEC, COMPOSANTS, ÉLECTRON tube fuse; ~ **à tube de verre** ALIM ÉLEC, COMPOSANTS, ÉLECTRON glass tube fuse

fût: ~ **de pylône** *m* ALIM ÉLEC tower body

FV *abrév* (*fréquence vocale*) GÉNÉRAL VF (*voice frequency*)

FWHM *abrév* (*étendue à mi-crête, largeur à mi-crête*) DONNÉES, ÉLECTRON, TRANSMIS FWHM (*full-width half-maximum*)

FX3 *abrév* (*télécopie du groupe 3*) TÉL FX3 (*Telefax Group 3*)

FX4 *abrév* (*télécopie du groupe 4*) TÉL FX4 (*Telefax Group 4*)

G

gabarit *m* COMPOSANTS jig, template, GÉNÉRAL formboard, jig, RADIO, TÉL mask, TESTS limit, TRANSMIS formboard, mask, template; **~ d'amplitude** TRANSMIS amplitude-response limit; **~ de câblage** TRANSMIS cable formboard, cable-forming board; **~ pour connexions intérieures** COMPOSANTS wiring fixture, wiring jig; **~ de coupe de fils** GÉNÉRAL cutting formboard; **~ de division** TRANSMIS *câbles* branch formboard; **~ d'écho** DONNÉES echo suppressor; **~ d'impulsion** TÉL, TRANSMIS pulse mask; **~ d'indice de réfraction à quatre cercles concentriques** TRANSMIS four-concentric-circle refractive-index template; **~ pour peigne** TRANSMIS cable-forming board, cable formboard; **~ à quatre cercles concentriques en champ proche** TRANSMIS four-concentric-circle near-field template; **~ du temps de propagation de groupe** TRANSMIS group-delay mask; **~ de traçage** COMPOSANTS scribing template

gâchette *f* ÉLECTRON gate

gain *m* ALIM ÉLEC, ÉLECTRON, RADIO, TÉL, TRANSMIS gain; **~ absolu** RADIO absolute gain, isotropic gain; **~ d'antenne** RADIO antenna gain; **~ des antennes d'émission du satellite** RADIO satellite transmit aerial gain, satellite transmit antenna gain; **~ en bord de couverture** RADIO edge-of-coverage gain; **~ en boucle ouverte** ALIM ÉLEC, CONTRÔLE open-loop gain; **~ composite** TRANSMIS composite gain; **~ de concentration** (*IG*) TRANSMIS interpolation gain (*IG*); **~ dans l'axe** RADIO on-axis gain; **~ dans l'axe du faisceau** RADIO beam-center gain (*AmE*), beam-centre gain (*BrE*); **~ différentiel en puissance** ÉLECTRON incremental power gain; **~ de directivité** RADIO directive gain; **~ de l'EMCN** TÉL DCME gain; **~ à l'émission** TRANSMIS transmit gain; **~ équivalent** TRANSMIS effective gain; **~ d'insertion** TRANSMIS insertion gain; **~ isotrope** RADIO isotropic gain, absolute gain; **~ logarithmique** CONTRÔLE logarithmic gain; **~ de masse sur les sources** RADIO feed mass saving; **~ de multiplication de circuits numériques** (*GMCN*) TÉL, TRANSMIS digital circuit multiplication gain (*DCMG*); **~ d'obstacle** RADIO obstacle gain; **~ optimal** ÉLECTRON optimum gain; **~ en petits signaux** ÉLECTRON small-signal gain; **~ en puissance** ALIM ÉLEC, ÉLECTRON power gain; **~ en puissance d'antenne** RADIO antenna power gain; **~ à la réception** RADIO receive gain; **~ réflexion** TRANSMIS reflection gain; **~ en régime de saturation** ÉLECTRON saturation gain; **~ en tension** ALIM ÉLEC voltage gain; **~ de transcodage** ÉLECTRON transcoding gain; **~ transductique** TÉL transducer gain; **~ à la transmission** TRANSMIS transmit gain; **~ de 1** ÉLECTRON unit gain, unity gain; **~ unitaire** ÉLECTRON unit gain, unity gain; **~ unité** ÉLECTRON unit gain, unity gain; **~ variable dans le temps** CONTRÔLE sensitivity time control (*STC*)

gainage *m* ALIM ÉLEC canning, sheathing, COMPOSANTS sheathing, jacket, jacketing, ÉLECTRON jacketing, buffering, sheathing, GÉNÉRAL sheathing, TRANSMIS buffer, jacket, sheathing; **~ d'épanouissement** ÉLECTRON breakout jacketing; **~ intermédiaire serré** TRANSMIS tight buffer; **~ de plomb** GÉNÉRAL lead sheathing; **~ rétractable** COMPOSANTS shrink tubing; **~ secondaire** TRANSMIS fiber buffer (*AmE*), fibre buffer (*BrE*); **~ serré** TRANSMIS tight jacket

gainé *adj* COMPOSANTS sheathed

gaine[1] *f* GÉNÉRAL sleeving, sheathing, tubing, sheath, jacket; **~ adaptée** TRANSMIS matched cladding; **~ d'ascension** TRANSMIS *pose de câbles* cable riser, cable runway; **~ de câble** TRANSMIS cable sheath; **~ de câble optique d'immeuble** ÉLECTRON lightguide building cable; **~ déprimée** TRANSMIS depressed cladding; **~ de dérivation** ÉLECTRON fan-out jacketing; **~ époxy** TRANSMIS epoxy buffer; **~ extérieure** ÉLECTRON jacket; **~ de fibres optiques** TRANSMIS fiber-optic cladding (*AmE*), fibre-optic cladding (*BrE*); **~ homogène** COMPOSANTS, TRANSMIS homogeneous cladding; **~ de plomb** GÉNÉRAL lead sheath; **~ de protection** ÉLECTRON protective sheath; **~ rétractable** COMPOSANTS shrink tube; **~ spirale** COMPOSANTS spiral wrap; **~ thermorétractable** COMPOSANTS, TRANSMIS heat-shrinkable sleeving, heat-shrink tubing

gaine[2]: **avec ~ PVC** *loc* ÉLECTRON PVC-jacketed

galette *f* COMPOSANTS, ÉLECTRON wafer

galvanomètre *m* TESTS galvanometer; **~ de zéro** TESTS null meter

gamme *f* DONNÉES scale, range, GÉNÉRAL range, INFORMAT gamut, range, scale, RADIO range, TÉL level, scale, range, TESTS, TRANSMIS range; **~ de claviers** INFORMAT keyboard range; **~ dynamique** TESTS dynamic range; **~ étendue** TV extended range; **~ des fréquences** RADIO, TRANSMIS frequency range; **~ de gris** DONNÉES gray scale (*AmE*), grey scale (*BrE*), TÉL gray level (*AmE*), grey level (*BrE*), gray scale (*AmE*), grey scale (*BrE*); **~ de mesure** TESTS measuring range; **~ tempérée** INFORMAT equal-tempered scale

gant: **~ capteur** *m* INFORMAT Dataglove; **~ sensitif** *m* INFORMAT Dataglove

garde: **en ~** *loc* TÉL on hold

gauchir *vt* COMPOSANTS, ÉLECTRON warp

gauchissement *m* COMPOSANTS, ÉLECTRON warping; **~ des bords** ÉLECTRON edge warping

gaufrage *m* COMPOSANTS embossing

gaz: **~ d'échappement** *m* CLIMAT exhaust, exhaust gas; **~ électronique** *m* GÉNÉRAL electron gas; **~ inerte** *m* CLIMAT inert gas; **~ rare** *m* CLIMAT noble gas, rare gas

GED *abrév* (*gestion électronique de documents*) DONNÉES electronic document management

gel *m* COMPOSANTS gel, DONNÉES, TRANSMIS freeze-out; **~ de l'image** TV freeze-picture

geler *vt* DONNÉES, INFORMAT, TV freeze

généralisation: **~ progressive** *f* GÉNÉRAL general phasing-in

générateur *m* GÉNÉRAL generator; **~ d'application** DONNÉES application generator; **~ d'arpèges** INFORMAT arpeggiator; **~ auxiliaire de fréquence porteuse** TRANSMIS redundant carrier-supply; **~ de base de temps** ÉLECTRON sawtooth generator, time-base generator; **~ de bruit** INFORMAT, TRANSMIS noise generator, noise-generating device; **~ de caractères** INFORMAT character generator; **~ de code** TÉL code generator; **~ de compilateurs** INFORMAT compiler generator; **~ de courant d'appel** PÉRIPH, TÉL ring-current set, ringing set, ringing-current set; **~ de courant de sonnerie** TÉL ringing machine; **~ de courbes** INFORMAT curve generator; **~ d'effets spéciaux** INFORMAT special-effects generator; **~ d'enveloppes** INFORMAT envelope generator; **~ de fonction** INFORMAT function generator; **~ de fréquence** ÉLECTRON, TÉL frequency generator; **~ de fréquence de balayage de ligne** TV line frequency generator; **~ de fréquence de ligne** TV line frequency generator; **~ de fréquence numérique** INFORMAT digital frequency generator; **~ de Hall** ÉLECTRON Hall generator; **~ d'harmoniques** TRANSMIS harmonic generator; **~ d'harmoniques numérique** INFORMAT digital harmonics generator; **~ d'impulsions** DONNÉES pulser, ÉLECTRON pulse generator, INFORMAT, TÉL, TRANSMIS pulser; ~ **d'impulsions de comptage** TÉL meter-pulse generator; ~ **d'impulsions d'horloge** ÉLECTRON clock oscillator; ~ **d'impulsions rectangulaires** ÉLECTRON square-pulse generator; ~ **d'impulsions de taxation** TÉL tariff generator; ~ **d'intervalles de temps** ALIM ÉLEC interval timer; ~ **de mire** TV pattern generator; ~ **de modulation** TRANSMIS modulation generator; ~ **multifréquence** SIGNAL, TÉL MF generator, multifrequency generator; ~ **d'onde pilote** TRANSMIS group pilot generation; ~ **d'ondes** ÉLECTRON, RADIO wave generator; ~ **de rythme** ÉLECTRON clock, TRANSMIS timing generator; ~ **de rythme de multiplexeur** COMMUT multiplex-timing generator; ~ **de rythme de régénérateur** DONNÉES regenerator-timing generator (*RTG*); ~ **de signal d'appel** TÉL ringing generator; ~ **de signal d'horloge** ÉLECTRON clock oscillator; ~ **de signalisation multifréquence** SIGNAL, TÉL MF generator, multifrequency generator; ~ **de signaux** SIGNAL, TESTS signal generator; ~ **de signaux d'appel** SIGNAL, TÉL tone-and-ringing set; ~ **de signaux carrés** ÉLECTRON, SIGNAL square-wave generator; ~ **de signaux d'interrogation** RADIO interrogation-signal generator (*ISG*); ~ **de signaux pseudo-aléatoires** DONNÉES pseudorandom-sequence generator; ~ **de son** DONNÉES acoustic generator; ~ **de tension d'appel** TÉL ringing-voltage generator; ~ **de tension de référence** ALIM ÉLEC reference-voltage generator; ~ **de Thévenin** ALIM ÉLEC Thévenin equivalent circuit; ~ **de tonalité** ÉLECTRON, INFORMAT tone generator; ~ **de tonalité d'intervention** TÉL tick-tone generator; ~ **de tonalité d'intrusion** TÉL tick-tone generator; ~ **de tonalité de retour d'appel** TÉL ring-tone generator; ~ **de tonalité de tic-tac** TÉL tick-tone generator; ~ **de trame** TRANSMIS frame generator

génération *f* GÉNÉRAL generation, INFORMAT, TÉL generator; ~ **d'impulsions de compteur** TÉL meter-pulse generation; ~ **d'impulsions d'ouverture** TÉL break-pulsing generator; ~ **de nombres aléatoires** TÉL random-number generation; ~ **d'ondes** ÉLECTRON wave generation; ~ **de paires d'électrons** COMPOSANTS pair generation, pair production; ~ **de parole** INFORMAT, RADIO, TÉL speech generation; ~ **de système** INFORMAT system generation; ~ **de tonalité** ÉLECTRON, INFORMAT tone generation; ~ **de tonalité FM numérique** INFORMAT digital FM-tone generator

générer *vt* GÉNÉRAL, INFORMAT generate

génériques *m pl* TV credits, production credits

génie *m* GÉNÉRAL engineering; ~ **cognitif** INFORMAT knowledge engineering; ~ **électrique** DONNÉES, GÉNÉRAL electrical engineering; ~ **électronique** DONNÉES electronic engineering; ~ **linguistique** INFORMAT linguistic engineering; ~ **logiciel** INFORMAT software engineering; ~ **logiciel assisté par ordinateur** INFORMAT computer-aided software engineering (*CASE*); ~ **des systèmes de contrôle** CONTRÔLE control engineering

géométrie: ~ **des fibres** *f* TRANSMIS fiber geometry (*AmE*), fibre geometry (*BrE*); ~ **solide constructive** *f* (*GSC*) INFORMAT constructive solid geometry (*CSG*)

géostationnaire *adj* RADIO geostationary

géosynchrone *adj* RADIO geosynchronous

gestion *f* DONNÉES management, GÉNÉRAL handling, operation, INFORMAT, PÉRIPH, RADIO, TÉL, TRANSMIS management; ~ **assistée par ordinateur** INFORMAT computer-assisted management; ~ **automatique de lignes** TRANSMIS automatic line management; ~ **automatisée** INFORMAT computer-assisted management; ~ **de base de données** DONNÉES, INFORMAT database management; ~ **du brouillage** DONNÉES scrambling control; ~ **de clés** DONNÉES, TÉL key management; ~ **de compression d'image** DONNÉES image compression management (*ICM*); ~ **des connexions** COMMUT switching complex control software machine; ~ **de couche** DONNÉES layer management; ~ **du document distant** TÉL remote-document management; ~ **de données** DONNÉES, INFORMAT data management (*DM*); ~ **électronique de documents** (*GED*) DONNÉES electronic document management; ~ **de facturation et taxation** TRANSMIS accounting management; ~ **des fréquences** RADIO frequency management; ~ **de localisations** RADIO location management; ~ **de la mémoire** DONNÉES storage management; ~ **d'objets** INFORMAT object management; ~ **périphérique** PÉRIPH peripheral management; ~ **du plan de trame** DONNÉES burst-time plan change control; ~ **de la qualité** APP SERV, TESTS quality management; ~ **de réseau** DONNÉES, GÉNÉRAL network management, TÉL network management, operations support, TRANSMIS network management; ~ **du réseau international** TÉL international network management; ~ **des situations de crise** TESTS crisis management; ~ **de système** DONNÉES system management; ~ **du système de télécommunications** GÉNÉRAL communications system management; ~ **totale de qualité** TESTS total quality management (*TQM*)

gestionnaire *m* COMMUT handler, DONNÉES manager, INFORMAT generator, manager, TÉL manager; ~ **de base de données** DONNÉES, INFORMAT database manager; ~ **de canal privé MCS** TÉL MCS private channel manager; ~ **de compression d'image** DONNÉES, TV image compression manager; ~ **de fichiers de disques compacts** INFORMAT compact disc file manager; ~ **de fichiers en réseau** (*NFM*) DONNÉES, INFORMAT network file manager (*NFM*); ~ **de fichiers séquentiels de caractères** INFORMAT sequential character file manager; ~ **de ligne de télécopie** DONNÉES *personne* facsimile line manager; ~ **de mémoire** INFORMAT memory manager; ~ **de mémoire de données** COMMUT, DONNÉES data store handler; ~ **de paquets** DONNÉES, TÉL packet handler (*PH*); ~ **de polices de caractères** INFORMAT font manager; ~ **de programme** INFORMAT program generator; ~ **de réseau** DONNÉES LAN manager; ~ **de serveur de terminaux** DONNÉES terminal server manager (*TSM*); ~ **des tests du client** DONNÉES *personne* client test manager

getter *m* ÉLECTRON getter

GFA *abrév* (*groupe fermé d'abonnés*) DONNÉES, TÉL closed subscriber group, CUG (*closed user group*)

GFU *abrév* (*groupe fermé d'usagers, groupe fermé d'utilisateurs*) DONNÉES, TÉL CUG (*closed user group*)

GHz *abrév* (*gigahertz*) ÉLECTRON, TESTS, TRANSMIS gc (*gigacycle*), GHz (*gigahertz*)

GIF *abrév* (*format d'échange de graphiques*) DONNÉES *Internet*, INFORMAT GIF (*graphics interchange format*)

gigabit *m* DONNÉES gigabit

gigahertz *m* (*GHz*) ÉLECTRON, TESTS, TRANSMIS gigacycle (*gc*), gigahertz (*GHz*)

gigue *f* ÉLECTRON, RADIO, TÉL, TRANSMIS, TV jitter, synchronization error; ~ **à l'entrée** TÉL input jitter; ~ **d'horloge** ÉLECTRON clock jitter; ~ **proportionnelle** RADIO proportional jitter; ~ **de rythme** TRANSMIS timing jitter

glissement *m* COMPOSANTS slide, DONNÉES slip, GÉNÉRAL glide, INFORMAT bend, slide, RADIO, TÉL, TRANSMIS slip; ~ **binaire** RADIO, TRANSMIS transmission bit slip; ~ **de bits** RADIO, TRANSMIS transmission bit slip; ~ **commandé** RADIO, TÉL controlled slip; ~ **de fréquence** RADIO frequency departure; ~ **horizontal de l'image** TV line slip; ~ **maîtrisable** RADIO controlled slip; ~ **de mode** ÉLECTRON moding; ~ **non commandé** RADIO, TÉL uncontrolled slip; ~ **non maîtrisable** RADIO, TÉL uncontrolled slip; ~ **de trame** DONNÉES frame slip

glisser: ~**-déplacer** *vt* INFORMAT drag and drop

GMCN *abrév* (*gain de multiplication de circuits numériques*) TÉL, TRANSMIS DCMG (*digital circuit multiplication gain*)

GO *abrév* (*guide d'ondes*) DONNÉES, RADIO, TRANSMIS waveguide

goniométrie *f* RADIO direction finding
gopher *m* DONNÉES, INFORMAT gopher
gorge *f* RADIO throat; **~ de raccordement** RADIO throat
goulot: **~ d'étranglement** *m* GÉNÉRAL bottleneck
goulotte *f* COMMUT chute, COMPOSANTS chute, duct, ÉLECTRON raceways, TÉL chute, TRANSMIS cable chute, chute; **~ latérale** TÉL side chute; **~ métallique** COMPOSANTS sheet metal duct; **~ de raccordement d'armoire** COMMUT cabinet connection chute; **~ de rangée** TÉL row chute; **~ secondaire** TÉL side chute; **~ transversale** COMPOSANTS cross chute
goutte: **~ d'époxy** *f* ÉLECTRON bead of epoxy
gouttière *f* TRANSMIS cabling duct
GPS *abrév* (*système de positionnement universel*) RADIO GPS (*global positioning system*)
gradation *f* TÉL grading
gradient: **~ d'indice** *m* ÉLECTRON, TRANSMIS graded index; **~ normal du coïndice** *m* RADIO standard refractivity vertical gradient; **~ de pesanteur** *m* RADIO gravity gradient; **~ de potentiel** *m* ALIM ÉLEC potential gradient
graduation *f* TESTS graduation, scale mark
graduer *vt* TESTS graduate
grain: **~ d'entrefer** *m* COMPOSANTS residual stud
grammaire: **~ abstraite** *f* DONNÉES, INFORMAT abstract grammar; **~ concrète** *f* DONNÉES concrete grammar; **~ graphique concrète** *f* DONNÉES concrete graphical grammar; **~ textuelle commune** *f* DONNÉES common textual grammar
grande: **à ~ distance** *loc* TÉL long-distance; **à ~ puissance** *loc* ALIM ÉLEC high-power
grandeur[1]: **~ de champ** *f* TESTS field strength; **~ de commande** *f* CONTRÔLE command variable; **~ commandée** *f* CONTRÔLE *commande automatique* controlled variable; **~ d'influence** *f* CONTRÔLE *commande automatique* actuating variable, TESTS influence quantity; **~ mesurable** *f* TESTS measurable quantity; **~ périodique** *f* GÉNÉRAL periodic quantity; **~ de perturbation** *f* CONTRÔLE disturbance variable; **~ de référence** *f* CONTRÔLE reference variable; **~ réglée** *f* CONTRÔLE *commande automatique* controlled variable; **~ de sortie** *f* CONTRÔLE output variable
grandeur[2]: **~ nature** *loc* GÉNÉRAL life-size
granularité *f* TV granularity
granule *m* DONNÉES cluster
granules *m pl* TV granules
graphe: **~ ET-OU** *m* DONNÉES AND-OR graph
graphie: **~ en mode coordonnées** *f* INFORMAT coordinate graphics; **~ en mode image** *f* TV image graphics
graphique *m* TV graphics; **~ de contrôle** TESTS inspection diagram; **~ en nuage de points** INFORMAT scatter graph
graphisme *m* TV graphics
grappe *f* GÉNÉRAL cluster; **~ d'adaptateur d'impédance** ÉLECTRON linking clusters of balun adaptors
gras: **en ~** *loc* INFORMAT *impression* in bold
gratuiciel *m* INFORMAT freeware package
gratuit *adj* GÉNÉRAL free of charge, TÉL toll-free
gravé: **~ numériquement** *loc* INFORMAT digitally-mastered
graver *vt* ÉLECTRON emboss, etch, INFORMAT etch; **~ chimiquement** GÉNÉRAL etch
graves *m pl* GÉNÉRAL bass, bass tones
gravité: **~ de la défaillance** *f* TESTS failure criticality
gravure: **~ chimique** *f* GÉNÉRAL etching; **~ de matrice** *f* INFORMAT digital mastering; **~ en relief** *f* ÉLECTRON embossing; **~ en retrait** *f* ÉLECTRON etch-back; **~ sous-jacente** *f* COMPOSANTS, ÉLECTRON undercut
grenailles: **~ de charbon** *f pl* COMPOSANTS *microphones* carbon granules
grillage *m* COMPOSANTS wire netting; **~ protecteur** COMPOSANTS grille
grille *f* COMPOSANTS grid, grating, grille, ÉLECTRON gate, grid, mesh, RADIO lattice; **~ alignée** ÉLECTRON shadow grid; **~ apparente** RADIO mesh effect; **~ d'arrêt** COMPOSANTS, ÉLECTRON suppressor grid; **~ de désionisation** ÉLECTRON de-ionization grid; **~-écran** ÉLECTRON screen grid; **~ à mailles** ÉLECTRON mesh electrode, field mesh; **~ de programme** TV program schedule (*AmE*), programme schedule (*BrE*); **~ de protection** ALIM ÉLEC shield grid, COMPOSANTS grille
grip: **~ de câble** *m* COMPOSANTS cable grip; **~ à chaussette** *m* COMPOSANTS steel wire braid; **~ de tirage** *m* COMPOSANTS, ÉLECTRON grip apparatus
gros: **~ ordinateur** *m* DONNÉES, INFORMAT mainframe
grossissement *m* INFORMAT zooming
groupe[1] *m* ALIM ÉLEC, COMMUT, CONTRÔLE group, set, DONNÉES group, segment, GÉNÉRAL group, INFORMAT bank, group, segment, TÉL group, cluster, segment; **~ actif** DONNÉES active group; **~ d'adaptateurs de signalisation** SIGNAL signaling adaptor group (*AmE*), signalling adaptor group (*BrE*); **~ amplificateur de puissance** ALIM ÉLEC, RADIO power cube; **~ d'attributs** INFORMAT attribute group; **~ d'attributs cumulatifs** INFORMAT accumulating-attribute group; **~ d'attributs mutuellement exclusifs** INFORMAT mutually exclusive attribute group; **~ de balais** COMMUT wiper set; **~ de base** TÉL basic group; **~ de câbles** GÉNÉRAL, TRANSMIS cable group; **~ de canaux sémaphores** SIGNAL signaling-link group (*AmE*), signalling-link group (*BrE*); **~ de comportements par défaut** DONNÉES default group; **~ de contacts de rupture** COMMUT break contact group, break contact spring set; **~ des demandés** DONNÉES invoked group; **~ de destinataires** DONNÉES multicast group; **~ détectable** INFORMAT detectable group; **~ de discussion** DONNÉES *Internet* newsgroup; **~ de données** DONNÉES, RADIO data group; **~ double** GÉNÉRAL twin group; **~ électrogène Diesel** ALIM ÉLEC Diesel generator set; **~ électrogène de relève** ALIM ÉLEC interruptible power supply; **~ d'entrées/sorties** COMMUT input/output group; **~ fermé d'abonnés** (*GFA*) DONNÉES, TÉL closed subscriber group, closed user group (*CUG*); **~ fermé d'usagers** (*GFU*) DONNÉES, TÉL closed user group (*CUG*); **~ fermé d'usagers avec accès sortant** DONNÉES, TÉL closed user group with outgoing access; **~ fermé d'usagers bilatéral** DONNÉES, TÉL bilateral closed user group (*BCUG*); **~ fermé d'usagers bilatéral avec accès sortant** DONNÉES, TÉL bilateral closed user group with outgoing access (*BCUGOA*); **~ fermé d'usagers vidéotex** INFORMAT, TV videotex closed user group; **~ fermé d'utilisateurs** (*GFU*) DONNÉES, TÉL closed user group (*CUG*); **~ frigorifique** CLIMAT cooling unit; **~ graphique** INFORMAT display group, display segment; **~ jumelé** GÉNÉRAL twin group; **~ de liaisons de signalisation** SIGNAL signaling-link group (*AmE*), signalling-link group (*BrE*); **~ de lignes** TRANSMIS line group; **~ de lignes réseau** TÉL group of matrix links; **~ de lignes à trafic équilibré** TÉL balanced grading group; **~ mixte d'experts en photographie** (*JPEG*) INFORMAT, TÉL, TV Joint Photographic Experts Group (*JPEG*); **~ mixte sur les images à deux niveaux** (*JBIG*) INFORMAT, TÉL, TV joint bi-level image group (*JBIG*); **~ de modules de test** TESTS test step group; **~ ouvert** DONNÉES open group; **~ PBX** TÉL PBX group; **~ primaire** TRANSMIS basic group; **~ de prises d'appel** TÉL call pick-up group; **~ quaternaire** TÉL line system, supermastergroup; **~ de remplacement** INFORMAT alternate group; **~ de ressorts bimétalliques** CONTRÔLE *dispositifs de réglage* bimetal springset; **~ secondaire** TÉL, TRANSMIS supergroup; **~ secondaire de base** TÉL, TRANSMIS basic supergroup; **~ spécial mobiles** (*GSM*) RADIO, TÉL global system for mobile communications (*GSM*); **~ tertiaire** TÉL mastergroup, TRANSMIS hypergroup, mastergroup (*AmE*); **~ de tests** DONNÉES, TESTS test group; **~ des tonalités et auxiliaires** TÉL *dans le CSN* tones and auxiliaries group; **~ d'unités administratives** (*GUA*) CONTRÔLE administrative unit

group; ~ **d'unités tributaires** DONNÉES tributary unit group (*TUG*); ~ **visuel** INFORMAT display group, display segment

groupe[2]: **en** ~ *loc* GÉNÉRAL by groups, in groups; **par** ~ *loc* GÉNÉRAL by groups, in groups

groupement *m* DONNÉES block, GÉNÉRAL grouping, TRANSMIS group, pool, *câbles* bunching

grouper *vt* DONNÉES block together, GÉNÉRAL group, TRANSMIS group, pool, bunch

GSC *abrév* COMMUT (*commutateur à autonomie d'acheminement*) GSC (*group switching center AmE, group switching centre BrE*), INFORMAT (*géométrie solide constructive*) CSG (*constructive solid geometry*)

GSM *abrév* (*groupe spécial mobiles*) RADIO, TÉL GSM (*global system for mobile communications*)

GUA *abrév* (*groupe d'unités administratives*) CONTRÔLE administrative unit group

guet *m* RADIO search

GUI *abrév* (*interface utilisateur graphique*) DONNÉES, INFORMAT GUI (*graphical user interface*)

guidage: ~ **orbital** *m* RADIO orbital guidance; ~ **vocal** *m* PÉRIPH voice prompt

guide *m* GÉNÉRAL guide, running bar; ~ **du cylindre** TÉL platen guide; **~-fils** COMPOSANTS wire guide; ~ **d'onde flexible** DONNÉES, RADIO, TRANSMIS flexible waveguide; ~ **d'onde à moulure** DONNÉES, RADIO, TRANSMIS ridge waveguide; ~ **d'onde nervuré** DONNÉES, RADIO, TRANSMIS ridge waveguide; ~ **d'onde optique** DONNÉES, RADIO, TRANSMIS optical waveguide; ~ **d'onde en titane diffusé** DONNÉES, RADIO, TRANSMIS titanium waveguide channel; ~ **d'onde torsadé** DONNÉES, RADIO, TRANSMIS twisted waveguide; ~ **d'onde unifilaire** COMPOSANTS waveguide wire, RADIO single-wire transmission line; ~ **d'ondes** (*GO*) DONNÉES, RADIO, TRANSMIS waveguide; ~ **d'ondes cloisonné** DONNÉES, RADIO, TRANSMIS septate waveguide; ~ **d'ondes en couche mince** DONNÉES, TRANSMIS thin-film waveguide; ~ **d'ondes fendu** DONNÉES, RADIO, TRANSMIS slot waveguide (*SWG*), slotted waveguide (*SWG*); ~ **d'ondes à fentes** DONNÉES, RADIO, TRANSMIS slot waveguide (*SWG*), slotted waveguide (*SWG*); ~ **d'ondes hélicoïdal** DONNÉES, RADIO, TRANSMIS spiral waveguide; ~ **d'ondes non réciproque** DONNÉES, RADIO, TRANSMIS nonreciprocal waveguide; ~ **d'ondes optiques** DONNÉES, RADIO optical waveguide, TRANSMIS optical waveguide, lightguide; ~ **d'ondes réctangulaires** DONNÉES, RADIO, TRANSMIS ridge waveguide; ~ **d'ondes à rubans** RADIO stripline; ~ **optique** TRANSMIS lightguide; ~ **parlant** TÉL intercept announcer, speaking guide; ~ **de référence** INFORMAT reference guide; ~ **rythmique** INFORMAT rhythmic guide; ~ **tridimensionnel** TRANSMIS three-dimensional guide; ~ **à trois dimensions** TRANSMIS three-dimensional guide; ~ **troposphérique** RADIO tropospheric radioduct

guider *vt* GÉNÉRAL guide, monitor, RADIO steer, guide

guillemets *m pl* INFORMAT inverted commas, quotation marks, quote marks, quotes

guipage *m* COMPOSANTS, ÉLECTRON, TRANSMIS *d'un câble* braiding, core wrap

guiper *vt* COMPOSANTS, ÉLECTRON, TRANSMIS *câble* braid

gyrateur *m* RADIO gyrator

gyrofréquence *f* RADIO gyrofrequency; ~ **transversale** RADIO transverse gyrofrequency

gyrotron *m* ÉLECTRON gyrotron

H

H *abrév* (*henry*) ÉLECTRON H (*henry*)
habillage: ~ **latéral** *m* TÉL side cover, side plate
hacheur *m* ÉLECTRON chopper
hachurer *vt* INFORMAT hatch
hachures *f pl* ÉLECTRON *circuits imprimés* crosshatching
halo *m* TV halation, halo
hampe *f* INFORMAT ascender
handover *m* RADIO handover
harmonique[1] *adj* ÉLECTRON harmonic
harmonique[2] *m* ÉLECTRON harmonic, INFORMAT overtone; ~ **inférieur** ÉLECTRON subharmonic
harmonisation: ~ **automatique** *f* INFORMAT automatic harmonization, automatic harmonizing
Hartley *m* GÉNÉRAL Hartley
hasard[1]: ~ **heuristique** *m* DONNÉES heuristic hazard
hasard[2]: **au** ~ *loc* GÉNÉRAL random
hauban *m* COMPOSANTS stay wire, guy wire, RADIO guy wire, support cable
haut-parleur *m* INFORMAT, TÉL loudspeaker, speaker; ~ **incorporé** INFORMAT built-in speaker, PÉRIPH internal speaker, built-in loudspeaker; ~ **de médiums** ÉLECTRON mid-range speaker; ~ **à membrane conique** COMPOSANTS cone loudspeaker; ~ **à oreiller** TÉL pillow loudspeaker; ~ **à volume réglable** PÉRIPH adjustable volume control
haute[1]: ~ **définition** *f* TV high definition; ~ **fréquence** *f* (*HF*) ÉLECTRON radio frequency (*RF*), RADIO high frequency, radio frequency (*RF*), TÉL radio frequency (*RF*); **très** ~ **fréquence** *f* (*VHF*) RADIO, TRANSMIS, TV very high frequency (*VHF*); ~ **résistance** *f* TESTS high resistance (*HR*); ~ **tension** *f* (*HT*) ALIM ÉLEC high tension, high voltage (*HV*), ÉLECTRON high voltage (*HV*)
haute[2]: **à** ~ **fréquence** *loc* GÉNÉRAL high-frequency; **à** ~ **impédance** *loc* ALIM ÉLEC, ÉLECTRON high-impedance; **à** ~ **résistance** *loc* ALIM ÉLEC, ÉLECTRON high-resistance; **à** ~ **résolution** *loc* INFORMAT, TV high-resolution (*hi-res*); **à** ~ **tension** *loc* ALIM ÉLEC high-voltage
hauteur *f* GÉNÉRAL height, INFORMAT, TÉL pitch; ~ **absolue** INFORMAT absolute pitch; ~ **apparente d'antenne** RADIO apparent height; ~ **de composant** ÉLECTRON component height; ~ **du conduit** RADIO duct height; ~ **effective** RADIO effective height; ~ **effective d'antenne** RADIO effective antenna height; ~ **effective au-dessus du sol moyen** (*HEASM*) RADIO effective height above average terrain (*EHAAT*); ~ **d'ensemble** INFORMAT overall pitch; ~ **équivalente d'antenne** RADIO effective antenna height; ~ **d'un fournisseur de service MCS** TÉL height of an MCS provider; ~ **générale** INFORMAT overall pitch; ~ **d'une impulsion** TÉL pulse magnitude; ~ **de l'œil** RADIO eye height; ~ **du son** INFORMAT pitch; ~ **standard** INFORMAT standard pitch; ~ **standard du clavier** INFORMAT standard keyboard pitch; ~ **virtuelle** RADIO virtual height; ~ **virtuelle de réflexion** RADIO virtual height
HCMOS *abrév* (*semiconducteur complémentaire à oxyde metallique à grande vitesse*) ÉLECTRON HCMOS (*high-speed complementary metal oxide semiconductor*)
HDB2 *abrév* (*code bipolaire à haute densité d'ordre 2*) DONNÉES, SIGNAL, TÉL HDB2 (*high-density bipolar 2 code*)
HDB3 *abrév* (*code bipolaire à haute densité d'ordre 3*) DONNÉES, SIGNAL, TÉL HDB3 (*high-density bipolar 3 code*)
HDR *abrév* (*adaptateur, adaptateur de mesure*) ÉLECTRON HDR (*header*)
HEASM *abrév* (*hauteur effective au-dessus du sol moyen*) RADIO EHAAT (*effective height above average terrain*)
hébergeur: ~ **de boîtes aux lettres** *m* DONNÉES *Internet* e-mail message store
hélice *f* ÉLECTRON, GÉNÉRAL, RADIO helix
hélicoïdal *adj* GÉNÉRAL helical
hélioalimenté *adj* ALIM ÉLEC solar-powered
HEMT *abrév* (*transistor à haute mobilité d'électrons*) ÉLECTRON HEMT (*high electron mobility transistor*)
henry *m* (*H*) ÉLECTRON henry (*H*)
heptode *f* ÉLECTRON heptode
herbe *f* INFORMAT, RADIO grass
hertz *m* (*Hz*) ÉLECTRON, TESTS, TRANSMIS hertz (*Hz*)
hertzien *adj* RADIO microwave
hétérochrone *adj* TÉL, TRANSMIS heterochronous
hétérodyne *adj* GÉNÉRAL heterodyne
hétérogène *adj* DONNÉES incompatible, GÉNÉRAL heterogeneous
hétérojonction *f* ÉLECTRON, TRANSMIS heterojunction; ~ **double** TRANSMIS double heterojunction; ~ **simple** TRANSMIS single heterojunction
heure: ~ **centrale** *f* COMMUT exchange time; ~ **chargée** *f* TÉL busy hour (*BH*); ~ **chargée moyenne** *f* TÉL mean busy hour, time-consistent busy hour; ~ **de l'échange** *f* DONNÉES interchange time; ~ **de pointe** *f* TÉL busy hour (*BH*), peak hour; ~ **de pointe variable** *f* TÉL bouncing busy hour (*BBH*); ~ **universelle** *f* GÉNÉRAL universal time (*UT*); ~ **universelle de coordonnée** *f* GÉNÉRAL universal time coordinated (*UTC*)
heures: ~ **croisées** *f pl* APP SERV off-peak tariff periods; ~ **d'écoute maximale** *f pl* TV prime time; ~ **de grande écoute** *f pl* TV prime time; ~ **pleines** *f pl* APP SERV peak tariff periods
heuristique *f* DONNÉES heuristics
hexode *f* ÉLECTRON hexode
HF *abrév* (*haute fréquence*) ÉLECTRON RF (*radio frequency*), RADIO high frequency, RF (*radio frequency*), TÉL RF (*radio frequency*)
hiérarchie *f* GÉNÉRAL hierarchy; ~ **croisée** INFORMAT tangled hierarchy; ~ **de multiplexage numérique** TÉL, TRANSMIS digital multiplex hierarchy; ~ **numérique** TÉL, TRANSMIS digital hierarchy; ~ **numérique plésiochrone** (*HNP*) TÉL, TRANSMIS plesiochronous digital hierarchy (*PDH*); ~ **numérique synchrone** (*HNS*) TÉL, TRANSMIS synchronous digital hierarchy (*SDH*)
hiérarchique *adj* GÉNÉRAL hierarchical
histogramme *m* ÉLECTRON histogram
HLDLC *abrév* (*commande de liaison de données à haut niveau*) DONNÉES, TÉL HLDLC (*high-level data link control*)
HLR *abrév* (*enregistreur de position de rattachement*) RADIO HLR (*home location register*)
HMOS *abrév* (*semiconducteur à oxide metallique à grande vitesse*) ÉLECTRON HMOS (*high-speed metal oxide semiconductor*)
HNP *abrév* (*hiérarchie numérique plésiochrone*) TÉL, TRANSMIS PDH (*plesiochronous digital hierarchy*)
HNS *abrév* (*hiérarchie numérique synchrone*) TÉL, TRANSMIS SDH (*synchronous digital hierarchy*)
HNS-FHN *abrév* (*faisceau hertzien numérique pour hiérarchie synchrone*) RADIO SDH-DRSS (*digital radio relay for synchronous hierarchy*)
holographique *f* COMMUT, DONNÉES, ÉLECTRON holograph

homochrone *adj* TÉL, TRANSMIS homochronous
homogène *adj* GÉNÉRAL homogeneous
homogénéifier *vt* GÉNÉRAL homogenize
homogénéiser *vt* GÉNÉRAL homogenize
homogénéité *f* GÉNÉRAL homogeneity
homojonction *f* ÉLECTRON homojunction
homologation *f* DONNÉES agreement, GÉNÉRAL certification of approval
homologué *adj* GÉNÉRAL commissioned
homologue *m* DONNÉES, RADIO, TRANSMIS peer
honoraires *m* GÉNÉRAL fee
horaire: ~ **de radiodiffusion** *m* RADIO broadcasting schedule
horizon: ~ **radar** *m* RADIO radar horizon; ~ **radio** *m* RADIO radio horizon; ~ **radioélectrique** *m* RADIO radio horizon; ~ **temporel** *m* INFORMAT, RADIO, TRANSMIS timescale
horizontale *f* COMMUT, COMPOSANTS selecting bar; ~ **numérique** COMMUT numerical horizontal, numerical selecting bar
horloge *f* GÉNÉRAL clock; ~ **asservie** ÉLECTRON slave clock; ~ **atomique** RADIO atomic standard clock; ~ **centrale** ÉLECTRON master clock; ~ **chien de garde** ÉLECTRON watchdog timer; ~ **coordonnée** RADIO coordinate clock; ~ **électronique** ÉLECTRON electronic clock; ~ **externe** INFORMAT external clock; ~ **interne** PÉRIPH internal clock; ~ **maîtresse** ÉLECTRON master clock; ~ **mère** ÉLECTRON master clock; ~ **de modem** DONNÉES modem clock; ~ **parlante** TÉL speaking clock; ~ **pilote** ÉLECTRON master clock; ~ **primaire** RADIO primary clock; ~ **principale** ÉLECTRON master clock; ~ **de référence** TRANSMIS reference clock; ~ **de synchronisation** INFORMAT sync clock, synch clock, synchronization clock; ~ **terrestre** TRANSMIS terrestrial standard clock
horodatage *m* PÉRIPH date-time stamping; ~ **vocal** PÉRIPH vocal date-time stamping
horodateur *m* COMMUT, INFORMAT, TESTS time stamp
hors: ~ **axe** *loc* RADIO off-axis; ~ **bande** *loc* ÉLECTRON, RADIO out-of-band (*OOB*); ~ **circuit** *loc* ALIM ÉLEC dead; ~ **code** *loc* DONNÉES shift-out; ~ **émission** *loc* RADIO off-air; ~ **faisceau** *loc* RADIO off-beam; ~ **ligne** *loc* DONNÉES, INFORMAT, TÉL off-line; ~ **portée** *loc* RADIO out-of-range, out-of-reach; ~ **service** *loc* ALIM ÉLEC out-of-service, TÉL busied out
hostile *adj* INFORMAT unfriendly
hôte *m* COMMUT, DONNÉES, ÉLECTRON, INFORMAT, RADIO host; ~ **de chargement** ÉLECTRON load host
HT *abrév* (*haute tension*) ALIM ÉLEC high tension, HV (*high voltage*), ÉLECTRON HV (*high voltage*)
HTML *abrév* (*hypertext markup language*) DONNÉES, INFORMAT *Internet* HTML (*hypertext markup language*)
hululement *m* GÉNÉRAL warble
hululer *vi* GÉNÉRAL warble
humidifuge *adj* GÉNÉRAL moisture-proof
humidité *f* GÉNÉRAL moisture
hurler *vi* TÉL, TRANSMIS howl
hurleur *m* TÉL, TRANSMIS howler
hybride *m* GÉNÉRAL hybrid
hydrofuge *adj* COMPOSANTS water-repellent
hydrofuger *vt* COMPOSANTS waterproof
hydrométéores *m pl* RADIO hydrometeors
hydrophone *m* PÉRIPH hydrophone
hyperbook *m* INFORMAT hyperbook
hypercours *m* INFORMAT hypercourse
hyperdata *m* INFORMAT hyperdata
hyperdocument *m* INFORMAT hyperdocument
hyperfréquence *f* RADIO, TV microwave frequency, superhigh frequency (*SHF*)
hypermédia *adj* DONNÉES, INFORMAT *intermédia* hypermedia
hyperobjet *m* INFORMAT hyperobject
hyperpiste *f* INFORMAT hypertrail
hypertexte *m* DONNÉES, INFORMAT hypertext
hyperviseur *m* TÉL hypervisor
hypsomètre *m* TESTS level meter, TRANSMIS level measuring set
hystérésis *f* ÉLECTRON hysteresis; ~ **magnétique** ÉLECTRON magnetic hysteresis
Hz *abrév* (*hertz*) ÉLECTRON, TESTS, TRANSMIS Hz (*hertz*)

I

I *abrév* (*implémenté*) ÉLECTRON, INFORMAT I (*implemented*)

IA *abrév* APP SERV, COMMUT, INFORMAT (*intelligence artificielle*) AI (*artificial intelligence*), TÉL (*interurbain automatique*) DDD (*AmE*) (*direct distance dialing*), STD (*BrE*) (*subscriber trunk dialling*)

IAA *abrév* (*identification des appels entrants*) DONNÉES, RADIO, TÉL ICI (*incoming-call identification*)

IAD *abrév* (*identification automatique du demandeur*) DONNÉES, RADIO, TÉL AIOD (*automatic identification of outward dialing AmE, automatic identification of outward dialling BrE*)

IAO *abrév* (*ingénierie assistée par ordinateur*) INFORMAT CAE (*computer-aided engineering*)

IBN *abrév* (*Institut belge de normalisation*) GÉNÉRAL BSI (*Belgian Standards Institute*)

IBO *abrév* (*recul d'entrée*) RADIO IBO (*input back-off*)

IBPT *abrév* (*Institut belge des services postaux et des télécommunications*) RADIO BIPT (*Belgian Institute for Post and Telephony*)

IC *abrév* TÉL (*indice de charge*) LN (*load number*), (*compteur des trames I*) IC (*I-frame counter*)

ICB *abrév* (*interdiction d'accès des appels entrants, interdiction des appels à l'arrivée*) DONNÉES, RADIO, TÉL incoming-call barring

icône *f* DONNÉES, INFORMAT icon; **~ grisée** INFORMAT dimmed icon; **~ d'images animées** INFORMAT pictorial icon (*picon*); **~ de liaison** INFORMAT link icon

iconoscope: **~ image** *m* ÉLECTRON image iconoscope

Id *abrév* (*identificateur*) GÉNÉRAL ID (*identifier*)

ID *abrév* (*identification*) GÉNÉRAL ID (*identification*)

IDA *abrév* (*accès numérique intégré*) DONNÉES IDA (*integrated digital access*)

identificateur *m* (*Id*) GÉNÉRAL identifier (*ID*); **~ d'appel** TÉL call identifier; **~ d'arbre** DONNÉES, INFORMAT tree identifier; **~ d'autorité et de format** DONNÉES, SIGNAL authority and format identifier (*AFI*); **~ de branche d'une transaction** DONNÉES transaction branch identifier; **~ de canal virtuel** DONNÉES, TRANSMIS virtual-channel identifier (*VCI*); **~ de commande** TÉL command identifier; **~ de communication** TÉL call identifier; **~ de conduit virtuel** DONNÉES, TRANSMIS virtual-path identifier (*VPI*); **~ de connexion de liaison de données** DONNÉES data-link connection identifier (*DLCI*); **~ de dépôt** DONNÉES, TÉL submission identifier; **~ de destination** DONNÉES, TÉL destination identifier; **~ de domaine initial** (*IDI*) DONNÉES, TÉL initial domain identifier (*IDI*); **~ d'enregistrement** DONNÉES registration identifier; **~ d'extrémité de connexion N** DONNÉES, TÉL N-connection end point identifier (*N-CEPI*); **~ d'extrémité de connexion de transport** DONNÉES, TÉL transport-connection end point identifier; **~ de format** (*FI*) ALIM ÉLEC, TÉL format identifier (*FI*); **~ de groupe** TÉL group identifier (*GI*); **~ de groupe de paramètres** INFORMAT parameter group identifier (*PGI*); **~ d'invocation d'entité d'application** DONNÉES application-entity invocation identifier; **~ d'invocation de processus d'application** DONNÉES application process invocation identifier; **~ de modem** DONNÉES data set identifier (*DSI*); **~ de module de test** TESTS test step identifier; **~ d'objet** DONNÉES object identifier; **~ d'origine** TÉL source identifier; **~ par défaut** DONNÉES default identifier; **~ de paramètre** TÉL parameter identifier (*PI*); **~ de plan de numérotage** DONNÉES, TÉL numbering plan identifier; **~ de point d'accès au réseau** DONNÉES service access point identifier (*SAPI*); **~ de point d'extrémité de terminal** DONNÉES terminal end-point identifier (*TEI*); **~ de protocole** DONNÉES protocol identifier (*PRT-ID*); **~ de réponse** TÉL response identifier (*RI*); **~ de syntaxe** DONNÉES, TÉL syntax identifier; **~ de test élémentaire** TESTS test case identifier; **~ de trajet virtuel** DONNÉES, TRANSMIS virtual-path identifier (*VPI*); **~ de transaction** DONNÉES, TÉL transaction identifier; **~ d'utilisateur du service MCS** DONNÉES, TÉL MCS user ID; **~ de voie virtuelle** DONNÉES, TRANSMIS virtual-channel identifier (*VCI*)

identification *f* (*ID*) GÉNÉRAL identification (*ID*); **~ de l'abonné appelant** (*CIG*) DONNÉES, TÉL calling-subscriber identification (*CIG*); **~ de l'abonné appelé** DONNÉES, TÉL called-subscriber identification (*CSI*); **~ de l'abonné demandé** DONNÉES, TÉL called-subscriber identification (*CSI*); **~ de l'abonné demandé-DIS** (*CSI-DIS*) DONNÉES, TÉL called-subscriber identification-DIS (*CSI-DIS*); **~ de l'abonné demandeur-DTC** (*CIG-DTC*) DONNÉES, TÉL calling-subscriber identification-DTC (*CIG-DTC*); **~ de l'abonné émetteur-DCS** (*TCI-DCS*) DONNÉES, TÉL transmitting subscriber identification-DCS (*TCI-DCS*); **~ ami-ennemi** DONNÉES, RADIO, TÉL identification friend or foe (*IFF*); **~ d'appel** DONNÉES, TÉL call identification, call trace; **~ de l'appel** TÉL calling-line identification; **~ d'appel malveillant** DONNÉES, TÉL malicious call identification; **~ des appels entrants** (*IAE*) DONNÉES, RADIO, TÉL incoming-call identification (*ICI*); **~ automatique** DONNÉES answer-back; **~ automatique du demandeur** (*IAD*) DONNÉES, RADIO, TÉL automatic identification of outward dialing (*AmE*) (*AIOD*), automatic identification of outward dialling (*BrE*) (*AIOD*); **~ de cible** RADIO target identification; **~ de couloir** RADIO lane identification; **~ d'une défaillance** TÉL failure recognition; **~ d'un dérangement** TÉL failure recognition; **~ du destinataire de l'échange** DONNÉES, TÉL interchange receiver identification; **~ d'échange** TÉL exchange identification; **~ d'empreinte spectrale** SIGNAL signal identification fingerprinting; **~ de l'expéditeur** DONNÉES, TÉL sender identification; **~ de l'expéditeur de l'échange** DONNÉES, TÉL interchange sender identification; **~ de fonction** TÉL function identification; **~ de la ligne appelante** (*ILA*) TÉL calling-line identification (*CLI*); **~ de la ligne appelée** TÉL called-line identification (*CDLI*); **~ de la ligne connectée** TÉL connected-line identification (*COLI*); **~ de la ligne demandée** TÉL called-line identification (*CDLI*); **~ de la ligne du demandeur** TÉL calling-line identification facility; **~ de multiplexage** TRANSMIS multiplexing identification (*MID*); **~ du numéro appelant** DONNÉES, RADIO, TÉL calling-number identification; **~ par défaut** DONNÉES default identification; **~ de paramètre** TÉL parameter identification (*PI*); **~ du plan de numérotage** DONNÉES, TÉL number plan identification (*NPI*), numbering plan identification; **~ du poste demandeur** TÉL calling station; **~ du réseau de transit** TRANSMIS transit network identification; **~ du risque** TESTS risk identification; **~ de service** APP SERV service identification; **~ du type ou du mode de la communication** TÉL mode or type of communication identification; **~ de l'usager du réseau** DONNÉES, TÉL network user identification (*NUI*); **~ de la voix** APP SERV, CONTRÔLE voice identification; **~ de zone de localisation** RADIO location area identification

identité: ~ **d'abonné mobile** *f* RADIO mobile subscriber identity; ~ **d'appel** *f* DONNÉES, TÉL call identity; ~ **d'appel de groupe** *f* DONNÉES, TÉL group call identity; ~ **de la ligne appelée** *f* DONNÉES, TÉL called-line identity; ~ **de la ligne du demandeur** *f* DONNÉES, TÉL calling-line identity; ~ **du réseau** *f* TÉL network identity; ~ **de réseau de transit d'EPR** *f* TÉL RPOA transit-network identity; ~ **de sélection** *f* TÉL selection identity; ~ **de la station côtière** *f* DONNÉES, RADIO, TÉL coast-station identity; ~ **de la station côtière terrienne** *f* DONNÉES, RADIO, TÉL coast-shore station identity; ~ **de la station demandée** *f* RADIO, TÉL called-station identity (*CSI*); ~ **de station de navire** *f* DONNÉES, RADIO, TÉL ship station identity; ~ **temporaire d'abonné mobile** *f* DONNÉES, RADIO, TÉL temporary mobile station identity

IDI *abrév* (*identificateur de domaine initial*) DONNÉES, TÉL IDI (*initial domain identifier*)

IEEE *abrév* (*Institut des ingénieurs électriciens et électroniciens*) ÉLECTRON IEEE (*Institute of Electrical and Electronics Engineers*)

IENT *abrév* (*Institut européen de normalisation en télécommunications*) GÉNÉRAL ETSI (*European Telecommunications Standards Institute*)

IFL *abrév* (*circuit entre installations de raccordement*) TRANSMIS IFL (*interfacility link*)

IG *abrév* (*gain de concentration*) TRANSMIS IG (*interpolation gain*)

IGD *abrév* (*message d'interrogation de groupe de circuits*) COMMUT CQM (*circuit-group query message*)

IGES *abrév* INFORMAT IGES (*initial graphics exchange specification*)

ignifuge *adj* COMPOSANTS fireproof, fire-resistant

ignitron *m* ÉLECTRON ignitron

ignorer *vt* INFORMAT bypass

IGR *abrév* (*message de réponse à une interrogation de groupe de circuits*) COMMUT CQR (*circuit-group query response message*)

ILA *abrév* (*identification de la ligne appelante*) DONNÉES, TÉL CLI (*calling-line identification*)

illimité *adj* GÉNÉRAL unrestricted

illumination: ~ **de l'ouverture** *f* RADIO aperture illumination

image *f* GÉNÉRAL, INFORMAT, TÉL, TV picture, image, frame; ~ **animée** GÉNÉRAL moving picture; ~**-b** (*image à codage prédictif bidirectionnel*) INFORMAT, TV b-picture (*bidirectionally-predictive coded picture*); ~**-b bi-trame** INFORMAT, TV b-frame picture; ~**-b monotrame** INFORMAT, TV b-field picture; ~ **binaire** DONNÉES binary image; ~ **brute d'indicateur** RADIO raw video display; ~ **de carte électronique** RADIO, TV video map; ~ **clé** INFORMAT keyframe; ~ **à codage intratrame** (*image-i*) INFORMAT, TV intracoded picture (*i-picture*); ~ **à codage prédictif bidirectionnel** (*image-b*) INFORMAT, TV bidirectionally-predictive coded picture (*b-picture*); ~ **codée MPEG** INFORMAT coded MPEG frame; ~ **en couleur** TÉL color image (*AmE*), colour image (*BrE*); ~ **dans l'image** INFORMAT inset picture, picture in picture (*PIP*); ~ **à deux niveaux** TÉL bi-level image; ~ **digitalisée** DONNÉES, TV digitized image; ~ **d'écran** INFORMAT screen image; ~ **fantôme** TV ghost; ~ **fixe** INFORMAT, TV still frame, stop frame; ~ **en fondu enchaîné** INFORMAT dissolved image; ~ **à haute définition** TV high-definition image; ~**-i** (*image à codage intratrame*) INFORMAT, TV i-picture (*intracoded picture*); ~**-i bi-trame** INFORMAT, TV i-frame picture; ~**-i monotrame** INFORMAT, TV i-field picture; ~ **incrustée** INFORMAT inset picture; ~ **intermédiaire** TV intermediate frame; ~ **limitée à quelques niveaux** TÉL limited multilevel image; ~ **mémoire** INFORMAT core image; ~ **mémoire exécutable** TÉL executable memory image; ~ **négative** INFORMAT negative effect; ~ **numérique fixe** INFORMAT digital still image; ~ **numérisée** TV digitized image; ~ **optique** TV optical image; ~ **orthicon** ÉLECTRON image orthicon; ~**-p** INFORMAT, TV p-picture (*predictive-coded picture*); ~ **à plusieurs niveaux** TÉL multilevel image; ~ **pour présentation visuelle** INFORMAT, TV soft-copy image; ~ **reconstruite** INFORMAT reconstructed picture; ~ **de référence** INFORMAT, TV reference picture; ~ **de référence antérieure** INFORMAT, TV past reference picture; ~ **de référence future** INFORMAT, TV future reference picture; ~ **en relief** TV three-dimensional image (*3D image*); ~ **de réseau** COMMUT, DONNÉES, TÉL network map; ~ **scannérisée** PÉRIPH scanned image; ~ **sortie sur support papier** TÉL hard-copy image; ~ **de synthèse** INFORMAT computer-generated image; ~ **de synthèse** DONNÉES, TV synthetic image; ~ **au trait** DONNÉES line drawing; ~ **tridimensionnelle** TV three-dimensional image (*3D image*); ~ **en trois dimensions** TV three-dimensional image (*3D image*); ~ **vidéo** INFORMAT, TV soft copy; ~ **virtuelle** DONNÉES, TV virtual image

imagette *f* INFORMAT inset

imageur/capteur: ~ **micro-ondes spécial** *m* ÉLECTRON special sensing microwave imager (*SSMI*)

imbriquer *vt* GÉNÉRAL, INFORMAT interleave, nest, TESTS nest

imitation *f* ÉLECTRON imitative deception

immobilisation *f* INFORMAT standstill

immunité: ~ **au brouillage** *f* TRANSMIS resilience to interference; ~ **au bruit** *f* TRANSMIS immunity to noise

IMP *abrév* GÉNÉRAL (*intermodulation passive*) PIM (*passive intermodulation*), DONNÉES, INFORMAT (*serveur de messages*) IMP (*interface message processor*)

impédance *f* GÉNÉRAL impedance; ~ **caractéristique** ÉLECTRON matching impedance; ~ **de charge** ÉLECTRON load impedance; ~ **en circuit ouvert** ALIM ÉLEC open-circuit impedance; ~ **en court-circuit** ALIM ÉLEC short-circuit impedance; ~ **effective d'entrée** ALIM ÉLEC effective input impedance; ~ **effective de sortie** ALIM ÉLEC effective output impedance; ~ **d'électrode** ÉLECTRON electrode impedance; ~ **d'entrée** ÉLECTRON, INFORMAT input impedance; ~ **d'entrée d'antenne** RADIO antenna input impedance; ~ **d'image** ÉLECTRON image impedance; ~ **inverse** ÉLECTRON inverse impedance; ~ **mutuelle** ÉLECTRON mutual impedance; ~ **d'onde** ÉLECTRON, RADIO wave impedance; ~ **d'onde caractéristique dans un guide** ÉLECTRON, RADIO guide characteristic wave impedance; ~ **série** ALIM ÉLEC series impedance; ~ **de sortie** ÉLECTRON, INFORMAT output impedance; ~ **de la source** ALIM ÉLEC source impedance; ~ **de terminaison** ÉLECTRON terminating impedance; ~ **terminale** ÉLECTRON terminal impedance; ~ **de transfert** ÉLECTRON transfer impedance

imperméable *adj* COMPOSANTS rainproof, waterproof, watertight

implantation *f* ÉLECTRON implantation, RADIO location; ~ **ionique** ÉLECTRON ion implantation

implanté *adj* ÉLECTRON implanted

implanteur *m* ÉLECTRON implanter

implémentation *f* ÉLECTRON, GÉNÉRAL, INFORMAT, TÉL implementation

implémenté *adj* (*I*) ÉLECTRON, INFORMAT implemented (*I*)

implémenter *vt* ÉLECTRON, GÉNÉRAL, INFORMAT, TÉL implement

implicite *adj* GÉNÉRAL implicit

import *m* DONNÉES import

importé *adj* DONNÉES imported

importer *vt* DONNÉES import

imposition: ~ **d'un seuil** *f* ALIM ÉLEC, ÉLECTRON thresholding

imprécision *f* TESTS inaccuracy

imprégner *vt* GÉNÉRAL impregnate

impression *f* ÉLECTRON *motif* pattern, GÉNÉRAL, INFORMAT printing, PÉRIPH print; ~ **à distance** INFORMAT remote printing; ~ **de journaux à distance** PÉRIPH teleprinting; ~ **laque thermographique** ÉLECTRON printing track area;

~ **laser** GÉNÉRAL laser printing; ~ **ligne par ligne** INFORMAT, TÉL line printing; ~ **multiple** ÉLECTRON multiple pattern; ~ **négative** ÉLECTRON negative pattern; ~ **non conductrice** ÉLECTRON nonconductive pattern

imprimante *f* DONNÉES, INFORMAT, PÉRIPH printer; ~ **à changement de phase** DONNÉES, PÉRIPH phase-change printer; ~ **à déposition d'ions** PÉRIPH ion deposition printer; ~ **à diodes électroluminescentes** DONNÉES, PÉRIPH LED printer; ~ **électrographique** PÉRIPH electrographic printer; ~ **électrostatique** PÉRIPH electrostatic printer; ~ **d'étiquettes** DONNÉES, PÉRIPH label printer; ~ **à fonctions multiples** PÉRIPH multifunction printer; ~ **à impact** DONNÉES, INFORMAT, PÉRIPH impact printer; ~ **à jet d'encre** DONNÉES, INFORMAT, PÉRIPH inkjet printer; ~ **laser** APP SERV, DONNÉES, INFORMAT, PÉRIPH laser printer; ~ **ligne** DONNÉES, INFORMAT, PÉRIPH line printer; ~ **ligne par ligne** DONNÉES, INFORMAT, PÉRIPH line printer; ~ **matricielle** INFORMAT, PÉRIPH matrix printer; ~ **matricielle par points** DONNÉES, INFORMAT, PÉRIPH dot matrix printer; ~ **à obturateur à cristal liquide** INFORMAT LCS printer; ~ **par ligne** DONNÉES, INFORMAT, PÉRIPH line printer; ~ **par page** DONNÉES, INFORMAT, PÉRIPH, TÉL page printer; ~ **portable** DONNÉES, INFORMAT, PÉRIPH portable printer; ~ **sans impact** DONNÉES, PÉRIPH, TÉL nonimpact printer; ~ **sérielle** DONNÉES, PÉRIPH serial printer; ~ **thermique** DONNÉES, PÉRIPH thermal printer; ~ **à la volée** TÉL on-the-fly printer

imprimer *vt* INFORMAT print out, PÉRIPH print, print out

improvisation *f* INFORMAT vamp

imp/s *abrév* (*impulsions par seconde*) RADIO, TÉL pps (*pulses per second*)

impulsion *f* GÉNÉRAL pulse, impulse; ~ **d'activation** ÉLECTRON strobe pulse; ~ **aléatoire** ÉLECTRON random pulse; ~ **de base de comptage** TÉL basic metering pulse; ~ **bipolaire** TÉL bipolar pulse; ~ **de commande** CONTRÔLE set pulse, *commande automatique* control pulse; ~ **de comptage** APP SERV charging impulse, TÉL meter pulse; ~ **cosinus** TRANSMIS cosine pulse; ~ **en cosinus carré** TRANSMIS cosine-squared pulse; ~ **de déblocage** ÉLECTRON gate pulse; ~ **décimale** TÉL decimal pulse (*DP*), dial pulse; ~ **de déclenchement** ÉLECTRON gate pulse, trigger pulse, triggering lead pulse; ~ **de demi-ligne** TV half-line pulse; ~ **en dents de scie** ÉLECTRON sawtooth pulse; ~ **d'échantillonnage** TESTS sample pulse; ~ **d'écho** RADIO reflected pulse; ~ **électromagnétique** ÉLECTRON electromagnetic pulse; ~ **d'encadrement** RADIO framing pulse; ~ **d'entrée** ÉLECTRON incoming surge, INFORMAT input pulse; ~ **d'erreur** DONNÉES error pulse; ~ **de fermeture** TÉL make pulse; ~ **gaussienne** TÉL, TRANSMIS Gaussian pulse; ~ **d'horloge** ÉLECTRON timing pulse, clock pulse, TÉL timing pulse; ~ **d'horloge maîtresse** ÉLECTRON master clock pulse; ~ **d'information** RADIO information pulse; ~ **d'inhibition** ÉLECTRON inhibiting pulse; ~ **d'interrogation** RADIO interrogating pulse; ~ **d'invalidation** ÉLECTRON inhibiting pulse; ~ **inverse** COMMUT revertible pulse; ~ **lumineuse** TV light pulse; ~ **de mise à un** CONTRÔLE set pulse; ~ **négative** ÉLECTRON negative pulse; ~ **non récurrente** GÉNÉRAL nonrecurring pulse; ~ **numérique** INFORMAT digital impulse; ~ **d'oscillation** ÉLECTRON, INFORMAT, TÉL sine-wave pulse; ~ **d'ouverture** TÉL break pulse; ~ **parasite** ÉLECTRON interference pulse; ~ **à plateau** GÉNÉRAL flat-topped pulse; ~ **rectangulaire** TÉL rectangular pulse; ~ **de remise à l'état initial** ÉLECTRON resetting pulse, restoring pulse; ~ **de remise à zéro** ÉLECTRON resetting pulse; ~ **de repérage de cap** RADIO course blip pulse; ~ **de restauration** ÉLECTRON resetting pulse, restoring pulse; ~ **de rétablissement** ÉLECTRON resetting pulse, restoring pulse; ~ **en sinus carré** ÉLECTRON, TÉL sine-squared pulse; ~ **sinusoïdale** ÉLECTRON, INFORMAT, TÉL sine-wave pulse; ~ **de sortie** INFORMAT output pulse; ~ **de strobage progressive** RADIO walking strobe pulse; ~ **stroboscopique** ÉLECTRON strobe pulse; ~ **de suppression du faisceau** TÉL *vidéo* blanking pulse; ~ **de synchronisation** ÉLECTRON timing pulse, GÉNÉRAL synchronization pulse, TÉL time measurement pulse, timing pulse; ~ **de transfert** INFORMAT strobe; ~ **triangulaire** ÉLECTRON triangular pulse; ~ **unidirectionnelle** TRANSMIS unidirectional pulse; ~ **unipolaire** TÉL unipolar pulse; ~ **de validation** COMPOSANTS enable pulse; ~ **de violation** TRANSMIS violation pulse

impulsions: ~ **par seconde** *loc* (*imp/s*) RADIO, TÉL pulses per second (*pps*)

impulsographe *m* TÉL, TRANSMIS pulse recorder

impureté *f* ÉLECTRON impurity

imputabilité *f* DONNÉES, GÉNÉRAL accountability

imputé *adj* GÉNÉRAL deducted

imputer *vt* APP SERV charge; ~ **l'appel à** APP SERV charge the call to

IN *abrév* (*indice de netteté*) RADIO, TÉL, TRANSMIS AI (*articulation index*)

inactif *adj* RADIO, SIGNAL, TÉL, TRANSMIS inactive

inactivité *m* GÉNÉRAL idleness

inauguration *f* RADIO inauguration

inaugurer *vt* GÉNÉRAL cut over, RADIO inaugurate

incapaciter *vt* INFORMAT disable

incertitude: ~ **de mesure** *f* TESTS uncertainty of measurement

incident *m* INFORMAT failure, glitch; ~ **de fonctionnement** TESTS malfunction, service incident, service irregularity

inclinaison *f* INFORMAT slant, RADIO inclination, tilt; ~ **axiale** TRANSMIS *optique de fibre* angular displacement; ~ **de caractères** APP SERV character skew; ~ **de l'onde de sol** RADIO wave tilt

INCLUSIF: ~ **OU** *f* INFORMAT, ÉLECTRON INCLUSIVE-OR

incohérence *f* INFORMAT inconsistency, TRANSMIS incoherence

incorporation *f* DONNÉES, ÉLECTRON, INFORMAT embedding

incrustation: ~**-couleur** *f* INFORMAT color separation overlay (*AmE*), colour separation overlay (*BrE*); ~ **douce** *f* INFORMAT color matting (*AmE*), colour matting (*BrE*); ~ **linéaire** *f* INFORMAT color matting (*AmE*), colour matting (*BrE*)

ind *abrév* (*indication*) GÉNÉRAL ind (*indication*)

IND *abrév* (*message de demande d'information*) DONNÉES, TÉL INR (*information request message*)

indépendance: ~ **du code utilisé** *f* TRANSMIS code transparency; ~ **des données** *f* DONNÉES, INFORMAT data independence; ~ **de la séquence des bits** *f* INFORMAT bit sequence independence

indépendant: ~ **de la charge** *adj* ALIM ÉLEC load-independent

index: ~ **de modulation optique** *m* TRANSMIS optical modulation index (*OMI*)

indexation *f* GÉNÉRAL, INFORMAT indexing

indicateur *m* GÉNÉRAL indicator, marker, PÉRIPH marker, RADIO display, SIGNAL indicator, meter, TÉL indicator, needle, pointer; ~ **d'abandon de bloc** DONNÉES block-dropping indicator; ~ **d'accusé de réception** TÉL acknowledgement indicator; ~ **d'acheminement détourné** TÉL alternative routing indicator; ~ **d'adresse de réacheminement** COMMUT, TÉL redirection address indicator; ~ **d'aiguillage** COMPOSANTS switch indicator; ~ **d'amplitude et d'azimut** RADIO range-bearing display; ~ **d'amplitude et de distance** RADIO range-amplitude display; ~ **d'appel** PÉRIPH calling indicator, TÉL call-indicating device, call indicator, ringing pilot indicator; ~ **d'appel d'essai** TESTS *système de signalisation R2* test call indicator; ~ **d'appel de groupe fermé d'usagers** TÉL closed user group call indicator; ~ **d'appel de groupe fermé d'usagers bilatéral** TÉL bilateral closed user group call indicator; ~ **d'appel réacheminé** COMMUT, TÉL redirected call indicator; ~ **d'arrêt** INFORMAT halt indicator; ~ **avec/sans**

taxation TÉL charge/no charge indicator; ~ **de battement** INFORMAT beat indicator; ~ **de bruit** TÉL noise indicator; ~ **de catégorie du demandeur** TÉL calling-party category indicator; ~ **de catégorie d'usagers** TÉL user class indicator; ~ **de choix de l'EPR** TÉL RPOA selection indicator; ~ **de commande de protocole** DONNÉES protocol control indicator; ~ **de communication nationale/internationale** TÉL national/international call indicator; ~ **de contrôle** INFORMAT check indicator, TV screening indicator (*SI*); ~ **de crédit** TÉL credit indicator; ~ **de crête** TESTS peak reading instrument; ~ **de débit** TÉL flow indicator; ~ **de demande** TÉL request indicator; ~ **de demande de catégorie du demandeur** TÉL calling-party category request indicator; ~ **de demande d'identification de la ligne du demandé** TÉL called-line identification request indicator; ~ **de demande d'identification du réseau d'origine** TÉL originating network identification request indicator; ~ **de demande d'identité de la ligne connectée** TÉL connected-line identity request indicator; ~ **de demande de taxation à l'arrivée** APP SERV, TÉL collect-calling request indicator (*AmE*), reverse-charging request indicator (*BrE*); ~ **de destination** TÉL destination indicator; ~ **de distance et gisement** RADIO type B display; ~ **de distance et de site** RADIO type E display; ~ **de domaine** TÉL field indicator; ~ **d'encombrement** RADIO congestion indicator; ~ **d'entrée** INFORMAT input indicator; ~ **d'essai** DONNÉES, TESTS test indicator; ~ **d'état** INFORMAT status flag; ~ **d'état de circuit** COMMUT circuit state indicator; ~ **de fin** TÉL clearing indicator; ~ **de force de réception** SIGNAL signal strength meter; ~ **de hauteur** RADIO height indicator; ~ **de hauteur et de distance** RADIO height-position indicator, range-height indicator; ~ **d'identité de la ligne appelée** TÉL called-line identity indicator; ~ **d'information de bout en bout** DONNÉES end-to-end information indicator; ~ **d'information demandée** DONNÉES solicited information indicator; ~ **d'interruption** INFORMAT interrupt indicator; ~ **de IPD/CIRD** SIGNAL DCC/DNIC indicator; ~ **de libération** DONNÉES, SIGNAL release indicator; ~ **de lien** INFORMAT link marker; ~ **de longueur** DONNÉES length indicator (*LI*); ~ **de longueur de domaine** TÉL field length indicator; ~ **de messages reçus** PÉRIPH message received indicator; ~ **de mesure** INFORMAT beat indicator; ~ **de méthode de bout en bout** DONNÉES end-to-end method indicator; ~ **de méthode SSCS** TÉL SCCP method indicator; ~ **de modification** DONNÉES modification indicator; ~ **national** TÉL national indicator; ~ **de la nature du circuit** TÉL nature-of-circuit indicator; ~ **de niveau de liquide** TÉL liquid-level indicator; ~ **de nouvelles données** DONNÉES new data flag (*NDF*); ~ **d'ordre des phases** ALIM ÉLEC phase-sequence indicator; ~ **d'origine** TÉL origin indicator; ~ **panoramique** RADIO plan-position indicator (*PPI*); ~ **de plan de numérotage** DONNÉES, TÉL numbering plan indicator; ~ **de plan de numérotage/TOA** DONNÉES, TÉL numbering plan indicator/TOA (*NPI/TOA*); ~ **de plan de numérotage/type d'adresse** DONNÉES, TÉL numbering and addressing plan indicator/type of address (*NAPI/TOA*); ~ **de polarité** ALIM ÉLEC polarity indicator; ~ **de présentation** DONNÉES presentation indicator (*PI*); ~ **prêt** TÉL ready indicator; ~ **de prises** TÉL seizure indicator; ~ **de prolongation d'appel possible** DONNÉES call-forwarding-may-occur indicator; ~ **de raison du réacheminement initial** TÉL original redirection reason indicator; ~ **de réacheminement** COMMUT, TÉL redirecting indicator; ~ **de recommandation** DONNÉES recommendation indicator; ~ **de rejet par le réseau** TESTS network discard indicator; ~ **de réponse à une demande de catégorie du demandeur** TÉL calling-party category response indicator; ~ **de restriction de divulgation d'événement** DONNÉES event presentation restriction indicator; ~ **de saturation** INFORMAT overflow indicator; ~ **sectoriel** RADIO sector display; ~ **de segmentation permise** DONNÉES segmentation permitted flag (*SPF*); ~ **de service** TÉL service indicator; ~ **de service supplémentaire** GÉNÉRAL facility indicator; ~ **de service utilisateur** TÉL service indicator; ~ **de signal** SIGNAL, TRANSMIS signal indicator; ~ **de signalisation d'usager à usager** DONNÉES user-to-user indicator; ~ **de sonnerie** TÉL ring indicator; ~ **de sortie** INFORMAT output indicator; ~ **de suppresseur d'écho** TÉL echo-suppressor indicator; ~ **de test** DONNÉES, TESTS test indicator; ~ **type A** RADIO type A display; ~ **visuel** TÉL star indicator; ~ **de vitesse** RADIO speed indicator, velocity indicator; ~ **de vitesse azimut** RADIO *radar* azimuth speed indicator

indicateurs: ~ **d'indicatifs de pays** *m pl* TÉL *système de signalisation R2* country-code indicators

indicatif *m* COMMUT code, DONNÉES answer-back code, code, GÉNÉRAL call letters, RADIO sign, SIGNAL code, TÉL number, answer-back, code, sign, *téléimprimeur* answer-back code; ~ **d'accès** COMMUT, TÉL service code; ~ **d'accès au standardiste** TÉL attendant access code; ~ **d'acheminement** TÉL routing code, routing number; ~ **à acheminer** TRANSMIS code to be routed; ~ **d'appel** GÉNÉRAL call letters, RADIO call sign, TÉL access code, call sign; ~ **interurbain** TÉL trunk code (*TC*); ~ **local** TÉL local index; ~ **national de destination** DONNÉES, TÉL national destination code (*NDC*); ~ **de numérotation** TÉL dialing code (*AmE*), dialling code (*BrE*); ~ **de pays** TÉL country code (*CC*), national code; ~ **de pays pour la transmission de données** DONNÉES data country code (*DCC*); ~ **régional** TÉL area code, numbering plan area (*NPA*); ~ **de service restreint** TÉL barred code; ~ **téléphonique de pays** TÉL telephone country code (*TCC*); ~ **de zone** TÉL area code

indication *f* (*ind*) GÉNÉRAL advice, information, display, indication (*ind*); ~ **d'alarme distante** TESTS remote alarm indication (*RAI*); ~ **d'alarme de retour sélective** RADIO selective back alarm indication; ~ **d'appel en attente** TÉL call-waiting indication; ~ **d'appel sans réponse** PÉRIPH, TÉL unanswered call indication; ~ **d'appels en instance** TÉL call waiting; ~ **d'attente de message** TÉL message waiting indication; ~ **audible** SIGNAL audible indication; ~ **audio active** TÉL audio indicate active; ~ **audio muette** TÉL audio indicate muted (*AIM*); ~ **automatique verbale des éléments de taxation** TÉL automatic ADC; ~ **de blocage** TÉL blocking indication; ~ **de compteur** TÉL meter indication; ~ **de concaténation** ÉLECTRON concatenation indication (*CI*); ~ **de coût** APP SERV charge advice, TÉL incremental charge advice; ~ **de coût en temps réel** APP SERV call-in-progress cost information; ~ **de déconnexion de réseau** DONNÉES N-disconnect indication; ~ **de défaillance** TÉL failure indication, fault indication; ~ **de défaillance en amont** TÉL upstream failure indication; ~ **de délai en transit** DONNÉES transit-delay indication (*TDI*); ~ **de demande de bloc de paramètres** TÉL parameter block request indication; ~ **à distance** DONNÉES remote indication, *télémétrie* remote sensing, remote reading, GÉNÉRAL remote reading, remote sensing, TESTS remote sensing; ~ **d'échappement** TÉL escape indication; ~ **d'état** INFORMAT, TESTS status indication; ~ **d'état de la console** TÉL attendant-status indicator; ~ **de l'état du poste** TÉL business set query busy station; ~ **de libération par l'ETCD** TÉL DCE clear indication; ~ **de la mesure** INFORMAT time signature; ~ **du mouvement absolu** RADIO true-motion display; ~ **multipoint de boucle** DONNÉES multipoint indication loop (*MIL*); ~ **de non-remise de datagramme** DONNÉES, TÉL datagram non-delivery indication; ~ **optique** INFORMAT, TÉL, TESTS visual indication; ~ **de panne** TÉL failure indication, fault indication; ~ **de pointeur zéro** DONNÉES null pointer indication (*NPI*); ~ **prêt** TÉL ready indication; ~ **primitive** DONNÉES primitive indication; ~ **de priorité** TÉL priority marking; ~ **pour rappel** TÉL reminder indication, reminder marking; **~ de réponse** TÉL answer indication; **~ de valeur de compteur** TÉL meter value indication; **~ verbale des éléments de taxation** TÉL advise duration and charge; **~ vidéo active** TV VIA (*video indicate active*); **~ vidéo prête à être activée** TV video indicate ready-to-activate (*VIR*); **~ vidéo supprimée** TV video indicate suppressed (*VIS*)

indice *m* DONNÉES subscript, ÉLECTRON figure, incidence, GÉNÉRAL index, INFORMAT subscript, RADIO figure, TÉL, TESTS, TRANSMIS rating; **~ de charge** (*IC*) TÉL load number (*LN*); **~ de correspondance spectrale** COMMUT pattern correspondence index (*PCI*); **~ d'état** GÉNÉRAL revision state; **~ de fuites globales** TRANSMIS cumulative leakage index (*CLI*); **~ de glissement amont** ÉLECTRON pushing figure; **~ de groupe** TRANSMIS group index; **~ de mérite** RADIO figure of merit; **~ de modulation** RADIO modulation index; **~ de netteté** (*IN*) RADIO, TÉL, TRANSMIS *d'un circuit téléphonique* articulation index (*AI*); **~ de netteté des logatomes** GÉNÉRAL logatom articulation index; **~ objectif global de volume des sons** TÉL overall objective loudness rating (*OOLR*); **~ objectif de volume des sons** TÉL objective loudness rating; **~ objectif de volume des sons pour l'effet local** TÉL side-tone objective loudness rating (*SOLR*); **~ objectif de volume des sons à l'émission** TÉL, TRANSMIS transmit objective loudness rating (*TOLR*); **~ objectif de volume des sons à la réception** TÉL receive objective loudness rating (*ROLR*); **~ de qualité** TÉL quality; **~ de qualité de transmission** TÉL, TESTS, TRANSMIS transmission performance rating; **~ de réfraction** ÉLECTRON, RADIO, TRANSMIS, TV refraction index, refractive index, index of refraction; **~ de réfraction modifié** RADIO modified refractive index; **~ à saut équivalent** TÉL equivalent step index (*ESI*); **~ de surcharge** ALIM ÉLEC overcurrent factor

indiquer *vt* COMMUT *une position* select, INFORMAT show

indisponibilité *f* DONNÉES, GÉNÉRAL, INFORMAT unavailability, TESTS outage, unavailability; **~ moyenne** TESTS mean unavailability

inductance *f* ALIM ÉLEC, COMPOSANTS, ÉLECTRON inductance, inductor; **~ additionnelle** ALIM ÉLEC series inductance; **~ d'aimantation** ÉLECTRON magnetizing inductance; **~-capacité** COMPOSANTS inductance capacity, ÉLECTRON inductance capacitance; **~ étalon** COMPOSANTS standard inductor, ÉLECTRON inductance standard, standard inductor; **~ mutuelle** ÉLECTRON mutual inductance; **~ propre** ALIM ÉLEC self-inductance; **~ série** ALIM ÉLEC series inductance

inductif *adj* ALIM ÉLEC, COMPOSANTS, ÉLECTRON inductive

induction *f* ALIM ÉLEC induction, ÉLECTRON flux density, induction, RADIO induction; **~ magnétique** ÉLECTRON flux density, magnetic induction; **~ mutuelle** ÉLECTRON mutual induction

induire *vt* GÉNÉRAL induce

induit: **~ fixe** *m* COMPOSANTS, ÉLECTRON stator

INF *abrév* (*message d'information*) GÉNÉRAL INF (*information message*)

infalsifiable *adj* TESTS foolproof, tamperproof

inférences: **~ logiques par seconde** *loc* INFORMAT, TESTS logical inferences per second (*LIPS*); **un million d'~ logiques par seconde** *loc* (*megalips*) INFORMAT, TESTS one million logical inferences per second (*megalips*)

infini: **à l'~** *adj* INFORMAT endless

infogérance *f* APP SERV facilities management (*FM*)

infographie *f* INFORMAT, TV computer graphics, computer imagery; **~ interactive** INFORMAT, TV interactive computer graphics

infographisme *m* INFORMAT, TV computer graphics

informaticien *m* GÉNÉRAL, INFORMAT computer scientist

information *f* GÉNÉRAL information; **~ d'adresse de protocole de réseau** DONNÉES network protocol address information; **~ annexe** INFORMAT side information; **~ d'appel** TÉL call information; **~ d'authentification** DONNÉES, SIGNAL authentication information; **~ de codage** INFORMAT *audio, vidéo* coding information; **~ de commande de position de paquet** DONNÉES burst position control information; **~ de commande de protocole de réseau** DONNÉES network protocol data unit; **~ de contrôle du protocole AAL** DONNÉES, TÉL AAL protocol control information (*AAL-PCI*); **~ de contrôle du protocole N** DONNÉES, TÉL N-protocol control information (*N-PCI*); **~ d'en-tête supplémentaire** INFORMAT additional header information; **~ d'entrée** DONNÉES entry information; **~ de fréquence vocale** TÉL tone message, tone-coded information; **~ de gestion** DONNÉES management information; **~ indirecte** INFORMAT side information; **~ de liaison défaillante** TÉL faulty link information; **~ de localisation** RADIO location information; **~ non-numérotée** DONNÉES unnumbered information (*UI*); **~ en retour** GÉNÉRAL feedback; **~ de rythme** DONNÉES, TRANSMIS timing information; **~ de service** DONNÉES, TÉL housekeeping information; **~ supplémentaire** TÉL additional information; **~ sur la qualité** GÉNÉRAL Q-information; **~ tarifaire** APP SERV advice of charge, charge advice; **~ de taxation** APP SERV charging information; **~ d'usager à usager** DONNÉES user-to-user information (*UUI*)

informations: **~ d'adressage du protocole N** *f pl* DONNÉES, TÉL N-protocol-addressing information (*N-PAI*); **~ d'autorisation** *f pl* DONNÉES, SIGNAL authorization information; **~ de commande** *f pl* INFORMAT control information; **~ de contrôle du protocole d'application** *f pl* DONNÉES, TÉL application protocol control information (*APCI*); **~ de fonctionnement** *f pl* DONNÉES, INFORMAT performance data; **~ numérisées** *f pl* DONNÉES digitized data; **~ de signalisation** *f pl* SIGNAL signaling information (*AmE*), signalling information (*BrE*); **~ significatives** *f pl* DONNÉES meaningful information; **~ supplémentaires sur l'instance de protocole destinées au test** *f pl* DONNÉES protocol implementation extra information for testing (*PIXIT*); **~ de synchronisation** *f pl* DONNÉES timing information, INFORMAT synchronization information, TRANSMIS timing information; **~ de verrouillage de trame** *f pl* TRANSMIS framing information

informatique *f* DONNÉES, INFORMAT computing, information technology (*IT*); **~ distribuée** DONNÉES, INFORMAT distributed computing; **~ domestique** DONNÉES, INFORMAT home computing; **~ de groupe** DONNÉES, INFORMAT collaborative computing; **~ répartie** DONNÉES distributed data processing, distributed processing, INFORMAT distributed processes; **~ tolérante aux fautes** DONNÉES, INFORMAT fault-tolerant computing; **~ de traitement d'alarmes** TESTS alarm management data processing; **~ de l'utilisateur** DONNÉES, INFORMAT end-user computing

informatisation *f* INFORMAT computerization

informatisé *adj* INFORMAT computerized

informatiser *vt* INFORMAT computerize

informer *vt* RADIO alert

informétrique *f* DONNÉES informetrics

inforoute *f* TRANSMIS information superhighway

infraréfraction *f* RADIO subrefraction

infrarouge[1] *adj* (*IR*) ÉLECTRON, GÉNÉRAL, INFORMAT, RADIO, TRANSMIS infrared (*IR*)

infrarouge[2] *m* ÉLECTRON, GÉNÉRAL, INFORMAT, RADIO, TRANSMIS infrared; **~ ondes courtes** RADIO shortwave infrared

infrason *m* GÉNÉRAL infrasound

infrastructure: **~ des réseaux** *f* TRANSMIS network infrastructure; **~ de télécommunications** *f* GÉNÉRAL telecommunications infrastructure; **~ unitaire industrielle** *f* (*IUI*) TV single industrial infrastructure

ingénierie: **~ assistée par ordinateur** *f* (*IAO*) INFORMAT computer-aided engineering (*CAE*); **~ de la**

connaissance *f* INFORMAT knowledge engineering; **~ de la fiabilité** *f* TESTS reliability engineering; **~ inverse** *f* ÉLECTRON, INFORMAT reverse engineering; **~ de la qualité** *f* TESTS quality engineering; **~ de réseau** *f* TÉL network engineering, networking; **~ de site** *f* GÉNÉRAL plant engineering; **~ du télétrafic** *f* TÉL teletraffic engineering

ingénieur: **~ de développement** *m* GÉNÉRAL development engineer; **~ électricien** *m* DONNÉES, GÉNÉRAL electrical engineer; **~ d'exploitation et de maintenance** *m* TÉL operation and maintenance engineer; **~ logiciel** *m* INFORMAT software engineer; **~ matériel** *m* GÉNÉRAL hardware engineer; **~ systèmes CFAO** *m* GÉNÉRAL, INFORMAT CAD/CAM systems engineer; **~ en télécommunications** *m* GÉNÉRAL telecommunications engineer

ingénieur-conseil *m* GÉNÉRAL consultant

inhiber *vt* ÉLECTRON inhibit

inhibition *f* ÉLECTRON, TESTS inhibition; **~ d'alarme de test** TESTS test alarm inhibition

ininflammable *adj* COMPOSANTS fireproof

initialisation *f* INFORMAT *de mémoire* initialization; **~ à distance** ÉLECTRON remote boot; **~ du tampon d'édition** INFORMAT edit buffer initialization

initialiser *vt* INFORMAT boot, boot up, bootstrap, initialize

initiateur: **~ N** *m* DONNÉES N initiator

initiation *f* RADIO, TÉL initiation

injecté *adj* TV injected

injection: **~ en orbite** *f* RADIO injection into orbit, orbit insertion

inopérabilité *f* ÉLECTRON, GÉNÉRAL inoperability

inopérable *adj* ÉLECTRON, GÉNÉRAL out of order

inopérant *adj* ÉLECTRON, GÉNÉRAL inoperative

inscription *f* DONNÉES, INFORMAT entry, RADIO registration, TÉL entry; **~ à l'annuaire** TÉL directory entry

inscrire *vt* APP SERV *demande* book, RADIO register, TÉL *demande* book

inscrit *adj* RADIO registered

insensibilité: **~ au brouillage** *f* RADIO immunity to interference

inséré *adj* ÉLECTRON hole-mounted, GÉNÉRAL inserted

insérer *vt* COMPOSANTS snap into, GÉNÉRAL insert, INFORMAT punch in, insert, SIGNAL, TRANSMIS insert

insertion *f* COMPOSANTS insert, GÉNÉRAL insertion, INFORMAT insert, punch-in, insertion, SIGNAL insertion, TRANSMIS insert, insertion; **~ et extraction** TRANSMIS drop and insert; **~ de mesures** INFORMAT measure insert; **~ de signaux** SIGNAL, TRANSMIS signal insertion

insonorisation *f* GÉNÉRAL, TÉL sound insulation, soundproofing

insonorisé *adj* GÉNÉRAL, TÉL sound-proof, sound-insulated

insonoriser *vt* GÉNÉRAL, TÉL soundproof

inspecter *vt* TESTS inspect

instabilité *f* ÉLECTRON instability, jitter, INFORMAT jitter, RADIO instability; **~ de la base de temps** ÉLECTRON time-base jitter; **~ de fréquence** RADIO frequency instability; **~ de mode** ÉLECTRON moding, TRANSMIS mode hopping; **~ thermique** COMPOSANTS, ÉLECTRON thermal instability

instable *adj* ÉLECTRON *éléments logiques* astable

installateur: **~-électricien** *m* GÉNÉRAL electrician

installation *f* ALIM ÉLEC system, GÉNÉRAL installation, plant, systems, facility, site, INFORMAT setting-up, PÉRIPH, RADIO, TÉL system; **~ d'abonné** DONNÉES, TÉL subscriber installation, subscriber's installation; **~ de commande par radio** RADIO radio control system; **~ échelonnée** TÉL echelon telegraphy; **~ d'interphone** TÉL intercom system; **~ matérielle** INFORMAT equipment installation; **~ de mise à la terre** ALIM ÉLEC earth system (*BrE*), ground system (*AmE*); **~ mixte** COMMUT, TRANSMIS hybrid system; **~ mobile** RADIO mobile installation; **~ murale** TÉL wall-mounted installation; **~ optique** TRANSMIS fiber cable plant (*AmE*), fibre cable plant (*BrE*); **~ de prise de terre** ALIM ÉLEC earth electrode installation (*BrE*), ground electrode installation (*AmE*); **~ privée** (*IP*) DONNÉES subscriber's installation, PÉRIPH private system, TÉL private installation; **~ privée d'abonné** DONNÉES, TÉL private subscriber installation; **~ de référence** TÉL reference installation; **~ de secours** APP SERV emergency installation; **~ souterraine** APP SERV underground installation; **~ téléphonique** TÉL telephone installation; **~ terminale d'abonné** TÉL, TV subscriber terminal installation; **~ terminale télétex à postes multiples** TÉL multistation teletex terminal installation; **~ terminale pour transmission de données** TÉL terminal installation for data transmission

installations: **~ extérieures** *f pl* TRANSMIS outside plant

installer *vt* GÉNÉRAL install

instance: **~ conforme** *m* DONNÉES conforming implementation; **~ sous test** *m* (*IUT*) DONNÉES implementation under test (*IUT*)

instant: **~ d'allumage** *m* ÉLECTRON firing time, ignition time; **~ d'amorçage** *m* ÉLECTRON ignition time; **~ de décision d'un signal numérique** *m* TÉL decision instant of a distant signal; **~ significatif** *m* SIGNAL, TÉL, TESTS significant instant; **~ significatif cohérent** *m* SIGNAL, TÉL, TESTS coherent significant instant

instants: **~ idéals** *m pl* TÉL ideal instants

instaurer *vt* INFORMAT set

Institut: **~ américain de normalisation** *m* (*ANSI*) GÉNÉRAL American National Standrads Institute (*ANSI*); **~ belge de normalisation** *m* (*IBN*) GÉNÉRAL Belgian Standards Institute (*BSI*); **~ belge des services postaux et des télécommunications** *m* (*IBPT*) RADIO Belgian Institute for Post and Telephony (*BIPT*); **~ européen de normalisation en télécommunications** *m* (*IENT*) GÉNÉRAL European Telecommunications Standards Institute (*ETSI*); **~ des ingénieurs électriciens et électroniciens** *m* (*IEEE*) ÉLECTRON Institute of Electrical and Electronics Engineers (*IEEE*)

instruction *f* ALIM ÉLEC instruction, INFORMAT instruction, order, TESTS, TRANSMIS instruction; **~ à une adresse** INFORMAT one address instruction; **~ d'affectation** DONNÉES assignment statement; **~ d'alarme** TÉL alarm statement; **~ d'arrêt** INFORMAT stop instruction; **~ d'arrêt facultatif** INFORMAT optional stop instruction; **~ de décision** INFORMAT decision instruction, discrimination instruction; **~ en direct** INFORMAT in-line instruction; **~ effective** INFORMAT effective instruction; **~ d'exécution** INFORMAT execute order, execute statement; **~ d'extraction** INFORMAT extract instruction; **~ de fin** TÉL end statement; **~ ineffective** INFORMAT NO OP, no operation instruction; **~ inopératoire** INFORMAT no operation instruction, NO OP; **~ logique** INFORMAT logic instruction; **~ machine** INFORMAT computer instruction; **~ à n plus une adresse** INFORMAT n-plus-one address instruction; **~ pas à pas** INFORMAT step-by-step instruction; **~ de pause** INFORMAT halt instruction, pause instruction; **~ de pause facultative** INFORMAT optional pause instruction; **~ à une plus une adresse** INFORMAT one-plus-one address instruction; **~ séquentielle** DONNÉES sequential statement

instructions: **~ de bobinage** *f pl* ALIM ÉLEC winding instructions; **~ concurrentes** *f pl* DONNÉES concurrent statements; **~ d'essai** *f pl* TESTS test instructions; **~ d'enroulement** *f pl* ALIM ÉLEC winding instructions; **~ d'ingénierie** *f pl* GÉNÉRAL engineering instructions; **~ d'installation** *f pl* GÉNÉRAL installation instructions; **~ de réglage** *f pl* GÉNÉRAL *documentation* adjusting instructions; **~ de réglage initial** *f pl* TRANSMIS line-up instructions

instrument: **~ à aiguille** *m* TÉL pointer instrument; **~ de mesure** *m* TESTS measuring instrument; **~ de mesure électrique** *m* TESTS electric measuring instrument; **~ de mesure indicateur** *m* TESTS indicating-measuring instrument; **~ de mesure de puissance optique** *m* TRANSMIS optical power meter; ~ **de mesure de la visibilité** *m* TESTS visibility meter; ~ **de musique numérique** *m* INFORMAT digital music instrument; ~ **de précision** *m* TESTS precision instrument; ~ **de vérification** *m* TESTS verifying instrument

insuffisance: ~ **de mémoire** *f* INFORMAT memory shortage

intégrateur *m* DONNÉES integrator, ÉLECTRON integrating circuit, RADIO integrator; ~ **sommateur** ÉLECTRON summing integrator; ~ **de systèmes** DONNÉES systems integrator

intégration: ~ **à échelle intermédiaire** *f* COMPOSANTS, ÉLECTRON, TÉL, TRANSMIS medium-scale integration (*MSI*); ~ **à grande échelle** *f* (*LSI*) COMPOSANTS, ÉLECTRON, TÉL, TRANSMIS large-scale integration (*LSI*); ~ **hypermédia** *f* INFORMAT hypermedia building; ~ **à moyenne échelle** *f* COMPOSANTS, ÉLECTRON, TÉL, TRANSMIS medium-scale integration (*MSI*); ~ **à petite échelle** *f* COMPOSANTS, ÉLECTRON, TÉL, TRANSMIS small-scale integration (*SSI*); ~ **de systèmes** *f* DONNÉES systems integration; ~ **à très grande échelle** *f* (*VLSI*) COMPOSANTS, ÉLECTRON, TÉL, TRANSMIS very large-scale integration (*VLSI*); ~ **voix-données** *f* DONNÉES voice-data integration

intégrité *f* DONNÉES, TÉL, TRANSMIS integrity; ~ **des données** DONNÉES data integrity; ~ **de la séquence des éléments numériques** TÉL digit sequence integrity; ~ **de la séquence des intervalles de temps** TÉL, TRANSMIS time slot sequence integrity

intelligence: ~ **artificielle** *f* (*IA*) APP SERV, COMMUT, INFORMAT artificial intelligence (*AI*)

intelligent *adj* INFORMAT smart, intelligent

intelligibilité: ~ **de la parole** *f* INFORMAT, RADIO, TÉL speech intelligibility

intensité *f* GÉNÉRAL intensity, strength, INFORMAT depth, TRANSMIS, TV intensity; ~ **d'appel** TÉL call intensity; ~ **de champ** ALIM ÉLEC, RADIO field strength, TESTS field intensity; ~ **de champ magnétique** ÉLECTRON magnetic-field intensity; ~ **du conduit** RADIO duct intensity; ~ **de courant** ALIM ÉLEC current intensity; ~ **de courant** ÉLECTRON ampere rating; ~ **de courant admissible** ALIM ÉLEC current-carrying capacity; ~ **d'un courant électrique** ALIM ÉLEC amperage; ~ **courant à vide** ALIM ÉLEC no-load current; ~ **de défaillance** TESTS failure intensity; ~ **énergétique** ÉLECTRON, TRANSMIS radiant intensity; ~ **instantanée de défaillance** TESTS instantaneous failure intensity; ~ **moyenne de défaillance** TESTS mean failure intensity; ~ **moyenne de dérangement** INFORMAT mean failure intensity; ~ **moyenne de trafic** TÉL average traffic intensity; ~ **de pluie** RADIO rainfall rate, precipitation rate; ~ **de précipitations** RADIO precipitation rate, rainfall rate; ~ **de rayonnement** ÉLECTRON, TRANSMIS radiant intensity; ~ **sonore** INFORMAT, TÉL loudness; ~ **de trafic** COMMUT, TÉL traffic intensity; ~ **du trafic acheminé** DONNÉES, TÉL traffic flow; ~ **de trafic équivalent** TÉL equivalent random traffic intensity

intention: ~ **d'appel** *f* TÉL call intent

interactif *adj* DONNÉES conversational, interactive, GÉNÉRAL interactive

interaction *f* DONNÉES, ÉLECTRON interaction; ~ **bidirectionnelle** DONNÉES two-way alternate interaction; ~ **de gestion** DONNÉES management interaction; ~ **ion-matière** ÉLECTRON ion-matter interaction; ~ **monologue** DONNÉES monologue interaction

interactivité *f* TV interactivity

interbande *adj* TÉL, TV interband

intercaler *vt* RADIO interleave

interception: ~ **d'appel** *f* TÉL call interception, call pick-up

interclasser *vt* INFORMAT collate

interclasseuse *f* INFORMAT collator

intercom *m* (*intercommunication*) TÉL intercom (*intercommunication*)

intercommunication *f* (*intercom*) TÉL intercommunication (*intercom*)

interconnectable *adj* GÉNÉRAL interfaceable

interconnecter *vt* ALIM ÉLEC, COMPOSANTS wire, ÉLECTRON cross-connect, GÉNÉRAL interconnect, PÉRIPH intercouple

interconnexion *f* ALIM ÉLEC, COMPOSANTS wiring, ÉLECTRON cross-strapping, GÉNÉRAL configuration, interconnection; ~ **de circuits** DONNÉES circuit interface; ~ **de composants périphériques** INFORMAT peripheral component interconnect (*PCI*); ~ **multisignaux** DONNÉES multisignal interconnect (*MSI*); ~ **de réseau** DONNÉES network interconnection; ~ **de réseaux** DONNÉES internetworking; ~ **de systèmes d'entreprise** INFORMAT enterprise systems connection (*ESCON*); ~ **de systèmes ouverts** (*ISO*) DONNÉES, INFORMAT, TÉL, TRANSMIS open systems interconnection (*OSI*)

interdiction *f* RADIO, TÉL barring, TRANSMIS denial; ~ **d'accès des appels entrants** (*ICB*) TÉL incoming-call barring; ~ **d'accès des appels sortants** DONNÉES, TÉL outgoing calls barred (*OCB*); ~ **d'appel** TÉL call barring; ~ **des appels d'arrivée** TÉL incoming-call barring; ~ **des appels de départ** DONNÉES, TÉL outgoing calls barred (*OCB*)

interdire *vt* ÉLECTRON inhibit, INFORMAT lock out, RADIO, TÉL bar, TRANSMIS deny

interdit *adj* ÉLECTRON inhibited, INFORMAT locked out, RADIO, TÉL barred, TRANSMIS denied

interface *f* GÉNÉRAL interface (*I/F*); ~ **d'adaptateur de périphérique** INFORMAT device adaptor interface; ~ **d'affichage graphique** INFORMAT graphic device interface (*GDI*); ~ **analogique** TRANSMIS analog interface; ~ **de bande** INFORMAT tape interface; ~ **cassette** INFORMAT cassette interface; ~ **CNE-UCN** TÉL UCN-CNE interface; ~ **de commande de support** (*MCI*) INFORMAT media control interface (*MCI*); ~ **de communication** INFORMAT, TRANSMIS transmission interface; ~ **de communication programmable** (*PCI*) TÉL programmable communication interface (*PCI*); ~ **contradirectionnelle** TRANSMIS contra-directional interface; ~ **de contrôle des médias** INFORMAT media control interface; ~ **du CSN avec les organes de connexion et de commande** TÉL interface of the CSN with the switching and control units; ~ **disque dur** INFORMAT hard-disk interface; ~ **de données avec distribution par fibres** DONNÉES, TRANSMIS fiber-distributed data interface (*AmE*), fibre-distributed data interface (*BrE*); ~ **graphique** INFORMAT graphical interface; ~ **homme-machine** INFORMAT man-machine interface; ~ **hors réseau** TÉL non-network interface; ~ **iconique** DONNÉES, INFORMAT icon interface; ~ **informatique** DONNÉES, INFORMAT computer interface; ~ **en langage naturel** DONNÉES natural language interface; ~ **de liaison réseau** TÉL matrix-link interface; ~ **de ligne numérique** TRANSMIS digital carrier module; ~ **de maintenance** TÉL maintenance interface; ~ **multimédia de l'utilisateur** DONNÉES multimedia user interface; ~ **multiplex** TÉL multiplex interface; ~ **de nœud de réseau** TRANSMIS network node interface (*NNI*); ~ **normalisée** GÉNÉRAL standard interface; ~ **numérique** DONNÉES digital interface; ~ **optique** TRANSMIS optical interface; ~ **d'ordinateur** DONNÉES, INFORMAT computer interface; ~ **d'ordinateur intelligente** DONNÉES, INFORMAT intelligent computer interface; ~ **ouverte de liaison de données** DONNÉES, ÉLECTRON open data link interface (*ODLI*); ~ **PABX-serveur** INFORMAT computer-PBX interface (*CPI*); ~ **parallèle** DONNÉES parallel interface; ~ **physique** DONNÉES physical interface (*PI*); ~ **physique de rythme de multiplexeur** TRANSMIS multiplexer timing physical interface (*MTPI*); ~ **de poste d'affichage** PÉRIPH display station interface (*DSI*); ~ **de programmation** DONNÉES, INFORMAT programming interface; ~ **de programmes d'application** (*IPA*) DONNÉES, INFORMAT application program interface (*API*); ~ **radio commune**

(*CAI*) RADIO common air interface (*CAI*); ~ **de réseau** DONNÉES, INFORMAT network interface; ~ **réseau** TÉL network interface; ~ **SCSI** (*SCSI*) INFORMAT small computer system interface (*SCSI*); ~ **en série** DONNÉES, ÉLECTRON serial interface; ~ **en série programmable** INFORMAT serial programmable interface; ~ **terrestre** RADIO terrestrial interface; ~ **de tranche de temps** TRANSMIS time slot interface; ~ **U** (*U*) DONNÉES, TÉL U interface (*U*); ~ **d'unité graphique** INFORMAT graphic device interface (*GDI*); ~ **universelle** DONNÉES, INFORMAT general-purpose interface; ~ **utilisateur graphique** (*GUI*) DONNÉES, INFORMAT graphical user interface (*GUI*); ~ **utilisateur-réseau** DONNÉES, TRANSMIS user-network interface (*UNI*); ~ **vocale** APP SERV vocal user interface

interférence: ~ **avec le canal adjacent** *f* RADIO, TRANSMIS co-channel interference; ~ **constructive** *f* TRANSMIS constructive interference; ~ **destructive** *f* TRANSMIS destructive interference; ~ **due à la propagation à trajets multiples** *f* RADIO multipath interference; ~ **électrique** *f* INFORMAT electrical interference; ~ **électromagnétique** *f* ÉLECTRON, TESTS electromagnetic interference (*EMI*); ~ **entre canaux voisins** *f* RADIO, TÉL, TRANSMIS sideband splatter; ~ **entre zones** *f* RADIO interzone interference; ~ **intersymbole** *f* INFORMAT, TRANSMIS, TV intersymbol interference (*ISI*); ~ **longitudinale** *f* INFORMAT, TRANSMIS, TV intersymbol interference (*ISI*); ~ **numérique** *f* TRANSMIS digital interference; ~ **d'ondes** *f* ÉLECTRON, RADIO wave interference; ~ **optique** *f* TRANSMIS optical interference; ~ **radioélectrique** *f* RADIO radio-frequency interference (*RFI*)

interféromètre *m* DONNÉES, TESTS interferometer; ~ **optique** DONNÉES light interferometer

interférométrie *f* RADIO slab interferometry, TRANSMIS axial interference microscopy, slab interferometry; ~ **axiale** RADIO, TRANSMIS axial slab interferometry; ~ **transversale** TRANSMIS transverse interferometry

interfonctionnement *m* DONNÉES interoperability, GÉNÉRAL interwork, interworking; ~ **entre catégories d'usagers du service** TÉL interworking between user classes of services; ~ **entre réseaux** TÉL interworking between networks; ~ **entre le service télétex et d'autres services** TÉL interworking between teletex and other services; ~ **par mise en correspondance de la commande des communications** DONNÉES interworking by call control mapping (*ICCM*); ~ **par point d'accès** (*IPA*) DONNÉES, INFORMAT interworking by port access (*IPA*); ~ **des plans de numérotage** DONNÉES, TÉL numbering plan interworking; ~ **des réseaux** DONNÉES internetworking; ~ **de service** TÉL service interworking; ~ **de la signalisation** SIGNAL signaling interworking (*AmE*), signalling interworking (*BrE*); ~ **vidéotex** TÉL, TV videotex interworking

interfonctionner *vi* GÉNÉRAL interwork

interlettrage *m* DONNÉES interletter spacing

interlignage *m* DONNÉES interline spacing

interligne *m* DONNÉES, INFORMAT line space, line spacing, TÉL blank line

interlocuteur *m* INFORMAT, TÉL speaker

intermittence *f* GÉNÉRAL intermittence, intermittency

intermodulation *f* TRANSMIS intermodulation; ~ **passive** (*IMP*) GÉNÉRAL passive intermodulation (*PIM*)

intermodule *adj* ÉLECTRON cross-module

internaute *m* DONNÉES surfer

Internet *m* DONNÉES, INFORMAT Internet (*Net*); ~ **protocol** (*IP*) DONNÉES, INFORMAT Internet protocol (*IP*)

interopérabilité *f* GÉNÉRAL interoperability, interworking, RADIO interworkability

interphone *m* PÉRIPH interphone, TÉL intercom, intercom telephone

interphonie *f* PÉRIPH interphone equipment

interpolateur *m* TRANSMIS interpolator

interpolation: ~ **différentielle** *f* INFORMAT differential interpolation; ~ **linéaire** *f* TRANSMIS linear interpolation; ~ **non linéaire** *f* TRANSMIS nonlinear interpolation; ~ **des signaux vocaux** *f* TRANSMIS interpolation of speech signals; ~ **temporelle des signaux vocaux** *f* (*système TASI*) SIGNAL, TRANSMIS time-assigned speech interpolation (*TASI*)

interposition *f* TÉL interposition

interprétabilité *f* GÉNÉRAL interpretability

interprétable *adj* GÉNÉRAL interpretable

interprétation *f* GÉNÉRAL interpretation; ~ **sémantique** DONNÉES semantic interpretation

interpréter *vt* GÉNÉRAL interpret

interpréteur *m* GÉNÉRAL interpreter

interréseau *adj* TRANSMIS cross-net

interrogateur *m* RADIO interrogator; ~**-répondeur** RADIO interrogator transponder

interrogation *f* GÉNÉRAL interrogation; ~ **à distance** GÉNÉRAL remote consultation, PÉRIPH remote interrogation; ~ **des navires** RADIO ship polling; ~ **par poste à télé** PÉRIPH remote interrogation by mf ph; ~ **par télécommande** PÉRIPH remote interrogation by telecommand; ~ **préalable** DONNÉES, INFORMAT, RADIO polling

interrogeable: ~ **à distance** *adj* PÉRIPH remotely-interrogable

interroger *vt* DONNÉES, GÉNÉRAL, INFORMAT, PÉRIPH, RADIO interrogate, poll

interrompre *vt* ALIM ÉLEC *courant* cut off, break, COMMUT, DONNÉES interrupt, disrupt, INFORMAT interrupt, TÉL interrupt, *une communication* pre-empt

interrompu *adj* ALIM ÉLEC cut-off, GÉNÉRAL broken, interrupted

interrupteur *m* ALIM ÉLEC, COMMUT, COMPOSANTS, ÉLECTRON switch, GÉNÉRAL cutoff switch, interrupter, INFORMAT, RADIO switch, TÉL interrupter; ~ **d'alimentation** ALIM ÉLEC, COMPOSANTS power on/off switch, power switch; ~ **à bascule** GÉNÉRAL tumbler switch, toggle; ~ **de clavier à membrane** ÉLECTRON membrane keyswitch; ~ **de défaut à la terre** ALIM ÉLEC earth-fault circuit breaker (*BrE*), ground-fault circuit breaker (*AmE*); ~ **de fin de course** ALIM ÉLEC limit switch; ~ **général** ALIM ÉLEC main circuit breaker, main switch, on-off switch, power switch; ~ **à membrane** ÉLECTRON membrane switch; ~ **de modulation** RADIO modulation switch; ~ **monostable** ALIM ÉLEC, COMMUT spring-loaded switch; ~ **à ressort** ALIM ÉLEC, COMMUT spring-loaded switch; ~ **rotatif** COMMUT, GÉNÉRAL, INFORMAT rotary switch; ~ **à rupture brusque** ALIM ÉLEC, COMMUT, ÉLECTRON, INFORMAT toggle switch; ~ **secteur** ALIM ÉLEC mains switch; ~ **de signal d'appel** TÉL ringing interruptor

interruption *f* COMMUT, DONNÉES interrupt, ÉLECTRON disruption, GÉNÉRAL break, cutoff, INFORMAT interrupt, TÉL interrupt, interruption, TESTS interruption, outage, TRANSMIS breakdown, failure; ~ **automatique** INFORMAT automatic interrupt; ~ **d'exploitation** TÉL operation interruption, service interruption, TESTS service interruption; ~ **imprévisible** TESTS unforeseen outcome; ~ **d'onde pilote de groupe primaire** TRANSMIS group pilot failure; ~ **de la procédure EOM** (*PRI-EOM*) DONNÉES, TÉL procedure interrupt EOM (*PRI-EOM*); ~ **de la procédure EOP** (*PRI-EOP*) DONNÉES, TÉL procedure interrupt EOP (*PRI-EOP*); ~ **de la procédure MPS** (*PRI-MPS*) DONNÉES, TÉL procedure interrupt MPS (*PRI-MPS*); ~ **de la procédure négative** (*PIN*) DONNÉES, TÉL, TRANSMIS procedure interrupt negative (*PIN*); ~ **de la procédure positive** (*PIP*) DONNÉES, TÉL, TRANSMIS procedure interrupt positive (*PIP*); ~ **de réseau** ÉLECTRON network disruption; ~ **du service** TÉL operation interruption, service interruption, TESTS service interruption; ~ **du service transmission** TÉL interruption of transmission service; ~ **totale des radiocommunications** RADIO radio blackout; ~ **de**

traçage COMMUT trace interrupt; **~ de transmission** TESTS transmission breakdown, TRANSMIS interruption of transmission, transmission breakdown
intersection *f* ÉLECTRON *éléments logiques* conjunction
intersupport *adj* ÉLECTRON cross-media
intertitre *m* INFORMAT subheading
interurbain[1] *adj* TÉL intercity
interurbain[2]: **~ automatique** *m* (*IA*) TÉL direct distance dialing (*AmE*) (*DDD*), subscriber trunk dialing (*AmE*) (*STD*), trunk dialing (*AmE*)
intervalle: **~ de confiance** *m* GÉNÉRAL confidence interval; **~ dans un enroulement** *m* ÉLECTRON unit interval (*UI*); **~ d'échantillonnage** *m* TRANSMIS sampling interval; **~ entre chiffres** *m* TÉL interdigital pause, interdigital time; **~ entre impulsions** *m* TÉL, TRANSMIS pulse interval; **~ d'impulsions de compteur** *m* TÉL meter pulse interval; **~ interchiffre** *m* TÉL interdigital pause, interdigital time; **~ minimum** *m* TÉL minimum interval; **~ minimum acceptable** *m* TÉL minimum acceptable interval; **~ de quantification** *m* TÉL quantizing interval; **~ de recouvrement** *m* APP SERV charging interval; **~ de réessai** *m* RADIO retry interval; **~ de relance** *m* RADIO retry interval; **~ de retour de ligne** *m* DONNÉES, TV horizontal blanking interval, line blanking interval; **~ de retour de trame** *m* TV field blanking interval (*FBI*), vertical blanking interval (*VBI*); **~ sans garde** *m* TÉL unguarded interval; **~ significatif** *m* SIGNAL, TÉL significant interval; **~ de silence** *m* TÉL silent period; **~ de suppression de trame** *m* TV field blanking interval (*FBI*), vertical blanking interval (*VBI*); **~ de suppression de ligne** *m* DONNÉES, TV horizontal blanking interval, line blanking interval; **~ temporel** *m* INFORMAT significant interval; **~ de temps** *m* (*IT*) COMMUT, RADIO, SIGNAL, TRANSMIS time slot; **~ de temps pour élément numérique** *m* TÉL digit time slot; **~ de temps pour élément numérique justifiable** *m* TÉL, TRANSMIS justifiable digit time slot; **~ de temps de signalisation** *m* SIGNAL signaling time slot (*AmE*), signalling time slot (*BrE*); **~ de temps du symbole de contrôle** *m* TRANSMIS check-symbol time slot; **~ de temps de verrouillage de trame** *m* TÉL, TRANSMIS frame-alignment time slot; **~ de temps de voie** *m* TÉL, TRANSMIS channel time slot; **~ unitaire** *m* GÉNÉRAL unit interval (*UI*)
intervallomètre *m* ALIM ÉLEC interval timer
intervention *f* TÉL barge-in, break-in, prompting; **~ ligne-réseau** COMMUT, TÉL trunk offer (*TKO*); **~ partielle** TÉL partial break-in; **~ de standardiste** TÉL attendant action; **~ totale** TÉL full break-in; **~ d'urgence** TÉL emergency action
intervertir *vt* GÉNÉRAL invert
interzonal *adj* ÉLECTRON interzonal
intitulé *m* DONNÉES destination details
intrabande *adj* SIGNAL, TRANSMIS in-band
intracodage *m* INFORMAT intracoding
intraficable *adj* TESTS tamperproof
intra-image *m* TV intraframe
Intranet *m* INFORMAT Intranet
introduction *f* INFORMAT input process, TÉL insertion; **~ progressive** RADIO phasing-in
introduire *vt* GÉNÉRAL input, insert; **~ en mémoire** INFORMAT usher in; **~ par clavier** GÉNÉRAL key in; **~ progressivement** GÉNÉRAL phase in
intrusion *f* ÉLECTRON, TÉL intrusion
inutilisable *adj* GÉNÉRAL out of order
invalidé *adj* ÉLECTRON inhibited, INFORMAT invalid, disabled
invalider *vt* ÉLECTRON inhibit, INFORMAT invalidate, disable
invariant: **~ dans le temps** *adj* TRANSMIS time-constant
inverser *vt* ALIM ÉLEC invert, reverse, ÉLECTRON invert, negate, reverse, INFORMAT negate, reverse, TV *image* invert
inverseur *m* ALIM ÉLEC changer, inverter, reversing switch, ÉLECTRON NOT circuit; **~ à bascule** ALIM ÉLEC, COMMUT, ÉLECTRON, INFORMAT toggle switch; **~ de phase** ALIM ÉLEC phase inverter; **~ de polarité** ALIM ÉLEC pole inverter, pole changer
inversion *f* ALIM ÉLEC inversion, reversal, CLIMAT, GÉNÉRAL inversion, ÉLECTRON shift, transposition, INFORMAT negation, PÉRIPH shift, RADIO inversion, TÉL reversal, shift, TRANSMIS inversion, reversal; **~ alternée des éléments numériques** TRANSMIS alternate digit inversion; **~ de bits alternée** TRANSMIS alternate bit inversion; **~ de charge en ligne** TÉL line reversal; **~ chiffres** PÉRIPH, TÉL *télex* figure shift; **~ de fréquence** RADIO frequency inversion; **~ lettres** TÉL letter shift (*LED*); **~ de lettres** INFORMAT letter shift (*LED*); **~ de ligne** TÉL *télex* line shift; **~ logique** ÉLECTRON negation, NOT operation; **~ partielle des nuances** TÉL partial tone reversal; **~ de phase** ALIM ÉLEC phase inversion; **~ de polarité** ALIM ÉLEC polarity inversion, polarity reversal, pole changing, pole reversal; **~ de polarité de ligne** TRANSMIS line polarity inversion, line polarity reversal; **~ de température** CLIMAT, RADIO temperature inversion
inversions: **~ de fils** *f pl* ÉLECTRON wire transpositions
investigation *f* GÉNÉRAL examination, TESTS investigation
invitation: **~ à continuer** *f* RADIO go-ahead (*GA*); **~ à émettre** *f* DONNÉES polling, invitation to transmit, INFORMAT, RADIO polling; **~ à numéroter** *f* TÉL proceed to select; **~ à recevoir** *f* COMMUT, DONNÉES, RADIO, TÉL selecting
invite *f* INFORMAT prompt
inviter: **~ à émettre** *vt* DONNÉES, INFORMAT, RADIO poll
invocation: **~ d'entité (N)** *f* DONNÉES (N)-entity invocation; **~ de processus d'application** *f* DONNÉES application process invocation
invoquer *vt* RADIO invoke
ion *m* ÉLECTRON, RADIO ion
ionisation *f* ÉLECTRON, RADIO ionization; **~ sporadique** RADIO sporadic ionization
ionogramme *m* RADIO ionogram
ionoscope *m* ÉLECTRON ionoscope
ionosonde *f* RADIO ionosonde
ionosphère *f* RADIO ionosphere; **~ supérieure** RADIO upper ionosphere
IP *abrév* DONNÉES (*installation privée*) subscriber's installation, (*Internet protocol, protocole Internet*) IP (*Internet protocol*), INFORMAT (*Internet protocol, protocole Internet*) IP (*Internet protocol*)
IPA *abrév* DONNÉES, INFORMAT (*interface de programmes d'applications*) API (*application program interface*), (*interfonctionnement par point d'accès*) IPA (*interworking by port access*)
IPX *abrév* COMMUT, DONNÉES IPX (*Internet packet exchange*)
IR *abrév* (*infrarouge*) ÉLECTRON, GÉNÉRAL, INFORMAT, RADIO, TRANSMIS IR (*infrared*)
IRA *abrév* (*alphabet international de référence*) TÉL IRA (*international reference alphabet*)
IRC *abrév* (*centre de recherche documentaire*) DONNÉES IRC (*information retrieval center, information retrieval centre*)
IRV *abrév* (*version internationale de référence*) TÉL IRV (*International reference version*)
ISA *abrév* DONNÉES, INFORMAT industry standard architecture (*ISA*)
ISO *abrév* DONNÉES (*interconnexion de systèmes ouverts*) OSI (*open systems interconnection*), GÉNÉRAL (*Organisme international de normalisation*) ISO (*International Standards Organization*), INFORMAT, TÉL, TRANSMIS (*interconnexion de systèmes ouverts*) OSI (*open systems interconnection*
Isoc *abrév* DONNÉES Isoc (*Internet society*)
isochrone[1] *adj* DONNÉES, SIGNAL, TÉL, TRANSMIS isochronous
isochrone[2]: **~ zéro** *f* RADIO zero isochrone
isochronisme: **~ interrompu** *m* TÉL burst isochronous; **~ par paquets** *m* TÉL burst isochronous
isolant[1] *adj* ALIM ÉLEC nonconducting, nonconductive, GÉNÉRAL insulating

isolant[2] *m* ALIM ÉLEC nonconductor, COMPOSANTS insulation, ÉLECTRON insulant, insulation, GÉNÉRAL, TRANSMIS insulation; **~ cellulaire** TRANSMIS cellular insulation

isolateur *m* ALIM ÉLEC, ÉLECTRON insulator, TRANSMIS isolator; **~ de borne à clip** ÉLECTRON small red plastic clip; **~ optique** TRANSMIS optical isolator; **~ de traversée** ALIM ÉLEC lead-in insulator

isolation *f* COMPOSANTS insulation, DONNÉES isolation, ÉLECTRON insulation, isolation, GÉNÉRAL insulation, TÉL isolation, TRANSMIS insulation; **~ à ceinture** TRANSMIS *câbles* belt insulation; **~ cellulaire** TRANSMIS cellular insulation; **~ de conducteur** TRANSMIS conductor insulation; **~ d'un défaut** TÉL fault isolation; **~ minérale** ÉLECTRON mineral insulation; **~ multicouche** ÉLECTRON multilayer isolation (*MLI*); **~ au papier préimprégné** TRANSMIS impregnated paper insulation; **~ par imprégnation dans la masse** ÉLECTRON mass-impregnated insulation; **~ en plastique extrudé massif** COMPOSANTS, GÉNÉRAL extruded solid insulation; **~ en plastique massif** COMPOSANTS, GÉNÉRAL solid insulation; **~ par polyoléfines** ÉLECTRON polyolefin insulation

isolé *adj* DONNÉES, INFORMAT stand-alone; **~ phoniquement** TÉL acoustically-muted; **~ de la terre** ALIM ÉLEC off earth (*BrE*), off ground (*AmE*), unearthed (*BrE*), ungrounded (*AmE*)

isolement *m* COMPOSANTS, ÉLECTRON, GÉNÉRAL, TRANSMIS insulation, RADIO isolation, TRANSMIS insulation; **~ d'air** ALIM ÉLEC insulation gap; **~ de polarisation** RADIO cross-polarization isolation

isoler *vt* ÉLECTRON insulate

isophone *m* TÉL isophon

isotrope *adj* ÉLECTRON, RADIO isotropic

IT *abrév* (*intervalle de temps*) COMMUT, RADIO, SIGNAL, TRANSMIS time slot

item: **~-filtre** *m* DONNÉES filter-item

itinéraire *m* COMMUT path, routing, TÉL path; **~ d'acheminement** TÉL, TRANSMIS route; **~ de parole** TÉL speech path; **~ de référence** COMMUT reference routing

itinérance *f* RADIO roaming

itinérant *adj* RADIO itinerant

IUCAF *abrév* (*Commission inter-unions pour l'attribution de fréquences à la radioastronomie et à la science*) RADIO IUCAF (*Inter-Union Commission on Frequency Allocations for Radio Astronomy*)

IUI *abrév* (*infrastructure unitaire industrielle*) TV single industrial infrastructure

IUT *abrév* (*instance sous test*) DONNÉES IUT (*implementation under test*)

IXP *abrév* (*protocole d'échange d'informations*) INFORMAT IXP (*information exchange protocol*)

J

jack *m* COMMUT, COMPOSANTS, ÉLECTRON, TÉL, TESTS jack; **~ de branchement** COMPOSANTS, ÉLECTRON branching jack, parallel jack; **~ pour casque téléphonique** ÉLECTRON banana jack; **~ coaxial** ÉLECTRON coaxial jack; **~ de dérivation** COMPOSANTS, ÉLECTRON branching jack, parallel jack; **~ d'écoute** TÉL listening jack; **~ d'essai** ÉLECTRON, TÉL test jack; **~ de parallèle** COMPOSANTS, ÉLECTRON parallel jack; **~ de réponse** TÉL answer jack; **~ à rupture** COMMUT *commutation manuelle* break jack; **~ en série** COMPOSANTS serial jack; **~ de service** COMPOSANTS, TÉL service jack; **~ de test** ÉLECTRON, TÉL test jack

jarretière *f* ALIM ÉLEC, COMMUT, COMPOSANTS, ÉLECTRON jumper, strap

jauge: **~ de contraintes** *f* COMPOSANTS strain gage (*AmE*), strain gauge (*BrE*); **~ d'épaisseur** *f* TESTS feeler gage (*AmE*), feeler gauge (*BrE*); **~ d'extension** *f* COMPOSANTS strain gage (*AmE*), strain gauge (*BrE*); **~ de fils** *f* COMPOSANTS wire gage (*AmE*), wire gauge (*BrE*); **~ de profondeur** *f* TESTS depth gage (*AmE*), depth gauge (*BrE*)

JBIG *abrév* (*groupe mixte sur les images à deux niveaux*) INFORMAT, TÉL, TV JBIG (*joint bi-level image group*)

jet: **~ d'air** *m* COMPOSANTS *tubes électroniques* air jet

jeton: **~ de commande multipoint d'assignation de jeton** *f* DONNÉES multipoint command assignment token (*MCA*)

jeux: **~ d'arcade** *m pl* INFORMAT arcade games; **~ ISO de caractères codés à 7 et à 8 éléments** *m pl* TÉL ISO 7- and 8-bit-coded character sets; **~ de matrices** *m pl* ÉLECTRON die sets

JFIF *abrév* (*format d'échange de fichiers JPEG*) INFORMAT JFIF (*JPEG file interchange format*)

jitter *m* ÉLECTRON jitter

joindre *vt* DONNÉES attach, GÉNÉRAL joint, TÉL reach, *abonné* obtain

joint *m* COMPOSANTS gasket, joint; **~ brasé** COMPOSANTS soldered joint; **~ caoutchouc** ÉLECTRON reusable rubber; **~ étanche** COMPOSANTS, ÉLECTRON seal; **~ d'étanchéité** COMPOSANTS closure gasket, seal, ÉLECTRON multirange grommet, seal; **~ de guide d'ondes** COMPOSANTS waveguide gasket; **~ hermétique** ÉLECTRON gas-tight seal; **~ torique** COMPOSANTS O-ring; **~ tournant** ALIM ÉLEC, RADIO, TÉL rotating joint; **~ d'usure** COMMUT, COMPOSANTS wear shim

jointif *adj* RADIO seamless

joncteur *m* COMMUT junctor, trunk, COMPOSANTS junctor, trunk-line relay set, DONNÉES junctor, ÉLECTRON trunk circuit, TÉL trunk, trunk circuit, trunk-line relay set, TRANSMIS junctor, trunk circuit; **~ d'abonné** COMMUT line circuit, TÉL line circuit, subscriber line circuit (*SLC*), subscriber line junctor; **~ d'arrivée** COMMUT, COMPOSANTS terminating junctor, TÉL terminating junctor, incoming trunk circuit, TRANSMIS incoming trunk circuit; **~ de circuit** TÉL circuit junctor; **~ de départ** COMMUT, COMPOSANTS originating junctor, TÉL, TRANSMIS outgoing trunk circuit; **~ entrant** TÉL, TRANSMIS incoming trunk circuit; **~ de ligne réseau** COMMUT exchange line junction set, exchange line junctor; **~ réseau départ** COMMUT outgoing trunk circuit; **~ sortant** TÉL, TRANSMIS outgoing trunk circuit

jonction *f* COMPOSANTS junction, DONNÉES interface (*I/F*), ÉLECTRON junction, GÉNÉRAL interface (*I/F*), RADIO, TÉL, TRANSMIS junction; **~ adaptée** RADIO matched junction; **~ de câbles** TRANSMIS cable trunk; **~ collecteur** ÉLECTRON collector junction; **~ émetteur-base** COMPOSANTS emitter junction; **~ par tirage** ÉLECTRON grown junction; **~ hybride** RADIO hybrid junction; **~ de modem** DONNÉES modem interface; **~ obtenue par tirage** ÉLECTRON grown junction; **~ par alliage** ÉLECTRON *semi-conducteurs* alloyed junction; **~ par diffusion** COMPOSANTS diffused junction; **~ PN** ÉLECTRON PN junction; **~ rentrante** TÉL re-entrant trunking; **~ au réseau terrestre** TRANSMIS terrestrial interface; **~ en T** RADIO T-junction; **~ tirée** ÉLECTRON grown junction; **~ urbaine** TRANSMIS local junction

jouer *vt* DONNÉES *les interrogations* handle, INFORMAT play

jour: **~ julien modifié** *m* RADIO modified Julian day

journal *m* DONNÉES log, report, newspaper, RADIO logbook, log, TÉL log; **~ d'appel** TÉL incoming-call log; **~ d'audit de sécurité** DONNÉES security audit trail; **~ de bord** INFORMAT log; **~ de communication** DONNÉES activity report, logbook; **~ des communications** RADIO, TÉL log; **~ de conformité** DONNÉES conformance log; **~ en ligne** DONNÉES on-line newspaper; **~ TV** TV current affairs programme

journaliser *vt* GÉNÉRAL log

journalisme: **~ électronique** *m* GÉNÉRAL, RADIO, TV electronic news gathering (*ENG*); **~ électronique par satellite** *m* RADIO, TV satellite news gathering (*SNG*)

JPEG *abrév* (*groupe mixte d'experts en photographie*) INFORMAT, TÉL, TV JPEG (*Joint Photographic Experts Group*)

justificatif: **~ d'identité** *m* DONNÉES credentials

justification *f* DONNÉES, INFORMAT justification, flush, TÉL, TRANSMIS justification, pulse stuffing; **~ négative** TÉL, TRANSMIS negative justification; **~ nulle** TÉL zero justification, TRANSMIS justification; **~ positive** TÉL, TRANSMIS positive justification; **~ positive-négative** TÉL, TRANSMIS positive-negative justification

justifier *vt* DONNÉES, INFORMAT, TÉL justify; **~ à droite** DONNÉES, INFORMAT right-justify; **~ à gauche** DONNÉES, INFORMAT left-justify

K

kaledia *m* INFORMAT kaledia
kHz *abrév* (*kilohertz*) ÉLECTRON, TESTS, TRANSMIS kc (*kilocycle*), kHz (*kilohertz*)
kilo: **~-instructions par seconde** *loc* INFORMAT kilo-instructions per second (*KIPS*); **~-operations par seconde** *loc* INFORMAT kilo-operations per second (*KOPS*)
kilobaud *m* INFORMAT kilobaud
kilohertz *m* (*kHz*) TESTS kilocycle (*kc*)
kiosque *m* DONNÉES kiosk; **~ Internet** INFORMAT Internet kiosk
kit: **~ antipoussière** *m* ÉLECTRON dustproofing kit; **~ de connexion** *m* DONNÉES connection kit; **~ d'installation sur véhicule** *m* RADIO *mobicom*, TÉL car installation kit; **~ d'outillage de réparation** *m* GÉNÉRAL, TÉL repair kit; **~ de pièces de rechange** *m* COMPOSANTS spare parts kit
klystron *m* ÉLECTRON, RADIO klystron; **~ multicavité** ÉLECTRON multicavity klystron; **~ reflex** ÉLECTRON reflex klystron; **~ à tube de glissement flottant** ÉLECTRON floating drift-tube klystron
KMS *abrév* (*système de gestion de connaissances*) INFORMAT KMS (*knowledge management system*)
krarupisation *f* GÉNÉRAL *de câbles*, TRANSMIS *de câbles* krarupization, continuous loading
krarupiser *vt* GÉNÉRAL, TRANSMIS krarupize, load continuously
krypton *m* COMPOSANTS krypton

L

LA *abrév* (*application locale*) TÉL LA (*local application*)
label: ~ **début** *m* INFORMAT HDR label, header label; ~ **interne** *m* DONNÉES internal label
laçage: ~ **de câble** *m* TRANSMIS cable forming
lacune *f* COMPOSANTS hole, vacancy, gap, ÉLECTRON hole, INFORMAT vacancy
lame: ~ **fixe de contact** *f* ALIM ÉLEC contact bar; ~ **d'inscription** *f* COMPOSANTS, TÉL writing bar, writing edge; ~ **prise modulaire** *f* ÉLECTRON modular plug blade; ~ **de rasoir** *f* ÉLECTRON razor blade; ~ **réversible** *f* ÉLECTRON reversible blade; ~ **d'usure** *f* COMPOSANTS wear blade
lamelle: ~ **de contact** *f* TÉL bank contact
laminage *m* ÉLECTRON lamination
lampe: ~ **d'appel** *f* PÉRIPH calling lamp, TÉL line lamp; ~ **de blocage** *f* TÉL blocking lamp; ~ **clignotante** *f* TV flickering light; ~ **à décharge** *f* COMPOSANTS, ÉLECTRON glow lamp; ~ **à décharge luminescente** *f* COMPOSANTS, ÉLECTRON glow discharge lamp; ~ **de fin de conversation** *f* TÉL clearing lamp; ~ **à incandescence** *f* CLIMAT filament lamp; ~ **luminescente** *f* COMPOSANTS, ÉLECTRON glow lamp; ~ **d'occupation** *f* TÉL busy lamp, seizure lamp; ~ **pilote** *f* TÉL signal lamp; ~ **portative** *f* GÉNÉRAL inspection lamp; ~ **de réponse** *f* TÉL answer lamp; ~ **témoin** *f* INFORMAT display light, pilot lamp, pilot light, PÉRIPH, RADIO pilot lamp, pilot light, TÉL indicating lamp, locating lamp, pilot lamp, pilot light; ~ **au xénon** *f* CLIMAT xenon lamp
lancement *m* DONNÉES, GÉNÉRAL, INFORMAT, RADIO launch, launching, TÉL calling, calling into execution, introduction, TRANSMIS launching; ~ **d'appel** RADIO call origination; ~ **de commande** TÉL transfer of control; ~ **de sauvegarde** RADIO backup launch; ~ **de secours** RADIO backup launch; ~ **de système** INFORMAT boot-up
lancer[1]: ~ **de rayons** *m* INFORMAT ray tracing
lancer[2] *vt* DONNÉES, GÉNÉRAL launch, INFORMAT launch, boot, boot up, bootstrap, RADIO initiate, launch, TÉL call into execution, initiate
lanceur *m* ÉLECTRON clamp; ~ **de la communication** DONNÉES call initiator; ~ **de sauvegarde** RADIO backup launcher; ~ **de secours** RADIO backup launcher
langage *m* INFORMAT language; ~ **algorithmique** (*ALGOL*) INFORMAT algorithmic language (*ALGOL*); ~ **artificiel** INFORMAT artificial language; ~ **d'assemblage** INFORMAT assembly language; ~ **assembleur** INFORMAT assembly language; ~ **de bases de données** DONNÉES, INFORMAT database language; ~ **à bas niveau** INFORMAT low-level language; ~ **clair** INFORMAT plain language; ~ **de commande** INFORMAT command language; ~ **de contrôle** ÉLECTRON, INFORMAT, TÉL control language; ~ **de contrôle de réseau** ÉLECTRON, TÉL network control language (*NCL*); ~ **conversationnel** INFORMAT conversational language; ~ **CPL/1** INFORMAT CPL/1 language; ~ **de description** DONNÉES description language; ~ **de description de circuits intégrés à très grande vitesse** (*langage VHDL*) INFORMAT very high-speed integrated circuit hardware description language (*VHDL*); ~ **de description de données** DONNÉES, INFORMAT data-description language; ~ **de description et de spécification** (*LDS*) INFORMAT specification and description language (*SDL*); ~ **de description de page** DONNÉES, INFORMAT page description language (*PDL*); ~ **de description structurelle de document** DONNÉES document description language; ~ **évolué** DONNÉES, ÉLECTRON, INFORMAT high-level language (*HLL*); ~ **de haut niveau** (*LHN*) DONNÉES, ÉLECTRON, INFORMAT high-level language (*HLL*); ~ **de haut niveau défini par le CCITT** INFORMAT, TÉL CCITT high-level language; ~ **homme-machine** INFORMAT man-machine language (*MML*); ~ **informatique** INFORMAT computer language; ~ **interactif** INFORMAT conversational language; ~ **d'interrogation** DONNÉES query language; ~ **d'interrogation structuré** (*SQL*) DONNÉES structured query language (*SQL*); ~ **machine** INFORMAT machine language, computer language; ~ **de manipulation de données** DONNÉES data-manipulation language; ~ **de modèles de pages en réalité virtuelle** INFORMAT virtual-reality modelling language (*VRML*); ~ **naturel** DONNÉES natural language; ~ **normalisé de balisage généralisé** INFORMAT standard generalized markup language (*SGML*); ~ **objet** INFORMAT object language; ~ **PL/M** INFORMAT PL/M language; ~ **de programme de gestion** INFORMAT control program language; ~ **source** INFORMAT source language; ~ **stratifié** INFORMAT stratified language; ~ **VHDL** (*langage de description de circuits intégrés à très grande vitesse*) INFORMAT VHDL (*very high-speed integrated circuit hardware description language*)
langue *f* DONNÉES, GÉNÉRAL, INFORMAT, TÉL language; ~ **cible** INFORMAT target language; ~ **naturelle** DONNÉES natural language
languette: ~ **de raccordement** *f* ÉLECTRON connecting tag
laque *f* COMPOSANTS lacquer; ~ **blanche** ÉLECTRON white lacquer
large: ~ **bande** *f* DONNÉES, INFORMAT, TÉL, TRANSMIS broadband
largeur *f* GÉNÉRAL width; ~ **angulaire** RADIO antenna width; ~ **de bande** RADIO, TRANSMIS bandwidth; ~ **de la bande de base** RADIO, TRANSMIS baseband bandwidth; ~ **de bande de cohérence** RADIO, TRANSMIS coherence bandwidth; ~ **de bande à la demande** INFORMAT bandwidth on demand; ~ **de bande effective globale de bruit** RADIO, TRANSMIS width of the effective overall noise band; ~ **de bande de fonctionnement** RADIO, TRANSMIS operating bandwidth; ~ **de bande à mi-puissance** RADIO, TRANSMIS half-power bandwidth; ~ **de bande nécessaire** RADIO, TRANSMIS necessary bandwidth; ~ **de bande nominale** INFORMAT, TV nominal bandwidth; ~ **de bande occupée** RADIO, TRANSMIS occupied bandwidth, occupied spectrum space; ~ **de bande à six décibels** ÉLECTRON nose bandwidth; ~ **de bande spectrale** RADIO, TRANSMIS spectral bandwidth; ~ **de bande du système d'alimentation** RADIO, TRANSMIS feed system bandwidth; ~ **du conducteur** ÉLECTRON conductor width; ~ **de faisceau** TRANSMIS beam width; ~ **d'impulsion** DONNÉES, ÉLECTRON, RADIO, TÉL, TRANSMIS pulse width (*PW*); ~ **à mi-crête** (*FWHM*) DONNÉES, ÉLECTRON, TRANSMIS full-width half-maximum (*FWHM*); ~ **nominale de canal** TV nominal channel width; ~ **de l'œil** RADIO eye width; ~ **de segment** INFORMAT stroke width; ~ **spectrale** TRANSMIS spectral width; ~ **de trait** ÉLECTRON line width
laryngophone *m* RADIO throat microphone, TÉL laryngophone, throat microphone, TRANSMIS throat microphone
laser *m* GÉNÉRAL laser; ~ **à diode** COMPOSANTS, ÉLECTRON, TRANSMIS diode laser, semiconductor laser; ~ **à double guide accordable** ÉLECTRON tunable twin-guide laser (*TTG laser*); ~ **à injection** COMPOSANTS, ÉLECTRON, TRANSMIS diode laser, semiconductor laser; ~ **à modes multiples**

ÉLECTRON, TRANSMIS multimode laser; ~ **monofréquence** ÉLECTRON, TRANSMIS single-frequency laser; ~ **monomodal** DONNÉES, ÉLECTRON, TRANSMIS monomode laser; ~ **multimodal** DONNÉES, ÉLECTRON, TRANSMIS multimode laser; ~ **rayons X** ÉLECTRON X-ray laser; ~ **à rétroaction répartie** DONNÉES, ÉLECTRON, TRANSMIS distributed feedback laser; ~ **semiconducteur** COMPOSANTS, ÉLECTRON, TRANSMIS diode laser, semiconductor laser; ~ **synchronisé par injection** ÉLECTRON injection-locked laser; ~ **ultrarapide** ÉLECTRON high-speed laser; ~ **unimodal** DONNÉES, ÉLECTRON, TRANSMIS monomode laser, single-mode laser; ~ **verrouillé par injection** ÉLECTRON injection-locked laser

LDS *abrév* (*langage de description et de spécification*) INFORMAT SDL (*specification and description language*)

lecteur *m* GÉNÉRAL reader, INFORMAT drive, reader, scanner; ~ **de badge** APP SERV badge reader; ~ **de bande** PÉRIPH tape recorder; ~ **de bande de contrôle** DONNÉES tally reader; ~ **de bande vidéo** TV videotape player (*VTP*); ~ **de caractères** APP SERV character reader; ~ **de carte** APP SERV, INFORMAT, PÉRIPH, TV card reader; ~ **de carte à mémoire** APP SERV smartcard reader; ~ **de cartouche** INFORMAT tape cartridge drive; ~ **de cassette vidéo** (*VCP*) TV videocassette player (*VCP*); ~ **CD** INFORMAT CD reader; ~ **de CD-ROM** INFORMAT CD-ROM drive, CD-ROM player; ~ **de code à barres** APP SERV, DONNÉES bar-code scanner; ~ **de courbe** INFORMAT curve follower; ~ **de disque** INFORMAT disk drive; ~ **de disque optique** DONNÉES optical disc drive; ~ **de disquette** INFORMAT diskette reader, diskette drive; ~ **laser portatif** INFORMAT Discman®; ~ **de nouvelles** INFORMAT newsreader; ~ **optique** DONNÉES, INFORMAT optical scanner, optical reader, TV optical reader; ~ **optique de code à barres** DONNÉES optical bar-code reader; ~**-perforateur de bande** INFORMAT tape station; ~ **de photo CD** INFORMAT photo CD player; ~**-reproducteur de microformes** DONNÉES microform reader/printer

lecture[1] *f* GÉNÉRAL read-out, reading, INFORMAT playback, read-out, reading, PÉRIPH, TV playback; ~ **arrière rapide** TÉL fast reverse playback; ~ **automatique** INFORMAT automatic playback; ~ **avant rapide** TÉL fast-forward playback; ~ **de banque** INFORMAT bank play; ~ **de chaîne** INFORMAT chain play; ~ **destructive** INFORMAT destructive read; ~ **d'une échelle de temps** INFORMAT, TRANSMIS timescale reading; ~ **linéaire** INFORMAT linear reading; ~ **de marques** GÉNÉRAL mark sensing; ~ **numérique** TESTS digital read-out; ~ **optique** DONNÉES, TV optical reading; ~ **optique de marques** DONNÉES mark scanning; ~ **par contact** TV mechanical contact reading; ~ **rapide** TÉL fast-forward playback; ~ **répétitive** INFORMAT repeat; ~ **vidéo par logiciel seul** INFORMAT, TV software-only video playback

légende *f* DONNÉES key to callouts, GÉNÉRAL marking, RADIO key

Lempel-Ziv-Welch *m* (*LZW*) INFORMAT Lempel-Ziv-Welch (*LZW*)

lent *adj* DONNÉES low-speed

lentille *f* INFORMAT lens; ~ **diélectrique** RADIO dielectric lens; ~ **E** RADIO E-plane lens; ~ **à échelons** RADIO zoned lens; ~ **à lames paralèlles** RADIO parallel-plate lens; ~ **multicellulaire** RADIO egg-box lens; ~ **radioélectrique** RADIO lens; ~ **à retard de phase** RADIO delay lens

LEOP *abrév* (*phase de lancement et de mise en orbite*) RADIO LEOP (*launch and early orbit phase*)

lettrage: ~ **blanc** *m* ÉLECTRON silk-screened lettering

lettre: ~ **inférieure** *f* DONNÉES, INFORMAT lower case; ~ **supérieure** *f* DONNÉES, INFORMAT upper case

lettres: ~ **espacées** *f pl* PÉRIPH spaced-out letters

levée: ~ **d'ambiguïté sur la phase** *f* TRANSMIS phase ambiguity resolution

levier *m* GÉNÉRAL lever; ~ **de dégagement** TÉL release lever; ~ **de frein** COMPOSANTS *génie mécanique* brake handle; ~ **de modulation** INFORMAT modulation lever

LHN *abrév* (*langage évolué, langage de haut niveau*) DONNÉES, ÉLECTRON, INFORMAT HLL (*high-level language*)

liaison *f* GÉNÉRAL link; ~ **d'accès** TÉL access link (*AL*); ~ **air-sol** RADIO air-ground link; ~ **ascendante** RADIO, TÉL uplink; ~ **en assemblage de quinze groupes secondaires** TÉL fifteen supergroup assembly link; ~ **asynchrone** DONNÉES asynchronous link; ~ **bidirectionnelle** INFORMAT two-way link; ~ **en cascade** INFORMAT cascade connection; ~ **commutée en mode circuit** DONNÉES circuit-switched connection; ~ **de connexion** RADIO, TRANSMIS feeder link; ~ **de contribution** RADIO contribution link; ~ **à courant porteur** TRANSMIS carrier-current connection; ~ **descendante** RADIO, TÉL down link; ~ **à deux bonds** RADIO second satellite hop; ~ **de distribution** RADIO distribution link; ~ **de distribution primaire** RADIO primary distribution link; ~ **de distribution secondaire** RADIO second distribution link, secondary distribution link; ~ **à diversité angulaire** DONNÉES angle diversity link; ~ **de données** DONNÉES, TRANSMIS data link; ~ **de données de signalisation** SIGNAL signaling data link (*AmE*), signalling data link (*BrE*); ~ **double attachement** ÉLECTRON dual homing; ~ **double interurbaine** ÉLECTRON long-distance dual-homing link; ~ **double sur longue distance** ÉLECTRON long-distance dual-homing link; ~ **entre satellites** RADIO, TV satellite-to-satellite link; ~ **à fibres optiques** ALIM ÉLEC, TRANSMIS optical fiber link (*AmE*) (*F/O link*), optical fibre link (*BrE*) (*F/O link*); ~ **à fréquence porteuse** TRANSMIS carrier-frequency connection; ~**-fusible** ALIM ÉLEC fuse link; ~ **à grande distance** TÉL long-distance connection; ~ **en groupe primaire** TÉL group link; ~ **en groupe quaternaire** TÉL supermastergroup link; ~ **en groupe secondaire** TÉL supergroup link; ~ **en groupe tertiaire** TÉL mastergroup link; ~ **hertzienne** RADIO microwave link; ~ **infrarouge** TRANSMIS infrared light link; ~ **d'interconnexion sur le site** TRANSMIS cross-site link; ~ **interface** ÉLECTRON interfacial connection; ~ **internationale** TÉL international link; ~ **internationale louée en groupe primaire** TÉL international leased group link; ~ **interprocesseur** DONNÉES interprocessor link; ~ **intersatellite** (*LIS*) RADIO intersatellite link (*ISL*); ~ **interurbaine** TÉL trunk connection, long-distance connection; ~ **en ligne** TÉL line link; ~ **de ligne numérique** TÉL, TRANSMIS digital line link; ~ **louée** TRANSMIS private circuit; ~ **MIC** TÉL PCM link; ~ **MIC interne** TÉL internal PCM link; ~ **de modulation numérique** INFORMAT, TRANSMIS digital modulation link; ~ **montante** RADIO, TÉL uplink; ~ **multiplex** TÉL multiplex link; ~ **multipoint** DONNÉES multipoint connection; ~ **multisatellite** RADIO, TV multisatellite link; ~ **non-transparente** TRANSMIS regenerative link; ~ **optique** TRANSMIS optical link; ~ **d'ordinateur** INFORMAT, TRANSMIS computer link; ~ **par câble** TRANSMIS *câbles* cable connection; ~ **par satellite** RADIO, TV satellite link; ~ **par transformateur** ALIM ÉLEC transformer coupling; ~ **permanente** GÉNÉRAL permanent link; ~ **à plusieurs circuits** RADIO multilink connection; ~ **porte-répéteur à fibres optiques** ÉLECTRON fiber-optic bridge-repeater link (*AmE*), fibre-optic bridge-repeater link; ~ **préprogrammée** TÉL preprogrammed line; ~ **de prestataire de service** TRANSMIS service provider link (*SPL*); ~ **primaire** DONNÉES, INFORMAT, TRANSMIS primary link (*PL*); ~ **primaire interactive** DONNÉES, INFORMAT interactive primary link (*IPL*); ~ **primaire pour les services de distribution** TRANSMIS distribution primary link (*DPL*); ~ **radio** RADIO radio link; ~ **radioélectrique** RADIO radio link; ~ **radiophonique internationale** TV international sound-program link (*AmE*), international sound-programme link (*BrE*); ~ **radiophonique internationale multiple** TÉL international multiple-destination sound-program link (*AmE*), international multiple-destination sound-programme link (*BrE*); **~ réseau** TÉL matrix link; **~ réseau entrante** TÉL incoming matrix link; **~ réseau sortante** TÉL outgoing matrix link; **~ satellite à plusieurs bonds** RADIO multihop satellite link; **~ satellite-terre** RADIO, TÉL down link; **~ à secret** TÉL privacy circuit, privacy connection; **~ sémaphore de données** SIGNAL signaling data link (*AmE*), signalling data link (*BrE*); **~ semi-permanente** TÉL semipermanent link; **~ de signalisation** SIGNAL, TÉL signalling link (*BrE*); **~ de signalisation régulière** SIGNAL, TÉL regular signaling link (*AmE*), regular signalling link (*BrE*); **~ spécialisée** (*LS*) DONNÉES direct line (*DL*), dedicated line, TÉL dedicated line, TRANSMIS direct line (*DL*), private wire (*PW*), dedicated line; **~ spécialisée à tous usages** TRANSMIS analog private wire, speech-grade private wire; **~ spécialisée transmission de données** DONNÉES, TRANSMIS data private wire; **~ studio-émetteur** (*LSE*) RADIO studio-transmitter link (*STL*); **~ synchrone** DONNÉES synchronous link; **~ téléphonique** TÉL telephone circuit, telephone connection, telephone link; **~ télévisuelle internationale** TV international television link; **~ télévisuelle internationale multiple** TV international multiple-destination television link; **~ terre-satellite sens montant** RADIO, TÉL uplink; **~ de transfert** TÉL transfer link; **~ de transmission** TÉL, TRANSMIS transmission link; **~ de transmission analogique** TRANSMIS analog transmission link; **~ de transmission de données par paquets** DONNÉES, TRANSMIS packet data link; **~ de transmission numérique** TRANSMIS digital transmission link; **~ transversale** TÉL transverse junction; **~ unidirectionnelle** INFORMAT one-way link; **~ vers réseau** TÉL matrix link; **~ vidéo** TV video link

liaisons *f pl* GÉNÉRAL flow

LIB *abrév* (*message de libération*) TESTS clear message, clearing message, REL (*release message*)

libération[1] *f* COMMUT, INFORMAT, TÉL clearing, disconnection, *de la connexion* release; **~ de la communication** COMMUT, INFORMAT, TÉL call clearing; **~ par l'abonné demandé** COMMUT, INFORMAT, TÉL called-party release, called-subscriber release; **~ par l'abonné demandeur** COMMUT, INFORMAT, TÉL calling-party release, calling-subscriber release; **~ par le dernier abonné** COMMUT, INFORMAT, TÉL last-party release; **~ par les deux abonnés** COMMUT, INFORMAT, TÉL both-party release; **~ par premier abonné** COMMUT, INFORMAT, TÉL first-party release, first-subscriber release; **~ prématurée** COMMUT, INFORMAT, TÉL premature release; **~ du secret** COMMUT, INFORMAT, TÉL privacy release; **~ sur abandon d'appel** COMMUT, INFORMAT, TÉL abandoned call disconnect

libération[2]: **~ d'alarme** *loc* TESTS alarm cleared

libérer *vt* COMMUT *circuits* release, RADIO free, TÉL free, clear, disconnect, TRANSMIS free

libre *adj* ÉLECTRON free-running, idle, GÉNÉRAL idle, free, unrestricted, TÉL free, idle, TRANSMIS idle

lien *m* INFORMAT link, linkage, TRANSMIS link; **~ bidirectionnel** INFORMAT two-way directionality; **~ chronologique** INFORMAT chronological link; **~ cible** INFORMAT destination link, target link; **~ créé par l'auteur** INFORMAT author-created link; **~ créé par l'utilisateur** INFORMAT user-created link; **~ de destination** INFORMAT destination link, target link; **~ explicite** INFORMAT explicit link; **~ de groupement** INFORMAT cluster link; **~ hiérarchique** INFORMAT hierarchical link; **~ hypermédia** INFORMAT hypermedia link; **~ hypertexte** INFORMAT hypertext link; **~ d'organisation** INFORMAT organizational link; **~ par mot-clé** INFORMAT keyword link; **~ primaire** INFORMAT, TRANSMIS primary link (*PL*); **~ de référence** INFORMAT referential link; **~ à sens unique** INFORMAT one-way directionality, one-way link; **~ source** INFORMAT source link; **~ unidirectionnel** INFORMAT one-way link

lier *vt* GÉNÉRAL link

lieu: **~ non-surveillé** *m* ÉLECTRON nonsupervised location; **~ de transfert** *m* CONTRÔLE loop transfer locus

LIFO *abrév* (*dernier entré, premier sorti*) INFORMAT LIFO (*last in, first out*)

ligne[1]: **en ~** *adj* DONNÉES, ÉLECTRON, INFORMAT, on-line, TÉL on-hook, on-line

ligne[2]: **en ~** *adv* DONNÉES, ÉLECTRON, INFORMAT on-line, TÉL on hook, on-line

ligne[3] *f* GÉNÉRAL line, INFORMAT row, TÉL line unit; **~ d'abonné** DONNÉES, INFORMAT, RADIO, TÉL, TRANSMIS subscriber loop, subscriber's line; **~ d'abonné à haut débit** TRANSMIS high-speed digital subscriber line (*HDSL*); **~ d'accès** DONNÉES, INFORMAT, RADIO, TÉL, TRANSMIS subscriber's line; **~ aérienne** ALIM ÉLEC, TRANSMIS overhead line; **~ d'alimentation** RADIO, TRANSMIS feeder; **~ d'annotateur** TÉL record line; **~ apériodique** RADIO flat line; **~ appelante** TÉL calling line; **~ artificielle** TRANSMIS artificial line, line simulator; **~ de balayage** INFORMAT scan line, TV scanning line; **~ de base** INFORMAT base line; **~ de branchement d'abonné** TÉL subscriber's lead-in; **~ de central** TÉL exchange line; **~ classique** TÉL plain old telephone line; **~ coaxiale** TRANSMIS coaxial line; **~ de commande** DONNÉES command line; **~ commune** TÉL party line; **~ commutée** DONNÉES dial-up line; **~ comportementale** DONNÉES behavior line (*AmE*), behaviour line (*BrE*); **~ de concentration de trafic** TÉL release link trunk (*RLT*); **~ directe** APP SERV nondialed connection (*AmE*), nondialled connection (*BrE*), DONNÉES direct line (*DL*), TÉL hot line, nondialed connection (*AmE*), nondialled connection (*BrE*), TRANSMIS direct line (*DL*); **~ directrice** GÉNÉRAL guideline; **~ de données** DONNÉES, RADIO data line; **~ double** APP SERV double-circuit line; **~ à double sens** TRANSMIS both-way line; **~ duplex** RADIO, TRANSMIS duplex line; **~ électrique** ALIM ÉLEC power line, power transmission line; **~ équipée** TRANSMIS line equipment; **~ entrante** TRANSMIS incoming line; **~ d'essai** TÉL test line; **~ d'extension** DONNÉES extension line; **~ extérieure privée** COMMUT, TÉL exclusive external line; **~ en faux appel** COMMUT, TÉL parked line; **~ à fente** RADIO slotted line; **~ en fils nus** TÉL open-wire line; **~ de fonctionnement** ÉLECTRON operating line; **~ formatrice d'impulsions** RADIO pulse-forming line; **~ à grande distance** TÉL, TRANSMIS long-distance line; **~ d'identification** DONNÉES identification line; **~ d'identification de l'appel** (*CIL*) TÉL call-identification line (*CIL*); **~ interautomatique** TÉL, TRANSMIS inter-PABX tie circuit; **~ internationale** TÉL, TRANSMIS international line; **~ interstandard** TÉL, TRANSMIS inter-switchboard line; **~ interurbaine** TÉL long-distance line, trunk line, TRANSMIS long-distance line; **~ d'invite** INFORMAT prompt line; **~ de jonction entre deux centraux** GÉNÉRAL, TÉL exchange line; **~ libre** TÉL available line, idle line, unoccupied line; **~ locale** TÉL local line; **~ de maintenance** TÉL line of maintenance; **~ médiane** TRANSMIS line of center (*AmE*), line of centre (*BrE*); **~ médiane de segment** INFORMAT stroke center line (*AmE*), stroke centre line (*BrE*); **~ de mesure à fente** RADIO slotted measuring line; **~ métallique** TRANSMIS physical line; **~ microruban** RADIO stripline; **~ mixte** TRANSMIS both-way line; **~ multipoint** TÉL party line; **~ nationale** TÉL national line; **~ non attribuée** TÉL spare line; **~ non résonnante** RADIO flat line; **~ ordinaire** TRANSMIS ordinary both-way line, ordinary line; **~ ouverte vingt-quatre heures sur vingt-quatre** TÉL twenty-four hour hotline; **~ partagée** TÉL party line, shared service line; **~ partagée entre deux abonnés** TÉL two-party line; **~ partagée entre deux postes** TÉL two-party line; **~ permanente** TÉL permanent line; **~ de pertes en surface** ALIM ÉLEC surface leakage path; **~ physique** TRANSMIS physical line; **~ principale de liaison** COMMUT, TÉL trunk flow, TRANSMIS main line; **~ de priorité de trafic**

TÉL priority-marked line; ~ **privée** COMMUT, TÉL, TRANSMIS direct line (*DL*), tie line; ~ **publique** TÉL public line; ~ **de rattachement** DONNÉES, INFORMAT, RADIO, TÉL subscriber's line; ~ **réseau** COMMUT central exchange trunk (*BrE*), central office trunk (*AmE*); ~ **à retard** DONNÉES delay line, ÉLECTRON delay circuit, RADIO *radar* pulse-forming line; ~ **à retard acoustique** DONNÉES *datacom* acoustic delay line; ~ **à retard magnétique** ÉLECTRON magnetic delay line; ~ **à retard numérique** INFORMAT digital delay line; ~ **à ruban** RADIO stripline; ~ **sandwich** TRANSMIS sandwich line; ~ **de service** TÉL service line; ~ **de service technique** TÉL engineering order wire (*EOW*); ~ **sortante** TÉL outgoing line; ~ **souterraine** TRANSMIS underground line; ~ **spécialisée** (*LS*) DONNÉES, TÉL, TRANSMIS dedicated line, leased line (*LS*); ~ **spécialisée arrivée** TÉL, TRANSMIS outgoing calls barred line, incoming-only line; ~ **spécialisée départ** TÉL, TRANSMIS incoming-calls-barred line; ~ **supplémentaire** COMMUT extension line; ~ **supraconductrice** TRANSMIS superconductor line; ~ **sur poteaux** TRANSMIS pole route; ~ **téléphonique** TÉL telephone link, telephone line; ~ **téléphonique principale** TRANSMIS main telephone line; ~ **téléphonique supplémentaire** COMPOSANTS telephone extension line; ~ **terrestre** TRANSMIS landline; ~ **de transfert** TÉL, TRANSMIS through line; ~ **de transmission** RADIO, TRANSMIS transmission line; ~ **de transmission avec transpositions** TÉL, TRANSMIS transposed line, transposed transmission line; ~ **à triple accès** COMMUT, TÉL triple-access line; ~ **à trois fils** COMMUT, TÉL three-wire line; ~ **trusquinée** COMPOSANTS scribed line, scribed mark; ~ **de visée** RADIO boresight; ~ **de vol** RADIO flight path

ligne[4]: ~ **occupée** *loc* TÉL trunk busy

lignes: ~ **par image** *f pl* TV lines per screen; ~ **mixtes** *f pl* TÉL mixed lines; ~ **multiples pour une même adresse** *f pl* TÉL multiple lines at the same address

limitation: ~ **d'accès** *f* TÉL access restriction; ~ **des collisions** *f* RADIO contention control; ~ **des conflits** *f* RADIO contention control; ~ **de courant** *f* ALIM ÉLEC current limiting; ~ **de crête** *f* TÉL peak limiting

limité: ~ **par l'affaiblissement** *adj* TRANSMIS attenuation-limited; ~ **par le bruit quantique** *adj* TRANSMIS quantum-noise-limited; ~ **par la dispersion** *adj* TRANSMIS dispersion-limited; ~ **par la distorsion** *adj* TRANSMIS distortion-limited; ~ **par la largeur de la bande** *adj* TRANSMIS bandwidth-limited

limite *f* COMPOSANTS, DONNÉES limit, ÉLECTRON *semiconducteurs* boundary, GÉNÉRAL limit, limitation, INFORMAT boundary; ~ **apparente d'élasticité** COMPOSANTS tensile yield limit; ~ **de bruit** RADIO noise limit; ~ **de caractères** APP SERV character boundary; ~ **d'une cellule** RADIO cell border; ~ **de confiance** GÉNÉRAL confidence limit; ~ **d'élasticité** COMPOSANTS *résistance*, TESTS breaking strength, ultimate strength, yield limit; ~ **de faisceau** RADIO beam edge; ~ **hors axe** RADIO off-axis limit; ~ **d'objet géré** DONNÉES managed-object boundary; ~ **de puissance** ÉLECTRON power limit; ~ **de qualité moyenne après contrôle** TESTS average outgoing quality limit (*AOQL*); ~ **de résolution au sol** ÉLECTRON ground resolution; ~ **statistique de couverture** TESTS statistical coverage limit; ~ **de surveillance** CONTRÔLE warning limit; ~ **de tolérance** TESTS tolerance limit

limites *f pl* RADIO zone boundaries; ~ **de contrôle** TESTS *qualité* control limits; ~ **d'exposition** GÉNÉRAL exposure limits; ~ **de maintenance** TÉL limits for maintenance purposes

limiteur *m* ALIM ÉLEC, ÉLECTRON, RADIO, TÉL limiter; ~ **de courant** ALIM ÉLEC current limiter; ~ **de durée de conversation** RADIO, TÉL conversation-time limiter; ~ **de parasites** TÉL noise limiter; ~ **de surtension** ALIM ÉLEC surge suppressor; ~ **de surtension transitoire** ALIM ÉLEC transient suppressor; ~ **de tension** ALIM ÉLEC voltage limiter

linéaire *adj* GÉNÉRAL linear

linéarisateur *m* TRANSMIS linearizer; ~ **d'amplificateur à haute puissance** ALIM ÉLEC, ÉLECTRON high-power amplifier linearizer

linéarité *f* GÉNÉRAL linearity

linguiste: ~**-informaticien** *m* INFORMAT computer linguist

linkway *m* INFORMAT linkway

liquidation: ~ **des comptes** *f* TÉL revenue-sharing arrangement

LIS *abrév* (*liaison intersatellite*) RADIO ISL (*intersatellite link*)

lisible *adj* GÉNÉRAL, INFORMAT readable, legible

lissage *m* GÉNÉRAL, INFORMAT smoothing, shading; ~ **constant** INFORMAT constant shading; ~ **de Gouraud** TV Gouraud shading; ~ **par interpolation bilinéaire** TV intensity interpolation shading; ~ **par interpolation de la couleur** INFORMAT color interpolation shading (*AmE*), colour interpolation shading (*BrE*); ~ **par interpolation des vecteurs de la normale** INFORMAT normal-vector interpolation shading; ~ **de Phong** INFORMAT Phong shading; ~ **polygonal** INFORMAT polygonal shading

lisse *adj* GÉNÉRAL smooth

lisser *vt* GÉNÉRAL smooth

listage *m* DONNÉES, INFORMAT listing; ~ **des appels** TÉL call log

liste[1]: **sur** ~ **rouge** *adj* TÉL ex-directory (*BrE*) (*XD*), unlisted (*AmE*)

liste[2]: ~ **des câbles** *f* TRANSMIS *documentation* cable list; ~ **des connexions enroulées** *f* COMPOSANTS wire-wrap connection list; ~ **de contrôle** *f* TESTS check list; ~ **de contrôle d'accès** *f* DONNÉES, TESTS access control list; ~ **de contrôle du client** *f* GÉNÉRAL client checklist; ~ **de contrôle de laboratoire de test** *f* TESTS test laboratory checklist; ~ **de diffusion** *f* DONNÉES, INFORMAT *pour la poste* distribution list, mailing list; ~ **directe** *f* DONNÉES, INFORMAT push-up list; ~ **de distribution** *f* DONNÉES, TÉL distribution list (*DL*); ~ **de groupes de câbles** *f* TRANSMIS *documentation* cable group list; ~ **inversée** *f* DONNÉES, INFORMAT push-down list; ~ **des mots-clés** *f* INFORMAT keyword list; ~ **refoulée** *f* DONNÉES, INFORMAT push-down list; ~ **de serveurs** *f* DONNÉES bookmark; ~ **des stations terriennes** *f* RADIO earth station record; ~ **des voies d'acheminement** *f* TÉL route list

listing *m* DONNÉES, INFORMAT listing

littéral *m* INFORMAT literal

livre: ~ **électronique** *m* DONNÉES electronic book; ~ **jaune** *m* INFORMAT yellow book; ~ **orange** *m* INFORMAT orange book; ~ **rouge** *m* INFORMAT red book

LLC-IE *abrév* (*élément d'information de compatibilité de couche inférieure*) COMMUT, TÉL LLC-IE (*low-layer compatibility information element*)

lm/W *abrév* (*lumen par watt*) TESTS lm/W (*lumen per watt*)

lobe *m* RADIO lobe; ~ **arrière** RADIO *antennes* back lobe; ~ **latéral** RADIO *antennes* side lobe; ~ **latéral d'antenne** RADIO antenna side lobe; ~ **latéral de grande ouverture** RADIO wide-angle side lobe; ~ **principal** RADIO main lobe; ~ **secondaire** RADIO side lobe

local[1]: **en** ~ *adj* DONNÉES off-line

local[2]: ~ **de câblage** *m* ÉLECTRON wiring closet

localisateur: ~ **de câble** *m* TRANSMIS *génie industriel extérieur* cable locater

localisation *f* ÉLECTRON pinpointing, INFORMAT locating, pinpointing, RADIO siting, whereabouts, *endroit* location, *repérage* localization, locating, TESTS locating; ~ **automatique** INFORMAT auto-locate, RADIO roaming; ~ **automatique de véhicules** APP SERV, RADIO automatic vehicle location (*AVL*); ~ **d'avaries** TÉL fault tracing; ~ **des câbles** TRANSMIS cable localization; ~ **des défauts** TÉL, TESTS fault localization; ~ **des dérangements** TÉL, TESTS fault localization; ~ **nominale** RADIO home location; ~ **de panne** TÉL, TESTS trouble shooting; ~ **par radio** RADIO radio location; ~ **sommaire** TESTS general localization

localiser *vt* ÉLECTRON pinpoint, INFORMAT pinpoint, locate, RADIO locate, site, TESTS locate; **~ une défaillance** TESTS trace a fault

log: **~ in** *vt* DONNÉES login, INFORMAT log in

logarithme *m* RADIO logarithm

logatome *m* RADIO, TÉL logatom

logement *m* INFORMAT slot; **~ d'antenne** RADIO antenna building

logements: **~ cerclés** *m pl* TV homes passed

loger *vt* COMPOSANTS house, GÉNÉRAL locate, RADIO site

logiciel *m* DONNÉES, INFORMAT computer program, software; **~ d'aide** DONNÉES, INFORMAT helpware; **~ d'antémémoire de disque** DONNÉES, INFORMAT disk caching software; **~ d'assistance** DONNÉES, INFORMAT helpware; **~ d'auteur** DONNÉES, INFORMAT authoring software; **~ de balayage** DONNÉES, INFORMAT browser; **~ de base** DONNÉES, INFORMAT system software, *du multiprocesseur de communication* operating system (*OS*); **~ de base et fonctions système de toute station SM** DONNÉES, INFORMAT, TÉL core functions control station software; **~ client** DONNÉES, INFORMAT client software; **~ de communication** DONNÉES, INFORMAT communication software; **~ complémentaire** DONNÉES, INFORMAT add-on software; **~ de compression de données** DONNÉES data-compression software; **~ didacticiel** DONNÉES, INFORMAT teachware; **~ du domaine public** DONNÉES, INFORMAT public-domain software; **~ d'émulation de terminal** DONNÉES, INFORMAT terminal emulation software; **~ d'essai** DONNÉES, INFORMAT testware; **~ expert** DONNÉES, INFORMAT knowledgeware; **~ d'extension** DONNÉES, INFORMAT add-on software; **~ fantôme** DONNÉES, INFORMAT vaporware (*AmE*), vapourware (*BrE*); **~ de filtrage** DONNÉES, INFORMAT firewall; **~ de gestion** DONNÉES, INFORMAT management software; **~ de groupe** DONNÉES, INFORMAT groupware; **~ d'initiation** DONNÉES, INFORMAT tutorial; **~ intégré** DONNÉES, INFORMAT integrated software; **~ en libre service** DONNÉES, INFORMAT shareware; **~ de maintien à poste** DONNÉES, INFORMAT on-station software; **~ de mise en pages** DONNÉES, INFORMAT page layout software; **~ navigateur** DONNÉES, INFORMAT navigation software; **~ de navigation** DONNÉES, INFORMAT steering software; **~ de pilotage** DONNÉES, INFORMAT steering software; **~ public** DONNÉES, INFORMAT freeware; **~ de reconnaissance optique de caractères** DONNÉES, INFORMAT optical character-recognition software; **~ serveur** DONNÉES, INFORMAT server software; **~ en shareware** DONNÉES, INFORMAT shareware package; **~ de survol** DONNÉES, INFORMAT browser; **~ de télécommunication** DONNÉES, INFORMAT communication software; **~ de test** DONNÉES, INFORMAT testware; **~ de traitement de texte** DONNÉES, INFORMAT word processing package

login *m* DONNÉES, INFORMAT login

logique *f* GÉNÉRAL logic; **~ bipolaire** ÉLECTRON bipolar logic; **~ à couplage par les émetteurs** (*ECL*) DONNÉES, ÉLECTRON emitter-coupled logic (*ECL*); **~ à décision majoritaire** ÉLECTRON majority decision logic; **~ de division verticale** TV vertical grating logic; **~ ECL positive** INFORMAT positive emitter-coupled logic (*PECL*); **~ d'édition de caractères** INFORMAT character-generation logic; **~ floue** INFORMAT soft logic, fuzzy logic; **~ formelle** INFORMAT formal logic; **~ de haut niveau** ÉLECTRON high-level logic; **~ à matrice programmable** (*PAL*) INFORMAT, TV programmable array logic (*PAL*); **~ multivalente** ÉLECTRON many-valued logic; **~ de noyaux magnétiques** INFORMAT core logic; **~ rapide** ÉLECTRON high-speed logic; **~ temporelle** DONNÉES temporal logic; **~ transistor-transistor** ÉLECTRON transistor-transistor logic (*TTL*)

logistique: **~ assistée par ordinateur** *f* INFORMAT computer-aided logistics (*CAL*); **~ de maintenance** *f* TÉL maintenance support performance, TESTS maintenance support logistics; **~ du service** *f* TÉL service support performance

logo *m* GÉNÉRAL logo

loi *f* GÉNÉRAL law; **~ A** TRANSMIS A-law; **~ de codage** TÉL, TRANSMIS coding law, encoding law; **~ de codage à segments** TÉL, TRANSMIS segmented coding law, segmented encoding law; **~ de composition des erreurs** TESTS law of combination of errors

loisirs: **de ~** *adj* DONNÉES *Internet* recreational (*rec.*)

longue[1]: **à ~ distance** *adj* TÉL long-distance

longue[2]: **~ ligne équipée** *f* TRANSMIS long-line equipment

longueur *f* GÉNÉRAL length; **~ d'appel** TÉL length of call; **~ du chemin optique** TRANSMIS optical path length; **~ de cohérence** TRANSMIS coherence length; **~ du contenu** DONNÉES content length; **~ de coupe** GÉNÉRAL cutting length; **~ de diffusion** COMPOSANTS diffusion length; **~ électrique** RADIO electrical length, electrical wavelength; **~ d'équilibrage** TRANSMIS equilibrium length; **~ de la file d'attente de datagrammes** DONNÉES, TÉL datagram queue length; **~ de groupe** TÉL group length (*GL*); **~ de kevlar** ÉLECTRON length of kevlar; **~ de lobe** ÉLECTRON lobe length; **~ de mot** DONNÉES word length; **~ d'onde** RADIO, TRANSMIS wavelength; **~ d'onde de coupure** RADIO, TRANSMIS cutoff wavelength; **~ d'onde dans le guide** RADIO, TRANSMIS guide wavelength; **~ d'onde du maximum** RADIO, TRANSMIS peak-intensity wavelength; **~ d'onde du maximum d'intensité** RADIO, TRANSMIS peak-intensity wavelength; **~ de paramètre** INFORMAT, TÉL parameter length (*PL*); **~ de portée** ALIM ÉLEC length of a span; **~ de rangée** TÉL row length; **~ de registre** DONNÉES register length; **~ de rupture** COMPOSANTS, TESTS *résistance* breaking length, length of fracture; **~ sans amplification** TV unamplified length; **~ totale de ligne** TÉL total scanning line length; **~ de trajet électrique** TRANSMIS electrical path length; **~ de trame** TRANSMIS frame length; **~ utile de ligne** TÉL usable scanning line length

lot *m* INFORMAT package, batch, TESTS batch; **~ de câble** TRANSMIS, TV cable set; **~ de livraison** TESTS delivery lot; **~ d'unités fonctionnelles de gestion des systèmes** INFORMAT systems management functional unit package

LS *abrév* (*ligne spécialisée*) DONNÉES, TÉL, TRANSMIS LS (*leased line*) dedicated line

LSE *abrév* (*liaison studio-émetteur*) RADIO STL (*studio-transmitter link*)

LSI *abrév* (*intégration à grande échelle*) COMPOSANTS, ÉLECTRON, TÉL, TRANSMIS LSI (*large-scale integration*)

LUF *abrév* (*fréquence minimale utilisable*) RADIO LUF (*lowest usable frequency*)

lumen[1] *m* TESTS lumen

lumen[2]: **~ par watt** *loc* (*lm/W*) TESTS lumen per watt (*lm/W*)

lumière *f* GÉNÉRAL light; **~ cohérente** TRANSMIS coherent light; **~ incohérente** TRANSMIS incoherent light; **~ visible** ÉLECTRON visible light

luminaire *m* ALIM ÉLEC lighting fixture, GÉNÉRAL light fitting, lighting fitting

luminance *f* ÉLECTRON luminance, radiance, INFORMAT luminance, TRANSMIS radiance, TV luminance; **~ énergétique** ÉLECTRON, TRANSMIS radiance; **~ énergétique spectrale** TRANSMIS spectral radiance; **~ spectrale** TRANSMIS spectral radiance

luminancemètre *m* TESTS luminancemeter

luminescence *f* TRANSMIS luminescence

lumineux *adj* ALIM ÉLEC illuminated, INFORMAT bright, illuminated

luminosité *f* GÉNÉRAL brilliance, brightness, lightness

luxmètre *m* GÉNÉRAL footcandle meter, illuminance meter

LXE *abrév* ÉLECTRON LXE (*lightguide express entry*)

LZW *abrév* (*Lempel-Ziv-Welch*) INFORMAT LZW (*Lempel-Ziv-Welch*)

M

M *abrév* (*module de réfraction*) RADIO M (*refractive modulus*)
mA *abrév* (*milliampère*) ÉLECTRON mA (*milliampere*)
MA *abrév* (*modulation d'amplitude*) DONNÉES, ÉLECTRON, INFORMAT, SIGNAL, TRANSMIS AM (*amplitude modulation*)
MAC: **~ amélioré** *m* TV enhanced MAC, E-MAC
machine: **~ pour connexions enroulées** *f* COMPOSANTS wrapping machine; **~ à dénuder** *f* COMPOSANTS stripping machine, GÉNÉRAL insulation stripper; **~ émettrice** *f* TÉL, TRANSMIS transmit machine; **~ logique** *f* INFORMAT software machine; **~ logique de gestion de commutateur** *f* COMMUT matrix switch-control software machine; **~ logique marqueur** *f* TÉL *distribution de messages à URM, ETA, COM* ML in charge of message distribution; **~ logique multi-enregistreur** *f* TÉL *traitement d'appel* call-handling software machine; **~ logique taxeur** *f* TÉL ML assigned to call charging; **~ parlante numérique** *f* PÉRIPH, TÉL digital announcement machine; **~ de protocole** *f* DONNÉES protocol machine (*PM*); **~ de protocole de canaux** *f* DONNÉES channel protocol machine (*CPM*); **~ de protocole de commande d'association** *f* DONNÉES, TÉL association control protocol machine (*ACPM*); **~ de protocole commun d'informations de gestion** *f* DONNÉES common management-information protocol machine (*CMIPM*); **~ de protocole de traitement transactionnel** *f* DONNÉES transaction-processing protocol machine (*TPPM*); **~ réceptrice** *f* TRANSMIS receive machine; **~ de traitement des données** *f* DONNÉES, INFORMAT data processor, **~ de traitement de texte** *m* DONNÉES, INFORMAT word processor; **~ virtuelle** *f* INFORMAT virtual machine
mâchoire *f* ÉLECTRON clamp
maclage *m* ÉLECTRON twinning
macro *f* DONNÉES, INFORMAT macro; **~-instruction** DONNÉES, INFORMAT macroinstruction
macrobloc *m* INFORMAT macroblock
macrocellule *f* RADIO macrocell
macrocourbure *f* GÉNÉRAL, TRANSMIS macrobending
macrogénérateur *m* INFORMAT macrogenerator, macrogenerating program
macrosegmentation *f* RADIO macrosegmentation
MAD *abrév* (*demande de modification d'appel*) TÉL CMR (*call-modification request*)
MADI *abrév* (*audio-numérique multicanaux*) INFORMAT MADI (*multichannel audio-digital interface*)
MAE *abrév* (*message de modification d'appel effectuée*) TÉL CMC (*call-modification completed message*)
magasin: **~ d'alimentation de cartes** *m* INFORMAT card hopper
magnéto: **~-optique** *adj* (*MO*) ÉLECTRON, INFORMAT magneto-optic (*MO*)
magnétorésistance *f* ÉLECTRON magnetoresistance
magnétoscope *m* ÉLECTRON magnetoscope, TV video recorder, videocassette recorder (*VCR*); **~ numérique** INFORMAT digital video tape recorder (*DVTR*)
magnétostriction *f* ÉLECTRON magnetostriction, magnetostrictive effect
magnétron *m* ÉLECTRON, RADIO magnetron; **~ accordé en tension** ÉLECTRON, RADIO injected-beam magnetron, voltage-tuned magnetron; **~ pulsé** ÉLECTRON, RADIO pulsed magnetron
mail *m* DONNÉES e-mail message
maillage *m* COMMUT linking, ÉLECTRON, INFORMAT meshing, RADIO mesh, TRANSMIS linking
maille *f* COMMUT link, ÉLECTRON, INFORMAT, RADIO mesh, TRANSMIS link
maillé *adj* COMMUT linked together, ÉLECTRON, INFORMAT, RADIO meshed, TRANSMIS linked together
maillon *m* TRANSMIS link
mains: **~ libres** *adj* GÉNÉRAL hands-free
maintenabilité *f* TÉL maintainability performance
maintenance: **~ active** *f* TÉL, TESTS *sécurité de fonctionnement* active maintenance; **~ affectant la fonction** *f* TÉL, TESTS function-affecting maintenance; **~-arrêt** *f* TÉL, TESTS function-preventing maintenance; **~ corrective** *f* TESTS fault maintenance, corrective maintenance; **~ corrective active** *f* TESTS active corrective maintenance; **~ corrective dirigée** *f* APP SERV, TESTS controlled corrective maintenance; **~ dégradant les fonctions** *f* TÉL function-degrading maintenance; **~ dirigée** *f* GÉNÉRAL controlled maintenance; **~ élémentaire** *f* TESTS first-line maintenance; **~ empêchant l'accomplissement de la fonction** *f* TÉL, TESTS function-preventing maintenance; **~ en exploitation** *f* TÉL, TESTS function-permitting maintenance; **~ facile** *f* ÉLECTRON trouble-free maintenance; **~ du flux lumineux** *f* TV lumen maintenance; **~ en fonctionnement** *f* TÉL, TESTS function-permitting maintenance; **~ hors site** *f* TESTS off-site maintenance; **~ des matériels de radiocommunication** *f* ALIM ÉLEC, RADIO radio equipment maintenance; **~ non programmée** *f* TÉL, TESTS unscheduled maintenance; **~ passive** *f* TESTS passive maintenance; **~ périodique** *f* TESTS scheduled maintenance; **~ préventive** *f* GÉNÉRAL, TESTS preventive maintenance; **~ programmée** *f* TESTS scheduled maintenance; **~ de réseau** *f* TV network maintenance; **~ sur site** *f* TESTS field maintenance, on-site maintenance; **~ systématique** *f* TESTS scheduled maintenance
maintien: **~ de la CAF** *m* RADIO AFC hold-in; **~ du niveau du noir** *m* INFORMAT, TÉL, TRANSMIS black-level retention; **~ par l'avant** *m* COMMUT forward hold; **~ en position** *m* RADIO station keeping; **~ à poste** *m* RADIO station keeping; **~ en séquence** *m* COMMUT, DONNÉES, INFORMAT, TÉL sequencing; **~ verrouillé** *m* ÉLECTRON latched detent position
maître: **~-oscillateur** *m* ÉLECTRON, TESTS master oscillator; **~-oscillateur stabilisé** *m* ÉLECTRON, TESTS stabilized master oscillator (*SMO*); **~ d'ouvrage** *m* GÉNÉRAL main contractor
maîtrise: **~ de la qualité** *f* APP SERV, TESTS quality control; **~ statistique de la qualité** *f* APP SERV, TESTS statistical quality control
maîtriser I> GÉNÉRAL oversee
majuscules *f pl* GÉNÉRAL capitals (*caps*), upper case
management *m* GÉNÉRAL management
manchette *f* INFORMAT banner
manchon *m* COMPOSANTS, ÉLECTRON sleeve, TRANSMIS cable sleeve, *de câble* cable distribution head; **~ de fixation** ÉLECTRON clamping sleeve; **~ de plomb** GÉNÉRAL lead sleeve; **~ de raccordement** COMPOSANTS jointing sleeve; **~ rétractable** COMPOSANTS shrink tube, shrink tubing; **~ thermorétractable** COMPOSANTS heat-shrinkable sleeve
manette *f* GÉNÉRAL lever, INFORMAT joystick; **~ de jeux** INFORMAT joystick
manipulateur *m* ALIM ÉLEC, TÉL key; **~ abaissé** TÉL key-down; **~ baissé** TÉL key-down; **~ Morse** TÉL Morse key
manipulation *f* GÉNÉRAL handling, operation; **~ d'amplitude**

TÉL amplitude keying; **~ corrélative par déplacement de phase** SIGNAL correlative phase-shift keying; **~ de document** TÉL document manipulation; **~ de documents** (*DM*) TÉL Document Manipulation (*DM*); **~ des données** DONNÉES data handling; **~ par tout ou rien** TÉL on-off keying (*OOK*); **~ par variation de fréquence à filtre gaussien** SIGNAL, TÉL Gaussian-filtered frequency-shift keying (*GFSK*); **~ simple courant** TÉL neutral keying

manivelle *f* PÉRIPH *postes téléphoniques* crank

manocontact *m* CLIMAT, COMPOSANTS pressure switch

manœuvre: **~ de correction** *f* TESTS active control

manomètre *m* TESTS manometer

manque *m* GÉNÉRAL lack, shortage, void, INFORMAT lack

manuel: **~ du central** *m* GÉNÉRAL exchange manual; **~ d'exploitation** *m* GÉNÉRAL operation manual; **~ d'installation** *m* GÉNÉRAL installation manual; **~ qualité** *m* TESTS quality manual; **~ de référence** *m* INFORMAT operation directory, reference manual; **~ de spécifications des sets de conformité** *m* TESTS conformance test specification manual; **~ de spécifications des tests de conformité** *m* TESTS conformance-testing specification manual (*CTSM*); **~ technique** *m* INFORMAT technical manual

mappage *m* INFORMAT mapping; **~ cognitif** INFORMAT cognitive mapping

mappe *f* INFORMAT map

mapper *vt* INFORMAT map

MAQ *abrév* (*modulation d'amplitude en quadrature*) DONNÉES, TRANSMIS QAM (*quadrature amplitude modulation*)

maquette *f* DONNÉES breadboard model, ÉLECTRON model, TESTS test rack; **~ électrique** ÉLECTRON electrical model; **~ sur table** ÉLECTRON breadboard model

MAR *abrév* (*message de refus de modification d'appel*) TÉL CMRJ (*call-modification reject message*)

marche *f* GÉNÉRAL operation, RADIO step strobe marker; **~ à suivre** GÉNÉRAL procedure; **~ à vide** GÉNÉRAL no-load operation

marge *f* RADIO, TÉL, TRANSMIS *de récepteur synchrone* margin; **~ applicable à une liaison** TRANSMIS link margin; **~ de décodage** RADIO decoding margin; **~ d'évanouissement uniforme** RADIO flat-fading margin; **~ nette** TÉL net margin; **~ de phase** CONTRÔLE phase margin; **~ de protection** RADIO protection margin; **~ de protection globale pour le canal adjacent** RADIO overall adjacent channel protection margin; **~ de protection globale pour le canal deuxième adjacent** RADIO overall second adjacent channel protection margin; **~ de protection globale dans le même canal** RADIO overall co-channel protection margin; **~ de protection globale équivalente** RADIO overall equivalent protection margin

marguerite *f* PÉRIPH daisy wheel

marquage *m* COMMUT marking, GÉNÉRAL labeling (*AmE*), labelling (*BrE*), designation; **~ de catégorie** TÉL category marking, class marking; **~ de classe** TÉL class marking; **~ de dérangement** TÉL failure marking, error marking, fault marking; **~ d'erreur** TÉL failure marking, error marking, fault marking; **~ interurbain** TÉL trunk marking; **~ de ligne vacante** TÉL vacancy marking; **~ de voies** RADIO channel marking; **~ de voie au repos** TÉL route-idle marking

marqué: **~ en attente** *adj* GÉNÉRAL standby-marked

marque[1]: **~-page** *m* DONNÉES page marker, INFORMAT bookmark, page marker

marque[2] *f* ÉLECTRON, INFORMAT brand, make; **~ de distance** RADIO range marker; **~ de groupe** DONNÉES group mark; **~ d'horodatage** COMMUT, INFORMAT, TESTS time stamp; **~ personnalisée** DONNÉES, INFORMAT personalized bookmark; **~ de répétition** INFORMAT rehearsal mark

marquer *vt* GÉNÉRAL designate; **~ occupé** TÉL mark busy

marqueur *m* GÉNÉRAL marker, INFORMAT software message distribution machine marker; **~ de chercheur d'enregistreur** COMMUT register-finder marker; **~ de début de bande** INFORMAT beginning-of-tape marker; **~ de parcours** INFORMAT trail marker, breadcrumbs; **~ de route** TÉL route marker; **~ stroboscopique** RADIO strobe marker

maser *m* ÉLECTRON maser

masquage *m* INFORMAT concealment, TÉL, TRANSMIS, TV masking; **~ par le bruit** TRANSMIS noise masking; **~ par compensation de mouvement** TV motion-compensated concealment; **~ à prédiction spatiale** INFORMAT spatial predictive concealment; **~ à prédiction temporelle** INFORMAT temporal predictive concealment

masque *m* ÉLECTRON mask; **~ photographique** ÉLECTRON photomask; **~ de touches** APP SERV, PÉRIPH *de poste téléphonique* key overlay

masquer *m* INFORMAT hide

masse *f* ALIM ÉLEC, ÉLECTRON earth (*BrE*), ground (*AmE*); **~ de données** DONNÉES data amount, INFORMAT bulk data, bulk information; **~ de l'électron** GÉNÉRAL electron mass

master *m* TV master tape

mastic *m* COMPOSANTS putty

mastiquer *vt* COMPOSANTS putty

mât: **~ d'antenne** *m* RADIO aerial mast

matelas *m* GÉNÉRAL pad, TRANSMIS *pose de câbles* bedding

matériau: **~ isolant** *m* ALIM ÉLEC, COMPOSANTS, ÉLECTRON insulating material; **~ magnétique doux** *m* COMPOSANTS, ÉLECTRON soft magnetic material; **~ magnétique dur** *m* COMPOSANTS, ÉLECTRON hard magnetic material

matériel *m* GÉNÉRAL equipment, hardware; **~ auxiliaire** ÉLECTRON, INFORMAT auxiliary equipment; **~ de canaux** TRANSMIS channel equipment (*CE*); **~ commun** COMMUT common equipment (*CE*); **~ de commutation** COMMUT, TÉL switching equipment; **~ de connectique** ÉLECTRON connecting devices; **~ de contrôle** CONTRÔLE control equipment; **~ de détection** RADIO *radar* detection equipment; **~ d'écoute** PÉRIPH tapping equipment; **~ électrique** DONNÉES electrical equipment; **~ électronique** DONNÉES electronic equipment; **~ de faible performance** INFORMAT low-performance equipment; **~ d'interconnexion** TÉL interconnection equipment; **~ de mesure** TRANSMIS trunk equipment; **~ de modules de lignes** COMMUT line-module equipment (*LME*); **~ périphérique** DONNÉES, GÉNÉRAL, INFORMAT peripheral equipment; **~ de protection** ALIM ÉLEC protection equipment; **~ de raccordement** GÉNÉRAL connection equipment; **~ de réserve** TÉL standby equipment; **~ de secours** TÉL standby equipment; **~ de servitude embarqué** RADIO airborne-support equipment; **~ de soutien au sol** RADIO ground-support equipment (*GSE*); **~ spécifique** COMPOSANTS specific material; **~ téléphonique** PÉRIPH telephone equipment; **~ test** TESTS test equipment; **~ traitement de chèques** APP SERV check-processing equipment (*AmE*), cheque-processing equipment (*BrE*); **~ de transmission de données** DONNÉES, INFORMAT, PÉRIPH data communications equipment (*DCE*); **~ tropicalisé** CLIMAT, COMPOSANTS tropicalized equipment; **~ de voies** TRANSMIS channel equipment (*CE*)

matière: **~ isolante** *f* ALIM ÉLEC, COMPOSANTS, ÉLECTRON insulating material; **~ plastique** *f* COMPOSANTS plastic, plastic material

matrice *f* GÉNÉRAL array; **~ active** INFORMAT active matrix (*AM*); **~ centrale de connexion** COMMUT central switching matrix; **~ de commutation** COMMUT, TÉL switching matrix; **~ de commutation de bande de base** COMMUT, TÉL baseband switch matrix (*BSM*); **~ de commutation numérique** COMMUT, TÉL digital switching matrix; **~ de commutation optique** COMMUT, TÉL optical switching matrix; **~ de commutation optique intégrée** COMMUT, TÉL integrated optical switching matrix; **~ de commutation optoélectronique** COMMUT, TÉL optoelectronic switching matrix; ~ **de commutation temporelle** COMMUT, TÉL symmetrical time matrix; ~ **de connexion** COMMUT connection matrix (*CM*); ~ **de ferrite** INFORMAT ferrite matrix; ~ **passive** INFORMAT passive matrix; ~ **de portes** ÉLECTRON gate array; ~ **de pressage** COMPOSANTS stamper, INFORMAT pressing master; ~ **de quantification** INFORMAT quantization matrix; ~ **en ruban** ÉLECTRON ribbon matrix; ~ **temporelle symétrique** COMMUT symmetrical time matrix; ~ **de tores** INFORMAT core matrix; ~ **à tores de ferrite** INFORMAT ferrite core matrix; ~ **de trafic** TÉL traffic matrix

matricé *adj* TÉL matrixed

Mayday: ~ **relay** *m* RADIO Mayday relay

Mbps *abrév* (*millions de bits par seconde*) DONNÉES, INFORMAT Mbps (*millions of bits per second*)

MCI *abrév* (*interface de commande de support*) INFORMAT MCI (*media control interface*)

MCSAP *abrév* (*point d'accès au service MCS*) TÉL MCSAP (*MCS service access point*)

MCSPDU *abrév* (*unité de données de protocole du service MCS*) TÉL MCSPDU (*MCS protocol data unit*)

MCT *abrév* (*modulation par codage en treillis*) TRANSMIS TCM (*trellis-coded modulation*)

MCV *abrév* (*centre mobile de transmissions*) TRANSMIS MCV (*message center vehicle AmE, message centre vehicle BrE*)

MCW *abrév* (*ondes entretenues modulées*) RADIO MCW (*modulated continuous waves*)

MD *abrév* (*modulation delta*) SIGNAL, TRANSMIS DM (*delta modulation*)

MDA *abrév* (*modulation par déplacement d'amplitude*) RADIO, SIGNAL ASK (*amplitude-shift keying*)

MDAQ *abrév* (*modulation par déplacement d'amplitude en quadrature*) RADIO, SIGNAL QASK (*quadrature amplitude-shift keying*)

MDF *abrév* (*modulation par déplacement de fréquence*) RADIO, SIGNAL FSK (*frequency-shift keying*)

MDM *abrév* (*modulation par déplacement minimal*) RADIO, SIGNAL MSK (*minimum-shift keying*)

MDP *abrév* (*modulation par déplacement de phase*) RADIO, SIGNAL PSK (*phase-shift keying*); ~ **quadrivalente** *f* RADIO, SIGNAL QPSK (*quadrature phase-shift keying, quadriphase-shift keying, quaternary phase-shift keying*)

MDPB *abrév* (*modulation par déplacement de phase bivalente*) RADIO, SIGNAL BPSK (*binary phase-shift keying*)

MDPC *abrév* (*modulation par déplacement de phase cohérente*) RADIO, SIGNAL CPSK (*coherent phase-shift keying*)

MDPQ *abrév* (*modulation par déplacement de phase quadratique*) RADIO, SIGNAL QPSK (*quadrature phase-shift keying*); ~ **avec filtrage** *f* RADIO filtered QPSK; ~ **avec mise en forme de phase** *f* RADIO, SIGNAL phase-shaped QPSK

ME *abrév* (*message enregistré*) DONNÉES stored message, TÉL RAN (*recorded announcement*), recorded message

méandres *m pl* DONNÉES ins and outs

MEB *abrév* (*microscope électronique à balayage*) ÉLECTRON SEM (*scanning electron microscope*), sweep microscope

mécanique: ~ **de central** *f* GÉNÉRAL exchange equipment practice; ~ **ondulatoire** *f* GÉNÉRAL wave mechanics

mécanisme: ~ **d'alimentation** *m* GÉNÉRAL feed mechanism; ~ **d'authentification** *m* DONNÉES authentication mechanism; ~ **d'avancement** *m* GÉNÉRAL feed mechanism; ~ **de commutation** *m* COMMUT, TÉL switching mechanism, selector mechanism; ~ **de déclenchement** *m* ÉLECTRON release mechanism, trigger mechanism, trip mechanism; ~ **de défaillance** *m* TESTS failure mechanism; ~ **d'entraînement de bande** *m* APP SERV, INFORMAT, TV tape transport unit (*TTU*); ~ **d'entraînement de disque** *m* INFORMAT disk storage drive; ~ **d'entraînement du générateur solaire** *m* (*MEGS*) ALIM ÉLEC bearing and power transfer assembly (*BAPTA*); ~ **d'exclusion** *m* TÉL queueing device; ~ **de mise en marche** *m* ÉLECTRON release mechanism, trigger mechanism, trip mechanism; ~ **de sélection** *m* COMMUT, TÉL selector mechanism, switching mechanism; ~ **de verrouillage** *m* COMPOSANTS locking mechanism

media: ~ **control interface** *m* INFORMAT media control interface

médial *m* ÉLECTRON mid-range speaker

médiatisé *adj* TV media-based

médium: ~ **interactif** *m* INFORMAT interactive medium

méga *m* DONNÉES meg of memory

megaflops *m* (*un million d'opérations en virgule flottante par seconde*) INFORMAT megaflops (*one million floating-point operations per second*)

megalips *m* (*un million d'inférences logiques par seconde*) INFORMAT megalips (*one million logical inferences per second*)

mégaoctet *m* (*Mo, Moctet*) DONNÉES, INFORMAT megabyte (*Mb, Mbyte*)

mégaoctets: ~ **de mémoire** *m pl* DONNÉES megabytes of memory

mégohmmètre *m* TESTS Megger®, megohmmeter

MEGS *abrév* (*mécanisme d'entraînement du générateur solaire*) ALIM ÉLEC BAPTA (*bearing and power transfer assembly*)

mélange *m* GÉNÉRAL blend; ~ **heuristique** DONNÉES heuristic mix; ~ **polyphonique** INFORMAT poly mix; ~ **réciproque** RADIO reciprocal mixing

mélangeur *m* GÉNÉRAL mixer; ~ **élévateur** DONNÉES, RADIO up-converter; ~ **équilibré** INFORMAT, RADIO balanced mixer; ~ **de modes** TRANSMIS mode mixer, mode scrambler

MEM *abrév* (*mémoire morte*) DONNÉES, INFORMAT read-only storage, ROM (*read-only memory*)

membrane *f* COMPOSANTS, RADIO diaphragm, membrane; ~ **basilaire** RADIO basilar membrane

mémo: ~ **appel** *m* GÉNÉRAL diary call; ~ **rappel** *m* APP SERV reminder call

mémoire *f* DONNÉES memory, storage, INFORMAT memory, storage, TÉL storage register; ~ **à accès aléatoire** (*SRAM*) INFORMAT static random-access memory (*SRAM, static RAM*); ~ **à accès direct** (*RAM*) DONNÉES, INFORMAT direct-access memory, random-access memory (*RAM*); ~ **à accès rapide** DONNÉES fast storage, INFORMAT fast-access memory, fast-access storage, quick-access memory; ~ **à accès sélectif** DONNÉES, INFORMAT direct-access memory; ~ **à accès séquentiel** DONNÉES, INFORMAT sequential-access storage; ~ **active** DONNÉES, INFORMAT active memory; ~ **additionnelle** INFORMAT additional memory; ~ **adressable par son contenu** DONNÉES data-addressable memory (*DAM*), INFORMAT content-addressable memory (*CAM*), data-addressable memory (*DAM*); ~ **d'appels** COMMUT call store, TÉL call memory; ~ **associative** DONNÉES content-addressable storage, INFORMAT associative storage, content-addressed storage; ~ **d'attribution de voies** TRANSMIS channel allocation store; ~ **autonome** INFORMAT off-line storage; ~ **auxiliaire** INFORMAT backup store; ~ **sur bande magnétique** DONNÉES, INFORMAT magnetic-tape storage; ~ **à bandes magnétiques** INFORMAT tape memory; ~ **à bulles magnétiques** DONNÉES magnetic-bubble memory; ~ **cache** INFORMAT cache memory, cache storage, cache store; ~ **à cartes magnétiques** DONNÉES magnetic-card storage; ~ **de catégories** TÉL category store; ~ **centrale** DONNÉES main storage, primary storage; ~ **de code de classe** TÉL class-mark store; ~ **commune** DONNÉES common store, INFORMAT common memory; ~ **de comptage d'appels** APP SERV call-metering store; ~ **de conservation de registre** COMMUT register-preserve memory; ~ **de contrôle** INFORMAT control memory, control storage; ~ **de destination** INFORMAT destination memory; ~ **directe** INFORMAT, PÉRIPH direct memory; ~ **disponible** INFORMAT available memory; ~ **sur disque magnétique**

DONNÉES, INFORMAT magnetic-disk storage; ~ **à disques** INFORMAT disk memory, disk storage, disk store; ~ **à dix numéros** PÉRIPH ten-number memory; ~ **de données** COMMUT, DONNÉES data store; ~ **dynamique** INFORMAT dynamic storage; ~ **effaçable** INFORMAT erasable store; ~ **d'effets** INFORMAT effect memory; ~ **d'éléments de taxation** APP SERV charging store; ~ **étendue** DONNÉES extended memory; ~ **d'exécution** INFORMAT performance memory; ~ **d'extension** INFORMAT extension store; ~ **externe** INFORMAT external memory; ~ **à ferrites** INFORMAT ferrite storage; ~**-fichier réelle** DONNÉES real file store; ~**-fichier virtuelle** DONNÉES virtual file store; ~ **fixe** (*ROM*) DONNÉES read-only memory (*ROM*), read-only storage, INFORMAT fixed store, read-only memory (*ROM*), read-only storage; ~ **de grande capacité** INFORMAT bulk store (*BrE*), bulk storage (*AmE*); ~ **holographique** DONNÉES holographic storage; ~ **hors ligne** INFORMAT off-line storage; ~ **d'images** CONTRÔLE picture store; ~ **inaltérable** (*ROM*) DONNÉES, INFORMAT read-only memory (*ROM*), read-only storage; ~ **indirecte** PÉRIPH indirect memory; ~ **individuelle** DONNÉES individual store; ~ **ineffaçable** INFORMAT nonerasable memory, nonerasable storage; ~ **intermédiaire** DONNÉES, INFORMAT buffer memory; ~ **interne** DONNÉES internal storage; ~ **des itinéraires** COMMUT path memory; ~ **lente** INFORMAT slow-access storage; ~ **à liste directe** DONNÉES, INFORMAT push-up storage; ~ **à liste inversée** DONNÉES push-down storage; ~ **magnétique** DONNÉES magnetic storage; ~ **magnéto-optique** DONNÉES magneto-optical storage; ~ **de masse** DONNÉES bulk store (*BrE*), bulk storage (*AmE*), mass storage, INFORMAT bank memory, bulk store (*BrE*), bulk storage (*AmE*); ~ **matricielle** INFORMAT matrix storage; ~ **de messagerie vocale** APP SERV, TÉL voice-messaging store (*VM-MS*); ~ **de messages** DONNÉES, TÉL message store (*MS*); ~ **de messages EDI** DONNÉES, TÉL EDI message store (*EDI-MS*); ~ **de messages de système de messagerie vocale** APP SERV, TÉL voice-messaging system message store (*VMGS-MS*); ~ **de morceaux de musique** INFORMAT song memory; ~ **morte** (*MEM*) DONNÉES, INFORMAT read-only memory (*ROM*), read-only storage; ~ **morte effaçable électriquement** ÉLECTRON electrically erasable programmable read-only memory (*EEPROM*); ~ **multiple** RADIO multiple memory; ~ **non effaçable** INFORMAT nonerasable memory, nonerasable storage; ~ **non rémanente** DONNÉES, INFORMAT volatile memory, volatile storage; ~ **non volatile** RADIO nonvolatile memory; ~ **numérique** DONNÉES digital storage; ~ **paginée** DONNÉES, INFORMAT expanded memory; ~ **sur papier numérique** DONNÉES, INFORMAT digital paper storage; ~ **préprogrammée** INFORMAT preset read-only memory, preset memory; ~ **principale** DONNÉES main storage, primary storage; ~ **de programme** INFORMAT program store; ~ **réelle** DONNÉES real storage; ~ **rémanente** DONNÉES, INFORMAT nonvolatile memory, nonvolatile storage; ~ **de sécurité** DONNÉES secure storage; ~ **à semiconducteurs** DONNÉES semiconductor storage; ~ **de séquence de commande suppléante** RADIO alternate control sequence memory; ~ **statique** DONNÉES static storage, INFORMAT static memory; ~ **à supraconduction** INFORMAT cryogenic storage; ~ **sur ruban optique** DONNÉES, INFORMAT optical tape storage; ~ **à tambour** COMMUT drum store; ~ **tampon** DONNÉES, INFORMAT buffer memory, buffer storage; ~ **tampon multitrame** DONNÉES, INFORMAT multiframe buffering; ~ **tampon temporaire** DONNÉES, INFORMAT temporary buffer memory; ~ **de textes** INFORMAT text memory; ~ **à tores** DONNÉES, INFORMAT core storage; ~ **à tores de ferrite** DONNÉES, INFORMAT ferrite-core storage; ~ **à tores magnétiques** DONNÉES, INFORMAT magnetic-core storage; ~ **de traduction** COMMUT translation store; ~ **de trame** TÉL speech memory; ~ **utilitaire** INFORMAT utility memory; ~ **vide statique** (*SRAM*) INFORMAT static RAM (*SRAM*), static random-access memory (*SRAM*); ~ **virtuelle** DONNÉES, INFORMAT virtual memory (*VM*), virtual storage; ~ **vive** (*RAM*) DONNÉES, INFORMAT random-access memory (*RAM*), read-write memory; ~ **vive dynamique** DONNÉES, INFORMAT dynamic RAM (*DRAM*); ~ **vocale** TÉL speech memory; ~ **volatile** DONNÉES, INFORMAT volatile memory, volatile storage

mémorisation *f* DONNÉES storage, memory storage, storing, INFORMAT storage, memory storage, storing; ~ **complète** DONNÉES, INFORMAT store-all; ~ **de données vocales** DONNÉES, INFORMAT storing of voice data

mémoriser *vt* DONNÉES, INFORMAT memorize, store

menace: ~ **active** *f* DONNÉES active threat; ~ **passive** *f* DONNÉES passive threat

menu *m* DONNÉES, INFORMAT menu; ~ **à bascule** INFORMAT toggle menu; ~ **déroulant** INFORMAT pull-down menu, scrolling menu; ~ **détachable** INFORMAT tear-off menu; ~ **détachant** INFORMAT tear-off menu; ~ **éclair** INFORMAT pop-up menu; ~ **d'essai** TESTS test menu; ~ **flottant** INFORMAT tear-off menu; ~ **fugitif** INFORMAT pop-up menu; ~ **incrusté** INFORMAT pop-up menu; ~ **des liens** INFORMAT links menu; ~ **local** INFORMAT pop-up menu; ~ **personnalisé** INFORMAT customized menu, tailored menu

mesa *m* ÉLECTRON mesa

mésochrone *adj* TÉL, TRANSMIS mesochronous

message[1]: ~ **d'acceptation d'appel** *m* DONNÉES, TÉL call-accepted message; ~ **d'acceptation de la demande d'annulation** *m* DONNÉES, TÉL facility cancellation request-accepted message; ~ **d'acceptation de la demande d'enregistrement** *m* DONNÉES, TÉL facility registration request-accepted message; ~ **d'acceptation de service supplémentaire** *m* (*SUAC*) DONNÉES, TÉL facility-accepted message (*FAA*); ~ **d'accueil** *m* INFORMAT banner; ~ **acquitté** *m* RADIO acknowledged message; ~ **d'adresse** *m* COMMUT, SIGNAL, TÉL address message; ~ **d'adresse initial** *m* COMMUT, SIGNAL, TÉL initial address message (*IAM*); ~ **d'adresse subséquent** *m* COMMUT, SIGNAL, TÉL subsequent address message; ~ **d'adresse ultérieur** *m* COMMUT, SIGNAL, TÉL subsequent address message; ~ **d'alarme** *m* TÉL alarm printout; ~ **alphanumérique** *m* INFORMAT text message; ~ **d'assignation** *m* DONNÉES assignment message; ~ **brouillé** *m* TÉL scrambled message; ~ **du canal de service** *m* TESTS service channel message; ~ **chiffré** *m* DONNÉES encoded message; ~ **de commande** *m* INFORMAT control message; ~ **de contrôle** *m* COMMUT supervisory tone, check print-out, CONTRÔLE supervisory tone, TÉL supervisory announcement; ~ **de demande d'adresse** *m* DONNÉES address prompt (*ADD*); ~ **de demande d'annulation de service** *m* TÉL facility cancellation request message; ~ **de demande d'enregistrement de service** *m* TÉL facility registration request message; ~ **de demande d'information** *m* (*IND*) DONNÉES, TÉL information request message (*INR*); ~ **de demande de service supplémentaire** *m* (*SUDM*) TRANSMIS facility request message (*FAR*); ~ **de dérangement** *m* DONNÉES fault message; ~ **à diffusion générale** *m* RADIO broadcast message; ~ **EDI** *m* DONNÉES, TÉL EDI message *(EDIM)*; ~ **enregistré** *m* (*ME*) DONNÉES stored message, TÉL recorded announcement (*RAN*), recorded message; ~ **d'erreur** *m* COMMUT error print-out, DONNÉES error message, INFORMAT fault print-out; ~ **d'état** *m* RADIO status message; ~ **d'état de circuit** *m* TÉL circuit state message; ~ **à faire passer** *m* DONNÉES pass-along message (*PAM*); ~ **de gestion de couche liaison consolidé** *m* DONNÉES consolidated link-layer management message (*CLLM*); ~ **de gestion de trame** *m* DONNÉES frame management message (*FMM*); ~ **d'incohérence** *m* DONNÉES confusion message; ~ **inconnu** *m* DONNÉES unknown message; ~ **d'information** *m* (*INF*) DONNÉES, GÉNÉRAL information message (*INF*);

~ **d'information d'usager à usager** *m* DONNÉES user-to-user information message; ~ **d'interrogation de groupe de circuits** *m* (*IGD*) COMMUT circuit-group query message (*CQM*); ~ **irrationnel** *m* TÉL unreasonable message; ~ **de libération** *m* (*LIB*) TESTS clear message, clearing message, release message (*REL*); ~ **de libération d'alarme** *m* TESTS alarm-cleared message; ~ **mémorisable** *m* PÉRIPH, RADIO, TÉL storable message; ~ **mémorisé** *m* PÉRIPH, RADIO, TÉL stored message; ~ **de modification d'appel effectuée** *m* (*MAE*) TÉL call-modification completed message (*CMC*); ~ **multiple** *m* TÉL multi-unit message (*MUM*); ~ **neutralisation-salve** *m* DONNÉES blank-and-burst message; ~ **de nouvelle affectation** *m* DONNÉES new assignment message; ~ **numérique** *m* RADIO numerical message; ~ **parlé** *m* PÉRIPH, TÉL spoken message; ~ **personnel** *m* PÉRIPH memo recording; ~ **de point de grille** *m* DONNÉES grid data message; ~ **préenregistré** *m* TÉL prerecorded message; ~ **de refus d'appel** *m* (*MRA*) DONNÉES, TÉL call-rejected message (*CRM*); ~ **de refus de la demande d'annulation** *m* TÉL facility cancellation request rejected message; ~ **de refus de la demande d'enregistrement** *m* TÉL facility registration request rejected message; ~ **de refus de modification d'appel** *m* (*MAR*) TÉL call-modification reject message (*CMRJ*); ~ **de refus de service supplémentaire** *m* (*SURF*) DONNÉES facility-rejected message (*FRJ*); ~ **de relève de dérangement** *m* TESTS fault-cleared message; ~ **de réponse avec taxation** *m* RADIO acceptance with charging message; ~ **de réponse à une interrogation de groupe de circuits** *m* (*IGR*) COMMUT circuit-group query response message (*CQR*); ~ **à segment unique** *m* DONNÉES single-segment message, SSM; ~ **à un seul segment** *m* DONNÉES single-segment message, SSM; ~ **de signalisation** *m* SIGNAL signal message, signaling message (*AmE*), signalling message (*BrE*); ~ **simple** *m* TÉL one-unit message; ~ **sortant** *m* PÉRIPH outgoing message (*OGM*); ~ **de supervision** *m* TÉL supervisory message; ~ **par télécopieur** *m* TÉL fax message; ~ **téléphonique** *m* TÉL telephone message; ~ **télex** *m* TÉL telex message; ~ **de transfert** *m* PÉRIPH transfer message; ~ **verbal** *m* TÉL verbal message; ~ **vocal** *m* (*MV*) DONNÉES, INFORMAT, TÉL vocal message, voice message (*VM*); ~ **vocal concerné** *m* DONNÉES subject voice message

message[2]: ~ **bien compris** *loc* RADIO roger; ~ **bien reçu** *loc* RADIO roger; ~ **compris** *loc* RADIO roger

messagerie *f* APP SERV message handling (*MH*), DONNÉES bulletin board system (*BBS*), message-handling system, INFORMAT mail, TÉL bulletin board system (*BBS*), message handling (*MH*); ~ **conviviale** DONNÉES, INFORMAT, TÉL bulletin board (*BB*); ~ **EDI** DONNÉES, INFORMAT, TÉL EDI messaging (*EDIMG*); ~ **électronique** APP SERV, DONNÉES, INFORMAT electronic mail (*e-mail*), electronic message service, electronic messaging; ~ **multimédia** DONNÉES, INFORMAT multimedia messaging; ~ **de personne à personne** TÉL interpersonal messaging (*IPM*); ~ **vidéo** INFORMAT, TV videomessaging; ~ **vocale** APP SERV, TÉL voice mail (*VM*), voice messaging (*VMG*)

mesurage: ~ **dynamique** *m* TESTS dynamic measurement

mesure *f* GÉNÉRAL *acte*, measurement, action, INFORMAT bar; ~ **acoustique** GÉNÉRAL acoustic measurement; ~ **d'affaiblissement** TRANSMIS attenuation measurement; ~ **de bruit** TÉL noise measurement; ~ **du bruit optique** DONNÉES optical noise measurement; ~ **de comparaison** TESTS comparison measurement; ~ **comparative** TESTS comparison measurement; ~ **de la disponibilité** TESTS *sécurité de fonctionnement* availability measure; ~ **de distance** RADIO ranging; ~ **échométrique** TÉL echometric measurement; ~ **de fiabilité** TESTS reliability measure; ~ **de la fréquence** TESTS frequency measurement; ~ **de grandeur de champ** TESTS field strength measurement; ~ **de l'information** GÉNÉRAL information measure; ~ **d'intensité de champ** ALIM ÉLEC, RADIO field strength measurement; ~ **d'isolement** ALIM ÉLEC insulation measurement; ~ **linéaire** TESTS linear measure; ~ **de longueur** TESTS linear measure; ~ **des pertes diélectriques** DONNÉES dielectric-loss measurement; ~ **de qualité** TESTS quality measure; ~ **de taux d'erreurs** DONNÉES error-rate measurement; ~ **de temps** TÉL, TESTS time measurement; ~ **de trafic** ÉLECTRON, TÉL, TESTS traffic metering; ~ **de transmission** TESTS, TRANSMIS transmission measurement; ~ **de la valeur du champ** TESTS field strength measurement

mesures: ~ **prises en cas d'alarme** *f pl* TÉL action taken upon alarm; ~ **sur le terrain** *f pl* TESTS field measurements; ~ **d'urgence** *f pl* GÉNÉRAL contingency plan

mesureur: ~ **d'intensité de champ** *m* TESTS field intensity meter, field strength meter; ~ **de puissance optique** *m* TRANSMIS optical power meter

métallisation *f* COMPOSANTS, ÉLECTRON metalization (*AmE*), metallization (*BrE*); ~ **des contacts** ÉLECTRON contact metalization (*AmE*), contact metallization (*BrE*); ~ **sous vide** COMPOSANTS vacuum metalization (*AmE*), vacuum metallization (*BrE*)

métaraisonnement *m* INFORMAT metareasoning

métastable *adj* ÉLECTRON metastable

météo *f* APP SERV weather line, TÉL weather forecast, weather line

méthode *f* GÉNÉRAL procedure, approach; ~ **d'accès** TÉL access method; ~ **à accès virtuel** COMMUT virtual access method (*VAM*); ~ **des attributs** TESTS attribute method; ~ **par cassette** COMPOSANTS splice-tray method; ~ **du champ réfracté** TRANSMIS refracted near-field technique; ~ **du champ réfracté proche** TRANSMIS refracted near-end method, refracted-ray method; ~ **du chemin critique** GÉNÉRAL critical-path method; ~ **comparative** TESTS comparison method; ~ **de correction d'erreur avec retransmission** TRANSMIS preventive cyclic retransmission error control method; ~ **de correction d'erreur de base** TÉL basic error correction method; ~ **des deux points directeurs** RADIO two-control-point method; ~ **de division des recettes de répartition** TÉL accounting revenue division procedure; ~ **d'échange** TÉL exchange method (*EM*); ~ **d'écho d'impulsion** TRANSMIS pulse echo method; ~ **d'essai** TESTS test engineering, test method, test technology, testing technique; ~ **d'exploration du champ réfracté proche** TRANSMIS refracted near-end method, refracted-ray method; ~ **de la fibre coupée** TRANSMIS cutback technique; ~ **heuristique** DONNÉES heuristic method; ~ **de masquage** INFORMAT concealment method; ~ **de mesure** TESTS measurement method; ~ **de mesure différentielle** TESTS differential measurement method; ~ **de mesure directe** TESTS direct method of measurement; ~ **de mesure indirecte** TESTS indirect method of measurement; ~ **de mesure par coïncidence** TESTS coincidence method of measurement; ~ **de mesure par comparaison** TESTS comparison method of measurement; ~ **de mesure par complément** TESTS complementary method of measurement; ~ **de mesure par contact** TESTS contact method of measurement; ~ **de mesure par déviation** TESTS deflection method of measurement; ~ **de mesure de référence** TESTS, TRANSMIS reference test method (*RTM*); ~ **de mesure de remplacement** TESTS practical test method, TRANSMIS alternative test method; ~ **de mesure par résonance** TESTS resonance method of measurement; ~ **de mesure par transposition** TESTS transposition method of measurement; ~ **des mesures** TESTS variable method; ~ **de modulation MA zéro** RADIO null AM modulation method; ~ **par nombre de défauts** TESTS defect-counting method; ~ **de pointage optique** RADIO boresight facility; ~ **de programmation** DONNÉES, INFORMAT programming method; ~ **de réflexion de Fresnel** TRANSMIS Fresnel reflection method; ~ **de réglage** TESTS, TRANSMIS line-up procedure; ~ **de rémunération par unité de trafic** TÉL traffic-unit price procedure; ~ **de rétrodiffusion** TÉL backscattering technique, TRANSMIS optical time domain reflectometry (*OTDR*), backscattering technique; ~ **statistique** APP SERV statistical method; ~ **structurelle** APP SERV structural method; ~ **de taxation** TÉL charging method; ~ **de test abstraite** DONNÉES, TESTS abstract test method (*ATM*); ~ **de test en boucle** TESTS loopback test method; ~ **de test coordonnée** TESTS coordinated test method; ~ **de test à distance** TESTS remote test method; ~ **de test encastré monocouche coordonnée** TESTS coordinated single-layer embedded test method (*CSE*); ~ **de test encastré monocouche locale** TESTS local single-layer embedded test method (*LSE*); ~ **de test locale** TESTS local test method; ~ **de test monocouche coordonnée** TESTS coordinated single-layer test method (*CS*); ~ **de test monocouche à distance** TESTS remote single-layer test method (*RS*); ~ **de test monocouche à distance encastrée** TESTS remote single-layer embedded test method (*RSE test method*); ~ **de test monocouche distribuée encastrée** (*DSE*) TESTS distributed single-layer embedded test method (*DSE*); ~ **de test monocouche locale** TESTS local single-layer test method (*LS test method*); ~ **de test monocouche répartie** TESTS distributed single-layer test method; ~ **de test répartie** TESTS distributed test method; ~ **de test transversale** TESTS transverse test method; ~ **de tests** TESTS test technology; ~ **à virgule fixe** INFORMAT fixed-point method; ~ **en visée directe au sol** RADIO boresight method; ~ **de zéro** TESTS null method, zero method

méthodes: ~ **administratives** *f pl* INFORMAT systems and procedures

méthodologie: ~ **de test abstraite** *f* DONNÉES, TESTS abstract testing methodology

métrologie: ~ **des fréquences** *f* TESTS frequency measurement

mettre *vt* GÉNÉRAL put; ~ **en application** GÉNÉRAL, TÉL implement; ~ **en attente** TÉL put on hold; ~ **au banc d'essai** TESTS benchtest; ~ **en charge** ALIM ÉLEC *une batterie* put on charge; ~ **en circuit** ALIM ÉLEC switch on, turn on; ~ **en communication** TÉL connect, put through; ~ **en communication avec** TÉL connect, put through to; ~ **en contact** TÉL put in touch; ~ **en décharge** ALIM ÉLEC *une batterie* put on discharge; ~ **en dégradé** INFORMAT shade; ~ **en dérivation** ÉLECTRON tap; ~ **sur écoute** TÉL wiretap, tap; ~ **en état d'alarme** TÉL alarm, alert; ~ **en état de prise** TÉL seize, occupy; ~ **en évidence** INFORMAT highlight; ~ **en exposant** DONNÉES, INFORMAT superscript; ~ **fin à** GÉNÉRAL, RADIO terminate; ~ **en forme** DONNÉES, INFORMAT format; ~ **en garde** COMMUT, TÉL *appel* put on hold; ~ **en garde tous postes** COMMUT, TÉL park; ~ **hors circuit** ALIM ÉLEC, COMMUT, ÉLECTRON, INFORMAT, PÉRIPH, RADIO switch off, turn off; ~ **hors service** TÉL remove from service, TESTS take out of service; ~ **hors tension** ALIM ÉLEC, COMMUT, ÉLECTRON, INFORMAT, PÉRIPH, RADIO switch off, turn off, power down; ~ **en indice** DONNÉES, INFORMAT subscript; ~ **à jour** ÉLECTRON upgrade, GÉNÉRAL update, bring up to date, INFORMAT upgrade, TÉL update; ~ **en lissage** INFORMAT shade; ~ **en marche** ALIM ÉLEC, COMMUT, ÉLECTRON, INFORMAT, PÉRIPH switch on, turn on; ~ **sur le marché** GÉNÉRAL launch; ~ **à la masse** ALIM ÉLEC, ÉLECTRON earth (*BrE*), ground (*AmE*); ~ **en mémoire** DONNÉES, INFORMAT, RADIO store; ~ **à niveau** ÉLECTRON retrofit, GÉNÉRAL upgrade; ~ **en occupation** TÉL busy out; ~ **en uvre** GÉNÉRAL, TÉL implement; ~ **en ombrage** TV shade; ~ **sur orbite** RADIO put into orbit; ~ **en page** INFORMAT lay out; ~ **en paquet** DONNÉES bundle; ~ **en phase** RADIO phase; ~ **en place** COMMUT *sélecteur* set, GÉNÉRAL, TÉL implement; ~ **en place progressivement** GÉNÉRAL phase in; ~ **au point** DONNÉES *programme* debug, GÉNÉRAL develop; ~ **en relation** TÉL put in touch; ~ **en relief** INFORMAT emboss; ~ **en retrait** INFORMAT indent; ~ **en route** ÉLECTRON switch on, INFORMAT run, start up, RADIO switch on; ~ **en service** GÉNÉRAL put into operation, put into service, INFORMAT enable, TÉL commission, cut over to service, put into operation, put into service; ~ **en surbrillance** INFORMAT highlight; ~ **sous tension** ALIM ÉLEC, COMMUT, ÉLECTRON, INFORMAT, PÉRIPH, RADIO switch on, turn on, power up; ~ **à la terre** ALIM ÉLEC, ÉLECTRON earth (*BrE*), ground (*AmE*); ~ **à un** ÉLECTRON set to one; ~ **à zéro** ÉLECTRON, TESTS adjust to zero, reset to zero, zero, zeroize

MF *abrév* (*modulation de fréquence*) RADIO, TRANSMIS, TV FM (*frequency modulation*)

MFA *abrév* (*modulation de fréquence asservie*) RADIO, TRANSMIS TFM (*tamed frequency modulation*)

MFBE *abrév* (*modulation de fréquence à bande étroite*) RADIO, TRANSMIS *satellite maritime* NBFM (*narrow-band frequency modulation*)

MHEG *abrév* INFORMAT MHEG (*multimedia and hypermedia information coding experts group*)

MHS-SE *abrév* (*élément de service de système de traitement de messages*) APP SERV, DONNÉES, RADIO, TÉL MHS-SE (*Message Handling System Service Element*)

MIA *abrév* (*modulation d'impulsions en amplitude*) TÉL, TRANSMIS PAM (*pulse amplitude modulation*)

MIC *abrév* (*modulation par impulsions codées*) DONNÉES, INFORMAT, TÉL, TRANSMIS PCM (*pulse code modulation*)

MICD *abrév* (*modulation par impulsions et codage différentiel*) DONNÉES, INFORMAT, TÉL, TRANSMIS DPCM (*differential pulse code modulation*)

MICDA *abrév* (*modulation par impulsions et codage différentiel adaptatif*) DONNÉES, INFORMAT, TÉL, TRANSMIS ADPCM (*adaptive differential pulse code modulation*)

MICR *abrév* (*reconnaissance magnétique des caractères*) APP SERV MICR (*magnetic ink character recognition*)

micro *m* (*microphone*) ÉLECTRON mike (*microphone*)

microaccordage *m* ÉLECTRON microtune, microtuning

microassemblage *m* ÉLECTRON micro-assembly

microcasque *m* TÉL lightweight headset; ~ **de téléphone** TÉL lightweight telephone headset

microcassette *f* PÉRIPH microcassette

microcircuit: ~ **à encapsulage plastique** *m* ÉLECTRON plastic-encapsulated microcircuit; ~ **intégré** *m* ÉLECTRON integrated microcircuit

microcommutateur *m* COMMUT microswitch

microcourbure *f* ÉLECTRON bending loss, TRANSMIS microbending

microcuvette *f* INFORMAT pit

microélectronique *f* ÉLECTRON microelectronics

microfente *f* TRANSMIS microslot

microforme *m* DONNÉES microform

microimage *f* DONNÉES micro-image

microinterrupteur *m* ÉLECTRON microswitch

micrologiciel *m* INFORMAT firmware

micromisation *f* INFORMAT downsizing

micromiser *vt* INFORMAT downsize

microonde *f* RADIO microwave

microordinateur *m* (*PC*) INFORMAT microcomputer, personal computer (*PC*)

microphone *m* (*micro*) DONNÉES transmitter (*XMTR*), ÉLECTRON microphone (*mike*), transmitter (*XMTR*), INFORMAT transmitter (*XMTR*), PÉRIPH, TÉL microphone (*mike*), transmitter (*XMTR*), TRANSMIS transmitter (*XMTR*); ~ **antibruit** TÉL noise-canceling microphone (*AmE*), noise-cancelling microphone (*BrE*); ~ **boutonnière** TÉL lapel microphone; ~ **à charbon** COMPOSANTS carbon microphone; ~ **à condensateur** TÉL capacitor microphone, condenser microphone, electrostatic microphone; ~ **à conducteur mobile** TÉL electrodynamic microphone;

~ **cravate** COMPOSANTS breastplate microphone; ~ **directionnel** TÉL directional microphone; ~ **électret** TÉL electret microphone; ~ **électrodynamique** TÉL electrodynamic microphone; ~ **électromagnétique** TÉL electromagnetic microphone; ~ **électrostatique** COMPOSANTS electrostatic microphone, TÉL capacitor microphone, condenser microphone, electrostatic microphone; ~ **plastron** COMPOSANTS *microphones* breastplate microphone; ~ **de proximité** COMPOSANTS close-talking microphone; ~ **à ruban** TÉL ribbon microphone; ~ **sonde** TÉL probe microphone; ~ **de table** TÉL desk microphone

microphonie *f* ÉLECTRON microphonic effect, microphony

microplaquette *f* ÉLECTRON microchip

microprocesseur *m* INFORMAT microprocessor

microprogrammation *f* INFORMAT microprogramming

microprogramme *m* INFORMAT firmware

microruban *m* ÉLECTRON microribbon

microscope: ~ **électronique** *m* ÉLECTRON electron microscope; ~ **électronique à balayage** *m* (*MEB*) ÉLECTRON scanning electron microscope (*SEM*), sweep microscope; ~ **optique** *m* TV light microscope

microscopie: ~ **axiale interférentielle** *f* ÉLECTRON axial interference microscopy, TRANSMIS axial interference microscopy, slab interferometry; ~ **électronique** *f* ÉLECTRON electron microscopy

microsegmentation *f* RADIO microsegmentation

microstripper *m* ÉLECTRON microstripper

microstructure *f* ÉLECTRON microcircuit

microterminal *m* TV very small aperture terminal (*VSAT*)

microtonal *adj* INFORMAT microtonal

MID *abrév* (*modulation d'impulsions en durée*) DONNÉES, RADIO, TÉL, TRANSMIS PDM (*pulse duration modulation*), PWM (*pulse width modulation*)

MIE *abrév* (*multiplexeur d'insertion-extraction*) TRANSMIS add-drop multiplexer

migration *f* GÉNÉRAL migration; ~ **électrique** ALIM ÉLEC electric migration; ~ **ionique** ÉLECTRON ion migration; ~ **d'ions** ÉLECTRON ion migration

migrer *vi* GÉNÉRAL migrate

milieu: ~ **de l'étendue** *m* GÉNÉRAL mid-range; ~ **laser** *m* ÉLECTRON, TRANSMIS laser medium; ~ **laser actif** *m* ÉLECTRON, TRANSMIS active laser medium

milliammètre: ~ **à rubans tendus** *m* TESTS taut band milliammeter

milliampère *m* (*mA*) ÉLECTRON milliamp (*mA*)

milliampèremètre *m* ÉLECTRON milliammeter

million: **un** ~ **d'inférences par seconde** *loc* (*megalips*) INFORMAT one million logical inferences per second (*megalips*); **un** ~ **d'opérations en virgule flottante par seconde** *loc* (*megaflops*) INFORMAT one million floating-point operations per second (*megaflops*)

millions: ~ **de bits par seconde** *loc* (*Mbps*) DONNÉES, INFORMAT millions of bits per second (*Mbps*)

milliwattmètre *m* ALIM ÉLEC, TESTS power meter

minidisque *m* INFORMAT mini disk

minidisquette *f* INFORMAT mini floppy, mini floppy disk

minienroulement *m* ÉLECTRON mini-wrapping

miniétoile *f* TV mini-star configuration

minimessagerie *f* SIGNAL user-to-user signaling (*AmE*) (*UUS*), user-to-user signalling (*BrE*) (*UUS*)

minimum: ~ **de courant** *m* ALIM ÉLEC undercurrent; ~ **de perception** *m* APP SERV minimum payable

miniprise *f* PÉRIPH mini jack

minuscules *f pl* INFORMAT lower case

minute: ~ **dégradée** *f* DONNÉES, RADIO degraded minute (*DM*)

minuter *vt* DONNÉES, INFORMAT, RADIO, TÉL, TESTS time

minuterie *f* ALIM ÉLEC interval timer, GÉNÉRAL timer; ~ **de compteur** ALIM ÉLEC meter changeover clock, COMMUT, DONNÉES time switch

minuteur *m* GÉNÉRAL timer

MIP *abrév* (*modulation d'impulsions en position*) DONNÉES, TÉL, TRANSMIS PPM (*pulse position modulation*)

mire *f* TV test pattern

miroir *m* DONNÉES, mirror; ~ **d'alignement** TRANSMIS alignment mirror; ~ **auxiliaire** RADIO subreflector; ~ **à couche secondaire** RADIO second-surface mirror; ~ **dichroïque** TRANSMIS dichroic mirror; ~ **grille** RADIO rod reflector; ~ **sphérique** TRANSMIS spherical mirror

mis: ~ **en attente** *adj* TÉL call-queued; ~ **en fonction NI** *adj* ÉLECTRON NORed; ~ **en fonction NON-ET** *adj* ÉLECTRON NANDed; ~ **en fonction OU** *adj* ÉLECTRON ORed; ~ **en garde** *adj* COMMUT parked, TÉL on-hold, parked; ~ **en garde tous postes** *adj* COMMUT, TÉL parked; ~ **hors tension** *adj* ALIM ÉLEC, ÉLECTRON powered-down; ~ **à la masse** *adj* ALIM ÉLEC, ÉLECTRON earthed (*BrE*), grounded (*AmE*); ~ **sous tension** *adj* ALIM ÉLEC, ÉLECTRON powered-up, connected to voltage; ~ **à la terre** *adj* ALIM ÉLEC, ÉLECTRON earthed (*BrE*), grounded (*AmE*)

MIS *abrév* (*semi-conducteur sur isolant métal, métal semi-conductuer sur isolant*) ÉLECTRON MIS (*metal insulator semiconductor*)

mise: ~ **en action** *f* INFORMAT actuation; ~ **en application** *f* GÉNÉRAL, TÉL implementation; ~ **en attente** *f* TÉL call queueing, call-queueing facility, call hold, call holding, call-holding facility; ~ **en attente du demandeur** *f* TÉL calling-party hold; ~ **en attente par le standardiste** *f* TÉL attendant camp-on; ~ **en attente sur poste occupé** *f* TÉL camp-on; ~ **en évidence** *f* INFORMAT highlighting; ~ **en file d'attente** *f* RADIO queueing, TÉL call-queuing facility, queue control; ~ **en file d'attente des appels** *f* TÉL call queueing; ~ **en file d'attente sur circuit de départ** *f* COMMUT, TÉL outgoing trunk queuing; ~ **en fonction d'opérateur** *f* INFORMAT operator enable; ~ **en forme** *f* DONNÉES, INFORMAT formatting; ~ **en forme de signal** *f* ÉLECTRON, SIGNAL signal shaping, signal transformation; ~ **en garde** *f* TÉL *d'un appel* call hold, call holding, call-holding facility; ~ **en garde d'appel** *f* RADIO queueing; ~ **en garde par indicatif** *f* TÉL *d'un appel* call park; ~ **en garde tous postes d'un appel** *f* TÉL call park; ~ **hors fonction d'opérateur** *f* INFORMAT operator disable; ~ **hors service** *f* INFORMAT disabling; ~ **hors service intempestive** *f* TESTS spurious switch-off (*SSO*); ~ **hors tension** *f* ALIM ÉLEC power-down; ~ **à jour** *f* ÉLECTRON upgrade, upgrading, GÉNÉRAL update, updating, INFORMAT upgrade, upgrading, PÉRIPH, TÉL update, updating; ~ **à jour automatique** *f* PÉRIPH automatic updating; ~ **à jour d'enregistrement autonome** *f* TÉL autonomous registration update; ~ **à jour de fichier** *f* INFORMAT file maintenance; ~ **à jour de localisation** *f* RADIO location updating; ~ **à jour rapide** *f* INFORMAT fast update; ~ **en marche dispositif** *f* TÉL device start; ~ **à la masse** *f* ALIM ÉLEC, ÉLECTRON earthing (*BrE*), grounding (*AmE*); ~ **en mémoire** *f* DONNÉES, INFORMAT, RADIO, TÉL, TV storage, storing; ~ **en mémoire cache** *f* DONNÉES caching; ~ **en mémoire de messages** *f* TÉL programmable announcements; ~ **en mémoire tampon** *f* DONNÉES buffering; ~ **en mémoire tampon des données vidéo** *f* DONNÉES, INFORMAT, TV video-data buffering; ~ **en mode asynchrone symétrique** *f* TÉL set asynchronous balanced mode (*SABM*); ~ **à niveau** *f* ÉLECTRON retrofit, GÉNÉRAL upgrade, upgrading; ~ **en occupation** *f* (*MO*) TÉL busy out (*BO*); ~ **en œuvre** *f* GÉNÉRAL, TÉL implementation; ~ **en orbite réussie** *f* RADIO successful injection; ~ **en page** *f* DONNÉES, INFORMAT layout, page layout; ~ **en paquet** *f* DONNÉES bundling; ~ **en parallèle des canaux** *f* TRANSMIS channelization; ~ **en phase** *f* RADIO, TÉL phasing; ~ **en place** *f* COMMUT setting, GÉNÉRAL setting-up, implementation, TÉL implementation, insertion; ~ **en place progressive** *f* RADIO phasing-in; ~ **au point** *f* GÉNÉRAL development; ~ **au point précise** *f* ÉLECTRON fine tuning; ~ **au point de**

programmes *f* DONNÉES, INFORMAT program debugging; **~ à poste** *f* RADIO positioning; **~ au rebut** *f* COMPOSANTS scrapping; **~ en relief** *f* INFORMAT embossing; **~ au repos** *f* COMPOSANTS release; **~ en route** *f* INFORMAT running, start-up; **~ en route à distance** *f* PÉRIPH remote playback; **~ en service** *f* GÉNÉRAL cut-over, putting into operation, putting into service, INFORMAT enabling, TÉL commissioning, introduction of service, putting into operation, putting into service; **~ sous contrôle de processus** *f* CONTRÔLE process control, TESTS process control, process quality control; **~ sous tension** *f* ALIM ÉLEC power-on; **~ sur écoute** *f* TÉL tapping, telephone tapping, wiretapping; **~ sur le marché** *f* INFORMAT, GÉNÉRAL launch, launching; **~ en surbrillance** *f* INFORMAT highlighting; **~ à la terre** *f* ALIM ÉLEC, ÉLECTRON earthing (*BrE*), grounding (*AmE*); **~ à la terre de protection** *f* ALIM ÉLEC, TESTS protective earthing (*BrE*), protective grounding (*AmE*); **~ à zéro** *f* ÉLECTRON, TESTS adjusting to zero, setting to zero, zeroing, zeroization, zeroizing

mixage: **~ de modes** *m* TRANSMIS mode mixing; **~ de pistes** *m* INFORMAT track mix

mixte *adj* GÉNÉRAL, RADIO, TÉL, TRANSMIS both-way, two-way

ML: **~ COM de branche A** *m* (*COMA*) TÉL ML COM of A branch (*COMA*); **~ COM de branche B** *m* (*COMB*) TÉL ML COM of B branch (*COMB*); **~ de traitement de protocole** *m* TÉL protocol-handling software machine; **~ unité de raccordement de multiplex** *f* TÉL *gestion des liaisons MIC* PCM link control software machine

MLM *abrév* (*modes longitudinaux multiples*) TÉL, TRANSMIS MLM (*multilongitudinal modes*)

MNP *abrév* DONNÉES MNP (*microcom network protocol*)

Mo *abrév* (*mégaoctet*) DONNÉES, INFORMAT Mb, Mbyte (*megabyte*)

MO *abrév* ÉLECTRON, INFORMAT (*magnéto-optique*) MO (*magneto-optic*), TÉL (*mise en occupation*) BO (*busy out*)

mobile *m* RADIO mobile radio station, TÉL mobile phone

mobilité *f* TESTS discrimination; **~ ionique** ÉLECTRON ionic mobility

mobilophone *m* TÉL mobile phone, mobile telephone

Moctet *m* (*mégaoctet*) DONNÉES, INFORMAT Mb, Mbyte (*megabyte*)

mode[1]: **en ~ non connecté** *adj* GÉNÉRAL connectionless (*CL*)

mode[2] *m* GÉNÉRAL mode, INFORMAT view; **~ d'accès à la mémoire** INFORMAT memory access mode; **~ d'accès multiple à répartition dans le temps** (*mode AMRT*) RADIO, SIGNAL, TÉL, TRANSMIS time-division multiple access mode (*TDMA mode*); **~ d'acquisition périodique** DONNÉES routine acquisition mode; **~ AMRT** (*mode d'accès multiple à répartition dans le temps*) RADIO, SIGNAL, TÉL, TRANSMIS TDMA mode (*time-division multiple access mode*); **~ analogique** TÉL, TRANSMIS analog control; **~ analogique linéaire** TRANSMIS linear analog control; **~ analogue linéaire** TÉL linear analog control; **~ associé** TÉL associated mode; **~ asymétrique** DONNÉES split baud rate; **~ asynchrone symétrique** TÉL asynchronous balanced mode, ABM; **~ de base** INFORMAT basic material; **~ bidirectionnel simultané** RADIO, TÉL, TV both-way communication, two-way communication; **~ bidirectionnel à vitesse multiple** DONNÉES split baud rate; **~ bivitesse** DONNÉES split baud rate; **~ de boucle de satellite** RADIO satellite loopback mode; **~ calcul** INFORMAT compute mode; **~ de chiffrement par adjonction de bits sur les trains** DONNÉES additive stream cipher mode; **~ de cœur** TRANSMIS *optique de fibre* core mode; **~ compression** DONNÉES compressed mode; **~ conditions initiales** INFORMAT initial condition mode, reset mode; **~ connexion** DONNÉES connection mode; **~ de contrôle partagé** CONTRÔLE, DONNÉES shared-control mode; **~ de contrôle polarisé** DONNÉES polarized-control mode; **~ conversationnel** INFORMAT conversational mode; **~ avec correction d'erreur** TÉL error-correction mode (*ECM*); **~ couplé** TRANSMIS coupled mode; **~ courant d'exploitation** RADIO background mode of operation; **~ déconnecté** (*DM*) TÉL disconnected mode (*DM*); **~ déconnecté asynchrone** TÉL asynchronous disconnected mode; **~ déconnexion** TÉL disconnected mode; **~ de dérangement** TÉL failure mode; **~ de destination multiple** DONNÉES multidestination mode; **~ destiné à identifier une entrée** DONNÉES mode dependency; **~ à deux canaux** INFORMAT dual-channel mode; **~ dialogue** INFORMAT conversational mode; **~ dialogué** INFORMAT interactive mode; **~ différé** APP SERV, DONNÉES store-and-forward mode; **~ E** RADIO E mode; **~ échoplex** TÉL echoplex mode; **~ électrique transversal** RADIO H mode; **~ électromagnétique** TRANSMIS electromagnetic mode; **~ d'emploi** INFORMAT operation manual; **~ d'essai** TESTS test mode; **~ d'exploitation avec attente** GÉNÉRAL delay working, delay-mode operation; **~ d'exploitation avec perte** TRANSMIS loss-mode operation, loss-mode working; **~ d'exploitation entièrement dissocié** TÉL fully-dissociated mode of operation; **~ d'exploitation non associé** TÉL nonassociated mode of operation; **~ d'exploitation quasi-associé** TÉL quasi-associated mode of operation; **~ d'extension** TÉL extension mode; **~ fonctionnel** TÉL functional mode; **~ de fonctionnement** TESTS functional mode; **~ de fonctionnement par ligne directe** TÉL hot-line mode of operation; **~ de fonctionnement simultané** TÉL simultaneous operation; **~ fondamental** TRANSMIS fundamental mode; **~ fréquences vocales** PÉRIPH MF mode; **~ de fuite** TRANSMIS tunneling mode (*AmE*), tunnelling mode (*BrE*), leaky mode; **~ de gaine** TRANSMIS cladding mode; **~ guidé** TRANSMIS *optique de fibre* bound mode, guided mode; **~ de guide d'ondes** RADIO, TRANSMIS waveguide mode; **~ H** RADIO H mode; **~ hybride** ÉLECTRON hybrid mode; **~ latéral** TRANSMIS side mode; **~ lié** TRANSMIS *optique de fibre* bound mode, guided mode; **~ local** TÉL local mode; **~ longitudinal unique** TRANSMIS single longitudinal mode, SLM; **~ LP** (*mode à polarisation rectiligne*) TRANSMIS LP mode (*linearly-polarized mode*); **~ magnétique transversal** RADIO, TRANSMIS transverse magnetic mode (*TM*); **~ de maintien** INFORMAT retain mode; **~ menu** TÉL menu mode; **~ message** DONNÉES message mode; **~ mixte** DONNÉES split baud rate, GÉNÉRAL mixed mode; **~ de modification** INFORMAT alter; **~ de montage** TÉL mounting mode; **~ multiclique** DONNÉES multiclique mode; **~ multidestination** DONNÉES multidestination mode; **~ non dupliqué** TÉL nonduplicated working mode; **~ non lié** TRANSMIS unbound mode; **~ normal** DONNÉES normal mode; **~ normal déconnecté** DONNÉES normal disconnected mode (*NDM*); **~ normal présentation** DONNÉES normal mode presentation; **~ normal de réponse** DONNÉES normal response mode (*NRM*); **~ d'oscillation** RADIO mode of oscillation; **~ d'ouverture de boucle** PÉRIPH loop-disconnect mode; **~ de panne** TESTS failure mode, fault mode; **~ par salve** RADIO burst mode; **~ pixel** INFORMAT pixel map; **~ point** DONNÉES bit mapping; **~ à polarisation rectiligne** (*mode LP*) TRANSMIS linearly-polarized mode (*LP mode*); **~ progressif hiérarchique** INFORMAT, TÉL hierarchical progressive mode; **~ progressif à résolution fixe** INFORMAT fixed-resolution progressive mode; **~ de propagation** TRANSMIS bound mode, *fibres optiques* propagation mode; **~ de propagation guidée** RADIO, TRANSMIS *optique de fibre* trapped mode; **~ à quantification d'amplitude** CONTRÔLE time-quantized control, TÉL amplitude-quantized control, time-quantized control; **~ quasi-associé** TÉL quasi-associated mode; **~ rafale** RADIO burst mode; **~ rapide** DONNÉES rapid mode; **~ rayonnant** ÉLECTRON, TRANSMIS radiation mode; **~ réception** RADIO receiving mode; **~ répétitif** TÉL continuation mode; **~ de représentation graphique** INFOR-

MAT graphics mode; ~ **routage source transparent** ÉLECTRON source rooting transparent (*SRT*); ~ **sans compression** DONNÉES, INFORMAT uncompressed operation; ~ **sans connexion** DONNÉES connectionless mode; ~ **secondaire** INFORMAT subjob; ~ **de secours** TESTS backup mode; ~ **séquentiel** CONTRÔLE, INFORMAT sequential control; ~ **de signalisation non associé** SIGNAL nonassociated mode of signaling (*AmE*), nonassociated mode of signalling (*BrE*), nonmode of signaling (*AmE*), nonmode of signalling (*BrE*); ~ **de signalisation quasi-associé** TÉL quasi-associated mode of signaling (*AmE*), quasi-associated mode of signalling (*BrE*); ~ **simplex** RADIO, TÉL, TRANSMIS simplex; ~ **de surcharge EMCN** TÉL DCME overload mode; ~ **de taxation** APP SERV charging mode, charging principle, TÉL charging mode; ~ **TE** TRANSMIS TE mode; ~ **TEM** TRANSMIS TEM mode; ~ **TM** RADIO, TRANSMIS TM mode; ~ **de transfert** INFORMAT transfer mode; ~ **de transfert asynchrone** (*MTA*) APP SERV *ISDN*, INFORMAT, TÉL, TRANSMIS asynchronous digital subscriber loop (*ADSL*), asynchronous transfer mode (*ATM*); ~ **de transfert synchrone** (*MTS*) DONNÉES, TRANSMIS synchronous transfer mode (*STM*); ~ **transparent** DONNÉES transparent mode; ~ **transversal** TRANSMIS transverse mode; ~ **troposphérique** RADIO tropospheric mode; ~ **tunnel** TRANSMIS leaky mode, tunneling mode (*AmE*), tunnelling mode (*BrE*); ~ **veille** RADIO monitoring mode, watching mode, TESTS standby mode; ~ **de vibration** ÉLECTRON vibration mode; ~ **virtuel** INFORMAT virtual mode

modèle *m* GÉNÉRAL model, INFORMAT template; ~ **à cases** TESTS compartmental model; ~ **compartimenté** TESTS compartmental model; ~ **de couleurs CMJ** INFORMAT CMY color model (*AmE*), CMY colour model (*BrE*); ~ **de couleurs RVB** TV RGB color model (*AmE*), RGB colour model (*BrE*); ~ **d'éclairement** INFORMAT lighting model; ~ **en fil de fer** COMPOSANTS wire-frame model; ~ **filaire** COMPOSANTS wire-frame model; ~ **général** TÉL general model; ~ **OSI** DONNÉES OSI model; ~ **OTAN de communication de données** RADIO standard NATO data model; ~ **de précipitations** RADIO rain model; ~ **psychoacoustique** INFORMAT psychoacoustic model; ~ **de référence** DONNÉES reference model; ~ **de référence de protocole** DONNÉES protocol reference model (*PRM*); ~ **du service à destinations multiples** DONNÉES multicast service model

modélisation: ~ **cognitive** *f* GÉNÉRAL cognitive modeling (*AmE*), cognitive modelling (*BrE*); ~ **des données** *f* DONNÉES data modeling (*AmE*), data modelling (*BrE*)

modem *m* (*modulateur/démodulateur*) DONNÉES, INFORMAT, TRANSMIS modem (*modulator/demodulator*); ~ **en bande de base** DONNÉES, TRANSMIS baseband modem; ~ **pour lignes téléphoniques** DONNÉES dial-up modem; ~ **en mode appel** DONNÉES calling modem; ~ **en mode réponse** DONNÉES answering modem; ~ **multimode** TRANSMIS multiple-mode modem; ~ **nul** DONNÉES modem eliminator; ~ **optique intelligent** DONNÉES intelligent optical modem; ~ **prêt** DONNÉES data-set ready (*DSR*); ~**-radio** RADIO modem radio; ~ **télécopieur** DONNÉES fax modem; ~ **de voie** TRANSMIS channel modem

modes: ~ **longitudinaux multiples** *m pl* (*MLM*) TÉL, TRANSMIS multilongitudinal modes (*MLM*)

modificateur *m* INFORMAT modifier; ~ **de données** DONNÉES, INFORMAT data modifier; ~ **d'instruction** INFORMAT instruction modifier

modification *f* INFORMAT alteration; ~ **d'annonce à distance** PÉRIPH remote announcement modification; ~ **d'appels** TÉL call modification; ~ **de capacité** COMMUT size alteration; ~ **de commande** CONTRÔLE control change; ~ **en cours d'appel** TÉL in-call modification; ~ **de fonction** GÉNÉRAL function change, function modification; ~ **des raccordements** COMPOSANTS wiring change; ~ **de volume** CONTRÔLE volume change

modulaire *adj* GÉNÉRAL, INFORMAT modular

modularisation *f* GÉNÉRAL, INFORMAT modularization

modularité *f* GÉNÉRAL, INFORMAT modularity

modulateur *m* ÉLECTRON, RADIO modulator, TÉL keyer, TRANSMIS modulator; ~ **en anneau** ÉLECTRON ring modulator; ~ **de Hall** ÉLECTRON Hall modulator; ~ **hyperfréquence** RADIO microwave modulator; ~ **optique** TRANSMIS optical modulator; ~ **optique spatial** ÉLECTRON spatial light modulator, SLM; ~ **de phase** TRANSMIS phase modulator; ~ **télégraphique** TÉL telegraphic modulator

modulateur/démodulateur *m* (*modem*) DONNÉES, INFORMAT, TRANSMIS modulator/demodulator (*modem*)

modulation *f* GÉNÉRAL, RADIO, TÉL, TRANSMIS modulation; ~ **d'amplitude** (*MA*) DONNÉES, ÉLECTRON, INFORMAT, SIGNAL, TRANSMIS amplitude modulation (*AM*); ~ **d'amplitude négative** ÉLECTRON, TV negative amplitude modulation; ~ **d'amplitude par absorption** RADIO absorption amplitude modulation; ~ **d'amplitude par la grille** TRANSMIS grid amplitude modulation; ~ **d'amplitude positive** ÉLECTRON, TV positive amplitude modulation; ~ **d'amplitude en quadrature** (*MAQ*) DONNÉES, TRANSMIS quadrature amplitude modulation (*QAM*); ~ **analogique** INFORMAT analog modulation; ~ **angulaire** TRANSMIS angle modulation; ~ **angulaire à grande excursion** TRANSMIS wide-deviation angle modulation; ~ **automatique** INFORMAT auto-bend; ~ **à balayage de tonalité sans cohérence** TRANSMIS noncoherent swept-tone modulation; ~ **à bas niveau** GÉNÉRAL low-level modulation; ~ **binaire** DONNÉES binary modulation; ~ **par codage en treillis** (*MCT*) TRANSMIS trellis-coded modulation (*TCM*); ~ **cohérente** TRANSMIS coherent modulation; ~ **de conductivité** ÉLECTRON conductivity modulation; ~ **delta** (*MD*) SIGNAL, TRANSMIS delta modulation (*DM*); ~ **delta avec compression et extension** SIGNAL, TRANSMIS companded delta modulation (*CDM*); ~ **delta MDP-2** (*modulation de phase delta 2*) SIGNAL delta modulation PSK-2 (*phase-shift keying 2 modulation*); ~ **à déplacement minimal avec filtrage gaussien** SIGNAL Gaussian-filtered minimum-shift keying (*GFSK*); ~ **par déplacement d'amplitude** (*MDA*) RADIO, SIGNAL amplitude-shift keying (*ASK*); ~ **par déplacement d'amplitude en quadrature** (*MDAQ*) RADIO, SIGNAL quadrature amplitude-shift keying (*QASK*); ~ **par déplacement de fréquence** (*MDF*) RADIO, TRANSMIS frequency-shift keying (*FSK*); ~ **par déplacement de fréquence rapide** SIGNAL fast frequency-shift keying; ~ **par déplacement minimal** RADIO, SIGNAL minimum-shift keying; ~ **par déplacement de phase** (*MDP*) RADIO, SIGNAL phase-shift keying (*PSK*); ~ **par déplacement de phase à bande étroite** SIGNAL narrow-band phase-shift keying (*NBPSK*); ~ **par déplacement de phase bivalente** (*MDPB*) RADIO, SIGNAL binary phase-shift keying (*BPSK*); ~ **par déplacement de phase cohérente** (*MDPC*) RADIO, SIGNAL coherent phase-shift keying (*CPSK*); ~ **par déplacement de phase avec décalage** DONNÉES, SIGNAL offset quarternary-phase-shift keying (*OQPSK*); ~ **par déplacement de phase quadratique** RADIO, SIGNAL (*MDPQ*) quadrature phase-shift keying (*QPSK*); ~ **différentielle en boucle** INFORMAT differential loop modulation; ~ **différentielle par impulsions et codage** INFORMAT, TÉL differential pulse-code modulation (*DPCM*); ~ **de données** DONNÉES, INFORMAT data modulation; ~ **double courant** TÉL polar keying; ~ **par étalement du spectre** RADIO spread-spectrum modulation; ~ **à états d'amplitude-phase** SIGNAL amplitude phase-shift keying (*APSK*); ~ **fragmentée** TÉL curbed modulation; ~ **de fréquence** (*MF*) RADIO frequency modulation (*FM*); ~ **de fréquence asservie** (*MFA*) RADIO, TRANSMIS tamed frequency modulation (*TFM*); ~ **de fréquence à bande étroite** (*MFBE*) RADIO, TRANSMIS *satellite maritime* narrow-band frequency modulation (*NBFM*); ~ **de fréquence négative** ÉLECTRON negative frequency modulation; ~ **en fréquence d'une sous-porteuse** TÉL subcarrier frequency modulation; ~ **de groupe primaire** TRANSMIS group translation; ~ **de hauteur** INFORMAT pitch modulation; ~ **horaire** DONNÉES modulated tariff time slot; ~ **de huit à dix** (*ETM*) INFORMAT eight to ten modulation (*ETM*); ~ **de huit à quatorze** (*ETF*) INFORMAT eight to fourteen modulation (*ETF*); ~ **d'impulsions** TÉL modulation of pulses, pulsed carrier modulation; ~ **d'impulsions en amplitude** (*MIA*) TÉL, TRANSMIS pulse amplitude modulation (*PAM*); ~ **par impulsions et codage différentiel** (*MICD*) DONNÉES, INFORMAT differential pulse code modulation (*DPCM*); ~ **par impulsions et codage différentiel adaptatif** (*MICDA*) DONNÉES, INFORMAT, TÉL, TRANSMIS adaptive differential pulse code modulation (*ADPCM*); ~ **par impulsions codées** (*MIC*) DONNÉES, INFORMAT, TÉL, TRANSMIS pulse code modulation (*PCM*); ~ **d'impulsions dans le temps** TÉL, TRANSMIS pulse time modulation (*PTM*); ~ **d'impulsions en durée** (*MID*) DONNÉES, RADIO, TÉL, TRANSMIS pulse duration modulation (*PDM*), pulse width modulation (*PWM*); ~ **d'impulsions échantillonnée** TRANSMIS quantized pulse modulation; ~ **d'impulsions en fréquence** TÉL pulse frequency modulation (*PFM*); ~ **d'impulsions en position** (*MIP*) DONNÉES, TÉL, TRANSMIS pulse position modulation (*PPM*); ~ **insuffisante** ÉLECTRON undermodulation; ~ **à interruption de porteuse** ALIM ÉLEC quiescent carrier modulation; ~ **des intervalles entre impulsions** TÉL pulse interval modulation (*PIM*); ~ **par inversion de phase** TÉL phase-inversion modulation; ~ **multi-état** RADIO multistate modulation; ~ **multiniveau** RADIO multilevel modulation; ~ **multiple** TRANSMIS multiple modulation; ~ **par mutation des fréquences** SIGNAL frequency-exchange signaling (*AmE*), frequency-exchange signalling (*BrE*); ~ **non cohérente** DONNÉES noncoherent demodulation; ~ **non essentielle** RADIO, TRANSMIS spurious modulation; ~ **numérique** INFORMAT, TRANSMIS digital modulation; ~ **numérique par déplacement de phase** TRANSMIS digital phase-shift modulation; ~ **optique** DONNÉES optical modulation; ~ **de phase** (*MP*) DONNÉES, TRANSMIS phase modulation (*PM*); ~ **de phase delta 2** (*modulation delta MDP-2*) SIGNAL phase-shift keying 2 modulation (*delta modulation PSK-2*); ~ **de phase à plusieurs niveaux** SIGNAL multilevel phase-shift keying (*MPSK*); ~ **de phase quadrivalente** (*MDPQ*) RADIO quadrature phase-shift keying (*QPSK*); ~ **à quatre états de phase avec décalage des trains binaires** DONNÉES, SIGNAL offset quaternary phase-shift keying (*OQPSK*); ~ **à référence fixe** TÉL modulation with a fixed reference; ~ **sans retour à zéro** TRANSMIS non-return to zero modulation; ~ **par séquences pseudo-aléatoires** TRANSMIS pseudonoise modulation; ~ **simple** RADIO simple modulation; ~ **d'une suite d'impulsions** TÉL modulation of pulses, pulsed carrier modulation; ~ **par suite d'impulsions** TRANSMIS modulation by pulses; ~ **sur l'étage final** RADIO high-level modulation; ~ **télégraphique** TÉL telegraphic modulation; ~ **télégraphique d'amplitude** TÉL amplitude keying; ~ **de timbre** INFORMAT timbre modulation; ~ **par tout-ou-rien** TÉL on-off keying (*OOK*); ~ **de vitesse** ÉLECTRON velocity modulation; ~ **vocale à bande étroite** TRANSMIS narrow-band voice modulation (*NBVM*)

modulats *m pl* TÉL modulation product

modulé: ~ **en amplitude** *adj* TRANSMIS amplitude-modulated

module *m* ÉLECTRON remote, GÉNÉRAL module; ~ **complémentaire** TÉL applique model; ~ **complémentaire de transmission de données** TÉL add-on data module; ~ **de compressibilité** COMPOSANTS *résistance* bulk modulus; ~ **concentrateur** (*ASM*) ÉLECTRON concentration module (*ASM*), asynchronous interface module (*ASM*); ~ **convertisseur** ÉLECTRON converter module; ~ **de coopération** RADIO, TÉL index of cooperation; ~ **de distribution** ALIM ÉLEC distribution module; ~ **d'échantillonnage** TESTS sampler module; ~ **d'élasticité longitudinale** COMPOSANTS modulus of elasticity, Young's modulus; ~ **électronique** INFORMAT electronic module; ~ **d'énergie de bâti** TÉL rack power module; ~ **d'extension** TÉL extension module; ~ **de fluage** TESTS creep modulus; ~ **fonctionnel** COMMUT function module; ~ **de gestion** DONNÉES, INFORMAT handler; ~ **de gestion des liaisons** DONNÉES, INFORMAT link-handling module; ~ **d'identité de l'abonné** RADIO subscriber-identity module, SIM; ~ **de jonction au réseau terrestre** TRANSMIS terrestrial interface module, TIM; ~ **de liens** DONNÉES web, INFORMAT link file, link map, web; ~ **de ligne** COMMUT line module; ~ **objet** INFORMAT object module; ~ **périphérique** PÉRIPH peripheral module; ~ **de protection contrôle surtension** ALIM ÉLEC, ÉLECTRON overvoltage protection unit; ~ **de raccordement** GÉNÉRAL interface module, IM; ~ **de réception optique** TRANSMIS fiberoptic receiver (*AmE*), fibreoptic receiver (*BrE*), optical receiver; ~ **de réfraction** (*M*) RADIO refractive modulus (*M*); ~ **de résistance à la flexion plastique** COMPOSANTS plastic-bending section modulus; ~ **sonore** INFORMAT sound module; ~ **standard** TÉL standard modular unit; ~ **de stockage de jarretière** ÉLECTRON jumper storage module (*JSM*); ~ **de test** TESTS test step; ~ **à touches et voyants** APP SERV, PÉRIPH key-and-lamp unit; ~ **de transfert synchrone** DONNÉES, TRANSMIS synchronous transfer module (*STM*); ~ **de transmission en mode semi-duplex** TÉL half-duplex transmission module (*HDTM*); ~ **de transport synchrone** DONNÉES, TRANSMIS synchronous transport module (*STM*); ~ **de transport synchrone-n** DONNÉES, TRANSMIS synchronous transport module-n (*STM-n*); ~ **de Young** COMPOSANTS modulus of elasticity, Young's modulus

moduler *vt* INFORMAT, SIGNAL, TRANSMIS, TV modulate

modulomètre: ~ **FM** *m* TESTS frequency deviation meter

moins: ~ **franc** *m* ALIM ÉLEC direct negative

moirage *m* TV moiré

moiré *m* TV moiré

moirure *f* TV moiré

molette *f* GÉNÉRAL thumbwheel; ~ **de hauteur** COMPOSANTS bender; ~ **de variation de hauteur** INFORMAT pitchbend wheel, pitchbender, pitch wheel

moment: ~ **cinétique** *m* RADIO angular momentum; ~ **de l'exécution** *m* INFORMAT run time

moments: **à** ~ **simultanés** *adj* TÉL coincident, sequential; **à** ~ **successifs** *adj* TÉL coincident

monaural *adj* GÉNÉRAL monaural

monétique *f* APP SERV automatic banking

moniteur *m* CONTRÔLE system monitor, monitor, INFORMAT system monitor, monitor, TV monitor; ~ **à balayage multiple** TV multiscan monitor; ~ **de trafic** ÉLECTRON, TÉL, TESTS traffic monitor

monocanal *adj* RADIO single-channel

monocellulaire *adj* RADIO monocellular

monochromateur *m* TRANSMIS monochromator

monochromatique *adj* TRANSMIS monochromatic

monocouche: ~ **répartie** *f* (*DS*) DONNÉES distributed single-layer (*DS*)

monocristal *m* ÉLECTRON single crystal

monofente *adj* ÉLECTRON single-slot

monofréquence *adj* ÉLECTRON single-frequency, RADIO one-frequency

monolithique *adj* ÉLECTRON monolithic

monomodal *adj* ÉLECTRON, TRANSMIS monomode

monophasé *adj* ALIM ÉLEC single-phase

monophonie *f* RADIO monophony

monophonique *adj* RADIO monophonic

monoscope *m* ÉLECTRON monoscope
monostable[1] *adj* ÉLECTRON monostable
monostable[2] *m* COMPOSANTS single-shot flip-flop, ÉLECTRON one-shot multivibrator, single-shot flip-flop
monovoie *adj* RADIO single-channel
montage *m* COMPOSANTS construction, jig, GÉNÉRAL mounting, jig, fitting, assembly, setting-up, INFORMAT setup, *d'un volume* software machine; **~ amplificateur** ALIM ÉLEC amplifier circuit; **~ asynchrone** ÉLECTRON ripple clocking arrangement; **~ en cascade** ÉLECTRON cascade connection; **~ cathodyne** ÉLECTRON cathode follower; **~ Darlington** ÉLECTRON Darlington amplifier, Darlington circuit; **~ en dérivation** ALIM ÉLEC shunting; **~ différentiel** TRANSMIS differential circuit; **~ à diodes** ÉLECTRON diode circuit, diode network, diode tree; **~ à émetteur commun** ÉLECTRON earthed emitter circuit (*BrE*), grounded emitter circuit (*AmE*); **~ émetteur suiveur** COMPOSANTS emitter follower; **~ émidyne** COMPOSANTS emitter follower; **~ encastré** GÉNÉRAL flush mounting; **~ d'essai** TESTS test setup; **~ en étoile** ALIM ÉLEC star connection, wye connection, Y-connection; **~ intégrateur** DONNÉES integrating circuit, integrator, ÉLECTRON integrating circuit, RADIO integrating circuit, integrator; **~ de mesure** TESTS test setup; **~ multimédia** INFORMAT, TV multimedia editing; **~ mural** PÉRIPH, TÉL wall mounting; **~ oscillateur** ALIM ÉLEC, ÉLECTRON oscillation circuit, oscillatory circuit; **~ en parallèle** ALIM ÉLEC parallel connection; **~ en pont** ALIM ÉLEC bridge connection; **~ push-pull** ALIM ÉLEC push-pull; **~ en saillie** GÉNÉRAL surface mounting; **~ en série** ALIM ÉLEC series connection, ÉLECTRON connection in series; **~ au sol** GÉNÉRAL floor mounting; **~ en support percé** COMPOSANTS, ÉLECTRON hole mounting; **~ sur platine** ALIM ÉLEC, COMPOSANTS strip-mounted set; **~ sur rubis** ALIM ÉLEC jewel bearing, jewel mount; **~ en surface** GÉNÉRAL surface mounting; **~ symétrique** ALIM ÉLEC push-pull, ÉLECTRON push-pull arrangement, push-pull circuit; **~ synchrone** ÉLECTRON parallel clocking arrangement; **~ vidéo numérique** INFORMAT digital video editing; **~ en Y** ALIM ÉLEC wye connection, Y-connection
montant *m* COMPOSANTS vertical, vertical iron; **~ de bâti** TÉL rack stanchion, rack upright; **~ extérieur** TÉL outer stanchion, outer vertical iron; **~ initial** TÉL first upright
monté: **~ à bord** *adj* RADIO vehicle-mounted; **~ en étoile** *adj* ALIM ÉLEC, COMPOSANTS star-connected; **~ en retrait** *adj* GÉNÉRAL recess-mounted; **~ en saillie** *adj* GÉNÉRAL surface-mounted; **~ en surface** *adj* GÉNÉRAL surface-mounted
montée: **sur la ~ d'horloge** *adj* ÉLECTRON on the up clock
monter *vt* COMPOSANTS house, GÉNÉRAL fit, install, INFORMAT set up; **~ en cascade** ÉLECTRON connect in cascade; **~ en dérivation** ALIM ÉLEC shunt; **~ en parallèle** ÉLECTRON connect in parallel; **~ en série** ÉLECTRON connect in series
monteur: **~ d'appareils électroniques et de télécommunication** *m* APP SERV, PÉRIPH telecommunications and electronic equipment technician; **~-électricien** *m* GÉNÉRAL electrical fitter; **~ de jarretières** *m* GÉNÉRAL jumper king
monture: **~ azimut-élévation** *f* RADIO azimuth elevation mount
moquette: **~ antistatique** *f* ÉLECTRON anti-static carpet
morceau: **~ de démonstration** *m* INFORMAT demo song; **~ de surface** *m* INFORMAT patch
morcellement *m* INFORMAT fragmentation
morphing *m* INFORMAT morphing
MOS *abrév* (*note moyenne d'opinion*) INFORMAT MOS (*mean opinion score*), (*metal-oxide semiconductor*) ÉLECTRON MOS (*metal-oxide semiconductor*)
mosaïque *f* ÉLECTRON mosaic
mot *m* GÉNÉRAL word; **~ activable** INFORMAT activable word; **~-clé** DONNÉES, INFORMAT keyword; **~-code** DONNÉES, INFORMAT code-word; **~-code de commande** DONNÉES, INFORMAT control code-word; **~ de commande** DONNÉES, INFORMAT control word; **~ d'index** DONNÉES, INFORMAT index word; **~-instruction** DONNÉES, INFORMAT instruction word; **~ à la mode** INFORMAT buzz word; **~ numérique** DONNÉES, INFORMAT numeric word; **~-paramètre** DONNÉES, INFORMAT parameter word; **~ de passe** DONNÉES, INFORMAT password; **~ de passe de référence du destinataire** DONNÉES, INFORMAT recipient's reference password; **~ réservé** DONNÉES, INFORMAT reserved word; **~ de synchronisation** DONNÉES sync word; **~ télégraphique** TÉL telegraphic word; **~ télégraphique conventionnel** TÉL telegraphic word; **~ unique de référence** DONNÉES reference unique word
moteur: **~ asynchrone** *m* ALIM ÉLEC asynchronous motor; **~ d'entraînement** *m* GÉNÉRAL driving motor; **~ d'entraînement d'antenne** *m* RADIO antenna drive motor; **~ à induction** *m* ALIM ÉLEC asynchronous motor; **~ d'inférence** *m* APP SERV inference engine; **~ pas à pas** *m* ALIM ÉLEC stepper motor; **~ de propulsion** *m* GÉNÉRAL driving motor; **~ de recherche** *m* DONNÉES search engine; **~ synchrone** *m* TÉL synchronous motor; **~ de traction** *m* GÉNÉRAL driving motor
motif *m* APP SERV, ÉLECTRON, GÉNÉRAL, INFORMAT pattern, RADIO cluster, pattern, TESTS, TRANSMIS pattern; **~ d'accompagnement** INFORMAT accompaniment pattern; **~ composite** ÉLECTRON multiple pattern; **~ de contrôle** TESTS test pattern; **~ de franges d'interférences** TRANSMIS interference fringe pattern; **~ en négatif** ÉLECTRON negative pattern; **~ de réutilisation** RADIO reuse pattern; **~ rythmique** INFORMAT rhythm pattern; **~ en zigzag** GÉNÉRAL zigzag pattern
mots: **~ par minute** *loc* TÉL words per minute; **~ supplémentaires à venir** *loc* TÉL additional words coming (*AWC*)
mou[1] *adj* COMPOSANTS slack
mou[2] *m* COMPOSANTS slack
mouillage *m* ÉLECTRON *des contacts* wetting; **~ au mercure** ÉLECTRON mercury wetting
moule: **~ à coins** *m* COMPOSANTS split mold (*AmE*), split mould (*BrE*); **~ à simple cavité** *m* COMPOSANTS single-cavity mold (*AmE*), single-cavity mould (*BrE*)
mouler: **~ sous pression** *vt* COMPOSANTS die-cast
mousse: **~ de polystyrène** *f* COMPOSANTS expanded polystyrene, expanded polystyrene foam
MOX *abrév* (*commande manuelle d'émission*) DONNÉES MOX (*manually-operated transmission*)
moyen: **~ de dissuasion** *m* INFORMAT deterrent; **~ économique** *m* GÉNÉRAL cost-effective method; **~ d'étiquetage** *m* ÉLECTRON, GÉNÉRAL labeling (*AmE*), labelling (*BrE*)
moyenne: **~ géométrique** *f* TESTS geometric mean; **~ des heures chargées en centaines de communications-secondes** *f* TÉL average busy hour in CCS; **~ des heures chargées de la saison de pointe** *f* TÉL average busy-season busy hour; **~ de temps de bon fonctionnement** *f* TESTS mean time between failures (*MTBF*); **~ des temps de fonctionnement avant la première défaillance** *f* TESTS mean time to first failure (*MTFF*); **~ de temps total de réparation** *f* TESTS mean time to repair (*MTTR*); **~ totale contrôlée** *f* TESTS *qualité* average total inspected (*ATI*); **~ de l'utilisation quotidienne** *f* TESTS average daily use
moyeu: **~ de bobine** *m* TÉL reel hub
MP *abrév* (*modulation de phase*) DONNÉES, TRANSMIS PM (*phase modulation*)
MPDU *abrév* (*unité de données de protocole de message*) DONNÉES MPDU (*message protocol data unit*)
MRA *abrév* (*message de refus d'appel*) DONNÉES, TÉL CRM (*call-rejected message*)
MRC *abrév* (*multiplexage par répartition en code*) COMMUT, RADIO, TÉL, TRANSMIS CDM (*code-division multiplexing*)

MRF *abrév* (*multiplexage par répartition en fréquence*) COMMUT, SIGNAL, TÉL, TRANSMIS FDM (*frequency-division multiplexing*)

MRL *abrév* COMMUT, TÉL, TRANSMIS (*multiplexage par répartition en longueur d'onde*) WDM (*wavelength-division multiplexing*), (*multiplexeur en longueur d'onde*) WDM (*wavelength-division multiplexer*)

MRT *abrév* (*multiplexage par répartition dans le temps*) COMMUT, SIGNAL, TÉL TDM (*time-division multiplex*)

MS-DOS® *abrév* (*système d'exploitation Microsoft®*) INFORMAT MS-DOS® (*Microsoft disk-operating system*)

MSSS *abrév* (*système de soutien multisatellite*) RADIO MSSS (*multisatellite support center, multisatellite support centre*)

MTA *abrév* (*mode de transfert asynchrone*) APP SERV *ISDN*, INFORMAT, TÉL, TRANSMIS ADSL (*asynchronous digital subscriber loop*), ATM (*asynchronous transfer mode*)

MTS *abrév* (*mode de transfert synchrone*) DONNÉES, TRANSMIS STM (*synchronous transfer mode*)

MUF *abrév* (*fréquence maximale utilisable*) RADIO MUF (*maximum usable frequency*); **~ d'exploitation** *m* RADIO operational MUF; **~ de référence** *m* RADIO basic MUF

muldex *m* TÉL muldex

multi[1]: **~-aires** *adj* RADIO multiarea; **~-relais** *adj* RADIO multirelay; **~-tâche** *adj* DONNÉES multitask, multitasking; **~-utilisateur** *adj* GÉNÉRAL multiuser

multi[2]: **~-adressage** *m* DONNÉES multiaddressing; **~-joncteur numérique** *m* TÉL digital multi-junctor; **~-sélecteur** *m* COMMUT cross-point switch

multibloc *m* TÉL multiblock

multibrin *adj* COMPOSANTS stranded, ÉLECTRON stranded, octopus, TRANSMIS multistranded

multicanal *adj* TRANSMIS multichannel

multicolonnage *m* DONNÉES multiple columns

multicommunication *f* DONNÉES multicast call

multiconstructeur *m* GÉNÉRAL multivendor

multicoupleur: **~ de récepteurs** *m* RADIO receiver multicoupler

multifilaire *adj* GÉNÉRAL, PÉRIPH multiline

multifréquence *f* SIGNAL multifrequency, TÉL multifrequency, multifrequency telephone, TRANSMIS multifrequency; **~ à deux tonalités** TÉL dual-tone multifrequency (*DTMF*)

multiligne *adj* GÉNÉRAL, PÉRIPH multiline, multiple-line

multimédia[1] *adj* INFORMAT, TV multimedia

multimédia[2] *m* INFORMAT, TV multimedia; **~ interactif** INFORMAT interactive multimedia

multimètre *m* TESTS universal meter, multimeter

multipaire *adj* TV multipair

multipiste *adj* DONNÉES multitrack

multiplage: **~ en fils nus** *m* TÉL bare-wire multiple; **~ partie asymétrique** *m* TÉL asymmetrical grading

multiple[1] *adj* ÉLECTRON, GÉNÉRAL multiple

multiple[2] *m* ÉLECTRON, GÉNÉRAL multiple; **~ de sélecteurs** COMMUT selector multiple, switch multiple, TÉL selector multiple, switch multiple; **~ téléphonique** COMMUT multiple switchboard; **~ à unités verticales** COMMUT vertical-unit multiple

multiplet *m* DONNÉES byte; **~ final** DONNÉES final character; **~ de signalisation** DONNÉES flag byte

multiplex: **~ d'accès aux stations de commande** *m* TÉL control-station access multiplex; **~ d'alarmes** *m* TÉL alarm link; **~ hétérogène** *m* TÉL heterogeneous multiplex; **~ homogène** *m* TÉL homogeneous multiplex; **~ interstation** *m* TÉL intercontrol station multiplex; **~ temporel de fréquences orthogonales codées** *m* TRANSMIS coded orthogonal frequency-division multiplex (*COFDM*)

multiplexage *m* COMMUT, DONNÉES, INFORMAT, RADIO multiplexing, TRANSMIS multiplexing, channeling (*AmE*), channelling (*BrE*); **~ continu** RADIO continuous multiplexing; **~ directionnel par division** TRANSMIS directional-division multiplexing (*DDM*); **~ en fréquence** COMMUT, SIGNAL, TÉL, TRANSMIS frequency multiplexing, frequency-division multiplexing (*FDM*); **~ en longueur d'onde** TÉL, TRANSMIS wavelength-division multiplex; **~ par paquets** RADIO packet multiplexing; **~ par répartition** COMMUT, SIGNAL, TÉL, TRANSMIS frequency multiplexing, frequency-division multiplexing (*FDM*); **~ par répartition en code** (*MRC*) COMMUT, TÉL, RADIO, TRANSMIS code-division multiplexing (*CDM*); **~ par répartition en fréquence** (*MRF*) COMMUT, SIGNAL frequency-division multiplexing (*FDM*), frequency-division multiple access (*FDMA*), TÉL, TRANSMIS frequency-division multiplexing (*FDM*), frequency-division multiple access (*FDMA*); **~ par répartition en longueur d'onde** (*MRL*) COMMUT, TÉL, TRANSMIS wavelength-division multiplex, wavelength-division multiplexing (*WDM*); **~ par répartition dans le temps** (*MRT*) COMMUT, SIGNAL time-division multiplex (*TDM*); **~ temporel** INFORMAT time multiplexing, TÉL time-division multiplexing, time multiplexing, TRANSMIS time-division multiplex (*TDM*); **~ temporel asynchrone** DONNÉES, INFORMAT asynchronous time-division multiplexing (*ATDM*); **~ dans le temps** INFORMAT, TÉL, TRANSMIS time multiplexing

multiplexer *vt* COMMUT, DONNÉES, INFORMAT, RADIO, TRANSMIS multiplex

multiplexeur *m* (*MUX*) COMMUT, DONNÉES, INFORMAT, RADIO, TRANSMIS multiplexer (*MUX*); **~ aiguilleur** COMMUT switching multiplexer, switching statistical multiplexer; **~-démultiplexeur** TÉL muldex; **~ distant** TRANSMIS remote multiplexer; **~ de données** DONNÉES, INFORMAT, TRANSMIS data multiplexer; **~ de données du deuxième ordre** DONNÉES, TRANSMIS second-data multiplexer; **~ de données du premier ordre** DONNÉES, TRANSMIS first-data multiplexer; **~ à guides d'ondes communs** RADIO common waveguide multiplexer; **~ d'insertion-extraction** (*MIE*) TRANSMIS add-drop multiplexer; **~ en longueur d'onde** (*MRL*) COMMUT, TÉL, TRANSMIS wavelength-division multiplexer (*WDM*); **~ de mémoire de données** COMMUT, DONNÉES data-store multiplexer; **~ numérique** TÉL, TRANSMIS digital multiplexer (*DM*); **~ optique** TV optical multiplexer; **~ du premier ordre** DONNÉES, TRANSMIS first-order multiplexer; **~ primaire** DONNÉES, TRANSMIS first-order multiplexer; **~ spectral** TÉL, TRANSMIS wavelength-division multiplexer (*WDM*); **~ synchrone** TRANSMIS synchronous multiplexer (*SM*); **~ temporel** COMMUT, SIGNAL time-division multiplexer (*TDM*); **~ de transfert** COMMUT transfer multiplexer

multiplicateur *m* ÉLECTRON, TÉL multiplicator, multiplier; **~ binaire** DONNÉES binary multiplier; **~ de fréquence** TRANSMIS frequency multiplier; **~ de Hall** ÉLECTRON Hall multiplier

multiplication: **~ d'erreurs** *f* TÉL error multiplication; **~ du gaz** *f* ÉLECTRON gas multiplication

multiplié *adj* ÉLECTRON, TÉL multiplied

multipoint *adj* GÉNÉRAL multipoint

multipolaire *adj* ALIM ÉLEC multipolar, multipole

multiprocesseur *m* DONNÉES, INFORMAT multiprocessor

multiprogrammation *f* INFORMAT multiprogramming

multirelève *f* INFORMAT multipolling

multisession *f* INFORMAT multisession

multisite *adj* RADIO multisite

multisynchrone *adj* TV multisynchronous

multitâche *adj* INFORMAT multitask, multitasking

multitimbral *adj* INFORMAT multitimbral

multitraitement *m* DONNÉES, INFORMAT multiprocessing

multitrame *f* DONNÉES, TÉL, TRANSMIS multiframe

multivibrateur *m* ÉLECTRON multivibrator (*MV*); **~ asservi** ÉLECTRON synchronized multivibrator; **~ instable** ÉLECTRON *éléments logiques* astable circuit, multivibrator (*MV*); **~ monostable** COMPOSANTS single-shot flip-flop,

ÉLECTRON one-shot multivibrator, single-shot flip-flop; **~ synchronisé** ÉLECTRON synchronized multivibrator
multivision *f* INFORMAT slide show
multivoie[1] *adj* TRANSMIS multichannel
multivoie[2] *f* TÉL *de transmission* multitone system
multivoix *adj* GÉNÉRAL multivoice
murmure: **~ confus** *m* TÉL babble
musique: **~ d'attente** *f* TÉL music on hold; **~ électronique** *f* INFORMAT electronic music
mutation *f* ÉLECTRON, TRANSMIS transfer
mutilation *f* GÉNÉRAL garbling, TÉL mutilation; **~ en début de bribe** TRANSMIS front-end clipping; **~ due à la saturation** TRANSMIS competitive clipping; **~ de la parole** TÉL, TRANSMIS clipping
mutiler *vt* GÉNÉRAL garble, TÉL mutilate
MUX *abrév* (*multiplexeur*) COMMUT, DONNÉES, INFORMAT, RADIO, TRANSMIS MUX
MV *abrév* (*message vocal*) DONNÉES, INFORMAT, TÉL VM (*voice message*); **~ concerné** *m* DONNÉES subject VM

N

N *abrév* (*newton*) TESTS N (*newton*)
NA *abrév* (*sans objet*) DONNÉES, GÉNÉRAL, TÉL N/A (*not applicable*)
N-A *abrév* (*numérique-analogique*) GÉNÉRAL D-A (*digital-to-analog*)
NACK *abrév* (*accusé de réception négatif*) RADIO NACK (*negative acknowledgement*)
NAI *abrév* (*numéro d'appel indirect*) TÉL, TRANSMIS IDN (*indirect directory number*)
nanoseconde *f* (*ns*) TESTS nanosecond (*ns*)
nappe: **~ de fils** *f* COMPOSANTS wiring form; **~ de multiple** *f* COMMUT multiple frame
navette *f* INFORMAT shuttle; **~ spatiale** GÉNÉRAL Earth-to-orbit shuttle, RADIO, TV space shuttle
navigateur *m* DONNÉES, INFORMAT navigator
navigation *f* DONNÉES navigation, INFORMAT navigation, browsing; **~ hypermédia** INFORMAT hypermedia navigation; **~ hypertexte** INFORMAT hypertext navigation; **~ multimédia** DONNÉES, INFORMAT multimedia navigation
naviguer *vt* DONNÉES, INFORMAT navigate
navire: **~ câblier** *m* TRANSMIS cable ship, cable-laying ship
NCSA *abrév* (*centre national des applications super-informatiques*) DONNÉES NCSA (*National Center for Supercomputing Applications AmE*)
nécessaire: **~ d'installation** *m* GÉNÉRAL installation kit; **~ de nettoyage** *m* GÉNÉRAL cleaning kit; **~ de réparation** *m* GÉNÉRAL, TÉL repair kit
NEF *abrév* (*trame inattendue*) TÉL NEF (*not-expected frame*)
négateur: **~ de paquets** *m* DONNÉES burst negator
négatif: **~ de distribution** *m* ALIM ÉLEC distribution negative; **~ franc** *m* ALIM ÉLEC direct negative
négation *f* ÉLECTRON NOT operation, negation, INFORMAT negation
négociation *f* ÉLECTRON negotiation; **~ de temps de transit de bout en bout** DONNÉES, TÉL end-to-end transit delay negotiation (*EETDN*)
néguentropie *f* DONNÉES negentropy
neige *f* RADIO snow
nématique[1]: **~ super-torsadé** *m* TRANSMIS supertwist nematic (*STN*); **~ torsadé supermonochrome** *m* ÉLECTRON monochrome supertwist nematic (*MSTN*)
nématique[2]: **~ en hélice** *f* ÉLECTRON twisted nematic
néoprène *m* COMPOSANTS neoprene
néper *m* ÉLECTRON neper (*np*)
NET *abrév* (*Norme européenne de télécommunications*) GÉNÉRAL ETS (*European Telecommunication Standard*)
NETBIOS *abrév* DONNÉES Network Basic Input Output System (*NETBIOS*)
netiquette *f* DONNÉES, INFORMAT netiquette
netsurfing *m* DONNÉES, INFORMAT netsurfing
netteté *f* DONNÉES cleanliness, sharpness; **~ phonétique** RADIO, TÉL, TRANSMIS articulation; **~ des syllabes** TÉL syllable articulation
neurones: **~ récursifs** *m pl* TRANSMIS recursive neurons
neutralisation *f* INFORMAT disablement, TÉL lockout; **~ de ligne** TÉL line lockout
neutraliser *vt* INFORMAT disable, TÉL lockout
neutrodynage *m* ÉLECTRON neutralization
news *m* DONNÉES news
News: **~ Watcher** *m* DONNÉES *Internet* News Watcher
newton *m* (*N*) TESTS newton (*N*)
NF *abrév* (*norme française*) GÉNÉRAL NF (*norme française*)
NFM *abrév* (*gestionnaire de fichiers en réseau*) DONNÉES, INFORMAT NFM (*network file manager*)
NFS *abrév* (*système de fichiers en réseau*) DONNÉES, INFORMAT NFS (*network file system*)
N_G *abrév* (*non-circularité de la gaine*) TÉL, TRANSMIS non-circularity of cladding
NI *m* ÉLECTRON NOR
nichrome *m* COMPOSANTS nichrome
nid: **~ d'abeilles** *m* RADIO cluster of cells
niveau *m* GÉNÉRAL level, TÉL level, shelf; **~ absolu de tension** TÉL absolute voltage level; **~ B** (*niveau bas*) ÉLECTRON L level (*low level*); **~ bas** (*niveau B*) ÉLECTRON low level (*L level*); **~ de base** COMMUT base level; **~ du blanc** INFORMAT, TÉL, TRANSMIS white level; **~ de blocage** TÉL blocking grade of service; **~ de brouillage** ALIM ÉLEC, RADIO interference level; **~ de bruit** ALIM ÉLEC, GÉNÉRAL, INFORMAT noise level; **~ de bruit d'un circuit** TÉL circuit noise level; **~ de bruit introduit par la masse** ÉLECTRON ground noise level; **~ de bruit perçu** TÉL perceived noise level; **~ de charge flottante** ALIM ÉLEC float charging level; **~ composite adapté** ALIM ÉLEC, TESTS terminated level, terminated reading; **~ de contrôle** TESTS inspection level; **~ de contrôle au casque** TÉL headphones monitoring level; **~ de coordination** DONNÉES coordination level; **~ de courant de maintien** ÉLECTRON holding current level; **~ de déclenchement** INFORMAT trigger level; **~ de décroissance** INFORMAT decay level; **~ de dépistage** COMMUT trace level; **~ de dureté aux rayonnements** COMPOSANTS radiation hardness level; **~ d'écho résiduel** TÉL residual echo level; **~ d'écoulement du trafic** DONNÉES, TÉL traffic-handling level; **~ d'écoute** INFORMAT listening level; **~ d'énergie** COMPOSANTS energy level; **~ d'ensemble** INFORMAT total level; **~ d'entrée** ÉLECTRON input level; **~ d'équilibre** INFORMAT balance level; **~ fantôme** TRANSMIS phantom level; **~ de fiabilité** GÉNÉRAL confidence level; **~ de fond** GÉNÉRAL *acoustique* background level; **~ H** ÉLECTRON high (*H*), high level (*H*), high-level state (*H*); **~ haut** ÉLECTRON high (*H*), high level (*H*), high-level state (*H*); **~ inchangé** TÉL unaffected level; **~ d'indentation** DONNÉES level of indentation; **~ d'intensité** TRANSMIS intensity level; **~ d'intensité sonore** TÉL loudness level; **~ kéranuique** RADIO keranuic level; **~ de limitation de courant** ALIM ÉLEC *électricité* current limit level; **~ de maintenance** TÉL level of maintenance; **~ de mauvais fonctionnement** TESTS malfunction level; **~ maximal admissible de brouillage** (*NMAB*) DONNÉES maximum permissible interference level (*MPIL*); **~ de mobilité** RADIO level of mobility; **~ moyen** RADIO mid-level; **~ neutre** INFORMAT flat level; **~ du noir** INFORMAT, TÉL, TRANSMIS black level; **~ non adapté** TESTS unterminated level; **~ de polarisation** INFORMAT bias level; **~ de pondération de clavier** INFORMAT keyboard scaling level; **~ de priorité** TÉL priority level; **~ de production** GÉNÉRAL production level; **~ de puissance** ALIM ÉLEC, INFORMAT power level; **~ de puissance acoustique** GÉNÉRAL *acoustique* acoustic power level; **~ de qualité** TESTS quality level; **~ de qualité acceptable** TESTS *qualité* acceptable quality level; **~ de référence** INFORMAT reference level; **~ de réglage** DONNÉES adjustment level; **~ relatif** TÉL relative level; **~ relatif nominal de puissance** TÉL nominal relative level; **~ relatif de puissance** TÉL relative power level; **~ relatif réel de puissance** ALIM ÉLEC actual relative level; **~ de retard** INFORMAT delay level; **~ de rétroaction** INFOR-

MAT feedback level; ~ **de réverbération** INFORMAT reverb level, reverberation level; ~ **du risque** TESTS risk level; ~ **de ronflement** ÉLECTRON hum level; ~ **de saturation** TÉL overload point; ~ **de signal** SIGNAL, TRANSMIS signal level; ~ **de signification** TESTS significance level; ~ **de sonie** TÉL loudness level; ~ **de sortie** ÉLECTRON, INFORMAT output level; ~ **de sortie moyen** ÉLECTRON, INFORMAT average output; ~ **de sortie nominal** ÉLECTRON, INFORMAT rated output level; ~ **de sortie d'opérateur** ÉLECTRON, INFORMAT operator output level; ~ **de suppression** TÉL *vidéo* blanking level; ~ **de tarification** APP SERV charge rate; ~ **de tension** ALIM ÉLEC voltage level; ~ **de traçage** COMMUT trace level; ~ **de trafic** TÉL traffic level; ~ **de traitement du trafic** DONNÉES, TÉL traffic-handling level; ~ **zéro** TESTS zero level

niveler *vt* GÉNÉRAL level

NMAB *abrév* (*niveau maximal admissible de brouillage*) DONNÉES MPIL (*maximum permissible interference level*)

NN *abrév* DONNÉES, TÉL (*notification négative*) NN (*negative notification*), (*numéro national*) NN (*national number*)

NNTP *abrév* (*protocole de transfert de chaîne d'informations*) DONNÉES NNTP (*network news transfer protocol*)

NO: ~ **OP** *m* INFORMAT no operation instruction (*NO OP*)

nodal *adj* INFORMAT nodal

nodecards *m pl* INFORMAT nodecards

nœud *m* ÉLECTRON branching point, GÉNÉRAL node, INFORMAT branching point, node, TÉL nodal point, node, TRANSMIS branching point, hub, *de réseau* node; ~ **d'accès** TÉL access node; ~ **actif** INFORMAT active node; ~ **d'aide à l'exploitation** TÉL operations support node; ~ **arborescent** DONNÉES, INFORMAT tree node; ~ **d'arbre** DONNÉES, INFORMAT tree node; ~ **ascendant** RADIO ascending node; ~ **cible** INFORMAT destination node, target node; ~ **de communication** COMMUT switching node; ~ **de commutation** COMMUT, SIGNAL switching node; ~ **à commutation de paquets** COMMUT, TRANSMIS packet-switching node (*PSN*); ~ **composite** INFORMAT composite node; ~ **de contrôle mobile** TÉL mobile control node (*MCN*); ~ **descendant** RADIO descending node; ~ **de destination** INFORMAT destination node, target node; ~ **feuille** DONNÉES leaf node; ~ **hypermédia** INFORMAT hypermedia node; ~ **hypertexte** INFORMAT hypertext node; ~ **inactif** TRANSMIS inactive node; ~ **de mise en mémoire des informations** TÉL information-storage node (*ISN*); ~ **racine** DONNÉES root node; ~ **de réseau** TRANSMIS network node; ~ **secondaire** TRANSMIS sub-hub; ~ **de service** TÉL service node; ~ **de service et de transfert** TÉL service-transfer node; ~ **source** INFORMAT source node; ~ **de tension** ALIM ÉLEC point of minimum voltage; ~ **de transfert** TÉL transfer node; ~ **de transit international** (*NTI*) COMMUT, TÉL, TRANSMIS international gateway node (*IGN*), international packet-switching gateway exchange, international transit node; ~ **de transmission** TRANSMIS transmission node

NOGAD *abrév* (*régulateur de niveau commandé par le bruit*) ÉLECTRON NOGAD (*noise-operated gain adjusting device*)

noir: ~ **de l'image** *m* INFORMAT picture black; ~ **nominal** *m* ALIM ÉLEC nominal black

nom: ~ **de combinaison exécution** *m* DONNÉES performance name; ~ **de connexion** *m* INFORMAT login name; ~ **descriptif** *m* DONNÉES descriptive name; ~ **distinctif** *m* DONNÉES distinguished name; ~ **distinctif relatif** *m* DONNÉES relative distinguished name; ~ **d'extension** *m* INFORMAT filename extension; ~ **générique** *m* DONNÉES generic name; ~ **hiérarchique d'enregistrement** *m* (*nom RH*) DONNÉES registration-hierarchical name; ~ **logique de fichier** *m* INFORMAT logical file name (*LFN*); ~ **de paramètre** *m* TÉL parameter name; ~ **d'un point de connexion** *m* RADIO login; ~ **primitif** *m* DONNÉES primitive name; ~ **qualifié** *m* DONNÉES qualified name; ~ **RH** *m* (*nom hiérarchique d'enregistrement*) DONNÉES registration-hierarchical name; ~ **spécifique** *m* DONNÉES distinguished name; ~ **spécifique relatif** *m* DONNÉES relative distinguished name; ~ **synonyme** *m* DONNÉES synonymous name; ~ **de syntaxe abstraite** *m* DONNÉES, INFORMAT abstract syntax name; ~ **de la touche** *m* INFORMAT key label; ~ **d'utilisateur** *m* INFORMAT user name

nomade *adj* RADIO itinerant

nombre[1]: ~ **d'atteintes du but** *m* RADIO hits per scan, hits per target; ~ **de base** *m* INFORMAT base number; ~ **binaire** *m* DONNÉES, INFORMAT binary numeral; ~ **de bits différents** *m* ALIM ÉLEC number of bit disagreements; ~ **de bits effectif** *m* DONNÉES effective number of bits (*ENOB*); ~ **composé au clavier** *m* INFORMAT keyed numeral; ~ **décimal** *m* INFORMAT decimal numeral; ~ **d'entités défectueuses** *m* TESTS number of defectives; ~ **hexadécimal** *m* INFORMAT hexadecimal numeral; ~ **de lectures maximal utilisable** *m* INFORMAT maximum-usable read number; ~ **non décimal** *m* TÉL nondecimal numeral; ~ **octal** *m* TÉL octal numeral; ~ **d'ondes** *m* TRANSMIS repetency, wave number; ~ **de points de bruit impulsif** *m* TÉL impulse noise count; ~ **de spires** *m* ALIM ÉLEC number of turns; ~ **d'unités défectueuses** *m* TESTS number of defectives

nombre[2]: ~ **moyen de bits par échantillon** *loc* DONNÉES average bits per sample

nominal *adj* ALIM ÉLEC rated

non[1]: ~ **agréé** *adj* PÉRIPH nonapproved; ~ **amplifié** *adj* ALIM ÉLEC, ÉLECTRON, TV unamplified; ~ **applicable** *adj* (*NA*) DONNÉES, GÉNÉRAL, TÉL not applicable (*N/A*); ~ **attribué** *adj* INFORMAT unassigned; ~ **autorisé** *adj* TÉL not permitted; ~ **chargé** *adj* INFORMAT, TRANSMIS unloaded; ~ **commuté** *adj* ALIM ÉLEC, COMMUT, ÉLECTRON, TÉL switchless; ~ **concluant** *adj* DONNÉES inconclusive; ~ **conducteur** *adj* ALIM ÉLEC insulating, nonconducting, nonconductive, ÉLECTRON insulating; ~ **confirmé** *adj* DONNÉES nonconfirmed; ~ **connecté** *adj* DONNÉES nonconnected, off-line; ~ **contigu** *adj* RADIO nonadjacent; ~ **convivial** *adj* GÉNÉRAL user-unfriendly, INFORMAT unfriendly; ~ **détecté** *adj* GÉNÉRAL undetected; ~ **entrelacé** *adj* INFORMAT noninterlaced; ~ **fiable** *adj* INFORMAT unreliable; ~ **initialisé** *adj* INFORMAT uninitialized; ~ **isolé** *adj* ALIM ÉLEC, ÉLECTRON uninsulated; ~ **masqué** *adj* INFORMAT unmasked; ~ **métallique** *adj* COMPOSANTS nonmetallic; ~ **perturbé** *adj* GÉNÉRAL disturbance-free; ~ **piloté** *adj* ÉLECTRON free-running; ~ **polarisé** *adj* ALIM ÉLEC nonpolarised; ~ **pondéré** *adj* INFORMAT unweighted; ~ **programmable** *adj* INFORMAT unprogrammable; ~ **protégé** *adj* INFORMAT unsecured; ~ **pupinisé** *adj* TRANSMIS unloaded; ~ **soudé** *adj* COMPOSANTS, ÉLECTRON unsoldered; ~ **structuré** *adj* INFORMAT nonstructured; ~ **synchronisé** *adj* ÉLECTRON free-running, INFORMAT off-beat; ~ **tendu** *adj* COMPOSANTS slack; ~ **testé** *adj* TESTS untested; ~ **translatable** *adj* INFORMAT unrelocatable; ~ **valable** *adj* GÉNÉRAL invalid; ~ **valide** *adj* DONNÉES *message* invalid; ~ **vocal** *adj* RADIO nonvoice; ~ **voisé** *adj* INFORMAT unvoiced

non[2]: **sur ~-inscription de l'abonné mobile** *loc* RADIO on mobile subscriber not registered; **sur ~-réponse** *loc* RADIO on no reply; **sur ~-réponse au message de recherche** *loc* RADIO, TÉL on no-paging response

non-brouillage *m* RADIO noninterface

non-circularité: ~ **du cœur** *f* TÉL, TRANSMIS noncircularity of core; ~ **de la gaine** *f* (N_G) TÉL, TRANSMIS noncircularity of cladding; ~ **de la surface de référence** *f* (N_R) TÉL noncircularity of reference surface

non-concordance *f* GÉNÉRAL, TRANSMIS mismatch; ~ **de caractères** INFORMAT type mismatch; ~ **de types** INFORMAT type mismatch

non-conformité *f* TESTS nonconformity

non-détection: ~ **du mot unique** *f* DONNÉES unique word-missed detection

non-échec *m* TESTS no-failure

NON-ET *m* ÉLECTRON not AND (*NAND*)
non-identification: **~ d'appel** *f* TÉL restriction of calling-line identification
non-inscription *f* RADIO nonregistration
non-juxtaposition *f* TÉL underlap
non-linéarités *f pl* TRANSMIS transients
NON-OU *m* ÉLECTRON nondisjunction, NOR
non-protection *f* RADIO nonprotection
non-rémanence *f* DONNÉES, INFORMAT volatility
non-répudiation *f* DONNÉES nonrepudiation
non-retour[1]: **~ à zéro inversé** *adj* DONNÉES nonreturn to zero inverted (*NRZI*)
non-retour[2]: **~ à zéro** *m* DONNÉES, TÉL nonreturn to zero (*NRZ*)
normalisation *f* GÉNÉRAL, INFORMAT normalization, standardization
normaliser *vt* GÉNÉRAL, INFORMAT normalize, standardize
norme *f* GÉNÉRAL standard; **~ de codage** TRANSMIS coding standard; **~ commune d'exploitation** GÉNÉRAL common operating standard; **~ DIN** GÉNÉRAL DIN standard; **~ de fait** DONNÉES de facto standard; **~ française** (*NF*) GÉNÉRAL norme française (*NF*); **~ d'interconnexion de systèmes ouverts** ÉLECTRON, TRANSMIS open-systems interconnection standard; **~ d'interface radio** RADIO air interface standard; **~ internationale** DONNÉES international standard (*IS*); **~ ISO** GÉNÉRAL ISO standard; **~ suisse** (*SN*) GÉNÉRAL Swiss Standard; **~ de test de conformité** DONNÉES, TESTS conformance testing standard
Norme: **~ européenne de télécommunications** *f* (*NET*) GÉNÉRAL European Telecommunication Standard (*ETS*)
NOS *abrév* (*système d'exploitation de réseau*) DONNÉES NOS (*network operating system*)
NOTAM *abrév* (*avis aux navigateurs*) RADIO NOTAM (*notice to airmen*)
notarisation *f* DONNÉES notarization
notation: **~ arborescente** *f* DONNÉES, INFORMAT tree notation; **~ d'arbre** *f* DONNÉES, INFORMAT tree notation; **~ combinée arborescente et tabulaire** *f* DONNÉES, INFORMAT tree and tabular combined notation (*TTCN*); **~ combinée arborescente et tabulaire en forme graphique** *f* DONNÉES, INFORMAT tree and tabular combined notation in graphical form (*TTCN,GR*); **~ hexadécimale** *f* INFORMAT hexadecimal notation; **~ de syntaxe abstraite numéro un** *f* DONNÉES, TÉL, TRANSMIS abstract syntax notation one (*ASN1*)
note: **~ de bas de page** *f* INFORMAT footnote, bottom note; **~ basse** *f* INFORMAT low note; **~ grave** *f* INFORMAT low-range note; **~ impaire** *f* INFORMAT odd note; **~ mobile** *f* INFORMAT sticky note; **~ moyenne d'opinion** *f* (*MOS*) INFORMAT mean opinion score (*MOS*); **~ paire** *f* INFORMAT even note; **~ repositionnable** *f* INFORMAT sticky note
NoteCards *f pl* INFORMAT NoteCards
notes: **toutes ~ relâchées** *loc* INFORMAT all notes off
notification *f* DONNÉES notification; **~ EDI** DONNÉES, TÉL EDI notification (*EDIN*); **~ d'encombrement explicite** COMMUT, DONNÉES, TÉL explicit congestion notification (*ECN*); **~ d'encombrement explicite émise vers l'arrière** COMMUT, DONNÉES, TÉL backward explicit congestion notification (*BECN*); **~ d'encombrement explicite émise vers l'avant** COMMUT, DONNÉES, TÉL forward explicit congestion notification (*FECN*); **~ de message vocal** APP SERV, DONNÉES, TÉL voice notification (*VN*); **~ de messagerie vocale** APP SERV, TÉL voice messaging notification; **~ négative** (*NN*) DONNÉES, TÉL negative notification (*NN*); **~ de personne à personne** APP SERV, DONNÉES, RADIO, TÉL interpersonal notification (*IPN*); **~ positive** DONNÉES, TÉL positive notification (*PN*); **~ de réception** DONNÉES, TÉL receipt notification (*RN*); **~ retransmise** DONNÉES, TÉL forwarded notification (*FN*); **~ de service** DONNÉES, TÉL service notification (*SN*)
nouveaux: **~ opérateurs** *m pl* GÉNÉRAL new common carriers (*NCC*)
nouvelle: **~ intervention** *f* TÉL re-entry
noyau *m* GÉNÉRAL core, INFORMAT kernel, nucleus; **~ de fer** ÉLECTRON iron core; **~ magnétique** ÉLECTRON magnetic core, iron core; **~ mobile** ALIM ÉLEC plunger; **~ plongeur** ALIM ÉLEC plunger; **~ à poudre de fer** ÉLECTRON iron-dust core, iron-powder core; **~ de relais** COMPOSANTS relay core; **~ torique** ÉLECTRON toroidal core
noyer *vt* DONNÉES, ÉLECTRON, INFORMAT embed
N_R *abrév* (*non-circularité de la surface de référence*) TÉL noncircularity of reference surface
ns *abrév* (*nanoseconde*) TESTS ns (*nanosecond*)
NSC-CIG-DTC *abrév* (*commande de fonction non-normalisée CIG-DTC*) DONNÉES, TÉL NSC-CIG-DTC (*nonstandard facility command CIG-DTC*)
NSF-CSI-DIS *abrév* (*CSI-DIS de fonction non-normalisée*) DONNÉES, TÉL NSF-CSI-DIS (*nonstandard facility CSI-DIS*)
NSS *abrév* (*établissement non-normalisé*) DONNÉES, TÉL NSS (*nonstandard setup*)
NTI *abrév* (*noeud de transit international*) COMMUT, TÉL, TRANSMIS IGN (*international gateway node*), international packet-switching gateway exchange, international transit node
nuance: **~ dans les clairs** *f pl* DONNÉES nuance of lightness; **~ dans les ombres** *f pl* DONNÉES nuance of darkness
nuancier *m* INFORMAT color gamut (*AmE*), colour gamut (*BrE*)
nul *m* RADIO null
NUL *abrév* (*caractère nul*) DONNÉES NUL (*null character*)
null: **~-modem** *adj* ÉLECTRON null-modem
numérique *adj* DONNÉES, GÉNÉRAL, INFORMAT digital, numeric, numerical; **~-analogique** (*N-A*) GÉNÉRAL digital-to-analog (*D-A*)
numériquement *adv* GÉNÉRAL digitally, numerically
numérisation *f* GÉNÉRAL digitalization, digitizing, digitization; **~ de signal** SIGNAL, TRANSMIS signal digitalization; **~ des signaux vocaux** INFORMAT, TÉL voice digitization; **~ de la voix** INFORMAT, TÉL voice digitization
numérisé *adj* GÉNÉRAL digitalized, digitized
numériser *vt* GÉNÉRAL digitalize, digitize
numériseur: **~ d'images vidéo** *m* INFORMAT, TV frame grabber, video digitizer; **~ vidéo** *m* INFORMAT, TV video digitizer, frame grabber
numéro *m* GÉNÉRAL number; **~ d'abonné** DONNÉES, TÉL subscriber number (*SN*); **~ d'abonné multiple** TÉL multiple subscriber number (*MSN*); **~ abrégé** TÉL abbreviated number; **~ absent** PÉRIPH unanswered call; **~ d'accès** TÉL access number; **~ d'annuaire** TÉL listed directory number (*LDN*); **~ d'appel** TÉL directory number (*DN*), subscriber's directory number, telephone number; **~ d'appel direct** TÉL through-dialing number (*AmE*), through-dialling number (*BrE*); **~ d'appel indirect** (*NAI*) TÉL, TRANSMIS indirect directory number (*IDN*); **~ d'appel sans frais** *Can* TÉL freephone number (*BrE*), toll-free number (*AmE*); **~ appelé** TÉL called number; **~ d'autorisation** TÉL authorization code; **~ de bande** TÉL band number; **~ demandé initial** TÉL original-called number; **~ du demandeur incomplet** TÉL calling-party number incomplete; **~ de déplacement de station mobile** TÉL mobile station roaming number (*MSRN*); **~ en dérangement** DONNÉES number out of order; **~ désaffecté** TÉL old number; **~ de désignation** TÉL designation number; **~ de dessin** GÉNÉRAL diagram number; **~ d'échantillon** TRANSMIS sample number; **~ d'équipement** COMMUT, TÉL equipment number; **~ inexistant** TÉL nonexistent number; **~ international** TÉL international number; **~ international pour la transmission de données** DONNÉES international data number; **~ interne** COMMUT extension number; **~ de jour**

julien RADIO julian day number; ~ **local** TÉL local number; ~ **de mémoire** PÉRIPH memory number; ~ **mobile** TÉL mobile number; ~ **national** (*NN*) DONNÉES, TÉL national number (*NN*); ~ **national significatif** TÉL national significant number (*NSN*); ~ **ne figurant pas à l'annuaire** TÉL ex-directory number (*BrE*); ~ **non attribué** DONNÉES number not allocated, TÉL spare number, vacant number; ~ **non-existant** TÉL nonexistent number; ~ **non-inscrit** TÉL ex-directory number (*BrE*); ~ **non-utilisé** TÉL vacant number; ~ **occupé** TÉL busy number; ~ **d'ordre** COMMUT, DONNÉES, INFORMAT sequence number (*SN*); ~ **d'ordre d'entrée maîtresse** COMMUT, DONNÉES, INFORMAT parent-sequence number; ~ **d'ordre d'une entrée secondaire** DONNÉES, INFORMAT child-sequence number; ~ **personnel d'identification** GÉNÉRAL personal identity number (*PIN*); ~ **de piste** INFORMAT track number; ~ **de poste** TÉL directory number (*DN*), subscriber's directory number; ~ **principal** TÉL listed directory number (*LDN*); ~ **programmé** PÉRIPH stored number; ~ **de réacheminement** RADIO roaming number; ~ **de renvoi** COMMUT redirection number; ~ **renvoyant l'appel** COMMUT, TÉL redirecting number; ~ **Roger** RADIO Roger number; ~ **de schéma** GÉNÉRAL diagram number; ~ **de séquence** COMMUT, DONNÉES, INFORMAT sequence number (*SN*); ~ **de séquence de commande** COMMUT command sequence number; ~ **de séquence vers l'arrière** DONNÉES, INFORMAT, TÉL backward sequence number; ~ **de séquence vers l'avant** COMMUT, DONNÉES forward sequence number; ~ **de série** GÉNÉRAL exchange code, TÉL serial number; ~ **de session** INFORMAT session number; ~ **de station de navire** DONNÉES, RADIO, TÉL ship station number; ~ **sur la liste rouge** TÉL ex-directory number (*BrE*); ~ **de téléphone** TÉL telephone number; ~ **télex national d'abonné** TÉL subscriber's national telex number; ~ **de terminal de réseau** DONNÉES, TÉL network terminal number (*NTN*); ~ **des terminaux de destination** DONNÉES destination terminal number; ~ **de touche** INFORMAT key number; ~ **universel** TÉL universal number, universal-access number; ~ **vert** (*cf numéro d'appel sans frais Can*) TÉL freephone number (*BrE*), toll-free number (*AmE*)

numérotage *m* GÉNÉRAL numbering; ~ **ouvert** TÉL variable-length numbering; ~ **de route** TÉL route numbering; ~ **uniforme** TÉL uniform numbering

numérotation *f* APP SERV keying, DONNÉES dialing (*AmE*), dialling (*BrE*), GÉNÉRAL numbering, PÉRIPH keying, RADIO numbering plan, TÉL dialing (*AmE*), dialling (*BrE*), pulsing; ~ **abrégée** DONNÉES abbreviated address calling, TÉL abbreviated address calling, abbreviated dialing (*AmE*), abbreviated dialling (*BrE*), short-code dialing (*AmE*), short-code dialling (*BrE*), speed calling; ~ **abrégée à un chiffre** TÉL single-digit dialing (*AmE*), single-digit dialling (*BrE*); ~ **automatique** TÉL automatic dialing (*AmE*), automatic dialling (*BrE*); ~ **avant décrochage** TÉL on-hook dialing (*AmE*), on-hook dialling (*BrE*); ~ **en boucle** PÉRIPH loop-disconnect dialing (*AmE*), loop-disconnect dialling (*BrE*); ~ **au clavier** APP SERV, PÉRIPH keying; ~ **combinée raccrochée** GÉNÉRAL, RADIO, TÉL on-hook dialing (*AmE*), on-hook dialling (*BrE*); ~ **en courant alternatif** TÉL AC pulsing, alternating current pulsing; ~ **dc/fv** PÉRIPH mixed-mode dialing (*AmE*), mixed-mode dialling (*BrE*); ~ **en fréquences vocales** TÉL tone dialing (*AmE*), tone dialling (*BrE*); ~ **FV** PÉRIPH, TÉL VF dialing (*AmE*), VF dialling (*BrE*); ~ **immédiate** TÉL immediate dialing (*AmE*), immediate dialling (*BrE*); ~ **incomplète** TÉL partial dial; ~ **mixte** PÉRIPH mixed-mode dialing (*AmE*), mixed-mode dialling (*BrE*); ~ **multifréquence** TÉL dual-tone multi-frequency dialing (*AmE*), dual-tone multi-frequency dialling (*BrE*); ~ **par commande vocale** TÉL voice dialing (*AmE*), voice dialling (*BrE*); ~ **par fréquence vocale** TÉL voice-frequency key sending; ~ **par ouverture de boucle** PÉRIPH loop-disconnect dialing (*AmE*), loop-disconnect dialling (*BrE*), TÉL loop dialing (*AmE*), loop dialling (*BrE*), loop pulsing; ~ **rapide** TÉL speed dialing (*AmE*), speed dialling (*BrE*); ~ **sans décrochage du combiné** GÉNÉRAL, RADIO, TÉL on-hook dialing (*AmE*), on-hook dialling (*BrE*); ~ **sans décrocher** GÉNÉRAL, RADIO on-hook dialing (*AmE*), on-hook dialling (*BrE*), TÉL on-hook dialing (*AmE*), on-hook dialling (*BrE*), preseizure dialing (*AmE*), preseizure dialling (*BrE*)

numéroter[1] *vt* GÉNÉRAL dial, TÉL *clavier* key

numéroter[2] *vi* GÉNÉRAL dial

numéroteur *m* TÉL dialer (*AmE*), dialler (*BrE*)

O

O *abrév* (*optionnel*) DONNÉES, INFORMAT, TÉL O (*optional*)
objet *m* DONNÉES, INFORMAT, TÉL object, TESTS scope; **~ d'association unique** DONNÉES single-association object; **~ d'entrée de discriminateur** DONNÉES discriminator input object; **~ géré** DONNÉES managed object (*MO*); **~ géré de couche N** DONNÉES N-layer managed object; **~ géré des systèmes** DONNÉES systems-managed object; **~ hypermédia** INFORMAT hypermedia object; **~ moulé** COMPOSANTS plastic molding (*AmE*), plastic moulding (*BrE*); **~ de service d'application** DONNÉES, TÉL application service object (*ASO*); **~ support de gestion** DONNÉES management support object
observation *f* GÉNÉRAL monitoring, TESTS measuring; **~ aberrante** GÉNÉRAL outlier; **~ automatique** TESTS *de qualité de service* automatic observation; **~ de charge** TÉL load observation; **~ manuelle** TESTS *de qualité de service* manual observation; **~ de qualité de service** TESTS service observation; **~ de radiovent** RADIO rawin flight, rawin observation, rawin sounding; **~ semi-automatique** TESTS *de qualité de service* semiautomatic observation; **~ de trafic** ÉLECTRON, TÉL, TESTS traffic measurement, traffic monitoring, traffic observation, traffic supervision
obstacles: **~ au sol** *m pl* RADIO ground clutter
obtenir: **~ la communication** *vt* TÉL get through
obturateur *m* TV shutter; **~ amovible** ÉLECTRON knockout blanking plate; **~ antipoussière** ÉLECTRON dust cover
occultation *f* RADIO shadowing; **~ log-normale** RADIO log-normal shadowing
occupation[1]: **sur ~** *adj* TÉL on-busy
occupation[2] *f* TÉL seizure; **~ à l'arrivée** TÉL random make busy (*RMB*); **~ par le système** (*OS*) TÉL system busy (*SB*); **~ d'une route** TÉL route seizure; **~ des sols** RADIO ground clutter; **~ du spectre** TRANSMIS spectral occupancy
occupé *adj* DONNÉES engaged, TÉL busy
octet *m* DONNÉES, INFORMAT, TÉL byte; **~ de service** DONNÉES service information
octode *f* ÉLECTRON octode
ODIF *abrév* (*format ouvert d'échange de documents*) DONNÉES ODIF (*office document interchange format*)
O-E *abrév* (*optoélectrique*) GÉNÉRAL, TV O-E (*optical-electrical, optoelectric*)
OECQ *abrév* (*Organisation européenne de contrôle de la qualité*) GÉNÉRAL EOQC (*European Organization for Quality Control*)
œil: **~ de caractère** *m* PÉRIPH *typographie* typeface
OEM *abrév* GÉNÉRAL OEM (*original equipment manufacturer*)
Office: **~ des normes générales du Canada** *m* (*ONGC*) GÉNÉRAL Canadian General Standards Board (*CGSB*)
offre *f* RADIO bid, TÉL offer; **~ d'appel** COMMUT, TÉL trunk offering
ohm *m* ÉLECTRON ohm
ohmmètre *m* ÉLECTRON, TESTS ohmmeter
ombrage *m* TV shade; **~ simple** TV flat shading
ombre: **~ de col** *f* ÉLECTRON neck shadow; **~ radar** *f* RADIO radar shadow; **~ radio** *f* RADIO radio shadow
OMT *abrév* (*ordre de modification technique*) COMPOSANTS ECO (*engineering change order*)
onde[1]: **~ moyenne** *adj* RADIO medium-wave (*MW*)
onde[2] *f* GÉNÉRAL wave; **~ acoustique de surface** ÉLECTRON, TRANSMIS surface acoustic wave (*SAW*); **~ carrée** ÉLECTRON, SIGNAL square wave; **~ centimétrique** TV super-high frequency; **~ de choc** GÉNÉRAL shock-wave; **~ complexe** RADIO complex wave; **~ composite** RADIO composite wave; **~ continue** TRANSMIS continuous wave (*CW*); **~ décamétrique** TRANSMIS decametric wave; **~ en dents de scie** ÉLECTRON, TRANSMIS sawtooth wave; **~ de dépolarisation piézoélectrique** ÉLECTRON piezoelectric depolarization; **~ directe** RADIO ground wave; **~ électromagnétique** DONNÉES, ÉLECTRON electromagnetic wave; **~ entretenue** TRANSMIS continuous wave (*CW*); **~ d'espace** RADIO space wave; **~ guidée** TRANSMIS guided wave; **~ hertzienne** ÉLECTRON, RADIO hertzian wave, radio wave; **~ hyperfréquence** RADIO microwave; **~ d'impulsion** INFORMAT, TÉL, TRANSMIS pulse wave; **~ ionosphérique** RADIO sky wave, ionospheric wave; **~ longitudinale** RADIO longitudinal wave; **~ métrique** TRANSMIS metric wave; **~ modulante** TRANSMIS modulating wave; **~ modulée** TRANSMIS modulated wave; **~ moyenne** RADIO medium wave; **~ myriamétrique** RADIO very low-frequency wave; **~ ordinaire** RADIO ordinary wave; **~ pilote** GÉNÉRAL pilot wave; **~ pilote émise du sol** RADIO ground-generated pilot, earth-generated pilot; **~ pilote de référence** RADIO, TRANSMIS reference pilot; **~ plane** RADIO plane wave; **~ porteuse** TRANSMIS carrier signal, carrier wave; **~ à porteuse complète** RADIO full-carrier wave; **~ porteuse couleur** TRANSMIS color carrier (*AmE*), colour carrier (*BrE*); **~ de pression** RADIO pressure wave; **~ progressive** ÉLECTRON outgoing wave, progressive wave, RADIO progressive wave, traveling wave (*AmE*), travelling wave (*BrE*), TRANSMIS progressive wave, traveling wave (*AmE*), travelling wave (*BrE*); **~ radioélectrique** ÉLECTRON, RADIO hertzian wave, radio wave; **~ rectangulaire** GÉNÉRAL rectangular wave; **~ sinusoïdale** ÉLECTRON, INFORMAT sine wave; **~ de sol** RADIO surface wave, ground wave; **~ stationnaire** ÉLECTRON stationary wave, RADIO, TRANSMIS standing wave; **~ de surface** ÉLECTRON surface wave; **~ de surface magnétostatique** ÉLECTRON magnetostatic surface wave (*MSSW*); **~ de tension** ALIM ÉLEC voltage surge; **~ de tension stationnaire** (*OTS*) RADIO, TRANSMIS voltage standing wave (*VSW*); **~ transversale** RADIO, TRANSMIS transverse wave, shear wave; **~ transversale électrique** RADIO, TRANSMIS transverse electric wave; **~ troposphérique** RADIO tropospheric wave; **~ ultracourte** RADIO ultrashort wave
ondemètre *m* TESTS wavemeter
ondes: **~ entretenues modulées** *f pl* (*MCW*) RADIO modulated continuous waves (*MCW*)
ondulateur *m* INFORMAT vibrator
ondulation *f* ÉLECTRON ripple; **~ de gain** ÉLECTRON gain ripple
ondulé *adj* COMPOSANTS *surface* undulating
onduleur *m* ALIM ÉLEC inverter, power supply, power supply unit
onduleurs *m pl* ALIM ÉLEC switched power supply
ONGC *abrév* (*Office des normes générales du Canada*) GÉNÉRAL CGSB (*Canadian General Standards Board*)
ongle: **~ de pouce** *m* INFORMAT thumbnail
ONI *abrév* (*ouverture numérique d'injection*) ÉLECTRON, RADIO, TRANSMIS LNA (*launch numerical aperture*)
OOR *abrév* ÉLECTRON, TÉL (*oscillateur à onde régressive, oscillateur à onde rétrograde*) BWO (*backward wave oscillator*)
OP *abrév* (*ordinateur personnel*) INFORMAT PC (*desktop computer, personal computer*); **~ multimédia** *m* (*ordinateur*

personnel multimédia) INFORMAT MPC (*multimedia personal computer*)

opérateur *m* GÉNÉRAL operator, switchboard operator, *exploitant* carrier, telephonist (*BrE*), telephone operator; **~ de réseau** TÉL network operating company, network operator; **~ de réseau câblé** TV cable network operating company, cable network operator; **~ du service de nuit** COMMUT night-service operator; **~ sonore** INFORMAT sound operator; **~ de SUT** DONNÉES SUT operator; **~ de télécommunications** DONNÉES, TÉL public operator; **~ de test** DONNÉES, TESTS test operator

opération: **~ abstraite d'avertissement** *f* DONNÉES alert abstract-operation; **~ abstraite de dépôt de message** *f* DONNÉES, INFORMAT message-submission abstract-operation; **~ abstraite d'enregistrement mémoire de messages** *f* DONNÉES register-MS abstract-operation; **~ abstraite extraction** *f* DONNÉES fetch abstract-operation; **~ abstraite de résumé** *f* DONNÉES summarize abstract-operation; **~ abstraite de suppression** *f* DONNÉES delete abstract-operation; **~ abstraite de listage** *f* DONNÉES list abstract-operation; **~ d'aménagement** *f* INFORMAT housekeeping operation; **~ auxiliaire** *f* INFORMAT auxiliary operation; **~ de base** *f* INFORMAT basic operation; **~ booléenne** *f* INFORMAT Boolean operation; **~ de chargement** *f* INFORMAT loading operation; **~ distante** *f* DONNÉES, TESTS remote operation (*RO*); **~ d'écriture** *f* INFORMAT write action; **~ d'effacement** *f* INFORMAT clear operation; **~ d'entrée/sortie** *f* INFORMAT input/output process; **~ erronée** *f* INFORMAT incorrect operation; **~ ET** *f* ÉLECTRON AND operation; **~ d'exclusion** *f* ÉLECTRON NOT-IF-THEN operation; **~ de gestion des systèmes** *f* INFORMAT systems management operation; **~ d'identité** *f* INFORMAT identity operation; **~ irréversible** *f* INFORMAT irreversible operation; **~ logique** *f* INFORMAT logic operation; **~ majoritaire** *f* ÉLECTRON majority operation; **~ NI** *f* ÉLECTRON nondisjunction, NOR operation; **~ NON-ET** *f* ÉLECTRON NAND operation, nonconjunction; **~ de non-identité** *f* ÉLECTRON nonidentity operation; **~ OU** *f* ÉLECTRON disjunction, OR operation; **~ OU exclusif** *f* ÉLECTRON EXCLUSIVE-OR operation, nonequivalence operation; **~ OU inclusif** *f* ÉLECTRON disjunction, OR operation; **~ de recherche et de sauvetage** *f* RADIO search and rescue operation (*SRR operation*); **~ de sauvegarde** *f* INFORMAT save operation; **~ de seuil** *f* ÉLECTRON, RADIO threshold operation; **~ à un seul pas** *f* INFORMAT one-step operation

opérations: **~ liées** *f pl* DONNÉES linked operations; **un million d'~ en virgule flottante par seconde** *loc* (*megaflops*) INFORMAT one million floating-point operations per second (*megaflops*)

opposition: **~ de phase** *f* ALIM ÉLEC *électricité* anti-phase

optimal *adj* GÉNÉRAL high, optimum,optimal

optimum *adj* GÉNÉRAL high

option[1]: **en ~** *adj* INFORMAT, TV optional

option[2] *f* GÉNÉRAL option; **~ d'accès** INFORMAT easy access; **~ de commande** TÉL command option; **~ normalisée** TÉL standardized option; **~ par défaut** INFORMAT default; **~ de partage** DONNÉES sharing option; **~ de présentation** TÉL layout option; **~ utilisateur** (*U*) DONNÉES, TÉL user option (*U*)

optionnel *adj* (*O*) DONNÉES, INFORMAT, TÉL optional (*O*)

options: **~ connectables** *f pl* ALIM ÉLEC strapping options; **~ d'impression** *f pl* INFORMAT print settings

optique *f* DONNÉES, GÉNÉRAL optics; **~ des fibres** TRANSMIS fiber optics (*AmE*), fibre optics (*BrE*); **~ géométrique** ÉLECTRON ray optics, TRANSMIS geometric optics (*GO*), ray optics; **~ intégrée** TRANSMIS integrated optics; **~ non linéaire** DONNÉES nonlinear optics; **~ ondulatoire** ÉLECTRON wave optics, TRANSMIS wave optics, physical optics; **~ physique** TRANSMIS physical optics; **~ des rayonnements** ÉLECTRON, TRANSMIS ray optics

optiquement: **~ actif** *adj* COMPOSANTS *optique de fibre*, TRANSMIS optically active

optocoupleur *m* DONNÉES, ÉLECTRON, TRANSMIS optocoupler

optoélectrique *adj* (*O-E*) GÉNÉRAL, TV optical-electrical (*O-E*), optoelectric (*O-E*)

optoélectronique[1] *adj* COMMUT, COMPOSANTS, ÉLECTRON, TRANSMIS optronic, optoelectronic

optoélectronique[2] *f* COMMUT, COMPOSANTS, ÉLECTRON, TRANSMIS optronics, optoelectronics

orage: **~ magnétique** *f* RADIO magnetic storm

orbite *f* RADIO orbit; **~ d'attente** RADIO parking orbit; **~ basse** RADIO low orbit, low earth orbit; **~ cimetière** RADIO graveyard orbit; **~ circulaire** RADIO circular orbit; **~ à déferlement lent** RADIO near-synchronous equatorial orbit; **~ directe** RADIO direct orbit; **~ elliptique** RADIO elliptical orbit; **~ elliptique très allongée** RADIO highly-eccentric orbit; **~ équatoriale** RADIO equatorial orbit; **~ équatoriale quasi-synchrone** RADIO near-synchronous equatorial orbit; **~ halo** RADIO halo orbit; **~ inclinée** RADIO inclined orbit; **~ non perturbée** RADIO unperturbed orbit; **~ polaire** RADIO polar orbit; **~ proche de la terre** RADIO closed earth orbit; **~ à projection fermée sur la surface terrestre** RADIO closed earth-track orbit; **~ des satellites géostationnaires** RADIO geostationary orbit, geostationary satellite orbit; **~ des satellites géosynchrones** RADIO geosynchronous satellite orbit; **~ très elliptique** RADIO highly elliptical orbit

orchestration *f* INFORMAT orchestration

ordinateur *m* INFORMAT computer; **~ analogique** INFORMAT analog computer; **~ asynchrone** INFORMAT asynchronous computer; **~ bloc-notes** INFORMAT clipboard, clipboard computer, notebook, notebook computer, notepad, notepad computer; **~ de bureau** (*PC*) INFORMAT desktop computer, personal computer (*PC*); **~ central** INFORMAT mainframe, mainframe computer; **~ de commande** INFORMAT controlling computer; **~ compatible** INFORMAT compatible computer; **~ d'entrée-sortie** INFORMAT front-end computer; **~ frontal** COMMUT front-end computer; **~ hôte** INFORMAT host computer; **~ hybride** INFORMAT hybrid computer; **~ incorporé** INFORMAT built-in computer; **~ individuel** (*PC*) INFORMAT personal computer (*PC*), desktop computer; **~ intégré** INFORMAT embedded computer; **~ à jeu d'instructions réduit** (*RISC*) DONNÉES, INFORMAT reduced-instruction-set computer (*RISC*); **~ neuronal** INFORMAT neural computer; **~ personnel** (*OP*) INFORMAT personal computer (*PC*); **~ personnel multimédia** (*OP multimédia*) INFORMAT multimedia personal computer (*MPC*); **~ de poche** INFORMAT palmtop computer, pocket computer; **~ portable** INFORMAT hand-held computer, notepad, laptop computer, notebook computer, notepad computer, notebook, portable computer; **~ RISC** (*RISC*) DONNÉES, INFORMAT reduced-instruction-set computer (*RISC*); **~ séquentiel** DONNÉES, INFORMAT sequential machine; **~ spécialisé** INFORMAT special purpose computer; **~ synchrone** INFORMAT synchronous computer

ordinogramme *m* INFORMAT programming flow chart

ordonnancement: **~ de paquets** *m* DONNÉES, TÉL packet sequencing

ordre *m* TÉL, TRANSMIS order; **~ d'affichage** INFORMAT display order; **~ de balayage en zigzag** INFORMAT zigzag scanning order; **~ codé** INFORMAT coded order; **~ de contrôle** INFORMAT control statement; **~ de diversité** RADIO order of diversity; **~ émis du sol** RADIO ground command; **~ d'interclassement** INFORMAT collating sequence; **~ de modification** GÉNÉRAL change order; **~ de modification technique** (*OMT*) COMPOSANTS engineering change order (*ECO*); **~ des phases** ALIM ÉLEC phase sequence; **~ de transmission des bits** INFORMAT bit-order of transmission

organe: **~ d'appel** *m* TÉL calling device; **~ de commande** *m*

CONTRÔLE control device, operating device; ~ **commun** *m* COMMUT common equipment (*CE*); ~ **de commutation** *m* ALIM ÉLEC, COMMUT, COMPOSANTS, TÉL switching component, switching device; ~ **de commutation numérique** *m* COMMUT digital-switching element (*DSE*); ~ **de contrôle** *m* CONTRÔLE monitoring unit, *aiguillage des messages relatifs à l'OM* operation and maintenance messaging software machine; ~ **de conversion et de mémorisation** *m* DONNÉES conversion and storage unit (*CSU*); ~ **du cycle d'acquisition** *m* DONNÉES acquisition-cycle unit (*ACU*); ~ **d'écoulement du trafic** *m* DONNÉES, TÉL traffic-carrying burst, traffic-carrying device; ~ **d'entrée** *m* INFORMAT input unit; ~ **d'entrée/sortie** *m* INFORMAT input/output unit; ~ **fictif** *m* GÉNÉRAL dummy; ~ **de gestion** *m* CONTRÔLE control unit, DONNÉES management unit; ~ **logiciel** *m* DONNÉES, INFORMAT software resource; ~ **moteur** *m* RADIO actuator; ~ **de signalisation** *m* SIGNAL signal device; ~ **de surveillance de paquets** *m* TESTS burst monitor (*BM*); ~ **de synthèse vocale** *m* (*OSV*) APP SERV, INFORMAT, TÉL voice synthesis unit; ~ **de verrouillage de phase** *m* ÉLECTRON phase lock unit (*PLU*)

organes: ~ **de commande communs** *m pl* COMMUT common control equipment

organigramme *m* ÉLECTRON chart, GÉNÉRAL, INFORMAT flow chart; ~ **de données** DONNÉES, INFORMAT data-flow chart, data-flow diagram; ~ **de fonction** GÉNÉRAL function flow chart; ~ **de programmation** INFORMAT programming flow chart

Organisation: ~ **arabe de communications par satellite** *f* RADIO Arab Satellite Communications Organization (*ARABSAT*); ~ **européenne de contrôle de la qualité** *f* (*OECQ*) GÉNÉRAL European Organization for Quality Control (*EOQC*); ~ **européenne de télécommunications par satellites** *f* RADIO European Telecommunications Satellite Organization (*EUTELSAT*); ~ **internationale de télécommunications maritimes par satellites** *f* (*INMARSAT*) RADIO International Maritime Satellite Organization (*INMARSAT*); ~ **internationale de télécommunications par satellites** *f* (*INTELSAT*) RADIO International Telecommunications Satellite Organization (*INTELSAT*); ~ **internationale des télécommunications spatiales** *f* (*INTERSPOUTNIK*) RADIO International Organization of Space Communications; ~ **maritime internationale** *f* TÉL International Maritime Organization

organisme *m* TESTS *organisation* body; ~ **certificateur** TESTS certification body; ~ **d'enregistrement international** DONNÉES international registration authority; ~ **de radiodiffusion** RADIO broadcasting organization, broadcasting authority

Organisme: ~ **international de normalisation** (*ISO*) GÉNÉRAL International Standards Organization (*ISO*)

orientabilité: ~ **d'antenne** *f* RADIO antenna steerability; ~ **du faisceau** *f* RADIO beam steerability

orientable *adj* RADIO steerable

orientation: ~ **d'antenne** *f* RADIO antenna steering

orienté: ~ **objet** *adj* INFORMAT object-oriented

orienter *vt* RADIO point, steer

originaire: ~ **de** *adj* RADIO uplinked from

origine[1]: ~ **d'une défaillance** *f* TESTS cause of failure, failure cause, failure source, source of failure, source of fault; ~ **d'un échec** *f* TÉL failure source; ~ **ETCD** *f* TÉL DCE source; ~ **d'une panne** *f* TESTS cause of failure, failure cause, failure source, source of failure, source of fault

origine[2]: **avoir pour** ~ *vt* INFORMAT originate in

orthicon *m* ÉLECTRON orthicon; ~ **à transfert d'image** ÉLECTRON image orthicon

OS *abrév* DONNÉES, INFORMAT (*système d'exploitation*) OS (*operating system*), TÉL (*occupation par le système*) SB (*system busy*), (*système d'exploitation*) OS (*operating system*), TESTS (*système d'exploitation*) OS (*operating system*)

oscillateur *m* ALIM ÉLEC, COMPOSANTS, ÉLECTRON, INFORMAT oscillator; ~ **en anneau** ÉLECTRON ring oscillator; ~ **de base de temps** ÉLECTRON time-base generator; ~ **basse fréquence** INFORMAT low-frequency oscillator; ~ **de battement de fréquence** (*BFO*) ÉLECTRON, TRANSMIS beat-frequency oscillator (*BFO*); ~ **cohérent** ÉLECTRON coherent oscillator; ~ **de commande de balayage** COMPOSANTS, ÉLECTRON sweep oscillator, time-base generator; ~ **à commande numérique** INFORMAT digital-controlled oscillator; ~ **à commande par tension** ÉLECTRON voltage-controlled oscillator (*VCO*); ~ **à décalage** ALIM ÉLEC phase-shift oscillator; ~ **étalon classique** COMPOSANTS, ÉLECTRON standard oscillator; ~ **excité par choc** ÉLECTRON, RADIO shock-excited oscillator, ringing oscillator; ~ **à extinction** COMPOSANTS, ÉLECTRON self-quenching oscillator; ~ **à fréquence variable** COMPOSANTS, ÉLECTRON variable-frequency oscillator; ~ **local** ÉLECTRON, RADIO local oscillator (*LO*); ~ **naturel** ÉLECTRON natural oscillator; ~ **à onde régressive** (*OOR*) ÉLECTRON, TÉL backward wave oscillator (*BWO*); ~ **à onde rétrograde** (*OOR*) ÉLECTRON, TÉL backward wave oscillator (*BWO*); ~ **d'ondes sinusoïdales** ÉLECTRON sinusoidal oscillator; ~ **à oscillation libre** ÉLECTRON free-running oscillator; ~ **paramétrique** ÉLECTRON parametric oscillator; ~ **paramétrique optique** ÉLECTRON optical parametric oscillator; ~ **pilote** ÉLECTRON master oscillator; ~ **à quartz** ÉLECTRON quartz oscillator; ~ **à quartz à commande numérique** ÉLECTRON digitally-compensated crystal oscillator (*DCXO*); ~ **à quartz contrôlé en tension** ÉLECTRON voltage-controlled crystal oscillator (*VCXO*); ~ **à relaxation** ÉLECTRON relaxation oscillator; ~ **à réseau en double T** COMPOSANTS, ÉLECTRON parallel-tee oscillator, twin-T oscillator; ~ **sinus-cosinus** ÉLECTRON sine-cosine oscillator; ~ **à transitron** COMPOSANTS, ÉLECTRON transitron oscillator

oscillation[1]: **à** ~ **propre** *adj* ÉLECTRON self-oscillating; **à** ~ **spontanée** *adj* ÉLECTRON self-oscillating

oscillation[2] *f* ALIM ÉLEC, ÉLECTRON, INFORMAT, TRANSMIS oscillation; ~ **amortie** ÉLECTRON damped oscillation; ~ **forcée** ALIM ÉLEC forced oscillation; ~ **harmonique** ÉLECTRON harmonic oscillation; ~ **d'intermodulation** TRANSMIS intermodulation component; ~ **libre** ÉLECTRON free oscillation, free-running frequency; ~ **linéaire** TRANSMIS linear oscillation; ~ **modulante** TRANSMIS modulating wave; ~ **non linéaire** TRANSMIS nonlinear oscillation; ~ **parasite** ÉLECTRON parasitic oscillation, RADIO ringing; ~ **propre** ÉLECTRON self-oscillation; ~ **spontanée** ÉLECTRON self-oscillation

oscillatoire *adj* ALIM ÉLEC, ÉLECTRON, INFORMAT, TRANSMIS oscillatory

osciller *vi* ALIM ÉLEC, ÉLECTRON, INFORMAT, TRANSMIS oscillate

oscilloscope *m* ÉLECTRON oscilloscope; ~ **à double faisceau** ÉLECTRON dual-beam oscilloscope

ossature *f* COMPOSANTS framework

OSV *abrév* (*organe de synthèse vocale*) APP SERV, INFORMAT, TÉL voice synthesis unit

OTS *abrév* (*onde de tension stationnaire*) RADIO, TRANSMIS VSW (*voltage standing wave*)

OU: ~ **exclusif** *m* ÉLECTRON EXCLUSIVE-OR operation, nonequivalence

oui *m* DONNÉES yes (*Y*)

outil: ~ **d'aide à la conception** *m* DONNÉES design aid; ~ **à cintrer** *m* COMPOSANTS bending tool; ~ **à cliquet à sertir** *m* ÉLECTRON crimp ratchet tool frame; ~ **de coupe** *m* GÉNÉRAL cutting tool; ~ **à couper et à dénuder** *m* GÉNÉRAL cutting and stripping tool; ~ **à dégainer** *m* COMPOSANTS sheath stripping tool; ~ **à dénuder** *m* COMPOSANTS stripping tool;

~ **dérouleur** *m* COMPOSANTS unwrapping tool; ~ **de diagnostic** *m* TESTS diagnostic aid; ~ **enrouleur** *m* COMPOSANTS wrapping tool; ~ **graveur** *m* COMPOSANTS scribe tool; ~ **logiciel** *m* DONNÉES, INFORMAT software tool; ~ **à mandrin** *m* TRANSMIS cleaving tool; ~ **de marquage** *m* COMPOSANTS scouring tool; ~ **de mise en place** *m* GÉNÉRAL insertion tool; ~ **de sertissage** *m* ÉLECTRON crimp tool; ~ **de supervision** *m* CONTRÔLE, TESTS supervisory aid; ~ **de tour** *m* COMPOSANTS turning tool

ouverture *f* ÉLECTRON cut-out, GÉNÉRAL, INFORMAT, RADIO *antennes* aperture; ~ **angulaire** RADIO antenna width; ~ **de faisceau** RADIO, TRANSMIS *optique de fibres* beam width; ~ **de faisceau à mi-puissance** RADIO, TRANSMIS half-power beamwidth; ~ **numérique** DONNÉES, ÉLECTRON, TRANSMIS numerical aperture (*NA*); ~ **numérique d'injection** (*ONI*) ÉLECTRON, RADIO, TRANSMIS launch numerical aperture (*LNA*); ~ **numérique théorique maximale** ÉLECTRON, TRANSMIS maximum theoretical numerical aperture; ~ **de l'œil** TRANSMIS eye opening; ~ **prioritaire du plus court chemin** ÉLECTRON, INFORMAT open shortest path first (*OSPF*); ~ **de session** DONNÉES, INFORMAT log in

ouvre-gaine: ~ **de câble** *m* COMPOSANTS sheath cutter

P

PA *abrév* DONNÉES, TÉL (*processeur d'application*) AP (*application process*), (*processus d'application*) AP (*application processor*)
PABX *abrév* (*autocommutateur privé*) APP SERV, COMMUT, CONTRÔLE, TÉL PABX (*private automatic branch exchange*); **~ à intégration de services** *m* APP SERV service integration PABX; **~ sans fil** *m* PÉRIPH cordless PABX
package *m* DONNÉES package
packetage: **~ standard** *m* DONNÉES standard packages
PAD *abrév* DONNÉES, TÉL (*assemblage-désassemblage de paquets*) PAD (*packet assembly-disassembly facility*), (*assembleur-désassembleur de paquets*) PAD (*packet assembler-disassembler*)
pagaille *f* ÉLECTRON chaos
page *f* DONNÉES, INFORMAT page, PÉRIPH type page; **~ d'accueil** DONNÉES opening screen, welcome page, INFORMAT front page, home page, welcome page; **~ à l'écran** DONNÉES screen page; **~-écran** TV screen page; **~ suivante** TÉL form feed (*FF*); **~ télétex** TÉL teletex page; **~ de titre** INFORMAT title page, cover page; **~ Web** DONNÉES, INFORMAT Web page
pages: **~ jaunes** *f pl* TÉL yellow pages
pagination *f* GÉNÉRAL, INFORMAT *traitement de texte* page numbering, pagination, paging
paginer *vt* GÉNÉRAL, INFORMAT number pages, page, paginate
paiement: **~ électronique** *m* APP SERV electronic payment; **~ à l'émission** *m* TV pay-per-view (*PPV*); **~ par émission** *m* TV pay-per-view (*PPV*)
pair[1] *adj* GÉNÉRAL, INFORMAT even
pair[2] *m* INFORMAT peer
pairage *m* TV pairing, twinning
paire *f* GÉNÉRAL pair; **~ de bornes** ALIM ÉLEC terminal pair; **~ en câble** TRANSMIS cable pair; **~ de service** TÉL administrative pair; **~ en simplex** TÉL simplex pair; **~ symétrique non-blindée** ÉLECTRON, TRANSMIS unshielded twisted pair (*UTP*); **~ torsadée** ÉLECTRON, TRANSMIS twisted pair; **~ torsadée blindée** ÉLECTRON, TRANSMIS shielded twisted pair (*STP*); **~ torsadée non-blindée** ÉLECTRON, TRANSMIS unshielded twisted pair (*UTP*)
PAL *abrév* (*logique à matrice programmable*) INFORMAT, TV PAL (*programmable array logic*)
palette *f* DONNÉES palette; **~ en bois** COMPOSANTS wooden paddle; **~ de couleurs** INFORMAT color scheme (*AmE*), colour scheme (*BrE*); **~ graphique** TV graphics package; **~ de jeu** INFORMAT paddle; **~ d'outils** DONNÉES, INFORMAT tool palette, toolbox
palier *m* DONNÉES, GÉNÉRAL level, RADIO plateau; **~ de comptage** TÉL metering rate
panel *m* GÉNÉRAL representative sample
paneuropéen *adj* GÉNÉRAL pan-European
panier: **~ de terminaison de ligne** *m* TRANSMIS line-terminating magazine
panne[1]: **en ~** *adj* TÉL, TESTS faulty; **~ facultative** *adj* ÉLECTRON fault-tolerant
panne[2] *f* GÉNÉRAL failure, hitch, breakdown, fault; **~ affectant les fonctions** TESTS function-disturbing fault; **~ d'alimentation** ALIM ÉLEC mains failure, power failure, power outage; **~ avec dégradation des fonctions** TESTS function-degrading fault; **~ complète** TÉL, TESTS function-preventing fault; **~ corrigeable** TESTS *sûreté de fonctionnement* correctable fault; **~ corrigée** TESTS *sûreté de fonctionnement* corrected fault; **~ de courant** ALIM ÉLEC mains failure, power failure, power outage; **~ due à la conception** TESTS design fault; **~ d'électricité** ALIM ÉLEC blackout; **~ d'équipement** INFORMAT equipment failure; **~ intermittente** TESTS intermittent fault, transient fault, volatile fault; **~ localisée** TESTS single-point failure (*SPF*); **~ de réseau** TESTS network breakdown; **~ du secteur** ALIM ÉLEC mains failure, power failure, power outage; **~ simple** TESTS individual fault; **~ temporaire** TÉL temporary fault, temporary malfunction, TESTS intermittent fault, temporary fault, temporary malfunction, transient fault, volatile fault
panneau *m* ALIM ÉLEC panel; **~ d'affichage** DONNÉES bulletin board (*BB*), notice board, INFORMAT, TÉL display panel, bulletin board (*BB*), notice board; **~ arrière** INFORMAT rear panel; **~ de bâti** TÉL rack panel; **~ à boutons-poussoirs** PÉRIPH push-button panel; **~ de cloison** ÉLECTRON bulkhead panel; **~ de commande** CONTRÔLE control panel, INFORMAT operator control panel; **~ en fibres de bois** COMPOSANTS particle board; **~ hôte** ÉLECTRON host panel; **~ indicateur** TÉL display panel; **~ de lampes** TÉL lamp display panel, lamp panel; **~ de lampes d'occupation** TÉL busy lamp panel, busy-state display, engaged state display; **~ latéral** TÉL side cover, side plate; **~ mobile** COMPOSANTS slide panel; **~ de montage encastrable** ÉLECTRON flush-mount panel; **~ de photopiles** ALIM ÉLEC solar array; **~ de répartition** RADIO, TÉL patch board, patch panel; **~ solaire** ALIM ÉLEC solar panel; **~ solaire de poupe** RADIO aft solar array; **~ solaire de proue** RADIO forward solar array; **~ supérieur** INFORMAT top panel; **~ de surveillance** CONTRÔLE observation panel, supervisory panel; **~ terre** RADIO Earth-facing panel; **~ de voyants** TÉL lamp display panel, lamp panel; **~ de voyants d'occupation** TÉL busy lamp field
panneaux: **~ de brassage** *m pl* ÉLECTRON cross-connect
panoramique: **~ stéréo** *f* INFORMAT stereo panning
panoramiquer: **~ droite-gauche** *vi* INFORMAT pan right-left; **~ gauche-droite** *vi* INFORMAT pan left-right
PAO *abrév* (*publication assistée par ordinateur*) APP SERV, DONNÉES DTP (*desktop publishing*), ÉLECTRON DTP (*desktop publishing system*), INFORMAT DTP (*desktop publishing*)
papier: **~ en continu** *m* PÉRIPH continuous form; **~ au format A4** *m* GÉNÉRAL *documentation* A4-sized paper
papillon *m* CLIMAT throttle valve, RADIO antenna butterfly
papillonnement *m* TV flicker, flickering
papillotement *m* TV flicker, flickering
papilloter *vi* TV flicker
paquet *m* DONNÉES bundle, packet, RADIO, TÉL packet; **~ d'appel** TÉL call setup packet; **~ d'appel entrant** DONNÉES, RADIO, TÉL incoming-call packet (*ICP*); **~ bref** DONNÉES short burst; **~ de demande d'appel** TÉL call request packet (CRP); **~ de demande de communication** TÉL call request packet; **~ de demande de libération** DONNÉES clear request packet; **~ de données** DONNÉES, RADIO data packet; **~ porteur de trafic** DONNÉES, TÉL traffic-carrying burst, traffic-carrying device; **~ de référence principal** DONNÉES primary reference burst (*PRB*); **~ de référence secondaire de secours** DONNÉES backup reference secondary burst; **~ unique de sous-paquets** DONNÉES single sub-burst burst
paquetage: **~ utilitaire sur type bit étendu** *m* INFORMAT bit-true arithmetic package

par *abrév* (*puissance apparente rayonnée*) RADIO erp (*effective radiated power*)
PAR *abrév* (*radiodétecteur d'approche de précision*) RADIO PAR (*precision approach radar*)
parabole *f* RADIO, TV parabola; ~ **à alimentation excentrée** RADIO offset dish; ~ **à alimentation latérale** RADIO front-fed dish
paraboloïde *m* RADIO, TV paraboloid
paradiaphonie *f* TRANSMIS near-end crosstalk
parafoudre *m* ALIM ÉLEC protector, lightning conductor, lightning arrester, COMPOSANTS surge diverter, surge suppressor, TESTS lightning arrester; ~ **à charbon** ALIM ÉLEC carbon lightning arrester; ~ **à gaz raréfié** ALIM ÉLEC gas-type arrester, vacuum lightning arrester
parallélisme: ~ **ET** *m* DONNÉES AND parallelism
paramètre: ~ **de codage** *m* INFORMAT coding parameter; ~ **conditionnel** *m* DONNÉES conditional parameter; ~ **contextuel** *m* DONNÉES context parameter; ~ **contextuel réel** *m* DONNÉES actual context parameter; ~ **défini par le nom de paramètre** *m* INFORMAT parameter-name-defined parameter; ~ **défini par la position** *m* TÉL position-defined parameter; ~ **de détachement abstrait** *m* DONNÉES abstract-unbind parameter; ~ **dynamique** *m* INFORMAT program-generated parameter; ~ **d'édition** *m* INFORMAT edit parameter; ~ **externe** *m* INFORMAT external program parameter; ~ **de fonction** *m* INFORMAT function parameter; ~ **de fonctionnement** *m* INFORMAT, TESTS performance parameter; ~ **d'identification** *m* RADIO keynote field; ~ **non-fondamental** *m* DONNÉES nonbasic parameter (*NBP*); ~ **obligatoire** *m* DONNÉES obligatory parameter; ~ **des paquets de référence de gestion** *m* DONNÉES controlling reference burst parameter; ~ **de profil** *m* TRANSMIS profile parameter; ~ **qualitatif** *m* INFORMAT, TESTS performance parameter; ~ **de rattachement abstrait** *m* DONNÉES abstract-bind parameter; ~ **réel** *m* DONNÉES actual parameter; ~ **de suite de tests** *m* TESTS test suite parameter; ~ **à tolérance** *m* TESTS tolerance parameter; ~ **de vitesse** *m* INFORMAT rate parameter
paramétrer *vt* GÉNÉRAL parameterize, INFORMAT set parameters to
parasitage *m* RADIO disturbance, interference
parasite[1] *adj* ALIM ÉLEC, INFORMAT, TÉL extraneous
parasite[2] *m* RADIO radio-frequency disturbance; ~ **électromagnétique** ÉLECTRON, TESTS electromagnetic disturbance, electromagnetic interference (*EMI*); ~ **industriel** ÉLECTRON, RADIO man-made noise
parasites *m pl* ALIM ÉLEC noise, RADIO disturbance, interference, TÉL background noise, extraneous noise
parasurtenseur *m* ALIM ÉLEC surge absorber, surge voltage protector, surge suppressor
paratonnerre *m* TESTS lightning conductor, lightning rod
parc: ~ **des abonnés** *m* RADIO, TÉL subscriber base; ~ **d'antennes** *m* RADIO antenna farm; ~ **automobile** *m* RADIO vehicle fleet
parcage *m* TÉL camp-on
parcourir *vt* INFORMAT scroll through, RADIO travel
parcours *m* INFORMAT path, RADIO travel; ~ **de câble** TRANSMIS *pose de câbles* cable route, cable way; ~ **libre** DONNÉES navigation, INFORMAT browsing, navigation; ~ **linéaire** INFORMAT linear path; ~ **de navigation hypermédia** INFORMAT hypermedia navigation path; ~ **de navigation hypertexte** INFORMAT hypertext navigation path
pare: ~**-étincelles** *m* ALIM ÉLEC spark quench, spark quencher, spark suppressor; ~**-vapeur** *m* CLIMAT vapor barrier (*AmE*), vapour barrier (*BrE*)
parité *f* DONNÉES, ÉLECTRON parity; ~ **d'entrelacement des bits** DONNÉES bit interleaved parity (*BIP*); ~ **X à entrelacement de bits** DONNÉES bit interleaved parity X (*BIP-X*)
parlante: ~ **numérique** *f* PÉRIPH digital talking machine
parole *f* RADIO, TÉL speech; ~ **artificielle** TÉL simulated speech
partage *m* DONNÉES sharing, ÉLECTRON, INFORMAT split, splitting, sharing, RADIO sharing, trunking; ~ **de charge** ÉLECTRON load sharing; ~ **de fréquences** RADIO trunking; ~ **de ligne** TÉL line sharing; ~ **de ressources** INFORMAT resource sharing; ~ **de temps** DONNÉES, INFORMAT time sharing; ~ **de trafic** RADIO traffic sharing
partager[1] *vt* DONNÉES share, ÉLECTRON, INFORMAT divide up, split, share, RADIO share; ~ **les fréquences** RADIO trunk
partager[2]: **se** ~ *v réfl* ÉLECTRON, INFORMAT split
partagiciel *m* DONNÉES, INFORMAT shareware, shareware package
particiel *m* DONNÉES, INFORMAT shareware
partie: ~ **de connaissance** *f* INFORMAT knowledge chunk; ~ **contraintes** *f* DONNÉES constraints part; ~ **du corps de données vocales** *f* DONNÉES, INFORMAT voice body part; ~ **déclarative** *f* DONNÉES declarations part; ~ **de domaine initial** *f* DONNÉES, TÉL initial domain part, IDP; ~ **dynamique** *f* DONNÉES dynamic part; ~ **présentation générale** *f* DONNÉES overview part; ~ **spécifique du domaine** *f* DONNÉES domain-specific part (*DSP*)
partiel *adj* DONNÉES, INFORMAT, TÉL partial
partiellement: ~ **obligatoire** *adj* (*PO*) TÉL conditionally mandatory (*CM*); ~ **supprimé** *adj* TV partially suppressed
parties: ~ **à défoncer** *f pl* ÉLECTRON knockouts
partition: ~ **des modes** *f* TRANSMIS mode partition
parv *abrév* (*puissance apparente rayonnée sur antenne verticale courte*) RADIO emrp (*effective monopole-radiated power*)
pas *m* GÉNÉRAL pitch, TV frequency step; ~ **d'assemblage** GÉNÉRAL length of lay; ~ **d'un câble** TRANSMIS cable pitch; ~ **d'entraînement** DONNÉES feed pitch; ~ **d'exploration** TÉL scanning pitch; ~ **d'hélice** RADIO helix pitch; ~ **longitudinal** DONNÉES row pitch; ~ **de masque** INFORMAT pitch; ~ **modulaire** TÉL spacing dimension; ~ **modulaire de bâti** TÉL rack spacing; ~ **de piste** INFORMAT track pitch; ~ **de régénération** TRANSMIS repeater span; ~ **de réglage** TESTS adjustment step; ~ **transversal** INFORMAT track pitch
PASC *abrév* INFORMAT PASC (*precision adaptive sub-band coding*)
passage[1]: **de** ~ *adj* GÉNÉRAL transient
passage[2] *m* INFORMAT run, RADIO pass; ~ **en connexion** TÉL through connection; ~ **de jeton** DONNÉES, TRANSMIS token passing; ~**-machine** INFORMAT run; ~ **au mode compression** (*PMC*) DONNÉES enter compressed mode; ~ **au mode transparent** (*PMT*) DONNÉES enter transparent mode; ~ **normal-secours** GÉNÉRAL changeover to standby; ~ **sous contrainte sur route de secours** TÉL forced rerouting; ~ **sur un satellite de secours** RADIO changeover to a spare satellite; ~ **d'urgence sur canal sémaphore de secours** TÉL emergency changeover; ~ **à zéro** ALIM ÉLEC, COMMUT zero crossing
passant *adj* ALIM ÉLEC conducting, ÉLECTRON conducting, conducting on, opening
passe: ~**-fil** *m* COMPOSANTS grommet; ~**-fils** *m* ÉLECTRON wireminder
passer[1]: ~ **une commande par télégramme** *vt* TÉL send an order by wire; ~ **commande des services** *vt* TV order services
passer[2] *vi* ALIM ÉLEC flow, TÉL *par opérateur* go; ~ **au niveau bas** ÉLECTRON go low; ~ **au niveau haut** ÉLECTRON go high
passerelle *f* DONNÉES gateway, INFORMAT bridge, TÉL, TRANSMIS gateway; ~ **entre réseaux** TRANSMIS gateway; ~ **de niveau application** TÉL application-level gateway; ~ **de réseau** DONNÉES network gateway
passivation *f* COMPOSANTS passivation; ~ **au nitrure de silicium** COMPOSANTS silicium nitride passivation
passive: ~ **double star** *m* TRANSMIS passive double star (*PDS*)
pastille *f* COMPOSANTS die, ÉLECTRON land, chip; ~ **d'entrefer** ÉLECTRON nonmagnetic armature shim; ~ **de microphone**

PÉRIPH transmitter cap; ~ **de soudure** COMPOSANTS solder pad; ~ **de traversée** ALIM ÉLEC feedthrough pad
pâte: ~ **thermoconductrice** *f* ALIM ÉLEC heat sink compound
patron *m* INFORMAT template
patte *f* ÉLECTRON pin
pause *f* ÉLECTRON hold
pavé: ~ **numérique** *m* INFORMAT numeric pad, numeric keypad, keypad
PAVI *abrév* (*point d'accès vidéotex*) TÉL, TV videotex access point, videotex gateway
pavillon *m* PÉRIPH *de l'écouteur* earcap, earpiece, receiver cap, TÉL *de l'écouteur* receiver cap, earcap, earpiece, *de microphone* funnel mouthpiece; ~ **de récepteur** PÉRIPH *de l'écouteur*, TÉL earcap, earpiece, receiver cap
pays: ~ **de destination** *m* GÉNÉRAL country of destination, TÉL administration of destination; ~ **d'origine** *m* GÉNÉRAL country of origin, TÉL administration of origin; ~ **terminal** *m* TÉL terminal country; ~ **de transit** *m* TÉL transit country; ~ **de transit en commutation** *m* TÉL switched-transit country; ~ **de transit direct** *m* TÉL direct-transit country
PB *abrév* (*point de branchement*) TRANSMIS drop point, TV branching point
PC *abrév* (*ordinateur de bureau, ordinateur individuel*) INFORMAT desktop computer, PC (*personal computer*); ~ **multimédia** *m* INFORMAT multimedia PC
PCI *abrév* TÉL (*interface de communication programmable*) PCI (*programmable communication interface*), TV (*porteuse cohérente incrémentielle*) ICC (*incrementally coherent carrier*)
PCL *abrév* DONNÉES PCL (*print command language*)
PCMCIA *abrév* (*Association internationale de normalisation des cartes à mémoire pour PC*) ÉLECTRON, INFORMAT PCMCIA (*Personal Computer Memory Card International Association*)
PCO *abrév* (*point de contrôle et d'observation*) DONNÉES PCO (*point of control and observation*)
PCS *abrév* COMMUT (*point de commande de services*) service command point, RADIO (*système de communications personnelles*) PCS (*personal communications system*)
PD *abrév* (*photodiode*) COMPOSANTS PD (*photodiode*), diode photodetector, DONNÉES PD (*photodiode*), TRANSMIS PD (*photodiode*), diode photodetector
PDA *abrév* (*photodiode à avalanche*) COMPOSANTS, DONNÉES, TRANSMIS *fibres optiques* APD (*avalanche photodiode*)
PDF *abrév* DONNÉES, INFORMAT PDF (*portable document file*)
PDU: ~ **de gestion de test** *m* DONNÉES, TESTS test management PDU (*TM-PDU*)
peau: ~ **avant** *f* RADIO active skin; ~ **de kapton revêtue de germanium** *f* DONNÉES germanium-coated kapton skin; ~ **en kevlar** *f* RADIO kevlar skin
PEB *abrév* (*puissance équivalente de bruit*) ÉLECTRON, TRANSMIS NEP (*noise equivalent power*)
pédale *f* INFORMAT pedal; ~ **d'expression** INFORMAT expression pedal; ~ **de portamento** INFORMAT portamento footswitch, portamento pedal; ~ **de soutenu** INFORMAT sustain pedal, sustain footswitch
peigné *adj* ÉLECTRON combed
peigne *m* GÉNÉRAL *câbles* laced cable fan; ~ **collecteur** TRANSMIS collecting comb; ~ **mobile** COMMUT lifting comb
pellicule *f* ALIM ÉLEC, COMPOSANTS film
pen: ~ **Collaboration Environment** *m* INFORMAT pen Collaboration Environment
pénalité *f* TRANSMIS penalty
pénétrer *vt* INFORMAT *illégalement* break into
pente *f* ÉLECTRON mutual conductance, slope, INFORMAT, RADIO, TESTS, TRANSMIS slope; ~ **de la courbe d'efficacité** TESTS discrimination ratio, slope of the OC curve; ~ **de la courbe du temps de propagation de groupe** TRANSMIS group-delay slope; ~ **de diminution** RADIO roll-off; ~ **de filtre** INFORMAT filter slope; ~ **aux frontières** RADIO attenuation slope; ~ **de gain** ÉLECTRON gain slope; ~ **du palier** RADIO plateau slope; ~ **du plateau** RADIO plateau slope; ~ **du sommet** TÉL pulse droop; ~ **de la trajectoire** RADIO flight-path angle
pentode *m* ÉLECTRON pentode
pénurie *f* RADIO *de fréquences* shortage
percée *f* INFORMAT breakthrough
percer *vt* COMPOSANTS bore, drill
percutant *adj* ÉLECTRON radical
perdre *vt* GÉNÉRAL lose; ~ **la connexion** TÉL drop out; ~ **le signal** TÉL drop out
père *m* DONNÉES, INFORMAT parent
pérenne *adj* GÉNÉRAL future use
perforation *f* ALIM ÉLEC puncture; ~ **d'entraînement** DONNÉES feed hole, INFORMAT feed hole, sprocket hole
perforations *f pl* ÉLECTRON punch-downs
perforatrice: ~ **à clavier** *f* INFORMAT keypunch
perforé *adj* GÉNÉRAL holed
perforer *vt* COMPOSANTS bore, drill
performance: ~ **opérationnelle** *f* TÉL operational performance; ~ **du réseau** *f* TESTS network performance (*NP*)
performances: **avoir des** ~ **supérieures à** *loc* INFORMAT outperform
périastre *m* RADIO periapsis
périgée *m* RADIO perigee
période *f* GÉNÉRAL time span; ~ **anomalistique** RADIO anomalistic period; ~ **d'arrêt** ÉLECTRON off period; ~ **d'attente** TESTS standby time; ~ **de charge** ALIM ÉLEC charge cycle, charge time; ~ **chargée** TÉL busy period; ~ **de conduction** ÉLECTRON conduction period; ~ **de décharge** ALIM ÉLEC discharge cycle, discharge time; ~ **de défaillance par usure** TESTS wear-out failure period; ~ **des défaillances précoces** TESTS early failure period; ~ **d'éclairement total** RADIO full sunlight period; ~ **erronée** DONNÉES erroneous period (*EP*); ~ **de fort trafic** TÉL busy period; ~ **d'installation** GÉNÉRAL installation time; ~ **d'intensité constante de dérangement** TÉL constant failure intensity period; ~ **de lignes** TV scanning line period; ~ **nodale** RADIO nodal period; ~ **non requise** TESTS nonrequired time; ~ **opérationnelle moyenne** TESTS *qualité* average run length; ~ **orbitale** RADIO orbital period, orbiting period, orbiting time, period of revolution; ~ **de pointe** TÉL busy period; ~ **préparatoire** TÉL preparatory period; ~ **de prolongation des essais** TESTS extension test period; ~ **de réglage** TÉL line-up period; ~ **de répétition des impulsions** (*PRI*) TÉL pulse repetition period (*PRP*); ~ **de repos** DONNÉES space period, ÉLECTRON, TÉL idle period; ~ **requise** TÉL, TESTS required time; ~ **de réserve** GÉNÉRAL backup period, contingency period; ~ **de révolution** RADIO orbital period, orbiting period, orbiting time, period of revolution; ~ **de révolution sidérale** RADIO sidereal period of revolution; ~ **de rotation sidérale** RADIO sidereal period of rotation; ~ **de taxation** TÉL charging period, metering period; ~ **de travail** TÉL mark period; ~ **vacante** TESTS free time, idle time
périodemètre *m* TESTS frequency counter, frequency meter
périodicité: ~ **de maintenance** *f* TESTS maintenance interval; ~ **moyenne de révisions** *f* TESTS mean time between overhauls (*MTBO*)
périphérie *f* INFORMAT edge, outer rim
périphérique *m* DONNÉES *traitement des données à distance* peripheral, peripheral device; ~ **à accès séquentiel** INFORMAT direct access service; ~ **de bureau** PÉRIPH tabletop device; ~ **d'entrée** INFORMAT input device; ~ **intelligent** (*PI*) PÉRIPH intelligent peripheral; ~ **de sortie** INFORMAT output device; ~ **téléphonique** PÉRIPH telephony-rated device; ~ **virtuel** INFORMAT virtual device
péritel *m* TV SCART socket television
péritéléphonie *f* PÉRIPH accessories, TÉL peritelephony, peripheral telephony

perl *m* DONNÉES perl
perle: ~ **isolante** *f* ÉLECTRON insulating bead
permanence[1]: ~ **par message enregistré** *f* TÉL after-hours recorded announcement
permanence[2]: **être de** ~ *vi* RADIO be on call
perméabilité *f* COMPOSANTS, ÉLECTRON permeability; ~ **initiale** ÉLECTRON initial permeability
perméable *adj* COMPOSANTS, ÉLECTRON permeable
perméamètre *m* TESTS permeameter
perméance *f* ÉLECTRON permeance
permittivité *f* ÉLECTRON permittivity; ~ **piézoélectrique de surface** ÉLECTRON surface piezoelectric permittivity; ~ **piézoélectrique de volume** ÉLECTRON bulk piezoelectric permittivity
permutation *f* INFORMAT swapping; ~ **de chaîne** COMMUT side shifting; ~ **circulaire** INFORMAT end-around shift; ~ **de groupe** COMMUT group switching; ~ **de liaison** COMMUT line switching; ~ **sous tension** ÉLECTRON hot swapping
persistance *f* ÉLECTRON lag, smearing; ~ **d'écran** TV persistence
personnaliser *vt* GÉNÉRAL customize
personnel: ~ **de maintenance** *m* GÉNÉRAL maintenance staff; ~ **de surveillance** *m* GÉNÉRAL manning
perte[1]: **avec** ~ *adj* INFORMAT, TÉL, TRANSMIS lossy
perte[2] *f* GÉNÉRAL loss; ~ **accidentelle** INFORMAT accidental loss; ~ **en aval de l'amplificateur** ÉLECTRON post-amplifier loss; ~ **de câble** TRANSMIS, TV cable loss; ~ **au connecteur** TRANSMIS connector loss; ~ **de connexion** TÉL drop-out; ~ **de couplage** ÉLECTRON coupling loss; ~ **de coupleur** ÉLECTRON coupler loss; ~ **en courbe** RADIO corner loss, TRANSMIS loss around a corner; ~ **de courbure** TRANSMIS bending loss; ~ **dans les câbles** TRANSMIS, TV cable loss; ~ **dans l'entrefer** TRANSMIS gap loss; ~ **dans le système d'alimentation** RADIO, TRANSMIS feeder loss; ~ **de données** DONNÉES data loss; ~ **due au pointage** RADIO pointing loss; ~ **due aux réflexions** ALIM ÉLEC, TRANSMIS reflection loss; ~ **effective** TRANSMIS effective loss; ~ **d'épissure** TRANSMIS splice loss; ~ **d'exploitation** GÉNÉRAL operating loss; ~ **extrinsèque de raccordement** TRANSMIS extrinsic joint loss; ~ **d'information** INFORMAT *vidéo* drop-out; ~ **d'insertion** TRANSMIS insertion loss; ~ **d'insertion des connecteurs** TRANSMIS connector insertion loss; ~ **d'insertion en fonctionnement** ÉLECTRON operating loss; ~ **d'interaction** TRANSMIS interaction loss; ~ **intrinsèque de raccordement** TRANSMIS intrinsic joint loss; ~ **d'isolement** TRANSMIS insulation loss; ~ **de ligne** TRANSMIS feed loss; ~ **du mot unique** DONNÉES unique word loss; ~ **de multitrame** TRANSMIS loss of multiframe (*LOM*); ~ **de niveau** TÉL drop-out; ~ **optique** DONNÉES, TRANSMIS optical loss; ~ **d'oxyde** INFORMAT particle shredding; ~ **par absorption** RADIO absorption loss; ~ **par couplage antenne-milieu** RADIO antenna-to-medium coupling loss, gain degradation; ~ **par couplage de polarisation** RADIO polarization coupling loss; ~ **par courants parasites** ALIM ÉLEC eddy current loss; ~ **par décalage latéral** TRANSMIS lateral offset loss; ~ **par défaut d'alignement** TRANSMIS misalignment loss; ~ **par déplacement longitudinal** TRANSMIS *fibres optiques* longitudinal offset loss; ~ **par déplacement transversal** TRANSMIS *fibres optiques* lateral offset loss, transverse offset loss; ~ **par désalignement angulaire** TRANSMIS *fibres optiques* angular misalignment loss; ~ **par espacement longitudinal** TRANSMIS gap loss; ~ **par inclinaison axiale** TRANSMIS *fibres optiques* angular misalignment loss; ~ **par macrocourbures** TRANSMIS macrobend loss; ~ **par microcourbures** TRANSMIS microbend loss; ~ **par réflexion optique** TRANSMIS optical return loss (*ORL*); ~ **de pointeur** TRANSMIS loss of pointer (*LOP*); ~ **aux points de raccordement** TRANSMIS splice loss; ~ **de puissance** ALIM ÉLEC, ÉLECTRON power dissipation; ~ **de puissance directe** ALIM ÉLEC, ÉLECTRON forward power dissipation, forward power loss; ~ **de puissance à l'état passant** COMPOSANTS, ÉLECTRON on-state power loss; ~ **au raccord** TRANSMIS splice loss; ~ **de signal** TÉL drop-out, TRANSMIS loss of signal (*LOS*); ~ **spécifique** TRANSMIS *de câbles coaxiaux* standard loss; ~ **en surface** ALIM ÉLEC surface leakage; ~ **de tension** ALIM ÉLEC voltage loss; ~ **de trame** TRANSMIS loss of frame (*LOF*); ~ **de transmission** TRANSMIS transmission loss; ~ **de verrouillage de multitrame** TRANSMIS loss of multiframe alignment; ~ **de verrouillage de trame** (*PVT*) TÉL, TRANSMIS loss of frame alignment (*LFA*)
pertes: ~ **dans les connecteurs** *f pl* TRANSMIS connector insertion loss; ~ **dues aux conditions locales** *f* RADIO site attenuation; ~ **en espace libre** *f pl* RADIO free-space basic loss; ~ **par hystérésis** *f pl* ÉLECTRON hysteresis loss; ~ **de pénétration dans les bâtiments** *f pl* RADIO building penetration loss
pertinence: ~ **rétroactive** *f* DONNÉES relevance feedback
perturbation *f* ALIM ÉLEC, ÉLECTRON, GÉNÉRAL, RADIO disturbance, interference, noise; ~ **électromagnétique** ÉLECTRON, RADIO electromagnetic disturbance, electromagnetic interference (*EMI*), TESTS electromagnetic disturbance, electromagnetic interference (*EMI*); ~ **industrielle** ÉLECTRON, RADIO man-made noise; ~ **ionosphérique** RADIO ionospheric disturbance; ~ **radioélectrique** RADIO radio-frequency interference (*RFI*), radio jamming, radio-frequency disturbance, static interference; ~ **temporaire** TÉL, TESTS temporary interruption, temporary perturbation
perturber *vt* GÉNÉRAL disturb
pervéance *f* ÉLECTRON perveance
phantastron *m* ÉLECTRON phantastron
phase[1]: ~ **de lancement et de mise en orbite** *m* (*LEOP*) RADIO launch and early orbit phase (*LEOP*)
phase[2] *f* GÉNÉRAL phase; ~ **d'appel** TÉL ringing period; ~ **ascensionnelle** RADIO ascent phase; ~ **d'attente** TESTS standby phase; ~ **de commande du réseau** ÉLECTRON, TÉL network control phase; ~ **de compilation** INFORMAT compilation phase, compiling phase; ~ **de données** DONNÉES, TRANSMIS data phase; ~ **d'établissement de la communication** TÉL call setup phase; ~ **d'exécution** INFORMAT execute phase; ~ **d'extension** TÉL extension step; ~ **nématique** ÉLECTRON nematic phase; ~ **pilote** TESTS field trial, field test; ~ **smectique** ÉLECTRON smectic phase; ~ **stationnaire** ÉLECTRON stationary phase; ~ **de transfert de données** DONNÉES, TÉL data-transfer phase; ~ **de transfert transparent de données** DONNÉES, TÉL, TRANSMIS transparent data-transfer phase; ~ **de transmission sur liaison montante** RADIO, TRANSMIS uplink transmission phase
phasemètre *m* ALIM ÉLEC phase meter
phases: ~ **d'un appel** *f pl* TÉL call phases; ~ **d'une communication** *f pl* TÉL call phases; ~ **successives d'une communication** *f pl* TÉL successive call phases
phénoplaste *m* COMPOSANTS phenolic plastic
phone *m* TÉL phon
Phong *m* INFORMAT Phong shading
phonie/graphie *f* TÉL voice/cw
phonique *adj* INFORMAT acoustical
photo: ~ **CD** *m* DONNÉES, INFORMAT photo CD
photocathode *f* ÉLECTRON photocathode
photocomposition *f* DONNÉES, TV photocomposition
photoconduction *m* ÉLECTRON, TRANSMIS internal photoelectric effect
photoconductivité *f* ÉLECTRON, TRANSMIS internal photoelectric effect, photoconductivity
photocoupleur *m* ÉLECTRON photocoupler
photocourant *m* ALIM ÉLEC, ÉLECTRON, TRANSMIS light current, photocurrent

photodétecteur *m* DONNÉES photodetector, ÉLECTRON light detector, TRANSMIS optical detector, photodetector
photodiode *f* (*PD*) COMPOSANTS photodiode; **~ à avalanche** (*PDA*) COMPOSANTS, DONNÉES, TRANSMIS *fibres optiques* avalanche photodiode (*APD*); **~ PIN** DONNÉES, TRANSMIS PIN photodiode
PhotoEdge *m* INFORMAT PhotoEdge
photoélectrique *adj* ÉLECTRON photoelectric
photoémission *f* ÉLECTRON external photoelectric effect, photoelectric emission, photoemission, TRANSMIS *fibres optiques* external photoelectric effect, photoemission
photolithographie *f* INFORMAT photolithography
photoluminescence *f* TRANSMIS photoluminescence
photomètre *m* TV exposure meter
photomultiplicateur *m* ÉLECTRON photomultiplier
photon *m* TRANSMIS photon
photonique *f* TRANSMIS photonics
photopile *f* COMPOSANTS, ÉLECTRON photovoltaic cell
photoréaliste *adj* INFORMAT photorealistic
photorécepteur *m* DONNÉES photodetector, ÉLECTRON light detector, TRANSMIS optical detector, photodetector
photosensible *adj* ÉLECTRON light-sensitive
photostyle *m* PÉRIPH light pen
phototélégraphie *f* TÉL facsimile telegraphy, photograph facsimile telegraphy
phototransistor *m* COMPOSANTS, TRANSMIS phototransistor
phototube: **~ à gaz** *m* ÉLECTRON gas-filled phototube
photovidéotex *m* DONNÉES photovideotex
PHS *abrév* (*service physique*) APP SERV, TÉL PhS (*physical service*)
PI *abrév* (*périphérique intelligent*) PÉRIPH intelligent peripheral
piano: **~ électronique** *m* PÉRIPH electronic piano
pianoter *vt* GÉNÉRAL key in
picoseconde *f* (*ps*) TESTS picosecond (*ps*)
picot *m* PÉRIPH pin
PICS *abrév* (*déclaration de conformité de mise en oeuvre du protocole*) DONNÉES PICS (*protocol implementation conformance statement*)
pictogramme *m* ÉLECTRON pictogram
pièce *f* COMPOSANTS part; **~ détachée** COMPOSANTS spare part; **~ d'écartement** GÉNÉRAL spacer; **~ polaire** ALIM ÉLEC field pole, pole, pole piece; **~ rapportée** COMPOSANTS insert, inset; **~ de rechange** COMPOSANTS spare part
pied: **~ à coulisse** *m* TESTS slide caliper (*AmE*), slide calliper (*BrE*)
piège *m* ALIM ÉLEC sliding frequency generator, choke, ÉLECTRON trap, trap circuit; **~ à ions** ÉLECTRON ion trap; **~ à ondes** RADIO wave trap
piégeage *m* DONNÉES, ÉLECTRON trapping
piéger *vt* DONNÉES, ÉLECTRON trap
pierre: **~ angulaire** *f* DONNÉES cornerstone
piézoélectricité *f* ÉLECTRON, GÉNÉRAL piezoelectricity
piézoélectrique *adj* ÉLECTRON, GÉNÉRAL piezoelectric
pigment *m* COMPOSANTS pigment
PILA *abrév* (*présentation de l'identification de la ligne appelante*) TÉL CLID (*calling-line identification display*), CLIP (*calling-line identification presentation*)
PILC *abrév* (*présentation de l'identification de la ligne connectée*) TÉL COLP (*connected-line identification presentation*)
pile *f* ALIM ÉLEC battery, cell, ÉLECTRON battery, INFORMAT stack; **~ déchargée** ALIM ÉLEC flat battery; **~ de disques** DONNÉES disk array; **~ FM** INFORMAT FM stack; **~ IP** INFORMAT IP stack; **~ longue durée** ALIM ÉLEC long-life battery; **~ primaire** ALIM ÉLEC primary battery; **~ rechargeable** ALIM ÉLEC rechargeable battery; **~ sèche** ALIM ÉLEC dry cell, dry battery; **~ solaire** ALIM ÉLEC solar battery; **~ de support flexible** ÉLECTRON flexible media stack
piles: **à ~** *adj* ALIM ÉLEC, ÉLECTRON battery-powered
pilotable *adj* DONNÉES steerable
pilotage *m* RADIO steering; **~ par signal horaire** PÉRIPH time-signal control; **~ de trafic** DONNÉES, TÉL traffic control
piloté: **~ par informatique** *adj* INFORMAT computer-controlled
pilote *m* DONNÉES driver, ÉLECTRON driver, master oscillator, GÉNÉRAL pilot, INFORMAT driver, RADIO, TRANSMIS pilot; **~ à diapason** RADIO tuning fork oscillator drive; **~ d'impression** DONNÉES, INFORMAT printer driver; **~ à ligne résonante** ÉLECTRON line-stabilized oscillator drive; **~ de périphérique** DONNÉES device driver; **~ de référence** RADIO, TRANSMIS reference pilot
piloter *vt* RADIO steer, TV drive
pilote/réserve *m* TÉL active/standby
PIN *abrév* (*interruption de la procédure négative*) DONNÉES, TÉL, TRANSMIS PIN (*procedure interrupt negative*)
pinceau *m* INFORMAT paintbrush, RADIO pencil beam
pinceur *m* PÉRIPH pinch roller
PIP *abrév* (*interruption de la procédure positive*) DONNÉES, TÉL, TRANSMIS PIP (*procedure interrupt positive*)
pipette *f* INFORMAT eyedropper
piquage: **~ de ligne** *m* TÉL line tapping
piquet: **~ de mise à la terre** *m* ALIM ÉLEC earthing rod (*BrE*), earthing probe (*BrE*), grounding rod (*AmE*)
piqûre *f* ÉLECTRON pinhole
piratage *m* DONNÉES piracy, GÉNÉRAL poaching, fraud, INFORMAT hacking, piracy, fraud, TV piracy; **~ informatique** INFORMAT computer fraud; **~ de logiciels** DONNÉES, INFORMAT software piracy
pirate[1] *adj* DONNÉES, GÉNÉRAL, INFORMAT, TV pirate
pirate[2] *m* DONNÉES, GÉNÉRAL, INFORMAT, TV pirate; **~ informatique** INFORMAT hacker
piraté *adj* DONNÉES, GÉNÉRAL, INFORMAT, TV pirated
pirater *vt* DONNÉES, GÉNÉRAL, INFORMAT, TV pirate
PIRE *abrév* (*puissance isotrope rayonnée équivalente*) RADIO EIRP (*equivalent isotropically-radiated power*)
piste *f* COMPOSANTS, DONNÉES, INFORMAT, RADIO, TV track; **~ audio** INFORMAT *de CD-ROM, de bande vidéo* audio track, soundtrack, TV soundtrack; **~ d'entraînement** COMPOSANTS sprocket track, DONNÉES feed track, INFORMAT feed track, sprocket hole; **~ magnétique** DONNÉES magnetic track, ÉLECTRON, TÉL magnetic strip; **~ de programmes** RADIO program tracking (*AmE*), programme tracking (*BrE*); **~ son** INFORMAT vocal track; **~ de son numérique** INFORMAT digital soundtrack; **~ vidéo** INFORMAT, TV video track
pistes: **~ par pouce** *loc* (*ppp*) DONNÉES, INFORMAT tracks per inch (*TPI*)
pistolet: **~ à air chaud** *m* ALIM ÉLEC heater gun
piston *m* RADIO plunger, shorting plunger, piston; **~ de court-circuit** ÉLECTRON shorting plunger, RADIO plunger, piston
pivot: **~ de barre de sélection** *m* COMMUT pivot for horizontal, pivot for selecting bar; **~ du générateur** *m* RADIO array hinge
pivoter *vt* COMPOSANTS swivel, INFORMAT swivel, rotate
pixel *m* ÉLECTRON, INFORMAT, TV pel, picture element, pixel
pixellisation *f* DONNÉES pixelation, pixelization (*AmE*), pixellization (*BrE*)
placage *m* COMPOSANTS veneer; **~ d'environnement** INFORMAT environment mapping; **~ métallique sélectif** ÉLECTRON pattern plating; **~ métallique total** ÉLECTRON panel plating; **~ de réflexion** INFORMAT reflection mapping; **~ de relief** INFORMAT bump mapping; **~ de rugosité** INFORMAT bump mapping; **~ de texture** INFORMAT pattern mapping, texture mapping
place: **sur ~** *adj* (*SP*) TÉL on-premises, on-scene (*ONS*)
placement *m* RADIO siting
placer *vt* GÉNÉRAL locate, RADIO site, TESTS locate; **~ sur orbite** RADIO put into orbit
plage *f* GÉNÉRAL range; **~ d'accord du dispositif mécanique**

ÉLECTRON mechanical tuning range; ~ **d'accord de fréquence** RADIO frequency tuning range; ~ **d'allongement** TRANSMIS expansion window; ~ **d'éclairement** ALIM ÉLEC illuminance range; ~ **de fonctionnement** TRANSMIS working range; ~ **des fréquences** RADIO, TRANSMIS frequency range; ~ **de gris** DONNÉES, TÉL gray scale (*AmE*), grey scale (*BrE*); ~ **horaire de tarifs** TÉL tariff time-slot; ~ **d'interaction** TRANSMIS interactive area; ~ **de maintien** ÉLECTRON holding range; ~ **de mesure** TESTS measuring range; ~ **de niveau du signal d'entrée** TRANSMIS input-signal level range; ~ **optimale d'accord du dispositif mécanique** ÉLECTRON optimum mechanical tuning range; ~ **d'orientation** RADIO steerability; ~ **de réglage** CONTRÔLE control range, GÉNÉRAL setting range; ~ **de réglage de la commande de souffle** INFORMAT breath-controller range; ~ **de réglage de la molette** INFORMAT modulation wheel range; ~ **de régulation** GÉNÉRAL range of regulation; ~ **de relèvement** RADIO swing; ~ **de sensibilité** RADIO lock-in range; ~ **de variation du contraste** TV contrast range

plan[1] *adj* GÉNÉRAL flat

plan[2] *m* GÉNÉRAL schedule, INFORMAT outline; ~ **d'acheminement du trafic** TÉL, TESTS traffic routing strategy; ~ **d'affectation des bornes** GÉNÉRAL terminal number drawing; ~ **de Babcock** RADIO, TÉL Babcock plan; ~ **de bits** TÉL bit plane; ~ **de câblage** COMPOSANTS wiring diagram, GÉNÉRAL *documentation* cable drawing; ~ **des câbles** TRANSMIS cable layout; ~ **de connexions** COMPOSANTS wiring drawing; ~ **coté** ÉLECTRON outline drawing; ~ **des couches** ÉLECTRON layer drawing; ~ **de crise** GÉNÉRAL contingency plan; ~ **directeur** RADIO master time plan (*MTP*); ~ **de disposition des voies à deux fréquences** TRANSMIS two-frequency channeling plan (*AmE*), two-frequency channelling plan (*BrE*); ~ **en double Babcock** RADIO, TÉL double Babcock plan; ~ **d'échantillonnage** TESTS sampling plan; ~ **d'émission en accès multiple à répartition dans le temps** RADIO, SIGNAL, TÉL, TRANSMIS time-division multiple access burst time plan; ~ **d'émission en AMRT** RADIO, SIGNAL, TÉL, TRANSMIS TDMA burst time plan; ~ **de fréquence** TRANSMIS frequency plan, TV frequency plan, spectrum plan; ~ **des fréquences OR** RADIO OR frequency plan; ~ **d'implantation** GÉNÉRAL floor plan; ~ **d'implantation des tables de commutation** TÉL switchboard floor plan; ~ **d'installation du central** GÉNÉRAL exchange installation drawing; ~ **maillé** RADIO lattice plan; ~ **de montage du central** GÉNÉRAL exchange assembly drawing; ~ **de numérotage** TÉL numbering plan, numbering scheme; ~ **de numérotage ouvert** TÉL variable-length numbering scheme; ~ **de numérotage privé** (*PNP*) TÉL private numbering plan (*PNP*); ~ **de numérotation** GÉNÉRAL numbering plan; ~ **de l'orbite** RADIO orbital plane; ~ **de perçage** COMPOSANTS, ÉLECTRON hole drawing; ~ **de polarisation** RADIO plane of polarization; ~ **qualité** TESTS quality plan; ~ **de référence** ÉLECTRON reference plane; ~ **de réseau** TÉL network chart, network plan; ~ **de secours** GÉNÉRAL contingency plan; ~ **de service** TV service plan; ~ **de sol** ALIM ÉLEC ground plane, RADIO ground system; ~ **temporel condensé et retransmis** DONNÉES retransmitted condensed burst time plan (*RCTP*); ~ **de trame** DONNÉES burst time plan (*BTP*); ~ **de trame simulé** DONNÉES simulator burst time plan (*SBTP*); ~ **de transmission** TRANSMIS transmission plan

plancher: ~ **de câblage** *m* GÉNÉRAL, TRANSMIS *mise à l'essai* cable floor; ~ **conducteur** *m* TESTS semiconductive floor

planification: ~ **de la maîtrise de la qualité** *f* TESTS quality-control planning; ~ **du réseau** *f* TÉL network planning; ~ **des systèmes** *f* INFORMAT systems planning; ~ **du trafic en situation réelle** *f* TÉL live-situation traffic planning

planifier *vt* GÉNÉRAL schedule

plaque: ~ **antirémanente** *f* COMPOSANTS residual plate; ~ **de base** *f* COMPOSANTS *structure*, ÉLECTRON base panel, base plate; ~ **de base de gabarit pour peigne** *f* TRANSMIS formboard plate; ~ **à bornes** *f* COMPOSANTS terminal strip; ~ **entretoise** *f* COMPOSANTS spacer plate; ~ **en fibre de bois** *f* COMPOSANTS wallboard; ~ **de fixation** *f* COMPOSANTS fastening plate, fixing plate; ~ **d'habillage** *f* COMPOSANTS cover plate; ~ **de Hall** *f* ÉLECTRON Hall plate; ~ **d'identification** *f* INFORMAT nameplate; ~ **isolante** *f* ALIM ÉLEC insulating plate; ~ **isolatrice** *f* ALIM ÉLEC insulating plate; ~ **latérale** *f* TÉL side cover, side plate; ~ **de montage** *f* COMPOSANTS mounting plate; ~ **de montage d'empilage** *f* COMPOSANTS spring-set mounting plate; ~ **murale** *f* COMPOSANTS, ÉLECTRON wallplate; ~ **de relais** *f* COMPOSANTS relay bar, relay mounting bar; ~ **de relais de concentration d'alarme** *f* TÉL alarm-assembling relay set; ~ **de relais de distribution** *f* COMPOSANTS distribution relay set; ~ **de relais de ligne d'annotateur** *f* TÉL record-line relay set; ~ **de relais de ligne de jonction** *f* TÉL junction-line relay set; ~ **de remplissage** *f* COMPOSANTS dummy strip; ~ **repère** *f* GÉNÉRAL marking plate; ~ **de séparation** *f* ALIM ÉLEC insulating plate; ~ **signalétique** *f* ALIM ÉLEC rating plate, INFORMAT nameplate; ~ **au sommet** *f* RADIO vertex plate; ~ **de support** *f* COMPOSANTS fastening plate, fixing plate, supporting plate; ~ **de terre** *f* ALIM ÉLEC earthing plate (*BrE*), grounding plate (*AmE*), earth plate (*BrE*), ground plate (*AmE*)

plaquette: ~ **de connexions enroulées** *f* COMPOSANTS wire-wrap plane; ~ **isolante** *f* ALIM ÉLEC insulating spacer

plasma: ~ **produit par laser** *m* ÉLECTRON laser plasma

plastifiant *m* COMPOSANTS plasticizer

plastique[1] *adj* COMPOSANTS plastic

plastique[2]: ~ **acétovinylique** *m* COMPOSANTS vinyl acetate plastic; ~ **acrylique** *m* COMPOSANTS acrylic plastic; ~ **antistatique** *m* ÉLECTRON anti-static plastic; ~ **chlorovinylique** *m* COMPOSANTS vinyl chloride plastic; ~ **époxyde** *m* COMPOSANTS epoxy plastic; ~ **estérique** *m* COMPOSANTS ester plastic, polyester plastic; ~ **éthénique** *m* COMPOSANTS polyethylene plastic; ~ **éthylénique** *m* COMPOSANTS polyethylene plastic; ~ **moulé** *m* ÉLECTRON molded plastic (*AmE*), moulded plastic (*BrE*); ~ **non-rigide** *m* COMPOSANTS nonrigid plastic; ~ **polyamidique** *m* COMPOSANTS polyamide plastic; ~ **polycarbonate** *m* COMPOSANTS polycarbonate plastic; ~ **polyestérique** *m* COMPOSANTS polyester plastic, ester plastic; ~ **renforcé** *m* COMPOSANTS reinforced plastic; ~ **rigide** *m* COMPOSANTS rigid plastic; ~ **semi-rigide** *m* COMPOSANTS semirigid plastic; ~ **stratifié** *m* COMPOSANTS laminated plastic; ~ **thermodurcissable** *m* COMPOSANTS thermosetting plastic

plastron: ~ **blindé** *m* ÉLECTRON shielded triple-faceplate outlet jack; ~ **à prise** *m* ÉLECTRON quadplex; ~ **rectangulaire** *m* ÉLECTRON rectangular faceplate; ~ **simple** *m* ÉLECTRON single-gang faceplate

plate: ~**-forme** *f* ÉLECTRON, TRANSMIS platform; ~**-forme de gestion de réseau** *f* TRANSMIS network-management platform; ~**-forme informatique banalisée ouverte** *f* DONNÉES open standards computing platform

plateau *m* INFORMAT platter, RADIO plateau; ~ **enrouleur de fibre** TRANSMIS fiber coiling plate (*AmE*), fibre coiling plate (*BrE*); ~ **d'impulsions** TÉL pulse top; ~ **de montage à amortisseurs** COMPOSANTS shock-absorbing tray

platine *f* COMPOSANTS mounting plate; ~ **avant** ÉLECTRON component front; ~ **encastrée** COMPOSANTS flush panel; ~ **mains libres** PÉRIPH hands-free tape deck; ~ **de relais** COMPOSANTS relay bar, relay mounting bar; ~ **de verticale** COMPOSANTS vertical unit base

plein[1]: ~ **écran** *adj* DONNÉES, INFORMAT full-screen

plein[2]: ~ **tarif** *m* APP SERV full rate

pleine: ~ **charge** *f* ALIM ÉLEC full load

plésiochrone *adj* TÉL, TRANSMIS plesiochronous

pleurage *m* ÉLECTRON wow

pliable *adj* GÉNÉRAL flexible

pliage: ~ **non guidé** *m* COMPOSANTS *ingénierie industrielle* air bending
plié: ~ **en accordéon** *adj* INFORMAT fan-folded
plier *vt* COMPOSANTS bend, INFORMAT fold
pliure *f* TRANSMIS *dans un câble* kink
plombage *m* COMPOSANTS sealing
plomber *vt* COMPOSANTS seal
plomberie *f* RADIO plumbing
plongeur *m* ALIM ÉLEC plunger, PÉRIPH switch hook, TÉL *de poste téléphonique* hook, plunger; ~ **de court-circuit** ÉLECTRON shorting plunger, RADIO plunger, piston
plot *m* ALIM ÉLEC eyelet, CLIMAT eyelet, contact plate, ÉLECTRON turret; ~ **de guidage** COMPOSANTS pilot stud; ~ **de positionnement** COMPOSANTS, ÉLECTRON tooling pin
plug[1]: ~**-in-and-play** *adj* INFORMAT plug-in-and-play
plug[2]: ~**-in** *m* DONNÉES plug-in unit (*PIU*)
plume *f* INFORMAT pen
pluralisme *m* ÉLECTRON pluralism
plus: ~ **franc** *m* ALIM ÉLEC direct positive
plusieurs: **à** ~ **valeurs** *adj* DONNÉES multivalued (*M*)
PLV *abrév* INFORMAT PLV (*production level video*)
PMC *abrév* (*passage au mode compression*) DONNÉES enter compressed mode
PMT *abrév* (*passage au mode transparent*) DONNÉES enter transparent mode
PNIC *abrév* (*code d'identification de réseau privé de données*) DONNÉES, TÉL PNIC (*private data-network identification code*)
PNP *abrév* (*plan de numérotage privé*) TÉL PNP (*private numbering plan*)
PO *abrév* (*partiellement obligatoire*) TÉL CM (*conditionally mandatory*)
poche: **de** ~ *adj* INFORMAT, RADIO pocket, pocket-sized
poignée: ~ **d'ajustement** *f* INFORMAT adjustment handle; ~ **du combiné** *f* PÉRIPH *de poste téléphonique* handset handle; ~ **de contrôle** *f* INFORMAT hand control; ~ **de redimensionnement** *f* INFORMAT resize handle
point[1]: ~ **à multipoint** *adj* RADIO point-to-multipoint; ~ **à point** *adj* COMPOSANTS, ÉLECTRON pin-to-pin, GÉNÉRAL point-to-point, *raccordement* direct, RADIO point-to-point
point[2] *m* INFORMAT dot, point, spot, TÉL point, TV dot; ~ **d'accès** DONNÉES access point, GÉNÉRAL access port, INFORMAT port, *de réseau* outlet, TÉL, TESTS access point; ~ **d'accès à un circuit** TÉL circuit access point; ~ **d'accès aux données** DONNÉES *d'ordinateur* data port; ~ **d'accès aux essais** TESTS test access point; ~ **d'accès Internet** DONNÉES, INFORMAT Internet access point; ~ **d'accès à la ligne** TÉL line access point; ~ **d'accès pour les mesures de circuit** TÉL circuit-test access point; ~ **d'accès public** TÉL public dial-up port; ~ **d'accès au service** DONNÉES service access point (*SAP*); ~ **d'accès au service de liaison de données** DONNÉES data-link service access point (*DLSAP*); ~ **d'accès au service MCS** (*MCSAP*) TÉL MCS service access point (*MCSAP*); ~ **d'accès au service N** DONNÉES N-service access point; ~ **d'accès au service physique** DONNÉES physical-service access point (*PhSAP*); ~ **d'accès au service de présentation** DONNÉES, TÉL presentation-service access point (*PSAP*); ~ **d'accès au service de réseau** DONNÉES, TÉL, TRANSMIS network service access point (*NSAP*); ~ **d'accès au service de session** DONNÉES session-service access point (*SSAP*); ~ **d'accès du service de transport** TÉL, TRANSMIS transport-service access point (*TSAP*); ~ **d'accès synchrone** INFORMAT, TRANSMIS synchronous port; ~ **d'accès vidéotex** (*PAVI*) TÉL, TV videotex access point, videotex gateway; ~ **actif** INFORMAT home spot; ~ **d'amorçage** TÉL singing point; ~ **d'amortissement** TRANSMIS damp point; ~ **d'ancrage** INFORMAT anchor point, anchor; ~ **d'appui** COMPOSANTS *génie mécanique* bearing surface; ~ **d'attache** RADIO hold-down point; ~ **de bifurcation** ÉLECTRON, INFORMAT, TRANSMIS branching point; ~ **de branchement** (*PB*) TRANSMIS drop point, TV branching point; ~ **de branchement optique** TRANSMIS optical tap; ~ **de brasage** COMPOSANTS soldering point; ~ **clignotant** INFORMAT flashing dot; ~ **de commande de services** (*PCS*) COMMUT service command point; ~ **de commutation auxiliaire** COMMUT auxiliary switching point; ~ **de commutation large bande** COMMUT broadband crosspoint; ~ **de comptage** APP SERV charging point; ~ **de connexion** COMMUT crosspoint, GÉNÉRAL connection point; ~ **de connexion international virtuel** TÉL virtual international connection point (*VICP*); ~ **de contact** ALIM ÉLEC, COMMUT contact point; ~ **de contrôle** INFORMAT checkpoint; ~ **de contrôle et d'observation** (*PCO*) DONNÉES point of control and observation (*PCO*); ~ **de convergence** GÉNÉRAL hub; ~ **de Curie** INFORMAT Curie point; ~ **de départ** INFORMAT initial point; ~ **de dérivation** ÉLECTRON, INFORMAT branching point, TRANSMIS drop point, branching point; ~ **de dérivation optique** TRANSMIS optical tap; ~ **de destination** TÉL destination point; ~ **de destination de copie** INFORMAT copy destination point; ~ **de détection** TESTS sensing point; ~ **de détermination de taxation** TÉL tariff determination point; ~ **de distribution** COMMUT operation point; ~ **de distribution secondaire** TRANSMIS pillar; ~ **d'échantillonnage** TESTS sample point; ~ **d'entrée** DONNÉES inlet access, INFORMAT entry point; ~ **d'entrée de baie** TÉL rack entry point; ~ **d'essai** TESTS test access point, test point; ~ **d'essai de ligne** TÉL line test point; ~ **d'essai de relais** COMPOSANTS relay test point; ~ **pour les essais de la signalisation de ligne** TRANSMIS testing point; ~ **pour les essais de la transmission** TRANSMIS testing point; ~ **de fonctionnement** ÉLECTRON operating point, working point; ~ **froid** RADIO point of minimum voltage; ~ **d'image** ÉLECTRON, TÉL, TV picture element (*pel, pixel*); ~ **d'indifférence** TESTS point of control, indifference point; ~ **d'insertion** INFORMAT insertion point, punch point; ~ **d'interconnexion entre systèmes** TÉL system interconnection point; ~ **d'interruption** INFORMAT breakpoint; ~ **d'intersection** COMMUT crosspoint; ~ **d'intersection CMOS** TÉL CMOS crosspoint; ~ **d'intersection de deux fils** COMMUT two-wire crosspoint; ~ **d'intersection à diode** COMMUT diode crosspoint; ~ **d'intersection électronique** COMMUT electronic crosspoint; ~ **d'intersection à maintien électrique** COMMUT electrically-held crosspoint; ~ **d'intersection à maintien magnétique** COMMUT magnetically-latched crosspoint; ~ **d'intersection métallique** COMMUT metallic crosspoint; ~ **d'intersection optique** COMMUT optical center (*AmE*), optical centre (*BrE*), optical-switching crosspoint; ~ **d'intersection optoélectronique** COMMUT optoelectronic crosspoint; ~ **d'intersection à quatre fils** COMMUT four-wire crosspoint; ~ **d'intersection à relais à lames souples** COMMUT reed-relay crosspoint; ~ **d'intersection remreed** COMMUT remreed crosspoint; ~ **d'intersection à semiconducteur** COMMUT semiconductor crosspoint; ~ **d'intersection à thyristor au silicium** ÉLECTRON SCR crosspoint, silicon-controlled rectifier crosspoint; ~ **isohypse** ALIM ÉLEC projected peak point; ~ **luminophore** INFORMAT phosphor dot; ~ **MCSAP de contrôle** TÉL control MCSAP; ~ **de mesure** TESTS measuring point, test access, test access point; ~ **milieu** GÉNÉRAL center point (*AmE*), centre point (*BrE*); ~ **Morse** TÉL Morse dot; ~ **neutre** ALIM ÉLEC neutral point, neutral; ~ **nodal** TÉL nodal point, node; ~ **d'origine** INFORMAT initial point; ~ **de partage** INFORMAT split point; ~ **de partage du clavier** INFORMAT keyboard split point; ~ **de pic** COMPOSANTS peak point; ~ **de polarisation** INFORMAT bias point; ~ **de raccordement** GÉNÉRAL connection point; ~ **de rattachement du sous-réseau** DONNÉES subnetwork point of attachment; ~ **de réception reconnu** RADIO genuine receive point; ~ **de référence** DONNÉES reference point,

GÉNÉRAL benchmark; ~ **de référence de bouche** TÉL mouth reference point; ~ **de référence interne** ÉLECTRON internal reference point (*IRP*); ~ **de référence oreille** TÉL ear reference point; ~ **de référence QOS** DONNÉES, GÉNÉRAL, TÉL, TESTS QOS reference point (*QRP*); ~ **de référence R** DONNÉES R-reference point; ~ **de référence S** DONNÉES S-reference point; ~ **de référence T** DONNÉES T-reference point; ~ **de référence de transmission** TÉL, TRANSMIS transmission reference point (*TRP*); ~ **de régulation** CONTRÔLE control point; ~ **de repère** TÉL checkpoint; ~ **de retour** INFORMAT re-entry point; ~ **de retournement** COMPOSANTS *de thyristor* breakover point; ~ **du risque du client** TESTS consumer risk point; ~ **du risque du fournisseur** TESTS producer's risk point (*PRP*); ~ **sémaphore** SIGNAL, TÉL signaling point (*AmE*), signalling point (*BrE*) (*SP*); ~ **sémaphore adjacent** SIGNAL adjacent signaling point (*AmE*), adjacent signalling point (*BrE*); ~ **sémaphore de destination** SIGNAL signaling destination point (*AmE*), signalling destination point (*BrE*); ~ **sémaphore d'origine** SIGNAL originating point, signaling originating point (*AmE*), signalling originating point (*BrE*); ~ **de signalisation** SIGNAL, TÉL signaling point (*AmE*) (*SP*), signalling point (*BrE*) (*SP*); ~ **de signalisation d'origine** SIGNAL originating point, signaling originating point (*AmE*), signalling originating point (*BrE*); ~ **de sortie** COMMUT outlet access; ~ **sub-satellite** RADIO subsatellite point; ~ **de taxation** APP SERV charging point; ~ **de test** TESTS test access point, test point; ~ **de test et d'opération** TÉL, TESTS test and operation point; ~ **de transfert de groupe quaternaire** TÉL through-group connection point; ~ **de transfert de groupe secondaire** TÉL through-supergroup connection point; ~ **de transfert de groupe tertiaire** TÉL through-mastergroup connection point; ~ **de transfert sémaphore** COMMUT, SIGNAL, TÉL, TRANSMIS signal transfer point (*STP*), signaling transfer point (*AmE*) (*STP*), signalling transfer point (*BrE*) (*STP*); ~ **de transfert de signaux** COMMUT, SIGNAL, TÉL, TRANSMIS signal transfer point (*STP*), signaling transfer point (*AmE*) (*STP*), signalling transfer point (*BrE*) (*STP*); ~ **de transfert d'assemblage de 15 groupes secondaires** TÉL through-15-supergroup assembly connection point; ~ **de vallée** COMPOSANTS valley point; ~ **zéro** TESTS zero point

pointage *m* INFORMAT pick, RADIO *d'antenne* positioning, steering, TESTS checking off; ~ **d'antenne** RADIO antenna positioning, antenna steering

pointe *f* COMPOSANTS point contact, ÉLECTRON peak, GÉNÉRAL transient; ~ **de bruit** TRANSMIS noise spike; ~ **de courant** ALIM ÉLEC current rush, current surge; ~ **de fiche** TÉL plug tip; ~ **de touche** TÉL tip, TESTS test probe, test prod; ~ **de tour** COMPOSANTS lathe center (*AmE*), lathe centre (*BrE*); ~ **de traçage** COMPOSANTS scriber; ~ **à tracer** COMPOSANTS scriber; ~ **de trafic** TÉL traffic peak

Pointel *m* ALIM ÉLEC, TÉL Telepoint

pointes: ~ **de bruit** *f pl* INFORMAT, RADIO grass

pointeur *m* DONNÉES, INFORMAT, TÉL pointer; ~ **optique** APP SERV, INFORMAT light pen; ~ **de pile** INFORMAT stack pointer

points[1]: ~ **d'entrée et de sortie pour la télégraphie** *m pl* TÉL points of telephony input and output; ~ **d'interconnexion internationale dans la bande de base** *m pl* TÉL points of international connection at baseband frequencies

points[2]: ~ **par pouce** *loc* (*ppp*) DONNÉES, INFORMAT dots per inch (*dpi*)

polaire *adj* INFORMAT, RADIO polar

polarisation[1]: **à ~ inverse** *adj* COMPOSANTS, RADIO reverse-biased

polarisation[2] *f* ALIM ÉLEC, ÉLECTRON, RADIO, TRANSMIS, TV bias, polarization; ~ **alternée** RADIO interleaved polarization; ~ **d'antenne** RADIO antenna polarization; ~ **circulaire** TRANSMIS circular polarization; ~ **dextrogyre** RADIO right-hand polarization, clockwise polarization; ~ **dextrorsum** RADIO right-hand polarization, clockwise polarization; ~ **directe** ÉLECTRON forward bias, forward biasing; ~ **électrique** ALIM ÉLEC electric polarization; ~ **elliptique** RADIO elliptical polarization; ~ **fixe** ÉLECTRON fixed bias; ~ **horizontale** TV horizontal polarization; ~ **inverse** RADIO reverse bias, reverse biasing; ~ **lévogyre** RADIO left-hand polarization, counter-clockwise polarization; ~ **linéaire** ALIM ÉLEC, RADIO, TRANSMIS linear polarization; ~ **des ondes** ÉLECTRON, RADIO wave polarization; ~ **optique** TRANSMIS optical polarization; ~ **orthogonale** RADIO orthogonal polarization; ~ **rectiligne** ALIM ÉLEC, RADIO, TRANSMIS linear polarization; ~ **senestrorsum** RADIO left-hand polarization, counter-clockwise polarization; ~ **verticale** TV vertical polarization

polarisations: **à ~ croisées** *adj* TRANSMIS cross-polarized

polarisé *adj* ALIM ÉLEC *électricité* biased

polariseur *m* ÉLECTRON, RADIO polarizer; ~ **en ondes guidées** ÉLECTRON guided-wave polarizer

polarité *f* ALIM ÉLEC, TV polarity; ~ **inverse** ALIM ÉLEC reverse polarity; ~ **de la modulation** TV modulation polarity; ~ **négative** ALIM ÉLEC minus polarity, negative polarity

pôle *m* ALIM ÉLEC pole; ~ **négatif** ALIM ÉLEC negative pole, negative terminal; ~ **positif** ALIM ÉLEC positive pole, positive terminal

police *f* DONNÉES, INFORMAT, PÉRIPH font; ~ **de base** DONNÉES basic font; ~ **de caractères** DONNÉES type font, INFORMAT character font, PÉRIPH font; ~ **de caractères optiques** DONNÉES optical type font; ~ **système** INFORMAT system font; ~ **vectorielle** INFORMAT outline font

polissage: ~ **électrolytique** *m* TRANSMIS electro-optic effect

politique *f* APP SERV, DONNÉES, TÉL, TESTS policy; ~ **qualité** TESTS quality policy; ~ **de sécurité fondée sur l'identité** DONNÉES identity-based security policy; ~ **de sécurité fondée sur des règles** DONNÉES rule-based security policy; ~ **tarifaire** APP SERV pricing policy, tariff policy, TÉL tariff policy

polyamide *m* COMPOSANTS polyamide

polyéthylène *m* COMPOSANTS polythene, polyethylene

polygone: ~ **de fréquence** *m* TESTS frequency polygon; ~ **des fréquences cumulées** *m* TESTS *qualité* cumulative frequency polygon

polymérisation *f* COMPOSANTS polymerization

polymériser *vt* COMPOSANTS polymerize

polyphonie *f* INFORMAT polyphony, simultaneous note capacity, simultaneous note output

polyphonique *adj* INFORMAT polyphonic

polystyrène: ~ **expansé** *m* COMPOSANTS expanded polystyrene

polythène *m* COMPOSANTS polythene

polyuréthane *m* COMPOSANTS polyurethane

pompage *m* ALIM ÉLEC hunting

pompe: ~ **à dessouder** *f* ÉLECTRON surplus solder collector

pondération: ~ **du bruit** *f* TRANSMIS noise weighting; ~ **de clavier** *f* INFORMAT keyboard scaling

pont *m* ALIM ÉLEC, DONNÉES, ÉLECTRON, GÉNÉRAL, INFORMAT, RADIO bridge; ~ **d'alimentation** ALIM ÉLEC feeding bridge, RADIO, TRANSMIS transmission bridge; ~ **autonome** ÉLECTRON stand-alone bridge; ~ **de conférence** APP SERV conference bridge; ~ **conforme** ÉLECTRON compliant bridge; ~ **de diodes** ÉLECTRON diode bridge; ~ **de distorsion** ÉLECTRON distortion bridge; ~ **de mesure** TESTS measuring bridge; ~ **de métal d'apport** COMPOSANTS solder bridge; ~ **redresseur** ALIM ÉLEC rectifier bridge; ~ **des répéteurs** SIGNAL repeater deck; ~**-routeur** DONNÉES bridge router; ~ **de transmission** RADIO, TRANSMIS transmission bridge; ~ **de Wheatstone** ALIM ÉLEC Wheatstone bridge

pontage *m* ALIM ÉLEC , COMPOSANTS strapping, TÉL bridging

ponté *adj* ALIM ÉLEC bridged, strapped, COMPOSANTS strapped, TÉL bridged

ponter *vt* ALIM ÉLEC, COMPOSANTS strap, TÉL bridge

pontet *m* ALIM ÉLEC, COMPOSANTS strap
pool: ~ **de concentration** *m* TRANSMIS interpolation pool
population *f* TESTS population, universe; ~ **type** TESTS standard population
poreux *adj* COMPOSANTS porous
porosité *f* COMPOSANTS porosity
port *m* DONNÉES, ÉLECTRON, INFORMAT port; ~ **d'accès asynchrone** DONNÉES asynchronous port; ~ **d'accès au réseau commuté** INFORMAT dial-up port; ~ **d'accès spécialisé** INFORMAT dedicated port; ~ **d'entrée** INFORMAT input port; ~ **imprimante** INFORMAT printer port; ~ **modem** INFORMAT modem port; ~ **parallèle** INFORMAT parallel port; ~ **série** INFORMAT series port, serial port; ~ **serveur** ÉLECTRON, INFORMAT serving port; ~ **de sortie** ÉLECTRON, INFORMAT output port; ~ **de transmission** INFORMAT communication port
portabilité *f* INFORMAT, TÉL portability; ~ **hypermédia** INFORMAT hypermedia portability; ~ **hypertexte** INFORMAT hypertext portability
portable[1] *adj* GÉNÉRAL portable, INFORMAT hand-held, lap-held, lap-size, laptop, RADIO, TÉL hand-portable
portable[2] *m* RADIO, TÉL hand-portable telephone; ~ **document file** (*PDF*) DONNÉES, INFORMAT portable document file (*PDF*)
portamento *m* INFORMAT portamento
portatif *adj* GÉNÉRAL portable, RADIO hand-held, hand-portable
porte[1]: ~**-étiquette** *m* GÉNÉRAL label holder; ~**-fusible** *m* ALIM ÉLEC fuse holder; ~**-puce enfichable** *m* ÉLECTRON leaded chip-carrier; ~**-puce non enfichable** *m* ÉLECTRON leadless chip-carrier (*LCC*); ~**-puce à souder** *m* ÉLECTRON leadless chip-carrier (*LCC*)
porte[2] *f* COMPOSANTS diode, ÉLECTRON gate, gate circuit, RADIO port, TÉL, TRANSMIS diode; ~ **amovible** ÉLECTRON lift-out door; ~ **collectrice ouverte** ÉLECTRON open collector gate; ~ **coulissante** COMPOSANTS sliding door; ~ **de distance** RADIO range gate; ~ **d'échantillonnage** TRANSMIS sampling gate; ~ **d'entrée** ÉLECTRON input gate; ~ **d'équivalence** ÉLECTRON, INFORMAT IF-AND-ONLY-IF gate; ~ **ET** ÉLECTRON AND gate; ~ **d'exclusion** ÉLECTRON NOT-IF-THEN element; ~ **à glissière** COMPOSANTS sliding door; ~ **d'identité** ÉLECTRON identity gate, identity element, TÉL identity element, ID; ~ **INCLUSIF OU** INFORMAT, ÉLECTRON INCLUSIVE-OR gate; ~ **logique** ÉLECTRON gate; ~ **NI** ÉLECTRON NOR element, NOR gate; ~ **NON** ÉLECTRON NOT element, NOT gate; ~ **NON-ET** ÉLECTRON NAND circuit, NAND gate; ~ **NON-OU** ÉLECTRON NOR element, NOR gate; ~ **OU** ÉLECTRON OR element, OR gate; ~ **à seuil** ÉLECTRON threshold gate; ~ **de sortie** ÉLECTRON output gate; ~ **de vitesse** RADIO speed gate, velocity gate; ~ **de voie** TÉL, TRANSMIS channel gate
portée *f* COMPOSANTS *génie mécanique* contact surface, bearing surface, DONNÉES span, range, GÉNÉRAL, INFORMAT range, RADIO *radar* range, TÉL range, TESTS scope, range; ~ **entre pôteaux** ALIM ÉLEC, TRANSMIS overhead span; ~ **en espace libre** RADIO free-space range; ~ **moyenne** RADIO medium range; ~ **optique** TRANSMIS line-of-sight, line-of-sight range; ~ **radar** RADIO distance, radar range
porter *vt* COMPOSANTS *sur génie mécanique* bear, lie
porteur *m* ÉLECTRON, TÉL, TV carrier; ~ **central** TRANSMIS central strength member, central strand; ~ **de charge** ÉLECTRON carrier, charge carrier; ~ **de charge libre** ÉLECTRON mobile charge carrier; ~ **libre** ÉLECTRON mobile charge carrier; ~ **majoritaire** ÉLECTRON majority carrier; ~ **minoritaire** ÉLECTRON minority carrier; ~ **de son de télévision** TV television sound carrier
porteuse[1]: **à** ~ **complète** *adj* RADIO full-carrier; **à** ~ **réduite** *adj* RADIO reduced-carrier
porteuse[2] *f* COMMUT, DONNÉES, RADIO, TRANSMIS, TV carrier; ~ **cohérente incrémentielle** (*PCI*) TV incrementally coherent carrier (*ICC*); ~ **commune** TRANSMIS common carrier; ~ **complète** RADIO full carrier; ~ **à destinations multiples** TRANSMIS multidestination carrier; ~ **de données** DONNÉES, TRANSMIS data carrier; ~ **de données avec commutation de paquets** COMMUT, TRANSMIS packet-switched data carrier; ~ **émise dans un faisceau de couverture globale** TRANSMIS global beam carrier; ~ **entrante** TRANSMIS in-band carrier; ~ **à fréquence vocale** RADIO tone-coded carrier; ~ **modulée en fréquence** TRANSMIS frequency-modulated carrier; ~ **non modulée** RADIO clean carrier, TRANSMIS continuous carrier; ~ **pilote** TRANSMIS pilot carrier; ~ **récupérée** TRANSMIS recovered carrier; ~ **réduite** RADIO reduced carrier; ~ **son** TRANSMIS audio carrier, TV sound carrier; ~ **supprimée** RADIO suppressed carrier; ~ **vision** TV visual carrier, video carrier
portier: ~ **automatique** *m* PÉRIPH door interphone, entry phone, hallway telephone, annunciator; ~ **téléphonique** *m* PÉRIPH door interphone, entry phone, hallway telephone; ~ **vidéo** *m* PÉRIPH door video
portillonner *vt* ÉLECTRON gate
pose *f* COMPOSANTS *de fil*, ÉLECTRON running, GÉNÉRAL fitting, mounting, TRANSMIS *d'un câble* laying, running; ~ **de câbles** TRANSMIS *génie industriel extérieur* cable laying, cable running; ~ **de repère** INFORMAT set mark
posemètre *m* TV exposure meter
poser *vt* COMPOSANTS run, ÉLECTRON *fil* connect, run, GÉNÉRAL *fil* run, TRANSMIS *câble* lay, run
positif: ~ **de distribution** *m* ALIM ÉLEC distribution positive; ~ **franc** *m* ALIM ÉLEC direct positive
position[1]: **en** ~ **verticale** *adj* TÉL erect
position[2] *f* GÉNÉRAL operating position, position, TÉL agent; ~ **A** TÉL position A; ~ **active** TÉL active position; ~ **angulaire** TÉL angular displacement; ~ **d'annotatrice** TÉL trunk record position; ~ **d'appel de départ** TÉL originating calling position; ~ **d'arrêt** COMPOSANTS *d'un relais* off position; ~ **auxiliaire** TÉL assistant operator position, auxiliary position; ~ **de barre de sélection** COMMUT selecting bar position; ~ **de caractère** TÉL character position; ~ **de commande** CONTRÔLE control position; ~ **de conversation** RADIO talking state, TÉL talking state, speaking position; ~ **à cran** INFORMAT click-stop position; ~ **demandé raccroché** SIGNAL clear-back condition; ~ **d'un élément numérique** SIGNAL digit position; ~ **d'un élément de signal** SIGNAL, TRANSMIS digit position; ~ **d'essai et de mesures** TÉL testing position, TESTS test position, testing position; ~ **de file d'attente** TÉL queue place; ~ **de flanc arrière** COMPOSANTS, TÉL trailing-edge position; ~ **de flanc avant** TÉL leading-edge position; ~ **de front arrière** COMPOSANTS, TÉL trailing-edge position; ~ **d'horizontale** COMMUT selecting bar position; ~ **d'impulsion** TÉL pulse position; ~ **initiale** INFORMAT initial position, RADIO home location; ~ **d'inscription** TÉL trunk record position; ~ **manuelle** TÉL manual position; ~ **nominale sur l'orbite** RADIO nominal orbital position; ~ **d'opérateur** GÉNÉRAL, TÉL operator position; ~ **d'opérateur d'assistance** TÉL assistant operator position, auxiliary position; ~ **d'opérateur de transit** COMMUT, TÉL traffic operator position; ~ **orbitale** TV orbital position; ~ **ouverte** COMMUT open position; ~ **d'ouverture** COMMUT open position; ~ **phototélégraphique internationale** (*PPI*) TÉL international telegraph position; ~ **de relais** TÉL relay position, relay space; ~ **de réponse** TÉL answering position; ~ **de repos** ALIM ÉLEC quiescent state, COMPOSANTS off position, release condition, ÉLECTRON idle position, rest position, INFORMAT home position, homing position; ~ **du satellite** RADIO satellite attitude; ~ **sélectionnée** TÉL selected position; ~ **de service de nuit** COMMUT night-service board; ~ **de sonnerie** TÉL ringing position; ~ **de symbole** DONNÉES symbol location; ~ **télex** TÉL telex position; ~ **télex internationale** TÉL international telex

position; ~ **terminale** APP SERV auxiliary service position; ~ **de tests et de mesures** TESTS testing position

positionnement *m* DONNÉES, GÉNÉRAL locating, moving, placing, positioning, setting, RADIO location, setting, siting; ~ **automatique** INFORMAT auto-locate; ~ **de départ** INFORMAT initial positioning

positionner *vt* DONNÉES, GÉNÉRAL move, locate, set, RADIO locate, set, site

positionneur *m* COMPOSANTS, GÉNÉRAL jig

positron *m* ÉLECTRON, GÉNÉRAL positive electron, positron

possibilité *f* APP SERV, INFORMAT facility; ~ **de choix** GÉNÉRAL option; ~ **d'émission en différé** APP SERV store-and-forward facility; ~ **d'extension** INFORMAT add-on facility, expandability, TÉL extendibility, expandability, extensibility; ~ **d'extension de mémoire** INFORMAT memory-expansion capacity; ~ **de maintien** INFORMAT hold facility; ~ **de stockage** TESTS storability; ~ **de traitement** TESTS process capability

possibilités: ~ **d'accord** *f pl* ÉLECTRON, INFORMAT tuning capability; ~ **de configuration** *f pl* ALIM ÉLEC strapping options; ~ **d'exécution** *f pl* INFORMAT playability

post[1]: ~**-amplificateur** *m* ALIM ÉLEC, ÉLECTRON post-amplifier; ~**-processeur** *m* INFORMAT postprocessor

post[2]: ~**-production** *f* INFORMAT postproduction pressing, TV postproduction

poste[1] *m* INFORMAT extension; ~ **d'abonné** TÉL subscriber's station; ~ **d'abonné de télécopie** TÉL subscriber's fax position; ~ **d'appel d'urgence** PÉRIPH emergency telephone; ~ **auxiliaire** COMMUT, TÉL additional telephone set; ~ **à câblage très limité** TÉL skinny wire set; ~ **à cadran** TÉL dial telephone, dial telephone set; ~ **à carte de crédit** TÉL creditphone; ~ **central** TÉL attendant's set; ~ **de charge** ALIM ÉLEC, ÉLECTRON charging point; ~ **à clavier** PÉRIPH push-button set, TÉL key-set telephone, keypad telephone, push-button set; ~ **à clavier mixte** TÉL dual-signaling telephone (*AmE*), dual-signalling telephone (*BrE*); ~ **de courrier** TÉL trader's turret; ~ **cryptophonique** DONNÉES key set, TÉL secure telephone (*STU*); ~ **cryptophonique à encryptage multiple** RADIO multiple-key set; ~ **dirigeur** TÉL attendant's set; ~ **d'essai** TESTS test position; ~ **d'exploitation déporté** PÉRIPH remote operating terminal; ~ **filaire** RADIO, TÉL wired handset; ~ **général de supervision** TÉL general supervisory station; ~ **à haut-parleur** TÉL speakerphone; ~ **hors réseau** TÉL off-net station; ~ **d'intercommunication** APP SERV, PÉRIPH, TÉL intercommunication telephone set, key telephone set; ~ **à interdiction d'accès au réseau** TÉL post-access-barred extension; ~ **interdit aux communications de départ** TÉL outgoing calls barred extension, outgoing calls barred line; ~ **intérieur** TÉL internal extension; ~ **d'interrogation** DONNÉES inquiry station; ~ **de lecture** INFORMAT sensing station; ~ **mains-libres** TÉL hands-free telephone set, speakerphone; ~ **mixte** RADIO, TÉL combi-telephone; ~ **mobile** TÉL table instrument, table set; ~ **mural** TÉL wall telephone set; ~ **d'opérateur** TÉL operator console, operator position, operator set, operator's telephone; ~ **d'opérateur de classe A** APP SERV, PÉRIPH key-per-line console (*BrE*), key-per-trunk console (*AmE*); ~ **d'opérateur de classe B** COMMUT switched-loop console; ~ **à prépaiement** PÉRIPH pay station, payphone, TÉL coin-operated telephone, pay station, payphone; ~ **principal** TÉL master set; ~ **principal d'abonné** TÉL subscriber's main station; ~ **public de télécopie** TÉL public facsimile station; ~ **radiotéléphonique mobile** RADIO, TÉL mobile telephone station; ~ **à réception amplifiée** TÉL loudspeaker telephone; ~ **résidentiel** PÉRIPH, TÉL residential telephone; ~ **à restriction d'accès au réseau** TÉL trunk-access-restricted extension; ~ **à restriction d'appel au départ** TÉL external calls barred extension; ~ **à restriction totale** COMMUT fully-restricted station; ~ **sans fil** PÉRIPH, TÉL cordless handset; ~ **sans limitation d'accès au réseau** TÉL trunk-unrestricted extension; ~ **secondaire** TÉL auxiliary set, auxiliary telephone set; ~ **de standardiste** TÉL attendant position; ~ **supplémentaire avec prise de ligne directe** TÉL unbarred extension, unrestricted extension; ~ **supplémentaire à lignes prioritaires** TÉL priority extension; ~ **de surveillance** TÉL observation telephone; ~ **de système numérique** TÉL digital telephone; ~ **téléphonique** PÉRIPH, TÉL telephone, telephone instrument, telephone set, telephone station; ~ **téléphonique à clavier** PÉRIPH, TÉL push-button set, push-button telephone, keypad telephone set; ~ **téléphonique à combiné** PÉRIPH, TÉL handset telephone; ~ **téléphonique dirigeur** TÉL attendant's set; ~ **téléphonique à écoute amplifiée** TÉL receive-amplified telephone; ~ **téléphonique à écoute amplifiée sur combiné** TÉL receive-amplified handset telephone; ~ **téléphonique électronique** TÉL electronic telephone set; ~ **téléphonique mains libres** TÉL hands-free telephone set; ~ **téléphonique à mémoire** PÉRIPH, TÉL telephone set with memory; ~ **téléphonique mobile** TÉL table instrument, table set; ~ **téléphonique mural** TÉL wall-mounted telephone set; ~ **téléphonique ordinaire** TÉL standard telephone set; ~ **téléphonique particulier à un utilisateur** TÉL proprietary telephone set; ~ **téléphonique de position à affichage numérique** TÉL agent position digit-display telephone; ~ **téléphonique à poussoir** *Can* APP SERV, PÉRIPH, TÉL key telephone set; ~ **téléphonique à prépaiement** PÉRIPH, TÉL coin-operated telephone, pay station, payphone; ~ **téléphonique public** (*PTP*) TÉL public telephone, public telephone booth; ~ **téléphonique sans cordon** PÉRIPH, TÉL cordless phone, cordless telephone (*CT*), cordless telephone set; ~ **téléphonique sans fil** PÉRIPH, TÉL cordless phone, cordless telephone (*CT*), cordless telephone set; ~ **téléphonique spécial** PÉRIPH, TÉL special telephone set; ~ **téléphonique supplémentaire** COMMUT, TÉL additional telephone set, extension set, extension telephone set; ~ **à touches musicales** TÉL MF telephone, multifrequency telephone; ~ **d'urgence** TÉL emergency telephone, emergency telephone set, emergency telephone station; ~ **virtuel** DONNÉES, INFORMAT, TÉL virtual terminal

poste[2]: ~ **occupé** *loc* TÉL position busy

PostScript® *m* INFORMAT PostScript®

postulant *m* GÉNÉRAL applicant

pot: ~ **de charge** *m* TRANSMIS loading-coil case, loading-coil pot

poteau *m* ALIM ÉLEC pole; ~ **télégraphique** TÉL telegraphic pole

potentiel: ~ **d'électrode** *m* ÉLECTRON electrode potential; ~ **du sol** *m* ALIM ÉLEC earth potential (*BrE*), ground potential (*AmE*); ~ **zéro** *m* ALIM ÉLEC neutral, zero potential

potentiomètre *m* ALIM ÉLEC, COMPOSANTS, ÉLECTRON pot, potentiometer; ~ **d'appoint** COMPOSANTS, ÉLECTRON trimming potentiometer; ~ **d'équilibrage** COMPOSANTS, ÉLECTRON trimming potentiometer; ~ **à glissière** INFORMAT slider, slider control; ~ **hélicoïdal** ALIM ÉLEC helical potentiometer; ~ **de panoramique** INFORMAT pan pot

poubelle *f* INFORMAT trash

poudrage *m* INFORMAT powdering

pourcentage: ~ **d'abonnés** *m* TÉL subscriber percentage; ~ **de débordement** *m* (*%DBM*) TÉL percentage overflow; ~ **de retour** *m* COMPOSANTS resetting percentage; ~ **d'utilisation** *m* RADIO duty cycle

%DBM *abrév* (*pourcentage de débordement*) TÉL percentage overflow

poursuite *f* INFORMAT, RADIO, TV tracking; ~ **automatique** RADIO auto-track, *radar* automatic tracking; ~ **et commande** TESTS tracking and command (*T&C*); ~ **continue** RADIO *radar* continuous tracking; ~ **en**

escalier RADIO step tracking; **~ en fréquence** RADIO frequency tracking; **~ d'objectif** RADIO target tracking; **~ pas à pas** RADIO step tracking; **~ programmée** RADIO program track (*AmE*), programme track (*BrE*); **~ radioélectrique** RADIO radio tracking; **~ du satellite** RADIO satellite tracking; **~, télémesure et télécommande** RADIO tracking, telemetry and command (*TT&C*)

poussoir *m* COMPOSANTS plunger

pouvoir: **~ d'alimentation** *m* ÉLECTRON supply current; **~ de coupure** *m* ALIM ÉLEC interrupting capacity; **~ isolant** *m* GÉNÉRAL insulating property; **~ de résolution** *m* ÉLECTRON resolving power, resolution, RADIO resolving power; **~ séparateur** *m* RADIO resolving power; **~ séparateur en distance** *m* RADIO range resolution; **~ séparateur radial** *m* RADIO range resolution, range discrimination; **~ séparateur en vitesse** *m* RADIO velocity resolution

pouvoirs *m pl* DONNÉES credentials

PPI *abrév* (*position phototélégraphique internationale*) TÉL international telegraph position

ppp *abrév* DONNÉES, INFORMAT (*points par pouce*) dpi (*dots per inch*), (*pistes par pouce*) TPI (*tracks per inch*)

PPP *abrév* (*protocole point à point*) DONNÉES *Internet* PPP (*point-to-point protocol*)

PPS-EOM *abrév* (*signal de page partielle EOM*) DONNÉES, TÉL PPS-EOM (*partial-page signal EOM*)

PPS-EOP *abrév* (*signal de page partielle EOP*) DONNÉES, TÉL PPS-EOP (*partial-page signal EOP*)

PPS-MPS *abrév* (*signal de page partielle MPS*) DONNÉES, TÉL PPS-MPS (*partial-page signal MPS*)

PPS-NULL *abrév* (*signal de page partielle NULL*) DONNÉES, TÉL PPS-NULL (*partial-page signal NULL*)

PPS-PRI-EOM *abrév* (*signal de page partielle PRI-EOM*) DONNÉES, TÉL PPS-PRI-EOM (*partial-page signal PRI-EOM*)

PPS-PRI-EOP *abrév* (*signal de page partielle PRI-EOP*) DONNÉES, TÉL PPS-PRI-EOP (*partial-page signal PRI-EOP*)

PPS-PRI-MPS *abrév* (*signal de page partielle PRI-MPS*) DONNÉES, TÉL PPS-PRI-MPS (*partial-page signal PRI-MPS*)

préaccentuation *f* ÉLECTRON pre-emphasis, pre-equalization, RADIO pre-emphasis

préaffecté *adj* GÉNÉRAL preassigned

préalimenter *vt* ALIM ÉLEC pre-energize

préambule *m* DONNÉES preamble; **~ d'essai** TESTS test preamble; **~ de test** TESTS test preamble

préampli *m* ÉLECTRON preamp

préamplificateur *m* ÉLECTRON preamplifier

préamplification *f* ÉLECTRON preamplification

préanalyse *f* TÉL preanalysis

préassemblage *m* TRANSMIS preassembly

préassigner *vt* GÉNÉRAL designate

précâblage *m* ÉLECTRON prewiring, TRANSMIS precabling

précâbler *vt* ÉLECTRON prewire, TRANSMIS precable

précâbles *adj* ÉLECTRON prewire

précautions: **~ électrostatiques** *f pl* ÉLECTRON, TESTS electrostatic precautions (*ESP*)

précision *f* GÉNÉRAL accuracy, precision; **~ absolue de fréquence de l'étalon** RADIO frequency-standard accuracy; **~ des échantillons** TÉL sample precision; **~ de mesure** TÉL *métrologie* accuracy of measurement; **~ de positionnement** GÉNÉRAL setting accuracy; **~ de réglage** GÉNÉRAL accuracy of adjustement; **~ du tracé** ÉLECTRON definition

précorrection *f* ÉLECTRON pre-equalization, TÉL precorrection

prédéterminer *vt* ÉLECTRON preset, GÉNÉRAL predetermine

prédicteur *m* INFORMAT predictor

prédiction: **~ anticipée** *f* TV forward prediction; **~ type** *f* TÉL typical prediction (*TP*)

prédiviseur *m* TESTS prescaler

préenregistré *adj* TV prerecorded

prééquipé *adj* ÉLECTRON preloaded

préférentiel *adj* INFORMAT discretionary

préfixe *m* GÉNÉRAL prefix; **~ de code de service** TÉL service code prefix; **~ de conduit** COMMUT, TÉL path overhead (*POH*); **~ international** TÉL international prefix; **~ interurbain** TÉL trunk prefix; **~ de numérotation abrégée** TÉL abbreviated dialing prefix (*AmE*), abbreviated dialling prefix (*BrE*)

préformaté *adj* DONNÉES, INFORMAT preformatted

préforme *f* GÉNÉRAL, TRANSMIS preform; **~ de verre** COMPOSANTS, TRANSMIS glass preform

préimprimé *adj* INFORMAT, PÉRIPH preprinted

préjugé *m* ÉLECTRON prejudice

prélèvement *m* TÉL, TESTS, TRANSMIS sample; **~ ponctuel d'un échantillon** TESTS spot sample

prélever *vt* TÉL, TESTS, TRANSMIS sample

prémarquage *m* TÉL *d'une bande magnétique* premarking

prématriçage *f* INFORMAT premastering

premier[1]: **~ commutateur n sept** *m* TÉL no. seven exchange - first; **~ plan** *m* TV foreground

premier[2]: **~ entré, premier sorti** *loc* GÉNÉRAL first in, first out (*FIFO*)

première: **~ phase de recette** *f* TESTS initial acceptance; **~ rupture** *f* ÉLECTRON first breakdown

premiers: **~ essais de mise en service** *m pl* TESTS initial full line-up (*IFLU*)

prendre: **~ en charge** *vt* GÉNÉRAL handle, INFORMAT deal with; **~ du son** *vt* TÉL pick up sound

préparation: **~ des données** *f* DONNÉES, INFORMAT data preparation; **~ de l'échantillon** *f* TESTS sample preparation

prépondérant *adj* GÉNÉRAL leading

préprocesseur *m* INFORMAT preprocessor

préprogrammé *adj* INFORMAT factory-preset, factory-programmed, preset

préréglage *m* GÉNÉRAL presetting; **~ des zéros** RADIO null placing technique

prérégler *vt* GÉNÉRAL preset

prescription *f* INFORMAT provision; **~ conditionnelle** GÉNÉRAL conditional requirement; **~ facultative** DONNÉES optional requirement; **~ obligatoire** DONNÉES mandatory requirement

présélecteur *m* TÉL, TV preselector; **~ de données** DONNÉES, INFORMAT data presetter

présélection *f* TÉL, TV preselection; **~ automatique** TÉL automatic preselection

présence: **~ de collisions** *f* ÉLECTRON occurrence of collisions; **~ obligatoire** *f* DONNÉES obligatory presence; **~ sur option** *f* DONNÉES optional presence

présentation *f* RADIO display, TÉL presentation; **~ à alignement automatique** RADIO azimuth-stabilized display; **~ de fichier** DONNÉES, INFORMAT file layout; **~ de l'identification de la ligne appelante** (*PILA*) TÉL calling-line identification display (*CLID*), calling-line identification presentation (*CLIP*); **~ de l'identification de la ligne connectée** (*PILC*) TÉL connected-line identification presentation (*COLP*); **~ en mode X.410-1984** DONNÉES X.410-1984 mode presentation; **~ panoramique stabilisée** RADIO *radar* azimuth-stabilized display; **~ synthétique** RADIO synthetic display; **~ visuelle** INFORMAT, TV soft copy

présenter *vt* INFORMAT exhibit

presse: **~-papiers** *m* INFORMAT clipboard

presseur *m* TÉL paper-winder bail

pression: **~ acoustique** *f* TÉL sound pressure; **~ après enfoncement** *f* INFORMAT after touch; **~ polyphonique** *f* INFORMAT polyphonic pressure; **~ positive** *f* GÉNÉRAL positive pressure; **~ sur une touche** *f* INFORMAT key depression; **~ de vapeur** *f* CLIMAT vapor pressure (*AmE*), vapour pressure (*BrE*)

pressurisation: **~ des câbles** *f* GÉNÉRAL cable pressurization

prestataire *m* DONNÉES provider; ~ **d'accès** DONNÉES access provider; ~ **de services** GÉNÉRAL, INFORMAT service provider; ~ **de services de maintenance** TESTS maintenance service provider (*MSP*)
prestations *f pl* TÉL services rendered
prêt: ~ **à émettre** *adj* DONNÉES ready for sending, clear to send (*CTS*), RADIO transmit-ready; ~ **à fonctionner** *adj* GÉNÉRAL serviceable; ~ **à fonctionner en ligne** *adj* DONNÉES ready on line; ~ **à fonctionner en secours** *adj* DONNÉES ready stand by; ~ **à recevoir** *adj* TÉL receive-ready (*RR*); ~ **pour la transmission de données** *adj* TRANSMIS ready for data
prétéléchargé *adj* ÉLECTRON preloaded
prétest *m* TESTS pretest
prétraitement *m* INFORMAT preprocessing
prévenir *vt* RADIO alert
prévention: ~ **de collision d'appels** *f* TÉL prevention of call collision; ~ **des désastres** *f* TESTS disaster prevention
prévision: ~ **de la couverture des cellules** *f* RADIO cell-coverage prediction; ~ **ionosphérique** *f* RADIO ionospheric prediction; ~ **de propagation radioélectronique** *f* RADIO radio-propagation prediction
prévisions: ~ **atmosphériques** *f pl* APP SERV weather line, TÉL weather forecast, weather line; ~ **météorologiques** *f pl* APP SERV weather line, TÉL weather forecast, weather line
prévisualisation *f* DONNÉES, INFORMAT preview
prévu *adj* RADIO prearranged
PRI *abrév* (*période de répétition des impulsions*) TÉL PRP (*pulse repetition period*)
PRI-EOM *abrév* (*interruption de la procédure EOM*) DONNÉES, TÉL PRI-EOM (*procedure interrupt EOM*)
PRI-EOP *abrév* (*interruption de la procédure EOP*) DONNÉES, TÉL PRI-EOP (*procedure interrupt EOP*)
primitive *f* DONNÉES, INFORMAT, TÉL primitive; ~ **de confirmation** (*conf*) DONNÉES, TÉL confirm primitive (*conf*); ~ **confirmation de service** DONNÉES N-service confirm primitive; ~ **de demande** DONNÉES request primitive (*req*); ~ **de demande associée** related-request primitive (*req*); ~ **demande de service N** DONNÉES N-service request primitive; ~ **d'indication** DONNÉES indication primitive; ~ **d'indication associée** DONNÉES related-indication primitive; ~ **indication de service N** DONNÉES N-service indication primitive; ~ **réponse à une demande de service** DONNÉES N-service response primitive; ~ **de service** TÉL service primitive; ~ **de service abstraite** (*ASP*) DONNÉES abstract service primitive (*ASP*); ~ **de service abstraite (N)** (*ASP(N)*) DONNÉES abstract (N) service primitive (*ASP(N)*)
PRI-MPS *abrév* (*interruption de la procédure MPS*) DONNÉES, TÉL PRI-MPS (*procedure interrupt MPS*)
principe: ~ **de conception modulaire** *m* COMMUT building-block principle; ~ **de construction modulaire** *m* COMMUT building-block principle; ~ **d'exploitation** *m* TÉL principle of operation; ~ **modulaire** *m* TÉL modular principle; ~ **de la récupération d'énergie** *m* ALIM ÉLEC flyback principle; ~ **de taxation** *m* APP SERV charging mode, charging principle
prioritaire *adj* GÉNÉRAL overriding
priorité[1] *f* INFORMAT, RADIO priority, TÉL override, priority facility, executive override; ~ **en cas d'alarme** TESTS alarm priority; ~ **inverse** INFORMAT reverse priority; ~ **de perte de cellule** RADIO cell loss priority (*CLP*); ~ **avec tonalité de notification** TÉL priority with notifying tone; ~ **de transfert de données MCS** TÉL MCS data transfer priority
priorité[2]: **avoir ~ sur** *vt* GÉNÉRAL override
prise *f* ALIM ÉLEC outlet (*AmE*), socket outlet (*BrE*), COMMUT jack, ÉLECTRON socket, INFORMAT jack, socket, RADIO socket, TÉL jack, seizure, TESTS jack, TRANSMIS socket, TV cable socket, socket; ~ **d'alimentation secteur** INFORMAT alternating current socket, AC socket; ~ **d'appel** TÉL call pick-up; ~ **d'appel par le standardiste** TÉL attendant call selection, call selection; ~ **de bobinage** ALIM ÉLEC winding tapping; ~ **de bouclage** DONNÉES, ÉLECTRON data wrap plug; ~ **casque** PÉRIPH headphones socket, headset jack, phones jack; ~ **centrale** ÉLECTRON center tap (*AmE*), centre tap (*BrE*), midpoint tap; ~ **en charge** GÉNÉRAL handling; ~ **de cordon secteur** INFORMAT alternating current power cord receptacle; ~ **de courant** ALIM ÉLEC current outlet, *électricité* current tap, INFORMAT alternating current outlet; ~ **DIN** INFORMAT DIN jack; ~ **directe du réseau** GÉNÉRAL, TÉL direct outward dialing (*AmE*), direct outward dialling (*BrE*) (*DOD*); ~ **double** TÉL double seizure; ~ **enfichable** ALIM ÉLEC socket outlet; ~ **d'enroulement** ALIM ÉLEC winding tapping; ~ **d'entrée** INFORMAT input socket; ~ **femelle modulaire** ÉLECTRON modular jack; ~ **informatique simple encastrée** ÉLECTRON flush-mounted single information outlet; ~ **de ligne** TÉL seizure; ~ **de ligne sans décrocher** GÉNÉRAL, RADIO, TÉL on-hook dialing (*AmE*), on-hook dialling (*BrE*); ~ **médiane** ÉLECTRON center tap (*AmE*), centre tap (*BrE*), midpoint tap; ~ **modulaire** ÉLECTRON modular jack; ~ **modulaire modifiée** ÉLECTRON modified modular jack (*MMJ*), modified modular plug (*MMP*); ~ **multiple** COMMUT multiple seizure; ~ **murale** ALIM ÉLEC wall outlet, wall socket, wall terminal box, ÉLECTRON faceplate, wall outlet, wall socket, wall terminal box, INFORMAT, TÉL wall outlet, wall socket, wall terminal box; ~ **péritel** TV AC mains socket wall, SCART socket television; ~ **d'une route** TÉL route seizure; ~ **S** TÉL S-universal access; ~ **secteur** ALIM ÉLEC, ÉLECTRON wall AC power socket, INFORMAT AC mains socket wall, AC power outlet, alternating current power outlet, wall AC power socket; ~ **secteur alternative** INFORMAT wall alternative current power socket; ~ **secteur murale** INFORMAT AC mains socket wall; ~ **séquentielle** TÉL sequential seizure; ~ **simultanée** TÉL dual seizure; ~ **de sortie** INFORMAT output socket; ~ **supplémentaire** GÉNÉRAL extension socket; ~ **de télécopieur** COMMUT faxswitch; ~ **téléphonique** COMMUT telephone socket, TÉL telephone connection socket, telephone socket; ~ **téléphonique standard** TÉL standard voice jack; ~ **de terre** ALIM ÉLEC earth electrode (*BrE*), ground electrode (*AmE*); ~ **de terre indépendante** ALIM ÉLEC independent earth electrode (*BrE*), independent ground electrode (*AmE*); ~ **twinaxiale métallique** ÉLECTRON metal twinax plug (*AmE*), metal twinaxe plug (*BrE*); ~ **universelle** TÉL S-universal access; ~ **de vue** INFORMAT, TV shot
prises[1]: ~ **raccordables** *f pl* TV homes passed
prises[2]: ~ **par circuit et par heure** *loc* TÉL seizures per circuit per hour (*SCH*)
prisme *m* ÉLECTRON prism; ~ **de déviation** TRANSMIS deviation prism; ~ **semi-réfléchissant** INFORMAT beam splitter
privé *adj* GÉNÉRAL private
privilège *m* ÉLECTRON favor (*AmE*), favour (*BrE*)
prix: ~ **d'abonnement** *m* DONNÉES, INFORMAT, RADIO, TÉL, TV subscription price; ~ **fixe** *m* GÉNÉRAL flat price; ~ **forfaitaire** *m* GÉNÉRAL flat price, inclusive price; ~ **horaire** *m* APP SERV hourly charge
probabilité: ~ **d'acceptation** *f* TESTS probability of acceptance; ~ **de défaillance** *f* TESTS probability of failure; ~ **de dépassement du temps d'attente** *f* DONNÉES probability of excess delay; ~ **de détection** *f* RADIO probability of detection; ~ **d'encombrement** *f* TÉL probability of congestion; ~ **d'erreur** *f* DONNÉES, SIGNAL error probability; ~ **d'exécution correcte du service** *f* TÉL probability of successful service completion; ~ **de fausse alarme** *f* GÉNÉRAL, RADIO false alarm probability; ~ **d'indication** *f* RADIO probability of indication; ~ **de rejet** *f* TESTS probability of rejection; ~ **de rupture de stock** *f* TESTS probability of shortage
problème: ~ **de chauffage** *m* ÉLECTRON heating fault; ~ **de conception** *m* TESTS design failure, design fault; ~ **d'erreur**

de manipulation *m* TESTS mishandling failure; ~ **de fabrication** *m* TESTS manufacturing failure, manufacturing fault; ~ **de fragilité** *m* TESTS weakness failure; ~ **de mauvaise utilisation** *m* TESTS misuse failure

procédé: ~ **du double creuset** *m* COMPOSANTS, TRANSMIS double-crucible technique; ~ **d'échange d'ions** *m* ÉLECTRON ion-exchange technique; ~ **d'échantillonnage** *m* TRANSMIS sampling process; ~ **krarup** *m* TRANSMIS krarup system; ~ **de modulation** *m* RADIO modulation technique; ~ **semi-additif** *m* ÉLECTRON semiadditive process; ~ **soustractif** *m* ÉLECTRON subtractive process; ~ **de la tige dans le tube** *m* ÉLECTRON, TRANSMIS rod-in-tube technique

procédure *f* GÉNÉRAL procedure; ~ **d'abandon** INFORMAT aborting procedure; ~ **d'acceptation** TESTS *qualité* acceptance procedure; ~ **d'accès à la liaison** DONNÉES line access procedure (*LAP*); ~ **d'accès à la liaison symétrique** TÉL link access procedure B (*LAPB*); ~ **d'acquisition et de synchronisation des trames d'émission** DONNÉES, TRANSMIS transmit-frame acquisition and synchronization procedure (*TFAS*); ~ **d'acquisition et de synchronisation des trames de transmission** DONNÉES, TRANSMIS transmit-frame acquisition and synchronization procedure (*TFAS*); ~ **d'acquisition en mode de recherche** DONNÉES search mode acquisition procedure, SMA procedure; ~ **d'acquisition par la méthode à créneaux** DONNÉES gated mode acquisition procedure, TRANSMIS gateway; ~ **d'arrivée** RADIO incoming procedure; ~ **de commande** TÉL control procedure; ~ **de commande d'appel** TÉL call-control procedure; ~ **de contrôle des voies** TESTS channel-check test procedure; ~ **de coordination des essais** TESTS test coordination procedure; ~ **de coordination des tests** TESTS test coordination procedure (*TCP*); ~ **de déconnexion** TÉL disconnection procedure; ~ **de départ** TÉL outgoing procedure; ~ **de dérivation** INFORMAT bypass procedure; ~ **détaillée** GÉNÉRAL detailed procedure (*DP*); ~ **de diagnostic** ÉLECTRON, INFORMAT, TÉL troubleshooting procedure, fault-tracing procedure; ~ **d'échantillonnage** TESTS sampling procedure; ~ **d'édition** GÉNÉRAL, INFORMAT editing procedure; ~ **d'essai** TESTS test procedure; ~ **d'essai générale** TESTS outline test procedure; ~ **d'exploitation** TÉL operational procedure; ~ **fondée sur les caractères** INFORMAT character-oriented procedure; ~ **de liaison** DONNÉES data-link protocol; ~ **à liaison simple** TÉL single-link procedure (*SLP*); ~ **de libération** SIGNAL clearing procedure; ~ **de mesure des sauts de cycle** TESTS cycle-skipping test procedure; ~ **de mise à jour** GÉNÉRAL, TÉL updating procedure; ~ **multiliaison** DONNÉES multilink procedure (*MLP*); ~ **de nouvelle conversation** DONNÉES new assignment procedure; ~ **opératoire** TÉL operational procedure; ~ **de programmation** DONNÉES, INFORMAT programming procedure; ~ **de rafraîchissement** DONNÉES, INFORMAT refreshment procedure; ~ **de réaffectation** RADIO reassignment procedure; ~ **de recherche binaire** DONNÉES binary search procedure; ~ **de recherche dichotomique** DONNÉES binary search procedure; ~ **de redressement** DONNÉES recovery procedure, recovery; ~ **de réglage** TESTS, TRANSMIS line-up procedure; ~ **de reprise** DONNÉES recovery procedure; ~ **de sélection en mode menu** TÉL menu selection procedure; ~ **de synchronisation de la trame de réception** TRANSMIS receive-frame synchronization procedure (*RFS*); ~ **de test** TESTS test procedure; ~ **du test de faisceau de routes sémaphores** SIGNAL signaling-route-set test procedure (*AmE*), signalling-route-set test procedure (*BrE*); ~ **de transfert autorisé** TÉL transfer-allowed procedure; ~ **de transfert interdit** TÉL transfer-prohibited procedure

processeur *m* COMMUT, DONNÉES, INFORMAT, TÉL processor; ~ **actif** INFORMAT active processor; ~ **analogique des appels** INFORMAT analog call processor; ~ **d'appels** TÉL call processor (*CP*); ~ **d'application** (*PA*) DONNÉES, TÉL application processor (*AP*); ~ **associatif** INFORMAT associative processor; ~ **central** COMMUT central processor; ~ **de centre d'exploitation et de maintenance** INFORMAT operation and maintenance center processor (*AmE*), operation and maintenance centre processor (*BrE*); ~ **de communications** COMMUT communications processor; ~ **de commutation** COMMUT, DONNÉES switching processor; ~ **de commutation de messages** PÉRIPH message-switching processor; ~ **de commutation de paquets** COMMUT, INFORMAT packet-switching processor; ~ **de gestion** INFORMAT administrative processor; ~ **de gestion de la trame** DONNÉES frame processor; ~ **graphique** INFORMAT graphics processor; ~ **hors service** TÉL processor outage; ~ **hybride** COMMUT hybrid processor; ~ **de langage** INFORMAT language processor; ~ **de maintenance** INFORMAT maintenance processor; ~ **maître** INFORMAT master processor; ~ **matriciel** DONNÉES array processor; ~ **numérique de signaux** (*DSP*) INFORMAT digital signal processor (*DSP*); ~ **périphérique** INFORMAT peripheral processor (*PP*); ~ **principal** INFORMAT principal processing unit (*PUP*); ~ **régional** COMMUT regional processor (*RP*); ~ **de réserve** COMMUT standby processor; ~ **de retard** INFORMAT delay processor; ~ **RISC** DONNÉES RISC processor; ~ **de signaux** SIGNAL signal processor (*SP*); ~ **de système de commutation** COMMUT, TÉL switching system processor; ~ **de traitement numérique des signaux** (*DSP*) INFORMAT digital signal processor (*DSP*); ~ **vectoriel** INFORMAT array processor; ~ **vidéo** INFORMAT video processor; ~ **à virgule flottante** INFORMAT floating-point processor

processeurs: ~ **fonctionnant en parallèle** *m pl* COMMUT parallel-working processors

processus *m* GÉNÉRAL process; ~ **additif** ÉLECTRON *circuits imprimés* additive process; ~ **aléatoire** ÉLECTRON random process; ~ **d'application** (*PA*) DONNÉES, TÉL application process (*AP*); ~ **d'application de gestion des systèmes** DONNÉES, INFORMAT systems management application process; ~ **de codage** INFORMAT, TÉL coding process; ~ **de commutation** COMMUT, TÉL switching process; ~ **de connexion** COMMUT, TÉL switching sequence; ~ **de décodage** INFORMAT decoding process; ~ **de mesure** TESTS measurement process; ~ **sous contrôle** TESTS process in control; ~ **stochastique** TRANSMIS stochastic process; ~ **suspendu** TÉL suspended process

procès-verbal: ~ **d'essai** *m* TESTS test record; ~ **de test** *m* TESTS test record

protection: ~ **de fichiers** *f* DONNÉES file protection

prodiffusion *f* RADIO forward scattering

ProDigi *f* (*PROfessional DIGItal*) INFORMAT ProDigi (*PROfessional DIGItal*)

producteur: ~ **de bases de données** *m* DONNÉES information provider

production *f* COMPOSANTS, GÉNÉRAL manufacturing; ~ **level video** (*PLV*) INFORMAT production level video (*PLV*); ~ **locale** TV local production; ~ **de paires** COMPOSANTS pair generation, pair production; ~ **de la parole** INFORMAT, RADIO, TÉL speech production; ~ **des pistes** RADIO track production

productique *f* INFORMAT computer-integrated manufacturing (*CIM*)

produit: ~ **de conversion de fréquence non essentiel** *m* RADIO spurious frequency-conversion product; ~ **d'étanchéité B** *m* ÉLECTRON B sealant; ~ **fini** *m* GÉNÉRAL end product; ~ **d'intermodulation** *m* TRANSMIS intermodulation product; ~ **d'intermodulation non essentiel** *m* RADIO, TRANSMIS spurious mixing product, spurious intermodulation product; ~ **logiciel** *m* INFORMAT software product; ~ **de modulation** *m* TÉL modulation product;

~ pâteux *m* ÉLECTRON paste-like material; **~ résistant à l'eau** *m* COMPOSANTS water-blocking compound

produits: **~ et services** *m pl* APP SERV products and services

PROfessional: **~ DIGItal** *m* (*ProDigi*) INFORMAT PROfessional DIGItal (*ProDigi*)

professionnel *adj* PÉRIPH professional-level

profil *m* ALIM ÉLEC, DONNÉES, ÉLECTRON profile, GÉNÉRAL *dessin technique* cross section, TRANSMIS profile; **~ du champ de mode** TRANSMIS mode-field profile; **~ à gradient exponentiel** ALIM ÉLEC, ÉLECTRON, TRANSMIS power law index profile; **~ à gradient d'indice** ÉLECTRON, TRANSMIS graded-index profile; **~ d'indice** ÉLECTRON, TRANSMIS *réfraction* index profile; **~ d'indice à loi en puissance** ALIM ÉLEC, ÉLECTRON, TRANSMIS power law index profile; **~ de l'indice de réfraction** TRANSMIS refractive-index profile; **~ à loi en puissance** ALIM ÉLEC, ÉLECTRON, TRANSMIS power law index profile; **~ normalisé** *f* TÉL standard profile; **~ parabolique** RADIO, TRANSMIS parabolic profile; **~ de paramètre** TESTS parameter profile; **~ à saut d'indice** ÉLECTRON, TRANSMIS step index profile; **~ à saut d'indice équivalent** ÉLECTRON, TRANSMIS equivalent-step-index profile (*ESI profile*)

profondeur: **~ de champ** *f* ÉLECTRON field depth; **~ de couleur** *f* DONNÉES depth of color (*AmE*), depth of colour (*BrE*); **~ de pénétration** *f* RADIO penetration depth

progiciel *m* DONNÉES, INFORMAT program package, package, software, software package, suite

programmable *adj* DONNÉES, INFORMAT, PÉRIPH alterable, programmable; **~ par l'utilisateur** INFORMAT user-programmable

programmateur *m* RADIO, TV programmer; **~ de PROM** INFORMAT blaster; **~ de rythmes** INFORMAT rhythm programmer; **~ de rythmes numérique** INFORMAT digital rhythm programmer

programmathèque *f* INFORMAT program library

programmation *f* DONNÉES, INFORMAT programming; **~ heuristique** DONNÉES heuristic programming; **~ logique** DONNÉES logic programming; **~ en nombres entiers** INFORMAT integer programming; **~ orientée accès** DONNÉES access-oriented programming; **~ orientée objet** INFORMAT object-oriented programming (*OOP*); **~ en parallèle** DONNÉES parallel programming; **~ paramétrique** INFORMAT chord intelligence, parametric programming

programmé: **~ en usine** *adj* INFORMAT factory-preset, factory-programmed

programme *m* DONNÉES computer program, program, GÉNÉRAL program (*AmE*), programme (*BrE*), schedule, INFORMAT computer program, program, RADIO, TV program (*AmE*), programme (*BrE*); **~ d'allocation d'itinéraire** TÉL path allocation program, routing program; **~ d'aménagement** INFORMAT housekeeping routine; **~ d'amorçage** INFORMAT boot, bootstrap loader; **~ d'amorce** INFORMAT boot, bootstrap loader; **~ d'analyse sélective** DONNÉES, INFORMAT snapshot program, snapshot trace program; **~ antivirus** INFORMAT anti-virus program; **~ d'assemblage** INFORMAT assembler, assembly program; **~ auto-organisateur** INFORMAT self-organizing program; **~ de balayage** INFORMAT scanning program; **~ de bibliothèque** INFORMAT library program; **~ de chargement** INFORMAT fetch program, boot; **~ clés en main** TV turnkey program; **~ de commande** INFORMAT control program; **~ de commande résident** INFORMAT nucleus of operating system, resident control program; **~ communal** TV community program (*AmE*), community programme (*BrE*); **~ de compilation** INFORMAT compiling program; **~ de contrôle** INFORMAT checking program; **~ de contrôle résident** INFORMAT nucleus of operating system, resident control program; **~ corésident** DONNÉES, INFORMAT, TÉL TSR program; **~ de démonstration** INFORMAT demo program; **~ de dépistage sélectif** DONNÉES snapshot program, INFORMAT snapshot program, snapshot trace program; **~ de diagnostic** INFORMAT diagnostic program; **~ d'échantillonnage** TESTS sampling scheme; **~ d'écoulement de trafic** DONNÉES, TÉL traffic-handling program; **~ encodé** INFORMAT coded program, output level; **~ d'enroulement** COMPOSANTS wrapping program (*AmE*), wrapping programme (*BrE*); **~ d'entrée/sortie** DONNÉES input/output program; **~ d'essai** TESTS testing program (*AmE*), testing programme (*BrE*); **~ d'essai de ligne** TÉL line test program (*AmE*), line test programme (*BrE*); **~ d'essai préliminaire** TESTS preliminary test program (*AmE*), preliminary test programme (*BrE*); **~ d'essais de recette** TESTS acceptance test plan; **~ d'essais sur orbite** RADIO orbital test program (*AmE*) (*OTP*), orbital test programme (*BrE*) (*OTP*); **~ d'exploration** INFORMAT scanning program; **~ en temps partagé** DONNÉES, INFORMAT time-sharing program; **~ générique** TÉL generic program; **~ de gestion** INFORMAT control program; **~ de gestion de mémoire** INFORMAT memory-management program; **~ de gestion de périphériques** INFORMAT handler; **~ de gestion de la qualité** TESTS quality program (*AmE*), quality programme (*BrE*); **~ indépendant** TV independent program (*AmE*), independent programme (*BrE*); **~ d'initialisation** INFORMAT initializing routine; **~ à installer** DONNÉES installation program; **~ interprétatif** INFORMAT interpreter; **~ d'interprétation** INFORMAT interpreter; **~ à ligne ouverte** RADIO, TÉL phone-in program (*AmE*), phone-in programme (*BrE*); **~ de maintenance** TESTS maintenance program (*AmE*), maintenance programme (*BrE*); **~ maître** TV master program (*AmE*), master programme (*BrE*); **~ moniteur** DONNÉES monitor program, monitoring program; **~ multitraitement** INFORMAT multiprocessing program; **~ objet** INFORMAT object program, TÉL target program (*AmE*), target programme (*BrE*); **~ par satellite** TV satellite program (*AmE*), satellite programme (*BrE*); **~ produit** INFORMAT package; **~ propre** TV own program (*AmE*), own programme (*BrE*); **~ radiophonique** RADIO radio program (*AmE*), radio programme (*BrE*); **~ de recherche d'itinéraire** TÉL path allocation program (*AmE*), path allocation programme (*BrE*), routing program (*AmE*), routing programme (*BrE*); **~ rentrant** INFORMAT re-entrant program; **~ résident** DONNÉES, INFORMAT, TÉL TSR program; **~ résultant** TÉL target program (*AmE*), target programme (*BrE*); **~ réutilisable** INFORMAT reusable program; **~ séquenceur** DONNÉES sequencer program; **~ séquentiel** CONTRÔLE sequential program; **~ de service** INFORMAT utility program, service program; **~ de sortie** INFORMAT output program; **~ source** INFORMAT source program; **~ superviseur** INFORMAT supervisory program, supervisor, TÉL executive program (*AmE*), executive programme (*BrE*); **~ télévisuel** TV television program (*AmE*), television programme (*BrE*); **~ temporel** INFORMAT time program; **~ de temporisation** INFORMAT time program, time-out program; **~ de test** TESTS test program (*AmE*), test programme (*BrE*); **~ de traçage sélectif** DONNÉES, INFORMAT snapshot program, snapshot trace program; **~ de traitement de texte** INFORMAT text editor; **~ de traitement du trafic** DONNÉES, TÉL traffic-handling program; **~ translatable** INFORMAT relocatable program; **~ TSR** INFORMAT TSR programme; **~ utilitaire** DONNÉES, INFORMAT service program, utility, utility program; **~ de vidage** INFORMAT dump routine

programmer *vt* DONNÉES program, GÉNÉRAL schedule, INFORMAT, PÉRIPH program

programmeur *m* INFORMAT programmer

progression: **~ d'appel** *f* TÉL call progress

progressivité: **~ de la coupure du filtre** *f* ÉLECTRON filter roll-off

projecteur *m* DONNÉES, TV projector; **~ vidéo** TV video projector

projection: ~ **américaine** *f* GÉNÉRAL third-angle projection
projet: ~ **du système de télécommunications par satellite** *m* TRANSMIS EHF satcom project
prologue: ~ **de procédure** *m* TÉL procedure prologue
prolongateur *m* ALIM ÉLEC, COMPOSANTS extension cord; ~ **téléphonique** COMPOSANTS telephone extension cable, telephone extension cord
prolongement: ~ **national** *m* TÉL national extension
prolonger *vt* RADIO extend
propagation *f* ÉLECTRON propagation, RADIO travel, propagation, TRANSMIS propagation; ~ **en espace libre** RADIO free-space propagation; ~ **ionosphérique** RADIO ionospheric propagation; ~ **de la lumière** TRANSMIS light propagation; ~ **normale** RADIO standard propagation; ~ **d'onde** ÉLECTRON, RADIO wave propagation; ~ **des ondes électromagnétiques** ÉLECTRON, RADIO electromagnetic wave propagation; ~ **des ondes sonores** ÉLECTRON, RADIO acoustic wave propagation; ~ **par diffusion ionosphérique** RADIO ionospheric-scatter propagation; ~ **par diffusion sur les précipitations** RADIO precipitation-scatter propagation; ~ **par impulsions météoriques** RADIO meteor-burst propagation; ~ **par salves météoriques** RADIO meteor-burst propagation; ~ **par trajets multiples** RADIO multipath propagation, multiple-path propagation; ~ **radioélectrique** RADIO radio propagation; ~ **sous-marine** RADIO underwater propagation; ~ **souterraine** RADIO subterranean propagation; ~ **suivant le mode des sifflements** RADIO whistler-mode propagation; ~ **transhorizon** RADIO transhorizon propagation; ~ **transionosphérique** RADIO transionospheric propagation; ~ **troposphérique** RADIO tropospheric propagation; ~ **troposphérique guidée** RADIO ducting; ~ **en visibilité** TRANSMIS line-of-sight propagation
prosodie *f* DONNÉES prosodic analysis
protection *f* DONNÉES safeguard, GÉNÉRAL protection, INFORMAT safeguard, TRANSMIS coating; ~ **acoustique** TÉL acoustic blanket; ~ **cathodique** COMPOSANTS cathodic protection; ~ **contre défauts à la terre** ALIM ÉLEC earth-fault protection (*BrE*), ground-fault protection (*AmE*); ~ **contre les dérangements** TÉL fail-safe; ~ **contre les échos** RADIO echo protection; ~ **contre l'entrée en ligne** TÉL protection against intrusion; ~ **contre les erreurs** DONNÉES error protection; ~ **contre les intrusions** TÉL protection against intrusion; ~ **contre rongeurs** ÉLECTRON rodent protection; ~ **contre les surcharges** ALIM ÉLEC, ÉLECTRON overload protection, overpower protection; ~ **contre les surtensions** ALIM ÉLEC, ÉLECTRON overvoltage protection; ~ **d'épissures** TRANSMIS splice protection; ~ **d'étiquette** TÉL label protection; ~ **de fichier** DONNÉES, INFORMAT file protection; ~ **de mémoire** DONNÉES storage protection, INFORMAT memory protection; ~ **en mémoire écrite** INFORMAT protect memory write; ~ **modem** DONNÉES modem protection equipment; ~ **de numéro de séquence** COMMUT, DONNÉES, INFORMAT sequence number protection (*SNP*); ~ **par fusible** ÉLECTRON fusing; ~ **par programme** INFORMAT software lockout; ~ **secondaire** TRANSMIS fiber buffer (*AmE*), fibre buffer (*BrE*); ~ **de section de multiplexage** TRANSMIS multiplex-section protection (*MSP*); ~ **sélective des champs** DONNÉES, RADIO selective field protection; ~ **télématique** DONNÉES communication protection
protégé *adj* GÉNÉRAL protected; ~ **contre les fausses manœuvres** GÉNÉRAL foolproof; ~ **contre les manipulations abusives** TESTS tamperproof; ~ **par fusible** ALIM ÉLEC, ÉLECTRON fuse-protected, fused
protocole *m* GÉNÉRAL protocol; ~ **d'accès** RADIO access protocol; ~ **d'accès à l'annuaire** DONNÉES directory access protocol (*DAP*); ~ **d'accès à la liaison** DONNÉES line access procedure (*LAP*); ~ **d'accord** GÉNÉRAL memorandum of understanding (*MOU*); ~ **d'application de gestion des systèmes** DONNÉES, INFORMAT systems management application protocol (*SMAP*); ~ **commun d'informations de gestion** DONNÉES common management-information protocol (*CMIP*); ~ **de communication** DONNÉES communication protocol; ~ **de contrôle de transmission/protocole Internet** (*TCP/IP*) DONNÉES transmission control protocol/Internet protocol (*TCP/IP*); ~ **de convergence sans connexion** DONNÉES connectionless convergence protocol (*CLCP*); ~ **de couche paquet** DONNÉES, TÉL packet layer protocol (*PLP*); ~ **de couche réseau en mode connexion** APP SERV connection-oriented network-layer protocol (*CONP*); ~ **de couche réseau sans connexion** APP SERV connectionless network-layer protocol (*CLNP*); ~ **d'échange d'informations** (*IXP*) INFORMAT information exchange protocol (*IXP*); ~ **entre homologues** TRANSMIS peer-to-peer protocol; ~ **de gestion** DONNÉES management protocol; ~ **de gestion de test** DONNÉES, TESTS test management protocol (*TMP*); ~ **d'informations de routage** DONNÉES routing information protocol (*RIP*); ~ **d'interfonctionnement** DONNÉES interworking protocol (*IWP*); ~ **Internet** (*IP*) DONNÉES, INFORMAT Internet protocol (*IP*); ~ **Internet de ligne en série** DONNÉES serial-line Internet protocol (*SLIP*); ~ **de maintenance de tables de routage** INFORMAT routing table maintenance protocol (*RTMP*); ~ **de messages de transmission de données numériques** DONNÉES digital data communications message protocol (*DDCMP*); ~ **N** DONNÉES N-protocol; ~ **point à point** (*PPP*) DONNÉES *Internet* point-to-point protocol (*PPP*); ~ **de réseau** DONNÉES, network protocol; ~ **de réservation rapide** DONNÉES fast-reservation protocol (*FRP*); ~ **sémaphore** SIGNAL signaling protocol (*AmE*), signalling protocol (*BrE*); ~ **de session** DONNÉES session protocol (*SP*); ~ **de signalisation** SIGNAL signaling protocol (*AmE*), signalling protocol (*BrE*); ~ **SNMP** (*SNMP*) DONNÉES simple network management protocol (*SNMP*); ~ **du système d'annuaire** (*DSP*) DONNÉES directory system protocol (*DSP*); ~ **de terminal standard international** ÉLECTRON international standard virtual terminal protocol; ~ **de transfert de chaîne d'informations** (*NNTP*) DONNÉES network news transfer protocol (*NNTP*); ~ **de transfert de fichiers** DONNÉES, INFORMAT file transfer protocol (*FTP*); ~ **de transfert hypertexte** DONNÉES, INFORMAT *Internet* hypertext transfer protocol (*HTTP*); ~ **de transfert du mts** DONNÉES mts transfer protocol; ~ **de transfert du mts 1984** DONNÉES mts transfer protocol 1984
provenance: ~ **d'un appel** *f* TÉL call origin; ~ **du message** *f* DONNÉES *Internet* origin of message
provision: ~ **de réseau ouvert** *f* TRANSMIS open network provision (*ONP*)
provoquer: ~ **un changement d'état** *vt* ÉLECTRON clock; ~ **la prise** *vt* TÉL seize
ps *abrév* (*picoseconde*) TESTS ps (*picosecond*)
pseudo[1]: ~-**aléatoire** *adj* DONNÉES, ÉLECTRON, TRANSMIS, TV pseudorandom
pseudo[2]: ~-**bruit** *m* DONNÉES, TRANSMIS pseudonoise; ~-**code** *m* INFORMAT pseudocode; ~-**événement** *m* DONNÉES pseudo-event
pseudo[3]: ~-**instruction** *f* INFORMAT directive, pseudo-instruction
psophomètre *m* TÉL, TESTS noise meter, noise-measuring device, noise-measuring set, psophometer
psophométrique *adj* TÉL psophometric
psychromètre *m* TÉL psychrometer
PTP *abrév* (*poste téléphonique public*) TÉL public telephone, public telephone booth
publication: ~ **assistée par ordinateur** *f* (*PAO*) APP SERV, DONNÉES desktop publishing (*DTP*), ÉLECTRON desktop publishing system, INFORMAT desktop publishing (*DTP*)
publicité: ~ **importune** *f* DONNÉES *Internet* junk mail
publiphone *m* PÉRIPH, TÉL coin box, coin-box set, pay station,

pay telephone, payphone, public telephone booth; **~ à cartes** TÉL cardphone

puce[1]: **sur ~** *adj* ÉLECTRON, INFORMAT on-chip

puce[2] *f* COMPOSANTS die, ÉLECTRON, INFORMAT chip; **~ décollée** INFORMAT lifted chip; **~ électronique** ÉLECTRON electronic chip; **~ intelligente** ÉLECTRON smart chip; **~ à micromiroirs** INFORMAT digital micro-mirror device (*DMD*); **~ à protubérances** ÉLECTRON flip chip

puissance *f* ALIM ÉLEC power, ÉLECTRON power, GÉNÉRAL power, *de bruit* loudness, RADIO, TRANSMIS power; **~ absorbée** ALIM ÉLEC power drain, ÉLECTRON input power; **~ acoustique** GÉNÉRAL acoustic power; **~ d'allumage** ÉLECTRON firing power; **~ apparente rayonnée** (*par*) RADIO effective radiated power (*erp*); **~ apparente rayonnée sur antenne verticale courte** (*parv*) RADIO effective monopole-radiated power (*emrp*); **~ d'attaque de grille** ÉLECTRON grid input power, grid driving power; **~ de calcul** DONNÉES, INFORMAT processing power; **~ de crête** ALIM ÉLEC peak power, GÉNÉRAL, RADIO *service maritime mobile* peak envelope power (*PEP*); **~ crête de la porteuse modulée** GÉNÉRAL, RADIO peak envelope power (*PEP*); **~ disponible** ALIM ÉLEC, TÉL, TRANSMIS available power; **~ dissipée** ALIM ÉLEC power dissipation; **~ dissipée en direct** ÉLECTRON forward power dissipation, forward power loss; **~ de distorsion de quantification** TÉL, TRANSMIS quantizing distortion power; **~ électrique** ALIM ÉLEC electrical power; **~ électromagnétique surfacique** TRANSMIS irradiance; **~ émise** ÉLECTRON radiated power; **~ d'émission** SIGNAL signal strength; **~ d'émission au satellite** RADIO satellite transmit power; **~ d'entrée** ÉLECTRON input power; **~ d'entrée anodique totale** ALIM ÉLEC, ÉLECTRON total anode-power input, total plate-power input; **~ d'entrée radiofréquence** RADIO radio-frequency input power, driving power; **~ équivalente de bruit** (*PEB*) ÉLECTRON, TRANSMIS noise equivalent power (*NEP*); **~ équivalente de crête** TÉL *d' un signal* equivalent peak envelope power of a telephone signal; **~ d'excitation** RADIO driving power, radio-frequency input power; **~ en fin de vie** ALIM ÉLEC end of life power (*EOL*); **~ fournie** ÉLECTRON input power; **~ de fuite pendant le palier** ÉLECTRON flat leakage power; **~ hors bande** RADIO out-of-band power; **~ hors bande admissible** RADIO permissible out-of-band power; **~ instantanée** ALIM ÉLEC instantaneous power; **~ isotrope rayonnée effective** RADIO effective isotropically-radiated power; **~ isotrope rayonnée équivalente** (*PIRE*) RADIO equivalent isotropically-radiated power (*EIRP*); **~ laser** ÉLECTRON laser power; **~ maximale admissible** TRANSMIS maximum admissible power; **~ moyenne** ALIM ÉLEC medium power, RADIO mean power, TRANSMIS average power; **~ nominale** ALIM ÉLEC rated power, rating; **~ optimale de sortie** ÉLECTRON optimum output power; **~ optique** TRANSMIS optical power; **~ optique de sortie** TRANSMIS optical power output; **~ porteuse** RADIO carrier power; **~ de porteuse requise** TRANSMIS carrier power requirement; **~ psophométrique** TÉL psophometric power; **~ rayonnante** ÉLECTRON, TRANSMIS radiant power; **~ rayonnée** ÉLECTRON radiated power; **~ réactive** ALIM ÉLEC reactive power; **~ requise** ÉLECTRON power requirement; **~ de saturation** ÉLECTRON saturation power; **~ de signal** SIGNAL signal power; **~ du signal** SIGNAL signal strength; **~ de sortie** ÉLECTRON output power; **~ de sortie en crête de modulation** ALIM ÉLEC, ÉLECTRON peak-pulse output power; **~ de sortie d'un émetteur** TRANSMIS transmitter output power; **~ en sortie d'impulsions** RADIO pulse output power; **~ de sortie nominale** ALIM ÉLEC rated output power; **~ de sortie utile** ALIM ÉLEC useful output power, load power, ÉLECTRON useful output power; **~ sur le trajet descendant** RADIO down-path power; **~ surfacique importante** RADIO high-power flux density; **~ surfacique limitée** RADIO low-power flux density; **~ surfacique minimale utilisable** RADIO minimum-usable power flux density; **~ surfacique moyenne** RADIO medium power flux density; **~ surfacique utilisable de référence** RADIO reference-usable field strength, reference-usable power flux density; **~ totale de fuite** ALIM ÉLEC, ÉLECTRON total leakage power; **~ utile** TÉL output; **~ à vide** ALIM ÉLEC no-load power

puits: **~ de chaleur** *m* ALIM ÉLEC, ÉLECTRON heat dissipator, heat sink

pulsion: **~ électromagnétique** *f* RADIO electromagnetic pulse

pulvérisateur *m* CLIMAT, COMPOSANTS spray nozzle

pupiniser *vt* TÉL load, pupinize, TRANSMIS load, coil-load, pupinize

pupitre *m* TV control desk; **~ d'annotateur** TÉL record desk; **~ de commande** CONTRÔLE control desk, observation desk; **~ de commutation** COMMUT *dirigeur*, TÉL, switchboard, telephone switchboard; **~ de commutation automatique** PÉRIPH auto-manual switchboard, automatic switchboard; **~ de conduite** CONTRÔLE control desk, observation desk; **~ d'essai de programme** INFORMAT program test panel; **~ d'exploitation** TÉL service supervision desk; **~ de mélange** INFORMAT mixing console, mixing desk; **~ d'observation de trafic** TÉL traffic observation desk, traffic supervision desk; **~ de surveillance** TÉL chief operator's desk, supervisor's desk, chief supervisor's desk; **~ de surveillance de trafic** TÉL traffic observation desk, traffic supervision desk

purgeur: **~ d'air** *m* COMPOSANTS *tubes électroniques* air valve

PVC *abrév* (*chlorure de polyvinyle*) ÉLECTRON PVC (*polyvinyl chloride*)

PVT *abrév* (*perte de verrouillage de trame*) TÉL, TRANSMIS LFA (*loss of frame alignment*)

pylône *m* RADIO radio mast; **~ hertzien** RADIO microwave tower

pyroélectricité *f* ALIM ÉLEC, ÉLECTRON pyroelectricity

Q

qdu *f* (*unité de distorsion de quantification*) TÉL qdu (*quantizing distortion unit*)
QFIC *abrév* (*quart du format intermédiaire commun*) INFORMAT QCIF (*quarter common intermediate format*)
QIC *abrév* (*cartouche en quart de pouce*) TV QIC (*quarter-inch cartridge*)
Q-mètre *m* TESTS Q-meter
quad: ~ **flatpack** *m* TRANSMIS quad flatpack (*QFP*)
quadrature: **en** ~ *adj* ALIM ÉLEC in quadrature
quadrillage *m* GÉNÉRAL gridiron pattern, INFORMAT grid, RADIO raster
quadripôle *m* ALIM ÉLEC, ÉLECTRON, TRANSMIS four-terminal network, two-terminal network, two-terminal pair; ~ **passif** ALIM ÉLEC, ÉLECTRON, TRANSMIS passive four-terminal network; ~ **symétrique** ALIM ÉLEC, ÉLECTRON, TRANSMIS passive four-terminal network, symmetrical two-terminal-pair network
qualificateur: ~ **d'entité d'application** *m* DONNÉES application-entity qualifier
qualificatif *m* DONNÉES, SIGNAL, TÉL qualifier; ~ **du code d'identification** TÉL identification-code qualifier; ~ **des informations d'autorisation** DONNÉES, SIGNAL authorization-information qualifier; ~ **de référence du destinataire** DONNÉES recipient-reference qualifier
qualification *f* TESTS qualification approval; ~ **de la conception** TESTS type approval; ~ **interne** TESTS type approval
qualimétrie *f* TESTS qualimetry
qualitatif *adj* ÉLECTRON quality
qualité *f* TESTS quality; ~ **assistée par ordinateur** INFORMAT computer-aided quality (*CAQ*); ~ **constante** INFORMAT constant quality; ~ **courrier** INFORMAT letter quality (*LQ*); ~ **d'écoulement du trafic** TESTS grade of service; ~ **de fonctionnement** DONNÉES modem performance; ~ **de fonctionnement d'une voie en polarisations croisées** TRANSMIS cross-polar performance; ~ **d'image** INFORMAT, TV picture quality; ~ **intolérable** TESTS unacceptable quality; ~ **moyenne après contrôle** TESTS average outgoing quality (*AOQ*); ~ **moyenne d'un processus** TESTS process average; ~ **numérique** RADIO digital quality; ~ **primaire de réception** RADIO primary grade of reception quality; ~ **secondaire de réception** RADIO secondary grade of reception quality; ~ **de service** DONNÉES, GÉNÉRAL, TÉL, TESTS quality of service (*QOS*), grade of service; ~ **de service dégradée** TESTS lower quality of service; ~ **de service inférieure** TESTS lower quality of service; ~ **sonore** INFORMAT, RADIO, TV sound quality; ~ **téléphonique** TÉL voice grade; ~ **de transmission** TÉL, TESTS, TRANSMIS transmission quality; ~ **des transmissions téléphoniques** TÉL telephony performance; ~ **vocale** ÉLECTRON, TÉL voice quality; ~ **de voies** TESTS channel performance
quantificateur *m* GÉNÉRAL, INFORMAT, TÉL, TRANSMIS quantizer
quantification *f* GÉNÉRAL, INFORMAT, SIGNAL, TÉL, TRANSMIS quantification, quantization, quantizing; ~ **adaptative** INFORMAT adaptive quantizing; ~ **des blocs** TRANSMIS block quantization; ~ **de données** DONNÉES, TÉL data quantization; ~ **non uniforme** TÉL, TRANSMIS nonuniform quantizing; ~ **de signal** SIGNAL signal quantization; ~ **solaire** SIGNAL solar quantization; ~ **uniforme** TÉL uniform quantizing, uniform encoding, TRANSMIS uniform encoding, uniform quantizing; ~ **vectorielle** DONNÉES, SIGNAL vector quantization; ~ **vectorielle avec un automate à états finis** SIGNAL finite-state vector quantization
quantifier *vt* GÉNÉRAL, INFORMAT, TÉL, TRANSMIS quantify, quantize
quantile: ~ **d'ordre de la durée de vie** *m* TESTS P-fractile life
quantité: ~ **de décisions** *f* INFORMAT decision content; ~ **d'information** *f* DONNÉES information content, negentropy; ~ **moyenne contrôlée** *f* TESTS *qualité* average amount of inspection
quantum *m* GÉNÉRAL quantum
quart: ~ **d'écran** *m* INFORMAT quarter screen; ~ **du format intermédiaire commun** *m* (*QFIC*) INFORMAT quarter common intermediate format (*QCIF*); ~ **de tour** *m* ÉLECTRON quarter turn
quarte *f* GÉNÉRAL, TRANSMIS quad; ~ **de câble** TRANSMIS cable quad
quartet *m* DONNÉES quartet
quartile: ~ **supérieur** *m* GÉNÉRAL upper quartile
quartz: ~ **de commande** *m* ÉLECTRON control crystal; ~ **fondu** *m* TRANSMIS fused quartz
quasi: ~-**instantané** *adj* TV near-instantaneous
quatre: ~ **prises modulaires** *f pl* ÉLECTRON quadplex
queue *f* INFORMAT queue; ~ **de cochon** ÉLECTRON flying lead, GÉNÉRAL pigtail; ~ **de cochon de fibres optiques** TRANSMIS optical fiber pigtail (*AmE*), optical fibre pigtail (*BrE*); ~ **de comète** INFORMAT comet tail; ~ **de raccordement** COMPOSANTS tab
queusot *m* ÉLECTRON tip
quicktime *m* INFORMAT quicktime
quintet *m* DONNÉES quintet
quitter[1] *vi* INFORMAT quit
quitter[2]: ~ **la cellule de localisation nominale** *loc* RADIO roam
quote-part: ~**de répartition** *f* TÉL accounting rate share; ~ **terminale** *f* TÉL terminal share; ~ **de transit** *f* TÉL transit share

R

R *abrév* (*réluctance*) ALIM ÉLEC R (*reluctance*)

raccord *m* COMPOSANTS joint, splice, TV in-between; **~ de fibres optiques** TRANSMIS fiber-optic splice (*AmE*), fibre-optic splice (*BrE*); **~ de guides d'ondes** RADIO, TRANSMIS waveguide connection; **~ multifibre** TRANSMIS multifiber joint (*AmE*), multifibre joint (*BrE*); **~ progressif** RADIO taper; **~ soudé** TRANSMIS fused joint; **~-union** COMPOSANTS screw coupling

raccordable *adj* TV passed

raccordé: **~ en permanence** *adj* GÉNÉRAL permanently connected; **~ au secteur** *adj* ALIM ÉLEC mains-connected

raccordement *m* ALIM ÉLEC, COMMUT, COMPOSANTS wiring, DONNÉES connection, ÉLECTRON connection, link, GÉNÉRAL connection, INFORMAT connection, hook-up, TÉL attachment, local line, drop; **~ automatique en cas de panne** ALIM ÉLEC, TÉL power failure feature, power failure transfer; **~ de câble** TRANSMIS cable termination; **~ à contacts autodénudants** ÉLECTRON insulation displacement termination; **~ de lobes enfichables** ÉLECTRON lobe attachment module; **~ manuel** TESTS manual repatching; **~ numérique** GÉNÉRAL digital connection; **~ par fiche et jack** ALIM ÉLEC plug-and-jack connection; **~ temporaire** TESTS patch out

raccorder *vt* ALIM ÉLEC wire, COMMUT patch, COMPOSANTS wire, DONNÉES attach, ÉLECTRON patch, GÉNÉRAL joint, INFORMAT connect, TÉL switch, connect

raccourci *m* INFORMAT shortcut; **~ clavier** INFORMAT keyboard shortcut; **~ de clavier** INFORMAT shorthand keying

raccrochage *m* TÉL hanging up, *du combiné* replacement; **~ du demandé** SIGNAL clear back

raccroché *adj* TÉL on the hook

raccrocher[1]: **~ le combiné** *vt* TÉL replace the handset, go on-hook, replace the receiver

raccrocher[2] *vi* TÉL hang up

racine *f* INFORMAT root top node, root, top node

rack: **~ d'équipement auxiliaire** *m* ÉLECTRON auxiliary equipment rack; **~ linéaire** *m* ÉLECTRON linear rack space

racon *m* (*radiophare*) RADIO racon (*radar beacon*), radio beacon

radar *m* RADIO radar set, radar; **~ d'acquisition** RADIO acquisition radar; **~ aéroporté** RADIO airborne radar; **~ aéroporté à visée latérale** RADIO side-looking airborne radar (*SLAR*); **~ aléatoire** RADIO frequency agility radar; **~ d'altimétrie** RADIO height-finding radar; **~ d'approche finale** (*PAR*) RADIO precision approach radar (*PAR*); **~ d'approche de précision** (*PAR*) RADIO precision approach radar (*PAR*); **~ d'autodirecteur** RADIO homing radar; **~ d'autopoursuite** RADIO homing radar; **~ à balayage latéral** RADIO noncoherent side-looking radar; **~ bistatique** RADIO bistatic radar; **~ diversité** RADIO diversity radar; **~ à diversité de fréquence** RADIO diversity radar; **~ Doppler** RADIO Doppler radar; **~ Doppler à impulsions** RADIO pulse-Doppler radar; **~ à effet Doppler** RADIO Doppler radar; **~ à émission continue** RADIO continuous-wave radar (*CW radar*); **~ IFF** DONNÉES, RADIO, TÉL identification friend or foe (*IFF*); **~ d'illumination** RADIO illuminating radar; **~ à impulsions** RADIO pulse radar, pulse-modulated radar; **~ à impulsions cohérentes** RADIO coherent-pulse radar; **~ à impulsions synchronisées** RADIO coherent-pulse radar; **~ météorologique** RADIO weather radar; **~ à modulation de fréquence** RADIO frequency-modulated radar; **~ mono-impulsion** RADIO monopulse radar; **~ monostatique** RADIO monostatic radar; **~ de navigation** RADIO navigational radar; **~ à ondes entretenues** RADIO continuous-wave radar (*CW radar*); **~ par impulsions à effet Doppler** RADIO pulse-Doppler radar; **~ parfaitement cohérent** RADIO fully-coherent radar; **~ de pistes** RADIO airport surface detection equipment (*ASDE*), ground movement radar (*GMR*); **~ de port** RADIO harbor radar (*AmE*), harbour radar (*BrE*); **~ primaire** RADIO primary radar; **~ secondaire** RADIO secondary radar; **~ de surveillance** RADIO surveillance radar; **~ de veille** RADIO search radar, surveillance radar; **~ à visée latérale non cohérent** RADIO noncoherent side-looking radar; **~ à visualisation des cibles mobiles** RADIO moving-target indicator radar (*MTI radar*)

radiateur *m* ÉLECTRON heat dissipator; **~ lambertien** ÉLECTRON, TRANSMIS Lambertian radiator, Lambertian source

radiation *f* RADIO deregistration; **~ électromagnétique** ÉLECTRON electromagnetic radiation; **~ infrarouge** ÉLECTRON infrared radiation, IR radiation; **~ de localisation** RADIO location deregistration

radio[1] *adj* RADIO radio

radio[2]: **par ~** *adv* RADIO over the air

radio[3]: **~-réveil** *m* PÉRIPH radio alarm

radio[4] *f* RADIO radio; **~ dans le réseau** RADIO radio in the loop (*RITL*); **~ maritime** RADIO maritime radio; **~ sol-train** RADIO truck-to-train radio

radioalignement: **~ à entrechevêtrement** *m* RADIO interlocking equisignal system

radioaltimètre *m* RADIO, TÉL radio altimeter

radioamateur *m* RADIO ham, radio ham, radio amateur

radioappel *m* DONNÉES, RADIO, TÉL, TRANSMIS radiopaging

radiobalise *f* RADIO radio beacon; **~ de détresse** RADIO emergency locater transmitter (*ELT*); **~ à impulsions** RADIO pulsed beacon; **~ de localisation de sinistre** RADIO emergency position-indicating radio beacon (*EPIRB*)

radioborne *f* RADIO marker beacon; **~ intérieure** RADIO inner marker beacon (*IM*); **~ intermédiaire** RADIO middle marker beacon

radiocanal *m* RADIO channel, radio channel, radio-frequency channel, RF channel

radiocommande *f* RADIO radio control

radiocommunication *f* RADIO radiocommunication; **~ intra-immeuble** RADIO indoor radio; **~ par paquets** RADIO packet radio; **~ de terre** RADIO terrestrial radiocommunication

radiocompas *m* RADIO radio compass

radiocotéléphonie: **~ à ressources partagées** *f* ALIM ÉLEC, RADIO trunked radio

radiodétecteur: **~ d'approche de précision** *m* (*PAR*) RADIO precision approach radar (*PAR*)

radiodétection: **~ primaire** *f* RADIO primary radar

radiodiffuseur *m* RADIO radio broadcaster

radiodiffusion *f* RADIO broadcasting, radio broadcasting, TÉL, TV broadcasting; **~ en différé** RADIO, TV recorded broadcast; **~ directe par satellite** (*RDS*) DONNÉES, RADIO, TV direct broadcasting by satellite (*DBS*); **~ numérique** RADIO digital radio broadcasting; **~ par satellite** RADIO, TV satellite broadcasting; **~ récréative** TV entertainment broadcasting; **~ restreinte** RADIO narrowcasting; **~ sonore** RADIO sound broadcasting; **~ sonore numérique** INFORMAT, TV digital audio broadcasting (*DAB*);

~ visuelle RADIO television (*TV*), radio broadcasting, TV television (*TV*), television broadcasting
radioélectrique *f* RADIO radio, radio-frequency interference (*RFI*)
radiofréquence *f* (*RF*) ÉLECTRON, RADIO, TÉL radio frequency (*RF*)
radiogoniomètre *m* RADIO direction finder, radiogoniometer; **~ Adcock à équilibreurs** RADIO balanced Adcock direction finder; **~ à antenne fixe** RADIO fixed direction finder, fixed-antenna direction finder; **~ à basculement du diagramme** RADIO switched-beam direction finder; **~ manuel** RADIO manual direction finder; **~ à tube cathodique** RADIO cathode-ray direction finder
radiogoniométrie: **~ automatique** *f* RADIO automatic direction finding (*ADF*); **~ visuelle à deux canaux de réception** *f* RADIO dual-channel visual direction finding
radioguidage *m* RADIO radio guidance
radiointerférométrie: **~ à très longue base** *f* (*RILB*) TESTS very long baseline interferometry (*VLBI*)
radiomessagerie *f* (*RM*) DONNÉES, RADIO, TÉL paging, radiopaging, TRANSMIS radiopaging; **~ avec accusé de réception** RADIO acknowledgement paging; **~ unilatérale** (*RMU*) DONNÉES, RADIO, TÉL, TRANSMIS display paging, radiopaging; **~ unilatérale urbaine** RADIO city paging; **~ à valeur ajoutée** RADIO value-added paging
radiomobile: **~ privé** *m* RADIO private mobile radio (*PMR*)
radionavigation *f* RADIO radio navigation; **~ hyperbolique** RADIO hyperbolic navigation system
radiopage *m* DONNÉES, RADIO, TÉL radiopager, pager
radiophare *m* (*racon*) RADIO radar beacon (*racon*), radio beacon; **~ à diagramme oscillant** RADIO sector-scanning beacon; **~ omnidirectionnel VHF** (*VOR*) RADIO VHF omnidirectional radio range (*VOR*)
radiophonie *f* RADIO, TÉL radiotelephony, wireless telephony
radioralliement *m* RADIO homing, radio homing
radiorecherche *f* DONNÉES, RADIO, TÉL, TRANSMIS radiopaging, paging
radiorepérage *m* RADIO radio location, radio determination
radiosité *f* INFORMAT radiosity
radiosonde *f* RADIO radiosonde
radiotélécopie *f* RADIO, TÉL radio facsimile
radiotélégraphie *f* RADIO, TÉL radiotelegraphy, wireless telegraphy
radiotéléimprimeur *m* RADIO radio teleprinter, radioteletype q.v.; **~ cinq moments** TÉL five-level RTTY
radiotélémètre *m* RADIO radar rangefinder; **~ à répondeur** RADIO Rebecca-Eureka system
radiotélémétrie *f* RADIO radio rangefinding
radiotéléphone *m* RADIO, TÉL mobile phone, mobile radio telephone, mobile telephone, mobile telephone station, *pour service mobile* radiophone, radiotelephone
radiotéléphonie *f* RADIO, TÉL mobile telephony, radiotelephony, wireless telephony; **~ cellulaire** RADIO cellular radiotelephony; **~ numérique** RADIO, TÉL digital radiotelephony
radiotélex *m* RADIO Telex Over Radio
radiothéodolite *m* RADIO radio theodolite
radiotransmission: **~ diplex** *f* DONNÉES, RADIO diplex radio transmission
radiovent *m* RADIO radio wind, rawin
radôme *m* RADIO radar dome, radome; **~ humide** RADIO wet radome
rafale: **~ d'impulsions** *f* DONNÉES, TRANSMIS pulse burst
raffinage *m* DONNÉES, INFORMAT refinement
raffiner *vt* DONNÉES, INFORMAT refine
rafraîchissement: **~ conditionnel** *m* DONNÉES conditional replenishment; **~ d'image** *m* TV image refreshing; **~ rapide** *m* INFORMAT fast update
RAG *abrév* (*rappel automatique*) PÉRIPH, TÉL *de répondeur automatique* call recall, automatic recall, automatic callback, automatic redialing (*AmE*), automatic redialling (*BrE*), RAG (*ring again*)
raidisseur *m* COMPOSANTS stiffening bar
raie: **~ spectrale** *f* TRANSMIS spectrum line, spectral line
rail *m* COMPOSANTS floor bar, floor rail; **~ de guidage** COMPOSANTS running bar
rainurage *m* COMPOSANTS scribing
rainure *f* GÉNÉRAL groove, RADIO ditch; **~ de clavette** GÉNÉRAL key groove, keyway
rainurer *vt* COMPOSANTS scribe
rainures[1]: **à ~ gravées** *adj* ÉLECTRON precision-etched
rainures[2]: **~ de ruban abouté** *f pl* ÉLECTRON butted-ribbon arrays
raisonnement *m* DONNÉES reasoning; **~ abductif** DONNÉES abductive reasoning; **~ analogique** DONNÉES analogical reasoning; **~ approximatif** DONNÉES approximate reasoning; **~ autoépistémique** DONNÉES auto-epistemic reasoning; **~ automatisé** DONNÉES automated reasoning; **~ à base de scénarios** DONNÉES script-based reasoning; **~ basé sur des cas** DONNÉES case-based reasoning; **~ basé sur la mémoire** DONNÉES, INFORMAT memory-based reasoning; **~ déductif** DONNÉES deductive reasoning; **~ formel** INFORMAT formal reasoning; **~ hybride** ÉLECTRON hybrid reasoning; **~ inductif** DONNÉES inductive reasoning; **~ monotone** DONNÉES monotonic reasoning; **~ par défaut** DONNÉES default reasoning; **~ plausible** DONNÉES plausible reasoning; **~ probabilistique** DONNÉES probabilistic reasoning; **~ qualitatif** DONNÉES qualitative reasoning; **~ quantitatif** DONNÉES quantitative reasoning; **~ de sens commun** DONNÉES common-sense reasoning; **~ temporel** DONNÉES temporal reasoning
rallonge: **~ de câble** *f* GÉNÉRAL extension cable
rallumage *m* ÉLECTRON reignition
RAM *abrév* DONNÉES, INFORMAT (*mémoire à accès direct*) RAM (*random-access memory*), (*mémoire vive*) read-write memory; **~ statique** *f* INFORMAT static RAM; **~ vidéo** *f* (*VRAM*) INFORMAT video RAM (*VRAM*)
ramark *m* RADIO ramark
ramener: **~ à la position initiale** *vt* COMMUT reset, return to home position, restore; **~ à la position de repos** *vt* COMMUT reset, return to home position, restore
rampe *f* TÉL ramp
RAN *abrév* (*retransmission audionumérique*) INFORMAT, TV DAB (*digital audio broadcast*)
random: **~ block file manager** *m* INFORMAT random block file manager
randomisation *f* GÉNÉRAL randomization
randomiser *vt* GÉNÉRAL randomize
rang: **~ d'un chiffre** *m* INFORMAT digit place
rangée *f* INFORMAT suite, TÉL rack row, row, suite; **~ arrière** TÉL rear row; **~ de broches** ÉLECTRON pin row; **~ de contacts** TÉL bank of contacts; **~ de pupitres d'opérateur** COMMUT, TÉL switchboard row, switchboard suite; **~ de sortie** ÉLECTRON output row; **~ de tableaux de commutation** COMMUT, TÉL switchboard row, switchboard suite; **~ de tables d'opérateur** COMMUT, TÉL switchboard row, switchboard suite
rapatrier *vt* DONNÉES transfer
rapidité: **~ d'apprêtage** *f* ÉLECTRON priming rate; **~ d'effacement** *f* INFORMAT erasing rate; **~ de modulation** *f* TÉL, TRANSMIS modulation rate, band rate
rappel *m* DONNÉES, INFORMAT recall, retrieval, TÉL recall, callback; **~ automatique** (*RAG*) PÉRIPH, TÉL *de répondeur automatique* ring again (*RAG*), call recall, automatic recall, automatic callback, automatic redialing (*AmE*), automatic redialling (*BrE*); **~ automatique du dernier numéro** PÉRIPH, TÉL *de répondeur automatique* last-number recall; **~ automatique sur libération des correspondants** TÉL ring again on idle EBS; **~ du dernier numéro** PÉRIPH, TÉL last-number recall; **~ du dernier son édité** INFORMAT edit

recall; ~ **d'enregistreur** TÉL, TRANSMIS register recall; ~ **de rendez-vous** PÉRIPH reminder call; ~ **du tampon d'édition** INFORMAT recall-edit buffer; ~ **temporisé** PÉRIPH, TÉL timed reminder, automatic timed recall

rappeler *vt* DONNÉES retrieve, INFORMAT recall, roll in, *à l'écran* retrieve, TÉL *numéro* recall, ring back, *personne* call back, phone back, telephone back; ~ **qn** TÉL return sb's call

rapport[1] *m* GÉNÉRAL ratio; ~ **d'affaiblissement dans le mode commun** TÉL common-mode rejection ratio; ~ **d'ajustement** COMPOSANTS setting ratio; ~ **avant-arrière** RADIO front-to-back ratio; ~ **des axes d'un pixel** INFORMAT, RADIO, TÉL pel aspect ratio, pixel aspect ratio; ~ **des charges financières annuelles** TÉL annual charge ratio; ~ **cœur-gaine** TRANSMIS core-cladding ratio; ~ **de compression** TRANSMIS compression ratio; ~ **des contrastes** TV contrast ratio; ~ **de conversion transversale** TÉL transverse conversion ratio, weighted terminal coupling loss (*TCLw*); ~ **cyclique** RADIO duty cycle; ~ **détaillé des appels** TÉL call-detail report; ~ **de discrimination** TESTS discrimination ratio, slope of the OC curve; ~ **de données dans la bande vocale** DONNÉES voice-band data ratio; ~ **de données numériques sans restriction** DONNÉES unrestricted digital data ratio; ~ **d'émission secondaire** ÉLECTRON secondary electron emission factor; ~ **énergie à densité de bruit** TRANSMIS energy-to-noise density; ~ **d'équilibre** TÉL *d'un récepteur de signaux sinusoïdaux* balance ratio; ~ **d'équilibre d'impédance** ÉLECTRON, TÉL impedance balance ratio; ~ **d'équilibre des signaux** SIGNAL *d'un générateur de signaux* signal balance ratio; ~ **d'étalement de la largeur de bande** RADIO, TRANSMIS bandwidth expansion ratio; ~ **d'extinction** ÉLECTRON, TRANSMIS extinction ratio (*EX*); ~ **fractionnaire** INFORMAT fractional ratio; ~ **de fréquence** TRANSMIS frequency ratio; ~ **global porteuse-brouillage** RADIO overall carrier-to-interference ratio; ~ **impulsionnel** RADIO impulsiveness ratio; ~ **d'impulsion-non-impulsion** TÉL pulse-no-pulse ratio; ~ **d'impulsions** COMMUT pulse ratio; ~ **d'incident** TÉL error report, trouble report, fault report; ~ **intrinsèque** ALIM ÉLEC intrinsic standoff ratio, standoff ratio; ~ **de luminance** TESTS luminance ratio; ~ **de mélange** RADIO mixing ratio; ~ **du nombre de spires** COMPOSANTS turn ratio; ~ **d'ondes de tension stationnaires** (*ROS*) RADIO, TRANSMIS standing wave ratio (*SWR*), voltage standing wave ratio (*VSWR*); ~ **porteuse-brouillage** TRANSMIS carrier-to-interference ratio (*C/I*); ~ **porteuse-bruit thermique du récepteur** TRANSMIS carrier-to-receiver thermal noise ratio; ~ **porteuse-densité de bruit** TRANSMIS carrier-to-noise density ratio; ~ **porteuse-densité de bruit d'intermodulation** TRANSMIS carrier-to-intermodulation noise density ratio; ~ **porteuse-parasites** TRANSMIS carrier-interference ratio; ~ **porteuse-puissance de bruit d'intermodulation** TRANSMIS carrier-to-intermodulation noise power ratio; ~ **porteuse-puissance de bruit thermique** TRANSMIS carrier-to-thermal noise power ratio; ~ **porteuse-puissance de bruit total** TRANSMIS carrier-to-total-noise-power ratio; ~ **de post-accélération** ÉLECTRON post-deflection acceleration ratio; ~ **de protection** RADIO protection ratio; ~ **de protection en audiofréquence** RADIO audio-frequency protection ratio; ~ **de protection en radiofréquence** RADIO radio-frequency protection ratio; ~ **de protection en vidéofréquence** RADIO, TV video-frequency protection ratio; ~ **de puissance de porteuse utile-porteuse brouilleuse** ÉLECTRON, RADIO, SIGNAL wanted-to-unwanted carrier power ratio; ~ **de réjection des canaux voisins** RADIO, TRANSMIS adjacent-channel rejection ratio; ~ **de reproduction** TÉL reproduction ratio; ~ **de retour** COMPOSANTS resetting ratio; ~ **S/B** (*rapport signal sur bruit*) ÉLECTRON, SIGNAL, TESTS, TRANSMIS SNR (*signal-to-noise ratio, signal-noise ratio*); ~ **signal bruit efficace pondéré** TRANSMIS signal-to-weighted rms noise ratio; ~ **signal sur brouillage** RADIO signal-to-interference ratio, signal interference ratio; ~ **signal sur bruit** (*rapport S/B*) ÉLECTRON, SIGNAL, TESTS, TRANSMIS signal-to-noise ratio (*SNR*); ~ **signal/brouillage en audiofréquence** RADIO audio-frequency signal-to-interference ratio; ~ **signal/brouillage en vidéofréquence** RADIO, TV video-frequency signal-to-interference ratio; ~ **signal/bruit normalisé** RADIO normalized signal-to-noise ratio; ~ **SINAD** GÉNÉRAL SINAD ratio; ~ **de test de conformité au protocole** DONNÉES, TESTS protocol conformance test report (*PCTR*); ~ **de test de conformité du système** DONNÉES, TESTS system conformance test report (*SCTR*); ~ **du trafic à l'heure chargée au trafic journalier** TÉL busy hour-to-day ratio; ~ **trafic à l'heure chargée/trafic journalier** TÉL busy hour-to-day ratio; ~ **du trafic journalier au trafic à l'heure chargée** TÉL day-to-busy-hour ratio; ~ **trafic journalier/trafic à l'heure chargée** TÉL day-to-busy-hour ratio; ~ **de transfert direct du courant** ÉLECTRON forward current transfer ratio; ~ **de transformation** COMPOSANTS ratio of transformation; ~ **de transmission** DONNÉES logbook; ~ **valeur de crête sur valeur moyenne** GÉNÉRAL peak-to-average ratio (*PAR*)

rapport[2]: ~ **énergie par bit d'information à densité de bruit** *loc* TRANSMIS energy per information bit to noise density ratio

rapporter *vt* GÉNÉRAL report

RAQ *abrév* (*règlement d'assurance qualité*) TESTS quality assurance ruling

raréfaction *f* RADIO rarefaction

raté: ~ **d'allumage** *m* ÉLECTRON misfire

râtelier *m* INFORMAT rack

ratio: ~ **de suppression de la bande latérale** *m* TRANSMIS side-mode suppression ratio (*SMSR*)

rattachement *m* DONNÉES bind; ~ **à un arbre** DONNÉES, ÉLECTRON, INFORMAT tree attachment; ~ **au point MCSAP** TÉL MCS attachment

rattrapage *m* COMPOSANTS retrofit

rattraper *vt* COMPOSANTS retrofit

rayage *m* COMPOSANTS scratching, scribing, RADIO deleting, TESTS scratching

rayer *vt* COMPOSANTS, TESTS scratch

rayon *m* GÉNÉRAL ray, INFORMAT ray, beam, RADIO range, TÉL shelf, TRANSMIS light ray; ~ **axial** RADIO, TRANSMIS *fibres optiques* axial ray; ~ **de courbure** ÉLECTRON bending radius, TRANSMIS *d'un câble* bend radius; ~ **de courbure minimal** TRANSMIS minimum bend radius; ~ **de fuite** TÉL leaky ray; ~ **lumineux** TRANSMIS light ray, ray; ~ **méridien** TRANSMIS meridional ray; ~ **non méridien** TRANSMIS skew ray; ~ **oblique** TRANSMIS skew ray; ~ **optique** TRANSMIS light ray, ray; ~ **d'outil** COMPOSANTS, ÉLECTRON tooling radius; ~ **paraxial** ÉLECTRON, TRANSMIS paraxial ray; ~ **de Pedersen** RADIO Pedersen ray; ~ **réfracté** TRANSMIS refracted ray; ~ **tangent** RADIO tangential wave path; ~ **terrestre équivalent** RADIO effective radius of the Earth; ~ **tunnel** TRANSMIS leaky ray, tunneling ray (*AmE*), tunnelling ray (*BrE*)

rayonnement *m* ÉLECTRON radiation, RADIO radio-frequency radiation, radiation, TRANSMIS radiation; ~ **ambiant** RADIO background radiation; ~ **d'antenne** RADIO antenna radiation; ~ **cohérent** RADIO coherent radiation; ~ **dans les lobes latéraux** RADIO side-lobe radiation; ~ **électromagnétique** ÉLECTRON, TRANSMIS electromagnetic radiation; ~ **des émissions hors bande** RADIO out-of-band emission (*OBE*); ~ **harmonique** RADIO harmonic emission; ~ **hors bande** RADIO out-of-band emission (*OBE*); ~ **incohérent** TRANSMIS incoherent radiation; ~ **infrarouge** ÉLECTRON infrared radiation, IR radiation; ~ **de la lumière visible** ÉLECTRON, TRANSMIS visible radiation; ~ **monochromatique** TRANSMIS monochromatic radiation; ~ **non désiré** RADIO unwanted emission; ~ **non**

essentiel RADIO, TRANSMIS spurious radiation, spurious emission; ~ **non-essentiel du satellite** RADIO spurious satellite emission; ~ **optique** TRANSMIS optical radiation; ~ **par le coffret** RADIO cabinet radiation; ~ **parasite** RADIO parasitic emissions, spurious emission, spurious radiation, TRANSMIS spurious radiation, spurious emission; ~ **parasite par conduction** ÉLECTRON conducted spurious emission; ~ **radioélectrique** ÉLECTRON, RADIO radiation; ~ **d'un récepteur** RADIO receiver radiation; ~ **secondaire** RADIO secondary radiation, reradiation; ~ **thermique** ALIM ÉLEC, ÉLECTRON heat radiation; ~ **ultraviolet** TRANSMIS ultraviolet radiation, UV radiation; ~ **visible** ÉLECTRON visible radiation, TRANSMIS light, visible radiation

rayonnements: ~ **divulgateurs** *m pl* RADIO compromising emanations

rayonner *vt* ÉLECTRON, GÉNÉRAL, TRANSMIS radiate

rayons *m pl* INFORMAT spoking

rayure *f* COMPOSANTS, TESTS scratch

RC *abrév* ÉLECTRON (*résistance-condensateur*) RC (*resistance-capacitance*), TÉL (*réactivation du compteur*) RC (*redrive counter*), TV (*répartiteur coaxial*) coaxial cable distributor

RCB *abrév* (*réactivation du compteur occupé*) TÉL RCB (*redrive counter busy*)

RCI *abrév* (*réseau conformateur d'impulsions*) ÉLECTRON PFN (*pulse-forming network*)

R-DAT *abrév* (*bande audio-numérique à tête rotative*) DONNÉES, INFORMAT R-DAT (*rotary-head digital audio tape*)

RDPCP *abrév* (*réseau de données public avec commutation par paquets*) DONNÉES PPSDN (*public packet-switched data network*)

RDS *abrév* (*radiodiffusion directe par satellite*) DONNÉES, RADIO, TV DBS (*direct broadcasting by satellite*)

réacheminement *m* COMMUT redirection, rerouting, DONNÉES rerouting, TÉL redirection, rerouting; ~ **automatique** RADIO roaming, TÉL automatic rerouting (*ARR*); ~ **et déviation des appels entre réseaux** DONNÉES internetwork call redirection and deflection (*ICRD*)

réacheminer *vt* COMMUT *trafic* redirect, reroute, DONNÉES reroute, TÉL *trafic* redirect, reroute

réactance *f* ALIM ÉLEC reactance; ~ **d'électrode** ÉLECTRON electrode reactance; ~ **monophase** ALIM ÉLEC single-phase reactance, single lashing

réaction *f* DONNÉES, ÉLECTRON, INFORMAT, RADIO feedback, positive feedback; ~ **acoustique** GÉNÉRAL acoustic feedback, acoustic regeneration, TÉL acoustic feedback; ~ **anodique** COMPOSANTS *métallurgie* anodic reaction; ~ **négative** ALIM ÉLEC negative feedback; ~ **positive** ÉLECTRON, INFORMAT, RADIO positive feedback; ~ **rapide** ÉLECTRON fast clamping

réactivation: ~ **du compteur** *f* (*RC*) TÉL redrive counter (*RC*); ~ **du compteur occupé** *f* (*RCB*) TÉL redrive counter busy (*RCB*)

réaffectation *f* GÉNÉRAL reallocate, RADIO reassign

réaffecté *adj* GÉNÉRAL reallocated, RADIO reassigned

réagir *vi* INFORMAT respond

réajustement: ~ **du rythme** *m* TÉL, TRANSMIS retiming

réalisable *adj* GÉNÉRAL feasible

réalisateur *m* GÉNÉRAL implementor; ~ **d'essai** TESTS test realizer; ~ **de tests** TESTS test realizer

réalisation *f* ÉLECTRON implementation, INFORMAT implementation; ~ **d'un essai** *f* TESTS test realization; ~ **d'un test** TESTS test realization

réalisé *adj* GÉNÉRAL implemented

réaliser *vt* GÉNÉRAL implement; ~ **une inversion** ÉLECTRON, INFORMAT negate; ~ **une liaison** TRANSMIS span

réalité: ~ **virtuelle** *f* (*RV*) DONNÉES, INFORMAT virtual reality (*VR*)

réamorçage *m* DONNÉES reboot, reignition, INFORMAT reboot, RADIO reignition

réarmer *vt* RADIO reset

réarrangement: ~ **automatique du trafic** *m* TRANSMIS automatic traffic rearrangement

réassemblage *m* TÉL reassembling

rebondir *vt* INFORMAT bounce

rebooter *vt* DONNÉES, INFORMAT reboot

reboucler *vt* ÉLECTRON loop

recadrage *m* INFORMAT reframe

recalage *m* RADIO, SIGNAL, TÉL, TRANSMIS *radar* aligning, alignment; ~ **automatique** RADIO *radar* automatic aligning, automatic alignment

recaler *vt* RADIO, SIGNAL, TÉL, TRANSMIS *radar* align

récepteur *m* DONNÉES, PÉRIPH, RADIO, TÉL radio receiver, receiver (*rcvr, RX*); ~ **alphanumérique d'appel de personnes** RADIO alphanumeric pager; ~ **d'appel** RADIO, TÉL pager; ~ **d'appel de personnes** RADIO, TÉL pager; ~ **à balayage** RADIO radio-frequency scanner, RF scanner; ~ **à bande étroite** RADIO narrow-band receiver; ~ **de code** TÉL code receiver; ~ **à couverture générale** RADIO general-coverage receiver; ~ **à cylindre** TÉL drum receiver; ~ **à déroulement continu** TÉL continuous receiver; ~ **différentiel de téléchauffage** ALIM ÉLEC torque differential receiver; **~-distributeur** RADIO receiver distributor; ~ **à double changement de fréquence** RADIO dual-conversion receiver; ~ **d'écoute** ÉLECTRON intercept receiver; ~ **à fibres optiques** TRANSMIS fiber-optic receiver (*AmE*), fibre-optic receiver (*BrE*), optoelectronic receiver, receive fiber-optic terminal device (*AmE*), receive fibre-optic terminal device (*BrE*); ~ **à fréquence réglable** RADIO floating receiver; ~ **générateur de fréquence** ÉLECTRON, TÉL frequency-generator receiver; ~ **intégré PIN-TEC** TÉL, TRANSMIS PIN-FET integrated receiver; ~ **d'interception** ÉLECTRON intercept receiver; ~ **à large bande** RADIO wide-band receiver; ~ **linéaire** RADIO linear receiver; ~ **de messages de tarification** TÉL tariff-message receiver; ~ **MF** SIGNAL MF receiver; ~ **de modem** DONNÉES modem receiver; ~ **multifréquence** SIGNAL multifrequency receiver; ~ **optique** TRANSMIS optical receiver; ~ **optoélectronique** TRANSMIS fiber-optic receiver (*AmE*), fibre-optic receiver (*BrE*), optoelectronic receiver, receive fiber-optic terminal device (*AmE*), receive fibre-optic terminal device (*BrE*); **~-perforateur** TÉL reperforator, receiving perforator; ~ **photoélectrique** DONNÉES photodetector, TRANSMIS optical detector, photodetector; ~ **de poche** RADIO pocket receiver; ~ **portable** RADIO portable receiver; ~ **portatif** RADIO hand-held receiver; ~ **à projection** TV projection television receiver; ~ **de prothèse auditive sans fil** RADIO wireless hearing aid receiver; ~ **radio modulation de fréquence** RADIO FM radio receiver; ~ **radioélectrique** PÉRIPH, RADIO, TÉL radio receiver; ~ **de radiomessagerie** (*RMU*) APP SERV Message-master, GÉNÉRAL message pager, RADIO message display pager, message pager, pager, TÉL pager, Message-master; ~ **à réaction** RADIO regenerative receiver; ~ **de réponse** RADIO responser, responsor; ~ **de signalisation à fréquence vocale** SIGNAL, TÉL voice-frequency signaling receiver (*AmE*), voice-frequency signalling receiver (*BrE*); ~ **de signaux** RADIO, SIGNAL signal receiver; ~ **de signaux horaires** RADIO, SIGNAL time-signal receiver; ~ **de télévision** TV television receiver; ~ **de télévision à projection** TV projection television receiver; ~ **de veille** RADIO guard receiver, watch-keeping receiver, watch receiver; ~ **de voie** TRANSMIS channel receiver; ~ **de voix** INFORMAT voice-input device

réception *f* DONNÉES receive (*REC*), INFORMAT, RADIO reception, accepting; ~ **à amplification directe** RADIO tuned radio frequency reception (*TRF reception*); ~ **communautaire** RADIO community reception; ~ **en diversité** RADIO diversity reception; ~ **en diversité d'espace** RADIO space-diversity reception; ~ **en diversité de fréquence** RADIO, TRANSMIS frequency-diversity reception; ~ **en diversité temporelle** DONNÉES, RADIO time-diversity reception; ~ **des données** DONNÉES received data; ~ **hétérodyne** RADIO heterodyne reception; ~ **individuelle** RADIO individual reception; ~ **à l'intérieur des bâtiments** RADIO in-building reception; ~ **de programmes** RADIO down-link of programs (*AmE*), down-link of programmes (*BrE*); ~ **de radiodiffusion** RADIO broadcast reception; ~ **avec régénération de porteuse** GÉNÉRAL reconditioned carrier reception; ~ **en régime permanent** RADIO steady-state reception (*SSR*); ~ **résiduelle** RADIO stray pick-up; ~ **superhétérodyne** RADIO superheterodyne reception; ~ **à superréaction** RADIO superregenerative reception; ~ **synchrone** RADIO homodyne reception

réception/émission *f* ÉLECTRON dial in/dial out

recette *f* TESTS *qualité* acceptance; ~ **sur site** RADIO on-site acceptance

recevoir *vt* INFORMAT download, TÉL receive; ~ **un courant de maintien** COMPOSANTS obtain holding current, be fed with holding current, receive holding current

rechargement *m* DONNÉES, INFORMAT reload, reloading, reboot

recharger *vt* DONNÉES, INFORMAT reboot, reload

réchauffage: ~ **préalable** *m* ÉLECTRON preheating

recherche *f* DONNÉES, INFORMAT search, RADIO research, search, TÉL hunting, TESTS research; ~ **d'accès** RADIO access hunting; ~ **appliquée** TESTS applied research; ~ **automatique de canal** RADIO automatic-channel tuning; ~ **binaire** INFORMAT binary search; ~ **booléenne** INFORMAT Boolean search; ~ **en chaîne** INFORMAT chaining search; ~ **commanditée** TESTS commissioned research; ~ **contextuelle** INFORMAT context search; ~ **dichotomique** INFORMAT dichotomizing search; ~ **en distance** RADIO range search; ~ **documentaire** DONNÉES, INFORMAT information retrieval; ~ **de fichier** DONNÉES, INFORMAT file search; ~ **fondamentale** TESTS basic research; ~ **dans un groupement de lignes** TÉL PBX hunting; ~ **heuristique** DONNÉES heuristic search; ~ **de l'information** DONNÉES, INFORMAT information retrieval; ~ **intuitive** INFORMAT intuitive searching; ~ **d'itinéraire** COMMUT path allocation; ~ **de ligne** TÉL line hunting; ~ **des messages** DONNÉES message retrieval; ~ **multiniveau** TÉL level hunt; ~ **de notes** INFORMAT note search; ~ **par décalage en fonction de la dernière ligne occupée** TÉL round-robin hunting; ~ **de personne** RADIO, TÉL paging; ~ **de personne téléphonique** RADIO loudspeaker paging, voice paging; ~ **préprogrammée** INFORMAT preset search; ~ **rapide** INFORMAT browsing; ~ **séquentielle** TÉL sequential hunting; ~ **simple niveau** TÉL rotary hunt, rotary hunting; ~ **subventionnée** TESTS commissioned research; ~ **de télétrafic** TÉL traffic research; ~ **textuelle** DONNÉES text search; ~ **en vitesse** RADIO speed search, velocity search

rechercher *vt* DONNÉES, INFORMAT find, RADIO, TÉL page

rechercher/remplacer *vt* INFORMAT find/replace, search and replace

réciprocité *f* ÉLECTRON reciprocity

reclassement *m* GÉNÉRAL upgrading, upgrade

reclasser *vt* GÉNÉRAL upgrade

recombinaison *f* COMPOSANTS recombination, TÉL recombining

recompilation *f* DONNÉES, INFORMAT translation

recompiler *vt* DONNÉES, INFORMAT translate

reconditionnement: ~ **négatif** *m* DONNÉES, TÉL retrain negative (*RTN*); ~ **positif** *m* DONNÉES, TÉL retrain positive (*RTP*)

reconfigurable *adj* ALIM ÉLEC restrappable, INFORMAT reconfigurable

reconfiguration *f* ALIM ÉLEC restrapping, INFORMAT reconfiguration, reconfiguring

reconnaissance *f* APP SERV, CONTRÔLE, DONNÉES, INFORMAT recognition, RADIO *de site* investigation, SIGNAL, TÉL, TRANSMIS *de signal* recognition; ~ **d'aptitude** TESTS *qualité* capability approval; ~ **de caractères** INFORMAT character recognition; ~ **de l'écriture manuelle** APP SERV handwriting recognition; ~ **de formes** APP SERV, CONTRÔLE, DONNÉES, TRANSMIS pattern recognition; ~ **d'impulsions** SIGNAL, TRANSMIS pulse recognition; ~ **du locuteur** APP SERV, INFORMAT, TÉL speaker recognition; ~ **magnétique des caractères** (*MICR*) APP SERV magnetic ink character recognition (*MICR*); ~ **monolocuteur de la parole** APP SERV, INFORMAT, TÉL single-voice speech recognition; ~ **multilangue** APP SERV, INFORMAT, TÉL multilingual recognition; ~ **multilocuteur de la parole** APP SERV, INFORMAT, TÉL multi-voice speech recognition; ~ **optique de caractères** (*ROC*) DONNÉES, INFORMAT optical character recognition (*OCR*); ~ **de la parole** APP SERV, INFORMAT, TÉL speech recognition, voice recognition; ~ **vocale** APP SERV, INFORMAT, TÉL speech recognition, voice recognition

reconnexion *f* GÉNÉRAL reconnection

recopie: ~ **d'écran** *f* INFORMAT screen printing

recouvrement *m* INFORMAT overlapping, overlay, RADIO, TÉL overlap, overlapping; ~ **étain-plomb** COMPOSANTS tin-lead coating

recouvrir *vt* INFORMAT overlap, overlay, RADIO, TÉL overlap

recueillir *vt* TÉL receive

recul: ~ **d'entrée** *m* (*IBO*) RADIO input back-off (*IBO*); ~ **de puissance** *m* RADIO back-off; ~ **de puissance de sortie** *m* TRANSMIS output back-off (*OBO*)

récupérateur *m* COMPOSANTS, ÉLECTRON spudger

récupération *f* DONNÉES, ÉLECTRON recovery, INFORMAT retrieval, RADIO recovery, recuperation, TÉL, TRANSMIS recovery; ~ **des fichiers** DONNÉES file recovery; ~ **inverse** ÉLECTRON reverse recovery; ~ **de rythme** TÉL, TRANSMIS timing recovery; ~ **du système** DONNÉES system recovery

récupérer *vt* DONNÉES, ÉLECTRON recover, INFORMAT retrieve, RADIO recover, recuperate, TÉL, TRANSMIS recover

récurrence *m* DONNÉES, INFORMAT recursiveness

récurrent *adj* DONNÉES, INFORMAT recursive

redémarrage *m* COMMUT restart, DONNÉES reactivation, reboot, rebooting, GÉNÉRAL restart, INFORMAT reboot, rebooting, restart, RADIO restart; ~ **restreint** COMMUT minor restart; ~ **système restreint** COMMUT minor system restart

redémarrer *vt* COMMUT restart, DONNÉES reactivate, reboot, GÉNÉRAL restart, INFORMAT reboot, restart, RADIO restart

redevance *f* GÉNÉRAL fee, licence (*BrE*), license (*AmE*), TÉL rental; ~ **forfaitaire** TÉL fixed charge

rediffuser *vt* RADIO, TRANSMIS, TV rebroadcast

rediffusion *f* RADIO, TRANSMIS, TV rebroadcast, rebroadcasting

rédiger *vt* GÉNÉRAL, INFORMAT edit

redimensionner *vt* ÉLECTRON right-size, INFORMAT resize

redirecteur *m* DONNÉES redirector

redondance *f* GÉNÉRAL, RADIO redundancy; ~ **active** TÉL, TESTS *sécurité de fonctionnement* active redundancy; ~ **en anneau** RADIO, TRANSMIS ring redundancy; ~ **en attente** TÉL standby redundancy; ~ **en mode annulaire** RADIO, TRANSMIS ring redundancy; ~ **partielle** TESTS partial redundance; ~ **passive** TESTS standby redundancy; ~ **relative** GÉNÉRAL relative redundancy

redressement *m* ALIM ÉLEC, ÉLECTRON rectification; ~ **à une alternance** ÉLECTRON half-wave rectification; ~ **à deux alternances** ÉLECTRON full-wave rectification

redresser *vt* ALIM ÉLEC rectify, COMPOSANTS straighten; ~ **à une alternance** ÉLECTRON half-wave rectify

redresseur *m* ALIM ÉLEC, COMPOSANTS, ÉLECTRON rectifier, rectifier unit; ~ **à une alternance** ÉLECTRON half-wave rectifier; ~ **de charge** ALIM ÉLEC charging rectifier; ~**-chargeur** ALIM ÉLEC power supply unit; ~ **connecté séquentiellement** ALIM ÉLEC, COMPOSANTS sequence-connected rectifier; ~ **à cristal** ALIM ÉLEC crystal rectifier; ~ **demi-onde** ÉLECTRON half-wave rectification, half-wave rectifier; ~ **à deux alternances** ÉLECTRON full-wave rectifier; ~ **pour élément de réduction** ALIM ÉLEC end-cell rectifier; ~ **sec** ÉLECTRON metallic rectifier; ~ **au sélénium** ALIM ÉLEC, COMPOSANTS selenium rectifier; ~ **simple alternance** ÉLECTRON half-wave rectification, half-wave rectifier; ~ **solide** ALIM ÉLEC solid-state rectifier, static rectifier, COMPOSANTS solid-state rectifier; ~ **à tension constante** ALIM ÉLEC constant-voltage rectifier; ~ **à thyristor au silicium** ALIM ÉLEC thyristor rectifier unit, COMPOSANTS, ÉLECTRON silicon-controlled rectifier (*SCR*); ~ **triphasé** ALIM ÉLEC, COMPOSANTS three-phase rectifier

réducteur: ~ **de bruit** *m* TRANSMIS noise reducer; ~ **de pression** *m* COMPOSANTS pressure-reducing valve; ~ **de souffle** *m* PÉRIPH, TÉL mouthpiece muffler

réduction *f* DONNÉES condensing, GÉNÉRAL cut, cutback, INFORMAT minimizing, reduction, TESTS *d'échantillon* reduction, scaling down; ~ **des données** DONNÉES, INFORMAT data reduction; ~ **de l'écho** RADIO, TÉL echo control; ~ **de puissance** RADIO back-off; ~ **de surface** COMPOSANTS reduction of area; ~ **de tarif** APP SERV reduction of tariff, tariff reduction; ~ **de la vitesse de groupe** RADIO group retardation

réduire *vt* DONNÉES condense, GÉNÉRAL cut, cut back, INFORMAT minimize, reduce, TESTS *échantillon* reduce, scale down

réduit *adj* DONNÉES condensed, GÉNÉRAL cut-back, INFORMAT minimized, reduced, TESTS reduced

réécouter *vt* PÉRIPH play back

réécrire *vt* DONNÉES rewrite

réel *adj* INFORMAT live

réémetteur *m* RADIO, TÉL, TRANSMIS, TV rebroadcasting transmitter, relay transmitter

réémettre *vt* RADIO, TÉL, TRANSMIS, TV retransmit

réémission *f* RADIO, TÉL, TRANSMIS, TV retransmission; ~ **des signaux d'adresse** SIGNAL, TÉL senderized operation

réenregistrable *adj* DONNÉES, INFORMAT rerecordable

réenregistrement *m* DONNÉES, INFORMAT reregistration

réenregistrer *vt* DONNÉES, INFORMAT rerecord

réentrant *m* INFORMAT re-entrant

rééquilibrage *m* GÉNÉRAL rebalancing

réétalonnage *m* GÉNÉRAL recalibration

réétalonner *vt* GÉNÉRAL recalibrate

réexécution *f* INFORMAT rerun

réexpédier *vt* GÉNÉRAL forward

référence[1]: **de** ~ *adj* RADIO baseline

référence[2]: ~ **alphanumérique** *f* INFORMAT reference designation; ~ **d'appel** *f* TÉL call reference; ~ **d'application** *f* DONNÉES, GÉNÉRAL application reference; ~ **de base de l'échange** *f* TÉL basic session reference; ~ **de commande** *f* TÉL command reference; ~ **de comportements par défaut** *f* DONNÉES default reference; ~ **aux contraintes** *f* DONNÉES constraints reference; ~ **de contrôle de l'échange** *f* DONNÉES interchange control reference; ~ **du destinataire** *f* DONNÉES recipient reference; ~ **élargie de l'échange** *f* TÉL expanded session reference; ~ **de groupe de comportements par défaut** *f* DONNÉES, TESTS default group inference; ~ **d'un groupe de modules de test** *f* DONNÉES, TESTS test step group reference; ~ **de groupe de tests** *f* DONNÉES, TESTS test group reference; ~ **d'orientation** *f* RADIO attitude reference

références: ~ **normatives** *f pl* TÉL normative references

réflectance *f* CLIMAT, ÉLECTRON, RADIO, TRANSMIS reflectance

réflecteur *m* CLIMAT, RADIO, TRANSMIS reflector; ~ **d'antenne à tiges** RADIO rod reflector; ~ **auxiliaire** RADIO subreflector; ~ **en D** RADIO cheese; ~ **lambertien** TRANSMIS Lambertian reflector; ~ **à mailles** RADIO mesh reflector; ~ **polyridique** RADIO corner reflector; ~ **radar** RADIO radar reflector; ~ **secondaire** RADIO subreflector; ~ **en segment** RADIO cheese; ~ **à tiges** RADIO rod reflector

réflectomètre: ~ **dans le domaine fréquenciel** *m* ALIM ÉLEC, ÉLECTRON, TESTS, TRANSMIS frequency-domain reflectometer; ~ **optique** *m* ALIM ÉLEC, ÉLECTRON, TESTS, TRANSMIS optical time domain reflectometer (*OTDR*); ~ **temporel** *m* ALIM ÉLEC, ÉLECTRON, TESTS, TRANSMIS reflectometer, time domain reflectometer (*TDR*)

réflectométrie: ~ **optique dans le domaine temporel** *f* ALIM ÉLEC, ÉLECTRON, TESTS, TRANSMIS optical time domain reflectometry (*OTDR*); ~ **temporelle** *f* ALIM ÉLEC, ÉLECTRON, TESTS, TRANSMIS time domain reflectometry (*TDR*)

réflexion *f* ÉLECTRON, INFORMAT, RADIO, TRANSMIS reflection; ~ **diffuse** INFORMAT diffuse reflection; ~ **de Fresnel** TRANSMIS Fresnel reflection; ~ **interne totale** ÉLECTRON, TRANSMIS total internal reflection; ~ **ionosphérique** RADIO ionospheric reflection; ~ **en M** RADIO M reflection; ~ **résiduelle** TRANSMIS residual reflectance; ~ **spéculaire** TRANSMIS specular reflection; ~ **totale** ÉLECTRON, TRANSMIS total reflection; ~ **troposphérique** RADIO tropospheric reflection

réflexions: ~ **par trajets multiples** *f pl* RADIO multipath reflections

réfraction *f* RADIO, TRANSMIS, TV refraction; ~ **atmosphérique** RADIO atmospheric refraction; ~ **normale** RADIO standard refraction; ~ **optique** TRANSMIS optical refraction

réfractomètre: ~ **portatif** *m* TRANSMIS hand-held refractometer

réfrigération *f* CLIMAT refrigeration, GÉNÉRAL cooling

refroidi: ~ **par air** *adj* COMPOSANTS air-cooled; ~ **par eau** *adj* CLIMAT water-cooled

refroidir *vt* CLIMAT, GÉNÉRAL cool

refroidissement *m* CLIMAT, GÉNÉRAL cooling; ~ **par eau** CLIMAT water cooling; ~ **en parallèle** CLIMAT parallel cooling; ~ **séparé** CLIMAT separate cooling

régénérateur *m* (*RG*) COMMUT, ÉLECTRON, SIGNAL, TÉL regenerator (*RG*); ~ **d'impulsions** SIGNAL, TÉL pulse regenerator; ~ **d'ondes lumineuses** DONNÉES light-wave regenerator; ~ **de synchronisation** TV synchronizing pulse regenerator

régénération: ~ **d'impulsions** *f* GÉNÉRAL, INFORMAT, SIGNAL, TÉL pulse regeneration; ~ **de signal** *f* SIGNAL signal regeneration

régie: ~ **d'abonné** *f* APP SERV network termination

régime *m* TÉL operating conditions, operational conditions; ~ **de charge** ALIM ÉLEC load range; ~ **de déclaration** TV declaration procedure; ~ **établi** GÉNÉRAL steady state; ~ **nominal** ALIM ÉLEC rating; ~ **permanent** GÉNÉRAL steady state

région: ~ **active** *f* TRANSMIS active region; ~ **de base** *f* ÉLECTRON *semiconducteurs* base region; ~ **de champ lointain** *f* RADIO, TRANSMIS far-field region; ~ **de champ proche** *f* RADIO, TRANSMIS near-field region; ~ **de charge d'espace** *f* COMPOSANTS, ÉLECTRON space-charge region; ~ **de collecteur** *f* COMPOSANTS drain region, ÉLECTRON collector region; ~ **de drain** *f* COMPOSANTS drain region; ~ **F** *f* RADIO F-region; ~ **de Fraunhofer** *f* RADIO, TRANSMIS Fraunhofer region, far-field region; ~ **de Fresnel** *f* RADIO, TRANSMIS Fresnel region, near-field region; ~ **de Geiger** *f* RADIO, TRANSMIS Geiger region; ~ **de grille** *f* ÉLECTRON gate region; ~ **d'interaction** *f* ÉLECTRON interaction region; ~ **de proportionnalité** *f* RADIO proportional region; ~ **de proportionnalité limitée** *f* RADIO region of limited proportionality; ~ **de rayonnement** *f* RADIO, TRANSMIS far-field region; ~ **de rayonnement proche** *f* RADIO, TRANSMIS near-field region; ~ **de source** *f* COMPOSANTS source region

registre *m* COMMUT, DONNÉES, ÉLECTRON, INFORMAT, TÉL register; ~ **d'adresse d'instructions** INFORMAT control instruction register, control register, instruction address register, program counter; ~ **d'appels** TÉL call register (*CR*); ~ **de comparaison de résultats** GÉNÉRAL result-comparing register; ~ **de consignation** DONNÉES log; ~ **de**

conversations TÉL call register; **~ à décalage** COMMUT, DONNÉES, ÉLECTRON shift register; **~ à décalage à boucle fermée** COMMUT, DONNÉES, ÉLECTRON feedback shift register; **~ à deux états** TÉL two-state register; **~ double** DONNÉES, INFORMAT double-length register; **~ d'index** INFORMAT index register; **~ d'instructions** INFORMAT instruction register; **~ de maintien** ÉLECTRON holding register; **~ quadrature** DONNÉES, INFORMAT quadruple-length register; **~ de translation** TÉL translation register

réglable *adj* GÉNÉRAL, INFORMAT adjustable, settable; **~ par paliers** GÉNÉRAL adjustable in steps, step-adjustable; **~ en pas à pas** GÉNÉRAL adjustable in steps, step-adjustable

réglage[1]: **~ en continu** *adj* CONTRÔLE infinitely variable

réglage[2] *m* CONTRÔLE, ÉLECTRON regulation, GÉNÉRAL adjustment, setting, INFORMAT regulation, setting; **~ d'accord par inducteurs en parallèle** ÉLECTRON inductive parallel tuning; **~ d'alarmes** TESTS alarm setting; **~ approché** GÉNÉRAL coarse adjustment; **~ approximatif** GÉNÉRAL coarse adjustment, coarse setting; **~ du déclenchement du répondeur** PÉRIPH, TÉL answer machine start-up adjustment; **~ de la dynamique** CONTRÔLE volume range control; **~ à échelonnement** COMMUT sequence adjustment; **~ fin** ÉLECTRON fine tuning, GÉNÉRAL fine adjustment, INFORMAT fine adjustment, fine setting, fine tuning; **~ de la fréquence** RADIO frequency tuning; **~ de fréquence de sonnerie** PÉRIPH ringing-period adjustment; **~ grossier** GÉNÉRAL coarse adjustment, coarse setting; **~ initial** TRANSMIS line-up; **~ de luminosité** PÉRIPH brightness control; **~ manuel** CONTRÔLE manual regulation, manual setting; **~ médian** INFORMAT dead center (*AmE*), dead centre (*BrE*); **~ du niveau sonore** APP SERV, PÉRIPH sound-level control; **~ par défaut** INFORMAT default setting; **~ de phase** ÉLECTRON phase control; **~ de phase généralisé** TESTS generalized phase control; **~ précis** GÉNÉRAL fine adjustment, INFORMAT fine adjustment, fine setting, fine tuning; **~ du rythme** TRANSMIS timing selection; **~ silencieux** RADIO squelch control, muting, squelch; **~ en usine** INFORMAT factory setting; **~ du zéro** ÉLECTRON, TESTS zero adjustment, zero setting, zeroing, zeroization, zeroizing

réglages *m pl* INFORMAT settings

règle *f* INFORMAT rule, provision; **~ liée au conditionnement** COMPOSANTS construction-bound rule

règlement: **~ d'assurance qualité** *m* (*RAQ*) TESTS quality assurance ruling; **~ simplifié** *m* GÉNÉRAL easy payment

réglementation *f* CONTRÔLE, ÉLECTRON, INFORMAT regulation

régler *vt* GÉNÉRAL adjust, *de nouveau* readjust, RADIO set; **~ à zéro** ÉLECTRON, TESTS zero, zeroize

règles: **~ de codage de base** *f pl* (*BER*) DONNÉES, INFORMAT, RADIO basic encoding rules (*BER*)

réglet *m* TESTS steel rule

réglette *f* ALIM ÉLEC, COMPOSANTS strip, ÉLECTRON connector, TÉL block terminal; **~ à broches** GÉNÉRAL terminal field; **~ câblée** ÉLECTRON prewired connector; **~ de coupure** ALIM ÉLEC fuse strip; **~ de distribution** ALIM ÉLEC fanning strip; **~ de fiches** ALIM ÉLEC plug bar; **~ de fusibles** ALIM ÉLEC fuse strip; **~ isolante** ALIM ÉLEC insulating strip; **~ de lampes** ALIM ÉLEC lamp strip; **~ de pontage** ALIM ÉLEC, COMPOSANTS strapping block; **~ porte-étiquette** GÉNÉRAL designation strip, label strip; **~ de raccordement** ALIM ÉLEC connection field, COMPOSANTS, ÉLECTRON strip connector, GÉNÉRAL terminal field; **~ de raccordement murale** ALIM ÉLEC wall terminal strip; **~ de répartition** ALIM ÉLEC fanning strip; **~ de strapage** ALIM ÉLEC, COMPOSANTS strapping block; **~ à vis** ALIM ÉLEC screw terminal block; **~-jauge** TÉL measuring rod

regroupement: **~ conceptuel** *m* INFORMAT conceptual clustering

regrouper *m* INFORMAT regroup

régulateur *m* ALIM ÉLEC regulator, CONTRÔLE *commande automatique* controller, regulator, ÉLECTRON, INFORMAT, TÉL regulator; **~ de batterie** ALIM ÉLEC battery regulator; **~ de charge de batterie** ALIM ÉLEC battery control unit; **~ de débit** CONTRÔLE flow regulator; **~ de niveau commandé par le bruit** (*NOGAD*) ÉLECTRON noise-operated gain adjusting device (*NOGAD*); **~ de pression d'évaporation** COMPOSANTS *tubes électroniques* back-pressure valve; **~ de tension** ALIM ÉLEC, ÉLECTRON voltage regulator; **~ de vitesse** TÉL speed regulator

régulation *f* CONTRÔLE feedback control, regulation, ÉLECTRON regulation, INFORMAT feedback control, regulation; **~ avec action anticipée** CONTRÔLE feedforward control; **~ avec action prévisionnelle** CONTRÔLE feedforward control; **~ en cascade** CONTRÔLE cascade control; **~ de correspondance** CONTRÔLE follow-up control; **~ de débit** DONNÉES flow control; **~ de fréquence** TRANSMIS frequency regulation; **~ de maintien** CONTRÔLE *commande automatique* control with fixed set-point; **~ par la fréquence** ALIM ÉLEC frequency control, frequency regulation; **~ de phase** ÉLECTRON phase regulation; **~ de précision** GÉNÉRAL fine positioning, fine regulation; **~ de la puissance d'émission** RADIO up-path power control; **~ de la température** CLIMAT temperature control; **~ du trafic** DONNÉES, TÉL traffic control

réimporter *vt* DONNÉES reimport

réinitialisation *f* DONNÉES, INFORMAT reinitialization, resetting, RADIO resetting

réinitialiser *vt* DONNÉES, INFORMAT reinitialize, reset, RADIO reset

réjection: **~ de la fréquence image** *f* RADIO image frequency rejection

rejet *m* TESTS nonacceptance, rejection; **~ à fréquence intermédiaire** RADIO intermediate frequency rejection, IF rejection; **~ sélectif** TÉL selective reject; **~ de trame** TÉL frame reject (*FRMR*); **~ de transaction d'entrée** DONNÉES input transaction rejected (*ITR*)

rejeter *vt* INFORMAT, TESTS reject

rejoindre *vt* ÉLECTRON butt

relâché *adj* COMPOSANTS released, slack, INFORMAT released

relâchement *m* COMPOSANTS, INFORMAT release; **~ de la touche** INFORMAT key-off, key release

relâcher *vt* COMPOSANTS, INFORMAT release; **~ en shuntant** COMPOSANTS release by shunting

relais *m* GÉNÉRAL relay; **~ d'alimentation en courant** ALIM ÉLEC current-feed relay; **~ d'alimentation en énergie** ALIM ÉLEC current-feed relay; **~ d'arrêt de sonnerie** TÉL ring-trip relay; **~ de cellules** TRANSMIS cell relay; **~ commun** RADIO trunk system; **~ de commutation d'antennes** RADIO antenna switching relay; **~ de commutation à tige** COMMUT coaxial reed relay changeover switch; **~ comparateur de phases** COMPOSANTS phase comparator relay; **~ de comptage** TÉL accounting relay; **~ à contact de mercure** ÉLECTRON mercury contact relay; **~ à contact sous vide** ALIM ÉLEC, COMPOSANTS reed relay; **~ à contacts mouillés au mercure** ÉLECTRON mercury-wetted relay, mercury-wetted contact relay; **~ de couche réseau** DONNÉES, TÉL, TRANSMIS network-layer relay (*NLR*); **~ de coupure** COMPOSANTS, ÉLECTRON break relay; **~ de coupure d'alarme** TÉL alarm cut-off relay; **~ de courant** ALIM ÉLEC current relay; **~ de courant fermé** ALIM ÉLEC closed-circuit relay; **~ de défaut à terre** COMPOSANTS, ÉLECTRON earth-fault relay (*BrE*), ground-fault relay (*AmE*); **~ à deux temps** COMPOSANTS two-step relay, two-stage relay; **~ différentiel** COMPOSANTS differential relay; **~ directionnel** COMPOSANTS directional relay; **~ de distribution** COMPOSANTS distribution relay; **~ diviseur d'impulsions** COMPOSANTS pulse-bisecting relay, bisecting relay; **~ des dizaines** COMPOSANTS, TÉL tens relay; **~ à double armature** COMPOSANTS double-armature relay; **~ électromagnétique** COMPOSANTS electromagnetic relay; **~ électromécanique** COMPOSANTS electromechanical relay;

~ électrostatique COMPOSANTS electrostatic relay; **~ d'émetteur** TRANSMIS transmitter relay; **~ d'émission** COMPOSANTS transmitter relay; **~ encapsulé** COMPOSANTS encapsulated relay, encased relay; **~ à enclenchement** ALIM ÉLEC latching relay; **~ enfichable** COMPOSANTS plug-in relay; **~ d'essai** COMPOSANTS, TESTS test relay; **~ de fin de train d'impulsions** COMPOSANTS, TÉL, TRANSMIS pulse-train relay; **~ d'impulsions** COMPOSANTS pulse relay; **~ indicateur de signal** SIGNAL, TESTS signal indicator relay; **~ à induction** ALIM ÉLEC induction relay; **~ intégrateur d'impulsions** COMPOSANTS notching relay, stepping relay; **~ à lames résonnantes** ALIM ÉLEC resonant-reed relay; **~ à lames souples** ALIM ÉLEC, COMPOSANTS reed relay; **~ à maximum** COMPOSANTS overcurrent relay, overpower relay, overvoltage relay; **~ à mercure** ÉLECTRON mercury relay; **~ à minimum** COMPOSANTS underpower relay, undervoltage relay; **~ à minimum de courant** COMPOSANTS undercurrent relay; **~ N** DONNÉES, TÉL N-relay; **~ nominal** RADIO mobile-to-base relay; **~ non-polarisé** ALIM ÉLEC, COMPOSANTS nonpolarized relay; **~ pas à pas** COMPOSANTS notching relay, stepping relay; **~ de passage** RADIO base-to-mobile relay; **~ passif** RADIO radar relay; **~ polarisé** ALIM ÉLEC, COMPOSANTS polarized relay, polar relay; **~ à pourcentage** COMPOSANTS percentage relay; **~ primaire** COMPOSANTS primary relay; **~ de prise** COMPOSANTS, TÉL seizure relay; **~ de protection** COMPOSANTS protection relay, protective relay; **~ de puissance** ALIM ÉLEC, COMPOSANTS power relay; **~ de radiocommunication** ALIM ÉLEC, RADIO radiocommunication relay; **~ de réactance** COMPOSANTS reactance relay; **~ de réception d'impulsions** COMPOSANTS pulse-receiving relay; **~ de réglage** COMPOSANTS regulation relay; **~ à rémanence** ALIM ÉLEC remanent relay; **~ de remise au repos** COMPOSANTS resetting relay, restoring relay; **~ réseau** DONNÉES network relay; **~ de ressort à fil** COMMUT wire-spring relay; **~ à retardement** COMPOSANTS, ÉLECTRON time-delay relay; **~ de retour** COMPOSANTS turning relay; **~ à retour de courant** ÉLECTRON reverse-current relay; **~ de route** TÉL route relay; **~ secondaire** COMPOSANTS secondary relay; **~ à semi-conducteurs** ALIM ÉLEC, COMPOSANTS, ÉLECTRON solid-state relay; **~ statique** ALIM ÉLEC, COMPOSANTS, ÉLECTRON solid-state relay, static relay; **~ sur shunt** COMPOSANTS shunt relay; **~ de surveillance** COMPOSANTS, TÉL supervisory relay; **~ à taux de variation** COMPOSANTS rate-of-change relay; **~ de téléalimentation** ALIM ÉLEC; **~ télégraphique** COMPOSANTS telegraphic relay; **~ temporisé** COMPOSANTS, ÉLECTRON time-delay relay, timer relay, timing relay; **~ à temps non-spécifié** COMPOSANTS nonspecified time relay; **~ de tension** ALIM ÉLEC, COMPOSANTS voltage relay; **~ de test** COMPOSANTS, TESTS test relay; **~ de traduction** COMPOSANTS, TÉL translation relay; **~ des trames** TRANSMIS frame relay; **~ à trois bobines** COMPOSANTS three-coil relay; **~ à zéro de tension** ALIM ÉLEC, COMPOSANTS zero-voltage relay

relance *f* COMMUT restart, DONNÉES reboot, GÉNÉRAL follow-up, restart, INFORMAT reboot, restart, RADIO restart

relancement *m* COMMUT restart, DONNÉES reboot, GÉNÉRAL follow-up, restart, INFORMAT reboot, restart, RADIO restart

relancer *vt* COMMUT restart, DONNÉES reboot, GÉNÉRAL follow-up, restart, INFORMAT reboot, restart, RADIO restart

relation *f* GÉNÉRAL relation, relationship, TÉL telex relation, relation; **~ homme-machine** (*RHM*) INFORMAT man-machine communications (*MMC*); **~ linéaire** GÉNÉRAL linear relation, linear relationship; **~ logique entre couches analogues** TRANSMIS peer-to-peer logical relationship; **~ sémaphore** SIGNAL signaling relation (*AmE*), signalling relation (*BrE*); **~ de service** DONNÉES service relationship; **~ source/collecteur** TÉL source/sink relationship; **~ téléphonique** TÉL telephone relation; **~ télex** TÉL telex relation; **~ temporelle** TRANSMIS timing relationship

relations: **~ de groupe** *f pl* DONNÉES group relationships; **~ homologues** *f pl* DONNÉES peer relationships; **~ publiques** *f pl* TESTS public relations (*PR*); **~ de sauvegarde** *f pl* DONNÉES backup relationships; **~ de substitution** *f pl* DONNÉES fall-back relationships

relevé *m* TÉL reading; **~ des communications** TÉL call logging; **~ du compteur** TÉL meter indication, meter-value indication; **~ détaillé** TÉL itemized statement; **~ radioélectrique** RADIO radio survey

relève *f* GÉNÉRAL takeover; **~ de dérangements** ÉLECTRON troubleshooting, GÉNÉRAL fault clearance, INFORMAT troubleshooting, TÉL fault tracing, troubleshooting

relèvement: **~ radio** *m* RADIO observed bearing; **~ radiogonométrique** *m* RADIO direction finding; **~ d'urgence** *m* TÉL emergency attention; **~ vrai** *m* RADIO *radar* true bearing

relever *vt* TÉL, TESTS clear; **~ un dérangement** DONNÉES *trouver* trace a fault, GÉNÉRAL, TÉL *réparer* clear a fault, *trouver* trace a fault, TESTS *réparer* clear a fault, *trouver* trace a fault

releveur: **~ de coordonnées** *m* INFORMAT locator

relier[1] *vt* ALIM ÉLEC, COMPOSANTS wire, DONNÉES link up, GÉNÉRAL interconnect, INFORMAT link, TÉL connect, TRANSMIS link up

relier[2]: **se ~** *v réfl* DONNÉES, TRANSMIS link up

relisez *m* TÉL read back

réluctance *f* (*R*) ALIM ÉLEC reluctance (*R*)

rémanence *f* ALIM ÉLEC remanence, ÉLECTRON retentivity, remanence, retention

remaniement: **~ des liaisons pendant la communication** *m* TÉL in-call rearrangement

remboursement: **~ complet** *m* TÉL full refund; **~ partiel** *m* TÉL partial refund

remède *m* GÉNÉRAL corrective measure

remettre: **~ à l'indicateur d'interruption** *vt* ÉLECTRON raise an interrupt; **~ au point** *vt* GÉNÉRAL realign, TESTS readjust, realign; **~ à zéro** *vt* ÉLECTRON, TESTS adjust to zero, reset to zero, zero, zeroize

remise: **~ différée** *f* TÉL delayed delivery; **~ en fonctionnement d'urgence** *f* TÉL emergency restart; **~ en forme** *f* ÉLECTRON reshaping; **~ à niveau** *f* ÉLECTRON, GÉNÉRAL, RADIO, TESTS reset; **~ physique** *f* DONNÉES, TÉL physical delivery (*PD*); **~ au point** *f* GÉNÉRAL, TESTS realignment; **~ en position initiale** *f* GÉNÉRAL reset, resetting; **~ à zéro** *f* ÉLECTRON zeroization, reset, zeroing, zeroizing, *d'une bande bascule* resetting, GÉNÉRAL, RADIO reset, TESTS reset, zeroing, zeroization, zeroizing; **~ à zéro de la mémoire** *f* INFORMAT core flush

remontée: **~ d'informations** *f* GÉNÉRAL feedback; **~ de pertinence** *f* DONNÉES relevance feedback

remorque: **~ pour touret de câble** *f* TRANSMIS *pose de câbles* cable trailer

remplacer *vt* INFORMAT replace

remplir *vt* DONNÉES, GÉNÉRAL, INFORMAT fill, pad; **~ de zéros** INFORMAT zero-fill

remplissage *m* DONNÉES, GÉNÉRAL, INFORMAT filler, filling, padding; **~ numérique** DONNÉES digital filling

rémunération: **~ pour utilisation en commun des circuits** *f* TÉL remuneration for shared use of circuits; **~ pour utilisation exclusive des circuits** *f* TÉL remuneration for exclusive use of circuits

rendement *m* GÉNÉRAL efficiency, RADIO *d'une antenne* radiation efficiency; **~ de couplage** ÉLECTRON coupling efficiency; **~ de l'émetteur** COMPOSANTS emitter efficiency, ÉLECTRON injection ratio; **~ énergétique** ALIM ÉLEC, TRANSMIS radiant efficiency, source power efficiency; **~ d'injection** COMPOSANTS emitter efficiency, ÉLECTRON injection ratio; **~ lumineux** TESTS light output; **~ quantique** ÉLECTRON, TRANSMIS quantum efficiency; **~ quantique différentiel** TESTS differential quantum efficiency; **~ de réponse** RADIO transponder reply efficiency;

~ **spectral** TRANSMIS spectral efficiency, spectrum efficiency; ~ **d'un tube** ÉLECTRON, TESTS tube efficiency

rendre *vt* INFORMAT render; ~ **compte** GÉNÉRAL report; ~ **égal** INFORMAT equate; ~ **inopérant** GÉNÉRAL disable, put out of order; ~ **inutilisable** GÉNÉRAL disable, put out of order; ~ **sûr** INFORMAT ensure

renforcement *m* CONTRÔLE, RADIO, TESTS enhancement, reinforcement, tightening; ~ **à gauche** ÉLECTRON left side of the front panel

renforcer *vt* CONTRÔLE, RADIO, TESTS tighten, enhance, reinforce

renfort *m* COMPOSANTS strain-relief mechanism, ÉLECTRON strength member; ~ **en kevlar** ÉLECTRON kevlar strength

renommer *vt* INFORMAT rename

renouveler *vt* GÉNÉRAL review; ~ **régulièrement** TESTS replace regularly

renouvellement: ~ **régulier** *m* TESTS regular replacement

renseignement *m* GÉNÉRAL information

renseignements: ~ **tarifaires** *m pl* TÉL billing information; ~ **téléphoniques** *m pl* APP SERV, TÉL directory assistance, directory enquiries (*BrE*) (*DQ*), directory information (*AmE*)

renvoi *m* GÉNÉRAL, TÉL forwarding, diversion; ~ **d'appel** TÉL call diversion, call forwarding; ~ **d'appel sur non-réponse** TÉL call diversion on no reply, diversion on no reply; ~ **d'appel sur occupation** TÉL call diversion on busy, diversion on busy; ~ **d'appel systématique** TÉL unconditional call forwarding; ~ **d'appels à téléprogrammations successives** TÉL follow-me call forward; ~ **automatique** SIGNAL automatic call transfer, TÉL call forward, call forwarding; ~ **automatique des appels** TÉL call forward, call forwarding; ~ **automatique intragroupe** TÉL call forward intra-group; ~ **automatique pas-de-réponse** TÉL call forward no answer, call forwarding no answer; ~ **automatique sur non-réponse** TÉL call forward no answer, call forwarding no answer; ~ **automatique sur occupation** TÉL call forward busy, call forwarding busy; ~ **automatique tout appel** TÉL call forward all calls, call forwarding all calls; ~ **automatique vers l'extérieur** TÉL call forward remote, call forwarding remote; ~ **en écho** DONNÉES echo return; ~ **de nuit** COMMUT night-service connection, night-service mode; ~ **au standardiste** TÉL attendant recall; ~ **systématique** TÉL unconditional call forwarding; ~ **temporaire** TÉL temporary transfer; ~ **temporaire d'appel** TÉL temporary call forwarding; ~ **temporaire de l'installation** TÉL installation call forwarding; ~ **temporaire de l'ITA** TÉL installation call forwarding; ~ **temporisé** RADIO holding call transfer; ~ **du terminal** APP SERV terminal call forwarding, TÉL call deflection; ~ **variable** TÉL diversion follow me, follow-me diversion

renvoyer *vt* GÉNÉRAL, TÉL divert, forward

repaginer *vt* INFORMAT repaginate

réparateur *m* GÉNÉRAL repairer, repairperson, TÉL engineer, repairer, repairperson

réparation *f* GÉNÉRAL, INFORMAT, TÉL repair, TESTS fault correction; ~ **par échange standard** TÉL swap repair; ~ **temporaire** TRANSMIS make good

réparer *vt* GÉNÉRAL fix

réparti *adj* ÉLECTRON fanned-out, GÉNÉRAL distributed, INFORMAT distributed, dispatched

répartir *vt* ÉLECTRON fan out, GÉNÉRAL distribute, INFORMAT dispatch, distribute

répartiteur *m* INFORMAT dispatcher, RADIO patch board, patch panel, TÉL cross-connecting unit, frame, patch board, patch panel, TRANSMIS distribution frame; ~ **coaxial** (*RC*) TV coaxial cable distributor; ~ **à double face** TRANSMIS double-sided distribution frame; ~ **de groupe primaire** TÉL, TRANSMIS group distribution frame; ~ **de groupe secondaire** TÉL, TRANSMIS supergroup distribution frame; ~ **de hautes fréquences** TRANSMIS high-frequency distribution frame; ~ **intermédiaire** (*RI*) COMMUT, TÉL intermediate distribution frame (*IDF*); ~ **de jonction** COMMUT, TRANSMIS junction distribution frame (*JDF*), trunk distribution frame (*TDF*); ~ **mixte** COMMUT combined distribution frame (*CDF*); ~ **monoface** COMMUT single-sided distribution frame; ~ **numérique** COMMUT digital cross-connect (*DXC*), digital distribution frame; ~ **optique** TRANSMIS optical-fiber frame (*AmE*) (*OFF*), optical-fibre frame (*BrE*) (*OFF*); ~ **principal** (*RP*) TÉL, TRANSMIS main distribution frame (*MDF*), mainframe; ~ **principal de station de répéteurs** TRANSMIS main repeater distribution frame; ~ **de puissance** ALIM ÉLEC, RADIO power divider

répartiteurs: ~ **optiques** *m pl* TRANSMIS optical control boards

répartition[1]: **à** ~ **dans le temps** *adj* TRANSMIS time-divided

répartition[2] *f* ÉLECTRON fanning, GÉNÉRAL distribution; ~ **adaptative du bruit** TRANSMIS adaptive noise allocation; ~ **du bruit** TRANSMIS noise allocation; ~ **de charge** ALIM ÉLEC load distribution, ÉLECTRON load sharing; ~ **en code** RADIO, TÉL code division; ~ **de défaillance** TÉL failure distribution; ~ **des durées d'occupation** TÉL holding-time distribution; ~ **des erreurs** TÉL error spread; ~ **dans l'espace** COMMUT, TÉL space division; ~ **exponentielle** GÉNÉRAL exponential distribution; ~ **en fréquence** RADIO, TÉL, TRANSMIS frequency division, code division; ~ **fréquentielle** RADIO, TRANSMIS frequency division; ~ **gaussienne** GÉNÉRAL Gaussian distribution; ~ **de mode à équilibre** TRANSMIS steady-state condition; ~ **des modes** TRANSMIS mode distribution; ~ **des modes à l'équilibre** ÉLECTRON steady-state condition, TRANSMIS equilibrium-mode distribution; ~ **des modes hors équilibre** TRANSMIS nonequilibrium-mode distribution; ~ **normale** GÉNÉRAL normal distribution; ~ **de probabilité des amplitudes** (*RPA*) SIGNAL amplitude probability distribution (*APD*); ~ **spatiale** COMMUT, TÉL space division; ~ **spectrale** TRANSMIS spectral distribution; ~ **des tâches** COMMUT job division; ~ **temporelle** COMMUT, RADIO, SIGNAL, TÉL, TRANSMIS time division; ~ **de temps** COMMUT, RADIO, SIGNAL, TÉL, TRANSMIS time division; ~ **du trafic** COMMUT, TÉL distribution of traffic, traffic distribution

repasser *vt* INFORMAT play, TÉL play back

repérage *m* ÉLECTRON labeling (*AmE*), labelling (*BrE*), GÉNÉRAL designation, labeling (*AmE*), labelling (*BrE*), RADIO, TESTS locating; ~ **automatique des déplacements** RADIO automatic roaming; ~ **des câbles** TRANSMIS cable localization; ~ **de l'information** DONNÉES, INFORMAT information retrieval; ~ **de l'information en ligne** DONNÉES on-line information retrieval; ~ **par radio** RADIO radio determination; ~ **de voies** RADIO channel marking; ~ **des voies au repos** COMMUT marked idle channel (*MIC*)

repère *m* INFORMAT pointer; ~ **de comparaison** INFORMAT compare mark; ~ **de distance** RADIO range marker; ~ **de fin** INFORMAT end mark; ~ **de graduation** TESTS mark, scale mark; ~ **de position de morceau de musique** INFORMAT song-position pointer; ~ **de recherche** INFORMAT search mark; ~ **de temps** RADIO time marker

repérer *vt* GÉNÉRAL designate, RADIO locate

répertoire *m* DONNÉES, INFORMAT directory, TÉL book, directory, repertoire; ~ **d'adresses** DONNÉES bookmark; ~ **des caractères graphiques télétex** TÉL teletex graphic character index; ~ **des caractères graphiques télétex de base** TÉL teletex basic graphic character index; ~ **des caractères télétex** TÉL teletex character index; ~ **de chaînes** INFORMAT chain directory; ~ **de code d'instruction** INFORMAT instruction repertoire; ~ **des fonctions de commande télétex** CONTRÔLE, TÉL teletex control function index; ~ **des fonctions de commande télétex de base** TÉL teletex basic control function index;

~ **racine** INFORMAT root directory; ~ **téléphonique** TÉL telephone directory, telephone book

répéter *vt* GÉNÉRAL, INFORMAT repeat

répéteur *m* GÉNÉRAL repeater, transponder; ~ **analogique** TÉL analog repeater; ~ **bidirectionnel** TRANSMIS two-wire repeater, two-way repeater; ~ **pour circuit à deux fils** TRANSMIS two-wire repeater, two-way repeater; ~ **correcteur d'impulsions** SIGNAL pulse-correcting register; ~ **en cuivre** ÉLECTRON shielded copper separator; ~ **gérable** TRANSMIS manageable repeater; ~ **immergé** TRANSMIS submerged repeater; ~ **d'impulsions** TÉL regenerative pulse repeater, pulse repeater; ~ **intermédiaire** TRANSMIS intermediate repeater; ~ **de liaison de transmission par impulsions** TÉL pulse link repeater (*PLR*); ~ **de ligne** TÉL line repeater; ~ **multipoint** TRANSMIS multipoint repeater; ~ **multiport Ethernet thin-wire** DONNÉES thin-wire Ethernet multiport repeater, ÉLECTRON DEMPR, thin-wire Ethernet multiport repeater; ~ **optique** TRANSMIS optical repeater; ~ **optique immergé** TRANSMIS submerged optical repeater; ~ **protégé en éclipse** RADIO eclipse-working transponder; ~ **à quatre fils** TRANSMIS four-wire repeater; ~ **régénérateur** TRANSMIS regenerative repeater; ~ **régénérateur optique** TRANSMIS optical repeater, optical regenerator, optical regenerative repeater; ~ **de satellite** RADIO satellite transponder; ~ **en service** RADIO active transponder; ~ **transparent** GÉNÉRAL transparent repeater

répétibilité *f* GÉNÉRAL repeatability

répétiteur: ~ **d'impulsions** *m* GÉNÉRAL pulse repeater

répétition *f* GÉNÉRAL repeat, repetition; ~ **automatique** TÉL automatic repetition; ~ **automatique du dernier numéro** PÉRIPH, TÉL last number called redial, last number dialed recall (*AmE*), last number dialled recall (*BrE*), last number redial; ~ **automatique des numéros** TÉL number redial; ~ **automatique des numéros choisis** TÉL selected number redial; ~ **automatique de tentative** TÉL automatic repeat attempt; ~ **automatique Verdan** TÉL automatic repetition; ~ **du dernier numéro** TÉL last number called redial, last number redial; ~ **des impulsions** ÉLECTRON, SIGNAL, TÉL, TRANSMIS pulse repetition; ~ **d'office** TÉL routine repetition; ~ **du réveil** PÉRIPH snooze function; ~ **sans fin** INFORMAT endless repeat; ~ **de transmission** TÉL rerun

rephasage *m* RADIO referencing

repli: ~ **automatique** *m* DONNÉES automatic fall-back

répondeur *m* DONNÉES, ÉLECTRON transponder, PÉRIPH recorded announcement machine, telephone answering machine, RADIO transponder, TÉL announcement machine, recorded announcement machine, telephone answering machine; ~ **d'association** DONNÉES association responder; ~ **de code** TÉL code-answering device, code-answering unit; ~ **pour contrôle de continuité** TÉL continuity check responder; ~ **enregistreur** TÉL answer-recording unit; ~ **interrogeable à distance** TÉL remote-access answering machine; ~ **à sélection d'appels** TÉL call-selector responder; ~ **simple et à interrogation à distance** TÉL answering device with voice synthesizer; ~ **téléphonique** PÉRIPH, TÉL announcement machine, recorded announcement machine, telephone answering machine

répondre[1]: ~ **à** *vt* TÉL *appel, téléphone* answer

répondre[2] *vi* DONNÉES answer, GÉNÉRAL answer, respond

réponse *f* DONNÉES *datacom* answering, GÉNÉRAL response, answer; ~ **en amplitude** ÉLECTRON, SIGNAL, TRANSMIS amplitude response, square-wave response; ~ **amplitude-fréquence** TRANSMIS amplitude-frequency response; ~ **automatique** DONNÉES *datacom*, TÉL automatic answering; ~ **en bande de base** TÉL baseband response; ~ **brouillée** RADIO garbled reply; ~ **de câble** TRANSMIS cable response; ~ **de clavier** INFORMAT keyboard response; ~ **codée** TÉL code answer; ~ **de comparateur** ÉLECTRON comparator response; ~ **pour continuer à corriger** (*CTR*) DONNÉES response to continue to correct (*CTR*); ~ **dynamique** GÉNÉRAL dynamic response; ~ **à l'échelon** TÉL step-function response; ~ **à un essai** TESTS test answer; ~ **pour fin de transmission** (*ERR*) DONNÉES, TÉL response for end of transmission (*ERR*); ~ **à la fonction échelon** TÉL step-function response; ~ **à la fonction rampe** CONTRÔLE ramp response; ~ **en fréquence** RADIO, TRANSMIS frequency response; ~ **impulsionnelle** CONTRÔLE unit impulse response, RADIO pulse response, TÉL impulse response, pulse response; ~ **impulsionnelle finie** (*RIF*) ÉLECTRON, TRANSMIS finite impulse response (*FIR*); ~ **impulsionnelle infinie** (*RII*) ÉLECTRON, TRANSMIS infinite impulse response (*IIR*); ~ **indicielle** CONTRÔLE step response; ~ **inintelligible** RADIO garbled reply; ~ **mains libres** TÉL hands-free answerback; ~ **manipulée** RADIO chopped response; ~ **manuelle** TÉL manual answering; ~ **neutre** ÉLECTRON flat response; ~ **parasite** RADIO spurious response; ~ **percussionnelle** RADIO pulse response, TÉL Dirac pulse response, impulse response, pulse response; ~ **en phase** TRANSMIS phase response; ~ **à la pression après enfoncement** INFORMAT after-touch response; ~ **primitive** DONNÉES primitive response, response primitive; ~ **sensitive** INFORMAT, PÉRIPH touch response; ~ **en signaux carrés** ÉLECTRON, SIGNAL, TRANSMIS square-wave response; ~ **temporelle** CONTRÔLE time response; ~ **à un test** TESTS test answer; ~ **uniforme des fréquences** ÉLECTRON flat frequency characteristic; ~ **vocale** TÉL voice response; ~ **vocale interactive** TÉL interactive voice response (*IVR*)

report: ~ **accéléré** *m* INFORMAT high-speed carry; ~ **anticipé** *m* ÉLECTRON look-ahead carry; ~ **circulaire** *m* INFORMAT end-around carry; ~ **négatif circulaire** *m* INFORMAT end-around borrow; ~ **à plat** *m* ÉLECTRON surface mounting; ~ **simultané** *m* INFORMAT high-speed carry

reportage: ~ **d'actualités** *m* RADIO, TV news gathering; ~ **électronique d'actualités** *m* GÉNÉRAL, RADIO, TV electronic news gathering (*ENG*); ~ **électronique d'actualités par satellite** *m* RADIO, TV satellite news gathering (*SNG*); ~ **électronique ENG** *m* GÉNÉRAL, RADIO, TV electronic news gathering (*ENG*); ~ **extérieur** *m* RADIO, TV outside broadcast; ~ **radiophonique** *m* RADIO audio news gathering; ~ **par satellite** *m* RADIO, TV satellite news gathering (*SNG*)

reporté: ~ **à plat** *adj* ÉLECTRON surface-mounted

repos[1]: **au** ~ *adj* ALIM ÉLEC, ÉLECTRON, TÉL, TRANSMIS idle

repos[2] *m* ALIM ÉLEC, ÉLECTRON, TÉL, TRANSMIS idle condition, rest state

repose: ~**-document** *m* DONNÉES document rest

reposer: ~ **sur** *vt* COMPOSANTS *génie mécanique* bear against, bear on, lie against

repositionnement: ~ **de transaction** *m* DONNÉES transaction roll-back

repositionner *vt* RADIO reset

reprendre *vt* INFORMAT repeat

représentation: ~ **analogique** *f* INFORMAT locater; ~ **codée** *f* INFORMAT coded representation; ~ **multifilaire** *f* GÉNÉRAL multiline representation; ~ **par limites** *f* INFORMAT boundary representation; ~ **par volumes** *f* INFORMAT bounding volumes; ~ **schématique** *f* GÉNÉRAL schematic representation; ~ **unifilaire** *f* GÉNÉRAL single-line representation; ~ **en virgule flottante** *f* INFORMAT floating-point representation

reprise *f* DONNÉES recovery, GÉNÉRAL takeover, INFORMAT rerun, restart, repeat, RADIO restart, recovery, TÉL call retrieval, *d'un téléphone* reconnection; ~ **après une catastrophe** DONNÉES disaster recovery; ~ **de dialogue** DONNÉES dialog recovery; ~ **de transmission** TÉL put-back

repriser *vt* DONNÉES recover, GÉNÉRAL take over, INFORMAT rerun, restart, repeat, RADIO restart, recover, TÉL reconnect, retrieve

reproducteur: ~ **de bande** *m* INFORMAT tape reproducer

reproductibilité *f* GÉNÉRAL reproducibility

reproductible *adj* GÉNÉRAL, TESTS reproducible
reproduction *f* GÉNÉRAL duplication, reproduction, INFORMAT playback, TESTS reproduction; ~ **par déclenchement** INFORMAT trigger play; ~ **rapide** INFORMAT quick playback
reproduire *vt* GÉNÉRAL reproduce, INFORMAT play, TESTS reproduce
reprographie *f* PÉRIPH copying device
répudiation *f* DONNÉES repudiation
requantification *f* INFORMAT requantization
requête *f* DONNÉES, INFORMAT, RADIO, TÉL request (*req*)
rerayonnement *m* RADIO secondary radiation, reradiation
rerouter *vt* COMMUT, DONNÉES redirect, reroute, TÉL *trafic* redirect, reroute; ~ **le trafic** COMMUT, TÉL redirect traffic
réseau[1]: **en** ~ *adj* DONNÉES, GÉNÉRAL networked
réseau[2] *m* ALIM ÉLEC mains, DONNÉES network, Net (*Internet*), array, ÉLECTRON lattice, GÉNÉRAL network, INFORMAT network, net, Net (*Internet*), TRANSMIS grating, backbone; ~ **d'accès** GÉNÉRAL access network; ~ **à accords décalés** ÉLECTRON stagger-tuned circuit; ~ **d'adaptation** ÉLECTRON matching network; ~ **aérien** ALIM ÉLEC, TRANSMIS overhead network; ~ **d'aiguillage** TRANSMIS combining network; ~ **en anneau** DONNÉES, TRANSMIS ring network; ~ **en anneau double arborescent** ÉLECTRON dual ring of trees; ~ **en anneau à passage jeton** DONNÉES, INFORMAT, TRANSMIS token ring network; ~ **arborescent** DONNÉES, INFORMAT tree and branch network, TRANSMIS tree and branch network, tree network; ~ **en arbre** DONNÉES, INFORMAT tree and branch network, TRANSMIS tree and branch network, tree network; ~ **asynchrone** TÉL asynchronous network; ~ **atténuateur** TRANSMIS pad; ~ **de base** TRANSMIS backbone network; ~ **bidirectionnel** TV bidirectional network; ~ **à blocage** COMMUT blocking network; ~ **de bornes relais** RADIO relay network; ~ **en boucle** TRANSMIS loop; ~ **de branchement** TV delivery network; ~ **en bus** TÉL bus network; ~ **en câble** TRANSMIS cable network; ~ **câblé** TV cable network; ~ **de câbles** TRANSMIS cable network; ~ **cellulaire** DONNÉES, RADIO, TÉL cellular network; ~ **cellulaire téléphonique mobile** RADIO, TÉL cellular mobile telephone network; ~ **classique** TRANSMIS ordinary network; ~ **client** TRANSMIS customer network (*CN*); ~ **de Clos** TÉL, TRANSMIS Clos network; ~ **de Clos imparfait** TÉL, TRANSMIS sub-Clos network; ~ **coaxial** TRANSMIS coaxial network; ~ **à commande de phase** RADIO multibeam phased array (*MPA*); ~ **communautaire** TV community network; ~ **de communication personnalisé** DONNÉES *traitement des données à distance*, RADIO personal communication network (*PCN*); ~ **de communication personnelle** DONNÉES *traitement des données à distance*, RADIO personal communication network (*PCN*); ~ **de commutation** COMMUT, TÉL switching network; ~ **à commutation de circuits** COMMUT circuit-switching network, TRANSMIS circuit-switched network; ~ **à commutation et transmission intégrées** TRANSMIS integrated transmission and switching network (*IST network*); ~ **à commutation de messages** COMMUT, SIGNAL store-and-forward switching network, TRANSMIS message-switching network; ~ **à commutation de paquets** COMMUT, TRANSMIS packet-switch network (*PSN*); ~ **commuté** COMMUT, TRANSMIS, TV switched network; ~ **de compensation** RADIO compensating network; ~ **de concentration** TRANSMIS concentration network; ~ **de conducteurs** ÉLECTRON conductive pattern; ~ **conformateur de faisceau** ÉLECTRON, RADIO beam-forming network; ~ **conformateur d'impulsions** (*RCI*) ÉLECTRON pulse-forming network (*PFN*); ~ **de connexion** COMMUT switching network, switching network complex, TÉL switching network; ~ **de connexion d'abonnés** COMMUT, TÉL subscriber switching network; ~ **de connexion à bande étroite** COMMUT, TRANSMIS narrow-band switching network; ~ **de connexion à large bande** COMMUT, TRANSMIS broadband switching network, wide-band switching network; ~ **de connexion non-bloquant** COMMUT, TRANSMIS nonblocking switching network; ~ **de connexion numérique** COMMUT digital switching network; ~ **de connexion optique** COMMUT, TRANSMIS optical switching network (*OSN*); ~ **de connexion sans blocage** COMMUT nonblocking switching network; ~ **de connexion temporelle** COMMUT, TRANSMIS time-division network; ~ **de coordination et de contrôle du système** TESTS system coordination and control network (*SCCN*); ~ **correcteur** ÉLECTRON equalizing network; ~ **correcteur de distorsion** TÉL signal-shaping network; ~ **cryptophonique** DONNÉES key net; ~ **décentralisé** INFORMAT distributed network; ~ **démocratique** TÉL *à synchronisation mutuelle* democratic network; ~ **dépliable** RADIO fold-out array; ~ **de désaccentuation** ÉLECTRON de-emphasis network; ~ **despotique** TÉL despotic network; ~ **de destination** TRANSMIS destination network (*DN*); ~ **à deux accès** ÉLECTRON, TRANSMIS two-terminal network; ~ **de diffraction** CLIMAT, TRANSMIS diffraction grating; ~ **à diffusion** RADIO broadcast network; ~ **de diodes** ÉLECTRON diode array; ~ **distribué** TV distributed network; ~ **de distribution** COMMUT distribution network, TÉL, TRANSMIS junction network, TV distribution network; ~ **de distribution de données pour groupes fermés d'usagers** DONNÉES, TRANSMIS closed user data-distribution network; ~ **domestique** DONNÉES domestic network; ~ **de données** DONNÉES, TRANSMIS data network; ~ **de données public avec commutation par paquets** (*RDPCP*) DONNÉES public packet-switched data network (*PPSDN*); ~ **de données synchrone** TÉL synchronous data network; ~ **dorsal** TRANSMIS backbone network; ~ **égaliseur** TRANSMIS equalizer circuit; ~ **électrique** ALIM ÉLEC household electricity; ~ **d'énergie électrique** ALIM ÉLEC electrical power system, electricity supply system; ~ **d'entreprise** TRANSMIS company network, enterprise network; ~ **équilibré** DONNÉES balanced network; ~ **à étages multiples** ÉLECTRON multistage network; ~ **en étoile** ALIM ÉLEC, TRANSMIS, TV star network; ~ **étoilé** ALIM ÉLEC, TRANSMIS, TV star network; ~ **en étoile commuté** ALIM ÉLEC, COMMUT, TRANSMIS, TV switched-star network; ~ **d'expansion** TRANSMIS expansion network; ~ **extérieur** TRANSMIS overseas network; ~ **à faible trafic** TRANSMIS thin-route network; ~ **fédérateur** INFORMAT backbone network; ~ **à fente** RADIO slot array; ~ **de fentes rayonnantes** RADIO slot array; ~ **filaire** RADIO, TÉL wired network, TRANSMIS fixed network, hard-wired network; ~ **de files d'attente** TÉL queueing network; ~ **G+** TV multiple-trunk distribution system; ~ **de gestion** TÉL, TRANSMIS management network; ~ **de gestion SDH** TRANSMIS SDH management network (*SMN*); ~ **de gestion des télécommunications** (*RGT*) DONNÉES, TÉL, TRANSMIS telecommunications management network (*TMN*); ~ **à grande distance** TÉL long-distance network; ~ **grappe** TÉL cluster; ~ **hertzien** RADIO microwave radio system; ~ **hiérarchisé** TÉL *à synchronisation mutuelle* hierarchic network; ~ **image** COMMUT bypath; ~ **indépendant** TRANSMIS independent network; ~ **d'information** DONNÉES information network; ~ **informatique** DONNÉES information network; ~ **d'installation d'abonné** TRANSMIS subscriber premises network (*SPN*); ~ **intégré** TRANSMIS integrated network; ~ **intelligent** (*RI*) APP SERV, INFORMAT, TRANSMIS intelligent network (*IN*), smart network; ~ **interactif** TRANSMIS interactive network; ~ **d'interconnexion** (*RI*) ALIM ÉLEC interconnection network, internetwork; ~ **d'interconnexion de multiprocesseur** DONNÉES multiprocessor interconnection network; ~ **international** TRANSMIS international network; ~ **international de transmission par paquets** TRANSMIS international packet-switched data network; ~ **interurbain** COMMUT trunk network, trunk-switching exchange area, TÉL trunk network,

long-distance telenetwork, long-distance network; ~ **JANET** TRANSMIS Joint Academic Network (*JANET*); ~ **à jeton** DONNÉES, INFORMAT, TRANSMIS token ring; ~ **de jonction** TÉL, TRANSMIS junction network; ~ **en L** ALIM ÉLEC L network; ~ **de liens** INFORMAT link network; ~ **de lignes** TÉL line network; ~ **local** DONNÉES, ÉLECTRON, INFORMAT, TÉL, TRANSMIS local area network (*LAN*); ~ **local d'entreprise** (*RLE*) DONNÉES, ÉLECTRON, INFORMAT, TÉL, TRANSMIS local area network (*LAN*); ~ **local étendu sans interruption** ÉLECTRON, INFORMAT, TÉL, TRANSMIS seamless extended local area; ~ **local de raccordement** TRANSMIS local distribution network; ~ **logique** DONNÉES logic array; ~ **longue distance** INFORMAT, TÉL wide-area network (*WAN*); ~ **longue portée** INFORMAT, TÉL wide-area network (*WAN*); ~ **à longue portée** TÉL long-distance network; ~ **maillé** TÉL polygon network, TRANSMIS meshed network, grid network; ~ **à mailles d'entraide sans blocage** TRANSMIS rearrangeable non-blocking network; ~ **métropolitain** TÉL city network, urban network, TRANSMIS metropolitan network, urban network; ~ **MIA** TÉL, TRANSMIS PAM network; ~ **de microcellules** RADIO microcell network; ~ **mobile public** RADIO, TRANSMIS public mobile network; ~ **mobile de transmission de données** RADIO mobile data-transmission network; ~ **de modulation d'impulsions en amplitude** TÉL, TRANSMIS pulse amplitude modulation network; ~ **mondial** TÉL worldwide network; ~ **multi-étage** ÉLECTRON multistage network; ~ **multimédia** TRANSMIS multimedia network; ~ **multiservice** TRANSMIS multiservice network; ~ **national de téléappel** TRANSMIS nationwide paging; ~ **neural** INFORMAT neural network; ~ **neuronique** INFORMAT neural network; ~ **de neurons** TRANSMIS neural network; ~ **à neutre isolé** ALIM ÉLEC isolated neutral system; ~ **de nœuds** INFORMAT nodes network; ~ **non-bloquant** COMMUT, TRANSMIS nonblocking network; ~ **non-synchronisé** TÉL nonsynchronized network; ~ **numérique** TRANSMIS digital network (*DN*); ~ **numérique à intégration de services** (*RNIS*) APP SERV, TÉL, TRANSMIS integrated services digital network (*ISDN*); ~ **numérique intégré** (*RNI*) TÉL, TRANSMIS integrated digital network (*IDN*); ~ **Odm** TV UHF network; ~ **oligarchique** TÉL oligarchic network; ~ **optique** TRANSMIS fiber-optic network (*AmE*), fibre-optic network (*BrE*), optical network unit; ~ **optique passif** TÉL, TRANSMIS passive optical network (*PON*); ~ **optique synchrone** INFORMAT, TRANSMIS synchronous optical network (*SONET*); ~ **d'ordinateurs** INFORMAT, TRANSMIS computer network; ~ **d'ordre supérieur** TRANSMIS high-order network; ~ **partagé de transmission de données** (*RPTD*) DONNÉES, TÉL, TRANSMIS shared data network (*SDN*); ~ **passe-partout** RADIO all-pass network; ~ **passif** ALIM ÉLEC passive network; ~ **de petites cellules** RADIO small-cell network; ~ **pilote** RADIO experimental network; ~ **plat d'antennes** RADIO planar-array antenna; ~ **pondérateur** TRANSMIS weighting network; ~ **de pondération** TRANSMIS weighting network; ~ **de portes** ÉLECTRON gate network; ~ **de préaccentuateur** ÉLECTRON pre-emphasis network; ~ **de préaccentuation** ÉLECTRON pre-emphasis network; ~ **primaire** TÉL primary network; ~ **privé** TÉL, TRANSMIS private branch network (*PBN*), private network; ~ **privé à intégration de services** TÉL, TRANSMIS private integrated-services network; ~ **à propagation arrière** DONNÉES back-propagation network; ~ **public** RADIO, TRANSMIS public network; ~ **public commuté** (*RPC*) COMMUT, DONNÉES, TÉL, TRANSMIS *télétransmissions* public switched network (*PSN*); ~ **public pour données** (*RPD*) DONNÉES, TÉL, TRANSMIS public data network (*PDN*); ~ **quadripôle** ALIM ÉLEC, ÉLECTRON, TRANSMIS two-terminal-pair network; ~ **quasi-associé** COMMUT semiassociated network; ~ **radio** RADIO radio network; ~ **radio à ressources partagées** (*3RP*) RADIO trunked radio network, trunked system, trunking system; ~ **de radiocommunications à interrogation** RADIO polling radio network; ~ **radiomobile privé** RADIO private mobile-radio network; ~ **de répartition** TRANSMIS distribution network; ~ **replié** TRANSMIS folded network; ~ **de résistances pondérées** ALIM ÉLEC variable-resistor network; ~ **de rétroréflecteurs** RADIO retroreflector array; ~ **rural** TRANSMIS rural network; ~ **sans blocage** COMMUT nonblocking network; ~ **à satellite** RADIO, TV satellite network; ~ **à satellite avec commutation de paquets** RADIO, TÉL packet satellite network (*PSN*); ~ **à satellite à réutilisation de fréquence** RADIO, TÉL frequency-reuse satellite network; ~ **secondaire** TÉL secondary network; ~ **de sécurité** TRANSMIS security network; ~ **sémantique** INFORMAT semantic network; ~ **sémaphore** COMMUT, SIGNAL, TRANSMIS signaling network (*AmE*), signalling network (*BrE*); ~ **de signalisation** COMMUT, SIGNAL, TRANSMIS signaling network (*AmE*), signalling network (*BrE*); ~ **au sol** TRANSMIS ground network; ~ **spatial-temporel-spatial** (*réseau STS*) COMMUT, TRANSMIS space-time-space network (*STS network*); ~ **STS** (*réseau spatial-temporel-spatial*) COMMUT, TRANSMIS STS network (*space-time-space network*); ~ **synchrone** TÉL synchronous network; ~ **de synchronisation** TRANSMIS synchronization network; ~ **à synchronisation despotique** TÉL despotic network; ~ **à synchronisation mutuelle** TÉL mutually-synchronized network; ~ **à synchronisation oligarchique** TÉL oligarchic network; ~ **synchronisé** TÉL synchronized network; ~ **en T** ALIM ÉLEC T network; ~ **de télécommunication privé** TÉL, TRANSMIS private branch network (*PBN*); ~ **de télécommunications** DONNÉES, GÉNÉRAL, RADIO, TÉL, TRANSMIS telecommunications network, telecommunications system; ~ **de télécommunications d'entreprises par satellite** TÉL, TRANSMIS satellite business telecommunications network; ~ **de télécommunications à grande distance** TÉL long-distance telenetwork; ~ **télégraphique** TÉL, TRANSMIS telegraphic network; ~ **télégraphique public** TÉL, TRANSMIS public telegraph network; ~ **de téléinformatique** INFORMAT, TRANSMIS teleprocessing network; ~ **téléphonique** TÉL, TRANSMIS telephone network, voice network; ~ **téléphonique classique** TÉL, TRANSMIS traditional telephone network; ~ **téléphonique commuté** (*RTC*) COMMUT, TÉL, TRANSMIS switched telephone network (*STN*); ~ **téléphonique commuté public** (*RTCP*) COMMUT, TÉL, TRANSMIS public switched telephone network (*PSTN*); ~ **téléphonique filaire** TÉL wired telephone network; ~ **téléphonique local** TÉL, TRANSMIS local telephone network; ~ **téléphonique privé** TÉL, TRANSMIS private telephone network; ~ **téléphonique public** TÉL, TRANSMIS public telephone network; ~ **téléphonique urbain** TÉL, TRANSMIS local telephone network; ~ **télésécurité** TRANSMIS remote security network; ~ **de télétraitement** INFORMAT, TRANSMIS teleprocessing network; ~ **de télévision par câbles** TV cable network, community network, cable television network; ~ **temporel-spatial-temporel** COMMUT, TRANSMIS time-space-time network; ~ **de terre** ALIM ÉLEC earthing network (*BrE*), grounding network (*AmE*); ~ **thermique équivalent** COMPOSANTS equivalent thermal network; ~ **transfrontalier** RADIO cross-border system; ~ **de transmission** TRANSMIS transmission network; ~ **de transmission de données** DONNÉES, TRANSMIS data-communication network; ~ **de transmission de données par paquets** DONNÉES, TRANSMIS packet data transmission network; ~ **de transmission local** TRANSMIS local communications network; ~ **de transport** TÉL transport network; ~ **de transport de données** DONNÉES, TRANSMIS data transport network; ~ **en treillis** ALIM ÉLEC lattice network; ~ **UHF** TV UHF network; ~ **unidirectionnel** TV unidirectional network; ~ **unifié** GÉNÉRAL consolidation; ~ **urbain** COMMUT, TÉL city network, urban network, local exchange area, TRANSMIS urban network, metropolitan network; ~ **à valeur ajoutée** (*RVA*) DONNÉES, TÉL, TRANSMIS, TV value-added network (*VAN*); ~ **VHF** TV VHF network; ~ **de vidéocommunication** TV videocommunication network; ~ **virtuel privé** INFORMAT, TÉL virtual private network; ~ **voix-données** DONNÉES, TRANSMIS speech-data network; ~ **VSAT** DONNÉES, TRANSMIS VSAT network; ~ **de zone urbaine** TRANSMIS metropolitan area network (*MAN*)

réseaux: ~ **d'immeubles raccordement** *m pl* TRANSMIS connection building networks

réserve[1]: **de** ~ *adj* DONNÉES backup, GÉNÉRAL backup, stand-by, INFORMAT, RADIO, TÉL, TESTS backup; **en** ~ *adj* GÉNÉRAL stand-by

réserve[2] *f* ÉLECTRON resist foil, TÉL stand-by; ~ **de puissance** ALIM ÉLEC reserve power; ~ **de soudure** COMPOSANTS solder resist

réserver *vt* APP SERV, TÉL *communication* book

réservoir *m* ÉLECTRON reservoir

résident[1] *adj* DONNÉES, INFORMAT resident

résident[2] *m* DONNÉES, INFORMAT, TÉL terminate and stay resident (*TSR*)

résidu: ~ **de connexion** *m* DONNÉES connection overhead; ~ **de section** *m* DONNÉES, TÉL section overhead (*SOH*); ~ **de trajet** *m* COMMUT, TÉL path overhead (*POH*)

résiliation *f* TÉL cancellation

résilience: ~ **du système** *f* INFORMAT system resilience

résilier *vt* TÉL cease; ~ **un abonnement** INFORMAT unsubscribe

résine: ~ **acrylique** *f* COMPOSANTS *matières plastiques* acrylate; ~ **époxyde** *f* COMPOSANTS epoxy resin; ~ **époxydique** *f* COMPOSANTS epoxy resin; ~ **à l'état B** *f* COMPOSANTS *matières plastiques* B-stage materials

résistance[1]: **à** ~ **fixe** *adj* ÉLECTRON fixed-resistance

résistance[2] *f* ALIM ÉLEC, COMPOSANTS resistor, ÉLECTRON, INFORMAT resistance; ~ **d'adaptation** COMPOSANTS, ÉLECTRON matching resistor; ~ **additionnelle** ALIM ÉLEC, COMPOSANTS series resistor; ~ **agglomérée** COMPOSANTS, ÉLECTRON composite resistor, molded composition resistor (*AmE*), moulded composition resistor (*BrE*); ~ **d'antenne** RADIO antenna resistance, aerial resistance; ~ **apparente directe** COMPOSANTS, ÉLECTRON forward slope resistance; ~ **apparente à l'état passant** COMPOSANTS, ÉLECTRON on-state slope resistance; ~ **ballast** ALIM ÉLEC ballast resistor; ~ **de base** ÉLECTRON *semi-conducteurs* base resistance; ~ **base-émetteur** ÉLECTRON *semi-conducteurs* base-emitter resistance; ~ **au bobinage** ALIM ÉLEC winding resistance; ~ **bobinée** ALIM ÉLEC, COMPOSANTS wire-wound resistor; ~ **au bouclage** ÉLECTRON loop series resistance; ~ **à brouillage** RADIO jamming resistance; ~**-capacité** (*RC*) ÉLECTRON resistance-capacitance (*RC*); ~ **de charge** ALIM ÉLEC, COMPOSANTS load resistor, plate resistor, plate-load resistor; ~ **de collecteur** ÉLECTRON collector resistance; ~ **à couche de carbone** COMPOSANTS, ÉLECTRON carbon-film resistor; ~ **à couche mince** COMPOSANTS, ÉLECTRON thin-film resistor; ~ **à couches diffusées** COMPOSANTS, ÉLECTRON diffused resistor; ~ **à couches épaisses** COMPOSANTS, ÉLECTRON thick-film resistor; ~ **pour courant continu** ALIM ÉLEC direct current resistance (*DC resistance*); ~ **à la déchirure** COMPOSANTS tearing resistance; ~ **de découplage** ALIM ÉLEC, COMPOSANTS decoupling resistor; ~ **différentielle** COMPOSANTS differential resistance; ~ **directe en courant continu** ÉLECTRON forward resistance; ~ **distribuée** ÉLECTRON spreading resistance; ~ **électrique** ÉLECTRON electrical resistance; ~ **d'électrode** ÉLECTRON electrode resistance; ~ **d'émetteur** COMPOSANTS emitter resistance; ~ **à l'enroulement** ALIM ÉLEC winding resistance; ~ **d'entrée** ÉLECTRON input resistance, transistor input resistance; ~ **à l'environnement** CLIMAT environmental endurance; ~ **d'équilibrage** ALIM ÉLEC, COMPOSANTS balancing resistor; ~ **équivalente** ALIM ÉLEC equivalent resistance; ~ **étalon** ALIM ÉLEC standard resistor, resistance standard, COMPOSANTS standard resistor; ~ **à étouffer les étincelles** COMPOSANTS spark-quenching resistor; ~ **extrinsèque de base** COMPOSANTS extrinsic base resistance; ~ **de faible dissipation** ALIM ÉLEC, COMPOSANTS low-wattage resistor; ~ **au feu** COMPOSANTS fire resistance; ~ **à film** ALIM ÉLEC film resistor; ~ **à la flexion** COMPOSANTS bending strength; ~ **de fuite** ALIM ÉLEC leakage resistance, *composant* bleeder resistor; ~ **de fuite de grille** COMPOSANTS, ÉLECTRON bleeder resistor, grid-leak resistor; ~ **de grille** ÉLECTRON gate resistance; ~ **Hall quantique** ÉLECTRON quantum Hall resistance; ~ **à l'humidité** GÉNÉRAL moisture resistance; ~ **à implantation ionique** COMPOSANTS, ÉLECTRON ion-implanted resistor; ~ **interne** ÉLECTRON internal resistance; ~ **inverse en courant continu** ALIM ÉLEC, ÉLECTRON reverse direct current resistance, reverse DC resistance; ~ **d'isolement** ALIM ÉLEC insulation resistance; ~ **de ligne** TRANSMIS line resistance; ~ **montée en série** ALIM ÉLEC, COMPOSANTS series resistor; ~ **non-linéaire** ÉLECTRON nonlinear resistance; ~ **en parallèle** ALIM ÉLEC shunt resistance; ~ **de perte** ALIM ÉLEC loss resistance; ~ **de plaque** ALIM ÉLEC, COMPOSANTS load resistor, plate resistor, plate-load resistor; ~ **de prise de terre** ALIM ÉLEC earth electrode resistance (*BrE*), ground electrode resistance (*AmE*); ~ **de protection** ALIM ÉLEC protecting resistor, COMPOSANTS protective resistor; ~ **de rappel à la masse** COMPOSANTS pull-down resistor; ~ **de rappel à la source** COMPOSANTS pull-up resistor; ~ **régulatrice** ALIM ÉLEC *composants* bleeder resistor; ~ **à la rupture** COMPOSANTS, TESTS breaking strength, ultimate strength; ~ **à la rupture en fluage** TESTS creep rupture stress; ~ **de saturation** ÉLECTRON saturation resistance; ~ **sensible à la tension** ALIM ÉLEC, COMPOSANTS voltage-dependent resistor; ~ **série** ALIM ÉLEC, COMPOSANTS series resistance; ~ **en série** ALIM ÉLEC, COMPOSANTS series resistor; ~ **série du collecteur** COMPOSANTS, ÉLECTRON collector series resistance; ~ **série de l'émetteur** COMPOSANTS, ÉLECTRON emitter series resistance; ~ **sous éclairement** ÉLECTRON resistance under illumination; ~ **spécifique** ALIM ÉLEC resistivity, specific resistance; ~ **superficielle** ALIM ÉLEC, ÉLECTRON surface resistance; ~ **de terre** ALIM ÉLEC earth resistance (*BrE*), ground resistance (*AmE*); ~ **thermique** COMPOSANTS, ÉLECTRON thermal resistance; ~ **à la traction** COMPOSANTS, GÉNÉRAL tensile strength; ~ **au vandalisme** APP SERV, COMPOSANTS, ÉLECTRON, GÉNÉRAL vandal-proofing; ~ **volumique** ALIM ÉLEC volume resistance

résistant: ~ **aux effets du rayonnement** *adj* COMPOSANTS radiation-hardened

résisteur: ~ **d'ajustage** *m* COMPOSANTS, ÉLECTRON trimming resistor; ~ **d'équilibrage** *m* COMPOSANTS, ÉLECTRON trimming resistor

résistivité *f* ALIM ÉLEC, ÉLECTRON resistivity, specific resistance; ~ **superficielle** ALIM ÉLEC, ÉLECTRON surface resistivity; ~ **volumique** ALIM ÉLEC, ÉLECTRON volume resistivity

résitol *m* COMPOSANTS *matières plastiques* B-stage materials

résolution *f* DONNÉES resolution, ÉLECTRON resolution, resolving power, INFORMAT, PÉRIPH, RADIO resolution, TESTS scale discrimination, TV resolution; ~ **d'affichage** INFORMAT display resolution; ~ **angulaire** RADIO *radar* angular resolution; ~ **maximale** TV maximum resolution; ~ **de notes** INFORMAT note resolution; ~ **optique** INFORMAT optical resolution; ~ **d'interpolation** PÉRIPH interpolation resolution; ~ **spatiale** INFORMAT, TÉL spatial resolution

résolveur: ~ **de noms de domaine** *m* DONNÉES domain name resolver (*DNR*)

résonance *f* ALIM ÉLEC, ÉLECTRON, GÉNÉRAL resonance; ~ **cyclotronique** ÉLECTRON cyclotronic resonance;

~ **parallèle** ALIM ÉLEC, ÉLECTRON parallel resonance; ~ **série** ALIM ÉLEC, ÉLECTRON series resonance

résonateur *m* ALIM ÉLEC, ÉLECTRON resonator, GÉNÉRAL resonator block, TRANSMIS resonator; ~ **acoustique** GÉNÉRAL acoustic resonator; ~ **à cavité** ÉLECTRON, TRANSMIS cavity resonator; ~ **diélectrique** ÉLECTRON dielectric resonator; ~ **Fabry-Pérot** TRANSMIS Fabry-Perot resonator; ~ **à hyperfréquences** RADIO microwave resonator; ~ **optique** TRANSMIS optical resonator

résonner *vi* GÉNÉRAL, INFORMAT resonate

Resource: ~ **Interchange File Format** *m* INFORMAT Resource Interchange File Format (*RIFF*)

responsabilité: ~ **EDI** *f* DONNÉES, TÉL EDI responsibility; ~ **EDIM** *f* DONNÉES, TÉL EDIM responsibility; ~ **du fait du produit** *f* TESTS product liability

ressaisie *f* DONNÉES, RADIO retrieval; ~ **de documents intégraux en temps réel** DONNÉES on-line full-text document retrieval

ressaisir *vt* DONNÉES, RADIO retrieve

resserrement *m* COMPOSANTS tightening

ressort[1]: **à** ~ *adj* COMPOSANTS spring-loaded

ressort[2]: ~ **Belleville** *m* COMPOSANTS *génie mécanique* Belleville spring; ~ **bimétallique** *m* CONTRÔLE *réglage* bimetal spring; ~ **de commande** *m* COMMUT lifting spring; ~ **à fil** *m* COMMUT wire spring; ~ **à lame** *m* COMPOSANTS, GÉNÉRAL leaf spring; ~ **de tension** *m* COMPOSANTS, GÉNÉRAL tension spring; ~ **de verrouillage** *m* COMPOSANTS locking spring

ressource: ~ **distante** *f* DONNÉES remote resource; ~ **documentaire** *f* DONNÉES documentary resource; ~ **d'interconnexion de systèmes ouverts** *f* (*ressource ISO*) DONNÉES, TÉL, TRANSMIS open systems interconnection resource (*OSI resource*); ~ **ISO** *f* (*ressource d'interconnexion de systèmes ouverts*) DONNÉES, TÉL, TRANSMIS OSI resource (*open systems interconnection resource*); ~ **locale** *f* DONNÉES local resource; ~ **partagée** *f* DONNÉES, INFORMAT shared resource

ressources: ~ **radioélectriques** *f pl* RADIO frequency resources; ~ **du réseau** *f pl* TÉL network resources; ~ **spectrales** *f pl* RADIO frequency resources, spectral resources, spectrum resources

restauration *f* GÉNÉRAL resetting, restoral, restoration

restaurer *vt* GÉNÉRAL reset, restore

rester *vi* ÉLECTRON remain; ~ **collé** ALIM ÉLEC hold in; ~ **en ligne** TÉL hold the line

restituer *vt* TÉL play back

restitution *f* ÉLECTRON, TÉL restitution; ~ **audionumérique** INFORMAT digital audio reproduction; ~ **du rythme** TRANSMIS clock recovery; ~ **sonore** INFORMAT, RADIO sound reproduction; ~ **du son** TV sound reproduction

restrapable *adj* ALIM ÉLEC restrappable

restrapage *m* ALIM ÉLEC restrapping

restreindre: ~ **l'accès à** *vt* RADIO, TÉL bar, restrict

restriction: ~ **d'accès au réseau** *f* TÉL trunk barring, trunk discrimination; ~ **d'appel** *f* TÉL call barring; ~ **d'appel interurbain** *f* TÉL toll-call barring (*AmE*), trunk-call barring (*BrE*); ~ **de l'identification de la ligne appelante** *f* (*RILA*) TÉL calling-line identification restriction (*CLIR*); ~ **de l'identification de la ligne connectée** *f* (*RILC*) TÉL connected-line identification restriction (*COLR*); ~ **partielle** *f* TÉL partial call barring; ~ **sur les zéros** *f* TRANSMIS zero restriction

restrictions: ~ **à l'extraction** *f pl* INFORMAT fetch restrictions

résultat *m* GÉNÉRAL, SIGNAL, TÉL response, TESTS result; ~ **d'essai** TESTS test outcome, test result; ~ **de la fonction OU de n signaux** INFORMAT OR of n signals; ~ **imprévu** TESTS *d'un test* unforeseen outcome; ~ **du mesurage** TESTS measured value; ~ **préliminaire** TESTS preliminary result; ~ **prévu d'un test** TESTS foreseen test outcome; ~ **de test** TESTS test outcome, test result; ~ **d'un test** DONNÉES observed test outcome

résumé *m* DONNÉES abstract, synopsis, GÉNÉRAL outline, INFORMAT abstract, synopsis

resynchronisation *f* ÉLECTRON reclocking, resynchronization, retiming

resynchroniser *vt* ÉLECTRON reclock, resynchronize, retime

rétablir[1] *vt* DONNÉES, ÉLECTRON restore, GÉNÉRAL reset, restore, INFORMAT, RADIO restore; ~ **le courant** ALIM ÉLEC restore supply

rétablir[2]: **se** ~ *v réfl* RADIO, TÉL recover

rétablissement *m* GÉNÉRAL restoral, restoration, INFORMAT restoration, RADIO recovery, restoral, restoration, TESTS recovery, restoration; ~ **d'un canal sémaphore** SIGNAL signaling-link restoration (*AmE*), signalling-link restoration (*BrE*); ~ **d'une liaison de signalisation** SIGNAL signaling-link restoration (*AmE*), signalling-link restoration (*BrE*); ~ **de niveau zéro** TÉL level shifting; ~ **du rythme** TRANSMIS clock recovery; ~ **de service** GÉNÉRAL restoral of service; ~ **du signal** SIGNAL signal restoration; ~ **de transaction** DONNÉES transaction recovery

retailler *vt* INFORMAT crop

retard *m* CONTRÔLE lead time, GÉNÉRAL delay; ~ **appliqué à l'émission** RADIO, TRANSMIS transmit delay; ~ **appliqué à la transmission** RADIO, TRANSMIS transmit delay; ~ **chrominance-luminance** TV chrominance-luminance delay inequality; ~ **à la croissance** COMPOSANTS delay time; ~ **à la décroissance** ÉLECTRON carrier storage time; ~ **dû à la formation de queues** TÉL queueing delay; ~ **d'exécution** DONNÉES, INFORMAT processing delay; ~ **de passage à l'ouverture** COMPOSANTS turn-off delay; ~ **du signal** SIGNAL signal delay; ~ **stéréo** INFORMAT stereo delay

retardateur *m* COMPOSANTS retarder

retarder *vt* ÉLECTRON lag

retenue: ~ **du renfort par vis sans sertissage** *f* ÉLECTRON crimpless-screw-type strength-member retention

retirer *vt* GÉNÉRAL phase out, INFORMAT roll out, withdraw

retouche *f* DONNÉES retouch; ~ **de brasage** COMPOSANTS soldering touch-up

retour: ~ **d'appel** *m* TÉL ring-back; ~ **arrière** *m* DONNÉES backtracking, INFORMAT backspace (*BS*); ~ **arrière intelligent** *m* DONNÉES intelligent backtracking; ~ **commun** *m* GÉNÉRAL common return; ~ **à l'exploitation** *m* TESTS return to service; ~ **de faisceau** *m* ÉLECTRON retrace; ~ **au fonctionnement normal** *m* DONNÉES error recovery; ~ **à la ligne** *m* GÉNÉRAL carriage return (*CR*); ~ **à la masse** *m* ÉLECTRON earth loop (*BrE*), ground loop (*AmE*); ~ **par la terre** *m* SIGNAL, TRANSMIS ground return; ~ **de pluie** *m* RADIO rain clutter; ~ **à la position initiale** *m* COMMUT restoring, return-to-home position; ~ **en position initiale** *m* INFORMAT homing; ~ **à la position de repos** *m* COMMUT restoring, return-to-home position; ~ **au repos** *m* COMMUT restoring, return-to-home position, ÉLECTRON, GÉNÉRAL, RADIO, TESTS reset; ~ **sous contrôle sur route normale** *m* TÉL controlled rerouting; ~ **du spot** *m* ÉLECTRON, INFORMAT retrace, TV flyback, retrace; ~ **sur canal sémaphore** *m* TÉL change back; ~ **sur la liaison normale** *m* TÉL change back; ~ **à zéro** *m* ÉLECTRON, INFORMAT, TÉL return to zero (*RZ*)

retournement: ~ **du modem** *m* DONNÉES modem back

retourner[1] *vt* COMPOSANTS reset, INFORMAT flip

retourner[2] *vi* INFORMAT revert

rétracter[1] *vt* COMPOSANTS shrink

rétracter[2]: **se** ~ *v réfl* COMPOSANTS shrink

rétraction *f* COMPOSANTS shrinkage, shrinking

retraire *vt* DONNÉES indent, retrieve, roll out, withdraw, RADIO retrieve, withdraw

retrait *m* INFORMAT *typographie* indent, *de données* retrieval, roll-out, withdrawal, RADIO retrieval, withdrawal

retraitable *adj* DONNÉES revisable

retraiter *vt* INFORMAT revise

retransmettre *vt* RADIO pass on, rebroadcast, relay, retrans-

mit, TÉL relay, retransmit, TRANSMIS, TV pass on, rebroadcast, relay, retransmit
retransmission *f* RADIO rebroadcast, rebroadcasting, relay, relaying, retransmission, TÉL relay, relaying, retransmission, TRANSMIS, TV rebroadcast, rebroadcasting, relay, relaying, retransmission; **~ audionumérique** (*RAN*) INFORMAT, TV digital audio broadcast (*DAB*); **~ automatique** DONNÉES auto-forward; **~ directe par satellite** TV direct satellite broadcasting (*DSB*); **~ EDI** DONNÉES, TÉL EDI-forwarding; **~ de VM** DONNÉES, TÉL VM forwarding
rétrécir[1] *vt* COMPOSANTS shrink
rétrécir[2] *vi* COMPOSANTS shrink
rétrécissement *m* COMPOSANTS shrinkage, shrinking
rétroaction *f* DONNÉES, ÉLECTRON, RADIO feedback; **~ de pertinence** DONNÉES relevance feedback; **~ répartie** ÉLECTRON distributed feedback (*DFB*)
rétrodiffusion *f* ALIM ÉLEC backscatter, backscattering, RADIO backscatter, backscattering, backward transmission, TÉL, TRANSMIS backscatter, backscattering; **~ directe** ALIM ÉLEC, RADIO short-distance backscatter; **~ indirecte** ALIM ÉLEC, RADIO indirect backscatter, long-distance backscatter
rétroéclairage *m* ALIM ÉLEC, ÉLECTRON, INFORMAT backlighting
rétroéclairé *adj* ALIM ÉLEC, ÉLECTRON, INFORMAT backlit
rétrograder *vt* INFORMAT demote
retrouver *vt* DONNÉES relocate
réunion *f* ÉLECTRON disjunction; **~ téléphone** TÉL meet-me call; **~ téléphonique** APP SERV, TÉL phone conference, phone conferencing
réutilisation: **~ de fréquence** *f* RADIO, TÉL frequency reuse, reuse of frequency channel; **~ de spectre de fréquences** *f* RADIO spectrum reuse
réveil: **~ automatique** *m* APP SERV alarm call, TÉL automatic wake-up (*AWU*), wake-up service; **~ téléphonique** *m* APP SERV alarm call, TÉL automatic wake-up (*AWU*), wake-up service
revendeur *m* RADIO reseller; **~ agréé** RADIO, TV accredited retailer
réverbération *f* ÉLECTRON, INFORMAT, TÉL reverberation (*reverb*); **~ numérique** INFORMAT digital reverb, digital reverberation
revêtement *m* GÉNÉRAL covering, TRANSMIS *d'un câble* coating; **~ antiréfléchissant** GÉNÉRAL anti-reflective coating; **~ atténuateur** TÉL acoustic blanket; **~ bicouche** COMPOSANTS double-layer coating; **~ ETFE** TRANSMIS epoxy buffer; **~ de fibre individuelle** COMPOSANTS single-fiber jacket (*AmE*), single-fibre jacket (*BrE*); **~ métallique** TRANSMIS metallic coating; **~ plastique** TRANSMIS plastic coating; **~ primaire** TRANSMIS epoxy buffer, primary coating; **~ protecteur** TRANSMIS fiber buffer (*AmE*), fibre buffer (*BrE*); **~ secondaire** COMPOSANTS, TRANSMIS fiber jacket (*AmE*), fibre jacket (*BrE*), secondary coating; **~ de silice** COMPOSANTS silica cladding, silica coating
revêtir *vt* GÉNÉRAL cover, TRANSMIS coat
révision *f* INFORMAT release
revu *adj* ÉLECTRON reset
revue: **~ de conception** *f* TESTS design review; **~ de conformité statique** *f* DONNÉES static conformance review; **~ critique de la conception** *f* GÉNÉRAL critical design review; **~ de direction du système qualité** *f* TESTS quality system review
RF *abrév* (*radiofréquence*) ÉLECTRON, RADIO, TÉL RF (*radio frequency*)
RG *abrév* (*régénérateur*) COMMUT, ÉLECTRON, SIGNAL, TÉL RG (*regenerator*)
RGT *abrév* (*réseau de gestion des télécommunications*) DONNÉES, TÉL, TRANSMIS TMN (*telecommunications management network*)
rhéostat *m* ALIM ÉLEC, COMPOSANTS rheostat; **~ de dérivation** COMPOSANTS shunt rheostat
RHM *abrév* (*relation homme-machine*) INFORMAT MMC (*man-machine communications*)
rhodiage *m* GÉNÉRAL rhodium plating
rhodier *vt* GÉNÉRAL rhodium-plate
RI *abrév* ALIM ÉLEC (*réseau d'interconnexion*) interconnection network, internetwork, APP SERV (*réseau intelligent*) IN (*intelligent network*), smart network, COMMUT (*répartiteur intermédiaire*), IDF (*intermediate distribution frame*), INFORMAT (*réseau intelligent*) IN (*intelligent network*), smart network, TÉL (*répartiteur intermédiaire*) IDF (*intermediate distribution frame*), TRANSMIS (*réseau intelligent*) IN (*intelligent network*), smart network
rich: **~ text format** *m* (*RTF*) INFORMAT rich text format (*RTF*)
rideau: **~ d'antennes** *m* RADIO aerial curtain, antenna curtain; **~ Chireix-Mesny** *m* RADIO Chireix array; **~ en dents de scie** *m* RADIO Chireix array; **~ de rétroflecteurs** *m* RADIO retroreflector array; **~ de Tchebychev** *m* RADIO Chebyshev array
RIF *abrév* (*réponse impulsionnelle finie*) ÉLECTRON, TRANSMIS FIR (*finite impulse response*)
rigidité: **~ diélectrique** *f* ALIM ÉLEC dielectric strength
RII *abrév* (*réponse impulsionnelle infinie*) ÉLECTRON, TRANSMIS IIR (*infinite impulse response*)
RILA *abrév* (*restriction de l'identification de la ligne appelante*) TÉL CLIR (*calling-line identification restriction*)
RILB *abrév* (*radiointerférométrie à très longue base*) TESTS VLBI (*very long baseline interferometry*)
RILC *abrév* (*restriction de l'identification de la ligne connectée*) TÉL COLR (*connected-line identification restriction*)
RISC *abrév* (*ordinateur à jeu d'instructions réduit, ordinateur RISC*) DONNÉES, INFORMAT RISC (*reduced-instruction-set computer*)
risque: **~ du client** *m* TESTS client risk (*CR*), consumer risk; **~ du fournisseur** *m* TESTS producer's risk (*PR*)
RLE *abrév* (*réseau local d'entreprise*) DONNÉES, ÉLECTRON, INFORMAT, TÉL, TRANSMIS LAN (*local area network*)
RM *abrév* (*radiomessagerie*) DONNÉES, RADIO, TÉL paging, radiopaging, TRANSMIS radiopaging; **~ avec accusé de réception** *f* RADIO acknowledgement paging; **~ à valeur ajoutée** *f* RADIO value-added paging
RMU *abrév* APP SERV (*récepteur de radiomessagerie*), Message-master, DONNÉES (*radiomessagerie unilatérale*) radiopaging, GÉNÉRAL message pager, RADIO, TÉL, TRANSMIS (*radiomessagerie unilatérale*) display paging, radiopaging (*récepteur de radiomessagerie*) message display pager, message pager, pager; **~ avec affichage numérique** *f* RADIO numerical display paging; **~ alphanumérique** *f* RADIO display paging, message display pager; **~ bip** *f* GÉNÉRAL tone pager; **~ urbaine** *f* RADIO city paging
RNI *abrév* (*réseau numérique intégré*) TÉL, TRANSMIS IDN (*integrated digital network*)
RNIS *abrév* (*réseau numérique à intégration de services*) APP SERV, TÉL, TRANSMIS ISDN (*integrated services digital network*); **~ à large bande** *m* (*RNIS-LB*) TÉL, TRANSMIS broadband ISDN (*B-ISDN*)
RNIS-LB *abrév* (*RNIS à large bande*) TÉL, TRANSMIS B-ISDN (*broadband ISDN*)
robinet: **~ à pression minimale constante** *m* COMPOSANTS *tubes électroniques* back-pressure valve; **~ de purge** *m* COMPOSANTS drain valve
robot: **~ d'appels** *m* TÉL call sender; **~ d'essai individuel** *m* TESTS routiner; **~ d'essais d'abonné** *m* TÉL, TESTS subscriber test responder
ROC *abrév* (*reconnaissance optique de caractères*) DONNÉES, INFORMAT OCR (*optical character recognition*)
rodage *m* INFORMAT burn-in
rodeuse *f* ALIM ÉLEC lapper
rôle *m* TV function; **~ d'agent** DONNÉES agent role; **~ de gestion** DONNÉES manager role; **~ de gestionnaire** DONNÉES manager role

ROM *abrév* (*mémoire à lecture seule*) INFORMAT ROM (*read only memory*), read-only storage
rompre[1] *vt* ALIM ÉLEC *courant* cut off, *électricité* break, open
rompre[2]: **se ~** *vi* ALIM ÉLEC break
rompu *adj* ALIM ÉLEC cut-off
ronflement *m* GÉNÉRAL hum, TÉL buzzer tone; **~ du secteur** ALIM ÉLEC mains hum
ronfleur *m* GÉNÉRAL, PÉRIPH, TÉL buzzer
ronronnement *m* GÉNÉRAL *d'une machine* hum
ROS *abrév* (*rapport d'ondes de tension stationnaires*) RADIO, TRANSMIS SWR (*standing wave ratio*), VSWR (*voltage standing wave ratio*); **~-mètre** *m* TESTS standing-wave meter
rosace: **~ de raccordement** *f* TÉL block terminal
rotor *m* TÉL rotor; **~ inductif** RADIO search coil
roue *f* INFORMAT wheel; **~ à caractères** PÉRIPH printwheel, type wheel; **~ codeuse** COMMUT, ÉLECTRON thumbwheel switch; **~ phonique** RADIO phonic motor
rouge: **~-vert-bleu** *adj* (*RVB*) TV red-green-blue (*RGB*)
rouleau *m* PÉRIPH pinch roller; **~ de pression** GÉNÉRAL pressure plug
rouleaux: **en ~** *loc* ÉLECTRON in rolls
roulements *m pl* TÉL reversals
routage *m* DONNÉES, INFORMAT, TÉL, TRANSMIS routing; **~ multiprotocole** DONNÉES, INFORMAT, TÉL, TRANSMIS multiprotocol routing
route *f* TÉL, TRANSMIS route; **~ détournée** TÉL alternative route; **~ de message** TRANSMIS message route; **~ des messages de signalisation** SIGNAL signaling-message route (*AmE*), signalling-message route (*BrE*); **~ sémaphore** SIGNAL signaling route (*AmE*), signalling route (*BrE*); **~ de signalisation** SIGNAL signaling route (*AmE*), signalling route (*BrE*)
router *vt* DONNÉES, ÉLECTRON, GÉNÉRAL, TÉL, TRANSMIS route
routeur *m* DONNÉES, ÉLECTRON, GÉNÉRAL, TÉL, TRANSMIS router
routine *f* DONNÉES, INFORMAT routine; **~ de diagnostic** INFORMAT diagnostic routine; **~ de service** INFORMAT utility routine; **~ utilitaire** INFORMAT utility routine
routinier *adj* ÉLECTRON repetitive
RP *abrév* (*répartiteur principal*) TÉL, TRANSMIS MDF (*main distribution frame*), mainframe
RPA *abrév* (*répartition de probabilité des amplitudes*) SIGNAL APD (*amplitude probability distribution*)
RPC *abrév* (*réseau public commuté*) COMMUT, DONNÉES, TÉL, TRANSMIS *télétransmissions* PSN (*public switched network*)
RPD *abrév* (*réseau public pour données*) DONNÉES, TÉL, TRANSMIS PDN (*public data network*)
RPTD *abrév* (*réseau partagé de transmission de données*) DONNÉES, TÉL, TRANSMIS SDN (*shared data network*)
RTC *abrév* (*réseau téléphonique commuté*) COMMUT, TÉL, TRANSMIS STN (*switched telephone network*)
RTCP *abrév* (*réseau téléphonique commuté public*) COMMUT, TÉL, TRANSMIS PSTN (*public switched telephone network*)
RTF *abrév* (*rich text format*) INFORMAT RTF (*rich text format*)
ruban *m* ALIM ÉLEC, COMPOSANTS, DONNÉES tape, ÉLECTRON tape, ribbon, GÉNÉRAL ribbon; **~ d'acier** COMPOSANTS steel tape; **~ adhésif isolant** ALIM ÉLEC insulating tape; **~ de contact en élastomère** COMPOSANTS elastomer contact-strip; **~ d'étiquetage** COMPOSANTS strip label, GÉNÉRAL designating tape; **~ isolant électrique** ALIM ÉLEC electrical tape; **~ magnétique** DONNÉES, INFORMATmagnetic tape; **~ perforé** TÉL perforated tape, punched tape
rubaner *vt* COMPOSANTS wrap, GÉNÉRAL lap
rubannage *m* COMPOSANTS wrapping, GÉNÉRAL lapping
rubrique: **~ professionnelle** *f* TÉL *dans l'annuaire* professional classification
rugosité *f* COMPOSANTS pulled surface, GÉNÉRAL roughness
rupture *f* ALIM ÉLEC *d'un conducteur* break, breakage, COMPOSANTS rupture, electric breakdown, fracture, ÉLECTRON breakdown, GÉNÉRAL break, cutoff, TÉL breakdown; **~ de câble** TRANSMIS cable break; **~ du diélectrique** ALIM ÉLEC, ÉLECTRON *semi-conducteurs* dielectric breakdown; **~ en fluage** TESTS creep rupture; **~ de séquence** INFORMAT jump
RV *abrév* (*réalité virtuelle*) DONNÉES, INFORMAT VR (*virtual reality*)
RVA *abrév* (*réseau à valeur ajoutée*) DONNÉES, TÉL, TRANSMIS, TV VAN (*value-added network*)
RVB *abrév* (*rouge-vert-bleu*) TV RGB (*red-green-blue*)
rythme *m* INFORMAT rhythm, swing; **~ d'appel** TÉL ringing periodicity; **~ automatique** INFORMAT automatic rhythm; **~ de batterie** INFORMAT drum rhythm; **~ binaire** DONNÉES, TRANSMIS bit timing; **~ de bits** DONNÉES, TRANSMIS bit timing; **~ de cerclage** TV leafleting periodicity; **~ de commande** TRANSMIS control timing; **~ diurne** GÉNÉRAL diurnal rhythm; **~ de multitrame émission** TRANSMIS transmit multiframe timing; **~ de multitrame transmission** TRANSMIS transmit multiframe timing; **~ de paquets acquis** DONNÉES burst-timing acquired (*BTA*); **~ de paquets non-acquis** DONNÉES burst-timing not acquired (*BTNA*); **~ récupéré** DONNÉES recovered clock
rythmeur *m* ÉLECTRON clock oscillator

S

SA *abrév* (*signal d'appel*) TÉL call signal, call waiting, calling signal

SAA *abrév* (*sélection automatique d'acheminement*) TÉL ARS (*automatic route selection*)

SAAFA *abrév* (*sélection automatique d'acheminement avec file d'attente*) TÉL ARSWQ (*automatic route selection with queueing*)

saccade *f* DONNÉES jerk; **~ d'images** INFORMAT, TV freeze update, strobe effect

saignée *f* GÉNÉRAL incision

saisie *f* DONNÉES, capture, keyboarding, INFORMAT keyboarding, TÉL seizure; **~ de données** DONNÉES data acquisition, data gathering; **~ sur clavier** DONNÉES, INFORMAT keyboarding

saisir *vt* GÉNÉRAL key in, INFORMAT grab, grip, feed

saison: **~ de pointe** *f* TÉL busy season

salle *f* ALIM ÉLEC, INFORMAT, TÉL, TESTS room; **~ des accumulateurs** ALIM ÉLEC battery room; **~ anéchoïque** TESTS *acoustique* anechoic chamber, anechoic room, dead room, free-field room; **~ de l'autocommutateur** TÉL switching room; **~ des batteries** ALIM ÉLEC battery room; **~ de commutation** TÉL switching room; **~ de contrôle** TESTS *O & M* control room; **~ dépoussiérée** TESTS dust-free room; **~ d'essai et mesures** TESTS test room; **~ d'exploitation** TESTS operations room, *O & M* control room; **~ des machines** TÉL engine room

salve *f* DONNÉES burst; **~ de signaux d'accès** DONNÉES access burst signal

sans: **~ amplification** *adj* ALIM ÉLEC, ÉLECTRON, TV unamplified; **~ armature** *adj* ÉLECTRON nonarmored (*AmE*), nonarmoured (*BrE*); **~ bruit** *adj* INFORMAT noise-free; **~ charge** *adj* ÉLECTRON free-running; **~ clavier** *adj* INFORMAT keyboardless; **~ compression** *adj* DONNÉES, INFORMAT uncompressed; **~ concentration numérique des conversations** *adj* (*SCNC*) TRANSMIS digital non-interpolated (*DNI*); **~ connexion** *adj* GÉNÉRAL connectionless (*CL*); **~ défaillance** *adj* TESTS fault-free; **~ entrée d'horloge** *adj* ÉLECTRON unclocked; **~ étiquette** *adj* INFORMAT unlabeled (*AmE*), unlabelled (*BrE*); **~ fil** *adj* GÉNÉRAL cordless, INFORMAT, RADIO, TÉL wireless; **~ fin** *adj* INFORMAT endless; **~ frais** *adj* TÉL toll-free; **~ halogène** *adj* ÉLECTRON nonhalogen; **~ perte** *adj* INFORMAT, TÉL lossless; **~ perturbation** *adj* GÉNÉRAL disturbance-free; **~ référence** *adj* INFORMAT unlabeled (*AmE*), unlabelled (*BrE*); **~ répéteur** *adj* GÉNÉRAL repeaterless; **~ restriction** *adj* GÉNÉRAL unrestricted; **~ ronflement** *adj* GÉNÉRAL hum-free; **~ signaux de synchronisation** *adj* ÉLECTRON free-running

SAOT *abrév* (*système d'analyse des options tarifaires*) APP SERV POAS (*payment option analysis system*)

satellisation *f* RADIO injection into orbit, putting into orbit

satelliser *vt* RADIO bring into orbit, inject into orbit, put into orbit

satellite *m* RADIO satellite, TÉL subordinate exchange, dependent exchange, TV satellite; **~ actif** RADIO active satellite; **~ d'application technologique** RADIO application technology satellite; **~ des artères principales** RADIO major-path satellite; **~ artificiel** DONNÉES, RADIO artificial satellite; **~ artificiel en orbite terrestre** RADIO earth-orbiting artificial satellite; **~ autonome** RADIO stand-alone satellite; **~ brouilleur** RADIO interfering satellite; **~ en chaîne** RADIO satellite in tandem; **~ de collecte de données** DONNÉES, RADIO data-collection satellite; **~ à commande d'orientation** RADIO attitude-stabilized satellite; **~ géostationnaire** RADIO geostationary satellite; **~ géostationnaire du service fixe** RADIO geostationary fixed satellite; **~ géosynchrone** RADIO earth-synchronous satellite, geosynchronous satellite; **~ hôte** RADIO host satellite; **~ laissé pour compte** RADIO cold bird; **~ loué à plein** RADIO hot bird; **~ maintenu en position** RADIO station-keeping satellite; **~ de navigation** RADIO beacon satellite; **~ occupant la même position orbitale** RADIO co-located satellite; **~ en orbite basse** RADIO low earth orbit satellite (*LEO satellite*); **~ pendant sa phase éclairée** RADIO sunlit satellite; **~ en phase** RADIO phased satellite, synchronized satellite; **~ à poste** RADIO on-station satellite; **~ à radiophare** RADIO beacon satellite; **~ RDS** TV DBS satellite; **~ réfléchissant** RADIO reflecting satellite; **~ réflecteur** RADIO reflecting satellite; **~ régénérateur** RADIO regenerative satellite; **~ relais de données** DONNÉES, RADIO data-relay satellite; **~ de relève** RADIO follow-on satellite; **~ de secours** RADIO spare satellite; **~ en service** RADIO hot satellite; **~ sous-synchrone** RADIO subsynchronous satellite; **~ stationnaire** RADIO stationary satellite; **~ super-synchrone** RADIO supersynchronous satellite; **~ sur orbite** RADIO orbiting satellite; **~ sur orbite à basse altitude** RADIO low earth orbit satellite (*LEO satellite*); **~ synchrone** RADIO synchronous satellite; **~ synchrone orienté vers la terre** RADIO earth-oriented geosynchronous satellite; **~ synchronisé** RADIO synchronized satellite, phased satellite; **~ de technologie des télécommunications** RADIO communications technology satellite (*CTS*); **~ de télécommunications** RADIO, TV telecommunications satellite; **~ de télécommunications civiles** RADIO civil communications satellite; **~ de télécommunications à corps stabilisé** RADIO body-stabilized communications satellite; **~ de télécommunications à grande capacité de transmission** RADIO high-capacity communication satellite; **~ de télédétection** RADIO remote sensing satellite; **~ de transmission** RADIO gap-filler satellite; **~ à vocation commerciale** RADIO business-oriented satellite

satisfaire: **~ aux exigences** *vi* GÉNÉRAL comply with requirements

saturation *f* CLIMAT temperature-limited state, ÉLECTRON saturation, saturation state, temperature-limited state, INFORMAT lightness

saturé *adj* ÉLECTRON saturated

saut *m* ALIM ÉLEC, INFORMAT jump, skip, ÉLECTRON hopping, RADIO hop, hopping, skip, TÉL skip, TRANSMIS jumping, skipping; **~ conditionnel** INFORMAT conditional jump; **~ de cycle** TRANSMIS cycle skipping; **~ de cycle d'horloge** TRANSMIS clock-cycle skipping; **~ de cycle de porteuse** TRANSMIS clock-carrier cycle skipping; **~ de fréquence** RADIO frequency hopping; **~ d'indice** ÉLECTRON, TRANSMIS step index profile; **~ de ligne** COMMUT, DONNÉES, INFORMAT, RADIO, TÉL line feed (*LF*); **~ de mode** TRANSMIS mode hopping, mode jumping; **~ de papier** INFORMAT slew, PÉRIPH, TÉL paper skip, paper throw; **~ de phase** RADIO phase hopping; **~ de répéteur** ÉLECTRON, RADIO transponder hopping; **~ de temps** RADIO time step; **~ de tension** ALIM ÉLEC voltage jump

sauter *vt* ALIM ÉLEC jump, ÉLECTRON hop, INFORMAT jump, skip, RADIO hop, skip, TÉL skip, TRANSMIS jump, skip

sautillement *m* RADIO jitter; ~ **d'image** RADIO jitter
sauvegarde[1]: **de** ~ *adj* DONNÉES, GÉNÉRAL, INFORMAT backup
sauvegarde[2] *f* DONNÉES backup, safeguard, saving, ÉLECTRON protected memory, INFORMAT backup, safeguard, save, PÉRIPH, TÉL save, saving; ~ **des comptes de taxe** TÉL meter account save; ~ **de données** DONNÉES data backup, data saving, INFORMAT data saving; ~ **des fichiers** DONNÉES, INFORMAT file backup; ~ **des messages** PÉRIPH, TÉL saving of messages; ~ **de programme** INFORMAT program save; ~ **sur cartouche** INFORMAT cartridge save, save to cartridge; ~ **sur cassette** INFORMAT cassette save, save to cassette; ~ **sur disquette** INFORMAT save to diskette; ~ **temporaire d'opérateur** INFORMAT save temporary operator
sauvegarder *vt* DONNÉES, INFORMAT back up, safeguard, save; ~ **sur cartouche** INFORMAT save to cartridge; ~ **sur cassette** INFORMAT save to cassette; ~ **sur disquette** INFORMAT save to diskette
SAVA *abrév* (*services à valeur ajoutée*) APP SERV, DONNÉES, TRANSMIS, TV VANS (*value-added network services*)
savoir: ~**-faire** *m* ÉLECTRON, INFORMAT know-how
scan *m* PÉRIPH, RADIO scan
scannage *m* PÉRIPH, RADIO scanning
scanner *m* DONNÉES, INFORMAT optical scanner, PÉRIPH, RADIO scanner; ~ **couleur** PÉRIPH color image scanner (*AmE*), colour image scanner (*BrE*); ~ **de diapositives** DONNÉES, INFORMAT slide scanner; ~ **d'images** PÉRIPH, TV image scanner; ~ **manuel** PÉRIPH hand-held scanner; ~ **à plat** PÉRIPH, TÉL flat-bed scanner
scannérisation *f* PÉRIPH, RADIO scanning
scannériser *vt* PÉRIPH, RADIO scan
SCAR *abrév* (*système de contrôle d'attitude et de roulis*) RADIO attitude and roll control system
sceau: ~ **électronique** *m* DONNÉES electronic key
scellé *adj* COMPOSANTS, GÉNÉRAL sealed; ~ **en configuration** ÉLECTRON butt-configurated
sceller *vt* COMPOSANTS, GÉNÉRAL seal
scénario *m* TESTS scenario; ~**-image électronique** INFORMAT electronic storyboard
schéma *m* APP SERV frame, framework, GÉNÉRAL diagram, scheme plan, drawing, pattern, sketch; ~ **d'affectation des bornes** GÉNÉRAL terminal number diagram; ~ **de l'autocommutateur** COMMUT exchange diagram; ~ **des bornes** GÉNÉRAL terminal diagram; ~ **de câblage** COMPOSANTS wiring diagram; ~ **de câbles** TRANSMIS cable scheme; ~ **du central** ALIM ÉLEC exchange diagram; ~ **des circuits** COMMUT, ÉLECTRON circuit diagram; ~ **de circuits et de câblage** ÉLECTRON circuit and wiring diagram; ~ **de commutation** COMMUT switching pattern; ~ **de connexions** COMPOSANTS wiring drawing, GÉNÉRAL connection diagram, INFORMAT plugging chart; ~ **de conséquence de défaillance** GÉNÉRAL failure consequence diagram; ~ **coté** ÉLECTRON outline drawing; ~ **de distribution d'énergie** ALIM ÉLEC power distribution diagram; ~ **d'écoulement du trafic** TÉL traffic pattern; ~ **de fonctionnement** INFORMAT system diagram; ~ **de groupement** GÉNÉRAL grouping diagram; ~ **d'implantation** ÉLECTRON layout sketch, GÉNÉRAL layout drawing; ~ **d'installation** GÉNÉRAL installation diagram, installation drawing; ~ **d'interfonctionnement** GÉNÉRAL interwork diagram; ~ **des liaisons** TÉL trunking diagram; ~ **logique** GÉNÉRAL logic diagram; ~ **de montage** INFORMAT setup diagram; ~ **de montage de câble** TRANSMIS *documentation* cable mounting drawing; ~ **de pontage** ALIM ÉLEC strapping diagram; ~ **de pose de fils** COMPOSANTS wire-running diagram; ~ **de principe** COMMUT, ÉLECTRON circuit diagram; ~ **de raccordement de câble** TRANSMIS *documentation* cable terminal drawing; ~ **de réflexion** TV reflectance diagram; ~ **de répartition des câbles** TRANSMIS *documentation* cable distribution diagram; ~ **de réseau** TÉL network diagram; ~ **de signalisation** GÉNÉRAL signaling diagram (*AmE*), signalling diagram (*BrE*); ~ **de strapage** ALIM ÉLEC strapping diagram; ~ **de trame** TRANSMIS alignment pattern; ~ **de verrouillage** TRANSMIS alignment pattern; ~ **de verrouillage de trame** DONNÉES frame pattern, framing pattern
scintillation *f* RADIO, TV glint, scintillation
scintillement *m* RADIO, TV flickering, flutter
SCNC *adj* (*sans concentration numérique des conversations*) TRANSMIS DNI (*digital non-interpolated*)
ScreenPlay *m* INFORMAT ScreenPlay
script *m* DONNÉES operations chart
Script: ~ **X** *m* INFORMAT Script X
scripter *vt* INFORMAT *programme SCRIPT* write
SCSI *abrév* (*interface SCSI*) INFORMAT SCSI (*small computer system interface*)
SDA *abrév* (*sélection directe à l'arrivée*) DONNÉES, TÉL DID (*direct inward dialing, direct inward dialling*), direct dialing-in (*AmE*), direct dialling-in (*BrE*), direct input dialing (*AmE*), direct input dialling (*BrE*)
SDL: ~ **de base** *m* DONNÉES basic SDL
SDLC *abrév* (*contrôle de liaison synchrone de données*) DONNÉES SDLC (*synchronous data link control*)
SDS *abrév* (*surdébit de section*) DONNÉES, TÉL SOH (*section overhead*)
SDSM *abrév* (*surdébit de section de multiplexage*) TRANSMIS multiplex-section overhead
SDSR *abrév* (*surdébit de section de régénération*) DONNÉES regenerator-section overhead
SE *abrév* DONNÉES, RADIO (*seconde avec erreurs, seconde entachée d'erreurs*) ES (*errored second*)
sébile: ~ **de restitution des pièces** *f* PÉRIPH coin-return cup, refund chute, return flap
SECAM *abrév* (*séquentiel couleur à mémoire*) TV SECAM (*séquentiel couleur à mémoire*)
second *m* INFORMAT runner-up; ~ **claquage** COMPOSANTS second breakdown
seconde *f* GÉNÉRAL second; ~ **avec erreurs** (*SE*) DONNÉES, RADIO errored second (*ES*); ~ **entachée d'erreurs** (*SE*) DONNÉES, RADIO errored second (*ES*); ~ **gravement erronée** (*SGE*) DONNÉES, RADIO severely-errored second (*SES*); ~ **d'indisponibilité** DONNÉES unavailable second (*UAS*); ~ **intercalaire** RADIO, TESTS leap second; ~ **de perte du verrouillage de trame** DONNÉES out-of-frame second (*OFS*); ~ **rupture** COMPOSANTS second breakdown; ~ **tonalité de numérotation** TÉL second dial tone, secondary dial tone
secours[1]: **de** ~ *adj* DONNÉES, GÉNÉRAL, INFORMAT, RADIO, TÉL, TESTS backup, stand-by
secours[2]: ~ **immédiat actif** *m* TESTS hot stand-by
secret *m* DONNÉES, PÉRIPH, TÉL privacy; ~ **des communications** TÉL call security; ~ **de conversation** TÉL call security; ~ **professionnel** GÉNÉRAL, TESTS professional secrecy
secteur *m* ALIM ÉLEC mains, GÉNÉRAL sector, INFORMAT mains, sector, RADIO sector, segment, TRANSMIS segment, TV sector; ~ **alternatif** INFORMAT AC mains; ~ **associatif** TV associated sector; ~ **audio CD-I** INFORMAT CD-I audio sector; ~ **de brouillage** RADIO interference sector; ~ **CD-I** INFORMAT CD-I sector; ~ **descripteur de piste** INFORMAT track-descriptor sector; ~ **de service** GÉNÉRAL, RADIO service sector; ~ **sol** TRANSMIS ground segment; ~ **spatial** RADIO space segment; ~ **vidéo CD-I** TV CD-I video sector
secteurs: ~ **par piste** *loc* (*spp*) INFORMAT sectors per track (*spt*)
section *f* DONNÉES section; ~ **d'adaptation** RADIO matching section, transforming section; ~ **du câble** TRANSMIS cable section; ~ **canaux** TRANSMIS channelized section; ~ **canaux d'un répéteur** TRANSMIS channelized stage of repeater; ~ **de circuit radiophonique** TV sound program circuit section (*AmE*), sound programme circuit section (*BrE*); ~ **de**

circuit télévisuel TV television circuit section; **~ d'un conducteur** ÉLECTRON conductor area; **~ creuse** ÉLECTRON air core; **~ élémentaire amplifiée** TÉL elementary repeater section; **~ élémentaire de câble** TÉL elementary cable section; **~ élémentaire régénérée** TÉL elementary regenerated section, elementary regenerator section; **~ de fil** ALIM ÉLEC wire area; **~ de gabarit pour peigne** TRANSMIS formboard section; **~ de groupe primaire** TÉL group section; **~ de groupe quaternaire** TÉL supermastergroup section; **~ de groupe secondaire** TÉL supergroup section; **~ de groupe tertiaire** TÉL mastergroup section; **~ hertzienne numérique** RADIO, TÉL digital radio section; **~ homogène** RADIO, TÉL homogeneous section; **~ internationale** TÉL international section; **~ internationale de circuit télévisuel multiple** TV international multiple-destination sound-program circuit (*AmE*), international multiple-destination sound-programme circuit (*BrE*), international multiple-destination television circuit; **~ internationale principale** TÉL international main section; **~ de ligne** TRANSMIS line section; **~ de ligne numérique** TÉL digital line section; **~ locale** TÉL extension circuit; **~ de multiplexage** TRANSMIS multiplex section (*MS*); **~ nationale** TÉL national section; **~ nationale principale** TÉL national main section; **~ numérique** TRANSMIS digital section (*DS*); **~ principale** TÉL main section; **~ principale d'essais** TÉL, TESTS principal test section; **~ de régénération** DONNÉES regenerator section (*RS*); **~ de régulation de ligne** TÉL regulated line section; **~ terminale nationale** TÉL terminal national section; **~ transversale** TRANSMIS transverse section; **~ unitaire** TV unit section

sectionneur *m* ALIM ÉLEC disconnector, isolator; **~ de terre** ALIM ÉLEC earthing switch (*BrE*), grounding switch (*AmE*)

séculaire *adj* ÉLECTRON age-old

sécurisation *f* COMMUT protection switching, GÉNÉRAL contingency planning, protection, TÉL protection switching

sécurisé *adj* GÉNÉRAL protected

sécurité[1]: **de ~** *adj* DONNÉES, GÉNÉRAL, INFORMAT, RADIO, TÉL, TESTS backup; **de ~ intrinsèque** *adj* TESTS fail-safe

sécurité[2] *f* GÉNÉRAL safeguard, safety, security; **~ d'exploitation** TÉL operational security; **~ des fichiers** DONNÉES, INFORMAT file security; **~ de fonctionnement** TESTS dependability, security arrangements; **~ d'information** TÉL information security; **~ informatique** DONNÉES computer security (*COMPUSEC*); **~ physique** DONNÉES, TESTS physical security; **~ de réseau** DONNÉES network security; **~ du système informatique** DONNÉES, INFORMAT data-processing system security; **~ de télécommunications** GÉNÉRAL telecommunications security; **~ du travail** CONTRÔLE industrial safety

segment *m* DONNÉES segment, INFORMAT group, section, segment, TÉL segment; **~ de caractère** COMPOSANTS stroke; **~ fédérateur** ÉLECTRON backbone; **~ graphique** INFORMAT display group, display segment; **~ ISA** DONNÉES, INFORMAT, TÉL ISA segment; **~ MHD** DONNÉES, INFORMAT, TÉL MHD segment; **~ ST** DONNÉES, INFORMAT, TÉL ST segment; **~ UNA** DONNÉES, INFORMAT, TÉL UNA segment; **~ UNB** DONNÉES, INFORMAT, TÉL UNB segment; **~ UNH** DONNÉES, INFORMAT, TÉL UNH segment

segmentation *f* DONNÉES, INFORMAT, TÉL segmentation, segmenting; **~ adaptative** INFORMAT adaptive segmentation; **~ et réassemblage** DONNÉES segmentation and reassembly (*SAR*); **~ fixe** INFORMAT fixed segmentation

segmenté *adj* DONNÉES, INFORMAT, TÉL segmented

segmenter *vt* DONNÉES, INFORMAT, TÉL segment

sélecter *vt* TÉL select

sélecteur *m* GÉNÉRAL selector, switch; **~ d'accès** COMMUT access switch, access selector; **~ d'adresse N** DONNÉES N-address selector; **~ de canal** RADIO channel switch, channel selector; **~ de canaux** TV channel selector; **~ du chercheur d'enregistreur** COMMUT register-finder selector; **~ à commande directe** COMMUT direct-driven selector; **~ de domaine MCS** TÉL MCS domain selector; **~ à double mouvement** TÉL two-motion selector, Strowger selector; **~ final** TÉL final selector; **~ de fréquence** RADIO, TRANSMIS frequency selector; **~ manuel** INFORMAT dial; **~ de mémoire de programme** INFORMAT program store switch; **~ N** DONNÉES N-address selector; **~ d'options** INFORMAT kill file; **~ pas à pas** COMMUT step-by-step selector; **~ de pondération de bruit** ÉLECTRON noise-weighing switch; **~ de programme** INFORMAT program selector, program switch; **~ rotatif** COMMUT rotary selector, rotary switch, uniselector, GÉNÉRAL, INFORMAT rotary switch; **~ Strowger** TÉL Strowger selector, two-motion selector; **~ tandem** COMMUT, TÉL tandem selector; **~ de tension** ALIM ÉLEC voltage adaptor switch, voltage selector; **~ de travail** INFORMAT job selector

sélection *f* GÉNÉRAL select, selecting, selection, TÉL *commutation* selection, selecting, *composition du numéro* dialing (*AmE*), dialling (*BrE*), TRANSMIS, TV select, selecting, selection; **~ aléatoire** TÉL random allotment, random selection; **~ automatique d'acheminement** (*SAA*) TÉL automatic route selection (*ARS*); **~ automatique d'acheminement avec file d'attente** (*SAAFA*) TÉL automatic route selection with queueing (*ARSWQ*); **~ conditionnelle** GÉNÉRAL conditional selection; **~ directe** INFORMAT direct select; **~ directe à l'arrivée** (*SDA*) DONNÉES, TÉL direct dialing-in (*AmE*), direct dialling-in (*BrE*), direct input dialing (*AmE*), direct input dialling (*BrE*), direct inward dialing (*AmE*) (*DID*), direct inward dialling (*BrE*) (*DID*); **~ directe d'un poste supplémentaire** TÉL direct dialing-in (*AmE*), direct dialling-in (*BrE*); **~ discontinue** INFORMAT discontinuous selection; **~ et amplification de branche** *f pl* TÉL branch selection and amplification; **~ et indication des paramètres de commande de flux** *f pl* TÉL flow-control parameter selection and indication; **~ et indication du délai en transit** DONNÉES, TÉL transit-delay selection and indication (*TDSAI*); **~ fréquentielle** RADIO, TV frequency selection; **~ de générateur d'enveloppes** INFORMAT envelope-generator select (*EG select*); **~ indirecte** INFORMAT indirect select; **~ d'information d'entrée** DONNÉES entry information selection; **~ intelligente de fréquence** RADIO smart frequency selection; **~ de ligne interurbaine spécifique** COMMUT specific trunk-line selection; **~ de longueur des paquets** DONNÉES, TÉL packet length selection; **~ d'opérateur** INFORMAT operator select; **~ par cadran numérique** INFORMAT dial switching; **~ rapide** TÉL fast select; **~ du rythme** TRANSMIS timing selection; **~ séquentielle** TÉL sequential selection; **~ de source** INFORMAT source select

sélectionner *vt* GÉNÉRAL, select, shuffle; **~ par une porte** ÉLECTRON gate

sélectivité *f* RADIO *d'un récepteur* TÉL, TRANSMIS selectivity; **~ effective d'un récepteur** RADIO, TRANSMIS effective selectivity; **~ proche** RADIO, TRANSMIS adjacent channel selectivity; **~ de voie adjacente** RADIO, TRANSMIS adjacent channel selectivity

selsyn *m* ÉLECTRON selsyn

sémantique: **~ d'image instantanée** *f* DONNÉES, INFORMAT snapshot semantics; **~ opératoire** *f* DONNÉES, INFORMAT operational semantics; **~ statique** *f* DONNÉES, INFORMAT static semantics

sémaphore *m* ÉLECTRON, TÉL semaphore

sématème *f* SIGNAL signal train

sémateur *m* TÉL semator

sémation *f* TÉL semation

semelle *f* COMPOSANTS *structure mécanique* base iron

semi-automatique *adj* COMMUT, TÉL semiautomatic

semiconducteur *m* ALIM ÉLEC, COMPOSANTS, ÉLECTRON semiconductor; **~ compensé** ALIM ÉLEC, COMPOSANTS,

ÉLECTRON compensated semiconductor; ~ **complémentaire à oxyde metallique à grande vitesse** (*HCMOS*) ÉLECTRON high-speed complementary metal oxide semiconductor (*HCMOS*); ~ **extrinsèque** ALIM ÉLEC, COMPOSANTS, ÉLECTRON extrinsic semiconductor; ~ **intrinsèque** ALIM ÉLEC, COMPOSANTS, ÉLECTRON intrinsic semiconductor; ~ **à oxyde metallique à grande vitesse** (*HMOS*) ÉLECTRON high-speed metal oxide semiconductor (*HMOS*); ~ **de puissance** ALIM ÉLEC, COMPOSANTS, ÉLECTRON power semiconductor; ~ **sur isolant métal** (*MIS*) ALIM ÉLEC, COMPOSANTS, ÉLECTRON metal insulator semiconductor (*MIS*); ~ **type intrinsèque** ALIM ÉLEC, COMPOSANTS, ÉLECTRON I-type semiconductor, intrinsic semiconductor; ~ **type N** ALIM ÉLEC, COMPOSANTS, ÉLECTRON N-type semiconductor; ~ **type P** ALIM ÉLEC, COMPOSANTS, ÉLECTRON P-type semiconductor

semi-duplex[1] *adj* RADIO, TÉL, TRANSMIS half-duplex, semi-duplex, simplex

semi-duplex[2] *m* RADIO, TÉL, TRANSMIS semiduplex, simplex

semi-graphique *adj* TESTS semigraphic

sens[1]: **à ~ unique** *adj* RADIO, TÉL one-way

sens[2] *m* ÉLECTRON, GÉNÉRAL, TÉL, TESTS, TRANSMIS direction; ~ **d'assemblage** GÉNÉRAL direction of lay; ~ **de blocage** ÉLECTRON *semi-conducteurs* blocking direction; ~ **de conduction** ÉLECTRON conducting direction; ~ **de conversation** TÉL speech direction; ~ **direct** ÉLECTRON forward direction; ~ **émission** TESTS, TRANSMIS transmit direction (*TX*); ~ **inverse** ÉLECTRON reverse direction; ~ **de liaison normal** GÉNÉRAL, TRANSMIS normal flow direction; ~ **de liaison secondaire** GÉNÉRAL, TRANSMIS secondary flow direction; ~ **longitudinal** GÉNÉRAL longitudinal direction; ~ **primaire de liaison** GÉNÉRAL, TRANSMIS primary flow direction; ~ **réception** TESTS, TRANSMIS receive direction (*RX*); ~ **de rotation** TÉL rotary direction; ~ **transmission** TESTS, TRANSMIS transmit direction

sensibilité *f* GÉNÉRAL sensitivity, susceptibility, responsivity, sensibility; ~ **d'accord** ÉLECTRON tuning sensitivity; ~ **de la commande de souffle** INFORMAT breath-controller sensitivity; ~ **au coude** ÉLECTRON knee sensitivity, knee-luminous flux; ~ **de détection** TRANSMIS detectivity; ~ **différentielle** TÉL differential sensitivity; ~ **énergétique** ÉLECTRON radiant sensitivity, responsivity, TRANSMIS responsivity; ~ **énergétique monochromatique relative** ÉLECTRON relative spectral sensitivity; ~ **d'entrée** TRANSMIS input sensibility; ~ **à l'erreur locale** RADIO site-error susceptibility; ~ **aux erreurs** DONNÉES error susceptibility; ~ **limite** TRANSMIS input sensibility; ~ **magnétique** ÉLECTRON magnetic sensitivity; ~ **maximale** GÉNÉRAL maximum sensitivity; ~ **maximale utilisable limitée par l'amplification** RADIO maximum usable sensitivity; ~ **à la modulation d'amplitude** INFORMAT amplitude modulation sensitivity; ~ **de la molette de modulation** INFORMAT modulation wheel sensitivity; ~ **à la pression après enfoncement** INFORMAT after-touch sensitivity; ~ **à la pression polyphonique** INFORMAT polyphonic-pressure sensitivity; ~ **au rayonnement** RADIO radiation sensitivity; ~ **du récepteur** RADIO receiver sensitivity (*RX sens*); ~ **de référence** RADIO, TRANSMIS reference sensibility, reference sensitivity; ~ **de la sonnerie** TÉL ringer sensitivity; ~ **spectrale** TRANSMIS spectral radiance, spectral responsivity; ~ **spectrale relative** ÉLECTRON relative spectral sensitivity; ~ **de synchronisation** GÉNÉRAL synchronization sensitivity; ~ **thermique** CLIMAT temperature sensitivity; ~ **à la vélocité de relâchement** INFORMAT release velocity sensitivity

sensible *adj* CLIMAT, DONNÉES, ÉLECTRON, GÉNÉRAL, INFORMAT, RADIO sensitive, responsive; ~ **aux décharges électrostatiques** TESTS electrostatic-discharge-sensitive; ~ **au point de vue stratégique** GÉNÉRAL commercially sensitive; ~ **à la pression** INFORMAT pressure-sensitive; ~ **à la vélocité de frappe** INFORMAT velocity-sensitive; ~ **à la vélocité du relâchement** INFORMAT release-velocity-sensitive

séparateur *m* GÉNÉRAL separator, TRANSMIS splitter; ~ **d'adresses** TÉL address separator; ~ **de blocs** TÉL block separator; ~ **de demandes de service complémentaire** TÉL facility request separator; ~ **d'éléments de données** DONNÉES data element separator; ~ **d'éléments de données constitutifs** DONNÉES component data element separator; ~ **de faisceau** INFORMAT, TRANSMIS *optique de fibres* beam splitter; ~ **de groupes** DONNÉES group separator (*GS*); ~ **de groupes de données** DONNÉES group mark; ~ **d'information** DONNÉES, ÉLECTRON information separator (*IS*); ~ **de rayon** INFORMAT, TRANSMIS *optique de fibres* beam splitter

séparation *f* ALIM ÉLEC, DONNÉES separation, ÉLECTRON separation, splitting, INFORMAT split, splitting, RADIO *fréquences* spacing, TÉL splitting, separation, TRANSMIS separation; ~ **angulaire** TRANSMIS angular separation; ~ **des appels** TÉL call splitting; ~ **des blocs** TÉL block separation

séparer[1] *vt* ÉLECTRON, INFORMAT divide up, split; ~ **des feuillets** INFORMAT burst

séparer[2]: **se ~** *v réfl* ÉLECTRON, INFORMAT split

septet *m* INFORMAT septet

séquence *f* GÉNÉRAL sequence; ~ **d'appel** TÉL call sequence, calling sequence; ~ **audio** INFORMAT audio sequence; ~ **de batterie** INFORMAT drum sequence; ~ **binaire** DONNÉES binary sequence; ~ **binaire pseudo-aléatoire** DONNÉES pseudorandom binary sequence (*PRBS*); ~ **câblée** CONTRÔLE, INFORMAT wired sequence controller; ~ **chaînée** DONNÉES chained sequence; ~ **du commutateur hyperfréquences** COMMUT microwave switch sequence; ~ **de commutation** COMMUT, TÉL switching sequence; ~ **de contrôle** TESTS check sequence; ~ **de contrôle de trame** DONNÉES, TÉL frame check sequence (*FCS*); ~ **directe** GÉNÉRAL direct sequence (*DS*); ~ **d'échappement** DONNÉES escape sequence; ~ **d'embrouillage** GÉNÉRAL scrambling sequence; ~ **d'entrée de paramètres avec format** INFORMAT format parameter-entry sequence; ~ **d'entrelacement** TV interlacing sequence; ~ **de fonctionnement dans le mode répétitif** TÉL continuation-mode operating sequence; ~ **de fonctionnement en mode menu** TÉL interactive-mode operating sequence, menu-mode operating sequence; ~ **d'images** TV image sequence; ~ **d'impulsions bidirectionnelles** TÉL bipolar sequence of pulses; ~ **d'introduction de bloc de paramètres** TÉL parameter block introduction sequence; ~ **de marquage** COMMUT marking sequence; ~ **de mesures à prendre en cas de panne** TESTS collapse action sequence; ~ **non chaînée** DONNÉES unchained sequence; ~ **nulle** DONNÉES invalid sequence; ~ **numérique** TRANSMIS digital sequence; ~ **numérique de pseudo-bruits** TRANSMIS digital pseudo-noise sequence; ~ **numérique de référence MIC** TÉL PCM digital reference sequence; ~ **paramétrable** INFORMAT skeletal code; ~ **de récupération du rythme binaire** DONNÉES, RADIO bit-timing recovery sequence; ~ **sémaphore** GÉNÉRAL signaling sequence (*AmE*), signalling sequence (*BrE*); ~ **de signalisation** GÉNÉRAL signaling sequence (*AmE*), signalling sequence (*BrE*); ~ **de symboles** DONNÉES symbol pattern; ~ **temporelle** GÉNÉRAL time sequence; ~ **unipolaire d'impulsions** TÉL unipolar sequence of pulses; ~ **de vérification de trame** DONNÉES, TÉL frame check sequence (*FCS*)

séquenceur *m* COMMUT, DONNÉES, INFORMAT, TÉL sequencer; ~ **fixe** DONNÉES fixed sequencer; ~ **microprogrammé** INFORMAT programmed sequence controller; ~ **multipistes** DONNÉES multitrack sequencer; ~ **programmable** COMMUT programmable sequencer; ~ **réparti** INFORMAT distributed-sequence controller

séquentiel[1] *adj* DONNÉES, ÉLECTRON, GÉNÉRAL, INFORMAT, TÉL sequential

séquentiel[2]: ~ **couleur à mémoire** *m* (*SECAM*) TV séquentiel couleur à mémoire (*SECAM*)

série *f* ALIM ÉLEC, ÉLECTRON series, INFORMAT, TÉL, TESTS strip, TRANSMIS series; ~ **de jacks d'essai** TESTS test jack strip; ~ **de jacks de test** TESTS test jack strip; ~ **de voyants lumineux** INFORMAT, TÉL, TESTS visual indicator strip

sérigraphie *f* COMPOSANTS screen printing

serre: **~-câble** *m* COMPOSANTS pull-relief clamp, ÉLECTRON cable clamp; **~-câbles** *m* ALIM ÉLEC strain relief, COMPOSANTS strain relief bracket; **~-joint** *m* COMPOSANTS screw clamp

serré *adj* COMPOSANTS tightened

serrer *vt* COMPOSANTS tighten

serrure: ~ **de sûreté** *f* TESTS safety lock

sertissage *m* ÉLECTRON crimp, crimping

servante: ~ **d'atelier** *f* COMPOSANTS shelved trolley

serveur *m* GÉNÉRAL server, hub, INFORMAT server, *service* on-line data service, *ordinateur* host computer; ~ **à accès multiples** TÉL S-universal interface; ~ **Audiotex** TÉL Audiotex server; ~ **à commande vocale** CONTRÔLE, PÉRIPH, TÉL voice command server; ~ **de communications** DONNÉES communications server; ~ **de destinations multiples** DONNÉES multicast server; ~ **de document** DONNÉES, INFORMAT, TESTS document server; ~ **de fichiers** DONNÉES, INFORMAT, TESTS file server; ~ **de fichiers multimédias** DONNÉES, INFORMAT, TESTS multimedia file server; ~ **de fichiers réparti** DONNÉES, INFORMAT, TESTS distributed file server; ~ **FTP** DONNÉES FTP server; ~ **de gestion** DONNÉES management database; ~ **d'imprimante** DONNÉES, INFORMAT printer server; ~ **d'information** DONNÉES information server; ~ **interactif** INFORMAT on-line host; ~ **de messagerie** DONNÉES, INFORMAT mail server; ~ **de messagerie vocale** DONNÉES, INFORMAT voice message server; ~ **de messages** (*IMP*) DONNÉES, INFORMAT interface message processor (*IMP*); ~ **multimédia** DONNÉES, INFORMAT multimedia server; ~ **de news** DONNÉES news server; ~ **de noms** DONNÉES name server; ~ **de réseau** DONNÉES network server; ~ **sans connexion** DONNÉES connectionless server (*CLS*); ~ **SDA virtuelle** TÉL virtual SDA server; ~ **de télécopie** DONNÉES, TÉL facsimile server; ~ **télématique** DONNÉES telematic server; ~ **de transfert fiable** INFORMAT reliable transfer server (*RTS*); ~ **vidéo** INFORMAT video server; ~ **vidéotex** INFORMAT, TÉL, TV videotex host, videotex server; ~ **vocal** DONNÉES, INFORMAT voice server; ~ **Web** DONNÉES, INFORMAT Web server; ~ **W3** DONNÉES, INFORMAT Web server

serveurs: ~ **d'alerte pour sites à hauts risques** *m pl* TESTS alert servers for high-risk sites; ~ **vocaux interactifs** *m pl* TÉL interactive voice data servers

servibilité *f* APP SERV, TÉL serveability, serveability performance, serviceability, serviceability performance

service[1]: **en** ~ *adj* RADIO hot

service[2] *m* GÉNÉRAL service; ~ **aux abonnés** DONNÉES, INFORMAT, RADIO, TÉL subscriber service; ~ **des abonnés absents** APP SERV, TÉL absent-subscriber service, answering service; ~ **abstrait mémoire de messages** DONNÉES MS abstract service; ~ **d'accès** DONNÉES access service; ~ **d'adaptation de syntaxe** DONNÉES syntax-matching service; ~ **d'affichage alphanumérique de renseignements** TÉL subscriber's alphanumerical display; ~ **d'amateur par satellite** RADIO amateur satellite service; ~ **amélioré** APP SERV enhanced service; ~ **des annotatrices** TÉL trunk ordering service; ~ **d'annuaire** (*DS*) DONNÉES directory services (*DS*); ~ **d'appel avec transfert automatique de taxe** TÉL automatic collect call (*AmE*), automatic reverse call (*BrE*), automatic transferred-charge call; ~ **d'appel direct** TÉL hot line; ~ **appel par appel** APP SERV call-by-call service, TÉL on-demand service; ~ **d'appel personnalisé** APP SERV custom-calling service; ~ **d'appels à destination fixe** TÉL fixed-destination call services; ~ **d'appels d'urgence** TÉL emergency call service; ~ **d'assistance annuaire** APP SERV, TÉL directory assistance, directory enquiries (*BrE*) (*DQ*), directory information (*AmE*); ~ **d'assurance qualité** APP SERV, TESTS quality assurance office; ~ **automatique** TÉL automatic service; ~ **automatique de cartes de crédit** APP SERV automatic credit card service; ~ **avec confirmation** GÉNÉRAL confirmed service; ~ **avec possibilité de rétablissement** RADIO restorable service; ~ **avec rétablissement possible** RADIO restorable service; ~ **de bas niveau** TESTS lower-level service; ~ **de base** TV basic service; ~ **bip** RADIO alert-only service; ~ **de boîte aux lettres électronique** INFORMAT electronic mailbox service; ~ **à bord de l'avion** RADIO in-flight telephone service; ~ **de bout en bout** TÉL end-to-end servicing; ~ **centralisateur** TV program booking center (*AmE*) (*PBC*), programme booking centre (*BrE*) (*PBC*); ~ **centrex** TÉL centrex service; ~ **de circuit commuté audio** APP SERV audio bearer service; ~ **de circuit commuté en canal B non transparent** APP SERV sixty-four kbps restricted service; ~ **de circuit commuté en canal B transparent** APP SERV clear sixty-four service; ~ **de circuit virtuel du RNIS** DONNÉES ISDN virtual-circuit bearer service; ~ **de circuit virtuel sur canal B** (*service CVB*) TÉL B-channel virtual-circuit service; ~ **commun d'informations de gestion** DONNÉES common management-information service (*CMIS*); ~ **de communication conférence** TÉL conference call service; ~ **de communication conférence unilatérale** TÉL lecture call; ~ **de communication multipoint** TÉL multipoint communication service; ~ **de communication personnel** TRANSMIS personal communications; ~ **de commutation des circuits** APP SERV circuit-switched service; ~ **de commutation par paquets** DONNÉES, TÉL packet-mode operation; ~ **commuté** COMMUT switched service; ~ **commuté de données multimégabit** DONNÉES switched multimegabit data service (*SMDS*); ~ **complémentaire** APP SERV, DONNÉES facility, TÉL connect facility; ~ **complémentaire d'assemblage de paquets pour la télécopie** DONNÉES facsimile packet assembly facility; ~ **complémentaire d'attente autorisée** TÉL waiting-allowed facility; ~ **complémentaire de désassemblage de paquets pour la télécopie** DONNÉES facsimile packet disassembly facility; ~ **complémentaire des enquêtes** TÉL information inquiry facility; ~ **complémentaire d'enregistrement de l'état ETTD** TÉL DTE-inactive registration facility; ~ **complémentaire d'enregistrement en ligne de service** TÉL on-line registration facility; ~ **complémentaire essentiel facultatif** TÉL essential optional user-facility; ~ **complémentaire de liaisons multipoints centralisées** DONNÉES, TÉL centralized multipoint facility; ~ **complémentaire de modification du bit D** INFORMAT D-bit modification facility; ~ **complémentaire offert en option à l'utilisateur** TÉL optional user-facility; ~ **complémentaire des renseignements** TÉL information facility; ~ **comportant un troisième correspondant** APP SERV, TÉL three-party service; ~ **de comptabilité sélective** TÉL selective accounting; ~ **confirmé** GÉNÉRAL confirmed service; ~ **confort** PÉRIPH star service; ~ **de consultation** INFORMAT retrieval service; ~ **conversationnel** INFORMAT conversational service; ~ **de la couche physique** TÉL, TESTS physical-layer service (*PLS*); ~ **de couche réseau sans connexion** APP SERV connectionless network-layer service (*CLNS*); ~ **CVB** (*service de circuit virtuel sur canal B*) TÉL B-channel virtual-circuit service; ~ **de datagramme** TÉL datagram service; ~ **à la demande** GÉNÉRAL demand service, TÉL on-demand service; ~ **de demande automatique d'une communication** TÉL automatic booked call; ~ **des demandes de secours** APP SERV emergency call service; ~ **de dépannage** TESTS fault reception center (*AmE*) (*FRC*), fault reception centre (*BrE*) (*FRC*); ~ **de destinations multiples** DONNÉES multicast service; ~ **de données à haut débit** APP SERV, DONNÉES connectionless broadband data service (*CBDS*); ~ **de données à large bande sans connexion** APP SERV, DONNÉES connectionless broadband data service (*CBDS*); ~ **EDA** (*service d'enregistrement des données d'appels*) APP SERV CDRS (*call-detail recording service*); ~ **EDC** (*service d'enregistrement des données de communications*) APP SERV CDRS (*call-detail recording service*); ~ **engendré par le fournisseur** DONNÉES provider-initiated service; ~ **d'enregistrement de communications d'arrivée** TÉL registration of incoming calls; ~ **d'enregistrement des données d'appels** (*service EDA*) APP SERV call-detail recording service (*CDRS*); ~ **d'enregistrement des données de communications** (*service EDC*) APP SERV call-detail recording service (*CDRS*); ~ **d'enregistrement écrit des éléments de taxation** APP SERV, TÉL printed record of duration and charge of calls service; ~ **d'exploitation internationale** APP SERV international operations service; ~ **fixe aéronautique** RADIO aeronautical fixed network (*AmE*), aeronautical fixed service (*BrE*), aeronautical fixed system (*AmE*); ~ **à frais virés** *Can* APP SERV reversed charging, TÉL collect-call service (*AmE*), reversed charging, transfer charging; ~ **des fréquences étalon par satellite** RADIO standard-frequency satellite service; ~ **de gestion des systèmes** INFORMAT systems management service; ~ **hors vacation** DONNÉES night service; ~ **d'identification d'appels malveillants** TÉL malicious call identification service; ~ **indépendant du support** TÉL bearer independent service (*BIS*); ~ **d'indicateur de taxe au domicile de l'abonné** TÉL subscriber call charge meter; ~ **d'indication automatique verbale des éléments de taxe** TÉL automatic verbal announcement of charges service; ~ **d'indication du numéro du demandeur** TÉL calling-number indication; ~ **d'information en ligne** DONNÉES on-line information service; ~ **d'informations enregistrées** TÉL recorded service; ~ **d'informations relatives à la disponibilité des systèmes** TÉL system availability information point; ~ **interactif** INFORMAT conversational service, TV interactive service; ~ **d'interception d'appels** TÉL intercept service; ~ **international de comptes télégraphiques transférés** TÉL international transferred-account telegraph service; ~ **international rapide** TÉL *exploitation manuelle* international manual demand service; ~ **inter-réseaux** TÉL network utility; ~ **d'interruption d'une conversation en cours** TÉL interruption of a call in progress; ~ **interurbain planifié** (*WATS*) TÉL wide-area telephone service (*WATS*); ~ **de liaison de données** APP SERV, DONNÉES data-link service (*DLS*); ~ **"libre-appel"** TÉL freephone service (*BrE*), toll-free service (*AmE*); ~ **en ligne** DONNÉES on-line service; ~ **de limitation des appels d'arrivée** TÉL incoming-call barring; ~ **de limitation des communications de départ** TÉL restriction in the outgoing direction; ~ **de location de circuits pour transmission de données** TRANSMIS leased-circuit data transmission service; ~ **maritime par satellite** RADIO maritime satellite service (*MSS*); ~ **de messagerie** APP SERV messaging service; ~ **de messagerie boîte aux lettres** INFORMAT mailbox messaging service; ~ **de messagerie vocale** APP SERV, TÉL voice messaging service; ~ **mixte parole** APP SERV alternate speech service; ~ **mobile** RADIO mobile service; ~ **mobile aéronautique** RADIO aeronautical mobile service; ~ **mobile aéronautique par satellite** RADIO aeronautical mobile satellite-service; ~ **mobile maritime** RADIO maritime mobile service; ~ **mobile maritime par satellite** RADIO maritime mobile-service satellite; ~ **mobile maritime de terre** RADIO maritime mobile terrestrial service; ~ **mobile terrestre par satellite** (*SMTS*) RADIO, TÉL, TRANSMIS land-mobile satellite service (*LMSS*); ~ **multimédia** TV multimedia service; ~ **N** TÉL N-service; ~ **ne pas déranger** TÉL do-not-disturb-service; ~ **de niveau élevé** APP SERV higher-level service; ~ **non confirmé** DONNÉES nonconfirmed service; ~ **de notification de gestion** DONNÉES management notification service; ~ **de nuit** DONNÉES night service; ~ **de nuit à renvoi fixe** TÉL fixed night-service; ~ **de nuit à renvoi sélectif** TÉL flexible night-service; ~ **de nuit universel** TÉL universal night-service; ~ **numérique européen de téléphonie sans fil** (*DECT*) ALIM ÉLEC digital European cordless telephone service (*DECT*); ~ **de numéro d'abonné permanent** TÉL permanent subscriber number; ~ **de numéros de téléphone doubles** TÉL dual telephone numbers; ~ **de numéros universels** TÉL universal-access number service; ~ **de numérotation abrégée** TÉL abbreviated dialing service (*AmE*), abbreviated dialling service (*BrE*); ~ **offert aux usagers** TÉL user facility, user service; ~ **d'opération de gestion** DONNÉES management operation service; ~ **optionnel** TV optional service, premium service; ~ **par satellite** RADIO, TV satellite service; ~ **particulier d'information enregistrée** TÉL customer recorded-information service; ~ **payable à l'arrivée** APP SERV, TÉL collect-call service (*AmE*), reversed charging; ~ **payant** GÉNÉRAL fee-based service; ~ **pays-direct** TÉL home-direct service; ~ **PCV** APP SERV, TÉL collect-call service (*AmE*), reversed charging, transfer charging; ~ **physique** (*PHS*) APP SERV, TÉL physical service (*PhS*); ~ **de présentation** DONNÉES presentation service; ~ **de présentation en mode sans connexion** DONNÉES connectionless-mode presentation service; ~ **de prise d'une communication établie** TÉL pick-up facility; ~ **public d'informations enregistrées** TÉL recorded public information service; ~ **public international de télécopie** TÉL international public facsimile service; ~ **public de transmission de données** TRANSMIS public data-transmission service; ~ **publiphone** APP SERV, TÉL payphone service; ~ **radio amateur** RADIO amateur radio service; ~ **de radiocommunication** RADIO radiocommunication service; ~ **de radiodiffusion par satellite** (*SRS*) RADIO, TV broadcasting-satellite service, satellite broadcasting service; ~ **radioélectrique d'appel unidirectionnel** RADIO radiopaging service; ~ **de radiolocalisation par satellite** RADIO radio-location satellite service; ~ **de radiorepérage par satellite** RADIO radio-determination satellite service (*RDSS*); ~ **radiotéléphonique mobile** (*SRM*) RADIO, TÉL mobile telephone service (*MTS*); ~ **radiotéléphonique mobile à courte portée** RADIO, TÉL limited roaming telephone service; ~ **radio-télex maritime automatique** TÉL automatic maritime radio-telex service; ~ **de rappel de date** TÉL diary service; ~ **de rappel de rendez-vous** TÉL reminder service; ~ **RDPCP** DONNÉES PPSDN service; ~ **de recherche de ligne libre dans un commutateur** COMMUT PBX line-hunting service; ~ **de recherche de lignes PBX** TÉL PBX hunting; ~ **de recherche de personnes** APP SERV, RADIO, TÉL paging service; ~ **de réclamation des dérangements** TÉL fault complaint service; ~ **à rémunération forfaitaire** APP SERV flat-rate service; ~ **de renseignements téléphoniques** APP SERV, TÉL directory enquiries (*BrE*) (*DQ*), directory information (*AmE*); ~ **de répartition** TRANSMIS dispatch service, dispatch system; ~ **de répétition de numéro** TÉL number repetition service; ~ **de réseau** DONNÉES, TÉL network service (*NS*); ~ **de réseau d'interconnexion de systèmes ouverts** (*service de réseau ISO*) APP SERV, DONNÉES, TÉL open systems interconnection network service (*OSI network service*); ~ **de réseau ISO** (*service de réseau d'interconnexion de systèmes ouverts*) APP SERV, DONNÉES, TÉL OSI network service (*open systems interconnection network service*); ~ **de réseau en mode connexion** APP SERV, DONNÉES, TÉL connection-oriented network service (*CONS*); **~ de réseau sans connexion** APP SERV, DONNÉES, TÉL connectionless-mode network service; **~ de réseau Xerox** (*XNS*) DONNÉES Xerox Network Service (*XNS*); **~ restreint** GÉNÉRAL restricted service, TÉL restricted service, call restriction; **~ restreint de transfert d'informations** APP SERV restricted information-transfer service; **~ de réveil** TÉL alarm call service; **~ de réveil automatique** TÉL automatic wake-up (*AWU*), wake-up service; **~ RNIS à large bande** APP SERV, TÉL B-ISDN service, broadband ISDN service; **~ sans confirmation** DONNÉES nonconfirmed service; **~ sans restriction** APP SERV, TÉL unrestricted service; **~ de session** DONNÉES session service; **~ de session en mode sans connexion** DONNÉES connectionless-mode session service; **~ de signal de retour d'appel d'un numéro particulier** SIGNAL private-number ringing service; **~ de signaux horaires par satellite** RADIO, SIGNAL time signal satellite service; **~ simplifié** DONNÉES night service; **~ de sous-réseau** DONNÉES subnet-work service; **~ spécial** TÉL special service; **~ spécial d'interception** TÉL special intercept; **~ supplémentaire** APP SERV supplementary service, DONNÉES facility, RADIO supplementary service, TÉL supplementary service, facility; **~ support** APP SERV, COMMUT, DONNÉES, INFORMAT, RADIO, SIGNAL, TÉL, TRANSMIS bearer service; **~ support audiofréquence** APP SERV audio bearer service; **~ support de circuit virtuel** DONNÉES, TRANSMIS virtual-circuit bearer service; **~ support à commutation de circuits** APP SERV circuit-mode bearer service; **~ support à commutation de paquets** APP SERV, COMMUT packet-switched bearer service; **~ support en mode circuit** APP SERV circuit-mode bearer service; **~ support en mode paquet** DONNÉES, TÉL packet-mode bearer service; **~ support en mode trame** APP SERV, TRANSMIS frame-mode bearer service (*FMBS*); **~ support non transparent** APP SERV, DONNÉES nontransparent bearer service; **~ support sans connexion** APP SERV connectionless bearer service; **~ de support sans restriction** APP SERV, TRANSMIS unrestricted bearer service; **~ de support de signalisation d'usager** DONNÉES, SIGNAL user-signaling bearer service (*AmE*), user-signalling bearer service (*BrE*); **~ support transparent** APP SERV, DONNÉES transparent bearer service; **~ support à 64 kbit/s sans restriction** APP SERV sixty-four kbps unrestricted bearer service; **~ support à 64 kbit/s avec restriction** APP SERV sixty-four kbps restricted bearer service; **~ à supports multiples** TÉL multibearer service; **~ de surveillance** APP SERV observation service; **~ de surveillance de la qualité des télécommunications** TÉL observation service; **~ à tarif de base** APP SERV basic rate service; **~ à tarif fixe** APP SERV flat-rate service; **~ de téléappel** APP SERV, RADIO, TÉL paging service; **~ de télécommunications** GÉNÉRAL telecommunications service; **~ de télécopie** DONNÉES, TÉL facsimile service; **~ télégraphique** TÉL telegraphic office (*TO*), telegraphic service; **~ télégraphique public** TÉL public telegram service; **~ téléimprimeur** APP SERV, PÉRIPH, TÉL teleprinter service (*BrE*), teletypewriter service (*AmE*); **~ de télématique** DONNÉES, INFORMAT, RADIO on-line service, telematics service; **~ télématique en ligne** DONNÉES, INFORMAT, RADIO on-line telematic service; **~ du téléphone en vol** RADIO in-flight telephone service; **~ téléphonique** TÉL telephone service (*TS*); **~ de téléphotographie** TÉL telephoto service; **~ de téléréunion** TÉL conference call service; **~ de télétexte** APP SERV teletext service; **~ temporaire** APP SERV short-term duty; **~ de transfert d'appel en cas de numéro occupé** TÉL automatic transferred debiting of charges, diversion-if-number-busy service; **~ de transfert d'informations avec accusé de réception** APP SERV, TÉL acknowledged information-transfer service (*AITS*); **~ de transfert d'informations sans accusé de réception** APP SERV, TÉL unacknowledged information transfer service (*UITS*); **~ de transfert d'informations sans restriction** APP SERV, TÉL unrestricted information-transfer service; **~ de transfert sans restriction** APP SERV, TÉL clear channel capability; **~ de transfert sur longueur d'onde privilégiée** APP SERV, TÉL clear channel capability; **~ de transmission de données** DONNÉES, TRANSMIS data transmission service; **~ de transmission de données avec commutation de circuits** DONNÉES, TRANSMIS circuit-switched data transmission service; **~ de transmission de données avec commutation de paquets** COMMUT, TRANSMIS packet-switched data transmission service; **~ de transmission d'un message verbal** TÉL, TRANSMIS verbal-message transmission service; **~ de transmission de messages** TRANSMIS message relay; **~ de transport** TÉL, TRANSMIS transport service (*TS*); **~ de type confirmé** GÉNÉRAL confirmed service; **~ d'urgence** APP SERV emergency service; **~ à valeur ajoutée de données** APP SERV, DONNÉES value-added data service; **~ vidéotex** APP SERV, TÉL, TV videotex service; **~ visiophonique** APP SERV, RADIO, TÉL, TV videophone service, visual telephone service; **~ vocal** APP SERV speech service; **~ vocal interactif** (*SVI*) APP SERV interactive voice service

Service: **~ de la réglementation des télécommunications** *m* (*SRT*) RADIO Telecommunications Regulatory Service (*TRS*); **~ Télématique** *m* TÉL bulletin board (*BB*)

services: **~ d'appels en instance** *m pl* TÉL call-waiting services; **~ de communication** *m pl* APP SERV communication services (*CS*); **~ de communications personnelles** *m pl* APP SERV personal communications services (*PCS*); **~ complémentaires** *m pl* DONNÉES add-on services; **~ facturés** *m pl* TÉL billed services; **~ d'intervention** *m pl* TV emergency service; **~ radiotéléphoniques mobiles** *m pl* RADIO, TÉL mobile radio telephone services; **~ d'urgence** *m pl* APP SERV emergency services; **~ à valeur ajoutée** *m pl* (*SAVA*) APP SERV, DONNÉES, TRANSMIS, TV value-added network services (*VANS*); **~ vocaux** *m pl* GÉNÉRAL voice services

servoamplificateur *m* ALIM ÉLEC, ÉLECTRON servoamplifier

servomécanisme *m* CONTRÔLE servomechanism

servosystème *m* CONTRÔLE servosystem

session *f* DONNÉES, TRANSMIS session

seuil *m* ALIM ÉLEC, ÉLECTRON, RADIO, SIGNAL, TÉL, TESTS, TRANSMIS threshold; **~ d'audition** RADIO, TÉL threshold of audibility, threshold of hearing, auditory threshold; **~ de bruit** TÉL noise threshold; **~ de décodage** RADIO decoding threshold; **~ de déploiement** RADIO deployment threshold; **~ de détection** ÉLECTRON, TÉL detection threshold, TRANSMIS sensitivity, detection threshold; **~ différentiel** TÉL differential threshold; **~ d'écoulement** GÉNÉRAL flow stress; **~ d'effet laser** ÉLECTRON, TRANSMIS lasing threshold; **~ d'émission laser** ÉLECTRON, TRANSMIS lasing threshold; **~ d'un flux lumineux** ÉLECTRON luminous-flux threshold; **~ de Geiger** RADIO, TRANSMIS Geiger threshold; **~ de gêne** RADIO, TÉL threshold of discomfort; **~ inférieur de brouillage** RADIO minimum interference threshold; **~ de longueur d'onde** ÉLECTRON, TRANSMIS threshold wavelength; **~ de mobilité** TESTS discrimination threshold; **~ de perception du courant** ALIM ÉLEC, ÉLECTRON, TRANSMIS threshold current; **~ du récepteur** RADIO receiver threshold level; **~ de sensibilité du dispositif avertisseur** TÉL alerting sensitivity; **~ de signal** SIGNAL signal threshold

seul: **d'un ~ bâti** *adj* ÉLECTRON single-bay

seule: **une ~ voie par porteuse** *loc* RADIO single channel per carrier (*SCPC*)

SGBD *abrév* (*système de gestion de bases de données*) DONNÉES, INFORMAT DBMS (*database management system*)

SGE *abrév* (*seconde gravement erronée*) DONNÉES, RADIO SES (*severely-errored second*)

shannon *m* GÉNÉRAL, TÉL shannon

shareware *m* INFORMAT shareware

SHARP *m* TRANSMIS SHARP (*stationary high-altitude relay platform*)
shell *m* DONNÉES shell
Shoebox *m* INFORMAT Shoebox
shunt *m* ALIM ÉLEC, COMPOSANTS shunt, shunt resistor
shuntage *m* ALIM ÉLEC, COMPOSANTS shunting
shunter *vt* ALIM ÉLEC, COMPOSANTS shunt, bypass
SIA *abrév* (*signal d'indication d'alarme*) TÉL, TESTS AIS (*alarm indication signal*)
SICT *abrév* (*système interurbain à communications tarifées*) TÉL MTS (*message toll service*)
siège: **~ de coussinet** *m* COMPOSANTS *génie mécanique* bearing seat
siemens *m* ALIM ÉLEC siemens
sifflement *m* COMPOSANTS singing, TÉL howling
SIGhyper *m* INFORMAT SIGhyper
sigle *m* GÉNÉRAL logo
signal *m* GÉNÉRAL signal; **~ d'abonné libre, sans taxation** TÉL *système de signalisation R2* subscriber-line free, no charge; **~ d'abonné occupé** SIGNAL electrical subscriber-busy signal, subscriber-busy signal, subscriber-line busy; **~ d'abonné transféré** SIGNAL subscriber-transferred signal; **~ d'acceptation** SIGNAL acceptance signal; **~ d'acceptation d'appel** SIGNAL, TÉL call-accepted signal, call acceptance signal; **~ d'acceptation d'enregistrement** SIGNAL registration-accepted signal; **~ d'accès au bus** COMMUT bus connection signal; **~ d'accès interdit** SIGNAL access barred signal; **~ d'accusé de réception** SIGNAL acknowledgement signal; **~ d'accusé de réception de blocage** SIGNAL blocking acknowledgement signal; **~ d'accusé de réception de commutation manuelle** SIGNAL manual-changeover acknowledgement signal; **~ d'accusé de réception de liaison de réserve prête** SIGNAL standby-ready acknowledgement signal; **~ d'accusé de réception de libération de circuit** SIGNAL circuit-released acknowledgement signal; **~ d'accusé de réception des multiblocs** SIGNAL multiblock acknowledgement signal, load-transfer acknowledgement signal; **~ d'accusé de réception de prise** SIGNAL seizing acknowledgement signal; **~ d'accusé de réception de réinitialisation** SIGNAL reset-band acknowledgement signal; **~ d'accusé de réception d'un signal d'autorisation de transfert** SIGNAL, TÉL transfer-allowed acknowledgement signal; **~ acoustique** DONNÉES acoustic signal; **~ d'action** ALIM ÉLEC forward signal; **~ d'activation de la porte d'entrée** ÉLECTRON, SIGNAL input-gate signal; **~ d'adresse** TÉL address signal (*AS*); **~ d'adresse abrégée** TÉL abbreviated address signal (*AAS*); **~ d'adresse complet** SIGNAL address-complete signal; **~ d'adresse incomplet** SIGNAL address-incomplete signal; **~ d'alarme** RADIO detector, TÉL alarm signal; **~ aléatoire** ÉLECTRON random signal; **~ d'alphabet** SIGNAL alphabetic signal; **~ alphabétique** SIGNAL alphabetic signal; **~ analogique** INFORMAT, SIGNAL analog signal; **~ d'analyse** SIGNAL scanner signal; **~ d'appel** (*SA*) TÉL call signal, call waiting, calling signal; **~ d'appel attente** TÉL call waiting; **~ d'appel audible** TÉL tone ringing; **~ d'appel cadencé** TÉL interrupted ringing signal; **~ d'appel en instance** TÉL call-waiting indication; **~ d'appel réacheminé** SIGNAL redirected call signal; **~ d'appel synchronisé** TÉL synchronized-ringing signal; **~ d'arrêt** DONNÉES, SIGNAL, TÉL stop signal, stop-send signal; **~ d'arrêt d'émission** DONNÉES, SIGNAL, TÉL stop-send signal; **~ audible** GÉNÉRAL *acoustique* audible signal; **~ audiofréquence** SIGNAL audio signal; **~ audio en mode d'attente** TÉL audio signal on hold; **~ d'autorisation de transfert** SIGNAL, TÉL transfer-allowed signal; **~ autosynchronisé** SIGNAL self-clocking signal; **~ d'avis aux navigateurs** RADIO navigation warning signal; **~ de balayage** INFORMAT sweep signal; **~ à bande étroite** ÉLECTRON, SIGNAL, TRANSMIS narrow-band signal; **~ de base** INFORMAT basic signal; **~ binaire** DONNÉES binary signal; **~ bip** SIGNAL tone signal; **~ bipolaire** SIGNAL, TÉL, TRANSMIS bipolar signal; **~ bipolaire alternant** SIGNAL, TÉL, TRANSMIS alternate mark inversion signal (*AMI*); **~ bipolaire modifié** SIGNAL, TÉL, TRANSMIS modified alternate mark inversion; **~ de blanc nominal** TÉL nominal white signal; **~ de blocage** TÉL blocking signal; **~ de bourrage DX** SIGNAL DX-stuffing signal; **~ de caractère** SIGNAL character signal; **~ carré** ÉLECTRON, SIGNAL square signal, square waveform; **~ de changement de canal** RADIO change channel signal; **~ de changement de numéro** SIGNAL changed-number signal; **~ de commande** INFORMAT actuating signal, SIGNAL control signal; **~ de commande du codage** DONNÉES coding control output; **~ de commande pour l'établissement des communications** SIGNAL call setup control signal; **~ de commande numérique** CONTRÔLE, DONNÉES digital command signal (*DCS*); **~ de commande temporelle** SIGNAL timing signal; **~ de commutation** SIGNAL changeover signal; **~ de commutation manuelle sur liaison de réserve** SIGNAL manual-changeover signal; **~ composé entièrement un** SIGNAL all-ones condition signal; **~ composite** ÉLECTRON composite waveform, SIGNAL aggregate signal, composite signal, compound signal; **~ de comptage** TÉL metering signal; **~ de compte rendu attendu** SIGNAL scheduled reporting signal; **~ de confusion** SIGNAL confusion signal; **~ de connexion** SIGNAL call-connected signal; **~ de connexion** SIGNAL connect signal; **~ de connexion sur le bus** COMMUT bus connection signal; **~ de connexion de transit** SIGNAL, TÉL transit through-connect signal; **~ de consigne** CONTRÔLE set point; **~ continu** DONNÉES continuous signal; **~ continuez** TÉL *télex* go-ahead signal; **~ de continuité** SIGNAL continuity signal; **~ de contrôle d'erreurs** RADIO error-check signal; **~ de correction linéaire** TV line tilt correction; **~ de correction linéaire de trame** TV field tilt correction; **~ de correction parabolique de ligne** TV line bend correction; **~ de correction parabolique de trame** TV field bend correction; **~ de correction de trapèze-ligne** TV line keystone correction; **~ de correction de trapèze-trame** TV field keystone correction, field keystone waveform; **~ de couleur** INFORMAT color signal (*AmE*), colour signal (*BrE*); **~ de couleur primaire** INFORMAT primary-color signal (*AmE*), primary-colour signal (*BrE*); **~ de coupure de l'alimentation de l'ETCD** SIGNAL DCE-power off signal; **~ de déblocage** SIGNAL unblocking signal; **~ de déblocage de l'émetteur** PÉRIPH, SIGNAL, TÉL, TRANSMIS transmitter turn-on signal; **~ de début de bloc** SIGNAL, TRANSMIS start-of-block signal; **~ de déclenchement de l'émetteur** PÉRIPH, SIGNAL, TÉL, TRANSMIS transmitter turn-on signal; **~ de déclenchement périodique** RADIO gating signal; **~ déclencheur** ÉLECTRON gate signal; **~ de déconnexion de l'enregistreur** TÉL register-disconnect signal; **~ de défaillance du réseau** SIGNAL network failure signal; **~ de défaillance du réseau sur la ligne d'abonné** SIGNAL network fault in local loop signal; **~ de dégradation du service** SIGNAL degraded-service signal; **~ de demande d'annulation** SIGNAL cancellation-request signal; **~ de demande d'appel** SIGNAL call request signal; **~ de demande d'enregistrement** SIGNAL registration request signal; **~ de demande de réacheminement** SIGNAL redirection request signal; **~ de départ** DONNÉES, SIGNAL start signal; **~ déphasé** GÉNÉRAL out-of-phase signal; **~ de dérangement** TÉL breakdown signal; **~ de détection de porteuse** DONNÉES carrier-sense signal; **~ déterministe à temps discret** INFORMAT, TRANSMIS discrete-time deterministic signal; **~ de détresse** TESTS danger signal, distress signal; **~ discret** INFORMAT, TRANSMIS discretely-timed signal; **~ de données** DONNÉES, SIGNAL data signal; **~ d'échantillonnage** GÉNÉRAL strobe; **~ échantillonné**

SIGNAL, TRANSMIS sampled signal; ~ **d'échec d'appel** SIGNAL call failure signal; ~ **électrique d'abonné occupé** SIGNAL electrical subscriber-busy signal; ~ **d'encombrement de l'équipement de commutation** COMMUT, SIGNAL switching-equipment congestion signal; ~ **d'encombrement du faisceau de circuits** DONNÉES, SIGNAL circuit group-congestion signal; ~ **d'encombrement du réseau** SIGNAL network congestion signal; ~ **d'encombrement sur le réseau national** SIGNAL national-network congestion signal; ~ **d'enregistrement effectué** SIGNAL registration-completion signal; ~ **d'enregistreur** SIGNAL, TÉL register signal; ~ **entrant** SIGNAL, TRANSMIS incoming signal; ~ **d'entrée** ÉLECTRON input signal; ~ **d'entrée interne** ÉLECTRON internal input signal; ~ **d'entrée vers l'arrière** SIGNAL, TÉL backward-input signal; ~ **d'entrée vers l'avant** SIGNAL *maritime mobile*, TÉL forward-input signal; ~ **envoyez le chiffre suivant** SIGNAL, TÉL send next digit signal; ~ **envoyez la tonalité spéciale d'information** SIGNAL, TÉL send special information tone signal; ~ **d'EPR hors service** SIGNAL, TÉL RPOA out-of-order signal; ~ **d'erreur** CONTRÔLE error signal; ~ **d'essai** SIGNAL, TESTS test signal; ~ **d'essai d'insertion** SIGNAL, TESTS insertion test signal (*ITS*); ~ **d'état du bus** COMMUT bus state signal; ~ **d'exécution d'annulation** SIGNAL cancellation-completed signal; ~ **extérieur d'interruption** COMMUT external interrupt signal; ~ **fantôme** TV ghost signal; ~ **FI** ÉLECTRON, RADIO IF signal; ~ **de fin** SIGNAL clear-forward signal, disconnect signal; ~ **de fin de bloc** SIGNAL end-of-block signal; ~ **de fin de communication** DONNÉES end-of-communication signal, TÉL clear-back signal; ~ **de fin de numérotation** SIGNAL *système de signalisation R2* end-of-pulsing signal; ~ **de fin de page** DONNÉES end-of-page signal; ~ **de fin de sélection d'enregistreur** TÉL register end-of-selection signal; ~ **à fréquence intermédiaire** ÉLECTRON, RADIO, SIGNAL intermediate frequency signal; ~ **de gestion** SIGNAL management signal; ~ **hertzien** RADIO, SIGNAL microwave signal; ~ **horaire** RADIO, SIGNAL time signal; ~ **d'horloge des octets** SIGNAL octet timing signal; ~ **d'identification de la ligne appelante** SIGNAL calling-line identification signal; ~ **d'identification de la ligne appelée** SIGNAL called-line identification signal; ~ **d'image** INFORMAT, TV picture signal; ~ **d'image vidéo en positif** INFORMAT positive video signal; ~ **d'incompatibilité de la catégorie d'usagers du service** SIGNAL incompatible service-user class signal; ~ **d'indicatif** DONNÉES, SIGNAL answer-back signal; ~ **d'indication d'alarme** (*SIA*) TÉL, TESTS alarm indication signal (*AIS*); ~ **d'indication d'alarme de section de multiplexage** (*SM-SIA*) TRANSMIS multiplex-section alarm indication signal; ~ **d'inoccupation** TÉL free-line signal (*FLS*); ~ **d'insertion de référence** SIGNAL insertion reference signal (*IRS*); ~ **intempestif de prise** SIGNAL false seizure; ~ **d'interdiction de transfert** SIGNAL, TÉL transfer-prohibited signal; ~ **interne** SIGNAL internal signal; ~ **d'interrogation** RADIO interrogating signal; ~ **d'intervention** SIGNAL forward transfer signal; ~ **d'invalidation** ÉLECTRON inhibiting signal; ~ **d'inversion chiffres** SIGNAL, TÉL figure-shift signal; ~ **d'inversion lettres** SIGNAL, TÉL letter-shift signal; ~ **invitant à différer la numérotation** SIGNAL *système de signalisation R1* delay-dialing signal (*AmE*), delay-dialling signal (*BrE*); ~ **d'invitation à émettre en transit** SIGNAL, TÉL transit proceed-to-send signal; ~ **d'invitation à transmettre** SIGNAL, TÉL *système de signalisation R1* TÉL proceed-to-send signal, start-dialing signal (*AmE*), start-dialling signal (*BrE*); ~ **KP** SIGNAL, TÉL *système de signalisation R1* KP signal; ~ **de lampe** TÉL lamp signal; ~ **à large bande** ÉLECTRON, SIGNAL, TRANSMIS wideband signal; ~ **de libération** DONNÉES disconnect signal, release signal; ~ **de libération en avant** TÉL clear-forward signal; ~ **de libération du circuit** SIGNAL circuit-released signal; ~ **de libération de l'enregistreur** SIGNAL register-release signal, TÉL register-clear signal, register-release signal; ~ **de libération de garde** SIGNAL, TÉL release-guard signal; ~ **de libération par le dernier abonné** TÉL last-party release signal; ~ **de libération par les deux abonnés** TÉL both party release signal; ~ **de ligne d'abonné en dérangement** TÉL *système de signalisation R2* subscriber-line out of order; ~ **de ligne hors service** SIGNAL line out-of-service signal, out-of-order signal; ~ **de luminance** TV luminance signal, brightness; ~ **lumineux** GÉNÉRAL light indicator; ~ **minimal détectable** (*SMD*) TRANSMIS minimum detectable signal (*MDS*); ~ **de mise à un** CONTRÔLE, SIGNAL set signal; ~ **de mise en phase** TÉL phasing signal; ~ **de mise en phase sur blanc** TÉL phase-white; ~ **de mise en phase sur noir** TÉL phase-black; ~ **de modulation** TRANSMIS modulating signal; ~ **modulé** SIGNAL, TRANSMIS modulated signal; ~ **monophonique** RADIO monophonic signal; ~ **n-aire redondant** SIGNAL redundant n-ary signal; ~ **de neutralisation d'alarme** (*SNA*) TESTS, TRANSMIS alarm inhibit signal; ~ **à niveau élevé** INFORMAT high-level signal; ~ **de noir nominal** TÉL nominal black signal; ~ **non filtré** SIGNAL unnotched signal; ~ **non prêt automatique** SIGNAL, TÉL uncontrolled not-ready signal; ~ **non prêt commandé** SIGNAL, TÉL controlled not-ready signal; ~ **de non-abonnement à l'acceptation de la taxation à l'arrivée** SIGNAL, TÉL reverse-charge acceptance not subscribed signal (*BrE*); ~ **de nouvelle réponse** SIGNAL reanswer signal; ~ **numérique** INFORMAT digital signal; ~ **numérique n-aire** SIGNAL n-ary digital signal; ~ **de numéro complet, ligne d'abonné libre, sans taxation** SIGNAL address-complete, subscriber-free signal, no charge; ~ **de numéro complet, sans taxation** SIGNAL address-complete signal, no charge; ~ **de numéro incomplet** SIGNAL address-incomplete signal; ~ **de numéro inutilisé** SIGNAL, TÉL unallocated-number signal; ~ **de numéro non accessible** SIGNAL unobtainable signal; ~ **de numéro non utilisé** SIGNAL, TÉL *système de signalisation R2* unallocated-number signal; ~ **de numéro occupé** SIGNAL number-busy signal; ~ **d'occupation de la ligne du terminal demandé** SIGNAL, TÉL terminal-engaged signal; ~ **OK** RADIO *service maritime mobile* OK signal; ~ **d'onde pilote** RADIO pilot signal; ~ **d'opérateur assistante** TÉL assistant operator signal; ~ **optique** INFORMAT, SIGNAL visual signal, TRANSMIS optical signal; ~ **orthogonal** RADIO orthogonal signal; ~ **de page partielle EOM** (*PPS-EOM*) DONNÉES, TÉL partial-page signal EOM (*PPS-EOM*); ~ **de page partielle EOP** (*PPS-EOP*) DONNÉES, TÉL partial-page signal EOP (*PPS-EOP*); ~ **de page partielle MPS** (*PPS-MPS*) DONNÉES, TÉL partial-page signal MPS (*PPS-MPS*); ~ **de page partielle NULL** (*PPS-NULL*) DONNÉES, TÉL partial-page signal NULL (*PPS-NULL*); ~ **de page partielle PRI-EOM** (*PPS-PRI-EOM*) DONNÉES, TÉL partial-page signal PRI-EOM (*PPS-PRI-EOM*); ~ **de page partielle PRI-EOP** (*PPS-PRI-EOP*) DONNÉES, TÉL partial-page signal PRI-EOP (*PPS-PRI-EOP*); ~ **de page partielle PRI-MPS** (*PPS-PRI-MPS*) DONNÉES, TÉL partial-page signal PRI-MPS (*PPS-PRI-MPS*); ~ **par impulsion** SIGNAL pulse signal; ~ **parasite** ALIM ÉLEC noise signal, perturbing signal, GÉNÉRAL interfering signal, SIGNAL spurious signal; ~ **de parole** SIGNAL, TÉL speech signal; ~ **périodique** ÉLECTRON periodic signal; ~ **perturbateur** ALIM ÉLEC noise signal, perturbing signal, interfering signal; ~ **de porte** ÉLECTRON gate signal, RADIO gating signal; ~ **de porte logique d'une largeur de symbole** DONNÉES one-symbol gate signal; ~ **porteur** INFORMAT carrier signal; ~ **pour plusieurs pages** DONNÉES, TÉL multipage signal (*MPS*); ~ **prioritaire** SIGNAL priority signal; ~ **de prise** SIGNAL, TÉL seizing signal, seizure signal; ~ **de prise de transit** SIGNAL, TÉL transit seizing signal, transit seizure signal;

~ de progression de l'appel SIGNAL, TÉL call progress signal (*CPS*); **~ pseudo-aléatoire** ÉLECTRON, TÉL pseudorandom signal; **~ pseudo-ternaire** ÉLECTRON, TÉL pseudoternary signal; **~ qui est là?** COMMUT, SIGNAL who-are-you signal (*WRU*); **~ de raccrochage** SIGNAL clear-back signal, hang-up signal, TÉL on-hook signal; **~ radar** RADIO radar signal; **~ de radiodiffusion** RADIO broadcasting signal; **~ de radiophare** SIGNAL beacon signal; **~ de rappel** SIGNAL *intervention*, TÉL ring-forward signal; **~ de rappel vers l'avant** TÉL forward recall signal; **~ RD** (*signal de réception de données*) DONNÉES RD (*receive data*); **~ de réaction** CONTRÔLE, SIGNAL feedback signal; **~ de réception de données** (*signal RD*) DONNÉES receive data (*RD*); **~ de reconstitution** GÉNÉRAL recovery signal; **~ de récupération** GÉNÉRAL recovery signal; **~ de refus de message** SIGNAL message refusal signal; **~ de réinitialisation de bande** SIGNAL reset-band signal; **~ de réinitialisation du circuit** SIGNAL reset-circuit signal; **~ de remise à zéro du registre** SIGNAL, TÉL register-reset signal; **~ de réponse** SIGNAL answer signal, TÉL answer signal, backward signal; **~ de réponse, avec taxation** SIGNAL answer signal, charge; **~ de réponse du demandé** TÉL B-answer signal; **~ de réponse, sans taxation** SIGNAL answer signal, no charge; **~ de repos** RADIO interval signal; **~ de réserve prête** SIGNAL stand by-ready signal; **~ retenu** SIGNAL retained signal; **~ de retour** TÉL backward signal; **~ de retour d'appel** TÉL audible ringing tone, ringing signal, ringing tone; **~ de rythme** SIGNAL timing signal; **~ de saisie** SIGNAL, TÉL seizing signal, seizure signal; **~ de sélection** SIGNAL selection signal; **~ de service** RADIO prosign, SIGNAL, TÉL service signal; **~ du service des abonnés absents** APP SERV, TÉL absent-subscriber service signal; **~ de service des renseignements pour les appels** SIGNAL call-information service signal; **~ à une seule fréquence** RADIO, SIGNAL, TÉL single-frequency signal; **~ simple** SIGNAL, TÉL simple signal; **~ sinusoïdal** ÉLECTRON, INFORMAT, SIGNAL, TÉL sinusoidal signal, sine-wave signal; **~ son** TV sound signal; **~ sonar** RADIO sonar signal; **~ sonore** DONNÉES audio cue, GÉNÉRAL *acoustique* audible signal, INFORMAT bleep, aural signal, alert sound, RADIO audible signal, SIGNAL sound signal, audible signal; **~ sonore bref** INFORMAT beep sound; **~ de sortie** ÉLECTRON, INFORMAT output signal; **~ à spectre étalé** RADIO spread-spectrum signal; **~ ST** SIGNAL ST signal; **~ de suppression** TÉL *vidéo* blanking signal; **~ de surveillance des multiblocs** SIGNAL multiblock monitoring signal; **~ de synchronisation** INFORMAT, SIGNAL, TV synchronizing signal, timing signal, synchronization signal; **~ de synchronisation de ligne** INFORMAT, SIGNAL, TV line-synchronizing signal; **~ de synchronisation de trame** INFORMAT, SIGNAL, TV field-synchronizing signal, field sync pulse; **~ télégraphique** SIGNAL, TÉL telegraphic signal; **~ téléphonique** SIGNAL, TÉL telephone signal; **~ temporel discret** INFORMAT, TRANSMIS discretely-timed signal; **~ circuits tous occupés** TÉL all-paths busy tone, all-trunks busy tone, fast-busy tone, reorder tone; **~ de transfert** INFORMAT strobe; **~ de transfert de charge** SIGNAL load-transfer signal; **~ de transfert d'urgence de la charge** SIGNAL emergency load-transfer signal; **~ transmis en retour** TV looped signal; **~ de transmission de numérotation** DONNÉES, SIGNAL wink signal, wink start signal; **~ de travail** RADIO marking wave; **~ d'urgence** RADIO urgency signal; **~ utile** ÉLECTRON, RADIO, SIGNAL wanted signal; **~ de validation** INFORMAT enabling signal; **~ de verrouillage de trame** SIGNAL, TRANSMIS frame-alignment signal (*FAS*); **~ de verrouillage de trame concentré** SIGNAL, TRANSMIS bunched frame-alignment signal; **~ de verrouillage de trame réparti** SIGNAL, TRANSMIS distributed frame-alignment signal; **~ vidéo** RADIO, SIGNAL, TV video signal; **~ vidéo synthétique** RADIO, SIGNAL, TV synthetic video signal; **~ vocal** SIGNAL, TÉL speech signal; **~ vocal codé** TÉL encrypted speech; **~ vocal numérisé** TÉL digitized speech

signaler *vt* GÉNÉRAL signal

signalisation *f* GÉNÉRAL report, signaling (*AmE*), signalling (*BrE*); **~ d'alarme** DONNÉES alarm report; **~ asservie** SIGNAL compelled signaling (*AmE*), compelled signalling (*BrE*); **~ associé** SIGNAL associated signaling (*AmE*), associated signalling (*BrE*); **~ pour l'attente** SIGNAL hold signaling (*AmE*), hold signalling (*BrE*); **~ automatique** TÉL automatic signaling (*AmE*), automatic signalling (*BrE*), auxiliary signaling (*AmE*), auxiliary signalling (*BrE*); **~ automatique de l'interurbain** TÉL trunk dialing (*AmE*), trunk dialling (*BrE*); **~ avec affectation à la demande** SIGNAL demand-assignment signaling (*AmE*), demand-assignment signalling (*BrE*); **~ de bout en bout** SIGNAL end-to-end signaling (*AmE*), end-to-end signalling (*BrE*); **~ de commande centralisée** SIGNAL centralized control signaling (*AmE*), centralized control signalling (*BrE*); **~ de commande décentralisée** SIGNAL decentralized-control signaling (*AmE*), decentralized-control signalling (*BrE*); **~ concentrée** TRANSMIS bunched signaling (*AmE*), bunched signalling (*BrE*); **~ de connexion et déconnexion** SIGNAL connect and disconnect signaling (*AmE*), connect and disconnect signalling (*BrE*); **~ à courant alternatif** ALIM ÉLEC, SIGNAL AC signaling (*AmE*), AC signalling (*BrE*), alternating-current signaling (*AmE*), alternating-current signalling (*BrE*); **~ en courant continu** ALIM ÉLEC, SIGNAL DC signaling (*AmE*), DC signalling (*BrE*), direct current signaling (*AmE*), direct current signalling (*BrE*); **~ dans la bande** SIGNAL in-band signaling (*AmE*), in-band signalling (*BrE*); **~ dans la bande avec codage** SIGNAL coded in-band signaling (*AmE*), coded in-band signalling (*BrE*); **~ dans l'intervalle de temps** SIGNAL in-slot signaling (*AmE*), in-slot signalling (*BrE*); **~ décimale** SIGNAL decadic signaling (*AmE*), decadic signalling (*BrE*); **~ de défaillances** ALIM ÉLEC fault reporting; **~ de dérangement** GÉNÉRAL, TÉL fault report; **~ à deux fréquences** SIGNAL, TÉL two-frequency signaling (*AmE*), two-frequency signalling (*BrE*); **~ à 2600 Hz** SIGNAL *système de signalisation R1* signaling at 2600 Hz (*AmE*), signalling at 2600 Hz (*BrE*); **~ d'enregistreur** SIGNAL, TÉL register signaling (*AmE*), register signalling (*BrE*); **~ entièrement dissociée** SIGNAL fully-dissociated signaling (*AmE*), fully-dissociated signalling (*BrE*); **~ entre enregistreurs** SIGNAL, TÉL register signaling (*AmE*), register signalling (*BrE*); **~ entre enregistreurs en bloc** SIGNAL *système de signalisation R2* en bloc register signaling (*AmE*), en bloc register signalling (*BrE*); **~ entre registreurs** SIGNAL, TÉL *système de signalisation R1* register signaling (*AmE*), register signalling (*BrE*); **~ et commutation avec affectation à la demande** SIGNAL demand-assignment signaling and switching (*AmE*) (*DASS*), demand-assignment signalling and switching (*BrE*) (*DASS*); **~ et commutation par affectation à la demande** SIGNAL demand-assigned switching and signaling (*AmE*), demand-assigned switching and signalling (*BrE*); **~ de l'état du demandeur** TÉL subscriber-state signaling (*AmE*), subscriber-state signalling (*BrE*); **~ à fréquence 500/20** TÉL 500/20 signaling (*AmE*), 500/20 signalling (*BrE*); **~ hors bande** SIGNAL out-band signaling (*AmE*), out-band signalling (*BrE*); **~ hors intervalle de temps** SIGNAL out-slot signaling (*AmE*), out-slot signalling (*BrE*); **~ inter-nœuds** SIGNAL interhub signaling (*AmE*), interhub signalling (*BrE*); **~ intrabande** SIGNAL in-band signaling (*AmE*), in-band signalling (*BrE*); **~ de ligne** SIGNAL line signaling (*AmE*), line signalling (*BrE*); **~ manuelle** GÉNÉRAL ring-down signaling (*AmE*), ring-down signalling (*BrE*); **~ en mode associée** GÉNÉRAL associated signaling (*AmE*), associated signalling (*BrE*); **~ multifréquence**

SIGNAL, TÉL tone signaling (*AmE*), tone signalling (*BrE*), dual-tone multifrequency signaling (*AmE*), dual-tone multifrequency signalling (*BrE*); **~ non associée** SIGNAL nonassociated signaling (*AmE*), nonassociated signalling (*BrE*); **~ par boucle de renvoi** GÉNÉRAL loopback configuration, loopback signaling (*AmE*), loopback signalling (*BrE*); **~ par canal sémaphore** SIGNAL common-channel signaling (*AmE*), common-channel signalling (*BrE*); **~ par éléments numériques vocaux** SIGNAL speech-digit signaling (*AmE*), speech-digit signalling (*BrE*); **~ par fréquences vocales** SIGNAL, TÉL voice-frequency code signaling (*AmE*), voice-frequency code signalling (*BrE*); **~ par impulsions** TÉL pulse signaling (*AmE*), pulse signalling (*BrE*); **~ par polarité** SIGNAL polarity signaling (*AmE*), polarity signalling (*BrE*); **~ par réassignation de bit** DONNÉES, SIGNAL robbed-bit signaling (*AmE*), robbed-bit signalling (*BrE*); **~ à point milieu** SIGNAL center-point signaling (*AmE*), centre-point signalling (*BrE*), midpoint signaling (*AmE*), midpoint signalling (*BrE*); **~ à point de symétrie** SIGNAL center-point signaling (*AmE*), centre-point signalling (*BrE*); **~ quasi-associée** SIGNAL quasi-associated signaling (*AmE*), quasi-associated signalling (*BrE*); **~ répartie** TRANSMIS distributed signaling (*AmE*), distributed signalling (*BrE*); **de ~ au réseau** TRANSMIS network addressing; **~ au réseau** TÉL network control signaling (*AmE*), network control signalling (*BrE*); **~ section par section** SIGNAL link-by-link signaling (*AmE*), link-by-link signalling (*BrE*), TRANSMIS link-by-link signalling (*BrE*), link-by-link signaling (*AmE*); **~ sur voie commune** SIGNAL common-channel signaling (*AmE*), common-channel signalling (*BrE*); **~ sur voie commune séparée** SIGNAL separate common-channel signaling (*AmE*), separate common-channel signalling (*BrE*); **~ TRON et RON** TÉL E&M signaling (*AmE*), E&M signalling (*BrE*); **~ d'usager à usager** (*UUS*) SIGNAL user-to-user signaling (*AmE*) (*UUS*), user-to-user signalling (*BrE*) (*UUS*); **~ vers l'avant** TÉL forward signaling (*AmE*), forward signalling (*BrE*); **~ voie par voie** SIGNAL channel-associated signaling (*AmE*) (*CAS*), channel-associated signalling (*BrE*) (*CAS*), common associated signaling (*AmE*), common associated signalling (*BrE*), TRANSMIS channel-associated signaling (*AmE*) (*CAS*), channel-associated signalling (*BrE*) (*CAS*)

signataire *m* TV signatory

signature: **~ de cible** *f* RADIO target signature; **~ du fond** *f* RADIO background signature; **~ numérique** *f* DONNÉES digital signature; **~ vocale** *f* INFORMAT, TÉL voice print

signaux: **~ de balise émis à partir de la terre** *m pl* RADIO earth-generated beacon; **~ de commande des appels** *m pl* SIGNAL call control signals; **~ de données anisochrones** *m pl* DONNÉES anisochronous data; **~ de données isochrones** *m pl* DONNÉES isochronous data; **~ effectivement transmis dans les transmissions** *m pl* SIGNAL effectively transmitted signals in sound-program (*AmE*), effectively transmitted signals in sound-programme (*BrE*); **~ d'égalisation** *m pl* TV equalizing pulse; **~ de gestion de réseau** *m pl* SIGNAL network management signals; **~ de gestion du réseau sémaphore** *m pl* SIGNAL signaling-network management signals (*AmE*), signalling-network management signals (*BrE*); **~ de gestion du réseau de signalisation** *m pl* SIGNAL signaling-network management signals (*AmE*), signalling-network management signals (*BrE*); **~ de maintenance du réseau** *m pl* SIGNAL network maintenance signals; **~ de sélection du réseau** *m pl* SIGNAL network selection signals; **~ vocaux numériques non linéaires** *m pl* TÉL nonlinear digital speech signals

signet *m* DONNÉES, INFORMAT marker, bookmark; **~ personnalisé** DONNÉES, INFORMAT personalized bookmark

significatif *adj* DONNÉES, INFORMAT, TÉL significant

silence: **~ écoute** *m* RADIO listening silence; **~ radio** *m* RADIO blackout, radio blackout, radio silence

silencieux[1] *adj* GÉNÉRAL hum-free, INFORMAT noiseless

silencieux[2] *m* GÉNÉRAL, TÉL, TRANSMIS deadener, silencer; **~ à commande par tonalités** RADIO tone squelch system; **~ de sous-porteuse** RADIO continuous tone-controlled squelch system (*CTCSS*)

silice *f* COMPOSANTS, TÉL, TRANSMIS silica; **~ fondue** TRANSMIS fused silica, vitreous silica; **~ recouverte de plastique** TÉL plastic-clad silica (*PCS*); **~ vitreuse** TRANSMIS fused silica, vitreous silica

silicium *m* COMPOSANTS, INFORMAT silicon

simple[1]: **~ étoile** *adj* TRANSMIS single-star (*SS*); **en ~ précision** *adj* INFORMAT single-precision

simple[2]: **~ mail transfer protocol** *m* DONNÉES simple mail transfer protocol (*SMTP*); **~ protocol for ATM network signalling** *m* TRANSMIS simple protocol for ATM network signaling (*AmE*) (*SPANS*), simple protocol for ATM network signalling (*SPANS*)

simplex[1] *adj* INFORMAT unidirectional, RADIO, TÉL simplex, unidirectional, TRANSMIS half-duplex, simplex

simplex[2] *m* RADIO, TÉL, TRANSMIS simplex; **~ bilatéral** TÉL two-way simplex system; **~ à deux fréquences** RADIO, TÉL two-frequency simplex

simplicité: **~ d'utilisation** *f* APP SERV, TÉL serviceability

simulateur *m* ALIM ÉLEC, INFORMAT, TÉL, TESTS, TRANSMIS simulator; **~ d'appels** TÉL, TESTS call simulator; **~ de câble** TRANSMIS artificial line, line simulator; **~ d'émetteur d'indicatif** TÉL answer-back unit simulator; **~ de tension** ALIM ÉLEC voltage simulator

simulation *f* ÉLECTRON, INFORMAT, TÉL, TESTS simulation; **~ de circuit** ÉLECTRON circuit simulation; **~ de dérangement** TESTS fault simulation; **~ de panne** TESTS fault simulation; **~ sur ordinateur** INFORMAT computer simulation; **~ en temps comprimé** INFORMAT fast-time simulation

simuler *vt* ÉLECTRON, INFORMAT, TÉL, TESTS simulate

sinusoïdal *adj* ÉLECTRON, SIGNAL sinusoidal

siphon *m* RADIO trap

sirène *f* TÉL siren

SITA *abrév (Société internationale de télécommunications aéronautiques)* RADIO SITA (*International Society for Aeronautical Telecommunications*)

site *m* GÉNÉRAL site, RADIO *radar* elevation, DONNÉES, ÉLECTRON, TV franchise area; **~ distant** ÉLECTRON remote site; **~ éloigné** TRANSMIS remote site; **~ d'installation** GÉNÉRAL installation site; **~ miroir** DONNÉES mirror site

situation *f* GÉNÉRAL siting, situation; **~ d'alarme** TÉL alarm situation, TESTS alarm condition; **~ de compte** TÉL account status

situé *adj* GÉNÉRAL, RADIO situated

situer *vt* GÉNÉRAL, RADIO locate, site, situate

skinner *m* TÉL skinner

SLS *abrév* (*suppression des lobes secondaires*) RADIO SLS (*side-lobe suppression*)

small: **~ office, home office** *loc* (*SOHO*) DONNÉES small office, home office (*SOHO*)

SMCN *abrév* (*système de multiplication de circuits numériques*) TÉL, TRANSMIS DCMS (*digital circuit multiplication system*)

SMD *abrév* (*signal minimal détectable*) TRANSMIS MDS (*minimum detectable signal*)

smectique *adj* ÉLECTRON smectic

S-mètre *m* TESTS S-meter

SM-SIA *abrév* (*signal d'indication d'alarme de section de multiplexage*) TRANSMIS multiplex-section alarm indication signal

SMT *abrév* RADIO, TÉL, TRANSMIS (*station mobile terrestre*) LMS (*land-mobile station*), (*système mondial de télécommunications*) GTS (*global telecommunications system*)

SMTS *abrév* (*service mobile terrestre par satellite*) RADIO, TÉL, TRANSMIS LMSS (*land-mobile satellite service*)
SMU *abrév* (*unité de mesure de source*) TESTS SMU (*source measure unit*)
SN *abrév* (*norme suisse*) GÉNÉRAL Swiss Standard
SNA *abrév* (*signal de neutralisation d'alarme*) TESTS, TRANSMIS alarm inhibit signal
SNMP *abrév* (*protocole SNMP*) DONNÉES SNMP (*simple network management protocol*)
société: **~ de dépannage** *f* GÉNÉRAL repair company; **~ exploitante de radiotélécommunications** *f* RADIO radio common carrier; **~ exploitante de réseau téléphonique** *f* TÉL Telephone Company (*telco*); **~ exploitante de télécommunications** *f* TRANSMIS common carrier, telecommunications common carrier; **~ exploitante du téléphone mobile** *f* TÉL mobile telephone operating company; **~ d'exploitation du réseau** *f* TÉL network operation company; **~ de services d'ingénierie en informatique** *f* INFORMAT software company
Société: **~ internationale de télécommunications aéronautiques** *f* (*SITA*) RADIO International Society for Aeronautical Telecommunications; **~ suisse des constructeurs de machines** *f* (*VSM*) GÉNÉRAL Swiss Association of Machinery Manufacturers (*VSM*)
socle: **~ d'armoire** *m* TRANSMIS cabinet base
sœur *f* INFORMAT sibling
sœurs *f pl* INFORMAT daughters, twins
SOHO *abrév* (*small office, home office*) DONNÉES SOHO (*small office, home office*)
sol[1]: **~-air** *adj* RADIO ground-to-air
sol[2]: **~ conducteur** *m* TESTS semiconductive floor
solénoïde *m* ALIM ÉLEC solenoid
sollicitation *f* RADIO invocation
sollicité *adj* DONNÉES looked-for
solliciter *vt* RADIO invoke
sommaire *m* GÉNÉRAL *résumé* outline
sommateur *m* ÉLECTRON analog adder, summer
somme: **~ de contrôle** *f* DONNÉES check sum; **~ numérique** *f* TÉL digital sum
sommeil: **en ~** *adj* INFORMAT dormant
sommet *m* GÉNÉRAL vertex
son[1]: **~ analogique** *m* INFORMAT analog sound; **~ atonal** *m* INFORMAT pitchless sound; **~ à attaque lente** *m* INFORMAT slow-attack sound; **~ audible** *m* GÉNÉRAL *acoustique* audible sound; **~ caverneux** *m* TÉL hollowness; **~ discordant** *m* INFORMAT jar; **~ fondamental** *m* TÉL fundamental sound; **~ grave** *m* GÉNÉRAL bass tone; **~ numérique** *m* INFORMAT digital audio; **~ préprogrammé** *m* INFORMAT preset voice; **~ vocal** *m* INFORMAT speech sound
son[2]: **~ sur son** *loc* INFORMAT sound on sound
sonar *m* RADIO sonar
sondage *m* GÉNÉRAL poll, polling, TESTS sensing; **~ ionosphérique** RADIO ionospheric sounding; **~ par téléphone** TÉL phone-in poll
sonde *f* GÉNÉRAL probe; **~ de Hall** ÉLECTRON Hall probe; **~ logique** ÉLECTRON logic probe; **~ microphonique** TÉL probe microphone; **~ de mise à la terre** ALIM ÉLEC earthing probe (*BrE*), grounding probe (*AmE*); **~ pyrométrique** CLIMAT, CONTRÔLE temperature sensor, thermometer probe; **~ spatiale** RADIO, TV space probe; **~ à thermistance** COMPOSANTS, TESTS thermistor mount
sonder *vt* GÉNÉRAL poll, TESTS sense
sondeur: **~ ionosphérique** *m* RADIO ionospheric recorder; **~ ionosphérique oblique** *m* RADIO oblique-incidence ionospheric recorder; **~ oblique** *m* RADIO oblique-incidence ionospheric recorder; **~ vertical** *m* RADIO ionosonde
sonie *f* INFORMAT pitch, TÉL loudness
sonique *adj* TÉL sonic
sonner *vi* TÉL ring, ring out
sonnerie *f* APP SERV ringer, ÉLECTRON buzzer, PÉRIPH bell, bellsets, TÉL *mécanisme* bell, ringer, *son* ringing; **~ d'alarme** TÉL alarm bell; **~ cadencée** TÉL interrupted ringing; **~ électrique polarisée** TÉL polarized electric bell; **~ électronique** PÉRIPH electronic call-signal generator, TÉL tone ringer; **~ manuelle** TÉL manual ringing; **~ monolithique** TÉL tone ringer; **~ par tonalité** TÉL *mécanisme* tone ringer, *son* tone ringing; **~ de poste** PÉRIPH extension bell; **~ de rappel** TÉL reminder ring; **~ réglable** APP SERV adjustable ringer; **~ sélective** TÉL selective ringing; **~ sélective en fréquence** TÉL frequency-selective ringing; **~ téléphonique** PÉRIPH telephone bell
sonomètre *m* TESTS sound-level meter
sonore *adj* GÉNÉRAL audible, INFORMAT acoustical, aural, audible, RADIO audible
sonorisation *f* INFORMAT sound system, RADIO public-address equipment
sonorité *f* GÉNÉRAL tone; **~ à intensité croissante** RADIO escalating loudness
sophistiqué *adj* ÉLECTRON sophisticated
sortance *f* ÉLECTRON fan-out
sortant *adj* GÉNÉRAL, TRANSMIS outgoing (*o/g*)
sortie *f* ALIM ÉLEC, APP SERV output, COMMUT outlet access, GÉNÉRAL *borne* terminal, *câble* lead, *réponse* output; **~ d'acceptation** TÉL acceptance output; **~ d'accusé de réception de sélection** TÉL selection acknowledgement output; **~ asymétrique** INFORMAT unbalanced output; **~ audio** ÉLECTRON, INFORMAT audio out, audio output; **~ de câble** TRANSMIS, TV cable outlet; **~ casque** ÉLECTRON, TÉL headphones output; **~ de collecteur** ÉLECTRON collector terminal; **~ de contrôle sur imprimante** COMMUT check print-out; **~ de demande** TÉL request output; **~ de distribution** ALIM ÉLEC distribution output; **~ des éléments de taxation** APP SERV charging output; **~ de format** TÉL format output; **~ hors dialogue** TÉL output outside dialog (*AmE*), output outside dialogue (*BrE*); **~ imprimée d'alarme** TÉL alarm print-out; **~ imprimée de contrôle** COMMUT check print-out; **~ d'instruction** TÉL guidance output; **~ d'intervention** TÉL prompting output; **~ de ligne** INFORMAT line-out, line output; **~ en mode menu** TÉL menu output; **~ mono** INFORMAT, PÉRIPH mono output; **~ mono/stéréo** INFORMAT, PÉRIPH mono/stereo output; **~ négociée** DONNÉES negotiated exit (*NE*); **~ de notes** INFORMAT note output; **~ numérique** INFORMAT digital output; **~ à plusieurs bits** DONNÉES multibit output; **~ de la porte OU recevant n signaux** INFORMAT OR of n signals; **~ de rejet** TÉL rejection output; **~ de réponse** TÉL response output; **~ spontanée** TÉL spontaneous output; **~ stéréo** INFORMAT, PÉRIPH stereo output; **~ sur papier** PÉRIPH hard-copy output; **~ sur support en papier** INFORMAT hard copy; **~ symétrique** INFORMAT balanced output; **~ du système** INFORMAT log-off; **~ de travail** COMPOSANTS output make-circuit
sorties *f pl* ALIM ÉLEC *d'un quadripôle* output; **~ radiales** ALIM ÉLEC radial leads
sortir[1] *vt* INFORMAT output, log-out; **~ sur imprimante** INFORMAT, PÉRIPH print out
sortir[2]: **~ de l'orbite** *vi* RADIO blast out of orbit
soudabilité *f* COMPOSANTS solderability
soudable *adj* COMPOSANTS solderable
soudage *m* COMPOSANTS bonding, soldering, welding, TRANSMIS welding, fusing, *avec fil à sonder* soldering; **~ par capillarité** COMPOSANTS capillary soldering; **~ par fils** COMPOSANTS wire bonding
soude: **~ caustique** *f* COMPOSANTS liquid caustic soda
souder *vt* COMPOSANTS *soudure autogène* weld, *avec fil à souder* solder together, solder, ÉLECTRON solder together, TRANSMIS weld
soudure *f* COMPOSANTS *action de souder, avec fil à souder* soldering, weld, *autogène* welding, *substance* solder, ÉLEC-

TRON soldering, TRANSMIS welding; **~ à l'étain** COMPOSANTS solder, soldering tin
soufflage: **~ magnétique** *m* ÉLECTRON magnetic blow-out
souffle *m* RADIO total continuous spectrum noise, TÉL hiss
soufflet: **~ de protection** *m* COMPOSANTS protective bellows
soufflette *f* ÉLECTRON air duster
soulager *vt* INFORMAT relieve
soumettre *vt* INFORMAT submit; **~ à une charge d'essai** TESTS test-load
soumission *f* RADIO bid; **~ par lots à distance** DONNÉES remote batch entry (*RBE*); **~ de travaux à distance** DONNÉES remote-job entry (*RJE*)
soupape *f* CLIMAT, COMPOSANTS, ÉLECTRON valve; **~ d'admission** CLIMAT throttle valve; **~ d'échappement** COMPOSANTS exhaust valve; **~ de purge** COMPOSANTS drain valve
souple *adj* GÉNÉRAL flexible
source *f* GÉNÉRAL source, RADIO feed; **~ acoustique** INFORMAT, TÉL sound source; **~ d'alimentation** ALIM ÉLEC, INFORMAT AC power supply, alternating-current power supply, power source; **~ d'antenne** RADIO antenna feed; **~ de brouillage** RADIO interfering source, TRANSMIS interference; **~ de bruit** ALIM ÉLEC noise source, source of interference; **~ de composition** TV graphics package; **~ de courant alternatif** ALIM ÉLEC, INFORMAT AC source, alternating current source; **~ d'énergie** ALIM ÉLEC, INFORMAT power source; **la ~ est la primitive de demande associée** DONNÉES the source is related-request primitive (*req*); **~ d'information stationnaire** GÉNÉRAL stationary information source; **~ isotrope** RADIO isotropic aerial, isotropic antenna, isotropic radiator; **~ lambertienne** ÉLECTRON, TRANSMIS Lambertian radiator, Lambertian source; **~ de lumière directionnelle** INFORMAT directional-light source; **~ de lumière dirigée** INFORMAT directional-light source; **~ de lumière étendue** INFORMAT distributed-light source; **~ de lumière lointaine** INFORMAT distant-light source; **~ de lumière répartie** INFORMAT distributed-light source, distant-light source; **~ lumineuse** TV light source; **~ de messages** TRANSMIS message source; **~ de messages stationnaire** GÉNÉRAL stationary message source; **~ de parasites** ALIM ÉLEC noise source, source of interference; **~ périscopique** RADIO beam waveguide feed; **~ de perturbation** ALIM ÉLEC noise source, source of interference; **~ primaire** RADIO feed, radiating device, radiating element, illuminating device, TV primary source; **~ primaire décalée** RADIO offset feed; **~ rayon X** ÉLECTRON X-ray source; **~ sonore** INFORMAT, TÉL sound source; **~ de synchronisation de multiplexeur** TRANSMIS multiplexer timing source (*MTS*); **~ à tonalité modulée incorporée** ÉLECTRON built-in warble tone source; **~ vidéo** TV video outlet, video source
sourdine: **en ~** *adj* INFORMAT mute
souris *f* INFORMAT mouse; **~ optique** INFORMAT optical mouse
sous-adressage *m* (*SUB*) APP SERV, DONNÉES, TÉL subaddressing (*SUB*); **~ des terminaux** APP SERV, DONNÉES, TÉL terminal subaddressing
sous-adresse *f* APP SERV, DONNÉES, TÉL subaddress (*SA*)
sous-arbre *m* DONNÉES, TÉL *de fournisseur de service MCS* subtree
sous-bande *f* COMMUT, INFORMAT, RADIO sub-band
sous-champ: **~ inférieur** *m* RADIO, TRANSMIS lower subfield; **~ supérieur** *m* RADIO, TRANSMIS upper subfield
sous-couche: **~ de convergence** *f* DONNÉES convergence sublayer (*CS*); **~ de convergence de transmission** *f* TRANSMIS transmission convergence sublayer; **~ de segmentation et réassemblage** *f* DONNÉES segmentation and reassembly sublayer; **~ de support physique** *f* TRANSMIS physical-medium sublayer
souscripteur *m* DONNÉES, INFORMAT, RADIO, TÉL, TV subscriber, underwriter
souscrire *vi* DONNÉES, INFORMAT, RADIO, TÉL, TV subscribe
sous-domaine: **~ de dénomination** *m* DONNÉES naming subdomain
sous-échantillonnage *m* INFORMAT, TEL subsampling
sous-ensemble *m* DONNÉES subset; **~ de domaine initial** DONNÉES, TÉL initial domain part (*IDP*); **~ de raffinement cohérent** DONNÉES consistent-refinement subset; **~ de servitude** INFORMAT housekeeping subsystem; **~ de subdivision cohérent** DONNÉES consistent partitioning subset
sous-en-tête *m* INFORMAT subheader
sous-faisceau *m* TÉL subunit, subroute, TRANSMIS cabling unit; **~ de circuits** TÉL circuit subgroup
sous-fenêtre *f* INFORMAT pane
sous-flotte *f* RADIO subfleet, secondary fleet
sous-groupe *m* TÉL subgroup, TRANSMIS pregroup, subgroup; **~ rationnel** *m* TESTS rational subgroup: **~ de recherche** *m* TÉL search subgroup
sous-harmonique *adj* ÉLECTRON subharmonic
sous-menu *m* DONNÉES submenu
sous-mode *m* INFORMAT subjob
sous-modulation *f* TRANSMIS undermodulation
sous-multiplexer *vt* TRANSMIS submultiplex
sous-multitrame *f* TÉL submultiframe (*SMF*)
sous-population *f* TESTS subpopulation
sous-porteuse *f* GÉNÉRAL subcarrier
sous-programme *m* DONNÉES, INFORMAT subprogram, subroutine; **~ fermé** INFORMAT closed subroutine; **~ ouvert** INFORMAT open subroutine, direct-insert subroutine; **~ rentrant** TÉL re-enterable subroutine, re-entrant program subroutine
sous-quantification *f* INFORMAT contouring, posteriation
sous-répartiteur *m* TRANSMIS subdistribution frame
sous-répertoire *m* INFORMAT subdirectory
sous-réseau *m* DONNÉES, TRANSMIS, TV subnetwork; **~ de gestion SDH** TÉL, TRANSMIS SDH management subnetwork (*SMS*); **~ réel** DONNÉES real subnetwork
sous-route *f* TÉL subroute
sous-statut *m* INFORMAT substatus
sous-structure: **~ de bloc** *f* DONNÉES block substructure; **~ de canal** *f* DONNÉES channel substructure
sous-système: **~ d'accès** *m* RADIO access subsystem; **~ d'alimentation électrique** *m* ALIM ÉLEC power subsystem (*PSS*); **~ de commande de connexions sémaphores** *m* (*SSCS*) SIGNAL signaling connection control part (*AmE*) (*SCCP*), signalling connection control part (*BrE*) (*SCCP*); **~ de contrôle de signalisation** *m* SIGNAL signaling control part (*AmE*), signalling control part (*BrE*); **~ de détermination et de correction d'attitude** *m* RADIO attitude determination control subsystem; **~ d'entrée/sortie** *m* COMMUT input/output subsystem; **~ exploitation et maintenance** *m* (*SSEM*) DONNÉES, TÉL, TESTS operations and maintenance subsystem (*OM subsystem*); **~ pour la gestion des transactions** *m* (*SSGT*) DONNÉES transaction capabilities application part, transaction management subsystem; **~ de localisation des véhicules** *m* RADIO vehicle localization subsystem; **~ radio** *m* RADIO radio subsystem; **~ de signalisation et commutation par affectation à la demande** *m* SIGNAL demand-assigned switching and signaling subsystem (*AmE*), demand-assigned switching and signalling subsystem (*BrE*); **~ de station de base** *m* TÉL base station subsystem (*BSS*); **~ de transfert** *m* COMMUT transfer subsystem; **~ de transfert de messages** *m* (*SSTM*) DONNÉES, TÉL message transfer part (*MTP*); **~ de transport de messages** *m* (*SSTM*) DONNÉES, TÉL message transfer part (*MTP*); **~ usager RNIS** *m* (*SSUR*) TÉL ISDN user part; **~ utilisateur** *m* TÉL user part; **~ utilisateur de données** *m* DONNÉES, SIGNAL data user part; **~ utilisateur**

pour le RNIS *m* (*SSUR*) TÉL ISDN user part; **~ utilisateur téléphonie** *m* (*SSUT*) DONNÉES, RADIO, TÉL telephone user part (*TUP*); **~ utilisateur téléphonique** *m* (*SSUT*) DONNÉES, RADIO, TÉL telephone user part (*TUP*)
sous-tension *f* ALIM ÉLEC undervoltage
sous-titrage *m* TV subtitling
sous-titre: **~-titre** *m* TV subtitle
sous-titrer: **sous-titrer** *vt* TV subtitle
soustracteur *m* ÉLECTRON subtracter; **~ complet** ÉLECTRON full subtracter
sous-type *m* DONNÉES subtype
sous-unité *f* ÉLECTRON subunit
sous-variable *f* INFORMAT subvariable
soutenu *m* INFORMAT sustain
soutien: **~ technique** *m* DONNÉES, GÉNÉRAL, INFORMAT backup
SP *abrév* (*sur place*) TÉL ONS (*on-scene*), on-premises
space: **~ bus** *m* DONNÉES space bus
spacistor *m* COMPOSANTS, ÉLECTRON spacistor
spécialisation: **~ des canaux** *f* TÉL channel dedication
spécialisé *adj* RADIO one-way
spécialiste: **~ du domaine** *m* INFORMAT, TRANSMIS domain expert
spécification *f* GÉNÉRAL requirement, specification, TESTS target specification; **~ de bloc fonctionnel** TÉL functional-block specification; **~ conditionnelle** GÉNÉRAL conditional requirement; **~ et conception des fonctions** COMMUT function specification and design; **~ facultative** DONNÉES optional requirement; **~ fonctionnelle** GÉNÉRAL *en LDS* functional specification (*FS*); **~ formelle** DONNÉES formal specification; **~ de gabarit pour peigne** TRANSMIS form-board specification; **~ d'instruction** TESTS inspection specification; **~ du matériel** COMPOSANTS specification of materials; **~ obligatoire** DONNÉES mandatory requirement; **~ relative à l'environnement** GÉNÉRAL environmental specification; **~ de la signalisation** GÉNÉRAL marshal document
spécifications: **~ de conformité** *f pl* TESTS conformity specifications; **~ d'interface du circuit de commande du réseau** *f pl* ÉLECTRON network drive interface specifications (*NDIS*)
spécifier *vt* TESTS specify
spécifique *f* ÉLECTRON specific
spécimen *m* TESTS specimen; **~ d'essai** TESTS test piece
spectre *m* DONNÉES, ÉLECTRON, INFORMAT, RADIO, TÉL, TRANSMIS spectrum; **~ acoustique** RADIO, TÉL, TRANSMIS sound spectrum; **~ de Brillouin** DONNÉES Brillouin spectrum; **~ de fréquences** RADIO, TRANSMIS broadcasting spectrum; **~ des fréquences** RADIO frequency spectrum; **~ harmonique** TRANSMIS harmonic spectrum; **~ d'harmoniques** INFORMAT spectrum of harmonics; **~ hors bande** RADIO out-of-band spectrum; **~ hors bande admissible** RADIO permissible out-of-band spectrum; **~ d'impulsions** TÉL, TRANSMIS pulse spectrum; **~ optique** TRANSMIS optical spectrum; **~ radioélectrique** ÉLECTRON radio spectrum; **~ de raies** ÉLECTRON, TRANSMIS line spectrum; **~ d'un son** RADIO, TÉL, TRANSMIS sound spectrum; **~ de train d'impulsions** TÉL pulse-train spectrum; **~ utile** RADIO desired spectrum
spectrogramme *m* RADIO, TÉL, TRANSMIS spectrogram; **~ acoustique** RADIO, TÉL, TRANSMIS sound spectrogram
spectrographie: **~ infrarouge à transformée de Fourier** *f* TESTS Fourier transform infrared spectography (*FTIR*)
spectromètre *m* TRANSMIS spectrometer
spectrométrie *f* TRANSMIS spectrometry; **~ d'absorption** RADIO absorption spectrometry
spectroscopie: **~ par résonance magnétique** *f* ÉLECTRON magnetic-resonance spectroscopy
spirale[1]: **en ~** *adj* GÉNÉRAL helical
spirale[2] *f* GÉNÉRAL helix; **~ de la qualité** TESTS quality loop, quality spiral
spire *f* COMPOSANTS *d'un enroulement* turn
spot *m* ÉLECTRON, INFORMAT, TV spot; **~ analyseur** INFORMAT scanning spot; **~ de balayage** TV scanning spot; **~ publicitaire** TV advertisement (*ad, advert*), commercial
spoule *m* INFORMAT spool
spp *abrév* (*secteurs par piste*) INFORMAT spt (*sectors per track*)
SQL *abrév* (*langage d'interrogation structuré*) DONNÉES SQL (*structured query language*)
SQS *abrév* (*Association suisse pour certificats d'assurance de la qualité*) GÉNÉRAL SQS (*Swiss Association for Quality Assurance Certificates*)
SRAM *abrév* (*mémoire vide statique*) INFORMAT SRAM (*static RAM, static random access memory*)
SRI *abrév* DONNÉES, TÉL (*station de réception d'informations*) IRS (*information receiver station*), (*système de référence intermédiaire*) IRS (*intermediate reference system*)
SRM *abrév* (*service radiotéléphonique mobile*) RADIO, TÉL MTS (*mobile telephone service*)
SRMA *abrév* (*système radiotéléphonique mobile automatique*) RADIO, TÉL AMTS (*automatic mobile telephone system*)
SRP *abrév* (*système de remise physique*) DONNÉES, TÉL PDS (*physical-delivery system*)
SRS *abrév* (*service de radiodiffusion par satellite*) RADIO, TV satellite broadcasting service
SSCS *abrév* (*sous-système de commande de connexions sémaphores*) SIGNAL SCCP (*signaling connection control part AmE, signalling connection control part BrE*)
SSEM *abrév* (*sous-système exploitation et maintenance*) DONNÉES, TÉL, TESTS OM subsystem (*operations and maintainance subsystem*)
SSGT *abrév* (*sous-système pour la gestion des transactions*) DONNÉES transaction capabilities application part, transaction management subsystem
SSTM *abrév* DONNÉES, TÉL (*sous-système de transfert de messages, sous-système de transport de messages*) MTP (*message transfer part*)
SSUR *abrév* (*sous-système usager RNIS, sous-système utilisateur pour le RNIS*) TÉL ISDN user part
SSUT *abrév* (*sous-système utilisateur téléphonie, sous-système utilisateur téléphonique*) DONNÉES, RADIO, TÉL TUP (*telephone user part*)
stabilisateur: **~ de tension** *m* ALIM ÉLEC, ÉLECTRON voltage stabilizer
stabilisation: **~ de tension** *f* ALIM ÉLEC, ÉLECTRON voltage regulation
stabilisé: **~ par montage en pont** *adj* TÉL bridge-stabilized
stabilité: **~ de circuit** *f* TRANSMIS circuit stability; **~ de couleur** *f* GÉNÉRAL colorfastness (*AmE*), colourfastness (*BrE*); **~ dimensionnelle** *f* COMPOSANTS dimensional stability; **~ dynamique** *f* GÉNÉRAL dynamic stability; **~ de fréquence** *f* RADIO frequency stability; **~ de phase** *f* ÉLECTRON phase stability
stable: **~ à la couleur** *m* GÉNÉRAL colorfast (*AmE*), colourfast (*BrE*)
stade: **~ d'assemblage** *m* INFORMAT assembling time; **~ d'exécution** *m* INFORMAT running time
standard *m* GÉNÉRAL operator console, operator's console, switchboard, telephone switchboard; **~ d'échange de documents numériques** DONNÉES digital document interchange standard (*DDIS*); **~ à lignes multipliées** COMMUT, TÉL multiple switchboard; **~ manuel à clés** GÉNÉRAL cordless switchboard; **~ privé** COMMUT, TÉL PBX switchboard; **~ téléphonique** COMMUT, TÉL telephone switchboard
standardisation *f* GÉNÉRAL standardization
standardiser *vt* GÉNÉRAL standardize
standardiste *m* COMMUT, TÉL attendant, operator, switchboard operator, telephone operator, telephonist (*BrE*)

stand-by: **être en ~** *loc* RADIO be on stand-by
station *f* GÉNÉRAL station; **~ d'accueil** RADIO host station; **~ aéronautique** RADIO aeronautical station; **~ d'aéronef** RADIO aircraft station; **~ d'alimentation** ALIM ÉLEC, TRANSMIS power-feeding station; **~ d'alimentation de répéteurs** ALIM ÉLEC, TRANSMIS power-feeding repeater station; **~ asservie** DONNÉES slave station; **~ aval** TRANSMIS downrange station; **~ de base** (*BS*) RADIO, TÉL base station (*BS*); **~ de base radio** RADIO *mobile* radio base station; **~ de base de rattachement** RADIO home base station; **~ centrale** RADIO hub station; **~ centrale de télésurveillance** TESTS central remote monitoring station; **~ de commande** INFORMAT control station; **~ de communication** RADIO communication station; **~ de contrôle** CONTRÔLE control station, TÉL *d'un faisceau* system control station; **~ côtière** RADIO, TÉL coast station, coastal earth station, coastal station, shore station; **~ couverte par une clause de droits acquis** RADIO grandfathered station; **~ à cuivre à raccordement double** ÉLECTRON dual-attaching copper station; **~ directrice** TÉL control station, controlling testing station; **~ directrice de circuit** TÉL circuit control station; **~ de distribution d'énergie** ALIM ÉLEC power supply station; **~ émettrice** RADIO broadcast station; **~ émettrice d'informations** DONNÉES information-sending station; **~ d'émission** RADIO broadcast station; **~ d'engin de sauvetage** RADIO survival craft station; **~ fixe aéronautique** RADIO aeronautical fixed station; **~ de fréquence étalon et de signaux horaires** RADIO standard-frequency and time-signal station; **~ hors réseau** TV off-net station; **~ hors service** TRANSMIS unserviceable station; **~ interdite** RADIO barred station; **~ maîtresse** RADIO master station; **~ mixte** INFORMAT combined station; **~ mixte d'émission et de réception** RADIO combined transmitting/receiving station; **~ mobile** RADIO, TÉL mobile station (*MS*), mobile unit; **~ mobile itinérante** RADIO roaming station; **~ mobile terrestre** (*SMT*) RADIO, TÉL, TRANSMIS land-mobile station (*LMS*); **~ multimédia** TV multimedia station; **~ multiprocesseur** TÉL control station; **~ multiprocesseur d'auxiliaires** TÉL auxiliaries control station; **~ multiprocesseur de commande** TÉL main control station; **~ multiprocesseur de connexion** TÉL matrix control station; **~ multiprocesseur de maintenance** TÉL maintenance station; **~ multiprocesseur de terminaison MIC** TÉL trunk control station; **~ de navire** RADIO ship station; **~ non surveillée** TRANSMIS unattended station; **~ passerelle** RADIO gateway station; **~ périphérique** DONNÉES, GÉNÉRAL remote site, remote station; **~ pourvue de personnel** GÉNÉRAL manned station; **~ primaire** DONNÉES primary station; **~ principale** RADIO master station; **~ principale de répéteurs** TÉL main repeater station; **~ à raccordement double** ÉLECTRON dual-attaching fiber station (*AmE*), dual-attaching fibre station (*BrE*); **~ à raccordement simple** ÉLECTRON single-attachment station (*SAS*); **~ radio** RADIO radio station; **~ radio relais** RADIO radio relay station; **~ de radiobalise de localisation des sinistres** RADIO emergency position-indicating radio beacon station; **~ de radiodiffusion** RADIO broadcasting station; **~ radioélectrique** RADIO radio station; **~ radiophare** RADIO radio range station; **~ de réception** TV receiver station; **~ de réception d'informations** (*SRI*) DONNÉES, TÉL information receiver station (*IRS*); **~ de référence à l'émission** RADIO, TÉL send reference station; **~ de référence principale** RADIO primary reference station; **~ relais** RADIO relay station; **~ de répéteurs** TÉL, TRANSMIS repeater station; **~ de répéteurs à alimentation indépendante** TÉL, TRANSMIS directly-powered repeater station; **~ de répéteurs autoalimentée** TÉL, TRANSMIS directly-powered repeater station; **~ de répéteurs téléalimentée** TÉL dependent repeater station; **~ sans personnel** TRANSMIS unattended station; **~ de secours** RADIO backup station; **~ de signaux horaires** RADIO standard time-signal emission station; **~ au sol** RADIO ground station; **~ sous-directrice** TÉL subcontrol station; **~ sous-directrice de circuit** TÉL circuit subcontrol station; **~ spatiale** RADIO, TV space station; **~ spatiale de radiodiffusion par satellite** RADIO, TV broadcasting-satellite space station; **~ subordonnée** DONNÉES tributary station; **~ de supervision externe** TÉL external supervision station; **~ téléalimentée** TÉL dependent station; **~ télécommandée** RADIO remotely-controlled station; **~ de temps et synchronisation** TÉL synchronization and time base station; **~ terminal de ligne** TRANSMIS line-terminal station; **~ terminale** TRANSMIS point station; **~ de terre** RADIO terrestrial station; **~ terrestre** RADIO, TÉL earth station, land station; **~ terrienne** GÉNÉRAL, RADIO earth station; **~ terrienne côtière** RADIO shore station; **~ terrienne homologuée** RADIO registered earth station; **~ terrienne maîtresse** RADIO master earth station; **~ terrienne terminale** RADIO ground terminal; **~ de travail** DONNÉES, INFORMAT, PÉRIPH work station; **~ de travail ATM** DONNÉES, INFORMAT, PÉRIPH ATM workstation; **~ de travail multimédia** DONNÉES, INFORMAT, PÉRIPH multimedia workstation; **~ tributaire** DONNÉES tributary station; **~ VSAT** RADIO VSAT terminal
stations *f pl* ÉLECTRON docking stations
statique *adj* GÉNÉRAL static
statistique *f* GÉNÉRAL statistic
statistiques: **~ d'exploitation** *f pl* TÉL operating statistics; **~ pluviométriques** *f pl* RADIO rainfall data; **~ de service** *f pl* TÉL operating statistics; **~ de taxation** *f pl* APP SERV charging statistics; **~ de trafic** *f pl* DONNÉES, RADIO, TÉL, TRANSMIS traffic statistics
stator *m* COMPOSANTS, ÉLECTRON stator; **~ inductif** RADIO field coils
stéréo[1] *adj* (*stéréophonique*) INFORMAT, PÉRIPH, TÉL, TV stereo (*stereophonic*)
stéréo[2] *f* (*stéréophonie*) INFORMAT, TÉL, TV stereo (*stereophony*); **~ MS** TV MS stereo
stéréophonie *f* (*stéréo*) INFORMAT stereophony (*stereo*), TÉL, TV stereo telephony; **~ à deux canaux** TÉL twin-channel stereo telephony
stéréophonique *adj* (*stéréo*) INFORMAT, PÉRIPH, TÉL, TV stereophonic (*stereo*)
stéréoscopie *f* INFORMAT, TV stereo vision
stéréoscopique *adj* INFORMAT, TV stereoscopic
stimulus *m* TÉL excitation; **~ de couleur** INFORMAT color stimulus (*AmE*), colour stimulus (*BrE*)
stipulation *f* INFORMAT provision
STM *abrév* APP SERV, DONNÉES, RADIO, TÉL (*système de transmission de messages*) MHS (*message handling system*), TRANSMIS (*système de télécommunications multipoint*) MCS (*multipoint communications system*)
stock[1]: **en ~** *adj* INFORMAT off-the-shelf
stock[2]: **~ de pièces de rechange** *m* COMPOSANTS spare parts inventory
stockage *m* COMPOSANTS, DONNÉES, INFORMAT, RADIO, TÉL storage, storing; **~ dans le réseau** TÉL storage within the network; **~ dynamique en mémoire tampon** INFORMAT dynamic buffering; **~ et repérage d'images** INFORMAT image storage and retrieval; **~ et retransmission** APP SERV, DONNÉES store and forward
stocké *adj* COMPOSANTS, DONNÉES, INFORMAT, RADIO, TÉL stored
stocker *vt* COMPOSANTS, DONNÉES, INFORMAT, RADIO, TÉL store
stoppé *adj* INFORMAT, TV stopped
storyboard: **~ live** *m* INFORMAT, TV storyboard live
strabisme *m* RADIO squint
strap *m* ALIM ÉLEC, COMPOSANTS strap
strapable *adj* ALIM ÉLEC, COMPOSANTS strappable

strapage *m* ALIM ÉLEC, COMPOSANTS strapping
straper *vt* ALIM ÉLEC, COMPOSANTS strap
strate *f* TESTS subpopulation
stratification *f* COMPOSANTS, ÉLECTRON laminating, lamination
stratifié *m* COMPOSANTS, ÉLECTRON laminate; **~ de papier** COMPOSANTS paper laminate
stripline *f* RADIO stripline
strobage *m* ÉLECTRON, GÉNÉRAL, TESTS strobing
strobe *f* ÉLECTRON, GÉNÉRAL, TESTS stroboscope (*strobe*)
stroboscope *m* ÉLECTRON, GÉNÉRAL, TESTS stroboscope (*strobe*)
structuration: **~ du contenu** *f* INFORMAT structuration organization
structure *f* GÉNÉRAL structure, TRANSMIS backbone; **~ d'antenne** RADIO antenna configuration; **~ arborescente** DONNÉES, ÉLECTRON, INFORMAT, TRANSMIS tree structure, tree and branch structure; **~ d'arbre** DONNÉES, ÉLECTRON, INFORMAT, TRANSMIS tree structure, tree and branch structure; **~ en arbre** DONNÉES, ÉLECTRON, INFORMAT, TRANSMIS tree structure, tree and branch structure; **~ de base** TRANSMIS backbone structure; **~ cellulaire** RADIO cellular structure; **~ de chaîne** INFORMAT chain structure; **~ de la commande** CONTRÔLE control structure; **~ de contrôle** CONTRÔLE control structure; **~ de la couche d'application** TRANSMIS application-layer structure (*ALS*); **~ en couches d'application étendue** TRANSMIS extended application-layer structure (*XALS*); **~ 2B+D** TRANSMIS 2B+D arrangement; **~ en étoile** TRANSMIS, TV star structure; **~ harmonique** TRANSMIS harmonic structure; **~ hiérarchique des données** DONNÉES, INFORMAT hierarchical data structure; **~ des informations de gestion** DONNÉES structure of management information (*SMI*); **~ mécanique de central** TÉL exchange mechanical structure; **~ mécanique de travée** TÉL row mechanical structure; **~ multitrame** DONNÉES multiframe structure; **~ parallèle** TESTS parallel structure; **~ du réseau** TÉL network structure; **~ de retard** ÉLECTRON slow-wave structure; **~ de suite de tests** TESTS test suite structure (*TSS*); **~ de suite de tests et objectifs de tests** *m pl* TESTS test suite structure and test purposes (*TSS&TP*); **~ syntaxique** INFORMAT syntactic structure; **~ de table** GÉNÉRAL table structure; **~ tarifaire** APP SERV, TÉL tariff structure; **~ de trame** RADIO time slot structure, TRANSMIS frame structure; **~ de la trame numérique** DONNÉES digital frame structure; **~ TS** TÉL TS network
stub *m* TÉL stub
studio: **~ local** *m* TV local studio
STX *abrév* (*caractère de début de texte*) DONNÉES, INFORMAT STX (*start of text, start-of-heading character*)
style *m* INFORMAT *instrument* pen, style, font style, stylus, TÉL stylus; **~ standard** INFORMAT plain style
stylo *m* INFORMAT stylus
SUAC *abrév* (*message d'acceptation de service supplémentaire*) DONNÉES, TÉL FAA (*facility accepted message*)
SUB *abrév* APP SERV (*sous-adressage*) SUB (*subaddressing*), DONNÉES, TÉL (*sous-adressage*) SUB (*subaddressing*), (*caractère de remplacement, caractère de substitution*) SUB (*substitute character*)
subdiviser[1] *vt* ÉLECTRON, INFORMAT split, TRANSMIS *échantillon* divide
subdiviser[2]: **se ~** *v réfl* ÉLECTRON, INFORMAT split
subdivision *f* DONNÉES, ÉLECTRON, INFORMAT partition, partitioning, splitting, TRANSMIS *d'un canal* division, partition, partitioning; **~ des données** DONNÉES, INFORMAT data partitioning; **~ de l'échantillon** TRANSMIS sample division
subir: **~ des essais** *vi* TESTS undergo trials
submicrostructure *f* GÉNÉRAL submicron structure
subminiature *f* ÉLECTRON subminiature
substance: **~ absorbante** *f* RADIO space cloth; **~ d'adaptation d'indice** *f* TRANSMIS index-matching material; **~ optiquement active** *f* COMPOSANTS optically active material
substrat *m* COMPOSANTS, ÉLECTRON substrate; **~ d'arséniure de gallium** COMPOSANTS gallium arsenide substrate
subtension *f* ALIM ÉLEC undervoltage
succès *m* DONNÉES pass verdict; **~ verdict** DONNÉES pass verdict
succession: **~ d'événements** *f* TESTS sequence of events
SUDM *abrév* (*message de demande de service supplémentaire*) DONNÉES, TRANSMIS FAR (*facility request message*)
suffixe: **~ de message** *m* INFORMAT message suffix
suite *f* DONNÉES continuation, suite, INFORMAT progression, suite, TESTS run, sequence, suite, TRANSMIS progression, *d'impulsions* train; **~ de chiffres** RADIO stream; **~ donnée** GÉNÉRAL action; **~ à donner** GÉNÉRAL action; **~ d'événements** TESTS sequence of events; **~ harmonique** TRANSMIS harmonic progression; **~ de message** DONNÉES continuation of message (*COM*); **~ de notes** INFORMAT note progression; **~ de tests** TESTS test suite; **~ de tests abstraite** DONNÉES, TESTS abstract test suite (*ATS*); **~ de tests abstraite normalisée** DONNÉES, TESTS standardized abstract test suite; **~ de tests abstraite normalisée de référence** DONNÉES, TESTS reference standardized abstract test suite; **~ de tests abstraits paramétrée** DONNÉES, TESTS parameterized abstract test suite (*PATS*); **~ de tests abstraits sélectionnés** DONNÉES, TESTS selected abstract test suite (*SATS*); **~ de tests de conformité** DONNÉES, TESTS conformance test suite; **~ de tests exécutables** DONNÉES, TESTS executable test suite (*ETS*); **~ de tests exécutables paramétrée** DONNÉES, TESTS parameterized executable test suite (*PETS*); **~ de tests exécutables sélectionnés** DONNÉES, TESTS selected executable test suite (*SETS*); **~ de tests génériques** DONNÉES, TESTS generic test suite; **~ de travaux** INFORMAT job stream, run stream
suivi *m* GÉNÉRAL follow-up, monitoring, tracking, INFORMAT follow, tracing, RADIO tracing, tracking, TV tracking; **~ automatique des déplacements** RADIO roaming; **~ des mobiles** RADIO roaming; **~ de rayons** INFORMAT ray tracing; **~ de touche** INFORMAT key follow
suivre *vt* ÉLECTRON track, GÉNÉRAL keep track of, follow up, monitor, track, RADIO trace, track, TV track
sujet *m* DONNÉES, INFORMAT topic
super: **~-autoroute de l'information** *f* TRANSMIS information superhighway; **~-iconoscope** *m* ÉLECTRON image iconoscope
superflu *adj* INFORMAT extraneous
superhétérodyne *adj* RADIO superheterodyne
supérieur *adj* DONNÉES, INFORMAT superscript
superluminescence *f* ÉLECTRON, TRANSMIS superluminescence, superradiance
superposable *adj* ÉLECTRON, INFORMAT piggyback
superposé *adj* ÉLECTRON, INFORMAT piggyback
superposer *vt* INFORMAT overlay, superpose, stack, superimpose, RADIO heterodyne, TÉL superimpose, superpose
superposition *f* INFORMAT overlay, stacking, superimposing, superposition, RADIO heterodyning, TÉL superimposing, superposition; **~ cartographique** RADIO, TV video mapping
superréfraction *f* RADIO superrefraction
superstructure: **~ cristalline** *f* GÉNÉRAL superlattice
superviser *vt* CONTRÔLE, GÉNÉRAL, INFORMAT, TÉL, TESTS monitor, supervise
superviseur *m* GÉNÉRAL supervisor, INFORMAT supervisor, supervisory program, TÉL, TESTS supervisor; **~ actif** DONNÉES active supervisor; **~ de réseau** TESTS network supervisor; **~ de secours** TÉL backup supervisor
supervision *f* ALIM ÉLEC, CONTRÔLE, GÉNÉRAL, INFORMAT, TÉL, TESTS monitoring, supervision; **~ de blocage de route** TÉL route lockout supervision; **~ de bruit** TÉL noise supervision, noise monitoring; **~ de faisceau** TÉL route

supervision; **~ des fusibles** ALIM ÉLEC fuse supervision; **~ d'itinéraire** TÉL route supervision; **~ des occupations** TÉL seizure supervision; **~ des perturbations** TÉL disturbance supervision; **~ de prise** TÉL seizure supervision; **~ de réseau** DONNÉES network supervision; **~ du réseau** TÉL network supervision

supplément: **~ d'affaiblissement** *m* TRANSMIS excess attenuation

supplémentaire *adj* INFORMAT, TÉL additional, extra

support *m* COMMUT hook switch, support, COMPOSANTS fixture, holder, support, tube socket, *de lampe* valve socket, CONTRÔLE medium, ÉLECTRON holder, socket, tube socket, *de lampe* valve socket, GÉNÉRAL medium, support, INFORMAT medium, stand, support, PÉRIPH *postes téléphoniques* bracket, cradle, cradle switch, hook switch, TÉL support, TRANSMIS bearer, bracket, medium, support, TV bearer, medium; **~ adaptable** DONNÉES, ÉLECTRON Data Grade Media; **~ d'alésage** COMPOSANTS boring fixture; **~ de bâti** TÉL rack support; **~ de câble** TRANSMIS cable bracket, cable support; **~ de câbles** TRANSMIS *pose de câbles* cable raceway; **~ de capot** COMPOSANTS case holder, cover holder, TÉL case support, cover support; **~ de circuit imprimé** COMPOSANTS, ÉLECTRON substrate; **~ de clavier** INFORMAT keyboard stand; **~ commutateur** PÉRIPH *postes téléphoniques* cradle switch, hook switch; **~ de composant** ÉLECTRON component holder; **~ de dispositif de sélection** COMMUT selecting finger base, selecting finger support; **~ en équerre** ÉLECTRON outrigger; **~ exploitable par machine** CONTRÔLE automated data medium, machine-readable medium; **~ hôte de chargement multiconstructeur** GÉNÉRAL multivendor host support; **~ isolant métallisé** ÉLECTRON metal-clad base material; **~ lisible par machine** CONTRÔLE automated data medium, machine-readable medium; **~ magnétique** INFORMAT magnetic media, magnetic support; **~ de mémoire** INFORMAT storage medium; **~ microphonique** PÉRIPH *postes téléphoniques* transmitter cup; **~ de montage** COMPOSANTS mounting holder; **~ mural** PÉRIPH wall bracket; **~ physique** TRANSMIS physical medium; **~ du poste** RADIO cradle; **~ de relais** TÉL relay support; **~ de roulement** COMPOSANTS, RADIO bearing; **~ secondaire de structures sur fibres** ÉLECTRON composite structure of grooved silicon chips; **~ de stockage numérique** INFORMAT digital storage media (*DSM*); **~ de transmission** DONNÉES bearer circuit, TRANSMIS, TV transmission bearer, transmission medium; **~ vierge** GÉNÉRAL virgin medium

supports: **~ optiques** *m pl* TV optical media

suppresseur *m* ALIM ÉLEC eliminator, quench, quencher, supressor, COMPOSANTS suppressor, ÉLECTRON canceler (*AmE*), canceller (*BrE*), suppressor, RADIO, TÉL, TRANSMIS suppressor; **~ d'écho** ÉLECTRON echo canceler (*AmE*), echo canceller (*BrE*), RADIO, TÉL, TRANSMIS echo suppressor (*ES*); **~ d'écho complet** RADIO, TÉL, TRANSMIS full echo suppressor; **~ d'écho différentiel** RADIO, TÉL, TRANSMIS differential echo suppressor; **~ d'écho à intervention adaptable** RADIO, TÉL, TRANSMIS adaptive break-in echo suppressor; **~ d'écho à intervention partielle** RADIO, TÉL, TRANSMIS partial break-in echo suppressor; **~ d'écho terminal** RADIO, TÉL, TRANSMIS terminal echo suppressor; **~ d'étincelles** ALIM ÉLEC spark quencher, spark suppressor, spark quench; **~ de perturbation** ALIM ÉLEC interference eliminator, noise eliminator; **~ de réaction** COMPOSANTS singing suppressor; **~ de surtension** ALIM ÉLEC surge suppressor; **~ de transitoires** ALIM ÉLEC transient suppressor

suppression *f* ÉLECTRON suppression, GÉNÉRAL deletion, INFORMAT detection, punch-out, RADIO cancellation, detection, suppression, TÉL suppression, *vidéo* blanking, TRANSMIS suppression; **~ de bruit** RADIO noise cancellation; **~ des codes zéro** TRANSMIS zero-code suppression; **~ d'interférence** TRANSMIS interference suppression; **~ des lobes secondaires** (*SLS*) RADIO side-lobe suppression (*SLS*); **~ des lobes secondaires de l'interrogateur** RADIO interrogator side-lobe suppression (*ISLS*); **~ de mode latéral** TRANSMIS side-mode suppression; **~ par étapes** RADIO phasing-out; **~ de la porteuse** TRANSMIS carrier suppression; **~ de trame** TV field blanking

supprimé *adj* ÉLECTRON, RADIO, TÉL suppressed

supprimer *vt* ÉLECTRON suppress, GÉNÉRAL delete, INFORMAT kill, punch out, RADIO suppress, TÉL suppress, *vidéo* blank; **~ la gigue** TRANSMIS dejitter; **~ progressivement** GÉNÉRAL phase out

supraconducteur *m* ÉLECTRON, TRANSMIS supraconductor, superconductor

supraphonique *adj* ÉLECTRON supraphonic

sur: **~-excursion** *f* RADIO overdeviation

suranné *adj* GÉNÉRAL out-of-date

surcapacité *f* RADIO redundancy

surchargé *adj* ALIM ÉLEC, COMMUT, ÉLECTRON, TÉL, TRANSMIS overloaded

surcharge *f* ALIM ÉLEC, COMMUT, ÉLECTRON, TÉL overload, TRANSMIS overload, overloading, overflow; **~ EMCN** TÉL DCME overload; **~ de transmission** TRANSMIS transmission overload

surcharger *vt* ALIM ÉLEC, COMMUT, ÉLECTRON, TÉL, TRANSMIS overload, overflow

surclassement *m* DONNÉES override

surcouverture *f* RADIO overcoverage

surdébit *m* DONNÉES, TÉL, TRANSMIS overhead; **~ de section** (*SDS*) DONNÉES, TÉL section overhead (*SOH*); **~ de section de multiplexage** (*SDSM*) TRANSMIS multiplex-section overhead; **~ de section de régénération** (*SDSR*) DONNÉES regenerator-section overhead

sûreté[1]: **à ~ intégrée** *adj* TESTS fail-safe

sûreté[2]: **~ de fonctionnement** *f* TESTS dependability

surexposer *vt* DONNÉES, TV overexpose

SURF *abrév* (*message de refus de service supplémentaire*) DONNÉES FRJ (*facility-rejected message*)

surface: **~ d'affichage** *f* INFORMAT display surface; **~ aléatoire** *f* GÉNÉRAL mixed terrain; **~ d'appui** *f* COMPOSANTS *génie mécanique* bearing surface, contact surface; **~ de captation** *f* RADIO effective area, *d'une antenne* absorption cross-section; **~ de contact** *f* COMPOSANTS *génie mécanique* bearing surface, contact surface; **~ de diffusion** *f* RADIO radar cross-section, radar echoing area; **~ dissipative** *f* RADIO radiator area; **~ effective** *f* RADIO effective area; **~ enregistrée** *f* INFORMAT signal surface; **~ équivalente** *f* RADIO effective area, radar cross-section, radar echoing area; **~ équivalente d'antenne** *f* RADIO antenna effective area; **~ lisse** *f* RADIO smooth surface; **~ de mémoire** *f* ÉLECTRON storage surface; **~ de miroir** *f* TRANSMIS mirror surface; **~ d'onde** *f* GÉNÉRAL wavefront; **~ d'ouverture** *f* RADIO aperture; **~ passive** *f* COMPOSANTS passivated surface; **~ portante** *f* COMPOSANTS *génie mécanique* bearing surface, contact surface; **~ de référence** *f* TRANSMIS reference surface; **~ rugueuse** *f* RADIO rough surface; **~ utile** *f* COMPOSANTS, ÉLECTRON *d'une électrode* working surface, RADIO useful area

surfer *vt* DONNÉES surf

surgir *vt* INFORMAT *à l'écran* pop up

surimpression *f* INFORMAT superimposition, overdub, overdubbing, TÉL superimposition

surintensité *f* ALIM ÉLEC excess current, overcurrent, surge current; **~ de courant** ALIM ÉLEC excess current, overcurrent, surge current

surluminance: **~ de bord** *f* INFORMAT edge flare

surmodulation *f* ÉLECTRON overexcitation, overmodulation

surnombre: **en ~** *adj* INFORMAT excess

surnommer *vt* GÉNÉRAL dub

suroscillation *f* RADIO ringing

surplomb *m* ÉLECTRON overhang
surtension *f* ALIM ÉLEC, ÉLECTRON excess voltage, overvoltage, surge, voltage spike, voltage surge; ~ **d'activation** ÉLECTRON *semi-conducteurs* activation overvoltage; ~ **positive** ALIM ÉLEC positive overvoltage
surveillance *f* GÉNÉRAL monitoring, supervision, surveillance; ~ **automatique du fonctionnement** TÉL automatic function supervision; ~ **de bruit** TÉL noise monitoring; ~ **des circuits de bout en bout** TESTS end-to-end circuit supervision; ~ **continue** TESTS continuous supervision; ~ **courante** TESTS routine monitoring; ~ **à distance** CONTRÔLE, TV remote supervision, remote surveillance; ~ **à distance du spectre** RADIO remote spectrum-monitoring (*RSM*); ~ **électronique** APP SERV electronic supervision; ~ **des erreurs sur un canal sémaphore** SIGNAL signaling-link error monitoring (*AmE*), signalling-link error monitoring (*BrE*); ~ **des erreurs sur une liaison de signalisation** SIGNAL signaling-link error monitoring (*AmE*), signalling-link error monitoring (*BrE*); ~ **et gestion des réseaux** TESTS network supervision and management; ~ **du fonctionnement** TÉL service supervision; ~ **d'indication de porteuse** TRANSMIS carrier-wave indication supervision; ~ **de l'onde pilote** TRANSMIS pilot supervision; ~ **par analyseur de spectre** RADIO spectrum-analyser monitoring (*BrE*) (*SAM*), spectrum-analyzer monitoring (*AmE*) (*SAM*); ~ **de la qualité** TESTS quality surveillance; ~ **de la qualité de fonctionnement** TÉL performance monitoring; ~ **du réseau** TÉL network supervision; ~ **secteur** ALIM ÉLEC mains monitoring; ~ **du service** TÉL service supervision; ~ **du système de télécommunications** TESTS communications system monitoring (*CSM*); ~ **du taux d'erreur d'alignement** TÉL alignment error rate monitoring; ~ **du taux d'erreur sur les trames sémaphores** SIGNAL, TÉL signal unit error rate monitoring; ~ **téléphonique** TÉL, TESTS telephone monitoring; ~ **de temporisation** DONNÉES, ÉLECTRON, RADIO, TÉL time-out supervision; ~ **de temps** TÉL time supervision; ~ **de la tension secteur** ALIM ÉLEC mains monitoring; ~ **de trafic** ÉLECTRON, TÉL, TESTS traffic supervision
surveillant *m* GÉNÉRAL supervisor
surveiller *vt* GÉNÉRAL monitor, supervise
survol *m* INFORMAT browsing
survoltage *m* ALIM ÉLEC step-up transformation
survolter *vt* ALIM ÉLEC step up
susceptibilité: ~ **au brouillage** *f* RADIO interference immunity
suspendre *vt* GÉNÉRAL cut off
suspendu *adj* GÉNÉRAL cut-off
suspension *f* ÉLECTRON hanger, GÉNÉRAL cutoff, INFORMAT hold
SVI *abrév* (*service vocal interactif*) APP SERV interactive voice service
S-vidéo *m* INFORMAT, TV S-video (*separated video*),
symbiose *f* ÉLECTRON symbiosis
symbole *m* GÉNÉRAL symbol, INFORMAT bullet, symbol; ~ **d'annotation** TÉL annotation symbol; ~ **de décision** TÉL decision symbol; ~ **d'entrée** TÉL input symbol; ~ **d'état** TÉL state symbol; ~ **graphique** DONNÉES, TÉL graphical symbol, graphic symbol; ~ **d'incertitude** TÉL uncertainty symbol; ~ **littéral** INFORMAT literal; ~ **logique** DONNÉES, ÉLECTRON logic symbol, logical symbol; ~ **de mise en réserve** INFORMAT save symbol; ~ **non terminal** TÉL nonterminal symbol; ~ **d'organigramme** INFORMAT flow-chart symbol; ~ **de poursuite** INFORMAT tracking symbol; ~ **de sortie** TÉL output symbol; ~ **de tâche** TÉL task symbol
symboles: ~ **identiques consécutifs** *m pl* DONNÉES consecutive identical digits (*CID*)
symétrique: ~ **par rapport à la terre** *adj* ALIM ÉLEC balanced-to-earth *BrE*, balanced-to-ground *AmE*
symétriseur *m* ALIM ÉLEC balanced-to-unbalanced transformer (*balun*)
symptôme: ~ **de panne** *m* TÉL failure symptom, fault symptom
SYN *abrév* (*caractère de synchronisation*) DONNÉES SYN (*synchronization character, synchronous idle character*)
synchro: ~**-résolveur** *m* ALIM ÉLEC selsyn; ~**-transmetteur** *m* RADIO synchro transmitter
synchrone *adj* ÉLECTRON, GÉNÉRAL, TÉL synchronous
synchronisateur: ~ **de trame** *m* TRANSMIS frame synchronizer
synchronisation *f* GÉNÉRAL synchronization (*sync, synch*), TÉL synchronizing, time measurement, TESTS time balance, time match, time measurement; ~ **de bande** INFORMAT tape synchronization; ~ **bilatérale** TÉL, TRANSMIS double-ended synchronization; ~ **en boucle fermée** ALIM ÉLEC, CONTRÔLE, TRANSMIS closed-loop synchronization; ~ **en boucle ouverte** ALIM ÉLEC, CONTRÔLE, TRANSMIS open-loop synchronization; ~ **de chrominance**® INFORMAT *Apple* ColorSync®; ~ **d'horloge** INFORMAT clock synchronization; ~ **image-son** TV image and sound synchronization; ~ **multimédia** DONNÉES multimedia synchronization; ~ **mutuelle** ÉLECTRON mutual synchronization; ~ **du rythme des paquets** DONNÉES burst-timing synchronization; ~ **de salve** DONNÉES burst synchronization; ~ **sur touche** INFORMAT key synchronization; ~ **de trame** TRANSMIS frame timing, frame synchronization, TRANSMIS frame synchronization; ~ **unilatérale** TÉL, TRANSMIS single-ended synchronization
synchronisé *adj* GÉNÉRAL synchronized (*sync, synch*)
synchroniser *vt* GÉNÉRAL synchronize (*sync, synch*)
synchroniseur: ~ **final** *m* DONNÉES postamble; ~ **initial** *m* DONNÉES preamble
synchronisme *m* RADIO, TV synchronism
synonyme *m* DONNÉES synonymous name
synopsis *m* DONNÉES, INFORMAT synopsis
syntactique *adj* DONNÉES, INFORMAT, TÉL syntactic
syntaxe: ~ **abstraite** *f* DONNÉES, INFORMAT abstract syntax; ~ **concrète** *f* DONNÉES, INFORMAT concrete syntax; ~ **graphique concrète** *f* DONNÉES, INFORMAT concrete graphical syntax; ~ **du nom de fichier** *f* DONNÉES, TÉL filename syntax; ~ **textuelle concrète** *f* DONNÉES, INFORMAT concrete textual syntax; ~ **de transfert** *f* TÉL transfer syntax
syntaxique *adj* DONNÉES, INFORMAT, TÉL syntactic
synthèse *f* GÉNÉRAL synthesis, SIGNAL, TÉL scanning, synthesis; ~ **additive** INFORMAT additive synthesis; ~ **additive trichrome** INFORMAT additive primary color (*AmE*), additive primary colour (*BrE*); ~ **arithmétique linéaire** INFORMAT linear arithmetic synthesis; ~ **FM** (*synthèse frequency modulation*) INFORMAT FM synthesis (*frequency modulation synthesis*); ~ **de fréquence numérique directe** ÉLECTRON direct digital frequency synthesis (*DDFS*); ~ **frequency modulation** (*synthèse FM*) INFORMAT frequency modulation synthesis (*FM synthesis*); ~ **d'hyperfréquences** RADIO microwave generation; ~ **du langage naturel** DONNÉES natural-language generation; ~ **numérique de fréquence** TÉL digital frequency synthesis; ~ **de la parole** APP SERV, INFORMAT, TÉL speech synthesis; ~ **de signal** DONNÉES, SIGNAL signal synthesis; ~ **sonore** INFORMAT sound synthesis; ~ **soustractive** INFORMAT substractive synthesis; ~ **vectorielle** INFORMAT vector synthesis; ~ **vocale** APP SERV, INFORMAT, TÉL speech synthesis, voice synthesis
synthétiser *vt* ÉLECTRON, INFORMAT, TÉL synthesize
synthétiseur *m* ÉLECTRON, INFORMAT, TÉL synthesizer; ~ **analogique** INFORMAT analog synthesizer; ~ **de fréquence** RADIO frequency synthesizer; ~ **linéaire** INFORMAT linear synthesizer; ~ **de son** INFORMAT sound synthesizer; ~ **sonore** INFORMAT sound synthesizer

syntonisateur *m* ÉLECTRON, INFORMAT, RADIO tuner
syntonisation *f* ÉLECTRON, INFORMAT, RADIO tuning
syntoniser *vt* GÉNÉRAL tune
systématique *f* TESTS systematics
système[1] *m* GÉNÉRAL system, INFORMAT device; ~ **d'abonné à courant porteur** TRANSMIS one plus one carrier system; ~ **d'accès** DONNÉES access system (*A/SYS*); ~ **d'accès et de mémorisation** DONNÉES access and storage system (*AS/SYS*); ~ **d'accès et de transfert** DONNÉES access and transfer system (*AT/SYS*); ~ **d'accès, de mémorisation et de transfert** DONNÉES access, storage and transfer system (*AST/SYS*); ~ **à accès multiple par assignation en fonction de la demande** RADIO demand-assigned multiple access system; ~ **d'accès au réseau** TRANSMIS network access system; ~ **adaptatif** DONNÉES adaptive system; ~ **d'adressage** DONNÉES, INFORMAT addressing; ~ **d'affichage de l'information opérationnelle** RADIO operational information display system; ~ **d'aide à la décision** DONNÉES decision support system; ~ **d'alignement de descente à deux fréquences** RADIO two-frequency glide path system; ~ **d'alignement de piste à deux fréquences** RADIO two-frequency localizer system; ~ **d'alimentation** ÉLECTRON feed assembly, INFORMAT *en papier* feeder; ~ **à alimentation directe** ALIM ÉLEC full-float system; ~ **d'alimentation feuille à feuille** INFORMAT cut-sheet feeder; ~ **alimenté par piles** ALIM ÉLEC, ÉLECTRON, INFORMAT battery system; ~ **ALOHA** RADIO, TÉL ALOHA system; ~ **ALOHA crénelé** RADIO, TÉL slotted ALOHA system; ~ **d'amplification** INFORMAT amplification system; ~ **analogique** GÉNÉRAL analog system; ~ **d'analyse des options tarifaires** (*SAOT*) APP SERV payment option analysis system (*POAS*); ~ **d'analyse en temps comprimé** INFORMAT fast-time analysis system (*FTAS*); ~ **à antennes-relais sans amplification** RADIO nonboosted antenna repeater system; ~ **d'appel de personnes** RADIO paging system, TÉL *télécom* paging system, staff locater system; ~ **d'appel sélectif** RADIO, TÉL selective calling system (*SELCAL*); ~ **d'appel unilatéral** RADIO, TÉL paging system, staff locater system; ~ **d'application** COMMUT, TÉL application system; ~ **d'arrêt d'urgence** CONTRÔLE emergency shutdown system, safety shutdown system; ~ **arythmique** TÉL start-stop system; ~ **asservi à boucle fermée** CONTRÔLE closed-loop control system; ~ **asservi à code multifréquence** SIGNAL compelled multi-frequency code; ~ **asymétrique** TRANSMIS unbalanced system; ~ **d'atterrissage hyperfréquence** RADIO microwave landing system (*MLS*); ~ **d'atterrissage aux instruments** RADIO instrument landing system (*ILS*); ~ **d'attribution de mémoire** DONNÉES storage allocation system (*EMS*); ~ **audio** GÉNÉRAL audio system; ~ **auteur** INFORMAT authoring system; ~ **auteur multimédia** INFORMAT multimedia authoring system; ~ **d'autocommutateur à quatre fils** COMMUT four-wire switching system; ~ **d'autocontrôle** CONTRÔLE, INFORMAT system monitor; ~ **automatique pas à pas** TÉL step-by-step automatic system; ~ **avec attente** TÉL delay system; ~ **avec perte** TÉL loss system; ~ **avec perte et attente** TÉL combined loss and delay system; ~ **de babillard électronique** (*BBS*) DONNÉES, TÉL bulletin board system (*BBS*); ~ **à banc de filtres** RADIO filter-bank system; ~ **à bande élargie** RADIO extended bandwidth system; ~ **binaire** INFORMAT binary system; ~ **bureautique intégré** PÉRIPH integrated office system; ~ **de bus** COMMUT bus system; ~ **de câblage BULL** (*BCS*) DONNÉES, INFORMAT BULL cabling system; ~ **de câblage IBM**® ÉLECTRON IBM cabling system® (*ICS*); ~ **de câblage de l'immeuble** ÉLECTRON premises-distribution system; ~ **de câblage polyvalent** ÉLECTRON multipurpose wiring system; ~ **de câbles** RADIO, TRANSMIS, TV cable system; ~ **de câblodistribution à courbe de réponse inclinée** TV tilted cable television (*tilted CATV*); ~ **de canalisations** GÉNÉRAL, TRANSMIS conduit system, duct system; ~ **à capacité de batterie répartie** ALIM ÉLEC divided-battery system, separate charging system; ~ **à cassettes** INFORMAT tape cassette drive system; ~ **CATV** TV CATV system (*cable television system*); ~ **cellulaire** RADIO cellular system; ~ **cellulaire numérique** RADIO digital cellular system (*DCS*); ~ **centralisé** COMMUT centralized system; ~ **centralisé d'enregistrement** TÉL recording central office; ~ **Centrex** TÉL Centrex system; ~ **de certification** TESTS certification system; ~ **à circuit unique** GÉNÉRAL single-circuit system; ~ **de codage à quatre signatures** RADIO *air mobile* four-signature coding system; ~ **de code des pays** GÉNÉRAL country-code system; ~ **à cœur réparti spécialisé** COMMUT function division system; ~ **à cœur réparti universel** COMMUT centralized traffic division system; ~ **collectif de prise de terre** ALIM ÉLEC collective earthing (*BrE*), collective grounding (*AmE*); ~ **de commande** CONTRÔLE control system; ~ **de commande d'attitude et d'orbite** RADIO attitude and orbit control system; ~ **de la commande du central** TÉL exchange control system; ~ **à commande centralisée** COMMUT, CONTRÔLE, INFORMAT centralized control system, common control system; ~ **à commande directe** COMMUT, INFORMAT direct control system; ~ **à commande indirecte** COMMUT, CONTRÔLE, INFORMAT indirect control system; ~ **de commande d'orientation** RADIO attitude control system; ~ **à commande par enregistreurs** COMMUT, TÉL register-controlled system; ~ **à commande par impulsions** COMMUT, TÉL pulse control system; ~ **à commande répartie** COMMUT distributed control system; ~ **à commande vocale** CONTRÔLE, PÉRIPH, TÉL voice command system; ~ **de commandement et conduite d'opérations** CONTRÔLE command and control system; ~ **de communication d'entreprises** APP SERV, RADIO, TRANSMIS business communication system, business system; ~ **de communication militaire** APP SERV, RADIO, TRANSMIS military communication system; ~ **de communication par câble rayonnant** TRANSMIS radiating-cable communication system; ~ **de communication personnel** TRANSMIS personal communications; ~ **de communication pour véhicules intelligents** RADIO vehicle intelligent communication system (*VICS*); ~ **de communications personnelles** (*PCS*) RADIO personal communications system (*PCS*); ~ **de commutation** COMMUT, TÉL switching system; ~ **de commutation analogique** COMMUT, TÉL analog switching system; ~ **à commutation de circuits** COMMUT, TÉL circuit-switching system; ~ **de commutation à commande centrale** COMMUT, TÉL common-control switching system; ~ **de commutation à distance** COMMUT, TÉL remote switching system; ~ **de commutation électronique** COMMUT, TÉL electronic switching system (*ESS*); ~ **de commutation intégrée** COMMUT, TÉL integrated digital exchange; ~ **de commutation manuelle** COMMUT, TÉL manual switching system; ~ **de commutation de messages** COMMUT, TÉL, TRANSMIS message-switching system; ~ **de commutation MIC** COMMUT, TÉL PCM switching system, PCM system; ~ **de commutation multidébit** COMMUT, TÉL multirate switching system; ~ **de commutation multiservice** COMMUT, TÉL, multiservice-switching system; ~ **de commutation numérique** COMMUT, TÉL digital switching system; ~ **de commutation optique** COMMUT, TÉL optical switching system; ~ **de commutation par répartition en fréquence** COMMUT, TÉL frequency-division switching system (*FDS system*); ~ **de commutation à programme câblé** COMMUT, TÉL hard-wired programmable switching system; ~ **de commutation à programme enregistré** COMMUT, CONTRÔLE, TÉL stored-program switching system (*AmE*), stored-programme switching system (*BrE*); ~ **de commutation de protection** COMMUT, TÉL *télécom* protection switching system; ~ **de commutation satellite**

COMMUT, TÉL remote switching system; ~ **de commutation semi-automatique** COMMUT, TÉL semiautomatic switching system; ~ **de commutation spatiale** COMMUT, TÉL spaced-division switching system; ~ **de commutation spatiale** COMMUT, TÉL spaced-division switching system; ~ **de commutation sur liaison d'éléments** ALIM ÉLEC cell switch system; ~ **de commutation sur secours** COMMUT, TÉL protection switching system; ~ **de commutation temporelle** COMMUT, DONNÉES, TÉL, TRANSMIS time-division switching system; ~ **de commutation tout optique** COMMUT, TÉL all-optical switching system; ~ **de contrôle** CONTRÔLE, DONNÉES, RADIO, TESTS monitor system, control system; ~ **de contrôle d'accès** DONNÉES, TESTS *sécurité* access control system; ~ **de contrôle d'attitude et de roulis** (*SCAR*) RADIO attitude and roll control system; ~ **de contrôle non linéaire** DONNÉES nonlinear control system; ~ **de contrôle à programme câblé** CONTRÔLE wired-program control system (*AmE*) (*WPC system*), wired-programme control system (*BrE*) (*WPC system*); ~ **conversationnel** INFORMAT conversational system; ~ **de coordonnées temporelles** INFORMAT time-coordinate system; ~ **correcteur d'erreurs** DONNÉES, TÉL error-correcting telegraph system; ~ **correcteur d'erreurs par retour d'information** DONNÉES, TÉL information feedback system; ~ **de correction d'erreurs** DONNÉES, TÉL error-correcting telegraph system; ~ **de couleurs de Munsell** TV Munsell color system (*AmE*), Munsell colour system (*BrE*); ~ **de couleurs d'Ostwald** INFORMAT Ostwald color system (*AmE*), Ostwald colour system (*BrE*); ~ **en cours de test** DONNÉES system under test (*SUT*); ~ **crénelé** RADIO, TÉL slotted system; ~ **crénelé à trame** RADIO, TÉL frame-slotted system; ~ **décentralisé** COMMUT decentralized system; ~ **décimal** INFORMAT decimal system; ~ **départ-arrêt** TÉL start-stop system; ~ **desservant des zones étendues** INFORMAT, TÉL, TRANSMIS wide-area system; ~ **détecteur d'erreurs avec demande de répétition** TÉL request-repeat system; ~ **détecteur d'erreurs sans répétition** TÉL error-detecting system; ~ **de détection** GÉNÉRAL sensor system; ~ **de détection de porteuse** DONNÉES carrier-sense system; ~ **de détresse radioélectrique** RADIO distress radio call system; ~ **à deux processeurs** COMMUT dual-processor system; ~ **à deux processeurs en partage de charge** COMMUT dual-processor load-sharing system; ~ **dialogué** INFORMAT conversational system; ~ **de diffusion de messages** DONNÉES message transmission system; ~ **discret** DONNÉES discrete system; ~ **de distribution optique** ÉLECTRON light-guide distribution system; ~ **à double processeur** COMMUT dual-processor system; ~ **d'échange de textes en mode message** DONNÉES message-oriented text interchange system (*MOTIS*); ~ **d'échantillonnage** TESTS sampling system; ~ **d'écriture** INFORMAT script system; ~ **EDA** (*système d'enregistrement des données d'appel*) TRANSMIS call-detail recording system; ~ **EDC** (*système d'enregistrement des données de communication*) TRANSMIS call-detail recording system; ~ **électronique à relais à tige** COMMUT reed-relay system; ~ **embarqué** RADIO airborne system; ~ **émetteur-récepteur** DONNÉES transponder system; ~ **d'émission de l'onde pilote** RADIO pilot system; ~ **d'enregistrement des données d'appel** (*système EDA*) TRANSMIS call-detail recording system; ~ **d'enregistrement des données de communication** (*système EDC*) TRANSMIS call-detail recording system; ~ **à entraînement mécanique** ALIM ÉLEC, RADIO, TÉL motor-driven system, power-driven system; ~ **d'essai** TESTS test system; ~ **à étalement du spectre** RADIO spread-spectrum system; ~ **expert** APP SERV, DONNÉES, INFORMAT expert system; ~ **expert-temps réel** COMMUT real-time expert system; ~ **d'exploitation** (*OS*) DONNÉES, INFORMAT operating system (*OS*), TÉL operating system (*OS*), executive program, TESTS operating system (*OS*); ~ **d'exploitation des entrées/sorties** (*BIOS*) TESTS basic input/output system (*BIOS*); ~ **d'exploitation Microsoft®** (*MS-DOS®*) INFORMAT Microsoft® disk operating system (*MS-DOS*); ~ **d'exploitation de réseau** (*NOS*) DONNÉES, ÉLECTRON network operating system (*NOS*); ~ **d'extrémité** DONNÉES end system (*ES*); ~ **d'extrémité réel** DONNÉES real end system; ~ **de fichiers en réseau** DONNÉES, INFORMAT network file system (*NFS*); ~ **à fréquence pilote** RADIO pilot-tone system; ~ **générique** COMMUT source system; ~ **géré** DONNÉES managed system; ~ **de gestion de bases de données** (*SGBD*) DONNÉES, INFORMAT database management system (*DBMS*); ~ **de gestion de connaissances** (*KMS*) INFORMAT knowledge management system (*KMS*); ~ **de gestion de flotte de taxis** RADIO taxi radio system; ~ **à grande échelle** DONNÉES large-scale system; ~ **graphique** DONNÉES, INFORMAT script; ~ **à groupement de voies duplex** RADIO duplex-channel group system; ~ **hertzien** RADIO microwave system; ~ **hexadécimale** INFORMAT hexadecimal system; ~ **hiérarchisé** GÉNÉRAL hierarchical system; ~ **hôte** DONNÉES host system; ~ **hybride** COMMUT, TRANSMIS hybrid system; ~ **hybride de commutation** COMMUT, TRANSMIS hybrid switching system; ~ **d'imagerie documentaire** DONNÉES document imaging system; ~ **à impulsions** COMMUT, TÉL pulse control system; ~ **indépendant du code utilisé** TÉL code-independent system; ~ **d'information** DONNÉES information system; ~ **d'information de région terminale en technique semi-conducteurs** INFORMAT solid-state automatic terminal information system (*SATIS*); ~ **informatique** INFORMAT computer system; ~ **intégré** DONNÉES, RADIO, TÉL, TRANSMIS integrated system; ~ **intégré de conduite du tir** RADIO integrated fire-control system; ~ **intégré de contrôle des communications** RADIO integrated communication control system (*ICCS*); ~ **intégré de téléphonie et de données** DONNÉES, TÉL voice and data integrated system (*VADIS*); ~ **interactif** TV interactive system; ~ **d'interception automatique** TÉL automatic intercept system; ~ **d'intercom** RADIO key system; ~ **d'intercommunication** TÉL intercom-PBX, intercommunication system; ~ **d'intercommunications électronique** PÉRIPH electronic key system; ~ **intermédiaire** DONNÉES, RADIO intermediate system (*IS*); ~ **d'interrogation séquentiel** INFORMAT polling system; ~ **interurbain** TÉL trunk system, intercity; ~ **interurbain à communications tarifées** (*SICT*) TÉL message toll service (*MTS*); ~ **de lancement double** RADIO dual-launch system; ~ **lié au code utilisé** TÉL code-dependent system; ~ **de ligne** TRANSMIS line system; ~ **à ligne coaxiale** TRANSMIS coaxial line system; ~ **de ligne à fibres optiques** TRANSMIS optical-fiber line system (*AmE*), optical-fibre line system (*BrE*); ~ **de ligne numérique** TÉL digital line system; ~ **de ligne sur fibre optique** TRANSMIS optical-fiber line system (*AmE*), optical-fibre line system (*BrE*); ~ **de limitation des possibilités de duplication** INFORMAT serial copy management system (*SCMS*); ~ **linéaire** GÉNÉRAL linear system; ~ **linéaire à temps discret** TRANSMIS discrete-time linear system; ~ **local maritime** TÉL maritime local system; ~ **de localisation** RADIO localization system; ~ **de localisation des véhicules** RADIO vehicle localization system; ~ **logique** INFORMAT logic system; ~ **à logique câblée** CONTRÔLE wired-logic system; ~ **loran** RADIO long-range navigation (*loran*); ~ **LRS** (*système de télémesures pour lignes d'abonnées*) TÉL LRS (*loop reporting system*); ~ **à mailles** TRANSMIS link system; ~ **de maintien alimenté par piles** PÉRIPH backup battery system; ~ **maritime** RADIO, TÉL maritime system; ~ **maritime par satellite** RADIO, TÉL maritime satellite system; ~ **de marquage** COMMUT marker control system, marker system; ~ **de mémorisation** DONNÉES storage system (*S/SYS*); ~ **de mémorisation et de transfert** DONNÉES storage and transfer system (*ST/SYS*);

~ **de messagerie** APP SERV, DONNÉES, RADIO, TÉL message handling system (*MHS*); ~ **de messagerie EDI** APP SERV, DONNÉES, RADIO, TÉL EDI messaging system (*EDIMS*); ~ **de messagerie électronique** APP SERV, DONNÉES, RADIO, TÉL electronic message system (*EMS*); ~ **de messagerie interpersonnelle** APP SERV, DONNÉES, RADIO, TÉL interpersonal messaging system (*IPMS*); ~ **de messagerie de personne à personne** APP SERV, DONNÉES, RADIO, TÉL interpersonal messaging system (*IPMS*); ~ **de messagerie vocale** APP SERV, RADIO, TÉL voice messaging system (*VMGS*); ~ **de mesures de fiabilité et de qualité** TESTS reliability and quality measurement system (*RQMS*); ~ **mixte local et transit** COMMUT combined local-toll system; ~ **mobile maritime par satellite** TÉL maritime mobile satellite system; ~ **mobile numérique** TÉL digital mobile system (*DMS*); ~ **mobile terrestre** RADIO land-mobile system; ~ **modulaire** TÉL modular system; ~ **à modulation avec étalement du spectre** RADIO spread-spectrum system; ~ **à modulation numérique** INFORMAT, TRANSMIS digital modulation system; ~ **à modulation polaire** RADIO polar modulation system; ~ **mondial de télécommunications** (*SMT*) RADIO, TÉL, TRANSMIS global telecommunications system (*GTS*); ~ **mondial de télécommunications par satellite** RADIO, TÉL, TRANSMIS global telecommunications satellite system; ~ **monoprocesseur** COMMUT single-processor system; ~ **multi-accès** DONNÉES multiaccess system; ~ **multidimensionnel** SIGNAL multidimensional signal; ~ **multimédia** TV multimedia system; ~ **à multiplexage spatial** COMMUT, TÉL spaced-division system; ~ **multiplex numérique T1** TRANSMIS T1 digital carrier (*T1C*); ~ **multiplex T1** TRANSMIS T1 carrier (*T1C*); ~ **de multiplication de circuits numériques** (*SMCN*) TÉL, TRANSMIS digital circuit multiplication system (*DCMS*); ~ **multiprocesseur** COMMUT multiprocessor system; ~ **multiservice par satellite** RADIO, TV satellite multiservice system; ~ **multitraitement** DONNÉES multiprocessing system; ~ **multi-utilisateur** DONNÉES multiuser system; ~ **national** TÉL national system; ~ **à niveaux multiples** TRANSMIS multilevel system; ~ **non dupliqué** TÉL nonduplicated system; ~ **non hiérarchique** TRANSMIS nonhierarchical system; ~ **non linéaire** DONNÉES nonlinear system; ~ **de numération** TÉL numbering system; ~ **numérique** INFORMAT digital system; ~ **à onduleurs** ALIM ÉLEC inverter system; ~ **à oscillateurs synchronisés** ALIM ÉLEC synchronized-oscillator system (*SOS*); ~ **ouvert** DONNÉES, TÉL, TRANSMIS open system; ~ **ouvert géré** DONNÉES, TÉL, TRANSMIS managed open system; ~ **ouvert réel** DONNÉES, TÉL, TRANSMIS real open system; ~ **panel** COMMUT panel system; ~ **en partage de charge** ÉLECTRON load-sharing system; ~ **en partage de fonctions** DONNÉES functionally divided system; ~ **de partage de trafic** COMMUT traffic division system; ~ **partagé du type dispatching** RADIO trunked dispatch system; ~ **pas à pas** COMMUT step-by-step system; ~ **pas-à-pas** COMMUT progressive switching magazine; ~ **de péage** APP SERV toll system; ~ **permettant d'utiliser efficacement la largeur de bande** TRANSMIS bandwidth-efficient system; ~ **à piles** ALIM ÉLEC, ÉLECTRON, INFORMAT battery system; ~ **en pipeline** INFORMAT pipeline system; ~ **plan** TRANSMIS planar system; ~ **à plusieurs antennes réparties** RADIO distributed multi-antenna; ~ **à plusieurs bonds** RADIO multihop system; ~ **point à multipoint** RADIO, TRANSMIS point-to-multipoint system; ~ **point à point** RADIO, TRANSMIS point-to-point system; ~ **des points de répartition primaire et secondaire** TRANSMIS cab and pillar distribution system; ~ **polyphase** ÉLECTRON polyphase system; ~ **à porteuse monovoie** RADIO single channel per carrier (*SCPC*); ~ **à porteuses cohérentes incrémentielles** TV incrementally coherent carrier (*ICC*); ~ **de porteuses en relation harmonique** TV harmonically related carrier (*HRC*); ~ **de porteuses en relation incrémentielle** TV incrementally related carrier (*IRC*); ~ **posé à même le sol** ÉLECTRON floor-standing system; ~ **de positionnement universel** RADIO global positioning system (*GPS*); ~ **de positions automatisées de telephonistes** TÉL operator system; ~ **de positions d'opératrices** TÉL operator system; ~ **de positions de téléphonistes** TÉL operator system; ~ **à postes embrochés** TÉL omnibus system; ~ **de poursuite** RADIO tracking system; ~ **de poursuite d'antenne** RADIO antenna TDRSS; ~ **de poursuite par échelons** RADIO step track system; ~ **privé local** RADIO private local system; ~ **de programmation** DONNÉES, INFORMAT programming system; ~ **en quadrature avec réponse partielle** DONNÉES quadrature partial-response system (*QPRS*); ~ **qualité** TESTS quality system; ~ **de raccordement d'abonnés** PÉRIPH, TÉL subscriber connection system; ~ **radio ferroviaire** RADIO railroad radio system (*AmE*), railway radio system (*BrE*), train radio system; ~ **radio pour gares de triage** RADIO shunting yard radio system; ~ **radio à large bande** RADIO broadband radio system; ~ **de radio numérique** DONNÉES, RADIO digital radio system; ~ **de radiobalises de localisation des sinistres** RADIO emergency position-indicating radio beacon system; ~ **de radiocommande** RADIO radio control system; ~ **de radiocommunication du service mobile** RADIO mobile radio system; ~ **de radiocommunications** RADIO radio system; ~ **de radiocommunications pour les chemins de fer** RADIO railroad radio system (*AmE*), railway radio system (*BrE*), train radio system; ~ **de radiocommunications pour les services des secours** RADIO rescue-service radio system; ~ **de radiocommunications pour taxis** RADIO taxi radio system; ~ **radioélectrique d'appel** RADIO radiopaging system; ~ **de radiomessagerie** RADIO radiopaging system; ~ **de radiomessagerie bilatérale** RADIO two-way paging system; ~ **de radiomessagerie par satellite** RADIO satellite paging system; ~ **de radiotéléphonie** RADIO, TÉL mobile telephony system; ~ **radiotéléphonique mobile** (*SRM*) RADIO, TÉL mobile telephone system (*MTS*); ~ **radiotéléphonique mobile automatique** (*SRMA*) RADIO, TÉL automatic mobile telephone system (*AMTS*); ~ **à rayonnement guidé** RADIO guided radiation system; ~ **de recherche documentaire** DONNÉES information retrieval system; ~ **de recherche de personnes** RADIO, TÉL paging system, staff locater system; ~ **réel** DONNÉES real system; ~ **de référence** GÉNÉRAL reference system; ~ **de référence intermédiaire** (*SRI*) DONNÉES, TÉL intermediate reference system (*IRS*); ~ **de refroidissement** CLIMAT cooling system; ~ **de régulation** CONTRÔLE, INFORMAT feedback control system; ~ **de régulation automatique** CONTRÔLE, INFORMAT automatic control system; ~ **de relais à deux fréquences** RADIO, TÉL two-frequency relay system; ~ **de relais par satellite** RADIO, TÉL satellite relay system; ~ **de remise physique** (*SRP*) DONNÉES, TÉL physical-delivery system (*PDS*); ~ **réparti** INFORMAT distributed system; ~ **de répartition** TRANSMIS dispatch system; ~ **à répartition dans l'espace** COMMUT, TÉL space-division system; ~ **en répartition géographique** DONNÉES segmented multiprocessor system; ~ **de repérage de l'information** DONNÉES information retrieval system; ~ **en réplique synchrone** INFORMAT parallel synchronous system; ~ **à réponse vocale** TÉL voice response system; ~ **à réseau image** COMMUT bypath system; ~ **de retour d'écoute avec intervention** TRANSMIS interrupt feedback; ~ **rotary** COMMUT rotary system; ~ **rotatif** COMMUT rotary system; ~ **rotatif à commutateurs rotatifs** COMMUT rotary system; ~ **à sangle** ÉLECTRON clamp-band separation system; ~ **à satellite** RADIO, TV satellite system; ~ **à satellite pour services inter-entreprises** RADIO satellite business system (*SBS*); ~ **de sauvegarde** DONNÉES backup system; ~ **SCAN**

de télécommande et de contrôle CONTRÔLE remote-control monitoring scan system (*RCMSS*); **~ SCPC avec affectation à la demande** RADIO demand-assigned SCPC system; **~ de secours** GÉNÉRAL stand-by system; **~ de secours immédiat** TESTS hot stand-by system; **~ de secours sous tension** TESTS hot stand-by system; **~ secouru** ALIM ÉLEC no-break system; **~ sémaphore** SIGNAL signaling system (*AmE*), signalling system (*BrE*); **~ semi-automatique** COMMUT, TÉL semiautomatic system; **~ séquentiel à mémoire** (*SECAM*) TV séquentiel couleur à mémoire (*SECAM*); **~ de signalisation** SIGNAL signaling system (*AmE*), signalling system (*BrE*); **~ de signalisation ferroviaire** CONTRÔLE, SIGNAL railroad signaling system (*AmE*), railway signalling system (*BrE*); **~ de signalisation de ligne** CONTRÔLE, SIGNAL line signaling system (*AmE*), line signalling system (*BrE*); **~ de signalisation des navires** RADIO, SIGNAL ship reporting system; **~ de signalisation de trafic** SIGNAL traffic signal system; **~ de signaux lumineux** RADIO, SIGNAL visual call system; **~ source** COMMUT source system; **~ de soutien multisatellite** (*MSSS*) RADIO multi-satellite support system; **~ spatial** RADIO space system; **~ Strowger** COMMUT step-by-step system; **~ à structure déconcentrée** COMMUT, CONTRÔLE fully-distributed control system; **~ à structure démocratique** TRANSMIS nonhierarchical system; **~ à structure éclatée** COMMUT, CONTRÔLE fully-distributed control system; **~ à structure hiérarchique** GÉNÉRAL hierarchical system; **~ à structure multimicroprocesseur** COMMUT multimicroprocessor system; **~ de studio virtuel** INFORMAT virtual-studio system; **~ de support** TÉL support system; **~ tacan** (*système tactical air navigation*) RADIO tacan (*tactical air navigation*); **~ tactical air navigation** (*système tacan*) RADIO tactical air navigation (*tacan*); **~ de tarification** APP SERV pricing scheme; **~ de tarification binaire** TÉL binary tariff system; **~ de tarification par mot** TÉL per-word tariff system; **~ TASI** (*interpolation temporelle des signaux vocaux*) SIGNAL, TRANSMIS TASI (*time-assigned speech interpolation*); **~ de taxation à perception unique** APP SERV single-billing system; **~ de télécommunication cryptophonique** RADIO remote secure voice system (*REMSEVS*); **~ de télécommunication mobile par satellite** RADIO mobile satellite telecommunication; **~ de télécommunications pour chemins de fer** RADIO railroad telecommunications system (*AmE*), railway telecommunications system (*BrE*); **~ de télécommunications multipoint** (*STM*) TRANSMIS multipoint communications system (*MCS*); **~ de télémesures pour lignes d'abonnées** (*système LRS*) TÉL loop reporting system (*LRS*); **~ téléphonique à batterie locale** TÉL local battery system; **~ téléphonique à poussoirs** APP SERV, PÉRIPH, TÉL key telephone system; **~ téléphonique rural** APP SERV, PÉRIPH, TÉL rural telephone system; **~ téléphonique à touches** APP SERV, PÉRIPH, TÉL key telephone system; **~ de télésurveillance** TESTS remote monitoring system; **~ de télévision** TV television system; **~ de télévision à antenne** TV master antenna TV system (*MATV system*); **~ de télévision par câbles** TV cable television system (*CATV system*); **~ de télévision par satellite à antenne principale** TV satellite master aerial television system (*SMATV*), satellite master antenna television system; **~ temporel** COMMUT, DONNÉES, TÉL, TRANSMIS time-division system; **~ à temps discontinu** DONNÉES discrete-time system; **~ à temps partagé** DONNÉES, INFORMAT time-sharing system; **~ de test** TESTS test system; **~ tout à relais** COMMUT, TÉL relay system; **~ de traitement des données** DONNÉES, INFORMAT, TÉL data-processing system; **~ de traitement de l'information** DONNÉES, INFORMAT, TÉL information processing system; **~ transactionnel** DONNÉES transactional system; **~ de transfert** DONNÉES transfer system (*T/SYS*); **~ de transfert de données** DONNÉES data transfer system; **~ de transfert de messages** DONNÉES, TÉL message transfer system (*MTS*); **~ de transfert de textes en mode message** DONNÉES message-oriented text interchange system (*MOTIS*); **~ de transmission de données** DONNÉES, INFORMAT data-communication system; **~ de transmission de messages** (*STM*) APP SERV, DONNÉES, RADIO, TÉL message handling system (*MHS*); **~ de transport universel** ÉLECTRON universal transport system (*UTS*); **~ universel de télécommunication mobile** RADIO, TÉL universal mobile telecommunication system (*UMTS*); **~ d'urgence** CONTRÔLE emergency stop; **~ vidéotex** APP SERV, TV videotex system; **~ Wheatstone** TÉL Wheatstone automatic system, Wheatstone system

système²: **~ exclusif** *loc* INFORMAT system-exclusive

Système: **~ international de satellites pour les recherches et le sauvetage** *m* (*COSPAS-SARSAT*) GÉNÉRAL International Satellite System for Search and Rescue (*COSPAS-SARSAT*)

systèmes: **~ asservis** *m pl* CONTRÔLE control engineering; **~ d'autoroute pour véhicules intelligents** *m pl* RADIO intelligent vehicle highway systems (*IVHS*)

T

T: ~ **parallèle** *m* RADIO shunt T

TA *abrév* (*téléphone d'affaires électronique*) TÉL BS (*electronic business set*)

table[1]: **de ~** *adj* INFORMAT desktop

table[2]: **~ d'allocation des fichiers** *f* DONNÉES, INFORMAT file allocation table (*FAT*); **~ d'annotateur** *f* TÉL record desk; **~ compacte des contraintes** *f* DONNÉES compact constraint table; **~ compacte des tests élémentaires** *f* DONNÉES compact test-case table; **~ de conversion** *f* INFORMAT conversion table; **~ de couleurs** *f* (*CLUT*) INFORMAT color look-up table (*AmE*) (*CLUT*), colour look-up table (*BrE*) (*CLUT*); **~ d'essai** *f* TESTS test desk; **~ de fonction** *f* GÉNÉRAL function table; **~ de fréquences** *f* TRANSMIS frequency table; **~ horizontale** *f* APP SERV, PÉRIPH key shelf; **~ de Karnaugh** *f* GÉNÉRAL Karnaugh map; **~ de mixage** *f* INFORMAT audio mixer, RADIO mixing console; **~ simple de contrainte** *f* DONNÉES single constraint table; **~ de test** *f* TESTS test desk; **~ traçante** *f* DONNÉES plotter, PÉRIPH, TÉL flat-bed plotter, plotter; **~ de vérité** *f* ÉLECTRON truth table

tableau *m* ALIM ÉLEC board, panel, table, COMMUT board, COMPOSANTS table, CONTRÔLE panel, DONNÉES array, board, ÉLECTRON panel, GÉNÉRAL list, panel, table, INFORMAT array, board, chart, panel, RADIO board, chart, list, panel, TÉL board, panel, table, TESTS panel, TRANSMIS list, table, TV panel; **~ d'affichage** DONNÉES, TÉL bulletin board (*BB*), notice board; **~ d'affichage électronique** DONNÉES, INFORMAT, TÉL bulletin board system (*BBS*); **~ de bord** INFORMAT control panel, RADIO dashboard, TV side panel; **~ de câblage** GÉNÉRAL *documentation* cabling table; **~-chargeur** ALIM ÉLEC charger, charging unit; **~ de code** TÉL code table; **~ de commande** CONTRÔLE control panel, INFORMAT operator control panel, TESTS instrument panel; **~ de commutation** COMMUT, TÉL switchboard, telephone switchboard; **~ de connexions** ALIM ÉLEC terminal table, terminal board, INFORMAT jack panel, pinboard, wiring board, TÉL terminal board; **~ de contingence** TRANSMIS frequency table; **~ de contrôle** CONTRÔLE control panel; **~ de correspondance** RADIO mapping list; **~ de correspondance des voies** TRANSMIS channel mapping list; **~ de diodes électroluminescentes** ÉLECTRON light-emitting diode panel; **~ de disposition des fils** GÉNÉRAL running list; **~ de distribution** ALIM ÉLEC distribution panel, power distribution board; **~ de distribution par câble** GÉNÉRAL *documentation* cable distribution table; **~ des fusibles** ALIM ÉLEC fuse panel; **~ de lot de câbles** GÉNÉRAL *documentation* cable-set table; **~ de prévision de fréquence** RADIO frequency prediction chart; **~ de réalisation** COMPOSANTS wiring table; **~ de répartition** RADIO, TÉL patch board, patch panel; **~ de répartition des câbles** GÉNÉRAL *documentation* cable distribution table; **~ de supervision des itinéraires** TÉL route supervision panel; **~ de surveillance** CONTRÔLE observation panel, supervisory panel; **~ de surveillance des voies** TÉL route supervision panel

tableaux: **~ de contrôle de vraisemblance** *m pl* TÉL reasonableness check tables

tables *f pl* RADIO register

tablette *f* INFORMAT tablet; **~ horizontale** APP SERV, PÉRIPH key shelf

tableur *m* INFORMAT spreadsheet

tabulation *f* INFORMAT tabulation

tabulatrice *f* INFORMAT tabulator (*tab*)

tabuler *vt* INFORMAT tabulate

tache *f* COMPOSANTS stain, ÉLECTRON shading, spot, TV ion burn; **~ ionique** *f* COMPOSANTS stain, TV ion burn

tâche *f* COMMUT job; **~ de fond** DONNÉES background operation, background task; **~ parallèle** *f* ÉLECTRON parallel tasking; **~ de traitement réparti de l'information** *f* DONNÉES distributed-information processing task

tacheture *f* DONNÉES, TRANSMIS speckle

tactile *adj* INFORMAT tactile

tagged: **~-image file format** *m* (*TIFF*) DONNÉES, INFORMAT tagged-image file format (*TIFF*)

TAI *abrév* (*temps atomique international*) RADIO IAT (*international atomic time*)

taille *f* DONNÉES, INFORMAT *du dossier* size; **~ du document** DONNÉES document size; **~ réelle** DONNÉES actual size

tambour: **~ d'impression** *m* PÉRIPH, TÉL print barrel, print drum; **~ magnétique** *m* DONNÉES magnetic drum

tamis *m* ÉLECTRON sieve

tampon *m* ÉLECTRON pad, swab, INFORMAT, TÉL, TRANSMIS buffer; **~ d'appel** TÉL call buffer; **~ audio** INFORMAT audio buffer; **~ d'édition** INFORMAT edit buffer; **~ d'édition de fonction** INFORMAT function edit buffer; **~ d'émission** INFORMAT, TÉL, TRANSMIS transmission buffer; **~ de réception** INFORMAT receive buffer; **~ de réémission** INFORMAT, TÉL, TRANSMIS retransmission buffer, transmit buffer; **~ de remise en ordre** INFORMAT reorder buffer; **~ de retransmission** INFORMAT, TÉL, TRANSMIS retransmission buffer; **~ temporaire** INFORMAT temporary buffer; **~ de transmission** INFORMAT, TÉL, TRANSMIS transmit buffer

tamponner *vt* COMPOSANTS stamp

tapis: **~ antistatique** *m* ÉLECTRON anti-static mat; **~ de sol** *m* RADIO grounding mat

tar *m* DONNÉES tar

tarif *m* APP SERV charge, rate, tariff, GÉNÉRAL rate, tariff RADIO rate, TÉL rate, tariff; **~ d'accès** GÉNÉRAL access charge rate; **~ blanc** *Fra* APP SERV, TÉL medium rate; **~ bleu** *Fra* APP SERV, TÉL cheap call rate; **~ bleu nuit** *Fra* APP SERV, TÉL cheap call rate; **~ fixe** APP SERV fixed rate; **~ inférieur** GÉNÉRAL lower tariff; **~ minimum** APP SERV minimum charge; **~ de raccordement** APP SERV initial connection charge; **~ réduit** APP SERV reduced rate, reduced tariff; **~ rouge** *Fra* APP SERV, GÉNÉRAL, TÉL peak rate; **~ supérieur** GÉNÉRAL higher tariff; **~ de taxe d'appel** APP SERV, TÉL call charge rate; **~ télégraphique** TÉL cable rate; **~ des terminaux** RADIO apparatus charge rate

tarifer *vt* APP SERV, TÉL charge

tarification *f* APP SERV pricing, tariff setting, TÉL tariff setting; **~ de base** APP SERV call charge rate; **~ à la durée** TÉL time charging, time-related charging, time-related tariffing

tas *m* COMPOSANTS bucker, dolly

tassement *m* TÉL packing

taux *m* GÉNÉRAL rate, ratio, INFORMAT depth, rate, ratio; **~ d'aboutissement des appels** TESTS call success rate; **~ d'annulation** RADIO *radar* cancellation ratio; **~ d'appel** TÉL calling rate; **~ d'appel des abonnés** TÉL subscriber calling rate; **~ d'appels** TÉL calling rate; **~ d'appels intempestifs** TÉL false calling rate; **~ de bande critique** APP SERV critical band rate; **~ de blocage** GÉNÉRAL blocking rate; **~ de bourrage** DONNÉES stuffing rate; **~ de codage** DONNÉES coding rate; **~ de compression** TÉL ratio of compression; **~ constant** DONNÉES, INFORMAT constant rate; **~ de croissance** GÉNÉRAL rate of increase; **~ de**

défaillance TESTS failure rate; **~ de dégradation** RADIO degradation figure; **~ de dérive** RADIO drift rate; **~ de désaccordage** INFORMAT detune depth; **~ de disponibilité** COMMUT availability figure; **~ d'échantillonnage** TRANSMIS sampling fraction, sampling frequency, sampling rate; **~ d'efficacité** TÉL completion ratio, efficiency ratio; **~ d'élimination d'écho parasite** RADIO clutter attenuation; **~ d'ellipticité** RADIO axial ratio; **~ d'erreur** TÉL error ratio, *d'une communication télégraphique* error rate; **~ d'erreur à long terme sur les bits** TÉL long-term bit error rate; **~ d'erreur d'une manipulation** TÉL error rate of keying; **~ d'erreur à moyen terme sur les bits** INFORMAT medium-term bit error rate; **~ d'erreur sur les bits** INFORMAT *binons* bit error code; **~ d'erreur sur les caractères** DONNÉES, INFORMAT character error rate, character error ratio; **~ d'erreur sur les éléments** TÉL element error rate; **~ d'erreurs** DONNÉES, SIGNAL error rate; **~ d'erreurs binaires** (*TEB*) DONNÉES, INFORMAT, TRANSMIS binary error rate (*BER*); **~ d'erreurs binaires résiduelles** (*TEBR*) DONNÉES, INFORMAT, RADIO residual binary error rate (*RBER*), residual bit error ratio (*RBER*); **~ d'erreurs résiduelles** DONNÉES, TÉL residual error rate (*RER*); **~ d'erreurs sur les bits** (*TEB*) DONNÉES, INFORMAT, RADIO bit error rate (*BER*); **~ d'erreurs sur les éléments unitaires** TÉL *sématèmes isochrones* unit-element error rate; **~ d'extension** TÉL ratio of expansion, rate of extension; **~ d'extinction** ÉLECTRON, TRANSMIS extinction ratio (*EX*); **~ de gel** DONNÉES freeze-out fraction; **~ d'harmoniques** ÉLECTRON distortion factor; **~ d'impulsions** APP SERV impulse rate, TÉL pulse rate; **~ instantané de défaillance** TESTS failure rate, instantaneous failure rate; **~ instantané de réparation** TÉL instantaneous repair rate; **~ de justification** TÉL, TRANSMIS justification ratio; **~ de modulation** TÉL modulation rate, TRANSMIS modulation depth, modulation factor, modulation percentage, modulation rate; **~ de modulation d'amplitude** TRANSMIS amplitude modulation depth; **~ de modulation de hauteur** INFORMAT pitch-modulation depth; **~ moyen de défaillance** TESTS mean failure rate; **~ moyen de dérangement** INFORMAT mean failure rate; **~ moyen de réparation** INFORMAT mean repair rate; **~ d'occupation** GÉNÉRAL filling rate, occupancy rate; **~ d'ondes de tension stationnaires** (*TOS*) RADIO, TRANSMIS standing wave ratio (*SWR*), voltage standing wave ratio (*VSWR*); **~ d'ondes stationnaires** (*TOS*) RADIO, TRANSMIS standing wave ratio (*SWR*), voltage standing wave ratio (*VSWR*); **~ de panne** PÉRIPH, TESTS failure rate; **~ de pénétration** TV penetration rate; **~ de perte de paquets** DONNÉES burst loss rate; **~ de pondération de clavier** INFORMAT keyboard scaling rate; **~ de prises avec réponse** TÉL answer-seizure ratio; **~ de puces bonnes** DONNÉES chip yield; **~ de réaction** ÉLECTRON feedback ratio; **~ de recombinaison** COMPOSANTS recombination rate; **~ de régénération** DONNÉES refresh rate, INFORMAT refresh rate, regeneration rate; **~ de réparation** TESTS repair rate; **~ de répétition** TÉL repetition rate; **~ de répétition des impulsions** TÉL pulse repetition rate (*PRR*); **~ de retour** ÉLECTRON feedback ratio; **~ de saut de cycle** TRANSMIS clock-skip rate; **~ de signalisations** GÉNÉRAL fault rate; **~ de transfert** DONNÉES, INFORMAT transfer rate; **~ de travail** TÉL marking percentage; **~ d'utilisation** TÉL usage

taxable *adj* APP SERV chargeable, TÉL subject to a charge

taxation[1] *f* APP SERV billing, charge rate, charging, PÉRIPH metering, RADIO charging, TÉL billing, charge metering, charging, metering, ticketing; **~ des appels** APP SERV *téléphonique* call charging; **~ à l'arrivée** TÉL *téléphonique* called-line charging, PÉRIPH per-second metering; **~ automatique** TÉL automatic billing; **~ centralisée** APP SERV centralized billing; **~ détaillée par destination** TÉL toll ticketing; **~ à la distance et à la durée** TÉL time and distance metering, time-zone metering; **~ individuelle** RADIO individual charging; **~ par impulsion périodique** TÉL random metering; **~ périodique** PÉRIPH periodic metering; **~ à l'unité** PÉRIPH per-unit metering; **~ à l'utilisation** PÉRIPH per-usage metering

taxation[2]: **sans ~** *loc* TÉL no charge

taxe *f* APP SERV charge, rate, GÉNÉRAL fee, RADIO charge, TÉL charge, fee, rate, tariff; **~ d'appel** APP SERV call charge; **~ de base** TÉL basic metered unit, basic charge; **~ de conversation** TÉL call fee; **~ forfaitaire** APP SERV flat rate; **~ de ligne** TÉL land-line charge; **~ locale** TÉL local rate, local tariff; **~ de mise en présence** TÉL initial charge; **~ de perception** TÉL collection charge; **~ de préavis** TÉL preavis charge, preavis fee; **~ de répartition** TÉL accounting rate; **~ de station mobile** TÉL mobile-station charge; **~ terrestre** RADIO, TÉL land-station charge; **~ unique** TÉL nonrecurrent charge, nonrecurrent fee; **~ unitaire** TÉL unit charge

taxer *vt* APP SERV, TÉL charge

taxeur *m* TÉL call accounting device

taxinomie: **~ croisée** *f* INFORMAT tangled taxonomy

TC *abrév* (*terminal de commutation*) COMMUT switching terminal

TCD *abrév* (*transformation en cosinus discret*) DONNÉES, ÉLECTRON, INFORMAT, TÉL DCT (*discrete cosine transform*)

TCI-DCS *abrév* (*identification de l'abonné émetteur-DCS*) DONNÉES, TÉL TCI-DCS (*transmitting subscriber identification-DCS*)

TCP *abrév* (*protocole de contrôle de transmission*) DONNÉES TCP (*transmission control protocol*)

T_{cs} *abrév* (*temps de transfert des messages aux points de transfert*) TRANSMIS message transfer time at signaling transfer points (*AmE*), message transfer time at signalling transfer points (*BrE*)

TD *abrév* (*traitement des données*) DONNÉES, INFORMAT DP (*data processing*)

TDEM *abrév* (*télédétection électromagnétique*) TESTS EMRS (*electromagnetic remote sensing*)

té: **~ hybride** *m* RADIO hybrid T; **~ magique** *m* RADIO magic T; **~ série** *m* RADIO series T

TE *abrév* (*transverse électrique*) TRANSMIS TE (*transverse electric*)

TEB *abrév* DONNÉES, INFORMAT, RADIO (*taux d'erreurs binaires*) BER (*binary error rate*), (*taux d'erreurs sur les bits*) BER (*bit error rate*)

TEBR *abrév* (*taux d'erreurs binaires résiduelles*) DONNÉES, INFORMAT, RADIO RBER (*residual binary error rate, residual bit error ratio*)

TEC *abrév* COMPOSANTS, ÉLECTRON (*transistor à effet de champ*) FET (*field effect transistor*), TÉL (*temps d'établissement de communication*) call setup time

technicien: **~ d'essai** *m* TESTS test clerk, test engineer; **~ réseau** *m* GÉNÉRAL network support engineer

technique: **~ cellulaire** *f* RADIO cellular technique; **~ de codage à fréquence synchrone** *f* DONNÉES synchronous-frequency encoding technique (*SFET*); **~ de commutation** *f* COMMUT switching technique; **~ des composants** *f* ÉLECTRON component engineering, component technology; **~ de comptage par impulsions** *f* RADIO *service maritime mobile* pulse-counting technique; **~ des courants faibles** *f* ALIM ÉLEC light-current engineering; **~ des courants forts** *f* ALIM ÉLEC heavy-current engineering, power-current engineering, power engineering; **~ de description formelle** *f* DONNÉES formal description technique (*FDT*); **~ de distorsion préalable** *f* ÉLECTRON predistortion technique; **~ d'échange ionique** *f* ÉLECTRON ion exchange technique; **~ d'essai** *f* TESTS test engineering, test method, test technology, testing technique; **~ des fibres optiques** *f* TRANSMIS fiber optics (*AmE*), fibre optics (*BrE*); **~ du froid** *f* CLIMAT refrigeration engineering; **~ de gestion de la qualité** *f* TESTS quality engineering; **~ des**

hyperfréquences *f* RADIO microwave engineering; **~ de mesure** *f* TESTS measurement technique; **~ de métamorphose** *f* INFORMAT morphing; **~ des micro-ondes** *f* RADIO microwave engineering; **~ numérique** *f* TÉL digital technology, digital engineering; **~ des ondes de sol** *f* RADIO surface-wave technology; **~ des ondes de surface** *f* ÉLECTRON surface-wave technology; **~ de perturbation** *f* TRANSMIS perturbation technique; **~ planar** *f* COMPOSANTS planar technique; **~ de rendu** *f* INFORMAT rendering technique; **~ des télécommunications** *f* TÉL communication engineering; **~ téléphonique** *f* TÉL telephone engineering; **~ de tests** *m* TESTS test technology; **~ de transmission** *f* TRANSMIS transmission technique; **~ Wiegand** *f* DONNÉES Wiegand effect

techniques: **~ d'extension de code** *f pl* TÉL code extension techniques

technologie[1]: **~ avancée** *f* COMPOSANTS highly advanced hardware; **~ de commutation** *f* COMMUT switching technology; **~ des composants** *f* ÉLECTRON component engineering, component technology; **~ électronique** *f* ÉLECTRON electronic technology; **~ de montage en surface** *f* COMPOSANTS, ÉLECTRON surface-mount technology (*SMT*); **~ de pointe** *f* GÉNÉRAL cutting-edge technology, leading-edge technology, INFORMAT advanced technology; **~ des salles de transaction** *f* PÉRIPH dealer room technology; **~ trois passes** *f* DONNÉES triple-pass technology; **~ vocale** *f* APP SERV, INFORMAT, RADIO, TÉL speech technology

technologie[2]: **à la pointe de la ~** *loc* GÉNÉRAL technologically advanced

TED *abrév* (*traitement électronique des données*) DONNÉES, INFORMAT EDP (*electronic data processing*)

teindre *vt* COMPOSANTS stain

teinte *f* INFORMAT lightness, TV hue; **~, luminosité, saturation** TV hue, lightness, saturation (*HLS*)

teinture *f* COMPOSANTS staining

tel: **~ écran, tel écrit** *loc* (*WYSIWYG*) INFORMAT what you see is what you get (*WYSIWYG*)

Tele: **~ 2x** *m* INFORMAT Tele 2x

téléachat *m* APP SERV teleshopping, GÉNÉRAL armchair shopping, telepurchase, TV armchair shopping, home shopping, teleshopping

téléalarme *f* DONNÉES alarm service, RADIO, TÉL remote alarm

téléalerte *f* APP SERV reminder alarm service

téléalimentation *f* ALIM ÉLEC current feed, power feed

téléappel *m* RADIO, TÉL, TRANSMIS paging; **~ national** TRANSMIS nationwide paging

téléblocage *m* CONTRÔLE remote blocking

télécarte *f* TÉL calling card, phonecard, prepaid card, telephone card

téléchargeable *adj* DONNÉES downloadable, ÉLECTRON down-line-loadable, downloadable, INFORMAT, TV downloadable

téléchargement *m* DONNÉES download, downloading, dumping, uploading, ÉLECTRON downloading, dumping, INFORMAT download, downloading, dumping, TESTS remote loading, TV downloading

télécharger *vt* DONNÉES download, dump, upload, ÉLECTRON, INFORMAT, TV download, dump

télécom *abrév* (*télécommunications*) GÉNÉRAL, RADIO, TÉL telecom (*telecommunications*)

télécommandable *adj* CONTRÔLE remote-control

télécommandé *adj* CONTRÔLE remote-controlled

télécommande *f* CONTRÔLE remote control, telecommand, telecontrol, GÉNÉRAL remote control, INFORMAT telecommand, telecontrol, PÉRIPH remote controller, telecommand, RADIO, TÉL telecommand, telecontrol; **~ d'alerte** CONTRÔLE remote control warning system; **~ d'antenne** RADIO antenna remote control; **~ et mesure de distance** TESTS telemetry command and ranging (*TCR*); **~ temporelle prédéterminée** CONTRÔLE time tag command; **~ par variation de fréquence** CONTRÔLE variable-frequency drive

télécommander *vt* CONTRÔLE, INFORMAT, PÉRIPH, RADIO, TÉL telecommand

Télécommunauté: **~ Asie-Pacifique** *f* (*APT*) RADIO Asia-Pacific Telecommunity (*APT*)

télécommunicateur *m* GÉNÉRAL carrier, TRANSMIS common carrier, telecommunications common carrier; **~ qui transmet uniquement l'information** TRANSMIS direct carrier

télécommunication *f* GÉNÉRAL telecommunication, RADIO, TÉL, TRANSMIS communication; **~ d'arrivée** RADIO, TÉL, TRANSMIS inbound communication; **~ de départ** RADIO, TRANSMIS outbound communication

télécommunications *f pl* APP SERV communications, GÉNÉRAL, RADIO, TÉL communications, telecommunications (*telecom*), TRANSMIS, TV communications; **~ à l'échelle mondiale** TÉL worldwide communications; **~ optiques** TRANSMIS optical communications; **~ par câble** APP SERV, TRANSMIS cable communications; **~ par salves météoriques** TRANSMIS burst meteor communications; **~ par satellite** RADIO, TÉL, TV satellite communications; **~ de secours** TÉL relief telecommunications; **~ spatiales** GÉNÉRAL space communications

télécommuniquer *vi* GÉNÉRAL telecommunicate

télécommutation *f* COMMUT remote switching

téléconférence *f* APP SERV phone conference, phone conferencing, teleconference, RADIO teleconference, TÉL conference calling, phone conference, phone conferencing, teleconference, TRANSMIS teleconference, TV conference; **~ multimédia** TÉL, TV multimedia conferencing; **~ à trois** APP SERV, TÉL three-way teleconference

téléconnecté *adj* TÉL remotely-connected

télécopie *f* GÉNÉRAL facsimile, fax, INFORMAT, TÉL telecopy, telefax; **~ contrastée** TÉL black-and-white reception; **~ fac-similée** TÉL document facsimile telegraphy; **~ du groupe 4** (*FX4*) TÉL Telefax Group 4 (*FX4*); **~ du groupe 3** (*FX3*) TÉL Telefax Group 3 (*FX3*); **~ noir sur blanc** TÉL black-and-white reception; **~ rapide** TÉL high-speed facsimile, fast facsimile; **~ sur réseaux privés** PÉRIPH facsimile on private networks

télécopier *vt* INFORMAT, TÉL telecopy

télécopieur *m* PÉRIPH, TÉL facsimile machine, facsimile apparatus, fax machine

télédémarrage *m* CONTRÔLE remote start

télédétection *f* DONNÉES *télémétrie*, GÉNÉRAL remote sensing, TESTS remote detection, remote sensing; **~ aérienne** TESTS airborne remote sensing; **~ aérospatiale** TESTS airborne and spaceborne remote sensing; **~ à distance** DONNÉES, GÉNÉRAL, TESTS remote sensing; **~ électromagnétique** (*TDEM*) TESTS electromagnetic remote sensing (*EMRS*)

télédiagnostique: **~ par circulation de courant** *m* TESTS current flow method

télédiaphonie *f* RADIO far-end crosstalk

télédiffuseur *m* TV television broadcaster

télédiffusion *f* RADIO broadcasting, TV television broadcasting

télédistribution *f* TRANSMIS cable distribution, TV cable television (*cable TV*), cabled distribution; **~ de programmes** TV program distribution (*AmE*), programme distribution (*BrE*)

télé: **~-écriture** *f* DONNÉES, INFORMAT, PÉRIPH, RADIO, TV telewriting

téléenseignement *m* TV instructional television (*ITV*), teletuition, distance learning

téléfax *m* INFORMAT, PÉRIPH, TÉL facsimile, facsimile apparatus, facsimile machine, fax, fax machine, telefax

télégestion *f* CONTRÔLE remote processing, DONNÉES remote processing, telemanagement, GÉNÉRAL remote control, INFORMAT telemanagement, TESTS remote management

télégramme *m* TÉL telegram; **~ d'état** TÉL government telegram; **~ fac-similé** TÉL document facsimile telegram; **~ en franchise officiel** TÉL official franking-privilege telegram; **~ phototélégraphique** TÉL photograph facsimile telegram; **~ de presse** TÉL press telegram; **~ de service** TÉL service telegram

télégraphie *f* RADIO, TÉL, TRANSMIS telegraphy; **~ alphabétique** TÉL alphabetic telegraphy; **~ arythmique** TÉL start-stop system; **~ diplex à quatre fréquences** TÉL four-frequency diplex telegraphy; **~ fac-similée** TÉL facsimile telegraphy; **~ à fréquences vocales** TÉL voice-frequency telegraphy (*VFT*); **~ harmonique** TÉL voice-frequency telegraphy (*VFT*); **~ infratéléphonique** TÉL sub-telephone telegraphy; **~ interbande** TÉL interband telegraphy; **~ intrabande** TÉL intraband telegraphy; **~ Morse** TÉL Morse telegraphy; **~ à mosaïque** TÉL mosaic telegraphy; **~ à courant porteur** TÉL carrier-current telegraphy; **~ par décomposition des signes** TÉL mosaic telegraphy; **~ sans fil** (*TSF*) RADIO, TÉL wireless telegraphy; **~ supratéléphonique** TÉL, TRANSMIS super-telephone telegraphy

téléguidage *m* RADIO teleguidance

téléimpression *f* INFORMAT remote printing

téléimprimeur *m* APP SERV, INFORMAT, PÉRIPH, RADIO, TÉL teleprinter (*BrE*), teletype (*TTY*), teletypewriter (*AmE*)

téléinformatique *f* DONNÉES remote data processing, teleinformatics, GÉNÉRAL teleprocessing, INFORMAT, RADIO remote data processing, teleinformatics

télélogiciel: **~ vidéotex** *m* INFORMAT, TÉL, TV videotex telesoftware

télémaintenance *f* DONNÉES, INFORMAT, TESTS, TV remote maintenance, telemaintenance

télémarketing *m* DONNÉES, INFORMAT telemarketing

télématique *f* DONNÉES, INFORMAT, RADIO telematics

télémesure *f* DONNÉES *télémétrie* remote metering, telemetering, telemetry, RADIO telemetering, telemetry, TESTS remote measuring, telemetering, telemetry, tracking and command (*T&C*); **~ de maintenance** TESTS housekeeping telemetry; **~ par variation de fréquence** TESTS variable-frequency telemetering

télémètre: **~ laser** *m* DONNÉES, TESTS laser rangefinder

télémétrie *f* DONNÉES telemetry, RADIO rangefinding, telemetry, TESTS telemetry; **~ laser** DONNÉES, TESTS laser telemetry

télépaiement *m* APP SERV electronic funds transfer (*EFT*), payment by telephone, DONNÉES, TÉL electronic funds transfer (*EFT*), TV remote payment

téléphone[1] *m* PÉRIPH, RADIO, TÉL telephone; **~ d'affaires électronique** (*TA*) TÉL electronic business set (*BS*); **~ appelant** TÉL calling telephone; **~ appelé** TÉL called telephone; **~ autogénérateur** TÉL sound-powered telephone; **~ à cadran** TÉL dial telephone; **~ de campagne** TÉL field telephone; **~ à carte de crédit** TÉL creditphone; **~ cellulaire** TÉL cellular, cellular telephone; **~ cellulaire numérique** RADIO digital cellular radio; **~ à clavier** PÉRIPH, TÉL push-button set, push-button telephone, touch-tone telephone; **~ à commande vocale** PÉRIPH, TÉL voice-activated phone; **~-fax** *m* PÉRIPH, TÉL fax phone; **~-fax-répondeur** *m* PÉRIPH, TÉL fax answerphone; **~ haut-parleur incorporé** TÉL speakerphone; **~ incorporé** PÉRIPH incorporated telephone; **~ mains-libres** TÉL speakerphone; **~ mobile** TÉL mobile telephone; **~ mobile mixte** RADIO, TÉL combi-telephone; **~ multifréquence** TÉL multifrequency telephone; **~ mural** TÉL wall telephone; **~ personnel** TÉL personal telephone; **~ de poche** RADIO pocket telephone, TÉL hand-portable, hand-portable telephone, pocket telephone; **~ de police secours** TÉL emergency call box; **~ portable** PÉRIPH, TÉL transportable telephone; **~ à poussoirs** *Can* APP SERV, PÉRIPH, TÉL key telephone set; **~ à prépaiement** PÉRIPH, TÉL pay telephone; **~ publique sans cordon** TÉL public cordless telephone; **~ publique sans fil** TÉL public cordless telephone; **~ à réception amplifiée** PÉRIPH amplified telephone; **~ répondeur-enregistreur** PÉRIPH, TÉL answerphone, telephone answering machine; **~ rouge** TÉL hot line; **~ sans fil** PÉRIPH, TÉL cordless telephone (*CT*); **~ sans fil de deuxième génération** (*CT2*) PÉRIPH, TÉL cordless telephone 2 (*CT2*); **~ sans fil de première génération** (*CT1*) PÉRIPH, TÉL cordless telephone 1 (*CT1*); **~ de secours** TÉL emergency telephone; **~ à touches** PÉRIPH, TÉL push-button set, push-button telephone, touch-tone telephone; **~ touch-tone** PÉRIPH, TÉL touch-tone telephone; **~ d'urgence** TÉL emergency telephone; **~ de voiture** RADIO, TÉL car phone, car telephone

téléphone[2]: **au ~** *loc* TÉL on the telephone, over the telephone

téléphoner[1] *vt* TÉL call, make a call, phone, ring (*BrE*); **~ à** TÉL call, phone, ring (*BrE*), telephone; **~ en PCV** APP SERV, TÉL call collect (*AmE*), reverse the charges (*BrE*), transfer the charges

téléphoner[2] *vi* TÉL phone, ring, telephone

téléphonie *f* RADIO, TÉL telephony; **~ en bande élargie** TÉL wideband telephony; **~ en bande normale** TÉL normal-band telephony; **~ à batterie centrale** TÉL central-battery system; **~ câble** TÉL cable telephony; **~ ferroviaire** RADIO, TÉL railroad telephony (*AmE*), railway telephony (*BrE*); **~ par câble** TÉL cable telephony; **~ publique sans fil** TÉL public wireless telephony; **~ rurale** TÉL rural network, rural telephony; **~ sans fil** RADIO cordless telephony; **~ spécialisée** PÉRIPH, TÉL specialized telephone

téléphoniste *m* COMMUT, TÉL operator, switchboard operator, telephone operator, telephonist (*BrE*)

téléport *m* GÉNÉRAL teleport, RADIO antenna farm

téléporter *vt* ÉLECTRON remote

téléprésence *f* INFORMAT telepresence

télépromotion *f* TV home shopping

télérelevé *m* DONNÉES, GÉNÉRAL remote reading, TV remote meter reading

téléréunion *f* TÉL conference call

téléscripteur *m* APP SERV teleprinter (*BrE*), teletypewriter (*AmE*), DONNÉES, INFORMAT telewriter, PÉRIPH teleprinter (*BrE*), teletype machine, teletypewriter (*AmE*), telewriter, RADIO telewriter, TÉL teleprinter (*BrE*), teletypewriter (*AmE*), TV telewriter

télésécurité *f* APP SERV telesecurity, TRANSMIS, TV remote security

téléservice *m* APP SERV teleservice

téléshopping *m* TV home shopping

télésignalisation: **~ d'alarme** *f* TESTS remote alarm indication (*RAI*)

télésondage *m* TV remote survey

télésouffleur *m* TV autocue (*BrE*), teleprompter (*AmE*)

télésoumission: **~ de lots** *f* DONNÉES remote batch entry (*RBE*); **~ de travaux** *f* DONNÉES remote job entry (*RJE*); **~ de travaux par lots** *f* DONNÉES remote batch entry (*RBE*)

téléspectateur *m* TV television viewer, viewer

télésurveillance *f* CONTRÔLE remote supervision, remote surveillance, RADIO telemonitoring, TESTS remote monitoring, telemonitoring, TV remote supervision, remote surveillance

télésurveillé *adj* TÉL remote-controlled

télétaxe *f* TÉL subscriber metering, telephone charge

télétex *m* INFORMAT, PÉRIPH, RADIO teletex

télétexte *m* (*TTX*) APP SERV, TV teletext (*TTX*)

télétrafic *m* COMMUT, TÉL (*trafic de télécommunications*) teletraffic (*telecommunications traffic*)

télétraitement *m* DONNÉES teleinformatics, GÉNÉRAL teleprocessing, INFORMAT, RADIO teleinformatics; **~ par lots** DONNÉES, INFORMAT remote batch processing

télétransmission *f* TRANSMIS remote transmission

télétravail *m* GÉNÉRAL telecommuting, teleworking

télétravailler *vt* APP SERV, DONNÉES, INFORMAT, TRANSMIS telecommute
télétravailleur *m* APP SERV, DONNÉES, INFORMAT, TRANSMIS teleworker
télétypewriter *m* INFORMAT, PÉRIPH, RADIO, TÉL teletype (*TTY*)
télévente *f* APP SERV telesales, TV home shopping, telesales
télévidéothèque *f* INFORMAT, TV video-on-demand (*VOD*)
téléviseur *m* TV television (*TV*), television set; **~ multistandard** TV multistandard television set
télévision *f (TV)* TV television (*TV*); **~ à accès conditionnel** TV pay television; **~ améliorée** TV enhanced television; **~ à analyse lente** TV slow-scan television; **~ à antenne collective** TV community-antenna television; **~ à antenne communautaire** TV community-antenna television (*CATV*); **~ à antenne principale** TV main-aerial television, main-antenna television, master antenna television (*MATV*); **~ à balayage lent** TV slow-scan television; **~ en circuit fermé** TV closed-circuit television (*CCTV*); **~ à la demande** TV pay-per-view television; **~ à exploration lente** TV slow-scan television; **~ à haute définition** (*TVHD*) TV high-definition television (*HDTV*); **~ à images fixes** TV still-picture television (*SPTV*); **~ industrielle** (*TVI*) TV industrial television (*ITV*); **~ interactive** (*TVI*) TV interactive television (*ITV*); **~ en noir et blanc** TV monochrome television; **~ par câble** TV cable television (*cable TV*); **~ par satellite à antenne principale** TV satellite master aerial television (*SMATV*), satellite master antenna television (*SMATV*); **~ payante** TV pay television; **~ à péage** TV pay television; **~ professionnelle** TV business television; **~ de qualité améliorée** TV enhanced-quality television; **~ de qualité conventionnelle** TV existing-quality television
télex *m* (*TLX*) DONNÉES, RADIO, TÉL telex (*TLX*); **~-plus** DONNÉES, RADIO, TÉL telex-plus
telnet *m* DONNÉES telnet
TEM *abrév* (*transverse électromagnétique*) TRANSMIS TEM (*transverse electromagnetic*)
témoin: **~ d'alimentation** *m* ÉLECTRON power indicator; **~ de défaillance** *m* TÉL failure alarm lamp, fault alarm lamp; **~ de dérangement** *m* TÉL failure alarm lamp, fault alarm lamp; **~ d'échec** *m* TÉL failure alarm lamp, fault alarm lamp; **~ lumineux** *m* INFORMAT pilot lamp, pilot light, PÉRIPH light indicator, pilot lamp, pilot light, RADIO pilot lamp, pilot light, TÉL indicating lamp, pilot lamp, pilot light; **~ de mise sous tension** *m* ÉLECTRON power indicator; **~ de panne** *m* TÉL failure alarm lamp, fault alarm lamp
température: **~ ambiante** *f* CLIMAT room temperature, INFORMAT ambient temperature; **~ du boîtier** *f* ÉLECTRON case temperature; **~ de bruit** *f* RADIO spot noise temperature, noise temperature; **~ de bruit du ciel** *f* RADIO sky noise temperature; **~ de bruit de la liaison par satellite** *f* RADIO satellite-link noise temperature; **~ de bruit du système** *f* TRANSMIS system noise temperature; **~ équivalente de bruit** *f* RADIO equivalent noise temperature, noise temperature; **~ de fonctionnement** *f* CLIMAT working temperature; **~ de jonction** *f* COMPOSANTS junction temperature; **~ moyenne de la plaque Hall** *f* ÉLECTRON mean Hall-plate temperature
tempête: **~ ionosphérique** *f* RADIO ionospheric storm
tempo: **~ d'enregistrement** *m* INFORMAT recording tempo; **~ par défaut** *m* INFORMAT default tempo
temporisateur *m* ÉLECTRON timer, interval timer, GÉNÉRAL timer; **~ actif** DONNÉES active timer; **~ de supervision** COMMUT *service maritime mobile* supervisory timer
temporisation *f* DONNÉES, ÉLECTRON time delay, time release, time-out, RADIO *service maritime mobile*, TÉL time release, time-out; **~ d'enregistrement** RADIO register time-out; **~ de protection inversement proportionnelle à la surcharge** ALIM ÉLEC inverse time overload protection
temporiser: **~ le déclenchement de** *vt* DONNÉES, ÉLECTRON, RADIO, TÉL time out
temps[1]: **en ~ différé** *adj* DONNÉES, INFORMAT, TÉL, TESTS nonreal time (*NRT*), NRT (*nonreal time*); **en ~ réel** *adj* DONNÉES, INFORMAT real-time; **~ réel** *adj* GÉNÉRAL on-line, TÉL, TESTS real-time
temps[2]: **en ~ réel** *adv* DONNÉES, INFORMAT, TÉL, TESTS in real time
temps[3] *m* GÉNÉRAL time; **~ d'accès** DONNÉES, INFORMAT access time, TÉL access delay (*AD*); **~ d'acquisition** DONNÉES acquisition time; **~ d'acquisition de régime stable de synthétiseur** TÉL synthesizer settling time; **~ d'amorçage par la gâchette** ÉLECTRON gate-controlled turn-on time; **~ d'arrêt moyen** TESTS mean downtime; **~ atomique international** (*TAI*) RADIO international atomic time (*IAT*); **~ d'attente** INFORMAT latency, rotational delay, waiting time; **~ d'attente de justification** TRANSMIS justification waiting time; **~ de balayage** ÉLECTRON sweep time; **~ blocable** COMPOSANTS turn-off time; **~ de bon fonctionnement** INFORMAT time between failures, TESTS uptime, time between failures; **~ de brasage** COMPOSANTS soldering time; **~ de commutation** COMMUT, TÉL, TESTS switching time; **~ de conduction** ÉLECTRON conduction time; **~ de convergence** TÉL convergence time; **~ de conversation** RADIO, TÉL talk time; **~-coordonné** RADIO coordinate time; **~ de correction de panne** TESTS fault correction time; **~ déblocable** COMPOSANTS turn-on time; **~ de déclin** TRANSMIS decay time; **~ de décroissance** ÉLECTRON pulse decay time, INFORMAT decay time, TÉL fall time, pulse decay time, pulse fall time; **~ de dépannage actif** TESTS *sécurité de fonctionnement* active corrective maintenance time; **~ de dépôt** DONNÉES submission time; **~ de désamorçage par la gâchette** ÉLECTRON gate-controlled turn-off time; **~ de descente** ÉLECTRON pulse decay time, TÉL decay time, fall time, pulse decay time, pulse fall time; **~ de descente de l'enveloppe des paquets** DONNÉES burst-envelope fall time; **~ de descente d'impulsion** ÉLECTRON, TÉL pulse decay time; **~ de développement** INFORMAT development time; **~ de diagnostic** TESTS diagnosis time; **~ de disponibilité** TÉL, TESTS uptime; **~ d'échantillonnage** TRANSMIS sampling time; **~ d'effacement minimal utilisable** ÉLECTRON minimum usable erasing time; **~ d'émission** RADIO airtime, on-air time; **~ d'émission du sous-système transport de messages** (T^{ms}) TÉL message transfer part sending time; **~ entre défaillances** INFORMAT, TESTS time between failures; **~ d'entretien** TÉL, TESTS *sécurité de fonctionnement* preventive maintenance time; **~ d'entretien correctif** TÉL, TESTS *sécurité de fonctionnement* active corrective maintenance time; **~ d'entretien préventif actif** TÉL, TESTS *sécurité de fonctionnement* active preventive maintenance time; **~ d'établissement** RADIO *d'un signal télégraphique* build-up time; **~ d'établissement d'appel** TÉL call setup time; **~ d'établissement de communication** (*TEC*) TÉL call setup time; **~ d'établissement de la communication dans le central** COMMUT exchange call set-up delay; **~ d'établissement relatif** RADIO *de signal télégraphique* relative build-up time; **~ d'exécution** INFORMAT execution time; **~ d'exploitation** GÉNÉRAL operating time; **~ d'extinction** ÉLECTRON fall time; **~ d'extraction** ÉLECTRON working function; **~ de fausse alarme** GÉNÉRAL, RADIO false alarm time; **~ de flashage** PÉRIPH flash time; **~ de fonctionnement** COMPOSANTS operate time, GÉNÉRAL operating time; **~ de fonctionnement avant défaillance** INFORMAT, TESTS time to failure; **~ de fonctionnement jusqu'à la première défaillance** INFORMAT, TESTS time to first failure; **~ de fonctionnement pour le blocage** TÉL suppression operate time; **~ de fonctionnement pour l'intervention partielle** RADIO, TÉL partial break-in operate time; **~ de fonctionnement pour l'intervention totale** RADIO, TÉL full break-in operate time; **~ fort** INFORMAT downbeat; **~ frappé** INFORMAT downbeat; **~ de fuite** TÉL

leak time; ~ **de garde** ÉLECTRON buffer time, RADIO guard time; ~ **de hold** INFORMAT hold time; ~ **d'illumination** RADIO illumination time; ~ **d'immobilisation** ÉLECTRON downtime, TESTS fault time; ~ **d'incapacité** TESTS disabled time, outage time; ~ **d'incapacité externe** TESTS external disabled time, external loss time; ~ **d'indisponibilité** DONNÉES, INFORMAT unavailability time (*UAT*), TESTS downtime; ~ **d'indisponibilité interne** TESTS internal disabled time; ~ **d'intervention** TESTS out-of-service time; ~ **d'inutilisation** GÉNÉRAL lost time; ~ **d'inversion** DONNÉES turnaround time; ~ **d'ionisation** ÉLECTRON ionization time; ~ **jusqu'à la saturation** TESTS time to saturation; ~ **de latence** TÉL stand-by time; ~ **de lecture maximal utilisable** INFORMAT maximum usable reading time; ~ **de libération** TÉL call-release time; ~ **de libération de la communication par le central** COMMUT exchange call-release delay; ~ **libre** TÉL free time, TESTS free time, idle time; ~ **de localisation de panne** TÉL, TESTS fault localization time; ~ **de maintenance** TÉL, TESTS maintenance time; ~ **de maintenance active** TÉL, TESTS *sécurité de fonctionnement* active maintenance time; ~ **de maintenance corrective** TÉL, TESTS *sûreté de fonctionnement* corrective maintenance time; ~ **de maintenance en hommes-heures** TESTS maintenance man-hours; ~ **de maintien** RADIO, TÉL hangover time; ~ **de maintien pour le blocage** TÉL suppression hangover time; ~ **de maintien pour l'intervention** TÉL break-in hangover time; ~ **de manipulation** TÉL user handling time; ~ **maximal de mémoire** ÉLECTRON maximum retention time; ~ **de mise en attente** TÉL queueing time; ~ **de mise en service de l'émetteur** PÉRIPH, TRANSMIS transmitter turn-on time; ~ **de mise sous tension** ALIM ÉLEC duty cycle; ~ **de montée** GÉNÉRAL rise time, build-up time, INFORMAT rise time, TÉL pulse rise time, rise time; ~ **de montée de l'enveloppe des paquets** DONNÉES burst-envelope rise time; ~ **mort** CONTRÔLE dead time, GÉNÉRAL delay, TESTS free time, idle time; ~ **moyen avant panne** TESTS mean time before failure; ~ **moyen de disponibilité** (*TMD*) TESTS mean uptime; ~ **moyen entre interruptions** TESTS mean time between interruptions (*MTBI*); ~ **moyen entre pannes** TESTS mean time between failures (*MTBF*); ~ **moyen de Greenwich** (*TMG*) GÉNÉRAL Greenwich Mean Time (*GMT*); ~ **moyen spécifique de réparation** TESTS specific mean repair time; ~ **de non-fonctionnement** TÉL, TESTS nonoperating time; ~ **d'occupation** TÉL holding time; ~ **opérateur** DONNÉES operating time; ~ **de panne latente** TESTS undetected fault time; ~ **partagé** DONNÉES, INFORMAT time sharing; ~ **de passage** TÉL, TESTS transit time; ~ **perdu** GÉNÉRAL lost time; ~ **pertinent** TESTS relevant time; ~ **de pose** TV exposure time; ~ **de positionnement** DONNÉES, INFORMAT positioning time; ~ **de préchauffage** ÉLECTRON preheating time; ~ **à prendre en compte** TESTS relevant time; ~ **de propagation** ALIM ÉLEC transit time, RADIO propagation delay, propagation time; ~ **de propagation différentiel** TRANSMIS differential delay time; ~ **de propagation électrique** TRANSMIS electrical path delay; ~ **de propagation de groupe** RADIO envelope delay, TRANSMIS group delay; ~ **de propagation moyen dans un sens** TÉL mean one-way propagation time; ~ **de propagation de phase** GÉNÉRAL, RADIO phase delay; ~ **de propagation sur le réseau** TÉL network transfer delay; ~ **de propagation sur la voie de données** (*Tp*) DONNÉES, TÉL data channel propagation time; ~ **propre** RADIO proper time; ~ **de réaction** GÉNÉRAL reaction time; ~ **de réception du sous-système transport de messages** (*Tmr*) INFORMAT message transfer part receiving time; ~ **de recouvrement inverse** COMPOSANTS recovery time; ~ **de récurrence des impulsions** (*TRI*) TÉL pulse repetition time (*PRT*); ~ **réel** DONNÉES real time, INFORMAT, TÉL, TESTS real time; ~ **de réémission** RADIO re-emission time (*RT*); ~ **de réglage** CONTRÔLE setting time; ~ **de relâchement** COMPOSANTS release time; ~ **de renversement** DONNÉES modem turnaround; ~ **de réparation** TÉL repair time, TESTS fault correction time; ~ **de réponse** GÉNÉRAL, INFORMAT, SIGNAL response time; ~ **de réponse à la prise d'un circuit d'arrivée** TÉL incoming response delay; ~ **de repos** ÉLECTRON idle period, off period, TÉL idle period; ~ **de reprise** DONNÉES recovery time, INFORMAT rerun time, TÉL reframing time; ~ **de reprise du verrouillage de trame** TRANSMIS frame-alignment recovery time; ~ **de réserve de batterie** ALIM ÉLEC battery storage time; ~ **de restitution** ÉLECTRON recovery time; ~ **de rétablissement** TRANSMIS restoring time; ~ **de rétablissement partiel** TRANSMIS partial restoring time; ~ **de retard** GÉNÉRAL delay time; ~ **de retombée** INFORMAT fall time; ~ **de réverbération** TÉL reverb time, reverberation time; ~ **de sauvegarde** INFORMAT backup time; ~ **scan image** PÉRIPH, TV image scanning time; ~ **de service** APP SERV service time; ~ **de sonnerie** TÉL ringing period; ~ **de stabilisation** ALIM ÉLEC settling time; ~ **de suppression d'une impulsion** GÉNÉRAL pulsed decay time; ~ **système** GÉNÉRAL overhead; ~ **total de transmission** TÉL, TRANSMIS total transmission time; ~ **de traitement pour le sous-système utilisateur de données** DONNÉES, SIGNAL data user part handling time; ~ **de transfert** ÉLECTRON, TÉL, TRANSMIS transfer time; ~ **de transfert dans le central** COMMUT cross-office transfer time (*Tcu*); ~ **de transfert à l'émission** TÉL sender transfer time; ~ **de transfert des messages aux points de transfert** (*Tcs*) TRANSMIS message transfer time at signaling transfer points (*AmE*), message transfer time at signalling transfer points (*BrE*); ~ **de transfert à la réception** TÉL receiver transfer time; ~ **de transit** TÉL, TESTS transit time; ~ **de transit cible** DONNÉES target transit delay (*TTD*); ~ **de transit cumulatif** DONNÉES cumulative transit delay (*CTD*); ~ **de transit maximal acceptable** DONNÉES maximum acceptable transit delay (*MATD*); ~ **de transit du signal** ÉLECTRON signal transit time; ~ **de transmission** TÉL, TRANSMIS transmission delay, transmission time; ~ **universel** (*TU*) GÉNÉRAL universal time; ~ **universel coordonné** GÉNÉRAL universal time coordinated (*UTC*); ~ **utilisable** GÉNÉRAL operable time, TESTS uptime; ~ **de vérification** TESTS check-out time; ~ **de vision maximal utilisable** TV maximum usable viewing time; ~ **zéro virtuel** TÉL virtual zero time

temps[4]: ~**-espace-temps** *loc* COMMUT time-space-time (*TST*); ~ **par minute** *loc* INFORMAT beats per minute

tenant: ~ **de câble** *m* TRANSMIS cable holder

tendeur *m* COMPOSANTS turnbuckle

tendre: ~ **en phase vapeur** *vt* COMPOSANTS solder using condensation technique

tendu *adj* TRANSMIS cleaved

teneur: ~ **en humidité** *f* GÉNÉRAL moisture content

tension[1] *f* ALIM ÉLEC voltage (*V*); ~ **d'alimentation** ALIM ÉLEC, ÉLECTRON supply voltage; ~ **d'allumage** ÉLECTRON ignition voltage, striking voltage; ~ **alternative** ALIM ÉLEC alternating voltage; ~ **alternative d'amorçage** ALIM ÉLEC, ÉLECTRON *d'un parafoudre* AC spark-over voltage; ~ **d'amorçage** ÉLECTRON ignition voltage, primer voltage, glow voltage, striking voltage, breakdown voltage; ~ **d'amorçage au choc** ÉLECTRON *d'un parafoudre* impulse spark-over voltage; ~ **de l'anode par rapport à la cathode** ÉLECTRON *semi-conducteurs* anode-cathode voltage; ~ **anodique** ÉLECTRON anode voltage, plate voltage; ~ **anodique directe de crête** ÉLECTRON peak forward-anode voltage; ~ **anodique négative de crête** ÉLECTRON peak negative-anode voltage; ~ **d'appel** TÉL ringing voltage; ~ **appliquée** ALIM ÉLEC applied voltage; ~ **d'arc** ALIM ÉLEC flashover voltage; ~ **assignée** ALIM ÉLEC rated voltage; ~ **d'attaque de grille** ÉLECTRON grid driving voltage, grid input voltage; ~ **d'attaque de grille-cathode** ÉLECTRON grid-cathode driving voltage, modulation voltage; ~ **auxiliaire** ALIM ÉLEC auxiliary voltage; ~ **d'avalanche**

ÉLECTRON *semi-conducteurs* avalanche voltage; ~ **base-émetteur** ÉLECTRON *semi-conducteurs* base-emitter voltage; ~ **de batterie** ALIM ÉLEC, ÉLECTRON battery voltage; ~ **de blocage** ÉLECTRON cutoff voltage; ~ **aux bornes** ALIM ÉLEC terminal voltage; ~ **de charge** ALIM ÉLEC charging voltage; ~ **de chauffage** ÉLECTRON filament voltage, heater voltage; ~ **à circuit ouvert** ALIM ÉLEC no-load voltage, open-circuit voltage; ~ **de claquage** ÉLECTRON *semi-conducteurs* breakdown voltage; ~ **collecteur-base** ÉLECTRON collector-base voltage; ~ **collecteur-émetteur** ÉLECTRON collector-emitter voltage; ~ **de commande** ALIM ÉLEC control voltage; ~ **de commande induite** ÉLECTRON induced control-voltage; ~ **composée** ALIM ÉLEC *d'un systèm polyphasé* voltage between lines; ~ **de concentration d'image** ÉLECTRON image focus voltage; ~ **continue** ALIM ÉLEC DC voltage, direct voltage; ~ **continue nominale d'amorçage** ALIM ÉLEC, ÉLECTRON *d'un parafoudre* DC spark-over voltage, nominal DC spark-over voltage; ~ **continue permanente** ÉLECTRON *thyristors* continuous direct off-state; ~ **continue pulsatoire** ALIM ÉLEC pulsating DC voltage; ~ **crénelée** ÉLECTRON, SIGNAL square-wave voltage; ~ **à croissance linéaire** ALIM ÉLEC linearly-increasing voltage, ramp voltage; ~ **de début du plateau** RADIO plateau-threshold voltage; ~ **de décalage** ÉLECTRON offset voltage; ~ **de décomposition** COMPOSANTS decomposition voltage; ~ **en dents de scie** ÉLECTRON sawtooth voltage; ~ **de désamorçage par la gâchette** ÉLECTRON gate turn-off voltage; ~ **directe** ÉLECTRON forward voltage; ~ **disruptive** ALIM ÉLEC disruptive voltage, flashover voltage; ~ **disruptive de choc** TESTS impulse spark-over voltage; ~ **d'écrêtage** ÉLECTRON clipping voltage; ~ **d'élément** ALIM ÉLEC cell voltage; ~ **d'émetteur** COMPOSANTS emitter voltage; ~ **émetteur-base** COMPOSANTS emitter-base voltage; ~ **entre les bornes** ALIM ÉLEC terminal voltage; ~ **d'entrée** ÉLECTRON input voltage; ~ **d'entretien** ÉLECTRON sustaining voltage; ~ **à l'état bloqué** COMPOSANTS, ÉLECTRON off-state voltage; ~ **à l'état passant** COMPOSANTS, ÉLECTRON on-state voltage; ~ **fantôme** ÉLECTRON phantom voltage; ~ **de filament** ÉLECTRON filament voltage, heater voltage; ~ **filtrée** ALIM ÉLEC filtered voltage, smoothed voltage; ~ **flottante** COMPOSANTS floating voltage; ~ **de fonctionnement** ALIM ÉLEC working voltage, ÉLECTRON operating voltage; ~ **de forme rectangulaire** ÉLECTRON, SIGNAL square-wave voltage; ~ **de Hall** ÉLECTRON Hall voltage; ~ **d'hélice** ÉLECTRON helix voltage; ~ **induite** ALIM ÉLEC induced voltage; ~ **induite résiduelle** ÉLECTRON inductive residual voltage; ~ **inverse** ÉLECTRON reverse voltage; ~ **inverse continue permanente** ÉLECTRON *diodes* continuous direct reverse-voltage; ~ **inverse de crête** ALIM ÉLEC, TÉL peak inverse voltage (*PIV*), peak reverse voltage (*PRV*); ~ **inverse de l'émetteur** COMPOSANTS emitter cutoff voltage, emitter reverse voltage; ~ **inverse initiale** ALIM ÉLEC initial inverse voltage; ~ **inverse moyenne** ALIM ÉLEC, ÉLECTRON mean reverse voltage; ~ **inverse de pointe non répétitive** ALIM ÉLEC, TÉL nonrepetitive peak reverse voltage; ~ **inverse de pointe répétitive** ALIM ÉLEC, TÉL repetitive peak reverse voltage; ~ **d'isolement** ALIM ÉLEC voltage rating; ~ **lissée** ALIM ÉLEC filtered voltage, smoothed voltage; ~ **de maintien** ÉLECTRON maintaining voltage; ~ **de modulation** ÉLECTRON modulation voltage; ~ **nominale** ALIM ÉLEC rated voltage, voltage rating; ~ **non lissée** ALIM ÉLEC unfiltered voltage, unsmoothed voltage; ~ **de non-amorçage par la gâchette** ÉLECTRON gate non-trigger voltage; ~ **d'ondulation** DONNÉES ripple voltage; ~ **parasite** ALIM ÉLEC extraneous voltage, RADIO radio influence voltage; ~ **de pénétration** ALIM ÉLEC punch-through voltage; ~ **perturbatrice** TESTS disturbance voltage, interference voltage, perturbation voltage; ~ **de phase** ALIM ÉLEC phase voltage; ~ **de pic** COMPOSANTS peak point voltage; ~ **de pincement** ALIM ÉLEC pinch-off voltage; ~ **de plafond** ALIM ÉLEC peak voltage; ~ **de point isohypse** ALIM ÉLEC projected peak point voltage; ~ **de point de vallée** ALIM ÉLEC, COMPOSANTS valley point voltage; ~ **de pointe répétitive** COMPOSANTS repetitive peak offstage voltage; ~ **de polarisation de grille** ÉLECTRON grid bias voltage; ~ **de pompage** ALIM ÉLEC pumping voltage; ~ **principale** ALIM ÉLEC main voltage, principal voltage; ~ **psophométrique** TÉL psophometric voltage; ~ **de quasi-crête** RADIO quasi-peak voltage; ~ **de rallumage** ÉLECTRON reignition voltage; ~ **en rampe** ALIM ÉLEC linearly increasing voltage, ramp voltage; ~ **de rappel** ÉLECTRON resetting voltage; ~ **de référence** ALIM ÉLEC reference voltage; ~ **réfléchie** TRANSMIS reflected voltage; ~ **résiduelle** ALIM ÉLEC residual voltage; ~ **de retournement** COMPOSANTS *thyristors* break-over voltage; ~ **de rupture** ÉLECTRON breakdown voltage; ~ **de saturation** ÉLECTRON saturation voltage; ~ **du secteur** ALIM ÉLEC mains voltage; ~ **secteur** INFORMAT AC voltage; ~ **de seuil** ALIM ÉLEC, ÉLECTRON threshold voltage; ~ **de seuil à l'état passant** COMPOSANTS, ÉLECTRON on-state threshold voltage; ~ **de signalisation** ALIM ÉLEC, SIGNAL signaling voltage (*AmE*), signalling voltage (*BrE*); ~ **de sonde** ÉLECTRON probe voltage; ~ **de sortie** ÉLECTRON output voltage; ~ **de synchronisme** ÉLECTRON synchronous voltage, synchronism voltage; ~ **thermique résiduelle** COMPOSANTS, ÉLECTRON thermal residual voltage; ~ **transversale** ALIM ÉLEC, TÉL transverse voltage; ~ **à vide** ALIM ÉLEC open-circuit voltage, no-load voltage

tension[2]: **sous ~** *loc* ALIM ÉLEC hot, live, ÉLECTRON hot

tension/fréquence *f* ALIM ÉLEC, ÉLECTRON voltage/frequency (*V/F*)

tentative *f* GÉNÉRAL attempt, RADIO, TÉL attempt, bid; ~ **d'accès frauduleuse** GÉNÉRAL fraudulent access attempt; ~ **d'appel** RADIO, TÉL call attempt; ~ **d'appel abandonnée** RADIO, TÉL abandoned call attempt; ~ **d'appel acheminée** RADIO, TÉL successful call attempt; ~ **d'appel efficace** RADIO, TÉL completed call attempt; ~ **d'appel perdue** RADIO, TÉL lost call attempt; ~ **d'appel répétée** RADIO, TÉL repeated call attempt; ~ **automatique** RADIO, TÉL automatic attempt; ~ **inefficace** RADIO, TÉL unsuccessful attempt; ~ **infructueuse** RADIO, TÉL unsuccessful attempt; ~ **manuelle** RADIO, TÉL manual attempt; ~ **de prise** GÉNÉRAL bid; ~ **répétée** TÉL reattempt

tentatives: ~ **de prise par circuit et par heure** *loc* TÉL bids per circuit per hour

tenue *f* INFORMAT sustain; ~ **à la flexion** GÉNÉRAL flexibility strength; ~ **au milieu ambiant** CLIMAT environmental endurance; ~ **aux radiations** COMPOSANTS radiation hardness; ~ **à la traction** COMPOSANTS tensile strength

téréphtalate: ~ **de polybutylène** *m* ÉLECTRON polybutylene terephthalate

terminaison *f* DONNÉES, ÉLECTRON, INFORMAT, RADIO, TÉL, TRANSMIS termination; ~ **adaptée** RADIO matched termination; ~ **anormale de test élémentaire** DONNÉES abnormal test termination; ~ **de câble** TRANSMIS cable termination; ~ **de collecteur** ÉLECTRON collector termination; ~ **de commutateur** TÉL exchange termination (*ET*); ~ **de conduit d'ordre inférieur** TÉL, TRANSMIS lower-order path termination (*LPT*); ~ **de conduit d'ordre supérieur** TÉL, TRANSMIS higher-order path termination (*HPT*); ~ **de ligne** TÉL, TRANSMIS line termination (*LT*); ~ **numérique d'abonné** TÉL subscriber digital termination; ~ **numérique de réseau** TÉL, TRANSMIS network digital termination (*NT1*); ~ **numérique réseau** TRANSMIS digital subscriber access unit; ~ **de réseau** TÉL, TRANSMIS network termination (*NT*); ~ **de réseau pour le RNIS-LB** TRANSMIS B-ISDN network termination (*NT-LB*); ~ **de réseau 2 pour le RNIS-LB** TRANSMIS B-ISDN network termination 2 (*NT2-LB*); ~ **de réseau 1 pour le RNIS-LB** TRANSMIS B-ISDN network termination 1 (*NT1-LB*); ~ **de section de**

multiplexage TRANSMIS multiplex-section termination (*MST*); ~ **de section de régénération** DONNÉES regenerator-section termination (*RST*); ~ **de segment** DONNÉES segment terminator

terminal *m* APP SERV, COMMUT, COMPOSANTS, CONTRÔLE, DONNÉES, INFORMAT, PÉRIPH, RADIO, TÉL, TRANSMIS, TV terminal; ~ **d'abonné** TÉL, TRANSMIS, TV subscriber terminal, user terminal; ~ **d'adaptation au réseau** TV set-top decoder; ~ **à analyse lent** TÉL slow-scan terminal; ~ **à balayage lent** TÉL slow-scan terminal; ~ **appelé** TÉL terminal called; ~ **de commande** CONTRÔLE control terminal (*CT*); ~ **commandé** CONTRÔLE controlled terminal; ~ **de commutation** (*TC*) COMMUT switching terminal; ~ **de destination** DONNÉES destination terminal; ~ **distant** DONNÉES remote station, remote terminal (*RT*); ~ **à écran** INFORMAT, TV screen terminal; ~ **à écran de visualisation** INFORMAT, TV display screen, video display terminal (*AmE*) (*VDT*), video display unit (*BrE*) (*VDU*); ~ **éloigné** PÉRIPH remote terminal (*RT*); ~ **d'exploitation** INFORMAT operating terminal; ~ **d'exploitation à distance** PÉRIPH remote operating terminal; ~ **à exploration lente** TÉL slow-scan terminal; ~ **à fibres optiques** TRANSMIS optical-line terminating equipment (*OLTE*), fiber-optic terminal device (*AmE*), fibre-optic terminal device (*BrE*); ~ **à fréquence porteuse** TRANSMIS carrier-frequency terminal; ~ **homme-machine** TÉL man-machine terminal; ~ **à liaison commutée** DONNÉES dial-up terminal; ~ **de ligne** TÉL line terminal; ~ **de ligne plésiochrone** TRANSMIS plesiochronous-line terminal; ~ **mains libres** PÉRIPH hands-free terminal; ~ **mixte** RADIO, TÉL combi-telephone; ~ **en mode paquet** TÉL packet-mode terminal; ~ **multimédia** TV multimedia terminal; ~ **de navire** RADIO shipboard terminal; ~ **numérique d'extrémité** TÉL end digital termination; ~ **numérique de ligne** TÉL line digital terminal; ~ **de paiement électronique** APP SERV electronic payment terminal; ~ **papier** TÉL printing terminal; ~ **passif** DONNÉES dumb terminal; ~ **de poche** RADIO hand-held terminal, pocket terminal; ~ **portatif de saisie de données** INFORMAT portable data entry terminal (*PDE terminal*); ~ **primaire** TÉL primary terminal; ~ **de référence principal** RADIO primary reference terminal (*PRT*); ~ **de RMU** RADIO, TÉL bleeper; ~ **RMU** RADIO, TÉL pager; ~ **RMU agile de fréquence** RADIO, TÉL frequency agile pager; ~ **RMU vibrateur** RADIO, TÉL silent alert vibrator; ~ **sous commande** CONTRÔLE controlled terminal; ~ **spécifique** TV dedicated terminal; ~ **de télécommunication de poche** DONNÉES hand-held message terminal; ~ **de télécommunications** RADIO, TÉL telecommunications terminal; ~ **de télécopie** PÉRIPH, TÉL facsimile, facsimile apparatus, facsimile machine, facsimile terminal, fax, fax machine; ~ **de télé-écriture** DONNÉES, INFORMAT, PÉRIPH, RADIO, TV telewriter; ~ **téléinformatique** DONNÉES data communication terminal; ~ **télématique** DONNÉES telematic terminal; ~ **de télépaiement** APP SERV home-banking terminal; ~ **téléphonique** PÉRIPH, TÉL telephone terminal; ~ **télétex** INFORMAT, PÉRIPH, RADIO teletex terminal; ~ **de télévision de réception seulement** TV TVRO; ~ **télex** DONNÉES, RADIO, TÉL telex terminal; ~ **de trafic** RADIO, TÉL traffic terminal; ~ **de traitement de données** DONNÉES, PÉRIPH, TÉL data terminal; ~ **de transmission de données** DONNÉES, PÉRIPH, TÉL data terminal; ~ **à très petite ouverture d'antenne** (*TTPO*) DONNÉES, PÉRIPH, TRANSMIS very small aperture terminal (*VSAT*); ~ **TV réception seulement** TV TVRO; ~ **TVRO** TV TVRO terminal; ~ **d'usager** TÉL, TRANSMIS, TV user terminal; ~ **d'usager local** PÉRIPH local user terminal (*LUT*); ~ **vidéotex** INFORMAT, TÉL, TV videotex terminal; ~ **virtuel** DONNÉES, INFORMAT, TÉL virtual terminal; ~ **WAN** INFORMAT WAN terminal

terminateur *m* PÉRIPH terminator

terminé *adj* RADIO over and out

terminer *vt* INFORMAT *une tâche* complete

termineur *m* COMPOSANTS terminating resistor, terminating set, DONNÉES terminating unit, ÉLECTRON, TRANSMIS terminating resistor, terminating set

terrain: ~ **d'essai d'antennes** *m* RADIO antenna range

terre *f* ALIM ÉLEC, ÉLECTRON earth (*BrE*), ground (*AmE*); ~ **de bâti** ALIM ÉLEC rack earth (*BrE*), rack ground (*AmE*); ~ **franche** ALIM ÉLEC direct positive; ~ **de protection** ALIM ÉLEC, TESTS protective earth (*BrE*), protective ground (*AmE*); ~ **de signalisation** SIGNAL, TRANSMIS signal earth (*BrE*), signal ground (*AmE*)

terrestre *adj* RADIO land-based

terrien *adj* ALIM ÉLEC, ÉLECTRON earth (*BrE*), ground (*AmE*)

tertiaire *m* GÉNÉRAL, RADIO service sector

test: ~ **assisté par ordinateur** *m* INFORMAT computer-aided testing (*CAT*); ~ **biaisé** *m* GÉNÉRAL *statistique* biased test; ~ **de capacité** *m* DONNÉES capability test; ~ **de comportement** *m* DONNÉES behavior test (*AmE*), behaviour test (*BrE*); ~ **de conformité** *m* DONNÉES, TESTS conformance testing; ~ **destructif** *m* TESTS destructive testing; ~ **élémentaire** *m* DONNÉES, TESTS test case; ~ **élémentaire abstrait** *m* DONNÉES, TESTS abstract test case; ~ **élémentaire abstrait paramétré** *m* DONNÉES, TESTS parameterized abstract test case; ~ **élémentaire exécutable** *m* DONNÉES, TESTS executable test case; ~ **élémentaire exécutable paramétré** *m* DONNÉES, TESTS parameterized executable test case; ~ **élémentaire générique** *m* DONNÉES, TESTS generic test case; ~ **d'évaluation des performances** *m* INFORMAT benchmark test; ~ **de fin** *m* TESTS completion test; ~ **final** *m* TESTS final test; ~ **d'interconnexion de base** *m* DONNÉES basic interconnection test (*BIT*); ~ **en laboratoire** *m* TESTS alpha trial, laboratory test; ~ **de liaison** *m* INFORMAT active sensing; ~ **à la limite** *m* TESTS marginal check; ~ **aux marges** *m* TESTS marginal test; ~ **d'occupation** *m* TÉL busy test; ~ **paramétrique** *m* TESTS parametric test; ~ **de performance** *m* INFORMAT benchmark; ~ **préliminaire** *m* TESTS preliminary test; ~ **progressif** *m* TESTS sequential test; ~ **de recette** *m* GÉNÉRAL acceptance test; ~ **de résolution de conformité** *m* DONNÉES, TESTS conformance-resolution test; ~ **de route** *m* TÉL route test; ~ **sélectif** *m* TESTS leapfrog test; ~ **de signification** *m* TESTS significance test; ~ **statistique** *m* TESTS statistical test; ~ **de transmission** *m* TESTS, TRANSMIS transmission test

testabilité *f* TESTS testability

testeur *m* TESTS tester; ~ **de câblage** TESTS cable tester; ~ **de circuit** TESTS circuit test set; ~ **d'équilibrage** TRANSMIS balance tester; ~ **inférieur** DONNÉES, TESTS lower tester (*LT*); ~ **de ligne** TESTS line tester; ~ **de modem** TESTS modem tester; ~ **multiple d'abonné** TÉL subscriber multiple test unit; ~ **rapide** TESTS high-speed tester; ~ **supérieur** DONNÉES, TESTS upper tester (*UT*); ~ **de taxation** APP SERV charging tester; ~ **de trafic** TÉL traffic tester, TESTS load tester, traffic tester

tests: ~ **de conformité** *m pl* DONNÉES, TESTS conformance testing; ~ **encastrés** *m pl* DONNÉES embedded testing

tête[1]: **en** ~ *m* INFORMAT header (*HDR*)

tête[2] *f* COMMUT head, COMPOSANTS head, termination, DONNÉES, ÉLECTRON, GÉNÉRAL head, INFORMAT head, root, root top node, top node, RADIO head, TÉL head, tip, TV head end; ~ **de câble** COMPOSANTS termination joint box, TRANSMIS cable distribution head, cable head, cable termination; ~ **au carbure** ÉLECTRON bend carbide; ~ **d'écriture** INFORMAT write head; ~ **de file** COMMUT head-of-the-queue; ~ **de lecture** COMPOSANTS read head, INFORMAT playback head; ~ **de lecture-écriture** COMPOSANTS read-write head; ~ **de ligne** TV head end; ~ **magnétique** ÉLECTRON magnetic head; ~ **magnéto-optique** DONNÉES magneto-optical head; ~ **de nettoyage**

TÉL cleaning head; ~ **optique** DONNÉES optical head; ~ **optique du capteur solaire** RADIO sun sensor head; ~ **radar** RADIO radar head; ~ **à radiofréquence** RADIO RF head; ~ **de réseau** (*TR*) GÉNÉRAL network head, TV head end, telephony head end; ~ **de réseau câblé** TV network head end

tétraphonie: ~ **à quatre canaux** *f* TÉL four-channel telephony

tétrode *m* COMPOSANTS, ÉLECTRON tetrode

texel *m* INFORMAT texel, texture element

texte *m* DONNÉES, INFORMAT, RADIO, TÉL, TV text; ~ **chiffré** DONNÉES, RADIO, TÉL, TV ciphertext; ~ **en clair** DONNÉES clear text; ~ **explicatif** TÉL clarifying text; ~ **intégral** INFORMAT full text; ~ **non formaté** INFORMAT draft text; ~ **par expansion** INFORMAT stretched text; ~ **variable** TÉL variable text

texture: ~ **par perturbation de la lumière** *f* INFORMAT bump mapping

TFR *abrév* (*transformation de Fourier rapide*) INFORMAT FFT (*fast Fourier transform*)

TFT *abrév* (*transfert de fichier télématique*) TÉL TFT (*telematic file transfer*)

théorème: ~ **de réciprocité** *m* ALIM ÉLEC reciprocity theorem; ~ **de Thévenin** *m* ALIM ÉLEC Thévenin's theorem

théorie: ~ **des automates** *f* DONNÉES automata theory; ~ **de la communication** *f* GÉNÉRAL communication theory; ~ **des ensembles flous** *f* DONNÉES fuzzy set theory; ~ **de l'information** *f* GÉNÉRAL information theory; ~ **des probabilités** *f* GÉNÉRAL probability theory; ~ **de rappel** *f* TRANSMIS background theory; ~ **de trafic** *f* TÉL traffic theory

thermistance *f* COMPOSANTS, ÉLECTRON thermistor

thermisteur *m* COMPOSANTS, ÉLECTRON thermistor

thermistor *m* COMPOSANTS, ÉLECTRON thermistor

thermocontact *m* CLIMAT temperature controller

thermocouple *m* COMPOSANTS, ÉLECTRON thermocouple

thermomètre: ~ **mouillé** *m* ÉLECTRON wet-bulb thermometer; ~ **à résistance** *m* TESTS resistance thermometer

thermorayonnance *f* ALIM ÉLEC, ÉLECTRON heat radiation

thermorétraction *f* ALIM ÉLEC, ÉLECTRON heat shrinking

thermostat *m* CLIMAT, COMPOSANTS, CONTRÔLE thermostat; ~ **de service** COMPOSANTS, CONTRÔLE operation thermostat

thyatron *m* COMPOSANTS, ÉLECTRON thyatron

thyristor *m* ALIM ÉLEC, COMPOSANTS, ÉLECTRON thyristor; ~ **blocable** COMPOSANTS turn-off thyristor; ~ **commandé au silicium** COMPOSANTS, ÉLECTRON silicon-controlled rectifier (*SCR*); ~ **diode bidirectionnel** COMPOSANTS, ÉLECTRON bidirectional diode thyristor; ~ **diode bloqué en inverse** COMPOSANTS, ÉLECTRON reverse blocking diode thyristor; ~ **diode passant en inverse** COMPOSANTS, ÉLECTRON reverse conducting diode thyristor; ~ **à gâchette éloignée** ÉLECTRON remote-gate thyristor; ~ **monté en diode bidirectionnel** COMPOSANTS, ÉLECTRON *semi-conducteurs* bidirectional diode thyristor; ~ **monté en triode bidirectionnel** COMPOSANTS, ÉLECTRON *semi-conducteurs* bidirectional triode thyristor; ~ **N** COMPOSANTS, ÉLECTRON N-gate thyristor; ~ **de puissance** ALIM ÉLEC, COMPOSANTS, ÉLECTRON power thyristor; ~ **triode bidirectionnel** COMPOSANTS, ÉLECTRON *semi-conducteurs* bidirectional triode thyristor

ticket: ~ **de taxation** *m* TÉL toll ticket

tiers *m* TÉL third party

TIFF *abrév* (*tagged-image file format*) DONNÉES, INFORMAT TIFF (*tagged-image file format*)

tige: ~ **d'ancrage** *f* COMPOSANTS stay iron, guy attachment; ~ **en delrin** *f* COMPOSANTS delrine rod; ~ **de gabarit pour peigne** *f* TRANSMIS formboard peg; ~ **de mise à la terre** *f* ALIM ÉLEC earthing rod (*BrE*), grounding rod (*AmE*); ~ **de paratonnerre** *f* TESTS lightning conductor, lightning rod; ~ **de soupape** *f* COMPOSANTS, ÉLECTRON valve spindle; ~ **de support** *f* COMPOSANTS supporting rod

tilde *m* INFORMAT *typographie* tilde

timbre *m* INFORMAT timbre, tone color (*AmE*), tone colour (*BrE*), TÉL timbre

tirade: ~ **électronique** *f* ÉLECTRON flame

tirage *m* INFORMAT hard copy, TRANSMIS *d'un câble* pulling; ~ **de câble** GÉNÉRAL *génie industriel extérieur* cable pulling, ~ **en conduite** TRANSMIS pulling in; ~ **à part** PÉRIPH offprint, reprint; ~ **préliminaire** INFORMAT, PÉRIPH preprint

titre[1]: **à** ~ **gratuit** *adv* GÉNÉRAL free of charge; **à** ~ **secondaire** *adv* RADIO on a secondary basis

titre[2] *m* DONNÉES title; ~ **de l'entité d'application** DONNÉES application-entity title; ~ **d'entité N** DONNÉES N-entity title; ~ **de processus d'application** DONNÉES application process title; ~ **de système** DONNÉES system title; ~ **de TPSU** DONNÉES TPSU title

titre[3]: **à** ~ **indicatif** *loc* GÉNÉRAL for your information (*FYI*)

titulaire: ~ **d'abonnement** *m* DONNÉES, INFORMAT, RADIO, TÉL, TV subscriber agreement holder

TLX *abrév* (*télex*) DONNÉES, RADIO, TÉL TLX (*telex*)

TM *abrév* (*transverse magnétique*) TRANSMIS TM (*transverse magnetic*)

TMD *abrév* (*temps moyen de disponibilité*) TESTS mean uptime

TMG *abrév* (*temps moyen de Greenwich*) GÉNÉRAL GMT (*Greenwich Mean Time*)

T_{mr} *abrév* (*temps de réception du sous-système transport de messages*) INFORMAT message transfer part receiving time

T_{ms} *abrév* (*temps d'émission du sous-système transport de messages*) TÉL message transfer part sending time

toc *m* COMPOSANTS dog, driver

tôle *f* COMPOSANTS sheet, sheet metal

tolérance *f* DONNÉES, ÉLECTRON, GÉNÉRAL, INFORMAT, RADIO, TESTS, TRANSMIS tolerance; ~ **d'affaiblissement** TRANSMIS attenuation tolerance; ~ **d'ajustement** GÉNÉRAL fitting tolerance; ~ **de bruit impulsif** RADIO impulsive noise tolerance; ~ **aux défaillances** DONNÉES, ÉLECTRON, INFORMAT, TESTS fault tolerance; ~ **aux fautes** DONNÉES, ÉLECTRON, INFORMAT, TESTS fault tolerance; ~ **de fréquence** RADIO frequency tolerance; ~ **aux pannes** DONNÉES, ÉLECTRON, INFORMAT, TESTS fault tolerance; ~ **de position** GÉNÉRAL positional tolerance; ~ **sur le diamètre du cœur** TRANSMIS core diameter tolerance

tolérant: ~ **aux pannes** *adj* DONNÉES, ÉLECTRON, INFORMAT, TESTS fault-tolerant

ton *m* GÉNÉRAL tone; ~ **absolu** INFORMAT absolute pitch

tonal *adj* INFORMAT, PÉRIPH, TÉL tonal

tonalité *f* COMMUT, CONTRÔLE, DONNÉES, ÉLECTRON, GÉNÉRAL, PÉRIPH, RADIO, SIGNAL tone, TÉL pitch, tone; ~ **d'accusé de réception** TÉL acknowledge tone; ~ **d'appel en attente** TÉL call-waiting tone; ~ **au-dessous de la bande** RADIO tone below band; ~ **au-dessus de la bande** RADIO tone above band; ~ **d'avertissement** TÉL burst of tone, warning tone; ~ **d'avertissement de neutralisation de ligne** TÉL line-lockout warning tone; ~ **dans la bande** RADIO tone in band; ~ **de débordement** TÉL overflow tone; ~ **dépistable** ÉLECTRON traceable tone; ~ **d'encombrement** TÉL congestion tone; ~ **d'enregistrement** TÉL record tone; ~ **d'entrée d'un tiers** TÉL third-party warning tone; ~ **d'entrée en tiers** TÉL busy override, override tone; ~ **de file d'attente** TÉL comfort tone; ~ **à fréquence vocale** TÉL buzzer tone, buzzing tone; ~ **de garde** TÉL tone on hold; ~ **d'identification de publiphone** TÉL payphone recognition tone; ~ **d'indication négative** TÉL negative indication tone; ~ **d'indication positive** TÉL positive indication tone; ~ **infrabande** RADIO tone below band; ~ **à l'intérieur de la bande** RADIO tone in band; ~ **interne de numérotation des autocommutateurs privés** TÉL PABX internal dial tone; ~ **d'intervention** TÉL barge-in tone, tick tone, ticker tone; ~ **intrabande** RADIO tone in band; ~ **d'intrusion** TÉL intrusion tone, tick tone, ticker tone; ~ **modulée incorporée** ÉLECTRON built-in warble tone;

~ **momentanée** SIGNAL burst of tone; ~ **de numérotation** DONNÉES, TÉL dial tone (*AmE*), dialling tone (*BrE*); ~ **d'occupation** GÉNÉRAL, TÉL busy signal (*AmE*), busy tone, engaged signal, engaged tone (*BrE*); ~ **d'option de file d'attente en mode de décrochage** TÉL off-hook queue offer tone; ~ **de paiement** TÉL pay tone; ~ **pilote** TRANSMIS pilot tone; ~ **de priorité** TÉL priority tone, override tone; ~ **de progression d'appel** TÉL call progress tone; ~ **de réponse** DONNÉES answer tone (*ANS*); ~ **de retour d'appel** TÉL ringing tone, audible ringing tone, ringing signal, ring tone; ~ **du ronfleur** TÉL buzzer tone; ~ **simple** GÉNÉRAL single tone; ~ **spéciale d'information** TÉL special information tone; ~ **spéciale de numérotation** DONNÉES, TÉL special dial tone (*AmE*), special dialling tone (*BrE*); ~ **spéciale de retour d'appel** TÉL special ringing tone; ~ **suprabande** RADIO tone above band; ~ **de surcharge** TÉL overflow tone; ~ **de surveillance** COMMUT, CONTRÔLE supervisory tone; ~ **de taxe de base** PÉRIPH end-of-period tone; ~ **de tic-tac** TÉL tick tone, ticker tone; ~ **circuits tous occupés** TÉL all-paths busy tone, all-trunks busy tone, fast-busy tone, reorder tone

tonie *f* TÉL pitch

tonique *m* INFORMAT keynote

tons: ~ **opposés** *m pl* TV *couleurs* contrasting colors (*AmE*), contrasting colours (*BrE*)

TOP *abrév* (*tube à ondes progressives*) ÉLECTRON, TRANSMIS TWT (*traveling wave tube, travelling wave tube*)

topologie *f* DONNÉES, ÉLECTRON, INFORMAT, TÉL topology; ~ **en anneau** DONNÉES ring topology; ~ **arborescente** DONNÉES, INFORMAT tree topology; ~ **en bus** DONNÉES bus topology; ~ **en étoile** DONNÉES star topology; ~ **du réseau** DONNÉES, TÉL network topology

tore *m* COMPOSANTS threading ring; ~ **de ferrite** INFORMAT ferrite core; ~ **magnétique** ÉLECTRON magnetic core

torsadé *adj* COMPOSANTS, ÉLECTRON twisted

torsade *f* COMPOSANTS, ÉLECTRON twist, twisting

torsader *vt* COMPOSANTS twine, twist, ÉLECTRON twist

torsion: ~ **transversale** *f* TRANSMIS *câbles* cross-twisting

TOS *abrév* RADIO, TRANSMIS (*taux d'ondes stationnaires, taux d'ondes de tension stationnaires*) SWR (*standing wave ratio*), VSWR (*voltage standing wave ratio*)

touche *f* APP SERV, CONTRÔLE, DONNÉES key, GÉNÉRAL button, INFORMAT, PÉRIPH button, key, RADIO button, TÉL button, key, TV button; ~ **d'accent** INFORMAT accent key; ~ **alphabétique** INFORMAT alphabetic key; ~ **alphanumérique** DONNÉES, PÉRIPH data character key; ~ **d'alternat** TÉL press-to-talk button; ~ **appel position** TÉL call agent key; ~ **d'arrêt** INFORMAT stop button; ~ **autobloquante** INFORMAT locking-type button; ~ **bis** PÉRIPH last number redial key, redial button; ~ **de coupe-circuit** TÉL blocking button, cut-out key; ~ **débordement** TÉL interflow key; ~ **de démonstration** INFORMAT demo button; ~ **de déverrouillage** INFORMAT unlock key; ~ **d'écoute de position** TÉL observe agent key; ~ **d'édition** INFORMAT edit button; ~ **à effleurement** DONNÉES electronic key, INFORMAT, PÉRIPH touch control; ~ **d'émission d'impulsions** TÉL pulsing key; ~ **enfoncée** INFORMAT key-on; ~ **envoi** RADIO send button; ~ **étoile** TÉL *postes téléphoniques* star button; ~ **d'exécution** INFORMAT execute button; ~ **de fonction** CONTRÔLE, INFORMAT function key; ~ **de fonction affichable** INFORMAT virtual pushbutton; ~ **à fonctions multiples** INFORMAT multifunction key; ~ **inopérante** INFORMAT invalid key; ~ **d'interruption** INFORMAT break key; ~ **d'inversion** INFORMAT reverse key; ~ **de liaison** INFORMAT tie button; ~ **majuscule** INFORMAT, PÉRIPH shift key; ~ **de marche-arrêt** INFORMAT start-stop button; ~ **modem** DONNÉES modem control key; ~ **à motion** INFORMAT, PÉRIPH shift key; ~ **multifonctions** INFORMAT multifunction key, multipurpose key; ~ **numérique** TÉL digit key; ~ **de position** TÉL agent key; ~ **de positionnement** INFORMAT, PÉRIPH shift key; ~ **de présélection** TV, RADIO key for direct selection, key for preset selection; ~ **prise de ligne** PÉRIPH line seizure button; ~ **de rappel** TÉL recall button; ~ **relâchée** INFORMAT key-off; ~ **à répétition** PÉRIPH typamatic key; ~ **répétitrice** INFORMAT repeat action key; ~ **de reprise** INFORMAT continue key; ~ **de reprise de tonalité** PÉRIPH redial button; ~ **secrète** PÉRIPH mute button, privacy button, secrecy button; ~ **de sélection** INFORMAT select button; ~ **de sortie** INFORMAT exit button; ~ **souple** INFORMAT soft key; ~ **tampon** INFORMAT soft key

touches *f pl* INFORMAT keyboard matrix, key matrix

tour[1]: ~ **à tourelle revolver** *m* COMPOSANTS turret lathe

tour[2]: ~ **hertzienne** *f* RADIO microwave tower

touret *m* COMPOSANTS lathe, TRANSMIS *de câble* cable drum, cable reel

tournage *m* COMPOSANTS turning; ~ **au tour à tourelle revolver** COMPOSANTS turret turning

tournant *m* COMPOSANTS rotary control

tourne: ~**-disque portatif** *m* INFORMAT Discman®

tourner[1] *vt* COMPOSANTS turn, DONNÉES, INFORMAT *logiciel* run; ~ **sur un axe** INFORMAT spin

tourner[2]: ~ **sur un axe** *vi* INFORMAT spin

tournevis: ~ **à clichet** *m* COMPOSANTS spiral ratchet screwdriver; ~ **coudé** *m* GÉNÉRAL offset screwdriver; ~ **pour écrous à fente** *m* COMPOSANTS slotted screwdriver; ~ **Torx** *m* COMPOSANTS Torx screwdriver, screwdriver for recessed head screws

T_p *abrév* (*temps de propagation sur la voie de données*) DONNÉES, TÉL data channel propagation time

TPA *abrév* (*architecture de protocole télématique*) TÉL TPA (*telematic protocol architecture*)

TPDU UN *abrév* (*TPDU d'utilisation de connexion de réseau*) DONNÉES UN TPDU (*use of network connection TPDU*); ~ **d'utilisation de connexion de réseau** *m* (*TPDU UN*) DONNÉES use of network connection TPDU (*UN TPDU*)

TR *abrév* DONNÉES, ÉLECTRON (*émetteur-récepteur*) TR (*transmitter-receiver*), GÉNÉRAL (*tête de réseau*) head end, network head, RADIO, TÉL, TRANSMIS (*émetteur-récepteur*) TR (*transmitter-receiver*), TV (*tête de réseau*) head end, network head

traçabilité *f* COMMUT, ÉLECTRON, TESTS traceability

traçage *m* COMMUT, ÉLECTRON, TESTS tracing

trace *f* COMMUT trace, ÉLECTRON scan, trace, TESTS trace; ~ **de fuite** RADIO tracking; ~ **rémanente** RADIO paint

tracé *m* INFORMAT *un graphique* plot

tracer *vt* COMMUT, ÉLECTRON trace, INFORMAT draw, TESTS trace

traceur *m* INFORMAT, TESTS plotter; ~ **de courbes** INFORMAT graph plotter, TESTS x-y plotter, x-y recorder; ~ **à tambour** INFORMAT drum plotter

traducteur: ~**-imprimeur sur bande** *m* PÉRIPH tape printer; ~ **de langage** *m* INFORMAT language translator; ~ **toroïdal** *m* TRANSMIS ring translator

traduction *f* DONNÉES, INFORMAT, TÉL, TV translation

traduire *vt* DONNÉES, INFORMAT, TÉL, TV translate

trafic *m* COMMUT, DONNÉES, ÉLECTRON traffic, GÉNÉRAL calls, traffic, RADIO traffic, TÉL calls, traffic, TESTS, TRANSMIS traffic; ~ **d'un abonné** TÉL subscriber traffic rate; ~ **aléatoire** TÉL random traffic; ~ **d'arrivée** TÉL incoming calls, terminating traffic; ~ **artificiel** TÉL artificial traffic; ~ **commandé par l'abonné** TÉL subscriber-dialed traffic (*AmE*), subscriber-dialled traffic (*BrE*); ~ **coordonné** RADIO coordinated traffic; ~ **à courte distance** TÉL short-distance traffic, short-haul traffic; ~ **de débordement** COMMUT, TÉL overflow traffic; ~ **de départ** TÉL originating traffic; ~ **à destination des services spéciaux** TÉL traffic to special services; ~ **direct** TÉL transverse traffic; ~ **écoulé** TÉL traffic carried, usage; ~ **efficace** COMMUT, GÉNÉRAL effective traffic, RADIO successful traffic; ~ **entrant** TÉL incoming traffic; ~ **d'essai** TÉL, TESTS, TRANSMIS test traffic; ~ **inefficace** RADIO, TÉL unsuccessful traffic; ~ **inter-entreprises** TÉL business traffic; ~ **interlocal** TÉL interlocal traffic; ~ **interne** TÉL internal traffic; ~ **interurbain** TÉL intertoll traffic, trunk traffic; ~ **de jour** TÉL day traffic; ~ **de ligne privée** COMMUT tie-line traffic, TÉL inter-PBX traffic, tie-line traffic, TRANSMIS tie-line traffic; ~ **local** TÉL local traffic; ~ **en mode duplex** RADIO duplex traffic; ~ **en mode simplex** RADIO simplex traffic; ~ **nonsélectif** RADIO nonselective traffic; ~ **perdu** TÉL lost traffic, waste traffic; ~ **de pointe** TÉL peak traffic; ~ **de Poisson** TÉL Poisson traffic, pure-chance traffic, random traffic; ~ **poissonien** TÉL Poisson traffic, pure-chance traffic, random traffic; ~ **de pur hasard** TÉL Poisson traffic, pure-chance traffic, random traffic; ~ **réacheminé** TÉL redirected traffic; ~ **en réception** TÉL incoming calls, terminating traffic; ~ **régularisé** TÉL smooth traffic; ~ **réseau de petite et moyenne envergure** ÉLECTRON light-to-medium network traffic; ~ **sortant** TÉL outgoing traffic; ~ **survariant** TÉL peaked traffic; ~ **de télécommunications** (*télétrafic*) COMMUT, TÉL telecommunications traffic (*teletraffic*); ~ **téléphonique** TÉL telephone traffic; ~ **télex** TÉL telex traffic; ~ **de transit** TÉL transit traffic

trafiquer *vt* TESTS tamper with

train: ~ **de bits** *m* DONNÉES bit stream; ~ **de bits en continu** *m* DONNÉES continuous bit oriented stream (*CBO stream*); ~ **de clés** *m* APP SERV, PÉRIPH key stream; ~ **de données numériques** *m* DONNÉES digital stream; ~ **d'impulsions** *m* SIGNAL pulse sequence, TÉL, TRANSMIS pulse train; ~ **d'impulsions numériques** *m* TÉL digital pulse stream; ~ **d'ondes** *m* DONNÉES trailer record, ÉLECTRON, RADIO wave train; ~ **porteur** *m* TRANSMIS pulse carrier

traînage *m* INFORMAT streaking, TÉL tailing, TV streaking; ~ **dynamique** INFORMAT comet tail

traînée *f* RADIO drag

traîner *vi* INFORMAT lag

trait *m* TÉL point; ~ **Morse** TÉL Morse dash; ~ **noir** ÉLECTRON black stroke

traité *adj* DONNÉES, INFORMAT processed

traitement *m* APP SERV, COMMUT processing, COMPOSANTS treatment, CONTRÔLE processing, DONNÉES handling, processing, ÉLECTRON processing, GÉNÉRAL handling, INFORMAT processing, RADIO processing, SIGNAL conditioning, processing, TÉL handling, processing, TRANSMIS, TV processing; ~ **adaptatif** RADIO adaptive processing; ~ **adapté aux signatures** RADIO signature-adapted processing; ~ **anodique** COMPOSANTS *métallurgie* anodic treatment; ~ **anti-rouille** COMPOSANTS coating with a rust preventive; ~ **d'appel** TÉL call control (*CC*); ~ **des appels** TÉL call handling, call processing; ~ **après production** INFORMAT postproduction processing; ~ **automatique des données** DONNÉES, INFORMAT automatic data processing, datamation; ~ **à bord** INFORMAT on-board processing (*OBP*); ~ **des communications** TÉL call handling; ~ **différé** DONNÉES, INFORMAT batch processing; ~ **direct** DONNÉES on-line processing; ~ **à distance** CONTRÔLE, DONNÉES remote processing; ~ **à distribution ouverte** INFORMAT open distribution processing (*ODP*); ~ **de documents** DONNÉES document processing; ~ **des données** (*TD*) DONNÉES, INFORMAT data processing (*DP*); ~ **électronique des données** (*TED*) DONNÉES, INFORMAT electronic data processing (*EDP*); ~ **des erreurs** DONNÉES error handling; ~ **et transmission de documents** DONNÉES document processing and interchange; ~ **ET-OU** DONNÉES AND-OR processing; ~ **de langage naturel** DONNÉES, INFORMAT natural language processing; ~ **de l'image** DONNÉES, TV image processing; ~ **de l'information** DONNÉES, INFORMAT data processing, TÉL information processing; ~ **intégré des données** DONNÉES integrated data processing; ~ **de la langue naturelle écrite** DONNÉES, INFORMAT natural language processing; ~ **des langues naturelles** DONNÉES, INFORMAT natural language processing; ~ **monomédia** TV monomedia processing; ~ **multimédia** INFORMAT multimedia processing; ~ **multitâche** DONNÉES, INFORMAT multitasking; ~ **non prioritaire** INFORMAT background processing; ~ **numérique** DONNÉES digital processing, INFORMAT full digital processing; ~ **optique** TRANSMIS optical processing; ~ **par lots** DONNÉES, INFORMAT batch processing; ~ **par lots à distance** DONNÉES, INFORMAT remote batch processing; ~ **en parallèle** DONNÉES, ÉLECTRON parallel processing; ~ **de la parole** APP SERV, TÉL speech processing; ~ **prioritaire** INFORMAT foreground processing; ~ **réel** INFORMAT live running; ~ **réparti** DONNÉES, INFORMAT distributing data processing; ~ **du signal** SIGNAL signal conditioning, signal processing; ~ **de signal acoustique** DONNÉES acoustic signal processing; ~ **de texte** DONNÉES word processing, INFORMAT text processing, word processing; ~ **thermique** ÉLECTRON tempering; ~ **du trafic** TÉL traffic handling; ~ **transactionnel** DONNÉES transaction processing (*TP*)

traiter *vt* DONNÉES handle, process, INFORMAT deal with, process

trajectoire *f* ÉLECTRON, INFORMAT, RADIO, TRANSMIS path; ~ **ascensionnelle** TRANSMIS climb path; ~ **sinusoïdale** ÉLECTRON sinusoidal path; ~ **de vol** RADIO flight path

trajet: ~ **de brouillage** *m* RADIO interference path; ~ **de câble** *m* TRANSMIS cable route, cable run, cable way; ~ **de canalisation** *m* GÉNÉRAL duct route, duct run, TRANSMIS conduit route, conduit run, duct route, duct run; ~ **de certification** *m* DONNÉES certification path; ~ **de communication** *m* DONNÉES communication path; ~ **de commutation** *m* COMMUT switching path; ~ **de conversation** *m* TÉL speech path; ~ **descendant de la liaison de connexion** *m* RADIO down-link feeder link; ~ **double** *m* TRANSMIS dual path; ~ **à grande impédance** *m* TRANSMIS high-impedance path; ~ **impraticable** *m* TRANSMIS untrafficable path; ~ **montant** *m* RADIO up-path; ~ **radioélectrique** *m* RADIO ray; ~ **radioélectronique** *m* RADIO radio path; ~ **de transmission** *m* RADIO, TRANSMIS transmission path; ~ **virtuel** *m* DONNÉES, TRANSMIS virtual path (*VP*)

tramage: ~ **aléatoire** *m* INFORMAT dithering

trame *f* DONNÉES field, frame, unit, ÉLECTRON raster, INFORMAT frame, RADIO raster, SIGNAL unit, TÉL frame, unit, TRANSMIS frame, TV frame, field; ~ **d'accusé de réception non-numéroté** TÉL unnumbered acknowledgement frame (*UA*); ~ **de déconnexion** TÉL disconnect frame; ~ **EMCN** TÉL DCME frame; ~ **de gestion** DONNÉES control frame; ~ **d'identification d'échange** TÉL exchange identification frame; ~ **d'identification numérique** DONNÉES digital identification frame; ~ **inattendue** (*NEF*) TÉL not-expected frame (*NEF*); ~ **inférieure** TV bottom field; ~ **d'information** TÉL information frame; ~ **en instance** DONNÉES outstanding frame; ~ **maîtresse** DONNÉES master frame; ~ **en mode déconnexion** TÉL disconnected mode frame; ~ **non prête à recevoir** TÉL receive-not-ready frame; ~ **périodique** INFORMAT periodic frame; ~ **prête à recevoir** TÉL receive-ready frame; ~ **de rejet** TÉL reject frame (*REJ*); ~ **de rejet sélectif** TÉL selective reject frame (*SREJ*); ~ **RNR** TÉL RNR frame; ~ **sémaphore** SIGNAL signal unit (*SU*); ~ **sémaphore d'état du canal sémaphore** SIGNAL link status signal unit; ~ **sémaphore de message** (*TSM*) DONNÉES, TÉL message signaling unit (*AmE*) (*MSU*), message signalling unit (*BrE*) (*MSU*); ~ **sémaphore de remplissage** TÉL fill-in signal unit; ~ **supérieure** TV top field; ~ **temporelle** INFORMAT, RADIO, TRANSMIS timescale

tranche *f* COMPOSANTS, ÉLECTRON wafer, TV *de câblage* tranche; ~ **horaire** COMMUT, RADIO, SIGNAL time slot, TÉL rate timeband, TRANSMIS time slot; ~ **de silicium** COMPOSANTS silicon wafer

transaction *f* DONNÉES transaction; **~ répartie** DONNÉES distributed transaction; **~ répartie assurée par l'application** DONNÉES application-supported distribution transaction; **~ répartie assurée par le fournisseur** DONNÉES provider-supported distributed transaction
transadmittance *m* ÉLECTRON transadmittance
transcodage *m* ÉLECTRON, TÉL, TESTS transcoding, TRANSMIS code conversion, transcoding
transcoder *vt* ÉLECTRON, TÉL, TESTS, TRANSMIS transcode
transcodeur *m* ÉLECTRON, TÉL, TESTS transcoder, TRANSMIS code converter, transcoder
transconductance *f* ÉLECTRON transconductance
transcripteur *m* TÉL transcriber
transducteur *m* ALIM ÉLEC, COMPOSANTS, ÉLECTRON, RADIO, TÉL, TESTS, TRANSMIS transducer, transductor; **~ électrique** ALIM ÉLEC electric signal transducer; **~ interdigité** TRANSMIS interdigital transducer (*IDT*); **~ de mesure** TESTS measuring transducer, measuring transductor; **~ orthomode** RADIO orthogonal mode transducer, orthomode transducer; **~ passif** ALIM ÉLEC passive transducer; **~ piézoélectrique** COMPOSANTS, ÉLECTRON piezoelectric transducer
transférer *vt* RADIO hand off (*AmE*), hand over (*BrE*), TÉL forward, transfer, TRANSMIS transfer; **~ une communication** TÉL transfer a call
transfert *m* APP SERV, COMMUT transfer, DONNÉES transfer, transmission, ÉLECTRON, GÉNÉRAL transfer, INFORMAT swapping, transfer, SIGNAL transfer, TÉL cut-over, extending, forwarding, through connection, transfer, TRANSMIS transfer, transmission; **~, accès et gestion de fichiers** (*FTAM*) DONNÉES, INFORMAT file transfer, access and management (*FTAM*); **~ d'alarme** TÉL alarm transfer; **~ d'appel** TÉL call extending, call transfer; **~ d'appel automatique** TÉL call forwarding; **~ d'appel en cas de numéro occupé** TÉL call transfer on "busy"; **~ d'appel en cas de retour d'appel et de non-réponse** TÉL call transfer on "ringing tone no reply"; **~ automatique des appels** SIGNAL automatic call transfer; **~ automatique intercellulaire** RADIO hand-off, hand-over; **~ autorisé** TÉL transfer allowed (*TFA*); **~ bidirectionnel** DONNÉES two-way; **~ de bloc** INFORMAT block transfer; **~ de catégorie** TÉL category transfer; **~ de charge** ÉLECTRON load transfer; **~ des communications réseau** TÉL auto-transfer; **~ de données** DONNÉES, TRANSMIS data transfer, data transmission, datacom; **~ de données demandé** DONNÉES, TRANSMIS data transfer requested; **~ électronique de fonds** APP SERV, DONNÉES, TÉL electronic funds transfer (*EFT*); **~ électronique de fonds au point de vente** APP SERV, DONNÉES, TÉL electronic funds transfer at point of sale (*EFTPOS*); **~ entre les cellules** COMMUT call-in-progress switching, RADIO intercell hand-off; **~ fiable** DONNÉES reliable transfer (*RT*); **~ de fichier** DONNÉES, INFORMAT file transfer; **~ de fichier télématique** (*TFT*) TÉL telematic file transfer (*TFT*); **~ de fichiers binaires** (*BFT*) DONNÉES, INFORMAT binary file transfer (*BFT*); **~ de groupe primaire** TRANSMIS through-group; **~ de groupe secondaire** TRANSMIS through super-group; **~ de groupe tertiaire** TRANSMIS through hypergroup, through mastergroup; **~ d'image** DONNÉES, TRANSMIS image transfer; **~ d'information** TÉL information transfer; **~ d'informations** TÉL information interchange; **~ intercellulaire** RADIO hand-off, hand-over; **~ intercellulaire courant** RADIO subsequent handover; **~ interdit** TÉL transfer prohibited (*TFP*); **~ magnétique** DONNÉES, INFORMAT print-through; **~ de masse de document** TÉL document bulk transfer; **~ de masse et manipulation de document** TÉL document bulk transfer and manipulation; **~ de mémoire** INFORMAT storage dumping; **~ de messages** DONNÉES, TÉL message transfer (*MT*); **~ mixte** DONNÉES two-way; **~ du mts** DONNÉES mts transfer; **~ multilatéral** DONNÉES n-way; **~ parallèle** INFORMAT parallel carry; **~ radial** INFORMAT radial transfer; **~ restreint** TÉL transfer restricted (*TFR*); **~ de secours** TÉL PBX failure transfer (*PFT*); **~ à sens unique** DONNÉES one-way; **~ série** DONNÉES serial carry, serial transfer; **~ de signaux** COMMUT, SIGNAL, TÉL, TRANSMIS signal transfer; **~ sous contrôle** TÉL transfer controlled (*TFC*); **~ sur panne secteur** ALIM ÉLEC, TÉL power failure feature, power failure transfer; **~ de technologie** INFORMAT technology transfer; **~ télégraphique** TÉL telegraphic transfer
transformable: **~ sur site** *adj* GÉNÉRAL field-upgradeable
transformateur *m* ALIM ÉLEC, COMPOSANTS, ÉLECTRON, RADIO, TÉL, TESTS, TV transformer; **~ abaisseur** ALIM ÉLEC step-down transformer; **~ d'adaptation** ÉLECTRON matching transformer; **~ d'alimentation** ALIM ÉLEC power transformer; **~ de courant d'appel** TÉL ringing-current transformer; **~ différentiel** ALIM ÉLEC differential transformer, ÉLECTRON bridge transformer; **~ différentiel équilibré** ALIM ÉLEC hybrid coil, ÉLECTRON hybrid transformer; **~ de distribution** ALIM ÉLEC distribution transformer; **~ d'entrée** ALIM ÉLEC input transformer; **~ d'isolation** ALIM ÉLEC isolating transformer; **~ de ligne** TV fly-back transformer; **~ de mesure** TESTS test transformer; **~ de mode** RADIO mode transformer, mode changer; **~ à point médian** ALIM ÉLEC center-tapped transformer (*AmE*), centre-tapped transformer (*BrE*); **~ relié au réseau** ALIM ÉLEC mains transformer; **~ Scott** ALIM ÉLEC, COMPOSANTS Scott-connected transformer; **~ sonde** RADIO probe transformer; **~ de sortie** ALIM ÉLEC output transformer; **~ suceur** ALIM ÉLEC draining transformer; **~ survolteur** ALIM ÉLEC step-up transformer; **~ symétrique/asymétrique** ALIM ÉLEC balanced-to-unbalanced transformer (*balun*); **~ symétriseur** ALIM ÉLEC balanced-to-unbalanced transformer (*balun*)
transformation *f* ALIM ÉLEC, DONNÉES, ÉLECTRON transform, INFORMAT alteration, transform, TÉL transform, TRANSMIS transform, transformation; **~ en cosinus discret** (*TCD*) DONNÉES, ÉLECTRON, INFORMAT, TÉL discrete cosine transform (*DCT*); **~ de Fourier** DONNÉES, INFORMAT, TESTS Fourier transformation (*FT*); **~ de Fourier discrète** DONNÉES, INFORMAT, TESTS discrete Fourier transform; **~ de Fourier rapide** (*TFR*) DONNÉES, INFORMAT, TESTS fast Fourier transform (*FFT*)
transformer *vt* ALIM ÉLEC, DONNÉES, ÉLECTRON transform, INFORMAT alter, transform, TÉL, TRANSMIS transform
transister *vi* TV travel
transistor *m* ALIM ÉLEC, COMPOSANTS, ÉLECTRON transistor; **~ à bascule** COMPOSANTS, ÉLECTRON transistor flip-flop; **~ bidirectionnel** ÉLECTRON *semi-conducteurs* bidirectional transistor; **~ bipolaire** COMPOSANTS, ÉLECTRON bipolar transistor; **~ bipolaire à hétérojonction** COMPOSANTS, ÉLECTRON heterojunction bipolar transistor (*HBT*); **~ de chute** COMPOSANTS series pass transistor, ÉLECTRON pass transistor, series pass transistor; **~ commandé en tension** ÉLECTRON voltage-controlled transistor; **~ en couche mince** COMPOSANTS, ÉLECTRON thin-film transistor (*TFT*); **~ à effet de champ** COMPOSANTS, ÉLECTRON (*FET, TEC*) field effect transistor (*FET*); **~ à effet de champ à jonction** COMPOSANTS, ÉLECTRON junction field effect transistor (*JFET*); **~ à effet de champ tétrode** COMPOSANTS, ÉLECTRON tetrode field effect transistor; **~ à effet de champ triode** COMPOSANTS, ÉLECTRON triode field effect transistor; **~ épitaxial** COMPOSANTS epitaxial transistor; **~ à faible puissance** COMPOSANTS low-power transistor; **~ à haute mobilité d'électrons** (*HEMT*) ÉLECTRON high electron mobility transistor (*HEMT*); **~ à hétérojonction** (*HEMT*) COMPOSANTS heterojunction transistor, ÉLECTRON heterojunction transistor, high electron mobility transistor (*HEMT*); **~ à jonction** ALIM ÉLEC, COMPOSANTS junction transistor; **~ à jonction par alliage** ÉLECTRON *semi-conducteurs* alloyed transistor, fused transistor; **~ à jonction par tirage** ÉLECTRON grown junction transistor; **~ à micro-**

alliage et diffusion ÉLECTRON micro-alloy diffused transistor (*MADT*); **~ monté en régulateur série** COMPOSANTS series pass transistor, ÉLECTRON pass transistor, series pass transistor; **~ optique** ÉLECTRON optical transistor; **~ de petits signaux** COMPOSANTS, ÉLECTRON small signal transistor; **~ planar** COMPOSANTS planar transistor; **~ à pointe** COMPOSANTS, ÉLECTRON point contact transistor; **~ de puissance** ALIM ÉLEC, COMPOSANTS power transistor; **~ tétrode** COMPOSANTS, ÉLECTRON tetrode transistor; **~ unijonction** (*TUJ*) ALIM ÉLEC, COMPOSANTS, ÉLECTRON unijunction transistor (*UJT*); **~ unijonction programmable** (*TUP*) ALIM ÉLEC, COMPOSANTS, ÉLECTRON programmable unijunction transistor (*PUT*); **~ unipolaire** COMPOSANTS, ÉLECTRON unipolar transistor

transistormètre *m* TESTS transistor analyser (*BrE*), transistor analyzer (*AmE*)

transit[1]: **de ~** *adj* TÉL tandem

transit[2]: **~ manuel par bande perforée** *m* TÉL, TRANSMIS torn-tape relay; **~ par bande perforée** *m* TÉL, TRANSMIS tape relay

transiter *vi* TRANSMIS pass through; **~ par** TÉL *opérateur* go via

transition *f* COMMUT, TÉL transition, TV crossfade; **~ descendante** ÉLECTRON high to low; **~ guide-coaxial** TRANSMIS waveguide-to-coax adaptor; **~ montante** ÉLECTRON low to high; **~ du niveau haut vers le niveau bas** ÉLECTRON high to low; **~ progressive** COMMUT progressive transition, RADIO taper; **~ progressive de fibres optiques** TRANSMIS tapered fiber (*AmE*), tapered fibre (*BrE*)

transitoire[1] *adj* GÉNÉRAL transient

transitoire[2]: **~ superposée** *f* TÉL pulse spike

transitron *m* COMPOSANTS, ÉLECTRON transitron

translater *vt* ÉLECTRON translate, INFORMAT relocate, translate, RADIO translate

translateur *m* TÉL repeat coil, repeating coil; **~ de niveau** ÉLECTRON level translator

translation *f* ÉLECTRON, RADIO translation, TÉL conversion, repeater; **~ pour conférence** TÉL conference repeater; **~ pour diffusion** TÉL broadcast repeater; **~ dynamique** INFORMAT dynamic relocation; **~ d'exploration** TÉL scanning shift; **~ de niveau** TÉL level shifting; **~ télégraphique** TÉL telegraphic conversion; **~ télégraphique pour conférence** TÉL conference telegraphic repeater

translitérer *vt* DONNÉES, INFORMAT transliterate

transmetteur: **~ automatique** *m* TÉL automatic transmitter; **~ à clavier** *m* TÉL keyboard transmitter; **~ à numéroteur automatique** *m* TÉL automatic numbering transmitter; **~ de synchro-détection** *m* TRANSMIS control transmitter; **~ télésignalisation** *m* APP SERV telesignaling transmitter (*AmE*), telesignalling transmitter (*BrE*)

transmettre *vt* DONNÉES, INFORMAT transmit, RADIO, TÉL relay, transmit, TRANSMIS relay, send, transmit, TV relay

transmis: **être ~ à** *vt* ÉLECTRON feed

transmission *f* ALIM ÉLEC, APP SERV transmission, DONNÉES communication, transmission, ÉLECTRON transfer, GÉNÉRAL communication, liaison, transfer, transmission, INFORMAT, RADIO, TÉL, TESTS transmission, TRANSMIS communication, transfer, transmission, TV transmission *f*; **~ d'alarme** TESTS alarm indication transmission; **~ d'alarmes** TESTS alarm transmission; **~ analogique** TRANSMIS analog transmission; **~ arythmique** TÉL start-stop transmission, TRANSMIS asynchronous transmission, start-stop transmission; **~ asynchrone** TÉL, TRANSMIS asynchronous transmission; **~ asynchrone de données** DONNÉES asynchronous data communication; **~ automatique** APP SERV automatic transmission; **~ en bande de base** DONNÉES, TÉL baseband transmission; **~ à bande latérale indépendante** RADIO, TRANSMIS independent sideband transmission; **~ à bande latérale unique** (*transmission BLU*) RADIO, TRANSMIS single sideband transmission (*SSB transmission*); **~ binaire synchrone** DONNÉES binary synchronous communication (*BSC*); **~ en bloc** TRANSMIS block transmission; **~ BLU** (*transmission à bande latérale unique*) RADIO, TRANSMIS SSB transmission (*single sideband transmission*); **~ caractère-série** TRANSMIS character-serial transmission; **~ en clair** TRANSMIS clear transmission; **~ de code** TRANSMIS code transmission; **~ codée** TRANSMIS coded transmission; **~ cohérente** TRANSMIS coherent transmission; **~ à courte distance** TRANSMIS short-haul transmission; **~ en différé** GÉNÉRAL deferred transmission; **~ différée** APP SERV, TRANSMIS store-and-forward transmission; **~ en direct** TRANSMIS, TV live transmission; **~ de documents** DONNÉES document transmission; **~ de données** DONNÉES, TRANSMIS data communication, data transfer, data transmission; **~ de données anisochrone** DONNÉES, TRANSMIS *datacom* anisochronous data transmission; **~ de données asynchrone** DONNÉES, TRANSMIS *datacom* asynchronous data transmission; **~ de données en différé** DONNÉES, TRANSMIS stored data transmission; **~ de données multimédia** DONNÉES, TRANSMIS multimedia transmission; **~ de données multiplexées** DONNÉES, TRANSMIS wideband data transmission; **~ de données de réseau en mode de connexion** DONNÉES, TRANSMIS network-connection-mode data transmission; **~ de données de réseau en mode sans connexion** DONNÉES, TRANSMIS network-connectionless-mode data transmission; **~ de données synchrone** DONNÉES, TRANSMIS synchronous data transmission; **~ à grande distance** TÉL long-haul transmission; **~ à grande puissance** RADIO, TRANSMIS high-power transmission; **~ à grande vitesse** DONNÉES high data rate; **~ d'image** DONNÉES, TRANSMIS image transmission; **~ isochrone** DONNÉES isochronous data transmission; **~ à large bande** TRANSMIS wideband transmission; **~ large bande** TRANSMIS broadband transmission; **~ en mode sans connexion** TRANSMIS connectionless-mode transmission; **~ en mode sans connexion N** DONNÉES, TRANSMIS N connectionless-mode transmission; **~ multiple** TRANSMIS multiple transmission; **~ multiplet-série** TRANSMIS byte-serial transmission; **~ numérique** TRANSMIS digital transmission; **~ numérique avec compression** TRANSMIS compressed digital transmission; **~ d'ondes** TRANSMIS wave transmission; **~ optique** TRANSMIS optical transmission; **~ du paquet d'accès** DONNÉES, TRANSMIS token passing; **~ par batterie centrale** ÉLECTRON, TÉL open-circuit working; **~ par bloc** TRANSMIS block transmission; **~ par courant continu** ALIM ÉLEC, TRANSMIS direct current transmission; **~ par courant simple** ALIM ÉLEC, TRANSMIS single current transmission; **~ par courants porteurs** ALIM ÉLEC, TRANSMIS carrier transmission; **~ par courroie** COMPOSANTS *génie mécanique* belt drive; **~ par double courant** TÉL *télex* double current working; **~ par double porteuse** ALIM ÉLEC, TRANSMIS dual carrier transmission; **~ par double-courant** TRANSMIS double current transmission; **~ par envoi de courant** ÉLECTRON, TÉL open-circuit working; **~ par envoi de fermeture** TRANSMIS closed-circuit working; **~ par fermeture de circuit** ÉLECTRON, TÉL open-circuit working, TRANSMIS closed-circuit working; **~ par fibres optiques** TRANSMIS fiber-optic transmission (*AmE*), fibre-optic transmission (*BrE*), optical fiber transmission (*AmE*), optical fibre transmission (*BrE*); **~ par interruption de courant** ÉLECTRON, TÉL open-circuit working; **~ par ouverture de circuit** ÉLECTRON, TÉL open-circuit working; **~ par paquets** TRANSMIS packet transmission; **~ par rafales** RADIO, TRANSMIS burst transmission; **~ par rupture de circuit** ÉLECTRON, TÉL open-circuit working; **~ par satellite** RADIO, TRANSMIS satellite transmission; **~ par trajets multiples** DONNÉES multipath transmission; **~ parallèle** INFORMAT, TRANSMIS parallel transmission; **~ en parallèle** INFORMAT, TRANSMIS parallel transmission; **~ radiophonique internationale** TV international sound-program transmission (*AmE*), international sound-programme transmission (*BrE*); **~ en rafales** ÉLECTRON broadcast storms; **~ semi-duplex** TRANSMIS half-duplex transmission; **~ sens montant** RADIO, TRANSMIS uplink transmission; **~ sens terre-satellite** RADIO, TRANSMIS uplink transmission; **~ séquentielle** RADIO, TRANSMIS sequential transmission; **~ série** TÉL, TRANSMIS serial transmission; **~ simple** TRANSMIS simple transmission; **~ simplex** TRANSMIS simplex transmission; **~ du son** APP SERV, TRANSMIS sound transmission; **~ sous incidence oblique** RADIO oblique-incidence transmission; **~ synchrone** DONNÉES, TRANSMIS synchronous transmission; **~ de téléphonie/données à l'alternat** TRANSMIS alternate voice-data transmission; **~ télévisuelle internationale** TV international television transmission; **~ télévisuelle multiple** TRANSMIS multiple television transmission; **~ en temps réel** INFORMAT real-time transmission; **~ vocale** TRANSMIS voice transmission

transmissions: **~ occasionnelles** *f pl* TRANSMIS occasional transmissions; **~ périodiques** *f pl* TRANSMIS regular transmissions

transmissomètre *m* TESTS, TRANSMIS transmissometer

transmodulation: **~ ionosophérique** *f* RADIO ionospheric cross-modulation

transmultiplexeur *m* ÉLECTRON, TÉL, TRANSMIS transmultiplexer (*TMUX*); **~ hiérarchique** TÉL hierarchical multiplexer

transordinateur *m* DONNÉES, INFORMAT transputer

transparence: **~ en code** *f* TRANSMIS code transparency; **~ du débit binaire** *f* DONNÉES, SIGNAL, TÉL, TRANSMIS data signaling rate transparency (*AmE*), data signalling rate transparency (*BrE*)

transparent[1] *adj* TRANSMIS nonregenerative

transparent[2] *m* DONNÉES, TÉL, TRANSMIS transparency

transphaseur *m* ÉLECTRON optical transistor

transpolarisation *f* RADIO cross-polarization

transport: **~ de charge acoustique** *m* (*ACT*) DONNÉES, TRANSMIS acoustic-charge transport; **~ d'énergie** *m* RADIO transmission

transportable *adj* GÉNÉRAL portable, RADIO hand-portable, portable

transporter *vt* TV carry

transposé: **~ en fréquence** *adj* RADIO frequency-translated

transposer *vt* DONNÉES, ÉLECTRON translate, INFORMAT, TÉL, TESTS, TRANSMIS transpose, TV convert

transposeur: **~ de fréquence** *m* DONNÉES up-converter, RADIO frequency translator, up-converter

transposition *f* DONNÉES, ÉLECTRON translation, INFORMAT transpose, transposition, RADIO translation, TÉL, TESTS, TRANSMIS transposition, TV conversion; **~ du clavier** INFORMAT key transpose; **~ de fréquence** RADIO frequency translation; **~ d'octave** INFORMAT octave transpose

transversale: **~ magnétique** *f* RADIO H line

transverse: **~ électrique** *adj* (*TE*) TRANSMIS transverse electric (*TE*); **~ électromagnétique** *adj* (*TEM*) TRANSMIS transverse electromagnetic (*TEM*); **~ magnétique** *adj* (*TM*) TRANSMIS transverse magnetic (*TM*)

travail *m* DONNÉES, INFORMAT work, job; **~ abandonné** INFORMAT aborted job; **~ d'accordage** INFORMAT tune job; **~ coopératif** DONNÉES cooperative work; **~ de désaccordage** INFORMAT detune job; **~ en série** ÉLECTRON daisy-chaining

travée *f* COMPOSANTS, INFORMAT suite, TÉL suite, rack row, row

traverse: **~ de bâti** *f* TÉL rack beam

traversée: **~ isolante** *f* ALIM ÉLEC bushing

traverser *vt* ALIM ÉLEC flow through, pass through, ÉLECTRON pass through

TRC *abrév* (*tube à rayon cathodique*) COMPOSANTS, ÉLECTRON CRT (*cathode ray tube*)

trébuchant *adj* DONNÉES tripping

treillis: **~ métallique** *m* COMPOSANTS wire netting

trémolo *m* INFORMAT tremolo

tressage *m* COMPOSANTS, ÉLECTRON, TRANSMIS *câbles* braiding

tresse *f* COMPOSANTS, ÉLECTRON, TRANSMIS *d'un câble* braid; **~ de blindage globale** ÉLECTRON overall braid screen; **~ métallique** ÉLECTRON metallic twist

tresser *vt* COMPOSANTS, ÉLECTRON, TRANSMIS *câble* braid

treuil: **~ de câble** *m* TRANSMIS cable winch; **~ pour pose de câbles** *m* TRANSMIS *génie industriel extérieur* cable-laying winch

tri *m* TESTS screening inspection; **~ par bloc** INFORMAT block sort

TRI *abrév* (*temps de récurrence des impulsions*) TÉL PRT (*pulse repetition time*)

triac *m* COMPOSANTS, ÉLECTRON *semi-conducteurs* triac

triax *m* COMPOSANTS triax, triax cable, triaxial cable

tributaire *m* DONNÉES, ÉLECTRON, RADIO, TÉL, TRANSMIS tributary (*TR*)

trier *vt* ALIM ÉLEC, INFORMAT sort; **~ à nouveau** INFORMAT resort; **~ par fusion** INFORMAT order by merging

trieur *m* ALIM ÉLEC, INFORMAT sorter

trigatron *m* ÉLECTRON trigatron

Trinitron *m* INFORMAT Trinitron

triode *f* COMPOSANTS, ÉLECTRON triode

triphasé *adj* ALIM ÉLEC three-phase

triple: **en ~ précision** *adj* DONNÉES triple-precision

trit *m* ÉLECTRON trit

troc *m* TV barter

trois: **à ~ états** *loc* ÉLECTRON tristate

3RP *abrév* (*réseau radio à ressources partagées*) RADIO trunked radio network, trunked system, trunking system

trombone *m* COMPOSANTS trombone

tronc: **~ commun** *m* COMMUT common highway

troncature *f* DONNÉES, INFORMAT truncation

tronçon *m* RADIO section, TRANSMIS length, run, section; **~ de câble** TRANSMIS cable run, cable length; **~ de circuits** TÉL circuit subgroup; **~ à section variable** RADIO squeeze section

tronquer *vt* DONNÉES, INFORMAT truncate

tropopause *f* RADIO tropopause

troposphère *f* RADIO troposphere

troquer *vt* TV barter

trou *m* COMPOSANTS, ÉLECTRON hole; **~ d'accès** ALIM ÉLEC *circuits imprimés* access hole; **~ de cadran** PÉRIPH *postes téléphoniques*, TÉL finger hole; **~ connexion** ÉLECTRON component hole; **~ de connexion transversale** ÉLECTRON through-connection hole; **~ de fixation** COMPOSANTS fixing hole; **~ d'indice** TRANSMIS index dip; **~ métallisé** ÉLECTRON plated through hole; **~ de passage** ÉLECTRON via hole, TRANSMIS cable hole; **~ de positionnement** ÉLECTRON locating hole, tooling hole; **~ sans pastille** ÉLECTRON landless hole

trouver *vt* INFORMAT locate

TRV *abrév* (*véhicule centre relais à bandes*) TV TRV (*tape-relay vehicle*)

TSF *abrév* (*télégraphie sans fil*) RADIO wireless telegraphy

TSM *abrév* (*trame sémaphore de message*) DONNÉES, TÉL MSU (*message signaling unit, message signalling unit*)

TTPO *abrév* (*terminal à très petite ouverture d'antenne*) DONNÉES, PÉRIPH, TRANSMIS VSAT (*very small aperture terminal*)

TTX *abrév* (*télétexte*) APP SERV, TV TTX (*teletext*)

tube *m* ALIM ÉLEC tube, COMPOSANTS, ÉLECTRON tube, valve, RADIO klystron, tube, TÉL, TESTS, TRANSMIS, TV tube; **~ acrylique transparent** ÉLECTRON clear acrylic tube; **~ amplificateur d'image** ÉLECTRON image intensifier tube; **~ analyseur à balayage** ÉLECTRON flying-spot scanner tube; **~ analyseur à électrons lents** ÉLECTRON low-velocity camera tube; ~ **analyseur à électrons rapides** ÉLECTRON anode-potential-stabilized camera tube, high-velocity camera tube; ~ **analyseur à mémoire** ÉLECTRON storage camera tube; ~ **analyseur photoconductif** TV photoconductive camera tube; ~ **analyseur photoémissif** ÉLECTRON photoemissive camera tube; ~ **analyseur à potentiel stabilisé sur l'anode** ÉLECTRON anode-potential-stabilized camera tube, high-velocity camera tube; ~ **analyseur à transfert d'image** ÉLECTRON image camera tube; ~ **à blocage éloigné** ÉLECTRON remote cutoff tube, variable mu tube; ~ **à blocage rapide** ÉLECTRON sharp cutoff tube; ~ **à cathode chaude** COMPOSANTS, ÉLECTRON filament; ~ **à cathode froide** ÉLECTRON cold cathode tube, cold cathode valve; ~ **cathodique à plusieurs canons** ÉLECTRON multiple-gun cathode ray tube; ~ **de champ** ALIM ÉLEC tube of force; ~ **à charge d'espace** COMPOSANTS space-charge-controlled tube; ~ **coaxial** TRANSMIS coaxial tube; ~**-compteur à aiguille** RADIO needle counter tube; ~**-compteur autocoupeur** RADIO self-quenched counter tube; ~**-compteur à cathode chaude** ÉLECTRON hot cathode stepping tube; ~**-compteur à circulation de gaz** ÉLECTRON gas flow counter tube; ~**-compteur Geiger-Müller** RADIO Geiger-Müller counter tube; ~**-compteur à halogène** COMPOSANTS, ÉLECTRON halogen-quenched counter tube; ~**-compteur à paroi mince** RADIO thin wall counter tube; ~**-compteur proportionnel** RADIO proportional counter tube; ~ **compteur de rayonnement** RADIO radiation counter tube; ~**-compteur à source gazeuse** ÉLECTRON gas sample counter tube; ~ **convertisseur d'image** COMPOSANTS, ÉLECTRON, TV image converter tube; ~ **convertisseur de fréquence** COMPOSANTS frequency converter tube; ~ **à décharge en arc à grille de commande** COMPOSANTS, ÉLECTRON grid-controlled arc discharge tube; ~ **à décharge luminescente** COMPOSANTS, ÉLECTRON glow discharge tube, glow discharge valve; ~ **déphaseur** ÉLECTRON phase-shifter tube; ~ **électronique** ÉLECTRON electronic tube, electronic valve; ~**-entretoise** COMPOSANTS distance tube; ~ **fluorescent** ÉLECTRON fluorescent lamp; ~ **de force** ALIM ÉLEC tube of force; ~ **à gaz** ÉLECTRON gas-filled tube, gas tube; ~ **de gaz rare** TESTS rare gas tube; ~ **générateur de bruit à plasma** ÉLECTRON noise generator plasma tube; ~ **à grilles alignées** TÉL aligned grid tube; ~ **image** ÉLECTRON picture tube, television tube, TV television tube; ~ **indicateur à luminescense** ÉLECTRON glow indicator tube; ~ **indicateur à néon** ÉLECTRON neon indicator tube; ~ **intensificateur d'image** ÉLECTRON image intensifier tube; ~ **isolant** ALIM ÉLEC insulating tube; ~ **en matière plastique** COMPOSANTS plastic hose, plastic tubing; ~ **à mémoire** ÉLECTRON storage tube; ~ **de mire** ÉLECTRON monoscope; ~ **multiple** ÉLECTRON multiple tube; ~ **à onde de charge** COMPOSANTS, ÉLECTRON space-charge wave tube; ~ **à onde de charge d'espace** COMPOSANTS, ÉLECTRON space-charge wave tube; ~ **à onde régressive** TÉL backward wave tube; ~ **à ondes progressives** (*TOP*) ÉLECTRON, TRANSMIS traveling wave tube (*AmE*) (*TWT*), travelling wave tube (*BrE*) (*TWT*), traveling wave valve (*AmE*), travelling wave valve (*BrE*), traveling waveguide (*AmE*), travelling waveguide (*BrE*); ~ **oscilloscope** ÉLECTRON oscilloscope tube, oscillograph tube; ~ **à pente réglable** ÉLECTRON remote cutoff tube, variable mu tube; ~ **préétiré rétractable à froid** ÉLECTRON prestretched cold-shrink tubing; ~ **pré-TR** ÉLECTRON pre-TR tube; ~ **de projection** ÉLECTRON projection tube; ~ **de puissance** ALIM ÉLEC power tube, ÉLECTRON power tube, power valve; ~ **à rayon cathodique** COMPOSANTS, ÉLECTRON (*TRC*) cathode ray tube (*CRT*); ~ **redresseur** ALIM ÉLEC rectifying tube, rectifying valve, ÉLECTRON rectifier tube; ~ **redresseur à cathode liquide** ÉLECTRON pool rectifier tube; ~ **de référence de tension** COMPOSANTS, ÉLECTRON voltage reference tube; ~ **régulateur de tension** ALIM ÉLEC, ÉLECTRON voltage regulator tube, voltage-stabilizing tube; ~ **relais** COMPOSANTS, ÉLECTRON trigger tube; ~ **souple en plastique** COMPOSANTS plastic hose, plastic tubing; ~ **stabilisateur de tension** ALIM ÉLEC, ÉLECTRON voltage regulator tube, voltage-stabilizing tube; ~ **stabilisateur de tension à plusieurs électrodes** ALIM ÉLEC, ÉLECTRON multielectrode voltage-stabilizing tube; ~ **TR** ÉLECTRON, RADIO TR tube; ~ **à trace sombre** ÉLECTRON skiatron; ~ **unité** ÉLECTRON unit tube; ~ **à vapeur de mercure** ÉLECTRON mercury vapor tube (*AmE*), mercury vapour tube (*BrE*); ~ **à vide** ÉLECTRON vacuum tube

TUJ *abrév* (*transistor unijonction*) ALIM ÉLEC, COMPOSANTS, ÉLECTRON UJT (*unijunction transistor*)

tuner *m* ÉLECTRON tuner, TV *du téléviseur* television tuner

TUP *abrév* (*transistor unijonction programmable*) ALIM ÉLEC, COMPOSANTS, ÉLECTRON PUT (*programmable unijunction transistor*)

tutoriel *m* DONNÉES, INFORMAT tutorial

tuyau: ~ **d'échappement** *m* COMPOSANTS exhaust pipe

TV *abrév* (*télévision*) TV TV (*television*); ~ **à accès conditionnel** *f* TV pay TV; ~ **en direct** *f* TRANSMIS, TV live TV; ~ **payante** *f* TV pay TV; ~ **à péage** *f* TV pay TV; ~ **professionnelle** *f* TV business TV

TVHD *abrév* (*télévision à haute définition*) TV HDTV (*high-definition television*)

TVI *abrév* TV (*télévision industrielle*) ITV (*industrial television*), (*télévision interactive*) ITV (*interactive television*)

type: ~ **d'acheminement** *m* TÉL routing case; ~ **d'action automatique** *m* DONNÉES auto-action type; ~ **d'adresse** *m* DONNÉES, TÉL type of address (*TOA*); ~ **d'attribut** *m* DONNÉES attribute type; ~ **de base** *m* DONNÉES base type; ~ **de base d'information** *m* DONNÉES information base type; ~ **de bloc** *m* DONNÉES block type; ~ **de caractère** *m* PÉRIPH typeface; ~ **chaîne binaire** *m* DONNÉES bistring type; ~ **chaîne de caractères** *m* DONNÉES character-string type; ~ **de chaîne d'octets** *m* DONNÉES octet-string type; ~ **de circuit radiophonique** *m* TÉL type of sound-program circuit (*AmE*), type of sound-programme circuit (*BrE*); ~ **de codage** *m* TRANSMIS coding type; ~ **de donnée abstrait** *m* DONNÉES, INFORMAT abstract data type; ~ **de données** *m* DONNÉES data type; ~ **d'ensemble** *m* DONNÉES set type; ~ **ensemble-de** *m* DONNÉES set-of type; ~ **d'entité N** *m* DONNÉES N-entity-type; ~ **d'entrée** *m* DONNÉES entry type; ~ **énuméré** *m* DONNÉES enumerated type; ~ **externe** *m* DONNÉES external type; ~ **d'information** *m* DONNÉES information type (*IT*); ~ **d'information codée d'origine** *m* DONNÉES original EITs; ~ **d'informations codées** *m* DONNÉES encoded information type (*EIT*); ~ **de modulation** *m* TRANSMIS modulation type; ~ **de notification** *m* DONNÉES notification type; ~ **de numéro** *m* DONNÉES, TÉL type of number (*TON*); ~ **de piste** *m* DONNÉES, INFORMAT, RADIO, TV track type; ~ **de processus d'application** *m* DONNÉES application process type; ~ **réel** *m* DONNÉES real type; ~ **de réverbération** *m* INFORMAT reverb type, reverberation type; ~ **de segment** *m* DONNÉES, INFORMAT segment type (*ST*); ~ **de sélection** *m* TÉL selection type; ~ **de séquence** *m* DONNÉES sequence type; ~ **séquence-de** *m* DONNÉES sequence-of type; ~ **structuré** *m* DONNÉES structured type; ~ **de transmission** *m* TÉL type of transmission

types: ~ **d'information codée retransmis** *m pl* DONNÉES delivered encoded-information types (*delivered EITs*)

typographie *f* APP SERV typography

U

U *abrév* DONNÉES, TÉL (*option utilisateur*) U (*user option*), (*interface U*) U (*U interface*)
UA *abrév* (*agent utilisateur*) APP SERV, DONNÉES, TÉL UA (*user agent*)
UAA *abrév* (*unité d'appel automatique*) TÉL automatic call unit
UAD *abrév* (*unité administrative*) CONTRÔLE AU (*administrative unit*)
UAL *abrév* (*unité arithmétique et logique*) INFORMAT ALU (*arithmetic and logic unit*)
UART *abrév* (*émetteur-récepteur asynchrone universel*) RADIO UART (*universal asynchronous receiver-transmitter*)
UC *abrév* COMMUT, INFORMAT, TÉL (*unité centrale de calcul, unité centrale de traitement*) CPU (*central processing unit*)
UCM *abrév* (*user communications manager*) INFORMAT UCM (*user communications manager*)
UER *abrév* (*Union européenne de radio-télévision*) RADIO EBU (*European Broadcasting Union*)
UHF *abrév* (*ultrahaute fréquence*) TV UHF (*ultrahigh frequency*)
UIT *abrév* (*Union internationale des télécommunications*) GÉNÉRAL ITU (*International Telecommunications Union*)
UL *abrév* (*unité logique*) DONNÉES, INFORMAT, TÉL LU (*logical unit, logic unit*)
ultrahaute: **~ fréquence** *f* (*UHF*) TV ultrahigh frequency (*UHF*)
ultrason *m* RADIO ultrasound
ultrasonique *adj* RADIO ultrasonic
ultrasonore *adj* RADIO ultrasonic
ultrasons: **à ~** *adj* RADIO ultrasonic
ultraviolet *adj* (*UV*) ÉLECTRON, TRANSMIS ultraviolet (*UV*)
U-Matic *m* TV U-Matic
un: **mis à ~** *f* COMPOSANTS, TESTS *d'une bascule* setting
unidirectionnel *adj* INFORMAT, RADIO unidirectional, TÉL unidirectional, one-way
unification *f* ÉLECTRON unification
uniformisation: **~ du temps de propagation** *f* TRANSMIS propagation time equalization
uniformité: **~ d'affichage** *f* ÉLECTRON uniformity of luminance
unilatéral *adj* INFORMAT, RADIO, TÉL unidirectional
unimodal[1] *adj* DONNÉES, ÉLECTRON, TRANSMIS monomode
unimodal[2] *m* DONNÉES, ÉLECTRON, TRANSMIS monomode laser, single-mode laser
union *f* ÉLECTRON disjunction
Union: **~ européenne de radiodiffusion** *f* RADIO European Union for Radio Broadcasting (*EURB*); **~ européenne de radio-télévision** *f* (*UER*) RADIO European Broadcasting Union (*EBU*); **~ européenne de recherches et d'études stratégiques en télécommunications** *f* (*EURESCOM*) GÉNÉRAL European Institute of Research and Strategic Studies in Telecommunications; **~ internationale des radio-amateurs** *f* RADIO International Amateur Radio Union (*IARU*); **~ internationale des télécommunications** *f* (*UIT*) GÉNÉRAL International telecommunications union (*ITU*); **~ panafricaine des télécommunications** *f* (*UPAT*) RADIO Pan-African Telecommunication Union (*PATU*); **~ postale universelle** *f* (*UPU*) TÉL Universal Postal Union (*UPU*); **~ de radiodiffusion Asie-Pacifique** *f* RADIO Asia-Pacific Broadcasting Union (*ABU*); **~ radio-scientifique internationale** *f* (*URSI*) RADIO International Union of Radio Science (*URSI*)
uniprogrammable *adj* INFORMAT one-time programmable (*OTP*)
unisson *m* INFORMAT unison
unité *f* GÉNÉRAL unit; **~ d'accès** GÉNÉRAL access unit (*AU*); **~ d'accès audio** INFORMAT audio access unit; **~ à accès contrôlé** TÉL controlled-access unit (*CAU*); **~ d'accès EDI** DONNÉES, TÉL EDI access unit (*EDI-AU*); **~ d'accès multiples** DONNÉES multistation access unit (*MAU*); **~ d'accès de remise physique** DONNÉES, TÉL physical delivery access unit (*PDAU*); **~ d'accès au service téléphonique** TÉL telephone service access unit (*TSAU*); **~ d'accès aux services télex** DONNÉES, RADIO, TÉL telex access unit (*TLXAU*); **~ d'accès au télétex** APP SERV teletex access unit (*TTXAU*); **~ d'accès télex** DONNÉES, RADIO, TÉL telex access unit (*TLXAU*); **~ d'adaptation** GÉNÉRAL interfacing unit, adaptation unit; **~ d'adaptation de bus** GÉNÉRAL bus adaptation unit; **~ additionnelle** GÉNÉRAL additional unit; **~ administrative** (*UAD*) CONTRÔLE administrative unit (*AU*); **~ d'adresse et de commande** TÉL address and control unit; **~ d'affichage** INFORMAT display unit; **~ d'alarme** TÉL alarm device, alarm unit; **~ d'alimentation** ALIM ÉLEC power supply; **~ d'alimentation de bâti** ALIM ÉLEC rack power unit; **~ d'appel automatique** (*UAA*) TÉL automatic call unit; **~ arithmétique et logique** (*UAL*) INFORMAT arithmetic and logic unit (*ALU*); **~ asservie** PÉRIPH remote terminal unit; **~ autonome** ÉLECTRON stand-alone, stand-alone unit; **~ auxiliaire** INFORMAT ancillary unit; **~ de canal** TRANSMIS channel modem; **~ de cartes magnétiques** DONNÉES magnetic card unit; **~ de cassettes** INFORMAT cassette drive; **~ centrale de calcul** (*UC*) COMMUT central processing unit (*CPU*), central processor, INFORMAT, TÉL central processing unit (*CPU*); **~ centrale de commande** COMMUT central control unit, central processor; **~ centrale de traitement** (*UC*) COMMUT central processing unit (*CPU*), central processor, INFORMAT, TÉL central processing unit (*CPU*); **~ de clés pour dicorde** COMMUT operator's switching set; **~ de commande** CONTRÔLE control system, control unit; **~ de commande de charge automatique** ALIM ÉLEC automatic charging control unit; **~ de commande et de connexion** COMMUT switching and control unit; **~ de commande numérique** TÉL digital control unit; **~ de commande principale** INFORMAT main control unit; **~ de commande de puissance** ALIM ÉLEC power control unit; **~ de commande de relais** COMPOSANTS relay contact unit; **~ de commutation** COMMUT operator's switching set, switching unit; **~ de compression de données** DONNÉES data compression processor; **~ de comptage d'appel** TÉL unit-charged marking; **~ de compteur** TÉL meter unit; **~ de concentration d'alarmes** TÉL alarm-assembling unit; **~ de conversion en différé** APP SERV store-and-forward conversion facility; **~ de conversion en temps réel** DONNÉES real-time conversion facility; **~ de correction d'impulsions** TRANSMIS pulse-correction unit; **~ décimale de quantité d'information** GÉNÉRAL Hartley; **~ de dépistage** COMMUT trace unit; **~ de détection** TESTS sensing unit; **~ de diagnostic et de commande** TESTS control diagnostic unit; **~ de disque dur** INFORMAT hard disk drive; **~ de disques** INFORMAT disk storage drive; **~ distante** COMMUT remote unit (*RU*); **~ de distorsion de quantification** TÉL quantizing distortion unit (*qdu*); **~ de distribution et de commande** CONTRÔLE distribution and control unit

(*DCU*); ~ **de données** DONNÉES, RADIO data unit; ~ **de données exprès du service de réseau** DONNÉES expedited network service data unit; ~ **de données d'interface MCS** DONNÉES, TÉL MCS interface data unit; ~ **de données de protocole** DONNÉES, TÉL protocol data unit (*PDU*); ~ **de données de protocole d'application** (*APDU*) DONNÉES, TÉL application protocol data unit (*APDU*); ~ **de données du protocole d'application A-ABANDON** (*ABRT*) DONNÉES, TÉL A-ABORT application protocol data unit (*ABRT*); ~ **de données du protocole d'application réponse** (*AARE*) DONNÉES, TÉL A-associate response (*AARE*); ~ **de données du protocole d'association demande** (*AARQ*) DONNÉES, TÉL A-associate request (*AARQ*); ~ **de données de protocole commun d'informations de gestion** DONNÉES, TÉL common management information protocol data unit (*CMIPDU*); ~ **de données de protocole de message** (*MPDU*) DONNÉES, TÉL message protocol data unit (*MPDU*); ~ **de données de protocole N** DONNÉES, TÉL N-protocol data unit; ~ **de données de protocole de présentation** DONNÉES, TÉL presentation protocol data unit (*PPDU*); ~ **de données de protocole de réseau** DONNÉES, TÉL network protocol data unit (*NPDU*); ~ **de données de protocole du service MCS** (*MCSPDU*) DONNÉES, TÉL MCS protocol data unit (*MCSPDU*); ~ **de données de protocole de session** DONNÉES, TÉL session protocol data unit (*SPDU*); ~ **de données de protocole de la sous-couche convergence** DONNÉES, TÉL convergence sublayer protocol data unit (*CSPDU*); ~ **de données de protocole de transport** DONNÉES, TÉL transport protocol data unit (*TPDU*); ~ **de données de service** DONNÉES, TÉL service data unit (*SDU*); ~ **de données de service AAL** DONNÉES, TÉL AAL service data unit (*AAL-SDU*); ~ **de données de service ATM** DONNÉES, TÉL ATM service data unit; ~ **de données du service MCS** DONNÉES, TÉL MCS service data unit; ~ **de données de service N** DONNÉES, TÉL N-service data unit; ~ **de données du service de présentation** DONNÉES, TÉL presentation service data unit; ~ **de données du service de réseau** DONNÉES, TÉL network service data unit; ~ **de données du service de transport** DONNÉES, TÉL, TRANSMIS transport service data unit (*TSDU*); ~ **de données synchrone** DONNÉES, ÉLECTRON synchronous data unit (*SDU*); ~ **d'échantillonnage** TESTS sampling unit; ~ **d'une échelle de temps** INFORMAT, RADIO, TRANSMIS timescale unit; ~ **d'effets spéciaux** INFORMAT effector; ~ **éloignée** COMMUT remote unit (*RU*); ~ **enfichable** COMPOSANTS, ÉLECTRON, INFORMAT, TÉL plug-in unit (*PIU*); ~ **d'enregistrement et de retransmission** APP SERV, COMMUT store-and-forward unit (*SFU*); ~ **d'entraînement du papier** PÉRIPH paper feed unit; ~ **d'entrée** INFORMAT input unit; ~ **d'entrelacement** RADIO, TRANSMIS interleaver; ~ **d'essai automatique** TESTS routiner; ~ **en état d'attente** GÉNÉRAL stand-by unit; ~ **d'extension** INFORMAT expander; ~ **d'extension FM** INFORMAT FM expander; ~ **extractible** COMPOSANTS slide-in unit, TÉL pull-out unit; ~ **fictive** TRANSMIS dummy unit; ~ **fonctionnelle** DONNÉES, TESTS *présentation* functional unit (*FU*); ~ **fonctionnelle de gestion des systèmes** INFORMAT systems management functional unit; ~ **fonctionnelle d'interconnexion** DONNÉES interworking functional unit (*IFU*); ~ **de fusibles** ALIM ÉLEC fuse unit; ~ **de gestion à distance** CONTRÔLE remote control unit (*RCU*); ~ **de gestion de mémoire** INFORMAT memory-control logic, memory management unit (*MMU*); ~ **d'incapacité** TESTS disablement unit, exclusion unit; ~ **d'indication** TÉL indicating unit; ~ **d'information** TÉL information unit; ~ **d'insertion de temps** DONNÉES time insertion unit; ~ **d'interconnexion de réseau local** ÉLECTRON local network interconnect; ~ **d'interfaçage** GÉNÉRAL adaptation unit, interfacing unit; ~ **d'interface** GÉNÉRAL adaptation unit, interfacing unit; ~ **d'interfonctionnement** DONNÉES interworking unit; ~ **intermédiaire** GÉNÉRAL stand-by unit; ~ **d'invalidation** TESTS disablement unit, exclusion unit; ~ **de jack de service** TÉL service jack unit; ~ **de jonction** GÉNÉRAL interfacing unit, adaptation unit; ~ **de ligne réseau** COMMUT exchange line unit; ~ **livrable** TÉL deliverable unit; ~ **logique** (*UL*) DONNÉES logical unit *m* (*LU*), logic unit (*LU*), logic device, ÉLECTRON logic device, INFORMAT, TÉL logical unit *m* (*LU*), logic unit (*LU*); ~ **M** RADIO M-unit; ~ **MCSPDU de contrôle** TÉL control MCSPDU; ~ **MCSPDU de domaine** TÉL domain MCSPDU; ~ **MCSPDU de données** DONNÉES, TÉL data MCSPDU; ~ **MCSPDU invalide** TÉL invalid MCSPDU; ~ **MCSPDU de type Connect** TÉL connect MCSPDU; ~ **MCSPDU valide** TÉL valid MCSPDU; ~ **de mémoire de données** COMMUT, DONNÉES data store unit; ~ **de mémoire de programme** INFORMAT program store unit; ~ **de mesure** TESTS unit of measurement; ~ **de mesure de source** (*SMU*) TESTS source measure unit (*SMU*); ~ **N** RADIO N-unit; ~ **numérique de traitement de signaux** DONNÉES digital signal processing chip; ~ **organisationnelle** TÉL organizational unit (*OU*); ~ **périphérique** DONNÉES, GÉNÉRAL, INFORMAT peripheral equipment; ~ **de présentation** INFORMAT presentation unit; ~ **processeur secondaire** INFORMAT secondary processor unit; ~ **processeur 32 bits** INFORMAT 32-bit processor unit; ~ **de protocole** DONNÉES protocol unit (*PU*); ~ **de raccordement** GÉNÉRAL interfacing unit, connection unit, TÉL *d'abonnés ou de circuits* connection unit; ~ **de raccordement d'abonné** TÉL remote subscriber stage, remote subscriber switch; ~ **de raccordement d'abonnés** COMMUT line interface module, PÉRIPH subscriber connection unit, TÉL subscriber access functional subsystem, subscriber connection unit; ~ **de raccordement des bornes** (*URB*) RADIO radio base station connection unit; ~ **de raccordement de bus** COMMUT bus connection unit; ~ **de raccordement de ligne** TRANSMIS line connection unit; ~ **de raccordement de multiplex** TÉL PCM link control software machine; ~ **de reconnaissance de la parole** APP SERV, INFORMAT, TÉL voice recognizer; ~ **de reconnaissance vocale** APP SERV, INFORMAT, TÉL voice recognizer; ~ **à réponse vocale** APP SERV, TÉL voice response unit; ~ **de réserve** GÉNÉRAL stand-by unit; ~ **de sélection** COMMUT selecting unit; ~ **sensible** TESTS sensing unit; ~ **de signalisation** SIGNAL signal unit (*SU*); ~ **de signalisation d'accusé de réception** SIGNAL acknowledgement signal unit; ~ **de signalisation pour la commande du système** SIGNAL system control signal unit (*SCU*); ~ **de signalisation initiale** SIGNAL initial signal unit (*ISU*); ~ **de signalisation solitaire** SIGNAL lone signal unit (*LSU*); ~ **de signalisation subséquente** SIGNAL subsequent signal unit; ~ **de signalisation de synchronisation** SIGNAL synchronization signal unit; ~ **de signalisation de synchronisation de multibloc** SIGNAL multiblock synchronization signal unit (*MBS*); ~ **de sortie** INFORMAT output device, output unit; ~ **de substitution** TRANSMIS dummy unit; ~ **de surveillance** CONTRÔLE, TÉL supervisory unit, TESTS monitor unit, supervisory unit; ~ **de surveillance du fonctionnement** COMMUT function supervisory unit; ~ **de synchronisation de trame** TRANSMIS frame synchronization unit; ~ **de synthèse** INFORMAT synthesizer unit; ~ **télécom** APP SERV charging unit, charge unit; ~ **de télécommande** CONTRÔLE remote control unit (*RCU*); ~ **de télégestion** CONTRÔLE remote control unit (*RCU*); ~ **temporelle** GÉNÉRAL time unit; ~ **de temps** GÉNÉRAL time unit; ~ **de traçage** COMMUT trace unit; ~ **de traitement** DONNÉES, INFORMAT processor unit; ~ **de traitement principal** DONNÉES, INFORMAT main processor unit, principal processing unit (*PUP*); ~ **de traitement de signaux** SIGNAL signal processor unit, signal processor (*SP*); ~ **de traitement de texte** DONNÉES, INFORMAT word processor; ~ **de trame** TRANSMIS frame unit; ~ **tributaire** DONNÉES tributary unit (*TU*); ~ **en virgule flottante** INFORMAT floating point unit (*FPU*); ~ **virtuelle** INFORMAT virtual unit

unités: ~ **constitutives** *f pl* COMPOSANTS constituent units; ~ **de données asynchrones** *f pl* DONNÉES, ÉLECTRON asynchronous data units (*ADU*); ~ **de stockage** *f pl* DONNÉES mass storage; ~ **utilisant le dB** *f pl* TÉL dB-related units

univocal *m* TÉL speech plus simplex, speech-plus-simplex equipment

Unix: ~**-to-Unix copy protocol** *m* (*UUCP*) DONNÉES *Internet* Unix-to-Unix copy protocol (*UUCP*)

uns: ~ **binaires non-brouillés** *m pl* DONNÉES unscrambled binary ones (*USB1*)

UPAT *abrév* (*Union panafricaine des télécommunications*) RADIO PATU (*Pan-African Telecommunication Union*)

UPS *abrév* (*alimentation sans coupure*) ALIM ÉLEC UPS (*uninterrupted power supply, uninterruptible power supply*)

UPU *abrév* (*Union postale universelle*) TÉL UPU (*Universal Postal Union*)

URB *abrév* (*unité de raccordement des bornes*) RADIO radio base station connection unit

urgence *f* APP SERV, CONTRÔLE, DONNÉES, GÉNÉRAL, PÉRIPH, RADIO, SIGNAL, TÉL emergency

URSI *abrév* (*Union radio-scientifique internationale*) RADIO URSI (*International Union of Radio Science*)

usage: ~ **national** *m* TÉL national usage (*NU*); ~ **parameter control** *m* APP SERV usage parameter control (*UPC*); ~ **privatif** *m* RADIO private use

usager *m* DONNÉES, GÉNÉRAL, INFORMAT, TÉL, TV user; ~ **itinérant** RADIO, TÉL roamer; ~ **du réseau** DONNÉES network user; ~ **à usager** *adj* GÉNÉRAL user-to-user

Usenet *m* INFORMAT Usenet

usinage: ~ **au tour** *m* COMPOSANTS turning

usiner: ~ **au tour** *vt* COMPOSANTS turn

usure: ~ **normale** *f* INFORMAT wear and tear

usurpation: ~ **d'identité** *f* DONNÉES masquerade

utilisateur *m* APP SERV, DONNÉES, GÉNÉRAL, INFORMAT, TÉL, TV user; ~ **DTAM** APP SERV, DONNÉES, TÉL DTAM user; ~ **EDI** APP SERV, DONNÉES, TÉL EDI user; ~ **final** GÉNÉRAL, TRANSMIS end user; ~ **mémoire de messages** APP SERV, DONNÉES, TÉL MS user; ~ **de la messagerie EDI** APP SERV, DONNÉES, TÉL EDI-messaging user, EDIMG user; ~ **de la messagerie vocale** APP SERV, DONNÉES, TÉL voice messaging user, voice-mail system user; ~ **référencé** INFORMAT registered user; ~ **de RTSE** APP SERV, DONNÉES, TÉL RTSE user; ~ **du service** APP SERV, DONNÉES, TÉL service user; ~ **du service abstrait mémoire de messages** APP SERV, DONNÉES, TÉL MS abstract service user; ~ **du service ACSE** APP SERV, DONNÉES, TÉL ACSE service user; ~ **du service CMISE** APP SERV, DONNÉES, TÉL CMISE service user; ~ **du service CMISE exécuteur** APP SERV, DONNÉES, TÉL performing CMISE service user; ~ **du service CMISE lanceur** APP SERV, DONNÉES, TÉL invoking CMISE service user; ~ **du service de contrôle d'association** APP SERV, DONNÉES, TÉL ACSE service user; ~ **du service MCS** APP SERV, DONNÉES, TÉL MCS user; ~ **du service de réseau** APP SERV, DONNÉES, TÉL network service user; ~ **du service de réseau demandé** APP SERV, DONNÉES, TÉL called network service user (*called NS user*); ~ **du service de réseau demandeur** APP SERV, DONNÉES, TÉL calling network service user (*calling NS user*); ~ **du service de traitement transactionnel** APP SERV, DONNÉES, TÉL transaction processing service user (*TPSU*); ~ **du système de messagerie de personne à personne** APP SERV, DONNÉES, TÉL interpersonal messaging system user; ~ **du système de messagerie vocale** APP SERV, DONNÉES, TÉL voice messaging system user

utilisateur-MIS *m* DONNÉES MIS user

utilisation *m* GÉNÉRAL use; ~ **commune de périphériques** *f* DONNÉES peripheral device sharing; ~ **étendue en période d'éclipse** *f* RADIO extended eclipse use; ~ **en répondeur simple** *f* PÉRIPH, TÉL announcement-only facility

utilitaire *m* DONNÉES, INFORMAT utility

UUCP *abrév* (*Unix-to-Unix copy protocol*) DONNÉES *Internet* UUCP (*Unix-to-Unix copy protocol*)

UUS *abrév* (*signalisation d'usager à usager*) SIGNAL UUS (*user-to-user signaling AmE, user-to-user signalling BrE*)

UV *abrév* (*ultraviolet*) ÉLECTRON, TRANSMIS UV (*ultraviolet*)

V

V *abrév* (*tension*) ALIM ÉLEC V (*voltage*)
va: **~-et-vient** *m* TÉL alternation on inquiry, inquiry reciprocation, double call
VAC *abrév* INFORMAT (*vitesse angulaire constante*) CAV (*constant angular velocity*), RADIO, TRANSMIS (*voie d'acheminement commune*) CRC (*common routing channel*)
vaccin *m* DONNÉES, INFORMAT vaccine
vacillement *m* TV jitter
vaisseau: **~ spatial** *m* RADIO space vehicle
valence *f* TÉL number of significant conditions
valeur[1]: **à ~ unique** *adj* DONNÉES single-valued (*S*)
valeur[2] *f* GÉNÉRAL value, TESTS result, value; **~ admissible** TESTS allowable value; **~ d'affaiblissement** TRANSMIS attenuation value; **~ ajoutée** DONNÉES added value; **~ d'ajustement** GÉNÉRAL setting value; **~ aléatoire** INFORMAT random value; **~ assignée** ALIM ÉLEC rated value; **~ d'attribut** DONNÉES attribute value; **~ d'authentification** DONNÉES authentication value; **~ autre que zéro** INFORMAT nonzero value; **~ de champ** TESTS field intensity, field strength; **~ choisie aux essais** TESTS select-on-test value; **~ de consigne** CONTRÔLE desired value; **~ de contrôle cryptographique** DONNÉES, RADIO, TÉL, TV cryptographic check value; **~ conventionnellement vraie d'une grandeur mesurable** TESTS *qualité* conventional true value of a quantity; **~ de crête** GÉNÉRAL, TESTS peak value; **~ de crête d'impulsion** TÉL pulse-peak value; **~ de décision** TRANSMIS decision value; **~ de départ** INFORMAT seed; **~ de données de présentation** DONNÉES presentation data value; **~ efficace** ÉLECTRON, GÉNÉRAL rms value, root-mean square value; **~ estimée** TESTS estimated value; **~ guide** GÉNÉRAL standard value; **~ instantanée** GÉNÉRAL instantaneous value; **~ mesurée** TESTS measured value; **~ modulée** DONNÉES modulated value; **~ moyenne d'affaiblissement** TRANSMIS mean loss value; **~ nominale** GÉNÉRAL, RADIO nominal value; **~ nominale de la puissance absorbée** COMPOSANTS rated burden; **~ normée** RADIO normalized value; **~ observée** TESTS observed value; **~ ordinale** DONNÉES ordinal value; **~ par défaut** GÉNÉRAL, INFORMAT default value; **~ de paramètre** TÉL, TESTS parameter value (*PV*); **~ de paramètre erronée** DONNÉES parameter out-of-range class; **~ de pointeur** COMMUT pointer value; **~ pouvant être prise par défaut** DONNÉES defaultable (*D*); **~ prédéterminée** TESTS preassigned value; **~ quadratique** ÉLECTRON, GÉNÉRAL root sum square; **~ quadratique moyenne** ÉLECTRON, GÉNÉRAL root mean square (*rms*); **~ de quantification** INFORMAT quantize value; **~ quantifiée** TRANSMIS quantized value; **~ réelle** CONTRÔLE *commande automatique* actual value; **~ de référence** GÉNÉRAL standard value, TESTS reference value; **~ de référence acceptée** TESTS *qualité* accepted reference value; **~ en régime établi** GÉNÉRAL steady state value; **~ en régime permanent** GÉNÉRAL steady state value; **~ de retour** COMPOSANTS resetting value; **~ spécifiée d'un caractère de qualité** TESTS specified quality measure; **~ spécifique** DONNÉES specific value; **~ type** GÉNÉRAL standard value
validation *f* DONNÉES validation, INFORMAT enabling; **~ des données d'entrée** DONNÉES input validation; **~ en entrée** DONNÉES input validation
valider *vt* INFORMAT enable
valise: **~ de tests** *f* TESTS test kit
valisette *f* GÉNÉRAL case
valorisation: **~ du risque** *f* TESTS risk quantification
valve: **~ de réduction** *f* COMPOSANTS reducing valve
vanne: **~ électromagnétique** *f* COMPOSANTS solenoid valve, CONTRÔLE magnetic valve
vantail *m* INFORMAT windowpane
VAO *abrév* (*vidéo assistée par ordinateur*) DONNÉES, INFORMAT DTV (*desktop video*)
variable[1] *adj* GÉNÉRAL changeable, variable
variable[2]: **~ aléatoire** *f* GÉNÉRAL random variable; **~ d'entrée** *f* CONTRÔLE input variable; **~ logique** *f* COMMUT switching variable; **~ de suite de tests** *f* TESTS test suite variable; **~ de test élémentaire** *f* TESTS test case variable
variante: **~ nationale** *f* TESTS national variant
variation: **~ aléatoire de la hauteur** *f* INFORMAT pitch randomize; **~ diurne** *f* TÉL day-to-day variation, diurnal variation; **~ de hauteur** *f* INFORMAT pitchbend, pitch variation; **~ de hauteur maximale** *f* INFORMAT pitchbend range; **~ de largeur** *f* INFORMAT width variation; **~ des lots** *f* TESTS *qualité* batch variation; **~ de niveau** *f* INFORMAT level variation; **~ d'orientation d'un faisceau** *f* RADIO *radar* beam switching; **~ rythmique** *f* INFORMAT fill-in; **~ de la somme numérique** *f* TÉL digital sum variation; **~ de température** *f* CLIMAT temperature difference; **~ de tension** *f* ALIM ÉLEC, TESTS voltage variation
varistance *f* COMPOSANTS, ÉLECTRON varistor
Vbox *m* INFORMAT Vbox
VCP *abrév* (*lecteur de cassette vidéo*) TV VCP (*videocassette player*)
vecteur *m* INFORMAT vector, RADIO vector line, TRANSMIS, TV vector; **~ chrominance** TV chrominance vector; **~ cinétique** INFORMAT, TV motion vector; **~ cinétique anticipé** INFORMAT, TV forward-motion vector; **~ cinétique différé** INFORMAT, TV backward-motion vector; **~ électrique** TRANSMIS electric vector; **~ de mouvement** INFORMAT, TV motion vector; **~ de mouvement anticipé** INFORMAT, TV forward-motion vector; **~ de mouvement différé** INFORMAT, TV backward-motion vector; **~ de réflexion** INFORMAT reflection vector
véhicule: **~ centre relais à bandes** *f* (*TRV*) TV tape-relay vehicle (*TRV*)
véhiculer *vt* DONNÉES *Internet*, TÉL carry
veille *f* RADIO monitoring, watching, search
veiller *vi* RADIO monitor
vélocité: **~ d'attaque** *f* INFORMAT attack velocity; **~ de relâchement** *f* INFORMAT release velocity
ventilateur *m* INFORMAT fan
ventilation *f* CLIMAT ventilation
ver: **~ automatique** *m* INFORMAT computer worm
verdict: **~ d'essai** *m* TESTS test verdict; **~ non concluant** *m* DONNÉES inconclusive verdict; **~ de test** *m* TESTS test verdict
vérificateur: **~ de câble** *m* TRANSMIS cable tester; **~ d'orthographe** *m* INFORMAT spellchecker
vérification *f* ALIM ÉLEC, COMMUT check, DONNÉES audit, verification, INFORMAT check, checking, test, testing, RADIO verification, SIGNAL test, TÉL test, testing, verification, verifying, TESTS test, testing, verification, verifying; **~ de l'acheminement du trafic** TÉL, TESTS traffic route testing (*TRT*); **~ d'affaiblissement** TRANSMIS attenuation check; **~ automatique des circuits** TÉL automatic trunk testing; **~ de la cohérence du circuit** DONNÉES circuit consistency verification; **~ des composants redondants** ALIM ÉLEC redundant component check; **~ de défaillance** TÉL failure verification; **~ et assistance des stations terriennes** *f pl*

RADIO earth station verification and assistance (*ESVA*); **~ de fonctionnement** TESTS function check-out; **~ des informations de service** TRANSMIS verification of overhead information; **~ d'itinéraire du trafic** TÉL, TESTS traffic route testing; **~ manuelle** INFORMAT desk checking; **~ de pile** INFORMAT check battery; **~ de présence** INFORMAT null data check; **~ de signal** SIGNAL, TESTS signal test; **~ du trajet dans le central** COMMUT cross-exchange check; **~ de la voie de conversation dans le central** COMMUT cross-office check

vérifier *vt* ALIM ÉLEC, COMMUT check, DONNÉES verify, INFORMAT check, test out, RADIO verify, SIGNAL test, TÉL, TESTS test, test out, verify, TRANSMIS check; **~ à l'oscilloscope** TESTS scope

vérifiez *loc* TÉL verify

vérité *f* TESTS truth

vernier *m* TESTS vernier

vernir: **~ au pistolet** *vt* COMPOSANTS spray-finish, spray-lacquer

vernis *m* COMPOSANTS, ÉLECTRON, TV lacquer; **~ épargne** COMPOSANTS solder-resist lacquer, ÉLECTRON resist lacquer; **~ au four** COMPOSANTS, TV stoving lacquer; **~ à la gomme laque** COMPOSANTS shellac; **~ isolant** ÉLECTRON insulating lacquer, insulating varnish; **~ soudable** COMPOSANTS soldering lacquer

verre: **~ renforcé** *m* COMPOSANTS reinforced glass

verrou *m* ÉLECTRON latch, RADIO, TESTS lock; **~ électronique** RADIO electronic lock; **~ d'émetteur** RADIO, TRANSMIS transmitter blocker cell (*TB cell*); **~ de récepteur** RADIO TR cell, transmit-receive cell, interlock switch; **~ de sûreté** TESTS safety lock

verrouillage *m* COMPOSANTS latch-up, DONNÉES alignment, ÉLECTRON alignment, coupling, locking, INFORMAT lock, locking, PÉRIPH lock, RADIO, TRANSMIS alignment, bolting, lock-on, TV locking; **~ de clavier** INFORMAT, PÉRIPH shift lock; **~ du clavier** INFORMAT key lock; **~ d'écran** INFORMAT screen locking; **~ à l'état passant** COMPOSANTS latch-up; **~ de fréquence** RADIO frequency alignment; **~ de guide d'ondes** RADIO, TRANSMIS waveguide bolting; **~ de multitrame** TRANSMIS multiframe alignment (*MFA*); **~ par baïonnette** ÉLECTRON bayonet coupling; **~ de phase** ÉLECTRON phase alignment, phase locking; **~ sur la cible** RADIO, TRANSMIS target lock-on; **~ de synchronisation** TV genlocking; **~ de trame** TÉL framing, TRANSMIS framing, frame alignment, frame synchronization; **~ de trame de base** TRANSMIS basic frame alignment (*BFA*)

verrouiller *vt* ÉLECTRON, RADIO lock on

version *f* DONNÉES version, INFORMAT release, version, TÉL version; **~ B** TÉL *système Alcatel E10* B version; **~ en cours** INFORMAT current version; **~ internationale de référence** (*IRV*) TÉL International reference version (*IRV*); **~ de syntaxe** DONNÉES, TÉL syntax version

vertical *adj* GÉNÉRAL vertical

VESA *abrév* ÉLECTRON, INFORMAT, TV VESA (*Video Electronics Standards Association*)

Veuillez: **~ accuser réception** *loc* APP SERV, DONNÉES, TÉL Acknowledge

veuve *f* INFORMAT *typographie* orphan, widow

VF *abrév* (*vidéofréquence*) TRANSMIS, TV VF (*videofrequency*)

VHF *abrév* (*très haute fréquence*) RADIO, TRANSMIS, TV VHF (*very high frequency*)

vibration: **~ acoustique** *f* TÉL acoustic vibration

vibrato: **~ par molette de modulation** *m* INFORMAT modulation wheel vibrato; **~ à retard** *m* INFORMAT delay vibrato

vibreur *m* ÉLECTRON, RADIO vibrator; **~ acoustique** PÉRIPH, TÉL buzzer; **~ sonore** GÉNÉRAL buzzer

vice: **~ de fabrication** *f* TESTS manufacturing defect

vidage: **~ après abandon** *m* INFORMAT abort dump; **~ après changement** *m* INFORMAT change dump; **~ global** *m* INFORMAT bulk dump; **~ de mémoire** *m* DONNÉES data dump, INFORMAT core dump, data dump, dump, memory dump, storage dump; **~ statique** *m* INFORMAT static dump

vide[1]: **à ~** *adj* ALIM ÉLEC unloaded, ÉLECTRON open-circuit, unloaded

vide[2] *m* GÉNÉRAL void

vidéo *f* APP SERV, DONNÉES, INFORMAT, TV video; **~ analogique** INFORMAT analog video; **~ assistée par ordinateur** (*VAO*) DONNÉES, INFORMAT desktop video (*DTV*); **~ composite** TV composite video; **~ à la demande** INFORMAT, TV video-on-demand (*VOD*); **~ interactif** TV interactive video; **~ inverse** INFORMAT reverse video, TV inverse video; **~ à large bande** TV wideband video; **~ linéaire** INFORMAT linear video; **~ numérique** INFORMAT digital video; **~ PAD** TV videotex PAD; **~ en temps réel** DONNÉES, INFORMAT real-time video

Video: **~ RISC Compression Processor** *m* INFORMAT Video RISC Compression Processor

vidéo-CD *m* TV CD-V

vidéocommunication *f* TV videocommunication

vidéoconférence *f* APP SERV, DONNÉES, INFORMAT, TÉL, TV videoconference, videoconferencing; **~ à images animées** TV full-motion videoconferencing; **~ à images fixes** TV slow-scan videoconferencing

vidéodisque *m* TV videodisk; **~ intéractif** INFORMAT, TV interactive videodisk

vidéofréquence *f* (*VF*) TRANSMIS, TV videofrequency (*VF*)

vidéographie *f* APP SERV, INFORMAT, RADIO videography, TV videography, video services; **~ diffusée** APP SERV, TV broadcast videography, teletext (*TTX*); **~ interactive** INFORMAT, RADIO videotex, TV interactive videography, videotex

vidéomessagerie *f* INFORMAT, TV videomessaging

vidéophone *m* APP SERV, PÉRIPH, RADIO, TÉL, TV picture phone, videophone, videotelephone, viewphone, visiophone, visual telephone

vidéophonie *f* APP SERV, INFORMAT, RADIO, TÉL, TV videophony, videotelephony, visual telephony

vidéotex *m* APP SERV, INFORMAT, RADIO, TÉL, TV interactive videography, videotex; **~ diffusé** APP SERV, TV broadcast videotex, teletext (*TTX*); **~ interactif** INFORMAT, RADIO videotex, TV interactive videotex, interactive videography, videotex; **~ syntaxique** TÉL syntax-based videotex (*SBV*)

vidéotique *f* DONNÉES, INFORMAT desktop video (*DTV*)

vidéotransmission *f* TRANSMIS, TV video transmission

vider *vt* DONNÉES empty, INFORMAT dump, empty

vidicon *m* ÉLECTRON vidicon; **~ à retour du faisceau** TÉL return-beam vidicon (*RBV*)

vie: **~ utile** *f* TÉL useful life, TESTS life, useful life

vieillir *vi* GÉNÉRAL *épreuves* age

vieillissement *m* GÉNÉRAL *épreuves* ageing; **~ artificiel** TESTS *épreuves* burn-in, *sécurité de fonctionnement* artificial ageing; **~ préalable** TESTS preageing

vignette *f* INFORMAT inset

vinyle *m* COMPOSANTS vinyl

violation: **~ de l'alternance des polarités** *f* TÉL alternate mark inversion violation; **~ de bipolarité** *f* TRANSMIS AMI violation, bipolar violation; **~ de code** *f* TRANSMIS code violation (*CV*); **~ du code de transmission** *f* TÉL transmission code violation; **~ de la règle de bipolarité** *f* TÉL alternate mark inversion violation; **~ de la règle de polarité** *f* TRANSMIS AMI violation

virement: **~ télégraphique** *m* TÉL cable transfer, telegraphic transfer

virgule *f* INFORMAT comma; **~ flottante** INFORMAT floating point; **~ flottante câblée** INFORMAT floating point hardware

virole *f* COMPOSANTS ferrule; **~ rotative** ÉLECTRON rotary ferrule

virtual: **~ DMA driver** *m* DONNÉES virtual DMA driver

virus *m* DONNÉES, INFORMAT virus; **~ informatique** DONNÉES computer virus

vis *f* ALIM ÉLEC, COMPOSANTS screw, ÉLECTRON jackscrew,

screw; ~ **à ailettes** COMPOSANTS, ÉLECTRON thumbscrew; ~ **d'arrêt de la palette** ALIM ÉLEC relay-armature stop; ~ **d'assemblage** COMPOSANTS fixing screw, mounting screw; ~ **à bois** COMPOSANTS woodscrew; ~ **de borne** ALIM ÉLEC terminal screw; ~ **de fixation** COMPOSANTS fixing screw, mounting screw; ~ **moletée** COMPOSANTS, ÉLECTRON thumbscrew; ~ **sans sertissage** ÉLECTRON crimpless screw; ~ **à six pans** COMPOSANTS screw with hexagonal head; ~ **de soupape** COMPOSANTS, ÉLECTRON valve spindle; ~ **de verrouillage** COMPOSANTS screwlock

viser *vt* RADIO, TÉL, TV target

visibilité: ~ **directe** *f* TRANSMIS line-of-sight; ~ **sous fouillis** *f* RADIO subclutter visibility

visiocasque *m* INFORMAT EyePhone

visioconférence *f* APP SERV, DONNÉES, INFORMAT, TÉL videoconference, videoconferencing, visioconference, TV confravision, videoconference, videoconferencing, visioconference

vision: ~ **assistée par ordinateur** *f* INFORMAT computer-assisted vision

visionneuse *f* INFORMAT, TV viewer; ~ **pour microforme** DONNÉES microform reader

visiophone *m* APP SERV, INFORMAT, PÉRIPH, RADIO, TÉL, TV picture phone, videophone, videotelephone, viewphone, visiophone, visual telephone

visiophonie *f* APP SERV, INFORMAT, RADIO, TÉL, TV videophony, videotelephony, visual telephony; ~ **à images fixes** RADIO, TÉL, TV still-picture videotelephony

visioréunion *f* TV satellite conference

visiteur: ~ **indiscret** *m* INFORMAT unauthorized visitor, unwelcome caller

visserie: ~ **de montage** *f* COMPOSANTS mounting hardware

visualisation *f* COMPOSANTS display, INFORMAT view, PÉRIPH, TESTS, TV display; ~ **des appels par LCD** PÉRIPH LCD calls display; ~ **de défaut** TESTS fault display

visualiser *vt* GÉNÉRAL display, INFORMAT view

visuel *m* INFORMAT display device

vitesse[1]: **à ~ moyenne** *adj* DONNÉES medium-speed

vitesse[2] *f* GÉNÉRAL rate, speed, velocity; ~ **angulaire constante** (*VAC*) INFORMAT constant angular velocity (*CAV*); ~ **d'apprêtage** ÉLECTRON priming speed; ~ **à l'attaque** INFORMAT attack rate; ~ **aveugle** RADIO *radar* blind speed; ~ **de balayage** ÉLECTRON sweep speed, RADIO scanning speed, sweep speed; ~ **de convergence** TESTS rate of approach; ~ **de décroissance** INFORMAT decay rate; ~ **efficace d'une particule** TÉL particle velocity; ~ **de groupe** TRANSMIS envelope velocity, group velocity; ~ **de l'image** INFORMAT, TV picture rate; ~ **d'inscription** PÉRIPH print capacity, printing speed; ~ **d'inscription maximale utilisable** ÉLECTRON maximum usable writing speed; ~ **de lecture** ÉLECTRON reading speed; ~ **de lecture minimale utilisable** ÉLECTRON minimum usable reading speed; ~ **de libération** RADIO escape velocity; ~ **linéaire constante** (*VLC*) INFORMAT constant linear velocity (*CLV*); ~ **de modulation** TRANSMIS modulation speed; ~ **moyenne de déplacement** COMPOSANTS drift velocity; ~ **normale de fonctionnement** DONNÉES free-running speed; ~ **de phase** RADIO phase speed, phase velocity; ~ **de propagation** ÉLECTRON, TRANSMIS propagation velocity; ~ **de relâchement** INFORMAT release rate; ~ **de repli** ÉLECTRON fall-back rate; ~ **de rotation** INFORMAT spin rate; ~ **de transfert** DONNÉES, INFORMAT transfer rate; ~ **de transfert de données** DONNÉES, TÉL data transfer rate; ~ **de transfert effective** DONNÉES *datacom* actual transfer rate; ~ **de transfert réelle** DONNÉES actual transfer rate; ~ **de transmission** INFORMAT baud rate, TRANSMIS transmission speed; ~ **de variation de phase** ÉLECTRON phase-change velocity

VLC *abrév* (*vitesse linéaire constante*) INFORMAT CLV (*constant linear velocity*)

VLSI *abrév* (*intégration à très grande échelle*) COMPOSANTS, ÉLECTRON, TÉL, TRANSMIS VLSI (*very large-scale integration*)

vobulation: ~ **du spot** *f* TV spot wobble

vocodeur *m* INFORMAT, RADIO, TÉL vocoder, voice coder, voice-operated coder; ~ **à canaux** RADIO channel vocoder; ~ **à formant** RADIO formant vocoder; ~ **PELP** RADIO pitch-excited vocoder

voguer *vt* DONNÉES surf

voie[1] *f* COMMUT channel, DONNÉES channel, path, ÉLECTRON channel, RADIO channel, circuit, path, route, SIGNAL channel, circuit, TÉL, TRANSMIS channel, circuit, path, route, TV channel, path; ~ **d'abonné** TÉL subscriber channel; ~ **d'accès** TRANSMIS, TV access channel, access path; ~ **d'accès de deuxième choix** TÉL alternate-access channel; ~ **d'accès de repli** TÉL alternate-access channel; ~ **d'accès de secours** TÉL alternate-access channel; ~ **d'acheminement** TÉL, TRANSMIS route; ~ **d'acheminement commune** (*VAC*) RADIO, TRANSMIS common routing channel (*CRC*); ~ **d'acheminement détournée** TÉL alternative route, alternate route; ~ **active** RADIO operating channel; ~ **d'affectation** TRANSMIS assignment channel; ~ **d'aller** DONNÉES, TÉL, TRANSMIS forward channel; ~ **d'appel** RADIO calling channel; ~ **d'appel unilatéral** TRANSMIS paging channel; ~ **d'assignation** TRANSMIS assignment channel; ~ **audiofréquence** TRANSMIS audio channel; ~ **avec concentration de la parole** TRANSMIS interpolated channel; ~ **binaire symétrique** ÉLECTRON, TÉL symmetrical binary channel; ~ **brouilleuse** TRANSMIS interfering channel; ~ **(circuit) mixte analogique-numérique** TÉL mixed analog-digital channel (circuit); ~ **de commande de chiffrement** TRANSMIS encryption control channel; ~ **commentaire** TRANSMIS commentary channel; ~ **de communication** RADIO, TÉL telecommunication circuit, TV communication channel; ~ **de communication bilatérale** TÉL telegraphic circuit; ~ **de communication télégraphique** TÉL telegraphic circuit; ~ **de contrôle du réseau** TRANSMIS network control channel; ~ **de conversation** DONNÉES traffic channel, RADIO traffic path, traffic channel, TÉL speech path, traffic path; ~ **de débordement** TRANSMIS overload channel; ~ **dédoublée** COMMUT split channel; ~ **à demi-débit** TRANSMIS half-rate channel; ~ **de départ** TRANSMIS outgoing channel; ~ **dérivée de fréquence** RADIO frequency-derived channel; ~ **dérivée en temps** TRANSMIS time-derived channel; ~ **descendante** TV down path; ~ **de données** TV data path; ~ **d'état** TÉL status channel; ~ **exclusive** TRANSMIS dedicated channel; ~ **exploitée** RADIO operating channel; ~ **à grande impédance** TRANSMIS high-impedance path; ~ **hors service** RADIO disabled channel; ~ **interpolée** TRANSMIS interpolated channel; ~ **d'interrogation-réponse** RADIO operating channel; ~ **de jonction numérique sans concentration** TRANSMIS digital non-interpolated channel (*DNI channel*); ~ **à large bande** TRANSMIS wideband channel, broadband channel; ~ **logique** TÉL logical channel; ~ **logique pour les signaux de progression de l'appel** TÉL call progress signal logical channel; ~ **de mémoire de données** COMMUT, DONNÉES data store channel; ~ **de mobile** TÉL reversal control channel; ~ **neutralisée** TRANSMIS blanked-off channel; ~ **à plein débit** TRANSMIS full-rate channel; ~ **porteuse** TÉL, TRANSMIS bearer channel (*BC*); ~ **porteuse d'information** TÉL information bearer channel; ~ **de poste hors lieux** TÉL off-premises extension (*OPX*); ~ **préassignée sans concentration** TRANSMIS preassigned noninterpolated channel; ~ **primaire** TÉL primary route; ~ **principale** RADIO operating channel; ~ **de qualité élevée** TRANSMIS high-grade channel; ~ **de qualité nominale** TRANSMIS standard-grade performance channel; ~ **radio** RADIO radio channel, radio path; ~ **radioélectronique** RADIO radio route;

~ radiophonique RADIO broadcast program channel (*AmE*), broadcast programme channel (*BrE*), TRANSMIS program sound circuit (*AmE*), programme sound circuit (*BrE*); **~ au repos** *f* TRANSMIS idle channel; **~ de retour** TÉL backward channel, TRANSMIS return path, backward channel, TV return path; **~ sans concentration de la parole** TRANSMIS noninterpolated channel; **~ satellite** RADIO, TV satellite channel; **~ secondaire** TÉL, TRANSMIS secondary route; **~ de service** TRANSMIS engineering service channel; **~ en service** RADIO operating channel; **~ de service par téléimprimeur** TRANSMIS teletype-order channel; **~ de service téléphonique** TÉL, TRANSMIS voice-order wire; **~ de signalisation** SIGNAL setup channel, signaling channel (*AmE*), signalling channel (*BrE*), TRANSMIS setup channel; **~ de signalisation non-spécialisée** TRANSMIS nondedicated signaling channel (*AmE*), nondedicated signalling channel (*BrE*); **~ de signalisation par satellite** SIGNAL satellite signaling channel (*AmE*), satellite signalling channel (*BrE*); **~ de son** TRANSMIS audio channel; **~ son** TRANSMIS audio path; **~ sortante** TRANSMIS outgoing channel; **~ support** TÉL, TRANSMIS bearer channel (*BC*); **~ symétrique** ÉLECTRON, TÉL symmetrical channel; **~ de téléphonie/données à l'alternat** TRANSMIS alternate voice data channel; **~ téléphonique** TÉL voice channel, TRANSMIS speech channel, voice channel; **~ téléphonique active** TÉL active telephone channel; **~ de trafic** TRANSMIS working channel; **~ de trafic au repos** RADIO idle working channel; **~ de transfert** COMMUT, TÉL transfer channel; **~ de transfert d'information** TÉL information channel; **~ de transmission** RADIO, TRANSMIS, TV transmission channel; **~ de transmission de données** DONNÉES, TRANSMIS data channel, data communication channel (*DCC*); **~ de transmission par satellite** RADIO, TRANSMIS, TV satellite transmission path; **~ de transmission télégraphique** TÉL telegraphic channel; **~ de transmultiplexeur** ÉLECTRON, TÉL, TRANSMIS transmultiplexer channel; **~ de type téléphonique** RADIO, TÉL telephone-type channel; **~ virtuelle** DONNÉES virtual path (*VP*), SIGNAL virtual channel (*VC*), TRANSMIS virtual channel (*VC*), virtual path (*VP*); **~ virtuelle diffusée de signalisation** TRANSMIS broadcast signaling virtual channel (*AmE*) (*BSVC*), broadcast signalling virtual channel (*BrE*) (*BSVC*); **~ virtuelle de signalisation** SIGNAL signaling virtual channel (*AmE*), signalling virtual channel (*BrE*) (*SVC*), switched virtual circuit (*SVC*)

voie[2]: **~ unique par porteuse** *loc* RADIO single channel per carrier (*SCPC*)

voies: **~ de secours** *f pl* TÉL emergency routes

voisé *adj* APP SERV, INFORMAT, TÉL voiced

voix *f* INFORMAT, PÉRIPH, TÉL voice; **~ artificielle** INFORMAT artificial voice; **~ synthétique** PÉRIPH synthetic voice

vol: **en ~** *adj* RADIO in-flight

volatilité *f* DONNÉES, INFORMAT volatility

volet *m* ÉLECTRON shutter, TV wipe; **~ pas à pas** TV trim wipe; **~ de protection** ÉLECTRON protective shutter

volt *m* ALIM ÉLEC, ÉLECTRON volt

voltampère *m* ALIM ÉLEC, ÉLECTRON volt-ampere

voltmètre: **~ de crête** *m* TESTS peak voltmeter; **~ efficace vrai** *m* TESTS true root mean square voltmeter (*true rms voltmeter*); **~ ferrodynamique** *m* TESTS iron-vane voltmeter, moving-vane voltmeter; **~ multicalibre** *m* TESTS multirange voltmeter; **~ numérique** *m* TESTS digital voltmeter; **~ à palette de fer doux** *m* TESTS iron-vane voltmeter, moving-vane voltmeter; **~ à redresseur** *m* TESTS rectifier meter; **~ sélectif** *m* TESTS frequency-selective voltmeter, tunable voltmeter

volume *m* ALIM ÉLEC, COMMUT, ÉLECTRON, GÉNÉRAL volume, INFORMAT loudness, volume, bulk, RADIO, TÉL, TRANSMIS volume; **~ de couverture** RADIO volumetric coverage; **~ des modes** TRANSMIS mode volume; **~ optique des modes** TRANSMIS effective mode volume; **~ de référence** TÉL reference telephonic power (*RTP*); **~ de trafic** COMMUT, TÉL traffic load, traffic loading, traffic volume; **~ utile** RADIO sensitive volume

volumineux *adj* COMPOSANTS bulky

VOR *abrév* (*radiophare omnidirectionnel VHF*) RADIO VOR (*VHF omnidirectional radio range*)

vous[1]: **à ~** *adv* RADIO over

vous[2]: **~ pouvez transmettre** *loc* RADIO you may transmit

voxel *m* INFORMAT, TV voxel

voyager *vi* INFORMAT travel

voyant *m* GÉNÉRAL indicator, INFORMAT display light, light indicator, pilot lamp, pilot light, visual indicator, PÉRIPH, RADIO pilot lamp, pilot light, TÉL pilot lamp, pilot light, visual indicator, TESTS visual indicator; **~ d'alarme** TÉL alarm lamp; **~ d'état** TESTS status lamp; **~ d'indication d'alarme** TESTS alarm indication lamp; **~ lumineux** GÉNÉRAL indicator lamp, PÉRIPH light-emitting diode (*LED*), TÉL indicating lamp, locating lamp; **~ d'occupation** TÉL busy lamp; **~ de réception de données** TÉL receive lamp, receive-data lamp

voyelle *f* GÉNÉRAL vowel

vrai *adj* ÉLECTRON, INFORMAT, RADIO, TESTS true

VRAM *abrév* (*RAM vidéo*) INFORMAT VRAM (*video RAM*)

vrillage *m* COMPOSANTS, ÉLECTRON twist

vrillé *adj* COMPOSANTS, ÉLECTRON twisted

vriller *vt* COMPOSANTS, ÉLECTRON twist

VSM *abrév* (*Société suisse des constructeurs de machines*) GÉNÉRAL VSM (*Swiss Association of Machinery Manufacturers*)

vu: **~ de l'arrière** *adj* TÉL rear-view; **~ de dos** *adj* TÉL rear-view

VU: **~-mètre** *m* TESTS VU-meter

vue *f* DONNÉES, GÉNÉRAL, INFORMAT view; **~ arborescente** DONNÉES, INFORMAT tree-structured view; **~ de dessous** GÉNÉRAL *dessin technique* bottom view; **~ éclatée** GÉNÉRAL cutaway view, exploded view

vulnérabilité: **~ au brouillage** *f* RADIO interference immunity

W

WAIS *abrév* INFORMAT, TÉL WAIS (*wide-area information server*)
WATS *abrév* (*service interurbain planifié*) TÉL WATS (*wide-area telephone service*)
wattmètre *m* ALIM ÉLEC, TESTS power meter, wattmeter; **~ à cloison** TESTS vane wattmeter; **~ à palette** TESTS vane wattmeter; **~ à pendule de torsion** TESTS torsion head wattmeter; **~ à tête de torsion** TESTS torsion head wattmeter; **~ à thermocouple** TESTS thermocouple wattmeter; **~ Thruline** TESTS Thruline wattmeter
Web *m* DONNÉES, INFORMAT Web
Whois *m* DONNÉES Whois
wilco *loc* RADIO, TÉL wilco
WMRA *abrév* INFORMAT WMRA (*write many read always*)
World: **~ Wide Web** *m* DONNÉES, INFORMAT (*W3, WWW*) Internet, World Wide Web (*W3, WWW*)
WORM *abrév* (*write once read many*) INFORMAT WORM (*write once read many*)
write: **~ once read many** *loc* (*WORM*) INFORMAT write once read many (*WORM*)
W3 *abrév* (*World Wide Web*) DONNÉES, INFORMAT Internet, W3 (*World Wide Web*), WWW (*World Wide Web*)
WWW *abrév* (*World Wide Web*) DONNÉES, INFORMAT Internet, W3 (*World Wide Web*), WWW (*World Wide Web*)
WYSIWYG *abrév* (*tel écran, tel écrit*) INFORMAT WYSIWYG (*what you see is what you get*)

X Y

XID *abrév* (*échange d'identifications*) TÉL XID (*exchange of identification*)
XNS *abrév* (*service de réseau Xerox*) DONNÉES XNS (*Xerox Network Service*)
YAG *abrév* (*grenat d'yttrium et d'aluminium*) ÉLECTRON YAG (*yttrium aluminum garnet AmE yttrium aluminium garnet BrE*)
YIG *abrév* (*grenat d'yttrium ferreux*) ÉLECTRON YIG (*yttrium iron garnet*)

Z

ZAA *abrév* (*zone à autonomie d'acheminement*) COMMUT, TÉL group switching center catchment area (*AmE*), group switching centre catchment area (*BrE*), group switching center exchange area (*AmE*), group switching centre exchange area (*BrE*), local exchange area, trunk switching exchange area

zénith *m* RADIO zenith

zéro *m* INFORMAT nought, TESTS zero point; **~ électrique** TESTS electrical zero, zero; **~ mécanique** TESTS mechanical zero

zéromètre *m* TESTS null meter

zigzag *m* ALIM ÉLEC, INFORMAT zigzag, zigzag pattern

zingage *m* COMPOSANTS zinc-plating; **~ électrolytique** COMPOSANTS zinc-electroplating

zinguer *vt* COMPOSANTS zinc-plate; **~ par électrolyse** COMPOSANTS zinc-electroplate

zone *f* GÉNÉRAL area, band, field, region, range, zone; **~ d'abonnement** DONNÉES, INFORMAT, RADIO, TÉL, TV subscription area; **~ active** INFORMAT hot zone, hot spot; **~ d'affaiblissement** TRANSMIS depressed cladding; **~ d'affichage** INFORMAT display field; **~ d'amorçage** INFORMAT lead-in area; **~ d'appels locaux** TÉL local calling area (*LCA*); **~ d'assemblage** GÉNÉRAL assembly area; **~ d'association** DONNÉES association area; **~ d'audibilité** TÉL range of audibility; **~ à autonomie d'acheminement** (*ZAA*) COMMUT, TÉL group switching center catchment area (*AmE*), group switching center exchange area (*AmE*), group switching centre catchment area (*BrE*), group switching centre exchange area (*BrE*), local exchange area, trunk switching exchange area; **~ aveugle** RADIO *radar* blind region; **~ de brouillage** GÉNÉRAL nuisance area; **~ de captage** RADIO capture area, catchment area; **~ de centre de transit principale** TÉL main trunk exchange area; **~ de centre de transit secondaire** COMMUT second-tier trunk exchange area; **~ cible** RADIO, TRANSMIS target zone; **~ de clics** INFORMAT clicking area; **~ de cœur** TRANSMIS core area; **~ de commande** CONTRÔLE *ordinateurs* control area; **~ commune** INFORMAT common field; **~ de contrôle de numéro** PÉRIPH n-field; **~ de contrôle du numéro** COMMUT check number field; **~ de correction** INFORMAT *ordinateurs* correction area; **~ de couverture** RADIO coverage area, footprint; **~ de couverture en absence de brouillage** RADIO interference-free coverage area; **~ de couverture nominale** RADIO nominal coverage area; **~ de couverture réelle** RADIO actual coverage area; **~ de couverture de la station de base** RADIO *mobile* base coverage area; **~ de couverture verticale** RADIO vertical coverage; **~ de début** INFORMAT lead-in area; **~ de dialogue** INFORMAT dialog box; **~ de dialogue de fichier standard** DONNÉES, INFORMAT standard file dialog box; **~ de l'échelle** TESTS scale range; **~ à l'émission** RADIO, TRANSMIS target zone, zone for transmission; **~ d'état** COMMUT status field; **~ étendue** COMPOSANTS overreaching, RADIO, TRANSMIS wide area, widespread area; **~ fonctionnelle de gestion des systèmes** DONNÉES, INFORMAT systems management functional area; **~ formante** TÉL formant; **~ de Fresnel** TRANSMIS Fresnel zone; **~ de glissement** INFORMAT drag region; **~ image** INFORMAT image storage space, coded image storage space; **~ imprimable** COMPOSANTS, INFORMAT printable area; **~ de lancement** RADIO launch site; **~ libre** APP SERV clear area, clear band, INFORMAT clear area; **~ de ligne de création** DONNÉES create-line area; **~ de localisation** RADIO location area; **~ de localisation nominale** RADIO home area; **~ M** (*zone de mode d'exploitation*) PÉRIPH M-field (*mode-of-operation field*); **~ de mémoire** INFORMAT storage block; **~ de mémoire de programmes** INFORMAT program memory area (*PMA*); **~ de mode d'exploitation** (*zone M*) PÉRIPH mode-of-operation field (*M-field*); **~ non occupée** APP SERV clear area, clear band; **~ de numérotage** TÉL routing number area, numbering area; **~ d'ombre** RADIO shadow region, shadow zone; **~ d'ombre crée par la terre** RADIO earth shadow; **~ la plus claire** PÉRIPH clearest zone; **~ la plus foncée** TV darkest zone; **~ de rayonnement** RADIO radiation zone; **~ de réception** RADIO reception area; **~ de recouvrement** RADIO overlap area; **~ réverbérée** TÉL reverberated field; **~ de saisie** INFORMAT text box, text entry field; **~ sautée** RADIO, TÉL skip area; **~ de service de liaison de connexion** RADIO, TRANSMIS feeder link service area; **~ de silence** RADIO blind region, silent zone, skip zone, TÉL skip zone, TRANSMIS shadow region, shadow zone; **~ d'une station de base** RADIO base station area; **~ de la station de base** RADIO base station area, *mobile* base area; **~ de taxation** TÉL tariff zone; **~ de taxe** TÉL charging area; **~ de télécommunications avancées** (*ZTA*) GÉNÉRAL teleport; **~ de texte** INFORMAT text space entry; **~ de tolérance** TESTS tolerance range, tolerance zone; **~ de traçage** INFORMAT plot area; **~ de trafic** RADIO traffic area; **~ de traitement d'appel** TÉL call-handling area; **~ de transition de l'émetteur** COMPOSANTS emitter depletion layer; **~ urbaine** TV urban area; **~ virtuelle** INFORMAT virtual area; **~ WATS** TÉL WATS zone

zoom *m* INFORMAT zooming; **~ arrière** INFORMAT zoom-out; **~ avant** INFORMAT zoom-in

zoomer *vt* INFORMAT zoom in

ZTA *abrév* (*zone de télécommunications avancées*) GÉNÉRAL teleport

ENGLISH–FRENCH
ANGLAIS–FRANÇAIS

A

A-ABORT: ~ **application protocol data unit** *n* (*ABRT*) DATA unité de données du protocole d'application A-ABANDON *f* (*ABRT*)

AAIC *abbr* (*accounting authority identification code*) TELEP code d'identification de l'autorité chargée de la comptabilité *m*

AAL *abbr* (*ATM adaptation layer*) DATA, GEN, TELEP, TRANS couche d'adaptation ATM *f*; ~ **protocol control information** *n* (*AAL-PCI*) DATA, TELEP information de contrôle du protocole AAL *f*; ~ **service data unit** *n* (*AAL-SDU*) DATA, TELEP unité de données de service AAL *f*

AAL-PCI *abbr* (*AAL protocol control information*) DATA, TELEP information de contrôle du protocole AAL *f*

AAL-SDU *abbr* (*AAL service data unit*) DATA, TELEP unité de données de service AAL *f*

AAR *abbr* (*automatic alternative routing*) SIG, TELEP acheminement détourné automatique *m*, acheminement automatique sur voie secondaire *m*

AARE *abbr* (*A-associate response*) DATA, TELEP AARE (*unité de données du protocole d'application réponse*)

AARQ *abbr* (*A-associate request*) DATA, TELEP AARQ (*unité de données du protocole d'association demande*)

AAS *abbr* (*abbreviated address signal*) TELEP signal d'adresse abrégée *m*

A-associate: ~ **request** *n* (*AARQ*) DATA, TELEP unité de données du protocole d'association demande *f* (*AARQ*); ~ **response** *n* (*AARE*) DATA, TELEP unité de données du protocole d'application réponse *f* (*AARE*)

abandoned: ~ **call attempt** *n* TELEP tentative d'appel abandonnée *f*; ~ **call disconnect** *n* COMP, TELEP libération sur abandon d'appel *f*

abbreviated: ~ **address calling** *n* DATA, TELEP numérotation abrégée *f*; ~ **address signal** *n* (*AAS*) TELEP signal d'adresse abrégée *m*; ~ **dialing** *AmE*, ~ **dialling** *n BrE* TELEP numérotation abrégée *f*; ~ **dialing prefix** *AmE*, ~ **dialling prefix** *n BrE* TELEP préfixe de numérotation abrégée *m*; ~ **dialing service** *AmE*, ~ **dialling service** *n BrE* TELEP service de numérotation abrégée *m*; ~ **number** *n* TELEP numéro abrégé *m*

abductive: ~ **reasoning** *n* DATA raisonnement abductif *m*

ability *n* COMP capacité *f*; ~ **concept** TEST *dependability* concept d'aptitude *m*

ABM *abbr* (*asynchronous balanced mode*) TELEP mode asynchrone symétrique *m*

abnormal: ~ **test termination** *n* DATA terminaison anormale de test élémentaire *f*

abort *n* COMP abandon *m*; ~ **dump** COMP vidage après abandon *m*

aborted: ~ **job** *n* COMP travail abandonné *m*

aborting: ~ **procedure** *n* COMP procédure d'abandon *f*

abrasive[1] *adj* PARTS abrasif

abrasive[2] *n* PARTS abrasif *m*

ABRT *abbr* (*A-ABORT application protocol data unit*) DATA ABRT (*unité de données du protocole d'application A-ABANDON*)

absent: ~**-subscriber service** *n* APPL, TELEP service des abonnés absents *m*; ~**-subscriber service equipment** *n* APPL, TELEP équipement de service des abonnés absents *m*; ~**-subscriber service signal** *n* APPL, TELEP signal du service des abonnés absents *m*

absolute: ~ **address** *n* COMP adresse absolue *f*; ~ **encoding** *n* DATA codage direct *m*; ~ **gain** *n* RADIO gain absolu *m*, gain isotrope *m*; ~ **pitch** *n* COMP hauteur absolue *f*, ton absolu *m*; ~ **voltage level** *n* TELEP niveau absolu de tension *m*

absorber: ~ **circuit** *n* RADIO circuit de compensation de charge *m*

absorption *n* GEN *fibre optics* absorption *f*; ~ **amplitude modulation** RADIO modulation d'amplitude par absorption *f*; ~ **circuit** RADIO circuit de compensation de charge *m*; ~ **cross-section** RADIO *aerials* surface de captation *f*; ~ **factor** RADIO facteur d'absorption *m*; ~ **loss** RADIO perte par absorption *f*; ~ **spectrometry** RADIO spectrométrie d'absorption *f*

absorptive: ~ **attenuator** *n* TRANS affaiblisseur à absorption *m*, atténuateur à absorption *m*

abstract *n* COMP, DATA résumé *m*; ~ **association** DATA association abstraite *f*; ~**-bind parameter** DATA paramètre de rattachement abstrait *m*; ~ **data type** COMP, DATA type de donnée abstrait *m*; ~ **grammar** COMP, DATA grammaire abstraite *f*; ~ **service primitive** (*ASP*) DATA primitive de service abstraite *f* (*ASP*); ~ **syntax** COMP, DATA syntaxe abstraite *f*; ~ **syntax name** COMP, DATA nom de syntaxe abstraite *m*; ~ **syntax notation one** (*ASN1*) DATA, TELEP, TRANS notation de syntaxe abstraite numéro un *f*; ~ **test case** DATA, TEST test élémentaire abstrait *m*; ~ **test case error** DATA, TEST erreur de test élémentaire abstraite *f*; ~ **test method** (*ATM*) DATA, TEST méthode de test abstraite *f*; ~ **test suite** (*ATS*) DATA, TEST suite de tests abstraite *f*; ~ **testing methodology** DATA, TEST méthodologie de test abstraite *f*; ~**-unbind parameter** DATA paramètre de détachement abstrait *m*

abstracting *n* DATA analyse documentaire *f*

ABU *abbr* (*Asia-Pacific Broadcasting Union*) RADIO Union de radiodiffusion Asie-Pacifique *f*

abusive: ~ **call** *n* TELEP appel importun *m*, appel malveillant *m*

AC *abbr* (*alternating current*) COMP, POWER, RADIO, SIG, TELEP CA (*courant alternatif*); ~ **generator** *n* POWER alternateur *m*; ~ **mains** *n* COMP secteur alternatif *m*; ~ **mains cord** *n* COMP cordon d'alimentation *m*; ~ **mains socket wall** *n* COMP prise secteur *f*, prise secteur murale *f*, TV prise péritel *f*; ~ **power** *n* COMP alimentation secteur *f*; ~ **power adaptor** *n* GEN adaptateur secteur *m*; ~ **power outlet** *n* COMP prise secteur *f*; ~ **power supply** *n* COMP source d'alimentation *f*; ~ **pulsing** *n* TELEP numérotation en courant alternatif *f*; ~ **signaling** *AmE*, ~ **signalling** *BrE n* POWER, SIG signalisation à courant alternatif *f*; ~ **socket** *n* COMP prise d'alimentation secteur *f*; ~ **source** *n* POWER source de courant alternatif *f*; ~ **spark-over voltage** *n* ELECT, POWER *of protector* tension alternative d'amorçage *f*; ~ **voltage** *n* COMP tension secteur *f*

accelerated: ~ **test** *n* TEST *dependability* essai accéléré *m*

accelerator *n* ELECT accélérateur *m*, PARTS *plastics* accélérateur promoteur *m*; ~ **board** DATA carte accélératrice *f*

accent: ~ **key** *n* COMP touche d'accent *f*

accentuated: ~ **contrast** *n* TV contraste accentué *m*

acceptable: ~ **interference** *n* RADIO brouillage tolérable *m*; ~ **quality level** *n* TEST niveau de qualité acceptable *m*

acceptance *n* COMP acceptation *f*, TEST recette *f*; ~ **angle** TRANS angle d'admission *m*, *fibre optics* angle d'ouverture *m*; ~ **with charging message** RADIO message de réponse avec taxation *m*; ~ **condition** TEST *quality* condition d'acceptation *f*; ~ **input** TELEP entrée d'acceptation *f*; ~ **inspection** TEST contrôle de réception *m*; ~ **number** TEST critère d'acceptation *m*; ~ **output** TELEP sortie d'acceptation *f*; ~ **procedure** TEST procédure d'acceptation *f*; ~ **sampling** TEST échantillonnage pour contrôle de réception *m*; ~ **signal** SIG signal d'acceptation *m*; ~ **test** GEN essai de qualification *m*, essai de réception *m*, essai de recette *m*, test de recette *m*; ~ **test plan** TEST programme d'essais de recette *m*

accepted: ~ **interference** *n* RADIO brouillage accepté *m*; ~ **reference value** *n* TEST valeur de référence acceptée *f*

accepting *n* COMP, RADIO réception *f*

acceptor *n* DATA, ELECT *semiconductors* accepteur *m*

access *n* GEN accès *m*; ~ **attribute** DATA attribut d'accès *m*; ~ **barred** TELEP accès interdit *m*; ~ **barred signal** SIG signal d'accès interdit *m*; ~ **burst signal** DATA salve de signaux d'accès *f*; ~ **channel** TELEP canal d'accès *m*, TRANS, TV voie d'accès *f*; ~ **charge** GEN frais d'accès au réseau *m*; ~ **charge rate** GEN tarif d'accès *m*; ~ **circuit** TELEP circuit d'accès *m*; ~ **code** DATA code d'accès *m*, TELEP indicatif d'appel *m*, TV code d'accès *m*; ~ **concentrator** TRANS concentrateur de voies asynchrones *m*; ~ **connection element** (*ACE*) DATA, TELEP élément de connexion d'accès *m*; ~ **control** DATA, TEST *security* contrôle d'accès *m*; ~ **control list** DATA, TEST *security* liste de contrôle d'accès *f*; ~ **control system** DATA, TEST système de contrôle d'accès *m*; ~ **delay** (*AD*) TELEP temps d'accès *m*; ~ **denied** TRANS accès interdit *m*; ~ **hole** POWER *printed circuits* trou d'accès *m*; ~ **hunting** RADIO recherche d'accès *f*; ~ **instruction** COMP commande d'accès *f*; ~ **link** (*AL*) TELEP liaison d'accès *f*; ~ **matrix** TELEP chercheur d'opérateur *m*; ~ **method** TELEP méthode d'accès *f*; ~ **network** GEN réseau d'accès *m*; ~ **node** TELEP nœud d'accès *m*; ~ **number** TELEP numéro d'accès *m*; ~ **on demand** DATA accès à la demande *m*; ~**-oriented programming** DATA programmation orientée accès *f*; ~ **path** COMP chemin d'accès *m*, TRANS, TV voie d'accès *f*; ~ **point** DATA, TELEP, TEST point d'accès *m*; ~ **port** GEN point d'accès *m*; ~ **protocol** RADIO protocole d'accès *m*; ~ **provider** DATA *Internet* prestataire d'accès *m*; ~ **restriction** TELEP limitation d'accès *f*; ~ **selector** SWG commutateur d'accès *m*, sélecteur d'accès *m*; ~ **service** DATA service d'accès *m*; ~ **and storage system** (*AS/SYS*) DATA système d'accès et de mémorisation *m*; ~**, storage and transfer system** (*AST/SYS*) DATA système d'accès, de mémorisation et de transfert *m*; ~ **subsystem** RADIO *paging* sous-système d'accès *m*; ~ **switch** SWG commutateur d'accès *m*, sélecteur d'accès *m*; ~ **system** (*A/SYS*) DATA système d'accès *m*; ~ **time** COMP, DATA temps d'accès *m*; ~ **to supplementary services** TELEP accès aux services supplémentaires *m*; ~ **and transfer system** (*AT/SYS*) DATA système d'accès et de transfert *m*; ~ **unit** (*AU*) GEN unité d'accès *f*

accessibility *n* GEN accessibilité *f*

accessible *adj* GEN consultable

accessories *n pl* PER APP péritéléphonie *f*

accessory *n* ~ **equipment** COMP équipement annexe *m*, TELEP équipement auxiliaire *m*; ~ **unit** COMP équipement annexe *m*

accidental: ~ **earthing** *n BrE* (*cf accidental grounding AmE*) POWER contact à la terre *m*; ~ **grounding** *n AmE* (*cf accidental earthing BrE*) POWER contact à la terre *m*; ~ **loss** *n* COMP perte accidentelle *f*

accompaniment: ~ **pattern** *n* COMP motif d'accompagnement *m*

account: ~ **code** *n* TELEP code de compte *m*; ~ **status** *n* TELEP situation de compte *f*

accountability *n* DATA, GEN imputabilité *f*

accounting: ~ **authority** *n* TELEP autorité chargée de la comptabilité *f*; ~ **authority identification code** *n* (*AAIC*) TELEP code d'identification de l'autorité chargée de la comptabilité *m*; ~ **counter** *n* TELEP compteur de règlement des comptes *m*; ~ **management** *n* TRANS gestion de facturation et taxation *f*; ~ **rate** *n* TELEP taxe de répartition *f*; ~ **rate share** *n* TELEP quote-part de répartition *f*; ~ **relay** *n* TELEP relais de comptage *m*; ~ **revenue division procedure** *n* TELEP méthode de division des recettes de répartition *f*

accreditation *n* TEST accréditation *m*

accredited: ~ **retailer** *n* RADIO, TV revendeur agréé *m*

accumulated: ~ **downtime** *n* TEST durée cumulée d'indisponibilité *f*; ~ **time** *n* TEST durée cumulée *f*

accumulating: ~**-attribute group** *n* COMP groupe d'attributs cumulatifs *m*

accumulation: ~ **effect** *n* TRANS *fibre optics* effet de cumul *m*; ~ **layer** *n* ELECT *semiconductors* couche d'accumulation *f*

accumulative: ~ **failure** *n* TEST défaillance cumulée *f*

accumulator *n* batterie d'accumulateurs *f*; ~ **battery** POWER batterie d'accumulateurs *f*; ~ **module** ELECT bloc accus *m*

accuracy *n* GEN précision *f*, exactitude *f*; ~ **of adjustment** GEN précision de réglage *f*; ~ **control** COMP contrôle d'exactitude *m*; ~ **of measurement** TELEP *of instrument* précision de mesure *f*

accurate *adj* GEN précis, exact

ACD *abbr* TELEP (*automatic call distributor*) DAA (*distributeur d'appels automatique*), (*automatic call distribution*) DAA (*distribution automatique d'appels*); ~ **in call** *n* TELEP appel d'arrivée DAA *m*

ACE *abbr* (*access connection element*) DATA, TELEP élément de connexion d'accès *m*

achromatic: ~ **color** *AmE*, ~ **colour** *BrE n* COMP, TV couleur achromatique *f*, couleur sans tonalité chromatique *f*

ACI *abbr* (*adjacent-channel interference*) RADIO, TRANS brouillage du canal adjacent *m*

ACID *abbr* (*atomicity, consistency, isolation and durability*) DATA ACID (*atomicité, cohérence, isolation et durabilité*)

ACK *abbr* (*acknowledgement*) DATA, SIG, TELEP AR (*accusé de réception*)

acknowledge[1]: ~ **tone** *n* TELEP tonalité d'accusé de réception *f*

acknowledge[2] *vt* DATA, SIG, TELEP accuser-réception de

Acknowledge *phr* DATA, SIG, TELEP Veuillez accuser réception

acknowledged: ~ **information-transfer service** *n* (*AITS*) APPL, TELEP service de transfert d'informations avec accusé de réception *m*; ~ **message** *n* RADIO message acquitté *m*

acknowledgement *n* (*ACK*) DATA, SIG, TELEP accusé de réception *m* (*AR*); ~ **indicator** TELEP indicateur d'accusé de réception *m*; ~ **paging** RADIO radiomessagerie avec accusé de réception *f*, RM avec accusé de réception *f*; ~ **request** DATA demande d'accusé de réception *f*; ~ **signal** SIG signal d'accusé de réception *m*; ~ **signal unit** SIG unité de signalisation d'accusé de réception *f*; ~ **window** TELEP fenêtre d'accusé de réception *f*

acoustic[1] *adj* GEN acoustique

acoustic[2]: ~ **absorption** *n* TELEP absorption acoustique *f*; ~ **antenna** *n* RADIO antenne acoustique *f*; ~ **blanket** *n* TELEP protection acoustique *f*, revêtement atténuateur *m*; ~**-charge transport** *n* (*ACT*) TRANS transport de charge acoustique *m* (*ACT*); ~ **coupler** *n* DATA, TELEP coupleur acoustique *m*; ~ **coupling** *n* DATA, TELEP couplage acoustique *m*; ~ **delay line** *n* DATA *datacom* ligne à retard acoustique *f*; ~ **feedback** *n* TELEP couplage acoustique interactif *m*, effet Larsen *m*, réaction acoustique *f*; ~ **field** *n* TELEP champ acoustique *m*; ~ **generator** *n* DATA générateur de son *m*; ~ **measurement** *n* GEN mesure acoustique *f*; ~ **noise** *n* DATA, TELEP bruit acoustique *m*; ~ **phase coefficient** *n* GEN déphasage linéique de propagation acoustique *m*; ~ **power** *n* GEN puissance acoustique *f*; ~ **power level** *n* GEN niveau de puissance acoustique *m*; ~ **regeneration** *n* GEN réaction acoustique *f*; ~ **resonator** *n* GEN résonateur acoustique *m*; ~ **shock** *n* ELECT choc acoustique *m*; ~ **signal** *n* DATA signal acoustique *m*; ~ **signal processing** *n* DATA traitement de signal acoustique

m; ~ **vibration** *n* TELEP vibration acoustique *f*; ~ **wave propagation** *n* ELECT, RADIO propagation des ondes sonores *f*
acoustical[1] *adj* COMP phonique, sonore
acoustical[2]: ~ **dispersion** *n* TELEP dispersion acoustique *f*
acoustically[1]: ~**-muted** *adj* TELEP isolé phoniquement
acoustically[2]: ~**-connected terminal equipment** *n* TELEP équipement terminal à raccordement acoustique *m*
acoustics *n* GEN acoustique *f*
acousto: ~**-electric effect** *n* ELECT effet électroacoustique *m*; ~**-optic effect** *n* TRANS *fibre optics* effet acousto-optique *m*
acousto-optical *adj* (*AO*) ELECT acousto-optique
ACPM *abbr* (*association control protocol machine*) DATA, TELEP machine de protocole de commande d'association *f*
acquisition *n* GEN acquisition *f*; ~**-cycle unit** (*ACU*) DATA organe du cycle d'acquisition *m*; ~ **radar** RADIO radar d'acquisition *m*; ~ **and steady-state reception** (*ASSR*) DATA acquisition et réception en régime permanent *f*; ~ **time** DATA temps d'acquisition *m*
acrylate *n* PARTS *plastics* résine acrylique *f*
acrylic: ~ **plastic** *n* PARTS plastique acrylique *m*
ACSE *abbr* (*association control service element*) DATA, TELEP élément de service de commande d'association *m*, élément de service de contrôle d'association *m*; ~ **provider** *n* DATA, TELEP fournisseur de l'ACSE *m*; ~ **service provider** *n* DATA, TELEP fournisseur du service ACSE *m*, fournisseur du service de contrôle d'association *m*; ~ **service user** *n* DATA, TELEP utilisateur du service ACSE *m*, utilisateur du service de contrôle d'association *m*
ACT *abbr* (*transverse conversion loss*) DATA, TELEP TCL (*affaiblissement de conversion transversale*)
action *n* GEN action *f*, mesure *f*, suite à donner *f*, suite donnée *f*; ~ **taken upon alarm** TELEP mesures prises en cas d'alarme *f pl*
activable: ~ **word** *n* COMP mot activable *m*
activate[1]: ~ **button** *n* COMP bouton de commande *m*
activate[2] *vt* COMP activer, déclencher, GEN, RADIO, TEST activer
activated: ~ **condition** *n* GEN état activé *m*
activating: ~ **unit** *n* RADIO activateur *m*
activation *n* COMP déclenchement *m*, GEN actionnement *m*, activation *f*, COMP, RADIO activation, TEST *alarms* activation *f*; ~ **energy** ELECT *semiconductors* énergie d'activation *f*; ~ **overvoltage** ELECT *semiconductors* surtension d'activation *f*
active[1] *adj* GEN actif
active[2]: ~ **application** *n* COMP application active *f*; ~ **component** *n* ELECT composant actif *m*; ~ **control** *n* TEST correction active *f*, manuvre de correction *f*; ~ **corrective maintenance** *n* TEST maintenance corrective active *f*; ~ **corrective maintenance time** *n* TELEP, TEST durée de maintenance corrective active *f*, TEST temps d'entretien correctif *m*, temps de dépannage actif *m*; ~ **expression** *n* DATA expression active *f*; ~ **group** *n* DATA groupe actif *m*; ~ **laser medium** *n* ELECT, TRANS milieu laser actif *m*; ~ **maintenance** *n* TELEP, TEST maintenance active *f*; ~ **maintenance time** *n* TELEP, TEST durée de maintenance active *f*, temps de maintenance active *m*; ~ **matrix** *n* (*AM*) COMP matrice active *f*; ~**-matrix flat screen** *n* TV écran plat à matrice active *m*; ~ **memory** *n* COMP, DATA mémoire active *f*; ~ **node** *n* COMP nœud actif *m*; ~ **position** *n* TELEP position active *f*; ~**-position addressing** *n* TELEP adressage de position active *m* (*PAP*); ~ **preventive maintenance time** *n* TELEP, TEST durée de maintenance préventive active *f*, temps d'entretien préventif actif *m*; ~ **processor** *n* COMP processeur actif *m*; ~ **redundancy** *n* TELEP, TEST *dependability* redondance active *f*; ~ **region** *n* TRANS *fibre optics* région active *f*; ~ **repair time** *n* TELEP durée de réparation active *f*; ~ **satellite** *n* RADIO satellite actif *m*; ~ **sensing** *n* COMP détection d'activité *f*, test de liaison *m*; ~ **sensor** *n* RADIO détecteur actif *m*; ~ **signaling link** *AmE*, ~ **signalling link** *BrE* *n* SIG canal sémaphore à l'état actif *m*, canal sémaphore actif *m*; ~ **skin** *n* RADIO peau avant *f*; ~ **supervisor** *n* DATA superviseur actif *m*; ~ **target** *n* RADIO cible répondeuse *f*; ~ **telephone channel** *n* TELEP voie téléphonique active *f*; ~ **threat** *n* DATA menace active *f*; ~ **timer** *n* DATA temporisateur actif *m*; ~ **transponder** *n* RADIO répéteur en service *m*; ~ **winding** *n* POWER enroulement de travail *m*
active/standby *n* TELEP pilote/réserve *m*
activity: ~ **report** *n* DATA journal de communication *m*
actual: ~ **context parameter** *n* DATA paramètre contextuel réel *m*; ~ **coverage area** *n* RADIO zone de couverture réelle *f*; ~ **deviation** *n* TEST écart réel *m*; ~ **dimension** *n* TEST cote effective *f*; ~ **key** *n* COMP clé absolue *f*; ~ **parameter** *n* DATA *SDL* paramètre réel *m*; ~ **relative level** *n* POWER niveau relatif réel de puissance *m*; ~ **size** *n* DATA *Internet* taille réelle *f*; ~ **transfer rate** *n* DATA vitesse de transfert effective *f*, vitesse de transfert réelle *f*; ~ **value** *n* CONTR valeur réelle *f*
actuating: ~ **signal** *n* COMP signal de commande *m*; ~ **variable** *n* CONTR grandeur d'influence *f*
actuation *n* COMP commande *f*, actionnement *m*, mise en action *f*
actuator *n* RADIO organe moteur *m*
ACU *abbr* (*acquisition-cycle unit*) DATA organe du cycle d'acquisition *m*
ad *abbr* (*advertisement*) TV spot publicitaire *m*
AD *abbr* (*access delay*) TELEP temps d'accès *m*
A/D *abbr* (*analog-to-digital*) COMP, DATA, ELECT, SIG, TELEP A/N (*analogique-numérique*); ~ **converter** *n* (*ADC*) COMP, DATA, ELECT, SIG, TELEP convertisseur A/N *m* (*CAN*)
adapt *vt* COMP, DATA, GEN adapter
adaptation: ~ **unit** *n* GEN unité d'adaptation *f*, unité de jonction *f*, unité d'interfaçage *f*, adaptation de jonction *f*, unité d'interface *f*
adaptative *see adaptive*
adapter *see adaptor*
adaptive: ~ **antenna** *n* RADIO antenne adaptative *f*; ~ **attenuation** *n* TELEP affaiblissement adaptable *m*; ~ **bit allocation** *n* COMP affectation binaire adaptative *f*; ~ **break-in echo suppressor** *n* RADIO, TELEP suppresseur d'écho à intervention adaptable *m*; ~ **coding** *n* SIG codage adaptif *m*; ~ **compression** *n* COMP compression adaptative *f*; ~ **differential pulse-code modulation** *n* (*ADPCM*) COMP, DATA, TELEP, TRANS modulation par impulsions et codage différentiel adaptatif *f* (*MICDA*); ~ **filtering** *n* TRANS filtrage adaptatif *m*; ~ **path-routing** *n* SIG, TELEP, TRANS acheminement adaptatif *m*; ~ **processing** *n* RADIO traitement adaptatif *m*; ~ **quantizing** *n* COMP quantification adaptative *f*; ~ **routing** *n* SIG, TELEP, TRANS acheminement adaptatif *m*; ~ **segmentation** *n* COMP segmentation adaptative *f*; ~ **system** *n* DATA système adaptatif *m*; ~ **transform acoustic coding** *n* (*ATRAC*) COMP codage par transformation acoustique adaptive *m*
adaptor *n* GEN adaptateur *m*
ADC *abbr* (*A/D converter, analog-to-digital converter*) COMP, DATA, ELECT, SIG, TELEP CAN (*convertisseur A-N, convertisseur analogique-numérique*)
add: ~ **in** *vt* GEN enficher
ADD *abbr* (*address prompt*) DATA message de demande d'adresse *f*
add-drop: ~ **multiplexer** *n* TRANS multiplexeur d'insertion-extraction *m* (*MIE*)
added: ~ **value** *n* DATA valeur ajoutée *f*
adder *n* DATA, ELECT additionneur *m*
add-in: ~ **board** *n* COMP carte d'extension *f*
additional[1] *adj* COMP, TELEP supplémentaire
additional[2]: ~ **function** *n* COMP fonction supplémentaire *f*; ~ **header information** *n* COMP information d'en-tête supplémentaire *f*; ~ **information** *n* TELEP information supplémentaire *f*; ~ **load** *n* POWER charge additionnelle *f*;

~ memory *n* COMP mémoire additionnelle *f*; **~ TC** *n* TELEP connexion TC additionnelle *f*; **~ telephone set** *n* TELEP, SWG poste téléphonique supplémentaire *m*, poste auxiliaire *m*; **~ unit** *n* GEN unité additionnelle *f*

additive: **~ color** *AmE*, **~ colour** *BrE n* COMP couleur additive *f*; **~ noise** *n* TELEP bruit additif *m*; **~ primary color** *AmE*, **~ primary colour** *BrE n* COMP couleur primaire additive *f*, synthèse additive trichrome *f*; **~ process** *n* ELECT *printed circuits* processus additif *m*; **~ stream cipher mode** *n* DATA mode de chiffrement par adjonction de bits sur les trains *m*; **~ synthesis** *n* COMP synthèse additive *f*

add-on[1] *adj* RADIO en frontal

add-on[2]: **~ board** *n* DATA carte d'extension *f*; **~ data module** *n* TELEP module complémentaire de transmission de données *m*; **~ facility** *n* COMP possibilité d'éxtension *f*; **~ services** *n pl* DATA services complémentaires *m pl*; **~ software** *n* COMP, DATA logiciel complémentaire *m*, logiciel d'extension *m*; **~ unit** *n* COMP élément additionnel *m*, élément supplémentaire *m*, DATA boîtier externe *m*

address *n* GEN adresse *f*; **~-complete signal** SIG signal d'adresse complet *m*; **~-complete signal, no charge** SIG signal de numéro complet, sans taxation *m*; **~-complete, subscriber-free signal, no charge** SIG signal de numéro complet, ligne d'abonné libre, sans taxation *m*; **~ and control unit** TELEP unité d'adresse et de commande *f*; **~ field** (*AF*) TELEP champ d'adresse *m*; **~ generation** COMP calcul d'adresse *m*; **~-incomplete signal** SIG signal d'adresse incomplet *m*, signal de numéro incomplet *m*; **~ message** SIG, SWG, TELEP message d'adresse *m*; **~ prompt** (*ADD*) DATA message de demande d'adresse *m*; **~ separator** TELEP séparateur d'adresses *m*; **~ signal** (*AS*) TELEP signal d'adresse *m*

addressable: **~ latch** *n* ELECT *logic elements* bascule adressable *f*

addressing *n* COMP, DATA adressage *m*, système d'adressage *m*

ADF *abbr* (*automatic direction finding*) RADIO radiogoniométrie automatique *f*

adhere *vi* ELECT *soldering, collage* adhérer

adhesion *n* ELECT *soldering, collage* adhérence *f*

adjacent: **~ channel** *n* RADIO, TELEP canal adjacent *m*; **~-channel interference** *n* (*ACI*) RADIO, TRANS brouillage du canal adjacent *m*; **~-channel rejection ratio** *n* RADIO, TRANS rapport de réjection des canaux voisins *m*; **~ channel selectivity** *n* RADIO, TRANS sélectivité de voie adjacente *f*, sélectivité proche *f*; **~ signaling point** *AmE*, **~ signalling point** *BrE n* SIG point sémaphore adjacent *m*

adjust *vt* GEN ajuster, régler; **~ to zero** ELECT, TEST mettre à zéro, remettre à zéro

adjustable[1] *adj* GEN ajustable, réglable; **~ in steps** GEN réglable en pas à pas, réglabie par paliers

adjustable[2]: **~ ringer** *n* APPL sonnerie réglable *f*; **~ volume control** *n* PER APP haut-parleur à volume réglable *m*

adjusting: **~ instructions** *n pl* GEN *documentation* instructions de réglage *f pl*; **~ to zero** *n* TEST mise à zéro *f*

adjustment *n* GEN ajustement, réglage *m*, TELEP ajustage *m*, ajustement *m*; **~ error** TELEP erreur d'ajustage *f*; **~ handle** COMP poignée d'ajustement *f*; **~ level** DATA niveau de réglage *m*; **~ step** TEST pas de réglage *m*

ADMD *abbr* (*administration management domain*) DATA, TELEP DGAD (*domaine de gestion d'administration*)

administration *n* GEN administration *f*; **~ of destination** TELEP administration de destination *f*, pays de destination *m*; **~ district** TELEP circonscription administrative *f*; **~ management domain** (*ADMD*) DATA, TELEP domaine de gestion d'administration *m* (*DGAD*); **~ of origin** TELEP administration d'origine *f*, pays d'origine *m*; **~ port** DATA accès d'administration *m*

administrative: **~ delay** *n* TEST délai administratif *m*; **~ pair** *n* TELEP paire de service *f*; **~ processor** *n* COMP processeur de gestion *m*; **~ time for corrective maintenance** *n* TELEP durée administrative pour la maintenance corrective *f*; **~ unit** *n* (*AU*) CONTR unité administrative *f* (*UAD*); **~ unit group** *n* CONTR groupe d'unités administratives *m* (*GUA*)

admittance *n* POWER admittance *f*

ADPCM *abbr* (*adaptive differential pulse-code modulation*) COMP, DATA, TELEP, TRANS MICDA (*modulation par impulsions et codage différentiel adaptatif*); **~ decoder** *n* TELEP décodeur MICDA *m*; **~ encoder** *n* TELEP codeur MICDA *m*

ADSL *abbr* (*asynchronous digital subscriber loop*) TRANS MTA (*mode de transfert asynchrone*)

adsorption *n* GEN adsorption *f*

ADU *abbr* (*asynchronous data units*) DATA, ELECT unités de données asynchrones *f pl*

advance: **~-feed tape** *n* COMP bande à entraînement décalé vers l'avant *f*, bande perforée à alignement frontal *f*

advanced: **~ statistical analysis** *n* ELECT analyse statistique *f*; **~ technology** *n* COMP technologie de pointe *f*

adverse: **~ state** *n* RADIO *scrambling process* état défavorable *m*

advert *n* (*advertisement*) TV spot publicitaire *m*

advertisement *n* (*advert*) TV spot publicitaire *m*

advice *n* GEN indication *f*; **~ of charge** APPL avis de taxation *m*, information tarifaire *f*

advise: **~ duration and charge** *n* TELEP indication verbale des éléments de taxation *f*

advisory: **~ committee** *n* GEN comité consultatif *m*

AE *abbr* COMP, RADIO (*aerial*) ant. (*antenne*), TELEP (*application entity*) entité d'application *f*

AEI *abbr* (*application-entity invocation*) DATA, RADIO, TELEP appel de l'entité d'application *m*

aerial *n* (*AE*) COMP, RADIO antenne *f* (*ant.*); **~ cable** POWER, TRANS câble aérien *m*; **~ current** RADIO courant d'antenne *m*; **~ curtain** RADIO antenne rideau *f*, rideau d'antennes *m*; **~ effect** RADIO effet d'antenne *m*; **~ feed** RADIO alimentation d'antenne *f*; **~ mast** RADIO mât d'antenne *m*; **~ resistance** RADIO résistance d'antenne *f*; **~ spacing** RADIO espacement d'antenne *m*, espacement entre les antennes *m*; **~-spacing error** RADIO erreur d'espacement d'antenne *f*

aeronautical: **~ fixed network** *n AmE* (*cf aeronautical fixed service BrE*) RADIO service fixe aéronautique *m*; **~ fixed service** *n BrE* (*cf aeronautical fixed system AmE, aeronautical fixed network AmE*) RADIO service fixe aéronautique *m*; **~ fixed station** *n* RADIO station fixe aéronautique *f*; **~ fixed system** *n AmE* (*cf aeronautical fixed service BrE*) RADIO service fixe aéronautique *m*; **~ mobile satellite-service** *n* RADIO service mobile aéronautique par satellite *m*; **~ mobile service** *n* RADIO service mobile aéronautique *m*; **~ station** *n* RADIO station aéronautique *f*

aerosol *n* RADIO aérosol *m*

AF *abbr* COMP (*audio frequency*) audiofréquence *f*, fréquence audio *f*, TELEP (*address field*) champ d'adresse *m*

AFC *abbr* (*automatic frequency control*) ELECT, TELEP, TRANS CAF (*commande automatique de fréquence, contrôle automatique de fréquence*); **~ hold-in** *n* RADIO accord de la CAF *m*, maintien de la CAF *m*; **~ pull-in** *n* RADIO établissement de la CAF *m*

AFI *abbr* (*authority and format identifier*) DATA, SIG identificateur d'autorité et de format *m*

AFNOR *abbr* (*French National Standards Association*) GEN AFNOR (*Association française de normalisation*)

A4: **~ size** *n* GEN *documentation* format A4 *m*; **~-sized paper** *n* GEN *documentation* papier au format A4 *m*

aft: **~ solar array** *n* RADIO panneau solaire de poupe *m*

after: **~-hours recorded announcement** *n* TELEP permanence par message enregistré *f*; **~ touch** *n* COMP pression après enfoncement *f*; **~-touch assign** *n* COMP affectation de la pression après enfoncement *f*; **~-touch response** *n* COMP réponse à la pression après enfoncement *f*; **~-touch**

sensitivity *n* COMP sensibilité à la pression après enfoncement *f*
AGC *abbr* (*automatic gain control*) ELECT, TELEP, TRANS CAG (*commande automatique de gain, contrôle automatique de gain*)
age[1]: **--old** *adj* ELECT séculaire
age[2] *vi* GEN vieillir
ageing *n* GEN vieillissement *m*; **~ failure** TEST défaillance par usure *f*, défaillance par vieillissement *f*
agent *n* GEN agent *m*, TELEP position *f*; **~ key** TELEP touche de position *f*; **~ position digit-display telephone** TELEP poste téléphonique de position à affichage numérique *m*; **~ role** DATA rôle d'agent *m*
aggregate: **~ interference** *n* RADIO brouillage global *m*; **~ signal** *n* SIG signal composite *m*
A-grade: **~ contour** *n* RADIO contour de classe A *m*
agreement *n* DATA agrément *m*, homologation *f*
AI *abbr* APPL, COMP (*artificial intelligence*) IA (*intelligence artificielle*), RADIO (*articulation index*) IN (*indice de netteté*), SWG (*artificial intelligence*) IA (*intelligence artificielle*), TELEP, TRANS (*articulation index*) IN (*indice de netteté*)
AIM *abbr* (*audio indicate muted*) TELEP indication audio muette *f*
aiming: **~ circle** *n* COMP champ de visée *m*; **~ field** *n* COMP champ de visée *m*; **~ symbol** *n* COMP champ de visée *m*
AIOD *abbr* (*automatic identification of outward dialing AmE, automatic identification of outward dialling BrE*) DATA, RADIO, TELEP IAD (*identification automatique du demandeur*)
air[1]: **--cooled** *adj* PARTS refroidi par air
air[2]: **over the ~** *adv* RADIO par radio
air[3]: **~ bending** *n* PARTS pliage non guidé *m*; **~ core** *n* ELECT section creuse *f*; **~ duster** *n* ELECT soufflette *f*; **~ gap** *n* POWER entrefer *m*; **--ground link** *n* RADIO liaison air-sol *f*; **~ interface standard** *n* RADIO norme d'interface radio *f*; **~ jet** *n* PARTS *valves* jet d'air *m*; **~ traffic control** *n* RADIO contrôle de la circulation aérienne *m*; **~ valve** *n* PARTS purgeur d'air *m*
airborne: **~ radar** *n* RADIO radar aéroporté *m*; **~ remote sensing** *n* TEST télédétection aérienne *f*; **~ and spaceborne remote sensing** *n* TEST télédétection aérospatiale *f*; **--support equipment** *n* RADIO matériel de servitude embarqué *m*; **~ system** *n* RADIO système embarqué *m*
airbrush *n* GEN aérographe *m*
aircraft: **~ station** *n* RADIO station d'aéronef *f*
airport: **~ surface detection equipment** *n* (*ASDE*) RADIO radar de pistes *m*
airtime *n* RADIO temps d'émission *m*, durée d'émission *f*
AIS *abbr* (*alarm indication signal*) TELEP, TEST SIA (*signal d'indication d'alarme*)
aisle *n* PARTS, SWG, TELEP allée *f*, couloir *m*
AITS *abbr* (*acknowledged information-transfer service*) APPL, TELEP service de transfert d'informations avec accusé de réception *m*
AJ *abbr* (*anti-jamming*) RADIO antibrouillage *m*
AL *abbr* (*access link*) TELEP liaison d'accès *f*, faisceau d'alarme *m*, faisceau de circuits d'alarme *m*
alarm[1] *n* DATA, TELEP, TEST alarme *f*; **--assembling relay set** TELEP plaque de relais de concentration d'alarme *f*; **--assembling unit** TELEP unité de concentration d'alarmes *f*; **~ bell** TELEP sonnerie d'alarme *f*; **~ call** APPL réveil automatique *m*, réveil téléphonique *m*; **~ call service** TELEP service de réveil *m*; **~ card** APPL carte d'alarme *f*; **~ center** *AmE*, **~ centre** *BrE* TELEP centre d'alarme *m*; **~ circuit** TELEP, TEST circuit d'alarme *m*; **--cleared message** TEST message de libération d'alarme *m*; **~ collection and display** TELEP collecte et visualisation d'alarmes *f pl*; **~ condition** TEST état d'alarme *m*, situation d'alarme *f*; **~ cut-off relay** TELEP relais de coupure d'alarme *m*; **~ device** TELEP unité d'alarme *f*; **~ equipment rack** TELEP bâti d'équipement d'alarme *m*; **~ function** TEST fonction d'alarme *f*; **~ history file** TELEP fichier historique des alarmes *m*; **~indication lamp** TEST voyant d'indication d'alarme *m*; **~ indication signal** (*AIS*) TELEP, TEST signal d'indication d'alarme *m* (*SIA*); **~ indication transmission** TEST transmission d'alarme *f*; **~ inhibit signal** TEST, TRANS signal de neutralisation d'alarme *m* (*SNA*); **~ inhibited** TRANS alarme neutralisée *f*; **~ lamp** TELEP voyant d'alarme *m*; **~ link** TELEP multiplex d'alarmes *m*; **~ management data processing** TEST informatique de traitement d'alarmes *f*; **~ print-out** TELEP message d'alarme *m*, sortie imprimée d'alarme *f*, TEST édition des alarmes *f*; **~ priority** TEST priorité en cas d'alarme *f*; **~ report** DATA signalisation d'alarme *f*; **~ satellite coupler** TELEP coupleur satellite d'alarmes *m*; **~ service** DATA téléalarme *f*; **~ setting** TEST réglage d'alarmes *m*; **~ signal** TELEP signal d'alarme *m*; **~ situation** TELEP situation d'alarme *f*; **~ statement** TELEP instruction d'alarme *f*; **~ transfer** TELEP transfert d'alarme *m*; **~ transmission** TEST transmission d'alarmes *f*; **~ transmitter** TELEP émetteur d'alarme *m*; **~ trunk** TELEP circuit d'alarme *m*; **~ trunk group** TELEP faisceau de circuits d'alarme *m*, faisceau d'alarme *m*; **~ unit** TELEP unité d'alarme *f*
alarm[2] *vt* TELEP alerter, mettre en état d'alarme
alarm[3]: **~ cleared** *phr* TEST libération d'alarme; **~ off** *phr* TRANS disparition d'alarme *f*; **~ on** *phr* TRANS déclenchement d'alarme *m*
A-law *n* TRANS loi A *f*
ALB *abbr* (*analog line board*) SWG carte de ligne analogique *f*
A-leg *n* TELEP conducteur de pointe *m*, fil de pointe *m*
alert[1] *n* TELEP avertissement *m*; **~ abstract-operation** DATA opération abstraite d'avertissement *f*; **--only pager** RADIO bipper *m*; **--only service** RADIO service bip *m*; **~ servers for high-risk sites** *n pl* TEST serveurs d'alerte pour sites à hauts risques *m pl*; **~ sound** COMP signal sonore *m*
alert[2] *vt* RADIO avertir, informer, prévenir, TELEP mettre en état d'alarme, alerter
alerting *n* RADIO, TEST alerte *f*; **~ detection device** TELEP circuit détecteur avertisseur *m*; **~ device** TELEP dispositif avertisseur *m*; **~ sensitivity** TELEP seuil de sensibilité du dispositif avertisseur *m*
algebraic: **~ decoding** *n* SIG décodage algébrique *m*
ALGOL *abbr* (*algorithmic language*) COMP ALGOL (*langage algorithmique*)
algorithm *n* COMP algorithme *m*
algorithmic[1] *adj* COMP algorithmique
algorithmic[2]: **~ language** *n* (*ALGOL*) COMP langage algorithmique *m* (*ALGOL*)
alias *adv* GEN alias
aliasing *n* COMP crénelage *m*, distorsion de repliement *f*, TRANS distorsion de repliement *f*
align *vt* COMP aligner, RADIO, SIG, TELEP, TRANS aligner, recaler
aligned[1] *adj* COMP, RADIO, SIG, TELEP aligné
aligned[2]: **~ grid tube** *n* TELEP tube à grilles alignées *m*
aligner *n* TELEP aligneur *m*
aligning *n* RADIO, SIG, TELEP, TRANS recalage *m*
alignment *n* COMP alignement *m*, DATA *of frame* verrouillage *m*, RADIO, SIG, TELEP, TRANS *fibre optics* alignement *m*, recalage *m*; **~ check** TRANS essai d'alignement *m*; **~ error** RADIO, TELEP erreur d'alignement *f*; **~ error-rate monitoring** TELEP surveillance du taux d'erreur d'alignement *f*; **~ fault** TRANS défaut d'alignement *m*; **~ mirror** TRANS miroir d'alignement *m*; **~ pattern** TRANS schéma de verrouillage *m*, schéma de trame *m*
alkaline: **~ cell** *n* POWER élément de pile alcaline *m*
all-glass: **~ fiber** *AmE*, **~ fibre** *BrE* *n* TRANS fibre de verre *f*
all-in-one *adj* DATA compréhensif
allocate *vt* COMP allouer, attribuer, GEN allouer, RADIO, TRANS attribuer
allocation *n* COMP, GEN affectation *f*, RADIO attribution *f*

all-ones: ~ **condition signal** *n* SIG signal composé entièrement un *m*
all-optical: ~ **switching system** *n* SWG système de commutation tout optique *m*
allot *vt* DATA allouer, ELECT attribuer, RADIO, TELEP attribuer
allotment: ~ **of capacity** *n* RADIO attribution de capacité *f*
allowable: ~ **value** *n* TEST valeur admissible *f*
alloyed: ~ **junction** *n* ELECT *semiconductors* jonction par alliage *f*; ~ **transistor** *n* ELECT *semiconductors* transistor à jonction par alliage *m*
alloying: ~ **constituent** *n* PARTS élément d'alliage *m*, composant d'alliage *m*
all-pass: ~ **network** *n* RADIO réseau passe-partout *m*
all-paths: ~ **busy** *n* (*APB*) TELEP circuits tous occupés *m*; ~ **busy tone** *n* TELEP signal circuits tous occupés *m*, tonalité circuits tous occupés *f*
all-plastic: ~ **fiber** *AmE*, ~ **fibre** *BrE n* DATA, PARTS, TRANS fibre plastique *f*
all-silica: ~ **fiber** *AmE*, ~ **fibre** *BrE n* DATA, PARTS, TRANS fibre tout silice *f*
all-trunks: ~ **busy** *n* (*ATB*) TELEP tous circuits occupés *m* (*TCO*); ~ **busy tone** *n* TELEP signal tous circuits occupés *m* (*signal TCO*), tonalité tous circuits occupés *f* (*tonalité TCO*)
alodine: ~ **treatment** *n* PARTS alodinage *m*
ALOHA: ~ **system** *n* RADIO, TELEP système ALOHA *m*
ALPD *abbr* (*axial lateral plasma deposition*) ELECT DALP (*dépôt axial latéral au plasma*)
alpha: ~ **dial** *n* COMP cadran alpha *m*, cadran de commande *m*; ~ **mosaic** *n* TEST alpha mosaïque *f*; ~ **trial** *n* TEST essai en laboratoire *m*, test en laboratoire *m*
alphabet *n* DATA alphabet *m*
alphabetic[1] *adj* COMP alphabétique
alphabetic[2]: ~ **code** *n* COMP code alphabétique *m*; ~ **data** *n pl* COMP données alphabétiques *f pl*; ~ **key** *n* COMP touche alphabétique *f*; ~ **signal** *n* SIG signal alphabétique *m*, signal d'alphabet *m*; ~ **telegraphy** *n* TELEP télégraphie alphabétique *f*
alphageometric[1] *adj* TV alphagéométrique
alphageometric[2]: ~ **display** *n* TELEP affichage alphagéométrique *m*
alphamosaic *adj* TV alphamosaïque
alphanumeric: ~ **display** *n* DATA, ELECT affichage alphanumérique *m*, TV afficheur alphanumérique *m*; ~ **keyboard** *n* COMP clavier alphanumérique *m*; ~ **pager** *n* RADIO récepteur alphanumérique d'appel de personnes *m*
alphaphotographic *adj* TV alphaphotographique
ALS *abbr* TRANS (*automatic laser shutdown*) coupure automatique du laser *f*, (*application-layer structure*) structure de la couche d'application *f*
alt. *abbr* (*alternative*) DATA *Internet* alt. (*alternative*)
alter[1] *n* COMP mode de modification *m*
alter[2] *vt* COMP changer, transformer
alterable *adj* COMP, DATA, PER APP programmable
alteration *n* COMP changement *m*, modification *f*, transformation *f*
alternate[1]: with ~ **handling** *adj* RADIO à l'alternat
alternate[2] *n* GEN débordement *m*; ~-**access channel** TELEP voie d'accès de deuxième choix *f*, voie d'accès de repli *f*, voie d'accès de secours *f*; ~ **bit inversion** TRANS inversion de bits alternée *f*; ~ **control sequence memory** RADIO mémoire de séquence de commande suppléante *f*; ~ **digit inversion** TRANS inversion alternée des éléments numériques *f*; ~ **final trunk group** ELECT, SWG faisceau de dernier choix *m*, TELEP faisceau de circuits de dernier choix *m*, faisceau de dernier choix *m*; ~ **group** COMP groupe de remplacement *m*; ~ **mark inversion code** SIG, TELEP, TRANS code AMI *m*; ~ **mark inversion signal** (*AMI*) SIG, TELEP, TRANS signal bipolaire alternant *m*; ~ **mark inversion violation** TELEP violation de l'alternance des polarités *f*, violation de la règle de bipolarité *f*; ~ **route** TELEP voie d'acheminement détournée *f*; ~ **route final trunk group** ELECT, SWG faisceau de dernier choix *m*, TELEP faisceau de circuits de dernier choix *m*, faisceau de dernier choix *m*; ~ **speech service** APPL *ISDN* service mixte parole *m*; ~ **transmission** TELEP alternat *m*; ~ **voice-data channel** TRANS voie de téléphonie/données à l'alternat *f*; ~ **voice-data circuit** (*AVD*) TRANS circuit de téléphonie/données à l'alternat *m*; ~ **voice-data transmission** TRANS transmission de téléphonie/données à l'alternat *f*
alternated *adj* RADIO alterné
alternating: ~ **current** *n* (*AC*) POWER courant alternatif *m* (*CA*); ~ **current coupler** *n* POWER coupleur alternatif *m*; ~ **current generator** *n* POWER alternateur *m*; ~ **current outlet** *n* COMP prise de courant *f*; ~ **current power cord receptacle** *n* COMP prise de cordon secteur *f*; ~ **current power outlet** *n* COMP prise secteur *f*; ~ **current power supply** *n* COMP, POWER source d'alimentation *f*; ~ **current pulsing** *n* TELEP numérotation en courant alternatif *f*; ~ **current signaling** *AmE*, ~ **current signalling** *BrE n* POWER, SIG signalisation à courant alternatif *f*); ~ **current socket** *n* COMP prise d'alimentation secteur *f*; ~ **current source** *n* POWER source de courant alternatif *f*; ~ **discharge current** *n* ELECT courant alternatif de décharge *m*; ~ **voltage** *n* POWER tension alternative *f*
alternation: ~ **on inquiry** *n* TELEP va-et-vient *m*
alternative *n* (*alt.*) DATA alternative *f* (*alt.*); ~ **route** SIG, TELEP, TRANS voie d'acheminement détournée *f*, route détournée *f*; ~ **routing** GEN débordement *m*, SIG, TELEP acheminement détourné *m*; ~ **routing indicator** TELEP indicateur d'acheminement détourné *m*; ~ **routing of signaling** *AmE*, ~ **routing of signalling** *BrE* SIG, TELEP, TRANS acheminement de signalisation de secours *m*; ~ **test method** TRANS méthode de mesure de remplacement *f*
altimeter *n* TELEP, TEST altimètre *m*
altitude: ~ **of the apogee** *n* RADIO altitude de l'apogée *f*
ALU *abbr* (*arithmetic and logic unit*) COMP UAL (*unité arithmétique et logique*)
AM *abbr* COMP (*active matrix*) matrice active *f*, (*amplitude modulation*) MA (*modulation d'amplitude*), DATA, ELECT, SIG, TRANS (*amplitude modulation*) MA (*modulation d'amplitude*)
amateur: ~ **radio service** *n* RADIO service radio amateur *m*; ~ **satellite service** *n* RADIO service d'amateur par satellite *m*
ambient: ~ **lighting** *n* COMP éclairage ambiant *m*; ~ **noise** *n* COMP, TELEP bruit d'ambiance *m*, bruit ambiant *m*; ~ **temperature** *n* COMP température ambiante *f*
American: ~ **National Standards Institute** *n* (*ANSI*) GEN Institut américain de normalisation *m* (*ANSI*); ~ **National Standards Institute committee 12** *n* (*ANSI X12*) GEN commission X12 de l'Institut américain de normalisation *f*; ~ **Standard Code for Information Interchange** *n* (*ASCII*) COMP, DATA American Code for Information Interchange *m* (*ASCII*)
AMI *abbr* (*alternate mark inversion signal*) SIG, TELEP, TRANS signal bipolaire alternant *m*, code AMI *m*; ~ **code** *n* TRANS code AMI *m*; ~ **violation** *n* TRANS violation de bipolarité *f*, violation de la règle de polarité *f*
aminoplastic *n* PARTS aminoplaste *m*
ammeter *n* TEST ampèremètre *m*
amorphous: ~ **channel** *n* TRANS canal amorphe *m*
amortization *n* GEN amortissement *m*
amortize *vt* GEN amortir
amperage *n* POWER ampérage *m*, intensité d'un courant électrique *f*
ampere *n* POWER ampère *m*; ~ **rating** ELECT intensité de courant *f*; ~ **turn** POWER ampère-tour *m*
amplification *n* COMP, ELECT, POWER, RADIO, TRANS amplification *f*; ~ **stage** ELECT étage d'amplification *m*; ~ **system** COMP système d'amplification *m*
amplified: ~ **circuit** *n* TRANS circuit amplifié *m*; ~ **handset** *n*

PER APP, TELEP combiné téléphonique à écoute amplifiée *m*; ~ **reception** *n* PER APP écoute amplifiée *f*; ~ **telephone** *n* PER APP téléphone à réception amplifiée *m*

amplifier: ~ **circuit** *n* POWER circuit amplificateur *m*, montage amplificateur *m*; ~ **stage** *n* ELECT, POWER étage d'amplificateur *m*

amplify *vt* COMP, ELECT, POWER, RADIO, TRANS amplifier

amplitude[1]: ~**-modulated** *adj* TRANS modulé en amplitude

amplitude[2] *n* GEN amplitude *f*; ~**-corrected echo** TELEP écho corrigé en amplitude *m*; ~ **curve** COMP courbe d'amplitude *f*; ~ **distortion** TRANS distorsion d'amplitude *f*; ~ **equalization** TRANS égalisation d'amplitude *f*; ~ **equalizer** COMP compensateur d'amplitude *m*, TRANS égaliseur d'amplitude *m*; ~ **error** COMP erreur d'amplitude *f*; ~ **frequency response** TRANS réponse amplitude-fréquence *f*; ~ **keying** TELEP modulation télégraphique d'amplitude *f*, manipulation d'amplitude *f*; ~ **modulation** (*AM*) COMP, DATA, ELECT, SIG, TRANS modulation d'amplitude *f* (*MA*); ~ **modulation depth** TRANS taux de modulation d'amplitude *m*; ~ **modulation sensitivity** COMP sensibilité à la modulation d'amplitude *f*; ~ **phase-shift keying** (*APSK*) SIG modulation à états d'amplitude-phase *f*; ~ **probability distribution** (*APD*) SIG distribution de probabilité des amplitudes *f* (*DPA*), répartition de probabilité des amplitudes *f* (*RPA*); ~**-quantized control** TELEP mode à quantification d'amplitude *m*; ~ **reflection factor** TRANS facteur de réflexion complexe *m*; ~ **response** ELECT, SIG, TRANS réponse en amplitude *f*; ~ **response limit** TRANS gabarit d'amplitude *m*; ~**-shift keying** (*ASK*) RADIO, SIG modulation par déplacement d'amplitude *f* (*MDA*)

AMTS *abbr* (*automatic mobile telephone system*) RADIO, TELEP SRMA (*système radiotéléphonique mobile automatique*)

AN *abbr* (*announcement trunk group*) TELEP faisceau de messages enregistrés *m*

anaerobic: ~ **adhesive** *n* ELECT adhésif anaérobique *m*

analog[1] *adj* GEN analogique

analog[2]: ~ **adder** *n* ELECT additionneur analogique *m*, sommateur *m*; ~ **call processor** *n* COMP processeur analogique des appels *m*; ~ **circuit** *n* COMP, TRANS circuit analogique *m*; ~ **color code** *AmE*, ~ **colour code** *BrE* *n* TELEP code couleur analogique *m*; ~ **computer** *n* COMP ordinateur analogique *m*; ~ **control** *n* TELEP, TRANS mode analogique *m*; ~ **converter** *n* COMP, DATA, ELECT, SIG, TELEP convertisseur analogique *m*; ~ **data** *n pl* COMP données analogiques *f pl*; ~ **device** *n* COMP dispositif analogique *m*, GEN appareil analogique *m*; ~ **display** *n* DATA, ELECT affichage analogique *m*; ~ **filter** *n* COMP, TRANS filtre analogique *m*; ~ **input** *n* COMP entrée analogique *f*; ~ **interface** *n* TRANS interface analogique *f*; ~ **line board** *n* (*ALB*) SWG carte de ligne analogique *f*; ~ **measuring instrument** *n* TELEP, TEST appareil de mesure analogique *m*; ~ **modulation** *n* COMP modulation analogique *f*; ~ **private wire** *n* TRANS liaison spécialisée à tous usages *f*; ~ **recording** *n* GEN enregistrement analogique *m*; ~ **repeater** *n* TELEP répéteur analogique *m*; ~ **signal** *n* COMP, SIG signal analogique *m*; ~ **sound** *n* COMP son analogique *m*; ~ **switching system** *n* SWG système de commutation analogique *m*; ~ **synthesizer** *n* COMP synthétiseur analogique *m*; ~ **system** *n* GEN système analogique *m*; ~ **transmission** *n* TRANS transmission analogique *f*; ~ **transmission link** *n* TRANS liaison de transmission analogique *f*; ~ **video** *n* COMP vidéo analogique *f*

analog-digital[1] *adj* (*A/D*) COMP, DATA, ELECT, SIG, TELEP analogique-numérique

analog-digital[2]: ~ **converter** *n* (*ADC*) COMP, DATA, ELECT, SIG, TELEP convertisseur analogique-numérique *m* (*CAN*, *convertisseur A/N*)

analogical: ~ **learning** *n* COMP apprentissage analogique *m*; ~ **reasoning** *n* DATA raisonnement analogique *m*

analog-to-digital[1] *adj* (*A/D*) COMP, DATA, ELECT, SIG, TELEP analogique-numérique (*A/N*)

analog-to-digital[2]: ~ **converter** *n* (*ADC*) COMP, DATA, ELECT, SIG, TELEP convertisseur analogique-numérique *m* (*CAN*, *convertisseur A-N*)

analyser *n* *BrE* ELECT, SIG analyseur *m*

analysis: ~ **filterbank** *n* COMP banc de filtres d'analyse *m*; ~ **for number length** *n* TELEP analyse de longueur de numéro *f*; ~ **meter** *n* TELEP analyseur de débordement *m*

analyzer *AmE see analyser BrE*

anamorphosis *n* GEN anamorphose *f*

anchor[1] *n* COMP point d'ancrage *m*; ~ **MSC** RADIO CCM de rattachement *m*, CCM nominal *m*; ~ **point** COMP point d'ancrage *m*

anchor[2] *vt* ELECT fixer

anchoring: ~ **bracket** *n* TRANS anneau d'ancrage *m*

ancillary[1] *adj* GEN annexe

ancillary[2]: ~ **unit** *n* COMP unité auxiliaire *f*

AND: **not** ~ *n* (*NAND*) ELECT NON-ET *m*; ~ **circuit** *n* ELECT circuit ET *m*; ~ **gate** *n* ELECT porte ET *f*; ~ **operation** *n* ELECT opération ET *f*; ~ **parallelism** *n* DATA parallélisme ET *m*

AND-OR: ~ **graph** *n* DATA graphe ET-OU *m*; ~ **processing** *n* DATA traitement ET-OU *m*

anechoic: ~ **chamber** *n* TEST chambre sourde *f*, salle anéchoïque *f*; ~ **room** *n* TEST chambre sourde *f*, salle anéchoïque *f*

angel *n* RADIO *radar* écho *m*, écho fantôme *m*

angle: ~ **of arrival** *n* TRANS angle d'arrivée *m*; ~ **from the axis** *n* RADIO angle par rapport à l'axe *m*; ~ **of azimuth** *n* RADIO angle d'azimut *m*, angle de gisement *m*; ~ **of departure** *n* TRANS angle de départ *m*; ~ **of depression** *n* RADIO angle de dépression *m*; ~ **diversity link** *n* DATA liaison à diversité angulaire *f*; ~ **of elevation** *n* RADIO, TRANS angle d'élévation *m*, angle de site *m*; ~ **encoder** *n* DATA codeur angulaire *m*; ~ **of incidence** *n* TRANS angle d'incidence *m*; ~ **modulation** *n* TRANS modulation angulaire *f*; ~ **of rotation** *n* TRANS angle de rotation *m*; ~ **of sight** *n* RADIO angle de visée *m*; ~ **of site** *n* TRANS angle de site *m*; ~ **of tilt** *n* RADIO angle d'inclinaison *m*

angular: ~ **displacement** *n* TELEP décalage angulaire *m*, déphasage angulaire *m*, position angulaire *f*, TRANS désalignement angulaire des axes de fibres *m*, inclinaison axiale *f*; ~ **misalignment loss** *n* TRANS perte par désalignement angulaire *f*, perte par inclinaison axiale *f*; ~ **momentum** *n* RADIO moment cinétique *m*; ~ **resolution** *n* RADIO résolution angulaire *f*; ~ **separation** *n* RADIO espacement angulaire *m*, TRANS séparation angulaire *f*

anisochronous[1] *adj* DATA, TRANS *digital transmission* anisochrone

anisochronous[2]: ~ **data** *n pl* DATA signaux de données anisochrones *m pl*; ~ **data transmission** *n* DATA, TRANS transmission de données anisochrone *f*

anisotropic *adj* TRANS *fibre optics* anisotrope

annealed: ~ **wire** *n* PARTS *heat treatment* fil recuit *m*

annotation *n* COMP, TELEP annotation *f*; ~ **button** COMP bouton de note *m*; ~ **symbol** TELEP symbole d'annotation *m*

announcement: ~ **machine** *n* PER APP répondeur téléphonique *m*, TELEP répondeur *m*; ~**-only facility** *n* PER APP *answer machine*, TELEP utilisation en répondeur simple *f*; ~ **trunk** *n* TELEP circuit d'annonces parlées *m*; ~ **trunk group** *n* (*AN*) TELEP faisceau de messages enregistrés *m*

annual: ~ **charge ratio** *n* TELEP rapport des charges financières annuelles *m*; ~ **subscription** *n* APPL, GEN abonnement annuel *m*

annunciator *n* PER APP portier automatique *m*

anode *n* ELECT anode *f*; ~ **battery** POWER batterie anodique *f*; ~**-cathode characteristics** ELECT *semiconductors* caractéristique tension-courant d'anode-cathode *f*; ~**-cathode voltage** ELECT *semiconductors* tension de l'anode par rapport à la cathode *f*; ~ **current** ELECT courant anodique *m*, courant plaque *m*; ~**-potential-stabilized camera tube** ELECT tube analyseur à électrons rapides *m*, tube analyseur à potentiel stabilisé sur l'anode *m*; ~ **terminal** ELECT *semiconductors* borne d'anode *f*; ~ **voltage** ELECT tension anodique *f*

anodic: ~ **reaction** *n* PARTS réaction anodique *f*; ~ **treatment** *n* PARTS traitement anodique *m*

anomalistic: ~ **period** *n* RADIO période anomalistique *f*

anomalous: ~ **dispersion** *n* TRANS *fibre optics* dispersion anomale *f*

ANS *abbr* (*answer tone*) DATA tonalité de réponse *f*

ANSI *abbr* (*American National Standards Institute*) GEN ANSI (*Institut américain de normalisation*); ~ **X12** (*American National Standards Institute committee 12*) GEN commission X12 de l'Institut américain de normalisation *f*

answer[1] *n* GEN réponse *f*; ~**-back** DATA identification automatique *f*, TELEP indicatif *m*; ~**-back code** DATA, TELEP *teleprinter* indicatif *m*; ~**-back signal** DATA, SIG signal d'indicatif *m*; ~**-back unit** DATA, TELEP émetteur automatique d'indicatif *m*; ~**-back unit simulator** TELEP simulateur d'émetteur d'indicatif *m*; ~ **code** TELEP code de réponse *m*; ~ **indication** TELEP indication de réponse *f*; ~ **jack** TELEP jack de réponse *m*; ~ **lamp** TELEP lampe de réponse *f*; ~ **machine function** PER APP, TELEP fonction répondeur *f*; ~ **machine start-up adjustment** PER APP, TELEP réglage du déclenchement du répondeur *m*; ~**-recording unit** TELEP répondeur enregistreur *m*; ~**-seizure ratio** TELEP taux de prises avec réponse *m*; ~ **signal** SIG, TELEP signal de réponse *m*; ~ **signal, charge** SIG signal de réponse, avec taxation *m*; ~ **signal, no charge** SIG signal de réponse, sans taxation *m*; ~ **tone** (*ANS*) DATA tonalité de réponse *f*

answer[2] *vt* TELEP *call* répondre à

answer[3] *vi* DATA, GEN, TELEP répondre

answering *n* DATA réponse *f*; ~ **delay** TELEP délai de réponse *m*; ~ **device with voice synthesizer** TELEP répondeur simple et à interrogation à distance *m*; ~ **equipment** TELEP équipement de réponse *m*; ~ **modem** DATA modem en mode réponse *m*; ~ **position** TELEP position de réponse *f*; ~ **service** APPL, TELEP service des abonnés absents *m*; ~ **time** TELEP délai de réponse *m*

answerphone *n* PER APP, TELEP téléphone répondeur-enregistreur *m*

ANT *abbr* (*antenna*) RADIO ant. (*antenne*)

antenna *n* (*ANT*) RADIO antenne *f* (*ant.*); ~ **building** RADIO logement d'antenne *m*; ~ **butterfly** RADIO papillon *m*; ~ **configuration** RADIO structure d'antenne *f*; ~ **current** RADIO courant d'antenne *m*; ~ **curtain** RADIO antenne rideau *f*, rideau d'antennes *m*; ~ **directivity diagram** RADIO diagramme de directivité d'antenne *m*; ~ **directivity factor** RADIO coefficient de directivité d'antenne *m*; ~ **discrimination** RADIO discrimination d'antenne *f*; ~ **drive motor** RADIO moteur d'entraînement d'antenne *m*; ~ **effect** RADIO effet d'antenne *m*; ~ **effective area** RADIO surface équivalente d'antenne *f*; ~ **factor** RADIO facteur d'antenne *m*; ~ **farm** RADIO complexe d'antenne *m*, parc d'antennes *m*, téléport *m*; ~ **fed out of phase** RADIO antenne à alimentation déphasée *f*; ~ **feed** RADIO alimentation d'antenne *f*, source d'antenne *f*; ~ **gain** RADIO gain d'antenne *m*; ~ **gain pattern** RADIO diagramme de gain d'antenne *m*, diagramme de gain de référence *m*; ~ **input impedance** RADIO impédance d'entrée d'antenne *f*; ~ **lead-in** RADIO, TELEP descente d'antenne *f*; ~**-loading coil** RADIO bobine de charge d'antenne *f*; ~**-pointing loss** RADIO affaiblissement dû au pointage d'antenne *m*; ~ **polarization** RADIO polarisation d'antenne *f*; ~ **positioning** RADIO pointage d'antenne *m*; ~ **power gain** RADIO gain en puissance d'antenne *m*; ~ **radiation** RADIO rayonnement d'antenne *m*; ~ **radiation pattern** RADIO diagramme de rayonnement d'antenne *m*; ~ **range** RADIO terrain d'essai d'antennes *m*; ~**-receiver unit** RADIO ensemble antenne-récepteur *m*; ~ **remote control** RADIO télécommande d'antenne *f*; ~ **resistance** RADIO résistance d'antenne *f*; ~ **side lobe** RADIO lobe latéral d'antenne *m*; ~ **spacing** RADIO espacement d'antenne *m*; ~ **spacing error** RADIO erreur d'espacement d'antenne *f*; ~ **steerability** RADIO orientabilité d'antenne *f*; ~ **steering** RADIO orientation d'antenne *f*, pointage d'antenne *m*; ~ **structure** RADIO bâti d'antenne *m*; ~ **supporting structure** RADIO bâti d'antenne *m*; ~ **swing off** RADIO dépointage d'antenne *m*; ~ **switching circuit** RADIO circuit de commutation d'antennes *m*; ~ **switching relay** RADIO relais de commutation d'antennes *m*; ~ **TDRSS** RADIO système de poursuite d'antenne *m*; ~**-to-medium coupling loss** RADIO perte par couplage antenne-milieu *f*; ~ **tuning** RADIO accord d'antenne *m*; ~ **tuning unit** (*ATU*) RADIO dispositif d'accord d'antenne *m*; ~ **width** RADIO largeur angulaire *f*, ouverture angulaire *f*

anti-aliasing[1] *adj* COMP, TRANS antirepliement

anti-aliasing[2]: ~ **filter** *n* TRANS filtre antirepliement *m*

anti-glare: ~ **filter** *n* DATA filtre antireflet *m*

anti-jamming *n* (*AJ*) RADIO antibrouillage *m*

anti-magnetic: ~ **shield** *n* DATA blindage antimagnétique *m*

anti-phase *n* POWER opposition de phase *f*

anti-radiation: ~ **filter** *n* DATA filtre antirayonnement *m*

anti-reflection[1] *adj* (*AR*) GEN antiréflexion (*AR*)

anti-reflection[2]: ~ **coating** *n* DATA antireflet *m*, TRANS *fibre optics* couche antireflets *f*

anti-reflective: ~ **coating** *n* GEN revêtement antiréfléchissant *m*

anti-shock: ~ **device** *n* TELEP dispositif anti-chocs *m*

anti-sidetone[1] *adj* TELEP antilocal

anti-sidetone[2]: ~ **circuit** *n* TELEP circuit antilocal *m*; ~ **device** *n* TELEP dispositif antilocal *m*

anti-sloshing *adj* RADIO anti-ballottant

anti-static[1] *adj* ELECT antistatique

anti-static[2]: ~ **carpet** *n* ELECT moquette antistatique *f*; ~ **mat** *n* ELECT tapis antistatique *m*; ~ **plastic** *n* ELECT plastique antistatique *m*; ~ **wristband** *n* ELECT bracelet antistatique *m*

anti-virus: ~ **program** *n* COMP programme antivirus *m*

anyvalue: ~ **expression** *n* DATA expression à valeur aléatoire *f*

AO *abbr* (*acousto-optical*) ELECT acousto-optique

AOQ *abbr* (*average outgoing quality*) TEST qualité moyenne après contrôle *f*

AOQL *abbr* (*average outgoing quality limit*) TEST limite de qualité moyenne après contrôle *f*

AP *abbr* DATA, TELEP (*application processor*) PA (*processeur d'application*), (*application process*) PA (*processus d'application*)

APB *abbr* (*all-paths busy*) TELEP circuits tous occupés; ~ **tone** *n* TELEP signal tous circuits occupés *m*, tonalité circuits tous occupés *f*

APC *abbr* (*automatic power control*) POWER commande automatique de puissance *f*

APCI *abbr* (*application protocol control information*) DATA, TELEP informations de contrôle du protocole d'application *f pl*

APD *abbr* DATA, PARTS (*avalanche photodiode*) PDA (*photodiode à avalanche*), SIG (*amplitude probability distribution*) DPA (*distribution de probabilité des amplitudes*), RPA (*répartition de probabilité des amplitudes*), TRANS (*avalanche photodiode*) PDA (*photodiode à avalanche*)

APDU *abbr* (*application protocol data unit*) DATA, TELEP APDU (*unité de données de protocole d'application*)

aperiodic[1] *adj* GEN *physics* apériodique

aperiodic[2]: ~ **aerial** *n* RADIO antenne apériodique *f*; ~ **antenna** *n* RADIO antenne apériodique *f*

aperture *n* COMP ouverture *f*, embouchure *f*, GEN ouverture *f*, RADIO surface d'ouverture *f*, *aerials* ouverture *f*; ~ **aerial** RADIO antenne à ouverture *f*, antenne à ouverture rayonnante *f*; ~ **antenna** RADIO antenne à ouverture *f*, antenne à ouverture rayonnante *f*; ~ **illumination** RADIO illumination de l'ouverture *f*

apex *n* COMP, GEN crête *f*

API *abbr* (*application program interface*) COMP, DATA IPA (*interface de programmes d'applications*)

apoapsis *n* RADIO apoastre *m*, apoapside *m*, apside *f*

apogee *n* RADIO apogée *m*

apparatus *n* ELECT, dispositif *m*, COMP device; ~ **charge rate** RADIO tarif des terminaux *m*

apparent: ~ **height** *n* RADIO hauteur apparente d'antenne *f*

append *vt* COMP ajouter

appliance *n* GEN dispositif *m*

applicable: **not** ~ *phr* (*N/A*) DATA, GEN, TELEP non applicable (*NA*)

applicant *n* GEN postulant *m*

application *n* GEN application *f*; ~ **appliance label** APPL étiquette verte *f*; ~ **association** DATA association d'application *f*; ~ **entity** (*AE*) TELEP entité d'application *f*; ~**-entity invocation** (*AEI*) DATA appel de l'entité d'application *f*; ~**-entity invocation identifier** DATA identificateur d'invocation d'entité d'application *m*; ~**-entity qualifier** DATA qualificateur d'entité d'application *m*; ~**-entity title** DATA appellation AE *f*, appellation d'entité d'application *f*, titre de l'entité d'application *m*; ~ **function** DATA fonction d'application *f*; ~ **generator** DATA générateur d'application *m*; ~ **layer** APPL *open systems interconnection*, DATA, TRANS couche application *f*; ~**-layer structure** (*ALS*) TRANS structure de la couche d'application *f*; ~**-level gateway** TELEP passerelle de niveau application *f*; ~ **process** (*AP*) DATA, TELEP processus d'application *m* (*PA*); ~ **process invocation** DATA invocation de processus d'application *f*; ~ **process invocation identifier** DATA identificateur d'invocation de processus d'application *m*; ~ **process title** DATA appellation AP *f*, appellation de processus d'application *f*, titre de processus d'application *m*; ~ **process type** DATA type de processus d'application *m*; ~ **processor** (*AP*) DATA, TELEP processeur d'application *m* (*PA*); ~ **program interface** (*API*) COMP, DATA interface de programmes appliqués *f*, interface de programmes d'applications *f* (*IPA*); ~ **protocol control information** (*APCI*) DATA, TELEP informations de contrôle du protocole d'application *f pl*; ~ **protocol data unit** (*APDU*) DATA, TELEP unité de données de protocole d'application *f* (*APDU*); ~ **reference** DATA, GEN référence d'application *f*; ~ **service element** (*ASE*) DATA, TELEP élément de service d'application *m* (*ASE*); ~ **service object** (*ASO*) DATA, TELEP objet de service d'application *m*; ~**-specific integrated circuit** (*ASIC*) ELECT, PARTS circuit intégré spécifique *m* (*ASIC*); ~**-supported distribution transaction** DATA transaction répartie assurée par l'application *f*; ~ **system** SWG, TELEP système d'application *m*; ~ **technology satellite** RADIO satellite d'application technologique *m*

applied: ~ **data bit** *n* DATA, TELEP bit de données appliqué *m*; ~ **data-bit scrambling process** *n* DATA, TELEP bit de données appliqué en brouillage *m*; ~ **research** *n* TEST recherche appliquée *f*; ~ **voltage** *n* POWER tension appliquée *f*

applique: ~ **model** *n* TELEP module complémentaire *m*

apply *vt* COMP, POWER appliquer, attribuer

appraisal: ~ **cost** *n* TEST coût d'évaluation *m*

approach *n* GEN méthode *f*

approved[1] *adj* GEN accepté

approved[2]: ~ **connecting agent** *n* TELEP dispositif de raccordement homologué *m*

approximate: ~ **reasoning** *n* DATA raisonnement approximatif *m*

approximation: ~ **error** *n* GEN erreur d'approximation *f*

APS *abbr* (*automatic protection switching*) SWG commutation de protection automatique *f*

APSK *abbr* (*amplitude phase-shift keying*) SIG modulation à états d'amplitude-phase *f*

APT *abbr* (*Asia-Pacific Telecommunity*) RADIO APT (*Télécommunauté Asie-Pacifique*)

AR *abbr* (*anti-reflection*) GEN AR (*antiréflexion*)

Arab: ~ **Satellite Communications Organization** *n* (*ARABSAT*) RADIO Organisation arabe de communications par satellite *f*

ARABSAT *abbr* (*Arab Satellite Communications Organization*) RADIO Organisation arabe de communications par satellite *f*

aramid: ~ **yarn** *n* ELECT fils d'aramide *m pl*

arc *n* POWER arc; ~ **current** ELECT courant d'arc *m*

arcade: ~ **games** *n pl* COMP jeux d'arcade *m pl*

Archie *n* DATA *Internet* Archie *m*

architecture *n* COMP architecture *f*

archivage *n* DATA archival *m*

archive[1]: ~ **file** *n* COMP, DATA fichier d'archives *m*; ~ **material** *n* TV archive *f*

archive[2] *vt* COMP, DATA, TV archiver

archives *n pl* COMP, DATA, TV archives *f pl*

archiving *n* COMP, DATA, TV archivage *m*

area *n* COMP, DATA zone *f*; ~ **code** TELEP indicatif de zone *m*, indicatif régional *m*; ~ **of propagation** ELECT, GEN *physics* aire de propagation *f*

argument *n* DATA argument *m*; ~ **of perigee** RADIO argument du périgée *m*

arithmetic: ~ **coding** *n* COMP, DATA codage arithmétique *m*; ~ **delimiter** *n pl* COMP délimiteur arithmétique *m*; ~ **expression** *n* COMP expression arithmétique *f*; ~ **and logic unit** *n* (*ALU*) COMP unité arithmétique et logique *f* (*UAL*)

armature: ~ **stick** *n* ELECT collage d'armature *m*; ~ **winding** *n* POWER enroulement d'induit *m*

armchair: ~ **shopping** *n* GEN, TELEP, TV achat à domicile *m*, téléachat *m*

armor *AmE see armour BrE*

armored *AmE see armoured BrE*

armoring *AmE see armouring BrE*

armour[1] *n BrE* PARTS *cables* armure *f*

armour[2] *vt BrE* TRANS *cables* armer

armoured[1] *adj BrE* GEN blindé

armoured[2]: ~ **cable** *n BrE* COMP, ELECT, TRANS câble armé *m*, câble blindé *m*

armouring *n BrE* TRANS *cables* armure *f*

arpeggiator *n* COMP arpégiateur *m*, générateur d'arpèges *m*

arpeggio *n* COMP arpège *m*

ARR *abbr* (*automatic rerouting*) TELEP réacheminement automatique *m*

arranger *n* COMP arrangeur *m*

array *n* COMP tableau *m*, DATA réseau *m*, tableau *m*, GEN matrice *f*; ~ **aerial** RADIO antenne réseau *f*; ~ **antenna** RADIO antenne réseau *f*; ~ **connection** ELECT connection rainuré *f*; ~ **connectors** *n pl* ELECT connecteurs rainurés *m pl*; ~ **hinge** RADIO pivot du générateur *m*; ~ **processor** COMP processeur vectoriel *m*, DATA processeur matriciel *m*

arrow *n* COMP, GEN flèche *f*

ARS *abbr* (*automatic route selection*) TELEP SAA (*sélection automatique d'acheminement*)

ARSWQ *abbr* (*automatic route selection with queueing*) TELEP SAAFA (*sélection automatique d'acheminement avec file d'attente*)

articulation *n* RADIO, TELEP, TRANS netteté phonétique *f*; ~ **index** (*AI*) RADIO, TELEP, TRANS indice de netteté *m* (*IN*); ~ **reference equivalent loss** TELEP affaiblissement équivalent pour la netteté *m* (*AEN*); ~ **score** RADIO appréciation de la netteté *f*

artificial: ~ **ageing** *n* TEST *dependability* vieillissement artificiel *m*; ~ **aerial** *n* RADIO antenne fictive *f*; ~ **antenna**

n RADIO antenne fictive *f*; ~ **intelligence** *n* (*AI*) APPL, COMP, SWG intelligence artificielle *f* (*IA*); ~ **language** *n* COMP langage artificiel *m*; ~ **line** *n* TRANS ligne artificielle *f*, simulateur de câble *m*; ~ **load** *n* RADIO charge fictive *f*; ~ **noise** *n* TEST bruit artificiel *m*; ~ **satellite** *n* DATA, RADIO satellite artificiel *m*; ~ **traffic** *n* TELEP trafic artificiel *m*; ~ **voice** *n* COMP voix artificielle *f*

artwork: ~ **master** *n* POWER *printed circuits* dessin modèle *m*

AS *abbr* (*address signal*) TELEP signal d'adresse *m*

ascender *n* COMP hampe *f*

ascending: ~ **node** *n* RADIO nœud ascendant *m*

ascent: ~ **phase** *n* RADIO phase ascensionnelle *f*

ASCII *abbr* (*American Standard Code for Information Interchange*) COMP, DATA, ASCII (*American Standard Code for Information Interchange*); ~ **data** *n pl* COMP données ASCII *f pl*

ASDE *abbr* (*airport surface detection equipment*) RADIO radar de pistes *m*

ASE *abbr* (*application service element*) DATA, TELEP ASE (*élément de service d'application*)

Asia-Pacific: ~ **Broadcasting Union** *n* (*ABU*) RADIO Union de radiodiffusion Asie-Pacifique *f*; ~ **Telecommunity** *n* (*APT*) RADIO Télécommunauté Asie-Pacifique *f* (*APT*)

ASIC *abbr* (*application-specific integrated circuit*) ELECT, PARTS ASIC (*circuit intégré spécifique*)

ASK *abbr* (*amplitude-shift keying*) RADIO, SIG MDA (*modulation par déplacement d'amplitude*)

ASM *abbr* (*concentration module*) ELECT module concentrateur *m*

ASN1 *abbr* (*abstract syntax notation one*) DATA, TELEP, TRANS notation de syntaxe abstraite numéro un *f*

ASO *abbr* (*application service object*) DATA, TELEP objet de service d'application *m*

ASP *abbr* (*abstract service primitive*) DATA ASP (*primitive de service abstraite*)

aspect: ~ **ratio** *n* RADIO, TELEP facteur d'aspect *m*

ASP(N) *abbr* (*abstract (N) service primitive*) DATA ASP(N) (*primitive de service abstraite (N)*)

assault: ~ **alarm device** *n* TEST *security* appareil d'alarme anti-agression *m*; ~ **alarm installation** *n* TEST *security* installation d'alarme anti-agression *f*

assembler *n* COMP assembleur *m*, programme d'assemblage *m*

assembling *n* GEN assemblage; ~ **time** COMP stade d'assemblage *m*

assembly *n* GEN assemblage *m*, montage *m*; ~ **area** GEN zone d'assemblage *f*; ~ **language** COMP langage assembleur *m*, langage d'assemblage *m*; ~ **program** COMP assembleur *m*, programme d'assemblage *m*

asserted *adj* TELEP actif

assign *vt* COMP allouer, attribuer, RADIO, TELEP attribuer

assigned: ~ **frequency** *n* RADIO fréquence assignée *f*; ~**-frequency band** *n* RADIO bande de fréquence assignée *f*

assignment *n* COMP affectation *f*, GEN affectation *f*, assignation *f*, TELEP *of classes of service* attribution *f*; ~ **channel** TRANS voie d'affectation *f*, voie d'assignation *f*; ~ **map** DATA carte d'affectation *f*; ~ **message** DATA message d'assignation *m*; ~ **of orbit** RADIO attribution de l'orbite *f*; ~ **statement** DATA instruction d'affectation *f*

assistant: ~ **operator position** *n* TELEP position auxiliaire *f*, position d'opérateur d'assistance *m*; ~ **operator signal** *n* TELEP signal d'opérateur assistant *m*

associated: ~ **connectors** *n pl* ELECT connecteurs associés *m pl*; ~ **mode** *n* TELEP *of operation* mode associé *m*; ~ **sector** *n* TV secteur associatif *m*; ~ **signaling** *AmE*, ~ **signalling** *BrE* *n* GEN signalisation en mode associée *f*, SIG signalisation associé *f*

association *n* DATA association; ~ **area** DATA zone d'association *f*; ~ **class** DATA classe d'associations *f*; ~ **control protocol machine** (*ACPM*) DATA, TELEP machine de protocole de commande d'association *f*; ~ **control service element** (*ACSE*) DATA, TELEP élément de service de commande d'association *m*, élément de service de contrôle d'association *m*; ~ **initiator** DATA demandeur d'association *m*; ~ **responder** DATA accepteur de l'association *m*, répondeur d'association *m*

associative: ~ **processor** *n* COMP processeur associatif *m*; ~ **storage** *n* COMP mémoire associative *f*

ASSR *abbr* (*acquisition and steady-state reception*) DATA acquisition et réception en régime permanent *f*

AS/SYS *abbr* (*access and storage system*) DATA système d'accès et de mémorisation *m*

astable[1] *adj* ELECT *logic elements* instable

astable[2]: ~ **circuit** *n* ELECT *logic elements* multivibrateur instable *m*

astatic *adj* GEN *physics* astatique

astrionics *n* ELECT électronique spatiale *f*

AST/SYS *abbr* (*access, storage and transfer system*) DATA système d'accès, de mémorisation et de transfert *m*

asymmetrical: ~ **distortion** *n* ELECT distorsion biaise *f*, TELEP distorsion biaise *f*, distorsion dissymétrique *f*; ~ **grading** *n* TELEP multiplage partie asymétrique *m*

asymptomatic: ~ **approximation** *n* GEN approximation asymptomatique *f*

asymptotic: ~ **availability** *n* TELEP, TEST *dependability* disponibilité asymptotique *f*; ~ **availability unavailability** *n* TELEP disponibilité indisponibilité asymptotique *f*

asynchronous[1] *adj* COMP, DATA, TRANS asynchrone

asynchronous[2]: ~ **balanced mode** *n* (*ABM*) TELEP mode asynchrone symétrique *m*; ~ **circuit** *n* DATA circuit asynchrone *m*; ~ **computer** *n* COMP ordinateur asynchrone *m*; ~ **counter** *n* TEST compteur asynchrone *m*; ~ **data communication** *n* DATA, TRANS transmission asynchrone de données *f*; ~ **data transmission** *n* DATA, TRANS transmission de données asynchrone *f*; ~ **data units** *n pl* (*ADU*) DATA, ELECT unités de données asynchrones *f pl*; ~ **digital subscriber loop** *n* (*ADSL*) TRANS mode de transfert asynchrone *m* (*MTA*); ~ **disconnected mode** *n* TELEP mode déconnecté asynchrone *m*; ~ **format** *n* DATA format asynchrone *m*; ~ **interface module** *n* (*ASM*) ELECT module concentrateur *m* (*ASM*); ~ **link** *n* DATA liaison asynchrone *f*; ~ **motor** *n* POWER moteur à induction *m*, moteur asynchrone *m*; ~ **network** *n* TELEP réseau asynchrone *m*; ~ **operation** *n* TELEP fonctionnement asynchrone *m*; ~ **port** *n* DATA port d'accès asynchrone *m*; ~ **time-division multiplexing** *n* (*ATDM*) COMP, DATA multiplexage temporel asynchrone *m*; ~ **transfer mode** *n* (*ATM*) APPL, COMP, TELEP, TRANS mode de transfert asynchrone *m* (*MTA*); ~ **transmission** *n* TELEP transmission asynchrone *f*, TRANS transmission asynchrone *f*, transmission arythmique *f*

A/SYS *abbr* (*access system*) DATA système d'accès *m*

ATB *abbr* (*all-trunks busy*) TELEP circuits tous occupés; ~ **tone** *n* TELEP signal circuits tous occupés *m*, tonalité circuits tous occupés *f*

ATDM *abbr* (*asynchronous time-division multiplexing*) COMP, DATA multiplexage temporel asynchrone *m*

ATI *abbr* (*average total inspected*) TEST *quality* moyenne totale contrôlée *f*

ATM *abbr* APPL, COMP (*asynchronous transfer mode*) MTA (*mode de transfert asynchrone*), DATA (*abstract test method*) méthode de test abstraite *f*, TELEP (*asynchronous transfer mode*) MTA (*mode de transfert asynchrone*), TEST (*abstract test method*) méthode de test abstraite *f*, TRANS (*asynchronous transfer mode*) MTA (*mode de transfert asynchrone*); ~ **adaptation layer** *n* (*AAL*) DATA, GEN, TELEP, TRANS couche d'adaptation ATM *f*; ~ **cell** *n* COMP, DATA cellule ATM *f*; ~ **connection** *n* COMP connexion ATM *f*; ~ **service data unit** *n* TELEP unité de données de service ATM *f*; ~ **switch** *n* SWG brasseur ATM *m*; ~ **virtual path** *n* APPL

conduit virtuel ATM *m*; ~ **workstation** *n* COMP, DATA, PER APP station de travail ATM *f*
ATME *abbr* (*automatic transmission-measuring equipment*) SWG AAMT (*appareil automatique de mesure de transmission*)
atmosphere *n* GEN atmosphère *f*
atmospheric: ~ **noise** *n* RADIO bruit atmosphérique *m*; ~ **refraction** *n* RADIO réfraction atmosphérique *f*
atomic: ~ **action data** *n pl* DATA données d'action atomique *f pl*; ~ **frequency standards** *n pl* ELECT étalons atomiques de fréquence *m pl*; ~ **standard clock** *n* RADIO horloge atomique *f*; ~ **time scale** *n* RADIO échelle de temps atomique *f*
atomicity *n* DATA atomicité *f*; ~, **consistency, isolation and durability** (*ACID*) DATA atomicité, cohérence, isolation et durabilité *f* (*ACID*)
atonal *adj* COMP atonal
at-once: ~ **recording** *n* COMP enregistrement unique *m*
ATRAC *abbr* (*adaptive transform acoustic coding*) COMP codage par transformation acoustique adaptive *m*
ATS *abbr* DATA, TEST (*abstract test suite*) suite de tests abstraite *f*
AT/SYS *abbr* (*access and transfer system*) DATA système d'accès et de transfert *m*
attach *vt* DATA *Internet* joindre, *scanner* raccorder
ATTACH: ~ **construct** *n* DATA construction ATTACH *f*
attachment *n* TELEP raccordement *m*
attack *n* GEN, COMP attaque *f*; ~ **rate** COMP vitesse à l'attaque *f*; ~ **velocity** COMP vélocité d'attaque *f*
attack *vt* COMP, GEN attaquer
attempt *n* GEN, RADIO, TELEP tentative *f*
attendant *n* TELEP opérateur *m*, standardiste *m*; ~ **access code** TELEP indicatif d'accès au standardiste *m*; ~ **action** TELEP intervention de standardiste *f*; ~ **busy override** TELEP entrée en tiers du standardiste *f*; ~ **call selection** TELEP prise d'appel par le standardiste *f*; ~ **camp-on** TELEP mise en attente par le standardiste *f*; ~ **conference** TELEP conférence établie par standardiste *f*; ~**-controlled conference** TELEP conférence établie par standardiste *f*; ~ **override** TELEP entrée en tiers du standardiste *f*; ~ **position** TELEP poste de standardiste *m*; ~ **recall** TELEP assistance du standardiste *f*, renvoi au standardiste *m*; ~**'s set** TELEP poste téléphonique dirigeur *m*, poste dirigeur *m*, poste central *m*; ~**-status indicator** TELEP indication d'état de la console *f*; ~**-to-attendant calling** TELEP appel entre standardistes *m*
attenuate *vt* COMP, POWER, RADIO, TRANS affaiblir, atténuer
attenuation[1]: ~**-limited** *adj* TRANS *fibre optics* limité par l'affaiblissement
attenuation[2] *n* COMP, POWER, RADIO, TRANS affaiblissement *m*, atténuation *f*; ~ **check** TRANS vérification d'affaiblissement *f*; ~ **coefficient** RADIO, TRANS coefficient d'affaiblissement *m*, coefficient d'atténuation *m*; ~ **constant** RADIO, TRANS affaiblissement linéique *m*, constante d'affaiblissement *f*; ~ **control** TRANS contrôle d'affaiblissement *m*; ~ **deviation** TRANS écart d'affaiblissement *m*; ~ **distortion** TELEP, TRANS distorsion d'affaiblissement *f*; ~ **factor** TRANS facteur d'affaiblissement *m*, facteur d'atténuation *m*; ~**-limited operation** TRANS fonctionnement limité par l'affaiblissement *m*; ~ **measurement** TRANS mesure d'affaiblissement *f*; ~ **slope** RADIO pente aux frontières *f*; ~ **tolerance** TRANS tolérance d'affaiblissement *f*; ~ **value** TRANS valeur d'affaiblissement *f*
attenuator *n* COMP, ELECT, POWER, RADIO, TRANS affaiblisseur *m*, atténuateur *m*
attitude: ~ **control** *n* RADIO commande d'attitude *m*, commande d'orientation *f*; ~ **control system** *n* RADIO chaîne de pilotage *f*, système de commande d'orientation *m*; ~ **control unit** *n* RADIO centrale d'orientation *f*; ~ **determination control subsystem** *n* RADIO sous-système de détermination et de correction d'attitude *m*; ~ **and orbit control system** *n* RADIO système de commande d'attitude et d'orbite *m*; ~ **reference** *n* RADIO référence d'orientation *f*; ~ **and roll control system** *n* RADIO système de contrôle d'attitude et de roulis *m* (*SCAR*); ~ **sensor** *n* RADIO détecteur d'orientation *m*; ~**-stabilized satellite** *n* RADIO satellite à commande d'orientation *m*
attribute *n* GEN attribut *m*; ~ **group** COMP groupe d'attributs *m*; ~ **method** TEST méthode des attributs *f*; ~ **type** DATA type d'attribut *m*; ~ **value** DATA valeur d'attribut *f*; ~**-value assertion** (*AVA*) DATA affirmation de valeur d'attribut *f*, assertion de valeur d'attribut *f*
ATU *abbr* (*antenna tuning unit*) RADIO dispositif d'accord d'antenne *m*
AU *abbr* CONTR (*administrative unit*) UAD (*unité administrative*), GEN (*access unit*) unité d'accès *f*
audibility *n* COMP, GEN, RADIO audibilité *f*; ~ **range** TELEP étendue d'audibilité *f*
audible[1] *adj* COMP, GEN, RADIO audible, sonore
audible[2]: ~ **alarm** *n* RADIO avertisseur sonore *m*, TEST alarme sonore *f*; ~ **indication** *n* SIG indication audible *f*; ~ **ringing tone** *n* TELEP signal de retour d'appel *m*, tonalité de retour d'appel *f*; ~ **signal** *n* GEN signal audible *m*, signal sonore *m*, RADIO, SIG signal sonore *m*; ~ **sound** *n* GEN son audible *m*
audio[1] *adj* COMP audio; ~**-digital** COMP audio-numérique
audio[2]: ~ **access unit** *n* COMP unité d'accès audio *f*; ~ **bearer service** *n* APPL service de circuit commuté audio *m*, service support audiofréquence *m*; ~ **buffer** *n* COMP tampon audio *m*; ~ **card** *n* COMP carte audio *f*; ~ **carrier** *n* TRANS porteuse son *f*; ~ **channel** *n* TRANS voie de son *f*, voie audiofréquence *f*, TV canal audio *m*; ~ **compression** *n* COMP compression audio *f*; ~ **compression board** *n* DATA carte de compression audio *f*; ~ **conference** *n* RADIO, TELEP audioconférence *m*; ~ **cue** *n* DATA signal sonore *m*; ~ **document** *n* DATA document sonore *m*; ~ **equipment** *n* COMP équipement acoustique *m*, TRANS équipement à fréquences vocales *m*; ~ **frequency** *n* (*AF*) RADIO, TELEP audiofréquence *f*, fréquence audio *f*; ~**-frequency circuit** *n* RADIO circuit basse fréquence *m*; ~**-frequency protection ratio** *n* RADIO rapport de protection en audiofréquence *m*; ~**-frequency signal-to-interference ratio** *n* RADIO rapport signal/brouillage en audiofréquence *m*; ~ **in** *n* COMP entrée audio *f*; ~ **indicate active** *n* TELEP indication audio active *f*; ~ **indicate muted** *n* (*AIM*) TELEP indication audio muette *f*; ~ **input** *n* COMP, ELECT entrée audio *f*; ~ **loop** *n* TELEP boucle audio *f*; ~**-messaging interchange specifications** *n* TELEP audio-messaging interchange specifications *m* (*AMIS*); ~ **mixer** *n* COMP table de mixage *f*; ~ **news gathering** *n* RADIO reportage radiophonique *m*; ~ **out** *n* COMP, ELECT sortie audio *f*; ~ **output** *n* COMP, ELECT sortie audio *f*; ~ **path** *n* TRANS voie son *f*; ~ **recording** *n* COMP enregistrement sonore *m*; ~ **sequence** *n* COMP séquence audio *f*; ~ **signal** *n* SIG signal audiofréquence *m*; ~ **splitter** *n* TRANS diviseur audiofréquence *m*; ~ **system** *n* GEN système audio *m*; ~ **tape** *n* COMP bande audio *f*, bande son *f*, bande sonore *f*, bande magnétique sonore *f*; ~ **track** *n* COMP *CD-ROM, video* piste audio *f*; ~**-videotex** *n* PER APP audio-vidéographie *f*, audio-vidéotexte *m*
audio[3]: ~ **signal on hold** *phr* TELEP signal audio en mode d'attente
Audio-Interchange: ~ **File Format** *n* COMP, DATA *Internet* Audio-Interchange File Format *m* (*AIFF*)
Audiotex: ~ **server** *n* TELEP serveur Audiotex *m*
audiovisual *adj* TV audiovisuel
audit *n* DATA vérification *f*, audit, GEN audit *m*
auditory: ~ **threshold** *n* TELEP seuil d'audition *m*
augmented: ~ **baseline** *n* RADIO base améliorée *f*
aural[1] *adj* COMP sonore
aural[2]: ~ **signal** *n* COMP signal sonore *m*
authentication *n* DATA, SIG authentification *f*; ~ **exchange**

DATA, SIG échange d'authentification *m*; ~ **function** DATA fonction d'authentification *f*; ~ **information** DATA, SIG information d'authentification *f*; ~ **mechanism** DATA mécanisme d'authentification *m*; ~ **procedure** SIG authentification *f*; ~ **value** DATA valeur d'authentification *f*

author: ~**-created link** *n* COMP lien créé par l'auteur *m*

authoring: ~ **software** *n* COMP, DATA logiciel d'auteur *m*; ~ **system** *n* COMP système auteur *m*

authority *n* GEN autorité *f*; ~ **and format identifier** (*AFI*) DATA, SIG identificateur d'autorité et de format *m*

authorization *n* GEN autorisation *f*; ~ **code** TELEP numéro d'autorisation *m*, code d'autorisation *m*, code d'habilitation *m*; ~ **information** DATA, SIG informations d'autorisation *f pl*; ~**-information qualifier** DATA, SIG qualificatif des informations d'autorisation *m*; ~ **key** TELEP clé d'habitation *f*, clé d'autorisation *f*

authorized: ~ **dealer** *n* GEN distributeur agréé *m*

auto: ~ **bass chord** *n* COMP accord de basse automatique *m*; ~ **power off** *n* COMP, POWER coupure automatique d'alimentation *f*

auto-action *n* DATA action automatique *f*; ~ **type** DATA type d'action automatique *m*

auto-alert *n* DATA avertissement automatique *m*

auto-arranger *n* COMP arrangeur automatique *m*

auto-bend *n* COMP modulation automatique *f*

auto-correct *n* COMP, DATA, SIG, TELEP correction automatique *f*

auto-correction *n* COMP, DATA, SIG, TELEP autocorrection *f*

auto-correlation: ~ **detector** *n* RADIO détecteur d'autocorrélation *m*

autocue *n BrE* (*cf teleprompter AmE*) TV télésouffleur *m*

auto-dialing *AmE see auto-dialling BrE*

auto-dialling *n BrE* TELEP appel automatique *m*

auto-epistemic: ~ **reasoning** *n* DATA raisonnement autoépistémique *m*

auto-forward *n* DATA retransmission automatique *f*

auto-installation *n* ELECT autoconfiguration *f*

autoload *n* COMP chargement automatique *m*

auto-locate *n* COMP localisation automatique *f*, positionnement automatique *m*

auto-manual: ~ **switchboard** *n* PER APP pupitre de commutation automatique *m*, SWG commutateur automatique *m*, TELEP commutateur automatique *m*

automata: ~ **theory** *n* DATA théorie des automates *f*

automate *vt* GEN automatiser

automated: ~ **data medium** *n* CONTR support exploitable par machine *m*, support lisible par machine *m*; ~ **reasoning** *n* DATA raisonnement automatisé *m*

automatic[1] *adj* COMP, GEN, RADIO, TELEP automatique

automatic[2]: ~ **acquisition** *n* RADIO accrochage automatique *m*; ~ **ADC** *n* TELEP indication automatique verbale des éléments de taxation *f*; ~ **aligning** *n* RADIO recalage automatique *m*; ~ **alignment** *n* RADIO recalage automatique *m*; ~ **alternative routing** *n* (*AAR*) SIG, TELEP, TRANS acheminement détourné automatique *m*, acheminement automatique sur voie secondaire *m*; ~ **answering** *n* DATA *datacom*, TELEP réponse automatique *f*; ~ **antenna-pattern recording** *n* RADIO enregistrement automatique de diagrammes de rayonnement *m*; ~ **arranger** *n* COMP arrangeur automatique *m*; ~ **attempt** *n* TELEP tentative automatique *f*; ~ **attenuation regulation** *n* TRANS contrôle automatique d'affaiblissement *m*; ~ **banking** *n* APPL monétique *f*; ~ **battery charging** *n* POWER charge automatique de batterie *f*; ~ **billing** *n* TELEP taxation automatique *f*; ~ **booked call** *n* TELEP service de demande automatique d'une communication *m*; ~ **bypass** *n* ELECT dérivation automatique *f*; ~ **call** *n* (*RAG*) DATA, TELEP appel automatique *m* (*RAG*); ~ **call acceptance** *n* TELEP acceptation automatique d'appels *f*; ~ **call distribution** *n* (*ACD*) TELEP distribution automatique d'appels *f* (*DAA*); ~ **call distributor** *n* (*ACD*) TELEP distributeur d'appels automatique *m* (*DAA*); ~ **call queue overflow** *n* TELEP débordement automatique des appels en file d'attente *m*; ~ **call transfer** *n* SIG renvoi automatique *m* (*Can*), transfert automatique des appels *m*; ~ **call transfer unit** *n* TELEP détourneur d'appels *m*; ~ **call transmitter** *n* TELEP émetteur automatique d'appels *m*; ~ **call unit** *n* TELEP unité d'appel automatique *f* (*UAA*); ~ **callback** *n* PER APP, TELEP rappel automatique *m* (*RAG*); ~ **calling** *n* DATA, TELEP appel automatique *m*; ~ **changeover** *n* TEST commutation automatique *f*; ~**-channel tuning** *n* RADIO recherche automatique de canal *f*; ~ **charging equipment** *n* APPL équipement de taxation automatique *m*; ~ **charging control unit** *n* POWER unité de commande de charge automatique *f*; ~ **collect call** *n AmE* (*cf automatic reverse call BrE*) TELEP service d'appel avec transfert automatique de taxe *m*; ~ **control** *n* CONTR commande automatique *f*; ~ **control system** *n* COMP, CONTR système de régulation automatique *m*; ~ **correction** *n* COMP, DATA, SIG, TELEP correction automatique *f*; ~ **credit card service** *n* APPL service automatique de cartes de crédit *m*; ~ **cutoff** *n* COMP arrêt automatique *m*; ~ **data processing** *n* COMP, DATA traitement automatique des données *m*; ~ **detection** *n* RADIO détection automatique *f*; ~ **dialer** *AmE*, ~ **dialler** *n BrE* TELEP composeur téléphonique *m*; ~ **dialing** *AmE*, ~ **dialling** *n BrE* TELEP numérotation automatique *f*; ~ **direction finding** *n* (*ADF*) RADIO radiogoniométrie automatique *f*; ~ **error correction** *n* COMP, DATA, SIG, TELEP correction automatique d'erreurs *f*; ~ **error detection** *n* RADIO, TELEP détection automatique d'erreurs *f*; ~ **exchange** *n* TELEP autocommutateur *m*, central automatique *m*; ~ **fall-back** *n* DATA repli automatique *m*; ~ **frequency control** *n* (*AFC*) ELECT, TELEP, TRANS commande automatique de fréquence *f* (*CAF*), contrôle automatique de fréquence *m* (*CAF*); ~ **function supervision** *n* TELEP surveillance automatique du fonctionnement *f*; ~ **gain control** *n* (*AGC*) ELECT commande automatique de gain *f* (*CAG*), contrôle automatique de gain *m* (*CAG*), TELEP, TRANS contrôle automatique de gain *m* (*CAG*), commande automatique de gain *f* (*CAG*); ~ **harmonization** *n* COMP harmonisation automatique *f*; ~ **harmonizing** *n* COMP harmonisation automatique *f*; ~ **identification of outward dialing** *AmE*, ~ **identification of outward dialling** *BrE n* (*AIOD*) DATA, RADIO, TELEP identification automatique du demandeur *f* (*IAD*); ~ **intercept system** *n* TELEP système d'interception automatique *m*; ~ **interrupt** *n* COMP interruption automatique *f*; ~ **laser shutdown** *n* (*ALS*) TRANS coupure automatique du laser *f*; ~ **level control** *n* COMP commande de niveau automatique *f*; ~ **line management** *n* TRANS gestion automatique de lignes *f*; ~ **location registration of ships** *n* APPL enregistrement automatique de la position des navires *m*; ~ **maritime radio-telex service** *n* TELEP service radio-télex maritime automatique *m*; ~ **measurement** *n* RADIO détermination automatique *f*; ~ **message switching center** *AmE*, ~ **message switching centre** *BrE n* TELEP centre automatique de commutation de messages *m*; ~ **mobile telephone system** *n* (*AMTS*) RADIO, TELEP système radiotéléphonique mobile automatique *m* (*SRMA*); ~ **numbering equipment** *n* TELEP émetteur à numérotation automatique *m*; ~ **numbering transmitter** *n* TELEP transmetteur à numéroteur automatique *m*; ~ **observation** *n* TEST *of service quality* observation automatique *f*; ~ **personal call** *n* TELEP conversation personnelle automatique *f*; ~ **playback** *n* COMP exécution automatique *f*, lecture automatique *f*; ~ **power control** *n* (*APC*) POWER commande automatique de puissance *f*; ~ **preselection** *n* TELEP présélection automatique *f*; ~ **protection switching** *n* (*APS*) SWG commutation automatique sur liaison de réserve *f*, commutation de protection automatique *f*; ~ **radio mute** *n* PER APP, RADIO coupure automatique de la radio *f*; ~ **recall** *n* PER APP, TELEP rappel automatique (*RAG*); ~ **redialing** (*AmE*), ~ **redialling** (*BrE*) PER APP, TELEP rappel automatique *m* (*RAG*); ~ **repeat attempt** *n* TELEP répétition automatique de tentative *f*; ~ **repetition** *n* TELEP répétition automatique Verdan *f*, répétition automatique *f*; ~ **rerouting** *n* (*ARR*) TELEP réacheminement automatique *m*; ~ **restart unit** *n* CONTR équipement de redémarrage automatique *m*; ~ **reverse call** *n BrE* (*cf automatic collect call AmE*) TELEP service d'appel avec transfert automatique de taxe *m*; ~ **rhythm** *n* COMP rythme automatique *m*; ~ **ring trip** *n* TELEP arrêt automatique de sonnerie *m*; ~ **ring tripping** *n* TELEP arrêt automatique de sonnerie *m*~ **roaming** *n* RADIO repérage automatique des déplacements *m*; ~ **route selection** *n* (*ARS*) TELEP sélection automatique d'acheminement *f* (*SAA*); ~ **route selection with queueing** *n* (*ARSWQ*) TELEP sélection automatique d'acheminement avec file d'attente *f* (*SAAFA*); ~ **sequential connection** *n* TELEP connexion automatique en séquence *f*; ~ **service** *n* TELEP service automatique *m*; ~ **shut-off** *n* ELECT extinction automatique *f*; ~ **signaling** *AmE*, ~ **signalling** *BrE n* TELEP signalisation automatique *f*; ~ **soldering** *n* ELECT brasage tendre à la machine *m*; ~ **stop** *n* COMP arrêt automatique *m*; ~ **switchboard** *n* PER APP pupitre de commutation automatique *m*, SWG, TELEP commutateur automatique *m*; ~ **switching** *n* TV *for circuits* commutation automatique *f*; ~ **switching equipment** *n* SWG, TELEP commutateur automatique *m*; ~ **telephone exchange** *n* SWG central téléphonique automatique *m*, TELEP autocommutateur *m*, central automatique *m*, central téléphonique automatique *m*; ~ **timed recall** *n* TELEP rappel temporisé *m*; ~ **tracking** *n* RADIO poursuite automatique *f*; ~ **traffic rearrangement** *n* TRANS réarrangement automatique du trafic *m*; ~ **transferred debiting of charges** *n* TELEP service de transfert d'appel en cas de numéro occupé *m*; ~ **transferred-charge call** *n* TELEP service d'appel avec transfert automatique de taxe *m*; ~ **transmission** *n* APPL transmission automatique *f*; ~ **transmission-measuring equipment** *n* (*ATME*) SWG appareil automatique de mesure de transmission *m* (*AAMT*); ~ **transmitter** *n* TELEP transmetteur automatique *m*; ~ **trunk testing** *n* TELEP vérification automatique des circuits *f*; ~ **trunk working** *n* PER APP, TELEP exploitation interurbaine automatique *f*; ~ **updating** *n* PER APP mise à jour automatique *f*; ~ **vehicle location** *n* (*AVL*) APPL, RADIO localisation automatique de véhicules *f*; ~ **verbal announcement of charges service** *n* TELEP service d'indication automatique verbale des éléments de taxe *m*; ~ **volume control** *n* (*AVC*) ELECT commande automatique de volume *f* (*CAV*), antifading *m*, TELEP, TRANS commande automatique de volume *f* (*CAV*); ~ **wake-up** *n* (*AWU*) TELEP réveil automatique *m*, réveil téléphonique *m*

automatically: ~ **allocated capacity** *n* TRANS capacité attribuée d'office *f*

automation *n* GEN automatisation *f*

automatization *n* CONTR automatisation *f*

automatize *vt* CONTR automatiser

auto-mode *n* DATA automode *m*

auto-modulation *n* TRANS automodulation *f*

autonomous[1] *adj* COMP, TELEP autonome

autonomous[2]: ~ **registration device** *n* TELEP dispositif d'enregistrement autonome *m*; ~ **registration update** *n* TELEP mise à jour d'enregistrement autonome *f*

auto-play *n* COMP écoute automatique des messages *f*

auto-stop *n* COMP arrêt automatique *m*

auto-synchronization *n* TRANS autosynchronisation *f*

auto-track *n* RADIO poursuite automatique *f*

auto-transfer *n* TELEP transfert des communications réseau *m*

auto-transformer *n* POWER autotransformateur *m*

AUTOVON: ~ **trunk access** *n* TELEP accès au réseau AUTOVON *m*

auxiliaries: ~ **control station** *n* TELEP station multiprocesseur d'auxiliaires *f*

auxiliary: ~ **board** *n* POWER carte auxiliaire *f*; ~ **circuit** *n* POWER circuit auxiliaire *m*; ~ **conditions** *n pl* TELEP états auxiliaires *m pl*; ~ **contact** *n* POWER contact auxiliaire *m*; ~ **equipment** *n* COMP, ELECT matériel auxiliaire *m*; ~ **equipment rack** *n* ELECT rack d'équipement auxiliaire *m*; ~ **management software machine** *n* TELEP équipement de tonalités et auxiliaires *m*; ~ **operation** *n* COMP opération auxiliaire *f*; ~ **position** *n* TELEP position auxiliaire *f*, position d'opérateur d'assistance *f*; ~ **service position** *n* APPL position terminale *f*; ~ **services trunk group** *n* TELEP faisceau de services auxiliaires *m*; ~ **set** *n* TELEP poste secondaire *m*; ~ **signaling** *AmE*, ~ **signalling** *BrE n* TELEP signalisation automatique *f*; ~ **switch** *n* COMP commutateur secondaire *m*, SWG commutateur auxiliaire *m*; ~ **switching point** *n* SWG point de commutation auxiliaire *m*; ~ **switching unit** *n* SWG commutateur auxiliaire *m*; ~ **telephone set** *n* TELEP poste secondaire *m*; ~ **voltage** *n* POWER tension auxiliaire *f*; ~ **winding** *n* POWER enroulement auxiliaire *m*

AVA *abbr* (*attribute-value assertion*) DATA affirmation de valeur d'attribut *f*, assertion de valeur d'attribut *f*

availability *n* DATA, GEN disponibilité *f*, TELEP accessibilité *f*, *of connection, leased circuit* disponibilité *f*, TEST disponibilité *f*; ~ **figure** SWG taux de disponibilité *m*; ~ **function** TEST disponibilité instantanée *f*; ~ **measure** TEST mesure de la disponibilité *f*; ~ **unavailability** TELEP disponibilité indisponibilité *f*

available: ~ **line** *n* TELEP ligne libre *f*; ~ **memory** *n* COMP mémoire disponible *f*; ~ **power** *n* POWER, TELEP, TRANS puissance disponible *f*

avalanche *n* ELECT *semiconductors* avalanche *f*; ~ **breakdown** ELECT claquage par avalanche *m*; ~ **diode** ELECT diode à avalanche *f*; ~ **effect** ELECT effet d'avalanche *m*; ~ **noise** ELECT bruit d'avalanche *m*; ~ **photodiode** (*APD*) DATA, PARTS, TRANS *fibre optics* photodiode à avalanche *f* (*PDA*); ~ **voltage** ELECT tension d'avalanche *f*

AVC *abbr* (*automatic volume control*) ELECT CAV (*commande automatique de volume*), antifading *m*, TELEP, TRANS (*automatic volume control*) CAV (*commande automatique de volume*)

AVD *abbr* (*alternate voice-data circuit*) TELEP, TRANS circuit de téléphonie-données à l'alternat *m*

average[1]: ~ **amount of inspection** *n* TEST *quality* quantité moyenne contrôlée *f*; ~ **busy hour in CCS** *n* TELEP moyenne des heures chargées en centaines de communications-secondes *f*; ~ **busy-season busy hour** *n* TELEP moyenne des heures chargées de la saison de pointe *f*; ~ **call duration** *n* TELEP durée moyenne de conversation *f*; ~ **cladding diameter** *n* TRANS diamètre moyen de la gaine *m*; ~ **core diameter** *n* TRANS diamètre moyen du cœur *m*; ~ **crossing rate** *n* GEN fréquence moyenne de passage à zéro *f*; ~ **daily use** *n* TEST moyenne de l'utilisation quotidienne *f*; ~ **operating time** *n* TELEP durée moyenne des manœuvres *f*; ~ **outgoing quality** *n* (*AOQ*) TEST qualité moyenne après contrôle *f*; ~ **outgoing quality limit** *n* (*AOQL*) TEST limite de qualité moyenne après contrôle *f*; ~ **output** *n* COMP, ELECT niveau de sortie moyen *m*; ~ **power** *n* TRANS puissance moyenne *f*; ~ **reference-surface diameter** *n* (*DRav*) TELEP diamètre moyen de la surface de référence *m*; ~ **run length** *n* TEST *quality* période opérationnelle moyenne *f*; ~ **sample number** *n* (*ASN*) TEST *quality* effectif moyen contrôlé *m*, effectif moyen de l'échantillon *m*; ~ **total inspected** *n* (*ATI*) TEST *quality* moyenne totale contrôlée *f*; ~ **traffic intensity** *n* TELEP intensité moyenne de traffic *f*

average[2]: ~ **bits per sample** *phr* DATA nombre moyen de bits par échantillon

AVL *abbr* (*automatic vehicle location*) APPL, RADIO localisation automatique de véhicules *f*
avoidance: ~ **angle** *n* RADIO angle d'évitement *m*
await: ~ **message** *n* TELEP attente de message *f*
AWU *abbr* (*automatic wake-up*) TELEP réveil automatique *m*, réveil téléphonique *m*, service de réveil automatique *m*
axial[1] *adj* GEN axial
axial[2]: ~ **interference microscopy** *n* ELECT microscopie axiale interférentielle *f*, TRANS interférométrie *f*, microscopie axiale interférentielle *f*; ~ **lateral plasma deposition** *n* (*ALPD*) ELECT dépôt axial latéral au plasma *m* (*DALP*); ~ **propagation coefficient** *n* ELECT, RADIO, TRANS exposant linéique de propagation longitudinale *m*; ~ **ratio** *n* RADIO taux d'ellipticité *m*; ~ **ray** *n* RADIO, TRANS *fibre optics* rayon axial *m*; ~ **slab interferometry** *n* RADIO, TRANS interférométrie axiale *f*
axiom *n* DATA axiome *m*
axis *n* COMP axe *m*
axle *n* COMP axe *m*, PARTS arbre *m*, axe *m*
azimuth *n* RADIO angle de gisement *m*, angle d'azimut *m*, azimut *m*; ~ **aerial** RADIO antenne d'azimut *f*; ~ **antenna** RADIO antenne d'azimut *f*; ~ **elevation mount** RADIO monture azimut-élévation *f*; ~ **speed indicator** RADIO indicateur de vitesse azimut *m*; ~-**stabilized display** RADIO présentation à alignement automatique *f*, présentation panoramique stabilisée *f*

B

B: ~ **sealant** *n* ELECT produit d'étanchéité B *m*; ~ **version** *n* TELEP *of Alcatel E10 system* version B *f*

babble *n* TELEP murmure confus *m*

Babcock: ~ **plan** *n* RADIO, TELEP plan de Babcock *m*

back[1]: ~ **dish** *n* RADIO coque arrière du réflecteur *f*; ~ **echo** *n* RADIO écho arrière *m*; **~-end processor** *n* DATA dorsal *m*; **~-end server** *n* COMP, DATA ante-serveur *m*; ~ **lobe** *n* RADIO *aerials* lobe arrière *m*; **~-off** *n* RADIO recul de puissance *m*, réduction de puissance *f*; ~ **panel** *n* PARTS fond de panier *m*; ~ **pressure** *n* GEN *physics* contre-pression *f*; **~-pressure valve** *n* PARTS régulateur à pression d'évaporation *m*, robinet à pression minimale constante *m*; **~-propagation network** *n* DATA réseau à propagation arrière *m*; **~-to-back parabolic aerials** *n pl* RADIO antennes paraboliques dos-à-dos *f pl*; **~-to-back parabolic antennas** *n pl* RADIO antennes paraboliques dos-à-dos *f pl*; **~-to-back scanners** *n pl* RADIO antennes de balayage dos-à-dos *f pl*; **~-to-back test** *n* TRANS essai en boucle *m*

back[2]: ~ **up** *vt* COMP *file* sauvegarder

backbone *n* ELECT segment fédérateur *m*, TRANS réseau *m*, structure *f*; ~ **circuit** ELECT, TRANS circuit principal *m*; ~ **flexibility** ELECT flexibilité du segment fédérateur *f*, TRANS flexibilité du réseau *f*, flexibilité de la structure *f*; ~ **network** COMP, DATA réseau fédérateur *m*, TRANS réseau dorsal *m*, réseau de base *m*; ~ **route** TELEP artère principale *f*; ~ **structure** TRANS structure de base *f*

backfire: ~ **aerial** *n* RADIO antenne à rayonnement vers l'arrière *f*; ~ **antenna** *n* RADIO antenne à rayonnement vers l'arrière *f*

background[1] *adj* COMP, GEN d'arrière-plan

background[2] *n* COMP, DATA fond *m*, ELECT contexte *m*; ~ **level** GEN *acoustics* niveau de fond *m*; ~ **mode of operation** RADIO mode courant d'exploitation *m*; ~ **mush** PER APP bruit de fond *m*; ~ **noise** COMP, GEN, PER APP bruit de fond *m*, TELEP bruit de fond *m*, bruit parasite *m*, parasites *m pl*; ~ **operation** DATA tâche de fond *f*; ~ **processing** COMP traitement non prioritaire *m*; ~ **radiation** RADIO rayonnement ambiant *m*; ~ **signature** RADIO signature du fond *f*; ~ **task** COMP tâche de fond *f*; ~ **theory** TRANS théorie de rappel *f*

backing: ~ **tape** *n* COMP bande de sauvegarde *f*

backlighting *n* COMP, ELECT éclairage par l'arrière *m*, rétroéclairage *m*

backlit *adj* ELECT éclairé par l'arrière, rétroéclairé

backlog *n* COMP arriéré *m*

backplane *n* ELECT configuration de fond *f*

backscatter *n* POWER, RADIO, TELEP, TRANS rétrodiffusion *f*

backscattering *n* POWER, RADIO, TELEP, TRANS rétrodiffusion *f*; ~ **technique** TELEP, TRANS *fibre optics* méthode de rétrodiffusion *f*

backspace[1] *n* (*BS*) COMP espace arrière *m*, retour arrière *m*; ~ **character** (*BS*) COMP caractère d'espacement arrière *m*, caractère de retour arrière *m*

backspace[2] *vt* COMP faire un espace arrière, faire un retour arrière

backtracking *n* DATA retour arrière *m*

backup[1] *adj* COMP, DATA de réserve, de sécurité, de sauvegarde, de secours, GEN auxiliaire, RADIO, TELEP, TEST de réserve, de sécurité, de sauvegarde, de secours

backup[2] *n* COMP, DATA, GEN copie de sécurité *f*, sauvegarde *f*, soutien technique *m*; ~ **battery** PER APP batterie de secours *f*; ~ **battery system** PER APP système de maintien alimenté par piles *m*; ~ **copy** COMP, DATA copie de sécurité; ~ **function** COMP, GEN fonction de sauvegarde *f*, fonction de secours *f*; ~ **information** COMP données de sauvegarde *f pl*; ~ **launch** RADIO lancement de secours *m*, lancement de sauvegarde *m*; ~ **launcher** RADIO lanceur de secours *m*, lanceur de sauvegarde *m*; ~ **mode** TEST mode de secours *m*; ~ **period** TEST période de réserve *f*; ~ **power supply** POWER alimentation de secours *f*; ~ **reference secondary burst** DATA paquet de référence secondaire de secours *m*; ~ **relationships** *n pl* DATA relations de sauvegarde *f pl*; ~ **station** RADIO station de secours *f*; ~ **store** COMP, DATA mémoire auxiliaire *f*; ~ **supervisor** TELEP superviseur de secours *m*; ~ **system** COMP, DATA système de sauvegarde *m*; ~ **tape** COMP, TEST bande de sauvegarde *f*; ~ **time** COMP, DATA temps de sauvegarde *m*

Backus: ~ **Normal form** *n* (*BNF*) DATA forme normale de Backus *f* (*forme BNF*); **~-Naur form** *n* (*BNF*) DATA forme de Backus-Naur *f* (*forme BNF*)

backward: ~ **alarm** *n* TEST alarme vers l'arrière *f*; ~ **attenuation** *n* TRANS atténuation vers l'arrière *f*; ~ **channel** *n* TELEP, TRANS voie de retour *f*; ~ **echo** *n* TELEP écho vers l'amont *m*; ~ **indicator bit** *n* COMP bit indicateur vers l'arrière *m*; **~-input signal** *n* SIG signal d'entrée vers l'arrière *m*; **~-interworking telephony event** *n* (*BITE*) TELEP événement téléphonique d'interfonctionnement vers l'arrière *m*; **~-motion vector** *n* COMP, TV vecteur cinétique différé *m*, vecteur de mouvement différé *m*; ~ **sequence number** *n* COMP, DATA, TELEP numéro de séquence vers l'arrière *m*; ~ **signal** *n* TELEP signal de retour *m*, signal de réponse *m*; ~ **transmission** *n* RADIO, TRANS rétrodiffusion *f*; ~ **wave oscillator** *n* (*BWO*) ELECT, TELEP oscillateur à onde régressive *m* (*OOR*), oscillateur à onde rétrograde *m* (*OOR*); ~ **wave tube** *n* TELEP tube à onde régressive *m*

badge: ~ **reader** *n* APPL lecteur de badge *m*

baffle *n* GEN *acoustics* écran acoustique *m*, enceinte acoustique *f*

bakelite *n* PARTS bakélite *f*

balance: ~ **level** *n* COMP niveau d'équilibre *m*; ~ **ratio** *n* SIG *for sinusoidal signal receiver* rapport d'équilibre *m*; **~-return loss** *n* TELEP affaiblissement d'équilibrage *m*; **~-return loss attenuation** *n* TRANS atténuation d'affaiblissement d'équilibrage *m*; ~ **tester** *n* TEST testeur d'équilibrage *m*; ~ **winding** *n* POWER enroulement d'équilibrage *m*

balanced[1] *adj* GEN compensé; **~-to-earth** *BrE* (*cf balanced-to-ground AmE*) POWER symétrique par rapport à la terre; **~-to-ground** *AmE* (*cf balanced-to-earth BrE*) POWER symétrique par rapport à la terre

balanced[2]: ~ **Adcock direction finder** *n* RADIO radiogoniomètre Adcock à équilibreurs *m*; ~ **error** *n* GEN *statistics* erreur centrée *f*; ~ **grading group** *n* TELEP *traffic* groupe de lignes à trafic équilibré *m*; ~ **input** *n* COMP entrée symétrique *f*; ~ **mixer** *n* COMP, RADIO mélangeur équilibré *m*; ~ **network** *n* DATA réseau équilibré *m*; ~ **opposing winding** *n* POWER enroulement bifilaire *m*; ~ **output** *n* COMP sortie symétrique *f*; **~-to-unbalanced transformer** *n* (*balun*) POWER symétriseur *m*, transformateur symétrique/asymétrique *m*, transformateur symétriseur *m*

balancing *n* GEN, TRANS équilibrage *m*, compensation *f*; ~ **resistor** PARTS, POWER résistance d'équilibrage *f*

ball: ~ **valve** *n* PARTS clapet à bille *m*

ballast *n* POWER ballast *m*; ~ **resistor** POWER résistance ballast *f*

balloon: ~ **help** *n* COMP bulles d'aide *f pl*
balun *n* (*balanced-to-unbalanced transformer*), POWER symétriseur *m*, transformateur symétrique/asymétrique *m*, transformateur symétriseur *m*; ~ **adaptor** ELECT adaptateur d'impédance *m*
banana: ~ **jack** *n* ELECT jack pour casque téléphonique *m*; ~ **plug** *n* ELECT, TEST fiche banane *f*
band *n* COMP bande *f*, GEN zone *f*, RADIO, TRANS *frequency* bande de fréquences *f*; ~ **elimination filter** TRANS filtre coupe-bande *m*; ~ **number** TELEP numéro de bande *m*; ~ **rate** TRANS rapidité de modulation *f*; ~ **rejection filter** TRANS filtre coupe-bande *m*, coupe-bande *m*; **out-~ signaling** *AmE*, **out-~ signalling** *BrE* SIG signalisation hors bande *f*
bandpass: ~ **filter** *n* (*BPF*) POWER, TRANS filtre passe-bande *m* (*FPB*)
bands *n pl* COMP bandeaux *m pl*
bandstop: ~ **filter** *n* TRANS coupe-bande *m*, filtre coupe-bande *m*
bandwidth[1]: ~-**limited** *adj* TRANS limité par la largeur de la bande
bandwidth[2] *n* DATA, ELECT *optical fibre* bande passante *f*, RADIO, TRANS largeur de bande *f*, TV bande passante *f*; ~ **compression** DATA compression de la largeur de bande *f*; ~-**efficient system** TRANS système permettant d'utiliser efficacement la largeur de bande *m*; ~ **expansion** DATA étalement de la largeur de bande *m*; ~ **expansion ratio** DATA rapport d'étalement de la largeur de bande *m*; ~-**limited operation** TRANS fonctionnement limité par la largeur de bande *m*; ~ **on demand** COMP bande passante à la demande *f*, largeur de bande à la demande *f*
bank *n* COMP banque *f*, groupe *m*, DATA, TV banque *f*; ~ **contact** TELEP lamelle de contact *f*; ~ **of contacts** GEN rangée de contacts *f*; ~ **memory** COMP mémoire de masse *f*; ~ **play** COMP lecture de banque *f*
banner *n* COMP énoncé *m*, manchette *f*, message d'accueil *m*
B-answer: ~ **signal** *n* SIG signal de réponse du demandé *m*
BAPTA *abbr* (*bearing and power transfer assembly*) POWER MEGS (*bloc support et transfert d'énergie du réseau solaire*)
bar[1] *n* COMP mesure *f*, PARTS *equipment practice*, TELEP profilé *m*; ~-**code** APPL, DATA code à barres *m*; ~-**code scanner** APPL, DATA lecteur de code à barres *m*; ~ **holder** PARTS attache de profilé *f*
bar[2] *vt* RADIO, SIG, TELEP bloquer, restreindre l'accès à, interdire
bare: ~ **wire** *n* ELECT, POWER *conductors* fil non isolé *m*, fil nu *m*; ~-**wire multiple** *n* TELEP multiplage en fils nus *m*; ~-**wire strap** *n* TELEP connexion à fil nu *f*; ~ **wiring** *n* TELEP câblage en fil nu *m*
barge: ~-**in** *n* TELEP intervention *f*; ~-**in tone** *n* TELEP tonalité d'intervention *f*
Bark *n* COMP Bark *m*
barred[1] *adj* RADIO, TELEP bloqué, interdit, restreint
barred[2]: ~ **code** *n* TELEP indicatif de service restreint *m*; ~ **station** *n* RADIO station interdite *f*
barrette: ~ **file** *n* TEST *tools* barrette *f*
barrier: ~ **layer** *n* ELECT couche barrière *f*, *semiconductors* couche de barrage *f*
barring *n* RADIO, TELEP blocage *m*, interdiction *f*
barter[1] *n* TV échange *m*, troc *m*
barter[2] *vt* TV échanger, troquer
barter[3] *vi* TV faire un échange, faire du troc
bartering *n* TV bartering *m*
base *n* ELECT *semiconductors* base *f*, PARTS *of structure* embase *f*; ~ **address** COMP adresse de base *f*; ~ **area** RADIO *mobicom* zone de la station de base *f*; ~ **band unit** (*BU*) RADIO coeur de chaîne *m*; ~ **charge** ELECT *semiconductors* charge de la base *f*; ~ **coverage area** RADIO zone de couverture de la station de base *f*; ~ **current** ELECT *semiconductors* courant de base *m*; ~-**emitter resistance** ELECT résistance base-émetteur *f*; ~-**emitter voltage** ELECT tension base-émetteur *f*; ~ **frequency** RADIO fréquence de base *f*; ~ **iron** PARTS profilé de base *m*, semelle *f*; ~ **level** SWG niveau de base *m*; ~ **line** COMP ligne de base *f*; ~ **load** POWER charge minimale *f*; ~ **metal** ELECT *soldering*; ~ **number** COMP nombre de base *m*; ~ **panel** ELECT, PARTS plaque de base *f*; ~ **plate** ELECT, PARTS *structure* plaque de base *f*; ~ **region** ELECT région de base *f*; ~ **resistance** ELECT résistance de base *f*; ~ **reverse current** ELECT courant inverse de la base *m*; ~ **station** (*BS*) RADIO, TELEP station de base *f*; ~ **station area** RADIO zone de la station de base *f*, zone d'une station de base *f*; ~ **station controller** RADIO contrôleur de station de base *m*; ~ **station separation** RADIO distance entre les stations de base *f*; ~ **station subsystem** (*BSS*) TELEP sous-système de station de base *m*; ~ **terminal** ELECT borne de base *f*; ~-**to-mobile relay** RADIO relais de passage *m*; ~ **type** DATA type de base *m*
baseband[1] *adj* RADIO, TRANS, TV en bande de base
baseband[2] *n* RADIO, TRANS, TV bande de base *f*; ~ **bandwidth** RADIO, TRANS largeur de la bande de base *f*; ~ **equipment** RADIO équipement aux fréquences de la bande de base *m*; ~ **modem** DATA, TRANS modem en bande de base *m*; ~ **response** TELEP réponse en bande de base *f*; ~ **response function** APPL, CONTR, DATA, TELEP, TRANS fonction de réponse en bande de base *f*; ~ **switch matrix** (*BSM*) SWG, TELEP matrice de commutation de bande de base *f*; ~ **transfer function** APPL, CONTR, DATA, TELEP, TRANS fonction de transfert en bande de base *f*; ~ **transmission** DATA, TELEP transmission en bande de base *f*
baseline[1] *adj* RADIO de base, de référence
baseline[2]: ~ **design** *n* RADIO conception de base *f*
basic: ~ **access** *n* TELEP, TRANS accès de base *m*; ~ **amplitude** *n* RADIO amplitude de base *f*; ~ **bit rate** *n* COMP, DATA, TELEP, TRANS débit binaire de base *m*; ~ **channel** *n* COMP canal de base *m*; ~ **charge** *n* TELEP taxe de base *f*; ~ **encoding rules** *n pl* (*BER*) COMP, DATA règles de codage de base *f pl* (*BER*); ~ **equipment rack** *n* PARTS bâti de matériel de base *m*; ~ **error correction method** *n* TELEP méthode de correction d'erreur de base *f*; ~ **font** *n* COMP police de base *f*; ~ **frame alignment** *n* (*BFA*) TRANS verrouillage de trame de base *m*; ~ **frequency** *n* ELECT fréquence fondamentale *f*, fondamentale *f*, fréquence de base *f*; ~ **group** *n* TELEP groupe de base *m*, TRANS groupe primaire *m*; ~ **input/output system** *n* (*BIOS*) COMP système d'exploitation des entrées/sorties *m* (*BIOS*); ~ **interconnection test** *n* (*BIT*) DATA test d'interconnexion de base *m*; ~ **loop** *n* COMP boucle simple *f*; ~ **loss** *n* TRANS affaiblissement de référence *m*; ~ **material** *n* COMP mode de base *m*; ~ **metered unit** *n* TELEP taxe de base *f*; ~ **metering pulse** *n* TELEP impulsion de base de comptage *f*; ~ **MUF** *n* RADIO MUF de référence *m*; ~ **noise** *n* TELEP bruit de fond *m*, bruit propre *m*; ~ **operation** *n* COMP opération de base *f*; ~ **rack** *n* PARTS bâti de base *m*; ~ **rate access** *n* (*BRA*) DATA, TELEP accès au tarif de base *m*; ~ **rate interface** *n* DATA, TELEP accès de base RNIS *m*; ~ **rate service** *n* APPL service à tarif de base *m*; ~ **research** *n* TEST recherche fondamentale *f*; ~ **SDL** *n* DATA *SDL* SDL de base *m*; ~ **service** *n* TV service de base *m*; ~ **session reference** *n* TELEP référence de base de l'échange *f*; ~ **signal** *n* COMP signal de base *m*; ~ **software package** *n* COMP bibliothèque de base *f*; ~ **subscription** *n* TELEP abonnement principal *m*; ~ **supergroup** *n* TELEP, TRANS groupe secondaire de base *m*; ~ **transmission loss** *n* RADIO, TRANS affaiblissement entre antennes isotropes *m*, affaiblissement de propagation *m*
basilar: ~ **membrane** *n* RADIO membrane basilaire *f*
basket *n* DATA corbeille *f*
bass *n* GEN *sound recording, reproduction* basses *f pl*, basses fréquences *f pl*, graves *m pl*; ~ **tone** GEN son grave *m*; ~ **tones** *n pl* GEN *sound recording, reproduction* graves *m pl*; ~ **unit** (*BU*) RADIO cellule de base *m*

batch *n* COMP, TEST lot *m*; **~ processing** COMP, DATA traitement différé *m*, traitement par lots *m*; **~ testing** PARTS essai par lots *m*; **~ variation** TEST *quality* variation des lots *f*

battery[1]: **~-powered** *adj* POWER à piles, alimenté par batterie, alimenté par piles

battery[2] *n* ELECT, POWER *combined cells* accumulateur *m*, batterie *f*, *single cell* pile *f*; **~ backup** PER APP batterie de secours *f*; **~ cabinets** *n pl* TRANS armoires batterie *f pl*; **~ cable** POWER câble de batterie *m*; **~ cart** POWER chariot porte-batteries *m*; **~ case** POWER coffret de batterie *m*; **~ cell** POWER élément de batterie *m*; **~ cell container** POWER bac d'éléments de batterie *m*; **~ charger** POWER chargeur *m*, chargeur de batterie *m*; **~ charging** POWER charge d'accumulateur *f*, charge de batterie *f*; **~ condition** POWER état des batteries *m*; **~ control unit** POWER régulateur de charge de batterie *m*; **~ fuse** PARTS, POWER fusible de batterie *m*; **~ fuse unit** PARTS, POWER ensemble coupe-circuit de batterie *m*; **~ monitor** POWER appareil de suivi de batterie *m*; **~ negative** POWER borne négative *f*; **~ positive** POWER borne positive *f*; **~ power supply** ELECT, POWER alimentation par batterie *f*, alimentation par piles *f*; **~ rack** POWER armoire à batterie *f*, bâti pour batterie *m*; **~ regulator** POWER régulateur de batterie *m*; **~ room** POWER salle des accumulateurs *f*, salle des batteries *f*; **~ storage time** ELECT, POWER temps de réserve de batterie *m*; **~ system** COMP, ELECT, POWER système à piles *m*, système alimenté par piles *m*; **~ tray** POWER caisse de groupement *f*, caisson *m*; **~ trolley** POWER chariot porte-batteries *m*; **~ voltage** ELECT, POWER tension de batterie *f*; **~ voltage terminal** POWER borne de tension de batterie *f*

baud *n* COMP baud *m*; **~ rate** COMP vitesse de transmission *f*

baudot: **~ code** *n* COMP code baudot *m*

bay *n* TELEP baie *f*

bayonet[1] *adj* ELECT, PARTS, POWER à baïonnette

bayonet[2]: **~ cap** *n* POWER *lighting* culot à baïonnette *m*; **~ coupling** *n* ELECT accouplement à baïonnette *m*, verrouillage par baïonnette *m*

BB *abbr* (*bulletin board*) COMP, DATA, TELEP babillard électronique *m* (*Can*), panneau d'affichage *m*, tableau d'affichage *m*, Service Télématique *m*; **~ Edit** *n* DATA *Internet* BB Edit *m*

BBH *abbr* (*bouncing busy hour*) TELEP heure de pointe variable *f*

BBS *abbr* (*bulletin board system*) COMP, DATA, TELEP BBS (*système de babillard électronique*), tableau d'affichage électronique *m* (*Fra*)

BC *abbr* GEN (*broadcast*) émission *f*, diffusion *f*, TELEP, TRANS (*bearer capability*) capacité support *f*, (*broadcast*) diffusion *f*, (*bearer channel*), voie support *f*, voie porteuse *f*

BCC *abbr* (*blind courtesy copy*) TELEP CCS (*copie conforme sélective*)

B-channel *n* DATA, TRANS *information channel* canal B *m*; **~ virtual-circuit service** TELEP service CVB *m* (*service de circuit virtuel sur canal B*), service de circuit virtuel sur canal B *m* (*service CVB*)

BCS *abbr* (*BULL cabling system*) COMP, DATA BCS (*système de câblage BULL*)

BCUG *abbr* (*bilateral closed user group*) TELEP groupe fermé d'usagers bilatéral *m*

BCUGOA *abbr* (*bilateral closed user group with outgoing access*) DATA, TELEP groupe fermé d'usagers bilatéral avec accès sortant *m*

beacon *n* GEN borne *f*, RADIO balise *f*; **~ satellite** RADIO satellite à radiophare *m*, satellite de navigation *m*; **~ signal** SIG signal de radiophare *m*; **~ transmitter** RADIO émetteur-balise *m*

bead: **~ of epoxy** *n* ELECT goutte d'époxy *f*

beam[1] *n* COMP rayon *m*, ELECT faisceau électromagnétique *m*, GEN faisceau *m*, RADIO, TRANS, TV faisceau électromagnétique *m*; **~ aerial** RADIO antenne directive *f*, antenne directionnelle *f*; **~ antenna** RADIO antenne directive *f*, antenne directionnelle *f*; **~ area** RADIO empreinte d'un faisceau *f*; **~ axis** RADIO axe de faisceau électromagnétique *m*, *aerials* axe de faisceau *m*; **~ center** *AmE*, **~ centre** *BrE* RADIO axe du faisceau *m*; **~-center gain** *AmE*, **~-centre gain** *BrE* RADIO gain dans l'axe du faisceau *m*; **~ diameter** TRANS diamètre de faisceau *m*; **~ divergence** RADIO, TRANS divergence de faisceau *f*; **~ edge** RADIO limite de faisceau *f*; **~ forming** ELECT, RADIO formage de faisceau *m*; **~-forming network** ELECT, RADIO réseau conformateur de faisceau *m*; **~ scanning** RADIO exploration *f*; **~ splitter** COMP prisme semi-réfléchissant *m*, séparateur de rayon *m*, TRANS séparateur de faisceau *m*; **~ steerability** RADIO orientabilité du faisceau *f*; **~ steering** RADIO commande d'orientation de faisceau *f*; **~ switching** SWG variation d'orientation d'un faisceau *f*; **~ waveguide** RADIO alimentation périscopique *f*; **~ waveguide feed** RADIO source d'alimentation périscopique *f*; **~ width** RADIO, TRANS diamètre d'un faisceau *m*, largeur de faisceau *f*, ouverture de faisceau *f*

beam[2]: **~ down** *vt* RADIO diffuser depuis un satellite

bear: **~ against** *vt* PARTS porter sur, reposer sur; **~ on** *vt* PARTS porter sur, reposer sur

bearer *n* TRANS *of circuits* support *m*, porteur *m*; **~ cable** TRANS câble porteur *m*; **~ capability** (*BC*) TELEP, TRANS capacité support *f*; **~ capacity** TRANS capacité du support *f*; **~ channel** (*BC*) TELEP, TRANS voie porteuse *f*, voie support *f*; **~ circuit** DATA support de transmission *m*; **~ identification code** DATA, TELEP code d'identification de support *m*; **~ independent service** (*BIS*) APPL, COMP, RADIO, TELEP, TRANS service indépendant du support *m*; **~ service** APPL, COMP, RADIO, TELEP, TRANS service support *m*; **~ unit** TELEP capacité support *f*

bearing *n* PARTS, RADIO *of antenna* support de roulement *m*; **~ housing** PARTS corps de palier *m*; **~ and power transfer assembly** (*BAPTA*) POWER bloc support et transfert d'énergie du réseau solaire *m* (*MEGS*); **~ seat** PARTS siège de coussinet *m*; **~ shell** PARTS corps de palier *m*; **~ surface** PARTS point d'appui *m*, portée *f*, surface d'appui *f*, surface de contact *f*, surface portante *f*

beat[1]: **~ frequency** *n* ELECT, TRANS fréquence de battement *f*; **~-frequency oscillator** *n* (*BFO*) ELECT, TRANS oscillateur de battement de fréquence *m* (*BFO*); **~ indicator** *n* COMP, SIG indicateur de mesure *m*, indicateur de battement *m*

beat[2]: **~ time** *vi* COMP battre la mesure

beats: **~ per minute** *phr* COMP temps par minute

bedding *n* TRANS *cable laying* matelas *m*

beep *n* COMP, RADIO bip *m*; **~ sound** COMP, RADIO bip sonore *m*, signal sonore bref *m*

beeper *n* COMP appareil de signalisation sonore *m*, RADIO bip *m*

beginning: **~ of call demand** *n* DATA début de la demande d'appel *m*; **~ of message** *n* (*BOM*) DATA, PER APP début de message *m*; **~-of-tape marker** *n* COMP marqueur de début de bande *m*; **~ of pulse** *n* COMP début d'impulsion *m*

behavior *AmE see behaviour BrE*

behavioral: **~ animation** *AmE see behavioural animation BrE*

behaviour *n BrE* DATA comportement *m*; **~ line** *BrE* DATA ligne comportementale *f*; **~ test** *BrE* DATA test de comportement *m*; **~ tree** *BrE* DATA arbre comportemental *m*

behavioural: **~ animation** *n BrE* COMP animation comportementale *f*

bel *n* RADIO *acoustics* Bel *m*

Beldfoil: **~ aluminium-polyester shield** *BrE*, **~ aluminum-polyester shield** *AmE n* ELECT blindage Beldfoil en aluminium polyester *m*

Belgian: **~ Institute for Post and Telephony** *n* (*BIPT*) RADIO Institut belge des services postaux et des télécommunications *m* (*IBPT*); **~ Standards Institute** *n* (*BSI*) GEN Institut belge de normalisation *m* (*IBN*)

bell *n* TELEP sonnerie *f*
Belleville: ~ **spring** *n* PARTS ressort Belleville *m*
bellsets *n* PER APP sonnerie *f*
belt: ~ **drive** *n* PARTS transmission par courroie *f*; ~ **insulation** *n* TRANS *cables* isolation à ceinture *f*; ~ **loop** *n* ELECT dragonne *f*
belted: ~ **cable** *n* TRANS câble à ceinture *m*
benchmark *n* COMP test de performance *m*, GEN point de référence *m*; ~ **test** TEST test d'évaluation des performances *m*; ~ **testing** TEST évaluation des performances *f*
benchtest *vt* TEST mettre au banc d'essai
benchtesting *n* TEST essai au banc *m*
bend[1] *n* PARTS cintrage *m*, coude *m*, TRANS *cables* courbe *f*, courbure *f*; ~ **aerial** RADIO antenne coudée *f*; ~ **carbide** ELECT tête au carbure *f*; ~ **radius** TRANS *cables* rayon de courbure *m*
bend[2] *vt* PARTS cintrer, plier, courber
bender *n* PARTS molette de hauteur *f*
bending: ~ **loss** *n* ELECT microcourbure *f*, TRANS affaiblissement de courbure *m*, perte de courbure *f*; ~ **radius** *n* ELECT rayon de courbure *m*; ~ **strength** *n* PARTS résistance à la flexion *f*; ~ **tool** *n* PARTS outil à cintrer *m*
BER *abbr* COMP, DATA, RADIO (*basic encoding rules*) BER (*règles de codage de base*), (*binary error rate*) TEB (*taux d'erreurs binaires*), (*bit error rate*) TEB (*taux d'erreurs sur les bits*)
best: ~ **possible circuit** *n* ELECT, TRANS circuit à impédance optimale *m*, circuit d'affaiblissement d'adaptation *m*
beta: ~-**trial** *n* TEST essai pilote *m*
Bézier: ~ **curve** *n* COMP courbe de Bézier *f*
BFA *abbr* (*basic frame alignment*) TRANS verrouillage de trame de base *m*
b-field: ~ **picture** *n* COMP, TV image-b monotrame *f*
BFO *abbr* (*beat-frequency oscillator*) ELECT, TRANS BFO (*oscillateur de battement de fréquence*)
b-frame: ~ **picture** *n* COMP, TV image-b bi-trame *f*
BFT *abbr* (*binary file transfer*) TELEP BFT (*transfert de fichiers binaires*); ~ **file attributes** *n pl* TELEP attributs de fichier BFT *m pl*
B-grade: ~ **contour** *n* RADIO contour de classe B *m*
BH *n* (*busy hour*) TELEP heure chargée *f*, heure de pointe *f*
bias *n* COMP biais *m*, POWER, RADIO polarisation *f*, TEST *quality* biais *m*; ~ **control** COMP commande de polarisation *f*; ~ **current** ELECT courant de polarisation *m*; ~ **distortion** ELECT, TELEP distorsion biaise *f*; ~ **level** COMP niveau de polarisation *m*; ~ **point** COMP point de polarisation *m*
biased[1] *adj* POWER polarisé
biased[2]: ~ **sample** *n* TEST *quality* échantillon biaisé *m*; ~ **test** *n* GEN *statistics* test biaisé *m*
biconic: ~ **connectors** *n pl* ELECT connecteurs biconiques *m pl*
biconical: ~ **aerial** *n* RADIO antenne biconique *f*; ~ **antenna** *n* RADIO antenne biconique *f*
bid *n* GEN tentative de prise *f*, RADIO offre *f*, soumission *f*, tentative *f*; ~ **for a line** TELEP demande de ligne *f*
bidirectional[1] *adj* TRANS, TV bidirectionnel
bidirectional[2]: ~ **diode thyristor** *n* ELECT, PARTS thyristor diode bidirectionnel *m*, thyristor monté en diode bidirectionnel *m*; ~ **network** *n* TV réseau bidirectionnel *m*; ~ **transistor** *n* ELECT, PARTS transistor bidirectionnel *m*; ~ **triode thyristor** *n* ELECT, PARTS thyristor monté en triode bidirectionnel *m*, thyristor triode bidirectionnel *m*
bidirectionally: ~-**predictive coded picture** *n* (*b-picture*) COMP, TV image à codage prédictif bidirectionnel *f* (*image-b*)
bids: ~ **per circuit per hour** *phr* TELEP tentatives de prise par circuit et par heure
bifilar: ~ **winding** *n* POWER enroulement bifilaire *m*
bilateral: ~ **closed user group** *n* (*BCUG*) TELEP groupe fermé d'usagers bilatéral *m*; ~ **closed user group call indicator** *n* TELEP indicateur d'appel de groupe fermé d'usagers bilatéral *m*; ~ **closed user group with outgoing access** *n* (*BCUGOA*) TELEP groupe fermé d'usagers bilatéral avec accès sortant *m*; ~ **control** *n* TELEP, TRANS commande bilatérale *f*
bi-level: ~ **image** *n* TELEP image à deux niveaux *f*
bilinear *adj* PER APP *scanner* bilinéaire
billed: ~ **services** *n pl* TELEP services facturés *m pl*
billing *n* APPL, DATA, RADIO facturation *f*, TELEP facturation *f*, taxation *f*; ~ **address** TELEP destination des factures téléphoniques *f*; ~ **center** *AmE*, ~ **centre** *BrE* APPL centre de facturation *m*; ~ **information** TELEP renseignements tarifaires *m pl*
bimetal *n* PARTS bimétal *m*; ~ **spring** CONTR *regulating device* ressort bimétallique *m*; ~ **springset** CONTR groupe de ressorts bimétalliques *m*
binary: ~ **coding** *n* SIG codage binaire *m*; ~ **counter** *n* ELECT *logic elements* compteur binaire *m*; ~ **data** *n pl* COMP, DATA, TRANS données binaires *f pl*; ~ **digit** *n* COMP, DATA, TRANS bit *m*, chiffre binaire *m*; ~ **error rate** *n* (*BER*) COMP, DATA, TRANS taux d'erreurs binaires *m* (*TEB*); ~ **figure** *n* COMP, TRANS chiffre binaire *m*; ~ **file** *n* COMP, DATA fichier binaire *m*; ~ **file transfer** *n* (*BFT*) COMP, DATA transfert de fichiers binaires *m* (*BFT*); ~ **image** *n* DATA image binaire *f*; ~ **modulation** *n* DATA modulation binaire *f*; ~ **multiplier** *n* DATA multiplicateur binaire *m*; ~ **numeral** *n* COMP, DATA nombre binaire *m*; ~ **phase-shift keying** *n* (*BPSK*) RADIO, SIG modulation par déplacement de phase bivalente *f* (*MDPB*); ~ **rate** *n* COMP, DATA, TELEP, TRANS débit binaire *m*; ~ **search** *n* DATA recherche binaire *f*; ~ **search procedure** *n* DATA procédure de recherche binaire *f*, procédure de recherche dichotomique *f*; ~ **sequence** *n* DATA séquence binaire *f*; ~ **serial signaling rate** *AmE*, ~ **serial signalling rate** *BrE* *n* SIG débit binaire série *m*; ~ **signal** *n* SIG signal binaire *m*; ~ **synchronous communication** *n* (*BSC*) DATA transmission binaire synchrone *f*; ~ **system** *n* COMP système binaire *m*; ~ **tariff system** *n* TELEP système de tarification binaire *m*
bind *n* DATA rattachement *m*
binomial: ~ **distribution** *n* GEN *statistics* distribution binomiale *f*
BIOS *abbr* (*basic input/output system*) DATA BIOS (*système d'exploitation des entrées/sorties*)
BIP *abbr* (*bit interleaved parity*) DATA parité d'entrelacement des bits *f*
biphase: ~ **code** *n* SIG code biphase *m*
bipolar: ~ **circuit** *n* ELECT circuit bipolaire *m*; ~ **code** *n* SIG code bipolaire *m*; ~ **code with three-zero substitution** *n* (*B3ZS*) DATA, SIG, TELEP code bipolaire avec substitution de trois zéros *m*; ~ **logic** *n* ELECT logique bipolaire *f*; ~ **pulse** *n* TELEP impulsion bipolaire *f*; ~ **sequence of pulses** *n* TELEP séquence d'impulsions bipolaires *f*; ~ **signal** *n* SIG signal bipolaire *m*; ~ **transistor** *n* ELECT, PARTS transistor bipolaire *m*; ~ **violation** *n* TRANS violation de bipolarité *f*
BIPT *abbr* (*Belgian Institute for Post and Telephony*) RADIO IBPT (*Institut belge des services postaux et des télécommunications*)
BIP-X *abbr* (*bit interleaved parity X*) DATA parité X à entrelacement de bits *f*
birefringence *n* DATA, TRANS *fibre optics* biréfringence *f*
BIS *abbr* (*bearer independent service*) TELEP service indépendant du support *m*
B-ISDN *abbr* (*broadband ISDN*) TELEP, TRANS RNIS-LB (*RNIS à large bande*); ~ **network termination** *n* (*NT-LB*) TRANS terminaison de réseau pour le RNIS-LB *f*; ~ **PBX** *n* TELEP autocommutateur privé pour le RNIS-LB *m*; ~ **service** *n* APPL, TELEP service RNIS à large bande *m*; ~ **terminal adaptor** *n* (*TA-LB*) TELEP adaptateur de terminal pour le RNIS-LB *m*; ~ **terminal equipment** *n* TELEP équipement terminal pour le RNIS-LB *m*; ~ **network termination 1** *n* (*NT1-LB*) TRANS terminaison de réseau 1

pour le RNIS-LB *f*; ~ **network termination 2** *n* (*NT2-LB*) TRANS terminaison de réseau 2 pour le RNIS-LB *f*

bisecting: ~ **relay** *n* PARTS relais diviseur d'impulsions *m*

bistable *adj* ELECT *logic elements* bistable

bistatic: ~ **radar** *n* RADIO radar bistatique *m*

bistring: ~ **type** *n* DATA type chaîne binaire *m*

bit[1]: ~**-by-bit** *adv* COMP bit à bit

bit[2] *n* COMP, DATA bit *m*, TRANS élément binaire *m* (*EB*), bit *m*; ~**-by-bit encoding** DATA codage bit à bit *m*; ~ **combination** COMP combinaison d'éléments binaires *f*, TELEP combinaison binaire *f*; ~ **disagreements** *n pl* DATA bits différents *m pl*; ~ **efficiency** DATA économie en mémoire *f*; ~ **error** TRANS erreur binaire *f*, erreur numérique *f*; ~ **error code** DATA, RADIO code d'erreur sur les bits *m*; ~ **error rate** (*BER*) COMP, DATA, RADIO taux d'erreurs sur les bits *m* (*TEB*); ~ **interleaved parity** (*BIP*) DATA parité d'entrelacement des bits *f*; ~ **interleaved parity X** (*BIP-X*) DATA parité X à entrelacement de bits *f*; ~ **mapping** DATA mode point *m*; ~**-order of transmission** COMP ordre de transmission des bits *m*; ~ **plane** TELEP plan de bits *m*; ~ **rate** COMP, DATA, TELEP, TRANS débit binaire *m*; ~**-rate compression** DATA compression des débits binaires *f*; ~ **sequence independence** COMP indépendance de la séquence des bits *f*; ~ **stream** DATA train de bits *m*; ~ **switch** SWG commutateur de bits *m*; ~ **timing** DATA, TRANS rythme binaire *m*, rythme de bits *m*; ~**-timing recovery sequence** DATA, RADIO séquence de récupération du rythme binaire *f*; ~**-true arithmetic package** COMP paquetage utilitaire sur type bit étendu *m*

BIT *abbr* (*basic interconnection test*) DATA test d'interconnexion de base *m*

BITE *abbr* (*backward-interworking telephony event*) TELEP événement téléphonique d'interfonctionnement vers l'arrière *m*

bi-timbral *adj* COMP bitimbral

Bitnet *n* DATA Bitnet *m*

bits: ~ **per color** *AmE*, ~ **per colour** *BrE phr* DATA *scanner* bits par couleur; ~ **per inch** *phr* (*bpi*) DATA bits par pouce (*bpp*); ~ **per second** *phr* (*bps*) DATA bits par seconde (*b/s*)

black: ~**-and-white reception** *n* TV télécopie contrastée *f*, télécopie noir sur blanc *f*; ~ **level** *n* COMP, TELEP, TRANS niveau du noir *m*; ~**-level retention** *n* COMP, TELEP, TRANS maintien du niveau du noir *m*; ~ **stroke** *n* ELECT trait noir *m*

blackout *n* RADIO disparition des signaux *f*, silence radio *m*, POWER panne d'électricité *f*

blade: ~ **contact** *n* ELECT, PARTS contact sabre *m*, contact mâle *m*; ~ **contact unit** *n* ELECT, PARTS élément à contact mâle *m*

blank[1] *n* ELECT coche *f*; ~**-and-burst message** DATA message neutralisation-salve *m*; ~ **entry** DATA entrée muette *f*; ~ **label** ELECT étiquette vierge *f*; ~ **line** TELEP interligne *m*

blank[2] *vt* TELEP supprimer

blanked: ~**-off channel** *n* TRANS voie neutralisée *f*

blanket: ~ **coverage** *n* RADIO couverture totale *f*

blanking *n* TV effacement *m*, suppression *f*; ~ **burr** PARTS bavure *f*; ~ **level** TV niveau de suppression *m*; ~ **pulse** TV impulsion de suppression du faisceau *f*; ~ **signal** TV signal de suppression *m*

blast: ~ **out of orbit** *vi* RADIO sortir de l'orbite

blaster *n* COMP programmateur de PROM *m*

bleeder: ~ **resistor** *n* ELECT, PARTS résistance de fuite de grille *f*, POWER résistance de fuite *f*, résistance régulatrice *f*

bleep *n* COMP signal sonore *m*, RADIO bip *m*, bipper *m*, avertissement sonore *m*

bleeper *n* RADIO bip *m*, bipper *m*, terminal de RMU *m*, avertisseur radio *m*

B-leg *n* TELEP conducteur de nuque *m*, fil de nuque *m*

blind[1] *adj* COMP aveugle

blind[2]: ~ **copy** *n* DATA copie aveugle *f*; ~ **courtesy copy** *n* (*BCC*) TELEP copie conforme sélective *f* (*CCS*); ~ **region** *n* RADIO zone de silence *f*, zone aveugle *f*; ~ **speed** *n* RADIO vitesse aveugle *f*

blink *vi* COMP, TELEP, TEST *indicating devices* clignoter

blinking *n* TELEP, TEST clignotement *m*

blip *n* RADIO écho *m*

block[1] *n* DATA groupement *m*, GEN *of parameters* bloc *m*; ~**-cancel character** COMP caractère d'annulation de bloc *m*; ~ **check** COMP contrôle de bloc *m*; ~ **code** COMP code complet *m*; ~ **coding** COMP codage par blocs *m*; ~ **companding** COMP compression-extension par blocs *f*; ~ **container** POWER *batteries* bac à compartiments *m*; ~**-dropping indicator** DATA indicateur d'abandon de bloc *m*; ~ **gap** DATA espace interbloc *m*, espace entre blocs *m*; ~ **interleaving** RADIO entrelacement des blocs *m*; ~ **quantization** TRANS quantification des blocs *f*; ~ **separation** TELEP séparation des blocs *f*; ~ **separator** TELEP séparateur de blocs *m*; ~ **sort** COMP tri par bloc *m*; ~ **substructure** DATA sous-structure de bloc *f*; ~ **terminal** TELEP boîte de raccordement *f*, rosace de raccordement *f*, réglette *f*, *on subscriber's telephone* broche de raccordement *f*; ~ **transfer** COMP transfert de bloc *m*; ~ **transmission** TRANS transmission en bloc *f*, transmission par bloc *f*; ~ **tree diagram** DATA diagramme d'arbre de blocs *m*; ~ **type** DATA type de bloc *m*

block[2] *vt* TELEP bloquer; ~ **together** DATA *data* grouper

blockage: ~ **effect** *n* GEN effet de blocage *m*

blocked *n* TELEP bloqué

blocking *n* DATA encombrement *m*, formation des blocs *f*, GEN, SIG, SWG, TELEP blocage *m*; ~ **acknowledgement signal** SIG signal d'accusé de réception de blocage *m*; ~ **alarm** TELEP alarme de blocage *f*; ~ **button** TELEP touche de coupe-circuit *f*; ~ **counter** TELEP compteur de blocages *m*; ~ **direction** ELECT sens de blocage *m*; ~ **grade of service** TELEP niveau de blocage *m*; ~ **indication** TELEP indication de blocage *f*; ~ **lamp** TELEP lampe de blocage *f*; ~ **network** SWG réseau à blocage *m*; ~ **rate** GEN taux de blocage *m*; ~ **signal** TELEP signal de blocage *m*

blow *vt* PARTS claquer, POWER *fuses* fondre, claquer

blowing *n* ELECT, POWER *of fuse* claquage *m*

blown: ~ **fiber** *AmE*, ~ **fibre** *BrE n* TRANS câble soufflé *m*; ~ **fuse** *n* PARTS, POWER fusible fondu *m*, fusible sauté *m*

blurred *adj* COMP flou, brouillé

BMLA *abbr* (*burst-mode link analyser*) DATA analyseur de liaison en mode paquet *m*

BNF *abbr* (*Backus-Naur form*) DATA forme BNF *f* (*forme de Backus-Naur, forme normale de Backus*)

BO *abbr* (*busy out*) TELEP MO (*mise en occupation*)

board *n* SWG *switchboard* tableau *m*; ~ **compartment** PARTS compartiment pour carte imprimée *m*; ~ **extraction tool** TEST extracteur de carte *m*; ~ **extractor** TEST *tools* extracteur de carte *m*; ~ **frame** PARTS châssis de cartes *m*; ~**-level testing** TEST essai au niveau carte *m*; ~ **position** PARTS emplacement de carte *m*; ~ **spacing** PARTS espacement de cartes *m*

bobbin *n* POWER bobine isolante *f*

Bode: ~ **diagram** *n* CONTR, POWER courbes de Bode *f pl*

body *n* GEN corps *m*, TEST *standardization purposes* organisme *m*; ~ **object** COMP élément corps *m*; ~ **placeholder** COMP espace réservé au corps *m*; ~**-stabilized communications satellite** RADIO satellite de télécommunications à corps stabilisé *m*

bold: **in** ~ *adj* COMP *printing font* en gras

bolting *n* RADIO, TRANS verrouillage *m*

BOM *abbr* DATA, PER APP (*beginning of message*) début de message *m*

bonding *n* ELECT collage *m*, PARTS soudage *m*, POWER collage *m*; ~ **sheet** ELECT, POWER feuille de collage *f*

book[1] *n* TELEP répertoire *m*

book[2] *vt* APPL, TELEP commander, inscrire, réserver

booked: ~ **call** *n* APPL, TELEP appel réservé *m*

bookmark *n* COMP marque-page *m*, signet *m*, DATA *Internet* signet *m*, répertoire d'adresses *m*, liste de serveurs *f*
Boolean[1] *adj* COMP, DATA booléen
Boolean[2]: ~ **operation** *n* COMP opération booléenne *f*; ~ **search** *n* COMP recherche booléenne *f*
boost: ~ **charge** *n* POWER *of battery* charge de biberonnage *f*, charge rapide *f*
booster *n* COMP *memory, capacity* accélérateur *m*; ~ **battery** POWER batterie auxiliaire *f*; ~ **converter** POWER convertisseur survolteur *m*
boot[1] *n* COMP programme d'amorçage *m*, programme d'amorce *m*, programme de chargement *m*; ~-**up** COMP amorçage *m*, chargement de système *m*, lancement de système *m*
boot[2] *vt* COMP amorcer, booter, charger, initialiser, lancer; ~ **up** COMP amorcer, booter, charger, initialiser, lancer
bootstrap[1] *n* COMP amorçage *m*, amorce *f*; ~ **loader** COMP programme d'amorçage *m*, programme d'amorce *m*
bootstrap[2] *vt* COMP amorcer, booter, charger, initialiser, lancer
border *n* GEN frontière *f*
bore *vt* PARTS aléser, percer, perforer
boresight *n* RADIO axe de visée *m*, dispositif de visée *m*, ligne de visée *f*; ~ **facility** RADIO méthode de pointage optique *f*; ~ **method** RADIO méthode en visée directe au sol *f*
boring *n* PARTS alésage *m*; ~ **fixture** PARTS support d'alésage *m*
both[1]: ~-**way** *adj* RADIO, TELEP, TRANS à double sens, mixte, bidirectionnel
both[2]: ~-**party release** *n* COMP, TELEP libération par les deux abonnés *f*; ~-**party release signal** *n* TELEP signal de libération par les deux abonnés *m*; ~-**way circuit** *n* TRANS circuit mixte *m*; ~-**way communication** *n* RADIO, TELEP, TV communication bidirectionnelle simultanée *f*; ~-**way group** *n* TRANS faisceau à double sens *m*, faisceau mixte *m*; ~-**way line** *n* TRANS ligne à double sens *f*, ligne mixte *f*
bottleneck *n* GEN goulot d'étranglement *m*
bottom *n* GEN fond *m*, bas *m*; ~ **field** TV trame inférieure *f*; ~ **note** COMP note de bas de page *f*; ~ **stud** PARTS butée inférieure *f*; ~ **view** GEN *technical drawing* vue de dessous *f*
bounce *vi* COMP rebondir
bouncing: ~ **busy hour** *n* (*BBH*) TELEP heure de pointe variable *f*
bound: ~ **data** *n pl* DATA données liées *f pl*; ~ **mode** *n* TRANS *fibre optics* mode de propagation *m*, mode guidé *m*, mode lié *m*
boundary *n* COMP, ELECT *semiconductors* limite *f*, GEN frontière *f*; ~ **representation** COMP représentation par limites *f*
bounding: ~ **volumes** *n* COMP représentation par volumes *f*
bow[1] *n* POWER *printed circuits* cintrage *m*, courbure *f*; ~-**tie aerial** RADIO antenne panneau *f*, antenne repliée *f*; ~-**tie antenna** RADIO antenne panneau *f*, antenne repliée *f*
bow[2] *vt* POWER cintrer
box *n* COMP *on screen* bouton *m*, coffret *m*, DATA boîtier *m*, coffret *m*
BPF *abbr* (*bandpass filter*) POWER, TRANS FPB (*filtre passe-bande*)
bpi *abbr* (*bits per inch*) DATA bpp (*bits par pouce*)
b-picture *n* (*bidirectionally-predictive coded picture*) COMP, TV image-b *f* (*image à codage prédictif bidirectionnel*)
bps *abbr* (*bits per second*) DATA b/s (*bits par seconde*)
BPSK *abbr* (*binary phase-shift keying*) RADIO, SIG MDPB (*modulation par déplacement de phase bivalente*)
BRA *abbr* (*basic rate access*) TELEP accès au tarif de base *m*
brace *n* GEN étai *m*
bracket *n* TRANS fixation *f*, support *m*
braid[1] *n* ELECT, PARTS *of cable*, TRANS tresse *f*
braid[2] *vt* ELECT, PARTS, TRANS guiper, tresser
braided: ~ **cable** *n* ELECT, PARTS, TRANS fil guipé *m*, fil tressé *m*
braiding *n* ELECT, PARTS, TRANS guipage *m*, tressage *m*
brake *n* GEN, PARTS frein *m*; ~ **handle** PARTS levier de frein *m*
branch *n* COMP branchement *m*, DATA, POWER branche *f*; ~ **flip-flop** SWG bascule de dérivation *f*; ~ **formboard** TRANS gabarit de division *m*; ~ **joint** TRANS branchement *m*, connexion *f*; ~-**off** TRANS dérivation *f*; ~ **selection and amplification** TELEP sélection et amplification de branche *f pl*
branching: ~ **block** *n* ELECT, PARTS *connecting devices* boîtier de dérivation *m*; ~ **bypass** *n* PARTS filtre d'aiguillage *m*; ~ **jack** *n* ELECT *connecting device*, PARTS jack de branchement *m*, jack de dérivation *m*; ~ **point** *n* COMP, ELECT, TRANS nud *m*, point de bifurcation *m*, point de dérivation *m*, TV point de branchement *m* (*PB*)
brand *n* COMP, ELECT marque *f*
breadboard: ~ **model** *n* DATA *Internet* maquette *f*, ELECT maquette sur table *f*
breadcrumbs *n pl* COMP marqueur de parcours *m*
break[1] *n* GEN coupure *f*, interruption *f*, rupture *f*, PARTS, POWER *electric conductors* rupture *f*, TRANS *cable* cassure *f*; ~ **contact group** SWG groupe de contacts de rupture *m*; ~ **contact spring set** SWG groupe de contacts de rupture *m*; ~-**in** TELEP intervention *f*; ~-**in hangover time** TELEP temps de maintien pour l'intervention *m*; ~ **jack** SWG jack à rupture *m*; ~ **key** COMP touche d'interruption *f*; ~-**out cable** TRANS câble de dérivation *m*; ~-**over voltage** PARTS *thyristors* tension de retournement *f*; ~ **pulse** TELEP impulsion d'ouverture *f*; ~-**pulsing generator** TELEP génération d'impulsions d'ouverture *f*; ~ **relay** ELECT, PARTS relais de coupure *m*
break[2] *vt* GEN couper, interrompre, PARTS, POWER couper, rompre, TRANS déchirer; ~ **down** ELECT claquer; ~ **into** COMP pénétrer
break[3] *vi* POWER se couper, se rompre; ~ **down** GEN tomber en panne
breakage *n* ELECT *conductors*, POWER rupture *f*
breakdown *n* ELECT *of insulator* claquage *m*, rupture *f*, GEN *mechanical failure* panne *f*, défaillance *f*, *of order* décomposition *f*, PARTS, claquage *m*, TELEP rupture *f*, TEST panne *f*, TRANS *of network* interruption *f*, coupure *f*; ~ **signal** SIG signal de dérangement *m*; ~ **voltage** ELECT, POWER tension d'amorçage *f*, tension de claquage *f*, tension de rupture *f*
breaking: ~ **length** *n* PARTS, TEST longueur de rupture *f*; ~ **strength** *n* PARTS, TEST charge de rupture *f*, limite d'élasticité *f*, résistance à la rupture *f*
breakout: ~ **box** *n* ELECT boîtier *m*; ~ **jacketing** *n* ELECT gainage d'épanouissement *m*
breakover: ~ **current** *n* ELECT, POWER *thyristors* courant de retournement *m*; ~ **point** *n* PARTS *thyristors* point de retournement *m*
breakpoint *n* COMP point d'interruption *m*
breakthrough *n* COMP percée *f*, RADIO *land mobile* brouillage *m*
breastplate: ~ **microphone** *n* PARTS *microphones* microphone cravate *m*, microphone plastron *m*
breath: ~ **control** *n* COMP commande de souffle *f*; ~ **controller** *n* COMP commande de souffle *f*; ~-**controller assign** *n* COMP affectation de la commande de souffle *f*; ~-**controller range** *n* COMP plage de réglage de la commande de souffle *f*; ~-**controller sensitivity** *n* COMP sensibilité de la commande de souffle *f*
Brewster: ~'**s angle** *n* TRANS *fibre optics* angle de Brewster *m*
bridge[1]: ~-**stabilized** *adj* TELEP stabilisé par montage en pont
bridge[2] *n* COMP pont *m*, passerelle *f*, DATA pont *m*, ELECT dérivation *f*, pont *m*; ~ **circuit** ELECT, POWER circuit de dérivation *m*, circuit en pont *m*; ~ **connection** POWER montage en pont *m*; ~ **control cord circuit** TELEP dicorde avec point *m*; ~ **converter** POWER convertisseur en pont *m*; ~ **duplex system** TELEP équipement duplex à pont *m*; ~ **router** DATA pont-routeur *m*; ~ **splice** TRANS épissure en dérivation *f*; ~ **transformer** ELECT transformateur différentiel *m*

bridge³ *vt* TELEP ponter
bridged¹ *adj* POWER ponté, TELEP en dérivation, ponté
bridged²: **~-relay configuration** *n* TELEP configuration de relais pontés *f*
bridging *n* TELEP pontage *m*
bright¹ *adj* GEN lumineux
bright²: **~-field detector** *n* TRANS détecteur de champ lumineux *m*; **~-field sensor** *n* TRANS capteur de champ lumineux *m*
brightening *n* COMP brillance *f*, TV luminance *f*
brightness *n* GEN luminosité *f*, TV signal de luminance *m*; **~ control** PER APP réglage de luminosité *m*
brilliance *n* COMP brillance *f*, GEN luminosité *f*; **~ control** COMP, TV commande de brillance *f*, commande de luminosité *f*
Brillouin: **~ spectrum** *n* DATA spectre de Brillouin *m*
Brinell: **~ test** *n* PARTS *durability* essai Brinell *m*
bring: **~ into orbit** *vt* RADIO satelliser; **~ up to date** *vt* GEN mettre à jour
British: **~ Standards Institution** *n* (*BSI*) GEN Institut britannique de normalisation (*BSI*)
broach *vt* PARTS aléser
broaching *n* PARTS alésage *m*
broadband *n* COMP, DATA, TELEP, TRANS large bande *f*; **~ amplifier** TRANS amplificateur à large bande *m*; **~ channel** TRANS voie à large bande *f*; **~ crosspoint** SWG point de commutation large bande *m*; **~ ISDN** (*B-ISDN*) TELEP, TRANS RNIS à large bande *m* (*RNIS-LB*); **~ ISDN PBX** APPL, CONTR, SWG, TELEP autocommutateur privé pour le RNIS-LB *m*; **~ ISDN service** APPL, TELEP service RNIS à large bande *m*; **~ noise** COMP bruit à large bande *m*; **~ radio system** RADIO système radio à large bande *m*; **~ switch** SWG commutateur à large bande *m*; **~ switching network** SWG, TRANS réseau de connexion à large bande *m*; **~ transmission** TRANS transmission large bande *f*
broadcast¹ *adj* GEN, TV diffusé; **in ~ mode** RADIO en diffusion; **~ by satellite** RADIO, TV diffusé par satellite, distribué par satellite
broadcast² *n* (*BC*) GEN émission *f*, TELEP *telex*, TRANS, TV diffusion *f*; **~ channel** RADIO canal de radiodiffusion *m*; **~ message** RADIO message à diffusion générale *m*; **~ network** RADIO réseau à diffusion *m*; **~ program channel** *AmE*, **~ programme channel** *BrE* RADIO voie radiophonique *f*; **~ reception** RADIO réception de radiodiffusion *f*; **~ repeater** TELEP translation pour diffusion *f*; **~ station** RADIO, TELEP, TV station émettrice *f*, station d'émission *f*; **~ storms** ELECT transmission en rafales *f*; **~ transmitter** PER APP émetteur de radiodiffusion *m*; **~ videography** APPL, TV vidéographie diffusée *f*; **~ videotex** APPL, TV vidéotex diffusé *m*
broadcast³ *vt* GEN, TV diffuser
broadcast⁴ *vi* GEN, TV diffuser
broadcaster *n* TV diffuseur *m*
broadcasting *n* DATA, ELECT, RADIO, TELEP, TV diffusion *f*, radiodiffusion *f*; **~ authority** RADIO organisme de radiodiffusion *m*; **~ band** RADIO bande de radiodiffusion *f*; **~ organization** RADIO, TV organisme de radiodiffusion *m*; **~-satellite service** RADIO, TV service de radiodiffusion par satellite *m*; **~-satellite space station** RADIO, TV station spatiale de radiodiffusion par satellite *f*; **~ schedule** RADIO horaire de radiodiffusion *m*; **~ signal** SIG signal de radiodiffusion *m*; **~ spectrum** RADIO, TRANS spectre de fréquences *m*; **~ station** RADIO, TV station de radiodiffusion *f*
broadside: **~ array aerial** *n* RADIO antenne à rayonnement transversal *f*; **~ array antenna** *n* RADIO antenne à rayonnement transversal *f*
broken *adj* GEN coupé, interrompu, PARTS, POWER rompu, TRANS cassé, déchiré
broker *n* COMP courtier *m*
B-router *n* DATA B-routeur *m*
brownout *n* POWER baisse de tension *f*
browser *n* COMP, DATA logiciel de balayage *m*, feuilleteur *m*, *software* logiciel de survol *m*, butineur *m*
browsing *n* COMP, DATA recherche rapide *f*, parcours libre *m*, consultation *f*, navigation *f*, survol *m*
brush: **~ discharge** *n* POWER aigrette *f*
BS *abbr* COMP (*backspace*) espace arrière *m*, retour arrière *m*, (*backspace character*) caractère d'espacement arrière *m*, caractère de retour arrière *m*, RADIO (*base station*) station de base *f* (*BS*), TELEP (*base station*) station de base *f* (*BS*), (*electronic business set*) TA (*téléphone d'affaires électronique*)
BSC *abbr* (*binary synchronous communication*) DATA transmission binaire synchrone *f*
BSI *abbr* GEN (*British Standards Institution*) BSI (*Institut britannique de normalisation*), (*Belgian Standards Institute*) IBN (*Institut belge de normalisation*)
BSS *abbr* (*base station subsystem*) TELEP sous-système de station de base *m*
B-stage: **~ materials** *n* PARTS *plastics* résine à l'état B *f*, résitol *m*
BSVC *abbr* (*broadcast signaling virtual channel AmE, broadcast signalling virtual channel BrE*) TRANS voie virtuelle diffusée de signalisation *f*
B3ZS *abbr* (*bipolar code with three-zero substitution*) DATA, SIG, TELEP code bipolaire avec substitution de trois zéros *m*
BTM *abbr* (*document bulk transfer and manipulation class*) TELEP BTM (*classe de transfert de masse et de manipulation de document*)
BTNA *abbr* (*burst-timing not acquired*) DATA rythme de paquets non-acquis *m*
BTP *abbr* (*burst time plan*) DATA plan de temps *m*
B-tree *n* COMP, DATA arbre balancé *m*
BU *abbr* RADIO (*bass unit*) cellule de base *f*, (*base band unit*) coeur de chaîne *m*
bucker *n* PARTS tas *m*
buffer *n* COMP tampon *m*, TRANS gainage *m*, gaine *f*; **~ amplifier** ELECT amplificateur séparateur *m*; **~ battery** POWER batterie tampon *f*; **~ circuit** COMP, DATA circuit tampon *m*; **~ memory** COMP, DATA mémoire intermédiaire *f*, mémoire tampon *f*; **~ state** ELECT état tampon *m*; **~ storage** COMP, DATA mémoire tampon *f*; **~ time** ELECT temps de garde *m*
buffering *n* DATA mise en mémoire tampon *f*, ELECT gainage *m*
bug¹: **~-free** *adj* COMP exempt d'erreur
bug² *n* DATA bogue *f*
build: **~-up time** *n* GEN temps de montée *m*, RADIO *of a telegraph signal* temps d'établissement *m*
building: **~-block principle** *n* SWG principe de conception modulaire *m*, principe de construction modulaire *m*; **~ penetration loss** *n* RADIO affaiblissement dû à l'effet d'écran du bâtiment *m*, pertes de pénétration dans les bâtiments *f pl*
built: **~-in check** *n* TEST contrôle incorporé *m*; **~-in computer** *n* COMP ordinateur incorporé *m*; **~-in filter** *n* ELECT filtre incorporé *m*; **~-in loudspeaker** *n* PER APP haut-parleur incorporé *m*; **~-in speaker** *n* COMP haut-parleur incorporé *m*; **~-in test equipment** *n* TELEP, TEST équipement de test incorporé *m*; **~-in warble tone** *n* ELECT tonalité modulée incorporée *f*; **~-in warble tone source** *n* ELECT source à tonalité modulée incorporée *f*; **~ slide** *n* COMP diapositive compilée *f*
bulk *n* COMP volume *m*; **~ data** *n pl* COMP masse de données *f*; **~ dump** COMP vidage global *m*; **~ erasing** COMP effacement global *m*; **~ information** COMP masse de données *f*; **~ modulus** PARTS module de compressibilité *m*; **~ piezoelectric permittivity** ELECT permittivité piézoélectrique de volume *f*; **~ sampling** TEST échantillonnage de matériaux en vrac *m*; **~ storage** *AmE* (*cf bulk store BrE*) COMP mémoire de masse *f*, mémoire de grande capacité *f*; **~ store** *BrE* (*cf*

bulk storage AmE) COMP mémoire de grande capacité *f*, mémoire de masse *f*

bulkhead: ~ **feed-through adaptor** *n* ELECT adaptateur à traversée de cloison *m*; ~ **panel** *n* ELECT panneau de cloison *m*

bulky *adj* PARTS encombrant, volumineux

BULL: ~ **cabling system** *n* (*BCS*) DATA système de câblage BULL (*BCS*)

bullet *n* COMP symbole *m*

bulletin *n* TELEP communiqué *m*, bulletin *m*; ~ **board** (*BB*) COMP, DATA, TELEP babillard électronique *m* (*Can*), messagerie conviviale *f*, panneau d'affichage *m*, tableau d'affichage *m*, Service Télématique *m*; ~ **board system** (*BBS*) COMP, DATA, TELEP babillard électronique *m* (*Can*), messagerie *f*, système de babillard électronique *m* (*BBS*), tableau d'affichage électronique *m*

bump *n* POWER *printed circuits* bosse *f*; ~ **mapping** COMP placage de relief *m*, placage de rugosité *m*, texture par perturbation de la lumière *f*; ~ **test** PARTS essai aux secousses *m*

bunch *vt* TRANS grouper

bunched: ~ **frame-alignment signal** *n* SIG, TRANS signal de verrouillage de trame concentré *m*; ~ **signaling** *AmE*, ~ **signalling** *BrE* *n* SIG, TRANS signalisation concentrée *f*

bunching *n* TRANS groupement *m*

bundle[1] *n* DATA paquet *m*, ELECT, TRANS faisceau *m*

bundle[2] *vt* DATA, GEN empaqueter, mettre en paquet

bundling *n* DATA mise en paquet *f*

burglar[1]: ~**-proof** *adj* TEST *security status* anti-effraction, à l'épreuve de l'effraction

burglar[2]: ~ **alarm** *n* TEST alarme anti-effraction *f*, système avertisseur anti-effraction *m*

burn: ~**-in** *n* COMP déverminage *m*, rodage *m*, TEST *testing components* vieillissement artificiel *m*

burr *n* PARTS bavure *f*

Burrus: ~ **diode** *n* PARTS, TELEP, TRANS DEL à émission frontale *f*, diode électroluminescente à émission frontale *f*, diode de Burrus *f*

burst[1] *n* DATA salve *f*; ~ **data section** DATA champ de données du paquet *m*; ~ **duration** DATA durée de salve *f*; ~**-envelope fall time** DATA temps de descente de l'enveloppe des paquets *m*; ~**-envelope rise time** DATA temps de montée de l'enveloppe des paquets *m*; ~ **isochronous** TELEP isochronisme interrompu *m*, isochronisme par paquets *m*; ~ **loss rate** DATA taux de perte de paquets *m*; ~ **meteor communications** *n pl* TRANS télécommunications par salves météoriques *f pl*; ~ **mode** RADIO, TRANS mode rafale *m*, mode par salve *m*; ~**-mode link analyser** *BrE*, ~**-mode link analyzer** *AmE* DATA analyseur de liaison en mode paquet *m*; ~**-mode tester** DATA, TEST dispositif d'essai en mode paquets *m*; ~ **negator** DATA annulateur de paquets *m*, négateur de paquets *m*; ~ **noise** TRANS bruit impulsif *m*; ~ **position control information** DATA information de commande de position de paquet *f*; ~ **rate** DATA fréquence des paquets de signaux *f*; ~ **synchronization** DATA synchronisation de salve *f*; ~ **time plan** (*BTP*) DATA plan de temps *m*; ~**-time plan change control** DATA gestion du plan de trame *f*; ~**-timing not acquired** (*BTNA*) DATA rythme de paquets non-acquis *m*; ~**-timing synchronization** DATA synchronisation du rythme des paquets *f*; ~ **of tone** SIG tonalité momentanée *f*, TELEP tonalité d'avertissement *f*; ~ **transmission** RADIO, TRANS transmission par rafales *f*

burst[2] *vt* COMP séparer des feuillets, PARTS faire crever, *cable* crever

bus *n* COMP, POWER, SWG, TELEP bus *m*, bus d'alimentation *m*; ~ **adaptation unit** SWG unité d'adaptation de bus *f*; ~ **arbitrator** COMP arbitre de bus *m*; ~ **cable** POWER câble omnibus *m*; ~ **cable chute** PARTS chemin de câble de bus *m*; ~ **configuration** COMP configuration par bus *f*; ~ **connection signal** SIG signal d'accès au bus *m*, signal de connexion sur le bus *m*; ~ **connection unit** SWG unité de raccordement de bus *f*; ~ **driver** ELECT *logic elements* circuit d'attaque de bus *m*, circuit de commande *m*; ~ **network** TELEP réseau en bus *m*, ~ **state signal** SWG signal d'état du bus *m*, ~ **system** SWG système de bus *m*; ~ **topology** DATA topologie en bus *f*

busbar *n* COMP, POWER, TELEP barre d'alimentation *f*, *connection purposes* barre omnibus *f*, barre de distribution d'énergie *f*, barre omnibus d'alimentation *f*, barre omnibus de distribution *f*

bushing *n* POWER traversée isolante *f*

busied: ~ **out** *adj* TELEP bloqué, hors service

business: ~ **communication system** *n* APPL, RADIO, TRANS système de communication d'entreprises *m*; ~**-oriented satellite** *n* RADIO, TRANS satellite à vocation commerciale *m*; ~ **set query** *n* TELEP indication de l'état du poste *f*; ~ **subscriber** *n* RADIO, TRANS abonné professionnel *m*; ~ **system** *n* APPL, RADIO, TRANS système de communication d'entreprises *m*; ~ **television** *n* TV télévision professionnelle *f*; ~ **traffic** *n* TELEP trafic inter-entreprises *m*; ~ **TV** *n* TV TV professionnelle *f*

busy[1] *adj* TELEP occupé

busy[2]: ~ **hour** *n* (*BH*) TELEP heure chargée *f*, heure de pointe *f*, heure de fort trafic *f*; ~ **hour-to-day ratio** *n* TELEP rapport du trafic à l'heure chargée au trafic journalier *m*; ~ **lamp** *n* TELEP lampe d'occupation *f*, voyant d'occupation *m*; ~ **lamp field** *n* TELEP panneau de lampes d'occupation *m*, panneau de voyants d'occupation *m*; ~ **lamp panel** *n* TELEP panneau de lampes d'occupation *m*, panneau de voyants d'occupation *m*; ~ **number** *n* TELEP numéro occupé *m*; ~ **occurrence** *n* TELEP appel sur ligne occupée *m*; ~ **out** *n* (*BO*) TELEP mise en occupation *f* (*MO*); ~ **override** *n* TELEP entrée en tiers *f*, tonalité d'entrée en tiers *f*; ~ **period** *n* TELEP période chargée *f*, période de fort trafic *f*, période de pointe *f*; ~ **season** *n* TELEP saison de pointe *f*; ~ **signal** *n* *AmE* (*cf engaged tone BrE*) TELEP tonalité d'occupation *f*; ~**-state display** *n* TELEP panneau de lampes d'occupation *m*; ~ **status** *n* TELEP état d'occupation *m*; ~ **study** *n* TELEP étude d'intensité de trafic *f*; ~ **test** *n* TEST test d'occupation *m*; ~ **tone** *n* TELEP tonalité d'occupation *f*

busy[3]: ~ **out** *vt* TELEP mettre en occupation

butt[1]: ~**-configurated** *adj* ELECT scellé en configuration

butt[2]: ~ **contact** *n* PARTS, SWG, TELEP contact à pression directe *m*

butt[3] *vt* ELECT rejoindre; ~**-joint** GEN abouter

butted: ~**-ribbon arrays** *n pl* ELECT rainures de ruban abouté *f pl*

button *n* COMP case de commande *f*, GEN touche *f*, bouton *m*

buzzer *n* ELECT *mechanism* sonnerie *f*, GEN, PER APP, TELEP ronfleur *m*, vibreur acoustique *m*, vibreur sonore *m*; ~ **tone** TELEP tonalité du ronfleur *f*, ronflement *m*, tonalité à fréquence vocale *f*

buzzing: ~ **tone** *n* TELEP tonalité à fréquence vocale *f*

buzzword *n* GEN mot à la mode *m*

BWO *abbr* (*backward wave oscillator*) ELECT, TELEP OOR (*oscillateur à onde régressive, oscillateur à onde rétrograde*)

bypass[1] *n* COMP bypass *m*, dérivation *f*, GEN, POWER contournement *m*; ~ **circuit** ELECT circuit de contournement *m*, procédure de dérivation *f*; ~ **procedure** COMP procédure de dérivation *f*

bypass[2] *vt* COMP ignorer, GEN contourner, POWER connecter en dérivation, contourner, shunter

bypassing *n* COMP, POWER dérivation *f*, contournement *m*

bypath *n* SWG réseau image *m*; ~ **system** SWG système à réseau image *m*

byte[1]: ~**-aligned** *adj* COMP aligné sur les octets

byte[2] *n* COMP octet *m*, DATA multiplet *m*, octet *m*, ELECT, TELEP octet *m*; ~ **switch** SWG commutateur d'octets *m*

C

C: ~ **component** *n* DATA élément conditionnel *m*

CA *abbr* DATA (*call accepted*) CA (*communication acceptée*), TELEP (*communication application*) CA (*application de communication*)

cab: ~ **and pillar distribution system** *n* TRANS système des points de répartition primaire et secondaire *m*

cabinet *n* COMP, DATA, PARTS, RADIO coffret *m*, SWG, TRANS armoire *f*, armoire de commande *f*; ~ **base** TRANS socle d'armoire *m*; ~ **connection chute** SWG goulotte de raccordement d'armoire *f*; ~ **radiation** RADIO rayonnement par le coffret *m*

cable *n* GEN câble *m*, TRANS fil *m*; ~ **address** TELEP adresse télégraphique *f* (*adr tél.*); ~ **armor** *AmE*, ~ **armour** *BrE* TRANS armure externe *f*, blindage *m*; ~ **assembly** GEN, PARTS, TRANS assemblage de câble *m*, câble équipé *m*; ~ **bend** TRANS courbe de câble *f*, courbure de câble *f*; ~ **box** ELECT, GEN, POWER, TRANS boîte de jonction *f*, boîte de division *f*, boîte de raccordement *f*; ~ **bracket** TRANS support de câble *m*, fixation de câble *f*; ~ **branch** TRANS branche de câble *f*; ~ **branching** TRANS division de câble *f*; ~ **break** TRANS rupture de câble *f*; ~ **butt** TRANS extrémité de gaine *f*; ~ **cabinet** TRANS armoire à câbles *f*; ~ **cap** TRANS embout de câble *m*; ~ **capacity** TRANS contenance d'un câble *f*; ~ **cellar** TRANS chambre souterraine *f*; ~ **chamber** TRANS chambre souterraine *f*; ~ **chute** TRANS goulotte *f*; ~ **clamp** ELECT serre-câble *m*, TRANS collier de câble *m*; ~ **code** TELEP code de câble *m*; ~ **communications** *n pl* APPL, TRANS télécommunications par câble *f pl*; ~ **connection** TRANS connexion par câble *f*, liaison par câble *f*; ~ **connection board** (*CCB*) TRANS carte de connexion de câble *f*; ~ **core** GEN, TRANS âme *f*; ~ **designation** TRANS désignation de câble *f*; ~ **distribution** TRANS télédistribution *f*, distribution par câble *f*; ~ **distribution cabinet** POWER armoire de dérivation *f*, armoire de distribution par câble *f*; ~ **distribution diagram** TRANS schéma de répartition des câbles *m*; ~ **distribution head** TRANS tête de câble *f*, manchon *m*; ~ **distribution table** GEN tableau de distribution par câble *m*, tableau de répartition des câbles *m*; ~ **drawing** GEN *documentation* dessin de pose *m*, plan de câblage *m*; ~ **drum** TRANS touret *m*; ~ **duct** TRANS caniveau de câble *m*; ~ **entry** TRANS entrée de câble *f*; ~ **factory** GEN, TRANS câblerie *f*; ~ **floor** GEN, TRANS faux plancher *m*, plancher de câblage *m*; ~ **formboard** TRANS gabarit de câblage *m*, *cables* gabarit pour peigne *m*; ~ **forming** TRANS façonnage de câble *m*, laçage de câble *m*; ~**-forming board** TRANS gabarit pour peigne *m*, gabarit de câblage *m*; ~ **grip** PARTS *plant engineering* grip de câble *m*; ~ **group** GEN, TRANS groupe de câbles *m*; ~ **group list** TRANS *documentation* liste de groupes de câbles *f*; ~ **head** TRANS tête de câble *f*; ~ **holder** TRANS tenant de câble *m*; ~ **hole** TRANS trou de passage *m*; ~ **inlet** TRANS entrée de câble *f*; ~ **jacket** GEN, TRANS gaine *f*; ~ **joining** TRANS épissure de câble *f*; ~ **joint** TRANS épissure de câble *f*; ~ **jointer** TRANS épisseur de câbles *m*, assembleur de câbles *m*; ~ **lacing** TRANS entrelacement de câblage *m*; ~ **landing** TRANS atterrissage de câble *m*; ~ **lasher** TRANS ficeleuse *f*; ~ **layer** TRANS *machine* charrue enfouisseuse *f*; ~ **laying** TRANS *outside plant engineering* pose de câbles *f*; ~**-laying ship** TRANS navire câblier *m*; ~**-laying winch** TRANS *outside plant engineering* treuil pour pose de câbles *m*; ~ **layout** TRANS plan des câbles *m*; ~ **length** TRANS tronçon de câble *m*; ~ **list** TRANS *documentation* liste des câbles *f*; ~ **localization** TRANS détection des câbles *f*, localisation des câbles *f*, repérage des câbles *m*; ~ **locater** TRANS détecteur de câble *m*, localisateur de câble *m*; ~ **loss** TRANS, TV perte dans les câbles *f*, perte de câble *f*; ~ **lug** TRANS attache de conducteur *f*, cosse de câble *f*; ~ **mounting drawing** TRANS schéma de montage de câble *m*; ~ **network** TRANS réseau en câble *m*, réseau de câbles *m*, TV réseau de télévision par câbles *m*, réseau câblé *m*; ~ **network operating company** TV opérateur de réseau câblé *m*; ~ **network operator** TV opérateur de réseau câblé *m*; ~ **operator** TV câblo-distributeur *m*, câblo-opérateur *m*; ~ **outlet** TRANS, TV sortie de câble *f*; ~ **pair** TRANS paire en câble *f*; ~ **pitch** TRANS pas d'un câble *m*; ~ **plough** *BrE*, ~ **plow** *AmE* GEN *outside plant engineering* charrue enfouisseuse *f*, charrue rigoleuse *f*, TRANS charrue enfouisseuse *f*, charrue sous-marin *f*; ~ **pressurization** GEN pressurisation des câbles *f*; ~ **pressurization equipment** GEN équipement de pressurisation de câbles *m*; ~ **program** *AmE*, ~ **programme** *BrE* TELEP émission sur chaîne câblée *f*; ~ **pulling** GEN tirage de câble *m*; ~ **quad** TRANS quarte de câble *f*; ~ **raceway** TRANS *cable laying* support de câbles *m*; ~ **rack** TRANS échelle de câbles *f*; ~ **rate** TELEP tarif télégraphique *m*; ~ **reel** TRANS touret *m*; ~ **response** TRANS réponse de câble *f*; ~ **riser** TRANS chemin de câble *m*, colonne montante *f*, gaîne d'ascension *f*; ~ **route** TRANS *cable laying* parcours de câble *m*, trajet de câble *m*; ~ **run** TRANS *cable laying* chemin de câble *m*, trajet de câble *m*, tronçon de câble *m*; ~**-run rung** TRANS barreau d'échelle *m*; ~ **running** TRANS pose de câbles *f*; ~ **runway** TRANS gaîne d'ascension *f*, chemin de câble *m*, cheminée d'ascension *f*; ~ **scheme** TRANS schéma de câbles *m*; ~ **section** TRANS, TV section du câble *f*; ~ **set** TRANS lot de câble *m*; ~**-set table** GEN tableau de lot de câbles *m*; ~ **sheath** TRANS gaine de câble *f*; ~ **shelf** TRANS étagère de câble *f*; ~ **ship** TRANS navire câblier *m*; ~ **sleeve** TRANS boîte de division *f*, manchon *m*; ~ **socket** RADIO, TRANS, TV prise *f*; ~ **splice** TRANS épissure de câble *f*; ~ **storage hold** TRANS cuve de stockage de câble *f*; ~ **support** TRANS support de câble *m*; ~ **system** RADIO, TRANS, TV système de câbles *m*; ~ **tag** TRANS étiquette de câble *f*; ~ **tank** TRANS cuve de stockage de câble *f*; ~ **telephony** TELEP téléphonie câble *f*, téléphonie par câble *f*; ~ **television** (*cable TV*) TV télédistribution *f*, télévision par câble *f*, câblo-diffusion *f*; ~ **television network** TV réseau de télévision par câbles *m*; ~ **television operator** TV exploitant de systèmes câbles *m*; ~ **television system** (*CATV system*) TV système CATV *m*, système de télévision par câbles *m*; ~ **terminal drawing** TRANS *documentation* schéma de raccordement de câble *m*; ~ **termination** TRANS raccordement de câble *m*, tête de câble *f*, terminaison de câble *f*; ~ **tester** TEST testeur de câblage *m*, TRANS vérificateur de câble *m*; ~ **tie** *n pl* PARTS attaches de câble *f pl*; ~ **trailer** TRANS remorque pour touret de câble *f*; ~ **transfer** TELEP virement télégraphique *m*; ~ **trimmer** TRANS coupe-gaine transversal *m*; ~ **trough** TRANS chemin de câbles *m*; ~ **trunk** TRANS jonction de câbles *f*; ~ **TV** (*cable television*) TV câble-diffusion *f*, câblo-diffusion *f*, télédistribution *f*, télévision par câble *f*; ~ **vault** TRANS chambre souterraine *f*; ~ **way** TRANS parcours de câble *m*, trajet de câble *m*; ~**-way drawing** GEN *documentation* dessin de parcours de câble *m*, dessin de trajet de câble *m*; ~ **winch** TRANS treuil de câble *m*

cabled: ~ **distribution** *n* TV câblodistribution *f*, télédistribution *f*

cabling *n* PARTS, TRANS câblage *m*; ~ **duct** TRANS gouttière *f*; ~ **element** TRANS élément de câblage *m*; ~ **plan** GEN dessin de câblage *m*; ~ **table** GEN tableau de câblage *m*; ~ **unit** TRANS faisceau *m*, sous-faisceau *m*

cache: ~ **memory** *n* COMP, DATA antémémoire *f*, mémoire cache *f*; ~ **storage** *n* COMP, DATA antémémoire *f*, mémoire cache *f*; ~ **store** *n* COMP mémoire cache *f*

caching *n* DATA mise en mémoire cache *f*

CAD *abbr* (*computer-aided design*) COMP CAO (*conception assistée par ordinateur*)

CAD/CAM *abbr* (*computer-aided design and computer-aided manufacturing*) COMP CFAO (*conception et fabrication assistées par ordinateur*); ~ **systems engineer** *n* COMP, GEN ingénieur systèmes CFAO *mf*

cadence *n* RADIO cadence *f*

CAE *abbr* (*computer-aided engineering*) COMP IAO (*ingénierie assistée par ordinateur*)

caesura *n* DATA césure *f*

CAI *abbr* (*common air interface*) RADIO CAI (*interface radio commune*)

CAL *abbr* COMP (*computer-assisted learning*) EAO (*enseignement assisté par ordinateur*), (*computer-aided logistics*) logistique assistée par ordinateur *f*

calibrate *n* TEST étalonner

calibrated: ~ **attenuator** *n* TRANS affaiblisseur étalonné *m*, atténuateur étalonné *m*

calibration *n* TEST étalonnage *m*; ~ **center** *AmE*, ~ **centre** *BrE* TEST centre d'étalonnage *m*; ~ **certificate** TEST certificat d'étalonnage *m*; ~ **error** TRANS erreur d'étalonnage *f*

call[1]: ~-**queued** *adj* TELEP mis en attente

call[2] *n* TELEP *incoming* appel *m*, *conversation* communication *f*, conversation *f*, conversation téléphonique *f*, coup de fil *m*; **in** ~ TELEP appel d'arrivée *m*; ~ **acceptance** TELEP acceptation d'appel *f*; ~ **acceptance signal** TELEP signal d'acceptation d'appel *m*; ~ **accepted** (*CA*) COMP, DATA, SIG, TELEP communication acceptée (*CA*), TELEP acceptation d'appel *f*; ~-**accepted condition** DATA, TELEP condition acceptation d'appel *f*; ~-**accepted message** DATA, TELEP message d'acceptation d'appel *m*; ~-**accepted signal** SIG signal d'acceptation d'appel *m*; ~ **accounting device** TELEP taxeur *m*; ~ **acknowledgement** TELEP accusé de réception d'appel *m*; ~ **address** RADIO adresse d'appel *f*; ~ **agent key** TELEP touche appel position *f*; ~ **allotter** TELEP distributeur d'appels *m*; ~ **attempt** RADIO, TELEP tentative d'appel *f*; ~ **barring** TELEP interdiction d'appel *f*, restriction d'appel *f*, *answer machine, call screening* discrimination des appels *f*; ~-**barring equipment** TELEP discriminateur *m*; ~-**binding posts** *n pl* TRANS bornes d'appels *f pl*; ~ **box** TELEP cabine publique *f*, cabine téléphonique *f*, cabine téléphonique publique *f*; ~ **buffer** TELEP tampon d'appel *m*; ~-**by-call service** APPL service appel par appel *m*; ~ **charge** APPL frais de communication *m pl*, taxe d'appel *f*; ~ **charge context** TELEP contexte de taxation *m*; ~ **charge counter** APPL compteur de taxes *m*; ~ **charge rate** APPL tarif de taxe d'appel *m*, tarification de base *f*, TELEP tarif de taxe d'appel *m*; ~ **charging** APPL taxation des appels *f*; ~-**charging analysis** APPL analyse taxation *f*; ~-**charging equipment** APPL équipement de facturation des appels *m*, équipement de taxation des appels *m*; ~ **clearing** COMP, TELEP libération de la communication *f*; ~ **collision** TELEP collision d'appels *f*; ~ **collision at the DTE/DCE interface** TELEP collision d'appel à l'interface ETTD/ETCD *f*; ~ **congestion** TELEP encombrement d'appel *m*; ~ **congestion loss** TELEP encombrement d'appel perte *m*; ~ **congestion probability of loss** TELEP encombrement d'appel probabilité de perte *m*; ~-**connected signal** SIG signal de connexion *m*; ~ **connection** TELEP établissement de communication *m*; ~ **control** (*CC*) TELEP traitement d'appel *m*; ~ **control agent** (*CCA*) DATA agent de traitement d'appel *m*; ~ **control character** TELEP caractère de commande d'appel *m*; ~-**control procedure** TELEP procédure de commande d'appel *f*; ~ **control signals** *n pl* SIG signaux de commande des appels *m pl*; ~ **counter** TELEP compteur d'appels *m*, compteur de conversations *m*; ~ **data recording** (*CDR*) APPL, DATA enregistrement des données d'appels *m* (*EDA*), enregistrement des données de communications *m* (*EDC*); ~ **deflection** TELEP renvoi du terminal *m*; ~ **destination** TELEP adresse *f*; ~-**detail recording service** (*CDRS*) APPL service d'enregistrement des données d'appels *m* (*service EDA*), service d'enregistrement des données de communications *m* (*service EDC*); ~-**detail recording system** APPL système d'enregistrement des données d'appel *m* (*système EDA*), système d'enregistrement des données de communication *m* (*système EDC*); ~-**detail report** TELEP rapport détaillé des appels *m*; ~ **discrimination** TELEP discrimination des appels *f*; ~ **display** TELEP affichage du numéro demandeur *m*; ~ **distributor** TELEP distributeur d'appels *m*; ~ **diversion** TELEP renvoi d'appel *m*; ~ **diversion on busy** TELEP renvoi d'appel sur occupation *m*; ~ **diversion on no reply** TELEP renvoi d'appel sur non-réponse *m*; ~ **diverter** TELEP détourneur d'appels *m*, déviateur d'appel *m*, dispositif de renvoi d'appels *m*; ~ **duration** TELEP *answer machine* durée de l'appel *f*, durée de la communication *f*, durée de la conversation *f*; ~ **establishment** TELEP établissement d'appel *m*; ~ **extending** TELEP transfert d'appel *m*; ~ **failure** TELEP échec d'appel *m*; ~ **failure signal** SIG signal d'échec d'appel *m*; ~ **fee** TELEP taxe de conversation *f*; ~ **filtering** TELEP filtrage *m*, filtrage des appels *m*, filtrage des communications *m*; ~ **finder** TELEP chercheur de ligne *m*, chercheur d'appel *m*; ~ **flow** TELEP flux du trafic *m*; ~ **forward** TELEP renvoi automatique des appels *m*, renvoi automatique *m*; ~ **forward all calls** TELEP renvoi automatique tout appel *m*; ~ **forward busy** TELEP renvoi automatique sur occupation *m*; ~ **forward intra-group** TELEP renvoi automatique intragroupe *m*; ~ **forward no answer** TELEP renvoi automatique pas-de-réponse *m*, renvoi automatique sur non-réponse *m*; ~ **forward reason display** TELEP affichage de la cause du renvoi automatique *m*; ~ **forward remote** TELEP renvoi automatique vers l'extérieur *m*; ~ **forwarder** TELEP déviateur d'appel *m*; ~ **forwarding** TELEP renvoi automatique *m* (*Can*), renvoi automatique des appels *m* (*Can*), renvoi d'appel *m*, transfert d'appel automatique *m*; ~ **forwarding all calls** TELEP renvoi automatique tout appel *m*; ~ **forwarding busy** TELEP renvoi automatique sur occupation *m*; ~ **forwarding-may-occur indicator** DATA indicateur de prolongation d'appel possible *m*; ~ **forwarding no answer** TELEP renvoi automatique pas de réponse *m*, renvoi automatique sur non-réponse *m*; ~ **forwarding remote** TELEP renvoi automatique vers l'extérieur *m*; ~ **handling** TELEP traitement des appels *m*, traitement des communications *m*; ~-**handling area** TELEP zone de traitement d'appel *f*; ~-**handling software machine** TELEP machine logique multi-enregistreur *f*; ~ **hold** TELEP mise en attente *f*, mise en garde *f*; ~ **holding** TELEP mise en attente *f*, mise en garde *f*; ~-**holding facility** TELEP mise en attente *f*, mise en garde *f*; ~ **identification** DATA, TELEP identification d'appel *f*; ~-**identification line** (*CIL*) TELEP ligne d'identification de l'appel *f* (*CIL*); ~ **identifier** TELEP identificateur d'appel *m*, identificateur de communication *m*; ~ **identity** DATA, TELEP identité d'appel *f*; ~ **in progress** TELEP communication en cours *f*, communication en cours d'établissement *f*, conversation en cours; ~-**in software** TELEP appel *m*; ~-**indicating device** TELEP indicateur d'appel *m*; ~ **indicator** TELEP indicateur d'appel *m*; ~ **information** TELEP information d'appel *f*; ~-**information service signal** SIG signal de service des renseignements pour les appels *m*; ~ **initiation** TELEP déclenchement d'un appel *m*; ~ **initiator** DATA lanceur de la communication *m*; ~-**in-progress cost information** APPL indication de coût en temps réel *f*; **~-in-progress switching** SWG commutation des appels en cours *f*, transfert entre les cellules *m*; **~ intensity** TELEP intensité d'appel *f*; **~ intent** TELEP intention d'appel *f*; **~ interception** TELEP interception d'appel *f*; **~ letters** *n pl* GEN indicatif *m*, indicatif d'appel *m*; **~ log** TELEP listage des appels *m*; **~ logging** TELEP enregistrement des appels *m*, relevé des communications *m*; **~ memory** TELEP mémoire d'appels *f*; **~ meter** TELEP compteur d'appels *m*, compteur de conversations *m*; **~ metering** TELEP comptage d'appels *m*, comptage de conversations *m*; **~-metering equipment** APPL équipement de comptage d'appels *m*; **~-metering store** APPL mémoire de comptage d'appels *f*; **~ modification** TELEP modification d'appels *f*; **~-modification completed message** (*CMC*) TELEP message de modification d'appel effectuée *m* (*MAE*); **~-modification reject message** (*CMRJ*) TELEP message de refus de modification d'appel *m* (*MAR*); **~-modification request** (*CMR*) TELEP demande de modification d'appel *f* (*MAD*); **~ not accepted** TELEP appel non accepté *m*; **~ origin** TELEP provenance d'un appel *f*; **~ origination** TELEP émission d'appel *f*, lancement d'appel *m*; **~ park** TELEP mise en garde tous postes d'un appel *f*, mise en garde par indicatif *f*; **~ phases** *n pl* TELEP phases d'un appel *f pl*, phases d'une communication *f pl*; **~ pick-up** TELEP interception d'appel *f*, prise d'appel *f*; **~ pick-up group** TELEP groupe de prises d'appel *m*; **~ processing** TELEP traitement des appels *m*; **~ processor** (*CP*) TELEP processeur d'appels *m*; **~ progress** TELEP progression d'appel *f*; **~ progress signal** (*CPS*) SIG, TELEP signal de progression de l'appel *m*; **~ progress signal logical channel** TELEP voie logique pour les signaux de progression de l'appel *f*; **~ progress tone** TELEP tonalité de progression d'appel *f*; **~ queue** TELEP file d'attente d'appels *f*; **~ queueing** TELEP mise en attente *f*, mise en file d'attente des appels *f*; **~-queueing facility** TELEP mise en attente *f*, mise en file d'attente *f*; **~-queueing system** TELEP exploitation avec mise en file d'attente des appels *f*; **~ recall** PER APP, TELEP rappel automatique *m* (*RAG*); **~ recorder** TELEP enregistreur de conversation *m*; **~ reference** TELEP référence d'appel *f*; **~ register** (*CR*) TELEP compteur d'appels *m*, compteur de conversations *m*, registre d'appels *m*, registre de conversations *m*; **~-rejected message** (*CRM*) TELEP message de refus d'appel *m* (*MRA*); **~-release time** TELEP temps de libération *m*; **~ released** DATA communication libérée *f*; **~ request** TELEP demande d'appel *f*, demande de communication *f*; **~ request packet** (*CRP*) TELEP paquet de demande d'appel *m*, paquet de demande de communication *m*; **~ request signal** SIG signal de demande d'appel *m*; **~ restriction** TELEP service restreint *m*; **~ retrieval** TELEP reprise *f*; **~ routine sequencer** TELEP distributeur séquentiel d'appels *m*; **~ screening** TELEP filtrage des appels *m*, filtrage des communications *m*; **~ security** TELEP secret de conversation *m*, secret des communications *m*; **~ selection** TELEP prise d'appel par le standardiste *f*; **~-selector responder** TELEP répondeur à sélection d'appels *m*; **~ sender** TELEP robot d'appels *m*; **~ sequence** TELEP séquence d'appel *f*; **~ sequencer** TELEP distributeur d'appels *m*, distributeur séquentiel d'appels *m*; **~ setup** TELEP établissement de communication *m*, établissement d'appel *m*; **~ setup control signal** SIG signal de commande pour l'établissement des communications *m*; **~ setup delay** (*CSD*, *CSUD*) TELEP durée d'établissement d'une communication *f*, délai d'établissement de la communication *m*, temps d'établissement de communication *m*; **~ setup packet** TELEP paquet d'appel *m*; **~ setup phase** TELEP phase d'établissement de la communication *f*; **~ setup time** GEN temps d'établissement d'appel *m*, temps d'établissement de communication *m* (*TEC*); **~ sign** RADIO, TELEP indicatif d'appel *m*; **~ signal** TELEP signal d'appel *m* (*SA*); **~ simulator** TELEP, TEST simulateur d'appels *m*; **~ splitting** TELEP consultation séparée des correspondants *f*, séparation des appels *f*; **~ store** SWG mémoire d'appels *f*; **~ string** TELEP chaîne d'appel *f*; **~ success rate** TEST taux d'aboutissement des appels *m*; **~ trace** DATA, TELEP identification d'appel *f*; **~ transfer** TELEP transfert d'appel *m*; **~ transfer on "busy"** TELEP transfert d'appel en cas de numéro occupé *m*; **~ transfer on "ringing tone no reply"** TELEP transfert d'appel en cas de retour d'appel et de non-réponse *m*; **~ waiting** (*CW*) TELEP appel en instance *m*, communication en attente *f*, indication d'appels en instance *f*, appel en attente *m*, signal d'appel *m* (*SA*), signal d'appel en attente *m* (*SA*); **~-waiting indication** TELEP indication d'appel en attente *f*, signal d'appel en instance *m*; **~-waiting services** *n pl* TELEP services d'appels en instance *m pl*; **~-waiting tone** TELEP tonalité d'appel en attente *f*; **~ wire** TELEP fil d'appel *m*

call³ *vt* GEN appeler, TELEP appeler, téléphoner, donner un coup de fil à, téléphoner à; **~ back** TELEP rappeler; **~ collect** *AmE* (*cf reverse the charges BrE*) APPL, TELEP téléphoner en PCV; **~ into execution** TELEP *subroutine* lancer; **~ over the intercom** TELEP appeler à l'interphone; **~ toll-free** TELEP appeler par numéro vert

call⁴ *vi* GEN appeler

call⁵: **be on ~** *phr* RADIO être de permanence; **~ connected** *phr* (*CC*) DATA communication établie

callback *n* TELEP rappel *m*; **~ queue** (*CBQ*) TELEP file d'attente avec rappel automatique *f*

called: **~ exchange** *n* TELEP central de destination *m*, commutateur de destination *m*; **~-line charging** *n* TELEP taxation à l'arrivée *f*; **~-line identification** *n* (*CDLI*) TELEP identification de la ligne appelée *f*, identification de la ligne demandée *f*; **~-line identification request indicator** *n* TELEP indicateur de demande d'identification de la ligne du demand*m*; **~-line identification signal** *n* SIG signal d'identification de la ligne appelée *m*; **~-line identity** *n* DATA, TELEP identité de la ligne appelée *f*; **~-line identity indicator** *n* TELEP indicateur d'identité de la ligne appelée *m*; **~-N address** *n* DATA adresse N d'appelé *f*; **~ network service user** *n* (*called NS user*) DATA utilisateur du service de réseau demandé *m*; **~ NS user** *n* (*called network service user*) DATA utilisateur du service de réseau demandé *m*; **~ number** *n* TELEP numéro appelé *m*; **~-number display** *n* TELEP affichage du numéro appelé *m*; **~ party** *n* TELEP appelé *m*, correspondant *m*, demandé *m*; **~-party release** *n* SWG, TELEP libération par l'abonné demandé *f*; **~-station identity** *n* (*CSI*) TELEP identité de la station demandée *f*; **~ subscriber** *n* TELEP demandé *m*; **~-subscriber identification** *n* (*CSI*) DATA, TELEP identification de l'abonné appelé *f*, identification de l'abonné demandé *f*; **~-subscriber identification-DIS** *n* (*CSI-DIS*) DATA, TELEP identification de l'abonné demandé-DIS *f* (*IADE*); **~-subscriber release** *n* SWG, TELEP libération par l'abonné demandé *f*; **~ telephone** *n* TELEP téléphone appelé *m*; **~ terminal** *n* TELEP équipement terminal demandé *m*

callee *n* TELEP demandé

caller *n* TELEP appelant *m*, demandeur *m*; **~ display** TELEP affichage du numéro demandeur *m*

calling *n* DATA *datacom* appel *m*, TELEP *for bids* lancement *m*; **~ card** TELEP télécarte *f*; **~ channel** RADIO voie d'appel *f*, *mobicom* canal d'appel *m*; **~ device** TELEP organe d'appel *m*; **~ indicator** PER APP indicateur d'appel *m*; **~ into execution** TELEP *of subroutine* lancement *m*; **~ lamp** PER APP lampe d'appel *f*; **~-line identification** (*CLI*) TELEP identification de l'appel *f*, identification de la ligne appelante *f* (*ILA*); **~-line identification display** (*CLID*) TELEP présentation de l'identification de la ligne appelante *f* (*PILA*); **~-line identification facility** TELEP identification de la ligne du demandeur *f*; **~-line identification presentation** (*CLIP*) TELEP présentation de l'identification de la ligne appelante *f* (*PILA*); **~-line identification restriction** (*CLIR*) TELEP restriction de l'identification de la ligne appelante *f* (*RILA*); **~-line identification signal** SIG signal d'identification de la ligne appelante *m*; **~-line identity** TELEP identité de la ligne du demandeur *f*; **~ modem** DATA modem en mode appel *m*; **~-N address** DATA adresse N d'appelant *f*; **~ network service user** (*calling NS user*) DATA utilisateur du service de réseau demandeur *m*; **~ NS user** (*calling network service user*) DATA utilisateur du service de réseau demandeur *m*; **~-number identification** DATA, RADIO, TELEP identification du numéro appelant *f*; **~-number indication** TELEP service d'indication du numéro du demandeur *m*; **~ party** TELEP appelant *m*, demandeur *m*; **~-party address** TELEP adresse de la ligne appelante *f*, adresse du demandeur *f*; **~-party category indicator** TELEP indicateur de catégorie du demandeur *m*; **~-party category request indicator** TELEP indicateur de demande de catégorie du demandeur *m*; **~-party category response indicator** TELEP indicateur de réponse à une demande de catégorie du demandeur *m*; **~-party data** *n pl* TELEP données d'appel *f pl*; **~-party hold** TELEP mise en attente du demandeur *f*; **~-party number incomplete** TELEP numéro du demandeur incomplet *m*; **~-party release** SWG, TELEP libération par l'abonné demandeur *f*; **~ rate** TELEP taux d'appel *m*, taux d'appels *m*; **~ sequence** TELEP séquence d'appel *f*; **~ signal** TELEP signal d'appel *m* (*SA*); **~ station** TELEP identification *f*, identification du poste demandeur *f*; **~ subscriber** TELEP demandeur *m*; **~-subscriber identification** (*CIG*) DATA, TELEP identification de l'abonné appelant *f*; **~-subscriber identification-DTC** (*CIG-DTC*) DATA, TELEP identification de l'abonné demandeur-DTC *f* (*CIG-DTC*); **~-subscriber release** SWG, TELEP libération par l'abonné demandeur *f*; **~ telephone** TELEP téléphone appelant *m*; **~ terminal** TELEP équipement terminal demandeur *m*; **~ tree** DATA arbre d'appel *m*

calls[1] *n pl* GEN trafic *m*

calls[2]: **~ barred** *phr* TELEP appels interdits; **~ per second** *phr* (*CS*) GEN appels par seconde; **~ waiting** *phr* PER APP *answer machine* appels en attente, appels en instance

CAM *abbr* (*computer-aided manufacturing*) COMP FAO (*fabrication assistée par ordinateur*)

CAMC *abbr* (*customer-access maintenance center AmE, customer-access maintenance centre BrE*) TEST centre de maintenance des accès clients *m*

camcorder *n* TV caméscope *m*

camera *n* TV caméra *f*

camp: **~-on** *n* TELEP mise en attente sur poste occupé *f*, parcage *m*

can *n* ELECT *semiconductors* boîtier *m*

Canadian: **~ General Standards Board** *n* (*CGSB*) GEN Office des normes générales du Canada *m* (*ONGC*); **~ Radio, Television and Telecommunications Commission** *n* (*CRTC*) RADIO, TELEP, TV Conseil de la radiodiffusion et des télécommunications canadiennes *m* (*CRTC*); **~ Standards Association** *n* (*CSA*) GEN Association canadienne de normalisation *f* (*ACN*)

cancellation *n* RADIO suppression *m*, TELEP *of telephone line* résiliation *f*; **~-completed signal** SIG signal d'exécution d'annulation *m*; **~ ratio** RADIO taux d'annulation *m*; **~-request signal** SIG signal de demande d'annulation *m*

canceler *AmE see canceller BrE*

canceller *n BrE* ELECT suppresseur *m*

canning *n* POWER gainage *m*

capability: **~ approval** *n* TEST *quality* reconnaissance d'aptitude *f*; **~ test** *n* DATA test de capacité *m*

capacitance *n* ELECT capacité *f*; **~ diode** ELECT *semiconductors*, PARTS diode à capacité variable *f*; **~ standard** ELECT étalon de capacité *m*

capacitive[1] *adj* ELECT capacitif

capacitive[2]: **~ coupling** *n* ELECT couplage capacitif *m*; **~ load** *n* ELECT charge capacitive *f*

capacitor *n* ELECT capacité *f*, condensateur *m*, PARTS, POWER condensateur *m*; **~ microphone** TELEP microphone électrostatique *m*, microphone à condensateur *m*; **~ plate** POWER armature d'un condensateur *f*

capacity *n* GEN capacité *f*; **~ decrease** (*CD*) GEN diminution de capacité *f*; **~ increase** GEN augmentation de capacité *f*; **~ top** RADIO *aerials* capacité terminale *f*

capillary: **~ soldering** *n* PARTS soudage par capillarité *m*

capitals *n pl* (*caps*) GEN majuscules *f pl*

caps *abbr* (*capitals*) GEN majuscules *f pl*

capture *n* DATA saisie *f*, ELECT accrochage *m*, GEN acquisition *f*, capture *f*; **~ area** RADIO zone de captage *f*; **~ effect** ELECT effet de capture *m*; **~ frequency** ELECT fréquence d'accrochage *f*

CAQ *abbr* (*computer-aided quality*) COMP qualité assistée par ordinateur *f*

car: **~ installation kit** *n* RADIO *mobicom*, TELEP kit d'installation sur véhicule *m*; **~ phone** *n* RADIO, TELEP téléphone de voiture *m*; **~ telephone** *n* RADIO, TELEP téléphone de voiture *m*

carbon: **~ copy** *n* (*Cc*) DATA copie carbone *m* (*Cc*); **~-film resistor** *n* ELECT, PARTS résistance à couche de carbone *f*; **~-granule chamber** *n* PARTS *microphones* chambre à grenailles *f*; **~ granules** *n pl* PARTS *microphones* grenailles de charbon *f pl*; **~ lightning arrester** *n* POWER *lightning protection* parafoudre à charbon *m*; **~ microphone** *n* PARTS microphone à charbon *m*; **~ microphone current feed** *n* POWER, TELEP courant d'alimentation du microphone à charbon *m*

carcinotron *n* ELECT carcinotron *m*

card *n* DATA, ELECT carte *f*; **~ edge** ELECT *printed circuits* bord de carte *m*; **~ follow-on** PER APP *payphone* enchaînement de carte *m*; **~ holder** TELEP détenteur d'une carte *m*; **~ hopper** COMP magasin d'alimentation de cartes *m*; **~-operated payphone** TELEP cabine à cartes *f*; **~ path** COMP chemin de cartes *m*; **~ reader** APPL, COMP, PER APP, TV lecteur de carte *m*

cardphone *n* TELEP cabine à cartes *f*, publiphone à cartes *m*

carriage: **~-control tape** *n* PER APP *printers* bande pilote *f*; **~ return** *n* (*CR*) retour à la ligne *m*

carrier *n* ELECT *semiconductors* porteur *m*, porteur de charge *m*, GEN télécommunicateur *m*, opératrice *f*, exploitant de réseaux interurbains *m*, *operator* opérateur *m*, TRANS porteuse *f*, courant porteur *m*; **~ circuit** TRANS circuit à courants porteurs *m*; **~ component** ELECT composant porteur *m*; **~-current connection** TRANS liaison à courant porteur *f*; **~-current device** TRANS dispositif à courants porteurs *m*; **~-current telegraphy** TELEP télégraphie à courant porteur *f*; **~ detect** (*CD*) TRANS détection de porteuse *f*; **~ frequency** TRANS fréquence porteuse *f*; **~-frequency connection** TRANS liaison à fréquence porteuse *f*; **~-frequency offset** RADIO, TRANS décalage de fréquence porteuse *m*; **~-frequency terminal** TRANS terminal à fréquence porteuse *m*; **~ group alarm** (*CGA*) TRANS alarme de groupe primaria *f* (*AGP*); **~-interference ratio** TRANS rapport porteuse-parasites *m*; **~-link interference** TRANS effet des résidus de porteurs *m*; **~-operated switch** (*COS*) RADIO commutateur commandé par porteur *m*; **~ power** RADIO puissance porteuse *f*; **~ power requirement** TRANS puissance de porteuse requise *f*; **~ rejection** TRANS affaiblissement de la porteuse *m*; **~-sense multiple access** (*CSMA*) RADIO, SIG, TELEP, TRANS accès multiple par détection de la porteuse *m* (*AMDP*); **~-sense multiple access with collision detection** (*CSMA/CD*) RADIO, SIG, TELEP, TRANS accès multiple avec écoute de la porteuse et détection de collisions *m* (*CSMA/CD*); **~-sense signal** DATA signal de détection de porteuse *m*; **~-sense system** DATA système de détection de porteuse *m*; **~ separation** TRANS écart entre porteuses *m*;

~ **signal** COMP signal porteur *m*, TRANS onde porteuse *f*; ~ **storage** ELECT, PARTS *semiconductors* accumulation des porteurs de charge *f*; ~ **storage time** ELECT *semiconductors* retard à la décroissance *m*; ~ **supply** TRANS alimentation des porteuses *f*; ~ **suppression** TRANS suppression de la porteuse *f*; ~**-to-interference ratio** (*C/I*) TRANS rapport porteuse-brouillage *m*; ~**-to-intermodulation noise density ratio** TRANS rapport porteuse-densité de bruit d'intermodulation *m*; ~**-to-intermodulation noise power ratio** TRANS rapport porteuse-puissance de bruit d'intermodulation *m*; ~**-to-noise density ratio** TRANS rapport porteuse-densité de bruit *m*; ~**-to-receiver thermal noise ratio** TRANS rapport porteuse-bruit thermique du récepteur *m*; ~**-to-thermal noise power ratio** TRANS rapport porteuse-puissance de bruit thermique *m*; ~**-to-total-noise-power ratio** TRANS rapport porteuse-puissance de bruit total *m*; ~ **transmission** TRANS transmission par courants porteurs *f*; ~ **wave** TRANS onde porteuse *f*; ~**-wave indication supervision** TRANS surveillance d'indication de porteuse *f*

carry *vt* DATA *Internet* véhiculer, TELEP *traffic* véhiculer, écouler, TV transporter; ~ **out** COMP, GEN exécuter

cartridge *n* GEN cartouche *f*; ~ **box** COMP coffret à cartouches *m*; ~ **fuse** PARTS, POWER fusible cartouche *m*; ~ **loading** COMP charge par cartouche *m*; ~ **save** COMP sauvegarde sur cartouche *f*

CAS *abbr* SIG, TRANS (*channel-associated signaling AmE, channel-associated signalling BrE*) signalisation voie par voie *f*

cascade[1]: ~ **connection** *n* COMP liaison en cascade *f*, ELECT montage en cascade *m*; ~ **control** *n* COMP commande en cascade *f*, CONTR régulation en cascade *f*

cascade[2] *vt* ELECT cascader

cascading: ~ **windows** *n pl* COMP fenêtres en cascade *f pl*

case *n* COMP étui *m*, ELECT *semiconductors* boîtier *m*, GEN valisette *f*, boîtier *m*, PARTS capot *m*, étui *m*, enveloppe *f*, coffret *m*, boîtier *m*, TELEP étui *m*; ~**-based reasoning** DATA raisonnement basé sur des cas *m*; ~ **capacitance** ELECT *semiconductors* capacité du boîtier *f*; ~ **holder** PARTS support de capot *m*; ~ **study** GEN étude de cas *f*; ~ **support** TELEP support de capot *m*; ~ **temperature** ELECT *semiconductors* température du boîtier *f*

CASE *abbr* (*computer-aided software engineering*) COMP génie logiciel assisté par ordinateur *m*

casing *n* COMP, PARTS, TELEP étui *m*

Cassegrain: ~ **antenna** *n* RADIO antenne à optique Cassegrain *f*

cassette *n* COMP cassette; ~ **drive** COMP unité de cassettes *f*; ~ **interface** COMP interface cassette *f*; ~ **save** COMP sauvegarde sur cassette *f*

CAT *abbr* (*computer-aided testing*) COMP test assisté par ordinateur *m*

catalog *AmE see catalogue BrE*

catalogue[1] *n BrE* COMP, DATA, TELEP catalogue *m*

catalogue[2] *vt BrE* COMP, DATA, TELEP cataloguer

catastrophic: ~ **degradation** *n* TELEP *fibre optics* dégradation catastrophique *f*; ~ **failure** *n* TEST défaillance catastrophique *f*

catchment: ~ **area** *n* RADIO zone de captage *f*

category: ~ **analyser** *BrE*, ~ **analyzer** *AmE* *n* TELEP analyseur de catégorie *m*; ~ **marking** *n* TELEP marquage de catégorie *m*; ~ **store** *n* TELEP mémoire de catégories *f*; ~ **transfer** *n* TELEP transfert de catégorie *m*

catenate *vt* COMP enchaîner

cathode *n* ELECT, PARTS cathode *f*; ~**-covering current** ELECT courant cathodique minimal *m*; ~ **follower** ELECT montage cathodyne *m*; ~**-ray direction finder** RADIO radiogoniomètre à tube cathodique *m*; ~**-ray screen** ELECT, PARTS écran cathodique *m*; ~**-ray tube** (*CRT*) ELECT, PARTS tube à rayon cathodique *m* (*TRC, TCR*); ~ **terminal** ELECT *semiconductors* borne de cathode *f*

cathodic: ~ **protection** *n* PARTS protection cathodique *f*

CATV *abbr* (*community-antenna television*) TV télévision à antenne collective *f*, télévision à antenne communautaire *f*; ~ **operator** *n* TV exploitant de réseaux câblés de télévision *m*; ~ **system** *n* TV système CATV *m*

CAU *abbr* (*controlled-access unit*) TELEP unité à accès contrôlé *f*

causality *n* DATA causalité *f*

cause: ~ **of failure** *n* TEST origine d'une défaillance *f*, origine d'une panne *f*

CAV *abbr* (*constant angular velocity*) COMP VAC (*vitesse angulaire constante*)

cavity *n* ELECT, TRANS *waveguides* cavité *f*; ~ **resonator** ELECT, TRANS *waveguides* cavité optique résonnante *f*, résonateur à cavité *m*

CB *abbr* (*citizens' band*) RADIO CB (*bande CB, bande publique*)

C-band *n* RADIO bande C *f*

CBDS *abbr* (*connectionless broadband data service*) APPL, DATA service de données à large bande sans connexion *m*

CBO: ~ **stream** *n* (*continuous bit oriented stream*) DATA train de bits en continu *m*

CBQ *abbr* (*callback queue*) TELEP file d'attente avec rappel automatique *f*

CBR *abbr* (*constant bit rate*) COMP, DATA, TELEP, TRANS débit binaire constant *m*

Cc *abbr* (*carbon copy*) DATA Cc (*copie carbone*)

CC *abbr* CONTR (*control channel*) canal de commande *m*, canal de contrôle *m*, DATA (*concatenated coding*) codage chaîné *m*, (*call connected*) communication établie, SIG (*connect confirm*) confirmation de connexion *f*, TELEP (*call control*) traitement d'appel *m*, (*country code*) indicatif de pays *m*

CCA *abbr* (*call control agent*) DATA agent de traitement d'appel *m*

CCB *abbr* (*cable connection board*) TRANS *equipment practice* carte de connexion de câble *f*

CCD *abbr* (*charge-coupled device*) ELECT DCC (*dispositif à couplage de charge*); ~ **detector** *n* PER APP *scanner* capteur CCD *m* (*captuer de dispositif à couplage de charge*)

CCITT: ~ **high-level language** *n* COMP, TELEP langage de haut niveau défini par le CCITT *m*

CCR *abbr* (*commitment, concurrency and recovery*) DATA engagement, concurrence et rétablissement *m*

CCTV *abbr* (*closed-circuit television*) TV *video* télévision en circuit fermé *f*

C^3 *abbr* (*C-cube*) COMP C^3 (*C-cube*)

C-cube *n* (C^3) COMP C-cube *m* (C^3)

CD *abbr* COMP (*compact disc*) CD (*disque compact*), DATA (*capacity decrease*) diminution de capacité *f*, ELECT (*collision detection*) détection de collision *f*, TRANS (*carrier detect*) détection de porteuse *f*; ~ **reader** *n* COMP lecteur CD *m*

CD-Audio *n* COMP CD-Audio *m*

CDB *abbr* (*command descriptor block*) DATA bloc de descripteur de commande *m*

CDE *abbr* (*convolutionally and differentially encoded*) DATA à codage convolutionnel différentiel

cdf *abbr* (*cumulative distribution function*) DATA fonction de distribution cumulative *f*

CDF *abbr* (*combined distribution frame*) SWG répartiteur mixte *m*

CD-I *abbr* (*compact disc interactive*) COMP CD-I (*disque compact interactif*); ~ **audio sector** *n* COMP secteur audio CD-I *m*; ~ **data** *n pl* COMP, DATA données CD-I *f pl*; ~ **sector** *n* COMP secteur CD-I *m*; ~ **video sector** *n* TV secteur vidéo CD-I *m*

CDLI *abbr* (*called-line identification*) DATA, TELEP identification de la ligne appelée *f*, identification de la ligne demandée *f*

CDM *abbr* SIG (*companded delta modulation*) modulation delta avec compression et extension *f*, RADIO, TELEP (*code-division multiplexing*) MRC (*multiplexage par répartition en code*), TRANS (*code-division multiplexing*) MRC (*multiplexage par répartition en code*), (*companded delta modulation*) modulation delta avec compression et extension *f*

CDMA *abbr* (*code-division multiple access*) RADIO, SIG, TELEP, TRANS AMDC (*accès multiple par différence de code*)

CDO *abbr* DATA (*connect data overflow*) débordement de données de connexion *m*, SWG (*community dial office*) central automatique local *m*

CDR *abbr* (*call data recording*) APPL EDA (*enregistrement des données d'appels*), EDC (*enregistrement des données de communications*)

CD-R *abbr* (*compact disc-read only*) COMP CD-R (*disque compacte à lecture seule*)

CD-ROM *abbr* (*compact disc read-only memory*) COMP CD-ROM (*disque optique numérique*), DON (*disque optique numérique*); **~ address** *n* COMP adresse CD-ROM *f*; **~ drive** *n* COMP lecteur de CD-ROM *m*; **~ player** *n* COMP lecteur de CD-ROM *m*; **~ real-time operating system** *n* (*CD-RTOS*) COMP CD-RTOS

CDRS *abbr* (*call-detail recording service*) APPL service EDA *m* (*service d'enregistrement des données d'appels*), service EDC *m* (*service d'enregistrement des données de communications*)

CD-RTOS *abbr* (*CD-ROM real-time operating system*) COMP CD-RTOS

CD-V *abbr* (*compact disc video*) TV vidéo-CD *m*

CD-WO *abbr* (*compact disc-write once*) COMP CD-WO (*disque compacte à écriture seule*)

CE *abbr* DATA (*connection element*) élément de connexion *m*, SWG (*common equipment*) organe commun *m*, TRANS (*channel equipment*) équipement de canaux *m*, matériel de canaux *m*, matériel de voies *m*

cease *vt* TELEP *service* résilier

CEB *abbr* (*consecutive error block*) DATA bloc erroné consécutif *m*

Ceefax® *n BrE* APPL Antiope® *m* (*Fra*)

cell *n* POWER *batteries* pile *f*, élément *m*, RADIO alvéole *f*, *land mobile* cellule *f*; **~ border** RADIO frontière d'une cellule *f*, limite d'une cellule *f*; **~ container** POWER *batteries* bac d'éléments *m*; **~-coverage prediction** RADIO prévision de la couverture des cellules *f*; **~ interconnection bar** POWER *batteries* barre d'interconnexion d'éléments *f*; **~ loss priority** (*CLP*) RADIO priorité de perte de cellule *f*; **~ relay** RADIO commutation par cellules *f*, TRANS relais de cellules *m*; **~ splitting** RADIO division de cellules *f*; **~ switch** POWER *batteries* commutateur d'éléments *m*; **~ switch system** POWER *batteries* système de commutation sur liaison d'éléments *m*; **~ switching** COMP commutation temporelle asynchrone *f*; **~ voltage** POWER *batteries* tension d'élément *f*; **~ voltage equalizer** POWER *batteries* égaliseur de tension d'élément *m*

cellular[1] *adj* ELECT, RADIO, TELEP cellulaire

cellular[2] *n* TELEP téléphone cellulaire *m*; **~ insulation** TRANS *cables* isolant cellulaire *m*, isolation cellulaire *f*; **~ mobile exchange** RADIO commutateur de mobiles *m*; **~ mobile telephone network** RADIO, TELEP réseau cellulaire téléphonique mobile *m*; **~ network** DATA, RADIO, TELEP réseau cellulaire *m*; **~ radio handset** RADIO combiné de radio mobile *m*; **~ radio telephony** RADIO radiotéléphonie cellulaire *f*; **~ rexolene conductor** ELECT diélectrique en rexolène cellulaire *m*; **~ structure** RADIO structure cellulaire *f*; **~ system** RADIO système cellulaire *m*; **~ technique** RADIO technique cellulaire *f*

CELP: **~ coder** *n* SIG codeur CELP *m*

cemf *abbr* (*counter-electromotive force*) POWER fcém (*force contre-électromotrice*)

CEN *abbr* (*European Committee for Standardization*) GEN CEN (*Comité européen de normalisation*)

CENELEC *abbr* (*European Committee for Electrotechnical Standardization*) GEN CENELEC (*Comité européen de normalisation électrotechnique*)

censored: **~ test** *n* TEST essai censuré *m*

center *AmE see centre BrE*

central: **~ alarm coupler** *n* CONTR coupleur central d'alarmes *m* (*CCAL*); **~ battery** *n* TELEP batterie centrale *f*; **~-battery system** *n* TELEP téléphonie à batterie centrale *f*; **~-charging equipment** *n* APPL équipement de taxation centralisée *m*; **~ conductor** *n* TRANS conducteur central *m*; **~ control unit** *n* SWG unité centrale de commande *f*; **~ defence** *BrE*, **~ defense** *AmE n* TELEP défense centrale *f*; **~ exchange** *n* SWG commutateur urbain *m*; **~ exchange switch** *n BrE* (*cf central office switch AmE*) SWG, TELEP commutateur public *m*; **~ exchange trunk** *n BrE* (*cf central office trunk AmE*) SWG ligne réseau *f*; **~ office** *n* SWG, TELEP central téléphonique *m*, commutateur téléphonique *m*; **~ office switch** *n AmE* (*cf central exchange switch BrE, cf public telephone exchange BrE*) SWG, TELEP commutateur public *m*; **~ office trunk** *n AmE* (*cf central exchange trunk BrE*) SWG ligne réseau *f*; **~ processing unit** *n* (*CPU*) COMP, SWG, TELEP unité centrale de calcul *f* (*UC*), unité centrale de traitement *f* (*UC*); **~ processor** *n* SWG processeur central *m*, unité centrale de calcul *f*, unité centrale de commande *f*, unité centrale de traitement *f* (*UC*); **~ pulse distributor** *n* SWG distributeur rapide *m*; **~ remote monitoring station** *n* TEST station centrale de télésurveillance *f*; **~ strand** *n* TRANS porteur central *m*; **~ strength member** *n* TRANS *cable* porteur central *m*; **~ suspension cable** *n* TRANS câble à porteur central *m*; **~ switching matrix** *n* SWG matrice centrale de connexion *f*; **~ switching office** *n* DATA autocommutateur *m*; **~ switching unit** *n* TELEP commutateur urbain *m*

centralized: **~ billing** *n* APPL taxation centralisée *f*; **~ control signaling** *AmE*, **~ control signalling** *BrE n* SIG signalisation de commande centralisée *f*; **~ control system** *n* SWG autocommutateur électronique à commande centralisée *m*, système à commande centralisée *m*; **~ DSA** *n* DATA DSA centralisé *m*; **~ multipoint facility** *n* DATA, TELEP service complémentaire de liaisons multipoints centralisées *m*; **~ system** *n* SWG système centralisé *m*; **~ traffic division system** *n* SWG système à cœur réparti universel *m*

centralograph *n* TELEP centralographe *m*

centre *n BrE* GEN centrale *f*; **~ frequency** *BrE* TRANS fréquence centrale *f*; **~ line** *BrE* RADIO axe de symétrie *m*; **~ point** *BrE* GEN point milieu *m*; **~-point signalling** *BrE* SIG signalisation à point de symétrie *f*, signalisation à point milieu *f*; **~ tap** *BrE* ELECT *transformers* prise centrale *f*, prise médiane *f*; **~-tapped transformer** *BrE* POWER *supply* transformateur à point médian *m*

centres: **between ~** *phr BrE* GEN entraxe

Centrex: **~ service** *n* TELEP service Centrex *m*; **~ system** *n* TELEP système Centrex

CEQ *abbr* (*customer equipment*) PER APP équipement d'usager *m*

ceramic[1] *adj* PARTS *materials* céramique

ceramic[2]: **~ capacitor** *n* ELECT, PARTS condensateur céramique *m*; **~ encapsulation** *n* ELECT *microcircuits* enveloppe céramique *f*; **~ encasement** *n* ELECT *microcircuits* enveloppe céramique *f*

ceramics *n* PARTS *materials* céramique *f*

certificate *n* GEN certificat *m*; **~ of approval** GEN certificat d'homologation *m*; **~ of compliance** TEST *quality* certificat de conformité *m*; **~ of conformance** TEST certificat de conformité *m*; **~ of conformity** TEST certificat de conformité *m*

certification *n* GEN *quality* certification *f*; **~ of approval** GEN homologation *f*; **~ authority** TEST autorité de certification *f*; **~ body** TEST organisme certificateur *m*; **~ path** DATA trajet de certification *m*; **~ system** TEST système de certification *m*

CF *abbr* (*control function*) CONTR, DATA, TELEP fonction de commande *f*, fonction de contrôle *f*
CFM *abbr* (*companding and frequency modulation*) TRANS CMF (*compression-extension avec modulation de fréquence*)
CFR *abbr* (*confirmation to receive*) DATA, TELEP confirmation de réception *f*
CGA *abbr* COMP (*color graphics adaptor AmE, colour graphics adaptor BrE*) CGA (*carte graphique couleur*), TRANS (*carrier group alarm*) AGP (*alarme de groupe primaire*); ~ **card** *n* COMP carte CGA *f*
CGSB *abbr* (*Canadian General Standards Board*) GEN ONGC (*Office des normes générales du Canada*)
chaff: ~ **echo** *n* RADIO écho de leurres *m*, écho de paillettes *m*; ~ **jamming** *n* RADIO brouillage par feuilles réfléchissantes *m*, brouillage par leurres *m*
chain[1] *n* SWG chaîne *f*; ~ **directory** COMP répertoire de chaînes *m*; ~ **edit** COMP édition de chaîne *f*; ~ **play** COMP lecture de chaîne *f*; ~ **sampling** TEST échantillonnage en chaîne *m*; ~ **structure** COMP structure de chaîne *f*
chain[2] *vt* COMP enchaîner
chained: ~ **sequence** *n* DATA séquence chaînée *f*
chaining *adj* DATA concaténation; ~ **search** *n* COMP recherche en chaîne *f*
change: ~ **channel signal** *n* RADIO signal de changement de canal *m*; ~ **dump** *n* COMP vidage après changement *m*; ~ **notice** *n* (*CN*) GEN avis de modification *m*; ~ **order** *n* GEN *documentation* ordre de modification *m*; ~ **request** *n* GEN demande de modification *f*
changeback *n* TELEP retour sur canal sémaphore *m*, retour sur la liaison normale *m*; ~ **code** SIG code de retour sur canal sémaphore normal *m*
changed: ~**-number signal** *n* SIG signal de changement de numéro *m*
changeover *n* GEN basculement *m*; ~ **signal** SIG signal de commutation *m*; ~ **switch** POWER commutateur *m*; ~ **to a spare satellite** RADIO passage sur un satellite de secours *m*; ~ **to standby** GEN passage normal-secours *m*
changer *n* POWER inverseur *m*
channel *n* DATA voie *f*, COMP canal *m*, RADIO radiocanal *m*, canal radioélectronique *m*, canal RF *m*, canal *m*, SIG voie *f*, TRANS canal *m*, voie *f*, TV canal *m*, chaîne *f*; ~ **allocation store** TRANS mémoire d'attribution de voies *f*; ~ **allocation time** TRANS durée d'affectation des canaux *f*; ~**-associated signaling** *AmE*, ~**-associated signalling** *BrE* (*CAS*) SIG, TRANS signalisation voie par voie *f*; ~ **capacity** TRANS capacité de canal *f*; ~**-check test procedure** TEST procédure de contrôle des voies *f*; ~ **coder** RADIO *mobile* codeur de canal *m*; ~ **concentrator** TRANS concentrateur de voies *m*; ~ **control** TRANS commande de voies *f*; ~ **dedication** TELEP spécialisation des canaux *f*; ~ **efficiency** TRANS efficacité des canaux *f*; ~ **equipment** (*CE*) TRANS équipement de canaux *m*, matériel de canaux *m*, matériel de voies *m*; ~ **extension equipment** TRANS équipement d'extension de canal *m*; ~ **filtering** TV filtrage des canaux *m*; ~ **gate** TELEP, TRANS porte de voie *f*; ~ **loading** RADIO capacité du canal *f*, charge du canal *f*, TRANS charge des canaux *f*; ~ **mapping list** TRANS tableau de correspondance des voies *m*; ~ **marking** RADIO marquage de voies *m*, repérage de voies *m*; ~ **modem** TRANS unité de canal *f*, modem de voie *m*; ~ **occupancy rate** RADIO degré d'occupation de voies *m*; ~ **performance** TEST qualité de voie *f*; ~ **pool** TRANS faisceau de voies *m*; ~ **protocol machine** (*CPM*) DATA machine de protocole de canaux *f*; ~ **receiver** TRANS récepteur de voie *m*; ~ **selector** RADIO sélecteur de canal *m*, TV sélecteur de canaux *m*; ~ **separation** RADIO espacement des canaux *m*, espacement entre canaux *m*; ~ **size** TRANS capacité des canaux *f*; ~ **spacing** RADIO espacement des canaux *m*, espacement entre canaux *m*; ~ **state** TRANS état de canal *m*; ~ **substructure** COMP, DATA sous-structure de canal *f*; ~ **switch** RADIO sélecteur de canal *m*; ~ **switching** SWG commutation de voies *f*; ~ **terminal** TRANS équipement terminal de voie *m*; ~ **time slot** TELEP intervalle de temps de voie *m*, TRANS créneau temporel de voie *m*, intervalle de temps de voie *m*; ~ **transmitter** TRANS émetteur de voie *m*; ~ **unit** TRANS équipement de voie *m*; ~ **vocoder** RADIO codeur à fréquences vocales à canaux *m*, vocodeur à canaux *m*
channeling *AmE see channelling BrE*
channelization *n* TRANS mise en parallèle des canaux *f*
channelized: ~ **section** *n* TRANS section canaux *f*; ~ **stage of repeater** *n* TRANS section canaux d'un répéteur *f*
channelling *n BrE* TRANS découpage en voies *m*, multiplexage *m*
chaos *n* ELECT diversification *f*, pagaille *f*
character *n* APPL, COMP, DATA, TELEP *specification and description language* caractère *m*; ~ **alignment** TELEP alignement de caractères *m*; ~ **boundary** APPL *character recognition* limite de caractères *f*; ~ **check** TELEP contrôle par caractère *m*; ~ **code** COMP code de caractère *m*; ~ **coupling switch** SWG commutateur caractère par caractère *m*; ~ **display** ELECT *indicating devices* afficheur de caractères *m*; ~ **error rate** COMP, DATA taux d'erreur sur les caractères *m*; ~ **error ratio** COMP, DATA taux d'erreur sur les caractères *m*; ~ **font** COMP police de caractères *f*; ~ **generation** COMP édition de caractères *f*; ~**-generation logic** COMP logique d'édition de caractères *f*; ~ **generator** COMP générateur de caractères *m*; ~ **instruction** COMP commande alphanumérique *f*; ~**-oriented procedure** COMP procédure fondée sur les caractères *f*; ~ **outline** APPL *character recognition* contour de caractère *m*; ~ **position** TELEP position de caractère *f*; ~ **reader** APPL lecteur de caractères *m*; ~ **recognition** COMP reconnaissance de caractères *f*; ~ **return** (*CR*) DATA caractère de retour à la ligne *m*; ~**-serial transmission** TRANS transmission caractère-série *f*; ~ **set** COMP jeu de caractères *f*; ~ **of signal** SIG caractéristiques du signal *f pl*; ~ **signal** SIG signal de caractère *m*; ~ **skew** APPL *character recognition* inclinaison de caractères *f*; ~**-spacing reference line** COMP axe de référence d'espacement *m*; ~**-string type** DATA type chaîne de caractères *m*; ~ **switch** SWG commutateur de caractères *m*
characteristic *n* TEST *quality* caractère *m*; ~ **distortion** DATA, TELEP distorsion caractéristique *f*; ~ **distortion compensation** DATA, TELEP compensation de la distorsion caractéristique *f*; ~ **frequency** RADIO fréquence caractéristique *f*
characters: ~ **per inch** *phr* (*cpi*) COMP caractères par pouce (*cpp*)
charge[1] *n* APPL *cost* taxe *f*, ELECT, POWER *electrical* charge *f*; ~ **advice** APPL affichage du coût de la communication *m*, avis de taxation *m*, indication de coût *f*, information tarifaire *f*; ~ **card** TELEP carte pastel *f*; ~ **carrier** ELECT *semiconductors* porteur de charge *m*; ~**-coupled device** (*CCD*) ELECT dispositif à couplage de charge *m* (*DCC*); ~**-coupled device detector** (*CCD detector*) PER APP capteur de dispositif à couplage de charge *m* (*capteur CCD*); ~**-coupled device circuit** DATA circuit à couplage de charge *m*; ~ **cycle** POWER période de charge *f*; ~ **injection device** (*CID*) POWER dispositif à injection de charge *m*; ~ **meter account** TELEP compte de taxe *m*; ~ **metering** TELEP taxation *f*; ~**-metering converter** APPL convertisseur de gestion *m* (*CG*); ~**/no charge indicator** *n* TELEP indicateur avec/sans taxation *m*; ~ **rate** APPL niveau de tarification *m*, taxation *f*; ~ **request** DATA demande de connexion *f*; ~ **time** POWER période de charge *f*; ~ **transfer device** (*CTD*) ELECT dispositif à transfert de charge *m* (*DTC*); ~ **unit** APPL unité télécom *f*
charge[2] *vt* APPL tarifer, charger, taxer, *ascribe costs* imputer, POWER *batteries* charger; ~ **the call to** APPL imputer l'appel à

chargeable *n* APPL taxable; ~ **duration** APPL, TELEP durée taxable *f*
charged: ~ **duration** *n* APPL, TELEP durée taxée *f*
charger *n* POWER *batteries* chargeur *m*, tableau-chargeur *m*
charging *n* APPL comptage *m*, taxation *f*, ELECT, POWER charge *f*; ~ **analysis** APPL analyse de taxation *f*; ~ **area** APPL circonscription tarifaire *f*, TELEP zone de taxe *f*; ~ **center** *AmE*, ~ **centre** *BrE* APPL centre de taxation *m*; ~ **current** ELECT, POWER courant de charge *m*; ~ **data** *n pl* DATA éléments de taxation *m pl*; ~ **equipment** APPL équipement de facturation *m*, équipement de taxation *m*; ~ **error** APPL erreur de facturation *f*, erreur de taxation *f*; ~ **impulse** APPL impulsion de comptage *f*; ~ **in progress PE** TELEP élément graphique de taxation en cours *m*; ~ **information** APPL information de taxation *f*, avis de taxation *m*; ~ **interval** APPL intervalle de recouvrement des *m*; ~ **method** TELEP méthode de taxation *f*; ~ **mode** APPL mode de taxation *m*, principe de taxation *m*, TELEP mode de taxation *m*; ~ **output** APPL sortie des éléments de taxation *f*; ~ **period** PER APP durée de la charge *f*, TELEP période de taxation *f*; ~ **point** APPL point de comptage *m*, point de taxation *m*, ELECT, POWER poste de charge *m*; ~ **principle** APPL mode de taxation *m*, principe de taxation *m*; ~**-recording center** *AmE*, ~**-recording centre** *BrE* TELEP centre d'enregistrement de la taxation *m*; ~ **rectifier** POWER redresseur de charge *m*; ~ **set** POWER *batteries* chargeur *m*, chargeur de batterie *m*; ~ **statistics** *n pl* APPL statistiques de taxation *f pl*; ~ **store** APPL mémoire d'éléments de taxation *f*; ~ **test** APPL essai de taxation *m*; ~ **tester** APPL, TEST dispositif d'essai de taxation *m*, testeur de taxation *m*; ~ **unit** APPL unité télécom *f*, POWER *batteries* tableau-chargeur *m*; ~ **voltage** POWER *batteries* tension de charge *f*
charstring *n* COMP, DATA chaîne de caractères *f*
chart *n* COMP tableau *m*, ELECT organigramme *m*
chassis *n* ELECT, TELEP châssis *m*
chat *n* TELEP bavardage *m*
chatter *n* COMP bruit parasite *m*
Chebyshev: ~ **approximation** *n* TRANS approximation de Tchebychev *f*; ~ **array** *n* RADIO rideau de Tchebychev *m*
check[1] *n* COMP, POWER, SWG vérification *f*; ~ **bit** DATA bit de contrôle *m*; ~ **box** COMP case d'option *f*; ~ **indicator** COMP indicateur de contrôle *m*; ~ **list** TEST *documentation* liste de contrôle *f*; ~ **loop** TELEP boucle pour contrôle de continuité *f*; ~**-number field** SWG zone de contrôle du numéro *f*; ~**-out equipment** TEST banc de contrôle *m*, équipement de contrôle *m*; ~**-out time** TELEP durée de vérification *f*, TEST *dependability* temps de vérification *m*; ~ **print-out** SWG message de contrôle *m*, sortie de contrôle sur imprimante *f*, sortie imprimée de contrôle *f*; ~**-processing equipment** *AmE see cheque-processing equipment BrE*; ~ **sequence** TEST séquence de contrôle *f*; ~ **sum** DATA somme de contrôle *f*; ~**-symbol time slot** TRANS intervalle de temps du symbole de contrôle *m*
check[2] *vt* COMP, POWER, SIG, SWG vérifier
checking *n* COMP vérification *f*; ~ **off** TEST pointage *m*; ~ **program** COMP programme de contrôle *m*
checkpoint *n* COMP point de contrôle *m*, TELEP point de repère *m*
cheese *n* RADIO réflecteur en D *m*, réflecteur en segment *m*; ~ **antenna** RADIO antenne en D *f*
cheque: ~**-processing equipment** *n BrE* APPL matériel traitement de chèques *m*
chief: ~ **operator** *n* TELEP chef de poste *m*, chef de surveillance *m*; ~ **operator's desk** *n* TELEP pupitre de surveillance *m*; ~ **supervisor** *n* TELEP chef de poste *m*, chef de surveillance *m*; ~ **supervisor's desk** *n* TELEP pupitre de surveillance *m*
child *n* COMP, DATA fils *m*; ~**-entry** COMP, DATA entrée secondaire *f*; ~**-sequence number** COMP, DATA numéro d'ordre d'une entrée secondaire *m*
childless: ~ **node** *n* COMP feuille *f*
chip *n* DATA élément *m*, bribe *f*, ELECT circuit intégré *m*, pastille *f*, puce *f*; ~ **card** ELECT carte à microcircuit *f*, carte à microprocesseur *f*, carte à puce *f*, carte à mémoire *f* (*CAM*); ~ **rate** DATA débit des éléments *m*, débit de bribes *m*; ~ **yield** DATA taux de puces bonnes *m*
Chireix: ~ **array** *n* RADIO rideau Chireix-Mesny *m*, rideau en dents de scie *m*
chirping *n* RADIO fluctuation de fréquence d'impulsion *f*, TRANS *fibre optics* fluctuation de longueur d'onde *f*
chloroprene *n* PARTS chloroprène *m*
choke *n* POWER bobine d'arrêt *f*, bobine de choc *f*, piège *m*
choking: ~ **coil** *n* POWER bobine de choc *f*
chopped: ~ **response** *n* RADIO réponse manipulée *f*
chopper *n* ELECT hacheur *m*; ~ **amplifier** ELECT amplificateur à découpage *m*
chord *n* COMP accord *m*; ~ **intelligence** COMP programmation paramétrique *f*
chosen: ~ **path** *n* COMP *Internet* chemin parcouru *m*
chromakey *n* COMP chromaclé *f*
chromatic[1] *adj* TRANS, TV chromatique
chromatic[2]: ~ **balance** *n* TV *scanner* équilibre chromatique *m*; ~ **color** *AmE*, ~ **colour** *BrE n* TV couleur chromatique *f*; ~ **dispersion** *n* TRANS *fibre optics* dispersion chromatique *f*; ~ **distortion** *n* TV distorsion chromatique *f*; ~ **space** *n* TV espace chromatique *m*, espace des couleurs *m*
chromaticity *n* TV chromaticité *f*
chrominance *n* GEN *optics*, TV chrominance *f*; ~ **axis** TV axe de chrominance *m*; ~ **component** TV composant de chrominance *m*; ~**-luminance delay inequality** TV retard chrominance-luminance *m*; ~ **vector** TV vecteur chrominance *m*
chronological: ~ **link** *n* COMP lien chronologique *m*
chunk *n* COMP fragment *m*
chute *n* PARTS, SWG, TELEP goulotte *f*, TRANS *cable laying* chemin de câbles *m*, goulotte *f*
CI *abbr* ELECT (*concatenation indication*) indication de concaténation *f*, RADIO (*conversation impossible*) CI (*conversation impossible*)
C/I *abbr* (*carrier-to-interference ratio*) TRANS rapport porteuse-brouillage *m*
CID *abbr* DATA (*consecutive identical digits*) symboles identiques consécutifs *m pl*, POWER (*charge injection device*) dispositif à injection de charge *m*
CIF *abbr* (*common intermediate format*) COMP FIC (*format intermédiaire commun*)
CIG *abbr* (*calling-subscriber identification*) DATA, TELEP CIG (*identification de l'abonné appelant*)
cigar: ~ **aerial** *n* RADIO antenne cigare *f*; ~ **antenna** *n* RADIO antenne cigare *f*
CIG-DTC *abbr* (*calling-subscriber identification-DTC*) DATA, TELEP CIG-DTC (*identification de l'abonné demandeur-DTC*)
CIL *abbr* (*call-identification line*) TELEP CIL (*ligne d'identification de l'appel*)
CIM *abbr* (*computer-integrated manufacturing*) COMP FIO (*fabrication intégrée par ordinateur*), productique *f*
cinching *n* COMP enroulement ondulé *m*
cipher *n* RADIO chiffre *m*
ciphering *n* RADIO chiffrement *m*
ciphertext *n* DATA, RADIO, SIG, TELEP, TV cryptogramme *m*, texte chiffré *m*
CIR *abbr* (*committed information rate*) DATA débit d'information garanti *m*
CIRC *abbr* (*cross-interleaved Reed-Solomon code*) COMP CIRC (*code Reed-Solomon à entrelacement croisé*)
circuit *n* GEN circuit *m*, RADIO, TRANS, TV voie *f*, SWG circuit de jonction *m*, TELEP circuit *m*; ~ **access point** TELEP point d'accès à un circuit *m*; ~ **availability** RADIO *maritime satellite*, TELEP disponibilité d'un circuit *f*; ~ **board** COMP

carte de circuit imprimé *f*, ELECT carte à circuits *f*; ~ **breaker** ELECT, POWER disjoncteur *m*; ~ **closer** POWER conjoncteur *m*; ~ **consistency verification** DATA vérification de la cohérence du circuit *f*; ~ **control station** TELEP station directrice de circuit *f*; ~ **design** GEN *technical drawing* conception des circuits *f*; ~ **diagram** ELECT, SWG schéma de principe *m*, schéma des circuits *m*; ~ **equipment** TELEP équipement de circuit *m*; ~ **group** GEN faisceau *m*; ~-**group query message** (*CQM*) SWG message d'interrogation de groupe de circuits *m* (*IGD*); ~-**group query response message** (*CQR*) SWG message de réponse à une interrogation de groupe de circuits *m* (*IGR*); ~ **group-congestion signal** DATA, SIG signal d'encombrement du faisceau de circuits *m*; ~ **interface** DATA interconnexion de circuits *f*; ~ **junctor** TELEP joncteur de circuit *m*; ~-**mode bearer service** APPL service support à commutation de circuits *m*, service support en mode circuit *m*; ~ **noise** TELEP bruit de ligne *m*; ~ **noise level** TELEP niveau de bruit d'un circuit *m*; ~-**released acknowledgement signal** SIG signal d'accusé de réception de libération de circuit *m*; ~-**released signal** SIG signal de libération du circuit *m*; ~ **route** TRANS faisceau de circuits *m*; ~ **simulation** ELECT *semiconductors* simulation de circuit *f*; ~ **stability** TRANS stabilité de circuit *f*; ~ **state indicator** SWG indicateur d'état de circuit *m*; ~ **state message** TELEP message d'état de circuit *m*; ~ **subcontrol station** TELEP station sous-directrice de circuit *f*; ~ **subgroup** TELEP tronçon de circuits *m*, sous-faisceau de circuits *m*, TRANS sous-faisceau de circuits *m*; ~ **switch** SWG commutateur de circuits *m*; ~ **switched** (*CS*) SWG commutation de circuits *f*; ~-**switched connection** DATA *datacom* liaison commutée en mode circuit *f*, SWG connexion par commutation de circuits *f*; ~-**switched data transmission service** DATA, TRANS service de transmission de données avec commutation de circuits *m*; ~-**switched exchange** SWG commutateur de circuits *m*; ~-**switched network** TRANS réseau à commutation de circuits *m*; ~-**switched service** APPL service de commutation des circuits *m*; ~ **switching** (*CS*) SWG, TELEP commutation de circuits *f*; ~-**switching center** *AmE*, ~-**switching centre** *BrE* SWG centre de commutation de circuits *m*; ~-**switching network** SWG réseau à commutation de circuits *m*; ~-**switching system** SWG système à commutation de circuits *m*; ~-**switching unit** SWG commutateur de circuits *m*; ~ **terminal** TELEP borne terminale *f*; ~-**test access point** TELEP point d'accès pour les mesures de circuit *m*; ~ **test set** TEST *testing* testeur de circuit *m*; ~ **and wiring diagram** ELECT schéma de circuits et de câblage *m*

circuitry *n* ELECT circuits *m pl*, ensemble des circuits *m*

circular: ~ **orbit** *n* RADIO orbite circulaire *f*; ~ **polarization** *n* TRANS polarisation circulaire *f*

circularize *vt* RADIO *orbit*, TRANS circulariser

circulator *n* RADIO, TRANS *waveguide* circulateur *m*

CIRM *abbr* (*International Marine Radio Association*) RADIO CIRM (*Comité international radio-maritime*)

CISE *abbr* (*complex instruction set equipment*) DATA CISE

citizens': ~ **band** *n* (*CB*) RADIO bande CB *f*

city: ~ **call** *n* TELEP appel local *m*; ~ **network** *n* TELEP réseau métropolitain *m*, réseau urbain *m*; ~ **paging** *n* RADIO radiomessagerie unilatérale urbaine *f*, RMU urbaine *f*

civil: ~ **communications satellite** *n* RADIO satellite de télécommunications civiles *m*

CL *abbr* (*connectionless*) GEN sans connexion

cladding *n* TRANS gaine *f*; ~ **center** *AmE*, ~ **centre** *BrE* TRANS centre de gaine *m*; ~ **diameter** TRANS diamètre de gaine *m*; ~ **mode** TRANS mode de gaine *m*; ~ **mode stripper** TRANS extracteur de modes de gaine *m*; ~ **surface diameter deviation** TELEP écart sur le diamètre de la gaine *m*; ~ **tolerance field** TRANS domaine de tolérance de gaine *m*, *fibre optics* champ de tolérance de gaine *m*

clamp *n* ELECT mâchoire *f*, lanceur *m*, circuit d'alignement *m*; ~-**band separation system** ELECT système à sangle *m*; ~ **diode** ELECT diode de fixation de niveau *f*, diode de niveau *f*; ~ **terminal** ELECT borne de fixation *f*

clamping: ~ **circuit** *n* ELECT circuit de verrouillage *m*, circuit d'alignement *m*; ~ **diode** *n* ELECT diode de fixation de niveau *f*, diode de niveau *f*; ~ **sleeve** *n* ELECT manchon de fixation *m*

clarifying: ~ **text** *n* TELEP texte explicatif *m*

clarity *n* GEN audibilité *f*

clashing *n* RADIO *land mobile* collision *f*

class *n* GEN classe *f*; ~ **of emission** RADIO classe d'émission *f*; ~ **of HDLC** TELEP classe de HDLC *f*; ~-**mark store** TELEP mémoire de code de classe *f*; ~ **marking** TELEP marquage de catégorie *m*, marquage de classe *m*

classification *n* GEN classification *f*; ~ **of failures** TEST classement des défaillances *m*

classify *vt* GEN classifier

CLCP *abbr* (*connectionless convergence protocol*) DATA protocole de convergence sans connexion *m*

clean[1]: ~ **carrier** *n* RADIO porteuse non modulée *f*

clean[2] *vt* COMP *screen* effacer

cleaning *n* COMP effacement *m*; ~ **head** TELEP tête de nettoyage *f*; ~ **kit** GEN nécessaire de nettoyage *m*

cleanliness *n* DATA *scanner* netteté *f*

clear[1]: ~ **to send** *adj* (*CTS*) DATA prêt à émettre

clear[2]: ~ **acrylic tube** *n* ELECT tube acrylique transparent *m*; ~ **area** *n* APPL bande de sécurité *f*, zone libre *f*, zone non occupée *f*, COMP zone libre *f*; ~ **back** *n* SIG raccrochage du demandé *m*; ~-**back condition** *n* SIG position demandé raccroché *f*; ~-**back signal** *n* SIG signal de raccrochage *m*, TELEP signal de fin de communication *m*; ~ **band** *n* APPL bande de sécurité *f*, zone libre *f*, zone non occupée *f*; ~ **channel capability** *n* APPL, TELEP service de transfert sans restriction *m*, service de transfert sur longueur d'onde privilégiée *m*; ~ **confirmation** *n* TELEP confirmation de libération *f*; ~-**forward signal** *n* SIG signal de fin *m*, TELEP signal de libération en avant *m*; ~ **message** *n* TEST message de libération *m* (*LIB*); ~ **operation** *n* COMP opération d'effacement *f*; ~ **request condition** *n* TELEP demande de libération *f*; ~ **request packet** *n* DATA paquet de demande de libération *m*; ~ **sixty-four service** *n* APPL service de circuit commuté en canal B transparent *m*; ~-**sky absorption loss** *n* RADIO affaiblissement dû à l'absorption par ciel clair *m*; ~ **speech** *n* RADIO conversation en clair *f*; ~ **text** *n* DATA texte en clair *m*; ~ **transmission** *n* TRANS transmission en clair *f*; ~ **weather attenuation** *n* RADIO affaiblissement atmosphérique par temps clair *m*

clear[3] *vt* COMP *screen* effacer, TELEP libérer, dégager, relever, TEST relever; ~ **a fault** TELEP, TEST relever un dérangement

clear[4]: **in** ~ *phr* TV en clair

clearance *n* COMP, CONTR, DATA, PARTS, TELEP *permission* autorisation *f*; ~ **hole** POWER *printed circuits* dégagement *m*

clearest: ~ **zone** *n* PER APP *scanner* zone la plus claire *f*

clearing *n* COMP, TELEP libération *f*; ~ **collision** TELEP collision de libération *f*; ~ **confirmation** TELEP confirmation de libération *f*; ~ **indicator** TELEP indicateur de fin *m*; ~ **lamp** TELEP lampe de fin de conversation *f*; ~ **message** TEST message de libération *m* (*LIB*); ~ **procedure** SIG procédure de libération *f*; ~ **request** TELEP demande de libération *f*

cleaved *adj* TRANS tendu

cleaver *n* TRANS fendon *m*

cleaving *n* TRANS fendage de la fibre *m*; ~ **tool** TRANS outil à mandrin *m*

CLI *abbr* DATA, TELEP (*calling-line identification*) ILA (*identification de la ligne appelante*), TRANS (*cumulative leakage index*) fuites globales *f pl*

click[1] *n* COMP clic *m*; ~-**stop position** COMP position à cran *f*

click[2] *vt* COMP cliquer; ~ **on** COMP cliquer sur

clicking: ~ **area** *n* COMP zone de clics *f*

CLID *abbr* (*calling-line identification display*) TELEP PILA (*présentation de l'identification de la ligne appelante*)
client *n* GEN client *m*; ~ **checklist** GEN liste de contrôle du client *f*; ~ **risk** (*CR*) TEST risque du client *m*; ~ **software** COMP, DATA logiciel client *m*; ~ **test manager** DATA gestionnaire des tests du client *mf*
client/server *n* COMP client/serveur *m*
climate: ~ **control equipment** *n* CLIM *environmental engineering* équipement régulateur d'ambiance *m*
climb: ~ **path** *n* TRANS trajectoire ascensionnelle *f*
clinched: ~-**wire through connection** *n* ELECT, POWER connexion transversale *f*
clip[1] *n* COMP coupure de page-écran *f*, ELECT clip *m*, PARTS attache *f*, clip *m*; ~-**art library** COMP bibliothèque d'images *f*
clip[2] *vt* ELECT attacher, SWG écrêter
CLIP *abbr* (*calling-line identification presentation*) TELEP PILA (*présentation de l'identification de la ligne appelante*)
clipboard *n* COMP ordinateur bloc-notes *m*, presse-papiers *m*; ~ **computer** COMP ordinateur bloc-notes *m*
clipping *n* COMP découpage *m*, fenêtrage *m*, SWG écrêtage *m*, TELEP distorsion *f*, mutilation de la parole *f*, distorsion de la parole *f*, *pulses* écrêtage *m*, TRANS mutilation de la parole *f*; ~ **voltage** ELECT *semiconductors* tension d'écrêtage *f*
CLIR *abbr* (*calling-line identification restriction*) TELEP RILA (*restriction de l'identification de la ligne appelante*)
CLNP *abbr* (*connectionless network-layer protocol*) TRANS protocole de couche réseau sans connexion *m*
CLNS *abbr* (*connectionless network-layer service*) APPL service de couche réseau sans connexion *m*
clock[1] *n* ELECT générateur de rythme *m*, GEN horloge *f*; ~-**carrier cycle skipping** TRANS saut de cycle de porteuse *m*; ~ **change** PER APP *answer machine* changement d'horaire *m*; ~ **cycle** COMP cycle d'horloge *m*; ~-**cycle skipping** TRANS saut de cycle d'horloge *m*; ~ **distribution** TRANS distribution de signaux horloge *f*; ~ **error** TRANS erreur de synchronisation *f*; ~ **jitter** ELECT gigue d'horloge *f*; ~ **oscillator** ELECT générateur de signal d'horloge *m*, générateur d'impulsions d'horloge *m*, rythmeur *m*, cadenceur *m*; ~ **phasing** TRANS décalage d'horloge *m*; ~ **pulse** ELECT impulsion d'horloge *f*; ~ **rate** TRANS fréquence d'horloge *f*, fréquence du rythme *m*; ~-**rate distribution** TRANS distribution de fréquence du rythme *f*; ~ **recovery** TRANS restitution du rythme *f*, rétablissement du rythme *m*; ~-**recovery circuit** ELECT circuit de rétablissement du rythme *m*; ~-**signal distribution** TRANS distribution de signaux de rythme *f*; ~-**skip rate** TRANS taux de saut de cycle *m*; ~ **synchronization** COMP synchronisation d'horloge *f*; ~ **time difference** GEN différence entre temps d'horloge *f*
clock[2] *vt* ELECT *logic elements* appliquer une impulsion d'horloge à, faire basculer, provoquer un changement d'état
clockwise: ~ **polarization** *n* RADIO polarisation dextrogyre *f*, polarisation dextrorsum *f*
clone[1] *n* COMP, DATA clone *m*
clone[2] *vt* COMP, DATA cloner
Clos: ~ **network** *n* TELEP, TRANS réseau de Clos *m*
close[1]: ~-**talking microphone** *n* PARTS microphone de proximité *m*
close[2] *vt* COMP, ELECT, PER APP, POWER, RADIO, SWG fermer
closed[1]: ~-**circuit** *adj* POWER, TV en circuit fermé, en courant fermé: ~-**loop** *adj* ELECT en boucle fermée
closed[2]: ~-**circuit current** *n* POWER courant en circuit fermé *m*; ~-**circuit relay** *n* POWER relais de courant fermé *m*; ~-**circuit television** *n* (*CCTV*) TV *video* télévision en circuit fermé *f*; ~-**circuit working** *n* TRANS transmission par envoi de fermeture *f*, transmission par fermeture de circuit *f*; ~ **coupler** *n* TELEP coupleur clos *m*; ~ **earth orbit** *n* RADIO orbite proche de la terre *f*; ~ **earth-track orbit** *n* RADIO orbite à projection fermée sur la surface terrestre *f*; ~ **loop** *n* ELECT boucle fermée *f*; ~-**loop circuit** *n* ELECT circuit en boucle fermée *m*; ~-**loop control** *n* CONTR commande asservissement *f*; ~-**loop control system** *n* CONTR système asservi à boucle fermée *m*; ~-**loop synchronization** *n* CONTR, POWER, TRANS synchronisation en boucle fermée *f*; ~ **subroutine** *n* COMP sous-programme fermé *m*; ~ **subscriber group** *n* DATA, TELEP groupe fermé d'abonnés *m* (*GFA*); ~ **user data-distribution network** *n* DATA, TRANS réseau de distribution de données pour groupes fermés d'usagers *m*; ~ **user group** *n* (*CUG*) DATA, TELEP groupe fermé d'usagers *m* (*GFU*), groupe fermé d'utilisateurs *m* (*GFU*); ~ **user group call indicator** *n* TELEP indicateur d'appel de groupe fermé d'usagers *m*; ~ **user group with outgoing access** *n* DATA, TELEP groupe fermé d'usagers avec accès sortant *m*
closure: ~ **gasket** *n* PARTS joint d'étanchéité *m*
cloud: ~ **return** *n* RADIO écho de nuage *m*
cloverleaf: ~ **antenna** *n* RADIO antenne en trèfle *f*
CLP *abbr* (*cell loss priority*) RADIO priorité de perte de cellule *f*
CLS *abbr* (*connectionless server*) DATA serveur sans connexion *m*
CLSF *abbr* (*connectionless service function*) APPL fonction de service sans connexion *f*
cluster *n* COMP *hardware* batterie *f*, DATA granule *m*, GEN grappe *f*, RADIO *of cells* motif *m*, TELEP *private branch exchange* ensemble *m*, groupe *m*, réseau grappe *m*; ~ **of cells** RADIO nid d'abeilles *m*; ~ **link** COMP lien de groupement *m*; ~ **sampling** GEN échantillonnage en grappes *m*
CLUT *abbr* (*color look-up table AmE, colour look-up table BrE*) COMP CLUT (*table de couleurs*)
clutter[1] *n* RADIO écho parasite *m*, fouillis d'écho *m*; ~ **attenuation** RADIO taux d'élimination d'écho parasite *m*; ~ **suppression** RADIO élimination des échos parasites *f*
clutter[2] *vt* COMP *hard disk* encombrer
CLV *abbr* (*constant linear velocity*) COMP VLC (*vitesse linéaire constante*)
CMB *abbr* (*CRC message bloc*) TELEP bloc de messages de CRC *m*
CMC *abbr* (*call-modification completed message*) TELEP MAE (*message de modification d'appel effectuée*)
cmf *abbr* (*cyclomotive force*) RADIO fcm (*force cyclomotrice*)
CMI *abbr* (*code-mark inversion*) TRANS code CMI *m*
CMIP *abbr* (*common management-information protocol*) DATA protocole commun d'informations de gestion *m*
CMIPDU *abbr* (*common management information protocol data unit*) DATA unité de données de protocole commun d'informations de gestion *f*
CMIPM *abbr* (*common management-information protocol machine*) DATA machine de protocole commun d'informations de gestion *f*
CMIS *abbr* (*common management-information service*) DATA service commun d'informations de gestion *m*, service commun de transfert d'informations de gestion *m*
CMISE *abbr* (*common management-information service element*) DATA élément de service commun d'informations de gestion *m*, élément de service commun de transfert d'informations de gestion *m*; ~ **service provider** *n* DATA, TELEP fournisseur du service CMISE *m*; ~ **service user** *n* DATA, TELEP utilisateur du service CMISE *m*
CMOS: ~ **crosspoint** *n* TELEP point d'intersection CMOS *m*
CMR *abbr* (*call-modification request*) TELEP MAD (*demande de modification d'appel*)
CMRJ *abbr* (*call-modification reject message*) TELEP MAR (*message de refus de modification d'appel*)
CMY: ~ **color model** *AmE*, ~ **colour model** *BrE* *n* COMP modèle de couleurs CMJ *m*
CN *abbr* DATA (*change notice*) avis de modification *m*, TRANS (*customer network*) réseau client *m*
C-n *abbr* (*container-n*) APPL, DATA, TELEP C-n (*conteneur-n*)
coalesce *vi* COMP se fondre
coarse: ~ **adjustment** *n* GEN réglage approché *m*, réglage

approximatif *m*, réglage grossier *m*; ~ **jumper** *n* ELECT cordon de raccordement coaxial *m*; ~ **setting** *n* GEN réglage approximatif *m*, réglage grossier *m*; ~ **sun sensor** *n* RADIO capteur solaire de première approximation *m*; ~ **switch** *n* ELECT bouton de réglage grossier *m*

coast: ~**-shore station identity** *n* DATA, RADIO, TELEP identité de la station côtière terrienne *f*; ~ **station** *n* RADIO, TELEP station côtière *f*; ~**-station identity** *n* DATA, RADIO, TELEP identité de la station côtière *f*

coastal: ~ **earth station** *n* RADIO station côtière *f*; ~ **station** *n* RADIO station côtière *f*

coat *vt* TRANS revêtir

coating *n* TRANS *fibre optics* protection *f*, revêtement *m*; ~ **with a rust preventive** PARTS traitement anti-rouille *m*

coax *n* PARTS coax *m*, TRANS câble coaxial *m*, coax *m*

coaxial[1] *adj* ELECT, RADIO, TRANS, TV coaxial

coaxial[2]: ~ **aerial** *n* RADIO antenne coaxiale *f*; ~ **antenna** *n* RADIO antenne coaxiale *f*; ~ **cable** *n* PARTS coax *m*, TRANS câble coaxial *m*, coax *m*; ~ **cable distributor** *n* TV répartiteur coaxial *m* (*RC*); ~ **jack** *n* ELECT *connecting devices* jack coaxial *m*; ~ **line** *n* TRANS ligne coaxiale *f*; ~ **line system** *n* TRANS système à ligne coaxiale *m*; ~ **network** *n* TRANS réseau coaxial *m*; ~ **plug** *n* ELECT *connecting devices* fiche coaxiale *f*; ~ **push-pull connector** *n* TRANS connecteur symétrique à lignes coaxiales *m*; ~ **reed relay changeover switch** *n* SWG relais de commutation à tige *m*; ~ **switched jumper** *n* TRANS commutateur coaxial *m*; ~ **tube** *n* TRANS *cables* tube coaxial *m*

co-channel *n* RADIO cocanal *m*, cofréquence *f*; ~ **interference** RADIO, TRANS brouillage dans le même canal *m*, brouillage sur la voie commune *m*, interférence avec le canal adjacent *f*; ~ **operation** RADIO *mobile* exploitation sur une même voie *f*; ~ **reuse distance** RADIO *land mobile* distance de réutilisation des fréquences dans la même voie *f*

code[1] *n* COMP code *m*, DATA, SIG, SWG code *m*, indicatif *m*, TELEP code télégraphique *m*, code *m*, code de données *m*, *dialling* indicatif *m*, TRANS code *m*, code télégraphique *m*; ~ **answer** TELEP réponse codée *f*; ~**-answering device** TELEP répondeur de code *m*; ~**-answering unit** TELEP répondeur de code *m*; ~ **block** DATA bloc codé *m*; ~ **character** TELEP caractère *m*, caractère télégraphique *m*; ~ **conversion** COMP, ELECT, TELEP, TRANS conversion de code *f*, transcodage *m*; ~ **converter** COMP, ELECT, TELEP, TRANS convertisseur de code *m*, transcodeur *m*; ~**-dependent system** TELEP système lié au code utilisé *m*; ~ **division** RADIO, TELEP répartition en code *f*, répartition en fréquence *f*; ~**-division multiple access** (*CDMA*) RADIO, SIG, TELEP, TRANS accès multiple par différence de code *m* (*AMDC*); ~**-division multiplexing** (*CDM*) RADIO, SWG, TELEP, TRANS multiplexage par répartition en code *m* (*MRC*); ~ **efficiency** SIG efficacité du codage *f*; ~ **element** COMP codet *m*; ~ **extension** TELEP extension de code *f*; ~ **extension techniques** TELEP techniques d'extension de code *f pl*; ~ **generator** TELEP générateur de code *m*; ~**-independent system** TELEP système indépendant du code utilisé *m*; ~ **key** RADIO clé de déchiffrement *f*; ~**-mark inversion** (*CMI*) TRANS code CMI *m*; ~ **memory board** RADIO carte mémoire de programme *f*; ~ **plug** RADIO fiche de code *f*; ~ **receiver** TELEP récepteur de code *m*; ~ **sender** TELEP émetteur de code *m*; ~**-sender finder** TELEP chercheur d'émetteur de code *m*; ~ **table** TELEP tableau de code *m*; ~ **to be routed** TRANS indicatif à acheminer *m*; ~ **transmission** TRANS transmission de code *f*; ~ **transparency** TRANS indépendance du code utilisé *f*, transparence en code *f*; ~ **violation** (*CV*) TRANS violation de code *f*; ~**-word** COMP, DATA mot-code *m*; ~ **x** TELEP chiffre x *m*

code[2] *vt* GEN coder

codec *abbr* (*coder/decoder*) COMP, TRANS codec (*codeur/décodeur*)

coded *adj* GEN codé; ~ **character data element** *n* DATA, TELEP élément de données en code caractère *m* (*élément de données CC*); ~**-data transmitter** *n* DATA émetteur d'informations *m*; ~ **image storage space** *n* COMP zone image *f*; ~ **in-band signaling** *AmE*, ~ **in-band signalling** *BrE n* SIG signalisation dans la bande avec codage *f*; ~ **MPEG frame** *n* COMP image codée MPEG *f*; ~ **order** *n* COMP ordre codé *m*; ~ **program** *n* COMP programme encodé *m*; ~ **representation** *n* COMP représentation codée *f*; ~ **ringing** *n* TELEP appel par sonnerie codée *m*, appel semi-sélectif *m*; ~ **transmission** *n* TRANS transmission codée *f*

coder *n* COMP, DATA, RADIO, SIG, TELEP, TRANS codeur *m*

coder/decoder *n* (*codec*) COMP, TRANS codeur/décodeur *m* (*codec*)

coding *n* COMP, DATA, SIG, TELEP, TRANS codage *m*, encodage *m*; ~ **circuit** SIG circuit de codage *m*; ~ **control output** DATA signal de commande du codage *m*; ~ **information** COMP information de codage *f*; ~ **law** TELEP, TRANS loi de codage *f*; ~ **parameter** COMP paramètre de codage *m*; ~ **process** COMP, TELEP processus de codage *m*; ~ **rate** DATA taux de codage *m*; ~ **standard** TRANS norme de codage *f*; ~ **type** TRANS type de codage *m*

coefficient *n* GEN coefficient *m*; ~ **of linear expansion** ELECT, PARTS coefficient de dilatation linéaire *m*

coercive: ~ **field** *n* ELECT *electromagnetism* champ cœrcitif *m*

coercivity *n* ELECT *electromagnetism* coercivité *f*

COF *abbr* (*cursor off*) COMP curseur arrêté

cognitive[1] *adj* GEN cognitif

cognitive[2]: ~ **mapping** *n* COMP mappage cognitif *m*; ~ **modeling** *AmE*, ~ **modelling** *BrE n* GEN modélisation cognitive *f*

coherence *n* DATA, GEN, TRANS cohérence *f*; ~ **area** TRANS aire de cohérence *f*; ~ **bandwidth** RADIO *land mobile*, TRANS largeur de bande de cohérence *f*; ~ **degree** TRANS *fibre optics* degré de cohérence *m*; ~ **of frequency** RADIO cohérence de fréquence *f*; ~ **length** TRANS longueur de cohérence *f*; ~ **time** TRANS durée de cohérence *f*

coherent: ~ **area** *n* TRANS aire de cohérence *f*; ~ **demodulation** *n* TRANS démodulation cohérente *f*; ~ **detection** *n* TRANS *fibre optics* détection cohérente *f*; ~ **light** *n* TRANS lumière cohérente *f*; ~ **modulation** *n* TRANS modulation cohérente *f*; ~ **oscillator** *n* ELECT oscillateur cohérent *m*; ~ **phase-shift keying** *n* (*CPSK*) RADIO, SIG modulation par déplacement de phase cohérente *f* (*MDPC*); ~**-pulse radar** *n* RADIO radar à impulsions cohérentes *m*, radar à impulsions synchronisées *m*; ~ **radiation** *n* RADIO rayonnement cohérent *m*; ~ **significant instant** *n* SIG, TELEP, TEST instant significatif cohérent *m*; ~ **transmission** *n* TRANS transmission cohérente *f*

coil[1] *n* POWER bobine *f*; ~ **former** POWER carcasse de bobine *f*

coil[2]: ~**-load** *vi* TRANS pupiniser

coin: ~ **box** *n* TELEP publiphone *m*; ~**-box set** *n* TELEP publiphone *m*; ~ **buffer** *n* TELEP caisse intermédiaire *f*; ~**-operated payphone** *n* TELEP cabine à pièces *f*; ~**-operated telephone** *n* PER APP, TELEP poste à prépaiement *m*, poste téléphonique à prépaiement *m*; ~**-return cup** *n* PER APP *payphones* sébile de restitution des pièces *f*; ~ **slot** *n* PER APP *payphones* fente d'introduction de pièces *f*

coincidence *n* TELEP *pulses* coïncidence *f*; ~ **method of measurement** TEST méthode de mesure par coïncidence *f*

coincident *adj* TELEP à moments simultanés, à moments successifs

cold: ~ **bird** *n* RADIO satellite laissé pour compte *m*; ~ **cathode tube** *n* ELECT tube à cathode froide *m*; ~ **cathode valve** *n* ELECT tube à cathode froide *m*

COLI *abbr* (*connected-line identification*) DATA, TELEP identification de la ligne connectée *f*

collaborative: ~ **computing** *n* COMP, DATA informatique de groupe *f*

collapse *n* GEN affaissement *m*; **~ action sequence** TEST séquence de mesures à prendre en cas de panne *f*
collate *vt* COMP interclasser
collating: **~ sequence** *n* COMP ordre d'interclassement *m*
collator *n* COMP interclasseuse *f*
collect: **~ call** *n AmE* (*cf reverse-charge call BrE*) TELEP appel à frais virés *m* (*Can*), appel en PCV *f* (*Fra*), communication en PCV *f* (*Fra*), communication payable à l'arrivée *f* (*Fra*); **~-call acceptance** *n AmE* (*cf reverse-charge acceptance BrE*) APPL, TELEP acceptation de la taxation à l'arrivée *f*, acceptation de la taxe à l'arrivée *f*; **~-call service** *n AmE* TELEP service à frais virés *m* (*Can*), service payable à l'arrivée *m*, service PCV *m*; **~-calling acceptance** *n AmE* (*cf reverse-charging acceptance BrE*) APPL, TELEP acceptation de la taxation à l'arrivée *f*, acceptation de la taxe à l'arrivée *f*; **~-calling request indicator** *n AmE* (*cf reverse-charging request indicator BrE*) APPL, TELEP indicateur de demande de taxation à l'arrivée *m*
collecting: **~ comb** *n* TRANS *cables* peigne collecteur *m*
collection: **~ charge** *n* TELEP taxe de perception *f*
collective: **~ aerial** *n* RADIO, TV antenne collective *f*; **~ earth system** *n BrE* (*cf collective ground system AmE*) POWER conducteur annulaire commun de mise à la terre *m*; **~ earthing** *n BrE* (*cf collective grounding AmE*) POWER système collectif de prise de terre *m*; **~ ground system** *n AmE* (*cf collective earth system BrE*) POWER conducteur annulaire commun de mise à la terre *m*; **~ grounding** *n AmE* (*cf collective earthing BrE*) POWER système collectif de prise de terre *m*
collector *n* ELECT, POWER collecteur *m*; **~-base voltage** ELECT tension collecteur-base *f*; **~ current** ELECT courant de collecteur *m*; **~ cutoff current** ELECT courant de coupure du collecteur *m*; **~-emitter voltage** ELECT tension collecteur-émetteur *f*; **~ junction** ELECT jonction collecteur *f*; **~ region** ELECT région de collecteur *f*; **~ resistance** ELECT résistance de collecteur *f*; **~ series resistance** ELECT, PARTS résistance série du collecteur *f*; **~ terminal** ELECT sortie de collecteur *f*; **~ termination** ELECT terminaison de collecteur *f*
collimation *n* TRANS *fibre optics* collimation *f*
collimator *n* TRANS collimateur *m*
collision *n* ELECT collision *f*; **~ detection** (*CD*) ELECT détection de collision *f*
collocated: **~ concentrator** *n* TELEP, TRANS concentrateur d'abonnés local *m*
co-located: **~ antenna** *n* RADIO antenne co-implantée *f*; **~ exchange concentrator** *n* TELEP concentrateur de central local *m*; **~ satellite** *n* RADIO satellite occupant la même position orbitale *m*
colon *n* COMP côlon *m*
colophony *n* ELECT *soldering* arcanson *m*, colophane *f*
color *AmE see colour BrE*
coloration *n* COMP coloration *f*
colored *AmE see coloured BrE*
colorfast *AmE see colourfast BrE*
colorimetric: **~ function** *n* COMP fonction colorimétrique *f*
colorimetry *n* GEN colorimétrie *f*
colorization *n* TV colorisation *f*
ColorSence® *n* COMP *Kodak multimedia* ColorSence® *m*
ColorSync® *n* COMP *Apple multimedia* synchronisation de chrominance® *f*
colour¹: **~-coded** *adj BrE* ELECT de code couleur
colour² *n BrE* GEN couleur *f*; **~ carrier** *BrE* TRANS onde porteuse couleur *f*; **~ code** *BrE* GEN *marking* code couleurs *m*; **~ component** *BrE* TELEP composant de couleur *m*; **~ gamut** *BrE* COMP nuancier *m*; **~ graphics adaptor** *BrE* (*CGA*) COMP carte graphique couleur *f* (*CGA*); **~ image** *BrE* TELEP image en couleur *f*; **~ image scanner** *BrE* PER APP scanner couleur *m*; **~ interpolation shading** *BrE* COMP lissage par interpolation de la couleur *m*; **~ look-up table** *BrE* (*CLUT*) COMP table de couleurs *f* (*CLUT*); **~ matting** *BrE* COMP incrustation douce *f*, incrustation linéaire *f*; **~ scheme** *BrE* COMP palette de couleurs *f*; **~ separation overlay** *BrE* COMP incrustation-couleur *f*; **~ signal** *BrE* COMP signal de couleur *m*; **~ slide** *BrE* GEN diapositive couleur *f*; **~ space** *BrE* TV espace chromatique *m*, espace des couleurs *m*; **~ stimulus** *BrE* COMP stimulus de couleur *m*
coloured: **~ noise** *n BrE* RADIO bruit coloré *m*
colourfast *adj BrE* GEN stable à la couleur
colourfastness *n BrE* GEN stabilité de couleur *f*
COLP *abbr* (*connected-line identification presentation*) TELEP PILC (*présentation de l'identification de la ligne connectée*)
COLR *abbr* (*connected-line identification restriction*) TELEP RILC (*restriction de l'identification de la ligne connectée*)
COM *abbr* (*continuation of message*) DATA suite de message *f*
combed *adj* ELECT peigné
combi: **~-telephone** *n* RADIO, TELEP poste mixte *m*, téléphone mobile mixte *m*, terminal mixte *m*
combination *n* COMP, DATA, TELEP combinaison *f*
combinational¹ *adj AmE* (*cf combinatorial BrE*) ELECT combinatoire
combinational²: **~ circuit** *n AmE* (*cf combinatorial circuit BrE*) ELECT circuit combinatoire *m*
combinatorial¹ *adj BrE* (*cf combinational AmE*) ELECT combinatoire
combinatorial²: **~ circuit** *n BrE* ELECT circuit combinatoire *m*, circuit mélangeur *m*
combine *vt* ELECT fédérer
combined: **~ distribution frame** *n* (*CDF*) SWG répartiteur mixte *m*; **~ local-toll system** *n* SWG système mixte local et transit *m*; **~ local and transit exchange** *n* SWG centre mixte urbain et de transit *m*; **~ loss and delay system** *n* TELEP système avec perte et attente *m*; **~ station** *n* COMP station mixte *f*; **~ transmitting/receiving station** *n* RADIO station mixte d'émission et de réception *f*
combiner *n* ELECT combineur *m*; **~ unit** ELECT combineur *m*
combining: **~ network** *n* TRANS réseau d'aiguillage *m*
comet: **~ tail** *n* COMP queue de comète *f*, traînage dynamique *m*; **~ tail effect** *n* COMP effet de queue de comète *m*, effet de traînage dynamique *m*
comfort: **~ tone** *n* TELEP tonalité de file d'attente *f*
comma *n* COMP virgule *f*
command *n* TELEP commande *f*; **~ antenna** RADIO antenne de télécommande *f*; **~ class** TELEP classe de commande *f*; **~ code** COMP, DATA, TELEP code de commande *m*; **~ and control system** CONTR système de commandement et conduite d'opérations *m*; **~ descriptor block** (*CDB*) DATA bloc de descripteur de commande *m*; **~ document start** DATA commande de début de document *f*; **~ identifier** TELEP identificateur de commande *m*; **~ interface unit** CONTR boîte de jonction de télécommande *f*; **~ language** COMP langage de commande *m*; **~ line** DATA *Internet* ligne de commande *f*; **~ option** TELEP option de commande *f*; **~ reference** TELEP référence de commande *f*; **~ sequence number** SWG numéro de séquence de commande *m*; **~ unit** CONTR bloc de commande *m*; **~ variable** CONTR grandeur de commande *f*
command/response: **~ bit** *n* (*C/R*) TELEP bit de commande/réponse *m*
commentary: **~ channel** *n* TRANS voie commentaire *f*
commercial *n* TV spot publicitaire *m*; **~ satellite telecast** TV émission de télévision commerciale par satellite *f*
commercially: **~ sensitive** *adj* GEN sensible au point de vue stratégique
commission *vt* TELEP mettre en service
commissioned¹ *adj* GEN homologué
commissioned²: **~ research** *n* TEST *development purposes* recherche commanditée *f*, recherche subventionnée *f*
commissioning *n* TELEP mise en service *f*; **~ test** TEST essai de mise en service *m*; **~ time** GEN délai de recette technique *m*

commitment *n* DATA engagement *m*; **~, concurrency and recovery** (*CCR*) DATA engagement, concurrence et rétablissement *m*; **~ coordinator** DATA coordinateur d'engagement *m*

committed: **~ information rate** *n* (*CIR*) DATA débit d'information garanti *m*

common: **~ aerial** *n* RADIO antenne collective *f*; **~ air interface** *n* (*CAI*) RADIO interface radio commune *f* (*CAI*); **~ ASEs** *n pl* DATA éléments de service d'application communs *m pl*; **~ associated signaling** *AmE*, **~ associated signalling** *BrE n* SIG signalisation voie par voie *f*; **~ base** *n* ELECT *transistors* base commune *f*; **~ base station working** *n* RADIO exploitation avec station de base commune *f*; **~ battery** *n* TELEP batterie centrale *f*; **~ branch** *n* TRANS branche commune *f*; **~ carrier** *n* GEN, TRANS entreprise de télécommunications *f*, exploitant de réseaux de télécommunications *m*, société exploitante de télécommunications *f*, télécommunicateur *m*, porteuse commune *f*; **~ channel exchange** *n* SWG centre utilisant un système de signalisation sur voie *m*; **~-channel signaling** *AmE see common-channel signalling BrE*; **~-channel signaling system defined by CCITT** *AmE see common-channel signalling system defined by CCITT BrE*; **~-channel signalling** *BrE n* SIG signalisation par canal sémaphore *f*, signalisation sur voie commune *f*; **~-channel signalling system defined by CCITT** *n BrE* SIG système de signalisation par canal sémaphore défini par le *m*; **~ collector** *n* ELECT *transistors* collecteur commun *m*; **~ control channel** *n* CONTR, TRANS canal de contrôle commun *m*; **~ control equipment** *n* SWG organes de commande communs *m pl*; **~-control switching system** *n* SWG système de commutation à commande centrale *m*; **~ control system** *n* SWG système à commande centralisée *m*; **~ emitter** *n* ELECT *transistors* émetteur commun *m*; **~ equipment** *n* (*CE*) SWG matériel commun *m*, organe commun *m*; **~ field** *n* COMP zone commune *f*; **~ highway** *n* SWG tronc commun *m*; **~ intermediate format** *n* (*CIF*) COMP format intermédiaire commun *m* (*FIC*); **~ management information protocol data unit** *n* (*CMIPDU*) DATA unité de données de protocole commun d'informations de gestion *f*; **~ management information protocol** *n* (*CMIP*) DATA protocole commun d'informations de gestion *m*; **~ management information protocol machine** *n* (*CMIPM*) DATA machine de protocole commun d'informations de gestion *f*; **~ management information service** *n* (*CMIS*) DATA service commun d'informations de gestion *m*; **~ management information service element** *n* (*CMISE*) DATA élément de service commun d'informations de gestion *m*; **~ memory** *n* COMP mémoire commune *f*; **~-mode rejection ratio** *n* TELEP rapport d'affaiblissement dans le mode commun *m*; **~ operating standard** *n* GEN norme commune d'exploitation *f*; **~ return** *n* GEN retour commun *m*; **~ routing channel** *n* (*CRC*) RADIO, TRANS voie d'acheminement commune *f* (*VAC*); **~-sense reasoning** *n* DATA raisonnement de sens commun *m*; **~ store** *n* DATA mémoire commune *f*; **~ textual grammar** *n* DATA grammaire textuelle commune *f*; **~ traffic terminal equipment** *n* (*CTTE*) TELEP équipement commun des terminaux de trafic *m*; **~ waveguide multiplexer** *n* RADIO multiplexeur à guides d'ondes communs *m*

commonality *n* TELEP communauté de conception *f*

communicate *vi* TV dialoguer

communication *n* DATA, GEN, RADIO transmission *f*, communication *f*; **~ application** (*CA*) DATA application de communication *f* (*CA*); **~ capability** GEN capacité de communication *f*; **~ card** DATA carte de communication *f*; **~ channel** TV voie de communication *f*; **~ console** DATA console télématique *f*; **~ control card** COMP carte PC de communication *f*; **~ engineering** TELEP technique des télécommunications *f*; **~ error** GEN erreur de communication *f*; **~ path** DATA trajet de communication *m*; **~ port** COMP port de transmission *m*; **~ protection** DATA protection télématique *f*; **~ protocol** DATA *Internet* protocole de communication *m*; **~ services** *n pl* (*CS*) APPL services de communication *m pl*; **~ software** COMP, DATA logiciel de communication *m*, logiciel de télécommunication *m*; **~ station** RADIO station de communication *f*; **~ theory** GEN théorie de la communication *f*

communications: **~ agreement** *n* GEN accord d'échange *m*; **~ centre** *n* TRANS centre de transmissions *m*; **~ processor** *n* SWG processeur de communications *m*; **~ research center** *AmE*, **~ research centre** *n BrE* GEN centre de recherche sur les communications *m*; **~ server** *n* DATA serveur de communications *m*; **~ spacecraft operation center** *AmE*, **~ spacecraft operation centre** *BrE n* RADIO centre d'opérations des engins spatiaux de télécommunications *m*; **~ system control** *n* (*CSC*) GEN commande du système de télécommunications *f*; **~ system management** *n* GEN gestion du système de télécommunications *f*; **~ system monitoring** *n* (*CSM*) TEST surveillance du système de télécommunications *f*; **~ technology satellite** *n* (*CTS*) RADIO satellite de technologie des télécommunications *m*

communicator *n* TELEP communicateur *m*

communiqué *n* TELEP communiqué

community: **~ aerial** *n* RADIO, TV antenne collective *f*, antenne communautaire *f*; **~ antenna** *n* TV antenne collective *f*, antenne communautaire *f*; **~-antenna television** *n* (*CATV*) TV télévision à antenne collective *f*, télévision à antenne communautaire *f*; **~ dial office** *n AmE* SWG central automatique local *m*; **~ network** *n* TELEP réseau de télévision par câbles *m*, TV réseau communautaire *m*; **~ program** *AmE*, **~ programme** *BrE n* TV programme communal *m*; **~ reception** *n* RADIO réception communautaire *f*; **~ of transponders** *n* RADIO communauté de répéteurs *f*

commutator *n* ELECT, POWER collecteur *m*

compact[1]: **~ constraint table** *n* DATA table compacte des contraintes *f*; **~ disc** *n* (*CD*), COMP disque compact *m* (*CD*); **~ disc file manager** *n* COMP gestionnaire de fichiers de disques compacts *m*; **~ disc interactive** *n* (*CD-I*) COMP disque compact interactif *m* (*CD-I, DCI*); **~ disc read-only memory** *n* (*CD-ROM*) COMP, DATA disque optique numérique *m* (*CD-ROM, DON*); **~ disc-read only** *n* (*CD-R*) COMP disque compacte à lecture seule (*CD-R*); **~ disc-write once** *n* COMP (*CD-WO*) disque compacte à écriture seule (*CD-WO*); **~ test-case table** *n* DATA table compacte des tests élémentaires *f*

compact[2] *vt* COMP compresser

compacted: **~ conductor** *n* TRANS âme rétreinte *f*

compaction *n* COMP compression *f*

compactor *n* COMP compacteur *m*

compand *vt* COMP, DATA, TELEP, TRANS comprimer-étendre

companded: **~ delta modulation** *n* (*CDM*) SIG, TRANS modulation delta avec compression et extension *f*

compander *n* DATA, SIG compresseur-extenseur *m*

companding *n* COMP, DATA, TELEP, TRANS compression-extension *f*; **~ and frequency modulation** (*CFM*) TRANS compression-extension avec modulation de fréquence *f* (*CMF*)

company: **~ code** *n* PER APP code d'identification de l'entreprise *m*; **~ identity code** *n* PER APP *PBX* code d'identification de l'entreprise *m*; **~ network** *n* TRANS réseau d'entreprise *m*

comparator *n* ELECT comparateur *m*; **~ response** ELECT *pulses* réponse de comparateur *f*

compare: **~ mark** *n* COMP repère de comparaison *m*

comparing: **~ element** *n* CONTR comparateur *m*

comparison: **~ circuit** *n* ELECT comparateur *m*; **~ measurement** *n* TEST mesure comparative *f*, mesure de comparaison *f*; **~ method** *n* TEST méthode comparative *f*;

~ **method of measurement** *n* TEST méthode de mesure par comparaison *f*
compartmental: ~ **model** *n* TEST modèle à cases *m*, modèle compartimenté *m*
compatible: ~ **computer** *n* COMP ordinateur compatible *m*
compelled: ~ **cycle** *n* SIG cycle asservi *m*; ~ **multi-frequency code** *n* SIG système asservi à code multifréquence *m*; ~ **signaling** *AmE*, ~ **signalling** *BrE* *n* SIG signalisation asservie *f*, asservissement *m*
compensated: ~ **semiconductor** *n* ELECT semiconducteur compensé *m*
compensating: ~ **element** *n* CONTR correcteur *m*; ~ **network** *n* RADIO *aerials* centreur d'impédance *m*, circuit de compensation *m*, réseau de compensation *m*
compensation: ~ **circuit** *n* ELECT circuit de compensation *m*
competitive: ~ **clipping** *n* TRANS mutilation due à la saturation *f*
compilation: ~ **phase** *n* COMP phase de compilation *f*
compile *vt* COMP, TELEP compiler
compiler *n* COMP, TELEP compilateur *m*; ~ **generator** COMP générateur de compilateurs *m*
compiling: ~ **phase** *n* COMP phase de compilation *f*; ~ **program** *n* COMP programme de compilation *m*
complementary: ~ **code** *n* TRANS code complémentaire *m*; ~ **method of measurement** *n* TEST méthode de mesure par complément *f*; ~ **service** *n* APPL complément du service *m*
complete[1]: ~ **charging period** *n* PER APP durée de la charge complète *f*; ~ **failure** *n* TEST *dependability* défaillance complète *f*; ~ **valid input-signal set** *n* DATA ensemble complet de signaux d'entrée valides *m*
complete[2] *vt* COMP *a task* terminer
completed: ~ **call** *n* TELEP appel efficace *m*; ~ **call attempt** *n* RADIO, TELEP tentative d'appel efficace *f*
completion: ~ **of calls** *n* TELEP aboutissement de l'appel *m*; ~ **ratio** *n* TELEP taux d'efficacité *m*; ~ **test** *n* TEST *testing* test de fin *m*
complex: ~ **instruction set equipment** *n* (*CISE*) DATA CISE; ~ **wave** *n* RADIO onde complexe *f*
compliance: ~ **test** *n* TEST essai de conformité *m*
compliant: ~ **bridge** *n* ELECT pont conforme *m*
comply: ~ **with requirements** *vi* GEN satisfaire aux exigences
component *n* DATA élément *m*, ELECT *discrete element* composant *m*, GEN élément *m*, élément constitutif *m*, PARTS composant *m*, TELEP élément *m*; ~ **data element separator** DATA séparateur d'éléments de données constitutifs *m*; ~ **engineering** ELECT *design technology* technique des composants *f*, technologie des composants *f*; ~ **front** ELECT *printed circuits* platine avant *f*; ~ **height** ELECT *printed circuits* hauteur de composant *f*; ~ **holder** ELECT support de composant *m*; ~ **hole** ELECT *printed circuits* trou connexion *m*; ~ **housing** ELECT armoire standard *f*, coffret de protection *m*; ~**-level testing** TEST *testing* essai au niveau composants *m*; ~ **part** GEN élément constitutif *m*; ~ **technology** ELECT *design* technique des composants *f*, technologie des composants *f*
composite: ~ **gain** *n* TRANS gain composite *m*; ~ **loss** *n* TELEP, TRANS affaiblissement composite *m*; ~ **node** *n* COMP nœud composite *m*; ~ **resistor** *n* ELECT, PARTS résistance agglomérée *f*; ~ **signal** *n* SIG signal composite *m*; ~ **signal coding** *n* TRANS codage de signal composite *m*; ~ **structure of grooved silicon chips** *n* ELECT support secondaire de structures sur fibres *m*; ~ **video** *n* TV vidéo composite *f*; ~ **wave** *n* RADIO onde composite *f*; ~ **waveform** *n* ELECT signal composite *m*
compound: ~ **parameter argument** *n* TELEP argument de paramètre composé *m*; ~ **signal** *n* SIG signal composite *m*
compress *vt* DATA compacter, compresser, comprimer, TELEP, TRANS compresser, comprimer
compressed[1] *adj* DATA, TELEP, TRANS avec compression
compressed[2]: ~ **data** *n pl* DATA données compactées *f pl*, données comprimées *f pl*; ~ **digital transmission** *n* TRANS transmission numérique avec compression *f*; ~ **mode** *n* DATA mode compression *m*
compression *n* COMP compression *f*, DATA compactage *m*, TELEP, TRANS compression *f*; ~ **of the luminance range** TELEP compression des luminances *f*; ~ **ratio** TRANS rapport de compression *m*
compressor *n* DATA, TELEP, TRANS compresseur *m*; ~**-expander** DATA, SIG compresseur-extenseur *m*
compromise: ~ **equalization** *n* TRANS égalisation fixe des modems *f*
compromising: ~ **emanations** *n pl* RADIO *security* rayonnements divulgateurs *m pl*
compulsory: ~ **clearing request** *n* SIG demande de libération forcée *f*
COMPUSEC *abbr* (*computer security*) DATA sécurité informatique *f*
compute: ~ **mode** *n* COMP mode calcul *m*
computer[1]: ~**-aided** *adj* COMP assisté par ordinateur; ~**-assisted** *adj* COMP assisté par ordinateur; ~**-controlled** *adj* COMP commandé par ordinateur, piloté par informatique
computer[2] *n* COMP ordinateur *m*; ~**-aided design** (*CAD*) COMP conception assistée par ordinateur *f* (*CAO*); ~**-aided design and computer-aided manufacturing** (*CAD/CAM*) COMP conception et fabrication assistées par ordinateur *f* (*CFAO*); ~**-aided engineering** (*CAE*) PARTS ingénierie assistée par ordinateur *f* (*IAO*); ~**-aided logistics** (*CAL*) PARTS logistique assistée par ordinateur *f*; ~**-aided manufacturing** (*CAM*) COMP fabrication assistée par ordinateur *f* (*FAO*); ~**-aided quality** (*CAQ*) COMP qualité assistée par ordinateur *f*; ~**-aided software engineering** (*CASE*) COMP génie logiciel assisté par ordinateur *m*; ~**-aided testing** (*CAT*) COMP test assisté par ordinateur *m*; ~**-assisted instruction** COMP *training* enseignement assisté par ordinateur *m*; ~**-assisted learning** (*CAL*) COMP enseignement assisté par ordinateur *m* (*EAO*); ~**-assisted management** COMP gestion automatisée *f*, gestion assistée par ordinateur *f*; ~**-assisted vision** COMP vision assistée par ordinateur *f*; ~ **bank** COMP batterie d'ordinateurs *f*; ~**-based learning** COMP éducation assistée par ordinateur *f*, éducation informatisée *f*; ~**-based training** APPL, COMP enseignement assisté par ordinateur *m* (*EAO*), enseignement informatisé *m*; ~ **bootstrapping** COMP amorçage *m*; ~ **buff** COMP fan d'ordinateur *m*; ~ **circuitry** COMP circuits informatiques *m pl*; ~ **fraud** COMP fraude informatique *f*, piratage informatique *m*; ~**-generated image** COMP image de synthèse *f*; ~ **graphics** COMP, TV infographie *f*, infographisme *m*; ~ **imagery** COMP infographie *f*; ~ **instruction** COMP instruction machine *f*; ~**-integrated manufacturing** (*CIM*) COMP fabrication intégrée par ordinateur *f* (*FIO*), productique *f*; ~ **interface** COMP, DATA interface d'ordinateur *f*, interface informatique *f*; ~ **language** COMP langage machine *m*, langage informatique *m*; ~ **linguist** COMP linguiste-informaticien *m*; ~ **link** COMP, TRANS liaison d'ordinateur *f*; ~ **network** COMP, TRANS réseau d'ordinateurs *m*; ~ **program** COMP, DATA logiciel *m*, programme *m*; ~ **scientist** GEN informaticien *m*; ~ **security** (*COMPUSEC*) DATA sécurité informatique *f*; ~ **simulation** COMP simulation sur ordinateur *f*; ~ **system** COMP système informatique *m*; ~ **virus** DATA virus informatique *m*; ~ **worm** COMP ver automatique *m*
computerization *n* COMP informatisation *f*
computerize *vt* COMP informatiser
computerized: ~ **power supply** *n* POWER alimentation informatisée *f*
computer-PBX: ~ **interface** *n* (*CPI*) COMP interface PABX-serveur *f*

computing *n* COMP, DATA informatique *f*; ~ **center** *AmE*, ~ **centre** *BrE* COMP centre de calcul *m*
COMSOC *n* RADIO centre d'opérations des engins spatiaux de télécommunications *m*
CON *abbr* COMP (*cursor on*) curseur en marche, TRANS (*concentrator*) concentrateur *m*
concatenate *vt* COMP enchaîner
concatenated: ~ **coding** *n* (*CC*) DATA codage chaîné *m*
concatenation *n* DATA concaténation *f*; ~ **indication** (*CI*) ELECT indication de concaténation *f*
concealment *n* COMP masquage *m*, RADIO dissimulation *f*; ~ **method** COMP méthode de masquage *f*
concentration *n* ELECT, GEN concentration *f*; ~ **in a switching stage** SWG concentration dans un étage de commutation *f*; ~ **module** (*ASM*) ELECT module concentrateur *m*; ~ **network** TRANS réseau de concentration *m*; ~ **stage** TRANS étage de concentration *m*
concentrator *n* (*CON*) ELECT, TRANS concentrateur *m*
concentric: ~ **conductor** *n* TRANS *cables* conducteur coaxial *m*; ~ **neutral cable** *n* TRANS câble à neutre périphérique *m*; ~ **stranding** *n* TRANS câblage en couches concentriques *m*
concept *n* TEST concept *m*
conceptual: ~ **clustering** *n* COMP regroupement conceptuel *m*
concession *n* TEST *quality* dérogation après production *f*
concrete: ~ **grammar** *n* DATA grammaire concrète *f*; ~ **graphical grammar** *n* DATA grammaire graphique concrète *f*; ~ **graphical syntax** *n* DATA syntaxe graphique concrète *f*; ~ **syntax** *n* DATA syntaxe concrète *f*; ~ **textual syntax** *n* DATA syntaxe textuelle concrète *f*
concurrent: ~ **statements** *n pl* DATA instructions concurrentes *f pl*
condensation: ~ **soldering** *n* ELECT brasage en phase vapeur *m*
condense *vt* DATA compacter, comprimer, réduire
condensed[1] *adj* DATA réduit
condensed[2]: ~ **sub-burst data** *n pl* DATA données condensées de sous-paquet *f pl*
condenser *n* ELECT condensateur *m*; ~ **microphone** TELEP microphone électrostatique *m*, microphone à condensateur *m*
condensing *n* DATA *of information* réduction *f*
condition *n* GEN état; ~ **A** TELEP condition A *f*; ~ **Z** TELEP condition Z *f*
conditional: ~ **access** *n* TV accès conditionnel *m*; ~ **access control** *n* RADIO commande de l'accès conditionnel *f*; ~ **component** *n* DATA élément conditionnel *m*, ELECT composant conditionnel *m*, élément conditionnel *m*, PARTS composant conditionnel *m*, TELEP élément conditionnel *m*; ~ **expression** *n* DATA *SDL* expression conditionnelle *f*; ~ **jump** *n* COMP saut conditionnel *m*; ~ **parameter** *n* DATA paramètre conditionnel *m*; ~ **replenishment** *n* DATA rafraîchissement conditionnel *m*; ~ **requirement** *n* GEN prescription conditionnelle *f*, spécification conditionnelle *f*; ~ **selection** *n* GEN sélection conditionnelle *f*
conditionally: ~ **toll-denied** *n* (*CTD*) SWG blocage partiel de l'accès à l'interurbain *m*
conditioning *n* RADIO traitement *m*
conductance *n* ELECT, POWER conductance *f*
conducted: ~ **effect** *n* ELECT effet propagé par les conducteurs *m*; ~ **spurious emission** *n* ELECT rayonnement parasite par conduction *m*
conducting[1] *adj* ELECT, POWER conducteur, passant; ~ **on** ELECT *transistors* passant
conducting[2]: ~ **direction** *n* ELECT sens de conduction *m*; ~ **layer** *n* ELECT couche conductrice *f*; ~ **state** *n* ELECT état conducteur *m*, état passant *m*
conduction *n* GEN conduction *f*; ~ **band** ELECT *semiconductors* bande de conduction *f*; ~ **electron** ELECT électron de conduction *m*, électron de valence *m*, électron libre *m*; ~ **period** ELECT période de conduction *f*; ~ **time** ELECT temps de conduction *m*; ~**-time control** ELECT commande du temps de conduction *f*
conductive[1] *adj* ELECT, POWER conducteur
conductive[2]: ~ **layer** *n* ELECT couche conductrice *f*; ~ **pattern** *n* ELECT *printed circuits* réseau de conducteurs *m*
conductivity *n* GEN conductibilité *f*; ~ **modulation** ELECT *semiconductors* modulation de conductivité *f*
conductor *n* GEN conducteur *m*, PARTS faisceau *m*, fil *m*, TELEP conducteur *m*; ~ **area** ELECT section d'un conducteur *f*; ~ **insulation** TRANS isolation de conducteur *f*; ~ **screen** TRANS *cables* écran sur âme *m*; ~ **spacing** ELECT distance entre conducteurs *f*; ~ **unit** TRANS faisceau d'âme *m*; ~ **width** ELECT *printed circuits* largeur du conducteur *f*
conduit *n* ELECT, PARTS, TRANS *cable laying* canalisation *f*, conduit *m*; ~ **route** TRANS trajet de canalisation *m*; ~ **run** TRANS trajet de canalisation *m*; ~ **system** TRANS système de canalisations *m*
cone: ~ **loudspeaker** *n* PARTS *acoustics* haut-parleur à membrane conique *m*
conference *n* TV téléconférence *f*; ~ **bridge** APPL pont de conférence *m*; ~ **call** TELEP conférence téléphonique *f*, téléréunion *f*; ~ **call service** TELEP service de communication conférence *m*, service de téléréunion *m*; ~ **calling** TELEP téléconférence *f*; ~ **circuit** TELEP circuit de conférence *m*; ~ **repeater** TELEP translation pour conférence *f*; ~ **telegraph repeater** TELEP translation télégraphique pour conférence *f*
conferencing: ~ **configuration** *n* TV configuration de type conférence *f*
confidence: ~ **coefficient** *n* GEN coefficient de fiabilité *m*; ~ **interval** *n* GEN intervalle de confiance *m*; ~ **level** *n* GEN niveau de fiabilité *m*; ~ **limit** *n* GEN limite de confiance *f*
configuration *n* GEN configuration *f*, interconnexion *f*; ~ **and data management** DATA configuration et gestion des données *f pl*
configurations: ~ **foldback** *n* ELECT épissures rabattues *f pl*
configure *vt* GEN configurer
confirmation *n* (*conf*) GEN confirmation *f* (*conf*); ~ **of receipt** (*COR*) DATA, TELEP confirmation de réception *f*
confirmed: ~ **service** *n* GEN service avec confirmation *m*, service confirmé *m*, service de type confirmé *m*
conformal: ~ **aerial** *n* RADIO antenne en trèfle *f*; ~ **antenna** *n* RADIO antenne en trèfle *f*
conformance: ~ **log** *n* DATA journal de conformité *m*; ~**-resolution test** *n* DATA, TEST test de résolution de conformité *m*; ~ **test specification manual** *n* TEST manuel de spécifications des tests de conformité *m*; ~ **testing** *n* DATA, TEST test de conformité *m*, tests de conformité *m pl*; ~ **testing specification document** *n* DATA document de spécification des tests de conformité *m*; ~**-testing specification manual** *n* (*CTSM*) DATA manuel de spécifications des tests de conformité *m*; ~ **testing standard** *n* DATA, TEST norme de test de conformité *f*; ~ **test suite** *n* DATA, TEST suite de tests de conformité *f*
conforming: ~ **implementation** *n* DATA instance conforme *m*
conformity: ~ **specifications** *n pl* TEST spécifications de conformité *f pl*
confravision *n* TV visioconférence *f*
confusion: ~ **jamming** *n* RADIO brouillage de diversion *m*; ~ **message** *n* DATA message d'incohérence *m*; ~ **signal** *n* SIG signal de confusion *m*
congest *vt* SWG, TELEP, TRANS encombrer
congested[1] *adj* SWG, TELEP, TRANS encombré
congested[2]: ~**-route counter** *n* SWG, TELEP compteur d'encombrements de faisceau *m*
congestion *n* SWG, TELEP, TRANS encombrement *m*; ~ **indicator** RADIO indicateur d'encombrement *m*; ~ **tone** TELEP tonalité d'encombrement *f*
conical[1] *adj* RADIO conique

conical[2]: **~ scan** *n* RADIO balayage conique *m*; **~ scanning with nutation** *n* RADIO balayage conique avec nutation *m*; **~ scanning with rotation** *n* RADIO exploration conique *f*

conjunction *n* ELECT conjonction *f*, intersection *f*

connect[1] *n* TELEP service complémentaire *m*; **~ confirm** (*CC*) SIG confirmation de connexion *f*; **~ data overflow** (*CDO*) DATA débordement de données de connexion *m*; **~ and disconnect signaling** *AmE*, **~ and disconnect signalling** *BrE* SIG signalisation de connexion et déconnexion *f*; **~ facility** TELEP service complémentaire *m*; **~ hold** TELEP attente connexion *f*; **~ MCSPDU** TELEP unité MCSPDU de type Connect *f*; **~ signal** SIG signal de connexion *m*; **~ time** COMP durée de connexion *f*

connect[2] *vt* DATA connecter, ELECT connecter, *strap* effectuer, poser, GEN, TELEP connecter, mettre en communication avec, relier à, mettre en communication, raccorder, SWG mettre en communication; **~ in cascade** ELECT monter en cascade; **~ in a loop** ELECT boucler un circuit, connecter en boucle; **~ in parallel** ELECT, POWER brancher en parallèle, connecter en parallèle, monter en parallèle; **~ in series** ELECT brancher en série, connecter en série, monter en série; **~ negative** POWER appliquer une polarité négative à

connect[3]: **~-data set to line** *phr* DATA connectez le poste de données sur la ligne

connectability *n* TV connectique *f*

Connect-Additional *n* TELEP Connect-Additional *m*

connected[1] *adj* COMP, DATA connecté, ELECT branché, TELEP connecté; **~ to** COMP *Internet* branché, TELEP en communication avec; **~ to voltage** POWER mis sous tension

connected[2]: **~-line identification** *n* (*COLI*) TELEP identification de la ligne connectée *f*; **~-line identification presentation** *n* (*COLP*) TELEP présentation de l'identification de la ligne connectée *f* (*PILC*); **~-line identification restriction** *n* (*COLR*) TELEP restriction de l'identification de la ligne connectée *f* (*RILC*); **~-line identity request indicator** *n* TELEP indicateur de demande d'identité de la ligne connectée *m*; **~ switching path PE** *n* SWG élément graphique de trajet de commutation établi *m*

connecting: **~ box** *n* TELEP boîte de raccordement *f*; **~ cable** *n* TRANS câble de liaison *m*, câble de raccordement *m*; **~ devices** *n pl* ELECT matériel de connectique *m*; **~ pin** *n* ELECT broche de raccordement *f*; **~ tag** *n* ELECT languette de raccordement *f*; **~ terminal** *n* COMP borne de connexion *f*; **~ wire** *n* ELECT *conductors* fil de connexion *m*

Connect-Initial *n* TELEP Connect-Initial *m*

connection *n* COMP, DATA, ELECT, GEN branchement *m*, connexion *f*, raccordement *m*, RADIO chaîne de connexion *f*, chaîne de connexion complète *f*, communication *f*, TELEP connexion *f*, *established call* communication *f*, chaîne de connexion *f*, TRANS connexion *f*; **~ box** TELEP boîte de raccordement *f*, TV boîtier de raccordement *m*; **~ building networks** *n pl* TRANS réseaux d'immeubles raccordement *m pl*; **~ cable** TRANS câble de raccordement *m*; **~ center** *AmE*, **~ centre** *BrE* TV centre de rattachement *m*; **~ charge** TELEP, TV droit d'accès *m*, droit de raccordement *m*; **~ communication path** TELEP connexion de trajet de communication *f*; **~ delay** TV délai de raccordement *m*; **~ diagram** GEN schéma de connexions *m*; **~ disk** COMP, DATA disquette de connexion *f*; **~ element** (*CE*) DATA élément de connexion *m*; **~ equipment** GEN matériel de raccordement *m*; **~ field** POWER champ de répartition *m*, réglette de raccordement *f*; **~ in progress** TELEP communication en cours d'établissement *f*; **~ in series** ELECT montage en série *m*; **~ kit** DATA kit de connexion *m*; **~ matrix** (*CM*) SWG matrice de connexion *f*; **~ mode** DATA mode connexion *m*; **~-oriented network service** (*CONS*) APPL, DATA service de réseau en mode connexion *m*; **~-oriented network-layer protocol** (*CONP*) TRANS protocole de couche réseau en mode connexion *m*; **~ overhead** DATA résidu de connexion *m*; **~ point** GEN point de connexion *m*, point de raccordement *m*; **~ provider** DATA fournisseur connexion *m*; **~-related function** (*CRF*) DATA, TRANS fonction liée à la connexion *f*; **~ request** DATA demande de connexion *f*, demande de modification *f*; **~ requirements** *n pl* DATA *Internet* besoins de connexion *m pl*; **~ strip** ELECT barrette à bornes *f*, barrette de connexion *f*; **~ terminal** COMP borne de raccordement *f*; **~ through an exchange** SWG connexion à travers un central *f*; **~ unit** GEN, TELEP *for subscribers or trunks* unité de raccordement *f*

connectionless[1] *adj* (*CL*) GEN en mode non connecté, sans connexion

connectionless[2]: **~ bearer service** *n* APPL service support sans connexion *m*; **~ broadband data service** *n* (*CBDS*) APPL, DATA service de données à large bande sans connexion *m*; **~ convergence protocol** *n* (*CLCP*) DATA protocole de convergence sans connexion *m*; **~ mode** *n* DATA mode sans connexion *m*; **~-mode network service** *n* APPL service de réseau sans connexion *m*; **~-mode presentation service** *n* DATA service de présentation en mode sans connexion *m*; **~-mode session service** *n* DATA service de session en mode sans connexion *m*; **~-mode transmission** *n* TRANS transmission en mode sans connexion *f*; **~ network-layer protocol** *n* (*CLNP*) APPL protocole de couche réseau sans connexion *m*; **~ network-layer service** *n* (*CLNS*) APPL service de couche réseau sans connexion *m*; **~ server** *n* (*CLS*) DATA serveur sans connection *m*; **~ service function** *n* (*CLSF*) APPL fonction de service sans connexion *f*

connectivity *n* COMP, TRANS, TV connectique *f*, connexité *f*

connector *n* GEN connecteur *m*; **out ~** GEN connecteur de sortie *m*; **~ insertion loss** TRANS *fibre optics* perte d'insertion des connecteurs *f*, pertes dans les connecteurs *f pl*; **~ loss** TRANS *fibre optics* perte au connecteur *f*; **~ terminal** ELECT borne de raccordement *f*

Connect-Response *n* TELEP Connect-Response *m*

Connect-Result *n* TELEP Connect-Result *m*

CONP *abbr* (*connection-oriented network-layer protocol*) TRANS protocole de couche réseau en mode connexion *m*

CONS *abbr* APPL, DATA (*connection-oriented network service*) service de réseau en mode connexion *m*

consecutive: **~ error block** *n* (*CEB*) DATA bloc erroné consécutif *m*; **~ identical digits** *n pl* (*CID*) DATA symboles identiques consécutifs *m pl*

consistency *n* DATA, GEN, TRANS cohérence *f*

consistent: **~ partitioning subset** *n* DATA sous-ensemble de subdivision cohérent *m*; **~-refinement subset** *n* DATA sous-ensemble de raffinement cohérent *m*

console *n* COMP console

consolidated: **~ link-layer management message** *n* (*CLLM*) DATA message de gestion de couche liaison consolidé *m*

consolidation *n* GEN réseau unifié *m*

constant: **~ angular velocity** *n* (*CAV*) COMP vitesse angulaire constante *f* (*VAC*); **~ bit rate** *n* (*CBR*) COMP, DATA, TELEP, TRANS débit binaire constant *m*; **~ failure intensity period** *n* TELEP période d'intensité constante de dérangement *f*; **~ linear velocity** *n* (*CLV*) COMP vitesse linéaire constante *f* (*VLC*); **~ quality** *n* COMP qualité constante *f*; **~ rate** *n* COMP, DATA taux constant *m*; **~ shading** *n* COMP lissage constant *m*; **~-voltage rectifier** *n* POWER redresseur à tension constante *m*

constituent: **~ units** *n pl* PARTS unités constitutives *f pl*

constraint *n* DATA contrainte *f*

constraints: **~ part** *n* DATA partie contraintes *f*; **~ reference** *n* DATA référence aux contraintes *f*

construction *n* PARTS montage *m*; **~-bound rule** PARTS règle liée au conditionnement *f*; **~ manager** PARTS chef de montage *m*

constructive: **~ interference** *n* TRANS *fibre optics* interférence constructive *f*; **~ solid geometry** *n* (*CSG*) COMP géométrie solide constructive *f* (*GSC*)

consultancy *n* GEN cabinet d'expertise *m*
consultant *n* GEN ingénieur-conseil *m*
consumer: ~ **risk** *n* TEST risque du client *m*; ~ **risk point** *n* TEST point du risque du client *m*
consumption: ~ **of a signal** *n* SIG absorption d'un signal *f*
contact *n* ELECT, PARTS, POWER contact *m*; ~ **bank** POWER banc de contacts *m*; ~ **bar** POWER lame fixe de contact *f*; ~ **metalization** *AmE*, ~ **metallization** *BrE* ELECT *semiconductors* métallisation des contacts *f*; ~ **method of measurement** TEST méthode de mesure par contact *f*; ~ **pin** ELECT *connecting devices* broche de contact *f*, broche de fiche mâle *f*; ~ **plate** CLIM *lighting* plot *m*; ~ **point** POWER, SWG *switchgear* point de contact *m*; ~ **spring adjuster** TEST cambreur de ressort porte-contact *m*; ~ **spring bender** TEST cambreur de ressort porte-contact *m*; ~ **surface** PARTS portée *f*, surface d'appui *f*, surface de contact *f*, surface portante *f*; ~ **unit** ELECT *connecting devices*, SWG contacteur *m*
contactor *n* ELECT, SWG *switchgear* contacteur *m*
container *n* (*CT*) APPL, DATA, TELEP conteneur *m*; ~**-n** (*C-n*) APPL, DATA, TELEP conteneur-n *m* (*C-n*)
content: ~**-addressable memory** *n* (*CAM*) COMP mémoire adressable par son contenu *f*; ~**-addressable storage** *n* DATA mémoire associative *f*; ~**-addressed storage** *n* COMP mémoire associative *f*; ~ **length** *n* DATA longueur du contenu *f*
contention *n* GEN conflit *m*, conflit d'accès *m*; ~ **control** RADIO limitation des collisions *f*, limitation des conflits *f*
context: ~ **parameter** *n* DATA paramètre contextuel *m*; ~ **search** *n* COMP recherche contextuelle *f*
continent: ~ **code** *n* TELEP *telex* code continental *m*
continental: ~ **circuit** *n* TELEP circuit continental *m*; ~ **connection** *n* TELEP communication continentale *f*; ~ **exchange** *n* SWG centre continental *m*
contingency: ~ **period** *n* GEN période de réserve *f*; ~ **plan** *n* GEN mesures d'urgence *f pl*, plan de crise *m*, plan de secours *m*; ~ **planning** *n* GEN sécurisation *f*
continuation *n* DATA suite *f*: ~ **character** *n* TELEP caractère de répétition *m*, caractère suite *m*; ~ **of message** *n* (*COM*) DATA suite de message *f*; ~ **mode** *n* TELEP mode répétitif *m*; ~**-mode operating sequence** *n* TELEP séquence de fonctionnement dans le mode répétitif *f*; ~ **rack** *n* PARTS bâti d'extension *m*
continue: ~ **key** *n* COMP touche de reprise *f*
continuity: ~ **check responder** *n* TELEP répondeur pour contrôle de continuité *m*; ~ **check transceiver** *n* TELEP émetteur-récepteur pour essais de continuité *m*; ~ **fault** *n* TEST défaut de continuité *m*; ~ **signal** *n* SIG signal de continuité *m*
continuous: ~ **action** *n* CONTR action progressive *f*; ~ **bit oriented stream** *n* DATA train de bits en continu *m*; ~ **carrier** *n* TRANS porteuse non modulée *f*; ~ **direct off-state** *n* ELECT tension continue permanente *f*; ~ **direct on-state current** *n* ELECT *thyristors* courant continu permanent à l'état passant *m*; ~ **direct reverse-voltage** *n* ELECT *diodes* tension inverse continue permanente *f*; ~ **form** *n* PER APP papier en continu *m*; ~ **loading** *n* GEN *cables* charge continue *f*, krarupisation *f*, TRANS krarupisation *f*; ~**-mode transmission monitor** *n* TEST appareil de surveillance de transmission en mode continu *m*; ~ **multiplexing** *n* RADIO multiplexage continu *m*; ~ **receiver** *n* TELEP récepteur à déroulement continu *m*; ~ **signal** *n* DATA signal continu *m*; ~ **supervision** *n* TEST surveillance continue *f*; ~ **tone-controlled squelch system** *n* (*CTCSS*) RADIO silencieux à commande par tonalités *m* (*CTCSS*); ~ **tracking** *n* RADIO poursuite continue *f*; ~ **wave** *n* (*CW*) TRANS onde continue *f*; ~**-wave radar** *n* (*CW radar*) RADIO radar à émission continue *m*, radar à ondes entretenues *m*

contour: ~ **correction** *n* COMP correction de contour *f*; ~ **extraction** *n* COMP extraction de contour *f*
contouring *n* COMP sous-quantification *f*
contradirectional: ~ **interface** *n* TRANS interface contradirectionnelle *f*
contrast *n* GEN contraste *m*; ~ **control** TV commande de contraste *f*; ~ **filter** TV *camerawork* filtre pour contrastes *m*; ~ **range** TV plage de variation du contraste *f*; ~ **ratio** TV rapport des contrastes *m*
contrasting: ~ **colors** *AmE*, ~ **colours** *BrE* *n pl* TV tons opposés *m pl*
contribution: ~ **link** *n* RADIO liaison de contribution *f*
control[1] *n* COMP, CONTR, GEN contrôle *m*, TELEP commande *f*; ~**-and-delay channel** TRANS canal de gestion et de retard *m*; ~ **area** CONTR zone de commande *f*; ~ **box** CONTR *equipment* boîte de commande *f*; ~ **cabinet** CONTR armoire de contrôle *f*, coffret de commande *m*; ~ **cable** CONTR, TRANS câble de commande *m* câble de contrôle *m*; ~ **card** CONTR carte de contrôle *f*; ~ **change** CONTR changement de commande *m*, modification de commande *f*; ~ **channel** (*CC*) CONTR canal de commande *m*, canal de contrôle *m*, TRANS canal de contrôle *m*; ~ **character** COMP, DATA, TELEP caractère de commande *m*; ~ **circuit** CONTR circuit de commande *m*; ~ **code-word** COMP, DATA mot-code de commande *m*; ~ **component** CONTR composant de régulation *m*; ~ **crystal** ELECT quartz de commande *m*; ~ **data** *n pl* COMP données de commande *f pl*; ~ **desk** CONTR pupitre de commande *m*, pupitre de conduite *m*, TV pupitre *m*; ~ **device** CONTR dispositif de commande *m*, organe de commande *m*; ~ **diagnostic unit** TEST unité de diagnostic et de commande *f*; ~ **element PE** TELEP élément graphique d'élément de commande *m*; ~ **engineering** CONTR *automatic* génie des systèmes de contrôle *m*, systèmes asservis *m pl*; ~ **equipment** CONTR équipement de commande *m*, matériel de contrôle *m*; ~ **with fixed set-point** CONTR régulation de maintien *f*; ~ **flow diagram** DATA diagramme de flux de contrôle *m*; ~ **frame** DATA trame de gestion *f*; ~ **frequency** CONTR fréquence de commande *f*; ~ **function** (*CF*) COMP, CONTR, DATA, TELEP fonction de commande *f*, fonction de contrôle *f*; ~ **information** COMP informations de commande *f pl*; ~ **instruction register** COMP registre d'adresse d'instruction *m*; ~ **knob** CONTR bouton de commande *m*, bouton de réglage *m*; ~ **language** COMP, ELECT, TELEP langage de contrôle *m*; ~ **limits** *n pl* TEST limites de contrôle *f pl*; ~ **loop** CONTR boucle d'asservissement *f*, boucle de régulation *f*; ~ **MCSAP** TELEP point MCSAP de contrôle *m*; ~ **MCSPDU** TELEP unité MCSPDU de contrôle *f*; ~ **memory** COMP mémoire de contrôle *f*; ~ **message** COMP message de commande *m*; ~ **operation** COMP fonction de commande *f*; ~ **panel** COMP tableau de bord *m*, CONTR panneau de commande *m*, tableau de commande *m*, tableau de contrôle *m*; ~ **point** CONTR point de régulation *m*; ~ **position** CONTR position de commande *f*; ~ **procedure** TELEP procédure de commande *f*; ~ **program** COMP programme de commande *m*, programme de gestion *m*; ~ **program language** COMP langage de programme de gestion *m*; ~ **pulse** CONTR impulsion de commande *f*; ~ **range** CONTR plage de réglage *f*; ~ **register** COMP registre d'adresse d'instruction *m*; ~ **room** TEST salle d'exploitation *f*, salle de contrôle *f*; ~ **signal** SIG signal de commande *m*; ~ **signaling code** *AmE*, ~ **signalling code** *BrE* SIG code de signalisation de commande *m*; ~ **statement** COMP ordre de contrôle *m*; ~ **station** COMP station de commande *f*, CONTR station de contrôle *f*, TELEP station directrice *f*, station multiprocesseur *f*; ~**-station access multiplex** TELEP multiplex d'accès aux stations de commande *m*; ~ **storage** COMP mémoire de contrôle *f*; ~ **structure** CONTR structure de contrôle *f*, structure de la commande *f*; ~ **system** CONTR système de commande *m*, système de contrôle *m*, unité de commande *f*;

~ **and system monitoring** TEST commande et surveillance du système *f*; ~ **terminal** (*CT*) CONTR terminal de commande *m*; ~ **timing** TRANS rythme de commande *m*; ~ **transmitter** TRANS transmetteur de synchro-détection *m*; ~ **unit** CONTR organe de gestion *m*, unité de commande *f*; ~ **voltage** POWER tension de commande *f*; ~ **word** COMP, DATA mot de commande *m*
control[2] *vt* GEN commander
controllability *n* CONTR contrôlabilité *f*
controlled: **~-access unit** *n* (*CAU*) TELEP unité à accès contrôlé *f*; **~-avalanche rectifier diode** *n* ELECT *semiconductors* diode de redressement à avalanche contrôlée *f*; ~ **corrective maintenance** *n* APPL, TEST maintenance corrective dirigée *f*; ~ **maintenance** *n* GEN maintenance dirigée *f*; ~ **not-ready signal** *n* SIG signal non prêt commandé *m*; ~ **rerouting** *n* TELEP retour sous contrôle sur route normale *m*; ~ **slip** *n* RADIO glissement maîtrisable *m*, glissement commandé *m*, TELEP glissement commandé *m*; ~ **terminal** *n* CONTR terminal sous commande *m*, terminal commandé *m*; ~ **variable** *n* CONTR grandeur commandée *f*, grandeur réglée *f*
controller *n* CONTR régulateur *m*, GEN contrôleur *m*
controlling: ~ **computer** *n* COMP calculateur pilote *m*, ordinateur de commande *m*; ~ **device** *n* CONTR dispositif de commande *m*; ~ **element** *n* CONTR élément de réglage *m*; ~ **equipment** *n* CONTR équipement de commande *m*; ~ **exchange** *n* SWG centre directeur *m*; ~ **reference burst parameter** *n* DATA paramètre des paquets de référence de gestion *m*; ~ **testing station** *n* TELEP *on circuit* station directrice *f*
convection *n* CLIM convection *f*
conventional: ~ **degree of distortion** *n* TELEP degré conventionnel de distorsion *m*; ~ **true value of a quantity** *n* TEST valeur conventionnellement vraie d'une grandeur mesurable *f*
convergence *n* DATA, TRANS convergence *f*; ~ **sublayer** (*CS*) DATA sous-couche de convergence *f*; ~ **sublayer protocol data unit** (*CSPDU*) DATA unité de données de protocole de la sous-couche convergence *f*; ~ **time** TELEP temps de convergence *m*
conversation: ~ **impossible** *n* (*CI*) RADIO conversation impossible (*CI*); ~ **time** *n* RADIO, TELEP durée de conversation *f*; **~-time limiter** *n* RADIO, TELEP limiteur de durée de conversation *m*
conversational[1] *adj* COMP conversationnel, interactif
conversational[2]: ~ **language** *n* COMP langage conversationnel *m*, langage interactif *m*; ~ **mode** *n* COMP mode conversationnel *m*, mode dialogue *m*; ~ **service** *n* COMP service conversationnel *m*, service interactif *m*; ~ **system** *n* COMP système conversationnel *m*, système dialogué *m*
conversion *n* DATA conversion *f*, TELEP translation *f*, TV transposition *f*; ~ **and storage unit** (*CSU*) DATA organe de conversion et de mémorisation *m*; ~ **table** COMP table de conversion *f*; ~ **to bipolar code** TRANS codage avec code bipolaire *m*, codage en bipolaire *m*
convert *vt* COMP, DATA convertir, TV transposer; ~ **to bipolar code** TRANS convertir en bipolaire
converted: ~ **encoded information types** *n pl* (*converted EITs*) DATA types d'information codée convertis *m pl*
converter *n* COMP, ELECT, GEN convertisseur *m*; ~ **cabinet** POWER armoire de conversion *f*; ~ **module** ELECT module convertisseur *m*; ~ **rack** POWER armoire de convertisseur *f*, bâti de convertisseur *m*
convolution: ~ **code** *n* COMP code de convolution *m*
convolutional: ~ **code** *n* COMP code de convolution *m*; ~ **coding** *n* COMP codage convolutif *m*; ~ **encoder** *n* COMP codeur convolutionnel *m*
convolutionally: ~ **and differentially encoded** *adj* (*CDE*) DATA à codage convolutionnel différentiel
convolutive: ~ **code** *n* COMP code convolutif *m*
cool *vt* CLIM, GEN refroidir
cooling *n* GEN réfrigération *f*, refroidissement *m*; ~ **flange** PARTS bride de dissipation de chaleur *f*; ~ **jacket** PARTS enveloppe de refroidissement *f*; ~ **system** CLIM système de refroidissement *m*, GEN circuit de refroidissment *m*; ~ **unit** CLIM groupe frigorifique *m*
cooperating: ~ **DSA** *n* DATA DSA coopérant *m*
cooperative: ~ **work** *n* DATA travail coopératif *m*
coordinate: ~ **clock** *n* RADIO horloge coordonnée *f*; ~ **graphics** *n* COMP graphie en mode coordonnées *f*; ~ **time** *n* RADIO temps-coordonné *m*
coordinated: ~ **single-layer embedded test method** *n* (*CSE*) TEST méthode de test encastré monocouche coordonnée *f*; ~ **single-layer test method** *n* (*CS*) TEST méthode de test monocouche coordonnée *f*; ~ **test method** *n* TEST méthode de test coordonnée *f*; ~ **time scale** *n* GEN échelle de temps coordonné *f*; ~ **traffic** *n* RADIO trafic coordonné *m*
coordination: ~ **level** *n* DATA niveau de coordination *m*
co-polar[1] *adj* RADIO copolaire
co-polar[2]: ~ **attenuation** *n* RADIO affaiblissement copolaire *m*, affaiblissement du signal copolaire *m*
co-polarized *adj* RADIO copolaire
copper: ~ **busbar** *n* POWER barre bus en cuivre *f*; ~ **cable** *n* TRANS câble de cuivre *m*; ~ **cabling** *n* TRANS câblage en cuivre *m*; ~ **conductor** *n* ELECT conducteur de cuivre *m*; ~ **wire** *n* ELECT, GEN, PARTS fil de cuivre *m*
coprocessor *n* COMP coprocesseur *m*
coproduction *n* TV coproduction *f*
copy[1] *n* DATA copie *f*; ~ **destination point** COMP point de destination de copie *m*
copy[2] *vt* COMP *file* copier
copying: ~ **device** *n* PER APP reprographie *f*
COR *abbr* (*confirmation of receipt*) DATA, TELEP confirmation de réception *f*
cord *n* COMP cordon *m*, GEN câble *m*, cordon *m*; ~ **circuit** TELEP, TEST dicorde *m*; ~ **set** TRANS cordon connecteur *m*
cordless[1] *adj* GEN sans fil
cordless[2]: ~ **handset** *n* PER APP, TELEP poste sans fil *m*; ~ **PABX** *n* PER APP PABX sans fil *m*; ~ **phone** *n* PER APP, TELEP poste téléphonique sans cordon *m*, poste téléphonique sans fil *m*; ~ **switchboard** *n* GEN *office* standard manuel à clés *m*, SWG, TELEP *switchroom* commutateur manuel sans cordons *m*, commutateur manuel à clés *m*; ~ **telephone** *n* (*CT*) PER APP, TELEP poste téléphonique sans cordon *m*, poste téléphonique sans fil *m*, téléphone sans fil *m*; ~ **telephone set** *n* PER APP, TELEP poste téléphonique sans cordon *m*, poste téléphonique sans fil *m*; ~ **telephony** *n* RADIO téléphonie sans fil *f*; ~ **telephone 1** *n* (*CT1*) TELEP téléphone sans fil de première génération *m* (*CT1*); ~ **telephone 2** *n* (*CT2*) PER APP, TELEP téléphone sans fil de deuxième génération *m* (*CT2*)
core *n* GEN noyau *m*, TRANS *cables* conducteur isolé *m*, *fibre optics* cœur *m*; ~ **area** TRANS *fibre optics* zone de cœur *f*; ~ **center** *AmE*, ~ **centre** *BrE* TRANS centre du cœur *m*; **~-cladding concentricity error** TELEP, TRANS *fibre optics* erreur de concentricité cœur-gaine *f*; **~-cladding ratio** TRANS rapport cœur-gaine *m*; ~ **diameter** TELEP, TRANS *cable* diamètre de l'âme *m*, diamètre du cœur *m*, diamètre du noyau *m*; ~ **diameter deviation** TELEP, TRANS écart sur le diamètre de l'âme *m*, écart sur le diamètre du cœur *m*, écart sur le diamètre du noyau *m*; ~ **diameter tolerance** TRANS tolérance sur le diamètre du cœur *f*; ~ **dump** COMP vidage de mémoire *m*; ~ **of exchange** DATA cœur de chaîne *m*; ~ **flush** COMP remise à zéro de la mémoire *f*; ~ **functions control station software** COMP, DATA, TELEP logiciel de base et fonctions système de toute station SM *m*; ~ **image** COMP image mémoire *f*; ~ **logic** COMP logique de noyaux magnétiques *f*; ~ **matrix** COMP matrice de tores *f*; ~ **mode**

TRANS mode de cœur *m*; **~-reference surface concentricity error** TELEP, TRANS erreur de concentricité cœur-surface de référence *f*; **~ screen** POWER *cables* écran sur enveloppe isolante *m*; **~ storage** COMP, DATA mémoire à tores *f*; **~ tolerance field** TRANS domaine de tolérance du cœur *m*; **~ wrap** ELECT, PARTS, TRANS *of cable* guipage *m*

corkscrew: **~ aerial** *n* RADIO antenne hélicoïdale *f*; **~ antenna** *n* RADIO antenne hélicoïdale *f*

corner: **~ aerial** *n* RADIO antenne dièdre *f*; **~ antenna** *n* RADIO antenne dièdre *f*; **~ loss** *n* RADIO perte en courbe *f*; **~ reflector** *n* RADIO réflecteur polyridique *m*

cornerstone *n* DATA pierre angulaire *f*

cornucopia *n* RADIO cornet *m*

corona *n* POWER effluve *f*; **~ arc testing** TEST essai corona *m*; **~ effect** POWER effet couronne *m*

corpus *n* COMP, DATA base de textes *f*

correctable: **~ fault** *n* TEST panne corrigeable *f*

corrected: **~ equivalent resistance error** *n* TELEP écart équivalent corrigé *m*; **~ fault** *n* TEST *dependability* panne corrigée *f*; **~ reference equivalent** *n* (*CRE*) TELEP équivalent de référence corrigé *m* (*ERC*)

correction *n* GEN correction; **~ area** COMP *computers* zone de correction *f*; **~ character** TELEP caractère de correction *m*; **~ factor** TEST facteur de correction *m*; **~ function** TEST fonction de correction *f*

corrective: **~ maintenance** *n* TEST maintenance corrective *f*; **~ maintenance time** *n* TELEP, TEST durée de maintenance corrective *f*, TEST temps de maintenance corrective *m*; **~ measure** *n* GEN remède *m*

correlation *n* GEN corrélation *f*

correlative: **~ phase-shift keying** *n* SIG manipulation corrélative par déplacement de phase *f*

correlator *n* ELECT, GEN corrélateur *m*

corridor *n* RADIO couloir *m*

corrugated: **~ horn feed** *n* RADIO cornet d'alimentation à surface cannelée *m*

COS *abbr* (*carrier-operated switch*) RADIO commutateur commandé par le porteur *m*

cosecant *n* RADIO cosécante *f*; **~-squared antenna** RADIO antenne à faisceau en cosécante carrée *f*

cosine: **~ emission** *n* TRANS cosinus de Lambert *m*; **~ pulse** *n* TRANS impulsion cosinus *f*; **~-squared pulse** *n* TRANS impulsion en cosinus carré *f*

COSPAS-SARSAT *abbr* (*International Satellite System for Search and Rescue*) GEN COSPAS-SARSAT (*Système international de satellites pour les recherches et le sauvetage*)

cost *n* GEN coût; **~-effective method** GEN moyen économique *m*; **~ of ownership** GEN coût de possession *m*; **~ of production** GEN coût de production *m*

Costas: **~ loop** *n* TELEP boucle de Costas *f*

cotton: **~ waste** *n* PARTS bourre de coton *f*

countdown: **~ timer** *n* PER APP *answer machine* compteur à rebours *m*, décompteur *m*

counter *n* GEN, TELEP compteur *m*; **~ cell** POWER élément d'opposition *m*; **~-clockwise polarization** RADIO polarisation lévogyre *f*, polarisation senestrorsum *f*; **~-electromotive force** POWER force contre-électromotrice *f* (*fcém*)

counteracting: **~ winding** *n pl* POWER enroulement antagoniste *m*, enroulement opposé *m*

counting: **~ circuit** *n* ELECT circuit de comptage *m*; **~ device** *n* ELECT dispositif de comptage *m*; **~ function** *n* ELECT fonction comptage *f*

countouring *n* COMP effet de sous-quantification *m*

country: **~ code** *n* (*CC*) TELEP indicatif de pays *m*; **~-code indicators** *n pl* TELEP indicateurs d'indicatifs de pays *m pl*; **~-code system** *n* GEN système de code des pays *m*; **~ of destination** *n* GEN administration de destination *f*, pays de destination *m*; **~ of origin** *n* GEN administration d'origine *f*, pays d'origine *m*

couple *vt* COMP coupler

coupled: **~ mode** *n* TRANS mode couplé *m*

coupler *n* ELECT coupleur *m*; **~ loss** ELECT perte de coupleur *f*

coupling *n* ELECT couplage *m*, verrouillage *m*; **~ capacitor** ELECT, POWER condensateur de couplage *m*; **~ efficiency** ELECT efficacité du couplage *f*, rendement de couplage *m*; **~ factor** ELECT facteur de couplage *m*; **~ loss** ELECT perte de couplage *f*

course: **~ blip pulse** *n* RADIO impulsion de repérage de cap *f*

courseware *n* COMP, DATA didacticiel *m*

courtesy: **~ music** *n* TELEP attente musicale *f*

cover[1] *n* PARTS boîtier *m*, capot *m*, couvercle *m*, enveloppe *f*; **~ holder** PARTS support de capot *m*; **~ page** COMP page de titre *f*; **~ plate** PARTS *equipment practice* plaque d'habillage *f*; **~ support** TELEP support de capot *m*

cover[2] *vt* GEN revêtir

coverage *n* RADIO couverture *f*; **~ area** RADIO aire de couverture *f*, zone de couverture *f*; **~ diagram** RADIO diagramme de couverture *m*; **~ factor** RADIO facteur de couverture *m*

covering *n* GEN *cables* revêtement *m*

CP *abbr* (*call processor*) TELEP processeur d'appels *m*, processeur de traitement des appels *m*

CPE *abbr* (*customer premises equipment*) PER APP équipement local d'abonnés *m*

cpi *abbr* (*characters per inch*) COMP cpp (*caractères par pouce*)

CPI *abbr* (*computer-PBX interface*) COMP interface PABX-serveur *f*

CPL/1: **~ language** *n* COMP langage CPL/1 *m*

CPM *abbr* (*channel protocol machine*) DATA machine de protocole de canaux *f*

CPS *abbr* (*call progress signal*) TELEP signal de progression de l'appel *m*

CPSK *abbr* (*coherent phase-shift keying*) RADIO, SIG MDPC (*modulation par déplacement de phase cohérente*)

CPU *abbr* (*central processing unit*) COMP, SWG, TELEP UC (*unité centrale de calcul, unité centrale de traitement*)

CQM *abbr* (*circuit-group query message*) SWG IGD (*message d'interrogation de groupe de circuits*)

CQR *abbr* (*circuit-group query response message*) SWG IGR (*message de réponse à une interrogation de groupe de circuits*)

CR *abbr* DATA (*character return*) caractère de retour à la ligne *m*, GEN (*carriage return*) retour à la ligne *m*, TELEP (*call register*) compteur d'appels *m*, registre d'appels *m*, TEST (*client risk*) *quality* risque du client *m*

C/R *abbr* (*command/response bit*) TELEP bit de commande/réponse *m*

crack: **~ detector** *n* TEST détecteur de fissures *m*

cracking *n* TEST criquage *m*; **~ test** TEST essai de criquage *m*, essai de fissuration *m*

crackle[1] *n* TELEP crachement *m*, friture *f*

crackle[2] *vi* TELEP *line* crépiter

crackling: **~ noise** *n* TELEP crachement *m*, crépitement *m*, friture *f*

cradle *n* PER APP support *m*, RADIO crochet commutateur *m*, support du poste *m*; **~ switch** PER APP support *m*, support commutateur *m*, RADIO, TELEP crochet commutateur *m*

crank *n* PER APP manivelle *f*

CRC *abbr* DATA (*cyclic redundancy check*) CRC (*contrôle de redondance cyclique*), (*cyclic redundancy code*) CRC (*code de redondance cyclique*) GEN (*communications research center AmE, communications research centre BrE*) centre de recherche sur les communications *m*, RADIO (*common routing channel*) VAC (*voie d'acheminement commune*), TELEP (*cyclic redundancy check*) CRC (*contrôle de redondance cyclique*), (*cyclic redundancy code*) CRC (*code de redondance cyclique*); **~ message block** *n* (*CMB*) TELEP bloc de messages de CRC *m*

CRE *abbr* (*corrected reference equivalent*) TELEP ERC (*équivalent de référence corrigé*)

create[1]: **~-line area** *n* DATA zone de ligne de création *f*
create[2] *vt* DATA créer
creation: **~ date** *n* COMP date de création *f*; **~ time** *n* DATA date-heure de création *f*
credentials *n pl* DATA pouvoirs *m pl*, justificatif d'identité *m*
credit: **~ card call** *n* TELEP appel sur carte de crédit *m*; **~ indicator** *n* TELEP indicateur de crédit *m*
creditphone *n* TELEP poste à carte de crédit *m*, téléphone à carte de crédit *m*
credits *n pl* TV génériques *m pl*
creep[1]: **~ modulus** *n* TEST *strength* module de fluage *m*; **~ rupture** *n* TEST *durability* rupture en fluage *f*; **~ rupture stress** *n* TEST *durability* résistance à la rupture en fluage *f*
creep[2] *vi* POWER fuire en surface
creepage *n* POWER fuite en surface *f*
creeping: **~ distance** *n* POWER distance de fuite *f*
crew: **~ factor** *n* TEST coefficient de pratique expérimentale *m*
CRF *abbr* (*connection-related function*) DATA, TRANS fonction liée à la connexion *f*
crimp *n* ELECT sertissage *m*; **~ ratchet tool frame** ELECT outil à cliquet à sertir *m*
crimping *n* ELECT sertissage *m*
crimpless: **~ screw** *n* ELECT vis sans sertissage *f*; **~-screw-type strength-member retention** *n* ELECT retenue du renfort par vis sans sertissage *f*
crisis: **~ management** *n* TEST gestion des situations de crise *f*
criterion *n* TEST critère
critical: **~ angle** *n* TRANS angle critique *m*; **~ band** *n* TELEP bande critique *f*; **~ band rate** *n* TELEP taux de bande critique *m*; **~ damping** *n* TELEP *acoustics* amortissement critique *m*; **~ design review** *n* GEN revue critique de la conception *f*; **~ failure** *n* TEST défaillance critique *f*; **~ frequency** *n* TRANS fréquence critique *f*, fréquence de seuil *f*; **~ path** *n* GEN *project management* chemin critique *m*; **~-path analysis** *n* GEN analyse du cheminement critique *f*; **~-path method** *n* GEN méthode du chemin critique *f*
CRM *abbr* (*call-rejected message*) DATA, TELEP MRA (*message de refus d'appel*)
crocus: **~ cloth** *n* ELECT chiffon non pelucheux *m*
crop *vt* COMP retailler, GEN élaguer
cross[1]: **~-media** *adj* ELECT intersupport; **~-module** *adj* ELECT intermodule; **~-net** *adj* TRANS interréseau; **~-polarized** *adj* TRANS à polarisations croisées
cross[2]: **~ assembler** *n* COMP assembleur croisé *m*; **~-band operation** *n* TRANS exploitation à bandes croisées *f*; **~-border system** *n* RADIO *land mobile* réseau transfrontalier *m*; **~ check** *n* COMP contrôle par balance carrée *m*, contre-vérification *f*; **~ chute** *n* PARTS *equipment practice* goulotte transversale *f*; **~-connect** *n* ELECT brassage *m*, panneaux de brassage *m pl*; **~-connect cabinet** *n* TELEP armoire de répartiteur *f*; **~-connecting unit** *n* TELEP répartiteur *m*; **~-exchange check** *n* SWG *cross-office* vérification du trajet dans le central *f*; **~-interleaved Reed-Solomon code** *n* (*CIRC*) COMP code Reed-Solomon à entrelacement croisé *m* (*CIRC*); **~-modulation noise** *n* RADIO bruit de transmodulation *m*; **~-office check** *n* SWG contrôle de continuité à travers un commutateur *m*, vérification de la voie de conversation dans le central *f*; **~-point switch** *n* SWG multi-sélecteur *m*; **~-polar channel** *n* TRANS canal contrapolaire *m*; **~-polar discrimination** *n* TRANS découplage de polarisations croisées *m*; **~-polar isolation** *n* TRANS discrimination par polarisation croisée *f*; **~-polar performance** *n* TRANS qualité de fonctionnement d'une voie en polarisations croisées *f*; **~-polarization** *n* RADIO transpolarisation *f*; **~-polarization canceler** *AmE*, **~-polarization canceller** *BrE* *n* RADIO annuleur de transpolarisation *m*; **~-polarization discrimination** *n* RADIO, TELEP discrimination de polarisation *f*, discrimination par polarisation croisée *f*; **~-polarization isolation** *n* RADIO isolement de polarisation *m*; **~ section** *n* GEN *technical drawing* coupe transversale *f*, profil *m*; **~-site link** *n* TRANS liaison d'interconnexion sur le site *f*; **~ state** *n* TRANS *fibre optics* état croisé *m*; **~-stranding** *n* TRANS câblage transversal *m*; **~-strapping** *n* ELECT interconnexion *f*; **~-twisting** *n* TRANS torsion transversale *f*
cross[3]: **~-connect** *vt* ELECT interconnecter
crossbar: **~ switch** *n* SWG commutateur crossbar *m*
crossed: **~ field** *n* TELEP champ croisé *m*; **~-field amplifier** *n* TRANS amplificateur à champs croisés *m*; **~-slot aerial** *n* RADIO antenne à fentes croisées *f*; **~-slot antenna** *n* RADIO antenne à fentes croisées *f*
crossfade *n* TV fondu enchaîné *m*, transition *f*
crosshatching *n* ELECT *on printed circuits* hachures *f pl*
cross-office: **~ transfer time** *n* (*Tcu*) SWG temps de transfert dans le central *m*
crossover: **~ patch cable** *n* ELECT câble de jonction *m*
crossplay: **~ sheath** *n* ELECT couches croisées *f pl*
crosspoint *n* SWG point de connexion *m*, point d'intersection *m*
crosstalk *n* TELEP, TRANS diaphonie *f*
CRP *abbr* DATA (*command repeat*) répéter la commande, TELEP (*call request packet*) paquet de demande d'appel *m*
CRT *abbr* (*cathode ray tube*) ELECT, PARTS TCR (*tube à rayon cathodique*)
CRTC *abbr* (*Canadian Radio, Television and Telecommunications Commission*) RADIO, TELEP, TV CRTC (*Conseil de la radiodiffusion et des télécommunications canadiennes*)
crush *vt* ELECT écraser
crushing *n* ELECT écrasement *m*
cryogenic: **~ current comparator** *n* ELECT comparateur de courant cryogénique *m*; **~ storage** *n* COMP mémoire à supraconduction *f*
cryogenics *n* GEN *physics* cryogénie *f*
cryotron *n* ELECT cryotron *m*
cryptanalysis *n* DATA, RADIO, TELEP, TV cryptographie *f*
cryptogram *n* DATA, RADIO, SIG, TELEP, TV cryptogramme *m*
cryptographic: **~ check value** *n* DATA, RADIO, TELEP, TV valeur de contrôle cryptographique *f*
cryptography *n* DATA, RADIO, TELEP, TV cryptographie *f*
crystal: **~ diode** *n* POWER diode à cristal *f*; **~ filter** *n* RADIO filtre à quartz *m*; **~ rectifier** *n* POWER redresseur à cristal *m*
CS *abbr* APPL (*communication services*) services de communication *m pl*, DATA (*convergence sublayer*) sous-couche de convergence *f*, GEN (*calls per second*) appels par seconde, SWG, TELEP (*circuit switching*) commutation de circuits *f*, TEST (*coordinated single-layer test method*) méthode de test mono-couche coordonnée *f*
CSA *abbr* (*Canadian Standards Association*) GEN ACN (*Association canadienne de normalisation*)
CSC *abbr* (*communications system control*) GEN commande du système de télécommunications *f*
CSD *abbr* (*call setup delay*) TELEP TEC (*temps d'établissement de communication*)
CSG *abbr* (*constructive solid geometry*) COMP GSC (*géométrie solide constructive*)
CSI *abbr* DATA, RADIO, TELEP (*called-station identity*) identité de la station demandée *f*, (*called-subscriber identification*) identification de l'abonné appelé *f*, identification de l'abonné demandé *f*
CSI-DIS *abbr* (*called-subscriber identification-DIS*) DATA, TELEP IADE (*identification de l'abonné demandé-DIS*)
CSM *abbr* (*communications system monitoring*) TEST surveillance du système de télécommunications *f*
CSMA *abbr* (*carrier-sense multiple access*) RADIO, SIG, TELEP, TRANS AMDP (*accès multiple par détection de porteuse*)
CSMA/CD *abbr* (*carrier-sense multiple access with collision detection*) DATA CSMA/CD (*accès multiple avec écoute de la porteuse et détection de collisions*)
CSPDU *abbr* (*convergence sublayer protocol data unit*) DATA

unité de données de protocole de la sous-couche convergence *f*

CSU *abbr* (*conversion and storage unit*) DATA organe de conversion et de mémorisation *m*

CSUD *abbr* (*call setup delay*) TELEP délai d'établissement de la communication *m*

CT *abbr* APPL (*container*) conteneur *m*, CONTR (*control terminal*) terminal de commande *m*, DATA (*container*) conteneur *m*, PER APP, TELEP (*cordless telephone*) poste téléphonique sans cordon *m*, poste téléphonique sans fil *m*, téléphone sans fil *m*, (*container*) conteneur *m*

CTCSS *abbr* (*continuous tone-controlled squelch system*) RADIO CTCSS (*silencieux à commande par tonalités*)

CTD *abbr* DATA (*cumulative transit delay*) temps de transit cumulatif *m*, ELECT (*charge transfer device*) DTC (*dispositif à transfert de charge*), SWG (*conditionally toll-denied*) blocage partiel de l'accès à l'interurbain *m*

CT1 *abbr* (*cordless telephone 1*) TELEP CT1 (*téléphone sans fil de première génération*)

CTR *abbr* (*response to continue to correct*) DATA CTR (*réponse pour continuer à corriger*)

CTS *abbr* DATA (*clear to send*) prêt à émettre, RADIO (*communications technology satellite*) satellite de technologie des télécommunications *m*

CTSM *abbr* (*conformance testing specification manual*) TEST manuel de spécifications des tests de conformité *m*

CTTE *n* (*common traffic terminal equipment*) TELEP équipement commun des terminaux de trafic *m*

CT2 *abbr* (*cordless telephone 2*) PER APP, TELEP CT2 (*téléphone sans fil de deuxième génération*)

CUG *abbr* (*closed user group*) DATA, TELEP GFA (*groupe fermé d'abonnés*), GFU (*groupe fermé d'utilisateurs, groupe fermé d'usagers*)

cumbersome *adj* PARTS encombrant

cumulative: ~ **distribution function** *n* (*cdf*) DATA fonction de distribution cumulative *f*; ~ **frequency polygon** *n* TEST *quality* polygone des fréquences cumulées *m*; ~ **leakage index** *n* (*CLI*) TRANS indice de fuites globales *m*; ~ **sum chart** *n* TEST *quality* carte à somme cumulée *f*; ~ **transit delay** *n* (*CTD*) DATA temps de transit cumulatif *m*

curbed: ~ **modulation** *n* TELEP modulation fragmentée *f*

Curie: ~ **point** *n* COMP point de Curie *m*

current *n* GEN courant *m*, flux *m*; ~ **affairs program** *AmE*, ~ **affairs programme** *BrE* TV journal TV *m*; ~ **amplification** POWER amplification de courant *f*; **~-carrying capacity** POWER intensité de courant admissible *f*; ~ **collector** POWER appareil de prise de courant *m*; ~ **density** POWER densité de courant *f*; ~ **distribution** POWER distribution de courant *f*; ~ **drift** POWER dérive de courant *f*; ~ **feed** POWER *power supply* alimentation en courant *f*, alimentation en énergie *f*, téléalimentation *f*; **~-feed relay** POWER *power supply* relais d'alimentation en courant *m*, relais d'alimentation en énergie *m* relais de téléalimentation *m*; ~ **flow method** TEST télédiagnostique par circulation de courant *m*; ~ **intensity** POWER intensité de courant *f*; ~ **limit level** POWER niveau de limitation de courant *m*; ~ **limiter** POWER limiteur de courant *m*; ~ **limiting** POWER limitation de courant *f*; ~ **load** POWER courant fourni *m*; ~ **loop** POWER boucle de courant *f*; ~ **outlet** POWER prise de courant *f*; ~ **relay** POWER relais de courant *m*; ~ **return conductor** POWER conducteur de courant de retour *m*; ~ **rush** POWER appel de courant *m*, pointe de courant *f*; ~ **state of the art** GEN état actuel de la technique *m*; ~ **surge** POWER appel de courant *m*, pointe de courant *f*; ~ **tap** POWER prise de courant *f*; ~ **version** COMP version en cours *f*

cursor[1] *n* COMP curseur *m*, RADIO disque d'azimut *m*

cursor[2]: ~ **off** *phr* (*COF*) COMP curseur arrêté; ~ **on** *phr* (*CON*) COMP curseur en marche

curvature *n of the earth* RADIO courbure *f*

curve *n* COMP, TRANS courbe *f*; ~ **follower** COMP lecteur de courbe *m*; ~ **generator** COMP générateur de courbes *m*

custom: **~-calling service** *n* APPL service d'appel personnalisé *m*

customer *n* GEN client *m*, donneur d'ordre *m*; **~-access maintenance center** *AmE*, **~-access maintenance centre** *BrE* (*CAMC*) TEST centre de maintenance des accès clients *m*; ~ **equipment** (*CEQ*) PER APP équipement d'usager *m*; ~ **network** (*CN*) TRANS réseau client *m*; ~ **premises equipment** (*CPE*) PER APP équipement local d'abonnés *m*; ~ **recorded-information service** TELEP service particulier d'information enregistrée *m*

customize *vt* GEN arranger sur demande, personnaliser

customized: ~ **menu** *n* COMP menu personnalisé *m*

cut[1]: **~-off** *adj* ELECT *transistors* bloqué, GEN suspendu, POWER *supply* coupé, interrompu, RADIO coupé

cut[2] *n* GEN coupure *f*, réduction *f*; ~ **face** TRANS *fibre optics* face clivée *f*; **~-out** ELECT ouverture *f*; **~-out key** TELEP touche de coupe-circuit *f*; **~-over** GEN mise en service *f*, TELEP basculement *m*, transfert *m*; **~-sheet feeder** COMP système d'alimentation feuille à feuille *m*

cut[3] *vt* COMP couper, GEN réduire; ~ **back** GEN réduire; ~ **off** ELECT *transistors* bloquer, GEN arrêter, *line* suspendre, POWER *supply* couper, interrompre, rompre, RADIO couper; ~ **over** GEN inaugurer; ~ **over to service** TELEP mettre en service; ~ **and paste** COMP copier-coller

cutaway: ~ **view** *n* GEN *technical drawing* vue éclatée *f*

cutback *n* GEN réduction *f*; ~ **technique** TRANS *fibre optics* méthode de la fibre coupée *f*

Cutler: ~ **feed** *n* RADIO alimentation par fentes *f*, alimentation Cutler *f*

cutoff *n* GEN arrêt *m*, coupure *f*, interruption *f*, rupture *f*, suspension *f*; ~ **current** ELECT *transistors* courant de coupure *m*, courant résiduel *m*; ~ **frequency** ELECT fréquence de coupure *f*; ~ **switch** GEN interrupteur *m*; ~ **voltage** ELECT *transistors* tension de blocage *f*; ~ **wavelength** RADIO, TRANS longueur d'onde de coupure *f*

cutting *n* GEN *wires* coupe *f*; **~-edge technology** GEN *research and development* technologie de pointe *f*; ~ **formboard** GEN *wires* gabarit de coupe de fils *m*; ~ **length** GEN *wires* longueur de coupe *f*; ~ **and stripping tool** GEN outil à couper et à dénuder *m*; ~ **tool** GEN outil de coupe *m*

CV *abbr* (*code violation*) TRANS violation de code *f*

CW *abbr* RADIO (*continuous wave*) émission continue *f*, TELEP (*call waiting*) appel en instance *m*, communication en attente *f*, TRANS (*continuous wave*) onde continue *f*, onde entretenue *f*; ~ **radar** *n* RADIO radar à émission continue *m*, radar à ondes entretenues *m*

cyber *adj* COMP cyber

cybercafé *n* COMP cybercafé *m*

cyberculture *n* COMP cyberculture *f*

cybernaut *n* COMP cybernaute *m*

cybernetics *n* COMP cybernétique *f*

cyberspace *n* COMP cyberspace *f*

cycle[1] *n* COMP, GEN cycle *m*; ~ **skipping** TRANS saut de cycle *m*; **~-skipping test procedure** TEST procédure de mesure des sauts de cycle *f*

cycle[2] *vt* TEST cycler

cycles: ~ **per second** *phr* COMP cycles par seconde

cyclic: ~ **alarm** *n* TEST alarme cyclique *f*; **~-block code** *n* DATA code de blocs cycliques *m*; ~ **code** *n* DATA code cyclique *m*; ~ **distortion** *n* TELEP distorsion cyclique *f*; ~ **redundancy check** *n* (*CRC*) DATA, TELEP contrôle de redondance cyclique *m* (*CRC*); ~ **redundancy code** *n* (*CRC*) DATA code de redondance cyclique *m* (*CRC*), contrôle de redondance cyclique *m* (*CRC*), TELEP contrôle de redondance cyclique *m* (*CRC*)

cyclical *adj* ELECT cyclique

cycling *n* TEST cyclage *m*

cyclomotive: ~ **force** *n* (*cmf*) RADIO force cyclomotrice *f* (*fcm*)
cyclotronic: ~ **resonance** *n* ELECT résonance cyclotronique *f*
cylindrical: ~ **reflector arial** *n* RADIO antenne à réflecteur cylindrique *f*; ~ **reflector antenna** *n* RADIO antenne à réflecteur cylindrique *f*

D

D *abbr* (*defaultable*) DATA valeur pouvant être prise par défaut *f*

D-A *abbr* (*digital-to-analog*) GEN N-A (*numérique-analogique*); **~ converter** *n* (*digital-to-analog converter, digital/analog converter*) COMP, DATA, ELECT, SIG, TELEP convertisseur N-A *m* (*convertisseur numérique-analogique*)

DAB *abbr* COMP, TV (*digital audio broadcast*) RAN (*retransmission audionumérique*), (*digital audio broadcasting*) radiodiffusion sonore numérique *f*

DAC *abbr* (*digital-to-analog converter, digital/analog converter*) COMP, DATA, ELECT, SIG, TELEP CNA (*convertisseur numérique-analogique*)

daisy[1]: **~-chaining** *n* ELECT travail en série *m*; **~ wheel** *n* PER APP *printing* marguerite *f*

daisy[2]: **~-chain** *vt* COMP, ELECT connecter en guirlande

DAM *abbr* (*data-addressable memory*) COMP, DATA mémoire adressable par son contenu *f*

DAMA *abbr* (*demand-assignment multiple access*) SIG accès multiple avec affectation à la demande *m*

damage *vt* TRANS endommager

damaged *adj* TRANS *cable* endommagé

damp[1]: **~ heat testing** *n* TEST essai de chaleur humide *m*; **~ point** *n* GEN, TRANS point d'amortissement *m*

damp[2] *vt* GEN, TRANS amortir, étouffer

damped: **~ oscillation** *n* ELECT *physics* oscillation amortie *f*

damping *n* TRANS amortissement *m*, *fibre optics* atténuation *f*

danger: **~ signal** *n* TEST *security* signal de détresse *m*

DAP *abbr* (*directory access protocol*) DATA protocole d'accès à l'annuaire *m*

dark: **~ current** *n* ELECT, TRANS courant d'obscurité *m*; **~ field** *n* TRANS *fibre optics* champ sombre *m*; **~-field detector** *n* TRANS *fibre optics* détecteur de champ sombre *m*

darken *vt* TV assombrir

darkest: **~ zone** *n* TV *scanner* zone la plus foncée *f*

Darlington: **~ amplifier** *n* ELECT montage Darlington *m*; **~ circuit** *n* ELECT montage Darlington *m*

dashboard *n* RADIO tableau de bord *m*

DASS *abbr* (*demand-assignment signaling and switching AmE, demand-assignment signalling and switching BrE*) SIG signalisation et commutation avec affectation à la demande *f*

DAT *abbr* (*digital audio tape*) COMP, TV DAT (*bande audionumérique*)

data[1]: **~-permissible** *adj* TRANS compatible ETD; **~-permissive** *adj* TRANS compatible ETD

data[2] *n pl* GEN données *f pl*; **~ above voice** (*DOV*) DATA, RADIO données supravocales *f pl*; **~ acquisition** *n* DATA saisie de données *f*; **~ adaptor unit** *n* COMP, DATA adaptateur de ligne *m*; **~-addressable memory** *n* (*DAM*) COMP, DATA mémoire adressable par son contenu *f*; **~ amount** *n* DATA masse de données *f*; **~ backup** *n* DATA sauvegarde de données *f*; **~ bank** *n* DATA banque de données *f*; **~ block** *n* DATA bloc de données *m*; **~ card** *n* COMP, DATA carte de données *f*; **~ carrier** *n* DATA, TRANS porteuse de données *f*; **~ carrier detection** *n* (*DCD*) DATA détection de porteuse de données *f*; **~ carrier failure detector** *n* DATA, TRANS détecteur d'interruption de la porteuse de données *m*; **~ cassette recorder** *n* (*DCR*) DATA, PER APP enregistreur de données à cassette *m*; **~ channel** *n* DATA *datacom* voie de transmission de données *f*; **~ channel failure detector** *n* DATA, TELEP détecteur d'interruption de la voie de données *m*; **~ channel propagation time** *n* DATA, TELEP temps de propagation sur la voie de données *m* (*Tᵖ*); **~ channel received line signal detector** *n* DATA, TRANS détecteur de signal de ligne reçu sur la voie de données *m*; **~ character key** *n* DATA, PER APP *keyboards* touche alphanumérique *f*; **~ circuit** *n* DATA circuit de données *m*, TRANS circuit de données *m*, circuit de transmission de données *m*; **~ circuit terminal equipment** *n* (*DCE*) COMP, DATA, PER APP, TRANS équipement de terminaison de circuit de données *m* (*ETCD*); **~ circuit terminating equipment** *n* (*DCTE*) COMP, DATA, PER APP, TRANS équipement de terminaison de circuit de données *m* (*ETCD*); **~ code** *n* DATA code de données *m*; **~ collection** *n* DATA collecte de données *f*; **~-collection center** *AmE*, **~-collection centre** *BrE* *n* DATA centre de collecte de données *m*; **~-collection equipment** *n* DATA équipement de collecte de données *m*; **~-collection satellite** *n* DATA, RADIO satellite de collecte de données *m*; **~ communication** *n* DATA, TRANS communication de données *f*, transmission de données *f*; **~ communication channel** *n* (*DCC*) DATA, TRANS canal de communication de données *m*, canal de transmission de données *m*, voie de transmission de données *f*; **~-communication network** *n* DATA, TRANS réseau de transmission de données *m*; **~-communication system** *n* COMP, DATA système de transmission de données *m*; **~ communication terminal** *n* DATA terminal téléinformatique *m*; **~ communications controller** *n* DATA contrôleur de communications *m*; **~ communications equipment** *n* (*DCE*) DATA, PER APP matériel de transmission de données *m*; **~ compression** *n* DATA compression des données *f*; **~-compression board** *n* DATA carte de compression de données *f*; **~-compression processor** *n* DATA unité de compression de données *f*; **~-compression software** *n* DATA logiciel de compression de données *m*; **~ concentrator** *n* DATA concentrateur de données *m*; **~ connection** *n* DATA chaîne de connexion de données *f*; **~ conversion** *n* DATA conversion de données *f*; **~ converter** *n* COMP, DATA convertisseur de données *m*; **~ country code** *n* (*DCC*) DATA indicatif de pays pour la transmission de données *m*; **~-description language** *n* COMP, DATA langage de description de données *m*; **~ distribution** *n* DATA, TELEP diffusion de données *f*; **~ dump** *n* COMP, DATA vidage de mémoire *m*; **~ editing** *n* COMP, DATA édition de données *f*; **~ element separator** *n* DATA séparateur d'éléments de données *m*; **~ entry** *n* COMP, DATA entrée de données *f*; **~ -flow** *n* COMP, DATA flux de données *m*; **~ -flow analysis** *n* DATA analyse du flux de données *f*; **~ -flow chart** *n* COMP, DATA organigramme de données *m*; **~ flow diagram** *n* COMP, DATA organigramme de données *m*; **~ gathering** *n* DATA saisie de données *f*; **~ group** *n* DATA, RADIO groupe de données *m*; **~ handling** *n* DATA manipulation des données *f*; **~ independence** *n* COMP, DATA indépendance des données *f*; **~ integrity** *n* DATA intégrité des données *f*; **~ interchange** *n* COMP, DATA échange de données *m*; **~ line** *n* DATA, RADIO ligne de données *f*; **~ link** *n* DATA, TRANS liaison de données *f*; **~-link connection** *n* (*DLC*) DATA connexion pour liaison de données *f*; **~-link connection identifier** *n* (*DLCI*) DATA identificateur de connexion de liaison de données *m*; **~-link control** *n* (*DLC*) DATA commande de liaison de données *f*; **~-link layer** *n* DATA, TRANS couche liaison de données *f*; **~-link protocol** *n* DATA procédure de liaison *f*; **~-link service** *n* (*DLS*) APPL, DATA service de liaison de données *m*; **~-link service access point** *n* (*DLSAP*) DATA point d'accès au service de liaison de données *m*; **~-link specialized access** *n* DATA accès spécialisé pour les liaisons de données *m*; **~-link switching** *n* (*DLS*) DATA, ELECT commutation de liaison de données *f*; **~ logger** *n* DATA enregistreur de données *m*; **~ loss** *n* DATA perte de données *f*; **~ management** *n* (*DM*) COMP, DATA gestion de données *f*; **~-manipulation language** *n* DATA langage de manipulation de données *m*; **~ MCSPDU** *n* DATA, TELEP unité MCSPDU de données *f*; **~ modeling** *AmE*, **~ modelling** *BrE* *n* DATA modélisation des données *f*; **~ modifier** *n* COMP, DATA modificateur de données *m*; **~ modulation** *n* COMP, DATA modulation de données *f*; **~ multiplexer** *n* COMP, DATA, TRANS multiplexeur de données *m*; **~ network** *n* DATA, TRANS réseau de données *m*; **~-network identification code** *n* (*DNIC*) DATA, TELEP code d'identification de réseau de données *m* (*CIRD*); **~-origin authentication** *n* DATA authentification de l'origine de données *f*; **~ packet** *n* DATA, RADIO paquet de données *m*; **~ partitioning** *n* COMP, DATA subdivision des données *f*; **~ path** *n* TV voie de données *f*; **~ phase** *n* DATA, TRANS phase de données *f*; **~ port** *n* DATA point d'accès aux données *m*; **~ preparation** *n* COMP, DATA préparation des données *f*; **~ presetter** *n* COMP, DATA présélecteur de données *m*; **~ private wire** *n* DATA, TRANS liaison spécialisée transmission de données *f*; **~ processing** *n* (*DP*) COMP, DATA traitement des données *m* (*TD*), traitement de l'information *m*; **~-processing center** *AmE*, **~-processing centre** *BrE* *n* DATA centre de traitement de l'information *m* (*CTI*); **~-processing connection** *n* DATA connection informatique *f*; **~-processing system** *n* COMP, DATA système de traitement des données *m*; **~-processing system security** *n* COMP, DATA sécurité du système informatique *f*; **~-processing terminal equipment** *n* DATA, PER APP équipement terminal de traitement de données *m* (*ETTD*); **~ processor** *n* COMP, DATA machine de traitement des données *f*; **~ quantization** *n* DATA, TELEP quantification de données *f*; **~ recording** *n* DATA enregistrement de données *m*; **~ reduction** *n* COMP, DATA réduction des données *f*; **~-relay satellite** *n* DATA, RADIO satellite relais de données *m*; **~ saving** *n* COMP, DATA sauvegarde de données *f*; **~ set identifier** *n* (*DSI*) DATA identificateur de modem *m*; **~-set ready** *n* (*DSR*) DATA modem prêt *m*; **~ sheet** *n* DATA *documentation* feuille de spécification *f*, fiche technique *f*; **~ signal** *n* DATA, SIG signal de données *m*; **~ signal quality detection** *n* DATA, SIG détection de la qualité du signal de données *f*; **~ signaling rate** *AmE see data signalling rate BrE*; **~ signaling rate transparency** *AmE see data signalling rate transparency BrE*; **~ signalling rate** *BrE* *n* COMP, DATA, TELEP, TRANS débit binaire *m*; **~ signalling rate transparency** *n* *BrE* DATA, SIG, TELEP, TRANS transparence du débit binaire *f*; **~ sink** *n* DATA collecteur de données *m*; **~ store** *n* DATA, SWG mémoire de données *f*; **~ store block** *n* COMP, DATA *code* bloc de mémoires de données *m*; **~ store channel** *n* DATA, SWG *code* voie de mémoire de données *f*; **~ store element** *n* DATA, SWG élément de mémoire de données *m*; **~ store handler** *n* DATA, SWG gestionnaire de mémoire de données *m*; **~-store multiplexer** *n* DATA, SWG *code* multiplexeur de mémoire de données *m*; **~ store unit** *n* DATA, SWG unité de mémoire de données *f*; **~ stream** *n* DATA flot de données *m*; **~ switch** *n* DATA, ELECT, SWG commutateur de données *m*; **~ switching** *n* DATA, ELECT, SWG commutation de données *f*; **~-switching exchange** *n* (*DSE*) DATA, ELECT, SWG centre de communication de données *m*, centre de commutation de données *m*; **~ terminal** *n* DATA, PER APP, TELEP terminal de traitement de données *m*, terminal de transmission de données *m*; **~ terminal equipment** *n* (*DTE*) DATA, PER APP, TELEP équipement terminal de données *m*; **~ terminal equipment address** *n* DATA, PER APP, TELEP adresse d'équipement terminal de traitement de données *f*; **~ transfer** *n* DATA, TRANS transfert de données *m*, transmission de données *f*; **~-transfer phase** *n* DATA, TELEP phase de transfert de données *f*; **~ transfer rate** *n* DATA, TELEP vitesse de transfert de données *f*; **~ transfer requested** *n* DATA transfert de données demandé *m*; **~ transfer system** *n* DATA système de transfert de données *m*; **~ transmission** *n* DATA, TRANS communication de données *f*, transfert de données *m*, transmission de données *f*; **~ transmission balloon** *n* DATA, TRANS ballon de transmission de données *m*; **~ transmission service** *n* DATA, TRANS service de transmission de données *m*; **~ transport network** *n* DATA, TRANS réseau de transport de données *m*; **~ type** *n* DATA type de données *m*; **~ under voice** (*DUV*) DATA, RADIO données infravocales *f pl*; **~ unit** *n* DATA, RADIO unité de données *f*; **~ user part** *n* DATA, SIG sous-système utilisateur de données *m*; **~ user part handling time** *n* DATA, SIG temps de traitement pour le sous-système utilisateur de données *m*; **~ wrap plug** *n* DATA, ELECT prise de bouclage *f*

Data: **~ Discman®** *n* COMP, DATA Data Discman® *m*; **~ Grade Media** *n* DATA, ELECT support adaptable *m*

database *n* COMP, DATA base de données *f*; **~ language** COMP, DATA langage de bases de données *m*; **~ management** COMP, DATA gestion de base de données *f*; **~ management system** (*DBMS*) COMP, DATA système de gestion de bases de données *m* (*SGBD*); **~ manager** COMP, DATA gestionnaire de base de données *m*

datacom *n* DATA, TRANS transfert de données *m*

Dataglove *n* COMP gant capteur *m*, gant sensitif *m*

datagram *n* DATA *Internet* datagramme *m*; **~ delivery data confirmation** DATA, SIG confirmation de remise de datagramme *f*; **~ non-delivery indication** DATA, TELEP indication de non-remise de datagramme *f*; **~ queue length** DATA, TELEP longueur de la file d'attente de datagrammes *f*; **~ service** DATA, TELEP service de datagramme *m*

datamation *n* COMP, DATA traitement automatique des données *m*

date *n* GEN date *f*; **~ or time of preparation** DATA date ou heure de préparation *f*; **~-time stamping** PER APP *answer machine* horodatage *m*; **~ and time of transmission** DATA date et heure de transmission *f*

daughterboard *n* COMP, ELECT *printed circuits* carte fille *f*

daughters *n pl* COMP sœurs *f pl*

day: **~-to-busy-hour ratio** *n* TELEP rapport du trafic journalier au trafic à l'heure chargée *m*, rapport trafic journalier/trafic à l'heure chargée *m*; **~-to-day variation** *n* TELEP variation diurne *f*; **~ traffic** *n* TELEP trafic de jour *m*

dB *abbr* (*decibel*) RADIO dB (*décibel*); **~-related units** *n pl* TELEP unités utilisant le dB *f pl*

D-bit: **~ modification facility** *n* COMP service complémentaire de modification du bit D *m*

dBm *abbr* (*decibels above one milliwatt*) TELEP dBm

DBMS *abbr* (*database management system*) COMP, DATA SGBD (*système de gestion de bases de données*)

DBS *abbr* (*direct broadcasting by satellite*) DATA, RADIO, TV RDS (*radiodiffusion directe par satellite*); **~ satellite** *n* TV satellite RDS *m*

DC *abbr* (*direct current*) POWER CC (*courant continu*); **~ bus** *n* POWER bus de courant continu *m*; **~ busbar** *n* POWER barre bus à courant continu *f*; **~ clamp diode** *n* RADIO diode de niveau *f*; **~ component** *n* TELEP composant continu *m*; **~ coupler** *n* POWER coupleur continu *m*; **~ generator** *n* POWER dynamo génératrice de courant continu *f*; **~ power** *n* POWER alimentation CC *f*; **~ resistance** *n* POWER résistance pour courant continu *f*; **~ restorer diode** *n* RADIO diode de niveau *f*; **~ signaling** *AmE*, **~ signalling** *BrE* *n* POWER, SIG signalisation en courant continu *f*; **~ spark-over voltage** *n* ELECT, POWER tension continue nominale d'amorçage *f*; **~ voltage** *n* POWER tension continue *f*

DCA *abbr* (*document content architecture*) DATA DCA (*architecture DCA*)

DCC *abbr* DATA (*data collection center AmE, data collection centre BrE*) centre de collecte de données *m*, (*data country code*) indicatif de pays pour la transmission de données *m*, (*data communication channel*) canal de communication de données *m*, canal de transmission de données *m*, voie de transmission de données *f*, TRANS (*data communication channel*) canal de communication de données *m*, canal de transmission de données *m*, voie de transmission de données *f*

DCC/DNIC: ~ **indicator** *n* SIG indicateur de IPD/CIRD *m*

DCD *abbr* (*data carrier detection*) DATA détection de porteuse de données *f*

DC-DC: ~ **converter** *n* DATA, POWER convertisseur de CC à CC *m*, *power supply* convertisseur continu-continu *m*

DCE *abbr* COMP, DATA, PER APP (*data circuit terminal equipment*) ETCD (*équipement de terminaison de circuit de données*), (*data communications equipment*) matériel de transmission de données *m*, TRANS (*data circuit terminal equipment*) ETCD (*équipement de terminaison de circuit de données*); ~ **clear indication** *n* TELEP indication de libération par l'ETCD *f*; ~**-power off signal** *n* SIG signal de coupure de l'alimentation de l'ETCD *m*; ~ **source** *n* TELEP origine ETCD *f*; ~ **waiting** *n* TELEP ETCD en attente *m*

D-channel *n* DATA, TELEP *ISDN* canal D *m*; ~ **virtual circuit** TELEP *ISDN* circuit virtuel sur canal D *m*

DCME *abbr* (*digital circuit multiplication equipment*) TELEP, TRANS EMCN (*équipement de multiplication de circuits numériques*); ~ **frame** *n* TELEP trame EMCN *f*; ~ **function** *n* TELEP fonction EMCN *f*; ~ **gain** *n* TELEP gain de l'EMCN *m*; ~ **overload** *n* TELEP surcharge EMCN *f*; ~ **overload mode** *n* TELEP mode de surcharge EMCN *m*

DCMG *abbr* (*digital circuit multiplication gain*) TELEP, TRANS GMCN (*gain de multiplication de circuits numériques*)

DCMS *abbr* (*digital circuit multiplication system*) TELEP, TRANS SMCN (*système de multiplication de circuits numériques*)

DCN *abbr* (*disconnect*) DATA DCN (*déconnexion*)

DCR *abbr* (*data cassette recorder*) DATA, PER APP enregistreur de données à cassette *m*

DCS *abbr* CONTR (*digital command signal*) signal de commande numérique *m*, DATA (*defined-context set*) ensemble de contextes définis *m*, (*digital command signal*) signal de commande numérique *m*, RADIO (*digital cellular system*) système cellulaire numérique *m*

DCT *abbr* (*discrete cosine transform*) COMP, DATA, ELECT, TELEP TCD (*transformation en cosinus discret*)

DCTE *abbr* (*data circuit terminating equipment*) COMP, DATA, PER APP, TRANS ETCD (*équipement de terminaison de circuit de données*)

DCU *abbr* (*distribution and control unit*) CONTR unité de distribution et de commande *f*

DCXO *abbr* (*digitally-compensated crystal oscillator*) ELECT oscillateur à quartz à commande numérique *m*

DD *abbr* (*double density*) COMP DD (*double densité*)

DDA *abbr* (*defined domain attribute*) TELEP attribut défini de domaine *m*

DDCMP *abbr* (*digital data communications message protocol*) DATA protocole de messages de transmission de données numériques *m*

DDD *abbr AmE* (*direct distance dialing*) TELEP IA (*interurbain automatique*)

DDE *abbr* (*dynamic data exchange*) DATA échange dynamique de données *m*

DDFS *abbr* (*direct digital frequency synthesis*) ELECT synthèse de fréquence numérique directe *f*

DDM *abbr* (*directional-division multiplexing*) TRANS multiplexage directionnel par division

de facto: ~ **standard** *n* DATA norme de fait *f*

deactivate *vt* COMP, DATA, ELECT, TELEP désactiver

deactivation *n* COMP, DATA, ELECT, TELEP désactivation *f*

dead[1] *adj* POWER hors circuit

dead[2]: ~ **center** *AmE*, ~ **centre** *BrE* *n* COMP réglage médian *m*; ~ **room** *n* TEST *acoustics* chambre sourde *f*, salle anéchoïque *f*; ~ **time** *n* CONTR retard *m*, temps mort *m*

deadener *n* GEN *ergonomics*, TELEP *acoustics* silencieux *m*

deadline *n* COMP date limite *f*

deal: ~ **with** *vt* COMP prendre en charge, traiter

dealer: ~ **room technology** *n* PER APP technologie des salles de transaction *f*

deblocking *n* RADIO, TELEP déblocage *m*

debug *vt* DATA dépanner, mettre au point

decade: ~ **counter** *n* ELECT *logic elements* compteur à décades *m*; ~ **resistance box** *n* POWER boîte de résistance à décades *f*

decadic: ~ **signaling** *AmE*, ~ **signalling** *BrE* *n* SIG signalisation décimale *f*

decametric: ~ **wave** *n* TRANS onde décamétrique *f*

decay[1] *n* COMP affaiblissement *m*, décroissance *f*, diminution *f*, amortissement *m*, déclin *m*; ~ **level** COMP niveau de décroissance *m*; ~ **rate** COMP vitesse de décroissance *f*; ~ **time** COMP temps de décroissance *m*, TELEP *pulses* temps de descente *m*, TRANS temps de déclin *m*

decay[2] *vt* COMP décroître, diminuer

decay[3] *vi* COMP, GEN diminuer

decentralize *vt* COMP, GEN décentraliser

decentralized: ~**-control signaling** *AmE*, ~**-control signalling** *BrE* *n* SIG signalisation de commande décentralisée *f*; ~ **operation** *n* SIG, SWG activité décentralisée *f*; ~ **system** *n* SWG système décentralisé *m*

deception: ~ **jamming** *n* RADIO brouillage de diversion *m*

DEChub: ~ **90 backplate** *n* ELECT, PARTS fond de panier DEChub 90 *m*

decibel *n* (*dB*) RADIO décibel *m* (*dB*)

decibels: ~ **above one milliwatt** *n pl* (*dBm*) TELEP dBm

decimal: ~ **numeral** *n* COMP nombre décimal *m*; ~ **pulse** *n* (*DP*) TELEP impulsion décimale *f*; ~ **system** *n* COMP système décimal *m*

decipher *vt* GEN déchiffrer

deciphering *n* GEN déchiffrement *m*

decipherment *n* GEN déchiffrement *m*

decision: ~ **circuit** *n* TELEP circuit de décision *m*; ~ **content** *n* COMP quantité de décisions *f*; ~ **feedback equalizer** *n* ELECT égaliseur avec rétroaction de décision *m*; ~ **function** *n* TEST *quality* fonction de décision *f*; ~ **instant of a distant signal** *n* TELEP instant de décision d'un signal numérique *m*; ~ **instruction** *n* COMP instruction de décision *f*; ~**-maker** *n* TV décideur *m*; ~ **support system** *n* DATA système d'aide à la décision *m*; ~ **symbol** *n* TELEP symbole de décision *m*; ~ **tree** *n* DATA arbre de décision *m*; ~ **value** *n* TELEP amplitude de décision *f*, TRANS valeur de décision *f*

declaration *n* COMP, DATA déclaration *f*; ~ **procedure** TV régime de déclaration *m*

declarations: ~ **part** *n* DATA partie déclarative *f*

decode *vt* COMP, RADIO, SIG, TV déchiffrer, décoder

decoded: ~ **stream** *n* COMP chaîne décodée *f*

decoder *n* COMP, RADIO, SIG, TV décodeur *m*

decoding *n* COMP, RADIO, SIG, TV déchiffrement *m*, décodage *m*; ~ **margin** RADIO marge de décodage *f*; ~ **process** COMP processus de décodage *m*; ~ **threshold** RADIO seuil de décodage *m*; ~ **time stamp** (*DTS*) COMP estampille de décodage *f*

decomposition *n* PARTS *electrochemistry* décomposition *f*; ~ **voltage** PARTS *electrochemistry* tension de décomposition *f*

decompressed[1] *adj Internet* décompacté

decompressed[2]: ~ **data** *n pl* COMP, DATA données décomprimées *f pl*

decompression *n* COMP, DATA décompression *f*

decoupling *n* ELECT, POWER découplage *m*; ~ **capacitor**

ELECT, POWER condensateur de découplage *m*; ~ **resistor** PARTS, POWER résistance de découplage *f*
decrease[1] *n* GEN diminution *f*, décroissance *f*
decrease[2] *vt* GEN diminuer, décroître; ~ **load** TELEP décharger, délester
decrease[3] *vi* COMP, GEN decrease, diminuer
decrement *n* COMP décrément *m*
decryption *n* COMP, DATA, GEN déchiffrage *m*
DECT *abbr* (*digital European cordless telephone service*) POWER DECT (*service numérique européen de téléphonie sans fil*)
dedicated: ~ **channel** *n* TRANS voie exclusive *f*; ~ **frequency** *n* TRANS fréquence réservée *f*; ~ **line** *n* DATA, TELEP, TRANS liaison spécialisée *f* (*LS*), ligne spécialisée *f* (*LS*); ~ **port** *n* COMP port d'accès spécialisé *m*, TRANS accès specialisé *m*; ~ **signaling channel** *AmE*, ~ **signalling channel** *BrE n* SIG canal spécialisé pour la signalisation *m*; ~ **terminal** *n* TV terminal spécifique *m*; ~ **terrestrial circuit** *n* TELEP, TRANS circuit terrestre spécialisé *m*
deducted *adj* GEN imputé
deductive: ~ **reasoning** *n* DATA raisonnement déductif *m*
de-emphasis *n* ELECT désaccentuation *f*; ~ **network** ELECT réseau de désaccentuation *m*
deep: ~ **space** *n* RADIO espace lointain *m*
default *n* COMP option par défaut *f*; ~ **behavior** *AmE*, ~ **behaviour** *BrE* DATA comportement par défaut *m*; ~ **button** COMP case par défaut *f*; ~ **context** DATA contexte par défaut *m*; ~ **group** DATA groupe de comportements par défaut *m*; ~ **group inference** DATA, TEST référence de groupe de comportements par défaut *f*; ~ **identification** DATA identification par défaut *f*; ~ **identifier** DATA identificateur par défaut *m*; ~ **library** DATA bibliothèque des comportements par défaut *f*; ~ **reasoning** DATA raisonnement par défaut *m*; ~ **reference** DATA référence de comportements par défaut *f*; ~ **setting** COMP réglage par défaut *m*; ~ **tempo** COMP tempo par défaut *m*; ~ **value** COMP, GEN valeur par défaut *f*
defaultable *n* (*D*) DATA valeur pouvant être prise par défaut *f*
defect *n* COMP, ELECT, POWER, TELEP, TEST défaut *m*; ~-**counting method** TEST méthode par nombre de défauts *f*
defective *adj* COMP, TELEP, TEST *quality* défectueux
deferrals *n pl* ELECT différés *m pl*
deferred[1] *adj* COMP, GEN différé
deferred[2]: ~ **addressing** *n* COMP adressage différé *m*; ~ **transmission** *n* GEN transmission en différé *f*
defined: ~-**context set** *n* (*DCS*) DATA ensemble de contextes définis *m*; ~ **domain attribute** *n* (*DDA*) TELEP attribut défini de domaine *m*
definition *n* COMP, DATA définition *f*, ELECT *printed circuits* définition *f*, précision du tracé *f*, TELEP, TRANS, TV définition *f*
deflection *n* ELECT, GEN, POWER, RADIO, TELEP, TEST, TRANS déviation *f*; ~ **method of measurement** TEST méthode de mesure par déviation *f*; ~ **yoke** ELECT, PARTS bobine de déviation *f*
deflector *n* RADIO déflecteur *m*
deframing *n* DATA *scanner* détramage *m*
defruiter *n* RADIO élagueur *m*, éliminateur de réponses parasites *m*
degradation *n* RADIO, SIG, TRANS dégradation *f*; ~ **figure** RADIO taux de dégradation *m*
degraded: ~ **minute** *n* (*DM*) DATA, RADIO minute dégradée *f*; ~-**service signal** *n* SIG signal de dégradation du service *m*
degree: ~ **of coherence** *n* TRANS degré de cohérence *m*; ~ **of distortion in service** *n* TELEP degré de distorsion en service *m*; ~ **of early anisochronous parallel distortion** *n* TELEP degré de distorsion parallèle anisochrone en avance *m*; ~ **of freedom** *n* GEN degré de liberté *m*; ~ **of gross start-stop distortion** *n* TELEP degré de distorsion arythmique globale *m*; ~ **of inherent distortion** *n* TELEP degré de distorsion propre *m*; ~ **of isochronous distortion** *n* TELEP degré de distorsion isochrone *m*; ~ **of late anisochronous parallel distortion** *n* TELEP degré de distorsion parallèle anisochrone en retard *m*; ~ **of modulation** *n* TRANS degré de modulation *m*; ~ **of standardized test distortion** *n* TELEP degré de distorsion d'essai normalisé *m*; ~ **of start-stop distortion** *n* TELEP degré de distorsion arythmique *m*; ~ **of synchronous start-stop distortion** *n* TELEP degré de distorsion arythmique au synchronisme *m*
de-ionization *n* ELECT désionisation *f*; ~ **grid** ELECT grille de désionisation *f*
dejitter *vt* TRANS dégiguer, supprimer la gigue
delamination *n* ELECT *printed circuits* décollement interlaminaire *m*
delay *n* GEN retard *m*, temps mort *m*; ~ **circuit** ELECT ligne à retard *f*, circuit à retard *m*; ~-**dialing signal** *AmE*, ~-**dialling signal** *BrE* SIG signal invitant à différer la numérotation *m*; ~ **effect** COMP effet de retard *m*; ~ **element** ELECT *logic elements* circuit à retard *m*, élément à retard *m*; ~ **equalizer** COMP compensateur de retard *m*; ~ **lens** RADIO lentille à retard de phase *f*; ~ **level** COMP niveau de retard *m*; ~ **line** DATA ligne à retard *f*; ~-**lock loop** ELECT boucle à retard de phase *f*; ~-**mode operation** GEN mode d'exploitation avec attente *m*; ~ **processor** COMP processeur de retard *m*; ~ **slug** PARTS *relays* bague de retard *f*; ~ **system** TELEP système avec attente *m*; ~ **time** GEN temps de retard *m*, PARTS retard à la croissance *m*; ~ **vibrato** COMP vibrato à retard *m*; ~ **working** GEN mode d'exploitation avec attente *m*
delayed: ~ **delivery** *n* TELEP remise différée *f*
delete[1]: ~ **abstract-operation** *n* DATA opération abstraite de suppression *f*; ~ **channel** *n* COMP effacement de canal *m*
delete[2] *vt* GEN annuler, éliminer, effacer, *score out* rayer, *suppress* supprimer
deleted: ~-**time sending** *n* DATA envoi en différé *m*
deleting *n* RADIO rayage *m*
deletion *n* GEN suppression *f*, annulation *f*, effacement *m*; ~ **character** COMP caractère d'effacement *m*
delimiter *n* GEN, TEST délimiteur *m*
deliverable: ~ **unit** *n* TELEP unité livrable *f*
delivered: ~ **EITs** *n pl* (*delivered encoded-information types*) DATA types d'information codée retransmis *m pl*; ~ **encoded-information types** *n pl* (*delivered EITs*) DATA types d'information codée retransmis *m pl*; ~-**message entry** *n* DATA entrée message remis *f*; ~-**report entry** *n* DATA entrée rapport remis *f*
delivery: ~ **confirmation** *n* TELEP confirmation de remise *f*; ~ **inspection** *n* TEST *quality* contrôle de livraison *m*; ~ **lot** *n* TEST *quality* lot de livraison *m*; ~ **network** *n* TV réseau de branchement *m*
de-load *vt* GEN *cables* dépupiniser
delrine: ~ **rod** *n* PARTS *plastics* tige en delrin *f*
delta: ~ **modulation** *n* (*DM*) SIG, TRANS modulation delta *f* (*MD*); ~ **modulation PSK-2** *n* (*phase-shift keying 2 modulation*) SIG modulation delta MDP-2 *f* (*modulation de phase delta 2*)
demagnetization *n* ELECT désaimantation *f*
demagnetize *vt* ELECT *electromagnetism* désaimanter
demagnetizing *n* ELECT désaimantation *f*
demand: ~-**assigned multiple access system** *n* RADIO système à accès multiple par assignation en fonction de la demande *m*; ~-**assigned SCPC system** *n* RADIO système SCPC avec affectation à la demande *m*; ~-**assigned switching and signaling** *AmE*, ~-**assigned switching and signalling** *BrE n* SIG signalisation et commutation par affectation à la demande *f*; ~-**assigned switching and signaling subsystem** *AmE*, ~-**assigned switching and signalling subsystem** *n BrE* SIG sous-système de signalisation et commutation par affectation à la demande *m*; ~ **assignment** *n* RADIO affectation à la demande *f*; ~-**assignment circuit** *n* ELECT circuit avec affectation à la

demande *m*; **~-assignment multiple access** *n* (*DAMA*) SIG accès multiple avec affectation à la demande *m*; **~-assignment signaling** *AmE see demand-assignment signalling BrE*; **~-assignment signaling and switching** *n AmE see demand-assignment signalling and switching BrE*; **~-assignment signalling** *n BrE* SIG, SWG signalisation avec affectation à la demande *f*; **~-assignment signalling and switching** *n BrE* (*DASS*) SIG, SWG signalisation et commutation avec affectation à la demande *f*; **~ operating** *n* TELEP exploitation en service rapide *f*; **~ service** *n* GEN service à la demande *m*; **~ working** *n* TELEP exploitation sans attente *f*
demo: **~ button** *n* COMP touche de démonstration *f*; **~ program** *n* COMP programme de démonstration *m*; **~ song** *n* COMP morceau de démonstration *m*
democratic: **~ network** *n* TELEP *mutually synchronized* réseau démocratique *m*
demodulate *vt* COMP, ELECT, RADIO, TRANS démoduler
demodulation *n* COMP, ELECT, RADIO, TRANS démodulation *f*
demodulator *n* COMP, ELECT, RADIO, TRANS démodulateur *m*
demote *vt* COMP rétrograder
demount *vt* GEN déposer
demultiplexer *n* COMP, RADIO, TELEP, TRANS, TV démultiplexeur *m*
demultiplexing *n* COMP, RADIO, TELEP, TRANS, TV démultiplexage *m*
denial *n* APPL *of service* déni *m*, TRANS interdiction *f*
denied *adj* TRANS interdit
dental: **~ pick** *n* ELECT crochet *m*
deny *vt* APPL dénier, TRANS *access* interdire
depacketizer *n* DATA, TELEP dépaqueteur *m*, désassembleur de paquets *m*
dependability *n* GEN fiabilité *f*, TEST sécurité de fonctionnement *f*, *quality* sûreté de fonctionnement *f*
dependable *adj* GEN *dependability* fiable
dependent: **~ conformance** *n* DATA conformité induite *f*; **~ exchange** *n* SWG central auxiliaire *m*, central tributaire *m*, TELEP central satellite *m*, satellite *m*; **~ repeater station** *n* TELEP station de répéteurs téléalimentée *f*; **~ station** *n* TELEP station téléalimentée *f*
depletion: **~ layer** *n* PARTS couche d'appauvrissement *f*; **~ mode operation** *n* PARTS fonctionnement en mode de déplétion *m*
deployment *n* GEN, RADIO déploiement *m*; **~ threshold** RADIO seuil de déploiement *m*
depointing *n* RADIO dépointage *m*
depolarization *n* RADIO dépolarisation *f*
deposit[1] *n* COMP dépôt *m*, TV dépôt *m*, dépôt de garantie *m*
deposit[2] *vt* GEN déposer
deposition *n* ELECT, PARTS dépôt *m*
depressed: **~ cladding** *n* TRANS gaine déprimée *f*, *fibre optics* zone d'affaiblissement *f*
depth *n* COMP amplitude *f*, intensité *f*, taux *m*; **~ of color** *AmE*, **~ of colour** *BrE* DATA profondeur de couleur *f*; **~ gage** *AmE*, **~ gauge** *BrE* TEST *measuring devices* jauge de profondeur *f*
dequantization *n* COMP déquantification *f*
deregistration *n* RADIO radiation *f*
deregulation *n* GEN dérégulation *f*, *economics* déréglementation *f*, TV dérégulation *f*
derivation: **~ path** *n* DATA chemin de dérivation *m*
derivative *n* GEN dérivé *m*, *mathematics* dérivée *f*
derive *vt* GEN *mathematics* dériver
descending: **~ node** *n* RADIO nœud descendant *m*
descramble *vt* DATA, RADIO, SIG, TELEP, TV désembrouiller
descrambler *n* RADIO décodeur *m*
descreening *n* DATA *scanner* détramage *m*
description: **~ language** *n* DATA langage de description *m*
descriptive: **~ name** *n* DATA nom descriptif *m*
descriptor *n* COMP descripteur *m*
desensitization *n* ELECT désensibilisation *f*
design: **~ aid** *n* DATA outil d'aide à la conception *m*; **~ failure** *n* TEST problème de conception *m*; **~ fault** *n* TEST panne due à la conception *f*, problème de conception *m*; **~ review** *n* TEST *quality assessment* revue de conception *f*
designate *vt* GEN désigner, préassigner, *marking* étiqueter, marquer, repérer, TELEP, TRANS désigner
designated: **~ frequency** *n* TRANS fréquence désignée *f*
designating: **~ tape** *n* GEN *marking* ruban d'étiquetage *m*
designation *n* GEN désignation *f*, *marking* repérage *m*, étiquetage *m*, marquage *m*, étiquette *f*, TELEP, TRANS désignation *f*; **~ drawing** GEN *documentation* dessin d'étiquetage *m*; **~ frame** GEN *marking* cadre d'étiquette *m*; **~ number** TELEP numéro de désignation *m*; **~ strip** GEN *marking* réglette porte-étiquette *f*
designer *n* GEN concepteur *m*
desired: **~ spectrum** *n* RADIO spectre utile *m*; **~ value** *n* valeur de consigne *f*
desk *n* COMP bureau *m*; **~ cassette** RADIO *mobile* console de table *f*; **~ checking** COMP vérification manuelle *f*, contrôle de programmation *m*; **~ microphone** TELEP *microphones* microphone de table *m*
desktop[1] *adj* COMP de table, de bureau
desktop[2] *n* COMP, DATA bureau *m*; **~ computer** COMP ordinateur individuel *m* (*PC*), ordinateur de bureau *m* (*PC*); **~ publishing** (*DTP*) APPL, COMP, DATA édition électronique *f*, publication assistée par ordinateur *f* (*PAO*), éditique *f*; **~ publishing system** ELECT publication assistée par ordinateur *f* (*PAO*); **~ video** (*DTV*) COMP, DATA vidéo assistée par ordinateur *f* (*VAO*), vidéotique *f*
desmoothing *n* RADIO délissage *m*
despotic: **~ network** *n* TELEP *synchronized* réseau à synchronisation despotique *m*, réseau despotique *m*
destaticize *vt* COMP appliquer un traitement antistatique à
destination *n* COMP, DATA, TELEP destination *f*; **~ address** TELEP adresse de destination *f*; **~ details** DATA *Internet* intitulé *m*; **~ equipment** COMP équipement destinaire *m*; **~ exchange** TELEP commutateur de destination *m*, central de destination *m*; **~ identifier** DATA, TELEP identificateur de destination *m*; **~ indicator** TELEP indicateur de destination *m*; **~ link** COMP lien cible *m*, lien de destination *m*; **~ memory** COMP mémoire de destination *f*; **~ network** (*DN*) TRANS réseau de destination *m*; **~ node** COMP nœud cible *m*, nud de destination *m*; **~ office** TELEP central de destination *m*, commutateur de destination *m*; **~ point** TELEP point de destination *m*; **~ point code** DATA, TELEP code du point de destination *m*; **~ terminal** DATA terminal de destination *m*; **~ terminal number** DATA numéro des terminaux de destination *m*
destruction: **~ characteristic of protector** *n* TELEP caractéristique de destruction *f*
destructive: **~ interference** *n* TRANS *fibre optics* interférence destructive *f*; **~ read** *n* COMP lecture destructive *f*; **~ testing** *n* TEST essai destructif *m*, test destructif *m*
detachment *n* GEN détachement *m*
detailed: **~ billing** *n* APPL, DATA, TELEP facturation détaillée *f*; **~ diagnosis** *n* GEN diagnostic détaillé *m*; **~ procedure** *n* (*DP*) GEN procédure détaillée *f*
detect *vt* GEN détecter
detectable: **~ element** *n* COMP élément détectable *m*; **~ group** *n* COMP groupe détectable *m*
detection *n* COMP détection *f*, suppression *f*, CONTR, DATA, ELECT détection *f*, GEN détection *f*, RADIO détection *f*, suppression *f*, TELEP, TRANS détection *f*; **~ equipment** RADIO matériel de détection *m*; **~ of marker** DATA détection du repère *f*; **~ threshold** ELECT, TELEP, TRANS seuil de détection *m*
detectivity *n* ELECT détectivité *f*, TRANS sensibilité de détection *f*, détectivité *f*
detector *n* GEN détecteur *m*, RADIO avertisseur *m*, signal

d'alarme *m*; ~ **power efficiency** PARTS *semiconductors* efficacité de détection en puissance *f*; ~ **voltage efficiency** PARTS *semiconductors* efficacité de détection en tension *f*

deter *vt* COMP décourager

deterioration: ~ **of quality** *n* TEST détérioration de la qualité *f*

determination: ~ **test** *n* TEST *equipment reliability* essai de détermination *m*

deterrent *n* COMP moyen de dissuasion *m*

detour *n* TELEP détournement *m*

detune[1] *n* COMP désaccord *m*, désaccordage *m*; ~ **depth** COMP taux de désaccordage *m*; ~ **job** COMP travail de désaccordage *m*

detune[2] *vt* COMP désaccorder

detuning *n* COMP déréglage *m*

Deutsche Industrienorm *n* (*DIN*) GEN Deutsche Industrienorm *m* (*DIN*)

develop *vt* GEN mettre au point, développer

development *n* GEN mise au point *f*, développement *m*; ~ **engineer** GEN ingénieur de développement *m*; ~ **time** COMP temps de développement *m*

deviation *n* GEN dérive *f*, écart *m*; ~ **measurement** GEN écartométrie *f*; ~ **permit** TEST dérogation avant production *f*; ~ **prism** TRANS prisme de déviation *m*

device[1] *n* COMP appareil *m*, dispositif *m*, système *m*, ELECT dispositif *m*, GEN dispositif *m*, TELEP dispositif *m*; ~ **adaptor** COMP adaptateur de périphérique *m*; ~ **adaptor interface** COMP interface d'adaptateur de périphérique *f*; ~ **control** COMP, DATA, ELECT contrôle des périphériques *m*; ~ **controller** COMP, DATA, ELECT contrôleur de périphériques *m*; ~ **description** TELEP description du dispositif *f*; ~ **driver** DATA pilote de périphérique *m*; ~ **start** TELEP mise en marche dispositif *f*; ~ **stop** TELEP arrêt dispositif *m*; ~ **under test** TEST dispositif en cours d'essai *m*

device[2]: ~ **waiting** *phr* (*DW*) TELEP dispositif en attente

devise *vt* COMP créer

dewetting *n* ELECT *soldering* démouillage *m*

DFB *abbr* (*distributed feedback*) ELECT rétroaction répartie *f*; ~ **laser diode** *n* TRANS *fibre optics* diode laser à rétroaction répartie *f*

diac *n* ELECT *semiconductors* diac *m*

DIAG *n* TELEP élément de diagnostic *m*

diagnosis *n* COMP, GEN, TEST diagnostic *m*; ~ **time** TEST temps de diagnostic *m*

diagnostic: ~ **aid** *n* TEST outil de diagnostic *m*; ~ **code** *n* TELEP *in Recommendation X.25* code de diagnostic *m*; ~ **element** *n* TELEP élément de diagnostic *m*; ~ **program** *n* COMP programme de diagnostic *m*; ~ **routine** *n* COMP routine de diagnostic *f*

diagnostics *n pl* COMP, GEN, TEST diagnostic *m*

diagram *n* GEN diagramme *m*, *documentation* schéma *m*; ~ **issue** GEN *documentation* édition de schéma *f*; ~ **number** GEN numéro de schéma *m*, *technical drawing* numéro de dessin *m*

dial[1] *n* COMP cadran *m*, sélecteur manuel *m*, PER APP cadran *m*, cadran d'appel *m*, TELEP cadran *m*; ~ **disk** *AmE* (*cf finger wheel BrE*) PER APP, TELEP disque mobile *m*, *telephone sets* disque du cadran d'appel *m*; ~ **in/dial out** ELECT réception/émission *f*; ~ **pulse** TELEP impulsion décimale *f*; ~ **switching** COMP sélection par cadran numérique *f*; ~ **telephone** TELEP téléphone à cadran *m*, poste à cadran *m*; ~ **telephone set** TELEP poste à cadran *m*; ~ **test** TEST contrôle des fréquences du clavier *m*, contrôle des impulsions du cadran *m*; ~ **tone** *AmE* (*cf dialling tone BrE*) DATA, TELEP tonalité de numérotation *f*; ~ **tone delay** TELEP durée d'attente de tonalité *f*; ~**-up** DATA *Internet*, ELECT connexion via modem *f*; ~**-up connection** DATA *internet* connexion à la demande *f*, connexion en dial-up *f*; ~**-up line** DATA ligne commutée *f*; ~**-up modem** DATA modem pour lignes téléphoniques *m*; ~**-up port** COMP port d'accès au réseau commuté *m*; ~**-up terminal** DATA terminal à liaison commutée *m*

dial[2] *vt* GEN numéroter, TELEP composer; ~ **up** DATA *Internet* établir une liaison

dial[3] *vi* GEN numéroter

Dial: ~**-up Internet Protocol** *n* DATA accès en mode ascendant *m*; ~**-up IP** *n* DATA accès en mode ascendant *m*

dialed: ~**-number display** *AmE see dialled-number display BrE*

dialer *AmE see dialler BrE*

dialing *AmE see dialling BrE*

dialled: ~**-number display** *n BrE* TELEP affichage du numéro composé *m*

dialler *n BrE* TELEP composeur *m*, numéroteur *m*

dialling *n BrE* DATA numérotation *f*, TELEP numérotation *f*, sélection *f*; ~ **code** *BrE* TELEP indicatif de numérotation *m*; ~ **error** *BrE* TELEP erreur de numérotation *f*; ~ **period** *BrE* TELEP durée de numérotation *f*; ~ **time** *BrE* TELEP durée de numérotation *f*; ~ **tone** *BrE* (*cf dial tone AmE*) DATA, TELEP tonalité de numérotation *f*

dialog: ~ **box** *n* COMP boîte de dialogue *f*, zone de dialogue *f*; ~ **recovery** DATA reprise de dialogue *f*

diaphragm *n* PARTS, RADIO diaphragme *m*, membrane *f*

diary: ~ **call** *n* GEN mémo appel *m*; ~ **service** *n* TELEP service de rappel de date *m*

dichotomizing: ~ **search** *n* COMP recherche dichotomique *f*

dichroic: ~ **filter** *n* ELECT, TRANS filtre dichroïque *m*; ~ **mirror** *n* TRANS *fibre optics* miroir dichroïque *m*

DID *abbr* (*direct inward dialing AmE, direct inward dialling BrE*) GEN SDA (*sélection directe à l'arrivée*)

die[1] *n* PARTS *semiconductors* pastille *f*, puce *f*; ~ **sets** *n pl* ELECT jeux de matrices *m pl*

die[2]: ~**-cast** *vt* PARTS couler sous pression, mouler sous pression

die[3]: ~ **away** *vi* POWER s'éteindre, s'affaiblir; ~ **down** *vi* POWER s'éteindre, s'affaiblir

dielectric[1] *adj* ELECT, PARTS diélectrique

dielectric[2] *n* ELECT, PARTS diélectrique *m*; ~ **aerial** RADIO antenne diélectrique *f*; ~ **antenna** RADIO antenne diélectrique *f*; ~ **breakdown** ELECT *semiconductors*, POWER claquage du diélectrique *m*, rupture du diélectrique *f*; ~ **charge** ELECT charge diélectrique *f*; ~ **constant** POWER constante diélectrique *f*; ~ **lens** RADIO lentille diélectrique *f*; ~**-loss measurement** DATA mesure des pertes diélectriques *f*; ~ **radiator** RADIO antenne diélectrique *f*; ~ **resonator** ELECT résonateur diélectrique *m*; ~ **rod radiator** RADIO antenne cierge *f*; ~ **strength** POWER rigidité diélectrique *f*; ~ **strength test** POWER essai de rigidité diélectrique *m*

Diesel: ~ **generator set** *n* POWER *power supply* groupe électrogène Diesel *m*

difference: ~ **of capacity** *n* TELEP écart de capacité *m*

differential[1] *adj* GEN différentiel

differential[2]: ~ **amplifier** *n* TRANS amplificateur différentiel *m*; ~ **circuit** *n* TRANS montage différentiel *m*; ~ **delay time** *n* TRANS temps de propagation différentiel *m*; ~ **echo suppressor** *n* RADIO, TELEP suppresseur d'écho différentiel *m*; ~ **encoding** *n* DATA codage par transition *m*; ~ **interpolation** *n* COMP interpolation différentielle *f*; ~ **loop modulation** *n* COMP modulation différentielle en boucle *f*; ~ **measurement method** *n* TEST méthode de mesure différentielle *f*; ~ **modal attenuation** *n* TRANS *fibre optics* affaiblissement modal différentiel *m*; ~**-mode attenuation** *n* TRANS *fibre optics* atténuation en mode différentiel *f*, affaiblissement en mode différentiel *m*; ~ **pulse-code modulation** *n* (*DPCM*) COMP, DATA, TELEP, TRANS modulation par impulsions et codage différentiel *f* (*MICD*), modulation différentielle par impulsions et codage *f*; ~ **quantum efficiency** *n* TEST rendement quantique différentiel *m*; ~ **relay** *n* PARTS relais différentiel *m*; ~ **resistance** *n* PARTS *semiconductors* résistance

différentielle *f*; ~ **sensitivity** *n* TELEP sensibilité différentielle *f*; ~ **threshold** *n* TELEP seuil différentiel *m*; ~ **transformer** *n* POWER transformateur différentiel *m*
differentiating: ~ **circuit** *n* POWER circuit de différenciation *m*
differentiation *n* TELEP *pulses* différentiation *f*
diffraction *n* TRANS *fibre optics* diffraction *f*; ~ **grating** CLIM, TRANS réseau de diffraction *m*; ~**-limited operation** TRANS exploitation limitée par diffraction *f*; ~ **pattern** RADIO, TRANS diagramme de diffraction *m*
diffuse: ~ **reflection** *n* COMP réflexion diffuse *f*; ~ **reflection coefficient** *n* RADIO, TRANS coefficient de réflexion diffuse *m*
diffused: ~ **junction** *n* PARTS jonction par diffusion *f*; ~ **resistor** *n* ELECT, PARTS résistance à couches diffusées *f*
diffuser *n* CLIM diffuseur *m*
diffusion *n* CLIM diffusion *f*; ~ **capacitance** PARTS *semiconductors* capacité de diffusion *f*; ~ **constant** PARTS constante de diffusion *f*; ~ **length** PARTS longueur de diffusion *f*
digit *n* DATA, TELEP, TRANS *display* élément numérique *m*, *number* chiffre *m*; ~ **display** TELEP *indicating devices* affichage numérique *m*; ~ **key** TELEP touche numérique *f*; ~ **place** COMP rang d'un chiffre *m*; ~ **plate** PER APP couronne numérotée *f*, disque numéroté *m*, TELEP couronne numérotée *f*; ~ **position** SIG position d'un élément de signal *f*, position d'un élément numérique *f*, TRANS position d'un élément de signal *f*; ~ **rate** TRANS débit numérique *m*; ~ **sequence integrity** TELEP intégrité de la séquence des éléments numériques *f*; ~ **storing** TELEP enregistrement de chiffres *m*; ~ **time slot** TELEP intervalle de temps pour élément numérique *m*
digital[1] *adj* COMP, DATA, GEN numérique; ~ **non-interpolated** (*DNI*) TRANS sans concentration numérique des conversations (*SCNC*); ~**-to-analog** (*D-A*) GEN numérique-analogique (*N-A*)
digital[2]: ~**-analog converter** *n* (*DAC*, *D-A converter*) COMP, DATA, ELECT, SIG, TELEP convertisseur numérique-analogique *m* (*CNA*, *convertisseur N-A*); ~ **announcement** *n* TELEP *answer machine* annonce numérique *f*; ~ **announcement machine** *n* PER APP, TELEP machine parlante numérique *f*; ~ **audio** *n* COMP son numérique *m*; ~ **audio broadcast** *n* (*DAB*) COMP, TV retransmission audionumérique *f* (*RAN*); ~ **audio broadcasting** *n* (*DAB*) COMP, TV radiodiffusion sonore numérique *f*; ~ **audio reproduction** *n* COMP restitution audionumérique *f*; ~ **audio tape** *n* (*DAT*) COMP, TV bande audionumérique *f* (*DAT*); ~ **block** *n* DATA, TELEP bloc numérique *m*; ~ **carrier module** *n* TRANS interface de ligne numérique *f*; ~ **cassette** *n* COMP cassette numérique *f*; ~ **cellular radio** *n* RADIO téléphone cellulaire numérique *m*; ~ **cellular system** *n* (*DCS*) RADIO système cellulaire numérique *m*; ~ **circuit** *n* TRANS circuit numérique *m*; ~ **circuit multiplication equipment** *n* (*DCME*) TELEP, TRANS équipement de multiplication de circuits numériques *m* (*EMCN*); ~ **circuit multiplication gain** *n* (*DCMG*) TELEP, TRANS gain de multiplication de circuits numériques *m* (*GMCN*); ~ **circuit multiplication system** *n* (*DCMS*) TELEP, TRANS système de multiplication de circuits numériques *m* (*SMCN*); ~ **circuitry** *n* COMP circuits numériques *m pl*; ~ **coding** *n* DATA codage numérique *m*; ~ **command signal** *n* (*DCS*) CONTR, DATA signal de commande numérique *m*; ~ **compact cassette** *n* (*DCC*) COMP, TV cassette compacte numérique *f*, cassette DCC *f*; ~ **compression** *n* TV compression numérique *f*; ~ **concentrator** *n* TELEP concentrateur numérique *m*; ~ **connection** *n* GEN raccordement numérique *m*, TRANS connexion numérique *f*; ~ **connectivity** *n* TRANS connexité numérique *f*; ~ **control** *n* COMP commande numérique *f*; ~ **control unit** *n* TELEP unité de commande numérique *f*; ~**-controlled oscillator** *n* COMP oscillateur à commande numérique *m*; ~ **converter** *n* COMP, DATA, SIG, TELEP convertisseur numérique *m*; ~ **cross-connect** *n* (*DXC*) SWG répartiteur numérique *m*; ~ **cross-connect equipment** *n* (*DXC*) SWG équipement d'interconnexion *m*; ~ **data** *n pl* COMP, DATA données numériques *f pl*; ~ **data communications message protocol** *n* (*DDCMP*) DATA protocole de messages de transmission de données numériques *m*; ~ **delay line** *n* COMP ligne à retard numérique *f*; ~ **demultiplexer** *n* TELEP démultiplexeur numérique *m*; ~ **display** *n* DATA, ELECT affichage numérique *m*; ~ **distribution frame** *n* SWG répartiteur numérique *m*; ~ **dynamic amplifier** *n* COMP amplificateur dynamique numérique *m*; ~ **dynamic filter** *n* COMP filtre dynamique numérique *m*; ~ **editing** *n* COMP édition numérique *f*; ~ **encoder-decoder** *n* COMP, TRANS codeur-décodeur numérique *m*; ~ **encoding** *n* DATA codage numérique *m*; ~ **engineering** *n* TELEP technique numérique *f*; ~ **error** *n* TELEP erreur numérique *f*; ~ **European cordless telephone service** *n* (*DECT*) POWER service numérique européen de téléphonie sans fil *m* (*DECT*); ~ **exchange** *n* SWG commutateur numérique *m*; ~ **feedback** *n* ELECT asservissement numérique *m*; ~ **filling** *n* DATA remplissage numérique *m*; ~ **filter** *n* COMP, DATA, TRANS filtre numérique *m*; ~ **filtering** *n* DATA, TRANS filtrage numérique *m*; ~ **FM-tone generator** *n* COMP génération de tonalité FM numérique *f*; ~ **frame structure** *n* DATA structure de la trame numérique *f*; ~ **frequency generator** *n* COMP générateur de fréquence numérique *m*; ~ **frequency synthesis** *n* TELEP synthèse numérique de fréquence *f*; ~ **harmonics generator** *n* COMP générateur d'harmoniques numérique *m*; ~ **hierarchy** *n* TELEP, TRANS hiérarchie numérique *f*; ~ **IC** *n* ELECT circuit intégré numérique *m*; ~ **identification frame** *n* DATA trame d'identification numérique *f*; ~ **impulse** *n* COMP impulsion numérique *f*; ~ **input** *n* COMP entrée numérique *f*; ~ **integrated circuit** *n* (*digital IC*) ELECT circuit intégré numérique *m*; ~ **interface** *n* DATA interface numérique *f*; ~ **interference** *n* TRANS interférence numérique *f*; ~ **line board** *n* (*DLB*) SWG carte de ligne numérique *f*; ~ **line link** *n* TELEP, TRANS conduit de ligne numérique *m*, liaison de ligne numérique *f*; ~ **line path** *n* TELEP, TRANS conduit de ligne numérique *m*; ~ **line section** *n* TELEP section de ligne numérique *f*; ~ **line system** *n* TELEP système de ligne numérique *m*; ~ **loop** *n* TRANS boucle numérique *f*; ~ **loss** *n* TRANS atténuation numérique *f*; ~ **main network switching center** *AmE*, ~ **main network switching centre** *BrE* *n* SWG centre de commutation numérique du réseau principal *m*; ~ **mastering** *n* COMP gravure de matrice *f*; ~ **mobile system** *n* (*DMS*) TELEP système mobile numérique *m*; ~ **modulation** *n* COMP, TRANS modulation numérique *f*; ~ **modulation link** *n* COMP, TRANS liaison de modulation numérique *f*; ~ **modulation system** *n* COMP, TRANS système à modulation numérique *m*; ~ **multi-junctor** *n* TELEP multi-joncteur numérique *m*; ~ **multiplex equipment** *n* TELEP, TRANS équipement de multiplexage numérique *m*, multiplexeur numérique *m*; ~ **multiplex hierarchy** *n* TELEP, TRANS hiérarchie de multiplexage numérique *f*; ~ **multiplexer** *n* (*DM*) TELEP, TRANS multiplexeur numérique *m*; ~ **music instrument** *n* COMP instrument de musique numérique *m*; ~ **network** *n* (*DN*) TRANS réseau numérique *m*; ~ **noise** *n* DATA *scanner* bruit numérisé *m*; ~ **non-interpolated channel** *n* (*DNI channel*) TRANS voie de jonction numérique sans concentration *f*; ~ **output** *n* COMP sortie numérique *f*; ~ **pad** *n* TRANS complément de ligne numérique *m*; ~ **paper storage** *n* COMP, DATA mémoire sur papier numérique *f*; ~ **path** *n* TELEP conduit numérique *m*, TRANS circuit numérique *m*, conduit numérique *m*; ~ **phase-shift modulation** *n* TRANS modulation numérique par déplacement de phase *f*; ~ **processing** *n* DATA traitement numérique *m*; ~ **pseudo-noise sequence** *n* TRANS séquence numérique de pseudo-bruits *f*;

~ **publishing** *n* APPL, COMP, DATA édition électronique *f*; ~ **pulse stream** *n* TELEP train d'impulsions numériques *m*; ~ **quality** *n* RADIO qualité numérique *f*; ~ **radio broadcasting** *n* RADIO radiodiffusion numérique *f*; ~ **radio concentrator** *n* RADIO concentrateur en radiocommunication numérique *m*; ~ **radio path** *n* RADIO, TELEP conduit hertzien numérique *m*; ~ **radio relay for synchronous hierarchy** *n* (*SDH-DRSS*) RADIO faisceau hertzien numérique pour hiérarchie synchrone *m* (*HNS-FHN*); ~ **radio section** *n* RADIO, TELEP section hertzienne numérique *f*; ~ **radio system** *n* DATA, RADIO système de radio numérique *m*; ~ **radio telephony** *n* RADIO, TELEP radiotéléphonie numérique *f*; ~ **read-out** *n* DATA, ELECT affichage numérique *m*, TEST lecture numérique *f*; ~ **recorder** *n* COMP, PER APP enregistreur numérique *m*; ~ **recording** *n* COMP, PER APP, TEST *answer machine* enregistrement digital *m*, enregistrement numérique *m*; ~ **reverb** *n* COMP réverbération numérique *f*; ~ **reverberation** *n* COMP réverbération numérique *f*; ~ **rhythm programmer** *n* COMP programmeur de rythmes numérique *m*; ~ **satellite concentrator** *n* RADIO concentrateur numérique satellite *m* (*CNS*); ~ **scale** *n* TEST échelle numérique *f*; ~ **section** *n* (*DS*) TRANS section numérique *f*; ~ **selective call** *n* (*DSC*) RADIO appel sélectif numérique *m*; ~ **sequence** *n* COMP séquence numérique *m*; ~ **signal** *n* COMP signal numérique *m*; ~ **signal processor** *n* (*DSP*) COMP processeur de traitement numérique des signaux *m* (*DSP*), processeur numérique de signaux *m* (*DSP*); ~ **signal processing chip** *n* DATA unité numérique de traitement de signaux *f*; ~ **signature** *n* DATA signature numérique *f*; ~ **soundtrack** *n* COMP piste de son numérique *f*; ~ **speech compression** *n* TELEP, TRANS concentration numérique de la parole *f* (*CNP*); ~ **speech interpolation** *n* (*DSI*) TELEP, TRANS concentration numérique des conversations *f* (*CNC*); ~ **still image** *n* COMP image numérique fixe *f*; ~ **storage** *n* DATA mémoire numérique *f*; ~ **storage media** *n* (*DSM*) COMP support de stockage numérique *m*; ~ **stream** *n* DATA train de données numériques *m*; ~ **subscriber access unit** *n* TRANS terminaison numérique réseau *f*; ~ **sum** *n* TELEP somme numérique *f*; ~ **sum variation** *n* TELEP variation de la somme numérique *f*; ~ **switch** *n* SWG autocommutateur numérique *m*; ~ **switching** *n* SWG commutation numérique *f*; ~ **switching center** *AmE*, ~ **switching centre** *BrE n* SWG centre de commutation numérique *m*; **~-switching element** *n* (*DSE*) SWG organe de commutation numérique *m*; ~ **switching matrix** *n* SWG, TELEP matrice de commutation numérique *f*; ~ **switching network** *n* SWG réseau de connexion numérique *m*; ~ **switching system** *n* SWG système de commutation numérique *m*; ~ **system** *n* COMP système numérique *m*; ~ **talking machine** *n* PER APP parlante numérique *f*; ~ **technology** *n* TELEP technique numérique *f*; ~ **telephone** *n* TELEP poste de système numérique *m*; **~-to-analog conversion** *n* COMP, DATA, SIG, TELEP conversion numérique-analogique *f*; **~-to-analog converter** *n* (*DAC*, *D-A converter*) COMP, DATA, ELECT, SIG, TELEP convertisseur numérique-analogique *m* (*CNA*, *convertisseur N-A*); ~ **transit command** *n* SIG commande de transit numérique *f*; ~ **transmission** *n* TRANS transmission numérique *f*; ~ **transmission link** *n* TRANS liaison de transmission numérique *f*; **~-transmit command** *n* (*DTC*) DATA commande d'émission numérique *f* (*DTC*); ~ **transmitters** *n pl* DATA émetteurs numériques *m pl*; ~ **video** *n* COMP vidéo numérique *f*; ~ **video editing** *n* COMP montage vidéo numérique *m*; ~ **video tape recorder** *n* (*DVTR*) COMP magnétoscope numérique *m*; ~ **voltmeter** *n* TEST *measuring devices* voltmètre numérique *m*

digitalization *n* GEN numérisation *f*, digitalisation *f*

digitalize *vt* GEN digitaliser, numériser

digitalized *adj* GEN numérisé, digitalisé

digitally¹: **~-mastered** *adj* COMP gravé numériquement

digitally² *adv* GEN numériquement

digitally³: **~-compensated crystal oscillator** *n* (*DCXO*) ELECT oscillateur à quartz à commande numérique *m*

digitization *n* GEN numérisation *f*, digitalisation *f*

digitize *vt* GEN digitaliser, numériser

digitized¹ *adj* GEN numérisé, digitalisé

digitized²: ~ **data** *n pl* DATA données numérisées *f pl*, informations numérisées *f pl*; ~ **image** *n* DATA *scanner*, TV image digitalisée *f*, image numérisée *f*; ~ **speech** *n* TELEP signal vocal numérisé *m*

digitizing *n* GEN numérisation *f*

digits: ~ **dialed** *AmE*, ~ **dialled** *BrE n pl* TELEP chiffres appelés *m pl*

dihedral: ~ **antenna** *n* TEST antenne dièdre *f*

dimension *n* COMP, DATA, GEN dimension *f*

dimensional: ~ **stability** *n* PARTS *materials* stabilité dimensionnelle *f*

dimensioning *n* COMP, GEN dimensionnement *m*

dimmed¹ *adj* COMP estompé

dimmed²: ~ **icon** *n* COMP icône grisée *f*

DIN *abbr* (*Deutsche Industrienorm*) GEN DIN (*Deutsche Industrienorm*); ~ **jack** *n* COMP prise DIN *f*; ~ **plug** *n* COMP fiche DIN *f*; ~ **standard** *n* GEN norme DIN *f*

diode *n* ELECT, PARTS, TELEP, TRANS diode *f*, porte *f*; ~ **array** ELECT réseau de diodes *m*; ~ **bridge** ELECT pont de diodes *m*; ~ **circuit** ELECT circuit à diodes *m*, montage à diodes *m*; ~ **crosspoint** SWG point d'intersection à diode *m*; ~ **laser** ELECT, PARTS, TRANS laser à diode *m*, laser à injection *m*, laser semi-conducteur *m*; ~ **network** ELECT montage à diodes *m*; ~ **photodetector** PARTS, TRANS photodiode *f*; ~ **tree** ELECT circuit à diodes *m*, montage à diodes *m*

dip: ~ **soldering** *n* ELECT brasage tendre au trempé *m*

DIP *abbr* (*dual-in-line package*) DATA boîtier DIP *m* (*boîtier à deux rangées de broches*)

diplex *n* PARTS, RADIO, TELEP, TRANS diplex *m*; ~ **radio transmission** DATA, RADIO radiotransmission diplex *f*

diplexer *n* PARTS, RADIO, TELEP, TRANS diplexeur *m*

dipole *n* RADIO doublet *m*; ~ **aerial** RADIO antenne de doublet *f*, antenne dipôle *f*, doublet *m*; ~ **antenna** RADIO antenne de doublet *f*, antenne dipôle *f*, doublet *m*

Dirac: ~ **function** *n* TELEP *pulses* fonction de Dirac *f*; ~ **pulse response** *n* TELEP réponse percussionnelle *f*

direct¹ *adj* GEN direct, point à point

direct²: ~ **access** *n* COMP, DATA, PER APP accès direct *m*; **~-access memory** *n* COMP, DATA mémoire à accès direct *f* (*RAM*), mémoire à accès sélectif *f*; ~ **access service** *n* COMP périphérique à accès séquentiel *m*; ~ **addressing** *n* COMP adressage direct *m*; ~ **beam** *n* RADIO faisceau latéral *m*; ~ **broadcasting by satellite** *n* (*DBS*) DATA, RADIO, TV radiodiffusion directe par satellite *f* (*RDS*); **~-buried cable** *n* ELECT câble directement enterré *m*; ~ **call** *n* DATA, TELEP appel direct *m*; ~ **carrier** *n* TRANS télécommunicateur qui transmet uniquement l'information *m*; ~ **comparison** *n* TEST comparaison directe *f*; ~ **control** *n* COMP, SWG commande directe *f*; ~ **control system** *n* COMP, SWG système à commande directe *m*; ~ **current** *n* (*DC*) POWER courant continu *m* (*CC*); ~ **current converter** *n* POWER convertisseur courant continu *m*; ~ **current coupler** *n* (*DC coupler*) POWER coupleur continu *m*; ~ **current signaling** *AmE*, ~ **current signalling** *BrE n* POWER, SIG signalisation en courant continu *f*; ~ **current transmission** *n* POWER, TRANS transmission par courant continu *f*; **~-dial call** *n* TELEP appel automatique *m*; ~ **dialing-in** *AmE*, ~ **dialling-in** *BrE n* DATA, TELEP accès direct à l'arrivée *m*, sélection directe à l'arrivée *f* (*SDA*), sélection directe d'un poste supplémentaire *f*; ~ **digital frequency synthesis** *n* (*DDFS*) ELECT synthèse de fréquence numérique directe *f*; ~ **digital input** *n* COMP entrée numérique directe *f*; ~ **distance dialing** *n AmE* TELEP interurbain automatique *m*

(*IA*); ~ **distance dialing access code** *n AmE* code d'accès à l'interurbain automatique *m*; ~ **distribution** *n* RADIO distribution directe *f*; ~**-driven selector** *n* SWG sélecteur à commande directe *m*; ~ **feed** *n* RADIO *aerials* alimentation directe *f*; ~ **function** *n* COMP fonction directe *f*; ~ **input dialing** *AmE*, ~ **input dialling** *BrE n* TELEP sélection directe à l'arrivée *f* (*SDA*); ~**-insert subroutine** *n* COMP sous-programme ouvert *m*; ~ **inward dialing** *AmE*, ~ **inward dialling** *BrE n* (*DID*) TELEP accès direct à un poste *m*, appel direct de l'extérieur *m*, sélection directe à l'arrivée *f* (*SDA*); ~ **line** *n* (*DL*) SWG ligne privée *f*, TELEP, TRANS liaison spécialisée *f*, ligne directe *f*, ligne privée *f*; ~ **memory** *n* COMP, PER APP mémoire directe *f*; ~ **memory access** *n* (*DMA*) COMP, PER APP accès direct à la mémoire *m*; ~ **method of measurement** *n* TEST méthode de mesure directe *f*; ~ **negative** *n* POWER moins franc *m*, négatif franc *m*; ~ **orbit** *n* RADIO orbite directe *f*; ~ **outward dialing** *AmE*; ~ **outward dialling** *BrE n* (*DOD*) TELEP accès direct au réseau *m*, prise directe du réseau *f*; ~ **positive** *n* POWER plus franc *m*, positif franc *m*, terre franche *f*; ~ **routing** *n* SIG, TELEP, TRANS acheminement direct *m*; ~ **satellite broadcasting** *n* (*DSB*) TV retransmission directe par satellite *f*; ~ **select** *n* COMP sélection directe *f*; ~ **sequence** *n* (*DS*) GEN séquence directe *f*; ~**-sequence spread spectrum** *n* RADIO étalement du spectre à séquence directe *m*; ~ **service circuit** *n* TELEP circuit de service direct *m*; ~**-transit country** *n* TELEP pays de transit direct *m*; ~ **voltage** *n* POWER tension continue *f*; ~ **wiring** *n* PARTS câblage direct *m*

direct[3] *vt* TELEP *traffic* aiguiller

directed: ~**-beam scan** *n* COMP balayage cavalier *m*; ~ **connection** *n* TELEP connexion dirigée *f*; ~ **scan** *n* COMP balayage cavalier *m*

direction *n* RADIO, TELEP *location* direction *f*, *of traffic* sens *m*; ~ **finder** RADIO radiogoniomètre *m*; ~ **finding** RADIO relèvement radiogonométrique *m*, goniométrie *f*; ~ **of lay** GEN *cables* sens d'assemblage *m*; ~ **of speech** TELEP direction de conversation *f*

directional: ~ **aerial** *n* RADIO antenne directive *f*; ~ **antenna** *n* RADIO antenne directive *f*; ~ **beam** *n* RADIO faisceau directionnel *m*; ~ **coupler** *n* DATA, SWG, TRANS *fibre optics* coupleur directif *m*, coupleur directionnel *m*; ~ **coupler optical switch** *n* SWG, TRANS commutateur optique à coupleur directif *m*; ~**-division multiplexing** *n* (*DDM*) TRANS multiplexage directionnel par division *m*; ~ **filter** *n* ELECT filtre d'aiguillage directionnel *m*; ~**-light source** *n* COMP source de lumière directionnelle *f*, source de lumière dirigée *f*; ~ **microphone** *n* TELEP microphone directionnel *m*; ~ **relay** *n* PARTS relais directionnel *m*; ~ **sample** *n* TEST *quality* échantillon d'orientation spécifiée *m*

directive *n* COMP déclaration *f*, pseudo-instruction *f*, DATA déclaration *f*; ~ **antenna** DATA antenne directive *f*; ~ **gain** RADIO gain de directivité *m*

directivity *n* RADIO directivité *f*

directly: ~**-heated cathode** *n* ELECT, PARTS cathode à chauffage direct *f*; ~**-powered repeater station** *n* TELEP, TRANS station de répéteurs à alimentation indépendante *f*, station de répéteurs autoalimentée *f*

directory *n* COMP répertoire *m*, DATA répertoire *m*, annuaire *m*, ELECT annuaire *m*, TELEP répertoire *m*; ~ **access protocol** (*DAP*) DATA protocole d'accès à l'annuaire *m*; ~ **assistance** APPL assistance annuaire *f* (*AA*), renseignements téléphoniques *m pl*, service d'assistance annuaire *m*, TELEP assistance annuaire *f* (*AA*); ~ **enquiries** *BrE* (*cf directory information AmE*) (*DQ*) APPL, TELEP *service* assistance annuaire *f* (*AA*), renseignements téléphoniques *m pl*, service d'assistance annuaire *m*, service de renseignements téléphoniques *m*; ~ **entry** DATA adresse *f*, adresse figurant dans l'annuaire *f*, TELEP inscription à l'annuaire *f*; ~ **information** *AmE* (*cf directory enquiries BrE*) APPL, TELEP assistance annuaire *f* (*AA*), renseignements téléphoniques *m pl*, service d'assistance annuaire *m*, service de renseignements téléphoniques *m*; ~ **information tree** (*DIT*) DATA, TELEP arbre d'informations d'annuaire *m* (*DIT*); ~ **number** (*DN*) TELEP numéro d'appel *m*, numéro de poste *m*; ~ **services** (*DS*) DATA service d'annuaire *m* (*DS*); ~ **store** TELEP fichier d'abonnés *m*; ~ **system agent** (*DSA*) DATA agent de système d'annuaire *m* (*DSA*); ~ **system protocol** (*DSP*) DATA protocole du système d'annuaire *m* (*DSP*); ~ **user agent** APPL, DATA, TELEP agent usager d'annuaire *m*, agent utilisateur d'annuaire *m*

disable *vt* COMP déconnecter, désactiver, incapaciter, invalider, neutraliser, GEN rendre inutilisable, rendre inopérant

disabled[1] *adj* COMP invalidé

disabled[2]: ~ **channel** *n* RADIO *mobile* voie hors service *f*; ~ **state** *n* TEST *dependability* état hors circuit *m*, état hors fonction *m*; ~ **time** *n* TEST *dependability* temps d'incapacité *m*

disablement *n* COMP neutralisation *f*; ~ **unit** TEST *dependability* unité d'incapacité *f*, unité d'invalidation *f*

disabling *n* COMP mise hors service *f*

disassemble *vt* GEN déposer

disassembly *n* GEN démontage *m*

disaster: ~ **prevention** *n* TEST *security* prévention des désastres *f*; ~ **recovery** *n* DATA reprise après une catastrophe *f*

disc: ~ **antenna** *n BrE* RADIO antenne à disque *f*

discharge[1] *n* POWER *batteries* décharge *f*; ~ **cycle** POWER période de décharge *f*; ~ **time** POWER période de décharge *f*

discharge[2] *vt* POWER *battery* décharger

disclosure *n* COMP divulgation *f*

Discman® *n* COMP lecteur laser portatif *m*, tourne-disque portatif *m*

disconnect[1] *n* (*DCN*) DATA déconnexion *f* (*DCN*); ~ **frame** TELEP trame de déconnexion *f*; ~ **signal** DATA signal de libération *m*, SIG signal de fin *m*

disconnect[2] *vt* COMP débrancher, déconnecter, ELECT débrancher, PER APP débrancher, TELEP déconnecter, libérer, TEST débrancher

disconnected *adj* COMP, ELECT, PER APP débranché, TELEP déconnecté, TEST débranché; ~ **mode** *n* (*DM*) TELEP mode déconnexion *m*, mode déconnecté *m* (*DM*); ~ **mode frame** *n* TELEP trame en mode déconnexion *f*

disconnection *n* COMP déconnexion *f*, libération *f*, TELEP libération *f*; ~ **procedure** TELEP procédure de déconnexion *f*

disconnector *n* POWER *switchgear* sectionneur *m*

discontinuous: ~ **selection** *n* COMP sélection discontinue *f*

discrete[1] *adj* GEN discret

discrete[2]: ~ **channel** *n* TRANS canal discret *m*; ~ **circuit** *n* ELECT circuit à composants discrets *m*; ~ **component** *n* PARTS composant discret *m*; ~ **cosine transform** *n* (*DCT*) COMP, DATA, ELECT, TELEP transformation en cosinus discret *f* (*TCD*); ~ **cosine transform coding** *n* COMP, DATA, ELECT, TELEP codage par transformée en cosinus discrète *m*; ~ **data collector** *n* COMP, DATA circuit collecteur de données discrètes *m*; ~ **Fourier transform** *n* DATA transformation de Fourier discrète *f*; ~ **system** *n* DATA système discret *m*; ~**-time deterministic signal** *n* COMP, TRANS signal déterministe à temps discret *m*; ~**-time linear system** *n* TRANS système linéaire à temps discret *m*; ~**-time system** *n* DATA système à temps discontinu *m*

discretely: ~**-timed signal** *n* COMP, TRANS signal discret *m*, signal temporel discret *m*

discretionary *adj* COMP préférentiel

discriminating: ~ **satellite exchange** *n* SWG central auxiliaire discriminateur *m*, central discriminateur *m*

discrimination *n* TEST *measuring devices* mobilité *f*;

~ **instruction** COMP instruction de décisions *f*; ~ **ratio** TEST *quality* pente de la courbe d'efficacité *f*, rapport de discrimination *m*; ~ **threshold** TEST seuil de mobilité *m*
discriminator *n* DATA, ELECT, SWG discriminateur *m*; ~ **input object** DATA objet d'entrée de discriminateur *m*
disengage *vt* GEN dégager
dish *n* RADIO antenne parabolique *f*; ~ **aerial** RADIO antenne à réflecteur parabolique *f*; ~ **antenna** RADIO antenne parabolique *f*
disjunction *n* ELECT opération OU *f*, union *f*, *logic elements* opération OU inclusif *f*, réunion *f*
disk: ~ **antenna** *AmE see disc antenna BrE*; ~ **array** *n* DATA pile de disques *f*; ~ **box** *n* COMP coffret à disquettes *m*; ~ **caching** *n* COMP, DATA antémémoire de disque *f*; ~ **caching software** *n* COMP, DATA logiciel d'antémémoire de disque *m*; ~ **change** *n* COMP changement de disque *m*; ~ **drive** *n* COMP lecteur de disque *m*; ~ **drive controller** *n* COMP contrôleur de lecteur de disque *m*; ~ **memory** *n* COMP mémoire à disques *f*; ~ **slot** *n* COMP fente d'insertion de disque *f*; ~ **space** *n* COMP *Internet* espace de disque *m*; ~ **storage** *n* COMP mémoire à disques *f*; ~ **storage drive** *n* COMP mécanisme d'entraînement de disque *m*, unité de disques *f*; ~ **store** *n* COMP mémoire à disques *f*
diskette *n* COMP disque souple *m*, disquette *f*
dismantle *vt* GEN déposer
dismantling *n* GEN démontage *m*
disparity *n* GEN, TELEP disparité *f*
dispatch[1]: ~-**received file** *n* (*DRF*) DATA distribution des fichiers reçus *f* (*DRF*); ~ **service** *n* TRANS service de répartition *m*; ~ **system** *n* TRANS système de répartition *m*
dispatch[2] *vt* COMP répartir, ELECT acheminer, GEN distribuer
dispatched *adj* COMP, GEN réparti
dispatcher *n* COMP répartiteur *m*, GEN expéditeur *m*
dispensing: ~ **tips** *n pl* ELECT canules d'application *f pl*
dispersed *adj* GEN *system* dispersé
dispersion[1]: ~-**limited** *adj* TRANS limité par la dispersion
dispersion[2] *n* ELECT, RADIO, TRANS dispersion *f*; ~ **coefficient** TRANS coefficient de dispersion *m*; ~-**flattened fiber** *AmE*, ~-**flattened fibre** *BrE* TRANS fibre à dispersion aplatie *f*
displacement *n* GEN déplacement *m*; ~ **amplitude** ELECT amplitude de déplacement *f*; ~ **from center frequency** *AmE*, ~ **from centre frequency** *BrE* RADIO, TRANS décalage par rapport à la fréquence centrale *m*
display[1] *n* COMP, DATA, ELECT affichage *m*, *screen* écran *m*, GEN indication *f*, PARTS visualisation *f*, PER APP affichage *m*, RADIO affichage *m*, présentation *f*, indicateur *m*, *screen* écran *m*, TELEP *screen* écran *m*, TV écran *m*, visualisation *f*; ~ **background** COMP fond d'image fixe *m*; ~ **board** DATA contrôleur d'écran *m*; ~ **console** COMP console d'affichage *f*, TV console de visualisation *f*; ~ **device** COMP dispositif d'affichage *m*, visuel *m*, DATA appareil d'affichage *m*; ~ **element** COMP élément graphique *m*; ~ **field** COMP zone d'affichage *f*; ~ **foreground** COMP élément d'image variable *m*; ~ **group** COMP groupe graphique *m*, groupe visuel *m*, segment graphique *m*; ~ **light** COMP lampe témoin *f*, voyant *m*; ~ **order** COMP ordre d'affichage *m*; ~ **paging** RADIO radiomessagerie unilatérale *f* (*RMU*), RMU alphanumérique *f*, TELEP, TRANS radiomessagerie unilatérale *f* (*RMU*); ~ **panel** TELEP *indicating devices* panneau d'affichage *m*, panneau indicateur *m*; ~ **resolution** COMP résolution d'affichage *f*; ~ **screen** COMP terminal à écran de visualisation *m*, PER APP, TV écran de visualisation *m*; ~ **segment** COMP segment graphique *m*, groupe visuel *m*, groupe graphique *m*; ~ **station interface** (*DSI*) PER APP interface de poste d'affichage *f*; ~ **surface** COMP surface d'affichage *f*; ~ **unit** COMP unité d'affichage *f*, TV console de visualisation *f*; ~ **window** COMP fenêtre d'affichage *f*
display[2] *vt* GEN *screen image* afficher, visualiser
Display: ~ **PostScript** *n* DATA Display PostScript *m*
disrupt *vt* DATA interrompre
disruption *n* ELECT interruption *f*
disruptive: ~ **breakdown** *n* POWER *electricity* claquage disruptif *m*; ~ **voltage** *n* POWER tension disruptive *f*
dissipation: ~ **factor** *n* POWER *supply* facteur de dissipation *m*
dissolved: ~ **image** *n* COMP image en fondu enchaîné *f*
distance *n* RADIO portée radar *f*; ~ **learning** TV téléenseignement *m*; ~ **tube** PARTS entretoise tubulaire *f*, tube-entretoise *m*
distant: ~-**light source** *n* COMP source de lumière lointaine *f*, source de lumière répartie *f*
distinguished: ~ **name** *n* DATA nom distinctif *m*, nom spécifique *m*
distortion[1]: ~-**limited** *adj* TRANS limité par la distorsion
distortion[2] *n* GEN distorsion *f*, déformation *f*; ~ **bridge** ELECT pont de distorsion *m*; ~ **factor** ELECT taux d'harmoniques *m*; ~-**limited operation** ELECT fonctionnement limité par la distorsion *m*
distress: ~ **alerting** *n* RADIO alerte de détresse *f*; ~ **radio call system** *n* RADIO système de détresse radioélectrique *m*; ~ **signal** *n* TEST *security* signal de détresse *m*
distribute *vt* COMP décentraliser, répartir, GEN décentraliser, répartir, distribuer, TELEP diffuser
distributed[1] *adj* COMP, GEN distribué, réparti, SWG distribué, TELEP diffusé
distributed[2]: ~ **application** *n* COMP application distribuée *f*; ~ **architecture** *n* SWG architecture distribuée *f*; ~ **computing** *n* COMP, DATA informatique distribuée *f*; ~ **control** *n* DATA contrôle réparti *m*; ~ **control system** *n* SWG système à commande répartie *m*; ~ **data processing** *n* DATA informatique répartie *f*; ~ **feedback** *n* (*DFB*) ELECT rétroaction répartie *f*; ~ **feedback laser** *n* DATA, ELECT, TRANS laser à rétroaction répartie *m*; ~ **feedback laser diode** *n* (*DFB laser diode*) TRANS diode laser à rétroaction répartie *f*; ~ **file** *n* DATA fichier réparti *m*; ~ **file server** *n* DATA serveur de fichiers réparti *m*; ~ **frame-alignment signal** *n* SIG, TRANS signal de verrouillage de trame réparti *m*; ~-**information processing task** *n* DATA tâche de traitement réparti de l'information *f*; ~-**light source** *n* COMP source de lumière étendue *f*, source de lumière répartie *f*; ~ **multi-antenna** *n* RADIO système à plusieurs antennes réparties *m*; ~ **network** *n* COMP réseau décentralisé *m*, TV réseau distribué *m*; ~ **PBX** *n* APPL, CONTR, SWG, TELEP autocommutateur privé réparti *m*; ~ **processes** *n* COMP informatique répartie *f*; ~ **processing** *n* DATA informatique répartie *f*; ~-**queue double system** *n* (*DQDBS*) DATA DQDBS; ~-**queue dual bus** *n* (*DQDB*) COMP DQDB; ~-**sequence controller** *n* COMP séquenceur réparti *m*; ~ **signaling** *AmE*, ~ **signalling** *BrE n* TRANS signalisation répartie *f*; ~ **single-layer** *n* (*DS*) DATA monocouche répartie *f* (*DS*); ~ **single-layer embedded test method** *n* (*DSE*) TEST méthode de test monocouche distribuée encastrée *f* (*DSE*); ~ **single-layer test method** *n* TEST méthode de test monocouche répartie *f*; ~ **system** *n* COMP système réparti *m*; ~-**system architecture** *n* (*DSA*) COMP architecture de systèmes distribués *f*; ~ **test method** *n* TEST méthode de test répartie *f*; ~ **transaction** *n* DATA transaction répartie *f*
distributing: ~ **data processing** *n* COMP traitement réparti *m*
distribution *n* GEN décentralisation *f*, distribution *f*, répartition *f*; ~ **block** ELECT, PARTS *connecting devices* boîtier de distribution *m*; ~ **box** POWER, TRANS boîte de dérivation *f*, boîte de distribution *f*; ~ **busbar** POWER barre omnibus de distribution *f*; ~ **cabinet** POWER armoire de dérivation *f*, TRANS coffret de distribution *m*, armoire de sous-répartiteur *f*; ~ **cable** POWER câble de distribution *m*; ~ **center** *AmE*, ~ **centre** *BrE* POWER, TV centre de distribution *m* (*CD*); ~ **channel** DATA *scanner* canal de distribution *m*; ~ **and control unit** (*DCU*) CONTR unité de distribution et de commande *f*; ~ **frame** TRANS répartiteur *m*; ~ **fuse** PARTS, POWER fusible de distribution *m*; ~ **link** RADIO liaison de distribution *f*; ~ **list** (*DL*) COMP *e-mail* liste

de diffusion *f*, DATA, TELEP liste de distribution *f*; ~ **module** POWER module de distribution *m*; ~ **negative** POWER négatif de distribution *m*; ~ **network** SWG, TRANS, TV réseau de distribution *m*, réseau de répartition *m*; ~ **output** POWER sortie de distribution *f*; ~ **panel** POWER tableau de distribution *m*; ~ **plug** POWER fiche de distribution *f*; ~ **point** (*DP*) TRANS branchement *m*; ~ **positive** POWER positif de distribution *m*; ~ **primary link** (*DPL*) TRANS liaison primaire pour les services de distribution *f*; ~ **rack** POWER armoire de distribution *f*, bâti de distribution *m*; ~ **relay** PARTS relais de distribution *m*; ~ **relay set** PARTS plaque de relais de distribution *f*; ~ **stage** TRANS étage de brassage *m*; ~ **terminal** POWER boîte de distribution *f*; ~ **of traffic** SWG, TELEP répartition du trafic *f*; ~ **transformer** POWER transformateur de distribution *m*

distributor *n* GEN distributeur *m*

disturb *vt* GEN perturber, déranger, *interfere* brouiller

disturbance[1]: ~**-free** *adj* GEN non perturbé, sans perturbation

disturbance[2] *n* GEN dérangement *m*, perturbation *f*, *interference* brouillage *m*, RADIO parasitage *m*, parasites *m pl*, perturbation *f*, TELEP perturbation *f*; ~ **counter** TELEP compteur de perturbations *m*; ~ **recording** TELEP enregistrement des perturbations *m*; ~ **supervision** TELEP supervision des perturbations *f*; ~ **suppression** TEST antibrouillage *m*, antiparasitage *m*; ~ **variable** CONTR grandeur de perturbation *f*; ~ **voltage** TEST tension perturbatrice *f*

DIT *abbr* (*directory information tree*) DATA, TELEP DIT (*arbre d'informations d'annuaire*)

ditch *n* RADIO rainure *f*

dithering *n* COMP tramage aléatoire *m*

diurnal: ~ **rhythm** *n* GEN *physiology* rythme diurne *m*; ~ **variation** *n* TELEP variation diurne *f*

diversion *n* GEN, TELEP renvoi *m*; ~ **follow me** TELEP renvoi variable *m*; ~**-if-number-busy service** TELEP service de transfert d'appel en cas de numéro occupé *m*; ~ **on busy** TELEP renvoi d'appel sur occupation *m*; ~ **on no reply** TELEP renvoi d'appel sur non-réponse *m*

diversity *n* GEN, RADIO diversité *f*; ~ **radar** RADIO radar à diversité de fréquence *m*, radar diversité *m*; ~ **reception** RADIO réception en diversité *f*

divert *vt* GEN, TELEP renvoyer; ~ **to** COMP dérouter sur

divide *vt* TRANS subdiviser; ~ **up** COMP, ELECT partager, séparer

divided: ~**-battery system** *n* POWER système à capacité de batterie répartie *m*

divider *n* COMP, ELECT diviseur *m*

division *n* TRANS subdivision *f*

DL *abbr* DATA (*distribution list*) liste de distribution *f*, SWG, TELEP, TRANS (*direct line*) liaison spécialisée *f*, ligne directe *f*, ligne privée *f*

DLB *abbr* (*digital line board*) SWG carte de ligne numérique *f*

DLC *abbr* DATA (*data-link connection*) connexion pour liaison de données *f*, (*data-link control*) commande de liaison de données *f*

DLCI *abbr* (*data-link connection identifier*) DATA identificateur de connexion de liaison de données *m*

DLL *abbr* (*dynamic-link library*) TELEP DLL (*bibliothèque de liens dynamiques*)

DLS *abbr* APPL, DATA (*data-link service*) service de liaison de données *m*, (*data-link switching*) commutation de liaison de données *f*, ELECT (*data-link switching*) commutation de liaison de données *f*

DLSAP *abbr* (*data-link service access point*) DATA point d'accès au service de liaison de données *m*

DM *abbr* COMP (*data management*) gestion de données *f*, DATA (*degraded minute*) minute dégradée *f*, (*data management*) gestion de données *f*, SIG (*delta modulation*) MD (*modulation delta*), TELEP (*disconnected mode*) DM (*mode déconnecté*), (*document manipulation*) DM (*manipulation de documents*), TRANS (*digital multiplexer*) multiplexeur numérique *m*, (*delta modulation*) MD (*modulation delta*)

DMA *abbr* (*direct memory access*) COMP accès direct à la mémoire *m*

DMS *abbr* (*digital mobile system*) TELEP système mobile numérique *m*

DN *abbr* TELEP (*directory number*) numéro d'appel *m*, numéro de poste *m*, TRANS (*destination network*) réseau de destination *m*, (*digital network*) réseau numérique *m*

DNI *abbr* (*digital non-interpolated*) TRANS SCNC (*sans concentration numérique des conversations*); ~ **channel** *n* (*digital non-interpolated channel*) TRANS voie de jonction numérique sans concentration *f*

DNIC *abbr* (*data-network identification code*) DATA, TELEP CIRD (*code d'identification de réseau de données*)

DNR *abbr* (*domain name resolver*) DATA résolveur de noms de domaine

DNTX: ~ **code** *n* SIG code DNTX *m*, code ne pas émettre *m*

do: ~**-not-disturb-service** *n* TELEP service ne pas déranger *m*; ~**-not-transmit code** *n* SIG code DNTX *m*, code ne pas émettre *m*

docking: ~ **stations** *n pl* ELECT stations *f pl*

document *n* GEN document *m*; ~ **bulk transfer** TELEP transfert de masse de document *m*; ~ **bulk transfer and manipulation** TELEP transfert de masse et manipulation de document *m*; ~ **content architecture** (*DCA*) DATA architecture DCA *f* (*DCA*); ~ **description language** DATA langage de description structurelle de document *m*; ~ **facsimile telegram** TELEP télégramme fac-similé *m*; ~ **facsimile telegraphy** TELEP télécopie fac-similée *f*; ~ **file** DATA fichier document *m*; ~ **format** DATA format de document *m*; ~ **imaging system** DATA système d'imagerie documentaire *m*; ~ **interchange format** DATA document interchange format *m*; ~ **loader** DATA *scanner* chargeur de documents *m*; ~ **manipulation** TELEP manipulation de document *f*; ~ **processing** DATA traitement de documents *m*; ~ **processing and interchange** DATA traitement et transmission de documents *m*; ~ **rest** DATA *scanner* repose-document *m*; ~ **server** DATA serveur de document *m*; ~ **size** DATA taille du document *f*; ~ **source** COMP, DATA document source *m*; ~ **transmission** DATA transmission de documents *f*

Document: ~ **Manipulation** *n* (*DM*) TELEP manipulation de documents *f* (*DM*)

documentary: ~ **resource** *n* DATA ressource documentaire *f*

documentation *n* GEN documentation *f*

DOD *abbr* (*direct outward dialing AmE, direct outward dialling BrE*) GEN, TELEP accès direct au réseau *m*, prise directe du réseau *f*

dog *n* PARTS entraîneur *m*, toc *m*

dolly *n* PARTS *device* tas *m*

domain *n* GEN domaine *m*; ~**-defined attribute** TELEP attribut défini de domaine *m*; ~ **expert** COMP expert du domaine *m*, spécialiste du domaine *mf*; ~ **MCSPDU** TELEP unité MCSPDU de domaine *f*; ~ **name resolver** (*DNR*) DATA résolveur de noms de domaine *m*; ~**-specific part** (*DSP*) DATA partie spécifique du domaine *f*

domestic: ~ **network** *n* DATA réseau domestique *m*

domotics *n pl* APPL domotique *f*

dongle *n* COMP clé électronique *f*

donor *n* PARTS donneur *m*

door: ~ **interphone** *n* PER APP portier automatique *m*, portier téléphonique *m*; ~ **video** *n* PER APP portier vidéo *m*

dope *vt* PARTS doper

doping *n*: ~ **compensation** *n* PARTS compensation par dopage *f*

Doppler: ~ **effect** *n* GEN *physics* effet Doppler *m*; ~ **radar** *n* RADIO radar à effet Doppler *m*, radar Doppler *m*

dormant *adj* COMP en sommeil

dot *n* COMP, TV point *m*; ~ **frequency** TV *video* fréquence de points *f*; ~ **interlace scanning** TV balayage découpé *m*; ~ **interlace sweep** TV balayage découpé *m*; ~ **matrix printer** COMP, PER APP imprimante matricielle par points *f*

dots: ~ **per inch** *phr* (*dpi*) COMP, DATA points par pouce (*ppp*)

double[1]: ~ **current** *adj* TELEP à double courant; ~**-precision** *adj* COMP en double précision

double[2]: ~**-armature relay** *n* PARTS relais à double armature *m*; ~ **Babcock plan** *n* RADIO, TELEP plan en double Babcock *m*; ~**-base diode** *n* PARTS diode à double base *f*; ~**-break contact** *n* PARTS *electric contacts* contact à double rupture *m*; ~ **call** *n* TELEP va-et-vient *m*; ~ **check** *n* COMP double vérification *f*; ~**-circuit line** *n* APPL *outside plant engineering* ligne double *f*; ~ **click** *n* COMP, DATA double clic *m*; ~ **conductor** *n* ELECT, PARTS faisceau double *m*; ~**-conversion converter** *n* TRANS abaisseur de fréquence à double changement de fréquence *m*; ~**-cord switchboard** *n* SWG commutateur à dicorde *m*; ~**-crucible technique** *n* PARTS, TRANS procédé du double creuset *m*; ~ **current** *n* TELEP *telex* double courant *m*; ~**-current code** *n* TELEP, TRANS code de câble bivalent *m*; ~**-current transmission** *n* TRANS transmission par double-courant *f*; ~**-current working** *n* TELEP transmission par double courant *f*; ~ **density** *n* (*DD*) COMP double densité *f* (*DD*); ~**-ended control** *n* CONTR, TRANS commande locale et distante *f*; ~**-ended synchronization** *n* TELEP, TRANS synchronisation bilatérale *f*; ~ **error** *n* TELEP erreur double *f*; ~ **heterojunction** *n* TRANS hétérojonction double *f*; ~**-layer coating** *n* PARTS revêtement bicouche *m*; ~**-length register** *n* COMP, DATA registre double *m*; ~**-make contact** *n* PARTS *electric contacts* contact à double fermeture *m*; ~**-phantom balanced telegraph circuit** *n* TELEP, TRANS circuit télégraphique superfantôme *m*; ~**-phantom circuit** *n* TRANS circuit superfantôme *m*; ~**-reflector aerial** *n* RADIO antenna à double réflecteur *f*; ~**-reflector antenna** *n* RADIO antenne à double réflecteur *f*; ~ **seizure** *n* TELEP prise double *f*; ~ **sideband** *n* (*DSB*) RADIO, TRANS double bande latérale *f* (*DBL*); ~**-sided distribution frame** *n* TRANS répartiteur à double face *m*; ~**-sided printed board** *n* ELECT *printed circuits* carte imprimée double face *f*; ~**-sided rack** *n* TELEP *equipment practice* bâti à double face *m*

doubler *n* ELECT doubleur *m*

DOV *abbr* (*data above voice*) RADIO données supravocales *f pl*

down: ~**-converter** *n* RADIO changeur abaisseur de fréquence *m*; ~ **lead** *n* RADIO *aerials*, TELEP descente d'antenne *f*; ~ **link** *n* RADIO, TELEP liaison descendante *f*, liaison satellite-terre *f*; ~**-link beam** *n* RADIO faisceau d'émission *m*; ~**-link feeder link** *n* RADIO trajet descendant de la liaison de connexion *m*; ~**-link of programmes** *BrE*, ~**-link of programs** *AmE* *n* RADIO réception de programmes *f*; ~ **path** *n* TV voie descendante *f*; ~**-path power** *n* RADIO puissance sur le trajet descendant *f*; ~**-range station** *n* TRANS station aval *f*; ~**-state** *n* TELEP état d'indisponibilité *m*; ~ **time** *n* TELEP durée d'indisponibilité *f*

downbeat *n* COMP temps fort *m*, temps frappé *m*

downgrade *vt* COMP abaisser, GEN déclasser

downline: ~**-loadable** *adj* ELECT téléchargeable

download[1] *n* COMP, DATA téléchargement *m*

download[2] *vt* COMP recevoir, télécharger, DATA, ELECT, TV télécharger

downloadable *adj* COMP, DATA, ELECT, TV téléchargeable

downloading *n* COMP, DATA, ELECT, TV téléchargement *m*

downsize *vt* COMP micromiser

downsizing *n* COMP micromisation *f*

downstream *adv* GEN en aval de

downtime *n* ELECT temps d'immobilisation *m*, TEST durée d'interruption *f*, temps d'indisponibilité *m*

downward[1] *adj* COMP décroissant

downward[2]: ~ **compatibility** *n* COMP compatibilité descendante *f*

DP *abbr* COMP, DATA (*data processing*) TD (*traitement des données*), GEN (*detailed procedure*) procédure détaillée *f*, TELEP (*decimal pulse*) impulsion décimale *f*, TRANS (*distribution point*) branchement *m*

DPCM *abbr* COMP, DATA, TELEP, TRANS (*differential pulse-code modulation*) MICD (*modulation par impulsions et codage différentiel*)

dpi *abbr* (*dots per inch*) COMP, DATA ppp (*points par pouce*)

DQ *abbr* (*directory enquiries*) APPL, TELEP AA (*assistance annuaire*)

DQDB *abbr* (*distributed-queue dual bus*) COMP DQDB

DQDBS *abbr* (*distributed-queue double system*) DATA DQDBS

draft: ~ **text** *n* COMP texte non formaté *m*

drag[1] *n* RADIO traînée *f*; ~ **region** COMP zone de glissement *f*

drag[2]: ~ **and drop** *vt* COMP glisser-déplacer, DATA drag and drop

drain: ~ **region** *n* PARTS *transistors* région de collecteur *f*, région de drain *f*; ~ **valve** *n* PARTS robinet de purge *m*, soupape de purge *f*; ~ **wire** *n* GEN *cables* fil de continuité d'écran *m*

drainage: ~ **coil** *n* POWER bobine d'absorption *f*

draining: ~ **transformer** *n* POWER transformateur suceur *m*

dramatic *adj* ELECT drastique

DRav *abbr* (*average reference-surface diameter*) TELEP diamètre moyen de la surface de référence *m*

draw *vt* COMP dessiner, tracer, GEN dessiner

drawing *n* GEN schéma *m*; ~ **process** TRANS *fibre optics* fibrage *m*

drawmap *n* COMP drawmap *m*

DRCS *abbr* (*dynamically-redefinable characters*) COMP caractères redéfinissables dynamiquement *m pl*

DRF *abbr* (*dispatch-received file*) DATA DRF (*distribution des fichiers reçus*)

drift[1] *n* GEN *tool* broche *f*, broche d'assemblage *f*, RADIO, TRANS dérive *f*; ~ **compensation** TELEP compensation de dérive *f*; ~ **failure** TEST défaillance par dérive *f*, défaillance progressive *f*; ~ **rate** RADIO taux de dérive *m*; ~ **velocity** PARTS *semiconductors* vitesse moyenne de déplacement *f*

drift[2] *vi* GEN dériver

driftpin *n* GEN broche *f*, broche d'assemblage *f*

drill *vt* PARTS percer, perforer

drive[1] *n* COMP lecteur *m*; ~ **distance** ELECT distance de propagation *f*

drive[2] *vt* TV piloter

driven: ~ **element** *n* RADIO élément excité *m*

driver *n* COMP pilote *m*, ELECT *circuit* circuit d'attaque *m*, circuit de commande *m*, driver *m*, pilote *m*, GEN dispositif d'entraînement *m*, dispositif de commande *m*, PARTS entraîneur *m*, toc *m*; ~ **stage** ELECT étage pilote *m*

driving: ~ **belt** *n* PARTS courroie d'entraînement *f*, courroie de commande *f*; ~ **circuit** *n* POWER circuit d'attaque *m*; ~ **current** *n* POWER courant d'attaque *m*, courant de commande *m*; ~ **motor** *n* GEN moteur d'entraînement *m*, moteur de propulsion *m*, moteur de traction *m*; ~ **power** *n* RADIO puissance d'entrée radiofréquence *f*, puissance d'excitation *f*; ~ **stage** *n* ELECT étage d'attaque *m*, circuit d'attaque *m*, circuit de commande *m*

drop[1]: ~ **and insert** *n* TRANS dérivation et insertion *f*, insertion et extraction *f*; ~**-out** *n* COMP perte d'information *f*, TELEP perte de connexion *f*, perte de niveau *f*, perte de signal *f*; ~ **point** *n* TRANS point de branchement *m* (*PB*), point de dérivation *m*; ~ **test** *n* TEST essai de chute *m*

drop[2]: ~ **out** *vi* TELEP perdre la connexion, perdre le signal

drum: ~ **factor** *n* TELEP facteur de cylindre *m*; ~ **machine** *n* COMP boîte à rythmes *f*, batterie électronique *f*; ~ **plotter** *n* COMP traceur à tambour *m*; ~ **receiver** *n* TELEP récepteur à cylindre *m*; ~ **rhythm** *n* COMP rythme de batterie *m*; ~ **sequence** *n* COMP séquence de batterie *f*; ~ **store** *n* SWG mémoire à tambour *f*; ~ **transmitter** *n* TELEP émetteur à cylindre *m*

dry: ~ **battery** *n* POWER pile sèche *f*; ~ **cell** *n* POWER pile sèche *f*; ~**-cell battery** *n* POWER batterie à pile sèche *f*

DS *abbr* DATA (*distributed single-layer*) DS (*monocouche répartie*), (*directory services*) DS (*service d'annuaire*), GEN (*direct sequence*) séquence directe *f*, TRANS (*digital section*) section numérique *f*

DSA *abbr* COMP (*distributed-system architecture*) architecture des systèmes distribués *f*, DATA (*directory system agent*) DSA (*agent de système d'annuaire*)

DSB *abbr* RADIO, TRANS (*double sideband*) DBL (*double bande latérale*), TV (*direct satellite broadcasting*) retransmission directe par satellite *f*

DSC *abbr* (*digital selective call*) RADIO appel sélectif numérique *m*

DSE *abbr* DATA (*distributed single-layer embedded test method*) DSE (*méthode de test monocouche distribuée encastrée*), (*data-switching exchange*) centre de communication de données *m*, ELECT, SWG (*data-switching exchange*) centre de commutation de données *m*, (*digital-switching element*) organe de commutation numérique *m*

DSI *abbr* DATA (*data set identifier*) identificateur de modem *m*, PER APP (*display station interface*) interface de poste d'affichage *f*, TELEP, TRANS (*digital speech interpolation*) CNC (*concentration numérique des conversations*)

DSM *abbr* (*digital storage media*) COMP support de stockage numérique *m*

DSP *abbr* COMP, DATA (*domain-specific part*) partie spécifique du domaine *f*, (*digital signal processor*) DSP (*processeur de traitement numérique des signaux, processeur numérique de signaux*)

D-star *n* ELECT, TEST, TRANS détectivité normée *f*, détectivité spécifique *f*

DTAM: ~ **user** *n* TELEP utilisateur DTAM *m*

DTC *abbr* (*digital-transmit command*) DATA DTC (*commande d'émission numérique*)

DTE *abbr* (*data terminal equipment*) DATA, PER APP, TELEP équipement terminal de données *m*; ~ **busy** *n* TELEP ETTD occupé *m*; ~ **clear request** *n* TELEP demande de libération émise par l'ETTD *f*; ~**-controlled not-ready** *n* TELEP ETTD non-prêt commandé *m*; ~**-inactive registration facility** *n* TELEP service complémentaire d'enregistrement de l'état ETTD *m*; ~**-uncontrolled not-ready** *n* TELEP ETTD non-prêt automatique *m*; ~ **waiting** *n* TELEP ETTD en attente *m*

DTMF *abbr* (*dual-tone multifrequency*) TELEP multifréquence à deux tonalités

DTP *abbr* (*desktop publishing*) APPL, COMP, DATA PAO (*publication assistée par ordinateur*), (*desktop publishing system*) ELECT PAO (*publication assistée par ordinateur*)

DTS *abbr* (*decoding time stamp*) COMP estampille de décodage *f*

DTV *abbr* (*desktop video*) COMP, DATA VAO (*vidéo assistée par ordinateur*), vidéotique *f*

dual[1] *adj* GEN double; ~**-ring** ELECT à anneau double; ~**-standard** GEN bistandard

dual[2]: ~**-attaching copper station** *n* ELECT station à cuivre à raccordement double *f*; ~**-attaching fiber station** *AmE*, ~**-attaching fibre station** *BrE* *n* ELECT station à raccordement double *f*; ~ **attachment** *n* ELECT double connexion *f*; ~**-beam antenna** *n* DATA antenne à double faisceau *f*; ~**-beam oscilloscope** *n* ELECT oscilloscope à double faisceau *m*; ~**-carrier algorithm** *n* COMP, RADIO, TRANS algorithme à deux porteurs *m*; ~**-carrier transmission** *n* TRANS transmission par double porteuse *f*; ~**-channel mode** *n* COMP mode à deux canaux *m*; ~**-channel visual direction finding** *n* RADIO radiogoniométrie visuelle à deux canaux de réception *f*; ~**-conversion receiver** *n* RADIO récepteur à double changement de fréquence *m*; ~**-frequency channel** *n* RADIO canal bifréquence *m*; ~ **front end** *n* RADIO entrée de réception double *f*; ~ **homing** *n* ELECT liaison double attachement *f*; ~**-in-line package** *n* (*DIP*) DATA boîtier à deux rangées de broches *m* (*boîtier DIP*); ~**-launch system** *n* RADIO système de lancement double *m*; ~ **path** *n* TRANS trajet double *m*; ~**-polarized antenna** *n* RADIO antenne à double polarisation *f*; ~**-processor load-sharing system** *n* SWG système à deux processeurs en partage de charge *m*; ~**-processor system** *n* SWG système à deux processeurs *m*, système à double processeur *m*; ~**-purpose connector** *n* ELECT connecteur à double usage *m*; ~**-purpose wire insertion cap** *n* ELECT bouchon d'insertion de câble double *m*; ~ **ring of trees** *n* ELECT réseau en anneau double arborescent *m*; ~ **seizure** *n* TELEP prise simultanée *f*; ~**-signaling telephone** *AmE*, ~**-signalling telephone** *BrE* *n* TELEP poste à clavier mixte *m*; ~ **telephone numbers** *n pl* TELEP service de numéros de téléphone doubles *m*; ~**-tone multifrequency** *n* (*DTMF*) TELEP multifréquence à deux tonalités *f*; ~**-tone multi-frequency dialing** *AmE*, ~**-tone multi-frequency dialling** *BrE* *n* TELEP numérotation multifréquence *f*; ~**-tone multi-frequency signaling** *AmE*, ~**-tone multi-frequency signalling** *BrE* *n* TELEP signalisation multifréquence *f*

dub *vt* GEN surnommer

dubbing *n* GEN doublage *m*

duct *n* ELECT alvéole *f*, canalisation *f*, conduit *m*, PARTS conduit *m*, goulotte *f*, TRANS alvéole *f*, canalisation *f*, conduit *m*; ~ **cable** GEN câble en canalisation *m*; ~ **height** RADIO hauteur du conduit *f*; ~ **intensity** RADIO intensité du conduit *f*; ~ **route** GEN, TRANS *cable laying* trajet de canalisation *m*; ~ **run** GEN, TRANS trajet de canalisation *m*; ~ **system** GEN, TRANS système de canalisations *m*

ducting *n* RADIO propagation troposphérique guidée *f*; ~ **layer** RADIO couche de guidage *f*

dumb: ~ **terminal** *n* DATA terminal passif *m*

dummy[1] *adj* GEN factice

dummy[2] *n* GEN organe fictif *m*; ~ **aerial** RADIO antenne fictive *f*; ~ **antenna** RADIO antenne fictive *f*; ~ **bit** DATA bit de bourrage *m*, bit factice *m*; ~ **capacitance** TRANS capacité fictive *f*; ~ **fiber** *AmE*, ~ **fibre** *BrE* TRANS fibre amorce *f*, fibre d'injection *f*; ~ **front** TELEP *equipment practice* faux panneau *m*; ~ **load** RADIO charge fictive *f*; ~ **panel** TELEP *equipment practice* faux panneau *m*; ~ **strip** PARTS fausse plaque *f*, plaque de remplissage *f*; ~ **unit** TRANS unité de substitution *f*, unité fictive *f*

dump[1] *n* COMP vidage de mémoire *m*; ~ **request** COMP demande de vidage *f*; ~ **routine** COMP programme de vidage *m*

dump[2] *vt* COMP télécharger, vider, clicher, DATA, ELECT, TV télécharger

dumping *n* COMP, DATA, ELECT téléchargement *m*

duplex[1] *adj* COMP bidirectionnel, GEN duplex, RADIO, TELEP bidirectionnel simultané, bilatéral simultané

duplex: ~**-channel group system** *n* RADIO système à groupement de voies duplex *m*; ~ **line** *n* RADIO ligne duplex *f*, TRANS ligne duplex *f*; ~ **traffic** *n* RADIO trafic en mode duplex *m*; ~ **working** *n* RADIO, TELEP fonctionnement en duplex *m*

duplexer *n* RADIO, SWG, TRANS duplexeur *m*

duplicate: ~ **equipment** *n* TELEP équipement dupliqué *m*, équipement en double *m*; ~ **operation** *n* TELEP fonctionnement en dupliqué *m*

duplication *n* GEN reproduction *f*

durability *n* GEN durabilité *f*

duration *n* COMP, GEN durée *f*

dust: ~ **cover** *n* ELECT obturateur antipoussière *m*, PARTS capot antipoussière *m*; ~**-free room** *n* TEST salle dépoussiérée *f*

dustproof *adj* GEN étanche à la poussière

dustproofing: ~ **kit** *n* ELECT kit antipoussière *m*

duty: ~ **cycle** *n* POWER coefficient d'utilisation *m*, temps de mise sous tension *m*, RADIO coefficient d'utilisation *m*, facteur d'utilisation *m*, pourcentage d'utilisation *m*, rapport cyclique *m*

DUV *abbr* (*data under voice*) DATA, RADIO données infravocales *f pl*

DVTR *abbr* (*digital video tape recorder*) COMP magnétoscope numérique *m*

DW *abbr* (*device waiting*) TELEP dispositif en attente

DX: **~-stuffing signal** *n* SIG signal de bourrage DX *m*

DXC *abbr* SWG (*digital cross-connect equipment*) équipement d'interconnexion *m*, (*digital cross-connect*) répartiteur numérique *m*

dynamic[1] *adj* GEN dynamique, RADIO agile en fréquence

dynamic[2] *n* GEN dynamique *f*; **~ allocation** COMP attribution dynamique des ressources *f*; **~ allocation of frequency** GEN allocation dynamique de fréquences *f*; **~ buffering** COMP stockage dynamique en mémoire tampon *m*; **~ chaining** DATA chaînage dynamique *m*; **~ channel allocation** RADIO allocation dynamique de fréquences *f*; **~ characteristic** COMP caractéristique dynamique *f*; **~ compression** COMP compression dynamique *f*; **~ conformance requirement** DATA condition de conformité dynamique *f*; **~ data exchange** (*DDE*) DATA échange de données dynamiques *m*; **~ keyboard** COMP clavier dynamique *m*; **~-link library** (*DLL*) TELEP bibliothèque de liens dynamiques *f* (*DLL*); **~ load control** SWG contrôle dynamique de la charge *m*; **~ measurement** TEST mesurage dynamique *m*; **~ noise** TELEP *acoustics* bruit dynamique *m*; **~ part** DATA partie dynamique *f*; **~ RAM** (*DRAM*) COMP, DATA mémoire vive dynamique *f*; **~ range** TEST gamme dynamique *f*; **~ relocation** COMP translation dynamique *f*; **~ resource allocation** COMP attribution dynamique des ressources *f*, affectation dynamique des ressources *f*; **~ response** GEN réponse dynamique *f*; **~ stability** GEN stabilité dynamique *f*; **~ storage** COMP mémoire dynamique *f*; **~ storage allocation** COMP attribution dynamique de mémoire *f*

dynamically: **~-redefinable characters** *n pl* (*DRCS*) COMP caractères redéfinissables dynamiquement *m pl*

E

E: ~ **mode** *n* RADIO mode E *m*

EA *abbr* (*external access equipment*) TELEP équipement d'accès externe *m*

E&M: ~ **signaling** *AmE*, ~ **signalling** *BrE* *n* TELEP signalisation TRON et RON *f*

ear: ~ **reference point** *n* TELEP point de référence oreille *m*

earcap *n* PER APP, TELEP pavillon *m*, pavillon de récepteur *m*

earlier: ~**-transmitted bits** *n* TRANS bits transmis précédemment *m*; ~**-transmitted bits scrambling process** *n* TELEP bits transmis précédemment en brouillage *m*

early: ~ **deployment** *n* GEN déploiement anticipé *m*; ~ **failure** *n* TEST défaillance précoce *f*; ~ **failure period** *n* TEST période des défaillances précoces *f*

earphone *n* COMP, PER APP écouteur *m*, TELEP écouteur téléphonique *m*, écouteur *m*

earpiece *n* COMP écouteur *m*, PER APP pavillon *m*, écouteur *m*, pavillon de récepteur *m*, TELEP pavillon *m*, écouteur *m*

earth[1] *adj BrE* (*cf ground AmE*) ELECT terrien; **off** ~ POWER isolé de la terre

earth[2] *n BrE* (*cf ground AmE*) ELECT, POWER masse *f*, terre *f*; ~ **acquisition** RADIO acquisition d'un repère terrestre *f*; ~ **bar** *BrE* (*cf ground bar AmE*) ELECT barre de masse *f*, POWER barre de mise à la terre *f*; ~ **button** *BrE* (*cf ground button AmE*) SWG bouton de mise à la terre *m*; ~ **cable** *BrE* (*cf ground cable AmE*) POWER câble de mise à la terre *m*; ~ **clamp** *BrE* (*cf ground clamp AmE*) POWER bride de masse *f*, collier de terre *m*; ~ **collector** *BrE* (*cf ground collector AmE*) POWER collecteur de terre *m*; ~ **collector bar** *BrE* (*cf ground collector bar AmE*) POWER collecteur de terre *m*; ~ **conductor** *BrE* (*cf ground conductor AmE*) POWER conducteur de masse *m*, conducteur de terre *m*; ~ **connection** *BrE* (*cf ground connection AmE*) POWER connexion de terre *f*; ~ **electrode** *BrE* (*cf ground electrode AmE*) POWER électrode de terre *f*, prise de terre *f*; ~ **electrode installation** *BrE* (*cf ground electrode installation AmE*) POWER installation de prise de terre *f*; ~ **electrode resistance** *BrE* (*cf ground electrode resistance AmE*) POWER résistance de prise de terre *f*; ~ **fault** *BrE* (*cf ground fault AmE*) POWER défaut à la masse *m*, défaut à la terre *m*; ~**-fault circuit breaker** *BrE* (*cf ground-fault circuit breaker AmE*) POWER interrupteur de défaut à la terre *m*; ~**-fault protection** *BrE* (*cf ground-fault protection AmE*) POWER protection contre défauts à la terre *f*; ~**-fault relay** *BrE* (*cf ground-fault relay AmE*) ELECT, PARTS relais de défaut à terre *m*; ~**-generated beacon** RADIO signaux de balise émis à partir de la terre *m pl*; ~**-generated pilot** RADIO onde pilote émise du sol *f*; ~ **lead** *BrE* (*cf ground lead AmE*) POWER fil de masse *m*; ~ **leakage** *BrE* (*cf ground leakage AmE*) POWER fuite à la terre *f*; ~ **leakage current** *BrE* (*cf ground leakage current AmE*) POWER courant de fuite à la terre *m*; ~ **leakage guard** *BrE* (*cf ground leakage guard AmE*) POWER détecteur de fuite à la terre *m*; ~ **loop** *BrE* (*cf ground loop AmE*) ELECT circuit de masse *m*, retour à la masse *m*; ~**-orbiting artificial satellite** RADIO satellite artificiel en orbite terrestre *m*; ~**-oriented geosynchronous satellite** RADIO satellite synchrone orienté vers la terre *m*; ~ **plane** *BrE* (*cf ground plane AmE*) POWER plan de sol *m*; ~ **plate** *BrE* (*cf ground plate AmE*) POWER plaque de terre *f*; ~ **potential** *BrE* (*cf ground potential AmE*) POWER potentiel du sol *m*; ~ **potential difference** *BrE* (*cf ground potential difference AmE*) POWER différence de potentiel entre deux terres *f*; ~ **radial termination** *BrE* (*cf ground radial termination AmE*) POWER connexion du réseau de terre radial *f*; ~ **resistance** *BrE* (*cf ground resistance AmE*) POWER résistance de terre *f*; ~ **return** *BrE* (*cf ground return AmE*) SIG, TRANS retour par la terre *m*; ~**-return circuit** *BrE* (*cf ground-return circuit AmE*) TRANS circuit à retour par la terre *m*; ~**-return double phantom circuit** *BrE* (*cf ground-return double phantom circuit AmE*) TELEP circuit approprié de fantôme *m*; ~**-return phantom circuit** *BrE* (*cf ground-return phantom circuit AmE*) TELEP circuit approprié *m*; ~ **ring collector** *BrE* (*cf ground ring collector AmE*) POWER collecteur annulaire de mise à la terre *m*; ~ **sensor** RADIO détecteur d'horizon terrestre *m*; ~ **shadow** RADIO zone d'ombre créée par la terre *f*; ~ **station** GEN station terrienne *f*, RADIO *satellite communications* station terrestre *f*, station terrienne *f*; ~ **station record** RADIO liste des stations terriennes *f*; ~ **station verification and assistance** (*ESVA*) RADIO vérification et assistance des stations terriennes *f pl*; ~**-synchronous satellite** RADIO satellite géosynchrone *m*; ~ **system** *BrE* (*cf ground system AmE*) POWER installation de mise à la terre *f*, RADIO plan de sol *m*; ~ **terminal** *BrE* POWER borne de mise à la terre *f*; ~ **wire** *BrE* (*cf ground wire AmE*) POWER câble de garde *m*, fil de masse *m*, fil de terre *m*

earth[3] *vt BrE* (*cf ground AmE*) ELECT, POWER mettre à la masse, mettre à la terre

Earth: ~**-facing panel** *n* RADIO panneau terre *m*; ~**-to-orbit shuttle** *n* GEN navette spatiale *f*

earthed[1] *adj BrE* (*cf grounded AmE*) ELECT, POWER mis à la masse, mis à la terre

earthed[2]: ~ **emitter circuit** *n BrE* (*cf grounded emitter circuit AmE*) ELECT circuit à émetteur commun *m*, montage à émetteur commun *m*; ~ **emitter stage** *n BrE* (*cf grounded emitter stage AmE*) ELECT *semiconductors* étage à émetteur commun *m*; ~**-grid amplifier** *n BrE* (*cf grounded-grid amplifier AmE*) ELECT amplificateur à grille à la masse *m*

earthing *n BrE* (*cf grounding AmE*) ELECT, POWER mise à la masse *f*, mise à la terre *f*; ~ **arrangement** *BrE* (*cf grounding arrangement AmE*) POWER dispositif de mise à la terre *m*; ~ **button** *BrE* (*cf grounding button AmE*) SWG *PBX* bouton de mise à la terre *m*; ~ **conductor** *BrE* (*cf grounding conductor AmE*) POWER conducteur de masse *m*, conducteur de terre *m*; ~ **indents** *n pl BrE* (*cf grounding indents AmE*) ELECT dentures de masse *f pl*; ~ **network** *BrE* (*cf grounding network AmE*) POWER réseau de terre *m*; ~ **plate** *BrE* (*cf grounding plate AmE*) POWER plaque de terre *f*; ~ **probe** *BrE* (*cf grounding probe AmE*) POWER piquet de mise à la terre *m*, sonde de mise à la terre *f*; ~ **rod** *BrE* (*cf grounding rod AmE*) POWER piquet de mise à la terre *m*, tige de mise à la terre *f*; ~ **switch** *BrE* (*cf grounding switch AmE*) POWER sectionneur de terre *m*

earthquake: ~**-resistant** *adj* TEST *construction* antisismique

easy[1]: ~**-to-use** *adj* COMP facile d'emploi

easy[2]: ~ **access** *n* COMP option d'accès *f*; ~ **payment** *n* GEN règlement simplifié *m*; **very** ~ **rodent-oriented Net-wide index to computerized archives** *n* (*Veronica*) DATA Veronica

eavesdropping *n* COMP écoute clandestine *f*, RADIO, TELEP écoute indiscrète *f*

EB *abbr* DATA (*erroneous block*) bloc erroné *m*, ELECT (*electron beam*) faisceau électronique *m*

EBCDIC *abbr* (*extended binary coded decimal interchange* *code*) DATA, TRANS EBCDIC (*code decimal codé binaire étendu*)

ebonite *n* PARTS ébonite *f*

EBU *abbr* (*European Broadcasting Union*) RADIO UER (*Union européenne de radio-télévision*)

EC *abbr* TELEP (*escape code*) code d'échappement *m*

ECC *abbr* COMP (*error-correcting code*) CCE (*code correcteur d'erreurs*), DATA (*embedded control channel*) canal de commande intégré *m*, (*error-correcting code*) CCE (*code correcteur d'erreurs*), SIG, TELEP (*error-correcting code*) CCE (*code correcteur d'erreurs*)

ECD *abbr* (*error-control device*) DATA dispositif de protection contre les erreurs *m*

echelon: ~ **telegraphy** *n* TELEP échelonné *m*, communication échelonnée *f*, faisceau échelonné *m*, installation échelonnée *f*

echo *n* GEN écho *m*; ~ **balance return loss** TELEP affaiblissement d'équilibrage pour l'écho *m*; ~ **box** RADIO boîte à écho *f*; ~ **cancelation** *AmE see echo cancellation BrE*; ~ **canceler** *AmE see echo canceller BrE*; ~ **canceling** *AmE see echo cancelling BrE*; ~ **cancellation** *BrE* ELECT annulation d'écho *f*, compensation d'écho *f*, RADIO annulation d'écho *f*, TRANS annulation d'écho *f*, compensation d'écho *f*; ~ **canceller** *BrE* ELECT annuleur d'écho *m*, compensateur d'écho *m*, suppresseur d'écho *m*, RADIO, TRANS annuleur d'écho *m*, compensateur d'écho *m*; ~ **cancelling** *BrE* ELECT, RADIO, TRANS annulation d'écho *f*; ~ **control** RADIO, TELEP réduction de l'écho *f*; ~-**control device** RADIO dispositif de protection anti-écho *m*, TELEP dispositif de réduction écho *m*; ~ **curve** TELEP courbe d'écho *f*; ~ **effect** COMP, TELEP effet d'écho *m*; ~ **loss** TELEP affaiblissement d'écho *m* (*AECHO*); ~ **protection** RADIO protection contre les échos *f*; ~ **return** DATA renvoi en écho *m*; ~ **suppression** ELECT, RADIO, TRANS annulation d'écho *f*; ~ **suppressor** (*ES*) DATA gabarit d'écho *m*, RADIO, TELEP, TRANS suppresseur d'écho *m*; ~-**suppressor indicator** TELEP indicateur de suppresseur d'écho *m*

echoless: ~ **chamber** *n* TEST chambre anéchoïque *f*

echometric: ~ **measurement** *n* TELEP mesure échométrique *f*

echoplex: ~ **mode** *n* TELEP mode échoplex *m*

ECL *abbr* (*emitter-coupled logic*) DATA, ELECT ECL (*logique à couplage par les émetteurs*)

eclipse: ~ **capability** *n* RADIO capacité de fonctionnement en période d'éclipse *f*; ~ **interference** *n* RADIO brouillage dû aux éclipses du soleil *m*; ~-**working transponder** *n* RADIO répéteur protégé en éclipse *m*

ECM *abbr* (*error-correction mode*) TELEP mode avec correction d'erreur *m*

ECMA *abbr* (*European Computer Manufacturers Association*) COMP ECMA (*Association européenne de constructeurs d'ordinateurs*)

ECN *abbr* (*explicit congestion notification*) DATA, SWG, TELEP notification d'encombrement explicite *f*

ECO *abbr* (*engineering change order*) PARTS OMT (*ordre de modification technique*)

economic: ~ **standard antenna** *n* RADIO antenne normale économique *f*

economy: ~ **of scale** *n* GEN économie d'échelle *f*

ECPT *abbr* (*European Conference of Postal and Telecommunications Administrations*) GEN CEPT (*Conférence européenne des administrations des postes et des télécommunications*)

EDA *abbr* (*electronic design automation*) PARTS CAO (*conception assistée par ordinateur*)

EDC *abbr* COMP, DATA, TELEP (*error-detecting code, error-detection code*) CDE (*code détecteur d'erreurs*)

eddy: ~ **current** *n* POWER courant de Foucault *m*; ~ **current loss** *n* POWER perte par courants parasites *f*

edge[1]: ~-**emitting** *adj* TRANS *fibre optics* à émission par la tranche

edge[2] *n* COMP périphérie *f*, ELECT *printed circuits* bord *m*, extrémité *f*, GEN *tools* fil *m*; ~-**board contact** ELECT contact d'extrémité de carte *m*; ~ **connector** ELECT connecteur de bord *m*, connecteur de bout de carte *m*, connecteur encartable *m*, GEN connecteur plat *m*; ~ **curl** COMP bord recourbé *m*; ~ **detection** DATA détection des contours *f*; ~ **distance** ELECT *printed circuits* distance de bord *f*; ~-**emitting diode** ELECT diode à émission par la tranche *f*, TRANS *fibre optics* diode à émission par la tranche *f*; ~-**emitting LED** TRANS *fibre optics* DEL à émission longitudinale *f*; ~ **flare** COMP surluminance de bord *f*; ~-**of-coverage gain** RADIO gain en bord de couverture *m*; ~-**triggered bistable** ELECT *logic elements* bascule déclenchée par front d'impulsion *f*; ~ **warping** ELECT *printed circuits* gauchissement des bords *m*

EDI *abbr* COMP, DATA, TELEP (*electronic data interchange*) EDI (*échange de données informatisées*); ~ **access unit** *n* (*EDI-AU*) TELEP unité d'accès EDI *f*; ~-**forwarding** *n* DATA, TELEP retransmission EDI *f*; ~ **message** *n* (*EDIM*) DATA, TELEP message EDI *m*; ~ **message store** *n* (*EDI-MS*) DATA, TELEP mémoire de messages EDI *f*; ~ **messaging** *n* (*EDIMG*) DATA, TELEP messagerie EDI *f*; ~-**messaging system** *n* (*EDIMS*) APPL, DATA, RADIO, TELEP système de messagerie EDI *m*; ~-**messaging user** *n* DATA, TELEP utilisateur de la messagerie EDI *m*; ~ **notification** *n* (*EDIN*) DATA, TELEP notification EDI *f*; ~ **responsibility** *n* DATA, TELEP responsabilité EDI *f*; ~ **user** *n* DATA, TELEP utilisateur EDI *m*; ~ **user agent** *n* (*EDI-UA*) APPL, DATA, TELEP agent utilisateur EDI *m*

EDI-AU *abbr* (*EDI access unit*) APPL, DATA, TELEP unité d'accès EDI *f*

EDIFACT *abbr* (*electronic data interchange for administration, commerce and transport*) DATA, TELEP EDIFACT (*échange de données informatisées pour l'administration, le commerce et le transport*)

EDIM *abbr* (*EDI message*) DATA, TELEP message EDI *m*; ~ **responsibility** *n* DATA, TELEP responsabilité EDIM *f*

EDIMG *abbr* (*EDI messaging*) DATA, TELEP messagerie EDI *f*; ~ **user** *n* TELEP utilisateur de la messagerie EDI *m*

EDIMS *abbr* (*EDI messaging system*) DATA, TELEP système de messagerie EDI *m*

EDIN *abbr* (*EDI notification*) DATA, TELEP notification EDI *f*

edit[1]: ~ **buffer** *n* COMP tampon d'édition *m*; ~ **buffer initialization** *n* COMP initialisation du tampon d'édition *f*; ~ **button** *n* COMP touche d'édition *f*; ~ **parameter** *n* COMP paramètre d'édition *m*; ~ **recall** *n* COMP rappel du dernier son édité *m*

edit[2] *vt* COMP, GEN éditer, rédiger

editing *n* COMP édition *f*; ~ **procedure** COMP, GEN procédure d'édition *f*

edition *n* GEN *documentation* édition *f*

editor *n* COMP, GEN éditeur *m*

EDI-UA *abbr* (*EDI user agent*) TELEP agent utilisateur d'EDI *m*

EDP *abbr* (*electronic data processing*) COMP, DATA TED (*traitement électronique des données*); ~ **analyst** *n* GEN analyste en informatique *m*

educational: ~ **software** *n* COMP, DATA logiciel éducatif *m*

EETDN *abbr* (*end-to-end transit delay negotiation*) DATA, TELEP négociation de temps de transit de bout en bout *f*

EF *abbr* (*entrance facility*) ELECT équipement d'entrée *m*

EFdS *abbr* (*error-free decisecond*) TELEP déciseconde sans erreur *m*

effect *n* COMP, DATA effet *m*; ~ **memory** COMP mémoire d'effets *f*

effective: ~ **address** *n* COMP adresse réelle *f*; ~ **antenna height** *n* RADIO hauteur effective d'antenne *f*, hauteur équivalente d'antenne *f*; ~ **area** *n* RADIO surface effective *f*, surface de captation *f*, surface équivalente *f*; ~ **call attempt** *n* RADIO tentative d'appel efficace *f*; ~ **capacity** *n*

TELEP capacité effective *f*; ~ **Earth-radius factor** *n* RADIO facteur multiplicatif du rayon terrestre *m*; ~ **gain** *n* TRANS gain équivalent *m*; ~ **height** *n* RADIO hauteur effective *f*; ~ **height above average terrain** *n* (*EHAAT*) RADIO hauteur effective au-dessus du sol moyen *f* (*HEASM*); ~ **input admittance** *n* POWER admittance effective d'entrée *f*; ~ **input capacitance** *n* POWER capacité effective d'entrée *f*; ~ **input impedance** *n* POWER impédance effective d'entrée *f*; ~ **instruction** *n* COMP instruction effective *f*; ~ **isotropically-radiated power** *n* RADIO puissance isotrope rayonnée effective *f*; ~ **loss** *n* TRANS perte effective *f*; ~ **mode volume** *n* TRANS volume optique des modes *m*; ~ **monopole-radiated power** *n* (*emrp*) RADIO puissance apparente rayonnée sur antenne verticale courte *f* (*parv*); ~ **number of bits** *n* (*ENOB*) DATA nombre de bits effectif *m*; ~ **output admittance** *n* POWER admittance effective de sortie *f*; ~ **output capacitance** *n* POWER capacité effective de sortie *f*; ~ **output impedance** *n* POWER impédance effective de sortie *f*; ~ **overall noise band** *n* TRANS bande effective globale de bruit *f*; ~ **radiated power** *n* (*erp*) RADIO puissance apparente rayonnée *f* (*par*); ~ **radius of the Earth** *n* RADIO rayon terrestre équivalent *m*; ~ **selectivity** *n* RADIO sélectivité effective d'un récepteur *f*; ~ **traffic** *n* GEN, SWG trafic efficace *m*

effectively: ~ **transmitted signals in sound-program** *AmE*, ~ **transmitted signals in sound-programme** *BrE n pl* SIG signaux effectivement transmis dans les transmissions *m pl*

effectiveness *n* GEN efficacité *f*

effector *n* COMP unité d'effets spéciaux *f*

efficiency *n* GEN rendement *m*, efficacité *f*; ~ **factor in time** TELEP facteur d'efficacité dans le temps *m*; ~ **ratio** TELEP taux d'efficacité *m*

E-field: ~ **antenna** *n* RADIO antenne champ E *f*

EFIR *abbr* (*equiripple finite-impulse response filter*) ELECT filtre réponse impulsionnelle finie ondulation *m*

EFT *abbr* (*electronic funds transfer*) APPL, DATA, TELEP télépaiement *m*, transfert électronique de fonds *m*

EFTPOS *abbr* (*electronic funds transfer at point of sale*) APPL, DATA, TELEP transfert électronique de fonds au point de vente *m*

EG: ~ **select** *n* (*envelope-generator select*) COMP sélection de générateur d'enveloppes *f*

EGA *abbr* (*enhanced graphics adaptor*) COMP EGA (*adaptateur EGA*)

egg: ~**-box lens** *n* RADIO lentille multicellulaire *f*

EHAAT *n* (*effective height above average terrain*) RADIO HEASM (*hauteur effective au-dessus du sol moyen*)

EHF: ~ **satcom project** *n* TRANS projet du système de télécommunications par satellite *m*

EID *abbr* (*escape in data*) DATA EDD (*échappement dans les données*)

eight: ~ **to fourteen modulation** *n* (*ETF*) COMP modulation de huit à quatorze *f* (*ETF*); ~ **to ten modulation** *n* (*ETM*) COMP modulation de huit à dix *f* (*ETM*)

EIRP *abbr* (*equivalent isotropically-radiated power*) RADIO PIRE (*puissance isotrope rayonnée équivalente*)

EISA *abbr* (*extended industry standard architecture*) COMP EISA (*architecture EISA*)

EIT *abbr* (*encoded information type*) DATA type d'informations codées *m*

eject[1]: ~ **button** *n* COMP, DATA bouton d'éjection *m*

eject[2] *vt* COMP, DATA *disk* éjecter

elasticity *n* GEN élasticité *f*

elastomer: ~ **contact-strip** *n* PARTS ruban de contact en élastomère *m*

elastometer *n* TEST élastomètre *m*

electret *n* PARTS, POWER électret *m*; ~ **microphone** TELEP microphone électret *m*

electric: ~ **arc** *n* POWER arc électrique *m*; ~ **breakdown** *n* ELECT *semiconductors* claquage *m*, PARTS *semiconductors* claquage *m*, rupture *f*, claquage du diélectrique *m*; ~ **cord** *n* COMP cordon électrique *m*; ~ **current** *n* POWER courant électrique *m*; ~ **field** *n* ELECT champ électrique *m*; ~ **measuring instrument** *n* TEST instrument de mesure électrique *m*; ~ **migration** *n* POWER migration électrique *f*; ~ **polarization** *n* POWER polarisation électrique *f*; ~ **power** *n* POWER énergie électrique *f*; ~ **power station** *n* POWER centrale électrique *f*; ~ **screen** *n* POWER écran électrique *m*; ~ **shock** *n* POWER choc électrique *m*; ~ **signal transducer** *n* POWER transducteur électrique *m*; ~ **vector** *n* TRANS *fibre optics* vecteur électrique *m*

electrical: ~ **admittance** *n* ELECT admittance électrique *f*; ~ **capacitance** *n* ELECT capacité électrique *f*; ~ **charge** *n* ELECT charge électrique *f*; ~ **circuit** *n* ELECT, POWER circuit électrique *m*; ~ **conductance** *n* ELECT conductance électrique *f*; ~ **data** *n pl* TRANS *fibre optics* caractéristiques électriques *f pl*; ~ **engineer** *n* DATA, GEN ingénieur électricien *m*; ~ **engineering** *n* DATA, GEN génie électrique *m*, POWER électrotechnique *f*; ~ **equipment** *n* DATA matériel électrique *m*; ~ **fitter** *n* GEN monteur-électricien *m*; ~ **integration unit** *n* ELECT dispositif d'intégration électrique *m*; ~ **interference** *n* COMP interférence électrique *f*; ~ **length** *n* RADIO longueur électrique *f*; ~ **model** *n* ELECT maquette électrique *f*; ~ **path delay** *n* TRANS temps de propagation électrique *m*; ~ **path equalization** *n* TRANS égalisation des trajets électriques *f*; ~ **path length** *n* TRANS longueur de trajet électrique *f*; ~ **power** *n* POWER puissance électrique *f*; ~ **power system** *n* POWER réseau d'énergie électrique *m*; ~ **resistance** *n* ELECT résistance électrique *f*; ~ **subscriber-busy signal** *n* SIG signal électrique d'abonné occupé *m*, signal d'abonné occupé *m*; ~ **tape** *n* POWER ruban isolant électrique *m*; ~ **wavelength** *n* RADIO longueur électrique *f*; ~ **zero** *n* TEST *measuring device* zéro électrique *m*

electrically: ~ **erasable programmable read-only memory** *n* (*EEPROM*) ELECT mémoire morte effaçable électriquement *f*; ~**-held crosspoint** *n* SWG point d'intersection à maintien électrique *m*

electrician *n* GEN installateur-électricien *m*

electricity: ~ **supply system** *n* POWER réseau d'énergie électrique *m*

electrification *n* TEST électrisation *f*

electro[1]: ~**-optic** *adj* ELECT, SWG, TRANS électro-optique; ~**-optical** *adj* ELECT, SWG, TRANS *fibre optics* électro-optique

electro[2]: ~**-optic effect** *n* TRANS *fibre optics* polissage électrolytique *m*, effet électro-optique *m*; ~**-optic switch** *n* SWG commutateur électro-optique *m*

electroacoustic *adj* TELEP électroacoustique

electroacoustics *n* TELEP électroacoustique *f*

electrochemical: ~ **capacitor** *n* PARTS condensateur électrochimique *m*; ~ **dissociation** *n* GEN dissociation électrochimique *f*

electrode *n* ELECT, POWER électrode *f*; ~ **admittance** ELECT admittance d'électrode *f*; ~ **capacitance** ELECT capacité d'électrode *f*; ~ **conductance** ELECT conductance d'électrode *f*; ~ **configuration** ELECT configuration d'électrode *f*; ~ **dark current** ELECT courant d'obscurité *m*; ~ **dissipation** ELECT dissipation d'électrode *f*; ~ **impedance** ELECT impédance d'électrode *f*; ~ **potential** ELECT potentiel d'électrode *m*; ~ **reactance** ELECT réactance d'électrode *f*; ~ **resistance** ELECT résistance d'électrode *f*

electrodynamic[1] *adj* ELECT, TELEP *electromagnetism* électrodynamique

electrodynamic[2]: ~ **microphone** *n* TELEP microphone à conducteur mobile *m*, microphone électrodynamique *m*

electroforming *n* ELECT *chemistry* électroformage *m*

electrographic: ~ **printer** *n* PER APP imprimante électrographique *f*

electroless: ~ **plating** *n* ELECT *printed circuits* dépôt par autocatalyse *m*, dépôt sans courant *m*

electroluminescence *n* ELECT, TRANS *fibre optics* électroluminescence *f*
electrolyte *n* ELECT électrolyte *m*
electrolytic: ~ **capacitor** *n* PARTS condensateur électrolytique *m*; ~ **cell** *n* ELECT cellule électrolytique *f*; ~ **dissociation** *n* GEN dissociation électrolytique *f*; ~ **dissociation constant** *n* GEN constante de dissociation électrolytique *f*
electromagnet *n* PARTS, POWER électroaimant *m*
electromagnetic[1] *adj* GEN électromagnétique
electromagnetic[2]: ~ **beam** *n* RADIO *aerials* faisceau électromagnétique *m*; ~ **disturbance** *n* ELECT parasite électromagnetique *m*, perturbation électromagnetique *m*, RADIO perturbation électromagnétique *f*, TEST brouillage électromagnétique *m*, parasite électromagnétique *m*, perturbation électromagnétique *f*; ~ **field** *n* ELECT champ électromagnétique *m*; ~ **force** *n* ELECT, POWER force électromagnétique *f*; ~ **interference** *n* (*EMI*) ELECT, TEST interférence électromagnétique *f*, brouillage électromagnétique *m*, parasite électromagnétique *m*, perturbation électromagnétique *f*; ~ **microphone** *n* TELEP microphone électromagnétique *m*; ~ **mode** *n* TRANS mode électromagnétique *m*; ~ **pulse** *n* ELECT impulsion électromagnétique *f*, RADIO pulsion électromagnétique *f*; ~ **radiation** *n* ELECT radiation électromagnétique *f*, TRANS rayonnement électromagnétique *m*; ~ **relay** *n* PARTS relais électromagnétique *m*; ~ **remote sensing** *n* (*EMRS*) TEST télédétection électromagnétique *f* (*TDEM*); ~ **wave** *n* DATA, ELECT onde électromagnétique *f*; ~ **wave propagation** *n* ELECT, RADIO propagation des ondes électromagnétiques *f*
electromechanical[1] *adj* GEN, PARTS, POWER électromécanique
electromechanical[2]: ~ **relay** *n* PARTS relais électromécanique *m*
electromechanics *n* GEN, PARTS, POWER électromécanique *f*
electromotive: ~ **force** *n* (*EMF*) ELECT, POWER force électromotrice *f*
electron *n* GEN électron *m*; ~ **beam** (*EB*) ELECT faisceau électronique *m*; ~ **charge** ELECT charge d'électron *f*; ~ **collision** ELECT collision d'électrons *f*; ~ **conduction** ELECT, PARTS *semiconductors* conduction par électrons *f*; ~ **current** ELECT courant électronique *m*, POWER courant électronique *m*, flux électronique *m*, flux d'électrons *m*; ~ **flow** ELECT courant électronique *m*, POWER courant électronique *m*, flux électronique *m*, flux d'électrons *m*; ~ **gas** GEN gaz électronique *m*; ~ **gun** ELECT canon à électrons *m*; ~ **mass** GEN masse de l'électron *f*; ~ **microscope** ELECT microscope électronique *m*; ~ **microscopy** ELECT microscopie électronique *f*; ~ **shell** GEN couche électronique *f*; ~-**volt** GEN électron-volt *m*
electronic[1] *adj* GEN électronique
electronic[2]: ~ **address** *n* DATA adresse électronique *f*, adresse E-mail *f*; ~ **automatic exchanges** *n pl* SWG autocommutateurs *m pl*; ~ **book** *n* DATA livre électronique *m*; ~ **business set** *n* (*BS*) TELEP téléphone d'affaires électronique *m* (*TA*); ~ **call-signal generator** *n* PER APP sonnerie électronique *f*; ~ **chip** *n* ELECT puce électronique *f*; ~ **circuit** *n* ELECT circuit électronique *m*; ~ **circuitry** *n* ELECT circuits électroniques *m pl*; ~ **clock** *n* ELECT horloge électronique *f*; ~ **commerce** *n* ELECT commerce électronique *m*; ~ **component** *n* COMP, ELECT, PARTS composant électronique *m*; ~ **countermeasure** *n* RADIO contre-mesure électronique *f*; ~ **crosspoint** *n* SWG point d'intersection électronique *m*; ~ **data interchange** *n* (*EDI*) COMP, DATA, TELEP échange de documents informatisés *m* (*EDI*), échange de données informatisées *m* (*EDI*); ~ **data interchange for administration, commerce and transport** *n* (*EDIFACT*) DATA échange de données informatisées pour l'administration, le commerce et le transport *m*; ~ **data processing** *n* (*EDP*) COMP, DATA traitement électronique des données *m* (*TED*); ~ **design automation** *n* (*EDA*) PARTS automatisation de conception électronique *f*, conception assistée par ordinateur *f* (*CAO*); ~ **directory** *n* DATA, TELEP annuaire électronique *m*; ~ **document interchange** *n* (*EDI*) COMP, DATA, TELEP échange de documents informatisés *m* (*EDI*); ~ **document management** *n* DATA gestion électronique de documents *f* (*GED*); ~ **engineering** *n* DATA génie électronique *m*; ~ **equipment** *n* DATA matériel électronique *m*, ELECT équipement électronique *m*; ~ **exchange** *n* COMP commutateur électronique *m*, SWG central électronique *m*, commutateur électronique *m*; ~ **funds transfer** *n* (*EFT*) APPL, DATA, TELEP télépaiement *m*, transfert électronique de fonds *m*; ~ **funds transfer at point of sale** *n* (*EFTPOS*) APPL, DATA, TELEP transfert électronique de fonds au point de vente *m*; ~ **key** *n* DATA sceau électronique *m*, touche à effleurement *f*; ~ **key system** *n* PER APP système d'intercommunications électronique *m*; ~ **lock** *n* RADIO verrou électronique *m*; ~ **mail** *n* (*e-mail*) APPL, COMP, DATA *Internet* courrier électronique *m*, messagerie électronique *f*; ~ **mailbox** *n* APPL, COMP, DATA *Internet* boîte aux lettres électronique *f*; ~ **mailbox service** *n* COMP service de boîte aux lettres électronique *m*; ~ **message service** *n* APPL, COMP, DATA messagerie électronique *f*; ~ **message switch** *n* SWG commutateur électronique de messages *m*; ~ **message system** *n* (*EMS*) APPL système de messagerie électronique *m*; ~ **messaging** *n* APPL, COMP, DATA messagerie électronique *f*; ~ **module** *n* COMP module électronique *m*; ~ **music** *n* COMP musique électronique *f*; ~ **news gathering** *n* (*ENG*) GEN, RADIO, TV journalisme électronique *m*, reportage électronique ENG *m*, reportage électronique d'actualités *m*; ~ **payment** *n* APPL paiement électronique *m*; ~ **payment terminal** *n* APPL terminal de paiement électronique *m*; ~ **pen** *n* DATA crayon électronique *m*; ~ **piano** *n* PER APP piano électronique *m*; ~ **publishing** *n* APPL, COMP, DATA édition électronique *f*; ~ **satellite concentrator** *n* RADIO concentrateur satellite électronique *m*; ~ **storyboard** *n* COMP conducteur visuel électronique *m*, scénario-image électronique *m*; ~ **supervision** *n* APPL surveillance électronique *f*; ~ **switch** *n* COMP, DATA, SWG commutateur électronique *m*; ~ **switching** *n* COMP, SWG commutation électronique *f*; ~ **switching system** *n* (*ESS*) SWG système de commutation électronique *m*; ~ **technology** *n* COMP technologie électronique *f*; ~ **telephone set** *n* TELEP poste téléphonique électronique *m*; ~ **time meter** *n* TEST compteur horaire électronique *m*; ~ **tube** *n* ELECT tube électronique *m*; ~ **tuning** *n* ELECT accord électronique *m*; ~ **valve** *n* ELECT tube électronique *m*
electronically: ~-**despun antenna** *n* RADIO antenne à compensation électronique *f*, antenne à contre-rotation électronique *f*
electronics *n* ELECT, GEN électronique *f*; ~ **technician** GEN électronicien *m*
electrostatic[1] *adj* ELECT, PARTS, PER APP, POWER, TELEP, TEST électrostatique; ~-**discharge-sensitive** TEST sensible aux décharges électrostatiques
electrostatic[2]: ~ **discharge** *n* (*ESD*) ELECT, TEST décharge électrostatique *f*; ~ **microphone** *n* PARTS microphone électrostatique *m*, TELEP microphone électrostatique *m*, microphone à condensateur *m*; ~ **precautions** *n pl* (*ESP*) ELECT, TEST précautions électrostatiques *f pl*; ~ **printer** *n* PER APP imprimante électrostatique *f*; ~ **relay** *n* PARTS relais électrostatique *m*
electrostatics *n* ELECT, PARTS, PER APP, POWER, TELEP, TEST électrostatique *f*
element *n* DATA, ELECT, GEN, TELEP élément *m*; ~ **error rate** TELEP taux d'erreur sur les éléments *m*
elementary: ~ **cable section** *n* TELEP section élémentaire de câble *f*; ~ **echo** *n* TELEP écho élémentaire *m*; ~ **regenerated section** *n* TELEP section élémentaire régénérée *f*; ~ **regenerator section** *n* TELEP section élémentaire régénérée *f*;

~ repeater section *n* TELEP section élémentaire amplifiée *f*; **~ stream** *n* COMP chaîne élémentaire *f*
eletromagnetic: **~ radiation** *n* ELECT rayonnement électromagnétique *m*
elevated: **~ duct** *n* RADIO conduit élevé *m*
elevation *n* RADIO site *m*; **~ angle** RADIO angle de site *m*
eliminator *n* POWER suppresseur *m*
elliptical: **~ orbit** *n* RADIO orbite elliptique *f*; **~ polarization** *n* RADIO polarisation elliptique *f*
ELT *abbr* (*emergency locater transmitter*) RADIO émetteur de localisation d'urgence *m*, radiobalise de détresse *f*
E-MAC *n* TV MAC amélioré *m*
e-mail *abbr* (*electronic mail*) APPL, COMP, DATA courrier électronique *m*, messagerie électronique *f*; **~ address** *n* DATA adresse électronique *f*, adresse E-mail *f*; **~ message** *n* DATA mail *m*; **~ message store** *n* DATA hébergeur de boîtes aux lettres *m*
embed *vt* COMP, DATA, ELECT noyer
embedded: **~ computer** *n* COMP ordinateur intégré *m*; **~ control channel** *n* (*ECC*) DATA canal de commande intégré *m*; **~ operations channel** *n* (*EOC*) DATA canal d'exploitation intégré *m*; **~ testing** *n* DATA tests encastrés *m pl*
embedding *n* COMP, DATA, ELECT incorporation *f*
embellish *vt* DATA *web page* agrémenter
emboss *vt* COMP mettre en relief, ELECT graver, GEN *printed circuits* embosser
embossing *n* COMP mise en relief *f*, ELECT *printed circuits* gravure en relief *f*, PARTS *of plastics* gaufrage *m*
emergency *n* GEN urgence *f*; **~ action** TELEP intervention d'urgence *f*; **~ attention** TELEP *fault maintenance* relèvement d'urgence *m*; **~ call box** TELEP téléphone de police secours *m*, borne d'appel d'urgence *f*; **~ call service** APPL service des demandes de secours *m*, TELEP service d'appels d'urgence *m*; **~ call stations** *n pl* PER APP bornes d'appels d'urgence *f pl*; **~ changeover** TELEP passage d'urgence sur canal sémaphore de secours *m*; **~ installation** APPL installation de secours *f*; **~ lighting** POWER éclairage de secours *m*; **~ load-transfer signal** SIG signal de transfert d'urgence de la charge *m*; **~ locater transmitter** (*ELT*) RADIO émetteur de localisation d'urgence *m*, radiobalise de détresse *f*; **~ position-indicating radio beacon** (*EPIRB*) RADIO radiobalise de localisation de sinistre *f*; **~ position-indicating radio beacon station** RADIO station de radiobalise de localisation des sinistres *f*; **~ position-indicating radio beacon system** RADIO système de radiobalises de localisation des sinistres *m*; **~ power supply** POWER alimentation électrique de secours *f*; **~ restart** TELEP remise en fonctionnement d'urgence *f*; **~ routes** *n pl* TELEP voies de secours *f pl*; **~ service** APPL service d'urgence *m*, TV services d'intervention *m pl*; **~ services** *n pl* APPL services d'urgence *m pl*; **~ shutdown system** CONTR système d'arrêt d'urgence *m*; **~ stop** CONTR système d'urgence *m*; **~ telephone** PER APP poste d'appel d'urgence *m*, TELEP poste d'urgence *m*, téléphone d'urgence *m*, téléphone de secours *m*; **~ telephone set** TELEP poste d'urgence *m*; **~ telephone station** TELEP poste d'urgence *m*; **~ transmitter** RADIO émetteur de secours *m*; **~ tripping device** GEN arrêt d'urgence *m*; **~ trunk group** GEN faisceau de secours *m*
emf *abbr* (*electromotive force*) ELECT, POWER fém (*force électromotrice*)
EMI *abbr* (*electromagnetic interference*) ELECT, TEST interférence électromagnétique *f*, brouillage électromagnétique *m*, parasite électromagnétique *m*, perturbation électromagnétique *f*
emission *n* ELECT, RADIO, TRANS émission *f*
emissivity *n* ELECT, RADIO, TRANS émissivité *f*
emit *vt* RADIO émettre
emittance *n* ELECT, RADIO, TRANS émittance *f*
emitter *n* GEN émetteur *m*; **~-base voltage** PARTS tension émetteur-base *f*; **~-coupled logic** (*ECL*) DATA, ELECT *logic elements* logique à couplage par les émetteurs *f* (*ECL*); **~ current** ELECT courant d'émetteur *m*; **~ cutoff current** ELECT courant résiduel de l'émetteur *m*; **~ cutoff voltage** PARTS tension inverse de l'émetteur *f*; **~ depletion layer** PARTS *semiconductors* zone de transition de l'émetteur *f*; **~ depletion-layer capacitance** PARTS capacité de jonction de l'émetteur *f*; **~ efficiency** PARTS rendement de l'émetteur *m*, rendement d'injection *m*; **~ follower** PARTS émettodyne *m*, montage émetteur suiveur *m*, montage émidyne *m*; **~ junction** PARTS jonction émetteur-base *f*; **~ resistance** PARTS résistance d'émetteur *f*; **~ reverse voltage** PARTS tension inverse de l'émetteur *f*; **~ series resistance** ELECT, PARTS résistance série de l'émetteur *f*; **~ terminal** PARTS borne de l'émetteur *f*; **~ voltage** PARTS tension d'émetteur *f*
empty[1]: **~ space** *n* TELEP *equipment practice* emplacement libre *m*
empty[2] *vt* COMP, DATA vider
emrp *abbr* (*effective monopole-radiated power*) RADIO parv (*puissance apparente rayonnée sur antenne verticale courte*)
EMRS *abbr* (*electromagnetic remote sensing*) TEST TDEM (*télédétection électromagnétique*)
EMS *abbr* APPL (*electronic message system*) système de messagerie électronique *m*
emulate *vt* COMP, ELECT émuler
emulation *n* COMP, ELECT émulation *f*
emulator *n* COMP, ELECT émulateur *m*
en bloc: **~ register signaling** *AmE*, **~ register signalling** *BrE* *n* SIG signalisation entre enregistreurs en bloc *f*
EN *abbr* (*Euronorm*) GEN EU (*Euronorm*)
enable[1]: **~ current** *n* POWER *semiconductors* courant de validation *m*; **~ dependency** *n* ELECT *logic elements* dépendance de validation *f*; **~ input** *n* ELECT *logic elements* entrée de validation *f*; **~ pulse** *n* PARTS *semiconductors* impulsion de validation *f*
enable[2] *vt* COMP mettre en service, valider
enabling *n* COMP mise en service *f*, validation *f*; **~ signal** COMP signal de validation *m*
enameled: **~ wire** *AmE see enamelled wire BrE*
enamelled: **~ wire** *n BrE* ELECT, GEN, PARTS fil émaillé *m*
encapsulated: **~ relay** *n* PARTS relais encapsulé *m*
encapsulating *n* COMP, ELECT, GEN, PARTS encapsulation *f*
encapsulation *n* COMP, ELECT encapsulation *f*
encased: **~ relay** *n* PARTS relais encapsulé *m*
encipher *vt* GEN chiffrer
encipherment *n* GEN chiffrement *m*
enclosed: **~ fuse-link** *n* POWER élément de remplacement à fusion *m*
encode *vt* COMP, DATA chiffrer, coder, encoder, RADIO, SIG, TELEP coder, encoder, TRANS, TV chiffrer, coder, encoder
encoded[1] *adj* COMP, DATA chiffré, codé, RADIO, SIG, TELEP codé, TRANS, TV codé, chiffré
encoded[2]: **~ information type** *n* (*EIT*) DATA type d'informations codées *m*; **~ message** *n* DATA message chiffré *m*
encoder *n* COMP, DATA, RADIO, SIG, TELEP, TRANS codeur *m*, encodeur *m*
encoding *n* COMP codage *m*, encodage *m*, DATA, RADIO, TELEP, TRANS, TV codage *m*; **~ law** TELEP, TRANS loi de codage *f*
encrypt *vt* DATA chiffrer, crypter, GEN, RADIO, SIG chiffrer, TELEP, TV crypter
encrypted: **~ speech** *n* TELEP signal vocal codé *m*
encrypting: **~ equipment** *n* DATA crypteur *m*, cryptographe *m*, GEN brouilleur *m*, dispositif de chiffrement *m*, dispositif de cryptage *m*, RADIO, TELEP, TV crypteur *m*, cryptographe *m*
encryption *n* DATA chiffrement *m*, cryptage *m*, RADIO cryptage *m*, SIG chiffrement *m*, cryptage *m*, TELEP cryptage *m*, TRANS chiffrement *m*, TV cryptage *m*; **~ control channel** TRANS voie de commande de chiffrement *f*; **~ key distribution**

RADIO distribution de clés de chiffrement *f*; ~ **unit** DATA, SIG équipement de chiffrement *m*

encryptor *n* DATA, RADIO, SIG, TELEP, TV crypteur *m*

end[1]: **~-to-end** GEN de bout en bout

end[2] *n* TELEP fin *f*; **~-around borrow** COMP report négatif circulaire *m*; **~-around carry** COMP report circulaire *m*; **~-around shift** COMP permutation circulaire *f*; ~ **cell** POWER *batteries* batterie terminale *f*; **~-cell rectifier** POWER redresseur pour élément de réduction *m*; ~ **of communication** (*EOC*) DATA, TELEP fin de communication *f*; ~ **of dialog** *AmE*, ~ **of dialogue** *BrE* TELEP fin de dialogue *f*; ~ **digital termination** TELEP terminal numérique d'extrémité *m*; ~ **of life power** (*EOL*) POWER puissance en fin de vie *f*; ~ **of line** PER APP *printers* fin de ligne *f*; ~ **mark** COMP repère de fin *m*; ~ **of message** (*EOM*) DATA, TELEP fin de message *f*; **~-of-block signal** SIG signal de fin de bloc *m*; **~-of-communication signal** DATA signal de fin de communication *m*; **~-of-page signal** DATA signal de fin de page *m*; **~-of-period tone** PER APP *payphones* tonalité de taxe de base *f*; **~-of-pulsing signal** SIG signal de fin de numérotation *m*; ~ **of output** DATA, TELEP fin de sortie *f*; ~ **part** TELEP *equipment practice* embout *m*; ~ **of procedure** (*EOP*) DATA, TELEP fin de procédure *f*; ~ **product** GEN produit fini *m*; ~ **of pulse** COMP fin d'impulsion *f*; ~ **of retransmission EOM** DATA, TELEP fin de retransmission EOM *f*; ~ **of retransmission EOP** DATA, TELEP fin de réémission EOP *f*, fin de retransmission EOP *f*; ~ **of retransmission MPS** DATA, TELEP fin de réémission MPS *f*, fin de retransmission MPS *f*; ~ **of retransmission NULL** DATA, TELEP fin de réémission NULL *f*, fin de retransmission NULL *f*; ~ **of retransmission PRI-EOM** DATA, TELEP fin de réémission PRI-EOM *f*, fin de retransmission PRI-EOM *f*; ~ **of retransmission PRI-EOP** DATA, TELEP fin de réémission PRI-EOP *f*, fin de retransmission PRI-EOP *f*; ~ **of retransmission PRI-MPS** DATA, TELEP fin de réémission PRI-MPS *f*, fin de retransmission PRI-MPS *f*; ~ **of selection** (*EOS*) DATA, TELEP fin de sélection *f*; ~ **statement** TELEP instruction de fin *f*; ~ **system** (*ES*) DATA système d'extrémité *m*; **~-to-end circuit supervision** TEST surveillance des circuits de bout en bout *f*; **~-to-end communication** TELEP communication poste à poste *f*; **~-to-end continuity test** TELEP essai de continuité de bout en bout *m*; **~-to-end digital connection** ELECT connexité numérique de bout en bout *f*; **~-to-end encipherment** DATA chiffrement de bout en bout *m*; **~-to-end information indicator** DATA indicateur d'information de bout en bout *m*; **~-to-end method indicator** DATA indicateur de méthode de bout en bout *m*; **~-to-end servicing** TELEP service de bout en bout *m*; **~-to-end signaling** *AmE*, **~-to-end signalling** *BrE* SIG signalisation de bout en bout *f*; **~-to-end test call** TELEP appel d'essai abonné à abonné *m*; **~-to-end transit delay negotiation** (*EETDN*) DATA, TELEP négociation de temps de transit de bout en bout *f*; ~ **of transaction** (*EOT*) DATA, TELEP fin de transaction *f*; ~ **of transmission** (*EOT*) DATA, TELEP fin de transmission *f*; ~ **user** GEN, TRANS *network* utilisateur final *m*; **~-user computing** COMP, DATA informatique de l'utilisateur *f*

endfire: ~ **array aerial** *n* RADIO antenne à rayonnement longitudinal *f*; ~ **array antenna** *n* RADIO antenne à rayonnement longitudinal *f*

endless[1] *adj* COMP à l'infini, sans fin

endless[2]: ~ **loop** *n* COMP boucle sans fin *f*; ~ **repeat** *n* COMP répétition sans fin *f*

endurance: ~ **test** *n* TEST essai d'endurance *m*

energizing: ~ **circuit** *n* POWER circuit d'excitation *m*

energy[1] *n* ELECT, GEN, PARTS, POWER, RADIO, TRANS énergie *f*; ~ **absorption** ELECT absorption d'énergie *f*; ~ **band** PARTS bande d'énergie *f*; ~ **converter** POWER convertisseur d'énergie *m*; ~ **density** GEN densité d'énergie *f*; ~ **dispersal** RADIO dispersion d'énergie *f*; ~ **gap** PARTS bande interdite *f*; ~ **level** PARTS niveau d'énergie *m*; ~ **saving** *n pl* GEN économie d'énergie *f*; **~-to-noise density** TRANS rapport énergie à densité de bruit *m*

energy[2]: ~ **per information bit to noise density ratio** *phr* TRANS rapport énergie par bit d'information à densité de bruit

ENG *abbr* (*electronic news gathering*) GEN, RADIO, TV journalisme électronique *m*, reportage électronique ENG *m*

engaged[1] *adj* DATA occupé

engaged[2]: ~ **signal** *n* TELEP tonalité d'occupation *f*; ~ **state display** *n* TELEP panneau de lampes d'occupation *m*; ~ **tone** *n BrE* (*cf busy signal AmE*) GEN, TELEP tonalité d'occupation *f*

engine: ~ **room** *n* TELEP salle des machines *f*

engineer *n* TELEP réparateur *m*

engineering: ~ **change order** *n* (*ECO*) PARTS ordre de modification technique *m* (*OMT*); ~ **data** *n pl* GEN *design technology* données techniques *f pl*; ~ **draftsman** *n* GEN dessinateur projeteur *m*; ~ **instructions** *n pl* GEN *documentation* instructions d'ingénierie *f pl*; ~ **order wire** *n* (*EOW*) TELEP ligne de service technique *f*; ~ **service channel** *n* TRANS voie de service *f*; ~ **service circuit** *n* (*ESC*) TRANS circuit de service *m*

enhance *vt* COMP enrichir, CONTR renforcer, GEN enrichir, RADIO, TEST renforcer

enhanced: ~ **graphics adaptor** *n* (*EGA*) COMP adaptateur EGA *m* (*EGA*); ~ **MAC** *n* TV MAC amélioré *m*; **~-quality television** *n* COMP télévision de qualité améliorée *f*; ~ **service** *n* APPL service amélioré *m*; ~ **television** *n* RADIO télévision améliorée *f*

enhancement *n* COMP augmentation *f*, CONTR renforcement *m*, enrichissement *m*, PARTS enrichissement *m*, RADIO enrichissement *m*, renforcement *m*, TEST renforcement *m*; ~ **mode operation** PARTS fonctionnement en mode d'enrichissement *m*

ENOB *abbr* (*effective number of bits*) DATA nombre de bits effectif *m*

ensure *vt* COMP rendre sûr

enter: ~ **compressed mode** *n* DATA passage au mode compression *m* (*PMC*); ~ **transparent mode** *n* DATA passage au mode transparent *m* (*PMT*)

enterprise: ~ **network** *n* TRANS réseau d'entreprise *m*; ~ **systems connection** *n* (*ESCON*) COMP interconnexion de systèmes d'entreprise *f*

entertainment: ~ **broadcasting** *n* TV radiodiffusion récréative *f*

entrance: ~ **facility** *n* (*EF*) ELECT équipement d'entrée *m*

entropy *n* GEN entropie *f*; ~ **coding** SIG codage entropique *m*

entry *n* COMP inscription *f*, DATA, SIG entrée *f*, TELEP entrée *f*, inscription *f*, TRANS entrée *f*; ~ **information** DATA information d'entrée *f*; ~ **information selection** DATA sélection d'information d'entrée *f*; ~ **level** DATA entrée de gamme *f*; ~ **phone** PER APP portier automatique *m*, portier téléphonique *m*; ~ **point** COMP point d'entrée *m*; ~ **status** DATA état de l'entrée *m*; ~ **type** DATA type d'entrée *m*

enumerated: ~ **type** *n* DATA type énuméré *m*

envelope *n* SWG enveloppe *f*; ~ **curve** COMP courbe d'enveloppe *f*; ~ **delay** RADIO temps de propagation de groupe *m*; ~ **generator** COMP générateur d'enveloppes *m*; **~-generator select** (*EG select*) COMP sélection de générateur d'enveloppes *f*; ~ **switch** SWG commutateur d'enveloppes *m*; ~ **velocity** TRANS vitesse de groupe *f*

environment *n* GEN environnement *m*; ~ **mapping** COMP placage d'environnement *m*

environmental: ~ **constraints** *n* CLIM contrainte d'environnement *f*; ~ **endurance** *n* CLIM *engineering* résistance à l'environnement *f*, tenue au milieu ambiant *f*; ~ **factor** *n* CLIM facteur ambiant *m*, facteur d'environnement *m*; ~ **requirements** *n pl* CLIM conditions ambiantes *f pl*; ~ **specification** *n* GEN spécification relative à l'environnement *f*; ~ **test** *n* CLIM essai d'ambiance *m*, essai sous conditions ambiantes *m*, GEN essai d'environnement *m*, TEST essai d'ambiance *m*; ~ **testing** *n* CLIM, TEST essais d'ambiance *m pl*, essais d'environnement *m pl*

EOC *abbr* DATA (*embedded operations channel*) canal d'exploitation intégré *m*, (*end of communication*) fin de communication *f*, TELEP (*end of communication*) fin de communication *f*

EOL *abbr* (*end of life power*) POWER puissance en fin de vie *f*

EOM *abbr* (*end of message*) DATA, TELEP fin de message *f*

EOP *abbr* (*end of procedure*) DATA, TELEP fin de procédure *f*

EOQC *abbr* (*European Organization for Quality Control*) GEN OECQ (*Organisation européenne de contrôle de la qualité*)

EOS *abbr* (*end of selection*) DATA, TELEP fin de sélection *f*

EOT *abbr* DATA, TELEP (*end of transmission*) fin de transmission *f*, (*end of transaction*) fin de transaction *f*

EOW *abbr* (*engineering order wire*) TELEP ligne de service technique *f*

EP *abbr* (*erroneous period*) DATA période erronée *f*

EPIRB *abbr* (*emergency position-indicating radio beacon*) RADIO radiobalise de localisation de sinistre *f*

epitaxial: ~ **layer** *n* PARTS couche épitaxiale *f*; ~ **transistor** *n* PARTS transistor épitaxial *m*

epitaxy *n* PARTS épitaxie *f*

E-plane: ~ **lens** *n* RADIO lentille E *f*

EPLD *abbr* (*erasable programmable logic device*) DATA, ELECT circuit logique programmable effaçable électriquement *m*

epoxide *n* PARTS *plastics* époxyde *m*

epoxy[1] *adj* ELECT, PARTS *plastics*, TRANS époxy

epoxy[2]: ~ **buffer** *n* TRANS gaine époxy *f*, revêtement ETFE *m*, revêtement primaire *m*; ~ **plastic** *n* PARTS plastique époxyde *m*; ~ **resin** *n* PARTS *plastics* résine époxyde *f*, résine époxydique *f*

EPROM *abbr* (*electrically programmable read-only memory*) ELECT EPROM (*mémoire morte programmable électroniquement*)

EQL *abbr* (*equalizer*) CONTR, ELECT, POWER, TELEP, TRANS égalisateur *m*, égaliseur *m*, correcteur *m*

equal: ~**-length code** *n* DATA, TELEP code à moments *m*; ~**-tempered scale** *n* COMP gamme tempérée *f*

equalization *n* ELECT, POWER, TRANS égalisation *f*

equalize *vt* ELECT, POWER, TRANS égaliser

equalizer *n* (*EQL*) CONTR, ELECT, POWER, TELEP, TRANS égalisateur *m*, égaliseur *m*, correcteur *m*; ~ **circuit** TRANS circuit égaliseur *m*, circuit d'égalisation *m*, réseau égaliseur *m*

equalizing: ~ **charge** *n* POWER charge d'égalisation *f*; ~ **network** *n* ELECT réseau correcteur *m*; ~ **pulse** *n* TV signaux d'égalisation *m pl*

equate *vt* COMP rendre égal

equatorial: ~ **orbit** *n* RADIO orbite équatoriale *f*

equilibrium: ~ **length** *n* TRANS longueur d'équilibrage *f*; ~ **mode distribution** *n* TRANS équilibre des modes *m*, répartition des modes à l'équilibre *f*; ~ **radiation pattern** *n* TRANS diagramme de rayonnement à l'équilibre *m*

equip *vt* COMP, TRANS équiper

equipment *n* COMP appareillage *m*, GEN équipement *m*, matériel *m*; ~ **failure** COMP panne d'équipement *f*; ~ **fuse** PARTS, POWER fusible d'appareil *m*; ~ **installation** COMP installation matérielle *f*; ~ **number** SWG, TELEP numéro d'équipement *m*

equiripple: ~ **finite-impulse response filter** *n* (*EFIR*) ELECT filtre à réponse impulsionnelle finie ondulation *m*

equivalent *n* ELECT, PARTS, POWER, RADIO, TELEP équivalent *m*; ~ **articulation loss** TELEP affaiblissement équivalent pour la netteté *m* (*AEN*); ~ **bit rate** TELEP débit binaire équivalent *m*; ~ **isotropically-radiated power** (*EIRP*) RADIO puissance isotrope rayonnée équivalente *f* (*PIRE*); ~ **noise temperature** RADIO température équivalente de bruit *f*; ~ **peak envelope power of a telephone signal** TELEP puissance équivalente de crête *f*; ~ **random circuit group** TELEP faisceau équivalent *m*; ~ **random traffic intensity** TELEP intensité de trafic équivalent *f*; ~ **resistance** POWER résistance équivalente *f*; ~ **resistance error** TELEP écart équivalent *m*; ~ **step index** (*ESI*) TELEP indice à saut équivalent *m*; ~**-step-index profile** (*ESI profile*) ELECT, TRANS profil à saut d'indice équivalent *m*; ~ **thermal network** PARTS réseau thermique équivalent *m*

erasable: ~ **programmable logic device** *n* (*EPLD*) DATA, ELECT circuit logique programmable effaçable électriquement *m*; ~ **store** *n* COMP mémoire effaçable *f*

erase *vt* COMP, PER APP, RADIO effacer

erasing *n* COMP effacement *m*; ~ **rate** COMP rapidité d'effacement *f*

erasure *n* PER APP, RADIO effacement *m*

erbium: ~**-doped fiber amplifier** *AmE*, ~**-doped fibre amplifier** *BrE* *n* TRANS amplificateur à fibre dopée erbium *m*

erect *adj* TELEP en position verticale

Ericsson: ~ **transport network architecture** *n* (*ETNA*) TRANS architecture de réseau de transport d'Ericsson *f*

Erlang *n* ELECT, TELEP erlang *m*

erp *abbr* (*effective radiated power*) RADIO par (*puissance apparente rayonnée*)

ERR *abbr* (*response for end of transmission*) DATA, TELEP ERR (*réponse pour fin de transmission*)

erroneous: ~ **bit** *n* COMP bit erroné *m*; ~ **block** *n* (*EB*) DATA bloc erroné *m*; ~ **period** *n* (*EP*) DATA période erronée *f*

error *n* GEN erreur *f*; ~ **analysis** DATA analyse des erreurs *f*; ~ **burst-correcting capability** DATA capacité de correction des paquets d'erreurs *f*; ~ **catalog** *AmE*, ~ **catalogue** *BrE* TELEP catalogue des pannes *m*, catalogue des erreurs *m*; ~ **check** RADIO détection d'erreur *f*; ~**-check character** RADIO caractère de contrôle d'erreurs *m*; ~**-check signal** RADIO signal de contrôle d'erreurs *m*; ~ **code** TELEP code d'erreur *m*; ~ **constant** CONTR constante d'erreur *f*; ~**-control device** (*ECD*) DATA dispositif de protection anti-erreur *m*; ~**-control loop** TELEP boucle de protection contre les erreurs *f*; ~**-correcting code** (*ECC*) COMP, DATA, SIG, TELEP code correcteur d'erreurs *m* (*CCE*); ~**-correcting telegraph system** DATA, TELEP système correcteur d'erreurs *m*, système de correction d'erreurs *m*; ~ **correction** COMP correction d'erreurs *f*, DATA correction d'erreurs *f*, correction de distorsion *f*, RADIO, SIG, TELEP correction d'erreurs *f*; ~ **correction coding** DATA codage correcteur d'erreurs *m*; ~**-correction mode** (*ECM*) TELEP mode avec correction d'erreur *m*; ~ **counter** TELEP compteur de défaillances *m*; ~ **density** DATA concentration d'erreurs *f*; ~**-detecting code** DATA code détecteur d'erreurs *m* (*CDE*); ~**-detecting system** TELEP système détecteur d'erreurs sans répétition *m*; ~ **detection** DATA détection d'erreur *f*; ~**-detection code** (*EDC*) COMP, DATA, SIG, TELEP code détection d'erreur *m* (*CDE*); ~**-detection coding** DATA codage de détection d'erreurs *m*; ~ **detector** DATA détecteur d'erreurs *m*; ~ **diagnosis** DATA diagnostic d'erreurs *m*; ~**-free decisecond** (*EFdS*) TELEP déciseconde sans erreur *m*; ~ **handling** DATA traitement des erreurs *m*; ~ **marking** TELEP marquage de dérangement *m*, marquage d'erreur *m*; ~ **message** DATA message d'erreur *m*; ~ **of method** TEST *metrology* erreur de méthode *f*; ~ **multiplication** TELEP multiplication d'erreurs *f*; ~ **multiplication factor** TELEP coefficient de multiplication d'erreurs *m*; ~ **pattern** DATA configuration d'erreurs *f*; ~ **print-out** SWG message d'erreur *m*; ~ **probability** DATA, SIG probabilité d'erreur *f*; ~**-protecting code** DATA code de protection contre les erreurs *m*; ~ **protection** DATA protection contre les erreurs *f*; ~ **protection code** DATA code de contrôle d'erreurs *m*, code de protection contre les erreurs *m*; ~ **pulse** DATA impulsion d'erreur *f*; ~ **rate** DATA, SIG taux d'erreurs *m*, TELEP *of* *telegraph communication* taux d'erreur *m*; ~ **rate of keying** TELEP taux d'erreur d'une manipulation *m*; ~-**rate measurement** DATA mesure de taux d'erreurs *f*; ~ **rate monitor** TELEP appareil de surveillance du taux d'erreur *m*; ~ **ratio** TELEP taux d'erreur *m*; ~ **recognition** TEST détection de panne *f*; ~ **recovery** DATA retour au fonctionnement normal *m*; ~ **report** TELEP fiche de dérangement *f*, rapport d'incident *m*; ~ **resilience** COMP élasticité aux erreurs *f*; ~ **signal** CONTR signal d'erreur *m*; ~ **spread** TELEP répartition des erreurs *f*; ~ **susceptibility** DATA sensibilité aux erreurs *f*

errored: ~ **second** *n* (*ES*) DATA, RADIO seconde avec erreurs *f* (*SE*), seconde entachée d'erreurs *f* (*SE*)

ES *abbr* DATA (*errored second*) SE (*seconde avec erreurs, seconde entachée d'erreurs*), (*end system*) système d'extrémité *m*, RADIO (*errored second*) SE (*seconde avec erreurs, seconde entachée d'erreurs*), TRANS (*echo suppressor*) suppresseur d'écho *m*

ESA *abbr* (*European Space Agency*) RADIO Agence spatiale européenne *f*

ESC *abbr* COMP, DATA, TELEP (*escape*) ESC (*échappement*), TRANS (*engineering service circuit*) circuit de service *m*

escalating: ~ **loudness** *n* RADIO sonorité à intensité croissante *f*

escape *n* (*ESC*) COMP, DATA, TELEP échappement *m* (*ESC*); ~ **character** DATA caractère d'échappement *m*; ~ **code** (*EC*) DATA, TELEP code d'échappement *m*; ~ **in data** (*EID*) DATA échappement dans les données *m* (*EDD*); ~ **indication** TELEP indication d'échappement *f*; ~ **sequence** DATA séquence d'échappement *f*; ~ **velocity** RADIO vitesse de libération *f*

ESCON *abbr* (*enterprise systems connection*) COMP interconnexion de systèmes d'entreprise *m*

ESD *abbr* (*electrostatic discharge*) ELECT, TEST décharge électrostatique *f*

ESI *abbr* (*equivalent step index*) TELEP indice à saut équivalent *m*; ~ **profile** *n* (*equivalent-step-index profile*) ELECT, TRANS profil à saut d'indice équivalent *m*; ~ **refractive index difference** *n* TELEP différence d'indice de réfraction *f*

esoteric[1] *adj* DATA ésotérique

esoteric[2]: ~ **marking** *n* DATA balisage ésotérique *m*

ESP *abbr* (*electrostatic precautions*) ELECT, TEST précautions électrostatiques *f pl*

ESS *abbr* (*electronic switching system*) SWG système de commutation électronique *m*

essential: ~ **optional user-facility** *n* TELEP service complémentaire essentiel facultatif *m*

establish *vt* DATA, GEN, RADIO *communication*, TELEP *connection, direct link* établir

establishment *n* GEN établissement *m*

ester: ~ **plastic** *n* PARTS plastique polyestérique *m*, plastique estérique *m*

estimate *vt* TEST estimer

estimated[1] *adj* TEST estimé

estimated[2]: ~ **value** *n* TEST valeur estimée *f*

ESVA *abbr* (*earth station verification and assistance*) RADIO vérification et assistance des stations terriennes *f pl*

e/t: ~ **switch** *n* RADIO commutateur ll/12 GHz *m*

ET *abbr* (*exchange termination*) TELEP terminaison de commutateur *f*

ETC *abbr* (*exchange terminal circuit*) SWG circuit terminal de central *m*

etch[1]: ~-**back** *n* ELECT gravure en retrait *f*; ~ **factor** *n* ELECT facteur de gravure *m*

etch[2] *vt* COMP, ELECT graver, GEN attaquer chimiquement, graver chimiquement

etching *n* GEN attaque chimique *f*, gravure chimique *f*

ETF *abbr* (*eight to fourteen modulation*) COMP ETF (*modulation de huit à quatorze*)

Ethernet *n* DATA Ethernet *m*; ~ **card** *n* DATA carte Ethernet *f*

ETM *abbr* (*eight to ten modulation*) COMP ETM (*modulation de huit à dix*)

ETNA *abbr* (*Ericsson transport network architecture*) TRANS architecture de réseau de transport de Ericsson *f*

ETS *abbr* DATA (*executable test suite*) suite de tests exécutables *f*, GEN (*European Telecommunication Standard*) NET (*Norme européenne de télécommunications*), TEST (*executable test suite*) suite de tests exécutables *f*

ETSI *abbr* (*European Telecommunications Standards Institute*) GEN IENT (*Institut européen de normalisation en télécommunications*)

EURB *abbr* (*European Union for Radio Broadcasting*) RADIO Union européenne de radiodiffusion *f*

EURESCOM *abbr* (*European Institute of Research and Strategic Studies in Telecommunications*) GEN EURESCOM (*Union européenne de recherches et d'études stratégiques en télécommunications*)

eurobeam: ~ **transmit aerial** *n* RADIO antenne d'émission du faisceau européen *f*; ~ **transmit antenna** *n* RADIO antenne d'émission du faisceau européen *f*

Euronorm *n* (*EN*) GEN Euronorm *m* (*EU*)

European: ~ **Broadcasting Union** *n* (*EBU*) RADIO Union européenne de radio-télévision *f* (*UER*); ~ **Committee for Electrotechnical Standardization** *n* (*CENELEC*) GEN Comité européen de normalisation électrotechnique *m* (*CENELEC*); ~ **Committee for Standardization** *n* (*CEN*) GEN Comité européen de normalisation *m* (*CEN*); ~ **Computer Manufacturers Association** *n* (*ECMA*) COMP Association européenne de constructeurs d'ordinateurs *f* (*ECMA*); ~ **Conference of Postal and Telecommunications Administrations** *n* (*ECPT*) GEN Conférence européenne des administrations des postes et des télécommunications *f*; ~ **Institute of Research and Strategic Studies in Telecommunications** *n* (*EURESCOM*) GEN Union européenne de recherches et d'études stratégiques en télécommunications *f*; ~ **Organization for Quality Control** *n* (*EOQC*) GEN Organisation européenne de contrôle de la qualité *f* (*OECQ*); ~ **Space Agency** *n* (*ESA*) RADIO Agence spatiale européenne *f*; ~ **Telecommunication Standard** *n* (*ETS*) GEN Norme européenne de télécommunications *f* (*NET*); ~ **Telecommunications Satellite Organization** *n* (*EUTELSAT*) RADIO Organisation européenne de télécommunications par satellites *f*, Organisation internationale de télécommunications par satellites *f* (*EUTELSAT, INTELSAT*); ~ **Telecommunications Standards Institute** *n* (*ETSI*) GEN Institut européen de normalisation en télécommunications *m* (*IENT*); ~ **Union for Radio Broadcasting** *n* (*EURB*) RADIO Union européenne de radiodiffusion *f*; ~ **Workshop for Open Systems** *n* (*EWOS*) GEN Atelier européen pour les systèmes ouverts *m* (*EWOS*)

EUTELSAT *abbr* RADIO EUTELSAT *abbr* (*Organisation internationale de télécommunications par satellites*)

evanescent: ~ **field** *n* ELECT, TRANS champ évanescent *m*

even[1] *adj* COMP, GEN pair

even[2]: ~ **note** *n* COMP note paire *f*; ~ **parity check** *n* DATA, TELEP contrôle de parité *m*; ~ **parity check odd parity check** *n* DATA, TELEP contrôle de parité ou d'imparité *m*

event[1]: ~-**driven** *adj* COMP commandé par événement

event[2] *n* COMP, DATA, ELECT, RADIO, TELEP, TEST événement *m*; ~-**forwarding discriminator** DATA discriminateur de transmission d'événement *m*; ~ **presentation restriction indicator** DATA indicateur de restriction de divulgation d'événement *m*

EWOS *abbr* (*European Workshop for Open Systems*) GEN EWOS (*Atelier européen pour les systèmes ouverts*)

ex[1]: ~-**directory** *adj BrE* (*XD*) (*cf unlisted AmE*) TELEP sur liste rouge

ex[2]: ~-**directory number** *n BrE* TELEP numéro ne figurant pas

à l'annuaire *m*, numéro non-inscrit *m*, numéro sur la liste rouge *m*
EX *abbr* (*extinction ratio*) ELECT, TRANS rapport d'extinction *m*, taux d'extinction *m*
examination *n* GEN examen *m*, investigation *f*
examine *vt* GEN examiner
excavation: ~ **damage** *n* APPL déterrement *m*
excavator: ~ **damage** *n* APPL déterrement *m*
excess[1] *adj* COMP en surnombre
excess[2]: ~ **attenuation** *n* TRANS supplément d'affaiblissement *m*; ~ **current** *n* POWER surintensité *f*, surintensité de courant *f*; ~ **solder** *n* ELECT excès de soudure *m*; ~ **voltage** *n* ELECT, POWER surtension *f*
exchange *n* COMP, GEN échange *m*, SWG, TELEP central de commutation *m* (*CC*), centre de commutation *m* (*CC*), commutateur téléphonique *m*; ~ **assembly drawing** GEN *documentation* plan de montage du central *m*; ~ **battery** TELEP batterie de central *f*; ~ **call set-up delay** SWG durée de sélection d'un commutateur *f*, temps d'établissement de la communication dans le central *m*; ~ **call-release delay** SWG temps de libération de la communication par le central *m*; ~ **code** GEN numéro de série *m*; ~ **concentrator** TELEP concentrateur de central *m*; ~ **control system** TELEP système de la commande du central *m*; ~ **diagram** POWER schéma du central *m*, SWG schéma de l'autocommutateur *m*; ~ **equipment practice** GEN mécanique de central *f*; ~ **identification** TELEP identification d'échange *f*; ~ **of identification** (*XID*) TELEP échange d'identifications *m* (*XID*); ~ **identification frame** TELEP trame d'identification d'échange *f*; ~ **installation drawing** GEN plan d'installation du central *m*; ~ **library** GEN documentation du central *f*; ~ **line** GEN ligne de jonction entre deux centraux *f*, TELEP ligne de jonction entre deux centraux *f*, ligne de central *f*; ~ **line junction set** SWG joncteur de ligne réseau *m*; ~ **line junctor** SWG joncteur de ligne réseau *m*; ~ **line unit** SWG unité de ligne réseau *f*; ~ **manual** GEN manuel du central *m*; ~ **mechanical structure** TELEP structure mécanique de central *f*; ~ **method** (*EM*) TELEP méthode d'échange *f*; ~ **terminal circuit** (*ETC*) SWG circuit terminal de central *m*; ~ **termination** (*ET*) TELEP terminaison de commutateur *f*; ~ **time** SWG heure centrale *f*
excitation *n* ELECT excitation *f*, TELEP stimulus *m*
excite *vt* ELECT exciter
exclusion *n* ELECT exclusion *f*; ~ **unit** TEST unité d'incapacité *f*, unité d'invalidation *f*
exclusive: ~ **external line** *n* SWG, TELEP ligne extérieure privée *f*
EXCLUSIVE-OR: ~ **element** *n* ELECT élément OU exclusif *m*; ~ **operation** *n* ELECT disjonction *f*, opération OU exclusif *f*, OU exclusif *m*
executable: ~ **memory image** *n* TELEP image mémoire exécutable *f*; ~ **test case** *n* DATA, TEST test élémentaire exécutable *m*; ~ **test case error** *n* DATA, TEST erreur de test élémentaire exécutable *f*; ~ **test suite** *n* (*ETS*) DATA, TEST suite de tests exécutables *f*
execute: ~ **button** *n* COMP touche d'exécution *f*; ~ **order** *n* COMP instruction d'exécution *f*; ~ **phase** *n* COMP phase d'exécution *f*; ~ **statement** *n* COMP instruction d'exécution *f*
execution: ~ **character** *n* TELEP caractère d'exécution *m*; ~ **time** *n* COMP temps d'exécution *m*
executive: ~ **override** *n* TELEP entrée en tiers *f*, priorité *f*; ~ **program** *AmE*, ~ **programme** *BrE* *n* TELEP programme superviseur *m*, système d'exploitation *m*
exhaust *n* CLIM gaz d'échappement *m*; ~ **gas** CLIM gaz d'échappement *m*; ~ **pipe** PARTS tuyau d'échappement *m*; ~ **valve** PARTS soupape d'échappement *f*
exhibit *vt* COMP présenter
existing: ~-**quality television** *n* COMP télévision de qualité conventionnelle *f*
exit: ~ **button** *n* COMP touche de sortie *f*
exocentric: ~ **angle** *n* RADIO angle exocentrique *m*
expandability *n* COMP, TELEP possibilité d'extension *f*, extensibilité *f*
expanded: ~ **memory** *n* COMP, DATA mémoire paginée *f*; ~ **polystyrene** *n* PARTS mousse de polystyrène *f*, polystyrène expansé *m*; ~ **polystyrene foam** *n* PARTS mousse de polystyrène *f*; ~ **session reference** *n* TELEP référence élargie de l'échange *f*; ~ **sweep** *n* RADIO balayage élargi *m*, balayage avec effet de loupe *m*, balayage dilaté *m*
expander *n* COMP expandeur *m*, unité d'extension *f*
expansion *n* COMP expansion *f*, ELECT détente *f*, dilatation *f*, GEN détente *f*, dilatation *f*, expansion *f*, PARTS dilatation *f*, SIG extension *f*; ~ **band** RADIO bande d'extension *f*; ~ **button** COMP bouton d'expansion *m*; ~ **card** COMP carte d'extension *f*; ~ **force** GEN force expansive *f*; ~ **network** TRANS réseau d'expansion *m*; ~ **slot** COMP connecteur d'extension *m*, emplacement libre pour extension *m*; ~ **stage** TRANS étage d'expansion *m*; ~ **window** TRANS plage d'allongement *f*
expected: **not-~ frame** *n* (*NEF*) TELEP trame inattendue *f* (*NEF*)
expedited: ~ **network service data-unit** *n* DATA unité de données exprès du service de réseau *f*
experimental: ~ **network** *n* RADIO réseau pilote *m*
expert: ~ **system** *n* APPL *artificial intelligence*, COMP, DATA système expert *m*
expiration: ~ **of timer** *n* *AmE* (*cf expiry of timer BrE*) RADIO expiration de la temporisation *f*
expiry: ~ **of timer** *n* *BrE* (*cf expiration of timer AmE*) RADIO *maritime* expiration de la temporisation *f*
explicit: ~ **congestion notification** *n* (*ECN*) DATA, SWG, TELEP notification d'encombrement explicite *f*; ~ **link** *n* COMP lien explicite *m*
exploded: ~ **view** *n* GEN *technical drawing* vue éclatée *f*
explosion: ~-**proof** *adj* TRANS antidéflagrant
exponential[1] *adj* COMP, ELECT exponentiel
exponential[2]: ~ **curve** *n* COMP courbe exponentielle *f*; ~ **distribution** *n* GEN *statistics* distribution exponentielle *f*, répartition exponentielle *f*; ~ **equation** *n* GEN équation exponentielle *f*; ~ **function** *n* GEN fonction exponentielle *f*
export *n* DATA export *m*, ELECT exportation *f*
exposure: ~ **limits** *n pl* GEN limites d'exposition *f pl*; ~ **meter** *n* TV photomètre *m*, posemètre *m*; ~ **time** *n* TV temps de pose *m*
expression *n* DATA, TELEP expression *f*; ~ **control** COMP commande d'expression *f*; ~ **pedal** COMP pédale d'expression *m*
extend *vt* COMP, DATA étendre, RADIO étendre, prolonger, TRANS, TV étendre
extended: ~ **application-layer structure** *n* (*XALS*) TRANS structure en couches d'application étendue *f*; ~ **bandwidth system** *n* RADIO *amateur radio* système à bande élargie *m*; ~ **binary coded decimal interchange code** *n* (*EBCDIC*) DATA, TELEP code decimal codé binaire étendu *m* (*EBCDIC*); ~ **BNF** *n* DATA forme BNF étendue *f*; ~ **eclipse use** *n* RADIO utilisation étendue en période d'éclipse *f*; ~ **industry standard architecture** *n* (*EISA*) COMP architecture EISA *f* (*EISA*); ~ **memory** *n* DATA mémoire étendue *f*; ~ **range** *n* TV gamme étendue *f*
extendibility *n* TELEP possibilité d'extension *f*
extensibility *n* COMP extensibilité *f*, TELEP extensibilité *f*, possibilité d'extension *f*
extension *n* (*X*) COMP poste *m*, DATA, GEN extension *f*, RADIO poste *m*, SIG extension *f*; ~ **bell** PER APP sonnerie de poste *f*; ~ **board** ELECT carte d'extension *f*; ~ **cable** GEN rallonge de câble *f*, POWER câble prolongateur *m*; ~ **circuit** TELEP section locale *f*; ~ **cord** POWER cordon prolongateur *m*, prolongateur *m*; ~ **line** DATA ligne d'extension *f*, SWG ligne supplémentaire *f*; ~ **mode** TELEP mode d'extension *m*; ~ **module** TELEP module d'extension *m*; ~ **number** SWG

numéro interne *m*; ~ **rack** TELEP bâti d'extension *m*; ~ **set** SWG, TELEP poste téléphonique supplémentaire *m*; ~ **socket** GEN prise supplémentaire *f*; ~ **step** TELEP étape d'extension *f*, phase d'extension *f*; ~ **store** COMP mémoire d'extension *f*; ~ **telephone set** SWG, TELEP poste téléphonique supplémentaire *m*; ~ **test period** TEST période de prolongation des essais *f*

external: ~ **access equipment** *n* (*EA*) TELEP équipement d'accès externe *m*; ~ **alarm** *n* RADIO *mobile* alarme externe *f*; ~ **blocking** *n* SIG blocage externe *m*; ~ **cable** *n* ELECT câble externe *m*; ~ **cabling** *n* ELECT câblage externe *m*; ~ **calls barred extension** *n* TELEP poste à restriction d'appel au départ *m*; ~ **clock** *n* COMP horloge externe *f*; ~ **device** *n* COMP appareil externe *m*, périphérique *m*; ~ **disabled time** *n* TEST temps d'incapacité externe *m*; ~ **equipment** *n* GEN équipement périphérique *m*; ~ **failure cost** *n* TEST *quality* coût des défaillances externes *m*; ~ **interrupt signal** *n* SWG signal extérieur d'interruption *m*; ~ **keyboard** *n* COMP clavier externe *m*; ~ **layer** *n* ELECT couche externe *f*; ~ **loss time** *n* TEST temps d'incapacité externe *m*; ~ **memory** *n* COMP mémoire externe *f*; ~ **photoelectric effect** *n* ELECT effet photoélectrique externe *m*, TRANS effet photoélectrique externe *m*, photoémission *f*; ~ **plant center** *AmE*, ~ **plant centre** *BrE n* GEN centre de construction de lignes *m* (*CCL*); ~ **program parameter** *n* COMP paramètre externe *m*; ~ **supervision station** *n* TELEP station de supervision externe *f*; ~ **type** *n* DATA type externe *m*

extinction: ~ **ratio** *n* (*EX*) ELECT, TRANS rapport d'extinction *m*, taux d'extinction *m*

extra[1] *adj* COMP, TELEP supplémentaire

extra[2]: ~ **digit** *n* TELEP chiffre supplémentaire *m*

extract[1]: ~ **instruction** *n* COMP instruction d'extraction *f*

extract[2] *vt* COMP extraire

extraneous[1] *adj* COMP superflu, parasite, POWER, TELEP parasite

extraneous[2]: ~ **noise** *n* TELEP bruit parasite *m*, parasites *m pl*; ~ **voltage** *n* POWER tension parasite *f*

extrinsic: ~ **base resistance** *n* PARTS *semiconductors* résistance extrinsèque de base *f*; ~ **conduction** *n* ELECT, PARTS conduction extrinsèque *f*; ~ **joint loss** *n* TRANS *fibre optics* perte extrinsèque de raccordement *f*; ~ **semiconductor** *n* ELECT, PARTS semiconducteur extrinsèque *m*

extruded: ~ **solid insulation** *n* GEN *cables*, PARTS isolation en plastique extrudé massif *f*

eye: ~ **diagram** *n* ELECT, TRANS diagramme en œil *m*; ~ **height** *n* RADIO hauteur de l'œil *f*; ~ **opening** *n* TRANS ouverture de l'œil *f*; ~ **pattern** *n* ELECT, TRANS diagramme en œil *m*; ~ **width** *n* RADIO largeur de l'œil *f*

eyedropper *n* COMP pipette *f*

eyelet *n* CLIM, POWER *lighting* plot *m*; ~ **bond strap** ELECT fils de liaison à illet *m pl*

EyePhone *n* COMP EyePhone *m*, visiocasque *m*

eyeshape: ~ **pattern** *n* ELECT, TRANS diagramme en œil *m*

F

F *abbr* (*farad*) POWER F (*farad*)
F1: **~-layer** *n* COMP, RADIO couche F1 *f*
F2: **~-layer** *n* COMP, RADIO couche F2 *f*
FAA *abbr* (*facility accepted message*) DATA, TELEP SUAC (*message d'acceptation de service supplémentaire*)
Fabry-Pérot: **~ laser diode** *n* (*FP laser diode*) TRANS *fibre optics* diode laser de Fabry-Pérot *f*; **~ resonator** *n* TRANS *fibre optics* résonateur Fabry-Pérot *m*
faceplate *n* ELECT prise murale *f*
facet *n* TRANS *fibre optics* facette *f*; **~ coating** TRANS enduit de facette *m*; **~ damage** TRANS endommagement de facette *m*
facial: **~ animation** *n* COMP animation faciale *f*; **~ warping** *n* COMP déformation faciale *f*
facilities: **~ management** *n* (*FM*) APPL infogérance *f*
facility *n* APPL possibilité, service complémentaire *m*, COMP possibilité *f*, DATA service complémentaire *m*, service supplémentaire *m*, GEN installation, TELEP service complémentaire *m*, service supplémentaire *m*; **~-accepted message** (*FAA*) DATA, TELEP message d'acceptation de service supplémentaire *m* (*SUAC*); **~ cancellation request-accepted message** DATA, TELEP message d'acceptation de la demande d'annulation *m*; **~ cancellation request message** TELEP message de demande d'annulation de service *m*; **~ cancellation request rejected message** TELEP message de refus de la demande d'annulation *m*; **~ indicator** GEN indicateur de service supplémentaire *m*; **~ registration request-accepted message** DATA, TELEP message d'acceptation de la demande d'enregistrement *m*; **~ registration request message** TELEP message de demande d'enregistrement de service *m*; **~ registration request rejected message** TELEP message de refus de la demande d'enregistrement *m*; **~-rejected message** (*FRJ*) DATA message de refus de service supplémentaire *m* (*SURF*); **~ request** DATA, TELEP demande de service complémentaire *f*, demande de service supplémentaire *m*; **~ request message** (*FAR*) TRANS message de demande de service supplémentaire *m* (*SUDM*); **~ request separator** TELEP séparateur de demandes de service complémentaire *m*
facsimile *n* GEN *copy* facsimile *m*, télécopie *f*, PER APP *machine* téléfax *m*, terminal de télécopie *m*; **~ apparatus** PER APP téléfax *m*, terminal de télécopie *m*, télécopieur *m*; **~ compression** TELEP compression de télécopie *f*; **~ information field** (*FIF*) COMP, DATA champ d'informations télécopie *m*; **~ line manager** DATA gestionnaire de ligne de télécopie *mf*; **~ machine** PER APP, TELEP télécopieur *m*, téléfax *m*, terminal de télécopie *m*; **~ message** TELEP message par télécopieur *m*; **~ on private networks** TELEP télécopie sur réseaux privés *f*; **~ packet assembly facility** DATA service complémentaire d'assemblage de paquets pour la télécopie *m*; **~ packet assembly-disassembly** DATA, TELEP assemblage-désassemblage de paquets pour la télécopie *m*; **~ packet disassembly facility** DATA service complémentaire de désassemblage de paquets pour la télécopie *m*; **~ server** DATA, TELEP serveur de télécopie *m*; **~ service** DATA, TELEP service de télécopie *m*; **~ telegraphy** TELEP phototélégraphie *f*, télégraphie fac-similée *f*; **~ terminal** PER APP télécopieur *m*, terminal de télécopie *m*; **~ transaction** TELEP envoi par télécopie *m*
factor *n* GEN facteur *m*; **~ of cooperation** TELEP facteur de coopération *m*
factorial *n* GEN factorielle *f*; **~ experiment** TEST expérience factorielle *f*
factoring *n* GEN *mathematics* factorisation *f*
factory[1]: **~-preset** *adj* COMP programmé en usine, préprogrammé; **~-programmed** *adj* COMP programmé en usine, préprogrammé
factory[2]: **~ setting** *n* COMP réglage en usine *m*
facts: **~ base** *n* COMP, DATA base de faits *f*
fade[1] *n* COMP évanouissement *m*, fondu *m*, fondu enchaîné *m*, RADIO disparition progressive *f*, TV fondu *m*; **~-in** COMP apparition progressive *f*, TV fondu à l'ouverture *m*; **~-out** COMP, RADIO évanouissement *m*, TV fondu à la fermeture *m*
fade[2] *vi* COMP, RADIO disparaître progressivement, s'évanouir, fondre
fader *n* COMP atténuateur *m*, commande de fondu sonore *f*
fading *n* COMP, RADIO évanouissement *m*, disparition progressive *f*, fondu *m*, TV fondu *m*; **~ rate** RADIO cadence des évanouissements *f*
fail[1]: **~-safe** *adj* TEST à sûreté intégrée, de sécurité intrinsèque
fail[2]: **~-safe** *n* TELEP protection contre les dérangements *f*; **~-safe concept** *n* TEST concept de sécurité intrinsèque *m*; **~ verdict** *n* DATA échec verdict *m*
fail[3] *vi* COMP, GEN, TELEP, TEST, TRANS échouer
failure *n* COMP incident *m*, panne *f*, ELECT panne *f*, GEN défaillance *f*, dérangement *m*, panne *f*, TELEP défaillance *f*, dérangement *m*, panne *f*, échec *m*, TEST défaillance *f*, dérangement *m*, échec *m*, TRANS interruption *f*, échec *m*; **~ alarm** TELEP alarme de dérangement *f*; **~ alarm circuit** TELEP circuit d'alarme de dérangement *m*; **~ alarm lamp** TELEP témoin de défaillance *m*, témoin de dérangement *m*, témoin de panne *m*, témoin d'échec *m*; **~ analysis** TEST *dependability* analyse de défaillance *f*; **~ catalog** *AmE*, **~ catalogue** *BrE* TELEP catalogue des défaillances *m*, catalogue des pannes *m*; **~ cause** TEST origine d'une défaillance *f*, origine d'une panne *f*; **~ consequence diagram** GEN *documentation* schéma de conséquence de défaillance *m*; **~ correction time** TELEP durée de relève d'un dérangement *f*; **~ counter** TELEP compteur de défaillances *m*; **~ criterion** TEST critère de défaillance *m*; **~ criticality** TEST gravité de la défaillance *f*; **~ diagnosis time** TELEP durée du diagnostic d'un dérangement *f*; **~ distribution** TELEP répartition de défaillance *f*; **~ indication** TELEP indication de panne *f*, indication de défaillance *f*; **~ intensity** TEST intensité de défaillance *f*; **~ marking** TELEP marquage de dérangement *m*, marquage d'erreur *m*; **~ mechanism** TEST mécanisme de défaillance *m*; **~ mode** TELEP mode de dérangement *m*, TEST mode de panne *m*; **~ occurrence** TELEP apparition d'un dérangement *f*; **~ rate** GEN, PER APP, TEST taux de panne *m*, taux de défaillance *m*, taux instantané de défaillance *m*; **~ recognition** TELEP identification d'un dérangement *f*, identification d'un échec *f*, identification d'une défaillance *f*; **~ source** TEST origine d'une défaillance *f*, origine d'une panne *f*, origine d'un échec *f*; **~ state** TELEP état de dérangement *m*; **~ symptom** TELEP symptôme de panne *m*; **~ to train** (*FTT*) DATA, TELEP échec du conditionnement *m* (*ECD*); **~ verification** TELEP vérification de défaillance *f*
fall: **~-back circuit** *n* TEST circuit de secours *m*; **~-back rate** *n* ELECT vitesse de repli *f*; **~-back relationships** *n pl* DATA relations de substitution *f pl*; **~ time** *n* COMP temps de retombée *m*, ELECT *of waveform* temps d'extinction *m*, TELEP *pulses* temps de décroissance *m*, temps de descente *m*
falling: **~ exponential curve** *n* COMP courbe exponentielle

descendante *f*; ~ **linear curve** *n* COMP courbe linéaire descendante *f*

false: ~ **alarm** *n* GEN fausse alarme *f*; ~ **alarm probability** *n* GEN probabilité de fausse alarme *f*; ~ **alarm time** *n* GEN temps de fausse alarme *m*; ~ **calling rate** *n* TELEP taux d'appels intempestifs *m*; ~ **echo** *n* RADIO faux écho *m*; ~ **seizure** *n* SIG signal intempestif de prise *m*; ~ **start** *n* GEN démarrage intempestif *m*, faux démarrage *m*

fan[1]: ~**-folded** *adj* COMP plié en accordéon

fan[2] *n* COMP ventilateur *m*; ~**-beam aerial** RADIO antenne à faisceau étalé *f*, antenne à faisceau fixe *f*; ~**-beam antenna** RADIO antenne à faisceau étalé *f*, antenne à faisceau fixe *f*; ~**-fold sheets** *n pl* ELECT feuilles d'étiquettes adhésives *f pl*; ~**-in** ELECT *electronics* entrance *f*; ~**-out** ELECT faisceau *m*, sortance *f*, facteur pyramidal *m*; ~**-out cable** ELECT câble de distribution *m*; ~**-out device** ELECT dispositif de sortance en pyramide *m*; ~**-out jacketing** ELECT gaine de dérivation *f*; ~**-shaped beam** POWER faisceau en éventail *m*

fan[3]: ~ **out** *vt* ELECT répartir

fanning *n* ELECT répartition *f*; ~ **strip** POWER plaquette de connexions *f*, réglette de distribution *f*, réglette de répartition *f*

far: ~**-end block error** *n* (*FEBE*) DATA erreur de bloc à l'extrémité distante *f*; ~**-end crosstalk** *n* RADIO télédiaphonie *f*; ~**-end receive failure** *n* (*FERF*) SIG défaut en réception à l'extrémité distante *m*, dérangement de réception à l'extrémité *m*; ~ **field** *n* RADIO, TRANS *fibre optics* champ lointain *m*; ~**-field analysis** *n* RADIO, TRANS analyse du champ lointain *f*; ~**-field diffraction pattern** *n* RADIO, TRANS diagramme de diffraction en champ lointain *m*; ~**-field radiation pattern** *n* RADIO, TRANS *fibre optics* diagramme de rayonnement de champ lointain *m*; ~**-field region** *n* RADIO, TRANS *fibre optics* région de champ lointain *f*, région de Fraunhofer *f*, région de rayonnement *f*

FAR *abbr* (*facility request message*) DATA, TRANS SUDM (*message de demande de service supplémentaire*)

farad *n* (*F*) POWER farad *m* (*F*)

Faraday: ~ **effect** *n* TRANS *fibre optics* effet Faraday *m*

FAS *abbr* SWG (*flexible access switch*) commutateur à accès flexible *m*, TELEP (*frame-alignment signal*), TRANS (*frame-alignment signal*) signal de verrouillage de trame *m*

fast: ~ **access** *n* COMP accès rapide *m*; ~**-access memory** *n* COMP mémoire à accès rapide *f*; ~**-access storage** *n* COMP mémoire à accès rapide *f*; ~**-busy tone** *n* TELEP signal circuits tous occupés *m*, tonalité circuits tous occupés *f*; ~ **circuit switch** *n* SWG commutateur à circuits rapides *m*; ~ **clamping** *n* ELECT reaction rapide *f*; ~ **facsimile** *n* TELEP télécopie rapide *f*; ~**-forward playback** *n* TELEP avance rapide *f*, lecture avant rapide *f*, lecture rapide *f*; ~ **Fourier transform** *n* (*FFT*) COMP transformation de Fourier rapide *f* (*TFR*); ~ **frequency-shift keying** *n* SIG modulation par déplacement de fréquence rapide *f*; ~ **packet switch** *n* (*FPS*) DATA, SWG, TELEP, TRANS commutateur de paquets rapide *m*; ~ **packet switching** *n* DATA, SWG, TELEP, TRANS commutation de paquets rapide *f*; ~**-reservation protocol** *n* (*FRP*) DATA protocole de réservation rapide *m*; ~ **reverse playback** *n* TELEP lecture arrière rapide *f*; ~ **select** *n* TELEP sélection rapide *f*; ~**-select acceptance** *n* TELEP acceptation de la sélection rapide *f*; ~ **storage** *n* DATA mémoire à accès rapide *f*; ~**-time analysis system** *n* (*FTAS*) COMP système d'analyse en temps comprimé *m*; ~**-time simulation** *n* COMP simulation en temps comprimé *f*; ~ **update** *n* COMP mise à jour rapide *f*, rafraîchissement rapide *m*; ~ **update request** *n* (*FUR*) COMP demande de rafraîchissement rapide *f*

fasteners *n pl* PARTS attaches *f pl*

fastening: ~ **plate** *n* PARTS plaque de fixation *f*, plaque de support *f*

FAT *abbr* (*file allocation table*) DATA table d'allocation des fichiers *f*

fatigue *n* PARTS *durability* fatigue *f*; ~ **test** TEST essai d'endurance *m*

fault[1]: ~**-free** *adj* TEST sans défaillance; ~**-tolerant** *adj* COMP, DATA, ELECT, TEST tolérant aux pannes

fault[2] *n* COMP, ELECT défaut *m*, POWER, TELEP, TEST dérangement *m*, *in machine* anomalie *f*, défaut *m*, *mistake* faute *f*, *breakdown* panne *f*; ~ **alarm circuit** TELEP circuit d'alarme de dérangement *m*; ~ **alarm lamp** TELEP témoin de défaillance *m*, témoin de dérangement *m*, témoin d'échec *m*, témoin de panne *m*; ~ **analysis** TEST *dependability* analyse des pannes *f*; ~ **center** *AmE*, ~ **centre** *BrE* TELEP centre de gestion des dérangements *m*; ~ **clearance** GEN, TEST relève de dérangements *f*; ~**-cleared message** TEST message de relève de dérangement *m*; ~ **code** TELEP code d'erreur *m*; ~ **complaint service** TELEP service de réclamation des dérangements *m*; ~ **condition** TELEP condition de dérangement *f*, état de panne *m*; ~ **correction** TEST réparation *f*; ~ **correction time** TEST temps de correction de panne *m*, temps de réparation *m*; ~ **current** ELECT courant de défaut *m*, courant de fuite *m*; ~ **detection** GEN détection des dérangements *f*, détection de défaut *f*; ~ **diagnosis** TEST diagnostic de panne *m*; ~ **display** TEST visualisation de défaut *f*; ~ **indication** TELEP indication de défaillance *f*, indication de panne *f*; ~ **isolation** TELEP isolation d'un défaut *f*; ~ **localization** TELEP, TEST localisation des défauts *f*, localisation des dérangements *f*; ~ **localization time** TELEP, TEST temps de localisation de panne *m*; ~ **maintenance** TELEP, TEST maintenance corrective *f*; ~ **marking** TELEP marquage de dérangement *m*, marquage d'erreur *m*; ~ **message** DATA message de dérangement *m*; ~ **mode** TEST mode de panne *m*; ~**-mode analysis** (*FMA*) TEST analyse des modes de panne *f*; ~ **pattern** TELEP cliché de panne *m*; ~ **print-out** COMP message d'erreur *m*; ~ **rate** GEN taux de signalisations *m*; ~ **reception center** *AmE*, ~ **reception centre** *BrE* TEST service de dépannage *m*; ~ **report** GEN signalisation de dérangement *f*, TELEP rapport d'incident *m*, signalisation de dérangement *f*, fiche de dérangement *f*; ~ **reporting** POWER signalisation de défaillances *f*; ~ **simulation** TEST simulation de dérangement *f*, simulation de panne *f*; ~ **state** TELEP état de panne *m*, état de dérangement *m*; ~ **symptom** TELEP symptôme de panne *m*; ~ **syndrome** TELEP cliché de panne *m*; ~ **time** TEST temps d'immobilisation *m*; ~ **tolerance** COMP, DATA, ELECT, TEST tolérance aux fautes *f*, tolérance aux pannes *f*, tolérance aux défaillances *f*; ~**-tolerant computing** COMP, DATA informatique tolérante aux fautes *f*; ~ **tracing** TELEP *general software name* localisation d'avaries *f*, détection de panne *f*, relève de dérangements *f*; ~**-tracing procedure** TELEP procédure de diagnostic *f*; ~ **tree** TEST arbre des pannes *m*, arbre de défaillance *m*; ~**-tree analysis** TEST analyse d'arbre de défaillance *f*, analyse d'arbre des pannes *f*

faulty[1] *adj* COMP défectueux, TELEP, TEST *quality* défectueux, en panne

faulty[2]: ~ **call** *n* TELEP, TEST communication défectueuse *f*; ~ **connection** *n* TELEP, TEST communication défectueuse *f*; ~ **link information** *n* TELEP information de liaison défaillante *f*; ~ **working** *n* TEST dysfonctionnement *m*; ~**-tracing procedure** *n* COMP, ELECT procédure de diagnostic *f*

favor *AmE see favour BrE*

favour *n BrE* ELECT privilège *m*

fax *n* GEN *copy* facsimile *m*, fax *m*, télécopie *f*, PER APP *machine* téléfax *m*, terminal de télécopie *m*; ~ **answerphone** PER APP, TELEP téléphone-fax-répondeur *m*; ~ **machine** PER APP télécopieur *m*, téléfax *m*, terminal de télécopie *m*, TELEP télécopieur *m*, terminal de télécopie *m*; ~ **modem** DATA *Internet* modem télécopieur *m*; ~ **phone** PER APP, TELEP téléphone-fax *m*; ~ **transmission** TELEP émission télécopie *f*

faxswitch *n* SWG prise de télécopieur *f*

FBI *abbr* (*field blanking interval*) DATA, TV intervalle de retour de trame *m*, intervalle de suppression de trame *m*
FC *abbr* (*functional class*) TELEP FC (*classe fonctionnelle*)
FCA *abbr* (*functional class A*) TELEP FCA (*classe fonctionnelle A*)
FCB *abbr* (*functional class B*) TELEP FCB (*classe fonctionnelle B*)
FCC *abbr AmE* (*Federal Communication Commission*) DATA FCC
FCS *abbr* (*frame check sequence*) DATA, TELEP séquence de contrôle de trame *f*, séquence de vérification de trame *f*
FD *abbr* (*functional description*) GEN description fonctionnelle *f*
FDHM *abbr* (*full-duration half-maximum*) DATA FDHM (*durée à mi-crête*)
FDM *abbr* SIG, SWG, TELEP, TRANS (*frequency-division multiplexing*) MRF (*multiplexage par répartition en fréquence*), multiplexage en fréquence *m*, multiplexage par répartition *m*
FDMA *abbr* (*frequency-division multiple access*) SIG, SWG, TELEP, TRANS AMRF (*accès multiple par répartition de fréquence*)
FDS: **~ system** *n* (*frequency-division switching system*) SWG commutateur en fréquence *m*, système de commutation par répartition en fréquence *m*
FDT *abbr* (*formal description technique*) DATA technique de description formelle *f*
FE *abbr* DATA, TELEP (*format effector*) caractère de mise en page *m*, caractère de présentation *m*
FEA *abbr* (*functional entity action*) DATA action d'entité fonctionnelle *f*
feasibility: **~ study** *n* GEN *systems development* étude de faisabilité *f*
feasible *adj* GEN faisable, réalisable
feature *n* COMP fonctionnalité *f*, GEN *function* fonction *f*, TELEP dispositif *m*
FEBE *abbr* (*far-end block error*) DATA erreur de bloc à l'extrémité distante *f*
FECC *abbr* COMP, DATA, RADIO, SIG, TELEP (*forward error correction codex*) correction d'erreurs sans voie de retour *f*
FECN *abbr* (*forward explicit congestion notification*) DATA, SWG, TELEP notification d'encombrement explicite émise vers l'avant *f*
fed: **be ~ with holding current** *vi* PARTS recevoir un courant de maintien
Federal: **~ Communication Commission** *n AmE* (*FCC*) DATA FCC (*Commission fédérale des communications*)
fee *n* GEN redevance *f*, taxe *f*, honoraire *m*; **~-based service** GEN service payant *m*; **~ digits** *n pl* TELEP chiffres de taxation *m pl*
feed[1] *n* COMP, ELECT, POWER *supply* alimentation *f*, RADIO source *f*, source primaire *f*; **~ assembly** ELECT système d'alimentation *m*; **~ hole** COMP, DATA perforation d'entraînement *f*; **~ horn** ELECT cornet d'alimentation *m*; **~ input** ELECT entrée de la source *f*; **~ loss** TRANS perte de ligne *f*; **~ mass saving** RADIO gain de masse sur les sources *m*; **~ mechanism** GEN mécanisme d'alimentation *m*, mécanisme d'avancement *m*; **~ pitch** DATA pas d'entraînement *m*; **~ system** ELECT circuit d'alimentation *m*; **~ system bandwidth** RADIO, TRANS largeur de bande du système d'alimentation *f*; **~ track** COMP, DATA piste d'entraînement *f*
feed[2] *vt* COMP alimenter, *computers* entrer, introduire, saisir, ELECT être transmis à, POWER, RADIO, TV alimenter; **~ back** COMP alimenter en retour
feedback *n* COMP réaction *f*, DATA, ELECT réaction *f*, feedback *m*, rétroaction *f*, *at controls* contre-réaction *f*, GEN *customer or audience response* information en retour *f*, feedback *m*, remontée d'informations *f*, RADIO réaction *f*, feedback *m*, rétroaction *f*, contre-réaction *f*; **~ AGC** TRANS CAG à contre-réaction *m*; **~ closed loop** ELECT boucle fermée de réaction *f*; **~ control** COMP, CONTR régulation *f*; **~ control system** COMP, CONTR système de régulation *m*; **~ decoding** ELECT décodage local *m*; **~ instability** GEN accrochage *m*; **~ level** COMP niveau de rétroaction *m*; **~ loop** COMP, ELECT boucle de réaction *f*, boucle de rétroaction *f*, RADIO boucle de retour de l'information *f*; **~ ratio** ELECT taux de réaction *m*, taux de retour *m*; **~ shift register** DATA, ELECT, SWG registre à décalage à boucle fermée *m*; **~ signal** CONTR, SIG signal de réaction *m*
feeder *n* COMP *paper* système d'alimentation *m*, dispositif d'alimentation *m*, PER APP dispositif d'alimentaion *m*, RADIO *aerials*, TRANS câble d'alimentation *m*, ligne d'alimentation *f*; **~ link** RADIO, TRANS liaison de connexion *f*; **~-link beam area** RADIO, TRANS empreinte d'un faisceau de liaison de connexion *f*; **~-link service area** RADIO, TRANS zone de service de liaison de connexion *f*; **~ loss** RADIO, TRANS perte dans le système d'alimentation *f*
feedforward: **~ AGC** *n* TRANS CAG avec précompensation *m*; **~ amplifier** *n* ELECT amplificateur à correction aval *m*; **~ control** *n* CONTR régulation avec action anticipée *f*, régulation avec action prévisionnelle *f*
feeding *n* COMP, PARTS, POWER *supply* alimentation *f*; **~ bridge** POWER pont d'alimentation *m*; **~ cable** TRANS aiguillage *m*; **~ current** POWER courant d'alimentation *m*; **~ device** COMP, PER APP dispositif d'alimentation *m*; **~ fuse** PARTS, POWER fusible d'alimentation *m*; **~ loop** POWER *supply* cercle fermé *m*
feedthrough: **~ capacitator** *n* POWER condensateur de traversée *m*; **~ pad** *n* POWER pastille de traversée *f*
feeler: **~ gage** *AmE*, **~ gauge** *BrE n* TEST *metrology* jauge d'épaisseur *f*
female: **~-female adapter** *n* COMP, ELECT coupleur femelle-femelle *m*; **~ plug** *n* COMP, ELECT, TEST connecteur femelle *m*
FERF *abbr* (*far-end receive failure*) SIG défaut en réception à l'extrémité distante *m*, dérangement de réception à l'extrémité *m*
ferrite *n* COMP, POWER ferrite *f*; **~ antenna** RADIO antenne à ferrites *f*, antenne à noyau magnétique *f*; **~ core** COMP tore de ferrite *m*; **~ core matrix** COMP matrice à tores de ferrite *f*; **~-core storage** COMP, DATA mémoire à tores de ferrite *f*; **~ matrix** COMP matrice de ferrite *f*; **~ storage** COMP mémoire à ferrites *f*
ferrule *n* PARTS bague *f*, ferrule *f*, virole *f*, TRANS *fibre optics* bague *f*
FET *abbr* (*field effect transistor*) ELECT *semiconductors*, PARTS FET (*transistor à effet de champ*), TEC (*transistor à effet de champ*)
fetch: **~ abstract-operation** *n* COMP opération abstraite extraction *f*; **~ program** *n* COMP programme de chargement *m*; **~ restrictions** *n pl* COMP restrictions à l'extraction *f pl*
FF *abbr* ELECT (*flip-flop*) *logic elements* bascule *f*, bascule bistable *f*, TELEP (*form feed*) page suivante *f*
FFD *abbr* (*freeform deformation*) COMP déformation de forme libre *f*
FFT *abbr* (*fast Fourier transform*) COMP TFR (*transformation de Fourier rapide*)
FI *abbr* (*format identifier*) POWER, TELEP FI (*identificateur de format*)
fiber *AmE see fibre BrE*
fiberglass *AmE see fibreglass BrE*
fibre[1]: **~-optic** *adj BrE* TRANS à fibres optiques, par fibres optiques, sur fibres optiques
fibre[2] *n BrE* GEN fibre *f*; **~ axis** *BrE* TRANS axe de la fibre *m*; **~ buffer** *BrE* TRANS gaînage secondaire *m*, protection secondaire *f*, revêtement protecteur *m*; **~ bundle** *BrE* TRANS faisceau de fibres *m*; **~ cable plant** *BrE* TRANS installation optique *f*; **~ coiling plate** *BrE* TRANS plateau enrouleur de fibre *m*; **~-distributed data interface** *BrE* DATA, TRANS interface de données avec distribution par fibre

f; ~ **end face** *BrE* TRANS face de la fibre *f*, face des fibres à raccorder *f*; ~ **geometry** *BrE* TRANS géométrie des fibres *f*; ~ **jacket** *BrE* PARTS, TRANS revêtement secondaire *m*; ~**-optic backbone** *BrE* ELECT épine dorsale en fibres optiques *f*; ~**-optic bridge-repeater link** ELECT liaison porte-répéteur à fibres optiques *f*; ~**-optic cladding** *BrE* TRANS gaine de fibres optiques *f*; ~**-optic loop** *BrE* TRANS boucle optique *f*; ~**-optic network** *BrE* TRANS réseau optique *m*; ~**-optic receiver** *BrE* TRANS récepteur à fibres optiques *m*, récepteur optoélectronique *m*, équipement optoélectronique de réception *m*, module de réception optique *m*; ~**-optic sensor** *BrE* TRANS détecteur optique *m*; ~**-optic splice** *BrE* TRANS raccord de fibres optiques *m*; ~**-optic terminal device** *BrE* TRANS dispositif d'extrémité de fibres optiques *m*, terminal à fibres optiques *m*; ~**-optic transmission** *BrE* TRANS transmission par fibres optiques *f*; ~ **optics** *BrE* TRANS optique des fibres *f*, technique des fibres optiques *f*; ~ **scattering** *BrE* TRANS diffusion de fibrage *f*

fibreglass *n BrE* ELECT fibre de verre *f*

fidelity *n* RADIO fidélité *f*

Fidonet *n* COMP *Internet* Fidonet *m*

field[1]: ~**-tested** *adj* GEN expérimenté sur le terrain, expérimenté en grandeur nature, piloté; ~**-upgradeable** *adj* GEN transformable sur site

field[2] *n* DATA trame *f*, zone *f*, GEN champ *m*, domaine *m*, TV *frame* trame *f*; ~ **bend correction** TV signal de correction parabolique de trame *m*; ~ **blanking** TV suppression de trame *f*; ~ **blanking interval** (*FBI*) TV intervalle de retour de trame *m*, intervalle de suppression de trame *m*; ~ **coils** *n pl* RADIO stator inductif *m*; ~ **depth** ELECT profondeur de champ *f*; ~ **effect** ELECT effet de champ *m*; ~**-effect emission** ELECT émission par effet de champ *f*; ~**-effect integrated circuit** ELECT circuit intégré à effet de champ *m*; ~ **effect transistor** (*FET*) ELECT *semiconductors*, PARTS transistor à effet de champ *m* (*FET*, *TEC*); ~ **frequency** TV fréquence de balayage vertical *f*; ~ **indicator** TELEP indicateur de domaine *m*; ~ **intensity** TEST valeur de champ *f*, intensité de champ *f*; ~ **intensity meter** TEST champmètre *m*, mesureur d'intensité de champ *m*; ~ **keystone correction** TV signal de correction de trapèze-trame *m*; ~ **keystone waveform** TV signal de correction de trapèze-trame *m*; ~ **length indicator** TELEP indicateur de longueur de domaine *m*; ~ **maintenance** TEST maintenance sur site *f*; ~ **measurements** *n pl* TEST mesures sur le terrain *f pl*; ~ **mesh** ELECT grille à mailles *f*; ~ **pole** POWER pièce polaire *f*; ~ **scan** COMP balayage de trames *m*; ~ **strength** POWER, RADIO intensité de champ *f*, TEST grandeur de champ *f*, valeur de champ *f*; ~ **strength measurement** POWER, RADIO mesure d'intensité de champ *f*, TEST mesure de grandeur de champ *f*, mesure de valeur de champ *f*; ~ **strength meter** TEST champmètre *m*, mesureur d'intensité de champ *m*; ~ **sweep** TV balayage de trames *m*; ~ **sync pulse** SIG, TV signal de synchronisation de trame *m*; ~**-synchronizing signal** SIG, TV signal de synchronisation de trame *m*; ~ **telephone** TELEP téléphone de campagne *m*; ~ **test** TEST essai en grandeur réelle *m*, essai pratique *m*, expérimentation *f*, expérimentation sur le terrain *f*, phase pilote *f*, essai en vraie grandeur *f*; ~ **tilt correction** TV signal de correction linéaire de trame *m*; ~**-time luminance distortion** COMP distorsion de luminance verticale *f*; ~**-time waveform distortion** COMP distorsion de l'ordre d'une trame *f*; ~ **trial** TEST essai pratique *m*, expérimentation sur le terrain *f*, phase pilote *f*, essai en grandeur réelle *m*, essai en vraie grandeur *m*, experimentation *f*; ~ **of vision** (*FOV*) RADIO angle de champ *m*, champ de visibilité *m*, champ de vision *m*, champ observé *m*

field[3]: ~**-test** *vt* GEN expérimenter sur le terrain, essayer en grandeur réelle, piloter

FIF *abbr* COMP, DATA (*facsimile information field*) champ d'informations télécopie *m*

FIFO *abbr* (*first in, first out*) GEN premier entré, premier sorti

fifteen: ~**-supergroup assembly link** *n* TELEP liaison en assemblage de quinze groupes secondaires *f*

figurative: ~ **constant** *n* COMP constante figurative *f*

figure *n* ELECT, RADIO indice *m*; ~ **of merit** RADIO coefficient de qualité *m*, facteur de mérite *m*, indice de mérite *m*; ~ **of merit meter** TEST appareil de mesure témoin *m*; ~**-of-eight diagram** RADIO diagramme en huit *m*; ~ **shift** PER APP, TELEP inversion chiffres *f*; ~**-shift signal** SIG signal d'inversion chiffres *m*

filament *n* ELECT, PARTS *plastics* filament *m*, tube à cathode chaude *m*; ~ **current** ELECT courant de chauffage *m*, courant de filament *m*; ~ **lamp** CLIM lampe à incandescence *f*; ~ **starting current** ELECT courant de chauffage au démarrage *m*, courant de démarrage du filament *m*, courant de filament au démarrage *m*; ~ **surge current** ELECT courant de chauffage au démarrage *m*, courant de filament au démarrage *m*, courant de pointe du filament *m*; ~ **voltage** ELECT tension de filament *f*, tension de chauffage *f*

file *n* COMP, DATA, TELEP, TEST fichier *m*; ~ **access** COMP, DATA accès aux fichiers *m*; ~ **allocation table** (*FAT*) COMP, DATA table d'allocation des fichiers *f*; ~ **attribute** COMP, DATA attribut de fichier *m*; ~ **backup** COMP, DATA sauvegarde des fichiers *f*; ~ **compression** COMP, DATA compression de fichier *f*; ~ **conversion** COMP, DATA conversion de fichier *f*; ~ **decompression** COMP, DATA décompression de fichier *f*; ~ **divider** TV diviseur de fréquence de trame *m*; ~ **layout** COMP, DATA définition de fichier *f*, disposition de fichier *f*, présentation de fichier *f*; ~ **maintenance** COMP, DATA mise à jour de fichier *f*; ~ **protection** COMP, DATA protection de fichier *f*; ~ **recovery** COMP, DATA récupération des fichiers *f*; ~ **search** COMP, DATA recherche de fichier *f*; ~ **security** COMP, DATA sécurité des fichiers *f*; ~ **server** COMP, DATA, TEST serveur de fichiers *m*; ~ **source** COMP, DATA fichier source *m*; ~ **transfer** COMP, DATA transfert de fichier *m*; ~ **transfer, access and management** (*FTAM*) COMP, DATA transfert, accès et gestion de fichiers *m* (*FTAM*); ~ **transfer access and management service element** (*FTAMSE*) COMP, DATA élément de service du transfert, accès et gestion de fichiers *m*; ~ **transfer protocol** (*FTP*) COMP, DATA protocole de transfert de fichiers *m*

filename: ~ **extension** *n* COMP nom d'extension *m*; ~ **syntax** *n* DATA, TELEP syntaxe du nom de fichier *f*

fill[1]: ~ **character** *n* DATA caractère de remplissage *m*; ~**-in** *n* COMP variation rythmique *f*; ~**-in signal unit** *n* TELEP trame sémaphore de remplissage *f*

fill[2] *vt* COMP, DATA, GEN remplir

filled: ~ **cable** *n* TRANS câble rempli *m*

filler *n* DATA caractère de remplissage *m*, GEN remplissage *m*, *cables* bourrage *m*, PARTS *plastics* charge *f*; ~ **character** DATA caractère de remplissage *m*

filling *n* COMP, DATA, GEN remplissage *m*; ~ **coefficient** DATA coefficient de remplissage *m*; ~ **compound** ELECT élément de remplissage *m*; ~ **rate** GEN taux d'occupation *m*

film *n* ELECT film *m*, PARTS, POWER film *m*, pellicule *f*, TV film *m*; ~ **resistor** PARTS, POWER résistance à film *f*

filter *n* COMP, DATA, ELECT, PARTS, POWER, RADIO, TRANS filtre *m*; ~ **amplifier** ELECT amplificateur de filtre actif *m*; ~**-and-sample detector** ELECT détecteur à filtre et échantillonnage *m*; ~**-bank system** RADIO *maritime satellite* système à banc de filtres *m*; ~ **capacitor** PARTS, POWER condensateur de filtrage *m*; ~ **coupler** ELECT filtre coupleur *m*; ~ **cutoff** COMP coupure de filtre *f*; ~ **flank** ELECT bordure du filtre *f*; ~ **fuse** PARTS, POWER fusible de filtre *m*; ~**-item** DATA item-filtre *m*; ~ **roll-off** ELECT progressivité de la coupure du filtre *f*; ~ **section** ELECT cellule de filtrage *f*; ~ **slope** COMP pente de filtre *f*

filterbank *n* COMP banc de filtrage *m*, banc de filtres *m*
filtered: ~ **QPSK** *n* RADIO *maritime satellite* MDPQ avec filtrage *f*; ~ **voltage** *n* POWER tension filtrée *f*, tension lissée *f*
filtering *n* COMP, DATA, ELECT, TRANS, TV filtrage *m*
final: ~ **amplifier** *n* ELECT amplificateur final *m*; ~ **bit** *n* TELEP bit de fin *m*; ~ **character** *n* DATA caractère final *m*, multiplet final *m*; ~ **circuit group** *n* TELEP faisceau final *m*; ~ **controlling element** *n* CONTR élément de réglage final *m*; ~ **inspection** *n* TEST contrôle final *m*; ~ **selection stage** *n* SWG étage de sélection final *m*; ~ **selector** *n* TELEP sélecteur final *m*; ~ **selector stage** *n* TELEP étage de sélecteur final *m*; ~ **stage** *n* RADIO, TELEP étage final *m*; ~ **state** *n* DATA état final *m*; ~ **test** *n* TEST essai final *m*, test final *m*
find/replace *vt* COMP rechercher/remplacer
fine: ~ **adjustment** *n* COMP, GEN réglage fin *m*, réglage précis *m*; ~ **diagnosis** *n* GEN diagnostic détaillé *m*; ~ **positioning** *n* GEN régulation de précision *f*; ~ **regulation** *n* GEN régulation de précision *f*; ~ **setting** *n* COMP réglage fin *m*, réglage précis *m*; ~ **switch** *n* ELECT bouton de réglage fin *m*; ~ **tuning** *n* COMP réglage fin *m*, réglage précis *m*, ELECT mise au point précise *f*, réglage fin *m*; ~**-wire fuse** *n* PARTS, POWER fusible pour faible intensité *m*
finger: ~ **hole** *n* PER APP, TELEP *on telephone* trou de cadran *m*; ~ **stock** *n* RADIO bande de contact *f*; ~ **stop** *n* PER APP, TELEP *on telephone* butée de cadran *f*; ~ **strip** *n* RADIO bande de contact *f*; ~ **wheel** *n BrE* (*cf dial disk AmE*) PER APP, TELEP *on telephone* disque du cadran d'appel *m*, disque mobile *m*
fingering *n* COMP doigté *m*
finish[1] *n* COMP finition *f*
finish[2]: ~ **off** *vt* GEN affiner
finishing off *n* GEN affinage *m*
finite: ~ **automation** *n* DATA automate à états finis *m*; ~ **field** *n* SIG corps fini *m*; ~ **impulse response** *n* (*FIR*) ELECT, TRANS réponse impulsionnelle finie *f* (*RIF*); ~ **impulse response filter** *n* ELECT, TRANS filtre à réponse impulsionnelle finie *m*; ~ **state** *n* DATA, SIG état fini *m*; ~**-state vector quantization** *n* SIG quantification vectorielle avec un automate à états finis *f*
FIR *abbr* (*finite impulse response*) ELECT, TRANS RIF (*réponse impulsionnelle finie*); ~ **filter** *n* ELECT, TRANS filtre RIF *m*
fire[1]: ~**-resistant** *adj* PARTS *materials* ignifuge
fire[2]: ~ **resistance** *n* PARTS résistance au feu *f*; ~ **valve** *n* PARTS clapet coupe-feu *m*
fireproof *adj* PARTS à l'épreuve du feu, ignifuge, ininflammable
firewall *n* COMP, DATA filtre sécuritaire *m*, logiciel de filtrage *m*
firing *n* ELECT allumage *m*; ~ **power** ELECT puissance d'allumage *f*; ~ **time** ELECT allumage *m*, instant d'allumage *m*
firmware *n* COMP microprogramme *m*, micrologiciel *m*
first[1]: ~**-adjacent channel** *n* RADIO canal premier-adjacent *m*; ~ **breakdown** *n* ELECT première rupture *f*; ~**-choice circuit group** *n* ELECT, SWG, TELEP faisceau de premier choix *m*; ~**-choice group** *n* ELECT, SWG, TELEP faisceau de premier choix *m*; ~**-data multiplexer** *n* DATA, TRANS multiplexeur de données du premier ordre *m*; ~**-line maintenance** *n* TEST maintenance élémentaire *f*; ~**-order multiplexer** *n* DATA, TRANS multiplexeur du premier ordre *m*, multiplexeur primaire *m*; ~**-party release** *n* COMP, TELEP libération par premier abonné *f*; ~**-subscriber release** *n* COMP, TELEP libération par premier abonné *f*; ~ **upright** *n* TELEP montant initial *m*
first[2]: ~ **in, first out** *phr* (*FIFO*) COMP premier entré, premier sorti
fishbone: ~ **antenna** *n* RADIO antenne en arête de poisson *f*
fishplate *n* PARTS ferrure d'assemblage *f*
fit *vt* DATA adapter, GEN adapter, monter
FITE *abbr* (*forward interworking telephony event*) TELEP ETAV (*événement téléphonique d'interfonctionnement vers l'avant*)
fitting *n* GEN montage *m*, pose *f*; ~ **tolerance** GEN tolérance d'ajustement *f*
fittings *n pl* PARTS ferrures *f pl*
five: ~**-level RTTY** *n* TELEP radiotéléimprimeur cinq moments *m*; ~**-point impairment scale** *n* TEST échelle de dégradation à cinq notes *f*; ~**-unit code** *n* TELEP code à cinq moments *m*
500/20: ~ **signaling** *AmE*, ~ **signalling** *BrE* *n* SIG signalisation à fréquence 500/20 *f*
fix *vt* GEN réparer
fixed[1]: ~**-resistance** *adj* ELECT à résistance fixe
fixed[2]: ~**-antenna direction finder** *n* RADIO radiogoniomètre à antenne fixe *m*; ~ **attenuator** *n* TRANS affaiblisseur fixe *m*, atténuateur fixe *m*; ~ **bias** *n* ELECT polarisation fixe *f*; ~ **blade connector** *n* PARTS *connecting devices* connecteur fixe multibroche *m*; ~ **charge** *n* TELEP redevance forfaitaire *f*; ~**-destination call services** *n pl* TELEP service d'appels à destination fixe *m*; ~ **direction finder** *n* RADIO radiogoniomètre à antenne fixe *m*; ~ **fork connector** *n* PARTS connecteur fixe à contacts lyre *m*; ~ **frequency** *n* RADIO fréquence fixe *f*; ~ **light** *n* TV *optics* feu fixe *m*; ~ **network** *n* TRANS réseau filaire *m*; ~ **night-service** *n* TELEP service de nuit à renvoi fixe *m*; ~ **packet recording** *n* COMP enregistrement par paquets fixes *m*; ~ **pin connector** *n* PARTS connecteur fixe multibroche *m*; ~**-point method** *n* COMP méthode à virgule fixe *f*; ~ **radio beacon** *n* RADIO borne fixe radio *f* (*BFR*); ~**-resolution progressive mode** *n* COMP mode progressif à résolution fixe *m*; ~ **segmentation** *n* COMP segmentation fixe *f*; ~ **sequencer** *n* DATA séquenceur fixe *m*; ~ **store** *n* COMP mémoire fixe *f* (*ROM*)
fixing: ~ **bracket** *n* PARTS bride de fixation *f*, étrier de fixation *m*, étrier de montage *m*; ~ **hole** *n* PARTS trou de fixation *m*; ~ **plate** *n* PARTS plaque de fixation *f*, plaque de support *f*; ~ **ring** *n* PARTS anneau de fixation *m*, bague de fixation *f*; ~ **screw** *n* PARTS vis d'assemblage *f*, vis de fixation *f*
flag *n* DATA drapeau *m*; ~ **bit** DATA bit de drapeau *m*; ~ **byte** DATA multiplet de signalisation *m*
flame *n* ELECT tirade électronique *f*
flap: ~ **attenuator** *n* RADIO affaiblisseur à lame *m*, atténuateur à lame *m*
flare: ~ **antenna** *n* RADIO antenne-cornet *f*, cornet *m*
flash[1] *n* GEN flash *m*; ~ **time** PER APP *scanner* temps de flashage *m*
flash[2] *vi* COMP, TELEP, TEST *indicating devices* clignoter
flashing *n* COMP, TELEP, TEST clignotement *m*; ~ **cursor** COMP curseur clignotant *m*; ~ **dot** COMP point clignotant *m*; ~ **light** TEST avertisseur lumineux *m*, TV *optics* feu clignotant *m*; ~ **rate** POWER fréquence des éclairs *f*
flashover *n* POWER contournement *m*; ~ **clearance** POWER distance de contournement *f*, distance d'isolement *f*; ~ **voltage** POWER tension d'arc *f*, tension disruptive *f*
flat[1] *adj* GEN plan, POWER dechargé
flat[2]: ~ **antenna** *n* RADIO antenne plane *f*; ~ **battery** *n* POWER pile déchargée *f*; ~**-bed plotter** *n* PER APP, TELEP table traçante *f*; ~**-bed scanner** *n* PER APP, TELEP scanner à plat *m*; ~**-bed transmitter** *n* PER APP, TRANS émetteur à plat *m*; ~ **cable** *n* ELECT câble plat *m*; ~ **fading** *n* RADIO évanouissement uniforme *m*, évanouissement par interférence *m*; ~**-fading margin** *n* RADIO marge d'évanouissement uniforme *f*; ~ **frequency characteristic** *n* ELECT réponse uniforme des fréquences *f*; ~ **leakage power** *n* ELECT puissance de fuite pendant le palier *f*; ~ **level** *n* COMP niveau neutre *m*; ~ **line** *n* RADIO ligne apériodique *f*, ligne non résonnante *f*; ~**-pack** *n* POWER boîtier plat *m*; ~ **panel display** *n* PER APP écran plat *m*; ~ **price** *n* GEN prix forfaitaire *m*, prix fixe *m*; ~ **rate** *n* APPL taxe forfaitaire *f*, tarif fixe *m*; ~**-rate service** *n* APPL service à rémunération forfaitaire *m*, service à tarif fixe *m*; ~ **response** *n* ELECT réponse neutre *f*; ~ **screen** *n* TV écran plat *m*; ~ **shading** *n* TV ombrage simple *m*; ~**-top antenna** *n*

RADIO antenne à nappe horizontale *f*, antenne en nappe *f*; **~-topped pulse** *n* GEN impulsion à plateau *f*
flatness: **~ of bearing** *n* RADIO flou *m*
F-layer *n* COMP, RADIO couche F *f*
FLCD *abbr* (*ferro-electric liquid crystal display*) COMP FLCD (*écran à cristaux liquides ferro-électriques*)
flex *n* ELECT flexion *f*, PARTS flexible *m*; **~-ferrule** ELECT coupleur *m*
flexibility *n* GEN flexibilité *f*; **~ strength** GEN tenue à la flexion *f*
flexible[1] *adj* GEN souple, flexible, pliable
flexible[2]: **~ access switch** *n* (*FAS*) SWG commutateur à accès flexible *m*; **~ cable** *n* POWER câble souple *m*; **~ conductor** *n* POWER *cables* conducteur souple *m*; **~ disk** *n* COMP disque souple *m*, disquette *f*; **~ hose** *n* PARTS flexible *m*; **~ media stack** *n* ELECT pile de support flexible *f*; **~ night-service** *n* TELEP service de nuit à renvoi sélectif *m*; **~ printed board** *n* ELECT *printed circuits* carte imprimée souple *f*; **~ waveguide** *n* DATA, RADIO, TRANS guide d'onde flexible *m*
flexstrip *n* POWER circuit souple *m*
flicker[1] *n* TV papillotement *m*, *optics* papillonnement *m*; **~ noise** TV bruit de scintillation *m*
flicker[2] *vi* TV *optics* papilloter
flickering *n* RADIO scintillement *m*, TV papillonnement *m*, scintillement *m*, papillotement *m*; **~ light** TV *optics* feu scintillant *m*, lampe clignotante *f*
flight: **~ path** *n* RADIO ligne de vol *f*, trajectoire de vol *f*; **~-path angle** *n* RADIO pente de la trajectoire *f*
flip[1]: **~ chip** *n* ELECT puce à protubérances *f*; **~-flop** *n* (*FF*) ELECT *logic elements* bascule *f*, bascule bistable *f*
flip[2] *vt* COMP basculer, retourner
float: **~ charging current** *n* POWER *batteries* courant de charge flottante *m*; **~ charging level** *n* POWER *batteries* niveau de charge flottante *m*
floating: **~ battery** *n* POWER batterie flottante *f*; **~ charge** *n* POWER *batteries* charge d'entretien *f*; **~ drift-tube klystron** *n* ELECT klystron à tube de glissement flottant *m*; **~ point** *n* COMP virgule flottante *f*; **~ point hardware** *n* COMP virgule flottante câblée *f*; **~ point processor** *n* COMP processeur à virgule flottante *m*; **~ point representation** *n* COMP représentation en virgule flottante *f*; **~ point unit** *n* (*FPU*) COMP unité en virgule flottante *f*; **~ receiver** *n* RADIO récepteur à fréquence réglable *m*; **~ voltage** *n* PARTS *semiconductors* tension flottante *f*
floor: **~ bar** *n* PARTS profilé de plancher *m*, rail *m*; **~ load** *n* GEN *construction* charge au sol *f*; **~ mounting** *n* GEN montage au sol *m*; **~ plan** *n* GEN *documentation* plan d'implantation *m*; **~ rail** *n* PARTS rail *m*, profilé de plancher *m*; **~-standing system** *n* ELECT système posé à même le sol *m*
floppy: **~ disk** *n* COMP disque souple *m*, disquette *f*; **~ disk drive** *n* COMP lecteur de disquette *m*; **~ disk reader** *n* COMP lecteur de disquette *m*
flow[1] *n* GEN écoulement *m*, flux *m*, *of flowchart* liaisons *f pl*; **~ chart** COMP, GEN organigramme *m*; **~-chart symbol** COMP symbole d'organigramme *m*; **~ control** CONTR contrôle de flux *m*, DATA commande de flux *f*, contrôle de flux *m*, régulation de débit *f*, TELEP commande de flux *f*, contrôle de flux *m*; **~-control parameter selection and indication** TELEP sélection et indication des paramètres de commande de flux *f pl*; **~ indicator** TELEP *indicating devices* indicateur de débit *m*; **~ regulator** CONTR *regulating devices* régulateur de débit *m*; **~ stress** GEN *strength* seuil d'écoulement *m*
flow[2]: **~ through** *vt* POWER traverser
flow[3] *vi* CONTR circuler, POWER circuler, s'écouler, passer
flowcharting *n* GEN *flowcharts* établissement d'un organigramme *m*
flowmeter *n* CONTR débitmètre *m*
FLS *abbr* (*free-line signal*) TELEP signal d'inoccupation *m*
fluctuation *n* POWER, TEST fluctuation *f*
fluorescence *n* RADIO, TRANS *fibre optics* fluorescence *f*
fluorescent: **~ lamp** *n* ELECT *electronics* tube fluorescent *m*
flush *n* COMP justification *f*; **~ antenna** RADIO antenne encastrée *f*; **~ conductor** ELECT conducteur affleurant *m*, conducteur encastré *m*; **~-mount panel** ELECT panneau de montage encastrable *m*; **~-mounted single information outlet** ELECT prise informatique simple encastrée *f*; **~ mounting** GEN montage encastré *m*; **~ panel** PARTS platine encastrée *f*
flutter *n* RADIO, TV scintillement *m*
flux *n* ELECT, GEN, TRANS flux *m*; **~ density** ELECT induction magnétique *f*, induction *f*, densité flux *f*, densité surfacique *f*; **~ pattern** ELECT diagramme des lignes de force *m*
fluxmeter *n* TEST fluxmètre *m*
fly: **~-back transformer** *n* TV transformateur de ligne *m*
flyback *n* TV retour du spot *m*; **~ principle** POWER *supply* principe de la récupération d'énergie *m*
flying: **~ lead** *n* ELECT queue de cochon *f*, *components* composant soudé *m*; **~-lead component** *n* ELECT composant soudé *m*; **~-spot scanner tube** *n* ELECT tube analyseur à balayage *m*
FM *abbr* APPL (*facilities management*), COMP (*facilities management*) infogérance *f*, RADIO (*frequency modulation*) MF (*modulation de fréquence*); **~ expander** *n* COMP unité d'extension FM *f*; **~ radio receiver** *n* RADIO récepteur radio modulation de fréquence *m*; **~ stack** *n* COMP pile FM *f*; **~ synthesis** *n* (*frequency-modulation synthesis*) COMP synthèse FM *f* (*synthèse frequency modulation*)
FMA *abbr* (*fault-mode analysis*) TEST *dependability* analyse des modes de panne *f*
FMBS *abbr* (*frame-mode bearer service*) APPL, TRANS service support en mode trame *m*
FN *abbr* (*forwarded notification*) DATA, TELEP notification retransmise *f*
F/O *abbr* (*fiber-optic AmE*, *fibre-optic BrE*) TRANS à fibre optique, par fibre optique, sur fibre optique; **~ link** *n* POWER, TRANS (*optical fiber link AmE*, *optical fibre link BrE*) liaison par fibre optique *f*
focal: **~-fed antenna** *n* RADIO antenne à alimentation focale *f*
focusing *n* ELECT concentration *f*; **~ coil** ELECT bobine de concentration *f*; **~ electrode** ELECT électrode de concentration *f*, électrode de focalisation *f*; **~ magnet** ELECT aimant de focalisation *m*
fog: **~ effect** *n* COMP effet de brouillard *m*
foil *n* PARTS feuille *f*, film *m*, POWER film *m*
fold[1]: **~-out array** *n* RADIO réseau dépliable *m*
fold[2] *vt* COMP plier
folded: **~-dipole antenna** *n* RADIO doublet replié *m*; **~ network** *n* TRANS réseau replié *m*
follow[1] *n* GEN suivi *m*; **~-me call forward** TELEP renvoi d'appels à téléprogrammations successives *m*; **~-me diversion** TELEP renvoi variable *m*; **~-on call** TELEP appel suivant *m*, PER APP *payphone* enchaînement de communication *m*; **~-on satellite** RADIO satellite de relève *m*; **~-up** GEN suivi *m*, relance *f*; **~-up control** CONTR *automatic* régulation de correspondance *f*
follow[2]: **~ up** *vt* GEN suivre
follow-on: **~ service advice** *n* DATA, TELEP avis de service à la suite *m*
font *n* DATA police *f*, PER APP police *f*, *printing* police de caractères *f*; **~ manager** COMP gestionnaire de polices de caractères *m*; **~ size** COMP corps *m*; **~ style** COMP style *m*
foolproof *adj* GEN protégé contre les fausses manœuvres, TEST *dependability* infalsifiable
footcandle: **~ meter** *n* GEN luxmètre *m*
footnote *n* COMP note de bas de page *f*
footprint *n* RADIO zone de couverture *f*, empreinte *f*, empreinte de faisceau du satellite *f*, empreinte de lobe d'antenne du satellite *f*

forced: ~ **oscillation** *n* POWER oscillation forcée *f*; ~ **rerouting** *n* TELEP passage sous contrainte sur route de secours *m*
foreground *n* TV premier plan *m*; ~ **processing** COMP traitement prioritaire *m*
foreseen: ~ **test outcome** *n* TEST résultat prévu d'un test *m*
fork: ~ **contact** *n* ELECT contact lyre *m*; ~ **contact unit** *n* ELECT barrette à contacts lyre *f*
forked: ~ **working** *n* TELEP bifurquée *f*, communication bifurquée *f*
form[1] *n* COMP, DATA, RADIO forme *f*; ~-**board section** TRANS *cables* section de gabarit pour peigne *f*; ~ **factor** TEST *EMC* facteur de forme *m*; ~ **feed** (*FF*) TELEP page suivante *f*
form[2]: ~ **charge** *vt* POWER *batteries* charger en formation; ~ **the fiber** *AmE*, ~ **the fibre** *BrE vt* TRANS fibrer
formal: ~ **description technique** *n* (*FDT*) DATA technique de description formelle *f*; ~ **logic** *n* COMP logique formelle *f*; ~ **reasoning** *n* COMP raisonnement formel *m*; ~ **specification** *n* DATA spécification formelle *f*
formant *n* TELEP *acoustics* formant *m*, zone formante *f*; ~ **vocoder** RADIO *maritime satellite* codeur à fréquences vocales à formant *m*, vocodeur à formant *m*
format[1] *n* COMP, DATA configuration *f*, format *m*, ELECT configuration *f*, PER APP, TELEP, TV format *m*; ~ **control** COMP commande d'édition *f*; ~ **effector** (*FE*) COMP caractère de présentation *m* DATA, TELEP caractère de mise en page *m*, caractère de présentation *m*; ~ **identifier** (*FI*) POWER, TELEP identificateur de format *m* (*FI*); ~ **output** TELEP sortie de format *f*; ~ **parameter input** COMP entrée de paramètres avec format *f*; ~ **parameter-entry sequence** COMP séquence d'entrée de paramètres avec format *f*; ~ **tape** DATA bande pilote *f*
format[2] *vt* COMP, DATA mettre en forme
formatting *n* COMP, DATA formatage *m*, mise en forme *f*
formboard *n* GEN, TRANS gabarit *m*; ~ **base** TRANS *cables* cadre pour peigne *m*; ~ **designation** TRANS *cables* désignation de gabarit pour peigne *f*; ~ **peg** TRANS *cables* tige de gabarit pour peigne *f*; ~ **plate** TRANS *cables* plaque de base de gabarit pour peigne *f*; ~ **specification** TRANS *cables* spécification de gabarit pour peigne *f*
fortuitous: ~ **distortion** *n* TELEP distorsion fortuite *f*
forum *n* DATA forum *m*
forward[1]: ~ **bias** *n* ELECT polarisation directe *f*; ~ **biasing** *n* ELECT polarisation directe *f*; ~ **chaining** *n* DATA chaînage avant *m*; ~ **channel** *n* DATA, TELEP, TRANS voie d'aller *f*; ~ **characteristic** *n* ELECT *diodes* caractéristique de conduction directe *f*, caractéristique directe *f*; ~ **current** *n* ELECT *diodes*, POWER courant direct *m*; ~ **current transfer ratio** *n* ELECT *semiconductors* rapport de transfert direct du courant *m*; ~ **direction** *n* ELECT sens direct *m*; ~ **echo** *n* TELEP écho vers l'aval *m*; ~ **error correction** *n* COMP, DATA, SIG, TELEP correction directe d'erreurs *f*; ~ **error correction codex** *n* (*FECC*) COMP, DATA, RADIO, TELEP correction d'erreurs sans voie de retour *f*; ~ **error correction coding** *n* COMP, TELEP codage correcteur d'erreurs sans circuit de retour *m*; ~ **explicit congestion notification** *n* (*FECN*) DATA, SWG, TELEP notification d'encombrement explicite émise vers l'avant *f*; ~ **hold** *n* SWG maintien par l'avant *m*; ~ **indicator bit** *n* TELEP bit indicateur vers l'avant *m*; ~-**input signal** *n* SIG signal d'entrée vers l'avant *m*; ~ **interworking telephony event** *n* (*FITE*) TELEP événement téléphonique d'interfonctionnement vers l'avant *m* (*ETAV*); ~-**motion vector** *n* COMP, TV vecteur cinétique anticipé *m*, vecteur de mouvement anticipé *m*; ~ **power dissipation** *n* ELECT *semiconductors*, POWER perte de puissance directe *f*, puissance dissipée en direct *f*; ~ **power loss** *n* ELECT *semiconductors*, POWER puissance dissipée en direct *f*, perte de puissance directe *f*; ~ **prediction** *n* TV prédiction anticipée *f*; ~ **recall signal** *n* TELEP signal de rappel vers l'avant *m*; ~ **resistance** *n* ELECT résistance directe en courant continu *f*; ~ **round-the-world echo** *n* RADIO écho avant tour de terre *m*; ~ **scattering** *n* RADIO prodiffusion *f*; ~ **sequence number** *n* DATA, SWG numéro de séquence vers l'avant *m*; ~ **signal** *n* POWER signal d'action *m*; ~ **signaling** *AmE*, ~ **signalling** *BrE n* TELEP signalisation vers l'avant *f*; ~ **slope resistance** *n* ELECT, PARTS résistance apparente directe *f*; ~ **solar array** *n* RADIO panneau solaire de proue *m*; ~ **transfer facility** *n* TELEP faculté d'intervention *f*; ~ **transfer signal** *n* SIG signal d'intervention *m*; ~ **voltage** *n* ELECT tension directe *f*; ~ **voltage drop** *n* ELECT chute de tension directe *f*
forward[2] *vt* GEN réexpédier, renvoyer, TELEP *calls* renvoyer, transférer
forwarded: ~ **notification** *n* (*FN*) DATA, TELEP notification retransmise *f*
forwarding *n* GEN, TELEP *calls* renvoi *m*; ~ **request** TELEP demande de renvoi *f*
four: ~-**channel telephony** *n* TELEP tétraphonie à quatre canaux *f*; ~-**concentric-circle near-field template** *n* TRANS gabarit à quatre cercles concentriques en champ proche *m*; ~-**concentric-circle refractive-index template** *n* TRANS gabarit d'indice de réfraction à quatre cercles concentriques *m*; ~-**frequency diplex telegraphy** *n* TELEP télégraphie diplex à quatre fréquences *f*; ~-**pole filter** *n* ELECT filtre quadripolaire *m*; ~-**signature coding system** *n* RADIO système de codage à quatre signatures *m*; ~-**terminal network** *n* ELECT, POWER, TRANS quadripôle *m*; ~-**wire cable** *n* ELECT câble à quatre fils *m*; ~-**wire chain** *n* TELEP chaîne à quatre fils *f*; ~-**wire circuit** *n* TELEP circuit à quatre fils *m*; ~-**wire crosspoint** *n* SWG point d'intersection à quatre fils *m*; ~-**wire repeater** *n* TRANS répéteur à quatre fils *m*; ~-**wire switch** *n* SWG autocommutateur à quatre fils *m*, commutateur à quatre fils *m*; ~-**wire switching** *n* SWG commutation à quatre fils *f*; ~-**wire switching system** *n* SWG commutateur à quatre fils *m*, système d'autocommutateur à quatre fils *m*; ~-**wire-type circuit** *n* TELEP circuit type quatre fils *m*
Fourier: ~ **transform infrared spectography** *n* (*FTIR*) TEST spectrographie infrarouge à transformée de Fourier *f*; ~ **transformation** *n* (*FT*) TEST transformation de Fourier *f*
FOV *abbr* (*field of vision*) RADIO angle de champ *m*, champ de visibilité *m*, champ de vision *m*, champ observé *m*
FP: ~ **laser diode** *n* (*Fabry-Pérot laser diode*) TRANS diode laser de Fabry-Pérot *f*
FPS *abbr* (*fast packet switch*) DATA, SWG, TELEP, TRANS commutateur de paquets rapide *m*
FPU *abbr* (*floating point unit*) COMP unité en virgule flottante *f*
fractal: ~ **dimension** *n* COMP dimension fractale *f*
fractional: ~ **ratio** *n* COMP rapport fractionnaire *m*
fracture *n* PARTS rupture *f*
fragmentation *n* COMP morcellement *m*
frame *n* APPL *artificial intelligence* cadre *m*, schéma *m*, COMP image *f*, DATA trame *f*, PARTS *rack* cadre *m*, *equipment* châssis *m*, TELEP répartiteur *m*, châssis *m*, TRANS trame *f*, TV cadre *m*, image *f*, trame *f*; ~ **acquisition and synchronization** TRANS acquisition et synchronisation de trame *f pl*; ~ **alignment** TRANS verrouillage de trame *m*; ~-**alignment recovery time** TRANS temps de reprise du verrouillage de trame *m*; ~-**alignment signal** (*FAS*) SIG signal de verrouillage de trame *m*; ~-**alignment state** TRANS état de verrouillage de trame *m*; ~-**alignment time slot** TELEP, TRANS intervalle de temps de verrouillage de trame *m*, créneau temporel de verrouillage de trame *m*; ~ **check sequence** (*FCS*) DATA, TELEP séquence de contrôle de trame *f*, séquence de vérification de trame *f*; ~ **code** DATA code de synchronisation de trame *m*; ~ **distributor** TRANS distributeur des trames *m*; ~ **format** TV format d'image *m*; ~ **frequency** TV fréquence de trame *f*, fréquence d'image *f*; ~ **generator** TRANS générateur de trame *m*; ~ **grabber** COMP numériseur d'images vidéo *m*, numériseur vidéo *m*, TV carte vidéo, numériseur d'images vidéo *m*, numériseur vidéo *m*;

~ **grabbing** TV capture vidéo *f*; ~ **length** TRANS durée de trame *f*, longueur de trame *f*; ~ **management message** (*FMM*) DATA message de gestion de trame *m*; ~**-mode bearer service** (*FMBS*) APPL, TRANS service support en mode trame *m*; ~ **overhead** DATA capacité de servitude de trame *f*; ~ **pattern** DATA schéma de verrouillage de trame *m*; ~ **processor** DATA processeur de gestion de la trame *m*; ~ **rate** DATA fréquence de trame *f*; ~ **reject** (*FRMR*) TELEP rejet de trame *m*; ~ **relay** TRANS commutation par trames *f*, relais des trames *m*; ~ **slip** DATA glissement de trame *m*; ~**-slotted system** RADIO système crénelé à trame *m*; ~ **structure** TRANS structure de trame *f*; ~ **synchronization** TRANS synchronisation de trame *f*, verrouillage de trame *m*; ~ **synchronization unit** TRANS unité de synchronisation de trame *f*; ~ **synchronizer** TRANS synchronisateur de trame *m*; ~ **timing** TRANS synchronisation de trame *f*; ~ **unit** TRANS unité de trame *f*; ~ **unit step** TRANS échelon unitaire de trame *m*

framed: ~ **2.048 Mbit/s channel** *n* DATA canal à 2,048 Mbit/s tramé *m*

framework *n* APPL *structure* cadre *m*, schéma *m*, COMP cadre de travail *m*, PARTS ossature *f*, cadre *m*, châssis *m*; ~ **agreement** GEN accord cadre *m*

framing *n* TELEP, TRANS verrouillage de trame *m*; ~ **bit** TELEP bit de verrouillage de trame *m*; ~ **information** TRANS informations de verrouillage de trame *f pl*; ~ **pattern** DATA schéma de verrouillage de trame *m*; ~ **pulse** RADIO impulsion d'encadrement *f*

franchise: ~ **area** *n* TV bassin d'audience *m*, site *m*

fraud *n* GEN piratage *m*

fraudulent: ~ **access attempt** *n* GEN tentative d'accès frauduleuse *f*

Fraunhofer: ~ **diffraction pattern** *n* TRANS diagramme de diffraction de Fraunhofer *m*; ~ **region** *n* RADIO, TRANS région de Fraunhofer *f*

frayed: ~ **wiring** *n* ELECT câblage effiloché *m*, fils effilochés *m pl*

FRC *abbr* (*fault reception center AmE*, *fault reception centre BrE*) TEST service de dépannage *m*

free[1] *adj* GEN, TELEP libre; ~ **of charge** GEN exempt de droits, gratuit; ~**-running** ELECT libre, non piloté, non synchronisé, sans charge, sans signaux de synchronisation

free[2]: ~ **of charge** *adv* GEN à titre gratuit

free[3]: ~ **blade connector** *n* ELECT, PARTS connecteur libre multibroche *m*, fiche multibroche *f*; ~ **connector** *n* ELECT, PARTS connecteur libre *m*; ~ **convection** *n* CLIM convection libre *f*, convection naturelle *f*; ~**-field room** *n* TEST *acoustics* chambre sourde *f*, salle anéchoïque *f*; ~ **fork connector** *n* ELECT, PARTS connecteur libre à contacts lyre *m*, fiche à contacts lyre *f*; ~ **form** *n* COMP forme libre *f*; ~ **format** *n* DATA format libre *m*; ~**-line signal** *n* (*FLS*) TELEP signal d'inoccupation *m*; ~ **oscillation** *n* ELECT oscillation libre *f*; ~ **pin connector** *n* ELECT fiche multibroche *f*, connecteur libre multibroche *m*, PARTS connecteur libre multibroche *m*; ~**-running frequency** *n* ELECT fréquence propre *f*, oscillation libre *f*; ~**-running oscillator** *n* ELECT oscillateur à oscillation libre *m*; ~**-running oscillator frequency** *n* ELECT fréquence d'oscillateur libre *f*; ~**-running speed** *n* DATA vitesse normale de fonctionnement *f*; ~ **sleeve connector** *n* ELECT, PARTS fiche à douilles *f*, *connecting devices* connecteur libre à douilles *m*; ~**-space antenna** *n* RADIO antenne en espace libre *f*; ~**-space attenuation** *n* RADIO, TRANS affaiblissement en espace libre *m*; ~**-space basic loss** *n* RADIO pertes en espace libre *f pl*; ~**-space basic transmission loss** *n* RADIO, TRANS affaiblissement d'espace libre *m*; ~**-space conditions** *n pl* RADIO conditions de propagation en espace libre *f pl*; ~**-space propagation** *n* RADIO propagation en espace libre *f*; ~**-space range** *n* RADIO *answer machine* portée en espace libre *f*; ~**-space transmission loss** *n* RADIO, TRANS affaiblissement d'espace libre *m*; ~ **state** *n* TEST état libre *m*, état vacant *m*; ~ **time** *n* TELEP temps libre *m*, TEST période vacante *f*, temps libre *m*, temps mort *m*

free[4] *vt* RADIO, TELEP, TRANS libérer

freeform: ~ **deformation** *n* (*FFD*) COMP déformation de forme libre *f*

freephone: ~ **call** *n BrE* (*cf toll-free call AmE*) TELEP appel avec tarification au demandé *m*, appel gratuit *m*, appel numéro vert *m* (*Fra*), appel sans frais *m* (*Can*); ~ **number** *n BrE* (*cf toll-free number AmE*) TELEP numéro d'appel sans frais *m* (*Can*), numéro vert *m* (*Fra*); ~ **service** *n BrE* (*cf toll-free service AmE*) TELEP service "libre-appel" *m*

freeware *n* COMP, DATA logiciel public *m*; ~ **package** COMP gratuiciel *m*

freeze[1]: ~**-frame** *n* TV arrêt sur image *m*; ~**-out** *n* DATA, TRANS gel *m*; ~**-out fraction** *n* DATA taux de gel *m*, TRANS fraction de gel *f*; ~**-picture** *n* TV gel de l'image *m*; ~**-picture request** *n* TV demande de gel de l'image *f*; ~ **update** *n* COMP, TV saccade d'images *f*

freeze[2] *vt* COMP, DATA, TV geler

F-region *n* RADIO région F *f*

French: ~ **National Standards Association** *n* (*AFNOR*) GEN Association française de normalisation *f* (*AFNOR*)

frequency[1]: ~**-translated** *adj* RADIO transposé en fréquence

frequency[2] *n* GEN fréquence *f*; ~ **agile pager** RADIO terminal RMU agile de fréquence *m*; ~ **agility radar** RADIO radar aléatoire *m*; ~ **alignment** RADIO verrouillage de fréquence *m*; ~ **allocation** RADIO allocation de fréquence *f*, attribution de fréquence *f*; ~ **allotment** RADIO attribution de fréquence *f*, allotissement de fréquence *m*; ~ **assignment** RADIO attribution de fréquence *f*, assignation de fréquence *f*, allocation de fréquence *f*; ~ **band** RADIO bande de fréquences *f*; ~ **band pairing** RADIO appariage des bandes de fréquence *m*; ~ **change signaling** *AmE*, ~ **change signalling** *BrE* RADIO formation des signaux par modulation de fréquence *f*; ~ **changer** RADIO changeur de fréquence *m*; ~ **changing** RADIO changement de fréquence *m*; ~ **channel** RADIO canal de fréquence *m*; ~ **characteristic** RADIO caractéristique fréquentielle *f*; ~ **control** RADIO régulation par la fréquence *f*, TRANS commande de fréquence *f*, contrôle de fréquence *m*; ~ **conversion** RADIO, TRANS, TV changement de fréquence *m*, conversion de fréquence *f*; ~ **converter** ELECT, POWER, TRANS, TV changeur de fréquence *m*, convertisseur de fréquence *m*; ~ **converter tube** PARTS tube convertisseur de fréquence *m*; ~ **counter** TEST compteur de fréquence *m*, fréquencemètre *m*, périodemètre *m*; ~ **curve** TEST *statistics* courbe de fréquence *f*; ~ **demodulation** TRANS démodulation de fréquence *f*; ~ **demodulator** TRANS démodulateur de fréquence *m*; ~ **density** TEST fréquence par largeur de classe *f*; ~ **departure** RADIO glissement de fréquence *m*, déviation de fréquence *f*; ~**-derived channel** RADIO voie dérivée de fréquence *f*; ~ **deviation** RADIO excursion de fréquence *f*; ~ **deviation meter** TEST excursiomètre *m*, modulomètre FM *m*; ~ **difference** RADIO différence de fréquence *f*; ~ **displacement** RADIO déplacement de fréquence *m*; ~ **distribution** GEN *statistics* distribution des fréquences *f*; ~ **diversity** RADIO, TRANS diversité de fréquence *f*; ~**-diversity reception** RADIO, TRANS réception en diversité de fréquence *f*; ~ **divider** RADIO, TRANS diviseur de fréquence *m*; ~ **division** RADIO, TRANS répartition en fréquence *f*, répartition fréquentielle *f*; ~**-division duplexing** RADIO, TRANS duplexage par division de fréquence *m*; ~**-division multiple access** (*FDMA*) RADIO, SIG, TELEP, TRANS accès multiple par répartition de fréquence *m* (*AMRF*); ~**-division multiplexing** (*FDM*) TELEP, TRANS multiplexage par répartition en fréquence *m* (*MRF*), multiplexage en fréquence *m*, multiplexage par répartition *m*; ~ **division switching** SWG, TRANS commutation par répartition en fréquence *f*; ~**-division**

switching system (*FDS system*) SWG commutateur en fréquence *m*, système de commutation par répartition en fréquence *m*; ~ **domain** ELECT, POWER, RADIO, TEST, TRANS *statistics* domaine fréquentiel *m*; ~**-domain analysis** ELECT, POWER, RADIO analyse dans le domaine fréquentiel *f*; ~**-domain characteristics** *n pl* RADIO caractéristiques dans le domaine fréquentiel *f pl*, caractéristiques fréquencielles *f pl*; ~**-domain reflectometer** ELECT, POWER, TEST, TRANS réflectomètre dans le domaine fréquenciel *m*; ~ **doubler** TRANS doubleur de fréquence *m*; ~ **down-conversion** TRANS abaissement de fréquence *m*; ~ **drift** ELECT, RADIO, TRANS dérive de fréquence *f*; ~ **encoding** RADIO codage de fréquence *m*; ~**-exchange signaling** *AmE*, ~**-exchange signalling** *BrE* SIG modulation par mutation des fréquences *f*; ~ **filter** ELECT, RADIO filtre de fréquence *m*; ~ **generating stage** ELECT, TRANS étage de génération de fréquence *m*; ~ **generator** ELECT, TELEP générateur de fréquence *m*; ~**-generator receiver** TELEP récepteur générateur de fréquence *m*; ~ **hopping** RADIO saut de fréquence *m*; ~**-hopping spread spectrum** RADIO étalement du spectre à sauts de fréquence *m*; ~ **instability** RADIO instabilité de fréquence *f*; ~ **inversion** RADIO inversion de fréquence *f*; ~ **lock** RADIO fréquence asservie *f*; ~ **management** RADIO gestion des fréquences *f*; ~ **measurement** TEST métrologie des fréquences *f*, mesure de la fréquence *f*; ~ **meter** TEST compteur de fréquence *m*, fréquencemètre *m*, périodemètre *m*; ~**-modulated carrier** TRANS porteuse modulée en fréquence *f*; ~**-modulated radar** RADIO radar à modulation de fréquence *m*; ~ **modulation** (*FM*) RADIO modulation de fréquence *f* (*MF*); ~**-modulation noise** TRANS bruit à modulation de fréquence *m*; ~ **modulation synthesis** (*FM synthesis*) COMP synthèse frequency modulation *f* (*synthèse FM*); ~ **monitor** RADIO contrôleur de fréquence *m*; ~ **multiplexing** (*FDM*) TELEP, TRANS multiplexage en fréquence *m*, multiplexage par fréquence *m*, multiplexage par répartition *m*, multiplexage par répartition en fréquence *m* (*MRF*); ~ **multiplier** TRANS multiplicateur de fréquence *m*; ~ **offset** RADIO, TRANS décalage de fréquence *m*; ~ **plan** TRANS, TV plan de fréquence *m*; ~ **polygon** TEST *statistics* polygone de fréquence *m*; ~ **prediction chart** RADIO tableau de prévision de fréquence *m*; ~ **pulling effect** ELECT *electronics* effet d'entraînement de fréquence *m*; ~ **range** RADIO, TRANS bande des fréquences *f*, plage des fréquences *f*, gamme de fréquences *f*; ~ **ratio** TRANS rapport de fréquence *m*; ~ **record** (*REC*) RADIO enregistrement de fréquence *m* (*ENR*); ~ **regulation** POWER régulation par la fréquence *f*, TRANS régulation de fréquence *f*; ~ **resources** *n pl* RADIO ressources spectrales *f pl*, ressources radioélectriques *f pl*; ~ **response** RADIO, TEST, TRANS réponse en fréquence *f*; ~ **response curve** ELECT courbe de réponse en fréquence *f*, RADIO courbe de réponse en fréquence *f*; ~ **reuse** RADIO, TELEP réutilisation de fréquence *f*; ~**-reuse satellite network** RADIO, TELEP réseau à satellite à réutilisation de fréquence *m*; ~ **scan** RADIO, TRANS balayage de fréquence *m*; ~ **scanner** RADIO, TRANS balayeur de fréquence *m*; ~ **selection** RADIO, TV sélection fréquentielle *f*; ~**-selective ringing** TELEP sonnerie sélective en fréquence *f*; ~**-selective sunshield** RADIO écran solaire sélectif en fréquence *m*; ~**-selective voltmeter** TEST voltmètre sélectif *m*; ~ **selector** RADIO, TRANS sélecteur de fréquence *m*; ~ **separation** RADIO espacement des fréquences *m*; ~ **shift** TRANS déplacement de fréquence *m*; ~**-shift keying** (*FSK*) TRANS modulation par déplacement de fréquence *f* (*MDF*); ~ **spacing** RADIO espacement des fréquences *m*; ~ **spectrum** RADIO spectre des fréquences *m*; ~ **stability** RADIO stabilité de fréquence *f*; ~ **staggering** RADIO, TRANS décalage de fréquence *m*; ~ **standard** RADIO étalon de fréquence *m*, fréquence étalon *f*; ~**-standard accuracy** RADIO précision absolue de fréquence de l'étalon *f*; ~ **step** TV pas *m*; ~ **sweep** RADIO, TRANS balayage de fréquence *m*; ~ **swing** GEN excursion de fréquence *f*; ~ **synthesizer** RADIO synthétiseur de fréquence *m*; ~ **table** TRANS table de fréquences *f*, tableau de contingence *m*; ~ **tolerance** RADIO tolérance de fréquence *f*; ~ **tracking** RADIO poursuite en fréquence *f*; ~ **translation** RADIO transposition de fréquence *f*; ~ **translator** RADIO transposeur de fréquence *m*; ~ **tuning** RADIO accord de fréquence *m*, réglage de la fréquence *m*; ~ **tuning range** RADIO plage d'accord de fréquence *f*; ~ **uncertainty** TELEP erreur de fréquence *f*; ~**-uncertainty band** TELEP bande d'incertitude des fréquences *f*

Fresnel: ~ **diffraction pattern** *n* TRANS diagramme de diffraction de Fresnel *m*; ~ **reflection** *n* TRANS réflexion de Fresnel *f*; ~ **reflection method** *n* TRANS méthode de réflexion de Fresnel *f*; ~ **region** *n* RADIO, TRANS région de Fresnel *f*; ~ **zone** *n* TRANS zone de Fresnel *f*; ~ **zone blockage** *n* TRANS blocage de la zone de Fresnel *m*

fringe: ~ **effect** *n* POWER effet de bords *m*

fringing *n* POWER effet de bords *m*

FRJ *abbr* (*facility-rejected message*) DATA SURF (*message de refus de service supplémentaire*)

front: ~**-end clipping** *n* DATA mutilation en début de bribe *f*; ~**-end computer** *n* COMP ordinateur d'entrée-sortie *m*, SWG frontal *m*, ordinateur frontal *m*; ~**-end equipment** *n* RADIO équipement RF associé à l'antenne *m*; ~**-fed dish** *n* RADIO parabole à alimentation latérale *f*; ~**-fed reflector antenna** *n* RADIO antenne à réflecteur et source frontale *f*; ~ **feed** *n* RADIO alimentation frontale *f*; ~ **loading** *n* GEN chargement frontal *m*; ~ **page** *n* COMP page d'accueil *f*; ~**-to-back ratio** *n* RADIO rapport avant-arrière *m*

FRP *abbr* (*fast-reservation protocol*) DATA protocole de réservation rapide *m*

frying *n* TELEP friture *f*

FS *abbr* DATA (*further study*) complément d'étude *m*, GEN (*functional specification*) spécification fonctionnelle *f*

FSK *abbr* (*frequency-shift keying*) RADIO, SIG MDF (*modulation par déplacement de fréquence*)

FT *abbr* (*Fourier transformation*) TEST transformation de Fourier *f*

FTAM *abbr* (*file transfer, access and management*) COMP *remote data processing*, DATA FTAM (*transfert, accès et gestion de fichiers*)

FTAMSE *abbr* (*file transfer access and management service element*) COMP, DATA élément de service du transfert, accès et gestion de fichiers *m*

FTAS *abbr* (*fast-time analysis system*) COMP système d'analyse en temps comprimé *m*

FTIR *abbr* (*Fourier transform infrared spectography*) TEST spectrographie infrarouge à transformée de Fourier *f*

FTP *abbr* (*file transfer protocol*) COMP, DATA protocole de transfert de fichiers *m*; ~ **server** *n* DATA serveur FTP *m*

FTT *abbr* (*failure to train*) DATA, TELEP ECD (*échec du conditionnement*)

FU *abbr* (*functional unit*) DATA, TEST unité fonctionnelle *f*

full[1]: ~**-carrier** *adj* RADIO à porteuse complète; ~**-duplex** *adj* TRANS bilatéral simultané; ~**-screen** *adj* DATA *Internet* plein écran

full[2]: ~ **adder** *n* ELECT additionneur complet *m*; ~ **availability** *n* TELEP disponibilité totale *f*; ~ **break-in** *n* TELEP intervention totale *f*; ~ **break-in operate time** *n* RADIO, TELEP temps de fonctionnement pour l'intervention totale *m*; ~ **carrier** *n* RADIO porteuse complète *f*; ~**-carrier emission** *n* RADIO émission à porteuse complète *f*; ~**-carrier wave** *n* RADIO onde à porteuse complète *f*; ~ **digital processing** *n* COMP traitement numérique *m*; ~ **duplex** *n* RADIO duplex *m*; ~**-duplex** *n* TRANS exploitation duplex *f*; ~**-duplex operation** *n* TRANS fonctionnement en duplex *m*; ~**-duration half-maximum** *n* (*FDHM*) DATA durée à mi-crête *f* (*FDHM*); ~ **echo suppressor** *n* RADIO, TELEP suppresseur d'écho complet *m*; ~**-float system** *n*

POWER *supply* système à alimentation directe *m*; ~ **load** *n* POWER charge maximale *f*, pleine charge *f*; ~-**motion videoconferencing** *n* TV vidéoconférence à images animées *f*; ~ **rate** *n* APPL plein tarif *m*; ~-**rate channel** *n* TRANS voie à plein débit *f*; ~ **refund** *n* TELEP remboursement complet *m*; ~-**scale test** *n* TEST essai en grandeur réelle *m*, essai en vraie grandeur *m*; ~-**screen** *n* COMP plein écran; ~-**steering capability aerial** *n* RADIO antenne orientable en tous sens *f*; ~-**steering capability antenna** *n* RADIO antenne orientable en tous sens *f*; ~ **subtracter** *n* ELECT *logic elements* soustracteur complet *m*; ~ **sunlight period** *n* RADIO période d'éclairement total *f*; ~ **text** *n* COMP texte intégral *m*; ~ **view** *n* COMP full view *f*; ~-**wave dipole** *n* RADIO *aerials* doublet en onde entière *m*; ~-**wave rectification** *n* ELECT *electronics* redressement à deux alternances *m*; ~-**wave rectifier** *n* ELECT *electronics* redresseur à deux alternances *m*; ~-**width half-maximum** *n* (*FWHM*) DATA, ELECT, TRANS étendue à mi-crête *f* (*FWHM*), largeur à mi-crête *f* (*FWHM*)

fully: ~-**automatic operation** *n* TELEP fonctionnement entièrement automatique *m*; ~-**charged battery** *n* POWER batterie chargée *f*; ~-**coherent radar** *n* RADIO radar parfaitement cohérent *m*; ~-**dissociated mode of operation** *n* TELEP mode d'exploitation entièrement dissocié *m*; ~-**dissociated signaling** *AmE*, ~-**dissociated signalling** *BrE* *n* SIG signalisation entièrement dissociée *f*; ~-**distributed control system** *n* CONTR, SWG système à structure éclatée *m*, système à structure déconcentrée *m*; ~-**provided circuit group** *n* TELEP faisceau totalement fourni *m*; ~-**restricted station** *n* SWG poste à restriction totale *m*

function[1]: ~-**dependent** *adj* GEN dépendant de la fonction

function[2] *n* GEN *operation* fonctionnement *m*, *mathematics* fonction *f*, TV rôle *m*; ~-**affecting maintenance** TELEP, TEST maintenance affectant la fonction *f*; ~ **block** SWG bloc fonctionnel *m*; ~ **change** GEN modification de fonction *f*; ~ **check-out** TEST vérification de fonctionnement *f*; ~ **data** *n pl* DATA données de fonction *f pl*; ~-**degrading failure** TELEP dérangement dégradant les fonctions *m*; ~-**degrading fault** TEST panne avec dégradation des fonctions *f*; ~-**degrading maintenance** TELEP maintenance dégradant les fonctions *f*; ~ **designation** GEN désignation de fonction *f*; ~-**disturbing fault** TEST panne affectant les fonctions *f*; ~ **division system** SWG système à cœur réparti spécialisé *m*; ~ **division system architecture** SWG architecture à cœur réparti spécialisé *f*; ~ **edit buffer** COMP tampon d'édition de fonction *m*; ~ **flow chart** GEN organigramme de fonction *m*; ~ **generator** COMP générateur de fonction *m*; ~ **identification** TELEP identification de fonction *f*; ~ **key** COMP, CONTR *equipment* touche de fonction *f*; ~ **keyboard** COMP clavier des fonctions *m*; ~ **modification** GEN modification de fonction *f*; ~ **module** SWG module fonctionnel *m*; ~ **parameter** COMP paramètre de fonction *m*; ~-**permitting failure** TELEP, TEST dérangement permettant l'accomplissement des fonctions *m*; ~-**permitting maintenance** TELEP, TEST maintenance en exploitation *f*, maintenance en fonctionnement *f*; ~-**preventing failure** TELEP, TEST dérangement empêchant l'accomplissement des fonctions *m*; ~-**preventing fault** TELEP, TEST panne complète *f*; ~-**preventing maintenance** TELEP maintenance-arrêt *f*; ~ **specification and design** SWG spécification et conception des fonctions *f*; ~ **supervisory unit** SWG unité de surveillance du fonctionnement *f*; ~ **switch** GEN commutateur de fonction *m*; ~ **table** GEN table de fonction *f*

functional: ~ **block** *n* TELEP *in SDL* bloc fonctionnel *m*; ~-**block boundary PE** *n* TELEP élément graphique de limite de bloc fonctionnel *m*; ~-**block description** *n* TELEP description de bloc fonctionnel *f*; ~-**block specification** *n* TELEP spécification de bloc fonctionnel *f*; ~ **characteristic** *n* GEN caractéristique fonctionnelle *f*; ~ **class** *n* (*FC*) TELEP classe fonctionnelle *f* (*FC*); ~ **class A** *n* (*FCA*) TELEP classe fonctionnelle A *f* (*FCA*); ~ **class B** *n* (*FCB*) TELEP classe fonctionnelle B *f* (*FCB*); ~ **description** *n* (*FD*) GEN description fonctionnelle *f*; ~ **entity action** *n* (*FEA*) DATA action d'entité fonctionnelle *f*; ~ **feature** *n* GEN caractéristique fonctionnelle *f*; ~ **mode** *n* TELEP mode fonctionnel *m*, TEST mode de fonctionnement *m*; ~ **specification** *n* (*FS*) GEN spécification fonctionnelle *f*; ~ **stress** *n* TEST contrainte de fonctionnement *f*; ~ **test** *n* TEST essai de fonctionnement *m*; ~ **testing** *n* TEST essais de fonctionnement *m pl*; ~ **unit** *n* (*FU*) DATA, TEST unité fonctionnelle *f*

functionality *n* DATA, TEST fonctionnalité *f*

functionally: ~ **divided system** *n* DATA système en partage de fonctions *m*

functor *n* COMP foncteur *m*

fundamental: ~ **component** *n* TEST composant fondamental *m*; ~ **mode** *n* TRANS mode fondamental *m*; ~ **sound** *n* TELEP son fondamental *m*

funnel: ~ **mouthpiece** *n* TELEP pavillon *m*

FUR *abbr* (*fast update request*) DATA demande de rafraîchissement rapide *f*

further: ~ **study** *n* (*FS*) DATA complément d'étude *m*

fuse[1]: ~-**protected** *adj* ELECT, POWER protégé par fusible

fuse[2] *n* ELECT, PARTS, POWER fusible *m*; ~ **alarm** TEST alarme de fusible *f*; ~ **box** POWER coffret à fusibles *m*; ~ **holder** POWER porte-fusible *m*; ~ **link** POWER liaison-fusible *f*; ~ **panel** POWER tableau des fusibles *m*; ~ **strip** POWER réglette de coupure *f*, réglette de fusibles *f*; ~ **supervision** POWER supervision des fusibles *f*; ~ **switch** POWER disjoncteur à fusible *m*, fusible interrupteur *m*; ~-**tapered fiber** *AmE*, ~-**tapered fibre** *BrE* TRANS fibre amincie par fusion *f*; ~ **unit** POWER coupe-circuit *m*, unité de fusibles *f*; ~ **wire** ELECT, PARTS, POWER fusible *m*

fused[1] *adj* ELECT, POWER protégé par fusible

fused[2]: ~-**fiber splice** *AmE*, ~-**fibre splice** *BrE* *n* TRANS épissure par fusion *f*, épissure par soudage *f*; ~ **joint** *n* TRANS *fibre optics* raccord soudé *m*; ~ **quartz** *n* TRANS quartz fondu *m*; ~ **silica** *n* TRANS silice fondue *f*, silice vitreuse *f*; ~ **transistor** *n* ELECT *semiconductors* transistor à jonction par alliage *m*

fusing *n* TRANS *fibre optics* soudage *m*

fusion: ~ **splice** *n* TRANS *fibre optics* épissure par fusion *f*, épissure par soudage *f*

future: ~ **reference picture** *n* COMP, TV image de référence future *f*; ~ **use** *n* GEN pérenne

fuzziness *n* TV *on scanner* flou *m*

fuzzy: ~ **logic** *n* COMP logique floue *f*; ~ **set theory** *n* DATA théorie des ensembles flous *f*

FWHM *abbr* (*full-width half-maximum*) DATA, ELECT, TRANS FWHM (*étendue à mi-crête, largeur à mi-crête*)

FX4 *abbr* (*Telefax Group 4*) TELEP FX4 (*télécopie du groupe 4*)

FX3 *abbr* (*Telefax Group 3*) TELEP FX3 (*télécopie du groupe 3*)

FYI *abbr* (*for your information*) GEN *Internet* à titre indicatif, pour information

G

GA *abbr* (*go-ahead*) RADIO invitation à continuer *f*
GaAs *abbr* (*gallium arsenide*) PARTS arséniure de gallium *f*
gage *AmE see gauge BrE*
gaging *AmE see gauging BrE*
gain *n* ELECT, POWER, RADIO, TELEP, TRANS gain *m*; ~ **calibration error** ELECT erreur d'étalonnage de gain *f*; ~ **control** ELECT commande de gain *f*; ~ **degradation** RADIO dégradation de gain *f*, perte par couplage antenne-milieu *f*; ~ **inequality** ELECT différence de gain *f*; ~ **ripple** ELECT ondulation de gain *f*; ~ **scale** ELECT axe des gains *m*, échelle des gains *f*; ~ **slope** ELECT pente de gain *f*
galactic: ~ **noise** *n* RADIO bruit galactique *m*
gallium: ~ **arsenide** *n* (*GaAs*) PARTS arséniure de gallium *m*; ~ **arsenide substrate** *n* PARTS *semiconductors* substrat d'arséniure de gallium *m*
galvanic: ~ **cell** *n* POWER *batteries* cellule galvanique *f*
galvanometer *n* TEST *measuring devices* galvanomètre *m*
gamma: ~ **curve** *n* PER APP *scanner* courbe de gamma *f*
gamut *n* COMP gamme *f*
gang: ~**-mountable adaptor** *n* ELECT, PARTS adaptateur coupleur *m*
ganged[1] *adj* ELECT couplé
ganged[2]: ~ **circuits** *n pl* ELECT circuits à commande unique *m pl*
gap: ~ **character** *n* DATA caractère de remplissage *m*; ~ **coding** *n* TRANS codage par découpage *m*; ~**-filler satellite** *n* RADIO satellite de transmission *m*; ~ **loss** *n* TRANS *longitudinal offset loss* perte par espacement longitudinal *f*, perte dans l'entrefer *f*
garble *vt* GEN *cryptology* altérer, mutiler
garbled: ~ **reply** *n* RADIO réponse brouillée *f*, réponse inintelligible *f*
garbling *n* GEN *cryptology* altération *f*, déformation *f*, mutilation *f*
gas: ~**-content factor** *n* ELECT facteur de vide *m*; ~ **current** *n* ELECT courant de gaz *m*; ~ **discharge** *n* ELECT décharge du gaz *f*; ~ **discharge display** *n* COMP, ELECT écran à plasma *m*; ~**-filled phototube** *n* ELECT phototube à gaz *m*; ~**-filled tube** *n* ELECT tube à gaz *m*; ~ **flow counter tube** *n* ELECT tube-compteur à circulation de gaz *m*; ~ **multiplication** *n* ELECT multiplication du gaz *f*; ~ **multiplication factor** *n* ELECT facteur de multiplication du gaz *m*; ~ **panel** *n* COMP, ELECT écran à plasma *m*; ~ **sample counter tube** *n* ELECT tube-compteur à source gazeuse *m*; ~**-tight seal** *n* ELECT joint hermétique *m*; ~ **tube** *n* ELECT tube à gaz *m*; ~**-type arrester** *n* POWER *protection* parafoudre à gaz raréfié *m*
gasket *n* PARTS joint *m*
gate[1] *n* ELECT porte *f*, porte logique *f*, *semiconductors* gâchette *f*, grille *f*; ~ **array** ELECT *logic elements* circuits prédiffusés *m pl*, matrice de portes *f*; ~ **array circuit** ELECT circuit à matrice de portes *m*; ~ **circuit** ELECT *logic elements* porte *f*; ~ **condition** ELECT *logic elements* état de porte *m*; ~**-controlled turn-off time** ELECT *thyristors* temps de désamorçage par la gâchette *m*; ~**-controlled turn-on time** ELECT *thyristors* temps d'amorçage par la gâchette *m*; ~ **cutoff current** ELECT *thyristors* courant de fuite d'une grille *m*; ~ **network** ELECT *logic elements* réseau de portes *m*; ~ **non-trigger current** ELECT *thyristors* courant de non-amorçage par la gâchette *m*; ~ **non-trigger voltage** ELECT *thyristors* tension de non-amorçage par la gâchette *f*; ~ **pulse** ELECT *logic elements* impulsion de déblocage *f*, impulsion de déclenchement *f*; ~ **region** ELECT *transistors* région de grille *f*; ~ **resistance** ELECT *thyristors* résistance de grille *f*; ~ **signal** ELECT *logic elements* signal déclencheur *m*, signal de porte *m*; ~ **terminal** ELECT *thyristors* borne de gâchette *f*; ~ **turn-off current** ELECT *thyristors* courant de désamorçage par la gâchette *m*; ~ **turn-off voltage** ELECT *thyristors* tension de désamorçage par la gâchette *f*; ~ **turn-on current** ELECT *thyristors* courant de gâchette de déblocage *m*
gate[2] *vt* ELECT *logic elements* portillonner, sélectionner par une porte
gated: ~ **amplifier** *n* ELECT *electronics* amplificateur à déclenchement periodique *m*; ~ **mode acquisition procedure** *n* DATA procédure d'acquisition par la méthode à créneaux *f*
gateway *n* DATA, TELEP passerelle *f*, TRANS passerelle *f*, centre de transit international *m*, passerelle entre réseaux *f*, procédure d'acquisition par la méthode à créneaux *f*; ~ **exchange** SWG centre d'accès *m*; ~ **MSC** SWG CCM d'entrées *m*; ~ **network element** TRANS élément passerelle de réseau *m*; ~ **station** RADIO station passerelle *f*
gating: ~ **signal** *n* RADIO signal de porte *m*, signal de déclenchement périodique *m*
gauge *n BrE* TEST *measuring devices* calibre *m*
gauging *n BrE* TEST *metrology* calibrage *m*
Gaussian: ~ **beam** *n* TRANS faisceau gaussien *m*; ~ **channel** *n* DATA canal de Gauss *m*; ~ **curve** *n* GEN *statistics* courbe gaussienne *f*; ~ **distribution** *n* GEN *statistics* distribution gaussienne *f*, répartition gaussienne *f*; ~**-filtered frequency-shift keying** *n* (*GFSK*) SIG manipulation par variation de fréquence à filtre gaussien *f*; ~**-filtered minimum-shift keying** *n* (*GMSK*) SIG modulation à déplacement minimal avec filtrage gaussien *f*; ~ **noise** *n* TELEP, TRANS bruit gaussien *m*; ~ **pulse** *n* TELEP, TRANS *fibre optics* impulsion gaussienne *f*
gc *abbr* (*gigacycle*) ELECT, TEST, TRANS GHz (*gigahertz*)
GDI *abbr* (*graphic device interface*) COMP interface d'affichage graphique *m*, interface d'unité graphique *m*
Geiger: ~ **region** *n* RADIO, TRANS région de Geiger *f*; ~ **threshold** *n* RADIO, TRANS seuil de Geiger *m*
Geiger-Müller: ~ **counter tube** *n* RADIO tube-compteur Geiger-Müller *m*
gel *n* PARTS *plastics* gel *m*; ~ **battery** POWER batterie à électrolyte gélifié *f*
general: ~ **attribute** *n* DATA attribut général *m*; ~ **auto-action** *n* DATA action automatique générale *f*; ~ **call** *n* TELEP appel général *m*; ~**-coverage receiver** *n* RADIO récepteur à couverture générale *m*; ~ **entertainment channel** *n* TV chaîne généraliste *f*; ~ **localization** *n* TEST localisation sommaire *f*; ~ **model** *n* TELEP modèle général *m*; ~ **negative-recorded announcement** *n* TELEP annonce enregistrée générale négative *f*; ~ **parameters** *n pl* COMP caractéristiques générales *f pl*; ~ **phasing-in** *n* GEN généralisation progressive *f*; ~ **positive-recorded announcement** *n* TELEP annonce enregistrée générale positive *f*; ~**-purpose interface** *n* COMP, DATA interface universelle *f*; ~**-purpose interface bus** *n* (*GPIB*) DATA bus d'interface universel *m*; ~ **recorded announcement** *n* TELEP annonce enregistrée générale *f*; ~ **supervisory station** *n* TELEP poste général de supervision *m*
generalized: ~ **phase control** *n* TEST réglage de phase généralisé *m*
generate *vt* COMP *report* générer, GEN engendrer, générer

generation *n* GEN génération *f*; ~ **rate** ELECT *semiconductors* cadence de création *f*

generator *n* COMP génération *f*, gestionnaire *m*, GEN générateur *m*, TELEP génération *f*

generic: ~ **address** *n* DATA adresse générique *f*; ~ **definition** *n* DATA définition générique *f*; ~ **flow control** *n* (*GFC*) CONTR, TRANS contrôle de flux générique *m*; ~ **name** *n* DATA nom générique *m*; ~ **program** *n* TELEP programme générique *m*; ~ **test case** *n* DATA, TEST test élémentaire générique *m*; ~ **test suite** *n* DATA, TEST suite de tests génériques *f*

genlocking *n* TV verrouillage de synchronisation *m*

genuine: ~ **receive point** *n* RADIO point de réception reconnu *m*

geocentric: ~ **angle** *n* RADIO angle géocentrique *m*

geographically: ~**-distributed exchange** *n* SWG centre géographiquement dispersé *m*

geometric: ~ **distortion** *n* TV distorsion géométrique *f*; ~ **mean** *n* TEST *quality* moyenne géométrique *f*; ~ **optics** *n pl* (*GO*) TRANS optique des rayons *f*, optique géometrique *f*; ~ **sample** *n* TEST *quality* échantillon d'orientation spécifiée *m*

geostationary[1] *adj* RADIO géostationnaire

geostationary[2]: ~ **fixed satellite** *n* RADIO satellite géostationnaire du service fixe *m*; ~ **orbit** *n* RADIO orbite des satellites géostationnaires *f*; ~ **satellite** *n* RADIO satellite géostationnaire *m*; ~ **satellite orbit** *n* RADIO orbite des satellites géostationnaires *f*

geosynchronous[1] *adj* RADIO géosynchrone

geosynchronous[2]: ~ **satellite** *n* RADIO satellite géosynchrone *m*; ~ **satellite orbit** *n* RADIO orbite des satellites géosynchrones *f*

germanium: ~**-coated kapton skin** *n* DATA peau de kapton revêtue de germanium *f*

get: ~ **connected** *vi* DATA *Internet* se connecter; ~ **through** *vi* TELEP obtenir la communication

getter *n* ELECT getter *m*

GFC *abbr* (*generic flow control*) CONTR, TRANS contrôle de flux générique *m*

GFSK *abbr* (*Gaussian-filtered frequency-shift keying*) SIG, TELEP manipulation par variation de fréquence à filtre gaussien *f*, modulation à déplacement minimal avec filtrage gaussien *f*

ghost *n* TV image fantôme *f*, effet d'écho *m*, fantôme *m*, double image *m*; ~ **signal** TV signal fantôme *m*

GHz *abbr* (*gigahertz*) ELECT, TEST, TRANS GHz (*gigahertz*)

GI *abbr* (*group identifier*) TELEP identificateur de groupe *m*

GIF *abbr* (*graphics interchange format*) COMP, DATA *Internet* GIF (*format d'échange de graphiques*)

gigabit *n* DATA gigabit *m*

gigacycle *n* (*gc*) ELECT, TEST, TRANS gigahertz *m* (*GHz*)

gigahertz *n* (*GHz*) ELECT, TEST, TRANS gigahertz *m* (*GHz*)

given: ~ **code** *n* DATA code considéré *m*

GL *abbr* (*group length*) TELEP longueur de groupe *f*

glare *n* TV éblouissement *m*

glass: ~ **capillary** *n* ELECT capillaire en verre *m*; ~ **fiber** *AmE*, ~ **fibre** *BrE* *n* PARTS, TRANS fibre de verre *f*; ~ **preform** *n* PARTS, TRANS *fibre optics* préforme de verre *f*; ~ **tube fuse** *n* ELECT, PARTS, POWER fusible à tube de verre *m*

glide *n* GEN glissement *m*; ~ **effect** COMP effet de glissement *m*

glint *n* RADIO scintillation *f*

glitch *n* COMP incident *m*

global: ~ **beam** *n* RADIO faisceau de couverture globale *m*; ~ **beam carrier** *n* TRANS porteuse émise dans un faisceau de couverture globale *f*; ~ **call** *n* RADIO appel global *m*; ~ **coverage** *n* RADIO couverture globale *f*; ~ **network addressing domain** *n* DATA domaine globale d'adressage du réseau *m*; ~ **positioning system** *n* (*GPS*) RADIO système de positionnement universel *m* (*GPS*); ~ **reference burst distribution** *n* DATA distribution du paquet de référence en mode global *f*; ~ **system for mobile communications** *n* (*GSM*) RADIO, TELEP groupe spécial mobiles *m* (*GSM*); ~ **telecommunications satellite system** *n* RADIO, TELEP système mondial de télécommunications par satellite *m*; ~ **telecommunications system** *n* (*GTS*) RADIO, TELEP, TRANS système mondial de télécommunications *m* (*SMT*)

glow: ~ **current of protector** *n* ELECT courant d'effluve *m*; ~ **discharge** *n* ELECT décharge luminescente *f*; ~ **discharge lamp** *n* ELECT, PARTS lampe à décharge luminescente *f*; ~ **discharge tube** *n* ELECT, PARTS tube à décharge luminescente *m*; ~ **discharge valve** *n* ELECT, PARTS tube à décharge luminescente *m*; ~ **indicator tube** *n* ELECT, PARTS tube indicateur à luminescense *m*; ~ **lamp** *n* ELECT, PARTS lampe à décharge *f*, lampe luminescente *f*; ~ **voltage** *n* ELECT tension d'amorçage *f*

GMR *abbr* (*ground movement radar*) RADIO radar de pistes *m*

GMSK *abbr* (*Gaussian-filtered minimum-shift keying*) SIG modulation à déplacement minimal avec filtrage gaussien *f*

GMT *abbr* (*Greenwich Mean Time*) GEN TMG (*temps moyen de Greenwich*)

go[1]: ~ **to return crosstalk** *n* TRANS diaphonie entre les deux sens de transmission *f*

go[2] *vi* TELEP *operator* passer, transiter par; ~ **high** ELECT *logic elements* passer au niveau haut; ~ **into circuit** TELEP entrer en ligne; ~ **low** ELECT *logic elements* passer au niveau bas; ~ **on-hook** TELEP raccrocher le combiné; ~ **out** POWER s'éteindre

GO *abbr* (*geometric optics*) TRANS optique des rayons *f*, optique géometrique *f*

go-ahead *n* (*GA*) RADIO invitation à continuer *f*; ~ **signal** TELEP *telex* signal continuez *m*

Golay: ~ **coding** *n* DATA codage Golay *m*

gold: ~**-bonded diode** *n* PARTS *semiconductors* diode à pointe d'or *f*

gooseneck: ~ **feed** *n* RADIO alimentation en col de cygne *f*

gopher *n* COMP, DATA *Internet* gopher *m*

Gouraud: ~ **shading** *n* TV lissage de Gouraud *m*

government: ~ **telegram** *n* TELEP télégramme d'état *m*; ~ **telex call** *n* TELEP communication télex d'État *f*

GPIB *abbr* (*general-purpose interface bus*) DATA bus d'interface universel *m*

GPS *abbr* (*global positioning system*) RADIO GPS (*système de positionnement universel*)

grab *vt* COMP saisir

grabber *n* COMP appareil de saisie *m*

graceful: ~ **degradation** *n* RADIO, TRANS dégradation progressive *f*, fonctionnement en mode dégradé *m*

grade[1] *n* TEST *quality* classe *f*; ~ **of service** TEST qualité d'écoulement du trafic *f*, qualité de service *f*

grade[2] *vt* TELEP classer

graded: ~ **index** *n* ELECT, TRANS gradient d'indice *m*; ~**-index fiber** *AmE*, ~**-index fibre** *BrE* *n* ELECT, PARTS, TRANS fibre à gradient d'indice *f*; ~**-index profile** *n* ELECT, TRANS profil à gradient d'indice *m*

grading *n* TELEP classement *m*, gradation *f*; ~**-coupling loss cable** TRANS câble à perte de couplage par gradation *m*; ~**-index fiber** *AmE*, ~**-index fibre** *BrE* ELECT, PARTS, TRANS fibre à gradient d'indice *f*

gradual: ~ **degradation** *n* RADIO, TRANS *fibre optics* dégradation progressive *f*; ~ **failure** *n* TEST *dependability* défaillance par dérive *f*, défaillance progressive *f*

graduate *vt* TEST *measuring devices* graduer

graduation *n* TEST *measuring devices* graduation *f*

grammatical: ~ **tagger** *n* COMP étiqueteur grammatical *m*

grandfathered: ~ **station** *n* RADIO station couverte par une clause de droits acquis *f*

grant *vt* RADIO, TV *licence* accorder

granular: ~ **noise** *n* TELEP bruit granulaire *m*

granularity *n* TV granularité *f*

granules *n pl* TV granules *m pl*

graph: ~ **plotter** *n* COMP traceur de courbes *m*

graphic: ~ **character** *n* DATA, TELEP caractère graphique *m*; ~ **code extension** *n* TELEP extension de code graphique *f*; ~ **device interface** *n* (*GDI*) COMP interface d'affichage graphique *f*, interface d'unité graphique *f*; ~ **programming aid** *n* COMP aide à la programmation graphique *f*; ~ **symbol** *n* DATA, TELEP symbole graphique *m*

graphical: ~ **character** *n* DATA, TELEP caractère graphique *m*; ~ **interface** *n* COMP interface graphique *f*; ~ **symbol** *n* DATA symbole graphique *m*; ~ **user interface** *n* (*GUI*) COMP, DATA interface utilisateur graphique *f* (*GUI*)

graphics *n* TV graphisme *m*, graphique *m*; ~ **database** COMP, DATA base de données graphiques *f*; ~ **interchange format** (*GIF*) COMP, DATA format d'échange de graphiques *m* (*GIF*); ~ **mode** COMP mode de représentation graphique *m*; ~ **package** TV palette graphique *f*, source de composition *f*; ~ **processor** COMP processeur graphique *m*; ~ **screen** COMP écran graphique *m*

grass *n* COMP, RADIO bruit de fond *m*, herbe *f*, pointes de bruit *f pl*

grating *n* PARTS grille *f*, TRANS *fibre optics* réseau *m*

graveyard: ~ **orbit** *n* RADIO orbite cimetière *f*

gravity: ~ **gradient** *n* RADIO gradient de pesanteur *m*

gray: ~ **level** *AmE see grey level BrE*; ~ **scale** *AmE see grey scale BrE*

Gray: ~ **code** *n* DATA code Gray *m*

Greenwich: ~ **Mean Time** *n* (*GMT*) GEN temps moyen de Greenwich *m* (*TMG*)

Gregorian: ~ **reflector aerial** *n* RADIO antenne à réflecteur grégorien *f*; ~ **reflector antenna** *n* RADIO antenne à réflecteur grégorien *f*

grey: ~ **level** *n BrE* TELEP gamme de gris *f*; ~ **scale** *n BrE* DATA, TELEP échelle de gris *f*, gamme de gris *f*, plage de gris *f*

grid *n* COMP quadrillage *m*, ELECT, PARTS grille *f*; ~ **amplitude modulation** TRANS modulation d'amplitude par la grille *f*; ~ **bias voltage** ELECT tension de polarisation de grille *f*; **~-cathode driving voltage** ELECT tension d'attaque de grille-cathode *f*; **~-controlled arc discharge tube** ELECT, PARTS tube à décharge en arc à grille de commande *m*; ~ **current** ELECT courant de grille *m*; ~ **data message** DATA message de point de grille *m*; ~ **driving power** ELECT puissance d'attaque de grille *f*; ~ **driving voltage** ELECT tension d'attaque de grille *f*; ~ **input power** ELECT puissance d'attaque de grille *f*; ~ **input voltage** ELECT tension d'attaque de grille *f*; **~-leak resistor** ELECT, PARTS résistance de fuite de grille *f*; ~ **network** TRANS réseau maillé *m*

gridded: ~ **reflector aerial** *n* RADIO antenne à réflecteur à grille *f*; ~ **reflector antenna** *n* RADIO antenne à réflecteur à grille *f*

gridiron: ~ **pattern** *n* GEN quadrillage *m*

grille *n* PARTS grille *f*, grillage protecteur *m*, grille de protection *f*

grip[1]: ~ **apparatus** *n* ELECT, PARTS grip de tirage *m*; **~-type fuse** *n* PARTS, POWER fusible à cartouche *m*

grip[2] *vt* COMP saisir

grommet *n* PARTS passe-fil *m*

groove *n* GEN rainure *f*

grooved: ~ **cable** *n* PARTS, TRANS *fibre optics* câble rainuré *m*

gross: ~ **bit rate** *n* COMP, DATA, TELEP, TRANS débit binaire composite *m*

ground[1] *adj AmE* (*cf earth BrE*) ELECT terrien; **~-to-air** RADIO sol-air

ground[2] *n AmE* (*cf earth BrE*) ELECT, POWER masse *f*, terre *f*; ~ **address** GEN adresse au sol *f*; ~ **bar** *AmE* (*cf earth bar BrE*) ELECT, POWER barre de masse *f*, barre de mise à la terre *f*; **~-based duct** RADIO conduit de surface *m*; **~-based pilot** RADIO onde pilote émise du sol *f*; ~ **button** *AmE* (*cf earth button BrE*) SWG bouton de mise à la terre *m*; ~ **cable** *AmE* (*cf earth cable BrE*) POWER câble de mise à la terre *m*; ~ **clamp** *AmE* (*cf earth clamp BrE*) POWER bride de masse *f*, collier de terre *m*; ~ **clutter** RADIO obstacles au sol *m pl*, occupation des sols *f*; ~ **clutter coefficient** RADIO coefficient d'occupation des sols *m*; ~ **collector** *AmE* (*cf earth collector BrE*) POWER collecteur de terre *m*; ~ **collector bar** *AmE* (*cf earth collector bar BrE*) POWER collecteur de terre *m*; ~ **command** RADIO ordre émis du sol *m*; ~ **conductor** *AmE* (*cf earth conductor BrE*) POWER conducteur de masse *m*, conducteur de terre *m*; ~ **connection** *AmE* (*cf earth connection BrE*) POWER connexion de terre *f*; ~ **control** RADIO commande du sol *f*; ~ **data** *n pl* RADIO données de terrain *f pl*; ~ **electrode** *AmE* (*cf earth electrode BrE*) POWER électrode de terre *f*, prise de terre *f*; ~ **electrode installation** *AmE* (*cf earth electrode installation BrE*) POWER installation de prise de terre *f*; ~ **electrode resistance** *AmE* (*cf earth electrode resistance BrE*) POWER résistance de prise de terre *f*; ~ **fault** *AmE* (*cf earth fault BrE*) POWER défaut à la masse *m*, défaut à la terre *m*; **~-fault circuit breaker** *AmE* (*cf earth-fault circuit breaker BrE*) POWER interrupteur de défaut à la terre *m*; **~-fault protection** *AmE* (*cf earth-fault protection BrE*) POWER protection contre défauts à la terre *f*; **~-fault relay** *AmE* (*cf earth-fault relay BrE*) ELECT, PARTS relais de défaut à terre *m*; **~-generated pilot** RADIO onde pilote émise du sol *f*; ~ **lead** *AmE* (*cf earth lead BrE*) POWER fil de masse *m*; ~ **leakage** *AmE* (*cf earth leakage BrE*) POWER fuite à la terre *f*; ~ **leakage current** *AmE* (*cf earth leakage current BrE*) POWER courant de fuite à la terre *m*; ~ **leakage guard** *AmE* (*cf earth leakage guard BrE*) POWER détecteur de fuite à la terre *m*; ~ **loop** *AmE* (*cf earth loop BrE*) ELECT circuit de masse *m*, retour à la masse *m*; ~ **movement radar** (*GMR*) RADIO radar de pistes *m*; ~ **network** TRANS réseau au sol *m*; ~ **noise** ELECT bruit de fond *m*; ~ **noise level** ELECT niveau de bruit introduit par la masse *m*; ~ **plane** RADIO plan de sol *m*; ~ **plane antenna** RADIO antenne à plan de sol *f*; ~ **plate** *AmE* (*cf earth plate BrE*) POWER plaque de terre *f*; ~ **potential** *AmE* (*cf earth potential BrE*) POWER potentiel du sol *m*; ~ **potential difference** *AmE* (*cf earth potential difference BrE*) POWER différence de potentiel entre deux terres *f*; ~ **radial termination** *AmE* (*cf earth radial termination BrE*) POWER connexion du réseau de terre radial *f*; ~ **resistance** *AmE* (*cf earth resistance BrE*) POWER résistance de terre *f*; ~ **resolution** ELECT limite de résolution au sol *f*; ~ **return** RADIO écho de sol *m*, SIG, TRANS retour par la terre *m*; **~-return circuit** *AmE* (*cf earth-return circuit BrE*) TRANS circuit à retour par la terre *m*; **~-return double phantom circuit** *AmE* (*cf earth-return double phantom circuit BrE*) TELEP circuit approprié de fantôme *m*; **~-return phantom circuit** *AmE* (*cf earth-return phantom circuit BrE*) TELEP circuit approprié *m*; ~ **ring collector** *AmE* (*cf earth ring collector BrE*) POWER collecteur annulaire de mise à la terre *m*; ~ **segment** TRANS secteur sol *m*; ~ **station** RADIO station au sol *f*; **~-support equipment** (*GSE*) RADIO équipement de soutien au sol *m*, matériel de soutien au sol *m*; ~ **system** POWER *AmE* (*cf earth system BrE*) installation de mise à la terre *f*, RADIO plan de sol *m*; ~ **terminal** POWER *AmE* (*cf earth terminal BrE*) borne de mise à la terre *f*, RADIO station terrienne terminale *f*; ~ **test** TRANS essai au sol *m*; ~ **wave** RADIO onde de sol *f*, onde directe *f*; ~ **wire** *AmE* (*cf earth wire BrE*) POWER câble de garde *m*, fil de masse *m*, fil de terre *m*

ground[3] *vt AmE* (*cf earth BrE*) ELECT, POWER mettre à la masse, mettre à la terre

grounded[1] *adj AmE* (*cf earthed BrE*) ELECT, POWER mis à la masse, mis à la terre

grounded[2]: ~ **emitter circuit** *n AmE* (*cf earthed emitter circuit BrE*) ELECT *semiconductors* circuit à émetteur commun *m*, montage à émetteur commun *m*; ~ **emitter stage** *n AmE* (*cf earthed emitter stage BrE*) ELECT *semiconductors* étage à émetteur commun *m*; **~-grid amplifier** *n AmE* (*cf earthed-grid amplifier BrE*) ELECT amplificateur à grille à la masse *m*

grounding *n AmE* (*cf earthing BrE*) ELECT, POWER mise à la masse *f*, mise à la terre *f*; ~ **arrangement** *AmE* (*cf earthing arrangement BrE*) POWER dispositif de mise à la terre *m*; ~ **button** *AmE* (*cf earthing button BrE*) SWG *PBX* bouton de mise à la terre *m*; ~ **conductor** *AmE* (*cf earthing conductor BrE*) POWER conducteur de masse *m*, conducteur de terre *m*; ~ **indents** *n pl AmE* (*cf earthing indents BrE*) ELECT dentures de masse *f pl*; ~ **mat** RADIO tapis de sol *m*; ~ **network** *AmE* (*cf earthing network BrE*) POWER réseau de terre *m*; ~ **plate** *AmE* (*cf earthing plate BrE*) POWER plaque de terre *f*; ~ **probe** *AmE* (*cf earthing probe BrE*) POWER sonde de mise à la terre *f*; ~ **rod** *AmE* (*cf earthing rod BrE*) POWER piquet de mise à la terre *m*, tige de mise à la terre *f*; ~ **switch** *AmE* (*cf earthing switch BrE*) POWER sectionneur de terre *m*

group[1] *n* COMP segment *m*, GEN groupe *m*, faisceau *m*, groupement *m*; ~ **call identity** DATA, TELEP identité d'appel de groupe *f*; ~ **calls** *n pl* TELEP appels destinés à des groupes de navires *m pl*; ~ **delay** TRANS temps de propagation de groupe *m*; ~**-delay distortion** TRANS distorsion du temps de propagation de groupe *f*; ~**-delay equalization** TRANS égalisation du temps de propagation de groupe *f*; ~**-delay mask** TRANS gabarit du temps de propagation de groupe *m*; ~**-delay slope** TRANS pente de la courbe du temps de propagation de groupe *f*; ~ **distribution frame** TELEP, TRANS répartiteur de groupe primaire *m*; ~ **identifier** (*GI*) TELEP identificateur de groupe *m*; ~ **index** TRANS *fibre optics* indice de groupe *m*; ~ **key** ELECT clé de groupe *f*; ~ **length** (*GL*) TELEP longueur de groupe *f*; ~ **link** TELEP liaison en groupe primaire *f*; ~ **listening** TELEP écoute collective *f*; ~ **mark** DATA marque de groupe *f*, séparateur de groupes de données *m*; ~ **of matrix links** TELEP groupe de lignes réseau *m*; ~ **monitoring** TELEP écoute collective sur postes distincts *f*; ~ **number call** TELEP appel de groupe *m*; ~ **pilot failure** TRANS interruption d'onde pilote de groupe primaire *f*; ~ **pilot generation** TRANS générateur d'onde pilote *m*; ~ **relationships** *n pl* DATA relations de groupe *f pl*; ~ **retardation** RADIO réduction de la vitesse de groupe *f*; ~ **section** TELEP section de groupe primaire *f*; ~ **separator** (*GS*) DATA séparateur de groupes *m*; ~ **switching** SWG permutation de groupe *f*; ~ **switching center** *AmE*, ~ **switching centre** *BrE* (*GSC*) SWG centre de commutation de groupement *m*, commutateur à autonomie d'acheminement *m* (*GSC*); ~ **switching centre catchment area** *BrE* SWG, TELEP zone à autonomie d'acheminement *f* (*ZAA*); ~ **switching centre exchange area** *BrE* SWG, TELEP zone à autonomie d'acheminement *f* (*ZAA*); ~ **testing** DATA essai par lots *m*; ~ **translating equipment** TELEP équipement de matériel de groupe primaire *m*, équipement de modulation de groupe primaire *m*; ~ **translation** TRANS modulation de groupe primaire *f*; ~ **velocity** TRANS vitesse de groupe *f*; ~ **velocity dispersion** (*GVD*) TRANS dispersion de vitesse de groupe *f*

group[2] *vt* GEN, TRANS grouper

grouping *n* GEN groupement *m*; ~ **diagram** GEN *documentation* schéma de groupement *m*

groups: **by** ~ *adv* GEN en groupe, par groupe; **in** ~ *adv* GEN en groupe, par groupe

groupware *n* COMP, DATA collectique *f*, logiciel de groupe *m*

grown: ~ **junction** *n* ELECT *semiconductors* jonction tirée *f*, jonction par tirage *f*; ~ **junction transistor** *n* ELECT *semiconductors* transistor à jonction par tirage *m*

GS *abbr* (*group separator*) DATA séparateur de groupes *m*

GSC *abbr* (*group switching center AmE*, *group switching centre BrE*) SWG GSC (*commutateur à autonomie d'acheminement*), centre de commutation de groupement *m*

GSE *abbr* (*ground support equipment*) RADIO équipement de soutien au sol *m*, matériel de soutien au sol *m*

GSM *abbr* (*global system for mobile communications*) RADIO, TELEP GSM (*groupe spécial mobiles*)

G/T *abbr* GEN facteur de mérite *m*, facteur G/T *m*

GTE *abbr* (*group translating equipment*) TRANS équipement de matériel de groupe primaire *m*, équipement de modulation de groupe primaire *m*

GTS *abbr* (*global telecommunications system*) RADIO, TELEP, TRANS SMT (*système mondial de télécommunications*)

guard: ~ **band** *n* RADIO bande de garde *f*; ~ **circuit** *n* TELEP circuit de garde *m*; ~ **receiver** *n* RADIO récepteur de veille *m*; ~ **ring** *n* TELEP anneau de garde *m*; ~ **space** *n* RADIO espacement de garde *m*; ~ **time** *n* RADIO temps de garde *m*

GUI *abbr* (*graphical user interface*) COMP, DATA GUI (*interface utilisateur graphique*)

guidance: ~ **output** *n* TELEP sortie d'instruction *f*

guide[1] *n* GEN guide *m*; ~ **characteristic wave impedance** ELECT, RADIO impédance d'onde caractéristique dans un guide *f*; ~ **finger** TELEP doigt de guidage *m*; ~ **spool** TELEP bobine de guidage *f*; ~ **wavelength** RADIO, TRANS longueur d'onde dans le guide *f*

guide[2] *vt* GEN guider

guided: ~ **mode** *n* TRANS *fibre optics* mode lié *m*, mode guidé *m*; ~ **radiation system** *n* RADIO système à rayonnement guidé *m*; ~ **wave** *n* TRANS *fibre optics* onde guidée *f*; ~**-wave demultiplexer** *n* (*GWD*) TRANS démultiplexeur d'onde guidée *m*; ~**-wave polarizer** *n* ELECT polariseur en ondes guidées *m*

guideline *n* GEN ligne directrice *f*

guillotine: ~ **attenuator** *n* RADIO affaiblisseur à guillotine *m*, atténuateur à guillotine *m*

guy *n* PARTS étai *m*; ~ **attachment** PARTS tige d'ancrage *f*; ~ **wire** PARTS, RADIO hauban *m*

GVD *abbr* (*group velocity dispersion*) TRANS dispersion de vitesse de groupe *f*

GWD *abbr* (*guided-wave demultiplexer*) TRANS démultiplexeur d'onde guidée *m*

gyrator *n* RADIO gyrateur *m*

gyrofrequency *n* RADIO gyrofréquence *f*

gyroscopic: ~ **drift** *n* RADIO dérive des gyroscopes *f*

gyrotron *n* ELECT gyrotron *m*

H

H *abbr* ELECT (*henry*) H (*henry*), (*high level*) *logic elements* niveau H *m* (*niveau haut*), (*high-level state*) état H *m* (*état haut*); **~ line** RADIO transversale magnétique *f*; **~ mode** RADIO mode électrique transversal *m*, mode H *m*
hacker *n* COMP pirate informatique *m*
hacking *n* COMP piratage *m*
halation *n* TV halo *m*
half[1]: **~-duplex** *adj* TRANS alternat, semi-duplex, simplex, à l'alternat; **~-wave** *adj* ELECT, POWER, RADIO, TRANS demi-onde
half[2]: **~-adder** *n* ELECT demi-additionneur *m*; **~-cycle** *n* POWER demi-alternance *f*, demi-période *f*; **~-duplex circuit** *n* TELEP circuit semi-duplex *m*; **~ duplex connection** *n* TELEP communication semi-duplex *f*; **~-duplex transmission** *n* TRANS transmission semi-duplex *f*; **~-duplex transmission module** *n* (*HDTM*) TELEP module de transmission en mode semi-duplex *m*; **~-echo suppressor** *n* TELEP demi-suppresseur d'écho *m*; **~-intensity** *n* TV demi-intensité *f*; **~-lattice filter** *n* ELECT, TRANS filtre en demi-treillis *m*; **~-line pulse** *n* TV impulsion de demi-ligne *f*; **~-power bandwidth** *n* RADIO, TRANS largeur de bande à mi-puissance *f*; **~-power beamwidth** *n* RADIO, TRANS ouverture de faisceau à mi-puissance *f*; **~-rate channel** *n* TRANS voie à demi-débit *f*; **~-rate FEC decoder** *n* SIG décodeur FEC de rapport 1/2 *m*; **~-size plug** *n* PARTS *connecting devices* fiche demi-grandeur *f*; **~-subtracter** *n* ELECT *logic elements* demi-soustracteur *m*; **~-tone distortion** *n* TV distorsion de demi-teinte *f*; **~-toning** *n* TELEP demi-ton *m*; **~-wave dipole** *n* RADIO doublet demi-onde *m*; **~-wave rectification** *n* ELECT redressement à une alternance *m*, redresseur demi-onde *m*, redresseur simple alternance *m*; **~-wave rectifier** *n* ELECT redresseur simple alternance *m*, redresseur à une alternance *m*, redresseur demi-onde *m*
half[3]: **~-wave rectify** *vt* ELECT redresser à une alternance
Hall: **~ angle** *n* ELECT angle de Hall *m*; **~ coefficient** *n* ELECT coefficient de Hall *m*; **~ effect** *n* ELECT effet Hall *m*; **~ effect device** *n* ELECT dispositif à effet Hall *m*; **~ generator** *n* ELECT générateur de Hall *m*; **~ integrated circuit** *n* ELECT circuit intégré à effet Hall *m*; **~ modulator** *n* ELECT modulateur de Hall *m*; **~ multiplier** *n* ELECT multiplicateur de Hall *m*; **~ plate** *n* ELECT plaque de Hall *f*; **~ probe** *n* ELECT sonde de Hall *f*; **~ terminal** *n* ELECT borne de Hall *f*; **~ voltage** *n* ELECT tension de Hall *f*
hallway: **~ telephone** *n* PER APP *for security purposes* portier automatique *m*, portier téléphonique *m*
halo *n* TV halo *m*; **~ orbit** RADIO orbite halo *f*
halogen: **~-quenched counter tube** *n* ELECT, PARTS tube-compteur à halogène *m*
halt: **~ indicator** *n* COMP indicateur d'arrêt *m*; **~ instruction** *n* COMP instruction de pause *f*
ham *n* RADIO radioamateur *m*
Hamming: **~ code** *n* DATA code de Hamming *m*; **~ distance** *n* COMP distance de Hamming *f*
hand[1]: **~-held** *adj* GEN portable, portatif, transportable; **~-portable** *adj* GEN portable, portatif, transportable
hand[2]: **~ control** *n* COMP poignée de contrôle *f*; **~-held computer** *n* COMP ordinateur portable *m*; **~-held message terminal** *n* DATA terminal de télécommunication de poche *m*; **~-held receiver** *n* RADIO récepteur portatif *m*; **~-held refractometer** *n* TRANS *optics* réfractomètre portatif *m*; **~-held scanner** *n* PER APP scanner manuel *m*; **~-held terminal** *n* RADIO terminal de poche *m*; **~-off** *n* RADIO transfert intercellulaire *m*, transfert automatique intercellulaire *m*; **~-over** *n* RADIO transfert intercellulaire *m*, transfert automatique intercellulaire *m*; **~-portable** *n* TELEP portable *m*, téléphone de poche *m*; **~-portable telephone** *n* TELEP portable *m*, téléphone de poche *m*
hand[3]: **~ off** *vt AmE* (*cf hand over BrE*) RADIO transférer; **~ over** *vt BrE* (*cf hand off AmE*) RADIO transférer
handle *vt* DATA traiter, GEN *control by hand* prendre en charge
handler *n* COMP programme de gestion de périphériques *m*, module de gestion *m*, DATA module de gestion *m*, SWG gestionnaire *m*
handling *n* GEN gestion *f*, prise en charge *f*, traitement *m*, manipulation *f*
handover *n* RADIO handover *m*
hands[1]: **~-free** *adj* GEN mains libres
hands[2]: **~-free answerback** *n* TELEP réponse mains libres *f*; **~-free conversation** *n* PER APP conversation mains libres *f*; **~-free operation** *n* RADIO fonctionnement mains libres *m*; **~-free tape deck** *n* PER APP platine mains libres *f*; **~-free telephone set** *n* TELEP poste mains-libres *m*, poste téléphonique mains libres *m*; **~-free terminal** *n* PER APP terminal mains libres *m*
handset *n* PER APP, TELEP combiné *m*, combiné téléphonique *m*; **~ cord** PER APP, TELEP cordon du combiné *m*; **~ handle** PER APP poignée du combiné *f*; **~ off hook** TELEP faux appel *f* (*FA*);**~ telephone** PER APP, TELEP poste téléphonique à combiné *m*
handshake *n* COMP, DATA colloque de reconnaissance *m*, établissement de liaison *m*, RADIO accusé de réception électronique *m*
handshaking *n* COMP, DATA colloque de reconnaissance *m*, établissement de liaison *m*, RADIO accusé de réception électronique *m*
handwriting: **~ recognition** *n* APPL reconnaissance de l'écriture manuelle *f*
hang[1]: **~ up** *vi* TELEP raccrocher
hang[2]: **~ on** *phr* TELEP ne quittez pas
hanger *n* ELECT suspension *f*
hanging: **~ up** *n* TELEP raccrochage *m*
hangover: **~ time** *n* RADIO, TELEP temps de maintien *m*
hang-up: **~ signal** *n* SIG signal de raccrochage *m*
Hann: **~ window** *n* TV fenêtre de Hann *f*
harbor: **~ radar** *AmE see harbour radar BrE*
harbour: **~ radar** *n BrE* RADIO radar de port *m*
hard: **~ copy** *n* COMP *printed version* sortie sur support en papier *f*, tirage *m*; **~-copy image** *n* TELEP image sortie sur support papier *f*; **~-copy output** *n* PER APP sortie sur papier *f*; **~ decision** *n* DATA, ELECT, SIG décision ferme *f*; **~-decision decoder** *n* DATA, ELECT, SIG dispositif de décodage à décision formelle *m*; **~-decision decoding** *n* DATA, ELECT, SIG décodage à décision formelle *m*; **~ disk** *n* COMP disque dur *m*; **~-disk card** *n* COMP disque dur sur carte d'extension *m*; **~ disk drive** *n* COMP unité de disque dur *f*; **~-disk interface** *n* COMP interface disque dur *f*; **~ magnetic material** *n* ELECT, PARTS matériau magnétique dur *m*; **~ shorting** *n* POWER court-circuit franc *m*; **~-wired network** *n* TRANS réseau filaire *m*; **~-wired programmable switching system** *n* SWG système de commutation à programme câblé *m*
hardening: **~ agent** *n* PARTS *plastics* agent de durcissement *m*

hardware *n* GEN matériel *m*; **~ engineer** GEN ingénieur matériel *m*
harmful: **~ interference** *n* RADIO brouillage préjudiciable *m*; **~ out-of-band components** *n pl* TELEP composants extra-bandes nuisibles *m pl*
harmless: **~ out-of-band components** *n pl* TELEP composants extra-bandes neutres *m pl*
harmonic[1] *adj* ELECT harmonique
harmonic[2] *n* ELECT harmonique *m*; **~ analysis** ELECT analyse harmonique *f*; **~ component** TRANS composant harmonique *m*; **~ distortion** TRANS distorsion harmonique *f*; **~ emission** RADIO rayonnement harmonique *m*; **~ generator** TRANS générateur d'harmoniques *m*; **~ interference** RADIO brouillage dû aux harmoniques *m*; **~ oscillation** ELECT oscillation harmonique *f*; **~ progression** TRANS suite harmonique *f*; **~ spectrum** TRANS spectre harmonique *m*; **~ structure** TRANS structure harmonique *f*
harmonically: **~ related carrier** *n* (*HRC*) TV système de porteuses en relation harmonique *m*
harness *n* PARTS faisceau *m*, câblage d'alimentation *m*
Hartley *n* GEN Hartley *m*, unité décimale de quantité d'information *f*
hash: **~ coding** *n* COMP, DATA adressage calculé *m*; **~ function** *n* COMP, DATA adressage calculé *m*
hashing *n* COMP, DATA adressage calculé *m*
hatch *vt* COMP hachurer
hazard *n* TEST aléa *m*; **~ analysis** TEST analyse de risque *f*
hazardous: **~ state** *n* TEST *dependability* état dangereux *m*
HBT *abbr* (*heterojunction bipolar transistor*) ELECT, PARTS transistor bipolaire à hétérojonction *m*
HCMOS *abbr* (*high-speed complementary metal oxide semiconductor*) ELECT HCMOS (*semiconducteur complémentaire à oxyde metallique à grande vitesse*)
H-corner *n* RADIO coude H *m*
HDB2 *abbr* (*high-density bipolar 2 code*) DATA, SIG, TELEP HDB2 (*code bipolaire à haute densité d'ordre 2*)
HDB3 *abbr* (*high-density bipolar 3 code*) DATA, SIG, TELEP HDB3 (*code bipolaire à haute densité d'ordre 3*)
HDR *abbr* (*header*) ELECT HDR (*adaptateur, adaptateur de mesure*); **~ label** *n* COMP label début *m*
HDSL *abbr* (*high-speed digital subscriber line*) TRANS ligne d'abonné à haut débit *f*
HDTM *abbr* (*half-duplex transmission module*) TELEP module de transmission en mode semi-duplex *m*
HDTV *abbr* (*high-definition television*) TV TVHD (*télévision à haute définition*)
head *n* COMP, DATA, ELECT, GEN, PARTS, SWG tête *f*; **~ end** GEN tête de réseau *f* (*TR*), TV tête de ligne *f*, tête de réseau *f* (*TR*), tête ; **~ gap** COMP entrefer *m*; **~ housing** TRANS boîtier des têtes optiques *m*; **~-of-the-queue** SWG tête de file *f*
header *n* (*HDR*) ELECT adaptateur *m* (*HDR*), adaptateur de mesure *m* (*HDR*); **~ card** COMP carte en-tête *f*; **~ error control** (*HEC*) COMP, DATA contrôle d'erreur sur en-tête *m*; **~ label** COMP label début *m*
headphones *n pl* TELEP casque *m*, casque d'écoute *m*; **~ monitoring** *n* TELEP contrôle au casque *m*; **~ monitoring level** *n* TELEP niveau de contrôle au casque *m*; **~ output** *n* ELECT, TELEP sortie casque *f*; **~ socket** *n* PER APP prise casque *f*
headset *n* TELEP casque *m*, casque téléphonique *m*; **~ jack** PER APP prise casque *f*
hearing: **~-aid compatible** *adj* APPL, PER APP *answer machine* compatible avec les appareils auditifs
heart *n* COMP cœur *m*
heat: **~ budget** *n* ELECT bilan thermique *m*; **~ coil** *n* POWER bobine thermique *f*; **~ detector** *n* TV capteur de chaleur *m*, capteur d'effraction *m*; **~ dissipation** *n* ELECT, POWER dissipation d'énergie *f*; **~ dissipator** *n* ELECT, POWER puits de chaleur *m*, radiateur *m*; **~ radiation** *n* ELECT, POWER rayonnement thermique *m*, thermorayonnance *f*; **~ shield** *n* ELECT écran thermique *m*; **~-shrink tubing** *n* PARTS, TRANS gaine thermorétractable *f*; **~-shrinkable sleeve** *n* PARTS manchon thermorétractable *m*; **~-shrinkable sleeving** *n* PARTS, TRANS gaine thermorétractable *f*; **~ shrinking** *n* ELECT thermorétraction *f*; **~ sink** *n* ELECT, POWER déperditeur de chaleur *m*, puits de chaleur *m*; **~ sink compound** *n* ELECT, POWER pâte thermoconductrice *f*; **~ strip** *n* ELECT, PARTS bande chauffante *f*; **~ trace cable** *n* TRANS câble chauffant pour tuyauterie *m*
heater: **~-cathode insulation current** *n* ELECT courant d'isolement entre filament et cathode *m*; **~ current** *n* ELECT courant de chauffage *m*, courant de filament *m*; **~ gun** *n* POWER pistolet à air chaud *m*; **~ starting current** *n* ELECT courant de chauffage au démarrage *m*, courant de démarrage du filament *m*, courant de filament au démarrage *m*; **~ surge current** *n* ELECT courant de chauffage au démarrage *m*, courant de filament au démarrage *m*, courant de pointe du filament *m*; **~ voltage** *n* ELECT tension de filament *f*, tension de chauffage *f*; **~ warm-up time** *n* ELECT durée de chauffage du filament *f*
heating: **~ cable** *n* POWER câble chauffant *m*; **~ fault** *n* ELECT *soldering* problème de chauffage *m*
heavy: **~ current** *n* POWER courant fort *m*; **~-current engineering** *n* POWER technique des courants forts *f*
HEC *abbr* (*header error control*) COMP, DATA contrôle d'erreur sur en-tête *m*
height *n* GEN hauteur *f*; **~-finding radar** RADIO radar d'altimétrie *m*; **~ indicator** RADIO indicateur de hauteur *m*; **~ of an MCS provider** TELEP hauteur d'un fournisseur de service MCS *f*; **~-position indicator** RADIO indicateur de hauteur et de distance *m*
helical[1] *adj* GEN hélicoïdal, en spirale
helical[2]: **~ aerial** *n* RADIO antenne en hélice *f*, antenne hélicoïdale *f*; **~ antenna** *n* RADIO antenne en hélice *f*, antenne hélicoïdale *f*; **~ potentiometer** *n* POWER potentiomètre hélicoïdal *m*
helix *n* GEN hélice *f*, spirale *f*; **~ aerial** RADIO antenne en hélice *f*, antenne hélicoïdale *f*; **~ antenna** RADIO antenne en hélice *f*, antenne hélicoïdale *f*; **~ pitch** RADIO pas d'hélice *m*; **~ spiral wrap** ELECT enroulement en spirale *m*; **~ voltage** ELECT tension d'hélice *f*
help: **~ file** *n* COMP fichier d'aide *m*
helpware *n* COMP, DATA logiciel d'aide *m*, logiciel d'assistance *m*
hemispherical: **~ coverage** *n* RADIO couverture hémisphérique *f*
HEMT *abbr* (*high electron mobility transistor*) ELECT HEMT (*transistor à haute mobilité d'électrons*)
henry *n* (*H*) ELECT henry *m* (*H*)
heptode *n* ELECT heptode *f*
hertz *n* (*Hz*) ELECT, TEST, TRANS hertz *m* (*Hz*)
hertzian: **~ wave** *n* ELECT, RADIO onde hertzienne *f*, onde radioélectrique *f*
heterochronous *adj* TELEP, TRANS hétérochrone
heterodyne[1] *adj* GEN hétérodyne
heterodyne[2]: **~ detection** *n* ELECT détection hétérodyne *f*; **~ frequency** *n* RADIO fréquence d'hétérodyne *f*; **~ frequency meter** *n* TEST fréquencemètre hétérodyne *m*; **~ reception** *n* RADIO réception hétérodyne *f*
heterodyne[3] *vt* RADIO superposer
heterodyning *n* RADIO superposition *f*
heterogeneous[1] *adj* GEN hétérogène
heterogeneous[2]: **~ multiplex** *n* TELEP multiplex hétérogène *m*
heterojunction *n* ELECT, TRANS hétérojonction *f*; **~ bipolar transistor** (*HBT*) ELECT, PARTS transistor bipolaire à hétérojonction *m*; **~ transistor** ELECT, PARTS transistor à hétérojonction *m*
heuristic: **~ decision** *n* DATA décision heuristique *f*; **~ hazard** *n* DATA hasard heuristique *m*; **~ method** *n* DATA méthode

heuristique *f*; ~ **mix** *n* DATA mélange heuristique *m*; ~ **programming** *n* DATA programmation heuristique *f*; ~ **search** *n* DATA recherche heuristique *f*

heuristics *n pl* DATA heuristique *f*

hexadecimal: ~ **digit** *n* COMP chiffre hexadécimale *m*; ~ **notation** *n* COMP notation hexadécimale *f*; ~ **numeral** *n* COMP nombre hexadécimal *m*; ~ **system** *n* COMP système hexadécimale *m*

hexode *n* ELECT hexode *f*

hidden: ~-**line removal** *n* COMP élimination des lignes cachées *f*

hide *vt* COMP cacher, masquer

hierarchical[1] *adj* GEN hiérarchique

hierarchical[2]: ~ **data structure** *n* COMP, DATA structure hiérarchique des données *f*; ~ **link** *n* COMP lien hiérarchique *m*; ~ **multiplexer** *n* TELEP transmultiplexeur hiérarchique *m*; ~ **network** *n* TELEP *mutually synchronized* réseau hiérarchisé *m*; ~ **progressive mode** *n* COMP, TELEP mode progressif hiérarchique *m*; ~ **routing** *n* SIG, TELEP, TRANS acheminement hiérarchique *m*; ~ **system** *n* GEN système à structure hiérarchique *m*, système hiérarchisé *m*

hierarchy *n* GEN hiérarchie *f*

high[1] *adj* GEN *setting* optimal, optimum; ~-**frequency** GEN à haute fréquence; ~-**impedance** ELECT, POWER à haute impédance; ~-**power** POWER à grande puissance; ~-**resistance** ELECT, POWER à haute résistance; ~-**resolution** (*hi-res*) COMP, TV à haute résolution; ~-**voltage** POWER à haute tension

high[2] *n* (*H*) ELECT *logic elements* niveau H *m*, niveau haut *m*, état H *m*, état haut *m*; ~-**BER alarm** DATA alarme de TEB élevé *f*; ~ **bit-rate ATM network mixer** SWG brasseur de réseau à haut débit ATM *m*; ~ **bit-rate private switch** SWG, TELEP commutateur privé haut débit *m*; ~-**capacity communication satellite** RADIO satellite de télécommunications à grande capacité de transmission *m*; ~ **data rate** DATA transmission à grande vitesse *f*; ~ **definition** TV haute définition *f*; ~-**definition image** TV image à haute définition *f*; ~-**definition television** (*HDTV*) TV télévision à haute définition *f* (*TVHD*); ~-**density binary code** COMP, DATA, ELECT code binaire à haute densité *m*; ~-**density bipolar 2 code** (*HDB2*) DATA, SIG, TELEP code bipolaire à haute densité d'ordre 2 *m* (*HDB2*); ~-**density bipolar 3 code** (*HDB3*) DATA, SIG, TELEP code bipolaire à haute densité d'ordre 3 *m* (*HDB3*); ~ **electron mobility transistor** (*HEMT*) ELECT transistor à haute mobilité d'électrons *m* (*HEMT*); ~ **frequency** RADIO haute fréquence *f* (*HF*); **very** ~ **frequency** (*VHF*) RADIO, TV très haute fréquence *f* (*VHF*); ~-**frequency distribution frame** TRANS répartiteur de hautes fréquences *m*; ~-**gain aerial** RADIO antenne à gain élevé *f*, antenne à grand gain *f*; ~-**gain antenna** RADIO antenne à gain élevé *f*, antenne à grand gain *f*; ~-**grade channel** TRANS voie de qualité élevée *f*; ~-**grade component** PARTS composant de haute qualité *m*; ~-**impedance circuit** ELECT circuit à grande impédance *m*; ~-**impedance path** TRANS trajet à grande impédance *m*, voie à grande impédance *f*; ~-**impedance state** ELECT état haute impédance *m*, état flottant *m*; ~-**layer compatibility** (*HLC*) DATA, SWG, TELEP compatibilité de couche supérieure *f*; ~ **level** (*H*) ELECT *logic elements* niveau haut *m* (*niveau H*); ~-**level data link control** (*HLDLC*) DATA, TELEP commande de liaison de données à haut niveau *f* (*HLDLC*); ~-**level language** (*HLL*) COMP, DATA, ELECT langage évolué *m*, langage de haut niveau *m* (*LHN*); ~-**level logic** ELECT logique de haut niveau *f*; ~-**level modulation** RADIO modulation sur l'étage final *f*; ~-**level signal** COMP signal à niveau élevé *m*; ~-**level state** (*H*) ELECT *logic elements* état haut *m* (*état H*); ~-**low action** CONTR action par tout ou peu *f*; ~-**order network** TRANS réseau d'ordre supérieur *m*; ~-**pass filter** POWER, TRANS filtre passe-haut *m*; ~-**power amplifier** (*HPA*) ELECT, POWER amplificateur à haute puissance *m*; ~-**power amplifier linearizer** ELECT, POWER linéarisateur d'amplificateur à haute puissance *m*; ~-**power flux density** RADIO puissance surfacique importante *f*; ~-**power transmission** RADIO, TRANS transmission à grande puissance *f*; ~ **resistance** (*HR*) TEST haute résistance *f*; ~-**speed carry** COMP report simultané *m*, report accéléré *m*; ~-**speed complementary metal oxide semiconductor** (*HCMOS*) ELECT semiconducteur complémentaire à oxyde metallique à grande vitesse *m* (*HCMOS*); ~-**speed data** *n pl* DATA données à grande vitesse *f pl*; ~-**speed digital subscriber line** (*HDSL*) TRANS ligne d'abonné à haut débit *f*; ~-**speed facsimile** TELEP télécopie rapide *f*; ~-**speed integrated circuit** ELECT circuit intégré à grande vitesse *m*; **very** ~-**speed integrated circuit** (*VHSIC*) ELECT circuit à très haute intégration *m*, circuit intégré à très grande vitesse *m* (*CITGV*); **very** ~-**speed integrated circuit hardware description language** (*VHDL*) COMP langage de description de circuits intégrés à très grande vitesse *m* (*langage VHDL*); ~-**speed laser** ELECT laser ultra-rapide *m*; ~-**speed logic** ELECT logique rapide *f*; ~-**speed metal oxide semiconductor** (*HMOS*) ELECT semiconducteur à oxyde metallique à grande vitesse *m* (*HMOS*); ~-**speed tester** TEST testeur rapide *m*; ~ **tension** POWER haute tension *f*; ~ **to low** ELECT transition descendante *f*, transition du niveau haut vers le niveau bas *f*; ~-**usage circuit group** TELEP faisceau à fort trafic *m*, faisceau débordant *m*; ~-**velocity camera tube** ELECT tube analyseur à électrons rapides *m*, tube analyseur à potentiel stabilisé sur l'anode *m*; ~-**velocity scanning** ELECT balayage par électrons rapides *m*; ~ **voltage** (*HV*) ELECT, POWER haute tension *f* (*HT*); ~-**voltage delay time** RADIO délai haute tension *m*; ~-**voltage surge** ELECT surtension *f*

higher: ~ **frequency** *n* RADIO fréquence supérieure *f*; ~-**level function** *n* (*HLF*) TRANS fonction de couche supérieure *f*; ~-**level service** *n* APPL service de niveau élevé *m*; ~-**order path adaptation** *n* (*HPA*) TELEP, TRANS adaptation de conduit d'ordre supérieur *f*; ~-**order path connection** *n* (*HPC*) TELEP, TRANS connexion de conduit d'ordre supérieur *f*; ~-**order path termination** *n* (*HPT*) TELEP, TRANS terminaison de conduit d'ordre supérieur *f*; ~ **tariff** *n* GEN tarif supérieur *m*

highlight *vt* COMP mettre en évidence, mettre en surbrillance

highlighting *n* COMP mise en évidence *f*, mise en surbrillance *f*

highly: ~ **advanced hardware** *n* PARTS technologie avancée *f*; ~-**eccentric orbit** *n* RADIO orbite elliptique très allongée *f*; ~ **elliptical orbit** *n* RADIO orbite très elliptique *f*

hi-res *abbr* (*high-resolution*) COMP, TV à haute résolution

hiss *n* TELEP souffle *m*

histogram *n* ELECT histogramme *m*

hit *n* RADIO atteinte de but *f*, coup de but *m*

hitch *n* COMP panne *f*

hi-tech: ~ **home** *n* TV foyer hi-tech *m*

hits: ~ **per scan** *phr* RADIO nombre d'atteintes du but *m*; ~ **per target** *phr* RADIO nombre d'atteinte du but *m*

HLC *abbr* (*high-layer compatibility*) DATA, SWG, TELEP compatibilité de couche supérieure *f*

HLDLC *abbr* (*high-level data link control*) DATA, TELEP HLDLC (*commande de liaison de données à haut niveau*)

HLF *abbr* (*higher-level function*) TRANS fonction de couche supérieure *f*

HLL *abbr* (*high-level language*) COMP, DATA, ELECT LHN (*langage de haut niveau*), langage évolué *m*

HLR *abbr* (*home location register*) RADIO HLR (*enregistreur de position de rattachement*)

HLS *abbr* (*hue, lightness, saturation*) TV teinte, luminosité, saturation *f*

HMOS *abbr* (*high-speed metal oxide semiconductor*) ELECT HMOS (*semiconducteur à oxyde metallique à grande vitesse*)

hoghorn *n* RADIO cornet parabolique *m*

hogtrough: ~ **antenna** *n* RADIO antenne en D *f*

hold[1] *n* COMP suspension *f*, ELECT arrêt de chronologie *m*, pause *f*; ~ **current** ELECT courant de maintien *m*, courant minimal de fonctionnement *m*; ~**-down point** RADIO point d'attache *m*; ~ **facility** COMP possibilité de maintien *f*; ~ **signaling** *AmE*, ~ **signalling** *BrE* SIG signalisation pour l'attente *f*; ~ **time** COMP temps de hold *m*
hold[2] *vt* TELEP *conversation* échanger; ~ **the line** TELEP rester en ligne
hold[3]: ~ **in** *vi* POWER rester collé
hold[4]: **on** ~ *phr* TELEP en garde
holder *n* PARTS fixation *f*, support *m*
holding *n* ELECT entretien *m*; ~ **call transfer** RADIO renvoi temporisé *m*; ~ **circuit** ELECT circuit de maintien *m*; ~ **coil** PARTS bobine de maintien *f*; ~ **contact** PARTS contact de maintien *m*; ~ **current** ELECT *semiconductors* courant de maintien *m*, courant minimal de fonctionnement *m*; ~ **current level** ELECT *semiconductors* niveau de courant de maintien *m*; ~ **range** ELECT plage de maintien *f*; ~ **register** ELECT registre de maintien *m*; ~ **time** TELEP durée d'occupation *f*, temps d'occupation *m*; ~**-time distribution** TELEP répartition des durées d'occupation *f*; ~ **winding** PARTS enroulement de maintien *m*
hole[1]: ~**-mounted** *adj* ELECT *printed circuits* inséré
hole[2] *n* ELECT, PARTS trou *m*, lacune *f*; ~ **drawing** ELECT, PARTS *printed circuits* plan de perçage *m*; ~ **mounting** ELECT, PARTS *printed circuits* montage en support percé *m*; ~ **pattern** ELECT, PARTS *printed circuits* configuration des perforations *f*; ~ **spacing** ELECT, PARTS *printed circuits* espacement des trous *m*; ~ **storage** ELECT, PARTS *semiconductors* accumulation des porteurs *f*, accumulation des trous *f*
holed *adj* GEN perforé
Hollerith: ~ **code** *n* COMP code Hollerith *m*
hollow: ~ **conductor** *n* TRANS *cables* conducteur creux *m*
hollowness *n* TELEP son caverneux *m*
holograph *n* DATA, ELECT, SWG holographique *f*
holographic: ~ **exchange** *n* SWG commutateur holographique *m*; ~ **optical element** *n* DATA élément optique holographique *m*; ~ **storage** *n* DATA mémoire holographique *f*
home: ~ **area** *n* RADIO zone de localisation nominale *f*; ~ **automation** *n* APPL domotique *f*; ~**-banking terminal** *n* APPL terminal de télépaiement *m*; ~ **base** *n* RADIO base de rattachement *f*; ~ **base station** *n* RADIO station de base de rattachement *f*; ~ **computing** *n* COMP, DATA informatique domestique *f*; ~**-direct service** *n* TELEP service pays-direct *m*; ~ **exchange** *n* SWG central d'origine *m*; ~ **location** *n* RADIO position initiale *f*, localisation nominale *f*; ~ **location register** *n* (*HLR*) RADIO enregistreur de position de rattachement *m* (*HLR*); ~ **page** *n* COMP *on Internet* page d'accueil *f*; ~ **position** *n* COMP position de repos *f*; ~ **record** *n* COMP enregistrement initial *m*; ~ **shopping** *n* TV téléachat *m*, télépromotion *f*, téléshopping *m*, télévente *f*; ~ **spot** *n* COMP point actif *m*; ~ **switching center** *AmE*, ~ **switching centre** *BrE n* SWG commutateur de rattachement *m*
homes: ~ **passed** *n pl* TV foyers cerclés *m pl*, logements cerclés *m pl*, prises raccordables *f pl*
homing *n* COMP retour en position initiale *m*, RADIO radioralliement *m*; ~ **position** COMP position de repos *f*; ~ **radar** RADIO radar d'autodirecteur *m*, radar d'autopoursuite *m*
homochronous *adj* TELEP, TRANS homochrone
homodyne: ~ **detection** *n* TRANS détection homodyne *f*; ~ **reception** *n* RADIO réception synchrone *f*
homogeneity *n* GEN homogénéité *f*
homogeneous[1] *adj* GEN homogène
homogeneous[2]: ~ **cladding** *n* PARTS, TRANS gaine homogène *f*; ~ **multiplex** *n* TELEP multiplex homogène *m*; ~ **section** *n* RADIO, TELEP section homogène *f*
homogenize *vt* GEN homogénéifier, homogénéiser
homojunction *n* ELECT homojonction *f*
hood *n* PARTS corps *m*, capot *m*
hook[1]: **off the** ~ *adj* TELEP décroché; **on the** ~ *adj* TELEP raccroché
hook[2]: **on** ~ *adv* COMP, TELEP en ligne
hook[3] *n* TELEP plongeur *m*; ~ **flash** SWG, TELEP actionnement du crochet commutateur *m*; ~ **switch** PER APP, SWG support *m*, support commutateur *m*; ~**-up** COMP connexion *f*, raccordement *m*; ~**-up wire** ELECT fil de connexion *m*
hook[4]: ~ **up** *vt* COMP brancher, connecter, TELEP accorder
hop[1] *n* RADIO bond *m*, saut *m*
hop[2] *vt* RADIO sauter
hopping *n* ELECT saut *m*; ~ **switch** RADIO, SWG commutateur des sauts *m*
horizontal: ~ **antenna** *n* RADIO antenne à polarisation horizontale *f*; ~ **blanking interval** *n* DATA, TV intervalle de retour de ligne *m*, intervalle de supression de ligne *m*; ~ **directivity pattern** *n* RADIO diagramme de directivité horizontal *m*; ~ **line frequency** *n* TV fréquence d'exploration horizontale *f*; ~ **polarization** *n* TV polarisation horizontale *f*; ~ **sweep** *n* TV balayage de ligne *m*, balayage horizontal *m*; ~ **sweep frequency** *n* TV fréquence de balayage horizontal *f*, fréquence de ligne *f*; ~ **tabulation character** *n* (*HT*) DATA, RADIO caractère de tabulation horizontale *m*
horn *n* RADIO cornet *m*; ~ **antenna** RADIO antenne à cornet *f*, antenne en cornet *f*
host *n* COMP, DATA, ELECT, RADIO, SWG hôte *m*; ~ **computer** COMP serveur *m*, ordinateur hôte *m*; ~ **concentrator** COMP concentrateur hôte *m*; ~ **exchange** RADIO central principal *m*, SWG commutateur de rattachement *m*; ~ **office** RADIO central principal *m*; ~ **panel** ELECT panneau hôte *m*; ~ **satellite** RADIO satellite hôte *m*; ~ **station** RADIO station d'accueil *f*; ~ **system** DATA système hôte *m*
hot[1] *adj* COMP *active* actif, ELECT, POWER *live* sous tension, RADIO *in service* en service
hot[2]: ~ **billing** *n* RADIO facturation à chaud *f*; ~ **bird** *n* RADIO satellite loué à plein *m*; ~ **carrier diode** *n* ELECT diode à porteurs chauds *f*; ~ **cathode** *n* ELECT, PARTS cathode chaude *f*; ~ **cathode stepping tube** *n* ELECT tube-compteur à cathode chaude *m*; ~ **conditions** *n pl* ELECT conditions de fonctionnement *f pl*; ~ **line** *n* TELEP appel sans numérotation *m*, ligne directe *f*, téléphone rouge *m*, service d'appel direct *m*; ~**-line mode of operation** *n* TELEP mode de fonctionnement par ligne directe *m*; ~ **satellite** *n* RADIO satellite en service *m*; ~ **spot** *n* COMP zone active *f*; ~ **stand-by** *n* TEST secours immédiat actif *m*; ~ **stand-by system** *n* TEST système de secours immédiat *m*, système de secours sous tension *m*; ~ **swapping** *n* ELECT permutation sous tension *f*; ~**-wire ammeter** *n* TEST ampèremètre thermique *m*; ~ **zone** *n* COMP zone active *f*
hot[3]: ~**-dip tin** *vt* PARTS étamer à chaud
hour: ~ **meter** *n* APPL, TEST compteur horaire *m*
hourly: ~ **charge** *n* APPL prix horaire *m*
house[1]: ~**-wiring cable** *n* POWER câble domestique *m*
house[2] *vt* PARTS loger, monter
household: ~ **electricity** *n* POWER réseau électrique *m*
housekeeping: ~ **bit** *n* DATA bit auxiliaire *m*, bit de service *m*; ~ **function** *n* DATA fonction auxiliaire *f*, fonction de service *f*; ~ **information** *n* DATA, TELEP information de service *f*; ~ **lifetime** *n* DATA durée de service *f*; ~ **operation** *n* COMP opération d'aménagement *f*; ~ **routine** *n* COMP programme d'aménagement *m*; ~ **subsystem** *n* COMP sous-ensemble de servitude *m*; ~ **telemetry** *n* TEST télémesure de maintenance *f*; ~ **unit** *n* DATA équipement de surveillance *m*
housing *n* PARTS boîtier *m*, coffret *m*, enveloppe *f*
how: ~ **do you read me?** *phr* RADIO comment me recevez-vous?
howl *vi* TELEP, TRANS heurler
howler *n* TELEP, TRANS heurleur *m*

howling *n* TELEP *acoustics* amorçage *m*, sifflement *m*, TRANS effet d'écho *m*
HPA *abbr* DATA (*higher-order path adaptation*) adaptation de conduit d'ordre supérieur, ELECT, POWER (*high-power amplifier*) amplificateur à haute puissance *m*, TELEP, TRANS (*higher-order path adaptation*) adaptation de conduit d'ordre supérieur *f*
HPC *abbr* (*higher-order path connection*) TELEP, TRANS connexion de conduit d'ordre supérieur *f*
HPT *abbr* (*higher-order path termination*) TELEP, TRANS terminaison de conduit d'ordre supérieur *f*
HR *abbr* (*high resistance*) TEST haute résistance *f*
hrc *abbr* RADIO, TELEP, TRANS (*hypothetical reference circuit*) CFR (*circuit fictif de référence*), TV (*harmonically related carrier*) système de porteuses en relation harmonique *m*
HRDP *abbr* (*hypothetical reference digital path*) RADIO, TELEP, TRANS CFNR (*conduit fictif numérique de référence*)
HT *abbr* (*horizontal tabulation character*) DATA, RADIO caractère de tabulation horizontale *m*
HTML *abbr* (*hypertext markup language*) COMP, DATA *Internet* HTML
HTTP *abbr* (*hypertext transfer protocol*) COMP, DATA *Internet* protocole de transfert hypertext *m*
hub *n* DATA, ELECT concentrateur *m*, GEN point de convergence *m*, serveur *m*, SWG centre de transit privé *m*, TRANS concentrateur *m*, nud *m*; **~ station** RADIO station centrale *f*
hue *n* TV teinte *f*; **~, lightness, saturation** (*HLS*) TV teinte, luminosité, saturation *f*
hum[1]: **~-free** *adj* GEN sans ronflement, silencieux
hum[2] *n* GEN bourdonnement *m*, bruit de fond *m*, ronflement *m*, ronronnement *m*; **~ level** ELECT niveau de ronflement *m*
human: **~ error** *n* TEST *dependability* erreur humaine *f*; **~-machine interface** *n* COMP interface homme-machine *f*
hundreds: **~ digit** *n* COMP chiffre des centaines *m*
hunting *n* POWER pompage *m*
HV *abbr* (*high voltage*) ELECT, POWER HT (*haute tension*); **~ delay time** *n* RADIO délai haute tension *m*
hybrid *n* GEN hybride *m*; **~ circuit** ELECT circuit hybride *m*; **~ circulator** RADIO anneau hybride *m*; **~ coil** ELECT, POWER transformateur différentiel équilibré *m*; **~ computer** COMP ordinateur hybride *m*; **~ connector** TRANS *fibre optics* connecteur hybride *m*; **~ integrated circuit** ELECT circuit intégré hybride *m*; **~ junction** RADIO jonction hybride *f*; **~ mode** ELECT mode hybride *m*; **~ processor** SWG processeur hybride *m*; **~ reasoning** ELECT raisonnement hybride *m*; **~ ring** RADIO anneau hybride *m*, dérivateur *m*; **~ spread spectrum** RADIO étalement du spectre hybride *m*; **~ switch** SWG commutateur hybride *m*; **~ switching system** SWG, TRANS système hybride de commutation *m*; **~ system** SWG, TRANS installation mixte *f*, système hybride *m*; **~ T** RADIO té hybride *m*; **~ transformer** ELECT, POWER transformateur différentiel équilibré *m*
hydrogen: **~ embrittlement** *n* TEST *strength* fragilisation par l'hydrogène *f*
hydrometeors *n pl* RADIO hydrométéores *m pl*
hydrophone *n* PER APP hydrophone *m*
hyperbolic: **~ navigation system** *n* RADIO radionavigation hyperbolique *f*
hyperbook *n* COMP hyperbook *m*
hypercourse *n* COMP hypercours *m*
hyperdata *n* COMP hyperdata *m*
hyperdocument *n* COMP hyperdocument *m*
hypergroup *n* TRANS groupe tertiaire *m*
hypermedia[1] *adj* COMP, DATA *Internet* hypermédia
hypermedia[2]: **~ application** *n* COMP application hypermédia *f*; **~ building** *n* COMP intégration hypermédia *f*; **~ conversion** *n* COMP conversion hypermédia *f*; **~ document** *n* COMP document hypermédia *m*; **~ function** *n* COMP fonction hypermédia *f*; **~ link** *n* COMP lien hypermédia *m*; **~ navigation** *n* COMP navigation hypermédia *f*; **~ navigation path** *n* COMP parcours de navigation hypermédia *m*, chemin de navigation hypermédia *m*; **~ node** *n* COMP nœud hypermédia *m*; **~ object** *n* COMP objet hypermédia *m*; **~ portability** *n* COMP portabilité hypermédia *f*
hyperobject *n* COMP hyperobjet *m*
hypertext *n* COMP, DATA hypertexte *m*; **~ application** COMP application hypertexte *f*; **~ conversion** COMP conversion hypertexte *f*; **~ document** COMP document hypertexte *m*; **~ link** COMP lien hypertexte *m*; **~ markup language** (*HTML*) COMP, DATA *Internet* HTML; **~ navigation** COMP navigation hypertexte *f*; **~ navigation path** COMP parcours de navigation hypertexte *m*; **~ node** COMP nud hypertexte *m*; **~ portability** COMP portabilité hypertexte *f*; **~ transfer protocol** (*HTTP*) COMP, DATA *Internet* protocole de transfert hypertexte *m*
hypertrail *n* COMP hyperpiste *f*
hypervisor *n* TELEP hyperviseur *m*
hypothetical: **~ reference circuit** *n* (*HRC*) RADIO, TELEP, TRANS circuit fictif de référence *m* (*CFR*); **~ reference circuit for sound program transmissions** *AmE*, **~ reference circuit for sound programme transmissions** *BrE* *n* RADIO circuit fictif de référence pour transmissions radiophoniques *m*; **~ reference circuit for telephony** *n* TELEP circuit fictif de référence pour la téléphonie *m*; **~ reference connection** *n* TELEP communication fictive de référence *f*; **~ reference digital path** *n* (*HRDP*) RADIO, TELEP, TRANS conduit fictif numérique de référence *m* (*CFNR*)
hysteresis *n* ELECT hystérésis *f*; **~ curve** ELECT courbe d'hystérésis *f*; **~ cycle** ELECT cycle d'hystérésis *m*; **~ loop** ELECT cycle d'hystérésis *m*; **~ loss** ELECT pertes par hystérésis *f pl*
Hz *abbr* (*hertz*) ELECT, TEST, TRANS Hz (*hertz*)

I

I *abbr* (*implemented*) COMP, ELECT I (*implémenté*)
IA *abbr* (*incoming access*) DATA accès entrant *m*
IAB *abbr* (*International Association of Broadcasting*) RADIO AIR (*Association internationale de retransmission*)
IA5 *abbr* (*international alphabet no. 5*) DATA, TELEP alphabet international n° 5 *m*
IAGC *abbr* (*instantaneous automatic gain control*) ELECT, TELEP, TRANS CAIG (*commande automatique instantanée de gain, contrôle automatique instantanée de gain*)
IALA *abbr* (*International Association of Lighthouse Authorities*) RADIO AISM (*Association internationale de signalisation maritime*)
IAT *abbr* (*international atomic time*) RADIO TAI (*temps atomique international*)
IB *abbr* (*intelligent building*) APPL édifice intelligent
IBM: ~ **cabling system**® *n* (*ICS*®) ELECT système de câblage IBM® *m*
IBO *abbr* (*input back-off*) RADIO IBO (*recul d'entrée*)
IC *abbr* DATA (*incoming call*) appel d'arrivée *m*, appel entrant *m*, ELECT, PARTS (*integrated circuit*) CI (*circuit intégré*), SIG (*intermediate cross-connect*) équipement de brassage intermédiaire *m*, RADIO (*incoming call*) appel d'arrivée *m*, appel entrant *m*, TELEP (*I-frame counter*) IC (*compteur des trames I*), (*interlock code*) code de verrouillage *m*, (*incoming call*) appel d'arrivée *m*, appel entrant *m*
I/C *abbr* (*incoming*) GEN entrant
ICC *abbr* (*incrementally coherent carrier*) TV PCI (*porteuse cohérente incrémentielle*)
ICCM *abbr* (*interworking by call control mapping*) DATA interfonctionnement par mise en correspondance de la commande des communications *m*
ICCS *n* (*integrated communication control system*) RADIO système intégré de contrôle des communications *m*
ICD *abbr* (*international code designator*) DATA désignateur de code international *m*
ICI *abbr* (*incoming-call identification*) DATA, RADIO, TELEP IAE (*identification des appels entrants*)
ICM *abbr* (*image compression management*) DATA, TV gestionnaire de compression d'image *m*
icon *n* COMP, DATA icône *f*; ~ **interface** COMP, DATA *Internet* interface iconique *f*
ICP *abbr* (*incoming-call packet*) DATA, RADIO, TELEP paquet d'appel entrant *m*
ICPA *n* (*integrated circuit panel assembly*) ELECT carte à circuits imprimés *f*
ICRD *abbr* (*internetwork call redirection and deflection*) DATA réacheminement et déviation des appels entre réseaux *m*
ICS *abbr* ELECT (*IBM cabling system*®) système de câblage IBM® *m*, GEN (*implementation conformance statement*) déclaration de conformité d'instance *f*
ID *abbr* GEN (*identification*) ID (*identification*), (*identifier*) Id (*identificateur*), SIG, TELEP (*identity element*) élément identité
IDA *abbr* (*integrated digital access*) DATA IDA (*accès numerique intégré*)
IDD *abbr* (*international direct dialing AmE, international direct dialling BrE*) TELEP automatique international *m*
IDDD *abbr* (*international direct distance dialing AmE, international direct distance dialing BrE*) TELEP automatique international *m*
ideal: ~ **instants** *n pl* TELEP instants idéals *m pl*; ~ **noise diode** *n* ELECT diode idéale de bruit *f*
identification *n* (*ID*) GEN identification *f* (*ID*); ~ **code** TELEP code d'identification *m*; ~-**code qualifier** TELEP qualificatif du code d'identification *m*; ~ **friend or foe** (*IFF*) DATA, RADIO, TELEP identification ami-ennemi *f*, radar IFF *m*; ~ **line** TELEP ligne d'identification *f*
identifier *n* (*ID*) GEN identificateur *m* (*Id*); ~ **circuit** ELECT circuit d'identification *m*, circuit de reconnaissance *m*, circuit identificateur *m*
identity: ~-**based security policy** *n* DATA politique de sécurité fondée sur l'identité *f*; ~ **element** *n* (*ID*) TELEP élément identité *m*; ~ **gate** *n* ELECT porte d'identité *f*, *logic elements* circuit d'identité *m*; ~ **operation** *n* COMP opération d'identité *f*
IDF *abbr* (*intermediate distribution frame*) SWG, TELEP RI (*répartiteur intermédiaire*)
IDI *abbr* (*initial domain identifier*) DATA, TELEP IDI (*identificateur de domaine initial*)
idle[1] *adj* ELECT, POWER, TELEP, TRANS libre, au repos
idle[2]: ~ **channel** *n* RADIO canal disponible *m*, TRANS voie au repos *f*; ~-**channel noise** *n* TRANS bruit d'une voie au repos *m*; ~ **circuit** *n* TRANS circuit libre *m*, circuit non utilisé *m*; ~ **condition** *n* ELECT repos *m*, POWER état de repos *m*, repos *m*, TELEP, TRANS repos *m*; ~ **line** *n* TELEP ligne libre *f*; ~ **period** *n* ELECT, TELEP période de repos *f*, temps de repos *m*; ~ **position** *n* ELECT position de repos *f*; ~ **state** *n* SWG, TELEP état de disponibilité *m*, *dependability* état inactif *m*, état libre *m*, TEST état vacant *m*; ~ **state test** *n* TEST état de test au repos *m*; ~ **testing state** *n* TEST état de test au repos *m*; ~ **time** *n* TEST temps mort *m*, temps libre *m*, période vacante *f*; ~ **voice channel** *n* RADIO canal vocal libre *m*; ~ **working channel** *n* RADIO *land mobile* voie de trafic au repos *f*
idleness *n* GEN inactivité *m*
idler: ~ **frequency** *n* RADIO fréquence idler *f*
IDN *abbr* TELEP, TRANS (*indirect directory number*) NAI (*numéro d'appel indirect*), (*integrated digital network*) RNI (*réseau numérique intégré*)
IDP *abbr* (*initial domain part*) DATA, TELEP partie de domaine initial *f*, sous-ensemble de domaine initial *m*
IDR *abbr* (*intermediate data rate*) DATA débit binaire intermédiaire *m*
IDSE *abbr* (*international data switching exchange*) DATA, SWG centre international de commutation de données *m*
IDT *abbr* (*interdigital transducer*) TRANS transducteur interdigité *m*
IE *abbr* (*information element*) TELEP élément d'information *m*
IEC *abbr* (*International Electrotechnical Commission*) GEN CEI (*Commission électrotechnique internationale*)
IEEE *abbr* (*Institute of Electrical and Electronics Engineers*) ELECT IEEE (*Institut des ingénieurs électriciens et électroniciens*)
IF *abbr* (*intermediate frequency*) RADIO, TRANS FI (*fréquence intermédiaire*); ~ **amplification** *n* RADIO, TRANS amplification FI *f*; ~ **amplifier** *n* RADIO, TRANS amplificateur FI *m*; ~ **filter** *n* RADIO, TRANS filtre FI *m*; ~ **rejection** *n* RADIO, TRANS atténuation à la fréquence intermédiaire *f*, rejet à la fréquence intermédiaire *m*; ~ **signal** *n* ELECT, RADIO signal FI *m*; ~ **stage** *n* ELECT, RADIO étage FI *m*
I/F *abbr* (*interface*) DATA, GEN interface *f*, jonction *f*
IF-AND-ONLY-IF: ~ **element** *n* COMP, ELECT *logic elements* circuit d'équivalence *m*; ~ **gate** *n* COMP, ELECT porte

d'équivalence *f*; ~ **operation** *n* COMP, ELECT équivalence logique *f*
IFF *abbr* (*identification friend or foe*) DATA, RADIO, TELEP identification ami-ennemi *f*, radar IFF *m*
i-field: ~ **picture** *n* COMP, TV image-i monotrame *f*
IFL *abbr* (*interfacility link*) TRANS IFL (*circuit entre installations de raccordement*)
IFLU *abbr* (*initial full line-up*) TEST premiers essais de mise en service *m pl*
i-frame: ~ **picture** *n* COMP, TV image-i bi-trame *f*
I-frame: ~ **counter** *n* (*IC*) TELEP compteur des trames I *m* (*IC*)
IF-THEN: ~ **element** *n* COMP, ELECT circuit d'inclusion *m*
IFU *abbr* (*interworking functional unit*) DATA unité fonctionnelle d'interconnexion *f*
IG *abbr* (*interpolation gain*) TRANS IG (*gain de concentration*)
IGES *abbr* (*initial graphics exchange specification*) COMP IGES
IGN *abbr* (*international gateway node*) SWG, TELEP, TRANS NTI (*nud de transit international*)
ignition *n* ELECT allumage *m*, amorçage *m*; ~ **time** ELECT instant d'allumage *m*, instant d'amorçage *m*; ~ **voltage** ELECT tension d'amorçage *f*, tension d'allumage *f*
ignitron *n* ELECT ignitron *m*
IH *abbr* (*intermediate high-usage trunk*) TELEP faisceau de circuits intermédiaire à débordement *m*, faisceau intermédiaire à débordement *m*
IIR *abbr* (*infinite impulse response*) ELECT, TRANS RII (*réponse impulsionnelle infinie*); ~ **filter** *n* ELECT, TRANS filtre RII *m*
ILD *abbr* (*injection laser diode*) TRANS diode laser à injection *f*
illuminance: ~ **meter** *n* GEN luxmètre *m*; ~ **range** *n* POWER plage d'éclairement *f*; ~ **uniformity ratio** *n* POWER facteur d'uniformité d'éclairement *m*
illuminate *vt* COMP, POWER éclairer
illuminated[1] *adj* COMP, POWER éclairé, RADIO lumineux
illuminated[2]: ~ **display** *n* RADIO affichage lumineux *m*; ~ **keyboard** *n* RADIO clavier lumineux *m*
illuminating: ~ **device** *n* RADIO source primaire *f*; ~ **radar** *n* RADIO radar d'illumination *m*
illumination: ~ **time** *n* RADIO temps d'illumination *m*
ILS *abbr* (*instrument landing system*) RADIO système d'atterrissage aux instruments *m*
IM *abbr* GEN (*interface module*) module de raccordement *m*, unité de raccordement *f*, RADIO (*inner marker beacon*) radioborne intérieure *f*, TELEP (*intermarker group trunk*) circuit entre groupes marqueurs *m*, (*intermarker trunk group*) faisceau de circuits entre groupes marqueurs *m*
IMA *abbr* (*input message acknowledgement*) DATA accusé de réception de message d'entrée *m*
image *n* GEN image *f*; ~ **analyser** *BrE* TV analyseur d'image *m*; ~ **analysis** TV analyse de l'image *f*; ~ **analyzer** *AmE see image analyser BrE*; ~ **attenuation coefficient** TRANS coefficient d'affaiblissement d'image *m*; ~ **attenuation constant** TRANS constante d'affaiblissement d'image *f*; ~ **bank** TV banque d'images *f*; ~ **camera tube** ELECT tube analyseur à transfert d'image *m*; ~ **coding** DATA codage d'image *m*; ~ **compression** DATA, TV compression d'image *f*; ~ **compression board** DATA, TV carte de compression d'images *f*; ~ **compression management** (*ICM*) DATA, TV gestion de compression d'image *f*; ~ **compression manager** DATA, TV gestionnaire de compression d'image *m*; ~ **conversion** DATA, TV conversion d'image *f*; ~ **converter tube** ELECT, PARTS, TV tube convertisseur d'image *m*; ~ **data** *n pl* DATA, TV données d'image *f pl*; ~ **enhancement** TV enrichissement d'image *m*; ~ **file** DATA fichier image *m*; ~ **focus voltage** ELECT tension de concentration d'image *f*; ~ **frequency** RADIO fréquence image *f*; ~ **frequency rejection** RADIO réjection de la fréquence image *f*; ~ **graphics** TV graphie en mode image *f*; ~ **iconoscope** ELECT iconoscope image *m*, super-iconoscope *m*; ~ **impedance** ELECT impédance d'image *f*; ~ **intensifier tube** ELECT tube intensificateur d'image *m*, tube amplificateur d'image *m*; ~ **orthicon** ELECT orthicon à transfert d'image *m*, image orthicon *f*; ~ **parameter filter** ELECT filtre de bandes *m*; ~ **phase-change coefficient** TRANS déphasage sur images *m*; ~ **processing** DATA, TV traitement de l'image *m*; ~**-processing equipment** DATA, TV equipement de traitement de l'image *m*; ~ **refreshing** TV entretien de l'image *m*, rafraîchissement d'image *m*; ~ **rejection ratio** RADIO affaiblissement sur la fréquence conjuguée *m*; ~ **retention fault** TV défaut de rémanence *m*; ~ **scanner** PER APP, TV scanner d'images *m*; ~ **scanning time** PER APP, TV temps scan image *m*; ~ **sensor** PER APP, TV détecteur d'image *m*, capteur d'images *m*; ~ **sequence** TV séquence d'images *f*; ~ **and sound synchronization** TV synchronisation image-son *f*; ~ **space** COMP espace image *m*; ~ **storage and retrieval** COMP stockage et repérage d'images *m*; ~ **storage space** COMP zone image *f*; ~ **transfer** DATA, TRANS transfert d'image *m*; ~ **transfer coefficient** DATA, TRANS exposant de transfert sur images *m*; ~ **transfer constant** DATA, TRANS exposant de transfert sur images *m*; ~ **transmission** DATA, TRANS transmission d'image *f*
IMC *abbr* (*international Morse code*) TELEP code Morse international *m*
imitative: ~ **deception** *n* ELECT imitation *f*
immediate: ~ **addressing** *n* COMP adressage immédiat *m*; ~ **dialing** *AmE*, ~ **dialling** *BrE* *n* TELEP numérotation immédiate *f*
immersed: ~ **gun** *n* ELECT canon immergé *m*
immunity: ~ **to noise** *n* TRANS immunité au bruit *f*
IMP *abbr* (*interface message processor*) COMP IMP (*serveur de messages*)
impact: ~ **printer** *n* COMP, DATA, PER APP imprimante à impact *f*; ~ **switch** *n* POWER commutateur à inertie *m*; ~ **test** *n* TEST essai de résistance *m*
impairment *n* RADIO, SIG, TRANS dégradation *f*
impedance *n* GEN impédance *f*; ~ **balance** ELECT, TELEP équilibre d'impédance *m*; ~ **balance ratio** ELECT, TELEP rapport d'equilibre d'impédance *m*; ~ **bond** POWER connexion inductive *f*, couplage inductif *m*; ~ **matching** RADIO adaptation d'impédance *f*; ~ **unbalance** ELECT, TELEP déséquilibre d'impédance *m*
imperfection *n* GEN défaut *m*
implantation *n* ELECT implantation *f*
implanted *adj* ELECT implanté
implanter *n* ELECT implanteur *m*
implement *vt* COMP, ELECT implémenter, GEN réaliser, TELEP implémenter, mettre en application, mettre en œuvre, mettre en place
implementation *n* COMP, ELECT implémentation *f*, réalisation *f*, GEN, TELEP implémentation *f*, mise en œuvre *f*, mise en application *f*, mise en place *f*; ~ **conformance statement** (*ICS*) GEN déclaration de conformité d'instance *f*; ~ **under test** (*IUT*) DATA instance sous test *m* (*IUT*)
implemented *adj* (*I*) COMP, ELECT implémenté (*I*), réalisé
implementor *n* GEN réalisateur *m*
implicit: ~ **send event** *n* DATA événement d'envoi implicite *m*
implied: ~ **addressing** *n* COMP adressage implicite *m*
import[1] *n* DATA import *m*
import[2] *vt* DATA importer
imported *adj* DATA importé
impregnate *vt* GEN *materials* imprégner
impregnated: ~ **paper insulation** *n* TRANS *cables* isolation au papier préimprégné *f*
impregnating: ~ **agent** *n* GEN *chemistry* agent d'imprégnation *m*
impulse *n* APPL, COMP, TELEP, TEST impulsion *f*; ~ **discharge current** ELECT *of protector* courant de choc de décharge *m*; ~ **excitation** ELECT excitation par choc *f*; ~ **noise** TELEP bruit impulsif *m*; ~ **noise count** TELEP nombre de points de bruit impulsif *m*; ~ **rate** APPL taux d'impulsions *m*, cadence

d'impulsions *f*; ~ **response** TELEP réponse impulsionnelle *f*, réponse percussionnelle *f*; ~ **sparkover voltage** ELECT *of protector* tension d'amorçage au choc *f*, TEST tension disruptive de choc *f*; ~ **spark-over voltage-time curve** TELEP courbe de tension d'amorçage au choc en fonction du temps *f*
impulsing *n* TELEP émission d'impulsions *f*
impulsive: ~ **noise tolerance** *n* RADIO tolérance de bruit impulsif *f*
impulsiveness: ~ **ratio** *n* RADIO rapport impulsionnel *m*
impurity *n* ELECT impureté *f*; ~ **center** *AmE*, ~ **centre** *BrE* ELECT centre d'impureté *m*
IN *abbr* (*intelligent network*) APPL, COMP, TRANS RI (*réseau intelligent*)
inaccuracy *n* TEST imprécision *f*
inaccurate *adj* GEN imprécis
inactive[1] *adj* RADIO, SIG, TELEP, TRANS inactif
inactive[2]: ~ **character** *n* TELEP caractère inactif *m*; ~ **node** *n* TRANS nœud inactif *m*; ~ **signaling link** *AmE*, ~ **signalling link** *BrE n* SIG canal sémaphore inactif *m*
inaugurate *vt* RADIO inaugurer
inauguration *n* RADIO inauguration *f*
in-band[1] *adj* SIG, TRANS dans la bande, intrabande
in-band[2]: ~ **carrier** *n* TRANS porteuse entrante *f*; ~ **data** *n pl* DATA données dans la bande *f pl*, données intrabandes *f pl*; ~ **signaling** *AmE*, ~ **signalling** *BrE n* SIG signalisation dans la bande *f*, signalisation intrabande *f*
in-between *n* TV raccord *m*
inbound: ~ **communication** *n* RADIO, TELEP, TRANS télécommunication d'arrivée *f*
in-building: ~ **coverage** *n* RADIO couverture à l'intérieur des bâtiments *f*; ~ **reception** *n* RADIO réception à l'intérieur des bâtiments *f*
in-call *n* TELEP communication en cours *f*; ~ **modification** TELEP modification en cours d'appel *f*; ~ **rearrangement** TELEP remaniement des liaisons pendant la communication *m*
incidence *n* ELECT indice *m*
incision *n* GEN saignée *f*
inclination *n* RADIO inclinaison *f*
inclined: ~ **orbit** *n* RADIO orbite inclinée *f*
inclusive: ~ **price** *n* GEN prix forfaitaire *m*
INCLUSIVE-OR *n* COMP, ELECT INCLUSIF OU *f*; ~ **circuit** COMP, ELECT circuit INCLUSIF OU *m*; ~ **element** COMP, ELECT circuit INCLUSIF OU *m*; ~ **gate** COMP, ELECT porte INCLUSIF OU *f*
incoherence *n* TRANS incohérence *f*
incoherent: ~ **light** *n* TRANS lumière incohérente *f*; ~ **radiation** *n* TRANS rayonnement incohérent *m*
incoming[1] *adj* (*I/C*) GEN entrant
incoming[2]: ~ **access** *n* (*IA*) DATA accès entrant *m*; ~ **call** *n* (*IC*) DATA, TELEP appel d'arrivée *m*, appel entrant *m*; ~-**call barring** *n* TELEP interdiction d'accès des appels entrants *f* (*ICB*), interdiction des appels d'arrivée *f*, service de limitation des appels d'arrivée *m*; ~-**call identification** *n* (*ICI*) DATA, RADIO, TELEP identification des appels entrants *f* (*IAA*); ~-**call log** *n* TELEP journal d'appel *m*; ~-**call packet** *n* (*ICP*) DATA, RADIO, TELEP paquet d'appel entrant *m*; ~ **calls** *n pl* TELEP trafic d'arrivée *m*, trafic en réception *m*; ~-**calls-barred line** *n* TELEP, TRANS ligne spécialisée départ *f*; ~ **circuit** *n* ELECT, SWG, TELEP circuit entrant *m*; ~ **group** *n* TRANS faisceau entrant *m*; ~ **inspection** *n* TEST contrôle d'entrée *m*; ~ **international R2 register** *n* TELEP enregistreur international R2 d'arrivée *m*; ~ **line** *n* TRANS ligne entrante *f*; ~ **matrix link** *n* TELEP liaison réseau entrante *f*; ~-**only line** *n* TELEP, TRANS ligne spécialisée arrivée *f*; ~ **procedure** *n* RADIO procédure d'arrivée *f*; ~ **R2 register** *n* TELEP enregistreur R2 d'arrivée *m*; ~ **register** *n* TELEP enregistreur d'arrivée *m*; ~ **response delay** *n* TELEP durée de présélection *f*, temps de réponse à la prise d'un circuit d'arrivée *m*; ~ **signal** *n* SIG, TRANS signal entrant *m*; ~ **surge** *n* ELECT impulsion d'entrée *f*; ~ **traffic** *n* TELEP trafic entrant *m*; ~ **trunk** *n* ELECT, SWG, TELEP circuit entrant *m*; ~ **trunk circuit** *n* (*ITC*) TELEP, TRANS circuit d'arrivée *m*, circuit entrant *m*, joncteur d'arrivée *m*, joncteur entrant *m*
incompatible[1] *adj* DATA *remote data processing* hétérogène
incompatible[2]: ~ **service-user class signal** *n* SIG signal d'incompatibilité de la catégorie d'usagers du service *m*
inconclusive *n* DATA non concluant; ~ **verdict** DATA verdict non concluant *m*
in-connector *n* RADIO connecteur d'entrée *m*
inconsistency *n* COMP incohérence *f*
incorporated: ~ **telephone** *n* PER APP *answer machine* téléphone incorporé *m*
incorrect: ~ **operation** *n* COMP opération erronée *f*
incremental: ~ **charge advice** *n* TELEP indication du coût *f*; ~ **power gain** *n* ELECT gain différentiel en puissance *m*; ~ **recording** *n* TV enregistrement incrémentiel *m*
incrementally: ~ **coherent carrier** *n* (*ICC*) TV porteuse cohérente incrémentielle *f* (*PCI*), système à porteuses cohérentes incrémentielles *m*; ~ **related carrier** *n* (*IRC*) TV système de porteuses en relation incrémentielle *m*
ind *abbr* (*indication*) GEN ind (*indication*)
indent[1] *n* COMP retrait *m*
indent[2] *vt* COMP mettre en retrait, DATA retraire
indentation *n* ELECT *printed circuits* enfoncement *m*
independent: ~ **alarm** *n* PER APP alarme indépendante *f*; ~ **earth electrode** *n BrE* (*cf independent ground electrode AmE*) POWER prise de terre indépendante *f*; ~ **ground electrode** *n AmE* (*cf independent earth electrode BrE*) POWER prise de terre indépendante *f*; ~ **network** *n* TRANS réseau indépendant *m*; ~ **program** *AmE*, ~ **programme** *BrE n* TV programme indépendant *m*; ~ **sideband** *n* (*ISB*) RADIO bande latérale indépendante *f* (*BLI*); ~ **sideband transmission** *n* RADIO, TRANS transmission à bande latérale indépendante *f*
index *n* GEN indice *m*; ~ **of cooperation** RADIO, TELEP module de coopération *m*; ~ **dip** TRANS creux central d'indice *m*, creux d'indice *m*, trou d'indice *m*; ~-**matching material** TRANS substance d'adaptation d'indice *f*; ~ **profile** ELECT, TRANS *refractive* profil d'indice *m*; ~ **of refraction** TRANS indice de réfraction *m*; ~ **register** COMP registre d'index *m*; ~ **word** COMP, DATA mot d'index *m*
indexed: ~ **address** *n* COMP adresse indexée *f*
indexing *n* COMP *software*, GEN indexation *f*
indicating: ~ **element** *n* TELEP élément d'indication *m*; ~ **lamp** *n* TELEP lampe témoin *f*, témoin lumineux *m*, voyant lumineux *m*; ~-**measuring instrument** *n* TEST appareil de mesure indicateur *m*, instrument de mesure indicateur *m*, appareil indicateur *m*; ~ **unit** *n* TELEP unité d'indication *f*
indication *n* (*ind*) GEN indication *f* (*ind*); ~ **of fracture** TEST amorce de rupture *f*; ~ **primitive** DATA primitive d'indication *f*
indicator *n* GEN indicateur *m*, voyant *m*; ~ **lamp** GEN voyant lumineux *m*
indifference: ~ **point** *n* TEST point d'indifférence *m*
indirect: ~ **address** *n* COMP adresse indirecte *f*; ~ **addressing** *n* COMP adressage indirect *m*; ~ **backscatter** *n* POWER, RADIO rétrodiffusion indirecte *f*; ~ **control system** *n* CONTR système à commande indirecte *m*; ~ **conversation** *n* DATA conversation en différé *f*; ~ **directory number** *n* (*IDN*) TELEP, TRANS numéro d'appel indirect *m* (*NAI*); ~ **distribution** *n* RADIO distribution indirecte *f*; ~ **memory** *n* PER APP mémoire indirecte *f*; ~ **method of measurement** *n* TEST méthode de mesure indirecte *f*; ~ **select** *n* COMP sélection indirecte *f*; ~ **stroke** *n* POWER coup de foudre indirect *m*; ~ **submission port** *n* DATA accès de dépôt indirect *m*
indirectly: ~-**heated cathode** *n* ELECT, PARTS cathode à chauffage indirect *f*

individual: ~ **aerial** *n* TV antenne individuelle *f*; ~ **antenna** *n* TV antenne individuelle *f*; ~ **charging** *n* RADIO taxation individuelle *f*; ~ **current feed** *n* POWER, TELEP alimentation en courant individuel *f*; ~ **distortion** *n* TELEP distorsion individuelle *f*; ~ **fault** *n* TEST dérangement simple *m*, panne simple *f*; ~ **reception** *n* RADIO réception individuelle *f*; ~ **store** *n* DATA mémoire individuelle *f*

indoor: ~ **radio** *n* RADIO radiocommunication intra-immeuble *f*

induce *vt* GEN induire

induced: ~ **control-voltage** *n* ELECT *semiconductors* tension de commande induite *f*; ~ **current** *n* POWER courant induit *m*; ~ **voltage** *n* POWER tension induite *f*

inductance *n* ELECT, PARTS, POWER inductance *f*; ~ **capacitance** ELECT inductance-capacité *f*; ~ **capacity** PARTS inductance-capacité *f*; ~ **coil** POWER bobine d'inductance *f*; ~ **standard** ELECT *electromagnetism* inductance étalon *f*

induction *n* ELECT, POWER, RADIO induction *f*; ~ **coil** ELECT bobine d'induction *f*; ~ **field** RADIO champ d'induction *m*; ~ **meter** POWER compteur à induction *m*; ~ **relay** POWER relais à induction *m*

inductive[1] *adj* ELECT inductif

inductive[2]: ~ **circuit** *n* ELECT circuit inductif *m*; ~ **load** *n* ELECT charge inductive *f*; ~ **parallel tuning** *n* ELECT réglage d'accord par inducteurs en parallèle *m*; ~ **reasoning** *n* DATA raisonnement inductif *m*; ~ **residual voltage** *n* ELECT tension induite résiduelle *f*

inductor *n* ELECT, PARTS inductance *f*, POWER bobine d'inductance *f*

industrial: ~ **safety** *n* CONTR sécurité du travail *f*; ~ **television** *n* (*ITV*) TV télévision industrielle *f* (*TVI*)

industry: ~ **standard architecture** *n* (*ISA*) COMP, DATA *remote data processing* ISA

ineffective: ~ **airtime** *n* RADIO durée d'émissions sans effet *f*; ~ **call** *n* TELEP appel inefficace *m*

inert: ~ **gas** *n* CLIM gaz inerte *m*

inertia: ~ **damper** *n* POWER frein d'inertie *m*; ~ **error** *n* TEST erreur due à l'inertie *f*

INF *abbr* (*information message*) GEN INF (*message d'information*)

inference: ~ **engine** *n* APPL *artificial intelligence* moteur d'inférence *m*

infinite: ~ **impulse response** *n* (*IIR*) ELECT, TRANS réponse impulsionnelle infinie *f* (*RII*); ~ **impulse response filter** *n* ELECT, TRANS filtre à réponse impulsionnelle infinie *m*

infinitely: ~ **variable** *n* CONTR réglage en continu

in-flight[1] *adj* RADIO en vol

in-flight[2]: ~ **telephone service** *n* RADIO service du téléphone en vol *m*, service à bord de l'avion *m*

influence: ~ **quantity** *n* TEST grandeur d'influence *f*

information[1] *n* GEN indication *f*, information *f*, renseignement *m*; ~ **base** DATA base d'information *f*; ~ **base type** DATA type de base d'information *m*; ~ **bearer channel** TELEP voie porteuse d'information *f*; ~ **bit** DATA bit d'information *m*, bit utile *m*, TELEP bit d'information *m*; ~ **channel** TELEP voie de transfert d'information *f*, TV chaîne d'information *f*; ~ **content** DATA quantité d'information *f*; ~ **element** (*IE*) TELEP élément d'information *m*; ~ **exchange** COMP, DATA échange d'information *m*; ~ **exchange protocol** (*IXP*) COMP protocole d'échange d'informations *m* (*IXP*); ~ **facility** TELEP service complémentaire des renseignements *m*; ~ **feedback system** DATA, TELEP système correcteur d'erreurs par retour d'information *m*; ~ **frame** TELEP trame d'information *f*; ~ **highway** TRANS autoroute de l'information *f*; ~ **inquiry facility** TELEP service complémentaire des enquêtes *m*; ~ **interchange** TELEP transfert d'informations *m*; ~ **measure** GEN mesure de l'information *f*; ~ **message** (*INF*) DATA, GEN message d'information *m* (*INF*); ~ **network** DATA réseau d'information *m*, *Internet* réseau informatique *m*; ~ **processing** TELEP traitement de l'information *m*; ~ **provider** DATA fournisseur d'information *m*, producteur de bases de données *m*; ~ **pulse** RADIO impulsion d'information *f*; ~ **receiver station** (*IRS*) DATA, TELEP station de réception d'informations *f* (*SRI*); ~ **request message** (*INR*) DATA, TELEP message de demande d'information *m* (*IND*); ~ **retrieval** COMP, DATA recherche de l'information *f*, recherche documentaire *f*, repérage de l'information *m*; ~ **retrieval center** *AmE*, ~ **retrieval centre** *BrE* (*IRC*) DATA centre de recherche documentaire *m* (*IRC*); ~ **retrieval system** DATA système de repérage de l'information *m*, système de recherche documentaire *m*; ~ **security** TELEP sécurité d'information *f*; ~-**sending station** DATA station émettrice d'information *f*; ~ **separator** (*IS*) DATA, ELECT séparateur d'information *m*; ~-**separator character** DATA caractère de séparation d'informations *m*; ~ **server** DATA serveur d'information *m*; ~-**storage node** (*ISN*) TELEP nœud de mise en mémoire des informations *m*; ~ **superhighway** TRANS inforoute *f*, super-autoroute de l'information *f*; ~ **system** DATA système d'information *m*; ~ **technology** (*IT*) COMP, DATA informatique *f*; ~ **theory** GEN théorie de l'information *f*; ~ **transfer** TELEP transfert d'information *m*; ~ **transfer rate** COMP, DATA débit de transfert d'information *m*; ~ **type** (*IT*) DATA type d'information *m*; ~ **unit** TELEP unité d'information *f*

information[2]: ~ **rate per character** *phr* DATA entropie moyenne par caractère; **for your** ~ *phr* (*FYI*) GEN à titre indicatif, pour information

informetrics *n pl* DATA informétrique *f*

in-frame: ~ **coding** *n* DATA codage intratrame *m*

infrared[1] *adj* (*IR*) COMP, ELECT, RADIO, TRANS infrarouge (*IR*)

infrared[2] *n* COMP, ELECT, RADIO, TRANS infrarouge *m*; ~ **detector** TRANS capteur infrarouge *m*; ~ **light link** TRANS liaison infrarouge *f*; ~ **radiation** ELECT radiation infrarouge *f*, rayonnement infrarouge *m*

infrasound *n* COMP infrason *m*

inherent: ~ **loss** *n* TRANS affaiblissement propre *m*

inhibit *vt* ELECT bloquer, inhiber, interdire, invalider

inhibited *adj* ELECT bloqué, interdit, invalidé

inhibiting: ~ **pulse** *n* ELECT impulsion d'inhibition *f*, impulsion d'invalidation *f*; ~ **signal** *n* ELECT signal d'invalidation *m*

inhibition *n* ELECT, TEST inhibition *f*

initial: ~ **acceptance** *n* TEST première phase de recette *f*; ~ **address message** *n* (*IAM*) SIG, SWG, TELEP message d'adresse initial *m*; ~ **alignment** *n* TELEP alignement initial *m*; ~ **attack slope** *n* COMP courbe d'attaque initiale *f*; ~ **charge** *n* TELEP taxe de mise en présence *f*; ~ **condition** *n* ELECT condition initiale *f*; ~ **condition mode** *n* COMP mode conditions initiales *m*; ~ **connection charge** *n* APPL tarif de raccordement *m*; ~ **domain identifier** *n* (*IDI*) DATA, TELEP identificateur de domaine initial *m* (*IDI*); ~ **domain part** *n* (*IDP*) DATA, TELEP partie de domaine initial *f*, sous-ensemble de domaine initial *m*; ~ **full line-up** *n* (*IFLU*) TEST premiers essais de mise en service *m pl*; ~ **graphics exchange specification** *n* (*IGES*) COMP IGES; ~ **inverse voltage** *n* POWER tension inverse initiale *f*; ~ **ionizing event** *n* RADIO événement ionisant initial *m*; ~ **metering** *n* APPL démarrage de la taxation *m*; ~ **permeability** *n* ELECT *electromagnetism* perméabilité initiale *f*; ~ **point** *n* COMP point d'origine *m*, point de départ *m*; ~ **position** *n* COMP position initiale *f*; ~ **positioning** *n* COMP positionnement de départ *m*; ~ **signal unit** *n* (*ISU*) SIG unité de signalisation initiale *f*; ~ **signal unit alignment** *n* SIG alignement initial des trames sémaphores *m*; ~ **signing unit alignment** *n* SIG alignement initial des trames sémaphores *m*; ~ **state** *n* DATA, ELECT état initial *m*; ~ **TC** *n* TELEP connexion TC initiale *f*; ~ **testing state** *n* DATA état de test initial *m*

initialization *n* COMP *of memory* initialisation *f*, GEN activation *f*

initialize *vt* COMP initialiser, GEN activer

initializing: ~ **routine** *n* COMP programme d'initialisation *m*

initiate *vt* RADIO, TELEP *call* émettre, appeler, effectuer, lancer

initiation *n* RADIO, TELEP initiation *f*

inject: ~ **into orbit** *vt* RADIO satelliser

injected[1] *adj* TV injecté

injected[2]: ~**-beam magnetron** *n* ELECT, RADIO magnétron accordé en tension *m*

injection: ~ **into orbit** *n* RADIO injection en orbite *f*, satellisation *f*; ~ **laser diode** *n* (*ILD*) TRANS diode laser à injection *f*; ~**-locked laser** *n* ELECT laser synchronisé par injection *m*, laser verrouillé par injection *m*; ~ **ratio** *n* ELECT rendement d'injection *m*, rendement de l'émetteur *m*

inkjet: ~ **printer** *n* COMP, DATA, PER APP imprimante à jet d'encre *f*

inland: ~ **call** *n* TELEP appel intérieur *m*

inlet: ~ **access** *n* DATA accès d'arrivée *m*, entrée *f*, point d'entrée *m*

in-line: ~ **instruction** *n* COMP instruction en direct *f*; ~ **recovery** *n* DATA correction immédiate *f*

inner: ~ **conductor** *n* TELEP conducteur central *m*, TRANS conducteur central *m*, *cables* conducteur intérieure *m*; ~ **duct** *n* ELECT conduit interne *m*; ~ **marker beacon** *n* (*IM*) RADIO radioborne intérieure *f*; ~ **shell electron** *n* ELECT électron interne *m*

inoperability *n* ELECT, GEN inopérabilité *f*

inoperative[1] *adj* ELECT, GEN inopérant

inoperative[2]: ~ **reference station code** *n* RADIO code de station de référence hors fonctionnement *m*

inopportune: ~ **test event** *n* DATA, TEST événement de test inopportun *m*

in-process: ~ **inspection** *n* TEST contrôle en cours de fabrication *m*; ~ **testing** *n* TEST essai en cours de fabrication *m*

input[1] *n* DATA, ELECT entrée *f*; ~ **admittance** POWER admittance d'entrée *f*; ~ **amplifier** ELECT amplificateur d'entrée *m*; ~ **back-off** (*IBO*) RADIO recul d'entrée *m* (*IBO*); ~ **capacitance** ELECT capacité d'entrée *f*; ~ **channel** COMP, ELECT canal d'entrée *m*; ~ **circuit** ELECT circuit d'entrée *m*; ~ **current** ELECT courant d'entrée *m*; ~ **data** *n pl* DATA données d'entrée *f pl*; ~ **device** COMP périphérique d'entrée *m*, DATA appareil d'entrée *m*; ~ **error** COMP, DATA erreur d'entrée *f*; ~ **filter** DATA filtre d'entrée *m*; ~ **gate** ELECT *logic elements* porte d'entrée *f*; ~**-gate signal** ELECT, SIG signal d'activation de la porte d'entrée *m*; ~ **impedance** COMP, ELECT impédance d'entrée *f*; ~ **indicator** COMP indicateur d'entrée *m*; ~ **jitter** TELEP *pulses* gigue à l'entrée *f*; ~ **level** ELECT niveau d'entrée *m*; ~ **message acknowledgement** (*IMA*) DATA accusé de réception de message d'entrée *m*; ~ **port** COMP port d'entrée *m*; ~ **power** ELECT puissance d'entrée *f*, puissance absorbée *f*, puissance fournie *f*; ~ **process** COMP introduction *f*; ~ **pulse** COMP impulsion d'entrée *f*; ~ **resistance** ELECT *semiconductors* résistance d'entrée *f*; ~ **sensibility** TRANS sensibilité d'entrée *f*, sensibilité limite *f*; ~ **signal** ELECT signal d'entrée *m*; ~**-signal level range** TRANS plage de niveau du signal d'entrée *f*; ~ **socket** COMP prise d'entrée *f*; ~ **stage** ELECT, RADIO étage d'entrée *m*; ~ **symbol** TELEP symbole d'entrée *m*; ~ **transaction accepted for delivery** (*ITD*) DATA acceptation de transaction pour remise *f*; ~ **transaction rejected** (*ITR*) DATA rejet de transaction d'entrée *m*; ~ **transformer** POWER transformateur d'entrée *m*; ~ **unit** COMP unité d'entrée *f*, organe d'entrée *m*, dispositif d'entrée *m*; ~ **validation** DATA validation en entrée *f*, validation des données d'entrée *f*; ~ **variable** CONTR variable d'entrée *f*; ~ **voltage** ELECT *electricity* tension d'entrée *f*

input[2] *vt* COMP, DATA introduire, entrer

input/output[1] *adj* (*I/O*) COMP, ELECT, TELEP entrée/sortie (*E/S*)

input/output[2] *n* (*I/O*) COMP, ELECT, TELEP entrée/sortie *f*; ~ **bus** SWG bus d'entrée/sortie *m*; ~ **channel** COMP canal d'entrée/sortie *m*; ~ **controller** COMP contrôleur d'entrée/sortie *m*; ~ **device** COMP dispositif d'entrée/sortie *m*; ~ **group** SWG groupe d'entrées/sorties *m*; ~ **process** COMP opération d'entrée/sortie *f*; ~ **program** DATA programme d'entrée/sortie *m*; ~ **subsystem** SWG sous-système d'entrée/sortie *m*; ~ **unit** COMP organe d'entrée/sortie *m*

inquiry: ~ **reciprocation** *n* TELEP va-et-vient *m*; ~ **station** *n* DATA poste d'interrogation *m*

INR *abbr* (*information request message*) DATA, TELEP IND (*message de demande d'information*)

in-register *adj* TV calé

inrush: ~ **current** *n* ELECT courant de surcharge *m*

ins: ~ **and outs** *n pl* DATA *Internet* méandres *m pl*

insert[1] *n* COMP insertion *f*, PARTS pièce rapportée *f*, insertion *f*, TRANS insertion *f*

insert[2] *vt* COMP, GEN insérer, TELEP *card, coin* introduire

insertion *n* COMP, GEN, SIG insertion, TELEP introduction *f*, mise en place *f*, TRANS insertion *f*; ~ **gain** TRANS gain d'insertion *m*; ~ **loss** TRANS perte d'insertion *f*; ~ **loss pad** TRANS affaiblisseur de perte d'insertion *m*, atténuateur de perte d'insertion *m*; ~ **point** COMP point d'insertion *m*; ~ **reference signal** (*IRS*) SIG signal d'insertion de référence *m*; ~ **test signal** (*ITS*) SIG signal d'essai d'insertion *m*; ~ **tool** GEN outil de mise en place *m*

inset *n* COMP imagette *f*, vignette *f*, PARTS insertion *f*, pièce rapportée *f*; ~ **picture** COMP image dans l'image *f*, image incrustée *f*

inside: ~ **call** *n* TELEP communication interne *f*; ~ **vapor phase oxidation** *AmE see inside vapour phase oxidation BrE*; ~ **vapour phase oxidation** *BrE n* (*IVPO*) ELECT dépôt en phase vapeur interne *m*

in-slot: ~ **signaling** *AmE*, ~ **signalling** *BrE n* SIG signalisation dans l'intervalle de temps *f*

inspect *vt* TEST contrôler, inspecter

inspection *n* TEST contrôle *m*; ~ **by attributes** TEST contrôle par attributs *m*; ~ **by variables** TEST contrôle par mesures *m*; ~ **diagram** TEST graphique de contrôle *m*; ~ **lamp** GEN baladeuse *f*, *electricity* lampe portative *f*; ~ **level** TEST niveau de contrôle *m*; ~ **specification** TEST spécification d'instruction *f*

instability *n* ELECT, RADIO instabilité *f*

install *vt* GEN installer, monter

installation *n* GEN installation *f*; ~ **address** TELEP adresse d'installation *f*; ~ **call forwarding** TELEP *ISDN* renvoi temporaire de l'installation *m*, renvoi temporaire de l'ITA *m*; ~ **data** *n pl* GEN données d'installation *f pl*; ~ **date** GEN date d'installation *f*; ~ **diagram** GEN schéma d'installation *m*; ~ **documentation** GEN documentation d'installation *f*; ~ **drawing** GEN schéma d'installation *m*; ~ **instructions** *n pl* GEN instructions d'installation *f pl*; ~ **kit** GEN nécessaire d'installation *m*; ~ **library** GEN bibliothèque d'installation *f*, documentation d'installation *f*; ~ **manual** GEN manuel d'installation *m*; ~ **program** DATA *Internet* programme à installer *m*; ~ **site** GEN chantier d'installation *m*, site d'installation *m*; ~ **site manager** GEN chef de chantier d'installation *m*; ~ **time** GEN période d'installation *f*

instantaneous: ~ **access** *n* COMP, TRANS accès instantané *m*; ~ **automatic gain control** *n* (*IAGC*) ELECT, TELEP, TRANS commande automatique instantanée de gain *f* (*CAIG*); ~ **availability** *n* TELEP disponibilité instantanée *f*; ~ **failure intensity** *n* TEST intensité instantanée de défaillance *f*; ~ **failure rate** *n* TEST taux instantané de défaillance *m*; ~ **frequency** *n* TRANS fréquence instantanée *f*; ~ **power** *n* POWER puissance instantanée *f*; ~ **repair rate** *n* TELEP taux instantané de réparation *m*; ~ **value** *n* GEN valeur instantanée *f*

Institute: ~ **of Electrical and Electronics Engineers** *n*

(*IEEE*) ELECT Institut des ingénieurs électriciens et électroniciens *m* (*IEEE*)

instruction *n* COMP, POWER, TEST, TRANS instruction *f*; ~ **address** COMP adresse d'instructions *f*; ~ **address register** COMP registre d'adresse d'instructions *m*; ~ **modifier** COMP modificateur d'instruction *m*; ~ **register** COMP registre d'instruction *m*; ~ **repertoire** COMP répertoire de code d'instruction *m*; ~ **word** COMP, DATA mot-instruction *m*

instructional: ~ **television** *n* (*ITV*) TV téléenseignement *m*

instrument: ~ **constant** *n* TEST constante d'un instrument de mesure *f*; ~ **cord** *n* PER APP, TÉL cordon flexible *m*; ~ **landing system** *n* (*ILS*) RADIO système d'atterrissage aux instruments *m*; ~ **panel** *n* TEST *indicating devices* tableau de commande *m*

instrumental: ~ **error** *n* TEST erreur instrumentale *f*

insulant *n* ELECT corps isolant *m*, isolant *m*

insulate *vt* ELECT isoler

insulated: ~ **cable** *n* ELECT, POWER câble isolé *m*; ~ **conductor** *n* ELECT fil isolé *m*, PARTS conducteur isolé *m*

insulating[1] *adj* ELECT, POWER non conducteur, isolant

insulating[2]: ~ **bead** *n* ELECT *printed circuits* perle isolante *f*; ~ **foil** *n* POWER film isolant *m*; ~ **lacquer** *n* ELECT vernis isolant *m*; ~ **layer** *n* POWER couche isolante *f*; ~ **material** *n* ELECT, PARTS, POWER matériau isolant *m*, matière isolante *f*; ~ **plate** *n* POWER plaque de séparation *f*, plaque isolante *f*, plaque isolatrice *f*; ~ **property** *n* GEN pouvoir isolant *m*; ~ **sheet** *n* POWER feuille isolante *f*; ~ **spacer** *n* POWER plaquette isolante *f*; ~ **strip** *n* POWER réglette isolante *f*; ~ **tape** *n* POWER ruban adhésif isolant *m*; ~ **tube** *n* POWER tube isolant *m*; ~ **varnish** *n* ELECT *printed circuits* vernis isolant *m*

insulation *n* ELECT, GEN, PARTS, TRANS isolant *m*, isolation *f*, isolement *m*; ~ **displacement termination** ELECT raccordement à contacts autodénudants *m*; ~ **fault** POWER défaut d'isolement *m*; ~ **gap** POWER espacement d'air *m*, isolement d'air *m*; ~ **loss** TRANS perte d'isolement *f*; ~ **measurement** POWER mesure d'isolement *f*; ~ **resistance** POWER résistance d'isolement *f*; ~ **screen** POWER écran sur enveloppe isolante *m*; ~ **stripper** GEN machine à dénuder *f*; ~ **test** POWER essai d'étanchéité *m*, essai d'isolement *m*

insulator *n* ELECT, POWER isolateur *m*

integer: ~ **programming** *n* COMP programmation en nombres entiers *f*

integral: ~ **antenna** *n* RADIO antenne incorporée *f*; ~ **sampling** *n* TRANS échantillonnage intégral *m*

integrated: ~ **access** *n* TELEP accès intégré *m*; ~ **adaptor** *n* COMP adaptateur intégré *m*; ~ **circuit** *n* (*IC*) ELECT, PARTS circuit intégré *m* (*CI*); ~ **circuit panel assembly** *n* (*ICPA*) ELECT carte à circuits imprimés *f*; ~ **communication control system** *n* (*ICCS*) RADIO système intégré de contrôle des communications *m*; ~ **data processing** *n* DATA traitement intégré des données *m*; ~ **digital access** *n* (*IDA*) DATA accès numérique intégré *m* (*IDA*); ~ **digital exchange** *n* SWG commutateur numérique intégré *m*, système de commutation intégrée *m*; ~ **digital network** *n* (*IDN*) TELEP, TRANS réseau numérique intégré *m* (*RNI*); ~ **digital services exchange** *n* SWG autocommutateur numérique à intégration de services *m*; ~ **electronics** *n pl* ELECT électronique intégrée *f*; ~ **fire-control system** *n* RADIO système intégré de conduite du tir *m*; ~ **microcircuit** *n* ELECT microcircuit intégré *m*; ~ **network** *n* TRANS réseau intégré *m*; ~ **office system** *n* PER APP circuit intégré logique *m*, système bureautique intégré *m*; ~ **optical circuit** *n* (*IOC*) ELECT, TRANS circuit d'optique intégré *m*, circuit optique intégré *m*; ~ **optical switch** *n* SWG, TRANS commutateur optique intégré *m*; ~ **optical switching matrix** *n* SWG, TELEP matrice de commutation optique intégrée *f*; ~ **optics** *n pl* TRANS optique intégrée *f*; ~ **optoelectronic circuit** *n* ELECT circuit optoélectronique intégré *m*; ~ **satellite test** *n* (*IST*) TEST essai de satellite intégré *m*; ~ **services digital network** *n* (*ISDN*) TELEP, TRANS réseau numérique à intégration de services *m* (*RNIS*); ~ **software** *n* COMP, DATA logiciel intégré *m*; ~ **system** *n* TRANS système intégré *m*; ~ **system test** *n* (*IST*) TEST essai de système intégré *m*; ~ **transmission and switching network** *n* (*IST network*) TRANS réseau à commutation et transmission intégrées *m*; ~ **voice-data PABX** *n* APPL, CONTR, SWG, TELEP autocommutateur privé voix-données *m*; ~ **voice-data switch** *n* SWG autocommutateur à intégration voix-données *m*

integrating: ~ **circuit** *n* ELECT intégrateur *m*, circuit d'intégration *m*, *electronics* montage intégrateur *m*; ~-**measuring instrument** *n* TEST appareil de mesure intégrateur *m*

integrator *n* DATA, RADIO intégrateur *m*, montage intégrateur *m*

integrity *n* DATA, TELEP, TRANS intégrité *f*

intelligent *n* COMP intelligent; ~ **backtracking** DATA retour arrière intelligent *m*; ~ **building** (*IB*) APPL édifice intelligent *m*; ~ **computer interface** COMP, DATA interface d'ordinateur intelligente *f*; ~ **network** (*IN*) APPL, COMP, TRANS réseau intelligent *m* (*RI*); ~ **optical modem** DATA modem optique intelligent *m*; ~ **peripheral** PER APP périphérique intelligent *m* (*PI*); ~ **tutoring systems** *n pl* (*ITS*) COMP enseignement intelligemment assisté par ordinateur *m* (*EIAO*); ~ **vehicle highway systems** *n pl* (*IVHS*) RADIO systèmes d'autoroute pour véhicules intelligents *m pl*

intelligible: ~ **crosstalk components** *n pl* TELEP composants de diaphonie intelligible *m pl*

intensity *n* GEN, TRANS, TV intensité *f*; ~ **control** TV commande d'intensité *f*; ~ **interpolation shading** TV lissage par interpolation bilinéaire *m*; ~ **level** TRANS niveau d'intensité *m*; ~ **sensor** TRANS capteur d'intensité *m*

interaction *n* ELECT interaction *f*; ~ **gap** POWER espace d'interaction *m*; ~ **loss** TRANS perte d'interaction *f*; ~ **region** ELECT région d'interaction *f*

interactive[1] *adj* COMP conversationnel, interactif, GEN interactif

interactive[2]: ~ **area** *n* TRANS *fibre optics* plage d'interaction *f*; ~ **audiovisual** *n* TV audiovisuel interactif *m*; ~ **computer graphics** *n pl* COMP, TV infographie interactive *f*; ~ **medium** *n* COMP médium interactif *m*; ~ **mode** *n* COMP mode dialogué *m*; ~-**mode operating sequence** *n* TELEP séquence de fonctionnement en mode menu *f*; ~ **multimedia** *n* COMP multimédia interactif *m*; ~ **network** *n* TRANS réseau interactif *m*; ~ **primary link** *n* (*IPL*) COMP, DATA liaison primaire interactive *f*; ~ **service** *n* TV service interactif *m*; ~ **system** *n* TV système interactif *m*; ~ **television** *n* (*ITV*) TV télévision interactive *f* (*TVI*); ~ **terminal** *n* TV borne interactive *f*; ~ **video** *n* TV vidéo interactif *f*; ~ **videodisk** *n* COMP, TV vidéodisque intéractif *m*; ~ **videography** *n* TV vidéographie interactive *f*, vidéotex interactif *m*; ~ **videotex** *n* TV vidéotex interactif *m*; ~ **voice data servers** *n pl* TELEP serveurs vocaux interactifs *m pl*; ~ **voice response** *n* (*IVR*) TELEP réponse vocale interactive *f*; ~ **voice service** *n* APPL service vocal interactif *m* (*SVI*)

interactivity *n* TV interactivité *f*

interband[1] *adj* TELEP, TV interbande

interband[2]: ~ **telegraphy** *n* TELEP télégraphie interbande *f*

interblock: ~ **gap** *n* DATA espace entre blocs *m*, espace interbloc *m*

intercell: ~ **hand-off** *n* RADIO transfert entre les cellules *m*; ~ **switching** *n* RADIO commutation des voies entre cellules *f*

intercept: ~ **announcer** *n* TELEP guide parlant *m*; ~ **equipment** *n* TELEP équipement d'interception *m*; ~ **receiver** *n* ELECT récepteur d'écoute *m*, récepteur d'interception *m*; ~ **service** *n* TELEP service d'interception d'appels *m*; ~ **trunk** *n* TELEP circuit d'interception *m*

interchange[1] *n* GEN échange *m*; ~ **circuit** TRANS circuit de jonction *m*, circuit de l'échange *m*; ~ **control reference** DATA référence de contrôle de l'échange *f*; ~ **date** DATA date de l'échange *f*; ~ **receiver ID** DATA, TELEP identification du destinataire de l'échange *f*; ~ **receiver identification** DATA, TELEP identification du destinataire de l'échange *f*; ~ **recipient** DATA, TELEP destinataire de l'échange *mf*; ~ **sender** DATA, TELEP expéditeur de l'échange *m*; ~ **sender ID** DATA, TELEP identification de l'expéditeur de l'échange *f*; ~ **sender identification** DATA, TELEP identification de l'expéditeur de l'échange *f*; ~ **time** DATA heure de l'échange *f*; ~ **trunk** TELEP circuit intracentral *m*; ~ **trunk group** *BrE* (*cf intraoffice trunk group AmE*) TELEP faisceau de circuits intracentral *m*, faisceau intracentral *m* (*BrE*); ~ **unit program** TELEP échangeur *m*

interchange[2] *vt* GEN échanger

interchangeable: ~ **connectors** *n pl* TRANS connecteurs interchangeables *m pl*

intercharacter: ~ **rest condition** *n* TELEP condition de repos intercalaire entre les caractères *f*

intercity *n* TELEP interurbain du réseau du gouvernement *m*, système interurbain *m*

intercom *n* TELEP interphone *m*, intercom (*intercommunication*); ~**-PBX** (*intercom-private branch exchange*) TELEP système d'intercommunication *m*; ~ **system** TELEP installation d'interphone *f*; ~ **telephone** TELEP interphone *m*

intercommunication *n* (*intercom*) TELEP intercommunication *f* (*intercom*); ~ **system** TELEP système d'intercommunication *m*; ~ **telephone set** TELEP poste d'intercommunication *m*

interconnect *vt* GEN connecter, interconnecter, relier

interconnection *n* GEN interconnexion *f*; ~ **circuit** TELEP circuit d'interconnexion *m*; ~ **equipment** TELEP matériel d'interconnexion *m*; ~ **network** POWER, TRANS réseau d'interconnexion *m* (*RI*)

intercontinental: ~ **circuit** *n* TELEP circuit intercontinental *m*; ~ **connection** *n* TELEP communication intercontinentale *f*; ~ **transit circuit** *n* TELEP circuit intercontinental de transit *m*; ~ **transit exchange** *n* TELEP centre intercontinental de transit *m*

intercontrol: ~ **station multiplex** *n* TELEP multiplex interstation *m*

intercouple *vt* PER APP coupler, interconnecter

interdigital: ~ **filter** *n* POWER filtre interdigital *m*; ~ **pause** *n* TELEP intervalle entre chiffres *m*, intervalle interchiffre *m*; ~ **time** *n* TELEP intervalle entre chiffres *m*, intervalle interchiffre *m*; ~ **transducer** *n* (*IDT*) TRANS transducteur interdigité *m*

interelectrode: ~ **capacitance** *n* ELECT capacité entre électrode *f*, capacité interélectrodes *f*

interexchange: ~ **trunk** *n BrE* (*cf interoffice trunk AmE*) TELEP circuit entre centres locaux *m*; ~ **trunk group** *n BrE* (*cf interoffice trunk group AmE*) TELEP faisceau de circuits entre centres locaux *m*, faisceau entre centres locaux *m*

interface *n* (*I/F*) DATA, GEN interface *f*, jonction *f*; ~ **adaptor** COMP, TELEP adaptateur d'interface *m*, adaptateur de jonction *m*; ~ **circuit** ELECT circuit d'interface *m*; ~ **coder** TRANS codeur d'interface *m*; ~ **connection** COMP branchement de liaison *m*; ~ **converter** DATA convertisseur d'interface *m*; ~ **of the CSN with the switching and control units** TELEP interface du CSN avec les organes de connexion et de commande *f*; ~ **decoder** TRANS décodeur d'interface *m*; ~ **message processor** (*IMP*) COMP serveur de messages *m* (*IMP*); ~ **module** (*IM*) GEN module de raccordement *m*

interfaceable *adj* GEN interconnectable

interfacial: ~ **connection** *n* ELECT liaison interface *f*

interfacility: ~ **link** *n* (*IFL*) TRANS circuit entre installations de raccordement *m* (*IFL*)

interfacing: ~ **unit** *n* GEN unité d'adaptation *f*, unité de jonction *f*, unité de raccordement *f*, unité d'interface *f*

interference *n* ELECT, GEN, POWER perturbation *f*, RADIO brouillage *m*, parasitage *m*, parasites *m pl*, perturbation *f*, TRANS source de brouillage *f*; ~ **canceler** *AmE*, ~ **canceller** *BrE* RADIO annuleur de perturbation *m*; ~ **eliminator** POWER suppresseur de perturbation *m*; ~ **filter** ELECT, TRANS filtre antibrouillage *m*, filtre antiparasite *m*, filtre d'antenne *m*, filtre interférentiel *m*; ~**-free coverage area** RADIO zone de couverture en absence de brouillage *f*; ~ **fringe pattern** TRANS *fibre optics* motif de franges d'interférences *m*; ~ **immunity** RADIO vulnérabilité au brouillage *f*, susceptibilité au brouillage *f*; ~ **level** POWER, RADIO niveau de brouillage *m*; ~ **level contribution** RADIO fraction de niveau de brouillage *f*; ~ **noise** RADIO bruit de brouillage *m*; ~ **path** RADIO trajet de brouillage *m*; ~ **pulse** ELECT impulsion parasite *f*; ~ **sector** RADIO secteur de brouillage *m*; ~ **suppression** ELECT, RADIO antibrouillage *m*, antiparasitage *m*, TRANS suppression d'interférence *f*; ~ **suppression equipment** PER APP équipement d'antiparasitage *m*; ~ **voltage** TEST tension perturbatrice *f*

interfering: ~ **channel** *n* TRANS voie brouilleuse *f*; ~ **power flux** *n* TRANS flux de puissance de brouillage *m*; ~ **satellite** *n* RADIO satellite brouilleur *m*; ~ **signal** *n* GEN signal parasite *m*, POWER signal perturbateur *m*; ~ **source** *n* RADIO source de brouillage *f*; ~ **transmitter** *n* RADIO émetteur brouilleur *m*

interferometer *n* DATA, TEST interféromètre *m*

interflow *n* TELEP débordement *m*; ~ **key** TELEP touche débordement *f*

interframe: ~ **coding** *n* DATA codage intertrame *m*; ~ **compression** *n* TV compression inter-image *f*

interhub: ~ **signaling** *AmE*, ~ **signalling** *BrE* *n* SIG signalisation inter-nœuds *f*

interlace *vt* RADIO, TELEP, TRANS, TV entrelacer

interlaced[1] *adj* RADIO, TELEP, TRANS, TV entrelacé

interlaced[2]: ~ **scanning** *n* TV balayage avec entrelacement *m*, balayage entrelacé *m*

interlacing *n* RADIO, TELEP, TRANS, TV entrelacement *m*; ~ **sequence** TV séquence d'entrelacement *f*

interlayer: ~ **connection** *n* ELECT *printed circuits* connexion entre couches *f*

interleave[1] *n* RADIO, TELEP, TRANS entrelacement *m*

interleave[2] *vt* COMP imbriquer, RADIO entrelacer, intercaler, TELEP, TRANS entrelacer

interleaved[1] *adj* RADIO, TELEP, TRANS entrelacé

interleaved[2]: ~ **carriers** *n pl* TRANS entrelacement de fréquences porteuses *m*; ~ **channel coding** *n* TRANS codage de canal entrelacé *m*; ~ **polarization** *n* RADIO polarisation alternée *f*

interleaver *n* RADIO, TRANS dispositif d'entrelacement *m*, unité d'entrelacement *f*

interleaving *n* COMP, GEN, TELEP, TRANS entrelacement *m*

interletter: ~ **spacing** *n* DATA crénage *m*, interlettrage *m*

interline: ~ **spacing** *n* DATA interlignage *m*

interlocal: ~ **traffic** *n* TELEP trafic interlocal *m*; ~ **trunk group** *n* TELEP faisceau local *m*, faisceau de circuits local *m*

interlock: ~ **code** *n* (*IC*) DATA, TELEP code de verrouillage *m*; ~ **switch** *n* RADIO verrou de récepteur *m*

interlocking: ~ **equisignal system** *n* RADIO radioalignement à entrechevêtrement *m*

intermarker: ~ **group trunk** *n* (*IM*) TELEP circuit entre groupes marqueurs *m*; ~ **trunk group** *n* (*IM*) TELEP faisceau de circuits entre groupes marqueurs *m*

intermateable: ~ **connector** *n* TRANS connecteur interconnectable *m*

intermediate: ~ **cross-connect** *n* (*IC*) SIG équipement de brassage intermédiaire *m*; ~ **data rate** *n* (*IDR*) DATA débit binaire intermédiaire *m*; ~ **distribution frame** *n* (*IDF*) SWG, TELEP répartiteur intermédiaire *m* (*RI*); ~ **equipment** *n* COMP équipement intermédiaire *m*, dispositif intermédiaire *m*, TELEP équipement intermédiaire *m*; ~ **frame** *n* TV image intermédiaire *f*; ~ **frequency** *n* (*IF*) RADIO, TRANS fréquence intermédiaire *f* (*FI*); ~ **frequency amplification** *n* RADIO, TRANS amplification à fréquence intermédiaire *f*; ~ **frequency amplifier** *n* RADIO, TRANS amplificateur à fréquence intermédiaire *m*; ~ **frequency filter** *n* RADIO, TRANS filtre à fréquence intermédiaire *m*; ~**-frequency rejection** *n* RADIO atténuation à fréquence intermédiaire *f*, rejet à fréquence intermédiaire *m*; ~ **frequency rejection ratio** *n* RADIO affaiblissement sur fréquence intermédiaire *m*; ~ **frequency signal** *n* ELECT, RADIO, SIG signal à fréquence intermédiaire *m*; ~ **frequency stage** *n* ELECT, RADIO étage à fréquence intermédiaire *m*; ~ **high-usage trunk** *n* (*IH*) TELEP faisceau de circuits intermédiaire à débordement *m*, faisceau intermédiaire à débordement *m*; ~ **reference system** *n* (*IRS*) DATA, TELEP système de référence intermédiaire *m* (*SRI*); ~ **repeater** *n* TRANS répéteur intermédiaire *m*; ~ **satellite band** *n* RADIO bande intermédiaire satellite *f*; ~ **system** *n* (*IS*) DATA, RADIO système intermédiaire *m*; ~ **trunk channel** *n* TRANS canal interurbain intermédiaire *m*

intermittence *n* GEN intermittence *f*

intermittency *n* GEN intermittence *f*

intermittent: ~ **failure** *n* TELEP dérangement intermittent *m*; ~ **fault** *n* TEST défaut intermittent *m*, panne intermittente *f*, panne temporaire *f*

intermodal: ~ **distortion** *n* TRANS *fibre optics* distorsion modale *f*

intermodulation *n* TRANS intermodulation *f*; ~ **attenuation** TRANS *electronics* affaiblissement d'intermodulation *m*; ~ **component** TRANS oscillation d'intermodulation *f*; ~ **distortion** TRANS distorsion d'intermodulation *f*; ~ **noise** TRANS *electronics* bruit d'intermodulation *m*; ~ **product** TRANS produit d'intermodulation *m*

internal: ~ **blocking** *n* SWG blocage interne *m*; ~ **circuitry** *n* ELECT circuit interne *m*; ~ **clock** *n* PER APP horloge interne *f*; ~ **connecter** *n* ELECT connecteur interne *m*; ~ **damage** *n* TEST défaillance interne *f*; ~ **disabled time** *n* TEST temps d'indisponibilité interne *m*; ~ **down time** *n* TELEP durée d'indisponibilité interne *f*; ~ **extension** *n* TELEP poste intérieur *m*; ~**-external downtime** *n* TELEP durée d'indisponibilité interne-externe *f*; ~ **failure** *n* TEST *quality* défaillance interne *f*; ~ **gas pressure cable** *n* TRANS câble sous pression gazeuse *m*; ~ **input signal** *n* ELECT signal d'entrée interne *m*; ~ **label** *n* DATA label interne *m*; ~ **layer** *n* ELECT *printed circuits* couche interne *f*; ~**-memory capacity** *n* COMP capacité de la mémoire interne *f*; ~ **PCM link** *n* TELEP liaison MIC interne *f*; ~ **photoelectric effect** *n* ELECT, TRANS effet photoélectrique interne *m*, photoconduction *m*, photoconductivité *f*; ~ **reference point** *n* (*IRP*) ELECT point de référence interne *m*; ~ **resistance** *n* ELECT résistance interne *f*; ~ **signal** *n* SIG signal interne *m*; ~ **speaker** *n* PER APP haut-parleur incorporé *m*; ~ **storage** *n* DATA mémoire interne *f*; ~ **traffic** *n* TELEP trafic interne *m*; ~ **wiring** *n* ELECT câblage interne *m*

international: ~ **alphabet no. 5** *n* (*IA5*) DATA, TELEP alphabet international n° 5 *m*; ~ **atomic time** *n* (*IAT*) RADIO temps atomique international *m* (*TAI*); ~ **automatic circuit** *n* TELEP circuit automatique international *m*; ~ **chain** *n* TELEP chaîne internationale *f*; ~ **circuit** *n* TELEP circuit international *m*; ~ **code designator** *n* (*ICD*) DATA désignateur de code international *m*; ~ **connection** *n* TELEP communication internationale *f*; ~ **data number** *n* DATA numéro international pour la transmission de données *m*; ~ **data number format** *n* DATA format du numéro international pour la transmission de données *m*; ~ **data switching exchange** *n* (*IDSE*) DATA, SWG centre international de commutation de données *m*; ~ **direct dialing** *AmE*, ~ **direct dialling** *BrE* *n* TELEP automatique international *m*; ~ **direct distance dialing** *AmE*, ~ **direct distance dialling** *BrE* *n* TELEP automatique international *m*; ~ **exchange** *n* SWG centre international *m*; ~ **gateway exchange** *n* SWG, TELEP commutateur de transit international *m*; ~ **gateway node** *n* (*IGN*) SWG, TELEP, TRANS nœud de transit international *m* (*NTI*); ~ **leased circuit** *n* TELEP circuit international loué *m*; ~ **leased group link** *n* TELEP liaison internationale louée en groupe primaire *f*; ~ **line** *n* TELEP, TRANS ligne internationale *f*; ~ **link** *n* TELEP liaison internationale *f*; ~ **main section** *n* TELEP section internationale principale *f*; ~ **manual demand service** *n* TELEP service international rapide *m*; ~ **Morse code** *n* (*IMC*) RADIO, TELEP code Morse international *m*; ~ **multiple-destination sound-program** *AmE see international multiple-destination sound-programme BrE*; ~ **multiple-destination sound-program circuit** *AmE see international multiple-destination sound-programme circuit BrE*; ~ **multiple-destination sound-program link** *AmE see international multiple-destination sound-program link BrE*; ~ **multiple-destination sound-programme** *BrE n* TV communication radiophonique internationale multiple *f*; ~ **multiple-destination sound-programme circuit** *n BrE* TV circuit radiophonique international multiple *m*, section internationale de circuit télévisuel multiple *f*; ~ **multiple-destination sound-programme link** *n BrE* TELEP liaison radiophonique internationale multiple *f*; ~ **multiple-destination television circuit** *n* TV circuit télévisuel international multiple *m*, section internationale de circuit télévisuel multiple *f*; ~ **multiple-destination television connection** *n* TV communication télévisuelle internationale multiple *f*; ~ **multiple-destination television link** *n* TV liaison télévisuelle internationale multiple *f*; ~ **network** *n* TRANS réseau international *m*; ~ **network management** *n* TELEP gestion du réseau international *f*; ~ **number** *n* TELEP numéro international *m*; ~ **operations service** *n* APPL service d'exploitation internationale *m*; ~ **packet-switched data network** *n* TRANS réseau international de transmission par paquets *m*; ~ **packet-switching gateway exchange** *n* SWG, TELEP, TRANS nœud de transit international *m* (*NTI*); ~ **prefix** *n* TELEP préfixe international *m*; ~ **public facsimile service** *n* TELEP service public international de télécopie *m*; ~ **reference alphabet** *n* (*IRA*) TELEP alphabet international de référence *m* (*IRA*); ~ **section** *n* TELEP section internationale *f*; ~ **sound-program center** *AmE see international sound-programme centre BrE*; ~ **sound-program circuit** *AmE see international sound-programme circuit BrE*; ~ **sound-program connection** *AmE see international sound-programme connection BrE*; ~ **sound-program link** *AmE see international sound-programme link BrE*; ~ **sound-program transmission** *AmE see international sound-programme transmission BrE*; ~ **sound-programme centre** *n BrE* (*ISPC*) RADIO centre radiophonique international *m* (*CRI*); ~ **sound-programme circuit** *n BrE* TV circuit radiophonique international *m*; ~ **sound-programme connection** *n BrE* TV communication radiophonique internationale *f*; ~ **sound-programme link** *n BrE* TV liaison radiophonique internationale *f*; ~ **sound-programme transmission** *n BrE* TV transmission radiophonique internationale *f*; ~ **standard** *n* (*IS*) DATA, ELECT, RADIO, TEST norme internationale *f*, TEST étalon international *m*; ~ **standard virtual terminal protocol** *n* ELECT protocole de terminal standard international *m*; ~ **subscriber dialling** *n BrE* (*cf international direct distance dialing AmE*) TELEP automatique international *m*; ~ **switching center** *AmE* , ~ **switching centre** *BrE n* (*ISC*) SWG centre international automatique *m*, centre international de commutation *m*, TELEP centre international de commutation *m*; ~ **switching maintenance center** *AmE*, ~ **switching maintenance centre** *BrE n* TEST centre international de maintenance de commutation *m* (*CIMC*); ~ **telegraph alphabet no. 2** *n* (*ITA2*) DATA, TELEP alphabet télégraphique international ň 2 *m*; ~ **telegraph position** *n* TELEP position phototélégraphique internationale *f* (*PPI*); **~ telephone call** TELEP communication téléphonique internationale; **~ telephone circuit** *n* TELEP, TRANS circuit téléphonique international *m*; **~ telephone connection** *n* TELEP communication téléphonique internationale *f*; **~ television center** *AmE*, **~ television centre** *BrE n* (*ITC*) TV centre télévisuel international *m*; **~ television circuit** *n* TV circuit télévisuel international *m*; **~ television connection** *n* TV communication télévisuelle internationale *f*; **~ television link** *n* TV liaison télévisuelle internationale *f*; **~ television program center** *AmE*, **~ television programme centre** *BrE n* (*ITPC*) TV centre télévisuel international *m*; **~ television transmission** *n* TV transmission télévisuelle internationale *f*; **~ telex position** *n* TELEP position télex internationale *f*; **~ transferred-account telegraph service** *n* TELEP service international de comptes télégraphiques transférés *m*; **~ transit exchange** *n* (*INTTR*) SWG, TELEP commutateur de transit international *m*; **~ transit node** *n* SWG, TRANS nœud de transit international *m* (*NTI*); **~ transmission-maintenance center** *AmE*, **~ transmission-maintenance centre** *BrE n* (*ITMC*) TEST centre international de maintenance de transmission *m* (*CIMT*); **~ X.121 format** *n* DATA format international X.121 *m*

International: **~ Amateur Radio Union** *n* (*IARU*) RADIO Union internationale des radio-amateurs *f*; **~ Association of Broadcasting** *n* (*IAB*) RADIO Association internationale de retransmission *f* (*AIR*); **~ Association of Lighthouse Authorities** *n* (*IALA*) RADIO Association internationale de signalisation maritime *f* (*AISM*); **~ Conference on Large High Voltage Electric Systems** *n* (*CIGRE*) RADIO Conférence internationale des grands réseaux électriques à haute tension *f*; **~ Electrotechnical Commission** *n* (*IEC*) GEN Commission électrotechnique internationale *f* (*CEI*); **~ Marine Radio Association** *n* (*CIRM*) RADIO Comité international radio-maritime *m*; **~ Maritime Organization** *n* TELEP Organisation maritime internationale *f*; **~ Maritime Satellite Organization** *n* (*INMARSAT*) RADIO Organisation internationale de télécommunications maritimes par satellites *f*; **~ Organization of Space Communications** *n* RADIO Organisation internationale des télécommunications spatiales *f*; **~ Radio Consultative Committee** *n* (*IRCC*) RADIO Comité consultatif international des radiocommunications *m* (*CCIR*); **~ reference version** *n* (*IRV*) TELEP version internationale de référence *f* (*IRV*); **~ Satellite System for Search and Rescue** *n* (*COSPAS-SARSAT*) GEN Système international de satellites pour les recherches et le sauvetage *m* (*COSPAS-SARSAT*); **~ Society for Aeronautical Telecommunications** *n* RADIO Société internationale de télécommunications aéronautiques *f* (*SITA*); **~ Special Committee on Radio Interference** *n* (*CISPR*) RADIO Comité international spécial des perturbations radioélectriques *m* (*CISPR*); **~ Standards Organization** (*ISO*) GEN Organisme international de normalisation (*ISO*); **~ Telecommunications Satellite Organization** *n* (*INTELSAT*) RADIO Organisation internationale de télécommunications par satellites *f* (*INTELSAT*); **~ Telecommunications Union** *n* (*ITU*) GEN Union internationale des télécommunications *f* (*UIT*); **~ Telegraph and Telephone Consultative Committee** *n* (*ITTCC*) TELEP Comité consultatif international télégraphique et téléphonique *m* (*CCITT*); **~ Union of Radio Science** *n* (*URSI*) RADIO Union radio-scientifique internationale *f* (*URSI*)

Internet *n* (*Net*) COMP, DATA Internet *m*, World Wide Web *m*, W3, WWW; **~ access point** COMP, DATA point d'accès Internet *m*; **~ account** COMP compte Internet *m*; **~ kiosk** COMP kiosque Internet *m*; **~ packet exchange** (*IPX*) DATA, SWG IPX; **~ protocol** (*IP*) COMP, DATA Internet protocol *m* (*IP*), protocole Internet *m* (*IP*); **~ service provider** COMP fournisseur d'accès Internet *m*; **~ society** DATA Isoc

internetwork *n* TRANS réseau d'interconnexion *m* (*RI*); **~ call redirection and deflection** (*ICRD*) DATA réacheminement et déviation des appels entre réseaux *m*; **~ protocol exchange** DATA IPX

internetworking *n* DATA interconnexion de réseaux *f*, interfonctionnement des réseaux *m*

interoffice: **~ trunk** *n AmE* (*cf interexchange trunk BrE*) TELEP circuit entre centres locaux *m*; **~ trunk group** *n AmE* (*cf interexchange trunk group BrE*) TELEP faisceau de circuits entre centres locaux *m*, faisceau entre centres locaux *m*

interoperability *n* DATA interfonctionnement *m*, GEN interopérabilité *f*

inter-PABX: **~ tie circuit** *n* TELEP, TRANS ligne interautomatique *f*

inter-PBX: **~ traffic** *n* TELEP trafic de ligne privée *m*

interpersonal: **~ message element of service** *n* (*IPMEOS*) APPL, DATA, TELEP élément de service de message de personne à personne *m*; **~ messaging** *n* (*IPM*) TELEP messagerie de personne à personne *f*; **~ messaging system** *n* (*IPMS*) RADIO, TELEP système de messagerie interpersonnelle *m*, système de messagerie de personne à personne *m*; **~ messaging system user** *n* TELEP utilisateur du système de messagerie de personne à personne *m*; **~ messaging system user agent** *n* (*IPMS UA*) APPL, DATA, RADIO, TELEP agent utilisateur de système de messagerie de personne à personne *m*; **~ notification** *n* (*IPN*) APPL, DATA, RADIO, TELEP notification de personne à personne *f*

interphone *n* PER APP interphone *m*; **~ equipment** PER APP interphonie *f*

interplay *n* GEN action réciproque *f*

interpolated: **~ channel** *n* TRANS voie avec concentration de la parole *f*, voie interpolée *f*

interpolation: **~ error** *n* TEST erreur d'interpolation *f*; **~ gain** *n* (*IG*) TRANS gain de concentration *m* (*IG*); **~ pool** *n* TRANS pool de concentration *m*; **~ resolution** *n* PER APP *scanner* résolution d'interpolation *f*; **~ of speech signals** *n* TRANS interpolation des signaux vocaux *f*

interpolator *n* TRANS interpolateur *m*

interposition *n* TELEP interposition *f*; **~ trunk** TELEP circuit interposition *m*, circuit d'intercommunication entre positions *m*; **~ trunk group** TELEP faisceau d'intercommunication entre positions *m*, faisceau de circuits d'intercommunication entre positions *m*

interpret *vt* GEN interpréter

interpretability *n* GEN interprétabilité *f*

interpretable *adj* GEN interprétable

interpretation *n* GEN interprétation *f*

interpreter *n* COMP interpréteur *m*, programme d'interprétation *m*, programme interprétatif *m*

interpretive: **~ code** *n* COMP code interprétatif *m*

interprocessor: **~ link** *n* DATA liaison interprocesseur *f*

interrecord: **~ gap** *n* DATA espace entre enregistrements *m*

interrogate *vt* GEN, PER APP interroger, RADIO consulter, interroger

interrogating: **~ pulse** *n* RADIO impulsion d'interrogation *f*; **~ signal** *n* RADIO signal d'interrogation *m*

interrogation *n* DATA demande d'identification *f*, GEN interrogation *f*, RADIO consultation *f*; **~-signal generator** (*ISG*) RADIO générateur de signaux d'interrogation *m*

interrogator *n* RADIO interrogateur *m*; **~ side-lobe suppression** (*ISLS*) RADIO suppression des lobes secondaires de l'interrogateur *f*; **~ transponder** RADIO interrogateur-répondeur *m*

interrupt[1] *n* COMP, DATA, SWG, TELEP interruption *f*; **~ feedback** TRANS système de retour d'écoute avec intervention *m*; **~ indicator** COMP indicateur d'interruption *m*

interrupt[2] *vt* COMP, DATA, SWG interrompre, TELEP couper, interrompre

interrupted: ~ **ringing** *n* TELEP sonnerie cadencée *f*, appel cadencé *m*; ~ **ringing signal** *n* TELEP signal d'appel cadencé *m*
interrupter *n* GEN, TELEP interrupteur *m*
interruptible: ~ **power supply** *n* POWER groupe électrogène de relève *m*
interrupting: ~ **capacity** *n* POWER pouvoir de coupure *m*
interruption *n* POWER coupure *f*, TELEP *by telephone operator* entrée en ligne *f*, interruption *f*, TEST interruption *f*; ~ **of a call in progress** TELEP service d'interruption d'une conversation en cours *m*; ~ **control** TELEP contrôle d'interruption *m*; ~ **of transmission** TRANS interruption de transmission *f*; ~ **of transmission service** TELEP interruption du service transmission *f*
intersatellite: ~ **link** *n* (*ISL*) RADIO liaison intersatellite *f* (*LIS*)
interscan *n* RADIO balayage intermédiaire *m*
intersite: ~ **cable** *n* TRANS câble inter-sites *m*
interspersed: ~ **frequency** *n* RADIO fréquence intercalée *f*
interstage: ~ **coupling** *n* ELECT couplage entre étages *m*
interswitchboard: ~ **line** *n* TELEP, TRANS ligne interstandard *f*
intersymbol: ~ **interference** *n* (*ISI*) COMP, TRANS, TV brouillage intersymbole *m*, interférence intersymbole *f*, interférence longitudinale *f*
intertandem: ~ **trunk** *n* TELEP circuit entre centres de transit *m*; ~ **trunk group** *n* TELEP faisceau entre centres de transit *m*, faisceau de circuits entre centres de transit *m*
intertoll: ~ **traffic** *n* TELEP trafic interurbain *m*; ~ **trunk** *n* TELEP circuit interurbain *m*; ~ **trunk group** *n* (*IT*) TELEP faisceau interurbain *m*, faisceau de circuits interurbain *m*
interval: ~ **signal** *n* RADIO signal de repos *m*; ~ **timer** *n* ELECT temporisateur *m*, POWER compteur d'intervalles de temps *m*, générateur d'intervalles de temps *m*, intervallomètre *m*, minuterie *f*
interwork[1] *n* GEN interfonctionnement *m*, dialogue *m*, échanges entre matériels *m pl*; ~ **description** GEN description d'interfonctionnement *f*; ~ **diagram** GEN schéma d'interfonctionnement *m*
interwork[2] *vi* GEN dialoguer, interfonctionner
interworkability *n* RADIO interopérabilité *f*
interworking *n* GEN échanges entre matériels *m pl*, dialogue *m*, interfonctionnement *m*, interopérabilité *f*; ~ **between networks** TELEP interfonctionnement entre réseaux *m*; ~ **between teletex and other services** TELEP interfonctionnement entre le service télétex et d'autres services *m*; ~ **between user classes of services** TELEP interfonctionnement entre catégories d'usagers du service *m*; ~ **by call control mapping** (*ICCM*) DATA interfonctionnement par mise en correspondance de la commande des communications *m*; ~ **by port access** (*IPA*) COMP, DATA interfonctionnement par point d'accès *m* (*IPA*); ~ **function** (*IWF*) TELEP fonction d'interfonctionnement *f* (*FIF*); ~ **functional unit** (*IFU*) DATA unité fonctionnelle d'interconnexion *f*; ~ **protocol** (*IWP*) DATA protocole d'interfonctionnement *m*; ~ **unit** DATA unité d'interfonctionnement *f*
interzonal *adj* ELECT interzonal
interzone: ~ **interference** *n* RADIO interférence entre zones *f*
intraband: ~ **telegraphy** *n* TELEP télégraphie intrabande *f*
intrabuilding: ~ **riser applications** *n pl* TRANS applications de câblage vertical en immeuble *f pl*
intracoded: ~ **picture** *n* (*i-picture*) COMP, TV image à codage intratrame *f* (*image-i*)
intracoding *n* COMP intra-codage *m*
intraframe *n* TV intra-image *m*; ~ **coding** TV codage intra-image *m*; ~ **compression** TV compression intra-image *f*
intramodal: ~ **distortion** *n* TRANS distorsion intramodale *f*
intranet *n* COMP intranet *m*
intraoffice: ~ **junctor circuit** *n* SWG circuit de connexion intracentral *m*; ~ **trunk** *n* TELEP circuit intracentral *m*; ~ **trunk group** *n AmE* (*cf interchange trunk group BrE*) TELEP faisceau de circuits intracentral *m*, faisceau intracentral *m*
intrinsic: ~ **conduction** *n* ELECT *semiconductors*, PARTS conduction intrinsèque *f*; ~ **error** *n* TEST erreur intrinsèque *f*; ~ **joint loss** *n* TRANS perte intrinsèque de raccordement *f*; ~ **junction loss** *n* TRANS affaiblissement intrinsèque de raccordement *m*; ~ **layer** *n* ELECT couche intrinsèque *f*; ~ **noise** *n* TRANS bruit interne *m*; ~ **semiconductor** *n* ELECT, PARTS semiconducteur intrinsèque *m*, semiconducteur type intrinsèque *m*; ~ **standoff ratio** *n* POWER rapport intrinsèque *m*
introduction *n* GEN, TELEP *product, system* lancement *m*; ~ **of service** TELEP mise en service *f*
intruder: ~ **presence detector** *n* APPL détecteur d'intrusion *m*
intrusion *n* ELECT, TELEP intrusion *f*; ~ **tone** TELEP tonalité d'intrusion *f*
INTTR *abbr* (*international transit exchange*) SWG, TELEP commutateur de transit international *m*
intuitive: ~ **searching** *n* COMP recherche intuitive *f*
invalid[1] *adj* COMP invalidé, DATA *message* non valide, GEN non valable
invalid[2]: ~ **cell** *n* COMP cellule invalidée *f*; ~ **command** *n* COMP commande non valide *f*, commande invalide *f*; ~ **key** *n* COMP touche inopérante *f*; ~ **MCSPDU** *n* TELEP unité MCSPDU invalide *f*; ~ **sequence** *n* DATA séquence nulle *f*; ~ **test event** *n* DATA, TEST événement de test invalide *m*
invalidate *vt* COMP invalider
inverse: ~ **electrode current** *n* ELECT courant inverse d'électrode *m*; ~ **function** *n* GEN fonction inverse *f*; ~ **impedance** *n* ELECT impédance inverse *f*; ~ **solidus** *n* COMP barre oblique inverse *f*; ~ **time overload protection** *n* POWER temporisation de protection inversement proportionnelle à la surcharge *f*; ~ **video** *n* TV vidéo inverse *f*
inversion *n* CLIM, GEN, POWER, RADIO, TRANS inversion *f*; ~ **layer** TRANS couche d'inversion *f*
invert *vt* ELECT inverser, GEN intervertir, POWER, TV inverser
inverted: ~ **background** *n* TV fond inversé *m*; ~ **commas** *n pl* COMP guillemets *m pl*
inverter *n* POWER inverseur *m*, onduleur *m*; ~ **circuit** ELECT circuit inverseur *m*, circuit NON *m*; ~ **system** POWER système à onduleurs *m*
inverting: ~ **amplifier** *n* POWER amplificateur inverseur *m*
investigate *vt* TEST examiner
investigation *n* TEST investigation *f*, examen *m*
invitation: ~ **to transmit** *n* DATA invitation à émettre *f*
invocation *n* RADIO sollicitation *f*
invoke *vt* RADIO solliciter, invoquer, TELEP appeler
invoked: ~ **group** *n* DATA groupe des demandés *m*
invoker *n* DATA demandeur *m*
invoking: ~ **CMISE service user** *n* DATA utilisateur du service CMISE lanceur *m*
I/O *abbr* (*input/output*) COMP, ELECT, TELEP E/S (*entrée/sortie*); ~ **channel** *n* COMP canal E/S *m*; ~ **controller** *n* COMP contrôleur d'E/S *m*; ~ **device** *n* COMP, TELEP dispositif d'entrée/sortie *m*
IOC *abbr* (*integrated optical circuit*) ELECT, TRANS circuit intégré optique *m*, circuit optique intégré *m*
ion *n* ELECT ion *m*; ~ **burn** TV tache ionique *f*, tache *f*; ~ **concentration** ELECT concentration ionique *f*, densité ionique *f*; ~ **deposition printer** PER APP imprimante à déposition d'ions *f*; ~ **exchange** ELECT échange ionique *m*; ~-**exchange technique** ELECT procédé d'échange d'ions *m*, technique d'échange ionique *f*; ~ **exchanger** ELECT échangeur d'ions *m*; ~ **implantation** ELECT *semiconductors* implantation ionique *f*; ~-**implanted resistor** ELECT, PARTS *semiconductors* résistance à implantation ionique *f*; ~-**matter interaction** ELECT interaction ion-matière *f*;

~ **migration** ELECT migration ionique *f*, migration d'ions *f*; ~ **noise** ELECT bruit ionique *m*; ~ **trap** ELECT piège à ions *m*
ionic: ~ **current** *n* ELECT courant ionique *m*; ~ **mobility** *n* ELECT *chemistry* mobilité ionique *f*
ionization *n* ELECT, RADIO ionisation *f*; ~ **energy** ELECT énergie d'ionisation *f*; ~ **time** ELECT durée d'ionisation *f*, temps d'ionisation *m*
ionizing: ~ **event** *n* ELECT événement ionisant *m*
ionogram *n* RADIO ionogramme *m*
ionoscope *n* ELECT ionoscope *m*
ionosonde *n* RADIO ionosonde *f*, sondeur vertical *m*
ionosphere *n* RADIO ionosphère *f*
ionospheric: ~ **cross-modulation** *n* RADIO transmodulation ionosophérique *f*; ~ **disturbance** *n* RADIO perturbation ionosphérique *f*; ~ **layer** *n* RADIO couche ionosphérique *f*; ~ **path error** *n* RADIO erreur de propagation ionosphérique *f*; ~ **prediction** *n* RADIO prévision ionosphérique *f*; ~ **propagation** *n* RADIO propagation ionosphérique *f*; ~ **recorder** *n* RADIO sondeur ionosphérique *m*; ~ **reflection** *n* RADIO réflexion ionosphérique *f*; ~ **scatter** *n* RADIO diffusion ionosphérique *f*; ~-**scatter propagation** *n* RADIO diffusion ionosphérique *f*, propagation par diffusion ionosphérique *f*; ~ **sounding** *n* RADIO sondage ionosphérique *m*; ~ **storm** *n* RADIO tempête ionosphérique *f*; ~ **wave** *n* RADIO onde ionosphérique *f*
IP *abbr* COMP, DATA (*Internet protocol*) IP (*protocole Internet*); ~ **stack** *n* COMP pile IP *f*
IPA *abbr* (*interworking by port access*) COMP, DATA IPA (*interfonctionnement par point d'accès*)
i-picture *n* (*intracoded picture*) COMP, TV image-i *f* (*image à codage intratrame*)
IPL *abbr* (*interactive primary link*) COMP, DATA liaison primaire interactive *f*
IPMEOS *abbr* (*interpersonal message element of service*) APPL, DATA, TELEP élément de service de message de personne à personne *m*
IPMS *abbr* (*interpersonal messaging system*) RADIO, TELEP système de messagerie interpersonnelle *m*, système de messagerie de personne à personne *m*; ~ **UA** (*interpersonal messaging system user agent*) APPL, DATA, RADIO, TELEP agent utilisateur de système de messagerie de personne à personne *m*
IPN *abbr* (*interpersonal notification*) APPL, RADIO, DATA, TELEP notification de personne à personne *f*
IPX *abbr* (*Internet packet exchange*) DATA, SWG IPX
IR *abbr* (*infrared*) COMP, ELECT, RADIO, TRANS IR (*infrarouge*); ~ **radiation** *n* ELECT *electromagnetism* radiation infrarouge *f*, rayonnement infrarouge *m*
IRA *abbr* (*international reference alphabet*) TELEP IRA (*alphabet international de référence*)
IRC *abbr* DATA (*information retrieval center AmE, information retrieval centre BrE*) IRC (*centre de recherche documentaire*), TV (*incrementally related carrier*) système de porteuses en relation incrémentielle *m*
IRCC *abbr* (*International Radio Consultative Committee*) RADIO CCIR (*Comité consultatif international des radiocommunications*)
iron: ~ **core** *n* ELECT *electromagnetism* noyau de fer *m*, noyau magnétique *m*; ~-**core coil** *n* ELECT bobine à noyau de fer *f*, bobine d'électro-aimant *f*; ~-**dust core** *n* ELECT noyau à poudre de fer *m*; ~-**powder core** *n* ELECT noyau à poudre de fer *m*; ~-**vane voltmeter** *n* TEST voltmètre ferrodynamique *m*, voltmètre à palette de fer doux *m*
IRP *abbr* (*internal reference point*) ELECT point de référence interne *m*
irradiance *n* TRANS éclairement énergétique *m*, flux lumineux *m*, puissance électromagnétique surfacique *f*
irradiation: ~-**saturation current** *n* ELECT, POWER courant de saturation par irradiation *m*
irrecoverable: ~ **error** *n* COMP erreur irréparable *f*
irregularity: ~ **reflection coefficient** *n* TELEP facteur de réflexion sur les irrégularités *m*
irreversible: ~ **operation** *n* COMP opération irréversible *f*
IRS *abbr* DATA (*information receiver station*) SRI (*station de réception d'informations*), (*intermediate reference system*) SRI (*système de référence intermédiaire*), SIG (*insertion reference signal*) signal d'insertion de référence *m*, TELEP (*information receiver station*) SRI (*station de réception d'informations*), (*intermediate reference system*) SRI (*système de référence intermédiaire*)
IRV *abbr* (*International reference version*) TELEP IRV (*version internationale de référence*)
IS *abbr* DATA, ELECT, RADIO, TEST (*international standard*) norme internationale *f*, (*information separator*) séparateur d'information *m*, (*intermediate system*) système intermédiaire *m*
ISA *abbr* (*industry standard architecture*) COMP ISA; ~ **segment** *n* COMP, DATA segment ISA *m*
ISB *abbr* (*independent sideband*) RADIO BLI (*bande latérale indépendante*)
ISC *abbr* SWG, TELEP (*international switching center AmE, international switching centre BrE*) centre international automatique *m*, centre international de commutation *m*
ISDN *abbr* (*integrated services digital network*) APPL, TELEP, TRANS RNIS (*réseau numérique à intégration de services*); ~ **access** *n* TELEP accès au RNIS *m*; ~ **adaptor** *n* TELEP adaptateur RNIS *m*; ~ **environment** *n* DATA environnement RNIS *m*; ~ **exchange** *n* SWG centre RNIS *m*; ~ **primary rate access** *n* TELEP accès du débit primaire du RNIS *m*; ~ **switch** *n* SWG autocommutateur RNIS *m*; ~ **user part** *n* TELEP sous-système usager RNIS *m* (*SSUR*), sous-système utilisateur pour le RNIS *m* (*SSUR*); ~ **virtual-circuit bearer service** *n* DATA service de circuit virtuel du RNIS *m*
ISG *abbr* (*interrogation signal generator*) RADIO générateur de signaux d'interrogation *m*
ISI *abbr* (*intersymbol interference*) COMP, TRANS, TV brouillage intersymbole *m*, interférence intersymbole *f*, interférence longitudinale *f*
ISL *abbr* (*intersatellite link*) RADIO LIS (*liaison intersatellite*)
ISLS *abbr* (*interrogator side-lobe suppression*) RADIO suppression des lobes secondaires de l'interrogateur *f*
ISMC *abbr* (*international switching maintenance center AmE, international switching maintenance centre BrE*) TEST CIMC (*centre international de maintenance de commutation*)
ISN *abbr* (*information-storage node*) TELEP nœud de mise en mémoire des informations *m*
ISO *abbr* (*International Standards Organization*) GEN ISO (*Organisme international de normalisation*); ~ **7- and 8-bit-coded character sets** *n pl* TELEP jeux ISO de caractères codés à 7 et à 8 éléments *m pl*; ~ **standard** *n* GEN norme ISO *f*
Isoc *abbr* (*Internet society*) DATA Isoc
isochronous[1] *adj* DATA, SIG, TELEP, TRANS isochrone
isochronous[2]: ~ **data** *n pl* DATA signaux de données isochrones *m pl*; ~ **data transmission** *n* DATA transmission isochrone *f*
isolated: ~ **neutral system** *n* POWER réseau à neutre isolé *m*
isolating: ~ **transformer** *n* POWER transformateur d'isolation *m*
isolation *n* DATA, ELECT isolation *f*, RADIO isolement *m*, TELEP isolation *f*
isolator *n* POWER sectionneur *m*, RADIO affaiblisseur non réciproque *m*, atténuateur unidirectionnel *m*, TRANS isolateur *m*
isophon *n* TELEP *acoustics* isophone *m*
isotropic[1] *adj* ELECT, RADIO isotrope
isotropic[2]: ~ **aerial** *n* RADIO antenne isotrope *f*, *aerials* source isotrope *f*; ~ **antenna** *n* RADIO antenne isotrope *f*, source isotrope *f*; ~ **gain** *n* RADIO gain absolu *m*, gain isotrope *m*;

~ **radiator** *n* RADIO *aerials* antenne isotrope *f*, source isotrope *f*
ISPC *abbr* (*international sound-program center AmE, international sound-programme centre BrE*) RADIO CRI (*centre radiophonique international*)
issue *n* GEN *documentation* édition *f*
IST *abbr* TEST (*integrated satellite test*) essai de satellite intégré *m*, (*integrated system test*), essai de système intégré *m*; ~ **network** *n* (*integrated transmission and switching network*) TRANS réseau à commutation et transmission intégrées *m*
ISU *abbr* (*initial signal unit*) SIG unité de signalisation initiale *f*
IT *abbr* COMP, DATA (*information technology*) informatique *f*, (*information type*), type d'information *m*, TELEP (*intertoll trunk group*) faisceau de circuits interurbain *m*, faisceau interurbain *m*
ITA2 *abbr* (*international telegraph alphabet no. 2*) DATA, TELEP alphabet télégraphique international n° 2 *m*
ITC *abbr* TRANS (*incoming trunk circuit*) circuit d'arrivée *m*, circuit entrant *m*, joncteur d'arrivée *m*, TV (*international television center AmE, international television centre BrE*) centre télévisuel international *m*
ITD *abbr* (*input transaction accepted for delivery*) DATA acceptation de transaction pour remise *f*
item *n* GEN entrée *f*
itemized: ~ **billing** *n BrE* TELEP facturation détaillée *f*; ~ **statement** *n* TELEP relevé détaillé *m*
itinerant *adj* RADIO itinérant, nomade
ITMC *abbr* (*international transmission-maintenance center AmE, international transmission-maintenance centre BrE*) TEST CIMT (*centre international de maintenance de transmission*)
ITPC *abbr* (*international television program center AmE, international television programme centre BrE*) TV centre télévisuel international *m*
ITR *abbr* (*input transaction rejected*) DATA rejet de transaction d'entrée *m*
ITS *abbr* COMP (*intelligent tutoring systems*) EIAO (*enseignement intelligemment assisté par ordinateur*), SIG (*insertion test signal*) signal d'essai d'insertion *m*
ITTCC *abbr* (*International Telegraph and Telephone Consultative Committee*) TELEP CCITT (*Comité consultatif international télégraphique et téléphonique*)
ITU *abbr* (*International Telecommunications Union*) GEN UIT (*Union internationale des télécommunications*)
ITV *abbr* TV (*instructional television*) téléenseignement *m*, (*industrial television*) TVI (*télévision industrielle*), (*interactive television*) TVI (*télévision interactive*)
I-type: ~ **semiconductor** *n* ELECT, PARTS semiconducteur type intrinsèque *m*
IUCAF *abbr* (*Inter-Union Commission on Frequency Allocations for Radio Astronomy*) RADIO IUCAF (*Commission inter-unions pour l'attribution de fréquences à la radioastronomie et à la science*)
IUT *abbr* (*implementation under test*) DATA IUT (*instance sous test*); ~ **functions** *n pl* DATA fonctions d'une IUT *f pl*
IVHS *abbr* (*intelligent vehicle highway systems*) RADIO systèmes d'autoroute pour véhicules intelligents *m pl*
IVPO *abbr* (*inside vapor phase oxidation AmE, inside vapour phase oxidation BrE*) ELECT dépôt en phase vapeur interne *m*
IVR *abbr* (*interactive voice response*) TELEP réponse vocale interactive *f*
IWF *abbr* (*interworking function*) TELEP FIF (*fonction d'interfonctionnement*)
IWP *abbr* (*interworking protocol*) DATA protocole d'interfonctionnement *m*
IXP *abbr* (*information exchange protocol*) COMP IXP (*protocole d'échange d'informations*)

J

jabber: ~ **protection** *n* ELECT brouillard protection *m*
jack *n* COMP prise *f*, ELECT, PARTS jack *m*, SWG, TELEP, TEST jack *m*, prise *f*; ~ **panel** COMP tableau de connexions *m*
jacket *n* ELECT gaine extérieure *f*, gaine *f*, PARTS, TRANS gainage *m*
jacketing *n* ELECT gainage *m*, PARTS gaine *f*
jackscrew *n* ELECT vis *f*
jam[1] *n* COMP bourrage *m*; ~ **inputs** *n pl* ELECT entrées de chargement parallèle direct *f pl*
jam[2] *vt* COMP, DATA *Internet link* embouteiller, ELECT charger directement en parallèle, brouiller, RADIO, TELEP brouiller
jammed *adj* COMP, DATA *Internet* brouillé, ELECT brouillé, chargé directement en parallèle, RADIO, TELEP brouillé
jammer *n* ELECT émetteur de brouillage *m*, brouilleur *m*, brouilleur intentionnel *m*, RADIO brouilleur intentionnel *m*, émetteur de brouillage *m*, brouilleur *m*
jamming *n* COMP, DATA, ELECT, RADIO brouillage *m*, brouillage intentionnel *m*; ~ **resistance** RADIO résistance à brouillage *f*; ~ **transmitter** RADIO émetteur de brouillage *m*, brouilleur intentionnel *m*
JANET *abbr* (*Joint Academic Network*) TRANS réseau JANET *m*
jar *n* COMP son discordant *m*
JBIG *abbr* (*joint bi-level image group*) COMP, TELEP, TV JBIG (*groupe mixte sur les images à deux niveaux*)
JDF *abbr* (*junction distribution frame*) SWG, TRANS répartiteur de jonction *m*
jelly: ~**-filled cable** *n* TRANS câble rempli de gelée imperméable *m*
jerk *n* DATA saccade *f*
jewel: ~ **bearing** *n* POWER montage sur rubis *m*; ~ **mount** *n* POWER montage sur rubis *m*
JFET *abbr* (*junction field effect transistor*) PARTS transistor à effet de champ à jonction *m*
JFIF *abbr* (*JPEG file interchange format*) COMP JFIF (*format d'échange de fichiers JPEG*)
jig *n* GEN, PARTS calibre *m*, gabarit *m*, montage *m*, positionneur *m*
jitter *n* COMP instabilité *f*, ELECT gigue *f*, instabilité *f*, jitter *m*, RADIO gigue *f*, sautillement *m*, sautillement d'image *m*, TELEP, TRANS gigue *f*, TV *onscreen* gigue *f*, fluctuation de phase *f*, vacillement *m*
job *n* COMP travail *m*, SWG tâche *f*; ~ **division** SWG répartition des tâches *f*; ~ **execution** COMP exécution de travail *f*; ~**-recovery control file** COMP fichier de gestion *m*; ~**-recovery file** COMP fichier de gestion *m*; ~ **selector** COMP sélecteur de travail *m*; ~ **stream** COMP flot de travaux *m*, suite de travaux *f*
Johnson: ~ **counter** *n* ELECT compteur de Johnson *m*
joint[1] *n* COMP articulation *f*, GEN *construction* épissure *f*, joint *m*, PARTS joint *m*, raccord *m*, épissure *f*; ~ **bi-level image group** (*JBIG*) COMP, TELEP, TV groupe mixte sur les images à deux niveaux *m* (*JBIG*); ~ **testing** TEST essais conjoints *m pl*; ~ **tests** *n pl* TEST essais conjoints *m pl*; ~ **user** TRANS co-usager *m*
joint[2] *vt* GEN raccorder, joindre
Joint: ~ **Academic Network** *n* (*JANET*) TRANS réseau JANET *m*; ~ **Photographic Experts Group** *n* (*JPEG*) COMP, TELEP, TV groupe mixte d'experts en photographie *m* (*JPEG*)
jointer *n* TRANS épisseur *m*, assembleur *m*
jointing *n* TRANS joint *m*; ~ **sleeve** PARTS manchon de raccordement *m*
joker *n* COMP caractère de générique *m*
Joule: ~ **effect** *n* GEN effet Joule *m*
joystick *n* COMP manette *f*, manette de jeux *f*
JPEG *abbr* (*Joint Photographic Experts Group*) COMP, TELEP, TV JPEG (*groupe mixte d'experts en photographie*); ~ **file interchange format** *n* (*JFIF*) COMP format d'échange de fichiers JPEG *m* (*JFIF*)
JSM *abbr* (*jumper storage module*) ELECT module de stockage de jarretière *m*
Julian: ~ **date** *n* RADIO date julienne *f*; ~ **day number** *n* RADIO numéro de jour julienne *m*
jump[1] *n* COMP saut *m*, rupture de séquence *f*
jump[2] *vt* COMP sauter
jumper[1] *n* COMP cavalier *m*, ELECT, PARTS, POWER bande de court-circuit *f*, bride de connexion *f*, *connection* connexion volante *f*, fil volant *m*, barrette *f*, jarretière *f*, SWG jarretière *f*; ~ **cable** TRANS câble de raccordement *m*; ~ **king** GEN monteur de jarretières *m*; ~ **ring** SWG anneau passe-jarretières *m*; ~ **storage module** (*JSM*) ELECT module de stockage de jarretière *m*; ~ **wire** TELEP fil jarretière *m*; ~**-wire eyelet** PARTS anneau guide-fils *m*; ~**-wire guide** PARTS anneau guide-fils *m*
jumper[2]: ~ **up** *vt* ELECT connecter les jarretières
jumpering *n* ELECT connection des jarretières *f*
jumping *n* TRANS saut *m*; ~ **allotter** TELEP distributeur de priorité aléatoire *m*
junction *n* ELECT, PARTS, RADIO, TELEP, TRANS jonction *f*; ~ **box** ELECT *cables*, GEN, POWER, TRANS boîte de raccordement *f*, boîte de jonction *f*; ~ **boxes and cabinet** *n pl* SWG boîtes et coffrets de raccordement *f pl*; ~ **capacitance** ELECT *semiconductors* capacité de jonction *f*; ~ **capacitor** ELECT condensateur à jonction *m*; ~ **circuit** TELEP circuit de jonction *m*; ~ **diode** PARTS diode à jonction *f*; ~ **distribution frame** (*JDF*) SWG, TRANS répartiteur de jonction *m*; ~ **field effect transistor** (*JFET*) PARTS transistor à effet de champ à jonction *m*; ~**-line relay set** TELEP plaque de relais de ligne de jonction *f*; ~ **network** TELEP, TRANS réseau de distribution *m*, réseau de jonction *m*; ~ **tandem exchange** SWG, TELEP commutateur de transit urbain *m*; ~ **temperature** PARTS température de jonction *f*; ~ **transistor** PARTS, POWER transistor à jonction *m*
junctor *n* DATA, PARTS, SWG, TELEP, TRANS circuit de connexion *m*, joncteur *m*; ~ **block** SWG bloc de joncteurs *m*
junk: ~ **mail** *n* DATA *Internet* publicité importune *f*
justifiable: ~ **digit** *n* TELEP, TRANS élément numérique justifiable *m*; ~**-digit time slot** *n* TELEP, TRANS intervalle de temps pour élément numérique justifiable *m*, créneau temporel élémentaire justifiable *m*
justification *n* TELEP, TRANS justification *f*, justification nulle *f*; ~ **ratio** TELEP, TRANS taux de justification *m*; ~ **service digit** TELEP, TRANS élément numérique de service de justification *m*; ~ **service unit** TELEP, TRANS élément numérique de service de justification *m*; ~ **waiting time** TRANS temps d'attente de justification *m*
justify *vt* COMP cadrer, justifier, TELEP justifier
justifying: ~ **digit** *n* TELEP élément numérique de justification *m*

K

kaledia *n* COMP kaledia *m*

Karnaugh: ~ **map** *n* GEN table de Karnaugh *f*

kc *abbr* (*kilocycle*) ELECT, TEST, TRANS kHz (*kilohertz*)

keep[1]: **~-alive circuit** *n* ELECT, POWER circuit d'entretien *m*; **~-alive electrode** *n* ELECT, POWER électrode d'entretien *f*

keep[2]: ~ **track of** *vt* GEN suivre

Kelvin: ~ **effect** *n* POWER effet de peau *m*, effet Kelvin *m*, effet pelliculaire *m*

keranuic: ~ **level** *n* RADIO niveau kéranuique *m*

kernel *n* COMP noyau *m*

Kerr: ~ **effect** *n* TRANS effet Kerr *m*

kevlar: ~ **skin** *n* RADIO peau en kevlar *f*; ~ **strength** *n* ELECT renfort en kevlar *m*

key[1]: **~-controlled** *adj* COMP commandé par touche

key[2] *n* APPL *on keyboard*, COMP, DATA clé *f*, touche *f*, ELECT détrompeur *m*, GEN clé *f*, PER APP *on keyboard* touche *f*, POWER détrompeur *m*, manipulateur *m*, RADIO clé *f*, légende *f*, TELEP clé *f*, manipulateur *m*, touche *f*; **~-and-lamp unit** APPL, PER APP module à touches et voyants *m*; ~ **assign** COMP affectation de touches *f*; ~ **click** RADIO, TELEP *telegraphy* claquement de manipulation *m*; ~ **depression** COMP frappe *f*, pression sur une touche *f*; **~-down** TELEP manipulateur abaissé *m*, manipulateur baissé *m*; ~ **factor** TV facteur clé *m*; ~ **filter** COMP filtre de touche *m*; ~ **follow** COMP suivi de touche *m*; ~ **for direct selection** RADIO, TV touche de présélection *f*; ~ **for preset selection** RADIO, TV touche de présélection *f*; ~ **groove** GEN rainure de clavette *f*; ~ **label** COMP nom de la touche *m*; ~ **lock** COMP verrouillage du clavier *m*; ~ **management** DATA, TELEP gestion de clés *f*; ~ **matrix** COMP touches *f pl*, clavier sans boîtier *m*, clavier nu *m*; ~ **net** DATA réseau cryptophonique *m*; ~ **number** COMP numéro de touche *m*; **~-off** COMP relâchement de la touche *m*, touche relâchée *f*; **~-on** COMP enfoncement de touche *m*, touche enfoncée *f*; ~ **overlay** APPL, PER APP masque de touches *m*; **~-per-line console** *BrE* (*cf key-per-trunk console AmE*) APPL, PER APP poste d'opérateur de classe A *m*; **~-per-trunk console** *AmE* (*cf key-per-line console BrE*) APPL, PER APP poste d'opérateur de classe A *m*; ~ **release** COMP relâchement de la touche *m*; ~ **set** DATA poste cryptophonique *m*; **~-set telephone** TELEP poste à clavier *m*; ~ **shelf** APPL, PER APP table horizontale *f*, tablette horizontale *f*; ~ **shift** COMP décalage de hauteur *m*, décalage de ton *m*; ~ **signature** COMP armature *f*, armure de clé *f*; ~ **split** COMP division du clavier *f*; ~ **stream** APPL, PER APP train de clés *m*; ~ **synchronization** COMP synchronisation sur touche *f*; ~ **system** RADIO système d'intercom *m*; ~ **telephone set** APPL, PER APP, TELEP poste téléphonique à poussoir *m* (*Can*), téléphone à poussoirs *m* (*Can*), poste d'intercommunication *m*; ~ **telephone system** APPL, PER APP, TELEP système téléphonique à poussoirs *m*, système téléphonique à touches *m*; ~ **to callouts** DATA légende *f*; ~ **touch force** COMP force de frappe *f*; ~ **transpose** COMP transposition du clavier *f*

key[3] *vt* TELEP *entry* numéroter; ~ **in** GEN saisir, pianoter, introduire par clavier

keyboard *n* GEN clavier *m*; ~ **amplifier** COMP amplificateur de clavier *m*; ~ **controller** COMP command de clavier *f*; ~ **matrix** COMP clavier nu *m*, clavier sans boîtier *m*, touches *f pl*; ~ **perforator** GEN clavier perforateur *m*; ~ **range** COMP gamme de claviers *f*; ~ **response** COMP réponse de clavier *f*; ~ **scaling** COMP pondération de clavier *f*; ~ **scaling level** COMP niveau de pondération de clavier *m*; ~ **scaling rate** COMP taux de pondération de clavier *m*; ~ **sender** APPL, PER APP émetteur à clavier *m*; ~ **shortcut** COMP raccourci clavier *m*; ~ **split point** COMP point de partage du clavier *m*; ~ **stand** COMP support de clavier *m*; ~ **transmitter** TELEP transmetteur à clavier *m*

keyboarding *n* COMP, DATA saisie sur clavier *f*, saisie *f*

keyboardless *adj* COMP sans clavier

keyed[2]: ~ **numeral** *n* COMP nombre composé au clavier *m*

keyer *n* TELEP modulateur *m*

keyframe *n* COMP image clé *f*; ~ **animation** COMP animation par images-clés *f*; ~ **rate** COMP fréquence d'images-clés *f*

keyframing *n* COMP animation par images-clés *f*

keying *n* APPL, PER APP numérotation *f*, numérotation au clavier *f*, TELEP détrompage *m*; ~ **error** APPL, PER APP erreur de numérotation *f*; ~ **filter** RADIO filtre de manipulation *m*

keynote *n* COMP tonique *m*; ~ **field** RADIO paramètre d'identification *m*

keypad *n* APPL cadran à clavier *m*, clavier numérique *m*, clavier téléphonique *m*, COMP bloc de touches *m*, pavé numérique *m*, clavier numérique *m*, PER APP clavier numérique *m*, cadran à clavier *m*, clavier téléphonique *m*, TELEP clavier de numérotation *m*; ~ **telephone** PER APP, TELEP, appareil téléphonique à clavier *m*, TELEP poste à clavier *m*; ~ **telephone set** TELEP poste téléphonique à clavier *m*

keypunch *n* COMP perforatrice à clavier *f*

keysender *n* TELEP clavier *m*

keyset *n* APPL, PER APP, TELEP clavier de numération *m*, cadran à clavier *m*, clavier numérique *m*, clavier téléphonique *m*; ~ **telephone** PER APP, TELEP appareil téléphonique à clavier *m*

keyswitch *n* TEST dispositif de sécurité à clé *m*

keyway *n* GEN rainure de clavette *f*

keyword *n* COMP, DATA mot-clé *m*; ~ **access** COMP accès par mot-clé *m*; ~ **link** COMP lien par mot-clé *m*; ~ **list** COMP liste des mots-clés *f*

kHz *abbr (kilohertz)* ELECT, TEST, TRANS kHz (*kilohertz*)

kick: **~-the-dog loop** *n* ELECT boucle à contrôle d'itération *f*

kill[1]: ~ **file** *n* COMP fichier de session *m*, sélecteur d'options *m*

kill[2] *vt* COMP supprimer, éliminer

kilo: **~-instructions per second** *phr* (*KIPS*) COMP kilo-instructions par seconde; **~-operations per second** *phr* (*KOPS*) COMP kilo-operations par seconde

kilobaud *n* COMP kilobaud *m*

kilocycle *n* (*kc*) TEST kilohertz *m* (*kHz*)

kilostream: ~ **circuit** *n* APPL circuit moyen débit *m*

kink *n* TRANS pliure *f*

kiosk *n* DATA kiosque *m*

KIPS *abbr* (*kilo-instructions per second*) COMP kilo-instructions par seconde

klystron *n* ELECT klystron *m*, RADIO klystron *m*, tube *m*

KMS *abbr* (*knowledge management system*) COMP KMS (*système de gestion de connaissances*)

knee: **~-luminous flux** *n* ELECT flux lumineux au coude *m*, sensibilité au coude *f*; ~ **point** *n* ELECT coude *m*; ~ **sensitivity** *n* ELECT flux lumineux au coude *m*, sensibilité au coude *f*

knife: ~ **switch** *n* POWER commutateur à couteau *m*, commutateur à lame *m*

knob *n* CONTR *control devices* bouton *m*

knockout: ~ **blanking plate** *n* ELECT obturateur amovible *m*

knockouts *n pl* ELECT parties à défoncer *f pl*

know-how *n* COMP, ELECT savoir-faire *m*
knowledge: ~ **acquisition** *n* APPL, COMP acquisition de connaissance *f*; ~ **base** *n* APPL, COMP *artificial intelligence* base de connaissances *f*; ~ **chunk** *n* COMP partie de connaissance *f*, fragment de savoir *m*; ~ **engineer** *n* COMP cogniticien *m*, ~ **engineering** *n* COMP génie cognitif *m*, ingénierie de la connaissance *f*, cognitique *f*; ~ **management system** *n* (*KMS*) COMP système de gestion de connaissances *m* (*KMS*)
knowledgeware *n* COMP, DATA logiciel expert *m*
KOPS *abbr* (*kilo-operations per second*) COMP kilo-operations par seconde
KP: ~ **signal** *n* TELEP signal KP *m*
krarup: ~ **cable** *n* TRANS câble à charge continue *m*, câble krarupisé *m*; ~ **loading** *n* TRANS charge continue *f*; ~ **system** *n* TRANS procédé krarup *m*
krarupization *n* GEN, TRANS krarupisation *f*
krarupize *vt* GEN *cables*, TRANS krarupiser
krypton *n* PARTS krypton *m*

L

L *abbr* (*low*) ELECT niveau bas *m*, état bas *m*, niveau B *m*, état B *m*; **~ network** *n* POWER réseau en L *m*
LA *abbr* (*local application*) TELEP LA (*application locale*)
label[1] *n* GEN étiquette *f*; **~ frame** GEN cadre d'étiquette *m*; **~ holder** GEN porte-étiquette *m*; **~ printer** DATA imprimante d'étiquettes *f*; **~ protection** TELEP protection d'étiquette *f*; **~ set** TELEP jeu d'étiquettes *m*; **~ strip** GEN réglette porte-étiquette *f*
label[2] *vt* GEN étiqueter
labeling *AmE see labelling BrE*
labelling *n BrE* GEN moyen d'étiquetage *m*, étiquetage *m*, marquage *m*, repérage *m*
laboratory: **~ test** *n* TEST essai en laboratoire *m*, test en laboratoire *m*
laced: **~ cable fan** *n* GEN peigne *m*
lack *n* COMP, GEN manque *m*
lacquer *n* PARTS vernis *m*, *paints and varnish* laque *f*
ladder: **~ on rollers** *n* PARTS échelle roulante *f*, échelle à roulettes *f*; **~-type filter** *n* RADIO filtre en échelle *m*
lag[1] *n* COMP décalage *m*, ELECT persistance *f*
lag[2] *vt* COMP décaler, ELECT retarder
lag[3] *vi* COMP traîner
lagging: **~ phase shift** *n* POWER déphasage arrière *m*
Lambertian: **~ emission** *n* TRANS émission lambertienne *f*; **~ radiator** *n* ELECT, TRANS radiateur lambertien *m*, source lambertienne *f*; **~ reflector** *n* TRANS réflecteur lambertien *m*; **~ source** *n* ELECT source lambertienne *f*, TRANS radiateur lambertien *m*, source lambertienne *f*
laminate *n* ELECT, PARTS *materials* stratifié *m*
laminated: **~ plastic** *n* PARTS plastique stratifié *m*
laminating *n* PARTS stratification *f*
lamination *n* ELECT laminage *m*, PARTS *plastics* stratification *f*
lamp: **~ base** *n* POWER culot de lampe *m*; **~ cap** *n* POWER culot *m*; **~ display panel** *n* TELEP panneau de lampes *m*, panneau de voyants *m*; **~ lens** *n* POWER cabochon *m*; **~ panel** *n* TELEP panneau de voyants *m*, panneau de lampes *m*; **~ signal** *n* TELEP signal de lampe *m*; **~ strip** *n* POWER réglette de lampes *f*
LAN *abbr* (*local area network*) COMP, DATA, ELECT, TELEP, TRANS RLE (*réseau local d'entreprise*), réseau local *m*; **~ manager** *n* DATA gestionnaire de réseau *m*
land[1]: **~-based** *adj* RADIO terrestre
land[2] *n* ELECT pastille *f*; **~ cable** TRANS câble terrestre *m*; **~-line charge** TELEP taxe de ligne *f*; **~-mobile satellite service** (*LMSS*) RADIO, TELEP, TRANS service mobile terrestre par satellite *m* (*SMTS*); **~-mobile station** (*LMS*) RADIO, TELEP, TRANS station mobile terrestre *f* (*SMT*); **~-mobile system** RADIO système mobile terrestre *m*; **~ return** RADIO écho de terre *m*; **~ station** RADIO, TELEP station terrestre *f*; **~-station charge** RADIO, TELEP taxe terrestre *f*
landless: **~ hole** *n* ELECT *printed circuits* trou sans pastille *m*
landline *n* TRANS ligne terrestre *f*
lane *n* RADIO couloir *m*; **~ identification** RADIO identification de couloir *f*
language *n* COMP langage *m*, GEN, TELEP langue *f*; **~ digit** TELEP chiffre de langue *m*; **~ processor** COMP processeur de langage *m*; **~ translator** COMP traducteur de langage *m*
lanthanide: **~ doped fiber** *AmE*, **~ doped fibre** *BrE n* TRANS fibre dopée lanthanide *f*
lap: **~-held** *adj* COMP portable; **~-size** *adj* COMP portable
LAP *abbr* (*line access procedure*) DATA procédure d'accès à la liaison *f*, protocole d'accès à la liaison *m*
LAPB *abbr* (*link access procedure B*) TELEP procédure d'accès à la liaison symétrique *f*
lapel: **~ microphone** *n* TELEP microphone boutonnière *m*
lapper *n* POWER rodeuse *f*
lapping *n* GEN *cables* rubannage *m*; **~ film** ELECT film de polissage *m*
laptop[1] *adj* COMP portable
laptop[2]: **~ computer** *n* COMP ordinateur portable *m*
large: **~-core multiple-mode optic fiber** *AmE*, **~-core multiple-mode optic fibre** *BrE n pl* TRANS fibre optique multimode à large cœur *m*; **~-scale integration** *n* (*LSI*) ELECT, PARTS, TELEP, TRANS intégration à grande échelle *f* (*LSI*); **very ~-scale integration** *n* (*VLSI*) ELECT, PARTS, TELEP, TRANS intégration à très grande échelle *f* (*VLSI*); **~-scale integration circuit** *n* ELECT, TELEP, TRANS circuit intégré à grande échelle *m*; **very ~-scale integration circuit** *n* ELECT, TELEP, TRANS circuit à très haut niveau d'intégration *m*, circuit intégré à très grande échelle *m*; **~-scale system** *n* DATA système à grande échelle *m*; **~ screen display** *n* DATA écran surdimensionné *m*
laryngophone *n* TELEP *microphones* laryngophone *m*
laser *n* GEN laser *m*; **~ beam** ELECT faisceau laser *m*; **~ beam scanning** ELECT balayage du rayon laser *m*; **~ diode** (*LD*) ELECT, TRANS diode laser *f*; **~ medium** ELECT, TRANS milieu laser *m*; **~ plasma** ELECT plasma produit par laser *m*; **~ power** ELECT puissance laser *f*; **~ printer** APPL, COMP, DATA, PER APP imprimante laser *f*; **~ printing** GEN impression laser *f*; **~ rangefinder** DATA, TEST télémètre laser *m*; **~ telemetry** DATA, TEST télémétrie laser *f*; **~ transmitter** TRANS émetteur laser *m*; **~ trimming** ELECT ajustage au laser *m*
lasing: **~ threshold** *n* ELECT, TRANS seuil d'émission laser *m*, seuil d'effet laser *m*
last[1]: **~-choice circuit group** *n* ELECT, SWG, TELEP faisceau de dernier choix *m*; **~ known location** *n* RADIO dernière localisation connue *f*; **~-number recall** *n* PER APP, TELEP rappel automatique du dernier numéro *m*; **~ number redial key** *n* PER APP touche bis *f*; **~-party release** *n* COMP, TELEP libération par le dernier abonné *f*; **~-party release signal** *n* TELEP signal de libération par le dernier abonné *m*; **~ try** *n* TELEP dernière tentative *f*
last[2]: **~ in, first out** *phr* (*LIFO*) COMP dernier entré, premier sorti (*LIFO*)
latch *n* ELECT *logic elements* verrou *m*; **~ circuit** ELECT circuit à verrouillage *m*; **~ lock** ELECT dispositif de verrouillage *m*; **~-up** PARTS *semiconductors* verrouillage à l'état passant *m*, verrouillage *m*
latched: **~ detent position** *n* ELECT maintien verrouillé *m*
latching: **~ current** *n* POWER *thyristors* courant d'accrochage *m*; **~ relay** *n* POWER relais à enclenchement *m*
latency *n* COMP temps d'attente *m*
lateral: **~ deviation** *n* RADIO déviation latérale *f*; **~ offset loss** *n* TRANS perte par décalage latéral *f*, *fibre optics* perte par déplacement transversal *f*; **~ plasma deposition** *n* ELECT dépôt latéral au plasma *m*
lathe *n* PARTS touret *m*; **~ center** *AmE*, **~ centre** *BrE* PARTS pointe de tour *f*
lattice *n* ELECT réseau *m*, RADIO grille *f*; **~ network** POWER réseau en treillis *m*; **~ plan** RADIO plan maillé *m*
launch[1] *n* COMP lancement *m*, mise sur le marché *f*, DATA, GEN

service lancement *m*; ~ **angle** TRANS angle de lancement *m*; ~ **complex** RADIO ensemble de lancement *m*; ~ **and early orbit phase** (*LEOP*) RADIO phase de lancement et de mise en orbite *m* (*LEOP*); ~ **numerical aperture** (*LNA*) ELECT, RADIO, TRANS ouverture numérique d'injection *f* (*ONI*); ~ **rocket** RADIO fusée de lancement *m*; ~ **site** RADIO zone de lancement *f*

launch[2] *vt* GEN lancer, mettre sur le marché

launching *n* GEN lancement *m*, mise sur le marché *m*; ~ **fiber** *AmE*, ~ **fibre** *BrE* TRANS fibre amorce *f*, fibre d'injection *f*

law *n* GEN loi *f*; ~ **of combination of errors** TEST loi de composition des erreurs *f*

lay[1]: ~ **person** *n* GEN profane *m*

lay[2] *vt* TRANS poser; ~ **out** COMP mettre en page

layer *n* GEN couche *f*; ~ **of cabling elements** GEN couche d'éléments de câblage *f*; ~ **diameter** GEN *cables* diamètre d'une couche *m*; ~ **drawing** ELECT *printed circuits* plan des couches *m*; ~ **management** DATA gestion de couche *f*; ~ **of units** GEN *cables* couche de faisceaux *f*

layered: ~ **architecture** *n* COMP architecture en couches *f*

laying *n* TRANS *of loss* pose *f*

layman *n* GEN profane *m*

layout *n* COMP, DATA mise en page *f*; ~ **character** COMP, DATA caractère de présentation *m*; ~ **drawing** GEN schéma d'implantation *m*; ~ **option** TELEP option de présentation *f*; ~ **sketch** ELECT croquis d'implantation *m*, schéma d'implantation *m*

LB *abbr* (*local battery*) TELEP batterie locale *f*

LBA *abbr* (*logical block address*) COMP adresse de bloc logique *f*

LCA *abbr* (*local calling area*) TELEP zone d'appels locaux *f*

LCC *abbr* (*leadless chip-carrier*) ELECT porte-puce à souder *m*, porte-puce non-enfichable *m*

LCD *abbr* (*liquid crystal display*) COMP, ELECT écran à cristaux liquides *m*, affichage à cristaux liquides *m*; ~ **calls display** *n* PER APP visualisation des appels par LCD *f*; ~ **projection panel** *n* DATA écran plat de projection à cristaux liquides *m*

LCS: ~ **printer** *n* COMP imprimante à obturateur à cristal liquide *f*

LD *abbr* (*laser diode*) ELECT, TRANS *fibre optics* diode laser *f*

LDN *abbr* (*listed directory number*) TELEP numéro d'annuaire *m*, numéro principal *m*

LE *abbr* (*local exchange*) SWG, TELEP central local *m*, central urbain *m*, commutateur d'abonnés *m*, commutateur local *m*

lead[1] *n* ELECT, GEN fil *m*, PARTS *semiconductors* fil *m*, sortie *f*, POWER, SWG, TELEP fil *m*; ~ **accumulator** POWER *batteries* accumulateur au plomb *m*, batterie au plomb *f*; **~-in** RADIO, TELEP descente d'antenne *f*; **~-in area** COMP zone de début *f*, zone d'amorçage *f*; **~-in insulator** POWER isolateur de traversée *m*; **~-peak ballast** POWER ballast capacitif à crête d'amorçage *m*; ~ **sheath** GEN *cables* gaine de plomb *f*; **~-sheathed cable** POWER câble sous plomb *m*; ~ **sheathing** GEN gainage de plomb *m*; ~ **sleeve** GEN manchon de plomb *m*; ~ **storage battery** POWER accumulateur au plomb *m*, batterie au plomb *f*

lead[2] *vt* ELECT *voltage* devancer

leaded: ~ **chip-carrier** *n* ELECT porte-puce enfichable *m*

leader *n* DATA amorce *f*, amorce de bande *f*, amorce de bande magnétique *f*

leading[1] *adj* GEN prépondérant

leading[2]: ~ **edge** *n* ELECT, GEN, PARTS, TELEP *pulses* front avant *m*; **~-edge position** *n* TELEP *pulses* position de flanc avant *f*; **~-edge technology** *n* GEN technologie de pointe *f*; ~ **phase shift** *n* POWER déphasage avant *m*

leadless: ~ **chip-carrier** *n* (*LCC*) ELECT porte-puce à souder *m*, porte-puce non-enfichable *m*

leaf *n* DATA feuille *f*; ~ **node** DATA nœud feuille *m*; ~ **spring** GEN, PARTS ressort à lame *m*

leafleting *n* TV cerclage *m*; ~ **periodicity** TV rythme de cerclage *m*

leak *n* TELEP fuite *f*; ~ **time** TELEP temps de fuite *m*

leakage *n* ELECT, POWER fuite *f*; ~ **current** ELECT courant de fuite *m*; ~ **field** ELECT champ de fuite *m*; ~ **resistance** POWER résistance de fuite *f*; ~ **test** POWER essai d'isolement *m*, essai d'étanchéité *m*

leaky: ~ **mode** *n* TRANS mode de fuite *m*, mode tunnel *m*; ~ **ray** *n* TELEP rayon de fuite *m*, TRANS rayon tunnel *m*

leap: ~ **second** *n* RADIO, TEST seconde intercalaire *f*

leapfrog: ~ **test** *n* TEST test sélectif *m*

leased: ~ **circuit** *n* TELEP circuit loué *m*, circuit spécialisé *m*, TRANS circuit loué *m*; **~-circuit data transmission service** *n* TRANS service de location de circuits pour transmission de données *m*; ~ **line** *n* (*LS*) DATA, TELEP, TRANS ligne spécialisée *f* (*LS*)

leasing *n* TV crédit-bail *m*

least: ~ **significant bit** *n* (*LSB*) DATA bit de plus faible poids *m*; ~ **significant digit** *n* (*LSD*) DATA élément numérique de poids faible *m*

lecture: ~ **call** *n* TELEP service de communication conférence unilatérale *m*

LED *abbr* COMP (*letter shift*) inversion de lettres *f*, ELECT, PARTS, PER APP, TELEP (*light-emitting diode*) DEL (*diode électroluminescente, diode émettrice de lumière*), (*letter shift*) *telegraphy* inversion lettres *f*, TRANS (*light-emitting diode*) DEL (*diode électroluminescente, diode émettrice de lumière*); ~ **display** *n* DATA, ELECT affichage à diodes électroluminescentes *m*; ~ **printer** *n* DATA imprimante à diodes électroluminescentes *f*; ~ **read-out** *n* COMP affichage à diodes électroluminescentes *m*

left[1]: **~-hand polarization** *n* RADIO polarisation senestrorsum *f*, polarisation lévogyre *f*; **~-right balance** *n* COMP balance droite-gauche *f*; ~ **side of the front panel** *n* ELECT renforcement à gauche *m*

left[2]: **~-justify** *vt* COMP, DATA cadrer à gauche, COMP *word processing* justifier à gauche

leg *n* TELEP fil *m*

legible *adj* GEN lisible

Lempel-Ziv-Welch *n* (*LZW*) COMP Lempel-Ziv-Welch *m* (*LZW*)

length *n* GEN longueur *f*, TRANS *of cable* tronçon *m*; ~ **of call** TELEP longueur d'appel *f*; ~ **of fracture** PARTS, TEST longueur de rupture *f*; ~ **indicator** (*LI*) DATA indicateur de longueur *m*; **~-indicator field** DATA champ d'indicateur de longueur *m*; ~ **of kevlar** ELECT longueur de kevlar *f*; ~ **of lay** GEN *cables* pas d'assemblage *m*; ~ **of a span** POWER longueur de portée *f*

lens *n* COMP lentille *f*, RADIO lentille radioélectrique *f*; ~ **aerial** RADIO antenne à lentille *f*; ~ **antenna** RADIO antenne à lentille *f*

LEO: ~ **satellite** *n* (*low earth orbit satellite*) RADIO satellite en orbite basse *m*, satellite sur orbite à basse altitude *m*

LEOP *abbr* (*launch and early orbit phase*) RADIO LEOP (*phase de lancement et de mise en orbite*)

letter: ~ **quality** *n* (*LQ*) COMP qualité courrier *f*; ~ **shift** *n* (*LED*) COMP inversion de lettres *f*, TELEP inversion lettres *f*; ~ **shift signal** *n* TELEP signal d'inversion lettres *m*

level[1] *n* DATA palier *m*, GEN niveau *m*, palier *m*, TELEP gamme *f*; ~ **attenuator** COMP affaiblisseur de niveau *m*, atténuateur de niveau *m*; ~ **control** COMP commande de niveau *f*; ~ **hunt** TELEP recherche multiniveau *f*; ~ **of indentation** DATA niveau d'indentation *m*; ~ **of maintenance** TELEP niveau de maintenance *m*; ~ **measuring set** TRANS hypsomètre *m*; ~ **meter** TEST hypsomètre *m*; ~ **of mobility** RADIO niveau de mobilité *m*; ~ **monitor** TRANS contrôleur de niveau *m*; ~ **monitoring** RADIO contrôle de niveau *m*; ~ **recorder** TEST, TRANS enregistreur de niveau *m*; **~-selecting bar** SWG barre de sélection de niveau de déclenchement *f*; **~-selecting horizontal bar** SWG barre de sélection d'aiguillage de niveau *f*; ~ **shifting** TELEP translation de niveau *f*, rétablissement de niveau zéro *m*; ~ **translator**

ELECT translateur de niveau *m*; ~ **variation** COMP variation de niveau *f*

level[2] *vt* GEN *construction* niveler

lever *n* GEN levier *m*, manette *f*

LF *abbr* COMP, DATA (*line feed*) changement de ligne *m*, saut de ligne *m*, RADIO, SWG (*low frequency*) BF (*basse fréquence*), TELEP (*line finder*) chercheur d'appel *m*, chercheur de ligne *m*, chercheur de ligne appelante *m*

LFA *abbr* (*loss of frame alignment*) TELEP, TRANS PVT (*perte de verrouillage de trame*)

LFC *abbr* (*local functional capabilities*) SWG CLF (*capacités de fonctionnement locales*)

LFN *abbr* (*logical file name*) COMP nom logique de fichier *m*

LI *abbr* (*length indicator*) DATA indicateur de longueur *m*

liaison *n* GEN liaison *f*, transmission *f*

LIB *abbr* (*line interface board*) SWG carte d'interface de ligne *f*

liberalization *n* TV déréglementation *f*

library *n* COMP, DATA bibliothèque *f*; ~ **of data** DATA bibliothèque de données *f*; ~ **program** COMP programme de bibliothèque *m*

LIC *abbr* (*line interface circuit*) SWG circuit d'interface de ligne *m*, circuit d'interface de transmission *m*, circuit de ligne *m*, TELEP circuit d'interface de transmission *m*, circuit de ligne *m*

licence *n BrE* GEN redevance *f*; ~ **restriction** *BrE* DATA *Internet* contrainte de licence *f*

license[1] *n AmE see licence BrE*

license[2] *vt* GEN autoriser

lid *n* COMP capot *m*, PARTS couvercle *m*

lie: ~ **against** *vt* PARTS porter sur, reposer sur

life[1]: ~**-size** *adj* GEN grandeur nature

life[2] *n* TEST *dependability* vie utile *f*; ~**-length distribution function** TEST fonction de répartition de la durée de vie *f*; ~ **testing** TEST essais d'endurance *m pl*, *dependability* essai de durée de vie *m*

lifespan *n* RADIO, TEST durée de vie *f*

lifetime *n* RADIO, TEST durée de vie *f*

LIFO *abbr* (*last in, first out*) COMP LIFO (*dernier entré, premier sorti*)

lift[1]: ~**-out door** *n* ELECT porte amovible *f*

lift[2] *vt* APPL, PER APP, TELEP décrocher

lifted: ~ **chip** *n* COMP puce décollée *f*; ~ **land** *n* ELECT décollement de pastille *m*

lifting *n* APPL, PER APP, TELEP décrochage *m*; ~ **bar** SWG barre de levage de verticale *f*; ~ **comb** SWG peigne mobile *m*; ~ **spring** SWG ressort de commande *m*

light[1]: ~**-sensitive** *adj* ELECT photosensible

light[2] *n* ELECT blanc *m*, GEN lumière *f*, TRANS rayonnement visible *m*; ~ **beam** ELECT faisceau lumineux *m*; ~ **current** ELECT, POWER, TRANS courant de faible puissance *m*, courant faible *m*, courant photoélectrique *m*, photocourant *m*; ~**-current engineering** POWER technique des courants faibles *f*; ~ **detector** ELECT photodétecteur *m*, photorécepteur *m*, TV capteur de lumière *m*; ~ **emitter** TV émetteur lumière *m*; ~**-emitting diode** (*LED*) ELECT, PARTS diode électroluminescente *f* (*DEL*), diode émettrice de lumière *f* (*DEL*), PER APP voyant lumineux *m*, TELEP, TRANS diode électroluminescente *f* (*DEL*), diode émettrice de lumière *f* (*DEL*); ~**-emitting diode panel** ELECT tableau de diodes électroluminescentes *m*; ~ **fitting** GEN luminaire *m*, appareil d'éclairage *m*; ~**-guide distribution system** ELECT système de distribution optique *m*; ~ **indicator** COMP voyant *m*, GEN signal lumineux *m*, PER APP témoin lumineux *m*; ~ **interferometer** DATA interféromètre optique *m*; ~ **microscope** TV *optics* microscope optique *m*; ~ **output** TEST rendement lumineux *m*; ~ **pen** APPL, COMP crayon optique *m*, pointeur optique *m*, PER APP photostyle *m*; ~**-pen detection** COMP détection par photostyle *f*, détection par crayon optique *f*; ~ **propagation** TRANS propagation de la lumière *f*; ~ **pulse** TV impulsion lumineuse *f*; ~ **ray** TRANS rayon lumineux *m*, rayon optique *m*, rayon *m*; ~**-signal transfer characteristic** ELECT caractéristique de transfert de signal lumineux *f*; ~ **source** TV *optics* source lumineuse *f*; ~**-to-medium network traffic** ELECT trafic réseau de petite et moyenne envergure *m*; ~**-wave regenerator** DATA régénérateur d'ondes lumineuses *m*

light[3]: ~ **up** *vi* COMP s'allumer

lighted: ~ **display** *n* COMP affichage lumineux *m*

lightguide *n* TRANS guide d'ondes optiques *m*, guide optique *m*; ~ **building cable** ELECT gaine de câble optique d'immeuble *f*; ~ **express entry** (*LXE*) ELECT LXE; ~ **interconnection unit** (*LIU*) ELECT boîte d'interconnexion optique *f*

lighting *n* POWER, TV éclairage *m*; ~ **fitting** CLIM appareil d'éclairage *m*, GEN luminaire *m*; ~ **fixture** POWER appareil d'éclairage *m*, luminaire *m*; ~ **model** COMP modèle d'éclairement *m*

lightness *n* COMP teinte *f*, saturation *f*, GEN luminosité *f*

lightning: ~ **arrester** *n* POWER, TEST parafoudre *m*; ~ **conductor** *n* POWER parafoudre *m*, TEST paratonnerre *m*, tige de paratonnerre *f*; ~ **impulse test** *n* POWER essai de tenue à la foudre *m*; ~ **rod** *n* TEST paratonnerre *m*, tige de paratonnerre *f*

lightwave: ~ **stream** *n* ELECT faisceau optique *m*

lightweight: ~ **headset** *n* TELEP microcasque *m*; ~ **telephone headset** *n* TELEP microcasque de téléphone *m*

limit *n* DATA, GEN, PARTS limite *f*, TEST gabarit *m*; ~ **switch** POWER fin de course *f*, interrupteur de fin de course *m*

limitation *n* GEN limite *f*

limited: ~ **access** *n* GEN accès limité *m*; ~**-access channel** *n* TV canal à code d'accès *m*; ~ **availability** *n* SWG disponibilité limitée *f*; ~ **bandwidth** *n* DATA bande passante limitée *f*; ~ **multilevel image** *n* TELEP image limitée à quelques niveaux *f*; ~ **roaming telephone service** *n* TELEP service radiotéléphonique mobile à courte portée *m*; ~ **scanning** *n* TELEP exploration réduite *f*; ~**-waiting queue** *n* TELEP file d'attente limitée *f*

limiter *n* ELECT, POWER, RADIO, TELEP limiteur *m*

limiting: ~ **current** *n* POWER courant coupé limité *m*; ~ **state** *n* TEST *dependability* état limite *m*

limits: ~ **for maintenance purposes** *n pl* TELEP limites de maintenance *f pl*

line[1]: **on the** ~ *adj* TELEP au bout du fil

line[2]: **off** ~ *adv* COMP, DATA, TELEP en différé; **on** ~ *adv* COMP, DATA, TELEP en ligne

line[3]: ~ **access point** *n* TELEP point d'accès à la ligne *m*; ~ **access procedure** *n* (*LAP*) DATA procédure d'accès à la liaison *f*, protocole d'accès à la liaison *m*; ~ **amplifier** *n* TV amplificateur de ligne *m*; ~ **analyser** *BrE*, ~ **analyzer** *AmE* *n* TEST analyseur de lignes *m*; ~ **balance** *n* TRANS équilibre de ligne *m*; ~ **bend correction** *n* TV signal de correction parabolique de ligne *m*; ~ **blanking interval** *n* DATA, TV intervalle de retour de ligne *m*, intervalle de supression de ligne *m*; ~ **of center** *AmE*, ~ **of centre** *BrE n* TRANS ligne médiane *f*; ~ **circuit** *n* SWG, TELEP circuit de ligne *m*, joncteur d'abonné *m*; ~ **circuit pack** *n* TELEP carte de ligne intérieure *f*; ~ **code** *n* DATA, TELEP code en ligne *m*, TRANS code de transmission *m*; ~ **concentrator** *n* TELEP, TRANS concentrateur d'abonnés *m*, concentrateur de lignes *m*; ~ **connection unit** *n* TRANS unité de raccordement de ligne *f*; ~ **connector cord** *n* COMP cordon secteur *m*; ~ **cord** *n* COMP, ELECT, POWER cordon d'alimentation *m*; ~ **current** *n* COMP courant de ligne *m*, courant secteur *m*; ~ **decoder** *n* TRANS décodeur en ligne *m*; ~ **digit rate** *n* TRANS débit numérique de ligne *m*; ~ **digital terminal** *n* TELEP terminal numérique de ligne *m*; ~ **drawing** *n* DATA *scanner* image au trait *f*; ~ **equipment** *n* TRANS ligne équipée *f*; ~ **fault** *n* TELEP dérangement d'une ligne *m*; ~ **feed** *n* (*LF*) COMP, DATA, RADIO, SWG, TELEP *telegraphy* changement de ligne *m*, chercheur de ligne *m*, saut de ligne *m*; ~**-feed character** *n*

DATA avancement d'un interligne *m*, caractère de changement de ligne *m*; ~ **filter** *n* COMP filtre antiparasite *m*, filtre secteur *m*, POWER filtre de ligne *m*; ~ **finder** *n* (*LF*) COMP, DATA, RADIO, SWG, TELEP chercheur d'appel *m*, chercheur de ligne *m*, chercheur de ligne appelante *m*; ~-**finder stage** *n* TRANS étage de chercheurs de ligne *m*; ~ **frequency** *n* COMP fréquence d'exploration horizontale *f*, TV fréquence de balayage horizontal *f*, fréquence de ligne *f*, fréquence du secteur *f*; ~ **frequency generator** *n* TV générateur de fréquence de balayage de ligne *m*, générateur de fréquence de ligne *m*; ~ **group** *n* SWG faisceau de lignes *m*, TRANS groupe de lignes *m*; ~ **hunting** *n* TELEP recherche de ligne *f*; ~-**in port** *n* PER APP accès d'entrée *m*, borne d'entrée de la ligne *f*; ~ **input** *n* COMP entrée de ligne *f*; ~ **interface board** *n* (*LIB*) SWG carte d'interface de ligne *f*; ~ **interface circuit** *n* (*LIC*) SWG circuit d'interface de transmission *m*, circuit de ligne *m*, circuit d'interface de ligne *m*, TELEP circuit de ligne *m*; ~ **interface module** *n* SWG unité de raccordement d'abonnés *f*; ~ **keystone correction** *n* TV signal de correction de trapèze-ligne *m*; ~ **lamp** *n* TELEP lampe d'appel *f*; ~ **lengthener** *n* RADIO extenseur de ligne à phase variable *m*; ~ **link** *n* TELEP liaison en ligne *f*; ~ **load control** *n* (*LLC*) SWG, TELEP contrôle de charge de ligne *m* (*CCL*); ~ **lockout** *n* TELEP neutralisation de ligne *f*; ~-**lockout warning tone** *n* TELEP tonalité d'avertissement de neutralisation de ligne *f*; ~ **of maintenance** *n* TELEP ligne de maintenance *f*; ~ **module** *n* SWG module de ligne *m*; ~-**module equipment** *n* (*LME*) SWG matériel de modules de lignes *m*; ~ **network** *n* TELEP réseau de lignes *m*; ~ **noise** *n* COMP bruit de ligne *m*; ~ **non-linearity** *n* TRANS distorsion de vitesse de balayage de ligne *f*, distorsion de balayage horizontal *f*; ~-**of-sight** *n* TRANS axe de visée *m*, portée optique *f*, visibilité directe *f*; ~-**of-sight propagation** *n* TRANS propagation en visibilité *f*; ~-**of-sight range** *n* TRANS portée optique *f*; ~-**out** *n* COMP sortie de ligne *f*; ~ **out-of-service signal** *n* SIG signal de ligne hors service *m*; ~ **output** *n* COMP sortie de ligne *f*; ~ **polarity inversion** *n* TRANS inversion de polarité de ligne *f*; ~ **polarity reversal** *n* TRANS inversion de polarité de ligne *f*; ~ **printer** *n* COMP, PER APP imprimante ligne *f*, imprimante ligne par ligne *f*, imprimante par ligne *f*; ~ **printing** *n* COMP, TELEP impression ligne par ligne *f*; ~ **rental** *n* PER APP *of telephone* abonnement *m*; ~ **repeater** *n* TELEP répéteur de ligne *m*; ~ **resistance** *n* TRANS résistance de ligne *f*; ~ **reversal** *n* TELEP inversion de charge en ligne *f*; ~ **scan** *n* COMP balayage par lignes *m*, balayage horizontal *m*; ~ **scratches** *n* TELEP crachement *m*, friture *f*; ~ **section** *n* TRANS section de ligne *f*; ~ **seizure button** *n* PER APP touche prise de ligne *f*; ~-**sequential drive** *n* ELECT excitation ligne par ligne *f*, excitation séquentielle *f*; ~ **sharing** *n* TELEP partage de ligne *m*; ~ **shift** *n* TELEP *telex* inversion de ligne *f*; ~ **signaling** *AmE*, ~ **signalling** *BrE* *n* SIG signalisation de ligne *f*; ~ **signaling code** *AmE*, ~ **signalling code** *BrE* *n* DATA, SIG (*LSC*) code de signalisation de ligne *m*; ~ **signaling equipment** *AmE*, ~ **signalling equipment** *BrE* *n* PER APP, TELEP équipement de signalisation de ligne *m*; ~ **signaling system** *AmE*, ~ **signalling system** *BrE* *n* CONTR, SIG système de signalisation de ligne *m*; ~ **signaling test** *AmE*, ~ **signalling test** *BrE* *n* TEST essai de signalisation de ligne *m*; ~ **simulator** *n* TRANS simulateur de câble *m*, ligne artificielle *f*; ~ **slip** *n* TV glissement horizontal de l'image *m*; ~ **space** *n* COMP *word processing* interligne *m*; ~ **spacing** *n* COMP *word processing* interligne *m*; ~ **spectrum** *n* ELECT, TRANS *fibre optics* spectre de raies *m*; ~-**stabilized oscillator drive** *n* ELECT pilote à ligne résonante *m*; ~ **stretcher** *n* RADIO extenseur de ligne à phase variable *m*; ~ **sweep** *n* TV balayage de ligne *m*; ~ **switch** *n* SWG commutateur de lignes *m*; ~ **switching** *n* SWG permutation de liaison *f*; ~-**synchronizing signal** *n* SIG signal de synchronisation de ligne *m*; ~ **system** *n* TELEP groupe quaternaire *m*, TRANS système de ligne *m*; ~ **tapping** *n* TELEP piquage de ligne *m*; ~ **terminal** *n* TELEP terminal de ligne *m*; ~-**terminal station** *n* TRANS station terminal de ligne *f*; ~-**terminating equipment** *n* PER APP, TELEP équipement de terminaison de ligne *m*; ~-**terminating magazine** *n* TRANS panier de terminaison de ligne *m*; ~ **termination** *n* (*LT*) TELEP, TRANS terminaison de ligne *f*; ~ **test program** *AmE*, ~ **test programme** *BrE* *n* TELEP programme d'essai de ligne *m*; ~ **test set** *n* TEST banc d'essai de ligne *m*; ~ **tester** *n* TEST testeur de ligne *m*; ~ **test-point** *n* TELEP point d'essai de ligne *m*; ~ **tilt correction** *n* TV signal de correction linéaire *m*; ~-**time luminance distortion** *n* COMP distorsion de luminance horizontale *f*; ~ **unit** *n* TELEP ligne *f*; ~-**up** *n* TRANS réglage initial *m*; ~-**up instructions** *n pl* TRANS instructions de réglage initial *f pl*; ~-**up period** *n* TELEP période de réglage *f*; ~-**up procedure** *n* TEST, TRANS méthode de réglage *f*, procédure de réglage *f*; ~-**up record** *n* TRANS formulaire de réglage *m*; ~ **width** *n* ELECT *printed circuits* largeur de trait *f*

linear[1] *adj* GEN linéaire

linear[2]: ~ **amplifier** *n* TRANS amplificateur linéaire *m*; ~ **amplitude distortion** *n* TRANS distorsion linéaire d'amplitude *f*; ~ **analog control** *n* TELEP, TRANS mode analogue linéaire *m*, mode analogique linéaire *m*; ~ **approximation** *n* ELECT approximation linéaire *f*; ~ **arithmetic synthesis** *n* COMP synthèse arithmétique linéaire *f*; ~-**array aerial** *n* RADIO antenne réseau rectiligne *f*; ~-**array antenna** *n* RADIO antenne réseau rectiligne *f*; ~ **bit density** *n* COMP densité d'enregistrement binaire *f*; ~-**block code** *n* SIG code de blocs linéaires *m*; ~ **channel** *n* TRANS canal linéaire *m*; ~ **circuit** *n* ELECT, TRANS circuit linéaire *m*; ~ **code** *n* DATA code linéaire *m*; ~ **control** *n* COMP, CONTR commande linéaire *f*; ~ **curve** *n* COMP courbe linéaire *f*; ~ **detection** *n* ELECT détection linéaire *f*; ~ **detector** *n* ELECT détecteur linéaire *m*; ~-**digital voice scrambler** *n* TELEP brouilleur numérique linéaire de la voix *m*; ~ **distortion** *n* TRANS distorsion linéaire *f*; ~ **expansion** *n* GEN *physics* dilatation linéaire *f*; ~ **expansion coefficient** *n* ELECT, PARTS coefficient de dilatation linéaire *m*; ~ **filter** *n* TRANS filtre linéaire *m*; ~ **filtering** *n* TRANS filtrage linéaire *m*; ~ **integrated circuit** *n* ELECT circuit intégré linéaire *m*; ~ **interpolation** *n* TRANS interpolation linéaire *f*; ~ **joint** *n* GEN épissure longitudinale *f*; ~ **measure** *n* TEST mesure de longueur *f*, mesure linéaire *f*; ~ **oscillation** *n* TRANS oscillation linéaire *f*; ~ **path** *n* COMP parcours linéaire *m*; ~ **polarization** *n* POWER, RADIO, TRANS polarisation linéaire *f*, polarisation rectiligne *f*; ~ **power amplifier** *n* TRANS amplificateur de puissance linéaire *m*; ~ **predicting coding** *n* (*LPC*) DATA codage prédictif linéaire *m*; ~ **predicting coding vocoder** *n* RADIO codeur à fréquences vocales LPC *m*; ~ **predictive vocoder** *n* TELEP codeur prédictif linéaire *m*; ~ **rack space** *n* ELECT rack linéaire *m*; ~ **reading** *n* COMP lecture linéaire *f*; ~ **receiver** *n* RADIO récepteur linéaire *m*; ~ **relation** *n* GEN *mathematics* relation linéaire *f*; ~ **relationship** *n* GEN *mathematics* relation linéaire *f*; ~ **scale** *n* TEST échelle linéaire *f*; ~-**scaling calculation** *n* DATA calcul d'échelle linéaire *m*; ~ **slider** *n* COMP commande linéaire *f*; ~ **synthesizer** *n* COMP synthétiseur linéaire *m*; ~ **system** *n* GEN système linéaire *m*; ~ **video** *n* TV vidéo linéaire *f*

linearity *n* GEN linéarité *f*; ~ **error** PARTS *semiconductors* erreur de linéarité *f*

linearizer *n* TRANS linéarisateur *m*

linearly: ~ **increasing voltage** *n* POWER tension en rampe *f*, tension à croissance linéaire *f*; ~-**polarized mode** *n* (*LP mode*) TRANS mode à polarisation rectiligne *m* (*mode LP*)

lines: ~ **per screen** *phr* TV lignes par image *f pl*

lingering: ~ **effect** *n* TV effet d'allongement *m*

linguistic: ~ **engineering** *n* COMP génie linguistique *m*

lining *n* PARTS doublure *f*

link[1] *n* COMP *between databases* lien *m*, ELECT *connection* raccordement *m*, RADIO *link-up* liaison *f*, SWG *in cable chain* maille *f*, TELEP, TRANS liaison *f*, lien *m*, maillon *m*; ~ **access procedure B** (*LAPB*) TELEP procédure d'accès à la liaison symétrique *f*; ~ **budget** RADIO, TRANS bilan de liaison *m*; ~**-by-link encipherment** DATA chiffrement de liaison par liaison *m*, chiffrement section par section *m*; ~**-by-link signaling** *AmE*, ~**-by-link signalling** *BrE* SIG, TRANS signalisation section par section *f*; ~**-by-link traffic routing** TELEP acheminement section par section *m*; ~ **coupling** RADIO couplage de liaison *m*; ~ **encryption** GEN *cryptology* chiffrement de liaison *m*; ~ **file** COMP fichier de liens *m*, module de liens *m*; ~**-handling module** COMP, DATA module de gestion des liaisons *m*; ~ **icon** COMP icône de liaison *f*; ~ **layer** DATA, TRANS couche liaison *f*; ~ **map** COMP fichier de liens *m*, module de liens *m*; ~ **margin** TRANS marge applicable à une liaison *f*; ~ **marker** COMP indicateur de lien *m*; ~ **network** COMP réseau de liens *m*; ~ **power budget** RADIO, TRANS bilan de puissance d'une liaison *m*; ~ **status signal unit** SIG trame sémaphore d'état du canal sémaphore *f*; ~ **system** TRANS système à mailles *m*; ~ **test** TEST essai de mailles *m*; ~ **test equipment** SWG, TEST dispositif d'essai de liaison *m*

link[2] *vt* COMP relier, GEN lier; ~ **up** DATA, TRANS relier, *Internet* se relier

linkage *n* COMP lien *m*; ~ **editor** COMP éditeur de liens *m*

linked[1]: ~ **together** *adj* SWG, TRANS maillé

linked[2]: ~ **offices** *n pl* TELEP bureaux connectés *m pl*; ~ **operations** *n pl* DATA opérations liées *f pl*

linker *n* COMP éditeur de liens *m*

linking *n* ELECT, TRANS maillage *m*; ~ **clusters of balun adaptors** *n pl* ELECT grappe d'adaptateur d'impédance *f*

links: ~ **menu** *n* COMP menu des liens *m*

linkway *n* COMP linkway *m*

lint: ~**-free cloths** *n pl* ELECT chiffons non-pelucheux *m pl*

lipping *n* COMP bords recourbés *m pl*

LIPS *abbr* (*logical inferences per second*) TEST inférences logiques par seconde

liquid: ~ **caustic soda** *n* PARTS soude caustique *f*; ~ **crystal display** *n* (*LCD*) DATA, ELECT écran à cristaux liquides *m*, affichage à cristaux liquides *m*; ~ **crystals** *n pl* ELECT, PARTS cristaux liquides *m pl*; ~**-level indicator** *n* TELEP indicateur de niveau de liquide *m*; ~**-sample counter** *n* RADIO compteur à source liquide *m*

LI/S *abbr* (*logical inferences per second*) TEST inférences logiques par seconde

list: ~ **abstract-operation** *n* DATA opération abstraite de listage *f*

listed: ~ **directory number** *n* (*LDN*) TELEP numéro d'annuaire *m*, numéro principal *m*

listen *vt* TELEP écouter à

listener: ~ **echo** *n* TELEP écho à la réception *m*; ~ **echo loss** *n* TELEP affaiblissement de l'écho à la réception *m*

listening *n* RADIO, TELEP, TEST écoute *f*; ~**-in** RADIO, TELEP, TRANS écoute *f*; ~ **jack** TELEP jack d'écoute *m*; ~ **level** COMP niveau d'écoute *m*; ~ **silence** RADIO silence écoute *m*; ~ **test** RADIO essai d'écoute *m*

listing *n* COMP, DATA listage *m*, listing *m*

lit *adj* COMP allumé

literal *n* COMP littéral *m*, symbole littéral *m*

LIU *abbr* (*lightguide interconnection unit*) ELECT boîte d'interconnexion optique *f*

live[1] *adj* COMP réel, DATA en direct, POWER sous tension

live[2]: ~ **conversation** *n* DATA *Internet* conversation en direct *f*; ~ **running** *n* COMP traitement réel *m*; ~**-situation traffic planning** *n* TELEP planification du trafic en situation réelle *f*; ~ **traffic test** *n* TEST essai de trafic opérationnel *m*, essai de trafic réel *m*; ~ **transmission** *n* TRANS, TV transmission en direct *f*; ~ **TV** *n* TRANS, TV TV en direct *f*; ~ **video coverage** *n* TV couverture vidéo en direct *f*

live[3]: ~ **wire** *phr* POWER *warning sign* conducteur sous tension

LLC *abbr* CONTR (*logical link control*) commande de liaison logique *f*, DATA (*low-layer compatibility*) compatibilité de couche inférieure *f*, SWG, TELEP (*line load control*) CCL (*contrôle de charge de ligne*), (*low-layer compatibility*) compatibilité de couche inférieure *f*

LLC-IE *abbr* (*low-layer compatibility information element*) SWG, TELEP LLC-IE (*élément d'information de compatibilité de couche inférieure*)

LME *abbr* SWG (*line module equipment*) matériel de modules de lignes *m*

LMS *abbr* (*land-mobile station*) RADIO, TELEP, TRANS SMT (*station mobile terrestre*)

LMSS *abbr* (*land-mobile satellite service*) RADIO, TELEP, TRANS SMTS (*service mobile terrestre par satellite*)

lm/W *abbr* (*lumen per watt*) TEST lm/W (*lumen par watt*)

LN *abbr* (*load number*) TELEP IC (*indice de charge*)

LNA *abbr* ELECT, RADIO, TRANS (*launch numerical aperture*) ONI (*ouverture numérique d'injection*), (*low-noise amplifier*) AFB (*amplificateur à faible bruit*)

LO *abbr* (*local oscillator*) ELECT, RADIO oscillateur local *m*

load[1]: ~**-independent** *adj* POWER indépendant de la charge

load[2] *n* GEN charge *f*; ~ **capacitance** ELECT capacité de charge *f*; ~ **capacity** ELECT capacité de charge *f*; ~ **circuit** ELECT circuit de charge *m*; ~ **current** POWER courant de charge *m*; ~ **curve diagram** POWER diagramme de courbe de charge *m*; ~ **cycle** POWER cycle de chargement *m*; ~ **distribution** POWER répartition de charge *f*; ~ **factor** POWER facteur d'utilisation *m*, facteur de charge *m*; ~ **function** POWER diagramme de charge *m*, caractéristiques en charge *f pl*; ~ **host** ELECT hôte de chargement *m*; ~ **impedance** ELECT impédance de charge *f*; ~ **number** (*LN*) TELEP indice de charge *m* (*IC*); ~ **observation** TELEP observation de charge *f*; ~ **power** POWER puissance de sortie utile *f*; ~ **range** POWER régime de charge *m*; ~ **resistor** PARTS, POWER résistance de charge *f*, résistance de plaque *f*; ~ **sharing** ELECT répartition de charge *f*, partage de charge *m*; ~**-sharing system** ELECT système en partage de charge *m*; ~ **tester** TEST testeur de trafic *m*; ~ **transfer** ELECT transfert de charge *m*; ~**-transfer acknowledgement signal** SIG signal d'accusé de réception des multiblocs *m*; ~**-transfer signal** SIG signal de transfert de charge *m*

load[3] *vt* COMP charger, TELEP, TRANS pupiniser; ~ **continuously** GEN *cables*, TRANS krarupiser

loaded[1] *adj* COMP chargé, TELEP en circuit fermé

loaded[2]: ~ **cable** *n* TELEP, TRANS câble blindé *m*, câble pupinisé *m*

loading *n* RADIO charge d'une ligne *f*, TRANS pupinisation *f*; ~ **cartridge** COMP chargement de cartouche *m*; ~ **coil** TRANS bobine de charge *f*, bobine Pupin *f*; ~**-coil case** TRANS boîte de bobine de charge *f*, pot de charge *m*; ~**-coil pot** TRANS boîte de charge *f*, pot de charge *m*; ~ **error** COMP erreur de chargement *f*; ~ **operation** COMP opération de chargement *f*; ~ **single** COMP chargement unitaire *m*

lobe *n* RADIO lobe *m*; ~ **attachment module** ELECT raccordement de lobes enfichables *m*; ~ **length** ELECT longueur de lobe *f*

local: ~ **action** *n* POWER décharge en circuit ouvert *f*; ~ **application** *n* (*LA*) TELEP application locale *f* (*LA*); ~ **area network** *n* (*LAN*) COMP, DATA, ELECT, TELEP, TRANS réseau local *m*, réseau local d'entreprise *m* (*RLE*); ~ **authority** *n* TV collectivité locale *f*; ~ **battery** *n* (*LB*) TELEP batterie locale *f*; ~ **battery system** *n* TELEP système téléphonique à batterie locale *m*; ~ **bus** *n* COMP bus local *m* (*BL*); ~ **bus card** *n* COMP carte sur bus local *f*; ~ **call** *n* TELEP appel local *m*, communication locale *f*, communication urbaine *f*; ~ **calling area** *n* (*LCA*) TELEP zone d'appels locaux *f*; ~ **central office** *n* TELEP centre de commutation local *m*, central local *m*; ~ **channel** *n* TV canal local *m*; ~ **charge-rate call** *n* TELEP appel à prix partagé *m*;

~ **communications network** *n* TRANS réseau de transmission local *m*; ~ **control** *n* CONTR, TRANS commande locale *f*; ~ **copy** *n* TRANS copie locale *f*; ~ **CSN** *n* TELEP CSN local *m*; ~ **digital concentrator** *n* TELEP concentrateur numérique local *m*; ~ **distribution cable** *n* TRANS câble de distribution secondaire *m*; ~ **distribution network** *n* TRANS réseau local de raccordement *m*; ~ **echo** *n* COMP écho local *m*; ~ **end** *n* TELEP ensemble terminal *m*; ~ **exchange** *n* (*LE*) SWG, TELEP central local *m*, central urbain *m*, commutateur d'abonnés *m*, commutateur local *m*; ~ **exchange area** *n* SWG, TELEP circonscription *f*, réseau urbain *m*, zone à autonomie d'acheminement *f* (*ZAA*); ~ **functional capabilities** *n* (*LFC*) SWG capacités de fonctionnement locales *f* (*CLF*); ~ **index** *n* TELEP indicatif local *m*; ~ **junction** *n* TRANS jonction urbaine *f*; ~ **line** *n* TELEP raccordement *m*, ligne locale *f*; ~ **line concentrator** *n* TELEP, TRANS concentrateur de lignes d'abonnés *m*; ~ **loop** *n* RADIO boucle d'abonné *f*; ~ **mode** *n* TELEP mode local *m*; ~ **network interconnect** *n* ELECT unité d'interconnexion de réseau local *f*; ~ **number** *n* TELEP numéro local *m*; ~ **operation** *n* CONTR exploitation locale *f*; ~ **oscillator** *n* (*LO*) ELECT, RADIO oscillateur local *m*; ~ **production** *n* TV production locale *f*; ~ **rate** *n* TELEP taxe locale *f*; ~ **record** *n* TELEP contrôle local *m*; ~ **register** *n* SWG enregistreur local *m*; ~ **resource** *n* DATA ressource locale *f*; ~ **sales office** *n* RADIO agence commerciale *f*; ~ **single-layer embedded test method** *n* (*LSE*) TEST méthode de test encastré monocouche locale *f*; ~ **single-layer test method** *n* (*LS test method*) TEST méthode de test monocouche locale *f*; ~ **studio** *n* TV studio local *m*; ~ **switch** *n* SWG, TELEP autocommutateur de rattachement *m*, commutateur local *m*; ~ **tariff** *n* TELEP taxe locale *f*; ~ **telephone network** *n* TELEP, TRANS réseau téléphonique local *m*, réseau téléphonique urbain *m*; ~ **test method** *n* TEST méthode de test locale *f*; ~ **traffic** *n* TELEP trafic local *m*; ~ **tree** *n* DATA arbre local *m*; ~ **user terminal** *n* (*LUT*) PER APP terminal d'usager local *m*

localization *n* RADIO localisation *f*; ~ **system** RADIO système de localisation *m*

locally[1]: **~-controlled** *adj* CONTR à commande locale, TELEP à commande locale, commandé localement

locally[2]: ~ **generated echo** *n* DATA écho local *m*

locate[1]: ~ **function** *n* COMP fonction de recherche *f*

locate[2] *vt* COMP localiser, trouver, DATA positionner, GEN loger, placer, situer, positionner, RADIO repérer, localiser, positionner, situer, placer, TEST localiser

locater *n* COMP représentation analogique *f*

locating *n* COMP localisation *f*, DATA, GEN positionnement *m*, RADIO, TEST localisation *f*, repérage *m*; ~ **hole** ELECT *printed circuits* trou de positionnement *m*; ~ **lamp** TELEP *indicating devices* lampe témoin *f*, voyant lumineux *m*; ~ **notch** ELECT *printed circuits* encoche de positionnement *f*

location *n* RADIO localisation *f*, implantation *f*, positionnement *m*, emplacement *m*; ~ **area** RADIO zone de localisation *f*; ~ **area identification** RADIO identification de zone de localisation *f*; ~ **cancelation** *AmE*, ~ **cancellation** *BrE* RADIO annulation d'une localisation *f*; ~ **deregistration** RADIO radiation de localisation *f*; ~ **information** RADIO information de localisation *f*; ~ **management** RADIO gestion de localisations *f*; ~ **register** RADIO enregistreur de localisations *m*, enregistreur de positions *m*; ~ **registration** RADIO enregistrement de la position *m*; ~ **updating** RADIO mise à jour de localisation *f*

locator *n* COMP releveur de coordonnées *m*

lock[1] *n* COMP, PER APP verrouillage *m*; **~-in range** RADIO bande de synchronisation *f*, plage de sensibilité *f*; **~-on** RADIO, TRANS verrouillage *m*

lock[2]: ~ **on** *vt* ELECT, RADIO *channel* verrouiller; ~ **out** *vt* COMP interdire

locked[1]: ~ **out** *adj* COMP interdit

locked[2]: ~ **file** *n* COMP fichier verrouillé *m*; ~ **loop** *n* ELECT boucle verrouillée *f*

locking *n* COMP, ELECT, TV verrouillage *m*; ~ **clip** ELECT clip de verrouillage *m*; ~ **device** PER APP dispositif de verrouillage *m*; ~ **mechanism** PARTS mécanisme de verrouillage *m*; ~ **spring** PARTS ressort de verrouillage *m*; **~-type button** COMP touche autobloquante *f*

lockout[1] *n* TELEP neutralisation *f*

lockout[2] *vt* COMP neutraliser

LOF *abbr* (*loss of frame*) TRANS perte de trame *f*

log[1] *n* COMP journal de bord *m*, DATA registre de consignation *m*, journal *m*, RADIO, TELEP journal *m*, journal des communications *m*; **~-commit record** DATA enregistrement de consignation d'engagement *m*; **~-damage record** DATA enregistrement de consignation de dommage *m*; **~-heuristic record** DATA enregistrement de consignation heuristique *m*; ~ **in** COMP, DATA ouverture de session *f*, commencement de session *m*, RADIO nom d'un point de connexion *m*; **~-in dialog** *AmE*, **~-in dialogue** *BrE* ELECT dialogue d'introduction *m*; **~-normal fading** RADIO *mobile* évanouissement log-normal *m*; **~-normal shadowing** RADIO évanouissement lent *m*, occultation log-normale *f*; **~-off** COMP fermeture de session *f*, fin de session *f*, sortie du système *f*; **~-periodic aerial** RADIO antenne log-périodique *f*, antenne à périodicité logarithmique *f*; **~-periodic antenna** RADIO antenne log-périodique *f*, antenne à périodicité logarithmique *f*; **~-ready record** DATA enregistrement de consignation prêt *m*; ~ **record** DATA enregistrement de consignation *m*

log[2] *vt* GEN journaliser; ~ **in** COMP login, DATA *Internet* login; ~ **on** RADIO se connecter à; ~ **onto** RADIO se connecter à; ~ **out** COMP *of system* sortir, clore

logarithm *n* RADIO logarithme *m*

logarithmic: ~ **gain** *n* CONTR *automatic control* gain logarithmique *m*

logatom *n* RADIO, TELEP logatome *m*; ~ **articulation index** GEN indice de netteté des logatomes *m*

logbook *n* DATA rapport de transmission *m*, journal de communication *m*, RADIO journal *m*

logger *n* COMP enregistreur automatique *m*

logging *n* COMP enregistrement *m*

logic *n* GEN logique *f*; ~ **analyser** *BrE*, ~ **analyzer** *AmE* DATA analyseur logique *m*; ~ **array** DATA réseau logique *m*; ~ **circuit** DATA, ELECT circuit logique *m*; ~ **design** DATA conception logique *f*, GEN étude logique *f*; ~ **device** DATA dispositif logique *m*, unité logique *f*; ~ **diagram** GEN schéma logique *m*; ~ **element** DATA, ELECT élément logique *m*, dispositif logique *m*; ~ **filtering** COMP filtrage logique *m*; ~ **instruction** COMP instruction logique *f*; ~ **operation** COMP opération logique *f*; ~ **probe** ELECT sonde logique *f*; ~ **programming** DATA programmation logique *f*; ~ **shift** COMP décalage logique *m*; ~ **state** ELECT état logique *m*; ~ **symbol** DATA, ELECT symbole logique *m*; ~ **system** COMP système logique *m*; ~ **unit** (*LU*) COMP, DATA, TELEP unité logique *f* (*UL*)

logical: ~ **block address** *n* (*LBA*) COMP adresse de bloc logique *f*; ~ **channel** *n* TELEP voie logique *f*; ~ **comparison** *n* COMP comparaison logique *f*; ~ **file** *n* DATA fichier logique *m*; ~ **file name** *n* (*LFN*) COMP nom logique de fichier *m*; ~ **inferences per second** *n pl* (*LIPS*) TEST inférences logiques par seconde *f pl*; ~ **link control** *n* (*LLC*) CONTR commande de liaison logique *f*; ~ **symbol** *n* DATA symbole logique *m*; ~ **unit** *n* (*LU*) COMP, DATA, TELEP unité logique *f* (*UL*)

login *n* COMP, DATA login *m*; ~ **name** COMP nom de connexion *m*

logistic: ~ **delay** *n* TELEP, TEST délai logistique *m*

logo *n* GEN logo *m*, sigle *m*

LOM *abbr* (*loss of multiframe*) TRANS perte de multitrame *f*

lone: ~ **signal unit** *n* (*LSU*) SIG unité de signalisation solitaire *f*

long[1]: ~**-distance** *adj* TELEP à grande distance, à longue distance

long[2]: ~**-address acceptance** *n* DATA acceptation d'adresses étendues *f*; **very ~ baseline interferometry** *n* (*VLBI*) TEST radiointerférométrie à très longue base *f* (*RILB*); ~ **circuit** *n* ELECT, TELEP circuit de grande longueur *m*; ~**-distance backscatter** *n* POWER, RADIO rétrodiffusion indirecte *f*; ~**-distance call** *n* TELEP appel longue distance *m*, communication longue distance *f*; ~**-distance carrier** *n* GEN exploitant de réseaux interurbains *m*; ~**-distance connection** *n* TELEP liaison à grande distance *f*, liaison interurbaine *f*; ~**-distance dual-homing link** *n* ELECT liaison double sur longue distance *f*, liaison double interurbaine *f*; ~**-distance line** *n* TELEP, TRANS ligne à grande distance *f*, ligne interurbaine *f*; ~**-distance network** *n* TELEP réseau à grande distance *m*, réseau à longue portée *m*, réseau interurbain *m*; ~**-distance telenetwork** *n* TELEP réseau interurbain *m*, réseau de télécommunications à grande distance *m*; ~**-haul transmission** *n* TELEP transmission à grande distance *f*; ~**-life battery** *n* POWER pile longue durée *f*, batterie de longue durée *f*; ~**-line equipment** *n* TRANS ligne longue équipée *f*; ~**-path bearing** *n* RADIO azimut du grand arc *m*; ~**-range navigation** *n* (*loran*) RADIO système loran *m*; ~**-term bit error rate** *n* TELEP taux d'erreur à long terme sur les bits *m*

longitudinal: ~ **balance** *n* POWER équilibre des circuits réels *m*; ~ **direction** *n* GEN sens longitudinal *m*; ~ **displacement** *n* GEN déplacement longitudinal *m*; ~ **judder** *n* TELEP broutage longitudinal *m*; ~ **offset loss** *n* TRANS perte par déplacement longitudinal *f*; ~ **redundancy check** *n* DATA contrôle longitudinal par redondance *m*; ~ **slot** *n* TRANS fente longitudinale *f*; ~ **time code** *n* (*LTC*) COMP code temporel longitudinal *m* (*CTL*); ~ **wave** *n* RADIO onde longitudinale *f*

look: ~**-ahead carry** *n* ELECT report anticipé *m*; ~**-through** *n* RADIO blanc *m*

looked: ~**-for** *adj* DATA sollicité

loop[1] *n* COMP, POWER boucle *f*, RADIO branchement d'abonné *m*, TRANS boucle *f*, réseau en boucle *m*; ~ **aerial** RADIO antenne cadre *f*; ~ **alignment error** TEST erreur de calage *f*; ~ **amplifier** TEST amplificateur de cadre *m*; ~ **antenna** RADIO antenne cadre *f*; ~ **circuit** POWER boucle *f*; ~ **dialing** *AmE*, ~ **dialling** *BrE* PER APP, TELEP émission d'impulsions par ouverture de boucle *f*, numérotation par ouverture de boucle *f*; ~**-disconnect dialing** *AmE*, ~**-disconnect dialling** *BrE* PER APP numérotation en boucle *f*, numérotation par ouverture de boucle *f*; ~**-disconnect mode** PER APP mode d'ouverture de boucle *m*; ~ **extender** TELEP dispositif de suralimentation *m*; ~ **pulsing** PER APP, TELEP numérotation par ouverture de boucle *f*, émission d'impulsions par ouverture de boucle *f*; ~ **reporting system** (*LRS*) TELEP système de télémesures pour lignes d'abonnées *m* (*système LRS*); ~ **series resistance** ELECT résistance au bouclage *f*; ~ **start** TELEP déclenchement par boucle *m*; ~**-start trunk** TELEP circuit à déclenchement par boucle *m*; ~ **station connector** (*LSC*) ELECT connecteur de station en boucle *m*; ~ **stop** COMP arrêt bouclé *m*; ~ **transfer locus** CONTR lieu de transfert *m*; ~ **wiring concentrator** (*LWC*) ELECT concentrateur de câblage en boucle *m*

loop[2] *vt* ELECT reboucler, *electricity* boucler, TRANS boucler

loopback *n* ELECT bouclage *m*, TRANS boucle *f*; ~ **command** TV commande de boucle *f*; ~ **configuration** GEN signalisation par boucle de renvoi *f*; ~ **signaling** *AmE*, ~ **signalling** *BrE* GEN signalisation par boucle de renvoi *f*; ~ **test method** TEST méthode de test en boucle *f*

looped[1] *adj* ELECT bouclé

looped[2]: ~ **signal** *n* TV signal transmis en retour *m*

looping *n* COMP, TELEP, TRANS bouclage *m*

loose: ~ **cable structure** *n* TRANS câblage libre *m*; ~ **connection** *n* POWER connexion lâche *f*; ~ **contact** *n* POWER contact lâche *m*; ~**-structure cable** *n* TRANS câble à structure à fibres libres *m*, câble à structure à fibres lisses *m*; ~**-tube cable** *n* TRANS câble à tubes assemblés *m*, câble classique à fibres libres *m*

LOP *abbr* (*loss of pointer*) TRANS perte de pointeur *f*

loran *n* (*long-range navigation*) RADIO système loran *m*

LOS *abbr* (*loss of signal*) TRANS perte de signal *f*

loss *n* COMP atténuation *f*, perte *f*, GEN *of power* affaiblissement *m*, RADIO affaiblissement *m*, atténuation *f*, TRANS affaiblissement *m*, perte *f*; ~ **around a corner** TRANS perte en courbe *f*; ~ **factor** POWER facteur de perte *m*, facteur d'affaiblissement *m*; ~ **of frame** (*LOF*) TRANS perte de trame *f*; ~ **of frame alignment** (*LFA*) TELEP, TRANS perte de verrouillage de trame *f* (*PVT*); ~ **of frame alignment detector** TELEP détecteur de perte de verrouillage de trame *m*; ~**-mode operation** TRANS mode d'exploitation avec perte *m*; ~**-mode working** TRANS mode d'exploitation avec perte *m*; ~ **of multiframe** (*LOM*) TRANS perte de multitrame *f*; ~ **of multiframe alignment** TRANS perte de verrouillage de multitrame *f*; ~ **of pointer** (*LOP*) TRANS perte de pointeur *f*; ~ **relative to free space** RADIO, TRANS affaiblissement par rapport à l'espace libre *m*; ~ **resistance** POWER résistance de perte *f*; ~ **of signal** (*LOS*) TRANS perte de signal *f*; ~ **system** TELEP système avec perte *m*

lossless *adj* COMP, TELEP sans perte

lossy *adj* COMP, TELEP avec perte

lost: ~ **call** *n* SWG, TELEP, TRANS appel perdu *m*; ~ **call attempt** *n* TELEP tentative d'appel perdue *f*; ~ **time** *n* GEN temps d'inutilisation *m*, temps perdu *m*; ~ **traffic** *n* TELEP trafic perdu *m*

loudness *n* COMP intensité sonore *f*, volume *m*, GEN puissance *f*, volume *m*, TELEP intensité sonore *f*, sonie *f*; ~ **contour** TELEP courbe isosonique *f*; ~ **level** TELEP niveau d'intensité sonore *m*, *acoustics* niveau de sonie *m*; ~ **scale** TELEP échelle de force sonore *f*

loudspeaker *n* COMP, TELEP haut-parleur *m*; ~ **paging** RADIO recherche de personne par haut-parleur *f*; ~ **telephone** TELEP poste à réception amplifiée *m*

low[1] *adj* GEN bas, faible; ~**-bit-rate** TV à faible débit; ~**-resistance** POWER à faible résistance; ~**-speed** DATA lent

low[2]: ~ **B to earth** *n* TEST *B-leg* basse résistance nuque/terre *f*; ~ **bit data communication** *n* DATA, RADIO communication de données bas débit *f*; ~ **dissipation** *n* ELECT faible dissipation *f*; ~ **earth orbit** *n* RADIO orbite basse *f*; ~ **earth orbit satellite** *n* (*LEO satellite*) RADIO satellite en orbite basse *m*, satellite sur orbite à basse altitude *m*; ~ **frequency** *n* (*LF*) RADIO, SWG basse fréquence *f* (*BF*); ~**-frequency amplifier** *n* ELECT amplificateur faible puissance *m*; ~**-frequency oscillator** *n* COMP oscillateur basse fréquence *m*; ~**-frequency sweep** *n* COMP balayage en basse fréquence *m*; **very ~-frequency wave** *n* RADIO onde myriamétrique *f*; ~**-grade code** *n* RADIO code à faible herméticité *m*; ~ **insertion loss** *n* TRANS faible perte d'insertion *f*; ~ **insulation loss** *n* TELEP faible perte d'isolement *f*; ~**-layer compatibility** *n* (*LLC*) DATA, SWG, TELEP compatibilité de couche inférieure *f*; ~**-layer compatibility information element** *n* (*LLC-IE*) SWG, TELEP élément d'information de compatibilité de couche inférieure *m* (*LLC-IE*); ~ **level** *n* (*L level*) ELECT état B *m*, état bas *m*, niveau bas *m* (*niveau B*); ~**-level language** *n* COMP langage à bas niveau *m*; ~**-level modulation** *n* GEN modulation à bas niveau *f*; ~**-level signaling device** *AmE*, ~**-level signalling device** *BrE* *n* GEN boîte de distribution et de traitement des signaux faibles *f*; ~ **loss** *n* TELEP faible atténuation *f*, faible perte *f*; ~**-loss connector** *n* TRANS connecteur à faible perte *m*; ~**-loss fiber connector** *AmE*, ~**-loss fibre connector** *BrE* *n* TRANS connecteur à fibre à faible perte *m*; ~ **medium** *n* COMP bas médium *m*; ~**-noise amplifier** *n* (*LNA*) ELECT,

RADIO, TRANS amplificateur à faible bruit *m* (*AFB*); ~ **note** *n* COMP note basse *f*; ~ **orbit** *n* RADIO orbite basse *f*; ~**-pass band** *n* TRANS bande basse *f*; ~**-pass filter** *n* POWER, TRANS filtre passe-bas *m*; ~**-performance equipment** *n* COMP matériel de faible performance *m*; ~**-power distress transmitter** *n* (*LPDT*) RADIO émetteur de détresse à faible puissance *m*; ~**-power flux density** *n* RADIO puissance surfacique limitée *f*; ~**-power laser diode** *n* ELECT diode laser à faible puissance *f*; ~**-power transistor** *n* PARTS *semiconductors* transistor à faible puissance *m*; ~**-power transmitter** *n* RADIO émetteur basse puissance *m*, émetteur faible puissance *m*; ~**-range note** *n* COMP note grave *f*; ~**-rate encoding** *n* (*LRE*) DATA codage à débit réduit *m*, codage à faible débit *m*; ~**-resistance distribution** *n* POWER *supply* distribution à basse impédance *f*; ~ **ring to ground** *n* TEST basse résistance nuque-terre *f*; ~ **speed** *n* GEN bas débit *m*, basse vitesse *f*; ~ **tip to ground** *n* TEST basse résistance pointe-terre *f*; ~ **tip to ring** *n* TEST basse résistance pointe-nuque *f*; ~ **A to B** *n* TEST basse résistance pointe-nuque *f*; ~ **A to B to earth** *n* TEST basse résistance pointe-nuque-terre *f*; ~ **A to earth** *n* TEST basse résistance pointe-terre *f*; ~ **to high** *n* ELECT transition montante *f*; ~**-velocity camera tube** *n* ELECT tube analyseur à électrons lents *m*; ~**-velocity scanning** *n* ELECT balayage par électrons lents *m*; ~ **voltage** *n* (*LV*) POWER basse tension *f*; ~**-wattage resistor** *n* PARTS, POWER résistance de faible dissipation *f*

lower[1]: ~ **case** *n* COMP *letters* minuscules *f pl*; ~ **frequency** *n* RADIO fréquence inférieure *f*; ~ **keyboard** *n* COMP clavier inférieur *m*; ~**-level service** *n* TEST service de bas niveau *m*; ~**-order path adaptation** *n* (*LPA*) TELEP, TRANS adaptation de conduit d'ordre inférieur *f*; ~**-order path connection** *n* (*LPC*) TELEP, TRANS connexion de conduit d'ordre inférieur *f*; ~**-order path termination** *n* (*LPT*) TELEP, TRANS terminaison de conduit d'ordre inférieur *f*; ~ **quality of service** *n* TEST qualité de service dégradée *f*, qualité de service inférieure *f*; ~ **sideband** *n* (*LSB*) TRANS bande latérale inférieure *f*; ~ **subfield** *n* RADIO, TRANS sous-champ inférieur *m*; ~ **tariff** *n* GEN tarif inférieur *m*; ~ **tester** *n* (*LT*) DATA, TEST testeur inférieur *m*

lower[2] *vt* GEN abaisser

lowest: ~ **usable frequency** *n* (*LUF*) RADIO fréquence minimale utilisable *f* (*LUF*)

LP: ~ **mode** *n* (*linearly-polarized mode*) TRANS mode LP *m* (*mode à polarisation rectiligne*)

LPA *abbr* (*lower-order path adaptation*) TELEP, TRANS adaptation de conduit d'ordre inférieur *f*

LPC *abbr* DATA (*linear predicting coding*) codage prédictif linéaire *m*, TELEP, TRANS (*lower-order path connection*) connexion de conduit d'ordre inférieur *f*

LPDT *abbr* (*low-power distress transmitter*) RADIO émetteur de détresse à faible puissance *m*

LPT *abbr* (*lower-order path termination*) TELEP, TRANS terminaison de conduit d'ordre inférieur *f*

LQ *abbr* (*letter quality*) COMP qualité courrier *f*

LRE *abbr* (*low-rate encoding*) DATA codage à débit réduit *m*, codage à faible débit *m*

LRS *abbr* (*loop reporting system*) TELEP système LRS *m* (*système de télémesures pour lignes d'abonnées*)

LS *abbr* (*leased line*) DATA, TELEP, TRANS LS (*liaison spécialisée, ligne spécialisée*)

LSB *abbr* DATA (*least significant bit*) bit de plus faible poids *m*, TRANS (*lower sideband*) bande latérale inférieure *f*

LSC *abbr* DATA (*line signaling code AmE, line signalling code BrE*) code de signalisation de ligne *m*, ELECT (*loop station connector*) connecteur de station en boucle *m*, SIG (*line signaling code, line signalling code*) code de signalisation de ligne *m*

LSD *abbr* (*least significant digit*) DATA élément numérique de poids faible *m*

L-section: ~ **filter** *n* ELECT filtre en L *m*

LSI *abbr* (*large-scale integration*) ELECT, PARTS, TELEP, TRANS LSI (*intégration à grande échelle*); ~ **circuit** *n* ELECT, TELEP, TRANS circuit LSI *m*

LSU *abbr* (*lone signal unit*) SIG unité de signalisation solitaire *f*

LT *abbr* DATA (*lower tester*) testeur inférieur *m*, TELEP (*line termination*) terminaison de ligne *f*, TEST (*lower tester*) testeur inférieur *m*, TRANS (*line termination*) terminaison de ligne *f*

LTC *abbr* (*longitudinal time code*) COMP CTL (*code temporel longitudinal*)

LU *abbr*(*logic unit, logical unit*) COMP, DATA, TELEP UL (*unité logique*)

LUF *abbr* (*lowest usable frequency*) RADIO LUF (*fréquence minimale utilisable*)

lumen[1] *n* TEST lumen *m*; ~ **maintenance** TV maintenance du flux lumineux *f*

lumen[2]: ~ **per watt** *phr* (*lm/W*) TEST lumen par watt (*lm/W*)

luminance *n* COMP, ELECT, TV luminance *f*; ~ **axis** COMP axe de luminance *m*; ~ **compensation** TV compensation de luminance *f*; ~ **component** COMP composant de luminance *m*; ~ **contrast** TV contraste de luminance *m*; ~ **factor** TV facteur de luminance *m*; ~ **ratio** TEST rapport de luminance *m*; ~ **signal** TV signal de luminance *m*

luminancemeter *n* TEST luminancemètre *m*

luminescence *n* TRANS luminescence *f*

luminescent: ~ **screen** *n* COMP, ELECT, TV écran fluorescent *m*

luminous: ~ **button** *n* CONTR bouton lumineux *m*; ~ **efficacy of a source of light** *n* DATA efficacité lumineuse d'une source de lumière *f*; ~**-flux threshold** *n* ELECT seuil d'un flux lumineux *m*

lurker *n* COMP *Internet* compagnon silencieux *m*

LUT *abbr* (*local user terminal*) PER APP terminal d'usager local *m*

LWC *abbr* (*loop wiring concentrator*) ELECT concentrateur de câblage en boucle *m*

LXE *abbr* (*lightguide express entry*) ELECT LXE

LZW *abbr* (*Lempel-Ziv-Welch*) COMP LZW (*Lempel-Ziv-Welch*)

M

M *abbr* DATA (*multivalued*) à plusieurs valeurs, RADIO (*refractive modulus*) M (*module de réfraction*); **~ component** *n* DATA élément obligatoire *m*; **~ reflection** *n* RADIO réflexion en M *f*
mA *abbr* (*milliampere*) ELECT mA (*milliampère*)
MA *abbr* (*medium adaptor*) ELECT, TELEP adaptateur de support *m*
MAC *abbr* (*medium access control*) CONTR commande d'accès au support *f*, contrôle d'accès au support *m*
MACF *abbr* (*multiple association control function*) DATA fonction de contrôle d'association multiple *f*
Mach: **-Zehnder interferometer** *n* TELEP coupleur Mach-Zehnder *m*
machine: **~ code** *n* COMP code machine *m*; **~ language** *n* COMP langage machine *m*; **~-readable medium** *n* CONTR support exploitable par machine *m*, support lisible par machine *m*; **~ soldering** *n* ELECT, PARTS brasage tendre à la machine *m*
macro *n* COMP, DATA macro *f*
macrobend: **~ loss** *n* TRANS *fibre optics* perte par macrocourbures *f*
macrobending *n* TRANS macrocourbure *f*
macroblock *n* COMP macrobloc *m*
macrocell *n* RADIO macrocellule *f*
macrogenerating: **~ program** *n* COMP macrogénérateur *m*
macrogenerator *n* COMP macrogénérateur *m*
macroinstruction *n* COMP, DATA macro-instruction *f*
macrosegmentation *n* RADIO macrosegmentation *f*
MADI *abbr* (*multichannel audio-digital interface*) COMP MADI (*audio-numérique multicanaux*)
MADT *abbr* (*micro-alloy diffused transistor*) ELECT, PARTS transistor à micro-alliage et diffusion *m*
MAF *abbr* (*management application function*) DATA fonction d'application de gestion *f*
magic: **~ T** *n* RADIO té magique *m*
magnet *n* ELECT, PARTS aimant *m*; **~ coil** ELECT, PARTS bobine d'électro-aimant *f*; **~ coupling** ELECT, PARTS couplage magnétique *m*; **~ wire** ELECT, PARTS fil à bobiner *m*, fil de bobinage *m*
magnetic: **~ blow-out** *n* ELECT soufflage magnétique *m*; **~-bubble memory** *n* DATA mémoire à bulles magnétiques *f*; **~ card** *n* DATA carte magnétique *f*; **~-card storage** *n* DATA mémoire à cartes magnétiques *f*; **~ card unit** *n* DATA unité de cartes magnétiques *f*; **~ circuit** *n* ELECT circuit magnétique *m*; **~ clutch** *n* PARTS embrayage électrique *m*; **~ core** *n* ELECT noyau magnétique *m*, tore magnétique *m*; **~-core storage** *n* COMP, DATA mémoire à tores magnétiques *f*; **~ deflection** *n* ELECT déviation magnétique *f*; **~ delay line** *n* ELECT ligne à retard magnétique *f*; **~ disk** *n* COMP, DATA, ELECT disque magnétique *m*; **~-disk storage** *n* COMP, DATA mémoire sur disque magnétique *f*; **~ drum** *n* DATA tambour magnétique *m*; **~ field** *n* ELECT champ magnétique *m*; **~ field intensity** *n* ELECT intensité de champ magnétique *f*; **~ flux density** *n* ELECT densité de flux magnétique *f*; **~ focusing** *n* ELECT concentration magnétique *f*; **~ head** *n* ELECT tête magnétique *f*; **~ hysteresis** *n* ELECT hystérésis magnétique *f*; **~ induction** *n* ELECT induction magnétique *f*; **~-induction current loop** *n* ELECT boucle d'induction magnétique *f*; **~ ink character recognition** *n* (*MICR*) APPL reconnaissance magnétique des caractères *f* (*MICR*); **~ media** *n* COMP support magnétique *m*; **~ potential difference** *n* ELECT différence de potentiel magnétique *f*; **~ recording** *n* ELECT enregistrement magnétique *m*; **~-resonance spectroscopy** *n* ELECT spectroscopie par résonance magnétique *f*; **~ sensitivity** *n* ELECT sensibilité magnétique *f*; **~ shield** *n* ELECT écran magnétique *m*; **~ storage** *n* DATA mémoire magnétique *f*; **~ storm** *n* RADIO orage magnétique *f*; **~ strip** *n* ELECT, TELEP feuillet magnétique *m*, piste magnétique *f*; **~-strip card** *n* ELECT carte à piste magnétique *f*; **~ support** *n* COMP support magnétique *m*; **~ tape** *n* COMP, DATA bande magnétique *f*, ruban magnétique *m*; **~-tape cartridge** *n* DATA cartouche de bande magnétique *f*; **~-tape drive** *n* DATA dérouleur de bande magnétique *m*; **~-tape leader** *n* DATA amorce de bande magnétique *f*; **~-tape storage** *n* COMP, DATA mémoire sur bande magnétique *f*; **~-tape trailer** *n* DATA fin de bande magnétique *f*; **~ track** *n* DATA piste magnétique *f*; **~ valve** *n* CONTR vanne électromagnétique *f*
magnetically: **~-latched crosspoint** *n* SWG point d'intersection à maintien magnétique *m*
magnetization *n* ELECT aimantation *f*
magnetizing: **~ current** *n* ELECT courant de magnétisation *m*, courant magnétisant *m*; **~ field** *n* ELECT champ magnétisant *m*; **~ inductance** *n* ELECT inductance d'aimantation *f*
magneto[1]: **~-optic** *adj* (*MO*) COMP, ELECT magnéto-optique (*MO*)
magneto[2]: **~-optic effect** *n* ELECT effet magnéto-optique *m*; **~-optical device** *n* DATA appareil magnéto-optique *m*; **~-optical disk** *n* COMP, DATA, ELECT disque magnéto-optique *m*; **~-optical head** *n* DATA tête magnéto-optique *f*; **~-optical storage** *n* DATA mémoire magnéto-optique *f*
magnetoresistance *n* ELECT magnétorésistance *f*
magnetoscope *n* ELECT magnétoscope *m*
magnetostatic: **~ surface wave** *n* (*MSSW*) ELECT onde de surface magnétostatique *f*
magnetostriction *n* ELECT magnétostriction *f*
magnetostrictive: **~ effect** *n* ELECT magnétostriction *f*
magnetron *n* ELECT, RADIO magnétron *m*; **~ injection gun** ELECT canon d'injection de magnétron *m*
mail *n* COMP, DATA messagerie *f*; **~ server** COMP, DATA serveur de messagerie *m*
mailbox *n* COMP, DATA boîte aux lettres *f* (*BAL*); **~ messaging service** COMP, DATA service de messagerie boîte aux lettres *m*
mailing: **~ list** *n* COMP, DATA liste de diffusion *f*
main: **~ aisle** *n* SWG allée principale *f*; **~-aerial television** *n* TV télévision à antenne principale *f*; **~-antenna television** *n* TV télévision à antenne principale *f*; **~ bus** *n* POWER barre d'alimentation principale *f*; **~ busbar** *n* POWER barre omnibus principale *f*; **~ chute** *n* SWG chemin de câble principal *m*; **~ circuit breaker** *n* POWER interrupteur général *m*; **~ contractor** *n* GEN maître d'ouvrage *m*; **~ control station** *n* TELEP station multiprocesseur de commande *f*; **~ control unit** *n* COMP unité de commande principale *f*; **~ current** *n* POWER courant principal *m*; **~ distribution frame** *n* (*MDF*) TELEP, TRANS répartiteur principal *m* (*RP*); **~ entry** *n* DATA entrée principale *f*; **~ event loop** *n* COMP boucle principale de gestion d'événements *f*; **~ exchange** *n* SWG commutateur de rattachement *m*, commutateur principal *m*, TELEP commutateur principal *m*; **~ fuse** *n* PARTS, POWER fusible principal *m*; **~ gap** *n* ELECT espace principal *m*; **~ international switching center** *AmE*, **~ international switching centre** *BrE* *n* SWG centre international de transit principal *m*; **~ international-trunk**

switching center *AmE*, ~ **international-trunk switching centre** *BrE n* SWG centre international de transit principal *m*; ~ **line** *n* TRANS ligne principale de liaison *f*; ~ **lobe** *n* RADIO lobe principal *m*; ~ **multiplex coupler** *n* TELEP coupleur multiplex principal *m*; ~ **processor unit** *n* COMP unité de traitement principal *f*; ~ **repeater distribution frame** *n* TRANS répartiteur principal de station de répéteurs *m*; ~ **repeater station** *n* TELEP station principale de répéteurs *f*; ~ **section** *n* TELEP section principale *f*; ~ **storage** *n* DATA mémoire centrale *f*, mémoire principale *f*; ~ **switch** *n* POWER interrupteur général *m*; ~ **telephone line** *n* TRANS ligne téléphonique principale *f*; ~ **trunk exchange area** *n* TELEP zone de centre de transit principale *f*; ~ **trunk switching center** *AmE*, ~ **trunk switching centre** *BrE n* SWG centre de transit principal *m*; ~ **voltage** *n* POWER tension principale *f*

mainframe *n* COMP gros ordinateur *m*, ordinateur central *m*, TELEP, TRANS répartiteur principal *m* (*RP*); ~ **computer** COMP ordinateur central *m*

mains[1]: ~**-connected** *adj* POWER branché sur le secteur, raccordé au secteur

mains[2] *n pl* POWER secteur *m*, *network* réseau *m*; ~ **connection** *n* POWER alimentation secteur *f*; ~ **failure** *n* POWER coupure secteur *f*, panne d'alimentation *f*, panne de courant *f*, panne du secteur *f*, coupure de courant *f*; ~**-failure alarm** *n* POWER alarme de panne de courant *f*; ~ **frequency** *n* POWER fréquence du réseau *f*, fréquence du secteur *f*; ~ **fuse** *n* PARTS, POWER fusible de secteur *m*; ~ **hum** *n* POWER ronflement du secteur *m*; ~ **monitor** *n* CONTR, POWER contrôleur secteur *m*; ~ **monitoring** *n* CONTR, POWER surveillance de la tension secteur *f*, surveillance secteur *f*; ~ **supply** *n* POWER alimentation par le secteur *f*; ~ **switch** *n* POWER interrupteur secteur *m*; ~ **transformer** *n* POWER transformateur relié au réseau *m*; ~ **voltage** *n* POWER tension du secteur *f*

maintainability: ~ **function** *n* TEST fonction de maintenabilité *f*; ~ **performance** *n* TEST maintenabilité *f*

maintaining: ~ **voltage** *n* ELECT, POWER tension de maintien *f*

maintenance: ~ **analysis** *n* TEST analyse de maintenance *f*; ~ **cell description** *n* (*MCD*) DATA description de cellule de maintenance *f*; ~ **channel** *n* TRANS canal de maintenance *m*; ~ **entity** *n* DATA cellule de maintenance *f*; ~ **function** *n* TEST fonction de maintenance *f*; ~ **interface** *n* TELEP interface de maintenance *f*; ~ **interval** *n* TEST périodicité de maintenance *f*; ~ **man-hours** *n pl* TEST temps de maintenance en hommes-heures *m*; ~ **processor** *n* COMP processeur de maintenance *m*; ~ **program** *AmE*, ~ **programme** *BrE n* TEST programme de maintenance *m*; ~ **service provider** *n* (*MSP*) TEST prestataire de services de maintenance *m*; ~ **staff** *n* TEST personnel de maintenance *m*; ~ **station** *n* TELEP station multiprocesseur de maintenance *f*; ~ **support logistics** *n pl* TEST logistique de maintenance *f*; ~ **support performance** *n* TEST logistique de maintenance *f*; ~ **time** *n* TELEP, TEST durée de maintenance *f*, temps de maintenance *m*

major[1]: ~ **defective** *adj* TEST défectueux majeur

major[2]: ~ **alarm** *n* TEST alarme majeure *f*; ~ **failure** *n* TEST défaillance majeure *f*; ~**-path concept** *n* RADIO concept des grandes artères *m*; ~**-path satellite** *n* RADIO satellite des artères principales *m*

majority: ~ **carrier** *n* ELECT porteur majoritaire *m*; ~ **decision logic** *n* ELECT logique à décision majoritaire *f*; ~ **element** *n* ELECT circuit majoritaire *m*; ~ **gate** *n* DATA, ELECT circuit logique majoritaire *m*, circuit majoritaire *m*; ~ **operation** *n* ELECT opération majoritaire *f*

make[1] *n* COMP, ELECT marque *f*; ~**-before-break contact** ELECT contact à chevauchement *m*; ~ **good** TRANS réparation temporaire *f*; ~ **pulse** TELEP impulsion de fermeture *f*

make[2]: ~ **a call to** *vt* TELEP appeler; ~ **a collect call to sb** *AmE vt* (*cf make a reverse-charge call to sb BrE*) TELEP appeler qn en PCV; ~ **a reverse-charge call to sb** *BrE vt* (*cf make a collect call to sb AmE*) TELEP appeler qn en PCV

make[3]: ~ **a call** *vi* TELEP appeler, effectuer un appel, téléphoner; ~ **make a collect call** *AmE vi* (*cf make a reverse-charge call BrE*) TELEP appeler en PCV; ~ **make a reverse-charge call** *BrE vi* (*cf make a collect call AmE*) TELEP appeler en PVC

male: ~ **plug** *n* GEN fiche mâle *f*, PARTS connecteur mâle *m*

malfunction *n* TEST incident de fonctionnement *m*; ~ **level** TEST niveau de mauvais fonctionnement *m*

malicious: ~ **call** *n* TELEP appel importun *m*, appel malveillant *m*; ~ **call identification** *n* TELEP identification d'appel malveillant *f*; ~ **call identification service** *n* TELEP service d'identification d'appels malveillants *m*

man: ~**-machine communications** *n pl* (*MMC*) COMP relation homme-machine *f* (*RHM*); ~**-machine dialog** *n* COMP dialogue homme-machine *m*; ~**-machine interface** *n* COMP interface homme-machine *f*; ~**-machine language** *n* (*MML*) COMP langage homme-machine *m*; ~**-machine terminal** *n* COMP terminal homme-machine *m*; ~**-made noise** *n* ELECT, RADIO bruit industriel *m*, parasite industriel *m*, perturbation industrielle *f*; ~**-pack transceiver** *n* RADIO émetteur-récepteur portable à dos d'homme *m*

MAN *abbr* (*metropolitan area network*) TRANS réseau de zone urbaine *m*

manageable: ~ **repeater** *n* TRANS répéteur gérable *m*

managed: ~ **object** *n* (*MO*) DATA objet géré *m*; ~**-object boundary** *n* DATA limite d'objet géré *f*; ~**-object class** *n* (*MOC*) DATA classe d'objet géré *f*; ~**-object conformance statement** *n* (*MOCS*) DATA déclaration de conformité d'objet géré *f*; ~ **open system** *n* DATA système ouvert géré *m*; ~ **system** *n* DATA système géré *m*

management *n* GEN gestion *f*, management *m*; ~ **application function** (*MAF*) DATA fonction d'application de gestion *f*; ~ **data** *n pl* DATA données de gestion *f pl*; ~ **database** DATA serveur de gestion *m*; ~ **domain** (*MD*) DATA, TELEP domaine de gestion *m*; ~ **information** DATA information de gestion *f*; ~ **information base** (*MIB*) COMP, DATA base de données de gestion *f*; ~ **interaction** DATA interaction de gestion *f*; ~ **network** TRANS réseau de gestion *m*; ~ **notification service** DATA service de notification de gestion *m*; ~ **operation service** DATA service d'opération de gestion *m*; ~ **protocol** DATA protocole de gestion *m*; ~ **signal** SIG signal de gestion *m*; ~ **software** COMP, DATA logiciel de gestion *m*; ~ **support object** DATA objet support de gestion *m*; ~ **unit** DATA organe de gestion *m*

manager *n* GEN gestionnaire *m*; ~ **role** DATA rôle de gestion *m*, rôle de gestionnaire *m*

Manchester[1]: ~**-coded** *adj* COMP codé Manchester

Manchester[2]: ~ **data** *n pl* COMP données Manchester *f pl*

mandatory: ~ **component** *n* DATA élément obligatoire *m*; ~ **requirement** *n* DATA spécification obligatoire *f*, prescription obligatoire *f*

manipulated: ~ **variable** *n* CONTR, PARTS écart *m*

manipulation: ~ **detection** *n* CONTR, DATA détection de modification *f*

manipulative: ~ **deception** *n* TELEP diversion par altération *f*

manned: ~ **station** *n* GEN station pourvue de personnel *f*

manning *n* GEN personnel de surveillance *m*

manometer *n* TEST manomètre *m*

manual: ~ **answering** *n* TELEP réponse manuelle *f*; ~ **attempt** *n* TELEP tentative manuelle *f*; ~ **calling** *n* TELEP appel manuel *m*; ~**-changeover acknowledgement signal** *n* SIG signal d'accusé de réception de commutation manuelle *m*; ~**-changeover signal** *n* SIG signal de commutation manuelle sur liaison de réserve *m*; ~ **control** *n* CONTR commande manuelle *f*; ~ **direction finder** *n* RADIO radiogoniomètre manuel *m*; ~ **marking** *n* DATA *Internet* balisage manuel *m*; ~ **observation** *n* TEST *of service quality* observation manuelle *f*; ~ **position** *n* CONTR position manuelle *f*; ~ **regulation** *n* CONTR réglage manuel *m*; ~ **repatching** *n* TEST raccordement manuel *m*; ~ **ringing** *n* TELEP appel manuel *m*, sonnerie manuelle *f*; ~ **setting** *n* CONTR réglage manuel *m*; ~ **switching system** *n* SWG système de commutation manuelle *m*

manually[1]: **~-controlled** *adj* CONTR à commande manuelle, commandé à la main

manually[2]: **~-operated transmission** *n* (*MOX*) DATA, TRANS émission à commande manuelle *f* (*MOX*), commande manuelle d'émission *f* (*MOX*)

manufacturing *n* PARTS fabrication *f*, production *f*; ~ **defect** TEST vice de fabrication *f*; ~ **failure** TEST problème de fabrication *m*; ~ **fault** TEST problème de fabrication *m*

many: **~-valued logic** *n* ELECT logique multivalente *f*

map[1] *n* COMP mappe *f*

map[2] *vt* COMP mapper, TELEP appliquer

mapping *n* COMP mappage *m*; ~ **list** RADIO tableau de correspondance *m*

MAR *abbr* (*miscellaneous apparatus rack*) PARTS, TELEP bâti pour appareils divers *m*

margin *n* TELEP *of synchronous receiver* marge *f*

marginal: ~ **check** *n* TEST test à la limite *m*, test aux marges *m*; ~ **test** *n* TEST test à la limite *m*, test aux marges *m*

maritime: ~ **account** *n* TELEP compte maritime *m*; ~ **center** *AmE seemaritime centre BrE*; ~ **center shore station** *AmE seemaritime centre shore station BrE*; ~ **centre** *BrE n* TELEP centre maritime *m*; ~ **centre shore station** *n BrE* TELEP centre maritime station côtière *m*, centre maritime station terriene côtière *m*; ~ **local system** *n* TELEP système local maritime *m*; ~ **mobile satellite system** *n* TELEP système mobile maritime par satellite *m*; ~ **mobile service** *n* RADIO service mobile maritime *m*; ~ **mobile terrestrial service** *n* RADIO service mobile maritime de terre *m*; ~ **mobile-service satellite** *n* RADIO service mobile maritime par satellite *m*; ~ **radio** *n* RADIO radio maritime *f*; ~ **satellite circuit** *n* RADIO circuit maritime par satellite *m*; ~ **satellite service** *n* (*MSS*) RADIO service maritime par satellite *m*; ~ **satellite switching center** *AmE*, ~ **satellite switching centre** *BrE n* (*MSSC*) RADIO, SWG centre de commutation du service maritime par satellite *m*; ~ **satellite system** *n* RADIO système maritime par satellite *m*; ~ **switching center** *AmE*, ~ **switching centre** *BrE n* (*MSC*) SWG, TELEP centre de commutation maritime *m*; ~ **system** *n* TELEP système maritime *m*; ~ **terrestrial circuit** *n* TELEP, TRANS circuit terrestre du système maritime *m*

mark[1] *n* TEST repère de graduation *m*; ~ **period** TELEP période de travail *f*; ~ **scanning** APPL lecture optique de marques *f*; ~ **scanning equipment** APPL équipement lecteur optique de marques *m*; ~ **sensing** APPL lecture de marques *f*

mark[2]: ~ **busy** *vi* TELEP marquer occupé

marked: ~ **idle channel** *n* (*MIC*) SWG repérage des voies au repos *m*

marker[1]: **~-controlled** *adj* CONTR, SWG à commande par marqueur

marker[2] *n* DATA *Internet* signet *m*, indicateur *m*, GEN indicateur *m*, marqueur *m*; ~ **beacon** RADIO radioborne *f*; ~ **control system** SWG système de marquage *m*; ~ **system** SWG système de marquage *m*

marking *n* GEN légende *f*, SWG marquage *m*, TELEP indication *f*; ~ **percentage** TELEP taux de travail *m*; ~ **plate** PARTS plaque repère *f*; ~ **sequence** SWG séquence de marquage *f*; ~ **wave** RADIO signal de travail *m*

marshal: ~ **document** *n* GEN spécification de la signalisation *f*

MASE *abbr* (*message administration service element*) DATA élément de service de gestion de messages *m*

maser *n* ELECT maser *m*

mask *n* ELECT, RADIO, TELEP, TRANS, TV gabarit *m* masque *m*

masking *n* ELECT, RADIO, TELEP, TRANS, TV masquage *m*, effect de masque *m*; ~ **curve** TELEP courbe d'effet de masque *f*

masquerade *n* DATA usurpation d'identité *f*

mass: **~-impregnated insulation** *n* ELECT, PARTS isolation par imprégnation dans la masse *f*; ~ **soldering** *n* ELECT, PARTS brasage tendre en grande série *m*, brasage tendre simultané *m*; ~ **storage** *n* DATA mémoire de masse *f*, unités de stockage *f pl*

master: ~ **antenna television** *n* (*MATV*) TV télévision à antenne principale *f*; ~ **antenna TV system** *n* TV système de télévision à antenne *m*; ~ **clear** *n* COMP effacement global *m*; ~ **clock** *n* ELECT horloge centrale *f*, horloge maîtresse *f*, horloge mère *f*, horloge pilote *f*, horloge principale *f*; ~ **clock pulse** *n* ELECT impulsion d'horloge maîtresse *f*; ~ **data** *n pl* DATA données permanentes *f pl*; ~ **document** *n* COMP document maître *m*; ~ **earth station** *n* RADIO station terrienne maîtresse *f*; ~ **frame** *n* DATA trame maîtresse *f*; ~ **oscillator** *n* ELECT maître oscillateur *m*, oscillateur pilote *m*; ~ **oscillator frequency** *n* ELECT fréquence d'oscillateur pilote *f*; ~ **processor** *n* COMP processeur maître *m*; ~ **program** *AmE*, ~ **programme** *BrE n* TV programme maître *m*; ~ **register** *n* DATA fichier de référence *m*; ~ **set** *n* TELEP poste principal *m*; **~-slave control** *n* CONTR asservissement maître-esclave *m*; **~-slave flip-flop** *n* ELECT bascule maître-esclave *f*, bistable maître-esclave *m*; ~ **station** *n* RADIO station maîtresse *f*, station principale *f*; ~ **tape** *n* TV master *m*, bande mère *f*; ~ **time plan** *n* (*MTP*) RADIO plan directeur *m*; ~ **volume control** *n* PER APP commande de volume général *f*; ~ **volume slider** *n* COMP curseur de volume général *m*

mastergroup *n* TELEP, TRANS groupe tertiaire *m*; ~ **link** TELEP liaison en groupe tertiaire *f*; ~ **section** TELEP section de groupe tertiaire *f*

matched: ~ **cladding** *n* PARTS gaine adaptée *f*; ~ **junction** *n* RADIO jonction adaptée *f*; ~ **load** *n* RADIO charge adaptée *f*; ~ **termination** *n* RADIO terminaison adaptée *f*

matching *n* DATA concordance *f*, ELECT, GEN adaptation *f*; ~ **device** ELECT dispositif d'adaptation *m*; ~ **impedance** ELECT impédance caractéristique *f*; ~ **network** ELECT réseau d'adaptation *m*; ~ **order** DATA confirmation correspondante *f*; ~ **resistor** ELECT, PARTS résistance d'adaptation *f*; ~ **section** RADIO section d'adaptation *f*; ~ **transformer** ELECT transformateur d'adaptation *m*

MATD *abbr* (*maximum acceptable transit delay*) DATA temps de transit maximal acceptable *m*

mate *vt* ELECT connecter

material: ~ **dispersion** *n* TRANS dispersion de matériaux *f*; ~ **dispersion coefficient** *n* TRANS coefficient de dispersion de matériaux *m*; ~ **scattering** *n* RADIO diffusion de matériaux *f*

mating: ~ **connectors** *n pl* PARTS connecteurs homologues *m pl*

matrix: ~ **circuit** *n* ELECT circuit matriciel *m*; ~ **configuration** *n* ELECT configuration en matrice *f*; ~ **control station** *n* TELEP station multiprocesseur de connexion *f*; ~ **detector** *n* PER APP *scanner* capteur matriciel *m*; ~ **display** *n* ELECT affichage matriciel *m*; ~ **link** *n* TELEP liaison réseau *f*, liaison vers réseau *f*; **~-link interface** *n* TELEP interface de liaison réseau *f*; ~ **printer** *n* COMP, PER APP imprimante matricielle *f*; ~ **remover** *n* ELECT extracteur de matrice *m*; ~ **storage** *n* COMP mémoire matricielle *f*; ~ **switch** *n* SWG commutateur matriciel *m*; **~-switch control software machine** *n* SWG machine logique de gestion de commutateur *f*

matrixed *adj* COMP, ELECT, SWG, TELEP matricé

MATV *abbr* (*master antenna television*) TV télévision à antenne principale *f*; ~ **system** *n* TV système de télévision à antenne *m*

MAU *abbr* (*multistation access unit*) DATA unité d'accès multiples *f*

maximum: ~ **acceptable transit delay** *n* (*MATD*) DATA temps de transit maximal acceptable *m*; **~ admissible power** *n* TRANS puissance maximale admissible *f*; **~ announcement duration** *n* PER APP, TELEP durée maximum de l'annonce *f*; **~ justification rate** *n* TELEP débit maximal de justification *m*; **~ likelihood decoder** *n* RADIO décodeur à vraisemblance maximale *m*; **~ memory capacity** *n* PER APP, TELEP capacité mémoire maximale *f*; **~ message duration** *n* PER APP, TELEP durée maximum du message *f*; **~ non-operate current** *n* ELECT courant inactif maximal *m*, courant maximal de non-attraction *m*; **~ permissible interference level** *n* (*MPIL*) DATA niveau maximal admissible de brouillage *m* (*NMAB*); **~ relative time interval error** *n* (*MRTIE*) DATA erreur relative maximum d'intervalle de temps *f*; **~ resolution** *n* TV résolution maximale *f*; **~ retention time** *n* ELECT temps maximal de mémoire *m*; **~ sensitivity** *n* GEN sensibilité maximale *f*; **~ theoretical numerical aperture** *n* ELECT, TRANS ouverture numérique théorique maximale *f*; **~ time interval error** *n* (*MTIE*) DATA erreur maximum d'intervalle de temps *f*; **~ usable frequency** *n* (*MUF*) RADIO fréquence maximale utilisable *f* (*MUF*); **~ usable reading time** *n* COMP temps de lecture maximal utilisable *m*; **~-usable read number** *n* COMP nombre de lectures maximal utilisable *m*; **~ usable sensitivity** *n* RADIO sensibilité maximale utilisable limitée par l'amplification *f*; **~ usable viewing time** *n* TV temps de vision maximal utilisable *m*; **~ usable writing speed** *n* ELECT vitesse d'inscription maximale utilisable *f*

Mayday: **~ relay** *n* RADIO Mayday relay *m*

Mb *abbr* (*megabyte*) COMP, DATA Mo, Moctet (*mégaoctet*)

M-bit *n* TELEP bit de continuation *m*, bit M *m*

Mbps *abbr* (*millions of bits per second*) COMP, DATA Mbps (*millions de bits par seconde*)

MBS *abbr* (*multiblock synchronization signal unit*) SIG unité de signalisation de synchronisation de multibloc *f*

Mbyte *abbr* (*megabyte*) COMP, DATA Mo, Moctet *m* (*mégaoctet*)

MCA *abbr* (*multipoint command assignment token*) DATA jeton de commande multipoint d'assignation *f*

MCD *abbr* (*maintenance cell description*) DATA description de cellule de maintenance *f*

MCF *abbr* DATA (*message communication function*), fonction de communication de messages *f*, (*message confirmation*) confirmation de message *f*

MCI *abbr* (*media control interface*) COMP MCI (*interface de commande de support*), interface de contrôle des média *f*

MCN *abbr* (*mobile control node*) TELEP nœud de contrôle mobile *m*

MCS *abbr* (*multipoint communications system*) TRANS STM (*système de télécommunications multipoint*); **~ attachment** *n* TELEP rattachement au point MCSAP *m*; **~ channel** *n* TELEP canal MCS *m*; **~ connection** *n* TELEP connexion MCS *f*; **~ data transfer priority** *n* TELEP priorité de transfert de données MCS *f*; **~ domain** *n* TELEP domaine MCS *m*; **~ domain selector** *n* TELEP sélecteur de domaine MCS *m*; **~ interface data unit** *n* DATA, TELEP unité de données d'interface MCS *f*; **~ private channel** *n* TELEP canal privé MCS *m*; **~ private channel manager** *n* TELEP gestionnaire de canal privé MCS *m*; **~ protocol data unit** *n* (*MCSPDU*) TELEP unité de données de protocole du service MCS *f* (*MCSPDU*); **~ provider** *n* DATA, TELEP fournisseur de service MCS *m*; **~ service access point** *n* (*MCSAP*) TELEP point d'accès au service MCS *m* (*MCSAP*); **~ service data unit** *n* DATA, TELEP unité de données du service MCS *f*; **~ user** *n* TELEP utilisateur du service MCS *m*; **~ user ID** *n* DATA, TELEP identificateur d'utilisateur du service MCS *m*

MCSAP *abbr* (*MCS service access point*) TELEP MCSAP (*point d'accès au service MCS*)

MCSPDU *abbr* (*MCS protocol data unit*) TELEP MCSPDU (*unité de données de protocole du service MCS*)

MCT *abbr* (*multipoint command token claim*) DATA commande multipoint de demande de jeton *f*

M-curve *n* RADIO courbe de M *f*

MCV *abbr* DATA (*multipoint command visualization forcing*) commande multipoint d'imposition de visualisation *f*, TRANS (*message center vehicle AmE*, *message centre vehicle BrE*) MCV (*centre mobile de transmissions*)

MCW *abbr* (*modulated continuous waves*) RADIO MCW (*ondes entretenues modulées*)

MD *abbr* COMP (*mediation device*) équipement de médiation *m*, dispositif de médiation *m*, DATA (*management domain*) domaine de gestion *m*

m-derived: **~ filter** *n* RADIO filtre dérivé en m *m*

MDF *abbr* (*main distribution frame*) TELEP, TRANS RP (*répartiteur principal*)

MDS *abbr* (*minimum detectable signal*) TRANS SMD (*signal minimal détectable*)

MDSE *abbr* (*message delivery service element*) DATA élément de service de remise de messages *m*

mean[1]: **~ availability** *n* TEST disponibilité moyenne *f*; **~ busy hour** *n* TELEP heure chargée moyenne *f*; **~ deviation** *n* TEST écart moyen *m*; **~ downtime** *n* TEST durée moyenne d'indisponibilité *f*, temps d'arrêt moyen *m*; **~ error** *n* TEST erreur moyenne *f*; **~ failure intensity** *n* TEST intensité moyenne de dérangement *f*, intensité moyenne de défaillance *f*; **~ failure rate** *n* TEST taux moyen de dérangement *m*, taux moyen de défaillance *m*; **~ forward current** *n* ELECT, POWER courant direct moyen *m*; **~ Hall-plate temperature** *n* ELECT température moyenne de la plaque Hall *f*; **~ holding time per seizure** TELEP durée d'occupation moyenne par prise *f*; **~ life** *n* TEST durée moyenne de vie *f*, durée moyenne de vie utile *f*; **~ loss value** *n* TRANS valeur moyenne d'affaiblissement *f*; **~ maintenance man-hour** *n pl* TEST durée moyenne de maintenance en homme-heure *f*; **~ one-way propagation time** *n* TELEP temps de propagation moyen dans un sens *m*; **~ opinion score** *n* (*MOS*) COMP note moyenne d'opinion *f* (*MOS*); **~ power** *n* RADIO puissance moyenne *f*; **~ repair rate** *n* TEST taux moyen de réparation *m*; **~ repair time** *n* TEST durée moyenne de réparation *f*; **~ re-use distance** *n* RADIO distance moyenne de réutilisation *f*; **~ reverse voltage** *n* ELECT, POWER tension inverse moyenne *f*; **~ time before failure** *n* TEST temps moyen avant panne *m*; **~ time between failures** *n* (*MTBF*) TEST moyenne des temps de bon fonctionnement *f*, temps moyen entre pannes *m*; **~ time between interruptions** *n* (*MTBI*) TEST temps moyen entre interruptions *m*; **~ time between overhauls** *n* (*MTBO*) TEST périodicité moyenne de révisions *f*; **~ time to first failure** *n* (*MTFF*) TEST moyenne des temps de fonctionnement avant la première défaillance *f*; **~ time to recovery** *n* (*MTTR*) TEST durée moyenne de panne *f*; **~ time to repair** *n* (*MTTR*) TEST moyenne de temps total de réparation *f*; **~ time to service restoration** *n* (*MTSR*) TEST délai moyen de rétablissement du service *m*; **~ unavailability** *n* TEST indisponibilité moyenne *f*; **~ uptime** *n* TEST durée moyenne de disponibilité *f*, temps moyen de disponibilité *m* (*TMD*); **~ waiting time** *n* COMP délai moyen d'attente *m*

mean[2]: **~ entropy per character** *phr* DATA entropie moyenne par caractère; **~ holding time per seizure** *phr* TELEP durée d'occupation moyenne par prise

meaningful: **~ information** *n* DATA informations significatives *f pl*

measurable: **~ quantity** *n* TEST grandeur mesurable *f*

measure: **~ insert** *n* COMP insertion de mesures *f*

measured: **~ value** *n* TEST résultat du mesurage *m*, valeur mesurée *f*

measurement: **~ error** *n* TEST erreur de mesure *f*; **~ method** *n* TEST méthode de mesure *f*; **~ process** *n* TEST processus de mesure *m*; **~ technique** *n* TEST technique de mesure *f*

measuring: ~ **amplifier** *n* TEST amplificateur de mesure *m*; ~ **bridge** *n* TEST pont de mesure *m*; ~ **circuit** *n* TEST circuit de mesure *m*; ~ **detector** *n* TEST détecteur de mesure *m*; ~ **equipment** *n* TEST équipement de mesure *m*; ~ **instrument** *n* TEST instrument de mesure *m* appareil de mesure *m*; ~ **loop** *n* TEST boucle de mesure *f*; ~ **plug** *n* TEST fiche de mesure *f*; ~ **point** *n* TEST point de mesure *m*; ~ **range** *n* TEST étendue de mesure *f*, gamme de mesure *f*, plage de mesure *f*; ~ **rod** *n* TEST réglette-jauge *f*; ~ **transducer** *n* TEST transducteur de mesure *m*; ~ **transductor** *n* TEST transducteur de mesure *m*

mechanical: ~ **contact reading** *n* TV lecture par contact *f*; ~ **life** *n* TEST durée de vie mécanique *f*; ~ **optical switch** *n* SWG commutateur optomécanique *m*; ~ **splice** *n* PARTS épissure mécanique *f*; ~ **spot displacement** *n* ELECT déplacement mécanique du spot *m*; ~ **spot misalignment** *n* ELECT excentrage mécanique du spot *m*; ~ **tuning range** *n* ELECT plage d'accord du dispositif mécanique *f*; ~ **zero** *n* TEST zéro mécanique *m*

media[1]: ~-**based** *adj* TV médiatisé

media[2]: ~ **control interface** *n* (*MCI*) COMP interface de commande de support *f* (*MCI*), interface de contrôle des média *f*; ~ **conversion** *n* DATA conversion de support *f*

mediation: ~ **device** *n* (*MD*) COMP équipement de médiation *m*, dispositif de médiation *m*; ~ **function** *n* (*MF*) COMP fonction de médiation *f*

medium[1]: ~-**speed** *adj* DATA à vitesse moyenne; ~-**wave** *adj* RADIO onde moyenne

medium[2] *n* COMP, CONTR, TV support *m*; ~ **access control** (*MAC*) CONTR commande d'accès au support *f*, contrôle d'accès au support *m*; ~ **adaptor** (*MA*) ELECT, TELEP adaptateur de support *m*; ~ **power** POWER puissance moyenne *f*; ~ **power flux density** RADIO puissance surfacique moyenne *f*; ~ **range** RADIO portée moyenne *f*; ~ **rate** APPL *call charge* tarif blanc *m* (*Fra*); ~-**scale integration** (*MSI*) ELECT, PARTS intégration à échelle intermédiaire *f*, intégration à moyenne échelle *f*; ~-**term bit error rate** COMP taux d'erreur à moyen terme sur les bits *m*; ~ **wave** (*MW*) RADIO onde moyenne *f*

meet: ~-**me bridge** *n* APPL dispositif de téléréunion *m*; ~-**me call** *n* TELEP réunion téléphone *f*; ~-**me conference** *n* APPL conférence rendez-vous *f*

meg: ~ **of memory** *n* DATA méga *m*

megabyte *n* (*Mb*, *Mbyte*) COMP, DATA mégaoctet *m* (*Mo*, *Moctet*)

megabytes: ~ **of memory** *n pl* DATA mégaoctets de mémoire *m pl*

megaflops *n* (*one million floating-point operations per second*) COMP megaflops *m* (*un million d'opérations en virgule flottante per seconde*)

megalips *n* (*one million logical inferences per second*) COMP megalips *m* (*un million d'inférences logiques par seconde*)

megastream: ~ **circuit** *n* TRANS circuit à haut débit *m*

Megger® *n* TEST mégohmmètre *m*

megohmmeter *n* TEST mégohmmètre *m*

melt *vi* POWER fondre

membrane *n* PARTS diaphragme *m*, membrane *f*; ~ **keyswitch** ELECT interrupteur de clavier à membrane *m*; ~ **switch** ELECT interrupteur à membrane *m*

memo: ~ **recording** *n* PER APP message personnel *m*

memorandum: ~ **of understanding** *n* (*MOU*) GEN accord cadre *m*, protocole d'accord *m*

memorize *vt* COMP, DATA mémoriser

memory *n* COMP, DATA mémoire *f*; ~ **access mode** COMP mode d'accès à la mémoire *m*; ~ **architecture** COMP architecture de mémoire *f*; ~ **bank** COMP, DATA bloc de mémoire *m*; ~-**based reasoning** COMP, DATA raisonnement basé sur la mémoire *m*; ~ **board** COMP, DATA carte mémoire *f*, carte à mémoire *f* (*CAM*); ~ **card** COMP, DATA carte mémoire *f*, carte à mémoire *f* (*CAM*); ~ **circuit** ELECT circuit de mémoire *m*; ~ **cleaning** COMP effacement mémoire *m*; ~ **configuration** COMP configuration mémoire *f*; ~ **dump** COMP vidage de mémoire *m*; ~ **erasure** COMP effacement mémoire *m*; ~-**expansion capacity** COMP possibilité d'extension de mémoire *f*; ~-**management program** COMP programme de gestion de mémoire *m*; ~ **management unit** (*MMU*) COMP unité de gestion de mémoire *f*; ~ **manager** COMP gestionnaire de mémoire *m*; ~ **number** PER APP numéro de mémoire *m*; ~ **protection** COMP protection de mémoire *f*; ~ **random access** COMP accès aléatoire à la mémoire *m*; ~ **shortage** COMP insuffisance de mémoire *f*; ~ **space** COMP espace mémoire *m*; ~ **storage** COMP, DATA mémorisation *f*

menu *n* COMP, DATA menu *m*; ~ **bar** COMP barre de menu *f*; ~ **mode** TELEP mode menu *m*; ~-**mode operating sequence** TELEP séquence de fonctionnement en mode menu *f*; ~ **output** TELEP sortie en mode menu *f*; ~ **selection procedure** TELEP procédure de sélection en mode menu *f*

mercury: ~ **contact** *n* ELECT contact au mercure *m*; ~ **contact relay** *n* ELECT relais à contact de mercure *m*; ~ **relay** *n* ELECT relais à mercure *m*; ~ **vapor tube** *AmE*, ~ **vapour tube** *BrE* *n* ELECT tube à vapeur de mercure *m*; ~-**wetted contact relay** *n* ELECT relais à contacts mouillés au mercure *m*; ~-**wetted relay** *n* ELECT relais à contacts mouillés au mercure *m*; ~ **wetting** *n* ELECT mouillage au mercure *m*

meridional: ~ **ray** *n* TRANS rayon méridien *m*

mesa *n* ELECT mesa *m*

mesh *n* COMP, ELECT, RADIO grille *f*, maille *f*, RADIO maillage *m*; ~ **analysis** ELECT analyse par boucles *f*; ~ **effect** RADIO grille apparente *f*; ~ **electrode** ELECT grille à mailles *f*; ~ **reflector** RADIO réflecteur à mailles *m*

meshed[1] *adj* COMP, ELECT, RADIO maillé

meshed[2]: ~ **network** *n* COMP, TRANS réseau maillé *m*

meshing *n* COMP, SWG maillage *m*

mesochronous *adj* TELEP, TRANS mésochrone

mesomorph: ~ **state** *n* ELECT état mésomorphe *m*

message: ~ **administration service element** *n* (*MASE*) DATA élément de service de gestion de messages *m*; ~ **announcement system** *n* TELEP émetteur d'annonces *m*; ~ **broadcaster** *n* PER APP diffuseur d'appels *m*, diffuseur de messages *m*; ~ **center vehicle** *AmE*, ~ **centre vehicle** *BrE* *n* (*MCV*) TRANS centre mobile de transmissions *m* (*MCV*); ~ **communication function** *n* (*MCF*) DATA fonction de communication de messages *f*; ~ **confirmation** *n* (*MCF*) DATA confirmation de message *f*; ~ **delivery service element** *n* (*MDSE*) DATA élément de service de remise de messages *m*; ~ **discrimination** *n* COMP discrimination des messages *f*; ~ **display pager** *n* RADIO radiomessagerie unilatérale *f* (*RMU*), RMU alphanumérique *f*; ~ **distribution** *n* TELEP distribution des messages *f*; ~ **feedback** *n* TELEP contrôle par retour de l'information *m*; ~ **handling** *n* (*MH*) APPL, TELEP messagerie *f*; ~-**handling system** *n* (*MHS*) APPL système de messagerie *m*, système de transmission de messages *m* (*STM*), DATA messagerie *f*, système de messagerie *m*, système de transmission de messages *m* (*STM*), RADIO système de messagerie *m*, système de transmission de messages *m* (*STM*), TELEP système de transmission de messages *m* (*STM*); ~ **handling system service element** *n* (*MHS-SE*) APPL, DATA, RADIO, TELEP élément de service de système de traitement de messages *m*; ~ **mode** *n* DATA mode message *m*; ~-**oriented text interchange system** *n* (*MOTIS*) DATA système d'échange de textes en mode message *m*, système de transfert de textes en mode message *m*; ~ **pager** *n* GEN, RADIO, TELEP, TRANS récepteur de radiomessagerie *m* (*RMU*); ~ **protocol data unit** *n* (*MPDU*) DATA unité de données de protocole de message *f* (*MPDU*); ~ **received counter** *n* PER APP, TELEP *answer machine* compteur de messages reçus *m*; ~ **received indicator** *n* PER APP *answer machine* indicateur de messages reçus *m*; ~ **refusal signal** *n*

SIG signal de refus de message *m*; ~ **register** *n* DATA compteur de taxes *m*; ~ **relay** *n* TRANS service de transmission de messages *m*; ~ **retrieval** *n* DATA recherche des messages *f*; ~ **retrieval service element** *n* (*MRSE*) DATA élément de service de recherche de messages *m*; ~ **route** *n* TRANS route de message *f*; ~ **signaling unit** *AmE*, ~ **signalling unit** *BrE n* DATA, TELEP trame sémaphore de message *f* (*TSM*); ~ **sink** *n* TRANS collecteur de messages *m*; ~ **source** *n* TRANS source de messages *f*; ~ **store** *n* (*MS*) DATA, TELEP mémoire de messages *f*; ~ **storing** *n* DATA archivage des messages *m*; ~**-submission abstract-operation** *n* DATA opération abstraite de dépôt de message *f*; ~ **submission service element** *n* (*MSSE*) DATA élément de service de dépôt de messages *m*; ~ **suffix** *n* COMP suffixe de message *m*; ~ **switch** *n* SWG commutateur de messages *m*; ~ **switching** *n* SWG commutation de messages *f*; ~**-switching center** *AmE*, ~**-switching centre** *BrE n* SWG centre de commutation de messages *m*; ~**-switching network** *n* TRANS réseau à commutation de messages *m*; ~**-switching processor** *n* PER APP processeur de commutation de messages *m*; ~**-switching system** *n* SWG, TRANS système de commutation de messages *m*; ~ **toll service** *n* (*MTS*) TELEP système interurbain à communications tarifées *m* (*SICT*); ~ **transfer** *n* (*MT*) DATA, TELEP transfert de messages *m*; ~ **transfer agent** *n* (*MTA*) DATA, TELEP agent de transfert de messages *m*; ~ **transfer part** *n* (*MTP*) DATA, TELEP sous-système de transfert de messages *m* (*SSTM*), sous-système de transport de messages *m* (*SSTM*); ~ **transfer part receiving time** *n* COMP temps de réception du sous-système transport de messages *m* (T_{mr}); ~ **transfer part sending time** *n* TELEP temps d'émission du sous-système transport de messages *m* (T_{ms}); ~ **transfer service element** *n* (*MTSE*) DATA élément de service de transfert de messages *m*; ~ **transfer system** *n* (*MTS*) DATA, TELEP système de transfert de messages *m*; ~ **transfer time at signaling transfer points** *AmE*, ~ **transfer time at signalling transfer points** *BrE n* TRANS temps de transfert des messages aux points de transfert *m* (*Tcs*); ~ **transmission system** *n* DATA système de diffusion de messages *m*; ~ **waiting indication** *n* TELEP indication d'attente de message *f*

Message: ~**-master** *n* APPL, TELEP récepteur de radiomessagerie *m*

messaging: ~ **service** *n* APPL service de messagerie *m*

messenger: ~ **cable** *n* TRANS câble porteur *m*; ~ **call** *n* TELEP communication avec avis d'appel *f* (*communication AVP*); ~ **wire** *n* TRANS câble porteur *m*

metal: ~**-clad base material** *n* ELECT support isolant métallisé *m*; ~**-clad conductor** *n* ELECT conducteur à revêtement métallique *m*, conducteur métallisé *m*; ~ **conductor cable** *n* TRANS câble métallique *m*; ~**-core board** *n* ELECT carte à cœur métallique *f*; ~**-film capacitor** *n* ELECT, PARTS condensateur à couche métallique *m*, condensateur à film métallique *m*; ~ **insulator semiconductor** *n* (*MIS*) ELECT semiconducteur sur isolant métal *m* (*MIS*); ~**-organic chemical vapor deposition** *AmE*, ~**-organic chemical vapour deposition** *BrE n* ELECT, TRANS dépôt en phase vapeur de matériaux métal-organiques *m*; ~**-organic vapor phase epitaxy** *AmE*, ~**-organic vapour phase epitaxy** *BrE n* ELECT épitaxie en phase vapeur de matériaux métal-organiques *f*; ~**-oxide semiconductor** *n* (*MOS*) ELECT metal-oxide semiconducteur *m*; ~ **twinax plug** *AmE*, ~ **twinaxe plug** *BrE n* ELECT prise twinaxiale métallique *f*

metalization *AmE see metallization BrE*

metalized *AmE see metallized BrE*

metallic: ~ **circuit** *n* TELEP circuit métallique *m*; ~ **coating** *n* TRANS revêtement métallique *m*; ~ **conductor cable** *n* TRANS câble à conducteurs métalliques *m*; ~ **crosspoint** *n* SWG point d'intersection métallique *m*; ~ **rectifier** *n* ELECT redresseur sec *m*; ~ **thin film** *n* ELECT couche mince métallique *f*; ~ **twist** *n* ELECT tresse métallique *f*

metallization *n BrE* ELECT, PARTS métallisation *f*

metallized: ~**-film capacitor** *n BrE* ELECT, PARTS condensateur à couche métallisée *m*, condensateur à film métallisé *m*; ~**-paper capacitor** *n BrE* POWER condensateur au papier métallisé *m*; ~ **screen** *n BrE* ELECT écran métallisé *m*

metareasoning *n* COMP métaraisonnement *m*

metastable *adj* ELECT métastable

meteor: ~**-burst communication** *n* RADIO communication par impulsions météoriques *f*, communication par salves météoriques *f*; ~**-burst propagation** *n* RADIO propagation par impulsions météoriques *f*, propagation par salves météoriques *f*

meter *n* GEN compteur *m*, SIG indicateur *m*; ~ **account save** TELEP sauvegarde des comptes de taxe *f*; ~ **changeover clock** POWER minuterie de compteur *f*; ~ **indication** TELEP indication de compteur *f*, relevé du compteur *m*; ~ **movement** TEST élément moteur *m*; ~ **pulse** TELEP impulsion de comptage *f*; ~**-pulse generation** TELEP génération d'impulsions de compteur *f*; ~**-pulse generator** TELEP générateur d'impulsions de comptage *m*; ~ **pulse interval** TELEP intervalle d'impulsions de compteur *m*; ~ **pulse sender** TELEP émetteur d'impulsions de compteur *m*; ~ **rack** TELEP bâti de compteurs *m*; ~ **unit** TELEP unité de compteur *f*; ~**-value indication** TELEP indication de valeur de compteur *f*, relevé du compteur *m*; ~ **wire** TELEP fil de compteur *m*

metering *n* TELEP comptage *m*, taxation *f*; ~ **period** TELEP période de taxation *f*; ~ **rate** TELEP palier de comptage *m*; ~ **signal** TELEP signal de comptage *m*

metric: ~ **wave** *n* TRANS onde métrique *f*

metropolitan: ~ **area network** *n* (*MAN*) TRANS réseau de zone urbaine *m*; ~ **network** *n* TRANS réseau métropolitain *m*, réseau urbain *m*; ~ **switch** *n* SWG commutateur urbain *m*

MF *abbr* COMP (*mediation function*) fonction de médiation *f*, GEN, SIG, TELEP (*multiple frequency*) FM (*fréquence multiple*); ~ **generator** *n* SIG, TELEP générateur de signalisation multifréquence *m*, générateur multifréquence *m*; ~ **mode** *n* PER APP *answer machine* mode fréquences vocales *m*; ~ **receiver** *n* SIG récepteur MF *m*; ~ **sender-receiver** *n* SIG, TELEP émetteur-récepteur multifréquence *m*; ~ **telephone** *n* TELEP poste à touches musicales *m*, téléphone multifréquence *m*

MFA *abbr* (*multiframe alignment*) TRANS verrouillage de multitrame *m*

m-factor *n* ELECT coefficient d'amplification *m*

M-field *abbr* (*mode-of-operation field*) PER APP zone M *f* (*zone de mode d'exploitation*)

MHD: ~ **segment** *n* COMP, DATA segment MHD *m*

MHEG *abbr* (*multimedia and hypermedia information coding experts group*) COMP MHEG

MHS *abbr* (*message handling system*) APPL, DATA, RADIO, TELEP STM (*système de messagerie, système de transmission de messages*)

MHS-SE *abbr* (*message handling system service element*) APPL, DATA, RADIO, TELEP MHS-SE (*élément de service de système de traitement de messages*)

MIB *abbr* (*management imformation base*) COMP, DATA base de donées de gestion *f*

MIC *abbr* ELECT (*microwave integrated circuit*) circuit intégré pour micro-ondes *m*, (*monolithic integrated circuit*) circuit intégré monolithique *m*, SWG (*marked idle channel*) repérage des voies au repos *m*

mica: ~ **capacitor** *n* ELECT, PARTS condensateur au mica *m*

MICR *abbr* (*magnetic ink character recognition*) APPL MICR (*reconnaissance magnétique des caractères*)

micro: ~**-alloy diffused transistor** *n* (*MADT*) ELECT transistor à micro-alliage et diffusion *m*; ~**-assembly** *n* ELECT microassemblage *m*; ~**-image** *n* DATA microimage *f*

microbend: ~ **loss** *n* TRANS perte par microcourbures *f*; ~ **sensor** *n* TRANS détecteur de microcourbures *m*
microbending *n* TRANS microcourbure *f*
microcassette *n* PER APP microcassette *f*
microcell: ~ **network** *n* RADIO réseau de microcellules *m*
microchip *n* ELECT microplaquette *f*
microcircuit *n* ELECT microstructure *f*
microcom: ~ **network protocol** *n* DATA MNP
microcomputer *n* COMP microordinateur *m* (*PC*)
microelectronics *n pl* ELECT microélectronique *f*
microform *n* DATA microforme *m*; ~ **reader** DATA visionneuse pour microforme *f*; ~ **reader/printer** DATA lecteur-reproducteur de microformes *m*; ~-**retrieval terminal** DATA appareil de repérage de microforme *m*
microphone *n* (*mike*) ELECT, PER APP, TELEP microphone *m* (*micro*); ~ **current feed** POWER, TELEP courant d'alimentation du microphone *m*; ~ **cutoff** PER APP *answer machine* coupure de microphone *f*; ~ **input** ELECT entrée du microphone *f*; ~ **noise** ELECT bruit microphonique *m*
microphonic: ~ **effect** *n* ELECT effet microphonique *m*, microphonie *f*; ~ **noise** *n* ELECT bruit microphonique *m*
microphonism *n* ELECT bruit microphonique *m*, effet microphonique *m*
microphony *n* ELECT effet microphonique *m*, microphonie *f*
microprocessor *n* COMP microprocesseur *m*; ~ **control** SWG commande par microprocesseurs *f*
microprogramming *n* COMP microprogrammation *f*
microribbon *n* ELECT microruban *m*
microsegmentation *n* RADIO microsegmentation *f*
microslot *n* TRANS microfente *f*
Microsoft®: ~ **disk-operating system** *n* (*MS-DOS®*) COMP DOS Microsoft® *m* (*MS-DOS*), système d'exploitation Microsoft® *m* (*MS-DOS*)
microstripper *n* ELECT microstripper *m*
microswitch *n* ELECT microinterrupteur *m*, SWG microcommutateur *m*
microtonal[1] *adj* COMP microtonal
microtonal[2]: ~ **scale** *n* COMP échelle microtonale *f*
microtune *n* ELECT microaccordage *m*
microtuning *n* ELECT microaccordage *m*
microwave[1] *adj* RADIO hertzien
microwave[2] *n* RADIO microonde *f*, onde hyperfréquence *f*; ~ **absorption** RADIO absorption hyperfréquence *f*; ~ **aerial** RADIO antenne hyperfréquence *f*; ~ **amplifier** RADIO amplificateur hyperfréquence *m*; ~ **antenna** RADIO antenne hyperfréquence *f*; ~ **attenuator** RADIO affaiblisseur hyperfréquence *m*, atténuateur hyperfréquence *m*; ~ **beam** RADIO faisceau hertzien *m*; ~ **circulator** RADIO, TRANS circulateur hyperfréquence *m*; ~ **communications** *n pl* RADIO communications hyperfréquence *f pl*, communications micro-ondes *f pl*; ~ **detector** RADIO détecteur d'hyperfréquences *m*; ~ **engineering** RADIO technique des hyperfréquences *f*, technique des micro-ondes *f*; ~ **frequency** RADIO, TV hyperfréquence *f*; ~ **generation** RADIO synthèse d'hyperfréquences *f*; ~ **integrated circuit** (*MIC*) ELECT circuit intégré pour micro-ondes *m*; ~ **landing system** (*MLS*) RADIO système d'atterrissage hyperfréquence *m*; ~ **link** RADIO liaison hertzienne *f*; ~ **link analyser** *BrE*, ~ **link analyzer** *AmE* TEST analyseur de faisceaux hertziens *m*, analyseur de liaisons hyperfréquences *m*; ~ **modulator** RADIO modulateur hyperfréquence *m*; ~ **phase changer** RADIO déphaseur hyperfréquence *m*; ~ **radio** RADIO faisceau hertzien *m*; ~ **radio system** RADIO réseau hertzien *m*; ~ **relay links** *n pl* TRANS faisceau hertzien *m*; ~ **resonator** RADIO résonateur à hyperfréquences *m*; ~ **signal** RADIO, SIG signal hertzien *m*; ~ **switch sequence** SWG séquence du commutateur hyperfréquences *f*; ~ **system** RADIO système hertzien *m*; ~ **tower** RADIO pylône hertzien *m*, tour hertzienne *f*
mid: ~-**level** *n* RADIO niveau moyen *m*; ~-**range** *n* GEN milieu de l'étendue *m*; ~-**range speaker** *n* ELECT haut-parleur de médiums *m*, médial *m*
MID *abbr* (*multiplexing identification*) TRANS identification de multiplexage *f*
middle: ~ **marker beacon** *n* RADIO radioborne intermédiaire *f*
midpoint: ~ **signaling** *AmE*, ~ **signalling** *BrE* *n* SIG signalisation à point milieu *f*; ~ **tap** *n* ELECT *of transformer* prise centrale *f*, prise médiane *f*
migrate *vi* GEN migrer
migration *n* GEN migration *f*
mike *n* (*microphone*) ELECT micro *m* (*microphone*), PER APP, TELEP microphone *m*
MIL *abbr* (*multipoint indication loop*) DATA indication multipoint de boucle *f*
military: ~ **communication system** *n* APPL, RADIO, TRANS système de communication militaire *m*
Miller: ~ **effect** *n* ELECT effet Miller *m*
milliammeter *n* ELECT milliampèremètre *m*
milliamp *n* (*mA*) ELECT milliampère *m* (*mA*)
milliken: ~ **conductor** *n* ELECT âme segmentée *f*
millions: ~ **of bits per second** *phr* (*Mbps*) COMP, DATA millions de bits par seconde (*Mbps*)
mime *vt* TV *song* chanter en playback
mineral: ~ **insulation** *n* ELECT isolation minérale *f*
mini: ~ **disk** *n* COMP minidisque *m*; ~ **floppy** *n* COMP minidisquette *f*; ~ **floppy disk** *n* COMP minidisquette *f*; ~ **jack** *n* PER APP mini-prise *f*; ~-**star configuration** *n* TV miniétoile *f*; ~-**wrapping** *n* ELECT mini-enroulement *m*
miniature: ~ **fuse** *n* PARTS, POWER fusible miniature *m*
minigage: ~ **loop** *AmE see minigauge loop BrE*
minigauge: ~ **loop** *n BrE* TELEP boucle minigauge *f*
minimize *vt* COMP réduire
minimizing *n* COMP réduction *f*
minimum: ~ **acceptable interval** *n* TELEP intervalle minimum acceptable *m*; ~ **bend radius** *n* TRANS rayon de courbure minimal *m*; ~ **cathode current** *n* ELECT courant cathodique minimal *m*; ~ **charge** *n* APPL tarif minimum *m*; ~ **detectable signal** *n* (*MDS*) TRANS signal minimal détectable *m* (*SMD*); ~ **interference threshold** *n* RADIO seuil inférieur de brouillage *m*; ~ **interval** *n* TELEP intervalle minimum *m*; ~ **net loss** *n* TRANS équivalent minimal net *m*, affaiblissement minimal admissible *m*; ~ **operate current** *n* ELECT courant minimal d'attraction *m*, courant minimal de commande *m*; ~ **payable** *n* APPL minimum de perception *m*; ~ **permissible basic transmission loss** *n* (*MPBTL*) TRANS affaiblissement de transmission de référence minimal admissible *m* (*ATRMA*); ~-**phase filter** *n* TRANS filtre à déphasage minimal *m*; ~-**shift keying** *n* (*MSK*) RADIO, SIG modulation par déplacement minimal *f* (*MDM*); ~-**standard aerial** *n* RADIO antenne normale minimale *f*; ~-**standard antenna** *n* RADIO antenne normale minimale *f*; ~-**usable erasing time** *n* ELECT temps d'effacement minimal utilisable *m*; ~ **usable field strength** *n* RADIO champ minimal utilisable *m*; ~-**usable power flux density** *n* RADIO champ minimal utilisable *m*, puissance surfacique minimale utilisable *f*; ~ **usable reading speed** *n* ELECT vitesse de lecture minimale utilisable *f*
minor[1]: ~ **defective** *adj* TEST défectueux mineur
minor[2]: ~ **alarm** *n* TEST alarme mineure *f*; ~ **defect** *n* TEST défaut mineur *m*; ~ **restart** *n* SWG redémarrage restreint *m*; ~ **system restart** *n* SWG redémarrage système restreint *m*
minority: ~ **carrier** *n* ELECT porteur minoritaire *m*
minus: ~ **polarity** *n* POWER polarité négative *f*
mirror[1] *n* DATA *Internet* miroir *m*; ~ **site** DATA site miroir *m*; ~ **surface** TRANS surface de miroir *f*; ~ **writing** DATA écriture miroir *f*
mirror[2] *vt* COMP créer une image miroir de
MIS *abbr* (*metal insulator semiconductor*) ELECT MIS (*semi-*

conducteur sur isolant métal); ~ **user** *n* DATA utilisateur-MIS *m*

misalignment: ~ **loss** *n* TRANS *fibre optics* perte par défaut d'alignement *f*

miscellaneous: ~ **apparatus rack** *n* (*MAR*) PARTS, TELEP bâti pour appareils divers *m*

misfire *n* ELECT raté d'allumage *m*

mishandling: ~ **failure** *n* TEST problème d'erreur de manipulation *m*

mismatch *n* ELECT discordance *f*, disharmonie *f*, GEN, TRANS adaptation défectueuse *f*, défaut d'adaptation *m*, désadaptation *f*, erreur d'adaptation *f*, erreur d'assortiment *m*, non-concordance *f*

mispointing *n* RADIO erreur de pointage *f*

misread *n* COMP erreur de lecture *f*

mistracking *n* COMP défaut d'asservissement *m*, erreur de guidage *f*

misuse: ~ **failure** *n* TEST problème de mauvaise utilisation *m*

mixed: ~ **analog-digital channel (circuit)** *n* TELEP voie (circuit) mixte analogique-numérique *f*; ~ **lines** *n pl* TELEP lignes mixtes *f pl*; ~ **mode** *n* GEN mode mixte *m*; **~-mode dialing** *AmE*, **~-mode dialling** *BrE n* PER APP numérotation dc/fv *f*, numérotation mixte *f*; ~ **terrain** *n* GEN surface aléatoire *f*

mixer *n* COMP, GEN, RADIO, TRANS mélangeur *m*; ~ **stage** RADIO étage mélangeur *m*

mixing: ~ **console** *n* COMP console de mélange *f*, console de mixage *f*, pupitre de mélange *m*, RADIO table de mixage *f*; ~ **desk** *n* COMP console de mélange *f*, console de mixage *f*, pupitre de mélange *m*; ~ **ratio** *n* RADIO rapport de mélange *m*; ~ **tray** *n* ELECT bac de mélange *m*

MJD *abbr* (*modified julian date*) RADIO DJM (*date julienne modifiée*)

ML: ~ **assigned to call charging** *n* TELEP machine logique taxeur *f*; ~ **in charge of message distribution** *n* TELEP machine logique marqueur *f*

MLI *abbr* (*multilayer isolation*) ELECT isolation multicouches *f*

MLM *abbr* (*multilongitudinal modes*) TELEP, TRANS MLM (*modes longitudinaux multiples*)

MLP *abbr* (*multilink procedure*) DATA procédure multiliaison *f*

MLS *abbr* (*microwave landing system*) RADIO système d'atterrissage hyperfréquence *m*

MMC *abbr* (*man-machine communications*) COMP RHM (*relation homme-machine*)

MMIC *abbr* (*monolithic microwave integrated circuit*) ELECT, PARTS circuit intégré monolithique hyperfréquence *m*

MMJ *abbr* (*modified modular jack*) ELECT prise modulaire modifiée *f*

MML *abbr* (*man-machine language*) COMP langage homme-machine *m*

MMP *abbr* (*modified modular plug*) ELECT prise modulaire modifiée *f*

MMU *abbr* (*memory management unit*) COMP unité de gestion de mémoire *f*

MNP *abbr* (*microcom network protocol*) DATA MNP

MO *abbr* COMP (*magneto-optic*) MO (*magnéto-optique*), DATA (*managed object*) objet géré *m*, ELECT (*magneto-optic*) MO (*magnéto-optique*); ~ **disk** *n* COMP, DATA, ELECT disque magnéto-optique *m*

mobile[1]: ~ **charge carrier** *n* ELECT porteur de charge libre *m*, porteur libre *m*; ~ **communications** *n pl* RADIO communications avec les mobiles *f pl*; ~ **control node** *n* (*MCN*) TELEP nœud de contrôle mobile *m*; ~ **data-processing terminal equipment** *n* RADIO équipement terminal de traitement de données mobile *m* (*ETTD-MT*); ~ **data-transmission network** *n* RADIO réseau mobile de transmission de données *m*; ~ **installation** *n* RADIO installation mobile *f*; ~ **location registration** *n* RADIO enregistrement de la position de la station mobile *m*; ~ **number** *n* TELEP numéro mobile *m*; ~ **office** *n* RADIO bureau mobile *m*; ~ **phone** *n* TELEP mobile *m*, mobilophone *m*, radiotéléphone *m*; ~ **radio fleet** *n* RADIO flotte *f*; ~ **radio station** *n* RADIO mobile *m*; ~ **radio system** *n* RADIO système de radiocommunication du service mobile *m*; ~ **radio telephone** *n* RADIO, TELEP radiotéléphone *m*; ~ **radio telephone services** *n pl* RADIO, TELEP services radiotéléphoniques mobiles *m pl*; ~ **satellite telecommunication** *n* RADIO système de télécommunication mobile par satellite *m*; ~ **service** *n* RADIO service mobile *m*; ~ **services switching center** *AmE*, ~ **services switching centre** *BrE n* SWG, TELEP centre de commutation de services mobiles *m*; ~ **station** *n* (*MS*) RADIO, TELEP station mobile *f*; **~-station charge** *n* TELEP taxe de station mobile *f*; ~ **station roaming number** *n* (*MSRN*) TELEP numéro de déplacement de station mobile *m*; ~ **subscriber identity** *n* RADIO identité d'abonné mobile *f*; ~ **switching center** *AmE*, ~ **switching centre** *BrE n* SWG, TELEP centre de commutation mobile *m*; ~ **telecommunication switching office** *n* (*MTSO*) RADIO centre de commutation de téléphonie mobile *m* (*CCTM*); ~ **telephone** *n* RADIO radiotéléphone *m*, TELEP téléphone mobile *m*, mobilophone *m*, radiotéléphone *m*; ~ **telephone operating company** *n* RADIO, TELEP société exploitante du téléphone mobile *f*; ~ **telephone service** *n* (*MTS*) RADIO, TELEP service radiotéléphonique mobile *m* (*SRM*); ~ **telephone station** *n* RADIO, TELEP poste radiotéléphonique mobile *m*, radiotéléphone *m*; ~ **telephone system** *n* RADIO, TELEP système radiotéléphonique mobile *m* (*SRM*); ~ **telephony** *n* RADIO radiotéléphonie *f*; ~ **telephony system** *n* RADIO, TELEP système de radiotéléphonie *m*; **~-to-base relay** *n* RADIO relais nominal *m*; ~ **unit** *n* TELEP station mobile *f*

mobile[2]: **on ~ subscriber not registered** *phr* RADIO sur non-inscription de l'abonné mobile

MOC *abbr* (*managed-object class*) DATA classe d'objet géré *f*

MOCS *abbr* (*managed-object conformance statement*) DATA déclaration de conformité d'objet géré *f*; ~ **proforma** *n* DATA formulaire MOCS *m*

modal: ~ **distortion** *n* TRANS distorsion modale *f*; ~ **noise** *n* TRANS bruit de tacheture *m*, bruit modal *m*

mode *n* GEN, COMP mode *m*; ~ **changer** RADIO transformateur de mode *m*; ~ **coupling** TRANS couplage de modes *m*; ~ **dependency** DATA mode destiné à identifier une entrée *m*; ~ **distortion** TRANS distorsion modale *f*; ~ **distribution** TRANS répartition des modes *f*; **~-field** TRANS champ de mode *m*; ~ **field diameter** TRANS diamètre du champ de mode *m*, diamètre du faisceau guidé *m*; **~-field profile** TRANS profil du champ de mode *m*; ~ **filter** TRANS filtre de mode *m*; ~ **hopping** TRANS instabilité de mode *f*, saut de mode *m*; ~ **jumping** PER APP changement de mode *m*, TRANS changement de mode *m*, saut de mode *m*; ~ **mixer** TRANS brasseur de modes *m*, brouilleur de modes *m*, mélangeur de modes *m*; ~ **mixing** TRANS brassage de modes *m*, mixage de modes *m*; **~-of-operation field** (*M-field*) PER APP zone de mode d'exploitation *f* (*zone M*); ~ **or type of communication identification** TELEP identification du type ou du mode de la communication *f*; ~ **of oscillation** RADIO mode d'oscillation *m*; ~ **partition** TRANS partition des modes *f*; **~-partition noise** TRANS bruit de partition des modes *m*; ~ **scrambler** TRANS brasseur de modes *m*, brouilleur de modes *m*, mélangeur de modes *m*; **~-selective coupler** TRANS coupleur à sélection de mode *m*; ~ **stripper** TRANS extracteur de modes de gaine *m*; ~ **transformer** RADIO transformateur de mode *m*; ~ **volume** TRANS volume des modes *m*

model *n* ELECT maquette *f*, GEN modèle *m*

modem *n* (*modulator/demodulator*) COMP, DATA, TRANS modem *m* (*modulateur/démodulateur*); ~ **back** DATA retournement du modem *m*; ~ **card** DATA carte modem *f*; ~ **clock** DATA horloge de modem *f*; ~ **control** DATA commande par modem *f*; ~ **control key** DATA touche modem *f*; ~ **eliminator** DATA modem nul *m*; ~ **holdover period** DATA durée de maintien de synchronisme de modem *f*; ~ **interface** DATA jonction de modem *f*; ~ **performance** DATA qualité de fonctionnement *f*; ~ **port** COMP port modem *m*; ~ **protection equipment** DATA protection modem *f*; ~ **radio** RADIO modem-radio *m*; ~ **receiver** DATA récepteur de modem *m*; ~ **tester** TEST testeur de modem *m*; ~ **transmitter** DATA émetteur de modem *m*; ~ **turnaround** DATA délai d'inversion *m*, temps de renversement *m*

modification: ~ **indicator** *n* DATA indicateur de modification *m*

modified: ~ **alternate mark inversion** *n* SIG, TRANS signal bipolaire modifié *m*; ~ **constraint** *n* DATA contrainte modifiée *f*; ~ **Julian date** *n* (*MJD*) RADIO date julienne modifiée *f* (*DJM*); ~ **Julian day** *n* RADIO jour julien modifié *m*; ~ **modular jack** *n* (*MMJ*) ELECT prise modulaire modifiée *f*; ~ **modular plug** *n* (*MMP*) ELECT prise modulaire modifiée *f*; ~ **refractive index** *n* RADIO indice de réfraction modifié *m*

modifier *n* COMP modificateur *m*

moding *n* ELECT glissement de mode *m*, instabilité de mode *f*

modular[1] *adj* COMP, GEN modulaire

modular[2]: ~ **jack** *n* ELECT prise modulaire *f*, prise femelle modulaire *f*; ~ **patching** *n* ELECT brassage modulaire *m*; ~ **plug blade** *n* ELECT lame prise modulaire *f*; ~ **principle** *n* TELEP principe modulaire *m*; ~ **spacing** *n* TELEP espacement modulaire *m*; ~ **system** *n* TELEP système modulaire *m*

modularity *n* GEN modularité *f*

modularization *n* COMP, GEN modularisation *f*

modulate *vt* COMP, SIG, TRANS, TV moduler

modulated: ~ **continuous waves** *n pl* (*MCW*) RADIO ondes entretenues modulées *f pl* (*MCW*); ~ **signal** *n* SIG, TRANS signal modulé *m*; ~ **tariff time slot** *n* DATA modulation horaire *f*; ~ **value** *n* DATA valeur modulée *f*; ~ **wave** *n* TRANS onde modulée *f*

modulating: ~ **action** *n* CONTR action par modulation *f*; ~ **frequency** *n* TELEP, TRANS fréquence de modulation *f*; ~ **signal** *n* TRANS signal de modulation *m*; ~ **wave** *n* TRANS onde modulante *f*, oscillation modulante *f*

modulation *n* GEN, RADIO, TELEP, TRANS modulation *f*; ~ **by pulses** TRANS modulation par suite d'impulsions *f*; ~ **coherence** TELEP cohérence d'un sématème *f*, cohérence d'une modulation *f*; ~ **depth** TRANS taux de modulation *m*; ~ **element** TELEP, TRANS élément de modulation *m*; ~ **factor** TRANS taux de modulation *m*; ~ **with a fixed reference** TELEP modulation avec référence fixe *f*; ~ **frequency** TELEP, TRANS fréquence de modulation *f*; ~ **generator** TRANS générateur de modulation *m*; ~ **index** RADIO indice de modulation *m*; ~ **lever** COMP levier de modulation *m*; ~ **noise** TRANS bruit de modulation *m*; ~ **percentage** TRANS taux de modulation *m*; ~ **polarity** TV polarité de la modulation *f*; ~ **product** TELEP modulats *m pl*, produit de modulation *m*; ~ **of pulses** TELEP modulation d'impulsions *f*, modulation d'une suite d'impulsions *f*; ~ **rate** TELEP, TRANS rapidité de modulation *f*, taux de modulation *m*; ~ **restitution signal element** SIG élément de modulation restitution signal *m*; ~ **speed** TRANS vitesse de modulation *f*; ~ **switch** RADIO interrupteur de modulation *m*; ~ **technique** RADIO procédé de modulation *m*; ~ **transfer function** ELECT fonction de transfert de modulation *f*; ~ **type** TRANS type de modulation *m*; ~ **voltage** ELECT tension d'attaque de grille-cathode *f*, tension de modulation *f*; ~-**wheel assign** COMP affectation de la molette de modulation *f*; ~ **wheel range** COMP plage de réglage de la molette *f*; ~ **wheel sensitivity** COMP sensibilité de la molette de modulation *f*; ~ **wheel vibrato** COMP vibrato par molette de modulation *m*

modulator *n* ELECT, RADIO, TRANS modulateur *m*; ~ **charger** ELECT chargeur de modulateur *m*

modulator/demodulator *n* (*modem*) COMP, DATA, TRANS modulateur/démodulateur *m* (*modem*)

module *n* GEN module *m*

modulo: ~-**n counter** *n* ELECT compteur modulo-n *m*

modulus: ~ **of elasticity** *n* PARTS module d'élasticité longitudinale *m*, module de Young *m*

moiré *n* TV moiré *m*, moirage *m*, moirure *f*

moisture[1]: ~-**proof** *adj* GEN étanche à l'humidité, humidifuge

moisture[2] *n* GEN humidité *f*; ~-**absorbing capacity** GEN capacité d'absorption d'humidité *f*; ~ **content** GEN état hygrométrique *m*, teneur en humidité *f*; ~ **resistance** GEN résistance à l'humidité *f*

molded *AmE see moulded BrE*

molecular: ~ **beam** *n* TV faisceau moléculaire *m*; ~ **electronics** *n* ELECT électronique moléculaire *f*

monaural *adj* GEN monaural

monitor[1] *n* COMP, CONTR moniteur *m*, TEST appareil de contrôle *m*, appareil de surveillance *m*, TV moniteur *m*; ~ **program** DATA programme moniteur *m*; ~ **system** CONTR système de contrôle *m*; ~ **unit** TEST unité de surveillance *f*

monitor[2] *vt* GEN contrôler, suivre, superviser, surveiller, guider

monitoring *n* GEN veille *f*, contrôle *m*, observation *f*, suivi *m*, supervision *f*, surveillance *f*; ~ **equipment** DATA équipement de supervision *m*; ~ **mode** RADIO mode veille *m*; ~ **program** DATA programme moniteur *m*; ~ **unit** CONTR organe de contrôle *m*

mono: ~ **input** *n* COMP, PER APP entrée mono *f*; ~ **output** *n* COMP, PER APP sortie mono *f*

monocellular *adj* RADIO monocellulaire

monochromatic[1] *adj* TRANS monochromatique

monochromatic[2]: ~ **radiation** *n* TRANS rayonnement monochromatique *m*

monochromator *n* TRANS monochromateur *m*

monochrome: ~ **supertwist nematic** *n* (*MSTN*) ELECT nématique torsadé super monochrome *m*; ~ **television** *n* TV télévision en noir et blanc *f*

monolithic[1] *adj* ELECT monolithique

monolithic[2]: ~ **circuit** *n* ELECT circuit monolithique *m*; ~ **filter** *n* ELECT filtre monolithique *m*; ~ **integrated circuit** *n* (*MIC*) ELECT circuit intégré monolithique *m*; ~ **microwave integrated circuit** *n* (*MMIC*) ELECT, PARTS circuit intégré monolithique hyperfréquence *m*

monologue: ~ **interaction** *n* DATA interaction monologue *f*

monomedia: ~ **communication** *n* TV communication monomédia *f*; ~ **processing** *n* TV traitement monomédia *m*

monomode[1] *adj* DATA, ELECT, TRANS monomodal, unimodal

monomode[2]: ~ **laser** *n* ELECT, TRANS laser monomodal *m*, laser unimodal *m*, unimodal *m*

monophonic[1] *adj* RADIO monophonique

monophonic[2]: ~ **signal** *n* RADIO signal monophonique *m*

monophony *n* RADIO monophonie *f*

monopole: ~ **aerial** *n* RADIO antenne monopôle *f*; ~ **antenna** *n* RADIO antenne monopôle *f*

monopulse: ~ **radar** *n* RADIO radar mono-impulsion *m*

monoscope *n* ELECT monoscope *m*, tube de mire *m*

monostable[1] *adj* ELECT monostable

monostable[2]: ~ **element** *n* ELECT élément monostable *m*

monostatic: ~ **radar** *n* RADIO radar monostatique *m*

mono/stereo: ~ **output** *n* COMP, PER APP sortie mono/stéréo *f*

monotonic: ~ **reasoning** *n* DATA raisonnement monotone *m*

monthly: ~ **subscription** *n* APPL, COMP, GEN abonnement mensuel *m*

morphing *n* COMP morphing *m*, technique de métamorphose *f*

Morse: ~ **code** *n* RADIO, TELEP code Morse *m*; ~ **dash** *n* TELEP trait Morse *m*; ~ **dot** *n* TELEP point Morse *m*; ~ **key** *n* TELEP manipulateur Morse *m*; ~ **space** *n* TELEP espace Morse *m*; ~ **telegraphy** *n* TELEP télégraphie Morse *f*

MOS *abbr* (*mean opinion score*) COMP MOS (*note moyenne d'opinion*)

mosaic *n* ELECT mosaïque *f*; ~ **effect** COMP effet mosaïque *m*;

~ **telegraphy** TELEP télégraphie à mosaïque *f*, télégraphie par décomposition des signes *f*
most: ~ **significant bit** *n* (*MSB*) DATA bit de plus fort poids *m*
motherboard *n* COMP carte mère *f*
motion: ~ **blur** *n* TV flou de mouvement *m*; ~**-compensated concealment** *n* TV masquage par compensation de mouvement *m*; ~ **compensation** *n* TV compensation de mouvement *f*; ~ **estimation** *n* TV estimation de mouvement *f*; ~ **picture experts group algorithm** *n* DATA algorithme MPEG *m*; ~ **picture experts group channel** *n* TV canal MPEG *m*; ~ **vector** *n* COMP, TV vecteur cinétique *m*, vecteur de mouvement *m*
MOTIS *abbr* (*message-oriented text interchange system*) DATA système d'échange de textes en mode message *m*, système de transfert de textes en mode message *m*
motor: ~**-driven system** *n* RADIO, POWER système à entraînement mécanique *m*; ~ **protector switch** *n* POWER disjoncteur du moteur *m*
MOU *abbr* (*memorandum of understanding*) GEN accord cadre *m*, protocole d'accord *m*
moulded: ~ **composition resistor** *n* *BrE* ELECT, PARTS résistance agglomérée *f*; ~ **plastic** *n* *BrE* ELECT plastique moulé *m*
mounting *n* GEN pose *f*, montage *m*; ~ **bar** PARTS profilé de montage *m*; ~ **base** PARTS embase *f*; ~ **bracket** PARTS étrier de montage *m*, bride de fixation *f*; ~ **cord** ELECT cordon de montage *m*; ~ **cut-out** ELECT découpe de montage *f*; ~ **frame** PARTS châssis de montage *m*, châssis *m*; ~ **hardware** PARTS visserie de montage *f*; ~ **holder** PARTS support de montage *m*; ~ **mode** TELEP mode de montage *m*; ~ **plate** PARTS plaque de montage *f*, platine *f*; ~ **screw** PARTS vis d'assemblage *f*, vis de fixation *f*
mouse *n* COMP souris *f*
mouth: ~ **reference point** *n* TELEP point de référence de bouche *m*
mouthpiece *n* PER APP embouchure *f*; ~ **muffler** PER APP, TELEP réducteur de souffle *m*
move *vt* COMP déplacer, DATA, GEN positionner; ~ **down** COMP déplacer vers le bas; ~ **up** COMP déplacer vers le haut
movie *n* TV film *m*; ~ **preview** TV avant-première *f*
moving *n* DATA, GEN positionnement *m*; ~**-coil instrument** POWER appareil à cadre mobile *m*; ~**-coil permanent magnet instrument** POWER appareil à cadre mobile et à aimant permanent *m*; ~ **fiber switch** *AmE*, ~ **fibre switch** *BrE* SWG commutateur à déplacement de fibre *m*; ~**-magnet instrument** POWER appareil à fer mobile *m*, appareil à aimant mobile *m*; ~ **picture** GEN image animée *f*; ~**-target indicator radar** (*MTI radar*) RADIO radar à visualisation des cibles mobiles *m*; ~**-vane voltmeter** TEST voltmètre à palette de fer doux *m*, voltmètre ferrodynamique *m*
MOX *abbr* (*manually-operated transmission*) DATA, TRANS MOX (*émission à commande manuelle, commande manuelle d'émission*)
MP: ~ **capacitor** *n* (*metallized-paper capacitor*) POWER condensateur au papier métallisé *m*
MPA *abbr* (*multibeam phased array*) RADIO réseau à commande de phase *m*
MPBTL *abbr* (*minimum permissible basic transmission loss*) TRANS ATRMA (*affaiblissement de transmission de référence minimal admissible*)
MPC *abbr* (*multimedia personal computer*) COMP OP multimédia *m* (*ordinateur personnel multimédia*)
MPDU *abbr* (*message protocol data unit*) DATA MPDU (*unité de données de protocole de message*)
MPEG: ~ **algorithm** *n* DATA algorithme MPEG *m*; ~ **channel** *n* TV canal MPEG *m*
MPIL *abbr* (*maximum permissible interference level*) DATA NMAB (*niveau maximal admissible de brouillage*)
MPS *abbr* (*multipage signal*) DATA, TELEP signal pour plusieurs pages *m*
MPSK *abbr* (*multilevel phase-shift keying*) SIG modulation de phase à plusieurs niveaux *f*
MRSE *abbr* (*message retrieval service element*) DATA élément de service de recherche de messages *m*
MRTIE *abbr* (*maximum relative time interval error*) DATA erreur relative maximum d'intervalle de temps *f*
ms: ~**-reliable-access** *n* DATA accès-fiable-ms *m*
MS *abbr* RADIO, TELEP (*mobile station*) station mobile *f*, TRANS (*multiplex section*) section de multiplexage *f*; ~ **abstract service** *n* DATA service abstrait mémoire de messages *m*; ~ **abstract service provider** *n* DATA, TELEP fournisseur du service abstrait mémoire de messages *m*; ~ **abstract service user** *n* DATA utilisateur du service abstrait mémoire de messages *m*; ~ **stereo** *n* TV stéréo MS *f*; ~ **user** *n* DATA utilisateur mémoire de messages *m*
MSB *abbr* (*most significant bit*) DATA bit de plus fort poids *m*
MSC *abbr* SWG, TELEP (*mobile switching center AmE, mobile switching centre BrE*) centre de commutation mobile *m*, (*maritime switching center AmE, maritime switching centre BrE*) centre de commutation maritime *m*, (*mobile services switching center AmE, mobile services switching centre BrE*), centre de commutation de services mobiles *m*
MS-DOS® *abbr* (*Microsoft disk operating system*®) COMP MS-DOS® (*DOS Microsoft, système d'exploitation Microsoft*)
MSI *abbr* DATA (*multisignal interconnect*) interconnexion multisignaux *f*, ELECT, PARTS (*medium-scale integration*) intégration à échelle intermédiaire *f*, intégration à moyenne échelle *f*; ~ **circuit** *n* TELEP circuit MSI *m*
MSK *abbr* (*minimum-shift keying*) RADIO, SIG MDM (*modulation par déplacement minimal*)
MSN *abbr* (*multiple subscriber number*) TELEP numéro d'abonné multiple *m*
MSP *abbr* TEST (*maintenance service provider*) prestataire de services de maintenance *m*, TRANS (*multiplex-section protection*) protection de section de multiplexage *f*
MSRN *abbr* (*mobile station roaming number*) TELEP numéro de déplacement de station mobile *m*
MSS *abbr* (*maritime satellite service*) RADIO service maritime par satellite *m*
MSSC *abbr* (*maritime satellite switching center AmE, maritime satellite switching centre BrE*) RADIO, SWG centre de commutation du service maritime par satellite *m*
MSSE *abbr* (*message submission service element*) DATA élément de service de dépôt de messages *m*
MSSS *abbr* (*multisatellite support center AmE, multisatellite support centre BrE*) RADIO MSSS (*système de soutien multisatellite*)
MSSW *abbr* (*magnetostatic surface wave*) ELECT onde de surface magnétostatique *f*
MST *abbr* (*multiplex-section termination*) TRANS terminaison de section de multiplexage *f*
MSTN *abbr* (*monochrome-super twist nematic*) ELECT nématique torsadé super monochrome *m*
MSU *abbr* DATA, TELEP (*message signaling unit AmE, message signalling unit BrE*) TSM (*trame sémaphore de message*)
MT *abbr* (*message transfer*) DATA, TELEP transfert de messages *m*
MTA *abbr* (*message transfer agent*) DATA, TELEP agent de transfert de messages *m*
MTBF *abbr* (*mean time between failures*) TEST moyenne des temps de bon fonctionnement *f*, temps moyen entre pannes *m*
MTBI *abbr* (*mean time between interruptions*) TEST temps moyen entre interruptions *m*
MTBO *abbr* (*mean time between overhauls*) TEST périodicité moyenne de révisions *f*
MTFF *abbr* (*mean time to first failure*) TEST moyenne des temps de fonctionnement avant la première défaillance *f*

MTI: ~ **radar** *n* (*moving-target indicator radar*) RADIO radar à visualisation des cibles mobiles *m*

MTIE *abbr* (*maximum time interval error*) TEST erreur maximum d'intervalle de temps *f*

MTP *abbr* DATA (*message transfer part*) SSTM (*sous-système de transfert de messages, sous-système de transport de messages*), RADIO (*master time plan*) plan directeur *m*, TELEP (*message transfer part*) SSTM (*sous-système de transfert de messages, sous-système de transport de messages*)

MTPI *abbr* (*multiplexer timing physical interface*) TRANS interface physique de rythme de multiplexeur *f*

mts: **~-forced-reliable-access** *n* DATA accès-fiable-imposé-mts *m*; **~-reliable-access** *n* DATA accès-fiable-mts *m*; ~ **transfer** *n* DATA transfert du mts *m*; ~ **transfer protocol** *n* DATA protocole de transfert du mts *m*; ~ **transfer protocol 1984** *n* DATA protocole de transfert du mts 1984 *m*

MTS *abbr* DATA (*message transfer system*) système de transfert de messages *m*, RADIO (*mobile telephone service*) SRM (*service radiotéléphonique mobile*), TELEP (*message toll service*) SICT (*système interurbain à communications tarifées*), (*mobile telephone service*) SRM (*service radiotéléphonique*), (*message transfer system*) système de transfert de messages *m*, TRANS (*multiplexer timing source*) source de synchronisation de multiplexeur *f*

MTSE *abbr* (*message transfer service element*) DATA, TELEP élément de service de transfert de messages *m*

MTSO *abbr* (*mobile telecommunication switching office*) RADIO CCTM (*centre de commutation de téléphonie mobile*)

MTSR *abbr* (*mean time to service restoration*) TEST délai moyen de rétablissement du service *m*

MTTR *abbr* TEST (*mean time to recovery*) durée moyenne de panne *f*, (*mean time to repair*) moyenne de temps total de réparation *f*

mu: ~ **factor** *n* POWER coefficient d'amplification *m*; ~ **law codec** *n* TRANS codec de loi mu *m*

MUF *abbr* (*maximum usable frequency*) RADIO MUF (*fréquence maximale utilisable*)

muldex *n* TELEP muldex *m*, multiplexeur-démultiplexeur *m*

multiaccess: ~ **system** *n* DATA système multi-accès *m*

multiaddress: ~ **calling** *n* TELEP adresses multiples *f pl*

multiaddressing *n* DATA multi-adressage *m*

multiarea *adj* RADIO multi-aires

multiband: ~ **filter** *n* TRANS filtre multibande *m*

multibeam: ~ **phased array** *n* (*MPA*) RADIO réseau à commande de phase *m*

multibearer: ~ **service** *n* TELEP *ISDN* service à supports multiples *m*

multibit: ~ **input** *n* DATA entrée à plusieurs bits *f*; ~ **output** *n* DATA sortie à plusieurs bits *f*

multiblock *n* TELEP multibloc *m*; ~ **acknowledgement signal** SIG signal d'accusé de réception des multiblocs *m*; ~ **monitoring signal** SIG signal de surveillance des multiblocs *m*; ~ **synchronization signal unit** (*MBS*) SIG unité de signalisation de synchronisation de multibloc *f*

multicast: ~ **call** *n* DATA multicommunication *f*; ~ **group** *n* DATA groupe de destinataires *m*; ~ **server** *n* DATA serveur de destinations multiples *m*; ~ **service** *n* DATA service de destinations multiples *m*; ~ **service model** *n* DATA modèle du service à destinations multiples *m*

multicavity: ~ **klystron** *n* ELECT klystron multicavité *m*

multichannel[1] *adj* TRANS multicanal, multivoie

multichannel[2]: ~ **audio-digital interface** *n* (*MADI*) COMP audio-numérique multicanaux *m* (*MADI*); ~ **filter** *n* TRANS filtre multivoie *m*; ~ **recording** *n* TV enregistrement multicanal *m*; ~ **rms deviation** *n* TRANS excursion efficace multivoie *f*; ~ **root mean square deviation** *n* TRANS excursion efficace multivoie *f*

multiclique: ~ **mode** *n* DATA mode multiclique *m*

multiconductor: ~ **cable** *n* TRANS câble multiconducteur *m*

multidestination: ~ **carrier** *n* TRANS porteuse à destinations multiples *f*; ~ **mode** *n* DATA mode de destination multiple *m*, mode multidestination *m*

multidimensional: ~ **data analysis** *n* DATA analyse des données multidimensionnelles *f*; ~ **digital filter** *n* DATA filtre numérique multidimensionnel *m*; ~ **filtering** *n* ELECT filtrage multidimensionnel *m*; ~ **signal** *n* SIG système multidimensionnel *m*

multielectrode: ~ **voltage-stabilizing tube** *n* ELECT tube stabilisateur de tension à plusieurs électrodes *m*

multielement: ~ **service charge** *n* APPL frais de service composés *m pl*

multifiber *AmE see multifibre BrE*

multifibre: ~ **cable** *n BrE* TRANS câble multifibre *m*; ~ **connector** *n BrE* TRANS connecteur multifibre *m*; ~ **joint** *n BrE* TRANS raccord multifibre *m*

multiframe *n* DATA, TELEP, TRANS multitrame *f*; ~ **alignment** (*MFA*) TRANS verrouillage de multitrame *m*; ~ **alignment code** TRANS code de verrouillage de multitrame *m*; ~ **buffering** COMP, DATA mémoire tampon multitrame *f*; ~ **structure** DATA structure multitrame *f*

multifrequency *n* SIG, TELEP, TRANS *signalling mode* multifréquence *f*; ~ **aerial** RADIO antenne multibande *f*, antenne multifréquence *f*; ~ **antenna** RADIO antenne multibande *f*, antenne multifréquence *f*; ~ **generator** SIG, TELEP générateur de signalisation multifréquence *m*, générateur multifréquence *m*; ~ **receiver** SIG récepteur multifréquence *m*; ~ **screen** TV écran multifréquence *m*; ~ **sender-receiver** SIG, TELEP émetteur-récepteur multifréquence *m*; ~ **signaling** *AmE*, ~ **signalling** *BrE* SIG signalisation multifréquence *f*; ~ **telephone** TELEP multifréquence *f*, téléphone multifréquence *m*

multifunction: ~ **key** *n* COMP touche à fonctions multiples *f*, touche multifonctions *f*; ~ **printer** *n* PER APP imprimante à fonctions multiples *f*, imprimante multifonction *f*

multihop: ~ **satellite link** *n* RADIO liaison satellite à plusieurs bonds *f*; ~ **system** *n* RADIO système à plusieurs bonds *m*

multilayer: ~ **CD** *n* DATA CD multicouche *m*; ~ **compact disc** *n* DATA disque compact multicouche *m*; ~ **isolation** *n* (*MLI*) ELECT isolation multicouche *f*; ~ **optical compact disc** *n* DATA disque compact multicouche *m*; ~ **printed circuit** *n* ELECT circuit imprimé multicouche *m*

multilevel: ~ **continuous sampling** *n* TEST échantillonnage continu multiniveau *m*; ~ **image** *n* TELEP image à plusieurs niveaux *f*; ~ **modulation** *n* RADIO modulation multiniveau *f*; ~ **phase-shift keying** *n* (*MPSK*) SIG modulation de phase à plusieurs niveaux *f*; ~ **system** *n* TRANS système à niveaux multiples *m*

multiline[1] *adj* GEN, PER APP multiligne, multifilaire

multiline[2]: ~ **representation** *n* GEN représentation multifilaire *f*

multilingual: ~ **recognition** *n* APPL, COMP, TELEP reconnaissance multilangue *f*

multilink: ~ **connection** *n* RADIO liaison à plusieurs circuits *f*; ~ **procedure** *n* (*MLP*) DATA procédure multiliaison *f*

multilobe: ~ **aerial** *n* RADIO antenne à lobes multiples *f*; ~ **antenna** *n* RADIO antenne à lobes multiples *f*

multilongitudinal: ~ **modes** *n pl* (*MLM*) TELEP, TRANS modes longitudinaux multiples *m pl* (*MLM*)

multimedia[1] *adj* COMP, TV multimédia

multimedia[2] *n* COMP, TV multimédia *m*; ~ **authoring system** COMP système auteur multimédia *m*; ~ **capabilities** *n pl* COMP capacités multimédia *f pl*; ~ **conferencing** TELEP, TV téléconférence multimédia *f*; ~ **configuration** COMP configuration multimédia *f*; ~ **database** COMP, DATA base de données multimédia *f*; ~ **document** DATA document multimédia *m*; ~ **editing** COMP, TV montage multimédia *m*; ~ **file server** COMP, DATA serveur de fichiers multimédias *m*; ~ **and hypermedia information coding experts group** (*MHEG*) COMP MHEG; ~ **kiosk** TV borne multimédia *f*; ~ **messaging** COMP, DATA messagerie multimédia *f*;

~ **navigation** COMP, DATA navigation multimédia *f*; ~ **network** TRANS réseau multimédia *m*; ~ **PC** COMP OP multimédia *m*; ~ **personal computer** (*MPC*) COMP ordinateur personnel multimédia *m* (*OP multimédia*); ~ **processing** COMP traitement multimédia *m*; ~ **server** COMP, DATA serveur multimédia *m*; ~ **service** TV service multimédia *m*; ~ **station** TV station multimédia *f*; ~ **synchronization** DATA synchronisation multimédia *f*; ~ **system** TV système multimédia *m*; ~ **terminal** TV terminal multimédia *m*; ~ **transmission** DATA, TRANS transmission de données multimédia *f*; ~ **user interface** DATA interface multimédia de l'utilisateur *f*; ~ **workstation** COMP, DATA, PER APP station de travail multimédia *f*

multimeter *n* TEST contrôleur universel *m*, multimètre *m*

multimetering *n* TELEP comptage multiple *m*

multimicroprocessor: ~ **system** *n* SWG système à structure multimicroprocesseur *m*

multimode: ~ **distortion** *n* TRANS distorsion modale *f*; ~ **fiber** *AmE*, ~ **fibre** *BrE* *n* DATA, PARTS, TRANS fibre multimodale *f*, fibre multimode *f*; ~ **laser** *n* ELECT, TRANS laser à modes multiples *m*, laser multimodal *m*

multipage: ~ **signal** *n* (*MPS*) DATA, TELEP signal pour plusieurs pages *m*

multipair[1] *adj* TV multipaire

multipair[2]: ~ **cable** *n* TRANS, TV câble multipaire *m*

multipath: ~ **fading** *n* RADIO évanouissement par trajets multiples *m*, évanouissement dû à la propagation par trajets multiples *m*; ~ **interference** *n* RADIO interférence due à la propagation à trajets multiples *f*; ~ **propagation** *n* RADIO propagation par trajets multiples *f*; ~ **reflections** *n pl* RADIO réflexions par trajets multiples *f pl*; ~ **time dispersion** *n* RADIO dispersion des temps par trajets multiples *f*; ~ **transmission** *n* DATA transmission par trajets multiples *f*

multipin: ~ **connector** *n* ELECT, PARTS connecteur multicontact *m*

multiple[1] *adj* ELECT, GEN multiple; **~-access** RADIO à accès multiple; **~-line** GEN, PER APP multiligne

multiple[2] *n* ELECT, GEN multiple *m*; ~ **access** GEN accès multiple *m*; ~ **association control function** (*MACF*) DATA fonction de contrôle d'association multiple *f*; ~ **beam** RADIO faisceau multiple *m*; **~-beam aerial** RADIO antenne multifaisceau *f*; **~-beam antenna** RADIO antenne multifaisceau *f*; ~ **cable** SWG câble multiple *m*; **~-code access** DATA accès code multiple *m*; ~ **columns** *n pl* DATA multicolonnage *m*; ~ **container** POWER *batteries* bac à compartiments *m*; ~ **diffraction** TRANS diffraction multiple *f*; **~-duct block** PARTS, TRANS bloc alvéolé *m*, bloc de canalisations multitubulaire *m*, conduit multitubulaire *m*; **~-filter chain** TRANS filtres en chaîne *m pl*; ~ **frame** SWG cadre multiplage *m*, nappe de multiple *f*; ~ **frequency** (*MF*) GEN, SIG, TELEP fréquence multiple *f* (*FM*); **~-frequency encoder** TRANS codeur multifréquence *m*; **~-gun cathode ray tube** ELECT tube cathodique à plusieurs canons *m*; ~ **image** TV effet d'écho *m*; **~-image production master** TV cliché de production à images multiples *m*; **~-key set** RADIO poste cryptophonique à encryptage multiple *m*; ~ **lines at the same address** *n pl* TELEP lignes multiples pour une même adresse *f pl*; ~ **memory** RADIO mémoire multiple *f*; **~-mode connectors** *n pl* TRANS connecteurs multimodes *m pl*; **~-mode modem** TRANS modem multimode *m*; ~ **modulation** TRANS modulation multiple *f*; **~-path propagation** RADIO propagation par trajets multiples *f*; ~ **pattern** ELECT impression multiple *f*, motif composite *m*; ~ **recorder** PER APP enregistreur multiple *m*; **~-reflector aerial** RADIO antenne à réflecteurs multiples *f*; **~-reflector antenna** RADIO antenne à réflecteurs multiples *f*; ~ **sampling** TRANS échantillonnage multiple *m*; ~ **seizure** SWG prise multiple *f*; **~-server queue** COMP, DATA file d'attente à serveurs multiples *f*; ~ **subsampling encoding** (*MUSE*) DATA codage d'échantillonnage multiple *m*; ~ **subscriber number** (*MSN*) TELEP numéro d'abonné multiple *m*; ~ **switchboard** SWG commutateur multiplex *m*, multiple téléphonique *m*, standard à lignes multipliées *m*; ~ **television transmission** TRANS transmission télévisuelle multiple *f*; ~ **transmission** TRANS transmission multiple *f*; ~ **transmitter** RADIO émetteur multiple *m*; **~-transmitter distributor** TELEP distributeur d'émetteur multiple *m*; **~-transponder operation** RADIO exploitation multirépéteur *f*; **~-trunk distribution system** TV réseau G+ *m*; ~ **tube** ELECT tube multiple *m*; ~ **twin cable** TRANS câble à quarte à paires combinables *m*

multiplex[1]: ~ **aggregate bit rate** *n* TELEP débit binaire cumulé d'un multiplex *m*; ~ **equipment** *n* TELEP, TRANS équipement multiplex *m*; ~ **interface** *n* TELEP interface multiplex *f*; ~ **link** *n* TELEP liaison multiplex *f*; ~ **section** *n* (*MS*) TRANS section de multiplexage *f*; **~-section alarm indication signal** *n* TRANS signal d'indication d'alarme de section de multiplexage *m* (*SM-SIA*); **~-section overhead** *n* TRANS surdébit de section de multiplexage *m* (*SDSM*); **~-section protection** *n* (*MSP*) TRANS protection de section de multiplexage *f*; **~-section termination** *n* (*MST*) TRANS terminaison de section de multiplexage *f*; ~ **space switch** *n* SWG commutateur spatial multiplex *m*; **~-timing generator** *n* SWG générateur de rythme de multiplexeur *m*

multiplex[2] *vt* COMP, DATA, RADIO, SWG, TRANS multiplexer

multiplexer *n* (*MUX*) COMP, DATA, RADIO, SWG, TRANS multiplexeur *m* (*MUX*); ~ **channel** TRANS canal multiplexeur *m*; ~ **timing physical interface** (*MTPI*) TRANS interface physique de rythme de multiplexeur *f*; ~ **timing source** (*MTS*) TRANS source de synchronisation de multiplexeur *f*

multiplexing *n* COMP, DATA, RADIO, SWG, TRANS multiplexage *m*; ~ **identification** (*MID*) TRANS identification de multiplexage *f*

multiplicator *n* ELECT, TELEP multiplicateur *m*

multiplied *adj* ELECT, TELEP multiplié

multiplier *n* ELECT, TELEP multiplicateur *m*

multipoint[1] *adj* GEN multipoint

multipoint[2]: ~ **command assignment token** *n* (*MCA*) DATA commande multipoint d'assignation de jeton *f*; ~ **command token claim** *n* (*MCT*) DATA commande multipoint de demande de jeton *f*; ~ **command visualization forcing** *n* (*MCV*) DATA commande multipoint d'imposition de visualisation *f*; ~ **communication service** *n* TELEP service de communication multipoint *m*; ~ **communications system** *n* (*MCS*) TRANS système de télécommunications multipoint *m* (*STM*); ~ **connection** *n* DATA liaison multipoint *f*; ~ **connector** *n* ELECT, PARTS connecteur multibroches *m*; ~ **indication loop** *n* (*MIL*) DATA indication multipoint de boucle *f*; ~ **repeater** *n* TRANS répéteur multipoint *m*

multipolar *adj* POWER multipolaire

multipole *adj* POWER multipolaire

multipolling *n* COMP multirelève *f*

multiprocessing *n* COMP, DATA multitraitement *m*; ~ **program** COMP programme multitraitement *m*; ~ **system** DATA système multitraitement *m*

multiprocessor *n* COMP, DATA multiprocesseur *m*; ~ **interconnection network** DATA réseau d'interconnexion de multiprocesseur *m*; ~ **station bus** TELEP bus de station multiprocesseur *m*; ~ **system** SWG système multiprocesseur *m*

multiprogramming *n* COMP multiprogrammation *f*

multiprotocol: ~ **routing** *n* COMP, DATA, TELEP, TRANS routage multiprotocole *m*; ~ **signaling coupler** *AmE*, ~ **signalling coupler** *BrE* *n* SIG coupleur de signalisation multiprotocole *m*

multipurpose: ~ **key** *n* COMP touche multifonctions *f*; ~ **wiring system** *n* ELECT système de câblage polyvalent *m*

multirange: ~ **grommet** *n* ELECT joint d'étanchéité *m*; ~ **voltmeter** *n* TEST voltmètre multicalibre *m*
multirate: ~ **filtering** *n* DATA filtrage multicadence *m*; ~ **switching system** *n* SWG système de commutation multidébit *m*
multiray: ~ **channel** *n* RADIO canal à ondes multiples *m*; ~ **diversity** *n* RADIO diversité à ondes multiples *f*
multirelay *adj* RADIO multi relais
multisampling *n* TRANS échantillonnage multiple *m*
multisatellite: ~ **link** *n* RADIO, TV liaison multisatellite *f*; ~ **support system** *n* RADIO système de soutien multisatellite *m* (*MSSS*)
multiscan: ~ **monitor** *n* TV moniteur à balayage multiple *m*
multiscreen: ~ **configuration** *n* TV configuration à écrans multiples *f*
multiservice: ~ **automatic exchange** *n* SWG autocommutateur multiservice *m*; ~ **network** *n* TRANS réseau multiservice *m*; ~**-switching system** *n* SWG système de commutation multiservice *m*
multisession *n* COMP multisession *f*
multisignal: ~ **interconnect** *n* (*MSI*) DATA interconnexion multisignaux *f*
multisite *adj* RADIO multisite
multislot: ~ **connection** *n* TELEP connexion à intervalles de temps multiples *f*
multispot: ~ **aerial** *n* RADIO antenne multifaisceau *f*; ~ **antenna** *n* RADIO antenne multifaisceau *f*
multistage: ~ **circuit** *n* ELECT circuit multi-étage *m*; ~ **network** *n* ELECT réseau à étages multiples *m*, réseau multi-étage *m*; ~ **sampling** *n* TEST échantillonnage multi-étage *m*
multistandard: ~ **television set** *n* TV téléviseur multistandard *m*
multistate: ~ **modulation** *n* RADIO modulation multi-état *f*
multistation: ~ **access unit** *n* (*MAU*) DATA unité d'accès multiples *f*; ~ **teletex terminal installation** *n* TELEP installation terminale télétex à postes multiples *f*
multistep: ~ **action** *n* CONTR action en plusieurs étapes *f*
multistrand: ~ **wire** *n* ELECT, PARTS conducteur multifilaire *m*
multistranded *adj* TRANS multibrin
multisynchronous *adj* TV multisynchrone
multitask *adj* COMP multitâche, DATA multi tâche
multitasking[1] *adj* COMP multitâche, DATA multi tâche
multitasking[2] *n* COMP, DATA traitement multitâche *m*
multiterminal: ~ **service circuit** *n* TELEP circuit de service multiterminal *m*
multitimbral *adj* COMP multitimbral
multitone: ~ **system** *n* TELEP multivoie *f*
multitrack[1] *adj* DATA multipiste
multitrack[2]: ~ **recorder** *n* DATA enregistreur multipiste *m*; ~ **sequencer** *n* DATA séquenceur multipistes *m*
multi-unit: ~ **message** *n* (*MUM*) TELEP message multiple *m*
multiuse: ~ **communication** *n* TELEP boucle de communication multi-usage *f*
multiuser[1] *adj* GEN multi-utilisateur
multiuser[2]: ~ **system** *n* DATA système multi-utilisateur *m*
multivalued[1] *adj* (*M*) DATA à plusieurs valeurs
multivalued[2]: ~ **attribute** *n* DATA attribut à plusieurs valeurs *m*
multivariable: ~ **control** *n* TEST contrôle à plusieurs variables *m*
multivendor *n* GEN multiconstructeur *m*; ~ **host support** GEN support hôte de chargement multiconstructeur *m*
multivibrator *n* (*MV*) ELECT bascule instable *f*, multivibrateur *m*, multivibrateur instable *m*
multivoice[1] *adj* GEN multivoix
multivoice[2]: ~ **speech recognition** *n* APPL, COMP, TELEP reconnaissance multilocuteur de la parole *f*
multiwire: ~ **conductor** *n* ELECT, PARTS conducteur multifilaire *m*
multizone: ~ **subscription** *n* RADIO abonnement multizone *m*
M-unit *n* RADIO unité M *f*
Munsell: ~ **color system** *AmE*, ~ **colour system** *BrE* *n* TV système de couleurs de Munsell *m*
MUSE *abbr* (*multiple subsampling encoding*) DATA codage d'échantillonnage multiple *m*
music: ~ **on hold** *n* TELEP musique d'attente *f*; ~**-on-hold system** *n* TELEP attente musicale *f*
mute[1] *adj* COMP en sourdine
mute[2]: ~ **button** *n* PER APP *answer machine, telephone* touche secrète *f*
mute[3] *vt* COMP assourdir
mutilate *vt* TELEP mutiler
mutilation *n* TELEP mutilation *f*
muting *n* RADIO, TELEP blocage automatique *m*, réglage silencieux *m*; ~ **device** RADIO dispositif de réglage silencieux *m*
mutual: ~ **branch** *n* POWER branche commune *f*; ~ **conductance** *n* ELECT pente *f*; ~ **coupling** *n* ELECT couplage mutuel *m*; ~ **impedance** *n* ELECT impédance mutuelle *f*; ~ **inductance** *n* ELECT inductance mutuelle *f*; ~ **induction** *n* ELECT induction mutuelle *f*; ~ **synchronization** *n* ELECT synchronisation mutuelle *f*
mutually: ~ **exclusive attribute group** *n* COMP groupe d'attributs mutuellement exclusifs *m*; ~**-synchronized network** *n* TELEP réseau à synchronisation mutuelle *m*
MUX *abbr* (*multiplexer*) TRANS MUX (*multiplexeur*)
MV *abbr* (*multivibrator*) ELECT bascule instable *f*, multivibrateur instable *m*
MW *abbr* (*medium wave*) RADIO onde moyenne *f*
mylar: ~ **capacitor** *n* ELECT condensateur au mylar *m*

N

N *abbr* (*newton*) TEST N (*newton*)

N: **~ address** *n* DATA adresse N *f*; **~-address selector** *n* DATA sélecteur N *m*, sélecteur d'adresse N *m*; **~-association** *n* DATA association N *f*; **~ connection** *n* TELEP connexion N *f*; **~-connection end point identifier** *n* (*N-CEPI*) DATA, TELEP identificateur d'extrémité de connexion (N) *m*; **~ connectionless-mode transmission** *n* DATA, TRANS transmission en mode sans connexion N *f*; **~ connect request** *n* DATA demande de connexion de réseau *f*; **~-directory-function** *n* DATA fonction de répertoire N *f*; **~-entity invocation** *n* DATA invocation d'entité N *f*; **~-entity title** *n* DATA titre d'entité N *m*; **~-entity-type** *n* DATA type d'entité N *m*; **~ function** *n* DATA fonction N *f*; **~ initiator** *n* DATA initiateur N *m*; **~-layer** *n* DATA, ELECT couche N *f*; **~-layer managed object** *n* DATA objet géré de couche N *m*; **~-protocol** *n* TELEP protocole N *m*; **~-protocol-addressing information** *n* (*N-PAI*) DATA, TELEP informations d'adressage du protocole N *f pl*; **~-protocol control information** *n* (*N-PCI*) DATA, TELEP information de contrôle du protocole N *f*; **~-protocol data-unit** *n* DATA unité de données de protocole N *f*; **~-recipient** *n* DATA destinataire N *m*; **~-relay** *n* DATA relais N *m*; **~-service** *n* TELEP service N *m*; **~-service data unit** *n* DATA, TELEP unité de données de service N *f*; **~-unit** *n* RADIO unité N *f*; **~-user data** *n pl* DATA, TELEP données utilisateur N *f pl*

NA *abbr* (*numerical aperture*) DATA, ELECT, TELEP, TRANS ouverture numérique *f*

N/A *abbr* (*not applicable*) DATA, GEN, TELEP NA (*sans objet*)

NACK *abbr* (*negative acknowledgement*) RADIO NACK (*accusé de réception négatif*)

NAD *abbr* (*noise amplitude distribution*) TRANS distribution d'amplitude de bruit *f*

NAE *abbr* (*network address extension*) DATA extension d'adresse de réseau *f*

name: **~ server** *n* DATA serveur de noms *m*

nameplate *n* COMP plaque d'identification *f*, plaque signalétique *f*

naming: **~ authority** *n* DATA autorité de dénomination *f*; **~ domain** *n* DATA domaine de dénomination *m*; **~ subdomain** *n* DATA sous-domaine de dénomination *m*; **~ tree** *n* DATA arbre de dénomination *m*

NAND *abbr* (*not AND*) ELECT NON-ET *m*; **~ circuit** *n* ELECT circuit NON-ET *m*, porte NON-ET *f*; **~ gate** *n* ELECT circuit NON-ET *m*, porte NON-ET *f*; **~ operation** *n* ELECT opération NON-ET *f*

NANDed *adj* ELECT en fonction NON-ET, mis en fonction NON-ET

nanosecond *n* (*ns*) TEST nanoseconde *f* (*ns*)

NAPI/TOA *abbr* (*numbering and addressing plan indicator/type of address*) DATA, TELEP indicateur de plan de numérotage/type d'adresse *m*

narrow: **~ band** *n* TRANS bande étroite *f*; **~-band frequency modulation** *n* (*NBFM*) RADIO, TRANS modulation de fréquence à bande étroite *f* (*MFBE*); **~-band noise** *n* TRANS bruit à bande étroite *m*; **~-band phase-shift keying** *n* (*NBPSK*) SIG modulation par déplacement de phase à bande étroite *f*; **~-band receiver** *n* RADIO récepteur à bande étroite *m*; **~-band signal** *n* ELECT, SIG, TRANS signal à bande étroite *m*; **~-band switch** *n* SWG commutateur à bande étroite *m*; **~-band switching network** *n* SWG, TRANS réseau de connexion à bande étroite *m*; **~-band voice modulation** *n* (*NBVM*) TRANS modulation vocale à bande étroite *f*; **~-beam transmit aerial** *n* RADIO antenne d'émission à faisceau étroit *f*; **~-beam transmit antenna** *n* RADIO antenne d'émission à faisceau étroit *f*; **~ shift** *n* RADIO faible déplacement *m*

narrowcasting *n* RADIO diffusion restreinte *f*, radiodiffusion restreinte *f*

n-ary: **~ digital signal** *n* SIG signal numérique n-aire *m*

national: **~ circuit** *n* TELEP circuit national *m*; **~ code** *n* TELEP indicatif de pays *m*; **~ control center** *AmE*, **~ control centre** *n BrE* TEST centre de coordination national *m* (*CCN*); **~ destination code** *n* (*NDC*) DATA, TELEP code national de destination *m*, indicatif national de destination *m*; **~ exchange** *n* SWG centre national *m*; **~ extension** *n* TELEP prolongement national *m*; **~ identification digit** *n* (*NID*) TELEP chiffre d'identification de nationalité *m*; **~ indicator** *n* TELEP indicateur national *m*; **~ line** *n* TELEP ligne nationale *f*; **~ main section** *n* TELEP section nationale principale *f*; **~-network congestion signal** *n* SIG signal d'encombrement sur le réseau national *m*; **~ number** *n* (*NN*) DATA, TELEP numéro national *m* (*NN*); **~ section** *n* TELEP section nationale *f*; **~ significant number** *n* (*NSN*) TELEP numéro national significatif *m*; **~ sound-program center** *AmE*, **~ sound-programme centre** *BrE n* RADIO centre radiophonique national *m* (*CRN*); **~ system** *n* TELEP système national *m*; **~ television center** *AmE*, **~ television centre** *n BrE* (*NTC*) TV centre télévisuel national *m*; **~ usage** *n* (*NU*) TELEP usage national *m*; **~ variant** *n* TEST variante nationale *f*

National: **~ Center for Supercomputing Applications** *n AmE* (*NCSA*) DATA centre national des applications super-informatiques (*NCSA*)

national/international: **~ call indicator** *n* TELEP indicateur de communication nationale/internationale *m*

nationwide: **~ paging** *n* TRANS réseau national de téléappel *m*, téléappel national *m*

natural: **~ frequency** *n* POWER fréquence propre *f*; **~ language** *n* DATA langue naturelle *f*, langage naturel *m*; **~-language analysis** *n* DATA analyse du langage naturel *f*; **~-language generation** *n* DATA synthèse du langage naturel *f*; **~ language interface** *n* DATA interface en langage naturel *f*; **~ language processing** *n* COMP, DATA traitement de la langue naturelle écrite *m*, traitement des langues naturelles *m*, traitement de langue naturel *m*; **~ noise** *n* ELECT bruit naturel *m*; **~ oscillator** *n* ELECT oscillateur naturel *m*

nature: **~-of-circuit indicator** *n* TELEP indicateur de la nature du circuit *m*

navigate *vt* COMP, DATA *Internet* naviguer

navigation *n* COMP, DATA navigation *f*, parcours libre *m*; **~ button** COMP, DATA bouton de navigation *m*; **~ path** COMP, DATA chemin de navigation *m*; **~ software** COMP, DATA *Internet* logiciel navigateur *m*; **~ warning signal** RADIO signal d'avis aux navigateurs *m*

navigational: **~ radar** *n* RADIO radar de navigation *m*

navigator *n* COMP, DATA navigateur *m*

NBFM *abbr* (*narrow-band frequency modulation*) RADIO, TRANS *maritime satellite* MFBE (*modulation de fréquence à bande étroite*)

NBP *abbr* (*nonbasic parameter*) DATA paramètre non-fondamental *m*

NBPSK *abbr* (*narrow-band phase-shift keying*) SIG modulation par déplacement de phase à bande étroite *f*

NBVM *abbr* (*narrow-band voice modulation*) TRANS modulation vocale à bande étroite *f*
NC *abbr* COMP, DATA (*numerical control*) commande numérique *f*, (*network connection*) connexion de réseau *f*, TELEP (*network connection*) connexion de réseau *f*
NCC *abbr* GEN (*new common carriers*) nouveaux opérateurs *m pl*, TEST (*network control center AmE, network control centre BrE*) centre de gestion de réseau *m*, (*national control center AmE, national control centre BrE*) CCN (*centre de coordination national*)
N-CEPI *abbr* (*N-connection end point identifier*) DATA, TELEP identificateur d'extrémité de connexion N *m*
NCL *abbr* (*network control language*) ELECT, TELEP langage de contrôle de réseau *m*
NCSA *abbr* (*National Center for Supercomputing Applications*) DATA *Internet* NCSA (*centre national des applications super-informatiques*)
ND *abbr* (*network default*) ELECT, TELEP DR (*défaut par le réseau*)
NDC *abbr* (*national destination code*) DATA, TELEP code national de destination *m*, indicatif national de destination *m*
NDF *abbr* (*new data flag*) DATA fanion de nouvelles données *m*, indicateur de nouvelles données *m*
NDIS *abbr* (*network drive interface specifications*) ELECT spécifications d'interface du circuit de commande du réseau *f pl*
N-disconnect: ~ **indication** *n* DATA indication de déconnexion de réseau *f*
NDN *abbr* (*nondelivery notification*) DATA, TELEP avis de non-remise *m*
NE *abbr* DATA (*negotiated exit*) sortie négociée *f*, TRANS (*network element*) élément de réseau *m*
near[1]: **~-instantaneous** *adj* TV quasi-instantané
near[2]: **at the near ~** *adv* TELEP *telex* chez l'émetteur
near[3]: **~-end crosstalk** *n* TRANS paradiaphonie *f*; ~ **field** *n* RADIO, TRANS champ proche *m*; **~-field analysis** *n* RADIO, TRANS analyse du champ proche *f*; **~-field diffraction pattern** *n* RADIO, TRANS diagramme de diffraction en champ proche *m*; **~-field radiation pattern** *n* RADIO, TRANS diagramme de rayonnement de champ proche *m*; **~-field region** *n* RADIO, TRANS région de champ proche *f*, région de Fresnel *f*, région de rayonnement proche *f*; **~-field scanning** *n* RADIO, TRANS exploration du champ proche *f*; **~-synchronous equatorial orbit** *n* RADIO orbite équatoriale quasi-synchrone *f*, orbite à déferlement lent *f*
near-Earth: ~ **space** *n* RADIO espace proche de la Terre *m*
necessary: ~ **bandwidth** *n* RADIO, TRANS largeur de bande nécessaire *f*
neck *n* ELECT col *m*; ~ **shadow** ELECT ombre de col *f*
needle *n* COMP aiguille *f*, TELEP indicateur *m*; ~ **counter tube** RADIO tube-compteur à aiguille *m*; ~ **deflection** TELEP déviation de l'aiguille *f*
NEF *abbr* (*not-expected frame*) TELEP NEF (*trame inattendue*)
negate *vt* COMP, ELECT complémenter, effectuer une inversion dans, inverser, réaliser une inversion
negation *n* COMP négation *f*, inversion *f*, ELECT inversion logique *f*, négation *f*
negative: ~ **acknowledge character** *n* TELEP caractère accusé de réception négatif *m*; ~ **acknowledgement** *n* (*NACK*) RADIO accusé de réception négatif *m* (*NACK*); ~ **amplitude modulation** *n* ELECT, TV modulation d'amplitude négative *f*; ~ **battery** *n* POWER accumulateur négatif *m*; ~ **effect** *n* COMP image négative *f*; ~ **feedback** *n* CONTR, DATA, ELECT contre-réaction *f*, POWER contre-réaction *f*, réaction négative *f*, RADIO contre-réaction *f*; **~-feedback amplifier** *n* ELECT amplificateur à contre-réaction *m*; ~ **frequency modulation** *n* ELECT modulation de fréquence négative *f*; ~ **indication tone** *n* TELEP tonalité d'indication négative *f*; ~ **justification** *n* TELEP, TRANS justification négative *f*; ~ **notification** *n* (*NN*) DATA, TELEP notification négative *f* (*NN*); ~ **pattern** *n* ELECT impression négative *f*, motif en négatif *m*; ~ **polarity** *n* POWER polarité négative *f*; ~ **pole** *n* POWER pôle négatif *m*; ~ **pulse** *n* ELECT impulsion négative *f*; ~ **terminal** *n* POWER *of battery* borne négative *f*, pôle négatif *m*; ~ **tilt** *n* TELEP chute du palier *f*; ~ **wire** *n* TELEP fil négatif *m*
negentropy *n* DATA néguentropie *f*, quantité d'information *f*
negotiated: ~ **exit** *n* (*NE*) DATA sortie négociée *f*
negotiation *n* ELECT négociation *f*
nematic: ~ **liquid crystals** *n pl* ELECT, PARTS cristaux liquides nématiques *m pl*; ~ **phase** *n* ELECT phase nématique *f*, état nématique *m*
neon: ~ **indicator tube** *n* ELECT tube indicateur à néon *m*
neoprene *n* PARTS néoprène *m*
NEP *abbr* (*noise equivalent power*) ELECT, TRANS PEB (*puissance équivalente de bruit*)
neper *n* (*np*) ELECT néper *m*
NERF *abbr* (*network element function*) TRANS fonction d'élément de réseau *f*
nest *vt* COMP *subroutine*, TEST emboîter, imbriquer
nested: ~ **sampling** *n* TEST échantillonnage imbriqué *m*
net *n* COMP réseau *m*; ~ **loss** TRANS équivalent net *m*; ~ **margin** TELEP marge nette *f*; ~ **radiometer** TEST bilanmètre *m*
Net *n* (*Internet*) COMP, DATA réseau *m* (*Internet*)
NETBIOS *abbr* (*Network Basic Input Output System*) DATA NETBIOS *m*
netiquette *n* COMP, DATA *Internet* netiquette *f*
netsurfing *n* COMP, DATA *Internet* netsurfing *m*
network *n* COMP, DATA réseau *m*; ~ **access** DATA, RADIO, TRANS accès au réseau *m*; ~ **access component** DATA, TELEP élément d'accès au réseau *m*; ~ **access system** TRANS système d'accès au réseau *m*; ~ **address** DATA adresse de réseau *f*; ~ **address extension** (*NAE*) DATA extension d'adresse de réseau *f*; ~ **addressing** TRANS de signalisation au réseau *f*; **~-addressing authority** DATA autorité d'adressage de réseau *f*; ~ **addressing domain** DATA domaine d'adressage du réseau *m*; ~ **administrator** COMP administrateur de réseau *m*; ~ **analyser** *BrE*, ~ **analyzer** *AmE* TEST analyseur de réseaux *m*; ~ **analysis point** TELEP centre d'analyse du réseau *m*; ~ **breakdown** TEST panne de réseau *f*; ~ **cable** TRANS câble de réseau *m*; ~ **carrier** TELEP entreprise de télécommunications *f*; ~ **chart** TELEP plan de réseau *m*; ~ **cluster** TELEP faisceau de réseaux *m*; ~ **computer** COMP analyseur de réseau *m*; ~ **configuration** TELEP configuration du réseau *f*; ~ **congestion** TELEP, TRANS encombrement du réseau *m*; ~ **congestion signal** SIG signal d'encombrement du réseau *m*; ~ **connection** (*NC*) DATA, TELEP connexion de réseau *f*; ~ **connection fee** TELEP frais de raccordement au réseau *m*; **~-connectionless-mode data transmission** DATA, TRANS transmission de données de réseau en mode sans connexion *f*; **~-connection-mode data transmission** DATA, TRANS transmission de données de réseau en mode de connexion *f*; ~ **control center** *AmE*, ~ **control centre** *BrE* TEST centre de gestion de réseau *m*; ~ **control channel** TRANS voie de contrôle du réseau *f*; ~ **control language** (*NCL*) ELECT, TELEP langage de contrôle de réseau *m*; ~ **control phase** ELECT, TELEP phase de commande du réseau *f*; ~ **control signaling** *AmE*, ~ **control signalling** *BrE* TELEP signalisation au réseau *f*; ~ **default** (*ND*) ELECT, TELEP défaut par le réseau *m*; ~ **diagram** TELEP schéma de réseau *m*; ~ **digital termination** (*NT1*) TELEP terminaison numérique de réseau *f*; ~ **discard indicator** TEST indicateur de rejet par le réseau *m*; ~ **disruption** ELECT interruption de réseau *f*; ~ **drive interface specifications** *n pl* (*NDIS*) ELECT spécifications d'interface du circuit de commande du réseau *f pl*; ~ **element** (*NE*) TRANS élément de réseau *m*; ~ **element function** (*NERF*) TRANS fonction d'élément de

réseau *f*; ~ **engineering** TELEP ingénierie de réseau *f*; ~ **establishment agreement** TV convention d'établissement *f*; ~ **failure** DATA dérangement du réseau *m*, TELEP défaillance du réseau *f*; ~ **failure signal** SIG signal de défaillance du réseau *m*; ~ **fault in local loop signal** SIG signal de défaillance du réseau sur la ligne d'abonné *m*; ~ **file manager** (*NFM*) COMP, DATA gestionnaire de fichiers en réseau *m* (*NFM*); ~ **file system** (*NFS*) COMP, DATA système de fichiers en réseau *m*; ~ **gateway** DATA passerelle de réseau *f*; ~ **head** GEN, TV tête de réseau *f* (*TR*); ~ **head end** TV tête de réseau câblé *f*; ~ **identity** TELEP identité du réseau *f*; ~ **infrastructure** TRANS infrastructure des réseaux *f*; ~ **interconnection** DATA *Internet* interconnexion de réseau *f*; ~ **interface** COMP, DATA interface de réseau *f*, TELEP interface réseau *f*; ~ **layer** (*NL*) DATA, TELEP, TRANS couche réseau *f*; ~**-layer relay** (*NLR*) DATA, TELEP, TRANS relais de couche réseau *m*; ~ **locking** DATA encombrement du réseau *m*; ~ **maintenance** TV maintenance de réseau *f*; ~ **maintenance signals** *n pl* SIG signaux de maintenance du réseau *m pl*; ~ **management** DATA, GEN gestion de réseau *f*, administration de réseau *f*, TELEP gestion de réseau *f*, TRANS gestion de réseau *f*, administration de réseau *f*; ~ **management application service element** (*NM-ASE*) DATA élément de service d'application de la gestion de réseau *m*; ~ **management architecture** TRANS architecture de la gestion de réseaux *f*; ~ **management center** *AmE*, ~ **management centre** *BrE* (*NMC*) DATA centre de gestion de réseau *m*; ~ **management data** *n pl* COMP données de gestion de réseau *f pl*; ~ **management function** TRANS fonction de la gestion de réseau *f*; ~**-management platform** TRANS plate-forme de gestion de réseau *f*; ~ **management point** TELEP centre de gestion du réseau *m*; ~ **management signals** *n pl* SIG signaux de gestion de réseau *m pl*; ~ **map** DATA, SWG *exchange data store*, TELEP image de réseau *f*; ~ **news transfer protocol** (*NNTP*) DATA *Internet* protocole de transfert de chaîne d'informations (*NNTP*); ~ **node** TRANS nœud de réseau *m*; ~ **node interface** (*NNI*) TRANS interface de nœud de réseau *f*; ~ **nonaddressing equipment** TELEP équipement sans possibilité d'adressage au réseau *m*; ~ **operating agreement** TV convention d'exploitation *f*; ~ **operating company** TELEP exploitant de réseau *m*, opérateur de réseau *m*; ~ **operating system** (*NOS*) DATA système d'exploitation de réseau *m* (*NOS*); ~ **operation company** TELEP société d'exploitation du réseau *f*; ~ **operation and maintenance center** *AmE*, ~ **operation and maintenance centre** *BrE* TEST centre d'exploitation et de maintenance de réseau *m*; ~ **operator** TELEP exploitant de réseau *m*, opérateur de réseau *m*; ~**-operators' maintenance channel** (*NOMC*) TRANS canal de maintenance des exploitants de réseau *m*; ~ **parameter** TELEP caractéristique du réseau *f*; ~ **performance** (*NP*) TEST performance du réseau *f*; ~ **plan** TELEP plan de réseau *m*; ~ **planning** TELEP planification du réseau *f*; ~ **protection device** (*NPD*) TELEP dispositif de protection du réseau *m* (*DPR*); ~ **protocol** COMP, DATA *Internet* protocole de réseau *m*, protocole réseau *m*; ~ **protocol address information** DATA information d'adresse de protocole de réseau *f*; ~ **protocol data unit** (*NPDU*) DATA information de commande de protocole de réseau *f*, unité de données de protocole de réseau *f*, TELEP unité de données de protocole de réseau *f*; ~ **relay** DATA relais réseau *m*; ~ **resources** *n pl* TELEP ressources du réseau *f pl*; ~ **routing** DATA, SIG, TELEP, TRANS acheminement réseau *m*; ~ **security** DATA sécurité de réseau *f*; ~ **selection signals** *n pl* SIG signaux de sélection du réseau *m pl*; ~ **server** DATA serveur de réseau *m*; ~ **service** (*NS*) DATA, TELEP service de réseau *m*; ~ **service access component** DATA, TELEP élément d'accès au réseau service *m*; ~ **service access point** (*NSAP*) DATA, TELEP, TRANS point d'accès au service de réseau *m*; ~ **service access-point address** DATA adresse de point d'accès au service de réseau *f*; ~ **service data unit** DATA unité de données du service de réseau *f*; ~ **service provider** DATA fournisseur du service de réseau *m*; ~ **service user** DATA utilisateur du service de réseau *m*; ~ **service utilization component** DATA, TELEP élément d'utilisation du réseau service *m*; ~ **structure** TELEP structure du réseau *f*; ~ **supervision** DATA supervision de réseau *f*, TELEP supervision du réseau *f*, surveillance du réseau *f*; ~ **supervision and management** TEST surveillance et gestion des réseaux *f*; ~ **supervisor** TEST superviseur de réseau *m*; ~ **support engineer** GEN technicien réseau *m*; ~ **terminal equipment** (*NTE*) TELEP, TRANS équipement terminal de réseau *m*; ~ **terminal number** (*NTN*) DATA, TELEP numéro de terminal de réseau *m*; ~ **terminating unit** (*NTU*) TRANS appareil terminal sur réseau *m*; ~ **termination** (*NT*) APPL régie d'abonné *f*, TELEP, TRANS terminaison de réseau *f*; ~ **topology** DATA, TELEP topologie du réseau *f*; ~ **transfer delay** TELEP temps de propagation sur le réseau *m*; ~ **user** DATA *Internet* usager du réseau *m*; ~ **user identification** (*NUI*) DATA, TELEP identification de l'usager du réseau *f*; ~ **utility** TELEP service inter-réseaux *m*; ~ **utility field** TELEP champ des services inter-réseaux *m*; ~ **utilization component** DATA, TELEP élément d'utilisation du réseau *m*

Network: ~ **Basic Input Output System** *n* (*NETBIOS*) DATA NETBIOS *m*

networked *adj* DATA, GEN communicant, coopérant, en réseau

networking *n* TELEP ingénierie de réseau *f*; ~ **operating system** ELECT système d'exploitation de réseau *m*

neural: ~ **computer** *n* COMP ordinateur neuronal *m*; ~ **network** *n* COMP réseau neural *m*, réseau neuronique *m*, TRANS réseau de neurons *m*

neutral *n* POWER potentiel zéro *m*, point neutre *m*; ~ **bar** POWER barre neutre *f*; ~ **conductor** GEN, POWER conducteur neutre *m*; ~ **keying** TELEP manipulation simple courant *f*; ~ **point** POWER point neutre *m*; ~ **terminal** POWER borne neutre *f*; ~ **wire** GEN, POWER conducteur neutre *m*

neutralization *n* ELECT neutrodynage *m*

neutralizing: ~ **capacitor** *n* ELECT condensateur de neutrodynage *m*

new: ~ **assignment message** *n* DATA message de nouvelle affectation *m*; ~ **assignment procedure** *n* DATA procédure de nouvelle conversation *f*; ~ **common carriers** *n pl* (*NCC*) GEN nouveaux opérateurs *m pl*; ~ **data flag** *n* (*NDF*) DATA fanion de nouvelles données *m*, indicateur de nouvelles données *m*; ~**-line character** *n* (*NL*) DATA caractère de retour à la ligne *m*

news *n* DATA actualités *f pl*, news *m*; ~ **gathering** RADIO, TV reportage d'actualités *m*; ~ **server** DATA *Internet* serveur de news *m*

News: ~ **Watcher** *n* DATA *Internet* News Watcher *m*

newsgroup *n* DATA *Internet* groupe de discussion *m*

newsletter *n* COMP bulletin d'information *m*

newspaper *n* DATA journal *m*

newsreader *n* COMP lecteur de nouvelles *m*

newton *n* (*N*) TEST newton *m* (*N*)

next: ~ **transmitted bit** *n* TRANS bit suivant transmis *m*

NF *abbr* (*norme française*) GEN NF (*norme française*)

N-facility *n* TELEP facilité N *f*

n-field *n* PER APP zone de contrôle de numéro *f*

NFM *abbr* (*network file manager*) COMP, DATA NFM (*gestionnaire de fichiers en réseau*)

NFS *abbr* (*network file system*) COMP, DATA NFS (*système de fichiers en réseau*)

N-gate: ~ **thyristor** *n* ELECT, PARTS thyristor N *m*

nichrome *n* PARTS nichrome *m*

nick *n* GEN encoche *f*

NID *abbr* (*national identification digit*) TELEP chiffre d'identification de nationalité *m*

night: ~ **effect** *n* RADIO effet nocturne *m*, effet de nuit *m*; ~ **error** *n* RADIO erreur de nuit *f*, erreur nocturne *f*; ~ **service** *n* DATA service de nuit *m*, service hors vacation *m*, service simplifié *m*; ~**-service board** *n* SWG position de service de nuit *f*; ~**-service connection** *n* SWG renvoi de nuit *m*; ~**-service key** *n* SWG commutateur pour renvoi sur le service de nuit *m*; ~**-service mode** *n* SWG renvoi de nuit *m*; ~**-service operator** *n* SWG, TELEP opérateur du service de nuit *m*, opératrice du service de nuit *f*

NL *abbr* DATA (*new-line character*) caractère de retour à la ligne *m*, (*network layer*) couche réseau *f*, TELEP, TRANS (*network layer*) couche réseau *f*

n-level: ~ **code** *n* TELEP code à n moments *m*

NLR *abbr* (*network-layer relay*) DATA, TELEP, TRANS relais de couche réseau *m*

NM-ASE *abbr* (*network management application service element*) DATA élément de service d'application de la gestion de réseau *m*

NMC *abbr* (*network management center AmE, network management centre BrE*) DATA centre de gestion de réseau *m*

NMSC *abbr* (*nonmetallic sheathed cable*) GEN câble sous gaine non-métallique *m*

NN *abbr* DATA, TELEP (*national number*) NN (*numéro national*), (*negative notification*) NN (*notification négative*)

NNE *abbr* (*non-SDH network element*) TRANS élément de réseau non-SDH *m*

NNI *abbr* (*network node interface*) TRANS interface de nud de réseau *f*

NNTP *abbr* (*network news transfer protocol*) DATA *Internet* NNTP (*protocole de transfert de chaîne d'informations*)

no.: ~ **seven exchange** *n* SWG commutateur ñ sept *m*; ~ **seven exchange - first** *n* TELEP premier commutateur ñ sept *m*; ~ **seven exchange - last** *n* SWG dernier commutateur ñ sept *m*

no: ~ **charge** *phr* TELEP sans taxation

NO: ~ **OP** *n* COMP instruction ineffective *f*, instruction inopératoire *f*, NO OP *m*

no-alarm *n* TEST absence d'alarme *f*

noble: ~ **gas** *n* CLIM gaz rare *m*

no-break: ~ **power** *n* POWER alimentation ininterrompue *f*; ~ **system** *n* POWER système secouru *m*

nodal[1] *adj* COMP nodal

nodal[2]: ~ **analysis** *n* TRANS analyse par nœud *f*, analyse nodale *f*; ~ **period** *n* RADIO période nodale *f*; ~ **point** *n* TELEP nœud *m*, point nodal *m*

node *n* COMP, GEN nœud *m*, TELEP nœud *m*, point nodal *m*, TRANS nœud *m*; ~ **attributes** *n pl* COMP attributs du nœud *m pl*

nodecards *n pl* COMP nodecards *m pl*

nodes: ~ **network** *n* COMP réseau de nuds *m*

no-failure *n* TEST absence d'échec *f*, non-échec *m*

NOGAD *abbr* (*noise-operated gain adjusting device*) ELECT NOGAD (*régulateur de niveau commandé par le bruit*)

noise[1]: ~**-free** *adj* COMP sans bruit, TRANS à contribution de bruit nul

noise[2] *n* ELECT perturbation *f*, GEN bruit *m*, POWER parasites *m pl*, perturbation *f*, RADIO bruit *m*, perturbation *f*, TELEP bruit *m*; ~ **allocation** TRANS attribution du bruit *f*, répartition du bruit *f*; ~ **amplitude distribution** (*NAD*) TRANS distribution d'amplitude de bruit *f*; ~**-canceling microphone** *AmE see noise-cancelling microphone BrE*; ~ **cancellation** RADIO suppression de bruit *f*; ~**-cancelling microphone** *BrE* TELEP microphone antibruit *m*; ~ **current** TELEP courant de bruit *m*; ~ **detector** TELEP détecteur de bruit *m*; ~ **eliminator** POWER suppresseur de perturbation *m*; ~ **equivalent power** (*NEP*) ELECT, TRANS puissance équivalente de bruit *f* (*PEB*); ~ **factor** RADIO, TEST, TRANS facteur de bruit *m*; ~ **field** TRANS champ de bruit *m*; ~ **figure** RADIO, TEST, TRANS facteur de bruit *m*; ~**-generating device** COMP, TRANS générateur de bruit *m*; ~ **generator** COMP, TRANS générateur de bruit *m*; ~ **generator diode** ELECT diode génératrice de bruit *f*; ~ **generator plasma tube** ELECT tube générateur de bruit à plasma *m*; ~ **indicator** TELEP indicateur de bruit *m*; ~ **level** COMP, GEN, POWER niveau de bruit *m*; ~ **level control** TELEP commande du niveau de bruit *f*; ~ **limiter** TELEP limiteur de parasites *m*; ~ **masking** TRANS masquage par le bruit *m*; ~ **measurement** TELEP mesure de bruit *f*; ~**-measuring amplifier** TELEP amplificateur psophométrique *m*; ~**-measuring device** TELEP psophomètre *m*; ~**-measuring set** TEST psophomètre *m*; ~ **meter** TELEP psophomètre *m*; ~ **monitor** TEST appareil de contrôle de bruit *m*, appareil de surveillance de bruit *m*; ~ **monitoring** TELEP analyse de bruit *f*, contrôle de bruit *m*, supervision de bruit *f*, surveillance de bruit *f*; ~ **1/f** ELECT bruit en 1/f *m*; ~ **limit** RADIO limite de bruit *f*; ~**-operated gain adjusting device** (*NOGAD*) ELECT régulateur de niveau commandé par le bruit *m* (*NOGAD*); ~ **reducer** TRANS réducteur de bruit *m*; ~ **signal** POWER signal perturbateur *m*, signal parasite *m*; ~ **source** POWER source de bruit *f*, source de parasites *f*, source de perturbation *f*; ~ **source enclosure** GEN coiffe insonorisante *f*; ~ **spike** TRANS pointe de bruit *f*; ~ **supervision** TELEP supervision de bruit *f*; ~ **suppression** RADIO élimination de bruit *f*; ~ **suppression component** TEST dispositif d'antiparasitage *m*; ~ **suppressor device** TEST dispositif d'antiparasitage *m*; ~ **temperature** RADIO température équivalente de bruit *f*, température de bruit *f*; ~ **test** TELEP essai de niveau de bruit *m*; ~ **threshold** TELEP seuil de bruit *m*; ~**-weighing switch** ELECT sélecteur de pondération de bruit *m*; ~ **weighting** TRANS pondération du bruit *f*

noiseless *adj* COMP exempt de bruit de fond, silencieux

no-load: ~ **current** *n* POWER intensité courant à vide *f*; ~ **operation** *n* GEN marche à vide *f*; ~ **power** *n* POWER puissance à vide *f*; ~ **voltage** *n* POWER tension à circuit ouvert *f*, tension à vide *f*

NOMC *abbr* (*network-operators' maintenance channel*) TRANS canal de maintenance des exploitants de réseau *m*

nominal: ~ **alternating discharge current** *n* POWER courant nominal alternatif de décharge nominal *m*; ~ **bandwidth** *n* COMP, TV largeur de bande nominale *f*; ~ **black** *n* POWER noir nominal *m*; ~ **black signal** *n* TELEP signal de noir nominal *m*; ~ **channel width** *n* TV largeur nominale de canal *f*; ~ **coverage area** *n* RADIO zone de couverture nominale *f*; ~ **DC spark-over voltage** *n* ELECT, POWER *of a protector* tension continue nominale d'amorçage *f*; ~ **impulse discharge current** *n* POWER courant nominal de choc de décharge *m*; ~ **justification rate** *n* TELEP débit nominal de justification *m*; ~ **load** *n* ELECT charge nominale *f*; ~ **orbital position** *n* RADIO position nominale sur l'orbite *f*; ~ **relative level** *n* TELEP niveau relatif nominal de puissance *m*; ~ **symbol rate** *n* DATA, TELEP débit de symboles nominal *m*; ~ **value** *n* GEN, RADIO valeur nominale *f*; ~ **white** *n* TELEP blanc nominal *m*; ~ **white signal** *n* TELEP signal de blanc nominal *m*

nonacceptance *n* TEST rejet *m*; ~ **number** TEST critère de rejet *m*

nonadjacent *adj* RADIO non contigu

nonapproved *adj* PER APP *apparatus* non agréé

nonarmored *AmE*, **nonarmoured** *BrE adj* ELECT sans armature

nonassociated: ~ **mode** *n* TELEP non associé *m*; ~ **mode of operation** *n* TELEP mode d'exploitation non associé *m*; ~ **mode of signaling** *AmE*, ~ **mode of signalling** *BrE n* SIG mode de signalisation non associé *m*; ~ **signaling** *AmE*, ~ **signalling** *BrE n* SIG signalisation non associée *f*

nonbasic: ~ **parameter** *n* (*NBP*) DATA paramètre non-fondamental *m*

nonbinary: ~ **code** *n* DATA code non binaire *m*

nonblocking: ~ **network** *n* SWG réseau non-bloquant *m*, réseau sans blocage *m*, TRANS réseau non-bloquant *m*; **~ switching network** *n* SWG réseau de connexion non-bloquant *m*, réseau de connexion sans blocage *m*, TRANS réseau de connexion non-bloquant *m*

nonboosted: **~ antenna repeater system** *n* RADIO système à antennes relais sans amplification *m*

noncircularity: **~ of cladding** *n* TELEP, TRANS non-circularité de la gaine *f* (N_G); **~ of core** *n* TELEP, TRANS non-circularité du cœur *f*; **~ of reference surface** *n* TELEP non-circularité de la surface de référence *f* (N_R)

noncoherent: **~ demodulation** *n* DATA modulation non cohérente *f*; **~ side-looking radar** *n* RADIO radar à balayage latéral *m*, radar à visée latérale non cohérent *m*; **~ swept-tone modulation** *n* TRANS modulation à balayage de tonalité sans cohérence *f*

nonconducting *adj* POWER isolant, non conducteur

nonconductive[1] *adj* POWER isolant, non conducteur

nonconductive[2]: **~ pattern** *n* ELECT impression non conductrice *f*

nonconductor *n* POWER corps non conducteur *m*, isolant *m*

nonconfirmed[1] *adj* DATA non confirmé

nonconfirmed[2]: **~ service** *n* DATA service non confirmé *m*, service sans confirmation *m*

nonconforming: **~ items** *n pl* TEST éléments non conformes *m pl*; **~ unit** *n* TEST dispositif non-conforme *m*

nonconformity *n* TEST non-conformité *f*

nonconjunction *n* ELECT opération NON-ET *f*

nonconnected *adj* DATA non connecté

nondecimal numeral *n* TELEP nombre non décimal *m*

nondedicated: **~ signaling channel** *AmE*, **~ signalling channel** *BrE* *n* TRANS voie de signalisation non-spécialisée *f*

nondelivery: **~ notification** *n* (*NDN*) DATA, TELEP avis de non-remise *m*

nondestructive: **~ testing** *n* DATA, TEST essai non destructif *m*

nondialed: **~ connection** *AmE see nondialled connection BrE*

nondialled: **~ connection** *n* *BrE* APPL, TELEP appel sans numérotation *m*, ligne directe *f*

nondisjunction *n* ELECT fonction de Pierce *f*, NON-OU *m*, opération NI *f*

nondissipative: **~ stub** *n* RADIO bras de réactance *m*

nondroppable: **~ blocks** *n pl* DATA blocs non supprimables *m pl*

nonduplicated: **~ equipment** *n* TELEP équipement non dupliqué *m*; **~ system** *n* TELEP système non dupliqué *m*; **~ working mode** *n* TELEP mode non dupliqué *m*

nonequilibrium: **~-mode distribution** *n* TRANS répartition des modes hors équilibre *f*

nonequivalence *n* ELECT disjonction *f*, OU exclusif *m*; **~ operation** ELECT disjonction *f*, opération OU exclusif *f*

nonerasable: **~ memory** *n* COMP mémoire ineffaçable *f*, mémoire non effaçable *f*; **~ storage** *n* COMP mémoire ineffaçable *f*, mémoire non effaçable *f*

nonexistent: **~ number** *n* TELEP numéro inexistant *m*, numéro non-existant *m*

non-Gaussian: **~ noise** *n* TELEP, TRANS bruit non gaussien *m*

nonhalogen *adj* ELECT sans halogène

nonhazardous: **~ state** *n* TEST état non dangereux *m*

nonhierarchical: **~ system** *n* TRANS système non hiérarchique *m*, système à structure démocratique *m*

nonidentity: **~ operation** *n* ELECT opération de non-identité *f*

nonimpact: **~ printer** *n* DATA, TELEP imprimante sans impact *f*

noninductive: **~ circuit** *n* POWER circuit non inductif *m*

noninterface *n* RADIO non-brouillage *m*

noninterlaced *n* COMP non entrelacé

noninterpolated: **~ channel** *n* TRANS voie sans concentration de la parole *f*

non-intra: **~ coding** *n* COMP codage non intratrame *m*

nonlinear: **~ amplification** *n* TRANS amplification non linéaire *f*; **~ circuit** *n* TRANS circuit non linéaire *m*; **~ control system** *n* DATA système de contrôle non linéaire *m*; **~ digital speech signals** *n pl* TELEP signaux vocaux numériques non linéaires *m pl*; **~ distortion** *n* ELECT, SIG, TRANS distorsion non linéaire *f*; **~ effect** *n* ELECT, TRANS effet non linéaire *m*; **~ filter** *n* DATA, TRANS filtre non linéaire *m*; **~ filtering** *n* TRANS filtrage non linéaire *m*; **~ interpolation** *n* TRANS interpolation non linéaire *f*; **~ optics** *n* DATA optique non linéaire *f*; **~ oscillation** *n* TRANS oscillation non linéaire *f*; **~ processing loss** *n* TELEP affaiblissement par traitement non linéaire *m* (*ATNL*); **~ recursive filter** *n* DATA, TRANS filtre récursif non linéaire *m*; **~ resistance** *n* ELECT résistance non linéaire *f*; **~ scale** *n* TEST échelle non linéaire *f*; **~ scattering** *n* RADIO, TRANS diffusion non linéaire *f*; **~ system** *n* DATA système non linéaire *m*

nonloaded: **~ cable** *n* TELEP câble non pupinisé *m*

nonmagnetic: **~ armature shim** *n* ELECT pastille d'entrefer *f*

nonmetallic[1] *adj* PARTS non métallique

nonmetallic[2]: **~ sheathed cable** *n* (*NMSC*) TRANS câble sous gaine non-métallique *m*

nonmode: **~ of signaling** *AmE*, **~ of signalling** *BrE* *n* SIG mode de signalisation non associé *m*

non-network: **~ interface** *n* TELEP interface hors réseau *f*

nonoperating: **~ state** *n* TELEP, TEST état de non-fonctionnement *m*; **~ time** *n* TELEP, TEST temps de non-fonctionnement *m*

nonpolarized[1] *adj* POWER non polarisé

nonpolarized[2]: **~ relay** *n* PARTS, POWER relais non-polarisé *m*

nonprecision: **~ frequency offset** *n* RADIO, TRANS décalage de fréquence de faible précision *m*

nonprotection *n* RADIO non-protection *f*

nonreal: **~ time** *adj* (*NRT*) COMP, DATA, TELEP, TEST en temps différé

nonreceipt: **~ notification** *n* (*NRN*) DATA, TELEP avis de non-réception *m*

nonreciprocal: **~ circuit** *n* ELECT circuit non réciproque *m*; **~ waveguide** *n* DATA, RADIO, TRANS guide d'ondes non réciproque *m*

nonrecurrent: **~ charge** *n* TELEP taxe unique *f*; **~ fee** *n* TELEP taxe unique *f*

nonrecurring: **~ pulse** *n* GEN impulsion non récurrente *f*

nonrecursive: **~ filter** *n* DATA, TRANS filtre non récursif *m*

nonregenerative *adj* TRANS transparent

nonregistration *n* RADIO non-inscription *f*

nonrelevant: **~ failure** *n* TEST défaillance non significative *f*

nonrepetitive: **~ peak reverse voltage** *n* PARTS, TELEP tension inverse de pointe non répétitive *f*

nonrepudiation *n* DATA non-répudiation *f*

nonrequired: **~ time** *n* TEST période non requise *f*

non-return[1]: **~ to zero inverted** *adj* (*NRZI*) DATA non-retour à zéro inversé

non-return[2]: **~ to reference recording** *n* DATA enregistrement non-retour à un niveau de référence *m*, enregistrement sans retour à un niveau de référence *m*; **~ to zero** *n* (*NRZ*) DATA, TELEP non-retour à zéro *m*; **~ to zero code** *n* (*NRZ code*) DATA, TRANS code non-retour à zéro *m* (*code NRZ*); **~ to zero modulation** *n* TRANS modulation sans retour à zéro *f*; **~ to zero recording** *n* DATA enregistrement sans retour à zéro *m*, enregistrement non-retour à zéro *m*

nonrigid: **~ plastic** *n* PARTS plastique non-rigide *m*

non-SDH: **~ network element** *n* (*NNE*) TRANS élément de réseau non-SDH *m*

nonselective: **~ traffic** *n* RADIO trafic nonsélectif *m*

non-self-maintained: **~ discharge** *n* ELECT décharge non-autonome *f*

non-sine-wave: **~ waveform** *n* COMP forme d'onde non sinusoïdale *f*

nonspecified: **~ time relay** *n* PARTS relais à temps non-spécifié *m*

nonstandard: **~ facilities setup** *n* (*NSS*) DATA, TELEP établissement de facilités non normalisées *m* (*NSS*);

~ **facility** *n* (*NSF*) DATA, TELEP fonction non normalisée *f*; ~ **facility command** *n* (*NSC*) DATA, TELEP commande de fonction non normalisée *f*; ~ **facility command CIG-DTC** *n* DATA, TELEP commande de fonction non normalisée CIG-DTC *f*; ~ **facility CSI-DIS** *n* DATA CSI-DIS de fonction non normalisée *f*

nonstructured *adj* COMP non structuré

nonsupervised: ~ **location** *n* ELECT lieu non-surveillé *m*

nonsynchronized: ~ **network** *n* TELEP réseau non synchronisé *m*

nonterminal: ~ **symbol** *n* TELEP symbole non terminal *m*

nontransparent: ~ **bearer service** *n* APPL, DATA service support non transparent *m*

nonuniform: ~ **encoding** *n* TELEP, TRANS codage non uniforme *m*; ~ **frequency response** *n* TRANS distorsion d'affaiblissement *f*; ~ **quantizing** *n* TELEP, TRANS quantification non uniforme *f*

nonuseful: ~ **d.c. component** *n* ELECT composant continu non significative *m*

nonverbal: ~ **communication** *n* DATA communication non verbale *f*

nonvoice *adj* RADIO non vocal

nonvolatile: ~ **memory** *n* COMP, DATA mémoire rémanente *f*, RADIO mémoire non volatile *f*; ~ **storage** *n* COMP, DATA mémoire rémanente *f*

nonzero: ~ **value** *n* COMP valeur autre que zéro *f*

noodle *n* COMP bruit de fond *m*

NOR *n* ELECT NI *m*, NON-OU *m*; ~ **circuit** ELECT circuit NON-OU *m*, circuit NI *m*; ~ **element** ELECT porte NON-OU *f*, porte NI *f*; ~ **gate** ELECT circuit NI *m*, porte NI *f*, circuit NON-OU *m*, porte NON-OU *f*; ~ **operation** ELECT fonction de Pierce *f*, fonction NON-OU *f*, opération NI *f*

NORed *adj* ELECT mis en fonction NI

normal: ~**-band telephony** *n* TELEP téléphonie en bande normale *f*; ~ **charging** *n* POWER charge de service *f*; ~ **disconnected mode** *n* (*NDM*) DATA mode normal déconnecté *m*; ~ **distribution** *n* GEN répartition normale *f*; ~ **flow direction** *n* GEN sens de liaison normal *m*; ~ **inspection** *n* TEST contrôle normal *m*; ~ **mode** *n* DATA mode normal *m*; ~ **mode presentation** *n* DATA mode normal présentation *m*; ~ **response mode** *n* (*NRM*) DATA mode normal de réponse *m*; ~ **routing** *n* SIG, TELEP, TRANS acheminement normal *m*; ~ **routing of signaling** *AmE*, ~ **routing of signalling** *BrE* *n* SIG, TELEP, TRANS acheminement normal de signalisation *m*; ~**-situation class** *n* DATA classe situation normale *f*; ~**-vector interpolation shading** *n* COMP lissage par interpolation des vecteurs de la normale *m*

normalization *n* COMP, GEN normalisation *f*

normalize *vt* COMP, GEN normalize

normalized: ~ **detectivity** *n* ELECT, TEST, TRANS détectivité spécifique, détectivité normée *f*, détectivité normalisée *f*; ~ **form** *n* COMP forme normalisée *f*; ~ **frequency** *n* RADIO, TRANS fréquence normalisée *f*, fréquence normée *f*; ~ **frequency departure** *n* RADIO écart de fréquence normée *m*; ~ **frequency difference** *n* RADIO différence de fréquence normée *f*; ~ **frequency drift** *n* RADIO dérive de fréquence normée *f*; ~ **frequency offset** *n* RADIO, TRANS décalage de fréquence normée *m*; ~ **offset** *n* RADIO, TRANS décalage normé *m*; ~ **signal-to-noise ratio** *n* RADIO rapport signal/bruit normalisé *m*; ~ **transmission loss** *n* (*NTL*) TRANS affaiblissement de transmission normalisé *m* (*ATN*); ~ **value** *n* RADIO valeur normée *f*

normative: ~ **references** *n pl* TELEP références normatives *f pl*

norme: ~ **française** *n* (*NF*) GEN *French standard* norme française *f* (*NF*)

NOS *abbr* (*network operating system*) DATA NOS (*système d'expoitation de réseau*)

nose: ~ **bandwidth** *n* ELECT largeur de bande à six décibels *f*

NOT: ~ **circuit** *n* ELECT circuit NON *m*, circuit inverseur *m*, inverseur *m*; ~ **element** *n* ELECT circuit NON *m*, porte NON *f*; ~ **gate** *n* ELECT circuit NON *m*, porte NON *f*; ~ **operation** *n* ELECT inversion logique *f*, négation *f*

NOTAM *abbr* (*notice to airmen*) RADIO NOTAM (*avis aux navigateurs*)

notarization *n* DATA notarisation *f*

notch *n* RADIO encoche *f*; ~ **filter** ELECT, TRANS filtre à coupure brusque *m*, filtre coupe-bande *m*, filtre coupe-bande à bande étroite *m*, filtre coupe-fréquence *m*, filtre de réjection *m*, filtre réjecteur *m*

notching: ~ **relay** *n* PARTS relais intégrateur d'impulsions *m*, relais pas à pas *m*

note: ~ **length** *n* COMP durée de notes *f*; ~ **output** *n* COMP sortie de notes *f*; ~ **progression** *n* COMP suite de notes *f*; ~ **resolution** *n* COMP résolution de notes *f*; ~ **search** *n* COMP recherche de notes *f*

notebook *n* COMP ordinateur bloc-notes *m*, ordinateur portable *m*; ~ **computer** COMP ordinateur bloc-notes *m*, ordinateur portable *m*

NoteCards *n pl* COMP NoteCards *f pl*

notepad *n* COMP ordinateur bloc-notes *m*, ordinateur portable *m*; ~ **computer** COMP ardoise électronique *f*, bloc-notes électronique *m*, ordinateur bloc-notes *m*, ordinateur portable *m*

notes: **all** ~ **off** *phr* COMP toutes notes relâchées

notice *n* GEN avis; ~ **board** COMP, DATA, TELEP panneau d'affichage *m*, tableau d'affichage *m*, babillard *m* (*Can*), babillard électronique *m* (*Can*); ~ **to airmen** (*NOTAM*) RADIO avis aux navigateurs m (*NOTAM*)

notification *n* DATA notification *f*, GEN avis *m*; ~ **type** DATA type de notification *m*

NOT-IF-THEN: ~ **element** *n* ELECT circuit d'exclusion *m*, porte d'exclusion *f*; ~ **operation** *n* ELECT opération d'exclusion *f*

nought *n* COMP zéro *m*

nozzle *n* CLIM, RADIO buse *f*

np *abbr* (*neper*) ELECT néper *m*

NP *abbr* (*network performance*) TEST performance du réseau *f*

NPA *abbr* (*numbering plan area*) TELEP indicatif régional *m*, NPA (*zone de plan de numérotage*)

N-PAI *abbr* (*N-protocol-addressing information*) DATA, TELEP informations d'adressage du protocole N *f pl*

N-PCI *abbr* (*N-protocol control information*) DATA, TELEP informations de contrôle du protocole N *f pl*

NPD *abbr* (*network protection device*) TELEP DPR (*dispositif de protection du réseau*)

NPDU *abbr* (*network protocol data unit*) DATA, TELEP unité de données de protocole de réseau *f*

NPI *abbr* DATA (*null pointer indication*) indication de pointeur zéro *f*, (*number plan identification*) identification du plan de numérotage *f*, TELEP (*number plan identification*) identification du plan de numérotage *f*

NPI/TOA *abbr* (*numbering plan indicator/TOA*) DATA, TELEP indicateur de plan de numérotage/TOA *m*

n-plus-one: ~ **address instruction** *n* COMP instruction à n plus une adresse *f*

NPO: ~**-type capacitator** *n* POWER condensateur NPO *m*

N-relay *n* TELEP relais N *m*

NRN *abbr* (*nonreceipt notification*) DATA, TELEP avis de non-réception *m*

NRT *abbr* (*nonreal time*) COMP, DATA, TELEP, TEST en temps différé

NRZ *abbr* (*nonreturn to zero*) DATA, TELEP non-retour à zéro *m*; ~ **code** (*nonreturn to zero code*) DATA, TRANS code NRZ *m* (*code non-retour à zéro*)

NRZI *abbr* (*nonreturn to zero inverted*) DATA, TELEP non-retour à zéro inversé

ns *abbr* (*nanosecond*) TEST ns (*nanoseconde*)

NS *abbr* (*network service*) DATA, TELEP service de réseau *m*

NSAP *abbr* (*network service access point*) DATA, TELEP, TRANS point d'accès au service de réseau *m*

NSC *abbr* (*nonstandard facility command*) DATA, TELEP commande de fonction non-normalisée *f*

NSC-CIG-DTC *abbr* (*nonstandard facility command CIG-DTC*) DATA, TELEP NSC-CIG-DTC (*commande de fonction non-normalisée CIG-DTC*)

N-service *n* DATA, TELEP service N *m*; **~ access point** (*NSAP*) DATA point d'accès au service N *m*; **~ access-point address** DATA adresse de NSAP *f*, adresse du point d'accès à des services N *f*; **~ confirm primitive** DATA primitive confirmation de service *f*; **~ indication primitive** DATA primitive indication de service N *f*; **~ request primitive** DATA primitive demande de service N *f*; **~ response primitive** DATA primitive réponse à une demande de service *f*

NSF *abbr* (*nonstandard facility*) DATA, TELEP fonction non-normalisée *f*

NSF-CSI-DIS *abbr* (*nonstandard facility CSI-DIS*) DATA, TELEP NSF-CSI-DIS (*CSI-DIS de fonction non-normalisée*)

NSN *abbr* (*national significant number*) TELEP numéro national significatif *m*

NSS *abbr* (*nonstandard setup*) DATA, TELEP établissement non-normalisé *m*

NT *abbr* (*network termination*) TELEP, TRANS terminaison de réseau *f*

NTC *abbr* (*national television center AmE, national television centre BrE*) TV centre télévisuel national *m*

NTE *abbr* (*network terminal equipment*) TELEP, TRANS équipement terminal de réseau *m*

nth: **~ choice group** *n* TELEP faisceau d'énième choix *m*; **~ order harmonic distortion** *n* TELEP affaiblissement de distorsion harmonique d'ordre n *m*

NTL *abbr* (*normalized transmission loss*) TRANS ATN (*affaiblissement de transmission normalisé*)

NT-LB *abbr* (*B-ISDN network termination*) TRANS terminaison de réseau pour le RNIS-LB *f*

NTN *abbr* (*network terminal number*) DATA, TELEP numéro de terminal de réseau *m*

NT1 *abbr* (*network digital termination*) TELEP terminaison numérique de réseau *f*

NT1-LB *abbr* (*B-ISDN network termination 1*) TRANS terminaison de réseau 1 pour le RNIS-LB *f*

NT2-LB *abbr* (*B-ISDN network termination 2*) TRANS terminaison de réseau 2 pour le RNIS-LB *f*

NTU *abbr* (*network terminating unit*) TRANS appareil terminal sur réseau *m*

N-type: **~ conduction** *n* ELECT *semiconductors*, PARTS conduction type N *f*; **~ semiconductor** *n* ELECT, PARTS semiconducteur type N *m*

NU *abbr* (*national usage*) TELEP usage national *m*

nuance: **~ of darkness** *n* DATA *scanner* nuance dans les ombres *f*; **~ of lightness** *n* DATA *scanner* nuance dans les clairs *f*

nucleus *n* COMP noyau *m*; **~ of operating system** COMP programme de commande résident *m*, programme de contrôle résident *m*

NUI *abbr* (*network user identification*) DATA, TELEP identification de l'usager du réseau *f*

nuisance: **~ area** *n* GEN zone de brouillage *f*; **~ call** *n* TELEP appel importun *m*

NUL *abbr* (*null character*) DATA NUL (*caractère nul*)

null[1]: **~-modem** *adj* ELECT null-modem

null[2] *n* RADIO évanouissement *m*, nul *m*; **~ AM modulation method** RADIO méthode de modulation MA zéro *f*; **~ character** (*NUL*) DATA caractère nul *m* (*NUL*); **~ data check** COMP contrôle de présence *m*, vérification de présence *f*; **~ detector** RADIO détecteur de zéro *m*; **~ meter** TEST zéromètre *m*, galvanomètre de zéro *m*; **~ method** TEST méthode de zéro *f*; **~ modem** DATA éliminateur de modem *m*; **~ placing technique** RADIO préréglage des zéros *m*; **~ pointer indication** (*NPI*) DATA indication de pointeur zéro *f*; **~-reference glide path** RADIO alignement de descente à amplitude de référence nulle *m*; **~ string** DATA chaîne vide *f*

number[1] *n* GEN numéro *m*, TELEP indicatif *m*; **~ of bit disagreements** DATA nombre de bits différents *m*; **~-busy signal** SIG signal de numéro occupé *m*; **~ of defectives** TEST nombre d'entités défectueuses *m*, nombre d'unités défectueuses *m*; **~ not allocated** DATA numéro non-attribué *m*; **~ out of order** DATA numéro en dérangement *m*; **~ plan identification** (*NPI*) DATA, TELEP identification du plan de numérotage *f*; **~ redial** TELEP répétition automatique des numéros *f*; **~ repetition service** TELEP service de répétition de numéro *m*; **~ of significant conditions** TELEP valence *f*; **~ of turns** POWER nombre de spires *m*

number[2] *vt* GEN numéroter; **~ pages** COMP, GEN paginer

numbering *n* GEN numérotage *m*, numérotation *f*; **~ and addressing plan indicator/type of address** (*NAPI/TOA*) DATA, TELEP indicateur de plan de numérotage/type d'adresse *m*; **~ area** TELEP zone de numérotage *f*; **~ drawing** GEN dessin de numérotage des broches *m*; **~ plan** GEN plan de numérotation *m*, RADIO numérotation *f*, TELEP plan de numérotage *m*; **~ plan area** (*NPA*) TELEP indicatif régional *m*, zone de plan de numérotage *f* (*NPA*); **~ plan identification** DATA, TELEP identification du plan de numérotage *f*; **~ plan identifier** DATA, TELEP identificateur de plan de numérotage *m*; **~ plan indicator** DATA, TELEP indicateur de plan de numérotage *m*; **~ plan indicator/TOA** (*NPI/TOA*) DATA, TELEP indicateur de plan de numérotage/TOA *m*; **~ plan interworking** DATA, TELEP interfonctionnement des plans de numérotage *m*; **~ scheme** TELEP plan de numérotage *m*; **~ system** TELEP système de numération *m*

numeric[1] *adj* COMP, DATA, GEN numérique

numeric[2]: **~ data** *n pl* COMP, DATA données numériques *f pl*; **~ keypad** *n* COMP clavier numérique *m*, pavé numérique *m*; **~ pad** *n* COMP clavier numérique *m*, pavé numérique *m*; **~ word** *n* COMP, DATA mot numérique *m*

numerical[1] *adj* COMP, DATA, GEN numérique

numerical[2]: **~ aperture** *n* (*NA*) DATA, ELECT, TRANS ouverture numérique *f*; **~ code** *n* DATA code numérique *m*; **~ control** *n* (*NC*) COMP commande numérique *f*; **~ data** *n pl* COMP, DATA données numériques *f pl*; **~ database** *n* COMP, DATA base de données numériques *f*; **~ display** *n* DATA, ELECT, RADIO affichage numérique *m*; **~ display paging** *n* RADIO RMU avec affichage numérique *f*; **~ horizontal** *n* SWG barre de sélection numérique *f*, horizontale numérique *f*; **~ message** *n* RADIO message numérique *m*; **~ selecting bar** *n* SWG barre de sélection numérique *f*, horizontale numérique *f*

numerically *adv* COMP, DATA, GEN numériquement

n-unit: **~ code** *n* TELEP code à n moments *m*, code à n éléments *m*, code à n éléments unitaires *m*

n-way *n* DATA transfert multilatéral *m*

nylon: **~ binder** *n* ELECT filin en nylon *m*

Nyquist: **~ diagram** *n* CONTR diagramme de Nyquist *m*; **~ sampling** *n* COMP échantillonnage de Nyquist *m*

O

O *abbr* (*optional*) COMP, DATA, TELEP O (*optionnel*); **~ component** *n* DATA élément facultatif *m*
OA *abbr* DATA (*outgoing access*) accès sortant *m*, SWG, TELEP (*overflow accept*) acceptation de débordement *f*
OAC *abbr* (*operator-assisted call*) GEN, TELEP appel assisté par l'opérateur *m*
OAMC *abbr* TELEP, TEST (*operation, administration and maintenance center AmE, operation, administration and maintenance centre BrE*) centre de gestion, d'exploitation et de maintenance *m*
OAMP *abbr* (*operation, administration, maintenance and provisioning*) TELEP, TEST exploitation, administration, maintenance et mise en service *f*
OBE *abbr* (*out-of-band emission*) RADIO émission hors bande *f*, rayonnement hors bande *m*
object[1]: **~-oriented** *adj* COMP orienté objet
object[2] *n* COMP, DATA élément *m*, objet *m*; **~ class** DATA classe d'objet *f*; **~ identifier** DATA identificateur d'objet *m*; **~ language** COMP langage objet *m*; **~ management** COMP gestion d'objets *f*; **~ module** COMP module objet *m*; **~-oriented programming** (*OOP*) COMP programmation orientée objet *f*; **~ program** COMP programme objet *m*; **~ space** COMP espace objet *m*
objective: **~ loudness rating** *n* TELEP indice objectif de volume des sons *m*
obligatory: **~ parameter** *n* DATA paramètre obligatoire *m*; **~ presence** *n* DATA présence obligatoire *f*
oblique: **~-incidence transmission** *n* RADIO transmission sous incidence oblique *f*; **~-incidence ionospheric recorder** *n* RADIO sondeur ionosphérique oblique *m*, sondeur oblique *m*; **~ stroke** *n* COMP barre oblique *f*
OBO *abbr* (*output back-off*) TRANS OBO (*recul de sortie*)
OBP *abbr* (*on-board processing*) COMP traitement à bord *m*
observation: **~ desk** *n* CONTR pupitre de commande *m*, pupitre de conduite *m*; **~ error** *n* TEST erreur d'observation *f*; **~ panel** *n* CONTR panneau de surveillance *m*, tableau de surveillance *m*; **~ service** *n* APPL service de surveillance *m*, TELEP service de surveillance de la qualité des télécommunications *m*; **~ telephone** *n* TELEP poste de surveillance *m*
observe *n* RADIO, TELEP écoute discrète *f*; **~ agent key** TELEP touche d'écoute de position *f*
observed: **~ bearing** *n* RADIO relèvement radio *m*; **~ deviation** *n* TEST écart observé *m*; **~ dimension** *n* TEST cote effective *f*; **~ test outcome** *n* DATA résultat d'un test *m*; **~ value** *n* TEST valeur observée *f*
obstacle: **~ effect** *n* TELEP effet de masque *m*; **~ gain** *n* RADIO gain d'obstacle *m*
obtain *vt* TELEP joindre; **~ holding current** PARTS recevoir un courant de maintien
OC *abbr* (*organization code*) COMP, DATA code d'organisation *m*; **~ curve** *n* (*operating characteristic curve*) GEN courbe d'efficacité *f*; **~ test** *n* (*open-circuit test*) ELECT essai à vide *m*
OCB *abbr* (*outgoing calls barred*) DATA, TELEP interdiction d'accès des appels sortants *f*, interdiction des appels au départ *f*
OCC *abbr* (*operations control center AmE, operations control centre BrE*) TEST centre de direction des opérations *m*
occasional: **~ transmissions** *n pl* TRANS transmissions occasionnelles *f pl*
occupancy: **~ rate** *n* GEN taux d'occupation *m*
occupation: **~ time** *n* GEN, TELEP durée d'occupation *f*
occupied: **~ band** *n* RADIO bande occupée *f*; **~ bandwidth** *n* RADIO, TRANS largeur de bande occupée *f*; **~ spectrum space** *n* RADIO, TRANS largeur de bande occupée *f*
occupy *vt* TELEP mettre en état de prise
occurrence: **~ of collisions** *n* ELECT présence de collisions *f*
OCR *abbr* (*optical character recognition*) COMP, DATA ROC (*reconnaissance optique de caractères*)
octal: **~ numeral** *n* TELEP nombre octal *m*
octantal: **~ error** *n* RADIO erreur octantale *f*
octave: **~ shift** *n* COMP décalage d'octave *m*; **~ transpose** *n* COMP transposition d'octave *f*
octet *n* COMP, DATA, ELECT byte *m*; **~ alignment** TELEP alignement d'octet *m*; **~-string type** DATA type de chaîne d'octets *m*; **~ timing signal** SIG signal d'horloge des octets *m*
octode *n* ELECT octode *f*
octopus *adj* ELECT multibrin
odd: **~ note** *n* COMP note impaire *f*
ODFMA *abbr* (*orthogonal frequency-division multiple access*) RADIO, SIG, TELEP, TRANS AMRFO (*accès multiple par répartition en fréquences orthogonales*)
ODIF *abbr* (*open document interchange format*) DATA ODIF (*format ouvert d'échange de documents*)
ODLI *abbr* (*open data link interface*) DATA, ELECT interface ouverte de liaison de données *f*
ODP *abbr* (*open distributed processing*) COMP ODP (*traitement à distribution ouverte*) *m*
O-E *abbr* GEN, TV (*optical-electrical, optoelectric*) O-E (*optoélectrique*)
OEM *abbr* (*original equipment manufacturer*) GEN OEM; **~ power supply** *n* POWER alimentation à incorporer *f*
off-air[1] *adj* RADIO hors émission
off-air[2]: **~ call setup** *n* RADIO établissement d'une connexion sans émission *m*
off-axis[1] *adj* RADIO en dehors de l'axe du faisceau principal, hors axe
off-axis[2]: **~ limit** *n* RADIO limite hors axe *f*
off-beam *adj* RADIO hors faisceau
offbeat *adj* COMP désynchronisé, non synchronisé
offer *n* TELEP offre *f*
off-hook[1] *adj* COMP débranché, déconnecté, ELECT, PER APP débranché, RADIO, TELEP déconnecté, décroché
off-hook[2]: **~ queue** *n* (*OHQ*) TELEP file d'attente d'appels au décrochage *f*; **~ queue offer tone** *n* TELEP tonalité d'option de file d'attente en mode de décrochage *f*
off-hook *vt* DATA décrocher
off-hooking *n* DATA décrochage *m*
office[1]: **~-level** *adj* DATA bureautique
office[2] *n* COMP bureau *m*; **~ automation** DATA, GEN bureautique *f*; **~ directory** DATA annuaire interne *m*; **~ document interchange format** DATA ODIF
official: **~ franking-privilege telegram** *n* TELEP télégramme en franchise officiel *m*
off-line[1] *adj* COMP autonome, hors ligne, DATA autonome, en local, hors ligne, non connecté, TELEP autonome, hors ligne, en différé
off-line[2]: **~ database** *n* COMP, DATA base de données hors ligne *f*, base de données off-line *f*; **~ storage** *n* COMP mémoire autonome *f*, mémoire hors ligne *f*; **~ working** *n* TELEP exploitation en autonome *f*
offload: **~ a satellite** *vt* RADIO décharger un satellite
off-net: **~ station** *n* TELEP poste hors réseau *m*, TV station

hors réseau *f*; **~ to on-net call** *n* TELEP appel en provenance de l'extérieur du réseau *m*
off-peak: **~ tariff periods** *n pl* APPL heures croisées *f pl*
offprint *n* PER APP tirage à part *m*
offset[1] *adj* COMP, RADIO décalé
offset[2] *n* COMP, RADIO décalage *m*; **~ current** ELECT courant de décalage *m*; **~ dish** RADIO parabole à alimentation excentrée *f*; **~ feed** RADIO source primaire décalée *f*; **~ Gregory antenna** RADIO antenne Gregory excentrée *f*; **~ key protection** ELECT clés de protection *f pl*; **~ parabolic aerial** RADIO antenne parabolique avec alimentation décalée *f*; **~ parabolic antenna** RADIO antenne parabolique avec alimentation décalée *f*; **~ quaternary phase-shift keying** (*OQPSK*) DATA, SIG modulation par déplacement de phase avec décalage *f*, modulation de phase de deux porteuses en quadrature avec décalage des trains binaires *f*; **~ screwdriver** GEN tournevis coudé *m*; **~ voltage** ELECT tension de décalage *f*
offset[3] *vt* COMP, RADIO décaler
off-site: **~ maintenance** *n* TEST maintenance hors site *f*
off-state *n* ELECT état inactivité *m*, état bloqué *m*, PARTS état bloqué *m*; **~ current** ELECT courant à l'état bloqué *m*; **~ voltage** ELECT, PARTS tension à l'état bloqué *f*
off-the-shelf *adj* COMP en stock
off-time *adj* COMP, RADIO décalé
OFS *abbr* (*out-of-frame second*) DATA seconde de perte du verrouillage de trame *f*
o/g *abbr* (*outgoing*) GEN, TRANS de départ, sortant
OGM *abbr* (*outgoing message*) PER APP annonce *f*
OHA *abbr* (*overhead access*) DATA accès au surdébit *m*, accès aux octets de surdébit *m*
ohm *n* ELECT ohm *m*
ohmic: **~ contact** *n* PARTS contact ohmique *m*
ohmmeter *n* TEST ohmmètre *m*
OHQ *abbr* (*off-hook queue*) TELEP file d'attente d'appels au décrochage *f*
oil: **~-filled capacitor** *n* ELECT condensateur à huile *m*
OK: **~ signal** *n* RADIO signal OK *m*
old: **~ code** *n* TELEP *code modification* ancien indicatif *m*; **~ number** *n* TELEP numéro désaffecté *m*
oligarchic: **~ network** *n* TELEP *synchronised* réseau à synchronisation oligarchique *m*, réseau oligarchique *m*
OLL *abbr* (*open-loop loss*) CONTR, POWER affaiblissement en boucle ouverte *m*
OLTE *abbr* (*optical-line terminating equipment*) TRANS terminal à fibre optique *m*
OM *abbr* (*operations and maintenance*) DATA, TELEP, TEST exploitation et maintenance *f*; **~ subsystem** *n* DATA, TELEP, TEST SSEM (*scus-système exploitation et maintenance*)
OMI *abbr* (*optical modulation index*) TELEP, TRANS indexe de modulation optique *m*
omnibus: **~ service circuit** *n* TELEP circuit de service omnibus *m*; **~ system** *n* TELEP système à postes embrochés *m*
omnidirectional: **~ aerial** *n* RADIO antenne omnidirectionnelle *f*; **~ antenna** *n* RADIO antenne omnidirectionnelle *f*
on: **~-demand testing** *n* TEST essais à la demande *m pl*
on-air: **~ time** *n* RADIO temps d'émission *m*
on-axis: **~ gain** *n* RADIO gain dans l'axe *m*
on-board[1] *adj* COMP embarqué
on-board[2]: **~ apparatus** *n* RADIO appareillage embarqué *m*; **~ beacon** *n* RADIO balise embarquée *f*; **~ equipment** *n* RADIO équipement à bord *m*, équipement embarqué *m*; **~ processing** *n* (*OBP*) COMP traitement à bord *m*; **~ subscriber** *n* RADIO abonné à bord *m*
on-busy *adj* TELEP sur occupation
on-chip *adj* COMP, ELECT sur puce
on-demand[1] *adj* TV à la demande
on-demand[2]: **~ service** *n* TELEP service à la demande *m*, service appel par appel *m*
one[1]: **~-frequency** *adj* RADIO monofréquence; **~-time programmable** *adj* (*OTP*) COMP uni-programmable; **~-way** *adj* RADIO spécialisé, TELEP à sens unique, unidirectionnel
one[2]: **~-way** *adv* RADIO, TELEP à sens unique
one[3]: **~ address instruction** *n* COMP instruction à une adresse *f*; **~-ahead addressing** *n* COMP adressage à progression automatique *m*; **~ million floating-point operations per second** *n pl* (*megaflops*) COMP un million d'opérations en virgule flottante par seconde *f pl* (*megaflops*); **~ million logical inferences per second** *n pl* (*megalips*) COMP un million d'inférences logiques par seconde *f pl* (*megalips*); **~ out of five decode** *n* ELECT décodage un sur cinq *m*; **~-piece cover** *n* TELEP couvercle monobloc *m*; **~-plus-one address instruction** *n* COMP instruction à une plus une adresse *f*; **~-plus-one carrier system** *n* TRANS système d'abonné à courant porteur *m*; **~-shot multivibrator** *n* ELECT multivibrateur monostable *m*, monostable *m*; **~-step operation** *n* COMP opération à un seul pas *f*; **~-symbol gate signal** *n* DATA signal de porte logique d'une largeur de symbole *m*; **~-unit message** *n* TELEP message simple *m*; **~-way** *n* DATA transfert à sens unique *m*; **~-way communication** *n* RADIO communication unidirectionnelle *f*, SWG, TELEP, TV communication unidirectionnelle *f*, communication unilatérale *f*; **~-way directionality** *n* COMP lien à sens unique *m*; **~-way link** *n* COMP liaison unidirectionnelle *f*, lien à sens unique *m*, lien unidirectionnel *m*
on-hold *adj* TELEP mis en garde
on-hook[1] *adj* COMP, TELEP en ligne
on-hook[2]: **~ condition** *n* TELEP état de raccrochage *m*; **~ dialing** *AmE*, **~ dialling** *BrE* *n* GEN, RADIO, TELEP numérotation sans décrochage du combiné *f*, numérotation avant décrochage *f*, numérotation combinée raccrochée *f*, numérotation sans décrocher *f*, *of answer machine* prise de ligne sans décrocher *f*; **~ signal** *n* TELEP signal de raccrochage *m*
on-line[1] *adj* COMP, DATA, ELECT en ligne, GEN temps réel, TELEP en direct, en ligne
on-line[2]: **~ catalog** *AmE*, **~ catalogue** *BrE* *n* DATA catalogue en ligne *m*; **~ data service** *n* COMP serveur *m*; **~ database** *n* COMP, DATA base de données en ligne *f*; **~ full-text document retrieval** *n* DATA ressaisie de documents intégraux en temps réel *f*; **~ host** *n* COMP serveur interactif *m*; **~ information retrieval** *n* DATA repérage de l'information en ligne *m*; **~ information service** *n* DATA *Internet* service d'information en ligne *m*; **~ newspaper** *n* DATA *Internet* journal en ligne *m*; **~ optical amplification** *n* TRANS amplification optique en ligne *f*; **~ processing** *n* DATA traitement direct *m*; **~ registration facility** *n* TELEP service complémentaire d'enregistrement en ligne de service *m*; **~ service** *n* DATA service de télématique *m*, service en ligne *m*; **~ telematic service** *n* DATA service télématique en ligne *m*
only-route: **~ circuit group** *n* TELEP faisceau d'acheminement unique *m*
on-net: **~ to off-net call** *n* TELEP appel en provenance du réseau privé *m*
on-off: **~ function** *n* PER APP fonction marche-arrêt *f*; **~ keying** *n* (*OOK*) TELEP manipulation par tout-ou-rien *f*, modulation par tout ou rien *f*; **~ switch** *n* COMP interrupteur général *m*
ONP *abbr* (*open network provision*) TRANS offre de réseau ouvert *f*
on-premises *adv* TELEP sur place (*SP*)
ONS *abbr* (*on-scene*) TELEP SP (*sur place*)
on-scene[1] *adj* (*ONS*) TELEP sur place

on-scene[2]: ~ **communications** *n pl* (*OSC*) RADIO communications sur place *f pl*
ONSD *abbr* (*optional network-specific digit*) TELEP chiffre facultatif spécifique au réseau *m*
on-site: ~ **acceptance** *n* RADIO recette sur site *f*; ~ **maintenance** *n* TEST maintenance sur site *f*
on-state *n* ELECT état conducteur *m*, état passant *m*, état activité *m*, PARTS état passant *m*, état activité *m*; ~ **current** ELECT courant à l'état passant *m*; ~ **power loss** ELECT, PARTS perte de puissance à l'état passant *f*; ~ **slope resistance** ELECT, PARTS résistance apparente à l'état passant *f*; ~ **threshold voltage** ELECT, PARTS tension de seuil à l'état passant *f*; ~ **voltage** ELECT, PARTS tension à l'état passant *f*
on-station: ~ **satellite** *n* RADIO satellite à poste *m*; ~ **software** *n* COMP, DATA logiciel de maintien à poste *m*
on-the-fly: ~ **printer** *n* TELEP imprimante à la volée *f*
OO *abbr* (*operator office*) TELEP accès bidirectionnel à l'opérateur *m*
OOB *abbr* (*out-of-band*) ELECT, RADIO hors bande
OOK *abbr* (*on-off keying*) TELEP manipulation par tout-ou-rien *f*, modulation par tout-ou-rien *f*
OOLR *abbr* (*overall objective loudness rating*) TELEP équivalent objectif global pour la sonie *m* (*EOGS*)
OOP *abbr* (*object-oriented programming*) COMP programmation orientée objet *f*
op: ~ **amp** *n* (*operational amplifier*) ELECT, TRANS ampliop *m* (*amplificateur opérationnel*)
OPC *abbr* (*originating point code*) DATA, TELEP code du point d'origine *m*
open[1]: ~**-circuit** *adj* ELECT à vide
open[2]: ~ **battery** *n* POWER batterie ouverte *f*; ~ **circuit** *n* POWER circuit ouvert *m*; ~**-circuit admittance** *n* POWER admittance en circuit ouvert *f*; ~**-circuit current** *n* POWER courant à vide *m*; ~**-circuit impedance** *n* POWER impédance en circuit ouvert *f*; ~**-circuit test** *n* (*OC test*) ELECT essai à vide *m*, essai en circuit ouvert *m*; ~**-circuit voltage** *n* POWER tension à vide *f*, tension à circuit ouvert *f*; ~**-circuit working** *n* ELECT, TELEP transmission par batterie centrale *f*, transmission par envoi de courant *f*, transmission par fermeture de circuit *f*, transmission par ouverture de circuit *f*, transmission par rupture de circuit *f*; ~ **collector-gate** *n* ELECT porte collectrice ouverte *f*; ~ **data link interface** *n* (*ODLI*) DATA, ELECT interface ouverte de liaison de données *f*; ~ **distributed processing** *n* (*ODP*) COMP traitement à distribution ouverte *m*; ~ **document interchange format** *n* (*ODIF*) DATA format ouvert d'échange de documents *m* (*ODIF*); ~ **forum** *n* DATA espace d'échange *m*; ~ **group** *n* DATA groupe ouvert *m*; ~ **loop** *n* COMP, POWER, TRANS boucle ouverte *f*; ~**-loop acquisition** *n* CONTR, POWER acquisition en boucle ouverte *f*; ~**-loop control** *n* CONTR, POWER commande en boucle ouverte *f*; ~**-loop gain** *n* CONTR, POWER gain en boucle ouverte *m*; ~**-loop loss** *n* (*OLL*) CONTR, POWER affaiblissement en boucle ouverte *m*; ~**-loop synchronization** *n* CONTR, POWER, TRANS synchronisation en boucle ouverte *f*; ~ **network architecture** *n* DATA architecture de réseau ouverte *f*; ~ **network provision** *n* (*ONP*) TRANS offre de réseau ouvert *f* (*ONP*); ~ **position** *n* SWG position d'ouverture *f*, position ouverte *f*; ~ **standards computing platform** *n* DATA plate-forme informatique banalisée ouverte *f*; ~ **subroutine** *n* COMP sous-programme ouvert *m*; ~ **system** *n* DATA, TELEP, TRANS système ouvert *m*; ~ **systems architecture** *n* (*OSA*) TELEP, TRANS architecture de systèmes ouverts *f*; ~ **systems interconnection** *n* (*OSI*) COMP, DATA, TELEP, TRANS interconnexion de systèmes ouverts *f* (*ISO*); ~ **systems interconnection layer** *n* COMP couche d'interconnexion de systèmes ouverts *f*; ~ **systems interconnection network layer** *n* (*OSI network layer*) DATA couche réseau d'interconnexion de systèmes ouverts *f* (*couche réseau ISO*); ~ **systems interconnection network service** *n* (*OSI network service*) DATA, TELEP service de réseau d'interconnexion de systesms ouverts *m* (*service de réseau ISO*); ~ **systems interconnection resource** *n* (*OSI resource*) DATA, TELEP, TRANS ressource d'interconnexion de systèmes ouverts *f* (*ressource ISO*); ~**-systems interconnection standard** *n* ELECT, TRANS norme d'interconnexion de systèmes ouverts *f*; ~ **systems interconnection transport layer** *n* (*OSI transport layer*) DATA couche transport d'interconnexion de systèmes ouverts *f* (*couche transport ISO*); ~ **wire** *n* GEN fil nu *m*, fil aérien *m*; ~**-wire line** *n* GEN fil nu aérien *m*, TELEP ligne en fils nus *f*
open[3] *vt* POWER couper, rompre
open[4]: ~ **shortest path first** *phr* (*OSPF*) COMP, ELECT ouverture prioritaire du plus court chemin *f*
opening[1] *adj* ELECT *transistors* passant
opening[2]: ~ **screen** *n* DATA page d'accueil *f*
operable: ~ **time** *n* GEN temps utilisable *m*
operate[1]: ~ **condition** *n* GEN état de marche *m*; ~ **current** *n* POWER courant d'appel *m*; ~ **time** *n* PARTS temps de fonctionnement *m*
operate[2] *vt* CONTR, GEN, TELEP exploiter
operated: ~ **position** *n* GEN état de marche *m*
operating *n* TELEP exploitation *f*; ~ **authorization** TV autorisation d'exploitation *f*; ~ **bandwidth** RADIO, TRANS largeur de bande de fonctionnement *f*; ~ **center** *AmE*, ~ **centre** *BrE* TELEP centre d'exploitation *m*; ~ **channel** RADIO voie d'interrogation-réponse *f*, voie active *f*, voie exploitée *f*, voie en service *f*, voie principale *f*; ~ **characteristic** TELEP caractéristique de fonctionnement *f*; ~ **characteristic curve** (*OC curve*) GEN courbe d'efficacité *f*; ~ **conditions** *n pl* ELECT conditions de fonctionnement *f pl*, TELEP conditions d'exploitation *f pl*, conditions de fonctionnement *f pl*, régime *m*; ~ **current** POWER courant actif *m*, courant de fonctionnement *m*, courant de travail *m*; ~ **curve** ELECT courbe de fonctionnement *f*; ~ **device** CONTR dispositif de commande *m*, organe de commande *m*; ~ **frequency** RADIO fréquence d'exploitation *f*, fréquence de fonctionnement *f*; ~ **line** ELECT courbe de fonctionnement *f*, ligne de fonctionnement *f*; ~ **loss** ELECT perte d'insertion en fonctionnement *f*, GEN perte d'exploitation *f*; ~ **point** ELECT point de fonctionnement *m*; ~ **position** PER APP position *f*; ~ **state** TEST état de fonctionnement *m*; ~ **statistics** *n pl* TELEP statistiques de service *f pl*, statistiques d'exploitation *f pl*; ~ **system** (*OS*) COMP *of communication multiprocessor*, DATA logiciel de base *m*, système d'exploitation *m* (*OS*), TELEP, TEST système d'exploitation *m* (*OS*); ~ **system function** (*OSF*) COMP, DATA, TELEP, TEST fonction de système d'exploitation *f*; ~ **terminal** COMP terminal d'exploitation *m*; ~ **time** DATA temps opérateur *m*, GEN temps d'exploitation *m*, temps de fonctionnement *m*; ~ **voltage** ELECT tension de fonctionnement *f*; ~ **winding** PARTS enroulement d'excitation *m*
operation *n* CONTR exploitation *f*, ELECT *of magnet* excitation *f*, GEN *of machine* fonctionnement *m*, gestion *f*, marche *f*, manipulation *f*, RADIO, SIG, TELEP, TEST, TRANS exploitation *f*; ~**, administration and maintenance center** *AmE*, ~**, administration and maintenance centre** *BrE* *n* TELEP, TEST centre de gestion, d'exploitation et de maintenance *m*; ~**, administration, maintenance and provisioning** *n* (*OAMP*) TELEP, TEST exploitation, administration, maintenance et mise en service *f*; ~ **class** *n* DATA classe d'opérations *f*; ~ **code** *n* COMP code opération *m*; ~ **directory** *n* COMP manuel de référence *m*; ~ **error** *n* COMP erreur de fonctionnement *f*; ~ **of horizontal** *n* SWG fonctionnement de barre de sélection *m*, fonctionnement d'horizontale *m*; **no** ~ **instruction** *n* COMP instruction ineffective *f*, instruction inopératoire *f*; ~ **interruption** *n* TELEP interruption d'exploitation *f*, interruption du service *f*; ~ **and maintenance center** *AmE*, ~ **and maintenance centre** *BrE* *n* TELEP, TEST centre d'exploitation et de

maintenance *m*; **~ and maintenance center processor** *AmE*, **~ and maintenance centre processor** *BrE n* COMP processeur de centre d'exploitation et de maintenance *m*; **~ and maintenance engineer** *n* TELEP ingénieur d'exploitation et de maintenance *m*; **~ and maintenance messaging software machine** *n* CONTR organe de contrôle *f*; **~ manual** *n* COMP mode d'emploi *m*, GEN manuel d'exploitation *m*; **~ point** *n* SWG point de distribution *m*; **~ of selecting** *n* SWG fonctionnement de barre de sélection *m*, fonctionnement d'horizontale *m*; **~ thermostat** *n* CONTR, PARTS thermostat de service *m*

operational: **~ amplifier** *n* (*op amp*) ELECT, TRANS amplificateur opérationnel *m* (*ampliop*); **~ availability factor** *n* TEST disponibilité effective *f*; **~ center** *AmE*, **~ centre** *BrE n* TELEP centre d'exploitation *m*; **~ conditions** *n pl* ELECT conditions de fonctionnement *f pl*, TELEP conditions d'exploitation *f pl*, conditions de fonctionnement *f pl*, régime *m*; **~ data** *n pl* DATA, TELEP données d'exploitation *f pl*; **~ description** *n* TELEP description d'exploitation *f*; **~ experience** *n* TELEP expérience acquise en exploitation *f*, expérience d'exploitation *f*; **~ information display system** *n* RADIO système d'affichage de l'information opérationnelle *m*; **~ lifetime** *n* TELEP, TEST durée de vie utile *f*; **~ MUF** *n* RADIO MUF d'exploitation *m*; **~ performance** *n* TELEP performance opérationnelle *f*; **~ procedure** *n* TELEP procédure d'exploitation *f*, procédure opératoire *f*; **~ security** *n* TELEP sécurité d'exploitation *f*; **~ semantics** *n* COMP, DATA sémantique opératoire *f*

operations: **~ center** *AmE*, **~ centre** *BrE n* GEN centre d'exploitation *m*, RADIO, TEST centre d'opérations *m*, TV centre d'exploitation *m*; **~ chart** *n* DATA script *m*; **~ control center** *AmE*, **~ control centre** *BrE n* (*OCC*) TEST centre de direction des opérations *m*; **~ and maintenance** *n pl* (*OM*) DATA, TELEP, TEST exploitation et maintenance *f*; **~ and maintenance subsystem** *n* (*OM subsystem*) DATA, TELEP, TEST sous-système exploitation et maintenance *m* (*SSEM*); **~ room** *n* TEST salle d'exploitation *f*; **~ support** *n* TELEP gestion de réseau *f*; **~ support node** *n* TELEP nœud d'aide à l'exploitation *m*

operator *n* GEN, RADIO opérateur *m*, exploitant *m*, SWG, TELEP opérateur *m*, *f*, téléphoniste *mf*, *private exchange* standardiste *mf*, TV câblo-opérateur *m*, exploitant *m*; **~-assisted call** (*OAC*) GEN appel assisté par opérateur *m*, TELEP appel avec assistance *m*, appel assisté par l'opérateur *m*; **~ authorization** TELEP autorisation d'accès par opérateur *f*; **~'s cashflow** ELECT compte de l'opérateur *m*; **~ center** *AmE*, **~ centre** *BrE* TELEP centre d'opérateurs *m*; **~ command** DATA commande opérateur *f*; **~-connected call** TELEP appel branché par l'opérateur *m*, appel connecté par l'opérateur *m*; **~ console** TELEP console d'exploitation *f*, poste d'opérateur *m*, poste d'opératrice *m*, standard *m*; **~ control panel** COMP tableau de commande *m*, *computers* panneau de commande *m*; **~ disable** COMP mise hors fonction d'opérateur *f*; **~ enable** COMP mise en fonction d'opérateur *f*; **~ factor** RADIO facteur d'opérateur *m*; **~ junctor trunk group** TELEP faisceau d'opératrice *m*; **~ output level** COMP, ELECT niveau de sortie d'opérateur *m*; **~ position** GEN, TELEP position d'opérateur *f*, poste d'opérateur *m*; **~ select** COMP sélection d'opérateur *f*; **~ service center** *AmE*, **~ service centre** *BrE* TELEP centre d'assistance d'opérateur *m*; **~ set** TELEP poste d'opérateur *m*; **~'s switching set** SWG unité de commutation *f*, unité de clés pour dicorde *f*; **~ system** TELEP système de positions automatisées de téléphonistes *m*, système de positions d'opérateurs *m*, système de positions de téléphonistes *m*; **~'s telephone** TELEP poste d'opérateur *m*, poste d'opératrice *f*

OPLL *abbr* (*optical phase-locked loop*) ELECT boucle à verrouillage de phase optique *f*

optic: **~ axis** *n* TRANS axe de la fibre *m*, axe optique *m*

optical[1]: **~-electrical** *adj* (*O-E*) GEN, TV optoélectrique (*O-E*)

optical[2]: **~ aberration** *n* TRANS aberration optique *f*; **~ activity** *n* TRANS activité optique *f*; **~ amplifier** *n* DATA, TRANS amplificateur optique *m*; **~ attenuator** *n* TRANS affaiblisseur optique *m*, atténuateur optique *m*; **~ axis** *n* TRANS axe de la fibre *m*, axe optique *m*; **~ bar-code reader** *n* DATA lecteur optique de code à barres *m*; **~ beam** *n* ELECT faisceau optique *m*; **~ bench** *n* TRANS banc optique *m*; **~ bistability** *n* TRANS bistabilité optique *f*; **~ branching** *n* TRANS branchement optique *m*, dérivation optique *f*; **~ cable** *n* TRANS câble optique *m*, câble à fibres optiques *m*; **~ cable assembly** *n* TRANS assemblage de câble optique *m*, câble optique assemblé *m*; **~ cavity** *n* ELECT cavité optique résonnante *f*, TRANS cavité optique *f*, cavité optique résonnante *f*; **~ center** *AmE*, **~ centre** *BrE n* SWG point d'intersection optique *m*; **~ character recognition** *n* (*OCR*) COMP, DATA reconnaissance optique de caractères *f* (*ROC*); **~ characteristic** *n* TRANS caractéristique optique *f*; **~ character-recognition software** *n* COMP, DATA logiciel de reconnaissance optique de caractères *m*; **~ circulator** *n* TRANS circulateur optique *m*; **~ coherence** *n* TRANS cohérence optique *f*; **~ combiner** *n* TRANS combinateur optique *m*; **~ communications** *n pl* TRANS télécommunications optiques *f pl*; **~ control boards** *n pl* TRANS répartiteurs optiques *m pl*; **~ cord** *n* TRANS cordon optique *m*; **~ coupler** *n* DATA, TRANS coupleur optique *m*; **~-data bus** *n* TRANS bus optique *m*; **~ demultiplexer** *n* TV démultiplexeur optique *m*; **~ detector** *n* TRANS détecteur de lumière *m*, détecteur optique *m*, détecteur photosensible *m*, photodétecteur *m*, photorécepteur *m*, récepteur photoélectrique *m*; **~ disk** *n* DATA disque optique *m*, TV disque optique numérique *m*; **~ disk drive** *n* DATA lecteur de disque optique *m*; **~ distribution** *n* TRANS distribution optique *f*; **~ exchange** *n* SWG, TRANS commutateur optique *m*; **~ fiber** *AmE see optical fibre BrE*; **~ fibre** *BrE n* COMP, DATA, PARTS, TRANS fibre optique *f*; **~ fibre cable** *n BrE* TRANS câble à fibres optiques *m*, câble optique *m*; **~ fibre connector** *n BrE* TRANS connecteur à fibres optiques *m*; **~ fibre coupler** *n BrE* DATA, TRANS coupleur optique *m*; **~-fibre line system** *n BrE* TRANS système de ligne à fibres optiques *m*, système de ligne sur fibres optiques *m*; **~ fibre link** *n BrE* (*F/O link*) POWER, TRANS liaison à fibres optiques *f*; **~ fibre pigtail** *n BrE* TRANS queue de cochon de fibres optiques *f*; **~ fibre sensor** *n BrE* TRANS capteur à fibres optiques *m*; **~ fibre splice** *n BrE* TRANS épissure de fibres optiques *f*; **~ fibre transmission** *n BrE* TRANS transmission par fibres optiques *f*; **~ filter** *n* TRANS filtre optique *m*; **~ head** *n* DATA tête optique *f*; **~ image** *n* TV image optique *f*; **~ interface** *n* TRANS interface optique *f*; **~ interference** *n* TRANS interférence optique *f*; **~ isolator** *n* TRANS isolateur optique *m*; **~-line terminating equipment** *n* (*OLTE*) TRANS terminal à fibre optique *m*; **~ link** *n* TRANS liaison optique *f*; **~ loss** *n* DATA, TRANS affaiblissement optique *m*, perte optique *f*; **~ media** *n pl* TV supports optiques *m pl*; **~ modulation** *n* DATA modulation optique *f*; **~ modulation index** *n* (*OMI*) TELEP, TRANS indexe de modulation optique *m*; **~ modulator** *n* TRANS modulateur optique *m*; **~ mouse** *n* COMP souris optique *f*; **~ multiplexer** *n* TV multiplexeur optique *m*; **~ network unit** *n* TRANS réseau optique *m*; **~ noise** *n* DATA bruit optique *m*; **~ noise measurement** *n* DATA mesure du bruit optique *f*; **~ parametric oscillator** *n* ELECT oscillateur paramétrique optique *m*; **~ path** *n* TRANS chemin optique *m*; **~ path length** *n* TRANS longueur du chemin optique *f*; **~ pencil** *n* APPL, COMP crayon optique *m*; **~ phase-locked loop** *n* (*OPLL*) ELECT boucle à verrouillage de phase optique *f*; **~ polarization** *n* TRANS polarisation optique *f*; **~ power** *n* TRANS puissance optique *f*; **~ power meter** *n* TRANS appareil de mesure de puissance optique *m*, instrument de mesure de puissance optique *m*, mesureur de puissance optique *m*; ~ **power output** *n* TRANS puissance optique de sortie *f*; ~ **processing** *n* TRANS traitement optique *m*; ~ **radiation** *n* TRANS rayonnement optique *m*; ~ **reader** *n* DATA, TV lecteur optique *m*; ~ **reading** *n* DATA, TV lecture optique *f*; ~ **receiver** *n* TRANS module de réception optique *m*, récepteur optique *m*; ~ **recording** *n* TRANS enregistrement optique *m*; ~ **refraction** *n* TRANS réfraction optique *f*; ~ **regenerative repeater** *n* TRANS répéteur régénérateur optique *m*; ~ **regenerator** *n* TRANS répéteur régénérateur optiquc *m*; ~ **repeater** *n* TRANS répéteur optique *m*, répéteur régénérateur optique *m*; ~ **resolution** *n* COMP résolution optique *f*; ~ **resonator** *n* TRANS résonateur optique *m*; ~ **return loss** *n* TRANS, perte par réflexion optique *f*; ~ **scanner** *n* COMP, DATA lecteur optique *m*, scanner *m*; ~ **shaft encoder** *n* ELECT codeur d'angle optique *m*; ~ **signal** *n* TRANS signal optique *m*; ~ **spectrum** *n* TRANS spectre optique *m*; ~ **splice** *n* TRANS épissure optique *f*; ~ **switch** *n* SWG, TRANS commutateur optique *m*; ~ **switching** *n* SWG, TRANS commutation optique *f*; **~-switching crosspoint** *n* SWG point d'intersection optique *m*; ~ **switching matrix** *n* SWG, TELEP matrice de commutation optique *f*; ~ **switching network** *n* (*OSN*) SWG, TRANS réseau de connexion optique *m*; ~ **switching system** *n* SWG système de commutation optique *m*; ~ **tap** *n* TRANS point de branchement optique *m*, point de dérivation optique *m*; ~ **tape storage** *n* COMP, DATA mémoire sur ruban optique *f*; ~ **tapping** *n* TRANS branchement optique *m*, dérivation optique *f*; ~ **time domain reflectometer** *n* (*OTDR*) ELECT, POWER, TEST, TRANS réflectomètre optique *m*; ~ **time domain reflectometry** *n* (*OTDR*) ELECT, POWER, TEST, TRANS réflectométrie optique dans le domaine temporel *f*; ~ **transistor** *n* ELECT transistor optique *m*, transphaseur *m*; ~ **transmission** *n* TRANS transmission optique *f*; ~ **transmitter** *n* TRANS émetteur optique *m*; ~ **tuning** *n* TRANS accord optique *m*; ~ **type font** *n* DATA police de caractères optiques *f*; ~ **waveguide** *n* DATA, RADIO, TRANS guide d'onde optique *m*, guide d'ondes optiques *m*

optically[1]: ~ **active** *adj* PARTS, TRANS optiquement actif

optically[2]: ~ **active material** *n* PARTS substance optiquement active *f*

optics *n* DATA, GEN optique *f*

optimal[1] *adj* GEN optimal

optimal[2]: ~ **control** *n* CONTR commande optimale *f*; ~ **path** *n* TRANS chemin optimal *m*; ~ **sampling** *n* TRANS échantillonnage optimal *m*

optimum[1] *adj* GEN optimum

optimum[2]: ~ **gain** *n* ELECT gain optimal *m*; ~ **load** *n* RADIO charge optimale *f*; ~ **mechanical tuning range** *n* ELECT plage optimale d'accord du dispositif mécanique *f*; ~ **output power** *n* ELECT puissance optimale de sortie *f*; ~ **scale range** *n* TEST étendue optimale de l'échelle *f*; ~ **traffic frequency** *n* (*OTF*) RADIO fréquence optimale de trafic *f* (*FOT*); ~ **working frequency** *n* (*OWF*) RADIO fréquence optimale de travail *f* (*FOT*)

option *n* GEN option *f*, possibilité de choix *f*

optional[1] *adj* (*O*) COMP optionnel (*O*), en option, DATA optionnel (*O*), TELEP optionnel (*O*), TV en option

optional[2]: ~ **component** *n* DATA élément facultatif *m*; ~ **network-specific digit** *n* (*ONSD*) TELEP chiffre facultatif spécifique au réseau *m*; ~ **pause instruction** *n* COMP instruction de pause facultative *f*; ~ **presence** *n* DATA présence sur option *f*; ~ **requirement** *n* DATA prescription facultative *f*, spécification facultative *f*; ~ **service** *n* TV service optionnel *m*; ~ **stop instruction** *n* COMP instruction d'arrêt facultatif *f*; ~ **user-facility** *n* TELEP service complémentaire offert en option à l'utilisateur *m*

optionally: **~-droppable blocks** *n pl* DATA blocs éventuellement supprimables *m pl*

optocoupler *n* DATA, ELECT, TRANS coupleur optique *m*, optocoupleur *m*

optoelectric *adj* (*O-E*) GEN, TV optoélectrique (*O-E*)

optoelectronic[1] *adj* ELECT optoélectronique

optoelectronic[2]: ~ **component** *n* PARTS composant optoelectronique *m*; ~ **crosspoint** *n* SWG point d'intersection optoélectronique *m*; ~ **equipment** *n* ELECT équipement optoélectronique *m*; ~ **receiver** *n* TRANS récepteur à fibre optique *m*, récepteur optoélectronique *m*; ~ **switching matrix** *n* SWG, TELEP matrice de commutation optoélectronique *f*

optoelectronics *n* ELECT, PARTS, SWG, TRANS optoélectronique *f*

optronic *adj* ELECT optoélectronique

optronics *n* ELECT optoélectronique *f*

OQPSK *abbr* (*offset quaternary-phase-shift keying*) DATA, SIG modulation par déplacement de phase à quatre états avec décalage *f*, modulation de phase de deux porteuses en quadrature avec décalage des trains binaires *f*

O/R *abbr* (*originator/recipient*) DATA, TELEP expéditeur/destinataire *m*

OR: ~ **circuit** *n* ELECT circuit OU *m*; ~ **element** *n* ELECT porte OU *f*; ~ **frequency plan** *n* RADIO plan des fréquences OR *m*; ~ **gate** *n* ELECT porte OU *f*, circuit OU *m*; ~ **of n signals** *n* COMP résultat de la fonction OU de n signaux *m*, sortie de la porte OU recevant n signaux *f*; ~ **operation** *n* ELECT *logic elements* opération OU *f*, opération OU inclusif *f*

orange: ~ **book** *n* COMP livre orange *m*; ~ **filter** *n* DATA *of scanner* filtre orangé *m*

orbit *n* RADIO orbite *f*; ~ **determination** RADIO détermination de l'orbite *f*; ~ **drift** RADIO dérive en longitude de l'orbite *f*; ~ **gap** RADIO écart angulaire en orbite *m*; ~ **insertion** RADIO injection en orbite *f*

orbital: ~ **decay** *n* RADIO dégénérescence d'une orbite *f*; ~ **elements** *n pl* RADIO éléments d'une orbite *m pl*; ~ **guidance** *n* RADIO guidage orbital *m*; ~ **inclination control** *n* RADIO correction d'inclinaison de l'orbite *f*; ~ **location** *n* RADIO emplacement orbital *m*; ~ **period** *n* RADIO période de révolution *f*, période orbitale *f*; ~ **plane** *n* RADIO plan de l'orbite *m*; ~ **position** *n* TV position orbitale *f*; ~ **slot** *n* RADIO créneau orbital *m*; ~ **test program** *AmE*, ~ **test programme** *BrE* *n* (*OTP*) RADIO programme d'essais sur orbite *m*

orbiting: ~ **period** *n* RADIO période de révolution *f*, période orbitale *f*; ~ **satellite** *n* RADIO satellite sur orbite *m*; ~ **time** *n* RADIO période de révolution *f*, période orbitale *f*

orchestration *n* COMP orchestration *f*

order[1]: **out of ~** *adj* GEN inopérable, inutilisable

order[2] *n* COMP instruction *f*, TELEP, TRANS ordre *m*; ~ **of diversity** RADIO ordre de diversité *m*

order[3] *vt* APPL, TELEP commander; ~ **by merging** COMP trier par fusion; ~ **services** TV passer commande des services

ordinal: ~ **date** *n* GEN date ordinale *f*; ~ **value** *n* DATA valeur ordinale *f*

ordinary: ~ **both-way line** *n* TRANS ligne ordinaire *f*; ~ **line** *n* TRANS ligne ordinaire *f*; ~ **network** *n* TRANS réseau classique *m*; ~ **private telex call** *n* TELEP communication télex privée ordinaire *f*; ~ **wave** *n* RADIO onde ordinaire *f*

ORed *adj* ELECT mis en fonction OU; ~ **with** ELECT en fonction OU

organization: ~ **code** *n* (*OC*) COMP, DATA code d'organisation *m*; ~ **and maintenance** *n* TELEP, TEST exploitation et maintenance *f*

organizational: ~ **link** *n* COMP lien d'organisation *m*; ~ **unit** *n* (*OU*) TELEP unité organisationnelle *f*

orientation: ~ **control** *n* CONTR commande d'orientation *f*

origin: ~ **indicator** *n* TELEP indicateur d'origine *m*; ~ **of message** *n* DATA provenance du message *f*

original: **~-called number** *n* TELEP numéro demandé initial *m*; ~ **document** *n* COMP, DATA document source d'origine *m*;

~ EITs *n* DATA type d'information codée d'origine *m*; **~ equipment manufacturer** *n* (*OEM*) GEN OEM; **~ inspection** *n* TEST contrôle en première présentation *m*; **~ production master** *n* ELECT cliché de production original *m*; **~ redirection reason indicator** *n* TELEP indicateur de raison du réacheminement initial *m*
originate[1]: **~ in** *vt* COMP avoir pour origine
originate[2]: **~ a call** *vi* TELEP appeler, effectuer un appel de départ
originating[1] *adj* TELEP de départ
originating[2]: **~ calling position** *n* TELEP position d'appel de départ *f*; **~ device** *n* TELEP dispositif émetteur *m*; **~ exchange** *n* SWG central de départ *m*; **~ feature code** *n* TELEP code de fonction utilisé pour le lancement d'un appel *m*; **~ junctor** *n* PARTS, SWG joncteur de départ *m*; **~ network identification request indicator** *n* TELEP indicateur de demande d'identification du réseau d'origine *m*; **~ point** *n* SIG point sémaphore d'origine *m*; **~ point code** *n* (*OPC*) DATA, TELEP code du point d'origine *m*; **~ register** *n* SWG enregistreur de départ *m*; **~ traffic** *n* TELEP trafic de départ *m*
originator *n* RADIO appelant *m*, TELEP demandeur *m*, expéditeur *m*
originator/recipient *n* (*O/R*) DATA, TELEP expéditeur/destinataire *m*
O-ring *n* PARTS joint torique *m*
orthicon *n* ELECT orthicon *m*
orthogonal: **~ co-channel** *n* RADIO cocanal orthogonal *m*; **~ frequency-division multiple access** *n* (*ODFMA*) RADIO, SIG, TELEP, TRANS accès multiple par répartition en fréquences orthogonales *m* (*AMRFO*); **~ mode transducer** *n* RADIO transducteur orthomode *m*; **~ polarization** *n* RADIO polarisation orthogonale *f*; **~ signal** *n* RADIO signal orthogonal *m*
orthomode: **~ transducer** *n* RADIO transducteur orthomode *m*
OS *abbr* (*operating system*) COMP, DATA, TELEP, TEST OS (*système d'exploitation*)
OSA *abbr* (*open systems architecture*) TELEP architecture de systèmes ouverts *f*
OSC *abbr* (*on-scene communications*) RADIO communications sur place *f pl*
oscillate *vi* COMP, ELECT, POWER osciller
oscillation *n* COMP, ELECT, POWER oscillation *f*; **~ circuit** ELECT, POWER montage oscillateur *m*; **~ frequency** ELECT fréquence d'oscillation *f*
oscillator *n* ELECT, POWER oscillateur *m*, RADIO bobine oscillatrice *f*; **~ frequency** ELECT fréquence d'oscillateur *f*
oscillatory[1] *adj* COMP, ELECT, POWER oscillatoire
oscillatory[2]: **~ circuit** *n* ELECT, POWER circuit oscillant *m*, montage oscillateur *m*
oscillograph: **~ tube** *n* ELECT tube oscilloscope *m*
oscilloscope *n* ELECT oscilloscope *m*; **~ tube** ELECT tube oscilloscope *m*
OSF *abbr* (*operating system function*) COMP, DATA, TELEP, TEST fonction de système d'exploitation *f*
OSI *abbr* (*open systems interconnection*) COMP, DATA, TELEP, TRANS ISO (*interconnexion de systèmes ouverts*) **~ layer** *n* COMP, DATA couche ISO *f*; **~ model** *n* DATA modèle OSI *m*; **~ model layer** *n* COMP, DATA couche ISO *f*; **~ network layer** *n* DATA couche réseau ISO *f*; **~ network service** *n* DATA, TELEP service de réseau ISO *m*; **~ resource** *n* DATA, TELEP, TRANS ressource ISO *f*; **~ transport layer** *n* DATA couche transport ISO *f*
OSN *abbr* (*optical switching network*) SWG, TRANS réseau de connexion optique *m*
OSPF *abbr* (*open shortest path first*) COMP, ELECT ouverture prioritaire du plus court chemin *f*
Ostwald: **~ color system** *AmE*, **~ colour system** *BrE* *n* COMP système de couleurs d'Ostwald *m*
OTDR *abbr* ELECT, POWER, TEST, TRANS (*optical time domain reflectometry*) réflectométrie optique dans le domaine temporel *f*, (*optical time domain reflectometer*) réflectomètre optique *m*
OTF *abbr* (*optimum traffic frequency*) RADIO FOT (*fréquence optimale de trafic*)
otherwise: **~ event** *n* DATA événement sinon *m*
OTP *abbr* COMP (*one-time programmable*) uni-programmable, RADIO (*orbital test program, orbital test programme*) programme d'essais sur orbite *m*
OU *n* (*organizational unit*) TELEP unité organisationnelle *f*
out: **over and ~** *phr* RADIO terminé
outage *n* TELEP défaillance *f*, TEST défaillance *f*, coupure *f*, interruption *f*, indisponibilité *f*; **~ time** TEST temps d'incapacité *m*
outbound: **~ communication** *n* RADIO, TRANS télécommunication de départ *f*
outdoor: **~ aerial** *n* RADIO antenne extérieure *f*; **~ antenna** *n* RADIO antenne extérieure *f*
outer: **~ conductor** *n* TELEP, TRANS conducteur extérieur *m*; **~ diameter** *n* GEN diamètre extérieur *m*; **~ electron** *n* ELECT électron périphérique *m*; **~ rim** *n* COMP périphérie *f*; **~ stanchion** *n* TELEP montant extérieur *m*; **~ vapor deposition** *AmE*, **~ vapour deposition** *BrE* *n* (*OVD*) PARTS dépôt par vaporisation *m*; **~ vertical iron** *n* TELEP montant extérieur *m*
outgoing[1] *adj* (*o/g*) GEN, TRANS de départ, sortant
outgoing[2]: **~ access** *n* (*OA*) DATA accès sortant *m*; **~ call** *n* DATA, RADIO, TELEP communication de départ *f*, appel sortant *m*; **~ calls barred extension** *n* TELEP, TRANS poste interdit aux communications de départ *m*; **~ calls barred line** *n* TELEP poste interdit aux communications de départ *m*, ligne spécialisée arrivée *f*; **~ channel** *n* TRANS voie sortante *f*, voie de départ *f*; **~ circuit** *n* ELECT, SWG, TELEP circuit sortant *m*; **~ group** *n* TELEP faisceau sortant *m*; **~ international R2 register** *n* TELEP enregistreur international R2 de départ *m*; **~ line** *n* TELEP ligne sortante *f*; **~ matrix link** *n* TELEP liaison réseau sortante *f*; **~ message** *n* (*OGM*) PER APP annonce *f*, message sortant *m*; **~ procedure** *n* TELEP procédure de départ *f*; **~ R2 register** *n* TELEP enregistreur R2 de départ *m*; **~ traffic** *n* TELEP trafic sortant *m*; **~ trunk** *n* ELECT, SWG, TELEP circuit de départ *m*, circuit sortant *m*; **~ trunk circuit** *n* SWG joncteur réseau départ *m*, TELEP, TRANS joncteur de départ *m*, joncteur sortant *m*; **~ trunk queuing** *n* SWG, TELEP mise en file d'attente sur circuit de départ *f*; **~ wave** *n* ELECT onde progressive *f*
outgoing[3]: **~ calls barred** *phr* (*OCB*) DATA, TELEP interdiction d'accès des appels sortants *f*, interdiction des appels au départ *f*
outgrowth *n* ELECT excroissance *f*
outlet *n* COMP *network* point d'accès *m*, POWER prise *f*; **~ access** SWG accès de départ *m*, point de sortie *m*, sortie *f*
outlier *n* GEN observation aberrante *f*
outline *n* COMP contour *m*, plan *m*, GEN contour *m*, croquis *m*, esquisse *f*, résumé *m*, sommaire *m*; **~ agreement** GEN accord cadre *m*; **~ drawing** ELECT croquis coté *m*, plan coté *m*, schéma coté *m*; **~ font** COMP police vectorielle *f*; **~ test procedure** TEST procédure d'essai générale *f*
out-of-action *adj* TELEP en dérangement
out-of-band[1] *adj* (*OOB*) ELECT, RADIO hors bande
out-of-band[2]: **~ emission** *n* (*OBE*) RADIO émission hors bande *f*; **~ power** *n* RADIO puissance hors bande *f*; **~ spectrum** *n* RADIO spectre hors bande *m*
out-of-channel: **~ filtering** *n* TRANS filtrage hors voie *m*
out-of-date *adj* GEN suranné
out-of-frame *n* DATA défaut de verrouillage trame *m* (*DVT*), DVT (*défaut de verrouillage trame*); **~ alignment time** TELEP durée de perte du verrouillage de trame *f*; **~ second** (*OFS*) DATA seconde de perte du verrouillage de trame *f*

out-of-order[1] *adj* TEST en dérangement
out-of-order[2]: **~ signal** *n* SIG signal de ligne hors service *m*
out-of-phase: **~ signal** *n* GEN signal déphasé *m*
out-of-range *adj* RADIO hors portée
out-of-reach *adj* RADIO hors portée
out-of-service[1] *adj* POWER hors service
out-of-service[2]: **~ time** *n* TEST temps d'intervention *m*
out-of-sync: **~ error** *n* TRANS décalage *m*
out-of-synchronization: **~ error** *n* TRANS décalage *m*
outperform *vt* COMP avoir des performances supérieures à, dépasser sur le plan des performances
outpulse *vi* GEN, SIG émettre des impulsions
output[1] *n* GEN sortie *f*, POWER *of a four-terminal network* sorties *f pl*, TELEP puissance utile *f*; **~ admittance** POWER admittance de sortie *f*; **~ angle** ELECT angle de sortie *m*, TRANS angle de sortie *m*, angle de rayonnement *m*; **~ assign** COMP affectation de sortie *f*; **~ back-off** (*OBO*) TRANS recul de sortie *m* (*OBO*); **~ capacitance** ELECT capacité de sortie *f*; **~ channel** COMP canal de sortie *m*; **~ characteristics** *n pl* ELECT caractéristiques de sortie *f pl*; **~ circuit** ELECT circuit de sortie *m*; **~ current** ELECT courant de sortie *m*; **~ data** *n pl* DATA données de sortie *f pl*; **~ device** COMP périphérique de sortie *m*, dispositif de sortie *m*, unité de sortie *f*; **~ divergence** ELECT divergence de sortie *f*; **~ gate** ELECT porte de sortie *f*; **~ impedance** COMP, ELECT impédance de sortie *f*; **~ indicator** COMP indicateur de sortie *m*; **~ jack** COMP borne de sortie *f*; **~ level** COMP programme encodé *m*, niveau de sortie *m*, ELECT niveau de sortie *m*; **~-level attenuation** COMP atténuation de niveau de sortie *f*; **~-level attenuator** COMP affaiblisseur de niveau de sortie *m*, atténuateur de niveau de sortie *m*; **~ load** POWER charge de sortie *f*, charge à la sortie *f*; **~ make-circuit** PARTS sortie de travail *f*; **~ outside dialog** *AmE*, **~ outside dialogue** *BrE* TELEP sortie hors dialogue *f*; **~ port** COMP, ELECT port de sortie *m*; **~ power** ELECT puissance de sortie *f*; **~ program** COMP programme de sortie *m*; **~ pulse** COMP impulsion de sortie *f*; **~ row** ELECT rangée de sortie *f*; **~ signal** COMP, ELECT signal de sortie *m*; **~ socket** COMP prise de sortie *f*; **~ stage** ELECT, RADIO étage de sortie *m*; **~ symbol** TELEP symbole de sortie *m*; **~ terminal** POWER borne de sortie *f*; **~ transformer** POWER transformateur de sortie *m*; **~ unit** COMP dispositif de sortie *m*, unité de sortie *f*; **~ variable** CONTR grandeur de sortie *f*; **~ voltage** ELECT tension de sortie *f*
output[2] *vt* COMP sortir
outrigger *n* ELECT support en équerre *m*
outside: **~ broadcast** *n* RADIO, TV reportage extérieur *m*; **~ location** *n* TV centre de reportage *m*; **~ party** *n* TELEP correspondant extérieur *m*; **~ plant** *n* TRANS installations extérieures *f pl*
out-slot: **~ signaling** *AmE*, **~ signalling** *n BrE* SIG signalisation hors intervalle de temps *f*
outstanding: **~ frame** *n* DATA trame en instance *f*
OVD *abbr* (*outer vapor deposition AmE*, *outer vapour deposition BrE*) PARTS dépôt par vaporisation *m*
over *n* RADIO à vous
overall: **~ adjacent channel protection margin** *n* RADIO marge de protection globale pour le canal adjacent *f*; **~ braid screen** *n* ELECT tresse de blindage globale *f*; **~ carrier-to-interference ratio** *n* RADIO rapport global porteuse-brouillage *m*; **~ co-channel protection margin** *n* RADIO marge de protection globale dans le même canal *f*; **~ equivalent protection margin** *n* RADIO marge de protection globale équivalente *f*; **~ objective loudness rating** *n* (*OOLR*) TELEP équivalent objectif pour la sonie *m* (*EOGS*); **~ pitch** *n* COMP hauteur d'ensemble *f*, hauteur générale *f*; **~ second adjacent channel protection margin** *n* RADIO marge de protection globale pour le canal deuxième adjacent *f*
overcoverage *n* RADIO surcouverture *f*
overcurrent *n* POWER surintensité *f*, surintensité de courant *f*; **~ factor** POWER indice de surcharge *m*; **~ relay** PARTS relais à maximum *m*
overdeviation *n* RADIO sur-excursion *f*
overdub *n* COMP surimpression *f*; **~ recording** COMP enregistrement par surimpression *m*
overdubbing *n* COMP surimpression *f*
overexcitation *n* ELECT surmodulation *f*
overexpose *vt* DATA, TV surexposer
overfading *n* RADIO affaiblissement supplémentaire *m*
overflow[1] *n* COMP dépassement de capacité *m*, ELECT surcharge *f*, GEN débordement *m*, POWER, SWG surcharge *f*, TELEP débordement *m*, surcharge *f*, TRANS surcharge *f*; **~ accept** (*OA*) SWG, TELEP acceptation de débordement *f*; **~ check** COMP contrôle de dépassement de capacité *m*; **~ indicator** COMP indicateur de saturation *m*; **~ meter** TELEP compteur d'appels non aboutis *m*, compteur de débordements *m*; **~ tone** TELEP tonalité de surcharge *f*, tonalité de débordement *f*; **~ traffic** SWG, TELEP trafic de débordement *m*
overflow[2] *vt* ELECT, POWER, SWG, TELEP, TRANS surcharger
overhang *n* ELECT surplomb *m*
overhead *n* DATA *transmission* surdébit *m*, GEN *time* temps système *m*; **~ access** (*OHA*) DATA accès au surdébit *m*, accès aux octets de surdébit *m*; **~ bit** DATA bit de servitude *m*, bit supplémentaire *m*, TELEP bit de service *m*; **~ cable** POWER, TRANS câble aérien *m*; **~ cabling** POWER, TRANS câblage aérien *m*; **~ line** POWER, TRANS ligne aérienne *f*; **~ network** POWER, TRANS réseau aérien *m*; **~ span** POWER, TRANS portée entre pôleaux *f*
overheads *n pl* GEN *costs* coût système *m*, frais généraux *m pl*
overlap[1] *n* TELEP chevauchement *m*, recouvrement *m*; **~ angle** ELECT angle d'empiétement *m*; **~ area** RADIO aire de recouvrement *f*, zone de recouvrement *f*; **~ operation** SIG, TELEP fonctionnement avec chevauchement *m*
overlap[2] *vt* COMP, RADIO, TELEP chevaucher, recouvrir
overlapping *n* COMP, RADIO chevauchement *m*, recouvrement *m*
overlay[1] *n* COMP recouvrement *m*, superposition *f*
overlay[2] *vt* COMP recouvrir, superposer
overload[1] *n* ELECT, POWER, SWG, TELEP, TRANS surcharge *f*; **~ channel** TRANS canal de surcharge *m*, voie de débordement *f*; **~ class field** TELEP champ de classe de surcharge *m*; **~ control** TELEP contrôle de surcharge *m*; **~ forward current** ELECT, POWER courant direct de surcharge prévisible *m*; **~ point** TELEP niveau de saturation *m*; **~ protection** ELECT, POWER protection contre les surcharges *f*
overload[2] *vt* ELECT, POWER, SWG, TELEP, TRANS surcharger
overloaded *adj* POWER, SWG surchargé
overloading *n* ELECT, POWER, SWG, TELEP, TRANS surcharge *f*
overmodulation *n* ELECT surmodulation *f*
overpower: **~ protection** *n* ELECT, POWER protection contre les surcharges *f*; **~ relay** *n* PARTS relais à maximum *m*
overreaching *n* PARTS zone étendue *f*
override[1] *n* DATA surclassement *m*, TELEP priorité *f*, entrée en tiers *f*; **~ circuit** TEST circuit de commande *m*; **~ feature** DATA, TELEP dispositif de priorité *m*; **~ tone** TELEP tonalité d'entrée en tiers *f*, tonalité de priorité *f*
override[2] *vt* GEN avoir priorité sur
overriding *adj* GEN prioritaire
overseas: **~ network** *n* TRANS réseau extérieur *m*
oversee *vt* GEN maîtriser
overshoot *n* TELEP dépassement *m*
overtone *n* COMP harmonique *m*
overview: **~ part** *n* DATA partie présentation générale *f*
overvoltage *n* ELECT, POWER surtension *f*; **~ alarm** POWER alarme de surtension *f*; **~ protection** ELECT, POWER protection contre les surtensions *f*; **~ protection unit**

ELECT, POWER module de protection contrôle surtension *m*; ~ **relay** PARTS relais à maximum *m*

overwrite *vt* DATA écraser

OWF *abbr* (*optimum working frequency*) RADIO FOT (*fréquence optimale de travail*)

own: **~-exchange supervisory circuit** *n* SWG circuit de connexion local *m*; ~ **program** *AmE*, ~ **programme** *BrE n* TV programme propre *m*

oxide: ~ **capacitor** *n* ELECT, PARTS condensateur à couche d'oxyde *m*; **~-coated cathode** *n* ELECT, PARTS cathode à couche d'oxydes *f*

P

PA *abbr* (*power amplifier*) COMP, DATA, POWER, TRANS amplificateur à gaine élevé *m*, amplificateur de puissance *m*

PABX *abbr* (*private automatic branch exchange*) APPL, CONTR, SWG, TELEP PABX (*autocommutateur privé*); **~ internal dial tone** *n* TELEP tonalité interne de numérotation des autocommutateurs privés *f*

pace *n* COMP, GEN cadence *f*

package *n* COMP lot *m*, programme produit *m*, progiciel *m*, DATA *Internet* package *m*

packet *n* DATA, RADIO, TELEP paquet *m*; **~ assembler-disassembler** (*PAD*) DATA, TELEP assembleur-désassembleur de paquets *m* (*PAD*); **~ assembly-disassembly facility** DATA, TELEP assemblage-désassemblage de paquets *m* (*PAD*); **~ data link** DATA, TRANS liaison de transmission de données par paquets *f*; **~ data transmission network** DATA, TRANS réseau de transmission de données par paquets *m*; **~ delay** DATA délai de transmission de paquets *m*; **~ disassembly** DATA, TELEP désassemblage de paquets *m*; **~ format** TELEP format de paquet *m*; **~ handler** (*PH*) DATA, TELEP dispositif de traitement de paquets *m*, gestionnaire de paquets *m*; **~ layer** DATA, TELEP couche paquet *f*; **~ layer protocol** (*PLP*) DATA, TELEP protocole de couche paquet *m*; **~ layer reference event** (*PE*) DATA, TELEP événement de référence de couche paquets *m*; **~ length selection** DATA, TELEP sélection de longueur des paquets *f*; **~-mode bearer service** DATA, TELEP service support en mode paquet *m*; **~-mode operation** DATA, TELEP commutation par paquets *f*, service de commutation par paquets *m*; **~-mode terminal** DATA, TELEP terminal en mode paquet *m*; **~ multiplexing** RADIO multiplexage par paquets *m*; **~ radio** RADIO radiocommunication par paquets *f*; **~ satellite network** (*PSN*) RADIO, TELEP réseau à satellite avec commutation de paquets *m*; **~ sequencing** DATA, TELEP ordonnancement de paquets *m*; **~ switch** (*PS*) SWG, TRANS commutateur de paquets *m* (*CP*); **~-switch network** (*PSN*) SWG, TRANS réseau à commutation de paquets *m*; **~-switched bearer service** SWG, TRANS service support à commutation de paquets *m*; **~-switched data carrier** SWG, TRANS porteuse de données avec commutation de paquets *f*; **~-switched data transmission service** SWG, TRANS service de transmission de données avec commutation de paquets *m*; **~ switching** (*PS*) SWG, TRANS commutation de paquets *f* (*CP*); **~-switching exchange** (*PSE*) SWG, TRANS centre de commutation de paquets *m*; **~-switching network** (*PSN*) SWG, TRANS réseau à commutation de paquets *m*; **~-switching node** (*PSN*) SWG nœud à commutation de paquets *m*; **~-switching processor** COMP, SWG processeur de commutation de paquets *m*; **~ transmission** TRANS transmission par paquets *f*

packing *n* TELEP tassement *m*; **~ density** DATA, ELECT densité d'enregistrement *f*, densité d'intégration *f*; **~ factor** DATA facteur de remplissage *m*; **~ fraction** DATA densité d'assemblage *f*

pad[1] *n* ELECT tampon *m*, GEN matelas *m*, POWER affaiblisseur fixe *m*, atténuateur *m*, atténuateur fixe *m*, TRANS affaiblisseur fixe *m*, atténuateur fixe *m*, réseau atténuateur *m*, atténuateur *m*

pad[2] *vt* COMP, DATA remplir

PAD *abbr* DATA, TELEP (*packet assembler-disassembler*) PAD (*assembleur-désassembleur de paquets*), (*packet assembly-disassembly facility*) PAD (*assemblage-désassemblage de paquets*)

padding *n* COMP, DATA remplissage *m*

paddle *n* COMP palette de jeu *f*; **~ board** POWER carte de connexion *f*; **~-shaped beam** RADIO faisceau plat *m*

page[1] *n* COMP, DATA page *f*; **~ description language** (*PDL*) COMP, DATA langage de description de page *m*; **~ frame** COMP, DATA cadre de page *m*; **~ layout** COMP, DATA mise en page *f*; **~ layout software** COMP, DATA logiciel de mise en pages *m*; **~ marker** COMP, DATA marque-page *m*; **~ numbering** COMP, GEN pagination *f*; **~ printer** DATA, TELEP imprimante par page *f*

page[2] *vt* COMP *word processing*, GEN paginer, RADIO, TELEP appeler

pager *n* RADIO, TELEP, TRANS récepteur d'appel *m*, récepteur d'appel de personnes *m*, récepteur de radiomessagerie *m* (*RMU*), radiopage *m*, terminal RMU *m*

paginate *vt* COMP *word processing*, GEN paginer

pagination *n* COMP, GEN pagination *f*

paging[1] *n* COMP pagination *f*, DATA radiomessagerie *f* (*RM*), GEN pagination *f*, RADIO téléappel *m*, appel de personne *m*, radiomessagerie *f* (*RM*), recherche de personne *f*, radiorecherche *f*, TELEP radiorecherche *f*, recherche de personne *f*, téléappel *m*, radiomessagerie *f* (*RM*); **~ channel** TRANS canal de recherche de personnes *m*, voie d'appel unilatéral *f*; **~ service** APPL, RADIO, TELEP service de recherche de personnes *m*, service de téléappel *m*; **~ system** RADIO, TELEP système d'appel de personnes *m*, système d'appel unilatéral *m*, système de recherche de personnes *m*

paging[2]: **on no-~ response** *phr* RADIO, TELEP sur non-réponse au message de recherche

paid: **~-service advice** *n* DATA, TELEP avis de service taxé *m*

paint *n* RADIO trace rémanente *f*

paintbrush *n* COMP pinceau *m*

pair *n* GEN paire *f*; **~ generation** PARTS génération de paires d'électrons *f*, production de paires *f*; **~ production** PARTS génération de paires d'électrons *f*, production de paires *f*

paired: **~ cable** *n* GEN câblage par paires *m*, câble à paires *m*; **~-disparity code** *n* TELEP code à disparité compensée *m*

pairing *n* TV pairage *m*

PAL *abbr* COMP, TV (*programmable array logic*) PAL (*logique à matrice programmable*)

palette *n* DATA palette *f*

palmtop: **~ computer** *n* COMP ordinateur de poche *m*

PAM *abbr* DATA (*pass-along method*) méthode du faire passer *m*, TELEP, TRANS (*pulse amplitude modulation*) MIA (*modulation d'impulsions en amplitude*); **~ network** *n* TELEP, TRANS réseau MIA *m*

PAMA *abbr* (*pulse-address multiple access*) RADIO AIMA (*accès multiple par impulsions d'adresse*)

pan[1]: **~ control** *n* COMP commande de panoramique *f*; **~ pot** *n* COMP potentiomètre de panoramique *m*

pan[2]: **~ left-right** *vi* COMP panoramiquer gauche-droite; **~ right-left** *vi* COMP panoramiquer droite-gauche

Pan-African: **~ Telecommunication Union** *n* (*PATU*) RADIO Union panafricaine des télécommunications *f* (*UPAT*)

pane *n* COMP sous-fenêtre *f*

panel *n* GEN tableau *m*, POWER panneau *m*; **~ plating** ELECT placage métallique total *m*; **~ system** SWG système panel *m*

pan-European *adj* GEN paneuropéen

panoramic: **~ space** *n* DATA espace panoramique *m*

Pantone: **~ scale** *n* ELECT échelle Pantone *f*

paper: **~ capacitor** *n* POWER condensateur au papier *m*; **~ directory** *n* TELEP annuaire papier *m*; **~ feed** *n* PER APP alimentation du papier *f*, entraînement du papier *m*; **~ feed unit** *n* PER APP unité d'entraînement du papier *f*; **~ format** *n* PER APP format du papier *m*; **~-insulated mass-impregnated cable** *n* TRANS câble à masse *m*; **~-insulated non-draining cable** *n* TRANS câble à matière non-migrante *m*; **~ laminate** *n* PARTS stratifié de papier *m*; **~-out switch** *n* PER APP contact de fin de papier *m*; **~ size** *n* PER APP format du papier *m*; **~ skip** *n* PER APP, TELEP saut de papier *m*; **~ throw** *n* PER APP, TELEP saut de papier *m*; **~ tray** *n* PER APP bac à papier *m*; **~ winder** *n* PER APP enrouleur de papier *m*; **~-winder bail** *n* TELEP presseur *m*

PAR *abbr* GEN (*peak-to-average ratio*) rapport valeur de crête sur valeur moyenne *m*, RADIO (*precision approach radar*) PAR (*radar d'approche de précision, radar d'approche finale, radiodétecteur d'approche de précision*)

parabola *n* RADIO, TV parabole *f*

parabolic: **~ aerial** *n* RADIO antenne parabolique *f*; **~ antenna** *n* RADIO antenne parabolique *f*; **~ cylinder aerial** *n* RADIO antenne cylindre parabolique *f*; **~ cylinder antenna** *n* RADIO antenne cylindre parabolique *f*; **~ profile** *n* RADIO, TRANS profil parabolique *m*; **~ reflector aerial** *n* RADIO antenne à réflecteur parabolique *f*; **~ reflector antenna** *n* RADIO antenne à réflecteur parabolique *f*

paraboloid *n* RADIO, TV paraboloïde *m*

parallax: **~ error** *n* TEST erreur de parallaxe *f*

parallel[1]: **~ adder** *n* ELECT additionneur parallèle *m*; **~ architecture** *n* TRANS architecture parallèle *f*; **~ carry** *n* COMP transfert parallèle *m*; **~ circuit** *n* ELECT circuit en parallèle *m*, circuit parallèle *m*; **~ clocking arrangement** *n* ELECT montage synchrone *m*; **~ connection** *n* POWER montage en parallèle *m*; **~ cooling** *n* CLIM refroidissement en parallèle *m*; **~ displacement** *n* RADIO, TRANS décalage parallèle *m*, déplacement parallèle *m*; **~ interface** *n* DATA interface parallèle *f*; **~ jack** *n* ELECT, PARTS jack de branchement *m*, jack de dérivation *m*, jack de parallèle *m*; **~ operation** *n* SWG fonctionnement en parallèle *m*; **~-plate lens** *n* RADIO lentille à lames paralèlles *f*; **~ port** *n* COMP port parallèle *m*; **~ processing** *n* DATA, ELECT traitement en parallèle *m*; **~ programming** *n* DATA programmation en parallèle *f*; **~ resonance** *n* ELECT, POWER résonance parallèle *f*; **~ resonant circuit** *n* ELECT, POWER circuit antirésonnant *m*, circuit résonnant parallèle *m*; **~ structure** *n* TEST structure parallèle *f*; **~ synchronous system** *n* COMP système en réplique synchrone *m*; **~ tasking** *n* ELECT tâche parallèle *f*; **~-tee oscillator** *n* ELECT oscillateur à réseau en double T *m*; **~-to-serial converter** *n* ELECT, TELEP convertisseur parallèle-série *m*; **~ transmission** *n* COMP, TRANS transmission en parallèle *f*, transmission parallèle *f*; **~-working processors** *n pl* SWG processeurs fonctionnant en parallèle *m pl*

parallel[2]: **~ up** *vt* POWER brancher en parallèle

parameter: **~ argument** *n* TELEP argument de paramètre *m*; **~ assign** *n* COMP affectation de paramètre *f*; **~ block** *n* TELEP bloc de paramètres spécifiques *m*; **~ block introduction sequence** *n* TELEP séquence d'introduction de bloc de paramètres *f*; **~ block request indication** *n* TELEP indication de demande de bloc de paramètres *f*; **~ group identifier** *n* (*PGI*) COMP identificateur de groupe de paramètres *m*; **~ identification** *n* (*PI*) TELEP identification de paramètre *f*; **~ identifier** *n* (*PI*) TELEP identificateur de paramètre *m*; **~ length** *n* (*PL*) COMP, TELEP longueur de paramètre *f*; **~ name** *n* TELEP nom de paramètre *m*; **~-name-defined parameter** *n* COMP paramètre défini par le nom de paramètre *m*; **~ out-of-range class** *n* DATA valeur de paramètre erronée *f*; **~ profile** *n* TEST profil de paramètre *m*; **~ value** *n* (*PV*) TELEP, TEST valeur de paramètre *f*; **~ word** *n* COMP, DATA mot-paramètre *m*

parameterize *vt* DATA, TEST paramétrer

parameterized: **~ abstract test case** *n* DATA, TEST test élémentaire abstrait paramétré *m*; **~ abstract test suite** *n* (*PATS*) DATA, TEST suite de tests abstraits paramétrée *f*; **~ executable test case** *n* DATA, TEST test élémentaire exécutable paramétré *m*; **~ executable test suite** *n* (*PETS*) DATA, TEST suite de tests exécutables paramétrée *f*

parametric: **~ amplifier** *n* (*paramp*) TRANS amplificateur paramétrique *m*; **~ oscillator** *n* ELECT oscillateur paramétrique *m*; **~ programming** *n* COMP programmation paramétrique *f*; **~ test** *n* TEST test paramétrique *m*

paramp *abbr* (*parametric amplifier*) TRANS amplificateur paramétrique *m*

paraphrase: **~ amplifier** *n* ELECT amplificateur déphaseur *m*

parasitic: **~ coupling** *n* ELECT couplage parasite *m*; **~ element** *n* RADIO élément parasite *m*; **~ emissions** *n pl* RADIO rayonnement parasite *m*; **~ error** *n* TEST erreur parasite *f*; **~ oscillation** *n* ELECT oscillation parasite *f*; **~ stopper** *n* GEN filtre antiparasite *m*

paraxial: **~ ray** *n* ELECT, TRANS rayon paraxial *m*

parent *n* COMP, DATA père *m*; **~ entry** DATA entrée maîtresse *f*; **~-sequence number** COMP, DATA, SWG numéro d'ordre d'entrée maîtresse *m*

Pareto: **~ analysis** *n* TEST analyse de Pareto *f*

parity *n* DATA, ELECT parité *f*; **~ bit** DATA bit de parité *m*, RADIO bit de contrôle *m*; **~ check** DATA, TELEP contrôle de parité *m*; **~ digit** COMP élément de parité *m*; **~ function** TELEP fonction de parité *f*; **~-generating circuit** SWG circuit générateur de parité *m*

park *vt* SWG, TELEP *call* mettre en garde tous postes

parked[1] *adj* SWG, TELEP en faux appel, mis en garde, mis en garde tous postes

parked[2]: **~ line** *n* SWG, TELEP ligne en faux appel *f*

parking: **~ orbit** *n* RADIO orbite d'attente *f*

part *n* PARTS pièce *f*; **~ sample** TEST échantillon partiel *m*; **~-time leased circuit** TELEP, TRANS circuit loué à temps partiel *m*

partial[1] *adj* COMP, DATA, TELEP partiel

partial[2]: **~ attribute request** *n* DATA demande partielle de valeurs d'attribut *f*; **~ break-in** *n* TELEP intervention partielle *f*; **~ break-in echo suppressor** *n* RADIO, TELEP suppresseur d'écho à intervention partielle *m*; **~ break-in operate time** *n* RADIO, TELEP temps de fonctionnement pour l'intervention partielle *m*; **~ call barring** *n* TELEP restriction partielle *f*; **~ coherence** *n* TRANS cohérence partielle *f*; **~ dial** *n* TELEP numérotation incomplète *f*; **~ failure** *n* TEST défaillance partielle *f*; **~-page request** *n* (*PPR*) DATA, TELEP demande de page partielle *f*; **~-page signal EOM** *n* (*PPS-EOM*) DATA, TELEP signal de page partielle EOM *m* (*SPP-EOM*); **~-page signal EOP** *n* (*PPS-EOP*) DATA, TELEP signal de page partielle EOP *m* (*SPP-EOP*); **~-page signal MPS** *n* (*PPS-MPS*) DATA, TELEP signal de page partielle MPS *m* (*SPP-MPS*); **~-page signal NULL** *n* (*PPS-NULL*) DATA, TELEP signal de page partielle NULL *m* (*SPP-NULL*); **~-page signal PRI-EOM** *n* (*PPS-PRI-EOM*) DATA, TELEP signal de page partielle PRI-EOM *m* (*SPP-PRI-EOM*); **~-page signal PRI-EOP** *n* (*PPS-PRI-EOP*) DATA, TELEP signal de page partielle PRI-EOP *m* (*SPP-PRI-EOP*); **~-page signal PRI-MPS** *n* (*PPS-PRI-MPS*) DATA, TELEP signal de page partielle PRI-MPS *m* (*SPP-PRI-MPS*); **~ redundance** *n* TEST redondance partielle *f*; **~ refund** *n* TELEP remboursement partiel *m*; **~ response code** *n* DATA code de réponse partielle *m*; **~ restoring time** *n* TRANS *of echo suppressor* temps de rétablissement partiel *m*; **~ test** *n* TEST essai partiel *m*; **~ tone reversal** *n* TELEP inversion partielle des nuances *f*

partially: **~ suppressed** *adj* TV partiellement supprimé

particle: **~ board** *n* PARTS panneau en fibres de bois *m*; **~ shredding** *n* COMP perte d'oxyde *f*; **~ velocity** *n* TELEP vitesse efficace d'une particule *f*

partition *n* COMP, ELECT, TRANS subdivision *f*; **~ noise** ELECT bruit de répartition *m*

partitioning *n* COMP, ELECT, TRANS *of channel* subdivision *f*

party: ~ **address** *n* DATA adresse du correspondant *f*; ~ **line** *n* TELEP ligne partagée *f*, ligne commune *f*, ligne multipoint *f*

pass[1] *n* RADIO *of satellite* passage *m*; ~**-along method** (*PAM*) DATA méthode à faire passer *m*; ~ **band** COMP, RADIO, TELEP, TRANS bande passante *f*; ~ **band attenuation** TRANS affaiblissement en bande passante *m*; ~ **transistor** ELECT transistor de chute *m*, transistor monté en régulateur série *m*; ~ **verdict** DATA succès *m*, succès verdict *m*

pass[2]: ~ **on** *vt* RADIO, TRANS, TV retransmettre; ~ **through** *vt* ELECT, POWER traverser, TRANS transiter

passed *adj* TV raccordable

passing: ~ **contact** *n* PARTS contact de passage *m*

passivated: ~ **surface** *n* PARTS surface passive *f*

passivation *n* PARTS passivation *f*

passive: ~ **alerting** *n* RADIO alerte passive *f*; ~ **bus** *n* ELECT bus passif *m*; ~ **circuit** *n* ELECT circuit passif *m*; ~ **component** *n* ELECT, PARTS composant passif *m*; ~ **filter** *n* ELECT filtre passif *m*; ~ **four-terminal network** *n* ELECT, POWER, TRANS quadripôle passif *m*, quadripôle symétrique *m*; ~ **intermodulation** *n* (*PIM*) GEN intermodulation passive *f* (*IMP*); ~ **maintenance** *n* TEST maintenance passive *f*; ~ **matrix** *n* COMP matrice passive *f*; ~ **network** *n* POWER réseau passif *m*; ~ **optical network** *n* (*PON*) TELEP, TRANS réseau optique passif *m*; ~ **sensor** *n* RADIO détecteur passif *m*; ~ **threat** *n* DATA menace passive *f*; ~ **transducer** *n* POWER transducteur passif *m*

password *n* COMP, DATA mot de passe *m*

past: ~ **reference picture** *n* COMP, TV image de référence antérieure *f*

paste[1]: ~**-like material** *n* ELECT produit pâteux *m*

paste[2] *vt* COMP coller

patch[1] *n* COMP morceau de surface *m*; ~ **board** RADIO, TELEP panneau de répartition *m*, répartiteur *m*, tableau de répartition *m*; ~ **cord** ELECT cordon de brassage *m*, TRANS cordon de raccordement *m*, fiche de connexion *f*; ~**-out** TEST aboutement, raccordement temporaire *m*; ~ **panel** RADIO, TELEP panneau de répartition *m*, répartiteur *m*, tableau de répartition *m*

patch[2] *vt* ELECT, SWG raccorder

patching *n* ELECT brassage *m*

patent: ~ **pending** *n* ELECT brevet en cours *m*

path *n* COMP, ELECT, RADIO trajectoire *f*, parcours *m*, SWG, TELEP itinéraire *m*, TRANS trajectoire *f*, voie *f*; ~ **allocation** SWG allocation d'itinéraire *f*, recherche d'itinéraire *f*; ~ **allocation program** COMP programme d'allocation d'itinéraire *m*, programme de recherche d'itinéraire *m*; ~ **a-t-b** RADIO *attenuation*, TELEP affaiblissement du trajet a-t-b *m*; ~ **attenuation** RADIO, TELEP affaiblissement sur un trajet *m*, atténuation de propagation *f*; ~ **memory** SWG mémoire des itinéraires *f*; ~ **name** COMP chemin d'accès *m*; ~ **overhead** (*POH*) SWG, TELEP préfixe de conduit *m*, résidu de trajet *m*

patrol: ~ **inspection** *n* TEST contrôle volant *m*

patron *n* TV mécène *m*

PATS *abbr* (*parameterized abstract test suite*) DATA, TEST suite de tests abstraits paramétrée *f*

pattern *n* APPL *of artificial intelligence* filtre *m*, forme *f*, motif *m*, COMP motif *m*, filtre *m*, forme *f*, ELECT impression *f*, motif *m*, GEN schéma *m*, motif *m*, RADIO, TEST, TRANS motif *m*; ~ **correspondence index** (*PCI*) SWG indice de correspondance spectrale *m*; ~ **generator** TV générateur de mire *m*; ~ **mapping** COMP placage de texture *m*; ~ **matching** APPL *of artificial intelligence*, COMP corrélation de formes *f*, filtrage *m*; ~ **plating** ELECT placage métallique sélectif *m*; ~ **recognition** APPL, CONTR, DATA, TRANS reconnaissance de formes *f*

PATU *abbr* (*Pan-African Telecommunication Union*) RADIO UPAT (*Union panafricaine des télécommunications*)

pause: ~ **instruction** *n* COMP instruction de pause *f*

pawl *n* SWG cliquet *m*

PAX *abbr* (*private automatic exchange*) APPL autocommutateur privé *m* (*PABX*), central automatique privé *m*

pay[1]: ~ **channel** *n* TV chaîne de télévision à péage *f*; ~ **console** *n* PER APP console à paiement *f*; ~**-per-view** *n* (*PPV*) TV paiement à l'émission *m*, paiement par émission *m*; ~**-per-view television** *n* TV télévision à la demande *f*; ~ **station** *n* PER APP, TELEP poste à prépaiement *m*, poste téléphonique à prépaiement *m*, publiphone *m*; ~ **telephone** *n* PER APP, TELEP publiphone *m*, téléphone à prépaiement *m*; ~ **television** *n* TV télévision à accès conditionnel *f*, télévision à péage *f*, télévision payante *f*; ~ **television network** *n* TV chaîne de télévision à péage *f*; ~ **tone** *n* TELEP tonalité de paiement *f*; ~ **TV** *n* TV TV à accès conditionnel *f*, TV à péage *f*, TV payante *f*

pay[2]: ~ **out** *vt* GEN *cable* dérouler

payload *n* ELECT, GEN charge utile *f*

payment: ~ **by telephone** *n* APPL télépaiement *m*; ~ **option analysis system** *n* (*POAS*) APPL système d'analyse des options tarifaires *m* (*SAOT*)

payphone *n* PER APP, TELEP poste à prépaiement *m*, poste téléphonique à prépaiement *m*, publiphone *m*; ~ **recognition tone** TELEP tonalité d'identification de publiphone *f*; ~ **service** APPL, TELEP service publiphone *m*

PBC *abbr* (*program booking centre, programme booking centre*) TV service centralisateur *m* (*PBC*)

PBN *abbr* (*private branch network*) TELEP, TRANS réseau de télécommunication privé *m*, réseau privé *m*

PBX *abbr* (*private branch exchange*) SWG, TELEP central d'abonné *m* (*PBX*), central privé *m*, commutateur privé *m*, commutateur téléphonique privé *m*; ~ **call** *n* TELEP appel de groupe *m*; ~ **group** *n* TELEP groupe PBX *m*; ~ **hunting** *n* TELEP service de recherche de lignes PBX *m*, recherche dans un groupement de lignes *f*; ~ **line-hunting service** *n pl* SWG service de recherche de ligne libre dans un commutateur *m*; ~ **subscriber** *n* TELEP abonné d'un commutateur privé *m*; ~ **switchboard** *n* TELEP standard privé *m*; ~ **trunk circuit** *n* TELEP circuit d'un commutateur privé raccordés à un circuit de transmission *m*

PBXIS-LB *abbr* TELEP autocommutateur privé pour le RNIS-LB *m*

PC *abbr* COMP (*personal computer*) OP (*ordinateur personnel*), PC (*micro-ordinateur, ordinateur de bureau, ordinateur individuel*), ELECT (*printed circuit*) CI (*circuit imprimé*), (*polarization controller*) contrôleur de polarisation *m*, PARTS (*printed circuit*) CI (*circuit imprimé*); ~ **support engineer** *n* GEN assistant en micro-informatique *m*

PCB *abbr* (*printed circuit board*) ELECT carte à circuits imprimés *f*

PCD: ~ **writer** *n* COMP enregistreur photo CD *m*

PCI *abbr* COMP (*peripheral component interconnect*) interconnexion de composants périphériques *f*, SWG (*pattern correspondence index*) indice de correspondance spectrale *m*, TELEP (*programmable communication interface*) PCI (*interface de communication programmable*)

PCIS: ~ **proforma** *n* DATA formulaire PCIS *m*

PCL *abbr* (*print command language*) DATA PCL

PCM *abbr* (*pulse-code modulation*) COMP, DATA, TELEP, TRANS MIC (*modulation par impulsions codées*); ~ **binary code** *n* DATA, ELECT, TELEP code binaire MIC *m*; ~ **digital reference sequence** *n* TELEP séquence numérique de référence MIC *f*; ~ **link** *n* TELEP liaison MIC *f*; ~ **link control software machine** *n* TELEP ML unité de raccordement de multiplex *f*, unité de raccordement de multiplex *f*; ~ **multiplex equipment** *n* TELEP, TRANS équipement de multiplexage MIC *m*; ~ **switching system** *n* TELEP système de commutation MIC *m*; ~ **system** *n* TELEP système de commutation MIC *m*; ~ **waveform** *n* COMP forme d'onde MIC *f*

PCMCIA *abbr* (*Personal Computer Memory Card Interna-*

tional Association) COMP, ELECT PCMCIA (*Association internationale de normalisation des cartes à mémoire pour PC*)

PCN *abbr* (*personal communication network*) DATA, RADIO réseau de communication personnelle *m*

PCO *abbr* (*point of control and observation*) DATA PCO (*point de contrôle et d'observation*)

PCS *abbr* APPL (*personal communications services*) services de communications personnelles *m pl*, RADIO (*personal communications system*) PCS *m* (*système de communications personelles*), TELEP (*plastic-clad silica*) silice recouverte de plastique *f*; **~-fiber** *AmE*, **~-fibre** *BrE n* DATA, PARTS, TRANS fibre silice-plastique *f*, fibre silice-silicone *f*

PCTR *abbr* (*protocol conformance test report*) DATA, TEST rapport de test de conformité au protocole *m*

PD *abbr* DATA (*physical delivery*) remise physique *f*, (*photodiode*) PD (*photodiode*), PARTS (*photodiode*) PD (*photodiode*), TELEP (*physical delivery*) remise physique *f*, (*protocol discriminator*) DP (*discriminateur de protocole*), TRANS (*photodiode*) PD (*photodiode*)

PDAU *abbr* (*physical delivery access unit*) DATA, TELEP unité d'accès de remise physique *f*

PDE: **~ terminal** *n* (*portable data entry terminal*) COMP terminal portatif de saisie de données *m*

PDF *abbr* COMP (*portable document file*) PDF (*portable document file*), DATA (*probability density function*) fonction de densité de probabilité *f*, (*portable document file*) PDF (*portable document file*)

PDH *abbr* (*plesiochronous digital hierarchy*) TELEP, TRANS HNP (*hiérarchie numérique plésiochrone*)

PDL *abbr* (*page description language*) COMP, DATA langage de description de page *m*

PDM *abbr* (*pulse duration modulation*) DATA, RADIO, TELEP, TRANS MID (*modulation d'impulsions en durée*)

PDN *abbr* (*public data network*) DATA, TELEP, TRANS RPD (*réseau public pour données*)

PDS *abbr* DATA (*physical-delivery system*), TELEP (*physical-delivery system*) SRP (*système de remise physique*)

PDU *abbr* (*protocol data unit*) DATA unité de données de protocole *f*

PE *abbr* COMP (*pictorial element*) élément graphique *m*, DATA (*packet layer reference event*) événement de référence de couche paquets *m*, (*protocol emulator*) émulateur de protocole *m*, TELEP (*packet layer reference event*) événement de référence de couche paquets *m*

peak[1]: **~-to-peak** *adj* POWER crête à crête

peak[2] *n* GEN crête *f*, pointe *f*; **~ amplitude** TELEP *of elementary echo* amplitude de crête *f*; **~ detector** GEN détecteur de crête *m*; **~ envelope power** (*PEP*) GEN, RADIO puissance crête de la porteuse modulée *f*, puissance de crête *f*; **~ forward-anode voltage** ELECT tension anodique directe de crête *f*; **~ frequency deviation** RADIO, TRANS excursion de fréquence maximale *f*, excursion de la fréquence de crête *f*; **~ hour** TELEP heure de pointe *f*; **~-intensity wavelength** RADIO, TRANS longueur d'onde du maximum *f*, longueur d'onde du maximum d'intensité *f*; **~ inverse voltage** (*PIV*) POWER, TELEP tension inverse de crête *f*; **~ limiting** TELEP limitation de crête *f*, TRANS écrêtage *m*; **~ limiting in quantizing** TRANS écrêtage *m*; **~ negative-anode voltage** ELECT tension anodique négative de crête *f*; **~ point** PARTS point de pic *m*; **~ point current** POWER courant de pic *m*; **~ point voltage** PARTS tension de pic *f*; **~ power** POWER puissance de crête *f*; **~-pulse output power** ELECT, POWER puissance de sortie en crête de modulation *f*; **~ rate** APPL, GEN tarif rouge *m*; **~ reading instrument** TEST indicateur de crête *m*; **~ reverse voltage** (*PRV*) POWER, TELEP tension inverse de crête *f*; **~ tariff periods** *n pl* APPL heures pleines *f pl*; **~-to-average ratio** (*PAR*) GEN rapport valeur de crête sur valeur moyenne *m*; **~-to-peak amplitude** GEN, RADIO amplitude crête à crête *f*; **~ traffic** TELEP trafic de pointe *m*; **~ value** GEN, TEST valeur de crête *f*; **~ voltage** POWER tension de plafond *f*; **~ voltmeter** TEST voltmètre de crête *m*

peaked: **~ traffic** *n* TELEP trafic survariant *m*

peakedness: **~ factor** *n* TELEP facteur d'irrégularité *m*

peaking: **~ coil** *n* POWER bobine de compensation *f*, bobine de crête *f*, bobine de relèvement *f*

pedal *n* COMP pédale *f*

Pedersen: **~ ray** *n* RADIO rayon de Pedersen *m*

peel: **~ strength** *n* ELECT force d'adhérence *f*

peer[1]: **~-to-peer** *adj* TRANS d'égal à égal

peer[2] *n* COMP pair *m*, DATA, RADIO, TRANS homologue *m*; **~-entity authentication** DATA authentification de l'entité homologue *f*; **~-identity authentication** DATA authentification de l'entité homologue *f*; **~ relationships** *n pl* DATA relations homologues *f pl*; **~-to-peer logical relationship** TRANS relation logique entre couches analogues *f*; **~-to-peer protocol** TRANS protocole entre homologues *m*

pel *n* (*picture element*) COMP, ELECT, TV élément d'image *m*, pixel *m*, point d'image *m*; **~ aspect ratio** COMP, RADIO, TELEP facteur d'aspect de pixel *m*, format du pixel *m*, rapport des axes d'un pixel *m*

pen *n* COMP plume *f*, style *m*, stylo *m*; **~ Collaboration Environment** COMP pen Collaboration Environment *m*

penalty *n* TRANS pénalité *f*

pencil: **~ beam** *n* RADIO faisceau filiforme *m*, faisceau ponctuel *m*, pinceau *m*; **~-beam aerial** *n* RADIO antenne à faisceau-crayon *f*; **~-beam antenna** *n* RADIO antenne à faisceau-crayon *f*

penetration: **~ depth** *n* RADIO profondeur de pénétration *f*; **~ rate** *n* TV taux de pénétration *m*

pentode *n* ELECT pentode *m*

PEP *abbr* (*peak envelope power*) GEN, RADIO puissance crête de la porteuse modulée *f*, puissance de crête *f*

per[1]: **on a ~-call basis** *adv* RADIO appel par appel

per[2]: **~-second metering** *n* PER APP taxation à l'arrivée *f*; **~-unit metering** *n* PER APP taxation à l'unité *f*; **~-usage metering** *n* PER APP taxation à l'utilisation *f*; **~-word tariff system** *n* TELEP système de tarification par mot *m*

perceived: **~ noise level** *n* TELEP niveau de bruit perçu *m*

percentage: **~ overflow** *n* TELEP pourcentage de débordement *m* (*%DBM*); **~ relay** *n* PARTS relais à pourcentage *m*

perforated: **~ tape** *n* TELEP ruban perforé *m*

perform *vt* COMP effectuer, exécuter

performance: **~ cartridge** *n* COMP cartouche d'exécution *f*; **~ characteristic** *n* TEST caractéristique de fonctionnement *f*; **~ combination** *n* COMP combinaison d'exécution *f*; **~ concept** *n* TEST concept de capacité *m*; **~ data** *n pl* COMP, DATA données d'exécution *f pl*, informations de fonctionnement *f pl*; **~ effect** *n* COMP effet d'exécution *m*; **~ index** *n* CONTR fonction de coût *f*; **~ memory** *n* COMP mémoire d'exécution *f*; **~ monitoring** *n* TELEP surveillance de la qualité de fonctionnement *f*; **~ name** *n* DATA nom de combinaison exécution *m*; **~ parameter** *n* COMP, TEST paramètre de fonctionnement *m*, paramètre qualitatif *m*; **~ test** *n* TEST essai de fonctionnement *m*

performing: **~ CMISE service user** *n* APPL, DATA utilisateur du service CMISE exécuteur *m*

periapsis *n* RADIO périastre *m*

perigee *n* RADIO périgée *m*

period: **off ~** *n* ELECT temps de repos *m*, période d'arrêt *f*; **~ of revolution** *n* RADIO période de révolution *f*, période orbitale *f*

periodic: **~ frame** *n* COMP trame périodique *f*; **~ metering** *n* PER APP taxation périodique *f*; **~ quantity** *n* GEN grandeur périodique *f*; **~ signal** *n* ELECT signal périodique *m*; **~ systematic sampling** *n* TEST échantillonnage systématique périodique *m*

peripheral *n* DATA périphérique *m*; **~ card** COMP carte périphérique *f*; ~ **component interconnect** (*PCI*) COMP interconnexion de composants périphériques *f*; ~ **controller** COMP, DATA, ELECT contrôleur de périphériques *m*; ~ **device** COMP dispositif périphérique *m*, DATA périphérique *m*, PER APP dispositif périphérique *m*; ~ **device sharing** DATA utilisation commune de périphériques *f*; ~ **equipment** COMP, DATA, GEN équipement périphérique *m*, matériel périphérique *m*, unité périphérique *f*; ~ **management** PER APP gestion périphérique *f*; ~ **module** PER APP module périphérique *m*; ~ **processor** (*PP*) COMP processeur périphérique *m*; ~ **telephony** TELEP péritéléphonie *f*

periscope: ~ **aerial** *n* RADIO antenne périscope *f*; ~ **antenna** *n* RADIO antenne périscope *f*

peritelephony *n* TELEP péritéléphonie *f*

perl *n* DATA perl *m*

permanent: ~ **circuit** *n* TELEP circuit permanent *m*; ~ **connection** *n* DATA connection permanente *f*; ~ **echo** *n* RADIO écho fixe *m*, écho permanent *m*; ~ **failure** *n* TELEP dérangement permanent *m*; ~ **line** *n* TELEP ligne permanente *f*; ~ **link** *n* GEN liaison permanente *f*; ~ **magnet** *n* ELECT, PARTS aimant permanent *m*; ~**-signal trunk group circuit** *n* TRANS faisceau de faux appels *m*, faisceau de circuits de faux appels *m*; ~ **subscriber number** *n* TELEP service de numéro d'abonné permanent *m*; ~ **virtual circuit** *n* (*PVC*) ELECT, SIG, SWG, TRANS circuit virtuel permanent *m*; ~ **wiring** *n* TELEP câblage fixe *m*

permanently[1]: ~ **connected** *adj* GEN connecté en permanence, raccordé en permanence

permanently[2]: ~**-locked envelope** *n* TELEP enveloppe verrouillée en permanence *f*

permeability *n* PARTS perméabilité *f*; ~ **tuning** RADIO accord par réluctance *m*, accord par variation de perméabilité *m*

permeable *adj* PARTS perméable

permeameter *n* TEST perméamètre *m*

permeance *n* ELECT perméance *f*

permissible: ~ **interference** *n* RADIO, TRANS brouillage admissible *m*; ~ **out-of-band power** *n* RADIO puissance hors bande admissible *f*; ~ **out-of-band spectrum** *n* RADIO spectre hors bande admissible *m*

permissive: ~ **circuit** *n* DATA circuit avec tolérance de transmission de données *m*

permitted: **not** ~ *adj* TELEP non autorisé

permittivity *n* ELECT permittivité *f*

persistence *n* TV persistance d'écran *f*

person: ~**-to-person call** *n* TELEP appel avec préavis *m*, communication avec préavis *f*, communication PAV *f*, communication personnelle *f*

personal: ~ **call** *n* TELEP appel avec préavis *m*, communication avec préavis *f* (*communication PAV*), communication personnelle *f*; ~ **communication network** *n* (*PCN*) DATA, RADIO réseau de communication personnalisé *m*, réseau de communication personnelle *m*; ~ **communications** *n pl* TRANS système de communication personnel *m*, service de communication personnel *m*; ~ **communications services** *n pl* (*PCS*) APPL services de communications personnelles *m pl*; ~ **communications system** *n* (*PCS*) RADIO système de communications personnelles (*PCS*); ~ **computer** *n* (*PC*) COMP microordinateur *m* (*PC*), ordinateur de bureau *m* (*PC*), ordinateur individuel *m* (*PC*), ordinateur personnel *m* (*OP*); ~ **digital assistant** *n* COMP assistant numérique personnel *m*; ~ **fax** *n* PER APP fax personnel *m*; ~ **icon** *n* DATA *Internet* devise personnelle *f*; ~ **identity number** *n* (*PIN*) GEN code confidentiel *m*, numéro personnel d'identification *m*; ~ **telephone** *n* TELEP téléphone personnel *m*

Personal: ~ **Computer Memory Card International Association** *n* (*PCMCIA*) COMP, ELECT Association internationale de normalisation des cartes à mémoire pour PC *f*

personalized: ~ **bookmark** *n* COMP, DATA marque personnalisée *f*, signet personnalisé *m*

perturbation: ~ **technique** *n* TRANS technique de perturbation *f*; ~ **voltage** *n* TEST tension perturbatrice *f*

perturbing: ~ **effect** *n* GEN effet perturbateur *m*; ~ **signal** *n* POWER signal parasite *m*, signal perturbateur *m*

perveance *n* ELECT pervéance *f*

PETS *abbr* (*parameterized executable test suite*) DATA, TEST suite de tests exécutables paramétrée *f*

P/F *n* TELEP bit d'invitation à émettre/fin *m*

PFM *abbr* (*pulse frequency modulation*) TELEP modulation d'impulsions en fréquence *f*

PFN *abbr* (*pulse-forming network*) ELECT RCI (*réseau conformateur d'impulsions*)

P-fractile: ~ **life** *n* TEST quantile d'ordre de la durée de vie *m*

PGI *abbr* (*parameter group identifier*) COMP identificateur de groupe de paramètres *m*

PH *abbr* (*packet handler*) DATA, TELEP dispositif de traitement de paquets *m* (*DTP*), gestionnaire de paquets *m*

phantastron *n* ELECT phantastron *m*

phantom: ~ **circuit** *n* TELEP, TRANS circuit combiné *m*, circuit fantôme *m*; ~ **level** *n* TRANS niveau fantôme *m*; ~ **telegraph circuit** *n* TELEP, TRANS circuit télégraphique fantôme *m*; ~ **voltage** *n* ELECT tension fantôme *f*

phantoming *n* TELEP, TRANS combinaison de circuits *f*

phase[1] *n* GEN phase *f*; ~ **advance** TRANS déphasage vers l'avant *m*; ~ **alignment** ELECT verrouillage de phase *m*; ~ **ambiguity resolution** TRANS levée d'ambiguïté sur la phase *f*; ~**-amplitude characteristic** TRANS caractéristique phase-amplitude *f*; ~**-amplitude distortion** ELECT, TRANS distorsion phase-amplitude *f*; ~ **angle** COMP, POWER angle de phase *m*; ~**-black** TELEP signal de mise en phase sur noir *m*; ~ **change coefficient** GEN *acoustics*, RADIO constante de phase *f*, déphasage linéique *m*; ~**-change coefficient** ELECT, TRANS constante de phase *f*, déphasage linéique *m*; ~**-change disk** DATA disque à changement de phase *m*; ~**-change printer** DATA imprimante à changement de phase *f*; ~**-change velocity** ELECT vitesse de variation de phase *f*; ~ **changer** ELECT, RADIO, SIG déphaseur *m*; ~ **coefficient** ELECT, GEN, RADIO, TRANS constante de phase *f*, déphasage linéique *m*; ~ **coherence** RADIO, TRANS cohérence de phase *f*; ~ **comparator relay** PARTS relais comparateur de phases *m*; ~ **compensation** ELECT, POWER compensation de phase *f*; ~ **conductor** POWER conducteur de phase *m*; ~ **constant** ELECT, GEN, RADIO, TRANS constante de phase *f*, déphasage linéique *m*; ~ **control** ELECT réglage de phase *m*, commande de phase *f*; ~**-corrected echo** RADIO, TELEP écho corrigé en phase *m*; ~**-corrected horn** RADIO, TELEP cornet à correction de phase *m*; ~ **delay** GEN temps de propagation de phase *m*; ~ **demodulation** ELECT démodulation de phase *f*; ~ **detector** ELECT détecteur de phase *m*; ~ **deviation** GEN décalage de phase *m*, déviation de phase *f*; ~ **difference** ELECT, POWER différence de phase *f*; ~ **distortion** ELECT, TRANS distorsion de phase *f*; ~ **encoding** DATA enregistrement en modulation de phase *m*; ~ **equalization** ELECT, POWER égalisation de phase *f*, compensation de phase *f*; ~ **equalizer** ELECT compensateur de phase *m*, POWER égaliseur de phase *m*, compensateur de distorsion de phase *m*, compensateur de phase *m*, correcteur de phase *m*; ~**-frequency distortion** ELECT, TRANS distorsion phase-fréquence *f*; ~ **hopping** RADIO saut de phase *m*; ~ **inversion** POWER inversion de phase *f*; ~**-inversion modulation** TELEP modulation par inversion de phase *f*; ~ **inverter** POWER inverseur de phase *m*; ~ **lock unit** (*PLU*) ELECT organe de verrouillage de phase *m*; ~**-locked filter** ELECT filtre à phase asservie *m*; ~**-locked loop** (*PLL*) ELECT boucle à phase asservie *f*, boucle à verrouillage de phase *f*; ~ **locking** ELECT verrouillage de phase *m*; ~ **margin** CONTR marge de phase *f*; ~ **meter** POWER phasemètre *m*; ~ **modulation** (*PM*) DATA, TRANS modulation de phase *f*

(*MP*); ~ **modulation recording** DATA enregistrement en modulation de phase *m*; ~ **modulator** TRANS modulateur de phase *m*; ~ **monitor** POWER dispositif de surveillance de phase *m*; ~ **plot** ELECT courbe de phase *f*; ~ **regulation** ELECT régulation de phase *f*; ~ **response** TRANS réponse en phase *f*; ~ **sensor** TRANS capteur de phase *m*; ~ **sequence** POWER ordre des phases *m*; ~-**sequence indicator** POWER indicateur d'ordre des phases *m*; ~-**sequence test** POWER essai d'ordre des phases *m*; ~-**shaped QPSK** RADIO, SIG MDPQ avec mise en forme de phase *f*; ~ **shift** ELECT, POWER, RADIO, TRANS décalage de phase *m*, déphasage *m*; ~-**shift distortion** TRANS distorsion de déphasage *f*; ~-**shift keying** (*PSK*) RADIO, SIG modulation par déplacement de phase *f* (*MDP*); ~-**shift keying 2 modulation** (*delta modulation PSK-2*) SIG modulation de phase delta 2 *f* (*modulation delta MDP-2*); ~-**shift oscillator** POWER oscillateur à décalage *m*; ~ **shifter** ELECT, RADIO, SIG déphaseur *m*; ~-**shifter tube** ELECT tube déphaseur *m*; ~-**shifting circuit** POWER circuit déphaseur *m*; ~-**shifting effect** COMP effet de décalage de phase *m*; ~ **speed** RADIO vitesse de phase *f*; ~ **stability** ELECT stabilité de phase *f*; ~ **tuning** ELECT accord en phase *m*; ~ **velocity** RADIO vitesse de phase *f*; ~ **voltage** POWER tension de phase *f*; ~-**white** TELEP signal de mise en phase sur blanc *m*; ~ **winding** PARTS enroulement de phase *m*

phase[2] *vt* RADIO mettre en phase; ~ **in** GEN introduire progressivement, mettre en place progressivement; ~ **out** GEN supprimer progressivement, éliminer, retirer

phased: ~-**array antenna** *n* RADIO antenne à balayage électronique *f*, antenne réseau à commande de phase *f*; ~ **satellite** *n* RADIO satellite synchronisé *m*, satellite en phase *m*

phasing *n* RADIO, TELEP mise en phase *f*; ~-**in** RADIO introduction progressive *f*, mise en place progressive *f*; ~-**out** RADIO élimination progressive *f*, suppression par étapes *f*; ~ **signal** TELEP signal de mise en phase *m*; ~-**type effect** COMP effet de déphasage *m*

PhC *abbr* (*physical connection*) TRANS connexion physique *f*

phenolic: ~ **plastic** *n* PARTS phénoplaste *m*

phon *n* TELEP phone *m*

phone[1]: ~ **box** *n* TELEP cabine téléphonique *f*; ~ **call** *n* TELEP appel téléphonique *m*, communication téléphonique *f*, coup de fil *m*; ~ **conference** *n* APPL, TELEP conférence téléphonique *f*, réunion téléphonique *f*, téléconférence *f*; ~ **conferencing** *n* APPL, TELEP conférence téléphonique *f*, réunion téléphonique *f*, téléconférence *f*; ~-**home feature** *n* DATA, TELEP fonction de rappel périodique *f*; ~-**in poll** *n* TELEP sondage par téléphone *m*; ~-**in program** *AmE*, ~-**in programme** *BrE n* RADIO, TELEP programme à ligne ouverte *m*; ~ **patch** *n* RADIO, TELEP coupleur téléphone-radio *m*

phone[2] *vt* TELEP *person* appeler, téléphoner à; ~ **back** TELEP rappeler

phone[3] *vi* TELEP appeler, téléphoner

phonecard *n* DATA carte pastel *f*, TELEP télécarte *f*, carte pastel *f*

phones: ~ **jack** *n* PER APP prise casque *f*

phonetic: ~ **alphabet** *n* RADIO code d'épellation *m*

Phong: ~ **shading** *n* COMP lissage de Phong *m*, Phong *m*

phonic: ~ **motor** *n* RADIO roue phonique *f*

phoning: ~ **comfort** *n* RADIO, TELEP confort téléphonique *m*

phosphor: ~ **dot** *n* COMP point luminophore *m*

photo: ~ **CD** *n* COMP, DATA photo CD *m*; ~ **CD catalog** *AmE*, ~ **CD catalogue** *BrE n* COMP catalogue photo CD *m*; ~ **CD player** *n* COMP lecteur de photo CD *m*; ~ **CD writer** *n* COMP enregistreur photo CD *m*

photocathode *n* ELECT photocathode *f*

photocomposition *n* DATA, TV photocomposition *f*

photoconductive: ~ **camera tube** *n* TV tube analyseur photoconductif *m*; ~ **cell** *n* ELECT cellule photoconductrice *f*; ~ **effect** *n* ELECT effet photoconductif *m*

photoconductivity *n* ELECT, TRANS effet photoélectrique interne *m*, photoconductivité *f*

photocoupler *n* ELECT photocoupleur *m*

photocurrent *n* ELECT, POWER, TRANS courant photoélectrique *m*, photocourant *m*

photodetector *n* DATA, TRANS détecteur de lumière *m*, détecteur optique *m*, détecteur photosensible *m*, photodétecteur *m*, photorécepteur *m*, récepteur photoélectrique *m*; ~ **diode** TRANS photodiode *f*

photodiode *n* (*PD*) DATA, PARTS, TRANS photodiode *f* (*PD*)

PhotoEdge *n* COMP PhotoEdge *m*

photoelectric[1] *adj* ELECT photoélectrique

photoelectric[2]: ~ **cell** *n* ELECT cellule photoélectrique *f*; ~ **current** *n* ELECT, POWER, TRANS courant photoélectrique *m*; ~ **effect** *n* ELECT, TRANS effet photoélectrique *m*; ~ **emission** *n* ELECT, TRANS émission photoélectrique *f*, effet photoélectrique externe *m*

photoemission *n* ELECT, TRANS photoémission *f*

photoemissive: ~ **camera tube** *n* ELECT tube analyseur photoémissif *m*; ~ **effect** *n* ELECT, TRANS effet de photoémission *m*, effet photoémissif *m*

photograph: ~ **facsimile telegram** *n* TELEP télégramme phototélégraphique *m*; ~ **facsimile telegraphy** *n* TELEP phototélégraphie *f*

photographic: ~ **coding** *n* COMP, TELEP codage photographique *m*

photolithography *n* COMP photolithographie *f*

photoluminescence *n* TRANS photoluminescence *f*

photomask *n* ELECT masque photographique *m*

photomultiplier *n* ELECT photomultiplicateur *m*

photon *n* TRANS photon *m*; ~ **noise** ELECT, TRANS bruit des photons *m*, bruit photonique *m*

photonics *n* TRANS photonique *f*

photorealistic *adj* COMP photoréaliste

photosensitive: ~ **cell** *n* DATA, ELECT cellule photosensible *f*

phototransistor *n* PARTS, TRANS phototransistor *m*

photovideotex *n* DATA photovidéotex *m*

photovoltaic: ~ **cell** *n* ELECT, PARTS cellule photovoltaïque *f*, photopile *f*; ~ **effect** *n* ELECT effet photovoltaïque *m*

phrase *n* COMP, DATA expression *f*

phrasing *n* COMP, DATA expression *f*

PhS *abbr* (*physical service*) APPL, TELEP PHS (*service physique*)

PhSAP *abbr* (*physical-service access point*) DATA PHSAP (*point d'accès au service physique*)

physical: ~ **address** *n* TELEP adresse physique *f* (*AP*); ~ **connection** *n* (*PhC*) TRANS connexion physique *f*; ~ **delivery** *n* (*PD*) DATA, TELEP remise physique *f*; ~ **delivery access unit** *n* (*PDAU*) DATA, TELEP unité d'accès de remise physique *f*; ~-**delivery system** *n* (*PDS*) DATA, TELEP système de remise physique *m* (*SRP*); ~ **interface** *n* (*PI*) DATA interface physique *f*; ~ **layer** *n* DATA, TRANS couche physique *f*; ~-**layer operations and maintenance cell** *n* (*PL-OAM cell*) RADIO cellule OAM de la couche physique *f*; ~-**layer service** *n* (*PLS*) TELEP, TEST service de la couche physique *m*; ~ **line** *n* TRANS ligne métallique *f*, ligne physique *f*; ~ **medium** *n* TRANS support physique *m*; ~-**medium sublayer** *n* TRANS sous-couche de support physique *f*; ~ **optics** *n* TRANS optique physique *f*, *fibre optics* optique ondulatoire *f*; ~ **security** *n* DATA, TEST sécurité physique *f*; ~ **service** *n* (*PhS*) APPL, TELEP service physique *m* (*PHS*); ~-**service access point** *n* (*PhSAP*) DATA point d'accès au service physique *m* (*PHSAP*)

PI *abbr* DATA (*physical interface*) interface physique *f*, (*presentation indicator*) indicateur de présentation *m*, TELEP (*parameter identification*) identification de paramètre *f*, (*parameter identifier*) identificateur de paramètre *m*

pick[1] *n* COMP pointage *m*

pick[2]: ~ **up** *vt* COMP copier, RADIO capter; ~ **up sound** *vt* TELEP prendre du son

picking: ~ **pliers** *n pl* GEN brucelles *f pl*

pick-up: ~ **facility** *n* TELEP service de prise d'une communication établie *m*
picon *n* (*pictorial icon*) COMP icône d'images animées *f*
picosecond *n* (*ps*) TEST picoseconde *f* (*ps*)
PICS *abbr* (*protocol implementation conformance statement*) DATA PICS (*déclaration de conformité d'une instance de protocole, déclaration de conformité de mise en oeuvre du protocole*); ~ **proforma** *n* DATA formulaire PICS *m*
pictogram *n* ELECT pictogramme *m*
pictorial: ~ **element** *n* (*PE*) COMP élément graphique *m*; ~ **icon** *n* (*picon*) COMP icône d'images animées *f*
picture *n* COMP, GEN, TELEP, TV image *f*; ~ **black** COMP noir de l'image *m*; ~ **blemish** TV défectuosité d'image *f*; ~ **channel** TV canal image *m*, chaîne image *f*; ~ **coding** COMP, TELEP codage d'image *m*; ~ **definition** TV définition d'image *f*; ~ **element** (*pel, pixel*) COMP, ELECT, TV élément d'image *m*, pixel *m*, point d'image *m*; ~ **frequency** COMP, TV fréquence d'image *f*; ~ **in picture** (*PIP*) COMP image dans l'image *f*; ~ **phone** PER APP, TELEP vidéophone *m*, visiophone *m*; ~ **quality** COMP, TV qualité d'image *f*; ~ **rate** COMP, TV vitesse de l'image *f*; ~ **signal** COMP, TV signal d'image *m*; ~**-sound desynchronization** COMP décalage image-son *m*, désynchronisation image-son *f*; ~**-sound lag** COMP décalage image-son *m*, désynchronisation image-son *f*; ~ **store** CONTR mémoire d'images *f*; ~ **tone** TV fréquence porteuse d'image *f*; ~ **tube** ELECT tube image *m*; ~ **white** COMP blanc de l'image *m*
pie: ~ **chart** *n* COMP camembert *m*
Pierce: ~ **gun** *n* ELECT canon de Pierce *m*
piezoelectric[1] *adj* ELECT piézoélectrique
piezoelectric[2]: ~ **crystal** *n* ELECT cristal piézoélectrique *m*; ~ **depolarization** *n* ELECT onde de dépolarisation piézoélectrique *f*; ~ **transducer** *n* ELECT, PARTS transducteur piézoélectrique *m*
piezoelectricity *n* GEN piézoélectricité *f*
piggyback[1] *adj* COMP, ELECT superposé, superposable
piggyback[2]: ~ **entry** *n* COMP accès clandestin *m*
pigment *n* PARTS pigment *m*
pigtail *n* ELECT câble amorce *m*, GEN queue de cochon *f*, TRANS fibre amorce *f*, amorce *f*, fibre d'injection *f*
pillar *n* TRANS point de distribution secondaire *m*
pillow: ~ **loudspeaker** *n* TELEP haut-parleur à oreiller *m*
pilot[1] *adj* GEN expérimental
pilot[2] *n* GEN, TRANS pilote *m*; ~ **carrier** TRANS porteuse pilote *f*; ~ **frequency** TRANS fréquence pilote *f*; ~ **interference** TRANS effet produit par les ondes pilotes *m*; ~ **lamp** COMP, PER APP, RADIO, TELEP *indicating device* lampe témoin *f*, témoin lumineux *m*, voyant *m*; ~ **light** COMP, PER APP, RADIO, TELEP *indicating device* lampe témoin *f*, témoin lumineux *m*, voyant *m*; ~ **operation** TELEP exploitation pilote *f*; ~ **signal** RADIO signal d'onde pilote *m*; ~ **stud** PARTS plot de guidage *m*; ~ **supervision** TRANS surveillance de l'onde pilote *f*; ~ **system** RADIO système d'émission de l'onde pilote *m*; ~ **tone** TRANS tonalité pilote *f*; ~**-tone monitoring** TRANS contrôle par tonalité pilote *m*; ~**-tone system** RADIO système à fréquence pilote *m*; ~ **trial** TEST, TRANS essai pilote *m*; ~ **wave** GEN onde pilote *f*
PIM *abbr* GEN (*passive intermodulation*) IMP (*intermodulation passive*), TELEP (*pulse interval modulation*) modulation des intervalles entre impulsions *f*
pin[1]: ~**-to-pin** *adj* ELECT, PARTS broche à broche, point à point
pin[2] *n* ELECT patte *f*, broche *f*, PARTS broche *f*, PER APP picot *m*; ~ **column** ELECT colonne de broches *f*; ~ **contact** ELECT, PARTS contact mâle *m*; ~ **contact unit** ELECT, PARTS *connecting device* élément à contact mâle *m*; ~ **feed** PER APP entraînement par picots *m*; ~**-out** ELECT brochage *m*; ~ **row** ELECT rangée de broches *f*; ~ **spacing** ELECT espacement des broches *m*; ~**-to-pin wiring** ELECT, PARTS câblage broche à broche *m*
PIN *abbr* DATA (*procedure interrupt negative*) PIN (*interruption de la procédure négative*), GEN (*personal identity number*) code confidentiel *m*, numéro personnel d'identification *m*, TELEP (*procedure interrupt negative*), TRANS (*procedure interrupt negative*) PIN (*interruption de la procédure négative*)
PIN: ~ **photodiode** *n* TELEP, TRANS photodiode PIN *f*
pinboard *n* COMP tableau de connexions *m*
pinch: ~ **effect** *n* PARTS effet de pincement *m*; ~**-off voltage** *n* POWER tension de pincement *f*; ~ **roller** *n* PER APP pinceur *m*, rouleau *m*
pincushion: ~ **distortion** *n* TV distorsion en coussin *f*
PIN-FET: ~ **integrated receiver** *n* TELEP, TRANS récepteur intégré PIN-TEC *m*
pinhole *n* ELECT piqûre *f*
pink: ~ **noise** *n* COMP bruit rose *m*
pinpoint *vt* COMP, ELECT localiser
pinpointing *n* COMP, ELECT localisation *f*
pip *n* TELEP bip *m*
PIP *abbr* COMP (*picture in picture*) image dans l'image *f*, DATA, TELEP, TRANS (*procedure interrupt positive*) PIP (*interruption de la procédure positive*)
pipe *n* COMP, DATA canal de communication *m*, ELECT, PARTS, TRANS conduit *m*; ~ **cable** GEN câble coaxial *m*
pipeline: ~ **system** *n* COMP système en pipeline *m*
piracy *n* COMP, TV piratage *m*
pirate[1] *n* COMP, TV pirate *m*
pirate[2] *vt* COMP, TV pirater
pirated *adj* COMP, TV contrefait, piraté
piston *n* RADIO piston *m*, piston de court-circuit *m*, plongeur de court-circuit *m*; ~ **attenuator** RADIO affaiblisseur à piston *m*, atténuateur à piston *m*
pit *n* COMP creux *m*, microcuvette *f*, *stone* alvéole *f*, PARTS cratère *m*, creux *m*
pitch *n* COMP hauteur *f*, hauteur du son *f*, pas de masque *m*, sonie *f*, GEN pas *m*, TELEP tonalité *f*, tonie *f*, hauteur *f*; ~ **bias** RADIO dépointage *m*; ~ **circle** GEN cercle primitif de référence *m*; ~ **diameter** GEN diamètre primitif *m*; ~**-excited vocoder** RADIO codeur à fréquences vocales activé par la hauteur du son *m*, vocodeur PELP *m*; ~ **modulation** COMP modulation de hauteur *f*; ~**-modulation depth** COMP taux de modulation de hauteur *m*; ~ **randomize** COMP variation aléatoire de la hauteur *f*; ~ **scale** TELEP échelle de hauteur *f*; ~**-to-MIDI converter** COMP convertisseur hauteur-MIDI *m*; ~**-to-voltage converter** COMP convertisseur hauteur-tension *m*; ~ **variation** COMP variation de hauteur *f*; ~ **wheel** COMP molette de variation de hauteur *f*
pitchbend *n* COMP variation de hauteur *f*; ~ **effect** COMP effet de variation de hauteur *m*; ~ **range** COMP variation de hauteur maximale *f*; ~ **wheel** COMP molette de variation de hauteur *f*
pitchbender *n* COMP molette de variation de hauteur *f*
pitchless: ~ **sound** *n* COMP son atonal *m*
PIU *abbr* (*plug-in unit*) COMP élément enfichable *m*, unité enfichable *f*, DATA *scanner* plug-in *m*, ELECT, PARTS, TELEP unité enfichable *f*, élément enfichable *m*
PIV *abbr* (*peak inverse voltage*) POWER, TELEP tension inverse de crête *f*
pivot: ~ **for horizontal** *n* SWG pivot de barre de sélection *m*; ~ **for selecting bar** *n* SWG pivot de barre de sélection *m*
pixel *n* (*picture element*) COMP, ELECT, TV élément d'image *m*, pixel *m*, point d'image *m*; ~ **aspect ratio** COMP, RADIO, TELEP facteur d'aspect de pixel *m*, format du pixel *m*, rapport des axes d'un pixel *m*; ~ **map** COMP mode pixel *m*
pixelation *n* COMP effet mosaïque *m*, DATA *of scanner* pixellisation *f*
pixelization *AmE see pixellization BrE*
pixellization *n BrE* COMP effet mosaïque *m*, DATA *of scanner* pixellisation *f*
PIXIT *abbr* (*protocol implementation extra information for*

testing) DATA PIXIT (*informations supplémentaires sur l'instance de protocole destinées au test*);~ **proforma** *n* DATA formulaire PIXIT *m*

PJE *abbr* (*pointer justification event*) DATA événement de justification de pointeur *m*

PL *abbr* COMP (*primary link, parameter length*) liaison primaire *f*, lien primaire *m*, longueur de paramètre *f*, TELEP (*parameter length*) longueur de paramètre *f*, TRANS (*primary link*) liaison primaire *f*, lien primaire *m*

placing *n* DATA, GEN positionnement *m*

plain: ~ **connector** *n* RADIO bride lisse *f*; ~ **coupling flange** *n* RADIO bride lisse *f*; ~ **language** *n* COMP langage clair *m*; ~ **old telephone line** *n* TELEP ligne classique *f*; ~ **style** *n* COMP style standard *m*

plan: ~**-position indicator** *n* (*PPI*) RADIO indicateur panoramique *m*

planar: ~**-array antenna** *n* RADIO antenne réseau planaire *f*, réseau plat d'antennes *m*, antenne réseau plan *f*; ~ **diode** *n* PARTS diode planaire *f*; ~ **system** *n* TRANS système plan *m*; ~ **technique** *n* PARTS technique planar *f*; ~ **transistor** *n* PARTS transistor planar *m*

plane: ~ **of polarization** *n* RADIO plan de polarisation *m*; ~ **wave** *n* RADIO onde plane *f*

plant *n* GEN installation *f*; ~ **engineering** GEN ingénierie de site *f*

plasma: ~ **diode** *n* ELECT diode au plasma *f*; ~ **display** *n* COMP, ELECT écran à plasma *m*; ~ **panel** *n* COMP, ELECT écran à plasma *m*

plastic[1] *adj* PARTS plastique; ~**-tipped** ELECT à extrémité en plastique

plastic[2] *n* PARTS matière plastique *f*; ~**-bending section modulus** PARTS module de résistance à la flexion plastique *m*; ~**-clad silica** (*PCS*) TELEP silice recouverte de plastique *f*; ~**-clad silica fiber** *AmE*, ~**-clad silica fibre** *BrE* DATA, PARTS, TRANS fibre silice-plastique *f*, fibre silice-silicone *f*; ~ **coating** TRANS revêtement plastique *m*; ~ **deformation** PARTS déformation plastique *f*; ~ **effect** COMP effet plastique *m*; ~**-encapsulated microcircuit** ELECT microcircuit à encapsulage plastique *m*; ~ **encapsulation** ELECT encapsulation plastique *f*; ~ **fiber** *AmE*, ~ **fibre** *BrE* DATA, PARTS, TRANS fibre plastique *f*; ~ **film** PARTS film plastique *m*; ~**-film capacitor** POWER condensateur à film plastique *m*; ~ **foil** PARTS film plastique *m*; ~**-foil capacitor** POWER condensateur à film plastique *m*; ~ **hose** PARTS tube en matière plastique *m*, tube souple en plastique *m*; ~**-insulated cable** TRANS câble à isolant plastique *m*, câble à isolant synthétique *m*; ~ **material** PARTS matière plastique *f*; ~ **molding** *AmE*, ~ **moulding** *BrE* PARTS *process* coulage *m*, *object* objet moulé *m*; ~ **tubing** PARTS tube en matière plastique *m*, tube souple en plastique *m*

plasticizer *n* PARTS plastifiant *m*

plate *n* ELECT anode *f*, plaque *f*, POWER *of capacitor* armature *f*; ~ **battery** POWER batterie anodique *f*; ~ **block** PER APP boîte à plaques *f*; ~ **current** ELECT courant anodique *m*, courant plaque *m*; ~ **finish** ELECT état de surface brut de presse *m*; ~**-load resistor** PARTS, POWER résistance de charge *f*, résistance de plaque *f*; ~ **resistor** PARTS, POWER résistance de plaque *f*, résistance de charge *f*; ~ **voltage** ELECT tension anodique *f*

plateau *n* RADIO plateau *m*, palier *m*; ~ **slope** RADIO pente du plateau *f*, pente du palier *f*; ~**-threshold voltage** RADIO tension de début du plateau *f*

plated: ~ **through hole** *n* ELECT trou métallisé *m*

platen: ~ **guide** *n* TELEP guide du cylindre *m*; ~ **knob** *n* TELEP bouton du cylindre *m*

platform *n* ELECT, TRANS plate-forme *f*

plating: ~ **bar** *n* ELECT barre de métallisation *f*

platter *n* COMP plateau *m*

plausible: ~ **reasoning** *n* DATA raisonnement plausible *m*

play *vt* COMP *audio* reproduire, *program, macro* repasser; ~ **back** PER APP *answer machine message* écouter, réécouter, TELEP repasser, restituer

playability *n* COMP possibilités d'exécution *f pl*

playback *n* COMP *disk* lecture *f*, *audio* reproduction *f*, *program, macro* exécution *f*, PER APP *of answer machine* lecture *f*, RADIO contrôle d'enregistrement *m*, TV lecture *f*; ~ **head** COMP tête de lecture *f*

PLC *abbr* (*programmable logic controller*) DATA contrôleur à logique programmable *m*

plenum: ~**-rated cable** *n* ELECT câble plenum *m*, câble anti-feu *m*

plesiochronous[1] *adj* TELEP, TRANS plésiochrone

plesiochronous[2]: ~ **digital hierarchy** *n* (*PDH*) TELEP, TRANS hiérarchie numérique plésiochrone *f* (*HNP*); ~**-line terminal** *n* TRANS terminal de ligne plésiochrone *m*; ~ **transmission equipment** *n* TRANS équipement de transmission plésiochrone *m*

PLL *abbr* (*phase-locked loop*) ELECT boucle à phase asservie *f*, boucle à verrouillage de phase *f*

PL/M: ~ **language** *n* COMP langage PL/M *m*

plot *n* COMP *graph* tracé *m*; ~ **area** COMP zone de traçage *f*

plotter *n* COMP traceur *m*, DATA, PER APP, TELEP table traçante *f*, TEST traceur *m*

PLP *abbr* (*packet layer protocol*) DATA, TELEP protocole de couche paquet *m*

PLS *abbr* (*physical-layer service*) TELEP, TEST service de la couche physique *m*

PLU *abbr* (*phase lock unit*) ELECT organe de verrouillage de phase *m*

plug[1]: ~**-in** *adj* COMP, POWER enfichable; ~**-in-and-play** *adj* COMP plug-in-and-play

plug[2] *n* COMP, ELECT, PARTS, POWER, RADIO, TEST fiche *f*; ~**-and-jack connection** POWER connexion par fiche et jack *f*, raccordement par fiche et jack *m*; ~ **bar** POWER réglette de fiches *f*; ~**-in board** ELECT carte enfichable *f*; ~**-in cable** ELECT câble enfichable *m*; ~**-in chip card** ELECT carte enfichable *f*; ~**-in connector** ELECT, PARTS connecteur enfichable *m*; ~**-in module** COMP, ELECT, PARTS élément enfichable *m*; ~**-in relay** PARTS relais enfichable *m*; ~**-in unit** (*PIU*) COMP élément enfichable *m*, unité enfichable *f*, DATA *scanner* plug-in *m*, ELECT, PARTS, TELEP unité enfichable *f*, élément enfichable *m*; ~ **pin** ELECT *connecting device*, PARTS broche de contact *f*, broche de fiche mâle *f*; ~ **tip** TELEP pointe de fiche *f*

plug[3]: ~ **in** *vt* ELECT brancher, GEN enficher

pluggable *adj* ELECT enfichable

plugging: ~ **chart** *n* COMP schéma de connexions *m*

plumbing *n* RADIO plomberie *f*

plunger *n* PARTS poussoir *m*, POWER plongeur *m*, noyau plongeur *m*, noyau mobile *m*, RADIO piston *m*, piston de court-circuit *m*, plongeur de court-circuit *m*, TELEP plongeur *m*; ~ **key** CONTR bouton-poussoir *m*

pluralism *n* ELECT pluralisme *m*

PM *abbr* DATA (*phase modulation*) MP (*modulation de phase*), (*protocol machine*) PM (*machine de protocole*), TRANS (*phase modulation*) MP (*modulation de phase*)

PMA *abbr* (*program memory area*) COMP zone de mémoire de programmes *f*

PMBX *abbr* (*private manual branch exchange*) SWG, TELEP commutateur privé manuel *m* (*PMBX*), commutateur manuel privé *m*

PMR *abbr* (*private mobile radio*) RADIO radiomobile privé *m*

PN *abbr* (*positive notification*) DATA, TELEP notification positive *f*; ~ **junction** *n* ELECT jonction PN *f*

PNIC *abbr* (*private data-network identification code*) DATA, TELEP PNIC (*code d'identification de réseau privé de données*)

PNP *abbr* (*private numbering plan*) TELEP PNP (*plan de numérotage privé*)

PNX *abbr* (*private network exchange*) SWG, TELEP commutateur de réseau privé *m*

PO *abbr* (*post office*) TELEP bureau de poste *m*

poaching *n* GEN *of software* piratage *m*

POAS *abbr* (*payment option analysis system*) APPL SAOT (*système d'analyse des options tarifaires*)

Pockels: ~ **effect** *n* TRANS effet Pockels *m*

pocket[1] *adj* COMP, RADIO de poche; ~-**sized** COMP, RADIO de poche

pocket[2]: ~ **computer** *n* COMP ordinateur de poche *m*; ~ **receiver** *n* RADIO récepteur de poche *m*; ~ **telephone** *n* RADIO, TELEP téléphone de poche *m*; ~ **terminal** *n* RADIO terminal de poche *m*

point[1]: ~-**to-multipoint** *adj* RADIO point à multipoint; ~-**to-point** *adj* GEN, RADIO point à point

point[2] *n* COMP point *m*, TELEP point *m*, trait *m*; ~ **code** DATA, TELEP code de point *m*; ~ **contact** PARTS contact ponctuel *m*, pointe *f*; ~-**contact diode** PARTS diode à pointe *f*; ~ **contact transistor** ELECT, PARTS transistor à pointe *m*; ~ **of control** TEST point d'indifférence *m*; ~ **of control and observation** (*PCO*) DATA point de contrôle et d'observation *m* (*PCO*); ~ **of minimum voltage** POWER nud de tension *m*, RADIO point froid *m*; ~ **station** TRANS station terminale *f*; ~-**to-area communication** RADIO, TRANS communication point à zone *f*; ~-**to-multipoint communication** RADIO, TRANS communication point à multipoint *f*; ~-**to-multipoint connection** APPL, RADIO, TRANS connexion point à multipoint *f*; ~-**to-multipoint system** RADIO, TRANS système point à multipoint *m*; ~-**to-point communication** RADIO, TRANS communication point à point *f*; ~-**to-point protocol** (*PPP*) DATA protocole point à point *m* (*PPP*); ~-**to-point system** TRANS système point à point *m*; ~-**to-point wiring** PARTS câblage point à point *m*

point[3] *vt* RADIO orienter

pointer *n* COMP pointeur *m*, repère *m*, DATA pointeur *m*, TELEP indicateur *m*, pointeur *m*; ~ **centering error** *AmE*, ~ **centring error** *BrE* RADIO erreur d'excentrement *f*; ~ **connection** SWG connexion de pointeur *f*; ~ **deflection** TELEP *of indicating device* déviation de l'aiguille *f*; ~ **instrument** TELEP instrument à aiguille *m*; ~ **justification count** (*PJC*) DATA compte de justification de pointeur *m*; ~ **value** SWG valeur de pointeur *f*

pointing: ~ **angle** *n* RADIO angle de pointage *m*; ~ **device** *n* COMP dispositif de pointage *m*, dispositif de désignation *m*; ~ **loss** *n* RADIO perte due au pointage *f*

points: ~ **of international connection at baseband frequencies** *n pl* TELEP points d'interconnexion internationale dans la bande de base *m pl*; ~ **of telephony input and output** *n pl* TELEP points d'entrée et de sortie pour la télégraphie *m pl*

Poisson: ~ **traffic** *n* TELEP trafic de Poisson *m*, trafic de pur hasard *m*, trafic poissonien *m*

polar[1] *adj* COMP, RADIO polaire

polar[2]: ~ **keying** *n* TELEP modulation double courant *f*; ~ **modulation system** *n* RADIO système à modulation polaire *m*; ~ **orbit** *n* RADIO orbite polaire *f*; ~ **relay** *n* POWER relais polarisé *m*

polarity[2] *n* TV polarité *f*; ~ **indicator** POWER indicateur de polarité *m*; ~ **inversion** POWER inversion de polarité *f*; ~ **reversal** POWER inversion de polarité *f*; ~ **signaling** *AmE*, ~ **signalling** *BrE* SIG signalisation par polarité *f*

polarization *n* POWER, RADIO polarisation *f*; ~ **controller** (*PC*) ELECT contrôleur de polarisation *m*; ~ **coupling loss** RADIO perte par couplage de polarisation *f*; ~ **current** POWER courant de polarisation *m*; ~ **error** RADIO erreur de polarisation *f*

polarized: ~-**control mode** *n* DATA mode de contrôle polarisé *m*; ~ **electric bell** *n* TELEP sonnerie électrique polarisée *f*; ~ **relay** *n* PARTS, POWER relais polarisé *m*

polarizer *n* ELECT, RADIO polariseur *m*

polarizing: ~ **isolation** *n* RADIO, TRANS discrimination de polarisation *f*; ~ **slot** *n* ELECT détrompeur *m*

pole *n* POWER pièce polaire *f*, poteau *m*, pôle *m*; ~ **changer** POWER inverseur de polarité *m*; ~ **changing** POWER inversion de polarité *f*; ~ **face** TELEP face polaire *f*; ~ **inverter** POWER inverseur de polarité *m*; ~ **piece** POWER pièce polaire *f*; ~ **reversal** POWER inversion de polarité *f*; ~ **route** TRANS ligne sur poteaux *f*; ~ **shoe** POWER épanouissement polaire *m*

policy *n* DATA politique *f*

polishing: ~ **jig** *n* ELECT disque à polir *m*

poll[1] *n* GEN sondage *m*; ~ **bit** DATA bit d'interrogation *m*, TELEP bit d'invitation *m*

poll[2] *vt* COMP, DATA interroger, inviter à émettre, GEN sonder, RADIO interroger, inviter à émettre

polling *n* COMP, DATA interrrogation préalable *f*, invitation à émettre *f*, appel sélectif *m*, GEN sondage *m*, RADIO interrogation préalable *f*, invitation à émettre *f*, appel sélectif *m*; ~ **key** COMP clé d'interrogation *f*; ~ **radio network** RADIO réseau de radiocommunications à interrogation *m*; ~ **system** COMP système d'interrogation séquentiel *m*

poly: ~ **mix** *n* COMP mélange polyphonique *m*

polyamide *n* PARTS polyamide *m*; ~ **plastic** PARTS plastique polyamidique *m*

polybutylene: ~ **terephthalate** *n* ELECT téréphtalate de polybutylène *m*

polycarbonate: ~ **plastic** *n* PARTS plastique polycarbonate *m*

polyester: ~ **plastic** *n* PARTS plastique estérique *m*, plastique polyestérique *m*

polyethylene *n* PARTS polyéthylène *m*; ~ **plastic** PARTS plastique éthénique *m*, plastique éthylénique *m*

polygon: ~ **network** *n* TELEP réseau maillé *m*

polygonal: ~ **shading** *n* COMP lissage polygonal *m*

polymerization *n* PARTS polymérisation *f*

polymerize *vt* PARTS polymériser

polyolefin: ~ **insulation** *n* ELECT insulation par polyoléfines *f*

polyphase: ~ **system** *n* ELECT système polyphase *m*

polyphonic[1] *adj* COMP polyphonique

polyphonic[2]: ~ **play** *n* COMP exécution polyphonique *f*; ~ **pressure** *n* COMP pression polyphonique *f*; ~-**pressure sensitivity** *n* COMP sensibilité à la pression polyphonique *f*

polyphony *n* COMP polyphonie *f*

polyrod *n* RADIO antenne cierge *f*

polythene *n* PARTS polythène *m*

polyurethane *n* PARTS polyuréthane *m*

polyvinyl: ~ **chloride** *n* (*PVC*) ELECT chlorure de polyvinyle *m* (*PVC*)

PON *abbr* (*passive optical network*) TELEP, TRANS réseau optique passif *m*

pool[1] *n* TRANS *of channels* groupement *m*; ~ **cathode** ELECT, PARTS cathode liquide *f*; ~ **rectifier tube** ELECT tube redresseur à cathode liquide *m*

pool[2] *vt* TRANS grouper

pop[1]: ~-**up menu** *n* COMP menu local *m*, menu éclair *m*, menu fugitif *m*, menu incrusté *m*

pop[2]: ~ **up** *vt* COMP *on screen* surgir

population *n* TEST population *f*

porosity *n* PARTS porosité *f*

porous *adj* PARTS poreux

port *n* COMP port *m*, point d'accès *m*, RADIO accès *m*, porte *f*

portability *n* COMP portabilité *f*, TELEP déplacement en cours de communication *m*, portabilité *f*

portable[1] *adj* COMP portable, portatif, GEN, RADIO portable, portatif, transportable

portable[2]: ~ **computer** *n* COMP ordinateur portable *m*; ~ **data entry terminal** *n* (*PDE terminal*) COMP terminal portatif de saisie de données *m*; ~ **document file** *n* (*PDF*) COMP, DATA portable document file (*PDF*); ~ **printer** *n* COMP, DATA, PER

APP imprimante portable *f*; ~ **receiver** *n* RADIO récepteur portable *m*

portamento *n* COMP portamento *m*; ~ **footswitch** COMP pédale de portamento *m*; ~ **pedal** COMP pédale de portamento *m*; ~ **time** COMP durée de glissade *f*

position[1] *n* GEN position; **off** ~ PARTS état de repos *m*, position d'arrêt *f*, position de repos *f*; ~ **A** TELEP position A *f*; ~**-defined parameter** TELEP paramètre défini par la position *m*; ~ **set** TELEP équipement de position *m*

position[2]: ~ **busy** *phr* TELEP poste occupé

positional: ~ **tolerance** *n* GEN tolérance de position *f*

positioning *n* DATA, GEN positionnement *m*, RADIO *of antenna* mise à poste *f*, pointage *m*; ~ **time** COMP, DATA temps de positionnement *m*

positive: ~ **amplitude modulation** *n* ELECT, TV modulation d'amplitude positive *f*; ~ **bar** *n* POWER barre positive *f*; ~ **battery** *n* POWER accumulateur positif *m*; ~ **electron** *n* ELECT, GEN électron positif *m*, positron *m*; ~ **emitter-coupled logic** *n* (*PECL*) COMP logique ECL positive *f*; ~ **feedback** *n* COMP, ELECT, RADIO réaction *f*, réaction positive *f*; ~ **glow** *n* ELECT colonne positive *f*; ~ **indication tone** *n* TELEP tonalité d'indication positive *f*; ~ **justification** *n* TELEP, TRANS justification positive *f*; ~**-negative justification** *n* TELEP, TRANS justification positive-négative *f*; ~ **notification** *n* (*PN*) DATA, TELEP notification positive *f*; ~ **overvoltage** *n* POWER surtension positive *f*; ~ **pole** *n* POWER pôle positif *m*; ~ **pressure** *n* GEN pression positive *f*; ~ **terminal** *n* POWER pôle positif *m*, *of battery* borne positive *f*; ~ **video signal** *n* COMP signal d'image vidéo en positif *m*; ~ **wire** *n* TELEP fil positif *m*

positron *n* ELECT, GEN électron positif *m*, positron *m*

post: ~**-amplifier** *n* ELECT, POWER post-amplificateur *m*; ~**-amplifier loss** *n* ELECT perte en aval de l'amplificateur *f*; ~ **code** *n* *BrE* (*cf zip code AmE*) COMP code postal *m*; ~**-deflection acceleration factor** *n* ELECT facteur de post-accélération *m*; ~**-deflection acceleration ratio** *n* ELECT rapport de post-accélération *m*; ~**-dialing delay** *AmE*, ~**-dialling delay** *BrE* *n* RADIO, TELEP attente après numérotation *f*, délai d'attente après numérotation *m*; ~ **office** *n* (*PO*) TELEP bureau de poste *m*

postamble *n* DATA synchroniseur final *m*, épilogue *m*

posteriation *n* COMP effet de sous-quantification *m*, sous-quantification *f*

postprocessor *n* COMP post-processeur *m*

postproduction *n* TV post-production *f*; ~ **pressing** COMP post-production *f*; ~ **processing** COMP traitement après production *m*

PostScript® *n* COMP PostScript® *m*

pot *n* ELECT potentiomètre *m*

potential: ~ **barrier** *n* PARTS barrière de potentiel *f*; ~ **difference** *n* POWER différence de potentiel *f*; ~ **function** *n* POWER fonction potentielle *f*; ~ **gradient** *n* POWER gradient de potentiel *m*; ~ **log record** *n* DATA enregistrement potentiel de consignation *m*

potentiometer *n* ELECT potentiomètre *m*

potting: ~ **boot** *n* GEN boîtier d'enrobage *m*

powdering *n* COMP poudrage *m*

power[1] *n* ELECT puissance *f*, GEN puissance *f*, énergie *f*, flux *m*, POWER puissance *f*, alimentation *f*, RADIO, TRANS puissance *f*; ~ **amplification** COMP, DATA, POWER, TRANS amplification de puissance *f*; ~ **amplifier** (*PA*) COMP, DATA, POWER, TRANS amplificateur à gaine élevé *m*, amplificateur de puissance *m*; ~ **budget** POWER, TRANS bilan de puissance *m*; ~ **busbar** POWER barre d'énergie *f*, barre omnibus d'alimentation *f*; ~ **cable** COMP, DATA, ELECT, POWER câble d'alimentation *m*, cordon d'alimentation *m*; ~ **cable chute** COMP, DATA, ELECT, POWER, TELEP chemin de câble d'alimentation *m*; ~ **control bus** POWER barre de régulation de puissance *f*; ~ **control unit** POWER unité de commande de puissance *f*; ~ **converter** COMP, DATA, POWER convertisseur d'alimentation *m*, convertisseur de courant *m*; ~ **cord** COMP, ELECT, POWER cordon d'alimentation *m*; ~ **cube** POWER, RADIO groupe amplificateur de puissance *m*; ~ **current** POWER courant fort *m*; ~**-current engineering** POWER technique des courants forts *f*; ~ **cut** POWER coupure de courant *f*, coupure secteur *f*; ~ **density** POWER densité de puissance *f*; ~ **dissipation** POWER perte de puissance *f*, puissance dissipée *f*; ~ **distribution** POWER distribution électrique *f*, distribution d'énergie *f*; ~ **distribution alarm** POWER alarme de distribution d'énergie *f*; ~ **distribution board** POWER tableau de distribution *m*; ~ **distribution diagram** POWER schéma de distribution d'énergie *m*; ~ **divider** POWER, RADIO diviseur de puissance *m*, répartiteur de puissance *m*; ~**-down** POWER mise hors tension *f*; ~ **drain** POWER puissance absorbée *f*; ~**-driven system** POWER, TELEP système à entraînement mécanique *m*; ~ **electronics** POWER électronique de puissance *f*; ~ **engineering** POWER technique des courants forts *f*; ~ **equipment** POWER équipement de puissance *m*; ~ **factor** POWER facteur de puissance *m*; ~ **failure** POWER panne d'alimentation *f*, panne de courant *f*, panne du secteur *f*; ~ **failure feature** POWER, TELEP transfert sur panne secteur *m*, raccordement automatique en cas de panne *m*; ~ **failure transfer** POWER, TELEP raccordement automatique en cas de panne *m*, transfert sur panne secteur *m*; ~ **feed** POWER téléalimentation *f*, alimentation en énergie électrique *f*; ~**-feeding repeater station** POWER, TRANS station d'alimentation de répéteurs *f*; ~**-feeding station** POWER, TRANS station d'alimentation *f*; ~ **fuse** PARTS, POWER fusible pour courants forts *m*; ~ **gain** ELECT, POWER gain en puissance *m*; ~ **indicator** ELECT témoin d'alimentation *m*, témoin de mise sous tension *m*; ~ **law index fiber** *AmE*, ~ **law index fibre** *BrE* ELECT, POWER, TRANS fibre à gradient exponentiel *f*; ~ **law index profile** ELECT, POWER, TRANS profil à gradient exponentiel *m*, profil à loi en puissance *m*, profil d'indice à loi en puissance *m*; ~ **level** COMP, POWER niveau de puissance *m*; ~ **limit** ELECT, POWER limite de puissance *f*; ~ **line** POWER ligne électrique *f*; ~ **meter** POWER, TEST wattmètre *m*, milliwattmètre *m*; ~**-off alarm** POWER alarme de coupure de courant *f*; ~**-off condition** POWER alimentation hors service *f*; ~**-on** POWER mise sous tension *f*; ~**-on display** PER APP, POWER affichage à la mise sous tension *m*; ~ **on/off switch** PARTS, POWER interrupteur d'alimentation *m*; ~ **outage** POWER panne d'alimentation *f*, panne de courant *f*, panne du secteur *f*; ~ **pack** POWER bloc d'alimentation *m*; ~ **plant** POWER centrale électrique *f*; ~ **rack** POWER armoire d'alimentation *f*, bâti d'énergie *m*; ~ **relay** PARTS, POWER relais de puissance *m*; ~ **requirement** ELECT puissance requise *f*; ~ **room** POWER atelier d'énergie *m*; ~ **semiconductor** PARTS, POWER semiconducteur de puissance *m*; ~ **source** POWER source d'alimentation *f*, source d'énergie *f*; ~ **stage** POWER, RADIO étage de puissance *m*; ~ **subsystem** (*PSS*) POWER sous-système d'alimentation électrique *m*; ~ **supply** POWER alimentation *f*, alimentation électrique *f*, bloc d'alimentation *m*, unité d'alimentation *f*, alimentation *f*, onduleur *m*; ~ **supply equipment** POWER équipement d'alimentation *m*; ~ **supply station** POWER station de distribution d'énergie *f*; ~ **supply unit** POWER bloc d'alimentation *m*, alimentation *f*, *backup* alimentation de secours *f*, *inverter* onduleur *m*, *rectifier* redresseur-chargeur *m*; ~ **switch** PARTS, POWER interrupteur d'alimentation *m*, interrupteur général *m*; ~**-test bus** POWER barre d'essai d'alimentation *f*; ~ **thyristor** ELECT, PARTS, POWER thyristor de puissance *m*; ~ **transformer** POWER transformateur d'alimentation *m*; ~ **transistor** PARTS, POWER transistor de puissance *m*; ~ **transmission line** POWER ligne électrique *f*; ~ **tube** ELECT, POWER tube de puissance *m*; ~ **unit** POWER alimentation *f*; ~ **valve** ELECT tube de puissance *m*

power[2] *vt* POWER alimenter; ~ **down** ELECT, POWER mettre hors tension; ~ **up** ELECT, POWER mettre sous tension
powered *adj* POWER alimenté; ~**-down** ELECT, POWER mis hors tension; ~**-up** ELECT, POWER mis sous tension
PP *abbr* (*peripheral processor*) COMP processeur périphérique *m*
PPDU *abbr* (*presentation protocol data unit*) DATA unité de données de protocole de présentation *f*
PPI *abbr* (*plan-position indicator*) RADIO indicateur panoramique *m*
p-picture *n* (*predictive-coded picture*) COMP, TV image-p *f*
PPM *abbr* (*pulse position modulation*) DATA, TELEP, TRANS MIP (*modulation d'impulsions en position*)
PPP *abbr* (*point-to-point protocol*) DATA PPP (*protocole point à point*); ~ **Internet connection** *n* DATA connexion Internet PPP *f*
PPR *abbr* (*partial-page request*) DATA, TELEP demande de page partielle *f*
pps *abbr* (*pulses per second*) RADIO, TELEP imp/s (*impulsions par seconde*)
PPSDN *abbr* (*public packet-switched data network*) DATA RDPCP (*réseau de données public avec commutation par paquets*); ~ **service** *n* DATA service RDPCP *m*
PPS-EOM *abbr* (*partial-page signal EOM*) DATA, TELEP SPP-EOM (*signal de page partielle EOM*)
PPS-EOP *abbr* (*partial-page signal EOP*) DATA, TELEP SPP-EOP (*signal de page partielle EOP*)
PPS-MPS *abbr* (*partial-page signal MPS*) DATA, TELEP SPP-MPS (*signal de page partielle MPS*)
PPS-NULL *abbr* (*partial-page signal NULL*) DATA, TELEP SPP-NULL (*signal de page partielle NULL*)
PPS-PRI-EOM *abbr* (*partial-page signal PRI-EOM*) DATA, TELEP SPP-PRI-EOM (*signal de page partielle PRI-EOM*)
PPS-PRI-EOP *abbr* (*partial-page signal PRI-EOP*) DATA, TELEP SPP-PRI-EOP (*signal de page partielle PRI-EOP*)
PPS-PRI-MPS *abbr* (*partial-page signal PRI-MPS*) DATA, TELEP SPP-PRI-MPS (*signal de page partielle PRI-MPS*)
PPV *abbr* (*pay-per-view*) TV paiement à l'émission *m*, paiement par émission *m*
PR *abbr* TEST (*public relations*) relations publiques *f pl*, (*producer's risk*) risque du fournisseur *m*
PRA *abbr* (*primary rate access*) TELEP accès au débit primaire *m*
practical: ~ **test method** *n* TEST méthode de mesure de remplacement *f*
practice: ~ **factor** *n* TELEP coefficient de pratique expérimentale *m*
PRB: ~ **code** *n* DATA code de paquet de référence principal *m*
PRBS *abbr* (*pseudorandom binary sequence*) DATA séquence binaire pseudo-aléatoire *f*
preageing *n* TEST vieillissement préalable *m*
preamble *n* DATA synchroniseur initial *m*, préambule *m*
preamp *n* ELECT préampli *m*
preamplification *n* ELECT préamplification *f*
preamplifier *n* ELECT préamplificateur *m*
preanalysis *n* TELEP préanalyse *f*
prearranged *adj* RADIO prévu
preassembled: ~ **cables** *n pl* TRANS câbles pré-équipés *m pl*
preassembly *n* TRANS préassemblage *m*
preassigned[1] *adj* GEN préaffecté
preassigned[2]: ~ **noninterpolated channel** *n* TRANS voie préassignée sans concentration *f*; ~ **value** *n* TEST valeur prédéterminée *f*
preavis: ~ **charge** *n* TELEP taxe de préavis *f*; ~ **fee** *n* TELEP taxe de préavis *f*
precable *vt* TRANS précâbler
precabling *n* TRANS précâblage *m*
precipitation: ~ **rate** *n* RADIO intensité de precipitation *f*, intensité de pluie *f*; ~ **return** *n* RADIO écho de précipitations atmosphériques *m*; ~ **scatter** *n* RADIO diffusion sur les précipitations *f*; ~**-scatter propagation** *n* RADIO propagation par diffusion sur les précipitations *f*; ~ **static** *n* TRANS brouillage atmosphérique *m*
precision[1]: ~**-etched** *adj* ELECT à rainures gravées
precision[2] *n* RADIO précision *f*, TEST fidélité *f*; ~ **approach radar** (*PAR*) RADIO radar d'approche de précision *m* (*PAR*), radar d'approche finale *m* (*PAR*), radiodétecteur d'approche de précision *m* (*PAR*); ~ **framing** DATA *of scanner* cadrage précis *m*; ~ **frequency offset** RADIO, TRANS décalage de fréquence de précision *m*; ~ **instrument** TEST instrument de précision *m*
precorrection *n* TELEP précorrection *f*
predefined: ~ **marker** *n* DATA balise prédéfinie *f*
predetermine *vt* GEN prédéterminer
predetermined *adj* GEN prédéterminé
prediction: ~ **error** *n* COMP erreur de prédiction *f*
predictive: ~**-coded picture** *n* (*p-picture*) COMP, TV image-p *f*; ~ **coding** *n* DATA codage prédictif *m*
predictor *n* COMP prédicteur *m*
predistortion: ~ **technique** *n* ELECT technique de distorsion préalable *f*
pre-emphasis *n* ELECT, RADIO préaccentuation *f*; ~ **network** ELECT réseau de préaccentuation *m*, réseau de préaccentuateur *m*
pre-empt *vt* TELEP *call* interrompre
pre-energize *vt* POWER préalimenter
pre-equalization *n* ELECT préaccentuation *f*, précorrection *f*
prefix *n* GEN préfixe *m*
preform *n* GEN, TRANS préforme *f*
preformatted *adj* COMP, DATA préformaté
preformed: ~ **fiber** *AmE*, ~ **fibre** *BrE* *n* TRANS baguette *f*
pregroup *n* TRANS sous-groupe *m*
preheating *n* ELECT réchauffage préalable *m*; ~ **time** ELECT temps de préchauffage *m*
prejudice *n* ELECT préjugé *m*
preliminary: ~ **result** *n* TEST résultat préliminaire *m*; ~ **test** *n* TEST essai préliminaire *m*, test préliminaire *m*; ~ **test program** *AmE*, ~ **test programme** *BrE* *n* TEST programme d'essai préliminaire *m*
preloaded *adj* ELECT prétéléchargé, prééquipé
premarking *n* TELEP *of magnetic tape* prémarquage *m*
premastering *n* COMP prématriçage *f*
premature: ~ **release** *n* COMP, TELEP libération prématurée *f*
premises: ~**-distribution system** *n* ELECT système de câblage de l'immeuble *m*; ~ **wiring** *n* TELEP circuits de câblage des locaux *m pl*, câblage des locaux *m*
premium: ~ **channel** *n* TV canal optionnel *m*, chaîne en option *f*; ~ **service** *n* TV service optionnel *m*
prepaid: ~ **card** *n* TELEP télécarte *f*
preparation: ~ **operating** *n* TELEP exploitation avec préparation *f*
preparatory: ~ **period** *n* TELEP période préparatoire *f*
preprint *n* COMP, PER APP tirage préliminaire *m*
preprinted *adj* COMP, PER APP préimprimé
preprocessing *n* COMP prétraitement *m*
preprocessor *n* COMP préprocesseur *m*
preprogrammed: ~ **line** *n* TELEP liaison préprogrammée *f*
prerecorded[1] *adj* TV préenregistré
prerecorded[2]: ~ **message** *n* TELEP message préenregistré *m*
prescaler *n* TEST prédiviseur *m*
preseizure: ~ **dialing** *AmE*, ~ **dialling** *BrE* *n* TELEP numérotation sans décrocher *f*
preselection *n* TELEP, TV présélection *f*
preselector *n* TELEP, TV présélecteur *m*
presentation *n* TELEP présentation *f*; ~ **address** DATA adresse de présentation *f*; ~ **connection** DATA connexion de présentation *f*; ~ **context** DATA contexte de présentation *m*; ~ **control function** TELEP fonction de commande pour la présentation *f*; ~ **data value** DATA valeur de données de présentation *f*; ~ **indicator** (*PI*) DATA indicateur de

présentation *m*; ~ **layer** DATA, TRANS couche présentation *f*; ~ **protocol data unit** (*PPDU*) DATA unité de données de protocole de présentation *f*; ~ **service** DATA service de présentation *m*; ~-**service access point** (*PSAP*) DATA, TELEP point d'accès au service de présentation *m*; ~-**service data unit** DATA unité de données du service de présentation *f*; ~ **time stamp** (*PTS*) COMP estampille de présentation *f*; ~ **unit** COMP unité de présentation *f*

preset[1] *adj* COMP préprogrammé

preset[2]: ~ **memory** *n* COMP mémoire préprogrammée *f*; ~ **read-only memory** *n* COMP mémoire préprogrammée *f*; ~ **search** *n* COMP recherche préprogrammée *f*; ~ **voice** *n* COMP son préprogrammé *m*

preset[3] *vt* ELECT prédéterminer, GEN prérégler

presetting *n* GEN préréglage *m*

press: ~ **telegram** *n* TELEP télégramme de presse *m*; ~-**to-talk button** *n* RADIO bouton de conversation *m*, TELEP touche d'alternat *f*

pressing: ~ **master** *n* COMP matrice de pressage *f*

pressure[1]: ~-**sensitive** *adj* COMP sensible à la pression

pressure[2]: ~ **contact** *n* PARTS contact à pression *m*; ~ **plug** *n* GEN rouleau de pression *m*; ~-**reducing valve** *n* PARTS détendeur *m*, réducteur de pression *m*; ~-**sensitive keyboard** *n* COMP clavier piézosensible *m*; ~ **switch** *n* CLIM, PARTS manocontact *m*; ~ **wave** *n* RADIO onde de pression *f*

pressurized: ~ **cable** *n* POWER, TRANS câble pressurisé *m*, câble sous pression *m*

prestretched: ~ **cold-shrink tubing** *n* ELECT tube préétiré rétractable à froid *m*

pretest *n* TEST essai préalable *m*, prétest *m*

pre-TR: ~ **tube** *n* ELECT tube pré-TR *m*

prevention: ~ **of call collision** *n* TELEP prévention de collision d'appels *f*; ~ **cost** *n* TEST coût de prévention *m*

preventive: ~ **cyclic retransmission error control method** *n* TRANS méthode de correction d'erreur avec retransmission *f*; ~ **inspection** *n* TEST contrôle préventif *m*; ~ **maintenance** *n* GEN, TEST maintenance préventive *f*; ~ **maintenance time** *n* TELEP, TEST durée de maintenance préventive *f*, temps d'entretien *m*

preview *n* COMP aperçu *m*, prévisualisation *f*, DATA *of scanner* prévisualisation *f*; ~ **channel** TV canal mosaïque *m*

prewire[1] *adj* ELECT précâblés

prewire[2] *vt* ELECT précâbler

prewired: ~ **connector** *n* ELECT réglette câblée *f*

prewiring *n* ELECT précâblage *m*

PRF *abbr* ELECT, SIG, TELEP, TRANS (*pulse recurrence frequency*) FRI (*fréquence de récurrence des impulsions*), (*pulse repetition frequency*) FRI (*fréquence de répétitions des impulsions*)

price: ~-**fixing** *n* APPL accords de prix *m pl*

pricing *n* APPL tarification *f*; ~ **policy** APPL politique tarifaire *f*; ~ **scheme** APPL système de tarification *m*

PRI-EOM *abbr* (*procedure interrupt EOM*) DATA, TELEP PRI-EOM (*interruption de la procédure EOM*)

PRI-EOP *abbr* (*procedure interrupt EOP*) DATA, TELEP PRI-EOP (*interruption de la procédure EOP*)

primary: ~ **access** *n* TELEP accès primaire *m*; ~ **battery** *n* POWER pile primaire *f*; ~ **block** *n* TELEP bloc primaire *m*; ~ **body** *n* RADIO corps principal *m*; ~ **center** *AmE*, ~ **centre** *BrE* *n* TELEP centre primaire *m*; ~ **clock** *n* RADIO horloge primaire *f*; ~ **coating** *n* TRANS revêtement primaire *m*; ~-**color signal** *AmE*, ~-**colour signal** *BrE* *n* COMP signal de couleur primaire *m*; ~ **distribution link** *n* RADIO liaison de distribution primaire *f*; ~-**electron emission** *n* ELECT émission d'électrons primaires *f*; ~ **failure** *n* TEST défaillance primaire *f*; ~ **flow direction** *n* GEN sens primaire de liaison *m*; ~ **frequency standard** *n* RADIO étalon primaire de fréquence *m*; ~ **function** *n* DATA fonction primaire *f*; ~ **grade of reception quality** *n* RADIO qualité primaire de réception *f*; ~ **high-usage trunk group** *n* TELEP faisceau de circuits à débordement *m*, faisceau direct à débordement *m*; ~ **interlock trunk** *n* TELEP circuit interurbain primaire *m*; ~ **intertoll trunk group** *n* TELEP faisceau de circuits interurbain primaire *m*, faisceau interurbain primaire *m*; ~ **link** *n* (*PL*) COMP, TRANS liaison primaire *f*, lien primaire *m*; ~ **network** *n* TELEP réseau primaire *m*, ~ **radar** *n* RADIO radar primaire *m*, radiodétection primaire *f*; ~ **rate access** *n* (*PRA*) TELEP accès au débit primaire *m*; ~ **rate interface** *n* DATA, TELEP accès primaire *m*; ~ **rate interface ISDN** *n* DATA, TELEP accès primaire RNIS *m*; ~ **reference burst code** *n* DATA code de paquet de référence principal *m*; ~ **reference station** *n* RADIO station de référence principale *f*; ~ **reference terminal** *n* (*PRT*) RADIO terminal de référence principal *m*; ~ **relay** *n* PARTS relais primaire *m*; ~ **route** *n* TELEP voie primaire *f*; ~ **source** *n* TV source primaire *f*; ~ **station** *n* DATA station primaire *f*; ~ **storage** *n* DATA mémoire centrale *f*, mémoire principale *f*; ~ **terminal** *n* TELEP terminal primaire *m*; ~ **winding** *n* PARTS enroulement primaire *m*

prime[1]: ~ **time** *n* TV heures d'écoute maximale *f pl*, heures de grande écoute *f pl*

prime[2] *vt* ELECT, GEN apprêter

primer *n* GEN apprêt *m*; ~ **electrode** ELECT électrode d'amorçage *f*; ~ **ignition** ELECT amorçage *m*; ~ **noise** ELECT bruit d'amorçage *m*; ~ **voltage** ELECT tension d'amorçage *f*

priming: ~ **rate** *n* ELECT rapidité d'apprêtage *f*; ~ **speed** *n* ELECT vitesse d'apprêtage *f*

primitive *n* COMP, DATA *statistics* primitive *f*; ~ **confirm** DATA, TELEP confirmation primitive *f*; ~ **indication** DATA indication primitive *f*; ~ **name** DATA nom primitif *m*; ~ **request** DATA demande primitive *f*; ~ **response** DATA réponse primitive *f*

PRI-MPS *abbr* (*procedure interrupt MPS*) DATA, TELEP PRI-MPS (*interruption de la procédure MPS*)

principal: ~ **processing unit** *n* (*PUP*) COMP processeur principal *m*, unité de traitement principal *f*; ~ **test section** *n* TELEP, TEST section principale d'essais *f*; ~ **voltage** *n* POWER tension principale *f*

principle: ~ **of operation** *n* TELEP principe d'exploitation *m*

print[1] *n* PER APP impression *f*; ~ **bar** TELEP barre d'impression *f*; ~ **barrel** PER APP tambour d'impression *m*; ~ **capacity** PER APP vitesse d'inscription *f*; ~ **command** COMP commande d'impression *f*; ~ **command language** (*PCL*) DATA PCL; ~-**control character** DATA caractère de commande d'impression *m*; ~ **drum** PER APP, TELEP tambour d'impression *m*; ~ **settings** *n pl* COMP configuration de l'impression *f*, options d'impression *f pl*; ~-**through** COMP, DATA transfert magnétique *m*

print[2] *vt* PER APP imprimer; ~ **out** COMP, PER APP imprimer, sortir sur imprimante

printable: ~ **area** *n* COMP zone imprimable *f*

printed: ~ **board** *n* ELECT carte imprimée *f*, carte nue *f*; ~ **board assembly** *n* ELECT carte équipée *f*, carte imprimée équipée *f*; ~ **circuit** *n* (*PC*) ELECT, PARTS circuit imprimé *m* (*CI*); ~ **circuit board** *n* (*PCB*) ELECT carte à circuits imprimés *f*; ~ **component** *n* ELECT composant imprimé *m*; ~ **conductor** *n* ELECT conducteur imprimé *m*; ~ **contact** *n* ELECT contact imprimé *m*; ~ **edge-board contact** *n* ELECT contact imprimé d'extrémité de carte *m*; ~ **edge-connector** *n* ELECT contact imprimé d'extrémité *m*; ~ **record of duration and charge of calls service** *n* APPL, TELEP service d'enregistrement écrit des éléments de taxation *m*; ~ **wiring** *n* ELECT, PARTS circuit imprimé *m* (*CI*)

printer *n* COMP, DATA imprimante *f*; ~ **controller** COMP, DATA contrôleur d'imprimante *m*; ~ **driver** COMP, DATA pilote d'impression *m*; ~ **port** COMP port imprimante *m*; ~ **server** COMP, DATA serveur d'imprimante *m*

printing *n* GEN impression *f*; ~ **block** PER APP cliché *m*; ~ **speed** PER APP vitesse d'inscription *f*; ~ **terminal** TELEP

terminal papier *m*; ~ **track area** ELECT impression laque thermographique *f*
printwheel *n* PER APP roue à caractères *f*
priority *n* COMP, RADIO priorité *f*; ~ **call** TELEP appel prioritaire *m*; ~ **classification** TELEP classification prioritaire *f*; ~ **disconnection** TELEP déconnexion de priorité *f*; ~ **extension** TELEP poste supplémentaire à lignes prioritaires *m*; ~ **facility** TELEP priorité *f*; ~-**indication bit** DATA bit d'identification de la priorité *m*; ~ **level** TELEP niveau de priorité *m*; ~-**marked line** TELEP ligne de priorité de trafic *f*; ~ **marking** TELEP indication de priorité *f*; ~ **with notifying tone** TELEP priorité avec tonalité de notification *f*; ~ **request** TELEP demande préalable *f*; ~ **signal** SIG signal prioritaire *m*; ~ **subscriber** TELEP abonné prioritaire *m*; ~ **tone** TELEP tonalité de priorité *f*
prism *n* ELECT prisme *m*
privacy *n* DATA, PER APP, TELEP exclusion des postes *f*, secret *m*; ~ **button** PER APP touche secrète *f*; ~ **circuit** TELEP liaison à secret *f*; ~ **connection** TELEP liaison à secret *f*; ~ **release** COMP, TELEP libération du secret *f*
private[1] *adj* DATA, PER APP, TELEP privé
private[2]: ~ **automatic branch exchange** *n* (*PABX*) APPL, CONTR, SWG, TELEP autocommutateur privé *m* (*PABX*); ~ **automatic exchange** *n* (*PAX*) APPL autocommutateur privé *m* (*PABX*), central automatique privé *m*; ~ **branch exchange** *n* (*PBX*) SWG, TELEP PBX (central d'abonné), central privé *m*, commutateur privé *m*, commutateur téléphonique privé *m*; ~ **branch network** *n* (*PBN*) TELEP, TRANS réseau de télécommunication privé *m*, réseau privé *m*; ~ **circuit** *n* TRANS liaison louée *f*; ~ **data-network identification code** *n* (*PNIC*) DATA, TELEP code d'identification de réseau privé de données *m* (*PNIC*); ~ **dial-up port** *n* SWG accès privé par réseau commuté *m*; ~ **exchange** *n* TELEP central privé *m*; ~ **installation** *n* TELEP installation privée *f*; ~ **integrated-services network** *n* TELEP, TRANS réseau privé à intégration de services *m*; ~ **local system** *n* RADIO système privé local *m*; ~ **management domain** *n* (*PRMD*) DATA, TELEP domaine de gestion privé *m*; ~ **manual branch exchange** *n* (*PMBX*) SWG, TELEP commutateur privé manuel *m* (*PMBX*), commutateur manuel privé *m*; ~ **manual exchange** *n* SWG, TELEP central manuel privé *m*; ~ **mobile radio** *n* (*PMR*) RADIO radiomobile privé *m*; ~ **mobile-radio network** *n* RADIO réseau radiomobile privé *m*; ~ **network** *n* TELEP, TRANS réseau privé *m*; ~ **network exchange** *n* (*PNX*) SWG, TELEP commutateur de réseau privé *m*; ~-**number ringing service** *n* SIG service de signal de retour d'appel d'un numéro particulier *m*; ~ **numbering plan** *n* (*PNP*) TELEP plan de numérotage privé *m* (*PNP*); ~ **subscriber installation** *n* DATA, TELEP installation privée d'abonné *f*; ~ **system** *n* PER APP *of answer machine* installation privée *f*; ~ **telephone network** *n* TELEP, TRANS réseau téléphonique privé *m*; ~ **use** *n* RADIO usage privatif *m*; ~ **wire** *n* (*PW*) TRANS liaison spécialisée *f*
PRMD *abbr* (*private management domain*) DATA, TELEP domaine de gestion privé *m*
probabilistic: ~ **reasoning** *n* DATA raisonnement probabilistique *m*
probability: ~ **of acceptance** *n* TEST probabilité d'acceptation *f*; ~ **of congestion** *n* TELEP probabilité d'encombrement *f*; ~ **density function** *n* (*PDF*) DATA fonction de densité de probabilité *f*; ~ **of detection** *n* RADIO probabilité de détection *f*; ~ **of excess delay** *n* DATA probabilité de dépassement du temps d'attente *f*; ~ **of failure** *n* TEST probabilité de défaillance *f*; ~ **of indication** *n* RADIO probabilité d'indication *f*; ~ **of rejection** *n* TEST probabilité de rejet *f*; ~ **of shortage** *n* TEST probabilité de rupture de stock *f*; ~ **of successful service completion** *n* TELEP probabilité d'exécution correcte du service *f*; ~ **theory** *n* GEN théorie des probabilités *f*
probe *n* GEN sonde *f*, TEST capteur *m*; ~ **current** ELECT courant de sonde *m*; ~ **microphone** TELEP microphone sonde *m*, sonde microphonique *f*; ~ **transformer** RADIO transformateur sonde *m*; ~ **voltage** ELECT tension de sonde *f*
procedural: ~ **error** *n* DATA erreur de procédure *f*
procedure *n* GEN procédure *f*, méthode *f*, marche à suivre *f*; ~ **call** RADIO appel de procédure *m*; ~ **epilogue** TELEP épilogue de procédure *m*; ~ **interrupt EOM** (*PRI-EOM*) DATA, TELEP interruption de la procédure EOM *f* (*PRI-EOM*); ~ **interrupt EOP** (*PRI-EOP*) DATA, TELEP interruption de la procédure EOP *f* (*PRI-EOP*); ~ **interrupt MPS** (*PRI-MPS*) DATA, TELEP interruption de la procédure MPS *f* (*PRI-MPS*); ~ **interrupt negative** (*PIN*) DATA, TELEP, TRANS interruption de la procédure négative *f* (*PIN*); ~ **interrupt positive** (*PIP*) DATA, TELEP, TRANS interruption de la procédure positive *f* (*PIP*); ~ **prologue** TELEP prologue de procédure *m*
proceed: ~ **to select** *n* TELEP invitation à numéroter *f*; ~-**to-send signal** *n* SIG, TELEP signal d'invitation à transmettre *m*
process[1] *n* GEN processus *m*; ~ **average** TEST qualité moyenne d'un processus *f*; ~ **capability** TEST aptitude au traitement *f*, aptitude du processus *f*, possibilité de traitement *f*; ~ **control** CONTR, TEST contrôle de processus *m*, mise sous contrôle de processus *f*; ~ **description** TELEP description de processus *f*; ~ **in control** TEST processus sous contrôle *m*; ~ **inspection** TEST contrôle de processus *m*; ~ **quality control** TEST mise sous contrôle de processus *f*
process[2] *vt* COMP, DATA traiter
processed *adj* COMP, DATA traité
processing *n* COMP fabrication *f*, COMP, DATA, SWG traitement *m*; ~ **card** COMP, DATA carte de traitement *f*; ~ **delay** COMP, DATA retard d'exécution *m*; ~ **power** COMP, DATA puissance de calcul *f*; ~ **priority code** COMP, DATA code de priorité de traitement *m*
processor *n* COMP, DATA processeur *m*; ~ **architecture** DATA architecture de processeur *f*; ~ **outage** TELEP processeur hors service *m*; ~ **unit** COMP, DATA unité de traitement *f*
ProDigi *n* (*PROfessional DIGItal*) COMP ProDigi *f* (*PROfessional DIGItal*)
producer: ~**'s quality cost** *n* TEST coût d'obtention de la qualité *m*; ~**'s risk** *n* (*PR*) TEST risque du fournisseur *m*; ~**'s risk point** *n* (*PRP*) TEST point du risque du fournisseur *m*
product: ~ **board** *n* ELECT carte de production *f*; ~ **liability** *n* TEST responsabilité du fait du produit *f*
production: ~ **cost** *n* GEN coût de fabrication *m*, coût de production *m*; ~ **credits** *n pl* TV génériques *m pl*; ~ **level** *n* GEN niveau de production *m*; ~ **master** *n* ELECT cliché de production *m*; ~ **permit** *n* TEST dérogation avant production *f*
products: ~ **and services** *n pl* APPL produits et services *m pl*
professional[1]: ~-**level** *adj* PER APP professionnel
professional[2]: ~ **classification** *n* TELEP *in directory* rubrique professionnelle *f*; ~ **secrecy** *n* GEN, TEST secret professionnel *m*
PROfessional: ~ **DIGItal** *n* (*ProDigi*) COMP PROfessional DIGItal *m* (*ProDigi*)
profile *n* DATA, TRANS profil *m*; ~ **dispersion** TRANS dispersion de profil *f*; ~-**dispersion parameter** TRANS facteur de dispersion de profil *m*; ~ **parameter** TRANS paramètre de profil *m*
program[1] *n* COMP, DATA programme *m*, GEN, RADIO, TV *AmE see programme BrE*; ~ **availability** *AmE see programme availability BrE*; ~ **booking center** *AmE see programme booking centre BrE*; ~ **cartridge** COMP cartouche de programmes *f*; ~ **counter** COMP registre d'adresse d'instruction *m*; ~ **debugging** COMP, DATA mise au point de programmes *f*; ~ **distribution** *AmE see programme distribution BrE*; ~-**generated parameter** COMP paramètre dynamique *m*; ~ **generator** COMP gestionnaire de

programme *m*; ~ **library** COMP bibliothèque de programmes *f*, programmathèque *f*; ~ **memory area** (*PMA*) COMP zone de mémoire de programmes *f*; ~ **package** COMP progiciel *m*; ~ **save** COMP sauvegarde de programme *f*; ~ **schedule** *AmE see programme schedule BrE*; ~ **selector** COMP sélecteur de programme *m*; ~ **sound circuit** *AmE see programme sound circuit BrE*; ~ **store** COMP mémoire de programme *f*; ~ **store block** COMP, DATA bloc de mémoires de programmes *m*; ~ **store bus** COMP, SWG bus de mémoire de programme *m*; ~ **store switch** COMP sélecteur de mémoire de programme *m*; ~ **store unit** COMP unité de mémoire de programme *f*; ~ **switch** COMP sélecteur de programme *m*; ~ **switch flip-flop** COMP bascule de changement de programme *f*; ~ **test panel** COMP pupitre d'essai de programme *m*; ~ **track** *AmE see programme track BrE*; ~ **tracking** *AmE see programme tracking BrE*

program[2] *vt* COMP, DATA, PER APP *equipment, apparatus* programmer

programmable[1] *adj* COMP, DATA, PER APP *equipment* programmable

programmable[2]: ~ **announcements** *n pl* TELEP mise en mémoire de messages *f*; ~ **array logic** *n* (*PAL*) COMP, TV logique à matrice programmable *f* (*PAL*); ~ **communication interface** *n* (*PCI*) TELEP interface de communication programmable *f* (*PCI*); ~ **control** *n* SWG commande programmable *f*; ~ **device** *n* ELECT appareil programmable *m*; ~ **keyboard** *n* COMP clavier programmable *m*; ~ **logic circuit** *n* ELECT circuit logique programmable *m*; ~ **logic controller** *n* (*PLC*) DATA contrôleur à logique programmable *m*; ~ **sequencer** *n* SWG séquenceur programmable *m*; ~ **unijunction transistor** *n* (*PUT*) ELECT, PARTS, POWER transistor unijonction programmable *m* (*TUP*)

programme *n BrE* GEN, RADIO, TV programme *m*; ~ **availability** *BrE* TV disponibilité des programmes *f*; ~ **booking centre** *BrE* TV service centralisateur *m*; ~ **distribution** *BrE* TV télédistribution de programmes *f*; ~ **schedule** *BrE* TV grille de programme *f*; ~ **sound circuit** *BrE* TRANS voie radiophonique *f*; ~ **track** *BrE* RADIO poursuite programmée *f*; ~ **tracking** *BrE* RADIO piste de programmes *f*

programmed: ~ **command** *n* TELEP commande programmée *f*; ~ **control** *n* CONTR commande programmée *f*; ~ **sequence controller** *n* COMP séquenceur microprogrammé *m*

programmer *n* COMP programmeur *m*, RADIO, TV programmateur *m*

programming *n* COMP, DATA programmation *f*; ~ **error** COMP, DATA erreur de programmation *f*; ~ **flow chart** COMP diagramme de programmation *m*, ordinogramme *m*, organigramme de programmation *m*; ~ **interface** COMP, DATA interface de programmation *f*; ~ **method** COMP, DATA méthode de programmation *f*; ~ **procedure** COMP, DATA procédure de programmation *f*; ~ **system** COMP, DATA système de programmation *m*

progression *n* COMP suite *f*

progressive: ~ **behavior** *AmE*, ~ **behaviour** *BrE n* COMP, TELEP comportement progressif *m*; ~ **coding** *n* COMP, TELEP codage progressif *m*; ~**-compatible-sequential coding** *n* COMP, TELEP codage séquentiel compatible avec le codage progressif *m*; ~ **scanning** *n* COMP balayage ligne par ligne *m*, balayage par lignes contiguës *m*, balayage sans entrelacement *m*; ~ **switching magazine** *n* SWG système pas-à-pas *m*; ~ **transition** *n* SWG transition progressive *f*; ~ **wave** *n* ELECT, TRANS onde progressive *f*

projected: ~ **peak point** *n* POWER point isohypse *m*; ~ **peak point voltage** *n* POWER tension de point isohypse *f*

projection: ~ **television receiver** *n* TV récepteur de télévision à projection *m*, récepteur à projection *m*; ~ **tube** *n* ELECT tube de projection *m*

projector *n* DATA, TV projecteur *m*

promotor *n* PARTS *plastics* accélérateur promoteur *m*

prompt *n* COMP invite *f*; ~ **line** COMP ligne d'invite *f*

prompting *n* TELEP intervention *f*; ~ **output** TELEP sortie d'intervention *f*

proof: ~ **test** *n* TEST, TRANS essai de rupture par traction *m*

propagation *n* ELECT, RADIO, TRANS propagation *f*; ~ **area** ELECT, GEN aire de propagation *f*; ~ **coefficient** ELECT, RADIO, TRANS constante de propagation *f*, exposant linéique de propagation *m*; ~ **constant** ELECT, RADIO, TRANS constante de propagation *f*, exposant linéique de propagation *m*; ~ **delay** RADIO temps de propagation *m*; ~ **error** RADIO erreur de propagation *f*; ~ **mode** TRANS mode de propagation *m*; ~ **performance** TELEP caractéristiques de propagation *f pl*; ~ **properties** *n pl* RADIO caractéristiques de propagation *f pl*; ~ **time** RADIO temps de propagation *m*; ~ **time equalization** TRANS égalisation du temps de propagation *f*, uniformisation du temps de propagation *f*; ~ **velocity** ELECT, TRANS vitesse de propagation *f*

proper: ~ **time** *n* RADIO temps propre *m*

proportional: ~ **action** *n* CONTR action proportionnelle *f*; ~ **counter tube** *n* RADIO tube-compteur proportionnel *m*; ~ **jitter** *n* RADIO gigue proportionnelle *f*; ~ **region** *n* RADIO région de proportionnalité *f*

proprietary: ~ **code** *n* DATA code natif *m*; ~ **system** *n* COMP système exclusif *m*; ~ **telephone set** *n* TELEP poste téléphonique particulier à un utilisateur *m*

prosign *n* RADIO signal de service *m*

prosodic: ~ **analysis** *n* DATA prosodie *f*

protect: ~ **memory write** *n* COMP protection en mémoire écrite *f*

protected[1] *adj* GEN protégé, sécurisé

protected[2]: ~ **contour** *n* TV contour de protection *m*; ~ **memory** *n* ELECT sauvegarde *f*

protecting: ~ **resistor** *n* POWER résistance de protection *f*

protection *n* GEN sécurisation *f*, protection *f*; ~ **against intrusion** TELEP protection contre l'entrée en ligne *f*, protection contre les intrusions *f*; ~ **cable** TRANS câble de protection *m*; ~ **circuit** TEST circuit de protection *m*; ~ **equipment** POWER matériel de protection *m*; ~ **margin** RADIO marge de protection *f*; ~ **ratio** RADIO rapport de protection *m*; ~ **relay** PARTS relais de protection *m*; ~ **switching** SWG, TELEP commutation sur liaison de réserve *f*, commutation sur liaison de secours *f*, sécurisation *f*; ~ **switching system** TELEP système de commutation de protection *m*, système de commutation sur secours *m*

protective: ~ **bellows** *n pl* PARTS soufflet de protection *m*; ~ **case** *n* PARTS boîtier de protection *m*, capot protecteur *m*; ~ **cover** *n* PARTS boîtier protecteur *m*, capot protecteur *m*; ~ **earth** *n BrE* (*cf protective ground AmE*) POWER, TEST terre de protection *f*; ~ **earthing** *n BrE* (*cf protective grounding AmE*) POWER, TEST mise à la terre de protection *f*; ~ **ground** *n AmE* (*cf protective earth BrE*) POWER, TEST terre de protection *f*; ~ **grounding** *n AmE* (*cf protective earthing BrE*) POWER, TEST mise à la terre de protection *f*; ~ **plastic coating** *n* COMP couche de plastique protectrice *f*; ~ **relay** *n* PARTS relais de protection *m*; ~ **resistor** *n* PARTS résistance de protection *f*; ~ **sheath** *n* ELECT gaine de protection *f*; ~ **shutter** *n* ELECT volet de protection *m*

protector *n* POWER parafoudre *m*

protocol *n* APPL, COMP, DATA, RADIO, SIG protocole *m*; ~ **analysis** DATA analyse de protocole *f*; ~ **class** DATA classe de protocole *f*; ~ **conformance test report** (*PCTR*) DATA, TEST rapport de test de conformité au protocole *m*; ~ **control indicator** DATA indicateur de commande de protocole *m*; ~ **converter** DATA convertisseur de protocole *m*; ~ **data unit** (*PDU*) DATA unité de données de protocole *f*; ~ **discriminator** (*PD*) DATA discriminateur de protocole *m* (*DP*); ~ **emulator** (*PE*) DATA émulateur de protocole *m*;

~ **error** TELEP erreur de protocole *f*; ~ **error class** DATA classe erreur de protocole *f*; ~**-handling software machine** TELEP ML de traitement de protocole *m*; ~ **identifier** (*PRT-ID*) DATA identificateur de protocole *m*; ~ **identifier field** DATA champ d'identificateur de protocole *m*; ~ **implementation conformance statement** (*PICS*) DATA déclaration de conformité d'une instance de protocole *f* (*PICS*), déclaration de conformité de mise en œuvre du protocole *f* (*PICS*); ~ **implementation extra information for testing** (*PIXIT*) DATA informations supplémentaires sur l'instance de protocole destinées au test *f pl* (*PIXIT*); ~ **machine** (*PM*) DATA machine de protocole *f*; ~ **reference model** (*PRM*) DATA modèle de référence de protocole *m*; ~ **unit** (*PU*) DATA unité de protocole *f*

provider *n* APPL, COMP fournisseur *m*, DATA fournisseur *m*, prestataire *m*, TELEP fournisseur *m*; ~**-initiated service** DATA service engendré par le fournisseur *m*; ~**-supported distributed transaction** DATA transaction répartie assurée par le fournisseur *f*

proving: ~ **stand** *n* TEST banc d'essai *m*

provision *n* APPL *of services* fourniture *f*, COMP *condition* prescription *f*, règle *f*, stipulation *f*, fourniture *f*, DATA, TRANS fourniture *f*

PRP *abbr* TELEP (*pulse repetition period*) PRI (*période de répétition des impulsions*), TEST (*producer's risk point*) point du risque du fournisseur *m*

PRR *abbr* (*pulse repetition rate*) TELEP taux de répétition des impulsions *m*

PRT *abbr* RADIO (*primary reference terminal*) terminal de référence principal *m*, TELEP (*pulse repetition time*) TRI (*temps de récurrence des impulsions*)

PRT-ID *abbr* (*protocol identifier*) DATA identificateur de protocole *m*

prune *vt* GEN élaguer

PRV *abbr* (*peak reverse voltage*) POWER, TELEP tension inverse de crête *f*

ps *abbr* (*picosecond*) TEST ps (*picoseconde*)

PS *abbr* DATA, SWG, TELEP, TRANS (*packet switch*) CP (*commutateur de paquets*), (*packet switching*) CP (*commutation de paquets*)

PSAP *abbr* (*presentation-service access point*) DATA, TELEP point d'accès au service de présentation *m*

PSE *abbr* (*packet-switching exchange*) DATA, SWG, TELEP, TRANS centre de commutation de paquets *m*

pseudo: ~**-event** *n* DATA pseudo-événement *m*; ~**-instruction** *n* COMP pseudo-instruction *f*

pseudocode *n* COMP pseudo-code *m*

pseudonoise *n* DATA, TRANS pseudo-bruit *m*; ~ **modulation** TRANS modulation par séquences pseudo-aléatoires *f*

pseudorandom[1] *adj* DATA, ELECT, TRANS, TV pseudo-aléatoire

pseudorandom[2]: ~ **binary sequence** *n* (*PRBS*) DATA séquence binaire pseudo-aléatoire *f*; ~ **bit** *n* DATA bit pseudo-aléatoire *m*; ~ **noise** *n* DATA, TRANS bruit pseudo-aléatoire *m*; ~**-noise code** *n* DATA, TRANS code de bruit pseudo-aléatoire *m*; ~**-sequence generator** *n* DATA générateur de signaux pseudo-aléatoires *m*; ~ **signal** *n* ELECT signal pseudo-aléatoire *m*

pseudoternary: ~ **signal** *n* TELEP signal pseudo-ternaire *m*

PSK *abbr* (*phase-shift keying*) RADIO, SIG MDP (*modulation par déplacement de phase*)

PSN *abbr* DATA (*public switched network*) RPC (*réseau public commuté*), RADIO (*packet satellite network*) réseau à satellite avec commutation de paquets *m*, réseau de transmission de données par paquets et par satellite *m*, SWG, TELEP, TRANS (*packet-switching network*) réseau à commutation de paquets *m*, (*public switched network*) RPC (*réseau public commuté*)

psophometer *n* TELEP psophomètre *m*

psophometric[1] *adj* TELEP psophométrique

psophometric[2]: ~ **electromotive force** *n* (*psophometric emf*) TELEP force électromotrice psophométrique *f* (*f.é.m. psophométrique*); ~ **emf** *n* (*psophometric electromotive force*) TELEP fém psophométrique *f* (*force électromotrice psophométrique*); ~ **power** *n* TELEP puissance psophométrique *f*; ~ **voltage** *n* TELEP tension psophométrique *f*

PSS *abbr* (*power subsystem*) POWER sous-système d'alimentation électrique *m*

PSTN *abbr* (*public switched telephone network*) SWG, TELEP, TRANS RTCP (*réseau téléphonique commuté public*)

psychoacoustic: ~ **model** *n* COMP modèle psychoacoustique *m*

psychrometer *n* TELEP psychromètre *m*

PTM *abbr* (*pulse time modulation*) TELEP, TRANS modulation d'impulsions dans le temps *f*

P-type: ~ **semiconductor** *n* ELECT, PARTS semiconducteur type P *m*

PU *abbr* (*protocol unit*) DATA unité de protocole *f*

public: ~**-address equipment** *n* RADIO sonorisation *f*; ~ **call box** *n* TELEP cabine publique *f*, cabine téléphonique publique *f*; ~ **cordless telephone** *n* TELEP téléphone publique sans fil *m*, téléphone publique sans cordon *m*; ~ **data network** *n* (*PDN*) DATA, TELEP, TRANS réseau public pour données *m* (*RPD*); ~ **data-transmission service** *n* TRANS service public de transmission de données *m*; ~ **dial-up port** *n* TELEP point d'accès public *m*; ~**-domain software** *n* COMP, DATA logiciel du domaine public *m*; ~ **exchange** *n* TELEP central public *m*, commutateur public *m*; ~ **facsimile bureau** *n* TELEP bureau public de télécopie *m*; ~ **facsimile station** *n* TELEP poste public de télécopie *m*; ~ **key** *n* RADIO clé publique *f*; ~**-key encryption** *n* RADIO chiffrement à clé publique *m*; ~ **line** *n* TELEP ligne publique *f*; ~ **mobile network** *n* RADIO, TRANS réseau mobile public *m*; ~ **network** *n* RADIO, TRANS réseau public *m*; ~ **operator** *n* DATA exploitant public *m*, opérateur de télécommunications *m*, TELEP exploitant public *m*, opérateur de télécommunications *m*; ~ **packet-switched data network** *n* (*PPSDN*) DATA réseau de données public avec commutation par paquets *m* (*RDPCP*); ~ **relations** *n* (*PR*) TEST relations publiques *f pl*; ~ **switched network** *n* (*PSN*) DATA, SWG, TELEP, TRANS réseau public commuté *m* (*RPC*); ~ **switched telephone network** *n* (*PSTN*) SWG, TELEP, TRANS réseau téléphonique commuté public *m* (*RTCP*); ~ **telegram service** *n* TELEP service télégraphique public *m*; ~ **telegraph network** *n* TELEP, TRANS réseau télégraphique public *m*; ~ **telephone** *n* TELEP poste téléphonique public *m* (*PTP*); ~ **telephone booth** *n* TELEP poste téléphonique public *m* (*PTP*), publiphone *m*; ~ **telephone exchange** *n BrE* (*cf central office switch AmE*) SWG, TELEP commutateur public *m*; ~ **telephone network** *n* TELEP, TRANS réseau téléphonique public *m*; ~ **wireless telephony** *n* TELEP téléphonie publique sans fil *f*

pull: ~**-down menu** *n* COMP menu déroulant *m*; ~**-down resistor** *n* PARTS résistance de rappel à la masse *f*; ~**-in** *n* POWER enclenchement *m*; ~**-off strength** *n* ELECT force d'arrachement *f*; ~**-out unit** *n* TELEP unité extractible *f*; ~**-relief clamp** *n* PARTS serre-câble *m*; ~**-up resistor** *n* PARTS résistance de rappel à la source *f*

pulled: ~ **surface** *n* PARTS rugosité *f*

pulling *n* TRANS *of cable* tirage *m*; ~ **figure** ELECT entraînement de fréquence *m*; ~ **force** TEST force d'arrachement *f*; ~ **in** TRANS tirage en conduite *m*; ~ **test** PARTS, TEST essai d'arrachement *m*

pulsating: ~ **DC current** *n* POWER courant continu pulsé *m*; ~ **DC voltage** *n* POWER tension continue pulsatoire *f*

pulse[1] *n* GEN impulsion *f*; ~**-address multiple access** (*PAMA*) RADIO accès multiple par impulsions d'adresse *m* (*AIMA*); ~ **amplitude** TRANS amplitude d'une impulsion *f*; ~ **amplitude modulation** (*PAM*) TELEP, TRANS modulation d'impulsions en amplitude *f* (*MIA*); ~ **amplitude modula-**

tion network TELEP, TRANS réseau de modulation d'impulsions en amplitude *m*; **~ base** TELEP base d'impulsion *f*; **~-bisecting relay** PARTS relais diviseur d'impulsions *m*; **~ broadening** ELECT, TELEP, TRANS élargissement d'impulsion *m*, étalement d'impulsion *m*, dispersion d'impulsion *f*; **~ burst** DATA, TRANS rafale d'impulsions *f*; **~ cam disc** *BrE*, **~ cam disk** *AmE* PER APP came d'impulsions *f*; **~ carrier** TRANS train porteur *m*; **~-code modulation** (*PCM*) COMP, DATA, TELEP, TRANS modulation par impulsions codées *f* (*MIC*); **~ coding** DATA, TRANS codage d'impulsions *m*; **~ compression** ELECT compression d'impulsions *f*; **~ control system** SWG, TELEP système à commande par impulsions *m*, système à impulsions *m*; **~-correcting register** SIG répéteur correcteur d'impulsions *m*; **~ correction** SIG, TRANS correction d'impulsions *f*; **~-correction circuit** TRANS circuit de correction d'impulsions *m*; **~-correction unit** TRANS unité de correction d'impulsions *f*; **~ corrector** TELEP filtre d'impulsions *m*; **~ counter** ELECT, SIG compteur d'impulsions *m*; **~-counting technique** RADIO technique de comptage par impulsions *f*; **~ decay time** ELECT, TELEP temps de décroissance *m*, temps de descente *m*, temps de descente d'impulsion *m*; **~ detection** SIG détection d'impulsions *f*; **~ disc** *BrE*, **~ disk** *AmE* PER APP came d'impulsions *f*; **~ dispersion** ELECT, TELEP, TRANS élargissement d'impulsion *m*, étalement d'impulsion *m*, dispersion d'impulsion *f*; **~ distortion** TELEP distorsion d'impulsion *f*; **~ droop** SIG déformation *f*, TELEP chute du palier *f*, déformation *f*, pente du sommet *f*; **~ duration** DATA, ELECT, RADIO, TELEP, TRANS durée d'impulsion *f*; **~ duration modulation** (*PDM*) DATA, RADIO, TELEP, TRANS modulation d'impulsions en durée *f* (*MID*); **~ duty factor** TELEP facteur de durée d'impulsions *m*; **~ echo attenuation** TELEP affaiblissement d'écho *m* (*AECHO*); **~ echo meter** TELEP échomètre à impulsions *m*; **~ echo method** TRANS méthode d'écho d'impulsion *f*; **~ echo return loss** TELEP affaiblissement d'écho *m* (*AECHO*); **~ fall time** TELEP temps de décroissance *m*, temps de descente *m*; **~-forming line** RADIO ligne à retard *f*, ligne formatrice d'impulsions *f*; **~-forming network** (*PFN*) ELECT réseau conformateur d'impulsions *m* (*RCI*); **~ frequency modulation** (*PFM*) TELEP modulation d'impulsions en fréquence *f*; **~ generator** ELECT générateur d'impulsions *m*; **~ interlacing** TELEP, TRANS entrelacement d'impulsions *m*; **~ interleaving** TELEP, TRANS entrelacement des impulsions *m*; **~ interval** TELEP, TRANS espacement entre impulsions *m*, intervalle entre impulsions *m*; **~ interval modulation** (*PIM*) TELEP modulation des intervalles entre impulsions *f*; **~ leading edge** ELECT front avant d'impulsion *m*; **~ length** DATA, ELECT, RADIO, TELEP, TRANS durée d'impulsion *f*; **~ length discriminator** RADIO discriminateur de durée d'impulsion *m*; **~ magnitude** TELEP amplitude d'une impulsion *f*, hauteur d'une impulsion *f*; **~ mask** TELEP, TRANS gabarit d'impulsion *m*; **~ meter** TELEP compteur d'impulsions *m*; **~-modulated radar** RADIO radar à impulsions *m*; **~ modulation** COMP, TELEP, TRANS modulation par impulsions *f*; **~-no-pulse ratio** TELEP rapport d'impulsion-non-impulsion *m*; **~ output power** RADIO puissance en sortie d'impulsions *f*; **~-peak value** TELEP valeur de crête d'impulsion *f*; **~ position** TELEP position d'impulsion *f*; **~ position modulation** (*PPM*) DATA, TELEP, TRANS modulation d'impulsions en position *f* (*MIP*); **~ radar** RADIO radar à impulsions *m*; **~ rate** TELEP taux d'impulsions *m*; **~ ratio** SWG rapport d'impulsions *m*; **~-receiving relay** PARTS relais de réception d'impulsions *m*; **~ recognition** SIG, TRANS reconnaissance d'impulsions *f*; **~ recorder** TELEP, TRANS enregistreur d'impulsions *m*, impulsographe *m*; **~ recurrence frequency** (*PRF*) ELECT, SIG, TELEP, TRANS fréquence de récurrence des impulsions *f* (*FRI*); **~ regeneration** COMP, GEN, SIG, TELEP régénération d'impulsions *f*; **~ regenerator** SIG, TELEP régénérateur d'impulsions *m*; **~ relay** PARTS relais d'impulsions *m*; **~ repeater** GEN répétiteur d'impulsions *m*, TELEP répéteur d'impulsions *m*; **~ repetition** ELECT, SIG, TELEP, TRANS répétition des impulsions *f*; **~ repetition frequency** (*PRF*) ELECT, SIG, TELEP, TRANS fréquence de répétition des impulsions *f* (*FRI*); **~ repetition period** (*PRP*) TELEP période de répétition des impulsions *f* (*PRI*); **~ repetition rate** (*PRR*) TELEP taux de répétition des impulsions *m*; **~ repetition time** (*PRT*) TELEP temps de récurrence des impulsions *m* (*TRI*); **~ response** RADIO, TELEP réponse impulsionnelle *f*, réponse percussionnelle *f*; **~ rise time** TELEP temps de montée *m*; **~ sequence** SIG train d'impulsions *m*; **~ shaper** TELEP conformateur d'impulsions *m*; **~-shaping circuit** ELECT circuit de mise en forme d'impulsions *m*; **~ signal** SIG signal par impulsion *m*; **~ signaling** *AmE*, **~ signalling** *BrE* TELEP signalisation par impulsions *f*; **~ spacing** TELEP, TRANS espacement des impulsions *m*, espacement entre impulsions *m*; **~ spectrum** TELEP, TRANS spectre d'impulsions *m*; **~ spike** TELEP transitoire superposée *f*; **~ spreading** ELECT, TELEP, TRANS élargissement d'impulsion *m*, étalement d'impulsion *m*, dispersion d'impulsion *f*; **~ stretcher** TELEP, TRANS élargisseur d'impulsions *m*; **~ stuffing** TELEP, TRANS justification *f*; **~ time modulation** (*PTM*) TELEP, TRANS modulation d'impulsions dans le temps *f*; **~ top** TELEP plateau d'impulsions *m*; **~ trailing edge** ELECT front arrière d'impulsion *m*; **~ train** TELEP, TRANS train d'impulsions *m*; **~-train relay** PARTS, TELEP, TRANS relais de fin de train d'impulsions *m*; **~-train spectrum** TELEP spectre de train d'impulsions *m*; **~ wave** COMP, TELEP, TRANS onde d'impulsion *f*; **~ widening** ELECT, TELEP, TRANS élargissement d'impulsion *m*; **~ width** (*PW*) DATA, ELECT, RADIO, TELEP, TRANS durée d'impulsion *f*, largeur d'impulsion *f*; **~ width modulation** (*PWM*) TRANS modulation d'impulsions en durée *f* (*MID*)

pulse[2]: **~ out** *vi* GEN, SIG émettre des impulsions

pulsed: **~ beacon** *n* RADIO radiobalise à impulsions *f*; **~ carrier modulation** *n* TELEP modulation d'impulsions *f*, modulation d'une suite d'impulsions *f*; **~ decay time** *n* GEN temps de suppression d'une impulsion *m*; **~ magnetron** *n* ELECT, RADIO magnétron pulsé *m*

pulse-Doppler: **~ radar** *n* RADIO radar Doppler à impulsions *m*, radar par impulsions à effet Doppler *m*

pulser *n* COMP, DATA, TELEP, TRANS générateur d'impulsions *m*

pulses: **~ per second** *phr* (*pps*) RADIO, TELEP impulsions par seconde (*imp/s*)

pulsing *n* TELEP numérotation *f*, émission d'impulsions *f*; **~ contact** TELEP contact de numérotation *m*; **~ key** TELEP touche d'émission d'impulsions *f*

pumping: **~ voltage** *n* POWER tension de pompage *f*

punch[1]: **~-down block** *n* ELECT bloc d'enfichage par impact *m*; **~-downs** *n pl* ELECT perforations *f pl*; **~-in** *n* COMP insertion *f*; **~-out** *n* COMP suppression *f*; **~ point** *n* COMP point d'insertion *m*; **~-through voltage** *n* POWER tension de pénétration *f*

punch[2]: **~ in** *vt* COMP insérer; **~ out** *vt* COMP supprimer

punched: **~ tape** *n* TELEP ruban perforé *m*

puncture *n* POWER perforation *f*

punctured: **~ convolutional code** *n* COMP code de convolution discontinu *m*

PUP *abbr* (*principal processing unit*) COMP processeur principal *m*, unité de traitement principal *f*

pupinize *vt* TELEP, TRANS pupiniser

pure: **~-chance traffic** *n* TELEP trafic de Poisson *m*, trafic de pur hasard *m*, trafic poissonien *m*

purely: **~-resistive circuit** *n* POWER circuit purement résistif *m*

push: **~-button** *n* CONTR bouton-poussoir *m*; **~-button dial** *n* PER APP clavier *m*; **~-button panel** *n* PER APP panneau à boutons-poussoirs *m*; **~-button set** *n* PER APP, TELEP appareil téléphonique à clavier *m*, poste à clavier *m*, poste téléphonique à clavier *m*, téléphone à clavier *m*, téléphone à touches *m*; **~-button telephone** *n* PER APP, TELEP appareil téléphonique à clavier *m*, poste téléphonique à clavier *m*, téléphone à clavier *m*, téléphone à touches *m*; **~-button terminal** *n* COMP borne à poussoir *f*; **~-down list** *n* COMP, DATA liste inversée *f*, liste refoulée *f*; **~-down storage** *n* DATA mémoire à liste inversée *f*; **~-on sneak current fuse** *n* ELECT fusible de courant de fuite enfichable *m*; **~-pull** *n* POWER montage push-pull *m*, montage symétrique *m*; **~-pull amplifier** *n* ELECT amplificateur push-pull *m*; **~-pull arrangement** *n* ELECT montage symétrique *m*; **~-pull circuit** *n* ELECT montage symétrique *m*; **~-pull operation** *n* ELECT fonctionnement en push-pull *m*; **~-to-talk button** *n* RADIO bouton de conversation *m*; **~-up list** *n* COMP, DATA liste directe *f*; **~-up storage** *n* COMP, DATA mémoire à liste directe *f*

pushing: **~ figure** *n* ELECT indice de glissement amont *m*

put[1]: **~-back** *n* TELEP reprise de transmission *f*

put[2]: **~ in touch** *vt* TELEP mettre en relation, mettre en contact; **~ into operation** *vt* GEN, TELEP mettre en service; **~ into orbit** *vt* RADIO mettre sur orbite, placer sur orbite, satelliser; **~ into service** *vt* GEN, TELEP mettre en service; **~ on charge** *vt* POWER mettre en charge; **~ on discharge** *vt* POWER mettre en décharge; **~ on file** *vt* TELEP archiver; **~ on hold** *vt* SWG mettre en garde, TELEP mettre en attente, mettre en garde; **~ out of order** *vt* GEN rendre inutilisable, rendre inopérant; **~ through** *vt* SWG, TELEP connecter directement, mettre en communication; **~ through a call** *vt* TELEP établir une communication; **~ through to** *vt* TELEP mettre en communication avec

PUT *abbr* (*programmable unijunction transistor*) ELECT, PARTS, POWER TUP (*transistor unijonction programmable*)

putting: **~ into operation** *n* GEN, TELEP mise en service *f*; **~ into orbit** *n* RADIO satellisation *f*; **~ into service** *n* GEN, TELEP mise en service *f*

putty[1] *n* PARTS mastic *m*

putty[2] *vt* PARTS mastiquer

PV *abbr* (*parameter value*) TELEP, TEST valeur de paramètre *f*

PVC *abbr* ELECT (*permanent virtual circuit*) circuit virtuel permanent *m*, (*polyvinyl chloride*) PVC (*chlorure de polyvinyle*), SIG, SWG, TRANS (*permanent virtual circuit*) circuit virtuel permanent *m*; **~-jacketed** *adj* ELECT avec gaine PVC

PW *abbr* DATA, ELECT, RADIO, TELEP (*pulse width*) durée d'impulsion *f*, largeur d'impulsion *f*, TRANS (*private wire*) liaison spécialisée *f*, (*pulse width*) durée d'impulsion *f*, largeur d'impulsion *f*

PWM *abbr* (*pulse width modulation*) ELECT, RADIO, TELEP, TRANS MID (*modulation d'impulsions en durée*)

pyroelectricity *n* ELECT, POWER pyroélectricité *f*

Q

Q: ~ **factor** *n* (*quality factor*) POWER facteur Q *m* (*facteur de qualité*)

QAM *abbr* (*quadrature amplitude modulation*) DATA, TRANS MAQ (*modulation d'amplitude en quadrature*)

QASK *abbr* (*quadrature amplitude-shift keying*) RADIO, SIG MDAQ (*modulation par déplacement d'amplitude en quadrature*)

QCIF *abbr* (*quarter common intermediate format*) COMP QFIC (*quart du format intermédiaire commun*)

qdu *abbr* (*quantizing distortion unit*) TELEP qdu *f* (*unité de distorsion de quantification*)

QED *abbr* (*quantum electrodynamics*) ELECT électrodynamique quantique *f*

QIC *abbr* (*quarter-inch cartridge*) TV QIC (*cartouche en quart de pouce*)

Q-information *n* GEN information sur la qualité *f*

Q-meter *n* TEST Q-mètre *m*

QOS *abbr* (*quality of service*) DATA, GEN, TELEP, TEST qualité de service *f*; ~ **reference point** *n* (*QRP*) DATA, GEN, TELEP, TEST point de référence QOS *m*

QPRS *abbr* (*quadrature partial-response system*) DATA système en quadrature avec réponse partielle *m*

QPSK *abbr* (*quadrature phase-shift keying*) RADIO, SIG MDPQ (*modulation par déplacement de phase quadratique*)

QRP *abbr* (*QOS reference point*) DATA, GEN, TELEP, TEST point de référence QOS *m*

quad *n* GEN, TRANS quarte *f*; ~ **cable** TRANS câble à quartes en étoile *m*; ~-**pair cable** TRANS câble à paires câblées en étoile *m*

quadplex *n* ELECT quatre prises modulaires *f pl*, plastron à prise *m*

quadrantal: ~ **error** *n* RADIO erreur quadrantale *f*

quadrature[1]: **in** ~ *adj* POWER en quadrature

quadrature[2]: ~ **amplitude modulation** *n* (*QAM*) DATA, TRANS modulation d'amplitude en quadrature *f* (*MAQ*); ~ **amplitude-shift keying** *n* (*QASK*) RADIO, SIG modulation par déplacement d'amplitude en quadrature *f* (*MDAQ*); ~ **component** *n* RADIO composant en quadrature *m*; ~ **mirror filter** *n* TRANS filtre miroir en quadrature *m*; ~ **partial-response system** *n* DATA système en quadrature avec réponse partielle *m*; ~ **phase-shift keying** *n* (*QPSK*) RADIO, SIG modulation par déplacement de phase quadratique *f* (*MDPQ*)

quadriphase: ~-**shift keying** *n* (*QPSK*) RADIO, SIG modulation par déplacement à quatre états *f* (*MDP4, MDPQ*), modulation par déplacement de phase quadratique *f* (*MDP4, MDPQ*), modulation par déplacement de phase quadrivalente *f* (*MDP4, MDPQ*)

quadruple: ~-**length register** *n* COMP, DATA registre quadrature *m*

qualification: ~ **approval** *n* TEST qualification *f*

qualified: ~ **event** *n* DATA événement qualifié *m*; ~ **name** *n* DATA nom qualifié *m*

qualifier *n* DATA qualificatif *m*

qualimetry *n* TEST qualimétrie *f*

qualitative: ~ **reasoning** *n* DATA raisonnement qualitatif *m*

quality[1] *adj* ELECT qualitatif

quality[2] *n* TELEP *of channel* indice de qualité *m*, TEST qualité *f*; ~ **assessment** APPL, TEST évaluation de la qualité *f*; ~ **assurance** APPL, TEST assurance qualité *f*; ~ **assurance office** APPL, TEST service d'assurance qualité *m*; ~ **assurance ruling** TEST règlement d'assurance qualité *m* (*RAQ*); ~ **audit** APPL, TEST audit de qualité *m*, contrôle de qualité *m*; ~ **certificate** TEST certificat d'assurance qualité *m*; ~ **chain** APPL chaîne de qualité *f*; ~ **circle** TEST cercle de qualité *m*; ~ **control** APPL, TEST contrôle de qualité *m*, maîtrise de la qualité *f*; ~-**control planning** TEST planification de la maîtrise de la qualité *f*; ~ **cost** TEST coût qualité *m*; ~ **degradation** TEST diminution de la qualité *f*; ~ **engineering** TEST ingénierie de la qualité *f*, technique de gestion de la qualité *f*; ~ **factor** (*Q factor*) POWER facteur de qualité *m* (*facteur Q*); ~ **level** TEST niveau de qualité *m*; ~ **loop** TEST boucle de la qualité *f*, spirale de la qualité *f*; ~ **management** APPL, TEST gestion de la qualité *f*; ~ **manual** TEST manuel qualité *m*; ~ **measure** TEST mesure de qualité *f*; ~ **plan** TEST plan qualité *m*; ~ **policy** TEST politique qualité *f*; ~ **program** *AmE*, ~ **programme** *BrE* TEST programme de gestion de la qualité *m*; ~ **requirement** TEST critère de qualité *m*, exigence de qualité *f*; ~ **of service** (*QOS*) DATA, GEN, TELEP, TEST qualité de service *f*; ~ **spiral** TEST boucle de la qualité *f*, spirale de la qualité *f*; ~ **surveillance** TEST surveillance de la qualité *f*; ~ **system** TEST système qualité *m*; ~ **system review** TEST revue de direction du système qualité *f*

quantification *n* COMP, GEN, TELEP, TRANS quantification *f*

quantify *vt* COMP, GEN, TELEP, TRANS quantifier

quantitative: ~ **reasoning** *n* DATA raisonnement quantitatif *m*

quantization *n* COMP, GEN, TELEP, TRANS quantification *f*; ~ **error** TRANS erreur de quantification *f*; ~ **matrix** COMP matrice de quantification *f*; ~ **noise** ELECT, TRANS bruit de quantification *m*; ~ **step** TRANS échelon de quantification *m*

quantize[1]: ~ **value** *n* COMP valeur de quantification *f*

quantize[2] *vt* COMP, GEN, TELEP, TRANS quantifier

quantized: ~ **pulse modulation** *n* TRANS modulation d'impulsions échantillonnée *f*; ~ **value** *n* TRANS valeur quantifiée *f*

quantizer *n* COMP, GEN, TELEP, TRANS quantificateur *m*

quantizing *n* COMP, GEN, TELEP, TRANS quantification *f*; ~ **distortion** TELEP, TRANS distorsion de quantification *f*; ~ **distortion power** TELEP, TRANS puissance de distorsion de quantification *f*; ~ **distortion unit** (*qdu*) TELEP unité de distorsion de quantification *f* (*qdu*); ~ **interval** TELEP intervalle de quantification *m*; ~ **noise** TRANS bruit de quantification *m*

quantum[1]: ~-**noise-limited** *adj* TRANS limité par le bruit quantique

quantum[2] *n* GEN quantum *m*; ~ **efficiency** ELECT, TRANS rendement quantique *m*; ~-**efficiency spectral characteristic** ELECT caractéristique spectrale de rendement quantique *f*; ~ **electrodynamics** (*QED*) ELECT électrodynamique quantique *f*; ~ **electronics** ELECT électronique quantique *f*; ~ **Hall effect** ELECT effet Hall quantique *m*; ~ **Hall resistance** ELECT résistance Hall quantique *f*; ~-**limited operation** ELECT fonctionnement limité par le bruit quantique *m*; ~ **noise** ELECT, TRANS bruit quantique *m*; ~-**noise-limited operation** ELECT fonctionnement limité par le bruit quantique *m*

quarter: ~ **common intermediate format** *n* (*QCIF*) COMP quart du format intermédiaire commun *m* (*QFIC*); ~-**inch cartridge** *n* (*QIC*) TV cartouche quart de pouce *f* (*QIC*); ~ **screen** *n* COMP quart d'écran *m*; ~ **turn** *n* ELECT quart de tour *m*; ~-**wave whip aerial** *n* RADIO antenne fouet quart d'onde *f*; ~-**wave whip antenna** *n* RADIO antenne fouet quart d'onde *f*

quartet *n* DATA quartet *m*
quartz: ~ **filter** *n* ELECT filtre à quartz *m*; ~ **oscillator** *n* ELECT oscillateur à quartz *m*
quasi: ~**-associated mode** *n* TELEP mode quasi-associé *m*; ~**-associated mode of operation** *n* TELEP mode d'exploitation quasi-associé *m*; ~**-associated mode of signaling** *AmE*, ~**-associated mode of signalling** *BrE* *n* TELEP mode de signalisation quasi-associé *m*; ~**-associated signaling** *AmE*, ~**-associated signalling** *BrE* *n* SIG signalisation quasi-associée *f*; ~**-impulsive interference** *n* RADIO brouillage quasi-impulsif *m*; ~**-omnidirectional antenna** *n* RADIO antenne quasi-équidirective *f*, antenne quasi-omnidirective *f*; ~**-peak voltage** *n* RADIO *air mobile* tension de quasi-crête *f*
quench *n* POWER suppresseur *m*; ~ **frequency** RADIO fréquence de découpage *f*
quencher *n* POWER suppresseur *m*
query: ~ **language** *n* DATA langage d'interrogation *m*
queue *n* GEN queue *f*, file d'attente *f*, TELEP, TV file d'attente *f*; ~ **control** TELEP mise en file d'attente *f*; ~ **file** TELEP file d'attente *f*; ~ **place** TELEP position de file d'attente *f*
queueing *n* RADIO mise en file d'attente *f*, mise en garde d'appel *f*, TELEP formation de files d'attente *f*; ~ **delay** TELEP retard dû à la formation de queues *m*; ~ **device** TELEP dispositif d'exclusion *m*, dispositif de mise en attente *m*, mécanisme d'exclusion *m*; ~ **equipment** TELEP équipement de mise en file d'attente *m*; ~ **network** TELEP réseau de files d'attente *m*; ~ **time** TELEP temps de mise en attente *m*
quick: ~**-access memory** *n* COMP mémoire à accès rapide *f*; ~**-acting fuse** *n* PARTS, POWER fusible à action instantanée *m*; ~ **charge** *n* POWER *of battery* charge de biberonnage *f*, charge rapide *f*; ~ **load** *n* COMP chargement rapide *m*; ~ **playback** *n* COMP reproduction rapide *f*
quicktime *n* COMP quicktime *m*
quiescent: ~ **carrier modulation** *n* POWER modulation à interruption de porteuse *f*; ~ **state** *n* POWER position de repos *f*
quiet: ~**-front** *adj* ELECT à faible encombrement
quintet *n* DATA quintet *m*
quit *vi* COMP quitter
quotation: ~ **marks** *n pl* COMP guillemets *m pl*
quote: ~ **marks** *n pl* COMP guillemets *m pl*
quotes *n pl* COMP guillemets *m pl*

R

R[1] *abbr* (*reluctance*) POWER R (*réluctance*)
R[2]: ~ **or T pads** *n pl* TELEP *in telephone extension* compléments de ligne R ou T plage du terminal *m pl*
raceway: ~ **cover** *n* ELECT couverture de chemin *f*
raceways *n* ELECT goulotte *f*
rack[1]: ~**-mounted** *adj* ELECT à câble connecté
rack[2] *n* COMP armoire *f*, râtelier *m*, POWER *for electronic equipment* bâti *m*, TELEP bâti *m*, baie *f*; ~ **beam** TELEP traverse de bâti *f*; ~ **cabling** TELEP câblage de bâti *m*; ~ **earth** *BrE* (*cf rack ground AmE*) POWER terre de bâti *f*; ~ **entry point** TELEP point d'entrée de baie *m*; ~ **ground** *AmE* (*cf rack earth BrE*) POWER terre de bâti *f*; ~ **panel** TELEP panneau de bâti *m*; ~ **power** POWER alimentation de bâti *f*; ~ **power module** TELEP module d'énergie de bâti *m*; ~ **power unit** POWER unité d'alimentation de bâti *f*; ~ **row** TELEP rangée *f*, travée *f*; ~ **spacing** TELEP pas modulaire de bâti *m*; ~ **stanchion** TELEP montant de bâti *m*; ~ **support** TELEP support de bâti *m*; ~ **upright** TELEP montant de bâti *m*
racon *n* (*radar beacon*) RADIO racon *m* (*balise de radiodétection, balise radar, radiophare*)
radar *n* RADIO radar *m*; ~ **aerial** RADIO antenne radar *f*; ~ **altitude control mode** RADIO contrôle automatique de l'altitude par radar de bord *m*; ~ **angel** RADIO ange *m*; ~ **antenna** RADIO antenne radar *f*; ~ **beacon** (*racon*) RADIO balise de radiodétection *f* (*racon*), balise radar *f* (*racon*), radiophare *m* (*racon*); ~ **camouflage** RADIO camouflage radar *m*, camouflage contre la radiodétection *m*; ~ **cross-section** RADIO surface de diffusion *f*, surface équivalente *f*; ~ **dome** RADIO capot d'antenne *m*, radôme *m*; ~ **duplexer** RADIO duplexeur de radar *m*; ~ **echo** RADIO écho radar *m*; ~ **echoing area** RADIO surface de diffusion *f*, surface équivalente *f*; ~ **equation** RADIO équation du radar *f*; ~ **head** RADIO tête radar *f*; ~ **horizon** RADIO horizon radar *m*; ~ **range** RADIO portée radar *f*; ~ **rangefinder** RADIO radiotélémètre *m*; ~ **reflector** RADIO réflecteur radar *m*; ~ **reflector buoy** RADIO balise passive *f*; ~ **relay** RADIO relais passif *m*; ~ **set** RADIO radar *m*; ~ **shadow** RADIO ombre radar *f*; ~ **signal** RADIO signal radar *m*; ~ **warning receiver** RADIO détecteur de radar *m*
radial: ~ **acceleration** *n* COMP accélération radiale *f*; ~**-field cable** *n* GEN câble à champ radial *m*; ~ **leads** *n pl* POWER sorties radiales *f pl*; ~ **transfer** *n* COMP transfert radial *m*
radiance *n* ELECT, TRANS luminance *f*, luminance énergétique *f*
radiant: ~ **efficiency** *n* TRANS rendement énergétique *m*; ~ **energy** *n* ELECT, TRANS énergie rayonnante *f*; ~ **flux** *n* ELECT, TRANS flux énergétique *m*; ~ **intensity** *n* ELECT, TRANS intensité énergétique *f*, intensité de rayonnement *f*; ~ **power** *n* ELECT, TRANS puissance rayonnante *f*; ~ **sensitivity** *n* ELECT sensibilité énergétique *f*
radiate *vt* ELECT, GEN, TRANS rayonner
radiated: ~ **interference** *n* ELECT brouillage transmis par rayonnement *m*; ~ **noise** *n* ELECT bruit transmis par rayonnement *m*; ~ **power** *n* ELECT puissance émise *f*, puissance rayonnée *f*
radiating: ~**-cable communication system** *n* TRANS système de communication par câble rayonnant *m*; ~ **device** *n* RADIO source primaire *f*; ~ **element** *n* RADIO source primaire *f*; ~ **horn** *n* RADIO cornet rayonnant *m*
radiation[1]: ~**-hardened** *adj* PARTS résistant aux effets du rayonnement
radiation[2] *n* ELECT, RADIO rayonnement *m*, rayonnement radioélectrique *m*; ~ **angle** ELECT, TRANS angle de rayonnement *m*, angle de sortie *m*; ~ **counter tube** RADIO tube compteur de rayonnement *m*; ~ **efficiency** RADIO *of antenna* rendement *m*; ~ **hardness** PARTS dureté aux rayonnements *f*, tenue aux radiations *f*; ~ **hardness level** PARTS niveau de dureté aux rayonnements *m*; ~ **mode** ELECT, TRANS mode rayonnant *m*; ~ **pattern** RADIO, TRANS diagramme de rayonnement *m*; ~ **sensitivity** RADIO sensibilité au rayonnement *f*; ~ **zone** RADIO zone de rayonnement *f*
radiator: ~ **area** *n* RADIO surface dissipative *f*
radical *adj* ELECT percutant
radio[1] *adj* RADIO radio; ~**-frequency** RADIO en radiofréquence
radio[2] *n* RADIO radioélectrique *f*, radio *f*; ~ **alarm** PER APP radio-réveil *m*; ~ **altimeter** RADIO, TELEP altimètre radioélectrique *m*, radioaltimètre *m*; ~ **amateur** RADIO radioamateur *m*; ~ **astronomical antenna** RADIO antenne radioastronomique *f*; ~ **base station** RADIO *mobile* station de base radio *f*; ~ **base station connection unit** RADIO unité de raccordement des bornes *f* (*URB*); ~ **beacon** RADIO radiobalise *f*, radiophare *m* (*racon*); ~ **blackout** RADIO interruption totale des radiocommunications *f*, silence radio *m*; ~ **broadcaster** RADIO radiodiffuseur *m*; ~ **broadcasting** RADIO radiodiffusion *f*, radiodiffusion visuelle *f*; ~ **button** COMP case d'option *f*; ~ **channel** RADIO canal radio *m*, radiocanal *m*, voie radio *f*; ~ **channel spacing** RADIO espacement des bandes de radio fréquences *m*, espacement des canaux radio *m*; ~ **circuit** RADIO circuit de radiocommunication *m*; ~ **common carrier** RADIO sociétés exploitantes de radiocommunications *f pl*; ~ **compass** RADIO radiocompas *m*; ~ **congestion** RADIO encombrement radio *m*; ~ **control** RADIO radiocommande *f*; ~ **control system** RADIO installation de commande par radio *f*, système de radiocommande *m*; ~ **coverage** RADIO couverture radioélectrique *f*; ~ **determination** RADIO repérage par radio *m*, radiorepérage *m*; ~**-determination satellite service** (*RDSS*) RADIO service de radiorepérage par satellite *m*; ~ **equipment maintenance** POWER, RADIO maintenance des matériels de radiocommunication *f*; ~ **exchange** RADIO autocommutateur radiotéléphonique *m*; ~ **facsimile** RADIO, TELEP radiotélécopie *f*; ~ **fade** RADIO affaiblissement *m*, évanouissement *m*; ~ **fade-out** RADIO évanouissement brusque *m*; ~ **frequency** (*RF*) ELECT, RADIO haute fréquence *f* (*HF*), radiofréquence *f* (*RF*), TELEP haute fréquence *f* (*HF*); ~**-frequency cable** RADIO câble haute fréquence *m*; ~**-frequency channel** RADIO canal radioélectrique *m*, canal radioélectronique *m*, radiocanal *m*; ~**-frequency disturbance** RADIO parasite *m*, perturbation radioélectrique *f*; ~**-frequency input power** RADIO puissance d'entrée radiofréquence *f*, puissance d'excitation *f*; ~**-frequency interference** (*RFI*) RADIO brouillage *m*, brouillage intentionnel *m*, interférence radioélectrique *f*, perturbation radioélectrique *f*, radioélectrique *f*; ~**-frequency protection ratio** RADIO rapport de protection en radiofréquence *m*; ~**-frequency radiation** RADIO rayonnement *m*; ~**-frequency scanner** RADIO récepteur à balayage *m*; ~ **guidance** RADIO radioguidage *m*; ~ **ham** RADIO radioamateur *m*; ~ **homing** RADIO radioralliement *m*; ~ **horizon** RADIO horizon radioélectrique *m*, horizon radio *m*; ~ **horizon distance** RADIO distance de l'horizon radioélectrique *f*; ~ **in the loop**

(*RITL*) RADIO radio dans le réseau *f*; ~ **influence voltage** DATA tension parasite *f*; ~ **jamming** RADIO brouillage des radiocommunications *m*, perturbation radioélectrique *f*; ~ **link** RADIO liaison radioélectrique *f*, liaison radio *f*; ~ **location** RADIO localisation par radio *f*, radiorepérage *m*; ~-**location satellite service** RADIO service de radiolocalisation par satellite *m*; ~ **mast** RADIO antenne *f* (*ant.*), pylône *m*; ~ **navigation** RADIO radionavigation *f*; ~ **network** RADIO réseau radio *m*; ~ **noise** RADIO bruit radioélectrique *m*; ~ **path** RADIO trajet radioélectronique *m*, voie radio *f*; ~ **program** *AmE*, ~ **programme** *BrE* RADIO programme radiophonique *m*; ~ **propagation** RADIO propagation radioélectrique *f*; ~-**propagation prediction** RADIO prévision de propagation radioélectronique *f*; ~ **range station** RADIO station radiophare *f*; ~ **rangefinding** RADIO radiotélémétrie *f*; ~ **receiver** PER APP, RADIO, TELEP récepteur *m*, récepteur radioélectrique *m*; ~ **relay station** RADIO station radio relais *f*; ~ **relay system** RADIO, TRANS faisceau hertzien *m*; ~ **route** RADIO voie radioélectronique *f*; ~ **shadow** RADIO ombre radio *f*; ~ **silence** RADIO silence radio *m*; ~ **spectrum** ELECT spectre radioélectrique *m*; ~ **station** RADIO station radio *f*, station radioélectrique *f*; ~ **subsystem** RADIO sous-système radio *m*; ~ **survey** RADIO relevé radioélectrique *m*; ~ **system** RADIO système de radiocommunications *m*; ~ **teleprinter** RADIO radiotéléimprimeur *m*; ~ **theodolite** RADIO radiothéodolite *m*; ~ **tracking** RADIO poursuite radioélectrique *f*; ~ **transmitter** RADIO émetteur radio *m*; ~ **unit** RADIO émetteur-récepteur *m*; ~ **warning** RADIO avertissement radio *m*, avertissement radioélectrique *m*; ~ **wave** ELECT, RADIO onde hertzienne *f*, onde radioélectrique *f*; ~ **wind** RADIO radiovent *m*

radiocommunication *n* RADIO radiocommunication *f*, communication radioélectrique *f*; ~ **relay** POWER, RADIO relais de radiocommunication *m*; ~ **service** RADIO service de radiocommunication *m*

radiogoniometer *n* RADIO chercheur radiogoniométrique *m*, radiogoniomètre *m*

radiographic: ~ **contrast** *n* RADIO contraste radiographique *m*

radiopager *n* DATA, RADIO radiopage *m*

radiopaging *n* DATA, RADIO, TELEP, TRANS radioappel *m*, radiomessagerie *f* (*RM*), radiomessagerie unilatérale *f* (*RMU*), radiorecherche *f*; ~ **service** RADIO service radioélectrique d'appel unidirectionnel *m*; ~ **system** RADIO système de radiomessagerie *m*, système radioélectrique d'appel *m*

radiophone *n* RADIO, TELEP radiotéléphone *m*

radiosity *n* COMP radiosité *f*

radiosonde *n* RADIO radiosonde *f*

radiotelegraphy *n* RADIO, TELEP radiotélégraphie *f*

radiotelephone *n* RADIO, TELEP radiotéléphone *m*

radiotelephony *n* RADIO, TELEP radiophonie *f*, radiotéléphonie *f*

radioteletype: ~ **q.v.** *n* (*RTTY*) RADIO radiotéléimprimeur *m* (*RTT, RTTY*)

radome *n* RADIO capot d'antenne *m*, radôme *m*

RAG *abbr* (*ring again*) PER APP, TELEP RAG (*rappel automatique*)

RAI *abbr* (*remote alarm indication*) TEST indication d'alarme distante *f*, télésignalisation d'alarme *f*

railings *n pl* ELECT brouillage en palissade *m*

railroad: ~ **radio system** *n AmE* (*cf railway radio system BrE*) RADIO système de radiocommunications pour les chemins de fer *m*, système radio ferroviaire *m*; ~ **signaling system** *n AmE* (*cf railway signalling system BrE*) CONTR, SIG système de signalisation ferroviaire *m*; ~ **telecommunications system** *n AmE* (*cf railway telecommunications system BrE*) RADIO système de télécommunications pour chemins de fer *m*; ~ **telephony** *n AmE* (*cf railway telephony BrE*) RADIO, TELEP téléphonie ferroviaire *f*

railway: ~ **radio system** *n BrE* (*cf railroad radio system AmE*) RADIO système de radiocommunications pour les chemins de fer *m*, système radio ferroviaire *m*; ~ **signalling system** *n BrE* (*cf railroad signaling system AmE*) CONTR, SIG système de signalisation ferroviaire *m*; ~ **telecommunications system** *n BrE* (*cf railroad telecommunications system AmE*) RADIO système de télécommunications pour chemins de fer *m*; ~ **telephony** *n BrE* RADIO, TELEP téléphonie ferroviaire *f*

rain: ~ **cell** *n* RADIO cellule de pluie *f*; ~ **clutter** *n* RADIO *on radar* retour de pluie *m*; ~ **fading** *n* RADIO évanouissement dû à la pluie *m*; ~ **model** *n* RADIO modèle de précipitations *m*

rainfall: ~ **data** *n pl* RADIO statistiques pluviométriques *f pl*; ~ **rate** *n* RADIO intensité de precipitation *f*, intensité de pluie *f*

raise: ~ **an interrupt** *vt* ELECT remettre à un l'indicateur d'interruption

RAM *abbr* (*random-access memory*) COMP, DATA RAM (*mémoire à accès direct, mémoire vive*); ~ **cartridge** *n* DATA cartouche RAM *f*; ~ **disk** *n* DATA disque RAM *m*, disque viruel en mémoire vive *m*

Raman: ~ **scattering** *n* RADIO, TRANS diffusion de Raman *f*

ramark *n* RADIO ramark *m*

ramp *n* TELEP rampe *f*; ~ **function** TELEP fonction rampe *f*; ~ **response** CONTR réponse à la fonction rampe *f*; ~ **voltage** POWER tension à croissance linéaire *f*, tension en rampe *f*

RAN *abbr* (*recorded announcement*) GEN ME (*message enregistré*), TELEP annonce enregistrée *f*, annonce parlée *f* (*Can*), ME (*message enregistré*)

RAND *n* TELEP élément de numéro aléatoire *m*

random[1] *adj* GEN aléatoire, au hasard

random[2]: ~ **access** *n* COMP accès aléatoire *m*; ~ **access device** *n* COMP dispositif à accès direct *m*; ~-**access memory** *n* (*RAM*) COMP, DATA mémoire à accès direct *f* (*RAM*), mémoire vive *f* (*RAM*); ~ **allotment** *n* TELEP affectation aléatoire *f*, attribution aléatoire *f*, sélection aléatoire *f*; ~ **allotter** *n* TELEP distributeur de priorité aléatoire *m*; ~ **block file manager** *n* COMP random block file manager *m*; ~ **error** *n* DATA, TELEP, TEST erreur aléatoire *f*; ~ **excitation** *n* ELECT excitation aléatoire *f*; ~ **factor** *n* COMP facteur aléatoire *m*; ~ **failure** *n* TEST défaillance aléatoire *f*; ~ **field** *n* RADIO champ aléatoire *m*; ~ **metering** *n* TELEP comptage aléatoire *m*, comptage au hasard *m*, taxation par impulsion périodique *f*; ~ **multiple access** *n* COMP accès multiple aléatoire *m*; ~ **noise** *n* ELECT, RADIO, TELEP bruit aléatoire *m*, bruit erratique *m*; ~ **number element** *n* TELEP élément de numéro aléatoire *m*; ~-**number generation** *n* TELEP génération de nombres aléatoires *f*; ~ **process** *n* ELECT processus aléatoire *m*; ~ **pulse** *n* ELECT impulsion aléatoire *f*; ~ **repetitive metering** *n* TELEP comptage multiple aléatoire *m*; ~ **sampling** *n* ELECT échantillonnage aléatoire *m*; ~ **selection** *n* TELEP sélection aléatoire *f*; ~ **signal** *n* ELECT signal aléatoire *m*; ~ **traffic** *n* TELEP trafic aléatoire *m*, trafic de Poisson *m*, trafic de pur hasard *m*, trafic poissonien *m*; ~ **value** *n* COMP valeur aléatoire *f*; ~ **variable** *n* GEN variable aléatoire *f*

randomization *n* GEN randomisation *f*

randomize *vt* GEN randomiser

range *n* GEN portée *f*, étendue *f*, gamme *f*, zone *f*, ~-**amplitude display** RADIO indicateur d'amplitude et de distance *m*; ~ **of audibility** TELEP étendue d'audibilité *f*, étendue de sensibilité auditive *f*, zone d'audibilité *f*; ~-**bearing display** RADIO indicateur d'amplitude et d'azimut *m*; ~ **chart** TEST carte de l'étendue *f*; ~ **discrimination** RADIO pouvoir séparateur radial *m*; ~ **gate** RADIO porte de distance *f*; ~-**height indicator** RADIO indicateur de hauteur et de distance *m*; ~ **marker** RADIO marque de distance *f*, repère de distance *m*; ~ **of regulation** GEN plage de régulation *f*; ~ **resolution** RADIO pouvoir séparateur radial *m*, pouvoir séparateur en distance *m*; ~ **search** RADIO recherche en distance *f*

rangefinding *n* RADIO télémétrie *f*

ranging *n* RADIO mesure de distance *f*

rank: ~ **of selectors** *n* TELEP étage de sélecteurs *m*

rapid: ~**-access loop** *n* DATA boucle d'accès rapide *f*; ~ **charge** *n* POWER *of battery* charge de biberonnage *f*, charge rapide *f*; ~ **mode** *n* DATA *of scanner* mode rapide *m*

rare: ~ **gas** *n* CLIM gaz rare *m*; ~ **gas tube** *n* TEST tube de gaz rare *m*

rarefaction *n* RADIO décompression *f*, raréfaction *f*

raster *n* ELECT trame *f*, RADIO quadrillage *m*, trame *f*; ~ **burn** ELECT brûlure de mire *f*

rasterize *vt* COMP convertir du vectoriel en bitmap, convertir en images point à point

rat: ~ **race** *n* RADIO anneau hybride *m*

ratchet: ~ **control** *n* ELECT contrôle par cliquet *m*

rate *n* APPL taxe *f*, GEN cadence *f*, taux *m*, vitesse *f*, TELEP taxe *f*; ~ **1/2 Golay code** DATA code Golay de rendement 1/2 *m*; ~ **of approach** TEST vitesse de convergence *f*; ~ **of extension** TELEP taux d'extension *m*; ~ **of increase** GEN taux de croissance *m*; ~ **of interruption** TELEP fréquence d'interruption *f*; ~**-of-change relay** PARTS relais à taux de variation *m*; ~ **parameter** COMP paramètre de vitesse *m*; ~ **timeband** TELEP tranche horaire *f*

rated[1] *adj* POWER nominal

rated[2]: ~ **burden** *n* PARTS valeur nominale de la puissance absorbée *f*; ~ **current** *n* POWER courant assigné *m*, courant nominal *m*; ~ **frequency** *n* POWER fréquence nominale *f*; ~ **output level** *n* COMP, ELECT niveau de sortie nominal *m*; ~ **output power** *n* POWER puissance de sortie nominale *f*; ~ **power** *n* POWER puissance nominale *f*; ~ **value** *n* POWER valeur assignée *f*; ~ **voltage** *n* POWER tension nominale *f*, tension assignée *f*

rating *n* POWER régime nominal *m*, puissance nominale *f*, TELEP, TEST, TRANS indice *m*; ~ **plate** POWER plaque signalétique *f*

ratio *n* GEN débit *m*, rapport *m*, taux *m*; ~ **circuit** ELECT circuit à rapport de résistances *m*; ~ **of compression** TELEP taux de compression *m*; ~ **of expansion** TELEP taux d'extension *m*; ~ **of transformation** PARTS rapport de transformation *m*

ratioless: ~ **circuit** *n* ELECT circuit sans diviseur *m*, circuit à précharge *m*

rational: ~ **subgroup** *n* TEST sous-groupe rationnel *m*

raw: ~ **data** *n pl* DATA données brutes *f pl*; ~ **video display** *n* RADIO image brute d'indicateur *f*

rawin *n* RADIO radiovent *m*; ~ **flight** RADIO observation de radiovent *f*; ~ **observation** RADIO observation de radiovent *f*; ~ **sounding** RADIO observation de radiovent *f*

ray *n* COMP, GEN rayon *m*, RADIO trajet radioélectrique *m*, TRANS rayon optique *m*, rayon lumineux *m*; ~ **optics** ELECT, TRANS optique des rayonnements *f*, optique géometrique *f*; ~**-path transmission loss** TRANS affaiblissement de transmission pour un trajet radioélectrique *m*; ~ **tracing** COMP suivi de rayons *m*, lancer de rayons *m*

Rayleigh: ~ **channel** *n* DATA, TELEP, TRANS canal de Rayleigh *m*; ~ **fading** *n* DATA, TELEP, TRANS évanouissement de Rayleigh *m*; ~ **scattering** *n* DATA, TELEP, TRANS diffusion de Rayleigh *f*

razor: ~ **blade** *n* ELECT lame de rasoir *f*

RBE *abbr* (*remote batch entry*) DATA soumission par lots à distance *f*, télésoumission de lots *f*, télésoumission de travaux par lots *f*

RBER *abbr* (*residual bit error ratio*) COMP, DATA, RADIO TEBR (*taux d'erreurs binaires résiduelles*)

RBV *abbr* (*return-beam vidicon*) TELEP vidicon à retour du faisceau *m*

RC *abbr* ELECT (*resistance-capacitance*) RC (*résistance-condensateur*), TELEP (*regional center AmE*, *regional centre BrE*) central régional *m*, centre régional *m*, (*réactivation du compteur*), (*redrive counter*) RC (*réactivation du compteur*); ~ **coupling** *n* ELECT couplage RC *m*

RCD: ~ **circuit** *n* ELECT circuit RCD *m*

RCP *abbr* (*restoration control point*) TELEP centre de commande de rétablissement du service *m*

RCU *abbr* CONTR (*remote control unit*) bloc de télécommande *m*, unité de gestion à distance *f*, unité de télécommande *f*, unité de télégestion *f*, RADIO (*remote concentration unit*) concentrateur satellite numérique *m*

rcvr *abbr* (*receiver*) DATA, PER APP, RADIO, TELEP récepteur *m*

RD *abbr* (*receive data*) DATA signal RD *m* (*signal de réception de données*)

RDA *abbr* (*remote database access*) DATA accès à des bases de données distantes *m*

R-DAT *abbr* (*rotary-head digital audio tape*) COMP, DATA R-DAT (*bande audio-numérique à tête rotative*)

RDB *abbr* (*relational database*) COMP, DATA BDR (*base de données relationnelles*)

RDSS *abbr* (*radio-determination satellite service*) RADIO service de radiorepérage par satellite *m*

RDU *abbr* (*remote data unit*) DATA équipement de télétraitement de l'information *m*

reach *vt* TELEP *subscriber* joindre

reactance *n* POWER réactance *f*; ~ **relay** PARTS relais de réactance *m*

reaction: ~ **time** *n* GEN temps de réaction *m*

reactivate *vt* DATA *scanner* redémarrer

reactivation *n* DATA redémarrage *m*

reactive: ~ **attenuator** *n* RADIO affaiblisseur réactif *m*, atténuateur réactif *m*; ~ **current** *n* POWER courant réactif *m*; ~ **load** *n* ELECT charge réactive *f*; ~ **power** *n* POWER puissance réactive *f*

read[1]: ~ **back** *n* TELEP relisez *m*; ~**-back check** *n* COMP contrôle par relecture *m*; ~ **head** *n* PARTS tête de lecture *f*; ~**-only memory** *n* (*ROM*) COMP, DATA mémoire fixe *f* (*ROM*), mémoire inaltérable *f* (*ROM*), mémoire morte *f* (*MEM*); ~**-only storage** *n* COMP, DATA mémoire fixe *f* (*ROM*), mémoire inaltérable *f* (*ROM*), mémoire morte *f* (*MEM*); ~**-out** *n* COMP, GEN affichage *m*, lecture *f*; ~**-write head** *n* PARTS tête de lecture-écriture *f*; ~**-write memory** *n* COMP, DATA mémoire vive *f* (*RAM*)

read[2]: ~ **out** *vt* COMP afficher

readable *adj* GEN lisible

reader *n* COMP, GEN lecteur *m*

reading *n* COMP, GEN lecture *f*, TELEP relevé *m*; ~ **error** COMP erreur de lecture *f*; ~ **gun** ELECT canon de lecture *m*; ~ **speed** ELECT vitesse de lecture *f*; ~ **wire** TELEP fil de lecture *m*

readjust *vt* GEN régler, TEST remettre au point

ready[1]: ~ **for data** *adj* TRANS prêt pour la transmission de données; ~ **for sending** *adj* DATA prêt à émettre; ~ **on line** *adj* DATA prêt à fonctionner en ligne; ~ **stand by** *adj* DATA prêt à fonctionner en secours

ready[2]: ~ **indication** *n* TELEP indication prêt *f*; ~ **indicator** *n* TELEP indicateur prêt *m*; ~**-to-change request** *n* DATA demande d'acceptation de changement *f*; ~**-to-commit state** *n* DATA état prêt à l'engagement *m*

real[1]: ~**-time** *adj* COMP, DATA, TELEP, TEST temps réel

real[2]: **in** ~ **time** *adv* COMP, DATA, TELEP, TEST en temps réel

real[3]: ~ **address** *n* COMP adresse réelle *f*; ~ **end system** *n* DATA système d'extrémité réel *m*; ~ **file store** *n* DATA mémoire-fichier réelle *f*; ~ **open system** *n* DATA système ouvert réel *m*; ~ **storage** *n* DATA mémoire réelle *f*; ~ **subnetwork** *n* DATA sous-réseau réel *m*; ~ **system** *n* DATA système réel *m*; ~ **time** *n* COMP, DATA, TELEP, TEST temps réel *m*; ~**-time control** *n* COMP commande en temps réel *f*; ~**-time conversion facility** *n* DATA unité de conversion en temps réel *f*; **~-time editing** *n* COMP édition en temps réel *f*; **~-time effect** *n* COMP effet en temps réel *m*; **~-time expert system** *n* SWG système expert-temps réel *m*; **~-time recording** *n* COMP enregistrement en temps réel *m*; **~-time transmission** *n* COMP transmission en temps réel *f*; **~-time video** *n* COMP, TV vidéo en temps réel *f*; **~-time writing** *n* COMP écriture en temps réel *f*; **~ type** *n* DATA type réel *m*

realign *vt* GEN, TEST remettre au point

realignment *n* GEN, TEST remise au point *f*

reallocated *adj* GEN réaffecté

ream *vt* PARTS aléser

reaming *n* PARTS alésage *m*

reanswer: **~ signal** *n* SIG signal de nouvelle réponse *m*

rear[1]: **~-view** *adj* TELEP vu de dos, vu de l'arrière

rear[2]: **~ panel** *n* COMP panneau arrière *m*; **~ row** *n* TELEP rangée arrière *f*

rearrangeable: **~ non-blocking network** *n* TRANS réseau à mailles d'entraide sans blocage *m*

reasonableness: **~ check tables** *n pl* TELEP tableaux de contrôle de vraisemblance *m pl*

reasoning *n* DATA raisonnement *m*

reassembling *n* TELEP réassemblage *m*

reassign *n* RADIO réaffectation *f*

reassigned *adj* RADIO réaffecté

reassignment: **~ procedure** *n* RADIO procédure de réaffectation *f*

reattempt *n* TELEP tentative répétée *f*

rebalancing *n* GEN rééquilibrage *m*

Rebecca-Eureka: **~ system** *n* RADIO radiotélémètre à répondeur *m*

reboot[1] *n* COMP, DATA réamorçage *m*, rechargement *m*, redémarrage *m*, relancement *m*, relance *f*

reboot[2] *vt* COMP, DATA rebooter, recharger, redémarrer, relancer

rebooting *n* COMP, DATA redémarrage *m*

rebroadcast[1] *n* RADIO, TRANS, TV rediffusion *f*, retransmission *f*

rebroadcast[2] *vt* RADIO, TRANS, TV rediffuser, retransmettre

rebroadcasting *n* RADIO, TRANS, TV rediffusion *f*, retransmission *f*; **~ transmitter** RADIO, TELEP, TRANS, TV réémetteur *m*

rec. *abbr* (*recreational*) DATA *Internet* de loisirs

REC *abbr* DATA (*receive*) réception *f*, RADIO (*frequency record*) ENR (*enregistrement de fréquence*)

recalibrate *n* GEN réétalonner

recalibration *n* GEN réétalonnage *m*

recall[1] *n* COMP, TELEP rappel *m*; **~ button** TELEP touche de rappel *f*; **~ edit buffer** COMP rappel du tampon d'édition *m*

recall[2] *vt* COMP, TELEP rappeler

receipt: **~ notification** *n* (*RN*) DATA, TELEP avis de réception *m*, notification de réception *f*

receive[1]: **~-ready** *adj* (*RR*) TELEP prêt à recevoir

receive[2] *n* (*REC*) DATA réception *f*; **~ aerial** RADIO antenne réceptrice *f*, antenne de réception *f*; **~-amplified handset telephone** TELEP poste téléphonique à écoute amplifiée sur combiné *m*; **~-amplified telephone** TELEP poste téléphonique à écoute amplifiée *m*; **~ antenna** RADIO antenne réceptrice *f*, antenne de réception *f*; **~ buffer** COMP tampon de réception *m*; **~ channel** COMP, TRANS canal de réception *m*; **~ characteristic** RADIO caractéristique de réception *f*; **~ conditions** *n pl* COMP conditions de réception *f pl*; **~ data** (*RD*) DATA signal de réception de données *m* (*signal RD*); **~-data lamp** TELEP voyant de réception de données *m*; **~ direction** (*RX*) TEST, TRANS sens réception *m*; **~ echo loss** TELEP affaiblissement de l'écho à la réception *m*; **~ end echo** TELEP écho à la réception *m*; **~ event** DATA événement de réception *m*; **~ fiber-optic terminal device** *AmE*, **~ fibre-optic terminal device** *BrE* TRANS récepteur à fibre optique *m*, récepteur optoélectronique *m*; **~-frame synchronization procedure** (*RFS*) TRANS procédure de synchronisation de la trame de réception *f*; **~ gain** RADIO gain à la réception *m*; **~ lamp** TELEP voyant de réception de données *m*; **~ loss** TELEP affaiblissement à la réception *m*; **~ machine** TRANS machine réceptrice *f*; **~-not-ready frame** TELEP trame non prête à recevoir *f*; **~-ready frame** TELEP trame prête à recevoir *f*

receive[3] *vt* TELEP recevoir, recueillir; **~ holding current** PARTS recevoir un courant de maintien

received: **~ data** *n pl* DATA réception des données *f*

receiver *n* DATA, PER APP, RADIO, TELEP (*rcvr*, *RX*) destinataire *m*, récepteur *m*; **~ cap** PER APP, TELEP pavillon *m*, pavillon de récepteur *m*; **~ case** PER APP, TELEP boîtier d'écouteur *m*, boîtier de récepteur *m*; **~ distributor** RADIO distributeur réception *m*, récepteur-distributeur *m*; **~ diversity** RADIO diversité en réception *f*; **~ headset** TELEP casque d'écoute *m*; **~ inset** PER APP, SIG, TELEP écouteur téléphonique *m*, capsule d'écouteur *f*, capsule réceptrice *f*; **~ multicoupler** RADIO multicoupleur de récepteurs *m*; **~-noise figure** RADIO facteur de bruit du récepteur *m*; **~ radiation** RADIO rayonnement d'un récepteur *m*; **~ rejection** RADIO affaiblissement d'un récepteur *m*; **~ sensitivity** (*RX sens*) RADIO sensibilité du récepteur *f*; **~ station** TV station de réception *f*; **~ threshold level** RADIO seuil du récepteur *m*; **~ transfer time** TELEP temps de transfert à la réception *m*; **~-transmitter isolation** RADIO découpage réception-émission *m*

receiving: **~ aerial** *n* RADIO antenne de réception *f*; **~ antenna** *n* RADIO antenne de réception *f*; **~ channel** *n* COMP canal de réception *m*; **~ device** *n* COMP, TELEP appareil récepteur *m*, dispositif récepteur *m*; **~ inspection** *n* TEST *quality* contrôle d'entrée *m*; **~ mode** *n* RADIO mode réception *m*; **~ objective loudness rating** *n* (*ROLR*) TELEP équivalent objectif pour la sonie à la réception *m*; **~ perforator** *n* TELEP récepteur-perforateur *m*; **~ sensitivity** *n* TELEP *of telephone system* efficacité à la réception *f*; **~ terminal** *n* TRANS extrémité réceptrice *f*

reception *n* COMP, RADIO réception *f*; **~ area** RADIO zone de réception *f*; **~ channel** COMP canal de réception *m*; **~ conditions** *n pl* COMP conditions de réception *f pl*; **~ congestion** TELEP encombrement de réception *m*; **~ definition** TELEP définition à la réception *f*

recess: **~-mounted** *adj* GEN monté en retrait

recessed: **~ access brackets** *n pl* ELECT cornières de montage en retrait *f pl*

rechargeable: **~ battery** *n* POWER accumulateur rechargeable *m*, batterie rechargeable *f*, pile rechargeable *f*

recipient *n* DATA, GEN destinataire *m*; **~ identification code** DATA, TELEP code d'identification du destinataire *m*; **~ reference** DATA référence du destinataire *f*; **~'s reference password** COMP, DATA mot de passe de référence du destinataire *m*; **~-reference qualifier** DATA qualificatif de référence du destinataire *m*

reciprocal: **~ bearing** *n* RADIO azimut inverse *m*; **~ circuit** *n* ELECT circuit réciproque *m*; **~ mixing** *n* RADIO mélange réciproque *m*

reciprocity *n* ELECT réciprocité *f*; **~ theorem** POWER théorème de réciprocité *m*

reclock *vt* ELECT resynchroniser

reclocking *n* ELECT resynchronisation *f*

recognition *n* APPL, COMP, CONTR, DATA, SIG, TELEP, TRANS *of signal* reconnaissance *f*; **~ circuit** ELECT circuit d'identification *m*, circuit de reconnaissance *m*, circuit identificateur *m*

recognized: **~ operating agency** *n* (*ROA*) APPL, DATA exploitation reconnue *f* (*ER*); **~ private operating agency** *n* (*RPOA*) APPL, DATA exploitation privée reconnue *f* (*EPR*); **~ private operation** *n* APPL, DATA exploitation privée reconnue *f* (*EPR*)

recombination *n* PARTS recombinaison *f*; **~ rate** PARTS taux de recombinaison *m*

recombining *n* TELEP recombinaison *f*
recommendation: ~ **indicator** *n* DATA indicateur de recommandation *m*
reconditioned: ~ **carrier reception** *n* GEN réception avec régénération de porteuse *f*
reconfiguration *n* COMP reconfiguration *f*
reconfiguring *n* COMP reconfiguration *f*
reconnect *vt* TELEP repriser
reconnection *n* GEN reconnexion *f*, TELEP reprise *f*
reconstructed: ~ **picture** *n* COMP image reconstruite *f*; ~ **sample** *n* TELEP échantillon reconstitué *m*
record *n* DATA enregistrement *m*; ~ **call** TELEP appel annoté *m*; ~ **cancel** COMP annulation d'enregistrement *f*; ~ **desk** TELEP pupitre d'annotateur *m*, table d'annotateur *f*; ~ **line** TELEP ligne d'annotateur *f*; ~**-line relay set** TELEP plaque de relais de ligne d'annotateur *f*; ~ **stand-by** COMP attente d'enregistrement *f*; ~ **tone** TELEP tonalité d'enregistrement *f*
recordable *adj* COMP, TELEP enregistrable
recorded[1] *adj* RADIO *broadcast* en différé, TELEP enregistré, TV *broadcast* en différé
recorded[2]: ~ **announcement** *n* (*RAN*) GEN message enregistré *m* (*ME*), TELEP annonce enregistrée *f*, annonce parlée *f* (*Can*), message enregistré *m* (*Can*) (*ME*); ~ **announcement machine** *n* PER APP, TELEP appareil à annonce enregistrée *m*, répondeur *m*, répondeur téléphonique *m*; ~ **announcement trunk** *n* TELEP circuit de messages enregistrés *m*; ~ **broadcast** *n* RADIO, TV radiodiffusion en différé *f*; ~ **message** *n* TELEP annonce enregistrée *f*, message enregistré *m* (*Can*) (*ME*); ~ **public information service** *n* TELEP service public d'informations enregistrées *m*; ~ **service** *n* TELEP service d'informations enregistrées *m*
recorder *n* APPL, PER APP enregistreur *m*
recording *n* PER APP enregistrement *m*, TELEP enregistrable; ~ **center** *AmE see recording centre BrE*; ~ **central office** TELEP système centralisé d'enregistrement *m*; ~ **centre** *BrE* SWG, TELEP centre d'enregistrement *m* (*CE*); ~ **channel** TELEP canal d'enregistrement *m*; ~**-completing trunk** TELEP circuit d'enregistrement et de raccordement *m*; ~**-completing trunk group** TELEP faisceau de circuits abonné-téléphoniste *m*, faisceau abonné-téléphoniste *m*; ~ **density** DATA densité d'enregistrement *f*; ~ **instrument** TEST appareil de mesure enregistreur *m*, appareil enregistreur *m*; ~ **keyboard** COMP clavier d'enregistrement *m*; ~ **operator** TELEP annotateur *m*; ~ **tempo** COMP tempo d'enregistrement *m*
recover[1] *vt* RADIO récupérer, repriser, TELEP récupérer
recover[2] *vi* RADIO, TELEP se rétablir
recovered: ~ **carrier** *n* TRANS porteuse récupérée *f*; ~ **clock** *n* DATA rythme récupéré *m*
recovery *n* DATA reprise *f*, procédure de redressement *f*, RADIO récupération *f*, reprise *f*, rétablissement *m*, TELEP récupération *f*, TEST rétablissement *m*; ~ **log** DATA consignation de reprise *f*; ~ **period** ELECT durée de désionisation *f*; ~ **procedure** DATA procédure de reprise *f*, procédure de redressement *f*; ~ **signal** GEN signal de reconstitution *m*, signal de récupération *m*; ~ **time** DATA temps de reprise *m*, ELECT temps de restitution *m*, PARTS temps de recouvrement inverse *m*
recreational *adj* (*rec.*) DATA *Internet* de loisirs
rectangular: ~ **faceplate** *n* ELECT plastron rectangulaire *m*; ~ **pulse** *n* TELEP impulsion rectangulaire *f*; ~ **wave** *n* GEN onde rectangulaire *f*
rectenna *n* RADIO antenne redresseuse *f*
rectification *n* POWER redressement *m*
rectifier *n* POWER redresseur *m*; ~ **bridge** POWER pont redresseur *m*; ~ **meter** TEST voltmètre à redresseur *m*; ~ **tube** ELECT tube redresseur *m*; ~ **unit** POWER redresseur *m*
rectify *vt* POWER redresser
rectifying: ~ **aerial** *n* RADIO antenne redresseuse *f*; ~ **antenna** *n* RADIO antenne redresseuse *f*; ~ **tube** *n* POWER tube redresseur *m*; ~ **valve** *n* POWER tube redresseur *m*
rectilinear: ~ **aerial** *n* RADIO antenne rectiligne *f*; ~ **antenna** *n* RADIO antenne rectiligne *f*
recuperate *vt* RADIO récupérer
recuperation *n* RADIO récupération *f*
recursion *n* COMP, DATA fonction de récurrence *f*
recursive[1] *adj* COMP, DATA récurrent
recursive[2]: ~ **digital filter** *n* DATA filtre numérique à mémoire infinie *m*; ~ **filter** *n* DATA, TRANS filtre récursif *m*; ~ **function** *n* COMP, DATA fonction de récurrence *f*, fonction récursive *f*; ~ **neurons** *n pl* TRANS neurones récursifs *m pl*
recursiveness *n* COMP, DATA récurrence *m*
red[1]: ~**-green-blue** *adj* (*RGB*) TV rouge-vert-bleu (*RVB*)
red[2]: ~ **book** *n* COMP livre rouge *m*; ~ **pass filter** *n* ELECT filtre rouge *m*
redial: ~ **button** *n* PER APP touche bis *f*, touche de reprise de tonalité *f*
redirect *vt* RADIO *vehicle* dérouter, SWG, TELEP réacheminer
redirected: ~ **call** *n* TELEP appel réacheminé *m*; ~ **call indicator** *n* SWG, TELEP indicateur d'appel réacheminé *m*; ~ **call signal** *n* SIG signal d'appel réacheminé *m*; ~ **traffic** *n* TELEP trafic réacheminé *m*
redirecting: ~ **counter** *n* SWG compteur de renvois *m*; ~ **indicator** *n* SWG, TELEP indicateur de réacheminement *m*; ~ **number** *n* SWG, TELEP numéro renvoyant l'appel *m*
redirection *n* SWG, TELEP réacheminement *m*; ~ **address** TELEP adresse de réacheminement *f*; ~ **address indicator** SWG, TELEP indicateur d'adresse de réacheminement *m*; ~ **number** SWG numéro de renvoi *m*; ~ **request signal** SIG signal de demande de réacheminement *m*
redirector *n* DATA redirecteur *m*
reduce *vt* COMP, TEST réduire
reduced[1]: ~**-carrier** *adj* RADIO à porteuse réduite
reduced[2]: ~ **carrier** *n* RADIO porteuse réduite *f*; ~**-carrier emission** *n* RADIO émission à porteuse réduite *f*; ~ **error** *n* TEST erreur réduite *f*; ~ **inspection** *n* TEST contrôle réduit *m*; ~**-instruction-set computer** *n* (*RISC*) COMP, DATA ordinateur à jeu d'instructions réduit *m* (*RISC*), ordinateur RISC *m* (*RISC*); ~ **rate** *n* APPL tarif réduit *m*, tarif bleu *m* (*Fra*), tarif bleu nuit *m* (*Fra*), TELEP tarif bleu *m* (*Fra*), tarif bleu nuit *m* (*Fra*); ~ **tariff** *n* APPL tarif réduit *m*
reducing: ~ **valve** *n* PARTS valve de réduction *f*
reduction *n* COMP, TEST *of sample* réduction *f*; ~ **of area** PARTS contraction *f*, réduction de surface *f*; ~ **of tariff** APPL réduction de tarif *f*
redundancy *n* GEN redondance *f*, RADIO redondance *f*, surcapacité *f*; ~ **check** DATA contrôle par redondance *m*; ~ **switching** TEST commutation normal-secours *f*, commutation redondante *f*, commutation sur les équipements de secours *f*
redundant: ~ **carrier-supply** *n* TRANS générateur auxiliaire de fréquence porteuse *m*; ~ **code** *n* SIG code redondant *m*; ~ **component check** *n* POWER vérification des composants redondants *f*; ~ **n-ary signal** *n* SIG signal n-aire redondant *m*
reed *n* PARTS contact scellé *m*; ~ **contact** PARTS contact à lames souples *m*; ~ **element** PARTS élément à lames souples *m*; ~ **frequency meter** TEST fréquencemètre à lames vibrantes *m*; ~ **relay** PARTS, POWER relais à contact sous vide *m*, relais à lames souples *m*; ~**-relay crosspoint** SWG point d'intersection à relais à lames souples *m*; ~**-relay electronic exchange** SWG central électronique à relais à lames souples *m*; ~**-relay system** SWG système électronique à relais à tige *m*
Reed-Solomon: ~ **code** *n* (*RS code*) COMP code de Reed-Solomon *m* (*code RS*)
reel *n* DATA bobine *f*; ~ **hub** TELEP moyeu de bobine *m*; ~ **of paper** PER APP bobine de papier *f*
re-emission: ~ **time** *n* (*RT*) RADIO temps de réémission *m*

re-enterable: ~ **subroutine** *n* TELEP sous-programme rentrant *m*

re-entrant *n* COMP réentrant *m*; ~ **program** COMP programme rentrant *m*; ~ **program subroutine** TELEP sous-programme rentrant *m*; ~ **trunking** TELEP jonction rentrante *f*

re-entry *n* TELEP nouvelle intervention *f*; ~ **point** COMP point de retour *m*

reference: ~ **aerial pattern** *n* RADIO diagramme de gain de référence *m*; ~ **antenna pattern** *n* RADIO diagramme de gain de référence *m*; ~ **atmosphere for refraction** *n* RADIO atmosphère de référence pour la réfraction *f*; ~ **clock** *n* TRANS horloge de référence *f*; ~ **conditions** *n pl* TEST conditions de référence *f pl*; ~ **designation** *n* COMP référence alphanumérique *f*; ~ **equivalent** *n* TELEP équivalent de référence *m*; ~ **frequency** *n* RADIO, TRANS fréquence de référence *f*; ~ **guide** *n* COMP guide de référence *m*; ~ **installation** *n* TELEP installation de référence *f*; ~ **level** *n* COMP niveau de référence *m*; ~ **manual** *n* COMP manuel de référence *m*; ~ **model** *n* DATA modèle de référence *m*; ~ **picture** *n* COMP, TV image de référence *f*; ~ **pilot** *n* RADIO, TRANS onde pilote de référence *f*, pilote de référence *m*; ~**-pilot generation** *n* RADIO, TRANS émission d'ondes pilotes de référence *f*; ~ **plane** *n* ELECT plan de référence *m*; ~ **point** *n* DATA point de référence *m*; ~ **publication format** *n* (*RPF*) DATA format de référence des publications *m*; ~ **range** *n* TEST domaine de référence *m*; ~ **routing** *n* SWG itinéraire de référence *m*; ~ **sensibility** *n* RADIO, TRANS sensibilité de référence *f*; ~ **sensitivity** *n* RADIO, TRANS sensibilité de référence *f*; ~ **standardized abstract test suite** *n* DATA, TEST suite de tests abstraite normalisée de référence *f*; ~**-status display** *n* RADIO affichage de l'état de la station de référence *m*; ~ **surface** *n* TRANS surface de référence *f*; ~**-surface center** *AmE*, ~**-surface centre** *BrE n* TELEP, TRANS centre de la surface de référence *m*; ~**-surface diameter** *n* TELEP, TRANS diamètre de la surface de référence *m*; ~**-surface diameter deviation** *n* TELEP, TRANS écart sur la surface de référence *m*; ~ **system** *n* GEN système de référence *m*; ~ **telephonic power** *n* (*RTP*) TELEP volume de référence *m*; ~ **terminal status interchange** *n* (*RTSI*) RADIO échange d'états du terminal de référence *m*; ~ **test method** *n* (*RTM*) TEST, TRANS méthode de mesure de référence *f*; ~ **unique word** *n* DATA mot unique de référence *m*; ~**-usable field strength** *n* RADIO champ utilisable de référence *m*, puissance surfacique utilisable de référence *f*; ~**-usable power flux density** *n* RADIO champ utilisable de référence *m*, puissance surfacique utilisable de référence *f*; ~ **value** *n* TEST valeur de référence *f*; ~ **variable** *n* CONTR grandeur de référence *f*; ~ **voltage** *n* POWER tension de référence *f*; ~**-voltage generator** *n* POWER générateur de tension de référence *m*

referencing *n* RADIO rephasage *m*

referential: ~ **link** *n* COMP lien de référence *m*

refine *vt* COMP, DATA *Internet* affiner, raffiner

refinement *n* COMP, DATA affinage *m*, raffinage *m*

reflectance *n* CLIM, ELECT, RADIO, TELEP, TRANS réflectance *f*, facteur de réflexion *f*; ~ **diagram** TV schéma de réflexion *m*

reflected: ~ **binary code** *n* DATA, ELECT code binaire réfléchi *m*; ~ **current** *n* TRANS courant réfléchi *m*; ~ **pulse** *n* RADIO impulsion d'écho *f*; ~ **voltage** *n* TRANS tension réfléchie *f*

reflecting: ~ **aerial** *n* RADIO antenne à réflecteur *f*; ~ **antenna** *n* RADIO antenne à réflecteur *f*; ~ **satellite** *n* RADIO satellite réflecteur *m*, satellite réfléchissant *m*

reflection *n* ELECT, TRANS réflexion *f*; ~ **coefficient** RADIO, TRANS coefficient de réflexion *m*; ~ **factor** CLIM, ELECT, RADIO, TELEP, TRANS facteur de réflexion *m*; ~ **gain** TRANS gain réflexion *m*; ~ **loss** POWER, TRANS perte due aux réflexions *f*; ~ **mapping** COMP placage de réflexion *m*; ~**-mode filter** RADIO filtre de mode à réflexion *m*; ~ **vector** COMP vecteur de réflexion *m*

reflectometer *n* ELECT, POWER, TEST, TRANS réflectomètre temporel *m*

reflector *n* CLIM réflecteur *m*; ~ **aerial** RADIO antenne à réflecteur *f*; ~ **antenna** RADIO antenne à réflecteur *f*

reflex: ~ **amplifier** *n* ELECT amplificateur reflex *m*; ~ **klystron** *n* ELECT klystron reflex *m*

reflow: ~ **soldering** *n* ELECT brasage tendre par refusion *m*

refracted: ~ **near-end method** *n* TRANS méthode d'exploration du champ réfracté proche *f*, méthode du champ réfracté proche *f*; ~ **near-field technique** *n* TRANS méthode du champ réfracté *f*; ~ **ray** *n* TRANS rayon réfracté *m*; ~**-ray method** *n* TRANS méthode d'exploration du champ réfracté proche *f*, méthode du champ réfracté proche *f*

refraction *n* RADIO, TRANS, TV réfraction *f*; ~ **index** TV indice de réfraction *m*

refractive: ~ **index** *n* ELECT, RADIO, TRANS, TV indice de réfraction *m*; ~**-index contrast** *n* TRANS contraste de l'indice de réfraction *m*, différence de l'indice de réfraction *f*; ~**-index distribution** *n* TRANS distribution de l'indice de réfraction *f*; ~**-index profile** *n* TRANS profil de l'indice de réfraction *m*; ~ **modulus** *n* (*M*) RADIO module de réfraction *m* (*M*)

refractivity *n* RADIO coïndice *m*

reframe *n* COMP recadrage *m*

reframing: ~ **time** *n* TELEP temps de reprise *m*

refresh: ~ **rate** *n* COMP, DATA fréquence de rafraîchissement *f*, taux de régénération *m*, fréquence verticale *f*

refreshment: ~ **procedure** *n* COMP, DATA procédure de rafraîchissement *f*

refrigerant *n* CLIM frigorigène *m*

refrigerate *vt* CLIM réfrigérer

refrigeration *n* CLIM réfrigération *f*; ~ **circuit** CLIM circuit de réfrigération *m*; ~ **engineering** CLIM technique du froid *f*

refund: ~ **chute** *n* PER APP *of payphone* sébile de restitution des pièces *f*

regeneration: ~ **rate** *n* COMP taux de régénération *m*

regenerative: ~ **link** *n* TRANS liaison non-transparente *f*; ~ **pulse repeater** *n* TELEP répéteur d'impulsions *m*; ~ **receiver** *n* RADIO récepteur à réaction *m*; ~ **repeater** *n* TRANS répéteur régénérateur *m*; ~ **satellite** *n* RADIO satellite régénérateur *m*

regenerator *n* (*RG*) ELECT, SIG, SWG, TELEP régénérateur *m* (*RG*); ~ **section** (*RS*) DATA section de régénération *f*; ~**-section overhead** DATA surdébit de section de régénération *m* (*SDSR*); ~**-section termination** (*RST*) DATA terminaison de section de régénération *f*

region *n* ELECT circonscription *f*; ~ **of limited proportionality** RADIO région de proportionnalité limitée *f*

regional: ~ **center** *AmE*, ~ **centre** *BrE n* TELEP central régional *m*, centre régional *m*; ~ **processor** *n* (*RP*) SWG processeur régional *m*

Regional: ~ **Commonwealth in the Field of Communications** *n* (*RCC*) RADIO Communauté régionale des communications *f*

register[1] *n* DATA registre *m*, RADIO tables *f pl*, SWG, TELEP enregistreur *m*, registre *m*; ~ **addressing** SWG adressage de l'enregistreur *m*; ~**-clear signal** TELEP signal de libération de l'enregistreur *m*; ~**-controlled system** SWG, TELEP système à commande par enregistreurs *m*; ~**-disconnect signal** TELEP signal de déconnexion de l'enregistreur *m*; ~ **end-of-selection signal** TELEP signal de fin de sélection d'enregistreur *m*; ~ **finder** SWG chercheur d'enregistrement *m*; ~**-finder marker** SWG marqueur de chercheur d'enregistreur *m*; ~**-finder selector** SWG sélecteur du chercheur d'enregistreur *m*; ~ **function** TELEP, TRANS fonction d'enregistreur *f*; ~ **length** DATA longueur de registre *f*; ~**-MS abstract-operation** DATA opération abstraite d'enregistrement mémoire de messages *f*; ~**-preserve memory** SWG mémoire de conservation de registre *f*; ~ **recall** TELEP, TRANS rappel d'enregistreur *m*;

~-release signal SIG, TELEP signal de libération de l'enregistreur *m*; **~-reset signal** SIG, TELEP signal de remise à zéro du registre *m*; **~ signal** SIG, TELEP signal d'enregistreur *m*; **~ signaling** *AmE*, **~ signalling** *BrE* SIG, TELEP signalisation d'enregistreur *f*, signalisation entre enregistreurs *f*, signalisation entre registreurs *f*; **~ time-out** RADIO temporisation d'enregistrement *f*; **~ translator** SWG enregistreur-traducteur *m*

register[2] *vt* RADIO inscrire

registered[1] *adj* RADIO enregistré, inscrit

registered[2]: **~ access** *n* COMP accès homologué *m*; **~ earth station** *n* RADIO station terrienne homologuée *f*; **~ user** *n* COMP utilisateur référencé *m*

registration *n* DATA enregistrement *m*, ELECT concordance *f*, RADIO enregistrement *m*, inscription *f*, TV coïncidence *f*; **~-accepted signal** SIG signal d'acceptation d'enregistrement *m*; **~ authority** DATA organisme d'enregistrement *m*; **~-completion signal** SIG signal d'enregistrement effectué *m*; **~-hierarchical name** DATA nom hiérarchique d'enregistrement *m* (*nom RH*); **~-hierarchical name tree** DATA arbre de nom hiérarchique d'enregistrement *m*; **~ identifier** DATA identificateur d'enregistrement *m*; **~ of incoming calls** TELEP service d'enregistrement de communications d'arrivée *m*; **~ request signal** SIG signal de demande d'enregistrement *m*

regression: **~ equation** *n* COMP, GEN équation de régression *f*

regroup *vt* COMP regrouper *m*

regular: **~ replacement** *n* TEST renouvellement régulier *m*; **~ signaling link** *AmE*, **~ signalling link** *BrE* *n* SIG canal sémaphore normal *m*, liaison de signalisation régulière *f*; **~ transmissions** *n pl* TRANS transmissions périodiques *f pl*

regularity: **~ loss** *n* TELEP affaiblissement de l'onde réfléchie sur les irrégularités *m*

regulated: **~ line section** *n* TELEP section de régulation de ligne *f*

regulating: **~ amplifier** *n* ELECT amplificateur régulateur *m*

regulation *n* COMP, CONTR, ELECT réglage *m*, régulation *f*, réglementation *f*; **~ relay** PARTS relais de réglage *m*

regulator *n* COMP, CONTR, ELECT régulateur *m*

rehearsal: **~ mark** *n* COMP marque de répétition *f*

reignition *n* DATA réamorçage *m*, ELECT rallumage *m*, RADIO réamorçage *m*; **~ voltage** ELECT tension de rallumage *f*

reimport *vt* DATA *Internet* réimporter

reinforce *vt* CONTR, RADIO renforcer

reinforced: **~ glass** *n* PARTS verre renforcé *m*; **~ plastic** *n* PARTS plastique renforcé *m*

reinforcement *n* CONTR, RADIO, TEST renforcement *m*

reinitialization *n* COMP, DATA réinitialisation *f*

reinitialize *vt* COMP, DATA réinitialiser

REJ *abbr* (*reject frame*) TELEP trame de rejet *f*

reject[1]: **~ character** *n* DATA caractère de rejet *m*; **~ frame** *n* (*REJ*) TELEP trame de rejet *f*

reject[2] *vt* COMP, TEST, rejeter

rejection *n* TEST rejet *m*; **~ number** TEST critère de rejet *m*; **~ output** TELEP sortie de rejet *f*

rejective: **~ condition** *n* TEST condition de refus *f*, condition de rejet *f*

REL *abbr* (*release message*) TEST LIB (*message de libération*)

related: **~-indication primitive** *n* DATA primitive d'indication associée *f*; **~-request primitive** (*req*) primitive de demande associée *f*

relation *n* TELEP relation *f*

relational: **~ database** *n* (*RDB*) COMP, DATA base de données relationnelles *f* (*BDR*)

relationship *n* COMP association *f*

relative: **~ address** *n* COMP adresse relative *f*; **~ amplitude** *n* TELEP *of elementary echo* amplitude relative *f*; **~ build-up time** *n* RADIO *of telegraph signal* temps d'établissement relatif *m*; **~ distinguished name** *n* DATA nom distinctif relatif *m*, nom spécifique relatif *m*; **~ error** *n* TEST erreur relative *f*; **~-intensity noise** *n* (*RIN*) ELECT bruit d'intensité *m*; **~ level** *n* TELEP niveau relatif *m*; **~ order** *n* COMP commande relative *f*; **~ power level** *n* TELEP niveau relatif de puissance *m*; **~ redundancy** *n* GEN redondance relative *f*; **~ spectral sensitivity** *n* ELECT sensibilité énergétique monochromatique relative *f*, sensibilité spectrale relative *f*; **~ spectral sensitivity characteristic** *n* ELECT, caractéristique de sensibilité énergétique monochromatique relative *f*, caractéristique de sensibilité spectrale relative *f*

relaxation: **~ oscillator** *n* ELECT oscillateur à relaxation *m*

relay[1] *n* RADIO, TELEP, TRANS, TV relais *m*, retransmission *f*; **~-armature stop** POWER vis d'arrêt de la palette *f*; **~ bar** PARTS plaque de relais *f*, platine de relais *f*; **~ chain** PARTS chaîne de relais *f*; **~ coil** POWER bobine de relais *f*; **~ contact** PARTS contact de relais *m*; **~ contact unit** PARTS unité de commande de relais *f*; **~ core** PARTS noyau de relais *m*; **~ driver** POWER amplificateur de relais *m*, attaqueur de relais *m*, excitateur de relais *m*; **~ frame** TELEP châssis de relais *m*; **~ magnet** ELECT, PARTS électroaimant de relais *m*, aimant de relais *m*; **~ mounting bar** PARTS platine de relais *f*, plaque de relais *f*; **~ network** RADIO réseau de bornes relais *m*; **~ position** TELEP position de relais *f*; **~ space** TELEP position de relais *f*; **~ spacing** TELEP espacement des relais *m*; **~ station** RADIO station relais *f*; **~ support** TELEP support de relais *m*; **~ system** SWG, TELEP commutateur tout à relais *m*, système tout à relais *m*; **~ test point** PARTS point d'essai de relais *m*; **~ transmitter** RADIO, TELEP, TRANS, TV réémetteur *m*; **~ tree** PARTS arborescence de relais *f*; **~-type switch** SWG, TELEP commutateur tout à relais *m*; **~ winding** PARTS enroulement de relais *m*; **~ yoke** ELECT, PARTS, TELEP culasse *f*

relay[2] *vt* RADIO, TELEP, TRANS, TV retransmettre, transmettre

relaying *n* RADIO, TELEP, TRANS, TV retransmission *f*; **~ and routing function** SWG fonction de relais et de routage *f*

release[1]: **~-velocity-sensitive** *adj* COMP sensible à la vélocité du relâchement

release[2] *n* COMP libération *f*, révision *f*, relâchement *m*, *software* version *f*, PARTS relâchement *m*, mise au repos *f*, TELEP *of connection* libération *f*; **~ condition** PARTS position de repos *f*; **~ current** POWER courant de libération *m*; **~-guard signal** SIG, TELEP signal de libération de garde *m*; **~ indicator** DATA, SIG indicateur de libération *m*; **~ lever** TELEP levier de dégagement *m*; **~ link trunk** (*RLT*) TELEP ligne de concentration de trafic *f*; **~ mechanism** ELECT mécanisme de déclenchement *m*, mécanisme de mise en marche *m*; **~ message** (*REL*) TEST message de libération *m* (*LIB*); **~ rate** COMP vitesse de relâchement *f*; **~ signal** DATA signal de libération *m*; **~ time** PARTS temps de relâchement *m*; **~ trunk** TELEP circuit de concentration de trafic *m*; **~ velocity** COMP vélocité de relâchement *f*; **~ velocity sensitivity** COMP sensibilité à la vélocité de relâchement *f*; **~ wire** SWG fil de libération *m*

release[3] *vt* COMP, PARTS relâcher; **~ by shunting** PARTS relâcher en shuntant

released *adj* COMP, PARTS relâché

relevance: **~ feedback** *n* DATA *Internet* contrôle de pertinence *m*, pertinence rétroactive *f*, rétroaction de pertinence *f*, remontée de pertinence *f*

relevant: **~ failure** *n* TEST défaillance significative *f*, défaillance pertinente *f*; **~ time** *n* TEST temps à prendre en compte *m*, temps pertinent *m*

reliability *n* COMP, ELECT, GEN, TELEP, TEST fiabilité *f*; **~ block diagram** TEST diagramme de fiabilité *m*; **~ engineering** TEST ingénierie de la fiabilité *f*; **~ measure** TEST mesure de fiabilité *f*; **~ performance** COMP, ELECT, GEN, TELEP, TEST fiabilité d'un équipement *f*; **~ and quality measurement system** (*RQMS*) TEST système de mesures de fiabilité et de qualité *m*

reliable[1] *adj* GEN fiable

reliable[2]: **~ transfer** *n* (*RT*) DATA transfert fiable *m*; **~ transfer server** *n* (*RTS*) COMP serveur de transfert fiable *m*
relief: **~ telecommunications** *n pl* TELEP télécommunications de secours *f pl*
relieve *vt* COMP soulager, TELEP décharger, délester
reload[1] *n* COMP rechargement *m*
reload[2] *vt* COMP recharger
reloading *n* COMP rechargement *m*
relocatable: **~ address** *n* COMP, TELEP adresse translatable *f*; **~ program** *n* COMP programme translatable *m*
relocate *vt* COMP translater, DATA *Internet* retrouver, RADIO déplacer
reluctance *n* (*R*) POWER réluctance *f* (*R*); **~ tuning** RADIO accord par réluctance *m*
remain *vt* ELECT rester
remanence *n* ELECT, POWER rémanence *f*
remanent: **~ relay** *n* POWER relais à rémanence *m*
reminder: **~ alarm service** *n* APPL téléalerte *f*; **~ call** *n* APPL appel aide-mémoire *m*, mémo rappel *m*, PER APP rappel de rendez-vous *m*; **~ indication** *n* TELEP indication pour rappel *f*; **~ marking** *n* TELEP indication pour rappel *f*; **~ ring** *n* TELEP sonnerie de rappel *f*; **~ service** *n* TELEP service de rappel de rendez-vous *m*
remote[1]: **~-control** *adj* CONTR télécommandable; **~-controlled** *adj* CONTR commandé à distance, télécommandé, TELEP télésurveillé
remote[2] *n* ELECT module *m*; **~ access** DATA, TELEP accès à distance *m*; **~-access answering machine** TELEP répondeur interrogeable à distance *m*; **~ alarm** RADIO, TELEP téléalarme *f*; **~ alarm indication** (*RAI*) TEST indication d'alarme distante *f*, télésignalisation d'alarme *f*; **~ announcement modification** PER APP modification d'annonce à distance *f*; **~ batch entry** (*RBE*) DATA soumission par lots à distance *f*, télésoumission de lots *f*, télésoumission de travaux par lots *f*; **~ batch processing** COMP, DATA télétraitement par lots *m*, traitement par lots à distance *m*; **~ blocking** CONTR blocage de ligne à distance *m*, téléblocage *m*; **~ boot** ELECT initialisation à distance *f*; **~ concentration unit** (*RCU*) RADIO concentrateur satellite numérique *m*; **~ concentrator** TRANS concentrateur distant *m*; **~ console** DATA console de télétraitement *f*; **~ consultation** GEN interrogation à distance *f*; **~ control** COMP commande à distance *f*, CONTR commande à distance *f*, télécommande *f*, GEN télécommande *f*, télégestion *f*, PER APP boîtier *m*, commande à distance *m*, RADIO, TELEP commande à distance *f*; **~ control unit** (*RCU*) CONTR bloc de télécommande *m*, unité de gestion à distance *f*, unité de télécommande *f*, unité de télégestion *f*; **~ control warning system** CONTR télécommande d'alerte *f*; **~-controlled exchange** TELEP central télésurveillé *m*; **~ controller** PER APP télécommande *f*; **~ CSE** TELEP CSE distant *m*; **~ CSN** TELEP CSN distant *m*; **~ cutoff tube** ELECT tube à blocage éloigné *m*, tube à pente réglable *m*; **~ data processing** COMP, DATA, RADIO téléinformatique *f*; **~ data unit** (*RDU*) DATA équipement de télétraitement de l'information *m*; **~ database access** (*RDA*) DATA accès à des bases de données distantes *m*; **~ detection** TEST télédétection *f*; **~ digital concentrator** TELEP concentrateur numérique éloigné *m* (*CNE*); **~ document access** DATA, TELEP accès du document distant *m*; **~-document management** TELEP gestion du document distant *f*; **~ error** DATA erreur distante *f*; **~-exchange concentrator** TELEP concentrateur de central distant *m*; **~-gate thyristor** ELECT thyristor à gâchette éloignée *m*; **~ indication** DATA indication à distance *f*; **~ interrogation** PER APP interrogation à distance *f*; **~ interrogation by mf ph** PER APP interrogation par poste à télé *f*; **~ interrogation by telecommand** PER APP interrogation par télécommande *f*; **~ job entry** (*RJE*) DATA soumission de travaux à distance *f*, télésoumission de travaux *f*; **~ loading** TEST téléchargement *m*; **~ log-in** ELECT connexion à distance *f*; **~ maintenance** COMP, DATA, TEST, TV télémaintenance *f*; **~ management** TEST télégestion *f*; **~ measuring** TEST télémesure *f*; **~ meter reading** TV télérelevé *m*; **~ metering** DATA télémesure *f*; **~ monitoring** TEST télésurveillance *f*; **~ monitoring system** TEST système de télésurveillance *m*; **~ multiplexer** TRANS multiplexeur distant *m*; **~ operating center** *AmE*, **~ operating centre** *BrE* RADIO centre d'enregistrement à distance *m* (*CED*); **~ operating terminal** PER APP poste d'exploitation déporté *m*, terminal d'exploitation à distance *m*; **~ operation** (*RO*) DATA, TEST exploitation à distance *f*, opération distante *f*; **~ payment** TV télépaiement *m*; **~ peripheral equipment** (*RPE*) COMP équipement périphérique distant *m* (*EPD*); **~ playback** PER APP mise en route à distance *f*; **~ printing** COMP impression à distance *f*, téléimpression *f*; **~ processing** CONTR, DATA télégestion *f*, traitement à distance *m*; **~ reading** DATA *telemetry*, GEN indication à distance *f*, télérelevé *m*; **~ resource** DATA ressource distante *f*; **~ security** TRANS, TV télésécurité *f*; **~ security network** TRANS réseau télésécurité *m*; **~ sensing** DATA, GEN, TEST indication à distance *f*, télédétection *f*, télédétection à distance *f*; **~ sensing satellite** RADIO satellite de télédétection *m*; **~ single-layer embedded test method** (*RSE test method*) TEST méthode de test monocouche à distance encastrée *f*; **~ single-layer test method** (*RS test method*) TEST méthode de test monocouche à distance *f*; **~ site** ELECT site distant *m*, GEN station périphérique *f*, TRANS site éloigné *m*; **~ spectrum-monitoring** (*RSM*) RADIO surveillance à distance du spectre *f*; **~ start** CONTR démarrage à distance *m*, télédémarrage *m*; **~ station** DATA station périphérique *f*, terminal distant *m*; **~ subscriber stage** TELEP unité de raccordement d'abonné *f*; **~ subscriber switch** TELEP unité de raccordement d'abonné *f*; **~ supervision** CONTR, TV surveillance à distance *f*, télésurveillance *f*; **~ surveillance** CONTR, TV surveillance à distance *f*, télésurveillance *f*; **~ survey** TV télésondage *m*; **~ switching** SWG télécommutation *f*; **~ switching point** SWG étage de commutation distante *m*; **~ switching stage** SWG étage de commutation distante *m*; **~ switching system** SWG système de commutation à distance *m*, système de commutation satellite *m*; **~ switching unit** (*RSU*) SWG centre satellite *m*; **~ terminal** (*RT*) PER APP terminal distant *m*; **~ terminal unit** PER APP unité asservie *f*; **~ test method** TEST méthode de test à distance *f*; **~ transmission** TRANS télétransmission *f*; **~ unit** (*RU*) SWG unité éloignée *f*, unité distante *f*
remote[3] *vt* ELECT téléporter
remotely[1]: **~-connected** *adj* TELEP connecté à distance, téléconnecté; **~-interrogable** *adj* PER APP *answer machine* interrogeable à distance
remotely[2]: **~-controlled exchange** *n* SWG centre télécommandé *m*; **~-controlled station** *n* RADIO station télécommandée *f*; **~-generated display** *n* DATA écho distant *m*
removable: **~ cartridge** *n* COMP cartouche amovible *f*; **~ keyed shroud** *n* ELECT coiffe à détrompeur amovible *f*; **~ magnetic disk** *n* COMP, DATA, ELECT disque magnétique amovible *m*
removal *n* ELECT dépose *f*, POWER déménagement *m*, démontage *m*
remove *vt* ELECT déposer, POWER déménager; **~ from service** TELEP mettre hors service; **~ the sheath from** GEN dégainer
remreed: **~ crosspoint** *n* SWG point d'intersection remreed *m*
remuneration: **~ for exclusive use of circuits** *n* TELEP rémunération pour utilisation exclusive des circuits *f*; **~ for shared use of circuits** *n* TELEP rémunération pour utilisation en commun des circuits *f*
rename *vt* COMP renommer
render *vt* COMP rendre
rendering: **~ technique** *n* COMP technique de rendu *f*
renewal: **~ function** *n* TEST fonction de renouvellement *f*

rental *n* TELEP redevance *f*; ~ **charge** GEN frais de location *m*
reorder: ~ **buffer** *n* COMP tampon de remise en ordre *m*; ~ **tone** *n* TELEP signal tous circuits occupés *m* (*signal TCO*), tonalité tous circuits occupés *f* (*tonalité TCO*)
repaginate *vt* COMP repaginer
repair[1] *n* COMP, GEN, TELEP réparation *f*, ~ **company** GEN société de dépannage *f*; ~ **kit** GEN, TELEP kit d'outillage de réparation *m*, nécessaire de réparation *m*; ~ **rate** TEST taux de réparation *m*; ~ **shop** TELEP atelier de réparation *m*; ~ **time** TELEP temps de réparation *m*
repair[2] *vt* GEN, TEST *equipment* dépanner
repairer *n* GEN, TELEP réparateur *m*
repairing: ~ **cables** *n pl* TRANS câbles réparation *m pl*
repairperson *n* GEN, TELEP réparateur *m*, réparatrice *f*
repeat[1] *n* COMP lecture répétitive *f*, répétition *f*, reprise *f*, GEN répétition *f*; ~ **action key** COMP touche répétitrice *f*; ~ **calls** *n pl* TELEP appels répétés *m pl*; ~ **coil** TELEP translateur *m*
repeat[2] *vt* COMP répéter, reprendre, GEN répéter
repeatability *n* DATA *of results* reproductibilité *f*, GEN répétibilité *f*
repeated: ~ **call attempt** *n* TELEP tentative d'appel répétée *f*
repeater *n* DATA, ELECT translation *f*, GEN répéteur *m*, SIG, TELEP, TRANS, TV translation *f*; ~ **deck** SIG pont des répéteurs *m*; ~ **fuse** POWER coupe-circuit à répétition *m*; **~-jammer** ELECT brouilleur-répéteur *m*; ~ **spacing** TRANS espacement des répéteurs *m*; ~ **span** TRANS espacement des répéteurs *m*, pas de régénération *m*; ~ **station** TELEP, TRANS station de répéteurs *f*
repeaterless *adj* GEN sans répéteur
repeating: ~ **coil** *n* TELEP translateur *m*
reperforator *n* TELEP récepteur-perforateur *m*
repertoire *n* TELEP répertoire *m*
repetency *n* TRANS nombre d'ondes *m*
repetition *n* GEN répétition *f*; ~ **frequency** TELEP fréquence de répétition *f*; ~ **rate** TELEP fréquence de répétition *f*, taux de répétition *m*
repetitive[1] *adj* ELECT routinier
repetitive[2]: ~ **metering** *n* TELEP comptage multiple *m*; ~ **peak forward current** *n* ELECT, POWER courant direct de pointe récurrent *m*; ~ **peak offstage voltage** *n* PARTS tension de pointe répétitive *f*; ~ **peak onstage current** *n* ELECT courant de pointe récurrent à passant *m*; ~ **peak reverse voltage** *n* PARTS, TELEP tension inverse de pointe répétitive *f*
replace *vt* COMP remplacer; ~ **the handset** TELEP raccrocher le combiné; ~ **the receiver** TELEP raccrocher le combiné; ~ **regularly** TEST renouveler régulièrement
replacement *n* TELEP *of handset* raccrochage *m*; ~ **board** ELECT carte de remplacement *f*, carte équivalente *f*
replay *n* RADIO contrôle d'enregistrement *m*
reply[1]: ~ **code** *n* RADIO code de réponse *m*
reply[2]: **on no** ~ *phr* RADIO sur non-réponse
report[1] *n* COMP état *m*, DATA journal *m*, signalisation *f*, GEN signalisation *f*
report[2] *vt* GEN rapporter, rendre compte
representative: ~ **sample** *n* GEN panel *m*, TEST échantillon représentatif *m*
reprint *n* PER APP tirage à part *m*
reproduce *vt* GEN, TEST reproduire
reproducibility *n* GEN, TEST reproductibilité *f*
reproducible[1] *adj* GEN, TEST reproductible
reproducible[2]: ~ **failure** *n* TEST défaillance reproductible *f*
reproductibility *n* GEN, RADIO reproductibilité *f*
reproduction *n* GEN reproduction *f*, RADIO contrôle d'enregistrement *m*; ~ **ratio** TELEP rapport de reproduction *m*
repudiation *n* DATA répudiation *f*
req *abbr* COMP (*request*) demande *f*, requête *f*, DATA (*request primitive*) primitive de demande, (*request*) demande *f*, requête *f*, (*related-request primitive*) primitive de demande associée, RADIO, TELEP (*request*) demande *f*, requête *f*, TV (*request*) demande *f*
requantization *n* COMP requantification *f*
request[1]: **~-response** *adj* RADIO demande-réponse
request[2] *n* (*req*) COMP, DATA, RADIO, TELEP demande *f*, requête *f*, TV demande *f*; ~ **data transfer** DATA, TELEP demande de transfert de données *f*; ~ **for proposal** (*RFP*) GEN appel d'offres *m*; ~ **for service** GEN demande de prestation *f*; ~ **indicator** TELEP indicateur de demande *m*; ~ **output** TELEP sortie de demande *f*; ~ **primitive** (*req*) DATA primitive de demande *f*; **~-repeat system** TELEP système détecteur d'erreurs avec demande de répétition *m*; ~ **to send** (*RTS*) DATA demande pour émettre *f*; ~ **transmission time** TRANS délai de transmission de la demande *m*
requestor *n* DATA demandeur *m*
required: ~ **time** *n* TELEP, TEST durée requise *f*, période requise *f*; ~ **unrequired time** *n* TELEP durée requise non-requise *f*
requirement *n* COMP exigence *f*, TEST exigence *f*
RER *abbr* (*residual error rate*) DATA, TELEP taux d'erreurs résiduelles *m*
reradiation *n* RADIO rerayonnement *m*, rayonnement secondaire *m*
rerecord *vt* COMP, DATA réenregistrer
rerecordable *adj* COMP, DATA réenregistrable
reregistration *n* COMP, DATA réenregistrement *m*
reroute *vt* DATA, SWG, TELEP *traffic* réacheminer
rerouting *n* DATA, SWG, TELEP *of traffic* réacheminement *m*
rerun *n* COMP réexécution *f*, reprise *f*, TELEP répétition de transmission *f*; ~ **time** COMP temps de reprise *m*
RES *abbr* (*reserved field*) ELECT champ réservé *m*
rescue: **~-service radio system** *n* RADIO système de radiocommunications pour les services des secours *m*
research *n* RADIO recherche *f*
reseller *n* RADIO revendeur *m*
reserve: ~ **circuit** *n* TELEP circuit de secours *m*; ~ **power** *n* POWER réserve de puissance *f*, alimentation de secours *f*; ~ **signaling link** *AmE*, ~ **signalling link** *BrE* *n* SIG canal sémaphore de secours *m*
reserved: ~ **field** *n* (*RES*) ELECT champ réservé *m*; ~ **switching path PE** *n* TELEP élément graphique de trajet de commutation réservé *m*; ~ **word** *n* COMP, DATA mot réservé *m*
reservoir *n* ELECT réservoir *m*
reset[1] *adj* ELECT revu
reset[2] *n* ELECT, GEN, RADIO, TEST remise à niveau *f*, remise à zéro *f*, retour au repos *m*; **~-band acknowledgement signal** SIG signal d'accusé de réception de réinitialisation *m*; **~-band signal** SIG signal de réinitialisation de bande *m*; ~ **button** CONTR bouton de remise à zéro *m*, bouton de remise en position initiale *m*, bouton de restauration *m*; **~-circuit signal** SIG signal de réinitialisation du circuit *m*; ~ **mode** COMP mode conditions initiales *m*
reset[3] *vt* COMP, DATA réinitialiser, GEN restaurer, rétablir, enclencher à zéro, PARTS retourner, RADIO réinitialiser, repositionner, réarmer, SWG ramener à la position de repos, ramener à la position initiale; ~ **to zero** ELECT, TEST mettre à zéro, remettre à zéro
resettability *n* RADIO fidélité *f*
resetting *n* COMP, DATA réinitialisation *f*, ELECT *flip-flop* remise à zéro *f*, GEN remise en position initiale *f*, restauration *f*, RADIO réinitialisation *f*; ~ **magnet** SWG aimant de réenclenchement *m*; ~ **percentage** PARTS pourcentage de retour *m*; ~ **pulse** ELECT impulsion de remise à l'état initial *f*, impulsion de remise à zéro *f*, impulsion de restauration *f*, impulsion de rétablissement *f*; ~ **ratio** PARTS rapport de retour *m*; ~ **relay** PARTS relais de remise au repos *m*; ~ **value** PARTS valeur de retour *f*; ~ **voltage** ELECT tension de rappel *f*
reshaping *n* ELECT remise en forme *f*
resident[1] *adj* COMP, DATA résident

resident[2]: ~ **control program** *n* COMP programme de contrôle résident *m*, programme de commande résident *m*
residential: ~ **telephone** *n* PER APP, TELEP poste résidentiel *m*
residual: ~ **binary error rate** *n* (*RBER*) COMP, DATA, RADIO taux d'erreurs binaires résiduelles *m* (*TEBR*); ~ **bit error ratio** *n* (*RBER*) COMP, DATA, RADIO taux d'erreurs binaires résiduelles *m* (*TEBR*); ~ **echo level** *n* TELEP niveau d'écho résiduel *m*; ~ **error** *n* COMP, DATA, TELEP erreur résiduelle *f*; ~ **error rate** *n* (*RER*) DATA, TELEP taux d'erreurs résiduelles *m*; ~ **magnetism** *n* ELECT aimantation résiduelle *f*; ~ **noise** *n* COMP bruit résiduel *m*; ~ **plate** *n* PARTS plaque antirémanente *f*; ~ **reflectance** *n* TRANS réflexion résiduelle *f*; ~ **sideband** *n* (*RSB*) RADIO, TRANS, TV bande latérale résiduelle *f* (*BLR*); ~ **stud** *n* PARTS grain d'entrefer *m*; ~ **voltage** *n* POWER tension résiduelle *f*
resilience: ~ **to interference** *n* TRANS immunité au brouillage *f*
resin: ~**-core solder** *n* ELECT fil à âme décapante *m*, fil à flux incorporé *m*; ~ **smear** *n* ELECT coulée de résine *f*
resist: ~ **foil** *n* ELECT réserve *f*; ~ **lacquer** *n* ELECT vernis épargne *m*
resistance *n* COMP, ELECT résistance *f*; ~**-capacitance** (*RC*) ELECT résistance-capacité *f* (*RC*); ~**-capacitance coupling** (*RC coupling*) ELECT couplage par résistance-capacité *m* (*couplage RC*); ~ **coupling** ELECT couplage par résistance *m*; ~ **standard** POWER résistance étalon *f*; ~ **thermometer** TEST thermomètre à résistance *m*; ~ **under illumination** ELECT résistance sous éclairement *f*
resistive: ~ **attenuator** *n* ELECT affaiblisseur résistif *m*, atténuateur résistif *m*; ~ **circuit** *n* ELECT circuit résistif *m*; ~ **cut-off frequency** *n* ELECT fréquence résistive de coupure *f*; ~ **load** *n* ELECT, POWER charge résistive *f*; ~ **reverse current** *n* ELECT courant inverse résistif *m*; ~ **temperature detector** *n* (*RTD*) CLIM détecteur de température résistif *m*
resistivity *n* ELECT résistivité *f*, POWER résistivité *f*, résistance spécifique *f*
resistor *n* PARTS résistance *f*
resize[1]: ~ **handle** *n* COMP poignée de redimensionnement *f*
resize[2] *vt* COMP redimensionner
resolution *n* COMP, DATA, ELECT définition *f*, résolution *f*, pouvoir de résolution *m*, TELEP, TRANS, TV définition *f*
resolving: ~ **power** *n* ELECT résolution *f*, pouvoir de résolution *m*, RADIO pouvoir de résolution *m*, pouvoir séparateur *m*
resonance *n* ELECT, GEN résonance *f*; ~ **frequency** ELECT, RADIO, TELEP fréquence de résonance *f*; ~ **method of measurement** TEST méthode de mesure par résonance *f*
resonant: ~ **absorption** *n* ELECT absorption résonnante *f*; ~ **cavity** *n* ELECT, TRANS cavité optique résonnante *f*; ~ **circuit** *n* ELECT, POWER circuit résonnant *m*; ~ **frequency** *n* ELECT, RADIO, TELEP fréquence de résonance *f*; ~**-mode filter** *n* RADIO filtre de mode résonnant *m*; ~**-reed relay** *n* POWER relais à lames résonnantes *m*
resonate *vt* COMP, GEN résonner
resonator *n* ELECT, POWER résonateur *m*; ~ **block** GEN résonateur *m*
re-sort *vt* COMP trier à nouveau
resource: ~ **sharing** *n* COMP partage de ressources *m*; ~**-unavailable class** *n* DATA classe ressource indisponible *f*
Resource: ~ **Interchange File Format** *n* (*RIFF*) COMP Resource Interchange File Format *m*
respond *vi* COMP réagir, GEN répondre
responding: ~**-(N) address** *n* DATA adresse (N) en réponse *f*
response *n* GEN réponse *f*, résultat *m*; ~ **for end of transmission** (*ERR*) DATA, TELEP réponse pour fin de transmission *f* (*ERR*); ~ **identifier** (*RI*) TELEP identificateur de réponse *m*; ~ **output** TELEP sortie de réponse *f*; ~ **primitive** DATA réponse primitive *f*; ~ **time** COMP temps de réponse *m*, GEN délai d'intervention *m*, temps de réponse *m*, SIG temps de réponse *m*; ~ **to continue to correct** (*CTR*) DATA réponse pour continuer à corriger *f* (*CTR*)
responser *n* RADIO récepteur de réponse *m*
responsive *adj* CLIM, COMP, DATA, ELECT, GEN, RADIO sensible
responsivity *n* ELECT, TRANS facteur de réponse *m*, sensibilité *f*, sensibilité énergétique *f*
responsor *n* RADIO récepteur de réponse *m*
rest: ~ **position** *n* ELECT position de repos *f*; ~ **state** *n* ELECT, POWER, TELEP, TRANS état de repos *m*, repos *m*
restart[1] *n* COMP, GEN, RADIO, SWG redémarrage *m*, relance *f*, relancement *m*, reprise *f*
restart[2] *vt* COMP, GEN redémarrer, RADIO redémarrer, relancer, repriser
restitution *n* ELECT, TELEP restitution *f*; ~ **delay** TELEP délai de restitution *m*
restorable: ~ **service** *n* RADIO service avec possibilité de rétablissement *m*, service avec rétablissement possible *m*
restoral *n* GEN restauration *f*, RADIO rétablissement *m*; ~ **of service** GEN rétablissement de service *m*
restoration *n* COMP rétablissement *m*, GEN restauration *f*, RADIO, TEST rétablissement *m*; ~ **control point** (*RCP*) TELEP centre de commande de rétablissement du service *m*
restore *vt* COMP, DATA, ELECT reconstituer, rétablir, GEN restaurer, RADIO reconstituer, rétablir, SWG ramener à la position initiale, ramener à la position de repos; ~ **supply** POWER rétablir le courant
restoring *n* SWG retour à la position de repos *m*, retour à la position initiale *m*, retour au repos *m*; ~ **button** GEN bouton de remise en position initiale *m*, bouton de remise à zéro *m*, bouton de restauration *m*; ~ **magnet** SWG aimant de remise au repos *m*; ~ **pulse** ELECT impulsion de rétablissement *f*, impulsion de remise à l'état initial *f*, impulsion de restauration *f*; ~ **relay** PARTS relais de remise au repos *m*; ~ **time** TRANS *echo suppressor* temps de rétablissement *m*
restrappable *adj* POWER reconfigurable, restrapable
restrapping *n* POWER reconfiguration *f*, restrapage *m*
restrict *vt* RADIO, TELEP restreindre l'accès
restricted: ~ **access** *n* TELEP accès restreint *m*; ~ **channel** *n* TV canal à code d'accès *m*; ~ **information-transfer service** *n* APPL service restreint de transfert d'informations *m*; ~ **service** *n* GEN, TELEP service restreint *m*
restriction: ~ **of calling-line identification** *n* TELEP non-identification d'appel *f*; ~ **in the outgoing direction** *n* TELEP service de limitation des communications de départ *m*
restrictor *n* ELECT discriminateur *m*
result *n* TEST *of measurement* résultat *m*, valeur *f*; ~**-comparing register** GEN registre de comparaison de résultats *m*
resynchronization *n* ELECT resynchronisation *f*
resynchronize *vt* ELECT resynchroniser
retain: ~ **mode** *n* COMP mode de maintien *m*
retained: ~**-data block** *n* DATA bloc de données conservé *m*; ~ **signal** *n* SIG signal retenu *m*
retarder *n* PARTS retardateur *m*
retention *n* ELECT rémanence *f*; ~ **period** RADIO délai de conservation *m*
retentivity *n* ELECT rémanence *f*
retest *n* TELEP, TEST contre-vérification *f*
retime *vt* ELECT resynchroniser
retiming *n* ELECT resynchronisation *f*, TELEP, TRANS réajustement du rythme *m*
retouch *n* DATA *scanner* retouche *f*
retrace *n* COMP retour du spot *m*, ELECT retour de faisceau *m*, retour du spot *m*, TV retour du spot *m*
retractable: ~ **aerial** *n* RADIO antenne escamotable *f*; ~ **antenna** *n* RADIO antenne escamotable *f*
retrain: ~ **negative** *n* (*RTN*) DATA, TELEP reconditionnement négatif *m* (*RCDN*); ~ **positive** *n* (*RTP*) DATA, TELEP reconditionnement positif *m*
retransmission *n* RADIO, TELEP, TRANS, TV réémission *f*,

retransmission *f*; ~ **buffer** COMP, TRANS tampon de réémission *m*, tampon de retransmission *m*
retransmit *vt* RADIO, TELEP, TRANS, TV réémettre, retransmettre
retrieval *n* COMP extraction *f*, rappel *m*, récupération *f*, retrait *m*, DATA rappel *m*, ressaisie *f*, RADIO ressaisie *f*, retrait *m*; ~ **port** DATA accès de recherche *m*; ~ **service** COMP service de consultation *m*
retrieve *vt* COMP récupérer, retraire, rappeler, DATA rappeler, ressaisir, retraire, RADIO ressaisir, retraire, TELEP repriser
retrofit[1] *n* ELECT mise à niveau *f*, PARTS rattrapage *m*
retrofit[2] *vt* ELECT mettre à niveau, PARTS rattraper
retroreflector: ~ **array** *n* RADIO réseau de rétroréflecteurs *m*, rideau de rétroflecteurs *m*
retry: ~ **interval** *n* RADIO intervalle de réessai *m*, intervalle de relance *m*
return[1]: ~-**beam vidicon** *n* (*RBV*) TELEP vidicon à retour du faisceau *m*; ~ **block** *n* TELEP blocage de retour *m*; ~ **conductor** *n* POWER conducteur de retour *m*; ~ **current** *n* TRANS courant réfléchi *m*; ~ **flap** *n* PER APP *of payphone* sébile de restitution des pièces *f*; ~ **loss** *n* TELEP, TRANS affaiblissement d'adaptation *m*, affaiblissement d'équilibrage *m*, affaiblissement de réflexion *m*; ~ **path** *n* TRANS, TV voie de retour *f*; ~-**service advice** *n* DATA, TELEP avis de service en retour *m*; ~ **to service** *n* TEST retour à l'exploitation *m*; ~ **to zero** *n* (*RZ*) COMP, ELECT, TELEP retour à zéro *m*; ~-**to-home position** *n* SWG retour à la position de repos *m*, retour à la position initiale *m*, retour au repos *m*; ~-**to-reference recording** *n* COMP enregistrement avec retour à un niveau de référence *m*; ~-**to-zero recording** *n* COMP enregistrement avec retour à zéro *m*; ~ **wire** *n* POWER fil de retour *m*
return[2]: ~ **sb's call** *vt* TELEP rappeler qn; ~ **to home position** *vt* SWG ramener à la position de repos, ramener à la position initiale
returned: ~-**content entry** *n* DATA entrée contenu renvoyé *f*
reusable: ~ **program** *n* COMP programme réutilisable *m*; ~ **rubber** *n* ELECT joint caoutchouc *m*
reuse: ~ **of frequency channel** *n* RADIO réutilisation de fréquence *f*; ~ **pattern** *n* RADIO motif de réutilisation *m*
revenue: ~-**sharing arrangement** *n* TELEP décompte *m*, liquidation des comptes *f*
reverb *n* (*reverberation*) COMP, ELECT réverbération *f*; ~ **level** COMP niveau de réverbération *m*; ~ **time** TELEP temps de réverbération *m*; ~ **type** COMP type de réverbération *m*
reverberated: ~ **field** *n* TELEP champ réverbéré *m*, zone réverbérée *f*
reverberation *n* (*reverb*) COMP, ELECT, TELEP réverbération *f*; ~ **level** COMP niveau de réverbération *m*; ~ **time** TELEP temps de réverbération *m*; ~ **type** COMP type de réverbération *m*
reversal *n* POWER, TELEP *of answer signal*, TRANS inversion *f*; ~ **control channel** TELEP voie de mobile *f*
reversals *n pl* TELEP roulements *m pl*
reverse[1]: ~-**biased** *adj* PARTS, RADIO à polarisation inverse
reverse[2]: ~ **bias** *n* RADIO polarisation inverse *f*; ~ **biasing** *n* RADIO polarisation inverse *f*; ~ **blocking current** *n* ELECT, POWER courant inverse de blocage *m*; ~ **blocking diode thyristor** *n* ELECT, PARTS thyristor diode bloqué en inverse *m*; ~ **blocking state** *n* ELECT, PARTS état bloqué dans le sens inverse *m*; ~-**charge acceptance** *n* *BrE* (*cf collect-call acceptance AmE*) APPL, TELEP acceptation de la taxation à l'arrivée *f*, acceptation de la taxe à l'arrivée *f*; ~-**charge acceptance not subscribed signal** *n* *BrE* (*cf collect-call acceptance not subscribed signal AmE*) SIG, TELEP signal de non-abonnement à l'acceptation de la taxation à l'arriv e *m*; ~-**charge call** *n* *BrE* (*cf collect call AmE*) TELEP appel à frais virés *m* (*Can*), appel en PCV *m* (*Fra*), communication en PCV *f* (*Fra*), communication payable à l'arrivée *f* (*Fra*); ~ **charging** *n* APPL, TELEP facturation à l'arrivée *f*; ~-**charging acceptance** *n* *BrE* (*cf collect-calling acceptance AmE*) APPL, TELEP acceptation de la taxation à l'arrivée *f*, acceptation de la taxe à l'arrivée *f*; ~-**charging request indicator** *n* *BrE* (*cf collect-calling request indicator AmE*) APPL, TELEP indicateur de demande de taxation à l'arrivée *m*; ~ **conducting diode thyristor** *n* ELECT, PARTS thyristor diode passant en inverse *m*; ~ **current** *n* ELECT courant inverse *m*; ~-**current circuit** *n* ELECT circuit à retour de courant *m*, circuit inverse *m*; ~-**current relay** *n* ELECT relais à retour de courant *m*; ~ **DC resistance** *n* ELECT, POWER résistance inverse en courant continu *f*; ~ **direction** *n* ELECT sens inverse *m*; ~ **display** *n* COMP affichage en inverse *m*; ~ **electrode current** *n* ELECT courant inverse d'électrode *m*; ~ **engineering** *n* COMP, ELECT ingénierie inverse *f*; ~ **key** *n* COMP touche d'inversion *f*; ~ **polarity** *n* POWER polarité inverse *f*; ~ **priority** *n* COMP priorité inverse *f*; ~ **recovery** *n* ELECT récupération inverse *f*; ~ **recovery current** *n* ELECT courant de recouvrement inverse *m*; ~ **routing** *n* SIG, SWG, TELEP, TRANS acheminement retour *m*; ~ **routing address** *n* DATA, SWG adresse d'acheminement retour *f*; ~ **video** *n* COMP vidéo inverse *f*; ~ **voltage** *n* ELECT tension inverse *f*
reverse[3] *vt* COMP, ELECT, POWER inverser; ~ **the charges** *BrE* (*cf call collect AmE*) APPL, TELEP téléphoner en PCV
reversed: ~ **charging** *n* *BrE* (*collect-call service AmE*) APPL, TELEP service à frais virés *m* (*Can*), service payable à l'arrivée *m*, service PCV *m*
reversible: ~ **blade** *n* ELECT lame réversible *f*; ~ **flexible disk** *n* COMP disquette double face *f*
reversing: ~ **switch** *n* POWER inverseur *m*
revert *vi* COMP retourner
revertible: ~ **pulse** *n* SWG impulsion inverse *f*
review *vt* GEN renouveler
revisable *adj* DATA retraitable
revise *vt* COMP retraiter
revision: ~ **state** *n* GEN état de révision *m*, indice d'état *m*
revolver *n* DATA boucle d'accès rapide *f*; ~ **loop** DATA boucle d'accès rapide *f*
rewritable: ~ **compact disc** *n* COMP, DATA disque compact réinscriptible *m*; ~ **optical disk** *n* COMP, DATA disque optique réinscriptible *m*
rewrite *vt* DATA réécrire
RF *abbr* (*radio frequency*) ELECT, RADIO, TELEP HF (*haute fréquence*), RF (*radiofréquence*); ~ **amplifier** *n* TELEP amplificateur à haute fréquence *m*, amplificateur de fréquences radioélectriques *m*, amplificateur RF *m*; ~ **channel** *n* RADIO canal radioélectronique *m*, canal RF *m*, radiocanal *m*, canal radioélectrique *m*; ~ **head** *n* RADIO tête à radiofréquence *f*; ~ **scanner** *n* RADIO récepteur à balayage *m*; ~ **test** *n* RADIO, TEST essai en radiofréquence *m*
RFI *abbr* (*radio-frequency interference*) RADIO brouillage *m*, interférence radioélectrique *f*, perturbation radioélectrique *f*; ~ **caulking** *n* RADIO calfeutrage antibrouillage RF *m*
RFP *abbr* (*request for proposal*) GEN appel d'offres *m*
RFS *abbr* (*receive-frame synchronization procedure*) TRANS procédure de synchronisation de la trame de réception *f*
RG *abbr* (*regenerator*) ELECT, SIG, SWG, TELEP RG (*régénérateur*)
RGB *abbr* (*red-green-blue*) TV RVB (*rouge-vert-bleu*); ~ **color model** *AmE*, ~ **colour model** *BrE* *n* TV modèle de couleurs RVB *m*
rheostat *n* PARTS, POWER rhéostat *m*
rhodium[1]: ~ **plating** *n* GEN rhodiage *m*
rhodium[2]: ~-**plate** *vt* GEN rhodier
rhombic: ~ **antenna** *n* RADIO antenne rhombique *f*, antenne en losange *f*
rhythm *n* COMP rythme *m*; ~ **pattern** COMP motif rythmique *m*; ~ **programmer** COMP boîte à rythmes *f*, composeur de rythmes *m*, programmeur de rythmes *m*; ~ **volume slider** COMP curseur de volume de rythme *m*
rhythmic: ~ **guide** *n* COMP guide rythmique *m*

RI *abbr* (*response identifier*) TELEP identificateur de réponse *m*
ribbon *n* ELECT ruban *m*, GEN bande *f*, ruban *m*; **~ cable** TRANS câble à structure ruban *m*, câble plat *m*, câble ruban *m*; **~ matrix** ELECT matrice en ruban *f*; **~ microphone** TELEP microphone à ruban *m*
rich: **~ text format** *n* (*RTF*) COMP rich text format *m* (*RTF*)
Rician: **~-fading channel** *n* ELECT canal avec évanouissement Rice *m*
ridge: **~ waveguide** *n* DATA, RADIO, TRANS guide d'onde à moulure *m*, guide d'onde nervuré *m*, guide d'ondes réctangulaires *m*
right[1]: **~-hand polarization** *n* RADIO polarisation dextrorsum *f*, polarisation dextrogyre *f*
right[2]: **~-justify** *vt* COMP, DATA cadrer à droite, justifier à droite; **~-size** *vt* ELECT redimensionner
rigid: **~ flex card** *n* ELECT carte flexo-rigide *f*; **~ plastic** *n* PARTS plastique rigide *m*
rigidize: **~-sensing grapple fixture** *n* CONTR dispositif de préhension détecteur de contact optimal *m*
RIN *abbr* (*relative-intensity noise*) ELECT bruit d'intensité *m*
ring[1] *n* DATA, ELECT anneau *m*, PARTS *of wire* conducteur de nuque *m*, fil de nuque *m*; **~-access control** ELECT contrôle d'accès à l'anneau *m*; **~ adaptor** DATA adaptateur d'anneau *m*; **~ again** (*RAG*) PER APP, TELEP rappel automatique *m* (*RAG*); **~ again on idle EBS** TELEP rappel automatique sur libération des correspondants *m*; **~ antenna** RADIO antenne en anneau *f*; **~-back** TELEP retour d'appel *m*; **~-back signal** TELEP signal de retour d'appel *m*; **~-back tone** TELEP tonalité de retour d'appel *f*; **~ configuration** TRANS configuration en anneau *f*, configuration en boucle *f*; **~ current** TELEP courant de sonnerie *m*; **~-current set** PER APP, TELEP générateur de courant d'appel *m*; **~-down signaling** *AmE*, **~-down signalling** *BrE* GEN signalisation manuelle *f*; **~-forward signal** SIG, TELEP signal de rappel *m*; **~ indicator** TELEP indicateur de sonnerie *m*; **~ main** POWER circuit bouclé *m*; **~-mode filter** RADIO filtre de mode à anneaux *m*; **~ modulator** ELECT modulateur en anneau *m*; **~ network** DATA, TRANS réseau en anneau *m*; **~ oscillator** ELECT oscillateur en anneau *m*; **~ redundancy** RADIO, TRANS redondance en anneau *f*, redondance en mode annulaire *f*; **~ start** TELEP déclenchement du signal d'appel *m*; **~ switch** RADIO commutateur à anneau *m*; **~ tone** TELEP tonalité de retour d'appel *f*; **~-tone generator** TELEP générateur de tonalité de retour d'appel *m*; **~ topology** DATA topologie en anneau *f*; **~ translator** TRANS traducteur toroïdal *m*; **~ trip** TELEP arrêt immédiat de l'appel *m*, arrêt de sonnerie *m*, arrêt de courant d'appel *m*; **~ tripping** TELEP arrêt immédiat de l'appel *m*, arrêt de sonnerie *m*, arrêt de courant d'appel *m*; **~-trip relay** TELEP relais d'arrêt de sonnerie *m*; **~ wire** (*R-wire*) TELEP fil de nuque *m* (*fil R*); **~ wrap** ELECT bouclage d'anneau *m*
ring[2] *vt* *BrE* (*cf call AmE*) TELEP appeler, téléphoner, téléphoner à; **~ back** *BrE* (*cf call back AmE*) TELEP *person* rappeler
ring[3] *vi* TELEP *BrE* (*cf call AmE*), téléphoner, *bell* sonner; **~ out** TELEP sonner
ringer *n* TELEP sonnerie *f*; **~ sensitivity** TELEP sensibilité de la sonnerie *f*
ringing *n* RADIO suroscillation *f*, oscillation parasite *f*, TELEP appel *m*, sonnerie *f*; **~ current** TELEP courant d'appel *m*; **~-current set** COMP, TELEP générateur de courant d'appel *m*; **~-current transformer** TELEP transformateur de courant d'appel *m*; **~ duration** TELEP durée de sonnerie *f*; **~ generator** TELEP générateur de signal d'appel *m*; **~ interruptor** TELEP interrupteur de signal d'appel *m*; **~ machine** TELEP générateur de courant de sonnerie *m*; **~ oscillator** RADIO oscillateur excité par choc *m*; **~ period** TELEP phase d'appel *f*, temps de sonnerie *m*; **~-period adjustment** PER APP réglage de fréquence de sonnerie *m*; **~ periodicity** TELEP rythme d'appel *m*; **~ pilot indicator** TELEP indicateur d'appel *m*; **~ position** TELEP position de sonnerie *f*; **~ set** COMP, TELEP générateur de courant d'appel *m*; **~ signal** TELEP signal de retour d'appel *m*, tonalité de retour d'appel *f*; **~ time** TV durée de sonnerie *f*; **~ tone** TELEP tonalité de retour d'appel *f*, signal de retour d'appel *m*; **~ voltage** TELEP tension d'appel *f*; **~-voltage generator** TELEP générateur de tension d'appel *m*
RIP *abbr* (*routing information protocol*) DATA protocole d'informations de routage *m*
ripcord *n* ELECT fil de déchirement *m*
ripple *n* ELECT ondulation *f*; **~ clocking arrangement** ELECT montage asynchrone *m*; **~ component** POWER composant d'ondulation *m*, composant ondulatoire *m*; **~-through counter** TEST compteur asynchrone *m*; **~ voltage** DATA tension d'ondulation *f*
RISC *abbr* (*reduced-instruction-set computer*) COMP, DATA RISC (*ordinateur à jeu d'instructions réduit, ordinateur RISC*); **~ processor** *n* DATA processeur RISC *m*
rise: **~ time** *n* COMP, GEN, TELEP temps de montée *m*; **~-time coefficient** *n* COMP, GEN, TELEP coefficient du temps de montée *m*
riser *n* ELECT colonne montante *f*; **~ shaft** ELECT colonne montante *f*
rising: **~ exponential curve** *n* COMP courbe exponentielle ascendante *f*; **~ linear curve** *n* COMP courbe linéaire ascendante *f*
risk: **~ analysis** *n* TEST analyse du risque *f*; **~ assessment** *n* TEST estimation du risque *f*; **~ economy** *n* TEST économie du risque *f*; **~ evaluation** *n* TEST évaluation du risque *f*; **~ identification** *n* TEST identification du risque *f*; **~ level** *n* TEST niveau du risque *m*; **~ quantification** *n* TEST valorisation du risque *f*
Rivest: **~, Shamir, Adleman algorithm** *n* (*RSA*) TELEP algorithme de Rivest, Shamir et Adleman *m*
RJE *abbr* (*remote job entry*) DATA soumission de travaux à distance *f*, télésoumission de travaux *f*
RLE *abbr* (*run-length encoding*) COMP codage par plages *m*
RLT *abbr* (*release link trunk*) TELEP ligne de concentration de trafic *f*
rms *abbr* (*root mean square*) ELECT valeur quadratique moyenne *f*, GEN écart type *m*; valeur quadratique moyenne *f*; **~ value** *n* ELECT, GEN valeur efficace *f*
RN *abbr* (*receipt notification*) DATA, TELEP avis de réception *m*, notification de réception *f*
RNR: **~ frame** *n* TELEP trame RNR *f*
RO *abbr* (*remote operation*) DATA, TEST exploitation à distance *f*, opération distante *f*
ROA *abbr* (*recognized operating agency*) APPL, DATA ER (*exploitation reconnue*)
roadside: **~ cabinet** *n* TV armoire quartier *f*; **~ furniture** *n* TV armoire de sous-répartition *f*
roam *vi* RADIO quitter la cellule de localisation nominale
roamer *n* RADIO, TELEP abonné itinérant *m*, usager itinérant *m*
roaming *n* RADIO itinérance *f*, localisation automatique *f*, réacheminement automatique *m*, suivi automatique des déplacements *m*, suivi des mobiles *m*; **~ capability** RADIO fonction d'enregistrement automatique *f*; **~ number** RADIO numéro de réacheminement *m*; **~ station** RADIO station mobile itinérante *f*; **~ subscriber** RADIO abonné itinérant *m*
robbed: **~-bit signaling** *AmE*, **~-bit signalling** *BrE* *n* DATA, SIG signalisation par réassignation de bit *f*
rocker: **~ button** *n* POWER bouton à bascule *m*
rocky: **~-point effect** *n* ELECT effet rocky-point *m*
rod *n* ELECT baguette *f*; **~ antenna** RADIO antenne-tige *f*; **~-in-tube technique** ELECT, TRANS *fibre optics* procédé de la tige dans le tube *m*; **~ reflector** RADIO réflecteur à tiges *m*, miroir grille *m*, réflecteur d'antenne à tiges *m*
rodent: **~ protection** *n* ELECT protection contre rongeurs *f*
Roger: **~ number** *n* RADIO numéro Roger *m*

roger *phr* RADIO message bien compris, message bien reçu, message compris

roll[1]: **~ axis** *n* RADIO axe de roulis *m*; **~-off** *n* RADIO décroissance *f*, pente de diminution *f*; **~-out** *n* COMP retraite *m*, RADIO déploiement *m*

roll[2]: **~ in** *vt* COMP rappeler; **~ out** *vt* COMP retirer, DATA retraire, RADIO déployer

rolling: **~ ball** *n* COMP boule de commande *f*; **~-code band splitting** *n* DATA découpage de la bande selon un code de brassage *m*; **~-track ladder** *n* PARTS *supply & distribution* échelle à roulettes *f*, échelle roulante *f*

rolls: **in ~** *adj* ELECT en rouleaux

ROLR *abbr* (*receiving objective loudness rating*) TELEP équivalent objectif pour la sonie à la réception *m*

ROM *abbr* (*read-only memory, read-only storage*) COMP, DATA MEM (*mémoire morte*), ROM (*mémoire fixe, mémoire inaltérable*)

rooftop: **~ aerial** *n* TV antenne de toit *f*; **~ antenna** *n* TV antenne de toit *f*

room *n* COMP salle *f*; **~ acoustics** TELEP acoustique de salle *f*; **~ noise** TELEP bruit de salle *m*; **~ temperature** CLIM température ambiante *f*

root *n* COMP racine *f*, tête *f*; **~ directory** COMP répertoire racine *m*; **~ mean square** (*rms*) ELECT valeur quadratique moyenne *f*, GEN écart type *m*, valeur quadratique moyenne *f*; **~-mean square value** ELECT, GEN valeur efficace *f*; **~ node** DATA nœud racine *m*; **~ sum square** ELECT, GEN valeur quadratique *f*; **~ top node** COMP racine *f*, tête *f*; **~ tree** DATA arbre racine *m*

rosin: **~ flux** *n* ELECT flux résineux *m*

rotary: **~ attenuator** *n* RADIO affaiblisseur rotatif *m*, atténuateur rotatif *m*; **~ control** *n* PARTS tournant *m*; **~ direction** *n* TELEP sens de rotation *m*; **~ exchange** *n* SWG central à cadran *m*; **~ ferrule** *n* ELECT virole rotative *f*; **~-head digital audio tape** *n* (*R-DAT*) COMP, DATA bande audionumérique à tête rotative *f* (*R-DAT*); **~ hunt** *n* TELEP recherche simple niveau *f*; **~ hunting** *n* TELEP recherche simple niveau *f*; **~ phase changer** *n* RADIO déphaseur rotatif *m*; **~ phase shifter** *n* RADIO déphaseur rotatif *m*; **~ selector** *n* SWG sélecteur rotatif *m*; **~ switch** *n* COMP commutateur rotatif *m*, interrupteur rotatif *m*, sélecteur rotatif *m*, GEN interrupteur rotatif *m*, commutateur rotatif *m*, sélecteur rotatif *m*, SWG interrupteur rotatif *m*, sélecteur rotatif *m*; **~ system** *n* SWG système rotary *m*, système rotatif *m*, système rotatif à commutateurs rotatifs *m*

rotatable: **~ aerial** *n* RADIO antenne orientable *f*; **~ antenna** *n* RADIO antenne orientable *f*

rotate *vt* COMP effectuer une rotation, pivoter

rotating: **~ field** *n* POWER, RADIO, TELEP champ tournant *m*; **~ field alternator** *n* POWER, RADIO, TELEP alternateur à champ tournant *m*; **~ joint** *n* POWER, RADIO, TELEP joint tournant *m*

rotational: **~ delay** *n* COMP temps d'attente *m*, délai rotationnel *m*

rotor *n* TELEP rotor *m*

rough: **~ surface** *n* RADIO surface rugueuse *f*

roughness *n* GEN rugosité *f*

round: **~-robin hunting** *n* TELEP recherche par décalage en fonction de la dernière ligne occupée *f*; **~-the-world echo** *n* RADIO écho tour de terre *m*

route[1] *n* ELECT axe *m*, RADIO voie *f*, TRANS faisceau *m*, itinéraire d'acheminement *m*, route *f*, voie d'acheminement *f*; **~ analysis** TELEP analyse de faisceau *f*, analyse de l'acheminement *f*; **~ blocking** TELEP blocage de faisceau *m*; **~-choice data** *n pl* TELEP données d'acheminement *f pl*; **~ congestion** TELEP encombrement d'itinéraire *m*, encombrement de faisceau *m*; **~ congestion counter** SWG, TELEP compteur d'encombrements de faisceau *m*; **~ determination** TELEP détermination de faisceau *f*, détermination d'itinéraire *f*; **~-idle marking** TELEP marquage de voie au repos *m*; **~ information digits** *n pl* TELEP chiffres d'identification de voie d'acheminement *m pl*; **~ list** TELEP liste des voies d'acheminement *f*; **~ lockout supervision** TELEP supervision de blocage de route *f*; **~ marker** TELEP marqueur de route *m*; **~ numbering** TELEP numérotage de route *m*; **~ relay** TELEP relais de route *m*, **~ seizure** TELEP occupation d'une route *f*, prise d'une route *f*; **~ supervision** TELEP supervision de faisceau *f*, supervision d'itinéraire *f*; **~ supervision panel** TELEP tableau de supervision des itinéraires *m*, tableau de surveillance des voies *m*; **~ test** TELEP test de route *m*; **~ test block** TELEP bloc de test de route *m*

route[2] *vt* GEN, TELEP, TRANS acheminer, router

router *n* DATA, ELECT, TELEP, TRANS routeur *m*

routine *n* COMP, DATA routine *f*; **~ acquisition mode** DATA mode d'acquisition périodique *m*; **~ monitoring** TEST surveillance courante *f*; **~ repetition** TELEP répétition d'office *f*; **~ task** COMP fonction courante *f*; **~ test** TEST essai de série *m*, essai préventif *m*, essai systématique *m*

routiner *n* TEST dispositif d'essai automatique *m*, unité d'essai automatique *f*, robot d'essai individuel *m*

routing *n* COMP, DATA routage *m*, SWG itinéraire *m*, SIG, TELEP acheminement *m*, routage *m*, TRANS acheminement *m*, *of circuit* routage *m*; **~ address** DATA, SWG adresse d'acheminement *f*; **~ case** TELEP type d'acheminement *m*; **~ code** TELEP indicatif d'acheminement *m*; **~ control** DATA contrôle de routage *m*; **~ data** *n pl* TELEP données d'acheminement *f pl*; **~ delay** DATA délai d'acheminement *m*; **~ device** DATA dérouteuse *f*; **~ digits** *n pl* SWG chiffres d'acheminement *m pl*; **~ diversity** TRANS diversité d'acheminement *f*; **~ information protocol** (*RIP*) DATA protocole d'informations de routage *m*; **~ label** TELEP étiquette d'acheminement *f*; **~ number** TELEP indicatif d'acheminement *m*; **~ number area** TELEP zone de numérotage *f*; **~ program** TELEP programme d'allocation d'itinéraire *m*, programme de recherche d'itinéraire *m*; **~ table maintenance protocol** (*RTMP*) COMP protocole de maintenance de tables de routage *m*; **~ wings** *n pl* ELECT colliers d'acheminement *m pl*

row *n* COMP, GEN ligne *f*, TELEP rangée *f*, travée *f*; **~ aisle** TELEP couloir intertravée *m*; **~ chute** TELEP goulotte de rangée *f*; **~-end cover** TELEP capot d'extrémité de rangée *m*; **~ length** TELEP longueur de rangée *f*; **~ mechanical structure** TELEP structure mécanique de travée *f*; **~ pitch** DATA pas longitudinal *m*; **~ spacing** TELEP espace intertravée *m*

RP *abbr* (*regional processor*) SWG processeur régional *m*

RPE *abbr* (*remote peripheral equipment*) COMP EPD (*équipement périphérique distant*)

RPF *abbr* (*reference publication format*) DATA format de référence des publications *m*

RPOA *abbr* (*recognized private operating agency*) APPL, DATA EPR (*exploitation privée reconnue*); **~ out-of-order signal** *n* TELEP signal d'EPR hors service *m*; **~ selection indicator** *n* TELEP indicateur de choix de l'EPR *m*; **~ transit-network identity** *n* TELEP identité de réseau de transit d'EPR *f*

RR *abbr* (*receive-ready*) TELEP prêt à recevoir

R-reference: **~ point** *n* DATA point de référence R *m*

RS *abbr* (*Reed-Solomon*) COMP RS (*Reed-Solomon*), (*regenerator section*) DATA section de régénération *f*, SIG RS (*Reed-Solomon*); **~ code** (*Reed-Solomon code*) COMP, SIG code RS *m* (*code de Reed-Solomon*); **~ test method** *n* TEST méthode de test monocouche à distance *f*

RSA *abbr* (*Rivest, Shamir, Adleman algorithm*) TELEP algorithme de Rivest, Shamir et Adleman *m*

RSB *abbr* (*residual sideband*) RADIO, TRANS, TV BLR (*bande latérale résiduelle*)

RSM *abbr* (*remote spectrum-monitoring*) RADIO surveillance à distance du spectre *f*

RST *abbr* (*regenerator-section termination*) DATA terminaison de section de régénération *f*

RSU *abbr* (*remote switching unit*) SWG centre satellite *m*

RT *abbr* DATA (*reliable transfer*) transfert fiable *m*, PER APP (*remote terminal*) terminal éloigné *m*, RADIO (*re-emission time*) temps de réémission *m*

RTD *abbr* (*resistive temperature detector*) CLIM détecteur de température résistif *m*

RTF *abbr* (*rich text format*) COMP RTF (*rich text format*)

RTMP *abbr* (*routing table maintenance protocol*) COMP protocole de maintenance de tables de routage *m*

RTN *abbr* (*retrain negative*) DATA, TELEP RCDN (*reconditionnement négatif*)

RTP *abbr* DATA (*retrain positive*) reconditionnement positif *m*, TELEP (*reference telephonic power*) volume de référence *m*, (*retrain positive*) reconditionnement positif *m*

RTS *abbr* COMP (*reliable transfer server*) serveur de transfert fiable *m*, DATA (*request to send*) demande pour émettre *f*

RTSE: **~ provider** *n* DATA, TELEP fournisseur de RTSE *m*; **~ user** *n* DATA utilisateur de RTSE *m*

RTSI *abbr* (*reference terminal status interchange*) RADIO échange d'états du terminal de référence *m*

RU *abbr* (*remote unit*) SWG unité éloignée *f*, unité distante *f*

rub *vt* COMP effacer; **~ out** COMP, RADIO décaler

rubber: **~ banding** *n* COMP étirement des lignes *m*

rule *n* COMP règle *f*; **~-based security policy** DATA politique de sécurité fondée sur des règles *f*

run[1] *n* COMP passage *m*, passage-machine *m*, TEST suite *f*, TRANS tronçon *m*; **~ duration** COMP durée d'exécution *f*; **~-length encoding** (*RLE*) COMP codage par plages *m*; **~ stream** COMP suite de travaux *f*, flot de travaux *m*; **~ time** COMP durée d'exécution *f*, moment de l'exécution *m*

run[2] *vt* COMP *program* mettre en route, tourner, CONTR exploiter, DATA *program* tourner, *at certain rate* cadencer, ELECT poser, GEN exploiter, poser

runner: **~-up** *n* COMP second *m*

running *n* COMP mise en route *f*, CONTR exploitation *f*, ELECT, PARTS pose *f*, TELEP exploitation *f*, TRANS pose *f*; **~ bar** GEN guide *m*, PARTS rail de guidage *m*; **~ diagram** COMP diagramme de fonctionnement *m*; **~ list** GEN tableau de disposition des fils *m*; **~ time** COMP exécution *f*, stade d'exécution *m*; **~-time meter** TELEP compte-temps *m*, compteur de temps de fonctionnement *m*

rupture *n* PARTS rupture *f*; **~ strain** PARTS allongement de rupture *m*

rural: **~ automatic exchange** *n* SWG central automatique rural *m*, commutateur automatique rural *m*, TELEP commutateur automatique rural; **~ exchange** *n* TELEP central rural *m*, commutateur rural *m*; **~ network** *n* TELEP téléphonie rurale *f*, TRANS réseau rural *m*; **~ switch** *n* SWG commutateur rural *m*; **~ telephone system** *n* TELEP système téléphonique rural *m*; **~ telephony** *n* TELEP téléphonie rurale *f*

R-wire *n* (*ring wire*) TELEP fil R *m* (*fil de nuque*)

RX *abbr* DATA, PER APP, RADIO, TELEP (*receiver*) récepteur *m*, TEST, TRANS (*receive direction*) sens réception *m*; **~ sens** *n* RADIO sensibilité du récepteur *f*

RZ *abbr* (*return to zero*) COMP, ELECT, TELEP retour à zéro *m*

S

S: **~ switch** *n* (*space switch*) SWG, TELEP commutateur S *m* (*commutateur spatial*)

SA *abbr* APPL (*subaddress*) sous-adresse *f*, DATA (*section adaptation*) adaptation de section *f*, (*subaddress*) sous-adresse *f*, TELEP (*subaddress*) sous-adresse *f*, (*system address*) AS (*adresse système*)

SABM *abbr* (*set asynchronous balanced mode*) TELEP fonctionnement en mode asynchrone symétrique *m*, mise en mode asynchrone symétrique *f*

SABME *abbr* (*set asynchronous balanced mode extended*) TELEP fonctionnement en mode asynchrone symétrique étendu *m*

safeguard[1] *n* COMP, DATA protection *f*, sauvegarde *f*, GEN sécurité *f*

safeguard[2] *vt* COMP, DATA sauvegarder

safety *n* GEN sécurité *f*; **~ lock** TEST serrure de sûreté *f*, verrou de sûreté *m*; **~ shutdown system** CONTR système d'arrêt d'urgence *m*

sag *n* PARTS flèche *f*

SAM *abbr* (*spectrum-analyser monitoring BrE*, *spectrum-analyzer monitoring AmE*) RADIO surveillance par analyseur de spectre *f*

sample[1] *n* GEN échantillon *m*, TELEP, TEST, TRANS prélèvement *m*; **~ division** TRANS subdivision de l'échantillon *f*; **~ and hold** CONTR échantillon et mémorisation *m*, échantillonnage et maintien *m*, échantillonneur-bloqueur *m*; **~ and hold action** CONTR action par échantillonnage et maintien *f*; **~ number** TRANS numéro d'échantillon *m*; **~ point** TEST point d'échantillonnage *m*; **~ precision** TELEP précision des échantillons *f*; **~ preparation** TEST préparation de l'échantillon *f*; **~ pulse** TEST impulsion d'échantillonage *f*; **~ space** TEST espace des résultats possibles *m*

sample[2] *vt* GEN échantillonner, TELEP, TEST, TRANS prélever

sampled: **~ signal** *n* SIG, TRANS signal échantillonné *m*

sampler *n* GEN échantillonneur *m*; **~ module** TEST module d'échantillonnage *m*

samples: **~ per second** *phr* TRANS échantillons par seconde

sampling *n* GEN échantillonnage *m*; **~ error** TRANS erreur d'échantillonnage *f*; **~ fraction** TRANS taux d'échantillonnage *m*; **~ frequency** TRANS fréquence d'échantillonnage *f*, taux d'échantillonnage *m*; **~ gate** TRANS porte d'échantillonnage *f*; **~ inspection** TEST contrôle par échantillonnage *m*; **~ interval** TRANS intervalle d'échantillonnage *m*; **~ plan** TEST plan d'échantillonnage *m*; **~ procedure** TEST procédure d'échantillonnage *f*; **~ process** TRANS procédé d'échantillonnage *m*; **~ rate** TRANS fréquence d'échantillonnage *f*, taux d'échantillonnage *m*; **~ scheme** TEST programme d'échantillonnage *m*; **~ system** TEST système d'échantillonnage *m*; **~ test** TEST essai par échantillonnage *m*, essai sur prélèvement *m*; **~ time** TRANS temps d'échantillonnage *m*; **~ unit** TEST unité d'échantillonnage *f*

sandwich: **~ line** *n* TRANS ligne sandwich *f*

SAP *abbr* (*service access point*) DATA point d'accès au service *m*

SAPI *abbr* (*service access point identifier*) DATA identificateur de point d'accès au service *m*

SAR *abbr* (*segmentation and reassembly*) DATA segmentation et réassemblage *f*

SAS *abbr* (*single-attachment station*) ELECT station à raccordement simple *f*

satellite *n* RADIO, TV satellite *m*; **~ aerial** RADIO, TV antenne satellite *f*; **~ antenna** RADIO, TV antenne satellite *f*; **~ attitude** RADIO position du satellite *f*; **~ band** RADIO bande de satellite *f*; **~ broadcast** RADIO, TV émission par satellite *f*; **~ broadcasting** RADIO, TV radiodiffusion par satellite *f*; **~ broadcasting service** RADIO, TV service de radiodiffusion par satellite *m* (*SRS*); **~ business system** (*SBS*) RADIO système à satellite pour services inter-entreprises *m*; **~ business telecommunications network** TELEP, TRANS réseau de télécommunications d'entreprises par satellite *m*; **~ channel** RADIO, TV voie satellite *f*; **~ closet** ELECT armoire satellite *f*; **~ communications** *n pl* RADIO, TELEP, TV télécommunications par satellite *f pl*; **~ conference** TV visioréunion *f*; **~ control center** *AmE*, **~ control centre** *BrE* (*SCC*) RADIO centre de contrôle des satellites *m*; **~ exchange** SWG central auxiliaire *m*, central satellite *m*; **~ in tandem** RADIO satellite en chaîne *m*; **~ link** RADIO, TV liaison par satellite *f*; **~-link noise temperature** RADIO température de bruit de la liaison par satellite *f*; **~ location** RADIO emplacement du satellite *m*; **~ loopback mode** RADIO mode de boucle de satellite *m*; **~ master aerial television** (*SMATV*) TV télévision par satellite à antenne principale *f*; **~ master aerial television system** TV système de télévision par satellite à antenne principale *m*; **~ master antenna television** (*SMATV*) TV télévision par satellite à antenne principale *f*; **~ master antenna television system** TV système de télévision par satellite à antenne principale *m*; **~ multiservice system** RADIO, TV système multiservice par satellite *m*; **~ network** RADIO, TV réseau à satellite *m*; **~ news gathering** (*SNG*) RADIO, TV contributions par satellite *f pl* (*SNG*), journalisme électronique par satellite *m*, reportage par satellite *m*, reportage électronique d'actualités par satellite *m*, reportage d'actualités par satellite *m*; **~ outage** RADIO, TV défaillance de satellite *f*; **~ paging system** RADIO système de radiomessagerie par satellite *m*; **~ program** *AmE*, **~ programme** *BrE* TV programme par satellite *m*; **~ range** RADIO distance du satellite *f*; **~ receiver aerial** RADIO antenne de réception par satellite *f*; **~ receiver antenna** RADIO antenne de réception par satellite *f*; **~ relay system** RADIO système de relais par satellite *m*; **~ service** RADIO, TV service par satellite *m*; **~ signaling channel** *AmE*, **~ signalling channel** *BrE* SIG voie de signalisation par satellite *f*; **~ spot beam** RADIO faisceau d'antenne du satellite *m*; **~ switch** RADIO, SWG commutateur satellite *m*; **~ switching** RADIO, SWG commutation satellite *f*; **~ switching center** *AmE*, **~ switching centre** *BrE* (*SSC*) RADIO, SWG centre de commutation à bord du satellite *m*; **~ system** RADIO, TV système à satellite *m*; **~ telecast** TV émission de télévision par satellite *f*; **~ television broadcasting** TV diffusion d'émissions de télévision par satellite *f*; **~ television channel** TV canal satellite de TV *m*; **~-to-satellite link** RADIO, TV liaison entre satellites *f*; **~ tracking** RADIO poursuite du satellite *f*; **~ transmission** RADIO, TRANS transmission par satellite *f*; **~ transmission path** RADIO, TRANS, TV voie de transmission par satellite *f*; **~ transmit aerial** RADIO antenne d'émission du satellite *f*; **~ transmit aerial gain** RADIO gain des antennes d'émission du satellite *m*; **~ transmit antenna** RADIO antenne d'émission du satellite *f*; **~ transmit antenna gain** RADIO gain des antennes d'émission du satellite *m*; **~ transmit power** RADIO puissance d'émission au satellite *f*; **~ transponder** RADIO répéteur de satellite *m*

SATS *abbr* (*selected abstract test suite*) DATA, TEST suite de tests abstraits sélectionnés *f*
saturated[1] *adj* ELECT saturé
saturated[2]: **~ state** *n* ELECT état saturé *m*
saturation *n* ELECT saturation *f*; **~ gain** ELECT gain en régime de saturation *m*; **~ power** ELECT puissance de saturation *f*; **~ resistance** ELECT résistance de saturation *f*; **~ state** ELECT saturation *f*; **~ voltage** ELECT tension de saturation *f*
save[1] *n* COMP sauvegarde *f*; **~ operation** COMP opération de sauvegarde *f*; **~ symbol** COMP symbole de mise en réserve *m*; **~ temporary operator** COMP sauvegarde temporaire d'opérateur *f*; **~ to cartridge** COMP sauvegarde sur cartouche *f*; **~ to cassette** COMP sauvegarde sur cassette *f*; **~ to diskette** COMP sauvegarde sur disquette *f*
save[2] *vt* COMP sauvegarder; **~ to cartridge** COMP sauvegarder sur cartouche; **~ to cassette** COMP sauvegarder sur cassette; **~ to diskette** COMP sauvegarder sur disquette
saving *n* DATA sauvegarde *f*
SAW *abbr* (*surface acoustic wave*) ELECT, TRANS onde acoustique de surface *f*; **~ device** *n* ELECT, PARTS, TRANS dispositif à ondes acoustiques de surface *m*
sawtooth: **~ current** *n* ELECT courant en dents de scie *m*; **~ generator** *n* ELECT générateur de base de temps *m*; **~ keyboard** *n* PER APP clavier à action directe *m*; **~ pulse** *n* ELECT impulsion en dents de scie *f*; **~ voltage** *n* ELECT tension en dents de scie *f*; **~ wave** *n* ELECT, TRANS onde en dents de scie *f*
SB *abbr* RADIO (*sideband*) BL (*bande latérale*), TELEP (*sideband*) BL (*bande latérale*), (*system busy*) OS (*occupation par le système*), TRANS (*sideband*) BL (*bande latérale*)
SB1 *abbr* (*scrambled binary ones*) DATA binaires brouillés *m pl*
SBS *abbr* (*satellite business system*) RADIO système à satellite pour services inter-entreprises *m*
SBV *abbr* (*syntax-based videotex*) TELEP vidéotex syntaxique *m*
S/C *abbr* (*short circuit*) COMP, ELECT, POWER court-circuit *m*
scalability *n* GEN évolutivité *f*
scale *n* COMP, DATA, TELEP gamme *f*; **~ of charges** APPL barème *m*; **~ discrimination** TEST résolution *f*; **~ division** TEST division de l'échelle *f*, échelon *m*; **~ mark** TEST repère de graduation *m*, graduation *f*; **~ numbering** TEST chiffraison *f*; **~ range** TEST zone de l'échelle *f*
scaling: **~ down** *n* TEST réduction *f*
scan[1] *n* COMP balayage *m*, ELECT balayage *m*, trace *f*, PER APP, RADIO balayage *m*, scan *m*; **~ converter** RADIO convertisseur de balayage *m*; **~ line** COMP ligne de balayage *f*
scan[2] *vt* COMP balayer, explorer, *character recognition* analyser, ELECT balayer, PER APP scannériser, RADIO balayer
scanned: **~ image** *n* PER APP image scannérisée *f*
scanner *n* COMP analyseur *m*, explorateur *m*, lecteur *m*, PER APP, RADIO appareil de balayage *m*, explorateur *m*, scanner *m*; **~ distributor** ELECT distributeur analyseur *m*; **~ signal** SIG signal d'analyse *m*; **~ unit** SWG analyseur *m*
scanning *n* COMP balayage *m*, exploration *f*, ELECT balayage *m*, GEN exploration *f*, PER APP balayage, *character recognition* scannérisation *f*, analyse *f*, scannage *m*, RADIO balayage *m*, scannage *m*, exploration *f*, SIG synthèse *f*, TELEP *fax* synthèse *f*, exploration *f*, TRANS exploration *f*; **~ aerial** RADIO antenne à balayage *f*; **~ antenna** RADIO antenne à balayage *f*; **~ capability** PER APP *character recognition* capacité d'acquisition *f*; **~-corrected aerial** RADIO antenne à correction par balayage *f*; **~-corrected antenna** RADIO antenne à correction par balayage *f*; **~ density** TELEP finesse d'exploration *f*; **~ electron microscope** (*SEM*) ELECT microscope électronique à balayage *m* (*MEB*); **~ field** TELEP *fax* champ lu *m*, champ exploré *m*; **~ interval** RADIO durée d'analyse *f*; **~ line** TV ligne de balayage *f*; **~ line frequency** TV fréquence de balayage horizontal *f*, fréquence de ligne de balayage *f*; **~ line period** TV période de lignes *f*; **~ pitch** TELEP pas d'exploration *m*; **~ program** COMP programme de balayage *m*, programme d'exploration *m*; **~ rate** PER APP *character recognition* capacité d'acquisition *f*; **~ shift** TELEP translation d'exploration *f*; **~ speed** RADIO vitesse de balayage *f*; **~ spot** COMP spot analyseur *m*, TV spot de balayage *m*
SCART: **~ socket television** *n* TV prise péritel *f*, péritel *m*
scatter: **~ diagram** *n* COMP *statistics* diagramme de corrélation *m*; **~ graph** *n* COMP graphique en nuage de points *m*
scattering: **~ coefficient** *n* RADIO coefficient de diffusion *m*; **~ loss** *n* RADIO affaiblissement de diffusion *m*
scatterometer *n* TEST diffusiomètre *m*
SCC *abbr* (*satellite control center AmE, satellite control centre BrE*) RADIO centre de contrôle des satellites *m*
SCCN *abbr* (*system coordination and control network*) TEST réseau de coordination et de contrôle du système *m*
SCCP *abbr* (*signaling connection control part AmE, signalling connection control part BrE*) SIG SSCS (*sous-système de commande de connexions sémaphores*); **~ method indicator** *n* TELEP indicateur de méthode SSCS *m*
scenario *n* TEST scénario *m*
SCH *abbr* (*seizures per circuit per hour*) TELEP prises par circuit et par heure
schedule[1] *n* GEN plan *m*, programme *m*, échéancier *m*, emploi du temps *m*, *specifications, requirements* cahier des charges *m*, RADIO emploi du temps *m*
schedule[2] *vt* GEN planifier, programmer
scheduled: **~ maintenance** *n* TEST maintenance périodique *f*, maintenance programmée *f*, entretien systématique *m*, maintenance systématique *f*; **~ operating time** *n* RADIO durée de fonctionnement prévue *f*; **~ reporting signal** *n* SIG signal de compte rendu attendu *m*; **~ stop** *n* TEST arrêt programmé *m*
schematic: **~ representation** *n* GEN représentation schématique *f*
scheme: **~ plan** *n* GEN schéma *m*
Schmitt: **~ limiter** *n* ELECT bascule de Schmitt *f*; **~ trigger** *n* ELECT bascule de Schmitt *f*
Schottky: **~ diode** *n* PARTS diode Schottky *f*; **~ effect** *n* ELECT effet Schottky *m*
scintillation *n* RADIO, TV scintillation *f*
scissoring *n* COMP détourage *m*
scope[1] *n* TEST étendue *f*, objet *m*, portée *f*
scope[2] *vt* TEST vérifier à l'oscilloscope
Scott: **~-connected transformer** *n* PARTS, POWER transformateur Scott *m*
scouring: **~ tool** *n* PARTS outil de marquage *m*
SCPC *abbr* (*single channel per carrier*) RADIO système par porteuse monovoie *m*, voie unique par porteuse, une seule voie par porteuse
SCR *abbr* ELECT (*silicon-controlled rectifier*), PARTS (*silicon-controlled rectifier*) redresseur à thyristor *m*, thyristor commandé au silicium *m*, TRANS (*signal-to-crosstalk ratio*) écart diaphonique *m*; **~ crosspoint** *n* ELECT point d'intersection à thyristor *m*
scramble *vt* COMP, DATA brouiller, crypter, GEN chiffrer, RADIO, TELEP, TV brouiller, crypter
scrambled *adj* COMP, DATA, RADIO, TELEP, TV brouillé
scrambled[2]: **~ binary ones** *n pl* (*SB1*) DATA binaires brouillés *m pl*; **~ message** *n* TELEP message brouillé *m*
scrambler *n* COMP, DATA brouilleur *m*, crypteur *m*, cryptographe *m*, cryptophone *m*, GEN dispositif de chiffrement *m*, RADIO, TELEP, TV brouilleur *m*, crypteur *m*, cryptographe *m*, cryptophone *m*
scrambling *n* COMP, DATA, RADIO, TELEP, TV brouillage *m*, cryptage *m*, cryptographie *f*; **~ control** DATA gestion du brouillage *f*; **~ sequence** GEN séquence d'embrouillage *f*
scrapping *n* PARTS mise au rebut *f*
scratch[1] *n* PARTS, TEST rayure *f*; **~ hardness** PARTS dureté par

rayage *f*; ~ **hardness test** PARTS, TEST essai à l'égratignure *m*, essai de résistance aux rayures *m*
scratch[2] *vt* PARTS, TEST rayer
scratching *n* PARTS, TEST rayage *m*; ~ **noise** TELEP bruit de surface *m*
screen *n* COMP *monitor*, ELECT, GEN, RADIO, TELEP, TV *monitor* écran *m*; ~ **background** DATA *scanner* fond d'écran *m*; ~ **burn** ELECT brûlure d'écran *f*; ~ **copy** DATA copie d'écran *f*; ~ **effect** GEN, RADIO, TV effet d'écran *m*; ~ **efficiency** ELECT efficacité de l'écran *f*; ~ **grid** ELECT grille-écran *f*; ~ **image** COMP image d'écran *f*; ~ **locking** COMP verrouillage d'écran *m*; ~ **page** DATA page à l'écran *f*, TV page-écran *f*; ~ **printing** COMP recopie d'écran *f*, PARTS sérigraphie *f*; ~ **terminal** COMP, TV terminal à écran *m*
screened: ~ **cable** *n* POWER, TELEP câble à écran *m*, câble blindé *m*
screening: ~ **effect** *n* GEN, RADIO, TV effet d'écran *m*; ~ **indicator** *n* (*SI*) TV indicateur de contrôle *m*; ~ **inspection** *n* TEST tri *m*; ~ **test** *n* TEST essai de sélection *m*
ScreenPlay *n* COMP ScreenPlay *m*
screensaver *n* COMP économiseur d'écran *m*
screw *n* PARTS vis *f*; ~ **base** PARTS culot à vis *m*; ~ **cap** PARTS culot à vis *m*; ~ **clamp** PARTS serre-joint *m*; ~ **coupling** PARTS raccord-union *m*; ~ **cover** PARTS couvercle à vis *m*; ~ **with hexagonal head** PARTS vis à six pans *f*; ~ **terminal** POWER borne à vis *f*; ~ **terminal block** POWER réglette à vis *f*; ~ **thread** PARTS filet *m*, filetage *m*; ~ **top** PARTS couvercle à vis *m*
screwdriver: ~ **for recessed head screws** *n* PARTS tournevis Torx *m*
screwed: ~ **connection** *n* POWER connexion à vis *f*
screwlock *n* PARTS vis de verrouillage *f*
scribe[1]: ~ **tool** *n* PARTS outil graveur *m*
scribe[2] *vt* PARTS rainurer
scribed: ~ **line** *n* PARTS ligne trusquinée *f*; ~ **mark** *n* PARTS ligne trusquinée *f*
scriber *n* PARTS pointe à tracer *f*, pointe de traçage *f*
scribing *n* PARTS rainurage *m*, rayage *m*; ~ **template** PARTS gabarit de traçage *m*
script *n* COMP, DATA système graphique *m*; ~**-based reasoning** DATA raisonnement à base de scénarios *m*; ~ **system** COMP système d'écriture *m*
Script: ~ **X** *n* COMP Script X *m*
scroll[1]: ~ **bar** *n* COMP barre de défilement *f*; ~ **box** *n* COMP ascenseur *m*, curseur de défilement *m*
scroll[2]: ~ **through** *vt* COMP parcourir
scrolling *n* COMP défilement *m*; ~ **menu** COMP menu déroulant *m*
SCS *abbr* (*system conformance statement*) DATA, TEST déclaration de conformité du système *f*
SCSI *abbr* (*small computer system interface*) COMP SCSI (*interface SCSI*)
SCTR *abbr* (*system conformance test report*) DATA, TEST rapport de test de conformité du système *m*
SCU *abbr* (*system control signal unit*) SIG SCU (*unité de signalisation pour la commande du système de signalisation*)
SD *abbr* (*signal degradation*) SIG dégradation du signal *f*
SDA *abbr* (*signal distortion analyser BrE, signal distortion analyzerAmE*) TEST analyseur de distorsion de signaux *m*
SDH *abbr* (*synchronous digital hierarchy*) TELEP, TRANS HNS (*hiérarchie numérique synchrone*); ~ **management network** *n* (*SMN*) TELEP, TRANS réseau de gestion SDH *m*; ~ **management subnetwork** *n* (*SMS*) TELEP, TRANS sous-réseau de gestion SDH *m*
SDH-DRSS *abbr* (*digital radio relay for synchronous hierarchy*) RADIO HNS-FHN (*faisceau hertzien numérique pour hiérarchie synchrone*)
S-distortion *n* ELECT distorsion en S *f*
SDL *abbr* (*specification and description language*) COMP LDS (*langage de description et de spécification*)
SDLC *abbr* (*synchronous data link control*) DATA SDLC (*contrôle de liaison synchrone de données*)
SDN *abbr* (*shared data network*) DATA, TELEP, TRANS RPTD (*réseau partagé de transmission de données*)
SDTNX: ~ **code** *n* (*selective do not transmit code*) DATA, SIG code SDTNX *m* (*code sélectif ne pas émettre*)
SDU *abbr* DATA (*service data unit*) unité de données de service *f*, (*synchronous data unit*) unité de données synchrone *f*, ELECT (*synchronous data unit*) unité de données synchrone *f*
sea: ~ **clutter** *n* RADIO écho de mer *m*, écho de vagues *m*; ~ **return** *n* RADIO écho de mer *m*, écho de vagues *m*
seal[1] *n* ELECT, PARTS joint étanche *m*, joint d'étanchéité *m*
seal[2] *vt* GEN sceller, PARTS étancher, plomber, sceller
sealant: ~ **applicator** *n* PARTS épandeur de matériel d'étanchéité *m*
sealed[1] *adj* GEN, PARTS scellé
sealed[2]: ~ **battery** *n* POWER batterie scellée *f*
sealing *n* PARTS plombage *m*
seamless[1] *adj* RADIO continu, jointif
seamless[2]: ~ **extended local area** *n* COMP, ELECT, TELEP, TRANS réseau local étendu sans interruption *m*
search[1] *n* RADIO guet *m*, recherche *f*, veille *f*; ~ **coil** RADIO rotor inductif *m*, bobine exploratrice *f*; ~ **condition** RADIO état de veille *m*, état de recherche *m*; ~ **engine** DATA moteur de recherche *m*; ~ **field** DATA champ de recherche *m*; ~ **key** COMP clé de recherche *f*; ~ **mark** COMP repère de recherche *m*; ~ **mode acquisition** (*SMA*) DATA acquisition par la méthode de recherche *f*; ~ **mode acquisition procedure** DATA procédure d'acquisition en mode de recherche *f*; ~ **radar** RADIO radar de veille *m*; ~ **and rescue operation** (*SRR operation*) RADIO opération de récherche et de sauvetage *f*; ~ **subgroup** TELEP sous-groupe de recherche *m*
search[2]: ~ **and replace** *vt* COMP, DATA rechercher/remplacer
SECAM *abbr* (*séquentiel couleur à mémoire*) TV SECAM (*système couleur à mémoire*)
second *n* GEN seconde *m*; ~ **adjacent channel** RADIO deuxième canal adjacent *m*; ~ **approximation diode** ELECT, PARTS diode selon la deuxième approximation *f*; ~**-attempt routing** SIG, TELEP, TEST, TRANS acheminement détourné *m*, acheminement par voie détournée *m*; ~ **breakdown** PARTS second claquage *m*, seconde rupture *f*; ~ **choice** TELEP deuxième choix *m*; ~**-data multiplexer** DATA, TRANS multiplexeur de données du deuxième ordre *m*; ~ **dial tone** TELEP seconde tonalité de numérotation *f*; ~ **distribution link** RADIO liaison de distribution secondaire *f*; ~ **harmonic distortion** (*2HD*) ELECT distorsion de la deuxième harmonique *f*; ~ **IF amplifier** POWER amplificateur à seconde FI *m*, deuxième amplificateur FI *m*; ~ **satellite hop** RADIO liaison à deux bonds *f*; ~**-surface mirror** RADIO miroir à couche secondaire *m*; ~**-tier trunk exchange area** SWG zone de centre de transit secondaire *f*; ~ **time around echo** RADIO deuxième écho *m*, écho hors cycle *m*
secondary[1]: **on a** ~ **basis** *adj* RADIO à titre secondaire
secondary[2]: ~ **coating** *n* PARTS, TRANS revêtement secondaire *m*; ~ **dial tone** *n* TELEP seconde tonalité de numérotation *f*; ~ **distribution link** *n* RADIO liaison de distribution secondaire *f*; ~ **electron emission** *n* ELECT émission d'électrons secondaires *f*; ~ **electron emission current** *n* ELECT courant d'émission électronique secondaire *m*; ~ **electron emission factor** *n* ELECT facteur d'émission secondaire d'électrons *m*, facteur d'émission électronique secondaire *m*, rapport d'émission secondaire *m*; ~ **failure** *n* TEST défaillance secondaire *f*; ~ **fleet** *n* RADIO sous-flotte *f*; ~ **flow direction** *n* GEN, TRANS sens de liaison secondaire *m*; ~ **frequency standard** *n* RADIO étalon secondaire de fréquence *m*; ~ **function** *n* DATA fonction secondaire *f*; ~ **grade of reception quality** *n* RADIO qualité sécondaire de réception *f*; ~ **ionic emission** *n* ELECT émission ionique secondaire *f*; ~ **multiplex coupler** *n* TELEP coupleur multiplex secondaire *m*; ~ **network** *n* TELEP réseau

secondaire *m*; ~ **processor unit** *n* COMP unité processeur secondaire *f*; ~ **radar** *n* RADIO radar secondaire *m*; ~ **radiation** *n* RADIO rerayonnement *m*, rayonnement secondaire *m*; ~ **reference burst code** *n* DATA code de paquet de référence secondaire *m*; ~ **refrigerant** *n* CLIM frigoporteur *m*; ~ **relay** *n* PARTS relais secondaire *m*; ~ **route** *n* TELEP, TRANS voie secondaire *f*; ~ **transmitter** *n* RADIO émetteur secondaire *m*; ~ **winding** *n* PARTS enroulement secondaire *m*

secrecy: ~ **button** *n* PER APP touche secrète *f*

secret: ~ **code** *n* RADIO code confidentiel *m*

section *n* COMP segment *m*, DATA section *f*, TRANS tronçon *m*; ~ **adaptation** (*SA*) DATA adaptation de section *f*; ~ **overhead** (*SOH*) DATA, TELEP surdébit de section *m* (*SDS*); ~ **spacing** TELEP espacement de section *m*; ~ **termination** TELEP extrémité de section *f*

sector *n* GEN secteur *m*; ~ **attributes** *n pl* COMP attributs du secteur *m pl*; ~ **cell** RADIO cellule sectorielle *f*; ~ **display** RADIO indicateur sectoriel *m*; ~ **scanning** RADIO balayage sectoriel *m*, exploration sectorielle *f*; ~-**scanning beacon** RADIO radiophare à diagramme oscillant *m*; ~-**shaped conductor** PARTS conducteur sectoral *m*

sectors: ~ **per track** *phr* (*spt*) COMP secteurs par piste (*spp*)

secure: ~ **access** *n* DATA accès sécurisé *m*; ~ **storage** *n* DATA mémoire de sécurité *f*; ~ **telephone** *n* (*STU*) TELEP poste cryptophonique *m*

security *n* GEN sécurité *f*; ~ **arrangements** *n pl* TEST sécurité de fonctionnement *f*; ~ **audit** DATA audit de sécurité *m*; ~ **audit trail** DATA journal d'audit de sécurité *m*; ~ **label** DATA étiquette de sécurité *f*; ~ **management information base** (*SMIB*) DATA base d'informations de gestion de sécurité *f*; ~ **network** TRANS réseau de sécurité *m*; ~-**related event** TEST événement relatif à la sécurité *m*

Seebeck: ~ **effect** *n* ELECT effet Seebeck *m*, effet thermoélectrique *m*

seed *n* COMP valeur de départ *f*; ~ **crystal** ELECT cristal germe *m*

S-effect *n* ELECT effet de charge de surface *m*, effet S *m*

segment[1] *n* COMP, DATA, TELEP groupe *m*, segment *m*, TRANS secteur *m*; ~ **terminator** DATA terminaison de segment *f*; ~ **type** (*ST*) COMP, DATA type de segment *m*

segment[2] *vt* COMP, DATA, TELEP segmenter

segmentation *n* COMP, DATA, TELEP segmentation *f*; ~ **permitted flag** (*SPF*) DATA indicateur de segmentation permise *m*; ~ **and reassembly** (*SAR*) DATA segmentation et réassemblage *f*; ~ **and reassembly sublayer** DATA sous-couche de segmentation et réassemblage *f*

segmented[1] *adj* COMP, DATA, TELEP segmenté

segmented[2]: ~ **coding law** *n* TELEP, TRANS loi de codage à segments *f*; ~ **encoding law** *n* TELEP, TRANS loi de codage à segments *f*; ~ **multiprocessor system** *n* DATA système en répartition géographique *m*

segmenting *n* COMP, DATA, TELEP segmentation *f*

seize *vt* TELEP provoquer la prise, mettre en état de prise

seizing: ~ **acknowledgement signal** *n* SIG signal d'accusé de réception de prise *m*; ~ **signal** *n* TELEP signal de prise *m*, signal de saisie *m*

seizure *n* TELEP saisie *f*, occupation *f*, prise *f*, prise de ligne *f*; ~ **circuit** TELEP circuit de prise *m*; ~ **condition** TELEP état de prise *m*; ~ **counter** TELEP compteur de prises *m*; ~ **indicator** TELEP indicateur de prises *m*; ~ **lamp** TELEP lampe d'occupation *f*; ~ **relay** PARTS, TELEP relais de prise *m*; ~ **signal** SIG, TELEP signal de prise *m*, signal de saisie *m*; ~ **supervision** TELEP supervision de prise *f*, supervision des occupations *f*; ~ **time** GEN, TELEP durée d'occupation *f*; ~ **wire** PARTS, TELEP fil de prise *m*

seizures: ~ **per circuit per hour** *phr* (*SCH*) TELEP prises par circuit et par heure

SELCAL *abbr* (*selective calling system*) RADIO, TELEP système d'appel sélectif *m*

select[1] *n* GEN, TRANS, TV sélection *f*; ~ **button** COMP touche de sélection *f*; ~-**on-test value** TEST valeur choisie aux essais *f*

select[2] *vt* COMP sélectionner, SWG indiquer, TELEP sélecter, sélectionner, TV sélectionner

selected: ~ **abstract test suite** *n* (*SATS*) DATA, TEST suite de tests abstraits sélectionnés *f*; ~ **executable test suite** *n* (*SETS*) DATA, TEST suite de tests exécutables sélectionnés *f*; ~ **number redial** *n* TELEP répétition automatique des numéros choisis *f*; ~ **position** *n* TELEP position sélectionnée *f*

selecting *n* DATA invitation à recevoir *f*, GEN sélection *f*, RADIO, SWG, TELEP invitation à recevoir *f*, TRANS, TV sélection *f*; ~ **bar** PARTS, SWG barre de sélection *f*, horizontale *f*; ~-**bar coil** SWG bobine d'électro de barre de sélection *f*; ~ **bar damper** SWG amortisseur de barre de sélection *m*, électroaimant de barre de sélection *m*; ~ **bar position** SWG position d'horizontale *f*, position de barre de sélection *f*; ~ **finger** PARTS, SWG doigt d'indication de barre de sélection *m*; ~ **finger base** SWG support de dispositif de sélection *m*; ~ **finger support** SWG support de dispositif de sélection *m*; ~ **unit** SWG unité de sélection *f*

selection *n* GEN, TRANS, TV sélection *f*; ~ **acknowledgement output** TELEP sortie d'accusé de réception de sélection *f*; ~ **allotter** TELEP distributeur de chercheurs *m*; ~ **allotter relay** TELEP distributeur de chercheurs *m*; ~ **allotter relay chain** TELEP distributeur de chercheurs *m*; ~ **box** COMP cadre de sélection *m*; ~ **code** TELEP code de sélection *m*; ~ **identity** TELEP identité de sélection *f*; ~ **input** TELEP entrée de sélection *f*; ~ **signal** SIG signal de sélection *m*; ~ **stage** SWG étage de sélection *m*; ~ **type** TELEP type de sélection *m*

selective: ~ **accounting** *n* TELEP service de comptabilité sélective *m*; ~ **back alarm indication** *n* RADIO indication d'alarme de retour sélective *f*; ~ **calling** *n* RADIO appel sélectif *m*; ~ **calling system** *n* (*SELCAL*) RADIO, TELEP système d'appel sélectif *m*; ~ **cancelation** *AmE*, ~ **cancellation** *BrE* *n* RADIO annulation sélective *f*; ~ **do not transmit code** *n* (*SDTNX code*) DATA, SIG code sélectif ne pas émettre *m* (*code SDTNX*); ~ **erasing** *n* DATA effacement sélectif *m*; ~ **field protection** *n* DATA, RADIO protection sélective des champs *f*; ~ **jamming** *n* ELECT, RADIO brouillage sélectif *m*; ~ **reject** *n* TELEP rejet sélectif *m*; ~ **reject frame** *n* (*SREJ*) TELEP trame de rejet sélectif *f*; ~ **ringing** *n* TELEP appel sélectif *m*, sonnerie sélective *f*

selectivity *n* RADIO *of receiver*, TELEP sélectivité *f*

selector *n* DATA, SWG, TELEP sélecteur *m*; ~ **mechanism** SWG, TELEP mécanisme de commutation *m*, mécanisme de sélection *m*; ~ **multiple** SWG, TELEP multiple de sélecteurs *m*; ~ **noise** SWG bruit de sélecteurs *m*; ~ **rack** SWG, TELEP bâti de commutateurs *m*, bâti de sélecteurs *m*; ~ **stage** TELEP étage de sélection *m*; ~ **switch** POWER, SWG, TELEP commutateur *m*

selenium: ~ **rectifier** *n* PARTS, POWER redresseur au sélénium *m*

self[1]: ~-**contained** *adj* COMP, ELECT autonome; ~-**oscillating** *adj* ELECT à oscillation propre, à oscillation spontanée

self[2]: ~-**boot** *n* ELECT auto-amorçage *m*; ~-**breaking contact** *n* PARTS contact auto-interrupteur *m*; ~-**carrying cable** *n* TRANS câble autoporté *m*, câble autoporteur *m*; ~-**cleaning contact** *n* PARTS contact autonettoyant *m*; ~-**clocking signal** *n* SIG signal autosynchronisé *m*; ~-**discharge** *n* POWER autodécharge *f*, décharge spontanée *f*; ~-**discharge current** *n* POWER courant d'autodécharge *m*; ~-**excited transmitter** *n* RADIO émetteur d'auto-excitation *m*; ~-**healing** *n* POWER autorégénération *f*; ~-**heating** *n* POWER échauffement propre *m*, auto-échauffement *m*; ~-**identification** *n* RADIO auto-identification *f*; ~-**inductance** *n* POWER inductance propre *f*, auto-inductance *f*; ~-**maintained discharge** *n* ELECT, POWER décharge auto-entretenue *f*, décharge autonome *f*; ~-**metering** *n* APPL autotarification *f*; ~-**neutralization**

frequency *n* ELECT fréquence d'autoneutrodynage *f*; **~-operating spring set** *n* PARTS, SWG empilage automoteur *m*; **~-optimization** *n* ELECT auto-optimisation *f*; **~-organizing program** *n* COMP programme auto-organisateur *m*; **~-orthogonal convolutional coding** *n* DATA codage auto-orthogonal par convolution *m*; **~-oscillation** *n* ELECT oscillation propre *f*, oscillation spontanée *f*; **~-phase modulation** *n* (*SPM*) TRANS automodulation de phase *f*; **~-quenched counter tube** *n* RADIO tube-compteur autocoupeur *m*; **~-quenching oscillator** *n* ELECT, PARTS oscillateur à extinction *m*; **~-registration** *n* DATA auto-enregistrement *m*; **~-relative address** *n* COMP adresse différentielle *f*; **~-resonant frequency** *n* RADIO fréquence propre *f*; **~-supporting antenna** *n* RADIO antenne autoporteuse *f*, antenne autoportante *f*; **~-supporting cable** *n* TRANS câble autoporté *m*, câble autoporteur *m*; **~-test** *n* TEST essai automatique *m*

Sellmeir: **~'s equation** *n* TRANS équation de Sellmeir *f*

selsyn *n* ELECT selsyn *m*, POWER synchro-résolveur *m*

SEM *abbr* (*scanning electron microscope*) ELECT MEB (*microscope électronique à balayage*)

semantic: **~ database** *n* COMP, DATA base de données sémantiques *f*; **~ interpretation** *n* DATA interprétation sémantique *f*; **~ network** *n* COMP réseau sémantique *m*

semantically: **~ invalid event** *n* TEST événement invalide sur le plan de la sémantique *m*

semaphore *n* ELECT, TELEP sémaphore *m*

semation *n* TELEP sémation *f*

semator *n* TELEP sémateur *m*

SEMF *abbr* (*synchronous equipment management function*) APPL fonction de gestion d'équipement synchrone *f*

semiadditive: **~ process** *n* ELECT procédé semi-additif *m*

semiassociated: **~ network** *n* SWG réseau quasi-associé *m*

semiautomatic[1] *adj* SWG, TELEP semi-automatique

semiautomatic[2]: **~ observation** *n* TEST observation semi-automatique *f*; **~ switching system** *n* SWG système de commutation semi-automatique *m*; **~ system** *n* SWG, TELEP système semi-automatique *m*

semicircular: **~ error** *n* RADIO erreur semi-circulaire *f*

semiconductive: **~ floor** *n* TEST plancher conducteur *m*, sol conducteur *m*

semiconductor *n* ELECT, PARTS semiconducteur *m*; **~ amplifier** ELECT amplificateur à semi-conducteurs *m*; **~ contact** PARTS contact à semiconducteur *m*; **~ crosspoint** SWG point d'intersection à semiconducteur *m*; **~ device** PARTS composant à semiconducteur *m*; **~ diode** ELECT, PARTS diode à semiconducteur *f*; **~ laser** ELECT, PARTS, TRANS laser à diode *m*, laser à injection *m*, laser semiconducteur *m*; **~ laser diode amplifier** (*SLDA*) ELECT amplificateur à laser semiconducteur *m*; **~ rectifier diode** ELECT, PARTS diode de redressement à semiconducteur *f*; **~ rectifier stack** PARTS batterie de redresseurs à semiconducteurs *f*; **~ storage** DATA mémoire à semiconducteurs *f*

semiduplex[1] *adj* TELEP semi-duplex

semiduplex[2] *n* TELEP semi-duplex *m*

semigraphic *adj* TEST semi-graphique

semiloop: **~ loss** *n* TELEP affaiblissement en demi-boucle *m*

semimajor: **~ axis** *n* RADIO demi-grand axe *m*

semipermanent: **~ access** *n* DATA accès semi-permanent *m*, accès à la demande *m*; **~ connection** *n* TELEP connexion semi-permanente *f*; **~ data** *n pl* TELEP données semi-permanentes *f pl*; **~ link** *n* TELEP liaison semi-permanente *f*

semirigid: **~ plastic** *n* PARTS plastique semi-rigide *m*

semitone *n* COMP demi-ton *m*

send[1]: **~ button** *n* RADIO touche envoi *f*; **~ event** *n* DATA événement d'envoi *m*; **~ next digit signal** *n* SIG, TELEP signal envoyez le chiffre suivant *m*; **~ reference station** *n* RADIO, TELEP station de référence à l'émission *f*; **~ special information tone signal** *n* SIG signal envoyez la tonalité spéciale d'information *m*

send[2] *vt* TRANS émettre, transmettre; **~ an order by wire** TELEP passer une commande par télégramme

sender *n* COMP, DATA, TELEP expéditeur *m*; **~ finder** TELEP chercheur d'émetteur de code *m*; **~ identification** DATA, TELEP identification de l'expéditeur *f*; **~ transfer time** TELEP temps de transfert à l'émission *m*

senderized: **~ operation** *n* SIG, TELEP réémission des signaux d'adresse *f*

sending *n* COMP, DATA émission *f*, GEN envoi *m*, RADIO, TRANS émission *f*; **~ sensitivity** TELEP *of local telephone circuit* efficacité à l'émission *f*

sense *vt* GEN détecter, TEST sonder

sensing *n* TEST sondage *m*; **~ element** TEST élément sensible *m*; **~ point** TEST point de détection *m*; **~ station** COMP poste de lecture *m*; **~ unit** TEST unité de détection *f*, unité sensible *f*

sensitive[1] *adj* CLIM, COMP, DATA, ELECT, GEN, RADIO sensible

sensitive[2]: **~ data** *n pl* COMP, DATA données confidentielles *f pl*; **~ volume** *n* RADIO volume utile *m*

sensitivity *n* COMP, DATA, ELECT, GEN, RADIO sensibilité *f*, TELEP *of artificial ear* efficacité *f*, TRANS seuil de détection *m*; **~ factor** RADIO facteur de sensibilité *m*; **~ time control** (*STC*) CONTR commande cyclique du gain *f*, gain variable dans le temps *m*

sensor *n* ELECT, GEN, RADIO capteur *m*, détecteur *m*; **~ system** GEN système de détection *m*; **~ wire** PARTS fil détecteur *m*

separate: **~ charging system** *n* POWER système à capacité de batterie répartie *m*; **~ common-channel signaling** *AmE*, **~ common-channel signalling** *BrE* *n* SIG signalisation sur voie commune séparée *f*; **~ cooling** *n* CLIM refroidissement séparé *m*

separated: **~ video** *n* (*S-video*) TV S-vidéo *m*

separately: **~ sheathed cable** *n* TRANS câble tréplomb *m*

separation *n* DATA, ELECT, POWER séparation *f*, RADIO écart *m*, TELEP séparation *f*

separator *n* COMP, GEN séparateur *m*

septate: **~ waveguide** *n* DATA, RADIO, TRANS guide d'ondes cloisonné *m*

septet *n* COMP septet *m*

septum *n* RADIO *waveguide* cloison *f*

sequence *n* GEN séquence *f*; **~ adjustment** SWG réglage à échelonnement *m*; **~ behavior** *AmE*, **~ behaviour** *BrE* COMP, TELEP comportement séquentiel *m*; **~ chart** TELEP diagramme de fonctionnement *m*, diagramme séquentiel *m*; **~-connected rectifier** PARTS, POWER redresseur connecté séquentiellement *m*; **~ data** *n pl* COMP, DATA données de séquence *f pl*; **~ of events** TEST succession d'événements *f*, suite d'événements *f*; **~ function** SWG fonctionnement échelonné *m*; **~ number** (*SN*) COMP, DATA, SWG numéro d'ordre *m*, numéro de séquence *m*; **~ number protection** (*SNP*) COMP, DATA, SWG protection de numéro de séquence *f*; **~-of type** DATA type séquence-de *m*; **~ recorder** COMP enregistreur de séquences *m*; **~ requirements** *n pl* SWG exigences de fonctionnement échelonné *f pl*; **~ spring set** PARTS, SWG empilage échelonné *m*; **~ switch** POWER, SWG commutateur séquentiel *m*; **~ type** DATA type de séquence *m*

sequencer *n* COMP, DATA, SWG, TELEP séquenceur *m*; **~ program** DATA programme séquenceur *m*

sequencing *n* COMP, DATA, SWG, TELEP maintien en séquence *m*; **~ system** DATA chaîne séquentielle *f*

sequential[1] *adj* COMP, DATA, ELECT, GEN séquentiel, TELEP séquentiel, à moments simultanés

sequential[2]: **~ access** *n* COMP, DATA accès séquentiel *m*; **~-access storage** *n* COMP, DATA mémoire à accès séquentiel *f*; **~ assembling** *n* COMP assemblage séquentiel *m*; **~ character file manager** *n* COMP gestionnaire de fichiers séquentiels de caractères *m*; **~ circuit** *n* DATA, ELECT circuit séquentiel *m*; **~ coding** *n* COMP, TELEP codage séquentiel *m*;

~ **control** *n* COMP, CONTR mode séquentiel *m*; ~ **decoding** *n* TELEP décodage séquentiel *m*; ~ **hunting** *n* TELEP recherche séquentielle *f*; ~ **logic element** *n* ELECT élément logique séquentiel *m*; ~ **machine** *n* COMP, DATA ordinateur séquentiel *m*; ~ **monitoring** *n* RADIO contrôle séquentiel *m*; ~ **operation** *n* COMP fonctionnement séquentiel *m*; ~ **program** *n* CONTR programme séquentiel *m*; ~ **reference burst distribution** *n* DATA distribution du paquet de référence en mode séquentiel *f*; ~ **sampling** *n* TEST échantillonnage progressif *m*; ~ **scanning** *n* COMP balayage ligne par ligne *m*, balayage sans entrelacement *m*, TELEP balayage séquentiel *m*, exploration séquentielle *f*; ~ **seizure** *n* TELEP prise séquentielle *f*; ~ **selection** *n* TELEP sélection séquentielle *f*; ~ **sensing** *n* TEST détection séquentielle *f*; ~ **statement** *n* DATA instruction séquentielle *f*; ~ **switching** *n* SWG commutation séquentielle *f*; ~ **test** *n* TEST test progressif *m*, essai séquentiel *m*; ~ **tone coding** *n* DATA codage à tonalités séquentielles *m*; ~ **transmission** *n* RADIO, TRANS transmission séquentielle *f*

séquentiel: ~ **couleur à mémoire** *n* (*SECAM*) TV système couleur à mémoire *m* (*SECAM*)

serial: ~ **access** *n* COMP accès séquentiel *m*; ~ **adder** *n* ELECT additionneur en série *m*; ~ **call** *n* TELEP appel en série *m*, communication en série *f*; **~-call key** *n* TELEP bouton d'appels en série *m*; ~ **carry** *n* DATA transfert série *m*; ~ **interface** *n* DATA, ELECT interface en série *f*; ~ **jack** *n* PARTS jack en série *m*; ~ **number** *n* TELEP numéro de série *m*; **~-parallel conversion** *n* DATA conversion série-parallèle *f*; ~ **port** *n* COMP connecteur série *m*, port série *m*; ~ **printer** *n* DATA imprimante sérielle *f*; ~ **programmable interface** *n* COMP interface en série programmable *f*; **~-to-parallel converter** *n* ELECT, TELEP convertisseur série-parallèle *m*; ~ **transfer** *n* DATA transfert série *m*; ~ **transmission** *n* TELEP, TRANS transmission série *f*

serializer *n* ELECT, TELEP convertisseur parallèle-série *m*

series[1]: **~-connected** *adj* GEN connecté en série

series[2] *n* ELECT, POWER, TRANS série *f*; ~ **circuit** ELECT circuit série *m*; ~ **connection** POWER montage en série *m*; ~ **impedance** POWER impédance série *f*; ~ **inductance** POWER inductance additionnelle *f*, inductance série *f*; ~ **pass transistor** ELECT, PARTS transistor de chute *m*, transistor monté en régulateur série *m*; ~ **port** COMP port série *m*; ~ **resistance** POWER résistance série *f*; ~ **resistor** PARTS, POWER résistance additionnelle *f*, résistance en série *f*, résistance montée en série *f*; ~ **resonance** ELECT, POWER résonance série *f*; ~ **resonant circuit** ELECT, POWER circuit résonnant série *m*; ~ **T** RADIO té série *m*

serve *vt* GEN desservir

serveability *n* APPL, TELEP servibilité *f*; ~ **performance** APPL, TELEP servibilité *f*

server *n* GEN serveur *m*, SWG, TELEP équipement de commutation *m*; ~ **command** DATA commande de serveur *f*; ~ **software** COMP, DATA logiciel serveur *m*

service[1] *n* GEN desserte *f*, service *m*; ~ **access point** (*SAP*) DATA point d'accès au service *m*; ~ **access point identifier** (*SAPI*) DATA identificateur de point d'accès au réseau *m*; ~ **access switch function** SWG, TELEP fonction commutateur d'accès aux services *f* (*CAS*); ~ **advice** TELEP avis de service *m*; ~ **alarm** TELEP alarme de service *f*; ~ **arc** RADIO arc de service *m*; ~ **availability performance** TELEP disponibilité du service *f*; ~ **bureau** COMP façonnier; ~ **call** TELEP appel de service *m*; ~ **channel message** TEST message du canal de service *m*; ~ **code** SWG, TELEP code de service *m*, indicatif d'accès *m*; ~ **code prefix** TELEP préfixe de code de service *m*; ~ **command point** SWG point de commande de services *m* (*PCS*); ~ **component** COMP composant de service *m*; ~ **conditions** *n pl* POWER conditions de fonctionnement *f pl*; ~ **data unit** (*SDU*) DATA unité de données de service *f*; ~ **digit** TELEP, TRANS élément numérique de service *m*; ~ **element** DATA, TELEP, TRANS élément de service *m*; ~ **identification** APPL identification de service *f*; ~ **incident** TEST anomalie de fonctionnement *f*, incident de fonctionnement *m*; ~ **indicator** TELEP indicateur de service *m*, indicateur de service utilisateur *m*; ~ **information** DATA *byte* octet de service *m*; ~ **integration PABX** APPL PABX à intégration de services *m*; ~ **interruption** TELEP, TEST interruption d'exploitation *f*, interruption du service *f*; ~ **interworking** TELEP interfonctionnement de service *m*; ~ **invocation component** TELEP élément de demande du service *m*; ~ **irregularity** TEST anomalie de fonctionnement *f*, incident de fonctionnement *m*; ~ **jack** PARTS, TELEP jack de service *m*; ~ **jack unit** TELEP unité de jack de service *f*; ~ **layer** TRANS couche de services *f*; ~ **line** TELEP ligne de service *f*; ~ **node** TELEP nœud de service *m*; ~ **notification** (*SN*) DATA, TELEP avis de service *m*, notification de service *f*; ~ **observation** TEST observation de la qualité de service *f*; **~-or-option class not-available** DATA classe service-ou-option indisponible *f*; **~-or-option class not-implemented** DATA classe service-ou-option non mis en œuvre *f*; ~ **plan** TV plan de service *m*; ~ **primitive** TELEP primitive de service *f*; ~ **program** COMP programme de service *m*, programme utilitaire *m*; ~ **provider** COMP prestataire de services *m*, GEN fournisseur de services *m*, prestataire de services *m*; ~ **provider link** (*SPL*) TRANS liaison de prestataire de service *f*; ~ **rate** COMP débit de service *m*; ~ **relationship** DATA relation de service *f*; ~ **sector** GEN, RADIO secteur de service *m*, tertiaire *m*; ~ **signal** SIG, TELEP signal de service *m*; ~ **string advice** DATA avis de chaîne de service *m*; ~ **supervision** TELEP surveillance du fonctionnement *f*, surveillance du service *f*; ~ **supervision desk** TELEP pupitre d'exploitation *m*; ~ **support performance** TELEP logistique du service *f*; ~ **telegram** TELEP télégramme de service *m*; ~ **telex call** PER APP communication télex de service *f*; ~ **time** APPL temps de service *m*; **~-transfer node** TELEP nœud de service et de transfert *m*; ~ **user** DATA, TELEP utilisateur du service *m*

service[2] *vt* GEN desservir

serviceability *n* APPL, TELEP aptitude au service *f*, facilité d'entretien *f*, servibilité *f*, simplicité d'utilisation *f*; ~ **performance** APPL, TELEP servibilité *f*

serviceable *adj* GEN prêt à fonctionner

services: ~ **rendered** *n pl* TELEP prestations *f pl*

servicing *n* GEN desserte *f*

serving *n* APPL *of subscribers* desserte *f*; ~ **exchange** SWG central serveur *m*, centre de rattachement *m*, commutateur de rattachement *m*; ~ **port** COMP, ELECT port serveur *m*

servoamplifier *n* ELECT, POWER servoamplificateur *m*

servomechanism *n* CONTR servomécanisme *m*

servosystem *n* CONTR servosystème *m*

SES *abbr* (*severely-errored second*) DATA, RADIO SGE (*seconde gravement erronée*)

session *n* DATA, TRANS session *f*; ~ **connection** DATA connexion de session *f*; ~ **layer** DATA, TRANS couche session *f*; ~ **level** DATA, TRANS couche session *f*; ~ **number** COMP numéro de session *m*; ~ **protocol** (*SP*) DATA protocole de session *m*; ~ **protocol data unit** (*SPDU*) DATA unité de données de protocole de session *f*; ~ **service** DATA service de session *m*; **~-service access point** (*SSAP*) DATA point d'accès au service de session *m*

set[1] *n* CONTR, POWER, SWG groupe *m*; ~ **of alternatives** DATA ensemble des propositions SI *m*, ensemble des propositions conditionnelles multiples *m*; ~ **asynchronous balanced mode** (*SABM*) TELEP fonctionnement en mode asynchrone symétrique *m*, mise en mode asynchrone symétrique *f*; ~ **asynchronous balanced mode extended** (*SABME*) TELEP fonctionnement en mode asynchrone symétrique étendu *m*; ~ **mark** COMP pose de repère *f*; **~-of type** DATA type ensemble-de *m*; ~ **point** CONTR signal de consigne *m*; ~ **pulse** CONTR impulsion de commande *f*, impulsion de mise à un *f*; ~ **signal** CONTR, SIG signal de mise à un *m*; ~ **status** COMP détermination de statut *f*; ~**-top decoder** TV *descrambling* terminal d'adaptation au réseau *m*; ~**-top unit** TV coffret d'abonné *m*; ~ **type** DATA type d'ensemble *m*; ~**-valued attribute** DATA attribut pour un ensemble de valeurs *m*

set[2] *vt* COMP instaurer, DATA cadencer, positionner, GEN positionner, RADIO armer, régler, positionner, SWG *selector* mettre en place; ~ **off** RADIO, TEST *alarm* déclencher; ~ **parameters to** COMP paramétrer; ~ **up** COMP *equipment* monter, DATA, RADIO *connection* établir, TELEP *call* établir; ~ **to one** ELECT mettre à un

SETS *abbr* (*selected executable test suite*) DATA, TEST suite de tests exécutables sélectionnés *f*

settable *adj* COMP, GEN réglable

setting *n* COMP réglage *m*, DATA positionnement *m*, GEN réglage *m*, positionnement *m*, PARTS *of a flip-flop* mise à un *f*, RADIO positionnement *m*, SWG mise en place *f*, TEST *of a flip-flop* mise à un *f*; ~ **accuracy** GEN précision de positionnement *f*; ~ **agent** PARTS *plastics* agent de durcissement *m*; ~ **range** GEN plage de réglage *f*; ~ **ratio** PARTS rapport d'ajustement *m*; ~ **time** CONTR temps de réglage *m*; ~ **to zero** TEST mise à zéro *f*; ~**-up** COMP installation *f*, GEN mise en place *f*, montage *m*, établissement *m*, RADIO *of connection*, TELEP *of call* établissement *m*; ~**-up time** CONTR durée de mise en œuvre *f*, TELEP durée d'établissement *f*; ~ **value** GEN valeur d'ajustement *f*

settings *n pl* COMP réglages *m pl*

settling: ~ **time** *n* POWER temps de stabilisation *m*

setup *n* COMP montage *m*, GEN établissement *m*, RADIO, TELEP établissement de communication *m*; ~ **channel** SIG, TRANS voie de signalisation *f*; ~ **diagram** COMP schéma de montage *m*

seven: ~**-segment display** *n* CONTR affichage à sept segments *m*; ~**-slot rack mounting assembly** *n* ELECT, PARTS armoire à sept logements *f*

severely: ~**-errored second** *n* (*SES*) DATA, RADIO seconde gravement erronée *f* (*SGE*)

SF *abbr* (*signal failure*) SIG défaillance de signal *f*

SFET *abbr* (*synchronous-frequency encoding technique*) DATA technique de codage à fréquence synchrone *f*

SFU *abbr* (*store-and-forward unit*) APPL, SWG unité d'enregistrement et de retransmission *f*

SGC *abbr* (*signaling grouping channel AmE, signalling grouping channel BrE*) SIG canal de groupement de signalisation *m*

SGML *abbr* (*standard generalized markup language*) COMP langage normalisé de balisage généralisé *m*; ~ **document** *n* COMP document SGML *m*

shade[1] *n* COMP, TV ombrage *m*

shade[2] *vt* COMP mettre en dégradé, mettre en lissage, TV mettre en ombrage

shading *n* COMP lissage *m*, ELECT tache *f*, GEN lissage *m*, TV effet de nuit *f*, effet d'ombrage *m*; ~ **correction** TV correction d'ombrage *f*

shadow: ~ **factor** *n* RADIO facteur d'ombre *m*; ~ **grid** *n* ELECT grille alignée *f*; ~ **offset** *n* COMP direction de l'ombre *f*; ~ **region** *n* RADIO zone d'ombre *f*, TRANS zone de silence *f*; ~ **zone** *n* RADIO zone d'ombre *f*, TRANS zone de silence *f*

shadowing *n* RADIO occultation *f*

shaft *n* PARTS arbre *m*, axe *m*; ~ **coupling** PARTS accouplement d'arbres *m*

shannon *n* GEN, TELEP shannon *m*

shape *n* COMP, RADIO forme *f*; ~ **conductor** PARTS conducteur profilé *m*

shaped: ~ **beam** *n* RADIO faisceau formé *m*; ~**-beam aerial** *n* RADIO antenne à faisceau formé *f*, antenne à faisceau modelé *f*; ~**-beam antenna** *n* RADIO antenne à faisceau formé *f*, antenne à faisceau modelé *f*; ~**-beam offset parabolic aerial** *n* RADIO antenne parabolique à faisceau formé *f*; ~**-beam offset parabolic antenna** *n* RADIO antenne parabolique à faisceau formé *f*; ~ **conductor** *n* PARTS âme en forme *f*

shaper *n* GEN conformateur *m*

shaping: ~ **filter** *n* DATA filtre de mise en forme *m*

share *vt* COMP, DATA, ELECT, RADIO partager

shared: ~**-control mode** *n* CONTR, DATA mode de contrôle partagé *m*; ~ **data network** *n* (*SDN*) DATA, TELEP, TRANS réseau partagé de transmission de données *m* (*RPTD*); ~ **disk** *n* COMP disque partagé *m*; ~ **frequency** *n* RADIO fréquence partagée *f*; ~ **resource** *n* COMP, DATA ressource partagée *f*; ~**-resource device** *n* COMP, DATA boîtier partageur de ressources *m*; ~ **service line** *n* TELEP ligne partagée *f*

shareware *n* COMP logiciel en libre service *m*, partagiciel *m*, particiel *m*, *Internet* shareware *m*, DATA partagiciel *m*, particiel *m*, logiciel en libre service *m*; ~ **package** COMP logiciel en shareware *m*, partagiciel *m*, DATA logiciel en shareware *m*, partagiciel *m*

sharing *n* COMP, DATA, ELECT, RADIO partage *m*; ~ **option** DATA option de partage *f*

sharp: ~ **cutoff tube** *n* ELECT tube à blocage rapide *m*

SHARP *abbr* (*stationary high-altitude relay platform*) TRANS SHARP *m*

sharpness *n* DATA *of scanner* netteté *f*

shear: ~ **wave** *n* RADIO onde transversale *f*

sheath *n* ELECT, GEN, TRANS *of cable* gaine *f*; ~ **current** TEST courant de gaine *m*; ~ **cutter** PARTS ouvre-gaine de câble *m*; ~ **retention clamp** ELECT bride de retenue *f*; ~ **stripping tool** PARTS outil à dégainer *m*

sheathed *adj* PARTS cuirassé, gainé

sheathing *n* ELECT gaine *f*, GEN, PARTS, POWER gainage *m*, gaine *f*, TRANS gaine *f*; ~ **layer** ELECT, PARTS couche de structure *f*

sheet *n* COMP, ELECT, GEN feuille *f*, PARTS *of metal* tôle *f*, POWER feuille *f*; ~ **feed** PER APP *of printer* avancement du papier *m*; ~ **metal** PARTS tôle *f*; ~ **metal duct** PARTS goulotte métallique *f*

shelf *n* TELEP étagère *f*, niveau *m*, rayon *m*; ~ **chassis** TELEP châssis d'alvéole *m*; ~ **fuse** PARTS, POWER, TELEP fusible d'alvéole *m*; ~ **fuse alarm** PARTS, POWER, TELEP alarme de fusible d'alvéole *f*; ~ **life** TEST durée de conservation *f*, durée de vie en stockage *f*

shell *n* DATA *Internet* shell *m*

shellac *n* PARTS vernis à la gomme laque *m*

shelved: ~ **trolley** *n* PARTS servante d'atelier *f*

SHF *abbr* (*superhigh frequency*) RADIO, TV hyperfréquence *f*

shield *n* POWER blindage *m*, écran de protection *m*; ~ **grid** POWER grille de protection *f*

shielded: ~ **cable** *n* POWER câble à écran *m*, câble blindé *m*; ~ **connector** *n* ELECT connecteur blindé *m*; ~ **copper separator** *n* ELECT répéteur en cuivre *m*; ~ **cord** *n* PARTS cordon blindé *m*; ~ **triple-faceplate outlet jack** *n* ELECT plastron blindé *m*; ~ **twisted pair** *n* (*STP*) ELECT, TRANS paire torsadée blindée *f*; ~ **wire** *n* PARTS fil blindé *m*

shielding *n* PARTS écran de protection *m*, blindage *m*; ~ **effect** GEN, RADIO, TV effet d'écran *m*

shift[1]: ~**-in** *adj* DATA en code; ~**-out** *adj* DATA hors code

shift[2] *n* COMP *keyboard* décalage *m*, ELECT déplacement *m*, inversion *f*, GEN déplacement *m*, PER APP inversion *f*, *keyboard* décalage *m*, TELEP *traffic* déplacement *m*, inversion *f*, TRANS déplacement *m*; ~ **clock** COMP décalage d'horloge *m*; ~ **function** COMP fonction décalage *f*; ~**-in character** DATA caractère de code normal *m*; ~ **key** COMP *keyboard*, PER APP touche à motion *f*, touche de positionnement *f*, touche majuscule *f*; ~ **lock** COMP, PER APP verrouillage de clavier *m*; ~**-lock keyboard** COMP, PER APP clavier à verrouillage *m*; ~**-out character** DATA caractère de code spécial *m*; ~ **register** DATA, ELECT, SWG registre à décalage *m*

ship: ~ **polling** *n* RADIO interrogation des navires *f*; ~ **reporting system** *n* RADIO, SIG système de signalisation des navires *m*; ~ **station** *n* RADIO station de navire *f*; ~ **station identity** *n* DATA, RADIO, TELEP identité de station de navire *f*; ~ **station number** *n* DATA, RADIO, TELEP numéro de station de navire *m*; **~-to-ship alerting** *n* RADIO alerte navire-navire *f*; **~-to-shore alerting** *n* RADIO alerte navire-terre *f*

shipboard: ~ **terminal** *n* RADIO terminal de navire *m*

shipping: ~ **sample** *n* TEST échantillon pour contrôle de livraison *m*

shock: **~-absorbing foam** *n* ELECT anti-chocs en mousse *m*; **~-absorbing tray** *n* PARTS plateau de montage à amortisseurs *m*; ~ **excitation** *n* ELECT, RADIO excitation par choc *f*; **~-excited oscillator** *n* ELECT, RADIO oscillateur excité par choc *m*; ~ **test** *n* TEST essai au choc *m*; **~-wave** *n* GEN onde de choc *f*

Shoebox *n* COMP Shoebox *m*

shore: ~ **station** *n* RADIO station côtière *f*, station terrienne côtière *f*; **~-to-ship alerting** *n* RADIO alerte terre-navire *f*

short[1]: **~-haul** *adj* TRANS à courte distance, à faible portée

short[2] *n* COMP, ELECT, POWER court-circuit *m*; ~ **burst** DATA paquet bref *m*; ~ **circuit** (*S/C*) COMP, ELECT, POWER court-circuit *m*; **~-circuit admittance** POWER admittance en court-circuit *f*; **~-circuit current** POWER courant de court-circuit *m*; **~-circuit impedance** POWER impédance en court-circuit *f*; **~-circuit test** ELECT essai en court-circuit *m*; ~ **code** TELEP *telex* code abrégé *m*; **~-code dialing** *AmE*, **~-code dialling** *BrE* TELEP numérotation abrégée *f*; **~-distance backscatter** POWER, RADIO rétrodiffusion directe *f*; **~-distance traffic** TELEP trafic à courte distance *m*; **~-haul traffic** TELEP trafic à courte distance *m*; **~-haul transmission** TRANS transmission à courte distance *f*; **~-range transmitter** RADIO émetteur à courte portée *m*; **~-term duty** APPL service temporaire *m*

short[3] *vt* COMP, ELECT, POWER court-circuiter; **~-circuit** COMP, ELECT, POWER court-circuiter

shortage *n* GEN manque *m*, RADIO *of frequencies* pénurie *f*

shortcut *n* COMP raccourci *m*

shortened: ~ **address** *n* TELEP adresse abrégée *f*

shorthand: ~ **keying** *n* COMP raccourci de clavier *m*

shorting: ~ **bars** *n pl* ELECT barres de court-circuit *f pl*; ~ **plunger** *n* ELECT plongeur de court-circuit *m*, piston de court-circuit *m*, RADIO piston *m*

shortwave: ~ **infrared** *n* RADIO infrarouge ondes courtes *m*

shot *n* COMP, TV prise de vue *f*; ~ **noise** ELECT, TELEP bruit de grenaille *m*

shoulder: **~-tap** *vt* COMP demander à une unité de prendre le relais

show *vt* COMP afficher, indiquer

shrink[1]: ~ **tube** *n* PARTS gaine rétractable *f*, manchon rétractable *m*; ~ **tubing** *n* PARTS gainage rétractable *m*, manchon rétractable *m*

shrink[2] *vt* PARTS rétrécir, rétracter

shrink[3] *vi* PARTS rétrécir, se contracter, se rétracter

shrinkage *n* PARTS contraction *f*, rétraction *f*, rétrécissement *m*

shrinking *n* PARTS contraction *f*, rétraction *f*, rétrécissement *m*

shroud *n* RADIO capot d'antenne *m*

shuffle *vt* COMP *audiovisual techniques and equipment* sélectionnner

shunt[1] *n* PARTS shunt *m*; ~ **capacitor** PARTS, POWER condensateur en parallèle *m*; ~ **relay** PARTS relais sur shunt *m*; ~ **resistance** POWER résistance en parallèle *f*; ~ **resistor** PARTS shunt *m*; ~ **rheostat** PARTS rhéostat de dérivation *m*; ~ **T** RADIO T parallèle *m*

shunt[2] *vt* POWER monter en dérivation, shunter

shunting *n* POWER montage en dérivation *m*, shuntage *m*; ~ **yard radio system** RADIO système radio pour gares de triage *m*

shutdown *n* POWER arrêt *m*, coupure *f*

shutter *n* ELECT volet *m*, TV obturateur *m*

shuttle *n* COMP navette *f*

SI *abbr* (*screening indicator*) TV indicateur de contrôle *m*

sibling *n* COMP sœur *f*

SIC *abbr* (*signal interface converter*) SIG convertisseur d'interface de signalisation *m*

side: ~ **aisle** *n* TELEP allée latérale *f*; ~ **chute** *n* TELEP goulotte latérale *f*, goulotte secondaire *f*; ~ **circuit** *n* TELEP circuit combinant *m*; ~ **cover** *n* TELEP habillage latéral *m*, panneau latéral *m*, plaque latérale *f*; ~ **echo** *n* RADIO écho latéral *m*; ~ **frequency** *n* RADIO fréquence latérale *f*; ~ **information** *n* COMP information annexe *f*, information indirecte *f*; ~ **lobe** *n* RADIO lobe secondaire *m*, lobe latéral *m*; **~-lobe performance** *n* RADIO caractéristiques des lobes latéraux *f pl*; **~-lobe radiation** *n* RADIO rayonnement dans les lobes latéraux *m*; **~-lobe suppression** *n* (*SLS*) RADIO suppression des lobes secondaires *f* (*SLS*); **~-looking airborne radar** *n* (*SLAR*) RADIO radar aéroporté à visée latérale *m*; ~ **mode** *n* TRANS mode latéral *m*; **~-mode suppression** *n* TRANS suppression de mode latéral *f*; **~-mode suppression ratio** *n* (*SMSR*) TRANS ratio de suppression de la bande latérale *m*; ~ **panel** *n* TV tableau de bord *m*; ~ **plate** *n* TELEP habillage latéral *m*, panneau latéral *m*, plaque latérale *f*; ~ **shifting** *n* SWG basculement *m*, permutation de chaîne *f*; **~-tone** *n* TELEP effet local *m*; **~-tone objective loudness rating** *n* (*SOLR*) TELEP indice objectif de volume des sons pour l'effet local *m*

sideband *n* (*SB*) RADIO, TELEP, TRANS bande latérale *f* (*BL*); ~ **with carrier** RADIO, TELEP, TRANS bande latérale avec porteuse *f*; ~ **interference** RADIO, TELEP, TRANS brouillage dû à une bande latérale *m*; ~ **splatter** RADIO, TELEP, TRANS interférence entre canaux voisins *f*

sidereal: ~ **period of revolution** *n* RADIO période de révolution sidérale *f*; ~ **period of rotation** *n* RADIO période de rotation sidérale *f*

siemens *n* POWER siemens *m*

sieve *n* ELECT tamis *m*

SIG *abbr* (*signature element*) SIG, TELEP élément signature *m*

SIGhyper *n* COMP SIGhyper *m*

sign[1] *n* RADIO, TELEP indicatif *m*; **~-binary character** COMP caractère binaire de signe *m*; ~ **bit** COMP bit de signe *m*; ~ **character** COMP caractère de signe *m*; ~ **digit** COMP chiffre de signe *m*

sign[2] ~ **on** *vt* RADIO se connecter à; ~ **onto** RADIO se connecter à

signal[1] *n* COMP, RADIO signal *m*, SIG, TELEP, TRANS *action* émission de signal *f*, *information* signal *m*; ~ **amplitude** SIG amplitude du signal *f*; ~ **analyser** *BrE*, ~ **analyzer** *AmE* COMP, SIG analyseur de signal *m*; ~ **analysis** SIG analyse de signaux *f*; ~ **attenuation** SIG affaiblissement du signal *m*; ~ **balance ratio** SIG *sinusoidal signal generators* rapport d'équilibre des signaux *m*; ~ **cable chute** TELEP chemin de câble de signalisation *m*; ~ **characteristic** SIG caractéristique du signal *f*; ~ **circuit** SIG, TELEP, TRANS circuit de signalisation *m*; ~ **code** SIG code de signal *m*, code de signalisation *m*; ~ **comparator** RADIO comparateur de signaux *m*; ~ **component** SIG composant de signal *m*; ~ **compression** SIG compression de signal *f*; ~ **conditioning** SIG traitement du signal *m*; ~ **conversion** SIG, TELEP conversion de signaux *f*; ~ **conversion equipment** SIG, TELEP équipement de conversion de signaux *m*; ~ **converter** SIG, TELEP convertisseur de signaux *m*; ~ **current** ELECT courant de signal *m*; ~ **degradation** (*SD*) SIG dégradation du signal *f*; ~ **delay** SIG retard du signal *m*; ~ **detection** SIG détection de signal *f*; ~ **detector** SIG détecteur de signal *m*; ~ **device** SIG organe de signalisation *m*; ~ **digitalization** SIG, TRANS numérisation de signal *f*; ~ **diode** ELECT diode de signal *f*; ~ **distance** SIG distance de Hamming *f*; ~ **distortion** GEN, SIG distorsion du signal *f*; ~ **distortion analyser** *BrE*, ~ **distortion analyzer** *AmE*

(*SDA*) TEST analyseur de distorsion de signaux *m*; **~ distribution** SIG distribution des signaux *f*; **~ duration** SIG durée du signal *f*; **~ earth** *BrE* (*cf signal ground AmE*) SIG, TRANS terre de signalisation *f*; **~ earth input** *BrE* (*cf signal ground input AmE*) SIG, TRANS entrée de terre de signalisation *f*, **~ electrode** ELECT électrode de signal *f*; **~ element** SIG, TELEP élément de signal *m*; **~-element timing** SIG, TELEP base de temps pour les éléments de signal *f*; **~ envelope** SIG enveloppe de signal *f*; **~ expansion** SIG extension de signal *f*; **~ extension** SIG extension de signal *f*; **~ extraction** SIG, TRANS extraction de signaux *f*; **~ failure** (*SF*) SIG défaillance de signal *f*; **~ flow diagram** SIG diagramme de fluence de signal *m*; **~ generator** SIG, TEST générateur de signaux *m*; **~ ground** *AmE* (*cf signal earth BrE*) SIG, TRANS terre de signalisation *f*; **~ ground input** *AmE* (*cf signal earth input BrE*) entrée de terre de signalisation *f*; **~ identification fingerprinting** SIG identification d'empreinte spectrale *f*; **~ indicator** SIG, TRANS indicateur de signal *m*; **~ indicator relay** SIG, TEST relais indicateur de signal *m*; **~ insertion** SIG, TRANS insertion de signaux *f*; **~ interface** SIG interface de signalisation *f*; **~ interface converter** (*SIC*) SIG convertisseur d'interface de signalisation *f*; **~ interference ratio** RADIO rapport signal sur brouillage *m*; **~ lamp** TELEP lampe pilote *f*; **~ level** SIG, TRANS niveau de signal *m*; **~ library** SWG bibliothèque de signaux *f*; **~ message** SIG message de signalisation *m*; **~ -noise ratio** (*SNR*) ELECT, SIG, TEST, TRANS rapport signal sur bruit *m* (*rapport S/B*); **~ path** SIG, TELEP, TRANS circuit de signalisation *m*; **~ power** SIG puissance de signal *f*; **~ processing** SIG traitement du signal *m*; **~ processor** (*SP*) SIG dispositif de traitement de signaux *m*, processeur de signaux *m*, unité de traitement de signaux *f*; **~ processor unit** SIG unité de traitement de signaux *f*; **~ quantization** SIG quantification de signal *f*; **~ receiver** RADIO, SIG récepteur de signaux *m*; **~ regeneration** SIG régénération de signal *f*; **~ restoration** SIG rétablissement du signal *m*; **~ sender** SIG, TRANS émetteur de signaux *m*; **~ shaping** ELECT, SIG mise en forme du signal *f*; **~-shaping filter** ELECT, SIG filtre de mise en forme du signal *m*; **~-shaping network** TELEP réseau correcteur de distorsion *m*; **~ state** GEN état du signal *m*; **~ and state recorder** SWG enregistreur de signal et d'état *m*; **~ strength** SIG force du signal *f*, puissance d'émission *f*, puissance du signal *f*; **~ strength meter** SIG indicateur de force de réception *m*; **~ surface** COMP surface enregistrée *f*; **~ synthesis** DATA, SIG synthèse de signal *f*; **~ test** SIG, TEST vérification de signal *f*; **~ test circuit** SIG, TEST circuit de vérification de signal *m*; **~ threshold** SIG seuil de signal *m*; **~-to-crosstalk ratio** (*SCR*) TRANS écart diaphonique *m*; **~-to-interference ratio** RADIO rapport signal sur brouillage *m*; **~-to-noise ratio** (*SNR*) ELECT, SIG, TEST, TRANS rapport signal sur bruit *m* (*rapport S/B*); **~-to-weighted rms noise ratio** TRANS rapport signal-bruit efficace pondéré *m*; **~ tracer** SIG détecteur de signal *m*; **~ train** SIG sématème *f*; **~ transfer** SIG, SWG, TELEP, TRANS transfert de signaux *m*; **~ transfer point** (*STP*) SIG, SWG, TELEP, TRANS point de transfert de signaux *m*, point de transfert sémaphore *m*; **~ transfer point switch** SIG, SWG, TELEP, TRANS commutateur de point de transfert de signaux *m*; **~ transformation** ELECT, SIG mise en forme de signal *f*; **~ transit time** ELECT temps de transit du signal *m*; **~ unit** (*SU*) SIG trame sémaphore *f* (*TS*), unité de signalisation *f*; **~ unit alignment** SIG alignement des trames sémaphores *m*; **~ unit error-rate monitoring** SIG, TELEP surveillance du taux d'erreur sur les trames sémaphores *f*; **~ weakening** SIG dégradation du signal *f*

signal[2] *vt* COMP, DATA, RADIO, SIG, TELEP signaler

signal[3] *vi* COMP, DATA, RADIO, SIG, TELEP émettre un signal

signaling *AmE see signalling BrE*

signalling *n BrE* COMP, DATA, RADIO, SIG, TELEP signalisation *f*; **~ adaptor group** *BrE* SIG groupe d'adaptateurs de signalisation *m*; **~ bit** *BrE* SIG, TRANS bit de signalisation *m*; **~ cable** *BrE* SIG, TRANS câble de signalisation *m*; **~ channel** *BrE* SIG canal sémaphore *m*, voie de signalisation *f*, TELEP canal sémaphore *m*; **~ conditions** *n pl BrE* SIG conditions de signalisation *f pl*; **~ connection control part** *BrE* (*SCCP*) SIG sous-système de commande de connexions sémaphores *m* (*SSCS*); **~ control part** *BrE* SIG sous-système de contrôle de signalisation *m*; **~ converter** *BrE* PER APP, SIG convertisseur de signaux *m*; **~ data link** *BrE* COMP, DATA, SIG liaison de données de signalisation *f*, liaison sémaphore de données *f*; **~ description** *BrE* SIG description de la signalisation *f*; **~ destination point** *BrE* SIG point sémaphore de destination *m*; **~ detector** *BrE* SIG détecteur de signalisation *m*; **~ diagram** *BrE* SIG schéma de signalisation *m*; **~ fault** *BrE* SIG faute de signalisation *f*; **~ grouping channel** *BrE* SIG canal de groupement de signalisation *m*; **~ information** *BrE* SIG informations de signalisation *f pl*; **~ information field** *BrE* SIG domaine d'informations de signalisation *m*; **~ interworking** *BrE* SIG interfonctionnement de la signalisation *m*; **~ link** *BrE* SIG, TELEP *channel* canal sémaphore *m*, *junction* liaison de signalisation *f*; **~-link blocking** *BrE* SIG blocage d'un canal sémaphore *m*, blocage d'une liaison de signalisation *m*; **~-link code** *BrE* SIG code de canal sémaphore *m*, code de liaison de signalisation *m*; **~-link error monitoring** *BrE* SIG surveillance des erreurs sur un canal sémaphore *f*, surveillance des erreurs sur une liaison de signalisation *f*; **~-link failure** *BrE* SIG défaillance d'un canal sémaphore *f*, défaillance d'une liaison de signalisation *f*; **~-link function** *BrE* RADIO, SIG fonction de canal sémaphore *f*, fonction de liaison de signalisation *f*; **~-link group** *BrE* SIG groupe de canaux sémaphores *m*, groupe de liaisons de signalisation *m*; **~-link management functions** *n pl BrE* SIG fonctions de gestion des canaux sémaphores *f pl*, fonctions de gestion de liaisons de signalisation *f pl*; **~-link restoration** *BrE* SIG rétablissement d'un canal sémaphore *m*, rétablissement d'une liaison de signalisation *m*; **~-link selection field** *BrE* SIG domaine de sélection de canal sémaphore *m*, domaine de sélection de liaison de signalisation *m*; **~-link set** *BrE* SIG faisceau de canaux sémaphores *m*, faisceau de liaisons de signalisation *m*, faisceau sémaphore *m*; **~-link unblocking** *BrE* SIG déblocage d'un canal sémaphore *m*, déblocage d'une liaison de signalisation *m*; **~ message** *BrE* SIG message de signalisation *m*; **~-message discrimination** *BrE* SIG discrimination des messages de signalisation *f*; **~-message distribution** *BrE* SIG distribution des messages de signalisation *f*; **~-message handling functions** *n pl BrE* SIG fonctions d'orientation des messages de signalisation *f pl*; **~-message route** *BrE* SIG route des messages de signalisation *f*; **~-message routing** *BrE* SIG, TELEP, TRANS acheminement des messages de signalisation *m*; **~ network** *BrE* SIG, SWG, TRANS réseau de signalisation *m*, réseau sémaphore *m*; **~-network functions** *n pl BrE* SIG fonctions du réseau sémaphore *f pl*, fonctions du réseau de signalisation *f pl*; **~-network management functions** *n pl BrE* SIG fonctions de gestion du réseau sémaphore *f pl*, fonctions de gestion du réseau de signalisation *f pl*; **~-network management signals** *n pl BrE* SIG signaux de gestion du réseau sémaphore *m pl*, signaux de gestion du réseau de signalisation *m pl*; **~ originating point** *BrE* SIG point sémaphore d'origine *m*, point de signalisation d'origine *m*; **~ point** *BrE* (*SP*) SIG, TELEP point sémaphore *m*, point de signalisation *m*; **~-point code** *BrE* SIG code d'un point sémaphore *m*, code d'un point de signalisation *m*; **~ protocol** *BrE* SIG protocole sémaphore *m*, protocole de signalisation *m*; **~ receiver PE** *BrE* SIG élément graphique de récepteur sémaphore *m*, élément graphique de récepteur de signalisation *m*; **~ relation** *BrE* SIG relation sémaphore *f*; **~ route** *BrE* SIG route sémaphore *f*, route de signalisation *f*; **~-route management functions** *n pl BrE* SIG fonctions de gestion des routes sémaphores *f pl*, fonctions de gestion des routes de signalisation *f pl*; **~-route set** *BrE* SIG faisceau de routes sémaphores *m*, faisceau de routes de signalisation *m*; **~-route-set test procedure** *BrE* SIG procédure du test de faisceau de routes sémaphores *f*; **~ routing** *BrE* SIG, TELEP, TRANS acheminement sémaphore *m*, acheminement de la signalisation *m*; **~ sender PE** *BrE* SIG élément graphique d'émetteur sémaphore *m*, élément graphique d'émetteur de signalisation *m*; **~ sequence** *BrE* SIG séquence sémaphore *f*, séquence de signalisation *f*; **~ state** *BrE* SIG état sémaphore *m*, état de signalisation *m*; **~ system** *BrE* SIG système sémaphore *m*, système de signalisation *m*; **~ terminal equipment** *BrE* (*STE*) TELEP équipement terminal sémaphore *m*, équipement terminal de signalisation *m*; **~ test** *BrE* SIG essai sémaphore *m*, essai de la signalisation *m*; **~ time slot** *BrE* SIG intervalle de temps de signalisation *m*; **~-traffic flow control** *BrE* DATA, SIG contrôle de flux de trafic sémaphore *m*; **~-traffic management functions** *n pl BrE* SIG fonctions de gestion de trafic sémaphore *f pl*; **~ transfer point** (*STP*) SIG, SWG, TELEP, TRANS point de transfert des signaux *m*, point de transfert sémaphore *m*; **~ virtual channel** *BrE* SIG, TRANS canal virtuel de signalisation *m* (*CVC*), circuit virtuel commuté *m* (*CVC*), voie virtuelle de signalisation *f*; **~ voltage** *BrE* POWER, SIG tension de signalisation *f*; **~ at 2600 Hz** *BrE* SIG signalisation à 2600 Hz *f*

signatory *n* TV signataire *m*

signature: **~-adapted processing** *n* RADIO traitement adapté aux signatures *m*; **~ element** *n* (*SIG*) SIG, TELEP élément signature *m*

signed: **~ response element** *n* TELEP élément de réponse signée *m*

significance: **~ level** *n* TEST niveau de signification *m*; **~ test** *n* TEST test de signification *m*

significant[1] *adj* COMP, DATA, TELEP significatif

significant[2]: **~ bit** *n* DATA bit significatif *m*; **~ condition** *n* SIG, TELEP, TEST état significatif *m*; **~ instant** *n* SIG, TELEP, TEST instant significatif *m*; **~ interval** *n* SIG, TELEP intervalle significatif *m*, TEST intervalle temporel *m*

silence: **~ elimination** *n* TEST, TRANS élimination de silence *f*

silencer *n* TELEP, TRANS amortisseur de bruit *m*, enceinte d'insonorisation *f*, silencieux *m*

silent: **~ alert vibrator** *n* RADIO terminal RMU vibrateur *m*; **~ period** *n* TELEP intervalle de silence *m*; **~ zone** *n* RADIO zone de silence *f*

silica *n* PARTS silice *f*; **~ cladding** PARTS revêtement de silice *m*; **~ coating** PARTS revêtement de silice *m*; **~ fiber** *AmE*, **~ fibre** *BrE* DATA, PARTS, TRANS fibre de silice *f*

silicium: **~ nitride passivation** *n* PARTS passivation au nitrure de silicium *f*

silicon *n* COMP, PARTS silicium *m*; **~-controlled rectifier** (*SCR*) ELECT, PARTS redresseur à thyristor au silicium *m*, thyristor commandé au silicium *m*; **~-controlled rectifier crosspoint** ELECT point d'intersection à thyristor au silicium *m*; **~ diode** PARTS diode au silicium *f*; **~ wafer** PARTS tranche de silicium *f*

silk: **~-screened lettering** *n* ELECT lettrage blanc *m*

silver[1]: **~-electroplating** *n* PARTS argenture électrolytique *f*; **~ mica capacitor** *n* ELECT, PARTS condensateur au mica argenté *m*; **~-plating** *n* PARTS argenture *f*

silver[2]: **~-electroplate** *vt* PARTS argenter par électrolyse; **~-plate** *vt* PARTS argenter

SIM *abbr* (*subscriber identity module*) RADIO module d'identité de l'abonné *m*

simple: **~ authentication** *n* DATA authentification simple *f*; **~ mail transfer protocol** *n* (*SMTP*) COMP *Internet* simple mail transfer protocol *m* (*SMTP*); **~ modulation** *n* RADIO modulation simple *f*; **~ multipoint circuit** *n* TELEP circuit multipoint simple *m*; **~ network management protocol** *n* (*SNMP*) DATA protocole SNMP *m* (*SNMP*); **~ parameter argument** *n* TELEP argument de paramètre simple *m*; **~ random sample** *n* TEST échantillon simple au hasard *m*; **~ signal** *n* SIG signal simple *m*; **~ transmission** *n* TRANS transmission simple *f*

simplex[1] *adj* RADIO à mode simplex, semi-duplex, simplex, TELEP, TRANS à l'alternat, semi-duplex, simplex

simplex[2] *n* RADIO, TELEP, TRANS mode simplex *m*, semi-duplex *m*, simplex *m*; **~ circuit** ELECT circuit simplex *m*, TELEP communication simplex *f*; **~ operation** TELEP exploitation simplex *f*; **~ pair** TELEP paire en simplex *f*; **~ traffic** RADIO trafic en mode simplex *m*; **~ transmission** TRANS transmission simplex *f*; **~ working** TELEP exploitation simplex *f*, fonctionnement en simplex *m*

simulate *vt* COMP, ELECT, TELEP, TEST simuler

simulated: **~ speech** *n* COMP parole artificielle *f*

simulation *n* COMP, ELECT, TELEP, TEST simulation *f*

simulator *n* COMP, ELECT, TELEP, TEST simulateur *m*

simulcast *n* RADIO, TV émission en synchronisme *f*

simultaneous: **~ call** *n* TELEP double appel *m*; **~-call terminal** *n* TELEP borne d'appels simultanés *f*; **~ note capacity** *n* COMP *sound reproduction* polyphonie *f*; **~ note output** *n* COMP *sound reproduction* polyphonie *f*; **~ operation** *n* TELEP exploitation simultanée *f*, mode de fonctionnement simultané *m*

SINAD: **~ ratio** *n* GEN rapport SINAD *m*

sine: **~-cosine oscillator** *n* ELECT oscillateur sinus-cosinus *m*; **~-squared pulse** *n* ELECT, TELEP impulsion en sinus carré *f*; **~ vibration test** *n* TEST essai aux vibrations en régime sinusoïdal *m*; **~ wave** *n* COMP, ELECT onde sinusoïdale *f*; **~-wave pulse** *n* COMP, ELECT, TELEP impulsion d'oscillation *f*, impulsion sinusoïdale *f*; **~-wave signal** *n* COMP, ELECT, TELEP signal sinusoïdal *m*

singing *n* PARTS amorçage *m*, sifflement *m*; **~ point** TELEP point d'amorçage *m*, point de sifflement *m*; **~ suppressor** PARTS suppresseur de réaction *m*

single[1]: **~-bay** *adj* ELECT d'un seul bâti; **~-channel** *adj* RADIO monocanal, monovoie; **~-frequency** *adj* ELECT monofréquence; **~-phase** *adj* POWER monophasé; **~ sideband** *adj* (*SSB*) RADIO, TRANS à bande latérale unique (*BLU*); **~-slot** *adj* ELECT monofente; **~-star** *adj* (*SS*) TRANS simple étoile

single[2]: **~-association control function** *n* DATA fonction de contrôle d'association unique *f*; **~-association object** *n* DATA objet d'association unique *m*; **~-attachment station** *n* (*SAS*) ELECT station à raccordement simple *f*; **~-billing system** *n* APPL système de taxation à perception unique *m*; **~-carrier algorithm** *n* COMP, RADIO, TRANS algorithme à un seul porteur *m*; **~-carrier operation** *n* COMP, RADIO, TRANS fonctionnement à un seul porteur *m*; **~-cartridge adaptor** *n* COMP adaptateur monocartouche *m*; **~-cavity mold** *AmE*, **~-cavity mould** *BrE* *n* PARTS moule à simple cavité *m*; **~-channel recorder** *n* TEST enregistreur monovoie *m*; **~-circuit system** *n* GEN système à circuit unique *m*; **~-compartment charger** *n* POWER chargeur monobatterie *m*, chargeur monocellulaire *m*; **~ conductor** *n* ELECT, PARTS conducteur unique *m*; **~-conductor cable** *n* ELECT câble unifilaire *m*; **~ constraint table** *n* DATA table simple de contrainte *f*; **~ contact** *n* PARTS contact simple *m*; **~-control bistable trigger circuit** *n* ELECT bascule bistable à une entrée *f*; **~-cord switchboard** *n* SWG, TELEP commutateur manuel à monocordes *m*; **~-core cable** *n* POWER câble monoconducteur *m*; **~ crystal** *n* ELECT monocristal *m*; **~ current** *n* POWER, TRANS courant simple *m*; **~-current transmission** *n* POWER, TRANS transmission par courant simple *f*; **~ device housing** *n* ELECT, PARTS composant unique autonome *m*; **~-digit dialing** *AmE*, **~-digit dialling** *BrE* *n* TELEP numérotation abrégée à un chiffre *f*; **~-duct block** *n* PARTS, TRANS bloc de canalisations monotubulaire *m*, bloc de conduite unitaire *m*, conduit monotubulaire *m*, TRANS conduit monotubulaire *m*; **~-ended amplifier** *n* POWER amplificateur asymétrique *m*, amplificateur non symétrique *m*; **~-ended control** *n* CONTR, TRANS commande locale *f*, commande unilatéral *f*; **~-ended synchronization** *n* TELEP, TRANS synchronisation locale *f*, synchronisation unilatérale *f*; **~ error** *n* TEST erreur simple *f*; **~-fee metering** *n* TELEP comptage simple *m*; **~-fiber cable** *AmE see single-fibre cable BrE*; **~-fiber jacket** *AmE see single-fibre jacket BrE*; **~-fibre cable** *n BrE* POWER câble monofibre *m*; **~-fibre jacket** *n BrE* PARTS revêtement de fibre individuelle *m*; **~-frequency interference** *n* RADIO, SIG, TELEP brouillage monofréquence *m*; **~-frequency laser** *n* ELECT, TRANS laser monofréquence *m*; **~-frequency operation** *n* RADIO, SIG, TELEP exploitation sur une fréquence *f*, fonctionnement à une seule fréquence *m*; **~-frequency return loss** *n* RADIO, SIG, TELEP affaiblissement de réflexion monofréquence *m*; **~-frequency signal** *n* RADIO, SIG, TELEP signal à une seule fréquence *m*; **~-gang faceplate** *n* ELECT plastron simple *m*; **~ heterojunction** *n* TRANS hétérojonction simple *f*; **~ industrial infrastructure** *n* TV infrastructure unitaire industrielle *f* (*IUI*); **~ in-line package** *n* (*SIP*) GEN boîtier à une rangée de connexions *m* (*boîtier SIP*); **~ lashing** *n* GEN réactance monophasée *f*; **~-level continuous sampling** *n* TEST échantillonnage continu uniniveau *m*; **~-line representation** *n* GEN représentation unifilaire *f*; **~-link procedure** *n* (*SLP*) TELEP procédure à liaison simple *f* (*PLU*); **~ longitudinal mode** *n* (*SLM*) TRANS mode longitudinal unique *m*; **~-mode coupler** *n* TRANS coupleur monomode *m*; **~-mode fiber** *AmE*, **~-mode fibre** *BrE n* DATA, PARTS, TRANS fibre monomodale *f*, fibre monomode *f*, fibre unimodale *f*; **~-mode laser** *n* DATA, ELECT, TRANS laser monomodal *m*, laser unimodal *m*, laser monomode *m*, unimodal *m*; **~-page feeding** *n* PER APP *of printer, scanner* alimentation feuille à feuille *f*; **~-pair cable box** *n* PARTS boîte de câble à une seule paire *f*; **~-phase alternating current** *n* POWER courant alternatif monophasé *m*; **~-phase reactance** *n* POWER réactance monophase *f*; **~-point failure** *n* (*SPF*) TEST défaillance au niveau élémentaire *f*, défaillance d'un élément vital *f*, panne localisée *f*; **~-processor system** *n* SWG système monoprocesseur *m*; **~-progression sequential coding** *n* COMP, TELEP codage séquentiel à progression unique *m*; **~ sampling** *n* TEST échantillonnage simple *m*; **~-segment message** *n* (*SSM*) DATA message à segment unique *m*, message à un seul segment *m*; **~-server queue** *n* COMP, DATA file d'attente à un seul serveur *f*; **~-sheet feeding** *n* PER APP *of printer, scanner* alimentation feuille à feuille *f*; **~-shot flip-flop** *n* ELECT, PARTS bascule monostable *f*, monostable *m*, multivibrateur monostable *m*; **~ sideband** *n* (*SSB*) RADIO, TRANS bande latérale unique *f* (*BLU*); **~ sideband emission** *n* RADIO, TRANS émission à bande latérale unique *f* (*émission BLU*), émission BLU *f* (*émission à bande latérale unique*); **~ sideband transmission** *n* RADIO, TRANS transmission à bande latérale unique *f* (*transmission BLU*), transmission BLU *f* (*transmission à bande latérale unique*); **~-sided distribution frame** *n* SWG répartiteur monoface *m*; **~-sided rack** *n* TELEP bâti à une face *m*; **~-stroke control key** *n* COMP commande monocartouche *f*; **~ sub-burst burst** *n* DATA paquet unique de sous-paquets *m*; **~ tone** *n* GEN fréquence unique *f*, tonalité simple *f*; **~-valued attribute** *n* DATA attribut à valeur unique *m*; **~-voice speech recognition** *n* APPL, COMP, TELEP reconnaissance monolocuteur de la parole *f*; **~-winding transformer** *n* POWER autotransformateur *m*; **~ wire** *n* ELECT, PARTS conducteur unique *m*; **~-wire circuit** *n* ELECT, PARTS circuit unifilaire *m*; **~-wire transmission line** *n* RADIO guide d'onde unifiliaire *m*

single[3]: **~ channel per carrier** *phr* (*SCPC*) RADIO système par porteuse monovoie *m*, voie unique par porteuse, une seule voie par porteuse

single-precision *adj* COMP en simple précision

sintering *n* PARTS frittage *m*

sinusoidal[1] *adj* ELECT, SIG sinusoïdal

sinusoidal[2]: **~ oscillator** *n* ELECT oscillateur d'ondes sinusoïdales *m*; **~ path** *n* ELECT trajectoire sinusoïdale *f*; **~ signal** *n* ELECT, SIG signal sinusoïdal *m*

SIP *abbr* (*single in-line package*) GEN boîtier SIP *m* (*boîtier à une rangée de connexions*)

siren *n* TELEP sirène *f*

SITA *abbr* (*International Society for Aeronautical Telecommunications*) RADIO SITA (*Société internationale de télécommunications aéronautique*)

site[1] *n* COMP emplacement *m*, site *m*, GEN installation *f*, RADIO emplacement *m*, site *m*; **~ attenuation** RADIO affaiblissement d'emplacement *m*, pertes dues aux conditions locales *f*; **~ error** RADIO erreur locale *f*; **~-error susceptibility** RADIO sensibilité à l'erreur locale *f*; **~ investigation** RADIO reconnnaissance du site *f*

site[2] *vt* GEN *antenna* situer, RADIO placer, loger, positionner, localiser

siting *n* GEN *of antenna* situation *f*, RADIO positionnement *m*, localisation *f*, placement *m*, situation *f*

situate *vt* RADIO situer

situated *adj* RADIO situé

situation *n* RADIO situation *f*

sixty: **~-four kbps restricted bearer service** *n* APPL *ISDN* service support à 64 kbit/s avec restriction *m*; **~-four kbps restricted service** *n* APPL *ISDN* service de circuit commuté en canal B non transparent *m*; **~-four kbps unrestricted bearer service** *n* APPL *ISDN* service support à 64 kbit/s sans restriction *m*

size[1] *n* COMP, DATA, GEN dimension *f*, taille *f*, SIG, SWG capacité *f*; **~ alteration** SWG modification de capacité *f*

size[2] *vt* COMP, GEN dimensionner

sizing *n* COMP, GEN dimensionnement *m*

skeletal: **~ code** *n* COMP séquence paramétrable *f*

sketch *n* GEN schéma *m*

skew: **~ ray** *n* TRANS *fibre optics* rayon non méridien *m*, rayon oblique *m*

skewness *n* GEN *of distribution* dissymétrie *f*

skiatron *n* ELECT tube à trace sombre *m*

skin: **~ effect** *n* POWER effet de peau *m*, effet Kelvin *m*, effet pelliculaire *m*; **~ target** *n* RADIO cible passive *f*

skinner *n* TELEP skinner *m*

skinny: **~ wire set** *n* TELEP poste à câblage très limité *m*

skip[1] *n* RADIO saut *m*; **~ area** RADIO, TELEP zone sautée *f*; **~ distance** RADIO, TELEP distance de saut *f*; **~ lot sampling** TEST échantillonnage successif partiel *m*; **~ zone** RADIO, TELEP zone de silence *f*

skip[2] *vt* COMP, RADIO, TELEP sauter

skipping *n* TRANS saut *m*

skirt *n* RADIO collerette *f*

sky: **~ noise temperature** *n* RADIO température de bruit du ciel *f*; **~ wave** *n* RADIO onde ionosphérique *f*

slab: **~ interferometry** *n* RADIO interférométrie *f*, TRANS interférométrie *f*, microscopie axiale interférentielle *f*

slack[1] *adj* PARTS mou, non tendu, relâché

slack[2] *n* PARTS *cable laying* mou *m*

slant *n* COMP *angle* inclinaison *f*; **~ range** RADIO distance de la cible *f*, distance radiale *f*

SLAR *abbr* (*side-looking airborne radar*) RADIO radar aéroporté à visée latérale *m*

slash *n* COMP barre oblique *f*

slave: **~ clock** *n* ELECT horloge asservie *f*; **~ flip-flop** *n* PARTS bascule esclave *f*; **~ station** *n* DATA station asservie *f*; **~ transmitter** *n* RADIO émetteur asservi *m*, émetteur esclave *m*

SLC *abbr* (*subscriber line circuit*) SWG circuit de ligne d'abonné *m*

SLD *abbr* (*superluminescent LED, superluminescent light-*

emitting diode) ELECT, PARTS, TRANS DSL (*diode superluminescente*)
SLDA *abbr* (*semiconductor laser diode amplifier*) ELECT amplificateur à laser semi-conducteur *m*
sleeve *n* ELECT, PARTS manchon *m*; ~ **contact** PARTS contact à douille *m*; ~ **contact unit** POWER barrette à douilles *f*
sleeving *n* ELECT, PARTS, POWER gaine *f*
slew *n* COMP *of printer* saut de papier *m*
slicer *n* ELECT, PARTS éminceur *m*
slide *n* COMP, PARTS *movement* glissement *m*, *photography* diapositive *f*; ~ **caliper** *AmE*, ~ **calliper** *BrE* TEST pied à coulisse *m*; ~ **contact** COMP contact glissant *m*; ~ **effect** COMP effet de glissement *m*; ~**-in unit** PARTS unité extractible *f*; ~ **panel** PARTS panneau mobile *m*; ~ **scanner** COMP, DATA scanner de diapositives *m*; ~ **show** COMP diaporama *m*, multivision *f*; ~ **switch** PARTS commutateur à coulisse *m*, commutateur à glissière *m*
slider *n* COMP commande linéaire *f*, curseur *m*, potentiomètre à glissière *m*; ~ **control** COMP commande linéaire *f*, curseur *m*, potentiomètre à glissière *m*
sliding: ~ **contact** *n* PARTS contact glissant *m*; ~ **door** *n* PARTS porte à glissière *f*, porte coulissante *f*; ~ **frequency generator** *n* POWER piège *m*; ~ **lid** *n* PARTS couvercle à glissière *m*, couvercle coulissant *m*; ~ **scale** *n* COMP échelle mobile *f*
slim: ~ **rack** *n* TELEP bâti étroit *m*
slip *n* DATA, RADIO, TELEP, TRANS glissement *m*; ~ **ring** POWER bague collectrice *f*
slipped: ~ **bank** *n* TELEP banc de broches décalé *m*
SLM *abbr* ELECT (*spatial light modulator*) modulateur optique spatial *m*, TRANS (*single longitudinal mode*) mode longitudinal unique *m*
slope *n* COMP, ELECT, RADIO, TEST, TRANS pente *f*; ~ **of the OC curve** TEST *quality* pente de la courbe d'efficacité *f*, rapport de discrimination *m*
sloping: ~ **V aerial** *n* RADIO antenne en V incliné *f*; ~ **V antenna** *n* RADIO antenne en V incliné *f*
slot[1]: ~**-based** *adj* ELECT, PARTS à emplacement interne
slot[2] *n* COMP créneau *m*, connecteur *m*, fente *f*, logement *m*, ELECT, PARTS emplacement *m*, fente *f*, PER APP, RADIO, TRANS fente *f*, TV créneau *m*; ~ **aerial** RADIO antenne à fente *f*, antenne rayonnante *f*; ~ **antenna** RADIO antenne à fente *f*, antenne rayonnante *f*; ~ **array** RADIO réseau de fentes rayonnantes *m*, réseau à fente *m*; ~**-fed dipole** RADIO doublet symétrisé par coaxial fendu *m*; ~ **in the hub** ELECT emplacement du hub *m*; ~ **radiator** RADIO *antenna* fente rayonnante *f*; **out-~ signaling** *AmE*, **out-~ signalling** *BrE* SIG signalisation hors intervalle de temps *f*; ~ **waveguide** (*SWG*) DATA, RADIO, TRANS guide d'ondes à fentes *m*, guide d'ondes fendu *m*
slotted: ~ **ALOHA system** *n* RADIO, TELEP système ALOHA crénelé *m*; ~**-core cable** *n* TRANS câble à cœur rainuré *m*, câble cylindrique rainuré *m*; ~**-core optical fiber cable** *AmE*, ~**-core optical fibre cable** *BrE* *n* TRANS câble optique à cœur rainuré *m*; ~ **line** *n* RADIO ligne à fente *f*; ~ **measuring line** *n* RADIO ligne de mesure à fente *f*; ~ **screwdriver** *n* PARTS tournevis pour écrous à fente *m*; ~ **system** *n* RADIO, TELEP système crénelé *m*; ~ **waveguide** *n* (*SWG*) RADIO, TRANS guide d'ondes à fentes *m*, guide d'ondes fendu *m*; ~**-waveguide aerial** *n* RADIO antenne à guide d'ondes fendu *f*; ~**-waveguide antenna** *n* RADIO antenne à guide d'ondes fendu *f*
slow: ~**-access storage** *n* COMP mémoire lente *f*; ~**-attack sound** *n* COMP son à attaque lente *m*; ~**-scan television** *n* TV télévision à analyse lente *f*, télévision à balayage lent *f*, télévision à exploration lente *f*; ~**-scan terminal** *n* TELEP terminal à balayage lent *m*, terminal à analyse lent *m*, terminal à exploration lente *m*; ~**-scan videoconferencing** *n* TV vidéoconférence à images fixes *f*; ~**-wave structure** *n* ELECT structure de retard *f*
SLP *abbr* (*single-link procedure*) TELEP PLU (*procédure à liaison unique*)
SLS *abbr* (*side-lobe suppression*) RADIO SLS (*suppression des lobes secondaires*)
slug *n* PARTS *relays* bague de retard *f*
SM *abbr* (*synchronous multiplexer*) TRANS multiplexeur synchrone *m*
SMA *abbr* (*search mode acquisition*) DATA acquisition en mode de recherche *f*; ~ **procedure** *n* DATA procédure d'acquisition en mode de recherche *f*
small: **very** ~ **aperture terminal** *n* (*VSAT*) DATA, PER APP, TRANS terminal à très petite ouverture d'antenne *m* (*TTPA*), TV microterminal *m*; ~**-cell network** *n* RADIO réseau de petites cellules *m*; ~ **computer system interface** *n* (*SCSI*) COMP interface SCSI *f* (*SCSI*); ~**-outline integrated circuit** *n* (*SOIC*) ELECT boîtier soie *m*, circuit intégré à connexions courtes *m*; ~ **red plastic clip** *n* ELECT isolateur de borne à clip *m*; ~**-scale integration** *n* (*SSI*) ELECT, PARTS, TELEP, TRANS intégration à petite échelle *f*; ~**-signal amplifier** *n* ELECT, SIG amplificateur bas niveau *m*, amplificateur de petits signaux *m*; ~**-signal capacitance** *n* ELECT capacité différentielle *f*; ~**-signal gain** *n* ELECT gain en petits signaux *m*; ~ **signal transistor** *n* ELECT, PARTS transistor de petits signaux *m*
SMAP *abbr* (*systems management application protocol*) DATA protocole d'application de gestion des systèmes *m*
smart[1] *adj* COMP intelligent
smart[2]: ~ **chip** *n* ELECT puce intelligente *f*; ~ **frequency selection** *n* RADIO sélection intelligente de fréquence *f*; ~ **network** *n* APPL, TRANS réseau intelligent *m* (*RI*)
smartcard *n* APPL, COMP carte à mémoire *f* (*CAM*), carte à microcircuit *f*, carte à microprocesseur *f*, carte intelligente *f*; ~ **access** APPL accès par carte intelligente *m*, accès par carte à mémoire *m*, accès par carte à microcircuit *m*, accès par carte à microprocesseur *m*; ~ **reader** APPL lecteur de carte à mémoire *m*
SMATV *abbr* (*satellite master aerial television*, *satellite master antenna television*) TV télévision par satellite à antenne principale *f*; ~ **system** *n* TV système de télévision par satellite à antenne principale *m*
SMC *abbr* (*surface-mounted component*) ELECT, PARTS CMS (*composant monté en surface*)
SMD *abbr* (*surface-mounted device*) ELECT, PARTS circuit à montage en surface *m*
SMDS *abbr* (*switched multimegabit data service*) DATA service commuté de données multimégabit *m*
SME *abbr* (*systems management equipment*) COMP équipement de gestion des systèmes *m*
smearing *n* ELECT persistance *f*
smectic[1] *adj* ELECT smectique
smectic[2]: ~ **liquid crystals** *n pl* ELECT, PARTS cristaux liquides en phase smectique *m pl*; ~ **phase** *n* ELECT phase smectique *f*; ~ **state** *n* ELECT état smectique *m*
S-meter *n* TEST S-mètre *m*
SMF *abbr* (*submultiframe*) TELEP sous-multitrame *f*
SMI *abbr* (*structure of management information*) DATA structure des informations de gestion *f*
SMIB *abbr* (*security management information base*) DATA base d'informations de gestion de sécurité *f*
SMN *abbr* (*SDH management network*) TELEP, TRANS réseau de gestion SDH *m*
SMO *abbr* (*stabilized master oscillator*) TEST maître-oscillateur stabilisé *m*
smooth[1] *adj* GEN lisse, POWER filtré
smooth[2]: ~ **surface** *n* RADIO surface lisse *f*; ~ **traffic** *n* TELEP trafic régularisé *m*
smooth[3] *vt* GEN lisser, POWER filtrer
smoothed: ~ **voltage** *n* POWER tension lissée *f*, tension filtrée *f*
smoothing *n* COMP finition *f*, lissage *m*, GEN lissage *m*; ~ **capacitor** PARTS, POWER condensateur de filtrage *m*,

condensateur de lissage *m*; ~ **filter** PARTS, POWER filtre de lissage *m*

SMPTE: ~ **time code** *n* COMP, TV code temporel SMPTE *m*

SMS *abbr* (*SDH management subnetwork*) TELEP, TRANS sous-réseau de gestion SDH *m*

SMSR *abbr* (*side-mode suppression ratio*) TRANS ratio de suppression de la bande latérale *m*

SMT *abbr* (*surface-mount technology*) ELECT, PARTS technologie de montage en surface *f*

SMTP *abbr* (*simple mail transfer protocol*) DATA SMTP (*simple mail transfer protocol*)

SN *abbr* COMP, DATA (*sequence number*) numéro d'ordre *m*, numéro de séquence *m*, (*service notification*) avis de service *m*, notification de service *f*, SWG (*sequence number*) numéro d'ordre *m*, numéro de séquence *m*, TELEP (*subscriber number*) numéro d'abonné *m*, (*service notification*) avis de service *m*, notification de service *f*

snap[1]: **~-action contact** *n* PARTS contact à action rapide *m*, contact à faible inertie *m*; ~ **connection** *n* ELECT connexion à encliquetage *f*; ~ **gage** *AmE*, ~ **gauge** *BrE n* TEST calibre à mâchoires *m*; ~ **locking** *n* PARTS fermeture à ressort *f*

snap[2]: ~ **in** *vt* PARTS enclencher; ~ **into** *vt* PARTS insérer

snap[3]: ~ **together** *vi* PARTS encliqueter

snapshot: ~ **program** *n* COMP, DATA programme d'analyse sélective *m*, programme de dépistage sélectif *m*, programme de traçage sélectif *m*; ~ **semantics** *n* COMP, DATA sémantique d'image instantanée *f*; ~ **trace program** *n* COMP, DATA programme d'analyse sélective *m*, programme de dépistage sélectif *m*, programme de traçage sélectif *m*

SNDCF *abbr* (*subnetwork dependent convergence function*) DATA fonction de convergence dépendant du sous-réseau *f*

sneak: ~ **current** *n* ELECT courant fantôme *m*

SNG *abbr* (*satellite news gathering*) RADIO, TV SNG (*contributions par satellite*), journalisme électronique par satellite *m*, reportage par satellite *m*, reportage électronique d'actualités par satellite *m*, reportage d'actualités par satellite *m*

SNMP *abbr* (*simple network management protocol*) DATA SNMP (*protocole SNMP*)

snooze: ~ **function** *n* PER APP répétition du réveil *f*

snow *n* RADIO neige *f*

SNP *abbr* (*sequence number protection*) COMP, DATA, SWG protection de numéro de séquence *f*

SNPA *abbr* (*subnetwork point of attachment*) DATA point de raccordement du sous-réseau *m*, point de rattachement du sous-réseau *m*

SNR *abbr* ELECT, SIG, TEST, TRANS (*signal-to-noise ratio, signal-noise ratio*) rapport S/B *m* (*rapport signal sur bruit*)

socket *n* COMP fiche femelle *f*, fixation *f*, prise *f*, ELECT broche *f*, prise *f*, support *m*, TV prise *f*; ~ **connector block** PARTS connecteur à vis *m*; ~ **outlet** POWER prise enfichable *f*, prise *f*

soft: ~ **case** *n* COMP étui souple *m*; ~ **copy** *n* COMP, TV image vidéo *f*, présentation visuelle *f*; **~-copy image** *n* COMP, TV image pour présentation visuelle *f*; ~ **decision** *n* DATA décision douce *f*, décision pondérée *f*, décision programmable *f*, décision souple *f*, ELECT, SIG décision douce; **~-decision decoding** *n* DATA décodage à décision douce *m*, décodage à décision programmable *m*, décodage à décision souple *m*, ELECT, SIG décodage à décision douce *m*; ~ **key** *n* COMP touche souple *f*, touche tampon *f*; ~ **logic** *n* COMP logique floue *f*; ~ **magnetic material** *n* ELECT, PARTS matériau magnétique doux *m*; ~ **shorting** *n* POWER court-circuit partiel *m*

software *n* COMP, DATA logiciel *m*, progiciel *m*; ~ **application** APPL, COMP, DATA application informatique *f*; ~ **company** COMP société de services d'ingénierie en informatique *f*; ~ **engineer** COMP ingénieur logiciel *m*; ~ **engineering** COMP génie logiciel *m*; ~ **lockout** COMP protection par programme *f*; ~ **machine** COMP machine logique *f*, montage *m*; ~ **message distribution machine marker** COMP marqueur *m*; **~-only video playback** COMP, TV lecture vidéo par logiciel seul *f*; ~ **package** COMP, DATA progiciel *m*; ~ **piracy** COMP, DATA piratage de logiciels *m*; ~ **product** COMP produit logiciel *m*; ~ **resource** COMP, DATA organe logiciel *m*; ~ **tool** COMP, DATA outil logiciel *m*

SOH *abbr* (*section overhead*) DATA, TELEP SDS (*surdébit de section*)

SOIC *abbr* (*small-outline integrated circuit*) ELECT boîtier soie *m*, circuit intégré à connexions courtes *m*

solar[1]: **~-powered** *adj* POWER à énergie solaire, alimenté par énergie solaire, hélioalimenté

solar[2]: ~ **array** *n* POWER panneau de photopiles *m*; ~ **battery** *n* POWER batterie solaire *f*, pile solaire *f*; ~ **interference** *n* RADIO brouillage dû à la présence du soleil *m*; ~ **panel** *n* POWER panneau solaire *m*; ~ **power** *n* POWER énergie solaire *f*; ~ **quantization** *n* SIG quantification solaire *f*

solder[1]: **~-plated** *adj* PARTS étamé

solder[2] *n* ELECT fil *m*, PARTS soudure *f*, soudure à l'étain *f*; ~ **bridge** PARTS pont de métal d'apport *m*; ~ **pad** PARTS pastille de soudure *f*; **~-plated quick clip** PARTS clip étamé à montage rapide *m*; ~ **resist** PARTS réserve de soudure *f*; **~-resist lacquer** PARTS vernis épargne *m*; ~ **spatter** PARTS éclaboussure de soudure *f*; ~ **splash** PARTS éclaboussure de soudure *f*

solder[3] *vt* ELECT braser, PARTS souder; ~ **together** ELECT, PARTS souder; ~ **using condensation technique** PARTS tendre en phase vapeur

solderability *n* PARTS brasabilité *f*, soudabilité *f*

solderable *adj* PARTS brasable, soudable

soldered: ~ **connection** *n* PARTS connexion soudée *f*, connexion brasée *f*; ~ **joint** *n* PARTS brasure *f*, joint brasé *m*

soldering *n* ELECT brasage *m*, brasage tendre *m*, PARTS soudage *m*, soudure *f*; ~ **lacquer** PARTS vernis soudable *m*; ~ **lug** PARTS broche à souder *f*, cosse à souder *f*; ~ **point** PARTS point de brasage *m*; ~ **tag** PARTS broche à souder *f*, cosse à souder *f*; ~ **time** PARTS temps de brasage *m*; ~ **tin** PARTS soudure à l'étain *f*; ~ **touch-up** PARTS retouche de brasage *f*; ~ **wire** PARTS fil à braser *m*, fil à souder *m*

solenoid *n* POWER solénoïde *m*; ~ **valve** PARTS vanne électromagnétique *f*

solicited: ~ **information indicator** *n* DATA indicateur d'information demandée *m*

solid[1]: **~-state** *adj* PARTS à état solide

solid[2]: ~ **conductor** *n* PARTS conducteur massif *m*; ~ **copperweld conductor** *n* PARTS conducteur monobrin en cuivre *m*; ~ **electrolyte capacitor** *n* PARTS condensateur à électrolyte solide *m*; ~ **insulation** *n* GEN, PARTS isolation en plastique massif *f*; **~-state amplifier** *n* ELECT amplificateur à transistors *m*; **~-state device** *n* ELECT dispositif transistorisé *m*, dispositif à semi-conducteurs *m*, PARTS dispositif à semi-conducteurs *m*, dispositif transistorisé *m*; **~-state disc** *BrE*, **~-state disk** *AmE n* ELECT disque à circuits intégrés *m* (*disque CI*); **~-state rectifier** *n* PARTS, POWER redresseur solide *m*; **~-state relay** *n* ELECT, PARTS, POWER relais à semi-conducteurs *m*, relais statique *m*

SOLR *abbr* (*side-tone objective loudness rating*) TELEP EOSEL (*équivalent objectif pour la sonie de l'effet local*)

son *n* COMP fils *m*

sonar *n* RADIO sonar *m*; ~ **signal** RADIO signal sonar *m*

S1 *abbr* (*unscrambled double dibit*) DATA double dibit non-embrouillé *m*; **~-field** *n* (*status field*) SWG zone d'état S1 *f*

SONET *abbr* (*synchronous optical network*) COMP, TRANS réseau optique synchrone *m*

song: ~ **memory** *n* COMP mémoire de morceaux de musique *f*; **~-position pointer** *n* COMP repère de position de morceau de musique *m*

sonic *adj* TELEP sonique

sophisticated *adj* ELECT sophistiqué

sort *vt* COMP, POWER trier

sorter *n* COMP, POWER trieur *m*

SOS *abbr* (*synchronized-oscillator system*) POWER système à oscillateurs synchronisés *m*

sound[1]: **~-insulated** *adj* GEN, TELEP insonorisé; **~ proof** *adj* GEN, TELEP insonorisé

sound[2]: **~ animation** *n* COMP *of scanner* animation sonore *f*; **~-attenuating cover** *n* PARTS, TELEP capot d'insonorisation *m*, couvercle d'insonorisation *m*; **~ attenuation** *n* TELEP affaiblissement des sons *m*; **~ board** *n* COMP carte audio *f*; **~ broadcast transmitter** *n* RADIO émetteur de radiodiffusion sonore *m*; **~ broadcasting** *n* RADIO radiodiffusion sonore *f*; **~ carrier** *n* TV porteuse son *f*; **~ carrier frequency** *n* TV fréquence porteuse son *f*; **~ channel** *n* TV canal son *m*; **~ density** *n* TELEP densité acoustique *f*; **~ effect** *n* COMP bruitage *m*, effet sonore *m*, TELEP effet sonore *m*; **~ fade-out** *n* COMP fondu sonore *m*; **~ file** *n* DATA fichier son *m*; **~ insulation** *n* GEN, TELEP insonorisation *f*; **~-level control** *n* APPL, PER APP réglage du niveau sonore *m*; **~-level meter** *n* TEST sonomètre *m*; **~ module** *n* COMP module sonore *m*; **~ operator** *n* COMP opérateur sonore *m*; **~-powered telephone** *n* TELEP téléphone autogénérateur *m*; **~ pressure** *n* TELEP pression acoustique *f*; **~-program circuit** *AmE*, **~-programme circuit** *BrE* *n* TV circuit radiophonique *m*; **~ program circuit section** *AmE*, **~ programme circuit section** *BrE* *n* TV section de circuit radiophonique *f*; **~ quality** *n* COMP, RADIO, TV qualité sonore *f*; **~ recording** *n* COMP, RADIO, TV enregistrement du son *m*; **~ reproduction** *n* COMP, RADIO restitution sonore *f*, TV restitution du son *m*; **~ signal** *n* SIG signal son *m*, signal sonore *m*; **~ source** *n* COMP, TELEP source acoustique *f*, source sonore *f*; **~ spectrogram** *n* RADIO, TELEP, TRANS spectrogramme acoustique *m*; **~ spectrum** *n* RADIO, TELEP, TRANS spectre acoustique *m*, spectre d'un son *m*; **~ synthesis** *n* COMP synthèse sonore *f*; **~ synthesizer** *n* COMP synthétiseur de son *m*, synthétiseur sonore *m*; **~ system** *n* COMP sonorisation *f*; **~ and television program connection** *AmE*, **~ and television programme connection** *BrE* *n* TV connexion radiophonique et télévisuelle *f*; **~ transmission** *n* APPL, TRANS transmission du son *f*

sound[3]: **~ on sound** *phr* COMP son sur son

soundproof *vt* GEN, TELEP insonoriser

soundproofing *n* GEN, TELEP insonorisation *f*

soundtrack *n* COMP, TV piste audio *f*, bande son *f*

source[1] *n* COMP, DATA, ELECT, POWER, SWG, TEST source *f*; **~ block** ELECT bloc-source *m*; **~ document** COMP, DATA document source *m*, document de base *m*; **~ of failure** TEST origine d'une défaillance *f*, origine d'une panne *f*; **~ of fault** TEST origine d'une défaillance *f*, origine d'une panne *f*; **~ identifier** TELEP identificateur d'origine *m*; **~ impedance** POWER impédance de la source *f*; **~ of interference** POWER source de bruit *f*, source de parasites *f*, source de perturbation *f*; **~ language** COMP langage source *m*; **~ link** COMP lien source *m*; **~ node** COMP nœud source *m*; **~ power efficiency** POWER, TRANS rendement énergétique *m*; **~ program** COMP programme source *m*; **~ region** PARTS *of field-effect transistor* région de source *f*; **~ rooting transparent** (*SRT*) ELECT mode routage source transparent *m*; **~ select** COMP sélection de source *f*; **~/sink relationship** TELEP relation source/collecteur *f*; **~ stream** COMP chaîne d'origine *f*; **~ system** SWG système source *m*, système générique *m*

source[2]: **the ~ is related-request primitive** *phr* (*req*) DATA la source est la primitive de demande associée *f*

SP *abbr* DATA (*space character, the source or sink is the ACPM, session protocol*) caractère d'espacement *m*, caractère espace *m*, la source ou le puits est l'ACPM, protocole de session *m*, SIG (*signalling point, signal processor*) point sémaphore *m*, unité de traitement de signaux *f*, TELEP (*signalling point*) point sémaphore *m*

space: **~ allocation** *n* GEN affectation d'espace *f*, attribution d'espace *f*; **~ bar** *n* COMP, PER APP barre d'espacement *f*; **~ bus** *n* DATA space bus *m*; **~ character** *n* (*SP*) DATA caractère d'espacement *m*, caractère espace *m*; **~ charge** *n* ELECT charge d'espace *f*; **~-charge debunching** *n* ELECT dégroupement de la charge d'espace *m*, dégroupement par répulsion mutuelle *m*; **~-charge limited current** *n* ELECT courant par charge d'espace *m*; **~-charge region** *n* ELECT, PARTS région de charge d'espace *f*; **~-charge wave tube** *n* ELECT, PARTS tube à onde de charge *m*, tube à onde de charge d'espace *m*; **~-charge-controlled tube** *n* PARTS tube à charge d'espace *m*; **~ cloth** *n* RADIO substance absorbante *f*; **~ communications** *n pl* GEN télécommunications spatiales *f pl*; **~-diversity reception** *n* RADIO réception en diversité d'espace *f*; **~ division** *n* TELEP répartition dans l'espace *f*, répartition spatiale *f*; **~-division switching** *n* SWG, TELEP commutation spatiale *f*; **~-division system** *n* SWG, TELEP système à répartition dans l'espace *m*; **~ period** *n* DATA période de repos *f*; **~ probe** *n* RADIO sonde spatiale *f*, TV sonde spaciale *f*; **~ radiocommunication** *n* RADIO, TV radiocommunication spatiale *f*; **~ segment** *n* RADIO secteur spatial *m*; **~ shuttle** *n* RADIO, TV navette spatiale *f*; **~ station** *n* RADIO, TV station spatiale *f*; **~ switch** *n* (*S switch*) SWG, TELEP commutateur spatial *m* (*commutateur S*); **~ system** *n* RADIO système spatial *m*; **~-time correlation** *n* DATA, SWG corrélation spatio-temporelle *f*; **~-time-space network** *n* (*STS network*) SWG, TRANS réseau spatial-temporel-spatial *m* (*réseau STS*); **~ vehicle** *n* RADIO vaisseau spatial *m*; **~ wave** *n* RADIO onde d'espace *f*

spaced: **~ carrier operation** *n* RADIO exploitation à porteuses distinctes *f*; **~-division switching system** *n* SWG, TELEP système de commutation spatiale *m*; **~-division system** *n* SWG, TELEP système à multiplexage spatial *m*; **~-out letters** *n pl* PER APP lettres espacées *f pl*

spacer *n* PARTS entretoise *f*, GEN pièce d'écartement *f*; **~ plate** PARTS plaque entretoise *f*

spacial: **~ multiplex switching** *n* SWG, TELEP commutation spatiale multiplexée *f*

spacing *n* COMP espacement *m*, RADIO espacement *m*, séparation *f*, TV écart *m*; **~ bar** PARTS profilé-entretoise *m*; **~ block** PARTS entretoise *f*; **~ dimension** TELEP pas modulaire *m*; **~ ring** PARTS bague d'espacement *f*

spacistor *n* ELECT, PARTS spacistor *m*

span[1] *n* DATA portée *f*

span[2] *vt* TRANS réaliser une liaison

spanning: **~-tree algorithm** *n* (*STA*) ELECT algorithme à arbre recouvrant *m*

spare: **~ line** *n* TELEP ligne non attribuée *f*; **~ number** *n* TELEP numéro non-attribué *m*; **~ part** *n* PARTS pièce de rechange *f*, pièce détachée *f*; **~ parts inventory** *n* PARTS stock de pièces de rechange *m*; **~ parts kit** *n* PARTS kit de pièces de rechange *m*; **~ satellite** *n* RADIO satellite de secours *m*

spark *n* POWER étincelle *f*; **~ gap** POWER éclateur *m*; **~-over** POWER *of protector* amorçage *m*; **~ quench** POWER étouffeur d'étincelles *m*, pare-étincelles *m*; **~ quench** POWER suppresseur d'étincelles *m*; **~ quencher** POWER étouffeur d'étincelles *m*, pare-étincelles *m*, suppresseur d'étincelles *m*; **~-quenching capacitor** PARTS condensateur étouffeur d'étincelles *m*, condensateur d'extinction des étincelles *m*, condensateur pare-étincelles *m*; **~-quenching resistor** PARTS résistance à étouffer les étincelles *f*; **~ suppressor** POWER étouffeur d'étincelles *m*, pare-étincelles *m*, suppresseur d'étincelles *m*; **~ transmitter** RADIO émetteur à étincelles *m*

sparking *n* POWER décharge disruptive *f*

spatial: **~ coherence** *n* TRANS cohérence spatiale *f*; **~ compression** *n* COMP compression spatiale *f*; **~ light modulator** *n* (*SLM*) ELECT modulateur optique spatial *m*; **~ predictive concealment** *n* COMP masquage à prédiction spatiale *m*; **~ resolution** *n* COMP, TELEP résolution spatiale *f*

SPC *abbr* (*stored-program control AmE, stored-programme* *control BrE*) CONTR, SWG, TELEP autocommutateur privé à programme enregistré *m*, commande programmable *f*, commande par programme enregistré *f*

SPDU *abbr* (*session protocol data unit*) DATA unité de données de protocole de session *f*

speaker *n* COMP, TELEP *person* interlocuteur *m*, *device* haut-parleur *m*; **~ circuit** TELEP circuit de service *m*; **~ recognition** APPL, COMP, TELEP reconnaissance du locuteur *f*

speakerphone *n* TELEP poste à haut-parleur *m*, poste mains-libres *m*, téléphone haut-parleur incorporé *m*, téléphone mains-libres *m*

speaking: **~ clock** *n* TELEP horloge parlante *f*; **~ guide** *n* TELEP guide parlant *m*; **~ position** *n* TELEP position de conversation *f*; **~ rate** *n* TELEP débit de la parole *m*

special: **~ character** *n* DATA caractère spécial *m*; **~ character combination** *n* DATA combinaison spéciale de caractères *f*; **~ connector** *n* PARTS connecteur spécial *m*; **~ dial tone** *n AmE* DATA, TELEP tonalité spéciale de numérotation *f*; **~ dialling tone** *n BrE* DATA, TELEP tonalité spéciale de numérotation *f*; **~-effects generator** *n* COMP générateur d'effets spéciaux *m*; **~ information tone** *n* TELEP tonalité spéciale d'information *f*; **~ intercept** *n* TELEP service spécial d'interception *m*; **~ purpose computer** *n* COMP ordinateur spécialisé *m*; **~ ringing tone** *n* TELEP tonalité spéciale de retour d'appel *f*; **~ service** *n* TELEP service spécial *m*; **~ telephone set** *n* PER APP, TELEP poste téléphonique spécial *m*

specialized: **~ telephone** *n* PER APP, TELEP téléphonie spécialisée *f*

specific *n* ELECT spécifique *f*; **~ detectivity** ELECT, TEST, TRANS détectivité spécifique *f*, détectivité normée *f*, détectivité normalisée *f*; **~ heat** GEN chaleur spécifique *f*; **~ material** PARTS matériel spécifique *m*; **~ mean repair time** TEST temps moyen spécifique de réparation *m*; **~ negative-recorded announcement without information** TELEP annonce enregistrée spécifique négative sans information *f*; **~ positive-recorded announcement with supplementary information** TELEP annonce enregistrée spécifique positive avec information supplémentaire*f*; **~ positive-recorded announcement without information** TELEP annonce enregistrée spécifique positive sans information *f*; **~ recorded announcement** TELEP annonce enregistrée spécifique *f*; **~ resistance** ELECT résistivité *f*, POWER résistivité *f*, résistance spécifique *f*; **~ trunk-line selection** SWG sélection de ligne interurbaine spécifique *f*; **~ value** DATA valeur spécifique *f*

specification *n* TEST spécification *f*; **~ and description language** (*SDL*) COMP langage de description et de spécification *m* (*LDS*); **~ of materials** PARTS spécification du matériel *f*

specified: **~ quality measure** *n* TEST valeur spécifiée d'un caractère de qualité *f*

specify *vt* TEST spécifier

specimen *n* TEST spécimen *m*

speckle *n* DATA, TRANS tacheture *f*; **~ noise** TRANS bruit de granulation *m*, bruit de tacheture *m*, bruit modal *m*

spectral: **~ analysis** *n* TEST analyse spectrale *f*; **~ bandwidth** *n* RADIO, TRANS largeur de bande spectrale *f*; **~ characteristic** *n* ELECT caractéristique spectrale *f*; **~ component** *n* PARTS, TRANS composant spectral *m*; **~ density** *n* DATA, TRANS densité spectrale *f*; **~ distribution** *n* TRANS répartition spectrale *f*; **~ efficiency** *n* TRANS rendement spectral *m*; **~ filter** *n* PARTS, TRANS filtre spectral *m*; **~ irradiance** *n* GEN densité spectrale d'éclairement *f*, densité spectrale d'éclairement énergétique *f*, éclairement énergétique spectral *m*, TRANS éclairement spectral *m*, éclairement énergétique spectral *m*; **~ line** *n* TRANS raie spectrale *f*; **~ occupancy** *n* TRANS occupation du spectre *f*; **~ radiance** *n* TRANS densité spectrale de luminance énergétique *f*, luminance énergétique spectrale *f*, luminance spectrale *f*, sensibilité spectrale *f*, densité spectrale de luminance *f*; **~ radiant emittance** *n* ELECT, TRANS émittance énergétique spectrale *f*; **~ resources** *n pl* RADIO ressources spectrales *f pl*; **~ responsivity** *n* TRANS sensibilité spectrale *f*, **~ width** *n* TRANS largeur spectrale *f*; **~ window** *n* TRANS fenêtre spectrale *f*

spectrogram *n* RADIO, TELEP, TRANS spectrogramme *m*

spectrometer *n* TRANS spectromètre *m*

spectrometry *n* TRANS spectrométrie *f*

spectrum *n* COMP, RADIO, TRANS spectre *m*; **~ allocation** RADIO allocation de spectre *f*; **~ amplitude** RADIO amplitude du spectre *f*; **~ analyser** *BrE*, **~ analyzer** *AmE* TEST analyseur de spectre *m*; **~-analyser monitoring** *BrE*, **~-analyzer monitoring** *AmE* (*SAM*) RADIO surveillance par analyseur de spectre *f*; **~ congestion** RADIO encombrement du spectre *m*; **~ efficiency** RADIO efficacité de l'emploi du spectre *f*, TRANS rendement spectral *m*; **~ expansion** RADIO élargissement spectral *m*; **~ of harmonics** COMP spectre d'harmoniques *m*; **~ line** TRANS raie spectrale *f*; **~ plan** TV plan de fréquence *m*; **~ resources** *n pl* RADIO ressources spectrales *f pl*; **~ reuse** RADIO réutilisation de spectre de fréquences *f*; **~ spreading** RADIO étalement du spectre *m* (*ES*)

specular: **~ reflection** *n* TRANS réflexion spéculaire *f*; **~ reflection coefficient** *n* RADIO, TRANS coefficient de réflexion spéculaire *m*

speech[1]: **~-controlled** *adj* CONTR, PER APP, TELEP à commande vocale, commandé par fréquence vocale, commandé par la voix

speech[2] *n* RADIO, TELEP parole *f*; **~ activity factor** TELEP facteur d'activité des signaux vocaux *m*; **~ analysis** TELEP analyse de la parole *f*; **~ band** TRANS bande des fréquences vocales *f*; **~ capture device** COMP dispositif de saisie vocale *m*; **~ channel** TRANS voie téléphonique *f*; **~ ciphering** COMP, TELEP cryptophonie *f*; **~ circuit** TELEP, TRANS circuit de conversation *m*, circuit de parole *m*; **~ code** TRANS code de signaux vocaux *m*; **~ coding** COMP codage de la parole *m*, DATA codage de la voix *m*, RADIO, TELEP, TRANS codage de la parole *m*, codage de la voix *m*; **~ compression** TELEP compression sonore *f*; **~ control** CONTR, PER APP, TELEP commande par la voix *f*, commande parlée *f*; **~-data network** DATA, TRANS réseau voix-données *m*; **~ detector** RADIO, TELEP détecteur de parole *m*; **~-digit signaling** *AmE*, **~-digit signalling** *BrE* SIG signalisation par éléments numériques vocaux *f*; **~ direction** TELEP sens de conversation *m*; **~ distortion** TELEP distorsion de la parole *f*; **~ encoder** COMP, RADIO, TELEP codeur de parole *m*; **~ encoding** COMP, RADIO, TELEP codage de la parole *m*; **~ generation** COMP, RADIO, TELEP génération de parole *f*; **~-grade private wire** TRANS liaison spécialisée à tous usages *f*; **~ intelligibility** COMP, RADIO, TELEP intelligibilité de la parole *f*; **~ memory** TELEP mémoire de trame *f*, mémoire vocale *f*; **~ path** TELEP trajet de conversation *m*, itinéraire de parole *m*, voie de conversation *f*; **~ plus duplex** TELEP bivocal *m*; **~ plus simplex** TELEP équipement univocal *m*, univocal *m*; **~-plus-duplex equipment** TELEP équipement bivocal *m*, bivocal *m*; **~-plus-simplex equipment** TELEP équipement univocal *m*, univocal *m*; **~ processing** APPL, TELEP traitement de la parole *m*; **~-processing card** APPL carte de traitement de la parole *f*; **~ production** COMP, RADIO, TELEP production de la parole *f*; **~ recognition** APPL, COMP, TELEP reconnaissance de la parole *f*, reconnaissance vocale *f*; **~ recording** TELEP enregistrement de la parole *m*; **~ sample** TRANS échantillon de parole *m*; **~ scrambler** RADIO, TELEP brouilleur de parole *m*; **~ service** APPL service vocal *m*; **~ signal** SIG, TELEP signal de parole *m*, signal vocal *m*; **~ sound** COMP son vocal *m*; **~ switchover** RADIO commutation parole *f*, commutation écoute *f*; **~ synthesis** APPL, COMP, TELEP synthèse de la parole *f*, synthèse vocale *f*;

~ technology APPL, COMP, RADIO, TELEP technologie vocale *f*; **~ wire** PARTS fil de conversation *m*
speed *n* GEN vitesse *f*; **~ calling** TELEP numérotation abrégée *f*; **~ dialing** *AmE*, **~ dialling** *BrE* TELEP numérotation rapide *f*; **~ gate** RADIO porte de vitesse *f*; **~ indicator** RADIO indicateur de vitesse *m*; **~ regulator** TELEP régulateur de vitesse *m*; **~ search** RADIO recherche en vitesse *f*; **~-up capacitor** PARTS, POWER condensateur d'accélération *m*
spellchecker *n* COMP vérificateur d'orthographe *m*
SPF *abbr* DATA (*segmentation permitted flag*) indicateur de segmentation permise *m*, TEST (*single-point failure*) défaillance au niveau élémentaire *f*, défaillance d'un élément vital *f*, panne localisée *f*
spherical: **~ aberration** *n* ELECT, TRANS aberration sphérique *f*; **~ aerial** *n* RADIO antenne sphérique *f*; **~ antenna** *n* RADIO antenne sphérique *f*; **~ mirror** *n* TRANS miroir sphérique *m*
spike: **~ leakage energy** *n* ELECT énergie de fuite de pointe *f*
spin[1]: **~ rate** *n* COMP vitesse de rotation *f*
spin[2] *vt* COMP tourner sur un axe
spin[3] *vi* COMP tourner sur un axe
spindle *n* PARTS axe *m*
spiral[1]: **~-wound** *adj* ELECT, PARTS enroulé en spirale
spiral[2]: **~ antenna** *n* RADIO antenne spirale *f*; **~ ratchet screwdriver** *n* PARTS tournevis à clichet *m*; **~ waveguide** *n* DATA, RADIO, TRANS guide d'ondes hélicoïdal *m*; **~ wrap** *n* PARTS gaine spirale *f*
SPITE *abbr* (*switching process interworking telephony event*) TELEP ETIC (*événement téléphonique de traitement pour les opérations de commutation à l'interface*)
SPL *abbr* (*service provider link*) TRANS liaison de prestataire de service *f*
splice[1] *n* GEN épissure *f*, PARTS épissure *f*, raccord *m*; **~ chamber** ELECT chambre d'épissage *f*; **~ closure** ELECT boîtier d'épissures *m*; **~ connector** PARTS connecteur bout-à-bout *m*; **~ loss** TRANS perte au raccord *f*, perte aux points de raccordement *f*, perte d'épissure *f*; **~ protection** TRANS protection d'épissures *f*; **~ tray** PARTS cassette d'épanouissement *f*; **~-tray method** PARTS méthode par cassette *f*
splice[2] *vt* GEN épissurer
splicer *n* PARTS épisseur *m*; **~ cable** TRANS câble de raccordement *m*
splicing: **~ facility** *n* COMP dispositif d'épissage *m*; **~ pigtail** *n* PARTS câble amorce d'épissage *m*
split[1] *n* GEN division *f*, COMP partage *m*, séparation *f*; **~ baud rate** DATA mode mixte *m*, mode asymétrique *m*, mode bivitesse *m*, mode bidirectionnel à vitesse multiple *m*; **~ channel** SWG voie dédoublée *f*; **~ connection** TELEP communication interrompue *f*; **~ feed** POWER, TELEP alimentation en courant individuel *f*; **~ headphones** *n pl* TELEP casque à deux écouteurs indépendants *m*; **~ mold** *AmE*, **~ mould** *BrE* PARTS moule à coins *m*; **~ point** COMP point de partage *m*
split[2] *vt* COMP, ELECT partager, séparer, subdiviser
split[3] *vi* COMP, ELECT se partager, se séparer, se subdiviser
splitter *n* COMP, ELECT diviseur *m*, TRANS séparateur *m*
splitting *n* GEN division *f*, partage *m*, séparation *f*, subdivision *f*, TELEP éclatement *m*, séparation *f*
SPM *abbr* (*self-phase modulation*) TRANS automodulation de phase *f*
SPN *abbr* (*subscriber premises network*) TRANS réseau d'installation d'abonné *m*
spoken: **~ message** *n* PER APP *of answer machine*, TELEP message parlé *m*
spoking *n* COMP rayons *m pl*
spontaneous: **~ emission** *n* ELECT, TRANS émission spontanée *f*; **~ output** *n* TELEP sortie spontanée *f*
spool *n* PARTS bobine *f*, spoule *m*
sporadic: **~ E layer** *n* RADIO couche E sporadique *f*; **~ failure** *n* TEST défaillance sporadique *f*; **~ ionization** *n* RADIO ionisation sporadique *f*
spot *n* COMP point *m*, ELECT spot *m*, tache *f*, TV spot *m*; **~ beam** RADIO faisceau étroit *m*, faisceau filiforme *m*, faisceau ponctuel *m*, TV faisceau étroit *m*; **~-beam coverage** RADIO couverture par faisceaux ponctuels *f*; **~ check** TEST contrôle par sondage *m*; **~ distortion** ELECT distorsion de spot *f*; **~ jamming** ELECT, RADIO brouillage sélectif *m*; **~ noise factor** RADIO facteur de bruit *m*; **~ noise figure** RADIO facteur de bruit *m*; **~ noise temperature** RADIO température de bruit *f*; **~ sample** TEST échantillon localisé *m*, prélèvement ponctuel d'un échantillon *m*; **~ size** TV diamètre du spot *m*; **~ wobble** TV vobulation du spot *f*
spotlight: **~ effect** *n* COMP effet de surbrillance *m*
spout *n* CLIM, PARTS buse *f*
spray[1]: **~ nozzle** *n* CLIM, PARTS buse de pulvérisation *f*, pulvérisateur *m*
spray[2]: **~-finish** *vt* PARTS vernir au pistolet; **~-lacquer** *vt* PARTS vernir au pistolet
spread *n* ELECT, RADIO étalement *m*, étendue *f*; **~ of bearings** RADIO étendue des relèvements *f*; **~ spectrum** (*SS*) RADIO étalement du spectre *m* (*ES*); **~-spectrum modulation** RADIO modulation par étalement du spectre *f*; **~-spectrum signal** RADIO signal à spectre étalé *m*; **~-spectrum system** RADIO système à étalement du spectre *m*, système à modulation avec étalement du spectre *m*
spreading: **~ loss** *n* RADIO affaiblissement géométrique *m*, atténuation géométrique *f*; **~ resistance** *n* ELECT résistance distribuée *f*
spreadsheet *n* COMP feuille de calcul *f*, tableur *m*
spring[1]: **~-loaded** *adj* PARTS à ressort
spring[2]: **~ adjuster** *n* PARTS, TEST cambreur de ressort *m*; **~ bender** *n* PARTS, TEST cambreur de lames de contact *m*, cambreur de ressort *m*; **~ contact** *n* PARTS contact élastique *m*; **~ holder** *n* PARTS attache-ressort *f*; **~-loaded switch** *n* POWER, SWG interrupteur à ressort *m*, interrupteur monostable *m*; **~ set** *n* SWG empilage *m*; **~-set load** *n* PARTS force d'empilage de lames de contact *f*; **~-set mounting plate** *n* PARTS plaque de montage d'empilage *f*
sprocket *n*: **~ hole** *n* COMP perforation d'entraînement *f*, piste d'entraînement *f*; **~ track** *n* PARTS piste d'entraînement *f*
spt *abbr* (*sectors per track*) COMP spp (*secteurs par piste*)
SPTV *abbr* (*still-picture television*) TV télévision à images fixes *f*
spudger *n* ELECT, PARTS récupérateur *m*
spurious: **~ count** *n* RADIO coup parasite *m*; **~ emission** *n* RADIO, TRANS rayonnement non essentiel *m*, rayonnement parasite *m*; **~ frequency-conversion product** *n* RADIO produit de conversion de fréquence non essentiel *m*; **~ intermodulation product** *n* RADIO, TRANS produit d'intermodulation non essentiel *m*; **~ mixing product** *n* TRANS produit d'intermodulation non essentiel *m*; **~ modulation** *n* RADIO, TRANS modulation non essentielle *f*; **~ radiation** *n* RADIO, TRANS rayonnement non essentiel *m*, rayonnement parasite *m*; **~ response** *n* RADIO réponse parasite *f*; **~ response rejection ratio** *n* RADIO affaiblissement sur la fréquence parasite *m*; **~ satellite emission** *n* RADIO rayonnement non-essentiel du satellite *m*; **~ signal** *n* SIG signal parasite *m*; **~ switch-off** *n* (*SSO*) TEST mise hors service intempestive *f*
SQL *abbr* (*structured query language*) DATA SQL (*langage d'interrogation structuré*)
SQS *abbr* (*Swiss Association for Quality Assurance Certificates*) GEN SQS (*Association suisse pour certificats d'assurance de la qualité*); **~ quality certificate** *n* TEST certificat d'assurance qualité SQS *m*, certificat SQS *m*
square: **~-law detection** *n* ELECT détection quadratique *f*; **~-law detector** *n* ELECT détecteur quadratique *m*; **~-pulse generator** *n* ELECT générateur d'impulsions rectangulaires *m*; **~ signal** *n* ELECT, SIG signal carré *m*; **~ wave** *n* ELECT, SIG

onde carrée *f*; **~-wave generator** *n* ELECT, SIG générateur de signaux carrés *m*; **~-wave response** *n* ELECT, SIG, TRANS réponse en amplitude *f*, réponse en signaux carrés *f*; **~-wave response characteristic** *n* ELECT, SIG, TRANS caractéristique de réponse à un échelon *f*; **~-wave voltage** *n* ELECT, SIG tension crénelée *f*, tension de forme rectangulaire *f*; **~ waveform** *n* ELECT, SIG signal carré *m*

squareax *n* TRANS coaxial carré *m*

squealing *n* RADIO accrochage *m*

squeeze[1]: **~ section** *n* RADIO tronçon à section variable *m*

squeeze[2] *vt* COMP comprimer

squelch *n* RADIO accord silencieux *m*, blocage automatique *m*, réglage silencieux *m*; **~ control** RADIO commande de blocage automatique *f*, réglage silencieux *m*

squint *n* RADIO erreur de directivité *f*, angle de strabisme *m*, strabisme *m*; **~ angle** RADIO angle de strabisme *m*, erreur de directivité *f*; **~ aerial** RADIO antenne à strabisme *f*; **~ antenna** RADIO antenne à strabisme *f*

SRAM *abbr* (*static random-access memory*) COMP SRAM (*mémoire à accés aléatoire, mémoire vide statique*)

SRD *abbr* (*superradiant diode*) ELECT, PARTS, TRANS *fibre optics* DSL (*diode superluminescente*)

S-reference: **~ point** *n* DATA point de référence S *m*

SREJ *abbr* (*selective reject frame*) TELEP trame de rejet sélectif *f*

SRR: **~ operation** *n* (*search and rescue operation*) RADIO opération de récherche et de sauvetage *f*

SRS *abbr* (*stimulated Raman scattering*) RADIO, TRANS diffusion Raman stimulée *f*

SRT *abbr* (*source rooting transparent*) ELECT mode routage source transparent *m*

SS *abbr* RADIO (*spread spectrum*) ES (*étalement du spectre*), TRANS (*single-star*) simple étoile

SSA *abbr* (*sun sensor assembly*) RADIO bloc capteur solaire *m*

SSAP *abbr* (*session service access point*) DATA point d'accès au service de session *m*

SSB *abbr* RADIO, TRANS (*single sideband*) BLU (*bande latérale unique*); **~ emission** *n* RADIO, TRANS émission BLU *f* (*émission à bande latérale unique*); **~ transmission** *n* RADIO transmission BLU *f* (*transmission à bande latérale unique*)

SSC *abbr* RADIO, SWG (*satellite switching center AmE, satellite switching centre BrE*) centre de commutation à bord du satellite *m*

SSI *abbr* (*small-scale integration*) ELECT, PARTS, TELEP, TRANS intégration à petite échelle *f*

SSM *abbr* (*single-segment message*) DATA message à segment unique *m*, message à un seul segment *m*

SSO *abbr* (*spurious switch-off*) TEST mise hors service intempestive *f*

SSR *abbr* (*steady-state reception*) RADIO réception en régime permanent *f*

S/SYS *abbr* (*storage system*) DATA système de mémorisation *m*

ST *abbr* (*segment type*) COMP, DATA type de segment *m*; **~ quality certificate** *n* TEST certificat d'assurance qualité SQS *m*, certificat SQS *m*

STA *abbr* (*spanning-tree algorithm*) ELECT algorithme à arbre recouvrant *m*

stability: **~ loss** *n* TELEP affaiblissement de stabilité *m*

stabilized: **~ master oscillator** *n* (*SMO*) TEST maître-oscillateur stabilisé *m*

stable: **~ state** *n* ELECT état stable *m*; **~ testing state** *n* TEST état de test stable *m*

stack[1] *n* COMP pile *f*; **~ pointer** COMP pointeur de pile *m*

stack[2] *vt* COMP superposer

stacking *n* COMP superposition *f*, PARTS empilage *m*

staff: **~ locater system** *n* RADIO système de recherche de personnes *m*, TELEP système d'appel de personnes *m*, système d'appel unilatéral *m*, système de recherche de personnes *m*

stage *n* GEN étage *m*

stagger: **~-tuned circuit** *n* ELECT réseau à accords décalés *m*

stain[1] *n* PARTS tache *f*, tache ionique *f*

stain[2] *vt* PARTS teindre

staining *n* PARTS teinture *f*

stamp *vt* PARTS tamponner, estampiller

stamper *n* PARTS matrice de pressage *f*

stand[1]: **~-alone** *adj* COMP, DATA autonome, isolé; **~-by** *adj* GEN de réserve, de secours, en attente, en réserve, POWER d'attente

stand[2]: **~-alone** *n* ELECT unité autonome *f*; **~-alone bridge** *n* ELECT pont autonome *m*; **~-alone DSA** *n* DATA DSA autonome *m*; **~-alone exchange** *n* SWG central autonome *m*; **~-alone rack** *n* TELEP bâti autonome *m*; **~-alone satellite** *n* RADIO satellite autonome *m*; **~-alone unit** *n* ELECT unité autonome *f*; **~-by** *n* COMP, RADIO attente *f*, TELEP réserve *f*; **~ set** *n* GEN appareil de réserve *m*, dispositif de réserve *m*; **~-by system** *n* GEN système de secours *m*; **~ by time** *n* TELEP temps de latence *m*; **~ by-ready signal** *n* SIG signal de réserve prête *m*

stand[3]: **~ by** *phr* TELEP attendez; **be on ~ by** *phr* RADIO être en standby

standard *n* GEN étalon *m*, norme *f*; **~ capacitor** POWER condensateur étalon *m*; **~ cell** ELECT cellule standard *f*; **~ cell circuit** ELECT circuit à cellules prédéfinies *m*; **~ component** ELECT, PARTS composant étalon *m*, composant normalisé *m*; **~ deviation** GEN écart type *m*; **~ error** GEN erreur type *f*; **~ file dialog box** COMP, DATA zone de dialogue de fichier standard *f*; **~ frequency** RADIO fréquence étalon *f*; **~-frequency emission** RADIO émission de fréquences étalon *f*; **~-frequency satellite service** RADIO service des fréquences étalon par satellite *m*; **~-frequency and time-signal station** RADIO station de fréquence étalon et de signaux horaires *f*; **~ generalized markup language** (*SGML*) COMP langage normalisé de balisage généralisé *m*; **~-grade performance channel** TRANS voie de qualité nominale *f*; **~ inductor** ELECT, PARTS inductance étalon *f*; **~ interface** GEN interface normalisée *f*; **~ keyboard pitch** COMP hauteur standard du clavier *f*; **~ loss** TRANS *of coaxial cable* perte spécifique *f*; **~ modular unit** TELEP module standard *m*; **~ NATO data model** RADIO modèle OTAN de communication de données *m*; **~ oscillator** ELECT, PARTS oscillateur étalon classique *m*; **~ packages** *n pl* DATA packetage standard *m*; **~ pitch** COMP hauteur standard *f*; **~ population** TEST population type *f*; **~ profile** TELEP profil normalisé *f*; **~ propagation** RADIO propagation normale *f*; **~ radio atmosphere** RADIO atmosphère radioélectrique normale *f*; **~ refraction** RADIO réfraction normale *f*; **~ refractivity vertical gradient** RADIO gradient normal du coïndice *m*; **~ resistor** PARTS, POWER résistance étalon *f*; **~ sheet** GEN feuille de norme *f*; **~ telephone set** TELEP poste téléphonique ordinaire *m*; **~ time-signal emission** RADIO émission des signaux horaires *f*; **~ time-signal emission station** RADIO station de signaux horaires *f*; **~ value** GEN valeur de référence *f*, valeur guide *f*, valeur type *f*; **~ voice jack** TELEP prise téléphonique standard *f*; **~ wave error** RADIO, TRANS erreur type de polarisation *f*

standardization *n* COMP normalisation *f*, GEN normalisation *f*, standardisation *f*

standardize *vt* COMP normaliser, GEN normaliser, standardiser

standardized: **~ abstract test suite** *n* DATA, TEST suite de tests abstraite normalisée *f*; **~ option** *n* TELEP option normalisée *f*

stand-by[1]: **~-marked** *adj* GEN marqué en attente

stand-by[2]: **~ conditions** *n pl* GEN conditions de veille *f pl*; **~ equipment** *n* TELEP équipement de réserve *m*, matériel de réserve *m*, matériel de secours *m*, équipement de secours *m*, équipement dupliqué *m*, équipement en attente *m*, équipement en double *m*; **~ mode** *n* TEST mode veille *m*;

~ phase *n* TEST phase d'attente *f*; **~ power** *n* POWER alimentation de secours *f*; **~ power supply** *n* POWER alimentation de secours *f*; **~ processor** *n* SWG processeur de réserve *m*; **~-ready acknowledgement signal** *n* SIG signal d'accusé de réception de liaison de réserve prête *m*; **~ redundancy** *n* TELEP redondance en attente *f*, TEST redondance passive *f*; **~ state** *n* TEST état d'attente *m*; **~ time** *n* TEST période d'attente *f*; **~ unit** *n* GEN unité de réserve *f*, unité en état d'attente *f*, unité intermédiaire *f*; **~ working** *n* TEST fonctionnement redondant *m*

standing: **~ current** *n* ELECT courant de repos *m*; **~ wave** *n* RADIO, TRANS onde stationnaire *f*; **~-wave meter** *n* TEST ROS-mètre *m*; **~ wave ratio** *n* (*SWR*) RADIO, TRANS rapport d'ondes stationnaires en tension *m* (*ROS*), taux d'ondes stationnaires *m* (*TOS*), taux d'ondes de tension stationnaires *m* (*TOS*)

standoff: **~ ratio** *n* POWER rapport intrinsèque *m*

standstill *n* COMP arrêt *m*, immobilisation *f*

star[1]: **~-connected** *adj* PARTS, POWER monté en étoile; **~-wired** *adj* POWER câblé en étoile

star[2]: **~ architecture** *n* TRANS, TV architecture en étoile *f*; **~ button** *n* TELEP touche étoile *f*; **~ configuration** *n* TRANS configuration en étoile *f*; **~ connection** *n* POWER montage en étoile *m*, TRANS connexion étoilée *f*; **~ coupler** *n* TRANS coupleur en étoile *m*; **~-delta starting** *n* POWER démarrage étoile-triangle *m*; **~ distribution** *n* TRANS distribution en étoile *f*; **~ indicator** *n* TELEP indicateur visuel *m*; **~ network** *n* POWER, TRANS, TV réseau étoilé *m*, réseau en étoile *m*; **~ service** *n* PER APP service confort *m*; **~ structure** *n* TRANS, TV structure en étoile *f*; **~ switch** *n* SWG commutateur en étoile *m*; **~ topology** *n* DATA topologie en étoile *f*

start[1]: **~-stop** *adj* GEN arythmique

start[2]: **~-dialing signal** *AmE*, **~-dialling signal** *BrE* *n* SIG, TELEP signal d'invitation à transmettre *m*; **~-of-block signal** *n* SIG, TRANS signal de début de bloc *m*; **~-of-heading character** *n* (*STX*) COMP, DATA caractère de début de texte *m* (*STX*); **~ signal** *n* DATA, SIG signal de départ *m*; **~-stop apparatus** *n* TELEP appareil arythmique *m*; **~-stop button** *n* COMP touche de marche-arrêt *f*; **~-stop distortion** *n* TELEP distorsion arythmique *f*; **~-stop format** *n* DATA format arythmique *m*; **~-stop system** *n* TELEP système arythmique *m*, système départ-arrêt *m*, télégraphie arythmique *f*; **~-stop transmission** *n* TELEP, TRANS transmission arythmique *f*; **~ of text** *n* (*STX*) COMP, DATA caractère de début de texte *m* (*STX*); **~-up** *n* COMP mise en route *f*; **~-up circuit** *n* POWER circuit de démarrage *m*; **~-up disk** *n* COMP, DATA disque de démarrage *m*; **~-up screen** *n* COMP écran de démarrage *m*

start[3]: **~ up** *vt* COMP mettre en route

starter: **~ gap** *n* ELECT espace de déclenchement *m*; **~ transfer current** *n* ELECT courant de transfert de l'électrode d'amorçage *m*

starting: **~ electrode** *n* ELECT électrode d'amorçage *f*

state *n* GEN état; **~ of alert** TELEP état d'alerte *m*; **~ symbol** TELEP symbole d'état *m*; **~ of transition** TELEP état de transition *m*; **~ transition diagram** TEST diagramme de transition d'état *m*

static[1] *adj* GEN statique

static[2] *n* TRANS brouillage atmosphérique *m*; **~ chaining** DATA chaînage statique *m*; **~ characteristic** ELECT caractéristique statique *f*; **~ conformance requirement** DATA condition de conformité statique *f*; **~ conformance review** DATA revue de conformité statique *f*, examen de conformité statique *m*; **~ dump** COMP vidage statique *m*; **~ field** ELECT champ statique *m*; **~ interference** RADIO perturbation radioélectrique *f*; **~ memory** COMP mémoire statique *f*; **~ RAM** (*static random-access memory*) COMP RAM statique *f*, mémoire vide statique *f* (*SRAM*); **~ random access memory** (*static RAM*) COMP mémoire à accès aléatoire *f* (*SRAM*), mémoire vide statique *f* (*SRAM*); **~ rectifier** POWER redresseur solide *m*; **~ relay** ELECT, PARTS relais statique *m*; **~ semantics** COMP, DATA sémantique statique *f*; **~ storage** DATA mémoire statique *f*; **~ switch** POWER, SWG commutateur statique *m*

station *n* GEN station *f*; **~ conference** RADIO conférence établie par un poste *f*; **~ keeping** RADIO maintien en position *m*, maintien à poste *m*; **~-keeping lifetime** RADIO durée de vie à poste *f*; **~-keeping satellite** RADIO satellite maintenu en position *m*

stationary: **~ high-altitude relay platform** *n* (*SHARP*) TRANS SHARP *m*; **~ information source** *n* GEN source d'information stationnaire *f*; **~ message source** *n* GEN source de messages stationnaire *f*; **~ phase** *n* ELECT phase stationnaire *f*; **~ satellite** *n* RADIO satellite stationnaire *m*; **~ wave** *n* ELECT onde stationnaire *f*

statistic *n* GEN statistique *f*

statistical: **~ coverage limit** *n* TEST limite statistique de couverture *f*; **~ delay of ignition** *n* ELECT durée statistique de préamorçage *f*; **~ frame traffic information** *n* ELECT données statistiques *f pl*; **~ method** *n* APPL méthode statistique *f*; **~ quality control** *n* TEST maîtrise statistique de la qualité *f*; **~ test** *n* APPL, TEST test statistique *m*

statistics: **~ counter** *n* TEST compteur de statistiques *m*

stator *n* ELECT, PARTS induit fixe *m*, stator *m*

status: **~ channel** *n* TELEP voie d'état *f*; **~ check** *n* RADIO contrôle d'état *m*; **~ and control code** *n* CONTR, DATA code d'état et de commande *m*; **~ data** *n pl* DATA, TEST données d'état *f pl*; **~ field** *n* SWG *first party connected* zone d'état *f*, TELEP domaine d'état *m*; **~ flag** *n* COMP indicateur d'état *m*; **~ indication** *n* COMP, TEST indication d'état *f*; **~ lamp** *n* TEST voyant d'état *m*; **~ message** *n* RADIO message d'état *m*; **~ monitoring** *n* CONTR, TEST contrôle d'état *m*

stay *n* PARTS étai *m*; **~ bar** PARTS barre de soutien *f*; **~ iron** PARTS tige d'ancrage *f*; **~ wire** PARTS hauban *m*

STC *abbr* (*sensitivity time control*) CONTR commande cyclique du gain *f*, gain variable dans le temps *m*

STD *abbr* *BrE* (*subscriber trunk dialling*) TELEP IA (*interurbain automatique*)

STE *abbr* (*signaling terminal equipment, signalling terminal equipment*) TELEP équipement terminal de signalisation *m*

steady: **~ state** *n* GEN régime établi *m*, régime permanent *m*, état d'équilibre *m*; **~-state condition** *n* ELECT condition de régime permanent *f*, répartition des modes à l'équilibre *f*, TRANS condition de régime permanent *f*, *fibre optics* répartition de mode à équilibre *f*, équilibre des modes *m*; **~-state launching conditions** *n pl* ELECT conditions d'injections d'équilibre *f pl*; **~-state reception** *n* (*SSR*) RADIO réception en régime permanent *f*; **~ state value** *n* GEN valeur en régime établi *f*, valeur en régime permanent *f*

steel: **~ armor** *AmE*, **~ armour** *BrE* *n* PARTS armure en acier *f*; **~ rule** *n* TEST réglet *m*; **~ tape** *n* PARTS ruban d'acier *m*; **~ tape armor** *AmE*, **~ tape armour** *BrE* *n* PARTS armure en feuillard métallique *f*; **~ wire** *n* ELECT, GEN, PARTS fil d'acier *m*; **~ wire braid** *n* PARTS grip à chaussette *m*

steer *vt* RADIO orienter, guider, piloter

steerability *n* RADIO plage d'orientation *f*

steerable[1] *adj* RADIO orientable, pilotable

steerable[2]: **~ aerial** *n* RADIO antenne orientable *f*; **~ antenna** *n* RADIO antenne orientable *f*; **~-beam aerial** *n* RADIO antenne à faisceau orientable *f*; **~-beam antenna** *n* RADIO antenne à faisceau orientable *f*

steered: **~ flip-flop** *n* ELECT, PARTS bascule à entrée de forçage *f*; **~ input** *n* ELECT entrée de forçage *f*

steering *n* RADIO pilotage *m*, pointage *m*; **~ input** ELECT entrée de forçage *f*; **~ software** COMP, DATA *Internet* logiciel de navigation *m*, *scanner* logiciel de pilotage *m*

step[1]: **~-adjustable** *adj* GEN réglable en pas à pas, réglable par paliers

step[2] *n* TRANS échelon; **~ action** CONTR action par étapes *f*; **~-by-step automatic system** TELEP système automatique pas à pas *m*; **~-by-step connection** POWER connexion pas à pas *f*, connexion progressive *f*; **~-by-step instruction** COMP instruction pas à pas *f*; **~-by-step office** TELEP central pas à pas *m*; **~-by-step operation** COMP exécution pas à pas *f*; **~-by-step selector** SWG commutateur pas à pas *m*, sélecteur pas à pas *m*, **~-by-step system** SWG système pas à pas *m*, système Strowger *m*; **~-down transformation** POWER abaissement de la tension *m*; **~-down transformer** POWER transformateur abaisseur *m*; **~ function** TELEP fonction échelon *f*; **~-function response** TELEP réponse à l'échelon *f*, réponse à la fonction échelon *f*; **~ index fiber** *AmE*, **~ index fibre** *BrE* DATA, PARTS, TRANS fibre à saut d'indice *f*; **~ index optical fiber** *AmE*, **~ index optical fibre** *BrE* DATA, PARTS, TRANS fibre optique à saut d'indice *f*; **~ index profile** ELECT, TRANS saut d'indice *m*, profil à saut d'indice *m*; **~ length** COMP durée de pas *f*; **~ operation** CONTR fonctionnement pas à pas *m*; **~ recording** COMP enregistrement pas à pas *m*; **~ response** CONTR réponse indicielle *f*; **~-stress test** TEST essai sous contrainte échelonnée *m*; **~ strobe marker** RADIO marche *f*; **~ track system** RADIO système de poursuite par échelons *m*; **~ tracking** RADIO poursuite en escalier *f*, poursuite pas à pas *f*; **~-up transformation** POWER survoltage *m*; **~-up transformer** POWER transformateur survolteur *m*; **~ write** COMP écriture par étapes *f*, écriture pas à pas *f*

step[3]: **~ down** *vt* POWER *circuit* dévolter, *voltage* abaisser par transformateur; **~ up** *vt* POWER *circuit* survolter, *voltage* augmenter par transformateur

stepper: **~ motor** *n* POWER moteur pas à pas *m*

stepping *n* GEN avance pas à pas *f*; **~ magnet** TELEP électroaimant d'ascension *m*, électroaimant de rotation *m*; **~ relay** PARTS relais intégrateur d'impulsions *m*, relais pas à pas *m*

stereo[1] *adj* (*stereophonic*) COMP stéréo (*stéréophonique*)

stereo[2] *n* (*stereophony*) COMP, TV stéréo *f* (*stéréophonie*); **~ delay** COMP retard stéréo *m*; **~ input** COMP, PER APP entrée stéréo *f*; **~ output** COMP, PER APP sortie stéréo *f*; **~ panning** COMP panoramique stéréo *f*; **~ telephony** TELEP, TV stéréophonie *f* (*stéréo*); **~ vision** COMP, TV stéréoscopie *f*

stereophonic[1] *adj* (*stereo*) COMP, PER APP, TELEP, TV stéréophonique (*stéréo*)

stereophonic[2]: **~ broadcast** *n* TV diffusion stéréo *f*; **~ circuit** *n* COMP circuit stéréo *m*

stereophony *n* (*stereo*) COMP, TV stéréophonie *f* (*stéréo*)

stereoscopic *adj* COMP, TV stéréoscopique

sticky: **~ note** *n* COMP note mobile *f*, note repositionnable *f*

stiffening: **~ bar** *n* PARTS raidisseur *m*

still: **~ frame** *n* COMP image fixe *f*, TV arrêt sur image *m*, image fixe *f*; **~-picture television** *n* (*SPTV*) TV télévision à images fixes *f*; **~-picture videotelephony** *n* RADIO, TELEP, TV visiophonie à images fixes *f*

stimulated: **~ emission** *n* ELECT, TRANS émission stimulée *f*; **~ Raman scattering** *n* (*SRS*) RADIO, TRANS diffusion Raman stimulée *f*

STL *abbr* (*studio-transmitter link*) RADIO LSE (*liaison studio-émetteur*)

STM *abbr* DATA (*synchronous transfer module*) module de transfert synchrone *m*, (*synchronous transfer mode*) MTS (*mode de transfert synchrone*), (*synchronous transport module*) module de transport synchrone *m*, TRANS (*synchronous transfer mode*) MTS (*mode de transfert synchrone*), module de transport synchrone *m*, (*synchronous transfer module*) module de transfert synchrone *m*

STM-n *abbr* (*synchronous transport module-n*) DATA, TRANS module de transport synchrone-n *m*

STN *abbr* SWG, TELEP, TRANS (*switched telephone network*) RTC (*réseau téléphonique commuté*)

stochastic: **~ process** *n* TRANS processus stochastique *m*

stock: **~ diameter** *n* TRANS diamètre du stockage *m*

stop: **~ band** *n* TRANS bande atténuée *f*; **~ bit** *n* DATA bit de stop *m*; **~ button** *n* COMP bouton d'arrêt *m*, touche d'arrêt *f*; **~ code** *n* SIG code d'arrêt *m*; **~ frame** *n* COMP image fixe *f*, TV arrêt sur image *m*, image fixe *f*; **~ instruction** *n* COMP instruction d'arrêt *f*; **~-send signal** *n* SIG signal d'arrêt *m*, TELEP signal d'arrêt d'émission *m*; **~ signal** *n* DATA, SIG, TELEP signal d'arrêt *m*

stopped *adj* COMP, TV stoppé

stopper: **~ circuit** *n* TELEP circuit de garde *m*

storability *n* TEST aptitude au stockage *f*, possibilité de stockage *f*

storable: **~ message** *n* PER APP, RADIO, TELEP message mémorisable *m*

storage *n* COMP *act*, DATA mémorisation *f*, mémoire *f*, mise en mémoire *f*, stockage *m*, GEN stockage *m*, RADIO, TELEP, TV mise en mémoire *f*; **~ allocation** COMP attribution de mémoire *f*, affectation de mémoire *f*; **~ battery** POWER accumulateur *m*; **~ block** COMP zone de mémoire *f*; **~ camera tube** ELECT tube analyseur à mémoire *m*; **~ capacity** DATA capacité de stockage *f*; **~ cell** DATA cellule de mémoire *f*; **~ control** DATA contrôle de la mémoire *m*; **~ dump** COMP vidage de mémoire *m*; **~ dumping** COMP transfert de mémoire *m*; **~ element** DATA élément de mémoire *m*, cellule de mémoire *f*; **~ file** DATA, TELEP archive *f*; **~ management** DATA gestion de la mémoire *f*; **~ medium** COMP support de mémoire *m*; **~ protection** DATA protection de mémoire *f*; **~ register** TELEP mémoire *f*; **~ surface** ELECT surface de mémoire *f*; **~ system** (*S/SYS*) DATA système de mémorisation *m*; **~ target** ELECT cible de mémoire *f*; **~ and transfer system** (*ST/SYS*) DATA système de mémorisation et de transfert *m*; **~ tube** ELECT tube à mémoire *m*; **~ within the network** TELEP stockage dans le réseau *m*

store[1]: **~-all** *n* COMP, DATA mémorisation complète *f*; **~-and-forward conversion facility** *n* APPL unité de conversion en différé *f*; **~-and-forward facility** *n* APPL possibilité d'émission en différé *f*; **~-and-forward mode** *n* APPL, DATA mode différé *m*; **~-and-forward switching network** *n* SIG, SWG réseau à commutation de messages *m*; **~-and-forward transmission** *n* APPL diffusion différée *f*, transmission différée *f*, TRANS transmission différée *f*, diffusion différée *f*; **~-and-forward unit** *n* (*SFU*) APPL, SWG unité d'enregistrement et de retransmission *f*; **~ and forward** *n* APPL, DATA stockage et retransmission *m*, enregistrement et retransmission *m*

store[2] *vt* COMP, DATA stocker, mémoriser, mettre en mémoire, RADIO stocker, mettre en mémoire

stored[1] *adj* COMP, DATA stocké, PER APP *answer machine* enregistré, RADIO stocké

stored[2]: **~ data transmission** *n* DATA, TRANS transmission de données en différé *f*; **~ message** *n* DATA message enregistré *m*, PER APP, RADIO, TELEP message mémorisé *m*; **~ number** *n* PER APP *answer machine* numéro programmé *m*; **~-program control** *AmE*, **~-programme control** *BrE* *n* (*SPC*) CONTR, SWG, TELEP autocommutateur privé à programme enregistré *m*, commande programmable *f*, commande par programme enregistré *f*; **~-program control exchange** *AmE*, **~-programme control exchange** *BrE* *n* CONTR, SWG, TELEP commutateur à programme enregistré *m*, commutateur privé de terminaison *m*; **~-programm switching system** *AmE*, **~-programme switching system** *BrE* *n* CONTR, SWG, TELEP système de commutation à programme enregistré *m*

storing *n* COMP, DATA mémorisation *f*, mise en mémoire *f*, stockage *m*, PARTS stockage *m*, RADIO mise en mémoire *f*, stockage *m*; **~ of digits** TELEP enregistrement de chiffres *m*; **~ of voice data** COMP, DATA mémorisation de données vocales *f*

storyboard: **~ live** *n* COMP, TV storyboard live *m*

stoving: **~ lacquer** *n* PARTS, TV vernis au four *m*

STP *abbr* ELECT (*shielded twisted pair*) paire torsadée blindée *f*, SIG, SWG, TELEP (*signal transfer point*, *signaling transfer point AmE*, *signalling transfer point BrE*) point de transfert sémaphore *m*, TRANS (*shielded twisted pair*) paire torsadée blindée *f*, (*signal transfer point*, *signaling transfer point AmE*, *signalling transfer point BrE*) point de transfert sémaphore *m*

straight: ~ **bank** *n* TELEP bancs de broches alignés *m pl*; ~ **binary code** *n* DATA, ELECT code binaire pur *m*; ~ **joint** *n* ELECT, PARTS épissure droite *f*; ~ **splice** *n* ELECT, PARTS épissure droite *f*

straighten *vt* PARTS dégauchir, redresser

straightforward: ~ **migration path** *n* ELECT canal de migration simple *m*

strain *n* PARTS allongement *m*; ~ **gage** *AmE*, ~ **gauge** *BrE* PARTS jauge de contraintes *f*, extensomètre *m*, jauge d'extension *f*; ~ **relief** POWER serre-câbles *m*; ~ **relief bracket** PARTS serre-câbles *m*; ~**-relief mechanism** PARTS renfort *m*; ~**-relief mounting bracket** PARTS bride de retenue *f*

strand *n* ELECT, PARTS brin *m*

stranded[1] *adj* ELECT, PARTS multibrin

stranded[2]: ~ **conductor** *n* ELECT, PARTS conducteur torsadé *m*, fil multibrin *m*, fil torsadé *m*; ~ **wire** *n* ELECT, PARTS fil multibrin *m*

stranding *n* ELECT, PARTS, TRANS câblage *m*

strap[1]: ~**-down** *adj* PARTS, POWER à composants liés

strap[2] *n* ELECT jarretière *f*, PARTS, POWER bande de connexion *f*, barrette *f*, bride de connexion *f*, cavalier *m*, connexion volante *f*, fil volant *m*, jarretière *f*, pontet *m*, SWG jarretière *f*

strap[3] *vt* PARTS, POWER ponter, straper

strappable *adj* PARTS strapable, POWER connectable par cavalier, strapable

strapped *adj* PARTS, POWER ponté

strapping *n* PARTS, POWER pontage *m*, strapage *m*; ~ **block** PARTS, POWER réglette de pontage *f*, réglette de strapage *f*; ~ **diagram** POWER schéma de pontage *m*, schéma de strapage *m*; ~ **options** *n pl* POWER options connectables *f pl*, possibilités de configuration *f pl*; ~ **plug** PARTS, POWER fiche de pontage *f*, fiche de strapage *f*

stratified: ~ **language** *n* COMP langage stratifié *m*; ~ **sample** *n* TEST échantillon stratifié *m*; ~ **sampling** *n* TEST échantillonnage stratifié *m*

stray: ~ **capacitance** *n* POWER capacité parasite *f*; ~ **current** *n* POWER courant vagabond *m*; ~ **emission** *n* ELECT émission parasite *f*; ~ **pick-up** *n* RADIO réception résiduelle *f*

streaking *n* COMP, TV traînage *m*

stream *n* GEN flux *m*, RADIO suite de chiffres *f*

streamer *n* COMP dérouleur *m*

streetside: ~ **cabinet** *n* TV armoire de sous-répartition *f*, armoire quartier *f*; ~ **furniture** *n* TRANS armoires de sous-répartition *f pl*

strength *n* GEN intensité *f*; ~ **member** ELECT renfort *m*; ~ **test** TEST essai de résistance *m*

stress *n* PARTS contrainte *f*, effort unitaire *m*; ~ **analysis** TEST analyse des contraintes *f*; ~ **crack** PARTS crique due à la tension *f*; ~ **relief loop** ELECT boucle anticontraintes mécanique *f*

stretched: ~ **text** *n* COMP texte par expansion *m*

striking *n* ELECT, POWER amorçage *m*; ~ **voltage** ELECT tension d'amorçage *f*, tension d'allumage *f*

string *n* DATA chaîne *f*

strip *n* PARTS réglette *f*, TELEP *of plug-sleeve indicators* série *f*; ~ **connector** ELECT, PARTS réglette de raccordement *f*; ~ **label** PARTS ruban d'étiquetage *m*; ~**-mounted set** PARTS, POWER montage sur platine *m*

stripline *n* RADIO guide d'ondes à rubans *m*, ligne à ruban *f*, ligne microruban *f*, stripline *f*

stripped: ~ **wire** *n* PARTS fil dénudé *m*

stripper *n* PARTS *knife* couteau à dénuder *m*

stripping *n* PARTS dénudage *m*; ~ **force** PARTS force d'arrachement *f*; ~ **knife** PARTS couteau à dénuder *m*; ~ **machine** PARTS machine à dénuder *f*; ~ **test** PARTS, TEST essai d'arrachement *m*; ~ **tool** PARTS outil à dénuder *m*

strobe[1] *n* (*stroboscope*) COMP impulsion de transfert *f*, signal de transfert *m*, ELECT strobe *f*, stroboscope *m*, GEN signal d'échantillonnage *m*, TEST strobe *f*, stroboscope *m*; ~ **effect** COMP, TV saccade d'images *f*; ~ **marker** RADIO marqueur stroboscopique *m*; ~ **pulse** ELECT impulsion d'activation *f*, impulsion stroboscopique *f*

strobe[2] *vt* ELECT activer par une impulsion

strobing *n* ELECT strobage *m*

stroboscope *n* (*strobe*) ELECT, TEST strobe *f*, stroboscope *m*

stroke *n* COMP bâtonnet *m*, *keystrike* frappe *f*, PARTS bâtonnet *m*, *typography* segment de caractère *m*; ~ **center line** *AmE*, ~ **centre line** *BrE* COMP ligne médiane de segment *f*; ~**-cleaving tool** PARTS fendoir *m*; ~ **edge** COMP bord de segment *m*; ~ **width** COMP largeur de segment *f*; ~ **writing** TV balayage aléatoire *m*

strong: ~ **authentication** *n* DATA authentification ferme *f*

Strowger: ~ **selector** *n* TELEP sélecteur Strowger *m*, sélecteur à double mouvement *m*

structural: ~ **method** *n* APPL méthode structurelle *f*

structuration: ~ **organization** *n* COMP structuration du contenu *f*

structure *n* GEN structure *f*; ~ **document** COMP document de structure *m*; ~ **flow chart** COMP diagramme de structure *m*; ~ **of management information** (*SMI*) DATA structure des informations de gestion *f*

structured: ~ **query language** *n* (*SQL*) DATA langage d'interrogation structuré *m* (*SQL*); ~ **type** *n* DATA type structuré *m*

STS: ~ **network** *n* (*space-time-space network*) SWG, TRANS réseau STS *m* (*réseau spatial-temporel-spatial*)

ST/SYS *abbr* (*storage and transfer system*) DATA système de mémorisation et de transfert *m*

STU *abbr* (*secure telephone*) TELEP poste cryptophonique *m*

stub *n* RADIO bras de réactance *m*, TELEP stub *m*; ~ **cable** POWER amorce *f*, câble amorce *m*

studio: ~**-transmitter link** *n* (*STL*) RADIO liaison studio-émetteur *f* (*LSE*)

stuffing *n* DATA bourrage *m*; ~ **character** DATA caractère de bourrage *m*; ~ **device** DATA dispositif de bourrage *m*; ~ **digit** DATA élément numérique de bourrage *m*, chiffre de bourrage *m*; ~ **rate** DATA taux de bourrage *m*

STX *abbr* COMP, DATA (*start of text*, *start-of-heading character*) STX (*caractère de début de texte*)

style *n* COMP style *m*

stylus *n* COMP style *m*, stylo *m*, TELEP style *m*

SU *abbr* (*signal unit*) SIG TS (*trame sémaphore*), unité de signalisation *f*

SUB *abbr* APPL (*subaddressing*) SUB (*sous-adressage*), DATA, TELEP (*subaddressing*) SUB (*sous-adressage*), (*substitute character*) SUB (*caractère de substitution*, *caractère de remplacement*)

subaddress *n* (*SA*) APPL, DATA, TELEP sous-adresse *f*

subaddressing *n* (*SUB*) APPL, DATA, TELEP sous-adressage *m* (*SUB*)

subaudio: ~ **frequency** *n* GEN fréquence infra-acoustique *f*

sub-band *n* COMP, RADIO, SWG sous-bande *f*; ~ **codec** COMP codec de sous-bande *m*; ~ **coding** SIG codage en sous-bandes *m*

subcarrier *n* GEN sous-porteuse *f*; ~ **frequency** RADIO, TELEP, TRANS fréquence de la sous-porteuse *f*, fréquence sous-porteuse *f*; ~ **frequency modulation** TELEP modulation en fréquence d'une sous-porteuse *f*

sub-Clos: ~ **arrangement** *n* TELEP, TRANS arrangement simplifié par rapport au modèle de Clos *m*; ~ **network** *n* TELEP, TRANS réseau de Clos imparfait *m*

subclutter: ~ **visibility** *n* RADIO visibilité sous fouillis *f*

subcontrol: ~ **station** *n* TELEP station sous-directrice *f*

subdirectory *n* COMP sous-répertoire *m*

subdistribution: ~ **cabinet** *n* TRANS coffret de sous-répartition *m*; ~ **cupboard** *n* TRANS armoire de sous-répartition *f*; ~ **frame** *n* TRANS sous-répartiteur *m*
subfleet *n* RADIO sous-flotte *f*
subgroup *n* TELEP, TRANS sous-groupe *m*
subharmonic[1] *adj* ELECT sous-harmonique
subharmonic[2] *n* ELECT harmonique inférieur *m*
subheader *n* COMP sous-en-tête *m*
subheading *n* COMP intertitre *m*
sub-hub *n* TRANS nœud secondaire *m*
subject[1]: ~ **to a charge** *adj* TELEP taxable
subject[2]: ~ **VM** *n* DATA MV concerné *m*; ~ **voice message** *n* DATA message vocal concerné *m*
subjective: ~ **test** *n* TEST essai subjectif *m*
subjob *n* COMP mode secondaire *m*, sous-mode *m*
submarine: ~ **cable** *n* TRANS câble sous-marin *m*
submenu *n* DATA *Internet* sous-menu *m*
submerged: ~ **optical repeater** *n* TRANS répéteur optique immergé *m*; ~ **repeater** *n* TRANS répéteur immergé *m*
submicron: ~ **structure** *n* GEN submicrostructure *f*
subminiature *n* ELECT subminiature *f*
submission: ~ **identifier** *n* DATA, TELEP identificateur de dépôt *m*; ~ **time** *n* DATA *duration* temps de dépôt *m*, *moment submitted* date et heure de dépôt *f*
submit *vt* COMP soumettre
submultiframe *n* (*SMF*) TELEP sous-multitrame *f*
submultiplex *vt* TRANS sous-multiplexer
subnetwork *n* DATA, TRANS, TV sous-réseau *m*; ~ **address** DATA adresse de sous-réseau *f*; ~ **dependent convergence function** (*SNDCF*) DATA fonction de convergence dépendant du sous-réseau *f*; ~ **functionality** DATA éléments fonctionnels du sous-réseau *m pl*; ~ **point of attachment** (*SNPA*) DATA point de rattachement du sous-réseau *m*; ~ **point of attachment address** DATA adresse du point de rattachement du sous-réseau *f*; ~ **service** DATA service de sous-réseau *m*
subordinate: ~ **exchange** *n* TELEP central satellite *m*, satellite *m*
subpopulation *n* TEST sous-population *f*, strate *f*
subprogram *n* COMP, DATA sous-programme *m*
subrack *n* TELEP bac *m*
subreflector *n* RADIO réflecteur auxiliaire *m*, miroir auxiliaire *m*, réflecteur secondaire *m*
subrefraction *n* RADIO infraréfraction *f*
subroute *n* TELEP sous-faisceau *m*, sous-route *f*
subroutine *n* COMP, DATA sous-programme *m*
subsampling *n* COMP, TELEP sous-échantillonnage *m*
subsatellite: ~ **point** *n* RADIO point sub-satellite *m*
subscribe *vi* GEN s'abonner, souscrire
subscriber[1]: ~-**controlled** *adj* TELEP commandé par l'abonné
subscriber[2] *n* COMP, DATA, RADIO, TELEP, TV abonné *m*, souscripteur *m*; ~ **access functional subsystem** TELEP unité de raccordement d'abonnés *f*; ~ **agreement holder** COMP, DATA, RADIO, TELEP, TV titulaire d'abonnement *m*; ~ **and analysis database** TELEP base de données et d'analyse *m*; ~ **base** RADIO, TELEP ensemble des abonnés *m*, parc des abonnés *m*; ~ **box** TV boîtier d'abonné *m*; ~-**busy signal** SIG signal d'abonné occupé *m*; ~ **call charge meter** TELEP service d'indicateur de taxe au domicile de l'abonné *m*; ~ **calling rate** TELEP taux d'appel des abonnés *m*; ~ **category PE** TELEP élément graphique de catégorie d'abonné *m*; ~ **channel** TELEP voie d'abonné *f*; ~ **circuit** TELEP circuit d'abonné *m*; ~ **connection system** PER APP, TELEP système de raccordement d'abonnés *m*; ~ **connection unit** PER APP, TELEP unité de raccordement d'abonnés *f*; ~-**dialed traffic** *AmE*, ~-**dialled traffic** *BrE* TELEP trafic commandé par l'abonné *m*; ~ **digital termination** TELEP terminaison numérique d'abonné *f*; ~ **digital-access unit** TELEP centre satellite numérique *m*; ~ **equipment** PARTS, TELEP équipement d'abonné *m*; ~ **equipment keeper** TELEP acquéreur *m*; ~-**identity module** (*SIM*) RADIO module d'identité de l'abonné *m*; ~ **installation** DATA, TELEP installation d'abonné *f*; ~-**line busy** SIG signal d'abonné occupé *m*; ~ **line circuit** (*SLC*) SWG, TELEP circuit de ligne d'abonné *m*; ~-**line free, no charge** TELEP signal d'abonné libre, sans taxation *m*; ~ **line junctor** TELEP joncteur d'abonné *m*; ~-**line out of order** TELEP signal de ligne d'abonné en dérangement *m*; ~ **line PE** TELEP élément graphique de ligne d'abonné *m*; ~ **loop** COMP, DATA, RADIO, TELEP, TRANS ligne d'abonné *f*; ~ **metering** TELEP télétaxe *f*; ~ **multiple test unit** TELEP testeur multiple d'abonné *m*; ~ **number** (*SN*) DATA, TELEP numéro d'abonné *m*; ~ **percentage** TELEP pourcentage d'abonnés *m*; ~ **premises network** (*SPN*) TRANS réseau d'installation d'abonné *m*; ~'**s alphanumerical display** TELEP service d'affichage alphanumérique de renseignements *m*; ~'**s call meter** TELEP compteur d'abonné *m*; ~'**s check meter** TELEP compteur télétaxe *m*; ~'**s dialing habits** *AmE*, ~'**s dialling habits** *BrE* *n pl* TELEP comportement de l'abonné *m*; ~'**s directory number** TELEP numéro d'appel *m*, numéro de poste *m*; ~ **service** COMP, DATA, RADIO, TELEP desserte aux abonnés *f*, service aux abonnés *m*; ~'**s fax position** TELEP poste d'abonné de télécopie *m*; ~'**s installation** DATA installation d'abonné *f*, installation privée *f* (*IP*), TELEP installation d'abonné *f*; ~'**s lead-in** TELEP ligne de branchement d'abonné *f*; ~'**s line** COMP, DATA, RADIO, TELEP, TRANS ligne d'abonné *f*, ligne d'accès *f*, ligne de rattachement *f*, TRANS ligne d'accès *f*; ~'**s line equipment** COMP, DATA, RADIO, TELEP, TV équipement de ligne d'abonné *m*; ~'**s main station** TELEP poste principal d'abonné *m*; ~'**s meter** PER APP, TELEP compteur de taxes d'abonné *m*, compteur d'abonné *m*; ~'**s national telex number** TELEP numéro télex national d'abonné *m*; ~'**s private meter** PER APP, TELEP compteur personnel *m*; ~'**s service center** *AmE*, ~'**s service centre** *BrE* TELEP centre du service d'abonnés *m*; ~'**s station** TELEP poste d'abonné *m*; ~ **stage** SWG étage de communication d'abonnés *m*; ~-**state signaling** *AmE*, ~-**state signalling** *BrE* TELEP signalisation de l'état du demandeur *f*; ~'**s store** SWG fichier d'abonnés *m*; ~ **switch** SWG étage d'abonnés *m*; ~ **switching network** SWG, TELEP réseau de connexion d'abonnés *m*; ~ **terminal** TELEP, TV terminal d'abonné *m*; ~ **terminal installation** TELEP, TV installation terminale d'abonné *f*; ~ **test responder** TELEP, TEST *for subscriber lines and sets* robot d'essais d'abonné *m*; ~ **traffic rate** TELEP trafic d'un abonné *m*; ~-**transferred signal** SIG signal d'abonné transféré *m*; ~ **trunk dialing** *AmE* TELEP interurbain automatique *m* (*IA*); ~ **trunk dialing access code** *AmE* TELEP code d'accès à l'interurbain automatique *m*; ~ **unit** TV coffret d'abonné *m*
subscript[1] *n* COMP indice *m*, *number* chiffre inférieur *m*, DATA indice *m*, *number* chiffre inférieur *m*
subscript[2] *vt* COMP, DATA mettre en indice
subscription *n* GEN, RADIO, TELEP, TV abonnement *m*; ~ **area** COMP, DATA, RADIO, TELEP, TV zone d'abonnement *f*; ~ **fee** COMP, DATA, RADIO, TELEP, TV frais d'abonnement *m pl*; ~ **price** COMP, DATA, RADIO, TELEP, TV prix d'abonnement *m*
subsequent: ~ **address message** *n* SIG, SWG, TELEP message d'adresse subséquent *m*, message d'adresse ultérieur *m*; ~ **handover** *n* RADIO transfert intercellulaire courant *m*; ~ **signal unit** *n* SIG unité de signalisation subséquente *f*
subset *n* DATA sous-ensemble *m*
substatus *n* COMP sous-statut *m*
substitute: ~ **character** *n* (*SUB*) DATA, TELEP caractère de substitution *m* (*SUB*), caractère de remplacement *m* (*SUB*); ~ **route** *n* TELEP faisceau de remplacement *m*
substractive: ~ **synthesis** *n* COMP synthèse soustractive *f*
substrate *n* ELECT, PARTS substrat *m*, support de circuit imprimé *m*; ~ **current** ELECT courant de substrat *m*
substring *n* DATA chaîne secondaire *f*

subsynchronous: **~ satellite** *n* RADIO satellite sous-synchrone *m*
subtelephone: **~ frequency** *n* TELEP fréquence infratéléphonique *f*; **~ telegraphy** *n* TELEP télégraphie infratéléphonique *f*
subterranean: **~ propagation** *n* RADIO propagation souterraine *f*
subtest *n* TEST essai partiel *m*
subtitle[1] *n* TV sous-titre *m*
subtitle[2] *vt* TV sous-titrer
subtitling *n* TV sous-titrage *m*
subtracter *n* ELECT soustracteur *m*
subtractive: **~ primary color** *AmE*, **~ primary colour** *BrE* *n* COMP couleur primaire soustractive *f*; **~ process** *n* ELECT procédé soustractif *m*
subtree *n* DATA, TELEP sous-arbre *m*
subtype *n* DATA sous-type *m*
subunit *n* ELECT sous-unité *f*, TELEP sous-faisceau *m*
subvariable *n* COMP sous-variable *f*
subzone: **~ center** *AmE*, **~ centre** *BrE* *n* SWG centre de transit *m*
successful: **~ call** *n* TELEP appel ayant abouti *m*; **~ call attempt** *n* RADIO, TELEP tentative d'appel acheminée *f*; **~ injection** *n* RADIO *of satellite* mise en orbite réussie *f*; **~ traffic** *n* RADIO trafic efficace *m*
successfully: **~-received** *adj* DATA bien reçu
successive: **~ call phases** *n pl* TELEP phases successives d'une communication *f pl*
sudden: **~ failure** *n* TEST défaillance soudaine *f*; **~ total failure** *n* TEST défaillance catastrophique *f*, défaillance soudaine et complète *f*
suite *n* COMP *of software* progiciel *m*, rangée *f*, suite *f*, travée *f*, DATA suite *f*, PARTS travée *f*, TELEP travée *f*, rangée *f*; **~ fuse** PARTS, POWER fusible de travée *m*; **~ fuse alarm** PARTS, POWER alarme de fusible de travée *f*; **~ supervision equipment** TELEP équipement terminal de travée *m*; **~ terminal equipment** TELEP équipement terminal de travée *m*
summarize: **~ abstract-operation** *n* DATA opération abstraite de résumé *f*
summation: **~ check** *n* COMP contrôle par totalisation *m*
summer *n* ELECT additionneur analogique *m*, sommateur *m*
summing: **~ integrator** *n* ELECT intégrateur sommateur *m*; **~-scaling amplifier** *n* POWER amplificateur de sommation et pondération *m*, amplificateur sommateur-pondérateur *m*
sun: **~ sensor assembly** *n* (*SSA*) RADIO bloc capteur solaire *m*; **~ sensor head** *n* RADIO tête optique du capteur solaire *f*
S-universal: **~ access** *n* TELEP prise universelle *f*, prise S *f*; **~ interface** *n* TELEP serveur à accès multiples *m*; **~ interface card** *n* TELEP carte d'interface S *f*
sunlit: **~ satellite** *n* RADIO satellite pendant sa phase éclairée *m*
sunshield *n* RADIO écran solaire *m*
super: **~-audio frequency** *n* GEN fréquence supra-acoustique *f*; **~-telephone frequency** *n* TELEP, TRANS fréquence supratéléphonique *f*; **~-telephone telegraphy** *n* TELEP, TRANS télégraphie supratéléphonique *f*
superconductor *n* ELECT, TRANS supraconducteur *m*; **~ cable** TRANS câble supraconducteur *m*; **~ line** TRANS ligne supraconductrice *f*
supergroup *n* TELEP, TRANS groupe secondaire *m*; **~ distribution frame** TELEP, TRANS répartiteur de groupe secondaire *m*; **~ link** TELEP liaison en groupe secondaire *f*; **~ section** TELEP section de groupe secondaire *f*; **~ translating equipment** TELEP équipement de transposition de groupe secondaire *m*, équipement de matériel de groupe secondaire *m*, équipement de modulation de groupe secondaire *m*
superheterodyne[1] *adj* RADIO superhétérodyne
superheterodyne[2]: **~ reception** *n* RADIO réception superhétérodyne *f*
superhigh: **~ frequency** *n* (*SHF*) RADIO hyperfréquence *f*, TV hyperfréquence *f*, onde centimétrique *f*
superimpose *vt* COMP, TELEP superposer
superimposing *n* COMP, TELEP superposition *f*
superimposition *n* COMP, TELEP surimpression *f*
superlattice *n* GEN superstructure cristalline *f*
superluminescence *n* ELECT, TRANS superluminescence *f*
superluminescent: **~ LED** *n* (*SLD*) ELECT, PARTS, TRANS diode superluminescente *f* (*DSL*); **~ light-emitting diode** *n* (*SLD*) ELECT, PARTS, TRANS diode superluminescente *f* (*DSL*)
supermastergroup *n* TELEP groupe quaternaire *m*; **~ link** TELEP liaison en groupe quaternaire *f*; **~ section** TELEP section de groupe quaternaire *f*
superphantom: **~ balanced telegraph circuit** *n* TELEP circuit télégraphique superfantôme *m*; **~ circuit** *n* TRANS circuit superfantôme *m*
superpose *vt* COMP, TELEP superposer
superposed: **~ circuit** *n* TELEP circuit superposé *m*
superposition *n* COMP, TELEP superposition *f*
superradiance *n* TRANS superluminescence *f*
superradiant: **~ diode** *n* (*SRD*) ELECT, PARTS, TRANS diode superluminescente *f* (*DSL*)
superrefraction *n* RADIO superréfraction *f*
superregenerative: **~ reception** *n* RADIO réception à superréaction *f*
superscript[1] *adj* COMP, DATA supérieur
superscript[2] *n* COMP, DATA, *number* chiffre supérieur *m*
superscript[3] *vt* COMP, DATA mettre en exposant
supersynchronous: **~ satellite** *n* RADIO satellite super-synchrone *m*
supervise *vt* COMP, CONTR, GEN, TELEP, TEST superviser, surveiller
supervision *n* COMP, CONTR, GEN, TELEP, TEST supervision *f*, surveillance *f*
supervisor *n* COMP, CONTR superviseur *m*, GEN surveillant *m*, TELEP, TEST superviseur *m*; **~'s desk** TELEP pupitre de surveillance *m*
supervisory: **~ aid** *n* CONTR, TEST outil de supervision *m*; **~ announcement** *n* TELEP message de contrôle *m*; **~ message** *n* TELEP message de supervision *m*; **~ panel** *n* CONTR panneau de surveillance *m*, tableau de surveillance *m*; **~ program** *n* COMP programme superviseur *m*, superviseur *m*; **~ relay** *n* PARTS, TELEP relais de surveillance *m*; **~ timer** *n* SWG temporisateur de supervision *m*; **~ tone** *n* CONTR, SWG message de contrôle *m*, tonalité de surveillance *f*; **~ unit** *n* CONTR, TELEP, TEST unité de surveillance *f*
supplementary: **~ service** *n* APPL, RADIO, TELEP service supplémentaire *m*
supplier: **~ evaluation** *n* TEST évaluation du fournisseur *f*, évaluation du vendeur *f*; **~ rating** *n* TEST cotation du fournisseur *f*; **~ valuation** *n* TEST évaluation du fournisseur *f*, évaluation du vendeur *f*
supply: **~ current** *n* ELECT pouvoir d'alimentation *m*, POWER courant d'alimentation *m*; **~ voltage** *n* ELECT, POWER tension d'alimentation *f*
support *n* PARTS *attachment* fixation *f*, *mounting* support *m*; **~ activities** *n pl* COMP fonctions complémentaires *f pl*; **~ cable** RADIO hauban *m*; **~ system** TELEP système de support *m*
supporting: **~ bar** *n* PARTS barre de support *f*; **~ comb** *n* PARTS, TELEP échelle fixe *f*; **~ plate** *n* PARTS plaque de support *f*; **~ rod** *n* PARTS tige de support *f*
suppress *vt* ELECT, RADIO, TELEP éliminer, supprimer
suppressed[1] *adj* ELECT, RADIO, TELEP supprimé
suppressed[2]: **~ aerial** *n* RADIO antenne encastrée *f*, antenne plaquée *f*; **~ antenna** *n* RADIO antenne encastrée *f*, antenne plaquée *f*; **~ carrier** *n* RADIO porteuse supprimée *f*; **~-carrier emission** *n* RADIO émission à porteuse supprimée *f*;

~ **demand** *n* RADIO demande latente *f*; ~ **frequency band** *n* TRANS bande atténuée *f*
suppression *n* ELECT, RADIO, TELEP élimination *f*, suppression *f*; ~ **hangover time** TELEP temps de maintien pour le blocage *m*; ~ **loss** TELEP affaiblissement de blocage *m*; ~ **operate time** TELEP temps de fonctionnement pour le blocage *m*
suppressor *n* ELECT suppresseur *m*; ~ **grid** ELECT, PARTS grille d'arrêt *f*
supraconductor *n* ELECT, TRANS supraconducteur *m*
supraphonic *adj* ELECT supraphonique
surf *vt* DATA *Internet* surfer, *scanner* voguer
surface[1]: ~**-mounted** *adj* ELECT reporté à plat, GEN monté en surface, monté en saillie
surface[2]: ~ **acoustic wave** *n* (*SAW*) ELECT, TRANS onde acoustique de surface *f*; ~ **acoustic wave device** *n* ELECT, PARTS, TRANS dispositif à ondes acoustiques de surface *m*; ~**-charge effect** *n* ELECT effet de charge de surface *m*, effet de surface *m*; ~ **connection** *n* ELECT connexion superficielle *f*; ~ **current** *n* ELECT courant superficiel *m*; ~ **duct** *n* RADIO conduit au sol *m*; ~**-emitting light-emitting diode** *n* PARTS, TELEP, TRANS diode électroluminescente à émission frontale *f*; ~ **finish** *n* PARTS fini de surface *m*; ~ **finish drawing** *n* PARTS dessin d'état de surface *m*; ~ **hardness** *n* PARTS dureté superficielle *f*; ~ **layer** *n* PARTS couche superficielle *f*; ~ **leakage** *n* POWER fuite en surface *f*, perte en surface *f*; ~ **leakage path** *n* POWER ligne de pertes en surface *f*; ~**-mount technology** *n* (*SMT*) ELECT, PARTS technologie de montage en surface *f*; ~**-mounted component** *n* (*SMC*) ELECT, PARTS composant monté en surface *m* (*CMS*); ~**-mounted device** *n* (*SMD*) ELECT, PARTS circuit à montage en surface *m*; ~ **mounting** *n* ELECT report à plat *m*, GEN montage en saillie *m*, montage en surface *m*; ~ **piezo-electric permittivity** *n* ELECT permittivité piézoélectrique de surface *f*; ~ **resistance** *n* ELECT, POWER résistance superficielle *f*; ~ **resistivity** *n* ELECT, POWER résistivité superficielle *f*; ~ **wave** *n* ELECT onde de surface *f*, RADIO onde de sol *f*; ~**-wave component** *n* ELECT composant d'onde de surface *m*; ~**-wave technology** *n* ELECT technique des ondes de surface *f*, RADIO technique des ondes de sol *f*
surfer *n* DATA *Internet* internaute *m*
surge *n* ELECT, POWER choc électrique *m*, surtension *f*; ~ **absorber** POWER parasurtenseur *m*; ~ **arrester** ELECT arrêt de surtension *m*; ~ **current** POWER courant de pointe *m*, surintensité *f*, surintensité de courant *f*, courant de surtension transitoire *m*; ~ **diverter** PARTS *lightning protection* parafoudre *m*, *supply line* divertisseur de surtension *m*; ~ **forward current** ELECT, POWER courant direct de pointe *m*; ~ **suppressor** POWER limiteur de surtension *m*, parasurtenseur *m*, suppresseur de surtension *m*, parafoudre *m*; ~ **voltage protector** POWER parasurtenseur *m*
surplus: ~ **solder collector** *n* ELECT pompe à dessouder *f*
surveillance TELEP, TEST surveillance*f*; ~ **radar** *n* RADIO radar de surveillance *m*, radar de veille *m*
survival: ~ **craft station** *n* RADIO station d'engin de sauvetage *f*
suspended: ~ **process** *n* TELEP processus suspendu *m*
suspension: ~ **strand** *n* TRANS câble horizontal de suspension *m*, POWER câble porteur *m*
sustain *n* COMP soutenu *m*, tenue *f*; ~ **footswitch** COMP pédale de soutenu *f*; ~ **pedal** COMP pédale de soutenu *f*; ~**-pedal assign** COMP affectation de la pédale de soutenu *f*; ~ **switch** COMP commutateur de soutenu *m*
sustaining: ~ **voltage** *n* ELECT tension d'entretien *f*
SUT *abbr* (*system under test*) DATA système en cours de test *m*; ~ **operator** *n* DATA opérateur de SUT *m*
SVC *abbr* ELECT (*switched virtual circuit*) CVC (*circuit virtuel commuté*), SIG (*signaling virtual channel AmE, signalling virtual channel BrE*) CVC (*canal virtuel de signalisation*), (*switched virtual circuit*) CVC (*circuit virtuel commuté*), SWG, TRANS (*switched virtual circuit*) CVC (*circuit virtuel commuté*)
S-video *n* (*separated video*) TV S-vidéo *m*
swab *n* ELECT tampon *m*
swan: ~ **feed** *n* RADIO alimentation en col de cygne *f*
swap: ~ **repair** *n* TELEP réparation par échange standard *f*
swapping *n* COMP permutation *f*, transfert *m*
sweep[1]: ~ **aerial** *n* RADIO antenne à balayage *f*; ~ **antenna** *n* RADIO antenne à balayage *f*; ~ **frequency** *n* ELECT, RADIO fréquence de balayage *f*; ~ **jamming** *n* ELECT, RADIO brouillage par balayage *m*; ~ **microscope** *n* ELECT microscope électronique à balayage *m* (*MEB*); ~ **oscillator** *n* ELECT, PARTS oscillateur de commande de balayage *m*; ~ **signal** *n* COMP signal de balayage *m*; ~ **speed** *n* ELECT, RADIO vitesse de balayage *f*; ~ **time** *n* ELECT durée de balayage *f*, temps de balayage *m*
sweep[2] *vt* ELECT, RADIO balayer
swept: ~ **frequency** *n* RADIO fréquence de balayage *f*
SWG *abbr* DATA, RADIO, TRANS (*slot waveguide, slotted waveguide*) guide d'ondes à fentes *m*, guide d'ondes fendu *m*
swing *n* COMP rythme *m*, cadence *f*, RADIO plage de relèvement *f*; ~**-arm feature** ELECT bras articulé *m*; ~ **error** RADIO erreur de plage *f*
swipe: ~ **card** *n* APPL, TELEP carte à mémoire *f* (*CAM*)
Swiss: ~ **Association for Quality Assurance Certificates** *n* (*SQS*) GEN Association suisse pour certificats d'assurance de la qualité *f* (*SQS*); ~ **Association of Machinery Manufacturers** *n* (*VSM*) GEN Société suisse des constructeurs de machines *f* (*VSM*); ~ **Standard** *n* GEN norme suisse *f* (*SN*); ~ **Standards Association** *n* GEN Association suisse de normalisation *f*
switch[1] *n* COMP interrupteur *m*, ELECT commutateur *m*, interrupteur *m*, PARTS, POWER, RADIO interrupteur *m*, SWG commutateur *m*, interrupteur *m*, TRANS duplexeur *m*, sélecteur *m*; ~ **driver** RADIO commutateur *m*; ~ **hook** PER APP plongeur *m*, RADIO, TELEP crochet commutateur *m*; ~ **indicator** GEN drapeau *m*, PARTS indicateur d'aiguillage *m*; ~ **multiple** SWG, TELEP multiple de sélecteurs *m*; ~**-off capability** POWER capacité de mise hors tension *f*; ~ **rack** SWG, TELEP bâti de commutateurs *m*, bâti de sélecteurs *m*; ~ **unit** SWG boîte de commutation *f*
switch[2] *vt* ELECT, POWER, RADIO connecter, SWG, TELEP connecter, raccorder; ~ **off** COMP *machine*, ELECT, GEN, PER APP, POWER, RADIO, SWG *machine* éteindre, fermer, *supply* arrêter, déconnecter, mettre hors circuit, mettre hors tension; ~ **on** COMP, ELECT, PER APP, POWER, RADIO, SWG *machine* brancher, *supply* allumer, connecter, mettre en circuit, mettre en marche, mettre sous tension, TELEP connecter
switchboard *n* SWG pupitre de commutation *m*, standard *m*, TELEP commutateur d'abonnés *m*, standard *m*, tableau de commutation *m*; ~ **equipment** SWG, TELEP équipement de commutateur *m*, équipement de standard *m*; ~ **floor plan** TELEP plan d'implantation des tables de commutation *m*; ~ **operator** GEN opérateur *m*, SWG, TELEP opérateur *m*, standardiste *mf*, téléphoniste *mf*; ~ **PE** SWG élément graphique de tableau commutateur *m*; ~ **row** SWG, TELEP rangée de pupitres d'opérateur *f*, rangée de tableaux de commutation *f*, rangée de tables d'opérateur *f*; ~ **suite** SWG, TELEP rangée de pupitres d'opérateur *f*, rangée de tableaux de commutation *f*, rangée de tables d'opérateur *f*
switched: ~**-beam direction finder** *n* RADIO radiogoniomètre à basculement du diagramme *m*; ~**-capacitor filter** *n* ELECT, PARTS filtre à capacités commutées *m*; ~ **circuit** *n* ELECT, TRANS circuit commuté *m*; ~**-loop console** *n* SWG poste d'opérateur de classe B *m*; ~ **multimegabit data service** *n* (*SMDS*) DATA service commuté de données multimégabit *m*; ~ **network** *n* SWG, TRANS, TV réseau commuté *m*; ~ **network layer** *n* SWG, TRANS, TV couche de réseau commuté *f*;

~ **power supply** *n* POWER onduleurs *m pl*; ~ **service** *n* SWG service commuté *m*; ~**-star network** *n* POWER, SWG, TRANS, TV réseau en étoile commuté *m*; ~ **telephone network** *n* (*STN*) SWG, TELEP, TRANS réseau téléphonique commuté *m* (*RTC*); ~**-transit country** *n* TELEP pays de transit en commutation *m*; ~ **virtual circuit** *n* (*SVC*) ELECT circuit virtuel commuté *m* (*CVC*), SIG circuit virtuel commuté *m* (*CVC*), voie virtuelle de signalisation *f*, SWG circuit virtuel commuté *m* (*CVC*), TRANS circuit virtuel commuté *m* (*CVC*), canal virtuel de signalisation *m*

switcher *n* SWG, TELEP autocommutateur *m*, TV aiguilleur *m*

switchgear *n* PARTS, POWER, SWG, TELEP dispositif de commutation *m*

switching *n* GEN commutation *f*; ~ **center** *AmE*, ~ **centre** *BrE* SWG, TELEP central de commutation *m* (*CC*), centre de commutation *m* (*CC*), commutateur téléphonique *m*; ~ **complex** SWG, TELEP chaîne centrale de connexion *f*; ~ **complex control software machine** SWG gestion des connexions *f*; ~ **component** PARTS, POWER, SWG, TELEP organe de commutation *m*; ~ **and control unit** SWG unité de commande et de connexion *f*; ~ **delay** SWG durée de sélection d'un commutateur *f*; ~ **device** PARTS, POWER, SWG, TELEP dispositif de commutation *m*, organe de commutation *m*; ~ **diode** ELECT, PARTS, SWG diode de commutation *f*; ~ **element** ELECT élément de commutation *m*; ~ **equipment** SWG, TELEP matériel de commutation *m*; ~**-equipment congestion signal** SIG, SWG signal d'encombrement de l'équipement de commutation *m*; ~ **exchange** SWG, TELEP central de commutation *m* (*CC*), central téléphonique *m*, centre de commutation *m* (*CC*), commutateur téléphonique *m*; ~ **function** SWG, TELEP fonction de commutation *f*; ~ **matrix** SWG, TELEP matrice de commutation *f*; ~ **mechanism** SWG, TELEP mécanisme de commutation *m*, mécanisme de sélection *m*; ~ **module PE** SWG élément graphique de module de commutation *m*; ~ **multiplexer** SWG multiplexeur aiguilleur *m*; ~ **network** SWG, TELEP réseau de commutation *m*, réseau de connexion *m*; ~ **network complex** SWG réseau de connexion *m*; ~ **node** SIG, SWG nœud de commutation *m*, nœud de communication *m*; ~ **path** SWG trajet de commutation *m*; ~ **path PE** SWG élément graphique de trajet de commutation *m*; ~ **pattern** SWG schéma de commutation *m*; ~ **point** SWG, TELEP étage de commutation *m*; ~ **process** SWG, TELEP processus de commutation *m*; ~ **process interworking telephony event** (*SPITE*) TELEP événement téléphonique de traitement pour les opérations de commutation à l'interface*m* (*ETIC*); ~ **processor** DATA, SWG processeur de commutation *m*; ~ **rate** SWG, TEST cadence de commutation *f*; ~ **room** TELEP salle de commutation *f*, salle de l'autocommutateur *f*; ~ **sequence** SWG, TELEP processus de connexion *m*, séquence de commutation *f*; ~ **stage** SWG, TELEP étage de commutation *m*; ~ **statistical multiplexer** SWG multiplexeur aiguilleur *m*; ~ **system** SWG, TELEP système de commutation *m*; ~ **system processor** SWG, TELEP processeur de système de commutation *m*; ~ **technique** SWG technique de commutation *f*; ~ **technology** SWG technologie de commutation *f*; ~ **terminal** SWG terminal de commutation *m* (*TC*); ~ **test** SWG, TEST essai de la commutation *m*; ~ **time** SWG, TELEP, TEST durée de commutation *f*, temps de commutation *m*; ~ **unit** SWG commutateur *m*, unité de commutation *f*; ~ **variable** SWG variable logique *f*

switchless *adj* ELECT, POWER, SWG, TELEP non commuté

switchover: ~ **diode** *n* ELECT, PARTS, SWG diode de commutation *f*

swivel[1]: ~ **stub** *n* ELECT câble amorce orientable *m*

swivel[2] *vt* COMP, PARTS pivoter

SWR *abbr* (*standing wave ratio*) RADIO, TRANS ROS (*rapport d'ondes stationnaires en tension*), TOS (*taux d'ondes stationnaires, taux d'ondes de tension stationnaires*)

syllabic: ~ **companding** *n* COMP, TELEP, TRANS compression-extension syllabique *f*

syllabization *n* APPL syllabization

syllable: ~ **articulation** *n* TELEP netteté des syllabes *f*; ~ **articulation test** *n* TELEP, TEST essai de netteté des syllabes *m*

symbiosis *n* ELECT symbiose *f*

symbol: ~ **location** *n* DATA position de symbole *f*; ~ **pattern** *n* DATA séquence de symboles *f*; ~ **rate** *n* DATA, TELEP débit de symboles *m*

symbolic: ~ **address** *n* COMP adresse symbolique *f*; ~ **addressing** *n* COMP adressage symbolique *m*

symmetrical: ~ **binary channel** *n* ELECT, TELEP voie binaire symétrique *f*; ~ **binary code** *n* DATA, ELECT, TELEP, TRANS code binaire symétrique *m*; ~ **channel** *n* ELECT, TELEP voie symétrique *f*; ~ **deflection** *n* ELECT déviation symétrique *f*; ~**-pair cable** *n* TRANS câble à paires symétriques *m*; ~ **short-circuit current** *n* POWER courant de court-circuit symétrique *m*; ~ **time matrix** *n* SWG matrice de commutation temporelle *f*, matrice temporelle symétrique *f*, TELEP matrice de commutation temporelle *f*; ~ **two-terminal-pair network** *n* ELECT, POWER, TRANS quadripôle symétrique *m*

SYN *abbr* (*synchronous idle character*) DATA SYN (*caractère de synchronisation*)

sync[1] *adj* GEN, TELEP synchronisé

sync[2] *n* GEN synchronisation *f*; ~ **clock** COMP horloge de synchronisation *f*; ~ **code** COMP, DATA code de synchronisation *m*; ~ **start** COMP départ synchro *m*; ~ **word** DATA mot de synchronisation *m*

sync[3] *vt* GEN synchroniser

synch[1] *adj* GEN (*synchronized*) synchronisé

synch[2] *n* GEN (*synchronization*) synchronisation *f*; ~ **clock** COMP horloge de synchronisation *f*; ~ **start** COMP départ synchro *m*

synch[3] *vt* GEN (*synchronize*) synchroniser

synchro: ~ **capacitor** *n* ELECT, PARTS compensateur synchrone *m*; ~ **transmitter** *n* RADIO synchro-transmetteur *m*

synchronism *n* RADIO, TV synchronisme *m*; ~ **voltage** ELECT tension de synchronisme *f*

synchronization *n* GEN (*sync*, *synch*) synchronisation *f*; ~ **bit** TELEP bit de synchronisation *m*; ~ **character** (*SYN*) DATA caractère de synchronisation *m* (*SYN*); ~ **clock** COMP horloge de synchronisation *f*; ~ **code** COMP, DATA code de synchronisation *m*; ~ **data** *n pl* COMP données de synchronisation *f pl*; ~ **error** ELECT, RADIO, TELEP, TRANS gigue *f*; ~ **information** COMP informations de synchronisation *f pl*; ~ **network** TRANS réseau de synchronisation *m*; ~ **pulse** GEN impulsion de synchronisation *f*; ~ **sensitivity** GEN sensibilité de synchronisation *f*; ~ **signal** SIG signal de synchronisation *m*; ~ **signal unit** SIG unité de signalisation de synchronisation *f*; ~ **and time base station** TELEP station de temps et synchronisation *f*

synchronize *vt* GEN (*sync*, *synch*) synchroniser

synchronized[1] *adj* GEN (*sync*, *synch*) synchronisé

synchronized[2]: ~ **multivibrator** *n* ELECT multivibrateur synchronisé *m*, multivibrateur asservi *m*; ~ **network** *n* TELEP réseau synchronisé *m*; ~**-oscillator system** *n* (*SOS*) POWER système à oscillateurs synchronisés *m*; ~ **playback** *n* COMP exécution synchronisée *f*; ~**-ringing signal** *n* TELEP signal d'appel synchronisé *m*; ~ **satellite** *n* RADIO satellite synchronisé *m*, satellite en phase *m*

synchronizing *n* GEN, TELEP synchronisation *f*; ~ **pulse regenerator** TV régénérateur de synchronisation *m*; ~ **signal** COMP, SIG signal de synchronisation *m*

synchronous[1] *adj* ELECT, GEN, TELEP synchrone

synchronous[2]: ~ **capacitor** *n* ELECT, PARTS compensateur synchrone *m*; ~ **circuit** *n* TRANS circuit synchrone *m*; ~ **computer** *n* COMP ordinateur synchrone *m*;

~ **condenser** *n* ELECT, PARTS compensateur synchrone *m*; ~ **correction** *n* TELEP correction de synchronisme *f*; ~ **data link control** *n* (*SDLC*) DATA contrôle de liaison synchrone de données *m* (*SDLC*); ~ **data network** *n* TELEP réseau de données synchrone *m*; ~ **data transmission** *n* DATA, TRANS transmission de données synchrone *f*; ~ **data unit** *n* (*SDU*) DATA, ELECT unité de données synchrone *f*; ~ **digital hierarchy** *n* (*SDH*) TELEP, TRANS hiérarchie numérique synchrone *f* (*HNS*); ~ **equipment management function** *n* (*SEMF*) APPL fonction de gestion d'équipement synchrone *f*; ~ **execution** *n* COMP exécution synchrone *f*; ~**-frequency encoding technique** *n* (*SFET*) DATA technique de codage à fréquence synchrone *f*; ~ **idle character** *n* (*SYN*) DATA caractère de synchronisation *m* (*SYN*); ~ **link** *n* DATA liaison synchrone *f*; ~ **motor** *n* TELEP moteur synchrone *m*; ~ **multiplexer** *n* (*SM*) TRANS multiplexeur synchrone *m*; ~ **network** *n* TELEP réseau synchrone *m*; ~ **optical network** *n* (*SONET*) COMP, TRANS réseau optique synchrone *m*; ~ **port** *n* COMP, TRANS point d'accès synchrone *m*; ~ **satellite** *n* RADIO satellite synchrone *m*; ~ **transfer mode** *n* (*STM*) DATA, TRANS mode de transfert synchrone *m* (*MTS*); ~ **transfer module** *n* (*STM*) DATA, TRANS module de transfert synchrone *m*; ~ **transmission** *n* DATA, TRANS transmission synchrone *f*; ~ **transport module** *n* (*STM*) DATA, TRANS module de transport synchrone *m*; ~ **transport module-n** *n* (*STM-n*) DATA, TRANS module de transport synchrone-n *m*; ~ **voltage** *n* ELECT tension de synchronisme *f*

synonymous: ~ **name** *n* DATA nom synonyme *m*, synonyme *m*

synopsis *n* COMP, DATA résumé *m*, synopsis *m*

syntactic[1] *adj* COMP, DATA, TELEP syntactique, syntaxique

syntactic[2]: ~ **analyser** *BrE*, ~ **analyzer** *AmE* *n* TELEP analyseur syntactique *m*; ~ **analysis** *n* TELEP analyse syntaxique *f*; ~ **structure** *n* COMP structure syntaxique *f*

syntactically: ~**-invalid test event** *n* DATA, TEST événement de test invalide sur le plan de la syntaxe *m*

syntax: ~**-based videotex** *n* (*SBV*) TELEP vidéotex syntaxique *m*; ~ **diagram** *n* TELEP diagramme syntaxique *m*; ~ **identifier** *n* DATA, TELEP identificateur de syntaxe *m*; ~**-matching service** *n* DATA service d'adaptation de syntaxe *m*; ~ **version** *n* DATA, TELEP version de syntaxe *f*

synthesis *n* COMP, ELECT, TELEP synthèse *f*

synthesize *vt* COMP, ELECT, TELEP synthétiser

synthesizer *n* COMP, ELECT, TELEP synthétiseur *m*; ~ **driver** COMP contrôleur de synthétiseur *m*; ~ **settling time** TELEP temps d'acquisition de régime stable de synthétiseur *m*; ~ **unit** COMP unité de synthèse *f*

synthetic: ~ **display** *n* RADIO présentation synthétique *f*; ~ **image** *n* DATA, TV image de synthèses *f*; ~ **video signal** *n* RADIO, SIG signal vidéo synthétique *m*; ~ **voice** *n* PER APP voix synthétique *f*

system[1]: ~**-exclusive** *adj* COMP système exclusif

system[2] *n* COMP appareil *m*, GEN installation *f*, système *m*, PER APP, POWER, RADIO installation *f*, TELEP exploitation *f*, installation *f*; ~ **address** (*SA*) TELEP adresse système *f* (*AS*); ~ **availability information point** TELEP service d'informations relatives à la disponibilité des systèmes *m*; ~ **configuration** COMP configuration du système *f*; ~ **conformance statement** (*SCS*) DATA déclaration de conformité du système *f*; ~ **conformance test report** (*SCTR*) DATA, TEST rapport de test de conformité du système *m*; ~ **console** TELEP console de système *f*; ~ **control** COMP contrôle de système *m*; ~ **control signal unit** (*SCU*) SIG unité de signalisation pour la commande du système de signalisation *f* (*SCU*); ~ **control station** TELEP station de contrôle *f*; ~ **coordination and control network** (*SCCN*) TEST réseau de coordination et de contrôle du système *m*; ~ **data** *n pl* COMP données système *f pl*; ~ **deviation** CONTR écart de régulation *m*; ~ **diagram** COMP schéma de fonctionnement *m*; ~ **flow chart** COMP diagramme de système *m*; ~ **font** police système *f*; ~ **generation** COMP génération de système *f*; ~ **interconnection point** TELEP point d'interconnexion entre systèmes *m*; ~ **loss** GEN affaiblissement du système *m*, RADIO affaiblissement entre bornes d'antennes *m*; ~ **management** COMP gestion de système *f*; ~ **monitor** COMP, CONTR moniteur *m*, système d'autocontrôle *m*; ~ **noise temperature** TRANS température de bruit du système *f*; ~ **provider** APPL fournisseur du système *m*; ~ **recovery** DATA récupération du système *f*; ~ **resilience** COMP résilience du système *f*; ~ **software** COMP, DATA logiciel de base *m*; ~ **testing** COMP, TEST essai de système *m*; ~ **title** DATA titre de système *m*; ~ **under test** (*SUT*) DATA système en cours de test *m*

system[3]: ~ **busy** *phr* (*SB*) TELEP occupation par le système *f* (*OS*)

systematic: ~ **effect** *n* TEST effet systématique *m*; ~ **error** *n* TEST erreur systématique *f*; ~ **failure** *n* TEST défaillance systématique *f*; ~ **sampling** *n* TEST échantillonnage systématique *m*; ~ **variation** *n* TEST écart systématique *m*

systematics *n* TEST systématique *f*

systems: ~ **analysis** *n* COMP, TEST analyse fonctionnelle *f*; ~ **integration** *n* DATA intégration de systèmes *f*; ~ **integrator** *n* DATA intégrateur de systèmes *m*; ~**-managed object** *n* DATA objet géré des systèmes *m*; ~ **management application process** *n* COMP, DATA processus d'application de gestion des systèmes *m*; ~ **management application protocol** *n* (*SMAP*) COMP, DATA protocole d'application de gestion des systèmes *m*; ~ **management application service element** *n* COMP, DATA élément de service d'application pour la gestion des systèmes *m*; ~ **management equipment** *n pl* (*SME*) COMP équipement de gestion des systèmes *m*; ~ **management function** *n* COMP, DATA fonction de gestion des systèmes *f*; ~ **management functional area** *n* COMP, DATA zone fonctionnelle de gestion des systèmes *f*; ~ **management functional unit** *n* COMP unité fonctionnelle de gestion des systèmes *f*; ~ **management functional unit package** *n* COMP lot d'unités fonctionnelles de gestion des systèmes *m*; ~ **management operation** *n* COMP opération de gestion des systèmes *f*; ~ **management service** *n* COMP service de gestion des systèmes *m*; ~ **planning** *n* COMP planification des systèmes *f*; ~ **and procedures** *n pl* COMP méthodes administratives *f pl*

systolic: ~ **architecture** *n* APPL, TRANS architecture systolique *f*

T

T: ~ **network** *n* POWER réseau en T *m*; ~-**reference point** *n* DATA point de référence T *m*; ~ **stage** *n* (*time stage*) SWG, TELEP étage T *m* (*étage temporel*); ~ **wire** *n* TELEP fil T *m*

TA *abbr* (*terminal adaptor*) DATA, PER APP, TELEP adaptateur de terminal *m*

tab *n* (*tabulator*) COMP tabulatrice *f*, PARTS queue de raccordement *f*

table: ~ **cassette** *n* RADIO console de table *f*; ~ **instrument** *n* PER APP, TELEP appareil téléphonique mobile *m*, poste mobile *m*, poste téléphonique mobile *m*; ~ **scanning** *n* GEN exploration de table *f*; ~ **set** *n* PER APP, TELEP appareil téléphonique mobile *m*, poste mobile *m*, poste téléphonique mobile *m*; ~ **structure** *n* GEN structure de table *f*

tablet *n* COMP tablette *f*

tabletop: ~ **device** *n* PER APP périphérique de bureau *m*

tabulate *vt* COMP, GEN tabuler

tabulation *n* COMP tabulation *f*

tabulator *n* (*tab*) COMP tabulatrice *f*

tacan *n* (*tactical air navigation*) RADIO système tacan *m* (*système tactical air navigation*)

tactical: ~ **air navigation** *n* (*tacan*) RADIO système tactical air navigation *m* (*système tacan*)

tactile[1] *adj* COMP tactile

tactile[2]: ~ **screen** *n* COMP écran tactile *m*

tag[1] *n* COMP *label* étiquette *f*, ELECT *of terminal spade* cosse *f*, PARTS *pin* broche *f*, TRANS *label* étiquette *f*; ~ **bit** DATA bit drapeau *m*; ~ **block** DATA, ELECT bloc de raccordement *m*; ~ **rash** ELECT éruption due au contact avec les blocs de raccordement *f*; ~ **strip** ELECT bande relais *f*, barrette à bornes *f*

tag[2] *vt* COMP étiqueter

tagged: ~-**image file format** *n* (*TIFF*) COMP, DATA tagged-image file format *m* (*TIFF*)

tagger *n* COMP étiqueteur *m*

tagging *n* COMP étiquetage *m*

tailing *n* TELEP traînage *m*

tailored: ~ **menu** *n* COMP menu personnalisé *m*

take: ~ **out of service** *vt* TEST mettre hors service

takeover *n* GEN relève *f*, reprise *f*

take-up: ~ **factor** *n* TELEP facteur de câblage *m*

TA-LB *abbr* (*B-ISDN terminal adaptor*) TELEP adaptateur de terminal pour le RNIS-LB *m*

talk[1] *n* DATA *Internet* dialogue *m*, RADIO, TELEP conversation *f*; ~ **battery** RADIO, TELEP batterie de conversation *f*; ~ **set** ELECT émetteur-récepteur *m*; ~ **spurt** RADIO, TELEP bribe de parole *f*; ~ **time** RADIO, TELEP temps de conversation *m*

talk[2] *vi* DATA *Internet* dialoguer, RADIO, TELEP converser

talking: ~ **state** *n* RADIO, TELEP position de conversation *f*

tally: ~ **reader** *n* DATA lecteur de bande de contrôle *m*

tamed: ~ **frequency modulation** *n* (*TFM*) RADIO, TRANS modulation de fréquence asservie *f* (*MFA*)

tamper: ~ **with** *vt* TEST trafiquer

tamperproof *adj* TEST infalsifiable, intraficable, protégé contre les manipulations abusives

T&C *abbr* (*tracking and command*) TEST poursuite et commande *f*

tandem[1] *adj* TELEP de transit

tandem[2]: ~ **completing trunk** *n* TRANS faisceau de transit local d'arrivée *m*, faisceau de circuits de transit local d'arrivée *m*; ~ **exchange** *n* DATA central de transit urbain *m*, centre nodal *m*, centre nodal automatique *m*, centre tandem *m*, SWG, TELEP central de transit urbain *m*, centre nodal *m*, centre nodal automatique *m*, centre tandem *m*, commutateur de transit *m*, commutateur de transit urbain *m*; ~ **private branch exchange** *n* SWG, TELEP, TRANS commutateur privé nodal *m*; ~ **PBX** *n* SWG, TELEP, TRANS commutateur privé nodal *m*; ~ **selector** *n* SWG, TELEP sélecteur tandem *m*

tangential: ~ **wave path** *n* RADIO rayon tangent *m*

tangled: ~ **hierarchy** *n* COMP hiérarchie croisée *f*; ~ **taxonomy** *n* COMP taxinomie croisée *f*

tangling *n* ELECT enchevêtrement *m*

tank: ~ **circuit** *n* ELECT circuit oscillant *m*, circuit bouchon *m*

tantalum: ~ **capacitor** *n* ELECT, PARTS condensateur au tantale *m*

tap *vt* ELECT *shunt* mettre en dérivation, TELEP *conversation* écouter clandestinement, *telephone* mettre sur écoute

tape[1] *n* APPL *audio, video*, COMP *audio* bande *f*, ELECT, PARTS *insulating* ruban *m*; ~ **cartridge drive** COMP lecteur de cartouche *m*; ~ **cassette drive system** COMP système à cassettes *m*; ~ **drive** COMP entraînement de bande magnétique *m*; ~ **interface** COMP interface de bande *f*; ~ **memory** COMP mémoire à bandes magnétiques *f*; ~ **printer** PER APP traducteur-imprimeur sur bande *m*; ~ **recorder** PER APP lecteur de bande *m*; ~ **relay** TELEP transit par bande perforée *m*; ~ **reproducer** COMP reproducteur de bande *m*; ~ **station** COMP lecteur-perforateur de bande *m*; ~ **synchronization** COMP synchronisation de bande *f*; ~ **transport unit** (*TTU*) APPL, COMP, TV dérouleur de bande *m*, mécanisme d'entraînement de bande *m*

tape[2] *vt* APPL, COMP enregistrer sur bande

taper *n* RADIO raccord progressif *m*, transition progressive *f*; ~ **curve** ELECT courbe de répartition de la résistance *f*

tapered[1] *adj* PARTS conique

tapered[2]: ~ **fiber** *AmE*, ~ **fibre** *BrE* *n* TRANS transition progressive de fibre optique *f*

tapping *n* TELEP *of telephone* écoute clandestine *f*, écoute téléphonique filaire *f*, branchement clandestin *m*, *of telephone line* mise sur écoute *f*; ~ **equipment** PER APP matériel d'écoute *m*

tar *n* DATA *file compression* tar *m*

target[1] *n* ELECT cible *f*, RADIO cible *f*; ~ **acquisition** RADIO accrochage de cible *m*; ~ **cell** RADIO cellule cible *f*; ~ **direction** RADIO direction d'objectif *f*; ~ **echo** RADIO écho de cible *m*, écho utile *m*; ~ **identification** RADIO identification de cible *f*; ~ **language** COMP langue cible *f*; ~ **link** COMP lien cible *m*, lien de destination *m*; ~ **lock-on** RADIO, TRANS verrouillage sur la cible *m*; ~ **node** COMP nud cible *m*, nud de destination *m*; ~ **program** *AmE*, ~ **programme** *BrE* TELEP programme résultant *m*; ~ **signature** RADIO signature de cible *f*; ~ **specification** TEST spécification *f*; ~ **tracking** RADIO poursuite d'objectif *f*; ~ **transit delay** (*TTD*) DATA temps de transit cible *m*; ~ **zone** RADIO, TRANS zone à l'émission *f*, zone cible *f*

target[2] *vt* RADIO, TELEP, TV destiner, viser

tariff *n* APPL tarif *m*, TELEP tarif *m*, taxe *f*; ~ **analysis** APPL analyse des tarifs *f*; ~ **area** TELEP circonscription de taxe *f*; ~ **class** TELEP classe de tarif *f*; ~ **determination** TELEP détermination de taxation *f*; ~ **determination point** TELEP point de détermination de taxation *m*; ~ **generator** TELEP générateur d'impulsions de taxation *m*; ~-**message receiver** TELEP récepteur de messages de tarification *m*; ~ **message sender** TELEP émetteur de messages de tarification *m*; ~ **policy** APPL, TELEP politique tarifaire *f*; ~ **pulse meter** TELEP compteur de taxes *m*; ~ **reduction** APPL réduction de tarif *f*; ~ **setting** APPL, TELEP fixation des tarifs *f*, tarification *f*; ~ **structure** APPL, TELEP structure tarifaire *f*; ~ **switching** SWG commutation de taxe *f*; ~-**switching class** SWG classe de changement de tarif *f*; ~ **time-slot** TELEP plage horaire de tarifs *f*; ~ **zone** TELEP zone de taxation *f*

TASI *abbr* (*time-assigned speech interpolation*) SIG, TRANS système TASI (*interpolation temporelle des signaux vocaux*)

task: ~ **description data** *n pl* (*TDD*) DATA données de description de tâches *f pl*; ~ **symbol** *n* TELEP symbole de tâche *m*

taut: ~ **band milliammeter** *n* TEST milliammètre à rubans tendus *m*

taxi: ~ **radio system** *n* RADIO système de gestion de flotte de taxis *m*, système de radiocommunications pour taxis *m*

TB: ~ **cell** *n* (*transmitter blocker cell*) RADIO, TRANS verrou d'émetteur *m*

T-branching: ~ **device** *n* ELECT, PARTS coupleur en T *m*

TC *abbr* ELECT (*telecommunications closet*) armoire de télécommunications *f*, TELEP (*trunk code*) indicatif interurbain *m*, (*transport connection*) connexion de transport *f*

TCC *abbr* (*telephone country code*) TELEP indicatif téléphonique de pays *m*

TCE *abbr* (*transit connection element*) TRANS élément de connexion de transit *m*

TCI-DCS *abbr* (*transmitting subscriber identification-DCS*) DATA, TELEP TCI-DCS (*identification de l'abonné émetteur-DCS*)

TCL *abbr* TELEP (*terminal coupling loss*) ACT (*affaiblissement de conversion transversale*), (*transverse conversion loss*) équivalent de couplage du terminal *m*

TCLw *abbr* (*weighted terminal coupling loss*) TELEP rapport de conversion transversale *m*

TCM *abbr* (*trellis-coded modulation*) TRANS MCT (*modulation par codage en treillis*)

T-coupler *n* ELECT, PARTS coupleur en T *m*

TCP *abbr* DATA (*transmission control protocol*) TCP (*protocole de contrôle de transmission*), TEST (*test coordination procedure*) procédure de coordination des tests *f*

TCR *abbr* (*telemetry command and ranging*) TEST télécommande et mesure de distance *f*

TCRF *abbr* (*transit connection-related function*) DATA, TRANS fonction liée à une connexion de transit *f*

Tcu *abbr* (*cross-office transfer time*) SWG *subtext* temps de transfert dans le central *m*

TDC *abbr* (*telex destination code*) DATA, TELEP code télex de destination *m*

TDD *abbr* TELEP (*telecommunications device for the deaf*) ATS (*appareil de télécommunications pour les sourds*), (*task description data*) données de description de tâches *f pl*

TDF *abbr* (*trunk distribution frame*) TRANS répartiteur de jonction *m*

TDI *abbr* DATA (*trade data interchange*) échange de données commerciales *m*, (*transit-delay indication*) indication de délai en transit *f*

TDM *abbr* SIG, SWG (*time-division multiplex*) MRT (*multiplexage par répartition dans le temps*), multiplexage temporel *m*, (*time-division multiplexer*) multiplexeur temporel *m*

TDMA *abbr* (*time-division multiple access*) RADIO, SIG, TELEP, TRANS AMRT (*accès multiple à répartition dans le temps*); ~ **burst time plan** *n* RADIO, SIG, TELEP, TRANS plan d'émission en accès multiple à répartition dans le temps *m*, plan d'émission en AMRT *m*; ~ **mode** *n* RADIO, SIG, TELEP, TRANS mode AMRT *m*

TDR *abbr* ELECT, POWER, TEST, TRANS (*time domain reflectometer*) réflectomètre temporel *m*, (*time domain reflectometry*) réflectométrie temporelle *f*

TDSAI *abbr* (*transit-delay selection and indication*) DATA, TELEP sélection et indication du délai en transit

TE *abbr* COMP, DATA, PER APP, TELEP (*terminal equipment*) équipement terminal *m*, TRANS (*transverse electric*) TE (*transverse électrique*); ~ **mode** *n* TRANS mode TE *m*

teachware *n* COMP, DATA didacticiel *m*, logiciel didacticiel *m*

tear: ~-**off menu** *n* COMP menu détachable *m*, menu détachant *m*, menu flottant *m*

tearing: ~ **resistance** *n* PARTS résistance à la déchirure *f*

technical: ~ **acknowledgement** *n* DATA accusé de réception technique *m*; ~ **breakdown** *n* TEST défaillance technique *f*; ~ **delay** *n* TEST délai technique *m*; ~ **delay time** *n* TEST durée de délai technique *f*; ~ **manual** *n* COMP manuel technique *m*; ~ **recommendations applications committee** *n* (*TRAC*) GEN comité d'application des recommandations techniques *m*

technologically: ~ **advanced** *adj* GEN à la pointe de la technologie

technology: ~ **transfer** *n* COMP transfert de technologie *m*

TEI *abbr* (*terminal end-point identifier*) DATA identificateur de point d'extrémité de terminal *m*

TELCO *abbr* (*telephone company*) TELEP société exploitante de réseau téléphonique *f*

telecom *abbr* (*telecommunications*) GEN, RADIO, TELEP télécom (*télécommunications*)

telecommand[1] *n* COMP, CONTR, PER APP, RADIO, TELEP télécommande *f*

telecommand[2] *vt* COMP, CONTR, PER APP, RADIO, TELEP télécommander, commander à distance

telecommunicate *vt* GEN télécommuniquer

telecommunication *n* GEN télécommunication *f*; ~ **circuit** RADIO, TELEP circuit de télécommunication *m*, voie de communication *f*

telecommunications *n pl* (*telecom*) GEN, RADIO, TELEP télécommunications *f pl*; ~ **administration** *n* TELEP administration des télécommunications *f*; ~ **board** *n* TELEP direction des télécommunications *f* (*DT*); ~ **cable** *n* TRANS câble de télécommunications *m*; ~ **closet** *n* (*TC*) ELECT armoire de télécommunications *f*; ~ **common carrier** *n* TRANS entreprise de télécommunications *f*, société exploitante de télécommunications *f*, télécommunicateur *m*; ~ **device for the deaf** *n* (*TDD*) TELEP appareil de télécommunications pour les sourds *m* (*ATS*); ~ **and electronic equipment technician** *n* APPL, PER APP monteur d'appareils électroniques et de télécommunication *m*; ~ **engineer** *n* GEN ingénieur en télécommunications *m*; ~ **equipment** *n* GEN équipement de télécommunications *m*; ~ **exchange** *n* TELEP central de télécommunications *m*, centre de télécommunications *m*; ~ **infrastructure** *n* GEN infrastructure de télécommunications *f*; ~ **management network** *n* (*TMN*) DATA, TELEP, TRANS réseau de gestion des télécommunications *m* (*RGT*); ~ **network** *n* DATA, TELEP, TRANS réseau de télécommunications *m*; ~ **operator** *n* GEN exploitant de télécommunications *m*; ~ **satellite** *n* RADIO, TV satellite de télécommunications *m*; ~ **security** *n* GEN sécurité de télécommunications *f*; ~ **service** *n* GEN service de télécommunications *m*; ~ **system** *n* DATA, RADIO, TELEP, TRANS réseau de télécommunications *m*; ~ **terminal** *n* RADIO, TELEP terminal de télécommunications *m*; ~ **traffic** *n* (*teletraffic*) SWG, TELEP trafic de télécommunications *m* (*télétrafic*)

Telecommunications: ~ **Regulatory Service** *n* (*TRS*) RADIO Service de la réglementation des télécommunications *m* (*SRT*)

telecommute *vi* APPL, COMP, DATA, TRANS télétravailler

telecommuting *n* APPL, COMP, DATA, TRANS télétravail *m*

teleconference *n* APPL, RADIO, TRANS téléconférence *f*

telecontrol *n* COMP, CONTR, RADIO, TELEP télécommande *f*

telecopy[1] *n* COMP, TELEP télécopie *f*, téléfax *m*
telecopy[2] *vt* COMP, TELEP télécopier
telefax *n* COMP, TELEP télécopie *f*, téléfax *m*
Telefax: ~ **Group 4** *n* (*FX4*) TELEP télécopie du groupe 4 *f* (*FX4*); ~ **Group 3** *n* (*FX3*) TELEP télécopie du groupe 3 *f* (*FX3*)
telegram *n* TELEP télégramme *m*
telegraphic: ~ **address** *n* TELEP adresse télégraphique *f* (*adr tél.*); ~ **alphabet** *n* TELEP alphabet télégraphique *m*; ~ **channel** *n* TELEP voie de transmission télégraphique *f*; ~ **circuit** *n* TELEP circuit télégraphique *m*, voie de communication bilatérale *f*, voie de communication télégraphique *f*, TRANS circuit télégraphique *m*; ~ **code** *n* TELEP, TRANS code télégraphique *m*; ~ **conversion** *n* TELEP translation télégraphique *f*; ~ **converter** *n* TELEP convertisseur télégraphique *m*; ~ **demodulator** *n* TELEP démodulateur télégraphique *m*; ~ **distortion** *n* TELEP distorsion télégraphique *f*; ~ **magnifier** *n* TELEP amplificateur télégraphique *m*; ~ **modulation** *n* TELEP modulation télégraphique *f*; ~ **modulator** *n* TELEP modulateur télégraphique *m*; ~ **network** *n* TELEP, TRANS réseau télégraphique *m*; ~ **noise** *n* TELEP bruit de télégraphe *m*; ~ **office** *n* (*TO*) TELEP bureau télégraphique *m*, service télégraphique *m*; ~ **pole** *n* TELEP poteau télégraphique *m*; ~ **relay** *n* PARTS relais télégraphique *m*; ~ **service** *n* TELEP service télégraphique *m*; ~ **service correspondence** *n* TELEP correspondance télégraphique de service *f*; ~ **signal** *n* SIG, TELEP signal télégraphique *m*; ~ **signal element** *n* SIG élément de signal télégraphique *m*; ~ **switchboard** *n* SWG, TELEP commutateur manuel télégraphique *m*; ~ **transfer** *n* TELEP transfert télégraphique *m*, virement télégraphique *m*; ~ **transmitter** *n* TELEP émetteur télégraphique *m*; ~ **word** *n* TELEP mot télégraphique conventionnel *m*, mot télégraphique *m*
telegraphy *n* RADIO, TELEP, TRANS télégraphie *f*
teleguidance *n* RADIO téléguidage *m*
teleinformatics *n* COMP, DATA, RADIO téléinformatique *f*, télétraitement *m*
telemaintenance *n* COMP, DATA télémaintenance *f*
telemanagement *n* COMP, DATA télégestion *f*
telemarketing *n* COMP, DATA télémarketing *m*
telematic: ~ **agent** *n* (*TLMA*) DATA, TELEP agent télématique *m*; ~ **file transfer** *n* (*TFT*) TELEP transfert de fichier télématique *m* (*TFT*); ~ **protocol architecture** *n* (*TPA*) TELEP architecture de protocole télématique *f* (*TPA*); ~ **server** *n* DATA serveur télématique *m*; ~ **terminal** *n* DATA terminal télématique *m*
telematics *n* COMP, DATA, RADIO télématique *f*; ~ **service** COMP, DATA, RADIO service de télématique *m*
telemetering *n* DATA, RADIO, TEST télémesure *f*; ~ **device** TEST dispositif de télémesure *m*
telemetry *n* DATA, RADIO, TEST télémétrie *f*, télémesure *f*; ~ **command and ranging** (*TCR*) TEST télécommande et mesure de distance *f*
telemonitoring *n* RADIO, TEST télésurveillance *f*
telephone[1] *n* PER APP, TELEP combiné téléphonique *m*, poste téléphonique *m*, téléphone *m*; ~ **answering machine** PER APP, TELEP téléphone répondeur-enregistreur *m*, répondeur téléphonique *m*, répondeur *m*; ~ **attachment** PER APP, TELEP appareil de téléphone *m*; ~ **bell** PER APP sonnerie téléphonique *f*; ~ **book** TELEP annuaire *m*, annuaire téléphonique *m*, bottin® *m*, répertoire téléphonique *m*; ~ **booth** TELEP cabine téléphonique *f*; ~ **box** TELEP cabine téléphonique *f*; ~ **call** TELEP appel téléphonique, conversation téléphonique *f*, coup de fil *m*; ~ **card** TELEP télécarte *f*; ~ **carrier** TELEP porteur de téléphonie *m*; ~ **charge** TELEP télétaxe *f*; ~ **charge analyser** *BrE*, ~ **charge analyzer** *AmE* TELEP analyseur de taxes téléphoniques *m*; ~ **charge printer** TELEP compteur imprimant télétaxe *m*; ~ **circuit** TELEP circuit téléphonique *m*, liaison téléphonique *f*, TRANS circuit téléphonique *m*; ~ **circuit loss** TELEP équivalent d'un circuit téléphonique *m*, équivalent d'une liaison téléphonique *m*; ~ **company** *n* (*TELCO*) TELEP société exploitante de réseau téléphonique *f*; ~ **conference** TELEP conférence téléphonique *f*; ~ **connection** TELEP liaison téléphonique *f*, circuit téléphonique *m*, TRANS circuit téléphonique *m*; ~ **connection socket** TELEP prise téléphonique *f*; ~ **conversation** TELEP conversation téléphonique *f*; ~ **country code** (*TCC*) TELEP indicatif téléphonique de pays *m*; ~ **density** TELEP densité téléphonique *f*; ~ **directory** TELEP annuaire *m*, annuaire téléphonique *m*, bottin® *m*, répertoire téléphonique *m*; ~ **engineering** TELEP technique téléphonique *f*; ~ **equipment** PER APP matériel téléphonique *m*; ~ **exchange** SWG, TELEP central téléphonique *m*, commutateur téléphonique *m*; ~ **extension cable** PARTS prolongateur téléphonique *m*; ~ **extension cord** PARTS prolongateur téléphonique *m*; ~ **extension line** PARTS ligne téléphonique supplémentaire *f*; ~ **extension reel** PARTS, PER APP enrouleur de prolongateur téléphonique *m*; ~ **frequency** TELEP, TRANS fréquence téléphonique *f*; ~ **hybrid** TELEP circuit différentiel *m*; ~ **installation** TELEP installation téléphonique *f*; ~ **instrument** PER APP, TELEP appareil téléphonique *m*, poste téléphonique *m*; ~ **instrument cord** PARTS, TELEP cordon d'appareil téléphonique *m*; ~ **interface equipment** TELEP équipement d'interface téléphonique *m*, équipement de raccordement téléphonique *m*; ~ **interference factor** TELEP facteur TIF *m*; ~ **jack** PARTS, PER APP, SWG, TELEP conjoncteur *m*; ~ **keypad** PER APP clavier téléphonique *m*; ~ **kiosk** TELEP cabine téléphonique *f*; ~ **line** TELEP ligne téléphonique *f*; ~ **link** TELEP liaison téléphonique *f*, ligne téléphonique *f*; ~ **message** TELEP message téléphonique *m*; ~ **monitoring** TELEP, TEST surveillance téléphonique *f*; ~ **network** TELEP, TRANS réseau téléphonique *m*; ~ **number** TELEP numéro d'appel *m*, numéro de téléphone *m*; ~ **operator** *AmE* (*cf telephonist BrE*) SWG, TELEP opérateur *m*, standardiste *m*, téléphoniste *m*; ~ **relation** TELEP relation téléphonique *f*; ~ **service** (*TS*) TELEP service téléphonique *m*; ~ **service access unit** (*TSAU*) TELEP unité d'accès au service téléphonique *f*; ~ **set** PER APP, TELEP appareil téléphonique *m*, combiné téléphonique *m*, poste téléphonique *m*; ~ **set with memory** PER APP, TELEP appareil téléphonique à mémoire *m*, combiné téléphonique à mémoire *m*, poste téléphonique à mémoire *m*; ~ **signal** SIG signal téléphonique *m*; ~ **socket** PARTS, PER APP, SWG, TELEP conjoncteur *m*, prise téléphonique *f*; ~ **station** PER APP, RADIO, TELEP poste téléphonique *m*; ~ **subscriber** TELEP abonné au téléphone *m*; ~ **subscription** TELEP abonnement téléphonique *m*; ~ **switchboard** SWG, TELEP commutateur d'abonnés *m*, commutateur manuel *m*, pupitre de commutation *m*, standard *m*, standard téléphonique *m*, tableau de commutation *m*; ~ **tapping** TELEP *of conversation* écoute clandestine *f*, écoute téléphonique filaire *f*, branchement clandestin *m*, *of telephone* mise sur écoute *f*; ~ **terminal** PER APP, TELEP terminal téléphonique *m*; ~ **traffic** TELEP trafic téléphonique *m*; ~ **traffic analyser** *BrE*, ~ **traffic analyzer** *AmE* TEST analyseur de trafic téléphonique *m*; ~**-type channel** RADIO, TELEP voie de type téléphonique *f*; ~**-type circuit** RADIO, TELEP circuit de type téléphonique *m*; ~**-type connection** RADIO, TELEP circuit de type téléphonique *m*; ~ **user part** (*TUP*) DATA, RADIO, TELEP sous-système utilisateur téléphonie *m* (*SSUT*), sous-système utilisateur téléphonique *m* (*SSUT*)
telephone[2] *vt* TELEP *person* appeler, téléphoner à; ~ **back** TELEP rappeler
telephone[3] *vi* TELEP téléphoner
telephone[4]: **on the** ~ *phr* TELEP au téléphone; **over the** ~ *phr* TELEP au téléphone

telephonist *n BrE* (*cf telephone operator AmE*) SWG, TELEP opérateur *m*, standardiste *m*, téléphoniste *m*
telephony *n* TELEP téléphonie *f*; **~ baseband test** TEST essai de téléphonie en bande de base *m*; **~ head end** TV tête de réseau *f*; **~ performance** TELEP qualité des transmissions téléphoniques *f*; **~-rated device** PER APP périphérique téléphonique *m*
telephoto: **~ service** *n* TELEP service de téléphotographie *m*
Telepoint *n* POWER, TELEP Pointel *m*
teleport *n* GEN téléport *m*, zone de télécommunications avancées *f* (*ZTA*)
telepresence *n* COMP téléprésence *f*
teleprinter *n BrE* (*cf teletypewriter AmE*) APPL, PER APP, TELEP téléimprimeur *m*, téléscripteur *m*; **~ service** *BrE* (*cf teletypewriter service AmE*) APPL, PER APP, TELEP service téléimprimeur *m*
teleprinting *n* PER APP *newspapers, magazines* impression de journaux à distance *f*
teleprocessing *n* GEN téléinformatique *f*, télétraitement *m*; **~ network** COMP, TRANS réseau de téléinformatique *m*, réseau de télétraitement *m*
teleprompter *n AmE* (*cf autocue BrE*) TV télésouffleur *m*
telepurchase *n* GEN téléachat *m*
telesales *n pl* APPL, TV télévente *f*
telesecurity *n* APPL télésécurité *f*
teleservice *n* APPL téléservice *m*
teleshopping *n* APPL, TV téléachat *m*
telesignaling: **~ transmitter** *AmE see telesignalling transmitter BrE*
telesignalling: **~ transmitter** *n BrE* APPL transmetteur télésignalisation *m*
teletex *n* COMP, PER APP, RADIO télétex *m*; **~ access unit** (*TTXAU*) APPL unité d'accès au télétex *f*; **~ basic control function index** TELEP répertoire des fonctions de commande télétex de base *m*; **~ basic graphic character index** TELEP répertoire des caractères graphiques télétex de base *m*; **~ call** TELEP communication télétex *f*; **~ character index** TELEP répertoire des caractères télétex *m*; **~ control function index** CONTR, TELEP répertoire des fonctions de commande télétex *m*; **~ document** TELEP document télétex *m*; **~ graphic character index** TELEP répertoire des caractères graphiques télétex *m*; **~ page** TELEP page télétex *f*; **~ terminal** COMP, PER APP, RADIO terminal télétex *m*
teletext *n* (*TTX*) APPL, TV télétexte *m* (*TTX*), vidéographie diffusée *f*, vidéotex diffusé *m*; **~ service** APPL service de télétexte *m*
teletrack *n* COMP effet de trace *m*, effet de traînée *m*
teletraffic *n* (*telecommunications traffic*) SWG, TELEP télétrafic *m* (*trafic de télécommunication, trafic de télécommunications*); **~ engineering** TELEP ingénierie du télétrafic *f*
teletuition *n* TV téléenseignement *m*
Tele: **~ 2x** *n* COMP Tele 2x *m*
teletype *n* (*TTY*) COMP, PER APP, RADIO, TELEP téléimprimeur *m*, télétypewriter *m*; **~ machine** PER APP téléscripteur *m*; **~-order channel** TRANS voie de service par téléimprimeur *f*
teletypewriter *n AmE* (*cf teleprinter BrE*) APPL, PER APP, TELEP téléimprimeur *m*, téléscripteur *m*; **~ service** *AmE* (*cf teleprinter service BrE*) APPL, PER APP, TELEP service téléimprimeur *m*
television *n* (*TV*) RADIO radiodiffusion visuelle *f*, TV radiodiffusion visuelle *f*, télévision *f*, *set* téléviseur *m*; **~ aerial** TV antenne classique *f*, antenne de télévision *f*; **~ antenna** TV antenne classique *f*, antenne de télévision *f* **~ broadcaster** TV télédiffuseur *m*; **~ broadcasting** TV radiodiffusion visuelle *f*, télédiffusion *f*; **~ channel** TV chaîne de télévision *f*; **~ circuit** TV circuit télévisuel *m*; **~ circuit section** TV section de circuit télévisuel *f*; **~ monitoring** TEST, TV contrôle de télévision *m*; **~ picture transmitter** TV émetteur d'images de télévision *m*; **~ program** *AmE*, **~ programme** *BrE* TV programme télévisuel *m*; **~ receive-only aerial** TV antenne de réception de télévision seulement *f*; **~ receive-only antenna** TV antenne de réception de télévision seulement *f*; **~ receiver** TV récepteur de télévision *m*, récepteur télévision *m*; **~ set** TV téléviseur *m*; **~ sound carrier** TV porteur de son de télévision *m*; **~ sound transmitter** TV émetteur de son de télévision *m*; **~ system** TV système de télévision *m*; **~ system converter** TV convertisseur du système de télévision *m*; **~ transmitter** TV émetteur de télévision *m*; **~ tube** ELECT, TV tube image *m*; **~ tuner** TV tuner *m*; **~ viewer** TV téléspectateur *m*
teleworker *n* GEN télétravailleur *m*
teleworking *n* GEN télétravail *m*
telewriter *n* COMP, DATA, PER APP, RADIO, TV téléscripteur *m*, terminal de télé-écriture *m*
telewriting *n* COMP, DATA, PER APP, RADIO, TV télé-écriture *f*
telex *n* (*TLX*) DATA, RADIO, TELEP télex *m* (*TLX*); **~ access unit** (*TLXAU*) DATA, RADIO, TELEP unité d'accès aux services télex *f*, unité d'accès télex *f*; **~ destination code** (*TDC*) DATA, TELEP code télex de destination *m*; **~ exchange** SWG, TELEP central télex *m*, commutateur télex *m*; **~ message** TELEP message télex *m*; **~-network identification code** DATA, TELEP code d'identification de réseau télex *m* (*CIRT*); **~ packet interworking function** (*TPIWF*) DATA fonction d'interfonctionnement télex-mode paquet *f* (*FITMP*); **~-plus** DATA, RADIO, TELEP télex-plus *m*; **~ position** TELEP position télex *f*; **~ relation** TELEP relation *f*, relation télex *f*; **~ switching center** *AmE*, **~ switching centre** *BrE* SWG, TELEP central télex *m*, commutateur télex *m*; **~ terminal** DATA, RADIO, TELEP terminal télex *m*; **~ traffic** TELEP trafic télex *m*
Telex: **~ Over Radio** *n* RADIO radiotélex *m*
telnet *n* DATA *Internet* telnet *m*
TEM *abbr* (*transverse electromagnetic*) TRANS TEM (*transverse électromagnétique*); **~ mode** *n* TRANS mode TEM *m*
temperature: **~ coefficient** *n* GEN coefficient de température *m*; **~ control** *n* CLIM régulation de la température *f*; **~ controller** *n* CLIM thermocontact *m*; **~ cycling** *n* CLIM, TEST cyclage thermique *m*; **~ difference** *n* CLIM variation de température *f*; **~ drop** *n* CLIM chute de température *f*; **~ error** *n* CLIM, TEST erreur due à la température *f*; **~ increase** *n* CLIM, GEN échauffement *m*; **~ inversion** *n* CLIM, RADIO inversion de température *f*; **~-limited state** *n* CLIM, ELECT saturation *f*; **~ rise** *n* CLIM, GEN échauffement *m*; **~ sensitivity** *n* CLIM sensibilité thermique *f*; **~ sensor** *n* CLIM, CONTR capteur de température *m*, sonde pyrométrique *f*
tempering *n* ELECT traitement thermique *m*
template *n* COMP *word processing* modèle *m*, patron *m*, PARTS, TRANS gabarit *m*
tempo: **~ control** *n* COMP commande de tempo *f*; **~ controller** *n* COMP appareil de commande de tempo *m*
temporal: **~ coherence** *n* TRANS cohérence temporelle *f*; **~ compression** *n* COMP, DATA compression temporelle *f*; **~ database** *n* COMP, DATA base de données temporelles *f*; **~ logic** *n* DATA logique temporelle *f*; **~ predictive concealment** *n* COMP masquage à prédiction temporelle *m*; **~ reasoning** *n* DATA raisonnement temporel *m*
temporary: **~ buffer** *n* COMP tampon temporaire *m*; **~ buffer memory** *n* COMP, DATA mémoire tampon temporaire *f*; **~ call forwarding** *n* TELEP renvoi temporaire d'appel *m*; **~ fault** *n* TELEP, TEST dérangement temporaire *m*, panne temporaire *f*; **~ interruption** *n* TELEP, TEST perturbation temporaire *f*; **~ malfunction** *n* TELEP, TEST dérangement temporaire *m*, panne temporaire *f*; **~ mobile station identity** *n* DATA, RADIO, TELEP identité temporaire d'abonné mobile *f*; **~ perturbation** *n* TELEP, TEST perturbation temporaire *f*; **~ transfer** *n* TELEP renvoi temporaire *m*; **~ window** *n* COMP fenêtre temporaire *f*

ten: **~-number memory** *n* PER APP mémoire à dix numéros *f*

tens: **~ counter** *n* ELECT compteur des dizaines *m*; **~ digit** *n* COMP chiffre des dizaines *m*; **~ relay** *n* PARTS, TELEP relais des dizaines *m*

tensile: **~ force** *n* PARTS effort de traction *m*; **~ strength** *n* GEN résistance à la traction *f*, PARTS résistance à la traction *f*, tenue à la traction *f*; **~ stress** *n* PARTS effort de tension *m*; **~ test** *n* PARTS, TEST essai de traction *m*; **~ yield limit** *n* PARTS limite apparente d'élasticité *f*

tension: **~ spring** *n* GEN, PARTS ressort de tension *m*

terminal *n* COMP, DATA terminal *m*, ELECT, GEN *semiconductor* sortie *f*, *contact* borne *f*, *device* terminal *m*, PER APP *contact* borne *f*, POWER borne *f*, borne de raccordement *f*, RADIO, TELEP terminal *m*; **~ adaptor** (*TA*) DATA, PER APP, TELEP adaptateur de terminal *m*; **~ administration** TELEP administration terminale *f*; **~ board** POWER, TELEP tableau de connexions *m*; **~ box** POWER boîte de distribution *f*, TELEP boîte de raccordement *f*; **~ call forwarding** APPL renvoi du terminal *m*; **~ called** TELEP terminal appelé *m*; **~ category PE** TELEP élément graphique de catégorie d'équipement *m*; **~ circuit** POWER, TELEP circuit terminal *m*; **~ clip** POWER borne d'élément *f*; **~ connector** ELECT cosse *f*; **~ country** TELEP pays terminal *m*; **~ coupling loss** (*TCL*) TELEP équivalent de couplage du terminal *m*; **~ diagram** GEN schéma des bornes *m*; **~ drawing** GEN dessin de raccordement *m*; **~ echo suppressor** RADIO, TELEP suppresseur d'écho terminal *m*; **~ emulation** DATA émulation de terminal *f*; **~ emulation software** COMP, DATA logiciel d'émulation de terminal *m*; **~ emulator** DATA émulateur de terminal *m*; **~ end-point identifier** (*TEI*) DATA identificateur de point d'extrémité de terminal *m*; **~-engaged signal** SIG, TELEP signal d'occupation de la ligne du terminal demandé *m*; **~ equipment** (*TE*) COMP, DATA, PER APP, TELEP équipement terminal *m*; **~ equipment package** TELEP ensemble d'équipements terminaux *m*; **~ equipment PE** PER APP élément graphique d'équipement terminal *m*; **~ exchange** TELEP, TRANS central terminal *m*, centre d'extrémité *m*, centre terminal *m*; **~ field** GEN réglette à broches *f*, réglette de raccordement *f*; **~ impedance** ELECT impédance terminale *f*; **~ installation for data transmission** TELEP installation terminale pour transmission de données *f*; **~ international center** *AmE*, **~ international centre** *BrE* TELEP centre terminal international *m*; **~ international exchange** TELEP centre terminal international *m*; **~ national center** *AmE*, **~ national centre** *BrE* TELEP centre terminal national *m*; **~ national section** TELEP section terminale nationale *f*; **~ node controller** (*TNC*) TRANS contrôleur de nœud terminal *m*; **~ number diagram** GEN schéma d'affectation des bornes *m*; **~ number drawing** GEN plan d'affectation des bornes *m*; **~ pair** POWER paire de bornes *f*; **~ screw** POWER vis de borne *f*; **~ server manager** (*TSM*) DATA gestionnaire de serveur de terminaux *m*; **~ share** TELEP quote-part terminale *f*; **~ strip** PARTS plaque à bornes *f*; **~ subaddressing** APPL, DATA, TELEP sous-adressage des terminaux *m*; **~ table** POWER tableau de connexions *m*; **~ voltage** POWER tension aux bornes *f*, tension entre les bornes *f*

terminate[1]: **~ and stay resident** *n* (*TSR*) COMP, DATA, TELEP corésident *m*, résident *m*; **~ and stay resident program** *n* COMP, DATA, TELEP programme corésident *m*, programme résident *m*, programme TSR *m*

terminate[2] *vt* GEN, RADIO mettre fin à, TRANS adapter, boucler

terminate[3] *vi* TELEP, TRANS *cable* aboutir; **~ to a station** TELEP, TRANS aboutir à un poste

terminated: **~ level** *n* POWER, TEST niveau composite adapté *m*; **~ reading** *n* POWER, TEST niveau composite adapté *m*

terminating: **~ equipment** *n* COMP, DATA, PER APP, TELEP équipement terminal *m*; **~ exchange** *n* SWG, TELEP central d'arrivée *m*, commutateur d'arrivée *m*; **~ impedance** *n* ELECT impédance de terminaison *f*; **~ junctor** *n* PARTS, SWG, TELEP joncteur d'arrivée *m*; **~ private branch exchange** *n* TELEP commutateur privé de terminaison; **~ PBX** *n* TELEP commutateur privé de terminaison *m*; **~ resistor** *n* ELECT, PARTS, TRANS termineur *m*; **~ set** *n* ELECT, PARTS, TRANS termineur *m*; **~ stage** *n* SWG étage terminal *m*; **~ traffic** *n* TELEP trafic d'arrivée *m*, trafic en réception *m*; **~ unit** *n* DATA termineur *m*

termination *n* DATA terminaison *f*, ELECT terminaison *f*, extrémité *f*, RADIO, TELEP, TRANS fin *f*, terminaison *f*; **~ joint** PARTS épissure de terminaison *f*; **~ joint box** PARTS boîte d'extrémité *f*, tête de câble *f*

terminator *n* PER APP terminateur *m*

terrain: **~ data** *n pl* COMP, DATA données relatives au terrain *f pl*; **~ and surface database** *n* COMP, DATA base de données des terrains et superficies *f*, base de données des terrains et surfaces *f*

terrestrial: **~ hypothetical reference circuit** *n* RADIO circuit fictif de référence pour système de terre *m*; **~ interface** *n* RADIO interface terrestre *f*, TRANS jonction au réseau terrestre *f*; **~ interface module** *n* (*TIM*) TRANS module de jonction au réseau terrestre *m*; **~ radiocommunication** *n* RADIO radiocommunication de terre *f*; **~ standard clock** *n* TRANS horloge terrestre *f*; **~ station** *n* RADIO station de terre *f*

test[1] *n* COMP, SIG vérification *f*, TEST essai *m*; **~ access** TEST point de mesure *m*; **~ access point** TEST point d'accès aux essais *m*, point d'essai *m*, point de mesure *m*, point de test *m*; **~ adaptor** TEST adaptateur d'essai *m*, adaptateur de test *m*; **~ alarm inhibition** TEST inhibition d'alarme de test *f*; **~ answer** TEST réponse à un essai *f*, réponse à un test *f*; **~ bar** SWG, TEST barre d'essai *f*; **~ bed** POWER, TEST banc d'essai *m*, banc de test *m*; **~ bench** POWER, TEST banc d'essai *m*, banc de test *m*; **~ block** TEST bloc d'essai *m*, bloc de test *m*; **~ board** ELECT, TEST carte d'essai *f*, carte de test *f*; **~ body** DATA, TEST corps d'essai *m*, corps de test *m*; **~ box** TEST boîte d'essai *f*, boîte de test *f*, boîtier de test *m*; **~ call indicator** TEST indicateur d'appel d'essai *m*; **~ call of the subscriber-to-subscriber type** TEST appel d'essai du type abonné à abonné *m*; **~ case** TEST *box* coffret d'essai *m*, *first attempt* test élémentaire *m*; **~ case error** TEST erreur de test élémentaire *f*; **~ case identifier** TEST identificateur de test élémentaire *m*; **~ case variable** TEST variable de test élémentaire *f*; **~ center** *AmE*, **~ centre** *BrE* TEST centre d'essai *m*, centre pour les essais *m*; **~ circuit** TELEP circuit d'essai *m*, TEST circuit d'essai *m*, circuit de test *m*; **~ clerk** TEST technicien d'essai *m*; **~ coordination procedure** (*TCP*) TEST procédure de coordination des essais *f*, procédure de coordination des tests *f*; **~ coupon** TEST éprouvette *f*; **~ customer** TEST client pilote *m*; **~ desk** TEST table d'essai *f*, table de test *f*; **~ duration** TEST durée d'essai *f*, durée de test *f*; **~ engineer** TEST technicien d'essai *m*; **~ engineering** TEST méthode d'essai *f*, technique d'essai *f*; **~ equipment** TEST équipement d'essai *m*, matériel test *m*; **~ event** DATA, TEST événement de test *m*; **~ group** DATA, TEST groupe de tests *m*; **~ group reference** DATA, TEST référence de groupe de tests *f*; **~ handset** TELEP, TEST combiné d'essai *m*, combiné de test *m*; **~ indicator** DATA, TEST indicateur d'essai *m*, indicateur de test *m*; **~ instructions** *n pl* TEST instructions d'essai *f pl*; **~ jack** ELECT, TELEP jack d'essai *m*, jack de test *m*; **~ jack strip** TEST série de jacks d'essai *f*, série de jacks de test *f*; **~ kit** TEST valise de tests *f*; **~ laboratory checklist** TEST liste de contrôle de laboratoire de test *f*; **~ line** TELEP ligne d'essai *f*; **~ load** TEST charge d'essai *f*; **~ loop** TEST boucle de mesure *f*; **~ management PDU** (*TM-PDU*) DATA, TEST PDU de gestion de test *m*; **~ management protocol** (*TMP*) DATA, TEST protocole de gestion de test *m*; **~ and measurement set** TEST banc de test et de mesure *m*; **~ menu** TEST menu d'essai *m*; **~ message** DATA, TEST

Internet envoi d'essai *m*; ~ **method** TEST méthode d'essai *f*, technique d'essai *f*; ~ **mode** TEST mode d'essai *m*; ~ **and operation bus** TELEP, TEST bus de test et d'exploitation *m*; ~ **and operation point** TELEP, TEST point de test et d'opération *m*; ~ **operator** DATA, TEST opérateur de test *m*; ~ **outcome** TEST résultat d'essai *m*, résultat de test *m*; ~ **pattern** TEST motif de contrôle *m*, TV *test card* mire *f*; ~ **piece** TEST éprouvette *f*, spécimen d'essai *m*; ~ **point** TEST point d'essai *m*, point de test *m*; ~ **position** TEST position d'essais et de mesures *f*, poste d'essai *m*; ~ **postamble** TEST épilogue d'essai *m*, épilogue de test *m*; ~ **preamble** TEST préambule d'essai *m*, préambule de test *m*; ~ **probe** TEST pointe de touche *f*; ~ **procedure** TEST procédure d'essai *f*, procédure de test *m*; ~ **prod** TEST pointe de touche *f*; ~ **rack** TEST bâti d'essai *m*, bâti de contrôle *m*, maquette *f*; ~ **realization** TEST réalisation d'essai *f*, réalisation d'un test *f*; ~ **realizer** TEST réalisateur d'essai *m*, réalisateur de tests *m*; ~ **record** TEST procès-verbal d'essai *m*, procès-verbal de test *m*; ~ **relay** PARTS, TEST relais d'essai *m*, relais de test *m*; ~ **result** TEST résultat d'essai *m*, résultat de test *m*; ~ **room** TEST salle d'essai et mesures *f*; ~ **setup** TEST montage de mesure *m*, montage d'essai *m*; ~ **signal** SIG, TEST signal d'essai *m*; ~ **stand** TEST banc d'essai *m*; ~ **step** TEST module de test *m*; ~ **step group** TEST groupe de modules de test *m*; ~ **step group reference** DATA, TEST référence d'un groupe de modules de test *f*; ~ **step identifier** TEST identificateur de module de test *m*; ~-**step library** TEST bibliothèque de modules de test *f*; ~ **suite** TEST suite de tests *f*; ~ **suite parameter** TEST paramètre de suite de tests *m*; ~ **suite structure** (*TSS*) TEST structure de suite de tests *f*; ~ **suite structure and test purposes** *n pl* (*TSS&TP*) TEST structure de suite de tests et objectifs de tests *m pl*; ~ **suite variable** TEST variable de suite de tests *f*; ~ **system** TEST système d'essai *m*, système de test *m*; ~ **technology** TEST méthode d'essai *f*, méthode de tests *f*, technique d'essai *f*, technique de tests *f*; ~ **traffic** TELEP, TEST, TRANS trafic d'essai *m*; ~ **transformer** TEST transformateur de mesure *m*; ~ **verdict** TEST verdict d'essai *m*, verdict de test *m*; ~ **wire** SWG, TELEP fil d'essai *m*, fil de test *m*

test[2] *vt* TEST *examine, investigate* vérifier; ~-**load** TEST soumettre à une charge d'essai; ~ **out** COMP, TEST vérifier

testability *n* TEST testabilité *f*

tester *n* TEST *equipment* appareil d'essai *m*, *person* testeur *m*

testing *n* TELEP, TEST vérification *f*; ~ **cycle** TEST cycle d'essai *m*; ~ **device** PER APP, TEST appareil d'essai *m*, dispositif d'essai *m*; ~ **equipment** PER APP, TEST appareil d'essai *m*, équipement d'essai *m*, équipement de test *m*; ~ **manager** TEST chef des essais *m*; ~ **point** SIG, SWG, TEST centre pour les essais de la signalisation de ligne *m*, centre pour les essais de la transmission *m*, TRANS point pour les essais de la signalisation de ligne *m*, point pour les essais de la transmission *m*; ~ **point line signaling** *AmE*, ~ **point line signalling** *BrE* SIG, TEST centre pour les essais de la signalisation de ligne *m*; ~ **point switching and interregister signaling** *AmE*, ~ **point switching and interregister signalling** *BrE* SWG, TEST commutation de point d'essai et signalisation inter-régistre *f*, commutation de point de test et signalisation inter-registre *f*; ~ **point transmission** TEST centre pour les essais de la transmission *m*, centre pour les tests de la transmission *m*; ~ **position** TELEP, TEST position d'essais et de mesures *f*, position de tests et de mesures *f*; ~ **program** *AmE*, ~ **programme** *BrE* TEST *series of tests* programme d'essai *m*, programme de test *m*; ~ **state** TEST état de test *m*; ~ **technique** TEST méthode d'essais *f*, technique d'essais *f*

testware *n* COMP, DATA logiciel d'essai *m*, logiciel de test *m*

tetrode *n* ELECT, PARTS tétrode *m*; ~ **field-effect transistor** ELECT, PARTS transistor à effet de champ tétrode *m*; ~ **transistor** ELECT, PARTS transistor tétrode *m*

texel *n* COMP texel *m*

text *n* COMP, DATA, TELEP texte *m*; ~ **block** COMP bloc de texte *m*; ~ **box** COMP zone de saisie *f*; ~ **editing** APPL, COMP édition de texte *f*; ~ **editor** COMP *program* programme de traitement de texte *m*, *person* éditeur de texte *m*; ~ **entry field** COMP zone de saisie *f*; ~ **file** DATA fichier texte *m*; ~ **mailbox** APPL, COMP boîte à lettres *f*; ~ **memory** COMP mémoire de textes *f*; ~ **message** COMP message alphanumérique *m*; ~ **processing** COMP traitement de texte *m*; ~ **search** DATA recherche textuelle *f*; ~ **space entry** COMP zone de texte *f*; ~ **string** TELEP chaîne de texte *f*

texture: ~ **element** *n* COMP texel *m*; ~ **mapping** *n* COMP placage de texture *m*

TFA *abbr* (*transfer allowed*) TELEP transfert autorisé *m*

TFAS *abbr* (*transmit-frame acquisition and synchronization procedure*) DATA, TRANS procédure d'acquisition et de synchronisation des trames d'émission *f*, procédure d'acquisition et de synchronisation des trames de transmission *f*

TFC *abbr* (*transfer controlled*) TELEP transfert sous contrôle *m*

TFEL *abbr* (*thin-film electroluminescent display*) ELECT, PARTS écran électroluminescent à couches minces *m*

TFM *abbr* (*tamed frequency modulation*) RADIO, TRANS MFA (*modulation de fréquence asservie*)

TFP *abbr* (*transfer prohibited*) TELEP transfert interdit *m*

TFR *abbr* (*transfer restricted*) TELEP transfert restreint *m*

TFT *abbr* ELECT (*thin-film transistor*), PARTS (*thin-film transistor*) transistor en couche mince *m*, TELEP (*telematic file transfer*) TFT (*transfert de fichier télématique*)

TG3 *abbr* (*telecopier group 3*) TELEP télécopieur du groupe 3 *m*

theme: ~ **channel** *n* TV canal thématique *m*

theoretical: ~ **duration** *n* TELEP durée théorique *f*; ~ **pitch diameter** *n* PARTS diamètre sur flanc *m*

thermal: ~ **agitation noise** *n* ELECT, PARTS bruit d'agitation thermique *m*, bruit thermique *m*; ~ **breakdown** *n* ELECT, PARTS claquage par emballement thermique *m*; ~ **capacitance** *n* ELECT, PARTS capacité thermique *f*; ~ **characteristic** *n* ELECT, PARTS caractéristique thermique *f*; ~ **conductivity** *n* ELECT, PARTS conductibilité thermique *f*; ~ **expansion** *n* ELECT, PARTS dilatation thermique *f*; ~ **expansion coefficient** *n* ELECT, PARTS coefficient de dilatation thermique *m*; ~ **instability** *n* ELECT, PARTS instabilité thermique *f*; ~ **noise** *n* ELECT, PARTS bruit thermique *m*; ~ **printer** *n* DATA, PER APP imprimante thermique *f*; ~ **residual voltage** *n* ELECT, PARTS tension thermique résiduelle *f*; ~ **resistance** *n* ELECT, PARTS résistance thermique *f*; ~ **runaway** *n* ELECT emballement thermique *m*

thermionic: ~ **emission** *n* ELECT émission thermionique *f*; ~ **emission efficiency** *n* ELECT efficacité d'émission thermionique *f*

thermistor *n* ELECT, PARTS thermistance *f*, thermisteur *m*, thermistor *m*; ~ **mount** PARTS, TEST sonde à thermistance *f*

thermocouple *n* ELECT, PARTS thermocouple *m*; ~ **wattmeter** TEST wattmètre à thermocouple *m*

thermoelectric: ~ **effect** *n* ELECT effet thermoélectrique *m*, effet Seebeck *m*

thermometer: ~ **probe** *n* CONTR capteur de température *m*, sonde pyrométrique *f*

thermosetting: ~ **plastic** *n* PARTS plastique thermodurcissable *m*

thermostat *n* CLIM, CONTR, PARTS thermostat *m*

Thévenin: ~ **equivalent circuit** *n* POWER circuit équivalent de Thévenin *m*, générateur de Thévenin *m*; ~**'s theorem** *n* POWER théorème de Thévenin *m*

thick: ~ **coax** *n* PARTS, TRANS coax gros *m*; ~ **coaxial cable** *n* TRANS câble coaxial gros *m*; ~-**film integrated circuit** *n* ELECT circuit intégré à couches épaisses *m*; ~-**film resistor** *n* ELECT, PARTS résistance à couches épaisses *f*; ~-**layer integrated circuit** *n* ELECT circuit intégré à couches épaisses *m*

thin: ~ **coax** *n* PARTS, TRANS coax fin *m*; ~ **coaxial cable** *n* TRANS câble coaxial fin *m*; ~ **Ethernet** *n* ELECT câble Ethernet noir *m*, Ethernet noir *m*; ~-**film capacitor** *n* ELECT, PARTS condensateur à couche mince *m*; ~-**film electroluminescent display** *n* (*TFEL*) ELECT, PARTS écran électroluminescent à couche mince *m*; ~-**film resistor** *n* ELECT, PARTS résistance à couche mince *f*; ~-**film transistor** *n* (*TFT*) ELECT, PARTS transistor en couche mince *m*; ~-**film waveguide** *n* DATA, TRANS guide d'ondes en couche mince *m*; ~-**layer capacitor** *n* ELECT, PARTS condensateur à couche mince *m*; ~-**route network** *n* TRANS réseau à faible trafic *m*; ~-**route TDMA** *n* RADIO AMRT à faible trafic *f*; ~ **wall counter tube** *n* RADIO tube-compteur à paroi mince *m*; ~-**wire Ethernet multiport repeater** *n* DATA, ELECT répéteur multiport Ethernet thin-wire *m*

thinnet: ~ **tap** *n* ELECT, PARTS auto-shunt *m*

third: ~-**angle projection** *n* GEN projection américaine *f*; ~ **harmonic distortion** *n* (*3HD*) ELECT distorsion de la troisième harmonique *f*; ~ **party** *n* TELEP tiers *m*; ~-**party billing** *n* TELEP facturation d'un tiers *f*; ~-**party billing call** *n* TELEP conversation payable par un tiers *f*; ~-**party warning tone** *n* TELEP tonalité d'entrée d'un tiers *f*

32: ~-**bit processor unit** *n* COMP unité processeur 32 bits *f*

Thomson: ~ **effect** *n* GEN effet Thomson *m*

thoriated: ~-**tungsten cathode** *n* ELECT, PARTS cathode en tungstène thorié *f*

thread *n* PARTS filet *m*, filetage *m*; ~ **gage** *AmE*, ~ **gauge** *BrE* TEST calibre de filetage *m*

threaded: ~ **bushing** *n* ELECT, PARTS bague filetée *f*; ~ **coupler** *n* ELECT, PARTS coupleur fileté *m*

threading: ~ **ring** *n* PARTS tore *m*

three[1]: ~-**phase** *adj* POWER triphasé

three[2]: ~-**coil relay** *n* PARTS relais à trois bobines *m*; ~-**condition cable code** *n* TELEP, TRANS code de câble trivalent *m*; ~-**condition telegraph code** *n* TELEP, TRANS code télégraphique trivalent *m*; ~-**conductor plug** *n* PARTS fiche tripolaire *f*; ~-**dimensional guide** *n* TRANS guide à trois dimensions *m*, guide tridimensionnel *m*; ~-**dimensional image** *n* (*3D image*) TV image en relief *f*, image en trois dimensions *f*, image tridimensionnelle *f*; ~-**gang condenser** *n* POWER condensateur à trois cages *m*; ~-**party conference** *n* APPL, TELEP conférence à trois *f*; ~-**party service** *n* APPL, TELEP service comportant un troisième correspondant *m*; ~-**phase circuit** *n* POWER circuit triphasé *m*; ~-**phase rectifier** *n* PARTS, POWER redresseur triphasé *m*; ~-**way call** *n* PER APP, RADIO, TELEP *m*, communication à trois *f*, conversation à trois *f*; ~-**way call chaining** *n* TELEP conférence à trois en chaîne *f*; ~-**way calling** *n* PER APP, RADIO, TELEP conversation à trois *f*; ~-**way conversation** *n* PER APP, RADIO, TELEP conversation à trois *f*; ~-**way teleconference** *n* APPL, TELEP téléconférence à trois *f*; ~-**wire line** *n* SWG, TELEP ligne à trois fils *f*

3D: ~ **image** *n* (*three-dimensional image*) TV image en relief *f*, image en trois dimensions *f*, image tridimensionelle *f*

3HD *abbr* (*third harmonic distortion*) ELECT distorsion de la troisième harmonique *f*

threshold *n* ELECT, POWER, RADIO, TRANS seuil *m*; ~ **of audibility** RADIO, TELEP seuil d'audition *m*; ~ **current** ELECT, POWER, TRANS courant de seuil *m*, seuil de perception du courant *m*; ~ **current density** ELECT, POWER, TRANS densité du courant de seuil *f*; ~ **decoding** DATA décodage à seuil *m*; ~ **of discomfort** RADIO, TELEP *hearing level* seuil de gêne *m*; ~ **element** ELECT élément à seuil *m*; ~ **extension demodulator** ELECT, RADIO démodulateur à abaissement de seuil *m*, démodulateur à seuil amélioré *m*; ~ **frequency** ELECT, TRANS fréquence critique *f*, fréquence de seuil *f*; ~ **function** ELECT fonction de seuil *f*; ~ **gate** ELECT porte à seuil *f*; ~ **of hearing** TELEP seuil d'audition *m*; ~ **operation** ELECT, RADIO opération de seuil *f*; ~ **voltage** ELECT, POWER tension de seuil *f*; ~ **wavelength** ELECT, TRANS seuil de longueur d'onde *m*

thresholding *n* ELECT, POWER imposition d'un seuil *f*

throat *n* RADIO gorge de raccordement *f*, gorge *f*; ~ **microphone** RADIO, TELEP, TRANS laryngophone *m*

throttle: ~ **valve** *n* CLIM papillon *m*, soupape d'admission *f*

through[1]: ~ **connection** *n* ELECT, POWER connexion transversale *f*, TELEP passage en connexion *m*, transfert *m*; ~-**connection delay** *n* TELEP durée d'établissement d'une connexion transversale *f*; ~-**connection hole** *n* ELECT trou de connexion transversale *m*; ~-**dialing number** *AmE*, ~-**dialling number** *BrE* *n* TELEP numéro d'appel direct *m*; ~-**group** *n* TRANS transfert de groupe primaire *m*; ~-**group connection point** *n* TELEP point de transfert de groupe quaternaire *m*; ~ **hypergroup** *n* TRANS transfert de groupe tertiaire *m*; ~ **line** *n* TELEP, TRANS ligne de transfert *f*; ~ **mastergroup** *n* TRANS transfert de groupe tertiaire *m*; ~-**mastergroup connection point** *n* TELEP point de transfert de groupe tertiaire *m*; ~ **supergroup** *n* TRANS transfert de groupe secondaire *m*; ~-**supergroup connection point** *n* TELEP point de transfert de groupe secondaire *m*; ~-**transmission loss** *n* TELEP, TRANS affaiblissement de transmission directe *m*

through[2]: ~-**connect** *vt* SWG connecter directement

through-15-supergroup: ~ **assembly connection point** *n* TELEP point de transfert d'assemblage de 15 groupes secondaires *m*

throughput *n* COMP, DATA débit *m*, RADIO capacité utile *f*; ~ **class** DATA classe de débit *f*

Thruline: ~ **wattmeter** *n* TEST wattmètre Thruline *m*

thruput *AmE see throughput*

thumbnail *n* COMP *graphics* ongle de pouce *m*

thumbscrew *n* ELECT, PARTS vis à ailettes *f*, vis moletée *f*

thumbwheel *n* ELECT molette *f*; ~ **switch** ELECT, SWG roue codeuse *f*

thyatron *n* ELECT, PARTS thyatron *m*

thyristor *n* ELECT, PARTS, POWER thyristor *m*; ~ **rectifier unit** POWER redresseur à thyristor *m*

tick: ~-**mark label** *n* COMP étiquette de graduation *f*; ~ **tone** *n* TELEP tonalité d'intervention *f*, tonalité d'intrusion *f*, tonalité de tic-tac *f*; ~-**tone generator** *n* TELEP générateur de tonalité de tic-tac *m*, générateur de tonalité d'intervention *m*, générateur de tonalité d'intrusion *m*

ticker: ~ **tone** *n* TELEP tonalité d'intervention *f*, tonalité d'intrusion *f*, tonalité de tic-tac *f*

ticketing *n* TELEP taxation *f*

tie: ~ **bar** *n* PARTS, TELEP profilé de jonction *m*; ~ **button** *n* COMP touche de liaison *f*; ~-**circuit interface** *n* TRANS équipement de ligne interautomatique *m*; ~ **line** *n* SWG, TELEP, TRANS ligne privée *f*; ~-**line traffic** *n* SWG, TELEP, TRANS trafic de ligne privée *m*; ~ **trunk** *n* TELEP circuit de jonction *m*; ~ **wrap** *n* PARTS, POWER attache autoblocante *f*, ceinture de câble *f*

TIFF *abbr* (*tagged-image file format*) COMP, DATA TIFF (*tagged-image file format*)

tight: ~ **buffer** *n* TRANS gainage intermédiaire serré *m*; ~ **jacket** *n* TRANS gainage serré *m*; ~-**jacketed cable** *n* TRANS câble à fibres enrobées *m*, câble à gainage serré *m*

tighten *vt* CONTR *conditions* renforcer, PARTS bloquer, serrer, RADIO, TEST *conditions* renforcer

tightened[1] *adj* PARTS bloqué, serré

tightened[2]: ~ **inspection** *n* CONTR, TEST contrôle renforcé *m*

tightening *n* CONTR renforcement *m*, PARTS resserrement *m*, RADIO, TEST *of conditions* renforcement *m*

tilde *n* COMP tilde *m*

tilt *n* RADIO inclinaison *f*

tilted: ~ **cable television** *n* (*tilted CATV*) TV système de câblodistribution à courbe de réponse inclinée *m*; ~ **CATV** *n* (*tilted cable television*) TV système de câblodistribution à courbe de réponse inclinée *m*; ~ **front** *n* ELECT, PARTS face avant en retrait *f*

TIM *abbr* TEST (*time interval meter*) chronomicromètre *m*, TRANS (*terrestrial interface module*) module de jonction au réseau terrestre

timbre *n* COMP, TELEP timbre *m*; ~ **envelope** COMP enveloppe de timbre *f*; ~ **modulation** COMP modulation de timbre *f*

time[1]: **~-constant** *adj* TRANS invariant dans le temps; **~-divided** *adj* TRANS à répartition dans le temps

time[2] *n* GEN temps *m*, durée *f*; **~-assigned speech interpolation** (*TASI*) SIG, TRANS interpolation temporelle des signaux vocaux *f* (*système TASI*); ~ **assignment speech concentration** TRANS concentration des conversations avec affectation temporelle *f*; ~ **axis** TEST, TRANS axe de temps *m*; ~ **balance** TEST synchronisation *f*; ~ **base** COMP, ELECT base de temps *f*, base temporelle *f*; **~-base generator** ELECT générateur de base de temps *m*, oscillateur de base de temps *m*, oscillateur de commande de balayage *m*; **~-base jitter** ELECT instabilité de la base de temps *f*; ~ **between failures** COMP, TEST temps de bon fonctionnement *m*, temps entre défaillances *m*; ~ **characteristic** COMP, ELECT, TEST, TRANS caractéristique temporelle *f*; ~ **charging** TELEP tarification à la durée *f*; ~ **code** COMP, DATA chronocode *m*, code temporel *m*, RADIO code horaire *m*; ~ **coefficient** TEST coefficient de temps *m*; ~ **comparison** RADIO, TRANS comparaison de temps *f*, comparaison temporelle *f*; ~ **compression** COMP, DATA compression de temps *f*, compression temporelle *f*; ~ **congestion** RADIO, TELEP congestion temporelle *f*; **~-consistent busy hour** TELEP heure chargée moyenne *f*; ~ **constant** COMP, ELECT, TEST, TRANS constante de temps *f*, constante temporelle *f*; **~-coordinate system** COMP système de coordonnées temporelles *m*; ~ **correlation** DATA, SWG corrélation temporelle *f*; **~-critical printing application** APPL application d'impression à durée critique *f*; ~ **delay** DATA, ELECT temporisation *f*; **~-delay circuit** ELECT circuit à retardement *m*, circuit temporisé *m*; **~-delay relay** ELECT, PARTS relais temporisé *m*, relais à retardement *m*; **~-derived channel** TRANS voie dérivée en temps *f*; ~ **dispersion** COMP, ELECT, TEST, TRANS dispersion des temps *f*, dispersion temporelle *f*; ~ **displacement** COMP, ELECT, TEST, TRANS écart de temps *m*; ~ **and distance metering** TELEP taxation à la distance et à la durée *f*; ~ **diversity** DATA, RADIO diversité temporelle *f*; **~-diversity reception** DATA, RADIO réception en diversité temporelle *f*; ~ **division** RADIO, SIG, SWG, TELEP, TRANS répartition temporelle *f*, répartition de temps *f*; **~-division exchange** SWG central temporel *m*; **~-division highway** SWG, TELEP canal à multiplexage dans le temps *m*; **~-division highway in switching** SWG, TELEP canal à multiplexage dans le temps en commutation *m*; **~-division multiple access** (*TDMA*) RADIO, SIG, TELEP, TRANS accès multiple à répartition dans le temps *m* (*AMRT*); **~-division multiple access burst time plan** RADIO, SIG, TELEP, TRANS plan d'émission en accès multiple à répartition dans le temps *m*, plan d'émission en AMRT *m*; **~-division multiple access mode** (*TDMA mode*) RADIO, SIG, TELEP, TRANS mode d'accès multiple à répartition dans le temps *m* (*mode AMRT*); **~-division multiplex** (*TDM*) SIG, SWG multiplexage par répartition dans le temps *m* (*MRT*), TRANS multiplexage temporel *m*; **~-division multiplex transmitter** COMP, TELEP, TRANS émetteur multiplex à répartition dans le temps *m*; **~-division multiplexer** (*TDM*) SIG, SWG multiplexeur temporel *m*; **~-division multiplexing** TELEP multiplexage temporel *m*; **~-division network** SWG, TRANS réseau de connexion temporelle *m*; **~-division switching** DATA, SWG, TELEP, TRANS commutation par répartition dans le temps *f*, commutation temporelle *f*; **~-division switching system** DATA, SWG, TELEP, TRANS système de commutation temporelle *m*; **~-division system** DATA, SWG, TELEP, TRANS système temporel *m*; ~ **domain** GEN dimension temporelle *f*; ~ **domain analysis** DATA analyse de la dimension temporelle *f*; **~-domain characteristics** *n pl* TEST caractéristiques dans le domaine temporel *f pl*; ~ **domain filter** ELECT filtre à réponse temporelle *m*, filtre temporel *m*; ~ **domain reflectometer** (*TDR*) ELECT, POWER, TEST, TRANS réflectomètre temporel *m*; ~ **domain reflectometry** (*TDR*) ELECT, POWER, TEST, TRANS réflectométrie temporelle *f*; ~ **insertion unit** DATA unité d'insertion de temps *f*; ~ **interval error** TELEP erreur sur la durée *f*; ~ **interval meter** (*TIM*) TEST chronomicromètre *m*; ~ **marker** RADIO repère de temps *m*; ~ **match** TEST synchronisation *f*; ~ **measurement** TELEP, TEST mesure de temps *f*, synchronisation *f*; ~ **measurement pulse** TELEP impulsion de synchronisation *f*; ~ **meter** APPL, TEST compteur horaire *m*; ~ **multiplex switching** SWG commutation temporelle multiplexée *f*; ~ **multiplexing** COMP, TELEP, TRANS multiplexage dans le temps *m*, multiplexage temporel *m*; **~-out** DATA, ELECT, RADIO, TELEP dépassement de temps imparti *m*, temporisation *f*; **~-out device** PARTS dispositif de temporisation *m*; **~-out disable facility** ELECT dispositif d'inhibition de temporisation *m*; **~-out event** DATA événement de fin de temporisation *m*; **~-out program** COMP programme de temporisation *m*; **~-out supervision** DATA, ELECT, RADIO, TELEP surveillance de temporisation *f*; ~ **program** COMP programme de temporisation *m*, programme temporel *m*; **~-quantized control** CONTR, TELEP mode à quantification d'amplitude *m*; ~ **queue** DATA, ELECT, RADIO, TELEP file d'attente temporelle *f*; **~-related charging** TELEP tarification à la durée *f*; **~-related tariffing** TELEP tarification à la durée *f*; ~ **relay** PARTS relais temporisé *m*; ~ **release** DATA, ELECT, RADIO, TELEP temporisation *f*; ~ **response** CONTR réponse temporelle *f*; ~ **sequence** GEN séquence temporelle *f*; ~ **sharing** COMP, DATA partage de temps *m*, temps partagé *m*; **~-sharing program** COMP, DATA programme en temps partagé *m*; **~-sharing system** COMP, DATA système à temps partagé *m*; ~ **signal** RADIO, SIG signal horaire *m*; **~-signal control** PER APP *on answer machine* pilotage par signal horaire *m*; ~ **signal receiver** RADIO, SIG récepteur de signaux horaires *m*; ~ **signal satellite service** RADIO, SIG service de signaux horaires par satellite *m*; ~ **signature** COMP indication de la mesure *f*; ~ **slicing** COMP découpage du temps *m*; ~ **slot** RADIO, SIG, SWG, TRANS intervalle de temps *m* (*IT*), tranche horaire *f*, créneau temporel *m*, TV *of programme* créneau horaire *m*; **~-slot code** SIG code d'intervalle de temps *m*, code de tranche horaire *m*, code de créneau temporel *m*; ~ **slot interchange** SWG échange entre intervalles de temps *m*; ~ **slot interface** TRANS interface de tranche de temps *f*; ~ **slot sequence integrity** TELEP, TRANS intégrité de la séquence des intervalles de temps *f*; ~ **slot structure** RADIO structure de trame *f*; **~-space-time network** SWG, TRANS réseau temporel-spatial-temporel *m*; ~ **span** GEN période *f*; ~ **stage** (*T stage*) SWG, TELEP étage temporel *m* (*étage T*); ~ **stamp** COMP, SWG, TEST *device* horodateur *m*, *marking* estampille *f*, marque d'horodatage *f*; ~ **standard** RADIO étalon de temps *m*; ~ **step** RADIO saut de temps *m*; ~ **supervision** TELEP surveillance de temps *f*; ~ **supervision flip-flop** PARTS, TELEP bascule de temporisation *f*; ~ **supervision of a process PE** TELEP élément graphique de temporisateur de contrôle d'un *m*; ~ **switch** SWG autocommutateur temporel *m*, commutateur temporel *m*, commutateur T , minuterie de compteur *f*; ~ **switching** DATA, SWG, TELEP, TRANS commutation temporelle *f*; ~ **tag command** CONTR télécommande temporelle prédéterminée *f*; ~ **to failure** TEST sans panne *f*, temps de fonctionnement jusqu'à la défaillance *m*; ~ **to first failure** COMP temps de fonctionnement avant la première défaillance *m*; ~ **to half value** TEST durée jusqu'à la mi-valeur *f*; ~ **to saturation** TEST temps jusqu'à la saturation *m*; ~ **unit** GEN unité de temps *f*, unité temporelle *f*; **~-variant amplifier** COMP amplificateur à variation temporelle *m*; **~-variant filter** COMP filtre à variation temporelle *m*; **~-zone metering** TELEP taxation à la distance et à la durée *f*

time[3] *vt* COMP, DATA, RADIO, TELEP, TEST chronométrer, minuter; **~ out** DATA, ELECT, RADIO, TELEP temporiser le déclenchement de

time[4]: **~-out** *vi* DATA, ELECT, RADIO, TELEP dépasser le temps imparti

time[5]: **~-space-time** *phr* (*TST*) SWG temps-espace-temps

timed: **~ call** *n* TELEP communication taxée à la durée *f*; **~ reminder** *n* PER APP, TELEP rappel temporisé *m*

timer *n* ELECT circuit de temporisation *m*, temporisateur *m*, GEN minuterie *f*, minuteur *m*, temporisateur *m*; **~ relay** ELECT, PARTS relais temporisé *m*

timescale *n* COMP, RADIO, TRANS échelle de temps *f*, échelle temporelle *f*, horizon temporel *m*, trame temporelle *f*; **~ difference** COMP, RADIO, TRANS différence entre échelles de temps *f*; **~ in synchronisation** RADIO échelle de temps en synchronisme *f*; **~ reading** COMP, TRANS lecture d'une échelle de temps *f*; **~ unit** COMP, RADIO, TRANS unité d'une échelle de temps *f*

timetable: **~ slot** *n* RADIO, TV créneau horaire *m*

timing: **~ chart** *n* GEN chronogramme *m*; **~ circuit** *n* ELECT circuit de temporisation *m*, GEN circuit d'horloge *m*; **~ control** *n* COMP, CONTR commande de synchronisation *f*; **~ correction device** *n* COMP dispositif de correction de synchronisation *m*; **~ diagram** *n* GEN chronogramme *m*; **~ generator** *n* TRANS générateur de rythme *m*; **~ information** *n* DATA, TRANS information de rythme *f*, informations de synchronisation *f pl*; **~ jitter** *n* TRANS gigue de rythme *f*; **~ pulse** *n* ELECT, TELEP impulsion d'horloge *f*, impulsion de synchronisation *f*; **~ recovery** *n* TELEP, TRANS récupération de rythme *f*; **~-recovery circuit** *n* ELECT, TRANS circuit de récupération de rythme *m*; **~ relationship** *n* TRANS relation temporelle *f*; **~ relay** *n* ELECT, PARTS relais temporisé *m*; **~ selection** *n* TRANS réglage du rythme *m*, sélection du rythme *f*; **~ signal** *n* SIG signal de commande temporelle *m*, signal de rythme *m*, signal de synchronisation *m*

tin[1]: **~-electroplating** *n* PARTS étamage électrolytique *m*; **~-lead coating** *n* PARTS recouvrement étain-plomb *m*; **~-plated conductor** *n* ELECT, PARTS conducteur étamé *m*; **~-plated copper braid shield** *n* ELECT, PARTS blindage à tresse en cuivre étamé *m*; **~-plated wire** *n* ELECT, GEN, PARTS fil étamé *m*; **~-plating** *n* PARTS étamage *m*; **~ stripping** *n* PARTS désétamage *m*

tin[2]: **~-electroplate** *vt* PARTS étamer par électrolyse; **~-plate** *vt* PARTS étamer

tinning *n* PARTS étamage *m*

tinsel: **~ conductor** *n* ELECT, PARTS fil rosette *m*

tip *n* ELECT queusot *m* PARTS *wire* fil de pointe *m*, TELEP pointe de touche *f*, tête *f*; **~ node** COMP feuille *f*

titanium: **~ waveguide channel** *n* DATA, RADIO, TRANS guide d'onde en titane diffusé *m*

title *n* COMP titre *m*; **~ bar** COMP barre de titre *f*; **~ page** COMP page de titre *f*

T-junction *n* RADIO jonction en T *f*

TKO *abbr* (*trunk offer*) SWG, TELEP intervention ligne-réseau *f*

TLMA *abbr* (*telematic agent*) DATA, TELEP agent télématique *m*

TLX *abbr* (*telex*) DATA, RADIO, TELEP TLX (*télex*)

TLXAU *abbr* (*telex access unit*) DATA, RADIO, TELEP unité d'accès aux services télex *f*, unité d'accès télex *f*

TM *abbr* TRANS (*transverse magnetic*) TM (*transverse magnétique*); **~ mode** *n* RADIO, TRANS mode TM *m*

TMN *abbr* (*telecommunications management network*) DATA, TELEP, TRANS RGT (*réseau de gestion des télécommunications*)

TMP *abbr* (*test management protocol*) DATA, TEST protocole de gestion de test *m*

TM-PDU *abbr* (*test management PDU*) DATA, TEST PDU de gestion de test *m*

TMUX *abbr* (*transmultiplexer*) ELECT, TELEP, TRANS transmultiplexeur *m*

TMUX-P *abbr* (*type P transmultiplexer*) TELEP transmultiplexeur de type P *m*

TMUX-S *abbr* (*type S transmultiplexer*) TELEP transmultiplexeur de type S *m*

TNC *abbr* (*terminal node controller*) TRANS contrôleur de nœud terminal *m*

TO *abbr* (*telegraphic office*) TELEP bureau télégraphique *m*, service télégraphique *m*

TOA *abbr* (*type of address*) DATA, TELEP type d'adresse *m*

toggle[1]: **~ menu** *n* COMP menu à bascule *m*; **~ switch** *n* COMP, ELECT, POWER, SWG, TELEP interrupteur à bascule *m*, interrupteur à rupture brusque *m*, inverseur à bascule *m*

toggle[2] *vt* COMP, ELECT, POWER, SWG, TELEP basculer

toggling *n* COMP, ELECT, POWER, SWG, TELEP basculement *m*

token: **~ passing** *n* DATA, TRANS passage de jeton *m*, transmission du paquet d'accès *f*; **~ ring** *n* COMP, DATA, TRANS anneau à jeton *m*, réseau à jeton *m*; **~ ring network** *n* COMP, DATA, TRANS réseau en anneau à passage jeton *m*

tolerance *n* TEST tolérance *f*; **~ limit** TEST limite de tolérance *f*; **~ parameter** TEST paramètre à tolérance *m*; **~ range** TEST zone de tolérance *f*; **~ zone** TEST zone de tolérance *f*

toll[1]: **~-free** *adj* COMP exempt de droits, TELEP gratuit, sans frais

toll[2]: **~ call** *n AmE* (*cf trunk call BrE*) TELEP appel à longue distance *m*, appel interurbain *m*, communication longue distance *f*, communication interurbaine *f*; **~-call barring** *n AmE* (*cf trunk-call barring BrE*) TELEP restriction d'appel interurbain *f*; **~-free call** *n AmE* (*cf freephone call BrE*) TELEP appel avec tarification au demandé *m*, appel gratuit *m*, appel numéro vert *m* (*Fra*), appel sans frais *m* (*Can*); **~-free number** *n AmE* (*cf freephone number BrE*) TELEP numéro d'appel sans frais *m* (*Can*), numéro vert *m* (*Fra*); **~-free service** *n AmE* (*cf freephone service BrE*) TELEP service "libre-appel" *m*; **~ switch** *n* SWG commutateur interurbain *m*; **~ system** *n* APPL système de péage *m*; **~ ticket** *n* TELEP ticket de taxation *m*; **~ ticketing** *n* TELEP taxation détaillée par destination *f*

TON *abbr* (*type of number*) DATA, TELEP type de numéro *m*

tonal *adj* COMP, PER APP, TELEP tonal

tone[1]: **~-pulse** *adj* PER APP, TELEP à fréquence vocale

tone[2] *n* GEN signal *m*, sonorité *f*, ton *m*, tonalité *f*, PER APP tonalité *f*, *of answer machine* bip *m*, bip sonore *m*, TELEP tonalité *f*; **~ above band** RADIO tonalité au-dessus de la bande *f*, tonalité suprabande *f*; **~-and-ringing set** SIG, TELEP générateur de signaux d'appel *m*; **~ in band** RADIO tonalité intrabande *f*, tonalité à l'intérieur de la bande *f*, tonalité dans la bande *f*; **~ below band** RADIO tonalité infrabande *f*, tonalité au-dessous de la bande *f*; **~ call** RADIO, TELEP appel par tonalité *m*; **~ code sender** TELEP contact d'interruption de tonalité *m*; **~-coded carrier** RADIO porteuse à fréquence vocale *f*; **~-coded information** TELEP information de fréquence vocale *f*; **~ color** *AmE*, **~ colour** *BrE* COMP timbre *m*; **~ control** COMP, CONTR commande de tonalité *f*; **~ dialing** *AmE*, **~ dialling** *BrE* TELEP numérotation en fréquences vocales *f*; **~ generation** COMP, ELECT génération de tonalité *f*; **~ generator** COMP, ELECT générateur de tonalité *m*; **~ message** TELEP information de fréquence vocale *f*; **~-off condition** COMP, ELECT, RADIO, TELEP état absence de fréquence *m*; **~-on condition** COMP, ELECT, RADIO, TELEP état présence de fréquence *m*; **~ on hold** TELEP tonalité de garde *f*; **~ pager** GEN RMU bip *f*, RADIO bip *m*; **~-plus-voice paging** RADIO bip vocal *m*; **~ ringer** TELEP sonnerie électronique *f*, sonnerie monolithique *f*, sonnerie par tonalité *f*; **~ ringing** TELEP signal d'appel audible *m*, sonnerie par tonalité *f*; **~ signal** SIG signal bip *m*; **~ signaling** *AmE*, **~ signalling** *BrE* SIG, TELEP signalisation multifréquence *f*; ~ **squelch system** RADIO silencieux à commande par tonalités *m*; ~ **wedge** TELEP échelle de nuances *f*

T1: ~ **carrier** *n* (*T1C*) TRANS système multiplex T1 *m*, entreprise de télécommunication de type 1 *f*; ~ **digital carrier** *n* TRANS système multiplex numérique *m*

T1C *abbr* (*T1 carrier*) TRANS système multiplex T1 *m*, système multiplex numérique T1 *m*, entreprise de télécommunication de type 1 *f*

tones: ~ **and auxiliaries group** *n* TELEP groupe des tonalités et auxiliaires *m*

tool: ~ **palette** *n* COMP, DATA palette d'outils *f*

toolbox *n* COMP, DATA palette d'outils *f*

tooling: ~ **hole** *n* ELECT trou de positionnement *m*; ~ **pin** *n* ELECT, PARTS plot de positionnement *m*; ~ **radius** *n* ELECT, PARTS rayon d'outil *m*

top: ~ **capacitive loading** *n* RADIO charge capacitive terminale *f*; ~ **field** *n* TV trame supérieure *f*; ~ **MCS provider** *n* DATA, TELEP fournisseur supérieur du service MCS *m*; ~ **node** *n* COMP racine *f*, tête *f*; ~ **panel** *n* COMP panneau supérieur *m*

topic *n* COMP, DATA *Internet search* sujet *m*

topocentric: ~ **angle** *n* RADIO angle topocentrique *m*

topology *n* ELECT topologie *f*

torn: ~**-tape relay** *n* TELEP, TRANS transit manuel par bande perforée *m*

toroidal: ~ **aerial** *n* RADIO antenne toroïdale *f*; ~ **antenna** *n* RADIO antenne toroïdale *f*; ~ **core** *n* ELECT noyau torique *m*

torque: ~ **differential receiver** *n* POWER récepteur différentiel de téléchauffage *m*

torsion: ~ **head wattmeter** *n* TEST wattmètre à tête de torsion *m*, wattmètre à pendule de torsion *m*

Torx: ~ **screwdriver** *n* PARTS tournevis Torx *m*; ~ **system** *n* GEN empreinte Torx *f*

total: ~ **anode-power input** *n* ELECT, POWER puissance d'entrée anodique totale *f*; ~ **charge** *n* APPL, COMP, RADIO, TELEP coût total de la communication *m*; ~ **configuration** *n* SWG configuration totale *f*; ~ **continuous spectrum noise** *n* RADIO souffle *m*; ~ **holding time** *n* TELEP durée totale d'occupation *f*; ~ **internal reflection** *n* ELECT, TRANS réflexion interne totale *f*; ~ **leakage power** *n* ELECT, POWER puissance totale de fuite *f*; ~ **level** *n* COMP niveau d'ensemble *m*; ~ **loss** *n* RADIO, TV *of link* affaiblissement global *m*; ~ **plate-power input** *n* ELECT, POWER puissance d'entrée anodique totale *f*; ~ **quality management** *n* (*TQM*) TEST gestion totale de qualité *f*; ~ **reflection** *n* ELECT, TRANS réflexion totale *f*; ~ **reflector current** *n* ELECT courant total de réflecteur *m*; ~ **scanning line length** *n* TELEP longueur totale de ligne *f*; ~ **starting time** *n* ELECT durée totale de démarrage *f*; ~ **transmission time** *n* TELEP, TRANS temps total de transmission *m*

touch: ~ **control** *n* COMP, PER APP touche à effleurement *f*; ~ **response** *n* COMP, PER APP réponse sensitive *f*; ~ **screen** *n* COMP écran tactile *m*; ~**-screen kiosk** *n* COMP borne interactive *f*; ~**-tone telephone** *n* PER APP, TELEP téléphone à clavier *m*, téléphone touch-tone *m*

tower: ~ **body** *n* POWER fût de pylône *m*

TP *abbr* DATA (*transaction processing*) traitement transactionnel *m*, TELEP (*typical prediction*) prédiction type *f*

TPA *abbr* (*telematic protocol architecture*) TELEP TPA (*architecture de protocole télématique*)

TPASE *abbr* (*transaction processing application service element*) DATA élément de service d'application pour le traitement des transactions *m*

TPDU *abbr* (*transport protocol data unit*) DATA unité de données de protocole de transport *f*

TPE *abbr* (*transmission path endpoint*) DATA, TRANS extrémité de voie de transmission *f*

TPI *abbr* (*tracks per inch*) COMP, DATA ppp (*pistes par pouce*)

TPIWF *abbr* (*telex packet interworking function*) DATA FITMP (*fonction d'interfonctionnement télex-mode paquet*)

TPSP *abbr* (*transaction processing service provider*) DATA fournisseur de service de traitement transactionnel *m*

TPSU *abbr* (*transaction processing service user*) DATA utilisateur de service de traitement transactionnel *m*; ~ **invocation** *n* DATA appel de TPSU *m*; ~ **title** *n* DATA titre de TPSU *m*

TPSUI *abbr* (*transaction processing service user invocation*) DATA appel de TPSU *m*

TQM *abbr* (*total quality management*) TEST gestion totale de qualité *f*

TR *abbr* DATA, ELECT, RADIO, TELEP, TRANS (*tributary*) tributaire *m*, (*transmitter-receiver*) TR (*émetteur-récepteur*); ~ **cell** *n* RADIO verrou de récepteur *m*; ~ **switch** *n* RADIO, SWG, TRANS duplexeur *m*; ~ **tube** *n* ELECT, RADIO tube TR *m*

TRAC *abbr* (*technical recommendations applications committee*) GEN comité d'application des recommandations techniques *m*

trace[1] *n* ELECT, SWG trace *f*; ~ **interrupt** SWG interruption de traçage *f*; ~ **level** SWG niveau de dépistage *m*, niveau de traçage *m*; ~ **unit** SWG unité de dépistage *f*, unité de traçage *f*

trace[2] *vt* ELECT tracer, RADIO suivre, SWG dépister, tracer, TEST tracer; ~ **a call** TELEP déterminer l'origine d'un appel, déterminer l'origine d'une communication; ~ **a fault** DATA, TELEP relever un dérangement, TEST localiser une défaillance, relever un dérangement

traceability *n* ELECT, SWG, TEST traçabilité *f*

traceable: ~ **tone** *n* ELECT tonalité dépistable *f*

tracer: ~ **bit** *n* SWG bit de traçage *m*

tracing *n* ELECT traçage *m*, RADIO suivi *m*, SWG traçage *m*, TEST traçage *m*; ~ **function** RADIO fonction de localisation *f*, fonction de suivi *f*, SWG fonction de dépistage *f*, fonction de traçage *f*, TEST fonction de traçage *f*

track[1] *n* COMP, DATA, RADIO, TV piste *f*; ~**-at-once recording** COMP enregistrement piste par piste *m*; ~ **density** COMP densité de piste *f*; ~**-descriptor sector** COMP secteur descripteur de piste *m*; ~ **mix** COMP mixage de pistes *m*; ~ **number** COMP numéro de piste *m*; ~ **pitch** COMP pas de piste *m*, pas transversal *m*; ~ **production** RADIO production des pistes *f*; ~ **type** COMP, DATA, RADIO, TV type de piste *m*

track[2] *vt* ELECT, GEN, RADIO, TV suivre

trackball *n* COMP boule de commande *f*, boule roulante *f*

tracker *n* ELECT circuit de poursuite *m*, RADIO antenne de poursuite *f*

tracking *n* COMP poursuite *f*, GEN suivi *m*, RADIO suivi *m*, trace de fuite *f*, *satellite* poursuite *f*, TV poursuite *f*, suivi *m*; ~ **aerial** RADIO antenne de poursuite *f*; ~ **antenna** RADIO antenne de poursuite *f*; ~ **and command** (*T&C*) TEST poursuite et commande *f*; ~ **error** RADIO erreur de poursuite *f*; ~ **symbol** COMP symbole de poursuite *m*; ~ **system** RADIO système de poursuite *m*; ~, **telemetry and command** (*TT&C*) RADIO poursuite, télémesure et télécommande *f*

tracks: ~ **per inch** *phr* (*TPI*) COMP, DATA pistes par pouce (*ppp*)

trade: ~ **data interchange** *n* (*TDI*) DATA échange de données commerciales *m*

trader: ~**'s turret** *n* TELEP poste de courrier *m*

traditional: ~ **telephone network** *n* TELEP, TRANS réseau téléphonique classique *m*

traffic *n* DATA, ELECT, RADIO, SWG, TELEP, TRANS trafic *m*; ~ **analyser** *BrE* DATA, TELEP, TEST analyseur de trafic *m*; ~ **analysis** DATA, TELEP, TEST analyse de trafic *f*; ~ **analyzer** *AmE see traffic analyser BrE*; ~ **area** RADIO zone de trafic *f*; ~ **bureau** RADIO bureau du trafic *m*; ~ **capacity** TELEP, TRANS capacité de trafic *f*; ~ **carried** TELEP trafic écoulé *m*; ~**-carrying burst** DATA, TELEP paquet porteur de trafic *m*, organe d'écoulement du trafic *m*; ~**-carrying device** DATA, TELEP organe d'écoulement du trafic *m*, paquet porteur de trafic *m*; ~ **channel** DATA, RADIO voie de conversation *f*, canal de trafic *m*; ~ **circuit** TELEP circuit de trafic *m*; ~ **control** DATA, TELEP pilotage de trafic *m*, régulation du trafic *f*; ~ **controller** DATA, TELEP contrôleur de trafic *m*; ~ **demand** RADIO demande de trafic *f*; ~ **density** RADIO, TELEP, TRANS densité de trafic *f*; ~ **distribution** SWG, TELEP répartition du trafic *f*; ~ **distribution imbalance** SWG, TELEP déséquilibre de la répartition du trafic *m*; ~ **division system** SWG système de partage de trafic *m*; ~ **flow** DATA écoulement de données *m*, écoulement du trafic *m*, flux du trafic *m*, intensité du trafic acheminé *f*, TELEP écoulement du trafic *m*, flux du trafic *m*, intensité du trafic acheminé *f*; ~ **flow control** DATA, TELEP contrôle de flux de trafic *m*; ~ **growth** TELEP accroissement du trafic *m*; ~ **handling** DATA, TELEP traitement du trafic *m*, écoulement du trafic *m*; ~-**handling capacity** DATA, TELEP capacité d'écoulement du trafic *f*, capacité de traitement du trafic *f*; ~-**handling level** DATA, TELEP niveau d'écoulement du trafic *m*, niveau de traitement du trafic *m*; ~-**handling program** DATA, TELEP programme d'écoulement de trafic *m*, programme de traitement du trafic *m*; ~ **imbalance** SWG, TELEP déséquilibre du trafic *m*; ~ **intensity** SWG, TELEP intensité de trafic *f*; ~ **intensity load** SWG, TELEP charge d'intensité de trafic *f*; ~ **level** TELEP niveau de trafic *m*; ~-**level counter** TELEP compteur de niveau des prises *m*; ~ **load** SWG, TELEP charge de trafic *f*, volume de trafic *m*; ~ **load imbalance** SWG, TELEP déséquilibre du trafic *m*; ~ **loading** SWG, TELEP charge de trafic *f*, volume de trafic *m*; ~ **matrix** TELEP matrice de trafic *f*; ~ **measurement** ELECT, TELEP, TEST observation de trafic *f*; ~ **meter** ELECT, TELEP, TEST compteur de trafic *m*, erlangmètre *m*; ~ **metering** ELECT, TELEP, TEST mesure de trafic *f*; ~ **monitor** ELECT, TELEP, TEST moniteur de trafic *m*; ~ **monitoring** ELECT, TELEP, TEST observation de trafic *f*; ~ **observation** ELECT, TELEP, TEST observation de trafic *f*; ~ **observation desk** TELEP pupitre d'observation de trafic *m*, pupitre de surveillance de trafic *m*; ~ **operator position** SWG, TELEP position d'opérateur de transit *f*; ~ **padding** DATA, SWG bourrage *m*; ~ **path** RADIO, TELEP voie de conversation *f*; ~ **pattern** TELEP schéma d'écoulement du trafic *m*; ~ **peak** TELEP pointe de trafic *f*; ~ **performance** TELEP capacité d'écoulement du trafic *f*; ~ **record** TELEP enregistrement de trafic *m*; ~ **recorder** TELEP enregistreur de trafic *m*; ~ **recording** TELEP enregistrement de trafic *m*; ~ **relation** TELEP flux de trafic *m*; ~ **research** TELEP recherche de télétrafic *f*; ~ **route** TELEP artère *f*; ~ **route tester** TELEP, TEST équipement d'essai d'itinéraire *m*, équipement d'essai de faisceau *m*, équipement de vérification de l'acheminement du trafic *m*; ~ **route testing** (*TRT*) TELEP, TEST vérification de l'acheminement du trafic *f*, vérification d'itinéraire du trafic *f*; ~ **routing** SIG, TELEP, TEST, TRANS acheminement du trafic *m*; ~ **routing strategy** TELEP, TEST plan d'acheminement du trafic *m*; ~ **sharing** RADIO partage de trafic *m*; ~ **signal system** SIG système de signalisation de trafic *m*; ~ **signaling equipment** *AmE see traffic signalling equipment BrE*; ~ **signaling installation** *AmE see traffic signalling installation BrE*; ~ **signalling equipment** *BrE* SIG équipement de signalisation routière *m*; ~ **signalling installation** *BrE* SIG équipement de signalisation routière *m*; ~ **statistics** *n pl* DATA, RADIO, TELEP, TRANS statistiques de trafic *f pl*; ~ **study** TELEP étude d'intensité de trafic *f*; ~ **superintendent** DATA, TELEP contrôleur de trafic *m*; ~ **supervision** ELECT, TELEP, TEST observation de trafic *f*, surveillance de trafic *f*; ~ **supervision desk** TELEP pupitre d'observation de trafic *m*, pupitre de surveillance de trafic *m*; ~ **terminal** RADIO, TELEP terminal de trafic *m*; ~ **tester** TELEP, TEST testeur de trafic *m*; ~ **theory** TELEP théorie de trafic *f*; ~ **to special services** TELEP trafic à destination des services spéciaux *m*; ~-**unit price procedure** TELEP méthode de rémunération par unité de trafic *f*; ~ **volume** SWG, TELEP volume de trafic *m*

trafficability: ~ **performance** *n* TELEP capacité d'écoulement du trafic *f*

trail: ~ **marker** *n* COMP marqueur de parcours *m*

trailer: ~ **record** *n* DATA train d'ondes *m*

trailing: ~ **contact** *n* PARTS, POWER contact à accompagnement *m*; ~ **edge** *n* ELECT, GEN, PARTS, TELEP front arrière *m*; ~-**edge position** *n* PARTS, TELEP position de flanc arrière *f*, position de front arrière *f*; ~ **noise** *n* RADIO bruit de traînage *m*

train *n* TRANS *of pulses* suite *f*; ~ **radio system** RADIO *on board* système radio ferroviaire *m*, système de radiocommunications pour les chemis de fer *m*

tranche *n* TV *of cable* tranche *f*

transaction *n* DATA transaction *f*; ~ **branch** DATA branche d'une transaction *f*; ~ **branch identifier** DATA identificateur de branche d'une transaction *m*; ~ **capabilities application part** DATA sous-système pour la gestion des transactions *m* (*SSGT*); ~ **commitment** DATA engagement de transaction *m*; ~ **identifier** DATA, TELEP identificateur de transaction *m*; ~ **logging** DATA enregistrement de transaction *m*; ~ **management subsystem** DATA sous-système pour la gestion des transactions *m* (*SSGT*); ~ **processing** (*TP*) DATA traitement transactionnel *m*; ~ **processing application service element** (*TPASE*) DATA élément de service d'application pour le traitement des transactions *m*; ~ **processing channel** DATA canal de traitement transactionnel *m*; ~-**processing protocol machine** (*TPPM*) DATA machine de protocole de traitement transactionnel *f*; ~ **processing service provider** (*TPSP*) DATA fournisseur de service de traitement transactionnel *m*; ~ **processing service user** (*TPSU*) DATA utilisateur du service de traitement transactionnel *m*; ~ **processing service user invocation** (*TPSUI*) DATA appel de TPSU *m*; ~ **recovery** DATA rétablissement de transaction *m*; ~ **roll-back** DATA repositionnement de transaction *m*

transactional: ~ **system** *n* DATA système transactionnel *m*

transadmittance *n* ELECT transadmittance *m*

transceiver *n* DATA, RADIO émetteur-récepteur *m*, émetteur-récepteur combiné *m*

transcode *vt* ELECT, TELEP, TEST, TRANS transcoder

transcoder *n* ELECT, TELEP, TEST, TRANS transcodeur *m*

transcoding *n* ELECT, TELEP, TEST, TRANS transcodage *m*; ~ **gain** ELECT gain de transcodage *m*

transconductance *n* ELECT transconductance *f*

transcriber *n* TELEP transcripteur *m*

transducer *n* ELECT, PARTS, POWER, RADIO, TELEP, TEST transducteur *m*; ~ **gain** TELEP gain transductique *m*; ~ **loss** TELEP affaiblissement transductique *m*

transductor *n* ELECT, PARTS, POWER, RADIO, TELEP, TEST transducteur *m*; ~ **amplifier** POWER amplificateur magnétique *m*

transfer[1] *n* ELECT mutation *f*, transfert *m*, transmission *f*, GEN transfert *m*, transmission *f*, TELEP *of call* transfert *m*; TRANS mutation *f*, transfert *m*, transmission *f*; ~-**allowed acknowledgement signal** SIG, TELEP signal d'accusé de réception d'un signal d'autorisation de transfert *m*; ~-**allowed procedure** TELEP procédure de transfert autorisé *f*; ~-**allowed signal** SIG, TELEP signal d'autorisation de transfert *m*; ~ **block** SWG bloc de transfert *m*; ~ **bus** SWG bus de transfert *m*; ~ **channel** SWG, TELEP voie de transfert *f*; ~ **characteristic** ELECT caractéristique de transfert *f*; ~-**charge call** TELEP communication en PCV *f* (*Fra*), communication payable à l'arrivée *f* (*Fra*); ~ **charging** TELEP service à frais virés *m* (*Can*), service payable à l'arrivée *m*, service PCV *m*; ~ **control** SWG commande de transfert *f*; ~ **of control** TELEP *to operator, supervisor* lancement de commande *m*; ~-**control block** SWG bloc de commande de transfert *m*; ~ **current** ELECT courant de transfert *m*; **~ function** APPL, CONTR, DATA, TRANS fonction de transfert *f*; **~ impedance** ELECT impédance de transfert *f*; **~ link** TELEP liaison de transfert *f*; **~ message** PER APP message de transfert *m*; **~ mode** COMP mode de transfert *m*; **~ multiplexer** SWG multiplexeur de transfert *m*; **~ node** TELEP nud de transfert *m*; **~-prohibited procedure** TELEP procédure de transfert interdit *f*; **~-prohibited signal** SIG, TELEP signal d'interdiction de transfert *m*; **~ rate** COMP, DATA débit de transfert *m*, taux de transfert *m*, vitesse de transfert *f*; **~ subsystem** SWG sous-système de transfert *m*; **~ syntax** TELEP syntaxe de transfert *f*; **~ system** (*T/SYS*) DATA système de transfert *m*; **~ time** ELECT, TELEP, TRANS temps de transfert *m*

transfer² *vt* DATA *Internet* rapatrier, TELEP, TRANS transférer; **~ a call** TELEP transférer une communication; **~ the charges** TELEP téléphoner en PCV

transfer³: **~ allowed** *phr* (*TFA*) TELEP transfert autorisé *m*; **~ controlled** *phr* (*TFC*) TELEP transfert sous contrôle *m*; **~ prohibited** *phr* (*TFP*) TELEP transfert interdit *m*; **~ restricted** *phr* (*TFR*) TELEP transfert restreint *m*

transform¹ *n* DATA, ELECT, POWER, TELEP, TRANS transformation *f*; **~ coding** SIG codage par transformée *m*

transform² *vt* ELECT, POWER transformer

transformation *n* TRANS transformation *f*

transformer *n* ELECT, PARTS, POWER, RADIO, TELEP, TEST, TV transformateur *m*; **~ coupling** POWER liaison par transformateur *f*

transforming: **~ section** *n* RADIO section d'adaptation *f*

transhorizon: **~ propagation** *n* RADIO propagation transhorizon *f*; **~ radio relay system** *n* RADIO faisceau hertzien transhorizon *m*

transient¹ *adj* GEN de passage, transitoire

transient² *n* ELECT clic *m*, GEN pointe *f*; **~ decay current** ELECT courant de déclin *m*, courant de traînage *m*; **~ fault** TEST panne intermittente *f*, panne temporaire *f*; **~ radiation effect on electronics** (*TREE*) ELECT effet des radiations transitoires sur les circuits électroniques *m*; **~ suppressor** POWER limiteur de surtension transitoire *m*, suppresseur de transitoires *m*; **~ testing state** DATA, TEST état de test transitoire *m*

transients *n pl* TRANS non-linéarités *f pl*

transionospheric: **~ propagation** *n* RADIO propagation transionosphérique *f*

transistor *n* ELECT, PARTS, POWER transistor *m*; **~ analyser** *BrE*, **~ analyzer** *AmE* TEST transistormètre *m*; **~ array IC** ELECT, PARTS CI à transistors *m*; **~ base** ELECT, PARTS base de transistor *f*; **~ collector** ELECT collecteur de transistor *m*; **~ emitter** ELECT émetteur à transistor *m*; **~ flip-flop** ELECT, PARTS transistor à bascule *m*; **~ input resistance** ELECT résistance d'entrée *f*; **~-transistor logic** (*TTL*) ELECT logique transistor-transistor *f*

transit: **~ administration** *n* TELEP, TRANS administration de transit *f*; **~ angle** *n* ELECT angle de transit *m*; **~ connection element** *n* (*TCE*) TRANS élément de connexion de transit *m*; **~ connection-related function** *n* (*TCRF*) DATA, TRANS fonction liée à une connexion de transit *f*; **~ country** *n* TELEP pays de transit *m*; **~-delay indication** *n* (*TDI*) DATA indication de délai en transit *f*; **~-delay selection and indication** *n* (*TDSAI*) DATA, TELEP sélection et indication du délai en transit *f*; **~ exchange** *n* DATA central de transit *m*, centre de transit *m*, SWG, TELEP central de transit *m*, centre de transit *m*, commutateur de transit *m*; **~ network identification** *n* TRANS identification du réseau de transit *f*; **~ proceed-to-send signal** *n* SIG, TELEP signal d'invitation à émettre en transit *m*; **~ routing** *n* SIG, TELEP, TRANS acheminement en transit *m*; **~ seizing signal** SIG, TELEP signal de prise de transit *m*; **~ seizure signal** *n* SIG, TELEP signal de prise de transit *m*; **~ share** *n* TELEP quote-part de transit *f*; **~ stage** *n* TELEP étage de transit *m*; **~ switching center** *AmE*, **~ switching centre** *BrE* *n* DATA centre de commutation de transit *m*, SWG, TELEP centre de commutation de transit *m*, commutateur de transit *m*; **~ through-connect signal** *n* SIG, TELEP signal de connexion de transit *m*; **~ time** *n* POWER temps de propagation *m*, TELEP, TEST temps de passage *m*, temps de transit *m*; **~ time jitter** *n* ELECT fluctuation du temps de transit *f*; **~ time spread** *n* ELECT étalement du temps de transit *m*; **~ traffic** *n* TELEP trafic de transit *m*

transition *n* TELEP transition *f*

transitional: **~ state** *n* TELEP état de transition *m*, état transitoire *m*

transitron *n* ELECT, PARTS transitron *m*; **~ oscillator** ELECT, PARTS oscillateur à transitron *m*

translate *vt* COMP *program code* translater, *language, image* recompiler, traduire, DATA recompiler, transposer, ELECT, RADIO translater, TELEP, TV traduire

translation *n* COMP *of program code* traduction *f*, *of language, image* recompilation *f*, traduction *f*, DATA recompilation *f*, ELECT *movement*, RADIO translation *f*, transposition *f*, TELEP, TV traduction *f*; **~ circuit** TELEP circuit de traduction *m*; **~ error** DATA erreur de transposition *f*; **~ register** TELEP registre de translation *m*; **~ relay** PARTS, TELEP relais de traduction *m*; **~ store** SWG mémoire de traduction *f*

transliterate *vt* COMP, DATA translitérer

transmission *n* DATA transfert *m*, GEN envoi *m*, RADIO émission *f*, transmission *f*, transport d'énergie *m*, TELEP, TRANS envoi *m*, transmission *f*, TV transmission *f*; **~ bearer** TRANS, TV support de transmission *m*; **~ bit rate** RADIO, TRANS débit binaire de transmission *m*; **~ bit slip** RADIO, TRANS glissement de bits *m*, glissement binaire *m*; **~ breakdown** TEST, TRANS interruption de transmission *f*; **~ bridge** RADIO, TRANS pont d'alimentation *m*, pont de transmission *m*; **~ buffer** TELEP, TRANS tampon d'émission *m*; **~ capability** DATA capacité de transmission *f*; **~ channel** RADIO, TRANS, TV canal de transmission *m*, voie de transmission *f*; **~ characteristic** RADIO, TRANS caractéristique de transmission *f*; **~ code violation** TELEP violation du code de transmission *f*; **~ coefficient** TRANS coefficient de transmission *m*; **~ conditions** *n pl* COMP, TRANS conditions de transmission *f pl*; **~ control protocol/Internet protocol** (*TCP/IP*) DATA protocole de contrôle de transmission/protocole Internet *m* (*TCP/IP*); **~ convergence** TRANS convergence de transmission *f*; **~ convergence sublayer** TRANS sous-couche de convergence de transmission *f*; **~ definition** TELEP, TRANS définition à l'émission *f*; **~ delay** DATA *Internet* délai d'acheminement *m*, TELEP *through digital exchange*, TRANS temps de transmission *m*; **~ equipment** TRANS équipement pour réseaux de transmission *m*; **~ error** TEST, TRANS erreur de transmission *f*; **~ highway** TRANS artère de transmission *f*; **~ interface** COMP, TRANS interface de communication *f*; **~ layer** TRANS couche réseau de transmission *f*; **~ line** RADIO, TRANS chaînon de voie de communication *m*, ligne de transmission *f*; **~ link** TELEP, TRANS chaîne de transmission *f*, chaînon de voie de communication *m*, chaînon de voie de transmission *m*, liaison de transmission *f*; **~ loss** TELEP affaiblissement de transmission *m*, TRANS affaiblissement de transmission *m*, perte de transmission *f*; **~ loss of a-t-b** TELEP, TRANS affaiblissement du trajet a-t-b *m*; **~ maintenance points for international line** *n pl* TELEP centres de maintenance de la transmission pour la ligne *m pl*; **~ measurement** TEST, TRANS mesure de transmission *f*; **~ medium** TRANS, TV support de transmission *m*; **~ network** TRANS réseau de transmission *m*; **~ node** TRANS nœud de transmission *m*; **~ overload** TRANS surcharge de transmission *f*; **~ path** RADIO, TRANS trajet de transmission *m*; **~ path endpoint** (*TPE*) DATA, TRANS extrémité de voie de transmission *f*; **~ performance** TELEP, TEST, TRANS fonctionnement de transmission *m*; **~ performance rating** TELEP, TEST, TRANS indice de qualité de transmission *m*;

~ **plan** TRANS plan de transmission *m*; ~ **priority code** DATA, TRANS code de priorité de transmission *m*; ~ **quality** TELEP, TEST, TRANS qualité de transmission *f*; ~ **recipient** DATA, TRANS destinataire *m*; ~ **reference point** (*TRP*) TELEP, TRANS point de référence de transmission *m*; ~ **sender** DATA, TELEP, TRANS expéditeur *m*; ~ **speed** TRANS vitesse de transmission *f*; ~ **technique** TRANS technique de transmission *f*; ~ **test** TEST, TRANS test de transmission *m*; ~ **test trunk** (*TTT*) TELEP, TEST, TRANS circuit d'essai de transmission *m*; ~ **time** TELEP, TRANS temps de transmission *m*; ~ **window** TELEP, TRANS fenêtre de transmission *f*

transmissometer *n* TEST, TRANS transmissomètre *m*

transmit[1]: ~-**ready** *adj* RADIO prêt à émettre

transmit[2]: ~ **aerial** *n* RADIO antenne émettrice *f*, antenne d'émission *f*; ~ **antenna** *n* RADIO antenne émettrice *f*, antenne d'émission *f*; ~ **buffer** *n* COMP, TRANS tampon de transmission *m*, tampon d'émission *m*; ~ **channel** *n* TRANS canal d'émission *m*, canal de transmission *m*; ~ **characteristic** *n* RADIO, TRANS caractéristique d'émission *f*, caractéristique de transmission *f*; ~ **data** *n pl* DATA, TRANS données de transmission *f pl*, données d'émission *f pl*; ~ **delay** *n* RADIO, TRANS retard appliqué à l'émission *m*, retard appliqué à la transmission *m*; ~ **direction** *n* (*TX*) TEST, TRANS sens émission *m*, sens transmission *m*; ~ **fiber optic terminal device** *AmE*, ~ **fibre optic terminal device** *BrE* *n* TRANS émetteur optique *m*; ~ **flow control** *n* TELEP, TRANS contrôle de flux d'émission *m*, contrôle de flux de transmission *m*; ~-**frame acquisition and synchronization procedure** *n* (*TFAS*) DATA, TRANS procédure d'acquisition et de synchronisation des trames d'émission *f*, procédure d'acquisition et de synchronisation des trames de transmission *f*; ~ **gain** *n* TRANS gain à l'émission *m*, gain à la transmission *m*; ~ **machine** *n* TELEP, TRANS machine émettrice *f*; ~ **multiframe timing** *n* TRANS rythme de multitrame émission *m*, rythme de multitrame transmission *m*; ~-**receive cell** *n* RADIO verrou de récepteur *m*; ~-**receive switch** *n* RADIO, SWG, TRANS duplexeur *m*

transmit[3] *vt* COMP, DATA, RADIO, TELEP, TRANS émettre, transmettre

transmit[4]: **you may** ~ *phr* RADIO vous pouvez transmettre; ~ **on** *phr* RADIO émission en cours

transmittance *n* ELECT, TRANS facteur de transmission *m*

transmitted: ~ **data** *n pl* DATA données émises *f pl*; ~ **phase advance** *n* TRANS avance de phase du signal transmis *f*, déphasage vers l'avant du signal transmis *m*

transmitter *n* (*XMTR*) GEN émetteur *m*, PER APP microphone *m*, microphone *m*; ~ **blocker cell** (*TB cell*) RADIO, TRANS verrou d'émetteur *m*; ~ **cap** PER APP capsule de microphone *f*, pastille de microphone *f*, SIG, TELEP capsule de microphone *f*; ~ **cup** PER APP support microphonique *m*; ~ **distortion** TELEP, TRANS distorsion à l'émission *f*; ~ **error** RADIO, TRANS erreur d'émetteur *f*; ~ **inset** PER APP capsule émettrice *f*; ~-**multiplexer** TRANS émetteur-multiplexeur *m*; ~ **output power** TRANS puissance de sortie d'un émetteur *f*; ~-**power control** RADIO commande de puissance d'émetteur *f*; ~-**receiver** (*TR*) DATA, ELECT, RADIO, TELEP, TRANS émetteur-récepteur *m* (*TR*); ~ **relay** PARTS relais d'émission *m*, TRANS relais d'émetteur *m*; ~ **turn-on signal** SIG, TELEP, TRANS signal de déblocage de l'émetteur *m*, signal de déclenchement de l'émetteur *m*; ~ **turn-on time** TELEP, TRANS temps de mise en service de l'émetteur *m*

transmitting: ~ **aerial** *n* RADIO antenne de transmission *f*; ~ **antenna** *n* RADIO antenne de transmission *f*; ~ **channel** *n* TRANS canal de transmission *m*; ~ **equipment** *n* PER APP, TRANS appareil émetteur *m*; ~ **objective loudness rating** *n* (*TOLR*) TELEP, TRANS équivalent objectif pour la sonie à l'émission *m* (*EOSE*); ~ **subscriber identification-DCS** *n* (*TCI-DCS*) DATA, TELEP identification de l'abonné émetteur-DCS *f* (*TCI-DCS*)

transmultiplexer *n* (*TMUX*) ELECT, TELEP, TRANS transmultiplexeur *m*; ~ **channel** ELECT, TELEP, TRANS voie de transmultiplexeur *f*

transparency *n* DATA *scanner for overhead projector*, TELEP, TRANS *network* transparent *m*

transparent: ~ **bearer service** *n* APPL, DATA service support transparent *m*; ~ **data-transfer phase** *n* DATA, TELEP, TRANS phase de transfert transparent de données *f*; ~ **designation labeling strip** *AmE*, ~ **designation labelling strip** *BrE* *n* ELECT, PARTS bande étiquetée de désignation transparente *f*; ~ **mode** *n* DATA mode transparent *m*; ~ **repeater** *n* GEN répéteur transparent *m*

transponder *n* DATA, ELECT, répondeur *m*, GEN répéteur *m*; ~ **allotment** DATA, ELECT, RADIO attribution de répéteurs *f*; ~ **hopping** ELECT, RADIO saut de répéteur *m*; ~ **reply efficiency** RADIO rendement de réponse *m*; ~ **system** DATA système émetteur-récepteur *m*

transport: ~ **connection** *n* (*TC*) TELEP connexion de transport *f*; ~-**connection endpoint identifier** *n* DATA, TELEP identificateur d'extrémité de connexion de transport *m*; ~ **layer** *n* DATA, TRANS couche transport *f*; ~ **locking device** *n* PER APP dispositif de blocage d'entraînement *m*, dispositif de verrouillage de défilement *m*; ~ **network** *n* TELEP réseau de transport *m*; ~ **protocol data unit** *n* (*TPDU*) DATA unité de données de protocole de transport *f*; ~ **service** *n* (*TS*) TELEP, TRANS service de transport *m*; ~-**service access point** *n* (*TSAP*) TELEP, TRANS point d'accès du service de transport *m*; ~-**service access-point address** *n* TELEP, TRANS adresse de point d'accès du service de transport *f*; ~ **service data unit** *n* (*TSDU*) TELEP, TRANS unité de données du service de transport

transportability *n* TEST aptitude au transport *f*

transportable: ~ **telephone** *n* PER APP, TELEP téléphone portable *m*

transpose *vt* COMP, TELEP, TRANS transposer

transposed: ~ **line** *n* TELEP, TRANS ligne de transmission avec transpositions *f*; ~ **transmission line** *n* TELEP, TRANS ligne de transmission avec transpositions *f*

transposition *n* COMP transposition *f*, ELECT inversion *f*, TELEP, TEST, TRANS transposition *f*; ~ **method of measurement** TEST méthode de mesure par transposition *f*

transputer *n* COMP, DATA transordinateur *m*

transversal: ~ **filter** *n* PER APP, TRANS filtre transversal *m*

transverse[1]: ~ **electric** *adj* (*TE*) TRANS transverse électrique (*TE*); ~ **electromagnetic** *adj* (*TEM*) TRANS transverse électromagnétique (*TEM*); ~ **magnetic** *adj* (*TM*) TRANS transverse magnétique (*TM*)

transverse[2]: ~ **conversion loss** *n* (*TCL*) DATA, TELEP affaiblissement de conversion transversale *m* (*ACT*); ~ **conversion ratio** *n* TELEP rapport de conversion transversale *m*; ~ **electric wave** *n* RADIO, TRANS onde transversale électrique *f*; ~ **gyrofrequency** *n* RADIO gyrofréquence transversale *f*; ~ **interferometry** *n* TRANS interférométrie transversale *f*; ~ **judder** *n* TELEP, TRANS broutage transversal *m*; ~ **junction** *n* TELEP liaison transversale *f*; ~ **magnetic mode** *n* (*TM*) RADIO, TRANS mode magnétique transversal *m*; ~ **mode** *n* TRANS mode transversal *m*; ~ **offset loss** *n* TRANS perte par déplacement transversal *f*; ~ **redundancy check** *n* DATA contrôle transversal par redondance *m*; ~ **reflection factor** *n* TELEP facteur de réflexion transversale *m*; ~ **return loss** *n* (*TRL*) TELEP affaiblissement de réflexion transversale *m*; ~ **route** *n* TELEP faisceau direct *m*; ~ **section** *n* TRANS section transversale *f*; ~ **slot** *n* TRANS fente transversale *f*; ~ **test method** *n* TEST méthode de test transversale *f*; ~ **traffic** *n* TELEP trafic direct *m*; ~ **voltage** *n* POWER, TELEP tension transversale *f*; ~ **wave** *n* RADIO, TRANS onde transversale *f*

trap[1] *n* ELECT filtre *m*, piège *m*, RADIO siphon *m*; ~ **circuit** ELECT piège *m*

trap[2] *vt* COMP débrancher, DATA piéger, ELECT filtrer, piéger

trapezoidal: ~ **distortion** *n* ELECT distorsion trapézoïdale *f*

trapped: ~ **mode** *n* RADIO, TRANS mode de propagation guidée *m*
trapping *n* COMP débranchement *m*, DATA *artificial intelligence* piégeage *m*, ELECT filtrage *m*, piégeage *m*
trash *n* COMP *icon* poubelle *f*
travel[1] *n* RADIO *of wave* parcours *m*, propagation *f*
travel[2] *vt* RADIO parcourir
travel[3] *vi* COMP voyager, TV transister
traveling *AmE see travelling BrE*
travelling: ~ **wave** *n BrE* RADIO, TRANS onde progressive *f*; **~-wave aerial** *n BrE* RADIO, TRANS antenne à ondes progressives *f*; **~-wave amplifier** *n BrE* (*TWA*) ELECT, TRANS amplificateur à ondes progressives *m*; **~-wave antenna** *n BrE* RADIO, TRANS antenne à ondes progressives *f*; ~ **wave tube** *n BrE* (*TWT*) ELECT, TRANS tube à ondes progressives *m* (*TOP*); ~ **wave tube amplifier** *n BrE* (*TWTA*) ELECT, TRANS amplificateur à tube à ondes progressives *m*; ~ **wave valve** *n BrE* ELECT tube à ondes progressives *m* (*TOP*); ~ **waveguide** *n BrE* ELECT, TRANS tube à ondes progressives *m* (*TOP*)
treatment *n* PARTS traitement *m*
treble *n* TRANS *sound recording, reproduction* aigus *m pl*
tree *n* COMP, DATA, ELECT arbre *m*; ~ **attachment** COMP, DATA, ELECT rattachement à un arbre *m*; ~ **and branch architecture** COMP, DATA architecture arborescente *f*, architecture en arbre *f*; ~ **and branch network** COMP, DATA, TRANS réseau arborescent *m*, réseau en arbre *m*; ~ **and branch structure** COMP, TRANS, DATA, ELECT arborescence *f*, structure en arbre *f*; ~ **concentrator** ELECT concentrateur arborescent *m*; ~ **distribution** TRANS distribution arborescente *f*, distribution en arbre *f*; ~ **expansion** COMP, DATA création de l'arborescence *f*, développement de l'arborescence *m*; ~ **identifier** COMP, DATA identificateur d'arbre *m*; ~ **leaf** COMP, DATA feuille d'arbre *f*; ~ **network** TRANS réseau arborescent *m*, réseau en arbre *m*; ~ **node** COMP, DATA nœud arborescent *m*, nœud d'arbre *m*; ~ **notation** COMP, DATA notation arborescente *f*, notation d'arbre *f*; ~ **structure** COMP, DATA, ELECT, TRANS arborescence *f*, structure arborescente *f*, structure d'arbre *f*; **~-structured view** COMP, DATA vue arborescente *f*; ~ **and tabular combined notation** (*TTCN*) COMP, DATA notation combinée arborescente et tabulaire *f*; ~ **and tabular combined notation in graphical form** (*TTCN,GR*) COMP, DATA notation combinée arborescente et tabulaire en forme graphique *f*; ~ **topology** COMP, DATA topologie arborescente *f*
TREE *abbr* (*transient radiation effect on electronics*) ELECT effet des radiations transitoires sur les circuits électroniques *m*
trellis: **~-coded modulation** *n* (*TCM*) TRANS modulation par codage en treillis *f* (*MCT*); ~ **coding** *n* DATA codage en treillis *m*
tremolo *n* COMP trémolo *m*; ~ **control** COMP commande de trémolo *f*
TRF: ~ **reception** *n* (*tuned radio frequency reception*) RADIO réception à amplification directe *f*
triac *n* ELECT *semiconductors*, PARTS triac *m*
trial *n* TEST essai technique *m*, expérimentation *f*
triangular: ~ **pulse** *n* ELECT impulsion triangulaire *f*; ~ **random noise** *n* ELECT, RADIO bruit erratique triangulaire *m*
triax *n* TRANS câble triax *m*, câble triaxial *m*, triax *m*; ~ **cable** TRANS câble triax *m*, câble triaxial *m*, triax *m*
triaxial: ~ **cable** *n* TRANS câble triax *m*, câble triaxial *m*, triax *m*
tributary[1] *adj* DATA, TELEP tributaire
tributary[2] *n* (*TR*) DATA, ELECT, RADIO, TELEP tributaire *m*, TRANS affluent *m*, tributaire *m*; ~ **station** DATA station tributaire *f*, station subordonnée *f*; ~ **unit** (*TU*) DATA unité tributaire *f*; ~ **unit group** (*TUG*) DATA groupe d'unités tributaires *m*
trickable: ~ **switch** *n* POWER, SWG commutateur déjouable *m*
trickle: ~ **charge** *n* POWER charge d'entretien *f*, charge de compensation *f*, charge de maintien *f*, charge lente *f*; ~ **charger** *n* PER APP, POWER chargeur d'entretien *m*, chargeur de compensation *m*, chargeur de maintien *m*, chargeur lent *m*
trigatron *n* ELECT trigatron *m*
trigger[1]: ~ **circuit** *n* ELECT circuit de déclenchement *m*; ~ **electrode** *n* ELECT, PARTS électrode de déclenchement *f*; ~ **gap** *n* ELECT espace de déclenchement *m*; ~ **input** *n* ELECT entrée de déclenchement *f*; ~ **level** *n* COMP niveau de déclenchement *m*; ~ **mechanism** *n* ELECT mécanisme de déclenchement *m*, mécanisme de mise en marche *m*; ~ **nozzle** *n* ELECT bec à déclic *m*; ~ **play** *n* COMP reproduction par déclenchement *f*; ~ **pulse** *n* ELECT impulsion de déclenchement *f*; ~ **tube** *n* ELECT, PARTS tube relais *m*
trigger[2] *vt* COMP, ELECT *alarm* déclencher, GEN activer, TEST *alarm* déclencher
triggered: ~ **flip-flop** *n* ELECT, PARTS bascule déclenchée *f*; ~ **spark gap** *n* ELECT éclateur déclenché *m*; ~ **time-base** *n* ELECT base de temps déclenchée *f*
triggering *n* COMP, ELECT, RADIO, TEST déclenchement *m*; ~ **lead pulse** ELECT impulsion de déclenchement *f*; ~ **waveform** RADIO forme d'onde de déclenchement *f*
trim[1]: ~ **wipe** *n* TV volet pas à pas *m*
trim[2] *vt* COMP, ELECT, POWER couper
trimmer: ~ **capacitor** *n* ELECT, PARTS condensateur ajustable *m*
trimming: ~ **potentiometer** *n* ELECT, PARTS potentiomètre d'équilibrage *m*, potentiomètre d'appoint *m*; ~ **resistor** *n* ELECT, PARTS résisteur d'ajustage *m*, résisteur d'équilibrage *m*
Trinitron *n* COMP Trinitron *m*
triode *n* ELECT, PARTS triode *f*; ~ **field effect transistor** ELECT, PARTS transistor à effet de champ triode *m*
trip[1] *n* TELEP arrêt; ~ **mechanism** ELECT mécanisme de déclenchement *m*, mécanisme de mise en marche *m*
trip[2] *vt* ELECT, POWER *circuit breakers* déclencher, TELEP arrêter; ~ **the ringing** TELEP arrêter le courant d'appel
triple[1]: **~-precision** *adj* DATA en triple précision
triple[2]: **~-access line** *n* SWG, TELEP ligne à triple accès *f*; **~-pass technology** *n* DATA *scanner* technologie trois passes *f*
tripping[1] *adj* DATA *Internet* trébuchant
tripping[2] *n* ELECT, POWER déclenchement *m*, TELEP *of ringing* arrêt *m*
tristate *adj* ELECT à trois états
trit *n* ELECT trit *m*
TRL *abbr* (*transverse return loss*) TELEP affaiblissement de réflexion transversale *m*
trombone *n* PARTS trombone *m*; ~ **connection** TELEP connexion en boucle *f*; ~ **loop connection** TELEP connexion en boucle *f*
tropicalized: ~ **equipment** *n* CLIM, PARTS matériel tropicalisé *m*
tropopause *n* RADIO tropopause *f*
troposphere *n* RADIO troposphère *f*
tropospheric: ~ **mode** *n* RADIO mode troposphérique *m*; ~ **propagation** *n* RADIO propagation troposphérique *f*; ~ **radioduct** *n* RADIO conduit troposphérique *m*, guide troposphérique *m*; ~ **reflection** *n* RADIO réflexion troposphérique *f*; ~ **scatter** *n* RADIO diffusion troposphérique *f*; ~ **wave** *n* RADIO onde troposphérique *f*
trouble: **~-free maintenance** *n* ELECT maintenance facile *f*; ~ **report** *n* TELEP fiche de dérangement *f*, rapport d'incident *m*
troubleshooting *n* COMP, ELECT, TELEP relève de dérangements *f*, TEST localisation de panne *f*; ~ **procedure** COMP, ELECT, TELEP procédure de diagnostic *f*

trough *n* ELECT creux *m*
troughing *n* ELECT caniveau *m*
TRP *abbr* (*transmission reference point*) TELEP, TRANS point de référence de transmission *m*
TRT *abbr* (*traffic route testing*) TELEP, TEST vérification de l'acheminement du trafic *f*
truck: **~-to-train radio** *n* RADIO radio sol-train *f*
true[1] *adj* COMP, ELECT, RADIO, TEST vrai
true[2]: **~ bearing** *n* RADIO relèvement vrai *m*; **~-motion display** *n* RADIO indication du mouvement absolu *f*; **~ rms voltmeter** *n* (*true root mean square voltmeter*) TEST voltmètre efficace vrai *m*; **~ root mean square voltmeter** *n* (*true rms voltmeter*) TEST voltmètre efficace vrai *m*
truncate *vt* COMP, DATA tronquer
truncation *n* COMP, DATA troncature *f*
trunk[1] *n* SWG, TELEP *circuit* circuit de jonction *m*, *exchange tie* joncteur *m*, *inter-switch tie* circuit de connexion *m*; **~-access-barred extension** TELEP poste à interdiction d'accès au réseau *m*; **~-access-restricted extension** TELEP poste à restriction d'accès au réseau *m*; **~ barring** TELEP restriction d'accès au réseau *f*; **~ cable** ELECT câble principal *m*, TRANS câble interurbain *m*; **~ call** *BrE* (*cf toll call AmE*) TELEP appel interurbain *m*; **~-call barring** *BrE* (*cf toll-call barring AmE*) TELEP restriction d'appel interurbain *f*; **~ channel** TRANS canal interurbain *m*; **~ circuit** ELECT circuit de jonction *m*, *exchange tie* joncteur *m*, SWG circuit interurbain *m*, TELEP circuit interurbain *m*, joncteur *m*, TRANS joncteur *m*, circuit interurbain *m*; **~ circuit block** ELECT, SWG, TELEP, TRANS bloc de circuits téléphoniques *m*; **~ code** (*TC*) TELEP indicatif interurbain *m*; **~ connection** TELEP liaison interurbaine *f*; **~ control station** TELEP station multiprocesseur de terminaison MIC *f*; **~ dialing** *AmE* TELEP accès direct à l'interurbain par numérotation *m*, interurbain automatique *m*, signalisation automatique de l'interurbain *f*; **~ discrimination** TELEP restriction d'accès au réseau *f*; **~ discrimination control** CONTR, TELEP contrôle de restriction d'accès au réseau *m*; **~ distribution frame** (*TDF*) TRANS répartiteur de jonction *m*; **~ equipment** TRANS équipement de circuit *m*, matériel de mesure *m*; **~ exchange** SWG, TELEP, TRANS central interurbain *m*, centre interurbain *m*; **~ flow** SWG, TELEP ligne principale de liaison *f*; **~ group** TELEP faisceau de circuits interurbain *m*, faisceau interurbain *m*, TRANS faisceau *m*; **~ line** TELEP ligne interurbaine *f*; **~-line relay set** PARTS, TELEP joncteur *m*; **~ marking** TELEP marquage interurbain *m*; **~ network** SWG, TELEP réseau interurbain *m*; **~ offer** (*TKO*) SWG, TELEP intervention ligne-réseau *f*; **~ offering** SWG, TELEP offre d'appel *f*; **~ ordering service** TELEP service des annotatrices *m*; **~ PE** TELEP élément graphique de circuit interurbain *m*; **~ prefix** TELEP préfixe interurbain *m*; **~ record position** TELEP position d'annotatrice *f*, position d'inscription *f*; **~-routed call** TELEP appel écoulé *m*; **~ switching** SWG commutation de transit *f*; **~ switching center** *AmE*, **~ switching centre** *BrE* SWG centre de commutation interurbaine *m*; **~ switching exchange area** SWG, TELEP réseau interurbain *m*, zone à autonomie d'acheminement *f* (*ZAA*); **~ switching stage** SWG étape de communication interurbaine *f*, TELEP étape de commutation interurbaine *f*; **~ system** RADIO relais commun *m*, TELEP système interurbain *m*; **~ telephone circuit** TELEP, TRANS circuit téléphonique interurbain *m*; **~ traffic** TELEP trafic interurbain *m*; **~ transit exchange** SWG centre de transit *m*; **~-unrestricted extension** TELEP poste sans limitation d'accès au réseau *m*
trunk[2] *vi* RADIO partager les fréquences
trunk[3]: **~ busy** *phr* TELEP ligne occupée; **~ free** *phr* TELEP circuit interurbain libre; **~ seized** *phr* TELEP circuit pris
trunked: **~ dispatch system** *n* RADIO système partagé du type dispatching *m*; **~ radio** *n* POWER, RADIO radiocotéléphonie à ressources partagées *f*; **~ radio network** *n* RADIO réseau radio à ressources partagées *m* (*3RP*); **~ system** *n* RADIO réseau radio à ressources partagées *m* (*3RP*)
trunking *n* RADIO partage *m*, partage de fréquences *m*; **~ diagram** TELEP schéma des liaisons *m*; **~ system** RADIO réseau radio à ressources partagées *m* (*3RP*)
trusted: **~ functionality** *n* DATA, TEST fonctionnalité de confiance *f*
truth *n* TEST vérité *f*; **~ table** ELECT table de vérité *f*
TS *abbr* TELEP (*telephone service*) service téléphonique *m*, (*transport service*) service de transport *m*, TRANS (*transport service*) service de transport *m*; **~ network** *n* TELEP structure TS *f*
TSAP *abbr* (*transport-service access point*) TELEP, TRANS point d'accès du service de transport *m*
TSAU *abbr* (*telephone service access unit*) TELEP unité d'accès au service téléphonique *f*
TSDU *abbr* (*transport service data unit*) DATA, TELEP, TRANS unité de données du service de transport *f*
TSM *abbr* (*terminal server manager*) DATA gestionnaire de serveur de terminaux *m*
TSR *abbr* (*terminate and stay resident*) COMP, DATA, TELEP TSR (*corésident*, *résident*); **~ program** *n* COMP, DATA, TELEP programme TSR *m*, programme corésident *m*, programme résident *m*
TSS *abbr* (*test suite structure*) TEST structure de suite de tests *f*
TSS&TP *abbr* (*test suite structure and test purposes*) TEST structure de suite de tests et objectifs de tests *m pl*
TST *abbr* (*time-space-time*) SWG temps-espace-temps
T/SYS *abbr* (*transfer system*) DATA système de transfert *m*
TT&C *abbr* (*tracking, telemetry and command*) RADIO poursuite, télémesure et télécommande *f*
TTCN *abbr* (*tree and tabular combined notation*) COMP, DATA notation combinée arborescente et tabulaire *f*; **~ statement** *n* DATA déclaration en notation combinée arborescente et tabulaire *f*
TTCN,GR *abbr* (*tree and tabular combined notation, graphical form*) COMP, DATA notation combinée arborescente et tabulaire, forme graphique *f*
TTD *abbr* (*target transit delay*) DATA temps de transit objectif *m*
TTG: **~ laser** *n* (*tunable twin-guide laser*) ELECT laser à double guide accordable *m*
TTL *abbr* (*transistor-transistor logic*) ELECT logique transistor-transistor *f*
TTT *abbr* (*transmission test trunk*) TELEP, TEST, TRANS circuit d'essai de transmission *m*
TTU *abbr* (*tape transport unit*) APPL, COMP, TV dérouleur de bande *m*, mécanisme d'entraînement de bande *m*
TTX *abbr* (*teletext*) APPL, TV TTX (*télétexte*), vidéotex diffusé *m*
TTXAU *abbr* (*teletex access unit*) APPL unité d'accès au télétex *f*
TTY *abbr* (*teletype*) COMP, PER APP, RADIO, TELEP téléimprimeur *m*, télétypewriter *m*
TU *abbr* (*tributary unit*) DATA unité tributaire *f*
tube *n* ELECT alvéole *f*, tube *m*, TRANS alvéole *f*; **~ base** ELECT, PARTS culot de tube *m*; **~ efficiency** ELECT, TEST rendement d'un tube *m*; **~ of force** POWER tube de champ *m*, tube de force *m*; **~ fuse** ELECT, PARTS, POWER fusible à tube *m*; **~ noise** ELECT bruit de tube *m*; **~ socket** ELECT, PARTS support *m*; **~ starting time** ELECT durée de démarrage d'un tube *f*; **~ voltage drop** ELECT chute de tension du tube *f*; **~ wrap** PARTS gaine *f*
tubing *n* ELECT, PARTS, POWER, TRANS gaine *f*
tubular: **~ capacitor** *n* PARTS, POWER condensateur tubulaire *m*
TUG *abbr* (*tributary unit group*) DATA groupe d'unités tributaires *m*
tumbler: **~ switch** *n* POWER interrupteur à bascule *m*
tunable: **~ twin-guide laser** *n* (*TTG laser*) ELECT laser à

double guide accordable *m*; ~ **voltmeter** *n* TEST voltmètre sélectif *m*
tune[1]: ~ **function** *n* COMP fonction d'accord *f*; ~ **job** *n* COMP travail d'accordage *m*
tune[2] *vt* GEN accorder, syntoniser
tuned: ~ **circuit** *n* POWER circuit accordé *m*; ~ **radio frequency reception** *n* (*TRF reception*) RADIO réception à amplification directe *f*
tuner *n* COMP syntonisateur *m*, ELECT, RADIO syntonisateur *m*, tuner *m*; ~ **circuit** ELECT circuit d'accord *m*
tuning *n* ELECT syntonisation *f*; ~ **capability** COMP, ELECT possibilités d'accord *f pl*; ~ **circuit** ELECT circuit d'accord *m*; ~ **fork** COMP, GEN diapason *m*; ~ **fork oscillator drive** RADIO pilote à diapason *m*; ~ **function** COMP fonction d'accord *f*; ~ **position meter** ELECT cadran de syntonisation *m*; ~ **sensitivity** ELECT sensibilité d'accord *f*
tunnel: ~ **action** *n* PARTS action tunnel *f*; ~ **diode** *n* ELECT, PARTS diode tunnel *f*; ~ **effect** *n* PARTS effet tunnel *m*
tunneling *AmE see tunnelling BrE*
tunnelling *n BrE* ELECT effet tunnel *m*; ~ **mode** *BrE* TRANS mode tunnel *m*, mode de fuite *m*; ~ **ray** *BrE* TRANS rayon tunnel *m*
TUP *abbr* (*telephone user part*) DATA, RADIO, TELEP SSUT (*sous-système utilisateur téléphonie, sous-système utilisateur téléphonique*)
turn[1] *n* PARTS spire *f*; **~-off delay** PARTS retard de passage à l'ouverture *m*; **~-off thyristor** PARTS thyristor blocable *m*; **~-off time** PARTS temps blocable *m*; **~-on time** PARTS temps déblocable *m*; ~ **ratio** PARTS rapport du nombre de spires *m*
turn[2] *vt* PARTS tourner, usiner au tour; ~ **off** GEN *device, machine* arrêter, *supply* mettre hors circuit, éteindre, mettre hors tension; ~ **on** GEN *supply* mettre en circuit, mettre en marche, mettre sous tension, *device, machine* allumer
turn[3]: ~ **on** *vi* ELECT devenir conducteur, devenir passant, se débloquer
turnaround: ~ **time** *n* COMP délai d'exécution *m*, DATA temps d'inversion *m*
turnbuckle *n* PARTS tendeur *m*
turning *n* PARTS façonnage au tour *m*, tournage *m*, usinage au tour *m*; ~ **relay** PARTS relais de retour *m*; ~ **tool** PARTS outil de tour *m*
turnkey: ~ **program** *n* TV programme clés en main *m*
turnstile: ~ **whip** *n* RADIO antenne fouet en tourniquet *f*
turret *n* ELECT plot *m*; ~ **lathe** PARTS tour à tourelle revolver *m*; ~ **turning** PARTS tournage au tour à tourelle revolver *m*
tutorial *n* COMP, DATA didacticiel *m*, logiciel d'initiation *m*, tutoriel *m*
TV *abbr* (*television*) TV TV (*télévision*)
TVRO *abbr* (*television receive-only*) TV TVRO (*terminal de télévision de réception seulement, terminal TV réception seulement*); ~ **aerial** *n* TV antenne de télévision de réception seulement *f*; ~ **antenna** *n* TV antenne de télévision de réception seulement *f*; ~ **terminal** *n* TV terminal TVRO *m*
TWA *abbr* ELECT, TRANS (*traveling-wave amplifier AmE, travelling-wave amplifier BrE*) amplificateur à ondes progressives *m*
twenty: **~-four hour hotline** *n* TELEP ligne ouverte vingt-quatre heures sur vingt-quatre *f*
twin: ~ **AX** *n* TRANS câble double *m*; ~ **cable** *n* ELECT, TRANS câble à paires symétriques *m*; **~-channel stereo telephony** *n* TELEP stéréophonie à deux canaux *f*; ~ **conductor** *n* ELECT, GEN, PARTS faisceau double *m*; ~ **conductor cable** *n* ELECT câble à paires *m*; ~ **contacts** *n pl* PARTS contacts doublés *m pl*, contacts jumelés *m pl*; ~ **group** *n* GEN groupe double *m*, groupe jumelé *m*
twine *vt* PARTS torsader
twinning *n* ELECT maclage *m*, TV pairage *m*
twins *n pl* COMP sœurs *f pl*
twin-T: ~ **oscillator** *n* ELECT, PARTS oscillateur à réseau en double T *m*
twist[1] *n* ELECT, PARTS torsade *f*, vrillage *m*
twist[2] *vt* ELECT, PARTS torsader, vriller
twisted[1] *adj* ELECT, PARTS torsadé, vrillé
twisted[2]: ~ **nematic** *n* ELECT nématique en hélice *f*; ~ **pair** *n* ELECT, TRANS paire torsadée *f*; **~-pair cable** *n* ELECT, TRANS cable à paires torsadées *m*; **~-pair cabling** *n* ELECT, PARTS, TRANS câblage à paires torsadées *m*; **~-pair hub** *n* ELECT, PARTS concentrateur à paires torsadées *m*; ~ **waveguide** *n* DATA, RADIO, TRANS guide d'onde torsadé *m*; ~ **wire** *n* ELECT, PARTS fil torsadé *m*
twisting *n* ELECT, PARTS torsade *f*
two[1]: **~-terminal-pair** *adj* ELECT, TRANS à deux paires de bornes; **~-way** *adj* GEN, RADIO, TRANS, TV bidirectionnel, à double sens, mixte; **~-wire** *adj* TV bifilaire
two[2]: **~-bit byte** *n* COMP, DATA doublet *m*; **~-condition cable code** *n* TELEP, TRANS code de câble bivalent *m*; **~-condition telegraph code** *n* TELEP, TRANS code télégraphique bivalent *m*; **~-conductor cable** *n* ELECT, TRANS câble à deux conducteurs *m*; **~-control-point method** *n* RADIO méthode des deux points directeurs *f*; **~-frequency channeling plan** *AmE*, **~-frequency channelling plan** *BrE n* TRANS plan de disposition des voies à deux fréquences *m*; **~-frequency glide path system** *n* RADIO, TRANS système d'alignement de descente à deux fréquences *m*; **~-frequency localizer system** *n* RADIO système d'alignement de piste à deux fréquences *m*; **~-frequency operation** *n* RADIO, SIG fonctionnement sur deux fréquences *m*; **~-frequency operation automatic repeater mode** *n* TELEP fonctionnement à deux fréquences avec relais automatique *m*; **~-frequency operation non-repeater mode** *n* TELEP fonctionnement à deux fréquences sans relais *m*; **~-frequency relay system** *n* RADIO, TELEP système de relais à deux fréquences *m*; **~-frequency signaling** *AmE*, **~-frequency signalling** *BrE n* SIG, TELEP signalisation à deux fréquences *f*; **~-frequency simplex** *n* RADIO, TELEP simplex à deux fréquences *m*; **~-gang capacitor** *n* PARTS, POWER condensateur à deux cages *m*; **~-gang condenser** *n* PARTS, POWER condensateur à deux cages *m*; **~-motion selector** *n* TELEP sélecteur Strowger *m*, sélecteur à double mouvement *m*; **~-out-of-five code** *n* DATA code deux sur cinq *m*, code quinaire *m*; **~-party line** *n* TELEP ligne partagée entre deux abonnés *f*, ligne partagée entre deux postes *f*; **~-piece cover** *n* PARTS, TELEP couvercle en deux parties *m*; **~-ray diversity** *n* RADIO diversité à deux ondes *f*; **~-stage contact spring set** *n* PARTS, SWG empilage à deux temps *m*; **~-stage relay** *n* PARTS relais à deux temps *m*; **~-state register** *n* TELEP registre à deux états *m*; **~-step action** *n* CONTR action en deux étapes *f*; **~-step contact spring set** *n* PARTS, SWG empilage à deux temps *m*; **~-step relay** *n* PARTS relais à deux temps *m*; **~-terminal network** *n* ELECT quadripôle *m*, réseau à deux accès *m*, POWER quadripôle *m*, TRANS quadripôle *m*, réseau à deux accès *m*; **~-terminal pair** *n* ELECT, POWER, TRANS quadripôle *m*; **~-terminal-pair network** *n* ELECT, POWER, TRANS réseau quadripôle *m*; **~-tone detector** *n* TEST détecteur bifréquence *m*; **~-way** *n* DATA transfert bidirectionnel, transfert mixte *m*, TELEP double sens *m*; **~-way alternate interaction** *n* DATA interaction bidirectionnelle *f*; **~-way circuit** *n* ELECT circuit bidirectionnel *m*; **~-way communication** *n* RADIO, TELEP, TV communication bidirectionnelle *f*, mode bidirectionnel simultané *m*; **~-way directionality** *n* COMP lien bidirectionnel *m*; **~-way link** *n* COMP liaison bidirectionnelle *f*; **~-way paging system** *n* RADIO système de radiomessagerie bilatérale *m*; **~-way repeater** *n* TRANS répéteur bidirectionnel *m*, répéteur pour circuit à deux fils *m*; **~-way simplex** *n* TELEP bilatérale *f*, communication bilatérale *f*; **~-way simplex connection** *n* TELEP bilatérale *f*, communication bilatérale *f*; **~-way simplex system** *n* TELEP exploitation par voies conjuguées *f*, simplex bilatéral *m*; **~-wire bus** *n* TV bifibus *m*; **~-wire circuit** *n* ELECT, TELEP,

TRANS circuit bifilaire *m*; **~-wire crosspoint** *n* SWG point d'intersection de deux fils *m*; **~-wire repeater** *n* TRANS répéteur bidirectionnel *m*, répéteur pour circuit à deux fils *m*; **~-wire switch** *n* SWG commutateur à deux fils *m*; **~-wire switching** *n* SWG commutation à deux fils *f*; **~-wire switching system** *n* SWG commutateur à deux fils *m*; **~-wire system** *n* SWG commutateur à deux fils *m*

2B+D: **~ arrangement** *n* TRANS structure 2B+D *f*

2HD *abbr* (*second harmonic distortion*) ELECT distorsion de la deuxième harmonique *f*

TWT *abbr* (*traveling wave tube AmE, travelling wave tube BrE*) ELECT, TRANS TOP (*tube à ondes progressives*)

TWTA *abbr* (*traveling-wave tube amplifier AmE, travelling-wave tube amplifier BrE*) ELECT, TRANS amplificateur à tube à ondes progressives *m*

TX *abbr* (*transmit direction*) TEST, TRANS sens émission *m*, sens transmission *m*

TY: **~ wrap** *n* PARTS, POWER attache autoblocante *f*, ceinture de câble *f*

typamatic: **~ key** *n* PER APP touche à répétition *f*

type: **~ of address** *n* (*TOA*) DATA, TELEP type d'adresse *m*; **~ A display** *n* RADIO indicateur type A *m*; **~ approval** *n* TEST agrément *m*, qualification de la conception *f*, qualification interne *f*; **~ bar** *n* TELEP barre d'impression *f*; **~-bar carriage** *n* TELEP chariot à jeu de barres *m*, chariot à barres *m*; **~ B display** *n* RADIO indicateur de distance et gisement *m*; **~ E display** *n* RADIO indicateur de distance et de site *m*; **~ exchange** *n* TELEP central type *m*; **~ font** *n* DATA police de caractères *f*; **~ mismatch** *n* COMP *printing* non-concordance de caractères *f*, *searching* non-concordance de types *f*; **~ of number** *n* (*TON*) DATA, TELEP type de numéro *m*; **~ page** *n* PER APP page *f*; **~ of sound-program circuit** *AmE*, **~ of sound-programme circuit** *BrE* *n* TELEP type de circuit radiophonique *m*; **~ test** *n* TEST essai de type *m*; **~ of transmission** *n* TELEP type de transmission *m*; **~ wheel** *n* PER APP *printing* roue à caractères *f*

typeface *n* PER APP *printing* œil de caractère *m*, type de caractère *m*

typical: **~ prediction** *n* (*TP*) TELEP prédiction type *f*

typing: **~ mistake** *n* COMP faute de frappe *f*

typo *n* COMP faute de frappe *f*

typography *n* APPL typographie *f*

U

U *abbr* DATA, TELEP (*user option*) U (*option utilisateur*), (*U interface*) U (*interface U*); **~ interface** *n* (*U*) DATA, TELEP interface U *f* (*U*)

UA *abbr* APPL, DATA (*user agent*) UA (*agent utilisateur*), TELEP (*user agent*) UA (*agent utilisateur*)

UAF *abbr* (*unnumbered acknowledgement frame*) TELEP trame d'accusé de réception non-numéroté *f*, accusé de réception non-numéroté *m*

UART *abbr* (*universal asynchronous receiver-transmitter*) RADIO UART (*émetteur-récepteur asynchrone universel*)

UAS *abbr* (*unavailable second*) DATA seconde d'indisponibilité *f*

UAT *abbr* (*unavailability time*) DATA temps d'indisponibilité *m*

UAX *abbr* (*unit automatic exchange*) SWG central automatique local *m*

UCM *abbr* (*user communications manager*) COMP UCM (*user communications manager*)

UCN-CNE: **~ interface** *n* TELEP interface CNE-UCN *f*

UDC *abbr* (*universal data connector*) ELECT connecteur de données universel *m*

UE *abbr* (*user element*) DATA élément utilisateur *m*, TELEP élément d'utilisateur *m*

UHF *abbr* (*ultrahigh frequency*) TV UHF (*ultrahaute fréquence*); **~ network** *n* TV réseau Odm *m*, réseau UHF *m*

UI *abbr* DATA (*unnumbered information*) information non-numérotée *f*, GEN (*unit interval*) intervalle unitaire *m*

UIH: **~ control field** *n* TELEP champ de commande UIH *m*

UITS *abbr* (*unacknowledged information-transfer service*) APPL service de transfert d'informations sans accusé de réception *m*

UJT *abbr* (*unijunction transistor*) ELECT, PARTS, POWER TUJ (*transistor unijonction*)

U-link *n* POWER cavalier *m*

ULSI: **~ circuit** *n* ELECT circuit ULSI *m*

ultimate: **~ strength** *n* PARTS, TEST charge de rupture *f*, limite d'élasticité *f*, résistance à la rupture *f*

ultrahigh: **~ frequency** *n* (*UHF*) RADIO, TV ultrahaute fréquence *f* (*UHF*)

ultralarge: **~ scale integration circuit** *n* ELECT circuit ULSI *m*

ultrashort: **~ wave** *n* RADIO onde ultracourte *f*

ultrasonic[1] *adj* RADIO à ultrasons, ultrasonique, ultrasonore

ultrasonic[2]: **~ frequency** *n* RADIO fréquence ultrasonore *f*

ultrasound *n* RADIO ultrason *m*

ultraviolet[1] *adj* (*UV*) ELECT, TRANS ultraviolet (*UV*)

ultraviolet[2]: **~ radiation** *n* TRANS rayonnement ultraviolet *m*

U-Matic *n* TV U-Matic *m*

umbrella: **~ cell** *n* RADIO cellule parapluie *f*

UMTS *abbr* (*universal mobile telecommunication system*) RADIO, TELEP système universel de télécommunication mobile *m*

UN: **~ TPDU** *abbr* (*use of network connection TPDU*) DATA TPDU UN (*TPDU d'utilisation de connexion de réseau*)

UNA: **~ segment** *n* DATA, TELEP segment UNA *m*

unabsorbed: **~ field strength** *n* RADIO champ en l'absence d'absorption *m*

unacceptable: **~ quality** *n* TEST qualité intolérable *f*

unacknowledged: **~ information transfer service** *n* (*UITS*) APPL service de transfert d'informations sans accusé de réception *m*

unaffected: **~ level** *n* TELEP niveau inchangé *m*

unallocated: **~-number signal** *n* SIG, TELEP signal de numéro inutilisé *m*, signal de numéro non utilisé *m*

unamplified[1] *adj* ELECT, POWER, TV non amplifié, sans amplification

unamplified[2]: **~ length** *n* TV longueur sans amplification *f*

unanswered: **~ call** *n* PER APP numéro absent *m*, TELEP appel sans réponse *m*; **~ call indication** *n* PER APP, TELEP indication d'appel sans réponse *f*

unassigned *adj* COMP non attribué

unattended: **~ exchange** *n* TELEP central automatique *m*, central sans personnel *m*; **~ station** *n* TRANS station non surveillée *f*, station sans personnel *f*

unauthorized: **~ visitor** *n* COMP *Internet* visiteur indiscret *m*

unavailability *n* COMP, DATA, TEST indisponibilité *f*; **~ time** (*UAT*) COMP, DATA temps d'indisponibilité *m*

unavailable: **~ second** *n* (*UAS*) DATA seconde d'indisponibilité *f*

UNB: **~ segment** *n* DATA, TELEP segment UNB *m*

unbalance *n* COMP asymétrie *f*, déséquilibre *m*, ELECT déséquilibre *m*, POWER asymétrie *f*, déséquilibre *m*, SWG, TELEP déséquilibre *m*

unbalanced[1] *adj* COMP asymétrique, déséquilibré, ELECT déséquilibré, POWER asymétrique, déséquilibré, SWG, TELEP déséquilibré

unbalanced[2]: **~ input** *n* COMP entrée asymétrique *f*; **~ operation normal response mode class** *n* (*UNC*) DATA classe d'exploitation asymétrique en mode réponse normal *f*; **~ output** *n* COMP sortie asymétrique *f*; **~ system** *n* TRANS système asymétrique *m*

unbarred: **~ extension** *n* TELEP poste supplémentaire avec prise de ligne directe *m*

unbind *vt* DATA détacher

unblanking *n* ELECT déblocage du faisceau *m*

unblock *vt* RADIO *channel*, TELEP débloquer

unblocking *n* RADIO *of channel*, TELEP déblocage *m*; **~ signal** SIG signal de déblocage *m*

unbound: **~ mode** *n* TRANS mode non lié *m*

UNC *abbr* (*unbalanced operation normal response mode class*) DATA classe d'exploitation asymétrique en mode réponse normal *f*

uncertainty: **~ of measurement** *n* TEST incertitude de mesure *f*; **~ symbol** *n* TELEP symbole d'incertitude *m*

unchained: **~ sequence** *n* DATA séquence non chaînée *f*

unclocked *adj* ELECT sans entrée d'horloge, asynchrone

uncoded *adj* TELEP en clair

uncompressed[1] *adj* COMP, DATA sans compression

uncompressed[2]: **~ operation** *n* COMP, DATA mode sans compression *m*

unconditional: **~ call forwarding** *n* TELEP renvoi systématique *m*, renvoi d'appel systématique *m*

uncontrolled: **~ not-ready signal** *n* SIG, TELEP signal non prêt automatique *m*; **~ slip** *n* RADIO, TELEP glissement non commandé *m*, glissement non maîtrisable *m*

undercurrent *n* POWER minimum de courant *m*; **~ relay** PARTS relais à minimum de courant *m*

undercut *n* ELECT, PARTS gravure sous-jacente *f*

undergo: **~ trials** *vt* TEST subir des essais

underground: **~ cable** *n* POWER câble enterré *m*, TRANS câble souterrain *m*; **~ chamber** *n* TRANS chambre souterraine *f*; **~ installation** *n* APPL installation souterraine *f*; **~ line** *n* TRANS ligne souterraine *f*

underlap *n* TELEP non-juxtaposition *f*

undermodulation *n* ELECT modulation insuffisante *f*, TRANS sous-modulation *f*
underpower: **~ relay** *n* PARTS relais à minimum *m*
undervoltage *n* POWER sous-tension *f*, subtension *f*; **~ alarm** POWER alarme de manque de tension *f*; **~ relay** PARTS relais à minimum *m*
underwater: **~ communication** *n* RADIO, TELEP communication sous-marine *f*; **~ propagation** *n* RADIO propagation sous-marine *f*
underwriter *n* TV souscripteur *m*
undetected[1] *adj* GEN non détecté
undetected[2]: **~ fault time** *n* TEST temps de panne latente *m*
undo *vt* COMP *command, operation* annuler
undulating *adj* PARTS *surface* ondulé
unearthed *adj BrE* (*cf ungrounded AmE*) POWER isolé de la terre
unfiltered: **~ voltage** *n* POWER tension non lissée *f*
unforeseen: **~ event** *n* TEST événement imprévu *m*; **~ outcome** *n* TEST interruption imprévisible *f*, résultat imprévu *m*
unfriendly *adj* COMP non convivial, hostile
unfurl *vt* RADIO *antenna* déployer
unfurlable *adj* RADIO *antenna* déployable
ungrounded *adj AmE* (*cf unearthed BrE*) POWER isolé de la terre
ungroup *vt* COMP dégrouper
unguarded: **~ interval** *n* TELEP intervalle sans garde *m*
UNH: **~ segment** *n* DATA, TELEP segment UNH *m*
unhook *vt* PER APP, TELEP décrocher
unhooking *n* PER APP, TELEP décrochage *m*
UNI *abbr* (*user-network interface*) DATA, TRANS interface utilisateur-réseau *f*
unidentified: **~ event** *n* TEST événement non identifié *m*
unidirectional[1] *adj* COMP, RADIO, TELEP simplex, unidirectionnel, unilatéral
unidirectional[2]: **~ aerial** *n* RADIO antenne directive *f*; **~ antenna** *n* RADIO antenne directive *f*; **~ communication** *n* RADIO, TELEP, TV communication unidirectionnelle *f*, communication unilatérale *f*; **~ connection** *n* SWG, TELEP, TV communication unidirectionelle *f*, communication unilatérale *f*; **~ network** *n* TV réseau unidirectionnel *m*; **~ pulse** *n* TRANS impulsion unidirectionnelle *f*
unification *n* ELECT unification *f*
uniform: **~ encoding** *n* TELEP, TRANS codage uniforme *m*, quantification uniforme *f*; **~ numbering** *n* TELEP numérotage uniforme *m*; **~ quantizing** *n* TELEP, TRANS quantification uniforme *f*; **~ resource locator** *n* (*URL*) COMP *Internet* uniform resource locator *m*; **~ spectrum random noise** *n* ELECT, RADIO, TELEP bruit aléatoire uniforme *m*; **~ wiring** *n* PARTS, RADIO, TELEP câblage standardisé *m*
uniformity: **~ of luminance** *n* ELECT uniformité d'affichage *f*
unigage *AmE see unigauge BrE*
unigauge: **~ loop** *n BrE* TELEP boucle unigauge *f*
unijunction: **~ diode** *n* ELECT, PARTS diode unijonction *f*; **~ transistor** *n* (*UJT*) ELECT, PARTS, POWER transistor unijonction *m* (*TUJ*)
unilateral: **~ control** *n* TELEP, TRANS commande unilatérale *f*
uninitialized *adj* COMP non initialisé
uninsulated[1] *adj* ELECT, POWER non isolé
uninsulated[2]: **~ conductor** *n* ELECT, POWER fil non isolé *m*, fil nu *m*
unintelligible: **~ crosstalk** *n* TELEP diaphonie inintelligible *f*; **~ crosstalk components** *n pl* TELEP composants de diaphonie inintelligible *m pl*
uninterrupted: **~ power supply** *n* (*UPS*) POWER alimentation sans coupure *f* (*UPS*)
uninterruptible: **~ power supply** *n* (*UPS*) POWER alimentation non interruptible *f* (*UPS*)
unipolar: **~ pulse** *n* TELEP impulsion unipolaire *f*; **~ sequence of pulses** *n* TELEP séquence unipolaire d'impulsions *f*; **~ transistor** *n* ELECT, PARTS transistor unipolaire *m*
unipole: **~ aerial** *n* RADIO antenne unipolaire *f*; **~ antenna** *n* RADIO antenne unipolaire *f*
unique: **~ word detection** *n* DATA détection du mot unique *f*; **~ word loss** *n* DATA perte du mot unique *f*; **~-word loss alarm** *n* DATA alarme de perte du mot unique *f*; **~ word window** *n* (*UW window*) DATA fenêtre du mot unique *f*; **~ word-missed detection** *n* DATA non-détection du mot unique *f*
uniselector *n* SWG sélecteur rotatif *m*
unison *n* COMP unisson *m*
unit *n* GEN unité *f*, fond *m*; **~ automatic exchange** (*UAX*) SWG central automatique local *m*; **~ charge** TELEP taxe unitaire *f*; **~-charge metering** TELEP comptage simple *m*; **~-charged marking** TELEP unité de comptage d'appel *f*; **~ design** PER APP, SWG conception d'unités *f*; **~-element error rate** TELEP *isochronous modulation* taux d'erreurs sur les éléments unitaires *m*; **~ gain** ELECT gain unitaire *m*, gain unité *m*, gain de 1 *m*; **~ impulse response** CONTR réponse impulsionnelle *f*; **~-indicating alarm** TELEP alarme indicatrice d'unité défectueuse *f*; **~ interval** (*UI*) GEN intervalle unitaire *m*; **~ of measurement** TEST unité de mesure *f*; **~ section** TV section unitaire *f*; **~ step** TELEP échelon unité *m*; **~ string** DATA chaîne unitaire *f*; **~ tube** ELECT tube unité *m*
units: **~ counter** *n* ELECT compteur des unités *m*; **~ input** *n* ELECT entrée des unités *f*
unitunnel: **~ diode** *n* ELECT, PARTS diode unitunnel *f*
unity: **~ gain** *n* ELECT gain de 1 *m*, gain unité *m*, gain unitaire *m*
universal: **~-access number** *n* TELEP numéro universel *m*; **~-access number service** *n* TELEP service de numéros universels *m*; **~ asynchronous receiver-transmitter** *n* (*UART*) RADIO émetteur-récepteur asynchrone universel *m* (*UART*); **~ data connector** *n* (*UDC*) ELECT connecteur de données universel *m*; **~ element** *n* COMP élément universel *m*; **~ meter** *n* TEST multimètre *m*; **~ mobile telecommunication system** *n* (*UMTS*) TELEP système universel de télécommunication mobile *m*; **~ night-service** *n* TELEP service de nuit universel *m*; **~ number** *n* TELEP numéro universel *m*; **~ switch** *n* SWG commutateur universel *m*; **~ time** *n* (*UT*) GEN temps universel *m* (*TU*); **~ time coordinated** *n* (*UTC*) GEN heure universel de coordonnée *f*, temps universel coordonné *m*; **~ transport system** *n* (*UTS*) ELECT système de transport universel *m*
Universal: **~ Postal Union** *n* (*UPU*) TELEP Union postale universelle *f* (*UPU*)
universe *n* TEST population *f*
Unix: **~-to-Unix copy protocol** *n* (*UUCP*) DATA Unix-to-Unix copy protocol *m* (*UUCP*)
unknown[2]: **~ message** *n* DATA message inconnu *m*
unlabeled *AmE see unlabelled BrE*
unlabelled *adj BrE* COMP sans étiquette, sans référence
unlisted *adj AmE* (*cf ex-directory BrE*) TELEP sur liste rouge
unloaded[1] *adj* COMP non chargé, ELECT à vide, POWER à circuit ouvert, à vide, TRANS non chargé, non pupinisé
unloaded[2]: **~ cable** *n* TRANS câble non chargé *m*, câble non pupinisé *m*
unloading *n* TELEP *volume* démontage *m*
unlock: **~ key** *n* COMP touche de déverrouillage *f*
unmanned: **~ exchange** *n* TELEP central automatique *m*, central sans personnel *m*
unmasked *adj* COMP non masqué
unnotched: **~ signal** *n* SIG signal non filtré *m*
unnumbered: **~ acknowledgement** *n* (*UA*) TELEP accusé de réception non-numéroté *m*; **~ acknowledgement frame** *n* (*UA*) TELEP trame d'accusé de réception non numéroté *f*;

~ **format** *n* TELEP format non numéroté *m*; ~ **information** *n* (*UI*) DATA information non numérotée *f*
unobtainable: ~ **signal** *n* SIG signal de numéro non accessible *m*
unoccupied: ~ **line** *n* TELEP ligne libre *f*
unpack *vt* COMP, DATA décondenser
unpacked *adj* COMP, DATA décondensé
unperturbed: ~ **orbit** *n* RADIO orbite non perturbée *f*
unplug *vt* COMP, ELECT, PER APP, TEST débrancher, déficher
unplugged *adj* COMP, ELECT, PER APP, TEST débranché, déficher
unprogrammable *adj* COMP non programmable
unqualified: ~ **event** *n* DATA événement non qualifié *m*
unreasonable: ~ **message** *n* TELEP message irrationnel *m*
unreliable *adj* COMP non fiable
unrelocatable *adj* COMP non translatable
unrestricted[1] *adj* GEN sans restriction, illimité, libre
unrestricted[2]: ~ **bearer service** *n* APPL, TRANS *ISDN* service de support sans restriction *m*; ~ **digital data ratio** *n* DATA rapport de données numériques sans restriction *m*; ~ **extension** *n* TELEP poste supplémentaire avec prise de ligne directe *m*; ~ **information-transfer service** *n* APPL, TELEP *ISDN* service de transfert d'informations sans restriction *m*; ~ **service** *n* APPL *ISDN*, TELEP service sans restriction *m*
unscheduled: ~ **maintenance** *n* TELEP, TEST maintenance non programmée *f*; ~ **stop** *n* TEST arrêt non programmé *m*
unscramble *vt* COMP décrypter
unscrambled: ~ **binary ones** *n pl* (*USB1*) DATA uns binaires non brouillés *m pl*; ~ **double dibit** *n* (*S1*) DATA double dibit non-embrouillé *m*
unsecured *adj* COMP non protégé
unserviceable: ~ **station** *n* TRANS station hors service *f*
unshielded: ~ **twisted pair** *n* (*UTP*) ELECT, TRANS paire symétrique non-blindée *f*, paire torsadée non-blindée *f*
unshift: ~**-on-space** *n* TELEP blanc des lettres automatique *m*
unsmoothed: ~ **voltage** *n* POWER tension non lissée *f*
unsolder *vt* ELECT, PARTS dessouder
unsoldered *adj* ELECT, PARTS non soudé
unstable: ~ **state** *n* GEN état instable *m*
unsubscribe *vi* COMP annuler une inscription, résilier un abonnement, se désabonner
unsuccessful: ~ **attempt** *n* RADIO, TELEP tentative inefficace *f*, tentative infructueuse *f*; ~ **call** *n* TELEP appel inefficace *m*, appel inefructueux *m*; ~ **traffic** *n* RADIO, TELEP trafic inefficace *m*
unterminated: ~ **level** *n* TEST niveau non adapté *m*
untested *adj* TEST non testé
untimed: ~ **call** *n* TELEP communication non taxée à la durée *f*
untrafficable: ~ **path** *n* TRANS trajet impraticable *m*
unvoiced *adj* COMP non voisé
unwanted: ~ **echo** *n* RADIO écho parasite *m*; ~ **emission** *n* RADIO émission indésirable *f*, rayonnement non desiré *m*
unweighted *adj* COMP non pondéré
unwelcome: ~ **caller** *n* COMP *Internet* visiteur indiscret *m*
unwind *vt* GEN dérouler
unwired: ~ **rack** *n* PARTS, TELEP bâti non câblé *m*
unwrap[1]: ~ **test** *n* GEN, TEST essai de déroulement *m*
unwrap[2] *vt* GEN, TEST dérouler
unwrapping *n* GEN, TEST déroulement *m*; ~ **test** TEST essai de déroulement *m*; ~ **tool** PARTS outil dérouleur *m*
unzip *vt* COMP, DATA décomprimer
unzipping *n* COMP, DATA décomprimer
up[1]: **on the ~ clock** *adj* ELECT sur la montée d'horloge
up[2]: ~**-converter** *n* DATA, RADIO changeur élévateur de fréquence *m*, convertisseur élévateur de fréquence *m*, mélangeur élévateur *m*, transposeur de fréquence *m*; ~**-down time** *n* TELEP durée de disponibilité-indisponibilité *f*; ~**-path** *n* RADIO trajet montant *m*; ~**-path power control** *n* RADIO régulation de la puissance d'émission *f*; ~**-state** *n* TELEP, TEST état de disponibilité *m*; ~ **time** *n* TELEP durée de disponibilité *f*
update[1] *n* GEN, PER APP, TELEP mise à jour *f*
update[2] *vt* GEN, PER APP, TELEP mettre à jour
updating *n* GEN, PER APP, TELEP mise à jour *f*; ~ **procedure** GEN, TELEP procédure de mise à jour *f*
upgrade[1] *n* COMP, ELECT mise à jour *f*, mise à niveau *f*; ~ **board** COMP carte d'extension *f*
upgrade[2] *vt* COMP, ELECT mettre à jour, mettre à niveau, reclasser, améliorer
upgrading *n* COMP, ELECT mise à jour *f* amélioration *f*, mise à niveau *f*, reclassement *m*
uplink *n* RADIO, TELEP liaison montante *f*, liaison ascendante *f*, liaison terre-satellite sens montant *f*; ~ **block** RADIO bloc de liaison montante *m*; ~ **fade** RADIO évanouissement sur le trajet montant *m*; ~ **transmission** RADIO, TRANS transmission sens terre-satellite *f*, transmission sens montant *f*; ~ **transmission phase** RADIO, TRANS phase de transmission sur liaison montante *f*
uplinked: ~ **from** *adj* RADIO originaire de
upload *vt* DATA télécharger
uploading *n* DATA téléchargement *m*
upper: ~ **bound** *n* TRANS borne supérieure *f*; ~ **case** *n* GEN *letters* majuscules *f pl*, PER APP motion haute *f*; ~ **ionosphere** *n* RADIO ionosphère supérieure *f*; ~ **keyboard** *n* COMP clavier supérieur *m*; ~ **quartile** *n* GEN quartile supérieur *m*; ~ **sideband** *n* (*USB*) RADIO, TRANS bande latérale supérieure *f* (*BL sup.*); ~ **subfield** *n* RADIO, TRANS sous-champ supérieur *m*; ~ **tester** *n* (*UT*) DATA, TEST testeur supérieur *m*
UPS *abbr* (*uninterruptible power supply*) POWER UPS (*alimentation non-interruptible*)
upstream: ~ **failure indication** *n* TELEP indication de défaillance en amont *f*
UPT *abbr* (*universal personal telecommunication*) APPL TPU (*télécommunications personnelles universelles*)
uptime *n* ELECT disponibilité *f*, GEN, TELEP temps de disponibilité *m*, TEST durée de bon fonctionnement *f*, temps de bon fonctionnement *m*, temps de disponibilité *m*, temps utilisable *m*
UPU *abbr* (*Universal Postal Union*) TELEP UPU (*Union postale universelle*)
upward: ~ **compatibility** *n* COMP compatibilité ascendante *f*
urban: ~ **area** *n* TV zone urbaine *f*; ~ **network** *n* TELEP, TRANS réseau métropolitain *m*, réseau urbain *m*
urgency: ~ **signal** *n* RADIO signal d'urgence *m*
urgent: ~ **alarm** *n* TELEP alarme majeure *f*, alarme urgente *f*; ~ **call** *n* TELEP appel urgent *m*
URL *abbr* (*uniform resource locator*) COMP *Internet* uniform resource locator *m*
URSI *abbr* (*International Union of Radio Science*) RADIO URSI (*Union radio-scientifique internationale*)
usable: ~ **field strength** *n* RADIO champ utilisable *m*; ~ **scanning line length** *n* TELEP longueur utile de ligne *f*
usage *n* TELEP taux d'utilisation *m*, trafic écoulé *m*
USB *abbr* (*upper sideband*) RADIO, TRANS BL sup. (*bande latérale supérieure*)
USB1 *abbr* (*unscrambled binary ones*) DATA uns binaires non-brouillés *m pl*
use *n* GEN utilisation *m*, *working* fonctionnement *m*; ~ **of network connection TPDU** (*UN TPDU*) DATA TPDU d'utilisation de connexion de réseau *m* (*TPDU UN*)
useful: ~ **area** *n* RADIO surface utile *f*; ~ **life** *n* TELEP, TEST durée de vie utile *f*, vie utile *f*; ~ **output power** *n* ELECT, POWER puissance de sortie utile *f*
Usenet *n* COMP Usenet *m*
user[1]: ~**-defined** *adj* COMP défini par l'utilisateur; ~**-friendly** *adj* COMP, GEN convivial, facile à exécuter, facile à utiliser; ~**-programmable** *adj* COMP programmable par l'utilisateur;

~-to-user *adj* GEN usager à usager; **~-unfriendly** *adj* COMP, GEN non convivial, difficile à exécuter, difficile à utiliser

user[2] *n* COMP, DATA, GEN, TELEP, TV usager *m*, utilisateur *m*, *subscriber* abonné *m*; **~ address** TELEP adresse d'usager *f*; **~ agent** (*UA*) APPL, DATA, TELEP agent utilisateur *m* (*UA*); **~-ASE** DATA, TELEP ASE d'utilisateur *m*; **~ class indicator** TELEP indicateur de catégorie d'usagers *m*; **~-created link** COMP lien créé par l'utilisateur *m*; **~ data** *n pl* COMP, DATA données d'usager *f pl*, données d'utilisateur *f pl*; **~ data field** DATA champ de données d'usager *m*; **~ element** (*UE*) DATA, TELEP élément d'utilisateur *m*; **~ facility** TELEP service offert aux usagers *m*; **~-friendliness** GEN convivialité *f*; **~ handling time** TELEP temps de manipulation *m*; **~ information transfer rate** COMP, DATA débit de transfert de l'information d'usager *m*; **~ name** COMP nom d'utilisateur *m*; **~-network interface** (*UNI*) DATA, TRANS interface utilisateur-réseau *f*; **~ option** (*U*) DATA, TELEP option utilisateur *f* (*U*); **~ part** TELEP sous-système utilisateur *m*; **~'s quality cost** TEST coût qualité supporté par l'utilisateur *m*; **~ service** TELEP service offert aux usagers *m*; **~-signaling bearer service** *AmE*, **~-signalling bearer service** *BrE* DATA, SIG service de support de signalisation d'usager *m*; **~ terminal** TELEP, TRANS, TV terminal d'abonné *m*, terminal d'usager *m*; **~-to-user indicator** DATA indicateur de signalisation d'usager à usager *m*; **~-to-user information** (*UUI*) DATA information d'usager à usager *f*; **~-to-user information message** DATA message d'information d'usager à usager *m*; **~-to-user signaling** *AmE*, **~-to-user signalling** *BrE* (*UUS*) SIG minimessagerie *f*, signalisation d'usager à usager *f* (*UUS*)

usher: **~ in** *vt* COMP introduire en mémoire

UT *abbr* DATA (*upper tester*) testeur supérieur *m*, GEN (*universal time*) heure universelle *f*, TEST (*upper tester*) testeur supérieur *m*

UTC *abbr* (*universal time coordinated*) GEN heure universel de cordonné *f*, temps universel coordonné *m*

utility *n* COMP, DATA utilitaire *m*, programme utilitaire *m*; **~ memory** COMP mémoire utilitaire *f*; **~ program** COMP programme utilitaire *m*, programme de service *m*; **~ routine** COMP routine utilitaire *f*, routine de service *f*

UTP *abbr* (*unshielded twisted pair*) ELECT, TRANS paire symétrique non-blindée *f*, paire torsadée non-blindée *f*

UTS *abbr* (*universal transport system*) ELECT système de transport universel *m*

utter *vt* COMP énoncer

utterance *n* COMP énoncé *m*

UUCP *abbr* (*Unix-to-Unix copy protocol*) DATA UUCP (*Unix-to-Unix copy protocol*)

UUI *abbr* (*user-to-user information*) DATA information d'usager à usager *f*

UUS *abbr* (*user-to-user signaling AmE*, *user-to-user signalling BrE*) SIG UUS (*signalisation d'usager à usager*), minimessagerie *f*

UV *abbr* (*ultraviolet*) ELECT, TRANS UV (*ultraviolet*); **~ radiation** *n* TRANS rayonnement ultraviolet *m*

UW: **~ window** *n* (*unique word window*) DATA fenêtre du mot unique *f*

V

V *abbr* (*voltage*) POWER V (*tension*)
vacancy *n* COMP *in memory*, PARTS *defect* lacune *f*; ~ **marking** TELEP marquage de ligne vacante *m*
vacant: ~ **number** *n* TELEP numéro non-attribué *m*, numéro non-utilisé *m*; ~ **position** *n* TELEP emplacement libre *m*
vaccine *n* COMP, DATA *virus protection* vaccin *m*
vacuum: ~ **factor** *n* ELECT facteur de vide *m*; ~ **lightning arrester** *n* POWER parafoudre à gaz raréfié *m*; ~ **metalization** *AmE*, ~ **metallization** *BrE* *n* PARTS métallisation sous vide *f*; ~ **suction** *n* ELECT aspiration à vide *f*; ~ **tube** *n* ELECT tube à vide *m*
VAD *abbr* (*vapor axial deposition AmE, vapour axial deposition BrE*) ELECT DAV (*dépôt axial en phase vapeur*)
valence: ~ **band** *n* PARTS bande de valence *f*
valid: ~ **cell** *n* COMP cellule valide *f*; ~ **event** *n* DATA, TEST événement valide *m*; ~ **MCSPDU** *n* TELEP unité MCSPDU valide *f*
validation *n* DATA validation *f*
validity: ~ **check** *n* CONTR, DATA contrôle de validité *m*
valley: ~ **point** *n* PARTS point de vallée *m*; ~ **point voltage** *n* PARTS, POWER tension de point de vallée *f*
valuation *n* GEN appréciation *f*
value: ~-**added data service** *n* APPL, DATA service à valeur ajoutée de données *m*; ~-**added network** *n* (*VAN*) DATA, TELEP, TRANS, TV réseau à valeur ajoutée *m* (*RVA*); ~-**added network services** *n pl* (*VANS*) APPL, DATA, TRANS, TV services à valeur ajoutée *m pl* (*SAVA*); ~-**added paging** *n* RADIO radiomessagerie à valeur ajoutée *f*, RM à valeur ajoutée *f*; ~ **assignment** *n* COMP affectation de valeur *f*, attribution de valeur *f*
valve *n* ELECT, PARTS *tube* tube *m*, *inlet, outlet* soupape *f*; ~ **base** ELECT, PARTS culot de tube *m*; ~ **socket** ELECT, PARTS support *m*; ~ **spindle** ELECT, PARTS tige de soupape *f*, vis de soupape *f*
VAM *abbr* (*virtual access method*) SWG méthode à accès virtuel *f*
vamp *n* COMP improvisation *f*
VAN *abbr* (*value-added network*) DATA, TELEP, TRANS, TV RVA (*réseau à valeur ajoutée*)
vandal: ~-**proofing** *n* APPL, ELECT, GEN, PARTS résistance au vandalisme *f*
vane: ~ **attenuator** *n* RADIO atténuateur à lame *m*; ~ **wattmeter** *n* TEST wattmètre à palette *m*, wattmètre à cloison *m*
VANS *abbr* (*value-added network services*) APPL, DATA, TRANS, TV SAVA (*services à valeur ajoutée*)
vapor *AmE see vapour BrE*
vaporware *AmE see vapourware BrE*
vapour: ~ **axial deposition** *n BrE* ELECT dépôt axial en phase vapeur *m* (*DAV*); ~ **barrier** *n BrE* CLIM barrière de vapeur *f*, pare-vapeur *m*; ~ **chemical deposition** *n BrE* ELECT dépôt chimique en phase vapeur *m* (*DCV*); ~ **concentration** *n BrE* CLIM concentration de vapeur *f*; ~-**phase soldering** *n BrE* ELECT brasage en phase vapeur *m*; ~ **pressure** *n BrE* CLIM pression de vapeur *f*
vapourware *n BrE* COMP, DATA logiciel fantôme *m*
varactor *n* ELECT, PARTS diode à capacité variable *f*, diode varicap *f*; ~ **tuning** ELECT accord par varactor *m*
variable: ~ **air capacitor** *n* ELECT, PARTS condensateur variable à air *m*; ~ **antenna coupling** *n* RADIO couplage d'antenne variable *m*; ~ **bit rate** *n* (*VBR*) COMP, DATA, TELEP, TRANS débit binaire variable *m* (*DBV*); ~ **capacitor** *n* ELECT, PARTS condensateur variable *m*; ~ **digital amplifier** *n* COMP amplificateur numérique variable *m*; ~ **digital filter** *n* COMP filtre numérique variable *m*; ~-**frequency drive** *n* CONTR télécommande par variation de fréquence *f*; ~-**frequency oscillator** *n* ELECT, PARTS oscillateur à fréquence variable *m*; ~-**frequency telemetering** *n* TEST télémesure par variation de fréquence *f*; ~-**length code** *n* COMP, DATA, TELEP code à longueur variable *m*; ~-**length coding** *n* (*VLC*) COMP, DATA, TELEP codage à longueur variable *m*; ~-**length numbering** *n* TELEP numérotage ouvert *m*; ~-**length numbering scheme** *n* TELEP plan de numérotage ouvert *m*; ~-**length OGM** *n* PER APP annonce de durée réglable *f*; ~-**length outgoing message** *n* PER APP annonce de durée réglable *f*; ~ **method** *n* TEST méthode des mesures *f*; ~ **mu tube** *n* ELECT tube à blocage éloigné *m*, tube à pente réglable *m*; ~-**packet recording** *n* COMP enregistrement par paquets variables *m*; ~ **queue** *n* TELEP file variable *f*; ~-**resistor network** *n* POWER réseau de résistances pondérées *m*; ~ **text** *n* TELEP texte variable *m*
varistor *n* ELECT, PARTS varistance *f*
VBI *abbr* (*vertical blanking interval*) DATA, TV intervalle de retour de trame *m*, intervalle de suppression de trame *m*
Vbox *n* COMP Vbox *m*
VBR *abbr* (*variable bit rate*) COMP, DATA, TELEP, TRANS DBV (*débit binaire variable*)
VC *abbr* APPL (*virtual container*) CTV (*conteneur virtuel*), DATA (*virtual channel*) canal virtuel *m*, voie virtuelle *f*, (*virtual container*) CTV (*conteneur virtuel*), (*virtual circuit*) circuit virtuel *m*, SIG (*virtual channel*) canal virtuel *m*, voie virtuelle *f*, TELEP (*virtual container*) CTV (*conteneur virtuel*), (*virtual call*) communication virtuelle *f*, TRANS (*virtual channel*) canal virtuel *m*, voie virtuelle *f*, (*virtual circuit*) circuit virtuel *m*
VCC *abbr* (*virtual-channel connection*) DATA, TRANS connexion de canal virtuel *f*, connexion de voie virtuelle *f*
VCCE *abbr* (*virtual-channel connection endpoint*) DATA, TRANS extrémité de connexion de canal virtuel *f*, extrémité de connexion de voie virtuelle*f*
VCCRF *abbr* (*virtual-channel connection-related function*) DATA, TRANS fonction liée à la connexion sur les canaux virtuels *f*, fonction liée à la connexion sur les voies virtuelles *f*
VCF *abbr* (*video command freeze-picture request*) TV commande vidéo de demande de gel de l'image *f*
VCI *abbr* (*virtual-channel identifier*) DATA, TRANS identificateur de canal virtuel *m*, identificateur de voie virtuelle *m*
VC-n *abbr* (*virtual container-n*) APPL, DATA, TELEP conteneur-n virtuel *m*
VCO *abbr* (*voltage-controlled oscillator*) ELECT oscillateur à commande par tension *m*
VCP *abbr* (*videocassette player*) TV VCP (*lecteur de cassette vidéo*)
VCR *abbr* (*videocassette recorder*) TV enregistreur de cassettes vidéo *m*, magnétoscope *m*
VCS *abbr* (*virtual-circuit switch*) SWG CCV (*commutateur à circuits virtuels*)
VCU *abbr* (*video command fast-update request*) TV commande vidéo de demande de rafraîchissement rapide *f*
VCXO *abbr* (*voltage-controlled crystal oscillator*) ELECT oscillateur à quartz contrôlé en tension *m*

VDT *abbr AmE* (*video display terminal, cf VDU BrE*) COMP, TV terminal à écran de visualisation *m*, écran de visualisation *m*
VDU *abbr BrE* (*video display unit, cf VDT AmE*) COMP, TV terminal à écran de visualisation *m*, écran de visualisation *m*
vector *n* COMP, TRANS, TV vecteur *m*; **~ document** DATA *Internet* document vectoriel *m*; **~ line** RADIO vecteur *m*; **~ quantization** DATA, SIG quantification vectorielle *f*; **~ scanning** COMP balayage cavalier *m*, balayage vectoriel *m*; **~ synthesis** COMP synthèse vectorielle *f*
vehicle[1]: **~-mounted** *adj* RADIO monté à bord, embarqué
vehicle[2]: **~ fleet** *n* RADIO flotte *f*, parc automobile *m*; **~ intelligent communication system** *n* (*VICS*) RADIO système de communication pour véhicules intelligents *m*; **~ localization subsystem** *n* RADIO sous-système de localisation des véhicules *m*; **~ localization system** *n* RADIO système de localisation des véhicules *m*
velocity[1]: **~-sensitive** *adj* COMP sensible à la vélocité de frappe
velocity[2] *n* GEN vitesse *f*; **~ gate** RADIO porte de vitesse *f*; **~ indicator** RADIO indicateur de vitesse *m*; **~ modulation** ELECT modulation de vitesse *f*; **~ resolution** RADIO pouvoir séparateur en vitesse *m*; **~ search** RADIO recherche en vitesse *f*; **~-sensitive keyboard** COMP clavier à réponse sensitive *m*
vendor: **~ appraisal** *n* TEST évaluation du fournisseur *f*, évaluation du vendeur *f*; **~ assessment** *n* TEST évaluation du fournisseur *f*, évaluation du vendeur *f*; **~ rating** *n* TEST cotation du fournisseur *f*
veneer *n* PARTS placage *m*
ventilation *n* CLIM ventilation *f*
verbal: **~ message** *n* TELEP message verbal *m*; **~-message transmission service** *n* TELEP, TRANS service de transmission d'un message verbal *m*
verification *n* TEST vérification *f*; **~ of overhead information** TRANS vérification des informations de service *f*
verify[1] *vt* TELEP, TEST vérifier
verify[2] *phr* TELEP vérifiez
verifying *n* TELEP, TEST vérification *f*; **~ instrument** TEST instrument de vérification *m*
vernier *n* TEST vernier *m*
version *n* COMP, DATA, TELEP version *f*
vertex *n* GEN sommet *m*; **~ angle** RADIO angle au sommet de l'alvéole *m*; **~ feed** RADIO alimentation axiale *f*; **~ plate** RADIO *of aerial reflector* plaque au sommet *f*
vertical[1] *adj* GEN vertical
vertical[2] *n* PARTS montant *m*; **~ blanking interval** (*VBI*) DATA, TV intervalle de retour de trame *m*, intervalle de suppression de trame *m*; **~ coverage** RADIO zone de couverture verticale *f*; **~ directivity pattern** RADIO diagramme de directivité vertical *m*; **~ displacement** GEN déplacement vertical *m*; **~ format** DATA, TRANS format vertical *m*; **~ grating logic** TV logique de division verticale *f*; **~ hold** TV commande de synchronisation verticale *f*, commande de stabilité verticale *f*; **~ interval time code** (*VITC*) COMP, DATA code temporel d'intervalle vertical *m*; **~ iron** PARTS montant *m*; **~ multiple cable** SWG câble multiple vertical *m*; **~ polarization** TV polarisation verticale *f*; **~ scan** COMP balayage de trames *m*; **~ tabulation character** (*VT*) COMP caractère de tabulation verticale *m*; **~ unit base** PARTS platine de verticale *f*; **~-unit multiple** SWG multiple à unités verticales *m*
VESA *abbr* (*Video Electronics Standards Association*) COMP, ELECT, TV VESA
vestigial: **~ sideband** *n* (*VSB*) RADIO, TRANS, TV bande latérale résiduelle *f* (*BLR*); **~-sideband emission** *n* RADIO, TRANS émission à bande latérale résiduelle *f*; **~-sideband transmission** *n* RADIO, TRANS émission à bande latérale atténuée *f*
VF *abbr* GEN (*voice frequency*) FV (*fréquence vocale*), TRANS (*videofrequency*), TV (*videofrequency*) VF (*vidéofréquence*); **~ dialing** *AmE*, **~ dialling** *BrE* *n* PER APP, TELEP numérotation FV *f*
V/F *abbr* (*voltage/frequency*) ELECT, POWER tension/fréquence *f*
VFT *abbr* (*voice-frequency telegraphy*) TELEP télégraphie à fréquences vocales *f*, télégraphie harmonique *f*
VGA *abbr* (*video graphics array*) COMP écran VGA *m*
VHDL *abbr* (*very high-speed integrated circuit hardware description language*) COMP langage VHDL *m* (*langage de description de circuits intégrés à très grande vitesse*)
VHF *abbr* (*very high frequency*) RADIO, TV VHF (*très haute fréquence*); **~ network** *n* TV réseau VHF *m*; **~ omnidirectional radio range** *n* (*VOR*) RADIO radiophare omnidirectif à ondes métriques VHF *m* (*VOR*)
VHSIC *abbr* (*very high-speed integrated circuit*) ELECT CITGV (*circuit intégré à très grande vitesse*)
via: **~ hole** *n* ELECT trou de passage *m*
VIA *abbr* (*video indicate active*) TV indication vidéo active
vibrating: **~-reed frequency meter** *n* TEST fréquencemètre à lames vibrantes *m*
vibration: **~ mode** *n* ELECT mode de vibration *m*; **~ test** *n* TEST essai de vibration *m*
vibrator *n* COMP ondulateur *m*, ELECT, RADIO vibreur *m*
VICP *n* (*virtual international connection point*) TELEP point de connexion international virtuel *m*
VICS *abbr* (*vehicle intelligent communication system*) RADIO système de communication pour véhicules intelligents *m*
video[1] *n* APPL, COMP, TV vidéo *f*; **~ acquisition** COMP acquisition vidéo *f*; **~ attribute** TV attribut vidéo *m*; **~ board** COMP carte vidéo *f*; **~ camera** TV caméra vidéo *f*; **~ capture** COMP capture vidéo *f*; **~ card** COMP carte vidéo *f*; **~ carrier** TV porteuse vision *f*; **~ cassette recorder** (*VCR*) COMP, TV enregistreur de cassettes vidéo *m*, magnétoscope *m*; **~ CD** TV CDV (*disque compact vidéo*); **~ codec** COMP codec vidéo *m*; **~ command fast-update request** (*VCU*) TV commande vidéo de demande de rafraîchissement rapide *f*; **~ command freeze-picture request** (*VCF*) TV commande vidéo de demande de gel de l'image *f*; **~ compression** DATA, TV compression vidéo *f*; **~ data** *n pl* COMP, DATA, TV données vidéo *f pl*; **~-data buffering** COMP, DATA, TV mise en mémoire tampon des données vidéo *f*; **~ digitizer** COMP carte vidéo *f*, numériseur vidéo *m*, numériseur d'images vidéo *m*, TV numériseur d'images vidéo *m*, numériseur vidéo *m*; **~ display terminal** *AmE* (*VDT, cf video display unit BrE*) COMP, TV terminal à écran de visualisation *m*, écran de visualisation *m*; **~ display unit** *BrE* (*VDU, cf video display terminal AmE*) COMP, TV terminal à écran de visualisation *m*, écran de visualisation *m*; **~ equipment** TV équipement de vidéo *m*; **~-frequency protection ratio** RADIO rapport de protection en vidéofréquence *m*; **~ graphics array** (*VGA*) COMP écran VGA *m*; **~ input** DATA, TV entrée vidéo *f*; **~ link** TV liaison vidéo *f*; **~ loop** TV boucle vidéo *f*; **~ mail** COMP courrier vidéo *m*; **~ map** RADIO, TV image de carte électronique *f*; **~ mapping** RADIO, TV superposition cartographique *f*; **~-on-demand** (*VOD*) COMP, TV télévidéothèque *f*, vidéo à la demande *f* (*VOD*); **~ outlet** TV source vidéo *f*; **~ processor** COMP processeur vidéo *m*; **~ projector** TV projecteur vidéo *m*; **~ RAM** (*VRAM*) COMP RAM vidéo *f* (*VRAM*); **~ recorder** TV enregistreur de cassettes vidéo *m*, magnétoscope *m*; **~ recording** TV enregistrement vidéo *m*; **~ server** COMP serveur vidéo *m*; **~ services** *n pl* TV vidéographie *f*; **~ signal** RADIO, SIG, TV signal vidéo *m*; **~ source** TV source vidéo *f*; **~ switch** SWG commutateur vidéo *m*; **~ tape** TV bande vidéo *f*; **~ track** COMP, TV piste vidéo *f*; **~ transmission** TRANS, TV vidéotransmission *f*
video[2]: **~ indicate active** *phr* (*VIA*) TV indication vidéo active; **~ indicate ready-to-activate** *phr* (*VIR*) TV indication vidéo prête à être activée; **~ indicate suppressed** *phr* (*VIS*) TV indication vidéo supprimée
Video: **~ Electronics Standards Association** *n* (*VESA*) COMP, ELECT, TV VESA; **~ RISC Compression Processor** *n* COMP Video RISC Compression Processor *m*

videocassette *n* TV cassette vidéo *f*; ~ **player** (*VCP*) TV lecteur de cassette vidéo *m* (*VCP*); ~ **recorder** (*VCR*) TV enregistreur de cassettes vidéo *m*, magnétoscope *m*
videocommunication *n* TV vidéocommunication *f*; ~ **equipment** TV équipement de vidéocommunication *m*; ~ **network** TV réseau de vidéocommunication *m*
videoconference *n* APPL, COMP, TV vidéoconférence *f*, visioconférence *f*
videoconferencing *n* APPL, COMP, TV vidéoconférence *f*, visioconférence *f*
videodisk *n* TV vidéodisque *m*
videofrequency *n* (*VF*) TRANS, TV vidéofréquence *f* (*VF*); ~ **band** TRANS, TV bande de fréquences vidéo *f*, bande vidéofréquences *f*; ~ **protection ratio** TV rapport de protection en vidéofréquence *m*; ~ **signal-to-interference ratio** RADIO, TV rapport signal-brouillage en vidéofréquence *m*
videography *n* COMP, RADIO, TV vidéographie *f*
videomessaging *n* COMP, TV messagerie vidéo *f*, vidéomessagerie *f*
videophone *n* APPL, TELEP, TV vidéophone *m*; ~ **service** APPL, RADIO, TELEP, TV service visiophonique *m*; ~ **switching system** SWG autocommutateur visiophonique *m*
videophony *n* RADIO, TELEP, TV vidéophonie *f*, visiophonie *f*
videotape *n* TV bande vidéo *f*; ~ **player** (*VTP*) TV lecteur de bande vidéo *m*; ~ **recorder** (*VTR*) TV enregistreur de bande vidéo *m*
videotelephone *n* APPL, RADIO, TELEP visiophone *m*, TV vidéophone *m*
videotelephony *n* RADIO, TELEP, TV vidéophonie *f*, visiophonie *f*
videotex *n* COMP, RADIO, TV vidéographie interactive *f*, vidéotex interactif *m*, vidéotex *m*; ~ **access point** TELEP, TV point d'accès vidéotex *m* (*PAVI*); ~ **closed user group** COMP, TV groupe fermé d'usagers vidéotex *m*; ~ **document** COMP document vidéotex *m*; ~ **gateway** TELEP, TV point d'accès vidéotex *m* (*PAVI*); ~ **host** TELEP, TV serveur vidéotex *m*; ~ **interworking** TELEP, TV interfonctionnement vidéotex *m*; ~ **PAD** TV vidéo PAD *f*; ~ **server** TELEP, TV serveur vidéotex *m*; ~ **service** APPL, TELEP, TV service vidéotex *m*; ~ **service center** *AmE*, ~ **service centre** *BrE* APPL, TELEP, TV centre de service vidéotex *m*; ~ **system** APPL, TV système vidéotex *m*; ~ **telesoftware** COMP, TELEP, TV télélogiciel vidéotex *m*; ~ **terminal** COMP, TELEP, TV terminal vidéotex *m*
vidicon *n* ELECT vidicon *m*
view[1] *n* COMP visualisation *f*, mode *m*, vue *f*; ~ **scale** COMP échelle d'affichage *f*
view[2] *vt* COMP visualiser
viewer *n* COMP *device* visionneuse *f*, TV *device* visionneuse *f*, *person* téléspectateur *m*
viewing: ~ **filter** *n* COMP filtre de visualisation *m*
viewphone *n* APPL, RADIO, TELEP visiophone *m*, TV vidéophone *m*
vinyl *n* PARTS vinyle *m*; ~ **acetate plastic** PARTS plastique acétovinylique *m*; ~ **chloride plastic** PARTS plastique chlorovinylique *m*
VIO *abbr* (*virtual input/output*) COMP entrée/sortie virtuelle *f*
violation: ~ **pulse** *n* TRANS impulsion de violation *f*
VIR *abbr* (*video indicate ready-to-activate*) TV indication vidéo prête à être activée
virgin: ~ **medium** *n* GEN support vierge *m*
virtual: ~ **access method** *n* (*VAM*) SWG méthode à accès virtuel *f*; ~ **address** *n* COMP adresse virtuelle *f*; ~ **addressing** *n* COMP adressage virtuel *m*; ~ **analog switching point** *n* SWG extrémité virtuelle analogique *f*; ~ **area** *n* COMP zone virtuelle *f*; ~ **call** *n* (*VC*) TELEP communication virtuelle *f*; ~-**call characteristic** *n* TELEP caractéristique de communication virtuelle *f*; ~ **call switched virtual circuit** *n* SWG communication virtuelle circuit virtuel commuté *f*; ~ **channel** *n* (*VC*) SIG, TRANS canal virtuel *m*, voie virtuelle *f*; ~-**channel connection** *n* (*VCC*) DATA, TRANS connexion de canal virtuel *f*, connexion de voie virtuelle *f*; ~-**channel connection endpoint** *n* (*VCCE*) DATA, TRANS extrémité de connexion de canal virtuel *f*, extrémité de connexion de voie virtuelle *f*; ~-**channel connection-related function** *n* (*VCCRF*) DATA, TRANS fonction liée à la connexion sur les canaux virtuels *f*, fonction liée à la connexion sur les voies virtuelles *f*; ~-**channel identifier** *n* (*VCI*) DATA, TRANS identificateur de canal virtuel *m*, identificateur de voie virtuelle *m*; ~ **circuit** *n* (*VC*) DATA, TRANS circuit virtuel *m*; ~-**circuit bearer service** *n* DATA, TRANS service support de circuit virtuel *m*; ~-**circuit switch** *n* (*VCS*) SWG commutateur à circuits virtuels *m* (*CCV*); ~-**circuit switching node** *n* SWG, TRANS centre de commutation de circuits virtuels *m*; ~ **communication** *n* DATA, TELEP communication virtuelle *f*; ~ **connection** *n* TELEP connection virtuelle *f*; ~ **container** *n* (*VC*) APPL, DATA, TELEP conteneur virtuel *m* (*CTV*); ~ **container-n** *n* (*VC-n*) APPL, DATA, TELEP conteneur-n virtuel *m*; ~ **decision value** *n* TELEP amplitude virtuelle de décision *f*; ~ **device** *n* COMP périphérique virtuel *m*; ~ **DMA driver** *n* DATA virtual DMA driver *m*; ~ **file store** *n* DATA mémoire-fichier virtuelle *f*; ~ **height** *n* RADIO hauteur virtuelle de réflexion *f*, hauteur virtuelle *f*; ~ **image** *n* DATA, TV image virtuelle *f*; ~ **input/output** *n* (*VIO*) COMP entrée/sortie virtuelle *f*; ~ **international connection point** *n* (*VICP*) TELEP point de connexion international virtuel *m*; ~ **machine** *n* COMP machine virtuelle *f*; ~ **memory** *n* (*VM*) COMP, DATA mémoire virtuelle *f*; ~ **mode** *n* COMP mode virtuel *m*; ~ **office** *n* GEN bureau virtuel *m*; ~ **path** *n* (*VP*) DATA, TRANS conduit virtuel *m*, trajet virtuel *m*, voie virtuelle *f*; ~-**path connection** *n* (*VPC*) DATA, TRANS connexion de conduit virtuel *f*, connexion de trajet virtuel *f*, connexion de voie virtuelle *f*; ~-**path connection endpoint** *n* (*VPCE*) DATA, TRANS extrémité de connexion de trajet virtuel *f*, extrémité de connexion de circuit virtuel *f*, extrémité de connexion de voie virtuelle *f*; ~-**path connection-related function** *n* (*VPCRF*) DATA, TRANS fonction liée à la connexion de conduit virtuel *f*; ~-**path identifier** *n* (*VPI*) DATA, TRANS identificateur de conduit virtuel *m*, identificateur de trajet virtuel *m*; ~ **private network** *n* COMP, TELEP réseau virtuel privé *m*; ~ **pushbutton** *n* COMP touche de fonction affichable *f*; ~ **reality** *n* (*VR*) COMP, DATA réalité virtuelle *f* (*RV*); ~-**reality modelling language** *n* (*VRML*) COMP langage de modèles de pages en réalité virtuelle *m*; ~ **SDA server** *n* TELEP serveur SDA virtuelle *m*; ~ **storage** *n* COMP, DATA mémoire virtuelle *f*; ~-**studio system** *n* COMP système de studio virtuel *m*; ~ **switching point** *n* SWG extrémité virtuelle *f*; ~ **terminal** *n* COMP, DATA, TELEP poste virtuel *m*, terminal virtuel *m*; ~ **unit** *n* COMP unité virtuelle *f*; ~ **zero time** *n* TELEP temps zéro virtuel *m*
virus *n* COMP, DATA virus *m*
VIS *abbr* (*video indicate suppressed*) TV indication vidéo supprimée
visibility: ~ **meter** *n* TEST instrument de mesure de la visibilité *m*
visible: ~ **arc** *n* ELECT arc de visibilité *m*; ~ **light** *n* ELECT lumière visible *f*; ~ **radiation** *n* ELECT, TRANS rayonnement de la lumière visible *m*, rayonnement visible *m*
visioconference *n* APPL, COMP, RADIO, TELEP, TV visioconférence *f*
visiophone *n* APPL, COMP, RADIO, TELEP, TV vidéophone *m*, visiophone *m*
visited: ~ **location register** *n* RADIO enregistreur de localisation visitée *m*
visual: ~ **alarm** *n* TEST alarme à indicateur visuel *f*, alarme visuelle *f*; ~ **call system** *n* RADIO, SIG système de signaux lumineux *m*; ~ **carrier** *n* TV porteuse vision *f*; ~ **indication** *n* COMP, TELEP, TEST indication optique *f*; ~ **indicator** *n* COMP,

TELEP, TEST voyant *m*; ~ **indicator strip** *n* COMP, TELEP, TEST série de voyants lumineux *f*; ~ **read-out** *n* COMP affichage *m*; ~ **signal** *n* SIG signal optique *m*; ~ **telephone** *n* APPL, COMP, PER APP, RADIO, TELEP visiophone *m*, TV vidéophone *m*; ~ **telephone service** *n* APPL, RADIO, TELEP, TV service visiophonique *m*; ~ **telephony** *n* RADIO, TELEP, TV vidéophonie *f*, visiophonie *f*

VITC *abbr* (*vertical interval time code*) COMP, DATA code temporel d'intervalle vertical *m*

Viterbi: ~ **decoding** *n* TELEP décodage de Viterbi *m*

vitreous: ~ **silica** *n* TRANS *fibre optics* silice vitreuse *f*, silice fondue *f*

VLBI *abbr* (*very long baseline interferometry*) TEST RILB (*radiointerférométrie à très longue base*)

VLC *abbr* (*variable-length coding*) COMP, DATA, TELEP codage à longueur variable *m*

VLSI *abbr* (*very large-scale integration*) ELECT, PARTS, TELEP, TRANS VLSI (*intégration à très grande échelle*); ~ **circuit** *n* PARTS, TELEP, TRANS circuit VLSI *m*

VM *abbr* APPL (*voice mail*) messagerie vocale *f*, COMP, DATA (*virtual memory*) mémoire virtuelle *f*, (*vocal message, voice message*) MV (*message vocal*), TELEP (*voice mail*) messagerie vocale *f*, (*vocal message, voice message*) MV (*message vocal*); ~ **forwarding** *n* DATA, TELEP retransmission de VM *f*

VMG *abbr* (*voice messaging*) APPL, TELEP messagerie vocale *f*

VMGS *abbr* (*voice messaging system*) APPL, TELEP système de messagerie vocale *m*

VMGS-MS *abbr* (*voice messaging system message store*) APPL, TELEP mémoire de messages de système de messagerie vocale *f*

VMGS-UA *abbr* (*voice messaging system user agent*) APPL, DATA, TELEP agent utilisateur de système de messagerie vocale *m*

VM-MS *abbr* (*voice messaging store*) APPL, TELEP mémoire de messagerie vocale *f*

VN *abbr* (*voice notification*) APPL, TELEP, DATA notification de message vocal *f*

vocal: ~ **date-time stamping** *n* PER APP *answer machine* horodatage vocal *m*; ~ **message** *n* COMP, TELEP message vocal *m* (*MV*); ~ **track** *n* COMP piste son *f*; ~ **user interface** *n* APPL interface vocale *f*

vocoder *n* COMP, RADIO, TELEP vocodeur *m*

vocoding *n* COMP, DATA, RADIO, TELEP codage de la voix *m*

VOD *abbr* (*video-on-demand*) COMP, TV télévidéothèque *f*, VOD (*vidéo à la demande*)

voice[1]: ~**-activated** *adj* CONTR, PER APP, TELEP à commande vocale, commandé par fréquence vocale, commandé par la voix; ~**-controlled** *adj* CONTR, PER APP, TELEP à commande vocale, commandé par fréquence vocale, commandé par la voix; ~**-operated** *adj* CONTR, PER APP, TELEP à commande vocale, commandé par fréquence vocale, commandé par la voix

voice[2] *n* COMP, PER APP, TELEP voix *f*; ~**-activated phone** PER APP, TELEP téléphone à commande vocale *m*; ~ **activation** CONTR, PER APP, TELEP commande par la voix *f*; ~ **amplifier** PER APP, TELEP amplificateur audio *m*, amplificateur basse fréquence *m*; ~ **annotation** COMP, TELEP annotation vocale *f*; ~ **band** DATA bande vocale *f*; ~**-band data detector** DATA détecteur de données dans la bande vocale *m*; ~**-band data ratio** DATA rapport de données dans la bande vocale *m*; ~ **body part** COMP, DATA partie du corps de données vocales *f*; ~ **channel** TELEP, TRANS voie téléphonique *f*; ~ **coder** COMP, RADIO, TELEP vocodeur *m*; ~ **command server** CONTR, PER APP, TELEP serveur à commande vocale *m*; ~ **command system** CONTR, PER APP, TELEP système à commande vocale *m*; ~ **control** CONTR, PER APP, TELEP commande par la voix *f*, commande parlée *f*, commande vocale *f*; ~**-controlled operation** CONTR, SWG, TELEP commutation à commande vocale *f*; ~**-data integration** DATA intégration voix-données *f*; ~**-data packet switch** DATA, SWG, TELEP, TRANS commutateur de paquets voix-données *m*; ~ **dialer** *AmE see voice dialler BrE*; ~ **dialing** *AmE see voice dialling BrE*; ~ **dialler** *BrE* TELEP composeur vocal *m*; ~ **dialling** *BrE* TELEP numérotation par commande vocale *f*; ~ **digitization** COMP, TELEP numérisation de la voix *f*, numérisation des signaux vocaux *f*; ~ **frequency** (*VF*) GEN fréquence vocale *f* (*FV*); ~**-frequency band** TRANS bande des fréquences vocales *f*; ~**-frequency code signaling** *AmE*, ~**-frequency code signalling** *BrE* SIG, TELEP signalisation par fréquences vocales *f*; ~**-frequency key sending** TELEP numérotation par fréquence vocale *f*; ~**-frequency signaling receiver** *AmE*, ~**-frequency signalling receiver** *BrE* SIG, TELEP récepteur de signalisation à fréquence vocale *m*; ~**-frequency telegraph** TELEP faisceau de système de télégraphie harmonique *m*; ~**-frequency telegraph circuit** TELEP, TRANS circuit de télégraphie harmonique *m*; ~**-frequency telegraphy** (*VFT*) TELEP télégraphie à fréquences vocales *f*, télégraphie harmonique *f*; ~ **grade** TELEP qualité téléphonique *f*; ~ **identification** APPL, CONTR identification de la voix *f*; ~**-input device** COMP dispositif d'entrée vocale *m*, récepteur de voix *m*; ~ **mail** (*VM*) APPL, TELEP messagerie vocale *f*; ~**-mail system user** APPL, TELEP utilisateur de la messagerie vocale *m*; ~ **mailbox** APPL, TELEP boîte à lettres vocale *f*; ~ **message** (*VM*) COMP, DATA, TELEP message vocal *m* (*MV*); ~ **message server** COMP, DATA serveur de messagerie vocale *m*; ~ **messaging** (*VMG*) APPL, TELEP messagerie vocale *f*; ~ **messaging notification** APPL, TELEP notification de messagerie vocale *f*; ~ **messaging service** APPL, TELEP service de messagerie vocale *m*; ~**-messaging store** (*VM-MS*) APPL, TELEP mémoire de messagerie vocale *f*; ~ **messaging system** (*VMGS*) APPL, TELEP système de messagerie vocale *m*; ~**-messaging system message store** (*VMGS-MS*) APPL, TELEP mémoire de messages de système de messagerie vocale *f*; ~ **messaging system user** APPL, DATA, TELEP utilisateur du système de messagerie vocale *m*; ~ **messaging system user agent** (*VMGS-UA*) APPL, DATA, TELEP agent utilisateur du système de messagerie vocale *m*; ~ **messaging user** APPL, TELEP utilisateur de messagerie vocale *m*; ~ **network** TELEP, TRANS réseau téléphonique *m*; ~ **notification** (*VN*) APPL, TELEP, DATA notification de message vocal *f*; ~**-operated coder** COMP, RADIO, TELEP codeur à fréquences vocales *m*, vocodeur *m*; ~**-operated relay** PARTS, SWG commutateur à commande vocale *m*; ~**-operated switch** PARTS, SWG commutateur à commande vocale *m*; ~**-operated switching** CONTR, SWG commutation à commande vocale *f*; ~**-order wire** TELEP, TRANS voie de service téléphonique *f*; ~**-order wire port** TELEP, TRANS accès de voie de service téléphonique *m*; ~ **paging** RADIO recherche de personne téléphonique *f*; ~ **print** COMP, TELEP empreinte vocale *f*, signature vocale *f*; ~ **prompt** PER APP *answer machine* guidage vocal *m*; ~ **quality** ELECT, TELEP qualité vocale *f*; ~ **recognition** APPL, COMP, TELEP reconnaissance de la parole *f*, reconnaissance vocale *f*; ~ **recognizer** APPL, COMP, TELEP unité de reconnaissance vocale *f*, unité de reconnaissance de la parole *f*; ~ **response** TELEP réponse vocale *f*; ~ **response system** TELEP système à réponse vocale *m*; ~ **response unit** APPL, TELEP unité à réponse vocale *f*; ~ **server** COMP, DATA serveur vocal *m*; ~ **services** *n pl* GEN services vocaux *m pl*; ~**-signal processing coupler** TELEP coupleur de traitement à signal vocal *m*; ~ **spurt** TELEP bribe de parole *f*; ~ **switching equipment** SWG appareil à commutation vocale *m*; ~ **synthesis** APPL, COMP, TELEP synthèse vocale *f*; ~ **synthesis unit** APPL, COMP, TELEP organe de synthèse vocale *m* (*OSV*); ~ **transmission** TRANS transmission vocale *f*; **~/cw** TELEP phonie/graphie *f*

voiced *adj* APPL, COMP, TELEP voisé

void *n* GEN manque *m*, vide *m*

volatile: ~ **fault** *n* TEST panne intermittente *f*, panne temporaire *f*; ~ **memory** *n* COMP, DATA mémoire non rémanente *f*, mémoire volatile *f*; ~ **storage** *n* COMP, DATA mémoire non rémanente *f*, mémoire volatile *f*

volatility *n* COMP, DATA non-rémanence *f*, volatilité *f*

volt *n* ELECT, POWER volt *m*; ~**-ampere** ELECT, POWER voltampère *m*

voltage[1]: ~**-controlled** *adj* CONTR, POWER commandé en tension, commandé par tension; ~**-dependent** *adj* CONTR, POWER dépendant de la tension

voltage[2] *n* (*V*) POWER différence de potentiel *f*, tension *f* (*V*); ~ **adaptor switch** POWER sélecteur de tension *m*; ~ **amplification factor** POWER facteur d'amplification en tension *m*; ~ **amplifier** POWER amplificateur de tension *m*; ~ **attenuation** POWER atténuation de tension *f*; ~ **between lines** POWER tension composée *f*; ~**-controlled amplifier** POWER amplificateur commandé par tension *m*; ~**-controlled crystal oscillator** (*VCXO*) ELECT oscillateur à quartz contrôlé en tension *m*; ~**-controlled filter** COMP, POWER filtre commandé par tension *m*; ~**-controlled oscillator** (*VCO*) ELECT oscillateur à commande par tension *m*; ~**-controlled transistor** ELECT transistor commandé en tension *m*; ~ **converter** POWER convertisseur de tension *m*; ~**-dependent resistor** PARTS, POWER résistance sensible à la tension *f*; ~ **discharge-current curve** TELEP courbe de tension en fonction du courant de décharge *f*; ~ **divider** POWER diviseur de tension *m*; ~ **doubler** POWER doubleur de tension *m*; ~ **drop** POWER chute de tension *f*; ~ **feed** POWER alimentation de tension *f*; ~ **fluctuation** POWER, TEST fluctuation de tension *f*; ~ **gain** POWER gain en tension *m*; ~ **jump** POWER saut de tension *m*; ~ **level** POWER niveau de tension *m*; ~ **limiter** POWER limiteur de tension *m*; ~ **loss** POWER perte de tension *f*; ~ **monitor** CONTR, POWER contrôleur de tension *m*, contrôleur secteur *m*; ~ **peak** POWER crête de tension *f*; ~ **rating** POWER tension d'isolement *f*, tension nominale *f*; ~ **reference diode** ELECT, PARTS diode de référence de tension *f*; ~ **reference tube** ELECT, PARTS tube de référence de tension *m*; ~ **regulation** ELECT, POWER stabilisation de tension *f*; ~ **regulator** ELECT, POWER régulateur de tension *m*; ~ **regulator diode** ELECT, PARTS diode régulatrice de tension *f*; ~ **regulator tube** ELECT, POWER tube régulateur de tension *m*, tube stabilisateur de tension *m*; ~ **relay** PARTS, POWER relais de tension *m*; ~ **saturation current** ELECT, POWER courant de saturation anodique *m*; ~ **selector** POWER sélecteur de tension *m*; ~ **simulator** POWER simulateur de tension *m*; ~ **spike** ELECT, POWER surtension *f*; ~ **stabilizer** ELECT, POWER stabilisateur de tension *m*; ~**-stabilizing circuit** ELECT circuit stabilisateur de tension *m*; ~**-stabilizing tube** ELECT, POWER tube régulateur de tension *m*, tube stabilisateur de tension *m*; ~ **standing wave** (*VSW*) RADIO, TRANS onde de tension stationnaire *f* (*OTS*); ~ **standing wave ratio** (*VSWR*) RADIO, TRANS rapport d'ondes de tension stationnaires *m* (*ROS*), taux d'ondes stationnaires *m* (*TOS*), taux d'ondes de tension stationnaires *m* (*TOS*); ~ **surge** ELECT, POWER onde de tension *f*, surtension *f*; ~ **test** POWER, TEST essai de tension *m*; ~**-tuned magnetron** ELECT, RADIO magnétron accordé en tension *m*; ~ **variation** POWER, TEST fluctuation de tension *f*, variation de tension *f*

voltage/frequency *n* (*V/F*) ELECT, POWER tension/fréquence *f*

volume *n* COMP, ELECT, GEN, POWER, RADIO volume *m*; ~ **change** CONTR modification de volume *f*; ~ **control** CONTR bouton de réglage de volume *m*, commande d'intensité sonore *f*, commande de volume *f*; ~ **density of charge** POWER densité volumique de charge *f*; ~ **envelope** COMP enveloppe de volume *f*; ~ **range control** CONTR réglage de la dynamique *m*; ~ **resistance** POWER résistance volumique *f*; ~ **resistivity** ELECT, POWER résistivité volumique *f*

volumetric: ~ **coverage** *n* RADIO volume de couverture *m*

VOR *abbr* (*VHF omnidirectional radio range*) RADIO VOR (*radiophare omnidirectif à ondes métriques VHF*)

vowel *n* GEN voyelle *f*

voxel *n* COMP, TV voxel *m*

VP *abbr* (*virtual path*) DATA, TRANS conduit virtuel *m*, trajet virtuel *m*, voie virtuelle *f*

VPC *abbr* (*virtual-path connection*) DATA, TRANS connexion de conduit virtuel *f*, connexion de trajet virtuel *f*, connexion de voie virtuelle *f*

VPCE *abbr* (*virtual-path connection endpoint*) DATA, TRANS extrémité de connexion de trajet virtuel *f*, extrémité de connexion de circuit virtuel *f*, extrémité de connexion de voie virtuelle *f*

VPCRF *abbr* (*virtual-path connection-related function*) DATA, TRANS fonction liée à la connexion de conduit virtuel *f*

VPI *abbr* (*virtual-path identifier*) DATA, TRANS identificateur de conduit virtuel *m*, identificateur de trajet virtuel *m*

VR *abbr* (*virtual reality*) COMP, DATA RV (*réalité virtuelle*)

VRAM *abbr* (*video RAM*) COMP VRAM (*RAM vidéo*)

VRML *abbr* (*virtual-reality modelling language*) COMP langage de modèles de pages en réalité virtuelle *m*

VSAT *abbr* (*very small aperture terminal*) DATA, PER APP, TRANS TTPO (*micro-station, micro-terminal*); ~ **network** *n* DATA, TRANS réseau VSAT *m*; ~ **terminal** *n* RADIO station VSAT *f*

VSB *abbr* (*vestigial sideband*) RADIO, TRANS, TV BLR (*bande latérale restante*)

V-shaped: ~ **aerial** *n* RADIO antenne à fils inclinés *f*; ~ **antenna** *n* RADIO antenne à fils inclinés *f*

VSM *abbr* (*Swiss Association of Machinery Manufacturers*) GEN VSM (*Société suisse des constructeurs de machines*)

VSW *abbr* (*voltage standing wave*) RADIO, TRANS OTS (*onde de tension stationnaire*)

VSWR *abbr* (*voltage standing wave ratio*) RADIO, TRANS ROS (*rapport d'ondes stationnaires en tension*), TOS (*taux d'ondes stationnaires, taux d'ondes de tension stationnaires*)

VT *abbr* (*vertical tabulation character*) DATA, RADIO caractère de tabulation verticale *m*

VTP *abbr* (*videotape player*) TV lecteur de bande vidéo *m*

VTR *abbr* (*videotape recorder*) TV enregistreur de bande vidéo *m*

VU: ~**-meter** *n* TEST VU-mètre *m*

W

wafer *n* ELECT *in semiconductor*, PARTS galette *f*, tranche *f*
WAIS *abbr* (*wide-area information server*) DATA WAIS
wait: **~-for-overhead message** *n* TELEP attente de message de service *f*
waiting: **~-allowed facility** *n* TELEP service complémentaire d'attente autorisée *m*; **~ call** *n* TELEP appel en attente *m*; **~ state** *n* COMP, TELEP état d'attente *m*; **~ time** *n* COMP, TELEP temps d'attente *m*
wake: **~-up service** *n* TELEP réveil automatique *m*, réveil téléphonique *m*, service de réveil automatique *m*
walking: **~ strobe pulse** *n* RADIO impulsion de strobage progressive *f*
wall: **~ AC power socket** *n* COMP, ELECT, POWER prise secteur *f*; **~ alternative current power socket** *n* COMP prise secteur alternative *f*; **~ bracket** *n* PER APP support mural *m*; **~ cabinet** *n* TELEP armoire murale *f*; **~-mountable interconnect center** *AmE*, **~-mountable interconnect centre** *BrE* *n* ELECT centre d'interconnexion mural *m*; **~-mountable splice center** *AmE*, **~-mountable splice centre** *BrE* *n* ELECT coffret d'épissure mural *m*; **~-mounted installation** *n* TELEP installation murale *f*; **~-mounted telephone set** *n* TELEP poste téléphonique mural *m*; **~ mounting** *n* ELECT élément mural *m*, PER APP, TELEP montage mural *m*; **~ outlet** *n* COMP, ELECT, POWER, TELEP prise murale *f*; **~ socket** *n* COMP, ELECT, POWER, TELEP prise murale *f*; **~ telephone** *n* TELEP téléphone mural *m*; **~ telephone set** *n* TELEP poste mural *m*; **~ terminal box** *n* COMP, ELECT, POWER, TELEP prise murale *f*; **~ terminal strip** *n* POWER réglette de raccordement murale *f*
wallboard *n* PARTS plaque en fibre de bois *f*
wallplate *n* ELECT, PARTS plaque murale *f*
WAN *abbr* (*wide-area network*) COMP, DATA, TELEP, TRANS réseau longue distance *m*, réseau longue portée *m*; **~ terminal** *n* COMP terminal WAN *m*
wand: **~ reader** *n* COMP *bar codes* crayon lecteur *m*
wanted: **~ emission** *n* ELECT, RADIO, SIG émission utile *f*; **~ signal** *n* ELECT, RADIO, SIG signal utile *m*; **~-to-unwanted carrier power ratio** *n* ELECT, RADIO, SIG rapport de puissance de porteuse utile-porteuse brouilleuse *m*
warble[1] *n* GEN hululement *m*
warble[2] *vi* GEN hululer
warming *n* CLIM, GEN échauffement *m*
warning: **~ lamp** *n* CONTR avertisseur lumineux *m*; **~ light** *n* CONTR avertisseur lumineux *m*; **~ limit** *n* CONTR limite de surveillance *f*; **~ tone** *n* TELEP tonalité d'avertissement *f*
warp *vt* ELECT, PARTS gauchir
warping *n* COMP déformation *f*, ELECT, PARTS gauchissement *m*
waste: **~ traffic** *n* TELEP trafic perdu *m*
watch: **~-keeping receiver** *n* PER APP écouteur témoin *m*, RADIO récepteur de veille *m*; **~ receiver** *n* PER APP écouteur témoin *m*, RADIO récepteur de veille *m*
watchdog: **~ timer** *n* ELECT horloge chien de garde *f*
watching *n* RADIO veille *f*; **~ mode** RADIO mode veille *m*
water[1]: **~-cooled** *adj* CLIM refroidi par eau; **~-repellent** *adj* PARTS hydrofuge
water[2]: **~-blocking compound** *n* PARTS produit résistant à l'eau *m*; **~ cooling** *n* CLIM refroidissement par eau *m*; **~ load** *n* RADIO charge à l'eau *f*, charge liquide *f*
waterproof *vt* PARTS imperméabiliser, hydrofuger
waterproofing *n* PARTS étanchéisation *f*
watertight *adj* PARTS étanche
WATS *abbr* (*wide-area telephone service*) TELEP WATS (*service interurbain planifié*); **~ zone** *n* TELEP zone WATS *f*
wattmeter *n* POWER, TEST wattmètre *m*
wave *n* GEN onde *f*; **~ amplification** ELECT amplification d'onde *f*; **~ analyser** *BrE*, **~ analyzer** *AmE* TEST analyseur d'ondes *m*; **~ coherence** ELECT cohérence d'onde *f*; **~ coupling** ELECT couplage d'ondes *m*; **~ diffraction** ELECT diffraction d'onde *f*; **~ dispersion** ELECT dispersion d'ondes *f*; **~ generation** ELECT génération d'ondes *f*; **~ generator** ELECT, RADIO générateur d'ondes *m*; **~ impedance** ELECT, RADIO impédance d'onde *f*; **~ interference** ELECT, RADIO interférence d'ondes *f*; **~ mechanics** GEN mécanique ondulatoire *f*; **~ number** TRANS nombre d'ondes *m*; **~ optics** ELECT, TRANS *fibre optics* optique ondulatoire *f*; **~ polarization** ELECT, RADIO polarisation des ondes *f*; **~ propagation** ELECT, RADIO propagation d'onde *f*; **~ shadowing effect** RADIO effet d'écran des ondes *m*; **~ shape** GEN forme d'onde *f*; **~ tilt** RADIO inclinaison de l'onde de sol *f*; **~ train** ELECT, RADIO train d'ondes *m*; **~ transmission** TRANS transmission d'ondes *f*; **~ trap** RADIO piège à ondes *m*, éliminateur de brouillage *m*, filtre antibrouillage *m*, filtre d'antenne *m*, circuit bouchon *m*
waveform *n* GEN forme d'onde *f*; **~ corrector** ELECT correcteur de forme d'onde *m*; **~ distortion** GEN, SIG distorsion du signal *f*; **~ monitor** CONTR écran de contrôle des signaux *m*
wavefront *n* GEN front d'onde *m*, surface d'onde *f*
waveguide *n* DATA, RADIO, TRANS guide d'ondes *m* (*GO*); **~ aerial** RADIO antenne à guide à fentes *f*; **~ antenna** RADIO antenne à guide à fentes *f*; **~ attenuator** RADIO, TRANS affaiblisseur en guide d'ondes *m*, atténuateur en guide d'ondes *m*; **~ bolting** RADIO, TRANS verrouillage de guide d'ondes *m*; **~ connection** RADIO, TRANS raccord de guides d'ondes *m*; **~ coupler** RADIO, TRANS coupleur de guides d'ondes *m*; **~ dispersion** RADIO, TRANS dispersion de guidage *f*, dispersion de guide d'ondes *f*; **~-fed aperture aerial** RADIO antenne à ouverture alimentée de guide d'ondes *f*; **~-fed aperture antenna** RADIO antenne à ouverture alimentée de guide d'ondes *f*; **~ gasket** PARTS joint de guide d'ondes *m*; **~ mode** RADIO, TRANS mode de guide d'ondes *m*; **~ phase changer** RADIO déphaseur de guide d'ondes *m*; **~ phase shifter** RADIO déphaseur de guide d'ondes *m*; **~ shim** PARTS, RADIO, TRANS cale de guide d'ondes *f*, cale mince de raccordement *f*; **~ switch** SWG commutateur de guide d'ondes *m*; **~-to-coax adaptor** TRANS transition guide-coaxial *f*; **~ wire** PARTS guide d'onde unifilaire *m*
wavelength *n* RADIO, TRANS longueur d'onde *f*; **~-division multiplex** SWG, TELEP, TRANS multiplexage en longueur d'onde *m*, multiplexage par répartition en longueur d'onde *m* (*MRL*); **~-division multiplex coupler** TELEP, TRANS coupleur par répartition en longueur d'onde *m*; **~-division multiplexer** (*WDM*) SWG, TELEP, TRANS multiplexeur en longueur d'onde *m* (*MRL*), multiplexeur spectral *m*; **~-division multiplexing** (*WDM*) SWG, TELEP, TRANS multiplexage par répartition en longueur d'onde *m* (*MRL*); **~-division switch** SWG commutateur en longueur d'onde *m*; **~ peak** GEN crête de longueur d'onde *f*; **~ switching** SWG commutation de longueur d'onde *f*
wavemeter *n* RADIO diffusiomètre vecteur vagues *m*, TEST ondemètre *m*
WDM *abbr* SWG, TELEP, TRANS (*wavelength-division*

multiplexing) MRL (*multiplexage par répartition en longueur d'onde*), (*wavelength-division multiplexer*) MRL (*multiplexeur en longueur d'onde*), multiplexeur spectral *m*
weakly: **~-guiding fiber** *AmE*, **~-guiding fibre** *BrE n* TRANS fibre à guidage faible *f*
weakness: **~ failure** *n* TEST problème de fragilité *m*
wear: **~ blade** *n* PARTS lame d'usure *f*; **~-out failure** *n* TEST *dependability* défaillance par usure *f*, défaillance par vieillissement *f*; **~-out failure period** *n* TEST période de défaillance par usure *f*; **~-out life** *n* TEST durée de vie jusqu'à l'usure *f*; **~ shim** *n* PARTS, SWG cale d'usure *f*, joint d'usure *f*; **~ and tear** *n* COMP usure normale *f*
weather: **~ forecast** *n* TELEP météo *f*, prévisions atmosphériques *f pl*, prévisions météorologiques *f pl*; **~ line** *n* APPL, TELEP météo *f*, prévisions atmosphériques *f pl*, prévisions météorologiques *f pl*; **~ radar** *n* RADIO radar météorologique *m*; **~-sealed cable** *n* POWER câble étanche *m*, câble intempérisé *m*
web *n* COMP, DATA module de liens *m*; **~ file** COMP, DATA fichier de liens *m*
Web *n* COMP, DATA *Internet* Web *m*; **~ browser** COMP explorateur Web *m*; **~ document** COMP, DATA *Internet* document Web *m*; **~ page** COMP, DATA *Internet* page Web *f*; **~ server** COMP, DATA serveur W3 *m*, serveur Web *m*
weighted: **~ keyboard** *n* COMP clavier pondéré *m*; **~ terminal coupling loss** *n* (*TCLw*) TELEP équivalent pondéré de couplage du terminal *m*
weighting: **~ filter** *n* ELECT filtre à pondération *m*, filtre pondérateur *m*, filtre psophométrique *m*; **~ network** *n* TRANS réseau de pondération *m*, réseau pondérateur *m*
welcome: **~ page** *n* COMP, DATA écran d'accueil *m*, page d'accueil *f*
weld[1] *n* PARTS soudure *f*
weld[2] *vt* PARTS souder
welding *n* PARTS, TRANS soudage *m*, soudure *f*
wet: **~-bulb thermometer** *n* ELECT thermomètre mouillé *m*; **~ radome** *n* RADIO radôme humide *m*, radome humide *m*
wetting *n* ELECT *of contacts* mouillage *m*; **~ current** ELECT, POWER courant de mouillage *m*
what: **~ you see is what you get** *phr* (*WYSIWYG*) COMP tel écran, tel écrit (*WYSIWYG*)
Wheatstone: **~ automatic system** *n* TELEP système Wheatstone *m*; **~ bridge** *n* POWER pont de Wheatstone *m*; **~ system** *n* TELEP système Wheatstone *m*
wheel *n* COMP roue *f*
whereabouts *n* RADIO localisation *f*
whip: **~ aerial** RADIO antenne fouet *f*; **~ antenna** *n* RADIO antenne fouet *f*
whistler: **~-mode propagation** *n* RADIO propagation suivant le mode des sifflements *f*
white *n* DATA, TELEP blanc *m*; **~ crushing** COMP, TELEP, TRANS écrasement des blancs *m*; **~ lacquer** ELECT laque blanche *f*; **~ level** COMP, TELEP, TRANS niveau du blanc *m*; **~ noise** RADIO, TELEP bruit blanc *m*
whitening: **~ filter** *n* ELECT filtre blanchissant *m*, filtre décolorant *m*
who: **~-are-you signal** *n* (*WRU*) SIG, SWG signal qui est là? *m*
Whois *n* DATA *Internet* Whois *m*
WIC *abbr* (*wall-mountable interconnect center AmE*, *wall-mountable interconnect centre BrE*) ELECT centre d'interconnexion mural *m*
wicking *n* ELECT, PARTS effet de mèche *m*
wide: **~-angle side lobe** *n* RADIO lobe latéral de grande ouverture *m*; **~ area** *n* RADIO, TRANS zone étendue *f*; **~-area information server** *n* (*WAIS*) COMP, TELEP WAIS; **~-area network** *n* (*WAN*) COMP, TELEP réseau longue distance *m*, réseau longue portée *m*; **~-area system** *n* COMP, TELEP, TRANS système desservant des zones étendues *m*; **~-area telephone service** *n* (*WATS*) TELEP service interurbain planifié *m* (*WATS*); **~-deviation angle modulation** *n* TRANS modulation angulaire à grande excursion *f*
wideband *n* TRANS bande large *f*; **~ amplifier** ELECT amplificateur à large bande *m*; **~ channel** TRANS voie à large bande *f*, canal à large bande *m*; **~ circuit** TRANS circuit à large bande *m*; **~ data transmission** DATA, TRANS transmission de données multiplexées *f*; **~ noise** DATA bruit à large bande *m*; **~ radio relay system** RADIO faisceau hertzien à large bande *m*; **~ receiver** RADIO récepteur à large bande *m*; **~ signal** ELECT, SIG, TRANS signal à large bande *m*; **~ switching network** SWG, TRANS réseau de connexion à large bande *m*; **~ telephony** TELEP téléphonie en bande élargie *f*; **~ transmission** TRANS transmission à large bande *f*; **~ video** TV vidéo à large bande *f*
widespread: **~ area** *n* RADIO, TRANS zone étendue *f*
widow *n* COMP veuve *f*
width *n* COMP, RADIO largeur *f*; **~ coding** RADIO codage par modulation de largeur *m*; **~ of the effective overall noise band** RADIO, TRANS largeur de bande effective globale de bruit *f*; **~ variation** COMP variation de largeur *f*
Wiegand: **~ effect** *n* DATA technique Wiegand *f*
wilco *phr* RADIO, TELEP wilco
wild: **~ connection** *n* COMP *Internet* connexion sauvage *f*
wildcard: **~ character** *n* COMP caractère générique *m*
wind *vt* PARTS, POWER bobiner, enrouler
winding *n* PARTS, POWER bobinage *m*, enroulement *m*; **~ data** *n pl* POWER caractéristiques d'enroulement *f pl*, caractéristiques de bobinage *f pl*; **~ instructions** *n pl* POWER instructions de bobinage *f pl*, instructions d'enroulement *f pl*; **~ layer** POWER couche à l'enroulement *f*, couche de bobinage *f*; **~ resistance** POWER résistance à l'enroulement *f*, résistance au bobinage *f*; **~ space** POWER espace d'enroulement *m*, espace de bobinage *m*; **~ tapping** POWER prise d'enroulement *f*, prise de bobinage *f*; **~ wire** POWER fil de bobinage *m*, fil d'enroulement *m*
window *n* COMP, DATA, RADIO, TELEP, TRANS, TV fenêtre *f*; **~ jamming** RADIO brouillage par leurres *m*, brouillage par paillettes *m*
windowpane *n* COMP vantail *m*
wink[1]: **~ signal** *n* DATA, SIG signal de transmission de numérotation *m*; **~ start signal** *n* DATA, SIG signal de transmission de numérotation *m*
wink[2] *vt* COMP, TELEP, TEST clignoter
winking *n* COMP, TELEP, TEST clignotement *m*
wipe[1] *n* TV volet *m*, enchaînement par volet *m*
wipe[2]: **~ out** *vt* COMP effacer
wiper *n* SWG balai *m*; **~ arm** SWG bras porte-balais *m*, bras porte-contacts *m*; **~ set** SWG groupe de balais *m*
wiping: **~ contact** *n* PARTS contact glissant *m*
wire[1] *n* GEN câble *m*, fil *m*, fil électrique *m*; **~ aerial** RADIO antenne à fil *f*, antenne filaire *f*; **~ antenna** RADIO antenne à fil *f*, antenne filaire *f*; **~ area** POWER section de fil *f*; **~ bonding** PARTS soudage par fils *m*; **~ end** PARTS extrémité de fil *f*; **~ fastener** PARTS attache-fil *f*; **~-frame model** PARTS modèle en fil de fer *m*, modèle filaire *m*; **~ fuse** PARTS, POWER fusible à fil *m*; **~ gage** *AmE*, **~ gauge** *BrE* PARTS jauge de fils *f*; **~ guide** PARTS guide-fils *m*; **~ guide beam** PARTS barre guide-fils *f*; **~-guide strip** PARTS profilé guide-fils *m*; **~-minder** ELECT passe-fils *m*; **~ netting** PARTS grillage *m*, treillis métallique *m*; **~-running diagram** PARTS schéma de pose de fils *m*; **~ spring** SWG ressort à fil *m*; **~-spring relay** SWG relais de ressort à fil *m*; **~ through connection** ELECT, POWER connexion transversale à fil *f*; **~ transpositions** *n pl* ELECT inversions de fils *f pl*; **~ tweezers** *n pl* PARTS brucelles *f pl*; **~-wound resistor** PARTS, POWER résistance bobinée *f*; **~-wrap connection** PARTS connexion enroulée *f*; **~-wrap connection list** PARTS liste des connexions enroulées *f*; **~-wrap plane** PARTS plaquette de connexions enroulées *f*; **~-wrap terminal**

PARTS borne pour connexion enroulée *f*; **~-wrapping** PARTS connexion par enroulement *f*

wire[2] *vt* PARTS, POWER *connect* raccorder, interconnecter, relier, connecter, TV câbler

wired: **~ handset** *n* RADIO, TELEP poste filaire *m*; **~-logic system** *n* CONTR système à logique câblée *m*; **~ network** *n* RADIO, TELEP réseau filaire *m*; **~-program control system** *AmE*, **~-programme control system** *BrE n* (*WPC system*) CONTR système de contrôle à programme câblé *m*; **~ rack** *n* PARTS, TELEP bâti câblé *m*; **~ sequence controller** *n* COMP, CONTR séquence câblée *f*; **~ telephone network** *n* TELEP réseau téléphonique filaire *m*

wireless[1] *adj* COMP, RADIO, TELEP sans fil

wireless[2]: **~ hearing aid receiver** *n* RADIO récepteur de prothèse auditive sans fil *m*; **~ PBX** *n* APPL, CONTR, SWG, TELEP autocommutateur privé sans fil *m*; **~ telegraphy** *n* RADIO, TELEP radiotélégraphie *f*, télégraphie sans fil *f* (*TSF*); **~ telephony** *n* RADIO, TELEP radiophonie *f*, radiotéléphonie *f*

wiretap *vt* TELEP *conversation* écouter clandestinement, *telephone* mettre sur écoute

wiretapping *n* TELEP écoute clandestine *f*, écoute téléphonique filaire *f*, branchement clandestin *m*, mise sur écoute *f*

wiring *n* PARTS *connecting* câblage, connexion *f*, interconnexion *f*, raccordement *m*, fils *m pl*, POWER *connecting* connexion *f*, interconnexion *f*, raccordement *m*, *laying wire* câblage *m*, TV câblage *m*; **~ board** COMP tableau de connexions *m*; **~ change** PARTS modification des raccordements *f*; **~ check** PARTS, TEST contrôle de câblage *m*, essai de câblage *m*; **~ closet** ELECT armoire de câblage *f*, local de câblage *m*; **~ concentrator** ELECT concentrateur de câblage *m*; **~ diagram** PARTS plan de câblage *m*, schéma de câblage *m*; **~ document** PARTS document de câblage *m*; **~ drawing** PARTS plan de connexions *m*, schéma de connexions *m*; **~ error** PARTS erreur de câblage *f*; **~ fixture** PARTS gabarit pour connexions intérieures *m*; **~ form** PARTS nappe de fils *f*; **~ harness** ELECT câblage électrique *m*; **~ jig** PARTS gabarit pour connexions intérieures *m*; **~ table** PARTS tableau de réalisation *m*; **~ test** TEST contrôle de câblage *m*, essai de câblage *m*; **~ test set** TEST équipement d'essai de câblage *m*; **~ tester** TEST équipement d'essai de câblage *m*

withdraw *vt* COMP retirer, DATA retraire, RADIO fermer, extraire, retraire

withdrawal *n* COMP retrait *m*, RADIO retrait *m*, *of service* fermeture *f*

WMRA *abbr* (*write many read always*) COMP WMRA

wooden: **~ paddle** *n* PARTS palette en bois *f*

woodscrew *n* PARTS vis à bois *f*

word *n* GEN mot *m*; **~ block** DATA bloc mot *m*; **~ length** DATA longueur de mot *f*; **~ processing** COMP, DATA traitement de texte *m*; **~ processing package** COMP, DATA logiciel de traitement de texte *m*; **~ processor** COMP, DATA unité de traitement de texte *f*, machine de traitement de texte *f*; **~ rate** DATA débit en mots *m*; **~ wrapping** COMP bouclage *m*

words: **~ per minute** *phr* TELEP mots par minute; **~ twice** *phr* RADIO chaque mot deux fois

work *n* COMP travail *m*; **~ file** COMP fichier de travail *m*; **~ sheet** COMP feuille de travail *f*, fiche de travail *f*; **~ station** COMP, DATA, PER APP station de travail *f*

workability *n* DATA fabricabilité *f*

working *n* PER APP exploitation *f*; **~ channel** TRANS voie de trafic *f*; **~ characteristics** *n pl* TELEP caractéristique de fonctionnement *f pl*; **~ data** *n pl* DATA données temporaires *f pl*; **~ diagram** COMP diagramme de fonctionnement *m*; **~ frequency** RADIO fréquence de travail *f*; **~ function** ELECT temps d'extraction *m*; **~ point** ELECT point de fonctionnement *m*; **~ range** TRANS plage de fonctionnement *f*; **~ standard** TEST étalon de travail *m*; **~ surface** ELECT, PARTS surface utile *f*; **~ temperature** CLIM température de fonctionnement *f*; **~ voltage** POWER tension de fonctionnement *f*

World: **~ Administrative Radiocommunications Conference** *n* GEN conférence administrative mondiale des radiocommunications *f* (*CAMR*); **~ Wide Web** *n* COMP (*W3*, *WWW*), DATA (*W3*, *WWW*) *Internet* World Wide Web *m* (*W3*, *WWW*)

worldwide: **~ communications** *n pl* TELEP télécommunications à l'échelle mondiale *f pl*; **~ network** *n* TELEP réseau mondial *m*

WORM *abbr* (*write once read many times*) COMP WORM (*disque inscriptible seule fois*)

wow *n* ELECT pleurage *m*

WPC: **~ system** *n* (*wired-programme control system*) CONTR système de contrôle à programme câblé *m*

wrap[1]: **~ post** *n* PARTS borne pour connexion enroulée *f*

wrap[2] *vt* PARTS *cable* rubaner, *connection* enrouler

wraparound *n* COMP bouclage *m*

wrapped: **~ connection** *n* GEN connexion enroulée *f*

wrapping *n* PARTS *of cables* rubannage *m*, POWER *coiling round* enroulement *m*; **~ fault** PARTS défaut d'enroulement *m*; **~ machine** PARTS machine pour connexions enroulées *f*; **~ pin** PARTS broche enrouleuse *f*; **~ program** *AmE*, **~ programme** *BrE* PARTS programme d'enroulement *m*; **~ tool** PARTS outil enrouleur *m*

write[1]: **~ action** *n* COMP opération d'écriture *f*; **~-back** *n* COMP écriture différée *f*; **~ head** *n* COMP tête d'écriture *f*; **~-through** *n* COMP double écriture *f*

write[2] *vt* COMP scripter

write[3]: **~ many read always** *phr* (*WMRA*) COMP WMRA; **~ once read many times** *phr* (*WORM*) COMP disque inscriptible seule fois (*WORM*)

writing: **~ bar** *n* PARTS, TELEP lame d'inscription *f*; **~ edge** *n* PARTS, TELEP lame d'inscription *f*

wrong: **~ connection** *n* faux numéro *m*; **~ number** *n* TELEP faux numéro *m*

WRU *abbr* (*who-are-you signal*) SIG, SWG signal qui est là? *m*

WSC *abbr* (*wall-mountable splice center AmE*, *wall-mountable splice centre BrE*) ELECT coffret d'épissure montable sur mur *m*

W3 *abbr* (*World Wide Web*) COMP, DATA *Internet* World Wide Web *m*, W3 (*World Wide Web*), WWW (*World Wide Web*)

WWW *abbr* (*World Wide Web*) COMP, DATA *Internet* W3 (*World Wide Web*), WWW (*World Wide Web*)

wye: **~ connection** *n* POWER montage en étoile *m*, montage en Y *m*

WYSIWYG *abbr* (*what you see is what you get*) COMP WYSIWYG (*tel écran, tel écrit*)

X

XALS *abbr* (*extended application-layer structure*) TRANS structure en couches d'application étendue *f*

XD *abbr* (*ex-directory*) TELEP sur liste rouge

xenon: ~ **lamp** *n* CLIM lampe au xénon *f*

Xerox: ~ **Network Service** *n* (*XNS*) DATA service de réseau Xerox (*XNS*)

X.410-1984: ~ **mode presentation** *n* DATA présentation en mode X.410-1984 *f*

XID *abbr* (*exchange of identification*) TELEP XID (*échange d'identifications*)

XMTR *abbr* (*transmitter*) GEN émetteur *m*

XNS *abbr* (*Xerox Network Service*) DATA XNS (*service de réseau Xerox*)

X-ray: ~ **laser** *n* ELECT laser rayon X *m*; ~ **source** *n* ELECT source rayon X *f*

x-y: ~ **plotter** *n* TEST traceur de courbes *m*; ~ **recorder** *n* TEST traceur de courbes *m*

Y

Y *abbr* (*yes*) DATA oui *m*

Yagi: **~ aerial** *n* RADIO antenne Yagi *f*; **~ antenna** *n* RADIO antenne Yagi *f*

Y-branching: **~ device** *n* ELECT, TRANS coupleur en Y *m*

Y-connection *n* POWER montage en étoile *m*, montage en Y *m*

Y-coupler *n* ELECT, TRANS coupleur en Y *m*

yellow: **~ book** *n* COMP livre jaune *m*; **~ pages** *n pl* TELEP pages jaunes *f pl*

yes *n* (*Y*) DATA oui *m*; **~ or no test** TELEP, TEST essai par "tout ou rien" *m*

yield: **~ limit** *n* PARTS, TEST limite d'élasticité *f*

yoke *n* ELECT, PARTS, TELEP culasse *f*; **~ assembly** ELECT, PARTS bloc de bobinage *m*, collier de déviation *m*; **~ coil** ELECT, PARTS bobine de déviation *f*

Young: **~'s modulus** *n* PARTS module d'élasticité longitudinale *m*, module de Young *m*

Z

z: ~ **time** *n* GEN fuseau z *m*
Z: ~ **component** *n* RADIO composant Z *m*
zap *vt* COMP effacer
ZBR *abbr* (*zone bit recording*) COMP enregistrement par zone *m*
Zener: ~ **diode** *n* ELECT diode Zener *f*
zenith *n* RADIO zénith *m*
zero[1] *n* TEST zéro; ~ **adjustment** TEST réglage du zéro *m*; ~-**code suppression** TRANS suppression des codes zéro *f*; ~ **crossing** POWER, SWG croisement au point nul *m*, passage à zéro *m*; ~ **crossing switch** SWG commutateur de passage à zéro *m*; ~ **error** TEST erreur de zéro *f*; ~ **isochrone** RADIO isochrone zéro *f*; ~ **justification** TELEP justification nulle *f*; ~ **level** TEST niveau zéro *m*; ~ **loss circuit** ELECT circuit sans perte *m*; ~ **method** TEST méthode de zéro *f*; ~ **point** TEST point zéro *m*, zéro *m*; ~ **point energy** GEN énergie à zéro absolu *f*; ~ **point entropy** GEN entropie à zéro absolu *f*; ~ **potential** POWER potentiel zéro *m*; ~ **restriction** TRANS restriction sur les zéros *f*; ~ **setting** TEST réglage du zéro *m*; ~-**voltage relay** PARTS, POWER relais à zéro de tension *m*
zero[2] *vt* ELECT, TEST mettre à zéro, remettre à zéro, régler à zéro; ~-**fill** COMP remplir de zéros
zeroing *n* ELECT, TEST mise à zéro *f*, réglage du zéro *m*, remise à zéro *f*
zeroization *n* ELECT, TEST mise à zéro *f*, réglage du zéro *m*, remise à zéro *f*
zeroize *vt* ELECT, TEST mettre à zéro, remettre à zéro, régler à zéro
zeroizing *n* ELECT, TEST mise à zéro *f*, réglage du zéro *m*, remise à zéro *f*
zigzag *n* COMP, POWER zigzag *m*; ~ **pattern** COMP, POWER zigzag *m*, GEN motif en zigzag *m*; ~ **scanning order** COMP ordre de balayage en zigzag *m*
zinc[1]: ~-**electroplating** *n* PARTS zingage électrolytique *m*; ~-**plating** *n* PARTS zingage *m*; ~-**stripping** *n* PARTS dézingage *m*
zinc[2]: ~-**electroplate** *vt* PARTS électrozinguer, zinguer par électrolyse; ~-**plate** *vt* PARTS zinguer
zip[1]: ~ **code** *n* *AmE* (*cf post code BrE*) COMP code postal *m*
zip[2] *vt* COMP *file* comprimer
zone *n* RADIO zone *f*; ~ **bit recording** (*ZBR*) COMP enregistrement par zone *m*; ~ **boundaries** *n pl* RADIO limites *f pl*; ~ **for transmission** RADIO, TRANS zone à l'émission *f*
zoned: ~ **lens** *n* RADIO lentille à échelons *f*
zoom[1]: ~ **box** *n* COMP case zoom *f*; ~-**in** *n* COMP zoom avant *m*; ~-**out** *n* COMP zoom arrière *m*
zoom[2]: ~ **in** *vi* COMP zoomer
zooming *n* COMP grossissement *m*, zoom *m*